FOUNDATIONS *of*

BIOGEOGRAPHY

FOUNDATIONS

Classic Papers with

Published in association with
THE INTERNATIONAL BIOGEOGRAPHY SOCIETY *and the*
NATIONAL CENTER FOR ECOLOGICAL ANALYSIS AND
SYNTHESIS (USA)

OF BIOGEOGRAPHY

Commentaries

EDITORS

Mark V. Lomolino, Dov F. Sax, and James H. Brown

THE UNIVERSITY OF CHICAGO PRESS
Chicago and London

The University of Chicago Press, Chicago 60637
The University of Chicago Press, Ltd., London
© 2004 by The University of Chicago
All rights reserved. Published 2004
Printed in the United States of America
25 24 23 22 21 20 19   5 6 7 8 9

ISBN-13: 978-0-226-49236-0 (cloth)
ISBN-10: 0-226-49236-2 (cloth)
ISBN-13: 978-0-226-49237-7 (paper)
ISBN-10: 0-226-49237-0 (paper)

Library of Congress Cataloging-in-Publication Data

Foundations of biogrography : classic papers with commentaries /
    editors, Mark V. Lomolino, Dov F. Sax, and James H. Brown
        p.   cm.
    "Published in association with the International
Biogeography Society and the National Center for Ecological
Analysis and Synthesis (USA)."
    Includes bibliographical references and index.
    ISBN 0-226-49236-2 (cloth : alk. paper) — ISBN 0-226-
    49237-0 (pbk. : alk. paper)
    1. Biogeography.   I. Lomolino, Mark V., 1953–   II. Sax,
Dov F.   III. Brown, James H., 1942 Sept. 25–
IV. International Biogeography Society.   V. National Center
for Ecological Analysis and Synthesis.
QH84 .F68   2004
578'.09—dc22
                                                    2003018272

Contents

Preface   xix
Introduction

James H. Brown   1

PART ONE

Early Classics
John C. Briggs and Christopher J. Humphries   5

1
Carolus Linnaeus (1781)
Excerpts from *Dissertation II, On the Increase of the Habitable Earth*

Translations by F. J. Brand from *Select Dissertations from the Amoenitates Academicae*
(London: G. Robinson and J. Robson, 1781; repr., New York: Arno, 1977)
14

2
Georges-Louis Leclerc, Compte de Buffon (1761)
Excerpts from *Natural History, General and Particular*

Translated by W. Smellie. London: W. Strahan and T. Cadell, 1791
16

3
Johann Reinhold Forster (1778)
Excerpts from *Observations Made during a Voyage Round the World,
on Physical Geography, Natural History, and Ethic Philosophy*

Edited by Nicholas Thomas, Harriet Guest, and Michael Dettelbach. Honolulu:
University of Hawai'i Press, 1966
19

4

Augustin de Candolle (1820)
Excerpt from *Essai Élémentaire de Géographie Botanique*
> Translation in A. P. Decandolle and K. Sprengel, *Elements of the Philosopy of Plants.*
> Edinburgh: Blackwood, 1821
> 28

5

Alexander von Humboldt (1805)
Excerpt from *Essay on the Geography of Plants*
> Translations by Francesca Kern and Philippe Janvier from *Essai sur la Géographie des
> Plantes.* Paris: Levrault, Schoell et Cie.
> 49

6

Edward Forbes (1844)
Excerpts from *Report on the Mollusca and Radiata of the Aegean Sea,
and on Their Distribution, Considered as Bearing on Geology*
> Reports of the British Association of Science for 1843, 130–93
> 58

7

James Dwight Dana (1853)
*On an Isothermal Oceanic Chart, Illustrating the Geographical
Distribution of Marine Animals*
> The American Journal of Science and Arts, 2d ser., 66:153–67, 325–27, 391–92
> 88

8

Sir Joseph Dalton Hooker (1853)
Excerpt from *The Botany of the Antarctic Voyage of H.M. Discovery Ships
Erebus and Terror in the Years 1839–1843*
> Part 2, *Flora Novae Zelandiae*, vol. 1: xix–xxvii
> 109

9

Philip Lutley Sclater (1858)
On the General Geographical Distribution of the Members of the Class Aves
> Journal of the Linnaean Society of London, Zoology 2:130–45
> 118

10
Asa Gray (1876)
Excerpt from *Darwiniana: Essays and Reviews Pertaining to Darwinism*
New York: D. Appleton
134

11
Charles Darwin (1859)
Excerpts from *On the Origin of Species by Means of Natural Selection,
or the Preservation of Favoured Races in the Struggle for Life*
London: John Murray
140

12
Alfred Russel Wallace (1876)
Excerpt from *The Geographical Distribution of Animals*
London: Macmillan
164

13
Ernst Haeckel (1876)
Excerpt from *The History of Creation, or the Development of the Earth
and Its Inhabitants by the Action of Natural Causes*
Translated by E. Ray Lankester. New York: D. Appleton, 1925
178

14
Hermann von Ihering (1900)
The History of the Neotropical Region
Science 12:857–64
194

15
Clinton Hart Merriam (1890)
Excerpt from *Results of a Biological Survey of the San Francisco
Mountain Region and Desert of the Little Colorado, Arizona*
North American Fauna, no. 3. U.S. Department of Agriculture, Washington, DC
202

16
William Diller Matthew (1915)
Excerpt from *Climate and Evolution*
2d Edition, 1939. Special Publications of the New York Academy of Sciences, vol. 1
234

17
Sven Ekman (1953)
Excerpt from *Zoogeography of the Sea*
> Translated by Elizabeth Palmer. London: Sidgwick and Jackson, 1953
> 245

18
Evgenii Vladimirovitch Wulff (1943)
Excerpt from *An Introduction to Historical Plant Geography*
> Translated by Elizabeth Brissenden. Waltham, MA: Chronica Botanica
> 249

PART TWO

Earth History, Vicariance, and Dispersal
Paul S. Giller, Alan A. Myers, and Brett R. Riddle   267

19
Alfred Wegener (1924)
Excerpt from *The Origin of Continents and Oceans*
> Translated by J. G. A. Skerl. 4th ed. London: Methuen, 1929
> 277

20
Lars Brundin (1966)
Excerpt from *Transantarctic Relationships and Their Significance,
as Evidenced by Chironomid Midges*
> Kungliga Svenska Vetenskapsakadamiens Handlingar, ser. 4, 11(1):437–56
> 295

21
Sherwin Carlquist (1966)
The Biota of Long-Distance Dispersal, I: Principles of Dispersal and Evolution
> The Quarterly Review of Biology 41:247–70
> 315

22
George Gaylord Simpson (1940)
Mammals and Land Bridges
> Journal of the Washington Academy of Sciences 30:137–63
> 339

23
Anthony Hallam (1967)
The Bearing of Certain Palaeozoogeographic Data on Continental Drift
        Palaeogeography, Palaeoclimatology, Palaeoecology 3:201–41
        366

24
Philip J. Darlington Jr. (1965)
Excerpt from *Biogeography of the Southern End of the World*
        Cambridge: Harvard University Press
        407

25
Larry G. Marshall, S. David Webb, J. John Sepkoski Jr., and David M. Raup (1982)
Mammalian Evolution and the Great American Interchange
        Science 215:1351–57
        419

26
Francis Dov Por (1971)
*One Hundred Years of Suez Canal—A Century of Lessepsian Migration:
Retrospect and Viewpoints*
        Systematic Zoology 20:138–59
        426

PART THREE

Species Ranges
        Robert Hengeveld, Paul S. Giller and Brett R. Riddle   449

27
Joseph Grinnell (1922)
The Role of the "Accidental"
        Auk 39:373–80
        456

28
Eric Hultén (1937)
Excerpts from *Outline of the History of Arctic and Boreal Biota
during the Quarternary Period*
        Stockholm: J. Cramer
        464

29
Evgenii Vladimirovitch Wulff (1943)
Excerpt from *An Introduction to Historical Plant Geography*
>   Translated by Elizabeth Brissenden. Waltham, Mass.: Chronica Botanica
>   513

30
Jeremy D. Holloway and Nicholas Jardine (1968)
Two Approaches to Zoogeography: A Study Based on the Distributions of Butterflies, Birds and Bats in the Indo-Australian Area
>   Proceedings of the Linnaean Society of London 179:153–88
>   536

31
Charles S. Elton (1958)
Excerpt from *The Ecology of Invasions by Animals and Plants*
>   London: Methuen and Co.
>   575

32
Daniel H. Janzen (1967)
Why Mountain Passes are Higher in the Tropics
>   American Naturalist 101:233–49
>   594

33
Philip V. Wells and Rainer Berger (1967)
Late Pleistocene History of Coniferous Woodland in the Mohave Desert
>   Science 155:1640–47
>   611

34
John R. Flenley (1979)
The Late Quaternary Vegetational History of the Equatorial Mountains
>   Progress in Physical Geography 3:488–509
>   619

35
Paul S. Martin (1973)
The Discovery of America
>   Science 179:969–74
>   641

PART FOUR

Revolutions in Historical Biogeography
Vicki A. Funk   647

36
Lars Brundin (1966)
Excerpt from *Transantarctic Relationships and Their Significance,
as Evidenced by Chironomid Midges*
> Kungliga Svenska Vetenskapsakadamiens Handlingar, ser, 4, 11(1):46–64
> 658

37
Willi Hennig (1966)
Excerpt from *Phylogenetic Systematics*
> Urbana: University of Illinois Press
> 679

38
Gareth J. Nelson (1969)
The Problem of Historical Biogeography
> Systematic Zoology 18:243–46
> 686

39
Leon Croizat (1962)
Excerpt from *Space, Time, Form: The Biological Synthesis*
> Caracas: Published by the author
> 690

40
Leon Croizat, Gareth J. Nelson and Donn Eric Rosen (1974)
Centers of Origin and Related Concepts
> Systematic Zoology 23:265–87
> 705

41
Gareth J. Nelson (1974)
Historical Biogeography: An Alternative Formalization
> Systematic Zoology 23:555–58
> 728

42
Norman I. Platnick and Gareth J. Nelson (1978)
A Method of Analysis for Historical Biogeography
Systematic Zoology 27:1–16
732

43
Donn E. Rosen (1978)
Vicariant Patterns and Historical Explanation in Biogeography
Systematic Zoology 27:159–88
748

PART FIVE

Diversification
Lawrence R. Heaney and Geerat Vermeij   779

44
Bernard Rensch (1960)
Excerpt from *Evolution above the Species Level*
New York: Columbia University Press
789

45
Ernst Mayr (1942)
Excerpt from *Systematics and the Origin of Species*
New York: Columbia University Press
811

46
David Lack (1947)
Excerpts from *Darwin's Finches*
Cambridge: Cambridge University Press, Cambridge
833

47
Philip J. Darlington Jr. (1959)
Area, Climate, and Evolution
Evolution 13:488–510
852

48
James W. Valentine (1969)
Patterns of Taxonomic and Ecological Structure of the Shelf Benthos during Phanerozoic Time
>Palaeontology 12:684–709
>875

49
David M. Raup (1972)
Taxonomic Diversity during the Phanerozoic
>Science 177:1065–71
>901

50
Jürgen Haffer (1969)
Speciation in Amazonian Forest Birds
>Science 165:131–37
>908

51
Guy L. Bush (1969)
Sympatric Host Race Formation and Speciation in Frugivorous Flies of the Genus Rhagoletis (Diptera, Tephritidae)
>Evolution 23:237–51
>915

PART SIX

The Importance of Islands
Robert J. Whittaker   931

52
Olof Arrhenius (1921)
Species and Area
>Journal of Ecology 9:95–99
>942

53
Edward O. Wilson (1959)
Adaptive Shift and Dispersal in a Tropical Ant Fauna
>Evolution, 13:122–44
>947

54

Robert H. MacArthur and Edward O. Wilson (1963)
An Equilibrium Theory of Insular Zoogeography
> Evolution 17:373–87
> 970

55

Daniel S. Simberloff and Edward O. Wilson (1970)
Experimental Zoogeography of Islands:
A Two-Year Record of Colonization
> Ecology 51:934–37
> 985

56

James H. Brown (1971)
Mammals on Mountaintops: Nonequilibrium Insular Biogeography
> The American Naturalist 105:467–78
> 989

57

Jared M. Diamond (1974)
Colonization of Exploded Volcanic Islands by Birds:
The Supertramp Strategy
> Science 184:803–6
> 1001

58

Jared M. Diamond (1975)
The Island Dilemma: Lessons of Modern Biogeographic Studies
for the Design of Natural Reserves
> Biological Conservation 7:129–46
> 1005

59

Storrs L. Olson and Helen F. James (1982)
Fossil Birds from the Hawaiian Islands: Evidence for Wholesale
Extinction by Man before Western Contact
> Science 217:633–35
> 1023

PART SEVEN

Assembly Rules
Nicholas J. Gotelli   1027

60
Philip J. Darlington Jr. (1957)
Excerpt from *Zoogeography: The Geographic Distribution of Animals*
New York: Wiley
1036

61
Charles S. Elton (1946)
Competition and the Structure of Ecological Communities
Journal of Animal Ecology 15:54–68
1041

62
Carrington Bonsor Williams (1947)
The Generic Relations of Species in Small Ecological Communities
Journal of Animal Ecology 16:11–18
1056

63
Robert H. Whittaker (1967)
Gradient Analysis of Vegetation
Biological Reviews 42:207–64
1064

64
Robert H. MacArthur (1972)
Excerpts from *Geographical Ecology: Patterns in the Distributions of Species*
New York: Harper and Row
1122

65
Jared M. Diamond (1975)
Excerpt from *Assembly of Species Communities*
In *Ecology and Evolution of Communities,* ed. M. L. Cody and J. M. Diamond.
Cambridge: Harvard University Press
1127

66
Edward F. Connor and Daniel S. Simberloff (1979)
The Assembly of Species Communities: Chance or Competition?
>Ecology 60:1132–40
>1135

PART EIGHT

*Gradients in Species Diversity: Why Are There So Many Species
in the Tropics?*
James H. Brown and Dov F. Sax   1145

67
Theodosius Dobzhansky (1950)
Evolution in the Tropics
>American Scientist 38:209–21
>1155

68
Alfred G. Fischer (1960)
Latitudinal Variations in Organic Diversity
>Evolution 14:64–81
>1168

69
George Gaylord Simpson (1964)
Species Density of North American Recent Mammals
>Systematic Zoology 13:57–73
>1186

70
Eric R. Pianka (1966)
Latitudinal Gradients in Species Diversity: A Review of Concepts
>The American Naturalist 100:33–46
>1203

71
Robert H. MacArthur (1972)
Excerpts from *Geographical Ecology: Patterns in the Distribution of Species*
>New York: Harper and Row
>1217

72
Robert H. Whittaker and William A. Niering (1975)
*Vegetation of the Santa Catalina Mountains, Arizona, V:
Biomass, Production and Diversity along the Elevation Gradient*
Ecology 56: 771–90
1254

References   1275
List of Contributors   1287
Index   1289

Preface

It is no longer necessary to justify publishing a collection of the classic papers in a scientific discipline. The value is attested by the enormous success of the first two "foundations" volumes published by University of Chicago Press: *Foundations of Ecology* (Real and Brown 1991) and *Foundations of Animal Behavior* (Houck and Drickamer 1996). Modeled on these successful precursors, *Foundations of Biogeography* is intended to play similar roles: as a text for advanced courses and graduate seminars on the history of the discipline, as a vetted collection of writings to provide students with an introduction to the theoretical and empirical foundations of the field, and as a handy source of the classic papers for practicing scientists.

As Brown and Lomolino (1998) observed, the field of biogeography has a long and distinguished history, and one interwoven with that of ecology and evolutionary biology. The seminal papers included in this volume should, therefore, be of interest to all scientists who study the origins, diversification, geography and conservation of biological diversity. Scientists have been working and making major contributions in biogeography for at least two centuries—as is attested by the earliest writings in this volume. Not until recently, however, did scientists identify themselves as biogeographers and view themselves as working in a discipline called biogeography. Only in the last few decades has biogeography come to be recognized as a science with its own conceptual and empirical foundations and with important practical applications to critical problems of global change and biological conservation. One of the primary goals of *Foundations of Biogeography*, therefore, is to foster the development of this emerging discipline by calling attention to its creative foundational works.

Like the preceding foundations volumes, this one was compiled by a committee. On October 1–4, 2000, biogeographers assembled at the National Center for Ecological Analysis and Synthesis (NCEAS) in Santa Barbara for an unprecedented meeting. This was the first time that such a large number of scientists representing the entire breadth of contemporary biogeography had assembled. The goal was to test whether they could overcome their divergent backgrounds, perspectives, and interests and undertake two important actions that would contribute to the unification and synthesis of the discipline of biogeography. One action was to found a new International Biogeography Society, which held its inaugural meeting in January of 2003 (www.biogeography.org).

The other action initiated at the meeting in Santa Barbara was to produce *Foundations of Biogeography* to demonstrate the common intellectual heritage of the discipline as reflected in its most important historical writings. A subset of the meeting participants have worked on this initiative—to discuss, assemble, and publish a collection of classic papers, with accompanying commentaries that place the writings in historical and contemporary perspective. This volume is the result of their labors.

We three editors are indebted to the many people who contributed to *Foundations of Biogeography*. These include, first and foremost, all of the people who attended the initial meeting in Santa Barbara: Julio Betancourt, John Briggs, James Brown, Robert Colwell, Michael Donoghue, Vicki Funk, Paul Giller, Nicholas Gotelli, Lawrence Heaney, Rob Hengeveld, Christie Henry, Chris Humphries, Mark Lomolino, Glen MacDonald, David Perault, Brett Riddle, Klaus Rohde, Dov Sax, Geerat Vermeij and Robert Whittaker. We are particularly grateful to those individuals who showed their dedication to producing this volume by serving as editor of one of the sections, presiding over the final choice of the selections to be reprinted, and writing the essays that introduce the featured papers.

Throughout the preparation of this book, we have received enthusiastic encouragement and expert assistance from University of Chicago Press. In particular, Christie Henry showed her interest in the book during the earliest discussions with the editors, she contributed invaluable help and advice throughout the preparation of the volume, and she worked with the editors, reviewers, and the press to insure rapid publication in accordance with the high standards set by previous foundations volumes.

The National Center for Ecological Analysis and Synthesis (NCEAS) provided invaluable support for producing this book, as well for the more general effort to unify biogeography by holding meetings and founding a society. NCEAS funded the initial meeting in Santa Barbara in 2000, and a second, smaller meeting on September 9–11, 2001 for the expressed purpose of completing most of the work on this volume. Both meetings were held at NCEAS headquarters in Santa Barbara. We are indebted to the staff and faculty of NCEAS for hosting these successful meetings and for other contributions that have led to the publication of this

book. In recognition of NCEAS support, royalties from this volume will go to NCEAS. We hope that these royalties will be used to further the development of biogeography, especially by contributing to the training of graduate students and postdoctoral scientists.

Finally, we thank all of the other people, too numerous to mention individually, who have contributed to the production of this volume. These include the scientists who published the selected writings, other scientists who have advanced the field of biogeography (some of whom are cited in the commentaries on each part), students and colleagues who have influenced our own thinking and development, the personnel at University of Chicago Press who have edited, and printed the book, and Chris Akios and Adam Harpster along with other helpful individuals in our home departments and libraries who have helped to compile, copy, and obtain permission to reprint the selected publications. Preparation of this volume has been a labor of love by many individuals dedicated to promoting the past accomplishments and future prospects of the exciting field of biogeography. We hope that new generations of students will learn from these foundational writings and be stimulated to make their own contributions.

Introduction

James H. Brown

This book is intended to provide a collection of foundational writings that will advance the science of biogeography. For more than two centuries, scientists have been working and publishing on topics that are easily recognized as biogeography. Until recently, however, the authors rarely considered themselves to be biogeographers, and they did not identify their field of endeavor as biogeography: only in the last few decades, has biogeography emerged as a single, coherent, important discipline. This coalescence has been facilitated by the increasingly interdisciplinary nature of much biogeographic research, by the growing interest in the spatial distributions and patterns of diversity of organisms, and by the increasing recognition of the extent to which our own species is changing environments and altering distributions of organisms.

Indeed, urgent calls to understand and address the many dimensions of "global change" have hastened the emergence of biogeography as a science with much to say about the present state and future trajectory of the earth and its biota. There is now a recognized need for biogeographers. There are journals, textbooks, university courses, and even job advertisements with the word "biogeography" prominently displayed in their titles. We hope that the publication of this compilation of foundational writings will further facilitate the development of biogeography by acquainting students and practicing scientists alike with the historical roots, conceptual and empirical content, and synthetic coherence of the discipline.

Our decision to produce this book was obviously spurred by the enormous success of earlier "foundations" volumes published by University of Chicago Press. More than a decade after its publication, *Foundations of Ecology* (Real and Brown 1991) continues to be widely used, both by students seeking to acquire a background in the historical development and seminal early works in their field, and by senior scientists wanting a convenient compilation of the classic papers. The subsequent book, *Foundations of Animal Behavior* (Houck and Drickamer 1996), enjoys similar success. These volumes, and others in the series, serve several important roles: they are used as texts or supplementary texts in courses and seminars, especially at the graduate level, and they are read by students seeking to improve their backgrounds and historical perspectives—including many graduate students studying for their comprehensive and qualifying exams! In a single, inexpensive, convenient format, the foundations volumes provide ready access to early works that are of timeless interest because they offer insights into the historical development, major conceptual and empirical advances, and synthetic unity of the field.

The editors of the first foundations volume, Les Real and James H. Brown, were stimulated to produce *Foundations of Ecology* because they "shared a love of the history of ecology and a study of the development of critical concepts in the field." Real and Brown had found that most students were unfamiliar with the historical roots of their discipline. The foundations volumes provide a convenient way to correct this deficiency. Left to their own devices, students feel compelled to keep up with the fast pace of modern science and tend to concentrate on reading the current literature. When the foundational writings are the subjects of graduate seminars, however, students find these classics to be not only of historical interest but also valuable for developing their own critical perspectives and original research programs. The value is not limited to students. Nearly all professional ecologists and animal behaviorists have a well-used copy of the respective foundations volume on their bookshelves.

1

A compilation of foundational publications should be even more valuable to students and practitioners of biogeography, because of that discipline's unusual history. On the one hand, it has a rich tradition extending back over two centuries. Many of the seminal contributions were made by some of the greatest scientists of their times. This volume reprints contributions of Alexander von Humboldt, Charles Darwin, Alfred Russel Wallace, Joseph Dalton Hooker, George Gaylord Simpson, Ernst Mayr, David Lack, Willi Hennig, Robert H. MacArthur, and Edward O. Wilson, among others. The writings of these authors speak for themselves.

On the other hand, recognition of biogeography as a coherent discipline in its own right, and the resulting synthesis and unification, has come only recently. If they had been asked "What do you call yourself?" none of the scientists named above would likely have answered "A biogeographer." The nineteenth-century scientists would have considered themselves to be naturalists. The twentieth-century scientists would have identified themselves with one of the recognized disciplines of their time, usually systematics, ecology, or paleontology. Such divergent specialization meant that biogeographers were trained in different departments, attended different meetings, and published in different journals. The unfortunate consequence was that biogeographic perspectives from ecology, systematics and phylogenetics, and paleobiology and earth history, instead of being integrated and synthesized, were often advanced without reference to each other, or even as competing viewpoints. This compartmentalization and isolation intensified as a result of taxonomic specialization. Throughout the first two-thirds of the twentieth century there was little interchange between phytogeographers, who studied plants, and zoogeographers, who specialized in animals. In fact, zoogeographers studying different animal groups like terrestrial vertebrates, marine invertebrates, or insects, had specialized audiences that were isolated from each other.

The result is that biogeography, despite its distinguished history, did not begin to coalesce into a recognized discipline until the last few decades. Major conceptual advances in the 1960s and 70s kindled intellectual excitement that spread rapidly beyond the specialized disciplines of that era. Especially noteworthy in this regard were the equilibrium theory of island biogeography of MacArthur and Wilson, the phylogenetic approach to reconstructing past distributions pioneered by Hennig, Rosen, Nelson, and Platnick, and the exploration by Brundin, Hallam, and others of the biogeographic implications of plate tectonics. It was in this period that the synthetic discipline called biogeography really began to emerge. It was signaled by the founding of the *Journal of Biogeography* in 1973, and by the publication of synthetic books with "Biogeography" in their titles (e.g., Cox and Moore 1973; Brown and Gibson, 1983; Myers and Giller 1988; and Hengeveld 1990).

These trends have accelerated in the last decade or so. Biogeography has emerged as well-recognized discipline with its own set of unifying perspectives, theoretical concepts, and supporting data. There are now senior scientists, including most of those who wrote the background commentaries that introduce the different parts of this book, who unhesitatingly call themselves biogeographers. Increasingly, major universities offer undergraduate courses and graduate training identified as biogeography. Numerous biogeographic resources are now available on the web, including Charles H. Smith's valuable bibliography and full-text archive, *Early Classics in Biogeography, Distribution, and Diversity Studies: To 1950* (http://www.wku.edu/~smithch/biogeog/). The most recent milepost is the formation of an International Biogeography Society (www.biogeography.org) to promote communication, integration, and synthesis across the discipline.

It is not coincidental that the Society's foundational meeting at the National Center for Ecological Analysis and Synthesis in Santa Barbara, California, in 2000 also led to the preparation of this book. That meeting brought together a wide spectrum of scientists who collectively represent much of the breadth of contemporary biogeography. These scientists faced a challenge. If biogeography is to be a coherent, viable discipline, with increasing interchange and integration of its historically specialized roots, then it should be possible to agree on a set of writings that have laid the foundation. As might be expected, the effort to achieve consensus on the works to be included in a *Foundations of Biogeography* volume engendered much spirited discussion. We wanted to give readers a sample of the long history of the field: to capture the breadth of the discipline by including selections that were both theoretical and empirical, represented important themes in both "historical" and "ecological" biogeography, and illustrated studies from different regions of the world and groups of organisms. We wanted to produce a volume that would be widely used in graduate courses and seminars, adopted as a "bible" by prospective biogeographers, and valued as a handy reference book by practicing scientists.

The book is organized into eight parts, each introduced by an essay describing the historical writings excerpted from published works. We have deliberately avoided organizing the parts to reflect obvious divisions like taxonomic group, environment or habitat, or region of the world. Instead, we have tried to organize the material around conceptual or synthetic themes. Our hope is that this scheme will juxtapose material in creative ways so as to break down the still existing barriers between specialized subdisciplines and research programs while emphasizing the ways that different perspectives contribute to the emerging unity of biogeography. The essays that introduce each part are intended to place the excerpted writings in historical and conceptual context by providing additional background information and citing other important works that either preceded or followed the reprinted publica-

tions. The introductory essays also try to make synthetic connections among the reprinted publications, both within and between parts.

In planning and assembling this volume, our primary difficulty was to select a severely limited number of writings that would nevertheless illustrate important advances that have laid the foundation across the entire breadth of modern biogeography. To produce a volume of reasonable size, many important contributions had to be omitted. We had to make a number of hard decisions: whether to reprint the first or the clearest publication on a topic; how to extract the most relevant sections from larger works; what weight to give to novel as opposed to synthetic contributions; and when to place the cutoff date for a "foundational" contribution (we chose 1975, with a few notable exceptions). We editors are indebted to all who have contributed, including especially the authors of the introductory essays, but also the others who attended the initial meeting and made valuable suggestions about what material to include and how to organize the book. The result is very much a book compiled by a committee. Most students and practitioners of biogeography probably would not have chosen exactly the same collection of readings. Approaching the discipline from their individualistic perspectives, they would have included other writings that they consider to be especially relevant and eliminated some that they consider less important. We have tried to achieve a reasonable compromise.

This book is intended primarily for students. Like all sciences, biogeography is a community endeavor that spans the generations. Major contributions by past scientists lay the foundation for creative breakthroughs and synthetic advances by future scientists. We hope that reading a collection of classic works will stimulate students to become biogeographers and make their own important contributions.

1 Early Classics

John C. Briggs and Christopher J. Humphries

In this part, we focus on those early works that we consider to be the most influential in advancing our knowledge of biogeography. Selecting them was not a simple task and, in so doing, we have omitted many works that might arguably have been included. Our list takes us up to 1953; more recent works are considered in subsequent parts of this volume. Here we will describe the broad historical background and provide a biographical sketch of each of the selected authors.

The origins of modern biogeography may be traced to the advent of the Renaissance in Europe. The opening of overland trade routes followed by the great voyages of exploration introduced a variety of exotic plants and animals. The Renaissance also introduced a gradual change in attitudes toward religion. The scriptures began to lose much of their allegorical meaning. By the middle of the seventeenth century, the practice of expounding the Bible's reality was widely established. As a result, the story of Noah and the Ark began to play an influential role in European philosophy (Browne 1983).

A Jesuit priest, Athanasius Kircher (1602–1680), undertook a detailed description of how the Ark must have been constructed in order to accommodate the 310 species of animals that he

recognized. No sooner was Kircher's book published, than it came under criticism from those who saw no room for the increasing number of exotic species. After that, most of the clergy ceased to discuss the size of the Ark and simply asserted that all modern species must have been passengers, otherwise they would not be in existence today. By the beginning of the eighteenth century, people informed on the subject of natural history generally abandoned the idea of the Ark. However, the concept of the Deluge was still strongly entrenched and the early naturalists still had to explain how thousands of species survived the floods.

The person considered to inaugurate our modern approach to systematics and biogeography (as distinct from the medieval) was the young Swede, CAROLUS LINNAEUS (Carl Linne) (1709–1778). He was a deeply religious man who felt that God spoke most clearly to mankind through the natural world. Linnaeus considered the earth to be a gigantic collection of natural objects given to him by God. It was his task to methodically describe and catalog all of the animals and plants into a series of descriptive compendia (see Nelson and Platnik 1981). Linnaeus (1781) solved the problem of the Ark by telescoping the story of the creation into that of the Deluge. Although Linnaeus was more deeply interested in systematics than in ecology and biogeography, he did have a profound interest in the geography of life and the different provenances of God's creations. He proposed that all living things had their origin on a high mountain in Paradise, an island situated under the equator at about the time that the primeval waters began to recede. Furthermore, he proposed that this Paradisiacal Mountain bore a variety of ecological conditions arranged in elevational and climatic zones so that each pair of animals was created in a particular habitat along with other species suited for that location. Linnaeus envisaged the various animals and plants migrating as the flood waters receded to their eventual homes where they remained unchanged for the rest of time.

Here we feature an excerpt from Linnaeus's "Dissertation II. On the Increase of the Habitable Earth" (paper 1), which clearly characterizes the struggles of scientists in the Age of Enlightenment to reconcile their growing appreciation of the complexity of nature with biblical scripture. Linnaeus's views on the dynamics of land and sea were antecedent to those of Lyell, Hooker and other nineteenth-century "extensionists," while Forster, von Humboldt, and Merriam would build upon Linnaeus's writings on elevation clines in climate and biotas.

Among Linnaeus's contemporaries, GEORGES-LOUIS LECLERC, COMPTE DE BUFFON (1707–1788), was most influential in persuading educated people to question the scientific validity of not just the fixity of the earth, but the Paradisiacal origins of plants and animals and the idea that species did not change through time. In his work *Histoire Naturelle, Générale et Particulière* [Natural History, General and Particular] published during the period 1749–1804 and excerpted here (paper 2), he compared the mammal faunas occurring in similar habitats in Africa and South America and discovered that those of the Old and New World tropics were exclusively confined to their own areas, with none being common to both continents. When Pierre Latreille (1762–1833) studied the insects and Georges Cuvier

(1769–1832) compared the reptiles of the two regions, Buffon's observations were found to be accurate and general—different regions, even those with very similar climates, were inhabited by distinct biotas. This became known as "Buffon's law."

This law would eventually be interpreted by some to mean that such animals had originated, or had been placed, *in situ* and had not migrated from Mt. Ararat, the Biblical landing place of the Ark. Nevertheless, Buffon still accepted Linnaeus's "center of origin" principle. He believed that life originated generally in the far north during a warmer period and had gradually moved south as the climate got colder. Because the New and Old Worlds were almost joined in the north, he concluded that the species in the two areas must originally have been identical. However, as the southward progression took place, the original populations were separated. In the New World, a kind of "structural degeneration" (evolution) took place that caused these species to depart from the primary type. While the specifics of his Northern Origins hypothesis was proven invalid, Buffon's law accurately emphasized the dynamics of not just Earth's climate, but its species as well. Just as important, the law became fundamental to all approaches for classifying biogeographic regions.

Buffon was struck by the similarities as well as the differences between biotas of the New and Old Worlds. In one of his most prescient statements, Buffon inferred that ."... the two Continents were formerly united, and that the species which inhabited the New World, because the soil and climate were most agreeable to their nature, were separated from the others by the irruption of the waters, when they divided Africa and America" (Buffon 1791: 451). This clearly is one of the earliest statements on what would come to be known as *vicariant events* (instances where species formation results from the imposition of a barrier dividing a formerly continuous population), either similar to those envisioned by the extensionists, or possibly by Wegener in his theory of continental drift (see parts 2, 3, and 4 of this volume).

Prussian-born JOHANN REINHOLD FORSTER (1729–1798) (accompanied by his son, Georg) was appointed to catalogue the floras and faunas for Captain Cook on his second expedition to the South Seas in 1772. Upon his return, Forster published his *Observations Made during a Voyage Round the World* ([1778] 1966). In this work, which is excerpted here (paper 3), he presented a worldwide view of the various natural regions and their biota. He explained how the different floras replaced one another as the physical characteristics of the environment changed, and also called attention to the way in which the type of vegetation determined the kinds of animals found in each region. Forster compared islands to the mainland and noted that the number of species in a given area was proportionate to the available physical resources. He remarked on the uniform decrease in floral diversity from the equator to the poles and attributed this phenomenon to the latitudinal change in the surface heating of the earth. These observations are fundamental to the fields of ecology and biogeography, and may be among the earliest descriptions of some of nature's most general patterns—the species-area and species-isolation relationships, latitudinal gradients in diversity, and (possibly) species-energy theory as well.

The German naturalist ALEXANDER VON HUMBOLDT (1769–1859) has often been called the "father of phytogeography" (Brown and Lomolino 1998). Like Linnaeus, von Humboldt felt that the study of geographical distribution was scientific inquiry of the highest order and that it could lead to the disclosure of fundamental laws of nature. He is perhaps best known for his explorations of Central and South America with the French naturalist Aimé Bonpland. The two men traveled along the Amazon and Orinoco Rivers exploring the Andes and parts of present-day Colombia, Ecuador, and Venezuela. By 1804, they had collected thousands of specimens of rocks and plants and had studied volcanoes, ocean currents, and the earth's magnetism. In 1829, Humboldt studied the vegetation and climatic conditions during his travels to the Ural Moun-

tains, the Caspian Sea, and parts of Siberia. In order to describe the precise climatic associations of plant communities, von Humboldt invented the *isobar* and *isotherm*—lines on a map joining points of equal atmosphere pressure and temperature, respectively. Included in the great 24-volume work *Voyage aux Régions Équinoxiales du Nouveau Continent* (1805–1837, with A. J. A. Bonpland), is von Humboldt's *Essai sur la Géographie des Plantes* [Essay on the Geography of Plants] (1805; paper 5). We have included an excerpt from this work because it is considered by many to be his best contribution to biogeography, and it illustrates his penchant for detailed analysis of physical environments and plant distributions and his passion for the beauty of nature. His theories of the geography of plants were strongly influenced by his climb of Mt. Chimborazo, an 18,000 foot peak in the Andes, where he observed a series of elevational, floristic belts equivalent to tropical, temperate, boreal, and arctic regions of the earth. This empirical observation laid the foundation for future comparisons of the relationships between vegetation, climate, and species distributions. Among his other works, von Humboldt's (1815) personal narrative of his travels in South America may have been most influential because it captivated a new generation of naturalists including Charles Darwin, Joseph Dalton Hooker and Alfred Russel Wallace.

A close friend of von Humboldt, the Swiss botanist AUGUSTIN DE CANDOLLE (1778–1841) provided important insights to both biogeography and ecology. Absorbed in the problems of plant distribution, he wrote extensively on what is now called ecological biogeography. In 1820, de Candolle published his *Essai élémentaire de géographie botanique* in the city of Strasbourg. In the same year, his work also appeared in the *Dictionnaire des Sciences Naturelles*, volume 18 (F. C. Levrault, Paris; see paper 4). In that work, de Candolle made a distinction between "stations" (habitats) and "habitations" (the major botanical provinces). This was the first real distinction to be made between ecological biogeography and historical biogeography. From this source, the young Charles Darwin learned his botanical geography, as did Joseph Hooker and Augustin's son Alphonse. In his "Geography of Plants," which we have excerpted here, Candolle respectfully but convincingly attacked Linnaeus's single origins hypothesis, and discussed the connections between plant physiology and what he called the history of plants—their "origin, diffusion, and gradual distribution."

While most biogeographical studies were, and continue to be focused on the terrestrial realm, nineteenth-century naturalists began exploring and comparing marine biotas from different regions. EDWARD FORBES (1815–1854), despite his short life, is considered to be one of the great philosophical naturalists, and his studies are fundamental to marine biogeography and dynamic biogeography, in general. When on board the survey ship *H.M.S. Beacon*, Forbes made a series of dredge hauls in the Mediterranean at depths between 100 and 250 fathoms. From this work, he presented the "Friday Evening Lecture" at the *Royal Institution* on Albermarle Street in London, an event that made him famous. His findings were published in his "Report on the Mollusca and Radiata of the Aegean Sea" (1844; paper 6), which included his important statement that "parallels in depth are equivalent to parallels in latitude" and announced his discovery of an "azoic," or lifeless zone at the greatest depths. According to Forbes (1844: 152), just as the distributions of terrestrial animals are influenced by three primary factors—"climate, mineral structure, and elevation"—the distributions of marine animals are determined by three great influences, namely climate, sea composition, and depth. Forbes went on to publish a book-length memoir, *On the Connection between the Distribution of the Existing Fauna and Flora of the British Isles, and the Geological Changes which Have Affected Their Area, Especially during the Epoch of the Northern Drift* (1846). This major study attempted to correlate present distribution patterns with past glaciations and land bridges, a topic that remains an active area of research for modern biogeographers.

In 1856, Forbes published the first worldwide survey of marine biogeography. This was in the form of a map of the distribution of marine life that was published, together with a descriptive text, in Alexander K. Johnston's *The Physical Atlas of Natural Phenomena*. In it, Forbes divided the world's oceans into 25 provinces located within a series of nine horizontal "homoizoic belts" and noted a series of five depth zones. The manuscript of Forbes's posthumous work, *The Natural History of European Seas,* was finished by his friend Robert Godwin-Austin and printed in 1859. In this book, Forbes asserted that, consistent with Buffon's law: (1) Each zoogeographic province is an area where there was a special manifestation of creative power, and the animals originally formed there were apt to become mixed with emigrants from other provinces; (2) each species was created only once, and individuals tended to migrate outward from their center of origin; and (3) provinces, to be understood, must be traced back, like species, to their origins in past time.

JAMES DWIGHT DANA (1813–1895) was a gifted American geologist, mineralogist, and naturalist who extended the work of Forbes and other early biogeographers who held that present distributions are somehow influenced by changes in sea levels. Dana provided fundamental insights in studies of mountain building, volcanic activity, sea life, and the origin of continents and ocean basins. Among his important works were *A System of Mineralogy* (1837) and *Manual of Mineralogy* (1848). When a young man, Dana joined the United States Exploring Expedition to the South Seas. He served four years (1838–42) as a geologist, but also undertook much of the zoological work. As the result of his discoveries on the distributions of corals and crustaceans, he was able to divide the surface waters of the world into several different zones based on temperature, and he used *isocrymes* (lines of mean minimum temperature) to separate them. He made the prophetic observation that: "The cause which limits the distribution of species northward or southward from the equator is the cold of winter rather than the heat of summer or even the mean temperature of the year" (1853; paper 7). Dana's observations would eventually be incorporated into twentieth-century explanations by Dobzhansky, MacArthur, and others attempting to explain the distributional limits and latitudinal diversity gradients of plants and animals (see part 8 of this volume).

SIR JOSEPH DALTON HOOKER (1817–1911) was an English naturalist noted for his botanical studies and for his friendship and encouragement of Charles Darwin when the latter was developing his theory of evolution. He was the younger son of William Jackson Hooker and, succeeding his father, became director of the Royal Botanical Gardens, Kew (from 1865 to 1885). He undertook many travels, initially as surgeon-botanist aboard *H.M.S. Erebus* on the Antarctic expedition of 1839–43. He discovered many species new to science and produced a stream of publications on plant geography. In his introductory essay to *The Botany of the Antarctic Voyage of H.M.S. Discover Ships "Erebus" and "Terror" in the Years 1839–1843* (paper 8), Hooker postulated the former existence of a great southern land mass connecting the South Pole with Australasia and, at a different time, a connection with Chile and Tierra del Fuego. In 1861, he published a paper, *Outline of the Distributions of Artic Plants*, which hypothesized the elevation of a great continent between eastern North America and northern Europe. In both works, Hooker challenged Darwin's explanation for disjunctions and other biogeographic "anomalies," namely dispersal; and instead proposed what modern biogeographers will recognize as a vicariance hypothesis for the origins of isolated floras. Hooker's views on land bridges were shared by the famous geologist Charles Lyell and by Edward Forbes, but Hooker never could convince Charles Darwin, who remained an advocate of long-distance dispersal by fortuitous means. Regardless of this, Hooker's theories were insightful and are recognized as being a precursor to modern vicariance biogeography.

PHILIP LUTLEY SCLATER (1829–1913), also a colleague and compatriot of Charles Darwin,

was an eminent ornithologist who, in the late nineteenth century, had described more than 1,000 new species of birds. In 1858 he published a small but truly seminal paper: *On the General Geographical Distribution of the Members of the Class Aves,* which we have featured here (paper 9). Just as Forbes had done for marine provinces, and Dana had for distributions of corals and crustaceans of the world's surface waters, Sclater applied Buffon's law to classify the world's terrestrial regions. Based on his knowledge of the distribution of bird species, he divided the world into six biogeographic regions. For the early zoologists, Sclater's regions provided the element missing from nineteenth-century biogeography—the idea that the interrelationships of areas can be defined by their groups of endemic species. The regions also called attention to those parts of the world that have had long and independent evolutionary histories (Ross 1974). Sclater acknowledged that his recognition of the six biogeographic regions was rudimentary and that much work was left to be done. Later, Alfred Russel Wallace (1876) included mammals and other animals in a much more comprehensive and sophisticated scheme of regions, with a detailed, hierarchical system of subregions and specific demarcations of geographic boundaries. However, Sclater's general arrangement of the principal biogeographic regions has stood the test of time, and many museum collections and textbook accounts remain faithful to his original scheme.

ASA GRAY (1810–1888) was an American botanist whose extensive studies did more than the work of any other scientist to unify the taxonomy of North American flora. His most widely used book, *Manual of the Botany of the Northern United States, from New England to Wisconsin and South to Ohio and Pennsylvania inclusive* (1848), commonly called *Gray's Manual,* remained a standard work through several editions. Gray's passion for biogeography grew when, at the age of 24, he joined the United States Exploring Expedition 1834–37. As a result, he became interested in the flora of Japan and East Asia. Gray and Charles Darwin maintained a frequent correspondence and

Gray's 1856 paper on plant distribution, "Statistics of the Flora of the Northern United States" was written partly in response to a request by Darwin. Darwin also inquired about the state of floral relationships between eastern Asia and North America, and Gray, thus encouraged, began to work on an extensive collection from Japan that had been made by Charles Wright. In 1859, Gray reported on that collection, and in so doing, noted the strong relationship to eastern North America and suggested that an interchange between the New and Old Worlds had taken place via Asia. Gray's hypothesis gave support to Darwin's theory of evolution, but refuted the theories of Louis Agassiz and others who believed in the fixity of species. The tenor of Gray's *Darwiniana: Essays and Reviews Pertaining to Darwinism* (1876), excerpted here (paper 10), provides an illuminating perspective on the great debates between the nineteenth-century Darwinians and creationists during this major paradigm shift in biogeography and evolutionary biology.

CHARLES DARWIN (1809–1882) was, of course, a great student of the early biogeographers, naturalists, and geologists. The observations of Buffon, Candolle, von Humboldt, Forbes, Dana, Gray, and especially Lyell no doubt influenced his development as a scientist. His most significant contribution to science was his theory of evolution by natural selection, which was published in 1858 in the *Journal of the Linnaean Society of London*, simultaneously with ALFRED RUSSEL WALLACE's (1823–1913) theory. Darwin's *Origin of Species by Means of Natural Selection* (1859) was a much more comprehensive treatise on the origin, spread, and diversification of species that quickly changed the thinking of the civilized world. When the young Darwin visited the Galapagos Islands in 1835, he was struck by the distinctiveness, yet basic similarity, of the fauna to that of mainland South America. Conversely, when Wallace traveled through the Indo Australian Archipelago, he was puzzled by the contrasting character of the island faunas, some with Australian relationships and others with South-East Asian affinities. After considerable thought about such matters, (many years on Darwin's part) each man arrived at a theoretical mechanism *(natural selection)* to account for the observed changes. The key for both Darwin and Wallace was the realization that distribution patterns had considerable evolutionary significance.

In Darwin's 1859 opus he dedicated two chapters to geographical distributions, which we have excerpted from here (paper 11), as they formed an integral part of his arguments for natural selection, and contributed directly to the field of biogeography. He explained his theory in such a way that he could account for all kinds of distributions. The finding of similar species in the British Isles and in Europe, which until relatively recent time had formed a continuous land surface, was easy to comprehend. At the same time, Buffon's earlier observations of the differences in the animals of Africa and South America, despite similar habitats, was understandable because the two continents had been separated for a long time. Thus, on the age of taxa, Darwin had a different opinion than that of Hooker in that plant and animal groups may be younger than the places they inhabit. Darwin thought that millions of chance dispersals over geologic time, especially to islands newly formed from the sea bed, would be sufficient to stock them with organisms for future evolution. Summarizing, Darwin made three points important to biogeography: (1) he emphasized that barriers to migration allowed time for the slow process of modification through natural selection; (2) he considered the concept of single centers of creation to be critical, that is, each species was first produced in one area only, and from that center it would proceed as far as its ability would permit; and (3) he noted that dispersal was a phenomenon of overall importance.

While Darwin went on to investigate many other aspects of evolutionary change, Wallace applied himself primarily to biogeography. During his travels in the Indo-Australian region, Wallace became particularly concerned about the location of the dividing line between the Australian and Oriental faunas. By 1863, he had decided that the line should run from east of the Philippines, south between Borneo and Celebes, and then between Bali and Lombok. Later to be known as "Wallace's Line," it was illustrated in his 1876 work and in his later book, *Island Life* (1880). In his 1910 book, *The World of Life*, Wallace changed his mind about the affiliation of Celebes (Sulawesi). It turned out that, for many groups of organisms, there is not a sharp demarcation but a zone of overlap between Australia-New Guinea and the large islands of Borneo and Java. This zone was eventually given the appropriate name of "Wallacea." Beyond his attention to this region of the world, Wallace contributed most broadly to the emerging field of biogeography with the publication of his monumental, two-volume work *The Geographical Distribution of Animals* (1876), which we have excerpted portions of here (paper 12). In this work, he reached a number of conclusions about biogeography that are still widely attested to today (and which are enumerated in part 4 of this volume).

While the origins of modern ecology may be traced to de Candolle's earlier writings, ERNST HEINRICH HAECKEL (1834–1919) is credited with coining the term *ecology* (or *oikos*), developing some of its most fundamental concepts, and urging the recognition of biogeography (which he called "chorology") as a distinguished field. He also proposed the principle which holds that ontogeny recapitulates phylogeny. Since Buffon's time, the central challenge of historical biogeography was to explain the origins, spread, and diversification of life. Haeckel's most significant contribution to this discipline was his curious work, *The History of Creation, or the Development of the Earth and its Inhabitants by the Action of Natural Causes*, which was published in 1876, and which we have featured here (paper 13). He placed the location of Linnaeus's Paradise in Lemuria, a continent (now sunk beneath the sea) in the center of the Indian Ocean. In a later edition of his book in 1907, he relocated Paradise to the north and east of India. The Lemuria concept strongly influenced subsequent writers and was resur-

rected by Hermann von Ihering as recently as 1927. By the time Haeckel wrote *The History of Creation*, it was clear to him that the theory of evolution must be incorporated into any explanation for the origin and distribution of nature; "the actual value and invincible strength of the Theory of Descent [is] that it explains *all* biological phenomena, that it makes *all* botanical and zoological series of phenomena intelligible in their relations to one another."

In a 1900 article in the journal *Science* reprinted here (paper 14), HERMANN VON IHERING (1850–1930) proposed a creative theory on the history of South America. He suggested that the continent formerly was divided in two parts (Archiplata and Archiamazonia) that were separated during most of the Tertiary. Archiplata was supposedly connected to Antarctica during the Cretaceous to the Eocene. Archiamazonia was supposed to have been connected to Africa through an ancient continent called Archiatlantica. Von Ihering noted that knowledge of the freshwater fauna was necessary in order to understand the zoogeography of the region. Moreover, von Ihering (1900) argued that historical reconstructions of the development and spread of biotas should be based on the zoogeography of ancient life forms.

In a later, more comprehensive work, *Die Geschichte des Atlantischen Ozeans* (1927), von Ihering used colored maps to depict the configurations of land and sea during the Cretaceous and Eocene. These show the Atlantic and Indian Oceans to be occupied by large continents, Archatlantis to the north, Archhelenis to the south, and Lemuria to the east. Rather than suggesting continents actually drifted across the globe, von Ihering attributed these changes in land and sea to vertical movements of ocean basins and sea levels, just as Lyell and Hooker had asserted during the nineteenth century. Although some of von Ihering's fossil data indicated useful relationships, biogeographers would eventually conclude that they resulted from continental movement rather than the rise and fall of ancient, trans-oceanic land bridges.

CLINTON HART MERRIAM (1885–1942) was a naturalist with broad interests in botany, mammalogy, ornithology, and anthropology. He conducted field surveys over many parts of the United States in which he interpreted historic patterns using ecological explanations. This lack of distinction between pattern and process was a common occurrence in his time. Although he published a myriad of papers, his most influential work for biogeographers was "Results of a Biological Survey of the San Francisco Mountain Region and the Desert of the Little Colorado, Arizona" (1890), featured here (paper 15). In that paper, he depicted a series of life zones based on the relationship between climate and vegetation, showing that elevational zonation of vegetation was, like latitudinal zonation, a response of species and communities to environmental gradients. In one sense, Merriam was following the lead of Forster and von Humboldt, but he had gathered more data and was able to analyze it to a much greater degree. In 1898, Merriam published a practical volume, *Life Zones and Crop Zones of the United States*. His discussions and detailed maps of "life zones" was antecedent to later syntheses on the distributions of biomes and, most recently, the ecoregions approach of R. G. Bailey (1996).

Also building on von Humboldt's earlier classic, along with those of Buffon and other earlier biogeographers, was WILLIAM DILLER MATTHEW (1871–1930)—a geologist and paleontologist who published an influential article *Climate and Evolution* (1915; paper 16). He was a dominant figure on the New York scene and was a great promoter of the idea that fossils, especially fossil mammals, provided a key to the past. His 1915 work emphasized the importance of centers of origin that were demonstrated by the discovery of fossil ancestors of modern forms. As Buffon had done a century earlier, Matthew advocated origins in the northern hemisphere (the "Holarctic Region") followed by gradual southern dispersals. Although his work was questioned by later biogeographers (e.g., Croizat 1962), the most enduring contribution turned out to be his

statement of evidence for the existence of centers of origin. Matthew said, "At any given period, the most advanced and progressive species of the race will be those inhabiting that [central] region; the most primitive and unprogressive species will be those most remote from this center. The remoteness is, of course, not a matter of geographic distance but of inaccessibility to invasion, conditioned by the habitat and facilities for migration and dispersal" (1915: 32).

Throughout the early development of the field, biogeographers continued to question the fixity of the continents and ocean basins. Between the years 1910 and 1912, Alfred Wegener (1880–1930), as well as Frederick B. Taylor and H. D. Baker, published what were then viewed as very radical ideas on continental drift (see the chapters from Wegener included in part 2). It was Wegener who showed the most tenacity in gathering evidence to defend his theory. Between 1915 and 1936, Wegener published six editions of his book *Die Entstehung der Kontinente und Ozeane*. English translations *(The Origins of Continents and Oceans)* were published in 1924 and 1936. These works created considerable controversy at a time when most geologists and geophysicists, especially in the United States, remained unconvinced. By the 1960s, research into paleomagnetism and marine geology began to offer overwhelming evidence for continental drift (plate tectonics). It had taken many years and well beyond his lifetime for Wegener's view to become accepted.

Despite the delayed acceptance of Wegener's theory, biogeographer's continued to advance the field during this period. In 1935, SVEN EKMAN (1876–1964) completed the huge task of analyzing all of the pertinent literature on marine animal distribution and published his results in a book entitled *Tiergeographie des Meeres*. This was the first comprehensive work on the worldwide distribution of marine animals. In 1953, Ekman published a revised edition in English called *Zoogeography of the Sea*, which is excerpted in this volume (paper 17). Although this book was published more recently than other works featured in this chapter, it nevertheless served as a classic volume for the emerging discipline of marine biogeography. In this book Ekman recognized a large number of faunal regions, characterized them in terms of their endemic species and genera, and discussed their interrelationships. The volume became the standard reference on the subject for more than twenty years and it is still useful to most students of marine biogeography.

In 1943, a significant book by Evgenii Vladimirovitch Wulff (1885–1941), *An Introduction to Historical Plant Geography*, was published in translation following the original Russian edition of 1932 (see paper 18). In this work, which has been described as perhaps the greatest work on botanical geography of the post-1940 period (Stott 1981), Wulff pointed out that some plant distribution patterns substantiated Wegener's theory of continental drift. Although he was attempting to locate major centers of origin for plants, he also recognized that present-day continents had formerly been joined, and he used the concept of vicariance to describe major disjunctions among the higher groups similar to those reported by Hooker and his colleagues during the nineteenth century. The key distinction here was that, whereas Hooker, Lyell and other "extensionists" proposed repeated emergence and submergence of oceanic-scale land bridges to explain disjunct distributions, Wulff correctly attributed these patterns to continental drift.

These early classics, as discussed here in a close-to-chronological sequence, include some of the most salient advances in the historical development of biogeography and the closely related fields of evolutionary biology and ecology. Naturalists were at first occupied with the problems of accommodation aboard the Ark and the means by which animals were able to disperse to various parts of the world following the Deluge. The Ark concept gave way to the Paradisiacal Mountain, which in turn yielded to the idea of creation in many different places. At the same time, the Linnaean axiom of the fixity of species through time was replaced by one of change under environmental influences. Then, naturalists began to study the associations of plants and animals in various parts of the world and, in so doing, began to appreciate the contrasts among the different regions.

The next major advance in biogeography took place when biologists began to analyze the distribution patterns of natural groups and found that such patterns could be used to predict past changes in the earth's surface, resulting in theories of land bridges and rising continents. A considerable advance occurred when Darwin and Wallace discovered the principle of natural selection from biogeographical evidence. Since that time, it has become apparent that almost every distribution pattern has evolutionary significance. Additional progress took place with the subsequent publication of important works by succeeding authors. Among them, Wegener was perhaps the most prescient, for he pointed the way toward the modern concept of plate tectonics. Together, this wealth of theories and insights from the early biogeographers formed the foundation for subsequent studies of the origin, spread, and diversification of life.

PAPER 1

From *Dissertation II. On the Increase of the Habitable Earth*
Carolus Linnaeus

Three years ago being at Fulleron, an estate of the senator Cronstedt, I saw a fragment of rock left by the water upon the shore where the inhabitants imbark: — Everybody can give evidence to this fact, and that such a stone had never been seen in that place before. If the Lake Melerus have force enough to move stones of a weight that many yokes of oxen are unable to draw, what must we expect from the ocean? — In the Dalic rocks, where Palmfjdllet borders upon the Lake Grusvelsjon, the eastern side of the mountain Wolen is annulated and worn away; a manifest indication of the effect of the waves once washing against it.

The sections of all rivers are wide at the surface, narrower at the bottom, and they every year excavate their beds deeper: the Simois and Xanthus which watered the meadows of Troy, so celebrated by the poets, are said by Bellonius now to be so diminished as not to be able to nourish the smallest fishes; they are quite dry in summer, and in winter "have scarce water enough to swim a goose." — At the Salmon leap at Luloa, in Lapmark, the height of the river has been every year constantly marked, and it appears annually decreasing.

From all these arguments I think we shall be justified in drawing the following conclusion, that the dry land is in a state of augmentation; that formerly it was much less than it now is, and at first a single, small island* in which, as in a compendium, all things were collected together which our gracious Creator had defined to the use of man.

We must now point out how it might have been effected, *"that vegetables in a small tract of land might find their proper soil, and every animal its proper climate."*

First let us conceive Paradise situated under the Equator; and nothing further is requisite to demonstrate the possibility of these two indispensable conditions, than supposing a very lofty mountain to have adorned its beautiful plains.

The higher a mountain rises into the middle regions of the air so much the more intense is the cold it is exposed to. The mountain of Ararat in Armenia preserves an eternal snow upon its summit, as well as the rocks of Lapland in the Artic circle: the same cold reigns on both; and on the tops and sides of such a mountain the same vegetables might grow, the same animals live, as in Lapland and the frigid zone; and in effect we find in the Pyrenean, Swiss, and Scotch mountains, upon Olympus, Lebanon, and Ida, the same plants which cover the Alps of Greenland and Lapland.

Tournefort, in his Journey to the East, makes one observation which deserves to be remembered on the present occasion: "I found," says he, "at the foot of Mount Ararat those plants which were common in Armenia, — a little further those which I had before seen in Italy; when I had ascended somewhat higher such vegetables as were common about Paris; the plants of Sweden possessed a more elevated region; but the highest tracts of the mountain, next the very summit, was occupied by the natives of the Swiss and the Lapland Alps." — By the plants growing upon the Alps of Dalecarlia, I was able to calculate how much lower these were than those of Lapland; for in Lapland I had accurately observed the heights at which I found every vegetable.

From the works of Cæsalpinus it is evident, that he reckoned all those plants that are common in the fields of Sweden under the title of Alpine plants, as he had found them growing in the mountains of Tuscany, which, however, are not Alps.

Hence we are authorized to conclude, that the height and elevation of the land exposes it to cold;

From Carl von Linné, *Select Dissertations from the Amoenitates Academicae*, trans. F. J. Brand (London: G. Robinson and J. Robson, 1781; repr., New York: Arno, 1977), 88–94, 113–15, 126–27.
  *See Note (D) at the end of this Essay.

and that the coldest winter and perpetual snow might be found under the line, upon a mountain of such a height that its top ascended above the region of the clouds. [. . .]

Thus you have heard with how exquisitely wise and attentive a care the great Artificer of Nature has provided that every seed shall find its proper soil, and be equally dispersed over the surface of the globe.

We have seen the Winds, the Rains, the Rivers, the Sea, Heat, Animals, Birds, the Structure of Seeds, and Seed Vessels, the peculiar Natures of Plants, and even Ourselves, contribute to this great work. — I have shewn, that any one single plant alone would have been able to have covered the face of the globe: — I have demonstrated that the dry land has always been increasing, and dilating itself; and therefore once was infinitely less than it is at this present: — I have traced back the orders of animals and vegetables, and found them terminate in individuals created by the hand of God.

If we add to this the proportion which subsists between carnivorous animals, and those subsisting upon vegetables, among birds, fishes, and insects, and between the Animal and Vegetable Kingdom; if besides we strengthen these inductions by the analogy of our Conclusion with the history of the Deluge, no one I believe will be able to say, that I have asserted without foundation, that one individual of every species of plant, and one sexual pair of every species of animal, was created at the beginning: — Thus the Garden of Paradise is rendered the most beautiful imagination can conceive, and the infinite glory of the Creator exalted, not depressed. Let those of this Assembly to whom the Parent of Nature has given a more happy genius, who possess an erudition more cultivated, and are more acute in the discovery of whatever the rules of genuine demonstration lead us to, let those with superior accuracy carry these enquiries to their ultimate end.—
Dixi.

(D) Under the Equator no snow can lay upon the mountains unless at very great heights above the level of the sea: the top of the mountain in this Paradisiacal island must be as much higher than the Pike of Teneriffe, as the Alps are above the level of the sea. And no such mountain is known to geographers. — But admitting such a one to be discovered, the difficulty is not removed; for let us suppose the seeds of its Alpine plants to be lodged upon the tops of other rocks as soon as they appeared above the water, being carried by the sea winds, or any other cause; they must remain lodged there, until by the subsiding of the ocean, the rock becomes so much elevated above its bed as to be covered with perpetual snow; as the surface of the sea falls (according to this theory) less than five feet in a century, such seeds must lay there with their vegetable virtues unimpaired at least a thousand ages before they spring. As the place where they are deposited becomes perpetually more and more elevated it is continually growing colder, their vegetable juices therefore must be supposed not to be put in motion or destroyed by the vicissitudes of moisture and superior heat for that whole period, and only beginning to act in the regions of a polar climate at the end of that time. — Many parts of the world, as the Cape of Good Hope, produce numerous tribes of very singular plants; how comes it that they are found in the intermediate tracts between the Cape and the Paradisiacal Mountains? in traveling along the continent they must have rooted, and produced seed from distance to distance. Many land animals also inhabit cold mountainous regions, separated from each other, which, whether they be of the same, or a different species, furnish very strong objections against this Hypothesis.

PAPER 2

Natural History, General and Particular
Georges-Louis Leclerc, Compte de Buffon

[. . .] The seal, or sea-calf, seems to be confined to northern countries, and is found equally on the coasts of Europe and of North America.

These are nearly all the animals which are common to the Old and New Worlds; and from this number, which is not considerable, we ought, perhaps, to retrench more than a third part, whose species, though apparently the same, may be different in reality. But, admitting the identity of all these species with those of Europe, the number common to the two Continents is very small, when compared with that of the species peculiar to each. It is farther apparent, that, of all these animals, it is those only which frequent the northern countries that are common to both Continents; and that none of those which cannot multiply but in warm or temperate climates are found in both worlds.

It is, therefore, no longer a doubtful point, that the two Continents either are, or have formerly been, contiguous towards the north, and that the animals common to both have passed from the one to the other by lands with which we have now no acquaintance. We are led to believe, especially since the discoveries made by the Russians to the north of Kamtschatka, that the lands of Asia are contiguous to those of America; for the north of Europe seems to have been always separated from the New World by seas too considerable to permit the passage of any quadruped. These animals, however, of North America, are not precisely the same with those of the north of Asia; but have a stronger resemblance to the quadrupeds of the north of Europe. It is the same with the animals which belong to the temperate climates. The argali or Siberian goat, the sable, the Siberian moles and the Chinese musk, appear not in Hudson's Bay, nor in any other north-west part of the New Continent; but, on the contrary, we find in the north-east parts of it, not only the animals common to the north of Europe and Asia, but likewise those which appear to be peculiar to Europe, as the elk, the rain-deer, &c. It must, however, be acknowledged, that the north-east parts of Asia are so little known, that we can have no certainty whether the animals of the north of Europe exist there or not.

We formerly remarked, as a singular phenomenon, that the animals in the southern province of the New Continent, are small in proportion to those in the warm regions of the Old. There is no comparison between the size of the elephant, the rhinoceros, the hippopotamus, the camelopard, the camel, the lion, the tiger, &c and the tapir, the cabiai, the ant-eater, the lama, the puma, the jaguar, &c, which are the largest quadrupeds of the New World: the former are four, six, eight, and ten times larger than the latter. Another observation brings additional strength to this general fact: All the animals which have been transported from Europe to America, as the horse, the ass, the ox, the sheep, the goat, the hog, the dog, &c. have become smaller; and those which were not transported, but went thither spontaneously, those, in a word, which are common to both Continents, as the wolf, the fox, the stag, the roebuck, the elk, &c. are also considerably less than those of Europe.

In this New World, therefore, there is some combination of elements and other physical causes, something that opposes the amplification of animated Nature: There are obstacles to the development, and perhaps to the formation of large germs. Even those which, from the kindly influences of another climate, have acquired their complete form and expansion, shrink and diminish under a niggardly sky and an unprolific land, thinly peopled with wandering savages, who, instead of using this territory as a master, had no property or empire; and, having subjected neither the animals nor the elements, nor conquered the seas, nor directed the motions of rivers, nor cultivated the earth, held only the first rank among animated beings, and existed as crea-

Excerpts are transcribed from pages 127–29, 149–52, and 450–52 of the 1791 translation by
W. Smellie of Buffon's *Histoire Naturelle, Générale et Particulière* (1761).

tures of no consideration in Nature, a kind of weak automations, incapable of improving or seconding her intentions. She treated them rather like a step-mother than a parent, [. . .]

I have said enough to guard the reader against errors both of general and particular kind, which are no where so numerous as in the works of nomenclators; because, being solicitous to comprehend every thing within the limits of their systems, they are obliged to associate all that they are ignorant of with the little that they know.

From what has been advanced, the following general conclusions may be drawn: That man is the only animated being on whom Nature has bestowed sufficient strength, genius, and ductility, to enable him to subsist and to multiply in every climate of the earth. No other animal, it is evident, has obtained this great privilege; for, instead of multiplying every where, most of them are limited to certain climates, and even to particular countries. Man is totally a production of heaven: But, the animals, in many respects, are creatures of the earth only. Those of one Continent are not found in another; or, if there are a few exceptions, the animals are so changed and contracted, that they are hardly to be recognised. Is any farther argument necessary to convince us, that the model of their form is not unalterable; that their nature, less fixed than that of man, may be varied, and even absolutely changed in a succession of ages; that, for the same reason, the least perfect, the least active, and the worst defended, as well as the most delicate and heavy species, have already, or will soon disappear; for their very existence depends on the form which man gives or allows to the surface of the earth?

The prodigious *mammouth*, whose enormous bones I have often viewed with astonishment and which were, at least, six times larger than those of the largest elephant, has now no existence; yet the remains of him have been found in many places remove from each other, as in Ireland, Siberia, Louisiana, &c. This species was unquestionably the largest and strongest of all quadrupeds; and, since it has disappeared, how many smaller, weaker, and less remarkable species must likewise have perished, without leaving any evidence of their past existence? How many others have undergone such changes, either from degeneration or improvement, occasioned by the great vicissitudes of the earth and waters, the neglect or cultivation of Nature, the continued influence of favourable or hostile climates, that

they are now no longer the same creatures? Yet the quadrupeds, next to man, are beings whose nature and form are the most permanent. Birds and fishes are subject to greater variations: The insect-tribes are liable to still greater vicissitudes: And, if we descent to vegetables, which ought not to be excluded from animated Nature, our wonder will be excited by the quickness and facility with which they assume new forms.

Hence, it is not impossible, that, without inverting the order of Nature, all the animals of the New World were originally the same with those of the Old, from whom they derived their existence; but that, being afterwards separated by immense seas, or impassable lands, they would, in the progress of time, suffer all the effects of a climate that had become new to them, and must have had its qualities changed by the very causes which produced the separation, and, consequently, degenerate, &c. But these circumstances should not prevent them from being now regarded as different species of animals. From whatever cause these changes, produced by the operation of time and the influence of climate, have originated, and though we should date them from the creation itself, they are not the less real. Nature, I allow, is in a perpetual state of fluctuation: But it is enough for man to seize her in his own age, and to look backward and forward, in order to discover her former condition, and what future appearances she may probably assume.

With regard to the utility of this mode of comparing animals, it is evident, that, independent of ascertaining names, of which some examples have been given, it extends our knowledge of the animal creation, and renders it more certain, and perfect; that it prevents us from ascribing, to American animals, properties which are peculiar to those of the East Indies, only because they have the same name; that, in examining the notices of foreign animals communicated by travellers, it will enable us to distinguish names and facts, and to refer each to its proper species; and, lastly, that it will render the history which I am now composing less defective, and perhaps more conspicuous and complete that they can be compared to none of them, and that it is impossible to refer them to any common origin, or to ascribe to the effects of degeneration the prodigious differences in their nature from that of any other animal.

Thus, of ten genera and four detached species, to which we have endeavoured to reduce all the animals peculiar to the New World, there are only two, namely, the genus of jaguars, ocelots, &c. and the species of the pecari, with their varieties, which can, with any degree of probability, be referred to the animals of the Old World. The jaguars and ocelots may be regarded as a species of the leopard or panther, and the pecari as a species of hog. There are also five genera and one detached species, namely, the species of the lama, the genera of sapajous, sagoins, moussettes, agoutis, and ant-eaters, which may be compared, though in an equivocal and very distant manner, with the camel, the monkeys, the pole-cat, the hare, and the scaly lizards: And, in line, there remain four genera and two detached species, namely, the opossums, the coaitis, the armadillos, the sloths, the tapir, and cabiai, which can neither be referred nor compared to any genera or species in the Old Continent. This seems to be a sufficient proof, that the origin of these animals peculiar to the New World cannot be attributed to degeneration alone: However powerful we may suppose the effects of degeneration, we can never suppose, with any appearance of reason, that these animals were originally the same with those of the Old Continent. It is more probable, that the two Continents were formerly united, and that the species which inhabited the New World, because the soil and climate were most agreeable to their nature, were separated from the others by the irruption of the waters, when they divided Africa from America. This is a natural cause; and similar causes might be conceived which would produce the same effect. For example, if the sea should make an irruption into Asia from east to west, and separate the southern regions of Africa and Asia from the rest of the Continent, all the animals peculiar to these countries, as the elephant, the rhinoceros, the giraffe, the zebra, the orang-outang, &c. would be in the same situation with those of South America. They would be entirely separated from those of the temperate regions, and could not be referred to an origin common to any of the species or genera which inhabit these countries, solely because some imperfect resemblances might be discovered between them.

Hence, to discover the origin of these animals, we must have recourse to the period when the two Continents were united, and tract the first changes which have happened on the surface of the earth. We must, at the same time, consider the two hundred species of quadrupeds as constituting thirty-eight families: And, though this is by no means the present state of nature, but, on the contrary, a state of much greater antiquity, which we can reach only by inductions and relations almost equally fugitive as time, that seems to have effaced their traces; we shall, however, endeavour to ascent, by facts and monuments still existing, to those first ages of Nature, and to exhibit those epochas which shall appear to be most clearly indicated.

PAPER 3

From *Observations Made during a Voyage Round the World*
Johann Reinhold Forster

CHAPTER V

Remarks on the Organic Bodies

OMNIS NATURA VULT ESSE CONSERVATRIX SUI, UT & IN GENERE CON-
SERVETUR SUO.
  M. Tullius Cicero *de Fin. Bon. & Mal.* l. 4[1]

The next article which demands our attention in the lands of the South Sea, is
the history of the organic bodies, which partly form, and partly dwell on their
immediate exterior surface. They constitute the vegetable and animal kingdoms
in the system of nature, the latter being distinguished from the first, by the powers
of perception, or the senses, the peculiar attributes of animal being.

ORGANIC BODIES

SECTION I

Vegetable Kingdom

The vegetation which cloaths our earth, VARIES considerably in every country we
met with during our circum-navigation, even as the appearances of the lands
themselves, are new and singular in almost every one of them. Between the
tropics, we met with the Low Islands, consisting of mere coral rocks, scarce
covered with sand. The Society Isles of vast height, surrounded by rich plains,
and included in coral reefs: and many other clusters of mountainous islands, destitute both of reefs and plains. We have observed how much the least attractive of
these tropical countries, surpasses the ruder scenery of New Zeeland: how much
more discouraging than this, are the extremities of America; and lastly, how
dreadful the southern coasts appear, which we discovered. In the same manner,
the plants that inhabit these lands, will be found to differ in number, stature,
beauty, and use.[2]

VEGETABLE
KINGDOM

113

Reprinted from *Observations Made during a Voyage Round the World*, ed. Nicholas Thomas,
Harriet Guest, and Michael Dettelbach (Honolulu: University of Hawai'i Press, 1966), 113,
119–23, 128, 135–36.

New Georgia

When we saw the barren side of Tierra del Fuego, we had scarce an idea of a more wretched country existing; but after standing sometime to the Eastward, we met with the isle of New-Georgia, which, though in the same latitude, appeared so much more dreadful, that before we came close up with it, it was suspected to be an island of ice. The shapes of its mountains are, perhaps, the most ragged and pointed on the globe; they are covered with loads of snow in the height of summer, almost to the water's edge; whilst here and there, the sun shining on points, which project into the sea, leaves them naked, and shews them craggy, black and disgustful. We landed in Possession-Bay, and found the whole Flora to consist of two species of plants, one a new plant* peculiar to the Southern hemisphere, the other a well-known grass; both which, by their starved appearance and low stature, denoted the wretchedness of the country.

However, as if nature meant to convince us of her power of producing something still more wretched, we found land about four degrees to the Southward of this, apparently higher than it, absolutely covered with ice and snow (some detached rocks excepted) and in all probability incapable of producing a single plant. Wrapt in almost continual fogs, we could only now and then have a sight of it, and that only of its lowest part, an immense volume of clouds constantly resting on the summits of the mountains, as though the sight of all its horrors would be too tremendous for mortal eyes to behold. The mind indeed, still shudders at the idea, and eagerly turns from so disgusting an object.

I. Number of Species

From what has been said, it appears, that the rigorous frost in the antarctic regions almost precludes the germination of plants; that the countries in the temperate zones, being chiefly uncultivated, produce a variety of plants, which only want the assistance of art to confine them within proper bounds; and lastly, that the tropical isles derive a luxuriance of vegetation from the advantage of climate and culture. But the number of vegetables is likewise commonly proportioned to the

---

* Ancistrum. Forster's *Nova Genera Plantarum*, p. 3, 4.

VEGETABLE
KINGDOM

extent of the country. Continents have therefore, at all times, been remarkable for their immense botanical treasures; and, among the rest, that of New-Holland, so lately examined by Messrs. BANKS and SOLANDER, rewarded their labors so plentifully, that one of its harbors obtained a name suitable to this circumstance, (Botany Bay). Islands only produce a greater or less number of species, as their circumference is more or less extensive. In this point of view, I think both New-Zeeland and the tropical isles rich in vegetable productions. It would be difficult to determine the number in the first with any degree of precision, from the little opportunities we had of examining its riches: our acquisitions of new species from thence amount to 120 and upwards; the known ones, recorded already in the works of Linnaeus, are only six, and consequently bear a trifling proportion to the new ones; but there is great reason to suppose that, including both the isles of New-Zeeland, a Flora of no less than 400 or 500 species, on a careful scrutiny, might be collected together; especially if the botanist should come at a more advanced season than the beginning of spring, or not so late as the beginning of winter; at which times we had the only opportunities of visiting this country.

In the tropical isles, the proportion of new and known species is very different. All our acquisitions of new ones from them amount to about 220 species; and the collection of the known or Linnaean, to 110, which gives the whole number 330; and shews, that one third were well known before. Cultivation contributes not a little towards this, because it probably contains such plants, as the first settlers of these isles brought with them from their original East-Indian seats, which of course are most likely to be known; and, with these cultivated ones, it is to be supposed there might come the seeds of many wild ones, also of East-Indian growth, and consequently known to the botanists. The new plants, therefore, can only be those which originally grew, peculiar to these countries, and such as have escaped the vigilance of the Europeans in India.

The number of individual species (330) which we found in the tropical isles, (old and new) is by no means to be considered as a perfect Flora, for which purpose, our opportunities of botanizing were greatly insufficient. On the contrary, I am rather inclined to think, that our number might almost be doubled on a more accurate search, which must be the work of years, not of a few days, as was the case with us. The greatest expectations are from the New-Hebrides, as they are large, uncultivated, but very fertile islands. The jealous disposition of their natives would not permit us to make many discoveries there; yet, from the outskirts of the country, we might form a judgment of the interior parts. As an instance, that we often have had indications of new plants, though we could never meet with the plants themselves, I shall only mention the wild nutmeg of the isle of Tanna, of which we obtained several fruits, without ever being able to find the tree. The first we met with was in the craw of a pigeon, which we had shot, (of that sort, which, according to Rumphius,[7] disseminates the true nutmegs in the East-India islands): it was still surrounded by a membrane of bright red, which was its mace; its color was the same as that of the true nutmeg, but its shape more oblong; its taste was strongly aromatic and pungent, but it had no smell. The natives afterwards brought us some of them. Quiros must have meant this wild nut, when he enumerates nutmegs among the products of his Tierra del Espiritù Santo.[8] This circumstance gives a strong proof (with many more of another nature) of the veracity of this famous navigator; and, as he likewise mentions silver, ebony, pepper, and cinnamon among the productions of Tierra del Espiritù Santo, and the isles in the neighbourhood, I am inclined to believe, that they are really to be met with there.

Another material obstacle to our compleating the Flora of the South-Seas, and which indeed is connected with the former, arises from the changes of seasons: for though, between the tropics, they be not strongly marked with the alternatives of heat and cold, yet, according to the approach or recess of the sun, vegetation is more or less active. This we experienced, by touching at some of the isles, two different times, after an interval of seven months. The first was in August (1773) or the height of the dry season; when we found every thing wearing a yellowish or exhausted colour; many trees had shed their leaves, and few plants were in flower. The second time, being in April (1774), soon after the rainy, or at the beginning of the dry season, we were surprized beyond measure by the lively hues which now appeared in those very objects, that had seemed as it were dead at our first visit: we found many plants which we have never seen before; observed many others in flower, and every thing covered with a thick foliage of a fresh and vivid green: and from this circumstance, and the longer time we spent at the Society Isles, our collections from thence are the most perfect. It is true, the difference of dry and rainy seasons is not so strongly marked as on continents, or in isles contiguous to them; especially as fruits of all kinds chiefly ripen during the wet months, which would be impossible, were the rains constant; and secondly, since even the dry months are not wholly exempted from showers: but the relative distinction holds notwithstanding, as the proportion of rain in one, is considerably greater than in another.

It is owing to the exceeding small size of the low isles, that their vegetable productions are so inconsiderable; though I must confess, we never landed on any one without meeting with something new. SAVAGE-ISLE, which is in fact no more than a low island, raised several feet above water, and clearly manifests its origin, by the bare coral rocks of which it consists, has some new plants, which, in the out-skirts of the isle, grew in the cavities of the coral without any the least soil. We might have made several acquisitions on this island, but the savage disposition of the natives forced us to abandon it.[9] As a contrast to the tropical isles, we

VEGETABLE
KINGDOM

ought to mention Easter-Island, which lies so little without the tropic, that is may well be classed with those isles which are actually included in it. This isle is either grossly misrepresented by the Dutch discoverers, or has since then been almost totally ruined.[10] Its wretched soil, loaded with innumerable stones, furnishes a Flora of only 20 species; among these, ten are cultivated; not one grows to a tree, and almost all are low, shrivelled and dry. In the opposite, or Westernmost part of the South-Sea, lies a small isle, which has obtained the name of Norfolk-Island: almost its whole vegetation corresponds with that of New-Zeeland, whose North end is not far distant from it; only some allowances must be made for the greater mildness of the climate, which gives every plant a greater luxuriance of growth. Peculiar to this isle, and to the Eastern end of Caledonia we found a species of coniferous tree, from the cones probably seeming to be a cypress: it grows here to a great size, and is very heavy but useful timber.

II. Stations

As the South-sea is bounded on one side by America, on the other by Asia, the plants, which grow in its isles, partly resemble those of the two continents; and the nearer they are either to the one or the other, the more the vegetation partakes of it. Thus the Easternmost isles contain a greater number of American, than of Indian plants; and again, as we advance farther to the West, the resemblance with India becomes more strongly discernible. There are, however, singular exceptions to this general rule: thus, for instance, we find the *gardenia* and *morus papyrifera*, both East-Indian plants, only in the Easterly groupes of the Friendly and Society Isles; the Tacca of Rumph, which is likewise an Indian species, is only found in the Society Isles.[11] On the other hand, some American species do not appear till we reach the Western Isles, called the Hebrides, which are however the farthest removed from that continent. Part of these exceptions are perhaps owing to the inhabitants, who, being of a more civilized nature in the Easterly isles, have brought several parts with them from India, for cultivation, which the others have neglected. The same circumstance also, accounts for the arrival of the spontaneous Indian species in these Easternmost isles; they being probably, as I have already observed, brought among the seeds of the cultivated sorts. In confirmation of which, it may be alledged, that the Indian species are commonly found on the plains in the Society Isles, and the spontaneous American species on the mountains.

A few plants are common to all the climates of the South Sea; among these is chiefly the celery, and a species of scurvy grass (Arabis) both which are generally found in the low islands between the tropics, on the beaches of New Zeeland,

and on the burnt islands of Tierra del Fuego. Several other species seem to have obviated the differences in the climate by a higher or lower situation: a plant, for instance, which occupies the highest summits of the mountains at O-Taheitèe, (or any of the Society Isles) and grows only as a shrub, in New Zeeland is found in the valley, and forms a tree of considerable height; nay the difference is sensible in different parts of New Zeeland itself: thus a fine shrubby tree at Dusky Bay or the Southern extremity, which there grows in the lowest part of the country, dwindles to a small inconsiderable shrub at Queen Charlotte's Sound, or the Northern end, where it is only seen on the highest mountains. A similarity of situation and climate sometimes produces a similarity of vegetation, and this is the reason why the cold mountains of Tierra del Fuego produce several plants, which in Europe are the inhabitants of Lapland, the Pyrenees, and the Alps.

III. Variety

The difference of soil and climate, causes more varieties in the tropical plants of the Southern isles, than in any other. Nothing is more common in the tropical isles, than two, three, four, or more varieties of the same plant, of which, the extremes sometimes, might have formed new species, if we had not known the intermediate ones, which connected them, and plainly shewed the gradation. In all these circumstances, I have always found that the parts most subject to variation, were the leaves, hairs, and number of flower stalks, (pedunculi) and that the shape and whole contents of the flower (partes fructificationis) were always the most constant. This however, like all other rules, is not without exceptions, and varieties arising from soil sometimes cause differences even there, but they are too slight to be noticed. A cold climate, or a high exposure shrinks a tree into a shrub, and vice versa. A sandy or rocky ground produces succulent leaves, and gives them to plants, which, in a rich soil have them thin and flaccid. A plant which is perfectly hairy in a dry soil, loses all its roughness, when it is found in a moister situation: and this frequently causes the difference between varieties of the same species in the Friendly Isles, and in the hills of the Society Isles: for the former, not being very high, are less moist than the hills of the latter, which are frequently covered with mists and clouds.

IV. Cultivation

That cultivation causes great varieties in plants, has been observed long since, and can no where be better seen than in the tropical South Sea isles, where the

SECTION II

Animal Kingdom

ANIMAL KINGDOM

The countries of the South Sea, and the Southern coasts, contain a considerable variety of animals, though they are confined to a few classes only. We have seen by what degrees nature descended from the gay enamel[1] of the plains of the Society Isles, to the horrid barrenness of the Southern SANDWICH LAND. In the same manner the animal world, from being beautiful, rich, enchanting, between the tropics; falls into deformity, poverty, and disgustfulness in the Southern coasts. We cannot help being in raptures, when we tread the paths of O-Taheitean groves, which at each step strike us with the most simple, and at the same time the most beautiful prospects of rural life; presenting scenes of happiness and affluence to our eyes, among a people, which, from our narrow prejudices we are too readily accustomed to call savage. Herds of swine are seen on every side; by every hut dogs lie stretched out at their ease, and the cock with his seraglio, struts about, displaying his gay plumage, or perches on the fruit trees to rest. An unintermitted chorus of small birds warbles on the branches all the day long, and from time to time, the pigeons cooe is heard with the same pleasure as in our woods. On the sea shore, the natives are employed in dragging the net, and taking a variety of beautiful fish, whose dying colours change every moment: or they pick some shells from the reefs, which, though well known to the naturalist, yet have a right to the philosopher's attention, who admires the wonderful elegance of nature alike, in her most common as in her rarest productions. To enhance the satisfaction we feel, this happy country is free from all noxious and troublesome insects; no wasps, nor mosquitoes, infest the inhabitants, as in other tropical countries; no beasts of prey, nor poisonous reptiles ever disturb their tranquility.*

Let us remove from hence to the temperate zone: what a falling off from the

---

* The common flies are, indeed, at some seasons troublesome, on account of their immense numbers, but they cannot be called *noxious* insects: the only disagreeable animal in O-Taheitee is the common black rat, which is very numerous there, and often does mischief by its voracity.

I. Number

The whole number of species in the greater classes of animals, *viz.* quadrupeds, cetacea, amphibia, birds, and fish, which we saw in the South-Sea, according to the above enumeration, amounts to between 260 and 270, of which about one third are well known. Let us allow, that this number comprehends two thirds of the animals of those classes, actually residing in the South-Sea, though we have reason to think, that the fauna is much more extensive, we shall have upwards of 400; and supposing the classes of insects and vermes to give only 150 species, the whole fauna of the South-Sea isles will consist of at least 550 species, a prodigious number indeed, when compared with that of the Flora.

II. Station

Though many of the birds in New-Zeeland are remarkable for the gay colors of their plumage; yet we found, when we came to Norfolk-Island, (which, as I have observed in my account of the plants, contains exactly the same species) that the same birds appeared there arrayed in far more vivid and burning tints, which must prove, that the climate has a considerable influence on colours. There is also a species of king-fisher common to all the South-Sea isles, of which the tropical varieties are much brighter than that of New-Zeeland. The plumage of birds is likewise adapted to the climate in another respect; for those of warm countries have a moderate covering, whilst those of the cold parts of the world, and such especially, as are continually skimming over the sea, have an immense quantity of feathers, each of which is double; and the pinguins, which almost constantly live in the water, have their short, oblong feathers lying as close above each other as the scales of fishes, being at the same time furnished with a thick coat of fat, by which they are enabled to resist the cold: the case is the same with the seals, the geese, and all other Southern aquatic animals. The land birds, both within and without the tropics, build their nests in trees, except only the common quail, which lives in New-Zeeland, and has all the manners of the European one. Of the water-fowl, some make their nests on the ground, such as the grallae, which breed only in pairs; whilst several species of shags, (*pelecani*) live gregarious in trees, and others in crevices of rocks; and some petrels (*procellariae*) by thousands together, burrow in holes under-ground close by each other, where they educate their young, and to which they retire every night. The most prolific species in the South Sea, are the ducks, which hatch several eggs at one brood, and though the shags, penguins, and petrels, do not hatch more than one or two, or at most three eggs at a time, yet by being never disturbed, and always keeping together in great flocks, they are become the most frequent and numerous. The most palatable spe-

ANIMAL KINGDOM   cies of fish are likewise the most prolific; but it must be observed, that there is no where such abundance of fish in the South-Sea, as at New-Zeeland, by which means they are become the principal nourishment of the natives, who have found that way of living to be attended with the least trouble, and consequently suited to that indolent disposition which they have in common with all barbarous nations.

III. Variety

It does not appear, that the individuals of the animal kingdom are so much subject to variety in the South-Seas, as those of the vegetable. Domestication, the great cause of degeneracy in so many of our animals, in the first place, is here confined to three species; the hog, dog, and cock: and secondly, it is in fact next to a state of nature in these isles: the hogs and the fowls run about at their ease the greatest part of the day; the last especially, which live entirely on what they pick up, without being regularly fed. The dog being here merely kept to be eaten, is not obliged to undergo the slavery, to which the varieties of that species are forced to submit in our polished countries; he lies at his ease all the day long, is fed at certain times, and nothing more is required of him: he is therefore not altered from his state of nature in the least; if probably inferior in all the sensitive faculties to any wild dog; (which may perhaps be owing to his food) and certainly, in no degree, partakes of the sagacity and quick perception of our refined variety. Among the wild birds, the varieties are very few: two species of pigeons, two of parrots, one of king-fishers, and one or two of fly-catchers, are the only I know of, that vary any thing in different isles; and it is much to be doubted, with regard to some of them, whether what we count varieties are not either distinct species, or only different sexes of one and the same; circumstances, which it is well known, require a long series of observations, not to be made on a cursory view. The varieties in other classes are still less considerable.

IV. Classification

The animals of the South-Seas, as we have already observed, are most of them new species. The known ones between the tropics, are chiefly such as are generally found all over the maritime parts of the torrid zone; those of the temperate zone being principally aquatic, are common to those latitudes in every sea; or consist of European species. Upon the whole, we found no more than two genera, which are distinct from those already known, and all the remaining species rank

PAPER 4

From *Essai Élémentaire de Géographie Botanique*
Augustin de Candolle

—◆—

CHAP. IV.

ON THE DISTRIBUTION OF PLANTS UPON THE EARTH.

Linné, Stationes plantarum, in Amœn. Acad. vol. iv.
Giraud Soulavie, Géographie physique de regne végétal.
F. Stromeyer, Historiæ vegetabilium geographicæ Specimen. ; Dissertatio.
G. R. Treviranus, Biologie, b. ii. s. 44, 137.
Humboldt et Bonpland, Essai sur la Geographie des Plantes.
Willdenow, im Magazin der Berlin, Gesellshaft naturforschender Freunde.
Wahlenberg, Flora Lapponica.
Dessen, Flora Carpathorum principalium.
Dessen, de Vegetatione et Climate Helvetiæ septentrionalis.
Brown's General Remarks, geographical and systematical, on the Botany of Terra australis.
Brown's Observations, systematical and geographical, on the Herbarium collected by Professor Smith in the vicinity of Congo.
Humboldt, Prolegomena ad nova genera plantarum.
Schouw, de sedibus plantarum originariis.
Jahrbucher der Gewachskunde.
Ritter's Sechs-Karten von Europa, mit erklarendem Text.
Titford's Sketches towards a Hortus botanicus Americanus.   Table of climates and habitats of plants.

390.

The geography of plants makes us acquainted with the present distribution of plants upon the earth and in the waters, and endeavours to refer their growth to external causes.   It is thus a part of the *Physiology of Plants*, since it investigates the laws according to which climate, temperature, soil, elevation above the surface of the sea, and distance from the equator, as also accidental external circumstances, operate upon the production of plants.   It is connected in some measure with the *History of Plants*, or with investigations respecting the origin, diffusion, and gradual distribution of

From *Essai Élémentaire de Géographie Botanique* (1820), translated in in A. P. Decandolle and K. Sprengel, *Elements of the Philosophy of Plants* (Edinburgh: Blackwood, 1821).

plants. Yet it must be distinguished from this; and when a sure foundation of facts is laid and arranged, it exerts an essential influence on the science of the cultivation of gardens and fields, on the rearing of forests and other civil occupations.

391.

We may investigate the laws of the distribution of the families and tribes of plants in different climates by two methods.

In the first place, we divide the surface of the Earth into certain zones, in which we seek for the plants that are produced, and thence draw general results. This method is indeed a laborious one, and is especially difficult on this account, that we are not yet completely acquainted with all the parts of every zone of the earth; while the lower families of plants have commonly been neglected by most travellers. Yet we can draw conclusions, with some probability, respecting unknown plants from those that are known; at least, this method leads to greater certainty than the following, on which only, however, most of our labours have been conducted. In this second method, we place the Floras of countries of different climates before us; we compare the plants which they contain, and in this way form conclusions respecting their distribution. But, as we are not in possession of complete Floras of all countries and their individual regions, it cannot but happen that false inferences and contradictions will arise, while we do not take into account the productions of neighbouring countries, or of those that lie between the districts which have been examined. Besides, we can only make use of the Floras of particular degrees of latitude and longitude, but not those of the whole zone; because most of the compilers of Floras have been acquainted with the products of vegetation only within a certain circle.

392.

Without entering, in this place, into the history of plants, we may state it as the fundamental law of the geography of plants, that the lower the organization of the body is, the more generally is it distributed.   As infusory animalculæ are produced in all zones, when the same conditions exist ; we find, in the same manner, that Fungi, Sponges, Algæ, and Lichens, and even Musci frondosi and hepatici, are distributed every where upon the earth, in the sea, and in the waters, when the same circumstances propitious to their production occur.   We have seen that the idea of genera and species can be applied with so much less strictness, the less perfect the vegetable is ; and hence, although the same or similar forms of Conyomici, Nematomici, Gastromici, and Sponges, are produced in all zones, we cannot pronounce in all cases respecting the identity of the species.   If travellers had not so much neglected the imperfect plants of foreign countries, this assertion might easily have been proved by innumerable testimonies.   But we must receive with caution and limitation, even what they have told us respecting the growth of common European cryptogamous plants in the most distant regions and waters of the earth, because many of these travellers had no exact knowledge of the cryptogamous plants.   The most distant countries of the earth, Europe and New Holland, the inhabitants of which are antipodes to each other, have, according to the testimony of Brown, a witness of the best information and highest credit, a considerable number of Lichens, almost indeed two-thirds of those that have hitherto been discovered in New Holland, of the same species with those that exist in Europe.   Of the Musci hepatici and frondosi, nearly one-third belong equally to New Holland and to Europe.   And, with respect to the Algæ, not only Confervæ, but Fuci, are common to the most distant seas. *Laminaria Agarum*, Lam., for instance, is found in Greenland, in Hudson's Bay, in Kamtschatka, and in the Indian Ocean. *Halidrys siliquosa*, Lyngb., *Sphærococcus ciliatus*, Ag., and many others, have a distribution equally extensive.

The Naiadæ and Rhizospermæ also are found in the same manner almost in all waters, as the *Marsilea quadrifolia*, *Zostera marina*, and the native Potamogetons and Lemnæ shew, which Brown found also in New Holland. Even the Grasses and Cyperoideæ share in this general distribution. A great many native members of these families, as the *Carex cæspitosa, Scirpus lacustris, Glyceria fluitans, Arundo phragmites, Panicum Crus Galli*, and so forth, grow also in New Holland. Humboldt has confirmed these observations, in respect to the growth of European Mosses, Grasses, and Cyperoideæ, in South America.

Higher and more perfect plants, on the contrary, are less generally distributed by nature, although, by cultivation, they also can be forced to vegetate in the most distant countries, provided favourable circumstances occur. Of these circumstances, a considerable number must always co-operate for the perfect evolution of plants of the higher orders. Yet there are some exceptions to this rule. *Verbena officinalis, Prunella vulgaris, Sonchus oleraceus, Hydrocotyle vulgaris, Potentilla anserina*, and some other common European plants, grow also, according to Brown, in New Holland. Almost the seventh part of the phanerogamous plants that vegetate in North America are found in Europe; yet we cannot deny that many of them have been transplanted hither, (401.) On Mascaren's Island, Bory St. Vincent found *Cladium Germanicum*, Schrad., *Cyperus fuscus, Potamogeton natans, Hydrocotyle vulgaris*, and some other European plants.

393.

The same distance from the equator, or the same degree of latitude, produces rather a resemblance in the forms,—an agreement in the families and genera,—than the same species, chiefly because, besides this geographical latitude, the height above the surface of the sea, the temperature during the growing season, the soil and constitution of the mountains, even the degree of longitude, and several other circumstances, have an influence on vegetation.

There are a great many perfect plants which exclusively belong to the tropics, which never pass beyond them, and which are found equally in Asia and Africa, in America and the South Sea Islands, and even in New Holland. Although, as we have said, these are rather families, as Palmæ, Scitamineæ, Museæ, Myrteæ, Sapindeæ, and Anoneæ; or genera, as *Epidendrum, Santalum, Olax, Cymbidium,* and so forth : yet there are particular species, which grow in all parts of the world only between the tropics, as, for instance, *Heliotropium Indicum, Ageratum conyzoides, Pistia stratiotes, Scoparia dulcis, Guilandina Bonduc, Sphenoclea zeylanica, Abrus precatorius, Boerhavia mutabilis,* and so forth.

But most commonly there are other species, which, under the same degree of latitude, supply in the new world the place of related species in the old. *Dryas octopetala,* indeed, grows equally upon the mountains of Canada, and in Europe ; but *Dryas tenella* of Pursh, which is very like the former, grows only in Greenland and Labrador. Instead of the *Platanus orientalis,* there grows in North America the *Platanus occidentalis ;* instead of *Thuia orientalis,* Thuia occidentalis ; instead of *Pinus Cembra,* in Europe and Asia, there grows in North America, *Pinus Strobus ;* instead of *Prunus Laurocerasus,* in Asia Minor, there grows under the same latitude in North America, the *Prunus Caroliniana.*

394.

There are many exceptions to this rule, however, depending on circumstances that have been already noticed. In the first place, countries are wont to share their Floras with neighbouring regions, especially islands lying under the same latitude, as the Azores possess the Floras of Europe and of Northern Africa, rather than those of America, because they are scarcely ten degrees of longitude from the coast of Portugal. Sicily, and still more Malta, possesses a Flora made up of those of the south of Europe and the north of Africa. The Aleutian Islands share their Flora with the north-west coast of America and the north-east of Asia. But the most distant countries, lying under the same latitude, may have the same, or a similar vegetation, while countries, or islands which lie between them, have not the least share in this particular Flora. The island of St Helena, which is scarcely eighteen degrees of longitude from the west coast of Africa, and which lies a little further south than Congo, has yet no plants, which are found in those last named regions; (Roxburgh's List of Plants seen in the Island of St Helena, Append. to Beatson's Island of St Helena.) Japan has a great many plants common to southern Europe, which, however, are not found in these regions of Asia that lie under the same latitude.

395.

We must further remark, that the eastern countries of the old world, and the eastern shores of America, as far as the Alleghany Mountains, have a much lower temperature than the western regions; or that it is always colder in Siberia and the north-east of Asia, than under the same latitude in Europe; and that even Petersburgh is colder than Upsal, and Upsal than Christiania, although they all three lie in the 60th degree of north latitude. In North America the difference is still greater, and there are commonly fifteen degrees of Fahrenheit's thermometer between the temperature of the east and west coast. It hence happens that many plants, which in Norway grow under the polar circle, scarcely reach the 60th degree on the limits between Asia and Europe. To this class belong the Silver-fir, Mountain-ash, Trembling Poplar, Black Alder, and Juniper. Even in the temperate zone, the vegetation of many trees ceases sooner in the east than in the west. In Lithuania and Prussia, under the 53d degree, neither Vines, nor Peaches, nor Apricots thrive; at least their fruit does not ripen, as also happens in the middle of England. The most remarkable example of this great difference of temperature is furnished by the *Mespilus Japonica*, which grows at Nanga sacki and Jeddo, under the 33d and 36th degrees of N. Lat., and which also grows in the open air in England, under the 52d degree of N. Lat., when it is planted against a wall, (Botanical Register, vol. v.)

396.

The same degrees of latitude, in the southern and northern hemisphere, are connected with very different temperatures, and produce a completely different vegetation. This, however, must be understood rather of the temperate and frigid zones, than of the tropical climates, which, as we have already noticed, are pretty much the same over all the earth. But the summer is shorter in the southern hemisphere, because the motion of the earth in her perigee is more rapid. The summer is there also colder, because the great quantity of ice over the vast extent of sea requires more heat for dissolving it than can be obtained ; as also, because the sun beams are not reflected in such quantity from the clear surface of the sea water, as to afford the proper degree of heat. It hence happens, that in the southern hemisphere, the Flora of the pole extends nearer the equator, than in the northern. Under the 53d and 54th degrees of south latitude, we meet with plants which correspond with the Arctic Flora. In Magellan's Land, and in Terra del Fuego, *Betula antarctica* corresponds with *Betula nana*, in Lapland ;—*Empetrum rubrum* with *Empetrum nigrum*,—*Arnica oporina* with *Arnica montana*,—*Geum Magellanicum* with *Geum rivale*, in England,— *Saxifraga Magellanica* with *Saxifraga rivularis*, in Finmark. Instead of *Andromeda tetragona* and *hypnoides*, of Lapland, Terra del Fuego produces *Andromeda myrsinites :* in place of *Arbutus alpina* and *Uva Ursi* of the Arctic polar circle, Terra del Fuego produces *Arbutus mucronata*, *microphylla*, and *pumila*. Aira antarctica reminds us of the *Holcus alpina* of Wahlenberg; and *Pinguicula antartica* recalls to our recollection *Pinguicula alpina*.

We must recollect, however, that in South America, the great mountain chains of the Andes stretch from the tropical regions, almost without interruption, to the Straits of Magellan, (from the 52d to the 53d degree of S. Lat.) ; and that, on this account, tropical forms are seen in that frigid southern zone, because, as we shall have occasion to remark more fully afterwards, the tract of mountains every where determines vegetation. It is hence that the Straits of Magellan are prolific of *Coronariæ*, *Onagræ*, *Dorsteniæ*, and *Heliotropiæ*, which in other parts of the world grow only within the tropics, or in their neighbourhood.

270   GEOGRAPHY OF PLANTS.

In general, the vegetation of the southern hemisphere is
very different from that of the northern, and there is a certain correspondence between the Floras of Southern Africa,
America, and New Holland. Most of the trees are woody,
with stiff leaves, blossoms sometimes magnificent, but fruit
of little flavour. In Southern Africa, as well as in New
Holland, it is the form of the Proteæ which prevails as if
appropriated to these regions. Instead of the South American *Ericæ*, we find the *Epacridæ* of New Holland. *Lobeliæ*,
Diosmeæ, and a great number of rare forms of compound
blossoms, and of *Umbellatæ*, are common to all these southern regions,

397.

For understanding the growth and distribution of plants,
we must also attend to the soil. Similar plants are found in
similar soils, completely separated from each other, and with
respect to which no supposition of interchange can be entertained, provided only the climate be not too different. Salt
soils produce almost every where particular Chenopodeæ,
species of Chenopodium, Atriplex, Salsola, Salicornia, and
Anabasis. Calcareous soils produce always the most numerous and distinguished forms of plants. Volcanic mountains,
particularly basalt, produce few forms, but those of a distinguished kind and very variable. Alluvial mountains, particularly in the neighbourhood of streams, usually, like marshes,
produce forms that are always the same. The primitive
mountains, on the other hand, almost every where separate
the Floras of countries. Thus the Pyrenees,—the Alps
which divide Italy from France and Switzerland,—those
which separate Upper Italy from the Tyrol and Carinthia,—
and the Carpathian mountains,—divide the Floras of the
southern from those of the northern countries.

It is hence of so much importance, along with Floras, to
describe the mountain rocks and the different soils, (258.)
The first example of a map of this kind was given by De
Candolle, in the second volume of his *Flore Francaise*, in

which, however, the mountains only, the southern shores, and the alluvial land, are distinguished, and the heights above the sea are given. Wahlenberg's Map of Lapland, in his *Flora Lapponica*, is much more correct. Maclure has given a Map of the constitution of the Mountains in the Free States of North America, (Geographische Ephemeriden). I know not that any person has given a more pleasant and instructive account of the soils and mountains of his country, in relation to their Floras, than the excellent Villar, respecting the Alps which divide Italy from Switzerland, in the Preface to the *Histoire des Plantes de Dauphiné*.

398.

This brings us to a very important circumstance, which must be taken into account in every examination of the causes of the growth of plants, namely, the height of their station above the level of the sea. As, upon the whole, the temperature on the highest mountain tops seems to be the same with the temperature at the polar circle, it is commonly believed, that under the snow-line, and near to it, the same vegetation is found as in the polar regions. The limit of perpetual snow, under the Equator, is at the height of 15,000 feet; in the 35th degree of N. Lat. it is at 10,800; in the 45th degree, at 8400; in the 50th degree, at 6000; in the 60th degree, at 3000; in the 70th degree, at from 1200 to 2000 feet above the level of the sea; and, at the 75th degree of N. Lat., the snow-line lies almost upon the ground In general, it has been ascertained by observations, that the same vegetation is produced at the same distance from the snow-line. It must be considered, however, that towards the pole, the summer is shorter, but hotter, than under the snow-line upon the tropical mountains, where winter and summer cause no change of temperature. On this account, a better vegetation must be produced in the polar regions during summer, especially as the plants are there exposed to the uninterrupted light of the sun : while, on the other hand, from the uniform temperature on the highest tropical mountains, throughout the year, a very different Flora is there produced from that which springs towards the pole. As yet we know only, from Humboldt's immortal labours, the vegetation upon the highest chains of the Andes, in South America; for the Floras of the much higher mountains in Northern India, which are called the Himalaya Mountains, and of the Mountains of the Moon, in Africa, are entirely unknown to us. The Flora of the Andes, at the height of 14,760 feet, is almost entirely of a peculiar kind; and if a pair of Ranunculi, a *Gentiana*, and a *Ribes*, remind us of the Flora of the poles, the remaining productions are completely peculiar, and prove that the height above the level of the sea is very far from producing universally the same forms. Yet we must add, that in Europe, at least, many northern and even polar forms appear under the snow-line of the Pyrenean and Helvetian Alps. Of this the Dwarf Willow, Dwarf Birch, Saxifrages, Ranunculi, Cerastia, and other genera, are striking proofs.

Of perfect plants, the *Daphne Cneorum* seems in Europe to hold the most elevated station, since, on Mont Blanc, it stands at 10,680 feet; and, on Mont Perdu, at 9036 feet high. The growth of woody plants ceases, on the Alps of central Europe, at the height of 5000 feet; and, on the Riesengebirge, at 3800. Oats grow on the Southern Alps at 3300, and on the Northern scarcely at 1800 feet. The Fir grows on Sulitelma, in Lapland (68 degrees N. Lat.), scarcely at the height of 600, and the Birch scarcely at the height of 1200 feet. On the other hand, upon the Alps which divide Italy from France and Switzerland, Oaks and Birches grow at 3600, Firs at 4800; and the same plants grow on the Pyrenees above the height of 600 feet.

In Mexico, the mountain chains, and, in particular, the Nevado of Toluca, are covered, above 12,000 feet high, with the occidental Pine (*Pinus occidentalis*); and, above 9000 feet, with the Mexican Oak (*Quercus Mexicana, spicata*); as also with the Alder of Jorullo (*Alnus Jorullensis.*)

On the Andes, Palms grow at the height of 3000 feet. The woody Ferns, (*Cyathea speciosa, Meniscium arborescens, Aspidium rostratum*), are found as high as 6600 feet; as are also the pepper species, Melastomeæ, Cinchonæ, Dorsteniæ, and

some Scitamineæ, rise to the same elevation. At the height
of 14,760 feet, we still find the Wax Palms, some Cinchonæ,
Winteræ, Escalloniæ, Espelettiæ, Culcitia, Joanneæ, *Vallea
stipularis*, *Bolax aretioides*, and some others.

399.

The growth of plants in society, or as individuals, is very
interesting. Many forms are so appropriated to certain regions, that amidst constant changes they still take in a great
tract of land, and are produced in exuberant abundance.
Others, on the contrary, stand quite insulated, and seem as if
they would utterly disappear, did not Nature, in a manner
which is often inexplicable, provide for their continuance.

While with us, *Polygonum aviculare*, *Erica vulgaris*,
Poa annua, *Aira canescens*, *Vaccinium Myrtillus*, grow always in society, and cover great tracts of country, we observe, on the contrary, that *Marrubium peregrinum*, *Carduus cyanoides*, *Stellera Passerina*, *Carex Buxbaumii*, *Cirsium eriophorum*, *Lathyrus Nissolia*, *Hypericum Kohlianum*,
Schœnus ferrugineus, and *Helianthemum Fumana*, are confined, in an insulated state, within a very narrow space, beyond which they never pass. The Cedar of Lebanon, *Forstera sedifolia* of New Zealand, *Melastoma setosum* on the
Volcano of Guadaloupe, and *Disa longicornis* on some spots
of the Table Mountain of the Cape, are examples of this
completely insulated growth, which renders the idea of the
migration of plants at least very doubtful.

400.

If we proceed through the separate families, we shall find
their geographical distribution pretty exactly ascertained, and
their increase or diminution determined according to the different zones. But, as many families consist but of single
groups, which are limited to certain zones or countries,—this
circumstance occasions always a variation in the account. If
we attend, for instance, to the Rubiaceæ, it is almost impossible to pronounce any general opinion respecting the geographical distribution of this family, because the first group of

this family, the Stellatæ, is almost peculiar to the temperate, and especially to the northern temperate zone, while the Spermacoceæ and Coffeaceæ are confined to the tropical zone. The Cinchoneæ, indeed, grow chiefly between the tropics, but always at a fixed height above the level of the sea, and they also pass beyond the tropics. Plants with compound flowers are indeed generally more abundant between the tropics, but the group, which I have named Perdicieæ (Labiatifloræ, De Cand.), is peculiar to South America, and descends even into the southern frigid zone.

Respecting Ferns, it is understood that, in the temperate zone, they constitute the sixtieth part of the whole vegetable kingdom. But how little certainty attends such conclusions, may be understood from this, that in New Zealand, the number of Ferns is to the number of other plants as 1 to 6; in Norfolk Island, and Tristan d'Acunha, as 1 to 3; in Otaheite as 1 to 4; in Mascaren's Island as 1 to 8; in Jamaica as 1 to 10; in St Helena as 1 to 2; and in Egypt we have as yet found but one species. The Grasses seem in all zones to maintain nearly the same proportion: They constitute the tenth or fifteenth part of the whole Flora. The Umbellatæ are evidently in greatest number in the temperate zone. They constitute about the thirtieth part of other plants. Towards the pole they diminish in number; and in the torrid zone there are scarcely any other Umbellatæ but some intermediate forms, which only appear at a very considerable height upon the mountains. The Cruciform plants exhibit a similar proportion. In the temperate zone, they are to the remaining plants perhaps as 1 to 20. Towards the pole they decrease in number, and between the tropics we find scarcely a trace of them. The reverse is the case with the Malvaceæ. Whilst these constitute, between the tropics, the fiftieth part of the other plants; in the temperate zone they bear to them the proportion of 1 to 200, and in the polar zone they fail entirely. The Leguminous plants are in greatest quantity between the tropics, where they form the twelfth part of the whole Flora. In the temperate regions they fail considerably, and in the polar zone they are to the other plants as 1 to 35.

The Primuleæ are almost the only plants that are common to the frigid and to the temperate zone. The Contortæ belong to the tropical region, where they form from the fortieth to the fiftieth part of the whole. In the temperate zone they diminish in number, until, towards the polar circle, they almost entirely cease.

401.

It is interesting also to know the limits within which the cultivation of the useful plants is confined. The cultivation of Cocoa, Coffee, Anatto, Cloves, and Ginger, is limited to inter-tropical regions. The Sugar Cane, Indian Figs, Dates, Indigo, and Battatas, pass the tropics as far as the 40th degree of N. Lat. Cotton, Rice, Olives, Figs, Pomegranates, Agrumæ, and Myrtles, grow in the open air, as far as the 45th and 46th degree. The Vine succeeds best with us within the 50th degree of N. Lat.; this, also, is the limit, especially in the West of Europe, of the cultivation of Maize, Chesnuts, and Almonds. Melons also succeed to the same latitude in the open air.

In the West of Europe, the cultivation of Plums, Peaches, Wheat, Flax, Tobacco, and Gourds, ceases at the 60th degree of N. Lat. In the East of Europe, the cultivation of Apples, Pears, Plums, and Cherries, terminates at the 57th degree; but Hops, Tobacco, Flax, Hemp, Buckwheat, and Pease, succeed there even under the 60th degree. Hemp, Oats, Barley, Rye, and Potatoes, are planted by the Norwegians under the polar circle; and the Strawberry flourishes, even at the North Cape, under the 68th degree.

—◆—

CHAP. V.

HISTORY OF THE DISTRIBUTION OF PLANTS.

Linné, de Telluris habitabilis incremento : in Amœn. Acad. vol. ii.
Zinn, Vom Ursprung der Pflanzen : im Hamburgishen Magazine.
Bergman, Jordklot. Phys. beskrifn. ii.
Zimmerman, Geographische Geschichte des Menschen.
Schouw, Diss. de Sedibus Plantarum originariis.

402.

We come now to answer the questions, in what manner plants have originated, and how they have distributed themselves. Are we to admit, that plants have been distributed from one point on the surface of the earth, to all its parts? or must we believe that they belong properly to every country in which they grow? The founder of Scientific Botany has defended the former of these opinions at a great expence of ingenuity, acuteness, and learning; but we apprehend that we must adopt the latter conclusion, with some limitations.

403.

When we examine the remains of the primeval world, we find the first traces of vegetable impressions in the slate formation. These remains of the former vegetable world belong almost entirely to the lower families : they consist, for the most part, of Grasses, Reeds, Palms, and Ferns,—the latter, however, being almost always destitute of fruit. But although these forms cannot be referred to any one of the species which are at present known, they have yet so much the appearance of tropical productions, that we are forced to admit a very high degree of heat at the surface of the earth during its former state, which heat must, at that time, have been diffused over all the zones, because we find the same productions in the slate formation of all parts of the earth *.   In order to explain this, it has been supposed, that the plane of the ecliptic, during the former state of the globe, was completely different in its position, and that, consequently, our planet had then another situation in respect to the sun.   But Bode, the worthy veteran of Prussian astronomers, has shewn, that the plane of the ecliptic has been, for 65,000 years, between the 20th and 27th degree; and that at present it is about 23 minutes less, and, consequently, the inclination of the axis of the earth as much greater, than in the time of Hipparchus, who lived about two thousand years ago.   The former solution must, therefore, be entirely abandoned; (Neue Schriften der Berlin Gesellschaft naturforschender Freunde).

Shall we then consider as satisfactory another explanation, which has been advanced by one of the most ingenious and learned investigators of the ancient state of the globe?   According to this author, the Earth, during its primeval state, was completely surrounded by water.   By slow degrees the sea retired; the highest mountain tracts were laid bare, and the lowest and densest atmospherical stratum, supported by the surface of the sea, now rested upon the highest primitive land, which, like an island, emerged but a little way above the ocean.   While the heat was chiefly generated in the lower strata of the air, it must also, at that time, have been equally diffused throughout all parts.   The naked summits of the mountains were gradually mouldered by the

---

* It seems, indeed, that all the carbonaceous matter of the more ancient slate formation ought to be considered as the oldest remains of plants which had been growing, but which had been stopped in their progress; and that all calcareous matter ought to be considered as the remains of a begun, but suppressed creation of animal bodies; (Steffen's Beyträge zur innern Naturgeschichte der Erde, s. 27. Dessen, Handbuch der Oryctognosie, b. ii. s. 353.) In what manner mineral substances are formed from corrupting vegetables, we perceive from the production of iron-pyrites, in our Peat Mosses, where it is found in layers, under the thin, broad, reed leaves, after they have become putrid.

influence of light, of aerial matters, and of moisture. To the primitive and transition rocks succeeded the horizontal flœtz formation. A multitude of bodies was now formed, which light elicited from the organizing water. These bodies decayed, and left behind them the original materials of carbon and lime, from which still more perfect forms were to spring, until, at length, Ferns, Grasses, and Palms, were produced, which, during the high level of the waters, enjoyed upon the declivities of the mountains, a high and equal temperature. With these forms, creative nature remained contented, till new revolutions of the surface of the earth gave the waters an opportunity of retreating still further, when new formations arose.

This hypothesis, which is favoured by the Geognosy, derives particular support from the concurring testimony of the most ancient inhabitants of the world, respecting universal inundations and floods; as also, more especially, from the Persian Cosmology, in which Albordsch, the highest primitive mountain, the hill of light, or navel of the earth, plays a principal part, while it is yet surrounded by water. The primitive light produces, upon this mountain, during an ever equal temperature, all living things; (Kanngiesser Altherthums wissenschaft, s. 8. u. 18.; Algemeine Encyclopædie, th. ii. s. 375.)

404.

What revolutions the surface of the earth underwent, before it assumed its present state, we know not. But with the period of the alluvial mountains begins the present vegetation of the earth's surface, and those forms which we now see in unvarying perpetuity, have been so for several thousand years, or since the earth assumed its present state, (142.)

Whether now, as Linnæus maintained, one example only of every genus of plant existed at the beginning: whether all these single genera were produced by the hand of Nature upon the highest mountain ridge of the earth, along with the single pair from which the human race has proceeded, and with similar individual pairs of all other animals: whether, at least, the highest mountain ridges may, in general, be regarded as the birth-places of the vegetable world, as Willdenow asserted; and whether, therefore, plants have been distributed from single stations, by their own migrations, and by other means, which nature provides,—these are questions which ought to be examined and answered with the greatest care.

405.

We are not disposed altogether to deny the migration of plants in similar climates, because we know from experience, that the *Datura Stramonium, Erigeron Canadense,* and *Æsculus Hippocastanum,* are not natives of Germany; but that the first was imported, as it is said, by the gypsies,—the Horse-Chesnut, by the Austrian embassy which was sent to Constantinople at the end of the sixteenth century,—and the *Erigeron Canadense,* in consequence of some commercial relations which we had with North America;—as this latter country has also probably received from us, in the course of commerce, the *Agrostis Spica Venti, Trichodium caninum, Anthoxanthum odoratum, Alopecurus pratensis, Poa trivialis, Bromus secalinus, mollis, Dactylis glomerata, Hordeum vulgare, Dipsacus sylvestris,* and many other plants, (392.) We know that a great many plants have passed from India into Italy, along with the cultivation of rice: we know that the West Indian negroes have introduced into the western world a great many plants from Africa, which at present grow wild there, as the *Cassia occidentalis,* and *Chrysobalanus Icaco.*

It is certain, that sea plants have been brought by ships, from the southern into the northern sea, as has been the case with *Fucus cartilagineus,* Turn., *Fucus natans,* and *bacciferus.* West Indian fruits are every year driven upon the coasts of Norway, and of the Faroe Islands, during storms from the south-west; as Cocoa Nuts, Gourds, the fruit of *Acacia scandens, Piscidia Erythrina,* and *Anacardium occidentale.* But after all this has been granted, we are still far from being in a condition to maintain the migration of plants and their distribution from one point over the surface of the earth.

406.

It betrays very limited ideas respecting the laws of vegetation to suppose, that all the plants upon the face of the earth, which require such different climates, such a different constitution of the mountain-rocks, so various a composition of soil and of water, could ever have been assembled upon one and the same high mountain ridge. All testimonies, indeed, confirm the supposition, that the human race, and the domestic animals, descended from the high mountain plains of central Asia, between the 27th and 44th degrees of N. Lat. We may also conjecture, that the different kinds of grain grow wild in these latitudes, as it is also probable that the domestic animals are there found in their native state. But the innumerable other wild growing plants, of all quarters of the globe, which are so frequently confined to a single island, or to a single circle of the continent, cannot possibly have their native seats in those regions; otherwise the remains of that vegetation which is now dispersed over the whole face of the earth, would be found in Northern India and Persia, in Thibet, and in the Mogul empire. It is physically impossible, that plants, which in Germany grow only upon calcareous soils, or upon basalt and other peculiar mountain rocks, could, at an early period, have flourished upon the primitive granite and gneiss of the Himalaya Mountains.

407.

As little can we assent to the opinion of those who consider the high mountain tracts as so many birth-places, or foci of vegetation, and of the neighbouring Floras. We admit, that when a particular mountain chain stretches into the level country beneath it, its peculiar plants will also appear in the low land. The flœtz limestone of central Germany confirms this conclusion in the strongest manner. But when this is not the case, the low country never partakes of the Flora of the neighbouring mountains. Otherwise, *Seseli Hippoma-*

rathrum, Teucrium montanum, Poa alpina, and *Stellera Passerina,* would soon diffuse themselves from our calcareous hills to the flat country, and even over our porphyry mountains.

It is true that mountain tracts commonly form the boundaries of Floras. But this happens, not because these mountains are the birth-places of the vegetable world, but because climate and temperature change with them. The Rhætian Alps separate Germany from Italy; on their southern declivities we observe Laurels, Pines, Beeches, Cypresses, Jasmins, and other similar plants, which do not grow on their northern sides. But the temperature on the opposite sides of these Alps is also completely different.

It must also be added, that the limits of Floras are not defined by mountain tracts alone, but that even in a great extent of level country the Floras have their proper boundaries. *Andropogon Ischæmum, Asperula cynanchica, glauca,* M. B. *Centunculus minimus, Lycopsis pulla, Bupleurum rotundifolium, Peucedanum officinale, Cnidium silaus, Silene noctiflora* and *conoidea,* and *Centaurea calcitrapa,* seem not to pass beyond the 52d degree North Lat. into central Germany. At that point, *Angelia Archangelica, Lonicera periclymenum, Andromeda polifolia, Arbutus uva ursi,* and other forms begin to appear. In completely level countries, the *Acer campestre, Pseudoplatanus, Populus alba, nigra,* and *Sambucus nigra,* cease to grow at the 56th degree N. Lat. The Myrtle, Mastick, Oak and Cork tree, the flowering Ash, and the Caper tree, pass not beyond the 44th degree North Lat., whether mountain tracts or level countries occur at this limit. The heights of the Wolga, or Alaunian Mountains of the ancients, are said to be the limit between the eastern and western Floras; but according to Pansner's recent examination, the entire Wolga heights are only alluvial land, covered with sea sand. Besides, the eastern Flora is seen a great way on this side of the Wolga heights, (Neue Geographische Ephemeriden, b. v. s. 141.) The Weichsel on the north, and the Oder on the south, seem better entitled to be considered as the limits of the western and eastern Flora. On the

2

farther side of these streams, we find *Plantago arenaria,* Kit. *Anchusa Barrelieri,* Vilm., *Flœrkia lilifolia, Angelica pratensis,* M. B., *Acer platanoides, Andromeda calyculata, Silene tatarica, Dianthus serotinus,* Kit. (east from Cracow), *Anemone patens, Ranunculus cassubicus, Teucrium Laxmanni, Dracocephalum Moldavica* (east from Grodno and Jaroslaw), *Bunias orientalis* (east from Lemberg), *Isatis tinctoria* (beyond Warsaw), *Astragalus Onobrychis, Melilotus polonica, Pentaphyllum Lupinaster, Hieracium collinum,* Bess., *Orchis cucullata* (beyond the Niemen.)

408.

Had plants been distributed from single, and, as it is thought, from elevated points on the earth's surface, the Floras of contiguous regions would necessarily have been confounded, and could not have been so distinctly appropriated, as we see them to be. It must be added, that winds and birds, rivers, and the waves of the sea, are far from being able completely to have effected the universal dispersion of plants. There can be no doubt, that the wind is able to diffuse to a certain extent some particular seeds, which are furnished with crowns, hairs, and other appendages. But it is not able to disperse to any distance the *Carduus cyanoides,* which grows on a single grassy hill near Halle, and on the steep banks of the Elb above Tochheim, although the seeds of this plant are furnished with a crown of bristly hairs. The Syngenesious plants, too, the seeds of which can be so easily transported by the wind, are by no means common in the greater number of countries. If the wind favoured the migration of plants, we might determine their correspondence in most countries from the distance. But we have already noticed (397.), that the most distant countries have common plants, whilst the most dissimilar Floras are found in neighbouring lands, and some plants grow quite insulated, (395.) The dispersion of plants over large tracts of country has also been ascribed to birds, because they devour the fruits, and often allow them to pass from them undigested. But no example of this can be produced except the Misletoe, and therefore this

3

assertion deserves no particular refutation.   Streams, indeed, can carry down seeds ; and plants from those higher regions through which the streams flow are accordingly often found growing on their banks.   But the Flora on the banks of one and the same stream, is very different in the different districts through which it passes, as is seen in the clearest manner upon the shores of the Elbe ; for in Bohemia very different plants appear from those which spring in the neighbourhood of Dresden,—others, again, make their appearance near Wittenberg and Barby,—and a yet different set near Lauenburgh and Hamburgh.

These considerations lead us to conclude, that the vegetable world has neither descended from one common birthplace, nor diffused itself from one country into another ; but that vegetation is in every case the product of the joint influence of temperature, soil, and the particular composition of the moisture of the earth.

Nor is the conclusion of Brown (on Congo, p. 50.), that the native country of a genus is always where the greatest variety of species is found, by any means to be admitted, since the example of Nicotiana shews the contrary.   The greatest number of its species are found in South America ; yet the *Nicotiana Chinensis*, Lehm. and *fruticosa* are certainly indigenous to Eastern Asia.

CHAP. VI.

ON MALFORMATIONS AND DISEASES OF PLANTS.

Linné, Philosophia botanica.   S. 119, 131.
Jäger, Uber die Missbildungen der Gewächse.
Gallesio, Theorie der vegetabilischen Reproduction.
Keith's System of Physiological Botany.

PAPER 5

From *Essay on the Geography of Plants*
Alexander von Humboldt

Foreword

Having been away from Europe these past five years, and having wandered through countries of which many had never been visited by naturalists, I should perhaps have hastened to publish the abridged account of my travel to the tropics and the series of phenomena that occurred successively before me over the course of my investigations. I would have been flattered by public appreciation of such haste as part of it has shown a generous interest in my own survival and in my expedition. However I thought that, before speaking about myself, and the obstacles I overcame during my expedition, I had better draw the attention of physicists to the major phenomena that Nature displays in the regions I visited. It is their whole that I consider in this essay where I present the results of the observations that will be published in greater detail in further works I am preparing for the public.

I include all the phenomena that one can observe on the surface of the globe, as well as in the surrounding atmosphere. The physicist familiar with the present state of science, primarily that of meteorology, will not be surprised to see such a large number of topics being touched upon in so few pages. Had I been able to work for a longer time on their drafting, my book would have been even shorter for an overview should only depict major views on physics, i.e., undisputed results that can be expressed as precise numbers.

I already conceived the idea of this work in my early youth and presented the first outline of a Geography of Plants in 1790 to the famous companion of Cook, M. Georges Forster, to whom friendship and gratitude attach me closely. Since then, my studies in several fields of the physical sciences have helped expand these initial ideas further. My travel to the tropics provided me with invaluable material for the physical history of the globe. It was when I actually saw the great objects I had to describe, at the foot of the Chimborozo, on the hillsides of the southern sea,

that I wrote most of this work. I thought that I had to retain the title "Essay on the Geography of Plants," because, when considering the imperfection of my work, any less modest title would have made it less worthy of public indulgence.

It is primarily for its style that I beg such indulgence. Having expressed myself for such a long time in several languages which are no more my mother tongue than is French, I dare not hope to now always maintain the purity of style that one would expect from a work written in my own language.

The overview presented here is based on my own observations and on those of M. Bonpland. United by the bonds of a most intimate friendship, having worked together for six years, and having shared the sufferings which travelers are necessarily exposed to in uneducated countries, we agreed that work on the results of our expedition will be co-authored under both our names.

It is in preparation of these works, which have been my main occupation since my return from Philadelphia, that I had to ask for help from renowned men who have honored me with every kindness. M. Laplace, whose name is far above my praise, was kind enough to show a most flattering interest in the work I brought back as well as in that completed since my return to Europe. Enlightening and refreshing, as if through the power of his wit and all that surrounds him, his help has become as useful to me as it generally is to all young men who approach him.

If it is a joy for me to pay friendship the tribute of my admiration and gratitude, this commits me to fulfill no less sacred duties. M. Biot was kind enough to honor me with his advice in the drafting of this work. Combining a physicist's perspicacity with a geometrician's profound knowledge, his company became a fertile source of instruction to me. Despite his numerous occupations, he was kind enough to calculate the tables of horizontal refraction and of light extinction in my table.

The facts I present on the history of fruit trees are

Translations by Francesca Kern and Philippe Janvier from Alexander von Humboldt, *Essai sur la Géographie des Plantes* (Paris: Levrault, Schoell et Cie., 1805).

taken from the work of M. Sickler, who combines—and this is rare—great erudition with much philosophical insight.

M. De Candolle provided me with interesting data on the Geography of Plants from the high Alps. M. Ramond provided me with information on the flora of the Pyrennées, and extracted data from the classical works of M. Willdenow. I regarded it as important to compare the phenomena of equinoctial vegetation with those of our European soil. M. Delambre was kind enough to enrich my overview with the measurements of summits never previously published. A large number of my barometric observations were calculated by M. de Prony (according to M. Laplace's formula and taking gravity into consideration). This honorable scientist was even kind enough to have four hundreds of my altitude measurements calculated under his supervision.

I am currently drafting the astronomical observations made during my expeditions, part of which have been submitted to the Bureau des Longitudes to have their accuracy checked.[1] It would have been imprudent to publish either the maps I made inland of the continent, or the account proper of my travels, before these were submitted because the position of the localities and their altitude have bearing on all the phenomena of the regions I visited. I am especially proud that the longitudinal observations I made during my trip on the Orinoco, the Cassiquiaré, and the Rio-Negro will be of interest to those concerned with the geography of southern America. Despite Father Caulin's precise description of the Cassiquiaré, more modern geographers doubt the accepted communication between the Orinoco and Amazon. Having worked there, I could not expect that some would reproach me bitterly[2] for discovering in Nature river courses and mountain directions completely at odds with what is shown on La Cruz's map; however, it is the fate of travelers to displease some when they observe facts that contradict received views.

After drafting the volume on astronomy, that of my other works will soon follow, and it will only be after my last voyage is published that I shall deal with a newly planned endeavor to throw the brightest light on meteorology and magnetic phenomena.

I cannot publish this first essay, this first result of my researches, without expressing my profound and respectful gratitude to the government which has honored me with such generous protection during my travels. Being granted a protection hitherto not granted to any individual, and having lived for five years in a frank and loyal nation, I was never confronted in the Spanish colonies with any other obstacle than those of physical Nature. The memory of this government's benevolence will remain in my soul, as deeply engraved as the expression of affection and interest that all classes of inhabitants honored me with during my stay in both Americas.

Alex. De Humboldt.

Essay on the Geography of Plants

Read to the physical sciences and mathematics classes of the Institut National, the 17 Nivôse[3] of the year 13.

Botanists' research is generally directed toward objects that only encompass a minute area of their science. They occupy themselves almost exclusively with the discovery of new plant species, their exterior structure, the characters that distinguish them, and the analogies that group them into classes and families

The knowledge of forms under which organisms present themselves is, without a doubt, the principal basis of descriptive natural history. One must view it as indispensable for the advancement of sciences which touch upon the medical properties of plants, their cultivation, or their application in the arts: if worthy of exclusively occupying a large number of botanists, although also capable of being considered philosophically, it is not any less important to target the Geography of Plants. This science as yet only exists as a name, and yet is an essential part of general physics.

The Geography of Plants considers plants by the relationships of their local association in different climates. As vast as the object it embraces, it broadly outlines the immense expanse covered by plants

---

1. The Bureau des Longitudes, which still exists in France, is a special office in charge of checking the accuracy of all geometrical measurements. TRANS.

2. *Géographie moderne,* by Pinkerton, translated by Walkenaer; vol. 6, pp. 174–77.

3. The fourth month of the French Republican calendar. (The equivalent date in the Gregorian calendar is 7 January 1805.) TRANS.

from regions with perpetual snow to the depths of the oceans, to the interior of the globe where—in obscure grottoes—cryptogams grow as unfamiliar to us as the insects which they nourish.

The superior limit of vegetation varies, like that of perpetual ice, according to its distance from a pole or the obliquity of solar rays. We ignore the extent of the inferior limit of plants: precise observations made in both hemispheres of subterranean vegetation prove that the interior of the globe is animated everywhere that organic germs have found the appropriate space for their development, and nourishment similar to their own constitution. The rocky and icy peaks which our eyes can hardly distinguish above the clouds are exclusively covered by mosses and lichenous plants; analogous cryptogams—sometimes studded, sometimes colored—ramify themselves in the vaults of mines and subterranean grottoes. In this way, two opposite limits of vegetation produce beings similar in structure, and whose physiology remain equally unfamiliar to us.

The Geography of Plants does not only arrange plants according to the zones and altitudes they are found in, nor is it content to consider plants according to the degrees of atmospheric pressure, temperature, humidity, and electric tension under which they live: it distinguishes among plants, as among animals, two classes which have very different ways of life and—dare one say it?—habits.

Some plants grow isolated and scattered: for example, in Europe *Solanum dulcamara, Lychnis dioica, Polygonum bistorta, Anthericum liliago, Crataegus aria, Weissia paludosa, Polytrichum piliferum, Fucus saccharinus, Clavaria pistillaris, Agaricus procerus*, or in the tropics *Theophrasta americana, Lysianthus longifolius, Cinchona, Hevea*. Other plants, gathered in societies like ants or bees, cover immense ground from which they exclude any hetergenous species: for example, strawberries [*Fragaria vesca*], bilberries [*Vaccinium myrtillus*], *Polygonum aviculare, Cyperus fuscus, Aria canescens, Pinus sylvestris, Sesuvium portulacastrum, Rhizophora mangle, Croton argenteum, Convolvulus brasiliensis, Brathys juniperina, Escallonia myrtilloides, Bromelia karatas, Sphagnum palustre, Polytrichum commune, Fucus natans, Sphaeria digitata, Lichen haematomma, Cladonia paschalis, Thelephora hirsuta*.

Plant associations are more common in temperate rather than tropical zones whose less uniform vegetation is all the more picturesque for this. From the banks of the Orinoco to those of the Amazon and the Ucayale, the entire surface of more than five leagues of soil is covered by thick forests; and if the rivers did not interrupt its continuity, monkeys—almost the only inhabitants of these solitudes—could swing from branch to branch from the boreal to the austral hemisphere. However, these immense forests do not offer a uniform view of social plants: every part produces diverse forms. Here we find mimosas, Psychotria or *Melastomas*, there baytrees, brasilettos, *Ficus, Carolinea*, and *Hevea*, all of which intertwine their branches with no plant exerting its authority over the others. However, this is not the case for the tropical region neighboring New Mexico and Canada. The plateau (1,500–3,000 m above sea level) from 17° to 22° latitude—the entire county of Anahuac [Chambers County, Texas]—is covered by oak trees and a species resembling *Pinus strobus*. On the eastern slope of the Cordillera, in the Xalapa valley, we find a vast forest of sweet gum: the soil, vegetation, and climate all take on the characteristics of temperate regions, a circumstance which has not been found elsewhere at the same altitudes in southern America.

The cause of this phenomenon appears to depend on the structure of the American continent: it stretches toward the North Pole, extending itself further in this direction than Europe. This makes the Mexican climate colder than it should be for its latitude and elevation from sea level. The plants of Canada and more northern regions have flooded back toward the south, and the volcanic mountains of Mexico are covered by the same fir trees which only appear to also be found at the sources of Gila and Missouri rivers.

By contrast, the great catastrophe in Europe that opened the Strait of Gibraltar and hollowed out the bed of the Mediterranean prevented African plants from migrating into southern Europe: we also find far fewer species to the north of the Pyrennées. However, the fir trees crowning the heights of the Tenochtitlan valley are species identical to those at the 45th parallel, and any painter traveling in the countries situated beneath the tropics to study the character of the vegetation there will not encounter the beauty and variety of forms found in equinoctial plants. In the parallel of Jamaica, they would find oak forests, fir trees, *Cupressus distichia*, and *Arbutus madronno*: indeed, forests that present all the characters and monotony of social plants in Canada, Europe, and northern Asia.

It would be interesting to designate on botanical maps the ground on which assemblages of same plants grow. They would appear as long bands whose compelling expansion diminished the population of states, separated neighboring nations, and put in place greater obstacles to their communication and commerce than mountains and seas. Heathland, this association of *Erica vulgaris* and *Erica tetralix*, of *Lichen icmadophila* and *Haematomma*, is scattered from the most northern extremity of Jutland, through Holstein and Lunebourg, to the 52nd parallel. From there, it turns toward the west, by the granitic sands of Munster and Breda, to the hillsides of the Ocean.

For many centuries, these plants caused widespread sterility of the soil, and thus exerted an absolute empire on these regions: despite his best efforts, man struggles against an almost untameable Nature and has secured only a small portion of ground for cultivation. These cultivated fields, small conquests of industry and charity for humanity, form small islands so to speak among the heathland: they remind the traveler of Lybya's oases, where ever-fresh greenery contrasts the desert sands.

A moss common to tropical and European swamps, *Sphagnum palustre*, covers a large part of Germania. This moss renders vast terrain inhabitable to the nomadic people whose customs Tacite described. A geological fact supports this phenomenon. The most ancient peat bogs mix MURIATE DE SOUDE [muriate of soda?] and marine shells and owe their origins to *Ulves* and *Fucus;* by contrast, more recently formed and more scattered peat bogs arose from *Sphagnum* and *Mnium serpillifolium*, and their existence proves to what extent cryptogams previously covered the globe. By cutting down forests, agricultural people diminished the humidity of the climate, dried swamps, and economic plants gradually conquered plains previously dominated by cryptogams adverse to cultivation.

Although the phenomenon of plant associations seems to appear predominantly in temperate zones, the tropics also offer several examples. At 3,000 m, on the back of the Andes range, there grow *Brathis juniperina, Jarava* (a grass genus neighboring *Papporophorum*), *Escallonia myrtilloides,* several species of *Molina*, and primarily *Tourrettia*, whose pith provides food for which indigenous Indians compete with bears. In the plains that separate the Amazon and Chinchipe rivers, we find a combination of *Croton argenteum, Bougainvillea,* and *Godoya*, and, as in the savannahs of Orinoco, the palm *Mauritia*, the herbaceous sensitive plants, and *Kyllingia*. In the kingdom of New Grenada, Bambusa and *Heliconia*

offer uniform bands uninterrupted by other plants: but these plant associations of same species are invariably less spread out and less numerous than in more temperate climates.

Geology pronounces on the ancient relationship of neighboring continents based on analogous hillsides, shallows of the Ocean, and the identity of animals which inhabit these. The Geography of Plants provides invaluable material for this sort of research: it can, to a certain extent, recognize the islands which—previously united—separated from one another, thus announcing that the separation of Africa from southern America occurred before the development of organisms. This science also shows which plants are common to both eastern Asia and the hillsides of Mexico and California. There are plants which exist in all zones and at every elevation above sea level. The Geography of Plants can help back up with some certitude the first physical state of the globe: it can decide whether, after the retreat of the waters whose abundant and agitated traces remain on conchiferous rocks, the whole surface of the world was immediately covered by a diversity of plants, or if—conforming to the traditions of different peoples—the globe at rest only produced plants in one region which sea currents then transported over the centuries to progressively more remote zones.

It is this science which examines if, across the immense variety of vegetative forms, one can recognize some primitive forms and if species diversity should be considered an effect of a degeneration which—with time—rendered the initially accidental forms constant.

If I dared to draw general conclusions on the phenomena I observed in the two hemispheres, the germs of cryptogams struck me as the only ones which Nature develops spontaneously in all climates. *Dicranum scoparium* and *Polytrichum commune, Verrucaria sanguinea* and *Verrucaria limitata* of Scopoli grow at all latitudes in Europe as at the equator, on the highest mountain ranges as at sea level: anywhere with shade and humidity.

On the banks of the Madeleine, between Honda and Egyptiaca, on a plain where the thermometer invariably indicates 28 to 30°C, at the base of *Macrocnemum* and *Ochroma*, mosses form a cover as beautiful and green as any found in Norway. If other travelers have asserted that cryptogams are very rare in the tropics, their assertions are doubtless based on the fact that they only visited the arid coasts and the cultivated islands without penetrating to the inland of continents. Lichenous plants of the same species can be found at all latitudes: their form appears as independent of the influence of climate as Nature is of the rocks it inhabits.

We do not yet know of a phanerogam whose organs are flexible enough to adjust to all zones and altitudes. In vain we pretended that *Alsine media*, *Fragaria vesca*, and *Solanum nigrum* enjoyed this flexibility reserved for man and the mammals surrounding him. The American and Canadian strawberry differs from the European. M. Bonpland and I believe we discovered several feet of the latter on the Andean cordillera, in passing the valley running from the Madeleine to the Cauca, by the snows of Quindiu. The solitude of the forests, composed of *Styrax*, passifloras growing on trees, wax palms, the absence of cultivation in the surroundings, and other circumstances seem to exclude any suspicion that these strawberry plants were spread by man's hand or by birds; but, had we perhaps discovered this plant in flower, we would have found it specifically different to *Fragaria vesca*, in the same way that subtle nuances differentiate *Fragaria elatior* from *Fragaria virginina*. During the five years that we botanized the two hemispheres, we—at the very least—never collected a European plant spontaneously produced by the southern American soil. We limit ourselves to believe that *Alsine media, Solanum nigrum, Sonchus oleraceus, Apium graveolens,* and *Portulaca oleracea* are plants which, like the Caucasian race, are very scattered in the northern part of the ancient continent. As we still know so little of the productions of the interior of soils, we must abstain from all general conclusions: I might add, otherwise, that we risk falling into the same trap as geologists who construct the entire globe according to the model of hills which surround them.

To decide on the big problem of plant migration, the Geography of Plants descends into the interior of the globe: it consults the ancient monuments Nature left behind as petrifications, in the fossil wood, and in the tombs of the first vegetation of our planet that are the carbon layers. It discovers the petrified fruits of the Indies, palms, tree ferns, Scitaminaea, and tropical bamboo all buried in the frozen ground of the north; it considers whether the equinoctial productions such the bones of elephants, tapirs, crocodiles, and marsupials recently unearthed in Europe could have been transported to temperate climates by the strength of submerged currents, or if these same climates previously sustained palms and tapirs, crocodiles and bamboo. We are inclined toward the latter opinion when we consider the local circumstances that accompanied the petrifications in the Indies. But can we admit to such great changes in atmospheric temperature without recourse to a shift in the stars, or a change in the axis of the Earth, which current astronomical knowledge claims unlikely? If the most striking geological phenomena tell us that in the past the entire crust of the planet was in a liquid state, and if stratification and differing rocks indicate to us that the formation of mountains and the crystallization of great masses around a common core were not completed at the same time on the entire surface of the globe, then we can conceive that the passage from a liquid to a solid state must have released an immense quantity of heat, and increased for a time the temperature of a region independently of solar heat: but would this local temperature increase have lasted Nature's required time to explain such phenomena?

Changes observed in the light of stars caused us to suspect that our sun endures analogous variations. Would an increase in the intensity of solar rays at particular times have spread tropical heat to the regions neighboring the poles? Are these variations which render Lapland habitable for equinoctial plants, elephants, and tapirs periodical? Or are they the effect of some passing and perturbing causes of our planetary system?

These discussions link the Geography of Plants to geology. It is its spread since the beginning of the primitive history of the globe that the Geography of Plants offers man's imagination so rich and interesting a field to cultivate.

Plants, if indeed analogues of animals by the irritability of their fibers and the stimulants which excite them, are essentially different in their mobility. The majority of animals do not leave their mother until adult. On the other hand, plants are fixed in place after their development and can only travel when still contained in an egg whose structure favors mobility. However, it is not only winds, currents, and birds that aid the migration of plants; man primarily takes care of this.

Once he abandoned the wandering life, he gathered around him animals and plants useful in clothing and feeding him. This transition from a nomadic to an agricultural lifestyle was belated with the people of the North. In equinoctial regions, between Orinoco and Amazon, the thickness of the forests prevents the savage from sustaining himself from hunting alone: he is forced to take care of some plants for subsistence such as several feet of *Jat-*

ropha, banana, and *Solanum*. Fishing, fruits, palms, and these small cultivated plots (if I dare call such a small collection of plants a plot), constitute the basis of the southern American's food. A savage's state is primarily modified by the Nature of the climate and soil he inhabits. It is these modifications alone that distinguish the first inhabitants of Greece from shepherd Bedouins, and from Canadian Indians.

Some plants which have been central to gardening and agriculture since the earliest times, have accompanied man from one end of the globe to another. In Europe, this is how the vine followed the Greeks, wheat the Romans, and cotton the Arabs. In America, the Tulteques [Toltecs, of Mexico] brought maize with them, and potatoes and quinoa are found everywhere where the inhabitants of the ancient Condinamarca passed. The migration of these plants is evident, but their homeland remains as unknown as that of the different races of man which we already find on every part of the globe since the beginning of their respective traditions. To the south and east of the Caspian Sea, to the banks of the Oxus [Amu Darya River], in the ancient Colchis [an area in Georgia, Asia], and particularly in the province of Curdistan [Kurdistan], where the high mountains are perpetually covered by snow as consequence of being 3,000 m above sea level, the soil is covered by lemon, pomegranate, cherry, pear, and all the other fruit trees found in our gardens. We ignore whether this is their natal site, or—if cultivated in the past—they have become wild and their existence confirms ancient agriculture of a region. It is these fertile countries between the Euphrates and the Indus, between the Caspian Sea, Pont-Euxine, and the Persian Gulf, which have provided the most precious produce in Europe. Persia sent us the walnut and peach; Armenia the apricot; Asia Minor the cherry and chestnut; Syria the fig, pear, pomegranate, olive, prune, and blackberry. In the time of Canto, the Romans knew not of cherries, peaches, or blackberries.

Hesiod and Homer already mentioned the cultivated olive in Greece and the island of the Archipelago. Under the reign of Tarquin the Elder, this tree only existed in Italy, Spain, and Africa. Under the consul of Appius Claudius olive oil was still very rare in Rome, but in the time of Pliny, olive trees had already reached France and Spain. The vine that we cultivate today did not originate in Europe: it appears wild on the hillsides of the Caspian Sea, Armenia, and Kerman. From Asia, it passed to Greece and, from there, to Sicily. The Phoenicians carried it to southern France, and the Romans planted it on the banks of the Rhine. The species of *Vites* that we find wild in northern America, and which gave the name of land of wine (Winenland) to the first part of the new continent the Europeans discovered, are very different to our *Vitis vinifera*.

A cherry-laden tree decorated the triumph of Lucullus; this was the first tree of this species to be seen in Italy. The dictator had removed it from the province of Pontus, and after the victory he carried it off to Mithridates. In less than a century, the cherry tree was common in France, Germany, and England. This is how man changes the surface of the globe to his liking, and gathers around him plants from the most remote climates. In the European colonies of both Indies, a small cultivated plot introduced coffee of Arabia, sugar cane of China, indigo of Africa, and a multitude of other plants belonging to both hemispheres. This variety of produce becomes all the more interesting when it recalls to the observer's imagination the series of events which spread the human race across the whole surface of the globe, and of which it appropriated all the produce.

This is how man—anxious and laboring, traveling to the various parts of the world—forced a certain number of plants to inhabit all climates and all altitudes; but this empire exerted on organisms has hardly de-Natured their primitive structure. The potato, cultivated in Chile at 3,000 m (1,936 fathoms) elevation, has the same flower of that plant introduced to the plains of Siberia. The barley that nourished the horses of Achilles was doubtless the same we sow today. The characteristic forms of plants and animals presented on the current surface of the globe do not appear to have been subjected to any changes since those ancient times. The ibis buried in the catacombs of Egypt, a bird whose antiquity goes almost as far back as the pyramids, is identical to that which fishes on the shores of the Nile today; its identity evidently proves that the enormous casts of fossil animals held in the bosom of the earth, not belonging to the variety of current species, in fact belong to a very different order of things than we currently live under, far too ancient for our traditions to include them.

Man, favoring the cultivation of newly introduced plants, has caused these to dominate over wild species; but this preponderance, which makes the appearance of European soil so monotone, and of which the botanist despairs during his excursions, only belongs to that small part of the globe where civilization has become perfect and, within which, by a necessary series of events, the population has increased the most. In the countries neighboring the equator, man is too weak to tame a vegetation which hides the soil from sight and leaves nothing free except the Ocean and rivers. Nature carries this wild and majestic characteristic before which all efforts at cultivation flounder.

The origin, the first homeland of the economic plants which have followed man since the most distant eras, is a secret and impenetrable as the first residence of all domestic animals. We ignore the homeland of grasses which provide the principal nourishment to the people of the Mongolian and Caucasian races; we do not know which region spontaneously produced the cereals, wheat, barley, oats, and rye. Rye does not even appear to have been cultivated by the Romans. We claimed to have found wild barley on the shores of Samara in Tartary, *Triticum spelta* in Armenia, rye in Crete, wheat in Baschiros in Asia: but these facts do not appear to be based on sufficient observation; it is very easy to mistake plants produced spontaneously, for plants— fleeing man's empire—have regained their former liberty. By devouring the grains of cereals, birds easily disseminate them in the woods. Plants which constitute the natural wealth of all inhabitants of the tropics, the banana tree, Caric *papaya,* Jatropha *manihot,* and maize have never been found in a wild state. I saw several feet of these on the shores of the Cassiquiaré and Rio-Negro, but the savage of these of regions, as melancholy as he is distrustful, cultivates small plots in the most solitary places; he abandons them shortly afterwards, and the plants left behind soon look natural to the soil that produced them. The potato, this beneficial plant upon which a large part of the population of the more barren countries in Europe subsists on, presents the same phenomenon as the banana tree, maize, and wheat. From the little research I could undertake in the field, I never learnt of any traveler who found potato in the wild, neither on the summit of the cordillera of Peru, nor in the kingdom of New Grenada where this plant is cultivated with *Chenopodium quinoa.*

These are some of the considerations agriculture presents, and its various produce depends on the latitude, origin, and needs of people. The influence of food, more or less stimulating the character and energy of passions, naval history, and wars undertaken for the dispute of produce of the vegetable kingdom; these all link the Geography of Plants to the political and moral history of man.

Without a doubt, these connections sufficiently demonstrate the area of science I am trying to de-limit; but man's sensitivity to the beauties of Nature also explains the influence vegetation's appearance has on the taste and imagination of people. Man would be advised to examine what the character of vegetation consists of, and the variety of sensations vegetation produces in the soul of those who contemplate it. These considerations are all the more important because they touch upon the means by which the arts of imitation and descriptive poetry act on us. The simple appearance of Nature, the sight of fields and woods, cause a rejoicing that differs essentially from the impression a particular study of the structure of an organized being provides. Here, it is the detail that interests us and excites our curiosity; there, it is the whole, whole masses, that agitate our imagination. What more differing impressions between the appearance of a vast prairie bordered by a few trees, and the appearance of a thick and sombre wood mixed of oak and fir trees? What more striking a contrast than that between the forests of temperate zones, and those of the equator, where the naked and entwined trunks of palms lift themselves above flowering mahogany to form majestic porticos in the air above? What is the moral cause of these sensations? Are they produced by Nature, by the grandeur of masses, the contour of forms, or the haven of plants? How can this haven, this view of Nature more or less rich, more or less pleasant, influence the mores and, primarily, the sensitivities of peoples? Of what consists the character of the vegetation of the tropics? What difference in physiognomy distinguishes plants from Africa from those of the New Continent? What analogy of forms unites Andean alpine plants with those found on the summits of the Pyrennées? These are questions little broached to at present, and doubtless deserve to occupy the physicist.

Among the variety of plants that cover the framework of our planet, we can distinguish without hesitation some general forms to which the majority of the others can be reduced, and which present as many analogous families or groups between them. I limit myself to naming fifteen of these groups whose physiognomy offer an important study to the landscape painter: 1. the form of the Scitaminae (*Musa, Heliconia, Strelitzia*); 2. those of palms; 3. tree ferns; 4. the form of *Arum, Pothos, and Dracontium;* 5. that of fir trees (*Taxus, Pinus*); 6. all the *Folia acerosa;* 7. that of tamarins (*Mimosa, Gleditsia, Porlieria*); 8. the form of the Malvaceae *(Sterculia, Hibiscus, Ochroma, Cavanillesia*); 9. those of lianas (*Vitis, Paullina*); 10. those of orchids (*Epidendrum,*

Serapias); 11. those of prickly pears (*Cactus*); 12. the Casuarinaceae (*Equisetum*); 13. those of grasses; 14. those of mosses; 15. and finally, those of lichens.

These physiognomic divisions have almost nothing in common with those that botanists make to this day along very different principles. Here, all we mean are the larger contours that determine the physiognomy of vegetation and the analogy of impressions on the observer of Nature, whereas descriptive botany groups plants according to the affinity presented by the different smaller, but most essential, parts for fructification. It would be an undertaking worthy of a distinguished artist to study the physiognomy of the plant groups in Nature I have enumerated here, and not in glasshouses and botanical volumes. What more interesting a subject for a painting than the trunk of a palm balancing its variegated leaves above a group of *Heliconia* and banana trees? What more picturesque a contrast than ferns in a tree surrounded by Mexican oak trees?

It is in the absolute beauty of forms, in harmony and contrast, that the assemblages are created of what we call the "natural character" of this or that region. Some forms, often the most beautiful (Scitaminae, palms, and bamboos), are entirely absent in temperate zones; others, for example trees with pinnate leaves, are very rare and less elegant. Arborescent species are low in number, smaller, and less weighed down by visually pleasing flowers. Also, the frequency of social plants mentioned earlier, and man's culture, make the soil's appearance less monotonous. By contrast, in the tropics Nature has contented itself to assemble all forms. Pines appear to be missing at first glance, but in the Andes of Quindiu, and in the temperate forests of Oxa and Mexico, there are cypress, fir, and juniper trees.

Plant forms closer to the equator are generally more majestic and imposing; the veneer of leaves is more brilliant, the tissue of the parenchyma more lax and succulent. The tallest trees are constantly adorned by larger, more beautiful and odoriferous flowers than in temperate zones. The weathered bark of their ancient trunks forms the most pleasant contrast against the young foliage of lianas, *Pothos*, and particularly orchids whose flowers imitate the form and plumage of the birds feeding on their nectar. However the tropics never offer our eyes the green expanse of prairies bordering rivers in the countries of the north: one hardly ever has the gentle sensation of spring awakening vegetation. Nature, beneficial to all beings, has reserved for each region particular gifts. A tissue of fibers more or less lax, vegetable colors more or less brash depending on the chemical mixture of elements and the stimulating strength of solar rays: these are just some of the causes that give each zone of the globe's vegetation its particular character. The great heights to which the soil near the equator elevates itself give the inhabitants of the tropics the curious spectacle of plants whose forms are the same as the plants in Europe.

The Andean valleys are adorned with banana and palm trees. Higher up grows that beneficial tree whose bark is a most efficient and rapid febrifuge. In this temperate region of the cinchona, and higher up toward that of the *Escallonia*, grow oak trees, fir trees, *Berberis*, *Alnus*, *Rubus*, and a mass of genera which we believe only belong to northern countries. Any inhabitant of the equinoctial regions also knows of all the plant forms Nature surrounds him with: the earth displays before his eyes as varied a spectacle as does the azure-colored vault of the sky which cannot hide any constellation from his sight.

Europeans do not enjoy this same advantage. Languid plants cultivated by love of science or refined luxury in glasshouses only hint at the shadow of the majesty of the equinoctial plants: many forms remain unknown to them forever, but perhaps the richness and perfection of their languages, the imagination and sensitivity of their poets and painters compensate for this. It is the arts of imitation which reproduce before our eyes the varied image of the equatorial regions. In Europe, an isolated man living on an arid coast can enjoy in his mind the scenery of the remote regions. If his soul is receptive to the production of art, and if his cultured mind is broad enough to reach for the great concepts of general physics, he is able—from the bottom of his loneliness, and without leaving his home—to appropriate all that the intrepid naturalist has discovered traveling through the air and ocean, penetrating subterranean caves, or climbing icy summits. It is probably in this way that the lights of civilization have the greatest influence on our individual happiness. They make us live in both past and present times, gathering around us all what Nature has produced in the various climates, bringing us into communication with all the peoples on Earth. Thanks to the discoveries we have already made, we can project ourselves into the future and, by foreseeing the consequences of the phenomena, we can erect forever the laws to which Nature submits. It is in undertaking these re-

	east of the Alleghanies		west of the Alleghanies
Aesculus flava can be found from	36° latitude	to	42° latitude.
Juglans nigra	41°		44°
Aristolochia sypho	38°		41°
Nelumbium luteum	40°		44°
Gleditsia triacanthos	38°		41°
Gleditsia monosperma	36°		39°
Glycine frutescens	36°		40°

searches that we prepare ourselves for an intellectual delight, a moral freedom that strengthens us against the blows of destiny, and which no external power could possibly destroy.

Additions to the Geography of Plants

In mentioning measures made by Spanish surveyors in this work, we made use of a reduction of vare of Castille in meters and fathoms which was not rigorous enough.[4] The vare is to the fathom 0.513074 : 1.196307, and instead of reducing by 2.3, one assumed one fathom = 2.3316 vares. Don Jorge Juan only accommodated for 2.32. However, consult the excellent work of M. Gabriel Ciscar *Sorba los nuevos pesos y medidas decimales* (1800). The 7,496 vares that the beautiful maps of Madrid's Deposito Hydrografico give to Chimborazo are, as a consequence, only 3,217 fathoms, which is the same number Bouguer published in his illustration of the world. The mountain from S. Elie is 6,507 vares, or 2,792 fathoms (5,441 m). That of Beau-Temps is 5,368 vares, or 2,304 fathoms (4,489 m). See *Viaje al Estrecho de Fuca hecho por las Goletas sutil y Mexicana* (1792), p. cxx–cxv.

II

In 1800, M. Barton read a memoir on the Geography of Plants of the United States to the Society of Philadelphia; it has not yet been published, but contains the most interesting ideas. He observed that *Mitchella repens* is the most scattered plant in North America. It occupies all the ground from 28° to 69° latitude. *Arbutus uva ursi* can also be found from New Jersey to 72° latitude where M. Hearne observed it.[5] On the contrary, Gordonia *Franklinia*, and *Dionaea muscipuls* are very isolated on a small terrain. M. Barton remarked that, in general, the same species of plants spread further north in countries situated to the west of the Alleghany Mountains than to the east where the climate is colder. Cotton is cultivated in Tennessee at a latitude that it is not found in North Carolina. The eastern hillsides of Hudson Bay are devoid of vegetation, whereas its western hillsides are covered. [See table above.]

M. Barton observed that: even *Crotalus horridus* (timber rattlesnake) can be found to the east of the Alleghanys until 44° latitude, whereas it spreads toward the north to the west of the mountains until 47° latitude. Also compare the excellent work of M. Volney on the soil and climate of the United States.[6]

4. The length of the meridian degree at latitude 45° was calculated to be 57,027 fathoms. EDS.

5. Here, Humboldt refers to the botanist, Benjamin Smith Barton (1766–1815) and Samuel Hearne, the author of *A Journey from Prince of Wales Fort in Hudson's Bay to the North Ocean . . . in the Years 1979, 1770, 1771, and 1772* (London: A. Strahan & T. Cadell, 1795). EDS.

6. Constantin François de Chasseboeuf, compte de Volney's *View of the Climate and Soil of the United States* (1804), is based on travels during 1795 and 1796. EDS.

next proceeded to describe the *Pholas dactylus* which he had found in clay-slate in Cornwall, and to describe particularly the form and actions of the animal, which he had kept alive in his house more than a month (there were fifteen or sixteen shells of all sizes), and although he marked the slab in which they were, he could not perceive that *they turned round for the purpose of boring.* In the same slab he also found *Pholas parva.*

Report on the Mollusca and Radiata of the Ægean Sea, and on their distribution, considered as bearing on Geology. By EDWARD FORBES, *F.L.S., M.W.S., Professor of Botany in King's College, London.*

THE British Association having done me the honour of requesting a report on the Mollusca and Radiata inhabiting the Ægean and Red Seas, considered more especially in their bearings on questions of distribution and of geology, I have now the pleasure of laying before this meeting such portion of it as relates to the eastern Mediterranean. The data upon which it is founded have been entirely derived from personal research during a voyage of eighteen months in the Ægean, when but few days passed by without being devoted to natural history observations. The calculations in the following pages have been based upon more than 100 fully recorded dredging operations in various depths, from 1 to 130 fathoms, and in many localities from the shores of the Morea to those of Asia Minor, besides numerous coast observations whenever opportunity offered. The circumstances under which these researches were made were peculiarly propitious. The merit of the results obtained is mainly due to Captain Graves in command of the Mediterranean Survey, at whose invitation the reporter joined H.M.S. Beacon as Naturalist, in April 1841, from which time, until his departure for England in October 1842, every possible assistance and means of observation were put at his disposal by that distinguished officer, and every cooperation afforded by the officers of the Survey. Without such aid it would have been quite impossible to have obtained the results now laid before the Association, which, from their having been made in connection with the Hydrographical Survey, may assume a value to which no private observations could lay claim*.

The Ægean Sea, although most interesting to the naturalist as the scene of the labours of Aristotle, has been but little investigated since his time. The partially-published observations of Sibthorpe, and the great French work on the Morea, include the chief contributions to its natural history. In the last-named work are contained catalogues of the Fishes and Mollusca, with notices of one or two Annelides. In all the marine tribes my lists greatly exceed the French catalogues, more than doubling the number of Fishes, and exceeding that of Mollusca by above 160 species, not to mention Radiata, Amorphozoa and Articulata. In the present report I propose to give an account of the distribution of the several tribes of Mollusca and Radiata in the eastern Mediterranean, exhibiting their range in depth, and the circumstances under which they are found; to inquire into the laws which appear

---

* A great portion of the observations among the Cyclades were made jointly with Lieut. Spratt, Assistant Surveyor of the Beacon, and of those relating to the coasts of Asia Minor with Mr. Hoskyn, late Master of the Beacon, and now Assistant Surveyor of H.M.S. Lucifer. Many independent observations of great value to the author were made by Lieut. Freeland, Lieut. Mansell, Mr. Chapman, and other officers of the Beacon, and he is desirous of recording his thanks to all the gentlemen named for their kindness in placing their collections at his disposal. He is happy to say that the Ægean researches have not ceased with his departure, Capt. Graves and his officers being actively engaged in natural history investigations in addition to their many scientific duties during the survey now in progress of the Island of Candia.

to regulate their distribution, and to show the bearings of the investigation on the science of geology.

I shall commence with an enumeration of the species of Mollusca and Radiata, prefacing the tabular view of each tribe with a few general remarks.

MOLLUSCA.

Cephalopoda.

Octopus vulgaris and *macropodius*, *Sepia officinalis* and *Sepiola rondeletii*, were the cuttle-fishes which I met with in the eastern Mediterranean. They are all inhabitants of the shallows, and are found in or near the littoral zone, where they are much sought after by the Greeks as articles of food. They are speared at night by torchlight when on their foraging excursions. The sandy shores of the island are thickly covered with the shell of the Sepia, sometimes forming beds of considerable thickness. In no instance did the shell occur when dredging, so that we may suppose that species to be confined to the littoral zone. The *Sepiola rondeletii* was taken on the coast of Asia Minor, as deep as 29 fathoms in a bottom of weed. *Octopus macropodius* only occurred once, and then among the rocks near watermark, in the Island of Cerigo, at the entrance of the Ægean. The *Argonauta* was much sought after, but never found. It is, however, a recorded inhabitant of the shores of Greece.

Pteropoda.

Eight species of Pteropoda, members of the genera *Hyalæa*, *Cleodora* and *Criseis*, inhabit the Ægean, and appear to be equally diffused in all parts of the eastern Mediterranean. The white mud which forms the sea bottom between 100 and 200 fathoms abounds with their remains, many hundreds coming up in a single dredge, chiefly *Criseis* and *Cleodora*. In the muddy deposits of upper regions they are scarce, in those of shallow water altogether absent. Though immense numbers of their dead shells were taken, comparatively few of these testacea occurred in a living state. Of the eight species four were taken alive, three of which were *Criseis*, and the fourth *Hyalæa tridentata*. The last was only observed once in the Bay of Cervi, at the entrance of the Ægean, in August 1841: the *Criseis* were abundant in the spring of the same year. They usually abound about three hours after noon and towards nightfall, sparkling in the water like needles of glass. Throughout the summer and autumn they were very seldom met with. It would appear that great flocks of Pteropoda live in the deeper parts of the sea, ascending to the surface only occasionally, and at definite seasons. That their range in depth is limited, is evident from the fact that their remains abound only between 100 and 200 fathoms, diminishing above and below that region.

Nucleobranchiata.

Seven species of undoubted Nucleobranchiata, with three probable members of that order, inhabit the Ægean, representatives of genera, four of which are shell-bearing and two naked. The observations regarding habitat and time of appearance apply equally to the members of this order and those of the last, with the exception of the *Firolæ*, which may be seen during most months of the year. Of the testaceous nucleobranes, the *Atlanta peronii* and two species of *Ladas* appear to be universally diffused in the Ægean. *Carinaria* is very rare, having only occurred twice, and then dead. A little shell of Bellerophon-like appearance is abundant in the mud of great depths, and from its resemblance to the young state of *Carinaria* I have placed it here. Two species of that very anomalous genus *Sagitta* were met with

K 2

[Pages 132–51 are omitted. Reading continues with page 152.]

lipora serpens in 20 to 40 fathoms. *Retepora* abundant between 15 and 30. *Alecto* incrusting shells in 150 fathoms. Four species of coral were taken, though dead, at 105 fathoms. *Eudendrium* was found at 20 fathoms. *Valkeria* and *Campanularia* at 30. *Crisia* at 20. *Actinia* ranged from the surface to 20 fathoms. *Alcyonium* as deep as 70.

Amorphozoa.

Sponges abound in the Ægean, inhabiting all depths of water between seamark, where the rocks are often of a brilliant scarlet with incrusting species, to nearly 200 fathoms, a sponge allied to *Grantia* having been dredged alive at 180 fathoms, and a small species of another genus at 185. The sponge of commerce is procured by divers from rocks in various depths between 7 and 30 fathoms. Most of the larger species are found at lesser depths, very large ones occurring in the second zone or region. The forms of the species do not appear to bear any relation to the depth in which they are found, tubular sponges, globular, incrusting and palmate species all inhabiting the littoral zone. I met with about twenty species of *Amorphozoa* in the eastern Mediterranean.

The distribution of marine animals is determined by three great primary influences, and modified by several secondary or local ones. The primary influences are climate, sea-composition and depth, corresponding to the three great primary influences which determine the distribution of land animals, namely climate, mineral structure and elevation. The first of these primary marine influences is uniform in the eastern Mediterranean. From Candia to Lycia, from Thessaly to Egypt, we find the same species of Mollusca and Radiata assembled together under similar circumstances. The uniformity of distribution throughout the Mediterranean is very surprising to a British naturalist, accustomed as we are to find distinct species of the same genera, *climatally representative* of each other, in the Irish and North seas, and on the shores of Devon and Zetland. The absence of certain species in the Ægean which are characteristic of the western Mediterranean, is rather to be attributed to sea-composition than to climate. The pouring in of the waters of the Black Sea must influence the fauna of the Ægean and modify the constitution of its waters. To such cause we must attribute the remarkable fact, that with few exceptions individuals of the same species are dwarfish compared with their analogues in the western Mediterranean. This is seen most remarkably in some of the more abundant species, such as *Pecten opercularis, Venerupis irus, Venus fasciata, Cardita trapezia, Modiola barbata,* and the various kinds of *Bulla, Rissoa, Fusus,* and *Pleurotoma,* all of which seemed as if they were but miniature representatives of their more western brethren.

To the same cause may probably be attributed the paucity of *Medusæ* and of corals and corallines. Sponges only seem to gain by it. The influence of depth is very evident in the general character of the Ægean fauna, in which the aborigines of the deeper recesses of the sea play an important part numerically, both as to amount of species and individuals.

The secondary influences which modify the distribution of animals in the Ægean are many. First in importance ranks the character of the sea-bottom, which, though uniform in the lowest explored region, is very variable in all the others. According as rock, sand, mud, weedy or gravelly ground prevails, so will the numbers of the several genera and species vary. The presence of the sponges of commerce often depends on the rising up of peaks of rock in the deep water near the coast. As mud forms by much the most extensive portion of the bottom of the sea, bivalve Mollusca abound more individually though not specifically than univalves. As the deepest sea-bottom is of fine mud, the delicate shells of Pteropoda and Nucleobranchiata are for the most part only preserved there. Where the bottom is weedy we find the naked Mollusca more numerous than elsewhere; where rocky, the strong-shelled Gasteropoda and active Cephalopoda. Few species either of Mollusca or Radiata inhabit all bottoms indifferently.

The nature of the sea-bottom is mainly determined by the geological structure of the neighbouring land. The general character of the fauna of the Ægean is in a great measure dependent on the great tracts of scaglia which border it, and of which so many of its islands are formed. The degradation of this cretaceous limestone fills the sea with a white chalky sediment, especially favourable to the development of Mollusca. Where the coast is formed of scaglia numerous marine animals abound which are scarce on other rocks. The genera *Lithodomus* and *Clavagella* among Mollusca, the *Cladocora cæspitosa* among Zoophytes, are abundant in such localities only.

In a report on the distribution of British terrestrial and fluviatile Mollusca, which I had the honour of presenting to the Association at Birmingham, I asserted that a remarkable negative influence was exercised by serpentine on the distribution of pulmoniferous Mollusca. This I have had peculiarly favourable opportunities of confirming in the Ægean, where whole islands being formed of serpentine, the almost total absence of those animals which are abundant on the islands of other mineral structure is most striking. But I found further, that not only does serpentine exercise a negative influence on air-breathing Mollusca, but also on marine species. An extensive tract on the coast of Lycia and Caria, indented with deep and land-locked bays, is formed of that rock. In such bays, with the exception of a few littoral species which live on all rocks, we find an almost total absence of Testacea; whilst in correspondent bays in the neighbouring districts, formed of scaglia, of saccharine marble, and even of slate, we find an abundance of Testacea, so that it can hardly be doubted that the absence or scarcity of shelled Mollusca in such case is owing to negative influence exercised by the serpentine. *The outline of the coast* is evidently an important element in such influences, or in modifying it.

Tides and *currents* in most seas are important modifying influences. In the Ægean the former are so slight as scarcely to affect the fauna; the latter, in places, must be powerful agents in the transportation of species and of the spawn of marine animals. Their action, however, like that of storms, appears materially to affect the upper regions only; the transportation of the species of one region into another seldom extending further than that of the regions immediately bounding that in which it is indigenous. Certain species, such as the *Rissoæ*, which live on sea-weed, may occasionally fall to the bottom region, of which they are not true natives, and may live for a time there, but such cases appear to be rare, and the sources of fallacy from *natural transportation* are fewer than might be imagined at first thought, and in most cases have arisen rather from the form of the coast than from currents. Thus where the coast-line is very steep, the sea suddenly deepening to 60 or 70 fathoms close to the rocks, limpets, littoral *Trochi* and other shells, when they die, fall to the bottom, and are found along with the exuviæ of the natural inhabitants of those depths. Several instances of this occurred during dredging.

The *influx of fresh water*, whether continual, or where a river empties itself into the sea, or temporary, as on the coast of Asia Minor during the rainy season, when every little ravine becomes suddenly filled with a raging torrent, bearing down trees and great masses of rock, and charged with thick mud, frequently modifies the marine fauna of certain districts very considerably. The first generates great muddy tracts, which present a fauna peculiar to themselves : the second, though of short duration, deposits detached patches of conglomerate, and by the sudden settling of the fluviatile mud forms thin strata at the bottom of the sea, often containing the remains of terrestrial and fluviatile animals, soon to be covered over by marine deposits with very different contents. From the influx of a great river we may have tropical or subtropical, terrestrial or fluviatile forms mingled with temperate marine. Thus among forty-six species of Testacea collected by Captain Graves and Mr. Hoskyn on the shore at Alexandria, there are four Egyptian land and freshwater Mollusca, three of which are of truly subtropical forms, viz. *Ampullaria ovata*, *Paludina unicolor*, and *Cyrena orientalis*. The marine associates of these are, however, noways more southern in appearance, and for the most part identical as species with the Testacea which strew the shore at Smyrna or at Toulon, in the former case mingled with Melanopsis, in the latter with characteristic European Pulmonifera.

When the sea washes the shores of Egypt, remains of vegetables of a subtropical character become mingled with similar associations of marine Mollusca with those in which the relics of more northern plants become imbedded in the waters of the Black Sea. The Nile may carry down the woods and animals of Upper Egypt, the Danube those of the Austrian Alps. Deposits presenting throughout similar organic contents of marine origin, may contain at one point the relics of marmots and mountain salamanders, at another those of ichneumons and crocodiles.

Vegetable remains are being imbedded in strata forming at very different depths. Thus olive leaves were scattered among the mud dredged from a depth of 30 fathoms on the coast of Lycia, at Symboli, and date stones and monocotyledonous wood from a depth of nine fathoms off Alexandria. Of course the associated Mollusca were very distinct in each instance, in the first being members of the fourth, in the second of the second region of depth.

Provinces of Depth.

There are eight well-marked regions of depth in the eastern Mediterranean, each characterised by its peculiar fauna, and when there are plants, by its flora. These regions are distinguished from each other by the associations of the species they severally include. Certain species in each are found in no other, several are found in one region which do not range into the next above, whilst they extend to that below, or *vice versâ*. Certain species have their maximum of development in each zone, being most prolific in individuals in that zone in which is their maximum, and of which they may be regarded as especially characteristic. Mingled with the true natives of every zone are stragglers, owing their presence to the action of the secondary influences which modify distribution. Every zone has also a more or less general mineral character, the sea-bottom not being equally variable in each, and becoming more and more uniform as we descend. The deeper zones are greatest in extent ; so that whilst the first or most superficial is but 12, the eighth, or lowest, is above 700 feet in perpendicular range. Each zone is capable of subdivision in smaller belts, but these are distinguished for the most part by negative characters derived from the cessation of species, the range of which is completed, and from local changes in the nature of the sea-bottom.

First Region, or Littoral Zone.

The first of the provinces in depth is the least extensive, and two fathoms may be regarded as its inferior limit. Its mineral nature is as various as the coast-line, and its living productions are influenced accordingly; sand, rock or mud presenting their several associations of species. Limited, too, as is its extent, it nevertheless presents well-marked subdivisions. That portion which forms the water-mark, and which (though in the Mediterranean the space be very small in consequence of the very slight tides) is left exposed to the air during the ebb, presents species peculiar to itself. Such on rock are *Littorina cœrulescens, Patella scutellaris, Kellia rubra, Mytilus minimus,* and *Fossarus adansoni*; on sand, *Mesodesma donacilla,* a bivalve which buries itself in great numbers immediately at the water's edge; in mud, a mineral character almost always derived from the influence of the influx of fresh water, *Nassa mutabile* and *neritoidea*; *Cerithium mammillatum* on all bottoms, usually under stones or weed; *Truncatella truncata* and *Auricula.* All these species are gregarious, most of them occurring in considerable numbers, and they are almost all Mollusca having a great geographic range; eight out of the eleven being widely distributed in the Atlantic, and one, the *Littorina cœrulescens,* extending from Tristan d'Acuna to the shores of Norway. The fuci of the coast-line, such as *Dictyota dichotoma* and *Corallina officinalis,* are also species of wide geographic diffusion. The bottomless barnacles (*Ochthosia*) are characteristic of this belt.

Immediately below this boundary line between the air and the water, we have a host of Mollusca of peculiar forms and often varied colours, associated with numerous Radiata and Articulata. In this under-belt we find the most characteristic Mediterranean forms, those which exhibit the action of the climatal influence most evidently. Boring in the sand live *Solen strigillatus, Lucina desmarestii, Amphidesma sicula, Venerupis decussata,* and various species of *Donax, Tellina* and *Venus*; in the mud abounds *Lucina lactea*; on the rocks we find *Cardita calyculata, Arca barbata, Chama gryphoides, Lithodomus, Chiton squamosus* and *cajetanus, Patella bonnardi, Fissurella costaria,* several species of *Vermetus, Haliotis,* numerous and peculiar *Trochi, Cerithium fuscatum, Fasciolaria tarentina, Fusus lignarius, Murex trunculus, Pollia maculosa, Columbella rustica, Cyprœa spurca,* and *Conus mediterraneus,* with various Radiata and Articulata, most of them peculiar forms. In this belt, in fact, we have the characteristic species of the Mediterranean fauna, those animals which give a subtropical aspect to the general assemblage of forms in that sea. It is worthy of note, that not only is the climatal influence evident in the colouring and size of the shells of Mollusca in this region, but also in that of the animals themselves, which often present the most varied combinations of brilliant hues, sources of well-marked specific character. This is especially the case with the Gasteropoda, and is equally true with the sublittoral forms of the Northern as of the Southern seas.

It is only in this subdivision of the highest zone that we see distinct instances of local distribution of species in the Ægean. This is especially the case with the genus *Trochus,* some of the species of which have a very limited distribution, though always abundant where they occur. It is also the case with the naked Mollusca and with Zoophytes. Among the last, the rocks of the first zone in Asia Minor are well distinguished from those in the islands, by the great abundance of a beautiful coral, *Cladocora cæspitosa,* which is found in large masses, but does not appear to live deeper than six or eight feet below the surface of the water. In the sheltered gulfs of Lycia and Caria, sponges (not the kinds used in commerce) of singular shapes and bright colours abound in this region, growing to a considerable size. In the Cyclades the beautiful *Actinea rubra* abounds in similar localities. *Padina pavonia* is the characteristic *Fucus* of the belt of the first region, and among its elegant fronds may be seen innumerable Crustacea prowling, whilst in the crevices of the rocks on which they grow live numerous fishes of the blenny and wrasse tribes, like all the other natives of this province, remarkable for the vivid painting of their skins.

The inhabitants of the lowest portion of this narrow but varied belt are equally characteristic, especially such as live on the sandy tracts covered with *Zostera.* The *Pinna squamosa* is most abundant here, and in rocky places the cuttle-fishes abound. On the Zostera live numerous *Rissoæ.*

Besides its true inhabitants, the littoral zone is continually receiving accessions to its fauna from the washing up of the exuviæ of the animals of the succeeding region, especially after storms, which strew the sandy shores with the remains of Mollusca. Mingled with these are the remains of freshwater animals carried into the sea by the streams. These are not necessarily found in the immediate neighbourhood of the streams by which they are brought down, but seem to be carried along the shore by eddies and currents, so that in a deep bay they may frequently be found at the opposite part of the shore to that where the stream which doubtless wafted them to the sea emptied itself, the depth of the intermediate gulf precluding the notion that they could have been washed across. Whilst the sea one day casts up numerous shells, Crustacea, &c., it often covers them up with silt the next, so that increasing alternations of organic bodies and sand or mud must be continually in process of formation in this region.

TESTACEA OF REGION I.

Lamellibranchiata.

Clavagella ——.*
Solen siliqua.
Solecurtus strigillatus.*
Ligula sicula.
Mactra stultorum.*
Kellia corbuloides.*
    rubra.*
Tellina donacina.
    fragilis.
    planata.
Lucina pecten.
    digitalis. ?
    lactea.*
    desmarestii.*
Venerupis irus.*
    decussata.*
Donax trunculus.*
    complanata.
    semistriata.

Mesodesma donacilla.*
Venus gallina.*
    decussata.*
    geographica. ?
Cardium rusticum.
    edule.*
Cardita calyculata.*
    trapezia.*
Arca barbata.*
    lactea.*
    noæ.*
Lithodomus lithophagus.*
Mytilus gallo-provincialis.*
    minimus.*
Pinna squamosa.*
Lima squamosa.*
    tenera. ?
Spondylus gadæropus.*
Ostrea plicatula.*

Gasteropoda.

Chiton squamosus.*
    cajetanus.*
    fascicularis.*

Patella scutellaris.*
    ferruginea.*
    bonnardi.*

Note.—The asterisk indicates that the species attains its maximum of development in that region; the note of interrogation implies that the species is probably a straggler.

Patella lusitanica.*
Gadinia garnoti. ?
Crepidula fornicata.
    unguiformis.*
Emarginula huzardi.*
Fissurella costaria.*
    gibba.*
Bullæa angustata. ?
    aperta. ?
Bulla striata.
    cornea. ?
    truncatula. ?
    truncata. ?
    striatula. ?
Eulima polita. ?
Parthenia elegantissima. ?
    humboldti.
Truncatella truncatulum.*
Rissoa desmarestii.*
    ventricosa.*
    oblonga.*
    violacea.*
    monodonta.*
    fulva.
    cancellata.
    granulata.
    montagui.*
    acuta.
    pulchella.
    conifera.
    cingilus.
    pulchra.
Littorina cœrulescens.*
Fossarus adansoni.*
Scalaria lamellosa. ?
Vermetus gigas.*
    subcancellatus.*
    arenarius.*
    glomeratus.*
    granulatus.*
Nerita viridis. ?
Haliotis lamellosus.*
Adeorbis subcarinata.
Trochus vielloti.*
    jussieui.*
    pallidus.*
    umbilicaris.*
    lyciacus.*
    richardi.*
    divaricatus.*
    articulatus.*

Trochus fragarioides.*
    therensis.*
    laugieri. ?
Phasianella pulla. ?
Ianthina nitens,* strag.
Cerithium fuscatum.*
    mammillatum.*
    lima. ?
    trilineatum. ?
Triforis adversum. ?
Pleurotoma albida. ?
    rude. ?
    purpurea. ?
    lævigata. ?
    lefroyi. ?
    fallax. ?
    linearis. ?
    lyciaca. ?
Fasciolaria tarentina.*
Fusus lyciacus. ?
    lavatus. ?
Murex brandaris. ?
    trunculus.*
    edwardsii.*
Ranella lanceolata. ?
Purpura hæmastoma.
Pollia maculosa.*
    candidissima. ?
Nassa reticulata.
    d'orbignii. ?
    variabile. ?
    cornicula.*
    mutabile.*
    gibbosula.*
    neritea.*
Columbella rustica.*
    linnæi.*
Mitra littoralis. ?
    cornea. ?
Marginella miliacea. ?
Ringuicula buccinea. ?
Cypræa lurida.
    rufa.*
    spurca.*
Conus mediterraneus.*
Dentalium 9-costatum. ?
    multistriatum. ?
    entalis. ?
    rubescens. ?
Auricula myosotis.*

158   REPORT—1843.

SECOND REGION.

The ground in the second region, which extends from two to ten fathoms, is most generally mud or sand, the former green with a beautiful Fucus, *Caulerpa prolifera*, abundant in the Archipelago, but I believe rare elsewhere, the latter abounding in *Zostera oceanica*. Great *Holothuriæ* are here found in abundance, and, among Mollusca, chiefly burying Conchifera. *Nucula margaritacea* and *Cerithium vulgatum* are the Testacea most generally distributed through this region. Those most prolific in individuals are, among Gasteropoda, *Cerithium vulgatum* and *lima*, *Trochus crenulatus* and *spratti*, *Rissoa ventricosa* and *oblonga*, and *Marginella clandestina*. Among Lamellibranchiata, *Tellina donacina*, *Lucina lactea*, *Nucula margaritacea*, and *Cardium exiguum*. Storms disturb this zone by washing up its inhabitants into the littoral region.

The smaller zoophytes, especially encrusting species and such as attach themselves to the leaves of *Zostera*, are frequent. *Caryophyllia cyathus* begins to appear here, ranging however through all the succeeding zones.

TESTACEOUS MOLLUSCA INHABITING THE SECOND REGION.

Lamellibranchiata.

Solen tenuis.*
  antiquatus.
Solecurtus strigillatus.
Ligula boysii.*
Solenomya mediterranea.*
Montacuta sp.
Byssomya guerinii.
Corbula nucleus.*
Pandora obtusa.
  rostrata.
Thracia phaseolina.
Psammobia vespertina.
Donax venusta.
Cytherea chione.
  lunata.
  apicalis.
Venus gallina.*
  verrucosa.*
  aurea.*
  geographica.*
Tellina donacina.*
  serrata.
  balaustina.
  distorta.*
Lucina flexuosa.
  pecten.
  lactea.*

Lucina rotundata.
  spinifera.
  transversa.
Cardium papillosum.*
  rusticum.
  exiguum.
Cardita sulcata.
  trapezia.
Arca barbata.
  lactea.*
Pectunculus glycimeris.*
Nucula emarginata.*
  nuclea.
Modiola barbata.*
  tulipa.*
  discrepans.*
  marmorata.*
Pinna squamosa.
Lima squamosa.
  tenera.
Pecten polymorphus.*
  hyalinus.*
  varius.
  sulcatus.
Spondylus gadæropus.
Ostrea plicatula.*
Chama gryphoides.

Palliobranchiata.

0.

Gasteropoda.

Chiton rissoi.*
  polii.*

Calyptræa sinense.*
Crepidula unguiformis.*

Emarginula huzardii.
Bulla hydatis.*
   cornea.
   ovulata.
   striatula.
   truncatula.*
   turgidula.
Natica valenciensii.*
   pulchella.
   olla.*
Eulima polita.*
   subulata.
Parthenia elegantissima.
Odostomia conoidea.
Rissoa desmarestii.*
   ventricosa.*
   oblonga.*
   violacea.*
   radiata.*
   cimicoides.*
   montagui.*
   buccinoides.*
   pulchella.*
   acuta.
Scalaria communis.
Turritella triplicata.
   terebra.*
Nerita viridis.*
Dentalium 9-costatum.*
   multistriatum.*
   entalis.*
   fissura.*
Trochus canaliculatus.*
   racketti.*
   spratti.*
   fanulum.*

Trochus adansoni.*
   conulus.*
   crenulatus.*
   gravesi.*
   exiguus.*
Turbo rugosus.*
Phasianella pulla.*
   intermedia.*
   vieuxii.*
Cerithium lima.*
   angustissimum.
Triforis adversum.*
Pleurotoma formicaria.*
   reticulata spinosa.*
   attenuata.*
   linearis.*
Fusus syracusanus.*
   lavatus.*
   lignarius.
Murex brandaris.*
   trunculus.*
   edwardsii.*
   fistulosus.*
Ranella gigantea.*
Nassa reticulata.*
   variabile.*
   musiva.
   granulata.*
   macula.*
   mutabile.*
Columbella rustica.*
   linnæi.*
Mitra obsoleta.*
Marginella clandestina.*
Ringuicula buccinea.*
Conus mediterraneus.*

THIRD REGION.

In this region, which extends from ten to twenty fathoms, the sea-bottom is very generally gravelly in places, great tracts of sand also being common. The *Caulerpa* and *Zostera* are still found, but cease towards its lower part. It may be regarded as a zone of transition presenting but few peculiarities. A very small and beautiful species of *Asterina* abounds on the fronds of *Zostera* here, and the large *Holothuriæ* are still abundant. *Aplysiæ* and the blue *Goniodoris* are the characteristic Mollusca. *Lucina lactea, Cardium papillosum, Tellina donacina,* and *Cerithium lima* are the Testacea most generally distributed. The species most prolific are *Cerithium lima, Cardium papillosum, Ligula boysii, Nucula margaritacea* and *emarginata, Lucina lactea* and *hiatelloides,* so that bivalves would appear to prevail.

TESTACEA OF REGION III.
Lamellibranchiata.

Solen tenuis. ? | Solen antiquatus.*

160   REPORT—1843.

Ligula boysii.*
Corbula nucleus.*
Neæra cuspidata.*
Pandora obtusa.
Thracia phaseolina.
Psammobia vesperti na.?
Tellina pulchella.*
    donacina.*
    serrata. ?
    balaustina.
Lucina flexuosa.*
    pecten.
    commutata.
    transversa.*
    lactea.*
    spinifera.*
Cytherea chione.
    lunata,
    apicalis.
Venus verrucosa.
    geographica.
    virginea.*
Cardium echinatum.
    papillosum.*
    exiguum.*

Cardium punctatum.*
Cardita sulcata.
    trapezia.*
Arca lactea.
Pectunculus glycimeris. ?
Nucula margaritacea.
    emarginata.
Chama gryphoides.
Modiola barbata.
    tulipa.
    discrepans.*
    marmorata.*
Pinna squamosa.
Lima squamosa. ?
    tenera. ?
    subauriculata.
Pecten jacobæus.
    polymorphus.
    hyalinus.
    opercularis.
    varius.
    pusio.
Spondylus gadæropus.
Ostrea plicatula.

Gasteropoda.

Calyptræa sinense.
Fissurella græca.
Bulla convoluta.
    ovulata.
    striatula.
    truncatula.
    truncata.
    akera.
Natica millepunctata.
    pulchella.
    guilleminii.
    valenciensii.
Eulima polita.
    subulata.
Parthenia elegantissima.
Odostomia conoidea. ?
Rissoa ventricosa.*
    violacea.*
    cimicoides.*
    montagui.
    acuta. ?
    conifera. ?
    pulchella.
Scalaria communis.
Turritella triplicata.
    terebra*.
Nerita viridis.

Trochus coutourii.
    canaliculatus.*
    racketti.*
    villicus.*
    spratti.*
    fanulum.
    adansoni.
    ziziphinus.*
    conulus.*
    crenulatus.*
    gravesi.*
    exiguus.
Turbo rugosus.
Phasianella pulla.
    vieuxii.*
Cerithium vulgatum.*
    lima.*
    angustum.*
Triforis adversum.*
Pleurotoma formicaria.
    bertrandi.
    reticulata spinosa.*
    gracilis.
    attenuata.
    ægeensis.*
    linearis.?
Fusus lignarius.

Fusus syracusanus.
　　lavatus.*
Murex brandaris.*
　　trunculus. ?
　　fistulosus. ?
Aporrhais pes-pelecani.*
Dolium galea. ?
Nassa prismatica.
　　variabile.*
　　granulata. ?

Nassa cornicula ?
Columbella rustica.*
　　linnæi.*
Mitra savignii.*
　　obsoleta.
Marginella clandestina.
Erato lævis.
Conus mediterraneus.?
Dentalium 9-costatum.*
　　multistriatum.

FOURTH REGION.

It extends through fifteen fathoms of length between twenty and thirty-five fathoms. The sea-bottom is very various, mud and gravel prevailing, sandy tracts being very rare. *Fuci* **are abundant, the** characteristic species being *Dictyomenia volubilis, Sargassum salicifolium, Codium bursa* and *flabelliforme,* and *Cystoceira.* The rare and curious *Hydrodictyon umbilicatum* was procured in this region on the coast of Asia Minor. Corallines are more frequent here than in the other zones. *Porites dædalea* occurs, but is very local. *Retepora cellulosa* is very abundant; several species of *Tubulipora* occur; *Myriapora truncata* and *Cellaria ceramioides* are characteristic species of this zone. Sponges abound, and some of the finest of those used in commerce grow here. Nullipore is abundant. *Echinidæ* are frequent, and *Comatula.* Crustacea are common, also *Annelides.*

Among Testacea the most generally distributed are *Nucula margaritacea* and *emarginata,* and *Dentalium 9-costatum:* those most prolific are *Nucula margaritacea, Arca lactea, Cardium papillosum, Corbula nucleus,* and *Ligula boysii; Dentalium 9-costatum* and *Cerithium lacteum. Mollusca tunicata* are common in this region.

TESTACEA OF REGION IV.

Lamellibranchiata.

Gastrochæna cuneiformis.
Solen tenuis.?
　　antiquatus.*
Ligula boysii.*
　　prismatica.
Kellia suborbicularis.*
Corbula nucleus.*
Neæra costellata.*
　　cuspidata.*
Pandora obtusa.*
Lyonsia striata.*
Thracia phaseolina.
Saxicava arctica.*
Psammobia discors.
　　ferroensis.
Tellina donacina.
　　serrata.
　　balaustina.
Lucina commutata.
　　digitalis.

Lucina transversa.
　　lactea.?
　　spinifera.
Astarte incrassata.
Cytherea apicalis.*
　　venetiana.
Venus verrucosa.
　　ovata.*
　　fasciata.
Cardium echinatum.
　　erinaceum.
　　lævigatum.
　　papillosum.*
　　exiguum.*
Cardita sulcata.*
　　squamosa.
　　trapezia.
Arca lactea.*
　　tetragona.*
　　noæ. ?

1843.   M

162   REPORT—1843.

Pectunculus glycimeris.
  pilosus.
  lineatus.
Nucula margaritacea.*
  emarginata.*
Chama gryphoides.
Modiola barbata.*
  tulipa.*
  discrepans.*
  marmorata.*
Pinna squamosa.
Avicula tarentina.
Lima squamosa.*
  tenera.

Lima fragilis.*
  subauriculata.
Pecten jacobæus.*
  polymorphus.*
  hyalinus.*
  testæ.*
  opercularis.*
  varius.*
  pusio.*
  similis.
Ostrea plicatula.
Anomia ephippium.*
  polymorpha.*

Palliobranchiata.

Terebratula detruncata.

Terebratula cuneata.*

Gasteropoda.

Chiton lævis.*
  freelandi.*
Calyptræa sinense.
Emarginula elongata.
Fissurella græca.*
Bullæa aperta.*
Bulla hydatis.
  cornea.*
  ovulata.*
  striatula.
  truncatula.
  truncata.
  convoluta.
Natica millepunctata.
  valenciensii.
  pulchella.
Eulima polita.
  nitida.
  subulata.*
Parthenia acicula.
  elegantissima.*
  scalaris.
  varicosa.
Odostomia conoidea.*
Rissoa ventricosa.*
  cimicoides.
  montagui.
  reticulata.
  acuta. ?
  pulchella.*
  striata.
  elongata. (?)
Turritella triplicata.*
  terebra.

Vermetus corneus.
Nerita viridis. ?
Trochus coutourii.
  magus.*
  spratti.*
  fanulum.
  adansoni.
  ziziphinus.*
  conulus.*
  gravesi.
  exiguus.*
Turbo sanguineus.*
  rugosus.*
Phasianella pulla.
  vieuxii.*
Cerithium vulgatum.*
  lima.*
  lacteum.
  angustissimum.
Triforis adversum.*
Pleurotoma formicaria.*
  reticulata var. spinosa.*
  maravignæ.*
  vauquelini.*
  gracilis.
  attenuata.*
  philberti.*
  turgida.*
  linearis.
Fusus lignarius. ?
  syracusanus.*
  lavatus.
Murex brandaris.*
  trunculus. ?

Murex cristatus.
   brevis.*
   fistulosus.
Aporrhais pes-pelecani.*
Nassa variabile.
   varicosa.
   granulata.
   prismatica.
Columbella rustica.*
   linnæi.
   gervillii.
Mitra ebenus.*

Mitra savignii.*
   obsoleta.*
   granum.*
Marginella clandestina.*
   secalina.*
   miliacea.
Erato lævis.
Tornatella fasciata.
Cypræa europæa.
Conus mediterraneus.?
Dentalium 9-costatum.*
   rubescens.*

FIFTH REGION.

From thirty-five to fifty-five fathoms, an extent of five fathoms more than the last, presents a well-marked fauna, and constitutes a fifth region. *Fuci* are much scarcer than in the last, but among its vegetable products are *Rytiphlœa tinctoria*, *Chrysimenia uvaria*, and *Dictyomenia volubilis*; the last, which gives a marked character to the preceding zone, being rare in this. Echinodermata are frequent here, Zoophytes not abundant. *Myriapora truncata* is frequent. The bottom is very generally nullipore and shelly. Muddy bottoms are scarce. The Testacea most generally distributed are *Nucula margaritacea*, *Pecten opercularis*, and *Turritella tricostata*. Those most abounding in individuals are *Nucula emarginata* and *striata*, *Cardium papillosum*, *Cardita aculeata*, and *Dentalium 9-costatum*.

TESTACEA OF REGION V.

Lamellibranchiata.

Solen tenuis.*
   antiquatus.*
Ligula boysii.
   prismatica.
Kellia suborbicularis.*
Corbula nucleus.*
   anatinoides.
Neæra cuspidata.*
   costellata.*
Pandora obtusa.
Lyonsia striata.?
Saxicava arctica.*
Psammobia discors.
   ferroensis.
Tellina donacina.
   serrata.
   balaustina.*
Lucina commutata.
   spinifera.*
Astarte incrassata.
Cytherea venetiana.
   apicalis.*
Venus verrucosa.
   ovata.
   fasciata.

Cardium echinatum.
   lævigatum.
   papillosum.
Cardita squamosa.
   trapezia.
Arca lactea.*
   imbricata.
   antiquata.
   tetragona.*
Pectunculus pilosus.
Nucula polii.
   margaritacea.*
   emarginata.*
   striata.*
Chama gryphoides.?
Modiola barbata.*
   tulipa.*
   discrepans.
   marmorata.
Lima squamosa.
   fragilis.*
   subauriculata.
   cuneata.
Pecten jacobæus.
   polymorphus.*

Pecten hyalinus.*
   testæ.*
   opercularis.*
   varius.*

Pecten pusio.*
   lævis.*
   fenestratus.
Anomia ephippium.

Palliobranchiata.

Terebratula detruncata. *
   cuneata. ?

Terebratula seminula.
Crania ringens.*

Gasteropoda.

Chiton lævis*
   freelandi.*
Lottia gussonii.
Calyptræa sinense.*
Emarginula capuliformis.*
   elongata.
Fissurella græca.*
Volva acuminata.
Bullæa aperta. ?
Bulla cornea.*
   utriculus.
   lignaria.
   ovulata.
   truncatula.*
   truncata.
Natica millepunctata. ?
   valenciensii. ?
   pulchella.*
Eulima distorta.
   nitida.*
Parthenia acicula.*
   elegantissima. ?
   pallida.
Odostomia conoidea.
Rissoa ventricosa.*
   cimicoides.
   reticulata.
Scalaria planicosta.
Turritella triplicata.*
   terebra. ?
Vermetus corneus.*
Siliquaria anguina.
Trochus coutourii.
   magus.*
   fanulum.
   ziziphinus.*
   gravesi.
   exiguus.
   millegranus.*
Turbo sanguineus.*

Turbo rugosus.*
Phasianella pulla. ?
Cerithium vulgatum.*
   lima.*
   angustum.*
Triforis adversum.*
Pleurotoma formicaria.*
   purpurea.
   reticulata.
   maravignæ.*
   vauquelini.
   gracilis.*
   attenuata.
   teres.
   philberti.
Fusus lavatus.
   muricatus.
   crispus.
   fasciolaria.
Murex brandaris.
   muricatus.
   distinctus.
   fistulosus.
Aporrhais pes-pelecani.*
Cassidaria tyrrhena.
Nassa intermedia.
Columbella rustica.
   linnæi.
Mitra ebenus.*
   obsoleta.
   phillippiana.
   granum.
Tornatella fasciata.
Marginella clandestina.*
   secalina.
Erato lævis.*
Cypræa europæa.*
Conus mediterraneus.?
Dentalium 9-costatum.

Sixth Region.

It extends through a range of twenty-four fathoms, between fifty-five and seventy-nine fathoms. Nullipore is the prevailing ground. *Fuci* have become extremely rare. *Cidaris histrix* is the characteristic Echinoderm. Several starfishes are not uncommon. *Venus ovata, Cerithium lima,* and *Pleurotoma maravignæ* are the most generally diffused species. *Turbo sanguineus, Emarginula elongata, Nucula striata, Venus ovata, Pecten similis,* and the various species of Brachiopoda those most prolific in individuals.

It will be observed, that although *Fuci* have become extremely scarce, and in the next zone altogether disappear, there are still a considerable number of Phytophagous Testacea. These are mostly found on "coral" ground, that is, on a clean bottom abounding in nullipore. Now that the observations of M. Decaisne, M. Kutzing and others have so clearly proved the vegetable nature of that singular production, so long regarded as a zoophyte, the source of the food of the Holostomatous Testacea in these deep regions is no longer problematical.

Testacea of Region VI.

Lamellibranchiata.

Ligula profundissima.
Kellia suborbicularis.*
Corbula nucleus.*
　　anatinoides.
Neæra cuspidata.
　　costellata.*
　　abbreviata.
Pandora obtusa.*
Lyonsia striata.*
Thracia pubescens.
Saxicava arctica.*
Kellia abyssicola.*
Lucina commutata.
　　bipartita.*
Astarte incrassata.
　　pusilla.
Cytherea apicalis.
Venus ovata.*
　　fasciata.
Cardium papillosum.
　　echinatum.
　　minimum.
Cardita squamosa.*
　　trapezia.

Arca lactea.*
　　scabra.
　　imbricata.
　　tetragona.*
Pectunculus pilosus.*
Nucula polii.
　　margaritacea.
　　striata.*
Modiola barbata.
Lima squamosa.
　　elongata.*
　　crassa.
Pecten jacobæus.
　　dumasii.
　　polymorphus.
　　hyalinus.
　　testæ.
　　varius.*
　　pusio.
　　pes felis.
　　similis.*
　　fenestratus.
　　concentricus.
Anomia polymorpha.

Palliobranchiata.

Terebratula truncata.*
　　detruncata.*
　　cuneata.*

Terebratula seminula.*
Crania ringens.

Gasteropoda.

Chiton lævis.
Lottia gussonii.

Lottia unicolor.*
Calyptræa sinense.

Emarginula elongata.
   capuliformis.
Fissurella græca.
Bullæa aperta. ?
Bulla cornea.*
   utriculus. ?
Coriocella perspicua.
Natica millepunctata.
   valenciensii.
   pulchella.
Eulima distorta.
   subulata.
   unifasciata.
Parthenia elegantissima. ?
Rissoa ventricosa. ?
   cimicoides.
   reticulata.*
   ovatella.
Turritella 3-plicata.*
   terebra.*

Siliquaria anguina.
Scissurella plicata.
Solarium stramineum.
Trochus coutourii.
   fanulum.
   exiguus.*
   millegranus.*
Turbo sanguineus.
   rugosus.*
Phasianella pulla.
Cerithium lima.*
   angustum.
Triforis adversum.
   perversum.*
Pleurotoma formicaria.*
   crispata.*
   reticulata var. spinosa.
   maravignæ.*
   vauquelini.

SEVENTH REGION.

The depths between 80 and 105 fathoms (an extent of 25), yield a characteristic fauna of their own. The sea-bottom is usually nullipore, more rarely sand or mud. Herbaceous *Fuci* have disappeared. *Echinodermata* are here not uncommon; *Zoophyta* and *Amorphozoa* scarce. Among the former are species of *Hornera*, *Lepralia* and *Cellepora*; among the latter a small round species of *Grantia* is frequent. *Echinus monilis*, *Cidaris histrix* and *Echinocyamus*, with some of the *Ophiuridæ*, are frequent alive: no *Asteriadæ* occur. *Mollusca tunicata* have ceased; as also *Nudibranchæa*. Crustacea are not unfrequent, as well as testaceous annelides, among which the glassy *Serpula* is very characteristic of this region.

The Testacea most generally distributed are *Lima elongata*, *Cardita aculeata*, *Rissoa reticulata*, and *Fusus muricatus*.

Those most prolific are *Rissoa reticulata*, *Turbo sanguineus*, *Venus ovata*, *Nucula striata*, *Pecten similis*, and the various species of Brachiopoda, which tribe abounds in this region.

TESTACEA OF REGION VII.

Lamellibranchiata.

Ligula profundissima.
Corbula nucleus.
Poromya anatinoides.
Neæra cuspidata.
   costellata.*
   abbreviata.
Pandora obtusa.
Saxicava arctica.*
Lucina commutata.
   bipartita.
Astarte incrassata.
   pusilla.
Cytherea apicalis.

Venus ovata.*
Cardium minimum.*
Cardita squamosa.*
Arca lactea.*
   scabra.
   imbricata.
   tetragona.
Nucula polii.
   margaritacea.
   striata.*
Modiola barbata.*
Lima elongata.
   crassa.

Pecten dumasii.
similis. ?
fenestratus. ?
concentricus. ?

Spondylus gussonii.*
Ostrea cochlear.
Anomia polymorpha.

Palliobranchiata.

Terebratula truncata.*
detruncata.*
lunifera.*
seminula.*

Terebratula vitrea.
appressa.*
Crania ringens.*

Gasteropoda.

Chiton lævis.*
Lottia unicolor.*
Pileopsis ungaricus.
Emarginula cancellata.
elongata.
capuliformis.
Fissurella græca.
Bullæa aperta. ?
Bulla utriculus.
Natica pulchella.
Eulima distorta.
subulata. ?
Parthenia elegantissima.
Rissoa ventricosa.*
reticulata.*
ovatella.
Turritella triplicata.
Scissurella plicata ?
Trochus tinei.
exiguus.*
millegranus.*

Turbo sanguineus.
rugosus.*
Phasianella pulla.*
Cerithium lima.*
Triforis adversum.
Pleurotoma formicaria. ?
crispata.*
reticulata.
maraviguæ.*
gracilis.*
Fusus muricatus.*
Murex cristatus.*
Nassa intermedia.
Mitra ebenus.*
phillippiana.
Tornatella fasciata.*
pusilla.
globulosa.
Marginella clandestina.
Dentalium 9-costatum.
5-angulare.

EIGHTH REGION.

The eighth region includes all the space explored below 105 fathoms, extending from that depth to 1380 feet beneath the surface of the sea, having a range of 125 fathoms, being more than twice the extent of all the other regions put together. Throughout this great, and I may say hitherto unknown province, for the notices we have had of it have been but few and fragmentary, we find an uniform and well-characterized fauna, distinguished from those of all the preceding regions by the presence of species peculiar to itself. Within itself the number of species and of individuals diminishes as we descend, pointing to a zero in the distribution of animal life as yet unvisited. It can only be subdivided according to the disappearance of species which do not seem to be replaced by others.

Sixty-five species of Testacea were taken in the eighth region, eleven of which were procured alive. Of the total number 22 were Univalves, 3 of which were found living; 30 Lamellibranchiate Bivalves, 8 living; 3 Palliobranchiate Bivalves, all dead, and possibly derived from the preceding region; and 10 Pteropoda and Nucleobranchiata, also dead. Of these, 17 Univalves, 23 Lamellibranchiata, and 3 Palliobranchiata occurred above 140 and under 180 fathoms; 4 Univalves, 11 Lamellibranchiata, and 1 Palliobranchiate Bivalve above 180 and under 200; and 1 Univalve, 4 Lamellibranchiate, and 1 Palliobranchiate Bivalve above 200 fathoms.

The Mollusca found alive at the greatest depths were *Arca imbricata* in 230 fathoms; accompanied by *Dentalium quinquangulare.* At 180 fathoms living examples of *Nucula ægeensis, Ligula profundissima, Neæra attenuata* and *costellata, Arca lactea,* and *Kellia abyssicola* occurred. *Trochus millegranus* was taken alive in 110 fathoms, along with the *Dentalium pusillum* of authors, which proved to be an annelide of the genus *Ditrupa,* and of which three species live in this region.

Pecten hoskynsii, Lima crassa, Nucula ægeensis, Scalaria hellenica, Parthenia fasciata and *ventricosa,* all new species, have been found in no other region. *Ligula profundissima, Pecten similis, Arca imbricata, Dentalium quadrangulare* and *Rissoa reticulata,* are more prolific of individuals in this region than in any other. *Ligula profundissima* and *Dentalium quinquangulare* are the most generally diffused species below 105 fathoms; the former being present in eleven localities, the latter in seven. The localities examined were eleven in number and far apart from each other, extending from Cerigo to the coast of Lycia.

The *Bullæa angustata, Rissoa acuta, Cerithium lima* and *Teredo* are probably only stragglers in this region.

Several *Ophiuridæ* are true inhabitants of the eighth region; as *Ophiura abyssicola, Amphiura florifera, Amphiura chiagi* and *Pectinura vestita,* all well adapted by their organisation to live in the white mud of great depths. The only other Echinoderm was *Echinocyamus* at 200 fathoms, which however was not taken alive. The Zoophytes are *Caryophyllia cyathus, Alecto* and an *Idmonea,* which occurs in very deep water. Small sponges of three genera were taken alive as deep as 180 fathoms. The deepest living Crustacea occurred at 140 fathoms, and the carapaces of small species are frequent. Besides the *Ditrupæ,* annelides of the genus *Serpula* were taken in the greatest depths explored. *Foraminifera* are extremely abundant through a great part of the mud of this region, and for the most part appear to be species very distinct from those in the higher zones. Representatives of the genera *Nodosaria, Textularia, Rotalia, Operculina, Cristellaria, Biloculina, Quinqueloculina* and *Globigerina* are among the number.

TESTACEA OF REGION VIII.

Lamellibranchiata.

Teredo.
Ligula profundissima.
Corbula anatinoides.
Neæra cuspidata.*
    costellata.*
    attenuata.
Pandora obtusa.
Thracia pholadomyoides.
Kellia abyssicola.*
    oblonga.
Astarte pusilla.
Venus ovata.
Lucina ferruginosa.
Cardium minimum.
Cardita squamosa.

Arca lactea.
    scabra.
    imbricata.
    tetragona.
Nucula polii.
    striata.*
    ægeensis.*
Lima elongata.
    crassa.
Pecten dumasii.
    similis.
    fenestratus.
    hoskynsi.
Ostrea cochlea.?
Anomia polymorpha.

Palliobranchiata.

Terebratula detruncata.
　vitrea.

Crania ringens.

Gasteropoda.

Lottia unicolor.
Bullæa aperta.
　angustata. ?
　alata.
Bulla utriculus.
　cretica.
Eulima subulata.
Parthenia ventricosa.
　turris.
　fasciata.
Rissoa reticulata.
　ovatella.

Rissoa acuta. ?
Scalaria hellenica.
Scissurella plicata.
Trochus millegranus.
Cerithium lima. ?
Pleurotoma abyssicola.
Fusus echinatus.
Nassa intermedia, var.
Marginella clandestina.
Dentalium quinquangulare.
　9-costatum ?

The following Diagram exhibits the comparative characters and relations of the several regions: —

DIAGRAM OF REGIONS OF DEPTH IN THE ÆGEAN SEA.

Sea-Bottom = deposits forming.	Region.	Depth in fathoms.	Characteristic Animals and Plants.
Extent—12 feet. Ground various. Usually rocky or sandy (conglomerates forming).	I.	 2	Littorina cœrulescens. Fasciolaria tarentina. Cardium edule. Plant :—Padina pavonia.
Extent—48 feet. Muddy. Sandy. Rocky.	II.	 10	Cerithium vulgatum. Lucina lactea. Holothuriæ. Plants :—Caulerpa and Zostera.
Extent—60 feet. Ground mostly muddy or sandy. Mud bluish.	III.	 20	Aplysiæ. Cardium papillosum.
Extent—90 feet. Ground mostly gravelly and weedy. Muddy in estuaries.	IV.	 35	Ascidiæ. Nucula emarginata. Cellaria ceramioides. Plants :—Dictyomenia volubilis. 　　　Codium bursa.
Extent—120 feet. Ground nulliporous and shelly.	V.	 55	Cardita aculeata. Nucula striata. Pecten opercularis. Myriapora truncata. Plant :—Rityphlœa tinctoria.
Extent—144 feet. Ground mostly nulliporous. Rarely gravelly.	VI.	 79	Venus ovata. Turbo sanguineus. Pleurotoma maravignæ. Cidaris histrix. Plant :—Nullipora.

170   REPORT—1843.

DIAGRAM OF REGIONS OF DEPTH IN THE ÆGEAN SEA (*continued*).

Sea-Bottom = deposits forming.	Region.	Depth in fathoms.	Characteristic Animals and Plants.
Extent—156 feet. Ground mostly nulliporous. Rarely yellow mud.	VII.		Brachiopoda. Rissoa reticulata. Pecten similis. —— Echinus monilis.
		105	Plant :—Nullipora.
Extent—750 feet. Uniform bottom of yellow mud, abounding for the most part in remains of Pteropoda and Foraminifera.	VIII.		Dentalium 5-angulare. Kellia abyssicola. Ligula profundissima. Pecten hoskynsi. —— Ophiura abyssicola. Idmonea.
		230	Alecto. Plants :—0.
Zero of Animal Life probably about 300 fathoms.			
Mud without organic remains.			

TRUE SCALE OF THE ABOVE DIAGRAM.

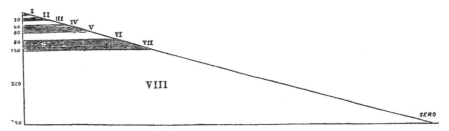

To all the eight regions only two species of Mollusca are common, viz. *Arca lactea* and *Cerithium lima*: the former a true native from first to last, the latter probably only a straggler in the lowest. Three species, namely, *Nucula margaritacea*, *Marginella clandestina* and *Dentalium 9-costatum*, are common to seven regions; the second possibly owing its presence in the lower ones to its having dropped off floating sea-weeds. Nine species are common to six regions.

Corbula nucleus.   Turritella 3-plicata.
Neæra cuspidata.   Triforis adversum.
Pandora obtusa.   Columbella linnæi.
Venus apicalis.   Cardita trapezia.
Modiola barbata.

Seventeen species are common to five regions.

Neæra costellata.   Pecten hyalinus.
Tellina pulchella.   varius.
Venus ovata.   Crania ringens.
Cardita squamosa.   Natica pulchella.
Arca tetragona.   Rissoa ventricosa.
Pecten polymorphus.   cimicoides.

ON ÆGEAN INVERTEBRATA.   171

Rissoa reticulata.	*Columbella rustica.*
Trochus exiguus.	*Conus mediterraneus.*

Terebratula detruncata.

When we inquire into the history of the species having such extensive ranges in depth, we find that more than one-half of them are such as have a wide geographic range, extending in almost every case to the British seas, and in some of those exhibiting the greatest range in depth, still further north ; many of them also ranging in the Atlantic far south of the gut of Gibraltar. If, again, we inquire into the species of Mollusca which are common to four out of the eight Ægean regions in depth, we find that there are 38 such, 21 of which are either British or Biscayan, and 2 are doubtfully British, whilst of the remaining 15, 6 are distinctly represented by corresponding speoics in the north.   Thus among the Testacea having the widest range in depth one third are Celtic or northern forms, whilst out of the remainder of Ægean Testacea, those ranging through less than four regions, only a little above a fifth are common to the British seas.   One-half of the Celtic forms in the Ægean which are not common to four or more zones in depth, are found among the cosmopolitan Testacea, inhabiting the uppermost part of the littoral zone. From these facts we may fairly draw a general inference, that the *extent of the range of a species in depth is correspondent with its geographical distribution.*

The proportion of Celtic forms in the faunæ of the zones varies in the several great families of Testacea. In the accompanying tables I have exhibited this variation conchologically, in order that they may be more useful to the geologist than if the unpreservable species were included.   It will be seen that there is a great disproportion in several of the regions between the number of Celtic forms of Univalves and of Bivalves, that whilst the Monomyaria and Dimyaria range as high as 35 and 30 per cent., the highest range of the Holostomatous univalve is only 13 and a fraction, and of the Siphonostomatous but 8, whilst the Aspiral species preserve a uniform per-centage of 6 in the three highest zones and of 3 in the three following.

Conchological Table, No. I.

Distribution of Shells in depth.

	Ægean total.	I.	II.	III.	IV.	V.	VI.	VII.	VIII.
Multivalves (molluscous).	7	3	2	0	2	2	1	1	0
Patelliform univalves ..	20	11	3	2	3	5	6	6	1
Tubular univalves (Dentalia)	6	4	4	2	2	1	1	2	2
Holostomatous spiral univalves (with Bullæ and Auricula)	115	50	40	40	44	35	28	17	15
Siphonostomat. and convolute spiral univalves.	104	40	27	30	41	36	30	16	5
Testaceous Pteropoda and Nucleobranchia.	12	1	0	0	0	0	0	3	12
Brachiopoda	8	0	0	0	2	4	5	7	3
Conchifera Lamellibranchiata.	135	38	53	52	68	58	48	34	28
	408	147	129	126	142	141	119	85	66

172   REPORT—1843.

Conchological Table, No. II.

Distribution of Celtic forms in the several zones.

	I.	II.	III.	IV.	V.	VI.	VII.	VIII.
Multivalves............	1	0	..	1	1	1	1	0
Patelliform univalves......	0	1	2	2	2	2	2	0
Tubular univalves........	1	1	0	0	0	0	0	0
Holostomatous spiral uni-valves...............	12	9	13	16	14	11	8	4
Siphonostomatous spiral uni-valves...............	4	5	7	8	9	6	5	2
Testaceous Pteropoda, and Nucleobranchia........	..	..	..	..	..	..	0	0
Brachiopoda	..	..	..	0	0	0	0	0
Conchifera Lamellibranchiata	16	25	28	39	33	19	11	7
	34	41	50	66	57	39	27	13
	‖	‖	‖	‖	‖	‖	‖	‖
	21 per cent.	36 per cent.	45 per cent.	43 per cent.	40 per cent.	35 per cent.	36 per cent.	20 per cent.

The importance of these results must be obvious to the geologist. The inductions as to climate or distribution which he may draw from his examination of the Testacea of a given stratum, will vary according to the depth in which those Testacea lived and the ground on which they lived ; for every zone of depth yields a different percentage ; and as the nature of the ground determines the tribe of Testacea which frequents it, and as every tribe yields a different per-centage, according to the variation of character of the sea-bottom, so will the conclusions of the geologist vary and become uncertain. The remedy is however obvious. By carefully observing the mineral character of the stratum in order to ascertain the nature of the former sea-bottom, by noticing the associations of species and the relative abundance of the individuals of each in order to ascertain the depth, and by calculating the percentage of northern or southern forms separately for each tribe, our conclusions will doubtless approximate very nearly to the truth.

A comparison of the Testacea and other animals of the lowest zones with those of the higher exhibits a very great distinction in the hues of the species, those of the depths being for the most part white or colourless, whilst those of the higher regions, in a great number of instances, exhibit brilliant combinations of colour. The results of an inquiry into this subject are as follows :

The majority of shells of the lowest zone are white or transparent : if tinted, rose is the hue ; a very few exhibit markings of any other colour. In the seventh region white species are also very abundant, though by no means forming a proportion so great as in the eighth. Brownish-red, the prevalent hue of the Brachiopoda, also gives a character of colour to the fauna of this zone : the Crustacea found in it are red. In the sixth zone the colours become brighter, reds and yellows prevailing, generally, however, uniformly colouring the shell. In the fifth region many species are banded or clouded with various combinations of colours, and the number of white species has greatly diminished. In the fourth, purple hues are frequent, and contrasts of colour common. In the third and second green and blue tints are met with, sometimes very vivid, but the gayest combinations of colour are seen in the littoral zone, as well as the most brilliant whites.

The animals of Testacea and the Radiata of the higher zones are much more brilliantly coloured than those of the lower, where they are usually white, whatever the hue of the shell may be. Thus the genus *Trochus* is an example of a group of forms mostly presenting the most brilliant hues both of shell and animal; but whilst the animals of such species as inhabit the littoral zone are gaily chequered with many vivid hues, those of the greater depth, though their shells are almost as brightly coloured as the coverings of their allies nearer the surface, have their animals for the most part of an uniform yellow or reddish hue, or else entirely white.

The chief cause of this increase of intensity of colour as we ascend is doubtless the increased amount of light above a certain depth. But the feeding grounds of the animals would appear to exert a modifying influence, and the reds and greens may be in many cases attributed to the abundance of nullipore and of the *Caulerpa prolifera*, a sea-weed of the most brilliant pea-green, the fronds of which the Mollusca of that colour, such as *Nerita viridis*, make their chosen residence.

The eight regions in depth are the scene of incessant change. The death of the individuals of the several species inhabiting them, the continual accession, deposition and sometimes washing away of sediment and coarser deposits, the action of the secondary influences and the changes of elevation which appear to be periodically taking place in the eastern Mediterranean, are ever modifying their character. As each region shallows or deepens, its animal inhabitants must vary in specific associations, for the depression which may cause one species to dwindle away and die will cause another to multiply. The animals themselves, too, by their over-multiplication, appear to be the cause of their own specific destruction. As the influence of the nature of sea-bottom determines in a great measure the species present on that bottom, the multiplication of individuals dependent on the rapid reproduction of successive generations of Mollusca, &c. will of itself change the ground and render it unfit for the continuation of life in that locality until a new layer of sedimentary matter, uncharged with living organic contents, deposited on the bed formed by the exuviæ of the exhausted species, forms a fresh soil for similar or other animals to thrive, attain their maximum, and from the same cause die off. This, I have reason to believe, is the case, from my observations in the British as well as the Mediterranean seas. The geologist will see in it an explanation of the phænomenon of interstratification of fossiliferous and non-fossiliferous beds.

Every species has three *maxima* of development,—in depth, in geographic space, in time. In depth we find a species at first represented by few individuals, which become more and more numerous until they reach a certain point, after which they again gradually diminish, and at length altogether disappear. So also in the geographic and geologic distribution of animals. Sometimes the genus to which the species belongs ceases with its disappearance, but not unfrequently a succession of similar species are kept up, representative as it were of each other. When there is such a representation the minimum of one species usually commences before that of which it is the representative has attained its correspondent minimum. Forms of representative species are similar, often only to be distinguished by critical examination. When a genus includes several groups of forms or subgenera, we may have a double or treble series of representations, in which case they are very generally parallel. The following examples from the Ægean fauna will serve to illustrate the representation in depth.

LIGULA ... { Ligula boysii.   Min. II. Max. III. Min. V.
         { Ligula profundissima.   Min. VI. Max. VIII.

NUCULA { Nucula margaritacea.   Min. II. Max. IV. Min. VI.
      { Nucula polii.   Min. V. Max. VIII.
      { Nucula emarginata.   Min. II. Max. IV. Min. V.
      { Nucula striata.   Min. IV. Max. VI. Min. VIII.

CARDIUM. { Cardium papillosum.   Min. II. Max. IV. Min. VI.
       { Cardium minimum.   Min. VI. Max. VIII.

CARDITA. { Cardita calyculata.   Max. I.
      { Cardita trapezia.   Min. I. Max. IV. Min. VI.
      { Cardita squamosa.   Min. IV. Max. VI. Min. VIII.

ARCA ... { barbata.   Max. I.
      { lactea.   Min. I. Max. IV. Min. VIII.
      { scabra.   Min. IV.? Max. VII. Min. VIII.
      { imbricata.   Min. V. Max. VIII.

TROCHUS { crenulatus.   Max. II. Min. III.
      { exiguus.   Min. II. Max. V. Min. VII.
      { ziziphinus.   Min. III. Max. IV. Min. V.
      { millegranus.   Min. V. Max. VII. Min. VIII.

NASSA ... { variabilis.   Min. I.? Max. II. Min. IV.
      { prismatica.   Max. IV.? Min. V.
      { intermedia.   Min. V. Max. VII. Min. VIII.

In cases equally evident, but where the maxima and minima are not so definite, the succession of representations may be exemplified thus :

LIMA ... { subauriculata.   III. IV. V.
     { cuneata.   V.
     { elongata.   VI. VII. VIII.

RISSOA . { granulata.   I. II.
      { cimicoides.   II. III. IV. V. VI.
      { reticulata.   V. VI. VII. VIII.

Genera like species have a fixed maximum of development in depth, not being irregularly distributed in the several zones, but presenting their greatest assemblage of species in some one, whilst the numbers fall away more or less gradually in the preceding and following zones. In making calculations of the *maxima* of genera in depth, we must be careful to exclude all stragglers from the zones in which they may occur, otherwise our figures will be untrue. In the following table I have exhibited the specific distribution in depth of such of the Ægean genera as present the greatest number of species.

	Ægean total.	I.	II.	III.	IV.	V.	VI.	VII.	VIII.
Cardium......	9	2	3	3	6	3	3	1	1
Pecten	14	0	4	6	8	9	11	4	5
Bulla	14	5	6	8	8	6	2	1	2
Rissoa........	21	14	10	7	7	3	3	3	2
Trochus	28	10	10	13	10	9	7	5	1
Pleurotoma....	24	3	5	7	10	11	9	5	1
Nassa	14	3	6	4	4	1	2	1	1

The consideration of the representation in space forms an important element in our comparisons between the faunas of distinct seas in the same or representative parallels. The analogies between species in the northern and southern, the eastern and western hemispheres, are instances. But there is another application of it which I would make here. The preceding tables and list afford indications of a very interesting law of marine distribution, probable *à priori*, but hitherto unproved. The assemblage of cosmopolitan species at the water's edge, the abundance of peculiar climatal forms in the highest zone, where Celtic species are scarce, the increase in the number of the latter as we descend, and when they again diminish the representation of northern forms in the lower regions, and the abundance of remains of Pteropoda in the lowest, with the general aspect of the associations of species in all, are facts which fairly lead to an inference *that parallels in latitude are equivalent to regions in depth*, correspondent to that law in terrestrial distribution which holds that *parallels in latitude are representative of regions of elevation*. In each case the analogy is maintained, not by identical species only, but mainly by representative forms; and accordingly, although we find fewer northern species in the faunas of the lower zones, the number of forms representative of northern species is so great as to give them a much more boreal or subboreal character than is presented by those regions where identical forms are more abundant.

The consideration of the law of *representation in time* illustrates importantly the history of the very few species hitherto known only as distinct, which were discovered during the course of these researches in the Ægean. They are either such species as have had their maxima during the tertiary æra and are now fast approaching extinction, or such as had their infancy in the latest præadamic formations and are now attaining their maxima. Of the first, *Nassa substriata*, hitherto regarded as a characteristic tertiary shell, is an instance. Abounding in all the latest tertiaries of the Archipelago and of Europe generally, apparently gregarious, half a dozen straggling individuals were all that occurred in above 150 dredgings throughout the Ægean, those too in a region below their usual habitation when the species was in its prime. Of the second, *Neæra costulata* is an example; a few specimens of which only had been derived from tertiary deposits.

The result of the examination of the Ægean fauna does not hold out much prospect of the discovery of any more important extinct forms in a living state. The very few which I have been so fortunate as to discover are not such as materially to disturb the calculations of the geologist, especially if he takes into consideration the relations of each species to others and to its own maximum and minimum in time and geographic distribution. To those who have looked forward to the finding of lost forms in the greater depths of the sea, the catalogues I here present to the Association must be unsatisfactory; for though two or three such have occurred, the majority of species in the great depths are either described existing forms, or altogether new. The zero of animal life in depth has been too nearly approached to hold out further hopes. The indefatigable researches of Captain Graves and his officers have supplied me, since my return, with a mass of new data from all depths and from many new localities; but the result of their examination has been to confirm the calculations I had made from my own observations, and to lead to the pleasing hope that the researches embodied in this report will form a safe base-line for future investigations in the same department of philosophic zoology.

Were the bottom of the Ægean sea, with its present inhabitants, to be elevated and converted into dry land, or even that sea be filled up by a long series of sedimentary depositions, the evidences of its fauna which would be presented may be summed up as follows :—

1. Of the higher animals, the marine Vertebrata, the remains would be scanty and widely scattered.

2. Of the highest tribe of Mollusca, the Cephalopoda, which though poor in species is rich in individuals, there would be but few traces, saving of the Sepia, the shell of which would be found in the sandy strata forming parts of the coast lines of the elevated sea-bed,

3. Of the Nudibranchous Mollusca there would not, in all probability, be a trace to assure us of their having been ; and thus, though we have every reason to suppose from analogy that those beautiful and highly character-istic animals lived in the tertiary periods of the earth's history, if not in older ages, as well as now, there is not the slightest remain to tell of their former existence.

4. Of the Pteropoda and Nucleobranchiata the shell-less tribes would be equally lost with the Nudibranchia, whilst of the shelled species we should find their remains in immense quantity characteristic of the soft chalky deposits derived from the lowest of our regions of depth.

5. The Brachiopoda we should find in deeply-buried beds of nullipore and gravel, and from their abundance we could at once predict the depth in which those beds were formed.

6. The Lamellibranchiate Mollusca we should find most abundant in the soft clays and muds, in such deposits generally presenting both valves in their natural position, whilst such species as live on gravelly and open bottoms would be found mostly in the state of single valves.

7. The testaceous Gasteropoda would be found in all formations, but more abundant in gravelly than in muddy deposits.   In any inferences we might wish to draw regarding the northern or southern character of the fauna, or on the climate under which it existed, whether from univalves or bivalves, our conclusions would vary according to the depth in which the particular stratum examined was found, and on the class of Mollusca which prevailed in the locality explored.

8. The Chitons would be found only in the state of single valves, and pro-bably but rarely, for such species as are abundant, living among disjointed masses of rock and rolled pebbles, which would afterwards go to form con-glomerate, would in all probability be destroyed, as would also be the case with the greater number of sublittoral Mollusca.

9. The *Mollusca tunicata* would disappear altogether, though now form-ing an important link between the Mediterranean and more northern seas.

10. Of the Arachnodermatous Radiata there would not be found a trace, unless the membranous skeleton of the *Velella* should under some peculiarly favourable circumstances be preserved in sand.

11. Of the Echinodermata certain species of *Echinus* would be found en-tire; species of *Cidaris*, on account of the depth at which that animal lives, would be not unfrequent, in certain strata, as the region in which it is found bounds the great lowermost region of chalky mud ; the spines would be found occasionally in that deposit, far removed from the bodies to which they be-longed.   Starfishes, saving such as live on mud or sand, would be only evi-denced by the occasional preservation of their ossicula.   Of the extent of their distribution and number of species no correct idea could be formed. Of the numerous *Holothuriadæ* and *Sipunculidæ* it is to be feared there would be no traces.   The single Crinoidal animal would be rarely preserved entire, but its ossicula and cup-like base would be found in the more shelly deposits.

ON ÆGEAN INVERTEBRATA.   177

12. Of the Zoophyta the corneous species might leave impressions resembling those of Graptolites in the shales formed from the dark muds on which they live. The Corals would be few, but perhaps plentiful in the shelly beds, mostly however fragmentary. The *Cladocora cæspitosa*, where present, would infallibly mark the bounds of the sea, and from the size of its masses, might be preserved in conglomerates where the Testacea would have perished. The *Actiniæ* would have disappeared altogether.

13. Of the Sponges, traces might be found of the more siliceous species when buried under favourable circumstances.

14. The Articulata, except the shelled Annelides, would be for the most part in a fragmentary state.

15. Foraminifera would be found in all deposits, their minuteness being their protection; but they would occur most abundantly in the highest and lowest beds, distinct species being characteristic of each.

16. Tracts would be found almost entirely deficient in fossils; some, such as the mud of the Gulf of Smyrna, containing but few and scattered, whilst similar muds in other localities would abound in organic contents. On sandy deposits formed at any considerable depth they would be very scarce and often altogether absent. Fossiliferous strata would generally alternate with such as contain few or no imbedded organic remains. Whilst at present the littoral zone presents the greatest number and variety of animal and vegetable inhabitants, including those most characteristic of the Mediterranean sea, when upheaved and consolidated, their remains would probably be imperfect as compared with those of the natives of deeper regions, in consequence of the vicissitudes to which they are exposed and the rocky and conglomeratic strata in which the greater number would be imbedded. A great part of the conglomerates and sandstones found would present no traces of animal life, which would be most abundant in the shales and calcareous consolidated muds.

Supposing such an elevation of the sea-bottom of the Ægean to have taken place, a knowledge of the associations of species in the Regions of Depth would enable us to form a pretty accurate notion of the depth of water in which each bed was deposited. This I had an opportunity of exemplifying at Santorin. During a visit to that remarkable volcanic crater, in company with Lieut. Spratt, we carefully examined the little island of Neokaimeni, which came up in 1707, with a view to ascertain, if possible, the depth at which the eruption took place from any portion of the sea-bottom which might be included in its substance. Our search was successful, for imbedded in the pumice was a thin stratum of sea-bottom with its testaceous inhabitants in beautiful preservation. The following were the species:—

Pectunculus pilosus, fine and double, the valves closed; *Arca tetragona, Cardita trapezia, Cytherea apicalis.*

Trochus ziziphinus, large and fine; *T. fanulum, T. exiguus,* and *T. coutourii; Turbo rugosus* and *sanguineus; Phasianella pulla, Turritella 3-costata, Rissoa cimicoides, Cerithium lima, Pleurotoma gracilis.*

A *Serpula,* fragments of *Cellepora* and *Millepora.*

Now there are only two of the regions in depth in which such an association of species would be met with,—the fourth and the fifth. Had it been the sixth, *Trochus ziziphinus* would have been replaced by its representative *Trochus millegranus.* In the third *Arca tetragona* has not commenced its range, but in the fourth and fifth we found all the species named. The state of the *Pectunculus* and the *Trochus ziziphinus* indicating their *maxima,* with the numbers taken of some of the others, refer us to the fourth region as the province in which the sea-bottom on which they lived was formed, *i. e.* in

1843.   N a depth between twenty and thirty-five fathoms. The thinness of the layer of organic remains resting in pumice indicated that no long period had past since a former disturbance of the bottom. The state of the bivalves, their shells double and their valves closed, with the epidermis remaining, indicated that they had been suddenly destroyed, for when *Pectunculi* and *Arcæ* die naturally the valves either separate or remain gaping. They had, doubtless, been smothered in the shower of pumiceous ash which now covers them. The Bay of Santorin, close to the island in question, afforded us no soundings with 150 fathoms line, so that either a high bank, on which lived the Mollusca enumerated, existed there in 1707, before the eruption, or the bottom was uniformly such as the association of animals on it certainly indicates, in which case a depression of more than 100 fathoms must have taken place in consequence of the convulsion.

A similar application may be made of the knowledge of associations of species in depth to the elucidation of the deposits of the tertiary and even of older periods. The determination of the depth by such means is of great importance, for we have already seen how calculations as to climate and northern or southern character of fauna may mislead, unless we attain a knowledge of the region in which the strata were deposited.

The bottom of the Ægean is probably gradually shallowing. The streams which pour into it are thickly charged with sediment. The lowest depth explored was 230 fathoms. Now when the sedimentary deposit shall have filled up that region and brought it to the lowest range of the region next above, it will present a thickness of 725 feet. We have seen that this lowest region had everywhere a bottom of yellowish mud, and that similar animal forms prevailed throughout its extent. Now the strata which shall have been formed by the filling up of that region will present throughout an uniform mineral character closely resembling that of chalk, and will be found charged with characteristic organic remains and abounding in Foraminifera. We shall in fact have an antitype of the chalk. But the Ægean is far deeper through a great portion of its extent than 230 fathoms. The depth below this point will doubtless be filled with a similar mineral deposit, in places perhaps several thousand feet in thickness. But we have seen that the diminution in the number of species and of individuals as we descend in this lowest region pointed to a not far distant zero; therefore the greater part of this immense under-deposit will in all probability be altogether void of organic remains. When indurated it would present the appearance of a great portion of the immense beds of scaglia or Apennine limestone which form such extensive districts in the South of Europe and West of Asia. This is supposing no change of level takes place during the deposition of the chalky mud. But any depression, rapid or gradual, will add to the extent of this great stratum, and by supposing such phænomenon to occur,—and the probability of its occurrence is attested by numerous examples of such in the Archipelago,— we may have a cretaceous formation produced of uniform mineral character and of indefinite thickness. On the other hand, any elevation, by raising the upper portions of the lower zone into the region next above it, will cause a correspondent change in its fauna, and if a depression ensue, we shall have an alternation of faunas, indicating very different depths and presenting very distinct zoological combinations.

Similar considerations respecting the other regions in depth must occur to the zoo-geologist who examines the facts embodied in the catalogues and tables of this report. I shall not swell its pages further by entering more at length into this attractive portion of my subject, which I leave to the consideration of more experienced inquirers, with the exception of calling attention to one other point in zoo-geology, which interested me in the course of my researches. It is this.

A very slight depression of land in the Gulf of Macri on the coast of Lycia, would now plunge below the sea muddy tracts, abounding in *Melania*, *Melanopsis*, *Neritina* and other freshwater Mollusca. Their successors in the first formed shallows would be *Cerithium mammillatum* and a few bivalves, the former mollusk in myriads. A drift of sand over this Cerithium mud would call into existence a new fauna, and every successive depression or elevation, however slight, would produce considerable zoological changes, for the subdivisions of the uppermost region are of small extent in depth, and very liable to be affected by secondary influences.

Now an inspection of the ancient monuments of the ruins of Telmessus proves that such elevations and depressions of small, but as regards animated nature, important extent, have occurred several times during the historical period; and a section of the great plain of Macri would doubtless exhibit such alternations of freshwater and marine strata with their characteristic organic contents.

In the preceding pages I have put forward several generalizations which to many may appear to be founded on inductions drawn from too limited a number of facts. The objection is, to a certain extent, true; though my data have been more numerous than would appear from this report, since the general conclusions embodied in it have not been founded only upon the observations in the Ægean, but also on a long series of researches previously conducted in the British seas. In the present state of the subject speculation is unavoidable, and indeed necessary for its advancement. If it be as important as the author believes, further researches are imperatively called for; and since this branch of inquiry, as at present conducted, may be said to have originated entirely with the British Association, he hopes that through encouragement afforded by that body, other and abler observers may be induced to enter the field, one in which the labourers require support, involving as it does time, expense and personal risk. Should the officers of the Navy and the members of Yacht Clubs take an interest in the subject, much might be done through their aid. To the surveying service the author from experience looks forward confidently for most valuable observations. Since questions of importance to navigation and commerce are intimately connected with this inquiry, it is not too much to look forward eventually to government for its support, the more so as the means of most naturalists—votaries of a science in which the pleasure of discovery is the only reward—do not warrant their adventuring privately in such researches.

Note.—In drawing up the tables of species embodied in this report, I have derived valuable assistance from several scientific friends, especially from Mr. Thompson of Belfast, who enabled me to compare my collections with a series of Mediterranean Testacea named by Michaud; from Mr. Cuming, in whose splendid collection is a series of Sicilian shells from Philippi; and from Mr. Harvey, who most kindly examined the Algæ necessary for the elucidation of the regions of depth.

THE

AMERICAN

JOURNAL OF SCIENCE AND ARTS.

[SECOND SERIES.]

Art. XVI.—*On an Isothermal Oceanic Chart, illustrating the Geographical Distribution of Marine animals;* by James D. Dana.*

THE temperature of the waters is well known to be one of the most influential causes limiting the distribution of marine species of life. Before therefore we can make any intelligent comparison of the species of different regions, it is necessary to have some clear idea of the distribution of temperature in the surface waters of the several oceans: and, if we could add also, the results of observations at various depths beneath the surface, it would enable us still more perfectly to comprehend this subject. The surface temperature has of late years been quite extensively ascertained, and the lines of equal temperature may be drawn with considerable accuracy. But in the latter branch of thermometric investigation almost everything yet remains to be done: there are scattering observations, but none of a systematic character, followed through each season of the year.

The Map which we present in illustration of this subject presents a series of lines of equal surface temperature of the oceans. The lines are isocheimal lines, or more properly, *isocrymal* lines; and where they pass, each exhibits the mean temperature of the waters along its course for the coldest thirty consecutive days of the year. The line for 68° F., for example, passes through the ocean where 68° F., is the mean temperature for extreme cold weather. January is not always the coldest winter

* From the Author's Expl. Exped. Report on Crustacea, p. 1451.

Second Series, Vol. XVI, No. 47.—Sept., 1853.   20

For the chart, see below, p. 103.

month in this climate, neither is the winter the coldest season in all parts of the globe, especially near the equator.   On this account, we do not restrict the lines to a given month, but make them more correctly the limit of the extreme cold for the year at the place.*   Between the line of 74° north and 74° south of the equator, the waters do not fall for any one month below 74° F. ; between 68° north and south, they do not fall below 68°.

There are several reasons why *isocrymal* are preferable to summer or *isotheral* lines.   The cause which limits the distribution of species northward or southward from the equator is the cold of winter, rather than the heat of summer or even the mean temperature of the year.   The mean temperature may be the same when the extremes are very widely different.   When these extremes are little remote, the equable character of the seasons, and especially the mildness of the winter temperature, will favor the growth of species that would be altogether cut off by the cold winters where the extremes are more intense.   On this account, lines of the greatest cold are highly important for a chart illustrating the geographical distributions of species, whether of plants or animals.   At the same time, summer lines have their value : but this is true more particularly for species of the land, and freshwater streams, and for sea-shore plants.   When the summer of a continent is excessive in its warmth, as in North America, many species extend far from the tropics that would otherwise be confined within lower latitudes. But in the ocean, the extremest cold in the waters, even in the Polar regions wherever they are not solid ice, (and only in such places are marine species found,) is but a few degrees below 32° Fahrenheit.   The whole range of temperature for a region is consequently small. The region which has 68° F. for its winter temperature, has about 80° for the hottest month of summer ; and the line of 56° F. in the Atlantic, which has the latitudes of the state of New York, follows the same course nearly as the summer line of 70° F.   In each of these cases the whole extent of the range is small, being twelve to fourteen degrees.†

In fresh-water streams, the waters, where not frozen, do not sink lower than the colder oceans, reaching at most but a few degrees below freezing.   Yet the extremes are greater than for the ocean ; for in the same latitudes which give for the ocean 56° and 70° F. as the limits, the land streams of America range in

---

* The word *isocrymal* here introduced is from the Greek ισος, *equal,* and κρυμος, *extreme cold,* and applies with sufficient precision to the lines for which it is used. These lines are not *isocheimal* lines, as these follow the *mean winter* temperature; and to use this term in the case before us, would be giving the word a signification which does not belong to it, and making confusion in the science.

† Moreover, the greatest range for all oceans is but 62° of Fahrenheit, the highest being 88°, and the lowest 26° ; while the temperature of the atmosphere of the globe has a range exceeding 150°.

temperature between 30° and 80° F., and the summer warmth in such a case, may admit of the development of species that would otherwise be excluded from the region.

While then both isocrymal and isotheral lines are of importance on charts illustrating distribution over the continents, the former are pre-eminently important where the geography of marine species is to be studied.

The lines of greatest cold are preferable for marine species to those of summer heat, because of the fact also that the summer range of temperature for thirty degrees of latitude either side of the equator is exceedingly small, being but three to four degrees in the Atlantic, and six to eight degrees in the Pacific. The July isothermal for 80° F. passes near the parallel of 30° ; and the extreme heat of the equatorial part of the Atlantic Ocean is rarely above 84°. The difficulty of dividing this space by convenient isothermals with so small a range is obvious.

It is also an objection to using the isotheres, that those towards the equator are much more irregular in course than the isocrymes. That of 80° for July, for example, which is given on our Map from Maury's Chart, has a very flexuous course. Moreover, the spaces between the isotheres fail to correspond as well with actual facts in geographical distribution. The courses of the cold water currents are less evident on such a chart, since the warm waters in summer to a great extent overlie the colder currents.

It is also to be noted that nothing would be gained by making the mean temperature for the year, instead of the extremes, the basis for laying down these lines, as will be inferred from the remarks already made, and from an examination of the chart itself.

The distribution of marine life is a subject of far greater simplicity than that of continental life. Besides the influence on the latter of summer temperature in connexion with that of the cold seasons, already alluded to, the following elements or conditions have to be considered :—the character of the climate, whether wet or dry ;—of the surface of the region, whether sandy, fertile, marshy, etc. ;—of the vegetation, whether that of dense forests, or open pasture-land, etc. ;—of the level of the country, whether low, or elevated, etc. These and many other considerations come in, to influence the distribution of land species, and lead to a subdivision of the Regions into many subordinate Districts. In oceanic productions, depth and kind of bottom have an important bearing: but there is no occasion to consider the moisture or dryness of the climate ; and the influence of the other peculiarities of region mentioned is much less potent than with continental life.

We would add here, that the data for the construction of this chart have been gathered, as regards the North Atlantic, from the isothermal chart of Lieutenant Maury, in which a vast amount of facts are registered, the result of great labor and study. For the rest of the Atlantic and the other oceans we have employed the Meteorological volume of Captain Wilkes of the Exploring Expedition Reports, which embraces observations in all the oceans and valuable deductions therefrom; also, the records of other travellers, as Humboldt, Duperey of the Coquille, D'Urville of the Astrolabe, Kotzebue, Beechey, Fitzroy, Vaillant of the Bonite, Ross in his Antarctic Voyage, together with such isolated tables as have been met with in different Journals. The lines we have laid down, are not however, those of any chart previously constructed, for the reason stated, that they mark the positions where a given temperature is the mean of the coldest month (or coldest thirty consecutive days) of the year, instead of those where this temperature is the mean annual or monthly heat; and hence, the apparent discrepancies, which may be observed, on comparing it with isothermal charts.

The isocrymal lines adopted for the chart are those of 80°, 74°, 68°, 62°, 56°, 50°, 44°, and 35° of Fahrenheit. The temperatures diminish by 6°, excepting the last, which is 9° less than 44°.

In adopting these lines in preference to those of other degrees of temperature, we have been guided, in the first place, by the great fact, that the isocryme of 68° is the boundary line of the coral-reef seas, as explained by the author in his Report on Zoophytes.* Beyond this line either side of the equator, we have no species of true Madrepora, Astræa, Meandrina or Porites; below this line, these corals abound and form extensive reefs. This line is hence an important starting point in any map illustrating the geography of marine life. Passing beyond the regions of coral reefs, we leave behind large numbers of Mollusca and Radiata, and the boundary marks an abrupt transition in zoological geography.

The next line below that of 68° F., is that of 74° F. The corals of the Hawaiian Islands, and the Mollusca also to a considerable extent, differ somewhat strikingly from those of the Feejees. The species of Astræa and Meandrina are fewer, and those of Porites and Pocillopora more abundant, or at least constitute a much larger proportion of the reef material. These genera of corals include the hardier species; for where they occur in the equatorial regions they are found to experience the greatest range in the condition of purity of the waters, and also the longest exposures out of water. Their abundance at the Hawaiian Islands, as at Oahu, is hence a consequence of their hardier character, and not a mere region peculiarity independent of temperature. There are grounds, therefore, for drawing a line between the Hawaiian Islands and the Feejees; and as the temperature at

---

* In the author's Report on Geology, 66° F. is set down as the limiting temperature of Coral-reef Seas; this, however, is given as the *extreme* cold. 68° appears to be the *mean* of the coldest month, and is therefore here used.

the Geographical Distribution of Marine Species.   157

the latter sinks to 74½° F. some parts of the year, 74° F. is taken as the limiting temperature. The Feejee seas are exceedingly prolific and varied in tropical species. The corals grow in great luxuriance, exceeding in extent and beauty anything elsewhere observed by the writer in the tropics. The ocean between 74° F., north of the equator, and 74° F. south, is therefore the proper tropical or torrid region of zoological life.

With respect to the line of 80° F., we are not satisfied that it is of much importance as regards the distribution of species. The range from the hottest waters of the ocean 88° to 74° F. is but fourteen degrees, and there are probably few species occurring within the region that demand a less range. Still, investigations hereafter made, may show that the hot waters limited by the isocryme of 80° include some peculiar species. At Sydney Island and Fakaafo, within this hot area, there appeared to be among corals a rather greater prevalence than usual of the genus Manopora, which as these are tender species, may perhaps show that the waters are less favorable for hardier corals than those of the Feejees, where the range of temperature is from 70° to 80° F.; but this would be a hasty conclusion, without more extended observations. The author was on these islands only for a few hours, and his collections were afterwards lost at the wreck of the Peacock, just as the vessel was terminating the voyage by entering the Columbia River.

It is unnecessary to remark particularly upon the fitness of the other isocrymals for the purposes of illustrating the geographical distribution of marine species, as this will become apparent from the explanations on the following pages.

The regions thus bounded require, for convenience of designation, separate names, and the following are therefore proposed. They constitute three larger groups : the *first*, the *Torrid* zone or *Coral-reef* seas, including all below the isocryme of 68° F.; the *second*, the *Temperate* zone of the oceans, or the surface between the isocrymes of 68° F. and 35° F.; the *third*, the *Frigid* zone, or the waters beyond the isocryme of 35° F.

I. TORRID OR CORAL-REEF ZONE.

Regions.				Isocrymal limits.
1. Supertorrid,	-	-	-	80° F. to 80° F.
2. Torrid, -	-	-	-	80° to 74°
3. Subtorrid,	-	-	-	74° to 68°

II. TEMPERATE ZONE.

1. Warm Temperate,	-	-	68°	to 62°
2. Temperate,	-	-	62°	to 56°
3. Subtemperate,	-	-	56°	to 50°
4. Cold Temperate, -	-	-	50°	to 44°
5. Subfrigid,	-	-	44°	to 35°

III. FRIGID ZONE.

1. Frigid, -	-	-	35°	to 26°

A ninth region—called the Polar—may be added, if it should be found that the distribution of species living in the Frigid zone requires it. There are organisms that occur in the ice and snow itself of the polar regions; but these should be classed with the animals of the continents; and the continental isotherms or iso-crymes, rather than the oceanic, are required for elucidating their distribution.

It seems necessary to state here the authorities for some of the more important positions in these lines, and we therefore run over the observations, mentioning a few of most interest. There is less necessity for many particulars with reference to the North Atlantic, as our facts are mainly derived from Lieut. Maury's Chart, to which the author would refer his readers.

1. NORTH ATLANTIC.—*Isocryme of* 74° F.—This isocryme passes near the reefs of Key West, and terminates at the northeast cape of Yucatan; it rises into a narrow flexure parallel with Florida along the Gulf Stream, and then continues on between the Little and Great Bahamas. To the eastward, near the African Coast, it has a flexure northward, arising from the hot waters along the coast of Guinea, which reach in a slight current upward towards the Cape Verde Islands. The line passes to the south of these islands, at which group, Fitzroy, in January of 1852, found the sea temperatures 71° and 72° F.

Isocryme of 68° F.—Cape Canaveral, in latitude 27° 30′, just north of the limit of coral reefs on the east coast of Florida, is the western termination of the line of 68°. The Gulf Stream oc-casions a bend in this line to 36° north, and the polar current, east of it, throws it southward again as far as 29° north. Westward it inclines much to the south, and terminates just south of Cape Verde, the eastern cape of Africa. Sabine found a temperature of 64° to 65° F. off Goree, below Cape Verde, January, 1822; and on February 9, 1822, he obtained $66\frac{1}{2}$° near the Bissao shoals. These temperatures of the cold season contrast strikingly with those of the warm season. Even in May (1831), Beechey had a temperature of 86° off the mouth of Rio Grande, between the parallels of 11° and 12° north.

Isocryme of 62° F.—This isocryme leaves the American coast at Cape Hatteras, in latitude $35\frac{1}{2}$° north, where a bend in the out-line of the continent prevents the southward extension of the polar currents close along the shores. It passes near Madeira, and bends southward reaching Africa nearly in the latitude of the Canaries.

Isocrymes of 56° *and* 50° F.—Cape Hatteras, for a like reason, is the limit of the isocrymes of 56° and 50° as well as of 62°, there being no interval between them on the American coast. The line of 56° F. has a deep northward flexure between the meridians of 35° and 40° west, arising from the waters of the

the Geographical Distribution of Marine Species.   159

Gulf Stream, which here (after a previous east and west course, occasioned by the Newfoundland Bank, and the Polar Current with its icebergs) bends again northeastward, besides continuing in part eastward. The Polar Current sometimes causes a narrow reversed flexure, just to the east of the Gulf Stream flexure. Towards Europe, the line bends southward, and passes to the southwest Cape of Portugal, Cape St. Vincent, or, perhaps to the north cape of the Straits of Gibraltar. Vaillant in the Bonite, found the sea-temperature at Cadiz in February, 49½° to 56° F. (9·7° to 13·4° C.), which would indicate that Cadiz, although so far south (and within sixty miles of Gibraltar), experiences at least as low a mean temperature as 56° F. for a month or more of the winter season. We have, however, drawn the line to Cape St. Vincent, which is in nearly the same latitude. Between Toulon and Cadiz, the temperature of the Mediterranean in February, according to Vaillant, was 55½° to 60¼° F. (13·1° to 15·7° C.), and it is probable, therefore, that Gibraltar and the portion of the Mediterranean Sea east and north to Marseilles, fall within the *Temperate* Region, between the isocrymes of 56° and 62° F., while the portion beyond Sardinia and the coast by Algiers is in the *Warm Temperate* Region, between the isocrymes of 62° and 68° F.

The line of 50° F., through the middle of the ocean, has the latitude nearly of the southern cape at the entrance of the British Channel; but approaching Europe it bends downward to the coast of Portugal. The low temperature of 49½° observed by Vaillant at Cadiz would carry it almost to this port, if this were the mean sea-temperature of a month, instead of an extreme within the bay. The line appears to terminate near latitude 42°, or six degrees north of the isocryme of 56°. This allows for a diminution of a degree Fahrenheit of temperature for a degree of latitude. A temperature as low as 61° F. has been observed at several points within five degrees of this coast in *July*, and a temperature of 52° F., in February. Vigo Bay, just north of 42° north, lies with its entrance opening westward, well calculated to receive the colder waters from the north; and at this place, according to Mr. R. Mac Andrew, who made several dredgings with reference to the geographical distribution of species, the Mollusca have the character rather of those of the British Channel than of the Mediterranean.*

Isocryme of 44° F.—This line commences on the west, at Cape Cod, where there is a remarkable transition in species, and a natural boundary between the south and the north. The cold waters from the north and the ice of Newfoundland Banks, press the line close upon those of 50° and 56° F. But after getting beyond these influences, it rapidly rises to the north, owing to the

---

* Rep. Brit. Assoc., 1850, p. 264.

expansion of the Gulf Stream in that direction, and forms a large fold between Britain and Iceland ; it then bends south again and curves around to the west coast of Ireland.

Isocryme of 35° F.—This line has a bend between Norway and Iceland like that of 44°, and from the same cause,—the influence of the Gulf Stream. But its exact position in this part has not been ascertained.

2. SOUTH ATLANTIC.—*Isocryme of* 74° F.—This line begins just south of Bahia, where Fitzroy found in August (the last winter month) a temperature of 74° to 75½° F. During the same month he had 75½° to 76½° F. at Pernambuco, five degrees to the north. Off Bahia, the temperature was two degrees warmer than near the coast, owing to the warm tropical current, which bends the isocryme south to latitude 17° and 18°, and the cold waters that come up the coast from the south. The line gradually rises northward, as it goes west, and passes the equator on the meridian of Greenwich. Sabine, in a route nearly straight from Ascension Island, in 8° south, to the African coast under the equator, obtained in June (not the *coldest* winter month) the temperatures 78°, 77°, 74°, 72·8°, 72·5°, 73°, the temperature thus diminishing on approaching the coast, although at the same time nearing the equator, and finally reaching it within a few miles. These observations in June show that the isocryme of 74° F. passes north of the equator. The temperatures mentioned in Maury's Chart afford the same conclusion, and lead to its position as laid down.

Isocryme of 68° F.—On October 23d to 25th, 1834, Mr. D. J. Browne, on board the U. S. Ship Erie, found the temperature of the sea on entering the harbor of Rio Janeiro, 67½° to 68½° F. Fitzroy on July 6, left the harbor with the sea temperature 70½° F. Beechey, in August, 1825, obtained the temperature 68·16° to 69·66° F. off the harbor. The isocryme of 68° F. commences therefore near Rio, not far south of this harbor. Eastward of the harbor, the temperature increases two to four degrees. In July, Fitzroy carried a temperature above 68° as far south as 33° 16′ south, longitude 50° 10′ west, the water giving at this time 68½ to 69½° F. Beechey in August obtained 68° F. in 31° south, 46° west. The isocryme of 68° F. thus bends far south, reaching at least the parallel of 30°. It takes a course nearly parallel with the line of 74° F., as different observations show, and passing just south of St. Helena, reaches the African coast, near latitude 7° south. Fitzroy, on July 10 (mid-winter), had a sea-temperature of 68½° near St. Helena ; and Vaillant, in the Bonite, in September found the sea-temperature 68·7° to 69·26° F.

Isocrymes of 56° *and* 50° F.—These two isocrymes leave the American coast rather nearly together. The former commences just north of the entrance of the La Plata. Fitzroy, in July 23 to 31, 1832, found the sea-temperature at Montevideo 56° to 58° F., and in August, 57° to 54½° F. These observations would lead to 56° F. as nearly the mean of the coldest month. The temperature 56 F. was also observed in 35° south, 53° west, and at 36° south, 56° 36′ west. But on July 10 and 13, 1833, at Montevideo, the sea-temperature was 46½° to 47½°, a degree of cold which. although only occasional, throws the line of 56° F. to the north of this place. The temperature near the land is several degrees of Fahrenheit lower than at sea three to eight degrees distant. East of the mouth of the La Plata, near longitude 30° west, Beechey, in July, 1828, found the temperature of the sea 61·86° F. So in April 23 to 29, Vaillant obtained the temperature 59·5° to 61·25° F. at Montevideo, while in 35° 5′ south, 49° 23′ west, on April 14, it was 66·2° F., and farther south, in 37° 42′ south, 53° 28′ west, April 30, it was 64·4° F. ; and in 39° 19′ south, 54° 32′ west, on May 1, it was 57¾° F. ; but a little to the westward, on May 2, in 40° 30′ south, 56° 54′ west, the temperature was 48° F., an abrupt transition to the colder shore waters. Beechey, in 39° 31′ south, 45° 13′ west, on August 28 (last of winter), found the temperature 57·25° F., and on the 29th, in 40° 27′ south, 45° 46′ west, it was 54·20° ; while on the next day, in 42° 27′ south, 45° 11′ west, the temperature fell to 47·83° F. These and other observations serve to fix the position of the isocryme of 56° F. It approaches the African coast, in 32° south, but bends upward, owing to cold waters near the land. On August 20, Vaillant, in 33° 43′ south, 15° 51′ east, found the temperature 56° F. ; while on the 22d, in the same latitude, and 14° 51′ east (or one degree farther to the westward), the temperature was 57·74° F., being nearly two degrees warmer. At Cape Town, in June (latitude 34°), Fitzroy found 55° to 61° F., while on August 16, farther south, in 35° 4′ south, and 15° 40′ west, one hundred and fifty miles from the Cape, Vaillant found the temperature 59·26° F. The high temperature of the last is due to the warm waters that come from the Indian Ocean, and which afford 61° to 64° F. in August, off the south extremity of Africa, west of the meridian of Cape Town.

The isocryme of 50° F. leaves the American coast just south of the La Plata ; after bending southwardly to the parallel of 41°, it passes east nearly parallel with the line of 56° F. It does not reach the African coast.

Isocrymes of 44° *and* 35° F.—Fitzroy in August (the last winter month) of 1833, found the sea-temperature at Rio Negro (latitude 41° south) 48½° to 50° F. But during the voyage from the La Plata to Rio Negro, a few days before, a temperature of 44½° to 46° was met with ; this was in the same month in which the low temperature mentioned above was found at Montevideo. The bend in the course north of the entrance to

162   *On an Isothermal Oceanic Chart, illustrating*

the La Plata, is to some extent, a limit between the warmer waters of the north, and the colder waters from the south; not an impassible limit, but one which is marked often by a more abrupt transition than occurs elsewhere along this part of the coast. The water was generally three or four degrees colder at Montevideo, than at Maldonado, the latter port being hardly sheltered from the influence of the tropical waters, while Montevideo is wholly so. The exact point where the line of 44° F. reaches the coast is somewhat uncertain; yet the fact of its being south of Rio Negro is obvious. After leaving the coast, it passes north of 47½° south, in longitude 53° west, where Beechey, in July, 1828, found the sea-temperature 40·70° F.

The line of 35° F. through the middle of the South Atlantic, follows nearly the parallel of 50°; but towards South America it bends southward and passes south of the Falklands and Fuegia. At the Falklands, Captain Ross, in 1842, found the mean temperature of the sea for July, 38·73°, and for August, 38·10°; while in the middle of the Atlantic, on March 24, latitude 52° 31′ south, and longitude 8° 8′ east, the temperature was down to 34·3° F., and in 50° 18′ south, 7° 15′ east, it was 37° F.; March 20, in 54° 7′ south on the meridian of Greenwich, it was 33·4° F. The month of March would not give the coldest temperature.

The temperature of the sea along the south coast of Fuegia sinks almost to 35°, if not quite, and the line of 35° therfore runs very near Cape Horn, if not actually touching upon Fuegia.

North Pacific Ocean.—*Isocryme of* 80° F.—The waters of the Atlantic in the warmest regions, sink below 80° F. in the colder season, and there is therefore no proper Supertorrid Region in that ocean. In the Gulf of Mexico, where the heat rises at times to 85° F., it sinks in other seasons to 74° and in some parts, even to 72° F.; and along the Thermal equator across the ocean, the temperature is in some portions of the year 78°, and in many places 74°.

But in the Pacific, where the temperature of the waters rises in some places to 88° F., there is a small region in which through all seasons, the heat is never below 80°. It is a narrow area, extending from 165° east to 148° west, and from 7½° north to 11° south. In going from the Feejees in August, and crossing between the meridians of 170° west and 180°, the temperature of the waters, according to Captain Wilkes, increased from 79° to 84° F., the last temperature being met with in latitude 5° south, longitude 175° west and from this, going northward, there was a slow decrease of temperature. The Ship Relief, of the Expedition, in October, found nearly the same temperature (83½°) in the same latitude and longitude 177° west.*   But the Peacock, in

---

* See, for these facts, Captain Wilkes's Report on the Meteorology of the Expedition.

January and February (*summer* months), found the sea-temperature 85° to 88° F., near Fakaafo, in latitude 10° south, and longitude 171° west. In latitude 5° south and the same longitude, on the 16th of January, the temperature was 84°; in 3° south, January 10th, it was 83° F.; on March 26th, in 5° south, and longitude 175° east, the temperature was 86° F.; on April 10th, in the same longitude, under the equator, at the Kingsmills, the temperature was 83½° F.; on May 2d, at 5° north, longitude 174° east, 83½° F.; May 5th, latitude 10°, longitude 169° east, 82° F. The fact that the region of greatest heat in the Middle Pacific is south of the equator, as it has been laid down by different authors, is thus evident; the limits of a circumscribed region of hot waters in this part of the Pacific, were first drawn out by Captain Wilkes.

Another Supertorrid region may exist in the Indian Ocean, about its northwestern portion; but we have not sufficient information for laying down its limits.

Isoeryme of 74° F.—At San Blas, on the Coast of Mexico, Beechey found the mean temperature of the sea for December, 1827, 74·63° F.; for January, 73·69° F.; for February, 72·40° F. The line of 74° F. commences therefore a degree or two south of San Blas. In the winter of 1827 on January 16 to 18, the temperature of 74·3° to 74·6° F. was found by Beechey, in 16° 4′ to 16° 15′ north, 132° 40′ to 135° west; and farther west, in the same latitude, longitude 141° 58′ west, the temperature was 74·83° F. West of the Sandwich Islands, near the parallel of 20° north, the temperature rises five degrees in passing from the meridian of 165° west to 150° east, and the isoeryme of 74° F., consequently trends somewhat to the north, over this part of the ocean. Between the meridians of 130° and 140° east, the temperature of the sea is quite uniform, indicating no northward flexure; and west of 130° east, nearing China, there is a rapid decrease of temperature, bending the line far south. Vaillant of the Bonite, found the sea of Cochin China, in latitude 12° 16′ north, 109° 28′ east, to have the temperature 74·12° F.; and even at Singapore, almost under the equator, the temperature on February 17 to 21, was 77·54° to 79·34° F. The isoeryme of 74° F. terminates therefore upon the southeastern coast of Cochin China.

Isoeryme of 68°—Off the Gulf of California, in 25° north, 117° west, Beechey obtained for the temperatue of the sea on December 13, 65° F.; on December 15, in 23° 28′ north (same latitude with the extremity of the peninsula of California), 115° west, a temperature of 69·41° F. The line of 68° will pass from the extremity of this peninsula, the temperature of the coast below, as it is shut off mostly from the more northern or cold waters, being much warmer. The temperature 69·41° in the middle of December, is probably two and a half degrees above the cold of the coldest month, judging from the relative temperatures of the latter half of December and the month of February at San Blas. Leaving California, the isocryme of 68° will therfore bend a little southerly to 22½°, in longitude 115° west. In 23° 56′ north, 128° 33′ west, Beechey, on January 11, found the temperature of the sea 67·83° F. The line of 68° passes north of the Sandwich Islands. The mean temperature of the sea at Oahu in February, 1827, was 69·69° F.

Near China, this isocryme is bent far south. At Macao, in winter, Vaillant found the sea-temperature, on January 4, 59° F.; on January 5 to 10, 52·7° to 50° F.; January 11, 12, 49·87° to 48·74° F.; January 13 to 16, 50·9° to 52·16° F.; and at Touranne in Cochin China, on February 6 to 24, the sea-temperature was 68° to 68½° F.; in 16° 22′ north, 108° 11′ east, on January 24, it was 67°; in 12° 16′ north, 109° 28′ east, it was 74·12° F. The very low Macao temperature is that of the surface of the Bay itself, due to the cold of the land, and not probably, as the other observations show, of the sea outside.

The line, before passing south, bends northward to the southeast shore of Niphon, which is far warmer than the southeast coast, along Kiusiu. In the Report of the Morrison's visit to Jeddo (Chinese Repository for 1837), a coral bottom is spoken of, as having been encountered in the harbor of Jeddo. According to Siebold (Crust. Faun. Japon., p. ix.), the mean winter temperature (air) of Jeddo is 57° F.; while that of Nagasaki, although farther south, is 44° F.

Isocryme of 62° F.—On January 8, 1827, Beechey found in 29° 42′ north, 126° 37′ west, the temperature 62·75° F.; while on the preceding day, 32° 42′ north, 125° 43′ west, the sea-temperature was 60·5° F. Again, on December 11, in 29° north, 120° west, the temperature was 62·58° F.

Isocryme of 56° F.—At Monterey, on January 1 to 5, the sea-temperature according to Beechey was 56°; but the mean temperature of the sea for November 1 to 17, was 54·91°. In the Yellow Sea, the January temperature is 50° to 56° F., and the line of 56° begins south of Chusan.

Isocryme of 50° F.—At San Francisco, from November 18 to December 5, 1826, Beechey found the mean sea-temperature to be 51·14° F., and off Monterey, in longitude 123° west, the temperature was 50·75° F., on December 6. But in December of 1826, the mean sea-temperature at San Francisco was 54·78° F.; and for November, 60·16° F. The line of 50° F. (mean of the coldest thirty consecutive days), probably leaves the coast at Cape Mendocino.

Isocrymes of 44° *and* 35° F.—Captain Wilkes found the temperature off the mouth of the Columbia River, through ten degrees of longitude, 48° to 49° F., during the last of April, 1841. The isocryme of 44° would probably reach the coast not far north of this place. The temperature on October 21, in the same latitude, but farther west, 147° west, was 52·08° F. On October 16, in 50° north, 169° west, the temperature was 44·91° F. According to some oceanic temperatures for the North Pacific, obtained from Lieutenant Maury, the sea-temperature off northern Niphon, in 41° north and 142½° east, was 44° F., in March, showing the influence of the cold Polar current; and in 42° north, and 149½° east, it was 43° F. The line of 44° hence bends southward as far as latitude 40° north, on the Japan coast.

Again in March, in 43° 50′ north, 151° east, the sea-temperature, was 41° F.; in 44° 50′ north, 152° 10′ east, 39° F.; in 46° 20′ north, 156° east, 33° F.; in 49° north, 157° east, 33° F.; and at the same time, west of Kamschatka, in 55° north, 153° east, 38° F.; in 55° 50′ north, 153° west, 38° F. The line of 35° consequently makes a deep bend, nearly to 45° north, along the Kurile Islands.

SOUTH PACIFIC.—*Isocrymes of* 74°, 68°, *and* 62° F.—The temperature of the sea at Guayaquil, on August 3d, was found by Vaillant, to be, in the river, from 70½° to 73½° F., and at the Puna anchorage, August 5 to 12, 74·7° to 75·2° F. But off the coast, August 15, in 2° 22′ south, 81° 42′ west, the temperature was 69·8° F.; and the next day, in 1° 25′ south, 84° 12′ west, it was 70° F.; on the 17th, 1° south, 87° 42′ west, it was 71·28° F.; and on the 14th, nearer the shore of Guayaquil, in 3° 18′ south, 80° 28′ west, it was 78° F. Again, at Payta, one hundred miles south of Guayaquil, in 5° south, the sea-temperature was found by Vaillant, July 26 to 31, to be 60·8° to 61½° F. The isocryme of 74° F., consequently leaves the coast just north of the bay of Guayaquil, while those of 68° and 62° F., both commence between Guayaquil and Payta. Payta is situated so far out on the western cape of South America that it receives the cold waters of the south, while Guayaquil is beyond Cape Blanco, and protected by it from a southern current. At the Gallapagos, Fitzroy found the temperature as low as 58½° F. on the 29th of September, and the mean for the day was 62°. The average for September was, however, nearer 66°. The Gallapagos appear therefore, to lie in the Warm Temperate Region, between the isocrymes of 62° and 68° F. Fitzroy, in going from Callao to the Gallapagos, early in September, left a sea-temperature of 57° F. at Callao, passed 62° F. in 9° 58′ north, and 79° 42′ west, and on the 15th, found 68½° F. off Barrington Island, one of the Gallapagos.

In the warm season, the cold waters about the Gallapagos have narrow limits; Beechey found a sea-temperature of 83·58° on the

30th of March, 1827, just south of the equator, in 100° west. But in October, Fitzroy, going westward and southward from the Gallapagos, found a sea-temperature of 66° F. at the same place; and in a nearly straight course from this point to 10° south, 120° west, found the sea-temperatures successively, 68°, 70°, 70½° 72½°, 73½°, 74°; and beyond this, 75½°, 76½°, 77½°F., the last on November 8, in 14° 24' south, 136° 51' west. These observations give a wide sweep to the cold waters of the colder seasons, and throw the isocrymes of 74° and 68° F., far west of the Gallapagos. Captain Wilkes, in passing directly west from Callao, found a temperature of 68° F., in longitude 85° west; 70° F., in 95° west; and 74° F., in 102° to 108° west. These and other observations lead to the positions of the isocrymes of 74°, 68°, and 62°, given on the Chart. The line of 74° passes close by Tahiti and Tongatabu, and crossing New Caledonia, reaches Australia in latitude 25° S.

In mid-ocean there is a bend in all the southern isocrymes.[*]

Isocrymes of 56° and 50° F.—The temperature at Callao, in July, averages 58½° or 59° F. At Iquique, near 20° south, Fitzroy had 58° to 60° F., on July 14, 1835; and off Copiapo, in the same month, 56½° F. At Valparaiso, Captain Wilkes found a sea-temperature of 52½° F., in May; and Fitzroy, in September, occasionally obtained 48° F., but generally 52° to 53°. Off Chiloe, Fitzroy found the temperature 48° to 51½° in July.

INDIAN OCEAN.—*Isocrymes of 74° and 68° F.*—Off the south extremity of Madagascar, in 27° 33' south, 47° 17' east, on August 4th, Vaillant found the temperature 69·26° F.; and in 29° 34' south, 46° 46' east, the temperature of 67·84° F.; off South Africa, August 12, in 34° 42' south, 27° 25' east, the temperature 63·5° F.; on August 14, in 35° 41' south, 22° 34' east, a temperature of 63·3° F.; while off Cape Town, two hundred miles to the west, the temperature was 50° to 54° F.

In the above review, we have mentioned only a few of the observations which have been used in laying down the lines, having selected those which bear directly on some positions of special interest, as regards geographical distribution.

The Chart also contains the *heat-equator*,—a line drawn through the positions of greatest heat over the oceans. It is a shifting line, varying with the seasons, and hence, there is some difficulty in fixing upon a course for it. We have followed mainly the Chart of Berghaus. But we have found it necessary to give it a much more northern latitude in the western Pacific, and also a flexure in the western Atlantic, both due to the currents from the south that flow up the southern continents.

Vaillant passing from Guayaquil to the Sandwich Islands, found the temperature, after passing the equator, slowly increase,

[*] See Observations by W. C. Cunningham, Am. J. Sci., [2] xv, 66.

from 76° F., August 19, in 2° 39′ north, 91° 58′ west (of Green-wich), to 81·9° F., in August 31, 11° 15′ north, 107° 3′ west, after which it was not above 80° F.   The same place in the ocean which gave Vaillant 76° F., in August, afforded Fitzroy (4° north, 96° west), on March 26 (when the sun had long been far north), 82½° F.   This fact shows the variations of temperature that take place with the change of season.

(To be continued.)

We proceed to give an enumeration of the several Zoological Provinces to which we are led by the temperature regions adopted. It should be again observed, that the isocryme of 68° is the grand boundary of coral reefs, and of the larger part of the zoological life connected with them, and that the *Torrid Zone* and *Coral-reef Zone* of oceanic temperature are synonymous terms.

We mention also the extent of the Provinces; and it will be found, that although seemingly numerous, few of them are under 500 miles in length, while some are full 4000 miles.

For zoological reasons which are explained in another place,* and which may be the subject of another communication to this Journal, we adopt for *Marine Zoological Geography*, three grand

* The Author's Report on Crustacea.

[This reading continues unabridged on page 325 of this issue of the *American Journal of Science and Arts*.]

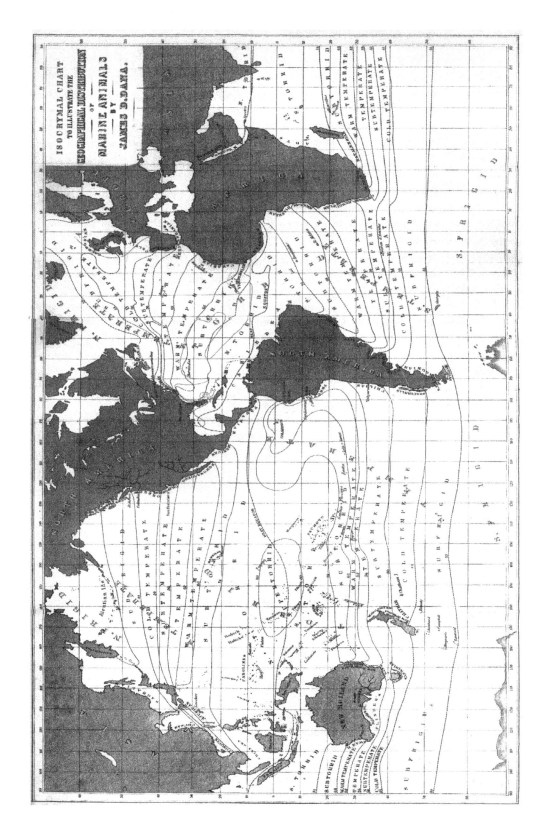

ISOTHERMAL CHART
TO ILLUSTRATE THE
GEOGRAPHICAL DISTRIBUTION
OF
MARINE ANIMALS
BY
JAMES D. DANA.

the Geographical Distribution of Marine Species.   325

divisions of the coasts of the globe. 1. The *American* or *Occidental* including East and West America; 2. The *Africo-European* including the coasts of Europe and Western Africa; and third, the *Oriental,* including the coasts of Eastern Africa, E. Indies, Eastern and Southern Asia, and Pacific. Besides these, there are the *Arctic* and *Antarctic* Kingdoms, including the coasts of the Frigid Zones, and in some places, as Fuegia, those of the extreme temperate zone. We add here, only in general terms, that there is a remarkable similarity in the genera of Eastern and Western America, and an identity of some few species; that the coast of Europe and Eastern Africa widely differ in Crustacea from either the American or Oriental; that the species of the Oriental division have a great similarity in genera, and that numerous species of Crustacea of Eastern Africa, are identical with those of the Pacific. We pass by, for the present, the details on these points.

We also omit the zoological characters of the Provinces here enumerated. Several of these Provinces are identical with those proposed by Milne Edwards, Prof. E. Forbes, and others; and as far as possible, the names heretofore used, are retained.

I. OCCIDENTAL KINGDOM.

A. WESTERN SECTION.

1. *Torrid or Coral-Reef Zone.*

Provinces.	Limits.	Length in Miles.
1. *Panamian,* (torrid) . . .	1° s. to 17½° N., . . .	1600
2. *Mexican,* Province, (N. subtorrid)	17½° N. to Californ. Penin., .	600
3. *Guayaquil,* " (S. subtorrid)	1° s. to Cape Blanco, 4¾° s., .	200

2. *North Temperate Zone.*

4. *Sonoran,* (warm temperate) .	Penin. Californ. to 28½° N., .	550
5. *Diego** or *Jacobian,* (temperate)	28½° N. to 34½° N., . .	450
6. *Californian,* (subtemperate) .	34½° N. to C. Mendocino, .	480
7. *Oregon,* (cold temperate) . .	C. Mendoc. to Puget's Sound, (?)	480
8. *Pugettian,* (subfrigid) .	Puget's Sound to 55° or 56°,	1200

3. *South Temperate Zone.*

9. *Gallapagos,* (warm temperate) .	Gallapagos.	
10. *Peruvian,* (temperate) . .	C. Blanco to Copiapo, 27½° s.,	1500
11. *Chilian,* (subtemperate) . .	27½° s. to 38° s., . .	700
12. *Araucanian,* (cold temperate) .	38° s. to 49° or 50° s., .	900
13. *South Patagonian,* (subfrigid) .	50° s. to Magellan Straits.	

B. EASTERN SECTION.

1. *Torrid Zone.*

1. *Caribbean,* (torrid) . . .	{ Key West & N. Yucatan } to 1° south of Bahia, } .	4000
2. *Floridan,* (N. subtorrid) . .	Key West to 27° N., . .	200
3. *Brazilian,* (S. subtorrid) . ·	15° s. to 24° s. . . .	600

2. *North Temperate Zone.*

4. *Carolinian,* (warm temperate) .	27° N. to C. Hatteras, . .	600
5. *Virginian,* (cold temperate) .	C. Hatteras to C. Cod,	650
6. *Acadian,* (subfrigid)† .	C. Cod to E. Cape of Newf'dl'd,	900

* May possibly be united conveniently to the Sonoran.
† Changed from Nova-Scotian in the Report on Crustacea.

326   *On an Isothermal Oceanic Chart, etc.*

3. *South Temperate Zone.*

Provinces.	Limits.	Length in Miles.
7. *St. Paul,** (warm temperate) .	24° s. to 30° s., . . .	480
8. *Uraguaian,* (temperate) . .	30° s. to north C. of La Plata,	360
9. *Platensian,* (subtemperate) .	Mouth of La Plata.	
10. *North Patagonian,* (cold temperate)	South C. of La Plata to 43° s.,	500
11. *South Patagonian,†* (subfrigid) .	43° s. to Magellan Straits, .	700

II. AFRICO-EUROPEAN KINGDOM.

1. *Torrid Zone.*

1. *Guinean,* (torrid) . . .	5° n. to 9° n., ...	1200
2. *Verdensian,* (N. subtorrid) .	{ 9° n. to 14½° n. including } { the Cape Verde Islands, }	1000
3. The *Biafrian,* (S. subtorrid) .	{ 5° n. to 7° or 8° s., includ- } { ing Ascension & St. Helena, }	900

2. *N. Temperate Zone.*

4. *Canarian,* (warm temperate) .	{ 14½° n. to 28° or 29° n. in- } { cluding the Canaries, }	1000
5. *Mediterranean,* (temperate) .	29° n. to C. St. Vincent with Mediterranean, excepting some of its northern coasts, and including Madeira and Azores.	
6. *Lusitanian,* (subtemperate) .	C. St. Vincent to 42° n., .	300
7. *Celtic,* (cold temperate) . .	42° n. to Scotland, . .	1000
8. *Caledonian,* (subfrigid) . .	N. Scotland, Shetland's Ferroe, etc.	

3. *South Temperate Zone.*

9. *Angolan,* (warm temperate) .	7° s. to 13° s., . . .	360
10. *Benguelan,* (temperate) . .	13° s. to 28° s., . .	900
11. *Capensian,* (subtemperate) .	28° s. to C. Agulhas, .	450
13. *Tristensian,* (cold temperate) .	Tristan d'Acunha.	

III. ORIENTAL KINGDOM.

I. AFRICAN SECTION, OR EAST COAST OF AFRICA AND NEIGHBORING ISLANDS.

1. *Abyssinian,* (torrid) . . .	{ 26¼° s. to 21° or 22° in Red } { Sea, including larger part of } { Madagascar and Islands north, }	3500
2. *Erythrean,* (N. subtorrid) .	Northern third of Red Sea, about	300
3. *Natalensian,* (S. subtorrid) .	26¼° s. to 31° s., with Southern Madagascar and Isle of France.	
4. *Algoan,* (warm temp. and temp.)	31° s. to C. Lagulhas, .	550

II. ASIATIC SECTION.

1. *Torrid Zone.*

1. *Indian,* (torrid)	East India Islands, N. Australia, Southern Asia, to 12½° n. on Cochin China.
2. *Liukiuan,* (N. subtorrid) .	12½° n. to 15° n., with Formosa, Loochoos, S. S.E. Shore of Japan.
3. *Endrachtian* or } W. Australian, } (S. subtorrid)	W. Australia 22° s. to 26½° s., 300

* The St. Paul province may perhaps be united with the Uraguaian.
† The South Patagonian is made to include both the eastern and western sides of this portion of the continent; but a division of the two may hereafter be found to be required.

2. *North Temperate Zone.*

Provinces.	Limits.
4. *Tonquin,* (warm temperate) . .	15° N. to 25° N., (Gulf Tonquin.)
5. *Chusan,* (subtemperate) . . .	25° N. into Japan Sea.
6. *Niphonensian,* (cold temp. & subtemp.)	East Coast of Niphon, to 40° N.
7. *Saghalian,* (subfrigid) . . .	Coast of Japan Sea, part of Western and Northern Niphon, Saghalian, Yeso, etc.

3. *South Temperate Zone.*

8. *Cygnian,* or Swan R. (warm temperate)	W. Australia, 26½° s. to S. W. Cape.
9. *Flinders,* (temperate) . . .	Southern Coast of Australia.
10. *Moreton,* (warm temp. and temp.) .	E. Australia, 26½° s. to 31° s.
11. *Bass,* (subtemperate) . . .	E. Australia, 31° or 32° s. to Van Diemens Land.
12. *Tasmanian,* (cold temperate) . .	Van Diemens Land.

III. PACIFIC SECTION.

1. *Torrid Zone.*

1. *Polynesian,* (torrid)	Pacific Ids. of Torrid Region.
2. *Hawaiian,* (N. subtorrid) . . .	Hawaiian range of Islands.
8. *Raratongan,* (S. subtorrid) . .	Hervey Islands and others of South Subtorrid Region.

2. *South Temperate Zone.*

4. *Kermadec,* (warm temp. and temp.) .	Kermadec Islands, etc.
5. *Wangaroan,* (subtemperate) . .	Northern New Zealand.
6. *Chatham,* (cold temperate) . .	Middle N. Z. to 46° S. and Chatham I.

The ARCTIC KINGDOM includes (1) the *Norwegian,* north of the Atlantic,—(2) the *Camtschatican,* north of the Pacific,—(3) the *North Polar;* the ANTARCTIC KINGDOM, includes (1) the *Fuegian,* Fuegia and Shetlands, etc.,—(2) the *Aucklandian,* Auckland and S. extremity of New Zealand,—(3) the *South Polar.*

ART. XXXII.—*The Coal Field of Bristol County and of Rhode Island;* by President E. HITCHCOCK.*

IN my Reports on the Geology of Massachusetts, I described a large tract in Bristol county, a part of Plymouth county, the whole of the island of Rhode Island, and a strip on the west side of Narraganset Bay, as underlaid by Graywacke; a rock older than the Coal Formation, and equivalent to the Silurian and Cambrian strata of late geological writers. While publishing my Final Report, however, in 1840, I became satisfied that a part of this region was a true coal formation, and so marked it on the map. But I now advance a step farther, and maintain that the whole of this tract, embracing not less than five hundred square miles, is a genuine coal field, that has experienced more than usual metamorphic action.

* From a Report to the Governor of Massachusetts, dated Feb. 23, 1853.

[Dana's companion article XLII, which begins on page 391 of the *Journal* follows in this reading.]

J. D. Dana on a supposed change of Ocean Temperature. 391

ART. XLII.—*On a change of Ocean Temperature that would attend a change in the level of the African and South American Continents ;* by JAMES D. DANA.

THE idea of a change of climate consequent upon a change in the distribution of land and water on the globe, brought forward by Sir Charles Lyell, has recently been discussed with much ability and precision, by Prof. Hopkins, especially with reference to the Northern Atlantic. As there is profit in this consideration of possibilities whether we can prove the actual occurrence of the supposed events or not, we briefly remark in this place upon another geological change that would affect the temperatures of both the Pacific and Atlantic Oceans.

Upon the oceanic isothermal chart issued with the last number of this Journal, and discussed in that and this number, it is observed that the whole western coast of South America is bordered by cold waters ; and that while in the Pacific, 80° F. is the coldest temperature of the year in mid-ocean, towards South America, even under the equator, the ocean temperature of 74° is not found, in the cold season, short of a distance of 2500 miles from the coast.

We have also remarked upon the evidence that a similar southern or extratropical current affects the temperature of the whole southern Atlantic, (see page 320) and makes this literally the cold ocean of the globe.

It is moreover evident from the temperature of the waters off western South America, that the extratropical or antarctic current has a vastly wider influence here than in the southern Atlantic ; the positions of the lines of 68° and 74° in the two regions make this sufficiently apparent. It is also obvious, that the South American Continent, by extending so far south,—22 degrees, or 1300 miles, beyond the south point of Africa,—should necessarily intercept to a large extent the antarctic current, and thus occasion in connection with other causes, the northern flow that influences so widely the temperature of the waters off this coast. The position of the isocryme of 35°, shows that this same current flows on, rising somewhat northward towards Cape of Good Hope ; yet the African continent lies so far to the north, that it can in fact intercept but a small part of the southern current, which consequently to a large extent passes on south of the Cape ; yet this small part produces the wonderful effects pointed out.*

* We find that at the recent meeting of the British Association, Mr. A. G. Findlay, in the course of a paper on the oceanic currents of the Atlantic and Pacific, takes the common view that the Lagulhas current is the origin of the current that flows up the West African coast, a view shown on page 322 to be untenable.

392 *J. D. Dana on a supposed change of Ocean Temperature.*

Suppose now, that by a change of level, America were to terminate in latitude 34° S., and Africa in latitude 56° S. : the relation of the two, and of the cold influences of the currents adjoining, would be entirely changed. The vast area in the South Pacific, embraced between the west South American coast and the isocryme of 74°,—which marks the influence in the colder season of the cold southern waters, though not by any means its extreme limit,—would, if transferred to the Atlantic equatorial regions, stretch nearly or quite across from Guinea to the East Cape of South America; and the line of 68° would sweep around north of the equator quite to mid-ocean. The actual extent of the change may be perceived with close accuracy if we transfer the isocrymal lines off this part of Western America to the Atlantic. In the Pacific, under the same circumstances, the line of 68° would nowhere reach within several degrees of the equator.

The distribution of marine life would be greatly changed. While now the west coast of South America is, as regards the ocean, one of the coldest regions for the latitude in the world, it would become very much moderated, and a considerable portion of coast would be bordered by tropical waters. Along by Lima, and far south, there might be coral reefs. In the Atlantic, on the contrary, the Gulf of Guinea now characterized by torrid waters, would be filled with the colder seas of the temperate zone, and true tropical life would be altogether excluded.

The influence also on the Gulf Stream would be very decided and the whole North Atlantic would feel the change.

It is a remarkable fact that while the west coast of America is bordered in the tropical part by cold waters, 10° to 12° below the mean of mid-ocean, and the marine zoology is hence extratropical, the temperature of the land is peculiarly torrid over the same latitudes. It is evident that in judging of the influence of the ocean temperature on the temperature of the land, the direction of the aerial currents for the year, should be considered as a most important element towards any just conclusions.

Although we cannot show that the supposed change of level in the continents has taken place, we may learn from the facts what vast changes in marine life, have happened in past ages, through such changes of level as have occurred in the earth's history. The changes on the land from this cause would be less marked; besides, these have had far less influence on the life of the rocks than those of the ocean, as the fossiliferous rocks are mainly of marine origin. We know that in the cretaceous and tertiary periods, the Andes were in part under water, or at a much lower level, and effects of the kind considered, cannot be altogether hypothetical.

From *The Botany of the Antarctic Voyage of H.M. Discovery Ships Erebus and Terror in the Years 1839–1843*

Sir Joseph Dalton Hooker

British Museum. When re-found in New Zealand it was described as new, and called *O. cataractæ*, and when found a third time in Tasmania, was called by still a third name, *O. lactea*. In this case a more important fact was smothered than that of the distribution of *O. corniculata*, namely, that of a very peculiar plant of the south temperate zone being common to these three widely sundered localities.

Many similar instances might be added, for there are several New Zealand plants (as *Pteris aquilina*) that have a different name in almost every country in the world, and, partly from changes in nomenclature, partly from the reduction of species, I have found myself obliged to quote 1500 names for the 720 New Zealand flowering-plants described, and I believe I might have doubled the number had my limits not obliged me to reduce the synonymy as much as possible; in many cases too much, I fear, for the requirements of working botanists in Europe.

§ 4.

The distribution of species has been effected by natural causes, but these are not necessarily the same as those to which they are now exposed.

Of all the branches of Botany there is none whose elucidation demands so much preparatory study, or so extensive an acquaintance with plants and their affinities, as that of their geographical distribution. Nothing is easier than to explain away all obscure phenomena of dispersion by several speculations on the origin of species, so plausible that the superficial naturalist may accept any of them; and to test their soundness demands a comprehensive knowledge of facts, which moreover run great risk of distortion in the hands of those who do not know the value of the evidence they afford. I have endeavoured to enumerate the principal facts that appear to militate against the probability of the same species having originated in more places (or centres) than one; but in so doing I have only partially met the strongest argument of all in favour of a plurality of centres, viz. the difficulty of otherwise accounting for the presence in two widely sundered localities of rare local species, whose seeds cannot have been transported from one to the other by natural causes now in operation. To take an instance: how does it happen that *Edwardsia grandiflora* inhabits both New Zealand and South America? or *Oxalis Magellanica* both these localities and Tasmania? The idea of transportation by aerial or oceanic currents cannot be entertained, as the seeds of neither could stand exposure to the salt water, and they are too heavy to be borne in the air. Were these the only plants common to these widely-sundered localities, the possibility of some exceptional mode of transport might be admitted by those disinclined to receive the doctrine of double centres; but the elucidation of the New Zealand Flora has brought up many similar instances equally difficult to account for, and has developed innumerable collateral phenomena of equal importance, though not of so evident appreciation. These, which all bear upon the same point, may be arranged as follows :—

1. Seventy-seven plants are common to the three great south temperate masses of land, Tasmania, New Zealand, and South America.

2. Comparatively few of these are universally distributed species, the greater part being peculiar to the south temperate zone.

3. There are upwards of 100 genera, subgenera, or other well-marked groups of plants entirely or nearly confined to New Zealand, Australia, and extra-tropical South America. These are represented by one or more species in two or more of these countries, and they thus effect a botanical relationship or affinity between them all, which every botanist appreciates.

4. These three peculiarities are shared by all the islands in the south temperate zone (including even Tristan d'Acunha, though placed so close to Africa), between which islands the transportation of seeds is even more unlikely than between the larger masses of land.

5. The plants of the Antarctic islands, which are equally natives of New Zealand, Tasmania, and Australia, are almost invariably found only on the lofty mountains of these countries.

Now as not only individual species, but groups of these, whether orders, genera, or their sub-divisions, are to a great degree distributed within certain limits or areas, it follows that the flora of every island or archipelago presents peculiarities of its own. Though an insular climate may favour the relative abundance of individuals, and even species of certain Natural Orders, there is nothing in the climate, or in any other attribute of insularity, which indicates the nature of the peculiarity of endemic species. The islands of each ocean contain certain botanically allied forms in common, which are more or less abundant in them, and rarely or never found on the neigh-bouring continents; thus there are curious genera peculiar to the North Atlantic islands, others to the North Pacific islands, others to those of the South Pacific, and others again to the Malayan Archipelago; just as there are still others peculiar to the Antarctic islands, and many to New Zealand, Fuegia, and Tasmania.

Each group of islands hence forms a botanical region, more or less definable by its plants as well as by its oceanic boundaries; precisely as a continuous area like Australia or South Africa does. There is however this difference, that whereas the Natural Orders that give a botanical character to a continuous area of a continent or to a large island (as the *Proteaceæ* in South Africa or in New Holland, and *Coprosma* in New Zealand) are numerous in species and often uniformly spread,—in clusters of small islands, distant from continents, they are few in species, and the individuals are scattered, appearing as if the vestiges of a flora which belonged to another epoch, and which is passing away: this is perhaps a fanciful idea, but one which I believe to contain the germ of truth; for no Botanist can reflect upon the destruction of peculiar species on small islands (such as is now going on in St. Helena amongst others), without feeling that, as each disappears, a gap remains, which may never be botanically refilled; that not only are those links breaking by which he con-nects the present flora with the past, but also those by which he binds the different members of the vegetable kingdom one to another. It is not true in every sense that all existing nature appears to the naturalist as an harmonious whole; each species combines by its own peculiarities two or more others more closely, and reveals their affinities more clearly, than any other does; just as the flora of an intermediate spot of land connects those of two adjacent areas better than any other locality does. It is often by one or a very few species that two large Natural Orders are seen to be related; just as by a few Chilian plants the whole flora of New Zealand is connected with that of South America. The destruction of a species must hence create an hiatus in our systems, and I believe that it is mainly through such losses that natural orders, genera, and species become isolated, that is, peculiar, in a naturalist's eyes.

To return to the distribution of existing species, I cannot think that those who, arguing for unlimited powers of migration in plants, think existing means ample for ubiquitous dispersion, suffi-ciently appreciate the difficulties in the way of the necessary transport. During my voyages amongst the Antarctic islands, I was led, by the constant recurrence of familiar plants in the most inaccessible spots, to reflect much on the subject of their possible transport; and the conviction was soon forced upon me, that, putting aside the almost insuperable obstacles to trans-oceanic migration between such islands as Fuegia and Kerguelen's Land, for instance (which have plants in common, not found elsewhere), there were such peculiarities in the plants so circumstanced, as rendered many of them the least likely of all to have availed themselves of what possible chances of transport there may have been. As species they were either not so abundant in individuals, or not prolific enough to have been the first to offer themselves for chance transport, or their seeds presented no facilities for migration*, or were singularly perishable from feeble vitality, soft or brittle integuments, the presence of oil that soon became rancid, or from having a fleshy albumen that quickly decayed†. Added to the fact that of all the plants in the respective floras of the Antarctic islands, those common to any two of them were the most unlikely of all to emigrate, and that there were plenty of species possessing unusual facilities, which had not availed themselves of them, there was another important point, namely, the little chance there was of the seeds growing at all, after transport. Though thousands of seeds are annually shed in those bleak regions, few indeed vegetate, and of these fewer still arrive at maturity. There is no annual plant in Kerguelen's Land, and seedlings are extremely rare there; the seeds, if not eaten by birds, either rot on the ground or are washed away; and the conclusion is evident, that if such mortality attends them in their own island, the chances must be small indeed for a solitary individual, after being transported perhaps thousands of miles, to some spot where the available soil is pre-occupied.

Beyond the bare fact of the difficulty of accounting by any other means for the presence of the same species in two of the islands, there appeared nothing in the botany of the Antarctic regions to support or even to favour the assumption of a double creation, and I hence dismissed it as a mere speculation which, till it gained some support on philosophical principles, could only be regarded as shelving a difficulty; whilst the unstable doctrine that would account for the creation of each species on each island by progressive development on the spot, was contradicted by every fact.

It was with these conclusions before me, that I was led to speculate on the possibility of the plants of the Southern Ocean being the remains of a flora that had once spread over a larger and more continuous tract of land than now exists in that ocean; and that the peculiar Antarctic genera and species may be the vestiges of a flora characterized by the predominance of plants which are now scattered throughout the southern islands. An allusion to these speculations was made in the 'Flora Antarctica' (pp. 210 and 368), where some circumstances connected with the distribution of the Antarctic islands were dwelt upon, and their resemblance to the summits of a submerged mountain chain was pointed out; but beyond the facts that the general features of the flora favoured such a view, that the difficulties in the way of transport appeared to admit of no other solution, and that there are no limits assignable to the age of the species that would make their creation posterior to such a series of geological changes as should remove the intervening land, there was nothing in the shape of evidence by which my speculation could be supported. I am indebted to the invaluable labours of Lyell and Darwin‡, for the facts that could alone have given countenance to such an hypothesis; the one showing that the necessary time and elevations and depressions of land

---

* Thus of the *Compositæ*, common to Lord Auckland's Group, Fuegia, and Kerguelen's Land, none have any pappus (or seed-down) at all! Of the many species *with* pappus, none are common to two of these islands!

† Of the seeds sent to England from the Antarctic regions, or transported by myself between the several islands, almost all perished during transmission.

‡ See Darwin's 'Journal of a Naturalist,' and 'Essays on Volcanic Islands and Coral Islands.' The proofs of the coasts of Chili and Patagonia having been raised continuously, for several hundred miles, to elevations varying between 400 and 1300 feet, since the period of the creation of existing shells, will be found in the first-named of these admirable works, which should be in the hands of every New Zealand Naturalist, if only from its containing

need not be denied; and the other, that such risings and sinkings are in active progress over large portions of the continents and islands of the southern hemisphere. It is to the works of Lyell* that I must refer for all the necessary data as to the influence of climate in directing the migration of plants and animals, and for the evidence of the changes of climate being dependent on geological change. In the 'Principles of Geology' these laws are proved to be of universal application, and amply illustrated by their being applied to the elucidation of difficult problems in geographical distribution. It follows from what is there shown, that a change in the relative positions of sea and land has occurred to such an extent since the creation of still existing species, that we have no right to assume that the plants and animals of two given areas, however isolated by ocean, may not have migrated over pre-existing land between them. This was illustrated by an examination of the natural history of Sicily (where land-shells, still existing in Italy, and which could not have crossed the Straits of Messina, are found imbedded on the flanks of Etna high above the sea-level), regarding which Sir Charles Lyell states that most of the plants and animals of that island are older than the mountains, plains, and rivers they now inhabit†.

It was reserved for Professor Edward Forbes, one of the most accomplished naturalists of his day, to extend and enlarge these views, and to illustrate by their means the natural history of an extensive area; which he did by applying a profound knowledge of geology and natural history to the materials he had collected during his arduous surveys of many of the shores of Europe and the Mediterranean. The result has been the enunciation of a theory, from which it follows that the greater part, if not all, of the animals and plants of the British Islands have immigrated at different periods, under very different climatic conditions; and that all have survived immense changes in the configuration of the land and seas of Northern Europe. The arguments which support this theory are based upon evidence derived from Zoology and Geology‡, and they receive addi-

important observations on his own islands. The fact of this accomplished Naturalist and Geologist having preceded me in the investigation of the Natural History of the Southern Ocean, has materially influenced and greatly furthered my progress; and I feel it the more necessary to mention this here, because Mr. Darwin not only directed my earliest studies in the subjects of the distribution and variation of species, but has discussed with me all the arguments, and drawn my attention to many of the facts which I have endeavoured to illustrate in this Essay. I know of no other way in which I can acknowledge the extent of my obligation to him, than by adding that I should never have taken up the subject in its present form, but for the advantages I have derived from his friendship and encouragement.

* To Sir Charles Lyell's works, indeed, I am indebted for the enunciation of those principles that are essential to the progress of every naturalist and geologist; those, I mean, that affect the creation and extinction, dispersion and subsequent isolation of organic beings; and though botanists still differ in opinion as to the views he entertains on the most speculative of subjects (the origin and permanence of species), there is, I think, but one as to the soundness and originality of his observations on all that relates to the strict dependence of organic beings on physical conditions in the state of the earth's surface. I feel that I cannot over-estimate the labours of this great philosopher, when I reflect that without them the science of geographical distribution would have been with me little beyond a tabulation of important facts; and that I am indebted to them, not only for having given a direction to my studies in this department, but for an example of admirable reasoning on the facts he has collected regarding the distribution of plants and animals. I have no hesitation in recommending the 'Principles of Geology' to the New Zealand student of Nature, as the most important work he can study.

† See the Principles of Geology, ed. 9. p. 702, and Address to the Geological Society of London by the President (Leonard Horner, Esq.), in 1847, p. 66.

‡ For the contents of the Essay itself, I must refer to the Records of the Geological Survey of Great Britain, vol. i. p. 336. This is the most original and able essay that has ever appeared on this subject, and though I cannot

tional weight from the fact that the distribution of British plants is in accordance with its principal features*.

The geographical distribution of British plants has been the subject of the most rigorous investigation by one of our ablest British botanists, Mr. H. C. Watson, who first drew attention to the various botanical elements of which the flora is composed, and grouped the species into botanical provinces. These provinces were intended for "showing the areas of plants, as facts in nature independent of all theoretical explanations and reasons." (Cybele Britannica, vol. i. p. 18.) An inspection of them shows the relations borne by the plants of England to those of certain parts of Europe and of the Arctic regions; and Professor Forbes, applying a modification of these botanical provinces to the illustration of his views of the original introduction of plants into the British Islands, proceeds to show that their migration took place at different periods, contemporary of course with the connection by land of each botanical region of Britain with that part of the continent which presents a similar association of plants.

To extend a theoretical application of these views to the New Zealand Flora, it is necessary to assume that there was at one time a land communication by which the Chilian plants were interchanged; that at the same or another epoch the Australian, at a third the Antarctic, and at a fourth the Pacific floras were added to the assemblage. It is not necessary to suppose that for this interchange there was a continuous connection between any two of these localities, for an intermediate land, peopled with some or all of the plants common to both, may have existed between New Zealand and Chili when neither of these countries was as yet above water†. To account, however, for the Antarctic plants on the lofty mountains, a new set of influences is demanded; no land connection between these islands and New Zealand could have effected this, for the climate of the intermediate area must necessarily have prevented it. But changes of relation between sea and land induce changes of climate, and the presence of a large continent connecting the Antarctic islands would, under certain circumstances, render New Zealand as cold as Britain was during the glacial epoch. Sir C. Lyell first demonstrated this, and showed what such conditions should be; and by consulting the 'Principles of Geology,' my reader will understand how such a climate would reign in the latitude of New Zealand, as that its flora should consist of what are now Antarctic forms of vegetation. The

subscribe to all its botanical details, I consider that the mode of reasoning adopted is sound, and of universal application. What I dissent from most strongly is, the origin of the gulf-weed, the peopling of Scotch mountains by iceberg transport of seeds, and the too great stress laid upon the west Irish flora, whose peculiarities appear to me to be considerably over-estimated.

* It may be well to state to the New Zealand student, that there are no reasons to suppose that Botany can ever be expected to give that direct proof of plants having survived geological changes of climate, sea, and land, which animals do; the cause is evident, for the bones of quadrupeds, shells of mollusca, and hard parts of many animals, afford an abundant means of specific identification, and such are preserved when the animals perish. In plants the case is widely different: their perishable organs of reproduction, which alone are available for systematic purposes, are seldom imbedded, even when other parts of the plants are.

† This disappearance of old land, and the migration of its flora and fauna to new, may be illustrated to a certain extent by the delta of any New Zealand river. A mud-bank on one shore, covered with mangroves, advances across the channel, the mangroves growing on the new land as it forms. The current changes, and the end of the bank (with its mangroves) is cut off, and becomes an island: another change of the river channel fills up that between the islet and the opposite shore, to which it hence becomes a peninsula, peopled by mangroves, whose parents grew on the opposite bank. Here, be it remarked, no subsidence is required, such as must have operated in the assumed isolation of New Zealand.

retirement of the plants to the summit of the New Zealand mountains*, would be the necessary consequence of the amelioration of climate that followed the isolation of New Zealand, and the replacement of the Antarctic continent by the present ocean.

The climate throughout the south temperate zone is so equable, and the isothermal lines are so parallel to those of latitude, that it is not easy for the New Zealand naturalist to realize the altered circumstances that would render the plains of his island suitable for the growth of plants that now inhabit its mountains only†; but if he glance at the map of the isothermal lines of the northern hemisphere, he will see how varied are the climates of regions in the same latitude; that London, with a mean temperature of 51°, is in the same latitude as Hudson's Bay, where the mean temperature is 30°, and the soil ever frozen: and he will further be able to understand by a little reflection, how a change in the relative positions of sea and land would, by isolating Labrador, raise its temperature 10°–15°, causing the destruction of all the native plants that did not retire to its mountain-tops, and favouring the immigration of the species of a more genial climate.

The first inference from such an hypothesis is that the Alpine plants of New Zealand, having survived the greatest changes, are its most ancient colonists; and it is a most important one in many respects, but especially when considered with reference to the mountain floras of the Pacific and southern hemisphere generally. These may be classed under three heads‡ :—

1. Those that contain identical or representative species of the Antarctic Flora, and none that are peculiarly Arctic; as the Tasmanian and New Zealand Alps§.

2. Those that contain, besides these, peculiarities of the Northern and Arctic Floras‖; as the South American Alps.

3. Those that contain the peculiarities of neither; as the mountains of South Africa and the Pacific Islands.

* With regard to the British mountains, Professor Forbes imagines that they were islets in the glacial ocean, and received their plants by transportation of seeds with soil, on ice from the Arctic regions. This appears to me to want support, and there is much in the distribution of Arctic plants especially, wholly opposed to the idea of ice transport being an active agent in dispersion. A lowering of 10° of mean temperature would render the greater part of Britain suitable to the growth of Arctic plants; it would give it the climate of Labrador, situated in the same latitude on the opposite side of the Atlantic. Britain is the warmest spot in its latitude, and a very slight geological change would lower its mean temperature many degrees.

† The New Zealand naturalist has probably a very simple means of determining for himself whether his island has been subject to a geologically recent amelioration of climate; to do which, let him examine the fiord-like bays of the west coast of the Middle Island, for evidence of the glaciers which there exist in the mountains having formerly descended lower than they now do. Glaciers to this day descend to the level of the sea in South Chili, at the latitude of Dusky Bay; and if they have done so in the latter locality, they will have left memorials, in the shape of boulders, moraines, and scratched and polished rocks.

‡ I need scarcely remind my reader that in thus sketching the characteristics of these Alpine floras, I make no allusion to exceptions that do not alter the main features. I am far from asserting that there are no peculiar Arctic or Antarctic forms in the Pacific Islands, nor any peculiarly Arctic ones in Tasmania and New Zealand: but if, on the one hand, future discoveries of such shall weaken the points of difference between these three mountain regions, on the other they might be very much strengthened by adducing the number of Arctic species common to the South American Alps, but not found in the others.

§ These Antarctic forms are very numerous; familiar ones are *Acæna, Drapetes, Donatia, Gunnera, Oreomyrrhis, Lagenophora, Forstera, Ourisia, Fagus, Callixene, Astelia, Gaimardia, Alepyrum, Oreobolus, Carpha, Uncinia.*

‖ *Berberis, Sisymbrium, Thlaspi, Arabis, Draba, Sagina, Lychnis, Cerastium, Fragaria, Lathyrus, Vicia, Hippuris, Chrysosplenium, Ribes, Saxifraga, Valeriana, Aster, Hieracium, Stachys, Primula, Anagallis, Pinguicula, Statice, Empetrum, Phleum, Elymus, Hordeum.*

We thus observe that the want of an Arctic or Antarctic Flora at all in the Pacific islands, and the presence of an Arctic one in the American Alps, are the prominent features; and I shall confine my remarks upon these to the fact that, with regard to the isolated islands of the Pacific, they are situated in too warm a latitude to have had their temperature cooled by changes in the relative position of land and ocean, so as to have harboured an Antarctic vegetation. With regard to the South American Alps, there is direct land communication along the Andes from Arctic to Antarctic regions; by which not only may the strictly Arctic genera and species have migrated to Cape Horn, but by which many Antarctic ones may have advanced northward to the equator*.

There is still another point in connection with the subject of the relative antiquity of plants, and in adducing it I must again refer to the 'Principles of Geology,' where it is said, "As a general rule, species common to many distant provinces, or those now found to inhabit many distant parts of the globe, are to be regarded as the most ancient . . . . their wide diffusion shows that they have had a long time to spread themselves, and have been able to survive many important changes in Physical Geography†." If this be true, it follows that, consistently with the theory of the antiquity of the Alpine flora of New Zealand, we should find amongst the plants common to New Zealand and the Antarctic islands, some of the most cosmopolitan; and we do so in *Montia fontana, Callitriche verna, Cardamine hirsuta, Epilobium tetragonum,* and many others.

On the other hand, it must be recollected that there are other causes besides antiquity and facility for migration, that determine the distribution of plants; these are their power, mentioned above, of invading and effecting a settlement in a country preoccupied with its own species, and their adaptability to various climates: with regard to the first of these points, it is of more importance than is generally assumed, and I have alluded to its effects under *Sonchus,* in the body of this work. As regards climates, the plants mentioned above seem wonderfully indifferent to its effects‡.

Again, even though we may safely pronounce most species of ubiquitous plants to have outlived many geological changes, we may not reverse the position, and assume local species to be amongst the most recently created; for whether (as has been conjectured) species, like individuals, die out in the course of time, following some inscrutable law whose operations we have not yet traced, or whether (as in some instances we know to be the case) they are destroyed by natural causes (geological or others), they must in either case become scarce and local while they are in process of disappearance.

In the above speculative review of some of the causes which appear to affect the life and range of species in the vegetable kingdom, I have not touched upon one point, namely, that which concerns the original introduction of existing species of plants upon the earth. I have assumed that they have existed for ages in the forms they now retain, that assumption agreeing, in my opinion, with the facts elicited by a survey of all the phenomena they present, and, according to the most eminent zoolo-

---

* Why these Antarctic forms have not extended into North America, as the Arctic ones have into South America, is a curious problem, and the only hypothesis that suggests itself is derived from the fact that though the Panama Andes are not now sufficiently lofty for the transit of either, there is nothing to contradict the supposition that they may have had sufficient altitude at a former period, and that one which preceded the advance of the Antarctic species to so high a northern latitude.

† Principles of Geology, ed. 9. p. 702.

‡ Mr. Watson (Cybele Britannica) gives the range of *Callitriche* in Britain alone as including mean temperatures of 40° to 52°, and as ascending from the level of the sea to nearly 2000 feet in the East Highlands of Scotland. *Montia,* according to the same authority, enjoys a range of 36° to 52°, and ascends to 3300 feet; *Epilobium,* a temperature of 40° to 51°, and ascends to 2000 feet; *Cardamine,* a temperature of 37° to 52°, and ascends to 3000 feet.

gists, with those laws that govern animal life also; but there is nothing in what is assumed above, in favour of the antiquity of species and their wide distribution, that is inconsistent with any theory of their origin that the speculator may adopt. My object has not so much been to ascertain what may, or may not, have been the original condition of species, as to show that, granting more scope for variation than is generally allowed, still there are no unassailable grounds for concluding that they now vary so as to obliterate specific character; in other words, I have endeavoured to show that they are, for all practical purposes of progress in botanical science, to be regarded as permanently distinct creations, which have survived great geological changes, and which will either die out, or be destroyed, with their distinctive marks unchanged. We have direct evidence of the impoverishment of the flora of the globe, in the extinction of many most peculiar insular species within the last century; but whether the balance of nature is kept up by the consequent increase of the remainder in individuals, or by the sudden creation of new ones, does not appear, nor have we any means of knowing: if the expression of an opinion be insisted on, I should be induced to follow the example of an eminent astronomer, who, when the question was put to him, as to whether the planets are inhabited, replied that the earth was so, and left his querist to argue from analogy. So with regard to species, we know that they perish suddenly or gradually, without varying into other forms to take their place as species, from which established premiss the speculator may draw his own conclusions.

And now that I have brought these desultory observations to a close, I cannot review them without fearing that I may incur the charges of, on the one hand, attempting to promote a spirit of theoretical inquiry amongst those naturalists of the distant colony whom I would fain instruct; and on the other, of giving way to it myself, and occupying the time of my readers with what is with too many the foundation of fruitless controversy. In answer to the first I would say, that the speculations which I have endeavoured to combat are becoming widely spread amongst superficial observers, and are quoted every day as objections to the devotion of time and labour to a systematic inquiry into any branch of Natural History. The very many aspirants to a knowledge of science whom I have had the pleasure of knowing in the Colonies, though well educated in the ordinary acceptation of the term, have never been trained to habits of observation, or of reasoning upon what they read in the book of nature, nor have they been grounded in the elements of natural science; they are hence prone to rely for information on these speculative subjects (which they seek with avidity) upon a class of works that are, with very few exceptions, by authors who have no practical acquaintance with the sciences they write about, or with the facts they so often distort. I have further had a more practical object in view—the offering of theoretical reasons for inculcating caution on the future botanists of New Zealand; I have endeavoured to make it clear to those who may read these remarks, that systematic botany is a far more difficult and important object than is generally supposed; that the progress the student will make himself, and hence that the science will make in his country, is not to be measured by the number of new species he may find, but by his manner of treating the old, and his desire to regard all as parts of the vegetable kingdom, and not of the New Zealand Flora only; and that there is no surer sign of his not appreciating the aim and scope of the science he cultivates, than a craving to load it with names, and to take contracted views of species, their variation and distribution.

To those who may accuse me of giving way to hasty generalization or loose speculation on the antiquity and dispersion of plants over parts of the Southern Hemisphere, I may answer, that no speculation is idle or fruitless, that is not opposed to truth or to probability, and which, whilst it co-ordinates a body of well established facts, does so without violence to nature, and with a due regard to the possible results of future discoveries. I may add, that after twelve years' devotion to the laborious accumulation and arrangement of facts in the field and closet, untrammelled by any theories to combat or vindicate, I have thought that I might bring forward the conclusions to which my studies have led me, with less chance of incurring such a reproach, than those would, who, with far better abilities and judgment, have not had my experience and opportunities.

CHAPTER III.

§ 1. ON THE PHYSIOGNOMY AND AFFINITIES OF THE NEW ZEALAND FLORA.

In the following remarks, the flowering plants alone of New Zealand are referred to, except when it is otherwise stated: my object being primarily to show the relation between the botany of New Zealand and that of the south temperate continents, I have, for several reasons, considered that the introduction of the Ferns even was not expedient:—1. Because they include only one family of *Cryptogamia*, and that the only one towards a knowledge of whose number and distribution in New Zealand we have even approximately accurate data.—2. Because the diffusion of their minute spores is so ubiquitous*, and their growth is so dependent on one climatic element, viz. humidity, that their geographic distribution does not harmonize with that of flowering plants in general.

The traveller from whatever country, on arriving in New Zealand, finds himself surrounded by a vegetation that is almost wholly new to him; with little that is at first sight striking, except the Tree-fern and *Cordyline* of the northern parts, and nothing familiar, except possibly the Mangrove; and as he extends his investigations into the Flora, with the exception of *Pomaderris* and *Leptospermum*, he finds few forms that remind him of other countries. Of the numerous Pines, very few recall by habit and appearance the idea attached either to trees of this family in the northern hemisphere, or to the *Callitris* of New Holland, or to the *Araucariæ* of that country and Norfolk Island; while of the families that on examination indicate the only close affinity between the New Zealand Flora and that of any other country, (the *Myrtaceæ, Epacrideæ,* and *Proteaceæ,*) few resemble in general aspect

---

* A most remarkable exemplification of this is found in the occurrence of *Lycopodium cernuum,* (a most universally distributed Fern in all warm climates) in the Azores, where it grows only around some hot springs. Within the last few months it has been also collected in St. Paul's Island (lat. 38° south), by the naturalists of Captain Denham's Expedition to the Pacific Islands: there, too, only where the ground is much heated by springs. These facts are most remarkable, for the *Lycopodium cernuum* does not inhabit Madeira or any spot in the Azores, except the vicinity of the hot springs, and St. Paul's Island is also far beyond its natural isothermal in that longitude of the southern hemisphere; and it is to be remarked, that in neither island is the *Lycopodium* accompanied by any other tropical plant, which would indicate the aerial transport of larger objects than the microscopic spores of *Lycopodia,* which are raised in clouds from large surfaces covered with the gregarious species.

Fig.

20. *Ceropachys oculatus.* 21. Head of the same. 22. Wing of the same.
    23. Antennæ of the same. 24. Abdomen of the same.
25. *Echinopla melanarctos.* 26. Section of the abdomen of the same,
    showing the styles, or blunt spines, with hairs on their summits,
    which cover the abdomen above. 27. Maxillary palpus of the same.
    28. Mandible of the same. 29. Labial palpus of the same.

Tab. II.

1. *Myrmosida paradoxa.* 1 *a*, antennæ; 1 *b*, wing.
2. *Crematogaster inflata.* 1 *b*, wing; 1 *c*, manble.
3. *Cataulacus horridus.*
4. *Cataulacus insularis.* 4 *a*, anterior wing.
5. *Meranoplus cordatus.* 6. *Meranoplus mucronatus.*
7. *Meranoplus castaneus.* 8. *Cataulacus reticulatus.*
9. Tongue of *Gayella pulchella.* 9 *a*, labial palpi; 9 *b*, paraglossæ. 10.
    Maxilla. 10 *a*, maxillary palpi.
11. Anterior wing of *Gayella pulchella.*

On the general Geographical Distribution of the Members of the
Class Aves.   By Philip Lutley Sclater, Esq., M.A.,
F.L.S.

[Read June 16th, 1857.]

An important problem in Natural History, and one that has
hitherto been too little agitated, is that of ascertaining the most
natural primary divisions of the earth's surface, taking the amount
of similarity or dissimilarity of organized life solely as our guide.
It is a well-known and universally acknowledged fact that we can
choose two portions of the globe of which the respective Faunæ
and Floræ shall be so different, that we should not be far wrong
in supposing them to have been the result of distinct creations.
Assuming then that there are, or may be, more areas of creation
than one, the question naturally arises, how many of them are
there, and what are their respective extents and boundaries, or
in other words, what are the most natural primary ontological divisions of the earth's surface?

In the Physical Atlases lately published, which have deservedly
attracted no small share of attention on the part of the public, too
little regard appears to have been paid to the fact that the divisions of the earth's surface usually employed are not always those

which are most natural when their respective Faunæ and Floræ are taken into consideration. The world is mapped out into so many portions, according to latitude and longitude, and an attempt is made to give the principal distinguishing characteristics of the Fauna and Flora of each of these divisions; but little or no attention is given to the fact that two or more of these geographical divisions may have much closer relations to each other than to any third, and, due regard being paid to the general aspect of their Zoology and Botany, only form one natural province or kingdom (as it may perhaps be termed), equivalent in value to that third. Thus in ' Johnston's Physical Atlas,' the earth is separated into sixteen provinces for Ornithology, solely according to latitude and longitude, and not after ascertainment of the amount of difference of ornithic life in the respective divisions. Six of these provinces are appropriated to America, one to Europe, and six to Asia, Australia, and the islands; a very erroneous division, according to my ideas, as I shall hereafter attempt to show. In Mr. Swainson's article in Murray's ' Encyclopedia of Geography,' and in Agassiz's introduction to Nott and Gliddon's 'Types of Mankind,' what I consider to be a much more philosophical view of this subject is taken. The latter author, in particular, attempts to show that the principal divisions of the earth's surface, taking zoology for our guide, correspond in number and extent with the areas occupied by what Messrs. Nott and Gliddon consider to be the principal varieties of mankind. The argument to be deduced from this theory, if it could be satisfactorily established, would of course be very adverse to the idea of the original unity of the human race, which is still strongly supported by many Ethnologists in this country. But I suppose few philosophical zoologists, who have paid attention to the general laws of the distribution of organic life, would now-a-days deny that, as a general rule, every species of animal must have been created within and over the geographic area which it now occupies. Such being the case, if it can be shown that the areas occupied by the primary varieties of mankind correspond with the primary zoological provinces of the globe, it would be an inevitable deduction, that these varieties of Man had their origin in the different parts of the world where they are now found, and the awkward necessity of supposing the introduction of the red man into America by Behring's Straits, and of colonizing Polynesia by stray pairs of Malays floating over the water like cocoa-nuts, and all similar hypotheses, would be avoided.

But the fact is, we require a far more extended knowledge

of zoology and botany than we as yet possess, before it can be told with certainty what *are* the primary ontological divisions of the globe. We want far more correct information concerning the families, genera, and species of created beings—their exact localities, and the geographical areas over which they extend—before very satisfactory conclusions can be arrived at on this point. In fact, not only families, genera, and species, but even local varieties must be fully worked out in order to accomplish the perfect solution of the problem. There is no reason, however, why attempts should not be made to solve the question, even from our present imperfect data, and I think the most likely way to make good progress in this direction, is for each inquirer to take up the subject with which he is best acquainted, and to work out what he conceives to be the most natural divisions of the earth's surface from that alone. Such being done, we shall see how far the results correspond, and on combining the whole, may possibly arrive at a correct solution of the problem—*to find the primary ontological divisions of the earth's surface.*

With these views, taking only the second group of the Order Vertebrata, the Class *Aves,* I shall attempt to point out what I consider to be the most natural division of the earth's surface into primary kingdoms or provinces, looking only to the geographical distribution of the families, genera, and species of this class of beings.

Birds, being of all the animated creation the class most particularly adapted for wide and rapid locomotion, would, at first sight, seem to be by no means a favourable part of Nature's subjects for the solution of such a problem. But, in fact, we know that there are many species, genera, and even families of this class, particularly amongst the *Passeres,* whose distribution is extremely local. The *Nestor productus,* confined to the little island called Philip Island; the several genera of Finches peculiar to the archipelago of the Galapagos; the gorgeous family *Paradiseidæ,* restricted to the Papuan territory, are familiar examples of this fact. Again, the migratory birds which traverse large districts of the earth's surface, how constant are they in returning only where they have been in former years! We do not find that the Nightingale extends its range farther to the west one year than another, nor that birds looked upon as occasional visitors to this country, grow more or less frequent. If the contrary be the case, it may always be accounted for by some external cause, generally referable to the agency of *man,* and not to any change in Nature's unvarying laws of distribution. It is, however, amongst the *Passeres* that we find *endemism* most normal; the *Accipitres*, *Anseres*, and, more than all, the *Grallæ* are ever disposed to be *sporadic*, and indeed some species belonging to the latter order may be denominated truly cosmopolitan.

Taking then the birds of the order *Passeres* (which I consider ought properly to include the *Scansores* or *Zygodactyli*) as the chief materials from which to derive our deductions, let us suppose a species of this group, but of doubtful form and obscure plumage, to be placed before the Ornithologist, from whom its name is required. The first thing he looks to is, whether it is from the Old World or the New; and this is a point which, as a general rule, a mere glance at the external appearance of the object is sufficient to settle. The most obvious geographical division of the birds of this order certainly corresponds with the usually adopted primary division of the earth's surface. In fact, taking Ornithology as our guide, we may at once pronounce that the Faunæ of the Old and New worlds may, to all appearance, have been the subjects of different acts of creation. There are very many natural families which are quite peculiar to one or the other of these great divisions of the earth's surface, more subfamilies, few genera really common to the two, and very few, if any, species*.

The appended Table will show some of the most noticeable of the natural families of birds which are confined to the Old and New worlds respectively.

Familiæ Neogeanæ, sive Novi Orbis.		Familiæ Palæogeanæ, sive Orbis Veteris.	
Todidæ.	Tyrannidæ.	Coraciidæ.	Promeropidæ.
Momotidæ.	Cotingidæ.	Eurylæmidæ.	Muscicapidæ.
Bucconidæ.	Rhamphastidæ.	Meropidæ.	Musophagidæ.
Galbulidæ.	Opisthocomidæ.	Upupidæ.	Coliidæ.
Trochilidæ.	Cracidæ.	Bucerotidæ.	Megapodidæ.
Icteridæ.	Tinamidæ.	Sturnidæ.	Pteroclidæ.
Cærebidæ.	Meleagrinæ.	Paradiseidæ.	Phasianidæ.
Formicariidæ.	Odontophorinæ.	Meliphagidæ.	Perdicinæ.
Dendrocolaptidæ.			

With regard to the genera of *Passeres*, common to the two worlds, when we have excepted the truly cosmopolitan forms *Turdus, Hirundo, Picus,* &c., the number will be found very small; and it will be observed that these are invariably genera

* There are now acknowledged only 8 species of the order *Passeres*, in

belonging to temperate regions, and such as extend themselves only through the northern portion of the New World, failing entirely before we reach Tropical and Southern America, the most really characteristic region of Neogean Ornithology.

Such is the case in the genera *Sitta, Certhia, Regulus, Parus, Lanius, Perisoreus, Pica, Corvus* and *Loxia.* No member of these genera (which are common to the temperate portions of both hemispheres) extends farther south in the New World than the Table-land of Mexico. They are all quite foreign to Neotropical (Tropical American) Ornithology, although in the Old World most of them reach the tropics.

Having, therefore, made our first territorial division that of the two worlds, agreeing so far with geographers, we will look at the great continent and Australia *en masse,* and see what are its most natural subdivisions.

Here we find ourselves at once at issue with ordinary geographers. Europe may be a very good continent of itself, in many ways, and in some respects worth all the rest of the world put together,—"*Better fifty years of Europe than a cycle of Cathay,*" says the Poet,—but it is certainly *not* entitled to rank as one of the primary zoological regions of the earth's surface, any more than as one of the physical divisions. Europe and Northern Asia are in fact quite inseparable. So far as we are acquainted with the ornithology of Japan—the eastern extremity of the temperate portion of the great continent, we there find no striking differences from the European *Avi-fauna,* but rather repetitions of our best-known European birds in slightly altered plumage,—representatives in fact of the European types. Temminck, indeed, has stated, that there are no less than 114 birds found in Japan, identical with European species. Some of these, however, have been since ascertained to be apparently distinct, but there can be no question as to the general strong resemblance of the Japanese Avi-fauna to that of Europe. How far south we are to extend the boundaries of this great temperate region of the Old World can

which no differences have, as yet, been detected in the comparison of specimens from the Old and New worlds, viz.:—

Cotyle riparia.	Linota linaria.
Ampelis garrula.	Plectrophanes nivalis.
Junco hyemalis.	Plectrophanes lapponica.
Linota borealis.	Loxia leucoptera.

The whole of these (with exception of *Cotyle riparia*) range to the extreme north, where the two worlds almost unite.

hardly be fairly ascertained, until the ornithology of Central Asia is much better worked out than is at present the case. While among the birds of the Himalayas we find many striking instances of the recurrence of European types, there is no doubt that the ornithology of the Indian Peninsula and the rest of Southern Asia, below the 30th parallel, is quite different from it.

Africa, north of the Atlas, along the southern shores of the Mediterranean, again appears to belong to Europe zoologically, and not to the continent to which it is physically joined. Such species of birds, foreign to Europe, as are found in Algeria and Morocco, are not usually connected with true African forms, but are again slightly modified representatives of Europæo-Asiatic species.

Such are the N. African species.	Representatives of the European.
Garrulus cervicalis.	Garrulus cristatus.
Pica mauritanica.	Pica caudata.
Fringilla spodiogenia.	Fringilla cælebs.
Parus ultramarinus.	Parus cæruleus.
Picus numidicus.	Picus major.

On the whole, therefore, I think we may consider Africa, north of the Atlas, Europe and Northern Asia, to form one primary zoological division of the earth's surface, for which the name Palæarctic or Northern Palæogean Region would be best applicable.

The great continent of Africa will form a second well-marked division, after cutting off the slice north of the Atlas, but including Madagascar (where the African type appears to have reached the height of its peculiar development) and Western Arabia, to the Persian Gulf; for in this latter region, so far as our information goes, the African type seems to predominate over the Indian. Although there are genera of *Passeres* common to Africa and India, and even a few species, yet there can be no question as to the generally dissimilar character of the *Avi-faunæ* of these two countries. This second African division may be called the Æthiopian or Western Palæotropical Region.

Another tropical region of the Old World seems to be constituted by Southern Asia and the islands of the Indian Archipelago. The Philippines, Borneo, Java, and Sumatra, certainly belong to this division, but it is of course not yet possible to decide where the line runs which divides the *Indian* zoology from the Australian. New Guinea presents probably only a more exaggerated production of the Australian type, and I should be inclined for the present not to separate New Zealand and the Pacific Islands generally from the Australian division. We should have, therefore, in the Old World one temperate region and three tropical; the eastern palæotropical or Australian advancing rather farther to the south than the others, the Indian or middle palæotropical being the most northern of the three.

In the New World we can simply divide the continent into northern and southern divisions; the northern, or Nearctic region, extending down the centre of the table-land of Mexico, and showing some indication of parallelism to the Palæarctic by the presence of certain temperate types; the Neotropical or southern (which embraces the whole of the rest of this great continent) being wholly free from any admixture of the sort, and in fact exhibiting, in my opinion (with the exception possibly of New Guinea), by far the richest and most peculiar *Avi-fauna* of the world's surface.

Having thus pointed out what I consider to be the primary divisions of the earth,—taking ornithology as our guide, I propose to devote a few lines to each region separately, noticing its apparent limits, its peculiarities, and most characteristic forms, and attempting to give an approximate estimate of the comparative abundance of ornithic species within its area.

The subjoined plan will serve to give at one view an illustration of my ideas as to the arrangement of these primary *Avi-faunæ* of the earth's surface. It must, however, be recollected that the calculations made as to the number of species to a square mile, can be only looked upon as mere attempts at approximations. Even in the whole general calculation, the presence of two variable elements—in the first place the number of square miles (about which geographers still give the most conflicting statements), and in the second place, the number of species of birds, concerning which ornithologists are as yet by no means agreed, greatly increases the uncertainty of the ratio deducible from them; and in working out the ratios in the respective regions, it is of course still more difficult to attain to any great degree of accuracy.

Taking however the whole number of square miles of dry land at 45,000,000, and the number of species of birds at 7500, which are both of them moderate estimates, we have on the average a single species to each 6000 square miles. In the different regions we shall attempt to show how far this ratio is departed from.

The zoological kingdoms or primary divisions are of course naturally separable into secondary divisions or provinces, but it would be extending the limits of this communication too far to attempt to go into these at the present time.

I. PALÆARCTIC REGION (*Regio Palæarctica*).

Extent.—Africa north of the Atlas, Europe, Asia Minor, Persia and Asia generally north of the Himalaya range, upper part of the Himalaya range?, northern China, Japan and the Aleutian Islands. Approximate area of 14,000,000 square miles.

Characteristic forms.—*Sylvia, Luscinia, Erythacus, Accentor, Regulus, Podoces, Fregilus, Garrulus, Emberiza, Coccothraustes, Tetrao.*

It cannot be denied that the ornithology of the Palæarctic or great temperate region of the Old World is more easily characterized by what it has not than by what it has. There are certainly few among the groups of birds occurring in this Region, which do not develope themselves to a greater extent elsewhere. For we must acknowledge that the most productive seats of animal life, where all the bizarre and extraordinary forms that the Naturalist best loves are met with, lie under the suns of the tropics, and far removed from temperate latitudes. The most prevalent forms among the *Passeres*, of the Palæarctic Region, are perhaps the plain dull-coloured *Sylviinæ*, distinguished rather for their melodious song than by any external beauty of plumage or singularity of form. Upwards of 35 species of this subfamily occur in the ornithology of Europe alone; and when Northern Africa and the whole North of Asia are taken into calculation, the number would be considerably increased, and this Region may be considered the true focus of the group.

The genus *Erythacus* would be perhaps as good a representative genus as any as a type of Palæarctic ornithology; a second species (*Erythacus akahige*) occurring at the eastern extremity of the Asiatic continent, and there beautifully representing our common Robin. True *Emberiza* is likewise very characteristic of the temperate portion of the Old World, nearly the whole of the known species being found in Europe or Northern Asia. *Accentor* is perhaps more strictly a northern Himalayan form, with several representatives within the Palæarctic Region; but *Fregilus, Podoces, Garrulus, Tetrao*, and numerous species of *Anatidæ* are likewise eminently noticeable as among the most typical forms of Palæarctic ornithology.

138   SCLATER ON THE GENERAL DISTRIBUTION OF AVES.

The most recent summary of the Birds of Europe gives—

1. Accipitres...... 57
2. Passeres 238
3. Scansores...... 12
4. Columbæ...... 7
5. Gallinæ........ 22   } 581 species.
6. Struthiones 0
7. Grallæ 101
8. Anseres........ 144

It is very difficult to say what additions should be made to this in order to give the approximate number of the birds of the whole Palæarctic Region; but a moderate calculation does not show more than 650 species truly belonging to this fauna: for it must be recollected that the number 581 contains many birds of rare occurrence in Europe, and which must be correctly reckoned as belonging to other divisions. As we have in the Palæarctic Region the enormous land area of probably upwards of 14,000,000 square miles, this will give us a species for each 21,000 square miles, speaking in round numbers; and it consequently follows (as might have been expected), that the Palæarctic is by far the least prolific region of ornithic life on the globe.   According to my ideas, therefore, the statement in Johnston's ' Physical Atlas,' that " *Europe possesses more species than any other zoological province,*" is exactly contrary to the fact.

II. Æthiopian or Western Palæotropical Region
(*Regio Æthiopica*).

Extent.—Africa, south of the Atlas range, Madagascar, Bourbon, Mauritius, Socotra and probably Arabia up to the Persian Gulf, south of 30° N. l.; an approximate area of 12,000,000 square miles.

Characteristic forms.—*Gypogeranus, Helotarsus, Polyboroides, Gypohierax, Melierax, Macrodipteryx, Irrisor, Fregilupus, Bucorvus, Apaloderma, Parisoma, Macronyx, Lioptilus, Sericolius, Malaconotus, Laniarius, Chaunonotus, Prionops, Sigmodus, Phyllastrephus, Lanioturdus, Vidua, Juida, Buphaga, Verreauxia, Læmodon, Indicator, Musophaga, Colius, Pæocephalus, Numida, Phasidus, Struthio, Balæniceps, Scopus.*

(Madagascar). *Euryceros, Falculia, Oriolia, Philipitta, Brachypteracias, Atelornis, Bernieria, Hartlaubius, Artamia, Vanga, Coua, Leptosomus, Vigorsia, Mesites, Biensis.*

The characteristic forms of African Ornithology are very numerous.   Several groups of birds, which seem clearly entitled to rank as distinct families, or at least as subfamilies, are wholly peculiar to this region, such as the *Coliidæ, Musophagidæ*, and *Buphaginæ.* There are also very many genera, of which the species are all confined to this continent; the principal of which I have enumerated in my List of Typical forms.   The island of Madagascar, however, is the locality where the African type seems pushed to its utmost degree of development.   There are many genera quite peculiar to this island, or which have a single representative or so upon the adjacent coast of the continent.   Such are *Oriolia, Atelornis, Brachypteracias, Vanga*, and others which I have mentioned above, not to mention the extinct gigantic *Æpyornis*.   Bourbon, Mauritius and the other Mascarene islands all belong to Africa zoologically, and have only recently lost the now extinct birds of the genera *Didus, Pezophaps* and their allies, which were, so far as we know, types quite peculiar to this locality.

Dr. G. Hartlaub's lately published *System der Ornithologie West-Africa's* gives as inhabitants of that part of the continent,—

Accipitres	56
Passeres	450
Scansores	69
Columbæ	17
Gallinæ	19
Struthiones	1
Grallæ	99
Anseres	42

753.

In the preface to Dr. Hartlaub's work will be found a *resumé* of all the most important facts known concerning African Ornithology.

For North-eastern Africa we have a List lately published by Dr. Heuglin, who mentions—

1. Accipitres......	95
2. Passeres	372
3. Scansores......	38
4. Columbæ......	14
5. Gallinæ	24
6. Struthiones	1
7. Grallæ	130
8. Anseres	80

754 species.

A correct catalogue of the Birds of S. Africa would probably be not less numerous in species.

10*

On the whole, therefore, I think we cannot allow for the Western Palæotropic region less than 1250 species, which, with an area of 12,000,000 square miles, gives one species to each 9600 square miles nearly.

III. INDIAN OR MIDDLE PALÆOTROPICAL REGION
(*Regio Indica*).

Extent.—India and Asia generally south of Himalayas, Ceylon, Burmah, Malacca and Southern China, Philippines, Borneo, Java, Sumatra and adjacent islands; an area of perhaps 4,000,000 square miles.

Characteristic forms.—*Harpactes, Colocalia, Calyptomena, Eurylæmus, Buceros, Garrulax, Liothrix, Malacocercus, Pitta, Timalia, Pycnonotus, Phyllornis, Pericrocotus, Analcipus, Acridotheres, Gracula, Sasia, Megalæma, Phænicophaus, Dasylophus, Palæornis, Pavo, Ceriornis, Polyplectron, Argus, Euplocamus, Rollulus, Casuarius.*

Mr. Swainson, in his article in H. Murray's 'Encyclopedia of Geography,' considers the mainland of Southern Asia and the larger Indian islands as belonging to two different zoological regions. But it is now generally acknowledged that this is not the case. There are so many generic forms which commence in Southern Asia and extend over the greater part of the Indian Archipelago, that it is not possible to look upon these countries as belonging to different regions, though they doubtless form distinct subkingdoms or provinces, in each of which will be found corresponding representative species. How far in an eastern direction we are to extend the boundaries of the Middle Palæotropical Region is a difficult question, which can hardly be answered until we know more of the Natural History of these great islands; but there is no doubt that Borneo, Sumatra and Java belong to this zoology, but probably not Celebes.

The most characteristic forms of the Indian region are without doubt the *Phasianidæ*, the whole of which magnificent group of birds may be said to be confined to this region,—one or two species only straying into the confines of Palæarctic zoology, and a single genus, *Meleagris*, representing them in America, and the few birds of the genera *Numida, Agelastus* and *Phasidus* in Africa.

If the number of species duly attributable to the Middle Palæotropical Region, be reckoned at about 1500, and its geographical area at nearly 4,000,000 square miles, we have a species to each 2600 miles nearly, which indicates a degree of intensity of species only surpassed by Tropical America.

IV. AUSTRALIAN OR WESTERN PALÆOTROPICAL REGION (*Regio Australiana*).

Extent.—Papua and adjacent islands, Australia, Tasmania and Pacific Islands ; an area of perhaps 3,000,000 square miles.

Characteristic forms.—

1. (Australia.) *Ægotheles, Falcunculus, Colluricincla, Grallina, Gymnorhina, Strepera, Cinclosoma, Menura, Psophodes, Malurus, Sericornis, Epthianura, Pardalotus, Chlamydera, Ptilonorhynchus, Struthidea, Licmetis, Calyptorhynchus, Platycercus, Euphema, Calopsitta, Climacteris, Scythrops, Myzantha, Talegalla, Leipoa, Pedionomus, Dromaius, Cladorhynchus, Tribonyx, Cereopsis, Anseranas, Biziura.*

2. (Papua.) *Sericulus, Melanopyrrhus, Ptiladela, Edoliosoma, Peltops, Rectes, Manucodia, Gymnocorvus, Astrapia, Paradisea, Epimachus, Nasiterna, Charmosyna, Cyclopsitta, Goura, &c.*

3. (New Zealand.) *Neomorpha, Prosthemadera, Anthornis, Acanthisitta, Mohoa, Certhiparus, Turnagra, Aplonis, Creadion, Nestor, Strigops, Apteryx, Ocydromus.*

4. (Pacific Islands.) *Moho, Hemignathus, Drepanis, Pomarea, Metabolus, Sturnoides, Leptornis, Tatare, Loxops, Coriphilus, Ptilonopus.*

New Guinea is in some respects so peculiar in its Ornithology, as far as we are acquainted with it, that it would at first sight appear as if it ought to form a zoological region of itself. But there are certainly many genera common to it and Australia (for example, *Podargus, Tanysiptera, Alcyone, Mimeta, Ptilorhis, Cracticus, Manucodia,* &c.) ; and for the present I am inclined to retain it as part of the Australian region. Both New Zealand and the Pacific islands have also some claims to stand alone as separate regions, their forms of ornithic life being in many cases extremely peculiar and local. If they can be attached anywhere, however, it is to Australia ; and I have included them temporarily in the same region. Mr. Gould's 'Birds of Australia' has made us well acquainted with the ornithology of that continent; but there still remains New Guinea and the multitudinous adjacent islands, which doubtless contain numbers of species as yet unknown to science. Mr. Gould, in his 'Birds of Australia,' enumerates—

1. Accipitres......	36	
2. Passeres	311	
3. Scansores......	36	
4. Columbæ	23	600.
5. Gallinæ........	16	
6. Struthiones	1	
7. Grallæ	78	
8. Anseres	99	

in all 600 species.

The most characteristic forms of this region are perhaps the *Paradiseidæ* and *Epimachidæ* (both peculiar to it) ; the *Meliphagidæ*, one or two genera only of which are found externally, and of which between 60 and 70 species occur in Australia alone ; the genera *Calyptorhynchus, Microglossa, Trichoglossus, Platycercus, Nestor, Strigops*, and many other forms amongst the *Psittacidæ*, besides a vast number of others.

Taking 3,000,000 of square miles as the amount of dry land in this region, and allowing 1000 species as peculiar to it, we have one species to every 3000 square miles, showing us that this is little inferior to the middle Palæotropical Region in intensity of species.

V. NEARCTIC or NORTH-AMERICAN REGION (*Regio Nearctica*).

Extent.—Greenland and North America down to centre of Mexico—area of perhaps 6,500,000 square miles.

Characteristic forms.—*Trochilus, Sialia, Toxostoma, Icteria, Vireo, Mniotiltinæ, Chamæa, Certhia, Sitta, Neocorys, Calamospiza, Zonotrichia, Picicorvus, Gymnocitta, Meleagris.*

As is the case in the Old World, most of the genera belonging to the northern part of the New World are better represented in its tropical than in its temperate portions. Northern America, however, produces *Sylvicolæ* and *Zonotrichiæ* in much greater abundance than southern America, and these genera (which are analogous to the *Sylviinæ* and *Emberizæ* of the Old World) are perhaps its most ordinary characteristic forms. I have already mentioned the chief genera common to the northern portions of both hemispheres. These are also characteristic of *Nearctic* in contrast to Neotropical zoology, as none of them extend into Southern America. The ornithology of the U. S. of America (which now embrace a very large proportion of the Nearctic region) contains upwards of 620 species.

Calculating the area of the Nearctic Region at six millions and a half of square miles, and the species peculiar to it at 660, we have about 9000 miles for each species, making this region, as might have been supposed, the least productive of ornithic life, after the Palæarctic.

VI. NEOTROPICAL or SOUTH-AMERICAN REGION (*Regio Neotropica*).

Extent.—West India Islands, Southern Mexico, Central America and whole of S. America, Galapagos Islands, Falkland Islands. Estimated area of about 5,500,000 square miles.

Characteristic forms.—1. (Continental.) *Sarcorhamphus, Ibycter, Milvago, Thrasaëtus, Cymindis, Herpetotheres, Steatornis, Nyctibius, Hydropsalis, Eleothreptus, Trogon, Bucco, Monasa, Galbula, Furnarius, Synallaxis, Anabates, Oxyrhamphus, Dendrocolaptes, Pteroptochos, Rhamphocænus, Campylorhynchus, Hylophilus, Lessonia, Agriornis, Formicarius, Formicivora, Grallaria, Tænioptera, Tityra, Conopophaga, Pipra, Rupicola, Phœnicercus, Cotinga, Gymnoderus, Cephalopterus, Vireolanius, Cyclorhis, Thamnophilus, Tanagra, Calliste, Saltator, Euphonia, Catamblyrhynchus, Phytotoma, Opisthocomus, Ramphastos, Picumnus, Celeus, Crotophaga, Cultrides, Penelope, Oreophasis, Crax, Thinocorus, Tinamus, Psophia, Cariama, Eurypyga, Parra, Palamedea, Chauna, Aramus, Merganetta, Heliornis.*

2. (Antilles.) *Todus, Priotelus, Cinclocerthia, Dulus, Loxigilla, Phœnicophilus, Spindalis, Glossiptila, Teretristis, Saurothera.*

3. (Galapagos.) *Certhidea, Cactornis, Camarhynchus, Geospiza.*

There can be no question, I think, that South America is the most peculiar of all the primary regions in the globe as to its ornithology. There are at least eight or nine distinct families of birds which are quite confined to this country, many of these embracing a multitude of different genera and species. The *Trochilidæ* (which are the distinguishing family of the new world *par emphase*) are now known to be more than 320 in number, and nearly the whole of them belong to tropical America, a few species only ranging into the northern portions of that continent. It is of course quite impossible to ascertain exactly the boundary between the northern and southern zoological regions of the New World; but many of the peculiar forms of the southern division appear to extend some way up the coast-line of Southern Mexico, even north of the isthmus of Tehuantepec; whilst northern forms range down the table-land quite into the Southern States of the Mexican Union. Thus we find one or two representatives of all the most characteristic South American groups occurring to the north of Panama,— *Galbula melanogenia* representing the *Galbulidæ*; *Pipra mentalis* and *Manacus Candæi*, the *Piprinæ*; *Calliste larvata*, the genus *Calliste*; *Cotinga amabilis*, the *Cotingæ*, and so on.

The Antilles seem to be a kind of debateable ground between the two regions, but are more properly referable, I suppose, or at least the greater portion of them, to the southern region. They furnish us, however, with several peculiar genera which do not occur elsewhere.

The Neotropical Region is without doubt, I think, rich in number of species beyond any other. A calculation which I made some short time ago of species occurring southwards of Panama gave me—

1. Accipitres	95	
2. Passeres	1360	
3. Scansores	230	
4. Columbæ	25	
5. Gallinæ	80	}2000 species;
6. Struthiones	2	
7. Grallæ	128	
8. Anseres	80	

and I am decidedly of opinion that, what with taking recent additions into consideration and adding on Central America, we cannot estimate the number of birds belonging to this region at less than 2250. Taking the approximate area at $5\frac{1}{2}$ millions of square miles, this will give a species to each 2400 square miles. It follows, therefore, that this region is more richly endowed with ornithic species than any other portion of the globe.

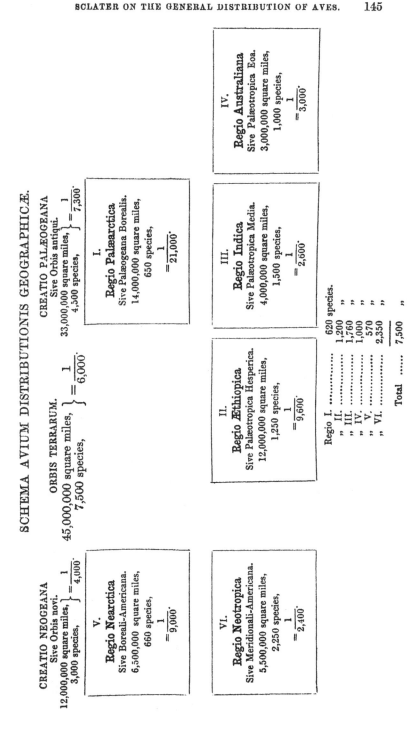

SCHEMA AVIUM DISTRIBUTIONIS GEOGRAPHICÆ.

ORBIS TERRARUM.
45,000,000 square miles, } = $\frac{1}{6,000}$.
7,500 species,

CREATIO PALÆOGEANA
Sive Orbis antiqui.
33,000,000 square miles, } $\frac{1}{7,300}$.
4,500 species,

I.
Regio Palæarctica.
Sive Palæogeana Borealis.
14,000,000 square miles,
650 species,
= $\frac{1}{21,000}$.

II.
Regio Æthiopica.
Sive Palæotropica Hesperica.
12,000,000 square miles,
1,250 species,
= $\frac{1}{9,600}$.

III.
Regio Indica.
Sive Palæotropica Media.
4,000,000 square miles,
1,500 species,
= $\frac{1}{2,600}$.

IV.
Regio Australiana.
Sive Palæotropica Eoa.
3,000,000 square miles,
1,000 species,
= $\frac{1}{3,000}$.

CREATIO NEOGEANA
Sive Orbis novi.
12,000,000 square miles, } = $\frac{1}{4,000}$.
3,000 species,

V.
Regio Nearctica.
Sive Boreali-Americana.
6,500,000 square miles,
660 species,
= $\frac{1}{9,000}$.

VI.
Regio Neotropica.
Sive Meridionali-Americana.
5,500,000 square miles,
2,250 species,
= $\frac{1}{2,400}$.

Regio I. ....   620 species.
,, II. 1,200   ,,
,, III. 1,760   ,,
,, IV. 1,000   ,,
,, V.   570   ,,
,, VI. 2,350   ,,
Total 7,500   ,,

From *Darwiniana: Essays and Reviews Pertaining to Darwinism*
Asa Gray

feres to guide the operation of physical causes." We italicize the word, for *interference* proves to be the keynote of Dr. Hodge's system. Interference with a divinely ordained physical Nature for the accomplishment of natural results! An unorthodox friend has just imparted to us, with much misgiving and solicitude lest he should be thought irreverent, his tentative hypothesis, which is, that even the Creator may be conceived to have improved with time and experience! Never before was this theory so plainly and barely put before us. We were obliged to say that, in principle and by implication, it was not wholly original.

But in such matters, which are far too high for us, no one is justly to be held responsible for the conclusions which another may draw from his principles or assumptions. Dr. Hodge's particular view should be gathered from his own statement of it:

> "In the external world there is always and everywhere indisputable evidence of the activity of two kinds of force, the one physical, the other mental. The physical belongs to matter, and is due to the properties with which it has been endowed; the other is the everywhere present and ever-acting mind of God. To the latter are to be referred all the manifestations of design in Nature, and the ordering of events in Providence. This doctrine does not ignore the efficiency of second causes; it simply asserts that God overrules and controls them. Thus the Psalmist says: 'I am fearfully and wonderfully made. My substance was not hid from Thee when I was made in secret, and curiously wrought (or embroidered) in the lower parts of the earth. . . . God makes the grass to grow, and herbs for the children of men.' He sends rain, frost, and snow. He controls the winds and the waves. He determines the casting of the lot, the flight of an arrow, and the falling of a sparrow" (pages 43, 44).

Far be it from us to object to this mode of conceiving divine causation, although, like the two other theistic conceptions referred to, it has its difficulties, and perhaps the difficulties of both. But, if we understand it, it draws an unusually hard and fast line between causation in organic and inorganic Nature, seems to look for no manifestation of design in the latter except as "God overrules and controls" second causes, and, finally, refers to this overruling and controlling (rather than to a normal action through endowment) all embryonic development, the growth of vegetables, and the like. He even adds, without break or distinction, the sending of rain, frost, and snow, the flight of an arrow, and the falling of a sparrow. Somehow we must have misconceived the bearing of the statement; but so it stands as one of "the three ways," and the right way, of "accounting for contrivances in Nature;" the other two being—1. Their reference to the blind operation of natural causes; and, 2. That they were foreseen and purposed by God, who endowed matter with forces which he foresaw and intended should produce such results, but never *interferes* to guide their operation.

In animadverting upon this latter view, Dr. Hodge brings forward an argument against evolution, with the examination of which our remarks must close:

"Paley, indeed, says that if the construction of a watch be an undeniable evidence of design, it would be a still more wonderful manifestation of skill if a watch could be made to produce other watches, and, it may be added, not only other watches, but all kinds of timepieces, in endless variety. So it has been asked, If a man can make a telescope, why cannot God make a telescope which produces others like itself? This is simply asking whether matter can be made to do the work of mind. The idea involves a contradiction. For a telescope to make a telescope supposes it to select copper and zinc in due proportions, and fuse them into brass; to fashion that brass into interentering tubes; to collect and combine the requisite materials for the different kinds of glass needed; to melt them, grind, fashion, and polish them, adjust their densities, focal distances, etc., etc. A man who can believe that brass can do all this might as well believe in God" (pp. 45, 46).

If Dr. Hodge's meaning is, that matter unconstructed cannot do the work of mind, he misses the point altogether; for original construction by an intelligent mind is given in the premises. If he means that the machine cannot originate the power that operates it, this is conceded by all except believers in perpetual motion, and it equally misses the point; for the operating power is given in the case of the watch, and implied in that of the reproductive telescope. But if he means that matter cannot be made to do the work of mind in constructions, machines, or organisms, he is surely wrong. *"Sovitur ambulando," vel scribendo;* he confuted his argument in the act of writing the sentence. That is just what machines and organisms are for; and a consistent Christian theist should maintain that is what all matter is for. Finally, if, as we freely suppose, he means none of these, he must mean (unless we are much mistaken) that organisms originated by the Almighty Creator could not be endowed with the power of producing similar organisms, or slightly dissimilar organisms, without successive interventions. Then he begs the very question in dispute, and that, too, in the face of the primal command, "Be fruitful and multiply," and its consequences in every natural birth. If the actual facts could be ignored, how nicely the parallel would run! "The idea involves a contradiction." For an animal to make an animal, or a plant to make a plant, supposes it to select carbon, hydrogen, oxygen, and nitrogen, to combine these into cellulose and protoplasm, to join with these some phosphorus, lime, etc., to build them into structures and usefully-adjusted organs. A man who can believe that plants and animals can do this (not, indeed, in the crude way suggested, but in the appointed way) "might as well believe in God." Yes, verily, and so he probably will, in spite of all that atheistical philosophers have to offer, if not harassed and confused by such arguments and statements as these.

There is a long line of gradually-increasing divergence from the ultra-orthodox view of Dr. Hodge through those of such men as Sir William Thomson, Herschel, Argyll, Owen, Mivart, Wallace, and Darwin, down to those of Strauss, Vogt, and Büchner.[1] To strike the line with telling power and good effect, it is necessary to aim at the right place. Excellent as the present volume is in motive and clearly as it shows that Darwinism may bear an atheistic as well as a theistic interpretation, we fear that it will not contribute much to the reconcilement of science and religion.

The length of the analysis of the first book on our list precludes the notices which we intended to take of the three others. They are all the production of men who are both scientific and religious, one of them a celebrated divine and writer unusually versed in natural history. They all look upon theories of evolution either as in the way of being established or as not unlikely to prevail, and they confidently expect to lose thereby no solid ground for theism or religion. Mr. St. Clair,[2] a new writer, in his "Darwinism and Design; or, Creation by Evolution," takes his ground in the following succinct statement of his preface:

"It is being assumed by our scientific guides that the design-argument has been driven out of the field by the doctrine of evolution. It seems to be thought by our theological teachers that the best defense of the faith is to deny evolution *in toto,* and denounce it as anti-Biblical. My volume endeavors to show that, if evolution be true, all is not lost; but, on the contrary, something is gained: the design-argument remains unshaken, and the wisdom and beneficence of God receive new illustration."

Of his closing remark, that, so far as he knows, the subject has never before been handled in the same way for the same purpose, we will only say that the handling strikes us as

[1] David Friedrich Strauss (1808–1874), German commentator on the Bible. Karl Vogt (1817–1895), German biologist who had to spend most of his career in Switzerland because of his political activity during the Revolution of 1848. Friedrich Karl Christian Ludwig Büchner (1824–1899), German writer on materialism, author of *Kraft und Stoff*.

It was only by such occasional comments as this that Gray indicated his awareness of the divergence of German thought from that of the Anglo-American world on the subject of Darwinism.

[2] George St. Clair, author of *Buried Cities and Bible Countries* as well as *Darwinism and Design*.

mainly sensible rather than as substantially novel. He traverses the whole ground of evolution, from that of the solar system to "the origin of moral species." He is clearly a theistic Darwinian without misgiving, and the arguments for that hypothesis and for its religious aspects obtain from him their most favorable presentation, while he combats the *dysteleology* of Häckel, Büchner, etc., not, however, with any remarkable strength.

Dr. Winchell,[3] chancellor of the new university at Syracuse, in his volume just issued upon the "Doctrine of Evolution," adopts it in the abstract as "clearly as the law of universal intelligence under which complex results are brought into existence" (whatever that may mean), accepts it practically for the inorganic world as a geologist should, hesitates as to the organic world, and sums up the arguments for the origin of species by diversification unfavorably for the Darwinians, regarding it mainly from the geological side. As some of our zoölogists and palæontologists may have somewhat to say upon this matter, we leave it for their consideration. We are tempted to develop a point which Dr. Winchell incidentally refers to—viz., how very modern the idea of the independent creation and fixity of species is, and how well the old divines got on without it. Dr. Winchell reminds us that St. Augustine and St. Thomas Aquinas were model evolutionists; and, where authority is deferred to, this should count for something.

Mr. Kingsley's[4] eloquent and suggestive "Westminster Sermons," in which he touches here and there upon many of the topics which evolution brings up, has incorporated into the preface a paper which he read in 1871 to a meeting of

[3] Alexander Winchell (1824–1891), American clergyman and evolutionary naturalist who was the principal in a celebrated if misunderstood academic freedom case at Vanderbilt.

[4] Charles Kingsley (1819–1875), famous British clergyman and author of *The Water Babies.* He had spoken highly of Gray's *Atlantic* articles (chapter III) in 1860 and subsequently visited Gray in the United States.

London clergy at Sion College, upon certain problems of natural theology as affected by modern theories in science. We may hereafter have occasion to refer to this volume. Meanwhile, perhaps we may usefully conclude this article with two or three short extracts from it:

"The God who satisfies our conscience ought more or less to satisfy our reason also. To teach that was Butler's mission; and he fulfilled it well. But it is a mission which has to be refulfilled again and again, as human thought changes, and human science develops. For if, in any age or country, the God who seems to be revealed by Nature seems also different from the God who is revealed by the then-popular religion, then that God and the religion which tells of that God will gradually cease to be believed in.

"For the demands of reason—as none knew better than good Bishop Butler—must be and ought to be satisfied. And, therefore, when a popular war arises between the reason of any generation and its theology, then it behooves the ministers of religion to inquire, with all humility and godly fear, on whose side lies the fault; whether the theology which they expound is all that it should be, or whether the reason of those who impugn it is all that it should be."

Pronouncing it to be the duty of the naturalist to find out the how of things, and of the natural theologian to find out the why, Mr. Kingsley continues:

"But if it be said, 'After all, there is no why; the doctrine of evolution, by doing away with the theory of creation, does away with that of final causes,' let us answer boldly, 'Not in the least.' We might accept all that Mr. Darwin, all that Prof. Huxley, all that other most able men have so learnedly and acutely written on physical science, and yet preserve our natural theology on the same basis as that on which Butler and Paley left it. That we should have to develop it I do not deny.

"Let us rather look with calmness, and even with hope and good-will, on these new theories; they surely mark a tendency toward a more, not a less, Scriptural view of Nature.

"Of old it was said by Him, without whom nothing is made, 'My Father worketh hitherto, and I work.' Shall we quarrel with Science if she should show how these words are true? What, in one word, should we have to say but this: 'We know of old that God was so wise that he could make all things; but, behold, he is so much wiser than even that, that he can make all things make themselves?' "

From *On the Origin of Species by Means of Natural Selection, or the Preservation of Favoured Races in the Struggle for Life*

Charles Darwin

CHAPTER XI

GEOGRAPHICAL DISTRIBUTION

Present distribution cannot be accounted for by differences in physical conditions – Importance of barriers – Affinity of the productions of the same continent – Centres of creation – Means of dispersal, by changes of climate and of the level of the land, and by occasional means – Dispersal during the Glacial period co-extensive with the world

IN considering the distribution of organic beings over the face of the globe, the first great fact which strikes us is, that neither the similarity nor the dissimilarity of the inhabitants of various regions can be accounted for by their climatal and other physical conditions. Of late, almost every author who has studied the subject has come to this conclusion. The case of America alone would almost suffice to prove its truth: for if we exclude the northern parts where the circumpolar land is almost continuous, all authors agree that one of the most fundamental divisions in geographical distribution is that between the New and Old Worlds; yet if we travel over the vast American continent, from the central parts of the United States to its extreme southern point, we meet with the most diversified conditions; the most humid districts, arid deserts, lofty mountains, grassy plains, forests, marshes, lakes, and great rivers, under almost every temperature. There is hardly a climate or condition in the Old World which cannot be paralleled in the New – at least as closely as the same species generally require; for it is a most rare case to find a group of organisms confined to any small spot, having conditions peculiar in only a slight degree; for instance, small areas in the Old World could be pointed out hotter than any in the New World, yet these are not inhabited by a peculiar fauna or flora. Notwithstanding this parallelism in the conditions of the Old and New Worlds, how widely different are their living productions!

344

GEOGRAPHICAL DISTRIBUTION

In the southern hemisphere, if we compare large tracts of land in Australia, South Africa, and western South America, between latitudes 25° and 35°, we shall find parts extremely similar in all their conditions, yet it would not be possible to point out three faunas and floras more utterly dissimilar. Or again we may compare the productions of South America south of lat. 35° with those north of 25°, which consequently inhabit a considerably different climate, and they will be found incomparably more closely related to each other, than they are to the productions of Australia or Africa under nearly the same climate. Analogous facts could be given with respect to the inhabitants of the sea.

A second great fact which strikes us in our general review is, that barriers of any kind, or obstacles to free migration, are related in a close and important manner to the differences between the productions of various regions. We see this in the great difference of nearly all the terrestrial productions of the New and Old Worlds, excepting in the northern parts, where the land almost joins, and where, under a slightly different climate, there might have been free migration for the northern temperate forms, as there now is for the strictly arctic productions. We see the same fact in the great difference between the inhabitants of Australia, Africa, and South America under the same latitude: for these countries are almost as much isolated from each other as is possible. On each continent, also, we see the same fact; for on the opposite sides of lofty and continuous mountain-ranges, and of great deserts, and sometimes even of large rivers, we find different productions; though as mountain-chains, deserts, &c., are not as impassable, or likely to have endured so long as the oceans separating continents, the differences are very inferior in degree to those characteristic of distinct continents.

Turning to the sea, we find the same law. No two marine faunas are more distinct, with hardly a fish, shell, or crab in common, than those of the eastern and western shores of South and Central America; yet these great faunas are separated only by the narrow, but impassable, isthmus of Panama. Westward of the shores of America, a wide space of open ocean extends,

THE ORIGIN OF SPECIES

with not an island as a halting-place for emigrants; here we have
a barrier of another kind, and as soon as this is passed we meet
in the eastern islands of the Pacific, with another and totally
distinct fauna. So that here three marine faunas range far northward and southward, in parallel lines not far from each other,
under corresponding climates; but from being separated from
each other by impassable barriers, either of land or open sea,
they are wholly distinct. On the other hand, proceeding still
further westward from the eastern islands of the tropical parts
of the Pacific, we encounter no impassable barriers, and we have
innumerable islands as halting-places, until after travelling over
a hemisphere we come to the shores of Africa; and over this vast
space we meet with no well-defined and distinct marine faunas.
Although hardly one shell, crab or fish is common to the above-
named three approximate faunas of Eastern and Western America and the eastern Pacific islands, yet many fish range from the
Pacific into the Indian Ocean, and many shells are common to
the eastern islands of the Pacific and the eastern shores of Africa,
on almost exactly opposite meridians of longitude.

A third great fact, partly included in the foregoing statements, is the affinity of the productions of the same continent
or sea, though the species themselves are distinct at different
points and stations. It is a law of the widest generality, and
every continent offers innumerable instances. Nevertheless the
naturalist in travelling, for instance, from north to south never
fails to be struck by the manner in which successive groups of
beings, specifically distinct, yet clearly related, replace each other.
He hears from closely allied, yet distinct kinds of birds, notes
nearly similar, and sees their nests similarly constructed, but
not quite alike, with eggs coloured in nearly the same manner.
The plains near the Straits of Magellan are inhabited by one
species of Rhea (American ostrich), and northward the plains
of La Plata by another species of the same genus; and not by a
true ostrich or emeu, like those found in Africa and Australia
under the same latitude. On these same plains of La Plata, we
see the agouti and bizcacha, animals having nearly the same
habits as our hares and rabbits and belonging to the same order
of Rodents, but they plainly display an American type of struc-

GEOGRAPHICAL DISTRIBUTION

ture. We ascend the lofty peaks of the Cordillera and we find an alpine species of bizcacha; we look to the waters, and we do not find the beaver or musk-rat, but the coypu and capybara, rodents of the American type. Innumerable other instances could be given. If we look to the islands off the American shore, however much they may differ in geological structure, the inhabitants, though they may be all peculiar species, are essentially American. We may look back to past ages, as shown in the last chapter, and we find American types then prevalent on the American continent and in the American seas. We see in these facts some deep organic bond, prevailing throughout space and time, over the same areas of land and water, and independent of their physical conditions. The naturalist must feel little curiosity, who is not led to inquire what this bond is.

This bond, on my theory, is simply inheritance, that cause which alone, as far as we positively know, produces organisms quite like, or, as we see in the case of varieties nearly like each other. The dissimilarity of the inhabitants of different regions may be attributed to modification through natural selection, and in a quite subordinate degree to the direct influence of different physical conditions. The degree of dissimilarity will depend on the migration of the more dominant forms of life from one region into another having been effected with more or less ease, at periods more or less remote; – on the nature and number of the former immigrants; – and on their action and reaction, in their mutual struggles for life; – the relation of organism to organism being, as I have already often remarked, the most important of all relations. Thus the high importance of barriers comes into play by checking migration; as does time for the slow process of modification through natural selection. Widely-ranging species, abounding in individuals, which have already triumphed over many competitors in their own widely-extended homes will have the best chance of seizing on new places, when they spread into new countries. In their new homes they will be exposed to new conditions, and will frequently undergo further modification and improvement; and thus they will become still further victorious, and will produce groups of modified descendants. On this principle of inheritance with modification, we can

THE ORIGIN OF SPECIES

understand how it is that sections of genera, whole genera, and even families are confined to the same areas, as is so commonly and notoriously the case.

I believe, as was remarked in the last chapter, in no law of necessary development. As the variability of each species is an independent property, and will be taken advantage of by natural selection, only so far as it profits the individual in its complex struggle for life, so the degree of modification in different species will be no uniform quantity. If, for instance, a number of species, which stand in direct competition with each other, migrate in a body into a new and afterwards isolated country, they will be little liable to modification; for neither migration nor isolation in themselves can do anything. These principles come into play only by bringing organisms into new relations with each other, and in a lesser degree with the surrounding physical conditions. As we have seen in the last chapter that some forms have retained nearly the same character from an enormously remote geological period, so certain species have migrated over vast spaces, and have not become greatly modified.

On these views, it is obvious, that the several species of the same genus, though inhabiting the most distant quarters of the world, must originally have proceeded from the same source, as they have descended from the same progenitor. In the case of those species, which have undergone during whole geological periods but little modification, there is not much difficulty in believing that they may have migrated from the same region; for during the vast geographical and climatal changes which will have supervened since ancient times, almost any amount of migration is possible. But in many other cases, in which we have reason to believe that the species of a genus have been produced within comparatively recent times, there is great difficulty on this head. It is also obvious that the individuals of the same species, though now inhabiting distant and isolated regions, must have proceeded from one spot, where their parents were first produced: for, as explained in the last chapter, it is incredible that individuals identically the same should ever have been produced through natural selection from parents specifically distinct.

GEOGRAPHICAL DISTRIBUTION

We are thus brought to the question which has been largely discussed by naturalists, namely, whether species have been created at one or more points of the earth's surface. Undoubtedly there are very many cases of extreme difficulty, in understanding how the same species could possibly have migrated from some one point to the several distant and isolated points, where now found. Nevertheless the simplicity of the view that each species was first produced within a single region captivates the mind. He who rejects it, rejects the *vera causa* of ordinary generation with subsequent migration, and calls in the agency of a miracle. It is universally admitted, that in most cases the area inhabited by a species is continuous; and when a plant or animal inhabits two points so distant from each other, or with an interval of such a nature, that the space could not be easily passed over by migration, the fact is given as something remarkable and exceptional. The capacity of migrating across the sea is more distinctly limited in terrestrial mammals, than perhaps in any other organic beings; and, accordingly, we find no inexplicable cases of the same mammal inhabiting distant points of the world. No geologist will feel any difficulty in such cases as Great Britain having been formerly united to Europe, and consequently possessing the same quadrupeds. But if the same species can be produced at two separate points, why do we not find a single mammal common to Europe and Australia or South America? The conditions of life are nearly the same, so that a multitude of European animals and plants have become naturalised in America and Australia; and some of the aboriginal plants are identically the same at these distant points of the northern and southern hemispheres? The answer, as I believe, is, that mammals have not been able to migrate, whereas some plants, from their varied means of dispersal, have migrated across the vast and broken interspace. The great and striking influence which barriers of every kind have had on distribution, is intelligible only on the view that the great majority of species have been produced on one side alone, and have not been able to migrate to the other side. Some few families, many sub-families, very many genera, and a still greater number of sections of genera are confined to a single region; and it has been observed by several

THE ORIGIN OF SPECIES

naturalists, that the most natural genera, or those genera in which the species are most closely related to each other, are generally local, or confined to one area. What a strange anomaly it would be, if, when coming one step lower in the series, to the individuals of the same species, a directly opposite rule prevailed; and species were not local, but had been produced in two or more distinct areas !

Hence it seems to me, as it has to many other naturalists, that the view of each species having been produced in one area alone, and having subsequently migrated from that area as far as its powers of migration and subsistence under past and present conditions permitted, is the most probable. Undoubtedly many cases occur, in which we cannot explain how the same species could have passed from one point to the other. But the geographical and climatal changes, which have certainly occurred within recent geological times, must have interrupted or rendered discontinuous the formerly continuous range of many species. So that we are reduced to consider whether the exceptions to continuity of range are so numerous and of so grave a nature, that we ought to give up the belief, rendered probable by general considerations, that each species has been produced within one area, and has migrated thence as far as it could. It would be hopelessly tedious to discuss all the exceptional cases of the same species, now living at distant and separated points; nor do I for a moment pretend that any explanation could be offered of many such cases. But after some preliminary remarks, I will discuss a few of the most striking classes of facts; namely, the existence of the same species on the summits of distant mountain-ranges, and at distant points in the arctic and antarctic regions; and secondly (in the following chapter), the wide distribution of freshwater productions; and thirdly, the occurrence of the same terrestrial species on islands and on the mainland, though separated by hundreds of miles of open sea. If the existence of the same species at distant and isolated points of the earth's surface, can in many instances be explained on the view of each species having migrated from a single birthplace; then, considering our ignorance with respect to former climatal and geographical changes and various occasional means of transport,

GEOGRAPHICAL DISTRIBUTION

the belief that this has been the universal law, seems to me incomparably the safest.

In discussing this subject, we shall be enabled at the same time to consider a point equally important for us, namely, whether the several distinct species of a genus, which on my theory have all descended from a common progenitor, can have migrated (undergoing modification during some part of their migration) from the area inhabited by their progenitor. If it can be shown to be almost invariably the case, that a region, of which most of its inhabitants are closely related to, or belong to the same genera with the species of a second region, has probably received at some former period immigrants from this other region, my theory will be strengthened; for we can clearly understand, on the principle of modification, why the inhabitants of a region should be related to those of another region, whence it has been stocked. A volcanic island, for instance, upheaved and formed at the distance of a few hundreds of miles from a continent, would probably receive from it in the course of time a few colonists, and their descendants, though modified, would still be plainly related by inheritance to the inhabitants of the continent. Cases of this nature are common, and are, as we shall hereafter more fully see, inexplicable on the theory of independent creation. This view of the relation of species in one region to those in another, does not differ much (by substituting the word variety for species) from that lately advanced in an ingenious paper by Mr Wallace, in which he concludes, that every species has come into existence coincident both in space and time with a pre-existing closely allied species.' And I now know from correspondence, that this coincidence he attributes to generation with modification.

The previous remarks on 'single and multiple centres of creation' do not directly bear on another allied question, – namely whether all the individuals of the same species have descended from a single pair, or single hermaphrodite, or whether, as some authors suppose, from many individuals simultaneously created. With those organic beings which never intercross (if such exist), the species, on my theory, must have descended from a succession of improved varieties, which will never have

THE ORIGIN OF SPECIES

blended with other individuals or varieties, but will have supplanted each other; so that, at each successive stage of modification and improvement, all the individuals of each variety will have descended from a single parent. But in the majority of cases, namely, with all organisms which habitually unite for each birth, or which often intercross, I believe that during the slow process of modification the individuals of the species will have been kept nearly uniform by intercrossing; so that many individuals will have gone on simultaneously changing, and the whole amount of modification will not have been due, at each stage, to descent from a single parent. To illustrate what I mean: our English racehorses differ slightly from the horses of every other breed; but they do not owe their difference and superiority to descent from any single pair, but to continued care in selecting and training many individuals during many generations.

Before discussing the three classes of facts, which I have selected as presenting the greatest amount of difficulty on the theory of 'single centres of creation,' I must say a few words on the means of dispersal.

[Pages 353–373 of chapter 11 are omitted. The reading continues with an excerpt from chapter 12, "Geographical Distribution—Continued."]

On the Inhabitants of Oceanic Islands. We now come to the last of the three classes of facts, which I have selected as presenting the greatest amount of difficulty, on the view that all the individuals both of the same and of allied species have descended from a single parent; and therefore have all proceeded from a common birthplace, notwithstanding that in the course of time they have come to inhabit distant points of the globe. I have already stated that I cannot honestly admit Forbes's view on continental extensions, which, if legitimately followed out, would lead to the belief that within the recent period all existing islands have been nearly or quite joined to some continent.

GEOGRAPHICAL DISTRIBUTION

This view would remove many difficulties, but it would not, I think, explain all the facts in regard to insular productions. In the following remarks I shall not confine myself to the mere question of dispersal; but shall consider some other facts, which bear on the truth of the two theories of independent creation and of descent with modification.

The species of all kinds which inhabit oceanic islands are few in number compared with those on equal continental areas: Alph. de Candolle admits this for plants, and Wollaston for insects. If we look to the large size and varied stations of New Zealand, extending over 780 miles of latitude, and compare its flowering plants, only 750 in number, with those on an equal area at the Cape of Good Hope or in Australia, we must, I think, admit that something quite independently of any difference in physical conditions has caused so great a difference in number. Even the uniform county of Cambridge has 847 plants, and the little island of Anglesea 764, but a few ferns and a few introduced plants are included in these numbers, and the comparison in some other respects is not quite fair. We have evidence that the barren island of Ascension aboriginally possessed under half-a-dozen flowering plants; yet many have become naturalised on it, as they have on New Zealand and on every other oceanic island which can be named. In St Helena there is reason to believe that the naturalised plants and animals have nearly or quite exterminated many native productions. He who admits the doctrine of the creation of each separate species, will have to admit, that a sufficient number of the best adapted plants and animals have not been created on oceanic islands; for man has unintentionally stocked them from various sources far more fully and perfectly than has nature.

Although in oceanic islands the number of kinds of inhabitants is scanty, the proportion of endemic species (i.e. those found nowhere else in the world) is often extremely large. If we compare, for instance, the number of the endemic land-shells in Madeira, or of the endemic birds in the Galapagos Archipelago, with the number found on any continent, and then compare the area of the islands with that of the continent, we shall see that this is true. This fact might have been expected on my

THE ORIGIN OF SPECIES

theory, for, as already explained, species occasionally arriving after long intervals in a new and isolated district, and having to compete with new associates, will be eminently liable to modification, and will often produce groups of modified descendants. But it by no means follows, that, because in an island nearly all the species of one class are peculiar, those of another class, or of another section of the same class, are peculiar; and this difference seems to depend on the species which do not become modified having immigrated with facility and in a body, so that their mutual relations have not been much disturbed. Thus in the Galapagos Islands nearly every land-bird, but only two out of the eleven marine birds, are peculiar; and it is obvious that marine birds could arrive at these islands more easily than landbirds. Bermuda, on the other hand, which lies at about the same distance from North America as the Galapagos Islands do from South America, and which has a very peculiar soil, does not possess one endemic land bird; and we know from Mr J. M. Jones's admirable account of Bermuda, that very many North American birds, during their great annual migrations, visit either periodically or occasionally this island. Madeira does not possess one peculiar bird, and many European and African birds are almost every year blown there, as I am informed by Mr E. V. Harcourt. So that these two islands of Bermuda and Madeira have been stocked by birds, which for long ages have struggled together in their former homes, and have become mutually adapted to each other; and when settled in their new homes, each kind will have been kept by the others to their proper places and habits, and will consequently have been little liable to modification. Madeira, again, is inhabited by a wonderful number of peculiar land-shells, whereas not one species of sea-shell is confined to its shores: now, though we do not know how sea-shells are dispersed, yet we can see that their eggs or larvae, perhaps attached to seaweed or floating timber, or to the feet of wading-birds, might be transported far more easily than landshells, across three or four hundred miles of open sea. The different orders of insects in Madeira apparently present analogous facts.

Oceanic islands are sometimes deficient in certain classes, and

GEOGRAPHICAL DISTRIBUTION

their places are apparently occupied by the other inhabitants; in the Galapagos Islands reptiles, and in New Zealand gigantic wingless birds, take the place of mammals. In the plants of the Galapagos Islands, Dr Hooker has shown that the proportional numbers of the different orders are very different from what they are elsewhere. Such cases are generally accounted for by the physical conditions of the islands; but this explanation seems to me not a little doubtful. Facility of immigration, I believe, has been at least as important as the nature of the conditions.

Many remarkable little facts could be given with respect to the inhabitants of remote islands. For instance, in certain islands not tenanted by mammals, some of the endemic plants have beautifully hooked seeds; yet few relations are more striking than the adaptation of hooked seeds for transportal by the wool and fur of quadrupeds. This case presents no difficulty on my view, for a hooked seed might be transported to an island by some other means; and the plant then becoming slightly modified, but still retaining its hooked seeds, would form an endemic species, having as useless an appendage as any rudimentary organ, – for instance, as the shrivelled wings under the soldered elytra of many insular beetles. Again, islands often possess trees or bushes belonging to orders which elsewhere include only herbaceous species; now trees, as Alph. de Candolle has shown, generally have, whatever the cause may be, confined ranges. Hence trees would be little likely to reach distant oceanic islands; and an herbaceous plant, though it would have no chance of successfully competing in stature with a fully developed tree, when established on an island and having to compete with herbaceous plants alone, might readily gain an advantage by growing taller and taller and overtopping the other plants. If so, natural selection would often tend to add to the stature of herbaceous plants when growing on an island, to whatever order they belonged, and thus convert them first into bushes and ultimately into trees.

With respect to the absence of whole orders on oceanic islands, Bory St Vincent long ago remarked that Batrachians (frogs, toads, newts) have never been found on any of the many islands with which the great oceans are studded. I have taken pains to verify this assertion, and I have found it strictly true. I have, however, been assured that a frog exists on the mountains of the great island of New Zealand; but I suspect that this exception (if the information be correct) may be explained through glacial agency. This general absence of frogs, toads, and newts on so many oceanic islands cannot be accounted for by their physical conditions; indeed it seems that islands are peculiarly well fitted for these animals; for frogs have been introduced into Madeira, the Azores, and Mauritius, and have multiplied so as to become a nuisance. But as these animals and their spawn are known to be immediately killed by sea-water, on my view we can see that there would be great difficulty in their transportal across the sea, and therefore why they do not exist on any oceanic island. But why, on the theory of creation, they should not have been created there, it would be very difficult to explain.

Mammals offer another and similar case. I have carefully searched the oldest voyages, but have not finished my search; as yet I have not found a single instance, free from doubt, of a terrestrial mammal (excluding domesticated animals kept by the natives) inhabiting an island situated above 300 miles from a continent or great continental island; and many islands situated at a much less distance are equally barren. The Falkland Islands, which are inhabited by a wolf-like fox, come nearest to an exception; but this group cannot be considered as oceanic, as it lies on a bank connected with the mainland; moreover, icebergs formerly brought boulders to its western shores, and they may have formerly transported foxes, as so frequently now happens in the arctic regions. Yet it cannot be said that small islands will not support small mammals, for they occur in many parts of the world on very small islands, if close to a continent; and hardly an island can be named on which our smaller quadrupeds have not become naturalised and greatly multiplied. It cannot be said, on the ordinary view of creation, that there has not been time for the creation of mammals; many volcanic islands are sufficiently ancient, as shown by the stupendous degradation which they have suffered and by their tertiary strata: there has also been time for the production of endemic species belonging to other classes; and on continents it is thought that mammals appear and disappear at a quicker rate than other and lower

GEOGRAPHICAL DISTRIBUTION

animals. Though terrestrial mammals do not occur on oceanic islands, aërial mammals do occur on almost every island. New Zealand possesses two bats found nowhere else in the world: Norfolk Island, the Viti Archipelago, the Bonin Islands, the Caroline and Marianne Archipelagoes, and Mauritius, all possess their peculiar bats. Why, it may be asked, has the supposed creative force produced bats and no other mammals on remote islands? On my view this question can easily be answered; for no terrestrial mammal can be transported across a wide space of sea, but bats can fly across. Bats have been seen wandering by day far over the Atlantic Ocean; and two North American species either regularly or occasionally visit Bermuda, at the distance of 600 miles from the mainland. I hear from Mr Tomes, who has specially studied this family, that many of the same species have enormous ranges, and are found on continents and on far distant islands. Hence we have only to suppose that such wandering species have been modified through natural selection in their new homes in relation to their new position, and we can understand the presence of endemic bats on islands, with the absence of all terrestrial mammals.

Besides the absence of terrestrial mammals in relation to the remoteness of islands from continents, there is also a relation, to a certain extent independent of distance, between the depth of the sea separating an island from the neighbouring mainland, and the presence in both of the same mammiferous species or of allied species in a more or less modified condition. Mr Windsor Earl has made some striking observations on this head in regard to the great Malay Archipelago, which is traversed near Celebes by a space of deep ocean; and this space separates two widely distinct mammalian faunas. On either side the islands are situated on moderately deep submarine banks, and they are inhabited by closely allied or identical quadrupeds. No doubt some few anomalies occur in this great archipelago, and there is much difficulty in forming a judgment in some cases owing to the probable naturalisation of certain mammals through man's agency; but we shall soon have much light thrown on the natural history of this archipelago by the admirable zeal and researches of Mr Wallace. I have not as yet had time to follow up this

THE ORIGIN OF SPECIES

subject in all other quarters of the world; but as far as I have gone, the relation generally holds good. We see Britain separated by a shallow channel from Europe, and the mammals are the same on both sides; we meet with analogous facts on many islands separated by similar channels from Australia. The West Indian Islands stand on a deeply submerged bank, nearly 1000 fathoms in depth, and here we find American forms, but the species and even the genera are distinct. As the amount of modification in all cases depends to a certain degree on the lapse of time, and as during changes of level it is obvious that islands separated by shallow channels are more likely to have been continuously united within a recent period to the mainland than islands separated by deeper channels, we can understand the frequent relation between the depth of the sea and the degree of affinity of the mammalian inhabitants of islands with those of a neighbouring continent, – an explicable relation on the view of independent acts of creation.

All the foregoing remarks on the inhabitants of oceanic islands, – namely, the scarcity of kinds – the richness in endemic forms in particular classes or sections of classes, – the absence of whole groups, as of batrachians, and of terrestrial mammals notwithstanding the presence of aërial bats, – the singular proportions of certain orders of plants, – herbaceous forms having been developed into trees, &c., – seem to me to accord better with the view of occasional means of transport having been largely efficient in the long course of time, than with the view of all our oceanic islands having been formerly connected by continuous land with the nearest continent; for on this latter view the migration would probably have been more complete; and if modification be admitted, all the forms of life would have been more equally modified, in accordance with the paramount importance of the relation of organism to organism.

I do not deny that there are many and grave difficulties in understanding how several of the inhabitants of the more remote islands, whether still retaining the same specific form or modified since their arrival, could have reached their present homes. But the probability of many islands having existed as halting-places, of which not a wreck now remains, must not be overlooked. I

GEOGRAPHICAL DISTRIBUTION

will here give a single instance of one of the cases of difficulty. Almost all oceanic islands, even the most isolated and smallest, are inhabited by land-shells, generally by endemic species, but sometimes by species found elsewhere. Dr Aug. A. Gould has given several interesting cases in regard to the land-shells of the islands of the Pacific. Now it is notorious that land-shells are very easily killed by salt; their eggs, at least such as I have tried, sink in sea-water and are killed by it. Yet there must be, on my view, some unknown, but highly efficient means for their transportal. Would the just-hatched young occasionally crawl on and adhere to the feet of birds roosting on the ground, and thus get transported? It occurred to me that land-shells, when hybernating and having a membranous diaphragm over the mouth of the shell, might be floated in chinks of drifted timber across moderately wide arms of the sea. And I found that several species did in this state withstand uninjured an immersion in sea-water during seven days: one of these shells was the Helix pomatia, and after it had again hybernated I put it in sea-water for twenty days, and it perfectly recovered. As this species has a thick calcareous operculum, I removed it, and when it had formed a new membranous one, I immersed it for fourteen days in sea-water, and it recovered and crawled away : but more experiments are wanted on this head.

The most striking and important fact for us in regard to the inhabitants of islands, is their affinity to those of the nearest mainland, without being actually the same species. Numerous instances could be given of this fact. I will give only one, that of the Galapagos Archipelago, situated under the equator, between 500 and 600 miles from the shores of South America. Here almost every product of the land and water bears the unmistakeable stamp of the American continent. There are twenty-six land birds, and twenty-five of those are ranked by Mr Gould as distinct species, supposed to have been created here; yet the close affinity of most of these birds to American species in every character, in their habits, gestures, and tones of voice, was manifest. So it is with the other animals, and with nearly all the plants, as shown by Dr Hooker in his admirable memoir on the Flora of this archipelago. The naturalist, looking at the

THE ORIGIN OF SPECIES

inhabitants of these volcanic islands in the Pacific, distant several hundred miles from the continent, yet feels that he is standing on American land. Why should this be so? why should the species which are supposed to have been created in the Galapagos Archipelago, and nowhere else, bear so plain a stamp of affinity to those created in America? There is nothing in the conditions of life, in the geological nature of the islands, in their height or climate, or in the proportions in which the several classes are associated together, which resembles closely the conditions of the South American coast: in fact there is a considerable dissimilarity in all these respects. On the other hand, there is a considerable degree of resemblance in the volcanic nature of the soil, in climate, height, and size of the islands, between the Galapagos and Cape de Verde Archipelagos: but what an entire and absolute difference in their inhabitants! The inhabitants of the Cape de Verde Islands are related to those of Africa, like those of the Galapagos to America. I believe this grand fact can receive no sort of explanation on the ordinary view of independent creation; whereas on the view here maintained, it is obvious that the Galapagos Islands would be likely to receive colonists, whether by occasional means of transport or by formerly continuous land, from America; and the Cape de Verde Islands from Africa; and that such colonists would be liable to modifications; – the principle of inheritance still betraying their original birthplace.

Many analogous facts could be given: indeed it is an almost universal rule that the endemic productions of islands are related to those of the nearest continent, or of other near islands. The exceptions are few, and most of them can be explained. Thus the plants of Kerguelen Land, though standing nearer to Africa than to America, are related, and that very closely, as we know from Dr Hooker's account, to those of America: but on the view that this island has been mainly stocked by seeds brought with earth and stones on icebergs, drifted by the prevailing currents, this anomaly disappears. New Zealand in its endemic plants is much more closely related to Australia, the nearest mainland, than to any other region: and this is what might have been expected; but it is also plainly related to South America, which,

GEOGRAPHICAL DISTRIBUTION

although the next nearest continent, is so enormously remote, that the fact becomes an anomaly. But this difficulty almost disappears on the view that both New Zealand, South America, and other southern lands were long ago partially stocked from a nearly intermediate though distant point, namely from the antarctic islands, when they were clothed with vegetation, before the commencement of the Glacial period. The affinity, which, though feeble, I am assured by Dr Hooker is real, between the flora of the south-western corner of Australia and of the Cape of Good Hope, is a far more remarkable case, and is at present inexplicable: but this affinity is confined to the plants, and will, I do not doubt, be some day explained.

The law which causes the inhabitants of an archipelago, though specifically distinct, to be closely allied to those of the nearest continent, we sometimes see displayed on a small scale, yet in a most interesting manner, within the limits of the same archipelago. Thus the several islands of the Galapagos Archipelago are tenanted, as I have elsewhere shown, in a quite marvellous manner, by very closely related species; so that the inhabitants of each separate island, though mostly distinct, are related in an incomparably closer degree to each other than to the inhabitants of any other part of the world. And this is just what might have been expected on my view, for the islands are situated so near each other that they would almost certainly receive immigrants from the same original source, or from each other. But this dissimilarity between the endemic inhabitants of the islands may be used as an argument against my views; for it may be asked, how has it happened in the several islands situated within sight of each other, having the same geological nature, the same height, climate, &c., that many of the immigrants should have been differently modified, though only in a small degree. This long appeared to me a great difficulty: but it arises in chief part from the deeply-seated error of considering the physical conditions of a country as the most important for its inhabitants; whereas it cannot, I think, be disputed that the nature of the other inhabitants, with which each has to compete, is at least as important, and generally a far more important element of success. Now if we look to those inhabitants of the

THE ORIGIN OF SPECIES

Galapagos Archipelago which are found in other parts of the
world (laying on one side for the moment the endemic species,
which cannot be here fairly included, as we are considering how
they have come to be modified since their arrival), we find a
considerable amount of difference in the several islands. This difference might indeed have been expected on the view of the
islands having been stocked by occasional means of transport –
a seed, for instance, of one plant having been brought to one
island, and that of another plant to another island. Hence when
in former times an immigrant settled on any one or more of
the islands, or when it subsequently spread from one island
to another, it would undoubtedly be exposed to different conditions of life in the different islands, for it would have to compete with different sets of organisms: a plant, for instance, would
find the best-fitted ground more perfectly occupied by distinct
plants in one island than in another, and it would be exposed
to the attacks of somewhat different enemies. If then it varied,
natural selection would probably favour different varieties in the
different islands. Some species, however, might spread and yet
retain the same character throughout the group, just as we see
on continents some species' spreading widely and remaining the
same.

The really surprising fact in this case of the Galapagos Archipelago, and in a lesser degree in some analogous instances, is that
the new species formed in the separate islands have not quickly
spread to the other islands. But the islands, though in sight of
each other, are separated by deep arms of the sea, in most cases
wider than the British Channel, and there is no reason to suppose that they have at any former period been continuously
united. The currents of the sea are rapid and sweep across the
archipelago, and gales of wind are extraordinarily rare; so that
the islands are far more effectually separated from each other
than they appear to be on a map. Nevertheless a good many
species, both those found in other parts of the world and those
confined to the archipelago, are common to the several islands,
and we may infer from certain facts that these have probably
spread from some one island to the others. But we often take, I
think, an erroneous view of the probability of closely allied

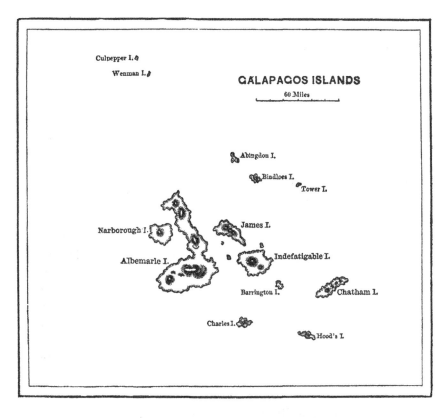

The wolf-like fox that inhabits the Falkland Islands.

GEOGRAPHICAL DISTRIBUTION

species invading each other's territory, when put into free inter-communication. Undoubtedly if one species has any advantage whatever over another, it will in a very brief time wholly or in part supplant it; but if both are equally well fitted for their own places in nature, both probably will hold their own places and keep separate for almost any length of time. Being familiar with the fact that many species, naturalised through man's agency, have spread with astonishing rapidity over new countries, we are apt to infer that most species would thus spread; but we should remember that the forms which become naturalised in new countries are not generally closely allied to the aboriginal inhabitants, but are very distinct species, belonging in a large proportion of cases, as shown by Alph. de Candolle, to distinct genera. In the Galapagos Archipelago, many even of the birds, though so well adapted for flying from island to island, are distinct on each; thus there are three closely-allied species of mocking-thrush, each confined to its own island. Now let us suppose the mocking-thrush of Chatham Island to be blown to Charles Island, which has its own mocking-thrush: why should it succeed in establishing itself there? We may safely infer that Charles Island is well stocked with its own species, for annually more eggs are laid there than can possibly be reared; and we may infer that the mocking-thrush peculiar to Charles Island is at least as well fitted for its home as is the species peculiar to Chatham Island. Sir C. Lyell and Mr Wollaston have communicated to me a remarkable fact bearing on this subject; namely, that Madeira and the adjoining islet of Porto Santo possess many distinct but representative land-shells, some of which live in crevices of stone; and although large quantities of stone are annually transported from Porto Santo to Madeira, yet this latter island has not become colonised by the Porto Santo species: nevertheless both islands have been colonised by some European land-shells, which no doubt had some advantage over the indigenous species. From these considerations I think we need not greatly marvel at the endemic and representative species, which inhabit the several islands of the Galapagos Archipelago, not having universally spread from island to island. In many other instances, as in the several districts of the same continent,

THE ORIGIN OF SPECIES

pre-occupation has probably played an important part in checking the commingling of species under the same conditions of life. Thus, the south-east and south-west corners of Australia have nearly the same physical conditions, and are united by continuous land, yet they are inhabited by a vast number of distinct mammals, birds, and plants.

The principle which determines the general character of the fauna and flora of oceanic islands, namely, that the inhabitants, when not identically the same, yet are plainly related to the inhabitants of that region whence colonists could most readily have been derived, – the colonists having been subsequently modified and better fitted to their new homes, – is of the widest application throughout nature. We see this on every mountain, in every lake and marsh. For Alpine species, excepting in so far as the same forms, chiefly of plants, have spread widely throughout the world during the recent Glacial epoch, are related to those of the surrounding lowlands; – thus we have in South America, Alpine humming-birds, Alpine rodents, Alpine plants, &c., all of strictly American forms, and it is obvious that a mountain, as it became slowly upheaved, would naturally be colonised from the surrounding lowlands. So it is with the inhabitants of lakes and marshes, excepting in so far as great facility of transport has given the same general forms to the whole world. We see this same principle in the blind animals inhabiting the caves of America and of Europe. Other analogous facts could be given. And it will, I believe, be universally found to be true, that wherever in two regions, let them be ever so distant, many closely allied or representative species occur, there will likewise be found some identical species, showing, in accordance with the foregoing view, that at some former period there has been inter-communication or migration between the two regions. And wherever many closely-allied species occur, there will be found many forms which some naturalists rank as distinct species, and some as varieties; these doubtful forms showing us the steps in the process of modification.

This relation between the power and extent of migration of a species, either at the present time or at some former period under different physical conditions, and the existence at remote

GEOGRAPHICAL DISTRIBUTION

points of the world of other species allied to it, is shown in another and more general way. Mr Gould remarked to me long ago, that in those genera of birds which range over the world, many of the species have very wide ranges. I can hardly doubt that this rule is generally true, though it would be difficult to prove it. Amongst mammals, we see it strikingly displayed in Bats, and in a lesser degree in the Felidae and Canidae. We see it, if we compare the distribution of butterflies and beetles. So it is with most fresh-water productions, in which so many genera range over the world, and many individual species have enormous ranges. It is not meant that in world-ranging genera all the species have a wide range, or even that they have on an *average* a wide range; but only that some of the species range very widely; for the facility with which widely-ranging species vary and give rise to new forms will largely determine their average range. For instance, two varieties of the same species inhabit America and Europe, and the species thus has an immense range; but, if the variation had been a little greater, the two varieties would have been ranked as distinct species, and the common range would have been greatly reduced. Still less is it meant, that a species which apparently has the capacity of crossing barriers and ranging widely, as in the case of certain powerfully-winged birds, will necessarily range widely; for we should never forget that to range widely implies not only the power of crossing barriers, but the more important power of being victorious in distant lands in the struggle for life with foreign associates. But on the view of all the species of a genus having descended from a single parent, though now distributed to the most remote points of the world, we ought to find, and I believe as a general rule we do find, that some at least of the species range very widely; for it is necessary that the unmodified parent should range widely, undergoing modification during its diffusion, and should place itself under diverse conditions favourable for the conversion of its offspring, firstly into new varieties and ultimately into new species.

In considering the wide distribution of certain genera, we should bear in mind that some are extremely ancient, and must have branched off from a common parent at a remote epoch; so

THE ORIGIN OF SPECIES

that in such cases there will have been ample time for great climatal and geographical changes and for accidents of transport; and consequently for the migration of some of the species into all quarters of the world, where they may have become slightly modified in relation to their new conditions. There is, also, some reason to believe from geological evidence that organisms low in the scale within each great class, generally change at a slower rate than the higher forms; and consequently the lower forms will have had a better chance of ranging widely and of still retaining the same specific character. This fact, together with the seeds and eggs of many low forms being very minute and better fitted for distant transportation, probably accounts for a law which has long been observed, and which has lately been admirably discussed by Alph. de Candolle in regard to plants, namely, that the lower any group of organisms is, the more widely it is apt to range.

The relations just discussed, – namely, low and slowly-changing organisms ranging more widely than the high, – some of the species of widely-ranging genera themselves ranging widely, – such facts, as alpine, lacustrine, and marsh productions being related (with the exceptions before specified) to those on the surrounding low lands and dry lands, though these stations are so different – the very close relation of the distinct species which inhabit the islets of the same archipelago, – and especially the striking relation of the inhabitants of each whole archipelago or island to those of the nearest mainland, – are, I think, utterly inexplicable on the ordinary view of the independent creation of each species, but are explicable on the view of colonisation from the nearest and readiest source, together with the subsequent modification and better adaptation of the colonists to their new homes.

From *The Geographical Distribution of Animals*
Alfred Russel Wallace

CHAPTER XXIII.

SUMMARY OF THE DISTRIBUTION, AND LINES OF MIGRATION, OF THE SEVERAL CLASSES OF ANIMALS.

HAVING already given summaries of the distribution of the several orders, and of some of the classes of land animals, we propose here to make a few general remarks on the special phenomena presented by the more important groups, and to indicate where possible, the general lines of migration by which they have become dispersed over wide areas.

MAMMALIA.

This class is very important, and its past history is much better known than that of most others. We shall therefore briefly summarise the results we have arrived at from our examination of the distribution of extinct and living forms of each order.

Primates.—This order, being pre-eminently a tropical one, became separated into two portions, inhabiting the Eastern and Western Hemispheres respectively, at a very early epoch. In consequence of this separation it has diverged more radically than most other orders, so that the two American families, Cebidæ and Hapalidæ, are widely differentiated from the Apes, Monkeys, and Lemurs of the Old World. The Lemurs were probably still more ancient, but being much lower in organisation, they became extinct in most of the areas where the higher forms of Primates became developed. Remains found in the Eocene formation indicate, that the North American and European

Primates had, even at that early epoch, diverged into distinct series, so that we must probably look back to the secondary period for the ancestral form from which the entire order was developed.

Chiroptera.—These are also undoubtedly very ancient. The most generalised forms—the Vespertilionidæ and Noctilionidæ— are the most widely distributed; while special types have arisen in America, and in the Eastern Hemisphere. Remains found in the Upper Eocene formation of Europe differ little from species still living in the same countries; so that we can form no conjecture as to the origin or migration of the group. Their power of flight would, however, enable them rapidly to spread over all the great continents of the globe.

Insectivora.—This very ancient group, now probably verging towards extinction, appears to have originated in the Northern continent, and never to have reached Australia or South America. It may, however, have become extinct in the latter country owing to the competition of the numerous Edentata. The Insectivora now often maintain themselves amidst more highly developed forms, by means of some special protection. Some burrow in the earth,—like the moles; others have a spiny covering,—as the hedgehogs and several of the Centetidæ; others are aquatic,—as the *Potamogale* and the desman; others have a nauseous odour,—as the shrews; while there are several which seem to be preserved by their resemblance to higher forms,—as the elephant-shrews to jerboas, and the tupaias to squirrels. The same need of protection is shown by the numerous Insectivora inhabiting Madagascar, where the competing forms are few; and by one lingering in the Antilles, where there are hardly any other mammalia.

Carnivora.—Although perhaps less ancient than the preceding, this form of mammal is far more highly organised, and from its earliest appearance appears to have become dominant in the world. It would therefore soon spread widely, and diverge into the various specialised types represented by existing families. Most of these appear to have originated in the Eastern Hemisphere, the only Carnivora occurring in North

American Miocene deposits being ancestral forms of Canidæ and Felidæ. It seems probable, therefore, that the order had attained a considerable development before it reached the Western Hemisphere. The Procyonidæ, now confined to America, are not very ancient; and the occurrence of a few allied forms in the Himalayas (*Ælurus* and *Æluropus*) render it probable that their common ancestors entered North America from the Palæarctic region during the Miocene period, but being a rather low type they have succumbed under the competition of higher forms in most parts of the Eastern Hemisphere. Bears and Weasels are probably still more recent emigrants to America. The aquatic carnivora (Seals, &c.) are, as might be expected, more widely and uniformly distributed, but there is little evidence to show at what period the type was first developed.

Ungulata.—These are the dominant vegetable-feeders of the great continents, and they have steadily increased in numbers and in specialisation from the oldest Tertiary times to the present day. Being generally of larger size and less active than the Carnivora, they have somewhat more restricted powers of dispersal. We have good evidence that their wide range over the globe is a comparatively recent phenomenon. Tapirs and Llamas have probably not long inhabited South America, while Rhinoceroses and Antelopes were once, perhaps, unknown in Africa, although abounding in Europe and Asia. Swine are one of the most ancient types in both hemispheres; and their great hardiness, their omnivorous diet, and their powers of swimming, have led to their wide distribution. The sheep and goats, on the other hand, are perhaps the most recent development of the Ungulata, and they seem to have arisen in the Palæarctic region at a time when its climate already approximated to that which now prevails. Hence they are pre-eminently a Temperate group, never found within the Tropics except upon a few mountain ranges.

Proboscidea.—These huge animals (the Elephants and Mastodons) appear to have originated in the warmer parts of the Palæarctic region, but they soon spread over all the great continents, even reaching the southern extremity of America. Their extinction has probably depended more on physical than on organic changes, and we can clearly trace their almost total disappearance to the effects of the Glacial epoch.

Rodentia.—Rodents are a very dominant group, and a very ancient one. Owing to their small size and rapid powers of increase, they soon spread over almost every part of the globe, whence has resulted a great specialisation of family types in the South American continent which remained so long isolated. They are capable of living wherever there is any kind of vegetable food, hence their range will be determined rather by organic than by physical conditions; and the occupation of a country by enemies or by competing forms, is probably the chief cause which has prevented many of the families from acquiring a wide range. The occurrence of isolated species of the South American families, Octodontidæ and Echimyidæ in the Ethiopian and Palæarctic regions, is an indication that the range of many of the families has recently become less extensive.

Edentata.—These singular and lowly-organised animals appear to have become almost restricted to the two great Southern lands—South Africa and South America—at an early period; and, being there free from the competition of higher forms, developed a number of remarkable types often of huge size, of which the Megatherium is one of the best known. The incursion of the highly-organised Ungulates and Carnivora into Africa during the Miocene epoch, probably exterminated most of them in that continent; but in America they continued in full force down to the Post-Pliocene period; and even now, the comparatively diminutive Sloths, Ant-eaters, and Armadillos, form a large and important portion of the fauna.

Marsupialia and Monotremata.—These are probably the representatives of the most ancient and lowly-organised types of mammal. They once existed in the northern continents, whence they spread into Australia; and being isolated, and preserved from the competition of the higher forms which soon arose in other parts of the world, they have developed into a variety of types, which, however, still preserve a general uniformity of organisation. One family, which continued to exist in Europe till the latter part of the Miocene period, reached America, and has there been preserved to our day.

Lines of Migration of the Mammalia.—The whole series of phenomena presented by the distribution of the Mammalia, looked at broadly, are in harmony with the view that the great continents and oceans of our own epoch have been in existence, with comparatively small changes, during all Tertiary times. Each one of them has, no doubt, undergone considerable modifications in its area, its altitude, and in its connection with other lands. Yet some considerable portion of each continent has, probably, long existed in its present position, while the great oceans seem to have occupied the same depressions of the earth's crust (varied, perhaps, by local elevations and subsidences) during all this vast period of time. Hence, allowing for the changes of which we have more or less satisfactory evidence, the migrations of the chief mammalian types can be pretty clearly traced. Some, owing to their small size and great vitality, have spread to almost all the chief land masses; but the majority of the orders have a more restricted range. All the evidence at our command points to the Northern Hemisphere as the birth-place of the class, and probably of all the orders. At a very early period the land communication with Australia was cut off, and has never been renewed; so that we have here preserved for us a sample of one or more of the most ancient forms of mammal. Somewhat later the union with South America and South Africa was severed; and in both these countries we have samples of a somewhat more advanced stage of mammalian development. Later still, the union by a northern route between the Eastern and Western Hemispheres appears to have been broken, partly by a physical separation, but almost as effectually by a lowering of temperature. About the same period the separation of the Palæarctic region from the Oriental was effected, by the rise of the Himalayas and the increasing contrast of climate; while the formation of the great desert-belts of the Sahara, Arabia, Persia, and Central Asia, helped to complete the separation of the Temperate and Tropical zones, and to render further intermigration almost impossible.

In a few cases—of which the Rodents in Australia and the pigs in Austro-Malaya are perhaps the most striking examples —the distribution of land-mammals has been effected by a sea-passage either by swimming or on floating vegetation; but, as a rule, we may be sure that the migrations of mammalia have taken place over the land; and their presence on islands is, therefore, a clear indication that these have been once connected with a continent. The present class of animals thus affords the best evidence of the past history of the land surface of our globe; and we have chiefly relied upon it in sketching out (in Part III.) the probable changes which each of our great regions has undergone.

Birds.

Although birds are, of all land-vertebrates, the best able to cross seas and oceans, it is remarkable how closely the main features of their distribution correspond with those of the Mammalia. South America possesses the low Formicaroid type of Passeres,—which, compared with the more highly developed forms of the Eastern Hemisphere, is analogous to the Cebidæ and Hapalidæ as compared with the Old World Apes and Monkeys; while its Cracidæ as compared with the Pheasants and Grouse, may be considered parallel to the Edentata as compared with the Ungulates of the Old World. The Marsupials of America and Australia, are paralleled, among birds, in the Struthionidæ and Megapodiidæ; the Lemurs and Insectivora preserved in Madagascar are represented by the Mascarene Dididæ; the absence of Deer and Bears from Africa is analogous to the absence of Wrens, Creepers, and Pheasants; while the African Hyracidæ and Chrysochloridæ among mammals, may well be compared with the equally peculiar Coliidæ and Musophagidæ among birds.

From these and many other similarities of distribution, it is clear that birds have, as a rule, followed the same great lines of migration as mammalia; and that oceans, seas, and deserts, have always to a great extent limited their range. Yet these barriers have not been absolute; and in the course of ages birds have been able to reach almost every habitable land upon the globe. Hence have arisen some of the most curious and interesting phenomena of distribution; and many islands, which are entirely destitute of mammalia, or possess a very few species, abound in birds, often of peculiar types and remarkable for some unusual character or habit. Striking examples of such interesting bird-faunas are those of New Zealand, the Sandwich Islands, the Galapagos, the Mascarene Islands, the Moluccas, and the Antilles; while even small and remote islets,—such as Juan Fernandez and Norfolk Island, have more light thrown upon their past history by means of their birds, than by any other portion of their scanty fauna.

Another peculiar feature in the distribution of this class is the extraordinary manner in which certain groups and certain external characteristics, have become developed in islands, where the smaller and less powerful birds have been protected from the incursions of mammalian enemies, and where rapacious birds—which seem to some degree dependent on the abundance of mammalia—are also scarce. Thus, we have the Pigeons and the Parrots most wonderfully developed in the Australian region, which is pre-eminently insular; and both these groups here acquire conspicuous colours very unusual, or altogether absent, elsewhere. Similar colours (black and red) appear, in the same two groups, in the distant Mascarene islands; while in the Antilles the parrots have often white heads, a character not found in the allied species on the South American continent. Crests, too, are largely developed, in both these groups, in the Australian region only; and a crested parrot formerly lived in Mauritius,—a coincidence too much like that of the colours as above noted, to be considered accidental.

Again, birds exhibit to us a remarkable contrast as regards the oceanic islands of tropical and temperate latitudes; for while most of the former present hardly any cases of specific identity with the birds of adjacent continents, the latter often show hardly any differences. The Galapagos and Madagascar are examples of the first-named peculiarity; the Azores and the Bermudas of the last; and the difference can be clearly traced to the frequency and violence of storms in the one case and to the calms or steady breezes in the other.

It appears then, that although birds do not afford us the same convincing proof of the former union of now disjoined lands as we obtain from mammals, yet they give us much curious and suggestive information as to the various and complex modes in which the existing peculiarities of the distribution of animals have been brought about. They also throw much light on the relation between distribution and the external characters of animals; and, as they are often found where mammalia are quite absent, we must rank them as of equal value for the purposes of our present study.

Reptiles.

These hold a somewhat intermediate place, as regards their distribution, between mammals and birds, having on the whole rather a wider range than the former, and a more restricted one than the latter.

Snakes appear to have hardly more facilities for crossing the ocean than mammals; hence they are generally absent from oceanic islands. They are more especially a tropical group, and have thus never been able to pass from one continent to another by those high northern and southern routes, which we have seen reason to believe were very effectual in the case of mammalia and some other animals. Hence we find no resemblance between the Australian and Neotropical regions, or between the Palæ-arctic and Nearctic; while the Western Hemisphere is comparatively poor as regards variety of types, although rich in genera and species. Deserts and high mountains are also very effectual barriers for this group, and their lines of migration have probably been along river valleys, and occasionally across narrow seas by means of floating vegetation.

Lizards, being somewhat less tropical than snakes, may have passed by the northern route during warm epochs. They are also more suited to traverse deserts, and they possess some unknown means of crossing the ocean, as they are not unfrequently found in remote oceanic islands. These various causes have modified their distribution. The Western Hemisphere is much richer in lizards than it is in snakes; and it is also very distinct from the Eastern Hemisphere. The lines of migration of lizards appear to have been along the mountains and deserts of tropical countries, and, under special conditions, across tropical seas from island to island.

Crocodiles are a declining group. They were once more generally distributed, all the three families being found in British Eocene deposits. Being aquatic and capable of living in the sea, they can readily pass along all the coasts and islands of the warmer parts of the globe. Tortoises are equally ancient, and the restriction of certain groups to definite areas seems to be also a recent phenomenon.

Amphibia.

The Amphibia differ widely from Reptiles in their power of enduring cold; one of their chief divisions, the Urodela or Tailed-Batrachia, being confined to the temperate parts of the Northern Hemisphere. To this class of animals the northern and southern routes of migration were open; and we accordingly find a considerable amount of resemblance between South America and Australia, and a still stronger affinity between North America and the Palæarctic continent. The other tropical regions are more distinct from each other; clearly indicating that, in this group, it is tropical deserts and tropical oceans which are the barriers to migration. The class however is very fragmentary, and probably very ancient; so that descendants of once widespread types are now found isolated in various parts of the globe, between which we may feel sure there has been no direct transmission of Batrachia. Remembering that their chief lines of migration have been by northern and southern land-routes, by floating ice, by fresh-water channels, and perhaps at rare intervals by ova being carried by aquatic birds or by violent storms,—we shall be able to comprehend most of the features of their actual distribution.

Fresh-water Fishes.

Although it would appear, at first sight, that the means of dispersal of these animals are very limited, yet they share to some extent the wide range of other fresh-water organisms. They are found in all climates; but the tropical regions are by far the most productive, and of these South America is perhaps the richest and most peculiar. There is a certain amount of identity between the two northern continents, and also between those of the South Temperate zone; yet all are radically distinct, even North America and Europe having but a small proportion of their forms in common. The occurrence of allied fresh-water species in remote lands—as the *Aphritis* of Tasmania and Patagonia, and the *Comephorus* of Lake Baikal, distantly allied to the mackerels of Northern seas— would imply that marine fishes are often modified for a life in fresh waters; while other facts no less plainly show that permanent fresh-water species are sometimes dispersed in various ways across the oceans, more especially by the northern and southern routes.

The families of fresh-water fishes are often of restricted range, although cases of very wide and scattered distribution also occur. The great zoological regions are, on the whole, very well characterized; showing that the same barriers are effectual here, as with most other vertebrates. We conclude, therefore, that the chief lines of migration of fresh-water fishes have been across the Arctic and Antarctic seas, probably by means of floating ice as well as by the help of the vast flocks of migratory aquatic birds that frequent those regions. On continents they are, usually, widely dispersed; but tropical seas, even when of small extent, appear to have offered an effectual barrier to their dispersal. The cases of affinity between Tropical America, Africa, Asia, and Australia, must therefore be imputed either to the survival of once widespread groups, or to analogous adaptation to a fresh-water life of wide-spread marine types; and these cases cannot be taken as evidence of any former land connection between such remote continents.

Insects.

It has already been shown (Vol. I. pp. 209-213 and Vol. II. pp. 44-48) that the peculiarities of distribution of the various groups of insects depend very much on their habits and general economy. Their antiquity is so vast, and their more important modifications of structure have probably occurred so slowly, that modes of dispersal depending on such a combination of favourable conditions as to be of excessive rarity, may yet have had time to produce large cumulative effects. Their small specific gravity and their habits of flight render them liable to dispersal by winds to an extent unknown in other classes of animals; and thus, what are usually very effectual barriers have been overstepped, and sometimes almost obliterated, in the case of insects. A careful examination will, however, almost always show traces of an ancient fauna, agreeing in character with other classes of animals, intermixed with the more prominent and often more numerous forms whose presence is due to this unusual facility of dispersal.

The effectual migration of insects is, perhaps more than in any other class of animals, limited by organic and physical conditions. The vegetation, the soil, the temperature, and the supply of moisture, must all be suited to their habits and economy; while they require an immunity from enemies of various kinds, which immigrants to a new country seldom obtain. Few organisms have, in so many complex ways, become adapted to their special environment, as have insects. They are in each country more or less adapted to the plants which belong to it; while their colours, their habits, and the very nature of the juices of their system, are all modified so as to protect them from the special dangers which surround them in their native land. It follows, that while no animals are so well adapted to show us the various modes by which dispersal may be effected, none can so effectually teach us the true nature and vast influence of the organic barrier in limiting dispersal.

It is probable that insects have at one time or another taken advantage of every line of migration by which any terrestrial organisms have spread over the earth, but owing to their small size and rapid multiplication, they have made use of some which are exclusively their own. Such are the passage along mountain ranges from the Arctic to the Antarctic regions, and the dispersal of certain types over all temperate lands. It will perhaps be found that insects have spread over the land surface in directions dependent on our surface zones—forests, pastures, and deserts;—and a study of these, with a due consideration of the fact that narrow seas are scarcely a barrier to most of the groups, may assist us to understand many of the details of insect-distribution.

Terrestrial Mollusca.

The distribution of land-shells agrees, in some features, with that of insects, while in others the two are strongly contrasted. In both we see the effects of great antiquity, with some special means of dispersal; but while in insects the general powers of motion, both voluntary and involuntary, are at a maximum, in land-molluscs they are almost at a minimum. Although to some extent dependent on vegetation and climate, the latter are more dependent on inorganic conditions, and also to a large extent on the general organic environment. The result of these various causes, acting through countless ages, has been to spread the main types of structure with considerable uniformity over the globe; while generic and sub-generic forms are often wonderfully localized.

Land-shells, even more than insects, seem, at first sight, to require regions of their own; but we have already pointed out the disadvantages of such a method of study. It will be far more instructive to refer them to those regions and sub-regions which are found to accord best with the distribution of the higher animals, and to consider the various anomalies they present as so many problems, to be solved by a careful study of their habits and economy, and especially by a search after the hidden causes which have enabled them to spread so widely over land and ocean.

The lines of migration which land-shells have followed, can

hardly be determined with any definiteness.   On continents they seem to spread steadily, but slowly, in every direction, checked probably by organic and physical conditions rather than by the barriers which limit the higher groups.   Over the ocean they are also slowly dispersed, by some means which act perhaps at very long intervals, but which, within the period of the duration of genera and families, are tolerably effective.   It thus happens that, although the powers of dispersal of land-shells and insects are so very unequal, the resulting geographical distribution is almost the opposite of what might have been expected,—the former being, on the whole, less distinctly localized than the latter.

CONCLUSION.

The preceding remarks are all I now venture to offer, on the distinguishing features of the various groups of land-animals as regards their distribution and migrations.   They are at best but indications of the various lines of research opened up to us by the study of animals from the geographical point of view, and by looking upon their range in space and time as an important portion of the earth's history.   Much work has yet to be done before the materials will exist for a complete treatment of the subject in all its branches; and it is the author's hope that his volumes may lead to a more systematic collection and arrangement of the necessary facts.   At present all public museums and private collections are arranged zoologically.   All treatises, monographs, and catalogues, also follow, more or less completely, the zoological arrangement; and the greatest difficulty the student of geographical distribution has to contend against, is the total absence of geographical collections, and the almost total want of complete and comparable local catalogues.   Till every well-marked district,—every archipelago, and every important island, has all its known species of the more important groups of animals catalogued on a uniform plan, and with a uniform nomenclature, a thoroughly satisfactory account of the Geographical Distribution of Animals will not be possible.   But more than this is wanted.   Many of the most curious relations between animal forms and their habitats, are entirely unnoticed, owing to the productions of the same locality *never* being associated in our museums and collections. A few such relations have been brought to light by modern scientific travellers, but many more remain to be discovered; and there is probably no fresher and more productive field still unexplored in Natural History. Most of these curious and suggestive relations are to be found in the productions of islands, as compared with each other, or with the continents of which they form appendages; but these can never be properly studied, or even discovered, unless they are visibly grouped together. When the birds, the more conspicuous families of insects, and the land-shells of islands, are kept together so as to be readily compared with similar associations from the adjacent continents or other islands, it is believed that in almost every case there will be found to be peculiarities of form or colour running through widely different groups, and strictly indicative of local or geographical influences. Some of these coincident variations have been alluded to in various parts of this work, but they have never been systematically investigated. They constitute an unworked mine of wealth for the enterprising explorer; and they may not improbably lead to the discovery of some of the hidden laws (supplementary to Natural Selection), which seem to be required, in order to account for many of the external characteristics of animals.

In concluding his task, the author ventures to suggest, that naturalists who are disposed to turn aside from the beaten track of research, may find in the line of study here suggested a new and interesting pursuit, not inferior in attractions to the lofty heights of transcendental anatomy, or the bewildering mazes of modern classification. And it is a study which will surely lead them to an increased appreciation of the beauty and the harmony of nature, and to a fuller comprehension of the complex relations and mutual interdependence, which link together every animal and vegetable form, with the ever-changing earth which supports them, into one grand organic whole.

From *The History of Creation, or the Development of the Earth and Its Inhabitants by the Action of Natural Causes*

Ernst Haeckel

(363)

CHAPTER XIV.

MIGRATION AND DISTRIBUTION OF ORGANISMS. CHOROLOGY AND THE ICE PERIOD OF THE EARTH.

Chorological Facts and Causes.—Origin of most Species in one Single Locality : " Centres of Creation."—Distribution by Migration.—Active and Passive Migrations of Animals and Plants.—Flying Animals.—Analogies between Birds and Insects.—Bats.—Means of Transport.—Transport of Germs by Water and by Wind.—Continual Change of the Area of Distribution by Elevations and Depressions of the Ground.—Chorological Importance of Geological Processes.—Influence of the Change of Climate.—Ice or Glacial Period.—Its Importance to Chorology.—Importance of Migrations for the Origin of New Species.—Isolation of Colonists.—Wagner's Law of Migration.—Connection between the Theory of Migration and the Theory of Selection.—Agreement of its Results with the Theory of Descent.

As I have repeatedly said, but cannot too much emphasize, the actual value and invincible strength of the Theory of Descent does not lie in its explaining this or that single phenomenon, but in the fact that it explains *all* biological phenomena, that it makes *all* botanical and zoological series of phenomena intelligible in their relations to one another. Hence every thoughtful investigator is the more firmly and deeply convinced of its truth the more he advances from single biological observations to a general view of the whole domain of animal and vegetable life.

Let us now, starting from this comprehensive point of view, survey a biological domain, the varied and complicated phenomena of which may be explained with remarkable simplicity and clearness by the theory of descent. I mean *Chorology*, or the theory of the *local distribution of organisms over the surface of the earth*. By this I do not only mean the *geographical* distribution of animal and vegetable species over the different parts and provinces of the earth, over continents and islands, seas, and rivers, but also their *topographical* distribution in a *vertical* direction, their ascending to the heights of mountains, and their descending into the depths of the ocean.

The strange chorological series of phenomena which show the horizontal distribution of organisms over parts of the earth, and their vertical distribution in heights and depths, have long since excited general interest. In recent times Alexander Humboldt[39] and Frederick Schouw have especially discussed the geography of plants, and Berghaus, Schmarda, and Wallace the geography of animals, on a large scale. But although these and several other naturalists have in many ways increased our knowledge of the distribution of animal and vegetable forms, and laid open to us a new domain of science, full of wonderful and interesting phenomena, yet Chorology as a whole remained, as far as their labours were concerned, only a desultory knowledge of a mass of individual *facts*. It could not be called a science as long as the *causes* for the explanation of these facts were wanting. These causes were first disclosed by the theory of selection and its doctrine of the *migrations* of animal and vegetable species, and it is only since Darwin that we have been able to speak of an independent *science*

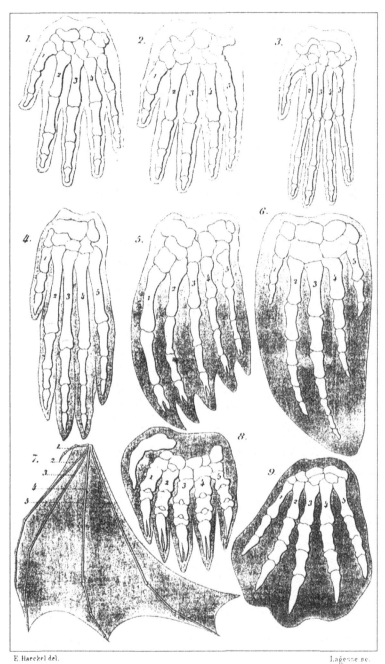

E. Haeckel del.   Lagesse sc.

1. Man, 2. Gorilla, 3. Orang, 4. Dog, 5. Seal
6. Porpoise, 7. Bat, 8. Mole, 9. Duck-bill.

Frontispiece to Ernst Haeckel, *The History of Creation*, volume 1 (8th edition, 1925).

of Chorology. Wallace and Moritz Wagner have done most, after Darwin, in this respect.

The first naturalist who clearly comprehended the theory of migration and correctly recognized its importance for the origin of new species, was the celebrated German geologist, Leopold Buch. In his " Physical Description of the Canary Island," as early as 1825—hence thirty-four years before the appearance of Darwin's work—he made those remarkable propositions which I have already quoted in my fifth chapter. He there states that the migration, distribution, and local separation of species are the three principal outward causes that effect the transformation of species; their influence, he thinks, is sufficient to produce new species by the internal interaction of variability and heredity. Buch, who was a great traveller and had made extensive observations himself, also discusses the great importance exercised by the local separation of animals and plants that have migrated to isolated islands. Unfortunately, this eminent geologist did not work this important idea out further, and was unable to convince his friend, Alexander Humboldt, of its great significance. Wagner, however, in his essay on Leopold Buch and Darwin (1883), has very justly pointed out that, with regard to the migration-theory, Buch must be looked upon as Darwin's greatest predecessor.

If all the phenomena of the geographical and topographical distribution of organisms are examined by themselves, without considering the gradual development of species, and if at the same time, following the customary superstition, the individual species of animals and plants are considered as forms independently created and independent of one another, then there remains nothing for us to do but to gaze at those phenomena as a confused collection of incomprehensible and inexplicable miracles. But as soon as we leave this low standpoint, and rise to the height of the theory of development, by means of the supposition of a blood-relationship between the different species, then all at once a clear light falls upon this strange series of miracles, and we see that all chorological facts can be understood quite simply and clearly by the supposition of a common descent of the species, and their passive and active migrations.

The most important principle from which we must start in chorology, and of the truth of which we are convinced by due examination of the theory of selection, is that, as a rule, every animal and vegetable species has arisen only *once* in the course of time and only in *one* place on the earth—its so-called "centre of creation"—by natural selection. I share this opinion of Darwin's unconditionally, in respect to the great majority of higher and perfect organisms, and in respect to most animals and plants in which the division of labour, or differentiation of the cells and organs of which they are composed, has attained a certain stage. For it is quite incredible, or could at best only be an exceedingly rare accident, that all the manifold and complicated circumstances—all the different conditions of the struggle for life, which influence the origin of a new species by natural selection—should have worked together in exactly the same agreement and combination more than once in the earth's history, or should have been active at the same time at several different points of the earth's surface.

On the other hand, I consider it very probable that

certain exceedingly imperfect organisms of the simplest structure, forms of species of an exceedingly indifferent nature, as, for example, many single-celled Protista (Algæ as well as Amœbæ and Infusoria), but especially the Monera, the simplest of them all, have several times or simultaneously arisen in their specific form in several parts of the earth. For the few and very simple conditions by which their specific form was changed in the struggle for life may surely have often been repeated, in the course of time, independently in different parts of the earth. Further, those higher specific forms also, which have not arisen by natural selection, but by *hybridism* (the previously mentioned hybrid species, pp. 150 and 151), may have repeatedly arisen anew in different localities. As, however, this proportionately small number of organisms does not especially interest us here, we may, in respect of chorology, leave them alone, and need only take into consideration the distribution of the great majority of animal and vegetable species in regard to which the *single origin of every species in a single locality*, in its so-called " central point of creation," can be considered as tolerably certain.

Every animal and vegetable species from the beginning of its existence has possessed the tendency to spread beyond the limited locality of its origin, beyond the boundary of its " centre of creation," or, in other words, beyond its *primæval home*, or its natal place. This is a necessary consequence of the relations of population and over-population. The more an animal or vegetable species increases, the less is its limited natal place sufficient for its sustenance, and the fiercer the struggle for life ; the more rapid the *over-population* of the natal spot, the more it leads to *emigration*. These *migrations* are common to all organisms, and are the real cause of the wide distribution of the different species of organisms over the earth's surface. Just as men leave over-crowded states, so all animals and plants migrate from their over-crowded primæval homes.

Many distinguished naturalists, especially Leopold Buch, Lyell,[11] and Schleiden, have before this repeatedly drawn attention to the great importance of these very interesting migrations of organisms. The means of transport by which they are effected are extremely varied. Darwin has discussed these most excellently in the eleventh and twelfth chapters of his work, which are exclusively devoted to "geographical distribution." The means of transport are partly active, partly passive; that is to say, the organism effects its migration partly by free locomotion due to its own activity, and partly by the movements of other natural bodies in which it has no active share.

It is self-evident that *active migrations* play the chief part in animals able to move freely. The more freely an animal's organization permits it to move in all directions, the more easily the animal species can migrate, and the more rapidly it will spread over the earth. *Flying* animals are of course most favoured in this respect, among vertebrate animals especially birds, and among articulated animals, insects. These two classes, as soon as they came into existence, can have more easily spread over the whole earth than any other animal, and this fact partly explains the extraordinary uniformity of structure which characterizes these two great classes of animals. For, although they contain an exceedingly large number of different species, and although the insect class alone is said to possess more different species than all other classes of animals together, yet all the innumerable species of insects, and in like manner, also, the different species of birds, agree most strikingly in all essential peculiarities of their organization. Hence, in the class of insects, as well as in that of birds, we can distinguish only a very small number of large natural groups or orders, and these few orders differ but very little from one another in their internal structure. The orders of birds with their numerous species are not nearly as distinct from one another as the orders of the mammalian class, containing much fewer species; and the orders of insects, which are extremely rich in genera and species, resemble one another much more closely in their internal structure than do the much smaller orders of the crab class. The general parallelism between birds and insects is also very interesting in relation to systematic zoology; and the great importance of their richness in forms, for scientific morphology, lies in the fact that they show us how, within the narrowest anatomical sphere, and without profound changes of the essential internal organization, the greatest variety in external bodily forms can be attained. The reason of this is evidently their flying mode of life and their free locomotion. In consequence of this, birds, as well as insects, have spread very rapidly over the whole surface of the earth, have settled in all possible localities inaccessible to other animals, and variously modified their specific form by superficial adaptation to particular local relations.

Of the flying vertebrates, *bats* are, moreover, of peculiar interest to chorology. For not a single island lying more than 300 miles from the nearest continent possesses other indigenous mammals from the mainland. On the other hand, numerous species of bat may be found on isolated islands, and many separate islands or groups of islands are distinguished for possessing quite peculiar species or even of peculiar species of bats. This remarkable fact is most easily accounted for by the theory of selection and migration, whereas it remains an unintelligible mystery without it. Land mammals, which cannot fly, are not able to wander across broad stretches of sea and to search far-off islands. This is possible only to bats, which can fly for some length of time, and are, moreover, easily carried hundreds of miles by storms. And when cast upon distant islands they have to adapt themselves to wholly different conditions of existence, and their descendants sooner or later become transformed into new species or even into new generic forms.

Next to the flying animals, those animals, of course, have spread most quickly and furthest which were next best able to migrate, that is, the best runners among the inhabitants of the land, and the best swimmers among the inhabitants of the water. However, the power of such active migrations is not confined to those animals which throughout life enjoy free locomotion. For the fixed animals also, such as corals, tubicolous worms, sea-squirts, lily encrinites, sea-acorns, barnacles, and many other lower animals which adhere to seaweeds, stones, etc., enjoy, at least at an early period of life, free locomotion. They all migrate before they adhere to anything. Their first free locomotive condition of early life is generally that of a "ciliated" larva, a roundish, cellular corpuscle, which, by

means of a garb of movable flimmer-hairs " (Latin, "cilia "),
swarms about in the water. All of these swimming ciliated
larvæ of the lower animals have developed out of the same
common germinal form, that is, out of the *Gastrula* (Plate V.,
Fig. 8, 18); and it too is originally capable of migrating,
owing to its garb of movable " flimmer-hairs."

But the power of free locomotion, and hence, also, of active
migration, is not confined to animals alone, but many plants
likewise enjoy it. Many lower aquatic plants, especially the
class of the Tangles (Algæ), swim about freely in the water
in early life, like the lower animals just mentioned, by
means of a vibratile hairy coat, a vibrating whip, or a
covering of tremulous fringes, and only at a later period
adhere to objects. Even in the case of many higher plants,
which we designate as creepers and climbing plants, we
may speak of active migration. Their elongated stalks and
perennial roots creep or climb during their long process of
growth to new positions, and by means of their widespread
branches they acquire new habitations, to which they
attach themselves by buds, and bring forth new colonies
of individuals of their species.

Influential as these active migrations of most animals
and many plants are, yet alone they would by no means
be sufficient to explain the chorology of organisms. *Passive
migrations* have ever been by far the more important, and
of far greater influence, in the case of most plants and in
that of many animals. Such passive changes of locality
are produced by extremely numerous causes. Air and
water in their eternal motion, wind and waves with their
manifold currents, play the chief part. The wind in all
places and at all times raises light organisms, small animals

and plants, but especially their young germs, animal eggs and plant seeds, and carries them far over land and seas. Where they fall into the water they are seized by currents or waves and carried to other places. It is well known, from numerous examples, how far in many cases trunks of trees, hard-shelled fruits, and other not readily perishable portions of plants are carried away from their original home by the course of rivers and by the currents of the sea. Trunks of palm trees from the West Indies are brought by the Gulf Stream to the British and Norwegian coasts. All large rivers bring down driftwood from the mountains, and frequently Alpine plants are carried from their home at the source of the river into the plains, and even further, down to the sea. Frequently numerous creatures live between the roots of the plants thus carried down, and between the branches of the trees thus washed away there are various inhabitants which have to take part in the passive migration. The bark of the tree is covered with mosses, lichens, and parasitic insects. Other insects, spiders, etc., even small reptiles and mammals, are hidden within the hollow trunk or cling to the branches. In the earth adhering to the fibres of the roots, in the dust lying in the cracks of the bark, there are innumerable germs of smaller animals and plants. Now, if the trunk thus washed away lands safely on a foreign shore or on a distant island, the guests who had to take part in the involuntary voyage can leave their boat and settle in the new country. A very remarkable kind of water-transport is formed by the floating icebergs which annually become loosened from the eternal ice of the Polar Sea. Although these cold regions are thinly peopled, yet many of their inhabitants, who were accidentally upon an iceberg while it was becoming loosened, are carried away with it by the currents, and landed on warmer shores. In this manner, by means of loosened blocks of ice from the northern Polar Sea, often whole populations of small animals and plants have been carried to the northern shores of Europe and America. Nay, even polar foxes and polar bears have been carried in this way to Iceland and to the British Isles.

Transport by air is no less important than transport by water in this matter of passive migration. The dust covering our streets and roofs, the earth lying on dry fields and dried-up pools, the light moist soil of forests, in short, the whole surface of the globe, contains millions of small organisms and their germs. Many of these small animals and plants can without injury become completely dried up, and awake again to life as soon as they are moistened. Every gust of wind raises up with the dust innumerable little creatures of this kind, and often carries them away to other places miles off. But even larger organisms, and especially their germs, may often make distant passive journeys through the air. The seeds of many plants are provided with light feathery processes, which act as parachutes and facilitate their flight in the air, and prevent their falling. Spiders make journeys of many miles through the air on their fine filaments, their so-called gossamer threads. Young frogs are frequently raised by whirlwinds into the air by thousands, and fall down in a distant part as a " shower of frogs." Storms may carry birds and insects across half the earth's circumference. They drop in the United States, having risen in England. Starting from California, they only come to rest in China. But, again, many other organisms may make the journey from one continent to another together with the birds and insects. Of course all parasites, the number of which is legion, fleas, lice, mites, moulds, etc., migrate with the organism upon which they live. In the earth which often remains sticking to the claws of birds there are also small animals and plants or their germs. Thus the voluntary or involuntary migration of a single larger organism may carry a whole small flora and fauna from one part of the earth to another.

Besides the means of transport here mentioned, there are many others which explain the distribution of animal and vegetable species over the large tracts of the earth's surface, and especially the general distribution of the so-called cosmopolitan species. But these alone would not be nearly sufficient to explain all chorological facts. How is it, for example, that many inhabitants of fresh water live in various rivers or lakes far away and quite apart from one another? How is it that many inhabitants of mountains, which cannot exist in plains, are found upon entirely separated and far-distant chains of mountains? It is difficult to believe, and in many cases quite inconceivable, that these inhabitants of fresh water should have in any way, actively or passively, migrated over the land lying between the lakes, or that the inhabitants of mountains in any way, actively or passively, crossed the plains lying between their mountain-homes. But here geology comes to our help, as a mighty ally, and completely solves these difficult problems for us.

The history of the earth's development shows us that the distribution of land and water on its surface is ever and continually changing. In consequence of geological changes of the earth's crust, *elevations* and *depressions* of the ground take place everywhere, sometimes more strongly marked in one place, sometimes in another. Even if they happen so slowly that in the course of centuries the seashore rises or sinks only a few inches, or even only a few lines, still they nevertheless effect great results in the course of long periods of time. And long—immeasurably long—periods of time have not been wanting in the earth's history. During the course of many millions of years, ever since organic life existed on the earth, land and water have perpetually struggled for supremacy. Continents and islands have sunk into the sea, and new ones have arisen out of its bosom. Lakes and seas have slowly been raised and dried up, and new water-basins have arisen by the sinking of the ground. Peninsulas have become islands by the narrow neck of land which connected them with the mainland sinking into the water. The islands of an archipelago have become the peaks of a continuous chain of mountains by the whole floor of their sea being considerably raised.

Thus the Mediterranean at one time was an inland sea, when, in the place of the Straits of Gibraltar, an isthmus connected Africa with Spain. England, even during the more recent history of the earth, when man already existed, has repeatedly been connected with the European continent and been repeatedly separated from it. Nay, even Europe and North America have been directly connected. The South Sea at one time formed a large Pacific continent, and the numerous little islands which now lie scattered in it were simply the highest peaks of the mountains covering that continent. The Indian Ocean formed a continent which extended from the Sunda Islands along the southern coast of Asia to the east coast of Africa. This large continent of former times Sclater, an Englishman, has called *Lemuria*, from the monkey-like animals which inhabited it, and it is at the same time of great importance from being the probable cradle of the human race, which in all likelihood here first developed out of anthropoid apes. The important proof which Alfred Wallace has furnished,[36] by the help of chorological facts, that the present Malayan Archipelago consists in reality of two completely different divisions, is particularly interesting. The western division, the Indo-Malayan Archipelago, comprising the large islands of Borneo, Java, and Sumatra, was formerly connected by Malacca with the Asiatic continent, and probably also with the Lemurian continent just mentioned. The eastern division, on the other hand, the Austro-Malayan Archipelago, comprising Celebes, the Moluccas, New Guinea, Solomon's Islands, etc., was formerly directly connected with Australia. Both divisions were formerly two continents separated by a strait, but they have now for the most part sunk below the level of the sea. Wallace, solely on the ground of his accurate chorological observations, has been able in the most acute manner to determine the position of this former strait, the south end of which passes between Bali and Lombok. And this deep strait, although only fifteen miles broad, still forms a sharp boundary between the islands of Bali and Lombok; the fauna of Bali belongs to further India, the fauna of the latter to Australia.

Thus, ever since liquid water existed on the earth, the boundaries of water and land have eternally changed, and we may assert that the outlines of continents and islands have never remained for an hour, nay, even for a minute,

exactly the same. For the waves eternally and perpetually
break on the edge of the coast, and whatever the land in
these places loses in extent, it gains in other places by the
accumulation of mud, which condenses into solid stone and
again rises above the level of the sea as new land.   Nothing
can be more erroneous than the idea of a firm and un-
changeable outline of our continents, such as is impressed
upon us in early youth by defective lessons in geography,
which are devoid of a geological basis.

I need hardly draw attention to the fact that these
geological changes of the earth's surface have ever been
exceedingly important to the migrations of organisms, and
consequently to their Chorology. From them we learn to
understand how it is that the same or nearly related species
of animals and plants can occur on different islands,
although they could not have passed through the water
separating them, and how other species living in fresh
water can inhabit different enclosed water-basins, although
they could not have crossed the land lying between them.
These islands were formerly mountain-peaks of a connected
continent, and these lakes were once directly connected
with one another. The former were separated by geological
depressions, the latter by elevations. Now, if we further
consider how often and how unequally these alternating
elevations and depressions occur on the different parts of
the earth, and how, in consequence of this, the boundaries
of the geographical tracts of distribution of species become
changed, and if we further consider in what exceedingly
various ways the active and passive migrations of organisms
must have been influenced by them, then we shall be in a
position to completely understand the great variety of the

SCIENCE

FRIDAY, DECEMBER 7, 1900.

CONTENTS :

The History of the Neotropical Region : DR. H. VON IHERING 857

A History of the Development of the Quantitative Study of Variation : PROFESSOR CHAS. B. DAVENPORT 864

Plant Geography of North America :— Composition of the Rocky Mountain Flora : DR. P. A. RYDBERG 870

A Tertiary Coral Reef near Bainbridge, Georgia : DR. T. WAYLAND VAUGHAN 873

Peach Yellows : A Cause Suggested : O. F. COOK.. 875

Scientific Books :—
Van't Hoff's Leçons de chimie physique : H. C. J. Meyer's Determination of Radicles in Carbon Compounds : PROFESSOR W. R. ORNDORFF 881

Scientific Journals and Articles 882

Societies and Academies :—
Geological Society of Washington : DR. F. L. RANSOME, DAVID WHITE. Section of Biology of the New York Academy of Sciences : PROFESSOR F. E. LLOYD. Torrey Botanical Club : PROFESSOR EDWARD S. BURGESS. Zoological Journal Club of the University of Michigan : DR. H. S. JENNINGS 884

Discussion and Correspondence :—
The Electrical Theory of Gravitation : PROFESSOR W. S. FRANKLIN. The Homing Instinct of a Turtle : PROFESSOR C. L. BRISTOL 887

Botanical Notes :—
Peach Leaf Curl ; A New Botanical Journal ; Engler's Pflanzenreich : PROFESSOR CHARLES E. BESSEY 890

Practical Results obtained from the Study of Earthquakes : PROFESSOR JOHN MILNE 891

Scientific Notes and News 982

University and Educational News....................... 896

MSS. intended for publication and books, etc., intended for review should be sent to the responsible editor, Professor J. McKeen Cattell, Garrison-on-Hudson, N. Y.

THE HISTORY OF THE NEOTROPICAL REGION.

IN No. 276 of SCIENCE, April, 1900, Dr. Henry F. Osborn published an article on the 'Geological and Faunal Relations of Europe and America during the Tertiary Period,' to which I may here refer, as it may be useful for science to discuss the different opinions to which our study has led us.

It is singular that Mr. Osborn has no knowledge at all of the numerous papers published by the writer on the history of the neotropical fauna, and consequently it is necessary to say at first some words on these papers, and the new discoveries and ideas published in them. Referring here only to those of my publications in which the geological and zoogeographical relations of South America were fully discussed, I name the following :

1. 'Die Geographische Verbreitung der Flussmuscheln.' *Das Ausland*, Stuttgart, 1890, Nos. 48–49 ; and translated 'The Geographical Distribution of the Freshwater Mussels.' The New Zealand *Journal of Science*, Vol. I., Dunedin, 1891, pp. 151-154.

2. 'Ueber die Beziehungen der chilenischen und suedbrasilianischen Suesswasser fauna.' *Verhandlungen des deutschen wissenschaftlichen Vereines zu Santiago*, Vol. II., 1891, p. 142–19.

3. 'Ueber die alten Beziehungen zwischen Neuseeland und Suedamerika.' *Ausland*, Stuttgart, 1891, No. 18. Translated, ' On

858   *SCIENCE.*   [N. S. Vol. XII. No. 310.

the Ancient Relations between New Zealand South America.' *Transactions* of the New Zealand Institute, Vol. XXIV., 1891, p. 431–445.

4. 'Die Palaeo-Geographie Suedamerikas.' *Ausland*, Stuttgart, 1893, Nos. 1–4.

5. 'Revision der von Spix in Brasilien gesammelten Najaden.' *Archiv. für Naturgeschichte*, 1890, p. 117–170, Taf. IX.

6. 'Najaden von S. Paulo und die geographische Verbreitung der Suesswasserfaunen von Suedamerika.' *Archiv. für Naturgeschichte*, 1893, pp. 45–140, Taf. III.–IV.

7. 'Das neotropische Florengebiet und seine Geschichte.' *Engler's Botanische Jahrbücher*, Vol. XVII., 1893, pp. 1–54.

8. 'Die Ameisen von Rio Grande do Sul.' *Berliner entomologische Zeitschrift*, Band 39, 1894, pp. 321–446.

9. 'Os molluscos dos terrenos terciarios da Patagonia.' *Revista do Museu Paulista*, Vol. II., 1898, pp. 217–382, Pl. III.–IX., with Conclusions in English, pp. 372–380.

The study of the fresh water fauna, and especially of the Unionidæ of South America, gave me as a practical result the separation of two sub-regions 'Archiplata' and 'Archamazonia.' The first contains Chili, Argentina, Uraguay and Southern Brazil, the second Central and Northern Brazil (Archibrazil) and Guyana, Venezuela, etc. (Archiguyana). Archiplata contains numerous genera of Mollusca, Crustacea, etc., that are common to Chili and the La Plata district, such as *Unio, Chilina, Parastacus, Aeglea*, etc., including many species and even their parasites (*Temnocephala*), which are identical on both sides of the Andes. This contrasts sharply with the Archamazonian fauna, as tropical genera extend to Rio Plata and Rio Negro which are completely wanting in Chili and Peru. In Ecuador, however, the Cordillere form no such zoogeographical division, due certainly to differences in the

geological history of both parts of the Andes. For example the Decapod Crustacea in Chili and in the whole of Archiplata are the Parastacidæ and Aegleidæ, but the Potamoninæ are in Archamazonia. Dr. Ortmann has opposed to my explanations the hypothesis that biological differences may be the true reason, exterminating the Potamoninæ that invaded Archiplata, but favoring the Parastacidæ. The observations made by the writer on the biology of these Crustacea emphatically annul the objection.' In Northern Argentine, Rio Grande do Sul and St. Catherina, both coexist in the same waters and while the Potamonidæ prefer rivers and brooks, living among aquatic plants, the *Parastacus* selects muddy territory where it can burrow.

That the explanation is a geographical one is proved also by the fact that the species of Unionidæ, Mutelidæ, Ampullariidæ, etc., which occur in the La Plata and in the Rio Paraguay, are almost all Amazonic species. Moreover the faunal relations of the Paraná River are totally different from those of the Paraguay River. In confirmation of these zoogeographical facts the geological ones indicate to us in the Entrerian-formation Unio of the Niæa group, Chilina, Strophocheilus, etc., that is to say, the pure Archiplata fauna. These facts point out that the invasion of the Archamazonian element into Archiplata is quite a recent one. The intrusion of the Archamazonian element is Pliocene or post-Tertiary, and the Andes formed a barrier insurmountable to fresh-water crabs and mussels as well as to fishes, chelonians and alligators.

It is evident that the two faunal elements of South America correspond to geographical districts which were separated by the ocean during the greater part of the Tertiary. The intermixture of the two elements, and especially the intrusion of Bolivian ants, land snails, etc., in Eastern Brazil is by no means finished, but is a fact which

DECEMBER 7, 1900.]   *SCIENCE.*   859

we observe to-day. It is highly probable that these conditions, which are decisive not only for the fresh-water fauna but also for the land gastropods, have determined also the history of the mammals, which may have reached Brazil only in the Pliocene.

Although these inferences concerning the different faunal elements of the neotropical fauna, based on the zoogeographical work of the writer, seem to be quite conclusive, the matter is more difficult and hypothetical if we turn to the ancient relations of Archiplata and Archamazonia to other regions of the globe.

The connection of Archiplata with a great antarctic continent during the Cretaceous and Eocene formation seems now to be generally accepted, but the historic data given on the matter by Osborn are very incomplete. The first to discuss the question was the eminent botanist, Sir William Hooker, but the work of Wallace, and especially his axiom of the permanence of the great oceanic depths, for a long time retarded further progress. Not until 1883 did Hutton, with reference to New Zealand, and in 1890 the writer, with reference to Archiplata, turn aside to publish new facts in favor of the hypothesis of Hooker, which was also confirmed by Fl. Ameghino.

In relation to the ancient connection of Africa and Archamazonia I have given arguments (1890) in favor of a Mesozoic 'archiatlantic continent,' which existed during the earlier Tertiary. At first because of some paleontological facts noted by Schlosser, I believed that this continent could have transmitted Eocene mammals from South Africa to Europe, an idea now defended by Ameghino and Osborn; but in 1893 I modified my opinion and set forth the hypothesis that no Eocene placental mammals had existed either in Archamazonia or in Æthiopian Africa. The ancient continent uniting Archamazonia with Af-

rica I named Archiatlantica in 1890, using in 1892 the term Helenis, and in 1893 that of Archhelenis, with the purpose of preventing confusion with the 'Atlantis' a hypothetical land bridge between South Europe and Central America proposed by Unger.

I will not repeat here what I have said elsewhere as to the intimate relation between the fresh water faunas of Brazil, Guiana, and of equatorial Africa, but I shall make some remarks on the geographical distribution of the fresh-water mussels. North America agrees in its Unionidæ with Eurasia, the genera *Unio*, *Margaritana* and *Anodonta* being predominant. The archiplatic element of South America is formed only by the genus *Unio*, and by a section of it which has no representatives in the holarctic region, forming the subgenus *Niæa*, which is found also in New Zealand and Australia. The numerous presumed genera of *Unio* now admitted in North America all agree in the characteristic sculpture of the beaks, which is quite different in *Niæa*. I consider, therefore, *Niæa* as a genus and the North American sections of *Unio* only as subgenera. In the archhelenic region we have representatives of *Unio* which are more intimately allied to *Niæa* than to *Unio*, no Anodontas, but very numerous representatives of the Mutelidæ. The South American 'Anodonta' are all *Glabaris*, a genus of Mutelidæ allied to *Spatha* of Africa.

Considering the geological history, we find the precursors of the actual North American Unionidæ as far back as the Jurassic period, and what we know of fossil mussels of New Zealand and Archiplata are only *Unios* of the *Niæa* section. On the other hand, Cretaceous deposits of Bahia show us representatives of *Glabaris* and *Mycetopus*. The actual conditions of distribution therefore predominated even in the Mesozoic time, and no explanation can be given of the intimate relation between the fresh-water faunas of tropical Africa and South America than the hypothesis of an ancient land bridge ; supposing that these faunas were only the remains of an ancient cosmopolitan tropical fauna, the paleontological evidence should be totally different.

In regard to the geological distribution of the mammals of South America, the opinions of the respective authors are very divergent. There is, however, one point of which there can be no doubt, *i. e.*, the Pliocene exchange of North and South American types. It must be decided by North American zoologists whether this interchange has commenced at the close of the Miocene or only in the Pliocene. We may therefore consider as Pliocene the Argentine Araucanian formation, where the northern Artiodactyla and other North American immigrants first appear ; the Entrerian formation, containing the neotropical forms must then be Miocene. This formation was considered by Ameghino Eocene (1889), or Oligocene (1898), and by the writer (1898), Miocene. In favor of his opinion Ameghino quotes the result of the study of G. Alessandri on the fossil selachian teeth of Entrerios which he believes to be Eocene. Mr. A. Smith-Woodward, to whom I have sent the material of our museum, writes me: "I conclude that the formation cannot be earlier than Miocene and is probably Pliocene." I have called attention to the fact that in the Entrerios deposits occurs *Monophora darwini*, a Scutellid with perforated disk which is common in the corresponding formation of the north Patagonian coast. No Scutellidæ with perforations of the disk are known earlier than the Miocene. On the other hand, the Mollusca of this formation are almost all extinct species and therefore I cannot believe it Pliocene.

Zittel in his 'Manual' has well explained the relations between the two American mammal faunas. I am, however, disposed to believe, contrary to him and to Ameghino, that the genus *Didelphys* in South America appears as a member of the North American immigration. If derived from the Patagonian Microbiotheriidæ, as suggested by Ameghino, this genus may have issued in the earlier Eocene time from Patagonia and Archinotis and, after having reached Europe in the Eocene and North America in the Miocene, turned to South America in the Pliocene. If Ameghino is right, the Proboscidia are derived from the Eocene Patagonian Pyrotheriidæ and, after having appeared in Europe and North America, returned to Argentina during the Pliocene in the form of the Mastodon.

If this migration is a relatively well established fact, it is quite doubtful in what manner Patagonia received its rich mammalian fauna in the Laramie period. Florentino Ameghino pointed out that this must have occurred by means of a landbridge which united both Americas at the beginning of the Tertiary period. On this matter there has been a discussion between Ameghino and the writer in the *Revista Argentina de Historia Natural*, Vol. I., Buenos Ayres, 1891, p. 122 ff. and p. 281 ff., in which I have combated this hypothesis. The fresh-water faunas of the two Americas, as I have shown, are so completely different that only a prolonged and absolute separation can explain the fact; the geological history of both North and South America demonstrate an enormous development of the Cretaceous Ocean, separating the two Americas, and in the Tertiary period the North American territory increased but slowly. This presumed primitive connection of the two Americas is not at all supported by facts, but only based on the predominance of wrong ideas of the history of the Australasian Territory. The Eocene mammals of Patagonia and North America certainly do not justify this hypothesis. The Eocene faunas of Reims and

DECEMBER 7, 1900.]	*SCIENCE.*	861

Puerco, although these localities are much more distant from each other than North and South America, correspond closely, but the characters of the earliest Tertiary Patagonian and North American mammalian faunas are quite divergent. We find nothing of the Toxodontia, Typotheriidæ and true Edentata in North America, and nothing of the Artiodactyla, Perissodactyla, Amblypoda, etc., in Patagonia. The orders and families which *are* represented in both Patagonia and North America may be such as were distributed over the whole area occupied by placental mammals in the Laramie period.

The third line of migration according to Ameghino and Osborn was determined by the land masses which connected Brazil and Africa. In my papers, and especially in the discussion with Ameghino, I have insisted upon the value of this Eocene land bridge, but I do not believe that it has served for the distribution of mammals, as I believe that Archamazonia was in the greater part of the Tertiary separated by the ocean from Archiplata. In this case Brazil has received mammals only in the Pliocene time, when the communication with Africa had long ago been interrupted. I have examined the deductions of Ameghino and Osborn with the purpose of verifying the facts proving their opinions, but these seem to be very insufficient. Osborn refers to the Pangolins and Aard Varks of the Æthiopian region as introduced from South America 'via Antarctica.' It must, however, be noted that these Edentata of the old world occur also in Asia and that they belong to the Nomarthra, while all the Patagonian representatives are Xenarthra. Both may be derived from a common Australasian ancestor, but if the South African Edentata had been derived from the Patagonian Eocene fauna, they should be Xenarthra. The genus *Orycteropus* occurs also in the Miocene of Samos, and

may have immigrated both to Samos and to Africa from its Indo-australian home. It may be observed here that I have shown that the claw of the Dasypodidæ develops in the form of a hoof, and it is wrong to classify the Xenarthra with the Unguiculata, as they are Ungulata. The Proboscidea and Hyracoidea are not Patagonian mammals at all, although in the Patagonian Laramie or Pyrotherium fauna the Pyrotheriidæ and Archæohyracidæ offer relations to the two above-mentioned living families. The case is the same with the sole Patagonian Insectivore, the genus *Necrolestes*, somewhat comparable with the Chrysochloridæ of South Africa. Evidently the few representatives in the Patagonian Eocene of the Insectivora, Prosimiæ and Hyracoidea are the isolated members of groups which were well represented in other regions then in connection with Patagonia. Thus the Chrysochloris argument for the Patagonian-South-African migration is not better than the hypothesis of a land-bridge uniting the Antilles with Madagascar, the sole localities where representatives are found to-day of the genus *Centetes*, which occurs also as Wallace affirms in the European Tertiary.

The intimate relations between the fresh water faunas of Africa and Brazil, and the colossal difference which exists between the fresh water faunas of Archamazonia and Archiplata, prove that both territories during the greater part of the Tertiary were separated quite as completely as the two Americas. In this case the mammalian fauna of Patagonia may have reached Ecuador or Colombia by means of the upheaval of the Andes, but not Brazil, and both Brazil and the Æthiopian region may have been without mammals and especially placental mammals, during the Eocene. When toward the close of the Eocene this land-bridge was submerged, there already existed many types that have been conserved until our time, and thus we find existing on the Central American and Brazilian coasts the same species of mangrove plants, and with them numerous identical forms of Crustacea, Mollusca, etc.; the distribution of Manatus must also be cited here.

We now turn to the relations of South America with Australia and New Zealand. As the views put forth on this point by Hutton and the writer seem to be now generally accepted, there is no reason for discussing the question here. It may be observed, however, that not only does the fresh-water fauna give evidence of an antarctic land bridge between Australia, New Zealand and Patagonia, but also numerous other zoological as well as botanical and paleontological facts. Osborn says only that this migration established the links with Australia, 'bringing in Marsupials, both polyprotodont and diprotodont.' Ameghino (Censo, p. 250) says that on this vast antarctic land was distributed the cretaceous mammalian fauna which he has described. No other conclusion is logically possible, and we cannot doubt that the Eocene fauna of the Australian region, though not at all known to-day, must have been very analogous to and in part identical with the Patagonian.

The different adaptive radiations of orders and families have given a very different aspect to the existing faunas of Australia and Patagonia, in Australia only Monotremates and Marsupials having survived, in Patagonia principally histricomorph Rodents and Edentata. The existing fauna of Australia, New Guinea and other allied islands has received by Miocene immigration some placental mammalia, as *Canis* and *Uromys* in Australia, *Sus* and *Uromys* in New Guinea, and other genera in the Moluccas. This proves that Australia and New Guinea, at least during the Miocene, continued to be connected with Asia as in the foregoing periods. There existed therefore in the earlier Tertiary a continuous land mass from Antarctica and Patagonia, via Australia and Asia, to Europe and North America. This enormous territory, my Eurygæa, was the birthplace of the placental mammals. The Stenogæa (or Archhelenis) extending from tropical South America to Africa, Madagascar and Bengal was in the Eocene without mammals.

It is certain that we have to-day no knowledge at all of the Eocene mammals of Australia, Brazil and Africa, but from the facts given it seems to be highly probable that future discoveries may confirm what we expect.

Paleophytical studies have given evidence of a great resemblance between the Cretaceous floras of North America and Eurasia. According to Fr. Kurtz, the same flora appears also at St. Cruz, Patagonia em Cerro Guido (*Revista Museu La Plata*, Vol. X., 1899, p. 43 ff.). According to the facts given above, this flora cannot have reached Patagonia from North America, as the two Americas were then separated and no South American continent existed. It is also impossible to admit that a land bridge formed by the Andes served for the migration, because these did not then exist, as the Cretaceous marine beds of the Andes prove. There must then have been a connection between the Antarctic Cretaceous continent, the Archinotis of the writer, and Asia. It may be observed that the genus *Quercus* was represented in the Cretaceous beds of both Patagonia and Australia, where to-day it has no representative. What has occurred in the case of *Quercus* and other genera in both Australia and Patagonia and what is observed in Patagonia with reference to mammals may have happened also in Australia to the earlier placental mammals. Further, it must be remembered that Australia, and South America also, developed by coalescence of different parts, each of which had its own history.

December 7, 1900.]   *SCIENCE.*   863

I may note here one more fact referring to the fresh water fauna: the dispersion of the cyprinoid fishes. These Holarctic fishes did not reach Australia, already isolated by the sea, but invaded Africa and Madagascar. Lemuria must therefore have persisted in connection with Asia, when the Australian region was already isolated. Thus Africa offers the same mixture of ancient indigenous elements and Neogene immigrants as Argentina and Southern Brazil, on account of the intrusion of archamazonic immigrants. Had this invasion occurred in the Eocene period, the Cyprinidæ would have reached Brazil ; supposing it to be Pliocene, these fishes would not have reached Madagascar. Probably Africa received its placental mammals at the same time that the invasion of Cyprinidæ into Africa took place, one of the most remarkable events in zoogeography.

We have no knowledge at all of the Cretaceous and Eocene mammals of Brazil, Guyana, Africa and Australia ; it is impossible to give a complete history of the mammals with incomplete materials. But combination of the known facts makes it probable that during the Cretaceous and Eocene period Archhelenis, or Stenogæa, was without placental mammals and that their origin was in Eurygæa. In regard to the flora the same holds good for many families of wide distribution, as for example the Cupuliferæ.

With reference to the terms used by Blandford, Lydekker and the writer, it must be said that the intention of the first two was to give names to *existing* zoogeographical regions, while the terms introduced by me refer to *supposed, ancient* zoogeographical and geographical regions. The two great Cretaceous continents Eurygæa and Stenogæa may have existed also during a part of the Eocene period and then dismembered. From Stenogæa, or Archhelenis, were separated first Bengal and then Madagascar, while Archamazonia after the loss of the connection with Africa consisted of Archiguyana and Archibrazil. Eurygæa split into (1) Archiboreas corresponding to the actual holarctic region and (2) Archinotis from which, in the Eocene, Archiplata was separated.

The comparison of the distribution of the mammals with the fresh water fauna makes especially evident the differences in the geographical conditions which must have determined their distribution. While the distribution of the existing types of mammals is a result of changes in geography during Tertiary time, the most fundamental facts in the distribution of the fresh-water fauna dates from the Mesozoic epoch. The fresh-water fauna of Chili preserved such a remnant of the Cretaceous fauna almost intact, and even the connection between the two Americas has not at all modified the South American fresh-water fauna. On the other hand, representatives of the Archamazonian fauna, in correlation with the geographical modifications of Central America and the Antilles have invaded the southern parts of the nearctic region. Thus in the Rio Usumacinta of Mexicao beside Cyprinidæ and Chromidæ we find also Characinidæ and Lepidosteus, also species of *Glabaris* intermediate between the northern Unios and Anodontas. There is a further difference in the distribution of mammals and fresh-water mussels. The former migrate on land bridges in both directions, the fresh-water fauna generally in only one, due to the opportunity given by the currents. Thus although there was an invasion of Cyprinid fishes into Africa there was no corresponding emigration of Æthiopian types. A similar fact is the sudden appearance of the Æthiopian faunal elements in the valley of the Nile, which occurred only at the close of the Pleistocene, as proved by paleontological facts. While the Pliocene connection of the two Americas was sufficient to mod-

SCIENCE. [N. S. Vol. XII. No. 310.

ify the distribution of the mammals in such a way that without paleontological researches it would be impossible to recognize the origin of the different faunal elements, the fresh-water faunas have resisted almost unchanged all modifications in the configuration of the continent.

The fresh-water fauna is not only older but also much more conservative than the distribution of the mammals. One of the most striking examples of this is given by the history of Africa. While the characteristic mammals are Neogene immigrants and Lydekker proceeds quite correctly in making Africa an annex only of the Holarctic region, thus establishing his Arctogæa, with relation to the fresh-water fauna, Africa is a part of South America, somewhat modified by the Neogene invasion of Cyprinid fishes. If as regards mammals Africa belongs to Arctogæa, with relation to the fresh-water fauna it belongs to the Archhelenic region.

This example demonstrates *the absurdity of the present system of construction of zoogeographical regions and maps. We can construct maps of the different classes and orders but not at all of the animal kingdom, because the geological history of the different groups is quite different.* When Osborn says that it is one problem 'to connect living distribution with distribution of past time,' he says only what had been the leading idea of Wallace and of Engler in their eminent works on zoo- and phytogeography, but when he continues 'and to propose a system which will be in harmony with both sets of facts,' he proposes a problem just as contradictory as would be the construction of descriptions and figures referring at the same time to egg, larva, nympha and imago of an insect. The works on 'zoogeography' are almost exclusively discussions of the distribution of mammals and birds, and the few words spent on other classes are only ornamental supplements. A wrong method cannot give valid results. For the exploration of the zoogeographical relations and regions of the beginning of the Tertiary and of the preceding Mesozoic epoch it is necessary to study and to discuss the more ancient classes and, as I have insisted for ten years, principally the fresh-water fauna.

H. von Ihering.

São Paulo, July 20, 1900.

A HISTORY OF THE DEVELOPMENT OF THE QUANTITATIVE STUDY OF VARIATION.[*]

The quantitative study of variation has for its object the investigation of evolution by exact, quantitative methods. The study demands a mathematical method as well as a biological subject matter; consequently the development of the science has proceeded along two main lines—the one biological and the other mathematical. Accordingly, the history of the development of the quantitative method involves a consideration of both the study of variation and the elaboration of the necessary method.

The fact of variation has been recognized since man began to think and to appreciate that in stature, color and mental capacity his fellow-men are diverse. The way for quantitative studies in biology was paved by the mathematical studies on the variation of measurements which engineers and astronomers found it necessary to make for their own purposes. These mathematical studies led to the discovery and elaboration of the law of error by Gauss and others—and this law is the corner-stone of the quantitative biological studies.

The application of the law of error to organic variation was, apparently, first made by an anthropological statistician, of the early part of the century, named Quetelet. In his book, entitled 'Lettres à Son Altesse Royale le Duc de Saxe-Coburg et

[*] Being part of the report of the Committee of the American Association for the Advancement of Science on the Quantitative Study of Variation.

PAPER 15

U. S. DEPARTMENT OF AGRICULTURE
DIVISION OF ORNITHOLOGY AND MAMMALOGY

NORTH AMERICAN FAUNA

No. 3

PUBLISHED BY AUTHORITY OF THE SECRETARY OF AGRICULTURE

[Actual date of publication, September 11, 1890]

Results of a Biological Survey of the San Francisco Mountain Region and Desert of the Little Colorado, Arizona

1. General Results, with special reference to the geographical and vertical distribution of species
2. Grand Cañon of the Colorado
3. Annotated List of Mammals, with descriptions of new species
4. Annotated List of Birds

> BY DR. C. HART MERRIAM

5. Annotated List of Reptiles and Batrachians, with descriptions of new species

> BY DR. LEONHARD STEJNEGER

WASHINGTON
GOVERNMENT PRINTING OFFICE
1890

FRONTISPIECE.

North American Fauna, No. 3.

SAN FRANCISCO PEAK.

AGASSIZ PEAK.

SAN FRANCISCO MOUNTAIN—FROM THE SOUTHWEST.

No. 3.   NORTH AMERICAN FAUNA.   August, 1890.

RESULTS OF A BIOLOGICAL SURVEY OF THE SAN FRANCISCO MOUNTAIN REGION AND DESERT OF THE LITTLE COLORADO IN ARIZONA.

By Dr. C. Hart Merriam.

PREFATORY NOTE.

Recent explorations in the west, conducted by the Division of Ornithology and Mammalogy of this Department, led to the belief that many facts of scientific interest and economic importance would be brought to light by a biological survey of a region comprehending a diversity of physical and climatic conditions, particularly if a high mountain were selected, where, as is well known, different climates and zones of animal and vegetable life succeed one another from base to summit. The matter was laid before the Assistant Secretary of Agriculture, the Hon. Edwin Willits, and I was authorized by the Secretary, the Hon. J. M. Rusk, to undertake such a survey of the San Francisco Mountain region in Arizona. San Francisco Mountain was chosen because of its southern position, isolation, great altitude, and proximity to an arid desert. The area carefully surveyed comprises about 13,000 square kilometers (5,000 square miles), and enough additional territory was roughly examined to make in all about 30,000 square kilometers (nearly 12,000 square miles), of which a biological map has been prepared. No less than twenty new species and subspecies of mammals were discovered, together with many new reptiles and plants; and the study of the fauna and flora as a whole led to unexpected generalizations concerning the relationships of the life areas of North America, necessitating a radical change in the primary and secondary divisions recognized.

The most important of the general results are :

(1) The discovery that there are but two primary life areas in North America, a northern (boreal) and a southern (subtropical), both extending completely across the continent and sending off long interpenetrating arms.

(2) The consequent abandonment of the three life areas commonly accepted by naturalists, namely: The Eastern, Central, and Western Provinces.

(3) The recognition of seven minor life zones in the San Francisco Mountain region, four of boreal origin, and three of subtropical or mixed origin.

(4) The correlation of the four boreal zones with corresponding zones in the north and east.

The present paper consists of five parts: (1) an announcement of the general results of the survey, with special reference to the geographic and vertical distribution of species; (2) results of a brief visit to the Grand Cañon of the Colorado; (3) an annotated list of the Mammals of the San Francisco Mountain region including the desert of the Little Colorado, with descriptions of new species; (4) an annotated list of the Birds; (5) an annotated list of the Reptiles and Batrachians, with descriptions of new species.

Prof. F. H. Knowlton, assistant paleontologist, U. S. Geological Survey, joined the party in the summer and collected the plants upon which many of my generalizations are based. He has placed me under great obligations by allowing me the unreserved use of this material and the privilege of announcing important results from the stand-point of the geographic distribution of species. I am indebted also to Mr. Frederick V. Coville, assistant botanist, U. S. Department of Agriculture, for the determination of many of the more difficult plants.

Dr. Leonhard Stejneger, curator of reptiles in the U. S. National Museum, joined the expedition in September. Though unable to visit the desert region, he made notes and colored sketches from the living animals collected by Mr. Bailey and myself, and has prepared the report on Reptiles and Batrachians which constitutes part four of the present bulletin.

My assistant, Mr. Vernon Bailey, deserves special recognition for the faithful and efficient performance of the duties assigned him, and it should be added that much of the success of the season's work is due to his zeal and intelligence.

It is proper also to acknowledge the assistance rendered by Mr. D. M. Riordan, and his brothers Thomas and M. J. Riordan, of Flagstaff, Arizona.

Much more would have been accomplished but for the insufficient fund available for the survey (only a little more than $600 to cover the total cost of transportation, outfitting, hire of animals and men, purchase of tents, supplies, etc.), thus permitting the employment of but one man as cook and general camp-hand; while the animals, both in number and quality, were far below the standard usually considered necessary for field work, which circumstance caused many annoying delays. All our traveling was done on horseback, and our packing on burros.

The altitudes given in the present paper were determined by means of aneroid barometers, and too much confidence must not be placed in their extreme accuracy.

The base maps made use of are those of the U. S. Geological Survey, for which I am indebted to the Director of the Survey, Maj. J. W. Powell, and to the chief geographer, Mr. Henry Gannett. The picture of San Francisco Mountain, which forms the frontispiece of this report, is from the sixth annual report of the U. S. Geological Survey.

The colored map of Arizona, showing the life areas of the Colorado plateau south of the Grand Cañon (map 1), is based upon the present survey, supplemented by information derived from the U. S. Geological Survey.

For the sake of convenience, the names employed to designate the various life areas are those in common use; and the author wishes to state that he does not commit himself to these names, or to the relative value of the terms indicating rank (Province, Region, Zone, etc.), all of which have been employed in diametrically opposite ways by different writers.

ITINERARY.

The following brief itinerary, in connection with the accompanying maps, will enable the reader to trace the routes of the expedition and determine the positions of the localities mentioned in the report.

Flagstaff, Arizona, a station on the Atlantic and Pacific Railway, is the point of departure for San Francisco Mountain. I reached Flagstaff July 26, 1889, and was joined next day by my assistant, Mr. Vernon Bailey. After spending three days in outfitting, we proceeded to Little Spring, at the north base of San Francisco Mountain, and pitched our tents in a grove of aspens and pines, on a knoll just northwest of the spring, at an altitude of 2,500 meters (8,250 feet). This was our base camp for two months, and from it numerous side-trips were made into the surrounding country. Three of these were of special importance, namely, two trips across the Painted Desert and one to the Grand Cañon of the Colorado. During these expeditions I crossed the Painted Desert and the Rio Colorado Chiquito four times, spending in all sixteen days on the desert. I visited also Walnut Cañon, about 9 kilometers (5½ miles) south of Elden Mountain; and local collections were made in the piñon and chaparral near a volcanic crater containing ruins of cliff dwellings 8 kilometers (5 miles) east of O'Leary Peak, and in various other directions. A branch camp was established just below timber line on the main peak of San Francisco Mountain, and the rocky summit above timber line was climbed several times. Kendrick and O'Leary Peaks also were ascended.

FIRST TRIP TO PAINTED DESERT, AUGUST 12 TO 19, INCLUSIVE.

The route followed skirted the north and east sides of San Francisco Mountain, passing through the pine forest by way of Partridge Spring, and along the edge of O'Leary Park, keeping west of O'Leary and

Sunset Peaks, and thence turning southeasterly to Turkey Tanks. The dry bed of the Little Colorado River was crossed at Grand Falls, and Tenebito Wash was followed to the high mesa on the east side of the desert; this mesa was ascended and a trail was taken northward to a point about 25 kilometers (16 miles) north, or a little west of north, of the Moki pueblo of Oraibi; an abrupt turn to the south was then made, and Grand Falls was reached by an Indian trail south of that taken on the outward journey, a short stop having been made at Oraibi, where water and goat's milk were obtained from the Indians. From Grand Falls the course lay across the lava beds direct to the north base of the mountain, instead of by way of Turkey Tanks, as on the outward journey. The total distance traveled was 370 kilometers (230 miles). The heat was intense and much suffering was occasioned by want of water.

SECOND TRIP ACROSS THE DESERT, SEPTEMBER 20 TO 27, INCLUSIVE.

A northeasterly course was taken from Little Spring to Black Tank, thence to the Little Colorado at Tanner's Crossing, following the Mormon trail and crossing the river about 56 kilometers (35 miles) north of Grand Falls, and continuing in a northeasterly direction to Moencopie Wash, which was followed to Echo Cliffs, and the southern point of Echo Cliffs mesa was crossed from Moa Ave to Tuba. Tanner's Gulch and the Pueblo of Moencopie were visited and Moencopie Wash was followed down to the point of departure for Echo Cliffs, whence the return to the mountain was made by nearly the same route as on the way out, the total distance traveled being about 280 kilometers (175 miles). The temperature was very much lower than during the former trip across the desert, and some of the nights were even cold. The recent heavy showers had left some water in the Little Colorado and in scattered alkaline pools in Moencopie Wash, and also in the gulches in the lava beds between San Francisco Mountain and the Little Colorado.

TRIP TO THE GRAND CAÑON OF THE COLORADO, SEPTEMBER 9 TO 16, INCLUSIVE.

The usual road was followed from Little Spring to Hull Spring and Red Horse Tank, and thence to the tank known as Cañon Spring on the Cocanini Plateau, close to the cañon, which is here about 1,800 meters (6,000 feet) in depth. Mr. Bailey and myself climbed down into the cañon and remained in it two days and two nights.

PART I.—GENERAL RESULTS OF A BIOLOGICAL SURVEY OF THE SAN FRANCISCO MOUNTAIN REGION IN ARIZONA, WITH SPECIAL REFERENCE TO THE DISTRIBUTION OF SPECIES.

By Dr. C. HART MERRIAM.

GENERAL PHYSICAL FEATURES OF ARIZONA.

Arizona as a whole may be readily divided into two very distinct physiographic areas—an elevated plateau area and a low desert area. A high cliff or escarpment, one of the best marked and most extensive in the North American continent, enters Arizona from Utah and completely crosses the Territory from northwest to southeast, marking the southern limit of the great Colorado Plateau. Though it does not everywhere maintain the form of a precipitous cliff, it has an average height of at least 1,200 meters (4,000 feet), and in some places its crest is more than 1,500 meters (5,000 feet) above the plain below. In its effects upon the life of the region it is an important faunal barrier. The region to the south is in the main an arid desert, interrupted by a few irregular ranges of mountains. The region to the north, beginning at the top of the cliff and occupying the northern part of Arizona, is a southward continuation of the Great Interior or Colorado Plateau, the plateau on which the Rocky Mountains rest.

GENERAL FEATURES OF THE SAN FRANCISCO MOUNTAIN REGION.

San Francisco Mountain is on this plateau, in the north-central part of the Territory (in latitude 35° 20′ N.; longitude 111° 41′ W.). It is a volcanic peak rising 3,900 meters (12,794 feet) above sea-level and rests on a lava base which is everywhere more than 2,130 meters (7,000 feet) in elevation, and overlies red sandstone and carboniferous limestone. This plateau comprises about 2,000 square kilometers (800 square miles), and measures about 72 kilometers (45 miles) from east to west by 53 kilometers (33 miles) from north to south.

Four other volcanic peaks (O'Leary, Kendrick, Sitgreaves, and Bill Williams), ranging in height from 2,750 to 3,200 meters (9,000 to 10,500 feet), together with many buttes, cones, and craters, some of which contain ' crater lakes,' occupy the same elevated base level. San Francisco Mountain proper, cut off from all surrounding and attached hills and

5

6   NORTH AMERICAN FAUNA.   [No. 3.

buttes at the height of 2,450 meters (8,000 feet), is about 19 kilometers (12 miles) in north and south diameter by 15 kilometers (9 miles) in east and west diameter, and covers about 180 square kilometers (70 square miles).

The lava plateau above 2,130 meters (7,000 feet) altitude is covered throughout by a beautiful forest of stately pines (*Pinus ponderosa*), which average at least 33 meters (100 feet) in height. There is no undergrowth to obstruct the view, and after the rainy season the grass beneath the trees is knee-deep in places, but the growth is sparse on account of the rocky nature of the surface. The pine forest extends up the mountain as high as 2,675 meters (8,800 feet), but loses its distinctive character at about 2,500 meters (8,200 feet), where it is replaced in the main by a forest of Douglas fir (*Pseudotsuga douglasii*), the same as that found from California to Puget Sound and British Columbia. The Douglas fir reaches an altitude of about 2,800 meters (9,200 feet), here giving place to Engelmann's spruce (*Picea engelmanni*), which covers the mountain sides between the altitude named and timber line (about 3,500 meters (11,500 feet). The fox-tail pine (*Pinus aristata*) begins a little lower down than Engelmann's spruce and accompanies it to the upper limits of tree growth, where both exist as depauperate forms scarcely more than a foot in height. The summit of the mountain above timber line consists of bare volcanic rock and is covered with snow about nine months of the year.

Again passing down to the plateau, and thence in an easterly direction to lower levels, a zone of cedar and piñon is first encountered—a belt varying in width from one to several miles according to the steepness of the slope. The only trees in this belt are junipers (locally known as 'cedars') and the piñon or nut pine (*Pinus edulis*), whose nut furnishes food to the Indians and the mammals and birds of the region. Descending still lower, the Desert of the Little Colorado is entered—an arid, treeless area whose upper limit may be set at the 1,800 meter (approximately 6,000 foot) contour or level. Parts of this desert are devoid of vegetation, while other parts support a scanty growth of cactus, greasewood, and a few other species.

In the foregoing account the general features of the several zones of the San Francisco Mountain region have been briefly outlined. Recapitulating, it may be said that in ascending from the hot and arid Desert of the Little Colorado to the cold and humid summit of the mountain no less than seven zones are encountered, each of which may be characterized by the possession of forms of life not found in the others. These zones, with their respective altitudes, are—first, the arid Desert region, below 1,800 meters (6,000 feet); second, the Piñon belt, from 1,800 to 2,100 meters (6,000 to 7,000 feet); third, the Pine, from 2,100 to 2,500 meters (7,000 to 8,200 feet); fourth, Douglas fir, from 2,500 to 2,800 meters (8,200 to 9,200 feet); fifth, Engelmann's spruce, from 2,800 to 3,500 meters (9,200 to 11,500); sixth a narrow zone of dwarf spruce; and seventh, the bare rocky summit, snow covered the greater part of the year.* These facts as isolated facts would be of comparatively little interest, but in their bearing on the problems of geographic distribution a very deep interest attaches to them. This will appear by passing in review the distinctive plants and animals of the several zones, and tracing their distribution in other parts of their ranges.

REMARKS ON THE GEOGRAPHIC DISTRIBUTION OF SPECIES CHARACTERISTIC OF THE SEVERAL ZONES OF THE SAN FRANCISCO MOUNTAIN REGION IN ARIZONA.

ALPINE ZONE.

[Approximate altitude : Above 3,500 meters, or 11,500 feet.]

Nine species of plants which grow on the bleak and storm-beaten summit of San Francisco Mountain were brought back from Lady Franklin Bay by Lieut. (now General) A. W. Greely. These species are :

Androsace septentrionalis	*Cystopteris fragilis*	*Saxifraga nivalis*
Arenaria verna	*Saxifraga cæspitosa*	*Oxyria digyna* ·
Cerastium alpinum	*Saxifraga flagellaris*	*Trisetum subspicatum*

One or more of them have been found at each of the following localities : British Columbia, Unalaska, Bering Strait, Kotzebue Sound, Point Barrow, Melville Island, Back's Great Fish River, Hudson Bay and Strait, Labrador, Baffin Bay, Greenland, Iceland, Spitzbergen, Newfoundland, Gulf of St. Lawrence, White Mountains of New Hampshire, Rocky Mountains, Selkirks, and Sierra Nevada. Several of them occur also in the arctic portions of the Old World, extending as far south along the coast as the island of Yeso, North Japan, and appearing again in the high mountains of Roumelia, in the Caucasus, the Carpathian Mountains, and the Alps.

Sibbaldia procumbens is another polar species inhabiting arctic America from the peninsula of Unalaska to Hudson Bay, Labrador, and Greenland, and flourishing also throughout the arctic regions of Asia. It comes south along the higher summits of the Cascade range, the Sierra Nevada, and the Rocky Mountains, and occurs in isolated colonies on the barren peaks of San Francisco Mountain in Arizona and Mount Washington in New Hampshire. In the same way it inhabits the mountains of Central Asia and Siberia, and also the Carpathian Mountains, the Apennines, the Alps, the Pyrenees, and the Himalaya.

Geum rossii belongs to the same category, growing from Greenland

---

* The normal altitudes here given for the various tree zones of San Francisco Mountain are averages for the northwest side of the mountain. Favorable southern and southwestern exposures carry the zones up a hundred meters or more above these limits, while similar northern and northeastern exposures, particularly in gulches and cañons, deflect the zones as much as two, or even three hundred meters. The normal average difference in altitude of the same zone on the southwest and northeast sides of San Francisco Mountain is about 275 meters (900 feet).

and the shores and islands of Hudson Strait to Melville Island and the coasts of Bering Strait and Unalaska, and also in the northern part of Siberia and Kamschatka. It comes southward in the Rocky Mountains, inhabiting the higher peaks of the Uintas and of Colorado, and is the most conspicuous plant above timber line on San Francisco Mountain, where it forms dense mats of green among the bare rocks—patches of such extent that they may be seen from the plateau level below.

Other arctic plants found above timber line on San Francisco Mountain, most of them circumpolar species, are:

Arenaria alpina	*Polemonium confertum*	*Silene acaulis*
Cerastium arvense	*Sagina linnæi*	*Stellaria umbellata*
Festuca brevifolia	*Saxifraga debilis*	*Thlaspi alpestre*

It appears from what has been said that many of the plants found on the high rocky summit of San Francisco Mountain occur on the higher peaks of the Rocky Mountains, the Sierra Nevada* and Cascade range, and the Appalachian chain; they occur along the arctic coasts of Alaska, Hudson Strait, North Labrador, Greenland, North Siberia, and Spitzbergen; they occur in the Alps of Europe, in the Altai and Ural Mountains, the Pyrenees, and some of them even in the Himalaya. In brief, they inhabit the arctic regions of the globe and extend far south on the summits of the higher mountain ranges. Plants and animals having such a distribution are termed *Arctic-Alpine Circumpolar species.*

We collected no insects at high altitudes on San Francisco Mountain, but butterflies and diptera from great elevations in Colorado have been shown to be identical with species from Mount Washington, Labrador, and Greenland.

Among birds, the Golden Eagle—a truly circumpolar species, though not confined to the arctic zone—rears its young on San Francisco Mountain.

There are no exclusively arctic mammals on the top of this high mountain, because such mammals could not exist long in so small an area. An Ermine Weasel (*Putorius* sp. ——?) inhabits the summit, and the Big-horn or Mountain Sheep, another truly circumpolar type, spends the summer there, descending in winter to lower levels.

SUB-ALPINE OR TIMBER-LINE ZONE.

[Approximate altitude, 3,200–3,500 meters, or 10,500–11,500 feet.]

Just below the barren arctic summit of the mountain is a narrow belt which may be named the Timber-line zone. Here the trees which reach timber line (in this case *Picea engelmanni* and *Pinus aristata*) lose the upright or arborescent habit and exist as stunted and prostrate trunks, whose gnarled and weather-beaten forms bear testimony to the severity of their struggle with the elements. In this narrow belt a number of

---

* Engler tells us that 26 per cent. of the plants found on the High Sierra Nevada are found also in the Alps and throughout arctic Europe.

hardy little plants attain their maximum development, decreasing rapidly in abundance both above and below.   Among these are :

Arenaria biflora carnulosa	*Gentiana barbellata*	*Potentilla dissecta*
Cerastium alpinum behring-	*Gentiana tenella*	*Primula parryi*
ianum	*Heuchera rubescens*	*Saxifraga debilis*
Corallorhiza multiflora	*Lazula spadicea parriflora*	*Sedum rhodanthum*
Draba aurea	*Pedicularis parryi*	*Veronica alpina*
Epilobium saximontanum	*Phleum alpinum*	

Many of them are circumpolar species found throughout the northern regions of America, and some of them throughout the northern regions of the world, coming south on high mountains and occurring in greatest perfection just at or near the edge of the northern limit of trees, and at timber-line on mountains further south. Such plants are known to botanists as '*Sub-Alpine species*,' and it would be well if the term *sub-alpine* were restricted to the characteristic species of this zone.

Among birds, the Titlark (*Anthus pensilvanicus*) was found at the top of the mountain, where it probably breeds. It breeds in grassy places on the high peaks of the Rocky Mountains, and at sea-level in Labrador, Greenland, and throughout arctic America; and birds congeneric with it are known to breed throughout the arctic portions of the Old World.

(CENTRAL) HUDSONIAN OR SPRUCE ZONE.

[Approximate altitude, 2,800–3,200 meters ; or 9,200–10,500 feet.]

Passing down into the next zone, the Spruce zone, a number of plants, birds, and mammals are encountered, which are characteristic of humid northern regions, but regions not quite so cold as those inhabited by the species which occur on the snowy summit and at timber-line. The characteristic trees of this zone are Engelmann's spruce (*Picea engelmanni*) and the fox-tail pine (*Pinus aristata*). Some of the small plants are :

Aquilegia chrysantha	*Pentstemon glaucus steno-*	*Solidago multiradiata*
Lathyrus arizonicus	*sepalus*	*Zygadenus elegans*
Mertensia paniculata	*Pyrola chlorantha*	
Moneses uniflora	*Ribes setosum*	

The fact of present interest is that many of the plants here enumerated as growing in the Spruce zone of this mountain are equally characteristic of the upper spruce belt of the higher Alleghanies, the Rocky Mountains, the Cascades, and the Sierra Nevada, and occur also in the great northern spruce forest of Canada. It is well known that the northernmost part of our own continent consists of bare rock and frozen tundras. There are no trees along the sea edge of Labrador or Hudson Strait, or along the coast region of arctic America from Boothia Felix to Alaska, but just south of this region a large forest begins which has been called the 'Great Pine Forest.' There is not a pine tree in it, but it is called pine because conifers in general are called pines by people who are not botanists. The tree that grows there is a species of spruce congeneric with the spruce which occurs high up on San Francisco Mountain, and many of the humbler plants are either identical or closely related representative forms.

Among the birds which breed in the Spruce belt on this mountain are the Goshawk, Dusky Horned Owl, Dusky Grouse, Evening Grosbeak, and Clark's Crow. The Goshawk and Dusky Horned Owl range throughout the spruce forests of the north, from Labrador to Alaska, and south in the mountains; while the others are confined to its western parts and outliers.

Of mammals, the Porcupine is the only one believed to be restricted to this belt during the season of reproduction, and, like the Big-horn, it comes down to lower levels during the winter. Bears (*Ursus*), Shrews (*Sorex*), Voles (*Arvicola*), and Red Squirrels (*Sciurus fremonti mogollonensis*) range throughout the spruce and fir zones but were not found below.

(CENTRAL) CANADIAN OR BALSAM FIR ZONE.

[Approximate altitude: 2,500–2,800 meters; or 8,200–9,200 feet.]

The distinctive tree of this zone is Douglas fir (*Pseudotsuga douglasii*), which ranges northward to British Columbia. Another tree of nearly coincident vertical distribution on the mountain is the lofty Rocky Mountain Pine (*Pinus flexilis macrocarpa*), which extends north to the Kootenai region and Calgary in Canada. Wherever the Douglas fir has been burned off, its place is taken by the aspen (*Populus tremuloides*), a species of wide distribution in the north, where it ranges from New England to Newfoundland and Labrador, and thence westward to Alaska, reaching its highest perfection along the southern part of the great coniferous forest of northern Canada, and coming south in the mountains.

Among the smaller plants of the Douglas fir zone are:

Actæa spicata	*Gentiana affinis*	*Potentilla fruticosa*
Berberis repens	*Gentiana heterosepala*	*Ribes rusbyi*
Ceanothus fendleri	*Geum triflorum*	*Viola canadensis scopulorum*

Nearly half of the above (namely, *Geum triflorum, Potentilla fruticosa, Actæa spicata,* and *Viola canadensis*) have a wide range in the Canadian flora of the East and North, or are representative forms of such species; and probably *Ceanothus fendleri* may be safely regarded as the western representative of *C. ovatus,* which ranges eastward from the Rocky Mountains to Vermont.

One batrachian, a Salamander of the genus *Amblystoma,* has been found in this zone. Allied species inhabit the Canadian fauna of the East.

A number of species of birds are characteristic of the Douglas fir zone. At least eight of these are either identical with or closely related representative forms of species which are well-known members of the *Canadian fauna* of the East, most of them breeding in northern New

England, the Adirondacks, and southward in the Alleghanies. These
are:

Three-toed Woodpecker (*Picoides americanus dorsalis*)

Olive-sided Flycatcher (*Contopus borealis*)

Crossbill (*Loxia currirostra stricklandi*)

Pine Linnet (*Spinus pinus*)

Audubon's Warbler (*Dendroica auduboni*)

Brown Creeper (*Certhia familiaris montana*)

Ruby-crowned Kinglet (*Regulus calen-
dula*)

Audubon's Thrush (*Turdus aonalaschkœ
auduboni*)

The following species which breed in the Douglas fir belt on San
Francisco Mountain do not occur in the East, though but one genus
(*Myadestes*) is unrepresented in the East:

Townsend's Solitaire (*Myadestes townsendii*)

Broad-tailed Humming-bird (*Trochilus
platycercus*)

Long-crested Jay (*Cyanocitta stelleri ma-
crolopha*)

Louisiana Tanager (*Piranga ludoviciana*)

Mountain Chickadee (*Parus gambeli*)

It is probable that *Parus gambeli* and *Myadestes townsendii* range up
from the Fir into the Spruce zone.

Of mammals, there are two species of Field Mice or Voles (*Arvicola
alticolus* and *A. mogollonensis*), one Shrew (*Sorex monticolus*), and one
Red Squirrel (*Sciurus fremonti mogollonensis*), all of which extend up
into the Spruce belt, but none of which were found below. It is evident
that the Spruce and Balsam zones are closely related.

NEUTRAL OR PINE ZONE.

[Approximate altitude: 2,100–2,500 meters, or 7,000–8,200 feet.]

The characteristic and only tree of the Pine zone is *Pinus ponderosa*,
which forms an unbroken forest over the whole of the lava plateau above
the altitude of 2,100 meters (about 7,000 feet) and extends up as high,
in some of the parks, as 2,675 meters (8,800 feet). As a distinctive
species, however, it loses its character at about 2,500 meters (8,200 feet)
where it is invaded, and soon after replaced, by *Pinus flexilis, Pseudotsuga douglasii*, and *Populus tremuloides*. *Pinus ponderosa* may be regarded as a tree of the middle elevations, occurring between the piñon
and cedar of the lower hills, and the firs and spruces of the higher
mountains. In such situations it ranges from the highlands of western
Texas and northern Mexico, northward along the Rocky Mountains and
Sierra Nevada to the dry interior of British Columbia, in latitude 51°,
30′, avoiding the region of excessive rain-fall along the coast from northern California northward.

Among the more conspicuous of the small plants occurring in the Pine
belt of San Francisco Mountain, and having a more or less coincident
distribution with that of *Pinus ponderosa* just cited, are:

Campanula parryi

Frasera speciosa

Gilia aggregata attenuata

Oxybaphus angustifolius

Oxytropis lamberti

Pentstemon barbatus torreyi

The one distinctive mammal of the Pine belt is Abert's Squirrel (*Sciurus aberti*) which ranges through the pine regions of Arizona, New

12   NORTH AMERICAN FAUNA.   [NO. 3.

Mexico, and Colorado, and has been reported from Durango, in Mexico.
Very little can be said with certainty as to the characteristic birds of
the Pine belt, the date of my arrival at the mountain being so late (end
of July) that the birds had finished breeding and were beginning to
wander. The following species, however, were nearly confined to the
pines at that date and are known to breed there:

Red-backed Junco (*Junco cinereus dor-* Western Flycatcher (*Empidonax difficilis*)
  salis)   Richardson's Flycatcher (*Contopus rich-*
Nuttall's Poor-will (*Phalænoptilus nut-*   *ardsoni*)
  tali)   Pigmy Nuthatch (*Sitta pygmæa*)

The only reptile found in the Pine belt is a handsome horned toad
(*Phrynosoma hernandesi*), which is abundant.

PIÑON ZONE.

[Approximate altitude, 1,800–2,100 meters, or 6,000–7,000 feet.]

The distinctive trees of this zone are the piñon, or nut pine (*Pinus
edulis*), and the so-called 'cedar' (*Juniperus occidentalis monosperma*) both
averaging about 5 meters (16½ feet) in height. The singular checker-
bark juniper (*Juniperus pachyphlœa*), a very handsome and conspicuous
species, occurs in two or three special localities, but is rare. Several
large shrubs not observed elsewhere are abundant in parts of this
belt, namely, *Berberis fremonti*, *Rhus aromatica trilobata*, and *Spiræa
discolor dumosa*. Near the Grand Cañon of the Colorado and again at
Walnut Cañon, where the lava rock gives place to limestone, these
shrubs are joined by *Cowania mexicana*, *Spiræa millifolium*, and *Robinia
neo-mexicana*; and *Yucca angustifolia* is replaced by *Yucca baccata*. *Juniperus californica utahensis* also grows at the Grand Cañon. A dense
chaparral (*Fallugia paradoxa*) forms extensive thickets east of O'Leary
Peak and occurs sparingly over most of the Piñon belt, even extending down into the desert in places. Both the piñon and cedar occupy
elevations of corresponding temperature in the arid lands from western Texas through New Mexico and Arizona and north to central Colorado, and the cedar reaches westward to southern California. Closely
related and strictly representative forms extend northward through the
Great Basin to the Plains of the Columbia. The other species mentioned occupy more or less of the same range, and some of them push
northward over the Great Plains as well as the interior basin.

The most conspicuous bird of the Piñon belt is the Piñon Jay (*Cyanocephalus cyanocephalus*). Other characteristic species are Woodhouse's
Jay (*Aphelocoma woodhousei*), the Gray Tufted Tit (*Parus inornatus
griseus*), the Gnatcatcher (*Polioptila cærulea*), and the Bush Tit (*Psal-
triparus plumbeous*). The range of these species, taken collectively, is
co-extensive with the distribution of the cedar belt above described.

The large Rock Squirrel (*Spermophilus grammurus*) is the most characteristic mammal of the Piñon belt, with which its range appears to be nearly coincident. It occurs in suitable localities from western Texas to the Great Basin in Utah and Nevada. Two or three small mammals, characterized by darkness of coloration, seem to be restricted to this belt, namely, *Spermophilus spilosoma obsidianus*, *Perognathus fuliginosus*, and *Onychomys fuliginosus*, which are here described for the first time. (See part III.)

Lizards abound in the Piñon belt, becoming more numerous toward the desert, but two species (*Sceloporus consobrinus* and *Uta ornata*) which abound in the Piñon belt were not found in the desert below.

THE DESERT AREA.

[Approximate altitude : 1,200–1,800 meters, or 4,000–6,000 feet.]

The Desert of the Little Colorado, sometimes known as the 'Painted Desert,' is a great basin about 1,000 meters (3,300 feet) in depth, situated on the top of the plateau. It was excavated, as its name indicates, by the drainage system of the Little Colorado River—the Colorado Chiquito of the Mexicans—and consequently is lowest at the north, its slope being *away* from the southern edge of the plateau. The river has cut its bed down to about 820 meters (2,700 feet) at the point where it empties into the Grand Cañon of the Colorado, and throughout the lower part of its course it flows through a cañon considerably below the level of the desert proper, the lowest part of which is but little less than 1,200 meters (approximately 4,000 feet) in altitude. Its upper limit may be set at 1,800 meters (6,000 feet). The term Painted Desert should be restricted, it seems to me, to that part of the basin which is below 1,500 meters (approximately 5,000 feet).*

The geology of the region is simple. The lowest stratum which comes to the surface is carboniferous limestone; above this is red sandstone, which in turn is overlaid by the so-called variegated marls or argillaceous clays, sometimes capped by a thin layer of impure coal or lignite. The limestone appears on the west side of the river only (?), where it is soon buried under the ancient lava floods from San Francisco Mountain and neighboring craters. The red sandstone is encountered everywhere, sometimes as surface rock, sometimes as high cliffs forming the escarpments of broad mesas, and sometimes as curiously sculptured tablets standing on the plain. The marls are widely distributed, and in many

---

* The area below 1,370 meters (4,500 feet) is about 120 kilometers (75 miles) in length, and that below 1,500 meters (5,000 feet), 200 kilometers (125 miles). The long axis of the desert, slightly crescentic in form, and curving from near the mouth of the Little Colorado in the northwest to New Mexico in the southeast, is 320 kilometers (200 miles) in length, with a transverse diameter of about 110 kilometers (70 miles) along the middle portion, and a total area of 29,800 square kilometers (11,500 square miles). Its eastern edge penetrates the boundary of New Mexico in two arms, following the usually dry courses of the Zuñi and the Carrizo, and nearly reaches the boundary along the Rio Puerco, the largest tributary of the Colorado Chiquito.

places, particularly south of the lower part of Moencopie Wash,* rise
from the surface level in the form of strangely eroded hills and ranges of
stratified cliffs whose odd shapes and remarkable combinations of colors
—red, white, blue, brown, yellow, purple, and green—have given the
area in which they occur the name 'Painted Desert.' There are hundreds of smoothly rounded, dome-shaped hills of bluish clay, utterly devoid of vegetation, and almost identical in appearance with the 'gumbo
hills,' of the Bad Lands bordering the Little Missouri in North Dakota.
Both the hills and the naked clayey flats between them abound in alkali
vents—miniature craterlets—where the alkali effloresces, crusting over
the surface in patches which resemble newly fallen snow. Many of the
hills are capped with fossil wood, and many of the flats and lower levels
east of the Little Colorado River are strewn with chips and pieces which
have tumbled down during the wearing away of the hill-sides. Logs 30
to 50 centimeters (roughly, a foot or a foot and a half) in diameter and
9 to 12 meters (30 or 40 feet) in length are still common, and several
sections were found, possibly from the same tree, which measured about
150 centimeters (5 feet) in diameter. There are pebble beds miles in extent, made up of agate, moss-agate, chalcedony, jasper, obsidian, and
fossil wood, with not so much as a spear of grass or bit of cactus between them. On the other hand, many of the mesas and plains are
covered with sand and decomposed marls which support a scanty growth
of cactus, yucca, grease-wood, and a few other forms of vegetation characteristic of arid regions.

The bed of the Little Colorado River contains the only running water
in this part of Arizona, and it 'goes dry' a large part of the year, a
little water remaining in scattered pools, which are strongly alkaline.
Some of the salt and alkali flats on the river-bottom support a luxuriant growth of a singular fleshy plant belonging to the genus *Salicornia*,
which at a little distance looks like a leafless bush with thick green
stems. During the rainy season, and whenever the river 'runs,' the
liquid which flows down its course is red alkaline mud, about the consistency of ordinary sirup. This is the case also with its tributaries, of
which Moencopie Wash and Tenebito Wash are the only ones which
cross the Painted Desert proper.

The physical and climatic features of the Painted Desert are peculiar
and striking, and result in the production of an environment hostile
alike to diurnal forms of animal life and to the person who traverses it.
The explorer is impressed with the unusual aspects of nature—the
strange forms of the hills, the long ranges of red and yellow cliffs, the
curiously buttressed and turreted buttes and mesas, the fantastic shapes

---

* The terms 'wash' and 'arroyo' are applied to the deep channels or ravines so
common in arid regions. "These arroyos are natural consequences of the unequal
manner in which the rain falls throughout the year. Sometimes not a drop falls for
several months; again, it pours down in a perfect deluge, washing deep beds in the
unresisting soil, leaving behind the appearance of the deserted bed of a great river."
—Emory, Mexican Boundary Survey, I, 1857, p. 57.

of the rocks carved by the sand-blast and rendered still more weird by the hazy atmosphere and steady glare of the southern sun, the sand-whirls moving swiftly across the desert, the extraordinary combination of colors exposed by erosion, the broad clayey flats whitened by patches of alkali and bare of vegetation, the abundance of fossil-wood, the extensive beds of shining pebbles, the unnatural appearance of the distant mountain sharply outlined against the yellow sky, the vast stretches of burning sand, the total absence of trees, the scarcity of water, the alluring mirage, the dearth of animal life, and the intense heat, from which there is no escape.*

The plant life of the desert is scattered and scanty, and consists of such characteristic arid land forms as grease-wood (*Atriplex canescens, A. confertifolia,* and *Sarcobatus vermiculatus*); weeds of the genera *Dicoria* and *Oxytænia* (*D. brandigei* and *O. acerosa*); a large brush-like shrub (*Tetradymia canescens*) with flowers suggesting the golden-rod; the singular *Ephedra,* which has no apparent foliage; the narrow-leaved yucca (*Yucca angustifolia*), and cactuses of several genera. But it must not be supposed that these rank and spiny forms of vegetation, whose gray or dull olive colors are in perfect harmony with the parched and barren aspects of the desert, are the only plants found there; for no sooner is the surface moistened by the passing showers of the so-called 'rainy season' than numerous plants spring into existence, and favored parts of the desert lose something of their usual desolate and dreary appearance. There are places where even the nutritious grama grass (*Bouteloua*) gains a precarious foot-hold, and where a dwarf lupine (*Lu-*

---

* Lieutenant Ives and Dr. Newberry attempted to cross this desert from the Little Colorado near Grand Falls, but were obliged to turn back the first day. After following up the river for three days they found an Indian trail leading north, and followed it to the Moki villages. The following quotation is from Ives's account of the first day on the desert: "The scene was one of utter desolation. Not a tree nor a shrub broke its monotony. The edges of the mesas were flaming red, and the sand threw back the sun's rays in a yellow glare. Every object looked hot and dry and dreary. The animals began to give out. We knew that it was desperate to keep on, but felt unwilling to return, and forced the jaded brutes to wade through the powdery impalpable dust for fifteen miles. The country, if possible, grew worse. There was not a spear of grass, and from the porousness of the soil and rocks it was impossible that there should be a drop of water. A point was reached which commanded a view twenty or thirty miles ahead, but the fiery bluffs and yellow sand, paled somewhat by distance, extended to the end of the vista. Even beyond the ordinary limit of vision were other bluffs and sand fields, lifted into view by the mirage, and elongating the hideous picture."

Woodhouse, in speaking of a somewhat similar desert which he crossed in western Arizona, states that a coyote, "becoming desperate, rushed to the spring, and was killed by one of the men with a stone." He says further: "The ravens were hovering over us while we remained here, eagerly watching our famished mules. Since we left Bill Williams's Fork there have been clouds seen every day, and anxiously did we watch for rain; but this seemed a thing impossible, to rain in this miserable country, where everything appears to be an enemy, and is armed with a thorn or a poisonous sting."

pinus capitatus) is abundant; and the higher levels are adorned by a kind of painted-cup (*Castellcia*) and scattered beds of a rather coarse plant (*Mirabilis multiflora*) which suggests the morning glory. The delicate pink blossoms of the graceful *Malvastrum*, and the more showy yellow and orange flowers of *Riddellia tagetina* and *Zinnia grandiflora* would attract attention anywhere, and their beauty is here heightened by contrast with their sombre surroundings.

Without going into details it may be said that these plants, taken collectively, occur in the arid parts of northern Mexico,* Texas, New Mexico, Arizona, and southern California, and some of them extend north in the Great Basin, even reaching the Plains of the Columbia; and a few spread northward over the Great Plains east of the Rocky Mountains.

Large black beetles of the genera *Eleodes* and *Asida* are common on the Painted Desert and are characteristic arid land forms, occurring also in Mexico.

Toads of the peculiar genus *Spea*, modified for life in desert regions, were found after rains in some of the arroyos or washes, which are dry the greater part of the year.

Lizards are the most conspicuous forms of animal life and many species abound throughout the desert. Among them are:

Crotaphytus baileyi	*Sceloporus elongatus*	*Holbrookia maculata flavilenta*
Crotaphytus wislizenii	*Uta stansburiana*	*Phrynosoma ornatissimum*
Sceloporus graciosus		

We saw only one rattlesnake, but others have been recorded. Several of the species and all of the genera of reptiles here mentioned occur also in Mexico.

Birds are scarce, both in species and individuals, and but few breed on the desert of the Little Colorado. The following species were observed there:

Black-throated Desert Sparrow (*Amphispiza bilineata*)
Nevada Sage Sparrow (*A. belli nevadensis*)
Boucard's Sparrow (*Peucæa ruficeps boucardi*)
Brewer's Sparrow (*Spizella breweri*)
Sage Thrasher (*Oroscoptes montanus*)
———— Thrasher (*Harporhynchus* sp. —?)
Burrowing Owl (*Speotyto cunicularia hypogæa*)

All of these are characteristic arid land birds, which come into the United States from Mexico and extend northward various distances. Boucard's Sparrow ranges north from the table-lands of Mexico to western Texas, New Mexico, and Arizona; the Black-throated Desert Sparrow, from Mexico and Texas westward to southern California and north in the Great Basin to Utah and Nevada; the Sage Sparrow, from Mexico north to the Plains of the Columbia; Brewer's Sparrow, from

---

* The number of Arizona plants which occur in the northern part of Mexico is very large. Hemsley, in the botanical part of Biologia Centrali-Americana, states that of the 560 genera of Arizona plants mentioned by Rothrock, no less than 402, or 72 per cent., occur also in northern Mexico.

Mexico north over the Great Plains and the Great Basin; the Sage Thrasher, from Mexico north through the Great Basin; and the Burrowing Owl, from southeastern Texas to California and northward to Canada wherever suitable localities exist. Another characteristic arid land bird, the Road Runner or Chaparral Cock (*Geococcyx californianus*), was not seen, but has been recorded from the Little Colorado, and, like the others, enters the United States from Mexico. It ranges from Texas to California and north to Colorado.

The characteristic mammals of the desert are small nocturnal forms, such as Kangaroo Rats (*Dipodomys*), Pocket Mice (*Chætodipus*, a subgenus of *Perognathus*), Big-eared Mice (*Hesperomys*—of the *eremicus* group), and Free-tailed Bats (*Nyctinomus*). All of these groups reach the United States from Mexico, and none of the species of the Painted Desert range much north of Arizona.

Thus it appears that most of the forms of life inhabiting the desert of the Little Colorado—its mammals, birds, reptiles, and plants—occur also in Mexico and extend northward as far as the arid lands are suited to their requirements; and some of its species range east into Texas and west into southern California.

In like manner it has been shown that the characteristic forms of life of the Piñon belt occur in similar areas in different parts of the arid lands from Mexico to the Plains of the Columbia; that lands which rise above the level of the Piñon belt are covered with forests of tall pines and in the main possess the same species from western Texas to British Columbia; that still higher elevations are clothed with balsam and spruce, and that the humbler plants, the birds, and the mammals of these balsam and spruce forests are essentially the same throughout the Rocky Mountains and the great northern forest of Canada from northern New England to Alaska; that the mountain peaks, if sufficiently high, are bare at the summit, or capped with snow and ice, and sustain the same species of plants that grow in the arctic regions of the world and come south on the high mountain ranges in all parts of the Northern Hemisphere; in brief, it has been found that the same species, or closely related representative species of animals and plants inhabit the remotest parts of these several zones that inhabit them on Sán Francisco Mountain.

INTERRELATIONS AND AFFINITIES OF THE SEVERAL ZONES.

The contemplation of the phenomena here described leads naturally to comparisons of similar areas throughout the country; to attempts to bring together these areas into natural biological zones and provinces, and to inquiries concerning their origin.

Without going into the history of the subject, it may be said that most zoologists recognize three primary zoo-geographical divisions in the United States—an '*Eastern*,' extending from the Atlantic Ocean to the Great Plains; a '*Central*,' from the eastern border of the Plains westward to the Sierra Nevada; and a '*Western*,' from the eastern

18   NORTH AMERICAN FAUNA.   [No. 3.

base of the Sierra Nevada to the Pacific. The arid region of the South.
west which enters the United States from Mexico has been recognized
as a distinct division by many naturalists, and has been named the
' *Chihuahuan*' or ' *Sonoran*' region.

The region east of the Great Plains was subdivided by Agassiz as
early as 1854 into three areas which he called Faunas, namely: (1) a
' *Canadian Fauna*,' (2) an ' *Alleghanian Fauna*,' or Fauna of the Middle States, and (3) a ' *Louisianian Fauna*,' or Fauna of the Southern
States. Subsequent writers, particularly Verrill and Allen, have circumscribed these Faunas, reduced their rank, and increased their number until at the present time ornithologists recognize eight faunal areas
in eastern North America, as follows: (1) Arctic; (2) Hudsonian; (3)
Canadian; (4) Alleghanian; (5) Carolinian; (6) Louisianian; (7) Floridian; and (8) Antillean. Cope, from a study of the reptiles and batrachians, united the Louisianian and Floridian Faunas into a district
of primary rank, which he named the '*Austroriparian*' region—the
exact equivalent of Agassiz's Louisianian Fauna. Passing over this
region as clearly of southern origin, there remain the Carolinian, Alleghanian, Canadian, Hudsonian, and Arctic Faunas. The three latter
are boreal in their affinities, while the Carolinian is suffused with southern forms, and the Alleghanian seems to be neutral ground.

In studying the several life-zones of the higher declivities of San Francisco Mountain it became apparent not only that each has its corresponding zone in the East, but that in many instances the zones of the
mountain may be recognized by the presence of the identical species
which characterize them in New England and Canada. In short, it was
found that the faunal and floral zones which go to make up the Boreal
Province in the East may be traced in a northwesterly direction around
the northern end of the Plains of the Saskatchewan and then south
along the sides of the Rocky Mountains, even to this isolated peak in
Arizona.* This has been pointed out somewhat in detail in the discussion under the head of each zone, and has been indicated further by the
headings themselves.

Each zone, while possessing throughout a certain number of common
or strictly representative species, undergoes a notable change in pass-

---

*This will be made clear by a glance at the accompanying map of North America
(map 5), on which the Boreal Province is represented in clear green.

Scudder, under the head of "Anomalies in the Geographical Distribution of our
Butterflies," mentions a number of cases in which northern species of butterflies
occur in supposed isolated colonies at remote points, all of which, it is significant to
observe, fall within the boundaries of the Boreal Province here defined. He cites the
brown elfin butterfly (*Incisalia augustus*) as a species throwing some light on this
' anomalous' distribution. It occurs, he states, in New England and New York, and
south in the Alleghanies to West Virginia. North of the United States it has been
found at Halifax, Quebec, Montreal, and thence westerly as far as Cumberland House
on the North Saskatchewan. In the West it again enters the United States along the
Rocky Mountains, and extends as far south as Colorado. A better example of a
typical boreal distribution could hardly be desired.

ing from the East to the West, each extreme being occupied by certain species not found in the other. It is necessary to recognize this difference in the names applied to the zones; hence the prefix 'central' has been used in each case to distinguish the Rocky Mountain arm from the eastern arm.

The several zones of the San Franciso Mountain region are interrelated in different degrees, some very closely and others very remotely. Many species and even genera which are common to two or more zones, and consequently of no value whatever in defining the single areas, become of the utmost importance in studying the interrelations of the several zones.   For instance, in the highest group of all—the mammalia— there are representatives of four distinct types, namely, Bears, Shrews, Voles, and Red Squirrels, which range from the top of the timber-line belt to the bottom of the Canadian or Douglas fir zone.*   All of these are circumpolar types, ranging over the boreal parts of the whole world and coming south in the mountains.   It is clear, therefore, that they are of boreal origin.   On the other hand, there are several very different types of mammals, among which may be mentioned the Kangaroo Rats, Pocket Mice, and Grasshopper Mice, which do not occur above the Piñon zone.   These are southern types reaching the United States from the table-lands of Mexico and extending northward over the arid lands as far as the conditions are suited to their requirements.   It is clear, therefore, that they are of southern origin.   In short, it may be stated, as a result of this biological survey of the San Francisco Mountain region, that all the forms of life inhabiting Arizona were derived from one of two directions—the north or the south.   And in extending these researches and generalizations so as to embrace the Great Interior Basin, the Rocky Mountain region, and the Great Plains, which together constitute the so-called 'Great Central Province,'† of naturalists, I was astonished to be forced into the belief that no such province exists.   Indeed, the present investigation demonstrates that there are but two primary life provinces in this country : a northern, which may be termed *Boreal*, and a southern, which, for our purposes, may be termed *Sonoran*, since it comes to us from Mexico through Sonora.   In attempting to arrange all the life zones of Arizona under these two headings the following conclusions have been reached: The Arctic-Alpine, Timber-line, Hudsonian, and Canadian zones, having been shown to be derived from the north, fall naturally under the *Boreal* division. The Desert and Piñon zones, having been shown to be derived from the south, fall naturally under the *Sonoran* division.   There remains but one area, namely, the Pine area, whose relationships are in any way obscure.   This area has

---

*Bears range over the lower levels at certain seasons of the year, but are not known to breed away from the spruce and fir forests.

†This province was outlined by Agassiz as long ago as 1854, and has been accepted so far as its essential features are concerned by LeConte, Baird, Wallace, Allen, Cope, Binney, Gray, Packard, and nearly all recent writers.

been shown to consist of a mixture of Boreal and Sonoran types, more or less modified by adaptation to environment. In other words, it is neutral territory. But since the number of its Sonoran types is greatly in excess of its Boreal types, it may be more properly referred to the Sonoran Province. Therefore, of the seven life-zones of the San Francisco Mountain region in Arizona, four may be referred to the Boreal Province and three to the Sonoran.

The zones composing each of these primary divisions are related to one another in different degrees. Thus, the Timber-line, Hudsonian, and Canadian zones are much more intimately related than the Timber-line and the Alpine; and the affinities of the Piñon and Desert are much closer than those of the Piñon and Pine. Hence it becomes possible to group the zones into categories of intermediate rank between the primary provinces and the tertiary zones or areas. These secondary divisions are here termed regions. Under the Boreal Province we may recognize two regions, an Arctic and a Boreal. The Arctic region contains but one zone, the Alpine. The Boreal region contains three zones, namely, the Timber-line, Hudsonian, and Canadian. The Sonoran or southern province may be likewise split into two regions, a Sub-Arid and an Arid. The Sub-Arid consists of a single zone, the Pine. The Arid region comprises two zones,* the Piñon and the Desert. The facts here set forth may be graphically represented by means of a table, thus:

Life Areas of the San Francisco Mountain Region in Arizona.

Provinces.	Regions.	Zones or Areas.
Boreal	Arctic	Alpine.
	Boreal	Timber-line.
		Hudsonian.
		Canadian.
Sonoran	Sub-Arid	Pine.
	Arid	Piñon.
		Desert.

The primary divisions are based on the possession of distinctive genera; the secondary and tertiary chiefly on distinctive species, though some of them possess distinctive genera also.

ORIGIN OF THE BOREAL FAUNA AND FLORA OF SAN FRANCISCO MOUNTAIN.

The Boreal zones of San Francisco Mountain are separated from corresponding areas elsewhere by a broad interval occupied by the upper faunas and floras of the Sonoran Province. The arctic summit of the mountain is distant more than 400 kilometers (250 miles) from the nearest peak of similar character in Colorado, and nearly 3,200 kilometers (2,000 miles) from the nearest point in the Arctic zone proper—all

---

* The Desert of the Little Colorado contains but two arid zones; further south a third is encountered.

the arctic areas within the United States being mere dots upon the map, and even the lower zones of the Boreal Province being widely separated from similar areas in the north. The question naturally arises as to the origin of these small colonies of arctic life which appear here and there over a great continent. It is perfectly evident that they could not have reached their present positions during existing climatic conditions; hence it is necessary to search the records of the past for the explanation. The period immediately preceding the present is known as the glacial age, because the northern parts of the globe were then buried in ice. This ice cap, which in places was several thousand feet in thickness, underwent two principal movements of advance and retreat, first crowding the life of the region far to the southward, then allowing it to return, to be again driven south by the next advance. The southern terminus of the great ice sheet extended from New Jersey to southern Illinois, and thence northwestward to British Columbia, and its effects upon the climate must have been felt throughout the United States and even into Mexico. The advance of the glacial period was so gradual that plants as well as animals had time to escape by extending their ranges southward, and during the return movement were enabled to keep pace with its slow retreat. Had either the process of refrigeration or the return of heat taken place more rapidly, most of the forms of life inhabiting the northern parts of the globe would have been exterminated. During the recession of the glacier many boreal plants and animals were stranded on mountains, where, by climbing upward as the temperature became warmer, they were able to find a final resting place with a climate sufficiently cool and moist for their needs; here they have existed ever since. This is the commonly accepted explanation of the presence of arctic forms on isolated mountain peaks widely removed from the southernmost limit of their continuous distribution.

Incidentally the ancient origin of arctic-alpine faunas leads to conclusions which might be of use to the geologist. For instance, San Francisco Mountain is a volcanic peak composed entirely of lava rock. Its summit is inhabited by species of animals and plants which could not have reached it since the recession of the glacial period. Hence the mountain itself can not be of more recent origin than this period. Here the living fauna and flora afford evidence of the age of a great mountain.

ORIGIN OF THE FAUNA AND FLORA OF THE PAINTED DESERT.

The Desert of the Little Colorado, it will be remembered, is a deep basin on top of the Great Colorado Plateau. It is wholly disconnected from the desert region of southern Arizona by the elevated and timber-covered highlands occupying the crest of the plateau escarpment. In fact the highest part of Arizona south of the Grand Cañon, except a few isolated mountains, is the edge of this plateau, which is nowhere below 2,130 meters (7,000 feet), and in places rises to the height of 2,740 meters (9,000 feet), as at the Mogollon Mesa. On the east, the desert is separated from the valley of the Upper Rio Grande by a broad area covered with cedar and piñon, through which the continental divide passes, at an elevation of upwards of 2,130 meters (7,000 feet). Therefore, the only possible channel through which the fauna and flora of the Painted Desert could have reached this desert during existing climatic conditions is by way of the Grand Cañon of the Colorado. At first thought it seems incredible that a fauna and flora should extend several hundred miles through a chasm of this character; but the evidence at hand indicates that it does. Our descent into the cañon from the Cocanini Plateau was made at a point about 25 kilometers (15 miles) below the mouth of the Little Colorado. Here the cañon is about 1,800 meters (more than a mile) in depth and nearly 25 kilometers (15 miles) wide at the top. Numerous side cañons cut into it, and there are many shelves and bottoms which support a flora of cactuses, yuccas, agaves, greasewoods, and other typical Sonoran forms. Pocket Mice of the sub-genus *Chætodipus*, Large-eared Mice of the *Hesperomys eremicus* group, and the Little Spotted Skunk (*Spilogale*) were secured, together with several birds (among them *Peucæa ruficeps boucardi*) and reptiles of the Sonoran fauna, some of which occur also on the Painted Desert.*

The inference is that the life of the Painted Desert is derived from the deserts of western Arizona, and that it came by the roundabout way of the Grand Cañon of the Colorado.

It might be urged that the climate of the Plateau region in the past may have been enough warmer than at present to admit of direct communication between the life of the Painted Desert and that of the deserts of southern Arizona; but Major C. E. Dutton, who has made a special study of the physiographic history of the Plateau region, assures me that its climate has not been warmer than now since glacial times.

GENERALIZATIONS CONCERNING THE DISTRIBUTION OF LIFE IN NORTH AMERICA.

OVERTHROW OF THE SO-CALLED 'CENTRAL PROVINCE' OF NATURALISTS.

The region almost universally recognized by recent writers as the 'Central Province' is made up of the Great Plains, the Rocky Mountains, and the Great Basin. A critical study of the life of the Rocky Mountains has shown it to consist of a southward extension of the Boreal Province, with an admixture of southern forms resulting from an intrusion or overlapping of representatives of the Sonoran Province, some of which, from long residence in the region, have undergone enough modification to be recognized as distinct subspecies or even species. A similar analysis of the life of the Great Plains and Great Basin has shown them to consist of northward extensions of the So-

---

* Among the reptiles found near the bottom of the cañon were two lizards (*Sceloporus clarkii* and *Uta symmetrica*) which belong to the torrid fauna of southern and western Arizona, and are not known to reach the Painted Desert.

noran Province, somewhat mixed with the southernmost fauna and flora of the Boreal Province.   Thus the whole of the so-called 'Great Central Province' disappears.

This explains a multitude of facts that are utterly incomprehensible under the commonly accepted zoological divisions of the country.   These facts relate particularly to the distribution of species about the northern boundaries of the supposed Central and Pacific Provinces, and to the dilemma we find ourselves in when attempting to account for the origin of so many primary life areas in a country where there are no impassable physical barriers to prevent the diffusion of animals and plants.

EVIDENCE ON WHICH THE 'CENTRAL PROVINCE' WAS BASED.

The conclusions here announced are so diametrically opposed to the long-accepted and current views of zoologists that it may be interesting to examine for a moment the evidence on which their generalizations were based.   This evidence, stated briefly, consists in the presence, in the region in question, of a large number of genera and species not found in the Eastern States.   It has just been shown that the vast majority of these forms were derived from the north or from the south. The remainder fall naturally into two categories: (1) Those so closely related to forms now living in adjoining regions as to leave no doubt that they are the immediate descendants of the same, modified by environment; and (2) isolated generic types, of which the number is small.

SIGNIFICANCE OF ISOLATED TYPES.

The presence of isolated types, however few, might be regarded as an obstacle to the acceptance of the views here advanced, but their significance becomes apparent as soon as an attempt is made to trace the life of the present back to the life of the past.   The colonies of big trees and redwoods of California (*Sequoia gigantea* and *S. sempervirens*) have no nearer relatives than the bald cypress (*Taxodium*) of the Gulf States and a related species from China (formerly recognized generically under the name *Glyptostrobus*).   This was pointed out many years ago by Dr. Asa Gray in connection with the circumstance that the ancestors of these trees once ranged throughout the boreal regions of the world.   A fossil species (*Sequoia langsdorfii*) closely related to the California redwood has been found in Spitzbergen, Iceland, Greenland, the north of Europe, Alaska, at the mouth of the Mackenzie River, and also in the Rocky Mountains, the Great Basin in Oregon, and the Bad Lands in Dakota.   Many parallel cases might be cited.   Thus the records of the rocks show that many of the types which have survived the perils incident to the successive shiftings of the fauna and flora during and subsequent to the ice age were formerly conspicuous over large areas in the north.   These facts are in complete accord with a general law which may be thus formulated :

When the physiographic conditions of a region are in process of change, those forms of life which are sufficiently plastic to adapt themselves to the rapidly changing conditions survive, while those which cannot so adapt themselves become extinct.

Isolated generic types are illustrations of this law and may be regarded as remnants of the past—the only living representatives of types once abundant and widely diffused. Such types are not confined to plants, but may be found in nearly every branch of the animal kingdom. Among North American mammals the genera *Neurotrichus* and *Aplodontia* may be cited as examples, both of them being confined to a narrow strip along the Pacific coast from northern California to British Columbia. The former has a near relative in Japan (*Urotrichus*), and the intermediate forms which connect it with the Shrews on the one hand and the Moles on the other are still living in eastern Asia (the genera *Scaptonyx* and *Uropsilus*). *Aplodontia* is a large rodent, the type and sole representative of an isolated family, and has no known living relative in any part of the world.

PRINCIPAL LIFE REGIONS OF NORTH AMERICA.

[See map 5.]

The most important generalization arrived at in the present investigation is that the whole of extratropical North America consists of but two primary life regions, a *Boreal* region, which is circumpolar; and a *Sonoran* or *Mexican table-land* region, which is unique.*

The *Boreal Province* [colored green on map 5] extends obliquely across the entire continent from New England and Newfoundland to Alaska, conforming in direction to the trend of the northern shores of the continental mass. It gives off three long arms or chains of islands which reach far south along the three great mountain systems of the United States—a western arm in the Cascades and Sierra Nevada, a central arm in the Rocky Mountains, and an eastern arm in the Alleghanies—and these arms interdigitate with northward prolongations of the Sono-

---

* Since the present paper was written (December, 1889) the author has been engaged in the preparation of an historical synopsis of the attempts that have been made to define the faunal and floral areas of North America. In the course of this investigation several important papers have been found which confirm, and in part anticipate, the general conclusions here announced, though none of them attempt to explain the significance of the areas recognized or to correlate them with the northern and southern origin of the life of the continent. For instance, the late Dr. Asa Gray stated that it is certain "that two types have left their impress upon the North American flora, and that its peculiarities are divided between these two elements. One we may call the *boreal-oriental element;* this prevails at the north, and is especially well represented in the Atlantic flora and in that of Japan and Manchuria; the other is the *Mexican-plateau element*, and this gives its peculiar character to the flora of the whole southwestern part of North America, that of the higher mountains excepted" (Bull. U. S. Geol. and Geog. Survey, VI, 1, Feb. 11, 1881, 62). At the same time, and in the same communication, Dr. Gray adopts the three great divisions usually recognized by zoologists—Eastern, Central, and Pacific.

ran Province, which latter completely surround the southern islands of the Boreal system.

The *Sonoran Province* [colored orange or yellow on map 5] comes into the United States from the south and is divisible into six subregions, namely : (1) an *Arid* or *Sonoran* subregion proper, occupying the table-land of Mexico and reaching north into western Texas, New Mexico, Arizona, and southern California ; (2) a *Californian* subregion, occupying the greater part of the State of that name ; (3) a *Lower Californian* subregion ; (4) a *Great Basin* subregion, occupying the area between the Rocky Mountains and the Sierra Nevada and extending as far north as the Plains of the Columbia ; (5) a *Great Plains* subregion, occupying the plains east of the Rocky Mountains and extending north to the Plains of the Saskatchewan ; and (6) a *Louisianian* or *Austroriparian* subregion, occupying the lowlands bordering the Gulf of Mexico and the Mississippi, and extending eastward, south of the Alleghanies, to the Atlantic seaboard, where it reaches as far north as the mouth of Chesapeake Bay.

The latter region requires a word of comment, since its true affinities have not been heretofore pointed out, though the region itself has been long recognized.*   That it is an offshoot of the Sonoran region is evident from the fact that most of its peculiar or distinctive animals and plants belong to Sonoran genera, and many of its species are identical with or closely related to Sonoran forms.   It contains no less than eight Sonoran genera of mammals, namely : *Spilogale, Urocyon, Neotoma, Sigmodon, Ochetodon, Geomys, Piecotus* (subgenus *Corinorhinus*), and *Nyctinomus*, most of which extend northward near the Atlantic seaboard as far as Norfolk, and at least one of them (*Urocyon*) considerably further.   It contains also a number of Sonoran genera of birds, reptiles, batrachians, and plants.   At the same time, it contains two Tropical American genera of mammals, namely, *Didelphys* and *Oryzomys ;* and perhaps *Urocyon, Sigmodon*, and *Nyctinomus* belong as much to one as to the other.   It contains also a number of Tropical genera of birds, reptiles, and plants.   Hence the *Austroriparian* subregion consists of a mixture of Sonoran and Tropical forms ; but since the number of its Sonoran types is greatly in excess of the Tropical, it may be fairly regarded as a subdivision of the former.

The *Tropical Province* [colored red on map 5], so far as North America is concerned, occupies Central America and the Antilles and pushes north along the lowlands on both sides of Mexico, reaching the mouth

---

*As early as 1817 the entomologist Latreille made it one of his circumpolar divisions.   In 1822 the botanist Schouw named it the *Realm of Magnolias ;* and in 1854, Agassiz named it the *Louisianian Fauna.*   These authors, and several other early writers (including Meyen, Martius, Berghaus, and Schmarda) regarded it as a region of primary rank.   More recent writers (including LeConte, Cooper, Binney, Baird, and Allen) looked upon it as a subdivision of the eastern forest region or *Eastern Province.*   Cope, in 1873, restored it to independent rank and named it the *Austroriparian region.*

of the Rio Grande on the Gulf of Mexico, and a little north of Mazatlan on the Pacific coast. It occupies also a narrow belt encircling the southern half of the peninsula of Florida. This tropical element in Florida is of comparatively recent origin, and consists mainly of a chain of island-like colonies of birds, insects, and plants which may easily have reached its shores and keys from the neighboring West Indies, as pointed out by Schwarz in an article on its peculiar Insect Fauna (Entomologica Americana, IV, No. 9, 1888). The interrelations of the Tropical and Sonoran Provinces are such as suggest that the chief difference may be due to humidity as much as temperature.

In the light of the general conclusions here announced, the only part of North America which is in any way obscure, so far as the relationships of its faunas and floras are concerned, is the so-called ' *Pacific Province;*' and, like the ' *Central Province*' already discussed, it is evidently made up of two distinct elements, a *mountain* element derived from the Boreal Province, and a *valley* element derived from the Sonoran; but owing to the peculiar physiographic conditions of the west coast it has undergone a greater amount of differentiation.

CAUSES WHICH DETERMINE DISTRIBUTION.

It is not the purpose of the present paper to discuss the causes that have to do with limiting the distribution of terrestrial animals and plants further than to point out a generalization which seems to have been overlooked. Omitting reference to the effects of physical barriers, which explain the differences in the life of disconnected continents, it may be stated that temperature and humidity are the most important causes governing distribution, and that temperature is more potent than humidity.* Authors differ as to the period during which temperature exerts the greatest influence, some maintaining that it is the temperature of the whole year, and others, that it is the temperature of a very brief period which determines the range of species. In the case of birds, it has been shown by Verrill and Allen that it is the temperature of the *breeding season.*

If this is true of birds, why is it not true of other forms of animal life and of plants as well? The season of reproduction for the plant, as for the animal, is the warm part of the year. After the period of reproduction the plant withers; after it flowers and fruits and matures its seed, it dies down or becomes physiologically inactive. And what the plant accomplishes in one way the animal accomplishes in another. To escape the cold of winter and its consequences the sensitive mammal hibernates; the bird migrates to a more southern latitude; the reptile and batrachian dig holes in the mud or sand and remain in a torpid condition; the insect sleeps in its cocoon or buries itself under leaves

---

*In arid districts humidity is an element of vastly more consequence that in regions of moderate or copious rain-fall, particularly in regard to the inception of the period of reproduction in plants.

or decomposing vegetation ; and none but the hardier forms of life are left to be affected by winter temperatures. Freezing does not hurt most plants when not in a state of reproductive activity. In the north, trees five and six feet in diameter freeze through to the heart every winter. It is obvious, therefore, that plants are not exceptions to the law that *the temperature during the season of reproductive activity* determines the distribution of life. In high arctic latitudes this period is very brief, while in the humid parts of the tropics it seems to extend over nearly if not quite the whole year.

Some eminent writers have assumed that plants and animals do not agree in distribution—that a faunal map (a map showing the distribution of an association of animals) must differ essentially from a floral map (a map showing the distribution of an association of plants). This assumption is illogical, for, as just stated, plants and animals are subjected to the same conditions during the season of reproduction—the season during which they are most affected by their surroundings. Furthermore, the field work on which the present paper is based, which was conducted with special reference to the determination of this point, demonstrated that complete coincidence exists in the limitation of the life-areas as defined independently by the study of the mammals, birds, reptiles, and plants of the San Francisco Mountain region.

Since the distribution of animals and plants depends primarily upon temperature, it follows that the physiographic conditions which influence temperature influence distribution also. In obedience to this law certain axioms of distribution may be thus expressed :

The distribution of species *in the same latitude* depends primarily on *altitude.*

The distribution of species in the same latitude and altitude is influenced notably by—

(*a*) Elevation above base-level.

(*b*) Slope-exposure.

(*c*) Proximity to and direction from large bodies of water.

(*d*) Meteorologic conditions affecting temperature.

In the case of mountains of equal altitude and low base-level:

(1) The number of faunal and floral zones (up to the limit of zones possible for the range of temperature) is inversely proportional to the distance from the equator.

(2) The width of the zones and the abruptness of the change from one to another is proportional to the steepness of the slope.

By elevation above base-level is meant the height of a given point above the plane it faces. This may be made clearer by an example. The mean altitude of base-level below the plateau rim in Arizona is less than 900 meters (3,000 feet), and above it more than 2,130 meters (7,000 feet). A mountain standing on the edge of the plateau will have a higher temperature at a given altitude on the north side than on the south side, because the plateau level (base-level) on the north side carries up the temperature. Many years ago Humboldt cited an instance of this kind in the Himalaya. The temperature on the north side of this lofty range is much higher than on the south side at the same elevation; or, to state it differently, the snow line and the timber line on the north side are about 900 meters (3,000 feet) higher than on the south side. This is due to the great height of the Thibetian Plateau as compared with the altitude of base level on the south side, and is in opposition to the influence of slope-exposure. By slope-exposure is meant the inclination of the surface of the earth in relation to the angle of reception of the sun's rays. The sun strikes the east side of a hill or mountain in the early part of the day, the south side a little later, the southwest and west sides in the afternoon, when its heat is greatest, and the northwest and north about sundown or not at all. But in case there is a high plateau on the north side, the heat from the plateau will force the timber line up. Therefore, of the influences under consideration, base-level is more powerful than slope exposure.

About half a century ago the elder Binney, in a work which he did not live to see published, made the following observation:

"The relations which the different levels of elevation bear to the parallels of latitude, although as interesting to the zoologist as to the botanist, have not yet been made the subject of examination in this country. But the Rocky Mountains * * * offer, in the great extent of their table-land and in the height to which they rise, a vast field of research to future naturalists, where they will be able to solve many of the most important questions connected with the geographical distribution of the terrestrial mollusks of our country."[*]

If the word 'mollusks' in the above quotation be changed to the more comprehensive word 'life,' Binney's remarks may be regarded as a prophecy fulfilled, in part at least, by the present Biological Survey of San Francisco Mountain. At the same time it should be remembered that the present report is little more than an announcement of the general conclusions resulting from a brief survey of a limited area, and that anything approaching a final discussion of the subject must be deferred until similar surveys of many regions result in the accumulation of a multitude of facts now unknown. As the late Leo Lesquereux once said of his favorite study:

"This science is in its infancy; and the childhood of science is marked, like that of man, by a series of trials and failures, from which strength and proficiency are derived. The first astronomers did not measure the distance from the earth to the fixed stars, nor weigh the planets by the diameter of their orbits."[†]

---

[*] Amos Binney, The Terrestrial Mollusks of the U. S., 1851, vol. I, 116–117.

[†] A Review of the Fossil Flora of North America. Bull. U. S. Geol. and Geog. Survey Terr., No. 5 (2d series) Jan., 1876, 248

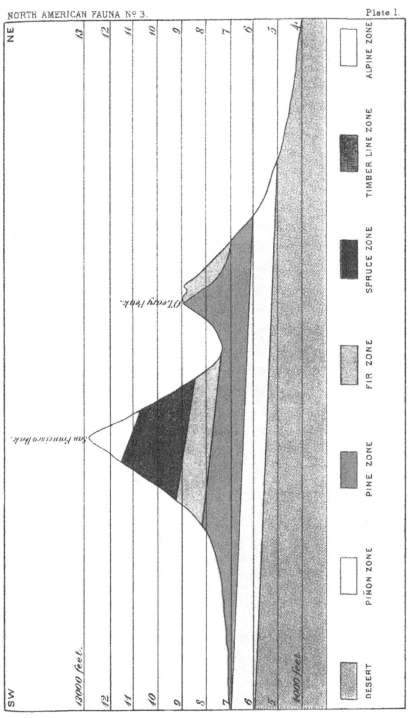

DIAGRAMMATIC PROFILE OF SAN FRANCISCO AND O'LEARY PEAKS FROM S. W. TO N. E... SHOWING THE SEVERAL LIFE ZONES AND EFFECTS OF SLOPE EXPOSURE.

DESERT   PIÑON ZONE   PINE ZONE   FIR ZONE   SPRUCE ZONE   TIMBER LINE ZONE   ALPINE ZONE

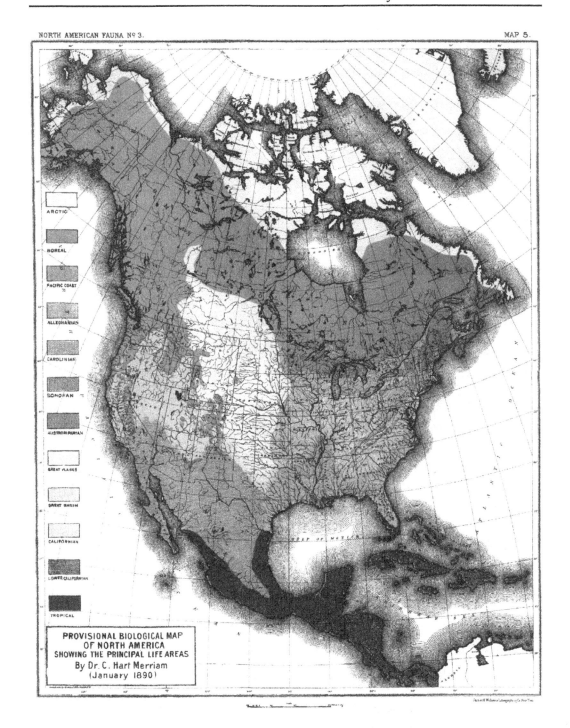

PROVISIONAL BIOLOGICAL MAP
OF NORTH AMERICA
SHOWING THE PRINCIPAL LIFE AREAS
By Dr. C. Hart Merriam
(January 1890)

PAPER 16

From *Climate and Evolution*
William Diller Matthew

So far as the correlation of the Pampean and Santa Cruz is concerned, their fossils agree wholly in preservation and degree of petrifaction with those preserved in similar Pleistocene and late Miocene formations, respectively, in the western Plains, and the degree of consolidation of the matrix is the same. We have in the West two fossiliferous formations, the Bridger (Eocene) and John Day (Oligocene), which are, like the Santa Cruz, composed of an andesitic volcanic ash, and similar ash strata are found in different levels of our Western Miocene formations. Now, the Santa Cruz matrix and fossils are very much less consolidated or thoroughly petrified than the Bridger and decidedly less so than the John Day, while they agree very well with the volcanic ash beds in the middle and upper Miocene. As there is no reason to suppose that the rock-making processes work at a different rate in different continents, this evidence is entitled to some consideration. On similar grounds, the Pampean fossils would be referred to middle Pleistocene, and the few fossils that I have seen from Monte Hermoso agree best with Pliocene fossil mammals from North America. I should place no weight on this kind of evidence except when, as in the present instance, the climatic conditions and the origin and method of deposition of the formations are substantially similar.

The foregoing digression is somewhat outside the limits of this discussion. It appears, however, to be necessary to show briefly the reasons on which the age assigned to the South American mammalian faunæ are based. It might, indeed, be logically objected that these correlations are based on the northern origin and migration of certain phyla and cannot, therefore, be used in support of the theories here advocated. But the phyla on which the demonstration rests are so universally admitted to have arisen in the north, and the evidence that they did so is so complete and conclusive, that there is no reasonable alternate to accepting them as such. And if so, the correlations of South American faunæ must be approximately as here stated, a conclusion supported by the wholly independent evidence of the degree of consolidation of the formation and of petrifaction of the fossils contained.

CENTERS OF DISPERSAL

Whether the evolution of a race be regarded as conditioned wholly by the external environment or as partly or chiefly dependent upon (unknown) intrinsic factors, it is admitted by everyone that it did not appear and progress simultaneously and *æquo pede* over the whole surface of the earth, or even over the whole area of a great continent. The successive steps in the progress must appear first in some comparatively limited region, and from that region the new forms must spread out, displacing the old and driving them before them into more distant regions. Whatever be the causes of evolution, we must expect them to act with maximum force in some one region; and so long as the evolution is progressing steadily in one direction, we should expect them to continue to act with maximum force in that region. This point then will be the center of dispersal of the race. At any given period, the most advanced and progressive species of the race will be those inhabiting that region; the most primitive and unprogressive species will be those remote from this center. The remoteness is, of course, not a matter of geographic distance but of inaccessibility to invasion, conditioned by the habitat and facilities for migration and dispersal.

If the environmental conditions in the center of dispersal pass the point of maximum advantage for the race-type that is being developed and become unfavorable to its progress, we should find its highest types arranged in a circle around a central region, which was the former point of dispersal, and the more primitive types arranged in concentric external circles. The central region will be unoccupied, or inhabited by specialized but not higher adaptations.

It would appear obvious that the present geographic distribution of a race must be interpreted in some such way as this by anyone who accepts the modern doctrine of evolution. Yet there are many high authorities on geographic distribution who proceed apparently upon a precisely opposite theory. According to these authors, the distribution center of a race is determined by the habitat of its most primitive species, and the highest and most specialized members of the race are most remote from its center of dispersal. This principle may be true enough so far as concerns the first appearance of a given race, *i. e.,* provided the most primitive species are also the oldest geologically; but it appears to me to be the direct reverse of fact as regards the present distribution, or the distribution at any one epoch of the past. The only ground on which it could be defended would be that the progress of the race is due to its migration, and those members which did not migrate did not progress. But this involves the view that its progressiveness up to the time that its geographical environment changed was due to staying at home, and the same progress after its environment changed was due to not staying at home. It seems to me that the prevalence of this view must be due to some fallacious notions about migration, unconsciously retained, involving a concept of it as analogous to travel in the individual. The successful business man, no doubt, may pack up his baggage and take to traveling, leaving home and going elsewhere and profiting much thereby. Nations have done the same thing, likewise to their advantage. But there is very little analogy here to the zoögeographic migration of species—which is a question of expansion or contraction of range, not directly of transference of habitat, although this may be the final result.

It seems obvious that the conditions which brought about the early progressiveness of the race in a particular locality would, so far as they were external, cause the continued progressiveness of those individuals which remained in that region; so far as they were intrinsic, they would affect the main bulk of the race, the center of its range, more than any outlying parts of it. The present writer is very thoroughly convinced that the whole of evolutionary progress may be interpreted as a response to external stimuli; and intends here to point out what he regards as the most important of these stimuli. It is therefore necessary to point out that these postulates regarding centers of dispersal and migration are not dependent upon the theories to be proved—we are not reasoning in a circle.

OCEANIC AND CONTINENTAL ISLANDS

Faunal Differences Between Oceanic and Continental Islands

One of the strongest arguments for the relative permanency of the deep oceans, especially during Cenozoic time, is afforded by the marked and striking contrast between the faunæ of those large islands which are, and those which are not, included within the continental shelf. The continental islands have the fauna of the continents to which they belong, large as well as small, differing only in the absence of types of recent evolution or of unsuitable adaptation and in the survival of primitive types which have disappeared from the mainland. But no question could be raised as to their former union with the mainland, no other possible solution would explain their fauna. We are compelled to assume the former connection of the British Isles with Europe, of Ceylon with India, of Japan with Korea or Siberia, of Sumatra, Java and Borneo with the Malayan mainland, of the Philippines with Borneo, of New Guinea and Tasmania with Australia, of Newfoundland and Cape Breton with Labrador and Nova Scotia. In each and all of these cases, the evidence is overwhelming, and, with the exceptions cited, the faunal identity is complete.

On the other hand, with all those islands which are separated by deep ocean from the mainland, we find that just that evidence is lacking which would afford convincing proof of former union with the mainland. Their faunæ are widely different from those of the adjoining mainland; they lack just those animals which could not possibly have reached there except by land bridges; they point often to long periods of independent evolution and expansion, and the primary elements of the faunæ of every one of them are such as might possibly at least have reached the island without continental union, whether by accidental transportation, by swimming or by other means.

Take for example the mammals of Sumatra, Java and Borneo. We cannot reasonably suppose that the rhinoceroses, tapirs, deer, wild dogs, felids and numerous other large animals common to them and the adjoining continents reached these islands except by land. They are too large for transportation on "rafts" of vegetation such as occasionally drift to sea from the mouths of tropical rivers. They are dry-land animals not given to swimming long distances. And we would not invoke the agency of man to account for a whole fauna. But most important is the fact that all the animals that we might fairly expect to find there in view of a former land connection are really present.

Contrast with this the fauna of Madagascar.[31] There are no ungulate mammals there, except for the bush-pig, possibly introduced by man (in accord with known customs of the Malays) and a pigmy hippopotamus (now extinct) which might have reached the island by swimming, as hippopotami are known to travel considerable distances by sea from one river mouth to another. The great majority of the unguiculate groups of the mainland are also absent. The only representatives are a few very peculiar carnivores of the family Viverridæ, a peculiar group of insectivores (Centetidæ) and a peculiar group of Cricetine rodents, each apparently evolved on the island from a single type introduced long ago, a species of shrew (Crocidura) of more recent introduction and a variety of bats. There are numerous lemurs and no monkeys there; and the lemurs appear to have radiated out from a single group[32] into a number of peculiar types, two of which, now extinct, paralleled the ungulates and the higher apes in several significant features. The fauna of the island does not resemble the present fauna of Africa, nor can it be derived from

---

31. A. R. WALLACE: Island Life. 381-412. 1881. See also Trouessart, Catalogus Mammalium and Suppl. Quinq.; Lydekker, Geog. Hit. Mam. 211-226. 1896. Lydekker's arguments for continental union are mostly invalidated by more recent discoveries.

32. See W. K. Gregory's studies upon the affinities of the Lemuroidea, forthcoming in Amer. Mus. Bulletin.

[W. K. GREGORY: Bull. Geol. Soc. Amer. 26: 419-446. 1915.—E.H.C.]

any one past fauna, known or inferential, of that continent. The attempt to derive it from the present or from any known or inferential past fauna of India involves still greater difficulties. On the contrary, the Malagasy mammals point to a number of colonizations of the island by single species of animals at different times and by several methods. Of these colonizations, the Centetidæ are the earliest, perhaps pre-Tertiary; the lemurs, rodents and viverrines are derivable from one or more middle Tertiary colonizations; and in both cases the "raft" hypothesis may reasonably be invoked.[33] The hippopotami may have arrived by swimming and the bush-pig and the shrew may have been introduced by man, while the bats may readily have arrived by flight. The extinct ground birds are easily derived from flying birds.

Dr. Arldt,[34] in his discussion of the Malagasy fauna, points out its composite character, derived from several successive invasions. This, I think, is clear enough; but it seems equally clear that these were not faunal invasions due to land connection but sporadic colonizations by a few species all at different times. The characters of the mammalian fauna, both negative and positive, practically exclude the theory of land connections during the Tertiary.

The West Indian islands afford another marked instance. In spite of its nearness to Florida, there are no North American mammals in Cuba, except the manatee,—analogous with the hippopotamus in Madagascar. Nor are the other islands richer in fauna. As also in Madagascar, we have a peculiar and very primitive insectivore *Solenodon* (Cuba and Haiti), a number of peculiar extinct ground-sloths, of which *Megalocnus* is the best known, and which although Pleistocene in age are derivable not from the Pliocene or Pleistocene ground-sloths of North or South America but from the Miocene ground-sloths of Patagonia, and evidently differentiated through a long-continued period of isolated evolution, and a couple of chinchillas—the hutias of the larger islands, the (extinct) *Amblyrhiza* in Anguilla. The *Solenodon* may be referred to a more ancient colonization, the ground-sloths probably arrived during the Miocene, the chinchillas more recently; and the direction of the prevalent ocean currents points out the reason why these are of South American derivation. Those who, like Dr. J. W. Spencer,[35] believe in gigantic elevation

33. The moist tropical conditions of early Tertiary times would favor the formation of such rafts, the small size and arboreal habits of the animals concerned would increase the chances of their being caught on such rafts and the uniform climate and consequently more placid seas would increase the distance over which the raft might be transported before it broke up.

34. THEODORE ARLDT: Entwicklung der Kontinente und ihrer Lebewelt. 119-142. 1907.

35. J. W. SPENCER: "Reconstruction of the Antillean Continent," Bull. Geol. Soc. Amer. 6: 103-140. 1895.

movements connecting the Antilles with the mainland in Pliocene and Pleistocene would account for the absence of the continental fauna by invoking a subsequent subsidence which drowned out everything else. The improbabilities involved in this hypothesis on stratigraphic and faunal grounds have been pointed out by W. H. Dall, R. T. Hill[36] and others.

Cuba, while near in actual distance to the North American continent, has been comparatively inaccessible to sporadic colonization from that source, on account of the direction of the ocean currents; but colonizations from South or (possibly Central) America have reached it. New Zealand is more remote and inaccessible, and, during the whole Mesozoic and Cenozoic eras, we have evidence of but two colonizations by land vertebrates, neither implying any necessary continental connection. The rock-lizard (*Sphenodon*) may, for aught we know to the contrary, be derived from a marine form; all its early Mesozoic relatives were aquatic, some apparently marine. The few other Reptilia may be best accounted for by sporadic colonizations of later date. The moas are probably derivatives from flying birds.

When we come to the smaller oceanic islands, their poverty of fauna is still more conspicuous. If their fauna is due to sporadic colonization, this should be expected, as the chances are reduced directly in proportion to the smaller length of coastline on which an immigrant might land, as well as by their effective distance from the mainland. The colonization of a group of islands one from another may be due to former land connection and subsequent isolation, or to the same method of accidental transport, subject to the same laws of chance.

It is quite possible that in certain instances the small size and unfavorable environment of islands formerly connected with the continent may account for non-survival of the continental fauna. The Falkland Islands are a case in point; but even here, we find the survivors closely allied to the continental fauna and including types which afford the conclusive proof of continental connection which is uniformly lacking in oceanic islands.[37]

The characteristics of continental and oceanic island faunæ have been very fully and ably elucidated by Wallace (Island Life), and it is intended

36. W. H. DALL: "Geological Results of the Study of the Tertiary Fauna of Florida," Trans. Wagn. Inst. 3(6). 1903.

R. T. HILL: "Geological History of the Isthmus of Panama and Portions of Costa Rica," Bull. Mus. Comp. Zoöl. 28: 151-285. 1898.

37. Introduction of *Canis antarcticus* by human agency in prehistoric times is, however, a possible explanation of its occurrence. It is the only alternate to a Pleistocene land connection.

here merely to assert that the progressive increase of our knowledge of the past life of the world tends only to emphasize the distinctions in the source of their faunæ which he has so clearly demonstrated and, so far as my acquaintance with the subject goes, to reduce still further the number of continental connections which he regarded as permissible.

To the argument so often advanced that the transportation of a species across a wide stretch of sea and its survival and success in colonizing a new country in this way is an exceedingly improbable accident, it may be answered that, if we multiply the almost infinitesimal chance of this occurrence during the few centuries of scientific record by the almost infinite duration of geological epochs and periods, we obtain a finite and quite probable chance, which it is perfectly fair to invoke, where the evidence against land invasion is so strong. Furthermore, the fact that continents have not in general been peopled in this way one from another is well accounted for by the fact that species already existed there which filled the place in the environment and by their competition prevented the new form from obtaining a foothold, or greatly reduced the chances thereof. In oceanic islands, however, the favorable environment existed without the animal to fill it. Very often, on account of this lack, some other type was evolved to fill its place; birds being widely distributed on account of their powers of flight have in many oceanic islands developed large terrestrial adaptations to take the place of the absent or scanty mammals.

Natural Rafts and the Probabilities of Over-sea Migration thereby

The following series of facts and assumptions may serve to give some idea of the degree of probability that attaches to the hypothesis of oversea transportation to account for the population of oceanic islands.

(1) Natural rafts have been several times reported as seen over a hundred miles off the mouths of the great tropical rivers such as the Ganges, Amazon, Congo and Orinoco.[38] For one such raft observed, a hundred have probably drifted out that far unseen or unrecorded before breaking up.

(2) The time of such observations covers about three centuries (I set aside the period of rare and occasional exploring voyages). The duration of Cenozoic time may be assumed at ((sixty)) million years (((Barrell's)) estimate).

(3) Living mammals have been occasionally observed in such records of natural rafts. Assume the chance of their occurrence (much greater than of their presence being noticed) at one in a hundred.

(4) Three hundred miles drift would readily reach any of the larger oceanic islands

---

38. Popular Science Monthly **79**: 303-307. Sept., 1911. Gives the recorded observations of the drift of a natural raft of this sort, covering over a thousand miles of travel.

except New Zealand. Assume as one in ten the probability that the raft drifted in such a direction as to reach dry land within three hundred miles.

(5) In case such animals reached the island shores and the environment afforded them a favorable opening, the propagation of the race would require either two individuals of different sex or a gravid female. Assume the probability of any of the passengers surviving the dangers of landing as one in three (by being drawn in at the mouth of some tidal river or protected inlet), of landing at a point where the environment was sufficiently favorable as one in ten, the chances of two individuals of different sexes being together might be assumed as one in ten, the alternate of a gravid female as one in five. The chance of one of the two happening would be $1/10 + 1/5 = 3/10$. The chance of the species obtaining a foothold would then be $3/10 \times 1/3 \times 1/10 =$ one in a hundred.

If then we allow that ten such cases of natural rafts far out at sea have been reported, we may concede that 1000 have probably occurred in three centuries and ((200,000,000)) during the Cenozoic. Of these rafts, only ((2,000,000)) will have had living mammals[39] upon them, of these only ((200,000)) will have reached land, and in only ((200)) of these cases will the species have established a foothold. This is quite sufficient to cover the dozen or two cases of Mammalia on the larger oceanic islands.

Few of these assumptions can be statistically verified. Yet I think that, on the whole, they do not overstate the probabilities in each case. ((The first is, I am confident, a gross understatement.)) They are intended only as a rough index of the degree of probability that attaches to the method, and to show that the populating of the oceanic islands through over-sea transportation, especially upon natural rafts, is not an explanation to be set aside as too unlikely for consideration.

I have considered the case only in relation to small mammals. With reptiles and invertebrates, the probabilities in the case vary widely in different groups, but in almost every instance they would be considerably greater than with mammals. The chance of transportation and survival would be larger and the geologic time limit in many instances much longer. Wind, birds, small floating drift and other methods of accidental transportation may have played a more important part with invertebrates, although they cannot be invoked to account for the distribution of vertebrates. The much larger variety and wider distribution of infra-mammalian life in oceanic islands is thus quite to be expected. And the extent and limits of such distribution are in obviously direct accord with the opportunities for over-sea transportation in different groups.

On the other hand, the transportation of very large animals in this way may fairly be regarded as a physical impossibility, which could not be multiplied into a probability by any duration of time. The only methods of accounting for such animals would be by evolution *in loco* from small

---

39. Small reptiles and invertebrates would only rarely be observed, if present.

ancestors, by swimming, by introduction through the agency of man and by actual continental union.

The first hypothesis would involve evolution in an isolated and more or less altered environment and would result in wide structural differences from any continental relatives. The second applies with greater probability to large than to small animals, but, except for animals of more or less aquatic habits and within certain limits of distance, it is an apparent physical impossibility. The third may be either intentional or accidental and should be considered in connection with the known custom among Malays and other races, of taming various captured animals and taking them along on sea-voyages. Its application is, of course, limited to distributional anomalies of late Pleistocene or modern origin. The last hypothesis, where it traverses the doctrine of the permanence of ocean basins, appears to me unnecessary, as I have failed to find a single instance of distribution which cannot reasonably be otherwise explained.

Considerations Affecting Probabilities of Over-sea Migration in Special Cases

The probabilities of over-sea transportation to an oceanic island will obviously be much greater if the island is large, and correspondingly reduced if it be of small size. The distance from the mainland will greatly reduce the chances of such rafts making a landing, for two reasons: first, the chances of survival of the animals are reduced proportionately to the length of their journey (or rather, in a varying relation, which for convenience we may consider as a direct proportion); second, most rafts will be carried out from one or more points along the coast, but not from all points equally (that is to say, from the mouths of one or more great rivers, where the conditions are favorable, seldom from any of the small rivers). If we disregard prevalent winds and currents and consider the rafts as drifting out in all directions the probability of their landing on a given island will be directly proportioned to its length opposite the mainland, inversely to the distance. The probabilities of survival of animals, so far as it depends on the raft holding together, will also be inversely as the number of days exposure to the sea, hence as the distance. Comparing Saint Helena, 1100 miles from Africa and 10 miles in diameter, with Madagascar, 200 miles from Africa and 1000 miles in length, we see that the probabilities of effecting a colonization would be $100 \times 3\frac{1}{2} \times 5\frac{1}{2}$, or 3025 times greater in the case of Madagascar. New Zealand, 800 miles long and 1200 miles from the Australian coast, will receive $8/10 \times 1/6$

$\times 1/6$, or $1/45$ as many colonizations as Madagascar, but $80 \times 11/12$ $\times 11/12$ or 67 times as many as Saint Helena.

I believe that it is to their small size rather than to unfavorable conditions for survival that the poverty of fauna, especially of higher vertebrates, in the smaller oceanic islands is due.

The oceanic currents and prevalent winds do, of course, profoundly modify the above generalities in each individual instance. They have prevented the populating of Cuba from North America, while facilitating invasions from South and Central America. The present set of currents reduces the probability of mammals reaching Madagascar from the African mainland, while increasing the chances of Oriental animals reaching it. It reduces materially the opportunities for Australian fauna to reach New Zealand.

We have no adequate data on which to base theories as to the former set of oceanic currents. A worldwide uniformity of climate would probably reduce the north and south movement of the waters; the east and west element of their motions is conditioned by the rotation of the earth, and its velocity would be reduced proportionately to the north and south movements; so that a more uniform climate would bring about a reduction of velocity rather than change in direction. The third principal conditioning element is the conformation of the continents, and doubtless the flooding of great areas and the opening up of broad though shallow passageways between seas now separated would profoundly modify the surface currents in many regions. The opening of a broad passage between North and South America would allow the Caribbean current to pass into the Pacific instead of being deflected northward and eastward along the shores of the Gulf of Mexico to find an outlet between Cuba and Florida. The absence of this initial part of the Gulf Stream would obviously be unfavorable to North or Central American animals reaching western Cuba. The great equatorial current would sweep across from Africa along the northern coast of South America, and uninterruptedly into the Pacific; transportation from Africa to South America or from South or Central America to the Galapagos Islands would thus be facilitated.

DISPERSAL OF MAMMALIA

Mankind

We may with advantage begin our review of the special evidence in support of our theory with the migration history of man. This is the most recent great migration; it has profoundly affected zoögeographic conditions; it is the one where our data are most complete and accurate; we can perceive its causes and conditions most clearly, and we have a great deal of corroborative evidence in history and tradition.

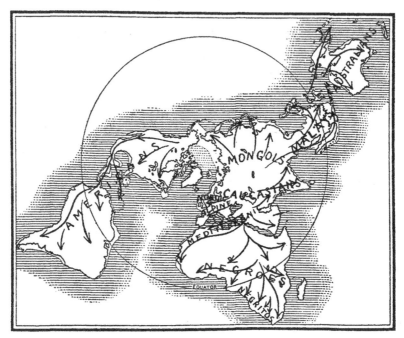

FIGURE 6. Dispersal and distribution of the principal races of man

No attempt is made to indicate anything beyond the broader lines of dispersal.

((Most)) authorities are to-day agreed in placing the center of dispersal of the human race in Asia.[39a] Its more exact location may be differently interpreted, but the consensus of modern opinion would place it probably in or about the great plateau of central Asia. In this region, now barren and sparsely inhabited, are the remains of civilizations perhaps more ancient than any of which we have record. Immediately around its borders lie the regions of the earliest recorded civilizations,—of Chaldea, Asia Minor and Egypt to the westward, of India to the south, of China to the east. From this region came the successive invasions which overflowed Europe in prehistoric, classical and mediæval times, each tribe pressing

---

[39a. Recent discoveries by Dr. Robert Broom show the presence of very primitive hominids in South Africa.—E.H.C.]

From *Zoogeography of the Sea*
Sven Ekman

pelagic types show a more pronounced benthal origin, thus the group Pelagica among the nemerteans, since epipelagic nemerteans are absent; the genera *Melanoteuthis* and *Cirrothauma* and the families Bolitænidæ, Amphitretidæ and Vampyroteuthidæ among the otherwise mainly benthal octopodid squids; the Pelagothuriidæ, which represent the only pelagic type among the echinoderms, and some fish families, namely the Ceratioidea, Saccopharyngidæ and the pelagic members of the Apoda. The nearest relatives of these fish are true bottom forms which partly belong to the deep sea and partly to the shelf.

We now turn from the biocœnotic to the regional origin. Because of the cold-water character of the bathypelagic fauna we might incline to the view that this fauna derives for the most part from the arctic and antarctic pelagic fauna. But this cannot be true since the polar pelagic faunas are considerably less rich in species than that of less cold regions. It is in these latter regions that we must look for the original home of the greater part of the bathypelagic fauna. This has been borne out by phylogenetic investigations on radiolarians[203], [417] and several metazoan groups. Theoretically, a polar origin is occasionally conceivable, for instance in connection with an equatorial submergence of a polar species.

*   *   *

CONCLUDING REMARKS

In the course of this work I have often had occasion to stress that as a characteristic of a zoogeographical region an endemic family is more important than an endemic genus, an endemic genus more important than an endemic species, and that endemic elements are more important than those which are common also to neighbouring regions. We must now devote further attention to these methodological questions.

The taxonomic (morphological) differences between two closely related genera are more comprehensive than the differences between two species of the same genus, and palæontological discoveries show that the generic differences have taken a longer time to develop than differences between species. In general, the endemic genus of a region has thus lived a longer time in the environmental conditions of this region than an endemic species in the environment of *its* region. The same is true of an endemic family as compared with an endemic genus. There is, therefore, in general a parallel between the taxonomic rank of an endemic element and the time it has lived in the environment in question. The "environment", of course, does not apply to regions with purely geographical delimitation, for instance by certain longitudes and latitudes, since these have often been subjected to climatical and other alterations, but to the ecological conditions to which the species (genus, family, etc.) is adapted.

The taxonomic scale contains, therefore, historical documents of great zoogeographical value. It is important in this connection that for close on 100 years taxonomy has operated mainly with phylogenetic, that is to say historical, concepts.

The surest basis for a historical zoogeography is clearly provided by palæontological, palæogeographical and palæoclimatological evidence. This is available to a certain extent and has been drawn upon in the preceding chapters. But as is well known, palæontology is by no means free from gaps and here taxonomy thus can provide a welcome complement.

Another factor which deserves special attention in the characterization of a fauna is endemism. Because of its exclusive occurrence within a environmental region a greater value must be attached to an endemic than to a non-endemic species, genus, etc., in the characterization of this region. We may imagine, for instance, an endemic genus with five species all of which are thus endemic in the same region. This genus will clearly be more characteristic for this region than another genus which likewise possesses five endemic species within the said region, but has some species in other regions as well. The same criterion must also be applied to an endemic family as compared with a non-endemic family. That an endemic species is more characteristic for a region than a species which is also found in other regions, is obvious.

The characterization of zoogeographical regions and the assessment of their relationship to one another results in a regional zoogeographical system with a graduated scale of super- and subregions. The parallel with the taxonomic system and its scale of classes, orders, families, etc., is clear. And just as the final aim of taxonomic research is not the graduated scale *per se* but the unravelling of the historical (phylogenetic) relationships between the taxonomic categories and thus the history of the animal kingdom, in the same way the final aim of zoogeography is not the graduated regional system in itself but the history which this system reflects, that is the history of the faunas. Zoogeography, like other sciences, strives to discover the ultimate causes. And the causal connections here are, as in so many other cases, to a great extent historical connections. Hence the importance of the parallelism between the rank of a region within the zoogeographical system and the position of its faunal constituents within the taxonomic scale.

To this must be added yet another factor. In the assessment of the

CONCLUDING REMARKS   373

position of a fauna within the zoogeographical system, that is to say its greater or lesser independence as a centre of development, we may introduce mathematical values for its various elements according to age within the environmental region, endemism or other facts, and these values may be combined into more comprehensive figures by summation, multiplication and division. This offers considerable advantages. For there is a fair number of partial values, which must be graded and combined so as to arrive at a reliable estimation: the value for families, genera and species in comparison to one another, the value for endemism and non-endemism in various taxonomic elements, the value for affinity with other regions as compared with independence, etc. It is advantageous to have a statistically expressed survey in order to be able to summarize the many combinations and so reach as far as possible an objective estimate instead of a more or less arbitrary and subjective one. The basis for such a statistical estimate is, of course, a fully adequate faunistical knowledge of the group or groups of animals with which this analysis is concerned. Space precludes a detailed description of the method. For this I must content myself with a reference to an earlier paper.[147]

*   *   *

Like all other biological phenomena, geographical distribution is the product of an interaction between two factors, namely the physiological properties of the living entity and the quality of the environment. The organisms must distribute themselves regionally in conformity with their own genotypical nature which is adapted to certain environmental conditions. But in the geographical distribution of the various species which thus comes about, there is not stagnation but change. And this change has its cause in the same two factors which we have mentioned above. The germ plasm may change; we call such changes mutations. Through these the organism becomes adapted ("pre-adapted") to new environmental conditions and is able to take possession of new regions. And the environment may change in various ways and thus give formerly useless mutations a "place in the sun". On the other hand changes in the germ plasm and environment may become unfavourable to the species and lead in time to its extinction. As far back as we have been able to trace life into ancient times, each geological period has shown examples in plenty of changes both in the organisms and in inanimate nature. Biologists have no difficulty in regarding time as a sort of fourth "dimension" in the whole of nature.

Throughout the phylogenetic evolution species, genera, families and so on and faunas have appeared, changed and disappeared. By the events in inanimate nature mountains and the deeps of the sea,

147. Ekman, S. (1940) Begrundung einer statischen Methode in der regionalen Tiergeografie. *Nova Acta Reg. Soc. Sci. Upsaliensis*, 12:2.

374   CONCLUDING REMARKS

ocean currents and climatic zones have appeared, changed and
disappeared and as a result of interactions of infinite complexity
between animate and inanimate nature the present biogeographical
conditions have emerged in the course of the ages. Time, which is in
reality nothing more than the succession of events, that is historical
happenings, is a factor of profound importance for all manifestations
of life. In other words: biogeography cannot confine itself simply to
describing the occurrence of living forms, arranging them regionally,
investigating the ecological causes of distribution. It must also
proceed historically.

From *An Introduction to Historical Plant Geography*

Evgenii Vladimirovitch Wulff

Chapter X

HISTORICAL CAUSES FOR THE PRESENT STRUCTURE OF AREAS AND THE COMPOSITION OF FLORAS

From the preceding chapters it is clear that in many cases the structure of the areas of species and the composition of floras cannot be explained by existing factors. The present distribution of any given species is a reflection of the geological revolutions and climatic changes that have occurred on our globe during the entire period of existence of that species. An elucidation of these great changes in the surface of our planet is the task of historical geology and paleogeography, a task as yet far from fulfillment. Consequently, the elucidation of the history of areas, the most difficult task of biogeography, likewise falls far short of achievement. In the present chapter we can, therefore, do no more than examine the chief theories that have been advanced and point out which give the most plausible and satisfactory explanation of the knotty problems of historical plant geography.

From very ancient times—at first without adequate foundation and later, with the development of geology and biogeography, on the basis of numerous data—the conviction has been held that the distribution of lands and seas was not always the same as now. For, if it had been, a considerable number of facts, both of a geological and biogeographical nature, would be inexplicable. The sedimentary character of the rocks covering extensive territories on the continents and the finding in these rocks of fossil marine animals testify to the fact that at one time seas covered these parts of the continents. That many islands formerly constituted a part of the mainland is shown by the geological structure of these islands, by their fossil and extant fauna and flora, and by the finding of submerged trees in various straits and channels, *e.g.*, in the English Channel. Furthermore, the outermost edges of continents and islands do not necessarily coincide with their shore lines, as their outer margins often lie submerged under so-called "shelf seas". The latter differ in extent, but their boundaries may be ascertained with considerable precision. The determination of these boundaries gives certain clues to the changes that have taken place in the distribution of lands and seas on the globe and on the probable existence in former times of connections between bodies of land now separated by the sea. Biogeographical data, in many cases, indicate that these changes occurred at a comparatively recent date.

1. *Theory of Land Bridges.*—We have already seen that a considerable number of cases of discontinuous areas of plants (and these might be supplemented by an equal number of instances of similarly distributed animals) cannot be regarded as accidental and require explanation. Often there seems to be only one possible explanation, *viz.*, that there formerly existed some sort of connection between the isolated habitats and that the now discontinuous areas were formerly continuous. Hence, many investigators have assumed that at one time

164

great "lost continents" or land-bridges of one form or another connected the continents now separated by oceans. On the basis of geological and paleontological data for different geological periods there have been postulated various connections between the continents, presumably later having sunk to the bottom of the sea and having been replaced by the upheaval of other land-bridges connecting other bodies of land. According to this theory, the distribution of land and sea has undergone constant change during the long history of the earth.

The complex conformation of these putative land-bridges clearly testifies to the artificial character of the hypotheses resting upon their probable rôle. This, together with the fact that there are a number of geophysical and geological arguments against this theory, has made it necessary to seek for new ways of explaining the former continuity of the now-discontinuous areas of organisms, all the more since the time when these land-bridges were supposed to exist is not always such as to be able to account for such continuity.

Even as regards biogeography, to which the creation of this cumbrous theory was a concession, it far from solves all the incomprehensible moments in the distribution of organisms. In particular, it leaves unclarified why plants in former geological periods grew in regions outside the climatic zones in which their present habitats are found, and also, which is very important, it gives no satisfactory explanation of discontinuous areas. If identical or closely related species are found on two continents separated by an ocean, we cannot explain this discontinuity of area merely by assuming the former existence of an intervening continent where the ocean now lies. It would likewise be necessary to assume that over the entire extent of this great "lost continent" there existed like ecological conditions, similar to those in the outlying portions of the area of the given species, these portions having been preserved in the form of isolated fragments of a once-continuous area. This can hardly be regarded as an acceptable hypothesis.

2. *Theory of the Permanence of Oceans and Continents.*—Against the theory of land-bridges, despite its partial fulfillment of the requirements of paleontology and biogeography and its acceptance by many geologists, there were advanced, beginning with the middle of the past century, a number of serious objections. On the basis of these objections a new theory was proposed, the theory of the permanence of oceans and continents. Among biologists this theory found many supporters, the most outstanding being DARWIN and WALLACE. The arguments advanced against the existence in the past of great continents that subsequently subsided to the bottom of the sea are so weighty that biogeographers cannot fail to take them seriously into account. These arguments, in brief, are as follows:

Upon the upheaval of continental masses in the area of our present oceans the great mass of water that had been in these oceans would have inundated all the continents, both old and new, except for the highest mountain peaks. Thus, the very continental connections that it was desired to create by the assumption of the existence of such "lost continents" would not have existed.

But if it be assumed that the position of the oceans and continents in former periods of the earth's history was not the same as at present, that where there are now continents, or at least over part of their territory, there were deep oceans and where the latter now are there were continents, then in the rocks of which our continents are composed there should be, over considerable expanses of territory, deep-sea deposits, attesting that at one time the given continent was at the bottom of the ocean. Such deep-sea deposits, however, are not to be found on our continents. This serves as grounds for concluding that our present continents were never covered by oceans and, in contrast to the ocean depths, always constituted elevated land-masses. On their surface in former geological periods, just as at present, there existed relatively shallow seas, but a large part (at least one-third) of their surface was always dry land (SOERGEL, 1917, p. 11). Slight variations in the level of the ocean, amounting to as much as several hundred meters, accompanied by a depression or elevation of the strand lines, may change considerably the contours of our continents, but they cannot affect their permanence as a whole nor the permanence of the ocean beds.

To geophysical objections to the land-bridge theory there may be added objections based on geological and paleontological data. These bear witness to the absence of transitional forms between the fossil marine faunas of the earlier geological periods and the sudden appearance of whole groups of species not connected with those of the formations preceding them in sequence of time, as would be required by the theory of evolution. These facts indicate that the successive phases of development of the inhabitants of the sea did not take place on the surface of our continents but must have taken place within the boundaries of the oceans as at present constituted. Hence, the following conclusions are drawn by the advocates of the theory of permanence (SOERGEL, 1917, p. 15): —

1. The great areas now occupied by oceans must have been thus occupied always, at least ever since the Pre-Cambrian era.

2. Fossil marine faunas have no roots on our present continents; they merely represent the repeated inland migrations of sea animals. (This is further confirmed by the absence in the faunas of present-day inland seas of elements that have preserved features of a deep-sea origin).

3. The territory occupied by our present continents has constituted a habitat of marine fauna always in contrast to the territory of the present oceans, a circumstance which is only understandable if one assumes that the former was an elevated territory, subject to alternate upheaval and depression, *i.e.*, a territory of a continental character. This means that the continents must have been permanent.

These conclusions, however, leave entirely unexplained the biogeographical data that indicate the need of assuming that the continents were at one time connected. It is true that now even advocates of the theory of permanence try to find a way out of the contradictions created. They state that, though we cannot concede the existence of large trans-oceanic land-bridges, we may concede the existence of some sort of connection between the continents, *e.g.*, narrow land-bridges between North America and Europe, Australia and South America,

Australia and Asia, Madagascar and Africa, the Antarctic continent
and South America (SOERGEL, 1917). But this concession to biogeography is not founded on any new facts. Moreover, conceding the
existence of such narrow land-bridges does not, by any means, solve all
the unexplained problems in the distribution of organisms; it is necessary, in order to explain the latter, to presume still other connections
not compatible with the theory of permanence. From this vicious
circle these early theories provide no exit. An exit is found, in our
opinion, only in WEGENER's theory of continental drift, which we shall
later discuss in detail.

4. *Pendulum Theory.*—An attempt to reconcile the contradictions
involved in the two theories above outlined was made by the advocates
of the so-called "pendulum theory", that aimed also to explain, on the
basis of climatic changes in former geological periods, those cases of
plant distribution that are not at all in accord with present-day climatic zones.

Changes in the position of the climatic zones on the globe—attested
to by data on the distribution of fossil plants, showing, for instance,
that at one time there was a rich flora within the area of the present
Arctic region—must presumably have been caused not by a different
location of the sun in relation to the earth but by a different location
of the continents in relation to the sun. The pendulum theory explains these phenomena by assuming that periodic changes have occurred in the position of the poles caused by their oscillating back and
forth like a pendulum. This theory was first advanced by the geologist
REIBISCH and later elaborated on the basis of biogeographical data by
SIMROTH (1914). These investigators start from the assumption that
the earth, besides an axis of rotation and poles of rotation located at
the north and south ends of this axis, has an axis on which it oscillates
like a pendulum. The two poles of this latter axis—known, in SIM-
ROTH's terminology, as "Schwingpolen"—are located one in Ecuador
and the other in Sumatra. Not subject to the oscillatory motion are
only this axis and its two poles, Ecuador and Sumatra. These points
alone remain fixed and under constant tropical conditions, while all
the other points on the earth's surface are subject to periodic changes
in climatic conditions induced by the oscillatory motion. Moreover,
the degree of these climatic changes is determined by the distance
from the equator. The greater this distance, the greater the deviations in climate suffered by any given point on the earth's surface.
Conversely, those regions of the earth nearest to the equator, particularly those nearest to the "Schwingpolen", are subject to the least
changes, climatic and physiographic, and, consequently, in these regions ancient plant and animal forms have been preserved to a much
greater extent than in those regions located on the outer arc of oscillation.

If the pendulum theory is accepted, there is no need for the theory
of land-bridges, since changes in the level of the sea induced by this
oscillation of the earth would suffice to cause the joining together and
disjoining of different parts of the earth's land surface and, moreover,
in such a way as serves to explain the distribution of plants and animals. SIMROTH's theory of the distribution of organisms is based on

those changes in the surface of the continents, particularly climatic changes, induced by the shifting of the position of the continents as a result of this oscillatory motion.

Every animal and plant, in the process of multiplication, naturally tends to spread within the limits of the same climatic zone within which it arose. Consequently, in the absence of obstacles to such distribution, their areas should encircle the globe in a band covering the territory occupied by the given climatic zone. This regularity in the distribution of organisms is upset by the oscillation of the earth. Thus, a plant or animal finding itself in the region of the outer arc of oscillation will be mechanically evicted from the climatic conditions to which it has been accustomed and will be forced to migrate to the west or east in order to return to its normal habitat conditions. This serves as an explanation of those discontinuous areas consisting of two halves symmetrically located on opposite sides of the arc of oscillation ("symmetrische Punkte" in SIMROTH's terminology). Moreover, if a species, as a result of such a change in habitat, does not undergo, in the process of adaptation to the new habitat conditions, vital changes in its morphological structure but acquires only a few, slight modifications, we shall find at these two points vicarious species. As an example of such "symmetrical points", we may take Japan and California. In case of the distribution of organisms farther inland in western North America and eastern Asia within the limits of the same latitudinal zone, we may speak of the *horizontal symmetry* of the distribution. This type of symmetry occurs chiefly during the polar phase of the oscillation. During its equatorial phase, on the other hand, the habitats shift to a very hot climatic zone, compelling marine animals, in order to attain more temperate climatic conditions, to go farther down into the sea and to adapt themselves to deep-water conditions, while terrestrial animals must ascend the mountains or migrate to the north or south, *i.e.*, in a meridional direction. As a result, we have distribution characterized by *meridional symmetry*, which is of most frequent occurrence in the outer arc of oscillation.

SIMROTH based his theory chiefly on carefully elaborated data on the geographical distribution of animals, but he also gave a number of examples of plant distribution that likewise agreed with his theory. Such agreement is expressed, first of all, in the actual existence of symmetry in the present-day distribution of certain species and in the fact that there are data establishing the existence in former times of habitats of these same species on corresponding parallels in the outer arc of oscillation. We shall here cite the most characteristic examples of those presented by SIMROTH and other investigators.

Let us first take a few instances from gymnosperm distribution. Out of fifteen species of the genus *Gnetum* seven grow in equatorial America, *i.e.*, in the region of the western "Schwingpole", one in Africa, and one on islands of the Pacific, while the six remaining species are grouped around the eastern "Schwingpole" in the eastern part of the Indian Ocean. Consequently, there is no doubt that we have here horizontal symmetry.

In the genus *Pinus* of most interest is the section *Taeda*, which comprises sixteen species. One group of these species grows in America

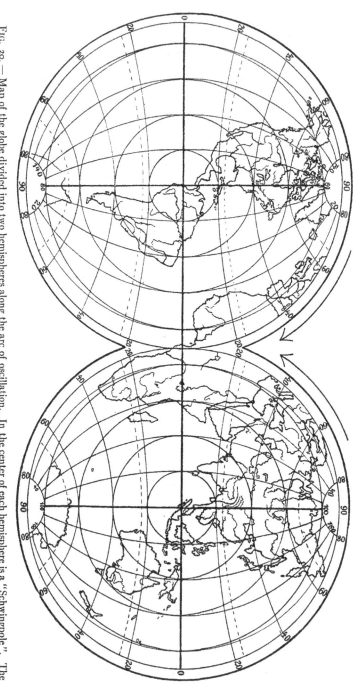

Fig. 20. — Map of the globe divided into two hemispheres along the arc of oscillation. In the center of each hemisphere is a "Schwingpole". The vertical meridian is the "culmination circle" (in the hemisphere at the left the 80th meridian, at the right, the 100th). The concentric circles represent the lines of motion of separate points during oscillation. (After Reibisch, from Simroth).

(in Florida and North and South Carolina, in the Rocky Mountains, in California, and in Mexico); the other group is found at a symmetrically opposite point in eastern Asia (the Himalayas, Tibet, and the Philippines). The area of one species, *P. canariensis*, growing in the mountains of the Canary Islands, seems to be an outlying spur of the former, more extensive range of this latter group.

Of two relic species of pine found on the Balkan peninsula, *Pinus Peuce* and *P. omorica*, the former is related to the eastern white pine, *P. Strobus*, of North America, and the latter to species of Manchuria and Japan. We have here, therefore, a clear case of the breaking up of a once-continuous area into two widely separated sections, one in the east and one in the west.

Analogous facts are likewise found in the distribution of angiosperms. For instance, the genus *Magnolia* has about 60 species, distributed in eastern Asia and in the Atlantic states of North America. *Liriodendron Tulipifera*, another member of the magnolia family, also grows both in China and in the Atlantic states of North America. The genus *Talauma*, also of the *Magnoliaceae*, has 32 species in India, Java, and the Philippines and 8 species in the New World (the West Indies, Mexico, Central America, and the northern part of South America). In a fossil state the *Magnoliaceae* are found in the outer arc of oscillation in deposits in Europe (beginning with the Cretaceous and ending with the Pliocene stage) and as far north as Greenland and Spitsbergen.

At first some biologists regarded the pendulum theory favorably, since they hoped to find in it a solution to those inexplicable features of the distribution of organisms about which we have already spoken. However, there are very serious objections to this theory which make it impossible to adopt it even as a working hypothesis. The chief objections are, first, that no cause can be found for such an oscillation, and, second, that geological data are not in accord with this theory. But even biologists advanced a number of serious objections, of which we shall note the most important. The assumption made by SIMROTH that Europe, as the region that has been most subjected to climatic changes, has constituted the chief center of origin of new forms is merely a hypothesis without any foundation. It is more logical to assume the exact contrary, since such great climatic changes as occurred in Europe during the Ice Age resulted in the creation of nothing essentially new. Europe, as compared with the other continents, is a comparatively small territory, and during the Tertiary period it was even smaller than now, so that it is difficult to believe that it constituted the place of origin of most of the flora and fauna of the world. Moreover, the existence of such centers of species-formation in other continents, *e.g.*, Asia, is beyond any doubt. Two other important objections are that many of the facts in the distribution of organisms cited by SIMROTH as caused by such oscillation of the earth may be explained, without assuming such oscillation, on the basis of ecological and edaphic data, and, lastly, that not all cases of discontinuous areas can be explained by the horizontal or meridional migration of species.

Nevertheless, the fact of symmetry in the discontinuous areas of many species and the fact that the break in these areas occurs precisely in the Euro-African sector were quite correctly established by

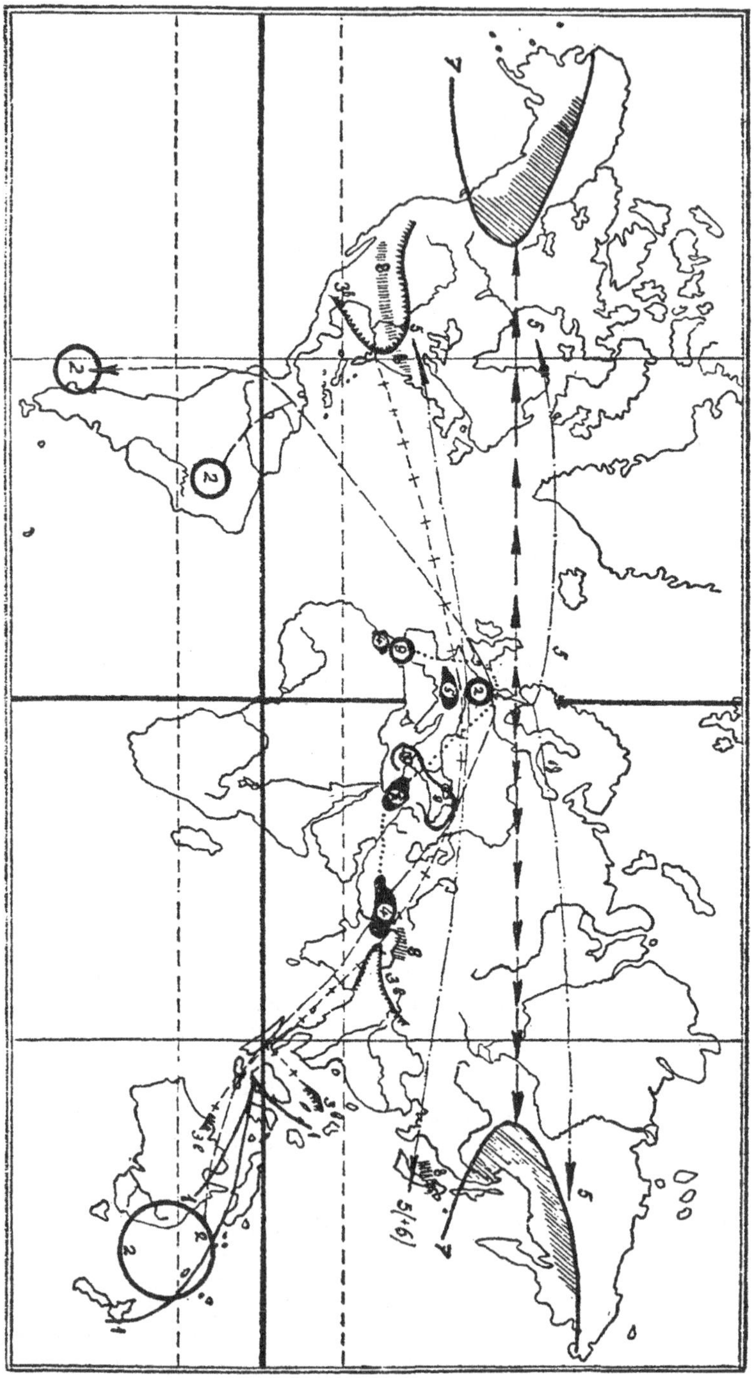

FIG. 21. — The Distribution of the *Coniferae*, according to SIMROTH: 1, *Dammara*; 2, *Araucaria*; 3, *Pinus Taeda*; 4, *Cedrus*; 5, *Larix*; 6, *Pseudo-larix*; 7, *Picea sitchensis*; 8, *Tsuga*; 9, *Abies pinsapo*.

SIMROTH; this symmetry, however, may be ascribed to other causes, as we shall show below.

5. *Theory of the Polar Origin of Floras.*—Another attempt to avoid the need of trans-oceanic land-bridges to explain the present distribution of floras is found in the theory of the origin of the latter from a single center lying in the north polar region, whence they spread radially toward the south in three directions—through Europe into Africa, through Asia and Malaysia into Australia, which were at that time connected, and through North America into South America. As a basis for this theory there served the investigations of HEER (1868) of the fossil flora of the Arctic.

HEER, in a number of papers, later collected in his "Flora Arctica", pointed out that in former geological epochs there grew in the Arctic woody plants now found only in temperate or subtropical regions. His data showed that climatic conditions in the Arctic during the Tertiary period were entirely different from now and supposedly confirmed the view that the polar region was the initial center of origin of floras.

This point of view was first advanced by FORBES and was later developed by DARWIN, BERRY, and others. It is based on data indicating that beginning with the Tertiary period the floras and faunas succeed one another in such a way that each successive one forces its predecessor to the south. This shifting of floras and faunas presumably began from the moment of the differentiation into climatic zones, which, according to the views then held, did not take place until the end of the Cretaceous or the beginning of the Tertiary period, prior to which the climatic conditions of the earth were allegedly uniform. Consequently, the lowering of temperature conditions, which reached its apogee during the Ice Age, must have had its effect on the vegetation of the entire globe and must have induced a migration of floras from north to south. On the basis of the foregoing, the proponents of this theory held that the flora that had inhabited the body of land encircling the North Pole, which body of land prior to the Ice Age was presumably even larger and more compact than now, constituted the initial flora from which arose all the vegetation at present inhabiting the earth, and that the steady decrease in temperature in the polar regions forced the vegetation ever farther and farther south (FÜRSTENBERG, 1909).

In order to explain the similarities in the floras of South America, Australia, and South Africa, and the affirmations of HOOKER as to the circumpolar nature of the antarctic flora, it was presumed that there was a trans-equatorial migration of the flora of the northern hemisphere into the southern, although DARWIN and HOOKER had themselves suggested that in the past there might have existed an antarctic continent embracing what are at present separate islands and the extremities of the continents of the southern hemisphere. Subsequently, the existence of such a continent was generally accepted, the remains of fossil flora found within the limits of present-day Antarctica having confirmed not only that the lands of the southern hemisphere had formerly been connected with one another by way of this Antarctic continent but also that there had occurred a change of climatic conditions in the Antarctic just as in the Arctic, in the sense of a decrease in temperature. Hence, there was created a basis for the presumption that there probably had been at one time a center of species-formation in the region around the South Pole, similar to that assumed for the North Polar region, and a migration of floras, under the influence of climatic changes, from south to north. Thus, the mono-boreal theory of the origin of life was replaced by the theory that this rôle was played by the lands encircling both poles.

The theory of the polar origin of floras, in the light of our present knowledge, cannot be accepted. It has now been established that climatic zones have existed during the entire history of the earth and that ice ages occurred not only in the Quaternary period but also during other geological periods, the glaciated regions, moreover, not being located in the present polar areas but in other parts of the globe.

The assumption that there were only these polar centers of development of floras is likewise contraverted by the fact that the existence of other centers of species-formation has been definitely established. The view advanced by HALLIER (1912), GOLENKIN (1925), IRMSCHER (1922, 1929), and others that the tropics served as a center of origin of the angiosperms, whence they at various times penetrated into temperate regions, has much data to support it, provided it is accepted that there took place shiftings of the tropical zone.

An approach to the solution of all these enigmatic moments in the history of the earth, making it seemingly impossible to explain the past distribution of its floras, has been provided by the theory of continental drift.

6. *Theory of Continental Drift.*—From the foregoing we have seen that the two chief theories as to the past history of the earth's surface —the theory of the permanence of the oceans and continents and the theory of trans-oceanic land-bridges—are mutually contradictory. A solution of this riddle may be found, if one accepts the permanence not of the separate oceans and continents as such but the permanence of the relative area of land and sea taken as a whole. Then, by assuming, as WEGENER (1929) does in his theory of continental drift, the possibility of a horizontal drifting of the continents, which, so to say, float on an underlying viscous substratum (the so-called "sima", composed of basic, igneous basalts), we are enabled, without departing from the principle of the permanence of oceans and continents, to explain the existence of connections between the continents, not on the assumption that there formerly existed additional continents where there is now sea, but on the assumption that our present continents were formerly in direct contact with one another.

According to this hypothesis, it is assumed that as late as the end of the Paleozoic era the continents were all united in one great continent, Pangaea, which only in the Mesozoic era begins to rift apart, two meridian lines of rupture being formed. These lines of rupture between Euro-Africa and America, on the one hand, and between Africa and India, on the other, lead to the creation of the Atlantic and Indian oceans. The separation of South America from Africa becomes wider and wider during the Cretaceous period, but even at the beginning of the Tertiary period there still persists a slight connection between the northeastern coast of South America and the west-central coast of

Africa, the complete separation of these two continents taking place only after the Eocene stage. At the same time, the sea separating America from Africa becomes wider and wider, due to the drifting of America westward.

The connection between Africa and India through Madagascar likewise persisted as late as the beginning of the Tertiary period, being broken only in the Eocene stage due to the drifting of India northward. This movement of India northward is confirmed, according to SAHNI (1936), by paleobotanical data. The Permo-Carboniferous flora of southern Asia belongs to the type of *Gigantopteris*, *i.e.*, it is a tropical flora, while the flora of India of the same period is of the *Glossopteris* type, undoubtedly adapted to a temperate climate. With India and Asia located as at present the contiguity of these two widely different floras would be entirely inexplicable. It must be presumed that India at that time lay considerably farther south and was separated from Asia by the Tethys Sea. Only later did India drift northward, resulting in the seeming juncture of these two floras.

At an earlier period, *i.e.*, in the Jurassic, Australia broke away from India and Ceylon and Antarctica from Africa. The latter, still retaining connection with South America, drifted in a southeast direction. Sometime during the Tertiary period Australia separated from Antarctica, which, however, preserved until the Quaternary period its connection with South America. Not until the Quaternary period, due to the westward drift of the Americas, did Antarctica break away from South America and drift farther and farther in the direction of the South Pole. At about the same time, during the Ice Age, Greenland broke away both from North America and Europe, thus causing the isolation of these two continents from each other.

From the foregoing we may conclude that the contact between Europe and North America persisted until the Quaternary period, between Africa and South America until the Eocene stage, between Africa and India also until the Eocene stage, between Australia and India until the Jurassic period, between Australia and Antarctica and South America until the middle of the Tertiary period, and, lastly, between Antarctica and South America until the Quaternary period.

DU TOIT (1937) gave in his recent book on "Our Wandering Continents" very important geological proofs of WEGENER's theory, at the same time introducing some modifications. According to his viewpoint, the continents were originally represented by two great land bodies. The southern, Gondwanaland, embraced Brazil, Guiana, Uruguay, Africa, Arabia, Madagascar, India, western and central Australia, and Antarctica; the northern, Laurasia, included central and eastern Canada, Greenland, Scandinavia, Finland, Siberia, and northern China. These two great continental masses were separated by the Tethys Sea. Regressions of the latter led for a short time to a connection between Gondwana and Laurasia (between Africa and Europe; between Indo-China and Australia).

The breaking up of Gondwanaland, according to DU TOIT, took place not earlier than the Cretaceous or, perhaps, the Tertiary period. India began its movement to the northeast, over a distance of 1,500 km., at the beginning of the Cretaceous period. During the Cre-

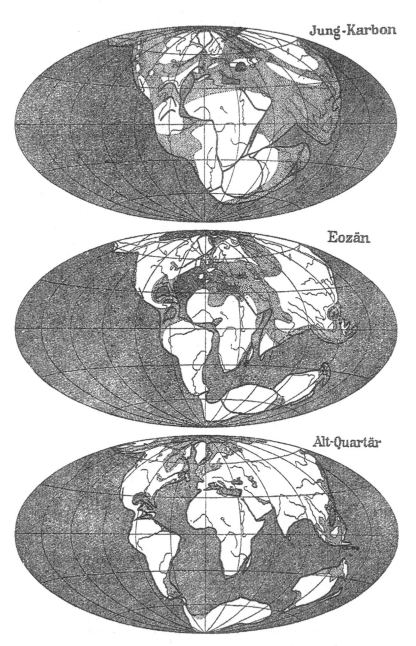

FIG. 22. — Reconstructions of the map of the globe according to data of the drift theory. Hatching = deep seas; stippling = shallow seas. Present-day rivers, contours of continents, etc. shown merely for purposes of orientation. (After WEGENER).

E. V. Wulff   —176—   Historical Plant Geography

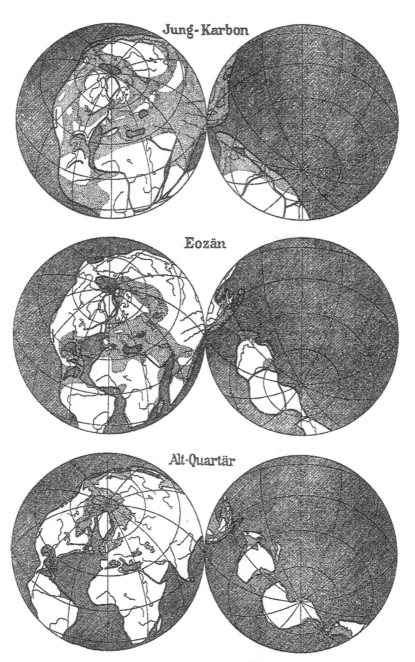

FIG. 23. — The same reconstructions as in FIG. 22, but projected in a different way. (After WEGENER).

taceous period New Guinea and New Zealand constituted the periphery of Australia. Not until the Tertiary period did they break away from Australia and drift off into the Pacific Ocean. The isolation of Australia, thus, presumably took place during the Tertiary period. At the beginning of the Cretaceous period it was still connected, by way of Madagascar and India, with southern Asia. The Andes of Antarctica constitute a continuation of the Andes of South America. The driftings of the continents induced changes in their position relative to the poles; this, in turn, led to changes in climatic zonation and, consequently, to changes in the distribution of living organisms.

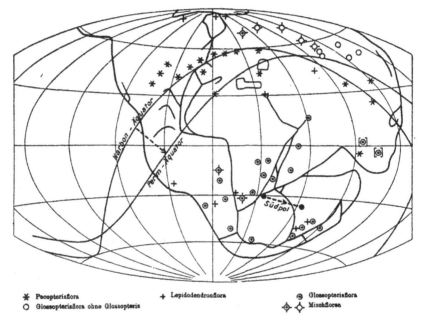

* Pecopterisflora   + Lepidodendronflora   ⊚ Glossopterisflora
O Glossopterisflora ohne Glossopteris   ◈ ◇ Mischfloren

FIG. 24. — Distribution of floras and location of the equator during the Carboniferous and Permian periods. (After KÖPPEN and WEGENER).

On the basis of the foregoing assumptions riddles that formerly seemed insoluble in the past and present geographical distribution of organisms are solved. Among such riddles we may mention, first, the distribution of plants and even of entire floras in zones that, as regards their present climatic conditions, are not suitable to these plants, and, second, uniformity of vegetation, indicating that formerly there existed uniform climatic conditions in regions where at present there are marked differences in climatic and vegetation zones.

Another such riddle that long seemed insoluble is the absence in the Carboniferous period of periodicity in plant growth. In the case of our present-day plants this periodicity is expressed by the alternation of active periods (*i.e.*, periods of intensive growth and development) and dormant periods (during that part of the year when climatic conditions are unfavorable). One of the characteristic features of such periodicity is, for instance, the possession by deciduous and coniferous trees and

E. V. Wulff   —178—   Historical Plant Geography

shrubs of dormant buds, which remain closed during unfavorable periods of the year and renew development with the onset of favorable climatic conditions. In plants of the Carboniferous period, despite the existence of large trees, such as *Lepidodendron, Cordaites, Calamites,* etc., no such buds are found. As another even more characteristic feature of periodicity we may mention the annual rings in the stems of woody plants. In Carboniferous plants, distributed over a considerable part of the northern hemisphere and, to a less extent, in the southern hemisphere, no annual rings are found. This circumstance and also the absence of dormant buds testify to continuous and uniform growth, which could have taken place only under uniform climatic conditions, characterized by the absence of alternating seasons. Such climatic conditions correspond to those found at present in the equatorial zone with its tropical vegetation. But the plants of the Carboniferous

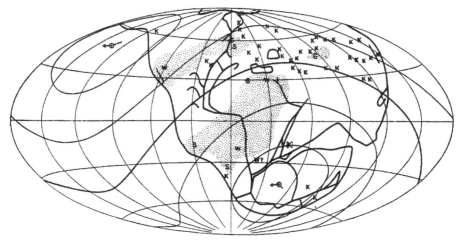

FIG. 25. — Distribution of swamps and deserts and location of the equator in the Jurassic period (K = coal, S = salt, G = gypsum, W = desert sandstone; stippling = arid regions. (After KÖPPEN and WEGENER).

period, though lacking periodicity, grew throughout the entire length and breadth of Europe.

In the flora of the Carboniferous period it is possible, according to POTONIÉ, to distinguish between elements of tropical origin, typified by the fern *Pecopteris,* and subtropical elements, typified by *Lepidodendron, Sigillaria,* etc. Whereas the former flora occupied in the Carboniferous period a quite limited, comparatively narrow strip, passing through North America and Europe and ending in eastern Asia, the subtropical (*Lepidodendron*) flora lay on both sides of this strip and had a considerably more extensive distribution. It is known as far north as Spitsbergen and as far south as the southern part of South America, occupying a latitudinal range of 120°.

Another riddle, which seemed no less difficult to solve, is the growing of trees near the poles as at present located. Moreover, as seems entirely incomprehensible, the long polar nights apparently had no effect on this vegetation.

Numerous attempts were made to solve these enigmatic peculiarities in the distribution and biology of fossil floras. It was assumed that plants of former periods possessed considerably greater ability, as compared with present-day plants, to adapt themselves to different climates and were considerably less sensitive to heat and cold. Or again, in order to explain the uniformity of climate over a considerable extent of the earth's surface, it was presumed that this uniformity of climate was due to the intense heat in the center of the earth and the insignificant amount of losses of this heat from irradiation owing to the fact that the earth was enveloped in a thick blanket of clouds. Lastly, it was considered possible to assume that the absence of annual rings was a peculiarity of plants of those times and that, consequently, this could not serve as a basis for conclusions regarding climatic conditions.

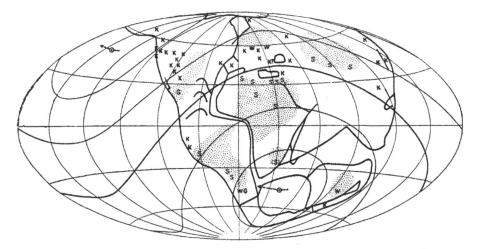

FIG. 26. — Same as in FIG. 25, but in the Cretaceous period. (After KÖPPEN and WEGENER).

These suppositions are refuted by the finding of traces of glaciers in the most ancient deposits of the earth and also by the fact that even in the Carboniferous period the climate was not everywhere uniform, since Carboniferous remains of trees having annual rings are known from the Falkland Islands and from Australia. In the floras of the succeeding geological stages periodicity in plant growth becomes of ever more widespread occurrence. All this indicates that climatic zones existed in past geological periods but that the location of these zones was undoubtedly entirely different from now.

It is likewise impossible to assume that there were any radical differences in the physiology and biology of plants of former geological periods. Paleobotanical data show that they were approximately the same as they are today. According to these data, fossil plants must have grown in climatic zones corresponding to the physiological peculiarities of these plants, whose requirements as regards light and heat must have been the same as those of their present-day descendants.

All the foregoing forces us to assume that the continents in past ages must have been differently situated with respect to the poles. A way out of all these difficulties in the geography of plants of former geological periods is provided, as we stated above, by WEGENER's theory of continental drift, in the light of which the climates of former geological periods receive an entirely different explanation. A detailed exposition of this viewpoint is given by KÖPPEN and WEGENER (1924) in a special work devoted to this problem. According to these investigators, in the Carboniferous period the equator passed through North America (from the southwest to the northeast), central Europe, the Caspian Sea, Asia, and the Sunda Isles. Central Europe was, therefore, included in the zone of equatorial rains, and it was, over much of its extent, submerged beneath the sea. Consequently, the Carboniferous deposits of central Europe must be of tropical origin; this is confirmed by the finding in such deposits of plants of the type of *Pecopteris*. Thus, the riddle of the finding in the Carboniferous deposits of central Europe of tropical flora lacking periodicity of growth is solved, and the assumption that the continents were formerly united explains the uniformity of climatic conditions over a considerable part of their territory. The theory of continental drift makes possible a new way of explaining problems involved in the paleogeography of plants, just as it aids in clearing up many formerly inexplicable facts in zoogeography.

The location of climatic zones in past geological periods was, according to WEGENER's theory, not at all the same as now; only gradually, beginning with the second half of the Tertiary period, did these zones begin to assume their present location. The shiftings of the climatic zones were accompanied by shiftings of floras, particularly on both sides of the Atlantic, in North America and Euro-Africa. Asia, on the other hand, suffered climatic changes to a considerably less degree, due to the fact that here the shiftings of the zones took place not symmetrically and in a circumpolar direction but asymmetrically. There are numerous facts both of a paleontological and of a floristic and faunistic character, long since noted by many investigators, that attest the relative constancy of the floras and faunas in the eastern part of the tropics of the Old World and, in particular, in the region of the Sunda Isles and eastern Asia. Thus, numerous Tertiary coal deposits have been found on the East-Asiatic coast and also on adjacent islands, *e.g.*, in the Soviet Far East (northern Sakhalin, Amur and Ussurian Regions), Manchuria, southern China, the Philippines, Java, Sumatra, and Borneo. According to data of Dutch geologists, there have been found on the Sunda Isles coal deposits from all stages of the Tertiary period. These facts, as well as the finding of species of palms in a fossil state, led KÖPPEN and WEGENER to the conclusion that throughout this region there has been a humid, tropical climate at least since the beginning of the Tertiary period.

IRMSCHER (1922, 1929), on the basis of ETTINGSHAUSEN's data, draws attention to the presence in the composition of the flora of this region not only of families but also of genera likewise found there in a fossil state. This has been confirmed by data of other paleobotanists, such as KUBART (1929) and KRÄUSEL (1929). MERRILL (1923), on the basis

Works Cited

du Toit, A. L. 1937. *Our Wandering Continents.* London.

Fürstenberg, v. Fürstenberg, A. 1909. "Die Polarregionen im Lichte geologischer und literarischer Forschung." *Naturw. Wochenschr.* 24.

Golenkin, M. I. 1927. *Victors in the Struggle for Existence* [in Russian]. Moscow.

Hallier, H. 1912. "Über frühere Landbrücken u. Völkerwanderungen zwischen Australien und Amerika." *Meded. 's Rijks Herbarium,* Leiden, nos. 8–14.

Heer, O. 1868–83. *Flora fossilis Arctica,* vols. 1–7. Zurich.

Irmscher, E. 1922, 1929. "Pflanzenverbreitung und Entwicklung der Kontinente," I and II. *Mitt. D. Inst. F. Allg. Bot. In Hamburg,* vols. 5 and 8.

Köppen, W., and A. Wegener. 1924. *Die Klimate der geologischen Vorzeit.* Berlin.

Sahni, B. 1936. "Wegener's theory of continental drift in the light of paleobotanical evidence." *J. Indian Bot.* 15, no. 5.

Simroth, H. 1914. *Die Pendulationstheorie.* 2d ed. Leipzig.

Soergel, W. 1917. *Das Problem der Permanenz der Ozeane und Kontinente.* Stuttgart.

2 Earth History, Vicariance, and Dispersal

Paul S. Giller, Alan A. Myers, and Brett R. Riddle

We cannot understand the present-day biogeographical distributions of animals and plants without an understanding of the history of the planet. This may seem a rather facile statement, but it is nevertheless true. The history of the earth—in terms of its geology, the changing location of land masses and oceans, changing sea levels, fluctuating climate, and indeed extraterrestrial impacts—forms the backdrop for understanding the modern distributions of species. Although ecological processes may be responsible for the control of local diversity or for influencing the fates of potential colonists, they also depend on the existing pool of available species—questions about how species accumulate and how biotas are assembled are different from questions about how local diversity is maintained (Vermeij 1978).

Historical biogeography attempts to reconstruct the origin of taxa; sequences of their dispersal, isolation, and extinction; and to explain how geological events have shaped present-day patterns of distribution (Myers and Giller 1988). It also uses both present and past biotic distributions to infer distributions of geological features in the past. In effect, historical biogeography addresses the how, when, and why of species distributions (Jablonski, Flessa, and

Valentine 1985). Important questions are why a taxon is absent from apparently suitable areas beyond its present range, and how taxa have become spatially separated (or disjunct). It is hypothesized that such patterns can be caused by the break-up of a once continuous range (vicariance), by long-distance dispersal, or through the separate origins of the taxon in two or more places. The latter is thought to be less likely, so the two processes of *dispersal* and *vicariance* underpin the key models to explain present day biogeography and form the cornerstones of the two schools of dispersal and vicariance biogeography (see also parts 4 and 5 of this volume).

Dispersal biogeography follows from the premise that species spread outward from a center of origin, sometimes across preexisting barriers, so that the present-day biota results from the accumulated dispersal of the various descendant lineages. Alternatively, recognition of an unstable earth stimulated the development of hypotheses that involved the splitting of biota (vicariance) into isolated populations by emerging barriers that resulted from the operation, often in concert, of so-called *TECO events* (tectonic [plate movements and associated orogeny], eustatic [changing sea levels], climatic [shifting climatic belts, ice ages], and oceanographic [oceanographic circulation changes]) (see Rosen 1984). In reality, large-scale disjunct distribution patterns can result from dispersal or vicariance or both, and distributional patterns are clearly modified over time by evolutionary and ecological processes discussed later in this volume.

The foundations of modern biogeography have been built largely upon the work of the two schools of dispersal and vicariance biogeography. Therefore, in part 2 we feature some of the major publications that have contributed much to the development of the ideas, concepts, and hypotheses in historical biogeography. There are, of course, many others that could also have a justifiable claim for inclusion here, and we will mention some of them in the discussion that follows.

Vicariance and the Unstable Earth

At the beginning of the twentieth century, most biogeographers still attempted to understand current animal and plant distributions by reference to a world on which, it was believed, the position of the various land masses had been relatively stable over the time scales associated with the distribution of the extant groups of organisms. To understand how organisms arrived at their present patterns of distribution it was necessary to consider range expansions, contractions, and migrations coupled with selective population extinctions, i.e., dispersal and centers of origin. Yet, as early as the middle of the nineteenth century, the botanist Joseph Hooker (1853–55) had commented that "... the botanical relationship [of the three great areas of land in the southern latitudes] is as strong as that which prevails throughout the lands within the Arctic and Northern Temperate zones, and which is not to be accounted for by any theory of transport or variation, but which is agreeable to the hypothesis of all being members of a once more extensive flora, which has been broken up by geological and climatic causes."

Hooker was far ahead of his time in envisaging a southern Gondwanaland before any hy-

potheses of continental drift had been put forward by geologists. However, Hooker held firm to the belief that the continents were fixed, but were connected at times by trans-oceanic land bridges. Five years later, Antonio Snider-Pellegrini (1858) suggested the idea of continental drift, but it was not until the influential publications of ALFRED WEGENER, culminating in the 1912 edition of his *Die Enstehung der Kontinente und Ozeane* (*The Origins of Continents and Oceans*; see below) that continental drift became an important matter of debate among geologists. The theory remained controversial for nearly half a century, during which time it had minimal impact on the science of biogeography. The reason for this was that the scientific community had difficulty testing the hypothesis rigorously with available techniques and also lacked a convincing explanatory mechanism. The far-reaching significance of continental drift was therefore not fully realized for many years. Indeed, as we will see later, even in the mid-twentieth century some of the foremost biogeographers were still questioning the hypothesis of drifting continents. They only later changed their views when the evidence, especially from marine geologists, became irrefutable.

The evidence emerged following the development of a range of new techniques. These included the plotting of past positions of the poles using paleomagnetic evidence from continental sedimentary and igneous rocks; the discovery of *aulacogens* (failed rift arms that extend from an oceanic margin into a continent) and of magnetic anomalies; the discovery of similar fossil assemblages across now-separated continents; and advances in seismology. Mapping of the ocean floor (much of the data for which was produced by the military during the Second World War) and discovery of mid-oceanic ridges separating strata with mirror-image magnetic signatures, suggested the mechanism—*sea-floor spreading* (Wilson 1963). Only then did biogeographers such as Darlington (1965) even consider taking up the challenge of trying to interpret biotic distributions in a world no longer viewed as topographically constant. Perhaps because of this hiatus between the hypotheses of continental drift and its modern acceptance as *plate tectonics*, those entering the field of analytical biogeography tend to commence their readings in the modern literature of plate tectonics. Wegener is rarely read, although frequently referenced. From a modern geological and biogeographic perspective, this is understandable, but from a historical perspective, it is valuable to read the seminal work of Wegener that was so instrumental in spawning the science of plate tectonics. We therefore include an extract from the 1966 reprint of the 1929 fourth edition of his translated book *The Origin of Continents and Oceans* in the current volume (paper 19).

The knowledge that the position of the continents of the earth had not always been constant gave an impetus to biogeographical studies and facilitated the development of analytical techniques, which began to replace the former narrative approaches. For the first time it became possible to understand many terrestrial distributions without the need to postulate trans-oceanic land bridges now sunk beneath the sea, or long-term range expansions coupled with local extinctions (see below). A classic paper in this mold was that of LARS BRUNDIN (1966), one of the earliest publications to employ Hennig's (1966) phylogenetic methods (cladistic biogeography) to demonstrate concordance between animal distributions on the one hand and plate tectonic movements on the other, and hence to link distributions with vicariant events (the origin of these techniques is explored in part 4 of this volume). We have included chapter 4 of Brundin's monograph, *Transantarctic Relationships and Their Significance, as Evidenced by Chironomid Midges* (paper 20), wherein he developed the fundamental rationale behind vicariance biogeography—that one must understand the general (i.e., vicariant) patterns in biogeography before attempting to decipher contingent (i.e., chance dispersal) patterns. Among his interpretations of resulting patterns were ". . that there once has been an important centre of evolution in the south and that Antarctica must have been a

vital part of that centre" (449), and ." . . the chironomid midges give clear evidence that their transantarctic relationships developed during periods when the southern lands were directly connected with each other" (451–52). In addition, Brundin highlighted the complex relationships of South American taxa, with ten distribution patterns connecting South America with New Zealand via West Antarctica, and ten patterns connecting South America with East Australia and Tasmania via East Antarctica. Much earlier, Wallace (1876) had also suggested that the closest relation of New Zealand was Magellanic South America. The details of the geological history of South America are still disputed, but whatever the ultimate conclusions, Brundin's 1966 paper was significant, not only as a milestone in the methodology of vicariance biogeography, but also because it highlighted the composite nature of some geographic areas and anticipated the possible role of allochthonous terranes in the formation of biotas. Most notably, Brundin resurrected and provided rigorous support for Hooker's (1853–55) conclusion that the southern continents shared a historically connected austral biota. But unlike Hooker, Brundin provided an explanation grounded in evidence for fragmentation of a once-united southern continent.

In contrast to terrestrial biogeographers, marine scientists, in the main, initially found little of interest in the burgeoning publications of plate tectonics and vicariance biogeography to help explain marine distributions. There were, however, some exceptions to this. For example, the endemism of the modern Atlantic biota could be explained by the suturing of Africa with Europe in the east and the rise of the Isthmus of Panama in the west, causing isolation of the tropical Atlantic from the Pacific and Indian oceans. Also, the realization that the Mediterranean was a remnant of the once extensive pan-tropical Tethys Sea could explain some odd disjunct taxa shared between Japan and the Mediterranean. These examples notwithstanding, the earlier biogeographers who had been working in the greater Indo-Pacific could be excused for believing that plate tectonics had left no biogeographic mark on this vast oceanic region. Most marine organisms, it has been argued, have planktonic larvae that disperse over large distances. The Indo-Pacific could, therefore, be considered as one large continuous ocean without substantial barriers, having remained relatively unaffected by the tectonic upheavals of the last 60 million years. As was described in part 1, however, Ekman (1935, 1953) in his two books on marine zoogeography recognized and characterized a large number of marine faunal regions. His scheme was further developed by Briggs in 1974 who divided the Indo-Pacific, as well as the rest of the marine realm, into provinces—each defined in terms of its endemic species (see also Jokiel 1990). Unfortunately, the alternative view that the oceans largely lack biogeographic barriers persists among some scientists even to today. As recently as 1995, the Committee on Biological Diversity in Marine Systems (CB-DMS 1995) termed marine ecosystems "open," implying that biogeographic barriers are all but absent (for discussion see Myers 1997).

This rather simplistic view of ocean biogeography was challenged by Springer's (1982) work on shore fishes, which suggested that the Pacific Plate, a tectonic plate lacking any continental land, should be considered a distinct biogeographic entity. Earlier, Thorne (1963) and van Balgooy (1971) had demonstrated a sharp break in distribution of terrestrial biota between the Indo-Malayan region and the oceanic islands of the Pacific, and Pacific Plate endemism has since been reported for shallow water marine amphipods which lack a planktonic dispersal phase (Myers 1994). Springer's hypothesis, has not been universally accepted, and Briggs (1995) has argued that the plate lacked sufficient overall endemism to merit designation as a distinct region. Regardless of one's view of the biogeographic status of the Pacific Plate, marine biogeographers have demonstrated that the constraints to dispersal previously invoked to explain the distributions of terrestrial and freshwater organisms can be equally important in explaining the endemic distributions of shallow-water marine organisms.

Many Pacific island archipelagos also exhibit high levels of endemism in their shallow-water marine organisms. Selective extinction of populations through a species range can result in the formation of *paleoendemics* (remnants of a previously more diverse and widely distributed fauna or flora). But most island endemics are attributed to the *founder principle:* rare stochastic arrival by propagules followed by establishment (thus colonization), spread, and evolution through phylogenesis. However, based on their work in the Hawaiian-Emperor chain, Rotondo et al. (1981) challenged this view by showing that continental lithospheric plate movements could result in volcanic islands being transported on the plate from their place of origin to another place where they may be integrated with other islands formed over hotspots. Other studies have shown that the extant islands of the Pacific are but a remnant of much more extensive islands and island archipelagos in the past (Schlanger and Premoli-Silva 1981, Nur and Ben Avraham 1982). *Island integration,* therefore, may well have been extensive through the late Cretaceous and Early Tertiary resulting in considerable reassortment of Pacific biotas. This is important in terms of our understanding of the nature and origination of island biota.

The fragmenting and suturing of land masses, the emergence of new lands (volcanoes and island arcs), the bringing of once distant lands into juxtaposition, are all dramatic rearrangements of the globe brought about by plate tectonics. The distributions of biota through space and time have also, however, been profoundly affected by eustatic, climatic, and oceanographic factors. The seminal paper by W. A. Berrgren and C. D. Hollister (1977), demonstrated the close link between changes in the earth's geography as a result of plate tectonics and ocean circulation over the last 200 million years. The authors paint a picture of tranquil Mesozoic seas under equable climates, followed by "commotion in the ocean" during Cenozoic times. The opening of the expanding ocean systems to Polar Regions, with consequent formation of cold deep bottom waters at high latitudes which then spread over the world's abyssal seas, allowed the development of thermospheric stratification that intensified circulation. The modern circulation pattern was completed by the opening of the Drake Passage and Tasman Sea, which allowed the development of circum-Antarctic circulation. Bergren and Hollister pointed out that because the world's oceans are in dynamic contact, and have been throughout most of their histories, it is virtually impossible to isolate the evolution of one from another.

Dispersal and Centers of Origin

Dispersal includes all types of geographic translocation of individuals leading to changes in the distribution of populations and species across a range of spatial scales. Two main dispersal processes are involved in biogeography: *range expansion* and *jump dispersal*. At the smallest scale, range expansion involves natural movement away from parents as a normal part of the life cycle. At a larger scale, it involves the gradual expansion of geographical ranges through the spread into areas beyond the boundaries of the initial range in response to habitat modifications, climatic shifts (e.g., global warming, ice retreat, etc.; see part 3 in this volume) or adaptations. The reverse, range contraction, can also occur, and may lead to disjunct distributions and isolation of closely related taxa in refugia (see Lomolino and Channell 1995; Channell and Lomolino 2000).

Jump dispersal is really a species-level process, carried out by individuals, that involves crossing some kind of barrier (e.g., mountain range, ocean, desert, or, in some cases, even quite small topographical or environmental features) through some chance or otherwise rare event. Jump dispersal has been considered the most important process shaping present-day distributions by some biogeographers (such as Croizat and Carlquist, see below). Here we concentrate on the role of dispersal in relation to longer-term earth history and defer the consideration of dispersal in relation to more recent events to part 3.

Linnaeus postulated that the world's biota originated from a single land mass, Paradise, and then spread out as more primordial land emerged from the sea—the "center of origin" hypothesis (Myers and Giller 1988) taken up later by Darwin. Elton (1958) pointed out that if we lived in a world with no barriers, we would have mostly pan-tropical and pan-temperate species distributions, bipolar forms, continental species reaching every island, and marine animals girding the world. The fact that most species occupy only a limited part of their potential global range, as is shown by the ability of exotic introduced species to colonize and spread into new areas, demonstrates the existence and effectiveness of barriers. The existence of large areas that contain sets of species unique to those areas (endemics) led Sclater and Wallace to identify the existence of six great faunal regions (Neotropical, Palaeartic, Nearctic, Oriental, Ethiopian, and Australasian). Wallace supposed that these regions had been left isolated for such long periods that they had evolved endemics, and we now know that many of the barriers that led to the creation of Wallace's regions were formed within the last 150 million years during the Cretaceous.

Despite apparent barriers, some species can achieve an almost cosmopolitan distribution, either because the barriers that affect other biota are ineffective for them, or because the species have good powers of jump dispersal. For example, the common blue damselfly *Enallagma cyathigerum* is found in a belt around the Northern hemisphere between 45°N and the Arctic Circle (Cox, Healy, and Moore 1973). The plantain *Plantago major* is found on all continents except Antarctica (Sagar and Harper 1964) (although man may have contributed to some extent to this distribution).

The colonization of oceanic islands can highlight the ability of species to cross barriers. An ideal study island would be one newly emerged from the sea, lacking any previous contact with land masses and devoid of life from its origin (Carlquist 1981). Krakatau is one such example, where the eruption in 1883 created pristine environments that were colonized by air and sea across 40 to 80 miles of water from Java and Sumatra by 271 plant species, 720 insect species, and more than 30 bird species, in just over 50 years (Docters van Leeuwen 1936; Dammerman 1948; Whittaker, Bush, and Richards 1989). The organisms that appeared all seemed to have good dispersal mechanisms, supporting the idea of jump dispersal. The closeness of Krakatau to nearby source pools might suggest that this is not really jump dispersal across a barrier, but rather *range expansion*. This alternate view, though, is less obviously applicable to other examples, like the Hawaiian chain. Here, colonizing land plants and animals have come from as far afield as North and South America, Australasia, New Guinea, and other Pacific islands. Although most of these colonists probably arrived by long-distance dispersal, some of them (especially some near-shore marine taxa) likely dispersed shorter distances via "stepping-stone" islands (Myers 1991).

SHERWIN CARLQUIST has been a strong proponent of long-distance dispersal, and his 1966 paper "The Biota of Long-distance Dispersal" (paper 21), provides a masterly review of dispersal to oceanic islands and subsequent evolutionary trends. In it, he points out that long-distance dispersal is unlikely to be achieved primarily by single introductions, but that repeated or simultaneous multiple introductions are generally needed for establishment of species. The idea of now-vanished high islands acting as possible stepping-stones or subsidiary source areas was also highlighted in this paper. In a later work, Carlquist (1981) called for acceptance of a major role for chance dispersal events, backing this up with a number of clear examples from a surprising range of organisms and detailing situations, as in Hawaii, where colonizing plant species have managed to hit small target areas. Caution should, however, be exercised when making assumptions about past dispersal based on the geography of the modern world. Much of the Hawaiian-Emperor chain exists today as seamounts that once were islands but have eroded and sunk, and are now submerged beneath the sea. This chain of is-

lands, which was formed over a hotspot, must have been more extensively exposed during periods of lowered sea levels, perhaps forming a larger "net" for sampling dispersing propagules.

Prior to acceptance of the theory of plate tectonics and sea-floor spreading, Matthew (1915) and his followers, Simpson (1940) and Darlington (1957, 1965), were the primary twentieth-century proponents of the Linnaean–Darwinian concept that geographic distributions are produced through dispersal away from a center of origin. This view of the world was based on the premise that continents had always been in their current positions (e.g., Simpson 1940; Darlington 1957), or that they had drifted to current positions prior to the time frame relevant to interpreting current biotic distributions (Darlington 1965). This assumption of movement away from a center of origin led to many attempts to explain distributions that invoked either highly improbable dispersal events, or the rapid emergence and submergence of transoceanic land bridges (Hooker 1853–55, Willis 1922, Simpson 1940). Much of this reasoning was soundly criticized for its ad hoc nature and lack of evidence (e.g., by Cain 1944). Even Charles Darwin, in a letter to Charles Lyell in 1856, felt compelled to criticize Lyell, one of his most respected and formidable mentors, for the "geological strides, which many of your disciples are taking" where land bridges were being created "as easy as a cook does pancakes." Darwin went on to warn Lyell, "If you do not stop this, if there be a lower region of punishment of geologists, I believe, my great master, you will go there." Nevertheless, it is instructive to read some of these early papers as the views espoused held sway for much of the first half of the twentieth century.

One of the classic dispersalist papers was GEORGE GAYLORD SIMPSON's 1940 essay, "Mammals and Land Bridges" (paper 22). Here, Simpson developed the biogeographic concepts of corridors, filter bridges, and sweepstakes routes, and presented a historical review of the ideas behind theoretical land bridges, migration routes, and the expansion from centers of origin. When major changes in climatic conditions occur, or when land masses come into contact, many species are likely to disperse along the same routes and more-or-less simultaneously (but probably at different rates), which is a very different mechanism from the rare independent events of jump dispersal. Udvardy (1969) used the term *corridor* to define the broad band of continuous habitat that allows potential dispersal over large distances without the need for crossing of major barriers, as in the case of the tree dispersal following the last ice age (see also Davis 1976, discussed in part 3 of this volume). Simpson had coined the term "corridor" to describe a route permitting the spread of many or most taxa from one area to another, across so-called land bridges. Aside from its relevance to the study of geographic distributions, Simpson's concept of corridors became a cornerstone of nature reserve design and more recent efforts to conserve biological diversity (see also part 6).

In his early work, Simpson played down the role of jump dispersal as an explanation of present biogeographic patterns (particularly of mammals), and was a strong advocate of the concept of land bridges. The underlying logic for the potential existence of land bridges is clear. The occurrence of animals now confined to one part of the globe but present as fossils in distant regions (regions from which it would now be impossible for them to migrate) implies that regions now separated by oceans must have been connected by land in the past. Simpson's concept does not suggest "lost continents" or broad trans-oceanic pathways, but rather relatively restricted (albeit extensive) links. Although the existence of most trans-oceanic land bridges has been falsified or discredited since, Simpson's interesting and well-written essay published in 1940 illustrates the thinking at the time and represents an important contribution to the development of the dispersal–vicariance debate.

In Simpson's later writings there seems to have been a shift in his opinion about the occurrence and importance of land bridges. A 1943 paper still challenged the idea of continental drift and supported the notion of stable continents, using dispersal across land bridges as a mechanism to explain disjunct distributions. However, in 1946, and contrary to his former opinions, Simpson argued against the notion of direct Atlantic bridges between Europe and North America and criticized zoologists postulating connections that violated geological principles; he likewise criticized geologists postulating continental connections that were not supported by zoological data. In fact, even in the 1940 paper you will see that Simpson discusses situations where insular and highly unbalanced faunas cannot be accounted for by land bridges, but are more likely to be the result of stochastic jump dispersal via "sweepstake routes." This is developed further in his later papers, including his 1952 article on the probability of accidental dispersal.

Plate tectonics provided the basis for development of an alternative to the prevailing view of fixed continents occasionally connected by trans-oceanic land bridges. In the 1967 paper, "The Bearing of Certain Palaeozoogeographic Data on Continental Drift," TONY HALLAM analyzed Mesozoic and Tertiary terrestrial vertebrate and marine mollusk fossils, and concluded that their palaeozoogeography provided convincing evidence of continental drift. This was a significant result because of "just how respectable . . . land bridges were to a former generation of thoughtful and erudite palaeontologists, to whom the idea of continental drift was too radical" (202). In the excerpt we have included here (paper 23), Hallam first argued against the existence of trans-oceanic land bridges in Atlantic and Indian oceans, pointing to the absence of any geological or oceanographic evidence for "sunken continents," but with clear evidence that the ocean floor is younger than would be required if it contained remnants of previous trans-continental connections. He then went on to develop a reconstruction of Gondwanaland and the sequence of its fragmentation, beginning with models developed by geologists (e.g. Wilson 1963), but revised with added insight provided by his analysis of the palaeozoogeographic evidence.

At the time of Hallam's synthesis, he was either unaware of or uninterested in the turmoil brewing among historical biogeographers elsewhere. Thus, while he discusses the contributions of dispersalists, including Matthew, Simpson, and Darlington, nowhere does he mention Leon Croizat, the iconoclastic father of modern vicariance biogeography. Through his analysis of distributional tracks, Croizat (1962) used repeated distributional patterns (generalized tracks) across very different taxa, such as birds and plants, to infer a causal relationship between biogeographic patterns and a historically dynamic earth. With his concept "panbiogeography," Croizat argued vehemently against the center of origin/dispersal scenarios popularized by the Matthewians. Instead, he advocated the view that ancestrally widespread taxa were isolated in different geographic regions at the time of their physical separation, and therefore that "life and earth did evolve together" (1962: 147). A sampling of Croizat's contribution is included in part 4 of this volume.

The dispersalists were not sufficiently convinced at first by either new support for continental drift or by Croizat's panbiogeography to abandon the center of origin/dispersal paradigm. By 1965, Philip J. Darlington had accepted the general fact that the southern continents had drifted apart over time, but still did not accept a causal association between continental drift and biogeography: "I have therefore become a Wegenerian, but not an extreme one. I doubt the former existence of a Pangaea or Gondwanaland, and I think that the movements of continents have been simpler and shorter than most Wegenerians suppose" (1965: 210). The excerpt we include here, from Darlington's 1965 book *Biogeography of the Southern End of the World* (see paper 24) deals with one of the more active questions in biogeography—why do plant and animal taxa in disjunct, cold temperate rainforests in higher latitudes of the southern hemisphere tend to be more similar to each other than they do to taxa distributed elsewhere on their respective continents? This question was formally addressed

much earlier by Hooker (1853–55), who concluded that plants endemic to these regions represented remnants of a once more extensive flora with a center of evolution in the southern hemisphere. Hooker's view was generally dismissed or ignored for the next century (Brundin 1966) in favor of one in which plants and animals originated in the northern hemisphere, dispersed to the southern hemisphere, and were driven to extinction in their northern homeland by newly emerging forms (Wallace 1876; Matthew 1915; Simpson 1940; Darlington 1957, 1965). Under this view, disjunct southern taxa resemble one another as a result of convergent evolution from separate northern stocks that had followed independent dispersal routes from the north. In the excerpt presented here, Darlington used carabid beetles as a case study illustrating the general model (see his figure 14) of successive waves of immigrants from the Northern into the Southern Hemisphere.

Much of the excitement generated by the acceptance of continental drift focused on its importance in vicariance biogeography. Nevertheless, plate tectonics also provided a mechanistic basis for interpreting significant dispersal events between historically isolated biotas. Embedded within Simpson's 1940 paper is a detailed description and analysis of a biogeographic pattern, recognized initially by Wallace (1876), that would occupy the attention of a generation of biogeographers (Stehli and Webb 1985). The intermingling of Nearctic and Neotropical biotas is called the "Great American Interchange," and resulted from emergence of the Panamanian land bridge, closing the Bolivar Trough between North America and South America about 3 million years ago. Simpson used this example to illustrate the action of a "filter-bridge." He summarized evidence that the South American fauna was very different in composition but not in richness before and after the interchange, but that the North American fauna changed little in composition following the interchange, hence, that the interchange was decidedly asymmetrical in magnitude and influence.

Simpson's primary explanation for the contraction and eventual extinction of a number of South American natives was their inferiority in competition with superior invaders from the north, which if true, provided an empirical example of Matthew's views on the superiority of species in northern faunas.

By the time LARRY G. MARSHALL, S. DAVID WEBB, J. JOHN SEPKOSKI JR., and DAVID M. RAUP published a detailed analysis of the Great American Interchange in a paper titled "Mammalian Evolution and the Great American Interchange" (Marshall et al. 1982; paper 25), the data base had been substantially improved through better sampling methods, improved taxonomies, and radioisotopic-based dates placed on mammal-bearing strata on both continents. Marshall et al. concluded that the interchange of mammals was symmetrical at the family level, but highly skewed in favor of inclusion of North American immigrants into the South American fauna at the generic level. They also compared aspects of faunal turnover with predictions derived from MacArthur–Wilson *equilibrium theory* (see part 6 in this volume), and suggested that this model could explain both the higher number of genus-level immigrants from North America to South America (with pre-interchange North America having a richer fauna), and the greater rate of post-interchange extinction of South American natives (relaxation of a supersaturated fauna). The one feature of interchange not predicted by equilibrium theory was the explosive adaptive radiation of North American taxa, particularly sigmodontine rodents, owing to speciation after immigration to South America. Important aspects of this interchange not examined by Marshall et al. are the patterns observed for other taxonomic groups, including birds, reptiles, and amphibians (Brown and Lomolino 1998). Today, details of the Great American Interchange continue to be actively explored, for example, the filtering role of glacial versus interglacial habitat configurations along the corridor (Webb 1991), the timing and generality of interchange (Bussing 1985, Engel et al., 1998, Vanzolini and Heyer 1985, Vuilleumier 1985), and the effect of the formation of the land bridge on the isolation of marine biotas (Lessios 1998).

The role of dispersal is perhaps the most difficult and controversial matter within historical biogeography owing to the deficiency of actual data (Brundin 1988) and the difficult issue of nonfalsifiability of dispersal processes. A number of distinguished biogeographers have suggested that the main patterns of global geographic distribution are orderly and thus unlikely to be the result of large-scale jump dispersal (Croizat 1962). This didn't stop the erection of various controversial hypotheses of jump dispersal across wide expanses of oceans or other large barriers to explain disjunct distributions. The concept of land bridges had similarly risen and then waned in popularity. What must not be forgotten, however, is that it is not dispersal alone but the entire colonization process (dispersal followed by establishment) that is the essential for a species to expand its geographical range. One of the most famous examples of this process is the colonization of species from the Red Sea to the Mediterranean Sea via the Suez Canal, with scarcely any evidence of colonization of species in the opposite direction (so-called Lessepsian Migration).

The opening of the Suez Canal in the mid 1800s provided the means of a great faunal interchange between the Atlantic and the Indo-Pacific Oceans, by way of the Red Sea and the Mediterranean. The first fifty years of the canal's existence were marked by a number of scientific investigations culminating in an expedition led by H. Munro Fox, which documented the fauna of the canal and the migration through it (Fox 1929). Because the occurrence of migration was taken more or less for granted, the second fifty years of the canal's existence passed by with diminished scientific interest, although distributional data were still accumulated. It was more than a hundred years after the opening of the canal that FRANCIS DOV POR (1971) published a review of Lessepsian migration that not only examined the distributional ecology of the fauna of the Canal and adjacent Mediterranean, but also reviewed what was known of the geological background and history of the Canal region. His explanation for the polarized nature of the interchange, was that the complex communities of the Red Sea prevented establishment of the species that dispersed in that direction. Colonization of the Mediterranean by Red Sea organisms has, on the other hand, been a stepwise process of range expansion through the canal. Por pointed out that migration was probably possible only for species that successfully settled in the Bitter lakes and that only a reduced number of species with a wide circumtropical range might have used the opportunity. Por has published extensively on this subject, but we have chosen to include his (1971) integrative review *One Hundred Years of Suez Canal—A Century of Lessepsian Migration* (paper 26), in which he discusses both possible pre-canal faunal interchange over a period of 2,000 years and the post-canal period of range expansion.

The emergence of modern vicariance biogeography and the theory of plate tectonics thus put an end to exaggerated hypotheses of jump dispersal and trans-oceanic land bridges. Now the majority of vicariance biogeographers accept that long-distance dispersal exists, but some still believe that it has played only a relatively minor role, superimposed on basic patterns dictated by geological history. However, new, isolated habitats like volcanic islands clearly do receive a biota from jump dispersal, and humans are effectively increasing the rate of such long distance dispersal (Elton 1958). The problem is not whether chance dispersal occurs, but more the extent to which patterns we see have been caused by it rather than by vicariance events. For example, recent studies on chameleon radiation using a combination of molecular and morphological approaches point to a greater role of oceanic dispersal than previously thought (Raxworthy, Forstner, and Nussbaum 2002). The challenge for the future is to evaluate further the influence and scale of dispersal activity, not only over an evolutionary time frame but also on an ecological one, and new and emerging techniques (such as molecular-based technologies) are already helping to make major advances on these age-old questions in biogeography.

From *The Origin of Continents and Oceans*
Alfred Wegener

CHAPTER 2

The Nature of the Drift Theory and Its Relationship to Hitherto Prevalent Accounts of Changes in the Earth's Surface Configuration in Geological Times

IT IS a strange fact, characteristic of the incomplete state of our present knowledge, that totally opposing conclusions are drawn about prehistoric conditions on our planet, depending on whether the problem is approached from the biological or the geophysical viewpoint.

Palæontologists as well as zoo- and phytogeographers have come again and again to the conclusion that the majority of those continents which are now separated by broad stretches of ocean must have had land bridges in prehistoric times and that across these bridges undisturbed interchange of terrestrial fauna and flora took place. The palæontologist deduces this from the occurrence of numerous identical species that are known to have lived in many different places, while it appears inconceivable that they should have originated simultaneously but independently in these areas. Furthermore, in cases where only a limited percentage of identities is found in contemporary fossil fauna or flora, this is readily explained, of course, by the fact that only a fraction of the organisms living at that period is preserved in fossil form and has been discovered so far. For even if the whole groups of organisms on two such continents had once been absolutely identical, the incomplete state of our knowledge would necessarily mean that only part of the finds in both areas would be identical and the other, generally larger, part would seem to display differences. In addition, it is obviously the case that even where the possibility of interchange was unrestricted, the organisms would not have been quite identical in both continents; even today Europe and Asia, for example, do not have identical flora and fauna by any means.

5

6   THE ORIGIN OF CONTINENTS AND OCEANS

Comparative study of *present-day* animal and plant kingdoms lead to the same result. The species found today on two such continents are indeed different, but the genera and families are still the same; and what is today a genus or family was once a species in prehistoric times. In this way the relationships between present-day terrestrial faunas and floras lead to the conclusion that they were once identical and that therefore there must have been exchanges, which could only have taken place over a wide land bridge. Only after the land bridge had been broken were the floras and faunas subdivided into today's various species. It is probably not an exaggeration to say that if we do not accept the idea of such former land connections, the whole evolution of life on earth and the affinities of present-day organisms occurring even on widely separated continents must remain an insoluble riddle.

Here is just one testimony amongst many: de Beaufort wrote [123]: "Many other examples could be given to show that it is impossible in zoogeography to arrive at an acceptable explanation of the distribution of animals if no connections between today's separate continents are assumed to have existed, and not only land bridges from which, as Matthew put it, only a few planks have been removed, but also such that joined land masses now separated by deep oceans."[1]

Obviously, there are many individual questions which are insufficiently explained by this theory. In many cases former land bridges have been assumed on the basis of very meagre evidence and have not been confirmed by the advance of research. In other cases there is still no complete agreement on the point in time when the connection was broken and the present-day separation began. However, in the case of the most important of these ancient land bridges, there does

1 Arldt [135] states: "Of course, there are today still some opponents of the land-bridge theory. Among them, G. Pfeffer is worth special mention. He starts from the point that various forms now restricted to the southern hemisphere are manifest as fossils in the northern hemisphere. This precludes any doubt, he says, that these forms were once more or less universally distributed. If this conclusion is not completely compelling, still less is the further conclusion that we should assume a universal distribution even in all cases where there is a discontinuous distribution in the south but no fossil evidence as yet in the north. If he wants to explain distribution anomalies solely by migrations between the northern continents and their mediterranean bridges, the assumption rests on a very uncertain footing." That the affinities found on the southern continents can be explained more *simply* and *completely* by direct land bridges than by parallel migrations from the common northern region will require no further comment, even though in individual cases the processs could have been the one that Pfeffer assumed.

already exist today a gratifying unanimity among specialists, whether they base their conclusions on geographical distribution of the mammals or earthworms, on plants or on some other portion of the world of organisms. Arldt [11], using the statements or maps of twenty scientists,[2] has drawn up a sort of table of votes for or against the existence of the different land bridges in the various geological periods. For the four chief bridges, I have presented the results graphically in Figure 1. Three curves are shown for each

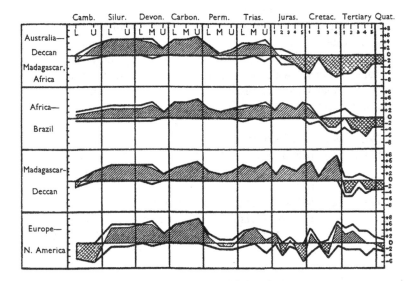

FIG. 1. The number of proponents (upper curves) and opponents (lower curves) of the existence of four land bridges since Cambrian times.

The difference (majority) is hatched, and crosshatched when the majority opposes.

bridge—the number of yeas, the number of nays and the difference between them, i.e., the strength of the majority vote, which is emphasised by hatching the appropriate area. Thus, the top section indicates that according to the majority of researchers the bridge between Australia on the one side and India, Madagascar and Africa

2 Arldt, Burckhardt, Diener, Frech, Fritz, Handlirsch, Haug, von Ihering, Karpinsky, Koken, Kossmat, Katzer, Lapparent, Matthew, Neumayr, Ortmann, Osborn, Schuchert, Uhlig and Willis.

(ancient "Gondwanaland") on the other lasted from Cambrian times
to the beginning of the Jurassic, but was then disrupted. The second
section shows that the old bridge between South America and Africa
("Arch-helenis") is considered by most to have broken in the Lower
to Middle Cretaceous. Still later, at the transition between Cretaceous
and Tertiary, the old bridge between Madagascar and the Deccan
("Lemuria") is assumed by the majority to have broken (see section
3 of Fig. 1). The land bridge between North America and Europe
was very much more irregular, as shown by section 4. But even here
there is a substantial measure of agreement in spite of the frequent
change in the behaviour of the curves. In earlier times the connection
was repeatedly disturbed, i.e., in the Cambrian, Permian and also
Jurassic and Cretaceous periods, but apparently only by shallow
"transgressions," which permitted subsequent re-formation. However, the final breach, corresponding now to a broad stretch of ocean,
can only have occurred in the Quaternary, at least in the north near
Greenland.

Many of the details of this will be treated later in the book. Only
one point is stressed here, so far not considered by the exponents of
the land-bridge theory, but of great importance: These former
land bridges are postulated not only for such regions as the Bering
Strait, where today a shallow continental-shelf sea, or floodwater
fills the gap, but also for regions now under ocean waters. All four
examples in Figure 1 involve cases of this latter type. They have
been chosen deliberately because it is precisely here that the new
concept of drift theory begins, as we have yet to show.

Since it was previously taken for granted that the continental
blocks—whether above sea level or inundated—have retained their
mutual positions unchanged throughout the history of the planet, one
could only have assumed that the postulated land bridges existed in
the form of intermediate continents, that they sank below sea level at
the time when interchange of terrestrial flora and fauna ceased and
that they form the present-day ocean floors between the continents.
The well-known palæontological reconstructions arose on the basis of
such assumptions, one example of them, for the Carboniferous, is
given in Figure 2.

This assumption of sunken intermediate continents was in fact the
most obvious so long as one based one's stand on the theory of the
contraction or shrinkage of the earth, a viewpoint we shall have to
examine more closely in what follows. The theory first appeared in

THE NATURE OF THE DRIFT THEORY   .9

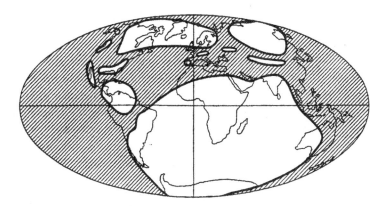

Fig. 2.   Distribution of water (hatched) and land in the Carboniferous, according to the usual conception.

Europe.   It was initiated and developed by Dana, Albert Heim and
Eduard Suess in particular, and even today dominates the fundamental
ideas presented in most European textbooks of geology.   The
essence of the theory was expressed most succinctly by Suess: "The
collapse of the world is what we are witnessing" [12, Vol. 1, p. 778].
Just as a drying apple acquires surface wrinkles by loss of internal
water, the earth is supposed to form mountains by surface folding as it
cools and therefore shrinks internally.   Because of this crustal
contraction, an overall "arching pressure" is presumed to act over
the crust so that individual portions remain uplifted as horsts. These
horsts are, so to speak, supported by the arching pressure.   In the
further course of time, these portions that have remained behind may
sink faster than the others and what was dry land can become sea floor
and vice-versa, the cycle being repeated as often as required.' This
idea, put forth by Lyell, is based on the fact that one finds deposits
from former seas almost everywhere on the continents.   There is no
denying that this theory provided historic service in furnishing an
adequate synthesis of our geological knowledge over a long period of
time.   Furthermore, because the period was so long, contraction
theory was applied to a large number of individual research results
with such consistency that even today it possesses a degree of attractiveness, with its bold simplicity of concept and wide diversity of
application.

Ever since our geological knowledge was made the subject of that impressive synthesis, the four volumes by Eduard Suess entitled *Das Antlitz der Erde*, written from the standpoint of contraction theory, there has been increasing doubt as to the correctness of the basic idea. The conception that all uplifts are only apparent and consist merely of remnants left from the general tendency of the crust to move towards the centre of the earth, was refuted by the detection of absolute uplifts [71]. The concept of a continuous and ubiquitous arching pressure, already disputed on theoretical grounds for the uppermost crust by Hergesell [124] has proved to be untenable because the structure of eastern Asia and the eastern African rift valleys have, on the contrary, enabled one to deduce the existence of tensile forces over large portions of the earth's crust. The concept of mountain folding as crustal wrinkling due to internal shrinkage of the earth led to the unacceptable result that pressure would have to be transmitted inside the earth's crust over a span of 180 great-circle degrees. Many authors, such as Ampferer [13], Reyer [14], Rudzki [15] and Andrée [16], among others, have opposed this quite rightly, claiming that the surface of the earth would have to undergo regular overall wrinkling, just as the drying apple does. However, it was particularly the discovery of the scale-like "sheet-fault structure" or overthrusts in the Alps which made the shrinkage theory of mountain formation, which presented enough difficulties in any case, seem more and more inadequate. This new concept of the structure of the Alps and that of many other ranges, which was introduced by the works of Bertrand, Schardt, Lugeon and others, leads to the idea of far larger compressions than did the earlier theory. Following previous ideas, Heim calculated in the case of the Alps a 50% contraction, but on the basis of the sheet-faulting theory, now generally accepted, contraction of $\frac{1}{4}$ to $\frac{1}{8}$ of the initial span [17]. Since the present-day width of the chain is about 150 km, a stretch of crust from 600 to 1200 km wide (5–10 degrees of latitude) must have been compressed in this case. Yet in the most recent large-scale synthesis on Alpine sheet-faults, R. Staub [18] agrees with Argand that the compression must have been even greater. On page 257 he concludes:

"The Alpine orogenesis is the result of the northward drift of the African land mass. If we smooth out only the Alpine folds and sheets over the transverse section between the Black Forest and Africa, then in relation to the present-day distances of about 1800 km, the original distance separating the two must have been about 3000 to 3500 km, which means an alpine compression (in the wider sense of the word Alpine) of around 1500 km. Africa must have been displaced relative to Europe by this amount. What is involved here is a true continental drift of the African land mass and an extensive one at that."[3]

Other geologists have put forward similar views, as for example F. Hermann [106], E. Hennig [19] or Kossmat [21], who states "that the formation of mountains must be explained by large-scale tangential movements of the crust, which cannot be incorporated in the scope of the simple contraction theory." In the case of Asia, Argand [20], especially, has developed an analogous theory in the course of a comprehensive investigation to which we shall return later. He and Staub have done the same for the case of the Alps. No attempt to relate these enormous compressions of the crust to a drop in temperature of the earth's core can be anything but a failure.

Moreover, even the apparently obvious basic assumption of contraction theory, namely that the earth is continuously cooling, is in full retreat before the discovery of radium. This element, whose decay produces heat continuously, is contained in measurable amounts everywhere in the earth's rock crust accessible to us. Many measurements lead to the conclusion that even if the inner portion had the same radium content, the production of heat would have to be incomparably greater than its conduction outwards from the centre, which we can measure by means of the rise of temperature with depth in mines, taking into account the thermal conductivity of rock. This would mean, however, that the temperature of the earth must rise continuously. Of course, the very low radioactivity of iron meteorites suggests that the iron core of the earth presumably contains much less radium than the crust, so that this paradoxical conclusion can

---

[3] It seems that estimations of the size of the Alpine compression are always on the increase. Staub wrote recently [214, similarly in 215]: "If we now, however, imagine these Alpine sheets, which are probably stacked twelvefold, to be smoothed out again . . ., the solid Alpine hinterland would necessarily lie much further south, and the original distance between foreland and hinterland would probably have been ten to twelve times greater than it is today." He adds: "Formation of a mountain range therefore originates quite clearly and certainly from independent drifting of larger blocks, surely continental blocks by their structure and composition; and thus, starting from Alpine geology and Hans Schardt's sheet theory, we arrive quite obviously and naturally at the acknowledgment of the basic principle of the great Wegener theory of continental drift."

perhaps be avoided.   In any case, it is no longer possible, as it once was, to consider the thermal state of the earth as a temporary phase in the cooling process of a ball that was formerly at a higher temperature. It should be regarded as a state of equilibrium between radioactive heat production in the core and thermal loss into space.   In fact, the most recent investigations into this question, which will be discussed in more detail later on, imply that actually, at least under the continental blocks, more heat is generated than is conducted away, so that here the temperature must be rising, though in the ocean basins conduction exceeds production.   These two processes lead to equilibrium between production and loss rate, taking the earth as a whole.   In any case, one can see that through these new views the foundation of the contraction theory has been completely removed.

There are still many other difficulties which tell against the contraction theory and its mode of thinking.   The concept of an unlimited periodic interchange between continent and sea floor, which was suggested by marine sediments on present-day continents, had to be strictly curtailed.   This is because more precise investigation of these sediments showed with increasing clarity that what was involved was coastal-water sediments, almost without exception.   Many sedimentary deposits formerly claimed as oceanic proved to be coastal; one example is chalk, as proved by Cayaux.   Dacqué [22] has given a good review of the problem.   Only in the case of a very few types of sediment, such as the low-lime Alpine radiolarites and certain red clays reminiscent of the red deep-sea clay, is formation in deep waters (4–5 km) still assumed today, particularly since sea water dissolves out lime only at great depths.   However, the area of these true deep-sea deposits on present-day continents is so tiny compared with the areas of the continents and the areas of coastal water sediments on them that the theory of the basically shallow-water nature of marine fossil deposits on present-day continents is unaffected.   For the contraction theory, however, a considerable difficulty arises.   Since coastal shallows must be counted, geophysically, as part of the continental blocks, the nature of these marine fossils implies that these blocks have been "permanent" throughout the history of the earth and have never formed ocean floors.   Are we then still to assume that today's sea floors were ever continents?   The justification for this conclusion is obviously removed by establishing that the marine sediments found on continents were formed in shallows.   But more than this, the conclusion now leads to an open contradiction.   If we reconstruct intercontinental bridges of the type shown in Figure 2, thus filling up a large part of today's ocean basins without having the possibility of compensating for this by submergence of present-day continental regions to the sea-floor level, there would be no room for the volume of the world's oceans in the now much reduced deep-sea basins. The water displacement of the intercontinental bridges would be so enormous that the level of the world's oceans would rise above that of the whole continental area of the earth and all would be flooded, today's continents and the bridges alike. The reconstruction would not therefore achieve the desired end, i.e., dry land bridges between continents. Figure 2 therefore represents an impossible reconstruction unless we introduce further hypotheses which are "ad hoc" improbabilities; for example, that the mass of ocean water was exactly the required amount less at the former period than it is today, or that the deep-sea basins remaining at that time were precisely the required amount deeper than today. Willis and A. Penck, among others, have brought up this peculiar difficulty.

Of the many objections to contraction theory, one more only will be emphasised; it has very special importance. Geophysicists have decided, mainly on the basis of gravity determinations, that the earth's crust floats in hydrostatic equilibrium on a rather denser, viscous substrate. This state is known as *isostasy*, which is nothing more than hydrostatic equilibrium according to Archimedes' principle, whereby the weight of the immersed body is equal to that of the fluid displaced. The introduction of a special word for this state of the earth's crust has some point because the liquid in which the crust is immersed apparently has a very high viscosity, one which is hard to imagine, so that oscillations in the state of equilibrium are excluded and the tendency to restore equilibrium after a perturbation is one which can only proceed with extreme slowness, requiring many millennia to reach completion. Under laboratory conditions, this "liquid" would perhaps scarcely be distinguishable from a "solid." However, it should be remembered here that even with steel, which we certainly consider a solid, typical flow phenomena occur, just before rupture, for example.

An example of perturbation of isostasy of the crust is shown by the load to which an inland icecap subjects it. The result is that the crust slowly sinks under this load and tends towards a new equilibrium position to correspond with the loading. When the icecap has melted, the original position of equilibrium is gradually resumed, and the shore lines formed during the process of depression are elevated along with the crust. The "isobase charts" of de Geer [23], drawn up from the shore lines, show for the last glaciation of Scandinavia a central depression of at least 250 m, gradually decreasing towards the perimeter; for the most extensive of the Quaternary glaciations still higher values must be assumed. In Figure 3 we reproduce a chart

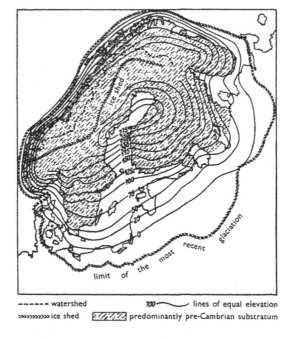

- - - - - watershed   100 ⌒ lines of equal elevation
≫≫≫≫≫≫ ice shed   ▨▨▨ predominantly pre-Cambrian substratum

FIG. 3. Post-glacial elevation contours (in metres) for Fennoscandia (according to Högbom).

of this post-glacial elevation of "Fennoscandia" (Finland, Sweden and Norway) according to Högbom (taken from Born [43]). The same phenomenon has been proved by de Geer to have occurred for the glaciated region of North America. Rudzki [15] has shown that, assuming isostasy, plausible values for the thickness of inland ice layers can be calculated, i.e., 930 m for Scandinavia and 1670 m for North America, where the depression amounted to 500 m. Because of the viscosity of the substrate the equilibration movements naturally lag far behind: the shore lines generally formed only after the melting of the ice, but before the elevation of the land, and even today Scandinavia is still rising by about 1 m in 100 years, as shown by tide-gauge readings.

Even sedimentary deposits result in a subsidence of the blocks, as Osmond Fisher was probably the first to recognise: every deposition from above leads to a subsidence of the block, somewhat delayed of course, so that the new surface occupies almost the same level as the old. In this way many kilometres' thickness of deposit can arise and yet all the layers are formed in shallow water.

Later on we shall examine the theory of isostasy more closely. Here we shall simply say that it has been established by geophysical observations over so wide a range that it is now part of the solid foundation of geophysics and its basic truth can no longer be doubted.[4]

One can see immediately that this result runs quite counter to the ideas of contraction theory and that it is very hard to combine one with the other. In particular, it seems impossible, in view of the isostatic principle, that a continental block the size of a land bridge of required size could sink to the ocean bottom without a load or that the reverse should happen. Isostasy is therefore in contradiction not only to contraction theory, but in particular also to the theory of sunken land bridges as derived from the distribution of organisms.[5]

[4] Americans, e.g., Taylor [101], sometimes mean by "isostasy" Bowie's theory of the origin of geosynclines and mountain ranges. According to Bowie [224], the initial elevation of sediment-filled basins, the geosynclines, comes from a rise in their isotherms, and hence a volumetric expansion. Once this has led to a land elevation, erosion sets in and a jagged mountain range is formed, whose substrate continually rises due to reduction in loading. Finally, the isotherms are raised to an abnormal height by this elevation, and begin to move slowly downwards; the block cools and contracts and the surface sinks; a depression is formed from the mountain region and renewed sedimentation occurs. This produces further depression or subsidence until the isotherms are abnormally low in level, then rise again, and so on over many cycles. This concept, which cannot be applied to the great folded ranges with their overthrusts, as Taylor and others have emphasised, does indeed make use of the principle of isostasy but should not be given the simple title of "the theory of isostasy."

[5] The objections to the contraction theory enumerated here are mainly directed against its typical earlier form. Very recently, attempts have been made to modernise the theory and to answer the objections, partly by restricting it and partly by adding hypotheses; various authors have been involved, such as Kober [24], Stille [25], Nölcke [26], and Jeffreys [102], among others. This is also true of the theory publicised by R. T. Chamberlin [160] which supposes contraction to be caused by "rearrangement" of material in the earth resulting from the planetesimal origin of the earth accepted by this author. Although one cannot deny

16   THE ORIGIN OF CONTINENTS AND OCEANS

In the foregoing, we deliberately reviewed the objections to contraction theory in some detail. This is because in one part of the train of thought discussed here another theory is rooted; this is known as the "theory of permanence" and is especially widespread among American geologists. Willis [27] formulated it as follows: "The great ocean basins constitute permanent features of the earth's surface, and have with little change in shape occupied the same positions as now since the ocean waters were first gathered." In fact, we have already referred above to the fact that the marine sediments on present-day continents were formed in shallow waters, and we deduced that the continental blocks as such have been permanent throughout the earth's history. Isostasy theory proves the impossibility of regarding present-day ocean floors as sunken continents, and this extends the scope of the result based on marine sediments to comprise a general permanence of deep-sea floors and continental blocks. Further, since here, too, the apparently obvious assumption was made that the continents have not changed their relative positions, Willis's formulation of the "permanence theory" appears to be a logical conclusion from our geophysical knowledge, disregarding, of course, the postulate of former land bridges, derived from the distribution of organisms. So we have the strange spectacle of two quite contradictory theories of the prehistoric configuration of the earth being held simultaneously—in Europe an almost universal adherence to the idea of former land bridges, in America to the theory of the permanence of ocean basins and continental blocks.

It is probably no accident that the permanence theory has its most numerous adherents in America: geology developed late there—thus simultaneously with geophysics—and this necessarily led to more rapid and complete adoption by geology of the results advanced by its sister science than in Europe. There was absolutely no temptation to make the contraction theory, which contradicts geophysics, one of the basic assumptions. It was quite otherwise in Europe, where geology already had a long period of development behind it before geophysics had produced its first results, and had, without benefit of geophysics, already arrived at an overall view of

---

that these attempts show a certain adroitness in pursuit of their aim, one cannot say that they really refute the objections, nor that they have brought the contraction theory into satisfactory agreement with new research, especially in the field of geophysics. A thorough discussion of this neo-contraction theory must, however, be dispensed with here.

the earth's evolution in the form of the contraction theory. It is quite understandable that it is difficult for many European scientists to free themselves completely from this tradition and that they view the results of geophysics with a mistrust that never completely fades.

However, where does the truth lie? The earth at any one time can only have had one configuration. Were there land bridges then, or were the continents separated by broad stretches of ocean, as today? It is impossible to deny the postulate of former land bridges if we do not want to abandon wholly the attempt to understand the evolution of life on earth. But it is also impossible to overlook the grounds on which the exponents of permanence deny the existence of sunken intermediate continents. There clearly remains but one possibility: there must be a hidden error in the assumptions alleged to be obvious.

This is the starting point of displacement or drift theory. The basic "obvious" supposition common to both land-bridge and permanence theory—that the relative position of the continents, disregarding their variable shallow-water cover, has never altered—must be wrong. The continents must have shifted. South America must have lain alongside Africa and formed a unified block which was split in two in the Cretaceous; the two parts must then have become increasingly separated over a period of millions of years like pieces of a cracked ice floe in water. The edges of these two blocks are even today strikingly congruent. Not only does the large rectangular bend formed by the Brazilian coast at Cape São Roque mate exactly with the bend in the African coast at the Cameroons, but also south of these two corresponding points every projection on the Brazilian side matches a congruent bay on the African, and conversely. A pair of compasses and a globe will show that the sizes are precisely commensurate.

In the same way, North America at one time lay alongside Europe and formed a coherent block with it and Greenland, at least from Newfoundland and Ireland northwards. This block was first broken up in the later Tertiary, and in the north as late as the Quaternary, by a forked rift at Greenland, the sub-blocks then drifting away from each other. Antarctica, Australia and India up to the beginning of the Jurassic lay alongside southern Africa and formed together with it and South America a single large continent, partly covered by shallow water. This block split off into separate blocks in the course of the Jurassic, Cretaceous and Tertiary, and the sub-blocks drifted

18   THE ORIGIN OF CONTINENTS AND OCEANS

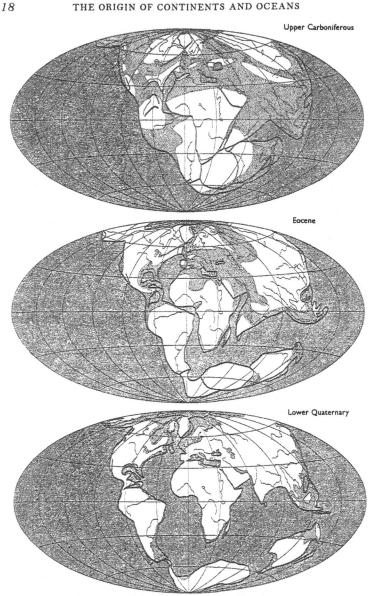

Fig. 4.   Reconstruction of the map of the world according to drift
theory for three epochs.

Hatching denotes oceans, dotted areas are shallow seas; present-day outlines
and rivers are given simply to aid identification.   The map grid is arbitrary
(present-day Africa as reference area; see Chapter 8).

THE NATURE OF THE DRIFT THEORY *19*

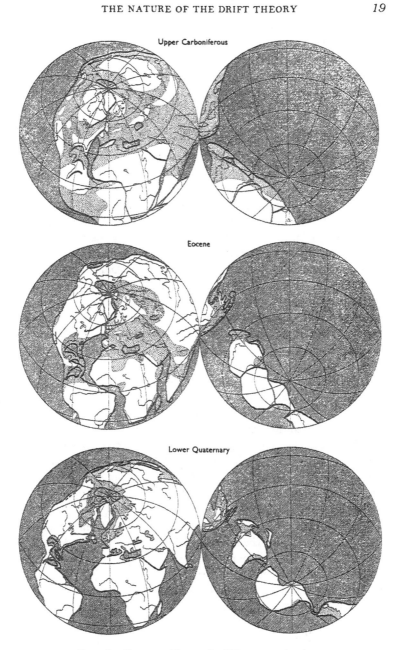

FIG. 5. Same as Fig. 4, in different projection.

away in all directions. Our three world maps (Figs. 4 and 5) for the Upper Carboniferous, Eocene and Lower Quaternary show this evolutionary process. In the case of India the process was somewhat different: originally it was joined to Asia by a long stretch of land, mostly under shallow water. After the separation of India from Australia on the one hand (in the early Jurassic) and from Madagascar on the other (at the transition from Tertiary to Cretaceous), this long junction zone became increasingly folded by the continuing approach of present-day India to Asia; it is now the largest folded range on earth, i.e., the Himalaya and the many other folded chains of upland Asia.

There are also other areas where the continental drift is linked causally with orogenesis. In the westward drift of both Americas, their leading edges were compressed and folded by the frontal resistance of the ancient Pacific floor, which was deeply chilled and hence a source of viscous drag. The result was the vast Andean range which extends from Alaska to Antarctica. Consider also the case of the Australian block, including New Guinea, which is separated only by a shelf sea: on the leading side, relative to the direction of displacement, one finds the high-altitude New Guinea range, a recent formation. Before this block split away from Antarctica, its direction was a different one, as our maps show. The present-day east coastline was then the leading side. At that time New Zealand, which was directly in front of this coast, had its mountains formed by folding. Later as a result of the change in direction of displacement, the mountains were cut off and left behind as island chains. The present-day cordilleran system of eastern Australia was formed in still earlier times; it arose at the same time as the earlier folds in South and North America, which formed the basis of the Andes (pre-cordilleras), at the leading edge of the continental blocks, then drifting as a whole before dividing.

We have just mentioned the separation of the former marginal chain, later the island chain of New Zealand, from the Australian block. This leads us to another point: smaller portions of blocks are left behind during continental drift, particularly when it is in a westerly direction. For instance, the marginal chains of East Asia split off as island arcs, the Lesser and Greater Antilles were left behind by the drift of the Central American block, and so was the so-called Southern Antilles arc (South Shetlands) between Tierra del Fuego and western Antarctica. In fact, all blocks which taper off towards

the south exhibit a bend in the taper in an easterly direction because
the tip has trailed behind: examples are the southern tip of Greenland, the Florida shelf, Tierra del Fuego, the Graham Coast and the
continental fragment Ceylon.

It is easy to see that the whole idea of drift theory starts out from
the supposition that deep-sea floors and continents consist of different
materials and are, as it were, different layers of the earth's structure.
The outermost layer, represented by the continental blocks, does not
cover the whole earth's surface, or it may be truer to say that it no
longer does so. The ocean floors represent the free surface of the
next layer inwards, which is also assumed to run under the blocks.
This is the geophysical aspect of drift theory.

If drift theory is taken as the basis, we can satisfy all the legitimate
requirements of the land-bridge theory and of permanence theory.
This now amounts to saying that there were land connections, but
formed by contact between blocks now separated, not by intermediate
continents which later sank; there is permanence, but of the area of
ocean and area of continent as a whole, but not of individual oceans or
continents.

Detailed substantiation of this new concept will form the chief part
of the book.

Works Cited

[1] WEGENER, A., "Die Entstehung der Kontinente," *Petermanns Mitteilungen*, 1912, pp. 185–195, 253–256,
305–309.

[2] WEGENER, A., "Die Entstehung der Kontinente," *Geologische Rundschau*, **3**, No. 4, 1912, pp. 276–292.

[3] WEGENER, A., *Die Entstehung der Kontinente und
Ozeane* (Sammlung Vieweg, No. 23, 94 pp.), Braunschweig, 1915; 2nd ed. (Die Wissenschaft, No. 66, 135
pp.), Braunschweig, 1920; 3rd ed., 1922.

[4] LÖFFELHOLZ VON COLBERG, Carl Freiherr, *Die Drehung
der Erdkruste in geologischen Zeiträumen* (62 pp.),
Munich, 1886. (2nd, much enlarged ed., 247 pp., Munich, 1895.)

[5] KREICHGAUER, D., *Die Äquatorfrage in der Geologie*
(304 pp.), Steyl, 1902; 2nd ed., 1926.

[6] WETTSTEIN, H., *Die Strömungen der Festen, Flüssigen
und Gasförmigen und ihre Bedeutung für Geologie,
Astronomie, Klimatologie und Meterologie* (406 pp.),
Zurich, 1880.

[7] SCHWARZ, E. H. L., *Geological Journal*, 1912, pp. 294–
299.

[8] PICKERING, *Journal of Geology*, **15**, No. 1, 1907; also
Gaea, 43, 1907, p. 385.

[9] COXWORTHY, W. FRANKLIN, *The Electrical Condition, or
How and Where Our Earth was Created*, London,
W. J. S. Phillips, 1890(?).

[10] TAYLOR, F. B., "Bearing of the Tertiary Mountain Belt
on the Origin of the Earth's Plan," *Bulletin of the Geological Society of America*, **21** (2), June 1910, pp. 179–
226.

[11] ARLDT, T., *Handbuch der Paläogeographie*, Leipzig, 1917.

[12] SUESS, E., *Das Antlitz der Erde*, **1**, 1885.

[13] AMPFERER, "Über das Bewegungsbild von Faltengebir-
gen," *Jahrbuch der k. k. Geologischen Reichsanstalt*,
56, Vienna, 1906, pp. 529–622.

[14] REYER, *Geologische Prinzipienfragen*, Leipzig, 1907.

[15] RUDZKI, M. P., *Physik der Erde*, Leipzig, 1911.

[16] ANDRÉE, K., *Über die Bedingungen der Gebirgsbil-
dung*, Berlin, 1914.

[17] HEIM, A., "Bau der Schweizer Alpen," *Neujahrsblatt
der Natur-forschung-Gesellschaft*, Zurich, 1908, Part
110.

[18] STAUB, R., "Der Bau der Alpen," *Beiträge zur geologischen Karte der Schweiz*, N.F., No. 52, Bern, 1924.

[19] HENNIG, E., "Fragen zur Mechanik der Erdkrusten-
Struktur," *Die Naturwissenschaften*, 1926, p. 452.

[20] ARGAND, E., "La tectonique de l'Asie," *Extrait du Compterendu du XIIIe Congrès géologique international 1922*, Liège, 1924.

[21] KOSSMAT, F., "Erörterungen zu A. Wegeners Theorie der Kontinentalverschiebungen," *Zeitschrift der Gesellschaft für Erdkunde zu Berlin*, 1921.

[22] DACQUÉ, E., *Grundlagen und Methoden der Paläogeographie*, Jena, 1915.

[23] GEER, G. DE, *Om Skandinaviens geografiska Utvekling efter Istiden*, Stockholm, 1896.

[24] KOBER, L., *Der Bau der Erde*, Berlin, 1921; *Gestaltungesgeschichte der Erde*, Berlin, 1925.

[25] STILLE, H., *Die Schrumpfung der Erde*, Berlin, 1922.

[26] NÖLCKE, F., *Geotektonische Hypothesen*, Berlin, 1924.

[27] WILLIS, B., "Principles of Palæogeography," *Science*, **31**, N.S., No. 790, 1910, pp. 241–260.

[43] BORN, A., *Isostasie und Schweremessung*, Berlin, 1923.

[63] GREEN, W. L., "The Causes of the Pyramidal Form of the Outline of the Southern Extremities of the Great Continents and Peninsulas of the Globe," *Edinburgh New Philosophical Journal*, **6**, n.s., 1857; also: *Vestiges of the Molten Globe*, 1875.

[86] MANTOVANI, R., "L'Antarctide," *Je m'instruis*, Sept. 19, 1909, pp. 595–597.

[101] TAYLOR, F. B., "Greater Asia and Isostasy," *American Journal of Science*, July 1926, pp. 47–67.

[102] JEFFREYS, H., *The Earth: Its Origin, History and Physical Constitution*, Cambridge University Press, 1924.

[106] HERMANN, F., "Paléogéographie et genèse penniques," *Eclogæ Geologicæ Helvetæ*, XIX, No. 3, 1925, pp. 604–618.

[123] DE BEAUFORT, L. F., "De beteekenis van de theorie van Wegener voor de zoögeografie," *Handlingen van het XXe Nederlandsch Natuur- en Geneeskundig Congress*, April 14/16, April 1925, Groningen.

[124] HERGESELL, H., "Die Abkühlung der Erde und die gebirgsbildenden Kräfte," *Beiträge zur Geophysik*, **2**, 1895, p. 153.

[160] CHAMBERLIN, ROLLIN T., "Objections to Wegener's Theory," 1928; in [228].

[214] STAUB, R., "Das Bewegungsproblem in der modernen Geologie," inaugural lecture, Zurich, 1928.

[215] STAUB, R., *Der Bewegungsmechanismus der Erde*, Berlin, 1928.

[224] BOWIE, W., *Isostasy* (275 pp.), New York, 1927.

From *Transantarctic Relationships and Their Significance,*
as Evidenced by Chironomid Midges
Lars Brundin

IV. THE NATURE OF TRANSANTARCTIC RELATIONSHIPS

CHIRONOMID MIDGES AS INDICATORS IN AUSTRAL BIOGEOGRAPHY

With the above monograph of three comparatively high-ranked midge groups as a background the way is prepared for a discussion of the true nature of transantarctic relationships and their general significance. Before that it seems appropriate, however, to raise the question: What about the general suitability of chironomid midges as indicators in biogeography? Is it, after all, advisable to try to answer intricate problems as to the history of the austral biotas and the existence or non-existence of former land connections on the basis of these small and fragile midges of the cool mountain streams? Are there not many groups which would give a more reliable evidence of general applicability?

We are here touching upon a subject which has been strongly coloured by misconceptions. There is a common belief that vertebrates and higher plants are the best organisms for biogeographic work because they are so "well known". But the knowledge of ecology, biology, and distribution, however detailed, is insufficient for biogeographic analyses. Many biologists realize, it is true, that a proper insight into the extrinsic and intrinsic relationships is absolutely essential for the interpretation of the history of a group, but only few seem to be fully aware of the fact that all endeavour to interpret relationships is futile if the arguments are based on typological thinking. Referring to our discussions in the foregoing chapters we are able to state that the present plant and animal systems still are largely typological, meaning that the primitive components of the different aggregates generally are paraphyletic. Only few have drawn the inescapable conclusion of this. Darlington is not among those. He stresses in his *Zoogeography* (1957) that the relationships of many groups are still unsettled and continues (p. 28):

"There is not much that zoogeographers can do about this until taxonomists improve the poor classification. In the meantime zoogeographers should work as much as possible with the best-known groups of animals, remember that classifications are imperfect, and proceed cautiously. Taxonomic experience is useful, almost essential, to a zoogeographer. An experienced taxonomist knows what classifications are and what other taxonomists mean when they talk about relationships. Also (but this is less important) a taxonomist can sometimes suspect unsound classification even of animals with which he is not familiar, much as an engineer might suspect an unsound bridge even if he has not built it himself."

"But this is the dark side of the matter. The bright side is that the facts now known of the geographical occurrence and classification at least of vertebrates are good enough."

It is significant of the confused situation in biogeography of today that the opinion of Darlington has encountered weak resistance. That the interpretation of the phylogenetic relationships is most unsatisfactory even among the mammals, was demonstrated above in the section on "The shortcomings of the typological method". A principal prerequisite for an understanding of the main trends in the early history of the placentals is of course the establishment of the sister group relationships of such high-ranked groups as the Insectivora, Primates, Edentata, Glires, Proboscidea, Hyracoidea, Tubulidentata, etc. These relationships are, however, very poorly known. But according to Darlington the classification of vertebrates is "good enough" for biogeographic analysis. This is fatal optimisim. When Darlington wrote his cited book

437

there was no major group among the biota of the world whose phylogenetic structure was so well known that a biogeographer could start reconstructing its history on a firm basis. There is, however, another aspect involved here, for if a scientist has been able to reconstruct the extrinsic and intrinsic phylogenetic relationships of a group, by application of proper arguments (including the rule of geographical vicariism of sister groups), then he has in fact at the same time demonstrated the main trends in the history of the group. In other words: under the given assumption there is simply no need for an intervention by a "professional" biogeographer. It seems that professionals like Darlington have not made this casual connection perfectly clear to themselves. Since the history of life and the history of phylogenies is the same thing, all scientists studying phylogenies are potential biogeographers in a truly progressive sense. Professional biogeographers in the sense of Darlington (1965), on the other hand, who are forced to compile and speculate, are doomed to remain outsiders.

That paleontology is able to give a direct information as to the phylogenetic relationships and history of a group, is a widely held belief. The untenability of that view has been demonstrated above in the chapter dealing with the principles of phylogenetic systematics.

The view that vertebrates and higher plants owing to better classification are more suitable for biogeographic work than insects and other invertebrates is mainly incorrect; and the general neglect of, or unwillingness to accept, the circumstance that the search for the sister group (Hennig's principle) is the essential task, still stands out as the stumbling-block of systematics and biogeography.

The problem as to the relative suitability of different groups for biogeographic work is thus primarily a question concerning their relative accessibility for proper phylogenetic analysis. Plants do not occupy a favoured position. Their comparatively simple construction combined with excessive parallelism and convergence must be serious obstacles to the botanical investigator. Many botanists centre their hope on cytogenetics, but the possible addition of phylogenetic arguments from that field will be limited, and they cannot be assigned any greater importance than other arguments.

Many animal groups are rich in complex structural systems facilitating the phylogenetic approach; but a group like the birds, with a very uniformly constructed body, will raise considerable difficulties. Comparing advantages and disadvantages of different groups we will find that few are better suited for a phylogenetic analysis than the holometabolous insect orders. In the possession of a developmental cycle comprising larva, pupa, and imago these orders offer three quite different types of organisation and adaptation within the limits of a species. We have thus the great advantage of being able to cross-check the phylogenetic conclusions drawn from one stage with the others, thus attaining a high degree of reliability by the establishment of monophyletic sister groups. The Nematocerous Diptera, which include the family Chironomidae, occupy a leading position among the Holometabola with regard to the number of available structural systems within the developmental cycle. This is a very important point. However, when stressing that the rheophil chironomid midges are better suited for a discussion of the true nature of the transantarctic relationships than most other groups among plants and animals, I have also other advantages in mind:

(1) The accessibility of the rheophil chironomid midges in the field because of the type of their habitats. It is an inestimable advantage that large and truly representative samples can be collected conveniently and in short time with the aid of the stream net method described above (p. 72).

(2) The austral rheophil chironomids are on the whole extremely cold-resistant but only moderately stenothermal, and they are more broadly adapted to different kinds of mountain streams and stream habitats than any other austral group outside Chironomidae. It is a remarkable fact that even the upper courses of large glacier torrents of the southern lands (southern South America, New Zealand) are inhabited by a chironomid fauna which is nearly quite as rich in species as that of clear, less wild and less cold mountain streams in the vicinity; and the composition of the chironomid fauna demonstrates that this phenomenon is the consequence of general adaptability and hardiness rather than specialisation. In the upper courses of torrents and rivers coming from glaciers the chironomids have to endure permanent water temperatures

L. BRUNDIN, *Transantarctic relationships* K. V. A. Handl. 11: 1

down to 0.5°C. It is significant in this connection that *Parochlus steineni*, the only fully winged insect and the largest permanent land resident of Antarctica, is a chironomid midge. Indeed, there are few other animal groups which have been less influenced by the Pleistocene glaciations in the south than the chironomids of the cool mountain streams.

At the other end of the temperature range, the power of resistance is rather different in various groups and species. As to the groups displaying transantarctic relationships, it seems reasonable to say that most species thrive well at water temperatures up to 15–17°C during the summer months; but I have found also several species, some in great abundance, at an upper range of 20–22°C. In other words, several groups are well represented also in mountain streams running through rain forests of the warm temperate zone, and some of the most tolerant species succeed even in subtropical environments. (For details see the monograph above.)

(3) To the advantages connected with the general hardiness of rheophil chironomid midges of the southern latitudes come in addition those of the smallness of the species whose wing length is only 0.8–3.7 mm. It is in general easier for smaller animals to maintain their populations above minimal limits during periods of climatic and geographic stress which commonly effects division into pockets of the old habitats and distribution areas.

(4) The larvae of the austral rheophil chironomids feed on diatoms, and obviously there has never been a problem of nourishment. The diatomaceous diet has permitted the midges to evolve and radiate independent of higher plants. This circumstance simplifies the interpretation of the history of the midges.

(5) The unchangeability of the habitats. From a geologic point of view any particular mountain stream or mountain range is ephemeral, but mountain streams in general are not. There are probably few habitats on land whose limiting abiotic factors have changed less through the ages than those of the upper reaches of the mountain streams and their springs.

(6) The absolute age of the chironomid midges. Tribes, families and orders are, unfortunately, completely incomparable concepts within the animal kingdom. It is important, therefore, to realize that the chironomid midges, though given only the rank of a family, obviously are fully comparable with the birds (even as to species number) and thus with a history which, very cautiously estimated, goes far back into the Jurassic. This view is supported by circumstantial fossil evidence (cf. Hennig 1954) and by the distribution patterns, especially the presence in Southern Africa of several austral disjunct groups displaying transantarctic relationships. In other words, the chironomid midges are old enough to perform the role of a key-group for the solution of the problem of transantarctic relationships.

(7) A great advantage when dealing with the austral rheophil chironomid groups is the circumstance that the distribution patterns so obviously are the result of processes and events beyond the action of man. The species of the different main areas are invariably endemic.

The advantages of the rheophil chironomid midges, as set out above, are obvious. There is little doubt, however, that many biologists will be inclined to attach a great importance to the fact that midges are small and winged creatures; the conclusion will be drawn that the midges are easily exposed to chance dispersal by strong winds and thus less well fitted as general indicators in austral biogeography. Such an opinion might seem all the more justified since chironomid midges have been established as members of the aerial plankton in connection with those large-scale trapping experiments which are carried out in Antarctic and subantarctic areas under the guidance of Gressitt. (Cf. Gressitt 1964 a, and references cited therein.)

It is not my intention to discuss the vast literature dealing with means of dispersal, "dispersal capacity", "spread potential" (Leston 1957), and winds and ocean currents as agents in the dispersal of different organisms. Enough opinions and suppositions have appeared in connection with the lengthy discussions on these matters. No one denies, I think, that the transporting capacity of gales, hurricanes, and tornados is impressive and that small insects can cross wide expanses of sea with the aid of air-borne dispersal. There is, further, reason to believe that long distance dispersal combined with successful establishment has occurred; and there has been ample time.

The extensive investigations made on chance dispersal and dispersal agents are in themselves interesting, but do they have some bearing on the biogeographical main problems? The truth seems to be that investigations along those lines are doomed to be beside the point because they can never answer the decisive question: What belongs to the main pattern, what is barely exceptional? The much-discussed problem as to the role played by chance dispersal can be solved or brought nearer solution only by additional detailed analyses of the phylogenetic structure of widespread disjunct groups with a theoretically high dispersal capacity. That role may have been different in different areas and during different epochs; as regards the circumantarctic lands and their disjunct groups of plants and animals it is evident that the answer given by a group like the chironomid midges will be of high general interest.

What has really happened? The answer, amazingly clear-cut, is contained in the connections between phylogenetic relationships and distribution patterns, as set out in detail in the above monograph. In the following I will first give a concentrated survey of the results of the phylogenetic analyses of the different groups and then a discussion of the performed evidence.

The evidence of the midges

According to the present investigation the cool mountain streams of temperate South America, Southern Africa, Tasmania-southeastern Australia, and New Zealand are inhabited by 600–700 chironomid species. South America possesses the largest number of species (more than 200), followed by the Tasmanian–Australian realm, New Zealand, and lastly Southern Africa.

Transantarctic relationship seems to be a widespread phenomenon within the family Chironomidae, and primarily it was my intention to analyse all the present cases. However, during the course of the work it became more and more clear that the demands for expert knowledge of the world fauna raised by the necessary phylogenetic analyses were so high that only cases exhibited by three of the seven major groups (subfamilies) could be considered here. Decisive for the evaluation of the subfamilies Podonominae, Aphroteniinae, and Diamesinae were (1) the reasonable size of these subfamilies, (2) their marked amphitropical distribution and complete absence from tropical lowlands, (3) their comparatively very strong involvement in transantarctic relationships, and (4) their possession of markedly diversified pupae rich in structural systems. At an advanced stage of the investigation it became apparent, moreover, that the austral disjunctive distribution pattern of all three subfamilies included also Southern Africa which is of course a fair assurance that the selected subfamilies can be employed for a treatment of the problem of transantarctic relationships in its whole scope.

The Podonominae

The subfamily Podonominae comprises 13 genera and 148 species. It has a markedly bipolar (amphitropical) distribution, but there is a pronounced imbalance in so far that the austral podonomine fauna consists of no less than 130 species (88 %) representing 9 genera, while in the northern continents there are only 5 genera and 18 species.

SOUTH AMERICA.—The present main centre of the austral component is South America whose podonomine fauna (inclusive that of Juan Fernandez) amounts to 6 genera and 98 species and makes up 66 % of the world fauna and 75 % of the southern hemisphere fauna of the subfamily. The largest genera are *Podonomus*, with 39 species, *Parochlus*, with 30 species, and *Podochlus*, with 21 species. The genera *Podonomopsis* and *Rheochlus* have only 5 and 2 species respectively, and the sixth genus, "Chile", known only in the pupal stage, is monotypic. The main part of the South American podonomine fauna is concentrated to the mountain streams of South Chile and Patagonia where the subfamily plays a considerably more prominent

L. BRUNDIN, *Transantarctic relationships* K. V. A. Handl. 11: 1

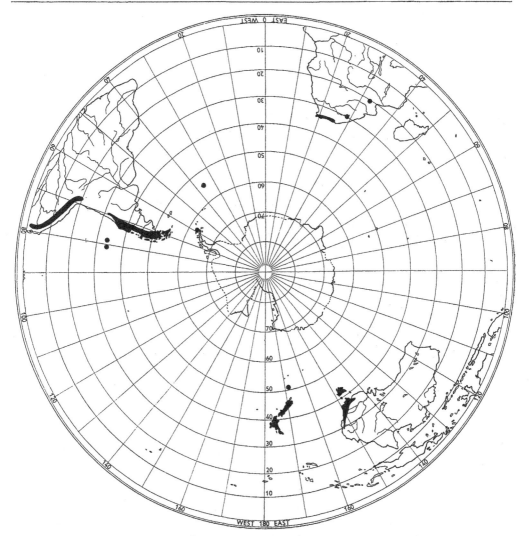

Fig. 633. Southern hemisphere distribution of the chironomid subfamilies Podonominae and Aphroteniinae.

role than in any other area of the world and makes up no less than 38 % of the chironomid species of the running waters, thus attaining a percentage which comes very close to that of Diamesinae + Orthocladiinae (42 %). The southern portion of South America, counted northwards to Santiago de Chile, and with includation of the Juan Fernandez Islands, is inhabited by 87 species. It is noteworthy that the podonomine fauna of the Magellanic region (here counted from Cerro Payne southwards to Cape Horn) is quite as rich in species as that of the Valdivian region (included here are the Lake District of South Chile, the adjoining parts of the Chilean and Argentinian Andes, and Chiloé Island), 42–43 species being known to me from each

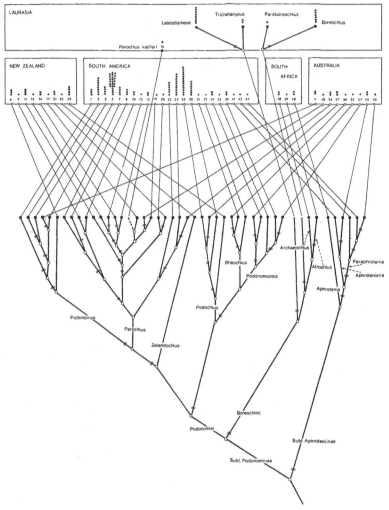

Fig. 634. Main features of the connection between geographic distribution and phylogenetic relationships in the midge aggregate Podonominae-Aphroteniinae. In the geographic sections the number of species in each group is indicated by small black dots. The apomorph groups of the phylogenetic diagram are indicated by two close-set transverse strokes. (From Brundin 1965, fig. 2, somewhat modified.)

The numbers at the bottom of the geographic sections indicate the following groups:

1. *Podonomus, maculatus* group, with 3 species in South America.
2. *Podonomus, albinervis* group, with 8 species in South America.
3. *Podonomus, nudipennis* group, with 4 species in South America.
4. *Podonomus, collessi* subgroup of the *decarthrus* group, with 2 species in Australia.
5. *Podonomus, decarthrus* group, with 23 species in South America.
6. *Podonomus, parochloides* group, with 3 species in New Zealand.
7. *Parochlus, chiloénsis* group, with 4 species in South America.

area. The high representation of the podonomine midges at the southernmost tip of the South American continent attests not only to the general hardiness of these forms, but also suggests that several podonomine species were able to survive the Pleistocene glaciations in ice-free areas of the extreme south, the most prominent of which were probably on the western coast. Further indication of survival in ice-free areas would be the persistence of endemism within the Magellanic region, and in fact quite a few podonomine species are known so far only from the extreme south. It is interesting to note that the evidence for ice-free sanctuaries in the Chilean Archipelago during the glacial maxima, as suggested by the midges, is in agreement with the opinion of Auer (1956, 1958), who has made extensive studies of Pleistocene stratigraphy in Fuegia and Patagonia.

The podonomine fauna of the JUAN FERNANDEZ ISLANDS comprises 3 *Parochlus* species and 3 *Podonomus* species. At least one species of the former genus and all species of the latter are probably endemic.

The present desert zone of northern Chile marks a break in the South American distribution area of the subfamily. But we find these midges again in the tropical Andes of Bolivia, Peru, and Ecuador, from 1700–1800 m above sea level and up to the lower limit of the high alpine block zone, i.e. to the uppermost limit of permanent surface water. The subfamily is represented here by 13 species, 10 of them belonging

8. *Parochlus, squamipalpis* group, with 3 species in South America.
9. *Parochlus conjungens* of the *conjungens* group, New Zealand.
10. *Parochlus subantarcticus* and *duséni* of the *conjungens* group, South America.
11. *Parochlus, novaezelandiae* subgroup of the *araucanus* group, with 3 species in New Zealand.
12. *Parochlus pauperatus* of the *pauperatus* subgroup, *araucanus* group, New Zealand.
13. *Parochlus crassicornis* of the *pauperatus* subgroup, *araucanus* group, South America.
14. *Parochlus ohakunensis* and *carinatus* of the *trigonocerus* subgroup, *araucanus* group, New Zealand.
15. *Parochlus trigonocerus* and *incaicus* of the *trigonocerus* subgroup, *araucanus* group, South America.
16. *Parochlus kiefferi* of the *araucanus* subgroup, *araucanus* group, Europe, Greenland, North America.
17. *Parochlus maorii* of the *araucanus* subgroup, *araucanus* group, New Zealand.
18. *Parochlus bassianus* of the *araucanus* subgroup, *araucanus* group, Tasmania.
19. *Parochlus araucanus* of the *araucanus* subgroup, *araucanus* group, South America.
20. *Parochlus patagonicus* of the *spinosus* subgroup of the *araucanus* group, South America.
21. *Parochlus spinosus* and *aotearoae* of the *spinosus* subgroup, *araucanus* group, New Zealand.
22. *Parochlus, steineni* group, with 6 species in South America, South Georgia, South Shetland Islands.
23. *Parochlus, nigrinus* group, with 10 species in South America.
24. *Parochlus, tonnoiri* group, with 2 species in Australia.
25. *Zelandochlus*, with 1 species in New Zealand.
26. *Podochlus, tenuicornis* group, with 13 species in South America.
27. *Podochlus, tasmaniensis* subgroup of the *tenuicornis* group, with 2 species in Australia.
28. *Podochlus, beschi* group, with 7 species in South America.
29. *Podochlus, grandis* group, with 4 species in New Zealand.
30. *Rheochlus wirthi*, Australia.
31. *Rheochlus prolongatus*, South America.
32. *Rheochlus insignis*, South America.
33. *Podonomopsis, discoceros* group, with 1 species in Australia.
34. *Podonomopsis, brevipalpis* group, with 2 species in South America.
35. *Podonomopsis andinus* of the *muticus* group, South America.
36. *Podonomopsis muticus* and *illiesi* of the *muticus* group, South America.
37. *Podonomopsis evansi* of the *muticus* group, Australia.
38. *Archaeochlus*, with 2 species in Southern Africa.
39. *Afrochlus*, with 1 species in Southern Africa.
40. *Aphrotenia*, with 2 species in Southern Africa.
41. *Aphroteniella*, "Peulla" group, with 1 species in South America.
42. *Aphroteniella, filicornis* group, with 2 species in Australia.
43. *Paraphrotenia excellens*, South America.
44. *Paraphrotenia multispinosa*, South America.
45. *Paraphrotenia fascipennis*, Australia.

to the genus *Podonomus*, the others to *Parochlus* (2 species) and *Podonomopsis* (1 species). All species of the tropical Andes are endemic there and stand out very clearly as apomorph offshoots of aggregates of the southern Andes. This is also the case with the endemic species of Juan Fernandez.

ANTARCTICA.—It is remarkable that one podonomine midge, *Parochlus steineni*, occurs on the South Shetland Islands off Graham Land. The species has been found also on South Georgia, but its main distribution is on the South American continent, from Tierra del Fuego to the Andes of Santiago de Chile. Remarkable also is the fact that the populations of the South Shetland Islands, South Georgia, and the southern Andes are fully winged, while the population at the northern distribution limit in the Santiago Andes is markedly short-winged and characterized also by other peculiarities, thus forming a subspecies of its own. Although these characteristics demonstrate a southern origin of the species, it is of course uncertain whether the *steineni* populations of South Georgia and the South Shetland Islands are preglacial relicts.

TASMANIA–AUSTRALIA.—The podonomine fauna of this realm is extremely similar to that of South America. All South American genera, except the peculiar, monotypic genus "Chile" of the Chilean Lake District, are represented, but the total species number is only 11. The strongholds of the members of the subfamily are the moist western part of Tasmania and the mountainous areas of Victoria and New South Wales. Only one species, *Podonomopsis evansi*, extends further north, into southern Queensland. Some species are known only from Tasmania, but whether this implies real endemism remains uncertain.

NEW ZEALAND.—Comprising 18 species the neozelandic podonomine fauna is larger than that of Tasmania–Australia. The composition of the fauna reveals a rather marked independency. Significant is the occurrence of the plesiomorph endemic genus *Zelandochlus* (monotypic), an inhabitant of typical glacier streams. Best represented is the genus *Parochlus*, with 10 species. The remaining genera, *Podochlus* and *Podonomus*, have only 4 and 3 species respectively. The genera *Rheochlus* and *Podonomopsis* are absent. The podonomine fauna of the South Island seems to be considerably richer than that of the North Island.

SOUTHERN AFRICA.—Of the podonomine faunas of the southern lands that of Southern Africa displays by far the strongest independency. Striking is also its paucity in species. Only two genera are represented, both endemic. The monotypic genus *Afrochlus* occurs in the mountains of Southern Rhodesia, while the two species of the genus *Archaeochlus* apparently are confined to the Great Escarpment of the Drakensberg.

The main features of the connection between phylogenetic relationships and geographic distribution in Podonominae (and Aphroteniinae) are demonstrated by the diagram in fig. 634.

The subfamily Podonominae consists of two major aggregates. One of them, the tribe PODONOMINI, comprises the whole podonomine fauna of South America, Tasmania–Australia, and New Zealand. This large austral group of 127 species has preserved, in its basic design, such a plesiomorph feature as double dististyles in the male hypopygium. The most prominent apomorph character of the basic design of the tribe appears to be the dominance of the dorsal prong of the dististyles (present in *Zelandochlus* and *Podochlus*).

The analysis of the apomorph characters indicates that one of the two groups making up the tribe Podonomini, the *Podonomus* group (subtribe Podonomina), comprises the genera *Zelandochlus*, *Parochlus*, and *Podonomus*. This group is the plesiomorph sister group of the other major group, the *Podochlus* group (subtribe Podochlina), which consists of the genera *Podochlus*, *Rheochlus*, *Podonomopsis*, and evidently also the monotypic genus "Chile" (known only in the pupal stage). The monotypic genus *Zelandochlus* is confined to New Zealand and forms the plesiomorph sister group of *Parochlus* + *Podonomus*, both of which occur in South America, Tasmania–Australia, and New Zealand.

Genus *Parochlus*.—Though a complete analysis of the phylogenetic relationships of the 45 species of the genus was not possible because of insufficient knowledge of the pupal stage of certain species, the main trends in the transantarctic phyletic connections could be firmly established. Although the numerous species of the *araucanus* group are generally inseparable in the imaginal stage they are fortunately readily distinguishable on the basis of excellent characters in the pupae.

The present main centre of *Parochlus* is South America where there are no less than 30 species. Most of these are confined to southern Chile–Patagonia. Of the 6 species groups present in South America the *araucanus-*, *conjungens-*, *chiloénsis-*, and *squamipalpis* groups constitute one unit, the *steineni-* and *nigrinus* groups another. All but the *araucanus-* and *conjungens* groups are exclusively South American.

Among the 10 neozelandic *Parochlus* species the species *conjungens* (No. 9 of the phylogenetic diagram in fig. 634) is the sole representative of the *conjungens* group (which further comprises 2 species in southern Chile), while the remaining 9 species are members of the large *araucanus* group. These two groups are comparatively plesiomorph. The structural systems of the pupae demonstrate that the neozelandic members of the *araucanus* group belong to 5 different subgroups (No. 11, 12, 14, 17, and 21 of the diagram referred to above). Of these subgroups 4 are represented also in temperate South America, while the comparatively apomorph *novaezelandiae* subgroup (No. 11) is confined to New Zealand. The above discussion (pp. 119–123) has demonstrated that there is strong evidence of complex transantarctic relationships between the *Parochlus* aggregates of South Chile–Patagonia on one hand and those of New Zealand on the other. Remarkable is the circumstance that we apparently are dealing with austral bicentricity even within the subgroups of the *araucanus* group, partly meaning the existence of austral bicentric species pairs.

But our survey of the *Parochlus* pattern is still far from complete: The *araucanus* subgroup of the *araucanus* group comprises not only the South American species *araucanus* (No. 19) and the New Zealand species *maorii* (No.17), but also *bassianus* (No. 18) of Tasmania and *kiefferi* (No. 16) of North America, Greenland and Europe. The Tasmanian species is more closely related to the plesiomorph *araucanus* than to any other species, and the holarctic *kiefferi* is clearly the most apomorph member of the subgroup. Thus the small *araucanus* subgroup displays not only austral tricentric distribution but also bipolarity.

The extraordinary complexity of the transantarctic relationships displayed by the genus *Parochlus* is emphasized by the presence of a further striking case. In the mountain streams of Tasmania and New South Wales occur the remarkable *Parochlus* species *tonnoiri* and *rieki*, both characterized by an extremely large "heel" on the dististyles of the male hypopygium. This *tonnoiri* group (No. 24) is doubtlessly the apomorph sister group of the *nigrinus* group (No. 23) of South America. These groups constitute the apomorph sister group of the South American *steineni* group (No. 22).

Genus *Podonomus*.—This genus comprises 44 species. In this respect it is comparable to *Parochlus*. However, the genus *Podonomus* displays a much simpler pattern of transantarctic relationships than its plesiomorph sister group. Alluded to is the fact that the small *parochloides* group of New Zealand (with 3 species) forms the plesiomorph sister group of an assemblage comprising all the other elements of the genus. The sister group of the *parochloides* group (No. 6) is formed by 4 subordinate groups: the *decarthrus* group (No. 4 + 5), the *nudipennis* group (No.3), the *albinervis* group (No. 2), and the *maculatus* group (No. 1). The latter small group is the apomorph sister group of a group comprising the three other groups. Groups No. 1–3 are exclusively South American, while the comparatively plesiomorph *decarthrus* group, comprising 24 South American species, has a bicentric distribution; the species *collessi* and *derwentensis* of Tasmania and southeastern Australia form a comparatively apomorph subgroup (No. 4) within the group.

In the subtribe Podochlina, the *Podochlus* group, we are confronted with the same types of connection between phylogenetic relationships and geographic distribution as in Podonomina. Of the four genera that constitute the subtribe Podochlina, only the comparatively plesiomorph genus *Podochlus* is represented in New Zealand. The *grandis* group of New Zealand (No. 29), comprising 4 species, is doubtlessly the plesiomorph sister group of an assemblage including all remaining species of the genus. The *beschi* group (No. 28), with 7 Chilean–Patagonian species, is the plesiomorph sister group of a group formed by the large, mainly South American *tenuicornis* group (No. 26 + 27). The species *australiensis* and *tasmaniensis* form a group (No. 27) which apparently is only a link in the sister group system made up by the South American members of the *tenuicornis* group. Precise identification of the sister group of the comparatively apomorph Australian group is, however, not possible at present because of lacking knowledge of the intrinsic relationships of the South American species cluster.

The 3 remaining podochline genera are all absent in New Zealand. The genus "Chile" is a peculiar mono-typic genus confined to the southern Andes. Since it is known only as pupa, its phylogenetic connections have not been determined precisely, but there is some indication that it is the apomorph sister group of a group that comprises all the other genera of the subtribe.

Genus *Podonomopsis*.—This comparatively apomorph genus displays twofold transantarctic relationships between groups of South America and Tasmania–Australia. One of the two major components of the genus, the *muticus* group, comprises 4 species. *P. andinus* of the Santiago Andes is the plesiomorph sister species (or sister group, No. 35) of a group comprising the 3 other species. Of these the Andean species *muticus* and *illiesi* form a group (No. 36) where the south Andean *muticus* constitutes the plesiomorph sister species of *illiesi* of the tropical Andes. *P. evansi* of Tasmania and southeastern Australia forms the apomorph sister group (No. 37) of *muticus + illiesi*. The other main component of *Podonomopsis*, i.e. the apomorph sister group of the *muticus* group, is composed of the south Andean *brevipalpis* group (No. 34), with 2 species, and the peculiar, monotypic *discoceros* group of Tasmania and southeastern Australia (No. 33). The latter is the apomorph sister group of the former.

The genus *Rheochlus* seems to be the plesiomorph sister group of *Podonomopsis*. Of the 3 *Rheochlus* species the comparatively plesiomorph *insignis* of the southern Andes is the sister group (No. 32) of a group including the two others. *R. prolongatus* of the southern Andes (No. 31) is the plesiomorph sister species of *wirthi* (No. 30) of New South Wales.

The above survey of transantarctic relationships displayed by the subfamily Podonominae has dealt only with genera belonging to the tribe Podonomini and with distribution patterns involving South America, Tasmania–Australia, and New Zealand. The only exception was the remarkable holarctic distribution of *Parochlus kiefferi*. The distribution patterns of the sister tribe BOREOCHLINI are entirely different. This interesting tribe is mainly boreal, and its austral occurrence is confined to southern Africa where the sister tribe is absent. The most plesiomorph genus is *Archaeochlus* of the Drakensberg whose larvae have preserved functional spiracles on the eighth abdominal segment. The genus comprises only 2 species. The monotypic genus *Afrochlus* of Southern Rhodesia is apparently a close relative of *Archaeochlus*. The absence of Boreochlini in the southwestern Cape Province and the apparent confinement of the two known genera to small temporary streams farther to the north suggests relic occurrence.

As *Afrochlus* is still incompletely known, analysis of the phylogenetic relationships of the African group was concentrated on *Archaeochlus*, the developmental cycle of which was described above. The analysis demonstrates that the pupae and larvae of *Archaeochlus* have no less than 5 unique apomorph characters in common with the boreal genera *Paraboreochlus*, *Boreochlus*, *Trichotanypus*, and *Lasiodiamesa*. Of these characters 3 are shared only with *Paraboreochlus* and *Boreochlus*. The presence of these synapomorphies demonstrates that *Archaeochlus* (and presumably also *Afrochlus*) is more closely related to the boreal genera than to any of the other austral podonomine genera, and, furthermore, that *Archaeochlus* (plus *Afrochlus*) is the sister group of a group consisting of the comparatively apomorph boreal genera *Paraboreochlus* and *Boreochlus*. It is also suggested that the African genera, together with the two boreal genera just mentioned, constitute the plesiomorph sister group of a group consisting of the remaining boreal genera, *Trichotanypus* and *Lasiodiamesa*.

There is strong reason to conceive the tribe Podonomini as the plesiomorph sister group of Boreochlini (cf. p. 103).

The Aphroteniinae

Though comprising only 8 species this exclusively austral group reveals itself as the sister group of Podonominae. Its relative apomorphy is evident. The aphroteniine midges prefer mountain streams running through temperate rain forests. The distribution area comprises the Valdivian region of South America, the narrow belt of moist mountain forests between Cape Town and Port Elizabeth, and southeastern

Australia northwards into southern Queensland. No species were found in the Magellanic region, and it appears probable that the members of the new subfamily are less cold-tolerant than most podonomine species.

Phylogenetic analysis shows that the South African genus *Aphrotenia*, with the two species *barnardi* and *tsitsikamae*, is the plesiomorph sister group (No. 40) of a group comprising the remaining elements of the subfamily. The genus *Aphroteniella* has one comparatively plesiomorph species (No. 41) in South America which is the sister group of a group consisting of the 2 Australian species *tenuicornis* and *filicornis* (No. 42). The genus *Paraphrotenia*, the apomorph sister group of *Aphroteniella*, is represented by 2 comparatively plesiomorph species in South America, *excellens* (No. 43) and *multispinosa* (No. 44). The less plesiomorph of these species, *multispinosa*, is the sister species of the apomorph Australian species *fascipennis* (No. 45).

The Diamesinae

In order to give a still greater weight to the documentation, I have chosen to monograph also the austral elements of the subfamily Diamesinae. This amphitropical subfamily is more closely related to the telmato-getonine and orthocladiine complexes than to Podonominae–Aphroteniinae. Most austral Diamesinae belong to the tribe Heptagyini, which includes 5 genera and 18 species. These comparatively large-sized midges play a prominent role in the cool mountain streams of most southern lands. They are, however, absent in Southern Africa.

Of the two major components making up the Heptagyini, the *Heptagyia* group in a strict sense, with the genera *Paraheptagyia* and *Heptagyia*, is comparatively plesiomorph. The latter monotypic genus is confined to South Chile and Patagonia, while its sister group, the plesiomorph genus *Paraheptagyia*, displays a distribution area which extends from Tierra del Fuego far northward into the tropical Andes and includes Tasmania and southeastern Australia. Within *Paraheptagyia* the 3 species making up the *cinerascens* group are comparatively plesiomorph and exclusively Andean. *P. nitescens* of the southern Andes is the plesiomorph sister species of *andina* of the tropical Andes. The subgroup formed by these two species is the apomorph sister group of the monotypic *cinerascens* subgroup.

The *Paraheptagyia semiplumata* group is the apomorph sister group of the *cinerascens* group and comprises two species pairs, one Andean, the other Australian. The latter, with the species *tasmaniae* and *tonnoiri*, is evidently the apomorph sister group of the South Andean species pair.

The other major component of Heptagyini is the comparatively apomorph *Araucania* group, comprising the Andean genera *Araucania* and *Limaya* and the New Zealand genus *Maoridiamesa*. In this assemblage *Araucania* constitutes the plesiomorph sister group of a group composed ot the two other genera. *Maoridiamesa* is on the whole more apomorph than *Limaya*. Of the 5 *Maoridiamesa* species 4 occur on the North and South Island of New Zealand, while the peculiar *insularis* is an endemic element of the fauna of Campbell Island. *M. insularis* is a strongly apomorph offshoot of the neozelandic *Maoridiamesa* complex.

New Zealand is, however, inhabited by another endemic diamesan genus, *Lobodiamesa*, with the single species *campbelli*. The larvae of this peculiar genus prefer slow-running streams. *Lobodiamesa* constitutes probably the apomorph sister group of the tribe Heptagyini and has thus to be accorded the rank of a tribe named Lobodiamesini. If we continue to ask for the sister group we arrive, as will always, sooner or later, be the case, in the northern hemisphere. The morphology of the highly specialized circumpolar genus *Boreoheptagyia* suggests that it is the apomorph sister group of the austral aggregate Heptagyini + Lobodiamesini. The peculiar larvae of the tribe Boreoheptagyini live in the splash zone on exposed rocks and boulders of the mountain streams of the Holarctic region and fix themselves by means of a pair of caudal suckers.

The way Southern Africa comes into the picture is significant. In the mountain streams of Transvaal and Southern Rhodesia occurs the primitive and monotypic genus *Harrisonina*. Procurement of the entire metamorphosis has demonstrated that this genus represents a tribe of its own. Harrisonini is the plesiomorph sister group of a group formed by the well-known holarctic tribes Diamesini and Protanypini. (It is possible

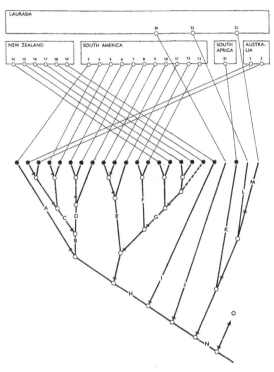

Fig. 635. Connection between geographic distribution and phylogenetic relationships in the austral tribes of the subfamily Diame-sinae. The comparatively apomorph groups and species of the phylogenetic diagram are indicated by two close-set transverse strokes. The letters and numbers at the bottom of the geographic sections and in the phylogenetic diagram indicate the following groups and species:

A, genus *Heptagyia*, with the single species *annulipes*; B, genus *Paraheptagyia*; C, the *semiplumata* group of *Paraheptagyia*; D, the *cinerascens* group of *Paraheptagyia*; E, genus *Araucania*; F, genus *Limaya*; G, genus *Maoridiamesa*; H, tribe Heptagyini; I, tribe Lobodiamesini; J, tribe Boreoheptagyini; K, tribe Harrisonini; L, tribe Diamesini; M, tribe Protanypini; N, subfamily Diamesinae; O, subfamily Telmatogetoninae.

1. *Heptagyia annulipes.*	14. *Maoridiamesa harrisi.*
2. *Paraheptagyia tonnoiri.*	15. *Maoridiamesa intermedia.*
3. *Paraheptagyia tasmaniensis.*	16. *Maoridiamesa stouti.*
4. *Paraheptagyia semiplumata.*	17. *Maoridiamesa glacialis.*
5. *Paraheptagyia umbraculata.*	18. *Maoridiamesa insularis.*
6. *Paraheptagyia andina.*	19. *Lobodiamesa campbelli.*
7. *Paraheptagyia nitescens.*	20. Genus *Boreoheptagyia*, with about 12 species in the Hol-
8. *Paraheptagyia cinerascens.*	arctic region.
9. *Araucania gelida.*	21. *Harrisonina petricola.*
10. *Araucania valdesiana.*	22. Here belong the holarctic genera *Diamesa*, *Onychodiamesa*,
11. *Araucania antiqua.*	*Sympotthastia*, *Potthastia*, *Pseudodiamesa*, and *Pagastia*,
12. *Limaya longitarsis.*	together with about 70 species.
13. *Limaya* sp. "Junín".	23. Genus *Protanypus*, with 6–7 species in the Holarctic region.

but still not strictly established that also the bipolar *Prodiamesa* group forms a part of this large assemblage; if so, the *Prodiamesa* group has to be interpreted as comparatively apomorph.) Within the subfamily Diamesinae we are thus faced with a major component comprising the tribes Heptagyini, Lobodiamesini, and Boreoheptagyini, which constitutes the plesiomorph sister group of another component made up by the tribes Harrisonini, Diamesini, and Protanypini. These components are both bipolar. See fig. 635.

It is evident from what has been set out above that the connection between phylogenetic relationships and distribution within Diamesinae confirms and further elucidates the results gained in Podonominae and Aphroteniinae. The important role played by Africa in conjunction with the development of old patterns of bipolar (amphitropical) distribution is evident.

The meaning of the phylogenetic connections and distribution patterns

The evidence presented in the preceding section is based on detailed analyses of the phylogenetic relationships and distribution patterns of 63 austral groups of chironomid midges, all of which are more or less directly involved in transantarctic relationships. This evidence is strengthened by the fact that no less than 55 of these austral groups are members of a monophyletic unit: Podonominae + Aphroteniinae. The exceedingly complex system of transantarctic relationships displayed by the disjunct elements of that major unit demonstrates in a conclusive way that there once has been an important centre of evolution in the south and that Antarctica must have been a vital part of that centre (fig. 636).

However, the relative plesiomorphy and apomorpy and the very nature of the relationships displayed by the involved groups, enable us to obtain a deeper insight into the matter. It is evident that Podonominae, Aphroteniinae and Diamesinae originated in the south; and the circumstance that the structure of the phylogenetic relationships and the distribution of the disjunct elements of the different austral groups form a major and regular pattern demonstrates the very important fact that the dispersal has been orderly throughout. The following observations are most noteworthy:

(1) The sister group of a New Zealand group lives always in South America, or in South America *and* Tasmania–Australia.

(2) There are no direct phylogenetic connections between a group of Tasmania–Australia and a group of New Zealand.

(3) A group of Tasmania–Australia is always an apomorph offshoot of the Chilean–Patagonian fauna.

Since there is no direct connection between the discussed groups of Tasmania–Australia and New Zealand, it seems obvious that Antarctica has offered separate connections between the two Australasian realms and South America. There can hardly be any uncertainty as regards the character of those transantarctic connections. Recent stratigraphic and tectonic research has confirmed the view that Antarctica consists of two very distinct geological provinces, the "Gondwanic" East Antarctica and the "Andean" West Antarctica (cf. Adie 1963, Hamilton 1964). Consequently we have good reason to suppose that the amphiantarctic elements of South America have been connected with those of Tasmania–Australia via East Antarctica, and with those of New Zealand via West Antarctica. The present concentration of plesiomorph groups in New Zealand and southern South America is strong evidence that West Antarctica has formed a very vital part of that austral centre of evolution whose former existence cannot be doubted any more. The general position of the Tasmanian–Australian groups as specialized sister groups of subordinate groups of major South American aggregates demonstrates on the other hand the subordinate evolutionary importance of East Antarctica and, especially, Tasmania-Australia. The constant position of Tasmania-Australia on the receiving side in respect to the cold-adapted disjunct elements stands out as one of the major features of transantarctic relationship.

Compared to the Tasmanian–Australian fauna the independent position of the New Zealand fauna of austral disjunct groups is strongly marked. This independency is stressed not only by the occurrence of several endemic genera like *Zelandochlus*, *Maoridiamesa*, and *Lobodiamesa*, but also by the strong relative plesiomorphy displayed by several neozelandic groups irrespective of their rank. Significant are also the negative characteristics. The comparatively apomorph (South American–Australian) podonomine genera *Rheochlus* and *Podonomopsis* are absent, as is the apomorph subfamily Aphroteniinae. Considering the plesiomorph austral Diamesinae the New Zealand fauna is, however, confined to the comparatively apomorph genera *Maoridiamesa* and *Lobodiamesa*, while the South American fauna exhibits 4 plesiomorph diamesan genera of its own (*Paraheptagyia*, *Heptagyia*, *Araucania*, *Limaya*), yet with the interesting exception that the most plesiomorph genus, *Paraheptagyia*, is represented by a (strongly apomorph) subgroup also in Tasmania–Australia. The positive and negative characteristics and systematic structure of the above mentioned faunal components of South America and New Zealand are strong evidence that the arc New Zealand–West Antarctica–South Chile has formed a very important centre of evolution and diversification, where the decisive evolutionary events in the history of different groups have taken place in different nodes (subcentres) along an axis with a mainly longitudinal extension. A parallel to this, through time and also in space, has been a far less dynamic axis connecting Patagonia with Tasmania–Australia via East Antarctica (see fig. 636).

It is most remarkable that as early as 1912 (p. 88) Hedley had a vision of the general state of matters that coincides exactly with the present results. He expresses himself as follows: "Whereas New Zealand in its relation with South America, via Antarctica, appears both as a giver and a receiver, Australia, on the contrary, seems to have made no return to South America, but to have received all and given nothing." There is no reason to change a single word.

But our reconstruction would be incomplete if we did not consider also the African sector. The negative characteristics of the southern African fauna of austral disjunct elements among the chironomid midges are quite as conspicuous and meaningful as the phylogenetic structure of the few austral groups occurring in that area. The circumstance that these groups represent just the most plesiomorph types among the apomorph main groups (tribe Boreochlini of Podonominae, subfamily Aphroteniinae, tribe Harrisonini of Diamesinae), demonstrates that Southern Africa once has been a vital part of the large austral evolutionary centre. As to general importance in the history of the austral disjunct chironomids the African subcentre was comparable with, but hardly equal to the arc New Zealand–West Antarctica–South Chile. The intimate connections between *Archaeochlus* of the Drakensberg and the holarctic podonomine genera on one hand, and between *Harrisonina* of Rhodesia–Transvaal and the holarctic tribes Diamesini and Protanypini on the other, indicate that old East African highlands have formed an important pathway for northward dispersal of progressive austral elements. It is also evident that the geographical connections between Southern Africa and the other southern lands were broken very early.

THE NATURE OF THE AUSTRAL CENTRE OF EVOLUTION

It has been stressed repeatedly in the present paper that the reliability of phylogenetic reasoning depends on the consistent use of synapomorph characters. The basic difficulty here is always the decision of whether a character is apomorph or plesiomorph. Having made a decision we are either quite right or quite wrong since there is nothing between. A complication is the circumstance that even apomorph characters in common only to two or three groups might be the result of parallelism. Though a misconception of one pair of characters may be neutralized by a correct interpretation of several others, there is of course never any absolute assurance that our conclusions concerning the phylogenetic relationships and relative apomorphy of one or two groups are correct. But there are gradations, and the situation undeniably becomes fairly promising if, faced with a certain biogeographic problem, we are able to demonstrate that in several

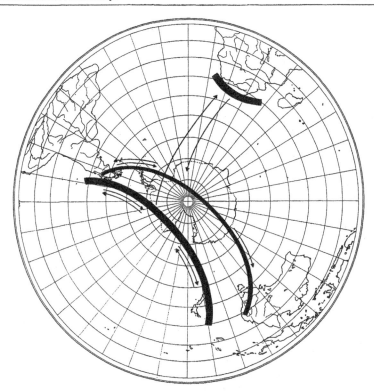

Fig. 636. Main pattern of transantarctic relationships, as evidenced by chironomid midges. The black arcs connect the inferred Mesozoic main nodes of evolution and diversification of the paleoaustral element. The arrows indicate directions of dispersal before the disruption of Gondwanaland.

pertinent groups, or pairs of groups, the connections between the inferred relationships and the distribution patterns are invariably the same and telling the same story.

In the present case long series of reconstructed phylogenies form an extensive network of mutual confirmation. Indeed, surveying the matter and finding all factual and inferred data fitting without any sort of disorder not only into a major austral pattern but also into a pattern of global extent, I feel confident that we at last have got on the track of the true nature of transantarctic relationships.

The 25 cases of transantarctic phyletic connections and their amphitropical ramifications analysed in the present paper tell a story of orderly dispersal which goes back to the middle Mesozoic. There is not the slightest evidence of chance dispersal over wide stretches of ocean or of any sort of hopping of ancestral forms from island to island between the southern continents. Indeed, there are factors of another kind behind such phenomena as the occurrence in Southern Africa of just *Archaeochlus*, *Aphrotenia* and *Harrisonina*, and in Tasmania-southeastern Australia of nothing but derivative members of the remote Chilean–Patagonian fauna, while in New Zealand there is a paleoaustral component of an originality well on a level with that of South Chile and Patagonia and with no direct relationships across the comparatively narrow Tasman Sea.

In spite of their theoretically high dispersal capacity the chironomid midges give clear evidence that their transantarctic relationships developed during periods when the southern lands were directly connected with each other. Moreover, it is plainly demonstrated by the nature of those relationships that the connections between the southern lands have been broken according to a certain sequence beginning with the separation of Southern Africa. The next event was the break in the connections between New Zealand and (West) Antarctica. The following separation between Tasmania–Australia and (East) Antarctica antedates, probably quite considerably, the break between southern South America and Antarctica.

Several scientists, and among them recently Axelrod (1960) and Florin (1963), have stressed that the dynamic fold-mountain belts with their rich variation of habitats and altitudinal climatic zones may have functioned as important centres of evolution and diversification of many groups, while the comparatively quiescent continental shields have been of secondary importance only. The basic soundness of these assumptions seems obvious. Since New Zealand, West Antarctica, and Chile stand out as fold-mountain areas, while Tasmania–Australia, Africa, East Antarctica, and (East) Patagonia all are continental shields, we are able to state that the geological nature of the southern lands corresponds exactly to the varying evolutionary role attached to them by the phylogenetical analysis of the austral disjunct chironomid groups. This remarkable coincidence can hardly be accidental because the causal connections are so obvious and so well in line with general evolutionary theory and the meaning of the speciation process.

The nature of the transantarctic relationships displayed by the chironomid midges indicates the presence in the Upper Jurassic, and not later, of a centre of origin in a continuous fold-mountain belt roughly corresponding to present New Zealand, West Antarctica, the Scotia Arc, and South Chile, but probably also including now sunken Pacific fore lands. This dynamic arc formed the southern margin of Gondwana and was situated in a zone of moist temperate climate.

In using the term Gondwana, I am touching upon the theory of continental drift and other lateral displacements of the earth's crust. Recent geologic, geophysic and paleomagnetic research has resulted in much new evidence, not least in the form of marine-stratigraphical data, in favour of an essentially modified theory of continental displacement including a comparatively rapid disruption of the old Gondwanaland. These processes are now interpreted as having taken place during the period Upper Jurassic-Upper Cretaceous. Since there is no clear evidence of continental drift during the Cenozoic, the continents appear to have reached their present positions at the end of Cretaceous (King 1962). This new time-table fits my biological data excellently. Referring to the geological data delivered by King (l.c.) we can assign the separation—relative to Antarctica—of Southern Africa to Upper Jurassic, of New Zealand to Lower Cretaceous, and of Australia to Middle Cretaceous.

After Wegener (1915) and Du Toit (1937) many attempts have been made to reconstruct Gondwanaland as it may have been in Permo-Carboniferous times. Reviewing the matter Holmes (1965, p. 1226) stresses that every attempt at re-assembly has so far encountered the same difficulty—"some item of seemingly favourable evidence has had to be sacrificed in order to achieve a reasonable 'fit' of all the parts". To Holmes the weakest link appears to be the very one that has always made the greatest appeal, i.e. the parallelism of the opposing shores of the Atlantic. He suggests that the first item to be sacrificed should be the traditional obsession with parallelism and points at the great improvement gained "without any geological loss" by swinging South America back so that the Brazilian bulge fits into the Gulf of Guinea, while the link with Antarctica is preserved (l.c., p. 1227).

Holmes remarks that equating the great concept of continental drift with Wegener's hypothesis "is an obvious source of unnecessary confusion", since a large number of hypotheses have been involved, some of them inadequate or wrong, while others "have turned out to be far more fruitful than even Wegener himself could have expected. In particular it should never be overlooked that in many respects his assemblage of the Carboniferous continents is in better accord with all modern evidence than several of the modifications proposed during the last fifty years" (Holmes, l.c., p. 1203).

It is noteworthy in this connection that the new biological data brought forward in the present paper are in a far better agreement with the reconstruction of Gondwanaland proposed by Wegener than with

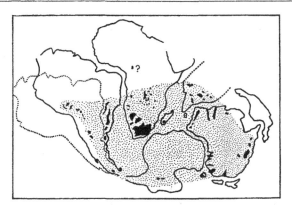

Fig. 637. Re-assembly of Gondwanaland, as suggested by Holmes, showing the distribution of the Permo-Carboniferous glaciations, with South America in the position indicated by a firm outline. The dotted outline, maintaining the parallelism of the opposing coasts, is rejected. (From Holmes 1965, fig. 875.)

those of, for example, Du Toit (1937), Carey (1958), and Tuzo Wilson (1963). I am inclined to attach a special importance to the circumstance that the re-assembly of Gondwanaland as suggested by Holmes (see fig. 637 of the present paper), the most modern known to me, is in very good accordance with the evidence delivered by the chironomid midges. We notice, it is true, that New Zealand is placed far from West Antarctica and very close to East Australia, but we need not care since Holmes always has looked upon New Zealand and West Antarctica as links in the peripheral orogenic ring of Gondwanaland (Holmes 1929; 1956, fig. 858). Referring to late advances in Antarctic geology King and Downard (1964, pp. 729–731) state:

"Break up of the Andean-west Antarctic orogenic belt occurred also in West Antarctica itself. Bentley and others (1960) have shown a gap in continuity between the Graham Land-Sentinel Range segment and the Kohler Range. The end of the west Antarctic part of the circum-Pacific girdle is at the Edsel Ford Range."

"All the broken disturbed pieces of the orogenic girdle may now be re-assembled in a continuous line upon the Gondwana re-assembly for the Cretaceous period when it is found that: beginning at the South Sandwich Islands, and connecting the various island parts of the south arm of the Scotia Arc to the Graham Land peninsula, adding thence the Kohler Range segment all in continuous linear relation, the end of the line comes with the Edsel Ford Range in almost exactly their present position on the globe."

"To complete the circum-Pacific girdle with New Zealand, the relations of Byrd Land, the width of the south-east Pacific submarine ridge at about lat. 66° S., long. 172° W. and the submarine platform south of New Zealand have to be taken into consideration. This has already been done (King 1958) when it is found that all the major fragments of the circum-Pacific girdle enumerated (plus the width of the submarine ridge specified) make a single, continuous feature spanning precisely from the South Sandwich Islands to New Zealand."

This opinion is in full accordance with the evidence furnished by the chironomid midges.

In all modern attempts to reconstruct Gondwanaland South America has been conceived as a homogenous mass of land. I am of the opinion that such a schematical and conventional course of action may have something to do with the difficulty in achieving a harmonic fit of southern South America, Southern Africa, and East Antarctica which is apparent also in the reconstruction of Holmes. Several geologists have suggested that the Brazil–Guiana shield and the Patagonian shield, the two fundamental units of South America, have operated in a different manner. King (1962, p. 75), for example, points out that in the Gondwana re-assembly the ancient massif of Patagonia "corresponds with the Weddell and Ross Sea areas of the Antarctic and to the low area between them thought by Admiral Byrd to be land. The flanks of the

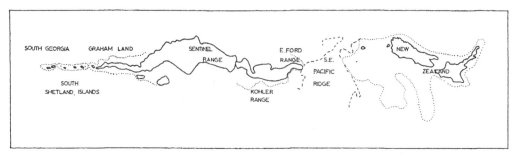

Fig. 638. Original continuity through West Antarctica of the Pacific circumvallation in the South Sandwich Islands–New Zealand sector. (From King and Downard 1964, fig. 4.)

Weddell Sea are possibly fractured on both east and west and the Ross Sea boundaries are certainly faulted so that these seas may represent subsided areas. ... Windhausen (1931) has indeed considered Patagonia as an element not belonging to the original framework of South America, a view that accords with a wide variety of data. ... A great part of this Patagonian mass is, indeed, now submerged in the broad continental shelf, directed towards the Falkland Islands."

The present hydrography of South America, with 90 % of the drainage directed to the Atlantic, is obviously a reversal of the original hydrography inherited from Gondwanaland when the rivers flowed westward from the eastern highlands. The intermittent rise of the Andes since the mid or late Jurassic has produced an impassable barrier in the west that diverted the outfall from westward to eastward. (King l.c., p. 76.)

On the whole the re-assemblage of the southern fragments of Gondwanaland does not seem to pose a too serious problem, and Holmes (1965, p. 1227) writes optimistically that "a few years' work should see the problem solved to general satisfaction".

There is no reason here to attempt to cite all new evidence in favour of continental drift which has appeared during the last few years and continues to appear, month by month. It will suffice to refer to the excellent review given by Arthur Holmes in his large work of 1965.

The theory of continental drift provides a background fitting all demands raised by the nature of the transantarctic relationships, as displayed by the chironomid midges. Indeed, the fit between the history told by the distribution patterns and reconstructed phylogenies on one hand, and the latest opinions concerning the geological nature and mutual connections of the Gondwana fragments, and the time-table affixed to the disruption of Gondwanaland on the other, is so close that there is agreement even in details.

Moreover, the theory of continental displacement is the only theory that is able to give a reliable background to the Permo-Carboniferous glaciations. Even many geologists seem to forget that objections against some details of the concept of continental displacement cannot shake its basic foundations as long as the critics fail to give another and more plausible explanation of the fact that land areas now more than 90 degrees of latitude apart (Antarctica, Peninsular India) were covered by ice sheets during the Permo-Carboniferous.

There seems also to be a deep biogeographical perspective involved here. The transantarctic and bipolar patterns of the chironomid midges form apparently only a fraction of similar and often much older patterns of the immense sister group system which makes up the holometabolous insects. Considering further the probable austral origin of many animal groups of greater age than the austral chironomid groups, one gets the impression that the development of austral disjunctive distribution (as a consequence of the disruption of Gondwanaland) was only an episode in a biotic history initiated by the Permo-Carboniferous glaciations. These lines of thought will be touched upon again in the following.

L. BRUNDIN, *Transantarctic relationships* K. V. A. Handl. 11: 1

BIPOLARITY AND AUSTRAL DISJUNCTIVE DISTRIBUTION

As I have alredy pointed out (p. 447), the search for the sister group of an austral group leads sooner or later to a group in the northern continents. The analyses have shown that *Parochlus kiefferi*, two groups of the tribe Boreochlini, the genus *Boreoheptagyia* (tribe Boreoheptagyini), and a major group comprising the tribes Diamesini and Protanypini are all boreal apomorph sister groups of austral plesiomorph groups. In these instances the bipolar distribution pattern is obviously a consequence of transtropic dispersal northwards. Among the chironomid midges there are also several examples of bipolarity caused by transtropic dispersal southwards, but among the comparatively plesiomorph groups especially studied here in connection with the problem of transantarctic relationships the only examples of this are displayed by the mainly boreal Diamesini (and the *Prodiamesa* group) which are represented by a few species in the southern Andes (cf. p. 336 above). That indicates a secondary double migration southwards along the eastern Pacific fold-belt of members of the primarily austral subfamily Diamesinae.

The subfamily Tanypodinae is a chironomid group of boreal origin according to the evidence delivered by Fittkau (1962). It is fairly well represented in the southern temperate zone, and there seems to be good reason to suppose that the austral disjunct distribution of a bipolar tanypodine genus like *Macropelopia* (South Chile–Patagonia, Tasmania–Australia, New Zealand) is a manifestation of dispersal via Antarctica. If so, the genus would fulfil the demands raised by Hennig's second criterion (cf. p. 56 above). Moreover, the subfamily Tanypodinae is obviously the apomorph sister group of the Podonominae and Aphroteniinae; and while the two latter subfamilies are typical rheophil and cold-adapted groups, the subfamily Tanypodinae is a mainly lenitic assemblage comprising many eurythermal and warm-adapted groups inhabiting the warm slow-running rivers and standing waters of the tropical lowlands.

The subfamilies Podonominae, Aphroteniinae and Tanypodinae form one of the two main components building up the family Chironomidae. In the other component, comprising the subfamilies Diamesinae, Telmatogetoninae, Orthocladiinae, and Chironominae, we encounter similar distribution patterns and adaptational trends as in the former main group, but in addition we meet a stronger representation in tropical lowlands, a more far-going adaptation to extreme lenitic habitats, a multiple adaptation on a rather large scale to terrestrial habitats, and, finally, a very remarkable adaptation of some groups (foremost Telmatogetoninae) to the life in the inter-tidal zone of rocky sea-coasts all around the world. By comparison with the former this major component stands out as derivative.

Among the chironomid midges the causal connection between a basically bipolar (amphitropical) distribution pattern and the history of the adaptations of the family is well traceable. A comparative study of larval and pupal ecology and of different structural systems of the young stages, especially the pupal leg sheath arrangements (cf. pp. 428–434 above), suggests that the family Chironomidae originated in cool running waters. There is strong reason to suppose that the basic adaptations and first steps in the phylogenetic evolution of the family took place in the upper courses of small, cool mountain streams and their springs, and that the present widespread occurrence of the chironomid midges in other types of habitats is due to secondary adaptive radiation. Indeed, there is a very wide array of habitats where the chironomid adaptations have led to a lasting success on a global scale.

Adaptations and evolving sister group systems are all intimately connected with physical translation (i.e. migration) and form-making in time and through space. The resulting patterns are essential manifestations of the history of life. With regards to the chironomid midges we are faced with the fact that the pronouncedly plesiomorph groups are confined to cool mountain streams of extra-tropical tracts of land. The evidence is indeed very strong that the cold-adaptation displayed by these amphitropical groups is a plesioec character. From the previous discussions it is clear, moreover, that the analysed cold-adapted chironomid groups are only forming fractions of a great system of sister groups which are all more or less directly involved in transantarctic relationships, bipolarity, and transtropic dispersal to the north and to the south. As the tropical groups of all altitudinal zones stand out as apomorph in relation to their cold-adapted austral and boreal sister groups, it is apparent that the main pattern of chironomid evolution has been of the bipolar type. The whole is thus the result of an intermittent interplay between two old main centres of evolution and diversification, one northern and one southern. The southern centre was broken up in the Mid Mesozoic and its huge Antarctic sector destroyed during the Pleistocene glaciations, but its profound influence on the faunas of both hemispheres is still clearly discernible.

At least as regards the chironomid midges we seem to have right to reverse the meaning of the first sentence of the present section and to say: Asking for the sister group of a boreal group we will sooner or later get into contact with an austral group; the latter will often prove to be involved in transantarctic relationships.

Works Cited

Adie, R. J. 1963. Geological evidence on possible Antarctic land connections. In *Pacific Basin Biogeoography*, ed. J. L. Gressitt, 455–63. Honolulu: Bishop Museum Press.

Axelrod, D. I. 1960. The evolution of flowering plants. In S. Tax, ed., *Evolution after Darwin*, vol. 1: *The Evolution of Life*, 227–305. Chicago: University of Chicago Press.

Bentley, C. R., A. P. Crary, N. A. Ostenso, and E. C. Thiel. 1960. Structure of West Antarctica. *Science* 131 (3394): 131–36.

Brundin, L. 1965. The bottomfaunistical lake type system and its application to the southern hemisphere. Moreover a theory of glacial erosion as a factor of productivity in lakes and oceans. *Verh. Int. Ver. Limnol.* 13: 288–97.

Carey, S. W. 1958. The tectonic approach to continental drift. In *Continental Drift: A Symposium*, 117–355. Hobart: University of Tasmania.

Du Toit, A. L. 1937. *Our Wandering Continents: An Hypothesis of Continental Drifting*. Edinburgh and London: Oliver and Boyd.

Fittkau, E. J. 1962. Die Tanypodinae (Diptera: Chironomidae). *Abhandl. Larvalsyst. Insekten* 6: 1–453. Berlin: Akademie-Verlag.

Florin, R. 1963. The distribution of conifer and taxad genera in time and space. *Acta Hort. Berg.* 20 (4): 121–312.

Gressitt, J. L. 1964. Ecology and biogeography of land arthropods in Antarctica. In *Biologie antarctique, premier symposium 1962*, 211–222. Paris.

Hamilton, W. 1964. Tectonic map of Antarctica: A progress report. In R. J. Adie, ed., *Antarctic Geology; Proceed-*

ings of the First International Symposium on Antarctic Geology, Cape Town, 16–21 September 1963, 676–79. Amsterdam: North-Holland.

Hennig, W. 1954. Flügelgeäder und System der Dipteren. *Beitr. Ent.* 4: 245–388.

Holmes, A. 1929. Radioactivity and earth movements. *Trans. Geol. Soc. Glasgow* 18: 559–606.

———. 1956. *Principles of Physical Geology*. London and Edinburgh: Thomas Nelson.

———. 1965. On the real nature of transantarctic relationships. *Evolution* 19: 496–505.

King, L. C. 1958. The origin and significance of the great sub-oceanic ridges. In *Continental Drift: A Symposium*, 62–102. Hobart: University of Tasmania.

———. 1962. *The Morphology of the Earth: A Study and Synthesis of World Scenery*. Edinburgh and London: Oliver and Boyd.

King, L. C., and T. W. Downard. 1964. Importance of Antarctica in the hypothesis of continental drift. In R. J. Adie, ed., *Antarctic Geology; Proceedings of the First International Symposium on Antarctic Geology, Cape Town, 16–21 September 1963*, 727–32. Amsterdam: North-Holland.

Leston, D. 1957. Spread potential and the colonization of islands. *Syst. Zool.* 6: 41–48.

Wegener, A. 1915. *Die Enstehung der Kontinente und Ozeane*. Braunschweig: Vieweg & Sohn.

Wilson, J. Tuzo. 1963. Continental drift. *Sci. American* 208 (4): 86–100.

Windhausen, A. 1931. *Geologia Argentina*. Buenos Aires.

VOL. 41 NO. 3   September, 1966

THE QUARTERLY REVIEW
of BIOLOGY

THE BIOTA OF LONG-DISTANCE DISPERSAL.
I. PRINCIPLES OF DISPERSAL AND EVOLUTION

By Sherwin Carlquist

Claremont Graduate School and Rancho Santa Ana Botanic Garden, Claremont, California

ABSTRACT

A growing consensus of biologists now favors the effectiveness of long-distance dispersal as a means of populating islands. The observational and experimental bases on which this opinion rests are strong, but additional work is needed. A clear understanding of long-distance dispersal is essential to an understanding of evolutionary trends on oceanic islands, because immigrant patterns are different from relict patterns. Since oceanic islands are short-lived, the evolutionary history of waif immigrants is also short. If a continental island maintains long isolation, arrivals by long-distance dispersal may show evolutionary patterns more completely, as is true on New Zealand, for example. The evolutionary patterns of waif biotas are influenced by isolation, by the broad range of available ecological opportunities, and, to a lesser extent, by the moderation characteristic of maritime climates. In addition to problems involved in becoming established, immigrants must overcome genetic disadvantages inherent in the fact that the number of original colonists is small. Increase of genetic variability may be governed by ecological diversity, and persistence of a phylad may be increased by maximizing outcrossing and hybridization. Among features which are exhibited by waif biotas are adaptive radiation, flightlessness in animals, loss of dispersal mechanism in plants, and development of new ecological habits and growth forms. Each of these adaptations is evidently governed by a wide variety of factors. "Weedy" groups seem to possess the greatest ability to disperse and become established; they also excel at sensitive adaptation to island conditions. The waif biota contains few relicts except for "recent relicts."

INTRODUCTION

THE faunas and floras of oceanic islands possess many distinctive characteristics which have long attracted attention. Interest in these lands as "evolutionary laboratories" was crystallized during the Darwin-Wallace era and has not diminished since. Floristic and faunistic studies of islands have been frequent (cf. Blake and Atwood, 1942; Darlington, 1957; Thorne, 1963), although more are needed and will undoubtedly be produced. Discussions of the possibility of dispersal to oceanic islands, and of whether particular islands are oceanic or continental, have occupied an inordinate number of pages and created considerable con-

247

troversy. Among the many discussions there have been a few outstanding contributions dealing with the problems of dispersal and patterns of insular evolution. Among those commendable for their comprehensiveness and rational outlook may be cited those of Gulick (1932), Zimmerman (1948), Darlington (1957), and Thorne (1963). To a surprising degree, the assessments which most appeal to the writer are those which confirm and extend the concepts of Darwin (1859) and Wallace (1880). Although adherents of alternative proposals may still be found, there seems to be a growing consensus among biologists regarding many of the ideas discussed below.

The reasons for presenting an additional essay at this time are several. Interest has shifted recently from a primarily floristic-faunistic outlook to a concern for evolutionary processes on islands. Principles which have emerged clearly, as well as those which can only be tentatively enunciated, are reviewed here because a foundation for further research seems needed and because a critical sifting of past hypotheses may prove helpful. An earlier generation of biologists stressed the amazing speciation which has occurred on oceanic islands. Today's biologists are inquiring into the background of this speciation: its origin and direction and its underlying mechanisms. The basic mechanisms of evolution on oceanic islands are the same as those on continental areas and no new "laws" are needed, yet the direction evolution takes and the products which result on islands are often quite distinctive, at least modally. Understanding of these evolutionary concepts must be based on sound principles of dispersal and biogeography. Therefore, these topics as they apply to the waif biota must be reviewed, however briefly. For example, the peculiar rosette trees of the Juan Fernandez Islands must be interpreted as relicts by those who see long-distance dispersal as an inadequate explanation for the Juan Fernandez flora. Those who can envisage these islands as recent and populated by means of long-distance dispersal can, on the contrary, regard the rosette trees as innovations —the products of recent evolution.

Literature on the evolution of island biotas is widely scattered among zoological, botanical, evolutionary, and other journals and books, and needs to be synthesized. Too often, pat-

terns of one group only, of animals only, or of plants only, are considered. There is a definite need for integrating the data from various disciplines. Information from one field not only often proves valuable to studies in other fields, but must eventually be coordinated into larger patterns.

The biota characteristic of oceanic islands is, ironically, not always found on such islands and may occur in other situations as well— hence the title of this paper. Oceanic islands near continents may show an essentially continental pattern. A completely glaciated continental island will bear a biota entirely oceanic in character (Fleming, 1963). A remote oceanic island may be too new or ecologically too poor to have shown the evolutionary patterns described below. Despite incongruities such as these, the concept that there are oceanic and continental islands, the differences between which reflect their geological history, still seems useful and is generally accepted (Fosberg, 1963). No scheme for classification of islands, however, can hope to reflect unexceptionably both geological history and the nature of the biota.

Most oceanic islands are volcanic, and volcanic islands are, on a geological scale short-lived. Therefore, true oceanic islands clearly display the early and middle stages in the phylesis of a waif biota; but such islands may vanish before evolutionary products are well advanced (for example, before they are differentiated to the level of distinct families). Ironically, later stages in the evolution of the long-distance-dispersed biota are often shown on old continental islands. For example, New Caledonia and New Zealand have been isolated for a very long time — since the Creataceous, if not much longer. Waif arrivals to these islands have had a much longer time in which to evolve than have arrivals to the Hawaiian Islands, which are probably no older than Miocene. The discerning biologist can, in most instances, detect which elements in the New Zealand biota are waif immigrants, and can follow their evolutionary patterns separately from those which probably are relicts from an era of continental interconnection or near-interconnection (such as conifers, primitive flowering plants, *Sphenodon* and *Leiopelma*). With an extended period of isolation and with a sizable land mass available for occupancy, the waif biota of New Zealand has been able to evolve extensively, to show waif evolutionary patterns better than such biotas on most oceanic islands. It is ironical to say that the evolutionary products of waif biotas are perhaps best studied on old continental or continental-like islands; nevertheless, it is possible to cite oceanic islands on which insular patterns are well fulfilled. Among these, the Hawaiian Islands surely deserve first place, although part of the evolutionary history of Hawaiian organisms may have taken place upon now-vanished island chains adjacent to the present Hawaiian Islands. The Hawaiian Islands are favored by ecological richness, extreme remoteness, and (for oceanic islands) relative oldness. The Galápagos Islands, although famed for their insular biota, are relatively young, relatively poor ecologically, and relatively close to continental areas. With the possible exception of Darwin's finches, the Galápagos biota best serves to show earlier stages in the evolution of insular groups. Completing this spectrum from old to young waif biotas, the earliest stages in evolution are perhaps best studied in animals and plants which have been introduced to islands by man. Mice have proved good examples (Berry, 1964).

Some persons will note that waif biotas are represented in non-insular but isolated situations as well: mountain-tops, caves, lakes, and the like. This is true. Islands differ from such situations, however, by offering a broad gamut of ecological opportunities (shore to alpine, aquatic to xeric) instead of a single extreme habitat (such as alpine). Some non-insular waif biotas can be informative, however, and are used among the examples to follow. Thus, a waif biota can occur wherever arrival is by means of long-distance dispersal and where isolation is high for a prolonged period of time. The assumption made here that waif arrivals on continental islands can be distinguished from preinsular arrivals will no doubt be challenged. Our methods of separating the two groups are inferential, but should improve as our knowledge of waif-biota characteristics become surer. I believe that most groups on an old continental island, and especially on a remote one, can be discriminated, and that we should aim for such discrimination. The basis

for this distinction is now, and will continue to be, largely a comparison with oceanic islands. For example, if conifers and primitive flowering plants are subtracted from the New Zealand flora, and if differences in climate are taken into account, the composition of that flora is remarkably similar to the Hawaiian flora with respect to families and even genera.

Hypotheses respecting the dispersal and evolution of waif biota are presented here as a series of principles, or if one prefers, topic sentences. The number of such principles could be enlarged or reduced, and ideas placed under some headings could just as well have been entered under others. Many of the ideas are closely interrelated, and lead me to conclude that there is an "insular syndrome" which derives (1) from difficulties of long-distance dispersal; (2) from isolation; (3) from ecological opportunity; and (4) to a much lesser extent, from climatic moderation of a maritime climate. The consequences of these factors are, however, quite manifold. Some of the principles stated below will appear restatements of the obvious, but others will represent views which have not hitherto been discussed to any appreciable extent in the literature on waif biotas. For definitions of which islands are continental and which ones are oceanic, the reader can consult Darlington (1957) or Carlquist (1965).

PRINCIPLES

1. *Disharmony in composition of an insular biota is considered a prime source of evidence for the occurrence of long-distance dispersal.*

The distinction between the concepts of harmonic and disharmonic biotas is a simple one, namely, that a harmonic flora or fauna contains a spread of forms with poor to excellent dispersal ability, whereas waif biotas will contain only the more easily dispersed end of the dispersibility spectrum. An excellent criterion for dispersibility is the maximum gap of salt water which may have been crossed by natural means by particular groups of plants or animals. A distance measured from a continental area to an oceanic island is the most reliable. Admittedly, the gap crossed may have been greater because the continental source population may have been inland, or narrower if a strait has widened since dispersal occurred. A rating of dispersibility was developed by Darlington (1957) for animals, and has been extended to plants by Carlquist (1965). Within taxonomic groups, dispersibility is related to size of the disseminule and to specific mechanisms which can vary widely even within a genus. The idea that particular groups have differential thresholds, or hurdle-values, roughly calculable in miles, is basic to the concept of disharmony, however.

A demonstration of disharmony is rendered more difficult by the fact that some regions, insular and otherwise, have biotic depauperation for ecological reasons. Poorer faunas or floras, however, are usually deprived of species with good dispersibility as well as those with poor dispersibility. Some poorly dispersing groups, by coincidence, also happen to be restricted to ecologically "good" situations (e.g., conifers). Also, good dispersal is often correlated with ability to occupy pioneer habitats — a wholly reasonable correlation because pioneer habitats are widely scattered (e.g., beaches) and often open up for occupancy suddenly, whereas stable habitats, such as rain forests, are limited in extent and do not change rapidly. The possibility of confusion between biotic depauperization owing to distance and that due to ecological factors is great, and has caused some workers, who are unwilling to make a distinction, to reject the concept of disharmony — usually in favor of the existence of hypothetical land-bridges or continental drift.

Strong support for the concept of disharmony comes from the fact that it occurs not once, but in as many replications as there are oceanic islands. The limit of dispersal ability of lizards is shown not only by their absence from the Society Islands, but from their absence on Easter Island, the Austral Islands, Samoa, etc. *Metrosideros* (Myrtaceae) is a tree which has reached not just Samoa, but all of the major groups of high islands of the Pacific, starting from an Indo-Malaysian source or sources.

An island can harbor both harmonic and disharmonic biotas if it is a continental island. New Zealand, Fiji, and New Caledonia are floristically notable for old harmonic elements (e.g., Araucariaceae) presumably derived from times when this arc was connected, or nearly connected, with an Antarctic route, perhaps in Cretaceous times (Thorne, 1963). Since that time, waif elements have also populated these islands; they bulk larger presumably both by virtue of continual immigration and the diversification of arrivals.

An oceanic island, by definition, will have only a disharmonic biota. Some biologists tend to overemphasize the difficulties of dispersal for groups with which they are most familiar. For oceanic islands, however, there can be no exceptions to over-water dispersal. Not generally appreciated is the tendency for loss of dispersal ability following arrival (see principles 16 and 17). Consequently, current dispersibility of island species cannot be used invariably as a basis for estimates of dispersibility. The pivotal controversy, as to which islands are continental and which are oceanic, seems to be dwindling. Only a few islands remain difficult to interpret, and the growing consensus regarding the interpretation of islands suggests that there is likewise increasing agreement on some of the principles discussed in this paper.

2. *Positive adaptations for long-distance dispersal and establishment are the key to disharmony, and disharmony is thus not a negative concept. All elements in a disharmonic biota are capable of long-distance dispersal or are derived from ancestors which were capable of it.*

Organisms which are clearly adapted to dispersal across very long or unlimited distances include spore-bearing plants, strand plants, and some birds and insects. Species with virtually unlimited dispersal (e.g., strand plants) are a virtually constant element on islands, and provide no problems in interpretation, nor do they develop patterns of evolutionary interest on islands. The strand flora usually remains a strand flora, and rarely evolves into montane species. Exceptions include *Erythrina* and *Acacia* (Rock, 1919b) on the Hawaiian Islands. Although dispersal by sea-water flotation has contributed little to montane floras of islands, one must remember that some plants for which sea-water dispersal may seem unlikely are, in fact, capable of it (e.g., *Gossypium:* Stephens, 1958a,b, 1963, 1964; *Lagenaria:* Whitaker and Carter, 1954, 1961).

Rafting is probably most effective over relatively short distances, but is doubtless responsible for cases of land-animal transport (for reviews and observed examples, see Wheeler, 1916; Darlington, 1938; Zimmerman, 1948; McCann, 1953).

For the remainder, birds or winds must be the vector. Gressitt and coworkers are to be congratulated for extensive aerial trapping experiments which demonstrate clearly not only that aerial transport of insects and spiders is likely, but that the species caught represent groups in the same proportions as they occur in insect faunas of oceanic islands (Gressitt, 1956, 1961a; Gressitt and Nakata, 1958; Gressitt, Sedlacek, Wise, and Yoshimoto, 1961; Gressitt and Yoshimoto, 1963; Yoshimoto and Gressitt, 1959, 1960, 1961, 1963, 1964; Yoshimoto, Gressitt, and Mitchell, 1962). Actual dispersal events can be witnessed with such rarity that Gressitt's experiments are as close to definitive proof as can be expected. Thus the contentions of such workers as Visher (1925), Ridley (1930), Setchell (1926, 1928, 1935), Andrews (1940), Zimmerman (1948), Myers (1953), and Fosberg (1963) that aerial dispersal is a reality wherever small or air-floated disseminules are formed seem justified.

Transport of fruits, seeds, and eggs externally, or of fruits and seeds internally by birds must be regarded as the dispersal mechanism of many species (Ridley, 1930; Gulick, 1932; Zimmerman, 1948; Falla, 1960; Holdgate, 1960; Wace, 1960). Observational and experimental evidence for these modes of transport is not as sufficient as one could wish, although morphology and texture (barbed fruit appendages or viscid surfaces of fruits or seeds, etc.) are highly suggestive. More experimental work needs to be done, because the striking array of species transported on intercepted birds demonstrates that birds definitely can function as vectors. Analysis of the montane flora and fauna of islands shows that the majority of species are adapted for transport externally or internally by birds. Detailed analyses of floras and faunas with respect to these mechanisms would be valuable; it has been done in only a few cases (Skottsberg, 1928; Lems, 1960b). Such analyses must be tempered by understanding that dispersal mechanisms change on islands, and may not be as good now as they once were (see principles 16 and 17). Continental disjunctions, such as the famed Chile–western

North American pattern (Constance, Heckard, Chambers, Ornduff, and Raven, 1963) can be as helpful in discovering means of long-distance transport by birds as can insular patterns.

High ability for long-distance dispersal is shown in the Pacific by the points at which genera terminate with reference to source areas. Smith (1955) has shown that 101 genera of flowering plants have their eastern terminus on Fiji, the most easterly major island within the line delimiting andesite rocks (which suggest the former presence of a larger land mass). Of the genera which do continue beyond Fiji — to Samoa, for example — many (or endemic genera derived from them) go all the way to Hawaii or the Society Islands, a fact which suggests that the flora of Polynesia is, in fact, the flora of long-distance dispersal. Careful analyses of distributions in Polynesia are needed, however.

3. *Long-distance dispersal is probably not achieved primarily by single introductions but by repeated or simultaneous introductions.*

There is a tendency to believe that introduction of a single seed can and will suffice to establish a new plant species. This possibility certainly cannot be denied, but several circumstances point to a different conclusion. First, animals require at least a breeding pair or a gravid female. A single gravid female would introduce less genetic material than a flock of adults and would be expected to be a poorer mode of introduction of an animal species. Second, this requirement holds true for dioecious or other obligately outcrossed plants (see principle 20). Third, a single individual seems less likely to yield progeny destined for long-term success because of its limited content of genetic variability (see principle 10).

In addition to these relatively demonstrable reasons one can suggest other cogent ones. The very frequent observation of stragglers among migratory birds (e.g., Munro, 1960) suggests repeated introduction. Repetitive stragglers succeeded in establishing a breeding colony of purple gallinules on Tristan da Cunha recently (Rand, 1955). Repeated arrival of straggling monarch butterflies on Canton Island was observed by Van Zwaluwenburg (1942). Shifts in ocean currents may result in unprecedented deposits on beaches of seeds of a previously absent species. Such populations of *Mucuna* seeds were observed on New Zealand beaches in 1956 (Mason, 1961). *Mucuna* seedlings were observed for the first time on not one but several of the Leeward Islands of the Hawaiian chain in 1964 by Charles Lamoureux (pers. commun.). Rafting would tend to bring not one individual, but a small population, as with the ants observed by Wheeler (1916). The pattern of distribution of some genera which occur on virtually all of the high islands of the Pacific, such as *Metrosideros*, suggests not introductions of single seeds but repeated dissemination of large quantities of seeds. Logic dictates that for every successful establishment, there are many unsuccessful arrivals.

Rapidity of colonization of the Krakatau islets by animals (Dammermann, 1948) and plants (Docters van Leeuwen, 1936) following the devastation of Krakatau, suggests introduction of many disseminules of a species. The equilibrium theory of MacArthur and Wilson (1963), a theory which seems well justified, depends on a continuous rate of immigration, rather than a few random accidents.

The majority of seed plants are cross-pollinated. This characteristic, together with a tendency to evolve toward self-pollination from cross-pollination but not the reverse, suggests that whenever there is a requirement for two or more individuals, they must not only grow and flower simultaneously, they must also grow within a short enough distance of each other for pollination to be possible. These requirements clearly favor simultaneous introductions for successful establishments. Such introductions may not be on a yearly or continual basis, of course, but rather the result of a violent occurrence such as an unusual storm that may deposit large quantities of alien trash on an island.

4. *Among organisms for which long-distance dispersal is possible, in the long run introduction is more probable than non-introduction to an island.*

This principle, together with appropriate calculations, has been clearly enunciated by Simpson (1952). Calculations demonstrating probabilities of this sort were offered by Matthew (1915) and Darlington (1938). The calculations of Fosberg (1948) on the origins of

the Hawaiian flora suggest the inevitability of the outcome. Attention is called to Fosberg's estimate that the Hawaiian species of no fewer than 23 genera of flowering plants have resulted from two independent introductions, whereas three or more than three introductions are hypothesized for Hawaiian species of 11 other genera. Similar calculations probably could be made of hypothetical immigrants in other groups of organisms and in various waif biotas.

Odd chance distributions do, of course, occur. *Lepinia* (Apocynaceae), which occurs on Tahiti and Ponape, is probably such an example (Fosberg, 1963). Peculiar distributions of this sort are discussed by Falla (1960), who suggests that habits of those birds which are possible vectors may be responsible. The distribution of *Bidens* in the Pacific suggests the operation of chance (Carlquist, 1965). Drastic disjunctions sometimes prove to be the result of incorrect taxonomic interpretation. The supposed occurrence of *Lipochaeta* (Compositae) on the Hawaiian Islands, Galápagos Islands, and New Caledonia now proves to be a case of three different genera which had not been distinguished because of insufficient study (Harling, 1962).

5. *Elements are present not only in proportion to dissemination ability, but also establishment ability.*

Difficulties of establishment seem much greater than those of transport. One can safely say that among successful introductions of disseminules in good condition, only a fraction manage to survive, reproduce, and establish a continuing colony. Animals with wide food preferences and plants of pioneer habitats and easily satisfied pollination requirements seem especially favored. Carnivores are less favored than herbivores, an example of the fact that primary elements in a food chain must become established on an island before later ones can be. This fact is shown on Tristan de Cunha, where the insect fauna contains an exceptionally high proportion of phytophagous species (Brinck, 1948). According to Gressitt (1961b), insects which inhabit plant debris or are leaf-miners or wood-borers are especially successful on oceanic islands.

Most forest trees or forest plants can become

established only when soils suitable for them are present. An exception is the chief forest tree of the Hawaiian Islands, *Metrosideros*, which can grow and eventually form a forest on new lava flows (Robyns and Lamb, 1939; Skottsberg, 1941). Lava pioneers are especially likely to succeed on oceanic islands, and by no coincidence, many plants of oceanic islands fall into this category. Rangitoto, a recent volcanic island in Auckland harbor, New Zealand, is vegetated by plants of the New Zealand flora which are capable of growth on bare lava. The floristic composition of Rangitoto contains few genera which are not also found on Hawaii and Tahiti (Carlquist, 1965).

Climatic requirements for establishment are numerous, and inadequate ecological conditions must screen out numerous potentially successful colonists. Examples on Canton Island described by Van Zwaluwenburg (1942) show this phenomenon clearly. Wilson (1959) has shown that in the Pacific, ants of open and marginal habitats have been the migrants to new island areas. Following arrival on islands they evolve into more specialized interior situations. More observations and experimental work on ecological requirements on island animals and plants need to be done.

Animals have the advantage of being able to seek suitable environments upon arrival, whereas plants, to survive, must be deposited upon locations suitable for their growth and maturation. Moreover, suitable pollination agents must be present. The claim that hermaphroditism or self-pollination is advantageous at the time of establishment (Baker, 1955) must be weighed against the undeniable value of outcrossing (see principle 19).

Perennial plants have been alleged to enjoy an advantage over annuals because longevity increases the likelihood for securing sufficient pollination to produce seeds to establish and maintain a species (Wallace, 1895), but alternative explanations for the lack of annuals on islands may take precedence (see principle 14).

6. *Migration to islands is governed by chance and probability, and ordinary concepts of migratory routes and biological provinces do not apply well to many islands.*

This statement is designed as a critique of the division of islands (chiefly those of the Pacific Ocean) into provinces, subprovinces, and the like. The concepts of biological provinces and migratory routes are primarily derived from, and therefore best applied to, continental situations. The best systems of biological provinces, such as those of Thorne (1963) and the workers he has cited, still remain unsatisfying to me for the following reasons:

(1) Groups of plants and animals on a particular island are present roughly in proportion to the distance from source areas. Many islands have acquired their biota from several source areas. Can the Hawaiian Islands logically be put into an Oriental province when an appreciable portion of their flora and fauna is American in origin (Fosberg, 1948)? The diverse sources of the New Zealand flora and fauna defy categorization into anything but multiple provinces and multiple routes (Dawson, 1958, 1963; McDowall, 1964). Darlington (1957) has shown that animals have entered the West Indies from many sources, and have used many ports of entry. These and other examples provide difficulties which in my opinion override whatever merits the province and route devices may have.

(2) Although there is merit in emphasizing major faunal regions, such as those delimited by Wallace's Line and Weber's Lines, these concepts apply to large land areas, not to oceanic islands. For example, the marsupials, which are a chief criterion of the Australo-Papuan region, do not reach oceanic islands or New Zealand.

(3) Criteria for one province differ from those for another. Placental mammals characterize the Malayan area, whereas biotic depauperation is the best criterion for Micronesia. In this connection, the Tuamotus must fall floristically into Micronesia, although geographically they are Polynesian.

(4) Each biologist will erect provinces and routes differently from others because, inevitably, each one will stress some groups not stressed by others, or will regard as of different importance historical, climatic, ecological, or geographical factors. No two biologists appear to be in agreement about the designation of provinces or routes, nor would agreement necessarily be desirable.

(5) The concepts of provinces and migration

routes result from an analysis of data but in their final expression do not of themselves yield information. They are a shorthand which cannot be translated. Because there is no substitute for knowledge of original data, the hypothetical construction of routes and provinces serves only their authors' ideas. Other persons would be well-advised to acquire the original data, if they would comprehend distribution patterns on islands.

7. *Guyots and other now-vanished high islands or lands more extensive formerly than now may have aided dispersal to oceanic islands as subsidiary source areas or "steppingstones," but not as dry land bridges.*

One can suppose, with Zimmerman (1948), that disseminules are abundant close to a source, and become progressively fewer with distance. In this case, presence of an island chain would attenuate the dispersal possibilities of a species. That this arrangement is indeed effective is shown by the Hawaiian Islands. Although the nearest area with any appreciable land surface is the North American continent, the Hawaiian biota is predominantly Indo-Malaysian. Moreover, the prevailing winds do not seem to favor immigration from Indo-Malaysia. However, the many small islands and atolls which lie west and south of Hawaii very likely aided, when they were larger high islands, in transmitting plants and animals to the Hawaiian Islands. Realizing the potential importance of these now-vanished lands, Zimmerman (1948) called for the development of paleontological information about the Pacific basin. Such data are now rapidly accumulating (Hamilton, 1953, 1956; Ladd, Ingerson, Townsend, Russell, and Stephenson, 1953; Ladd, 1958; Cloud, 1956; Cloud, Schmidt, and Burke, 1956; Menard, 1956; Menard and Hamilton, 1963; Stark and Schlanger, 1956; Durham, 1963). These authors have contributed to a picture of large former archipelagos, some as early as the Eocene, lying in the mid-Pacific. The maps by Menard and Hamilton (1963) strongly suggest a dispersal potential for now-vanished islands and should serve greatly to reduce skepticism concerning long-distance dispersal as a means for populating the Hawaiian Islands.

"Steppingstone" islands are potentially less effective as a source for dispersal than large land masses, because dispersal may be expected to be in proportion to the number of individuals in a source area (Johnson, 1960a, b). Moreover, autochthonous loss of dispersal mechanisms may lower the dispersal function of "steppingstone" islands. That such islands may function as staging areas in dispersal cannot be denied, however.

Within archipelagos redispersal occurs, often with interesting consequences as a result of successive events of isolation and reinvasion (Hamilton and Rubinoff, 1963). Many theoretical possibilities of how archipelagos may affect dispersal have been summarized by Carlquist (1965). One can say that any species with dispersal ability sufficient to bring it to an archipelago can be expected to succeed in dispersing (but not, perhaps, in establishing) itself throughout the archipelago. A species may lose pioneering ecological characteristics or dispersal mechanisms, however (Wilson, 1959).

8. *The size and systematic composition of particular insular biotas are determined by many factors which differ in relative importance from island to island.*

To attempt to say which factor is of prime importance is difficult. Even for particular islands factors are not easily ranked, although attempts have been made (Hamilton and Rubinoff, 1963; for a critique, see Carlquist, 1965). Among factors which have been claimed to be of importance are island area and altitude (Darlington, 1957); size, nearness, and richness of source (Zimmerman, 1948); latitude, climate, age, and geological events. Archipelago effects (size of island; size of neighbor island, nearness to neighbor island, altitude, altitude of neighbor island) are manifold (Hamilton and Rubinoff, 1963).

Factors which are influential in an island's biotic richness are measures of (1) ecological opportunity, (2) the degree to which barriers to dispersal to an island can be overcome, (3) the number of barriers within an island which can serve for isolation during speciation, and (4) the requirement for a certain minimal area for maintenance and evolution of a population (see principle 10).

9. *Relicts in the strictest sense are few or absent on oceanic islands, although every immigrant group has a history, and one can designate more primitive island autochthones as "recent relicts."*

Good dispersal ability is not entirely restricted to phylogenetically advanced groups. If primitive forms migrated to oceanic islands while the mainland remnants have become extinguished recently, they appear as relicts. This is probably the case with the primitive flowering plant *Lactoris* (Carlquist, 1964) and the fern *Thyrsopteris* on the Juan Fernandez Islands. These islands do not appear to be ancient, and the remainder of their flora and fauna contains no relicts. In the West Indies, the cycad *Microcycas* and the insectivore *Solenodon* (together with its fossil relative *Nesophontes*) may be considered relicts (McDowell, 1958; Darlington, 1957). Other than the above, no spectacular phylogenetic relicts can be found on oceanic islands.

Insular groups may tend to have a rapid cycle of speciation and extinction, and the latter stages may be said to contain, in a sense, "relict" species (see principle 18).

Why are there so few real relicts among the waif biota, whereas continental islands, such as Tasmania, are notable for relictism? One answer may lie in the fact that the most successful groups in the waif biota appear to be those which are evolutionarily "upgrade" or "weedy" and which have a greater degree of genetic "momentum." Most antique groups make relatively poor immigrants, both because of their generally poorer adaptability, and also because they often have poorer dispersal mechanisms.

10. *Immigrant species must overcome the restriction of genetic material related to the very small size of the initial population.*

The small size of populations with which waif groups begin is one of the unique features of insular existence. With few exceptions (such as the strand flora), an immigrant population will receive no new genetic material from the parent population. Indeed, if such infusions occurred with frequency, endemism on islands would be much lower than it is. Such infusions do occur with relatively great frequency among the strand flora, so endemism among those species is, in fact, relatively low (Fosberg, 1963).

Genetic variability must be manufactured following migration. The degree to which mutations are retained depends not only upon the mutability of a species but upon the ecological opportunity. Chromosomal polymorphism in West Indian populations of *Drosophila willistoni* is greater on islands which are ecologically diverse (Dobzhansky, 1957). *Metrosideros polymorpha*, the chief forest tree of the Hawaiian Islands, is extraordinarily diverse and has proved difficult for taxonomists. This multiform tree ranges from bare lowland lava to high bogs, where it is a shrub (Rock, 1917). Polymorphism in morphology characterizes many island species. Notable examples of variable island species in stages of expansion were demonstrated in the land shell *Partula* by Crampton (1916, 1925, 1932). There is every reason to believe that physiological and ecological opportunities play a controlling role. Small area and ecological poverty are cited by MacArthur and Wilson (1963) as reasons why extinction is high on some oceanic islands. The requirements of populations of a species for a land area large enough for maintenance of genetic variability, and thus, for the maintenance of the species itself are cited by various authors. Darlington (1957) believes that this explains the extinction of large mammals on Ceylon.

"Weedy" immigrants would be expected to have the advantage of high mutability, in my opinion. Moreover, the broader ecological tolerances of "weedy" plants and animals would permit occupancy by more numerous individuals, thus increasing potential genetic variability. Smaller body size would also favor development of larger populations, which similarly could be genetically more viable.

A species which immigrates as a flock of individuals rather than as one or two would seem to enjoy a great genetic advantage. The genetic disadvantage of self-pollination in an immigrant population of plants seems considerable.

The colonizing individuals represent only a portion of the genetic content of a species, and this portion will influence the characteristics of the eventual island population and its distinctions from the remnant mainland populations (Berry, 1964).

11. *Rapid evolution of island immigrants is not only possible, it is frequent. Change following arrival is inevitable.*

"Explosive" evolution is demonstrated by various groups which have been afforded good ecological opportunities. Among outstanding examples may be cited the Hawaiian Drosophilidae (perhaps 400 species according to Elmo Hardy, pers. commun.) or the Hawaiian species of *Cyrtandra* (Gesneriaceae): about 130 species of *Cyrtandra* on Oahu alone are claimed by Harold St. John (1966).

Actual times have been estimated for some cases of rapid insular evolution. Five species of endemic Hawaiian banana-constant moths have evolved in the approximately 1,000 years since human introduction of the banana there (Zimmerman, 1960). Freshwater lakes can be considered insular situations. In Lake Lanao in the Philippines, four endemic genera of fish with 18 species have evolved from a probable single species in 10,000 years or less (Myers, 1960). In Lake Baikal, endemism has reached the familial level in the case of the fish family Comephoridae (Kozhov, 1963).

Biologists are increasingly aware that geological time is short on volcanic islands. Evolution must proceed within this time span, although to this length of time can be added that available on an archipelago as a whole, as well as that on "steppingstone" islands, if any.

Factors favoring rapid evolution on oceanic islands include the lack of competitors and predators, presence of a wide spectrum of ecological opportunities, and numerous potential barriers. Geographical isolation is abundantly provided by the sharp relief of deep valleys and narrow ridges formed as volcanic islands erode. This may, in part, have been responsible for the remarkable speciation of *Partula* (Crampton, 1916, 1925, 1932). Lava flows may subdivide populations and provide isolation (Zimmerman, 1948), and catastrophic episodes of volcanism may well spur evolution. The barriers among islands of an archipelago seem important in the speciation of insular groups (Darlington, 1957). Other forms of reproductive isolation among insular species are

possible, however, and should not be overlooked (Bailey, 1956).

Where distinct endemic species and genera evolve, one may expect changes in any part of a plant or animal. For example, it is illogical to expect that the dispersal mechanism in the Hawaiian silversword, *Argyroxiphium*, is the same as that of the ancestors of this endemic genus. Possibilities that genes may have pleiotropic effects should not be overlooked.

12. *Situations on islands new to immigrants will dictate their courses of evolution. Adaptive radiation is the inevitable result on an island or archipelago where a small number of immigrant groups is faced with a broad spread of ecological opportunities.*

The "genus-and-family-poor but species-rich" condition of oceanic islands is one reflection of adaptive radiation. What is not reflected in taxonomic terms is the tendency for island species to evolve into ecological niches that would be occupied by a member of an entirely different group on a comparable mainland area. The consequence of the absence of mammals on most oceanic islands is the assumption of mammalian roles by birds and reptiles. For example, the dodo of Mauritius, a large terrestrial herbivore, represented a sort of avian rabbit. Phases of adaptive radiation are related to topics discussed under principles 14, 15, 16, and 17 below, in which diversification of habits within groups is cited.

Well-known examples of adaptive radiation such as Darwin's finches (Lack, 1945, 1947; Bowman, 1961, 1963) or the Hawaiian honeycreepers (Amadon, 1950) are only a few of the excellent ones which could be named. Other instances, particularly among plants, deserve greater currency. Among these are the Hawaiian tarweeds (Carlquist, 1959) and the Canary Island Aeoniums (Lems, 1960a). Descriptions of these and others and a catalogue of instances of adaptive radiation are offered elsewhere (Carlquist, 1965).

Time is a requisite for completion of a cycle of adaptive radiation. On some islands, time (as well as other conditions) has been insufficient for achievement of spectacular radiation. In some cases, secondary cycles of radiation can begin. For example, the genus *Psittacirostra* (Drepanididae) not only represents one product of adaptive radiation, it contains species which differ in bill size and shape and in food sources (Amadon, 1950).

13. *An immigrant group which is not confronted with, or cannot take advantage of, a broad spectrum of ecological opportunities on an island may evolve into one or a few niches.*

This statement is a way of saying that in addition to "definitive" adaptive radiation, islands may bear portions of a gamut of adaptive radiation. Examples of "incomplete" or "lopsided" adaptive radiation are abundant on islands. These examples usually demonstrate entry into a few new habits or mechanisms. The Hawaiian species of *Viola* and *Bidens* range from herbs to small shrubs, but not large shrubs or trees.

Islands may play a role in fostering peculiar adaptations, such as unusual diets (seaweed is eaten by the Galápagos iguanid *Amblyrhynchus*), specialized pollination mechanisms (ornithophily has been developed by Hawaiian lobeliads), peculiar dispersal mechanisms (the Hawaiian lobeliad *Trematolobelia*), new food-getting mechanisms (the tool-using Galápagos finch *Cactospiza*), etc. Unique developments in these general categories do occur, of course, on mainland areas as well as on islands. One can speculate that the specific examples cited in parentheses not only represent responses to opportunities on islands, they also may be preserved in the less competitive island situation (Carlquist, 1965).

14. *New growth forms evolve among plants on oceanic islands. Most conspicuously, there is a tendency toward increased stature.*

The chief changes which seem to occur are from herb to rosette-tree or rosette-shrub, herb to shrub, shrub to rosette-tree, shrub to "true" tree. Examples of these tendencies are described elsewhere (Carlquist, 1965). The reasons for these alterations in growth form are several, and one or more of these may serve to explain the habits of particular species:

(1) A high volcanic island potentially could support the same growth forms in the same taxonomic groups as a continental area with similar climate. Dispersal to such an island tends to favor not the entire spectrum of plants,

but mostly herbaceous and shrubby plants, chiefly those of pioneer or "open" habitats. Plants of stable forest areas, many of which are often large-seeded, are disadvantaged as waif immigrants not only because of poor dispersal but because of specialized ecological requirements. From the limited spectrum of growth forms which dispersal brings, the entire gamut tends to be evolved on islands.

(2) Herbs which follow a strongly seasonal regime on continents are "released" on tropical, subtropical, and even temperate islands where latitude and the tempering effect of maritime surroundings moderate climate. Under these conditions, the growing season lasts throughout the year and a rosette-plant which would ordinarily die down to the ground each winter forms elongate stems. Moreover, this is a more efficient growth form where climate permits, for vegetative growth and flowering stems can be produced continuously, by-passing seedling stages required by annuals each year. These seem to be major factors in the development of rosette shrubs such as the large *Plantago* species which have evolved independently on St. Helena, the Juan Fernandez, Hawaiian, and Canary Islands. Similar considerations apply to the Hawaiian lobeliads. Climatic factors have been cited as responsible for development of rosette-shrubs in the Canary Islands (Schenck, 1907; Rikli, 1912; Johnston, 1953). Climatic considerations also seem basic to evolution of rosette trees in equatorial-alpine regions on continents (*Espeletia* and *Puya* in the Andes; *Senecio* and *Lobelia* in the African Alps).

(3) A relative lack of herbivores may permit survival of succulent and large-leaved shrubs and trees, once they have evolved on islands, although this circumstance may not be basic to their evolution.

(4) Competition among individuals of an herbaceous species was claimed by Darwin (1859) to be a selective force. Larger individuals which overtopped their neighbors were interpreted by him as leading to shrubby habits. This seems rather unlikely in the example selected by Darwin, *Scalesia*, shrubby species of which grow mostly not in dense stands, but in a scattered fashion on rather bare lava areas of the Galápagos Islands (Harling, 1962).

Reasons (1) and (2) above are considered most significant in alteration of growth forms. The tendency for typically herbaceous groups — such as the Compositae — to develop into rosette-trees, rosette-shrubs, or true trees is represented by the abundance of such Compositae on the Hawaiian Islands (Rock, 1913), the Juan Fernandez Islands (Skottsberg, 1953), St. Helena (Melliss, 1875), the Canary Islands (Schenck, 1907; Børgesen, 1924), and other islands (Hemsley, 1885; Carlquist, 1965). Compositae have been successful at these innovations both because they are, as a group, rapid-evolving, adaptable, well-represented on islands because of good dispersibility, and because, although many are herbs, most are at least somewhat woody and can presumably evolve increased woodiness easily.

The alternative proposition — that strange growth forms such as those cited above are relicts (Skottsberg, 1956; Lems, 1960a, 1961) — seems clearly contradicted by data referable to the above concepts. Where comparative studies of the island forms and their mainland relatives are undertaken, the interpretation of insular rosette-trees and the like as innovations, rather than relicts, becomes inescapable. Anatomical evidence on the nature of insular rosette trees shows that they are juvenilistic; the stem and wood structure suggests an herb pattern the ontogeny of which never, or only slowly, reaches adult characteristics (Carlquist, 1962).

Particular ecological conditions undoubtedly play a role in determining the nature of the growth form evolved — e.g., whether a large-leaved rosette-tree or a rosette-shrub, whether a stem-succulent or a wiry fountain-shaped shrub (Carlquist, 1965). Further anatomical studies are in progress in an effort to elucidate the nature of changes in growth forms on oceanic islands.

A corollary to insular change in growth form is that mainland relatives will have different growth forms. This truism must be remembered when attempts are made to establish the relationships of "anomalous" insular endemics. One must often investigate groups with rather different growth forms. For example, Keck (1936) thought that the Hawaiian silversword *Argyroxiphium* might be related to a Juan Fernandez rosette-tree with similar habits, *Robinsonia*, but the closest affinities actually seem to be with the California tarweeds (Carlquist, 1959), which are relatively non-woody annuals and perennials.

15. Changes in form, size, and color of animals often occur on islands: gigantism, dwarfism, changes in body proportions, and melanism are among the changes represented.

The animals which show these trends well are those which are best represented on islands and which are evolutionarily plastic. Insects, birds, reptiles, and land molluscs are chief among these. The former two groups, being mostly volant, often have alterations in shape, size, and proportions which are related to flight (and are considered under principle 17 below). Reptiles offer particularly good examples of change in form and size, and the review by Mertens (1934) of island reptiles is a landmark.

Gigantism is common among insular animals as compared with their mainland relatives (Berland, 1924; Mertens, 1934; Steven, 1953; Hill, 1959; Cook, 1961; Berry, 1964). The reasons for gigantism are rather speculative at present, and well-designed studies would be very welcome. Among the explanations for insular gigantism which can be offered at present are the following:

(1) Lack of predators. Larger sizes may be achieved if predator pressure is reduced. Increased longevity and thereby increased size are possible under these circumstances.

(2) A new diet. Especially for herbivorous and to some extent insectivorous animals, food supply may be more abundant. This abundance may not take the form of a greater quantity, for an animal population would soon enlarge to take advantage of the supply. On an island, however, three or four food items might be taken by one animal species, whereas on a continental area, several animal species might compete for them. On some islands, food supplies might be present throughout the year, owing to milder climatic conditions. More likely, however, an animal can adapt to new, and in particular to larger, articles of diet. The quantity and types of nutrition available might be expected to be correlated with body

size, and a species faced with increase in nutrition could be "released" from some factor limiting size and could evolve into a larger race.

(3) Lack of competitors except for members of their own species. If competition for food is *within* a species rather than between or among species, one might expect gigantism to have a selective advantage. Moreover, mating struggles might favor larger individuals; this would be true on mainland areas, too, but predation and food supply might cancel out this trend on the continents.

Dwarfism is also a notable tendency on islands. The following may be the reasons:

(1) Smaller forms on islands may reflect the small size of immigrant species, a fact related to their greater ease in dispersal.

(2) Smaller size may be an adaptation to pressures of predators, if greater on a particular island (Hecht, 1952).

(3) Smaller size may be an adaptation to smaller food articles or smaller food supplies, and thus is the reverse of (1) under gigantism above. A smaller animal could maintain a viable population size in a situation where food is scarce.

Changes in proportion which are evident on islands include:

(1) Fatter bodies, fatter tails, and thicker limbs (Mertens, 1934). These changes, shown especially well by lizards, may be forms of gigantism, and to that extent they are explained above. Lack of need for locomotion may also underlie these trends, as may need for storage during dry seasons.

(2) Shorter limbs and tails are characteristic of many island reptiles. A good instance of this is reported in a well-designed study by Kramer (1951). Stubby legs and clumsy habits characterize some island insects, such as certain ones in Hawaii (Perkins, 1913) or the New Zealand wetas. Adaptation to a terrestrial way of life where food supplies are secured easily by crawling may be responsible. Change in location of available foods may also explain alteration of body proportions in reptiles. The instances studied by Kramer suggest that lack of predators and smaller spaces on small islets lower the selective value of long, agile limbs suited to evasive tactics.

(3) Change to new habits may result in leglessness. This trend is shown independently

in two genera of skinks: *Brachymeles* (Philippines) and *Grandidierina* (Madagascar). These genera show adaptation to life underground or in forest litter.

Melanism is widespread in island reptiles (Mertens, 1934). This is not a protective color except in a few cases. Among the most appealing of the explanations offered by Mertens is that of thermoregulation. Cooler maritime climates of islands might render heat-absorptive coloration valuable, especially for reptiles whose size is greater owing to gigantism. Coloration in the Galápagos ground finches is related to the proportion of time spent on or near the ground in feeding (Bowman, 1963), and may be related to protection from predators.

16. Dispersal mechanisms and dispersal ability may be lost during the evolution of plants on oceanic islands.

Autochthonous loss or alteration of dispersal mechanisms on islands seems not to have been recognized. Study of floras of oceanic islands — particularly in the Pacific — has convinced me that it is just as real a phenomenon as flightlessness in insular birds and insects. Reasons for this tendency are as follows:

(1) As noted above (under principle 12), islands which can support a forest must, to a greater or lesser extent, reconstitute a forest from a waif flora, which consists mostly of nonforest species. Large fruit and seed size characterizes many true forest species, a fact related to the requirement of seedlings in deep forest for greater food supplies as they advance to upper, better-lighted levels of the forest. Such fruits and seeds are poorly suited to dispersal, whereas the best-dispersed seeds and fruits are usually small — a condition antithetical to an abundance of storage materials. Increase of seed and fruit volume, an evolutionary trend on many oceanic islands, has the net effect of reducing dispersal ability.

(2) Dispersibility may acquire a neutral or negative value in island situations. A plant with excellent dispersal ability — as with plants typical of pioneering habitats — disperses its seeds widely. If such a species reaches an island and adapts to an ecological situation of limited geographical extent (as have many island species), good dispersibility actually acquires a negative value. The area favorable for a

particular island species is usually much smaller than the island itself — often only a ridge or valley. Although some persons have noted that a dispersal mechanism which permits disseminules to be swept off an island is unfavorable, the limits beyond which disseminules are "wasted" are, in fact, much narrower than the island itself. Precinctiveness is thus linked to poor dispersal; this may operate on mainland areas, but it does so to a greater extent on islands.

(3) In some cases, a dispersal mechanism takes the form of an appendage or some other special formation. Economy would dictate that such structures be gradually lost if they no longer have any selective value. Even if it is assumed that a dispersal mechanism has a neutral value, it may disappear as a result of changes within the fruit — by means of pleiotropic gene action, for example. Island plants still need a certain dispersal ability, although it may be much less than that of continental plants. A dispersal mechanism may dwindle to the point where it is equal to the dispersal required for success of a species.

Among the more salient examples of loss of dispersal on oceanic islands are *Bidens* (Polynesia, including Hawaii) and *Dendroseris* (Juan Fernandez Islands), both Compositae. Loss or alteration of dispersal mechanisms is especially prominent in Polynesian floras and will be detailed in a forthcoming paper by the writer.

Another change which may occur is the tendency for seeds of island species to lose prolonged viability. One would expect that as pioneer species develop into forest species, the shorter viability typical of seeds of rain-forest trees might be acquired. Data on this point are very much needed.

17. *Flightlessness may evolve in volant groups of animals in response to insular conditions.*

The number of flightless birds and insects among waif faunas on islands is quite startling. Early noted by Darwin (1855), it has received comment from various authors such as Perkins (1913), Darlington (1943), Brinck (1948), Zimmerman (1948, 1957), Hagen (1952), Gressitt, Leech, and Wise (1963), Holloway (1963), and Gressitt (1964). A listing of flightless insular birds has been assembled (Carlquist, 1965).

Loss of flight occurs most frequently in groups which normally are airborne only a small proportion of the time, such as rails (Rallidae), a family that is exceptionally well represented by non-flying as well as flying species on islands (Hagen, 1952; Carlquist, 1965). An explanation for the frequency of rails on islands might be that because they are infrequent and perhaps poor fliers, they are easily swept out to sea, cannot fight against storm winds, and thus cannot return home as easily as birds which are stronger fliers. Because of limited flying ability, some birds seem more likely to stay on islands than others, even before loss of flight.

Flightlessnes probably does not have a single cause (Darlington, 1943), but one or more of the following may operate in particular cases:

(1) Absence on islands of ground-feeding animals — such as most mammals — permits birds and insects to enter a terrestrial way of life, where flight becomes unnecessary as an adjunct to food-getting. As flight becomes infrequent, a bird or insect species easily and imperceptibly crosses the threshold from ability to inability to fly. Volant species which have a near-terrestrial mode of life (such as rails) would cross this threshold more easily.

(2) Entry into the ground-feeding habit is associated with winglessness in insects, and stable forests of tropical islands have an abundance of flightless species (Darlington, 1943). In birds, a mimicking of the habits of some mammals may be accompanied by increase in body size without increase in wing size and musculature. Change in feeding habits may be accompanied by increased precinctiveness if food sources occur within a limited area; in this case, restriction of flight pattern might acquire a positive selective value. In New Zealand species of Lucanidae (Coleoptera), flightless species occupy smaller ranges than do volant species (Holloway, 1963). Entry into new ecological situations is related to winglessness in certain non-insular situations also: insects of mountaintops (Darlington, 1943), caves (May, 1963), and some groups of parasitic insects, such as those flies that parasitize bats (Streblidae, Nycteribiidae).

(3) Increase in leg size and musculature may force a decline in size of other body parts. If wings are no longer used, they are the most expendable structures. In this concept, the body would be viewed as a closed economy or budget, in which increase in one item forces decrease in another. Of the rails native to Tasmania, the best at running is the native hen (*Tribonyx mortieri*), which is also the only one unable to fly (Sharland, 1958).

(4) Ability to fly does not continue without maintenance by positive selective pressure. If this pressure lessens or vanishes, flight will, in time, disappear. Flight is an ability dependent on a complex of components, just as is flight of an aircraft, and decrease in selective pressure will permit mutation of only one or a few genes to cancel flying ability.

(5) Lack of predators or enemies lowers the value of flight as an evasive tactic.

(6) Flight becomes unfavorable if an appreciable proportion of a population could be blown away or straggle away from an island or even away from suitable portions of an island. This is the most likely explanation among the smaller volant animals and on locations where wind pressure is often high, as on the subantarctic islands. Experimental evidence that wind, and possibly also straggling, are selective factors in flightlessness has been obtained in *Drosophila* by L'Heritier, Neefs, and Teissier (1937). Gressitt (1964) has shown that in the insect fauna of Campbell Island, at least 40 per cent which belong to normally winged groups' show a reduction in wing size. This reduction varies from moderate to complete absence of wings. Subantarctic islands are particularly notable for their high proportion of flightless moths (Enderlein, 1909; Viette, 1948, 1952a,b, 1954, 1959; Salmon and Bradley, 1956; Gressitt, 1964). On the basis of this explanation, one can also understand why males are more fully winged than females (which would be less expendable) within many of these moth species. The flightless moths of subantarctic islands seemingly have compensated for lack of flight by developing grasshopper-like springing ability (Salmon and Bradley, 1956), as have flightless flies of these islands (Gressitt, 1964). This development may have contributed correlatively to wing reduction.

(7) Development of winglessness may be permitted on islands which are relatively free from flooding (Darlington, 1943).

Darlington (1943) views flightlessness as a habit controlled by a number of factors, and predominance of one or more of these may shift the proportion of winged to wingless. Clearly, wingless species are not restricted to islands, and thus these factors operate in many situations.

As soon as the threshold to flightlessness has been crossed, wings seem to be rapidly reduced phylogenetically. At various stages in their reduction, wings may function for climbing or as brakes during hopping (Hagen, 1952). Because insect wings are less complex than bird wings and are governed by fewer genes, one may expect wingless insects to evolve more frequently and rapidly on islands than do wingless birds. Of the flightless birds on oceanic islands, all have some wing vestiges; later stages in wing loss are shown on old islands such as New Zealand (e.g., *Apteryx*, the kiwi). Selective pressure for loss of wings would be maximal at the time of and immediately following the loss of flying ability, and would decrease as wings become mere vestiges that neither have any function nor require the expenditure of much energy in formation or maintenance. Darlington (1943) offers an alternative explanation, that pleiotropic genes are responsible for retention of wing vestiges.

Flightlessness may be a matter of habit, as well as morphology. According to Gressitt and Weber (1959), wingless or partially winged antarctic and subantarctic insects are active, whereas winged species are "sluggish." When these two groups are added together, the insect fauna of far-southern islands has a strongly sedentary aspect.

18. *Competitive ability is often decreased slightly to markedly among endemics of oceanic islands.*

This statement has an intentional vagueness which hinges on the inexactness of the concept "competition." The phenomenon is real enough, certainly, and aspects have been noticed by many biologists who have witnessed the rapidity with which indigenous island species yield to continental species introduced by man (cf. Elton, 1958). Competitive ability would seem to depend upon constant selective pressure, and is thus abundantly represented on continents, where numerous aggressive groups are evolving simultaneously. This pressure vanishes or is lowered when a species migrates to an island where there are fewer species, and where many ecological opportunities which are not preempted exist. Each immigrant group would, in this view, lose competitive ability following arrival, so that the total flora or fauna at any time would be less competitive.

In the Hawaiian Islands, almost any introduced continental species of plant seems capable of replacing autochthonous species of comparable ecological requirements. Even if it is admitted that soils have been disturbed by man and his domesticates, native island species show poorer self-replacement after disturbance than do native species in a comparably disturbed continental area. The rain forests of Kauai now host a remarkable variety of weeds, including many garden flowers, few of which would be noxious — if they were weedy at all — in continental areas. Even high bogs are not exempt from weeds (e.g., *Rubus* is now covering the bogs of Mt. Kaala, Oahu). This situation augurs poorly for attempts at conservation of island endemics. Not only are weeds well entrenched in many areas of the Hawaiian Islands, but also efforts to remove them would very likely only renew and widen the areas of disturbance and encourage more weedy growth than before. Many plants which are now weeds in the Hawaiian Islands can hardly be kept out of areas because of their good dispersal mechanisms (e.g., *Schinus* and *Psidium* are spread by frugivorous birds). Only a few native Hawaiian plants, such as *Scaevola*, *Pipturus*, and *Acacia*, seem capable of occupying disturbed sites.

Loss of competitive ability may in part be due to genetic depauperation, a fate which seems common to most island autochthones. One might expect that even in prehuman times, there would have been genera that "lost momentum." In fact, there do seem to be such genera in the Hawaiian biota. These include the genera of land shells *Carelia* (Cooke, 1931) and (to a lesser extent) *Achatinella* (Cooke and Kondo, 1960). Among plant genera, the lobeliad *Delissea* has been labelled "decadent" by

Rock (1962), and other Hawaiian genera which could be so described include *Hesperomannia*, *Hibiscadelphus*, *Isodendrion*, *Kokia*, *Munroidendron*, *Pteralyxia*, *Remya*, *Rollandia*, and, to some extent, *Cyanea* and *Clermontia*.

Irreversible adaptation to excessively specialized locations may be a key to these vanishing genera. These specializations would be irreversible if loss of variability occurred. An insular phylad would be expected to be either in a state of expansion and speciation as it draws on genetic variability, retained or acquired; or in a state of decadence, as adaptability to new situations dwindles. The upgrade groups would be expected to replace the downgrade taxa, so that groups in both categories could be found at any given time. Paucity of individuals would seem to play a key role in loss of genetic variability. Adaptation to an ecological zone of limited extent (and on oceanic islands, any zone would be limited in extent) would result in a smaller number of individuals per species. Excessive specialization is exemplified among Hawaiian animals by the land shell *Achatinella*, which is capable of eating, not foliage, but only epiphytic algae. The Hawaiian drosophilids seem to subsist only on leaves of the endemic arborescent Araliaceae and on rotting lobeliads (Elmo Hardy, pers. commun.), an unusual diet in this family of flies. Some Hawaiian insect species may be so restricted that they occur only on a single tree (Zimmerman, 1948).

An interesting manifestation of loss of competitiveness is seen in genera and even families which are widespread on islands but which have, with few exceptions, not managed to gain or retain a foothold on continents. Such genera of land shells as *Tornatellides* and *Elasmias* (Cooke and Kondo, 1960) or *Partula* (Germain, 1934) have almost unbelievable distributions which span wide stretches of the Pacific, even reaching offshore islands of continents but not the continents themselves. Such distributions would suggest a kind of relictism, except that the genera and the islands they occupy are doubtless relatively recent in geological terms. Possibly these genera range the Pacific by means of good dispersal, establishing themselves only in situations which are both suitable ecologically and in which competitiveness is rather low. Further observational and, if possible, experimental evidence is needed to demonstrate the nature and causes of "incompetent" insular species.

19. *Means for outcrossing become highly developed in waif floras. Species without potential for outcrossing are probably doomed to a short tenure.*

Baker (1955) hypothesized that self-compatibility is advantageous for the establishment of plant immigrants. The obvious advantages of this habit for initial establishment may well be overwhelmingly outweighed by the long-term disadvantages of inbreeding. So disadvantageous is inbreeding on islands that self-pollinated species do not bulk large in insular floras, and, where they are present, one often suspects that they are declining species. Where population size is limited, as it necessarily is on islands, and where inflow of new genetic material is cut off by the sea barrier, maximizing of outcrossing seems a necessity. All genetic variability can be dispersed throughout a population by means of outcrossing. A species with flowers that will self-pollinate if cross-pollination does not occur within the first day or two after opening, would possess an advantage. This habit characterizes the Compositae, a family notable for its success on islands. The long-term mode toward which insular groups evolve is cross-pollination, however. Baker (1955) has claimed that the characteristics of the Plumbaginaceae offer evidence of a relationship between island habitats and the occurrence of self-pollination, but calculations based on his data on one genus from this family, *Limonum* (Baker, 1953), proves the reverse. In the genus *Limonium* as a whole, 79.6 per cent have dimorphic flowers, a condition which would make outcrossing a virtual necessity. Of the species of *Limonium* endemic to small or medium-sized oceanic islands, 94.5 per cent have dimorphic flowers. There are, however, instances reported in which speciation on islands is accompanied by selfing, as in the Galápagos tomatoes, *Lycopersicon* (Rick, 1963). Autogamy may be expected more in the strand flora, less in the montane flora of islands. Re-introduction of new genetic material is possible in the strand flora, whereas the montane flora is more effectively cut off from source areas. Autogamy may also be expected in relatively recent, up-

grade groups on islands, less in the older species of the forest flora. Many species capable of autogamy may experience sufficient outcrossing to maintain genetic variability.

The importance of some means for outcrossing on islands has not been generally appreciated. Outcrossing may be enforced simply by one or several gene pairs, or it may be required by deep-seated and visible means, such as unisexual flowers. Dioecism makes outcrossing mandatory. In the New Zealand flora, 14.5 per cent of the species are dioecious, whereas in a continental flora, such as that of the British Isles, only 2 per cent are dioecious (Rattenbury, 1962). New Zealand may not seem a small enough island for the limitation of population size to make outcrossing highly valuable. In the Pleistocene, however, many New Zealand species were doubtless very much restricted by the advance of ice sheets. Families which elsewhere are represented mostly by normal bisexual flowers have dimorphic or unisexual flowers in their New Zealand representatives (Godley, 1955; Franklin, 1962; Dawson, 1964).

Flowers suited for outcrossing by the various means listed below are particularly abundant in the floras of the Juan Fernandez Islands (Skottsberg, 1928, 1938), the Desventuradas Islands (Skottsberg, 1963), and the Hawaiian Islands (Hillebrand, 1888; Skottsberg, 1936a,b, 1944a,b, 1945; Fosberg, 1956; St. John and Frederick, 1949; Carlquist, 1965). Because this phenomenon has been insufficiently appreciated, the writer is preparing an account of outcrossing mechanisms in the Hawaiian flora.

Mechanisms which aid or insure outcrossing in the floras of the above islands are as follows: (a) dioecism; (b) gynodioecism; (c) monoecism; (d) heterostyly, floral dimorphism, and trimorphism; (e) wind pollination; (f) pollen sterile on some plants, ovules infertile on others; (g) protandry or protogyny; and (h) massing of flowers into large, conspicuous inflorescences.

20. *Natural hybridization acquires a positive value in evolution of the waif biota.*

In a sense, this is merely an extension of outcrossing, for exchange of genetic material is involved. With hybridization, however, a greater degree of difference between, or a barrier between, species (or other taxa) is implied. Species on oceanic islands are notoriously "unstable" or "variable." This condition does not seem to be wholly the result of the lack of extinction of intermediates. A variable species is probably a successful species on an island, and as long as it can retain or attain a degree of polymorphism, it would seem to have an evolutionary future.

Does hybridization play an important role on islands? Only a few cases of variability in island species have so far been traced to hybridization. There has been a tendency to overlook hybridization, however, partly because taxonomic studies of island organisms are often based upon museum specimens rather than field studies. One can safely say that hybrids would prove to be more frequent if students of particular groups would entertain this possibility. The Hawaiian species of *Scaevola* (Goodeniaceae), hitherto regarded as a series of easily defined endemics, now prove to be highly polymorphic, mostly linked to each other by a series of hybridizations, current and old (George W. Gillett, pers. commun.). "Cylic hybridization" has been claimed as a mechanism by which many species in the New Zealand flora survived radical climatic changes of the Pleistocene (Rattenbury, 1962).

Reports of hybridization among waif floras include those of Rock (1919a) in *Clermontia*, Skottsberg (1939) in *Viola*, Fosberg (1956) in *Gouldia*, Lems (1958) in *Adenocarpus*, Dawson (1960) in *Acaena*, and Franklin (1962, 1964) in *Gaultheria* and *Pernettya*. Many reports of hybrids in the New Zealand flora are summarized by Cockayne and Allen (1934) and Allan (1961).

Hybrids might be said to occur between races (although some will argue that this violates the meaning of the term "hybrid") as well as between species on islands. A successful curriculum (at least in plants) would seem to be that of a population which takes advantage of reproductive isolation in exploring (speciating into) new ecological territory, but which retains fertility with other populations. As conditions change, gene flow among a series of semi-separate populations could maintain a high level of adaptability. Instances, modes, and extent of natural hybridization remain rewarding avenues for investigation by students of insular botany. These investigations are particularly

urgent, because they can be pursued only while insular floras are relatively intact.

21. *Pollination relationships correspond to and change with respect to availability of insects and other pollinating agents on islands.*

Wallace (1895) noted the paucity of conspicuous flowers in the floras of New Zealand and the Galápagos Islands, a fact he correlated with poverty of insects on those islands. This insect poverty is best described, however, in terms of the absence or scarcity of particular groups, such as butterflies in the Galápagos (Wallace, 1895) or long-tongued bees in New Zealand (Rattenbury, 1962). The Hawaiian forest flora is also abundant in small green or white flowers which are poor in scent. Smaller flowers have probably established preferentially or evolved in these islands to suit the smaller size, and the habits and preferences of available pollinators. As will be noted, the waif insect fauna consits of immigrants belonging to smaller size classes, and this may have promoted the evolution of smaller floral sizes. That island conditions do influence pollination mechanisms is suggested by Hagerup (1950a,b, 1951). Flowers in floras of oceanic islands often have simple open forms, suitable to entry by a wide variety of potential pollinators.

Massing of flowers may serve to attract pollinators, and is theoretically advantageous if pollinators are scarce. This factor might or might not help to explain the occurrence of insular monocarpic rosette trees, such as *Wilkesia* and *Trematolobelia* (Hawaiian Islands) or *Centaurodendron*, *Yunquea*, and *Phoenicoseris* (Juan Fernandez Islands). Other plants on these islands with massive inflorescences could easily be cited. Wallace (1895) has claimed that the perennial habit aids in securing pollinators. The value of this habit seems questionable, because a few additional flowering seasons will probably not serve to secure pollination if suitable pollinators are not present in sufficient numbers during any one season. Probability seems to favor either the presence of a suitable insect in sufficient numbers during any given year, or else its complete absence from an island. The high proportion of perennials on oceanic islands probably is primarily related to climatic factors (see principle 14, 2).

The attractiveness of Compositae to a variety of insects is alleged by Wallace (1895) as a reason for the abundance of Compositae on islands. If operative, this reason is probably subsidiary to the good dispersibility, weediness, and adaptability characteristic of this family. Compositae, by congestion of flowers into a head, may be said to fulfil the floral aggregation cited above as a possible advantage in securing pollination.

Change partly or wholly to wind-pollination solves the problem of a scarcity of suitable insect pollinators. Autochthonous evolution into anemophily may have occurred on a few islands (e.g., *Rhetinodendron* on the Juan Fernandez Islands; Skottsberg, 1928). The proportion of anemophilous flowers may be higher in waif floras, but calculations are apparently not yet available.

Speculations such as the above suggest that a study of pollination relationships in the waif biota is very much needed. Few definite statements can be made or particular flower-insect relationships cited.

22. Some mutations which would be lethal or disadvantageous in continental environments have a more nearly neutral value in the less competitive environment of an oceanic island.

This statement, easy to make, is difficult to demonstrate, although many of the changes described under headings above might qualify as examples. One feature which clearly does qualify is the fearlessness of island animals. Many descriptions of fearlessness in island animal — particularly birds — have been offered (Lönnberg, 1920; Beebe, 1924; Rand, 1938; Greenway, 1958; Rice, 1964). Where predators are absent, evasive action would seem to be wasteful, because it would interfere with other activities, such as feeding.

Development of conspicuous color patterns may be a visual equivalent of fearlessness, reflecting, as does fearless behavior, the absence of predators. This characteristic occurs prominently among certain lizards (Mertens, 1934; Carlquist, 1965).

Various forms of reproductive inefficiency characterize some island animals, although this subject is as yet poorly explored. Possibly, careful field studies would reveal more instances of the sort reported by Hagen (1952) on Tristan

da Cunha: birds with smaller clutches of eggs than characteristic of mainland relatives, and with prolonged sexual immaturity. Vivipary, as in New Zealand geckos, and abundance of intersexes, as in the viper *Bothrops insularis* on the Brazilian island, Queimada Grande (Anon., 1959), would probably be disadvantageous on continental areas. Natural selection might favor a low rate of reproduction in a species with restriction to a small land area, as would such factors as lack of competitors, or greater longevity, all of which conditions do occur on islands.

23. Endemism, although high on oceanic islands, is not of itself a criterion for identification of an island as oceanic; the nature of the endemism may be indicative, however.

Oceanic islands tend to have endemism restricted to lower categories, particularly species; to a lesser extent, genera. Old continental islands may be likely to possess endemic families, even orders, as relicts from continents. Setchell (1928) hints at this.

Endemism is a constant byproduct of evolutionary change, and the percentage of endemics is more a measure of degree of isolation in time and space than of mode of origin of islands. If island biotas contain phylogenetically primitive forms with poor dispersal ability, a continental origin for the island may be suspected. For example, the presence of araucarias strongly suggests that New Caledonia is an old, continental-like island.

24. Evolutionarily plastic groups will be sensitive indicators of directions of evolution in the biota of long-distance dispersal.

Groups such as insects or composites may be expected to fit themselves rapidly and closely to the templates provided by the island environment. As one example of the usefulness of this concept, one can discount Skottsberg's (1956) claim that peculiar growth forms on the Juan Fernandez Islands are relicts. One would expect such growth forms, if they are relicts, to be representatives of relatively primitive or slowly evolving groups. Instead, the Juan Fernandez rossette-trees and rosette-shrubs occur in upgrade, predominantly herbaceous groups, characterized by weediness and rapid evolution: the Plantaginaceae, the Umbelliferae, and three tribes of the Compositae.

266   *THE QUARTERLY REVIEW OF BIOLOGY*

LIST OF LITERATURE

ALLAN, H. H. 1961. *Flora of New Zealand.* Vol. I. R. E. Owen, Gov't Printer, Wellington.

AMADON, DEAN. 1950. The Hawaiian honeycreepers (Aves, Drepaniidae). *Bull. Amer. Mus. Natur. Hist.,* 95:151–262.

ANDREWS, E. C. 1940. Origin of the Pacific insular floras. *Proc. 6 Pacific Sci. Congr.,* 4:613–620.

ANONYMOUS 1959. Queer vipers. *Time,* October 19, 1959: 53.

BAILEY, D. W. 1956. Re-examination of the diversity in *Partula taeniata. Evolution,* 10:360–366.

BAKER, H. G. 1953. Dimorphism and monomorphism in the Plumbaginaceae. II. Pollen and stigmata in the genus *Limonium. Ann. Bot.,* 17:433–445.

——. 1955. Self-compatibility and establishment after "long-distance" dispersal. *Evolution,* 9: 347–349.

BEEBE, W. 1924. *Galápagos: World's End.* G. P. Putnam's Sons, New York.

BERLAND, L. 1924. Araignées de l'île de Pâques et des îles Juan Fernandez. *Natur. Hist. Juan Fernandez and Easter I.,* 3:419–437.

BERRY, R. J. 1964. The evolution of an island population of the house mouse. *Evolution,* 18:468–483.

BLAKE, S. F., and A. C. ATWOOD. 1942. *Geographical Guide to Floras of the World. Part. I.* U. S. Dept. Agr. Misc. Pub. 401, Washington, D.C.

BØRGESEN, F. 1924. Contributions to the knowledge of the vegetation of the Canary Islands. *Mém. Acad. Roy. Sci. Danemark, Sect. Sci.,* Ser. 8, 6:285–398.

BOWMAN, R. I. 1961. Morphological differentiation and adaptation in the Galápagos finches. *Univ. Calif. Pub. Zool.,* 58:1–302.

——. 1963. Evolutionary patterns in Darwin's finches. *Occasional Papers Calif. Acad. Sci.,* 44:107–140.

BRINCK, P. 1948. Coleoptera of Tristan da Cunha. *Results Norwegian Sci. Exp. Tristan da Cunha,* 17:1–121.

CARLQUIST, S. 1959. Studies on Madinae: anatomy, cytology, and evolutionary relationships. *Aliso,* 4:171–236.

——. 1962. A theory of paedomorphosis in dicotyledonous woods. *Phytomorphology,* 12:30–45.

——. 1964. Morphology and relationships of Lactoridaceae. *Aliso,* 5:421–435.

——. 1965. *Island Life.* Natural History Press, New York.

CLOUD, P. E., JR. 1956. Provisional correlation of selected Cenozoic sequences in the western and central Pacific. *Proc. 8. Pacific Sci. Congr.,* 2:555–573.

CLOUD, P. E. JR., R. G. SCHMIDT, and H. W. BURKE. 1956. Geology of Saipen, Mariana Islands. Part 1. General Geology. *U. S. Geol. Surv. Profess. Papers,* 208A:1–126.

COCKAYNE, L., and H. H. ALLAN. 1934. An annotated list of groups of wild hybrids in the New Zealand flora. *Ann. Bot.,* 48:1–55.

CONSTANCE, L., L. HECKARD, K. L. CHAMBERS, R. ORNDUFF, and P. R. RAVEN. 1963. Amphitropical relations in the herbaceous flora of the Pacific coast of North and South America: a symposium. *Quart. Rev. Biol.,* 38:109–177.

COOK, L. M. 1961. The edge effect in population genetics. *Amer. Natur.* 95:295–307.

COOKE, C. M., Jr. 1931. The land snail genus *Carelia. Bishop Mus. Bull.,* 85:1–97.

COOKE, C. M., and Y. KONDO. 1960. Revision of Tornatellidae and Achatinellidae (Gastropoda, Pulmonata). *Bishop Mus. Bull.,* 221:1–303.

CRAMPTON, H. E. 1916. Studies on the variation, distribution and evolution of the genus *Partula. Carnegie Inst. Wash. Pub.,* 228:1–311.

——. 1925. Contemporaneous differentiation in the species of *Partula* living on Moorea, Society Islands. *Amer. Natur.* 59:5–35.

——. 1932. Studies on the variation, distribution and evolution of the genus *Partula.* The species inhabiting Moorea. *Carnegie Inst Wash. Pub.,* 410:1–335.

DAMMERMANN, K. W. 1948. The fauna of Krakatau 1883–1933. *Verh. kon. Ned. Akad. Wetensch. Afd. Natuurk.,* Ser. 2, 44:1–594.

DARLINGTON, P. J., JR. 1938. The origin of the fauna of the Greater Antilles, with discussion of dispersal of animals over water and through the air. *Quart. Rev. Biol.,* 13:274–300.

——. 1943. Carabidae of mountains and islands: data on the evolution of isolated faunas and on atrophy of wings. *Ecol. Monogr.,* 13:37–61.

——. 1957. *Zoogeography.* John Wiley & Sons, New York.

DARWIN, C. 1855. Letter to J. D. Hooker, March 7, 1855. *In* F. Darwin (ed.), *Life and Letters of Charles Darwin.* Basic Books, New York.

——. 1859. *On the Origin of Species by Means of Natural Selection.* (Reprint of first edition, 1950.) Watts & Co., London.

DAWSON, J. W. 1958. Interrelationships of the Australasian and South American floras. *Tuatara,* 7:1–6.

——. 1960. Natural *Acaena* hybrids growing in the vicinity of Wellington. *Trans. Roy. Soc. N. Z.,* 88:13–27.

——. 1963. Origins of the New Zealand alpine flora. *Proc. N. Z. Ecol. Soc.,* 10:1–4.

——. 1964. Unisexuality in the New Zealand Umbelliferae. *Tuatara,* 12:67–68.

DOBZHANSKY, T. 1957. Genetics of natural populations. XXVI. Chromosomal variability in island and continental populations of *Drosophila willistoni* from Central America and the West Indies. *Evolution,* 11:280–293.

DOCTERS VAN LEEUWEN, W. M. 1936. Krakatau, 1883–1933. *Ann. Jard. Bot. Buitenzorg,* 46–47: xii + 506.

DURHAM, J. W. 1963. Paleogeographic conclusions in light of biological data. *In,* J. L. Gressitt (ed.), *Pacific Basin Biogeography,* p. 355–365. Bishop Museum Press, Honolulu.

ELTON, C. S. 1958. *The Ecology of Invasions by Animals and Plants.* John Wiley & Sons, New York.

ENDERLEIN, G. 1909. Die Insektfauna der Insel St. Paul. Die Insektfauna der Insel Neu-Amsterdam. *Deut. Südpolar-Exped., Zool.,* 2 (4):481–492.

FALLA, R. A. 1960. Oceanic birds as dispersal agents. *Proc. Roy. Soc., Lond.,* B, 152:655–659.

FLEMING, C. A. 1963. Paleontology and southern biogeography. *In* J. L. Gressitt (ed.), *Pacific Basin Biogeography,* p. 369–382. Bishop Museum Press, Honolulu.

FOSBERG, F. R. 1948. Derivation of the flora of the Hawaiian Islands. *In* E. C. Zimmerman (ed.), *Insects of Hawaii. Introduction,* p. 107–119. Univ. of Hawaii Press, Honolulu.

——. 1956. Studies in Pacific Rubiaceae: I–IV. *Brittonia,* 8:165–178.

——. 1963. Plant dispersal in the Pacific. *In* J. L. Gressitt, (ed.), *Pacific Basin Biogeography,* p. 273–281. Bishop Museum Press, Honolulu.

FRANKLIN, D. A. 1962. The Ericaceae in New Zealand (*Gaultheria* and *Pernettya*). *Trans. Roy. Soc. N. Z. (Bot.),* 1:155–173.

——. 1964. *Gaultheria* hybrids on Rainbow Mountain. *N. Z. J. Bot.,* 2:34–43.

GERMAIN, L. 1934. Études sur les faunes malacologiques de l'océan Pacifique. *Mém. Soc. Biogéogr.,* 4:89–153.

GODLEY, E. J. 1955. Breeding systems in New Zealand plants. I. *Fuchsia. Ann. Bot.,* 19:549–559.

GREENWAY, J. C., JR. 1958. *Extinct and Vanishing Birds of the World.* Amer. Com. Int. Wild Life Protection, New York.

GRESSITT, J. L. 1956. Some distribution patterns of Pacific Island faunas. *Syst. Zool.,* 5:11–32, 47.

——. 1961a. Problems in the Zoogeography of Pacific and Antarctic Insects. *Pacific Insects Monogr.,* 2:1–94.

——, 1961b. Zoogeography of Pacific Coleoptera. *Verh. XI. Int. Kong. Entomol. (Wien),* 1:463–465.

——. 1964. Insects of Campbell Island. Summary. *Pacific Insects Monogr.,* 7:531–600.

GRESSITT, J. L., R. E. LEECH, and K. A. WISE. 1963. Entomological investigations in Antarctica. *Pacific Insects,* 5:287–304.

GRESSITT, J. L., and S. NAKATA. 1958. Trapping of air-borne insects on ships on the Pacific. *Proc. Hawaiian Entomol. Soc.,* 16:363–365.

GRESSITT, J. L., J. SEDLACEK, K. A. J. WISE, and C. M. YOSHIMOTO. 1961. A high speed airplane trap for air-borne organisms. *Pacific Insects,* 3:549–555.

GRESSITT, J. L., and N. A. WEBER. 1959. Bibliographic introduction to Antarctic-Subantarctic entomology. *Pacific Insects,* 1:441–480.

GRESSITT, J. L., and C. M. YOSHIMOTO. 1963. Dispersal of animals in the Pacific. *In* J. L. Gressitt (ed.), *Pacific Basin Biogeography,* p. 283–292. Bishop Museum Press, Honolulu.

GULICK, A. 1932. Biological peculiarities of oceanic islands. *Quart. Rev. Biol.,* 7:405–427.

HAGEN, Y. 1952. Birds of Tristan da Cunha. *Results Norweg. Sci. Exped. Tristan da Cunha,* 20:1–248.

HAGERUP, O. 1950a. Thrips pollination in *Calluna. Kong. Danske Videnskab. Selskab. Biol. Medd.,* 18 (4):1–16.

——. 1950b. Rain pollination. *Kong. Danske Videnskab. Selskab. Biol. Medd.,* 18 (5):1–19.

——. 1951. Pollination in the Faroes—in spite of rain and poverty of insects. *Kgl. Danske Videnskab. Selskab, Biol. Medd.,* 18 (15):1–48,

HAMILTON, E. L. 1953. Upper Cretaceous, Tertiary, and Recent planktonic Foraminifera from mid-Pacific flat-topped seamounts. *J. Paleontol.*, 27:204–237.

——. 1956. Sunken island of the mid-Pacific mountains. *Geol. Soc. Amer., Mem.*, 64:1–97.

HAMILTON, T., and I. RUBINOFF. 1963. Isolation, endemism and multiplication of species in the Darwin finches. *Evolution*, 17:388–403.

HARLING, G. 1962. On some Compositae endemic to the Galápagos Islands. *Acta Horti Bergiani*, 20:63–120.

HECHT, M. K. 1952. Natural selection in the lizard genus *Aristelliger*. *Evolution*, 6: 122–124.

HEMSLEY, W. B. 1885. Endemic and arborescent Compositae in oceanic islands. *Rep. Sci. Results Voyage H. M. S. Challenger, Bot.*, 1:19–24.

HILL, J. E. 1959. Rats and mice from the islands of Tristan da Cunha and Gough, South Atlantic Ocean. *Results Norweg. Sci. Exped. Tristan da Cunha*, 46:1–5.

HILLEBRAND, W. 1888. *Flora of the Hawaiian Islands.* Privately published, Heidelberg.

HOLDGATE, M. W. 1960. The fauna of the mid-Atlantic islands. *Proc. Roy. Soc., Lond.*, B. 152:550–567.

HOLLOWAY, B. A. 1963. Wing development and evolution of New Zealand Lucanidae (Insecta: Coleoptera). *Trans. Roy. Soc. N. Z., Zool.*, 3:99–116.

JOHNSON, C. G. 1960a. A basis for a genera system of insect migration and dispersal by flight. *Nature*, 186:348–350.

——. 1960b. Present position in the study of insect dispersal and migration. *Rep. 7. Commonw. Entomol. Conf., Lond.*: 140–145.

JOHNSTON, I. M. 1953. Studies in the Boraginaceae. A revaluation of some genera in the Lithospermeae. *J. Arnold Arboretum*, 34:258–299.

KECK, D. D. 1936. The Hawaiian silverswords: systematics, affinities and phytogeographic problems of the genus *Argyroxiphium*. *Occasional Papers Bishop Mus.*, 11 (19):1–38.

KOZHOV, M. 1963. *Lake Baikal and Its Life.* Dr. W. Junk, The Hague.

KRAMER, G. 1951. Body proportions in mainland and island lizards. *Evolution*, 5:193–206.

LACK, D. 1945. The Galápagos finches (Geospizinae). *Occasional Papers. Calif. Acad. Sci.*, 21:1–151.

——. 1947. *Darwin's Finches.* Cambridge University Press, Cambridge.

LADD, H. S. 1958. Fossil land shells from western Pacific atolls. *J. Paleontol.*, 32:183–198.

LADD, H. S., E. INGERSON, R. C. TOWNSEND, M. RUSSELL, and H. K. STEPHENSON. 1953. Drilling on Eniwetok Atoll, Marshall Islands. *Bull. Amer. Asso. Petrol. Geol.*, 37:2257–2280.

LEMS, K. 1958. Botanical notes on the Canary Islands. I. Introgression among the species of *Adenocarpus*, and their role in the vegetation of the islands. *Bol. Inst. Nac. Invest. Agron. (Madrid)*, 39:351–360.

——. 1960a. Botanical notes on the Canary Islands. II. The evolution of plant forms in the islands: *Aeonium*. *Ecology*, 41:1–17.

——. 1960b. Floristic botany of the Canary Islands. *Sarracenia*, 5:1–94.

——. 1961. Botanical notes on the Canary Islands. III. The life form spectrum and its interpretation. *Ecology*, 42:569–572.

L'HERITIER, P., Y. NEEFS, and G. TEISSIER. 1937. Aptérisme des insectes et selection naturelle. *Compt. Rend. Acad. Sci. Paris*, 204:907–909.

LÖNNBERG, E. 1920. The birds of the Juan Fernandez Islands. *Natur. Hist. Juan Fernandez and Easter I.*, 3:1–17.

MACARTHUR, R. H., and E. O. WILSON. 1963. An equilibrium theory of insular zoogeography. *Evolution*, 17:373–387.

MASON, R. 1961. Dispersal of tropical seeds by ocean currents. *Nature*, 191:408–409.

MATTHEW, W. D. 1915. Climate and evolution. *Ann. N. Y. Acad. Sci.*, 24:171–318.

MAY, B. M. 1963. New Zealand Cave Fauna. II. The limestone caves between Port Waikato and Piopio Districts. *Trans. Roy. Soc. N. Z. Zool.*, 3: 181-204.

MCCANN, C. 1953. Distribution of the Gekkonidae in the Pacific area. *Proc. 7. Pacific Sci. Congr.* 4:27–32.

MCDOWALL, R. M. 1964. The affinities and derivation of the New Zealand fresh-water fish fauna. *Tuatara*, 12:59–67.

MCDOWELL, S. B., JR. 1958. The Greater Antillean insectivores. *Bull. Amer. Mus. Natur. Hist.*, 115:113–214.

MELLISS, J. C. 1875. *St. Helena.* L. Reeve & Co., London.

MENARD, H. W. 1956. Recent discoveries bearing on linear tectonics and seamounts in the Pacific basin. *Proc. 8. Pac. Sci. Cong.*, 2a:809.

MENARD, H. W., and E. L. HAMILTON. 1963. Paleogeography of the tropical Pacific. *In* J. L. Gressitt (ed.), *Pacific Basin Biogeography*, p. 193–217. Bishop Museum Press, Honolulu.

MERTENS, R. 1934. Die Insel-reptilien, ihre Ausbreitung, Variation, und Artbildung. *Zoologica*, 32:1–209.

MUNRO, G. C. 1960. *Birds of Hawaii*. Ed. 2. The Ridgeway Press, Rutland, Vermont.

MYERS, G. S. 1953. Ability of amphibians to cross sea barriers, with especial reference to Pacific zoogeography. *Proc. 7. Pacific Sci. Congr.*, 4:19–26.

——. 1960. The endemic fish fauna of Lake Lanao, and the evolution of higher taxonomic categories. *Evolution*, 14:232–333.

PERKINS, R.C. L. 1913. *Fauna Hawaiiensis. Introduction*. Cambridge University Press, Cambridge.

RAND, A. L. 1938. Results of the Archbold Expedition No. 22. On the breeding habits of some birds of paradise in the wild. *Amer. Mus. Natur. Hist. Novit.*, 993:1–8.

——. 1955. The origin of the land birds of Tristan da Cunha. *Fieldiana, Zool.*, 37:139–163.

RATTENBURY, J. A. 1962. Cyclic hybridization as a survival mechanism in the New Zealand forest flora. *Evolution*, 16:348–363.

RICE, D. W. 1964. The Hawaiian monk seal. *Natur. Hist.*, 73 (2):48–55.

RICK, C. M. 1963. Biosystematic studies on Galápagos tomatoes. *Occasional Papers. Calif. Acad. Sci.*, 44:59–77.

RIDLEY, H. N. 1930. *The Dispersal of Plants Throughout the World*. L. Reeve & Co., Ashford, Kent.

RIKLI, M. 1912. *Lebensbedingungen und Vegetationsverhältnisse der Mittelmeerländer und der Atlantischen Inseln*. Verlag Kramer, Jena.

ROBYNS, W., and S. H. LAMB. 1939. Preliminary ecological survey of the Island of Hawaii. *Bull. Jard. Bot. Bruxelles*, 15:241–293.

ROCK, J. F. 1913. *The Indigenous Trees of the Hawaiian Islands*. Privately published, Honolulu.

——. 1917. The ohia lehua trees of Hawaii. *Terr. Hawaii Bd. Agr. Forest. Bull.*, 4:1–76.

——. 1919a. A monographic study of the Hawaiian species of the tribe Lobelioideae, family Companulaceae. *Mem. Bishop Mus.*, 7 (2):1–394.

——. 1919b. The arborescent indigenous legumes of Hawaii. *Terr. Hawaii Bd. Agr. Forest. Bot. Bull.*, 5:1–153.

——. 1962. Hawaiian lobelioids. *Occasional Papers Bishop Mus.*, 23:65–75.

ST. JOHN, H. 1966. Monograph of *Cyrtandra* (Gesneriaceae) on Oahu, Hawaiian Islands. *Bishop Mus. Bull.*, 229:1–466.

ST. JOHN, H., and L. FREDERICK. 1949. A second species of *Alectryon* (Sapindaceae): Hawaiian plant studies 17. *Pacific Sci.*, 3: 296–301.

SALMON, J. T., and J. D. BRADLEY. 1956. Lepidoptera from the Cape Expedition and Antipodes Islands. *Rec. Dominion Mus.* (N. Z.) ,3:61–81.

SCHENCK, H. 1907. Beiträge zur Kenntniss der Vegetation der Kanarischen Inseln. *Wiss. Ergebn. deut. Tiefsee-Exped. "Valdivia"*, 2 (1: 2) :225–406.

SETCHELL, W. A. 1926. Les migrations des oiseaux et la dissemination des plantes. *Compt. Rend. Soc. Biogéogr.*, 3:54–57.

——. 1928. Migration and endemism with reference to Pacific insular floras. *Proc. 3. Pan-Pacific Sci. Congr.*, 1:869–875.

——. 1935. Pacific insular floras and Pacific paleogeography. *Amer. Natur.*, 69:289–310.

SHARLAND, M. 1958. *Tasmanian Birds*. Ed. 3. Angus & Robertson, Sydney.

SIMPSON, G. C. 1952. Probabilities of dispersal in geologic time. *Bull. Amer. Mus. Natur. Hist.*, 99:163–176.

SKOTTSBERG, C. 1928. Pollinationsbiologie und Samenverbreitung auf den Jan Fernandez Inseln. *Natur. Hist. Juan Fernandez and Easter I.*, 2:503–547.

——. 1936a. The arboreous Nyctaginaceae of Hawaii. *Svensk Bot. Tidskr.* 30:722–743.

——. 1936b. Vascular plants from the Hawaiian Islands. II. *Medd. Göteborgs Bot. Trädg.*, 10:97–193.

——. 1938. On Mr. C. Bock's collection of plants from Masatierra (Juan Fernandez), with remarks on the flowers of *Centaurodendron*. *Medd. Göteborgs Bot. Trädg.*, 12:361–373.

——. 1939. A hybrid violet from the Hawaiian Islands. *Bot. Notis.*, 1939:805–812.

——. 1941. Plant succession on recent lava flows in the island of Hawaii. *Göteborgs Kgl. Vetenskaps. Vitterhets-Samhäll Handl.*, Ser. b, 1 (8) :1–32.

——. 1944a. On the flower dimorphism in Hawaiian Rubiaceae. *Ark. Bot.*, 31a (4):1–28.

——. 1944b. Vascular plants from the Hawaiian Islands. IV. *Medd. Göteborgs Bot. Trädg.*, 15:275–531.

——. 1945. The flower of *Canthium*. *Ark. Bot.*, 32a(5):1–12.

——. 1953. The vegetation of the Juan Fernandez Islands. *Natur. Hist. Juan Fernandez and Easter I.*, 2:793–960.

——. 1956. Derivation of the flora and fauna of Juan Fernandez and Easter Island. *Natur. Hist. Juan Fernandez and Easter I.*, 1:193–438.

——. 1963. Zur Naturgeschichte der Insel San Ambrosio (Islas Desventuradas, Chile). 2. Blütenpflanzen. *Ark. Bot.*, 4:465–488.

SMITH, A. C. 1955. Phanerogam genera with distributions terminating in Fiji. *J. Arnold Arboretum*, 36:273–292.

STARK, J. T., and S. O. SCHLANGER. 1956. Stratigraphic succession on Guam. *Proc. 8. Pacific Sci. Congr.*, 2:262–266.

STEPHENS, S. G. 1958a. Factors affecting seed dispersal in *Gossypium*. *N. Carolina Agr. Exp. Sta. Tech. Bull.*, 131:1–32.

——. 1958b. Salt water tolerance of seeds of *Gossypium* species as a possible factor in seed dispersal. *Amer. Natur.*, 92:83–92.

——. 1963. Polynesian cottons. *Ann. Missouri Bot. Gard.*, 50:1–22.

——. 1964. Native Hawaiian cotton (*Gossypium tomentosum* Nutt. *Pacific Sci.*, 18:385–398.

STEVEN, D. M. 1953. Recent evolution in the genus *Clethrionomys*. *Symp. Soc. Exp. Biol.*, 7:310–319.

THORNE, R. F. 1963. Biotic distribution patterns in the tropical Pacific. *In* J. L. Gressitt (ed.), *Pacific Basin Biogeography*, p. 311–354. Bishop Museum Press, Honolulu.

VAN ZWALUWENBURG, R. H. 1942. Notes on the temporary establishment of insect and plant species on Canton Island. *Hawaiian Planter's Rec.*, 46:49–52.

VIETTE, P. 1948. Croisière du Bougainville aux îles australes Françaises. 20. Lepidoptera. *Mus. Nat. Hist. Natur. (Paris), Mém.*, n.s., 27(1):1–28.

—— 1952a. Lepidoptera. *Results Norweg. Sci. Exped. Tristan da Cunha*, 23:1–19.

——. 1952b. Lepidoptera. *Sci. Res. Norweg. Antarct. Exped.*, 33:1–4.

——. 1954. Une nouvelle éspèce de Lépidoptère de l'île Campbell. *Entomol. Med., Copenhagen*, 27:19–22.

——. 1959. Lépidoptères de l'île d'Amsterdam (récoltes de Patrice Paulian, 1955–1956). *Bull. Soc. Entomol. France*, 64:22–29.

VISHER, S. S. 1925. Tropical cyclones and the dispersal of life from island to island in the Pacific. *Amer. Natur.*, 59:70–78.

WACE, N. M. 1960. The botany of the southern oceanic islands. *Proc. Roy. Soc. Lond.*, B, 152:475–490.

WALLACE, A. R. 1880. *Island Life*. Macmillan & Co., London.

——. 1895. *Natural Selection and Tropical Nature*. Macmillan & Co., London.

WHEELER, W. N. 1916. Ants carried in a floating log from the Brazilian coast to San Sebastian Island. *Psyche*, 28:180.

WHITAKER, T. W., and G. F. CARTER. 1954. Oceanic drift of gourds—experimental observations. *Amer. J. Bot.*, 41:697–700.

—— and ——. 1961. A note on the longevity of seed of *Lagenaria siceraria* (Mol.) Standl. after floating in sea water. *Bull. Torrey Bot. Club*, 88:104–106.

WILSON, E. O. 1959. Adaptive shift and dispersal in a tropical ant fauna. *Evolution*, 13:122–144.

YOSHIMOTO, C. M., and J. L. GRESSITT. 1959. Trapping of air-borne insects on ships on the Pacific. II. *Proc. Hawaiian Entomol. Soc.*, 17:150–155.

——, and ——. 1960. Trapping of air-borne insects on ships on the Pacific. III. *Pacific Insects*, 2:239–243.

——, and ——. 1961. Trapping of air-borne insects on the Pacific. IV. *Pacific Insects*, 3:556–558.

——, and ——. 1963. Trapping of air-borne insects in the Pacific-Antarctic area, 2. *Pacific Insects*, 5:873–883.

——, and ——. 1964. Dispersal studies in Aphididae, Agromyzidae and Cynipoidea. *Pacific Insects*, 6:525–531.

YOSHIMOTO, C. M., J. L. GRESSITT, and C. J. MITCHELL. 1962. Trapping of air-borne insects in the Pacific area, 1. *Pacific Insects*, 4:847–858.

ZIMMERMAN, E. C. 1948. *Insects of Hawaii. Introduction*. University of Hawaii Press, Honolulu.

——. 1957. *Insects of Hawaii. 6. Ephemeroptera — Neuroptera — Trichoptera*. University of Hawaii Press, Honolulu.

——. 1960. Possible evidence of rapid evolution in Hawaiian moths. *Evolution*, 14:137–138.

JOURNAL

OF THE

WASHINGTON ACADEMY OF SCIENCES

VOL. 30   APRIL 15, 1940   No. 4

PALEONTOLOGY.—*Mammals and land bridges.*[1] GEORGE GAY-
LORD SIMPSON, American Museum of Natural History, New
York. (Communicated by C. LEWIS GAZIN.)

It was well known to the ancients that different regions of the earth
were characterized by different sorts of animal life. The Roman em-
perors seeking all manners of beasts for their diversions knew that
they must send to various countries each inhabited by characteristic
animals. Later, when European travelers began to penetrate the far
reaches of the earth, among the first questions asked them was what
peculiar creatures inhabited the deserts of Tartary or the jungles of
Ethiopia. Cartographers delighted in putting pictures of native animals on their maps, and their efforts to amaze and to embellish produced the first zoogeographic charts. Generations secure in the belief
in the creation of things as they are seldom sought any explanation of
the differences in fauna between one region and another, and few
men obscurely guessed that this might be the outcome of a shifting
history rather than the static result of divine command.

The rise of science in its modern form found here a whole series of
fascinating problems ready to hand. From a descriptive point of view
the main outlines of the present distribution of mammals were long
since correctly sketched, and now almost all the details are also
known. Confident that the processes of nature are orderly and can be
summarized by general theories and explained by general principles,
the students of the nineteenth century began the attempt to deduce
from the present faunal distribution the historical sequence that led
to it. In this new field of inference many blunders were made (and
we are surely still making some) because of the lack of historical
documents. On this basis alone, the history really can not be de-
ciphered, any more than one could reconstruct the political history of
Europe from the present boundaries of its nations if all actual records
of the past were destroyed. Here the paleontologist came to the rescue. His discoveries are the historical documents of animal dis-

[1] Address delivered before the Washington Academy of Sciences, February 15,
1940. Received February 9, 1940.

137

138 JOURNAL OF THE WASHINGTON ACADEMY OF SCIENCES VOL. 30, NO. 4

tribution. They have solved many problems in this field and at the same time they have revealed many others as yet unsolved.

Among the plainest inferences from the study of recent mammals is the fact that some of them have been able to cross regions that are now impassable to them. Aquatic animals have somehow traversed areas now dry land, and land animals have gone from one area to another now isolated by a barrier of water. The paleontologist was called on to reveal how such movements were possible and when and under what conditions they occurred.

Now geologists became vitally interested. Caring nothing about the distribution of animals as such, they care a great deal about the past distribution of land and sea, the evolution of climates, the rise and fall of connections between the continents, and other problems that are involved in or that depend on paleontological studies of distribution. Research in this field constantly assumes new aspects and touches new fields of knowledge until from being a curiously specialized and abstruse detail it has become vital for work in several different sciences and has acquired importance and meaning for anyone who takes any intellectual interest in the world in which he lives.

When I undertook to discuss this subject, it was my first intention to take up the various theoretical land bridges from one continent to another and to summarize the evidence for and against each one in order to produce a historical account of where and when such bridges have existed. It soon became apparent that such an account, if it were to have any value, would involve a mass of detail that would, indeed, be of interest only to specialists in this field. It also became evident that relatively few such specialists have risen above this mass of detail to make a conscious survey of the general principles involved and of the basic assumptions underlying their studies. Such a general survey is, then, not only of wider interest but also fresher and more needed in the present stage of study.

To review all the broad problems and principles in one paper is a manifest impossibility, and attention will be directed to two aspects on which it now seems possible and useful to make some suggestions. The first is the broadest problem of all in this field, the general way in which land mammals tend to become distributed and in which their distribution tends to change in time. The second is more particular: the different types of migration routes between major land areas, the way in which one type or another can be inferred from the faunal evidence, and the effect that a given type has on the faunas that use it. In order to lend reality to these abstractions and to point out some further promising leads for research, one specific example of

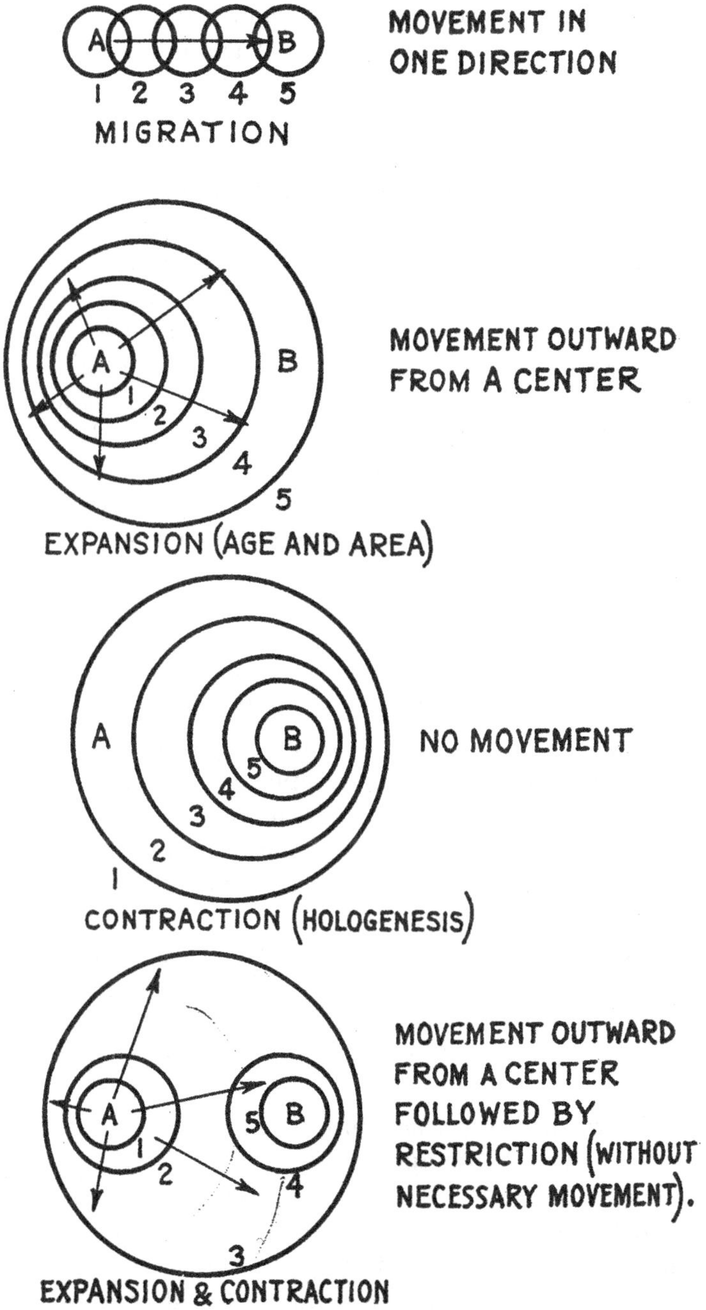

APR. 15, 1940   SIMPSON: MAMMALS AND LAND BRIDGES   139

Fig. 1.—Diagram showing various theoretical explanations of the spread of a group of mammals from one place to another. The given facts are that the group occurs at both A and B and is known at B later than at A. The numbered circles represent the limits of distribution of the group at successive times, from 1 to 5.

140   JOURNAL OF THE WASHINGTON ACADEMY OF SCIENCES   VOL. 30, NO. 4

the rise of a migration route between continents is then taken and what happened to the continental faunas as a consequence is briefly considered.

TEMPORAL PATTERNS OF MAMMALIAN DISTRIBUTION

We commonly speak of changes in mammalian distribution as being caused by migration and extinction. "Migration" suggests a trek from one area into another or periodic movement back and forth between two regions, both rare and unimportant phenomena in dealing with the broader outlines of mammalian distribution. It would be more accurate to substitute "expansion and contraction" for "migration and extinction."[2] However the words be used, it is clear that mammals do not as a rule acquire new territory simply by traveling into it but by a less purposeful peripheral expansion in all possible directions. Similarly, they do not usually lose territory simply by traveling away from it, but by a complex sequence of attenuation and local extinction that can be called contraction. (Fig. 1.)

Regarding the usual relationship of spatial distribution to time, there are two extreme theories, that of "age and area," expounded by Willis (1922), and that of "hologenesis," advanced by Rosa (1931) and supported in its zoogeographic implications by Fraipont and Leclerq (1932). Willis is a botanist and bases his theory mainly on plants but believes it probably also applicable to mammals. His basic postulate is that new forms of life originate in definite, limited regions from which as centers they expand slowly and steadily as time goes on. Then, as a rule with exceptions, at any given point in time, the area occupied by a form of life should be directly proportional to the age of that form of life. The theory involves various interesting corollaries, such as the belief that endemics or isolated forms of life with narrow distribution are usually young forms that originated where they are found and are just starting on their careers of expansion.

Rosa's theory of hologenesis, on the contrary, has the basic postulate that a new form of life appears simultaneously over a great area, over the entire range occupied by an ancestral form or predecessor. There is, then, no such thing as a center of distribution or a cradle of any form of life. The distribution, as a rule with unimportant exceptions, is at the beginning as wide as it will ever be. Migration (in any sense, or expansion), if it occurs at all, is so insignificant that

---

[2] Although, since usage makes meaning, I am not prepared to grant that "migration" can not mean what nine zoogeographers out of ten use it to mean.

the broad features of distribution are about what they would be if migration never occurred. The area covered by a form of life tends always to decrease, not to increase. Hence as a rule area is inversely, not directly, proportional to age. The corollary regarding isolated forms of narrow distribution is that they are necessarily the relicts of old groups once more widely distributed.

To a more or less orthodox zoologist Rosa's theory seems at first sight so fantastic as hardly to warrant serious discussion. This is still more true of some of the nongeographic aspects of the theory of hologenesis not pertinent here. It seems so obvious that most of the essential geographic implications of the theory are incorrect that I shall not devote time to disproving them, but it is necessary to recognize considerable merit in the work of Rosa, especially as supplemented by Fraipont and Leclerq, less on the theoretical side than in the description and emphasis of real sequences of geographic events. From this limited point of view both the age and area theory and that of hologenesis give true but incomplete pictures. One theory reaches an unsatisfactory conclusion, as far as mammals are concerned, and the other departs from an unsatisfactory postulate, but the combination of the less disputable parts of the two gives a satisfactory result.

One of the many moderate opinions intermediate between the extreme views of Willis and of Rosa is that of Matthew (1915, 1939). Matthew's main thesis, now well known, is that groups tend to spread from centers, that the marginal forms are generally conservative and the central forms progressive, and that most of the main, primary centers of such spreading have, for mammals at least, been in the Northern Hemisphere, most southern mammals being relatively primitive types pushed away from the north by peripheral expansion about these centers. This thesis is not under discussion in the present paper, but the general type of geographic history assumed by Matthew to be typical for mammals is that here more explicitly supported. Writing in 1915, before the recent denials of the existence of centers of dispersal, Matthew took these as universally admitted. His work is full of examples of contracting phases in mammalian geographic history, and it was mainly on a consideration of these that he built his theory.

As a concrete example of expansion and contraction, the distribution of the mastodonts is enlightening and was chosen by Fraipont and Leclerq as one item of evidence for hologenesis. Their map shows a Tertiary distribution essentially world-wide except for Australia

142   JOURNAL OF THE WASHINGTON ACADEMY OF SCIENCES   VOL. 30, NO. 4

and a Quaternary distribution including all North America, a spot in Ecuador, one in India, and one in Java. It is not fatal to their theory that their facts are not straight. Mastodonts entered South America only at the end of the Tertiary and were typically Quaternary all over that continent. They died out in the Old World near the beginning of the Quaternary and are typically Tertiary, only, in those continents.

The fatal flaw in the hologenetic presentation of mastodont history is not factual but in the method of generalization. Lumping the Tertiary as if it were a single point in time, they make it appear that mastodonts arose *in situ* everywhere, which is their thesis but which is certainly contrary to fact. In the Oligocene mastodonts are known only from northern Africa. Many great Oligocene faunas from other continents are known, and it is inconceivable that mastodonts or any possible ancestors of mastodonts would be (as they are) entirely unknown in them if these then already had anything comparable to their maximum distribution. Similarly it is as nearly certain as such conclusions can ever be that mastodonts were present in Eurasia (known in the Lower Miocene) earlier and thence spread to North America (not known until Upper Miocene) and that they were in North America long before they reached South America (not known until the end of the Pliocene) and spread from North America to South America. These facts are consistent with the age and area idea of expansion from a center and are radically inconsistent with the hologenetic idea of simultaneous appearance throughout the whole range.

On the other hand, as mastodonts declined it is evident that their area greatly diminished until only one or a few relicts were left in relatively limited regions. This part of the history, if taken alone, is consistent with hologenesis. It is not, in itself, inconsistent with age and area, which admits the reality of such cases as exceptions, but it becomes inconsistent if shown to be usual rather than exceptional, and this can, I think, be shown.

The accompanying map (Fig. 2) epitomizes what is known of mastodont distribution in space and time. I hold no brief for the accuracy of this map in detail: there are great gaps in knowledge, and later discoveries will necessitate changes in the distribution boundaries of the map, which are time contours or isochrones of mastodont expansion and contraction. These isochrones are, however, consistent with what is now known (which the map of Fraipont and Leclerq is not), and I venture to predict that later changes of detail will not much affect the general character of their pattern.

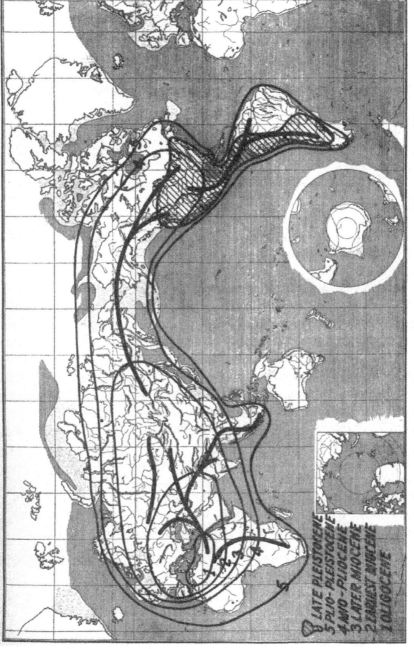

Fig. 2. –Approximate known distribution of mastodonts at various times. The numbered lines, 1 to 5, represent stages in the expanding phase, the shaded area a nearly terminal stage of the more rapid contracting phase. The heavy lines roughly represent some of the major lines of travel, or so-called migration routes.

144   JOURNAL OF THE WASHINGTON ACADEMY OF SCIENCES   VOL. 30, NO. 4

This I believe to be the type of pattern that would be shown by almost any form of life[3] that had run its entire course from origin to extinction. A form appears in some center or "cradle," not an exact spot that could be marked with a monument but, say, a single biotic district or province. Thence it tends to spread steadily in all directions until it encounters insuperable barriers. After a time it begins to contract, possibly but not usually toward its center of origin and often splitting into disjunctive spots as it contracts. Finally it disappears. (Fig. 3.)

The expansion of a group of animals involves actual motion. Individual animals must move from place to place, and some of them must travel where their immediate ancestors had never been. The population as a whole must move outward along its periphery. Contraction does not, or need not, involve any motion. It does not necessarily mean and in reality very seldom means a contraction of the population in the sense that there is predominant inward motion along the periphery. It involves rather a process of disappearance or extinction, commonly preceded by a general lowering or attenuation of the population. A population may decrease greatly and actually be well along in its contraction phase before it loses any significant amount of its range by local complete extinction. This phase of contraction can not be simply represented by contours as in the accompanying diagrams, and this essential difference between expansion and contraction must be understood if the diagrammatic representation is not to be misleading.

An excellent descriptive analogy is provided by the expansion and contraction of ice caps. In their expanding phase there is actual movement outward from a center. They may begin to contract even while the movement is still outward, but their definitive contraction is accompanied by stagnation, with thinning of the ice (attenuation of the animal population) before any considerable regression is obvious. Commonly parts of the ice mass will be isolated and remain *in situ* until they melt entirely, just as relicts of once widespread animal groups may be isolated in one or in several separate regions before they become extinct.

It is tempting to go into many of the details and corollaries of this history, but I must limit myself to mention of only one or two. As

[3] The expression "form of life" is intentionally vague for the purpose of generalization. Of course, it is not supposed that a single race or species goes through the whole course of such a cycle unchanged. General racial evolution, modification in local environments, and many other factors greatly complicate the issue. It would be impossible in limited space to attempt consideration of such modifications and it would merely confuse the broader trends that are believed to be real despite these complications.

APR. 15, 1940   SIMPSON: MAMMALS AND LAND BRIDGES   145

regards age and area, it is evident that this pattern is partly in agreement with that theory, but the theory is unbalanced in tending to stress the expansive phase as usual or normal and to consider the diminishing phase, which seems really to be an inevitable, integral part of the whole process, as unusual or abnormal. Whether a majority of animals at any given time were really distributed in accordance with age and area would depend on whether more were then in

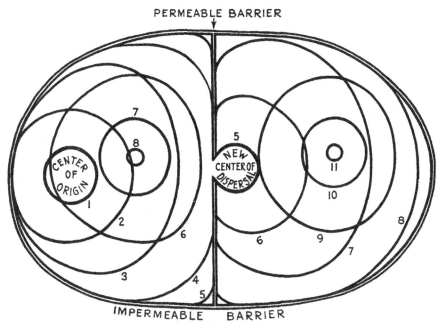

PERMEABLE BARRIER

IMPERMEABLE   BARRIER

Fig. 3.—Diagram of a common type of mammalian expansion and contraction, exemplified in varying detail by the mastodonts and other groups. The numbered lines 1 to 11, represent limits of distribution at various times. 1 to 5 represent the primary expansion of the group on the land-mass where it originated, 5 to 8 contracting phases here. At time 5 it crosses a barrier and from 5 to 8 expands on a second land-mass, contracting there from 8 to 11. From 6 to 8 the group has discontinuous (disjunctive) distribution in two areas. After 8, it is extinct in its home-land but survives abroad. After 11, it is everywhere extinct.

the expanding or in the contracting phase and on the relative speed of these phases. It seems probable that at the present time, including the recent past, more mammals are actually in the contracting phase, so that age and the area is a poor guide to the recent distribution of this particular group of animals.

One other striking detail is that we can as yet seldom follow the actual expansion of a group of mammals within its set of barriers. Sometimes related mammals do really seem to appear all at once over the whole of a great area inhabitable by them, and subsequent ex-

146   JOURNAL OF THE WASHINGTON ACADEMY OF SCIENCES VOL. 30, NO. 4

pansion, if it occurs, is not such in the simple age and area sense but is by flooding through a broken barrier: this is in part true of the spread of mastodonts into North America in the Miocene.[4] It is these cases of sudden widespread appearance that are used to support hologenesis, but the support is spurious.[5] One reason for the apparent widespread simultaneous appearance is the imperfection of the record and of our interpretation of it. For any one age we are lucky to get one good fossil deposit on a continent and almost never have deposits so placed all over a land mass that the expansion could be recorded. And even if we did have such ideal data, our usual methods of correlation would very seldom permit so precise a following of the real sequence. We usually establish a theoretical sequence by assuming (doubtless contrary to fact but as a workable approximation) that given types of mammals did appear simultaneously over the whole area—obviously it is then ridiculous to expect this sequence to show that they did not. Only when different faunas cross barriers and impinge on one another is it easy to show that expansion has occurred. The other reason is that the expansive phase of mammals is normally very rapid unless definite obstacles slow it down. Once a group of mammals gains access to a land mass, it tends to spread over it in the wink of an eye, geologically speaking. A century or a millennium may suffice, and in most cases such periods are imperceptibly short to the paleontologist.[6] Only in dealing with recent mammals is one likely really to see expansion taking place on a smaller scale.

Mammalian distribution as the paleontologist sees it is thus seldom concerned with the spread of any group on a single land mass. Relatively local differences are usually to be assigned to environmental or facial causes, while differences between larger areas are usually to be interpreted not primarily from the age and area viewpoint of simple time elapsed but more from the point of view of the rise, fall, and character of intervening barriers. The paleontologist's

---

[4] But I do not doubt that they would have passed the barrier earlier if they had reached it earlier, so that this is only a modification, not a contradiction, of the age and area type of expansion.

[5] To mention only one of several cogent reasons, because adequate data always show that new forms appear first only on one side and never on both sides of a barrier.

[6] Willis foresaw that the great mobility of mammals might vitiate the application to them of his age and area theory, which as a matter of practical observation demands that spread should be very slow, as it commonly is among plants and some animals but rarely among mammals. His thought that the theory might, after all, apply to mammals was based largely on the fact that some of them gave a "hollow curve" for number of genera plotted against number of species, but such a curve seems to me inevitable either in the expanding "age and area" phase or in the contracting phase. Perhaps in an intermediate relatively stable maximum phase it would not occur, but even this is doubtful.

cardinal principles (open to exception) are (a) that strong differences between approximately contemporaneous mammalian faunas of similar facies imply an intervening barrier and (b) that strong resemblances between such faunas denote an intervening connection.

Thus the paleontologist would seldom conclude that a given sort of mammal occurred on one continent but not on another simply because it had not had time to reach the second, but because there was no likely way for it to get there. If, then, this sort of animal did later appear on the second continent, he would normally conclude that something had happened to provide the means of getting there, and not that the animal only then got around to using the means that existed all along. These interpretive principles are widely accepted, so much so that real exceptions to them have greatly confused zoogeographers.[7] Generally true, they are the basis on which the paleontologist and zoogeographer collaborate with the geologist in establishing the probable presence or absence of land connections between the continents in past times.

TYPES AND EFFECTS OF MIGRATION ROUTES

Corridors

If no barrier at all exists between two areas, it is to be expected that their faunas will be very similar, or as far as genera or larger groups are concerned practically identical. Such radical differences as exist will be mainly or wholly caused by the survival or development of local forms in some narrow environment, that is, will be facial and not geographic in a broader sense.

As an example, a comparison of the living mammals of Florida and New Mexico (Simpson, 1936) shows the degree of similarity attained by areas in which there is no significant geographic barrier but where the local climates and facies are almost completely different in the two areas. For various reasons not pertinent here, the mammalian fauna of Florida is relatively small, with only a quarter as many species as in New Mexico, but of the orders of mammals present in Florida, all occur in New Mexico, of the families over nine-tenths, of the genera two-thirds, and of the species nearly one-fifth. If these were fossil faunas resemblance this great (or, as is often the case, greater) would warrant the conclusion that no barrier did exist between the two. This criterion can be applied in close parallel. It was formerly sometimes supposed that when Florida first definitively

[7] Cases of spread over "sweepstakes routes," discussed on a later page, are the most confusing of these real exceptions.

148   JOURNAL OF THE WASHINGTON ACADEMY OF SCIENCES   VOL. 30, NO. 4

appeared as dry land in the mid-Tertiary it was not yet connected with North America. Now we have from there Middle Miocene mammalian faunas with ten genera surely and six others doubtfully identified. Of these, all but one are common in contemporaneous beds in western North America. The conclusion that there was no sea or other notable barrier between Florida and these States is inescapable. Such evidence suggests not merely that a bridge existed but that none was needed; that the two areas were part of a single land mass.

Filter-Bridges

When two regions are separated by a strong barrier, they develop quite different faunas, the differences being roughly proportional to the lapse of time since the regions were connected. If now some means of passing the barrier appears, the two faunas intermingle, but usually the result is not the production of a single fauna even in the sense that Florida and New Mexico have one fauna. Several factors are concerned in the usual fact that such regions tend indefinitely after they are united still to have distinctive faunas, despite their sharing of some faunal elements. From this point of view the fact that the regions often are different environmentally exerts a profound effect, but one not of primary importance in the phenomena here considered because the effect might have been analogous even if the regions had always been united. A more important factor is that biological pressure of immigrant forms may inhibit the expansion of some groups in one region without being sufficient to cause rapid extinction, although in such cases extinction usually follows sooner or later. Equilibrium does occur but is seldom or never permanent.

Another and for the present subject a more important reason for the continued distinction of two faunas between which a barrier-crossing has been established is the character (including the position) of that crossing. Its approaches may be inaccessible for some animals, and of course they can not use a bridge that they can not reach. From the animals that do expand into a new land mass, it is sometimes possible to infer where the bridge was. Thus when North America and Asia had a great faunal interchange in the Pleistocene, no mammals then confined to southern North America reached Asia and none then confined to southern Asia reached North America. Obviously the bridge was in the north and exclusively southern animals could not reach it. It is also noteworthy that none of the mammals that had come into North America from South America reached Asia. To reach North America they had to come through the Tropics, and none was sufficiently adaptive also to pass over a relatively cold bridge.

Here the character of the mammals themselves is a determining factor. What is a barrier for one is not for another, and conversely what is an open route for one is not for another. The Asia–North America bridge opened the barrier for elephants (mammoths) but not for gazelles. The North America–South America bridge opened the barrier for horses but not for bison. This strongly selective action depending on the position and character of the bridge and the consequent environmental conditions of it and of its approaches is a rule with few exceptions. Another way of putting this would be to say that the true barrier in such cases was not the presence of a stretch of sea but some less obvious environmental factor, such as climate or vegetation, and that for these animals the apparent bridging of a barrier had no meaning because the true barrier remained untouched. (Fig. 4.)

In the inference of intercontinental land connections from faunal relationships it is, therefore, wrong to demand that anything like a complete faunal interchange be adduced as evidence of the existence of the connection. A wide-open, nonselective connection, a corridor, is the only sort that could approach such a result, and these are rare.[8] In the whole history of mammals there are exceedingly few cases (e.g., Lower Eocene between Europe and North America) where the evidence really warrants the inference of a wide-open corridor between two now distinct continental masses. The usual sort of connection is selective, not acting as a corridor or open door but as a sort of filter, permitting some things to pass but holding back others. From the probable mechanism of such filtering of faunas, it follows that these connections were usually of narrow environmental scope and their continental abutments limited, drawing only on one faunal zone of the continent, not on its fauna as a whole. In other words, the usual evidence for such connections does not suggest "lost continents" comprising parts of two or more as they exist today, or even broad transoceanic pathways, but relatively restricted links. The analogy of a bridge for such selective or filtering connections is fairly good, and it is to them that the term "land bridge" most properly applies.

From the point of view of paleogeography, the sort of bridge that

[8] Europe and Asia are now connected by a corridor, but zoogeographically they are not distinct continents. One of the many arguments against the Wegener hypothesis, at least in any application to mammals, is that the connections that it provides are corridors, but the faunal relationships on which it depends for evidence would not be produced by corridors.

150   JOURNAL OF THE WASHINGTON ACADÉMY OF SCIENCES   VOL. 30, NO. 4

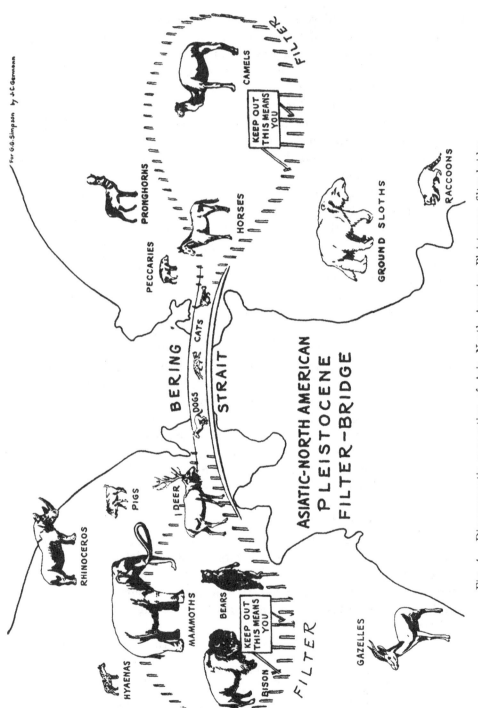

Fig. 4.—Diagrammatic conception of Asia–North America Pleistocene filter-bridge.

best fits the zoological evidence in such cases of extensive but filtered faunal interchange is an isthmian link in the sense of Bailey Willis (1932). The broad land bridges of many paleogeographers should be corridors from a faunal point of view, but isthmian links, more nearly than any other geologically postulated connections, fill the requirements of a filter-bridge, which the faunal evidence shows to be the usual type of intercontinental connection although, of course, by no means the only type.

When it is recognized that a filter-bridge does not lead to an integral transfer of continental faunas, it is a practical problem to determine what sort and degree of resemblance does indicate such a bridge. There have been students who did not hesitate to build extensive individual bridges in all directions to account for peculiarities of distribution in single forms of life. Thus, to mention only a few of his many connections,[9] Joleaud (1924 and elsewhere) has an individual Late Oligocene route from Haiti to west-central Africa for insectivores, one diagonally across this from Brazil to northwestern Africa in the Late Eocene for certain rodents, one in the Early Miocene straight across the Atlantic from the United States to Spain for a genus of horses, *Anchitherium*, one at the same time parallel to but south of this from northern Africa to Florida for the mastodonts, and so on. Similarly, von Ihering built a special bridge across the Pacific from South America to Asia for raccoons and bears,[10] and examples could be multiplied. Aside from geological considerations, which in themselves are almost enough to exclude these particular bridges at these places and times, and aside from what are now known to be errors in the factual data adduced for them, such individual, self-service bridges are supposed to have acted in a way in which no surely established bridge is known to have acted, and I can not believe in their reality.

One good criterion of the reality of a bridge is that it should have acted in both directions. Provided that both areas had land faunas, there seems to be no proved case in which a bridge has conducted animals only from one to the other and not in both directions. This is true even when one fauna was decidedly dominant and tended as a

[9] Postulated not necessarily as bridges but possibly as connections of similar effect but a different sort by an "accordion" motion of the continents on Wegenerian lines.
[10] Such a rapid summary is hardly fair either to Joleaud or to von Ihering, who adduced considerable evidence for their views (although some of the evidence has since been shown to be erroneous), but it is necessary to mention one or two instances as briefly as possible in order to demonstrate that I have not set up a straw man. Citation of the vagaries of less distinguished men would not warrant mentioning the point in so general a review.

152   JOURNAL OF THE WASHINGTON ACADEMY OF SCIENCES   VOL. 30, NO. 4

general rule to suppress the other or to inhibit its expansion. For instance, the South American ground sloths were doomed to extinction when they came in contact with the North American fauna, but first they penetrated far into North America. The armadillos, also archaic animals such as might be expected to contract in distribution, have gained an even more enduring foothold in North America and are now (for at least the second time) expanding there. One of the best arguments against the disputed derivation of South American marsupials by land bridge from Australia (direct or via Antarctica) is that the evidence favors migration only from Australia to South America, with none in the reverse direction even though the South American mammals must have been at least as capable of expansion as the Australian.[11] This and other evidence regarding this particular hypothetical migration route have been discussed elsewhere (Simpson, 1940). (Fig. 5.)

The second and perhaps the best criterion of the reality of a land bridge is that even though it rarely transports whole faunas, it does tend to transport integrated faunules. It does not transport all the genera of a continent, but neither does it transport one genus all by itself. For instance, it is improbable that only herbivores or only carnivores would cross such a bridge (although they need not both cross in the same direction). Where herbivores go, carnivores can and will accompany them, and carnivores can not go where there are no herbivores. The postulation of land bridges on the basis of one or a few mammals is thus very uncertain. Unless there is reasonable possibility that their companions have not been discovered, a theoretical bridge based on such evidence is probably unreal.

Sweepstakes Routes

There are, however, instances of migrations of single groups of mammals or of unbalanced faunas that did occur but that do not meet these criteria for filter-bridge connections and, of course, still less those for corridors. Many insular faunas are of this type, as a whole. Madagascar and the West Indies are classic examples. As carnivores, Madagascar has only peculiar viverrids, relatives of the civets, although nearby Africa is abundantly provided with cats

---

[11] It is conceivable that a bridge might function in one direction by a sort of lock or storm-door action, an otherwise uninhabited region receiving a fauna first from one source, losing that connection, and only then being united with a second continent, so that animals would be transported from the first to the second but not in the other direction. There is, however, no good evidence that such a peculiar sequence of events ever actually happened and it should hardly be postulated except in the absence of any acceptable alternative hypothesis.

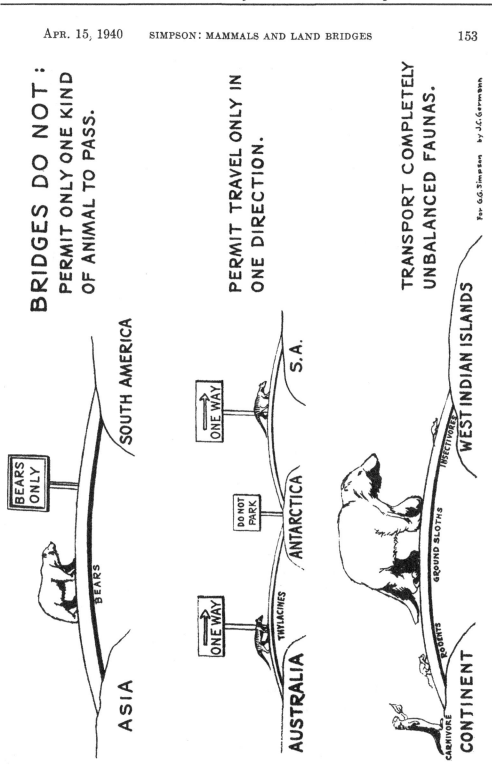

BRIDGES DO NOT:
PERMIT ONLY ONE KIND OF ANIMAL TO PASS.

PERMIT TRAVEL ONLY IN ONE DIRECTION.

TRANSPORT COMPLETELY UNBALANCED FAUNAS.

For G.G.Simpson   by J.C.Germann

ASIA

BEARS ONLY

BEARS

SOUTH AMERICA

AUSTRALIA

THYLACINES

ONE WAY

DO NOT PARK

ANTARCTICA

ONE WAY

S.A.

CONTINENT

CARNIVORE

RODENTS

GROUND SLOTHS

INSECTIVORES

WEST INDIAN ISLANDS

Fig. 5.—Diagrammatic conception of filter-bridge incompetence.

154   JOURNAL OF THE WASHINGTON ACADEMY OF SCIENCES   VOL. 30, NO. 4

large and small and various other carnivores. Madagascar's insectivores and rodents are also peculiar and each group is related to only one of many African types. Madagascar has many primitive primates and lemurs, but no apes or monkeys. These are all ancient forms and constitute a very unbalanced fauna that must have entered (whether together or separately) by the middle Tertiary at latest. The only ungulates are a pigmy hippopotamus (now extinct) and a bush-pig, both of which must have reached Madagascar much later than its other mammals and which are, again, an example of migration that can not possibly be explained by an ordinary filter-bridge. In the West Indies the Pleistocene land mammals included only peculiar rodents, insectivores, and ground sloths, without any of the ungulates, carnivores, and other groups abundant on all adjacent continental areas. This fauna, too, is inexplicable as a result of normal filtering on a land bridge such as is here envisioned. I am aware that some excellent authorities do maintain that these faunas arrived over bridges (see general summary in Schuchert, 1935), but I can not feel that they have clearly seen or considered the conditions that could give such a result. (Fig. 6.)

There are also instances of the appearance of isolated immigrants on continental masses. A curious and relatively neglected example, among many that might be cited, is that of the sudden appearance in South America of small relatives of the North American raccoon. These procyonids appear as fossils in the Late Miocene or Early Pliocene of Argentina definitely before any of the other carnivores or any of the abundant North American ungulates reached there. Since in this case a filter-bridge certainly existed at a later time, it is usual to assume that the procyonids came on this bridge and that their appearance dates the formation of the bridge as a practicable migration route or true and complete filter-bridge. If, however, we consider only the time when the procyonids did appear, disregarding our knowledge of what was destined to happen later, such a conclusion is not warranted. If my previous remarks as to filter-bridges are true, or are acceptable as a theory of general tendencies, then it is wrong to conclude that a bridge can account for the appearance of this one group of small carnivores and no other animals of similar geographic origin at that time, unless the bridge was then so nearly impassable as not to warrant the name in its usual accepted sense.

The late W. D. Matthew, who was probably the most distinguished and best informed student of problems like this, concluded that insular and highly unbalanced faunas were probably to be accounted for

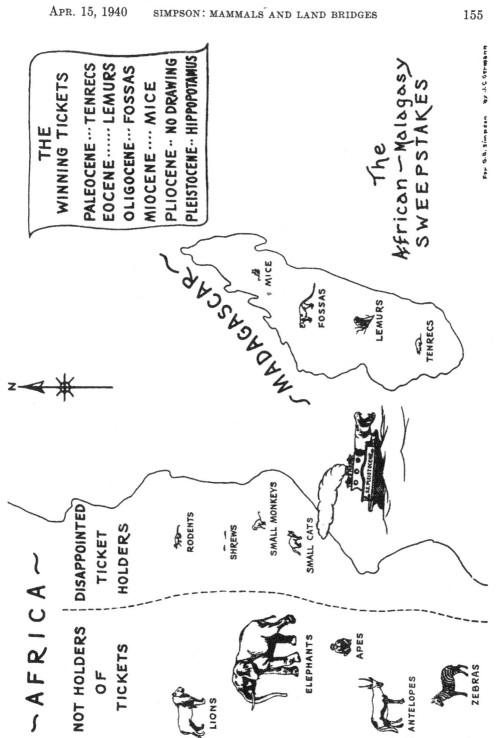

Fig. 6.—Diagrammatic conception of the "sweepstakes" route between Africa and Madagascar.

156   JOURNAL OF THE WASHINGTON ACADEMY OF SCIENCES   VOL. 30, NO. 4

by sporadic transportation of land animals on natural rafts, without the existence of a dry-land route (Matthew, 1918, 1939). This opinion has been severely criticized in some quarters. It has been claimed or felt, even by some adherents of Matthew's general thesis of "Climate and Evolution," that this sort of adventitious migration is dragged in when necessary to explain away any facts that contradict the main thesis.

It has not been sufficiently emphasized even by Matthew that the role of such a theory may be positive and primary, not merely negative and supplementary. Adventitious migration has indeed been used and sometimes abused simply to get inconvenient facts out of the way of a favored hypothesis, but there are instances in which adventitious migration is itself the most probable hypothesis and the most economical theory. In the cases of the faunas of Madagascar and the West Indies, for instance, I strongly favor this explanation, and I do so not at all in order to explain away data for a land bridge where I do not want to believe in one—as Matthew has, quite incorrectly, been accused of doing. It is to be favored because it does explain, simply and completely, facts that the land-bridge theory does not explain.

This sort of migration can be extended to include cases other than those of transportation by natural rafts, although doubtless these provide the most common instances. Any barrier, whether of water, climate, biota, or other, may or will be involved in such migration if its crossing at any one time is highly improbable but is not impossible. The action is not merely like that of a relatively less permeable filter but is different in kind as well as in intensity. A filter-bridge permits some animals to pass and holds others back, but in general those that can cross it do cross it and do so fairly soon after the bridge becomes available to them. It is relatively deterministic as to the fact of crossing, as to the animals that do or do not cross, and as to the time of crossing. An adventitious route, which I call "a sweepstakes route" to emphasize this characteristic, is indeterministic. Its use depends purely on chance and is therefore unpredictable and, except in a broad way, can not be clearly correlated with other events in time and space, as filter-migration can.

If a sweepstakes route exists, it depends on chance whether a given type of animal that can cross it will really do so, which of two types of animals will cross first, and when any particular types will cross it. It is, for instance, my belief that such a sweepstakes route for land mammals now exists between Asia and Australia, that it has existed since toward the end of the Mesozoic, and that no more tangible route, such as a filter-bridge, has existed there during that time. Certain Asiatic mammals can not follow such a route, and in this sense it, too, has a filtering action. It is not really a route for such mammals: they do not hold tickets in the sweepstakes. Other mammals, particularly small arboreal types, can. All these have tickets in the sweepstakes, some types holding more tickets than others and so having more chances, and any of them might win at any time, but any one is unlikely to do so at any one time and the less likely can win before the more likely do. A given sort of mammal might have crossed at once, might have crossed at any time from Cretaceous to Recent, or might never have crossed. Whether it crossed and when it crossed were matters of chance, in a sense almost exactly analogous to the chance of throwing a given point with dice. (Fig. 6.)

This is, I think, the only theory yet advanced that really is capable of explaining all the peculiarities of the Australian fauna and many similar but less extreme peculiarities of land faunas in other parts of the world. That such theories have not received much attention and that they are uncongenial to many zoogeographers are perhaps a reflection of the mechanistic scientific philosophy dominant in the Victorian age, from which zoogeography has not fully emerged. Land-bridge migration seems more mechanistic because it is often more simply predictable. In fact, of course, it too depends on chance, but here on the chances of a probable event, whereas sweepstakes migration depends on the chances of an improbable event.[12] The viewpoint involved is, I believe, new, and it merits detailed consideration, but this can not be given it here. Among other points, the physical nature of such sweepstakes routes needs study. It is not to be supposed that they are invariably island stepping-stones or that natural rafts are the sole means of transport involved.

A FILTER-BRIDGE IN ACTION

As an example of what actually happens when two continents are united by a filter-bridge, the case of North and South America is one of the most interesting and the facts about it are now fairly well known. These continents were separated (except, probably, for a sweepstakes route) almost throughout the Tertiary. Toward the end of the Pliocene they were united by an isthmian link antecedent and

[12] Students of statistics will recognize a relationship with the binomial of probability approaching forms like the normal distribution when chances are about equal and approaching forms like the very different Poisson distribution when chances are very unequal.

158   JOURNAL OF THE WASHINGTON ACADEMY OF SCIENCES   VOL. 30, NO. 4

similar to that now existing, the Isthmus of Panama. For the mammalian faunas this was and is a filter-bridge.

Just before the two continents were united, South America had about 29 families of land mammals and North America about 27.[13] With two doubtful exceptions,[14] they did not then have any families in common. Shortly after the union of the continents, in the Pleistocene, they had 22 families in common, 7 of South American origin, 14 North American, and 1 doubtful. Some extinction already having taken place, South America then had 17 native families still confined to it and North America 9.[15] With further extinction and some further migration, the Recent faunas of these continents have 14 families of land mammals in common and there are 15 families confined to South America (not all native) and 9 confined to North America. There was thus a great faunal interchange but one that never produced even approximately identical faunas, involving many mammals from each continent but never all or even the majority— a typical picture of the action of a filter-bridge on the continental faunas at each end of it.

In passing, there are various interesting facts involved in these summary figures. The South American fauna is now about as rich as it was before the interchange, but very different. North America has a decidedly poorer fauna than before the interchange, but its general composition has not changed so much as in South America. Both faunas reached their maximum in variety soon after the interchange and later declined.

The broad outlines of what actually happened can be seen by summary of the histories of the various major groups of mammals involved.

Certain groups expanded into the other continent and became permanently at home there, without losing much of their former range, the "age and area" type of expansion. The groups of which this is true were almost entirely of North American origin and include some rodents, especially the cricetids, most of the carnivores, and, among ungulates, the deer. Among South American mammals only the porcupine can unquestionably be placed in this category although there are one or two other less clear or more complex cases, e.g. the peba armadillo.

[13] The exact figures depend on the classification used and are not important except as they express relative values.

[14] Didelphidae and Procyonidae, possible exceptions for different reasons not affecting the basic situation here described.

[15] Some of these, on each side, did manage to spread slightly beyond the isthmus, but not to colonize the other continent widely.

Some groups seem to have been almost unaffected. It must be supposed of these that they had not reached a contracting phase in their history, that a barrier continued to exist for them (the filter-bridge filtered them out of the flow of mammals), and that the contact of new types of mammals was not lethal. This is true of a few more South American than North American mammals, but the difference is not significant. In North America the moles, pocket gophers, beavers, kangaroo-rats, prongbucks, bison, and a few others belong here. In South America the (tree) sloths, anteaters, most of the armadillos, the monkeys, and most of the native rodents (eight families out of twelve, and lesser groups in the other families) may be mentioned. Some of these managed to get onto the bridge (for instance several sorts of monkeys), but none really succeeded in crossing it.

I do not know of any single unified theory that would account well for the fact that these animals did not cross the bridge and yet did not markedly contract. The age and area theory demands that they (or most of them) be new groups that have not yet had time for this expansion, but this is clearly false. Most of these are ancient types of animals in their own continent. All certainly have had ample time to cross the bridge if they were going to do so. They are not inconsistent with Matthew's "Climate and Evolution" theory, but neither does it explain them; these data are outside that field of theory. The reasons are probably too varied to be reduced to a formula more specific than that of general filter action. For many of these animals, such as the monkeys, the absence of necessary environmental conditions beyond the bridge is an evident reason for their stopping where they did. Others, like the bison, were evidently kept by analogous environmental barriers from reaching the bridge. In some cases, for instance many of the rodents, it is hard to believe that the physical, climatic, or floral environment can have sufficed to prevent their spread and the most reasonable inference seems to be that these animals were able to maintain their places in the shifting fauna around them, in the region where they were well acclimated, but not quite able to invade the same ecologic niches where these were already occupied under somewhat different conditions, even though these conditions would not have been deterrent if there were no competition. The explanation is vague and not very satisfactory because it seems unlikely that so delicate an equilibrium could long be maintained.

Some groups began to contract at or soon after the time of faunal interchange. Doubtless some would have contracted anyway, but it

160   JOURNAL OF THE WASHINGTON ACADEMY OF SCIENCES   VOL. 30, NO. 4

can not be coincidence that so many did so just at this time. In North America there were few examples of this. Some North American groups have contracted since the connection with South America, but in these cases there is little doubt that the contraction had quite different causes and would have occurred regardless of the rise of the land bridge in question. In South America, however, the sharp contraction and eventual extinction of all the native carnivores (the borhyaenid marsupials) and all the native ungulates (notoungulates and litopterns) undoubtedly were related to this event and so, probably, was the contraction, with or without extinction as yet, of various native rodents and of the caenolestid marsupials.

It is highly improbable that none of these animals could have crossed the bridge successfully as far as most environmental factors go. Some of the notoungulates and borhyaenids, for instance, were ecologically similar to animals that did cross the bridge and they lived in environments abundantly available in North America. The only probable explanation is that these animals were biologically inferior to immigrants from North America. The impact of the latter not only prevented the expansion of these South American groups but also started or hastened their contraction. The contraction was slow in some cases, occupying a million years or more, but it effectively prevented acquisition of new territory and in most cases has now ended in extinction. In this instance, and probably this is the rule for mammals, expansion of groups that did expand was plainly more rapid than the contraction of those that did contract.

A final category is provided by the various sorts of mammals that expanded when the continents were united but that later contracted again. This was true of about as many North American as South American mammals. It is a phenomenon still more complex than those already mentioned and the land mammals so affected may be placed in three categories:

1. Those that expanded into the other continent and then became extinct in both:
   (a) Of South American origin: Glyptodonts, ground sloths (several families).
   (b) Of North American origin: Gomphotheres (bunodont mastodonts),[16] horses.

2. Those that expanded into the other continent and then contracted (or in one case became extinct) there but were not much restricted in their original home:

---

[16] These were not ultimately of North American origin, but those involved in this interchange were. Throughout this discussion North American origin means simply not South American, only these two continents being considered.

(a) Of South American origin: Capybaras, armadillos.
(b) Of North American origin: None.
3. Those that expanded into and survived in the other continent and became restricted or extinct in their continent of origin:
    (a) Of South American origin: None.
    (b) Of North American origin: Tapirs, camels, peccaries, short-faced bears.[17]

The first of these three categories can be dismissed (although hardly explained) as including groups that would have become extinct in any case but that happened to share in this last expansive movement before fatal restrictions set in. The last two are complementary and show an interesting relationship. No North American groups became extinct in South America and not in North America. If they became extinct in South America they were, so to speak, slated for extinction anyway and the new environment did not save them. On the other hand several North American groups became extinct at home but not in South America.[18] These were, then, contracting groups, for which extinction was postponed by the change of environment. South America was an asylum for them in their retreating phase and the preceding expansion was rather an incident than an indication of potency against their old environment. Here again both the age and area and the hologenetic theories are far beside the point when confronted by the actual facts. On the other hand, this particular class of facts is broadly consistent with Matthew's views, especially when details here omitted are considered.

South American groups that were contracting, or were destined soon to contract, in that continent either were unable to reach or, in rarer cases, did reach but could not survive in the northern continent. Even some animals that remained potent and at least did not markedly contract in South America were unable to maintain themselves in North America after reaching there. Generally speaking, the faunal interchange was far from equal. In the long run the two faunas did not mingle as much as one invaded the other. The North American mammals were on the whole definitely more potent and more expansive than the South American, both in their ability to migrate and in their ability to survive, a generalization supported by the following tabulations, in which the figures are numbers of families of land mammals known to have existed in the two continents at about the time when the bridge arose. (Doubtful cases are omitted.)

[17] These have also contracted considerably in South America.
[18] I include peccaries in this group because they contracted greatly in North America (also, but to far less extent, in South America) and in all probability would have become extinct in the north if no asylum had been offered them, and may indeed still become extinct first in the north.

162 JOURNAL OF THE WASHINGTON ACADEMY OF SCIENCES VOL. 30, NO. 4

TABLE 1.—ASSOCIATION OF MIGRATION AND SURVIVAL WITH GEOGRAPHIC ORIGIN

	Of South American origin	Of North American origin	Ratio of ratios, favoring North America
Migrated to other continent............	7	14	
Did not migrate to other continent......	21	11	
Ratio................................	.33	1.27	3.8
Now surviving........................	17	21	
Now extinct..........................	11	4	
Ratio................................	1.55	5.25	3.4

These differences are statistically significant, the first surely, the second probably.

A priori it would be expected that the ability to accomplish such a migration, an indication of expansive power at the time, would be related to ability to survive. It is possible that there is a relationship here, but if so it is more complex and involves other factors. Simple tabulation of the same families shows no such tendency:

TABLE 2.—ASSOCIATION OF MIGRATION WITH SURVIVAL

	Migrated to other continent	Did not migrate to other continent
Now surviving....................	15	23
Now extinct......................	6	9
Ratio...........................	2.50	2.56

The difference is far from significant. As far as these figures show, a family capable of spreading to the other continent was no more likely to survive than one that did not spread.[19] Thus in the final outcome of the interchange, as far as yet reached, the ability of these faunas to expand and their ability to survive are both associated with geographic origin, or biologically with the general character of the historically northern, Holarctic, as opposed to the historically southern, Neotropical, fauna. But ability to expand and ability to survive are two different faunal characteristics in this instance with no apparent relationship to each other.

Like so many phases of this great subject on which I have barely been able to touch in passing, this unexpected conclusion has far-reaching implications and merits much more detailed consideration than can now be given it. An enormous amount of work has been done to unearth the facts of faunal distribution in the past and present.

[19] Use of smaller taxonomic units, such as genera, gives larger figures but obscures the conclusion sought. Commonly the act of spreading from one continent to the other was accompanied by evolution of generic rank. The use of actual phyla would be ideal but is impractical because these are not sufficiently well known in many cases.

Far less progress has yet been made in finding the broad interpretive principles that may be revealed by these facts. Here an effort has been made to indicate what a few of these principles may prove to be and, more particularly, to suggest a few of the lines of attack that may lead to clearer grasp of these and to the discovery of others.

LITERATURE CITED

FRAIPONT, C., and S. LECLERQ. *L'évolution. Adaptations et mutations. Berceaux et migrations.* Actualités Scientifiques et Industrielles **47**: 1–38. 1932.

IHERING, H. VON. *Die Umwandlungen des amerikanische Kontinentes während der Tertiärzeit.* Neues Jahrb. Min. Geol. Pal., Beil-Bd. **32**: 134–176. 1911.

JOLEAUD, L. *L'histoire biogéographique de l'Amérique et la théorie de Wegener.* Journ. Soc. Amér. Paris, n.s., **16**: 325–360. 1924.

MATTHEW, W. D. *Climate and evolution.* Ann. New York Acad. Sci. **24**: 171–318. 1915. (Reprinted, with additions, in Matthew, 1939.)

—— *Affinities and origin of the Antillean mammals.* Bull. Geol. Soc. Amer. **29**: 657–666. 1918. (Reprinted in Matthew, 1939.)

—— *Climate and evolution*, ed. 2, revised and enlarged. New York Acad. Sci., Special Publ. **1**: i–xiii, 1–223. 1939. (Also several other papers bearing on this subject.)

ROSA, D. *L'ologénèse. Nouvelle théorie de l'évolution et de la distribution géographique*, pp. i–xii, 1–368. 1931. Félix Alcan, Paris. (Revised and translated from the Italian edition, 1918.)

SCHUCHERT, C. *Historical geology of the Antillean-Caribbean region or the lands bordering the Gulf of Mexico and the Caribbean Sea.* John Wiley & Sons, New York, 1935.

SIMPSON, G. G. *Data on the relationships of local and continental mammalian faunas.* Journ. Pal. **10**: 410–414. 1936.

—— *Antarctica as a faunal migration route.* Proc. Pacific Congress of 1939, in press. 1940.

WILLIS, B. *Isthmian links.* Bull. Geol. Soc. Amer. **43**: 917–952. 1932.

WILLIS, J. C. *Age and area. A study of geographical distribution and origin of species*, pp. i–x, 1–259. University Press, Cambridge (England), 1922.

PALEOBOTANY.—*The Pliocene Esmeralda flora of west-central Nevada.*[1] DANIEL I. AXELROD. (Communicated by RoLAND W. BROWN.)

One of the results of recent collections of later Tertiary floras over the Great Basin province has been the discovery that the Esmeralda flora described by Knowlton (1900) from the northern end of the Silver Peak Range, Esmeralda County, Nev., is distinct from the Coal Valley flora reported by Berry (1927), which lies in the drainage of the East Walker River 75 miles northwest. A well-preserved flora of approximately 50 plants has been collected at Coal Valley and will form the basis of a subsequent paper. The present brief report adds six species to the Esmeralda flora and includes an analysis of previously collected material now at the United States National Museum. Acknowledgement is made to Dr. Roland W. Brown for assistance in examining the collections, to the Carnegie Institution of Washington under whose auspices the collections were made, and

[1] Received November 20 1939.

Palaeogeography, Palaeoclimatology, Palaeoecology
Elsevier Publishing Company, Amsterdam – Printed in The Netherlands

THE BEARING OF CERTAIN PALAEOZOOGEOGRAPHIC DATA ON CONTINENTAL DRIFT

ANTHONY HALLAM

Grant Institute of Geology, University of Edinburgh, Edinburgh (Great Britain)

(Received November 1, 1966)

SUMMARY

Data concerning the distribution of land vertebrates and marine molluscs from the Triassic to the present are utilised in an investigation of former continental relationships. Factors controlling the distribution of living representatives of these groups are discussed and it is indicated how regions of deep ocean restrict the distribution of the predominantly neritic molluscan faunas almost as effectively as they do land animals. Complementarity in distribution patterns in the continental and marine faunas resulting from the emergence of land bridges proves an especially useful tool in interpreting fossil distributions.

The information derived from living (together with recently extinct) faunas is applied to the study of older faunas. The distribution of many Mesozoic faunas appears to demand the existence of land and/or shelf sea connections between continents where none exists today. The old concept of trans-oceanic land bridges being considered unsatisfactory, an interpretation based on Late Mesozoic and Early Tertiary continental drift is preferred, taking into account other geological and geophysical data. None of the palaeozoogeographic data surveyed is obviously incompatible with drift, and much is difficult to explain without it.

Palaeozoogeographic data are thought to be especially useful in establishing a timetable for continental drift. In the hypothesis proposed it is argued that Gondwanaland did not begin to disintegrate substantially until well into the Cretaceous. The opening of the North Atlantic could possibly have taken place earlier but data are as yet inadequate to decide. The best evidence so far comes from the North Pacific region and suggests that a Bering land bridge was established by Middle or Late Cretaceous time. During the Early Tertiary the Australasian continent is considered to have moved northwards towards Indonesia and the African–Arabian continent rotated slightly anticlockwise, impinging upon Asia Minor and hence dividing the Tethyan seaway into two. The onset of a significant amount of drifting in the Middle and Late Cretaceous might be bound up with the growth of a major oceanic ridge system, which also had the effect of causing sea level to rise substantially.

A. HALLAM

INTRODUCTION

Palaeontological evidence of the former distribution of animals and plants played a notable role in Wegener's hypothesis (1924) of continental drift. A separate chapter of his famous book, *The Origin of Continents and Oceans*, was devoted to this evidence, which led him to the conclusion that the various components of his "Pangaea" had remained united until the Late Mesozoic or ever later.

Nevertheless, palaeobiogeography has been conspicuously neglected in the many discussions provoked in the last few years by the resurgence of interest in continental drift. Nowhere is this better illustrated than in the recent Royal Society symposium (BLACKETT et al., 1965). Not one of the numerous articles in this publication is devoted to this topic, though it does receive a cursory mention in one of them (by T. S. Westoll) and the following comment in E. C. Bullard's concluding remarks to the symposium: "The most troublesome differences of opinion about the interpretation of the facts relate to the distribution of plants and animals in the past." Evidently palaeobiogeography is currently regarded as the poor relation of geology and geophysics.

The reasons for this state of affairs are not far to seek. Westoll remarked, in the article referred to above, that "Perhaps more dubious and unconvincing "evidence" has been published in this field by enthusiastic "drifters" than in all the other fields put together." Unfortunately it is hard to disagree with this statement. Too much work has been of a low critical standard and so has tended to bring the whole subject into disrepute. Certain biologists have made extravagant claims about the past distributions of land and sea without reference to geology, and certain geologists have made use of biogeographic data without an adequate knowledge of the factors controlling the distribution of organisms.

A second reason is that, even in those instances where the evidence for past land connections appeared to be good, resort was made to the alternative of transoceanic land bridges. It is perhaps not sufficiently appreciated today just how respectable such land bridges were to a former generation of thoughtful and erudite palaeontologists, to whom the idea of continental drift was too radical.

Finally, there are a number of difficulties intrinsic to the subject. Many organisms leave a poor fossil record, and important discoveries in all groups are inevitably subject to a considerable element of chance. Factors controlling distribution differ significantly with different groups of organisms, so that broad generalisations need treating with especial caution. Good biogeographic research depends upon good taxonomy, which is not always available and anyhow involves a considerable subjective element.

Despite these difficulties the author believes that pessimism about the validity of palaeobiogeographic evidence is premature and that the time is ripe for a further appraisal of some of this evidence. Within such a vast field individual workers must severely restrict the scope of their investigations. In this paper attention is confined

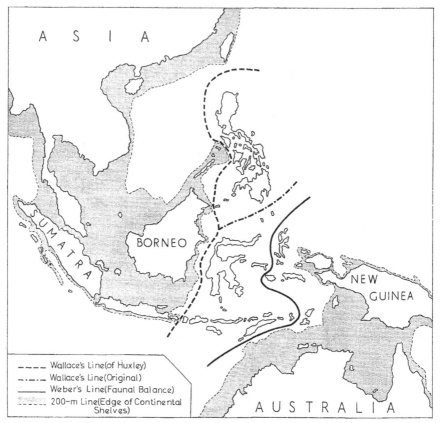

Fig.1. Zoogeographic boundaries in the Indonesian–Australian region. Data from MAYR (1944).

to certain pertinent data concerning continental vertebrates and marine molluscs, Mesozoic or younger in age. This choice is justified on the following grounds.

The subject of this study is continental drift in the sense understood by WEGENER (1924), which he claimed took place since the Palaeozoic. Vertebrates have been studied far more intensely than any other group that leaves a relatively good fossil record. The distribution of continental vertebrates is controlled more by the presence or absence of land connections than by climate, whereas the reverse is true for plants (SIMPSON, 1952). Thus there are no pan-tropical mammals, as there are pan-tropical plants. MAYR (1953) gave a good illustration of this difference. The plants of New Guinea are like those of Malaysia, which has a similar climate, and unlike those of Australia, which has a different climate. The New Guinea vertebrates, however, are similar to the Australian but strikingly different from those of Malaysia, Java, Sumatra, and Borneo, which lie on the Sunda Shelf, separated from the Sahul Shelf of Australia and New Guinea by deep marine straits (Fig.1). It is apparent that vertebrates are likely to be more reliable indicators of past land connections than plants.

The marine molluscs are among the best-studied of invertebrates and the shell-bearing species have an excellent chance of preservation in the fossil record. Most of them are confined to shelf seas and cannot readily cross wide oceanic barriers. Information on vertebrates and molluscs can be used for purposes of cross-checking in a way that will be explained later.

Inevitably the author has had to rely upon the opinions of experts for the various groups concerned. Much of the older literature bearing on past continental relationships, as summarised for instance by ARLDT (1919–1922), has required or still requires critical revision in the light of modern knowledge and the more rigorous standards of today. Attention is therefore confined in this study to the work of modern specialists. While this is of course no guarantee of definitive statements there is the considerable advantage that none of these specialists has been an advocate of continental drift (quite the reverse in some cases). This diminishes the likelihood that awkward facts unfavourable to the drift hypothesis have been suppressed. Many of the data discussed in this paper have not in fact been utilised previously in discussions about continental drift.

The procedure to be followed will be to outline the principal factors controlling the distribution of continental vertebrates and marine molluscs as understood by zoogeographers at the present time, and then to apply this knowledge to the understanding of fossil distributions in terms of land and sea connections in the past. Consideration of the broader and more controversial question of former continental relationships will await a systematic account, period by period, of faunal affinities which is largely descriptive, or rather involves interpretation at a comparatively humble level.

FACTORS CONTROLLING FAUNAL DISTRIBUTION

Continental vertebrates

Though much can be deduced from the distribution of living animals, Recent zoogeographic data need to be supplemented by evidence from fossils for an adequate understanding of the factors controlling distribution. This is well illustrated by the case of the so-called primitive or relict vertebrates occurring in the southern continents. These animals include (among the mammals) the monotremes and marsupials of Australia and New Guinea, the marsupial opossums of South America and the lemurs of Madagascar; also the giant ratite birds and lungfish of South America, Africa and Australasia, and a number of amphibians and reptiles.

The distribution of these undoubtedly primitive types has led to the common belief of former times that the southern continents had direct land connections. This view has been criticised on the grounds that fossils of these groups are known also in the Northern Hemisphere. Thus fossil lungfish, closely related to the living three genera, indicate a world-wide distribution in Mesozoic times, and fossil marsupials are known from the Cretaceous of the Northern Hemisphere. MATTHEW (1915)—concerning himself strictly with mammals—was a leading exponent of the view that relict organisms have a peripheral distribution with respect to the continents, well away from the main centres of evolution in the Northern Hemisphere (Holarctic Realm) from which they have been driven, and that the present distribution of animals can be adequately accounted for by a distribution of continents much like the present one. These ideas have become widely accepted, though with inevitable modifications (e.g., DARLINGTON, 1948).

Besides competition from other animals, vertebrate distribution is controlled primarily by climate and geography. In all classes, diversity of types is greater in low than in high latitudes. Only the mammals and birds, being warm-blooded, have representatives that are well adapted to the cold polar regions. The reptiles are the least tolerant of cold, though they disperse more readily than the amphibia and fresh-water fish (DARLINGTON, 1948). It should be borne in mind however that the strong climatic zonation of the present time is untypical of earth history and that during the Mesozoic and Early Tertiary the restraining influence of low temperature may have been appreciably less than today in most parts of the world.

A much more potent restraint is imposed by geographic barriers of one kind or another. The most important of these in the present context are marine barriers. MYERS (1938, 1953b) has distinguished two groups among the so-called true freshwater fish. The *primary* group consists of families rigorously restricted to fresh water and includes the lungfish. The *secondary* group includes forms that have been able to cross short marine barriers and reach, for instance, the Greater Antilles and Madagascar, though not New Zealand. Myers considered that the primary fresh-water fish are the most useful of all groups in vertebrate zoogeography.

Amphibia are also severely restricted by barriers of sea water, though frogs are comparatively mobile, one group even having reached New Zealand (MYERS, 1953a). In the case of reptiles and mammals, the chance of successfully negotiating a marine barrier varies greatly according to the type of animal. Thus small arboreal rodents stranded on drifting logs have a vastly greater chance than, say, elephants.

The American palaeontologist Simpson has made notable contributions to post-Mesozoic mammalian zoogeography and the principles he has outlined have considerable relevance to other vertebrate classes as well.

SIMPSON (1940a) argued that strong differences between approximately contemporaneous faunas of similar facies imply the presence of an intervening barrier, while strong resemblances between such faunas denote an intervening connection. SIMPSON (1947) made an attempt to give quantitative expression to the degree of faunal resemblance between two such faunas. If N_1 represents the smaller fauna, N_2 the larger and C the number of taxa in common, then the degree of resemblance can be expressed as: $100 \ C/N_1$, which gives a percentage.

206 A. HALLAM

While the desire for quantitative expression is commendable in itself there is perhaps a danger of a spurious air of precision being introduced by such indices as the one quoted above, since the basic data consist of subjectively-determined taxa. Qualitative expressions of degrees of resemblance, preferably by more than one authority, may therefore be the best one can hope for in the present state of taxonomic knowledge.

Land animals can migrate between areas partly or wholly separated by sea by three distinct means.

(*1*) Corridors. These are land bridges which allow free communication of a large proportion of the fauna in both directions. A good example is the land bridge that developed in the Late Pliocene linking the two American subcontinents (SIMPSON, 1940a, b). In Earlier Tertiary time South America was isolated by sea in the Central American region, as shown by geological evidence. A highly distinctive endemic mammalian fauna developed there undisturbed by outside competition; it was dominated by borhyaenids (marsupial carnivores), notoungulates and liptoterns (placental ungulates) and glyptodonts. Immediately prior to the union with North America there were about twenty-nine known families of mammals in South America and about twenty-seven different ones in North America. After union, the Pleistocene faunas show twenty-two families in common. There was a major invasion of North American forms, including mastodonts, tapirs, camels and dogs. The native ungulates and carnivores of South America became extinct and were replaced by northern types with a similar mode of life. The traffic was not one-way, however, since porcupines, armadillos and ground sloths moved northwards.

Another example of a corridor is the Bering land bridge in Eocene times. The interchange of mammals across this land bridge involved the major part of the whole fauna, and North America and Asia formed essentially a single zoogeographic province at this time (SIMPSON, 1947; Fig.2).

(*2*) Filter bridges. These differ from corridors in that climate or some other factor "filters out" certain elements among the potential migrants. Thus although there was a pronounced interchange of mammals across the Bering land bridge during the Pleistocene it only involved a minority of the total fauna of the two continents. Warmth-loving forms were excluded because of the cold climate (SIMPSON, 1947).

(*3*) Sweepstakes routes. Unlike corridors and filter bridges, these involve chance migration across the sea on natural rafts. Only a small minority of the available fauna obtains winning tickets in the sweepstakes and the result might be the migration of an unbalanced fauna, i.e., herbivores will not be accompanied normally by their natural predators. It is in this way that oceanic islands are colonised. Primitive groups may flourish in the absence of competition and come to occupy a wide variety of ecological niches.

New Zealand provides an outstanding example of a land area that has been colonised by this process. It lacks indigenous mammals, snakes, tortoises and fresh-

BEARING OF PALAEOZOOGEOGRAPHIC DATA ON CONTINENTAL DRIFT   207

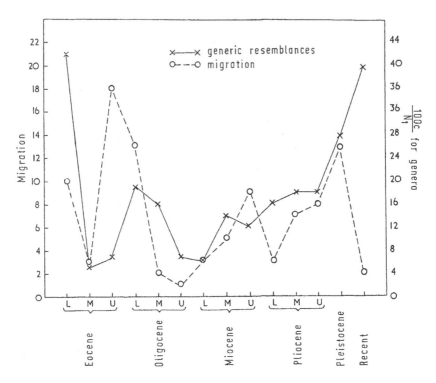

Fig.2. Generic resemblance and migration curves for Eurasian and North American Tertiary mammals. The generic resemblance index is discussed in text. Simpson's migration index is proportional to the presumed freedom of migration. Adapted from SIMPSON (1947, fig.4).

water fish, though it has a wide variety of unusual birds. There are a few groups of frogs and lizards of primitive types, of which the best known is the rhynchocephalian "living fossil" *Sphenodon*. The evidence clearly points against a land connection since the beginning of the Tertiary and possibly for much longer.

The case of Madagascar is more controversial, and both WEGENER (1924) and DU TOIT (1937) argued that its mammalian fauna demanded a connection with Africa during mid-Tertiary times. This contention has been strongly disputed by both MATTHEW (1915) and SIMPSON (1940a, 1943). Simpson maintained that the nature of the Madagascan mammalian fauna is precisely what is to be expected of sweepstakes route colonisation.

The fauna is of limited variety. Among insectivores there are only tenrecs and one shrew, among primates only lemuroids and among carnivores only viverrids. Rodents are limited to cricetids and artiodactyls to a bush pig and pigmy hippopotamus (now extinct). Large animals such as flourish on the African mainland are absent. If there had been a land bridge with Africa at any time during the Tertiary there should be a far more diverse and better-balanced fauna.

208   A. HALLAM

A similar pattern is discernible in other vertebrate classes. Thus there are no venomous snakes, caecilian and urodele amphibians, and chameleons have flourished in the absence of competition in a way comparable to the lemurs. The affinities of the fauna are clearly with Africa and not with the Orient (MILLOT, 1952).

SIMPSON (1952) has pointed out that migrations across the sea should be thought of in terms of probability and has stressed the importance of geological time in this respect. Thus, if the chance that any individual animal will cross a geographic barrier in any one year is, say, one in a million, or $p = 0.000001$, then in a million years $p = 0.99995$. Clearly, even with a low probability of dispersal, a dozen or less successful sweepstakes colonisations of Madagascar by mammals within about seventy million years seem distinctly plausible.

Another important instance which has been quoted in the continental drift controversy is that of Australia and New Guinea. The vertebrate fauna, birds included, is strikingly different from that of the Oriental Province, which includes Malaysia, Sumatra, Java, and Borneo. The mammals, for instance, are restricted to monotremes and marsupials, with the exception of a few bats and rodents which immigrated in the Late Tertiary or Pleistocene. It is now generally agreed that the evidence indicates that the Australian region has been isolated by sea since the beginning of the Tertiary or the Late Cretaceous (the age of the first fossil marsupials) but there has been dispute about the geographical position of Australia and New Guinea with respect to the islands of the Sunda Shelf. This is a matter which will be taken up later, but it is desirable at this point to indicate where modern zoogeographers would place the boundary of the Australian and Oriental vertebrate faunal provinces. The contrast between the two faunas was first brought to general notice by the "father of zoogeography" A. R. Wallace and the line of separation he proposed, between Bali and Lombok and between Borneo and the Celebes, was termed "Wallace's Line" by Huxley, who made a slight modification (Fig.1). MAYR (1944) has proposed that Weber's Line of Pelseneer would be a better zoogeographic boundary, since this corresponds to the line of faunal balance (Fig.1). To the east the fauna is dominantly Australo–Papuan, to the west dominantly Oriental. Wallace and earlier zoogeographers were misled because the Celebes and neighbouring islands have greatly impoverished insular faunas. The significance of Wallace's Line is that it marks a deep marine strait which must have remained a seaway even during the maximum Pleistocene lowering of sea level.

Marine molluscs

The vast majority of marine molluscs are inhabitants of shelf seas or the neritic zone. As with vertebrates their distribution is controlled primarily by climate and geography.

There is a strong correlation between faunal diversity and latitude, with the most varied faunas occupying the tropics (FISCHER, 1961) and it is evident that low temperatures restrict the spread of a wide variety of forms. Once more, it should be borne in mind that latitudinal climatic zonation in the Mesozoic and Lower Tertiary was apparently weaker than today.

Climatic barriers are gradational and only perceptible over a large distance. In this respect they tend to differ from geographic barriers. The latter are of two sorts.

(*1*) Land barriers. These may impose sharp restrictions within very limited areas, so that contrasted faunas may evolve separately within only a short distance of each other. A good example is provided by the faunas on the Pacific and Atlantic sides of the Central American land bridge. These faunas differ widely at the specific but are closely related at the generic level; there are many so-called geminate or twin species suggesting derivation from a common ancestor (EKMAN, 1953). A direct marine connection must have existed until the recent geological past, that is, prior to the end-Tertiary emergence of the land bridge. Another striking example concerns the faunas of the Mediterranean and Red Seas. These have only eight species of molluscs in common and most of these can be traced all around the coast of Africa (DAVIES, 1934).

(*2*) Deep sea barriers. The eastern and western shores of the Atlantic Ocean have only a very small percentage of warm-water species in common. The East Pacific, which has the world's greatest expanse of uninterrupted deep sea, corresponds with the most pronounced break in the world's warm-water fauna, perceptible even at the generic level (EKMAN, 1953). The reason for this lies in the limited dispersal potentialities of the larvae, a subject gone into thoroughly by Thorson.

THORSON (1961) studied data concerning the larvae of about 200 invertebrate species and found that about 80% settle after less than 6 weeks of pelagic life; only 5% remain in the plankton for more than three months. Even if these times are doubled, to allow for especially favourable conditions, it remains true that only a minute fraction qualify as long-distance larvae. This is made clear by reference to the average velocities of marine surface currents. Thus it takes an average of 22–23 weeks for water to travel from Cape Hatteras to the Azores and 19–20 weeks to travel from Somalia to the west coast of India. Thorson concluded that the present pattern of distribution of coastal faunas must depend partly on what he called "transport miracles". Even so, only a fairly small percentage of the total Indo-West Pacific coastal fauna has reached Hawaii despite the abundance of intermediate islands.

Especially significant for the present purpose is Thorson's conclusion that the Bivalvia (or lamellibranchs) are the least capable of long-distance transportation of all the invertebrate groups studied. This is because they mostly have a very short larval life and because the heavy larval shells sink readily upon slight disturbance. The same is true of the shell-bearing prosobranch gastropods, with the exception of a few specialised groups whose larvae might live in the plankton for several months. These include genera such as *Cassis* and *Cypraea* and it is precisely among such forms that wide-ranging circumtropical species are known.

A restricted number of neritic animals can cross deep oceans by attaching themselves to floating seaweed or driftwood, and the importance of this factor in the colonisation of Hawaii is stressed by EKMAN (1953, p.21). Among the mollusca, many of the anisomyarian bivalves have the potential to travel in this way by means of byssal attachments. The possible importance of this process in the past is signified by the discovery of numerous *Inoceramus* associated with driftwood in the Liassic *Posidonienschiefer* of southern Germany, generally thought to be a euxinic deposit (HAUFF, 1953). This particular group apart, it would seem to be a fair generalisation that strong affinities between bivalve faunas of different areas signify shelf sea connections, or at least deep-water barriers of negligible width.

In the case of the ammonites, which have figured largely in palaeozoogeographic literature, one is limited to speculation. It is probable that many ammonites could tolerate a wide range of depth like the living *Nautilus*. This, together with their undoubted swimming ability, must have rendered them much more capable of wide dispersal than most benthonic molluscs, though it must be doubted whether they were capable of crossing wide oceanic barriers. It is thought unlikely that post-mortal transport of shells was of more than limited importance. Empirical data suggest, of course, that many ammonite species were able to disperse widely. While this renders them excellent for purposes of correlation it reduces their importance for palaeozoogeography.

Complementarity in the distribution of vertebrate and molluscan faunas

Consideration of the evidence of both continental and marine faunas allows a two-pronged attack, as it were, on problems of former land and sea relationships. This is especially well seen in those instances where the making or breaking of a land bridge has had contrasting effects on the two types of fauna.

Once again, the Central American land bridge provides an excellent example. As already mentioned, its creation at the end of the Tertiary had the joint effect of allowing free migration of land animals between two formerly isolated landmasses, and dividing into two a marine fauna that had formerly been united. Another example is the Bering land bridge in the Early Tertiary. The evidence of fossil vertebrates from Eurasia and North America suggests that these two regions became isolated during Middle Eocene times, following extensive interchange in the Early Eocene (SIMPSON, 1947). The Middle Eocene is precisely the time when there was a major unification and spread of marine faunas (DAVIES, 1934) which appears to be related to a eustatic rise of sea level (HALLAM, 1963a). It will be seen subsequently how useful a tool is this complementary behaviour of continental and marine faunas.

DATA BEARING ON FORMER CONTINENTAL RELATIONSHIPS

The following account makes no pretence to being exhaustive, but study of the literature has produced enough data for a reasonably coherent synthesis which, it is hoped, will in turn provide a stimulus for further work.

Triassic

Vertebrates

The best evidence relates to tetrapods. The fossil record of fresh-water fish during the whole of the Mesozoic is regarded as too poor to have a decisive bearing on the problem of land connections across the Atlantic (SCHAEFFER, 1952) but the claim has been made that the Triassic fish fauna of Australia has an endemic character (HILLS, 1958).

Two important general points are made by COLBERT (1952); firstly, that the Triassic tetrapod fauna of the world is a comparatively homogeneous one, with no striking intercontinental differences like today; secondly, that the presence of closely-related, rather than identical, genera in different continents with land connections should be regarded as the normal condition.

There has been little dispute that the tetrapod evidence favours free communication during the Triassic between the northern and southern continents of the Old World (HAUGHTON, 1956; OLSON, 1957; COLBERT, 1958). Land connections across the Atlantic at this time have been more controversial. HAUGHTON (1956), for the Late Palaeozoic, and DIENER (1915), for the Triassic, favoured land bridges across the North Atlantic to account for the close similarities in the tetrapod faunas of North America and Europe. Both, however, were sceptical of a South Atlantic land bridge connecting Africa with South America. This is the more crucial connection, since it could be argued that an emergent Bering land bridge allowed free migration between the northern continents. This matter has been gone into a few years ago by two leading authorities on fossil tetrapods.

Perhaps the most striking feature of the faunas of Brazil and South Africa is the presence of cynodont reptiles, which are not known anywhere else. COLBERT (1952) used Simpson's index of faunal affinity to make an intercontinental comparison of the whole fauna at the familial and subordinal levels. The results are shown in Table I.

Colbert considered that this evidence is slightly in favour of a South Atlantic land connection. The strongest evidence against a North American link with the Old World is the apparent absence there of cynodonts.

A. S. Romer (in: MAYR, 1952, pp.250–254) observed that it was meaningless to group together the very different Early and Late Triassic faunas as Colbert had done. It was necessary to compare faunas of a similar age, and this could be done between the Middle Triassic Santa Maria Beds of Brazil and the Manda Beds of

212 A. HALLAM

TABLE I

INTERCONTINENTAL FAUNAL COMPARISON BASED ON SIMPSON'S INDEX OF FAUNAL AFFINITY
(EARLY AND LATE TRIASSIC FAUNAS)

	Percentage in common	
	Suborders	Families
South America–Africa	100	75
South America–Europe	86	75
South America–North America	71	63
South America–Asia	43	37
South America–Australia	14	12

Tanzania. The two tetrapod faunas are found by Romer to be remarkably similar and the differences are no greater than may be expected in two parts of the same continental area. Applying Simpson's index to Middle Triassic faunas only, the results differ appreciably from Colbert's (see Table II).

Romer concluded decisively that the burden of proof rests on those who would deny a South Atlantic land connection.

The reptiles of the Late Triassic Maleri fauna of India have strong affinities with those of Europe and North America, implying free land communication (COLBERT, 1958).

Molluscs

Generally speaking, Triassic marine molluscs, like the tetrapods, have a cosmopolitan character. This is most evident in the ammonites and pterioid bivalves (*Daonella, Halobia, Monotis, Rhaetavicula*) which have many wide-ranging species and are, with ammonites, the best means of correlation between continents. It is significant that the pterioid species are so widespread, since they are anisomyarians with byssi that could have attached themselves to drifting seaweeds and logs.

TABLE II

INTERCONTINENTAL FAUNAL COMPARISON BASED ON SIMPSON'S INDEX OF FAUNAL AFFINITY
(MIDDLE TRIASSIC FAUNAS)

	Percentage in common	
	Suborders	Families
South America–Africa	83	72
South America–North America	33	?14

DIENER (1915), in a major zoogeographic study of Triassic faunas, thought he could distinguish from the ammonite evidence a separate Andine Realm for the Americas, though he recognised its very close affinities to his Mediterranean Realm, with many species in common. Such a distinction is not upheld by modern work (a similar fate has befallen UHLIG's (1911) Andine Realms of the Later Mesozoic).

The best evidence concerning the problem of the Atlantic comes from gastropods and bivalves excluding pterioids. Both in South America (JAWORSKI, 1923) and the western United States (MULLER and FERGUSON, 1939) these groups show very strong affinities to the Mediterranean region and the great majority of species appear to be identical. This evidence weighs strongly against a wide separation of deep ocean between these different regions, as Jaworski fully recognised, and a continuous shelf sea connection seems much more likely.

In contrast there are many endemic molluscs in the Triassic of New Zealand, which appears to have been somewhat isolated from the main Tethyan Realm, though ammonites and pterioids were seemingly able to migrate freely (FLEMING, 1962). A deep ocean barrier therefore quite possibly existed in this instance.

Jurassic

Vertebrates

The fossil tetrapod record is generally poor compared with the Cretaceous and Triassic. Rich dinosaur faunas are known only from the Upper Jurassic. Like the other Mesozoic faunas they signify world-wide uniformity of type. The famous Tendaguru fauna of Tanzania is remarkably similar to that of the Morrison Formation in the western United States and suggests that there was free land communication between the two areas (COLBERT, 1962).

The Australian tetrapod fauna is impoverished compared with other parts of the world, as in the rest of the Mesozoic. The dinosaur *Bothriospondylus*, also known from Europe, has been found in Madagascar. No dinosaurs of any age have been found in New Zealand.

Molluscs

As in the Triassic, the faunas of the Lower Jurassic have a cosmopolitan character. In both North and South America endemic elements among the gastropods and bivalves are greatly outnumbered by species apparently identical with European species, while endemic ammonites are virtually absent (JAWORSKI, 1914; HALLAM, 1965). As in the Triassic, a shelf sea connection seems probable.

This close relationship between faunas on the two sides of the Atlantic persists into the Later Jurassic and the main zoogeographic change, recognisable in both the Old and New Worlds, was the differentiation of Tethyan and Boreal Realms among the ammonites (ARKELL, 1956) and belemnites (STEVENS, 1965). This differentiation began during the Middle Jurassic, becoming clearly established

214

A. HALLAM

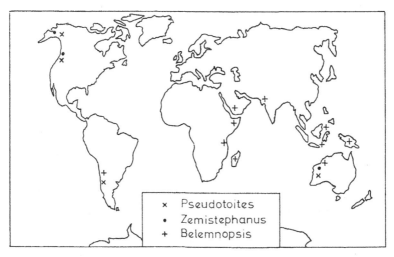

Fig.3. World distribution of the Middle Jurassic ammonite genera *Pseudotoites* and *Zemistephanus* and the Late Jurassic belemnite *Belemnopsis*.

during the Callovian and persisted into the Cretaceous. As there is a gradual transition rather than a sharp boundary between the two realms, and as it is roughly parallel to present-day latitudinal climatic belts, modern workers follow NEUMAYR (1883) and UHLIG (1911) in attributing the differentation to climate.

ARKELL (1956) also proposed a third, Pacific, realm, based on many fewer data. The best evidence related to the occurrence of the Bajocian ammonite genera *Pseudotoites* and *Zemistephanus* in Western Australia, Argentina, Canada, Alaska, and nowhere else (Fig.3). Somewhat weaker supporting evidence comes from the Kimmeridgian. *Idoceras* species of the *durangense* Burckhardt group are abundant in Central America and occur in New Zealand and Indonesia. As Arkell admitted, this instance is not conclusive since a single species is known also in the Mediterranean region. The Kimmeridgian genus *Epicephalites* is known only from Mexico and New Zealand. STEVENS (1965) found no evidence of a separate Pacific realm among the belemnites.

ARKELL (1956, chapter 27) argued strongly in favour of an Atlantic Ocean comparable to the present during the Late Jurassic on the grounds that Mexico and Cuba possessed endemic ammonites not known in Europe. However, as he himself admitted, the majority of the ammonite fauna is strikingly similar at all stratigraphical levels on both sides of the Atlantic. The existence of subordinate endemic ammonites could be due to a variety of causes apart from separation by a wide oceanic barrier (for instance, the existence of partly isolated marine gulfs). Indeed, Arkell recognised a number of endemic elements in the faunas around the Indian Ocean whose direct communication with the European Tethys was not questioned.

Whereas no marine Triassic rocks are known on the margins of the so-called "Gondwana" continents around the Indian Ocean important marine Jurassic sequences occur in northwest India, east Africa and Madagascar (where the oldest beds are Toarcian in age) and Western Australia (starting with Bajocian). The faunas belong to the Tethyan Realm, but from Late Oxfordian times onwards a separate Indo-Pacific province is recognisable from both ammonite and belemnite evidence.

The Upper Oxfordian ammonite family Mayaitidae serves to characterise the province well, since it is confined to Indonesia, the Himalayas, Kutch, east Africa and Madagascar (ARKELL, 1956, chapter 27). Marine communication with the Mediterranean region is indicated, however, by abundant ammonites common to the two areas. The appearance of endemic species of *Belemnopsis*, whose development reached its acme during the Kimmeridgian, marks a complete separation of the Indo-Pacific belemnites from the Mediterranean region (dominated by *Hibolithes*), according to STEVENS (1965).

To account for the more drastic isolation of the Indo-Pacific belemnites, Stevens drew upon evidence from facies distributions to suggest that most belemnites were confined during life to shelf seas and were restricted by deep oceanic barriers more readily than ammonites. Accordingly he suggested the development of a deep sea zone in Late Jurassic times somewhere in the Middle Eastern region as the isolating factor (Fig.4). In support of this, the facies of this age suggests deposition in deep water on a variety of grounds (HÖLDER, 1964).

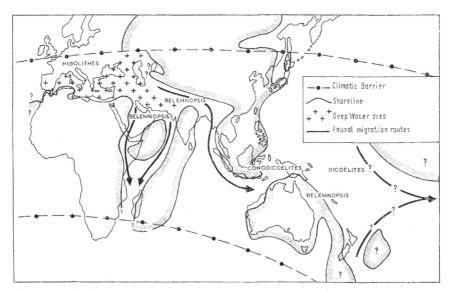

Fig.4. Stevens' palaeogeographic reconstruction for the Lower Kimmeridgian to show how the migration of belemnite faunas could have been controlled by areas of deep water and land barriers. (After STEVENS, 1965, fig.35, with slight modifications.)

216   A. HALLAM

More pertinent to the subject of continental drift is the development of distinct subprovinces within the Indo-Pacific province. The Callovian ammonite genus *Obtusicostites* is confined to Kutch, east Africa and Madagascar and the Toarcian genus *Bouleiceras* is abundant in the latter two areas though it also occurs more rarely in the Iberian Peninsula. The affinities of the whole ammonite fauna between Kutch and the African region are strikingly close (ARKELL, 1956). The Oxfordian and Kimmeridgian belemnites of these regions are also closely similar and different from those of the Himalayas, suggesting partial isolation from the main Tethyan belt (STEVENS, 1965). Stevens suggested furthermore that the affinities of the faunas of Kutch and Madagascar are actually closer than between Kutch and Somalia-east Africa. Endemic elements are also recognisable in the rest of the molluscan fauna, for instance, the Late Jurassic–Early Cretaceous bivalve subgenus *Indotrigonia* (genus *Trigonia* s.s.), known only from Kutch and east Africa (COX, 1965).

The rich Middle Bajocian fauna of Western Australia, with abundant *Fontannesia, Pseudotoites* and *Zemistephanus*, is very different from that in beds of the same age in the east African region or, for that matter, anywhere else in the Indo-Pacific. There are also signs of isolation in the bivalve and nautiloid faunas of this age. The Australian Oxfordian and Tithonian fauna of ammonites, belemnites and bivalves has strong Indonesian affinities, however (ARKELL, 1956, pp.457–458).

Both Arkell and Stevens have argued that the differences between the faunas of the two regions discussed above indicate that no direct marine communication existed between them. A deep sea barrier is implausible because shelf faunas could have migrated around the coastlines and both authors favoured a landmass occupying the site of the central part of the Indian Ocean (Fig.4).

The Lower Jurassic zoogeographic picture for New Zealand resembles that of the Triassic, with the occurrence of a number of endemic bivalves (*Clavigera, Otapiria, Pseudaucella*) suggesting a degree of isolation (FLEMING, 1964). This is also suggested in Stevens' opinion by the absence of belemnites, which were considered, unlike the ammonites, unable to cross a deep sea barrier. During the Later Jurassic there was a strong influx of Tethyan elements in the fauna, and endemic forms were reduced to a very small percentage.

Cretaceous

Vertebrates

Rich dinosaur faunas are confined to the Upper Cretaceous. They have in general a cosmopolitan character suggesting the existence of free land communication between the continents (COLBERT, 1952, 1962). The theropod *Genyodectes* has a world-wide distribution. Since this was probably an upland form there must have been adequate avenues for dry-land dispersal (COLBERT, 1952). Other wide-

TABLE III

INTERCONTINENTAL FAUNAL COMPARISON BASED ON SIMPSON'S INDEX OF FAUNAL AFFINITY
(JURASSIC–CRETACEOUS)

	Percentage in common	
	Suborders	Families
South America–Africa	80	33
South America–Europe	90	58
South America–North America	90	66
South America–Asia	70	42
South America–Australia	10	8

ranging genera include *Laplatosaurus*, known from South America, east Africa, Madagascar, and India. A land link between Madagascar and the African mainland seems required at some time during the Cretaceous for this large animal, as also for *Bothriospondylus* in the Jurassic.

COLBERT (1952) calculated Simpson's index of faunal affinity between the various continents and obtained the results which are shown in Table III.

The closest affinities of the South American Late Cretaceous fauna are apparently with North America and Europe and the intimate relationship with Africa that had existed during the Triassic seems to have been lost. This foreshadows the Tertiary picture, when South American and African mammals evolved in complete isolation from each other.

Dinosaur evidence also has some bearing on the possible existence of a Bering Straits land bridge during the Late Cretaceous. The rich faunas of Asia and North America appear to be somewhat more closely related to each other than to those of Europe, Africa and South America, notably in the occurrence of ceratopsians (COLBERT, 1962, pp.246–247). This suggests the likelihood of an emergent land bridge. E. H. Colbert (personal communication, 1966) has kindly offered the following remarks: "The Upper Cretaceous reptilian faunas of northeastern Asia are certainly closer to those of North America than to any others. Indeed, there is a close identity of faunas between these two areas, indicating a free interchange of faunal elements back and forth, presumably across a Bering land bridge. This is particularly noticeable in the composition of the dinosaurian elements in the fauna with such genera as *Tyrannosaurus*, *Ornithomimus*, *Saurolophus*, and others, common to both regions. But the resemblances extend to other reptiles as well. This has been brought out by the work of the Central Asiatic Expeditions done by the American Museum of Natural History through the past few decades, and more recently by the work of the Russians and Poles in Mongolia".

The fossil record of other vertebrate groups is too poor to warrant deductions about land relationships, but the distribution of living cold-blooded vertebrates is significant in one highly relevant respect, namely that certain primitive and, by general consent, closely related groups are found only in Africa and South America. These include the pelomedusid tortoises (also known in Madagascar), pipid frogs and several families of freshwater fish including characins (primary), osteoglossids and cyprinodonts (secondary). DARLINGTON (1948) recognised that there are only three ways of accounting for this distribution. Either the various animals were able to migrate across the Atlantic Ocean, or by a South Atlantic land connection, or by a long land journey via Asia and North America. The first possibility was considered highly implausible and the second rejected because it departs from orthodox geological views. Darlington was left with the hypothesis of a migration across the Holarctic Realm which has, however, left no trace in the indigenous fauna, living or fossil. The connection was presumed to have been during the Cretaceous. This subject is of sufficient importance to warrant a few quotations from Darlington's writings, since they express with admirable clarity the dilemma in which he found himself.

South America was (p.115) "certainly connected with the rest of the world at or before the beginning of the Tertiary, but it is a question whether the connection was with North America. If so, the old fauna which came over it should in a general way resemble the new fauna which has come via North America." This new fauna dates from the emergence of the Pliocene land bridge.

A few lines later we read: "The old fauna, then, apparently lacks a distinctive North American element, while the new fauna has one." Nevertheless there follows an involved and not very convincing argument in favour of a North American connection for the old fauna as well. Though the South American animals are derivatives of a fraction of Old World families, North America has left them untouched. "That is, the most characteristic North American stocks may have been the least persistent in South America."

Finally (p.116): "All the strictly freshwater fish of South America may be derived from African groups; apparently none, from any North American group; not a single fossil has yet been found in North America to prove the former presence there of any of the fishes concerned except the Osteoglossidae, which may have been salt-tolerant rather than strictly fresh-water forms; and enough other animals parallel the distribution of the fishes to make it unlikely that the latter's African–South American relationships are due to dispersal through the sea."

An uncommitted observer might well conclude that the most simple explanation of the facts was that a direct land connection existed between South America and Africa at some time during the Cretaceous. Darlington's interpretation is perhaps symptomatic of an over-reaction on the part of modern zoogeographers against hypotheses involving subsided land bridges or continental drift.

Molluscs

It has long been known that the Lower Cretaceous ("Neocomian") trigoniids of South America and southern Africa are remarkably similar to each other and distinct from those in other parts of the world. Their distribution was the principle reason why UHLIG (1911) included the Uitenhage Beds of Natal in his South Andine faunal province, and deduced that shelf sea communication must have existed between the two continents. Modern work has upheld the close similarities, but it is now recognised that the faunal province also includes Madagascar, east Africa and northwest India. This is indicated, for example, by the distribution of the subgenus *Megatrigonia* of the genus of the same name and by the so-called pseudoquadrate species of the trigoniid genus *Yaadia*. The subgenus *Iotrigonia* of the genus *Megatrigonia* is also common throughout the province but extends also to New Zealand and, possibly, British Columbia (COX, 1951). Cox added in his paper that other bivalve groups confirm the existence of a distinct faunal province embracing western South America and the eastern side of Africa.

The characteristic ammonite of the province is *Rogersites*, a subgenus of *Olcostephanus*. A number of endemic genera (e.g., *Favrella*) occur in the Hauterivian of Patagonia, and *Pulchellia* is abundant, though it also occurs in other parts of the world. The belemnites of the east African and Madagascan Lower Cretaceous contain apparent relicts, which suggests some degree of isolation from the Tethys as in the Late Jurassic (STEVENS, 1965).

Free communication between the Old and New Worlds is also suggested by the distribution of another Lower Cretaceous trigoniid genus, *Buchotrigonia*, known only from the northern Andes and the Iberian peninsula.

Whereas during Lower Cretaceous times belemnites of the *Hibolithes-Duvalia* assemblage migrated freely into Australasia from the Tethys, isolation of this region took place in about the Aptian, as indicated from this time onwards by the endemic subfamily Dimitobelinae. This contains four distinctive genera, confined to Australia, New Guinea and New Zealand except for one species known in south India (STEVENS, 1965). After Cenomanian times belemnites apparently died out in Australia and New Guinea.

Ammonites and *Inoceramus* species were apparently able to migrate freely into the area however so that, for instance, Cenomanian ammonite species in New Zealand compare closely with species in India and Europe. Stevens accounted for this difference by invoking a deep-water barrier separating Australasia from the main Tethyan shelf seas. It may be noted here that the apparent free migration of *Inoceramus* species might be the result of a pseudoplanktonic mode of life of some of the adults, as probably with the Triassic pterioids.

In the Late Cretaceous the Indo-Pacific faunas of ammonites and *Inoceramus* are clearly distinguishable from those of Europe and North America (excluding the west coast) and may signify a major zoogeographic realm (MATSUMOTO, 1959). This is well indicated by the Kossmaticeratidae, which family has a dominantly austral distribution in Campanian times (Chile, Antarctica, New Zealand, Madagascar) but is present less abundantly also in the North Pacific region (BASSE, 1953; MATSUMOTO, 1959).

A separate Austral province or realm may be distinguishable in Late Senonian times. Free communication of shelf faunas at this time between New Zealand, Antarctica and Chile is suggested by the occurrence there (and nowhere else) of a number of bivalves and gastropods including *Lahillea, Pacitrigonia* and *Stuthioptera*. Faunal distribution was probably in part the result of climate. The austral seas were evidently warmer than today but cooler than in the Tethyan belt, as signified by the absence of rudists and hermatypic corals (FLEMING, 1962).

There is a limited amount of information available on the Late Cretaceous isolation of Madagascar. COLLIGNON (1934) studied a typical neritic fauna in the Turonian which was characterised by an abundance of bivalves and gastropods. He recognised that four fifths of the species distinguished were identical with forms occurring in Europe, north Africa and India, which implies free shelf-sea communication. The sea lapped around the eastern shores of Madagascar for the first time in the Campanian. By Maastrichtian times, according to ROGER (1949), faunal affinities with India and Africa had diminished appreciably from what they were earlier in the Late Cretaceous, but no details were given.

The Cretaceous faunas on the two sides of the North Atlantic have strong affinities with each other (COBBAN and REESIDE, 1952) but quantitative comparisons using something like Simpson's index would be premature in the present state of taxonomic knowledge. Thus in such monographs as those of WADE (1926) and STEPHENSON (1941, 1952) virtually all the molluscan species described are North American forms. A study of these works suggests that many of the species have been interpreted on a narrow typological basis and that no serious attempt has been made to compare them with the European.

Middle and Late Cretaceous faunas on the two sides of the South Atlantic are less well known than those further north. REYMENT (1965) has recently listed five ammonite and one bivalve species common to west Africa and Brazil and thereby deduced that the Late Cretaceous marine separation of these two regions was much narrower than at present, implying that there has been considerable drift since the Cretaceous. Such a conclusion seems unwarranted by the evidence of so few species, especially since they are not well chosen for the problem in question. Ammonites have wider distributions than most other contemporary molluscs and the solitary bivalve is an anisomyarian which could have attached itself to drifting plant debris by means of its byssus.

Of especial relevance to the subject of continental drift are the Cretaceous molluscan faunas of the lands bordering the North Pacific. The Late Cretaceous ammonite and *Inoceramus* faunas of the west coast of North America, from California to southern Alaska, have much in common with those of Japan but very little with those of the North American Western Interior and northern Alaska and clearly belong to different faunal provinces (MATSUMOTO, 1959; JONES, 1963;

JONES and GRYC, 1960). The two provinces seem distinguishable in the Albian (IMLAY, 1960b, 1961) but before this, in the Hauterivian and Valanginian, the west coast faunas of North America have abundant affinities both with Europe and Japan, suggesting free marine connections to the north, west and south (IMLAY, 1960a).

The sharp differences between the faunas of the west coast and the Western Interior, which cut across the (climatically controlled?) Tethyan-Boreal belts, are readily explained by the land barrier portrayed in all palaeogeographic reconstructions of the Cretaceous. It is evident from the fauna alone that this barrier must have persisted into central Alaska to separate southern and northern seas. A further deduction is that the Bering Strait must have been an emergent land bridge in Middle and Upper Cretaceous times, since otherwise the faunas of northern and southern Alaska would have intermingled freely.

So far attention has been confined to molluscs that had powers of wide dispersal. From what has been said earlier in this section, it is apparent that the many Cretaceous trigoniids with restricted geographic distribution suggest that the group had limited powers of dispersal.

According to NAKANO (1960) the affinities of the Japanese Neocomian–Turonian trigoniids are with India and Africa rather than with the west coast of North America, as exemplified by *Pterotrigonia*, *Rutitrigonia* and *Acanthotrigonia*. The only North American west coast genera reported in the Neocomian are *Yaadia* and *Iotrigonia*, neither of them known from the West Pacific. By Senonian times, however, the situation changed, by the establishment of affinities between North America and Japan. The subgenus *Steinmanniella* (*Yeharalla*) has been found only in the North Pacific borderlands and may be taken as characteristic of the North Pacific province.

Tertiary

Vertebrates

The zoogeographic picture for the Tertiary is clearer than for earlier times because of the comparative excellence of the fossil mammal record and the relatively good geological control.

An impressive account of mammal migrations between Eurasia and North America has been given by SIMPSON (1947). Major faunal interchange apparently took place in the Early Eocene, Late Eocene, Early Oligocene, Late Miocene, Middle–Late Pliocene, and Pleistocene; very little or none took place in Middle Eocene and in Middle and Late Oligocene times. These conclusions are based on degrees of generic resemblance (Fig.2). The only land link required is across the Bering Straits. This was certainly operative from Late Eocene times onwards, since the Asian faunas resemble the North American much more closely than the European. Previous to this the Holarctic mammal faunas were more uniform. It has been noted earlier that the character of the Bering land bridge changed from a corridor to a filter bridge during the course of the Tertiary, as a result of climatic deterioration. KURTÉN (1966) has, however, disputed Simpson's conception of a unified Palaearctic zoogeographic region in earliest Tertiary times and claims direct intermigration of mammals by a land bridge between Europe and North America in the Late Palaeocene and Early Eocene.

South America evidently received mammals from North America at about the beginning of Tertiary times but land links with other continents were subsequently severed and only renewed by the uplift in Pliocene times of the Central American land bridge, as previously discussed (SIMPSON, 1940a, b). The African and South American faunas of land mammals are totally dissimilar, with the exception of hystricomorph rodents, which appear without known antecedents in the Oligocene of Argentina. This isolated case could possibly represent a rare "transport miracle" involving driftwood floating across the Atlantic, but WOOD (1950) claimed that the African and South American hystricomorphs have evolved separately from a more widely-distributed Eocene ancestor, so that no migration across the ocean need be invoked.

The relationships of Madagascar and Australia to neighbouring continents during the Tertiary have received attention from many people, including Wegener. The case of Madagascar has already been discussed, and reasons put forward in favour of its isolation by sea.

WEGENER (1924) claimed that the common occurrence of marsupials in South America and Australia signified a land link via Antarctica which was not broken until about Eocene times. This view has been strongly contested by SIMPSON (1939, 1943), who pointed out that the faunal resemblance is not especially close, being confined to one marsupial family and suborder. As fossil marsupials are known from Cretaceous deposits in the Northern Hemisphere, the existing forms can be considered as southern relicts which have evolved independently. If a continuous Antarctic land bridge had existed during the Early Tertiary there should be a much closer resemblance between the South American and Australian faunas, with abundant placentals in the latter region. STIRTON (1958) thought that marsupials probably arrived on driftwood rafts in the Late Cretaceous and have been isolated ever since. If this were so, one wonders why no placentals did the same until the end of the Tertiary, bearing in mind SIMPSON's (1952) study on the influence of time on probability of dispersal.

Similar views to those of Simpson have been expressed for other vertebrates by MAYR (1953) and DARLINGTON (1948). Mayr considered that the strongly endemic living bird population of Australia and New Guinea demands drastic isolation at some time in the past, which he dated as Early Tertiary. Palaeontological support for this dating is not available. If there had been free land connection with South America, one would expect relict elements from there whereas in fact they are absent. Darlington (p.117) found "no good reason to think that any cold-blooded vertebrate has crossed an Antarctic land bridge, but that salt-tolerant fish and giant land turtles may have dispersed across the water gaps of an Antarctic archipelago under conditions presumably more favourable than now." It is known, in fact, that the land turtle *Testudo* has been able to cross sea barriers several hundred miles wide.

Finally, the relationships of African and Eurasian mammalian faunas deserve a comment. The proboscids and anthropoids appear to be groups which evolved in isolation in Africa. Land connection with Eurasia was not established until mid-Tertiary times, as indicated by the sudden appearance there of these groups in Burdigalian (Lower Miocene) deposits (DAVIES, 1934, p.37). SIMPSON (1940a) has noted that fossil mastodont remains of Oligocene age have been found only in Africa. The earliest-known mastodonts in Eurasia are Lower Miocene and in North America Upper Miocene.

Molluscs

As in the case of the vertebrates much more is known about Tertiary molluscs than about their Mesozoic ancestors. Amidst a maze of detail, two major changes may be discerned during the course of the Tertiary. Firstly, a gradual lowering in temperature caused a progressive contraction of the belts with climates like those occurring today in the tropics. Secondly, the old Tethyan seaway, which had extended continuously across the Old and New Worlds, became divided into two in the mid Tertiary, separating an Indo-West Pacific from an Atlantic-East Pacific Realm (EKMAN, 1953). The present-day Atlantic fauna is but an impoverished remnant of a rich Lower Tertiary fauna which closely resembled that surviving still in the Indo-West Pacific region.

Relationships of the fauna on the two sides of the Atlantic have an important bearing on the problem of continental drift, as recognised by DURHAM (1952). At the present time it appears that about 16 % of the bivalve and 8 % of the gastropod species occurring on the eastern seaboard of the United States are found also in Europe and less than half of these have a discontinuous distribution, i.e., they have not evidently migrated via the North Atlantic, where the deep sea barrier is narrower. Although it has been widely believed that the corresponding early Tertiary faunas had closer affinities to each other than at present (DAVIES, 1934; EKMAN, 1953) Durham quoted the taxonomic work of specialists (HARRIS and PALMER, 1946–1947) who recognised only one molluscan species out of 352 as common to the Eocene of the Gulf Coast and Europe. Durham cited this as good evidence that the Atlantic Ocean must have been as wide in the Eocene as at present, so that no drift could have occurred during this time. Nevertheless, bearing in mind the data for living species, it is extra-ordinary that so little resemblance was recognised by Harris and Palmer, such that the suspicion is aroused that these workers may have interpreted species in an unduly parochial way. The subject is of sufficient interest to warrant thorough revision by palaeontologists on both sides of the Atlantic.

The splitting up of the Tethys is dated by DAVIES (1934, p.105) as Oligocene. From this time onwards the faunas of the Indo-Pacific and the Atlantic (including the Mediterranean) underwent separate evolution and today there is little detailed resemblance between them. The separation is clearly due to the appearance of a land barrier in the Middle East. This was not completely efficient at first, as there is evidence of two brief marine incursions into Egypt during the Miocene, bringing an Indo-Pacific fauna. Subsequently the progressive lowering of sea level during the Late Tertiary (HALLAM, 1963a) ensured that separation was more effective.

Although the Tertiary and Recent molluscan faunas of Australia and New Zealand are dominated by warm-water Indo-Pacific elements there are a number of relicts of austral affinities, of which perhaps the best-known are the gastropod *Struthiolaria* and the bivalve *Neotrigonia*. As FLEMING (1964) pointed out, the latter is the sole surviving genus of a flourishing Mesozoic family and seems to have evolved directly from trigoniids, themselves apparently relicts, that occurred in the Austral Realm in the Late Cretaceous. *Neotrigonia* is a bivalve with comparatively primitive morphological features and is a veritable "living fossil". Its several species are restricted to the Australian coastal seas and exhibit wide tolerance of temperature and depth. The genus apparently owes its survival, like the marsupials, to the Late Mesozoic–Early Tertiary (?) isolation of the Australian region. A resistant strain (the *Eotrigonia-Neotrigonia* lineage) was evidently able to adapt to later competition from Indo-Pacific invaders.

Another feature of interest concerning the Australasian faunas is that they signify a temperature maximum in mid-Tertiary times. FLEMING (1962) put this maximum in the Lower Miocene for New Zealand, GILL (1961) in the Oligocene and Miocene for southern Australia, as signified by such tropical and subtropical elements as giant cowries, volutes and cone-shells. This is confirmed by oxygen isotope palaeotemperature determinations (DORMAN and GILL, 1959). Analyses of shells of *Chlamys*, *Ostrea* and *Glycimeris* indicate temperatures rising from Early to mid-Tertiary and then falling to present-day values. These results are abnormal compared with those obtained in other well-studied parts of the world. Both the plant and animal evidence for Europe and North America (DURHAM, 1959; SCHWARZBACH, 1963) signify clearly a more or less gradual decline of temperature from the Early Tertiary. The same picture is revealed by oxygen isotope studies of calcareous benthonic foraminifers from the floor of the equatorial Pacific (EMILIANI, 1954, 1961).

INTERPRETATION OF THE PALAEOZOOGEOGRAPHICAL DATA

Inadequacy of the concept of trans-oceanic land bridges

In order to account for certain features of ammonite distribution NEUMAYR

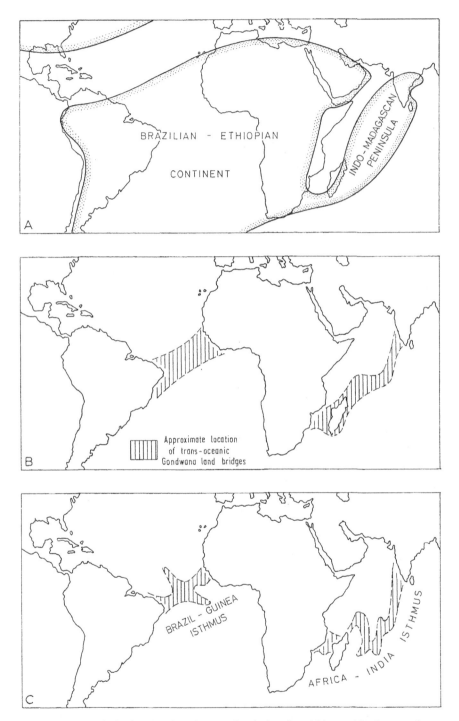

Fig.5. Trans-oceanic land connections between South America, Africa and India according to (A) NEUMAYR (1887), (B) SCHUCHERT (1932), and (C) WILLIS (1932).

(1887) proposed that the southern continents had been linked in the Jurassic by major landmasses occupying the site of part of the present South Atlantic and Indian Oceans. South America was joined to Africa by the "Brazilian–Ethiopian continent" and Madagascar to India by a continental landmass which came to be known as "Lemuria" (Fig.5A). Such was the need felt for land links to account for a whole range of faunal and floral similarities that Neumayr's interpretation was widely accepted by European geologists during the succeeding few decades, including some of the leading figures of the time (e.g., BLANFORD, 1890; HAUG, 1900; SUESS, 1906; UHLIG, 1911). It was generally accepted furthermore that these landmasses had disappeared by the beginning of the Tertiary.

A major difficulty of this hypothesis, fully appreciated by Wegener, is the isostatic problem involved in sinking an extensive "sialic" continent without trace. This posed a dilemma for the leading American student of the subject, Schuchert. On the one hand he was convinced after many years of study that the palaeo-biogeographic evidence was overwhelmingly in favour of a "Gondwana" land bridge until the end of the Cretaceous (SCHUCHERT, 1932, p.878). On the other hand he did not want to disregard the isostatic problem. His solution (Fig.5B) was to pare down the size of the landmasses in the two oceans. This was not sufficiently radical for WILLIS (1932), since the landmasses envisaged by Schuchert were still considered to be composed of normal continental crustal material. Willis' "isthmian links", which roughly followed the same lines as Schuchert's land bridges, were conceived of as basic rock thrown up from beneath the sea bed by volcanic activity; by the close of the Mesozoic they had subsided. While Willis' interpretation, which achieved a fair measure of popularity among biogeographers, appears to solve or at least significantly diminish the isostatic problem, it runs into the difficulty that the volcanic islands and submarine ridges that he invoked as evidence are now widely believed to be geologically youthful features, largely or entirely post-Mesozoic in age. He also, inconsistently, included the ancient granitic Seychelles in his construction (Fig.5C). Willis' isthmian links are thereby diminished to the rank of an ad hoc hypothesis, for which independent evidence is lacking.

In more recent years oceanographic research has, of course, fully confirmed the absence of any trace of "sunken continents" in the Atlantic and Indian oceans and suggested on a variety of grounds that the ocean floor is young. For instance, the sedimentary layer is surprisingly thin, no sediments older than the Cretaceous have been discovered and, at least in the Atlantic, there is no properly-developed oceanic crustal layer, unlike in the Pacific (MENARD, 1964, p.150). Unless one is to deny the validity of the palaeozoogeographic evidence there is only one plausible alternative to Mesozoic land bridges, continental drift.

A hypothesis involving continental drift

The cumulative effect of a wide range of geological and geophysical data, especially that concerned with palaeoclimates, palaeomagnetism, oceanic rift structures and the excellence of "fit" of the Atlantic continents, has at last made continental drift in the Wegenerian sense a respectable hypothesis, which is beginning to achieve wide acceptance (BLACKETT et al., 1965). There remains considerable uncertainty about details and especially about the timing of the events. It is hoped to show how the palaeozoogeographic data can contribute in these respects, on the assumption, here adopted, that continental drift has actually happened.

The indications of free "intercontinental" communication of land and neritic faunas as deduced from strong faunal affinities suggest that the continents may not have begun to drift apart until the Cretaceous. As far as the opening of the South Atlantic is concerned this in an unsurprising conclusion. It accords with Wegener's own views and has been generally accepted by geological supporters of drift, who have been impressed by the absence around the South Atlantic of marine sediments older than mid-Cretaceous. The case needs arguing more fully, however, for the Indian Ocean, since it has been proposed that this began to open up in the Jurassic. The ground for this belief is the occurrence of marine Jurassic sediments in parts of the bordering lands.

As has been mentioned earlier, differences in the Jurassic molluscan faunas appear to demand a land separation between the east African–northwest Indian

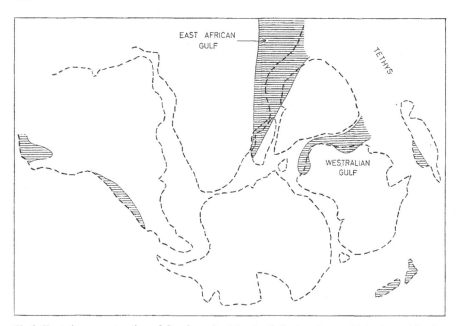

Fig.6. Tentative reconstruction of Gondwanaland for the Callovian. Areas with horizontal-lined ornament signify seas extending over regions of continental crust. Data on land and sea distribution for this (and Fig.7 and 8) based primarily on FURON (1963), TEICHERT (1958), and TERMIER and TERMIER (1960).

region and Western Australia. This landmass has been interpreted by ARKELL (1956) and STEVENS (1965), neither of them supporters of continental drift, as corresponding to the familiar "Lemuria" of many authors (Fig.4). This interpretation is here rejected in favour of one based upon a reassembly of "Gondwanaland" (Fig.6).

The reconstruction favoured represents something of a comprise between those of WILSON (1963, Fig.6) and HOLMES (1965, fig.875). The relationships of India and Australia are based on evidence from both geology (HOLMES, 1965, p.1223) and palaeomagnetism (IRVING, 1964, p.271). Australia and Antarctica are closed up towards the line of the Pacific–Antarctic Ridge and India is placed closer to Madagascar than by Holmes, on the ground that the Late Jurassic belemnites of Kutch and Madagascar show closer affinities to each other than to east Africa and Somalia. Allowing for an area of continental crust which is supposed, according to CAREY (1958) and HOLMES (1965), to have underridden the Himalayan region, it will be seen that the general shape of India fits well with that of east Africa in this reconstruction. Something must be sacrificed in all Gondwanaland reconstructions, and it will be seen that the South America–western Antarctica relationship is rather awkward, though not implausible. At least it avoids the overlap of Wilson's reconstruction. Holmes solved this problem by abandoning the apparent close fit of South America and Africa. Such a solution is felt to be too radical.

The Jurassic faunas and sediments of the areas under consideration give no indication of being other than neritic. Fig.6 shows how they may be interpreted as the deposits of two shallow marine gulfs, whose only communication was via the Tethyan seaway. This accounts both for the general Tethyan affinities of the two regions, and for the occurrence of endemic elements suggestive of partial isolation. Although the earliest Mesozoic marine deposits are Jurassic both the east African and Westralian gulfs have a history of intermittent subsidence and marine deposition extending back into the Palaeozoic (FURON, 1963; TEICHERT, 1958). Both probably marked areas of relatively weak and thin crust which would be the obvious loci for subsequent rifting.

Reconstruction of the southern continents along Wegenerian lines also helps to explain certain otherwise extremely puzzling features of Jurassic ammonite and belemnite distribution. ARKELL (1956) thought that the distribution of *Pseudotoites*, *Zemistephanus* and certain Kimmeridgian ammonites demanded that they had migrated across the Pacific (Fig.3). Trans-Pacific migration of benthonic or nektonic molluscs during the Jurassic would have been remarkable enough; it seems incredible that animals so mobile and adaptable that they could nevertheless achieve this, should at the same time have been inhibited from penetrating other marine regimes. A similar problem is presented by the distribution of Indo-Pacific species of *Belemnopsis*, which STEVENS (1965) thought must have been able to migrate across the Pacific from Australia and Indonesia to Argentina (Fig.3). Yet elsewhere in his monograph Stevens argued that belemnites were confined to a neritic habitat and

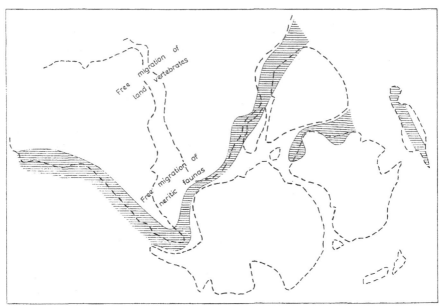

Fig.7. Tentative reconstruction of Gondwanaland for the Neocomian. Ornamentation as in Fig.6.

Belemnopsis species, in particular, were prevented from entering the Mediterranean region by a deep-water barrier.

The faunal individuality of the east African region persisted into the Early Cretaceous, when, however, there is evidence of a marine link-up with the Andean region, allowing free migration of neritic faunas, which must have passed south of Neumayr's "Brazilian–Ethiopian" continent. The interpretation tentatively proposed here is that this marks the beginning of rifting between Africa, South America and Antarctica, so allowing the sea to penetrate between these continents (Fig.7). There are still no indications at this time of direct marine communication between the east African and Westralian gulfs.

Active drifting apart of the components of Gondwanaland was probably well under way by mid-Cretaceous times, as indicated for instance by the development of an endemic belemnite subfamily in the Australian region, suggesting isolation from Tethyan shelf faunas by a deep-water barrier. The isolation of the marsupials could well date from this time also. Judging from fossil evidence this was quite probably the dominant mammalian group in the Cretaceous. Extensive Early Tertiary radiation of the more advanced placentals subsequently led to the decline of marsupials through competition in other parts of the world where they were not protected by marine barriers. Isolation of the bird fauna might also have taken place in the Cretaceous, rather than the Early Tertiary, as MAYR (1953) presumed.

TEICHERT (1958) has argued that the impoverished Mesozoic tetrapod fauna of Australia indicates that there have been negligible land connections with other

230   A. HALLAM

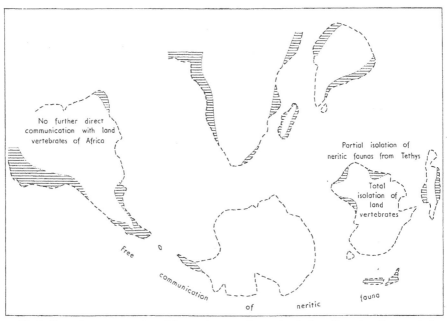

No further direct
communication with land
vertebrates of Africa

Partial isolation of
neritic faunas from Tethys

Total
isolation of
land
vertebrates

Free

Communication

of   neritic   fauna

Fig.8. Tentative reconstruction of the southern continents for the Late Senonian. Ornamentation as in Fig.6.

continents during this era, perhaps only island chain connections. Assuming that this impoverishment is genuine and not due to collection failure, an alternative explanation is available. IRVING and BROWN (1964) studied the faunal diversity of Permo-Triassic labyrinthodont amphibians over the whole world and were able to correlate the low diversity of the Australian fauna with high palaeomagnetic latitude, thus establishing a relationship which is valid for most animals at the present day. Palaeomagnetic data indicate that Australia remained at a high latitude throughout the Mesozoic (IRVING, 1964). Whether or not this latter explanation proves acceptable, something more than island chain links is required to account for the presence in Australia of the lungfish *Neoceratodus* and large Jurassic and Cretaceous dinosaurs.

By Late Senonian times Madagascar apparently had become isolated from India and Africa and a distinct austral faunal province was established, linking Australasia with South America and allowing free communication of shelf faunas. A tentative attempt at a possible configuration of the southern continents at this time is given in Fig.8. The main points that this figure is meant to bring out are that Australia was closer to Antarctica than India, and that continental drift, though well under way, continued into the Tertiary.

Shelf sea connections between South America and Africa are indicated by the trigoniid and associated bivalve fauna and a land connection at some time during the Cretaceous seems the most plausible explanation of the primitive cold-blooded vertebrate distribution discussed by DARLINGTON (1948). By the close of the Cretaceous, however, and probably earlier, judging by the dinosaur evidence, South America was isolated from Africa by an ocean barrier. The problem of the North Atlantic appears more difficult and seems best tackled indirectly at present, by considering the North Pacific region.

A Bering land bridge must have been established by Late Cretaceous times,

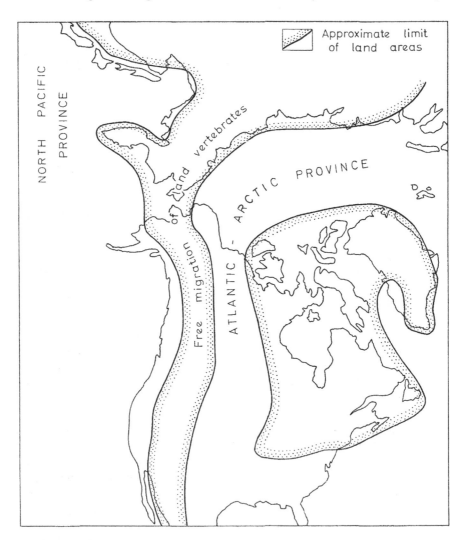

Fig.9. Tentative interpretation of the geographical relationships of North America and Asia in the Late Cretaceous.

to account both for the separation of the North Pacific and Atlantic-Arctic mollus-can provinces and for the apparent free communication of dinosaurs between Asia and North America (Fig.9). In terms of continental drift theory this can be accounted for in two different ways.

According to CAREY (1958) there has been no westward translation of Alaska, but instead a clockwise rotation of North America about the Alaskan orocline. Thus, Greenland formerly occupied a site in the Arctic Ocean north of Spitsber-gen and a Bering land bridge or shelf connection existed even before drift (Fig.9). It is interesting to note that Wegener anticipated this interpretation in attempting to counter a criticism by Diener concerning this same supposed connection (WEGE-NER, 1924, p.78).

However, E. C. BULLARD and co-authors (in: BLACKETT et al., 1965) have recently demonstrated the excellence of fit of the Greenland shelf to that of Europe in a position southwest of Spitsbergen, and Harland, in a contribution to the Royal Society symposium, has pointed out the abundance of close geological correspondences if this fit is adopted. Therefore the evidence seems to favour the view that there has been westward translation of the whole of North America and that accordingly a Bering shelf connection could not have existed before continental drift took place.

A suggestive hint about the timing of this movement is provided by the distribution of trigoniids on the North Pacific borderlands. The fact that a close relationship of Japanese and North American forms was not established until the Senonian could be an indication that before this time a deep-sea barrier existed between Alaska and Siberia. On the other hand the distribution of ammonites and *Inoceramus* suggests that a Bering land bridge might have been established by Albian times, if not earlier. Clearly, many more data on North Pacific faunas are required. Meanwhile the possibility must be allowed for that the North Atlantic began to open up before Gondwanaland split apart, perhaps in the Late Jurassic, as WILSON (1963) has proposed on other grounds.

There is no reason, of course, for assuming that drift ceased suddenly at the close of the Cretaceous, but the palaeozoogeographic data give no indication as yet of the possible extent of westward movement of the Americas during the Ter-tiary. Perhaps after the Siberian and Alaskan shelves closed up the continued operation of subcrustal forces caused the more southerly part of North America to rotate clockwise slightly about the Alaskan orocline. This would be consistent with Carey's view that North America has sheared westwards along sinistral faults with respect to South America.

It is possible to draw more positive conclusions about the Australian region. The anomalous climatic changes deduced from the molluscan data, suggesting a temperature rise to a mid-Tertiary maximum, followed by a decline, can readily be accounted for by northward drift of both Australia (with New Guinea) and New Zealand towards Indonesia. From a high latitude with temperate climate this region,

BEARING OF PALAEOZOOGEOGRAPHIC DATA ON CONTINENTAL DRIFT   233

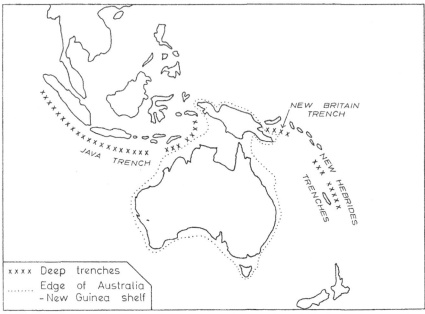

Fig.10. The relationship of the Australia–New Guinea continental block to the belt of deep trenches extending from Indonesia towards New Zealand. Based on MENARD (1964, fig.1,10, 1,11 and 5,6).

moving presumably as one crustal block, would have passed into a subtropical zone. During the Late Tertiary, drift having slowed down or ceased, Australasia would have experienced with other continents the pronounced climatic cooling which led on to the Quaternary Ice Age.

Support for northward movement in the Tertiary comes from both structural geology and palaeomagnetism. Fig.10 portrays in a simplified way the relationship of Australia and New Guinea with the belt of deep oceanic trenches and associated negative gravity anomalies that runs from Indonesia towards New Zealand. It will be seen that this belt gives the appearance of having been violently bent and disrupted by the northward movement of New Guinea. This major disturbance could have taken place in the Late Miocene, a time of intense folding and thrusting of the Banda Arc islands (UMBGROVE, 1947, p.181). Palaeomagnetic data suggest that a locality in South Australia moved northwards from a latitude of about 60°S at the beginning of the Tertiary to about 28°S at the present day (IRVING, 1964, p.269).

This interpretation accords essentially with that proposed by Wegener, except that a land connection with Antarctica and South America during the Early Tertiary is rejected for reasons discussed earlier. It is clearly incompatible with the hypothesis of CAREY (1958), who suggested that Australia had moved eastwards from India along a shear belt and had only become close to New Guinea at a late

234 A. HALLAM

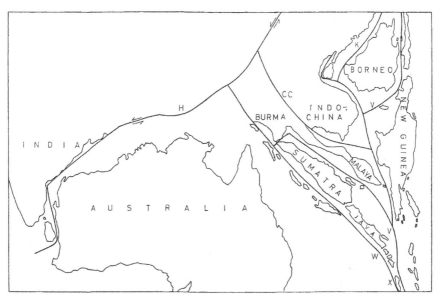

Fig.11. Relationship of Australia to India and New Guinea before continental drift, according to CAREY (1958, simplified from fig.40c).

stage in the Tertiary (Fig.11). The vertebrate evidence suggests on the contrary that New Guinea has been an integral part of the Australasian landmass since the Mesozoic.

Both SIMPSON (1943) and MAYR (1953) thought that the relationships of the Oriental and Australian vertebrate faunas could be accounted for without invoking continental drift. It might be felt that they hardly did justice to the peculiarities of the latter fauna, especially bearing in mind the presumed multiple colonisations of Madagascar in the Tertiary. In Wegener's words, Australia is, faunally speaking, "like a foreign body from another world."

Invertebrate palaeontologists might raise the objection that the Mesozoic faunas of Australasia have strong affinities with those of Indonesia, which implies close geographic proximity. This is not necessarily true. The admitted similarities of the two faunas could merely reflect the homogeneity of freely-communicating faunas on both sides of an equatorial Tethyan belt lacking a wide oceanic barrier in its western part (Fig.12).

The mid-Tertiary isolation of the Mediterranean and Atlantic marine fauna from that of the Indo-West Pacific coincides closely in time with the first evidence of free migration of African mammals into Eurasia. It is here suggested that this complementary behaviour of the land and sea faunas was the result of the sealing off of the Mediterranean in the Middle East by the anticlockwise rotation of the African–Arabian continent.

It is evident from the location of the mid-Atlantic Ridge and the apparent "fit" of the Atlantic continents (BLACKETT et al., 1965, and Fig.12) that Africa must have undergone some anticlockwise rotation in addition to eastward translation during drift (see also IRVING's reconstruction (1964, fig.10, 17) based on palaeomagnetic data). This movement was presumably only halted when Arabia impinged upon Asia Minor. The result was drastic not only on the fauna but on the palaeogeography, with the creation of a series of brackish water inland seas where formerly there had been a continuous seaway. In more recent times the opening of the Gulf of Aden and Red Sea rifts has come close to restoring the old marine connection between the eastern and western "Tethys".

DISCUSSION

The timing of the initiation of continental drift deduced in this paper is similar to that proposed by DU TOIT (1937) and KING (1962) insofar as these authors thought that the components of Gondwanaland did not begin drifting apart substantially until the Cretaceous. A different interpretation is put on the evidence of marine Jurassic sediments in the Indian Ocean borderlands, however. Both Du Toit and King thought that rifting began in the Jurassic.

KING (1962, pp.57–60) actually proposed a time sequence of rifting based on the age of the oldest marine deposits. Thus disruption is supposed to have started along the eastern side of Africa, while not until well into the Cretaceous had the South Atlantic begun to open up.

Both of these South African geologists seem to have underestimated the importance of marine transgressions resulting from eustatic rises of sea level. The first marine sediments of the east African (Lower Toarcian) and Westralian (mid-

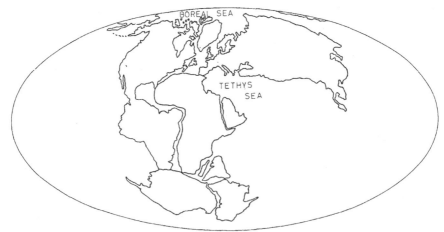

Fig.12. Wilson's reconstruction of the continents in mid-Mesozoic time. Adapted from WILSON (1963, fig.6).

Bajocian) gulfs correspond to the times of major transgressions which appear to be world-wide (HALLAM, 1963b). Periodic flooding of subdued continental areas by shallow epeiric seas need not necessarily have anything to do with rifting and independent evidence is required, such as changing faunal distributions. Similarly, the first marine sediments of the South Atlantic shores correlate with an extensive mid-Albian transgression which also appears to have been world-wide. Relationship with continental drift could in part be an indirect one, as will be discussed later.

Du Toit and King differed in their estimate of the importance of continental drift during the Tertiary. Whereas the former presumed active drifting, with the production of major fold belts and separation of Australia, Antarctica and South America delayed until well after the close of the Cretaceous, King thought that there was no geological evidence favouring drifting during this time. The present interpretation inclines towards Du Toit's view, while differing in detail.

Divergence from the views expressed by palaeomagnetists is more radical. RUNCORN (1962) showed that a separation of pole positions for Europe and North America, of a sort compatible with continental drift, existed until after the Triassic. Although there are no data available for post-Triassic times, he proposed that the North Atlantic continents began to drift apart about $2 \cdot 10^8$ years ago, that is, in the Middle Triassic. CREER (1964) based his interpretation on the apparent divergence of polar wandering curves for different continents and concluded that the North Atlantic began to open up in the Late Carboniferous and Gondwanaland started to disintegrate in the Late Permian. This latter interpretation is clearly incompatible with the one favoured in this paper.

If the continents (North America possibly excepted) did not commence drifting apart to any significant extent until well into the Cretaceous one may reasonably expect to find evidence of other spectacular events at this time. One such event is the greatest marine transgression since the Palaeozoic, starting in the mid-Cretaceous and reaching a peak in the Senonian (probably Campanian–Maastrichtian).

The relationship that must exist between the growth and collapse of oceanic rises and eustatic changes of sea level has been outlined by HALLAM (1963a) and MENARD (1964, chapter 11). The case of the Darwin Rise in the West Pacific is dealt with by Menard in some detail. This huge rise, occupying an area of over $4 \cdot 10^7$ km, began to subside about 10^8 years ago following presumed uplift in the Earlier Mesozoic. The effect of this subsidence since about mid-Cretaceous times would be to lower sea level about 100 m. Menard calculated that uplift of other, younger oceanic rises within the same period would cause a sea level rise of about 300 m.

Now, if this uplift had been confined to the Tertiary, there would have been a net rise of sea level since the Cretaceous, whereas in fact there has been a progressive lowering (HALLAM, 1963a; HOLMES, 1965, fig.719). The East Pacific Rise appears to be a genuine Tertiary structure and has a number of characteristics interpreted by Menard as youthful compared with the Mid-Atlantic–Mid-Indian–Pacific–Antarctic Ridge system. Guyots, indicating subsidence, have indeed been found on the flanks of the Mid-Atlantic Ridge south of the Azores (HEEZEN et al., 1959). Guyots in the Gulf of Alaska signify Tertiary subsidence of an older part of the East Pacific Rise (MENARD, 1964, p.135).

The evidence appears to be compatible with the view that at least a large part of the Mid-Atlantic and associated ridge system commenced uplift in the latter part of the Cretaceous and has subsequently begun to subside, although the loss of relief has been compensated to some extent by continued outpourings of volcanic rock. Adapting the familiar model of convection cells in the mantle, as used by Menard, it is suggested that growth of this ridge system had two main consequences: (*a*) the initiation of continental drift, and (*b*) a major marine transgression as a result of the displacement of sea water.

It is possible that the pronounced marine regression at the close of the Cretaceous (HALLAM, 1963a) is related to the beginning of widespread subsidence of the ridge system. Alternatively there might at this time have been a significant collapse of the Darwin Rise. At any rate, it seems a reasonable presumption that the oceanic ridge system, the East Pacific Rise excepted, has undergone a decline during the Tertiary. The implication of this is that continental drift has slowed down, perhaps ceasing altogether in some areas. In support of this, palaeomagnetic data give no indication of drift since the Early Tertiary (IRVING, 1964, p.256). Also, the present rate of dyke formation in Iceland signifies dilation across the line of the Mid-Atlantic Ridge of from 3 to 6 mm per year (BODVARSSON and WALKER, 1964). This amount is less by an order of magnitude than that required, for instance, by WILSON (1963) for drifting since the Late Mesozoic.

The hypothesis propounded in this paper is amenable to testing in a variety of ways. The most decisive independent test will come from borehole cores through the older sediments of the Atlantic and Indian Ocean floors and by dating lavas dredged from the sea bed. Meanwhile there is an enormous amount of important palaeozoogeographic work to be done. Of especial value would be a detailed study of Cretaceous neritic faunas on the two sides of the North Pacific. Thorough analyses, stage by stage, of Cretaceous and Lower Tertiary neritic faunas on both sides of the Atlantic should also yield results of great interest. India, according to prediction, was isolated by sea during Late Cretaceous and Early Tertiary times and hence should reveal endemism in its vertebrate faunas of this age. These are just a few of the fascinating projects awaiting investigation.

ACKNOWLEDGEMENTS

The author has benefited from stimulating discussions with a number of

238   A. HALLAM

colleagues and is indebted to Professor F. H. Stewart and Drs. G. Y. Craig and
E. N. K. Clarkson for their critical reading of the manuscript.

REFERENCES

ARKELL, W. J., 1956. *Jurassic Geology of the World*. Oliver and Boyd, Edinburgh, 806 pp.
ARLDT, T., 1919–1922. *Handbuch der Paläogeographie*. Bornträger, Leipzig, 1 (1919): 679 pp.;
    2 (1922): 967 pp.
BASSE, E., 1953. L'extension des *Kossmaticeras* dans les Mers Antarctico–Indo-Pacifiques au
    Néocrétacé. *Proc. Pacific Sci. Congr. Pacific Sci. Assoc., 7th, New Zealand, 1949*, 2: 130–135.
BLACKETT, P. M. S., BULLARD, E. C. and RUNCORN, S. K. (Editors), 1965. A symposium on
    continental drift. *Phil. Trans. Roy. Soc. London, Ser. A*, 258: 1–323.
BLANFORD, W. T., 1890. The permanence of ocean basins. *Proc. Geol. Soc. London, 1889–1890*:
    59–109.
BODVARSSON, G. and WALKER, G. P. L., 1964. Crustal drift in Iceland. *Geophys. J.*, 8: 285–300.
CAREY, S. W., 1958. A tectonic approach to continental drift. *Geol. Dept. Univ. Tasmania, Symp.*,
    5: 177–355.
COBBAN, W. A. and REESIDE, J. B., 1952. Correlation of the Cretaceous formations of the Western
    Interior of the United States. *Bull. Geol. Soc. Am.*, 63: 1011–1044.
COLBERT, E. H., 1952. The Mesozoic tetrapods of South America. *Bull. Am. Museum Nat. Hist.*,
    99: 237–249.
COLBERT, E. H., 1958. Relationships of the Triassic Maleri fauna. *J. Palaeontol. Soc. India*, 3:
    68–81.
COLBERT, E. H., 1962. *Dinosaurs: Their Discovery and Their World*. Hutchinson, London, 288 pp.
COLLIGNON, M., 1934. Fossiles turoniens d'Antantiloky. *Ann. Geol. Serv. Mines (Madagascar)*,
    4: 7–59.
COX, L. R., 1951. Notes on the Trigoniidae, with outlines of a classification of the family. *Proc.
    Malac. Soc. London*, 29: 45–70.
COX, L. R., 1965. Jurassic Bivalvia and Gastropoda from Tanganyika and Kenya. *Bull. Brit.
    Museum, Geol. Suppl.*, 1: 1–213.
CREER, K. M., 1964. A reconstruction of the continents for the Upper Palaeozoic from palaeo-
    magnetic data. *Nature*, 203: 1115–1120.
DARLINGTON, P. J., 1948. The geographical distribution of cold-blooded vertebrates. *Quart.
    Rev. Biol.*, 23: 1–26, 105–123.
DAVIES, A. M., 1934. *Tertiary Faunas*. Murby, London, 2: 252 pp.
DIENER, C., 1915. Die marinen Reiche der Trias. *Denkschr. Akad. Wiss. Wien, Math. Naturw. Kl.*,
    92: 405–549.
DORMAN, F. H. and GILL, E. D., 1959. Oxygen isotope paleotemperature determinations of
    Australian Cainozoic fossils. *Science*, 130: 1576.
DURHAM, J. W., 1952. Early Tertiary marine faunas and continental drift. *Am. J. Sci.*, 250: 321–
    343.
DURHAM, J. W., 1959. Palaeoclimates. In: L. H. AHRENS (Editor), *Physics and Chemistry of the
    Earth*. Pergamon, London, 3: 1–16.
DU TOIT, A., 1937. *Our Wandering Continents*. Oliver and Boyd, Edinburgh, 366 pp.
EKMAN, S., 1953. *Zoogeography of the Sea*. Sidgwick and Jackson. London, 418 pp.
EMILIANI, C., 1954. Temperature of Pacific bottom waters and polar superficial waters during the
    Tertiary. *Science*, 119: 853–855.
EMILIANI, C., 1961. The temperature decrease of surface seawater in high latitudes and of abyssal-
    hadal water in open oceanic basins during the past 75 million years. *Deep-Sea Res.*, 8:
    144–147.
FISCHER, A. G., 1961. Latitudinal variations in organic diversity. *Am. Scientist*, 49: 50–74.
FLEMING, C. A., 1962. New Zealand biogeography; a paleontologist's approach. *Tuatara*, 10:
    53–108.

FLEMING, C. A., 1964. History of the bivalve family Trigoniidae in the southwest Pacific. *Australian J. Sci.*, 26: 196–204.

FURON, R., 1963. *Geology of Africa*. Oliver and Boyd, Edinburgh, 377 pp.

GILL, E. D., 1961. The climates of Gondwanaland in Kainozoic time. In: A. E. M. NAIRN (Editor), *Descriptive Palaeoclimatology*. Interscience, New York, N.Y., pp. 332–353.

HALLAM, A., 1963a. Major epeirogenic and eustatic changes since the Cretaceous, and their possible relationship to crustal structure. *Am. J. Sci.*, 261: 397–423.

HALLAM, A., 1963b. Eustatic control of major cyclic changes in Jurassic sedimentation. *Geol. Mag.*, 100: 444–450.

HALLAM, A., 1965. Observations on marine Lower Jurassic stratigraphy of North America, with special reference to United States. *Bull. Am. Assoc. Petrol. Geologists*, 49: 1485–1501.

HAUFF, B., 1953. *Das Holzmadenbuch*. Rau, Öhringen, 80 pp.

HAUG, E., 1900. Les geosynclinaux et les aires continentales; contribution à l'étude des transgressions et des régressions marines. *Bull. Soc. Géol. France*, 3(38): 617–711.

HAUGHTON, H. S., 1956. Gondwanaland and the distribution of early reptiles. *Trans. Geol. Soc. S. Africa—Du Toit Mem. Lecture*, 56: 1–30.

HEEZEN, B. C., THARP, M. and EWING, M., 1959. The floors of the oceans. 1. The North Atlantic. *Geol. Soc. Am., Spec. Papers*, 65.

HILLS, E. S., 1958. A brief review of Australian fossil vertebrates. In: T. S. WESTOLL (Editor), *Studies on Fossil Vertebrates*. Univ. Press, London, pp.86–107.

HÖLDER, H., 1964. *Jura*. (Band IV of *Handbuch der stratigraphischen Geologie*). Enke, Stuttgart, 603 pp.

HOLMES, A., 1965. *Principles of Physical Geology*, 2nd ed. Nelson, London, 1288 pp.

IMLAY, R. W., 1960a. Ammonites of Early Cretaceous age (Valanginian and Hauterivian) from the Pacific coast states. *U.S., Geol. Surv., Profess. Papers*, 334-F: 167–228.

IMLAY, R. W., 1960b. Early Cretaceous (Albian) ammonites from the Chitina Valley and Talkeetna Mountains, Alaska. *U.S., Geol. Surv., Profess. Papers*, 354-D: 87–114.

IMLAY, R. W., 1961. Characteristic Lower Cretaceous megafossils from northern Alaska. *U.S., Geol. Surv. Profess. Papers*, 335: 1–74.

IRVING, E., 1964. *Paleomagnetism and its Applications to Geological and Geophysical Problems*. Wiley, New York, 399 pp.

IRVING, E. and BROWN, D. A., 1964. Abundance and diversity of the labyrinthodonts as a function of paleolatitude. *Am. J. Sci.*, 262: 689–708.

JAWORSKI, E., 1914. Beiträge zur Kenntnis des Jura in Südamerika. *Neues Jahrb. Mineral., Geol. Paläontol.*, 37: 285–342, 40: 364–456.

JAWORSKI, E., 1923. Die marine Trias in Südamerika. *Neues Jahrb. Mineral., Geol. Paläontol.*, 47: 93–200.

JONES, D. L., 1963. Upper Cretaceous (Campanian and Maestrichtian) ammonites from southern Alaska. *U.S., Geol. Surv., Profess. Papers*, 432: 1–53.

JONES, D. L. and GRYC, G., 1960. Upper Cretaceous pelecypods of the genus *Inoceramus* from northern Alaska. *U.S., Geol. Surv., Profess. Papers*, 334-E: 149–165.

KING, L. C., 1962. *Morphology of the Earth*. Oliver and Boyd, Edinburgh, 699 pp.

KURTÉN, B., 1966. Holarctic land connexions in the Early Tertiary. *Comm. Biol. Soc. Sci. Fenn.*, 29: 1–5.

MATSUMOTO, T., 1959. Zonation of the Upper Cretaceous of Japan. *Mem. Fac. Sci., Kyushu Univ., Ser. D*, 9: 55–93.

MATTHEW, W. D., 1915. Climate and evolution. *Ann. N.Y. Acad. Sci.*, 24: 171–318.

MAYR, E., 1944. Wallace's Line in the light of recent zoogeographic studies. *Quart. Rev. Biol.*, 19: 1–14.

MAYR, E. (Editor), 1952. The problem of land connections across the South Atlantic, with special reference to the Mesozoic. *Bull. Am. Museum Nat. Hist.*, 99: 85–258.

MAYR, E., 1953. Fragments of a Papuan ornithography. *Proc. Pacific. Sci. Congr. Pacific Sci. Assoc., 7th, New Zealand, 1949*, 4: 11–19.

MENARD, H. W., 1964. *Marine Geology of the Pacific*. McGraw Hill, New York, N.Y., 271 pp.

MILLOT, J., 1952. La faune malgache et le mythe gondwanien. *Mem. Inst. Sci. Madagascar, Sér. A*, 7: 1–36.

MULLER, S. W. and FERGUSON, H. G., 1939. Mesozoic stratigraphy of the Hawthorne and Tonopah quadrangles, Nevada. *Bull. Geol. Soc. Am.*, 50: 1572–1624.

MYERS, G. S., 1938. Fresh-water fishes and West Indian zoogeography. *Smithsonian Inst., Ann. Rept.*, 1937: 339–364.

MYERS, G. S., 1953a. Ability of amphibians to cross sea barriers, with especial reference to Pacific zoogeography. *Proc. Pacific Sci. Congr. Pacific Sci. Assoc., 7th, New Zealand, 1949*, 4: 19–27.

MYERS, G. S., 1953b. Palaeogeographical significance of fresh-water fish distribution in the Pacific. *Proc. Pacific Sci. Congr. Pacific Sci. Assoc., 7th, New Zealand, 1949*, 4: 38–48.

NAKANO, M., 1960. Stratigraphic occurrences of the Cretaceous trigoniids in the Japanese Islands and their faunal significances. *J. Sci. Hiroshima Univ., Ser. C*, 3: 215–280.

NEUMAYR, M., 1883. Über klimatische Zonen während der Jura- und Kreidezeit. *Denkschr. Akad. Wiss. Wien, Math. Naturw. Kl.*, 47: 277–310.

NEUMAYR, M., 1887. *Erdgeschichte*. Bibliogr. Inst., Leipzig, 1: 653 S.; 2: 879 S.

OLSON, E. C., 1957. Catalogues of localities of Permian and Triassic terrestrial vertebrates of the territories of the U.S.S.R. *J. Geol.*, 65: 196–226.

REYMENT, R. A., 1965. Upper Cretaceous fossil molluscs in South America and west Africa. *Nature*, 207: 1384.

ROGER, J., 1949. Paléobiologie des invertébrés fossiles de Madagascar. *Mem. Inst. Sci. Madagascar, Sér. D*, 1: 97–115.

RUNCORN, S. K., 1962. Palaeomagnetic evidence for continental drift and its geophysical cause. In: S. K. RUNCORN (Editor), *Continental Drift*. Acad. Press, New York, N.Y.

SCHAEFFER, B., 1952. The problem of the fresh-water fishes. *Bull. Am. Museum Nat. Hist.*, 99: 227–235.

SCHUCHERT, C., 1932. Gondwana land bridges. *Bull. Geol. Soc. Am.*, 43: 875–916.

SCHWARZBACH, M., 1963. *Climates of the Past*. Van Nostrand, London, 340 pp.

SIMPSON, G. G., 1939. Antarctica as a faunal migration route. *Proc. Pacific Sci. Congr. Pacific Sci. Assoc. 6th, 1949*, pp.755–768.

SIMPSON, G. G., 1940a. Mammals and land bridges. *J. Wash. Acad. Sci.*, 30: 137–163.

SIMPSON, G. G., 1940b. Review of the mammal-bearing Tertiary of South America. *Proc. Am. Phil. Soc.*, 83: 649–710.

SIMPSON, G. G., 1943. Mammals and the nature of continents. *Am. J. Sci.*, 241: 1–31.

SIMPSON, G. G., 1947. Holarctic mammalian faunas and continental relationships during the Cenozoic. *Bull. Geol. Soc. Am.*, 58: 613–688.

SIMPSON, G. G., 1952. Probabilities of dispersal in geologic time. *Bull. Am. Museum Nat. Hist.*, 99: 163–176.

STEPHENSON, L. W., 1941. The larger invertebrates of the Navarro Group, Texas. *Texas, Univ. Publ.*, 4101: 1–641.

STEPHENSON, L. W., 1952. The larger invertebrate fossils of the Woodbine Formation (Cenomanian) of Texas. *U.S., Geol. Surv., Profess. Papers*, 242: 1–226.

STEVENS, G. R., 1965. The Jurassic and Cretaceous belemnites of New Zealand and a review of the Jurassic and Cretaceous belemnites of the Indo-Pacific region. *New Zealand Geol. Surv. Palaeont. Bull.*, 36: 1–283.

STIRTON, R. A., 1958. The relationships and origin of Australian monotremes and marsupials. *Geol. Dept. Univ. Tasmania, Symp.*, 5: 172–174.

SUESS, E., 1906. *The Face of the Earth*, 2. Clarendon Press, Oxford, 556 pp.

TEICHERT, C., 1958. Australia and Gondwanaland. *Geol. Rundschau*, 47: 562–590.

TERMIER, H. and TERMIER, G., 1960. *Atlas de Paléogéographie*. Masson, Paris, 99 pp.

THORSON, G., 1961. Length of pelagic larval life in marine bottom invertebrates as related to larval transport by ocean currents. In: M. SEARS (Editor), *Oceanography—Am. Assoc. Advan. Sci.*, 67: 455–474.

UHLIG, V., 1911. Die marinen Reiche des Jura und der Unterkreide. *Mitt. Geol. Ges. Wien*, 3: 329–448.

UMBGROVE, J. H. F., 1947. *The Pulse of the Earth*, 2nd ed. Nijhoff, The Hague, 358 pp.

WADE, B., 1926. The fauna of the Ripley Formation on Coon Creek, Tennessee. *U.S., Geol. Surv., Profess. Papers*, 137: 1–272.

WEGENER, A., 1924. *The Origin of Continents and Oceans*, 3rd ed. Methuen, London, 212 pp.
WILLIS, B., 1932. Isthmian links. *Bull. Geol. Soc. Am.*, 43: 917–952.
WILSON, J. T., 1963. Hypothesis of earth's behaviour. *Nature*, 198: 925–929.
WOOD, A. E., 1950. Porcupines, paleogeography and parallelism. *Evolution*, 4: 87–98.

From *Biogeography of the Southern End of the World*
Philip J. Darlington Jr.

4. Southern distributions in relation to the world: four groups of carabid beetles

To illustrate the fourth and fifth principles listed at the beginning of Chapter 1, that situations in the southern cold-temperate zone should be considered in relation to the whole world and that it is important not only to investigate single cases in detail but also to compare cases, I shall take a set of four groups of Carabidae.

Migadopini. The distribution of this tribe is described in sufficient detail in the preceding chapter. The tribe itself is confined to the south temperate zone and most of the species are southern *cold-temperate* or subantarctic in distribution (Fig. 9). However, the closest relatives of the Migadopini are apparently the Elaphrini, and the latter are confined to the cooler parts of the north temperate zone, in temperate Eurasia and North America. This suggests (as one possibility but not the only one) the existence a long time ago, perhaps in the Mesozoic, of a common ancestral stock that was adapted primarily to temperate climates but that somehow crossed the tropics, in one direction or the other, and then evolved separate tribes in northern and southern temperate areas.

Broscini. Occurring with Migadopini in the wet forests of southwestern South America and Tasmania are certain members of the tribe Broscini, a tribe of medium-sized or large, ground-living Carabidae (Fig. 10A). Some of the South Chilean and Tasmanian forest-living forms are very much alike, although they belong to different genera, *Cascellius* in Chile and *Promecoderus* in Tasmania. These two genera may actually be related, but probably less closely than their external similarity suggests. Certain particular species of these two genera resemble each other remarkably, perhaps as a result of convergence in similar niches in the South Chilean and Tasmanian forests. For example, *Cascellius* in Chile and *Promecoderus* in Tasmania each include, living in wet *Nothofagus* forest, a medium-sized bright-green species and a smaller brown one. An observer who knew only the Chilean and Tasmanian faunas might be overimpressed by these similarities.

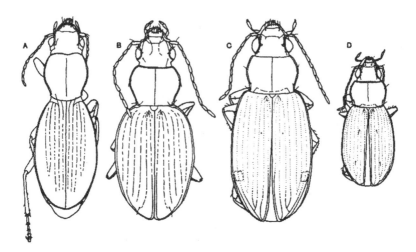

Fig. 10. Subantarctic carabid beetles from Puerto Williams, South Chile: *A, Cascellius gravesi* Curt. (Broscini); *B, Trechisibus antarcticus* Dej. (Trechini); *C, Bembidion kuscheli* Jean.; *D, Bembidion nitidum* Jean.; *A*, ca. 4.2×; others, ca. 10×.

However, other broscines exist, and their distribution as a whole is amphitropical (Fig. 11). They occur around the world both north and south of the tropics but are absent in the tropics themselves, except that primarily temperate genera apparently enter the edges of the tropics in Asia and Australia. The geographic relationships of broscine genera have been discussed several times recently (Jeannel 1941; Britton 1949; Ball 1956). I shall be concerned here mainly with the southern forms. Of the northern ones I need say only that they do exist and that some of them are winged, as the ancestor of the tribe must have been. The wings have atrophied in all the southern forms, so far as I can determine.

In the Southern Hemisphere, Broscini occur in the same three general areas as Migadopini: southern South America, southern Australia with Tasmania, and New Zealand. However, details are different. Broscines are more widely distributed than migadopines in open steppe as well as forest in southern and western South America and in open as well as forested parts of southern Australia, while migadopines are more widely distributed on the southernmost moorland. For example, migadopines are represented by endemic genera on subantarctic moorland in extreme southern South America and the Falkland Islands and also on the Auckland Islands south of New Zealand, while broscines are undifferentiated

Relation to the World   41

on the southernmost moorland in South America and absent on
the islands in question. In South Chile several species of broscines
of the genus *Cascellius* do extend onto subantarctic moorland, but
they are all primarily forest-living species and the genus as a whole
is primarily forest-living.

On the whole, although Migadopini and Broscini occur together in many places, the distribution of the Migadopini within
the south-temperate zone is more southern than that of the Broscini, and the Migadopini are more differentiated in the southernmost habitats. This difference suggests that migadopines have
been in the southern cold-temperate zone longer than broscines,

Fig. 11. Distribution of carabid beetles of the tribe Broscini. Areas of occurrence
in the Southern Hemisphere are bounded (except that northern limits in Australia and South America are not determined), and southern limits in the Northern
Hemisphere are shown (approximately) by heavy lines. Double-ended arrows indicate the geographic relationships of the broscids of different areas.

long enough for generic differentiation on the moorland, while
the broscines have moved south more recently and are just beginning to invade the southernmost moorland. However, broscines
have evidently been in the Southern Hemisphere for a considerable time. All the southern genera are different from northern
ones, and no genus occurs in more than one of the three main
areas of distribution in the south.

Several genera of Broscini occur on New Zealand. They are
all endemic and all may be derived from a single ancestor (Ball
1956:44, 47). This hypothetical ancestor of all New Zealand Broscini presumably reached New Zealand from Australia. Related
forms still occur in Australia, and also in the Northern Hemisphere. However, the New Zealand broscines are *not* directly related to the small, forest-living forms (*Cascellius* and *Promecoderus*)
that are so similar in Tasmania and southern Chile. The result
is that, while there are or may be direct relationships between
some Australian and South American broscine genera, and between other Australian and New Zealand genera, there are no
direct relationships between the broscines of South America and
New Zealand (see arrows on Fig. 11). If the existing Australian
relatives of the New Zealand broscines should become extinct,
some surviving Australian and South American genera would be
(or seem to be) related, while the New Zealand forms would
seem to be derived from a separate, northern stock. This pattern
of relationship does occur in some other groups of Carabidae,
including the Trechini (below; and see p. 64).

Trechini. A third tribe of Carabidae important in the southern
hemisphere is the Trechini. It is a large tribe of small carabids
(Fig. 10*B*). Some are winged; others are flightless, with atrophied
wings. Some of the winged forms live beside running water. Most
of the flightless ones live on the ground in forest, on moorland,
on mountains, or in caves. A few trechines live on the seashore
between tide lines and may disperse on ocean drift, but they
form a separate taxonomic group and are not concerned in the
history of the terrestrial members of the tribe.

Jeannel (1926–1928) has published a classic monograph of the
tribe Trechini (which he considered a subfamily). The tribe as
a whole is almost amphitropical in distribution (Fig. 12), best
represented in the north temperate and south temperate zones,

Fig. 12. Distribution of carabid beetles of the two principal terrestrial subtribes of the tribe Trechini. Areas of occurrence in the Southern Hemisphere are bounded (the occurrence in southwestern Australia is based on the discovery of an undescribed species there), and southern limits in the Northern Hemisphere are shown (approximately) by heavy lines. Triangles indicate geographic relationships: Homaloderina occur in Australia (but not New Zealand), South America, and Spain; Trechina, in the Northern Hemisphere and New Zealand.

although a few water-loving genera occur in or across the tropics. However, the tropical forms are mostly winged. Flightless trechines are almost all either northern or southern, not tropical. Most of the flightless species belong to two principal subtribes (called tribes by Jeannel), one mainly northern, the other mainly southern. Although most species of both subtribes are now flightless, both include some winged species and both are evidently derived from winged ancestors presumably capable of dispersal by flight. The northern subtribe (Trechina) is dominant across temperate

Eurasia and North America. This subtribe is absent in South America and in Australia and Tasmania but is represented on New Zealand by a group of 3 interrelated endemic genera (all, I should think, derived from one ancestor) and at least 9 species, most of them described by Britton (1962 and other papers). The other principal subtribe (Homaloderina) includes most of the terrestrial trechines of southern South America and southern Australia and Tasmania, and they occur in the southernmost wet forest and on moorland in both places. However, the South American and Australian-Tasmanian genera are all different. And this subtribe is not represented on New Zealand but is represented in the Northern Hemisphere by one blind, presumably relict, monotypic genus in a cave in Spain! This pattern of distribution is indicated by triangles in Fig. 12. The geographic pattern is so surprising that I doubted the reality of it until I had personally confirmed the characters on which the classification is based.

The classification of subtribes of Trechini is based on structural characters including mandibular teeth, and the characters do seem to be valid. The structural classification is partly but not wholly supported by an apparent difference in adaptive potential of the subtribes. Members of the northern subtribe commonly enter caves and become blind and otherwise highly modified there (in Europe, Japan, eastern North America, and elsewhere), and some of the supposedly related New Zealand forms are cave dwelling too (Britton 1962). On the other hand, the southern subtribe has apparently produced no true cave dwellers in South America or Australia-Tasmania, while the supposedly related relict in Spain is a blind cave dweller, and this fact may throw some doubt on its supposed relationship.

However, regardless of details, the tribe Trechini (Jeannel's subfamily Trechinae) surely has two principal zones of dominance, north and south of the tropics, and several different members of the tribe surely have crossed or are crossing the tropics, in one direction or the other. Each of the two principal subtribes discussed above has crossed at least once, and three additional, primarily winged groups extend part or all of the way across the tropics now. Those that do so in the Old World are *Perileptus*, which ranges almost continuously from Japan and southeastern Asia to southeastern Australia (and from Europe to South Africa), and *Trechodes*, which occurs very discontinuously in southeastern

Asia, the Philippines, and Australia (and in Africa and Madagascar), with a related genus, *Cyphotrechodes*, on mountain moorland in Tasmania. And, in the New World, trechines of the genus *Cnides* range from part of Central America to part of Chile.

Bembidion. The fourth and last group of Carabidae to be discussed now is the genus *Bembidion*. This is a very large genus of small carabids (Fig. 10C, D) which live on the ground and are comparable to trechines in size and habits, except that *Bembidion* rarely enters caves. Some *Bembidion* are winged and some have atrophied wings, and some live beside water and others away from it. A few live on the seashore or between tide lines, but these are not concerned in the distribution of the dominant terrestrial forms. The genus *Bembidion* as a whole has not been revised, but fortunately Jeannel (1962) has recently reviewed the known South American species, and I (1962), the Australian ones. The Eurasian ones have been put in basic order by Netolitzky (1942–1943) and most of the North American ones by Lindroth (1963a). Netolitzky, Lindroth, and I have treated *Bembidion* as one large genus with many subgenera, while Jeannel has split it into many separate genera, but this difference in usage does not affect the distribution of the group and need not concern present readers. In any case, *Bembidion* in the broad sense is a much more compact group than the tribe Trechini and is therefore probably more recent in evolution and dispersal.

The distribution of this single genus (Fig. 13) roughly parallels that of the whole tribe Trechini. *Bembidion*, like the Trechini, is bizonally dominant, with hundreds of species around the world north of the tropics in Eurasia and North America, and also fair numbers in southern South America, a few in southern Australia and Tasmania, and some on New Zealand. The genus is very poorly represented in the tropics, although a few species do occur there in both Old and New Worlds. Most of the few tropical species live at high altitudes on mountains or, if at low altitudes, beside fresh or salt water, the presence of which seems to give some protection against tropical climate. Most of the few lowland tropical species are winged as well as water-loving. For some further discussion of the general distribution of *Bembidion* see Darlington 1953 and 1959c.

The distribution of *Bembidion* in southern South America has

Fig. 13. Distribution of carabid beetles of the genus *Bembidion*. Terrestrial forms occur on all habitable continents and principal islands of the world, except New Guinea, but are concentrated principally north and south of the tropics, in areas indicated by heavier hatching. Arrows show apparent direction of dispersal of 7 stocks of the genus into the Southern Hemisphere.

been noted in the discussion of southward diminution of the insect fauna (p. 22). About 70 species of the genus are known in South America, most of them along the Andes or at low altitudes in the warmer part of the south temperate zone. Some (all?) of them are related to and apparently derived from North American forms. In fact, the South American species of *Bembidion* include at least 3 northern subgenera (called genera by Jeannel), indicating at least 3 crossings of the American tropics by different members of the genus. At least 8 species range south to the Straits of Magellan on the opener, warmer, eastern side of South America, but only 2 reach the south side of the Beagle Channel, and they occur there only in special habitats that receive full sun (see p. 24). No species of *Bembidion* occurs in the southernmost forest or moorland in South America, although Trechini do so. However, one of the two southernmost South American species of *Bembidion* has entered another habitat characteristic of the far south, sphagnum bogs, and has lost or is losing its wings there. I have short-winged specimens of it from Puerto Williams.

Five native species of *Bembidion* occur in Australia, chiefly in the southern part of the continent (Darlington 1962). They are not directly related to South American forms but seem to have been derived independently from the north, from Eurasia or the Old World tropics, by three successive invasions of Australia. Four of the five species extend to Tasmania and, although three of them occur at only low altitudes there, the fourth reaches mountain moorland. Also, one of the species (not the one that reaches the mountain moors) is undergoing wing atrophy: most Tasmanian specimens of it are short-winged. There is thus a remarkable convergence between unrelated stocks of *Bembidion* in southernmost South America and Tasmania. In each place members of the genus, independently derived from the north, are now beginning to invade far-southern habitats and to evolve flightless forms.

The *Bembidion* of New Zealand have not been critically studied, but superficial examination suggests that they are derived not from any existing Australian or South American stock but from a separate ancestor which, if it came by way of Australia, has disappeared there. New Zealand possesses more, and more diverse, species of *Bembidion* than Australia does. This suggests a longer period of evolution in New Zealand, which is consistent with derivation of the New Zealand forms from a separate, older ancestor.

Apparent dispersal cycle of Carabidae. The geographic history of southern Carabidae must be deduced from their present distributions and from various indirect clues, for pertinent fossils are lacking.

When the four groups of Carabidae discussed above are compared in reverse order—*Bembidion* first and Migadopini last—it is seen that they may represent successive stages in a common, world-wide cycle of evolution and dispersal. The apparent cycle is: rise on the large land masses in the Northern Hemisphere, or possibly in the tropics; dispersal southward into southern South America and southern Australia by separate routes, and to New Zealand probably from Australia; disappearance of the tropical or tropics-crossing forms, leaving an amphitropical pattern; and finally disappearance from the Northern Hemisphere, leaving survivors on the three main pieces of land in the southern cold-temperate zone. (Other groups might disappear in the Southern Hemisphere and survive only in the Northern.)

All four groups of Carabidae discussed above still include winged forms. All the groups are therefore derived from winged ancestors probably capable of dispersing rapidly, for long distances, and across barriers. All four groups can tolerate cold climates (some Carabidae apparently cannot), and all now occur principally in cool- or cold-temperate areas. But all four groups have also a few representatives either actually in the tropics (*Bembidion* and Trechini) or in warm-temperate parts of southern South America and Australia (Migadopini and Broscini). All the groups are therefore apparently cold-tolerant rather than strictly cold-adapted. Moreover, the tropical forms, when they exist, are usually winged and presumably able to make the long, multiple dispersals required by the hypothetical cycle. In most cases the existing tropical forms cannot be the direct ancestors of existing south-temperate groups, but they show how the latters' ancestors may have crossed the tropics.

If these four groups of Carabidae do represent stages in a common cycle of dispersal, *Bembidion* is just beginning it, or is nearest the beginning. This is suggested by the relatively slight differentiation of the genus in different parts of the world. It is neither well represented nor much differentiated in the *colder* parts of the south temperate zone. It is apparently just beginning to invade and become adapted to far-southern habitats, and is doing so, independently, in both southern South America and Tasmania.

The Trechini seem to have moved somewhat further than *Bembidion* through the hypothetical cycle. The dominant tribes have formed an essentially bizonal pattern of distribution in the temperate zones, although a few winged forms still extend across the tropics. Also, the trechines as a group have diversified much more than *Bembidion*, and have invaded, differentiated in, and radiated in far-southern habitats much more than *Bembidion* has done.

The Broscini now have a strictly bizonal distribution. They

Relation to the World   49

must have crossed the tropics, but the tropics crossers have disappeared. To this extent at least Broscini have gone further than Trechini through the hypothetical cycle.

Finally the Migadopini are nearest the end of the cycle. If they ever existed in the tropics and the Northern Hemisphere, they have disappeared there and survive as relicts only in the

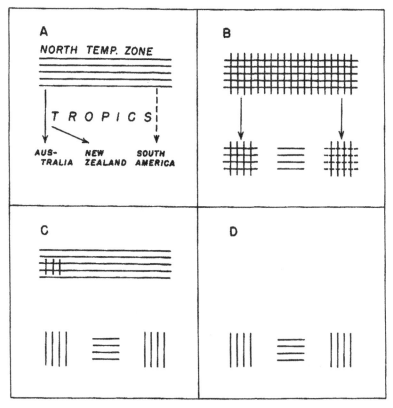

Fig. 14. Diagram of the hypothetical pattern of evolution and dispersal suggested by carabid beetles. *A,* an initial stock evolves in the north and disperses southward to Australia, New Zealand, and perhaps South America. *B,* a second stock evolves in the north and disperses southward to Australia and South America, but not New Zealand. *C,* the *first* stock dies out in Australia and (if it existed there) in South America, and the *second* stock dies out in the north except for local relicts. *D,* both stocks disappear completely in the north, leaving relicts in the three principal areas in the southern cold-temperate zone. Note that *C* is the actual pattern of distribution of the principal subtribes of Trechini now (Fig. 12). Many variations of this general pattern are possible.

south temperate zone. The differentiation, localization, and adaptation of different genera to different far-southern habitats (indicating long residence in the south) and the fact that their relatives, so far as they have relatives, are northern and tribally distinct (suggesting a crossing of the tropics a long time ago) are at least consistent with the Migadopini having passed through the suggested dispersal cycle long ago. But even if they have done so, they may have undergone further dispersal, later, in the far south.

The hypothetical dispersal pattern suggested by Carabidae is diagrammed in Fig. 14, and the complexity of it is further discussed, in connection with New Zealand, in Chapter 6.

Works Cited

Ball, G. E. 1956. Notes . . . on the classification of the tribe Broscini . . . . *Coleopterists' Bull.* 10: 33–52.

Britton, E. B. 1949. The Carabidae (Coleoptera) of New Zealand. Part III: A revision of the tribe Broscini. *Trans. Royal Soc. New Zealand* 77: 533–81.

———. 1962. New genera of beetles (Carabidae) from New Zealand. *Ann. Mag. Nat. Hist.* 13 (4): 665–72.

Darlington, P. J., Jr. 1953. A new *Bembidion* (Carabidae) of zoogeographic importance from the Southwest Pacific. *Coleopterists' Bull.* 7: 12–16.

———. 1959c. The *Bembidion* and *Trechus* (Col.: Carabidae) of the Malay Archipelago. *Pacific Insects,* 1:331–44. Honolulu: Bishop Museum.

———. 1962. Australian carabid beetles X. *Bembidion. Breviora* (Mus. Comp. Zool.), no. 162, 1–12.

Jeannel, R. 1926–28. Monographie des Trechinae. *L'Abeille* 32 (3): 33, 35.

———. 1941. *Coléoptères carabiques, première partie.* Vol. 39 of *Faune de France.* Paris: Lechevalier.

———. 1962. Les trechides de la Paléantarctide Occidentale. *Biologie de l'Amérique Australe,* vol. 1: *Études sur la faune du sol,* 527–655. Paris: CNRS.

Lindroth, C. H. 1963a. *The Ground-beetles of Canada and Alaska,* parts 2 and 3 (*Opuscula Entomol.,* suppl. 20 and 24, Lund, Sweden).

Netolitzky, F. 1942–43. Bestimmungstabellen . . . Gattung *Bembidion* Latr. *Koleopterologische Rundschau,* Band 28, Heft 1/3, 3/6; Band 29, Heft 1/3.

12 March 1982, Volume 215, Number 4538

SCIENCE

Mammalian Evolution and the Great American Interchange

Larry G. Marshall, S. David Webb
J. John Sepkoski, Jr., David M. Raup

Biogeographers have long recognized the late Cenozoic mingling of the previously separated American continental biotas as a monumental natural experiment, the Great American Interchange. Comparison of the "wonderful extinct fauna . . . discovered in North America, with what was previously known from South America" allowed Wallace to first recognize the existence of this event in 1876 (1). However, the direction in which representatives of the various animal groups dispersed was not well understood by Wallace (2). It took another 15 years of intense paleontological exploration and study by Cope and Marsh in North America, and by Carlos and Florentino Ameghino in South America before sufficient data existed to permit clarification of many of the basic issues in this event (3). By 1891 a comprehensive and balanced overview of the interchange existed:

. . . not only did North American taxa cross the newly opened land bridge, greatly expanding their ranges, but also South American autochthons began ranging into North America, and thus toward the end of the Pliocene epoch took place one of the most remarkable faunal exchanges that Geology has known [Karl A. von Zittel (4)].

Continued research in the present century has resulted in elaboration of the histories of the participant and nonparticipant taxa. This cumulative knowledge has been periodically summarized by Matthew (5), Scott (6), Simpson (7), Patterson and Pascual (8), and others (9–12).

Despite the wealth of accumulated knowledge, many finer details of the

interchange have remained obscure. Recent improvements in paleontological sampling (especially screen-washing for taxa of small body size); refined taxonomic studies spanning both continents; and availability of an array of radioisotopic age determinations interpolated within the late Cenozoic land-mammal bearing strata on each continent permit clarification of unresolved earlier prob-

Mammals (8, 13). This isolation ended about 3 million years ago with the disappearance of the Bolivar Trough Marine Barrier in the area of northwestern Colombia and southern Panama, and the total emergence of the Panamanian land bridge (8). Thereafter the fossil record documents a reciprocal intermingling of the long-separated North and South American terrestrial biotas. Since the Bolivar Trough served as the final geographic barrier separating these biotas, the area to the north of it is here referred to as North America and the area south of it as South America. The area of the former Bolivar Trough is thus the "gateway" for the Great American Interchange.

On the basis of the timing and the means of dispersal, the participants in the Great American Interchange can be divided into two groups. The first group includes late Miocene waif immigrants, which are believed to have dispersed along island arcs before the final emergence of the land bridge (7). This group

Summary. A reciprocal and apparently symmetrical interchange of land mammals between North and South America began about 3 million years ago, after the appearance of the Panamanian land bridge. The number of families of land mammals in South America rose from 32 before the interchange to 39 after it began, and then back to 35 at present. An equivalent number of families experienced a comparable rise and decline in North America during the same interval. These changes in diversity are predicted by the MacArthur-Wilson species equilibrium theory. The greater number of North American genera (24) initially entering South America than the reverse (12) is predicted by the proportions of reservoir genera on the two continents. However, a later imbalance caused by secondary immigrants (those which evolved from initial immigrants) is not expected from equilibrium theory.

lems. There now exists sufficient knowledge of these aspects of the interchange to permit quantitative, rather than simply qualitative, examination of patterns of faunal dispersal and evolution. In this article we have compiled such quantitative data, and we use them to examine both empirical patterns of faunal interchange and correspondence to models of equilibrial diversities and biogeography.

Qualitative Aspects of the Interchange

South America was isolated from other continents during most of the Age of

includes members of two families of North American origin: (i) procyonids (racoons and allies), which are first recorded in beds of late Miocene (Huayquerian) age in Argentina (11, 14), and (ii) cricetid rodents (New World rats and mice) of the tribe Sigmodontini (Hesper-

L. G. Marshall is an assistant curator of fossil mammals in the Department of Geology, Field Museum of Natural History, Chicago, Illinois 60605. S. D. Webb is curator of fossil vertebrates, Florida State Museum and professor of Zoology, University of Florida, Gainesville 32611. J. John Sepkoski, Jr., is an assistant professor of paleontology in the Department of the Geophysical Sciences, University of Chicago, Chicago, Illinois 60637. D. M. Raup is dean of science, Field Museum of Natural History, Chicago, Illinois 60605 and professor in the Department of the Geophysical Sciences, University of Chicago, Chicago, Illinois 60637.

0036-8075/82/0312-1351$01.00/0 Copyright © 1982 AAAS

1351

419

omyini), which are first recorded in beds of early Pliocene (Montehermosan) age (15) in Argentina. It also includes members of the extinct South American ground sloth families Megalonychidae and Mylodontidae, which are first recorded in North America in beds of late Miocene (Hemphillian) age (11).

Included within the second group of participants are those taxa that walked across the bridge after its final emergence. The North American immigrants to South America include members of the families (i) Mustelidae (skunks and allies) and Tayassuidae (peccaries), which first appear in the late Pliocene (Chapadmalalan Age) (8); (ii) Canidae (dogs, wolves, foxes), Felidae (cats), Ursidae (bears), Camelidae (camels, llamas), Cervidae (deer), Equidae (horses), Tapiridae (tapirs), and Gomphotheriidae

(mastodonts) which appear in the early Pleistocene (Uquian Age) (8); and (iii) Heteromyidae (kangaroo rats and allies), Sciuridae (squirrels), Soricidae (shrews), and Leporidae (rabbits) which are known only from Holocene or Recent (or both) (8, 16, 17). The South American immigrants to North America include members of the families (i) Dasypodidae (armadillos), Glyptodontidae (glyptodonts), Hydrochoeridae (capybaras), and Erethizontidae (porcupines), which appear in the late Pliocene (late Blancan Age) (9, 11); (ii) Didelphidae (opossums) and Megatheriidae (ground sloths), which appear in the early and middle Pleistocene (Irvingtonian Age) (9, 11); (iii) Toxodontidae (toxodonts), which are recorded in the late Pleistocene (Rancholabrean Age) (8); and (iv) Callitrichidae (marmosets and tamarins), Cebidae

(New World monkeys), Choleopodidae (tree sloths), Bradypodidae (tree sloths), Cyclopidae (anteaters), Myrmecophagidae (anteaters), Dasyproctidae (agoutis, pacas), and Echimyidae (spiny rats), which are known only in Recent faunas (17).

The late Cenozoic record of fossil mammals in North and South America is relatively well documented. The one great deficiency of the South American record is that it is largely restricted to Argentina (8) and to a lesser extent Bolivia (18). Both the North and South American records are deficient for mammals from tropical latitudes. Nevertheless, inferences gleaned from these records yield generalities that are probably valid for the continents as a whole (7). Furthermore, the majority of faunas sampled appear to represent savanna-

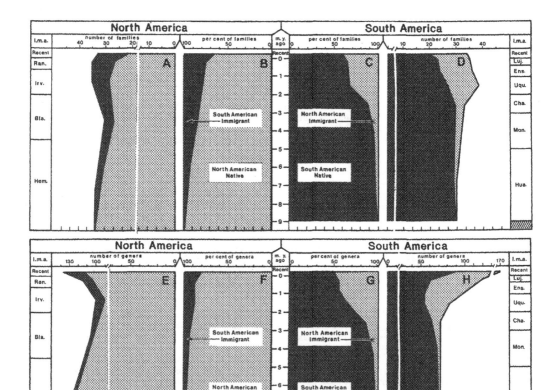

Fig. 1. Numbers (or "diversities") of known families (top) and genera (bottom) in successive late Cenozoic land mammal ages in North (left) and South (right) America. Graphs show total number of native and immigrant taxa and their percentage contribution to each land mammal age fauna. Note that scales are different for cumulative numbers of families and genera. The land mammal ages for North America are Ran., Rancholabrean; Irv., Irvingtonian; Bla., Blancan; and Hem., Hemphillian; for South America they are Luj., Lujanian; Ens., Ensenadan; Uqu., Uquian; Cha., Chapadmalalan; Mon., Montehermosan; and Hua., Huayquerian. Abbreviations: l.m.a., land mammal age; m.y., million years.

grassland habitats. We are thus dealing primarily with the evolutionary history of ecologically similar faunas during the interchange period, and this feature lessens the potential for sampling bias.

Aspects of Faunal Dynamics

The late Cenozoic stratigraphic ranges of families and genera of terrestrial mammals can be used to analyze simple aspects of taxonomic evolution [that is, measurements of changes in total numbers of taxa and changes of taxa within clades through time (19–21)]. For North America we analyze these data for the last 12 million years (divided into five standard land mammal ages) (22), and for South America we analyze these data for the last 9 million years (divided into six land mammal ages) (18) (Table 1). The

perspective given by this temporal framework permits establishment of a pre–land bridge "basal metabolism" to which post–land bridge faunal dynamics can be compared.

Table 1 lists summations for each land mammal age of (i) familial and generic diversity, (ii) numbers of first and last fossil occurrences (that is, observed originations and extinctions) of genera of native and immigrant taxa, and (iii) aspects of faunal dynamics and taxonomic evolution (21) based on data in (22) for North America and (18) for South America. Numbers of families and genera in each land mammal age on each continent are shown in Fig. 1 (top and bottom, respectively), with shading indicating the continent of origin of the taxa (23) (Fig. 1). Below we consider aspects of the taxonomic evolution first of families and then of genera.

Families. The total number of known families remained relatively constant throughout the late Cenozoic on both continents; the average diversity from late Miocene to Recent in both North and South America was about 34. Today, the familial diversities remain similar, with 35 in South America and 33 in North America (17).

In South America the peak in familial diversity (39 families) followed the appearance of the land bridge and the arrival of members of eight North American families in the Uquian, raising the number of immigrant families to 12; the total number of families then dropped to 36 before Lujanian time (Table 1). For North America the record likewise indicates a sharp rise in South American immigrant families after emergence of the isthmus: representatives of a total of six new families appeared in the late

Table 1. Faunal dynamics (genera per million years) of late Cenozoic land mammal genera in South America (left) and North America (right). The number of families represented are listed in brackets. The land mammal ages for South America are (from oldest to youngest) H, Huayquerian; M, Montehermosan; C, Chapadmalalan; U, Uquian; E, Ensenadan; and L, Lujanian. For North America they are C, Clarendonian; H, Hemphillian; B, Blancan; I, Irvingtonian, and R, Rancholabrean.

Indices (21)	South American land mammal age						North American land mammal age				
	H	M	C	U	E	L	C	H	B	I	R
a. Durations (million years)	4.0	2.0	1.0	1.0	0.7	0.3	2.5	5.0	2.5	1.3	0.7
b. Number of genera											
North American	1[1]	4[2]	10[4]	29[12]	49[12]	61[12]	92[33]	128[33]	99[25]	90[26]	102[26]
South American	72[29]	68[30]	62[29]	55[27]	58[24]	59[24]	0[0]	3[2]	8[6]	11[8]	12[9]
Total	73[30]	72[32]	72[33]	84[39]	107[36]	120[36]	92[33]	131[35]	107[31]	101[34]	114[35]
c. Originations (No.)											
North American	1	3	6	21	26	13	43	75	54	28	22
South American	55	32	14	25	19	8	0	3	7	4	1
Total	56	35	20	46	45	21	43	78	61	32	23
d. Extinctions (No.)											
North American	0	0	2	6	1	20	37	81	40	9	23
South American	36	20	32	16	7	25	0	2	1	0	9
Total	36	20	34	22	8	45	37	83	41	9	32
e. Running means											
North American	0.5	2.5	6.0	15.5	35.5	44.5	52	50.0	65.0	71.5	79.5
South American	26.5	42.0	39.0	34.5	45.0	42.5		0.5	4.0	9.0	7.0
Total	27.0	44.5	45.0	50.0	80.5	87.0	52	50.5	69.0	80.5	86.5
f. Origination rates											
North American	0.3	1.5	6.0	21.0	37.0	43.3	17.2	15.0	21.6	21.5	31.4
South American	13.8	16.0	14.0	25.0	27.0	26.7		0.6	2.8	3.1	1.4
Total	14.1	17.5	20.0	46.0	64.0	70.0	17.2	15.6	24.4	24.6	32.9
g. Extinction rates											
North American	0	0	2.0	6.0	1.5	66.7	14.8	16.2	16.0	6.9	32.9
South American	9.0	10.0	32.0	16.0	10.0	83.3		0.4	0.4	0	12.9
Total	9.0	10.0	34.0	22.0	11.5	150.0	14.8	16.6	16.4	6.9	45.8
h. Turnover rates											
North American	0.1	0.8	4.0	13.5	19.3	55.0	16.0	15.6	18.8	14.2	32.2
South American	11.4	13.0	23.0	20.5	18.5	55.0		0.5	1.6	1.6	7.2
Total	11.5	13.8	27.0	34.0	37.8	110.0	16.0	16.1	20.4	15.8	39.4
i. Per-genus turnover	0.4	0.3	0.6	0.7	0.5	1.3	0.3	0.3	0.3	0.2	0.5
j. Breakdown estimate of immigrants											
Total number											
Primary	1	1	2	10	18	20	0	2	6	8	9
Secondary	0	3	8	19	31	41	0	1	2	3	3
Originations											
Primary	1	0	1	8	9	2	0	2	6	3	1
Secondary	0	3	5	13	17	11	0	1	1	1	0
Extinctions											
Primary	0	0	0	1	0	7	0	2	1	0	7
Secondary	0	0	2	5	1	13	0	0	0	1	2

Blancan (four families) and early Irvingtonian (two families). Both continents experienced a notable decline in familial diversity at the end of the Pleistocene (24).

A sharp rise in the number of known immigrant families between the late Pleistocene and Recent occurs on both continents. This rise is due to our ignorance of late Cenozoic tropical faunas (17). Today, members of 14 North American families occur in South America and contribute 40 percent to the familial diversity of that continent, whereas members of 12 South American families occur in North America and account for a nearly equivalent 36 percent of that continent's familial diversity.

In summary, the data show that at the family level the interchange was balanced (12). The fact that total familial diversity on both continents is virtually the same today as it was in pre–land bridge times might be construed as indicative of symmetrical replacement of native by immigrant taxa at high taxonomic levels (Fig. 1).

Genera. The known late Cenozoic diversity of fossil genera is, on the average, greater in North America than in South America (Table 1, row b); this is in contrast to the Recent fauna which is slightly more diverse in South America

(170 genera) than in North America (141 genera) (17). In South America known diversity remained near 72 genera for at least 6 million years prior to the land bridge (Fig. 1, column H), suggesting that an equilibrium was established at about that level, at least in the environments sampled. After the appearance of the land bridge, generic diversity rose rapidly to 84 in the Uquian, 107 in the Ensenadan, and 120 in the Lujanian (Table 1, row b). During this time the North American immigrants increased sharply but steadily in number and progressively contributed a larger part of the South American land mammal fauna. Today, 85, or 50 percent, of the mammal genera in South America are derived from members of immigrant North American families (17).

In North America, observed numbers of genera drop from 131 in the Hemphillian to 101 in the Irvingtonian and then rise to 114 in the Rancholabrean (Table 1). The known South American immigrants rose from 3 to 12 during this period and came to contribute only 11 percent to the total North American land mammal fauna in the Rancholabrean. Today, 29 (21 percent) of the land mammal genera in North America are derived from immigrant South American families (17, 25, 26). Most of the genera not

sampled as fossils in North America live in subtropical to tropical latitudes in the Neotropical Realm (8).

The generic diversities of Recent and pre–land bridge faunas in North America are similar. However, in South America a major increase in generic diversity followed the appearance of the land bridge, the result of adding immigrant taxa. At the same time, the number of native South American genera declined by 13 percent between pre–land bridge and Lujanian faunas, a percentage reduction comparable to the 11 percent decline among native genera in North America. Thus, as in the case of families, the percentage decline of native genera was virtually identical on both continents, and in this regard the interchange was balanced. But the increase in both numbers and percentages of immigrant taxa was much greater in South America, as is discussed further below.

Rarefaction Analysis

Before considering the dynamics of the Great American Interchange further, some aspects of the quality and robustness of the taxonomic data must be analyzed in more detail. The diversity values for land mammal age faunas shown in Table 1 and Fig. 1 are somewhat higher than the numbers of taxa actually recorded. Some taxa occur in preceding and succeeding land mammal age faunas, and their presence in the intervening fauna or faunas is inferred. This usage of inferred ranges is conventional and legitimate for general analysis of diversity patterns, although an alternative analysis by rarefaction methods incorporates only the genera and families actually recorded in a particular land mammal age (27).

A taxonomic rarefaction curve is computed on the basis of the frequency distribution of genera within families in each land mammal age (Fig. 2). The distal end of each curve represents the number of genera and families actually recorded, and the curves provide estimates of the number of families that would have been recorded had fewer genera been found. Thus, rarefaction analysis provides answers to two basic questions: (i) Is the nature of sampling and taxonomic treatment consistent throughout the data set? and (ii) Did familial diversity differ significantly among the time intervals (land mammal ages) sampled?

All of the rarefaction curves in Fig. 2 have approximately the same shape, and crossing of curves is minimal, suggesting minimal overall differences in sampling

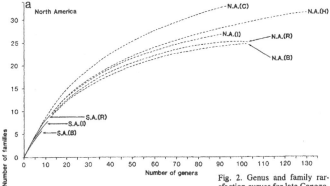

Fig. 2. Genus and family rarefaction curves for late Cenozoic land mammal age faunas in North (a) and South (b) America. (a) *S.A.*, genera that evolved from families of South American origin (solid lines); *N.A.*, genera that evolved from families of North American origin (dashed lines); (b) *S.A.*, genera that evolved from families of South American origin (solid lines); *N.A.*, genera that evolved from families of North American origin (dashed lines). For key to abbreviations of land mammal ages, see legend of Table 1.

both among land mammal ages and between continents. A sharp drop in North American native families (statistically significant, $P < .05$) occurs between the Clarendonian and Hemphillian (Fig. 2a), although no drop is seen in the number of actual or inferred families shown in Table 1. This drop could represent either a true decrease in diversity or a generic radiation among existing families. The diversity of native North American families within North America continues to drop after initiation of the interchange, but the changes are not statistically significant.

Diversity histories for South American natives in South America (Fig. 2b) reflect effects of the interchange. The post-interchange faunas (Uquian, Ensenadan, Lujanian) show significantly lower familial diversities than the pre–land bridge faunas, and the post-interchange curves for South American natives show decreasing diversity in chronostratigraphic order, the youngest being the lowest.

Thus, analysis of the data by rarefaction confirms the analysis of the raw family data presented above. The rarefaction work also lessens the possibility that the patterns observed in Fig. 1 are artifacts of sampling.

Patterns and Rates of Generic Evolution

The Great American Interchange has played a primary role in the development of basic biogeographic principles regarding tempo and mode of large-scale dispersal. Although these principles were formulated under a model of stationary continents, most can be applied to the dynamic paradigm with logical modifications and extensions (28). With the documentation and acceptance of plate tectonic theory, the Great American Interchange has become a classic example for studying the biological consequences of continental suturing.

Equilibrial biogeographic models have been suggested as applicable to the Great American Interchange (9, 19, 29) and have been used to test the extension of the island biogeographic theory of MacArthur and Wilson (30) to continental scales. This theory was first developed to explain biogeographic patterns on oceanic islands (30, 31) and only later was applied to continental and global systems (19, 32–36). The fundamental prediction of equilibrium models is that species diversity in any restricted area (that is, island, continent) will, under constant conditions, eventually attain a dynamic equilibrium maintained by balanced rates of origination (or immigra-

tion, or both) and extinction (Fig. 3). The resultant diversity and the time span required to attain equilibrium are largely dependent on the size of the area; thus continents will have a higher species diversity and lower per-species turnover rate and will require a longer time span to attain equilibrium, as compared to oceanic islands (37). Monte Carlo simulations and empirical studies of fossil data indicate that genera and families show patterns of diversification commensurate with diversification of their constituent species so long as large numbers of higher taxa are involved (29, 34, 36). This relationship permits study of patterns of diversification at higher taxonomic levels even though species are the real units of evolution.

As discussed above, North and South American land mammal faunas each appear to have attained equilibrial diversity

prior to the Great American Interchange. These equilibria were dynamic, with diversity remaining steady despite continuous origination and extinction of taxa (Table 1). In South America, per-genus turnover rate averaged 0.4 genera per genus per million years from 9 to 2 million years ago, while in North America the per-genus turnover rate averaged only 0.3 genera per genus per million years over the same period. The greater generic diversity and lower overall turnover rates of mammals in North America prior to the interchange are consistent with its greater total area (24×10^6 square kilometers compared to 18×10^6 square kilometers for South America), as predicted by equilibrium models (38).

The emergence of the Panamanian land bridge ended the phase of simple equilibrium for both continents and made each a potential source of immi-

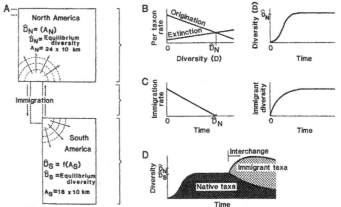

Fig. 3. Components of a hypothetical equilibrium model for the Great American Interchange. (A) Geographic constraints. North and especially South America were isolated continents through much of the Tertiary Period. As such, each should have supported a unique equilibrium diversity ($\hat{D}$), which may have been proportional, to a first approximation, to the area (A) of the continent (29, 30). Interconnection of the continents across the Panamanian land bridge terminated the equilibrium phase and permitted immigration of taxa between the two continents. (B) Closed-system diversification. Prior to interconnection, per taxon rates of origination (speciation) may have been high and rates of extinction low when diversity was low; with increasing diversity (and hence crowding of ecosystems), origination rates may have decreased, and extinction rates increased until becoming approximately equal at the equilibrium diversity $\hat{D}$ (33–35). This "diversity dependence" of evolutionary rates would result, in a simple deterministic system, in diversity increasing sigmoidally from some initial low to the equilibrium and then maintaining that equilibrium so long as the system remained closed (43). (C) Open-system immigration. Complete or partial interconnection of a large fauna at equilibrium with an area lacking fauna will initiate a flow of immigrants to the new area. The rate of immigration will be high at first and decline as the fauna in the newly colonized area becomes a larger and larger subset of the source fauna (30, 37). As a result, diversity in the new area should increase rapidly at first but later asymptotically approach an equilibrium determined by the equilibrium of the source area and by the local immigration and extinction rates. (D) Combined models. If taxa immigrate into a large area containing a native fauna, such as occurred in South America, the addition of immigrant taxa will, in essence, supersaturate the fauna of the new area. Extinction rates of both native and immigrant taxa will increase as diversity exceeds the equilibria of both faunal components. This will slow the increase in immigrant diversity and cause an exponential decline in native diversity. If the diversity of the source fauna is greater than the equilibrium of the native fauna, native diversity will eventually dwindle to zero in this simple model. More realistic constraints in the model (which could slow or prevent extinction of native taxa) would include backflow of immigrants north into North America and autochthonous evolution of taxa of immigrant ancestry, such as seen in the actual fossil records of North and South America.

grant taxa for the other. However, the tropical areas of North and South America seem to have acted as a barrier to dispersal of some representatives of genera and families. Only families with at least some constituent species distributed in tropical or subtropical areas took part in the interchange, whereas families with only temperate species did not disperse. Of all the families with part or all of their distribution in tropical areas, 17 South American families and 16 North American families dispersed; six South American and seven North American families did not. Thus, with regard to the potential family participants, the interchange was balanced (*10, 12, 17*).

Island biogeographic theory predicts that the effect of a source fauna on another fauna receiving immigrants should be proportional to the size (or diversity) of the source fauna (Fig. 3). In an attempt to apply this prediction to the interchange we divided the immigrant genera in Table 1 into primary immigrants (those with members that came directly from the other continent) and secondary immigrants (those whose founding species apparently evolved from primary immigrants after their arrival on the other continent) (*39*). During the interchange representatives of 1 to 11 percent of known available native genera in North America immigrated to South America in any given land mammal age, while representatives of 2 to 7 percent of known available native genera in South America immigrated to North America. These proportions are generally statistically indistinguishable (*40*). Ultimately, representatives of more North American genera immigrated to South America than vice versa, but this apparently represents a simple consequence of North America having a 60 percent greater average generic diversity than South America during the late Cenozoic. Thus, as predicted, the number of primary immigrants appears proportional to the size of the respective source faunas.

However, the subsequent evolutionary histories of the primary immigrants are significantly different. Various members of the 12 South American primary immigrants in North America gave rise to three secondary genera, whereas the 21 North American primary immigrants in South America gave rise to 49 secondary genera, derived subequally from members of five immigrant groups: cricetine rodents, carnivorans, proboscideans, perissodactyls, and artiodactyls (Table 1, row j). This difference in evolutionary histories represents nearly an order of magnitude difference in per-genus rates of origination between the

respective primary immigrants (*26*). The Recent record further emphasizes this trend as demonstrated by the remarkable secondary diversity (more than 40 genera) of cricetine rodents.

The resulting faunal dynamics of native taxa on the two continents are predictable. North America, with a proportionately small "input" of primary immigrants, exhibits no detectable change in per-genus turnover rate; on the other hand, South America, where generic diversity eventually exceeded previous equilibrium levels by more than 50 percent, exhibits an increase of nearly 70 percent in per-genus extinction rates among native taxa (Table 1). The observed gradual decline in diversity of native South American genera subsequent to the land bridge (Fig. 1, column H and Table 1, row b) is consistent with patterns expected for a supersaturated biogeographic system (Fig. 3) [figure 6 in (*35*)]. North American immigrant genera in South America exhibit a marked increase in extinction rates over genera remaining in North America (per-genus extinction rate averages 0.3 genera per genus per million years for immigrants, and 0.2 genera per genus per million years for native North American taxa), but the North American immigrants in South America maintained lower average extinction rates (0.3 genera per genus per million years) than South American natives (0.5 genera per genus per million years) (*41*). These differences in per-genus rates reflect the continued diversification of North American immigrants in South America, their documented replacement of South American natives, and the significant increase in South American generic diversity and faunal enrichment on a continent-wide basis.

Conclusions

Some aspects of faunal dynamics of the Great American Interchange (that is, prior equilibrium, difference in turnover rates, importance of source faunas, and increased extinction with supersaturation) are predicted from elementary considerations of equilibrium theory. However, the significant and apparently rapid diversification of North American secondary immigrants within South America is not predicted by simple extrapolation of equilibrium models into evolutionary time frames. This radiation is thus the unique aspect of the interchange story; it alone seems to account for the long-observed asymmetry in interchange dynamics between the two continents

and for the great change in taxonomic composition of the post–land bridge mammal fauna in South America (Fig. 1).

A possible but speculative explanation for the post–land bridge history of the South American fauna exists. During the late Cenozoic, a phase of orogeny beginning about 12 million years ago resulted in a significant elevation of the Andes Mountain range (*8*). A major phase of these orogenic movements occurred between 4.5 and 2.5 million years ago with a rise of from 2000 to 4000 meters (*42*). The newly elevated Andes served as a barrier to moisture-laden Pacific winds (*8*), and a rain shadow was created on the eastern (leeward) side. The southern South American habitats changed from primarily savanna-woodland to drier forests and pampas, and precocious pampas environments and desert and semidesert systems came into prominence at about that time. Many subtropical savanna-woodland animals retreated northward (*8*), and new opportunities favoring higher generic diversity arose for those animals able to adapt to these new ecologies.

The greater diversification of North American genera after they had reached South America is evident in such different groups as cricetid rodents, canid carnivores, gomphotheres, horses, llamas, and peccaries. If the relative success of northern groups is attributed to competitive displacement of equivalent southern groups, it becomes necessary to develop a number of complex scenarios with a great deal of uncertainty concerning which groups of species compete and on which adaptive bases (*10*). Perhaps it is more reasonable to attribute the success of the North American groups to some general ability inherent in their previous history to insinuate themselves into narrower niches (*8*). In any event, their success in South America is a clear pattern not predicted by simple equilibrium theory.

References and Notes

1. A. R. Wallace, *The Geographical Distribution of Animals* (Macmillan, London, 1876).
2. For example, Wallace (*1*) regarded *Galera* (a skunk), *Tapirus*, and *Lama* as moving from South America to North America, but the reverse is now known to be true.
3. It is somewhat ironic that these workers neither discussed, nor even believed in, the Great American Interchange as such (G. G. Simpson, personal communication, 11 June 1981); see also (*6*).
4. Translated from K. A. von Zittel, in *Handbuch der Palaeontologie*, vol. 4, *Band Vertebrata (Mammalia)* (Ouldenberg, Munich, 1891–1893), pp. 754–755. "Aber nicht nur nordamerikanische Typen benützen die neuröffnete Bahn, um ihr Verbreitungsgebiet zu vergrössen, sondern auch die südlichen Autochthonen begannen nach Norden zu wandern, und so vollzog sich am Schluss der Pliocaenzeit eine der merkwürdigsten Faunenüberschliebungen, welche die Geologie zu verzeichnen hat."

5. W. D. Matthew, *Ann. N.Y. Acad. Sci.* **24**, 171 (1915); *Spec. Publ. N.Y. Acad. Sci.* **1**, 1 (1939).
6. W. B. Scott, *A History of Land Mammals in the Western Hemisphere* (Hafner, New York, 1937).
7. G. G. Simpson, *J. Wash. Acad. Sci.* **30**, 137 (1940); *The Geography of Evolution* (Chilton, New York, 1965); in *Biogeography and Ecology in South America*, E. J. Fittkau *et al.*, Eds. (Junk, The Hague, 1969), p. 879; *Splendid Isolation: The Curious History of South American Mammals* (Yale Univ. Press, New Haven, Conn., 1980).
8. B. Patterson and R. Pascual, in *Evolution, Mammals and Southern Continents*, A. Keast, F. C. Erk, B. Glass, Eds. (State Univ. of New York Press, Albany, 1972), p. 274.
9. S. D. Webb, *Paleobiology* **2**, 216 (1976).
10. ——, *Annu. Rev. Ecol. Syst.* **8**, 355 (1977); *ibid.* **9**, 393 (1978); *Paleobiology* **4**, 206 (1978); L. G. Marshall and M. K. Hecht, *ibid.*, p. 203; L. G. Marshall, *ibid.* **5**, 126 (1979); in *Biotic Crises in Ecological and Evolutionary Time*, M. Nitecki, Ed. (Academic Press, New York, 1980), p. 133; I. Ferrusquía-Villafranca, *Univ. Nac. Auton. Mex. Inst. Geol. Bol.* **101**, 193 (1978); R. Hoffstetter, *Acta Geol. Hisp.* **16**, 71 (1981).
11. L. G. Marshall, R. F. Butler, R. E. Drake, G. H. Curtis, R. H. Tedford, *Science* **204**, 272 (1979).
12. J. M. Savage, *Nat. Hist. Mus. Los Angeles Cty. Contrib. Sci.* **260**, 1 (1974).
13. D. H. Tarling, in *Evolutionary Biology of the New World Monkeys and Continental Drift*, R. L. Ciochon and A. B. Chiarelli, Eds. (Plenum, New York, 1980), p. 1; M. C. McKenna, in *ibid.*, p. 43.
14. O. J. Linares, thesis, University of Bristol (1978).
15. O. A. Reig, *Publ. Mus. Munic. Cienc. Nat. Lorenzo Scaglia* **2**, 164 (1978).
16. P. Hershkovitz, in *Evolution, Mammals and Southern Continents*, A. Keast, F. C. Erk, B. Glass, Eds. (State Univ. of New York Press, Albany, 1972), p. 311.
17. S. D. Webb and L. G. Marshall, in *South American Mammalian Biology*, H. Genoways and M. Mares, Eds. [Special Publication Series of the Pymatuning Laboratory of Ecology, University of Pittsburgh (Univ. of Pittsburgh Press, Pittsburgh, Pa., in press)].
18. The range distributions of South American taxa are drawn from L. G. Marshall, R. Hoffstetter, R. Pascual, *Fieldiana Geol.*, in press; and L. G. Marshall *et al.*, *ibid.*, in press.
19. S. D. Webb, *Evolution* **23**, 688 (1969).
20. C. W. Harper, Jr., *J. Paleontol.* **49**, 752 (1975); H. R. Lasker, *Paleobiology* **4**, 135 (1978).
21. Indices used for measuring aspects of mammalian faunal dynamics in Table 1 are defined and computed as follows: (i) Durations of each land mammal age are given in millions of years (to the nearest tenth) and are based on all available radioisotopic, paleomagnetic, and biostratigraphic data, much of which is summarized in (*11*). (ii) The number of genera is the total number (or "diversity") of terrestrial mammal genera known for each land mammal age in each continent. We exclude *Homo*, and aquatic and volant mammals such as sea cows and bats. (iii) Originations are number of first appearances of a particular taxon (genus) in a given time interval (land mammal age) on each continent. This category combines three different kinds of originations: (a) New native autochthons (taxa whose members evolved in situ); (b) new immigrant allochthons (taxa with members that immigrated from outside the continent, or at least outside the area previously sampled); and (c) pseudo-originations produced by taxonomists when an evolving lineage changes enough to warrant a new name. Sufficient data are not available to consistently make distinctions among these alternatives. (iv) Extinctions are last appearances of a taxon on a given continent. These may not be "true extinctions" since the same taxon may continue to live on another continent, as with the Rancholabrean "extinctions" of *Tapirus* and *Equus* in North America, or the same population evolves to the point where it receives a new name, thus producing a taxonomic or pseudoextinction. As in the case of originations, the data do not consistently permit discrimination between these alternatives; however, pseudoextinctions appear to be of minor importance in this data set. (v) Running means (R_m) are expressions of the standing crop of a taxon (*19*, *20*). This statistic compensates for time intervals of unequal duration by subtracting the average of originations (O_i) and extinctions (E_i) for a given age from the number of genera (S_i) for that age; thus, $R_m = S_i - (O_i + E_i)/2$. (vi and vii) Origination rates (O_r) are indices adjusted for time intervals of unequal length by dividing the total number of originations (O_i) of taxa occurring during a given time interval by the duration (d) of that interval; thus, $O_r = O_i/d$. By similar reasoning, the extinction rate, $E_r = E_i/d$. (viii) Turnover rates (T) are the average number of taxa of a given rank that either originate or go extinct during a given time interval (that is, rates of first and last appearances). Turnover rates represent the average of origination rates and extinction rates for a given time interval; thus, $T = (O_r + E_r)/2$. (ix) The per-genus turnover rate is the turnover rate adjusted for average diversity calculated by dividing the total turnover rate (T) by the total running mean (R_m).
22. The range distributions of North American genera are drawn from S. D. Webb, in *Pleistocene Extinctions, the Search for a Cause*, P. Martin and R. Klein, Eds. (Univ. of Arizona Press, Tucson, ed. 2, in press).
23. We have counted the trans-Beringian immigrants with the native North American genera, while freely recognizing the fact that North America was not a "closed system" as was South America. Three trans-Beringian genera (*Pseudocyon, Pseudoceras, Torynobelodon*) appear in the North American Clarendonian. After the Clarendonian, incursions into North America from the Old World grew steadily; trans-Beringian origins account for 18 genera in the Hemphillian, 20 genera in the Blancan and Irvingtonian, and 27 genera in the Rancholabrean. Most Hemphillian and Blancan immigrants were cricetid rodents and diverse Carnivora; by Irvingtonian and Rancholabrean time the principal taxa were arvicoline rodents and bovid and cervid ruminants. In fact, the balance of generic exchange between North America and Eurasia shifted strongly in favor of the latter during the late Cenozoic [C. A. Repenning, in *The Bering Land Bridge*, D. M. Hopkins, Ed. (Stanford Univ. Press, Stanford, Calif., 1967)], p. 288. Nonetheless, virtually all Irvingtonian and Rancholabrean immigrants from trans-Beringea were steppe-tundra grazers. Their ecological impact was surely concentrated in north temperate latitudes and considerably removed from major impact on the Great American Interchange in tropical North America.
24. Of the 36 families of mammals recorded in beds of Lujanian age in South America, 8 (22 percent) are now extinct. These include 6 of 24 (25 percent) native and 2 of 12 (17 percent) immigrant groups. All eight families were present in North America at about that time, and became extinct there as well. Of the 35 families of mammals recorded in beds of Rancholabrean age in North America, 11 (31 percent) are now extinct. These include 6 of 26 (23 percent) native families and 5 of 9 (56 percent) immigrant families.
25. Of the 120 genera known from the Lujanian of South America, 45 (40 percent) became extinct. Included were 25 of 59 (42 percent) native South American genera and 20 of 61 (33 percent) immigrant genera. These differences are not statistically significant (*26*). Of the 114 genera known from the Rancholabrean of North America, 32 (28 percent) became extinct. Included are 23 of 102 (23 percent) native North American genera, and 9 of 12 (75 percent) immigrant genera.
26. Tests for differences in proportions: $z = 1.08$, $P > .05$ for South American; $z = 3.82$, $P < .01$ for North American.
27. D. M. Raup, *Paleobiology* **1**, 333 (1975). The mathematical technique used here was that applied by Raup to higher taxa.
28. M. C. McKenna, in *Implications of Continental Drift to the Earth Sciences*, D. H. Tarling and S. K. Runcorn, Eds. (Academic Press, New York, 1973), p. 21; R. H. Tedford, in *Paleogeographic Provinces and Provinciality*, C. A. Ross, Ed. (Soc. Econ. Paleontol. Mineral. Spec. Publ. 21, 1974), p. 109.
29. K. W. Flessa, *Paleobiology* **1**, 189 (1975).
30. R. H. MacArthur and E. O. Wilson, *Evolution* **17**, 373 (1963); *The Theory of Island Biogeography* (Princeton Univ. Press, Princeton, N.J., 1967).
31. D. S. Simberloff, *Annu. Rev. Ecol. Syst.* **5**, 161 (1974); *Science* **194**, 572 (1976); in *Biotic Crises in Ecological and Evolutionary Time*, M. Nitecki, Ed. (Academic Press, New York, 1980).
32. R. H. MacArthur, *Biol. J. Linn. Soc.* **1**, 19 (1969).
33. M. L. Rosenzweig, in *Ecology and Evolution of Communities*, M. L. Cody and J. M. Diamond, Eds. (Belknap, Cambridge, Mass., 1975), p. 121.
34. J. J. Sepkoski, Jr., *Paleobiology* **4**, 223 (1978).
35. ——, *ibid.* **5**, 222 (1979).
36. D. S. Simberloff, *J. Geol.* **82**, 267 (1974).
37. ——, in *Models in Paleobiology*, T. J. M. Schopf, Ed. (Freeman, Cooper, San Francisco, 1972), p. 160.
38. Alternatively, the higher average turnover rates for South America may simply represent the need for a refined synthetic systematic review of these faunas [D. M. Raup and L. G. Marshall, *Paleobiology* **6**, 9 (1980)], and rate differences between the continental faunas may be due to different approaches to taxonomic treatment (lumpers versus splitters). We consider only potential mammal-mammal interactions yet recognize that nonmammalian groups may be involved as well [J. H. Brown and D. W. Davidson, *Science* **196**, 880 (1977)].
39. This subdivision of immigrants into primary and secondary groups has never before been formally attempted, although many workers have noted existence of these categories. Several caveats should accompany such an interpretive subdivision, the first being our wholly inadequate knowledge of late Cenozoic mammalian evolution in the American tropics, regarded as an undocumented source area for many of the immigrants (*8*). Nevertheless, our criteria for subdivision are simple and can be easily tested by further fossil discoveries. Primary immigrant genera are native genera (those that belong to native families) with members that occur on the other continent or are genera with members so closely related to known genera on the other continent that further taxonomic studies will probably show them to be congeneric (a criterion based on the observations of L.G.M. and S.D.W.). Secondary immigrant genera are those that belong to families native to the other continent but are unknown on that continent and apparently lack possible congeneric forms. By these criteria, the primary South American immigrants to North America are *Didelphis, Kraglievichia, Dasypus, Glyptotherium, Pliometanastes, Nothrotheriops, Eremotherium, Thinobadistes, Glossotherium, Hydrochoerus, Neochoerus*, and *Mixotoxodon*; and the primary North American immigrants to South America are *Calomys, Canis, Felis, Leo, Smilodon, Conepatus, Galera, Lutra, Mustela, Cyonasua, Nasua, Arctodus, Hemiauchenia, Odocoileus, Dicotyles, Platygonus, Equus, Hippidion, Tapirus, Cuvieronius*, and *Stegomastodon*. The remaining immigrant taxa listed in (*22*) for North America and in (*18*) for South America are regarded as secondary.
40. A z-test for differences in proportions [G. W. Snedecor and W. G. Cochron, *Statistical Methods* (Iowa State Univ. Press, Ames, ed. 6, 1967, p. 220)] applied to total numbers of primary immigrants moving north and south as proportions of the total size of their respective native faunas over the whole of the last 9 million years reveals no significant difference between North and South America ($z = 0.719$, $P > .05$). However, significant differences do occur within the Ensenadan-Lujanian interval (≈ Rancholabrean), when approximately 2 percent of available South American genera move north compared to 11 percent of available North American genera which move south ($z = 2.278$, $P < .05$).
41. It is possible that the comparatively low extinction rate for immigrant taxa in Table 1, row g may involve multiple immigrations rather than any special quality of the immigrants. If all members of a taxon endemic to continent A died out on that continent, it is for all practical purposes, regarded extinct. (Members of this taxon may have dispersed to continent B before their extinction on A, and then may have been able to reinvade A from B after the population on A died out, but the chance of this happening is regarded as having a very low probability.) If, on the other hand, an immigrant from continent B to continent A died out on A, its extinction would not be universal so long as it survived on continent B. This population from B could then reestablish itself on A by subsequent and repeated invasions from B and create the impression that the original population in A never really "became extinct." Such reestablishments of populations would result in lower extinction rates for immigrants relative to native taxa (that is, Table 1, row g) even if actual extinction rates in both groups were the same on each continent.
42. B. S. Vuilleumier, *Paleobiology* **1**, 273 (1975).
43. For applicability to mammalian evolution, see J. A. Lillegraven, *Taxon* **21**, 261 (1972).
44. We thank W. Burger, J. Cracraft, G. McGhee, K. Luchterhand, J. M. Savage, G. G. Simpson, S. Stanley, and W. D. Turnbull for reading the manuscript. Supported in part by NSF grant EAR 7909515 and DEB 7901976 (to L.G.M.), NSF grant DEB 7810672 (to S.D.W.), and NSF grant EAR 75-03870 (to D.M.R.).

ONE HUNDRED YEARS OF SUEZ CANAL—A CENTURY OF LESSEPSIAN MIGRATION: RETROSPECT AND VIEWPOINTS

F. D. POR

Abstract

Por, F. D. (Dept. of Zoology, Hebrew Univ., Jerusalem) 1971. One hundred years of Suez Canal—a century of Lessepsian migration: Retrospect and Viewpoints. Syst. Zool., 20:138–159. One hundred years has passed since one of the biggest biogeographical experiments started with the opening of the Suez Canal. The successful migration along the Canal of a great many Red Sea species into the Mediterranean is seen as a process to which most of these species were preadapted in the littoral environment of the Red Sea. The repeated transgressions of this sea into the basin of the Bitter Lakes and the existence—presently and in the past—of several slightly hypersaline lagoons along its shores, resulted in the creation of a special stock of species able to perform this migration. Progress through the Canal is either a step-by-step one, or a result of active swimming (or transport). Neither the role of planktonic spread by means of larvae and/or currents, nor the obstacles put by the salinity, should be overemphasized. The main factors in helping or hindering migration are considered to be the presence of suitable substrates and water transparencies in the Canal.

The Levant Basin of the Mediterranean was also "preadapted" to receive the immigrants because of its high salinity and temperature and the resulting biological undersaturation with temperate fauna. Most probably this explains the one-way stream of the migrants into the Mediterranean. The distribution limits of the immigrants in the Mediterranean are most probably set by the increasing competitive capacities of the aboriginal species. The westward advance of the immigrants in the Mediterranean will probably become more pronounced after the damming of the Nile waters behind the newly erected Asswan dam. [Suez Canal: Lessepsian migration: Red Sea: Mediterranean.]

". . The waves intermingled friendly
and Ocean regained hold of the
lands he once owned".

Stephen, 1872 (describing the
opening of the Suez Canal).

INTRODUCTION

In 1865, four years before the opening of the Suez Canal, Louis Vaillant expressed his conviction that the new channel "will doubtlessly bring about an interchange of the species" and he foresaw many of the ecological and evolutionary results which might ensue. A century has elapsed since the intermingling of Mediterranean and Red Sea waters, and it can be said, in retrospect, that much of Vaillant's prophecy has come true. But the fact that the biological consequences of Ferdinand de Lesseps' great achievement have been foreseen, gives only a meager satisfaction if we consider how poor is our understanding today of this unique and one-hundred-year-old process.

For about 50 years, Suez Canal research was in the center of the scientific interest. One is still compelled to admire the scientific works of Lesseps, Fuchs, Keller, Tillier and, last but not least, of Munro Fox, leader of the 1924 Cambridge University Expedition to the Suez Canal. It seems that the two volumes of results of this latter expedition, published in the Transactions of the Zoological Society, gave the false impression that everything was already known. The fact that there is migration through the Suez Canal came safely down into the textbooks without any further attempt to investigate how this migration became possible and what are the ecological, zoogeographical and evolutionary lessons to be learned. Since the Cambridge Expedition there have

138

been only episodic visits by isolated biologists to the canal area. W. Steinitz's passionate outcry in 1919, has remained unheard. He wrote: "The Suez Canal . . . is the only place on earth where two totally separated faunal provinces are freely interpenetrating and there, of all places, no research institute is found."

The ever-troubled political atmosphere around the Suez Canal cannot serve as a justification for this neglect. As early as 1882 (during Arabi Pasha's uprising), Keller wrote: "The present political troubles . . . . may even endanger the Canal's very existence and interrupt the migration." Unfortunately, not too much has changed, in this respect during the 86 years which have elapsed.

Today, the hypothetical parallel foreseen by Nourse (1870) may become true, and a similar sea level channel may be dug across the central American isthmus. How can we use the century old experience of the lessepsian migration?

FIRST ACHIEVEMENTS OF THE SMITHSONIAN INSTITUTION—HEBREW UNIVERSITY PROJECT

The faunal interchange through the Suez Canal is presently being studied by a joint Smithsonian-Hebrew University team. The considerations included in this paper are partly based on new information obtained in the first two years of the project. Not too much data could be obtained in the Suez Canal itself, but the Bitter Lakes were investigated to a certain extent. The Sirbonian lagoons (Sabkhat Bardawil), waterbodies never investigated before, have been surveyed, as well as the open shores off northern Sinai. Along the Mediterranean coast of Israel, previous investigations have been followed up with special emphasis on the migration problem and considerable knowledge has also been obtained of the littoral and offshore fauna of the island of Cyprus.

At the Red Sea end, the most important progress was made in investigating the almost unknown fauna of the Gulf of Suez, which is a water body differing considerably in its living world from both the open Red Sea and the Gulf of Elat.

THE PRE-LESSEPSIAN CONTACTS

From the very beginning it was clear to the learned world that the achievement of Ferdinand de Lesseps was not something new, but only a repetition of an old situation, performed by modern techniques. Fuchs (1878) showed that fossiliferous deposits of the Red Sea fauna, of recent aspect, extend as far north as Lake Timsah (84 km north of Suez). Fossiliferous layers of present-day Mediterranean type, reach southwards to the sill of El Guisr, near the opening of the present Canal into Lake Timsah, 75 km south of the present shoreline. This leaves a gap of only about 9 km which is covered by fossils of fresh to brackish water type ("pseudosarmatic" in Fuchs' words).

The recognition by Napoleon's scientists participating in the Egyptian Campaign that, in olden times, there had been a channel (or rather several channels) built across the Isthmus of Suez led to the idea of the modern Canal.[1] Lesseps already knew that there had existed a connection, by various means and ways, between the Gulf of Suez and the Mediterranean, for a period of over 2000 years (from Ramses II in the 13th century BC till the Calif Al Mansur, late in the 8th century AD).

An additional fact gained more and more recognition in later years: the existence of considerable eustatic changes in sea level during the Pleistocene period. At high interglacial levels, the sea undoubtedly covered the present day Isthmus.

Let us now discuss these three historical aspects separately and in their chronological sequence.

1. There is not much doubt that during interglacial periods contact existed between the two seas. It is, however, also evident that during the following glacial periods,

---

[1] In fact, the idea of rebuilding the Sea-route of the antiquity was first suggested in a memorandum to Louis XIV by the German philosopher Leibnitz.

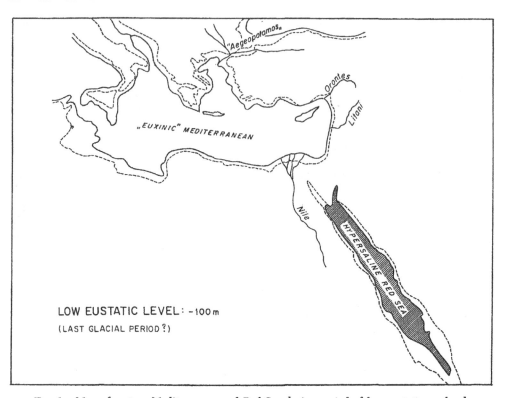

Fig. 1.—Map of eastern Mediterranean and Red Sea during period of low eustatic sea level.

with low eustatic levels, the Eastern Mediterranean became a brackish basin with saline stratification, and the Red Sea a hypersaline lake (Fig. 1). If organic forms succeeded in crossing the Isthmus during times of high sea level, they were most probably exterminated by the brackish or the hypersaline conditions during the following glacial event.

The Gulf of Suez still bears evidence to show that the Gulf lay dry during the Riss glaciation, which involved a sea level decrease of 60 m. At –40 m, the hypersaline Red Sea could already transgress into the Gulf as far as Ras Sudr. At –30 m, the sea reached as far as Ras Misalla, but area north of Ras Sudr was probably a semi-closed hypersaline lagoon. At successively increasing postglacial levels, up to a level of –10 m, the area north of Ras Misalla became another semi-closed lagoon. Under

these conditions a special biotic association evolved in the Red Sea, capable of living at increased salinities of 50–60‰. As will be further shown, this biotic association is especially important in recent lessepsian migration. (See Fig. 2).

On the Mediterranean side, where brackish conditions existed during the glacial periods, a euryhaline fauna lived, in many aspects identical to that of other brackish areas of the Mediterranean. Many species of this fauna still exist in the euryhaline media of the present day isthmus.

2. In postglacial, prehistoric times (perhaps also during the early pharaonic dynasties), with slightly higher (Flandrian) eustatic levels, a continuous, though peculiar, waterway connected the two seas (Fig. 3). At first shown by Fuchs (op. cit.) and by Krukenberg (1888), conditions were

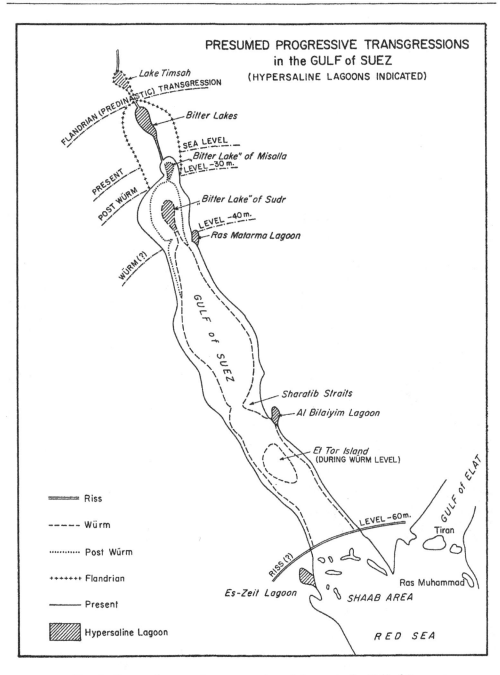

PRESUMED PROGRESSIVE TRANSGRESSIONS
in the GULF of SUEZ
(HYPERSALINE LAGOONS INDICATED)

FIG. 2—Presumed progressive transgressions of the sea in the Gulf of Suez.

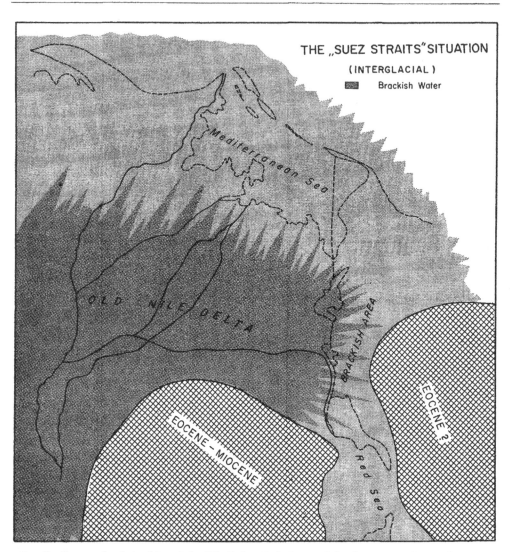

FIG. 3.—Presumed relationships of the Nile Delta, Red Sea and Mediterranean during late inter-glacial times.

perhaps similar to those of the Straits of Tartary, between the Island of Sakhalin and the Asiatic mainland, where the inflowing freshwaters of the river Amur create a brackish barrier across the straits. Then the Red Sea reached as far north as Serapeum (about half the distance of the present width of the isthmus). The Nile Delta was situated much more to the south and had

two, then very active, branches: one corresponding to Wadi Tumilat, flowing eastward into the basin of the present lake Timsah, and the other, the Pelusiac branch which reached the Mediterranean at the head of the present Gulf of Pelusium (Et Tineh). Both branches created a barrier of fresh and brackish waters at the opening of the Straits into the Mediterranean. The

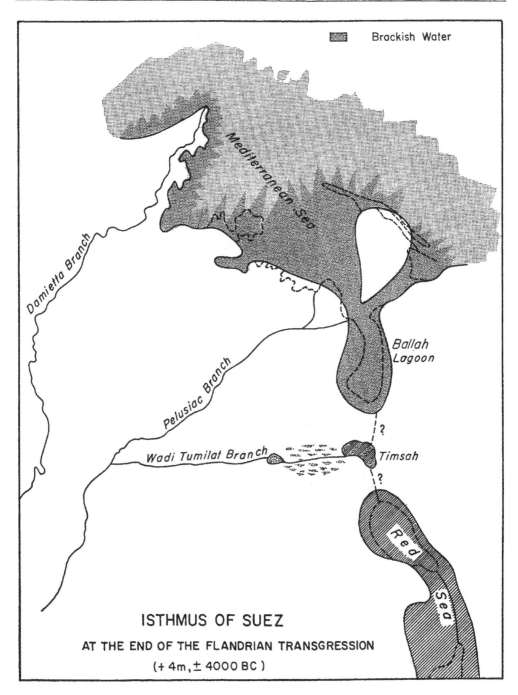

ISTHMUS OF SUEZ

AT THE END OF THE FLANDRIAN TRANSGRESSION

(+ 4m, ± 4000 BC)

FIG. 4.—Presumed relationships of Red Sea and Mediterranean during Flandrian transgression showing Isthmus of Suez.

area of Lake Timsah is particularly interesting in this connection; here, freshwater deposits are laminated with marine Red Sea deposits showing an alternation of conditions (Fuchs op. cit.) or a more probable, contemporaneous, stratification of brackish and marine faunas (W. Steinitz, 1927).

It appears therefore that contact between the Red Sea and the Mediterranean fauna was difficult, if at all possible, during this last presumed period in which a continuous and natural waterway existed.

3. The pharaonic channel, was in fact, a stubborn attempt to maintain, by artificial means, the above mentioned natural waterway, following the gradual widening of the land gap between the two seas (Fig. 4). In the north, the Nile Delta advanced further and further north, the Ballah lagoon became a shallow salt marsh, and the branch of Wadi Tumilat became less and less active. In the south, the Red Sea retreated beyond the rising Shallufa ridge (presently at +3 m) spilling over into the Bitter Lake basin progressively less frequently. The southern tract of the pharaonic channel was thus cut through the Shallufa ridge. From Lake Timsah, the channel used the artificially maintained flow of Wadi Tumilat. From Wadi Tumilat, ships could enter the Mediterranean through the Nile. Marine organisms had to cross the freshwater section of the Nile, a feat which they could probably only rarely accomplish (Fig. 5). But, at least in Ptolemaic times (4th century BC) and perhaps even under the late period of pharaoh Nechao (end of 7th century BC) a channel existed leading from Lake Timsah directly to the north, and cutting through the deltaic lakes Ballah and Menzala, which were left behind by the advancing Nile Delta. This channel used the other old branch of the delta, the Pelusiac, as outlet from the lagoons into the Mediterranean. Animals and plants could perhaps have

more readily followed this second artificial waterway, since it probably implied only the crossing of brackish lagoons instead of freshwater areas.

It follows that the lessepsian channel, the modern Suez Canal, and the consequent opportunities offered for faunal advance into the Isthmus waters and onward into the "other" sea, are not essentially new. The Red Sea fauna had repeated opportunities to invade the Bitter Lakes which had existed for a long time at least as a swampy area. North of these swamps, a system of more or less brackish lagoons persisted, the southernmost being Lake Timsah. Passage from one lagoon to the other was at times possible by natural means, but on occasion was artificially enhanced by ship channels. In this system of more or less interconnected basins of various salinities, the potential for faunal movement existed all the time. It is possible that a few of the faunal elements might have reached the Mediterranean and successfuly settled there having crossed the Isthmus from the Red Sea.

The presence of pre-lessepsian Red Sea elements in the Mediterranean has been admitted by several authors, though always only a few exceptional cases were involved. (Steinitz, 1926, Kosswig, 1955, Gohar, 1954, Peres, 1958.) Even if we disallow some reports which cannot be substantiated (as that of the mantis shrimp *Gonodactylus chiragra* by Milne Edwards), or delete some which have proved to be wrong (e.g., the porcelain crab *Petrolisthes boscii*), there still remain a small number of organisms which may be suspected of being pre-lessepsians. However, as knowledge of the Mediterranean fauna was still in its infancy in 1869, most of the judgments have to be made post-factum, and are therefore much open to criticism.

A useful criterion might be found in the very early appearance of a Red Sea organism at considerable distances from the Canal

→

FIG. 5.—Old canals across the Isthmus of Suez.

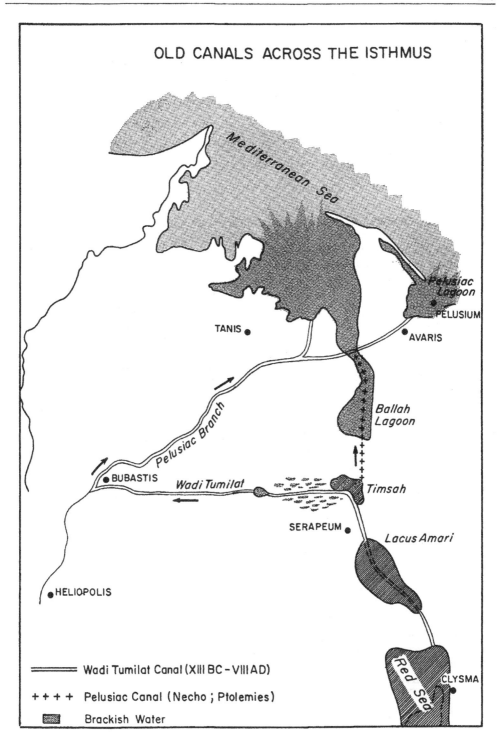

OLD CANALS ACROSS THE ISTHMUS

Wadi Tumilat Canal (XIII BC – VIII AD)

++++ Pelusiac Canal (Necho ; Ptolemies)

Brackish Water

and in well established populations outside ports. In such cases, the existence of the organism in the Mediterranean, prior to the opening of the Suez Canal, might be a more acceptable explanation than rapid immigration after 1869. The pearl oyster *Pteria occa* (Plate I, Fig. 8) was reported from the Gulf of Gabes in 1895 by Bavay (1897); the first report of this organism in the Suez Canal was only twelve years earlier (Keller 1882). The marine eel-grass *Halophila stipulacea* (Plate III, Fig. 4) was reported by Fritsch in 1895 from the port of Rhodes. It is hard to believe that the plant reached this far coast on a ship hull, although such a mechanism may have been used by *Pteria*. Peres (op. cit.) considers, in fact, that *Halophila* is a tropical relic in the Mediterranean. The ascidian *Metrocarpa nigra*, also frequently listed as a Red Sea immigrant, is likewise considered by Peres (op. cit.) as a tropical relic in the Mediterranean, as it occurs today on the Tunisian coast. A few fishes may also be include in the category of pre-lessepsian elements. Such are the shark *Carcharius brevippinna* known from the coast of Tripoli, *Parexocoetus mento* found in 1880 at Naples and recently by Gilat (Ben-Tuvia, 1966) in the Gulf of Sidra and *Leiognathus klunzingeri* (Plate III, Fig. 2) found near Lampedusa and Rhodes and considered pre-lessepsian by Kosswig (op. cit.). The Polychaeta *Cirriformia semicincta*, *Branchiosyllis uncinigera* and *Spirobranchus giganteus*, considered by Laubier (1966) to be lessepsian immigrants, are more probably circumtropical species, which are warm water relics along the Levant coast, where Laubier found them, and not necessarily immigrants through the Suez Canal.

In all these cases, and in several ones not mentioned, it would be difficult to distinguish whether the species are tropical relics, in the geological time sense, Red Sea immigrants through a pre-lessepsian waterway, or just circumtropical elements which came through Gibraltar.

It might, however, be worthwhile to investigate the fauna of the gulfs of Gabes and Sirta more thoroughly. These areas could probably have served as a refuge to tropical elements during the glacial periods, much better than the Levant coast. In some cases, even a recent repopulation of the Levant basin with a "Gabes" fauna, proceeding simultaneously with the Suez Canal immigration, may be assumed.

THE ISTHMUS FAUNA

All the geological and historical information indicates that the Isthmus of Suez and adjacent areas (the Sinai shore and the Nile valley) were covered during the Pleistocene by water bodies of fluctuating dimensions and salinities. More or less ephemeral contacts existed between these water bodies and with the neighbouring seas. The fauna inhabiting these water bodies—extremely euryhaline species, mostly of marine origin —did not need an historically well-circumscribed occasion, such as the opening of the lessepsian canal, in order to spread from basin to basin. This fauna was consequently "in situ," when in 1869, the modern Suez Canal filled and connected the different salty swamps and lakes of the Isthmus. Species belonging to this biotic complex have, at times, been considered recent Red Sea or Mediterranean migrants. In fact, their presence in the Isthmus area goes back to earlier and perhaps also Pleistocene times.

This fauna can be studied at its best in the Sirbonian lagoons (Sebkhat el Bardwil), hypersaline basins of 50 to 100S‰, situated east of Port Said, and connected with the Menzaleh lagoon through the periodically flooded swamps of the Romani area. The lagoons can be considered to be outside the area directly influenced by the recent lessepsian changes. If therefore, an animal is living today in the Sirbonian lagoons, with their high salinities, it can be assumed that it could have existed in the water bodies of the Isthmus before the opening of the Canal.

Among these organisms we may find cosmopolites, circummediterranean lagoon species, species of Red Sea origin, but also

endemics. The following is a first and preliminary list of this fauna, which might be
called the *Ruppia maritima* fauna, after the
flowering plant which is one of its characteristic shelters.

Mollusca: *Cardium edule* (Plate I, Fig. 9)
            Mytilus variabilis (Plate I,
            Fig. 10)
            Pirenella conica (Plate I, Fig. 5)
            Pirenella caillaudi
Polychaeta: *Augeneriella lagunari* (Gitay,
            1970)
Copepoda: *Canuella perplexa*
            Canuellina insignis
            Robertsonia salsa
            Paralaophonte quinquespinosa
            Neocyclops salinarum
            Euryte sp.
            Pseudodiaptomus salinus
Ostracoda: *Cyprideis torosa*
            Aglaiocypris sp.
Cirripedia: *Balanus amphitrite*
Isopoda: *Sphaeroma serratum*
            Sphaeroma walkeri
Pisces: *Aphanius dispar.* (Plate III, Fig.
            1)

It seems that some other organisms may also
be included in this fauna, such as the fish
Pranesus pinguis and perhaps species of
Mugil, the chironomide *Cricotopus mediterraneus* and a few hydrozoans. This fauna
(and flora) will always remain a restricted
group of species with all the additions
which can be expected in the future—the
same case as is met with in every environment of changing salinities known in the
world. The existence of this "third fauna"
(i.e., neither a Mediterranean nor a Red Sea
fauna) has to be taken into account in every
zoogeographical analysis of the lessepsian
migration.

THE LITTORAL NECTOBENTHOS

The most active and, in fact, the pioneer
ecological group to settle and cross the
Suez Canal into the Mediterranean, were
littoral fishes, portunide swimming crabs
and penaeid shrimps. All these animals live

inshore, at shallow depths, more or less confined by their feeding and behaviour to
shallow level bottoms. This ecological group
everywhere forms the most obvious part of
the sterile population of the hypersaline lagoons. The latter are the feeding grounds
of the adults; reproduction occurs outside
the lagoon in normal sea water. The high
osmotic capacities of the adult specimens
of this ecological group, combined with
their ability to cover wide distances by
swimming, enabled them very early to inhabit different parts of the Canal and to be
the first to appear in the Mediterranean. It
is however a questionable point whether
these animals reproduce in canal waters,
whether they are real inhabitants of the
waterway or only quick-moving transients.

THE "HALOPHILA FAUNA"

The speed with which the Bitter Lakes
were populated with benthic animals, in the
very first years following the opening of the
Suez Canal, is outstanding. The shell *Mactra olorina* (Plate I, Fig. 7) reached El
Qantara in the north as early as 1876 and
by the nineties *Murex tribulus* (Plate I, Fig.
4) annoyed the European bathers in Lake
Timsah (Bavay, 1879). The shell *Gafrarium
pectinatum*, the conch *Strombus tricornis*
(Plate I, Fig. 1), the medusa *Cassiopea
andromeda* and the small sea urchin *Nudechinus scotiopremnus* (Plate III, Fig. 8)
were frequent in the Bitter Lakes by 1880
(Keller, 1882.) *Cassiopea andromeda* reached
Lake Timsah a few years later (Krukenberg,
1888). By the end of the century, the lamellibranch *Malleus regula* (Plate I, Fig. 6)
and the gastropod *Diodora ruppeli* had
spread all over the Canal (Rillier et Bavay,
1905). The eel-grass *Halophila stipulacea*,
although not reported in the Canal in the
early eighties by Keller, probably became
dominant in the Bitter Lakes shortly afterwards.

This fauna which first entered and settled
the main water body of the Canal, the Bitter
Lakes, is one of shallow sandy-muddy bottoms, associated all over the Red Sea with
Halophila stipulacea. It also inhabits the

EXPLANATION OF PLATES

Elements of the fauna and flora of the Red Sea, Suez Canal and Mediterranean.

PLATE I

1. *Strombus tricornis* L., an early immigrant into the Suez Canal, did not reach the Mediterranean; 2. *Fusus marmoratus* Phil., a dominant mollusc of the Bitter Lakes, successful immigrant in the Mediterranean; 3. *Murex (Chicoreus) anguliferus* Lamarck, dominant species in the Bitter Lakes, of Red Sea origin, unknown in the Mediterranean; 4. *Murex tribulus* L., an early and massive inhabitant of the Suez Canal, successful immigrant in the Mediterranean; 5. *Pirenella conica* a hypersaline "isthmus" species; 6. *Malleus regula* Forsk., wide-spread Red Sea immigrant in the Mediterranean; 7. *Mactra olorina* Phil., a wide-spread Red Sea immigrant in the Suez Canal; 8. *Pteria occa* Reeve, the indopacific pearl oyster reported in the Mediterranean a few years after the opening of the Suez Canal; 9. *Cardium edule* Lam., a hypersaline "isthmus" species of Mediterranean origin; 10. *Mytilus variabilis* Krauss, a euryhaline species, possibly belonging to the isthmus fauna.

PLATE II

1. *Charybdis hellerii* (Milne Edwards), an Indopacific portunide crab, reported from Haifa in 1929; 2. *Myra fugax* Fabricius, a Red Sea species now wide-spread in the Eastern Mediterranean; 3. *Leucosia signata* Paulson, a dominant Red Sea species of the Bitter Lakes, recently reported also in the Mediterranean; 4. *Atergatis roseus* Ruppell, a recent Red Sea immigrant, now wide-spread along the Israeli coast; 5. *Squilla massawensis* Kossmann, a most successful Red Sea immigrant in the Eastern Mediterranean.

PLATE III

1. *Aphanius dispar* Rüpp., a species probably of Red Sea origin, wide-spread in hypersaline lagoons; 2. *Leiognathus klunzingeri* Steindehn., Red Sea immigrant in the Mediterranean; 3. *Upeneus moluccensis* Blkr., a now dominant Mediterranean fish of Red Sea origin; 4. *Halophila stipulacea*, Indopacific flowering plant wide-spread in the Eastern Mediterranean; 5. *Ophiactis savignyi* Müll. Trosch., Indopacific brittle star in the Eastern Mediterranean; 6. *Astropecten polyacanthus*, Müll. Trosch., Red Sea species, among the early inhabitants of the Bitter Lakes; 7. *Asterina wega* Perrier, a recent Red Sea immigrant undergoing massdevelopment along the Levant coasts; 8. *Nudechinus scotiopremnus* H. L. Clark, a mass form in the Bitter Lakes which did not reach the Mediterranean.

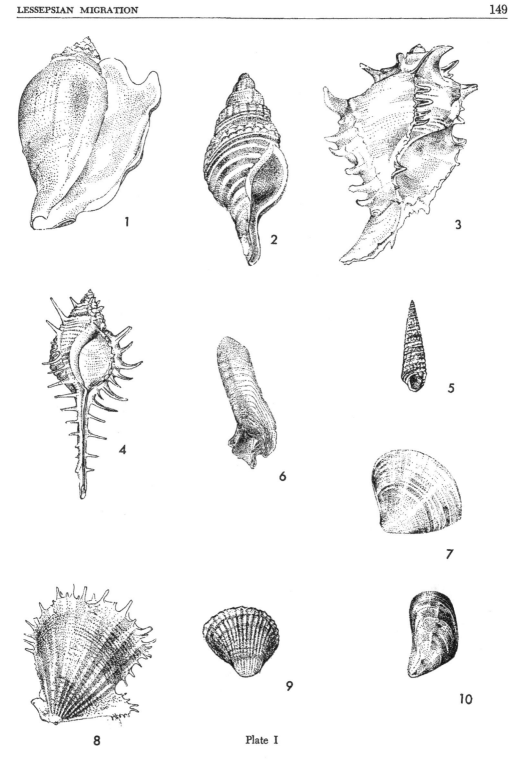

Plate I

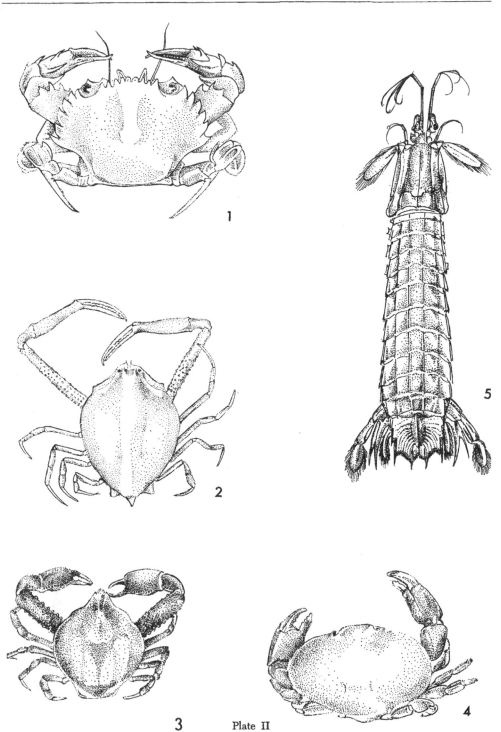

1

2

5

3   Plate II   4

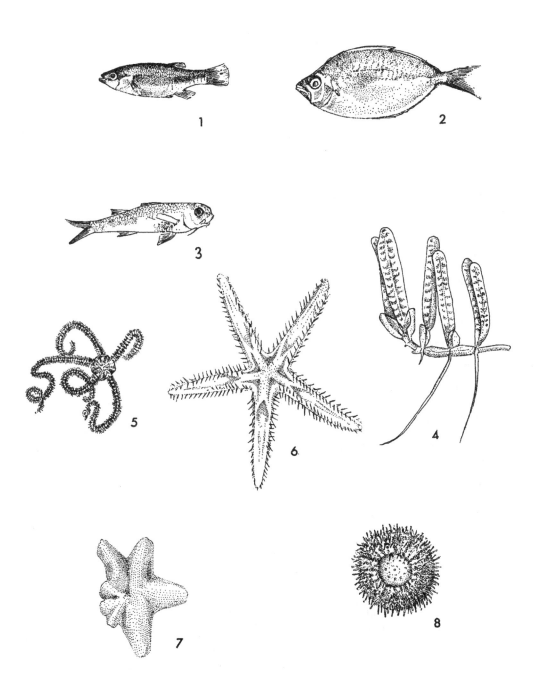

Plate III

slightly hypersaline "Khor's" and "Gubbet's"—semiclosed lagoons along the shores of the Red Sea. The importance of this *Halophila* association becomes particularly great in the northern part of the Gulf of Suez. It was only natural therefore that such a fauna of shallow level bottoms, adapted to high salinities, should invade the Bitter Lakes after the opening of the Canal. The *Halophila* fauna of the Bitter Lakes can be compared with that of such hypersaline lagoons along the Gulf of Suez as Khor El Bilaiyim near Ras Durba or the lagoons of Ras Matarma in the north. It is well understood that only part of the *Halophila* fauna could settle the Bitter Lakes, since the planktonic larval stages of species reproducing by this means, have difficulty in reaching the lakes.

THE LITTORAL BOULDER FAUNA

Another faunal element which settled the Bitter Lakes and made quite an early advance in the waters of the Canal is the fauna (or rather representatives of the fauna) of the littoral boulders of the Red Sea. The Bitter Lakes are very poor in rocky shores and even in littoral boulders. There are only a few localities with this bottom type, e.g., around Kabret, near Fayed on the African shore and at the Deversoir. The combined biological activity of *Mytilus*, some sponges and ascidians and probably also an admixture of bituminous residues is presently building up a peculiar beachrock, consisting of sand and a large amount of shells. The big *Murex* and *Strombus* shells half buried in the sandy sediment also serve as a pebblelike substrate for hard-bottom fauna.

Elements of the fauna, which inhabit the littoral boulders and pebbles of the Red Sea, exposed both to the permanent danger of being buried or obturated by sand and to the possibility of increased salinities in the shallow littoral, invaded the Canal at an early date and settled the Bitter Lakes and even Lake Timsah. This fauna includes the snails *Nerita forskali* (appearing in Timsah by the beginning of the century), *Clan-*

culus pharaonis (in the Bitter Lakes by 1880), the opisthobranch *Berthella*, the small sea star *Asterina wega* (Plate III, Fig. 7), and even the small gorgonarian *Akabaria*, found in the Bitter Lakes by the Cambridge Expedition in 1924.

RICHNESS AND CONSTANCY OF THE FAUNA IN THE BITTER LAKES

There is a tendency in scientific opinion to consider the Bitter Lakes as a barren brine-pool. In fact, it is only the bottom below the 10 m isobath, in the center of the Great Bitter Lake and overlying the sub-recent salt deposits, which at times has been highly hypersaline, more or less anaerobic and, frequently, abiotic. The shallow bottom areas around this central basin have probably never had a salinity much above 50‰ and were, from the first years inhabited, by a qualitatively and quantitatively, rich fauna. As shown above, this fauna is chiefly composed of representatives of three ecological groups of the Red Sea: the littoral nectobenthos, the *Halophila* fauna and the fauna of the littoral boulders. As early as 1882, Keller considered the fauna of the Bitter Lakes as richer than that of the Suez shore in the open Gulf of Suez. Munro-Fox (1927) reported as follows on the shallow fauna of the Bitter Lakes: ". . . the fauna is rich both as regards numbers of individuals and species. Thus, the high salinity is not unfavourable to at any rate certain forms of marine life."

The rich littoral fauna which is presently found in the Bitter Lakes does not differ basically from that which inhabited the lakes eighty years ago. The same mollusks, the same echinoderms (*Nudechinus, Astropecten polyacanthus* (Plate III, Fig. 6) *Ophiactis savignyi* (Plate III, Fig. 5), and the various sea cucumbers) dominate now, as before. The biota of the lake is thus a quite equilibrated and stable association. The mollusks are also much the same as those found by Fuchs (op. cit.) in subfossil form in the partly dry basin of the future Bitter Lakes, before the opening of the

Canal. Thus, the lessepsian canal did not change anything essential: it gave to a certain faunal complex the opportunity to reconquer grounds which it had settled in the past under natural circumstances.

ROLE OF PASSIVE TRANSPORT

Transport on ship hulls brought about several well known and spectacular changes in the distribution of marine species. It was only natural, therefore, to suppose that lessepsian migration should be greatly enhanced by passive transport, particularly because the Suez Canal used to be one of the busiest waterways in the world. Munro-Fox (1929) makes, in fact, the only reference to passive transport, noting that "evidence was obtained that Polyzoa and Isopods have spread in this manner" (i.e., on ship hulls). But neither Omer Cooper listing the Isopoda of the Cambridge Expedition, nor Hastings dealing with the Polyzoa collected, mention such passive transport. W. Steinitz (1927) rightly expected the appearance of Indopacific immigrants in the Mediterranean ports of call of ships which traversed the Suez Canal. Very few such cases, however, have been reported as yet (for example the ephemerous appearance of the rock lobster, *Thennus orientalis* in Ryeka [Fiume]).

New Indopacific immigrants are discovered every year along the Israeli coast (among them, lately, four species of Indopacific Polyzoa [Powell, 1969]) notwithstanding the fact that in the past twenty years no ship which had crossed the Suez Canal called at Israeli ports. In Cyprus, however, which lies across the Canal opening and with a number of ports frequently visited by ships a few hours out of Port Said, the influence of Indopacific immigrants seems to be reduced. The recent Hebrew-University Smithsonian Expedition to Cyprus, led by Dr. M. Tsurnamal did not find the Indopacific organisms here which are now so obvious along the Israeli coast. Preliminary analyses of the Mollusca (Barash, pers. comm.), of the Polyzoa (Eytan, pers. comm.) and of littoral fish (H. Stein-

itz, pers. comm.), strongly support this view.

Passive transport cannot entirely be ruled out, but it is suggested that only a reduced number of species, with wide circumtropical range, have used this opportunity (as for example some of the Polyzoa). In this case, although it would be immigration in the strict sense, yet as shown by H. Steinitz (1968), it is immigration only in the quantitative sense, and as such it would be difficult to prove. It might however be that species with very wide, presumably prelessepsian, distribution ranges in the Mediterranean, such as *Pteria occa* or *Metrocarpa nigrum* (see above) have been transported passively through the waterways of antiquity.

MEIOBENTHOS AND LESSEPSIAN MIGRATION

The term meiobenthos is used for small metazoans which live preferably in the level bottom and have no planktonic larval stages. Nematodes, copepods, ostracods, halacarids form the most important elements of this ecological group.

Meiobenthic organisms have not been investigated in this area until recently. Only Gurney (1927) listed a fairly large number of benthic copepods from the Suez Canal. The present author, studying the same group of animals, found very little Indopacific influence on the fauna of the Israeli coast (1964). For the Ostracoda too (Ruth Lerner-Seggev, pers. comm.) the Indopacific influence seems to be minimal in this area. Nevertheless, in a recent discussion of the Copepod fauna of the Sirbonian lagoons, which is very rich despite the high salinity, and in a study of the distribution of the family Canuellidae, the present author (Por 1968; 1969) has reached the following conclusions, which might apply to the whole meiobenthos: The immigration advance of the meiobenthic species is a very slow one, which, in the short time of 100 years, possibly did not influence the pre-existent distributional patterns in open sea. The meiobenthos, however, has a much higher degree of resistance to high and changing salinities than the macrobenthos and has, therefore, a much higher potential capacity to settle biotopes along the Suez Canal itself. This property results in an increased ability to use pre-lessepsian contacts.

Another positive factor in the migration of meiobenthos might be the turbidity resulting from the stirring up of the soft bottoms caused by passing ships or by dredgers working in the Canal. The meiobenthos might thereby be swept along passively for some distance, and be the only ecological group which is thus favored by the turbidity, the most limiting of the environmental factors affecting other Suez Canal animals.

ECOLOGICAL GROUPS EXCLUDED IN
PRINCIPLE FROM LESSEPSIAN MIGRATION

1. The sublittoral and much more, the bathyal fauna of all sorts. These are naturally excluded from the migration process since the deepest bottoms in the Canal are about 18 m.

2. The fauna of the rocky bottoms and transparent waters. Not only has the Suez Canal turbid waters and very few hard bottoms, either natural or artificial, but at both ends of the Canal, in the Gulf of Suez and off Port Said, the bottoms are primarily sediments. Therefore, the richest of the faunal associations of the Red Sea, the coral reef with all the associated fauna and flora, is excluded from migration. As rightly pointed out by Keller (1882), corals should not be expected in the Canal since in the adjacent areas of Suez there are no actively growing reefs.

3. The holoplanktonic species and the benthic species with meroplanktonic larvae. Since there are no throughgoing currents in the Suez Canal and no currents at all in the 36 kilometers of the Bitter Lakes, only the animals which are able to live in the Bitter Lakes for at least one reproductive generation have any chance of crossing the Canal. In the turbid waters of the Canal, and of the Lakes, the planktonic organisms

and larvae which feed on phytoplankton are the first to suffer. The shallow Bitter Lakes with their saline stratification behave as every other brackish or hypersaline lake: they have a much impoverished plankton, since vertical migration, essential to many of the planktonic organisms, is impossible.

Several holoplanktonic groups such as, Siphonophora, Pteropoda and Heteropoda, Hyperiidea, Appendiculariidae and Salpidae have not been reported from Canal waters. Benthic groups with planktonic larvae are very sparsely represented in the Canal, e.g., the Echinodermata. Several littoral brachyuran crabs and hermit crabs, which are frequent in the *Halophila* bottoms and the boulder littoral of the Red Sea—such as *Cryptodromia, Calappa, Grapsus* and *Metopograpsus, Coenobita* and *Dardanus*, all with planktonic larvae—are unknown in the Suez Canal.

BRIEF CONCLUSIONS ABOUT THE
PAST MIGRATION PROCESS

1. Although repeated pre-lessepsian faunal interchange was physically possible, only a restricted number of euryhaline species could have made good use of it. Considerable barriers of varied and rapidly changing salinities, always existed along the old waterways.

2. A highly euryhaline biotic association became established in these waterbodies of the Isthmus. This fauna is still found "in situ" but does not play an important numerical role in the migration process.

3. The Bitter Lakes have a key role in the migration; migration was probably possible only for species which successfully settled in the lakes. The three ecological groups of Red Sea fauna—the littoral nectobenthos, the *Halophila* fauna and the littoral boulder fauna—have repeatedly inhabited the Bitter Lakes in the past and reoccupied them shortly after the opening of the modern canal.

4. The migrants into the Mediterranean are essentially those species which presently

inhabit the Bitter Lakes and Lake Timsah. Their presence in these Lakes is perhaps the most concrete proof that they in fact migrated through the Suez Canal. There is not much evidence to support the fact that species previously present in the Canal have disappeared since. On the other hand, there seems to be no sizeable increase in the number of species settling in the Canal for the last few decades.

5. The high salinities in the Bitter Lakes served as an obstacle only for the holoplanktonic and meroplanktonic organisms, not chiefly by virtue of the high salinity itself, but due to the existence of a saline stratification in the lakes.

6. The edaphic conditions, such as the lack of hard bottoms and, primarily, the high turbidity due to the continuous stirring up of the soft bottoms, constitute perhaps the chief obstacle to migration in the Suez Canal. The meiobenthos might be the only group favored by this latter circumstance.

7. The peculiar conditions found off Suez, with prevailing sediment bottoms, low temperatures and increased salinities, form the first sieve in the immigration path into and through the Suez Canal. The coral reef fauna, which is the most spectacular and richest fauna of the Red Sea, is already stopped there.

8. The edaphic conditions are again the chief limiting factor at the receiving end of the migration route. Outside Port Said, along the Sinai coast, the shore is alluvial and the next rocky littoral bottom is hundreds of kilometers away (south of Jaffa). It follows that at the entrance to the Mediterranean the fauna of the littoral level bottom is again at an advantage.

The importance of the "brackish barrier" (Thorson, 1968) at the northern end of the Suez Canal should not be overemphasized. The salinity outside Port Said probably never drops below 25‰, a value which is well above typical brackish values. The littoral fauna is usually able to withstand such slightly lowered salinities, especially the fauna living in the *Halophila* meadows of the alluvial fans of the Red Sea wadis, where sudden floods during the winter may sometimes reduce the salinity to a considerable extent. A true brackish barrier exists west of Port Said at the mouth of the Nile Delta—and this especially is responsible for the main eastward direction of the immigrants in the Mediterranean.

It should also be made clear that the section of the Canal which crosses Lake Menzala is artificially almost completely isolated from the brackish waters of the lake. The brackish barrier was of course much more effective in the pre-lessepsian waterways which crossed the open lagoons of the Ballah and Menzala area.

THE UNIDIRECTIONAL CHARACTER OF THE LESSEPSIAN MIGRATION

The expectations, first expressed by Vaillant, became only partly true; there occurred no faunal interchange, in the strict meaning of the word. The lessepsian migration is a one-way movement of fauna from the Red Sea into the Mediterranean.

The most recent short list of Mediterranean immigrants (Steinitz, 1968) into the Gulf of Suez indicates that there are extremely few convincing cases—and all of these are fishes. On the other hand, the number of the Red Sea immigrants into the Mediterranean is most probably running into the hundreds. How can this practically unidirectional sense of the migration be explained?

Previous authors were almost unanimous in trying to explain this situation by invoking the prevailing northward directed current in the Canal. But as shown above, there is no such throughgoing current—at least in the early 40 km segment of the Bitter Lakes. On the other hand, the outgoing current stops at the pier of Port Said, where the most complicated problem for the immigrants—that of conquering ecological niches in the "new sea"—commences.

Ben-Tuvia (1966) was among the first to

make reference to the scarcity of the East Mediterranean fish-fauna as compared to that of the Red Sea: 550 species against 800 species. Since Le Danois (1925),[3] it is universally recognized that the fauna of the Eastern Mediterranean is an impoverished fauna of the Western basin. Several hydrological and historical reasons, which need not be discussed here, result in numerous species and genera, and even a few higher taxa—such as the Gorgonaria—not reaching the Levant coast. Many of the species which settled the area are represented by small-sized specimens. The extremely low nutritive capacity of the Eastern Mediterranean probably does not allow the few well-adapted species to compensate for this qualitative scarcity with a quantitative flourishing. It can be said that the Eastern Mediterranean is a zoogeographical "cul de sac," a tropical sea, undersaturated with an Atlantic-temperature fauna.

By reopening artificially the contact with the Red Sea—a typical tropical sea—Lesseps helped unknowingly to reestablish a zoogeographical equilibrium and to fill this peculiar ecological vacuum in the Eastern Mediterranean.

In support of this reasoning, it is important to emphasize that most of the presently known Red Sea immigrants are very successful and widespread in their new home along the Levant coast. Ben-Tuvia (op. cit.) points out that of the 24 species of immigrant fish, 13 are already very common and 11 are even of commercial value. Other examples of mass development of immigrants are furnished by the decapods *Myra fugax* (Plate II, Fig. 2) and *Atergatis roseus* (Plate II, Fig. 4), the mantis shrimp *Squilla massawensis* (Plate II, Fig. 5), the gastropod *Murex tribulus* and the sea star *Asterina wega*. The successful settling of the newcomers may be explained by the fact that they either occupied empty ecological niches or replaced the less successful local species. There are several well known examples of the last case.

The facts discussed above give also a partial answer to the second aspect of this problem: why did the "Mediterraneans" not suceed in establishing themselves in the Gulf of Suez. The Red Sea is a tropical sea with a population "at equilibrium" with its environment.

The 70 km long narrow artificial channel which starts immediately at Port Said constitutes another very difficult barrier for the "Mediterraneans." Until they reach the quiet waters of Lake Timsah, southward migrating species have to cover a long distance, quite incomparable to the distance which separates the Bitter Lakes from Suez (less than 30 km). In these stretches of artificial channel, the importance of the prevailing currents (northward for about 10 months, between Timsah and Port Said) plays an important, though probably not essential, role. This becomes evident when considering the fact that a fair number of Mediterranean species succeeded in reaching Lake Timsah and the Bitter Lakes. Immigration into the Suez Canal waters resulted therefore in a real mixing of the two faunas (or three if we consider the Isthmus fauna). The unidirectional character of the lessepsian migration—i.e., in the unilateral Indopacific invasion of an old Atlantic realm, appears only in its zoogeographical "end product" in the open sea.

THE FUTURE OF THE LESSEPSIAN MIGRATION

Two recent environmental changes may exert their influence upon the lessepsian migration in the immediate future.

The first is the interruption of shipping and of maintenance of the Suez Canal, following the Six Day War. For two years, because many tens of thousands of ships have failed to pass through, the waters of the Canal have become clear and quiet as never before. Never has the bottom of the Canal been so undisturbed by dredgers and baggers. Numerous piers and port installations have not been maintained and cleaned from algae and sessile fauna. It seems, that this rather unexpected side-product of re-

---

[3] Quoted by Tortonese (1951).

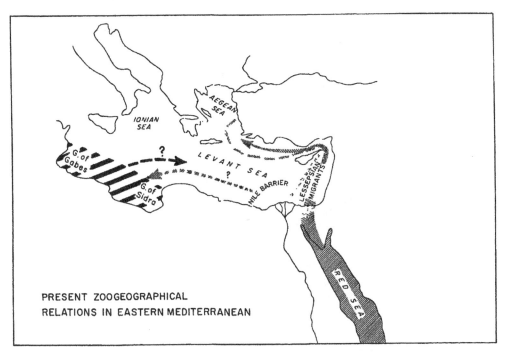

PRESENT ZOOGEOGRAPHICAL
RELATIONS IN EASTERN MEDITERRANEAN

Fig. 6.—Present zoogeographical relations in the eastern part of the Mediterranean.

cent hostilities should have a positive influence on the lessepsian migration.

The second "man-made" change is the definitive damming of the Nile by the new high-dam at Aswan. Since 1966, only one fourth of the previous amount of Nile waters reaches the Mediterranean. The sea off Port Said has, therefore, an almost constant salinity of 39‰, very similar to that of normal Levantine sea water. The considerable salinity increase of the barrier in front of the Nile Delta will perhaps enhance the ability of the immigrants to choose also the westward route along the African coast. However, since the Nile was the main supplier of nutrient salts for the excessively poor Levant basin, the fact that the future fauna of this basin will have to live within an even narrower trophic frame has also to be taken into account. Although this is slightly beside the point of the present discussion—a general decrease in the total amount of life in the basin will be a negative phenomenon; it remains an

open question whether the increased competition for food will favor the old inhabitants or the new immigrants.

How far will the qualitative enrichment and the faunal replacement proceed? Will the Levant basin and the Eastern Mediterranean finally become another province of the Indo-Pacific realm, not unlike the Honshu region on the Japanese coast or the Natal region?

As impressive as the lessepsian migration might appear, the Suez Canal will never duplicate a natural way of contact between two faunas. It might be reopened to navigation or slowly decay, but the problems which face the immigrants on their way through the Canal will not change radically for the better. Scleractinians and other reef organisms are unlikely to be able to cross into the Mediterranean. The same holds true for holoplankton and for benthic organisms with pelagic larvae.

Devoid of the reef fauna and of many other elements of the Indopacific benthos and plankton, it is improbable that the Levant basin will be changed into an appendix of the Red Sea, provided no artificial transport takes place, either irresponsibly or made for scientific or economic reasons.

It seems that the next step in the immigration will be for Bitter Lake species, which have not yet managed to invade the Mediterranean, to successfully complete their migration. There seems to be no reason to assume that immigration into the Canal will qualitatively increase at the Red Sea end. The lessepsian migration is, therefore, a phenomenon with a rather clearly set frame which is rapidly approaching its fulfillment.

Where will the geographical limits of the lessepsian immigration be set? The migration in the Mediterranean is clearly a longshore advance of littoral species along the Levant coast (Fig. 6). For the time being, it seems that the limit of the influence is in the area of Rhodes. The colder areas of the Aegean Sea probably constitute an environment in which the Atlantic fauna is much better represented and better fitted to withstand the competition of the tropical immigrants. The case of the Island of Cyprus, where the Indopacific species do not seem to have too much weight, indicates that direct migration across the open Mediterranean is taking place at a very much slower rate.

To the west, the question of the original fauna of the North African gulfs still remains incompletely answered. A considerable effort should be made to investigate the fauna there in the immediate future, because the disappearance of the Nile Delta barrier will probably lead to an invasion of that region by lessepsian immigrants.

REFERENCES

BAVAY, A. 1897. Au sujet du passage d'un mollusque de la Mer Rouge dans la Mediterranee. Bull. Soc. Zool. France, 5(23):161.

BEN-TUVIA, A. 1966. Red Sea fishes recently found in the Mediterranean. Copeia, 2:255-275.

FOX, H. M. 1926. Cambridge Expedition to the Suez Canal, 1924. I General Part. Trans. Zool. Soc. London, 22/1:1-64.

FOX, H. M. 1929. Cambridge Expedition to the Suez Canal. Summary of Results. Trans. Zool. Soc. London, 22/6:843-863.

FUCHS, TH. 1878. Die geologische Beschaffenheit der Landenge von Suez. Denkschr. Akad. Wiss. Wien, Math.-nat., Kl. 38:25.

GITAY, A. 1970. A review of *Augeneriella* (Pol: Sab.) and a new species from northern Sinai. Israel J. Zool. 19:105-110.

GOHAR, H. A. F. 1954. The place of the Red Sea between the Indian Ocean and the Mediterranean. Hidrobiologi. Istanbul, B. 2:47-82.

GRUVEL, A. 1936. Contribution a l'etude de la bionomie generale et de l'exploitation de la faune du Canal de Suez. Mem. Inst. d'Egypte, 29:1-255.

GURNEY, R. 1927a. Cambridge Expedition to the Suez Canal. VIII. Report on the Crustacea—Copepoda and Cladocera of the Plankton. Trans. Zool. Soc. London, 22:139-171.

GURNEY, R. 1927b. Cambridge Expedition to the Suez Canal XXXIII. Report on the Crustacea: Copepoda (Littoral and Semiparasitic). Trans. Zool. Soc. London, 22/6:451-475.

HASTINGS, A. B. 1927. Report on the Polyzoa. Cambridge Expedition to the Suez Canal. Trans. Zool. Soc. London, 22/3:331-354.

KELLER, C. 1882. Die Fauna in Suez Canal und die Diffusion der mediterraneen und erythraischen Tierwelt. Denkschriften der Schweiz. Ges. für die gesamten Naturwiss., 28:39.

KLAUSEWITZ, W. 1959. Systematisch-evolutive Untersuchungen uber die Abstammung einiger Fische des Roten Meeres. Verh. Deut. Zool. Ges. Leipzig, 175-182.

KOSSWIG, C. 1956. Beitrag zur Faunengeschichte des Mittelmeeres. Publ. Staz. Zool. Napoli, 28:78-88.

KRUKENBERG, C. F. W. 1888. Die Durchfluthung des Isthmus von Suez in chorologischer, hydrographischer und historischer Beziehung. Heidelberg.

LAUBIER, L. 1966. Sur quelques Annelides Polychetes de la region de Beyrouth. Am. Univ. Beyrouth Misc. Papers. Nat. Sc., 5:9-23.

LERNER-SEGGEV, RUTH. 1964. Preliminary notes on the Ostracoda of the Mediterranean coast of Israel. Israel J. Zool., 13:145-176.

LESSEPS, F. DE. 1867-8. Les lacs amers. Comp. Rend. Acad. Sci. 78 and 82.

MORKOS, S. A. 1960. Die Verteilung des Salzgehaltes im Suez Canal. Kieler Meeres Forsch., 16:133-154.

NOURSE, J. E. 1870. The Maritime Canal of Suez: brief memoir of the enterprise and comparison of its probable results with those of a ship canal across Darien. Washington. 57 p.

OMER-COOPER, J. 1927. Cambridge Expedition

to the Suez Canal. XII. Report on the Crustacea Isopoda and Tanaidacea. Trans. Zool. Soc. London, 22/8:201–209.

OREN, O. H. 1957. Changes in temperature of the Eastern Mediterranean in relation to the catch of the Israel Trawl fishery during the years 1954–55 and 1955–56. Bull. Inst. Oceanogr., Monaco, 1102:1–10.

PERES, J. M. 1958. Ascidies de la Baie de Haifa collectees par E. Gottlieb. Bull. Res. Council Israel, B7:151–164.

PERES, J. M. 1967. The mediterranean-benthos. Oceanogr. Mar. Biol. Ann. Rev., 5:449–553.

POR, F. D. 1964. A study of the Levantine and Pontic Harpacticoida (Crustacea, Copepoda), Zool. Verh. Leiden, 64:128.

POR, F. D. (in press). The zoobenthos of the Sirbonian lagoons. Rapports et Proc.-verbaux des reunions de la C.I.E.S.M.M.

POR, F. D. 1969. The Canuellidae (Copepoda, Harpacticoida) in the waters around the Sinai peninsula and the problem of lessepsian migration in this family. Israel J. Zool., 18:169–178.

SCHLEIDEN, M. J. 1858. Die Landenge von Suez. Zur Burtheilung des Canal-projects und des Auszugs der Israeliten aus Aegypten. Leipzig. 202.

SEYMOUR-SEWELL, R. B. 1948. The Free-swimming Planktonic Copepoda Geographical Distribution. John Murray Exp. Sci. Rep., 8/3:317–595.

STEINITZ, H. 1967. A tentative list of immigrants via the Suez Canal. Letter to the editor. Israel J. Zool., 16:166–169.

STEINITZ, H. 1969. Remarks on the Suez Canal as pathway and as habitat. Rapp. Comm. int. Mer. Medit., 19:139–141.

STEINITZ, W. 1919. Denkschrift zur Begründung einer zoologischen Meerestation an der Küste Palästinas, Breslau, 24 p.

STEINITZ, W. 1929. Die Wanderung indopazifischer Arten ins Mittelmeer seit Beginn der Quartarperiode Int. Revue ges. Hydrobiol. Hydrogr., 22:1–90.

THORSON, G. 1968. Animal Migrations through the Suez Canal in the Past. Recent Years and the Future. III-e Symposium· European de Biologie Marine Arachon. (mimeographed).

TILLIER, J. B. 1902. Le Canal de Suez et sa faune ichthyologique. Mem. Soc. Zool. France, 15:297–318.

TILLIER, L. AND A. BAVAY. 1905. Les mollusques testaces du Canal de Suez, Bull. Soc. Zool. France, 30:70.

TORTONESE, E. 1951. I Caratteri biologici del Mediterraneo orientale e i problemi relativi. Attual. Zool., 7:207–251.

(*Received November 5, 1969*)

3 Species Ranges

Robert Hengeveld, Paul S. Giller, and Brett R. Riddle

Each species lives in a geographical area—the *species range*—which is sometimes limited by topographical or climatic barriers, sometimes by the distribution of specific habitat conditions in relation to the dispersal ability or physiology of the species, or sometimes by interactions with other species. A species can reach the limits of its geographic distribution abruptly at a sharp topographical boundary, or more subtly at the edge of a particular habitat transition. The environment, though, is not uniform in either space or time. As the spatial configuration of their living conditions changes, the distribution of individuals over the species range will shift accordingly, and so will the range boundaries. How a species responds to these changes depends on its dispersal ability, habitat plasticity, and other biological attributes. Because of the variation in habitat conditions over space and time, the species needs to have some capacity to move or to be moved.

Thus, geographical ranges are defined by the relationship between attributes of species and the spatial variation in their environment, a spatial pattern that itself shifts over time. Principal to this spatially dynamic view is the notion that temporal persistence of a species across its range is the biological product of response mechanisms of each individual organism to environmental variation in space.

Spatial analysis is difficult, particularly when it concerns large-scale dynamics of habitats and ranges. We, therefore, tend to look at spatial dynamics from a local and short-term viewpoint. In contrast, a more comprehensive understanding of geographic range dynamics requires that we scale-up our studies to include patterns and processes operating over much larger temporal and spatial scales. Only recently have we begun to analyze range dynamics as specific habitat-tracking responses to long-term and large-scale shifts in spatial variation in the environment.

In contrast, classical approaches to biogeography considered species ranges to be spatially static entities. At the turn of the twentieth century, Hugo de Vries (1901–5) formulated his "mutation" theory. Willis (1922) later elaborated in biogeographical terms a centrifugal geographical spread from a center of origin after mutation, rather like the ripples created from a stone dropped into a pond. Variations on Willis' (1922) theme included the picture developed in Grinnell's (1943) book on the dynamic structure of ranges whereby individuals spread centrifugally from central, high-density regions to those of lower density at the margins, and

Hultén's (1937) equiformal progressive areas (to be discussed below).

In 1922, JOSEPH GRINNELL, who perhaps is better known for his early contributions to community ecology, published an essay "The Role of the 'Accidental,'" which we have included in this part (paper 27). This paper was among the first to highlight the importance of long-range dispersal of individuals to range expansion and persistence of species. Working on bird species in California, Grinnell proposed that occurrences of so-called *accidentals* (species recorded only once in an area or region outside their normal range) is a "regular thing to be expected." He also pointed out that although accidentals were usually bird species with good dispersal abilities, some were more sedentary in habit. Based on the rate of additions of occurrences of species to the list of accidentals for the 35 years prior to publication of the paper, and all else being equal, he predicted it would take 410 years (until the year 2331) for all the birds of North America to be seen, at least once, in California.

Grinnell (1922) suggested that some range expansion results from pioneers—individuals in the periphery of the species' range—testing out adjoining areas, with a few of them locating new and suitable habitats. The ultimate outcome in this case would be gradual extension of the species' habitat and range limits. Other individuals (accidentals) may cross barriers (i.e., exhibit "jump dispersal"), thus providing a means by which species can spread across barriers to new sites in response to shifting conditions. For example, recent range expansion by some butterflies and bush crickets in Britain appears to have been associated with evolution of increased range of habitats colonized by these species (Thomas et al. 2001). Still other species can increase the rate of range expansion by increasing the fraction of dispersive individuals in the population.

Grinnell's (1943) approach for studying range expansion started from a local viewpoint, considering the dynamic build-up of ranges as an overflow of individuals from high-density populations in central regions toward lower densities at the range periphery. He assumed that the distribution of favorable and unfavorable conditions, causing differences in densities within the range, remained more or less the same. He highlighted, almost prophetically as we sit in the current grasp of global warming, that in order to persist in a world of changing conditions it is necessary that a close match be maintained between a species and the distribution of its preferred living conditions. The emergence of new dispersing phenotypes may underlie the responses of many species to past and future climatic change.

Although Grinnell's scenario contained the dynamic element of overflowing populations, most twentieth-century biogeographers continued to assume that geographic ranges were static, at least since the end of the last ice age, some 10,000 years ago (see Udvardy 1969). ERIC HULTÉN's *Outline of the History of Arctic and Boreal Biota during the Quarternary Period* (1937; paper 28) assumes the relatively static location of centers of origin from which plants disperse. In the absence of dispersal barriers or climatic changes of the Pleistocene, the geographic ranges would be circular, as Willis (1922) had originally suggested. Overall, however, the shapes and sizes of the various ranges were thought to reflect, not the relative ages of species, but differences in dispersal rates among species. For example, plant species with bulbs or thick rhizomes are not able to disperse as quickly as annuals and, therefore, should have smaller geographic ranges. Thus, assuming that speciation takes place in particular areas, Hultén's classification of geographic ranges of Holarctic plant species included groups of species whose ranges overlap, are similar in shape but different in size; in his terms, "equiformal progressive areas."

This classification was intuitive rather than statistical. Moreover, Hultén's equiformal progressive areas, categorized in terms of location and shape, contained not only species of a single taxon, but also those of many unrelated taxa. Central to this classification was his view that the location of the various ranges is static; they would not have shifted since the species originated. Hultén made no connection between the ecological history of an area and the taxonomy of its species. Similar to Grinnell's (1922) explanations, Hultén's included processes in ecological, rather than geological time, thus representing early examples of ecological biogeography.

"Areas, Their Centers and Boundaries," chapter 3 of EVGENII VLADIMIROVITCH WULFF's *An Introduction to Historical Plant Geography* (trans. 1943; see paper 29) gives an exemplary overview of the ways in which species ranges had been studied up to 1943. As such, it also formed the basis of many approaches and concepts in later developments of biogeography. Wulff's chapter described the whole complexity of ranges as both historical and ecological entities. The historical entities, representing time lags to ecological adaptation (Hengeveld 1990), were discussed in terms of the spread from some presumed center of origin. At the same time, Wulff emphasized the impact of ecological conditions, with, for example, highest abundances occurring at the center because that is where optimal conditions are found. Interestingly, he also realized that the present location of a range may differ greatly from that in the past, so that at least in some respects, ranges may not be static, thus confusing attempts to locate centers of origin. Unfortunately, Wulff did not define the time scale over which such shifts in range location may have taken place.

One of the advantages of assuming geographic ranges to be static was that it made it much easier to classify them, as Hultén did in 1937; the classes thus obtained were interpreted either historically and evolutionarily, or ecologically. The latter case assumed that the classes form closed systems that would resist invasion of species from outside. Within this ecological context of species ranges, much emphasis was placed on the existence of biogeographical barriers between geographical regions. Another consequence of this eco-evolutionary interpretation and classification of floras and faunas was that eco-geographical units were assumed to be separated by sharp boundaries. If species somehow managed to cross these boundaries, the invaded system would remain out of balance for some time, only gradually resuming a new state of community equilibrium (see, e.g., Elton 1958).

One of the best-known biogeographic boundaries is Wallace's Line, separating the Oriental and Australian realms, which lie on two adjacent tectonic plates, and which partly bump into and partly shift past each other (see, e.g., Whitmore 1981). The line is, therefore, also generally consistent with the deepwater zones that separated the ice-age mainland of Malaysia (which included Sumatra, Java, and Borneo) from the islands of eastern Indonesia and Sahul—the combined landmasses of New Guinea, Australia, and Tasmania (see Brown and Lomolino 1998: fig. 7.9). The position of the biogeographic demarcation was debated, leading to a multitude of lines depending on the focal taxa and various approaches employed (Simpson 1977). In relation to this debate, the analysis of JEREMY D. HOLLOWAY and NICHOLAS JARDINE (1968) became significant because it was also among the first studies to use multivariate statistical techniques to analyze biogeographical patterns. Applying the same technique to data sets concerning three taxa, their paper "Two Approaches to Zoogeography: A Study Based on the Distributions of Butterflies, Birds and Bats in the Indo-Australian Area" (paper 30), showed that the members of these taxa trespassed across the tectonic plates differently according to the dispersal capacities of their species. Oriental bats advanced well beyond Wallace's Line, while butterflies fell well short of it. Later, Flessa (1975) extended these ideas to consider the effects of dispersal on the similarity of mammals in different biogeographic regions.

Thus, it appeared that, to varying degrees, dispersal could easily blur differences in distributions that originated in the geological past, with the extent of the shifts depending on the dispersal power of the species of the taxon concerned. In earlier geographical classifications like Hultén's (1937), the species' ranges and the patterns that they formed had no obvious, unique interpretation in terms of either taxonomic composition or evolution. Both the geological history and dispersal processes were involved. Moreover, in addition to these two factors, climate is also involved in partitioning biotas within a region, again affecting different taxa differentially (Lincoln 1975).

Despite the strength and effectiveness of barriers and limits to the dispersal capacities of species, the past few hundred years have seen perhaps some of the largest-scale dispersal events in the history of the world. Humans are notoriously successful dispersal agents, effectively increasing the rate of range expansion by eroding barriers to dispersal or, more directly, by accidental or purposeful introductions of exotic species. This long-distance, short-time transport mechanism is now having an unprecedented influence on distributions of flora and fauna. We may pity the biogeographers of the future as they try to unravel the mysterious mechanisms (natural or anthropogenic) that led to the patterns they will study!

The role of humans in the distribution of biotas is a central theme of CHARLES S. ELTON's classic book *The Ecology of Invasions by Animals and Plants* (1958). Here he brought together ideas from three different streams of thought from his studies over the previous 30 years: faunal history, ecology, and conservation. Using many diverse examples of the impact of man through introductions, Elton's book provides a stark picture of what actually happened in ecological time under a situation of enhanced dispersal across heretofore impassable barriers. For example, by 1954, the European Starling, *Sturnus vulgaris*, had dispersed across almost the entire United States from a stock of 80 birds introduced in 1891 into Central Park, New York. Similarly, the North American Muskrat, *Ondatra zibethicus*, expanded its range in within about 50 years to almost the whole of Europe from five individuals introduced in 1905 into what was then Czechoslovakia. These examples from Elton's book illustrate how, through the action of humans, a Palearctic (starling) or a Nearctic (muskrat) species could attain Holarctic distributions within just half a century. Although

Elton's book should be read in its entirety as a wonderful insight into the influence of man on nature, we have selected chapter 4, "The Fate of Remote Islands," for inclusion here (paper 31).

Under natural circumstances, the ability to follow changing environmental conditions, and thereby to cross barriers, depends not only on physiology, but also on the behavior and ecological tolerance (niche breadth) of the focal species. In his classic 1967 essay "Why Mountain Passes are Higher in the Tropics" (paper 32), DANIEL H. JANZEN postulated that topographic barriers may be more effective in the tropics than in the temperate regions, owing to the relationship between climatic variability and niche breadth of resident species. In the tropics, terrestrial organisms are exposed to relatively stable conditions and, as a result, are presumed to have narrow niches, often limiting their travels to one watershed and elevational zone (for examples, see Flenley 1979, featured in this part, or Wolda 1978). In contrast, in order to survive the substantial variation in climate of temperate regions, resident species must have relatively broad niches and the capacity to cross a wide range of habitats and life zones. Indeed, Janzen showed that if a temperate species dispersed across a mountain pass in summer, it would be unlikely to encounter temperatures lower than it regularly experiences in the winter. This would definitely not be true for a tropical lowland species attempting to cross an equally high pass; that is, mountain passes *are* higher in the tropics. Quite obvious in retrospect, but no one had articulated this before Janzen!

As was discussed in the introduction to part 2, although much of the earlier biogeographical analysis was based on dispersal biogeography, the acceptance of plate tectonics and the gradual appreciation of its potential impact on the distribution of biota provided a key mechanism for vicariance biogeography of the 1960s and 1970s. Also during this period, MacArthur and Wilson's watershed work, *The Theory of Island Biogeography* (1967; see also MacArthur and Wilson 1963), challenged the static paradigm, at least among ecologists, emphasizing the role of dispersal and dynamic turnover of communities over ecological time. Among historical biogeographers, however, the dispersal versus vicariance debate waged on. Quaternary ecologists studying biotas of the Pleistocene were allied more with the dispersal biogeographers, basing most of their analyses on the distributions of fossil mammals and insects, and of plants from pollen and macrofossils. Previously, though, the palynologists had been applying a local approach in the interpretation of pollen diagrams. As information on the changing distributions of Pleistocene glaciers and climates began to accumulate, however, they started to frame their results within a geographical setting, showing maps with colonization patterns.

The pollen and macrofossils from arid regions of the world, like southwestern North America, were relatively rare and unexploited as indicators of Quaternary plant distributions and community assemblages (Betancourt, Van Devender, and Martin 1990). However, the seminal papers of Wells and Jorgensen (1964) and Wells and Berger (1967) showed that a paleovegetation record from the Mojave Desert, spanning the past 40,000 years, was represented in exquisite detail in the middens of woodrats of the genus *Neotoma*. Woodrat middens consist of a temporally stratified accumulation of vegetation harvested by foraging rats, preserved within an indurated matrix of urine and debris. The well-preserved leaves and seeds permit precise identification of plant taxa as well as accurate radiocarbon dating of the material.

The featured contribution by PHILIP V. WELLS and RAINER BERGER, "Late Pleistocene History of Coniferous Woodland in the Mohave Desert" (1967; paper 33), revealed dramatic shifts in the distributions of species and the compositions of communities across the Pleistocene–Holocene threshold approximately 10,000 years ago. For example, at many sites throughout the Mojave Desert below about 1,500-meter elevation, Wells and Berger documented a regional pattern of rapid transition from pygmy conifer woodlands to xeric

shrub assemblages. Subsequently, the analysis of middens of small herbivorous mammals has expanded throughout arid regions of western North America, and to some extent onto other continents. These studies serve as an important bridge between historical and ecological biogeography by providing a means to trace directions and rates of range dynamics and to document shifts in community structure and species associations during and after the most recent glacial–interglacial cycle (see Betancourt, Van Devender, and Martin 1990 for a thorough review).

Margaret Davis was another pioneer of phytogeography during the Quaternary, analyzing the colonization history of North American tree species during the late-Pleistocene and early Holocene. In her 1976 paper, "Pleistocene Geography of Temperate Deciduous Forests," Davis illustrated how dispersal characteristics of particular species influenced their responses to the changing environments since the last glacial period. This work, based on pollen records in lake and soil sediments of North America and Europe, revealed how dispersal and range expansion of trees varied among regions as well as among species. It also showed how these differences in dispersal abilities led to continuously changing community composition over time, such that many past assemblages have no present-day analogues (see also Huntley 1980; Premoli, Kitzberger, and Veblen 2000).

Other Quaternary biogeographers described marked elevational shifts in geographic ranges in response to the last glacial recession, echoing the earlier hypotheses of montane origins and range shifts proposed by Linnaeus and Karl Willdenow. For example, JOHN FLENLEY's 1979 article, "The Late Quaternary Vegetational History of the Equatorial Mountains," included in this part (paper 34), compares the elevational distributions of plant communities along mountain slopes in East Africa, New Guinea, and Colombia (see also Walker and Flenley 1979). It appears that the tree line shifted upward as much as 1,500 meters in the Colombian mountains over a few thousands of years. Also, the shifts in tree lines for these three tropical regions appear to have occurred at the same time, hinting at worldwide effects of global climate variation. Finally, these and other studies reveal that warm conditions of the present time are exceptional, as interglacial periods existed for less than 10 percent of the total, two-million-year period of the Pleistocene.

These accounts of range dynamics and continuous turnover of plant communities shocked the proponents of climax theory (Clements 1916), as well as those who had related stability and equilibria with the complexity of communities (MacArthur 1955). These ideas and their proponents were still dominating community theory in the 1970s. Ramensky's ([1924] 1965) and Gleason's (1926) *individualistic*, and Curtis's (Curtis and McIntosh, 1951) *continuum* approaches appeared more applicable to plant communities and were becoming more widely accepted (e.g., Whittaker 1967, in part 7, and Whittaker and Niering 1975, in part 8 of this volume). Up to that time, some ecologists and biogeographers not only believed that communities and biogeographic provinces existed as entities, well-balanced and impenetrable to invaders, but also that they could even be compared with organisms, originating and developing to maturity. In ecology, this type of holistic reasoning was basic to Clement's (1916) theory of community succession and much of continental vegetation science. In biogeography, Meusel (1943) considered the range as an organism leading its own life. Similar to ideas in the 1920s and 1930s in embryology, philosophy, and societal organization, some special force was assumed to shape the organization and development of such entities, although certainly not all ecologists or biogeographers adhered to such ideas (see Anker 2001). Implicitly, these views deny the relevance of individualistic behaviors of organisms and species to range dynamics and community assembly (Diamond 1975a and b—featured in parts 6 and 7 of this volume; Whittaker and Niering 1975; Hengeveld and Walter 1999; Walter and Hengeveld 2000).

Responding to the trends in population biology at large, paleoclimatologists started mapping changes in climate patterns at continental and global scales, concluding that not only are the high global temperatures of the Holocene exceptional, but so are those of the last millennium. This paradigmatic shift from theories of a static world to those of biogeographic dynamics and disequilibria paralleled advances in Quaternary climatology, which then began to emphasize variability in climate at all temporal and spatial scales (e.g., Bryson 1966; Lamb 1972, 1977)—a scientific revolution shattering any ideas of stability and permanency (Imbrie and Imbrie 1979, Coope 1975, Cushing 1982, Fritts 1976, Rainey 1963; see also Hengeveld and Walter 1999, Walter and Hengeveld 2000).

This new trend leading to a more genuine, dynamic biogeography effectively started in the late 1960s and early 1970s and continues today. As mentioned above, it considers species to behave individualistically, rather than as interdependent members of communities. Moreover, it was not only the particular species, but their constituent individuals that were beginning to be considered to be the basic units in the dispersal process. This actually is a fundamental assumption of the models on range expansion based on reaction-diffusion processes, as suggested by Skellam (1951), more fully described by mathematical ecologists like Mollison (1977), Okubo (1980), or Van den Bosch, Metz, and Diekmann (1990), and subsequently promoted by some Quaternary ecologists (e.g. Dexter, Banks, and Webb 1987) and ecologists (Van den Bosch, Hengeveld, and Metz 1992). This conceptual development mirrored that concerning the more-or-less contemporaneous analyses of the spread of post-Neolithic humans (e.g., Ammerman and Cavalli-Sforza 1971), their inventions (e.g., Hägerstrand 1967, Brown 1981) and diseases (Cliff et al. 1981, Van der Plank 1963).

The individual-based approach, in fact, links present analytical methods with ones formulated much earlier, and concerns range shifts and the spread of genes. Yet, as mentioned above, this spatially dynamic approach was uncommon, in fact only being promoted in a couple of early papers, one by the plant geographer Good (1931), and the other one by the population geneticist Fisher (1937). Similarly, the very general tendency for population densities to be highest near the center of a species' geographic range (Hengeveld and Haeck 1981, Brown 1995), was largely ignored by most biogeographers, even though it was initially proposed by the early plant geographers (e.g., Livington and Shreve 1921), and also applied long ago in agronomy (Klages 1942, Wilsie 1962). To a large extent, biogeographers remained reluctant to consider the influence of physiologic responses of individuals and population dynamics of species at broad spatial scales.

Thus, although the seeds of a more dynamic and biologically informed species range were sown many years ago (see Cain 1944, Wulff 1943), the real breakthroughs had to await more recent developments. Meanwhile, the study of species ranges remained concentrated on qualitative, static distribution patterns, usually indicated solely by the range limit. It took more than half a century to fill the void and to introduce a dynamic aspect to biogeography, one that included the more mechanistic and individualistic hypothesis that geographic patterns often result from ecological responses of individuals busily tracking their preferred living conditions.

We conclude this section with a paper that nicely characterizes the state of dynamic biogeography in the 1970s: a subdiscipline on the verge of paradigmatic shifts, rich with bold, insightful, yet provocative and controversial opinions. PAUL MARTIN's "The Discovery of America" (1973; paper 35), challenged both the static view of species ranges and the nearly universal reverence of "noble savages" who somehow lived in harmony with their environment (see also various chapters in Martin and Wright 1967). Instead, Martin argued that invasions and subsequent ecological impacts of our own species were not fundamentally different from those described in Elton's (1958) *Ecology of Invasions* (also excerpted in part 3, paper 31). According to Martin (1973), the collapse of geo-graphic ranges of North America's Pleistocene megafauna was the inevitable consequence of invasions by early humans: exponential population growth of these novel predators in a land where the prey were ecologically naïve to what Darwin once referred to as the "strangers' craft of power," and subsequent and surprisingly swift decimation of the native fauna.

Martin's paper was, and remains today, a flashpoint for insights and controversy. His general thesis, and especially its particulars (colonization dates, estimates of population sizes and biomass), continue to draw criticism from those proposing a climatic rather than anthropogenic cause for megafaunal extinctions (e.g., Axelrod 1967, Guilday 1967, Slaughter 1967, Graham 1986, Graham et al. 1996), and from archaeologists claiming earlier residence of human societies in the Americas (e.g., Meltzer 1997). While Martin's views will continue to draw criticism for some time, his "overkill" thesis was nonetheless seminal to both dynamic biogeography and the early development of conservation biology; distributions of many contemporary species, and the decline and loss of many more species in historic and prehistoric times may have been strongly influenced by the biogeographic dynamics of humans (Olson and James 1982, 1984; Flannery 1994, 2001; Steadman 1993, 1995; MacPhee and Marx 1997; Lomolino and Channell 1995).

Overall, the studies featured in part 3 chronicle the debate and eventual replacement of a static paradigm with a more dynamic view of species' ranges and ecological assemblages. The new, and now generally accepted paradigm of dynamic biogeography benefited from fundamental insights of Quaternary biogeographers studying the marked and seemingly continuous range shifts of plants and animals, including those of our own species. Modern biogeography can no longer be considered a discipline concerned just with patterns, but as one in which a rich diversity of biogeographic phenomena are to be explained as spatially dynamic, ecological, and evolutionary responses of biotas.

PAPER 27

pair breeding at Lake Forest. They seem, like the Cardinal, to be gradually extending their range to the north.

22. **Sitta canadensis.** RED-BREASTED NUTHATCH.—Very common during the falls of 1915, 1916 and 1921. Other years only a few have been seen.

23. **Polioptila c. caerula.** BLUE-GRAY GNATCATCHER.—Some years a very common migrant during May. A male was taken May 31, 1920 in the oak scrub and later the same day and throughout June a female was seen in the pines, generally in the same place and always uttering a plaintive call. I spent a great deal of time watching her on different occasions but if there was a nest it was never found. Dr. Eifrig reported seeing several May 30, 1921, and I intended making a search for a nest but my trip was delayed until July 24. On this date, I had just started through the oak scrub when I heard the call of the Gnatcatcher and found five in a small oak, two adults and three young. The young were almost fully grown and were catching insects for themselves but I saw the parents feed them a number of times. When the old birds approached the young opened their mouths and quivered their wings. This is a rare breeder in northeastern Illinois but is more common in the Sand Dunes in Indiana.

24. **Hylocichla g. guttata.** ALASKA HERMIT THRUSH. Mr. Coale took this bird November 5, 1916, and has already reported it in 'The Auk' (Vol. XXXIV, No. 1) being the first record east of the Rockies.

25. **Planesticus m. migratorius.**—ROBIN. Robins sometimes spend the winter in this region; seen December 31, 1914 and February 18, 1917.

P. O. Box 55, Chicago, Ill.

THE ROLE OF THE "ACCIDENTAL."[*]

BY JOSEPH GRINNELL.

The total number of species and subspecies of birds recorded upon definite basis from California amounts at the present moment to 576. Examination of the status of each species, and classification of the whole list according to frequency of observation, show that in 32 cases out of the 576 there is but one occurrence known. In 10 cases the presence of the species has been ascertained twice, in 6 cases three times, and for all the rest there are 4 or more

[*] Contribution from the Museum of Vertebrate Zoology of the University of California.

records of occurrence. Some 500 species can be called regularly migrant or resident.

Examination of the records for the past 35 years shows that the proportion of one-occurrence cases is continually increasing. In other words, the state is so well known ornithologically that regular migrants and residents have all or nearly all of them been discovered, and *their* number now remains practically constant, while more and more non-regulars are coming to notice. This might be explained on the ground that there are continually more and better-trained observers on the lookout for unusual birds. This is probably correct, partially. But also, I believe, there is indicated a continual appearance, within the confines of the state, through time, of additional species of extra-limital source.

In published bird lists generally, species which have been entered upon the basis of one occurrence only, are called "accidentals." This is true of county lists, of state lists, and, quite patently, of the American Ornithologists' Union's 'Check-list of North American Birds'. The idea in the adoption of the word "accidental" seems to have been that such an occurrence is wholly fortuitous, due to some unnatural agency (unnatural as regards the behavior of the bird itself) such as a storm of extraordinary violence, and that it is not likely to be repeated. This understanding of the word "accidental" is borne out by the explicit meaning given it in the 'Century Dictionary,' for instance, which is "taking place not according to the usual course of things," "happening by chance or accident, or unexpectedly." Now the way in which the word is used by ornithologists is really a misapplication of the term; for, as I propose to show, the occurrence of individual birds a greater or less distance beyond the bounds of the plentiful existence of the species to which they belong is the *regular thing, to be expected*. There is nothing really "accidental" about·it; the process is part of the ordinary evolutionary program.

However, as I have intimated, the word is firmly fixed in distributional literature. We had better continue to use it; but let us do so with the understanding that it simply means that any species so designated has occurred in the locality specified on but one known occasion. No special significance need to be implied.

Vol. XXXIX]
1922   GRINNELL, *The Role of the "Accidental."*   375

Accidentals are recruited mostly from those kinds of birds which are strong fliers. It is true that the majority belong to species of distinctly migratory habit. But some of our accidentals exemplify the most sedentary of species. Examples of one-instance occurrences, in other words "accidentals," are as follows: the Western Tanager in Wisconsin, the Louisiana Water-Thrush in southern California, Townsend's Solitaire in New York, the Catbird on the Farallon Islands, the Tennessee Warbler in southern California, and Wilson's Petrel on Monterey Bay. In the North American list some of the accidentals come from South America, some even from Asia and from Europe.

I would like to emphasize the point now that there is no species on the entire North American list, of some 1250 entries, that is not just as likely to appear in California sooner or later as some of those which are known to *have* occurred. Expressing it in another way, it is only a matter of time theoretically until the list of California birds will be identical with that for North America as a whole. On the basis of the rate for the last 35 years, $1\frac{3}{5}$ additions to the California list per year, this will happen in 410 years, namely in the year 2331, if the same intensity of observation now exercised be maintained. If observers become still more numerous and alert, the time will be shortened.

It will be observed that there are now many more one-occurrence, "accidental," cases than there are two-occurrence cases, and that there are more of the two-occurrence group than there are three-occurrence, and so forth, there being a regular reduction in the intervals so that, if we just had enough observations, a smooth curve would probably result. If the one-instance occurrences should continue to accumulate without any modification of the process, in the course of about 300 years there would be more of these "accidentals" in California than of regularly resident species, and the other groups would grade down in a steeper curve. I attempted to carry out the figures, which seem to behave according to some mathematical formula; but when I came to deal with $\frac{3}{5}$ of an occurrence I decided it was profitless to go on!

It is evident, however, that another process takes place, of quite opposite effect. With the lapse of time second-occurrence cases replace one-occurrence cases, to be followed by a third order of accretion, and this by a fourth; and the process might continue on ad infinitum, until theoretically, sometime after the full number of the North American list had been reached, our state list would no longer contain any accidentals at all. To cite an example, the eastern White-throated Sparrow was recorded first from California, and then as an accidental, on December 23, 1888; in 1889 a second specimen was taken; in 1891, a third was taken; and so on until in 1921, 19 occurrences have been recorded. This species is considered now simply as rare, certainly not accidental, casual, or even specially noteworthy save from a very local standpoint.

It comes to the mind here that if observations could be carried on so comprehensively as to bring scrutiny each year of every one of the 200,000,000 birds in California, this being the estimated minimum population maintained within the state from year to year, a great many more accidentals would be detected than are now known, and in addition some birds now known from but a few records, or even as accidentals, would come to be considered of frequent, though not necessarily regular, occurrence. With the White-throated Sparrow it is not impossible that a thousand of the birds have wintered in California in certain years.

Some of the considerations in the preceding paragraphs, while of interest in themselves perhaps, have confessedly been rather beside the issue. For the definite question which I wished to ask and which I will now briefly discuss is as to the function or role played by accidentals. Are they a mere by-product of species activity or do they in themselves constitute part of a mechanism of distinct use to the species?

The rate of reproduction in all birds, as with other animals, is so great that the population rapidly tends toward serious congestion except as relieved by death of individuals from various causes or else by expansion of the area occupied. The individuals making up a given bird species and occupying a restricted habitat may be likened to the molecules of a gas in a container which are continually beating against one another and against the confining walls, with resulting pressure outwards. But there is an essential difference in the case of the bird in that the number of individual

Vol. $\frac{XXXIX}{1922}$]   Grinnell, *The Role of the "Accidental."*   377

units is being augmented 50 per cent, 100 per cent, in some cases even 500 per cent, at each annual period of reproduction, with correspondingly reinforced outward pressure.

The force of impingement of the species against the barriers which operate to hem it in geographically, results in the more than normally rapid death of those individuals which find themselves under frontier conditions. There follows, through time and space both, a continual flow of the units of population from the center or centers toward the frontiers.

The common barriers which delimit bird distribution are as follows: Land to aquatic species and bodies or streams of water to terrestrial species, the climatic barriers of temperature, up or down beyond the limits to which the species may be accustomed, and of atmospheric humidity beyond critical limits of percentage; the limits of occurrence of food as regards amount and kind with respect to the inherent food-getting and food-using equipment of the species concerned; and the limits of occurrence of breeding places and safety refuges of a kind prescribed by the structural characters of the species requiring them.

An enormous death rate results from the process of trial and error where individuals are exposed wholesale to adverse conditions. This can be no less, on an average, than the annual rate of increase, if we grant that populations are, on an average, maintaining their numbers from year to year in statu quo. But before the individuals within the metropolis of a species succumb directly or indirectly to the results of severe competition, or those at the periphery succumb to the extreme vicissitudes of unfavorable conditions of climate, food or whatnot obtaining there, the latter have served the species invaluably in *testing out* the adjoining areas for possibly new territory to occupy. These *pioneers* are of exceeding importance to the species in that they are continually being centrifuged off on scouting expeditions (to mix the metaphor), to seek new country which may prove fit for occupancy. The vast majority of such individuals, 99 out of every hundred perhaps, are foredoomed to early destruction without any opportunity of breeding. Some few individuals may get back to the metropolis of the species. In the relatively rare case two birds comprising a pair, of greater hardihood, possibly, than the average, will find themselves a little beyond the confines of the metropolis of the species, where they will rear a brood successfully and thus establish a new outpost. Or, having gone farther yet, such a pair may even stumble upon a combination of conditions in a new locality the same as in its parent metropolis, and there start a new detached colony of the species.

It is this rare instance of success that goes to justify the prodigal expenditure of individuals by the species. Such instances, repeated, result in the gradual extension of habitat limits on the part especially of species in which the frontier populations are in some degree adaptable—in which they can acquire modifications which make them fit for still farther peripheral invasion against forbidding conditions.

Incidentally, the great majority of these pioneers are, I believe, birds-of-the-year, in the first full vigor of maturity; such birds are innately prone to wander; and furthermore it is the autumnal season when the movement is most in evidence, a period of food-lessening when competitive pressure is being brought to bear upon the congested populations within their normal habitats. The impetus to go forth is derived from several sources.

The "accidentals" are the exceptional individuals that go farthest away from the metropolis of the species; they do not belong to the ordinary mob that surges against the barrier, but are among those individuals that cross through or over the barrier, by reason of extraordinary complement of energy, in part by reason of hardihood with respect to the particular factors comprising the barrier, and in part of course, sometimes, through merely fortuitous circumstances of a favoring sort.

Geologists tell us that barriers of climate are continually moving about over the earth's surface, due to uplift and depression, changes in atmospheric currents, and a variety of other causes. Animal populations are by them being herded about, as it were, though that is too weak a word. The encroaching barrier on the one side impinges against the population on that side; the strain may be relieved on the opposite side, *if* the barrier on that side undergoes parallel shifting, with the result that the species

Vol. XXXIX]
1922   GRINNELL, *The Role of the "Accidental."*   379

as a whole may, through time, flow in a set direction. If anything should happen that a barrier on one side impinged on a species without corresponding retreat of the barrier on the other side, the habitat of the species would be reduced like the space between the jaws of a pair of pliers, and finally disappear: the species would be extinct.

But, in the case of persistence, it is the rule for the population, by means of those individuals and descent lines on the periphery of the metropolis of the species, to keep up with the receding barrier and not only that but to press the advance. I might picture the behavior of the population of a given bird as like the behavior of an active amoeba. This classic animal advances by means of outpushings here and there in reaction to the environment or along lines of least resistance. The whole mass advances as well. The particles of protoplasm comprising the amoeba may be likened to the individuals comprising the entire population of the animal in question, the mass of the amoeba to the aggregate of the population.

It is obvious that the interests of the individual are sacrificed in the interests of the species. The species will not succeed in maintaining itself except by virtue of the continual activity of pioneers, the function of which is to seek out new places for establishment. Only by the service of the scouts is the army as a whole able to advance or to prevent itself being engulfed: in the vernacular, crowded off the map—its career ended.

The same general ideas that I have set forth with regard to birds, who happen to be endowed with means of easy locomotion, hold, I believe, also for mammals, and probably in greater or less degree for most other animals. I can conceive of a snail in the role of an "accidental," an individual which has wandered a few feet or a few rods beyond the usual confines of the habitat of its species. Given the element of time (and geologists are granting this element in greater and greater measure of late), the same processes will hold for the slower moving creatures as they seem to do for those gifted with extreme mobility.

Migration, by the way, looks to me to be just a phase of distribution, wherein more or less regular seasonal shifting of popula-

380   McAtee, *Food Habits of the Shoveller Duck.*   [Auk
[July

tions takes place in response to precisely the same factors as hem in the ranges of sedentary species.

The continual wide dissemination of so-called accidentals, has, then, provided the mechanism by which each species as a whole spreads, or by which it travels from place to place when this is necessitated by shifting barriers. They constitute sort of sensitive tentacles, by which the species keeps aware of the possibilities of areal expansion. In a world of changing conditions it is necessary that close touch be maintained between a species and its geographical limits, else it will be cut off directly from persistence, or a rival species, an associational analogue, will get there first, and the same fate overtake it through unsuccessful competition—supplantation.

Museum Vert. Zool., Univ. of California, Berkeley, Calif.
(September 7, 1921.)

NOTES ON FOOD HABITS OF THE SHOVELLER OR SPOONBILL DUCK (SPATULA CLYPEATA).

BY W. L. MCATEE.

REPORTS have been made by the Biological Survey upon the food habits of all of the shoal-water ducks of the United States except the Shoveller. McAtee wrote the accounts[1] of the Mallard, Black Duck and Southern Black Duck, and Mabbott those[2] of the Gadwall, Baldpate, European Widgeon, Green-winged, Blue-winged and Cinnamon Teals, Pintail and Wood Duck. The Shoveller would have been included in the latter report had the author returned from war. However, design as well as fate had to do with the omission; truth is that the food habits of the Spoonbill duck are more difficult to study than those of any other anatine species yet investigated and the work, therefore, was postponed to the last. Even so the pioneer analyses by McAtee have not yet been supplemented and these are here reported upon so that some data on the food habits of the Shoveller will be available, and the

[1] U. S. Dept. Agr. Bull. 720, 35 pp., 1 Pl. Dec. 1918.
[2] U. S. Dept. Agr. Bul. 862, 67 pp., 7 pls. Dec. 1920.

From *Outline of the History of Arctic and Boreal Biota during the Quaternary Period; their evolution during and after the glacial period as indicated by the equiformal progressive areas of present plant species*

Eric Hultén

Materials for the present study

When writing the Kamtchatka flora the author of this paper ventured to quote as accurately as possible the t o t a l areas of distribution of each species treated. The intention was to supplement that flora with a study on the phytogeography of the Bering Sea region. This task, however, proved impracticable, as the Aleutian Islands formed a practically unknown region and as their central position in the Bering Sea area rendered any discussion of the phytogeography of that area, without a detailed knowledge of the conditions in the Aleutians, worthless. I therefore undertook an expedition to the Aleutian Islands, and the results will be found in the Flora of the Aleutian Islands, which is to be published in a short time. The investigations into the total areas of the Bering Sea plants were continued during the work with this flora. The Kamtchatka Flora and the Aleutian Flora therefore form the main basis of the present study.

Kamtchatka and the Aleutian Islands constitute, however, the centre of what has up to the present been the least known district in the boreal belt, viz. the area between Lena R. in the west and Mackenzie R. in the east and between the Arctic Ocean in the north and southern Alaska and the middle Kuriles in the south, cutting out a strip of the arctic and boreal belts approximately from 125° E. long. to 130° W. long, and consequently comprising somewhat more than one fourth of these belts (= area of fig. 4). A number of botanical investigations have been made within this vast area, but their results are very inaccessible, as they are scattered amongst numerous small Russian, Japanese and American papers and have not been brought together in summary form. I have therefore ventured to summarize all existing published information about that vast region and to record it on dot-maps, one map for each species present. These dot-maps thus represent a summary of my knowledge of the occurrence and distribution of vascular plants within the above-mentioned area. The basis of the present study is thus t h e t o t a l a r e a s o f e a c h s p e c i e s known to me to occur within the area of fig. 4. The species concerned number about 2000. It would naturally have been preferable to base a study of this nature upon a detailed knowledge of the areas of all plants in the arctic and boreal belts. This, however, would involve so much work that it seems beyond the capacity of a single investigator. In the following pages I hope to be able to show that the evolution of the arctic and boreal plants can be well outlined on the basis of the material to which I have had access and that a similar study of the arctic and boreal plants of the entire circumference of the globe will only supplement the results at which I have arrived.

5

It must be emphasized that the results of this work concern only the arctic and boreal plants, that is, the plants whose areas were destroyed or appreciably disturbed by the Pleistocene glacial periods. The conditions south of that belt might be different and are not dealt with in this paper.

The total geographical areas of different biota naturally exhibit the most profuse diversity. Practically speaking the areas of two different forms hardly ever cover one another completely. Different history, different specific properties, causing different reactions to the ecological conditions, and other circumstances account for this ample variation which is met with in the distribution of living things.

The variation is, however, by no means an irregular one; rather the contrary. The geographical areas of the biota can be divided up into more or less distinctly delimited groups, the interpretation and study of which will doubtless contribute substantially to our knowledge and conception of the origin and development of life on the earth.

HOOKER (1862) and CHRIST (1867) were the pioneers in the study of geographical distributions. Since then a considerable body of literature has grown up around the subject, possessing in certain respects a quite complicated terminology, developed by JEROSCH (1903), DIELS (1910), BROWN-BLANQUET (1919, 1928), KULCZYNSKI (1924), STEFFEN (1924, 1925), WANGERIN (1932) and others. As on the whole the present study proceeds along other lines than those followed by the above-mentioned authors, I find it unnecessary to enter into any discussion on these terminological questions.

With but few exceptions earlier authors have dealt with partial areas only. The reason for this is apparently the difficulty of determining with some degree of accuracy the total area of the more widespread species. I must emphasize, however, that a knowledge of the total area of a species essentially entails the possession of much vital information concerning the plant in question. This is perhaps not so commonly recognized by botanists as it should be. Not infrequently one comes across floras and systematical works in which the geographical distribution of the species dealt with is very much neglected or even omitted altogether. Even students of the problems of the evolution, migration and development of the floras are sometimes found to declare that the knowledge of the "rough area" itself "naturally" does not give any clue to the history of the plant in question. More rarely one finds the importance of the knowledge of the total areas clearly emphasized. Yet ENGLER (1879 p. 183) wrote: "Alle Untersuchungen über Pflanzenverbreitung in Mitteleuropa, welche nicht mit einer Kenntniss der Verhältnisse in Südeuropa und im ostlichen Asien gemacht sind, sind für die Entscheidung unserer Fragen (migration, origin) von geringem Werth..." Recently WANGERIN expressed much the same view when he wrote (1932 p. 517): "... weil für alle weiteren einwandrungsgeschichtlichen, genetischen u. s. w. Fragen die Kenntniss der gegenwärtigen Verbreitung und der auf Grund dieser zusammengehörigen Florenbestandtheile die unverrückbare Grundlage bilden muss". That false conclusions are likely to be drawn when only part of the area is taken into consideration will be clear from the following examples:

6

A plant that is found in the steppes of southern Europe is likely to be referred by students of the Europaean flora to a group of southern steppe plants. A knowledge of the total area, however, might show that in Eastern Siberia it occurs from the arctic shore down to China (compare Pl. 3—8). It can thus hardly be correct to classify it as a southern steppe plant.

A plant found in Europe exclusively in the most continental parts is very likely to be classed as a markedly continental species. A knowledge of the total area might show that it occurs on the shores of the Pacific and thus has no specially marked continental character. In Europe it might be protruding westwards and might not yet have reached its climatic limits. If it migrates from Asia it might naturally appear as a continental plant in Europe.

In Ireland is found an orchidaceous plant *Spiranthes Romanzoffiana*. It is customary to refer it to the "Atlantic" group and some authors have even included it in still more oceanic groups, as it only occurs in the most oceanic part of Europe. TROLL (1925 p. 310—11), for instance, refers it to a "hyperoceanic" group, which he considers to have a climatically limited area. *Spiranthes Romanzoffiana* also occurs, however, throughout North America from the Aleutian Islands in the west to the Atlantic coast in the east. It is by no means an oceanic plant, although it happens to be restricted in Europe to a small area on the Atlantic coast.

Similar mistakes are frequently made by students of phytogeography who have a knowledge of and deal with only a fraction of the total areas of the species they discuss. Recently some authors (EIG 1931, WANGERIN 1932, STEFFEN 1935) have proposed to separate the "rough area" of a species into two parts, "the compact area" and "the radiations" ("Ausstrahlungen"). When they discuss the matters they lay stress chiefly upon the "compact area". This is bound to lead to false conclusions. The area of a species can only mean the t o t a l area, and nothing else. It is just the stations found outside the "compact area" that are likely to be the most valuable ones, which can give a clue as to how the development has taken place. They are so to speak "the living fossils" of the species in question. In connection with these considerations another question should be mentioned, viz. that of the "mass centre" ("Massencentrum"). CHRIST and other authors considered that a plant is likely to have originated in a district where its most numerous individuals are now found. HEER already opposed this view. It is natural that if a plant at the border of its perhaps wide original area should find favourable conditions and multiply freely, so that numerous individuals are developed, such a phenomenon will afford no indication of the earlier history of the species. It must also be unsafe to assume that a plant originates in the place where it has its most numerous relatives. In most cases such a consideration will perhaps be correct, but in others it must be misleading. This point will be more fully discussed later on.

The difficulties that arise in determining the total area of a species are many and serious, and it is not to be expected that they can be completely overcome in a work such as this study. On the other hand, when so many species are dealt with as here, and when most species have been subjected individually to careful taxonomical treatment, one may be justified in expecting that the mistakes will

7

be reduced to a minimum and will have little influence on the general picture arrived at.

It is evident, however, that the possibilities of arriving at tolerably correct total areas have existed only during the last two decades. Before that there had been large gaps in our knowledge of the circumboreal flora and no unbroken survey had been obtainable of the distribution of the arctic and boreal floras. A short review is given below of those modern works, which have enabled us to gain a comparatively complete picture of the present distribution of these plants.

In Europe, perhaps the best-known section of the boreal belt, a great many comprehensive works on the flora and phytogeography of different countries — too well known to need enumeration here — facilitate the work. Among modern treatments the Icelandic flora by OSTENFELD and GRÖNTVED (1934) and the Flora of Nova Zembla by LYNGE (1923) are specially worthy of mention. TOL-MATCHEV's lists of the flora of Vajgatch (1919), Nova Zembla (1926) and Kolgujev (1930), as well as ANDREJEV's paper on northern Kanin (1931) are also of importance. The vast country of Siberia has always offered serious handicaps to the student of the distribution of boreal plants, since practically the only source accessible to the botanist not acquainted with the Russian language has been LEDEBOUR's admirable Flora Rossica, which, however, having been printed in 1842—1853, can hardly be expected to furnish information as detailed as is required today or to be up to date in its taxonomy. Above all, the late Dr. KRYLOV's capital work "Flora Sibiriae Occidentalis", started in 1927, has greatly facilitated present-day research on the flora of this important section of the boreal belt. KOMAROV's "Introduction à l'étude de la Flore de l'Iakoutie" (1926), as well as KOMAROV and KLOBUKOVA-ALISÒVA's "Key to the plants of the Far Eastern region of the USSR" (1931), although not critical floras, give valuable information concerning the regions in question. TOLMATCHEV has recently made valuable contributions to the knowledge of areas in arctic Siberia (the mouth of the Yenisej and the Tajmyr Penins.), and through the works of KOMAROV and the author of this paper the knowledge of the flora of the Kamtchatka Peninsula may be said to meet modern requirements. KUDO's work "The vegetation of Yezo" (1925) and KUDO and MIYABE's "Flora of Hokkaido and Saghalin", started in 1930, make the flora of northern Japan and Saghalin easily accessible for such studies, and TATEWAKI's recent works on the flora and vegetation of the Kuriles fill the gap between Japan and Kamtchatka. Lastly, the comprehensive work of Fl. SSSR, which is now being rapidly issued, affords a valuable survey of the entire flora of the vast territory of Soviet Russia, although when the areas of the species are given the districts mentioned are so chosen, that they often give no adequate idea of the real distribution of the plant in question, especially when the Arctic is concerned.

In America, ABRAM's "Illustrated Flora of the Pacific States" (1923), of which only the first part has appeared so far, RYDBERG's "Flora of the Rocky Mountains" (1917) and "Flora of the Prairies and Plains of Central North America" by the same author (1932) give us valuable manuals for the districts in question. RAUP's work on the plants in Wood Buffalo Park (1935) and his

"Phytogeographic studies in the Athabaska — Great Slave Lake region" (1936), SIMMONS's "A Survey of the Phytogeography of the Arctic American Archipelago" (1913), MARIE VICTORIN's "Flore Laurentienne" (1935), the new edition of BRITTON & BROWN, "Illustrated Flora of the Northern States and Canada" (1913), the tables in FERNALD's "Persistence of plants in unglaciated areas of Boreal America" (1925), as well as the studies on the Newfoundland Flora (1933) by the same author, and lastly what is published of "North American Flora", afford us possibilities of phytogeographically handling with some degree of accuracy the arctic and boreal belts of North America. Concerning Greenland, OSTENFELD gave a review in his "The Flora of Greenland and its origin" (1926), later supplemented by several authors. Numerous recent smaller papers, which it would take too long to enumerate here, contribute materially to our knowledge of boreal floras and have been used in the preparation of the maps in this work. Without access to the recent works enumerated above it would not have been worth while attempting a differentiation of the geographical areas of all plants from our area into distinct groups. This is the reason why our knowledge of the distributional types in the flora of the boreal countries has up to the present been so limited. It must be admitted that vast areas of the boreal belt are still but incompletely known in regard to their flora and are very much in need of further botanical investigation.

With the aid of the floristic material now available, supplemented with information derived from monographs and other taxonomical works, and after the study of a comprehensive herbarium material from all parts of the boreal belt in several leading herbaria, I think it has been possible for me to gain a fairly correct view of the total areas of our plants. Further investigations will, however, doubtless add important details and correct mistakes.

It should perhaps be added that the total area of each species was at first mapped out on an outline-map in LAMBERT's surface-true polar projection of the northern hemisphere, the same as used in the plates of this paper. The entire work was then based on a comparison of these maps.

It is of importance to chose a map in suitable projection for such studies. Maps in MERKATOR's projection are impossible to use as the proportions especially in the north are too distorted.

Already in 1931 I had arranged the plants found in Kamtchatka into groups exactly corresponding to those presented in this paper. I even gave a lecture about these groups in botanical societies both in Stockholm and in Uppsala but neither I myself nor any of the botanists hearing the lectures were then able to give a probable interpretation of the presented facts.

The theory of the progressive areas

When tackling the problems offered by the geographical areas of plants, many authors have chosen to discuss peculiar or singular types having a distribution out of the common, in the hope that they will suddenly give a clue to the solution of the problem. This is merely appealing to the imagination, it is, so to speak, a romantic method of investigation. On the other hand, it is obvious that considerable difficulties in interpretation are bound to arise if the chief consideration is given to such singular types. It is surely more rational to start the investigation with the simplest types, those that show the least possible peculiarities. When they have been interpreted, the complicated and often strongly interrupted areas of the singular or peculiar types are likely to be better understood. This is a more prosaic, but probably also a safer method of procedure.

All plants can be divided according to the character of their area into two series, which might be called the Oceanic and the Continental series. To the Oceanic series should be referred all plants that occur exclusively in territories with an insular or distinctly coastal climate, but avoid the interior of the continents, while to the Continental series should be taken those that exclusively inhabit the interior of the Continents or are found both there and in the coastal districts. The difference must be assumed to be caused by the continental climate, especially by its low humidity, low winter temperatures and wide range of variation between the extreme temperatures. It is reasonable to assume that the possibilities to stand such conditions must be characteristic properties, which have only to an inconsiderable degree changed during the late history of a plant, and that due consideration to the oceanity, respectively continentality, of the plants always must be taken. The reasons why some plants sometimes have a continental appearance in one part of their area and an oceanic in another depends upon that their areas are reduced and will be later discussed.

The simplest distributional group seems to me to be the continental Eurasiatic, that is, the plants confined to the Eurasiatic Continent which are not decidedly coastbound. I have therefore chosen it as the first group to tackle in order to see whether a close study of their areas can reveal anything concerning their origin and history. Some general considerations concerning migrations of plants might, however, be in place before we start that study.

The first that strikes one when trying to group geographical areas is that only very rarely do the areas of two species completely cover one another. This has long been recognized and considered to be a drawback in all such attempts. If this were not the case, the plants would inevitably exhibit a degree of conformity that could hardly be expected, not only in regard to their centre of dispersal but also to their reaction to the ecological conditions, rate and means of dispersal, and other circumstances. The wide variation in the total areas of plants is indeed a very natural phenomenon, and it would be very hard to explain the reasons for any contrary behaviour. The question is to find the lines along which the dispersal of the plants take place, to understand the nature and rate of that process and to arrange the empirically found areas in such a way as to give us natural groups.

Let us consider what might be expected to happen if a number of different plants were placed within a large area of suitable soil cleared from all vegetation and allowed to develop under equable and favorable conditions without competition from other plants. Each species would spread in all directions around the original centre, but it could hardly be expected that all plants would spread with the same rapidity. Some of them, possessing better means of propagation — vegetative or sexual — would doubtless be able in a given time to cover an area several times larger than others with less effective propagative organs. The result would doubtless be a p p r o x i m a t e l y   c i r c u l a r   a r e a s   o f   d i f- f e r e n t   s i z e   a r o u n d   t h e   c e n t r e. If the experiment were carried on long enough and the available space were limited, the final result would be that a l l   p l a n t s   w o u l d   u l t i m a t e l y   r e a c h t h e b o u n d a r y o f t h e   a r e a   a t   d i s p o s a l   a n d   t h u s   a l l   f i n a l l y   o c c u p y   t h e s a m e   a r e a. In principle this is what happens under natural conditions. In nature the areas doubtless rarely attain the theoretically circular form, although in cer- tain districts approximations to such circular areas are found. (Compare, for in- stance, the map of 13 deciduous trees in S. E. United States in LIVINGSTON & SHREVE 1921 p. 54). Theoretical cases rarely materialize in nature. Usually the natural conditions are not so equal and schematic as in the theoretical case. The plants are subject to many influences that are apt to disturb the picture. For instance, the climatic belts stretching around the globe must give the theoreti- cally circular areas a more oval form and barriers in form of open seas, deserts, inland-ice or other agencies put a stop to distribution in certain directions. There still remains, however, the chief feature of comparatively recent areas, t h e i r   c o n c e n t r i c i t y   a r o u n d   t h e   p l a c e   f r o m   w h e r e   t h e y   r a- d i a t e d. In other words: those plants radiate from the same centre, which have equiformal areas of different size, and this centre can be found if the areas of as many species as possible belonging to the group are compared.

Let us now return to the Eurasiatic continental plants and see how this theory, which might be called t h e   t h e o r y   o f   t h e   e q u i f o r m a l   p r o- g r e s s i v e   a r e a s, works out practically.

The continental Eurasiatic plants

A superficial survey of the Eurasiatic-continental plants shows that they can be divided into two widely differing groups. One of them is confined to the mountain system stretching from Eastern Siberia through Central Asia to Persia, the Caucasus, the Alps and the Pyrenees, having another lobe of distribution along the shore of the Arctic Ocean. They might be called the Arctic-montane plants. (The term Arctic-alpine is avoided, as plants belonging to other distri- butional types have often been classed as Arctic-alpine, and furthermore the word "alpine" has two meanings, viz. living in the Alps and living in the moun- tains.) For the moment this large and interesting group might be left out of consideration. The other chief group comprises chiefly lowland plants, distri- buted between the Pacific and the Atlantic, and possessing more or less exten- sive areas. If the above principles of the equiformal progressive areas are applied to that group, we will find that they can be arranged in different series, certain of which should be more closely discussed here, as they seem to represent the most direct approximation in the boreal belt to the theoretical case of progressive areas.

RUBUS HUMULIFOLIUS
SMILACINA DAHURICA
CAREX GLOBULARIS
STELLARIA RADIANS
MAJANTHEMUM BIFOLIUM

Fig. 1. Representation of equiformal progressive areas corresponding to the figure in Plate 1.

The first of them comprises plants having a fairly northern distribution in Eurasia (Pl. 1), the second one (Pl. 2) plants having a similar area but reaching still further to the north in Eastern Asia, protruding even into the Kolyma valley. Those with smaller areas occupy eastern Siberia and the Amur Prov., those with larger areas penetrate progressively towards northern Europe (all of them occur in *northern* Scandinavia*), and, finally, those with the widest area cover a broad belt from the Pacific to the Atlantic. They are northern but not really

---

* Scandinavia is in this paper taken to include Sweden, Norway, Finland and Kola penins.

arctic plants, and the area covered by them coincides approximately with the boreal coniferous belt. Only the species with the widest distribution extend a considerable way outside that belt, some few even reach Iceland and Greenland. It should be noted that those that reach Iceland and Greenland also have the widest distribution as well to the north as to the south in Siberia and central Asia, that is, that the condition of conformity is filled.

The group as illustrated in Pl. 1 and enumerated below does not comprise all the plants that should belong to it. Part of its core consists of plants found in Amur Prov., which do not reach the area covered by this study and therefore are not yet known to me. If the area of investigation were extended to comprise that country also, some further species with both smaller and larger areas could certainly be added. There can hardly be any doubt that the plants in question spread from a centre in eastern Siberia and Amur westwards towards Europe. It would not be possible to arrange in any other way the Eurasiatic continental plants of our area in a series such as the above discussed theoretical case indicates. The difference between the natural and the theoretical case is that the areas are oval owing to the pressure of the arctic climate in the north and the warm and dry climate to the south, to which this group of plants is not accustomed. The centre is placed asymmetrical owing to the barrier formed by the Pacific. Not a few have approximately reached the limits of their dispersion under present conditions within the space at disposal (Eurasia). They are plants that long ago reached Europe and whose present area is limited by the climate and by their ability to transgress the barriers rising in all directions at their borders. Five of the plants in Pl. 1 are mapped separately in Fig. 1.

In order to facilitate the discussion I propose to call plants radiating from a centre and occupying areas of different extension around that centre *radiants* and those that have not moved considerably from the centre *centrants*. The plants belonging to the groups discussed can thus be called in brief A m u r-r a d i a n t s and E a s t e r n S i b e r i a n radiants.

The species included in Pl. 1, b e g i n n i n g w i t h t h o s e h a v i n g t h e s m a l l e s t a r e a a n d p r o c e e d i n g a s t h e a r e a s g r o w w e s t-w a r d , a r e :

Not reaching beyond Lake Bajkal:

Ribes Warzewiczii	Minuartia laricina
Trigonotis radicans	Juniperus dahurica
„ myosotideum	Trollius Ledebourii
Smilacina davurica	Cardamine prorepens
Glyceria spiculosa	Cimicifuga simplex
Aconitum Kuznetzoffii	Rosa davurica

Reaching beyond Lake Bajkal but not the Urals:

Carex Meyeriana	Carex amblyolepis
„ pseudocuraica	Stellaria radians
Ribes dikuscha	Viola Komarovii

Hierochloë Bungeana

Potentilla flagellaris

Carum buriaticum

Tanacetum boreale

Reaching to about the Urals:

Alyssum sibiricum

Arabis Stelleri

Viola uniflora

Stellaria Bungeana

Reaching beyond the Urals but not Sweden:

Poa sibirica

Rubus humulifolius

Reaching Sweden but not western Europe:

Sagittaria natans

Agrostis clavata

Actaea erythrocarpa

Platanthera parvula

Carex laxa

Athyrium crenatum

Ribes rubrum (sens. lat.)

Sparganium Friesii

Carex pediformis

Glyceria lithuanica

Carex globularis

Sparganium glomeratum

Reaching western Europe:

Nuphar pumilum

Majanthemum bifolium

The plants on which Pl. 2 was founded (as always further on in this paper beginning with those having the smallest area and proceeding as the area grows from the centre) are:

Reaching to about Lake Bajkal:

Saxifraga Merckii

Carex amgunensis

Populus suaveolens

Pinus pumila

Reaching west of Lake Bajkal to the Urals:

Carex Schmidtii

Petasites Gmelini

Zygadenus sibiricus

Heracleum dissectum

Cardamine macrophylla

Gentiana barbata

Chrysanthemum sibiricum

Reaching west of the Urals but not western Europe:

Cacalia hastata

Androsace filiformis

Ledum palustre (excl. subsp. decumbens)

Mulgedium sibiricum

Carex laevirostris

Viola epipsila (incl. subsp. repens)

Galium ruthenicum

Reaching western Europe:

Prunus Padus

Tanacetum vulgare (incl. var. boreale)

Solidago virgaurea (incl. var. leio-
carpa)

Reaching Iceland:

Sorbus aucuparia (sens. lat.)

Valeriana officinalis

Erigeron acris

14

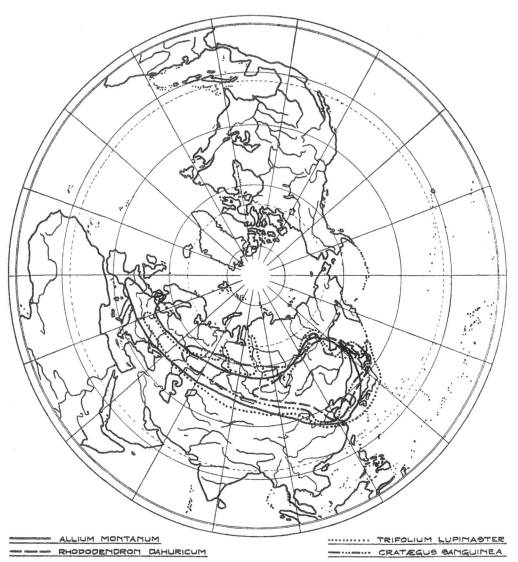

ALLIUM MONTANUM

RHODODENDRON DAHURICUM

TRIFOLIUM LUPINASTER

CRATÆGUS SANGUINEA

Fig. 2. Representation of equiformal progressive areas corresponding to the figure in Plate 3.

Reaching Greenland:

Rubus saxatilis   Thymus Serpyllum

Vicia Cracca

Most of these plants are lowland plants. Some few are, however, montane plants, and these have small areas, which is to be expected, as there are mountains in Amur and eastern Siberia and the plants found in them cannot continue along the same route as the lowland plants, as there are no mountains further to the west in northern Siberia. A kind of barrier was thus placed in the way of their dispersion.

15

Let us now turn to some Eurasiatic continental plants with a distribution of a different type, those, namely, that are found to have a wide distribution in eastern Asia just like the first two groups, but which, contrary to them, penetrate towards Europe on a southern and far narrower route. They form a large and fairly distinct group and can, I think, be treated as a separate unit; but, in order to avoid misinterpretations, it might be suitable to divide them into several subdivisions. The first of these is the group occurring in Amur and Ussuri (and frequently also in the islands off the Pacific coast) and in southern Jakutsk province, and which are not found south of the desert belt of central Asia (Pl. 3). The following plants are included in that map:

Not reaching west of Lake Bajkal:

Saxifraga Sieversiana
Oxytropis ajanensis
Gypsophila violacea
Rhododendron Redowskyanum
Rumex jacutensis
Brachypodium villosum
Avenastrum Krylovii
Agrostis jacutica
Plantago depressa

Sorbaria grandiflora
Fragaria orientalis
Plantago canescens
Artemisia jacutica
Cassiope ericoides
Cirsium pendulum
Senecio palmatus
Alnus hirsuta

Reaching west of Lake Bajkal but not the line Tobol R. — Irtysh R.:

Viola dactyloides
Gentiana triflora
Ribes fragrans
Conioselinum cenolophioides
Aster dahuricus
Sedum Stephani
Iris laevigata
Festuca pseudosulcata
Festuca jacutica
Hierochloe glabra
Dontostemon pectinatus
Lilium dahuricum
Poa botryoides
Rhododendron dahuricum
Patrinia rupestris
Selaginella sanguinolenta
Rumex Gmelini
Carex Sedakovii
Bupleurum triradiatum
Festuca lenensis
Hypochaeris grandiflora
Ephedra monosperma
Gagea pauciflora
Carex eleusinoides

Orchis salina
Poa subfastigiata
Anandria Bellidiastrum
Delphinium grandiflorum
Viola brachyceras
Gentiana macrophylla
Aconitum barbatum
Scutellaria scordiifolia
Ribes procumbens
Artemisia commutata
Scorzonera radiata
Stellaria Cherleriae
Viola dissecta
Hypericum Gebleri
Vicia amoena
Salix rorida
Braya siliquosa
Selaginella borealis
Smelowskia alba
Campanula pilosa
Allium Ledebourianum
Hypericum Ascyron
Vicia unijuga

16

Reaching the line Tobol — Irtysh but not the Volga:

Nepeta lavendulacea
Aegopodium alpestre
Sorbaria sorbifolia
Polygonatum humile
Viola Mauritii
Eritrichium pectinatum
Halenia corniculata

Aconitum volubile
Patrinia sibirica
Hordeum secalinum breviaristulatum
Thermopsis lanceolata
Iris ruthenica
Schizachne callosa
Chamaerhodos erecta

Reaching the Volga but not Switzerland:

Crataegus sanguinea
Artemisia latifolia
Artemisia sericea
Arabis pendula

Veronica spuria
Iris flavissima
Anthriscus nemorosa

Reaching central and western Europe:

Trifolium lupinaster
Scorzonera austriaca

Allium montanum

The areas of four of these species are represented in Fig. 2 in order to exemplify individual areas.

Pl. 4 comprises plants also occuring south of the Central Asiatic desert belt. They are:

Not reaching west of Lake Bajkal:

Nasturtium globosum
Taxus cuspidata
Carex drymophila

Cacalia auriculata
Sanguisorba tenuifolia
Filipendula palmata

Reaching Lake Bajkal but not the Volga:

Potentilla tanacetifolia
Allium tenuissimum
Allium odorum
Isopyrum fumarioides

Aster altaicus
Pedicularis resupinata
Lathyrus humilis

Reaching the Volga but not western Europe:

Stipa sibirica
Axyris amaranthoides
Cypripedium macranthum
Potentilla bifurca

Lepidium apetalum
Geranium sibiricum
Artemisia scoparia

Reaching western Europe:

Allium victorialis
Calamagrostis pseudophragmites

Rosa pimpinellifolia

The fifth group includes plants occurring besides in the Jakutsk — Amur districts, also further to the northeast at Kolyma or even Anadyr in eastern Siberia, but not south of the desert belt in Central Asia (Pl. 5). The following plants are referred to this group:

17

Not reaching west of Lake Bajkal:

Asperella sibirica   Delphinium cheilanthum
Aquilegia parviflora   Potentilla Sanguisorba
Galium davuricum   Salix macrolepis
Astragalus adsurgens

Reaching Lake Bajkal but not the Urals:

Polygonum divaricatum   Artemisia macrobotrys
Rhododendron chrysanthum   Potentilla fragarioides
Agropyron Turczanninovii

Reaching the Urals but not western Europe:

Vicia multicaulis   Salix Gmelini
Chenopodium aristatum   Sisymbrium polymorphum
Puccinellia Hauptiana

Reaching western Europe:

Polygonum undulatum   Allium strictum
Carex alba

The group represented in Pl. 6 comprises plants with much the same distribution as group 5, but with the area extending also south of the Central Asiatic deserts. These are:

Not reaching the Urals:

Urtica angustifolia   Leontopodium Palibinianum
Oxytropis strobilacea

Reaching the Urals but not western Europe:

Cotyledon spinosus   Artemisia sacrorum
Potentilla viscosa   Saxifraga sibirica
Sedum aizoon   Linum perenne
Elymus sibiricus   Thalictrum foetidum
Artemisia Sieversiana   Artemisia laciniata

Reaching western Europe:

Androsace villosa   Ribes petraeum

Lastly, two groups should be considered in this connection, which differ from all the foregoing in being absent or practically absent from the Pacific littoral, that is, in having a still more continental type of distribution than the others.

The first of them (Pl. 7) does not occur south of the Central Asiatic desert belt, while the second (Pl. 8) is found also in the mountains south of that belt. The plants on which Pl. 7 is based are the following:

Reaching to about Lake Bajkal:

Bupleurum dahuricum
Lysimachia dahurica
Saussurea amurensis
Plantago paludosa
Saussurea Karoi

Potentilla asperrima
Hypericum attenuatum
Lactuca versicolor
Corydalis paeoniaefolia
Pedicularis rubens

Reaching west of Lake Bajkal but not the Urals:

Orobanche ammophila
Limnas Stelleri
Astragalus angarensis
Petasites saxatilis
Pedicularis altaica
Crepis Bungei
Erigeron armeriaefolius
Galium densiflorum
Salix Kochiana
Puccinellia tenuiflora
Dracocephalum pinnatum
Trifolium eximium
Axyris sphaerosperma
Agropyron geniculatum
Pedicularis tristis
Claytonia Joanneana
Aquilegia sibirica
Spiraea alpina

Astragalus fruticosus
Erysimum altaicum
Potentilla evestita
Iris tigridia
Arenaria formosa
Draba ochroleuca
Astragalus multicaulis
Mertensia davurica
Rheum Rhaponticum
Elymus dasystachys
Arabidopsis salsuginea
Armoracia sisymbrioides
Cirsium serratuloides
Salix minutiflora
Peucedanum baicalense
Peucedanum vaginatum
Vicia megalotropis
Gentiana decumbens

Reaching the Urals but not western Europe:

Oxytropis glabra
Dianthus ramosissimus
Thesium refractum
Statice speciosa
Potentilla sibirica
Gypsophila Patrini
Stipa Krylovii
Oxytropis uralensis
Geranium pseudosibiricum
Sedum hybridum
Artemisia macrantha
Urtica cannabina
Carex curaica

Clausia aprica
Serratula nitida
Hedysarum Gmelini
Saussurea amara
Artemisia armeniaca
Veronica incana
Allium albidum
Rumex ucrainicus
Phlomis tuberosa
Adenophora liliifolia
Lycopus exaltatus
Carex secalina

Reaching western Europe:

Allium angulosum
Plantago Cornuti
Aconitum Lycoctonum
Lilium Martagon

Polygonum patulum
Pedicularis comosa
Androsace maxima

Pl. 8 includes the following species:

Not reaching west of Aral Lake:

Braya rosea	Caragana jubata
Melandryum brachypetalum	Dracocephalum nutans
Agropyron cristatum	Cobresia capillifolia (incl. var. filifolia)
Crepis tenuifolia	

Reaching Aral Lake but not Central Europe:

Potentilla cericea	Draba lanceolata
Suaeda corniculata	Primula nivalis
Ranunculus subsimilis	Cobresia schoenoides
Carex pseudofoetida	Elymus junceus

Reaching beyond the Black Sea:

Hieracium virosum	Stipa capillata
Artemisia dracunculus	Euphrasia tartarica
Saussurea pygmaea	Delphinium elatum

It must be admitted that the centrants (i. e. the plants with a fairly small area around the countries from which all plants in the above 8 groups radiate) are somewhat arbitrarily divided up into the different groups. They are so divided that their area such as it is known to me at present conforms as far as possible to that of the group in which they were placed. (It should be remembered that most of these plants are endemic in the country east and northeast of Lake Bajkal of which no modern Floras exist). If their areas and vertical distribution were better known and if ecological requirements and other facts concerning their occurrence were known, a better and more exact division could probably be made. For the present discussion, however, the question of a possible incorrect division of these centrants is of little or no importance.

Rigid and plastic areas

Let us now examine groups 1 and 2 and contrast them with groups 3 — 8. Very conspicuous differences at once strike the eye. Both first mentioned groups spread evenly from a centre in Eastern Asia westwards to Europe, following the Arctic Coast though sometimes at some distance from it. Only a few plants, namely those with the widest area, reach down to the steppe and desert zone or go beyond it. The latter six groups, on the other hand, spread from the same Eastern Asiatic centre but further west occupy a narrower and more southernly area, which in Europe i s d e c i d e d l y s o u t h e r n and f a r r e - m o t e f r o m t h e A r c t i c c o a s t. It is hardly reasonable to assume that both these types of distribution are caused by climatic conditions. In Eastern Asia they occupy one and the same area, but in Europe strikingly different

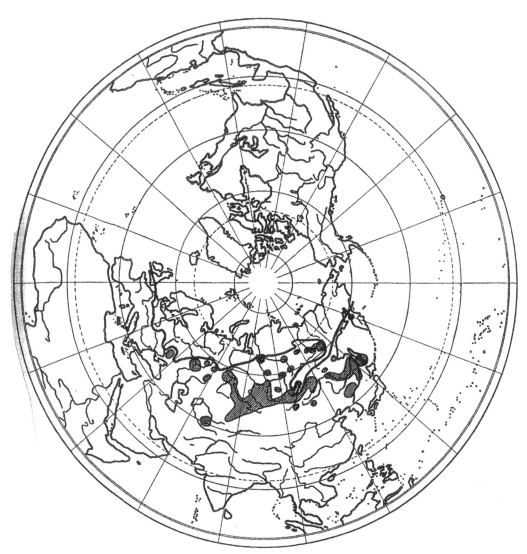

Fig. 3. Distribution of *Arabis pendula* (chiefly after Busch). ————— Border of maximum
glaciation (chiefly after Shaparenko).

areas. If they were accustomed to climates that forced them to occupy such
widely separated areas in Europe, it is hard to understand how these climatic
requirements can allow them to unite in Eastern Asia. Even if the theory
of the equiformal progressive areas were not accepted and the above groups
were thus not recognized as natural groups, the area of each of the more
widespread species belonging to groups 1 and 2 could be compared with that
of any of the more widespread species belonging to groups 3 — 8, and the
same striking contrast would have to be explained. The correct interpretation

21

is doubtless that g r o u p s 3 — 8 f o r m e r l y o c c u p i e d a r e a s simi-
l a r to those of the g r o u p s 1 — 2, but these a r e a s were
p a r t i a l l y d e s t r o y e d by the ice that c o v e r e d northern
E u r o p e and n o r t h w e s t e r n S i b e r i a d u r i n g the P l e i s t o c e n e
g l a c i a t i o n. The plants belonging to groups 3 — 8 thereby lost either wholly
or for the most part their capability of occupying new areas and t h e y r e-
m a i n t o d a y m o r e o r l e s s in the s a m e s t a t i o n s as when
the ice had its maximal extension.

I propose to call species with such properties r i g i d s p e c i e s and such
areas r i g i d a r e a s. The others, group 1—2, were not deprived of their
capacity to spread but have occupied in postglacial time the earlier glaciated area.
They thus, contrary to the rigid species, have performed considerable migrations
in postglacial time. I propose to call such species p l a s t i c s p e c i e s. A striking example of a completely rigid species is *Arabis pendula* (see fig. 3).

As it is out of the question that most of these plastic plants, belonging to the group 1 — 2, originated after the maximum glaciation the rigid
and the plastic plants must have been reacting in different ways to the pro-
bationships of the glacial period. How this could be thought to have taken
place will be discussed a little here. A possible explanation is the following:
Within a given species there is always a certain potential variation. Under
the influence of a catastrophe such as the glacial period, when large parts of
the area of the species in question are covered with ice and the rest is severely
exposed to unfavorable conditions, a reduction of this variability is inevitable,
as all biotypes, being more sensitive to the hardships, are exterminated (compare HAGELDOORN 1921). Yet it was those very biotypes which occured in
the n o r t h e r n part of the area and which might be expected to be hardier
than those in the southern part of it, that were completely exterminated, while
only part of the biotypes of the s o u t h e r n area survived. The population
close to the s o u t h e r n boundary of an area was presumably selected as to
comprise types that can stand better than others the warm and dry climate
there. When just such a population is exposed to the pressure of the severe
climate in the neighbourhood of the ice it is natural that only a few biotypes,
and lastly perhaps only one single biotype, can survive, while the others, which
are more sensitive to the cold climate, perish. A very strong reduction of the
biotypes is bound to ensue and, parallel to it, a strong reduction of the potential
variability and thereby also presumably of the spreading capacity.

A further condition must enhance the difficulty which eventual remaining biotypes have in following the retreating ice to the north: those specimens which during
the maximum glaciation formed the northern boundary of the area of a certain species will necessarily comprise that part of the surviving population which includes
the least number of biotypes. The biotypes found south of them will, however, meet
with considerable difficulties in reaching the new country when the ice retreats,
as, the border biotypes having occupied most of the habitats that are suitable for
the species, they must enter into competition with them in the border zone
before they can reach the open country laid bare by the retreating ice. In other
words: the border population will form a kind of barrier which may be

expected — sometimes at least — to be difficult to pass for the more biotype-rich population behind it.

Obviously vicissitudes of the kind discussed above are sometimes bound to lead to the complete extermination of a species even if its area is not completely covered by the ice. Especially if a species has been exposed to the climatic hardships of more than one glaciation it is natural that its biotypes and variability will be so depauperated that it will almost completely lose its ability to spread. An example of a recent plant, which existed in Europe before the last glaciation but has been exterminated from that region, is *Dulichium spathaceum* (Cyperaceae), which now occurs only in E. America, whereas before the last glaciation it was found both in Germany and in Denmark.

Other plants had during the glacial period a geographical area so situated that they were not so hard pressed by the vicissitudes of the conditions (that is to say they possessed considerable areas relatively far from the ice) and their biotype supply was not so depauperated. They thus had larger possibility of spreading again when favorable conditions returned. Others again may have properties that make them less sensitive to the above-mentioned vicissitudes and may thus be able to reoccupy their former area.

It should be noted that many of the above plants showing distinctly rigid areas have bulbs or thick rhizomes. Such plants apparently have greater difficulty than others in moving from the place where they live. They must also be expected to have a better chance to survive hard climatic conditions as they can subsist large parts of the year sheltered by the earth.

All this reasoning is in contradiction to commonly held views. It is often presumed that the ice, as it proceeded, "chased" or forced the plants south. In my opinion this was not so, at any rate not in most cases. The migration of plants seems to proceed far more slowly, at least as far as perennial plants and plants with bulbs and rootstocks are concerned. They must have had very little chance of escaping the ice. Anyone who has observed a recent glacier on the move, as I have in Alaska, must have been impressed by the fact that the ordinary vegetation, woods with their complete undergrowth, grows close to the foot of the ice, and as the ice proceeds the trees are overthrown and the entire plant society destroyed. It is not the case that vast tundras surround the advancing ice and the plants retreat in front of it. WOLDSTEDT (1929) has pointed out that no conclusions could be drawn as to the conditions during glacial periods from the fact that forests grow on the moraines of the Malaspina glacier in Alaska. This may be true to a certain extent, seeing that no anticyclones are formed over the relatively small glaciers in Alaska as they must have been formed over the icesheets during the glacial periods. The dry winds blowing from these anticyclones must have been one of the most severe factors which the plants close to the ice had to withstand. The differences should, however, not be overestimated and the above facts hardly affect the present argument as they do not make it any more probable that the plants escaped the severe conditions close to the ice by migrations southwards, they only indicate that many species were exterminated by these severe conditions on the spot where they stood in front of the ice and not by the ice itself.

Between the Scandinavian ice and that of the Alps there was undoubtedly, as the fossils indicate, a forestless tundra during the period of maximum glaciation, but conditions there were still worse than at the border of the ice in other places, as the vegetation was severely influenced by the t w o   icesheets, one north and one south of the district in question. The present conditions in Greenland seem to indicate that a luxuriant vegetation with high shrubs or even trees can stand the conditions close to an inland ice. It is true, however, that these borders of the ice are along the seacoast and thus the climate along them is more genial than it would have been if these borders were situated inland.

Annual plants may have had a better chance of escaping the ice. However, they play a very insignificant part in the present arctic and boreal flora. Mountain plants naturally had a greater chance of escaping the ice, especially in large mountain ranges, by moving a short distance in altitudinal direction. The climatic conditions there, however, were probably very unfavourable, so that only the more hardy species could survive.

On the other hand, it is evident that when the ice retreats plants follow comparatively close to its edge. The virgin soil is doubtless very congenial to the plants and the climate must have undergone considerable amelioration before the ice of a glacial period starts to recede. In Alaska I have seen receding glaciers (for instance the Mendenhall glacier near Juneau), where plants follow so close to the glacier that the area of the moraines a few hundred metres from the ice is covered with vegetation. The vegetation naturally consists of only a few species, in this case chiefly *Lupinus nootkatensis*. After the glacial period similar migrations seem to have taken place. Then, too, it seems that only a few species accompanied the ice on its retreat. As early as in 1870 this was proved by NATHORST, when he found remains of a few arctic plants in the glacial clay of Alnarp in southern Sweden. NANNFELT (1935 p. 77) arrived at a similar conclusion. When the ice edge had receded to central Sweden, the climate was so genial that the flora in its neighbourhood was no longer arctic, as was pointed out by GUNNAR ANDERSSON (1906, p. 59 — 60).

Position of the centres of the equiformal progressive areas

We have now discussed in some details two major progressive groups of Eurasiatic plants, both of which proved to be radiants from Eastern Asia, the first group, the plastic ones, possessing areas in which primarily no influence of the ice of the glacial period could be traced, the second group, the rigid ones, having areas which are apparently only remnants of earlier larger areas destroyed by the maximum glaciation. Let us now see if the rest of the flora within the area discussed can be divided up into other similar groups and if they too will show any reference to the glacial periods.

If an attempt is made to arrange all plants of the area investigated in similar equiformal progressive groups, it will soon become evident that, with the exception of a few species with very much split-up and complicated areas, this will be quite possible. The species can quite freely be arranged in a number of progressive series radiating from a centre with "accumulative", or as I have chosen to call them equiformal, progressive areas. Some of these groups comprise a large number of species, others fewer. It is characteristic of all of them that they contain plants from all parts of the system, seemingly without any regularity. These groups will be dealt with one by one further on in this paper and it will perhaps suffice for the present to state where the centres of the progressive groups are situated, that is to say, the regions from which the groups have radiated. (Compare the plates.) Such centres are found, as we have already seen, in North-Eastern Siberia and in the Amur — Manchuria region. Another occurs in the Altai — Sajan region, sending out radiants towards the Arctic shore. A third centre is northern Japan, whence numerous plants radiate to the north and to the coast of the Asiatic Continent. A centre of great importance it the region around the northern part of the Bering Sea. It sends out progressive radiants reaching symetrically as well to the west into arctic Asia and Europe as to the east to Eastern America, and also often extends arms along both the Asiatic and American Pacific coast. In America radiants proceed from the Yukon valley along the Arctic American coast, others centre around the Arctic Archipelago, and others again have the centre of their progressive figures in the State of Washington and radiate along the American coast or along the Rocky Mts. to Alaska. Of the plants discussed in this paper no groups could be formed having their centres in northern Europe or western Siberia, or in North-Eastern America or in the country between Yukon Valley and the Great Lakes. The conclusion must be drawn that no plants of the region spread from these latter districts. The reason for this is apparently that t h e s e   d i s t r i c t s   w e r e   c o v e r e d   w i t h   i c e   d u r i n g   t h e   m a x i m u m   P l e i s t o c e n e   g l a c i a t i o n.   A l l   p l a n t s   o f   o u r   a r e a   r a d i a t e   f r o m   d i s t r i c t s   t h a t   w e r e   n o t   c o m p l e t e l y   b u r i e d   u n d e r   t h e   i c e - s h e e t   o f   t h e   m a x i m u m   g l a c i a t i o n. I n   o t h e r   w o r d s :   t h e   p l a n t s   h a v e   s p r e a d   o v e r   t h e   a r c t i c   a n d   b o r e a l   b e l t   f r o m   t h e   r e f u g i a   c l o s e   t o   t h e   i c e ,   w h e r e   t h e y   w e r e   l e f t   i n   p o s s e s s i o n   o f   a   s m a l l   p a r t   o f   t h e i r   e a r l i e r   a r e a ,   a n d   w h e r e   t h e y   w e r e   a b l e   t o   s u r v i v e   t h e   s e v e r e   c o n - d i t i o n s   o f   t h e   m a x i m u m   g l a c i a t i o n.

In the light of this statement we no longer have any difficulty in accepting the fact that the progressive groups are built up of species without systematical relationship, which now seems only natural. S u c h   a   g r o u p   c o n s t i t u t e s   a n   e v o l u t i o n   o f   a   f r a g m e n t   o f   t h e   e a r l i e r   v e g e t a t i o n , w h i c h   w a s   m o r e   o r   l e s s   i s o l a t e d   w i t h i n   a   c e r t a i n   d i s t r i c t   b y   t h e   i c e   o f   t h e   m a x i m u m   g l a c i a t i o n. It must thus be expected to comprise forest plants, meadow plants, mountain plants, bog plants, Cryptogams and Angiospermes that survived the ice in the district in question. Even animals, which more or less directly subsist on plants, must be expected, on the whole. to have survived in the same districts and to have followed the same routes in their migrations from the refugia.

In order to be able to discuss this theory in greater detail it will be necessary to know as accurately as possible the extension of the Pleistocene ice-sheets as well as the conditions that presumably prevailed during the glacial period. These questions will be discussed in the next chapter.

The Pleistocene glaciation

The evolution of the Pleistocene glaciation is still very incompletely known. Within the last few years, however, a tendency has been noticeable to recognize the results at which PENCK and BRÜCKNER (1909) arrived after studying the glaciation in the Alps as having a more general bearing and as being applicable throughout the glaciated area of the globe. PENCK and BRÜCKNER stated that four different glaciations had taken place in the Alps, separated by more or less distinctly warmer interglacial periods. The four glacial periods were named by PENCK and BRÜCKNER Güntz, Mindel, Riss and Würm. The two first of these were, however, recently by BECK (1933) placed back in the Pliocene, and he sets in their place two other early Pleistocene glaciations, "Kander" and "Glütsch". The number is thus the same. The interglacials between all six glaciations were by BECK called the A, B, C, D and E interglacials respectively, and for the sake of shortness I shall in the following use these names.

Already in 1898 KRAUSE had pointed out the possibility that the loess formations found in Europe were formed during the glacial periods. This view has gradually been accepted and confirmed, and it is now clear, thanks to the investigations of SOERGEL (1919), that this must have been the case. Close to the ice, where a cold climate was prevalent and where dry winds were blowing from the high pressures formed above the ice, the vast fields of fine deposits ("sandr" in Icelandic) were whirled up by the storms, especially during wintertime, and accumulated further from the ice. When the ice borders the open sea, as was the case during the receding stages of Würm ice over Scandinavia, no loess is formed; but it cannot be too much emphasized that the condition for loess formation is not only the arid climate but also the nival one, if sufficiently large fields of fine deposits without vegetation are present. Compare the investigations by C. SAMUELSSON on Iceland and Spitsbergen (see SAMUELSSON 1925).

In the Ukraine KROKOS (1927) found four loess layers, corresponding to four glaciations (the first indistinct), and three layers containing humus of the "black earth" type, separating them and corresponding to the three last interglacials. (Compare also BUBNOFF 1930). In southern Europe the interglacials are marked as arid periods, while the pluvial periods there correspond to the glaciations in the northern area. According to BROOKS (1922), the pluvial periods in Egypt and Syria are clearly parallel to the glacials, although the desert-phase of the C-interglacial is not very clearly distinguishable. According to the same author four glacials have been noted in Corsica. According to KESSLER (1925) river-valleys that are still visible were formed during the last glacial period in the now dry parts of N. Africa. BLANC found the following layers in a cave on the Apulian Coast:

1) Layer corresponding to the last part of the D-interglacial, containing remains of *Elephas antiquus, Rhinoceros Merckii* and *Hippopotamus;*
2) Stalagmite layers corresponding to the Riss glacial with remains of hare and fox;
3) Layers of the E-interglacial with a fauna resembling that of 1) mixed with some steppe animals;
4) Stalagmite layers of early Würm glacial with remains of steenbok;
5) Loess layers with cold continental fauna corresponding to part of Würm ("Aurignac");
6) Loess layers with cold fauna, for instance *Alca impennis*, corresponding to late Würm.

On the Wershojansk and Tscherski mountains in E. Siberia three glaciations can be distinguished, of which the middle (Riss) had the largest extension (GRIGORIEV 1927).

According to WOLDSTEDT (1929) the first Pleistocene glaciation in Europe (Günz of PENCK and BRÜCKNER, Kander of BECK) corresponds to the Nebraskan and Yersyan driftsheets in N. America. In the late Pliocene deposits of the "Apscheron" layers in southern Russia a fauna is found pointing to a slow deterioration of the climate, interpreted as the initial of that glacial.

The C-interglacial corresponds to the "Aftonian" interglacial of American geologists. PENCK and BRÜCKNER estimated, by a comparison of the weathering of deposits from that period and the post-glacial deposits, that it must have had an extention of about 60,000 years. *Magnolia Kobus* and *Vitis vinifera* were at that period growing in the lower Rhine valley. In Central Europe the Sabre-toothed tiger *(Michairodus)* and a zebra like horse *(Euquus stenonis)* were found, in North America the giant sloth *(Megatherium).*

The following glaciation, Mindel by earlier authors, Glütsch by BECK, corresponds to the "Elster" and in America to the "Kansan" drift. Its extention was on the whole greater than earlier glaciations and in Europe about the same as Riss.

The lower layers of fossil ice on the new Siberian Islands are considered by WOLDSTEDT (1929) to have been formed during that early glaciation.

The D-interglacial is often called the great interglacial. It was estimated by PENCK and BRÜCKNER to have lasted about 240,000 years and thus to have been considerably more than ten times longer than the postglacial time. GOESEN-HAGEN, GISTEL and DEWALL have counted 10—11,000 year varves in rock meal (kiselguhr) on the Lüneburg heath from this period. BECK, however, considers the interglacial between Glütsch and Riss to be only about 70,000 years. If, as he presumes, the Glütsch and Kander glaciations had no correspondence in the British Isles and the Netherlands, but were severely felt only in the continental parts of Europe, then perhaps the Elster still is to be considered as paralell to

27

the Mindel, as assumed by PENCK and BRÜCKNER. In that case the D-interglacial retains its formerly presumed length except in some continental regions, where it was interrupted by comparatively small glaciations. In any case the D-interglacial must have been longer than the E-interglacial. It corresponds according to GAMS to the Dürntenien of GEIKE, and in America to the "Yarmouth" interglacial, to which, according to ANTEVS (1929) the Toronto interglacial beds should be referred. These beds possess remnants of a flora, which according to BRAUN (1928) now grows 3°—5° of latitude further south. The Cromer forest beds are also considered to belong to this interglacial. WOLDSTEDT (1929) and other authors also consider the climate of the D-interglacial to have been more genial than the present. In the *Quercus* forest of Europe *Elephas antiquus, Rhinoceros Merckii* and *Hippopotamus* were present, perhaps also *Homo heidelbergensis. Trapa* and *Ficus* have been found in the deposits and *Rhododendron ponticum* grew in the Alps, where it was exterminated by the Riss ice, and never returned.

On the New Siberian Islands the D-interglacial is shown as a humus stratum between the two ice layers. In these layers fruiting specimens of *Alnus fruticosa* were found (TOLL 1895) 4° north of the northernmost station for this shrub today. GAMS (1934) considers that towards the end of the Great interglacial the climate in the Mediterranean region was so dry that no Tertiary woods could grow there and that such woods could only occur towards the boundary of the districts that had a cold climate during the glacial periods.

After the D-interglacial came the Riss glaciation, corresponding to the Saale, the Polandian and in America to the Illinoian (and ? Iowan) driftsheets. To this glaciation the Dnieper lobe doubtless belongs and most authors, for instance ANTEVS (1929), consider the maximum glaciation in Siberia to belong to this period. Its maximum is estimated by MILANKOVITCH to have been reached about 116,000 years ago. The upper ice of the New Siberian Islands is considered to be the ice of the Riss glaciation (compare WOLDSTEDT 1929). On Kanin Peninsula and on Kolgujev Island erratics originating from the Timan Mts. and from the Urals are found, having been transported to these places during Riss (ANTEVS 1929). It has been estimated that the water stored on the land at the maximum of Riss must have caused the water level of the oceans to sink 93 m. (ANTEVS 1929 p. 82, DALY 1925 p. 10). It is evident that a large part of the continental shelf at that period must have been above water level. Thus, for instance, it has been shown that the valley of the Hudson R. in eastern N. America can be followed down to a depth of 79 m. below sealevel, 173 km. off the coast of Sandy Hook. At the end of the Riss glaciation the so called "Chosarian transgression" of the Caspian Sea took place (BUBNOFF 1930).

During the Riss glaciation the mammoth *(Elephas primigenius)* and the woolly Rhinoceros *(R. tichorhinus)* replaced the above mentioned, warmth-loving pachyderms at least in Central Europe.

The E-interglacial, or the so-called last interglacial, corresponds according to GAMS (1936) to the Eem and further to the Atelian, the Sangamon and the Peorian interglacial. It is estimated by PENCK and BRÜCKNER to have lasted for about 60,000 years. According to WOLDSTEDT (1929) and others, the climate approximated to that of the present time. A colder climate seems to have prevailed during a period at its middle. Some consider the Iowan drift in America to have been caused by this climatic deterioration, but it seems more probable that the Iowan drift represents a minor advance of the Riss glaciation. *Quercus* forests, *Juglans regia, Brasenia purpurea* and *Trapa natans* were found in Central Europe and the elephants, the wild horses, the cave bear and the lion coexisted with Neanderthal man.

The last glaciation, called Würm by PENCK and BRÜCKNER, Mecklenburgian by GEIKE, Weichsel in northern Germany, and Wisconsin in America (Vashon in W. America), is considered to have reached its maximum about 35,000 years ago (ANTEVS 1929). MILANKOVICH (1930) on theoretical grounds arrived at the conclusion that its beginning took place at least 71,000 years ago. Würm had a considerably smaller extension than the foregoing. In Europe it did not reach eastern Russia (Scandinavian erratics only reach the western half of Kanin Peninsula). Its drift lines are found considerably north of those belonging to Saale and Elster in Germany. In America the Rocky Mts. were hardly totally glaciated. Würm is naturally the best known of the glacials and its later part has been surveyed in detail through the investigations of the clay varves by DE GEER. It made at least four major advances and several smaller ones. The summer temperature in central Europe is estimated to have been about 10° lower than at present. *Betula* and *Pinus silvestris* were growing in the upper Rhine valley, in Bohemia and Hungaria, in the latter place also *Larix* and *Pinus cembra*. During the maximum of the Würm glaciation trees were not growing in the area between the northern ice sheet and that of the Alps (GAMS 1936). *Dryas,* of which earlier remains were not found in the Alps, immigrated there during Würm (GAMS 1936).

The musk ox, lemming, reindeer, bison and mammoth occurred far south in Europe together with primitive *Homo sapiens* who pictured them inside the caves, probably for magic purposes. Perhaps northernmost Norway was also the home of men living under conditions very similar to those of the "polar Eskimoes" of N. Greenland.

It seems fairly evident that districts free from ice occured in Scandinavia during the Würm glaciation. RAMSAY has presumed that Fiskar Penins. and Kildin I. in western Kola Penins. were not glaciated during Würm. Already in 1904 A. M. HANSEN had emphasized the existence of the ice-free refugia in western Norway, and NORDHAGEN (1935) has shown that most probably ice-free districts have existed along the coast of Norway. Refugia for the plants must also have been present in Scotland. In the Alps the glaciation was far from complete. (Compare the map in GAMS 1936 p. 11).

In Eastern America large areas were unglaciated, viz. Gaspé Peninsula, part of the islands and shores of St. Lawrence Bay, and the Long Range Mts. in Newfoundland. Earlier, DALY and COLEMAN had considered that the Thorngat Mts. in northern Labrador were not glaciated above 700 m., but according to ABBE (1931) ODELL showed that they actually were glaciated, while districts close to those mountains at sea-level have possibly been left unglaciated. Large parts of the Rocky Mts. apparently were also not glaciated at this period. High peaks in the Rocky Mts. seem, on the whole, never to have been glaciated. RAUP (1934), for instance, found places on the Caribou Mountain plateau (W. of Lake Athabasca) which had apparently never been covered by ice. The northern end of Queen Charlotte Islands was never glaciated (ANTEVS 1929).

On the whole early authors seem to have considered that the ice did not leave any areas within its limits free, but with our growing detailed knowledge it is becoming more and more clear that such areas must have occurred in many places.

From the last stage of the Würm glaciation in Scandinavia, the so called finiglacial period, SMITH (1920) found the following plants in middle Sweden: *Empetrum nigrum, Dryas, Asplenium viride, Primula stricta, Salix herbacea, reticulata* and *polaris, Saxifraga oppositifolia, Scirpus caespitosus* and (?) *Hippophae rhamnoides.* He also found the characteristic pollen of two species which could not be identified with the pollen of any existing Scandinavian plant.

The time that has elapsed since the year when the last ice in northern Scandinavia had melted so far that it was divided into two separate parts (the year zero in the chronology of DE GEER) up to the present year (1937) is according to DE GEER 8,677 years, undoubtedly a very accurate figure. On the basis of surveys of the quantity of helium in the mineral Zircon, KÖNIGSBERGER estimated the Quarternary period to have lasted (half a million to) one million years.

In postglacial time a distinct amelioration, with subsequent deterioration, of the climate took place. In many places evidences have been found of the warm postglacial period between these vicissitudes. So for instance ANDERSSON (1902) stated that the boundary of the *Corylus avellana* in Sweden at that time was found to be considerably further to the north (3° at the coast, 1° inland) than now. In Greenland marine layers from this period have been found at an elevation of about 30 m. containing remains of more southern species than those existing there at the present (GELTING 1934). On Iceland *Betula pubescens* is found in peatbogs north of its present boundary. In the south, *Betula verrucosa,* which does not now live on Iceland is found in the peat. (LINDROTH 1931). According to LINDROTH many of the insects of Iceland are such as chiefly inhabit woods, although Iceland is now for the most part destitute of woods. In E. America the peatbogs indicate according to BROOKS (1922) a drier climate than the present, and in the New Siberian Islands marine deposits are found above the last ice, containing molluscs indicating a warmer climate. On the coast of northern Siberia lastly postglacial (sometimes interglacial?) deposits containing *Larix* and *Alnus* together with bones of *Mammoth* are often found considerably north of the present northernmost outpost of the forest. The tundra belt thus widened 1° — 2° there in postglacial time.

The maximum of the postglacial warm period seems to have been reached about 4 — 2000 years B. C.

It should be mentioned that during the melting of the last ice-sheet the water-level of the oceans must naturally have risen again, although the differences between the water-levels of the last glaciation and the postglacial time may

30

have been smaller than during and after the Riss glaciation. The Caspian sea increased in size and the so called Chwalynsk transgression took place.

To sum up, we know that, since the Tertiary period, at least three, probably four glacial periods, separated by warmer interglacial intervals, have passed over the boreal belt of the globe. There can hardly be any doubt that the glacial periods were synchronous, at least all over the northern hemisphere. The earlier theory of American geologists, that the American centra of glaciation during the Pleistocene gradually moved eastwards, was opposed by ANTEVS and is not, as far as I can find, now current among the leading American glaciologists. It has instead been taken for granted by those scientists, who plead for the theory of migrations of the Poles, and constitutes one of their most beloved arguments (compare below).

The extension of the ice-sheets of each separate glaciation is not yet known with accuracy, and it is likewise uncertain whether the ice vanished completely during all the interglacials. The extension of the last glaciation in Europe is well known, as also the rate, time and mode of its retreat. Another very important feature is also approximately known, viz. t h e   m a x i m a l   e x t e n s i o n   o f   t h e g l a c i a t i o n. That one warm, long interglacial has existed before the maximal glaciation is also clear. The knowledge of these groups of facts is the most important information we have to rely on in dealing with the problems of the origin and development of the boreal flora. I have therefore tried to re-present the maximal glaciation so far as it is known today, in the map in Pl. 43. As I have already said, most probably it corresponds to the Riss glaciation. In Europe its extension is known in detail, but in regard to northern Asia there is some uncertainty. As shown on maps by ANTEVS, OSBORN & REEDS, TOL-MATCHEW and others, the ice was thought to have reached only western Siberia, stretching a tongue up to Tajmyr Peninsula, with an isolated centre north and northwest of the Ochotsk Sea. OBRUTCHEW and others, however, have assumed that the glaciation had a far larger extension in northern and eastern Siberia. The compilatory work of OBRUTCHEW is admirable, and if his written word instead of his map is taken into consideration it is possible to get a fairly good idea of the glaciation. His arguments show that at any rate there cannot have been any covering icesheets in this part of Siberia, as such leave neither lateral nor terminal moraines in the mountain valleys, which, however, are his most prominent evidence of glaciation. The covering ice-sheets of the map correspond sometimes to such statements in the text as the following: "Nachrichten über Eiszeitspuren N vom Tscherskigebirge fehlen noch — — —. Wenn man aber die hohe Breite (67—71°) des Gebietes und die Existenz von 600—1000 m. er-reichenden Gebirgszügen berücksichtigt, so scheint es notwendig, eine Vereisung wenigstens einzelner Zentren anzunehmen, von denen die Gletschern nach allen Seiten herabflossen und sich vielleicht zu einer ununterbrochenen Eisdecke ver-einigten." TOLMACHEV, who himself visited at least part of the region, strongly opposes the absurd exaggerrations of OBRUTCHEW. (Compare map in TOL-MATCHEV 1932 p. 45.)

The extravagance of the views held on the subject of glaciation in Eastern Siberia is, I find, most in evidence when OBRUTCHEW discusses

Kamtchatka, where I myself have had an opportunity of forming a personal opinion of the conditions. Southern Kamtchatka has certainly never been covered by an ice-sheet. No erratic blocks from other regions are found there and the *Betula* forests and meadows covering its lowlands with their extraordinarily distinct associations were doubtless established there already before the time of the maximum glaciation. There are, however, very distinct signs that the present feeble glaciation was in earlier periods considerably stronger. In the mountains numerous hanging glaciers have existed and in some places large tongues of ice once flowed down also in the lowland region. In the mountains of central Kamtchatka a large covering ice-sheet seems to have existed at one time.

The study of the river terraces, so common along the Siberian rivers, will certainly contribute much to our knowledge of the number and length of the Siberian glaciations. Probably the level surface of these terraces was formed by the accumulation of the sand-loaded rivers, when the sea-level rose at the end of the glaciations, and the steep slopes between them by the eroding rivers at the initial stages of the glaciations with their sinking sea-level. When I visited Kamtchatka I was not aware of the significance of these terraces, so prominent along all large rivers, and did not investigate them specially. From my notes it is clear that almost all rivers have three terraces separated by steep slopes. These slopes may possibly correspond to the Riss glaciers and some earlier glaciation. The traces of the Würm glaciation may be expected to be less marked, as the great ice-sheets of that glaciation were very distant from Kamtchatka and its influence probably not so much felt in the mountains there. On the lower Balaganchik R., the Koseten R. and the Unkanakchek R., however, I noted four terraces with three intermediate slopes, possibly corresponding to the "Mindel", Riss and Würm glaciations. Especially at the sources of the two last-mentioned rivers the traces of former glaciations are fresher than in other parts of South Kamtchatka, and the recent snowline is lower and more pronounced. An investigation of these terraces would doubtless yield most interesting results; in particular, as the synchronous pumice layers in Kamtchatka, which have spread from the volcanoes at different epochs and are recognizable from their different colour and structure, offer rare opportunities for the study of chronological questions.

The conditions in other parts of Eastern Siberia would appear from the descriptions to have been very similar to those in Kamtchatka. In other words: in Eastern Siberia a strong local glaciation occurred during at least two (and in places three) glacial periods. The glaciation in the mountains there may have been of the same type as the present glaciation of the ranges on the northern and northeastern shores of the Gulf of Alaska. The conditions in northern Siberia thus seem to be similar to the much better known conditions on the eastern side of the Bering Sea, maps of which showing the maximal glaciation in considerable detail are available.

In a paper by SHAPARENKO (Acta Inst. Bot. Acad. Sc. URSS Ser. 1 fasc 2, 1936 opposite p. 26) a map is published of the glaciation in Siberia, so far as it is known at present. However this author, apparently in order to avoid the difficulties in differentiating between solid ice-sheets and strong local glaciations marks out on the map "the ancient glaciations of mountains and main-lands". His map thus shows the districts that were completely free from ice throughout the Pleistocene time, contrasted with those once covered with solid ice-sheets and those more or less locally glaciated. From a botanical point of view, however, the latter differentation is necessarily of great importance. Further-more easternmost Asia is omitted. In the map Pl. 43 I have tried to summarize as correctly as possible our present knowledge of the Siberian maximal glacia-tion. In America the boundary of the maximal glaciation is, as already mentioned, fairly well known.

It is evident from the map that large parts of eastern Siberia were never covered with ice. It should be noted that the Lena-Aldan valley, although practically surrounded by glaciers, was left unglaciated, and that the country west of Lake Baikal, the Amur-Manchuria district, and at least the western slopes of the Stanovoj Mts., were also left intact. The country bordering the Bering Sea was only locally glaciated, and the Yukon valley was free from ice.

Perhaps the most remarkable feature is that the Arctic American Archipelago does not show any traces of glaciation and the conditions there during the glacial period must have been similar to those prevailing in northernmost Greenland today.

It was not only these large areas, however, that were free from ice during the maximal glaciation. It seems very probable that more or less extensive ice-free districts occurred in practically all places where the maximum ice-sheet reaches the coasts. In Eastern America, as already mentioned, it has long been known through the works of FERNALD and other recent authors that the mountains of the Gaspé Penins., the Long Range Mts. in Newfoundland, and other districts were never glaciated. Driftless areas have also been found south of the Great Lakes not so far from the southern boundary of the maximum glaciation as well as on islands in Lake Superior. In Western America unglaciated areas are known at the northern end of the Queen Charlotte Islands. Unglaciated mountains occur in E. Greenland (GELTING 1934). LINDROTH (1931 p. 440) presumes that part of southern Iceland was probably free from ice during the maximum glaciation. AHLMANN (1919) and NORDHAGEN (1931, 1935) have produced evidence that makes it very probable that ice-free districts occurred, even during maximal glaciation, in Western and Northern Norway.

It seems to me that on the whole the conditions during maximum glaciation must have been similar to the present conditions close to the great ice-sheets of Greenland and the Antarctic. Greenland has an ice-free border and in the Antarctic there are numerous nunataks, as well as ice-free shores and islands.

There is the further point that the storage of water in the great ice-sheets of the maximum glaciation lowered the sea-level very considerably. Districts along the coast, for the present only to a very limited extent accessible for investigation, must have been above the sea at that time. As I have mentioned earlier, the sea-level is estimated to have sunk during maximum glaciation by 90 — 100 m., which might be a minimum figure, as in its calculation only the storage of water was taken into consideration and not the isostatic changes, which must be expected to work in the same direction, forcing up the margins and the levels adjacent to the ice-sheets. A more probable figure for the m a x i m u m depth, to which the continental shelf may have been laid bare, seems to be 150 m. To what extent these continental shelves were glaciated is for the most part unknown, but there can be no doubt that they must have played a very important rôle as refugia for the vegetation during the maximum glaciation.

FERNALD has discussed the banks off the coast of eastern N. America, which prove to have been dry during the maximum glaciation, and their significance for the vegetation.

The Doggers Bank in the North Sea is known to have been above the sea, populated and forestclad in glacial or early postglacial time.

It is apparent that there must actually have been a connection between Asia and America over the Bering Straits during the glacial periods. The continental shelf there is very shallow. A lowering of the sea-level even by 50 m. would result in a connection between Asia and America 300 km. broad. The 100 and 200 m. depth curves follow one another fairly closely and mark the steep escarpment of the continental shelf towards the very deep southern part of the Bering Sea. A water-level 100 m. lower than at present would result in the entire northern half of the Bering Sea becoming dry, the present islands low mountains. The two continents would be connected by a land-bridge about 1200 km wide. (Compare fig. 4). In fact the distribution of the plants indicate that a connection of about that extension actually did exist during Quarternary time.

If not only the lowering of the waterlevel but also, as is indeed probable, an isostatic pressure brought this broad bridge into being, it may be expected to have been in existence also for some time a f t e r the glacial periods. This land-mass, which I shall hereinafter call B e r i n g i a , must have been a good refugium for the biota during the glacial period. The connection with the cold Arctic Ocean was closed and its southern shores were washed by the comparatively warm water of the Pacific. It did not possess any high mountains and there is hardly any reason to believe that it was more strongly glaciated than were the present shores of Bering Sea. In the following we shall see that during maximum glaciation Beringia was the home of a large portion of the coast-bound flora of the northern Pacific and the Arctic Ocean. The Aleutians, however, as I myself have had an opportunity of establishing, were heavily glaciated, probably owing to their high mountains and the great precipitation from southernly winds. It seems not improbable that the middle part of that chain of islands was glaciated to such an extent that only a very small area was left unglaciated.

* *

*

The origins of the ice ages are still obscure. They are, however, not without interest to the biologist. For instance, the biological conditions must presumably have differed fairly considerably if, as is commonly supposed, the

34

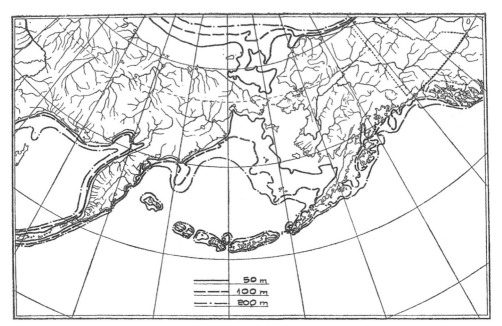

Fig. 4. Depth of the sea in the Northern Pacific area.

ice ages were caused by a lowering of the average temperature over the entire earth, or if, as SIMPSON recently proposed, they were caused by an i n c r e a s e of the temperature. SIMPSON thinks that a rising temperature must cause an increase of the circulation in the atmosphere and a higher precipitation and that the temperature is low enough in the interglacials to allow ice-sheets to be formed, provided that the moisture and precipitation were considerably increased. This seems quite probable as far as regards the ice-sheets originating from countries near or inside the Polar Circle, but, as PENCK and BRÜCKNER have estimated, the precipitation must have been about 20 times as much as it is now in order to cause the glaciations of Mindel or Riss time in such a southern region as the Alps. The lowering of the snow line in the tropical mountains all over the globe, which undoubtedly took place simultaneously to the ice ages, cannot be reconciled with SIMPSON's theory.

Another theory was recently developed by MILANKOVITCH, who deduces from climatic variations, caused by the variable eccentricity of the ecliptic, the displacement of the vernal equinox and the variations in the declination of the ecliptic, the four Pleistocene glaciations and their interglacials. As this theory seems to be the most plausible of all and has been payed much attention I think that a short account for it may be at place here.

The declination of the ecliptic varies within a period of 40,000 years. When it has its smallest value, the differences in climate between the Equator and the Poles are greater than they are now, as the former receives more and the latter less radiation from the sun than now, and these differences, as com-

35

pared with present conditions, are greater in summer than in winter. Thus a decrease in the angle between the ecliptic and the equator diminishes the difference between summer and winter in high latitudes, but widens the differences between the climatic belts. As mentioned above, the latter circumstance will cause an increase of the circulation in the atmosphere and thus a greater precipitation. Thus cold and wet summers, at least in high latitudes, are results of a decrease in the declination of the ecliptic. These vicissitudes are felt equally in both hemispheres. So is also the variation in the eccentricity of the ecliptic, which has a period of 92000 years and whose influence is such that when it has its maximal value the difference between the summer and winter halves of the year is great. This difference also varies, and still more so, according to the wandering of the equinoxes around the ecliptic (the precession). These move around the ecliptic in 21000 years. When they are in perihelion and aphelion, respectively, the hemispheres have the same climate, and the summer and winter halves of the year are equal. At the present time, the earth is in the perihelion at the beginning of January, and the summer half of the year is one week longer than the winter half. The difference in climate, caused by the precession, is felt alternately in the two hemispheres.

At certain periods of the past all the three variations have worked in the same direction, causing cool and wet summers alternately in the two hemispheres, and thus inducing the different glaciations. The periods of cool and wet summers often occur two or three in comparatively close succession, and as an ice-sheet itself causes a considerable deterioration of the climate, two or three periods, even if separated by a warmer period of 30 — 35000 years, can quite naturally together form one single glaciation, at least in the central parts of the ice-sheet. In the marginal parts, especially in the Alps, SŒRGEL (1925) and EBERL (1930) have found oscillations in the glaciations, which are interpreted as corresponding to the periods of MILANKOVITCH.

The displacement of the poles, regarded by some authors as causing the glacials, was recently disputed by WOLDSTEDT on the strength of glacial observations in the South American Andes, and it seems quite impossible to place the Pole in such a position that the simultaneous heavy glaciation of N. America, Europe and, especially, Western Siberia can be explained by the shorter distance of those regions from it without also accepting the theory of continental drift (KÖPPEN & WEGENER 1924).

It would doubtless be of great help to the biologists if a rational cause of the glaciations could be found, but for the present it seems necessary to work without that help.

The deposits formed during the glaciations and the interglacials do, however, give some information as to the conditions during the periods in question. Reference to these deposits has already been made above, when the different periods were reviewed. In any case, the initial climatic change that caused a glaciation must either have been a lowering of the summer temperature or an increase in the precipitation, or both. Thus the primary result in the landscape was, as already pointed out by NANNFELDT in 1935, an enormous extension of the peat bogs, due in the first place to a decrease in the evaporation.

36

These conditions must have formed an impassable barrier for the mountain plants, especially the more pretentious ones.

Still further to the south, in the present steppe and desert belts, the change must have led to a moister climate, either directly by the higher precipitation or indirectly through the shifting of the climatic zones in a southernly direction, which is bound to follow from the lowered temperature in the north. The fact that even a comparatively slight deterioration of the climate in the north must have such an influence is known from the historic time. That cold period which had its maximum in 800 — 500 B. C. (the "fimbulvinter" of the sagas) was characterized in the Mediterranean countries by a moister climate, allowing wheat to be cultivated in places where it now does not grow, and thus facilitating the high culture in those countries (BROOKS 1922).

The deterioration of the climate, initiating a glacial period, thus means a certain amelioration of the climate further to the south. The plants already found there, at least most of them, must have had a good chance of retaining their position. They must thus be expected to have offered a considerable resistance in the competition to the flora that eventually migrated southwards in front of the ice. These facts likewise contradict the idea of the flora being "chased" south.

It seems clear that during the glacials loess formations were formed in the border districts of the ice, while a pluvial climate was prevailing further south of it. The high pressure of air, forming above the ice, must have sent out strong, comparatively dry winds chiefly blowing f r o m the ice. These winds doubtless were very trying for the vegetation in districts close to the border of the ice, and many species certainly could not have withstand them. The fact that many of the species which apparently survived close to the ice have bulbs or rhizomes is perhaps due to this phenomenon. Their stems, which are partly subterranean, were better sheltered from the dry and cold annihilating winds.

Towards the end of the glaciations, when the ice recedes, the climate must be considerably more genial than during their growing and maximum stages, as the retreat of the ice takes place much later than the start of the amelioration of the climate. During the interglacials a "normal" climate set in, steppe and desert conditions spread northwards, the climatic zones approximated to their present positions or even surpassed them. As the interglacials were many times longer than post-glacial time, the area of the former ice was doubtless completely covered by a vegetation that had had an opportunity of attaining stable conditions corresponding to the new climatic belts.

The glacial periods and their climatic consequences have apparently played the most prominent part in the development of present arctic and boreal biota, and unless these features are studied parallel to the variation and present area of different species the problems of their origin and evolution will remain unsolved. In the rest if this paper I am going to examine the flora of our area from this point of view, in order as far as possible to ascertain the influence of the maximum glaciation on the origin, migration and variation of present arctic and boreal plants.

Bibliography

Taxonomical papers and works, from which information concerning the distribution of the plants were gathered, are not quoted here. For information concerning them see my papers on the Flora of Kamtchatka and the Flora of the Aleutian Islands.

ABBE, E. C. Botanical results of the Grenfell-Forbes Northern Labrador Expedition, 1931. — Rhodora Vol. 38 (1936) p. 102—161.

ABRAMS, LE ROY. Endemism and its significance in the Californian flora. — Proc. Internat. Congr. Pl. Sc. Ithaca 1926 (1929) p. 1520—1524.

AHLMANN, H. Geomorphological Studies in Norway. — Geogr. Annaler Bd. 1 (1919) p. 1—148.

ANDERSSON, G. Hasseln i Sverige fordom och nu. — Sv. geol. unders. Ser. Ca. n:r 3 (1902) (168 p.).

ANTEVS, E. The last glaciation. — Amer. Geogr. Soc. Research Series No. 17, New York 1928 (292 p.).

— Maps of the Pleistocene glaciations. — Bull. Geol. Soc. Amer. Vol. 40 (1929) p. 631—720.

ARLDT, TH. Handbuch der Palaeogeographie Bd. 1—2 1919—1922 (679 + 1647 p.).

BECK, P. Über das schweizerische und europäische Pliozän und Pleistozän. — Eclog. geol. Helv. Bd. 26 (1933) p. 335—437.

BERRY, E. W. Former land connection between Asia and North America as indicated by the distribution of fossil trees. — Proceed. of the Fifth Pacif. Sc. Congr. Canada 1933 Vol. 4 (1934) p. 3093—3106.

BRAUN, E. L. Glacial and Post-Glacial Plant Migrations indicated by relic Colonies of Southern Ohio. — Ecology Vol. 9 (1928) p. 284—302.

BRAUN-BLANQUET, J. Essai sur les notions d' "élément" et de "territoire" phytogeographiques. — Arch. Sc. Phys. et Nat. Vol. 1, 5 sér. (1919) p. 497—512.

— Über die Genesis der Alpenflora. — Verhandl. der Naturforsch. Ges. in Basel Bd. 35 (1923—24) p. 243—261.

— Über die pflanzengeographischen Elemente Westdeutschlands. — Der Naturforscher Bd. 5 (1928) p. 297—306.

BROOKS, C. E. P. The evolution of climate. London 1922 (173 p.).

— Climate through the ages. London 1926 (439 p.).

BUBNOFF, S. v. Das Quartär in Russland. — Geol. Rundschau Bd. 21 (1930) p. 177—200.

CHANDLER, M. E. J. The upper Eocene flora of Hordle, Hants. — Palaeontogr. Soc. Part I (1925) p. 1—32 and II (1926) p. 33—52.

CHODAT, R. L'Endémisme alpin et les réimmigrations post-glaciaires. — Verhandl. der Naturforsch. Ges. in Basel Bd. 35 (1923—1924) p. 69—82.

CHRIST, H. Die Geographie der Farne. Jena 1910 (357 p.).

COLEMAN, A. P. Interglacial periods in Canada. — 10 sess. Congr. Géol. Internat. Mexico 1907 p. 1237—1258.

— Ice Ages Recent and Ancient. 1926 (296 p.).

DALY, R. A. Pleistocene changes of level. — Amer. Journ. Sc. Ser. 5 Bd. 10 (1925) p. 281—313.

DIELS, L. Genetische Elemente in der Flora der Alpen. — Engler Bot. Jahrb. Bd. 44 (1910) p. 7—46.

EBERL, B. Die Eiszeitenfolge in nördlichen Alpenvorlande. Augsburg 1930.

EGGLER, J. Arealtypen in der Flora und Vegetation der Umgebung von Graz. — Mitteil. des Naturwissenschaftl. Ver. für Steiermark Bd. 71 (1934) p. 18 —32.

EIG, A. Les éléments et les groupes phytogéographiques auxiliaires dans la flore palestinienne. I Texte, II Tableaux analytiques. — FEDDE, Repert. spec. nov. regni veget. Beihefte Bd. 68 (1931) (201 p.).

EKMAN, S. Djurvärldens utbredningshistoria på Skandinaviska halvön. Stockholm 1922 (614 p.).

ELFSTRAND, M. Var hava fanerogama växter överlevat sista istiden i Skandinavien? — Sv. Bot. Tidskr. 1927 p. 269—284.

ENGLER, A. Versuch einer Entwicklungsgeschichte der Extratropischen Florengebiete der nördlichen Hemisphäre. Leipzig 1879 (386 p.).

— Beiträge zur Entwicklungsgeschicte der Hochgebirgsfloren erläutert an der Verbreitung der Saxifragen. — Abhandl. der Königl. Preuss. Akad. des Wissensch. Phys. Math. Klasse N:r 1 (1916) p. 1—113.

FERNALD, M. L. The geographic affinities of the vascular floras of New England, the maritime Provinces and Newfoundland. — Journ. Amer. Bot. 5 No. 5 (1918) p. 219—236.

— Isolation and endemism in Northeastern America and their relation to the age- and-area hypothesis. — Am. Journ. of Botany. Vol. 11 (1924) p. 558—572.

— Persistence of plants in unglaciated areas of boreal America. — Mem. Amer. Acad. of Arts and Sc. Vol. 15 (1925) p. 241—342.

— The antiquity and dispersal of vascular plants. — The Quarterly Review of Biology Vol. 1 (1926) p. 213—245.

— Some relationships of the floras of the Northern Hemisphere. — Proceed. of the Internat. Congr. of Plant Sc. Vol. 2 (1926) p. 1487—1507.

— Specific Segregations and Identities in some floras of Eastern North America and the old world — Rhodora Vol. 33 (1931) p. 25—63.

— Recent discoveries in the Newfoundland Flora. — Rhodora Vol. 35 (1933) p. 1—403.

— Some beginnings of specific differentiation in plants. — Science Vol. 79 (1934) p. 573—578.

FRIES, TH. C. E. Botanische Untersuchungen in Nördlichsten Schweden. Ein Beitrag zur Kenntnis der alpinen und subalpinen Vegetation in Torne Lappmark. — Vetenskapl. och prakt. undersökn. i Lappland. Uppsala 1913 (361 p.).

GAMS, H. Das Alter des alpinen Endemismus. — Bericht des Schweizer. Bot. Ges. Bd. 42 Heft 2 (1933) p. 407—483.

— Zur Geschichte, klimatischen Begrenzung und Gliederung der immergrünen Mittelmeer-stufe. — Ergebn. Int. Pflanzen-geogr. Exkurs. durch Mittel-italien 1934. Verröffentl. Geobot. Inst. Rübel Zürich Heft. 12 (1934) (42 p.).

— Beiträge zur Mikrostratigraphie and Paläontologie der Pliozäns und Pleisto-zäns von Mittel- und Osteurope und Westsibirien. — Eclogae geol. Helv. Vol. 28, No. 1 (1935) (31 p.).

— Der Einfluss der Eiszeiten auf die Lebewelt der Alpen. — Jahrb. des Ver. zum Schutze der Alpenpflanzen und -Tiere. Jahrg. 8 (1936) p. 7—29.

GELTING, P. Studies on the vascular plants of East Greenland between Franz Joseph Fjord and Dove Bay. — Meddel. om Grønland Bd. 101 No. 2 (1934) (340 p.).

GRIGORJEV, A. A. Geomorphologische Skizze von Jakutien. 1927. The work "Jakutia" published by the Russian Academy of Science (48 p.).

HAGEDOORN, A. L. & A. C. The relative value of the process causing evolution. The Hague 1921 (294 p.).

HANSEN, A. M. Bre og Biota. Skr. Det Norske Videnskaps-Akademi i Oslo. Bd. 3 (1903) (255 p.).

HOFMAN, E. Isoporien der europäischen Tagfalter. 52 p. Inaug. Diss. Jena 1873.

HOFSTEN, N. v. Die Echinodermen des Eisfjords. — Kungl. Sv. Vet. Akad. Handl. Bd. 54 No. 2 (1915) (282 p.).

HOOKER, J. D. Outlines of the Distribution of Arctic Plants. — Transact. Linn. Soc. Vol. 23 (1862) p. 251—348.

HULTÉN, E. Eruption of a Kamtchatka Volcano in 1907 and its atmospheric Consequences. — Geol. Fören. Förh. 46, 1924 p. 407—417.

— Flora of Kamtchatka and the adjacent Islands. — Kungl. Sv. Vet. Akad. Handl. Ser. 3 Bd. 5 and Bd. 8. Stockholm 1927—1930. 1134 p.

— On the American component in the flora of Eastern Asia. — Sv. Bot. Tidskr. 22 (1928) p. 220—229.

— Studies on the origin and distribution of the flora in the Kurile Islands. — Bot. Not. 1933 p. 325—345.

— Flora of the Aleutian Islands including westernmost Alaska Penins. with notes on the Commander Islands. (In press).

JEROSCH, M. CH. Geschichte und Herkunft der schweizerischen Alpenflora. Leipzig 1903 (223 p.).

KESSLER, P. Das eiszeitliche Klima. Stuttgart 1925 (210 p.)

KNOPF, A. The probable Tertiary land connection between Asia and North America. — Univ. Calif. publ. Bull. Dept. Geol. Vol. 5 (1906—10) p. 413—420.

KROKOS, W. Materialien zum Studium der Böden der Ukraine. Heft 5 (1927) Charkow 1927.

KRYSHTOFOVICH, A. N. Evolution of the Tertiary flora in Asia. — The New Phytologist Vol. 28 (1929) p. 303—312.
— Materialien zur Tertiärflora des fernen Ostens von Asien. Mat. Geol. Fernen Ostens Nr. 18 (1921) Wladivostok 1923. In Russian.
— Die amerikanische Butternuss aus den Süsswasserablagerungen des Jakutskgebietes. — Tr. Geol. Kom. N. S. Bd. 124 (1915). In Russian.
— The tertiary flora of the Korf Gulf, Kamtchatka. — Transact. Far East Geol. Prosp. trust. Fasc. 62, 1934. 32 p.
— A final link between the Tertiary Floras of Asia and Europe. — The New Phytologist Vol. 34 No. 4 (1935) p. 339—344.

KULCZYNSKI, S. Das boreale und arktisch-alpine Element in der Mittel-Europäischen Flora. — Bull. Internat. de l'Acad. Polonaise des Sc. No. 1—10 (1923) p. 127—214.

KÖPPEN, W. & WEGENER, A. Die Klimate der geologischen Vorzeit. Berlin 1924 (255 p.).

LAVRENKO, E. M. Les centres de la conservation des relictes sylvestres tertiaire entre les Carpathes et l'Altai. — Journ. Soc. Bot. Russ. T. 15 (1930) p. 351—363.

LEVERETT, F. Pleistocene glaciations of the Northern Hemisphere. — Bull. Biol. Soc. Amer. Vol. 40 (1928) p. 745—760.

LINDROTH, C. H. Die Insektenfauna Islands und ihre Probleme. — Zoologiska Bidrag från Uppsala Bd. 13 (1931) p. 105—599.

LIVINGSTON, B. E. & SHREVE, F. The distribution of vegetation in the United States, as related to climatic conditions. Washington 1921 (590 p.).

MILANKOVITCH, M. Matematische Klimalehre und Astronomische Theorie der Klimaschwankungen. — Handbuch der Klimatologie Bd. 1 (1930) Teil A p. 1—176.

NANNFELDT, J. A. Taxonomical and plant-geographical studies in the Poa laxa group. A contribution to the history of the North European mountain floras. — Symbolae botanicae Upsalienses Vol. 5 (1935) (113 p.).

NATHORST, A. G. Om några arktiska växtlämningar i en sötvattenslera vid Alnarp i Skåne. — Lunds Univ. Årsskr. Bd. 7 (1871) 2 Avd. f. Math. o. Nat. No. 9.

NORDHAGEN, R. Studien über die skandinavischen Rassen des Papaver radicatum Rottb. sowie einige mit denselben verwechselte neue Arten. — Bergens Museums Årsbok 1931 N:r 2 Naturvidenskaplig rekke, p. 1—50.
— Om Arenaria humifusa Wg. og dens betydning for utforskningen av Skandinavias eldste floraelement. — Bergens Museums Årbok 1935 Naturvidenskapelig rekke N:r 1 (183 p.).
— Skandinavias fjellflora og dens Relasjoner til den siste Istid. — 19:de Skand. Naturforskarmötet i Hälsingfors 1936, p. 93—124.
— Versuch einer neuen Einteilung der subalpinen—alpinen Vegetation Norwegens. — Bergens Museums Årsbok 1936 n:r 7 Naturvidenskapelig rekke, p. 1—88.

OBRUTSCHEW, W. A. Geologie von Sibirien. — Fortschritte der Geologie und Palaeontologie Heft 15 (1926) (572 p.).

— Die Verbreitung der Eiszeitspuren in Nord- und Zentralasien. — Geol. Rundschau Bd. 21 (1930) p. 243—283.

OLBRICHT, K. Die diluviale Eiszeit in Ostasien. — Centralblatt für Mineralogie, Geologie und Paläontologie Jahrg. 1923 p. 726—730.

OSBORN, H. F. Influence of the glacial age on the evolution of Man. — Bull. Geol. Soc. Amer. Vol. 40 (1929) p. 589—596.

OSBORN, H. F. & REEDS, C. A. Old and new standards of Pleistocene division in relation to the prehistory of Man in Europe. — Bull. Geol. Soc. Amer. Vol. 33 (1922) p. 411—490.

OSTENFELD, C. H. The flora of Greenland and its origin. — Det Kgl. Danske Vidensk. Selsk. Biol. Med. 6 (1926) (71 p.).

PAWLOWSKI, B. Die geographischen Elemente und die Herkunft der Flora der subnivalen Vegetationsstufe im Tatra-Gebirge. — Bull. Acad. Polonaise des Sc. Nat. (1928) p. 161—202.

PENCK, A. & BRÜCKNER, E. Die Alpen im Eiszeitalter. Leipzig 1909 Bd. I, II & III (1199 p.).

PENNELL, F. W. Castilleja in Alaska and Northwestern Canada. — Proceed. Acad. Nat. Sc. Philad. Vol. 85 (1934) p. 517—540.

RAMSAY, W. Über die geologische Entwicklung der Halbinsel Kola in der Quartärzeit. — Fennia 16 (1898) Helsingfors.

RAUP, H. M. The distribution and affinities of the vegetation of the Athabasca—Great Slave Lake Region. — Rhodora Vol. 32 (1930) p. 187—208.

— Phytogeographic studies in the Peace and upper Liard River Regions, Canada. — Contr. from the Arnold Arb. of Harv. Univ. 6 (1934) (230 p.).

REID, C. & REID, E. M. The fossil flora of Tegelen-Sur-Meuse, near Venloo, in the province Limburg. — Verhandl. Koninklijke Akad. van Wetenschappen. Deel 13 No. 6 (1907) (26 p.).

— The Pliocene floras of the Dutch-Preussian Border. — Mededeel. van de Rijksopsporing van Delfstoffen No. 6 (1915) (178 p.).

REID, E. M. Recherches sur quelques graines pliocènes de Pont-de Gail (Cantal). — Bull. Soc. Géol. de France T. 20 (1920) p. 49—53.

— On two preglacial Floras from Castle Eden. — Quarterly Journ. Geol. Soc. of London Vol. 76 (1920) p. 104—144.

REID, E. M. & CHANDLER, M. E. J. The London Clay Flora. London 1933 (561 p.).

REID, E. M., BOSWELL, P. G. H., CHANDLER, M. E. J., GODWIN, H., WILMOTT, A. J., SALISBURY, E. J., RAISTRICK, A., DU RIETZ, E. Discussion on the Origin and Relationship of the British Flora and Open Discussion. — Proceed. Roy. Soc. London Vol. 118 (1935) p. 197—241.

REINHARD, A. v. Über die eiszeitliche Vergletscherung Kamtschatkas. — Zeitschr. der Ges. für Erdkunde. Berlin 1915, p. 180—183.

REINIG, W. F. Die Holarktis. Ein Beitrag zur diluvialen und alluvialen Geschichte der Zirkumpolaren Faunen und Florengebiete. Jena 1937 (124 p.).

RIDLY, H. N. The dispersal of plants throughout the world. 1930 (744 p.).

RIKLI, M. Alpin-arktische Arten und einige Bemerkungen über die Beziehungen der Flora unserer Alpen mit derjenigen der Polarländer. — Veröffentl. des Geobot. Inst. Rübel in Zürich Heft 3 (1925) (Festschrift Carl Schröter) p. 96—108.

RYDBERG, P. A. A short phytogeography of the prairies and great plains of central North America. — Brittonia Vol. 1 (1931) p. 57—66.

SAMUELSSON, C. Några studier över erosionsföreteelserna på Island. — Ymer 1925, p. 339—355.

SARASIN, P. Zur Frage von der prähistorischen Besiedelung von Amerika. — Denkschriften der Schweiz. Naturforsch. Ges. Bd. 64 (1928) p. 235—273.

SCHARFF, R. F. Distribution and Origin of Life in America. 1911 (497 p.).

SCHMID, E. Über Florenelemente. — Ber. der Schweizer. Bot. Ges. Bd. 44 (1935) p. 444—445.

SCHULTZ, A. Über die Entwicklungsgeschichte der gegenwärtigen Phanerogamen Flora und Pflanzendecke der skandinavischen Halbinsel und der benachbarten schwedischen und norwegischen Inseln. — Abhandl. der Naturforsch. Ges. zu Halle Bd. 22 (1900) (316 p.).

SHAPARENKO, K. K. The nearest ancestors of Ginkgo biloba L. — Fl. et Syst. Pl. Vasc. Fasc. 2, 1926 p. 5—32.

SIMPSON, G. C. Past Climates. — Mem. and Proceed. Manch. Literary & Philosophical Soc. Vol. 74 (1930) p. 1—34.

— Discussion on Geological Climates. — Proceed. Roy. Soc. of London Vol. 106 Ser. B (1930) p. 301—317.

SMITH, J. P. Climate Relations of the Tertiary and Quarternary faunas of the California region. — Proceed. of the California Acad. of Sc. Vol. 9 No. 4 (1919) p. 123—173.

SMITH, H. Vegetationen och dess utvecklingshistoria i det centralsvenska högfjällsområdet. — Norrländskt Handbibliotek 9. Uppsala 1920. (238 p.)

SOERGEL, W. Lösse, Eiszeiten und Paläolithische Kulturen. Jena 1919.

— Die Gliederung und absolute Zeitrechnung des Eiszeitalters. — Fortschr. Geol. Palaeontol. Heft 13, Berlin 1925 (251 p.).

STEFFEN, H. Versuch einer Gliederung der arktischen Flora in geographische bzw. genetische Florenelemente. — Bot. Archiv Bd. 6 Heft 1—3 (1924) p. 7—49.

— Weitere Beiträge zur Gliederung der arktischen Flora. — Bot. Archiv Bd. 10 Heft 5—6 (1925) p. 335—349.

— Beiträge zur Begriffsbildung und Umgrenzung einiger Florenelemente Europas. — Beihefte zum Bot. Centralblatt Bd. 53 Abt. B Heft 2—3 (1935) p. 330—404.

— Gedanken zur Entwicklungsgeschichte der arktischen Flora. — Beih. Bot. Centralbl. Bd. 56 Abt. B, Heft. 3, 1937, p. 409—447.

STERNER, R. The Continental Element in the flora of South Sweden. — Geogr. Ann. H. 3—4 (1922) p. 221—441.

SUKATCHEW, W. Einige Angaben zur vorglazialen Flora Nordens von Sibirien. — Tr. Geol. Mus. Peter des Grossen IV (1910) L. 4 (In Russian).

— Sur la trouvaille de la flore arctique fossile sur la rive de fleuve Irtyche près du village Demianskóe, gouv. Tobolsk. — Bull. de l'Acad. Impér. des Sc. de St. Petersbourg T. 4 (1910) p. 457—464.

TOLL, E. v. Die fossilen Eislager und ihre Beziehung zu den Mammutleichen. — Mém. Ac. Sc. St. Petersb. Sér. 7, 42 (1895) Nr. 13.

TOLMATCHEV, A. P. Flora tsentralnoj tchasti vostotchnovo Tajmyra (Flora of the central part of eastern Tajmyr). — Trudy Poljarnoj Komissii Vyp. 8, 13 and 25 (1932—1935). Together 281 p. (In Russian).

TROLL, K. Ozeanische Züge im Pflanzenkleid Mitteleuropas. — Freie Wege Vergleichender Erdkunde. Festgabe Erich von Drygalski 1925 p. 307—335.

TURESSON, G. Die Genenzentrumtheorie und das Entwicklungszentrum der Pflanzenart. — Kungl. Fysiogr. Sällsk. i Lund Förhandl. Bd. 2 n :r 6 (1932) (11 p.).

— Rassenökologie und Pflanzengeographie. Einige kritische Bemerkungen. — Bot. Not. 1936 p. 420—437.

WANGERIN, W. Florenelemente und Arealtypen. — Beih. Bot. Centralbl. Bd. 49 Erg. Bd. (Drudefestschrift) p. 515—566.

WEGENER, A. Die Entstehung der Kontinente und Ozeane. — Die Wissenschaft Bd. 66 (1922) (144 p.).

WILLE, N. The flora of Norway and its immigration. — Ann. Mo. Bot. Gard. Vol. 2 (1915) p. 59—108.

VOSS, J. Postglacial Migration of Forest in Illinois, Wisconsin and Minnesota. — The Botanical Gazette. — Vol. 96 : 1 (1934) p. 3—43.

WOLDSTEDT, P. Das Eiszeitalter. Stuttgart 1929. (406 p.).

Index to the Plates

The different shades in the maps are limited by lines going through stations where equal numbers of species belonging to the group in question are known to occur. The corresponding numbers are indicated in the index found at the foot af each plate.

List of species is found on page:	Number of Plate:	Characteristic of groups shown in the plates.
13	1	Boreal Eurasiatic plants, plastic in Europe, not reaching the (lower) Kolyma valley to the North-East.
14	2	Boreal Eurasiatic plants, plastic in Europe, reaching also the Kolyma valley to the North-East.
16	3	Boreal Eurasiatic plants, rigid in Europe, not reaching lower Kolyma valley to the North-East.
17	4	Boreal Eurasiatic plants, rigid in Europe, not reaching the Kolyma valley to the North-East but reaching south of the Central Asiatic desert belt.
17	5	Boreal Eurasiatic plants, rigid in Europe, reaching the Kolyma valley to the North-East, but not south of the Central Asiatic desert belt.
18	6	Boreal Eurasiatic plants, rigid in Europe, reaching both Kolyma valley to the North-East and south of the Central Asiatic desert belt.
18	7	Boreal Eurasiatic plants, rigid in Europe, of more continental character than the preceding, not reaching south of the Central Asiatic desert belt.
20	8	Boreal Eurasiatic plants, rigid in Europe, of continental character, reaching south of the Central Asiatic desert belt.

Pl. 1.

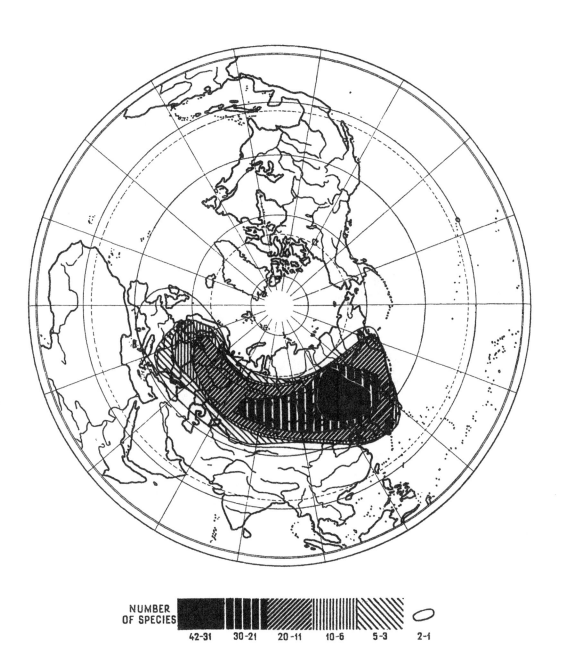

NUMBER
OF SPECIES

42-31   30-21   20-11   10-6   5-3   2-1

Pl. 2.

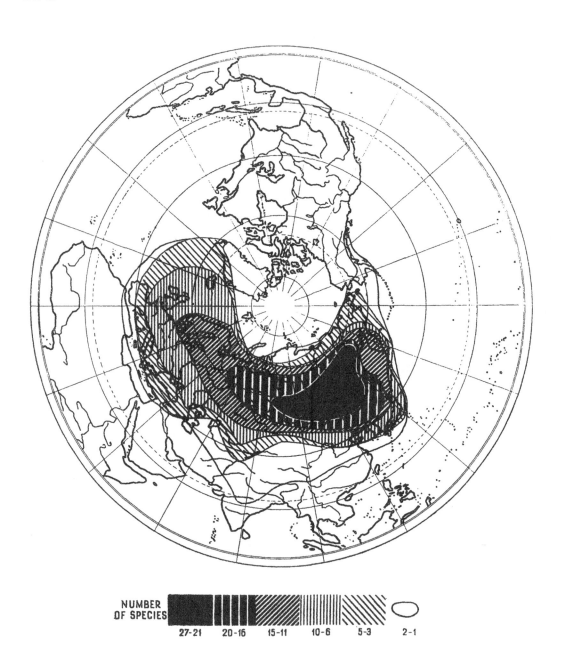

Pl. 3.

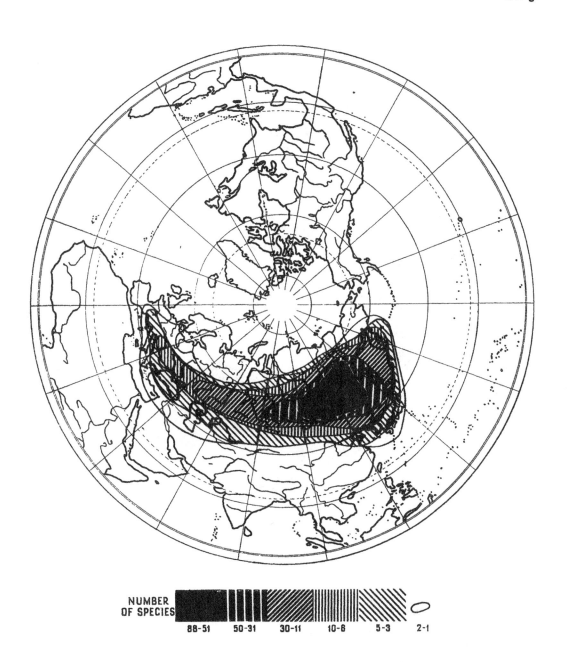

NUMBER
OF SPECIES

88-51   50-31   30-11   10-6   5-3   2-1

Pl. 4.

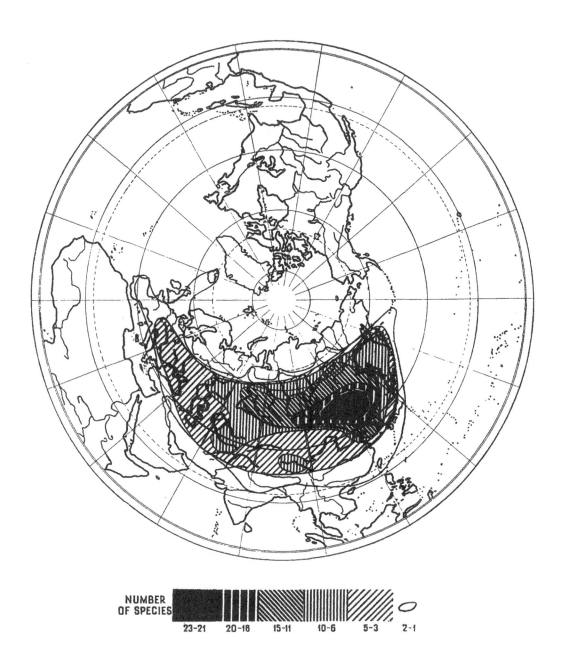

NUMBER
OF SPECIES

23-21   20-16   15-11   10-6   5-3   2-1

Pl. 5.

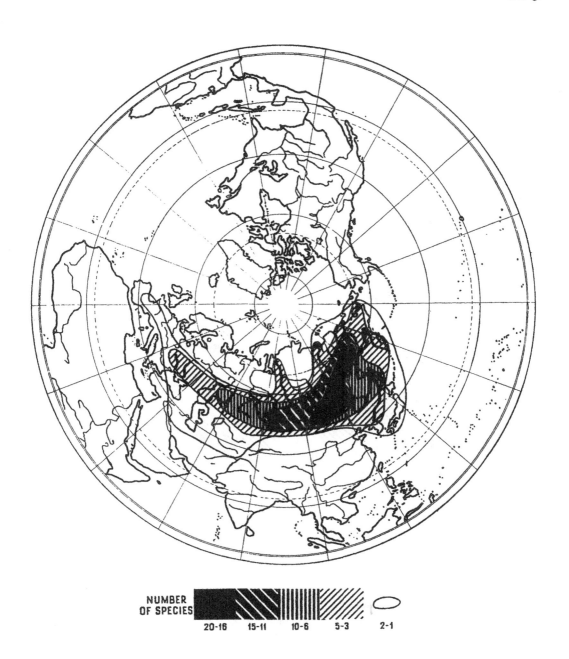

NUMBER
OF SPECIES
20-16   15-11   10-6   5-3   2-1

Pl. 6.

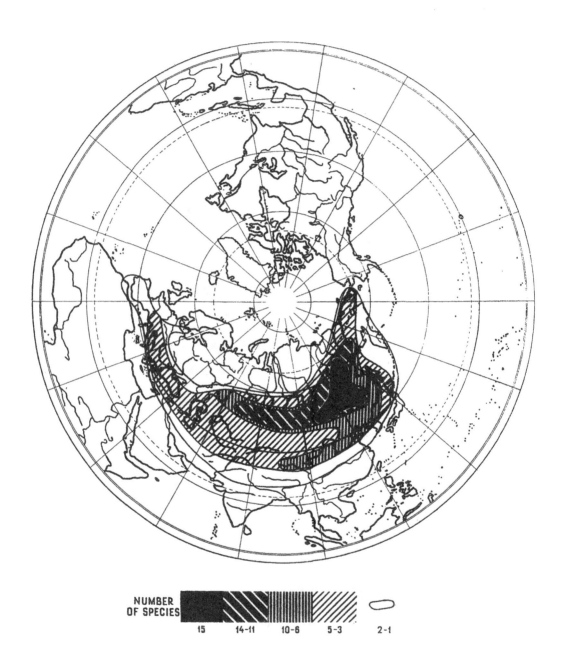

NUMBER OF SPECIES   15   14-11   10-6   5-3   2-1

Pl. 7.

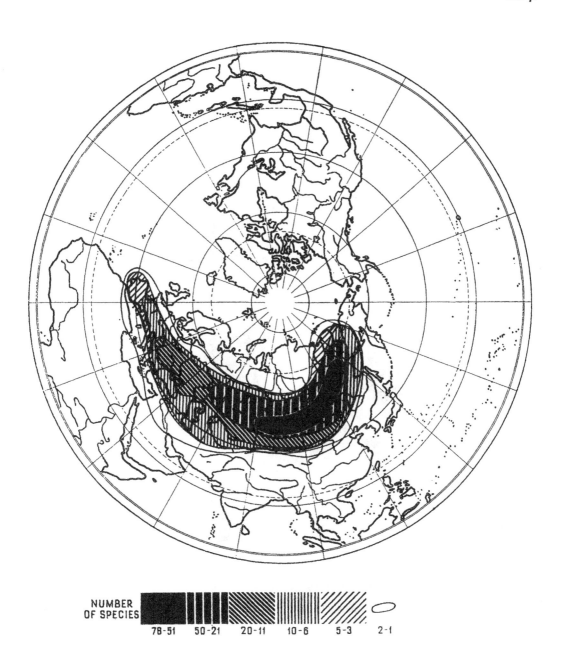

NUMBER
OF SPECIES

78-51   50-21   20-11   10-6   5-3   2-1

Pl. 8.

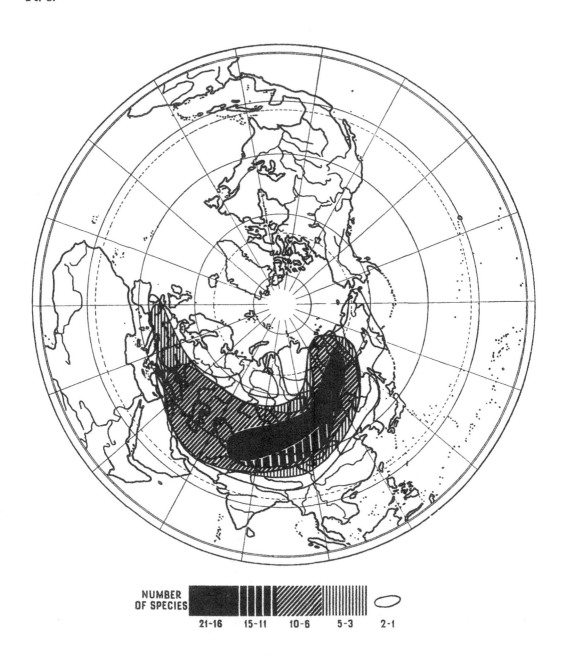

NUMBER
OF SPECIES

21-16   15-11   10-6   5-3   2-1

From *An Introduction to Historical Plant Geography*
Evgenii Vladimirovitch Wulff

Chapter III

AREAS, THEIR CENTERS AND BOUNDARIES

Area Concept Defined: — By area (*area geographica*) is understood, taking the Latin meaning of this word, the region of distribution of any taxonomic unit (species, genus, or family) of the plant (or animal) world. A distinction is made between *natural areas*, occupied by a plant as a result of its dispersal caused by the combined action of various natural factors, and *artificial areas*, arising as a consequence of the intentional or accidental introduction of a plant by man.

Within the limits of its area a plant does not occupy the entire surface of the earth but leaves smaller or larger intervening spaces unoccupied. This is due to the biological peculiarities of plants and to their adaptation to local habitat conditions, which even within the

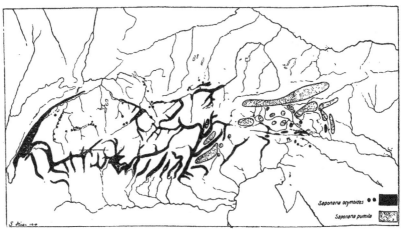

FIG. 1. — Example of a discontinuous area. Areas of *Saponaria ocymoides* and *S. pumila*, restricted to mountain-tops in the Alps. (After HEGI).

limits of a small territory may vary to a considerable extent. Among such local conditions are: physical and chemical properties of the soil and its humidity, micro-relief, micro-climate, geographical location with reference to the countries of the world, influence of animals and man, interrelations with other plants, etc. The character of the distribution of a plant within the limits of its area or, in other words, the local distribution of a plant, is known as its *topography* (DE CANDOLLE, 1855).

The area of distribution of a plant is best pictured by maps, on which all its known habitats are indicated by dots. Connecting by a line all the outer points of the distribution of a given plant, we are able to judge as to the *shape* of its area. The shape of an area depends on the combined effect of the biological peculiarities of the plant and

25

the physico-geographical conditions of the country, the latter usually playing the predominant rôle. The configuration of an area depends, to a large extent, on the latitude. In the frigid and temperate zones, as DE CANDOLLE pointed out, the diameter of most areas from west to east is much greater than that from north to south, due to the considerably greater variation in climatic, particularly temperature, conditions, in the case of the latter direction. Such areas, therefore, have the shape of an ellipse extending from west to east. The areas of species in the torrid zone have a relatively longer (as compared to the preceding case) diameter from north to south. Cases in which both diameters are of the same length, the area being roughly circular in shape, are of very rare occurrence.

The establishment of regularities in the formation of areas and a study of the areas themselves lead to the elucidation of their history and origin, a basic task of historical plant geography, since it enables us to arrive at conclusions as to the history and origin of floras. The carrying out of this task is by no means easy. Difficulties arise from a number of sources. The chief is our insufficient knowledge regarding the flora of many regions of the globe and, hence, regarding the present geographical distribution of species, a factor which taxonomists have begun only recently to take into account in a general way. Our knowledge of the former distribution of species, since the finding of well-preserved fossil remains is of exceptionally rare and chance occurrence, is even more meager, often practically negligible. And it happens, as we shall see below, that the structure of an area in many cases, particularly of a discontinuous area, may be explained only on the basis of its conformation in former times and not on the basis of natural causes now in force.

Another obstacle to an elucidation of the distribution of a species is our inadequate knowledge of the peculiarities of its ecology. Among such peculiarities we may mention: ability to grow only within certain restricted temperature limits—*stenothermy;* adaptability to specific habitat conditions—soil, humidity, light, cohabitation with other organisms (symbiosis, mycorrhiza, parasitism), presence of special insect pollinators—*stenotopy.* Lastly, in most cases we do not know whether the given area represents the limit of possible distribution of the species or whether the area is still in the process of expansion.

As an example of the close relation between plant distribution and definite edaphic conditions we may mention plants found only on serpentine soils (LÄMMERMAYR, 1926, 1927; NOVÁK, 1928). However, as early as 1865 NAEGELI pointed out that the character of the distribution of a plant cannot be explained only by the physical or chemical nature of the soil, since the latter factor acts in combination with climatic and biotic factors. When growing apart from one another, species may be indifferent to soil conditions, while when growing together, the same species, due to mutual competition, may show preference for definite and different soils. Thus, *Rhododendron hirsutum* and *R. ferrugineum,* when found apart, grow both on soils rich and on soils poor in lime, but when found together, the former is adapted to calcareous and the latter to non-calcareous soils. Hence, edaphic conditions themselves constitute in this case only an indirect cause of the "adaptation" of these species to different soils, such "adaptation" being determined by competition between the two species.

Center of an Area: — Of vital importance for the study and understanding of an area is the determination of that initial territory whence a genus or species began its dispersal whereby it reached the present boundaries of its area. This initial territory where an area originated is known as the center of the area. One of the first definitions of the concept of the center of an area was given by Robert Brown (1869),

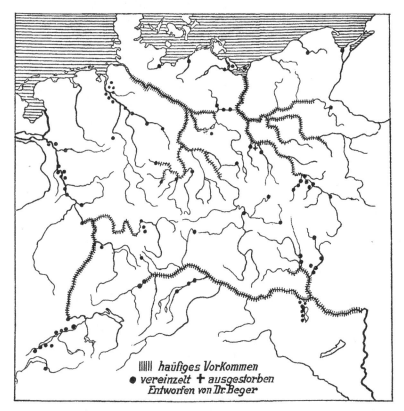

FIG. 2. — Example of a discontinuous area. Localities of *Euphorbia palustris*, restricted to the river basins of central Europe. (After Hegi).

who formulated it approximately as follows: Each genus seems to have arisen in that center in which the greater number of its species is found; these centers have doubtless undergone many modifications as a result of geological changes, and many anomalies in the distribution of plants may be thus explained.

There are no grounds for presuming that a new species will not extend its area beyond the limits of the region of its origin. It will, without any doubt, begin to spread in all directions open to it, and the region of its origin will constitute the center of the area being formed. Further on we shall discuss more in detail the origin of an area, and we shall then refer again to the problem of the initial center of an area. We shall now examine the methods of locating the center of an area.

The determination of the location of the center of an area is closely connected with the establishment of the habitats of the species or other taxonomic unit whose area is under study. Consequently, the historico-geographical study of an area should be based on a monographic study of the given taxonomic unit—species, genus, or family—and the elucidation of its kinship to closely related taxonomic units of the same or different rank.

In order to establish the center of formation of an area and the successive stages of its development, it is necessary to know its past history, which paleobotany alone is in a position to give. Unfortunately, its data are very incomplete and rarely enable one, on their basis, to establish the entire past history of an area. Nevertheless, there are very few cases when the fossil remains of any given taxonomic unit are found exclusively within the boundaries of the present area of that unit, *i.e.*, cases when we might consider the present area to be the place of origin and habitat of the given taxonomic unit during the entire period of its existence. Usually fossil remains are found outside the present area, sometimes embracing an area of distribution considerably larger than the present one and occupying entirely different regions. In such cases even very incomplete paleobotanic data give us indications as to the past area and guard us from falsely interpreting facts of the present distribution and from determining the center of the area only on the basis of contemporary data. Paleobotanic data have in many cases shown that the habitats of a taxonomic unit and, consequently, also the initial center of its area, may have been situated outside its present boundaries, a circumstance occurring both in the geography of plants and animals. Hence, in most cases only the center of the present-day distribution of a given unit may be found within its present area but not the center of its origin, *i.e.*, not the center of the area itself.

Theoretically we can distinguish between two kinds of centers of areas: the first, the region where there is accumulated the greatest number of habitats of the given taxonomic unit—the *center of frequency* (Frequenzcentrum—Samuelsson, 1910); the second, the region where there is concentrated the greatest diversity and wealth of forms—the *mass center* or *center of maximum variation* (Turrill, 1939). The latter center of an area, taking into account our present, insufficiently detailed knowledge of wild species, may be located, for the most part, only for units of the higher ranks. There is no doubt, however, that for the elucidation of the origin of an area it is of more importance to locate the center of maximum variation of the taxonomic unit whose area is under study than to locate its center of frequency, which depends more on ecological than on historical causes.

Both as regards the variety and frequency of stations and as regards the concentration of diverse forms, we may consider that a variety or species newly arising from an initial form will be found in greatest numbers not far from the place of its origin, its representatives gradually decreasing in number as one proceeds from this center of the area toward its periphery. At the time of its origin a species naturally finds itself in favorable conditions, since otherwise there would be no occasion for its arising there. If a species should accidentally arise in unsuitable conditions, it would be immediately destroyed as a result of natural selection. A new species is highly variable, reacting to all the micro- and macro-conditions of its habitat, and, hence, gives rise to a large number of forms.

PACHOSKY (1921) considers that the above-noted decrease in number of representatives of a species or variety toward the periphery of its area is closely connected with adaptability to definite habitat conditions, particularly to a definite type of soil. In the center of an area, where, as a rule, the habitat conditions of a given species most nearly approximate the optimum, it can grow under fairly diverse conditions, even on different soils. On the other hand, farther from this region, *i.e.*, nearer to the periphery of the area, not only is an optimum combination of factors of more rare occurrence but often there is lacking

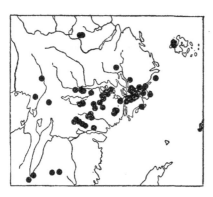 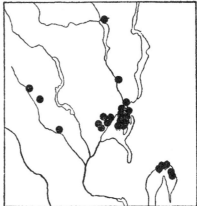

FIG. 3. — Centers of frequency in areas of species of the genus *Hieracium* on the Scandinavian peninsula (Swedish West Coast): *left, Hieracium meticeps; right, H. chloroleucum.* (After SAMUELSSON).

even that minimum of conditions required for the normal existence of a species. For example, the beech tree, ordinarily capable of growing on a variety of soils, on the periphery of its area is confined solely to lime soils. Relic species and species becoming extinct likewise prefer localities with lime or chalk soils for their habitats. Apparently, the physical conditions of the substrata of these soils provide for such species more favorable conditions as regards competition with other species, thus allowing them to maintain themselves, despite the fact that the habitat conditions as a whole deviate considerably from those normal for them.

As regards the effect of climatic factors on the distribution of a species within the limits of an area, GRAEBNER (1910) distinguishes between the "region of compact distribution", within which asp ecies finds itself in optimum conditions as regards climatic factors and the "absolute limit of distribution", where the stations of a species are confined to certain localities having specific habitat conditions.

From the foregoing it is clear that the farther from the center of an area the more rarely do conditions suitable for the growth of a given species occur, which results in the peripheral regions of an area being more sparsely inhabited by the species than the center. Moreover, plants growing under conditions unsuitable for them will quite naturally find themselves subjected to competition and crowding out by closely related species for which these same conditions are more suitable.

This, however, can by no means be taken as an unconditional and universal proposition. We can assume also the occurrence of such cases—and they actually do occur—where a species, spreading in the direction of the periphery of its area, encounters, often far from the place of its origin, favorable habitat conditions, perhaps even more favorable than existed in the center of the area, which give an impetus to new form-genesis. But such form-genesis, leading even to the origin of new species on the periphery of the area of the initial species, may also be due to unfavorable conditions. We shall discuss this in more detail later. We should, therefore, distinguish between the center of origin of an area and the center of its development, the latter in such cases being necessarily regarded as a *secondary mass center* (or centers, since there may be several of them) of the area.

Hence, when a species arises at any point of its future area there is created, first of all, a center of propagation of this species, the center of frequency of its area, and then there develops its differentiation into a number of forms of different taxonomic rank, the creation of a center of maximum variation, or a mass center of the area. In young species the latter center may not exist at all, or it may coincide with the center of frequency. Later on, such coincidence will be broken, since the primary mass center will remain in its original place, while the initial center of frequency may disintegrate and arise anew at one or another point in the migration of forms issuing from the primary center of formation of a species. A species during the course of its dispersal may, under especially favorable conditions, enter into a phase of new form-genesis, as a result of which there will arise a secondary mass center of the area, which in contrast to the primary center of its origin constitutes a center of the subsequent development of the area and will be characterized by the presence of younger forms. The secondary nature of such a center may be established by a combination of various methods of botanical study, by which it would be shown that in the center of origin there is a concentration of more ancient and primitive forms, as compared with those concentrated in the secondary center of development.

From the foregoing we may conclude that the region of frequency of stations may ordinarily be expected to coincide with the region of maximum variation, *i.e.*, with the center of origin of the area, in those cases where the distribution of the given taxonomic unit has not yet been subjected to any later influences inducing alterations in the character of the area.

As regards the area of a species, correspondence of the center of frequency with the center of the area is characteristic, as we noted above, for areas of *young species*. To illustrate this point we may present data obtained by SAMUELSSON (1910) from a study of the areas of several species endemic to the Scandinavian peninsula. His maps of the distribution of *Hieracium meticeps* and *H. chloroleucum* give a clear picture of the character of the areas of these species and the centers of greatest frequency of their stations. These centers of frequency are at the same time the centers of origin of the areas of these species. SAMUELSSON considers these species to be of comparatively recent origin, having arisen during the Ice Age after the end of the last (Mecklenburg) glaciation. Each arose—possibly by mutation—at a certain point, which became the center of its area, whence dispersal proceeded in various directions. The present boundaries of these areas, therefore, cannot be regarded as climatic boundaries. They merely mark in each case the limits of that territory which the given species has succeeded in occupying at this stage of its dispersal, a territory which in the future will continue to expand.

With respect to the terminology of the concepts of the center of an area, ARWIDSSON (1928) has proposed that those areas entirely included within the limits of one well-defined region be called *unicentric*, in contrast to *bicentric* (embracing two regions), *tricentric*, etc., or *polycentric* areas (CHRIST), if there are many such centers.

As an example of how the center of maximum variation of a genus may be located—without giving, however, any indication as to the primary or secondary nature of this center—we may take the data of SHIRJAEV (1932) on the area of distribution of the genus *Ononis*. The species and subspecies of this genus are distributed throughout the world as follows:

Country	No. of species and subspecies	Country	No. of species and subspecies
Morocco	52	Austria	6
Algeria	44	Istria	6
Spain	44	Rhodes	6
Italy	24	Carpathians	6
Portugal	20	Albania	5
Syria and Palestine	19	Egypt	5
Tunis	18	Central Europe	4
Sicily	18	Hungary and Rumania	4
Asia Minor	17	Bulgaria	4
France	16	Arabia	4
Sardinia	14	Armenia and Transcaucasia	3
Cyprus	13	Crimea	3
Tripolitania and Libya	13	Madeira	3
Canary Islands	11	Greece	2
Mesopotamia and Kurdistan	10	England	2
Crete	10	Turkestan	2
Corsica	10	Afghanistan	2
Jugoslavia	10	Caucasus (excl. Transcaucasia)	1
Islands of Aegean Sea	9	So. European U.S.S.R.	1
Dalmatia	9	Southern Siberia	1
Iran	7	Mongolia	1
Balearic Islands	7	Northwestern India	1
Tyrol	7	Eritrea	1
Thrace	7	Abyssinia	1
Switzerland	6		

From a study of these data SHIRJAEV draws the conclusion that the center of origin and development of the area of the genus *Ononis* comprised the Iberian Peninsula (Spain, 44 species; Portugal, 20 species), Morocco (52 species), and Algeria (44 species), which at one time formed a united region. This conclusion has been visualized by a map drawn by SZYMKIEWICZ (1933) on the basis of these same data.*

Another example is provided by a table given by SZYMKIEWICZ himself, where the number of species are given for regions located at approximately the same latitude from the Iberian peninsula to Japan. For each genus the figures in bold-faced type indicate the region or regions in which the genus is represented by the greatest diversity of species.

GENUS	Iberian Penin-sula (so. part)	Italy	Greece	Asia Minor	Armenia & Caucasus	Iran	Soviet Central Asia	Altais	Far Eastern Region	Japan
Armeria	**37**	16	3	2	0	0	0	0	1	0
Genista	**47**	34	13	12	5	0	1	0	0	0
Helianthemum	**27**	16	10	11	7	5	1	0	0	0
Trifolium	54	**98**	64	61	45	15	14	7	2	1
Lotus	20	**24**	17	14	10	6	4	3	1	1
Coronilla	9	**11**	8	6	5	2	1	0	0	0
Silene	58	65	**86**	73	65	41	49	14	10	10
Alyssum	13	16	20	**40**	27	14	11	3	0	0
Gypsophila	3	3	7	**24**	23	16	19	7	3	0
Onobrychis	10	8	7	21	**27**	22	13	1	0	0
Astragalus	43	28	37	146	253	317	**328**	55	6	6
Ferula	5	4	3	4	9	14	**35**	3	0	0
Artemisia	20	17	5	13	20	23	**68**	30	30	17
Saussurea	0	0	0	0	1	2	**41**	23	24	19

Similarly, we may take the distribution of wild species of *Nicotiana*, a genus including, according to data of a study of the geography of this genus made by GRABOVETSKAYA (1937), a total of 76 species. They are distributed as follows: North America, 12 (of which 7 are endemic); Central America, 14 (7 endemic); South America, 43 (39 endemic); Australia, 14 (all endemic). On the basis of these data the center of the area of this genus must be regarded as South America.

This method of determining the location of the centers of origin of genera, and on the basis of the latter, the centers of development of floras, is at the present time generally accepted, but it should, nevertheless, be emphasized that this method is only relatively reliable, in many cases leading undoubtedly to incorrect conclusions. Most genera of angiosperms originated in the Cretaceous period, some probably even earlier, in the Jurassic. Having attained at the end of the Cretaceous and beginning of the Tertiary periods a very wide distribution, most of them had at that time a considerably more limited intrageneric differentiation than at present. Intensive processes of species-formation and the initiation of geographical series of species (see below) were not begun until later, in the second half of the Tertiary period. These were induced by climatic changes, particularly decrease in humidity, and geomorphological changes—mountain-forming processes (*e.g.*, the

---

* Discrepancies between the figures on the map and those in SHIRJAEV's table arise from the fact that for the map SZYMKIEWICZ used only the number of species, whereas SHIRJAEV included subspecies as well.

uplifting of the Alps and the Himalayas, which radically altered the climatic conditions of the regions lying north of these mountain ranges), shifting of sea basins (formation of deserts where formerly there were seas and the formation of seas where formerly there was dry land), separation from the continents of archipelagos and islands, which formerly constituted a united whole, etc.

Consequently, the present-day concentration of species only in rare instances can reflect the actual center of origin of the genus; usually it indicates the center not of the past but of the present development of the genus. In view of this, conclusions made on the basis of data on the migration of a genus from such a center of development are founded on incorrect premises and in many cases are utilized for broad generalizations, which cannot be accepted without reservations, except upon further verification.

This explains why newer methods of determining the center of origin of a genus are being sought, but it may be taken for granted that these methods will be able to give full assurance as to the reliability of the results obtained only in case they are confirmed by paleobotanic data. In the absence of the latter, conclusions drawn solely on the basis of the present distribution of species will evoke doubts as to their validity. SZYMKIEWICZ (1934, 1936, 1937) has in recent years made intensive studies with the aim of finding new methods of locating the centers of areas and of tracing the development of floras. The methods proposed by him may be summarized as follows:

If we take as the center of origin of a genus that region where the greatest number of its species are concentrated, we do not take into account differences in the character of the areas of the various species, as a result of which we compare figures that are phytogeographically of unequal value. SZYMKIEWICZ divides species, as regards the character of their areas, into three categories: (1) endemic and subendemic, the latter meaning species whose areas extend only slightly beyond the boundaries of their primary natural regions; (2) species whose areas embrace, in addition, a second natural region phytogeographically identical to the first; (3) widely distributed species, in whose areas the primary natural region occupies only an inconsiderable part. These three categories of species, in judging as to the center and origin of the area of a genus, provide data of unequal value, the first being of greater significance than the second and the second greater than the third. SZYMKIEWICZ (1937) proposes, therefore, that the center of the area of a genus should be established not on the basis merely of data as to the total number of species but of data as to the number of species in each of the three above-mentioned categories, and he points out that by the latter method it is easier to detect a second center of concentration of species, in case there are two such centers. By way of illustration are given below the data obtained by SZYMKIEWICZ for the genus *Carex*:

E. V. Wulff   —34—   Historical Plant Geography

Carex

REGIONS OF DISTRIBUTION	NO. OF SPECIES IN EACH CATEGORY	TOTAL NO. OF SPECIES
Europe	27–101–14	142
Siberia	8– 70–16	94
Mediterranean Basin	31– 67–22	120
Eastern Asia	259– 70–18	347
No. America, Pacific Coast .	61– 77– 5	143
– – , Atlantic –	86– 75– 6	167
Mexico	9– 11– 3	23
Andes	41– 16– 3	60
Neotropical region	12– 5– 2	19
Tropical Africa	28– 6– 0	34
Malaysia	51– 17– 8	76
South Africa	5– 9– 0	14
Australia	40– 11– 1	52

This table shows that, taking the greatest concentration of species as a basis, Eastern Asia would appear to be the chief center of origin of the genus *Carex* and the Atlantic Coast of North America the next most important center. Judging by the concentration of each of the three categories of species, the conclusion that these are the two most important centers is confirmed, for they contain the greatest number of endemic species.

But in thus locating the center of an area it is necessary also to take into account the fact, well known to every author of a botanical monograph, that species themselves are not uniform, not of equal value. This may usually be compensated for by the grouping together of closely related species into sections, subgenera, and other such units. Consequently, for the purpose of checking the conclusions as to the center of the area of a genus arrived at on the basis of a calculation of the number of species in the three categories, SZYMKIEWICZ proposes that analogous calculations be made of the number of sections or subgenera in each of the same three categories. Thus, one first calculates the number of species of each of the three categories in each section, and then, on this basis, determines the number of endemic sections, sections distributed in two regions, and sections having a wide distribution. As an example let us take the data obtained for the genus *Sisymbrium:* —

Chapter III   —35—   On Areas

Sisymbrium

REGIONS OF DISTRIBUTION	No. OF SECTIONS IN EACH CATEGORY	TOTAL NO. OF SECTIONS	No. OF SPECIES IN EACH CATEGORY	TOTAL NO. OF SPECIES
Europe	0–1–4	5	5–5–0	10
Siberia	0–0–2	2	0–3–0	3
Mediterranean Basin . . .	5–2–1	8	15–6–1	22
Eastern Asia.	0–0–2	2	1–1–0	2
No. America, Pacific Coast	0–0–2	2	2–0–0	2
Mexico	0–0–2	2	3–0–0	3
Andes	5–1–0	6	25–0–0	25
Neotropical region	0–0–1	1	1–0–0	1
Tropical Africa	0–1–1	2	2–0–0	2
South Africa	2–0–0	2	10–0–0	10

In this case by all three methods we get the same result: the existence of two centers, one in the Mediterranean Basin and one in the Andes. This connection between the Mediterranean Basin and Central America and adjoining territories of North and South America is characteristic of a number of genera, such as *Draba, Eryngium, Centaurea, Astragalus, Trifolium, Lupinus,* etc. But such an exact agreement of results from all three methods of investigation does not, by any means, always hold true. As an illustration of this, we may take the distribution of sections in the genus *Carex,* for each of its subgenera separately, and compare the data thus obtained by SZYMKIEWICZ with those given above for species.

Carex

REGIONS OF DISTRIBUTION	Subgenus *Primocarex* No. OF SECTIONS IN EACH CATEGORY	Subgenus *Vignea* No. OF SECTIONS IN EACH CATEGORY	Subgenus *Indo-Carex* No. OF SECTIONS IN EACH CATEGORY	Subgenus *Eu-Carex* No. OF SECTIONS IN EACH CATEGORY
Europe	0–3–4	2–2–11	0–0–0	2–1–18
Siberia	0–1–5	0–0–11	0–0–0	0–1–15
Mediterranean Basin . .	0–1–4	1–1–11	0–1–0	0–0–20
Eastern Asia	2–1–0	2–2–10	2–0–3	8–4–10
North America — Pacific Coast	3–3–3	2–4– 8	0–0–0	1–3–16
North America — Atlantic Coast . . .	2–3–4	2–4– 6	0–0–0	7–5–10
Mexico	0–0–0	0–0– 3	0–1–1	0–1– 8
Andes	1–0–2	2–0– 6	0–0–2	1–1–11
Neotropical region	1–0–0	1–0– 1	0–0–2	0–0– 4
Tropical Africa	1–0–0	0–1– 0	0–0–1	1–0– 4
Malaysia	0–0–2	0–0– 5	3–0–0	0–0–10
South Africa	0–0–0	0–0– 2	0–0–1	0–0– 6
Australia	0–0–1	2–1– 4	0–0–0	1–1– 7

In this table there does not stand out any definite center of the area of the genus, the East-Asiatic center so prominent in the other table for the genus *Carex* not being in evidence here at all. We thus see how complicated is the problem of determining the location of the center of an area, and that for its solution statistical calculations alone can have only a very limited significance. It is necessary to assemble data of an all-sided study of species—as, for instance, was done by Ko-marov (1908)—disclosing the phylogenetically most primitive types, the direction of their evolution, the centers of concentration of these primitive, initial types, and the direction of their further distribution. Only such a monographic study, based also on paleobotanic data, can give a more or less correct idea as to the initial center of the area of a genus and of the secondary centers of its development.

In this respect cytological data may prove of great value. It has now been established that in some cases species belonging to the same genus differ in chromosome number, and that many of the polyploid species originated, apparently, as a result of chromosome mutations induced by the action of external factors. Species, in dispersing from the center of their origin, often extend their area beyond the boundaries of optimum conditions for their existence. As a result of the action of ecological factors to which a species is not accustomed there occur irregularities at meiosis in the sex cells, which result in the phenomenon of polyploidy. This connection between the origin of polyploid species and definite ecological conditions is the reason why such species have in many cases quite specific geographical areas, differing from the areas of the initial species (see below; also, in more detail, Wulff, 1937).

It is now possible, therefore, theoretically to advance the proposition that floras of those regions of the globe characterized by extremely low temperature, such as arctic regions and mountain peaks, or by very high temperature and low humidity, such as deserts, are distinguished by an exceptionally large number of polyploid species. It must also be presumed that such chromosome mutations have occurred in nature not only under present-day conditions, as a result of species having become widely distributed and having penetrated into localities with ecological conditions differing from those normal for them, but also as a result of the great climatic changes that took place in former geological times and of the migrations of species in those times.

Arranging the species of a genus in order according to chromosome number, we obtain so-called polyploid series of species, which at the same time reflect the direction of evolution of the genus and also the direction of its dispersal. Starting from the premise that the species having the smallest chromosome number in a polyploid series usually is the initial species, we may consider that the areas of species charac-terized by such chromosome numbers are more ancient than areas of species with larger chromosome numbers and that, consequently, in the regions occupied by these ancient areas one must seek for the initial center of the area of a genus. As an illustration we may cite the data of a cytological investigation of the genus *Iris* carried out by Simonet (1932). This genus is widely distributed throughout the entire north-ern hemisphere. Species having rhizomes occupy the largest areas, practically identical to that of the genus, while tuber-bearing species are considerably more restricted in their distribution. The area of the latter is confined to the Mediterranean Basin—from the Iberian Penin-sula to Soviet Central Asia, inclusive. Not only all four sections of tuber-bearing irises but also three of the sections of rhizome-bearing irises are limited in their distribution to the Mediterranean Basin. Moreover, almost all the other sections have representatives here. These circumstances force one to presume that the center of origin of the area of the genus *Iris* must be located in the Mediterranean Basin. This conclusion is confirmed by cytological data. Precisely in the Mediterranean Basin are concentrated those species with the lowest chromosome numbers ($n = 8, 9, 10, 11$), whereas American species with the highest chromosome numbers have areas located at the greatest distance from this center. Moreover, the tuber-bearing species of irises, being the most ancient species and having an area confined to the Mediterranean Basin, have the lowest chromosome numbers. Hence, if, in determining the location of the centers of areas, we utilize cytological data for those genera the species composing which have different chromosome numbers, we acquire an additional method facilitating, in combination with other methods, the solution of this difficult problem.

If, after a species has died out over a considerable portion of its area, favorable conditions should reoccur, the species may renew its dispersal from those retreats where it preserved its habitats. For example, many species lived through the Ice Age in restricted localities, which served as retreats for them and whence, in inter- and post-glacial periods, they renewed their dispersal. In such cases these retreats are known as *centers of dispersal* (centres de dispersion—JEANNEL and JOLEAUD, 1924) or *centers (regions) of preservation* (Erhaltungsgebiete—IRMSCHER, 1929).

In determining the location of the center of an area great caution must be observed, since if any factors whatsoever favoring or hindering the distribution of species are not taken into account, entirely incorrect conclusions may be drawn. For instance, PALMGREN (1927), on the basis of the character of the distribution of species on the Åland Islands, draws conclusions as to the extent of their penetration into the territory of these islands and the general direction of their migration. He considers that, in case a species is distributed *uniformly* within the limits of a given territory, we cannot obtain any facts as regards its former migrations from its present distribution. By uniform distribution he means the approximately equal frequency of occurrence of a species in *all* parts of the given territory and, consequently, the absence of any perceptible concentration of stations in any one part. If, on the other hand, the frequency of occurrence of a species grows clearly less or greater in some definite direction, this indicates, in his opinion, the direction of migration of the species. Thus, by his investigation of the Åland Islands PALMGREN found that, in addition to uniformly distributed species, there are three other categories of species, which, in contradistinction to the former, shed some light on their origin on the islands. The first category embraces species with a clear decrease in the frequency of their occurrence toward the east, which gives grounds for considering that they migrated from the west. This group of species is the largest in point of numbers in the flora of the Åland Islands. To the second category belong a few species, the frequency of occurrence of which decreases toward the west, and which, consequently, migrated presumably from the east. Lastly, the third category embraces species distributed within the limits of the islands in two isolated areas—western and eastern. From these data PALMGREN draws the conclusion that the first category of species, constituting the great bulk of the flora of the islands, migrated to the islands from the Scandinavian peninsula, from Sweden; the second category from Finland or the eastern section of the Baltic seacoast; and, lastly, the third category from both directions.

But, in opposition to the foregoing, EKLUND (1931) shows that in southwestern Finland there is found a very great diversity of habitat conditions, particularly of edaphic conditions. This diversity is very clearly reflected in the distribution of plants. In the western part of this region there are the best soil conditions, shown, first of all, in the fairly high content of lime in the soil. Here the flora is richest. From this locality in all directions the soil grows poorer, accompanied by an impoverishment of the flora. Hence, the decrease in the frequency of occurrence of species from west to east is to be explained not by the greater distance from the place from which they migrated, as PALM-GREN assumed, but by ecological causes, expressed in this case by the indicated differences in soil conditions. EKLUND remarks that in the Åland Islands there may be observed a decrease in the frequency of occurrence of species and an impoverishment of the flora from west to east, while in Uppland there is just the reverse—an impoverishment from east to west. In both cases this impoverishment is to be explained by one and the same cause, by a decrease in the content of lime in the soil. Consequently, PALMGREN's conclusions with respect to the direction of migration and the chief country from which the Åland Islands derived their flora, based on the decrease in frequency of occurrence of species from west to east, are in the given case incorrect, since he did not take ecological conditions into account. EKLUND comes to the conclusion that the islands were populated with species from an entirely different direction than PALMGREN supposed.

The present areas of many species do not constitute the maximum territory that they may possibly be capable of occupying. The further expansion of these areas has been curtailed by obstacles that have up to the present prevented these species from continuing to spread. By the artificial introduction of plants into new habitats outside their natural area it is frequently found that a species may grow under a considerably wider range of ecological and geographical conditions. This shows that each species has, besides its actual area, a "potential area" (GOOD, 1931). This circumstance is of exceptional practical importance in the introduction and regional allocation of new crops.

Boundaries of an area: — The limits of distribution of a species, the *boundaries of its area*, formerly very ineptly termed "vegetation lines", are determined by the reaction of the species to any of numerous factors or combinations of these factors. Among the most obvious causes hindering the dispersal of a species are purely mechanical obstacles, such as mountains, seas, deserts, etc. Only in rare cases does a plant, by the mere dissemination of its seeds and their transport by chance agents, succeed in overcoming such obstacles and extending its area beyond them.

Among other factors limiting the extent of an area—and for the plant usually insurmountable—are climatic conditions. The latter, creating the *climatic boundaries* of an area, may limit the distribution of a species both horizontally (to the north, south, east, and west) and vertically (altitudinally). Climatic boundaries are not determined by any one climatic factor but by all of them taken together, in consequence of which a study of the climatic boundaries of an area and the elucidation of the rôle of individual climatic factors meet with very great difficulties. The latter are all the greater because the reaction of species to climatic phenomena is closely linked with their biological characteristics, as a result of which their climatic boundaries are characterized by extraordinary diversity. Nevertheless, a study of the areas of plants provides a basis for determining the most important climatic factors affecting their distribution.

Altitudinal climatic boundaries are the result of a particularly complex combination of causes, often very difficult to fathom, the most important of all being insufficient warmth (inadequate sum of temperatures above the minimum temperature required for the given species). In addition to the latter, insufficient humidity, intensity of the sun's rays at high altitudes, strong heat radiation, eternal snow or late melting of snow, height above sea level depending on the latitude of the locality, and other factors also play a part in determining the altitudinal climatic boundaries of an area.

Until the end of the nineteenth century evaluations of climate and also of the climatic boundaries of the areas of plants were made chiefly on the basis of temperature data. At the beginning of the twentieth century it became clear that atmospheric humidity, as a factor determining the boundaries of the distribution of plants, was of predominant importance. In many cases, as, for example, for most evergreen plants, temperature plays only an indirect rôle in the limitation of the distribution of a species, the chief factor being humidity conditions. Hence, in determining the boundaries of areas, both these factors should be taken into account (GAMS, 1931).

The boundaries of an area may be determined not only by climatic causes but also by edaphic causes or by a combination of edaphic, climatic, and geographical causes. Lastly, competition with other plants may create an insurmountable obstacle to the further distribution of a species.

In many cases the boundaries of areas cannot be explained by any cause at present in force, due to the fact that these areas were formed under the influence of conditions in past epochs, often in other geological periods. A study of such areas and the establishment, on the basis of such a study, of the history of distribution of a given species constitute one of the chief tasks of historical plant geography.

The boundaries of areas may, then, be subdivided into three main types: first, boundaries set by physical barriers impassable for the given species, such as seas, straits, rivers, mountains, deserts, etc.; second, boundaries determined by ecological conditions; and, third, boundaries determined by competition among species. Moreover, an area may be in a state of expansion, in case the dispersal of the species is still in progress, or, on the other hand, it may be in a state of contraction, in case of retrogressive distribution. In the latter case the contraction of an area may for a time be in abeyance and the boundaries of the area remain temporarily without change, but subsequently they may either continue to contract or, in case of the onset of more favorable conditions, begin again to expand.

Areas vary greatly in size, depending on a combination of factors, among which the history of the given species plays an important rôle. If we assume that an area has a center of origin, from which there took place the gradual dispersal of a species or other taxonomic unit in different directions, it seems necessary likewise to assume that the size of the area occupied, in case of unhindered dispersal, would depend in part on the duration of such dispersal, which may be designated as the "age" of the species. Thus, SCHULZ (1894) considers that only very few species of the flora of central Europe have succeeded in attaining in post-glacial times, and these only in a few places, their natural boundaries as set by their edaphic and climatic requirements and by their ability to spread.

In botanico-geographical literature age as a factor in plant distribution has long been recognized. As early as 1853 LYELL in his "Principles of Geology", in chapters on the distribution of plants and animals, wrote that, if we assume that a species arises only in one place, it must have considerable time to become distributed over an extensive area. If this hypothesis is accepted, it follows that restricted distribution may, in the case of some species, be due to their recent origin and, in the case of others, to the fact that the area they once occupied has been greatly contracted as a result of climatic changes. The former are young, local species that have not existed long enough to have had the possibility for widespread dispersal, while the latter are no doubt of considerable age.

HOOKER, in his "Flora Novae Zelandiae" (1853), writes that "consistently with the theory of the antiquity of the alpine flora of New Zealand, we should find amongst the plants common to New Zealand and the Antarctic Islands some of the most cosmopolitan, and we do so". But, at the same time, HOOKER, fully conceding that all the diversity in the geographical distribution of plants cannot be explained by age alone, goes on to say that ". . . though we may safely pronounce most species of ubiquitous plants to have outlived many geological changes, we may not reverse the position, and assume local species to be among the most recently created, for species, like individuals, die out in the course of time; whether following some inscrutable law whose operations we have not yet traced, or whether . . . they are destroyed by natural causes (geological or other) they must in either case become scarce and local while they are in process of disappearance" (p. xxv).

An equally clear exposition of the significance of age as a botanico-geographical factor may be found in BENTHAM's "Notes on the Classification, History, and Geographical Distribution of the Compositae" (1873).

SCHRÖTER (1913) points out that the degree of disruption may also be utilized in determining the age of an area, an extensive and much disrupted area indicating its considerable age. POHLE (1925) proposes that ancient species be called "senior species" and young species—"junior species".

These citations fully suffice to show that the significance of age was never lost sight of by botanical geographers, although, no doubt, as compared with other biological factors, it was given too little attention and its importance underrated.

Age—or the length of time during which the dispersal of species and, hence, also the formation of floras have taken place—is for historical plant geography a factor of just such prime importance as the duration of geological periods established by LYELL was for the theory of evolution. Just as a necessary premise for the evolution of organisms is the duration of time taken by the latter for their development, so all the regularities in historical biogeography may be understood only by taking into account the length of time during which they have existed.

Consequently, we cannot fail to give serious consideration to WILLIS's treatise on "Age and Area" (1922), devoted to a study of it as a botanico-geographical factor and constituting a summary of his many investigations on this problem begun in 1907. The works of WILLIS are not mere armchair theorizing but are based on twenty years of field work devoted to the geographical study of plants in nature, in tropical South America and particularly in tropical Asia, where for a long time he was director of the Botanical Garden at Ceylon. He verified his conclusions by comparison with the data published by him in his "Flora of Ceylon" and other floristic works. At the same time, however, the very fact that he used the floras of these tropical regions as the basis for his conclusions constitutes the cause of the onesidedness of his conclusions, on account of which numerous criticisms were directed against him. The floras of the tropical regions of America and Asia are the only floras on the globe that since the Cretaceous period have not been subjected to great climatic changes. Consequently, laws established with respect to the formation of areas of species of these floras are applicable only to such floras as are characterized by unhampered development. They are not of universal significance and cannot be applied to all the floras of the earth. The criticisms of WILLIS's book were directed chiefly on this flaw in the propositions advanced by him.

Studying the flora of Ceylon, WILLIS was struck by the great differences in the size of areas occupied by different species of the same genus, some of which were endemic to the island and others not. This led him to the conclusion that "the endemic species occupied, on the *average*, the smallest areas in the island, those found also in Peninsular India (but not beyond) areas rather larger, and those that ranged beyond the peninsula the largest areas of all (again on the average)" (WILLIS, 1922, p. 65). At the same time, the number of species in each class was found to vary, increasing or diminishing depending on the size of the area. This may be clearly illustrated by his data on the flora of New Zealand. Taking the extent of the areas of species in this flora along the north and south diameter of the island, the following gradations are obtained (*ibid.*, p. 64): —

	Range in N. Z. (miles)	Endemics	Wides
1.	881–1080	112	201
2.	641– 880	120	77
3.	401– 640	184	53
4.	161– 400	190	38
5.	1– 160	296	30*

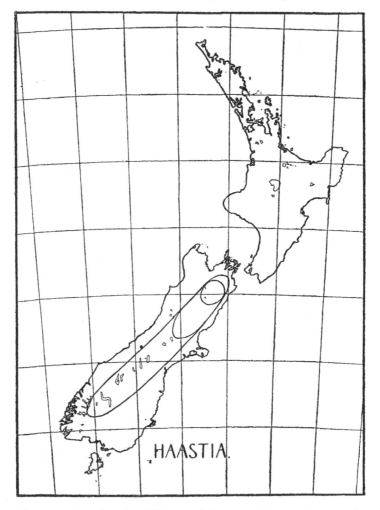

FIG. 4. — Areas of species of different ages, the most ancient species occupy the largest areas: species of the genus *Haastia* (*Compositae*) in New Zealand. (After WILLIS).

The widely held view that endemic species are either relics approaching extinction or species that have arisen as a result of adaptation to local conditions cannot explain the fact of gradual gradation in the areas occupied both by endemics and by wides, the first from many

* Largely undoubted introductions of recent years.

small areas to few large and the second in the reverse order.   To
explain this regularity it is necessary to concede the significance of age
as a factor in distribution.   The older species with extensive areas of
distribution reached New Zealand prior to its separation from Australia
and had enough time to become widely distributed there.   Hence, it is
clear why in the zone of areas of least extent we find the smallest

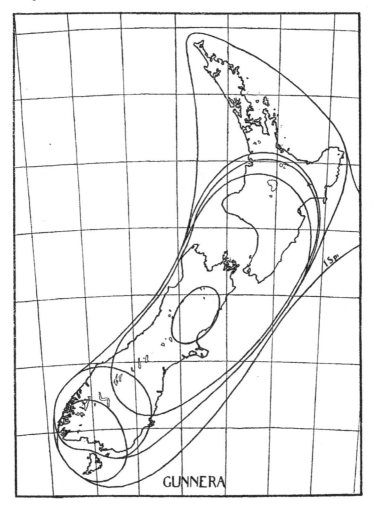

Fig. 5. — Areas of species of different ages: the genus *Gunnera* (*Halorrha-
gaceae*) in New Zealand. (After Willis).

number of widely distributed species.   On the other hand, endemic
species—regarded by Willis as young species that had their origin
at a later time, after New Zealand had become an island—become more
and more rare, the greater the distance from the place of their origin.
Consequently, it was to be expected, and investigations confirmed this,
that the islands surrounding New Zealand would have a flora consist-

ing of the oldest species, those most widely distributed in New Zealand. To quote WILLIS (1922, p. 75): "In fact it was found that on the average its species ranged nearly 300 miles more in New Zealand than did those that did not reach the islands".

As regards the way in which a country is peopled by invasions of plants, WILLIS gives the following rule based on his general hypothesis (*ibid.*, p. 83): "If a species enter the country and give rise casually to new (endemic) species, then, if the country be divided into equal zones, it will generally occur that the endemic species occupy the zones in numbers increasing from the outer margins to some point near the centre at which the parent entered". Applying this to New Zealand, he found that all the genera in its flora adhered to this rule.

Supplementary to his central "Age and Area" hypothesis, WILLIS proposes a second principle called by him "Size and Space", which he formulates as follows (*ibid.*, p. 118): "If species spread in a country mainly in accordance with their age, then it is clear that on the average some of those in the genera represented by most species will have arrived before the first of those in the genera represented by few; . . . on the whole, keeping to the same circle of affinity, a group of large genera will occupy more space than a group of small. The space occupied will vary more or less with the number of species".

From the foregoing follows also the final implication of WILLIS's theory, *viz.*, that monotypic genera, that is, genera with one species only, like endemic genera and species, are "young beginners" that have just commenced their geographical distribution. Here, however, one must make the reservation (which WILLIS himself fails to make) that this conclusion is not applicable to genera that have acquired their monotypic character as a result of the dying out of most of their species nor to ancient endemic genera and species.

All of WILLIS's views may be summarized in this basic hypothesis— the area of a species is proportional to its age. If this proposition could be universally applied, it would simplify the solution of many problems of botanical geography. But, as his critics pointed out, this proposition is applicable only to certain genera and species, and so it cannot serve as a general rule for determining the age of an area. For instance, paleobotanic data show that some genera now occupying small areas were widely distributed in the past and are often older than genera now having extensive areas. The same applies to those endemic species that are the descendants of species at one time widely distributed but whose areas were much contracted (BERRY, 1917). In his later papers and in his book WILLIS, in answer to the deluge of criticisms, reformulated his Age and Area hypothesis, qualifying it by so many reservations that it became very complicated and practically unworkable. In its latest version it read: "The area occupied at any given time, in any given country, by any group of allied species at least ten in number, depends chiefly, so long as conditions remain reasonably constant, upon the ages of the species of that group in that country, but may be enormously modified by the presence of barriers such as seas, rivers, mountains, changes of climate from one region to the next, or other ecological boundaries, and the like, also by the action of man, and by other causes" (WILLIS, 1922, p. 63).

The areas of species of a number of floras (of England—GUPPY. 1925; MATTHEWS, 1922; of North America—FERNALD, 1924; of South Africa—SCHONLAND, 1924; and others) have been studied with the aim of testing WILLIS's theories, and the investigators came to the conclusion that these theories were not applicable to the cases studied by them. The size of the areas studied depended not so much on their age as on the adaptability of the given species and on whether or not ecological conditions favored dispersal. Species that had migrated at a later period often had larger areas than older species, parts of the areas of which had been destroyed during the Ice Age. The region of greatest concentration of endemics did not coincide with the place of origin of the genera within the limits of the given flora. Moreover, it was shown that WILLIS entirely ignored those changes in the composition of floras induced by man's activities (RIDLEY, 1923). Likewise studies of various families and genera (*e.g.*, *Magnoliaceae*—GOOD, 1925; *Passerina*—THODAY, 1925) also revealed a number of data disagreeing with WILLIS's theory. On the other hand, several investigators have presented data that agree with the regularities established by him.

All this indicates that the size of an area does not depend solely on the age of a species. The latter constitutes only one of a combination of factors on which area-formation depends. Nevertheless, the study of age as a factor in plant distribution, the significance of which was first emphasized by WILLIS, should be continued.

References:

ARBER, AGNES, 1919: On the law of age and area in relation to the extinction of species (Ann. Bot., Vol. 33, pp. 211–213).

ARWIDSSON, TH., 1928: Bizentrische Arten in Skandinavien—eine terminologische Erörterung (Bot. Notiser, No. 1).

BENTHAM, G., 1873: Note on the classification, history and geographical distribution of the *Compositae* (Linn. Soc. Jour., Vol. 13).

BERRY, E. W., 1917: A note on the age and area hypothesis (Science, Vol. 46, pp. 539–40).

BERRY, E. W., 1924: Age and area as viewed by the paleontologist (Amer. Jour. Bot., Vol. 11, No. 9).

BROCKMAN-JEROSCH, H., 1913: Der Einfluss des "Klimacharakters" auf die Grenzen der Pflanzenareale (Vierteljahrsschr. Naturf. Ges. Zürich, Vol. 58).

BROCKMAN-JEROSCH, H., 1913: Der Einfluss des "Klimacharakters" auf die Verbreitung der Pflanzengesellschaften (Engler's Bot. Jahrb., Vol. 49, Suppl. 109).

BROWN, R., 1869: On the geographical distribution of the *Coniferae* and *Gnetaceae* (Trans. Bot. Soc., Vol. 10).

BUNGE, S., 1874: Weite und enge Verbreitungsbezirke einiger Pflanzen (Sitzungsber. Dorp. Naturf. Ges. Jurjev, Vol. 11).

CHRIST, H., 1913: Über das Vorkommen des Buchsbaumes (*Buxus sempervirens*) in der Schweiz und weiterhin durch Europa und Vorderasien (Verhandl. Naturf. Ges. Basel, Vol. 24).

EKLUND, O., 1931: Über die Ursachen der regionalen Verteilung der Schärenflora Südwest-Finnlands (Acta Bot. Fenn., Vol. 8).

EKLUND, O., 1937: Klimabedingte Artenareale (Acta Soc. pro Fauna et Flora Fennica, Vol. 60).

FERNALD, M., 1924: Isolation and endemism in northeastern America and their relation to the age and area hypothesis (Amer. Jour. Bot., Vol. 11, No. 9).

FERNALD, M., 1926: The antiquity and dispersal of vascular plants (Quart. Rev. Biol., Vol. 1).

GAMS, H., 1931: Die klimatische Begrenzung der Pflanzenareale (Zeitschr. d. Ges. f. Erdkunde, No. 9/10).

GLEASON, H. A., 1924: Age and area from the viewpoint of phytogeography (Amer. Jour. Bot., Vol. 11).

GOOD, R. D'O., 1925: The past and present distribution of the *Magnoliae* (Ann. Bot., Vol. 39, No. 154).

E. V. Wulff   —46—   Historical Plant Geography

GOOD, R. D'O., 1931: A theory of plant geography (New Phytol., Vol. 30, No. 3).
GRABOVETSKAYA, A. N., 1937: A contribution to our knowledge of the genus *Nicotiana* L. (In Russian, Eng. summary; Bull. Appl. Bot., Gen. and Plantbr., Ser. 1, No. 2).
GREENMAN, J. M., 1925: The age and. area hypothesis with special reference to the flora of Tropical America (Amer. Jour. Bot., Vol. 12, No. 3).
GRIGGS, R. F., 1914: Observations on the behavior of some species on the edges of their ranges (Bull. Torrey Bot. Club, Vol. 41).
GUPPY, H. B., 1910: Die Verbreitung der Pflanzen und Tiere (Peterm. Mitt., Vol. 56).
GUPPY, H. B., 1921: The testimony of the endemic species of the Canary Islands in favour of the age and area theory of Dr. WILLIS (Ann. Bot., Vol. 35).
GUPPY, H. B., 1925: A side issue of the Age and Area hypothesis (Ann. Bot., Vol. 39, No. 156).
HULTÉN, E., 1937: Outline of the history of arctic and boreal biota (Stockholm).
IRMSCHER, E., 1929: Pflanzenverbreitung und Entwicklung der Kontinente, Parts I and II (Mitt. aus d. Inst. f. allg. Bot., Hamburg, Vol. 8).
JEANNEL, R. et JOLEAUD, L., 1924: Centres de dispersion (C. R. Soc. Biogéogr., No. 2).
KRAŠAN, FR., 1880: Über gewisse extreme Erscheinungen aus der geographischen Verbreitung der Pflanzen (Zeitschr. Oest. Ges. f. Meteor., Vol. 15).
KRAŠAN, FR., 1882: Die Erdwärme als Pflanzengeographischer Factor (Engler's Bot. Jahrb., Vol. 11).
LÄMMERMAYR, L., 1926: Materialien zur Systematik und Oekologie der Serpentinflora (Sitz. Ber. Ak. Wiss. Wien, Abt. I, Vol. 135, No. 9).
LANGLET, O., 1935: Über den Zusammenhang zwischen Temperatur und Verbreitungsgrenzen von Pflanzen (Meddel. Stat. Skogsförsöksanst., Stockholm, Vol. 4, No. 28).
MATTHEWS, J. R., 1922: The distribution of plants in Perthshire in relation to Age and Area (Ann. Bot., Vol. 36, No. 143).
NAEGELI, K., 1865: Über die Bedingungen des Vorkommens von Arten und Varietäten innerhalb ihres Verbreitungsbezirkes (Sitzungsber. d. math.-phys. Klasse d. Bayr. Akad., Vol. 2, No. 4).
NOVÁK, F., 1928: Quelques remarques relatives au problème de la végétation sur les terrains serpentiques (Preslia, Vol. 6).
PALMGREN, A., 1927: Die Einwanderungswege der Flora nach den Ålandsinseln (Acta Bot. Fennica, Vol. 2).
POHLE, R., 1925: Drabae asiaticae (Repert. sp. nov. Beih., Vol. 32).
RIDLEY, H. N., 1923: The distribution of plants (Ann. Bot., Vol. 37, No. 1).
RIDLEY, H. N., 1925: Endemic plants (Jour. Bot., Vol. 63).
SALISBURY, E. J., 1926: The geographical distribution of plants in relation to climatic factors (Geogr. Jour.).
SAMUELSSON, G., 1910: Über die Verbreitung einiger endemischen Pflanzen (Arkiv f. Bot., Vol. 9, No. 12).
SCHONLAND, S., 1924: On the theory of Age and Area (Ann. Bot., Vol. 38, No. 151).
SCHULZ, A., 1894: Grundzüge einer Entwicklungsgeschichte der Pflanzenwelt Mitteleuropas seit dem Ausgange der Tertiärzeit (Jena).
SHIRJAEV, G., 1932: Generis *Ononis* revisio critica (Beih. z. Bot. Centralbl., Vol. 49, Abt. 2).
SIMONET, M., 1932: Nouvelles recherches cytologiques et génétiques chez les *Iris* (Ann. Sci. Nat. Bot., sér. 10).
SINNOTT, E., 1917: The Age and Area hypothesis and the problem of endemism (Ann. Bot., Vol. 31).
SINNOTT, E., 1924: Age and Area and the history of species (Amer. Jour. Bot., Vol. 11, No. 9).
SZYMKIEWICZ, D., 1933: Contributions à la géographie des plantes, I–III (Kosmos, Vol. 53).
SZYMKIEWICZ, D., 1934: Une contribution statistique à la géographie floristique (Acta Soc. Bot. Polon., Vol. 11, No. 3).
SZYMKIEWICZ, D., 1936: Seconde contribution statistique à la géographie floristique (Acta Soc. Bot. Polon., Vol. 13, No. 4).
SZYMKIEWICZ, D., 1937: Contributions à la géographie des plantes, IV. Une nouvelle méthode pour la recherche des centres de distribution géographique des genres (Kosmos, Vol. 61).
SZYMKIEWICZ, D., 1939: Une nouvelle méthode . . . (Chron. Bot. 5:201).
THODAY, D., 1925: The geographical distribution and ecology of *Passerina* (Ann. Bot., Vol. 30, No. 153).
TURRILL, W. B., 1939: The principles of plant geography (Bull. Misc. Information, Kew, No. 5, pp. 208–237).
WILLIS, J. C., 1917: Further evidence for age and area (Ann. Bot., Vol. 31).
WILLIS, J. C., 1922: Age and Area. A study in geographical distribution and origin of species (Cambridge).
WILLIS, J. C., 1923: Age and Area: a reply to criticism, with further evidence (Ann. Bot., Vol. 37, No. 146).

WILLIS, J. C., 1923: The origin of species by large, rather than by gradual change and by GUPPY's method of differentiation (Ann. Bot., Vol. 37, No. 148).

WILLIS, J. C., 1936: Some further studies in endemism (Proc. Linn. Soc. London, Session 148, Part 2).

WULFF, E. V., 1936: Area y Edad (Rev. Argent. de Agronom., Vol. 3, No. 1; Pub. in 1927 in Russian in Bull. Appl. Bot., Vol. 17, No. 4).

WULFF, E. V., 1937: Polyploidy and the geographical distribution of plants [In Russian; Achievements Mod. Biol. (Uspekhi Sovremennoy Biologii), Vol. 7, No. 2].

153

Proc. Linn. Soc. Lond., **179**, 2, *pp.* 153–188
With 21 figures
Printed in Great Britain
June, 1968

Two approaches to zoogeography: a study based on the distributions of butterflies, birds and bats in the Indo-Australian area

By J. D. HOLLOWAY and N. JARDINE*

Department of Zoology and Department of History and Philosophy of Science,
Cambridge

(*Accepted for publication September*, 1967)

Communicated by John Smart, F.L.S.

In this paper the recorded distributions of butterfly, bird and bat taxa in the Indo-Australian area are subjected to various kinds of numerical analysis. Numerical methods are used to assign primary areas to zoogeographic regions and to assign taxa to faunal elements. An attempt is made to relate these classifications to the past geographical history of the area, and hence to infer something about past trends of spread and speciation in each group.

CONTENTS

1. Introduction	153
2. Numerical methods	154
(i) Derivation of zoogeographic regions	154
(ii) Non-metric multidimensional scaling	157
(iii) Derivation of faunal elements	158
(iv) The problem of weighting	158
3. Zoogeographic regions in the Indo-Australian area	159
(i) Introduction	159
(ii) Sources of data	159
(iii) Results	159
4. Non-metric multidimensional scaling of primary areas	165
5. Faunal elements in the Indo-Australian area	171
(i) Introduction	171
(ii) Outline of the geographical history of the Indo-Australian area . .	171
(iii) Sources of data	174
(iv) Results	175
Conclusion	185
Acknowledgements	186
References	186

1. INTRODUCTION

In this paper we use numerical methods for the analysis of distributional data to illustrate and contrast two approaches to zoogeography. The first approach is based upon the faunal affinities of primary areas with respect to taxa from particular groups. Numerical taxonomic methods are used to group primary areas into *zoogeographic regions*, and the technique of non-metric multidimensional scaling is used to relate the pattern of faunal affinities to geographical disposition. The second approach is based upon the geographical coincidences of taxa from particular groups. Numerical taxonomic methods are used to group taxa, into *faunal elements*, that is sets of taxa having similar patterns of distribution.

* The work was carried out during the tenure of S.R.C. Research Studentships by the authors.

The first questions to be asked about any method of classification are: What is its purpose? What kinds of inference and prediction can be made from it (cf. Gilmour, 1940, 1961)? The two kinds of classification used here are contrasted by Voous (1963) as follows:—

> 'The geographical method is static; it tries to define the borders of zoogeographical regions, districts, or provinces. It is part of the classical zoogeography of Philip Lutley Sclater and Alfred Russell Wallace. The faunal method is dynamic; it tries to detect and to describe the far-reaching intergradation of separate faunas throughout the continents . . . This method starts from the conception that there are distinct faunas but no distinct zoogeographic regions. This concept is no less arbitrary and no less disputable than that of the zoogeographic regions.'

We shall demonstrate that the two approaches are complementary, and that neither approach need be arbitrary. The classification into zoogeographic regions, together with the results of non-metric multidimensional scaling, shows up discontinuities in faunal distribution, and may, by revealing something about past trends of spread, help to explain the origins of the faunal elements revealed by the second approach. The classification into faunal elements may, by revealing something about past centres of radiation, help to explain the discontinuities in distribution shown up by the first approach. The two approaches, together with what is known of the past geographical history of the region, may enable us to make generalizations about the past history of each group in the Indo-Australian area.

The arbitrariness pointed out by Voous (1963) may have arisen, in part, from a tendency to confuse the inferences which may be made from a classification with the method of classification itself. It was partly to avoid this kind of confusion that numerical taxonomic methods were originally developed. However, we do not follow Sokal & Sneath (1963) in rejecting the traditional methods of classification in which there is an interplay between method and inference, for, as shown by Hull (1967), such an interplay need not result in logical circularity, but is perhaps better described as a process of 'reciprocal illumination' (Hennig, 1950). Instead, we regard the numerical methods simply as extensions of the traditional methods of classification in zoogeography. If rational use is to be made of numerical taxonomic methods, the well-established results of traditional methods of classification must be used as a check on the validity of applying a given numerical method to particular problems; otherwise there is a danger, embarrassingly exposed in the recent literature of numerical taxonomy, that numerical methods which are 'objective' and mathematically sophisticated will be applied in such a way as to yield results that are useless. The value of numerical taxonomic methods lies in their ability to handle consistently much larger quantities of data than can be dealt with by intuitive methods. But a method may be consistently right or wrong; that is, it may yield consistent results that are useful or useless as a basis for further inference.

2. NUMERICAL METHODS

The raw data for each of the methods described here are simply the recorded distributions of taxa, but the ways in which this information is handled are various.

(i) *Derivation of zoogeographic regions*

There are six stages in this process.

(1) A selection of taxa of a given rank from a particular group of organisms is made.

(2) Primary areas are defined.

(3) The distributions of the taxa selected in (1) amongst the primary areas defined in (2) are tabulated.

(4) From (3) a coefficient of faunal dissimilarity between each pair of primary areas is calculated.

(5) The matrix of dissimilarity coefficients is subjected to cluster-analysis to obtain a dendrogram (see Figs 1, 2 and 3), that is, informally expressed, a hierarchy with numerically defined levels.

(6) A hierarchic classification into discrete regions, sub-regions, etc. is derived from the dendrogram.

In stage (1), if the selection of taxa is to be regarded as adequate, the taxonomy of the groups concerned must be adequate (cf. Van Tyne & Berger, 1959). In particular, it is important that the taxonomic treatment of the group concerned should be as uniform as possible; that is, that taxa of a particular rank should not have been delimited by 'lumpers' in some parts of a group and by 'splitters' in other parts. Ideally, we suggest that quantitative zoogeographic studies should be preceded by numerical taxonomic studies of the groups concerned, but for such groups as the birds and butterflies this would be a colossal undertaking. In the first stage we must decide the rank at which we are to select taxa. In general the lower the rank the more recent the past trends of spread and centres of radiation that may be inferred from the resultant classification. This assumption does not rest upon a premise of constant evolutionary rate, but on the weaker premise that 'on average' the greater the phenetic dissimilarity between taxa the more ancient their evolutionary divergence.

Serious sampling problems arise in stage (1). The taxonomic and distributional data that would be needed for a study based upon all the species of the butterflies, birds and bats is not available, and much of the information that is available is unreliable. Even to handle all the reliable information would be impracticable, hence we must sample. If we had detailed knowledge of the past history of each group in the area we might be able to predict, to some extent, the effect that different methods of sampling would have on the resultant classification, but it is precisely this kind of information that we wish to infer from the resultant classification. We can, however, test the extent to which the resultant classification varies with different samples from each group.[*] We found that for the butterflies and birds different samples gave very similar results. For the bats, however, the agreements were not so good and this is probably accounted for by the fact that the two sub-orders of the bats, the Microchiroptera and the Megachiroptera, have had very different histories in the area (cf. Tate, 1946).

In stage (2) it is assumed that the primary areas should be delimited by geographical criteria rather than ecological or climatic criteria, since we wish to derive a zoogeographic classification. The choice of primary areas in this study was dictated to a large extent by the way in which the distributions of taxa have been recorded in the literature. Given precise distributional data it is possible to devise precise methods for selecting primary areas of relative faunal homogeneity. Such methods are given by Webb (1950), Ryan (1963), Hagmeier & Stults (1964), Huheey (1965) and Hagmeier (1966). What is important in the selection of primary areas is that areas which are faunally heterogeneous should not be lumped together; division of homogeneous areas into two or more primary areas matters little for they will be grouped together in the cluster-analysis. For example, in Hagmeier (1966) it is shown that the zoogeographic regions derived from the distributions of North American mammals in Hagmeier & Stults (1964) are inconsistent with those derived using the same numerical methods but a finer selection of primary areas. Our selection of primary areas for the mainland regions is arbitrary. Of the islands both New Guinea and Java are probably faunally heterogeneous, the former consisting of

---

[*] A statistical method for comparing dendrograms is given by Sokal and Rohlf (1962), and a measure of the difference between dendrograms is suggested in Jardine and Sibson (in press, *a*).

geologically distinct regions which may have had very different origins (Cheesman, 1951), the latter being divisible into a western region where rain forest predominates and an eastern region where savannah conditions predominate. The distributional data used were not precise enough to enable us to recognize more than one primary area for each island.

Numerous coefficients of dissimilarity which might be used in stage (4) have been devised. Some of these are reviewed by Dagnelie (1960), Long (1963) and by Hagmeier & Stults (1964). The only coefficient that we have seen that has a sound statistical basis is that of Preston (1948, 1962), and this we have used.* Preston's coefficient of faunal dissimilarity, z, between two primary areas is given by the 'resemblance equation':

$$x^{1/z} + y^{1/z} = 1,$$

where x is the proportion of the joint fauna found in one primary area and y is the proportion of the joint fauna found in the other. A table for the solution of the resemblance equation is given by Preston (1962: 419), and a programme to obtain more precise solutions to the equation has been written for use on the Titan computer.

Methods of cluster-analysis that might be used in stage (5) are reviewed in Sokal & Sneath (1963), Williams & Dale (1965) and Lance & Williams (1967). In Jardine, Jardine & Sibson (1967) it is shown that only the single-link (nearest-neighbour) method satisfies certain obvious requirements, for example the requirement that continuous variation in the data should produce continuous variation in the resultant dendrogram. The standard objection to this method is that it groups together basic taxa (in this case primary areas) that are linked by chains of intermediates. In Jardine & Sibson (in press, a) it is suggested that this is an inherent defect of hierarchic classification, and that the kinds of information concealed by chaining, for example information about the relative homogeneity of the groups recognized, can be revealed only by cluster-methods leading to non-hierarchic systems. An informal account of these methods is given in Jardine & Sibson (in press, b). Within the model for taxonomy given in Jardine & Sibson (in press, a) measures of the degree of isolation, and of the relative homogeneity, of groups in a taxonomic system are derived, and a method for determining the extent to which an array of dissimilarity coefficients shows 'hierarchic structure' is given. The measure of isolation for a group in a hierarchic classification is given by the length of the 'stalk' of the corresponding cluster in the dendrogram from which the hierarchy is derived; hence the isolation of groups may be observed by inspecting the dendrograms. This measure is suggested also by Wirth, Estabrook & Rogers (1966). It is valid only for dendrograms derived by a single-link method. The measure of homogeneity and the method for finding out the extent to which given data have hierarchic structure are rather complex; some use is made of the latter in the following sections, but details of calculation would be out of place here: the reader is referred to Jardine & Sibson (in press, a) for a detailed discussion.

Stage (6) is the derivation of a hierarchic classification from the dendrogram. This is done by choosing levels in the dendrogram at which the various categories in the hierarchy, in this case region, sub-region and province, are to be recognized. Here it seems reasonable to fix the first level, the level at which we recognize faunal provinces, near 0·27, for as shown by Preston (1962) a value of z lying between 0·27 and 1 for two areas indicates that 'there is some interaction, but it is incomplete, and there is, and long has been some degree of

---

* Preston's theoretical model is based on the assumption that the distribution of individuals amongst taxa of a given rank is a particular kind of lognormal distribution. Several alternative hypotheses about the distribution of individuals amongst taxa of a given rank have been suggested. Williams (1947, 1949) suggested that the distribution is a logarithmic series; MacArthur (1957) suggested that the distribution is a Barton-Davis distribution (often referred to as the 'broken-stick' distribution). Preston (1962) re-examined some of the data upon which these alternative hypotheses were based and showed that they fit his hypotheses at least as well. Part of the difficulty in deciding between these alternative hypotheses arises from the fact that, as shown by Goodall (1952), it is difficult to find data which will discriminate between them. Preston's lognormal distribution seems to be indirectly verified by the accuracy of the predictions obtained in his model.

genuine isolation'. In practice we have fixed the level at which faunal provinces are recognized at 0·30 to allow for sampling errors; since a single-link cluster method is used the faunal provinces recognized will be such that there is significant isolation between all basic areas belonging to different faunal provinces. A similar argument is used by Hagmeier & Stults (1964) to decide the level at which faunal provinces should be recognized, but is invalid since an average-link cluster method was used. Groups which are clustered at a given level, z, by a single-link method will be such that basic areas belonging to different groups cannot be linked by a coefficient of dissimilarity less than, or equal to, z. In an average-link method this is not the case. Various criteria for fixing the other levels might be devised; in a zoogeographic classification it would be reasonable to choose the levels so as to obtain zoogeographic regions that are as distinct as possible, and hence we might fix the levels in such a way as to maximize the average of the measures of isolation of the groups of primary areas clustering at or below the levels. As indicated above, the isolation of each of the regions recognized is given by the length of the 'stalk' of the corresponding cluster in the dendrogram. This method does not, however, work well where the number of basic areas is small. Instead we have fixed the levels in such a way as to yield zoogeographic regions that correspond as closely as possible to those already described by previous authors; this is justifiable since we are primarily concerned to check the validity of the various zoogeographic regions and faunal boundaries described in the Indo-Australian area.

In one respect the method of classification described here departs from the method of classification used in classical zoogeographic studies. The method described here is based upon the distributions of taxa at a single taxonomic level within a group, whereas classical studies were based upon a composite assessment of the distributions of taxa at all taxonomic levels within a group. One of the advantages of using a single taxonomic level is that the comparison between the classifications obtained at different taxonomic levels within a group may itself be informative, for the higher the taxonomic level used the more ancient the trends of spread and radiation that may be inferred from the resultant classification (see p. 155). An alternative approach would be to employ some kind of differential weighting of the distributions of taxa, as described below.

(ii) *Non-metric multidimensional scaling*

Given as data measures of dissimilarity between all pairs in a set of objects, non-metric multidimensional scaling may be used to obtain a representation in a given number of dimensions in which the distances between the points representing the objects 'distort' the data as little as possible. A full account of the theoretical background to this technique is given in Kruskal (1964a). In this case the objects are the primary areas, and we wish to find out the extent to which the pattern of faunal dissimilarities, given by the faunal dissimilarity coefficients, may be related to the present geographical disposition of the primary areas. This may be determined by computing a two-dimensional representation and comparing it after suitable scaling and orientation with the geographical map.

The two-dimensional representation, R, is scaled and orientated with respect to the geographical map, as follows. First the geographical map is replaced by a map, G, showing the disposition of points representing the primary areas. R and G are then centred by identifying their centroids. They are reduced to the same scale by equalizing the sums of the squared distances of points from the centroid. R and G are orientated by rotating about the centroid so as to minimize the sum of the squared distances of points in R from corresponding points in G. R is reflected if this leads to a better optimum fit on rotation. This kind of scaling and orientation is discussed at length in Sneath (1967a). We may use $\Delta^{\frac{1}{2}}$, the standardized root mean square distance between points in G and corresponding points in the scaled and orientated R, as a measure of 'goodness-of-fit' between G and R. Precise tests of statistical significance for comparing values of $\Delta^{\frac{1}{2}}$ cannot be devised since we do not know the sampling distribution of Δ^i. Some approximate tests for comparing values of Δ^i are given in Sneath (1967a, Appendix 3.5). Differences between R and G may yield information about past trends of spread in the area, and this may in turn be related to the past geographical history of the area. In this study we have judged the patterns of deviation of R from G by eye. Trend-surface analysis (Miller 1956, Krumbein 1959, Sneath 1967a) may provide more precise techniques for estimating such patterns of deviation. But in this case, where the number of primary areas is small and the patterns of deviation are fairly obvious, to use trend-surface analysis would be to use a sledge-hammer to crack a nut.

The method of computing used in non-metric multidimensional scaling is described in Kruskal (1964b). In this study the programme (NJ 867/PROG O/KRUSKAL) was used on the computer Titan at the Cambridge University Mathematical Laboratory.

A similar method has been used by Sneath (1967b) to test the relation between the floristic affinities of primary areas and their past and present geographical distributions. The study was based on the world distributions of conifer taxa mapped by Florin (1963). His general approach is close to that described here. However, we prefer the method of non-metric multidimensional scaling to that of principal component analysis on the grounds that the use of principal component analysis to obtain a meaningful representation of primary areas in a reduced number of dimensions depends upon an assumption of linear regression which is questionable for data of this kind (cf. Cattell, 1965). Where world distributions of taxa are considered the best fitting to a sphere of points representing primary areas is required, and the technique of non-metric multidimensional scaling could be modified to obtain such a fitting. Non-metric multidimensional scaling differs from principal component analysis in making no assumptions about linearity of regression.

(iii) *Derivation of faunal elements*

The first stages in the derivation of faunal elements are the same as stages (1)-(3) for the derivation of zoogeographic regions. In stage (2) it seems reasonable to group together as single primary areas those primary areas which cluster below 0·30 in the dendrogram obtained in the study of zoogeographic regions.

Stage (4) is the calculation of a coefficient of dissimilarity in distribution between all pairs of taxa. We have used the coefficient $1 - m/n$, where m is the number of primary areas in which both taxa occur, and n is the total number of primary areas.

In stage (5) the matrix of dissimilarity coefficients is subjected to single-link cluster-analysis to obtain a dendrogram (see Figs 9, 10 and 11). There is little point in deriving a hierarchic classification of faunal elements from the dendrogram. Instead, we may simply recognise as faunal elements the clusters in the dendrogram. The most suggestive way of displaying the results is to tabulate the number of taxa from each faunal element occurring in each primary area. Particular faunal elements may then be shown on a map by means of contours, each contour embracing those primary areas in which occur more than a given number of taxa belonging to the faunal element (see Figs 12–19).

(iv) *The problem of weighting*

One obvious criticism of all these methods which must be answered is that they give equal weight to the distribution of each taxon selected, regardless of whether the taxon is considered to be of ancient or of relatively recent origin. In this study equal weighting is justifiable, for little is known about the details of the phylogeny of the butterflies, birds and bats of the Indo-Australian area. In a group of organisms with a good fossil record some kind of prior weighting may be reasonable. Even where little is known of past evolutionary history some kind of weighting may be possible. Intuitively, we should attach greater weight to coincidences in distribution between highly dissimilar taxa than to coincidences in distribution between closely similar taxa. This is based on the assumption that there is likely, at least on average, to be some connexion between phenetic similarity and recency of common ancestry. Where a study of the distribution of taxa in an area is preceded by a numerical taxonomic study, it may be possible to incorporate some such differential weighting. The danger of differential weighting of the distributions of taxa is simply that greater weight may be given to just those distributions that support some particular hypothesis about the past geographical history of an area.

3. ZOOGEOGRAPHIC REGIONS IN THE INDO-AUSTRALIAN AREA

(i) *Introduction*

The numerical method for zoogeographic classification, described in 2(i), is an extension of the classical approach to zoogeography pioneered by Sclater (1858) and Wallace (1860). This approach has been criticized, for example by Peters (1955), on the grounds that no two taxa have identical distributions, and hence that the imposition of faunal boundaries is merely conventional. An excellent defence of zoogeographical classification is given by Darlington (1957, p. 419 *et seq.*). The method used here shows where the statistical discontinuities in the distributions of taxa lie, and enables us to investigate the extent to which these discontinuities coincide for different groups.

(ii) *Sources of data*

(a) *Rhopalocera*. 870 species from 130 genera were used in this study. Except for *Danaus* and part of *Euploea*, for which distributions were taken from Seitz (1927), all distributional data were taken from recent revisions as follows: *Tenaris*, Brooks (1950); *Euploea*, Carpenter (1953); Hesperidae, Evans (1949); *Ixias*, Gabriel (1943); *Curetis*, Evans (1954); *Arhopala*, Evans (1957); *Delias*, Talbot (1928–1937); *Troides*, Zeuner (1943); *Eurema, Symbrenthia, Chersonesia, Cyrestis*, revisions in preparation by J. D. H.

(b) *Chiroptera*. 124 species from 61 genera were used. All distributional data were taken from Tate (1946).

(c) *Aves*. 876 species from 10 orders were used. Distributional data were taken from the *Check List of the Birds of the World* (Peters, 1931–1951; Mayr & Greenway, 1951–1963). All migrants and sea birds were excluded.

(iii) *Results*

The coefficients of faunal dissimilarity are given in Tables 1 and 2. The primary areas used for the Chiroptera which differ from those used for the Rhopalocera and Aves, were taken directly from Tate (1946). The dendrograms obtained by single-link cluster-analysis for the Rhopalocera, Aves and Chiroptera respectively are shown in Figs 1, 2 and 3. In each case the levels at which we have recognized faunal regions, sub-regions and provinces are indicated by heavy horizontal lines. As pointed out in 2(i), the level at which faunal provinces are recognized is suggested by the nature of the coefficient of dissimilarity used, but the other levels are chosen so as to give a classification conforming as well as possible with the zoogeographic regions and faunal boundaries recognized by previous authors. In Figs 5A, 6A and 7A the boundaries of the regions and sub-regions are shown on maps of the Indo-Australian area. These boundaries may be contrasted with the faunal boundaries proposed by various authors, shown in Fig. 4.

Classical zoogeographic studies have generally been based on the assumption that the present faunal distributions in the Indo-Australian area are the product of interaction between distinct Oriental and Australasian faunas, and have attempted to find the best

160   *Proceedings of The Linnean Society of London*   [179,

Table 1. *Coefficients of faunal dissimilarity between primary areas, calculated from the distributions of selected taxa of the Aves and Rhopalocera in the Indo-Australian area*

AVES (upper triangle) — **RHOPALOCERA** (lower triangle); diagonal cells carry the area index (1–23).

Area	1	2	3	4	5	6	7	8	9	10	11	12	13	14	15	16	17	18	19	20	21	22	23
Burma (1)	**1**	0·395	0·696	0·530	0·591	0·637	0·736	0·743	0·727	0·848	0·862	0·473	0·682	0·734	0·816	0·882	0·926	0·943	0·971	0·967	0·957	0·934	0·524
Indo-China (2)	0·321	**2**	0·672	0·813	0·864	0·848	0·887	0·862	0·862	0·853	0·878	0·864	0·884	0·901	0·936	0·941	0·888	0·941	0·968	0·971	0·969	0·929	0·627
Formosa (3)	0·838	0·790	**3**	0·813	0·864	0·440	0·689	0·862	0·730	0·775	0·853	0·304	0·616	0·711	0·816	0·823	0·888	0·862	0·943	0·970	0·957	0·942	0·862
Malaya (4)	0·482	0·568	0·926	**4**	0·226	0·424	0·654	0·730	0·758	0·812	0·862	0·266	0·639	0·730	0·810	0·882	0·904	0·942	0·929	0·962	0·918	0·943	0·551
Sumatra (5)	0·655	0·658	0·922	0·212	**5**	0·391	0·654	0·730	0·758	0·763	0·862	0·266	0·639	0·750	0·816	0·882	0·963	0·926	0·971	0·966	0·916	0·943	0·622
Java (6)	0·599	0·662	0·934	0·524	0·226	**6**	0·312	0·440	0·730	0·758	0·812	0·304	0·467	0·711	0·786	0·861	0·936	0·936	0·971	0·936	0·942	0·971	0·695
Bali (7)	0·711	0·677	0·971	0·594	0·571	0·499	**7**	0·391	0·568	0·667	0·756	0·467	0·662	0·730	0·786	0·861	0·915	0·888	0·971	0·965	0·071	0·943	0·836
Lombok (8)	0·825	0·810	0·943	0·747	0·639	0·474	0·413	**8**	0·334	0·334	0·467	0·862	0·897	0·832	0·743	0·888	0·861	0·834	0·912	0·926	0·940	0·943	0·761
Sumbawa (9)	0·788	0·804	0·915	0·839	0·672	0·474	0·474	0·334	**9**	0·329	0·204	0·862	0·882	0·874	0·752	0·886	0·861	0·834	0·910	0·888	0·916	0·943	0·777
Flores (10)	0·832	0·816	0·916	0·848	0·763	0·562	0·474	0·204	0·329	**10**	0·512	0·817	0·882	0·874	0·764	0·886	0·888	0·836	0·881	0·862	0·915	0·943	0·777
Timor (11)	0·897	1·000	0·943	0·883	0·839	0·730	0·534	0·512	0·204	0·512	**11**	0·888	0·836	0·832	0·700	0·913	0·888	0·888	0·836	0·825	0·887	0·915	0·777
Borneo (12)	0·662	0·746	0·929	0·415	0·282	0·529	0·685	0·763	0·763	0·817	0·888	**12**	0·541	0·644	0·752	0·812	0·965	0·965	0·971	0·967	0·943	0·886	0·730
Palawan (13)	0·788	0·777	0·943	0·717	0·695	0·788	0·802	0·889	0·836	0·836	0·836	0·541	**13**	0·644	0·780	0·804	0·971	0·943	1·000	0·918	0·934	0·887	0·853
Philippines (14)	0·786	0·784	0·871	0·782	0·780	0·761	0·740	0·722	0·722	0·804	0·832	0·758	0·897	**14**	0·707	0·794	0·936	0·843	0·941	0·908	0·904	0·777	0·836
Celebes (15)	0·859	0·862	0·910	0·880	0·878	0·807	0·834	0·752	0·752	0·764	0·700	0·853	0·816	0·577	**15**	0·400	0·856	0·774	0·906	0·895	0·963	0·804	0·832
Sula Is. (16)	0·918	0·963	0·941	0·904	0·904	0·834	0·875	0·859	0·859	0·884	0·913	0·864	0·887	0·794	0·461	**16**	0·824	0·769	0·884	0·895	0·834	0·862	0·887
N. Moluccas (17)	0·968	0·940	1·000	0·965	0·963	0·936	0·915	0·861	0·861	0·888	0·884	0·965	0·971	0·936	0·856	0·824	**17**	0·134	0·001	0·682	0·889	0·688	0·940
S. Moluccas (18)	0·968	0·940	1·000	0·965	0·926	0·936	0·888	0·834	0·834	0·836	0·888	0·943	0·943	0·843	0·774	0·769	0·515	**18**	0·734	0·734	0·723	0·709	0·939
New Guinea (19)	0·970	1·000	1·000	1·000	0·971	0·971	0·971	0·912	0·912	0·881	0·836	0·971	1·000	0·941	0·906	0·941	0·001	0·734	**19**	0·807	0·636	0·719	0·912
Australia (20)	1·000	1·000	1·000	1·000	0·965	1·000	0·971	0·926	0·888	0·862	0·825	0·825	0·887	0·908	0·895	0·957	0·001	0·734	0·807	**20**	0·847	0·689	0·967
Bismarck Is. (21)	0·969	0·970	1·000	0·965	0·963	0·960	0·943	0·942	0·913	0·915	0·887	0·967	1·000	0·936	0·886	0·878	0·737	0·811	0·639	0·875	**21**	0·554	0·970
Solomon Is. (22)	1·000	1·000	0·887	0·063	0·878	1·000	1·000	0·943	0·971	0·971	0·915	0·964	0·971	0·777	0·804	0·910	0·688	0·737	0·734	0·901	0·554	**22**	0·502
Andaman & Nicobar Is. (23)	0·969	0·970	0·887	0·725	0·725	0·734	0·740	0·836	0·835	0·834	0·888	0·747	0·836	0·836	0·825	0·862	0·942	0·942	1·000	0·915	0·970	0·502	**23**

RHOPALOCERA

Table 2. *Coefficients of faunal dissimilarity between primary areas, calculated from the distributions of the taxa of the Chiroptera listed by Tate (1946)*

1	S.E. China, Hainan and Formosa												
0·368	2 Indo-China												
0·532	0·240	3 Malay Peninsula											
0·530	0·395	0·240	4 Sumatra										
0·694	0·534	0·360	0·453	5 Borneo, Palawan and Calamianes									
0·441	0·433	0·515	0·324	0·464	6 Java and Bali								
0·658	0·572	0·534	0·529	0·554	0·534	7 Philippines							
0·804	0·811	0·730	0·713	0·666	0·681	0·568	8 Celebes						
0·888	0·864	0·768	0·758	0·696	0·768	0·476	0·499	9 Sanghir Talaut					
0·943	0·850	0·734	0·794	0·816	0·672	0·843	0·764	0·867	10 Lesser Sundas				
0·807	0·786	0·780	0·786	0·736	0·780	0·732	0·556	0·476	0·740	11 Moluccas			
0·807	0·862	0·836	0·836	0·836	0·836	0·786	0·711	0·654	0·752	0·529	12 New Guinea		
0·915	0·888	0·856	0·862	0·808	0·830	0·780	0·736	0·636	0·856	0·554	0·407	13 Solomon Is.	
0·886	0·889	0·859	0·836	0·810	0·780	0·808	0·737	0·725	0·824	0·666	0·487	0·642	14 Australia

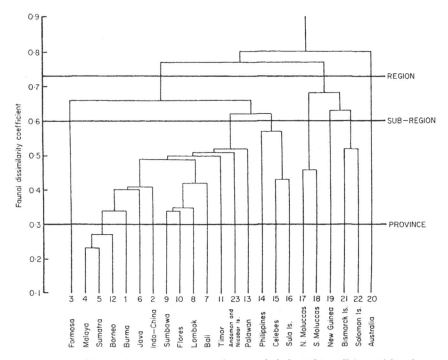

Fig. 1. Dendrogram derived by single-link cluster-analysis from the coefficients of faunal dissimilarity between primary areas for the Rhopalocera (see Table 1). Heavy horizontal lines indicate the levels at which we recognise Regions, Sub-regions and Provinces.

dividing-line between these faunas. A detailed review of the various proposed faunal boundaries is given by Burkill (in Scrivenor *et al.*, 1942). In the account given here we summarize this and review more recent work.

Wallace (1860) demarcated the line named after him on the basis of material, mainly of butterflies, birds and mammals, collected during his travels in the area, described in Wallace (1869). This line was later modified by Huxley (1868) to exclude the Philippines from the Oriental region, largely on the basis of the distribution of the Megapodiidae (brush-turkeys).* Kloss (1929) pointed out that this modification of Wallace's line coincides with the limit of the Sundaland continental area that was exposed at times of low sea-level during the Pleistocene, and hence that the Wallace–Huxley line may indicate a barrier to recent faunal spread. The limit of the Sunda-Oriental region in the Pleistocene

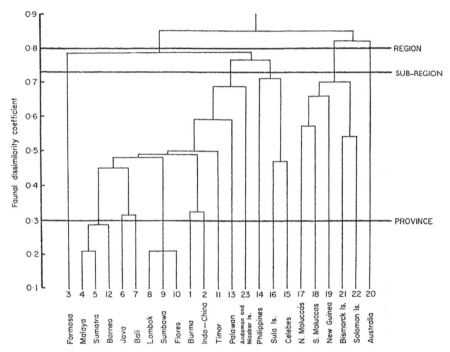

Fig. 2. Dendrogram derived by single-link cluster-analysis from the coefficients of faunal dissimilarity between primary areas for the Aves (see Table 1). Heavy horizontal lines indicate the levels at which we recognize Regions, Sub-regions and Provinces.

may correspond approximately to the present Asiatic 100 fathom line. Kloss suggested that the Australian 100 fathom line, the boundary of the Sahul shelf, may similarly correspond to the limit of the Australian continental area at times of low sea-level in the Pleistocene. We should expect that the Pleistocene continental limits would be most clearly indicated in the distributions of groups depending solely upon land connexion for dispersal. This is confirmed by the studies by Raven (1935) and Brongersma (1936) of mammal distributions (excluding bats and murid rodents), and by the studies of the distributions of freshwater fish by Darlington (1948, 1957). For such groups as the Aves, Chiroptera and Rhopalocera which may be dispersed by flight and wind, and the murid rodents which may be dispersed by rafting, these limits should be less clear.

* In section 5(iv) it is shown that the distribution of the genus *Megapodius* is in fact anomalous.

Attempts to delimit the Australian and Oriental faunas fall into two main categories. One school, following Pelseneer (1904), has attempted to find a single line separating the region in which the fauna is predominantly Australian from the region in which it is predominantly Oriental. The line most commonly supported (e.g. by Mayr, 1944) is Weber's line, originally proposed by Pelseneer, and often called the 'line of faunal balance'. Kloss pointed out that this line coincides with a zone of negative gravimetric anomaly and vulcanism.

Another school, exemplified by Lydekker (1896), Dickerson (1928) and Tate (1946), takes the Wallace-Huxley line and the line delimiting the Sahul Shelf as the limits of the

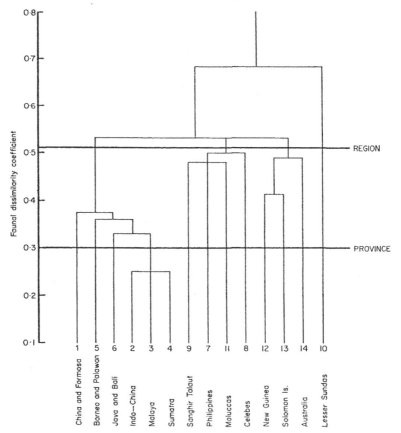

Fig. 3. Dendrogram derived by single-link cluster-analysis from the coefficients of faunal dissimilarity between primary areas for the Chiroptera (see Table 2). Heavy horizontal lines indicate the levels at which we recognize Regions and Provinces.

main Oriental and Australian regions and treats the intervening region variously as a 'transition' zone or as a region of distinct intercontinental speciation.

Before discussing the various proposed regions and faunal boundaries in the light of the numerical classifications derived in this study some general points must be made. First, calculations using the method for determining the extent to which given data shows 'hierarchic structure', described in Jardine & Sibson (in press), show that the dissimilarity

11

coefficients have relatively poor hierarchic structure. Poorer, for example, than the dissimilarity coefficients used in the study of faunal elements and poorer than dissimilarity coefficients calculated for species in the butterfly genera *Cyrestis* Bsd. and *Eurema* Hbn. This does *not* mean that the hierarchic classification is arbitrarily imposed: it does mean, however, that variation in the data will lead to relatively large (though continuous) variation in the resultant dendrogram and hence to changes in the resultant classification. The effect of varying the sample of taxa selected from each group was tested as described in section 2(i), and it was shown that this leads to little change in the resultant dendrograms. But, in the absence of sufficiently precise distributional data we were unable to test the results of different selections of primary areas. In particular we suspect that Java, New Guinea and Australia are faunally heterogeneous. Both our treatment of these heterogeneous areas as single primary areas, and inaccuracies and incompleteness in the

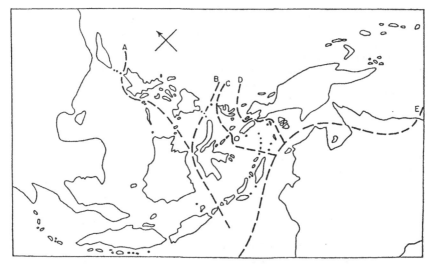

Fig. 4. Faunal boundaries suggested within the Indo-Australian area. A, Huxley, 1868; B, Wallace, 1860; C, Pelseneer, 1904 (Weber's line of faunal balance); D, Lydekker, 1896; E, Gressitt, 1956b.

distributional data used, might lead to significant changes in the resultant classification. Hence we must be cautious in interpreting the numerically derived classifications.

One conclusion is immediate. The zoogeographic classifications derived from the study of distributions of taxa from the Rhopalocera and Aves differ widely from that derived for the Chiroptera. It is clear that zoogeographic classifications, at least in this area, should be considered relative to particular taxonomic groups.

We discuss first the zoogeographic classification derived for the Chiroptera since this differs markedly from that derived for the Rhopalocera and Aves. The conclusion drawn by Tate (1946) that there is a distinct intercontinental or 'Wallacean' region is confirmed, but the Lesser Sunda Islands are excluded from this, appearing as a distinct region. As Tate pointed out, the bat fauna of the Lesser Sundas is imperfectly known and small. Further information might well alter the position of the Lesser Sundas in the classification. We have called Tate's Australian region the Papuan region since as shown in section 5, most of the bats found in this region belong to a faunal element strongly centred on New Guinea.

The classifications derived from the distributions of the Rhopalocera and Aves are closely comparable. Inspection of the relevant dendrograms, Figs 1 and 2, shows that in each case Weber's line is confirmed as a boundary between the Oriental region and the Papuan and Australian regions. The results conflict with the classical zoogeographic regions in strongly suggesting that a Papuan region, distinct from the Australian region, should be recognized. For the Rhopalocera, this region clusters with the Oriental region rather than the Australian region at the highest level.

The recognition of a separate Papuan region for the Rhopalocera and Aves correlates well with studies by Gressitt (1956a, 1956b, 1958, 1959, 1960, 1961) of the distribution of the Coleoptera, from which he concluded that the fauna of Melanesia (including the Moluccas, New Guinea, the Bismarck Archipelago and the Solomon Is.) is of Oriental rather than Australian origin. He suggested that Melanesia, together with north-east Queensland, should be treated as a distinct Papuan region. This view was endorsed by Gupta (1962) on the basis of the taxonomy and distribution of the hymenopteran genus *Theronia*, and by Metcalf (1949) from a study of the distribution of the Homoptera. Both Gupta (1962) and Zimmerman (1942) supposed that the main spread of insect fauna has been along the Melanesian arc into Oceania, with only minor spread into Australia from New Guinea.

We obtain more detailed information about past trends of spread and centres of radiation, and about the relation between the present distributions of taxa and the past geographical history of the area, from the results of non-metric multidimensional scaling described in section 4, and from the study of faunal elements described in section 5.

4. NON-METRIC MULTIDIMENSIONAL SCALING OF PRIMARY AREAS

We should expect some kind of connexion between the faunal dissimilarities of primary areas and their dispositions. Non-metric multidimensional scaling of the array of dissimilarity coefficients to obtain a two-dimensional representation provides us with a useful test of this. Further, we may reasonably hope to be able to infer from the disposition of points representing primary areas something about past trends of spread; hence we may be able to relate to the past geographical history of the area the deviations of the two dimensional representation from the geographical map. The 2-D representations obtained by computing are shown in Figs 5B, 6B and 7B. For purposes of comparison they have been scaled and orientated with respect to the geographical map, as described, and the boundaries of the regions and sub-regions derived in section 3 have been indicated. The *stress*, a measure of the goodness-of-fit of the two-dimensional representation to the coefficients of faunal dissimilarity, lies near 10% in each case, a value which Kruska (1964a) suggests can be verbally evaluated as 'fair'. Computation of the best fitting 1-D representation and of the best fitting 3-6D representations showed that in each case the stress increases evenly down to 2-D and then shows a marked increase. This may indicate that use of a 2-D representation of the data is appropriate. What conclusions can be drawn from these results?

(1) The measures of 'goodness-of-fit' between R and G are as follows: Rhopalocera $\Delta^{\frac{1}{2}} = 0.78$; Aves, $\Delta^{\frac{1}{2}} = 0.59$; Chiroptera, $\Delta^{\frac{1}{2}} = 0.58$. The fit is in each case significantly better than that expected for a random disposition of points. The fact that $\Delta^{\frac{1}{2}}$ is greater for the Rhopalocera than for the Aves and the Chiroptera may arise from the fact that the Rhopalocera, being at a relatively high taxonomic level (see p. 169), and having relatively low spread potential, reveal more ancient trends of spread. These may in turn be related to relatively more ancient dispositions of the primary areas.

We conclude that there is an intimate connexion between present geographic disposition and the pattern of faunal affinities. In fact it is the very closeness of the overall fit that entitles us to infer something from the differences between the 2-representations and the geographical map.

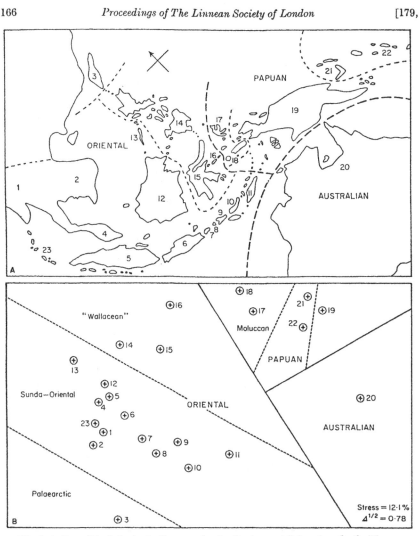

Fig. 5, A, Map of the Indo-Australian area showing Regions and Sub-regions for the Rhopalocera derived from the dendrogram (Fig. 1). B, Two-dimensional representation of primary areas obtained from the coefficients of faunal dissimilarity by non-metric multidimensional scaling. The representation has the same scale and orientation as the geographical map and the regional and sub-regional boundaries are marked. The stress is a measure of the 'goodness-of-fit' to the dissimilarity coefficients, and $\Delta^{\frac{1}{2}}$ is a measure of the 'goodness-of-fit' to the map.

(2) In each case the primary areas lying on the Sunda Shelf are bunched together on the 2-D representation relative to their geographical dispositions. This indication that the faunal isolation of these primary areas is less than would be expected from their geographical separations is supported by the fact that, as shown in the previous section, Borneo, Malaya and Sumatra cluster near 0·27 in the zoogeographical classification, a value which is shown by Preston (1962) to indicate that areas are behaving as samples of a

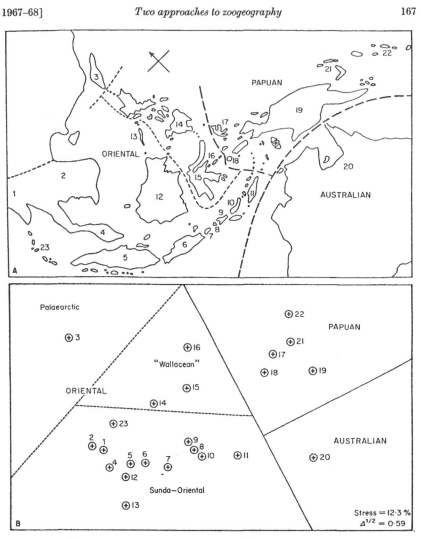

Fig. 6. A, Map of the Indo-Australian area showing Regions and Sub-regions for the Aves derived from the dendrogram (Fig. 2). B, Two-dimensional representation of primary areas obtained from the coefficients of faunal dissimilarity by non-metric multidimensional scaling. The representation has the same scale and orientation as the geographical map and the regional and sub-regional boundaries are marked. The stress is a measure of the 'goodness-of-fit' to the dissimilarity coefficients, and $\Delta^{\frac{1}{2}}$ is a measure of the 'goodness-of-fit' to the map.

single homogeneous area of faunal equilibrium. It is known that these areas were united as a single land-mass, Sundaland, at times of low sea-level during the Pleistocene (van Bemmelen, 1949). We may conclude that there was little barrier to faunal spread in this period and that relatively little speciation has taken place in more recent times.

(3) In contrast to the bunching together of areas lying on the Sunda Shelf, New Guinea and Australia, lying on the Sahul Shelf bounded by the Australian 100-fathom line, are

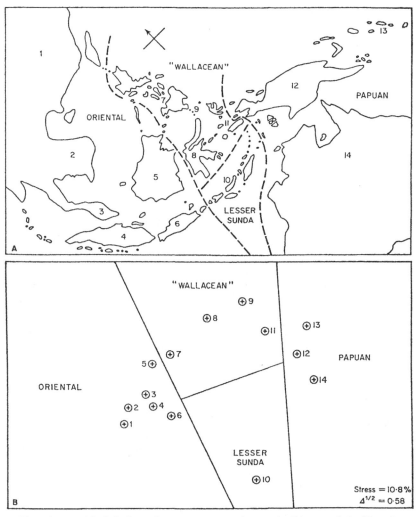

Fig. 7. A, Map of the Indo-Australian area showing Regions for the Chiroptera derived from the dendrogram (Fig. 3). B, Two-dimensional representation of primary areas obtained from the coefficients of faunal dissimilarity by non-metric multidimensional scaling. The representation has the same scale and orientation as the geographical map and the regional boundaries are marked. The stress is a measure of the 'goodness-of-fit' to the dissimilarity coefficients, and $\Delta^{\frac{1}{2}}$ is a measure of the 'goodness-of-fit' to the map.

relatively widely separated in the 2-D representations, though rather less so for the Chiroptera than for the Aves and Rhopalocera. This must be interpreted with caution since we have treated both New Guinea and Australia, which are faunally heterogeneous, as single primary areas. Both the Rhopalocera and the Aves, but not the Chiroptera, are thought to have reached Australia from mainland Asia before the subsidence of inter-mediate land areas at the beginning of the Tertiary. Secondary spread in more recent times has taken place from the Sunda-Oriental region to New Guinea and Australia, New Guinea being a strong centre of secondary radiation and spread. Routes of spread and centres of radiation will be discussed in the following subsection and in section 5. It seems that for the Rhopalocera and Aves there has been a strong barrier to spread to and from Australia since the end of the Cretaceous. This is supported by the fact that, as shown in the previous section, distinct Papuan and Australian regions can be recognized for these groups. For the Chiroptera the barrier seems to have been less strong, and this is supported by the fact that Australia and New Guinea form a single region in the zoogeographic classification. At first sight this is surprising, for as shown by Kloss (1929) and David (1950) on the basis of present geomorphology, and by Raven (1935) on the basis of marsupial distribution, there is strong evidence for a late-Pleistocene land connexion between Australia and New Guinea. Two tentative explanations may be offered. First, it may be supposed that at the time of the appearance of New Guinea in the Miocene the Australian continent lay further south than it does today. Evidence for such a continental drift of Australia is summarized in Darlington (1965) and recent palaeomagnetic evidence is given in Cox & Doell (1960) and Briden (1967). Secondly, the present ecological barrier to faunal spread between tropical forest and savannah zones may be of long standing. This might explain, in part, why the spread of the Chiroptera into Australia has been more extensive, for the spread of many tropical Aves and Rhopalocera may be limited by available forest cover, whereas of the Chiroptera only the fructivorous Megachiroptera are limited in this way.

(4) We may expect to discover something about past routes of spread for each group from the 2-D representations. As pointed out in section 2, the lower the rank of the taxa selected in a given group the more recent the patterns of spread that will be revealed; similarly, the greater the average spread potential of individuals in a given group the more recent the patterns of spread that will be revealed.

Comparison of taxonomic level between different groups is risky since in the absence of common characters there is no 'morphological yardstick'. However, the situation here is extreme. The birds are generally agreed to have been much 'split' by taxonomists. This is supported by the finding that birds from different taxa of a given rank are much more similar to one another serologically than, for example, mammals from different taxa of the same rank (De Falco, 1942). Tate's treatment of the bats was, as he admitted, that of a 'lumper'. The treatment of species in the Rhopalocera is intermediate between these two extremes. The spread potential is certainly greatest for the Aves, and probably least for the Rhopalocera. We conclude that non-metric multidimensional scaling should reveal more recent trends of spread for the Aves than for the Rhopalocera and Chiroptera.

For the Aves and Rhopalocera two well-marked routes of spread are indicated by chaining of the primary areas in the 2-D representation. The northern (Melanesian) route includes the Philippines, Celebes and the Melanesian arc; the southern (Lesser Sunda) route includes Java, Sumatra and the Lesser Sunda Is. (see Fig. 22). For the Chiroptera only the northern chain is well-marked, though further information about the small bat fauna of the Lesser Sundas might show up a southern route. Geological evidence suggests that the southern route may be more recent (see pp. 171–2) and this is consistent with the fact that it is most clearly indicated for the Aves.

No information about the direction of spread can be obtained from the multidimensional scaling representation. We may, however, obtain such information from the original faunal dissimilarity coefficients by observing the coefficients between successive islands in the chain and the primary areas terminal to the chain. In Fig. 8 these coefficients are plotted for the southern chain. The result indicates clearly that for the Rhopalocera spread has been predominantly from the Sunda-Oriental region with relatively little spread from Australia. For the Aves, also spread has been mainly from the Sunda-Oriental region, but there has been rather more spread from Australia. This may consist of recent spread from the Australian savannah into the savannah habitats of the Lesser Sundas and East Java. The faunal dissimilarity coefficients for the northern chain give little information about the

170 *Proceedings of The Linnean Society of London* [179,

direction of spread, probably because this route of spread is more ancient, and, as shown in the next section, extensive speciation and secondary spread has taken place in the islands which comprise the northern route.

An alternative approach to the analysis of the relation between the distribution of taxa in a group and the geographical disposition of the areas amongst which they are distributed is given in MacArthur and Wilson (1963). A theoretical model, based on the Barton-Davis distribution (see footnote on p. 155), is set up, and it is shown that a 'distance effect' can be derived within the model. The 'distance effect' is the generalization that there is an inverse relation between the distance of islands from a faunal source and the degree of

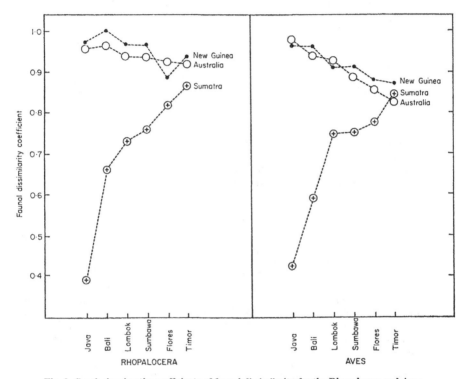

Fig. 8. Graph showing the coefficients of faunal dissimilarity for the Rhopalocera and Aves, between islands in the southern route of spread and Sumatra, Australia and New Guinea.

faunal saturation of islands (cf. Mayr, 1944). The same effect can be derived within the alternative model suggested by Preston (1962). The main difficulty in this approach is that it depends upon the assumption that we can decide *a priori* what is the faunal source for an area. For example, in MacArthur & Wilson (1963) it is assumed that New Guinea is the faunal source for the Melanesian area. In section 5(i) we discuss the possibility of relating faunal elements, determined by present day coincidences in the distributions of taxa, to past centres of radiation. We suggest that it may in some cases be possible to reach conclusions about past centres of radiation; for example in section 5(iv) we reach definite conclusions about the origin of the fauna of New Guinea in relation to its past geographical history. But *a priori* judgements about faunal sources seem to us to be unjustified.

5. FAUNAL ELEMENTS IN THE INDO-AUSTRALIAN AREA

(i) *Introduction*

In this study the geographical coincidences in the distributions of taxa form the sole basis for the recognition of faunal elements. Concepts of floristic elements or types have been used, for example, by Stapf (1913), Matthews (1937) and Hultén (1937), and of faunal elements or types, for example, by Dunn (1931), Stegmann (1938), Mayr (1946), Raup (1947) and Udvardy (1958, 1963). In some cases (Hultén (1937), Stegmann (1938)) common geographical origin has been considered to be part of the definition of a floristic or faunal element or type. Thus Hultén's floral types are explicitly related to centres of isolation in the Pleistocene glaciation. This seems to us to be a mistake. Common geographical origin is but one of several factors which may be used to explain present day coincidences in distribution, just as common ancestry is but one of the factors which may be used to explain present-day morphological resemblances between taxa. Other factors will include common climatic limitations and common ecological requirements. Thus, in the Indo-Australian area the faunal elements in the birds and bats which we have called 'Aru Merauke' elements (see pp. 183–4) are more convincingly explained in terms of common preferences for savannah habitats by the taxa concerned than in terms of their common geographical origin. It is clear that present-day coincidences in the geographical distributions of taxa require explanation, and that no single factor can be used to provide an explanation. To define faunal elements or types in terms of common geographical origin must lead to confusion, for it conceals the fact that the inference from present geographical distribution to past geographical origin is subject to uncertainty.

(ii) *Outline of the geographical history of the Indo-Australian area*

Before presenting and interpreting the results of the numerical derivation of faunal elements we summarize the past geographical and geological history of the area. The main source consulted was van Bemmelen (1949). Reviews by Zeuner (1943), Umbgrove (1943), Mayr (1944), David (1950) and Collenette (1958) were also consulted. Detailed accounts of the geological structure of New Guinea are given in Zeuner (1943) and Cheesman (1951).

In the Late Cretaceous it is thought that there were extensive land connexions between Asia and Australia, and that there was a uniform climate, flora, and fauna throughout. In the Eocene large-scale subsidence occurred with the formation of a geosyncline between the stable blocks of Australia and mainland Asia, the latter extending to Malaya and a large part of Borneo. In the Miocene orogenesis occurred in the unstable area (see Fig. 20).

Initially there was massive downfolding along a line extending from the southern edge of the Sunda block through to the present location of Timor and thence northwards between the Sula Islands and the Moluccas. This downfold is detectable today as a negative gravimetric anomaly. Subsequently land emerged in the Philippines, probably in two separate localities in Celebes (first in the Minahassa and then in the south), and in northern Borneo. To the east of the Philippines, through the Talaud Islands, the North Moluccas and the northern ranges of New Guinea (including the Gautier Mountains, Torricelli Mountains and Finisterre Mountains) to the Bismarck Islands, and possibly the Solomon Islands, a 'Melanesian arc' of islands arose. Towards the end of the Miocene low-lying land, which now constitutes the central range of New Guinea (including the Nassau Mountains, the Oranje Mountains, the Bismarck Range and the Owen Stanley Range) appeared, though this was well isolated from the Melanesian arc.

In the Late Miocene, following the downfold along its length from Malaya to Timor, the two Banda anticlines were pushed up. The outer chain (Andamans, Nias, Timor, Tenimber, Kei, South Moluccas) is thought to have emerged first. The inner chain (Sumatra, Java, the northern Lesser Sunda Islands via Wetar to the Banda Islands), though probably initiated at the same time is thought to have been no more than a chain of small volcanic islands until the Pliocene–Pleistocene period, when further orogenesis uplifted these islands to roughly their present state. The outer chain was further uplifted in this period, especially at its eastern end. This latter orogenesis produced changes in other regions, causing the Sunda Shelf to attain its present form. The Philippines and Celebes also came to approximate to their present geomorphology.

But the major changes in this period were in the New Guinea region. Today New Guinea consists of an old northern range of mountains, a younger central range and a slight ridge in the very south. The areas in between these consist of young marine and alluvial sediments. They are thought to have been covered by sea until very recently (early Pleistocene). The northern range is part of the old Melanesian arc (cf. above). The central range existed in the early Tertiary as a shallow marine basin between the incipient Melanesian arc and North Australia. A first uplift occurred in the Miocene as indicated above, and during the Pliocene–Pleistocene orogeny the range gradually arose further, and became conjoined with the Melanesian part of New Guinea. The southern ridge is probably of late Pleistocene origin.

During the Pleistocene glaciations there were large scale fluctuations in sea-level ranging from c. 100 m higher than at present (during the longest, warmest interglacial) to c. 200 m lower during the glacial periods. The figures are approximate and maximum. During periods of high sea-level the Sundaland area was very much reduced in size. During periods of maximum fall the exposed area may have been as shown in Fig. 21, and much of the Sahul Shelf may have been exposed. If the orogeny in New Guinea was still in progress during the Pleistocene, then it is probable that during the early glaciations southern New Guinea was still submerged. Evidence of the effects of past glaciations in the Snowy Mountains suggests that they were at a much lower level even during the most recent glaciations (Scrivenor, in Scrivenor *et al.*, 1942).

The Indo-Australian area has been cited in support of theories of continental drift (see for example Du Toit, 1937). Evidence for continental drift was reviewed by Darlington (1965) from a biogeographical standpoint. Briefly, he concluded that most biogeographic evidence held to support the theory is open to alternative explanation. He was convinced that the evidence of Palaeozoic glaciation, and the associated *Glossopteris* flora, shows the 'Gondwana' lands to have been separated by smaller water gaps than at present. He was convinced, also, that palaeomagnetic evidence (Cox & Doell, 1960) indicates drift of land masses; in particular it indicates that India and Australia have drifted northwards throughout the late Mesozoic and Tertiary. Recent reviews of the evidence for continental drift are; Blackett (ed., 1965), Good (1966), Runcorn (1966), Harland (1967) and Briden (1967). Drift of Australia in the latter period is relevant to this study.

The geological events outlined above are compatible with a northwards drift of Australia. Such drift may have contributed to the initial geosyncline in the unstable zone, followed by downfolding, giving rise to the negative gravimetric anomaly. Australian drift may also have contributed to the uplift of the Melanesian arc, and, together with drift of the Indian block, to the ruckling of the sea bed against the relatively stable blocks of Sundaland, South Celebes and Melanesia into the folds of the Banda arcs. Du Toit drew attention to the 'crustal swirls' of the eastern ends of the Banda arcs and of the Bismarck Archipelago, and to the way Celebes and Halmaheira appear to have been pushed west. He suggested that the northward force due to drift of Australia, resisted by the Melanesian arc, was resolved into easterly and westerly components. The east-west orientation of the mountain chains of New Guinea may similarly be explained; for to 'flatten out' the folds of these mountains would push the Australian block well to the south of its present position. Such arguments from present geomorphology to past geological events are highly speculative.

Past climate is even less well known. It is thought that in early Tertiary times the climate was much warmer, the tropical zones extending far to the north. Throughout the Tertiary there was a gradual cooling of climate, culminating in the present periods of glaciation. Scrivenor (in Scrivenor *et al.*, 1942) discussed the climate during the Pleistocene, and concluded that the periods of glaciation were cooler, despite certain evidence to the

Table 3. *Taxa used in the computing of faunal elements for the Rhopalocera, numbered as in the dendrogram (Fig. 9). Numbers following generic names refer to unnamed subgeneric and sectional groupings by the author of the revision of the genus (see p. 158)*

1	Bibasis	45	Oerane, Narathura sect. Agaba
2	Coladenia	46	Quedara, Unkana, Narathura sect. Amphi-
3	Allora, Ornithoptera		muta
4	Hasora sect. Lizetta	47	Isma
5	Hasora sect. Myra, Astictopterus, Pithauria,	48	Plastingia
	Iambrix, Scobura, Symbrenthia 3,	49	Lotongus
	Panchala	50	Zela, Narathura sect. Perimuta
6	Hasora sect. Chromus	51	Gangara, Acerbas
7	Hasora sect. Celaenus, Eurema 1, Noto-	52	Erionota
	crypta, Piccarda	53	Ilma, Symbrenthia 1
8	Hasora sect. Thridas	54	Ge, Hidari, Narathura sect. Epimuta
9	Hasora sect. Hasora	55	Taractrocera
10	Badamia, Tagiades sect. Japetus	56	Ocybadistes
11	Choaspes	57	Suniana
12	Euschemon, Trapezites, Hesperilla, Signeta,	58	Potanthus
	Pasma, Motasingha	59	Arrhenes
13	Chaetocneme, Delias 11	60	Telicota
14	Capila, Pintara, Arnetta, Iton	61	Cephrenes
15	Lobocla, Sarangesa, Ctenoptilum, Cartero-	62	Pirdana
	cephalus, Ochus, Baracus, Sovia, Pedesta	63	Prusiana
16	Charmion	64	Ptychandra
17	Celaenorrhinus	65	Sabera
18	Netrocoryne, Toxidia	66	Parnara
19	Tapena, Eetion, Cyrina, Narathura sect.	67	Borbo, Pelopidas
	Rama	68	Polytremis
20	Darpa	69	Caltoris
21	Odina	70	Symbrenthia 2
22	Satarupa	71	Chersonesia
23	Seseria, Ampittia	72	Eurema 2
24	Daimio	73	Ixias
25	Tagiades sect. Nestus	74	Narathura sect. Anthelus
26	Mooreana, Baoris	75	Narathura sect. Camdeo
27	Abraximorpha, Onryza, Thoressa, Ochlodes,	76	Narathura sect. Agesias, Narathura sect.
	Narathura sect. Belphoebe		Vihara, Aurea
28	Exometoeca, Anisyntoides, Anisyntes,	77	Narathura sect. Abseus
	Oreisplanus, Dispar, Mesodina, Croitara	78	Narathura sect. Theba
29	Odontoptilum, Matapa	79	Narathura sect. Hercules
30	Caprona	80	Narathura sect. Cleander
31	Vlasta, Hewitsoniella, Prada, Tiacellia,	81	Narathura sect. Atrax
	Pastria, Banta, Kobrona, Delias 5–10,	82	Narathura sect. Nobilis
	Delias 14–16	83	Narathura sect. Wildei
32	Felicena, Mimene	84	Narathura sect. Acetes
33	Neohesperilla	85	Narathura sect. Eumolphus
34	Aeromachus	86	Narathura sect. Centaurus
35	Sebastonyma, Pudicitia	87	Narathura sect. Fulla
36	Stimula	88	Delias 1
37	Halpe	89	Delias 2
38	Isoteinon	90	Arhopala, Delias 19
39	Koruthaialos, Suada, Hyarotis, Narathura	91	Delias 4, Schoenbergia
	sect. Agelastus	92	Delias 12
40	Psolos, Cupitha, Narathura sect. Aedias,	93	Delias 13
	Narathura sect. Democritus, Flos	94	Delias 17
41	Ancistroides	95	Delias 18
42	Udaspes	96	Trogonoptera
43	Suastus	97	Troides
44	Zographetus, Oriens		

contrary (for example, Dickerson, 1928). Climate on the Sunda Shelf must have fluctuated from maritime to continental with change in sea-level. At times of low sea-level there are thought to have been two large river basins in Sundaland, one flowing between Borneo and Java, the other between Borneo and Indo-China; oceanographic evidence for this is cited in van Bemmelen (1949). At present most of the lowland areas from Burma to north-east Queensland are covered by tropical forest of Asian origin (Lam, 1934; Burbidge, 1960) The age of the tropical forests of north-east Queensland is disputed, and hence the extent to which the faunal isolation of Australia can be accounted for by the existence of a long standing ecological barrier is uncertain. In the Lesser Sundas from East Java to Tenimber, and in parts of Celebes, the influence of the Australian continent is evident and conditions resulting in savannah predominate. The phytogeography of the area is reviewed by Lam (1934), van Steenis (1934–1936) and Keast (ed., 1959).

(iii) *Sources of data*

(a) *Rhopalocera*. The same sources of information about distribution were used as in the derivation of zoogeographic regions. The taxa selected were of subgeneric rank. Those taxa whose recorded distributions are entirely coincident were treated as single 'units' for computational purposes. The taxa are listed in Table 3, numbered as they appear in the dendrogram.

(b) *Aves*. Taxa were selected at the subgeneric level from the *Check List of the Birds of the World*. The taxa used are listed in Table 4, numbered as they appear in the dendrogram.

(c) *Chiroptera*. The same taxa as were used in the derivation of zoogeographic regions were used in this study, and are listed in Table 5, numbered as they appear in the dendrogram.

Table 4. *Taxa used in the computing of faunal elements for the Aves, numbered as in the dendrogram (Fig. 10)*

1 Zanclostomus, Platylophus, Hemicircus	16 Saxicola
2 Blythipicus, Micropternus	17 Actinodura, Paradoxornis, Spizixos
3 Harpactes, Chrysocolaptes, Copsychus, Stachyris	18 Geomalia, Cataponera, Malia, Enodes, Scissirostrum, Cryptophaps,
4 Amalocichla, Androphobus, Ptilorrhoa, Melampitta, Ifrita, Campochaera, Chaetorhynchus, Archboldia, Amblyornis, Loria, Pardigalla, Epimachus, Drepanornis, Astrapia, Loboparadisea, Cnemophila, Macgregoria, Manucodia, Seleucidis, Paradisea, Lophorina, Parotia, Pteridophora, Cicinnurus, Diphyllodes, Rallicula, Megacrex, Talegalla, Anurophasis, Clytoceyx, Melidora, Garritornia, Trugon, Otidiphaps, Goura, Psittrichas, Pseudeos, Oreopsittacus, Neopsittacus, Psittaculirostris, Psittacella, Caliechthrus, Microdynamis	Macrocephalon, Aramidopsis, Cittura, Meropogon, Diopezus
	19 Pycnonotus
	20 Ceyx, Halcyon, Centropus, Tyto, Artamus, Corvus, Zoothera, Coracina, Oriolus
	21 Criniger, Aegithina, Dinopium
	22 Turdus
	23 Setornis, Platysaurus, Melanoperdix, Houppifer, Lophura, Argusianus
	24 Opopsitta, Probosciger, Dacelo, Ailuroedus, Sericulus, Phonygammus, Orthonyx
	25 Chloropsis, Irena
	26 Corcorax, Struthidia, Cracticus, Strepera, Leipoa, Eulabornis, Tribonyx, Choriotis, Erythrogonys, Recurvirostra, Ocyphaps, Phaps, Lophophaps, Geophaps, Histriophaps, Lathamus, Calypsorhynchus, Lophochroa, Eolophus, Platycercus, Northiella, Psephotus, Melopsittacus, Psophodes
5 Geopelia, Mirafra	
6 Chlamydochaera, Haematortyx, Lobiophasis	
7 Brachypteryx	
8 Lalage	
9 Rhyacornis	
10 Cinclidium, Pavo	27 Hypsipetes
11 Pericrocotus	28 Sphenostoma, Glossopsitta, Zanda, Callocephalon, Licmetis, Nymphicus, Polytelis, Spathopterus, Purpureicephalus, Neopherna, Pezoporus, Geopsittacus, Zonifer, Elseyornis, Peltolycus, Pedionomus
12 Enicurus, Myiophoneus, Pomatorhinus	
13 Hemipus	
14 Cochoa	
15 Pellorneum, Tephrodornis, Cissa, Lacedo	

Table 4—*continued*.

29	Aploornis	72	Hydrophasianus
30	Grallina, Gymnorhina, Calamydera, Cinclosoma, Pomatostomus	73	Nesoclopeus, Microgoura, Coryphoenas
		74	Metopidius
31	Leucopsar	75	Gymnocrex
32	Acridotheres	76	Habropteryx
33	Terricolumba, Mino	77	Rogibyx
34	Berenicornis, Anorrhinus, Eupetes, Kenopia, Psittinus, Rhinortha, Rollulus	78	Podargus
		79	Lobibyx
35	Basilornis	80	Eurylaimus, Batrachostomus
36	Trichastoma	81	Scolopax
37	Streptocitta	82	Sphenurus
38	Mimizuku, Pseudoptynx, Leonardina, Micromacroum, Phapiteron, Darylophus, Lepidogrammus, Bolbopsittacus, Sarcops	83	Treron
		84	Aegotheles
		85	Leucoteron, Penelopides
39	Gracula	86	Hemiprocne
40	Malacopteron	87	Ptilinopus
41	Aprosmictus, Sphecotheres	88	Pelargopsis
42	Rimator	89	Megaloprepia
43	Alectura, Leucosarcia, Lopholaimus, Petrophassa, Scenopoeetes, Prionodura, Ptylonorhynchus, Ptiloris	90	Tanysiptera
		91	Ducula
		92	Collocalia
44	Ptilochichla	93	Domicella, Charmosyna, Micropsitta, Gymnophaps
45	Habroptila, Lycocorax, Semioptera		
46	Napothera, Buceros, Sasia	94	Aceros
47	Arborophila, Pteruthius, Alcippe, Garralux, Dendrocitta	95	Macropygia
		96	Reinwardtoena
48	Heterophasia	97	Anthracoceros
49	Crypsirina	98	Chalcophaps
50	Yuhina, Ketupa, Picus	99	Serilophus
51	Megapodius	100	Henicophaps
52	Eulipoa, Eos	101	Megalaima
53	Phodilus, Cypsiurus	102	Gallicolumba
54	Coturnix	103	Chalcopsitta
55	Synoicus	104	Mulleripicus
56	Otus	105	Dryocopus
57	Excalfactoria	106	Dendrocopos
58	Poliolimnas	107	Trichoglossus
59	Tropicoperdix, Rhopodytes, Nyctyornis, Calyptomena, Corydon, Calorhamphus, Cymbirhynchus, Psarisomus	108	Psitteuteles
		109	Kakatoe
		110	Ducorpsius
60	Tribonyx	111	Lorius
61	Caloperdix	112	Geoffroyus
62	Gallicrex	113	Prioniturus
63	Bambusicola	114	Tanygnathus
64	Chalcurus, Psilopogon	115	Alisterus
65	Gallinula	116	Loriculus
66	Polyplectron	117	Penthoceryx
67	Porphyrio	118	Cacomantis
68	Turnix	119	Surniculus
69	Rallus	120	Eudynamys
70	Hypotaenidia	121	Rhamphococcyx
71	Irediparra		

(iv) *Results*

The lists of coefficients of dissimilarity in distribution for each group are too bulky to be given here. The dendrograms derived by a single-link method are shown in Figs 9, 10 and 11. On each dendrogram the clusters corresponding to the faunal elements recognised are numbered. The distributions of taxa in each faunal element amongst the primary areas are given in Tables 6, 7 and 8. Some of the more important faunal elements are displayed in contour form on maps of the Indo-Australian areas in Figs 12–19.

Table 5. *Taxa used in the computing of faunal elements for the Chiroptera, numbered as in the dendrogram (Fig. 11)*

1 Roussettus leschenaulti	42 Pipistrellus affinis
2 Roussettus amplexicaudatus	43 Pipistrellus circumdatus
3 Boneia bidens, Styloctenium wallacei,	44 Pipistrellus tasmaniensis, Macroderma,
Eonycteris rosenbergi, Pipistrellus	Rhinolophus megaphyllus, Rhinonycteris
minahassae	45 Chalinolobus. 4 spp.
4 Pteropus hypomelanus	46 Pipistrellus joffrei
5 Pteropus mariannus	47 Glischropus
6 Pteropus conspicillatus	48 Tylonycteris pachypus
7 Pteropus caniceps	49 Tylonycteris robustulus
8 Pteropus melanopogon	50 Eptesicus
9 Pteropus rayneri, Dobsonia viridis,	51 Scoteinus 2 spp., Taphozous australis,
Syconycteris crassa, Emballonura	Hipposideros muscinus
raffrayana, Aselliscus 2 spp.	52 Scotophilus temminckii
10 Pteropus lombocensis, Dobsonia peroni	53 Scotophilus heathii
11 Pteropus samoensis, Notopteris macdonaldi,	54 Miniopteris schreibersii
Lasiurus semotus, Emballonura spp.,	55 Miniopteris tristis
Mystacops	56 Miniopteris australis
12 Pteropus pselaphon	57 Murina
13 Pteropus temmincki	58 Harpiocephalus
14 Pteropus vampyrus	59 Kerivoula hardwickii
15 Pteropus alecto	60 Kerivoula pusilla
16 Pteropus neohibernicus, Melonycteris	61 Kerivoula papillosa
17 Pteropus macrotis, Pteropus scapulatus,	62 Kerivoula picta
Syconycteris australis	63 Phoniscus
18 Acerodon celebensis	64 Nyctophilus
19 Acerodon jubatus	65 Chaerephon johorensis
20 Pteralopex 2 spp., Nesonycteris,	66 Chaerephon plicatus
Anamygdon, Anthops	67 Chaerephon 2 spp., Saccolaimus flaviventris
21 Dobsonia minor, Paranyctimene, Philetor	68 Mormopterus
22 Dobsonia magna, Nyctimene papuanus	69 Emballonura monticola
23 Cynopterus sphinx, Pipistrellus abramus	70 Mosia
24 Cynopterus brachyotis	71 Taphozous longimanus, Nycteris javanica
25 Niadius	72 Taphozous melanopogon
26 Ptenochirus, Haplonycteris, Saccolaimus	73 Saccolaimus saccolaimus
pluto	74 Megaderma
27 Megaerops	75 Rhinolophus simplex
28 Dyacopterus, Hesperopterus	76 Rhinolophus borneënsis
29 Balionycteris, Penthetor	77 Rhinolophus affinis
30 Chironax	78 Rhinolophus lepidus
31 Thoöpterus	79 Rhinolophus arcuatus
32 Nyctimene cephalotes	80 Rhinolophus euryotis
33 Eonycteris spelaea, Cheiromeles	81 Rhinolophus philippinensis
34 Macroglossus minimus, Rhinolophus	82 Rhinolophus macrotis
trifoliatus	83 Rhinolophus luctus
35 Macroglossus lagochilus	84 Hipposideros bicolor
36 Harpionycteris 2 spp.	85 Hipposideros calcaratus
37 Selysius	86 Hipposideros galeritus
38 Chrysopteron	87 Hipposideros diadema
39 Leuconoë	88 Hipposideros speoris
40 Pipistrellus coromandra	89 Aselliscus
41 Pipistrellus tenuis	90 Coelops

In the following sections the past history of each group is discussed in the light of the faunal elements recognized and the past geographical history of the area. It may be useful in reading these accounts to refer to Figs 20 and 21 in which routes of spread are related, speculatively, to the Miocene and Pleistocene geography of the Indo-Australian area.

(a) *Rhopalocera*

Five main faunal elements may be recognized; a Sundaland based element (Fig. 12), a Melanesian element (Fig. 13), an element endemic to all Australia, an element endemic to

Table 6. *The distributions amongst the primary areas of taxa in the faunal elements recognized for the Rhopalocera. The numbers assigned to the elements are the same as those assigned to the corresponding clusters in the dendrogram (Fig. 9). The distributions of elements 1 and 2 are shown in contour form in Figs 12 and 13 respectively*

Faunal Elements	Formosa	Indo-China	Burma	Sunda Is.	Java	Lesser Sunda Is.	Timor	Palawan	Phillipines	Celebes	Sula Is.	N. Moluccas	S. Moluccas	N. Australia	S. Australia	Bismarck Is.	Solomon Is.	New Guinea
1 Sunda-based	34	85	110	98	80	36	21	54	70	54	23	28	28	15	2	16	15	23
a	12	29	35	36	35	9	2	24	31	20	5	1						
b	14	18	21	21	21	21	15	16	21	21	14	21	21	12		14	14	19
2 Melanesian						2	1			2		12	13	10	1	12	6	18
3 Australian					1									7	7			
Endemics			2	3						2		1	7					16

Table 7. *The distributions amongst the primary areas of taxa in the faunal elements recognized for the Aves. The numbers assigned to the elements are the same as those assigned to the corresponding clusters in the dendrogram (Fig. 10). The distribution of element 1 is shown in contour form in Fig. 14, the distribution of elements 1b and 2 in Fig. 15, the distribution of elements 3 and 7 in Fig. 16, and the distribution of elements 5, 6 and 8 in Fig. 17*

Faunal Elements	Formosa	China	Indo-China	Burma	Malaya and Sumatra	Palawan	Borneo	Java	Bali	Lesser Sunda Is.	Timor	Phillipines	Celebes	Sula Is.	N. Moluccas	S. Moluccas	New Guinea	Bismarck Is.	Solomon Is.	N. Australia	S. Australia
1 Sunda-based	37	49	96	112	118	44	106	87	54	36	28	56	40	24	24	28	27	27	21	22	14
a	18	25	46	50	52	37	50	52	37	27	22	38	25	19	18	19	21	21	18	19	12
b	13	16	16	16	16	1	14	8	3			1									
c			3	3	3		3	2	3			3	3	2	3	3	3	3			
2 Melanesian										1					8	8	8	8	6	3	
3 Australian											2		1	3	1	18	2			43	33
4 Melanesian																	4	3	3		
5 ⎫ Aru-Merauke										1	2				1	2				2	2
6 ⎭										1	2	1			1	2				2	2
7 Wallacean												4	4	1	2						
8 Aru-Merauke							2	1	1	2	1				2	2				2	1
9 Palaearctic	2	2										1									
Endemics				1	2						1	9	11		3		43	1	3	8	16

Table 8. *The distribution amongst the primary areas of taxa in the faunal elements recognized for the Chiroptera. The numbers assigned to the elements are the same as those assigned to the corresponding clusters in the dendrogram (Fig. 11). The distributions of elements 1 and 2 are shown in Figs 18 and 19*

Faunal Elements	India	Africa	China	Indo-China	Malaya	Sumatra	Borneo	Java and Bali	Phillipines	Celebes	Sanghir Talaut	Lesser Sunda Is.	Moluccas	New Guinea	Solomon Is.	Australia	Oceania
1 Sunda-based	14	2	17	27	36	31	26	32	24	15	6	5	11	9	7	8	1
a	7	5	5	8	8	7	8	7	5	4	3	7	6	6	7	1	
b	2	1	7	9	9	9	4	9	5	1	1	1					
c				7	10	9	8	6	1	1				2			
d					3	1	3	3	2								
2 Melanesian									2	5		5	17	25	14	15	
a													2	9	12	9	7
b									1	2			2	2	2	2	
c											3	2	3	2			
3 Wallacean										3	3	1	2				
4 Aru-Merauke						1				2		2	1	2		2	
5 Oceanic													3		1	1	3
Endemics						2	1	3	4	2		1	3	4		4	5

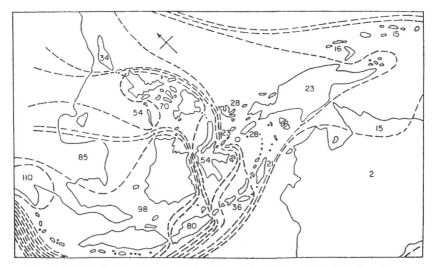

Fig. 12. The distribution of taxa in the Sunda-based faunal element (1) of the Rhopalocera.

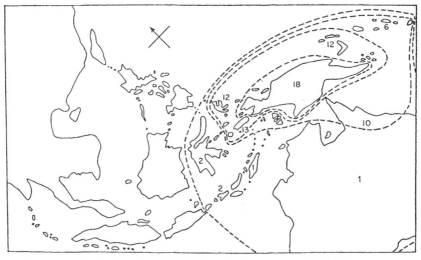

Fig. 13. The distribution of taxa in the Melanesian faunal element (2) of the Rhopalocera.

southern and western Australia, and an element endemic to New Guinea. Other smaller elements, clustering at high levels in the dendrogram, are either endemic to a single primary area or consist of one or two genera having anomalous distributions.

The Australian endemics considered in this study are all members of one taxon of higher rank, the Trapezitinae, also centred on Australia and showing only minor spread from the area. (Two genera of the Trapezitinae occur in New Guinea and one occurs in both New Guinea and Australia.) It seems that the Australian endemic element is very old. If the

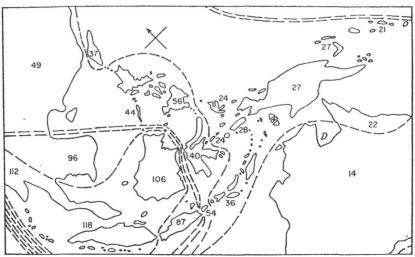

Fig. 14. The distribution of taxa in the Sunda-based element (1) of the Aves.

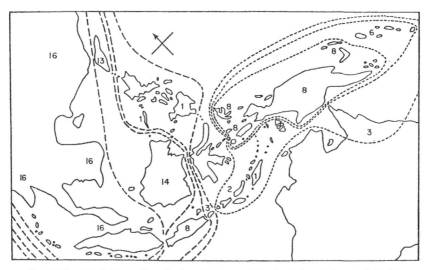

Fig. 15. The distributions of taxa in the sub-element of the Sunda-based element limited by Weber's line (1b), and in the Melanesian element (2), of the Aves.
————, Element 1b; - - - -, element 2.

Trapezitinae and other Australian endemics had arrived in the Miocene when land between Asia and Australia reappeared, we should expect to find related forms in Melanesia. We conclude that the Australian endemic element reached Australia in the early Tertiary. It follows that the faunal elements centred on New Guinea and Melanesia are of Asian origin, having spread into the area when the land area became sufficient for their establish-

12

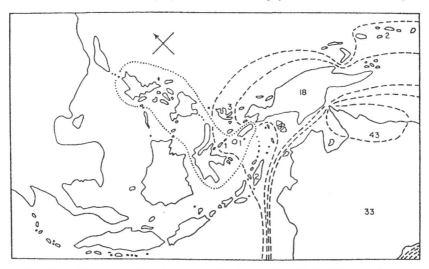

Fig. 16. The distributions of taxa in the Australian element (3), and in the Wallacean element (7), of the Aves. The numbers of taxa are given only for the Australian element.
————, Element 3; . . . . , element 7.

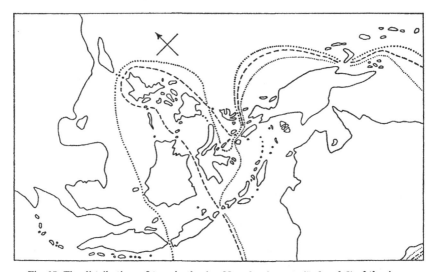

Fig. 17. The distributions of taxa in the Aru-Merauke elements (5, 6 and 8) of the Aves.
. . . . , Element 5; ————, element 6; . . . . , element 8.

ment and further radiation in the Miocene. This is supported by Zeuner's (1943) study of the *Troides* aggregate of genera, where the initial isolation of the four genera involved is dated approximately at the Miocene–Pliocene. Thus in the Miocene the general picture emerges of spread into the Melanesian area and subsequent radiation of groups there. This has probably continued until Recent times.

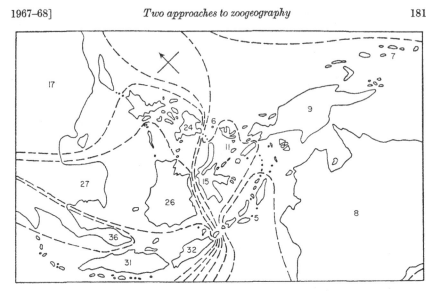

Fig. 18. The distribution of taxa in the Sunda-based faunal element (1) of the Chiroptera.

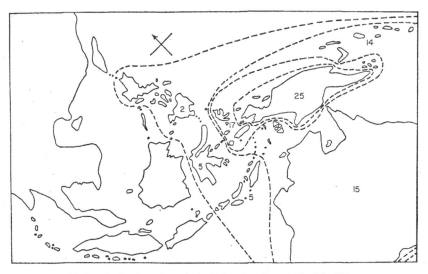

Fig. 19. The distribution of taxa in the Melanesian element (2) of the Chiroptera.

The study of zoogeographic regions indicates a barrier to spread at Weber's line, which also marks the western boundary of the Melanesian element. The Sunda-based element shows spread into Melanesia, but the contours are aggregated between the Sunda and Wallacean sub-regions, and along the northern part of Weber's line, suggesting that there have been checks to spread. Taxa that have crossed Weber's line in recent times may still be under isolative pressure in Melanesia. An interesting example of this has emerged in studies at present being carried out (J. D. H.) on the genus *Symbrenthia*. *Symbrenthia*

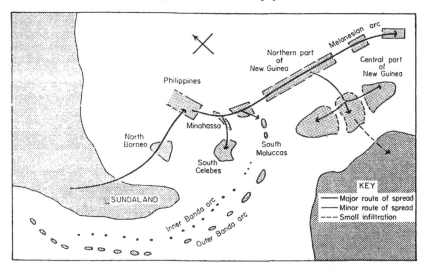

Fig. 20. Diagrammatic representation of Miocene geography (partly after Umbgrove, 1943) illustrating suggested routes of spread.

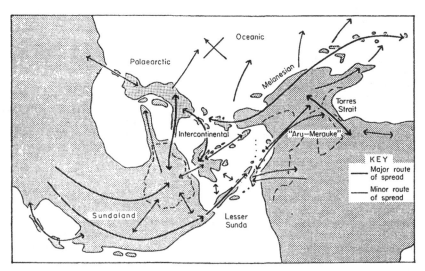

Fig. 21. Diagrammatic representation of Pleistocene geography at periods of low sea-level, illustrating and naming the suggested routes of spread.

hippoclus Cramer, with a distribution from Sundaland to the Philippines, Celebes, Sula Isles, Lesser Sundas to Timor, and to the Moluccas, New Guinea and the Louisiade Archipelago, should be divided into two taxa on the basis of the morphology of the genitalia and other characters. One of these forms replaces the other to the east of Weber's line. Similar replacement at Weber's line is seen in the subspecies of *Appias paulina* Moore, and *A. nero* Butler. There are many other cases where taxonomic differentiation within genera coincides with the faunal elements recognized. Such coincidences give powerful support to the interpretation of the faunal elements derived by numerical methods as representing past centres of isolation and speciation (see p. 169).

The butterfly fauna of New Guinea belongs to three distinct faunal elements; about one third are endemic, one-third belong to a pandemic sub-element of the Sunda-based element, and about one-third are Melanesian. This heterogeneity of the New Guinea butterfly fauna shows up well in Talbot's (1928–37) analysis of *Delias*, and is noted also by Carpenter (1953) for the distribution of species of *Euploea*, and Lieftinck (1949) for the distribution of dragonflies. We conclude that this faunal heterogeneity can be related directly to the geological history of New Guinea (see p. 171 and Fig. 20), the Melanesian and endemic elements having arisen separately as a result of spread from Asia into the Melanesian arc and into the formerly separate southern part of New Guinea since the Miocene. That this isolation has continued until quite recently was suggested by Toxopeus (1950) in studies of the subspecific variation of butterflies in the Papuan region.

The distribution of the Sunda-based element (Fig. 12) and the results of multidimensional scaling (Fig. 5B) suggest that relatively little spread into the Papuan region or Australian has taken place along the southern (Lesser Sunda) route of spread, *Udaspes* and *Ixias* (42 and 73) being exceptions. Similarly there is little spread from Australia into the Lesser Sundas, though *Eurema smilax* (treated here as a section of *Eurema*), centred in Australia, is found in Java, and *Ocybadistes*, a hesperid genus, shows an 'Aru Merauke' distribution comparable with that found in the Aves and Chiroptera (see pp. 183–4 and Fig. 17). Since the southern (Lesser Sunda) route of spread, revealed by multidimensional scaling, is thought to be more recent than the northern (Melanesian) route, we should expect it to be more strongly indicated in a study of faunal elements in the Rhopalocera at the specific level. Preliminary study indicates that this is in fact the case.

(b) *Aves*

The faunal elements derived for the Aves match well those derived for the Rhopalocera. However subgenera of the Aves are, in general, at a lower taxonomic level than those of the Rhopalocera (see p. 169), and the dispersive powers of birds are greater. Hence we should expect the faunal elements derived to yield information about more recent spread and speciation. This is confirmed by preliminary study (J. D. H.) of the faunal elements derived from the distribution of Rhopalocera at the specific level which match the faunal elements derived for the Aves even better than those derived at the subgeneric level. It is further confirmed by the high degree of endemism in the primary areas compared with that for the Rhopalocera and Chiroptera (the numbers of taxa endemic to each primary area for the Rhopalocera, Aves and Chiroptera are shown in Tables 6, 7 and 8 respectively).

The faunal elements recognized for the birds are shown in Table 7, and the main elements are displayed in contour form in Figs 14, 15, 16, and 17. The situation is more complex than for the Rhopalocera, but the same general pattern emerges. There is a large Sunda-based element (1) shown in Fig. 14, with several sub-elements. Among the sub-elements are a large evenly spread 'pandemic' element, (1a), a mainland-centred element bounded by Weber's line (1b) shown in Fig. 15, and an element widespread, but absent from China, Formosa and Australia (1c). Element 1b clearly indicates a barrier to faunal spread at Weber's line.

Australia shows less faunal isolation for the Aves than for the Rhopalocera. Element 3, shown in Fig. 16, is a north-east Australian centred element, including the 24 genera occurring in both north and south Australia but not elsewhere. It includes also 17% of the New Guinea genera. As for the Rhopalocera many of the endemic Australian genera are certainly of ancient Asian origin, but the spread between Australia and New Guinea, and the 'Aru Merauke' spread, are probably fairly recent (Pleistocene). New Guinea is again shown to be faunally heterogeneous. 42% of the genera are endemic, probably having spread into the area from Asia in the Miocene. 26% of the genera belong to the widespread sub-elements, 1a and 1c, of the Sunda-based element, and some of these may be relatively recent immigrants. A small part of the bird fauna (10%) is referable to the two Melanesian elements, 2 (shown in Fig. 15) and 4·6% belong to the 'Aru Merauke' elements (5, 6 and 8) shown in Fig. 17. The heterogeneity revealed in the bird and butterfly fauna of New Guinea contrasts with the conclusions drawn by Mayr (1953), who based an attack on the continental drift hypothesis on the assumption that there is a uniform Australo-Papuan fauna.

As for the Rhopalocera, the existence of a northern route of spread, shown in the results of multidimensional scaling, is supported by the presence of Melanesian elements. This Melanesian route of spread, as suggested for the Rhopalocera, probably arose in the Miocene (see Fig. 20).

Element 7 (shown in Fig. 16) a Philippine-Celebian or 'Wallacean' element, is of particular interest. Both the zoogeographic classification and the results of multidimensional scaling show a faunal link between Celebes and the Philippines. The Philippines and Celebes form a 'Wallacean' sub-region of the Sunda-Oriental region for the Aves and Rhopalocera and, together with the Moluccas, a 'Wallacean' region for the Chiroptera. In all the multidimensional scaling representations (Figs 5B, 6B, 7B) Celebes appears between the Philippines and the Moluccas and Sula Islands as part of the northern route of spread, and is widely separated from both Borneo and the Lesser Sundas. This ties up well with what is known of the past history of Celebes. The faunal links with the Philippines are consistent with the emergence in the Miocene of land in the Minahassa peninsula earlier than, and separated from, land in the south of Celebes. The absence of faunal links with Borneo is consistent with the hypothesis that Celebes has been 'pushed' westwards.

More recent spread in the southern route revealed by multidimensional scaling (see Figs 6B and 8) is shown up by the 'Aru Merauke' elements (5, 6 and 8) shown in Fig. 17. They probably represent recent spread of savannah habitat forms from Australia via the geologically most recent southern part of New Guinea and Aru to the Lesser Sundas, Philippines and Celebes.

The pattern of distribution of birds at the subgeneric level is thus seen to be more complex that that for butterflies at the subgeneric level, more weight being given to relatively recent spread. Comparison of the dendrogram for the Aves, Fig. 10, with the dendrogram for the Rhopalocera, Fig. 9, shows that there are many more single genera or small groups of genera clustering at a high level. Some of these are endemic genera; the rest are genera having anomalous or disjunct distributions. For example, the genus *Megapodius*, used by Huxley (1868) in support of his proposed demarcation of the Oriental and Australian faunas, is anomalous. One of these small elements, 9, is of interest for it indicates spread of Palaearctic montane forms into the area. This would have been more clearly shown up if the study had been carried out at the specific level, or if we had made preliminary segregation of taxa into lowland and montane forms. (cf. Moreau, 1966, who showed that the faunal affinities of lowland and montane birds in tropical Africa differ widely.)

(c) *Chiroptera*

The results for the Chiroptera must be interpreted with caution. The primary areas used are taken from Tate (1946) and are rather unsatisfactory, the Lesser Sundas for instance being treated as a single primary area. The taxa used are also taken from Tate and are mostly 'aggregates' of closely related but vicarious species. The distributions of the elements are shown in Table 8, and the two main elements, a Sunda-based element, 1, and a Melanesian element, 2, are shown in Figs 18 and 19 respectively.

The patterns of distribution revealed differ markedly from those for the Aves and Rhopalocera. This almost certainly arises from the fact that there can have been no spread of the Chiroptera into Australia from Asia before the subsidence of intermediate land areas at the beginning of the Tertiary, the earliest known Chiroptera being of Eocene origin (Darlington, 1957: 319). This is confirmed by the fact that about half of the Australian bat fauna falls into the Melanesian element centred on New Guinea, and hence must represent Miocene or post-Miocene spread from New Guinea. It is confirmed also by the fact that the percentage of endemic taxa in areas to the east of Weber's line is much lower than for the Rhopalocera and Aves.

Within the Sunda-based element we may note a uniform pandemic sub-element, 1a, and three sub-elements largely restricted to Sundaland (1b, 1c and 1d). Within the Melanesian element three sub-elements may be recognized (2a, 2b and 2c), 2a being centred on New Guinea. Element 5, which includes the Oceanic taxa, is represented also in the Philippines and Melanesia. This correlates with the conclusion of Zimmerman (1942) and Gressitt (1960) that the insect fauna of the Pacific Islands originated in the Philippines and Melanesia. Element 4 shows an 'Aru Merauke' distribution, and element 3 is an intercontinental, or 'Wallacean' element.

The faunal elements recognized here correlate badly with those derived by Tate (1946) from the same data, probably because Tate based his faunal elements upon presumed centres of origin of the taxa concerned, an approach criticised on p. 169. However, the results confirm Tate's suggestion that there were two waves of invasion from Asia, an early (Miocene) one in which the Megachiroptera predominated, and a later one in which the Microchiroptera predominated, the Megachiroptera being most strongly represented in the New Guinea centred sub-element of the Melanesian element, 2a. The paucity of the Megachiroptera in Australia may result from the fact that, being fructivorous, they are more limited to tropical forest.

CONCLUSION

This study was undertaken primarily as an exercise in method. In other words, the main aim was to test the extent to which numerical methods can be used as adjuncts to intuitive methods in zoogeography. Our application of numerical methods has been rather crude; we have used distributional data that are probably in many cases incomplete; our selection of primary areas is imprecise being determined by the way in which various authors have described the distributions of taxa in the Indo-Australian area; and we have relied upon other authors' taxonomic treatment of the groups concerned. Nevertheless, the results seem promising.

Some of the ways in which the methods described here might be refined are: (a) Where precise distributional data are available, quantitative methods for delimiting regions of relative faunal homogeneity as primary areas may be devised (see p. 155). (b) Ideally we suggest that a taxonomic revision of each of the groups of organisms studied should preceed the application of numerical methods for the analysis of their distributions. As has been mentioned (see p. 158), a numerical taxonomic revision might be used as a basis for differential weighting of the distributions of taxa. (c) The methods of classification used in this study are hierarchic. Recent work suggests that the numerical methods for non-hierarchic classification developed by Jardine & Sibson (in press, *a* and *b*) may prove to be a more powerful tool for use in zoogeographic studies. Application of non-hierarchic methods of classification to the study of the distribution in Australia of endemic Rhopalocera and Blattidae of the tribe of the Polyzosterinae is at present being undertaken.

From the study of zoogeographic regions and from the study of faunal elements, we conclude that all generalizations made on the basis of analysis of distributional data should be considered to be *strictly relative* to the particular area studied and to the particular taxonomic group studied. Only experiment can determine the extent to which classifications into zoogeographic regions and faunal elements will coincide with those obtained in studies covering a wider area, or based upon different taxonomic groups. As shown in this study numerical methods are most powerful when used on a comparative basis, for the results may then be used to infer something about past trends of spread and speciation of forms referable to particular taxonomic groups in relation to the past geographical and geological history of an area.

We conclude that both intuitive and numerical methods for grouping primary areas into zoogeographic regions are in themselves mainly of use for descriptive purposes. It is only when such methods are used in conjunction with such more refined methods as multi-dimensional scaling of primary areas and grouping of taxa into faunal elements that detailed inferences about the history of particular taxonomic groups can be made.

We feel that studies based upon the distributions of large groups over a wide area should be regarded only as a first stage in the zoogeographic study of a region. The second stage should consist of detailed studies of the patterns of variation and distribution of taxonomic groups of lower rank and attempts to relate these in detail to past and present geographical and ecological factors.

ACKNOWLEDGEMENTS

Our thanks are due to Dr John Smart, Dr W. B. Harland, Mr P. D. Sell, Mr C. W. Benson and Dr A. D. McLaren for advice and encouragement, to Mr C. J. Jardine and Miss J. Champion who helped in the preparation of the manuscript and in the computing and to Mrs V. Cole who typed the manuscript.

REFERENCES

BEMMELEN, R. W. VAN, 1949. *The Geology of Indonesia*. The Hague: Government Printing Office.
BLACKETT, P. M. S. (ed.), 1965. Symposium on continental drift. *Phil. Trans. R. Soc.* **1088**: 1–323.
BRIDEN, J. C., 1967. Recurrent continental drift of Gondwanaland. *Nature, Lond.* **215**: 1334–9.
BRONGERSMA, L. D., 1936. Some comments upon H. C. Raven's paper: 'Wallace's line and the distribution of Indo-Australian mammals'. *Archs néerl. Zool.* **2**: 240–56.
BROOKS, C. J., 1950. A revision of the genus *Tenaris* Hübner (Lepidoptera: Amathusiidae). *Trans. R. ent. Soc. Lond.* **101**: 179–238.
BURBIDGE, N. T., 1960. The phytogeography of the Australian region. *Aust. J. Bot.* **8**: 75–211.
CARPENTER, G. D. H., 1953. The genus *Euploea* (Lep. Danaidae) in Micronesia, Melanesia, Polynesia and Australia. A zoogeographical study. *Trans. zool. Soc. Lond.* **28**: 1–165.
CATTELL, R. B., 1965. Factor analysis: an introduction to essentials. 1. The purpose and underlying models. *Biometrics*, **21**: 190–215.
CHEESMAN, L. E., 1951. Old mountains of New Guinea. *Nature, Lond.* **168**: 597.
COLLENETTE, P., 1958. *Geology of the Kinabalu and Jesselton area*. Sarawak Govt. Press, Mem. No. 6.
COX, A. G. & DOELL, R. R., 1960. Review of palaeomagnetism. *Bull. geol. Soc. Am.* **71**: 645–768.
DAGNELIE, P., 1960. Contribution à l'étude des communautés végétales par l'analyse factorielle. *Bull. Serv. Carte phytogéogr.* (Sér. B), **5**: 7–71, 93–195.
DARLINGTON, P. J. JR., 1948. The geographical distribution of cold-blooded vertebrates. *Q. Rev. Biol.* **23**: 1–26, 105–23.
DARLINGTON, P. J. JR., 1957. *Zoogeography: the geographical distribution of animals*. New York: John Wiley.
DARLINGTON, P. J. JR., 1965. *Biogeography of the Southern End of the World*. Cambridge, Mass.: Harvard University Press.
DAVID, T. W. E., 1950. *The geology of the Commonwealth of Australia*. London: Edward Arnold, 3 vols.
DE FALCO, R. J., 1942. A serological study of some avian relationships, *Biol. Bull. mar. biol. Lab., Woods Hole*, **83**: 205–18.
DICKERSON, R. E., 1928. The distribution of life in the Philippines. *Monogr. Philipp. Bur. Sci.* **21**:1–322.
DUNN, E. R., 1931. The herpetological fauna of the Americas. *Copeia*, 1931: 106–19.
DU TOIT, A. L., 1937. *Our wandering continents*. Edinburgh: Oliver and Boyd.
EVANS, W. H., 1949. *A catalogue of the Hesperiidae from Europe, Asia and Australia in the British Museum (Natural History)*. London: British Museum (Nat. Hist.).
EVANS, W. H., 1954. A revision of the genus *Curetis* (Lepidoptera: Lycaenidae) *Entomologist*, **87**: 190, 212.
EVANS, W. H., 1957. A revision of the *Arhopala* group of oriental Lycaenidae (Lepidoptera: Rhopalocera). *Bull. Br. Mus. nat. Hist. (Entomology)*, **5**: 85–141.
FLORIN, R., 1963. The distribution of conifer and taxad genera in time and space. *Acta Horti Bergiani*, **20**: 122–312.
GABRIEL, A. G., 1943. A revision of the genus *Ixias* Hübner (Lepidoptera: Pieridae). *Proc. R. ent. Soc. Lond.* (ser. B), **12**: 55–70.
GILMOUR, J. S. L., 1950. Taxonomy and philosophy. In J. S. Huxley (ed.) *The New Systematics*, pp. 461–74. Oxford: Clarendon Press.

GILMOUR, J. S. L., 1961. Taxonomy. In A. M. Macleod & A. S. Cobley (eds), *Contemporary botanical thought*, pp. 27–43. Edinburgh: Oliver and Boyd.

GOOD, R., 1966. The botanical aspects of continental drift. *Sci. Prog., Oxf.* 54: 315–24.

GOODALL, D. W., 1952. Quantitative aspects of plant distribution. *Biol. Rev.* 27: 194–245.

GRESSITT, J. L., 1956a. Zoogeography of insects. *A. Rev. Ent.* 3: 207–30.

GRESSITT, J. L., 1956b. Some distribution patterns of Pacific Island faunae. *Syst. Zool.* 5: 11–32, 47.

GRESSITT, J. L., 1958. New Guinea and insect distribution. *Proc. 10th Int. Congr. Ent., Montreal,* 1956: 1–431.

GRESSITT, J. L., 1959. The Wallace Line and insect distribution. *Proc. 15th Int. Congr. Zool., London,* 1958: 66–8.

GRESSITT, J. L., 1960. Zoogeography of Pacific Coleoptera. *Proc. 11th Int. Congr. Ent., Wien, 1960:* 463–5.

GRESSITT, J. L., 1961. Problems in the zoogeography of Pacific and Antarctic insects. *Pacif. Insects, Monogr.* 2: 1–127.

GUPTA, V. K., 1962. Taxonomy, zoogeography, and evolution of Indo-Australian *Theronia* (Hymenoptera: Ichneumonidae). *Pacif. Insects, Monogr.* 4: 1–142.

HAGMEIER, E. M., 1966. A numerical analysis of the distributional patterns of North American mammals. II. Re-evaluation of the provinces. *Syst. Zool.* 15: 279–99.

HAGMEIER, E. M. & STULTS, C. D., 1964. A numerical analysis of the distributional patterns of North American mammals. *Syst. Zool.* 13: 125–55.

HARLAND, W. B., 1967. Tectonic aspects of continental drift. *Sci. Prog., Oxf.* 55: 1–14.

HENNIG, W., 1950. *Grundzüge einerTheorie der phylogenetischen Systematik.* Berlin: Deutsche Zentralverl

HUHEEY, J. E., 1965. A mathematical method of analysing biogeographical data. I. Herpetofauna of Illinois. *Am. Nat. Midl. Nat.* 73: 490–500.

HULL, D. L., 1957. Certainty and circularity in evolutionary taxonomy. *Evolution,* 21: 174–89.

HULTÉN, E., 1937. *Outline of the history of the Arctic and boreal biota during the Quaternary period* Stockholm: Bokforlags Aktiebolaget Thule.

HUXLEY, T. H., 1868. On the classification and distribution of the Alectoromorphae and Hetero morphae. *Proc. zool. Soc. Lond.* 1868: 294–319.

JARDINE, C. J., JARDINE, N. & SIBSON, R., 1967. The structure and construction of taxonomic hier archies. *Mathematical Biosciences,* 1: 173–9.

JARDINE, N. & SIBSON, R. (In press, a). A model for taxonomy. *Mathematical Biosciences.*

JARDINE, N. & SIBSON, R. (In press b). The construction of hierarchic and non-hierarchic classifications *Computer Journal.*

KEAST, A. L., (ed.), 1959. Biogeography and ecology in Australia. *Monographiae biol.* 8: 1–640.

KLOSS, C. B., 1929. The zoogeographical boundaries between Asia and Australia and some Orienta subregions. *Bull. Raffles Mus.* 2: 1–10.

KRUMBEIN, W. C., 1959. Trend-surface analysis of contour-type maps with irregular control-poin spacing. *J. geophys. Res.* 64: 823–34.

KRUSKAL, J. B., 1964a. Multidimensional scaling by optimising goodness-of-fit to a nonmetri hypothesis. *Psychometrika,* 29: 1–27.

KRUSKAL, J. B., 1964b. Nonmetric multidimensional scaling: a numerical method. *Psychometrik* 29: 115–29.

LAM, H. J., 1934. Material towards a study of the flora of the island of New Guinea. *Blumea,* 1: 115–5

LANCE, G. N. & WILLIAMS, W. T., 1967. A general theory of classificatory sorting strategie I. Hierarchical systems. *Comput. J.* 9: 373–80.

LIEFTINCK, M. A., 1949. Synopsis of the odonate fauna of the Bismarck Archipelago and the Solomo Islands. *Treubia,* 20: 319–74.

LONG, C. A., 1963. Mathematical formulas expressing faunal resemblance. *Trans. Kans. Acad. Sc* 66: 138–40.

LYDEKKER, R., 1896. *A geographical history of mammals.* Cambridge University Press.

MACARTHUR, R. H., 1957. On the relative abundance of bird species. *Proc. natn. Acad. Sci. U.S.A* 43: 293–4.

MACARTHUR, R. H. & WILSON, E. O., 1963. An equilibrium theory of insular zoogeography. *Evolutio* 17: 373–87.

MATTHEWS, J. R., 1937. Geographical relationships of the British flora. *J. Ecol.* 25: 1–90.

MAYR, E., 1944. Wallace's line in the light of recent zoogeographic studies. *Q. Rev. Biol.* 19: 1–14.

MAYR, E., 1946. History of the North American bird fauna. *Wilson Bull.* 58: 3–41.

MAYR, E., 1953. Fragments of a Papuan ornithogeography. *Proc. 7th Pacific Sci. Congr., New Zealan* 1949, 4: 11–19.

METCALF, Z. P., 1949. Zoogeography of the Homoptera. *Proc. 13th Int. Congr. Zool., Paris, 194* 538–44.

MILLER, R. L., 1956. Trend surfaces: their application to analysis and description of environments sedimentation. *J. Geol.* 64: 425–46.

MOREAU, R. E., 1966. *Bird faunas of Africa and its islands.* London: Academic Press.

188   *Proceedings of The Linnean Society of London*

PELSENEER, P., 1904. La ligne de Weber, limite zoologique de l'Asie et de l'Australie. *Bull. Acad. r. Belg. Cl. Sci.* **1904**: 1001–22.

PETERS, J. A., 1955. Use and misuse of the biotic province concept. *Am. Nat.* **89**: 21–8.

PETERS, J. L., 1931–1951 (ed. vols. 1–7), MAYR, E. & GREENWAY, J. C. JR., 1951–1963 (eds vols 8–10, 15). *Check List of the Birds of the World.* Cambridge, Mass.: Harvard University Press.

PRESTON, F. W., 1948. The commonness, and rarity, of species. *Ecology*, **29**: 254–83.

PRESTON, F. W., 1962. The canonical distribution of commonness and rarity. *Ecology*, **43**: 185–215, 410–32.

RAUP, H. M., 1947. Some natural floristic areas in boreal America. *Ecol. Monogr.* **16**: 221–34.

RAVEN, H. C., 1935. Wallace's line and the distribution of Indo-Australian mammals. *Bull. Am. Mus. nat. Hist.* **68**: 179–283.

RUNCORN, S. K., 1966. Rock magnetism. *Sci. Prog., Oxf.* **54**: 467–482.

RYAN, R. M., 1963. The biotic provinces of Central America as indicated by mammalian distribution. *Acta zool. mex.* **6**: 1–55.

SCLATER, P. L., 1858. On the general geographical distribution of the members of the class Aves. *J. Linn. Soc. (Zool)*, **2**: 130–45.

SCRIVENOR, J. B., BURKILL, I. H., SMITH, M. A., CORBET, A. S., AIRY SHAW, H. K., RICHARDS, P. W. & ZEUNER, F. E., 1942. Symposium: Biogeography of the Indo-Australian Archipelago. *Proc. Linn. Soc. Lond.* **154**: 120–65.

SEITZ, A. (ed.), 1927. *Macrolepidoptera of the World. Vol. 9. Oriental and Australian Rhopalocera.* Stuttgart: Alfred Kernen.

SNEATH, P. H. A., 1967a. Trend-surface analysis of transformation grids. *J. Zool., Lond.* **151**: 65–122.

SNEATH, P. H. A., 1967b. Conifer distributions and continental drift. *Nature, Lond.* **215**: 467–70.

SOKAL, R. R., & ROHLF, T. J., 1962. The comparison of dendrograms by objective methods. *Taxon*, **11**: 33–40.

SOKAL, R. R. & SNEATH, P. H. A., 1963. *Principles of numerical taxonomy.* San Francisco and London: W. H. Freeman.

STAPF, O., 1913. The southern element in the British flora. *Engler's Bot. Jb.*, **1**: 509–25.

STEENIS, C. G. G. T. VAN, 1934–1936. The origins of the Malaysian mountain flora. *Bull. Jard. bot. Buitenz* (Ser. III), **13**: 135–262, 289–417; **14**: 56–72.

STEGMANN, B., 1938. Les Oiseaux. Principes généraux des subdivisions ornithogéographiques de la région paléarctique. In *Fauna de L'URSS* (n.s.19) *Vol.* 1, *no.* 2, 1–156. Moscow and Leningrad: Acad. Sci. U.R.S.S.

TALBOT, G., 1928–1937. *Monograph of the Pierine genus Delias.* London: John Bale, Danielsson and British Museum (Nat. Hist.).

TATE, G. H. H., 1946. Geographical distribution of the bats in the Australian Archipelago. *Am. Mus. Novit.* **1323**: 1–21.

TOXOPEUS, L. J., 1950. The geological principles of species evolution in New Guinea. *8th Int. Congr. Ent., Stockholm,* 1948: 508–22.

UDVARDY, M. D. G., 1958. Ecological and distributional analysis of North American birds. *Condor,* **60**: 50–66.

UDVARDY, M. D. F., 1963. Bird faunas of North America. *Proc. 13th Int. orn. Congr. Ithaca,* 1962, **2**: 1147–67.

UMBGROVE, J. H. F., 1943. *Structural History of the East Indies.* Cambridge University Press.

VAN TYNE, J. & BERGER, A. J., 1959. *Fundamentals of Ornithology.* New York: Wiley.

VASWANI, S., 1950. Assumptions underlying the use of the tetrachoric correlation coefficient. *Sankhya,* **10**: 269–76.

VOOUS, K. H., 1963. The concept of faunal elements or faunal types. *Proc. 13th Int. orn. Congr., Ithaca,* 1962, **2**: 1104–8.

WALLACE, A. R., 1860. On the zoological geography of the Malay Archipelago. *J. Linn. Soc. (Zool.),* **4**: 172–184.

WALLACE, A. R., 1869. *The Malay Archipelago.* 2 vols. London: Macmillan.

WALLACE, A. R., 1876. *The Geographical Distribution of Animals.* London: Macmillan.

WEBB, W. L., 1950. Biogeographic areas of Texas and Oklahoma. *Ecology,* **31**: 426–33.

WILLIAMS, C. B., 1947. The logarithmic series and the comparison of island floras. *Proc. Linn. Soc. Lond.* **158**: 104–8.

WILLIAMS, C. B., 1949. Jaccard's generic coefficient and coefficient of floral community, in relation to the logarithmic series and the index of diversity. *Ann. Bot.,* (N.S.), **13**: 53–8.

WILLIAMS, W. T. & DALE, M. B., 1965. Fundamental problems in numerical taxonomy. *Adv. bot. Res.* **2**: 35–68.

WIRTH, M., ESTABROOK, G. F. & ROGERS, D. J., 1966. A graph theory model for systematic biology, with an example from the Oncidiinae (Orchidaceae). *Syst. Zool.* **15**: 59–69.

ZEUNER, F. E., 1943. Studies in the systematics of *Troides* Hbn. and its allies. *Trans. zool. Soc. Lond.* **25**: 107–84.

ZIMMERMAN, E. C., 1942. Distribution and origins of some eastern oceanic insects. *Am. Nat.* **76**: 280–307.

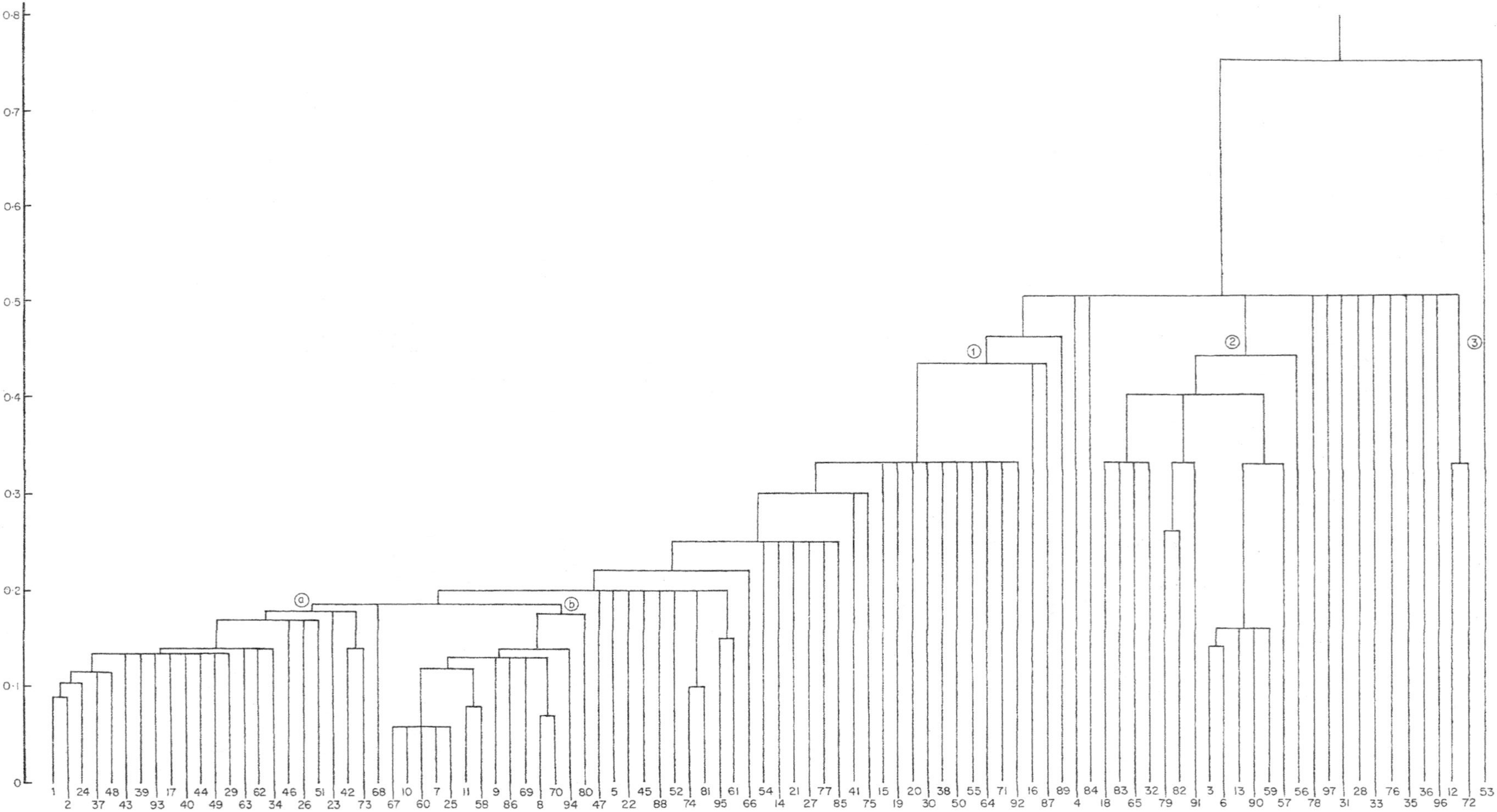

Fig. 9. Dendrogram derived by single-link cluster-analysis from the coefficients of coincidence in distribution between subgenera of the Rhopalocera. The taxa are numbered as in Table 3. Ringed numbers and letters indicate the clusters that are recognized as faunal elements and sub-elements respectively.

(Facing p. 172)

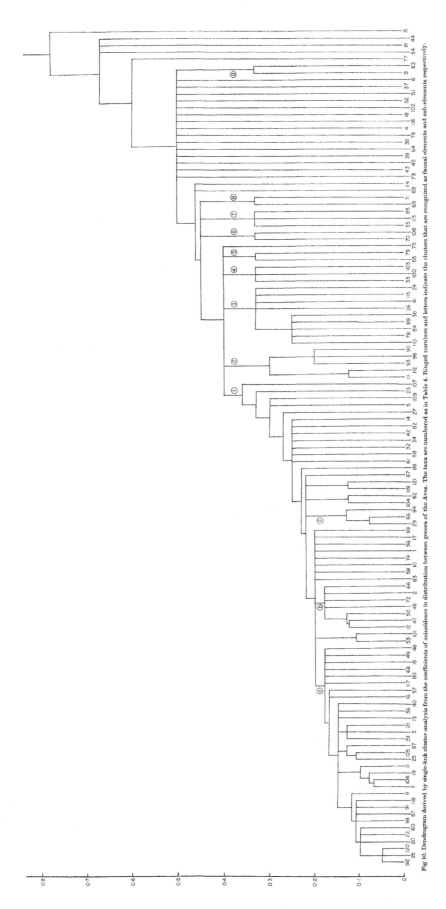

Fig 10. Dendrogram derived by single-link cluster-analysis from the coefficients of coincidence in distribution between genera of the Aves. The taxa are numbered as in Table 4. Ringed numbers and letters indicate the clusters that are recognized as faunal elements and sub-elements respectively.

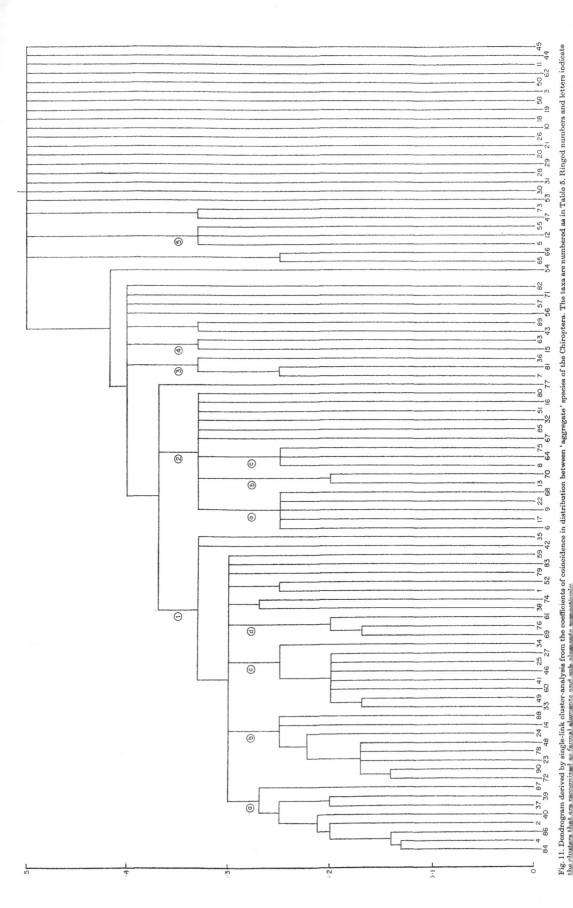

Fig. 11. Dendrogram derived by single-link cluster-analysis from the coefficients of coincidence in distribution between 'aggregate' species of the Chiroptera. The taxa are numbered as in Table 5. Ringed numbers and letters indicate the clusters that are recognized as faunal elements and sub-elements respectively.

From *The Ecology of Invasions by Animals and Plants*
Charles S. Elton

CHAPTER FOUR

The Fate of Remote Islands

When Captain Cook anchored off Easter Island in March 1774, he noted that 'Nature has been exceedingly sparing of her favours to this spot'.[122] This was exactly true, for Nature had only with great difficulty managed to get there at all. The nearest continent is South America, 2,280 miles away, and even the nearest vegetated Pacific island (Ducie Island) a thousand miles. This bit of volcanic rock (from which the famous hatted statues were carved out), covered with hills and grassy downlands, is only about a third the size of the Isle of Wight (Pl. 23). This is about 9 per cent. of the combined areas of the Marquesas Islands—one of the remotest of the Pacific mountain island archipelagos; these in turn are about 8 per cent. of the combined area of the Hawaiian Islands (which amounts to 6,400 square miles). The Hawaiian group is the largest, most varied and richest in life of the truly oceanic islands of the central Pacific: the area of Africa is nearly twelve million square miles!

Plants and animals have managed not only to reach these remote archipelagos and islands without the help of man, but in some have evolved luxuriant tropical vegetation and sometimes, though not very often, unique and peculiar groups of plants and animals. The island of Krakatau, which blew off its head in 1883 and absolutely destroyed all life under a rain of hot volcanic ash that lay more than a hundred feet deep on some of the slopes, was recolonized by plants and animals from the nearest land, and after fifty years had already a rich and maturing jungle of forest inhabited by epiphytic plants and many kinds of animals. By 1933 there were at least 720 species of insects, 30 kinds of resident birds, and a few species of reptiles and mammals, though no frogs or toads. But these species only had to cross by various means from the adjacent

77

Reprinted from Charles S. Elton, *The Ecology of Invasions by Animals and Plants* (London: Methuen, 1958), with kind permission from Kluwer Academic Publishers.

THE ECOLOGY OF INVASIONS

tropical lands of Java and Sumatra, a mere twenty-five miles over the sea.[123]

When the first white man made collections there Easter Island had extremely few native plants and animals compared with Krakatau, though this has only 12 per cent. of the area of the former. The Swedish Expedition under Skottsberg that visited Easter Island for a short time in 1917 has published a very good series of reports on the place, as well as upon Juan Fernandez—Robinson Crusoe's island.[144-6] But there are two things that have to be considered besides the remoteness of the island. One is that it has been a great deal modified by human activities, especially grazing sheep and cattle and the removal of timber (Pl. 24). It seems likely that the original condition was a sort of forest savannah with grass. So some of the indigenous plants and animals may have died out before they could be collected by biologists. The only tree, *Sophora toromiro*, is nearly extinct now. The second thing is that no absolutely complete collection of insects and other small animals has been done, and even the Swedish party only spent a fortnight there. Nevertheless, there are certainly no native earthworms at all, only one introduced species; and no land birds or other vertebrates except a few introduced by man. In the flora there are 31 species of flowering plants, apart from cultivated plants like the plantain, sweet potato, and sugar-cane; 15 kinds of fern, of which four are endemic; 14 kinds of moss, of which nine are endemic. That is, less than fifty species that may originally have been native to the island—a tiny flora. The number of animal species that seem to be native is almost absurd. There are so far known to be only five endemic: a green lacewing,[139] a fly,[125] a weevil,[113] a water beetle, and a land snail (Pl. 25).[135] The water beetle, *Bidessus skottsbergi*, was found among algae in the crater lake of Rano Kao, where there are also some kinds of endemic aquatic mosses.[157] No other fresh-water animals have yet been found, though probably microscopic life would be rich enough, because it is easily air-borne even to distant lands. Practically all the rest of the land animals are either known to have been introduced, or else this can be supposed from their cosmopolitan man-borne distribution: about 44 kinds of insects, spiders, and other invertebrates, of which one (a dragonfly) probably arrived under its own power; two introduced

THE FATE OF REMOTE ISLANDS

lizards; two kinds of birds brought from Chile; and rats. The surviving native animals are therefore outnumbered in species by about 10 to 1, and far more so in populations.

There are thousands and thousands of small remote islands that have, like Easter Island, been too far from the busy evolutionary centres of the continents to acquire more than a sprinkling of accidental immigrants before the arrival of man began to make this process of dispersal so much easier and faster. The small atoll of Palmyra Island in the equatorial Pacific had only fourteen species of native plants. The insect fauna of Midway Island, lying at the western extremity of the Hawaiian chain, was also minute: of beetles there were only six species, of flies only nine.[128]

Before considering some of the more catastrophic invasions of oceanic islands, it is worth examining what is happening on three very remote islands in the South Atlantic—the Tristan da Cunha group. Here a very thorough biological survey was made by the Norwegian Scientific Expedition to Tristan da Cunha in 1937–8, under the leadership of Christophersen.[121] There are also earlier records, especially for the plants and birds. These three islands (Tristan, Nightingale, and Inaccessible) are even farther away from the nearest continent than Easter Island—2,900 miles from South Africa; 3,200 from Brazil and 4,500 from Cape Horn. Tristan itself is the upper part of a volcano risen over 12,000 feet from the sea bottom, and having about half of this exposed above the sea. On the top is an ancient crater with a lake inside it. Down the sides there grows fairly rich vegetation, with only one kind of small tree but with tree ferns, and there is much heavy tussock grass and rather wet heath. On a shelf of land above the shore three miles long lives the small human community.

There are some fifty native and seventy alien species of flowering plants on these islands, besides nearly 300 of ferns, mosses, and liverworts.[120] Animal life, other than sea-birds, is very poor. Five species of birds, some of which have evolved differences on the separate islands— a member of the flycatcher family that looks like a thrush, two kinds of finch, a flightless rail and coot (the rail on Inaccessible Island, the coot on Tristan Island, though now almost extinct).[129] The insect life can be illustrated by some examples. Of the twelve recorded moths and

79

THE ECOLOGY OF INVASIONS

butterflies, the Expedition took only eight. Only five of the twelve appear to be native, the rest brought by man.[150] Only four kinds of plant-lice, of which two at least have come in with man.[141] The beetles were analysed with special thoroughness. Of the twenty species thought to be indigenous, only two are predatory and the rest herbivorous, many of the latter being weevils which are one of the widespread kinds of beetle in oceanic islands elsewhere (the fifteen Tristan ones are all peculiar, and are flightless). Besides these native beetles there are six or seven that are or seem to have been brought in by man.[116]

Consider how extremely little traffic has gone between Tristan da Cunha and other places. Yet the fauna will soon contain as many invaders as there used to be native fauna. In 1882 a ship was wrecked and a few Norway rats got ashore from it. The pastor urged that these should be destroyed, but they were allowed to get in, and now infest many parts of the island of Tristan, eating potatoes (the people's most important crop on land) and killing nesting birds in the wilder parts of the mountain.[114] They are supposed to have destroyed the Tristan coot, perhaps assisted by feral cats, and it is fortunate that rats have not reached the two other islands yet. There are no records so far of great outbreaks among the introduced insect populations, in fact the chief enemy of the potato is a native moth: it has no parasites at all. But one of the invading plant-lice, *Myzus persicae*, is able in other countries to carry two of the worst potato viruses, and one of the moths is a well-known eater of cruciferous plants. It is to be noticed that some introduced species are still confined almost entirely to the limited shelf of settlement, with its pasture and gardens and potato crops; but that others like the rat have spread to the natural habitats as well. An introduced staphylinid beetle, *Quedius mesomelinus*, and a species of European millipede, *Cylindroiulus latestriatus* (that has also been spread by man to North America, South Africa, and the Azores), has colonized a very wide range in tree-fern ground, bogs, the sea-shore drift line and other places. But another European millipede, *Blaniulus guttulatus*, was found on cultivated land.[130]

Here then is an oceanic island in which man has carved out a small patch for himself, leaving the rest of it wild. Except for exploitation of wood and of seabirds, his new influence in the wilder parts is through

THE FATE OF REMOTE ISLANDS

invading species brought on ships. To see the same process acting on a much larger island group, we may turn to Hawaii.

No need to describe the Hawaiian Islands: remote, mountainous, volcanic, tropical, rich, and until modern times holding within their archipelago one of the most extraordinary island floras and faunas ever known. Few people now deny that these islands are truly oceanic, like Tristan da Cunha and Easter Island, and that the endemic species there are descended from rare immigrants that had to cross several thousand miles of ocean. America and Australia are over four thousand miles away; the nearest continental islands are Japan—3,400 miles. Fiji—near or on the edge of the old sunken outlier of the Australasian continent—is about 2,800 miles away. Before the Polynesian canoes reached Hawaii in about the twelfth century A.D. or earlier the islands were probably covered with luxuriant forest, except in places where the climate is locally dry or there were recent lava flows. Since then the forest line has retreated from the coast until it now covers only a quarter of its former extent, on the mountains mostly. Fire and wild cattle, sheep, goats and horses, and the clearing of land for crops, have all contributed to this retreat. But from a quite rich percentage of surviving forms and the earlier records a pretty good stock-taking has been made, though many species may have died out before white men came, and it has even been suggested that as many as a third of the original insect fauna had disappeared unrecorded. Over nine-tenths of the 1,729 species of flowering plants are found nowhere outside these islands. Zimmerman, to whose *Insects of Hawaii* I am indebted for much critical information, estimates that 3,722 of the 6,000 or so species of insects known there are also endemic; the rest being comprised of species also living elsewhere, naturally or artificially introduced.[158] There are two large families of land snails. The Achatinellidae with 215 species are unique to the islands; the Amastridae with 294 almost so (Pl. 26). The former family lives entirely in trees. The shells are gaily marked and coloured. The Amastridae show a great deal of evolution into different ecological forms, both in trees and on the ground. Of the 77 kinds of endemic birds ever found on the whole Hawaiian chain, about 43 species and sub-species belong to the Drepaniidae, a family within which more than a dozen ecological ways of life have been

F   81

THE ECOLOGY OF INVASIONS

evolved within Hawaii—honey-suckers, wood-insect hunters, other insect-eaters, seed-eaters, nut-eaters, fruit-eaters—differences that would in a continent be developed in separate orders, not just genera of birds. It is not surprising that Captain King was puzzled when he saw one. In 1779 he wrote: 'A bird with a yellow head, which, from the structure of its beak, we called a perroquet, is likewise very common. It, however, by no means belongs to that tribe, but greatly resembles the yellow cross-bill, *Loxia flavicans* of Linnaeus.'[131] (Pl. 27). This was the Drepanid *Psittacirostra psittacea* (Pl. 27). Taxonomists have tried to calculate how many ancestors these Hawaiian groups may have had; that is, how many original ancestors arriving by various routes to the islands. For flowering plants it is about 272 species; for insects between 233 and 254; for Achatinellidae only one; for land snails from 22 to 24;[158] and for birds about 14—the Drepaniidae coming only from one of these forms.[133]

What has been the fate of this marvellous flora and fauna? First of all the list has been enormously added to by introduction, partly on purpose and partly by mistake. The full roll-call for insects has not yet been finished, but out of the 1,100 or so species given in the first five volumes of the *Insects of Hawaii*, 420 are thought to be adventive. This is a rough estimation and there still are nine orders of insects to be assessed, including such predominantly important ones as moths, beetles, flies, and Hymenoptera. In 1953 a list was published of 49 'economic' insects found to have become established since 1939.[137] The geographical sources of these species are very mixed. They came from California, Mexico, the Philippines, Samoa, Fiji, Guam, Saipan, and New Guinea, and for some the origin is unknown. Among these immigrants was the Argentine ant (probably from California) in 1940. When Wheeler compiled a list of Hawaiian ants in 1934 he mentioned that this species had been intercepted by quarantine and had not by then invaded the islands.[153] He also recorded that the leaf-cutting ant *Pheidole megacephala* was then displacing a previously introduced ant, *Solenopsis rufa*, on the island of Oahu. Zimmerman wrote in 1948: 'The voracious *Pheidole megacephala* alone has accounted for untold slaughter. One can find few endemic insects within the range of that scourge of native insect life. It is almost ubiquitous from the seashore to the beginnings of damp forest. Below

82

THE FATE OF REMOTE ISLANDS

about 2,000 feet few native insects can be found today.' It was known that the leaf-cutting species had invaded the Canary Islands and Madeira, to be followed at a later date by the Argentine ant, which not only wiped out *Pheidole* but also practically all the native ants below 3,000 feet.[153] Perhaps the same thing will now happen in Hawaii.

Every new insect pest may cause a train of operations with foreign counterpests. A very recent report on the annual increment of counterpests to Hawaii in 1953-5 gives quite a vivid notion of this process.[152] For control of the shrub *Lantana*, a Mexican longicorn beetle that bores in the stems, also a Central American chrysomelid beetle and a phalaenid moth from California whose grubs and caterpillars respectively eat the leaves. There are already an introduced seed-eating fly, and some other insects for this job. Then from Mexico a fly whose larvae eat the flower heads of *Eupatorium glandulosum*, a relative of our own hemp agrimony that has become a tropical weed in Hawaii, as well as in the Philippines and elsewhere. A moth from Brazil to eat the leaves of another locally troublesome plant, the Christmas berry tree. Four kinds of Mexican dung-beetles, whose grubs might help in controlling the maggots of some kinds of flies. Two Scoliid wasps from Guam, to attack various kinds of scarabaeid beetles. Two parasites from Arizona to try again (after a failure to establish them ten years earlier) in the control of a moth that attacks the flowers of the mesquite tree. A Mexican ladybird to feed on aphids in the sugar-cane fields. Finally, two carnivorous snails, one from the Mariana Islands and one from Florida, to try against giant African snails. It is quite an exchange and bazaar for species, a scrambling together of forms from the continents and islands of the world, a very rapid and efficient breaking down of Wallace's Realms and Wallace's *Island Life*!

Most of the herbivorous insects have followed in the wake of earlier plant introductions (as field crops, fruit trees, forest trees and garden plants) and it is usually some years before the animals catch up with their plant hosts, as has been seen recently in the case of *Leucaena glauca*.[143] This large leguminous shrub is probably an original native of South America, though it has been spread to other parts of the world, including Hawaii since 1888. It is valued there as a forage crop that is full of protein, and its seeds were collected and sown, and also later on became used in

83

THE ECOLOGY OF INVASIONS

the island manufacture of seed jewellery. It became a thriving additional crop on the islands, but in 1954 a small anthribid beetle from the region of Indo-China and the Philippines was discovered to be living in the seeds and, on the island of Oahu, sometimes destroying the complete seed crop. Hitherto its control has not been achieved.

The native life is not just retreating with the forest, keeping its forces intact though on a smaller area. It is true that a good deal of the forest, and of some other upland habitats, survives because it is impossible to cultivate. Yet roaming cattle and other feral animals have done much harm. And ship rats, whose violent influence is a frequent refrain in the modern history of islands, have also gone into the forest. The native moths have diminished greatly from their former strength and some have died out. This Zimmerman attributes to the invasion of forest by ichneumonid parasites brought in as counterpests on agricultural land, and he cites especially three species, *Casinaria infesta*, *Cremastus flavoorbitalis*, and *Hyposoter exiguae*, that have a very wide range of hosts. For Hawaiian insects were not naturally parasitized or adapted against parasites and insect enemies to the extent that continental insects are. Furthermore, this decrease in native moths may be the reason why some species of *Odynerus*, a genus of hunting wasps that has many species in Hawaii, have also declined in numbers; for they depend on caterpillars for stocking larders in which their own young grow up.

Amongst the many invaders of Hawaii none can have had such a long and steady progress across the Indo-Pacific world before its arrival as the giant snail *Achatina fulica*. This genus is otherwise entirely Ethiopian in distribution, with over 65 species that live chiefly in tropical forests. It contains the largest living land snails, the biggest of all, *Achatina achatina*, being still confined to West Africa, where it is a favourite food of the natives. *A. fulica*, though rather smaller (Pl. 28), is something to be considered if you have to collect 400 of them every night in a small garden, as a resident of Batavia in Java was doing in 1939.[78] This is an East African species that may have been introduced to Madagascar long ago. It began to spread to the outlying islands of Mauritius (by 1800), Reunion (by 1821), the Seychelles (by 1840) and the Comoro Islands (by 1860). Some were released in Calcutta in 1847, and Bequaert, who

84

THE FATE OF REMOTE ISLANDS

has documented its travels, as well as the systematics of the whole group, says that 'at first, the spread of the snail in southern Asia was very slow'.[115] It was in Ceylon by 1900, Malay Peninsula certainly by 1922 and probably twelve years earlier; Borneo by 1928; Siam in 1937–8; and Hong Kong in 1941. It moved through the Netherlands East Indies in the nineteen-twenties and thirties. It was in Japan, though not doing very successfully, by 1925, reached the Palau Islands in 1938 and on to the Marianas, soon becoming a major agricultural problem in many of the Micronesian islands. In Guam especially it became a plague, having been brought there from Saipan in the Marianas in 1946. By 1948 it had a foothold at three points in and near New Guinea. Meanwhile a few got into California about 1947; but it is not thought likely that it can ever become established in the United States, because the climate is unsuitable for a tropical snail. This majestic spread was accomplished by partly accidental dispersal on transported stores and plant materials; and also to quite a large extent because of its value for food. Their size, voracity, and abundance give a remarkable atmosphere to these invasions. There can be few invading species which become such a menace to motor traffic that they cause cars to skid on the roads! The whole extended snail may measure nine inches, not counting the projection of its shell.

The giant snail reached Hawaii in 1936, and in spite of great efforts for its control, it now inhabits Oahu where it damages crops—an invasion thought to have started from only two individuals brought from Formosa. The trial of enemy snails to kill *Achatina* is already in full swing, and it may be noted that these have been brought not only from the original home of the species in East Africa, but also from one of the Mariana Islands and from Florida.[154] Meanwhile many of the beautiful native Amastrid snails have become scarce in the lowlands and some species extinct. One factor bringing this about seems to have been the attacks of foreign rats (Pl. 26);[147] the original Polynesian rat having rather different food habits, cannot have been decisive in causing this decline.

The birds also display the consolidation of alien species and, on the whole, the diminution of native ones. In 1940, when E. H. Bryan summarized the position in Hawaii, there were already about half as many introduced as original native forms: 94 kinds of foreign birds had been

85

THE ECOLOGY OF INVASIONS

tried, and only 41 found wanting.[117] The 53 established ones show the varied pattern of origins that is becoming familiar in this book. The ring-necked pheasant, *Phasianus colchicus*, derived from Europe; the green Japanese pheasant, *Phasianus versicolor*; the California quail, *Lophortyx californica*; the painted quail, *Coturnix coturnix*, from Japan; the lace-necked dove, *Streptopelia chiensis*, from Eastern Asia; the barred dove, *Geopelia striata*, from the Malay Archipelago—to mention only some game-birds that have now got a firm hold in the islands.[142] The two pheasants have not only spread widely, but in some places hybridized (Pl. 29). The wild jungle fowl, *Gallus gallus*, must have been brought from Malaya by the Polynesian voyagers themselves, as these birds existed in Hawaii when Captain Cook discovered the islands. The rock pigeon, *Columbia livia*, instead of being only a town bird as it is in the United States, has become wild on the cliffs. The Indian mynah, *Acridotheres tristis*, brought from India in 1865, is well known to ecologists because of the part it played in originally spreading the seeds of *Lantana*.

According to Munro, who had known the Hawaiian birds for a lifetime, and was with Perkins and other early naturalists when they explored and collected at the end of the nineteenth century, there are several introduced birds that have penetrated more or less deeply into the forests.[134] The babbler or Pekin nightingale, *Leiothrix lutea*, a Chinese species brought over in 1918–20, is now on most of the main islands, and it is on record that bird malaria has been found in this species in Japan. There are also the Chinese thrush, *Trochalopterum canorum*, from about 1900, a bird of the scrub layer that has gone deeper into the forests than any other species, but in 1944 was reported to be diminishing locally; and the Japanese tit, *Parus varius*, from 1890 onwards, which has made itself quite at home in the forest. It has been suggested, though without direct proof, that species like these might carry diseases of birds from the lowlands into the upper zones, and possibly harm the Drepanids. These wonderful birds have practically all become reduced in numbers, even in their remaining natural haunts. Only about a third of the species and island sub-species have a good chance of survival in the future. Some which were thought probably to be extinct have yet been found to persist, but have been missed through the physical difficulty of searching for them:

86

THE FATE OF REMOTE ISLANDS

thus the crested honey-eater, *Palmeria dolei*, had not been seen since 1907, yet was telephotographed on the high mountains of Maui as recently as 1950. And *Pseudonestor xanthophrys*, not seen for half a century, was observed on Maui at 6,400 feet in 1950.[140]

Captain King's 'perroquet', the Ou, *Psittacirostra psittacea*, used to be abundant on the main islands, but Perkins said in 1903 that though it was still widespread on other islands the bird had become practically extinct on Oahu. This he thought might be caused by competition from the ship rats, *Rattus rattus*, that had spread in the forests there: 'Now over extensive areas it is often difficult to find a single red Ieie fruit, which the foreign rats have more or less eaten and befouled, and they may thus have indirectly brought about the extinction of the Ou.'[138] The bill of this bird is indeed especially useful for picking out the fruits of the Ieie, a bright-flowered liana, *Freycinetia arnotti*, that climbs on forest trees; and the Ou's chief habitat on all these islands was in the zone where the liana occurred. But the birds have decreased also on the other islands, they will eat other fruits and also caterpillars (on which they feed their young), and have for some time been known to feed on the fruits of introduced plants like the guava, which itself is a wide-spreading invader in Hawaii. It is also known that domestic birds have brought in diseases; Drepanids have been seen with at least one of these; and bird malaria has come in to the islands, and may possibly be spread by mosquitoes.[112] The mosquitoes are apparently all introduced species, the islands originally being free from them. W. A. Bryan in his *Natural History of Hawaii* says that the one that bites at night, *Culex fatigans*, was thought to have arrived on a ship from Mexico as early as 1826. There are also some day-biting species of *Stegomyia*.[118] All these possibilities only serve to suggest the tangle of influences that are likely to be at work: no one has sorted them out in a thorough way. For example, what of the scarcity of native caterpillars affecting young forest birds? The Ou still survives, though in 1944 it was thought to be dangerously near extinction. In 1950–1 some were seen in a mountain forest reserve and in the National Park on the island of Hawaii, at a height of several thousand feet. It mitigates the fate of remote islands in this century if some of their species are saved from the wreckage, like the few survivors that clamber out of

87

FIG. 33. Introduced birds and mammals that have established populations in New Zealand, with their countries of origin (the percentage from each shown by the black in the circles). (From K. A. Wodzicki, 1950.)

THE FATE OF REMOTE ISLANDS

a smashed aircraft. This bird especially, of which Perkins wrote: 'Sometimes it sings as it flies, and when a small company are on the wing together they not infrequently sing in concert, as they sometimes do at other times, and in a very pleasing manner.' [138]

The only rival to Hawaii among remote islands is New Zealand. Here is a country that looks like part of a continent, yet was probably never joined to one directly, and has been isolated for an immensely long period. It has therefore partly the environment of a continent but the history of an oceanic island. No place in the world has received for such a long time such a steady stream of aggressive invaders, especially among the mammals—successful in the short run, though often affecting the future of their own habitats in a decisive manner. Originally there were no native mammals except bats. In 1950 Wodzicki could write an ecological monograph upon some twenty-nine kinds of 'problem animals', among them being eleven kinds of ungulates: four from Japan, four from North America and three from Europe (including England), and to these must be added feral (and domestic stock), not neglecting pigs.[156] A list of such successfully established mammals and birds compiled by Wodzicki is set out in Fig. 33. Red deer, *Cervus elaphus*, were liberated between 1851 and 1910, and quickly multiplied in both islands of New Zealand (Pl. 30). Now spreading patches from different centres have merged within each Island, but the greatest occupation is still in the South Island (Figs. 34–5). The red deer have already made a profound impact upon native forests, especially in the drier types of woodland; but it is in the wetter regions that forest damage leads to most serious soil erosion. It is likely that on many watersheds the deer, helped by domestic stock, have tipped the scale towards a cycle of catastrophic soil erosion, which is felt not only in the mountains but also in those parts of the lowland valleys that receive the extra load of silt washed from above.

To follow the story of invading insects in New Zealand would only repeat what has already been indicated for other countries. Taking only populations that had arrived and become noticeable, there were fourteen species between 1929 and 1939, and another nine by 1949.[124] Among the last was the common ground-nesting wasp of Europe, *Vespula germanica*.[136] In 1945 beekeepers at one place in the North Island observed some strange

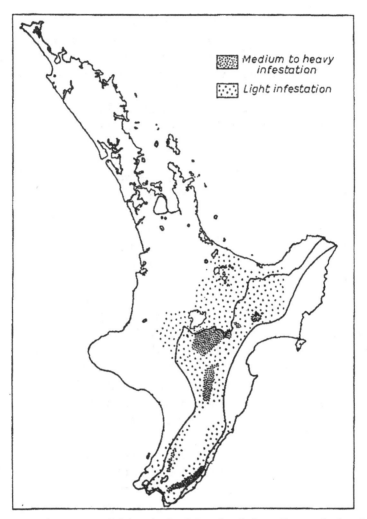

Fig. 34. Areas occupied by the introduced red deer, *Cervus elaphus*, in the North Island of New Zealand, 1947. The southern beech, *Nothofagus*, forests are enclosed by the black line. (From K. A. Wodzicki, 1950. Forest areas mapped by C. M. Smith and A. L. Poole.)

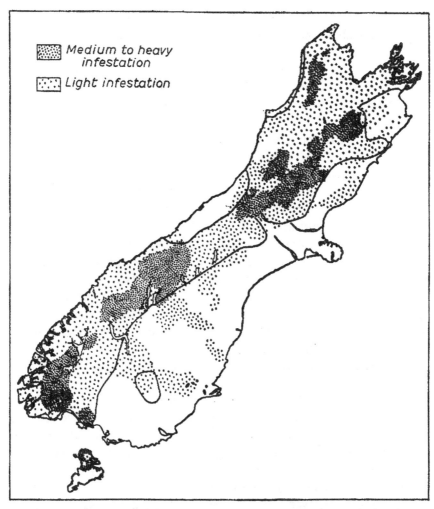

FIG. 35. Areas occupied by the introduced red deer, *Cervus elaphus*, in the South Island of New Zealand, 1947. The southern beech, *Nothofagus*, forests are enclosed by the black line. (From K. A. Wodzicki, 1950. Forest areas mapped by C. M. Smith and A. L. Poole.)

THE ECOLOGY OF INVASIONS

wasps flying about their hives. After this date the wasps spread and increased at a great rate, probably aided by the absence of winter cold to check their breeding. That year seven nests were destroyed, in 1946, 140, and by 1948 over 3,000—and this did not stop the spread. When a bounty was offered for resting queens, one schoolboy brought in 2,400 in less than a week!

The fate of remote islands is rather melancholy, even after one has made allowances for all the human excellence that has remained or developed again in some of them after our invading civilizations settled down. The reconstitution of their vegetation and fauna into a balanced network of species will take a great many years. So far, no one has even tried to visualize what the end will be. What is the full ecosystem on a place like Guam or Kauai or Easter Island? How many species can get along together in one place? What is the nature of the balance amongst them? Can we combine the simple culture of crops with the natural complexity of nature, especially when there is an almost inexhaustible reservoir of continental species that may send new colonists to disturb the scene? All these questions are much nearer than the horizon, though most ecologists have not looked at them with any enthusiasm, or if they have glanced at them, shuddered and turned away towards the already tedious and difficult task of understanding the biology of a single species, dead or alive.

I would like, however, to leave the subject with a back-glance at a more pleasant and balanced ecological world, before Atlantic civilized man crashed into this remote galaxy of island communities. In that age, when the numbers of human beings were regulated by customs, often harsh enough, but meeting the end desired, a great many of the Pacific Islands were inhabited by quite large numbers of a small species of rat, derived from a Malayan form, and evidently brought by the Polynesians in their great migrations eastwards and southwards some hundreds of years ago.[149] *Rattus exulans* (with closely similar forms like *Rattus hawaiiensis* (Pl. 31)) is a small rat, much gentler in habits and less aggressive than the larger ship and Norway rats: it has been found in New Zealand for example that the Maori rat does not do harm in bird island sanctuaries. On many of the islands of the Pacific the native rat was exterminated either by cats or by the arrival of these larger species. For a long time it

THE FATE OF REMOTE ISLANDS

was believed that they were extinct in Hawaii, until it was discovered that they had been confused with the young of the grey form of the ship rat, and are actually living there with the other two foreign species of rat.[148] From early missionary books we learn that these little rats were an important part of civilized life. On the island of Raratonga, in Mid-Pacific, they were highly prized for sport. 'In those days—ere the cat had been introduced—rats were very plentiful. Rat hunting was the grave employment of bearded men, the flesh being regarded as delicious.' [127] And on the Tonga Islands about 1806, the King and court used to go out and shoot the rats along the forest paths, using huge bows and arrows six feet long: 'Whichever party kills ten rats first, wins the game. If there be plenty of rats, they generally play three or four games.'[132] There were elaborate rules, as we should now have for football or hunting: precedence, offside, and—a wonderful dispensation—if you shot a bird you could count it as a rat! Even late in the nineteenth century 'the proverb "sweet as a rat" survives in Mangaia'.[126] Von Hochstetter, writing in 1867 about the small Maori rat in New Zealand, relates that 'this indigenous rat was so scarce already at the time of the arrival of the first Europeans, that a chief, on observing the large European rats on board one of the vessels, entreated the captain to let these rats run ashore, and thus enable the raising of some new and larger game'.[151] Returning to Captain Cook at Easter Island in 1774: 'They also have rats, which it seems they eat; for I saw a man with some dead ones in his hand, and he seemed unwilling to part with them, giving me to understand they were for food.' [122] Can we still find a remote island where people will be unwilling to part with the new rats that have arrived there in the last 180 years? Perhaps we could bear in mind the story told by Buxton and Hopkins, about the arrival of the human flea in one small Pacific Island more than a hundred years ago: 'The placid natives of Aitutaki, observing that the little creatures were constantly restless and inquisitive, and even at times irritating, drew the reasonable inference that they were the souls of deceased white men.' [119] We may hope that this same restless curiosity in the form of research will find out how the broken balance can be restored and protected.

Works Cited

111. Abbott, R. T. 1949. March of the giant African snail. *Nat. Hist.*, N.Y. **80:** 68–71.

112. Amadon, D. 1950. The Hawaiian honeycreepers (Aves, Drepaniidae). *Bull. Amer. Mus. Nat. Hist.* **95:** 155–262.

113. Aurillius, C. 1926. Coleoptera-Curculionidae von Juan Fernandez und der Oster-Insel. In *The natural history of Juan Fernandez and Easter Island* (Ed. C. Skottsberg). Uppsala. **3:** 461–77.

114. Barrow, K. M. 1910. *Three years in Tristan da Cunha.* London.

115. Bequaert, J. C. 1950. Studies in the Achatinae, a group of African land snails. *Bull. Mus. Comp. Zool.* **105:** 1–216.

116. Brinck, P. 1948. Coleoptera of Tristan da Cunha. *Results of the Norwegian Sci. Exped. to Tristan da Cunha 1937–1938,* No. 17: 1–121.

117. Bryan, E. H. 1940. A summary of the Hawaiian birds. *Proc. 6th Pacif. Sci. Congr.,* 1939, **4:** 185–9.

118. Bryan, W. A. 1915. *Natural history of Hawaii.* Honolulu.

119. Buxton, P. A., and Hopkins, G. H. E. 1927. *Researches in Polynesia and Melanesia . . .* Parts I–IV. (*Relating principally to medical entomology.*) *Mem. Lond. Sch. Hyg. Trop. Med.* **1:** 1–260.

120. Christophersen, E. 1939. Problems of plant geography in Tristan da Cunha. *Norsk Geogr. Tidsskr.* **7:** 106–12. (And personal communication.)

121. Christophersen, E. 1940. *Tristan da Cunha: the lonely isle.* London.

122. [Cook, J.] [*c.* 1890 ed.] *The three famous voyages of Captain James Cook round the world . . .* London and New York.

123. Dammerman, K. W. 1948. *The fauna of Krakatau 1883–1933.* Amsterdam.

124. Dumbleton, L. J. 1953. Entomological aspects of insect quarantine in New Zealand. *Proc. 7th Pacif. Sci. Congr.,* 1949, **4:** 331–4.

125. Enderlein, G. 1938. Die Dipterenfauna der Juan-Fernandez-Inseln und der Oster-Insel. In *The Natural History of Juan Fernandez and Easter Island* (Ed. C. Skottsberg). Uppsala. **3:** 634–80.

126. Gill, W. W. 1876. *Life in the Southern Isles.* London.

127. Gill, W. W. 1880. *Historical sketches of savage life in Polynesia.* Wellington.

128. Gulick, A. 1932. Biological peculiarities of oceanic islands. *Quart. Rev. Biol.* **7:** 405–27.

129. Hagen, Y. 1952. Birds of Tristan da Cunha. *Results of the Norwegian Sci. Exped. to Tristan da Cunha 1937–1938,* No. 20: 1–248.

130. Jeekel, C. A. W. 1954. Diplopoda. *Results of the Norwegian Sci. Exped. to Tristan da Cunha 1937–1938,* No. 2: 5–9.

131. [King, J. in Cook, J., p. 1010.]

132. Martin, J. 1818. *An account of the natives of the Tonga Islands in the South Pacific Ocean . . . compiled from the extensive communications of Mr William Mariner.* London. 2 vols. **1:** Ch. 9.

133. Mayr, E. 1943. The zoogeographic position of the Hawaiian Islands. *Condor,* **45:** 45–8.

134. Munro, G. C. 1944. *Birds of Hawaii.* Honolulu.

135. Odhner, N. H. 1926. Mollusca from Juan Fernandez and Easter Island. In *The natural history of Juan Fernandez and Easter Island* (Ed. C. Skottsberg). Uppsala. **3:** 219–54.

136. Paterson, C. R. 1953. The establishment and spread in New Zealand of the wasp *Vespa germanica.* Proc. 7th Pacif. Sci. Congr., 1949, **b:** 358–62.

137. Pemberton, C. E. 1953. Economic entomology in Hawaii. *Proc. 7th Pacif. Sci. Congr., 1949,* **4:** 91–4.

138. Perkins, R. C. L. 1903. *Fauna Hawaiiensis or the Zoology of the Sandwich (Hawaiian) Islands.* Vol. I. Part IV. Vertebrata. 363–466. Cambridge.

139. Petersen, P. Esben-. 1924. More Neuroptera from Juan Fernandez and Easter Island. In *The natural history of Juan Fernandez and Easter Island* (Ed. C. Skottsberg). Uppsala. **3:** 309–13.

140. Richards, L. P., and Baldwin, P. H. 1953. Recent records of some Hawaiian honeycreepers. *Condor,* **55:** 221–2.

141. Ris Lambers, D. H. 1955. Aphididae of Tristan da Cunha. *Results of the Norwegian Sci. Exped. to Tristan da Cunha 1937–1938,* No. 34: 1–5.

142. Schwartz, C. W. and E. R. 1949. *A reconnaissance of the game birds in Hawaii.* Hilo, Hawaii.

143. Sherman, M., and Tamashiro, M. 1956. Biology and control of *Araecerus levipennis* Jordan (Coleoptera: Anthribidae). *Proc. Hawaii. Ent. Soc.* **16:** 138–48.

144. Skottsberg, C. 1920. Notes on a visit to Easter Island. In *The natural history of Juan Fernandez and Easter Island* (Ed. C. Skottsberg). Uppsala. **1:** 1–20.

145. Skottsberg, C. 1927. The vegetation of Easter Island. In *The natural history of Juan Fernandez and Easter Island* (Ed. C. Skottsberg). Uppsala. **2:** 487–502.

146. Skottsberg, C. 1956. Derivation of the flora and fauna

of Juan Fernandez and Easter Island. In *The natural history of Juan Fernandez and Easter Island* (Ed. C. Skottsberg). Uppsala. **1:** 193–438.

147. Stokes, J. F. G. 1917. Notes on the Hawaiian rat. *Occ. Pap. Bishop Mus.* **3** (4): 11–21.

148. Svihla, A. 1936. The Hawaiian rat. *Murrelet,* **17:** 3–14.

149. Tate, G. H. H. 1935. Rodents of the genera *Rattus* and *Mus* from the Pacific Islands. *Bull. Amer. Mus. Nat. Hist.* **68:** 145–78.

150. Viette, P. E. L. 1952. Lepidoptera. *Results of the Norwegian Sci. Exped. to Tristan da Cunha 1937–1938*, No. 23: 1–19.

151. von Hochstetter, F. 1867. *New Zealand: its physical geography, geology and natural history.* Stuttgart.

152. Weber, P. W. 1956. Recent introductions for biological control in Hawaii. 1. *Proc. Hawaii. Ent. Soc.* **16:** 162–4.

153. Wheeler, W. M. 1934. Revised list of Hawaiian ants. *Occ. Pap. Bishop. Mus.* **10** (21): 1–21.

154. Williams, F. X. 1953. Some natural enemies of snails of the genus *Achatina* in East Africa. *Proc.7th Pacif. Sci. Congr., 1949,* **4:** 277–8.

155. Wilson, S. B., and Evans, A. H. 1890–99. *Aves Hawaiiensis: the birds of the Sandwich Islands.* London.

156. Wodzicki, K. A. 1950. Introduced mammals of New Zealand: an ecological and economic survey. *Bull. D.S.I.R., N.Z.* **98:** 1–255.

157. Zimmermann, A. 1924. Coleoptera-Dytiscidae von Juan Fernandez und der Oster-Insel. In *The natural history of Juan Fernandez and Easter Island* (Ed. C. Skottsberg). Uppsala. **3:** 298–304.

158. Zimmerman, E. C. 1948. *Insects of Hawaii.* **1.** *Introduction.* Honolulu.

Vol. 101, No. 919   The American Naturalist   May–June, 1967

WHY MOUNTAIN PASSES ARE HIGHER IN THE TROPICS*

DANIEL H. JANZEN

Department of Entomology, The University of Kansas, Lawrence

INTRODUCTION

This paper is designed to draw attention to the relation between tropical climatic uniformity at a given site and the effectiveness of topographic barriers adjacent to the site in preventing movements of plants and animals. This is not an attempt to explain tropical species diversity (see Pianka, 1966, for a review of this subject), but rather to discuss a factor that should be considered in any discussion of the relation between topographic and climatic diversity, and population isolation. Simpson (1964) states that "Small population ranges and numerous barriers against the spread and sympatry of related populations would therefore tend to increase density of species in a region as a whole. It will be suggested below that this is a factor in the increase of species densities in regions of high topographic relief. I do not, however, know of any evidence that it is more general or more effective in the tropics." I believe that the climatic regimes discussed below, and the reactions of organisms to them, indicate that topographic barriers may be more effective in the tropics. Mountain barriers and their temperature gradients in Central America, as contrasted to those in North America, are used as examples; but it is believed that the central idea equally applies to other tropical areas, types of barriers, and physical parameters.

There are three thoughts central to the argument to be developed: (1) in respect to temperature, it is the temperature gradient across a mountain range which determines its effectiveness as a barrier, rather than the absolute height; (2) in Central America, terrestrial temperature regimes are generally more uniform than North American ones, and differ in their patterns of overlap across geographic barriers; and (3) it can be assumed that animals and plants are evolutionarily adapted to, and/or have the ability to acclimate to, the temperatures normally encountered in their temporal and geographic habitat (or microhabitat).

MOUNTAIN TEMPERATURE GRADIENTS

Animals and plants encounter a mountain barrier as, among other things, different temperature regime from that to which they are acclimated or evolutionarily adapted. In general, and granting other environmental factors to be similar, this different temperature regime could occur as a band across a flat plain and still be just as impassable. The problem is the usual one of

*Contribution No. 1337 of the Department of Entomology, University of Kansas, Lawrence, Kansas. This paper is a by-product of National Science Foundation Grant No. GB-5206.

234 THE AMERICAN NATURALIST

how much overlap there is between the temperature regime at the top of the pass and the valley below. This overlap bears two major considerations: (1) the number of hours, days, or months when the temperatures on the pass are similar to those in the valley, and (2) the amount of time and degree to which the organism can withstand temperatures different from those in the valley while it is crossing the pass. The outcomes of such crossings are theoretically measurable at all levels, from the single individual which establishes a new population but is never joined by further immigrating members, to varying rates of individual organism flow (sometimes resulting in gene flow); there will be some level where the overlap is so great that for a given population the barrier no longer exists. However, for some other population with less ability to withstand previously unexperienced temperatures, the same region may be an absolute barrier.

CLIMATIC UNIFORMITY AND OVERLAP

There are areas of great temperature stability in temperate North America (e.g., coastal bays backed by low mountain ranges such as the area of San Francisco Bay, California), and areas of large temperature fluctuation in Central America (e.g., areas of well marked dry and wet seasons, and under the influence of cold air masses off the Caribbean during the northern winter, e.g., Veracruz, Mexico). However, contrast of weather records from a Central American country with those of any state in the United States will quickly show that, in general, the Central American temperature regime at a given site is more uniform on a monthly and daily basis at any altitude of distance from the seas than that of a geographically comparable site in the United States (Fig. 1; representative monthly means of the daily means, maxima and minima, for six sites in Costa Rica and in the United States). It is clear from Fig. 1 that from site to site there may be large differences in monthly temperatures; but at a given site, relative uniformity is the rule in a representative tropical country such as Costa Rica.

In respect to the impassability of mountain barriers, and intimately related to the relative uniformity of temperature regimes, the amount of overlap between the weather regime at the top of the pass and the valley below is of utmost importance; the more overlap, the less of a temperature barrier the mountain presents, and the greater the difference between monthly mean maxima and minima in the two adjacent regimes, the more overlap there is likely to be. Six representative overlap patterns in the United States and Costa Rica are exemplified in Fig. 1. For Costa Rica, the weather records were extracted from the Annuario Meteorologico for 1961 and 1964. For the United States, they were taken from U.S. Department of Commerce Weather Bureau reports (1959) and Marr (1961). Both sets were chosen on the basis of availability of maxima and minima weather data, and their position on altitudinal transects. In Fig. 1, patterns of overlap are presented from virtually no overlap (a: Costa Rica—Palmar Sur to Villa Mills, 16 to 3096 meters; d: Colorado—Grand Junction to the top of the front range behind Boulder, 1616 to 4100 meters), to high containment of the mountaintop regime within that of the valley below (c: Costa Rica—Palmar Sur to San Isidro del General, 16 to 703 meters; f: California—Fresno to Bishop, 110 to 1369 meters).

The form of the Costa Rican temperature regimes and their overlaps in Fig. 1 is clearly not the same as that of the United States' regimes. The following traits of these representative patterns are of importance to organisms living in one regime and confronted with the problem of moving through the other temperature regime to get to another area, or merely into the other regime for short-term activities.

1. The temperate regimes involve much greater changes over the year than the tropical, in respect to monthly values, and daily values (not shown in the figures, but clear from weather records).

2. The variation in difference between the monthly mean maxima and minima, across the change of seasons, is greater in the temperate examples than the tropical ones.

3. The time of maximum difference between the monthly mean maxima and minima is the summer (growing season) in the temperate examples and is the dry season (dormancy season) in the tropical examples.

4. The absolute differences between monthly mean maxima and minima are greater in the temperate examples during the summer than in the tropical examples during the rainy season.

5. The greatest amount of overlap between temperature regimes occurs during the growing season in the temperate examples but during the dry season in the tropical examples.

It is hypothesized that the amount of overlap between two temperature adjacent regimes should be greater in the temperate region than in the tropics, for any given elevational difference between the sites of the two adjacent regimes, because of the greater distances between the extremes for the temperate regimes as contrasted to tropical. To test this, overlap values between the temperature regimes of 15 pairs of sites in Costa Rica and 15 pairs of sites in the continental United States (Appendix) were calculated by the following formula:

$$\text{overlap value} = \sum_{i=1}^{12} \frac{d_i}{\sqrt{R_{1i}R_{2i}}}$$

where d_i is the amount (in degrees) of one regime that is included within the other, for the ith month. If one regime is not included within the other, d_i is considered negative and has the value of the number of degrees separating the regimes. R_{1i} is the difference in degrees between the monthly mean maximum and minimum for the ith month of the higher elevational regime and R_{2i} is the equivalent value for the lower elevational regime. Overlap is being considered in units of the geometric mean between R_{1i} and R_{2i}; hence it is the *relative* overlap, called hereafter simply "overlap." The overlap value has the property that if the monthly mean maxima of the higher elevation regime are equal to the monthly mean minima of the lower elevation regime for all 12 months, the overlap value is zero (a case intermediate

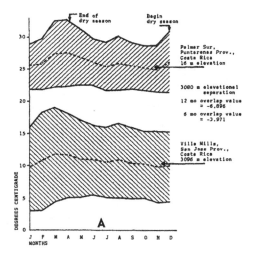

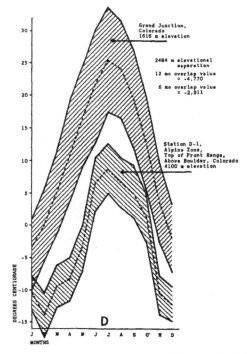

FIG. 1. Representative temperature regimes of three tropical (A, B, C) sites and three temperate (D, E, F) sites. Each graph figures two regimes. The dotted lines trace the monthly temperature means; in all cases the mean of the lower elevation regime is above that of the higher elevation mean. Solid lines trace the monthly means of daily maxima and minima; in all cases the two lower solid lines in each

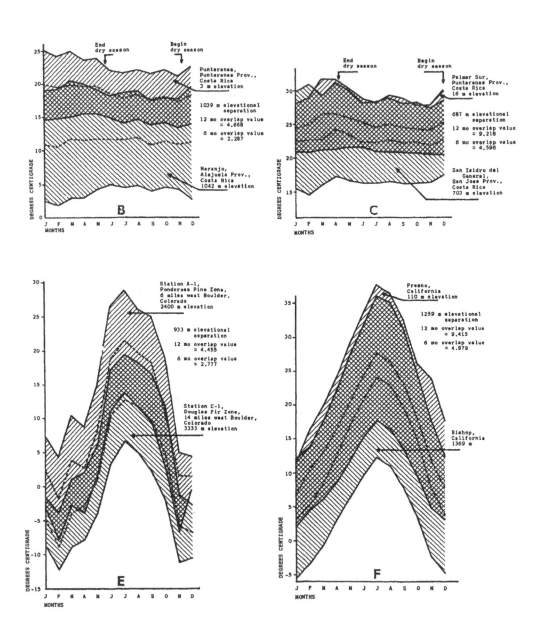

graph represent the monthly mean minima of the two regimes, with the lowermost representing the higher elevation site. Overlap between temperature regimes of adjacent sites is portrayed by cross-hatching, and increases from A and D to C and F. All graphs are to the same scale.

between that exemplified in Fig. 1a and 1b). As overlap increases, the overlap value increases to a value of 12 at the point where the two temperature regimes are congruent. Complete inclusion of one regime in the other does not guarantee maximal overlap (e.g., Fig. 1c, 1f). There is no lower limit to negative overlap values (e.g., Fig. 1a, 1d) as elevational separation between two regimes become progressively greater.

It is clear that the amount of overlap between the temperature regime of the valley bottom and the temperature regime on the mountainside above, at a specific geographic area, is primarily a function of the distance in elevation between the two weather stations. Thus the overlap between a pair of adjacent regimes in the tropics can only be compared with a temperate example of overlap where the elevational separation is similar in the two geographic areas. In Fig. 2 and 3, overlap values for each of the 30 pairs of adjacent regimes in Appendix A are plotted against the amount of elevational difference between the regimes of each pair. In Fig. 2, the overlap values are calculated for the entire 12 months of the year, while in Fig. 3, they are calculated for the six months of the growing season in the temperate examples (April through September) and for the first six months of the rainy season in the Costa Rican examples (May through October). Ideally, all 30 elevational transects should have been chosen from areas of similar rainfall patterns, but these weather data are not available.

From examination of these scattergrams and their regression lines, a number of statements have been generated that have a bearing on the effectiveness of temperature barriers in restricting the movements of organisms.

1. The overlap values of all 30 pairs of regimes for both 12 and 6 month periods, show an apparently linear relation of decreasing overlap with increasing elevational separation between the lowland and highland temperature regimes. It is this relation that leads to the classical feeling that effectiveness of mountain barriers is roughly related to their height.

2. The 12 month regression line (line I) for the Costa Rican sites has the steepest slope of the three lines in Fig. 2, but is not significantly different in slope from the 12 month regression line (line II) for all the United States' sites (line I, $b = -149$; line II, $b = -114$, $t_{26 d.f.} = 1.084$). However, the t value is high enough to suspect a relation that is obscured by the small sample size and high variance. If line I and line II are really representative of two different populations, as they appear to be following the manipulation described under 3 below, then it can be said that the overlap values demonstrated across the tropical elevational separations are less than those of the temperate examples, and become proportionally less as the elevational separation becomes less.

3. When the regression lines for the six months values are compared (Fig. 3) their slopes are highly significantly different (line IV, $b = -299$; line V, $b = -208$; $t_{26 d.f.} = 3.1531$) indicating that there are two separate populations of overlap values. This indicates that there is more dissimilarity between the overlaps of the Costa Rican paired regimes and the overlaps of the United States' paired regimes during the growing season than during the dormancy

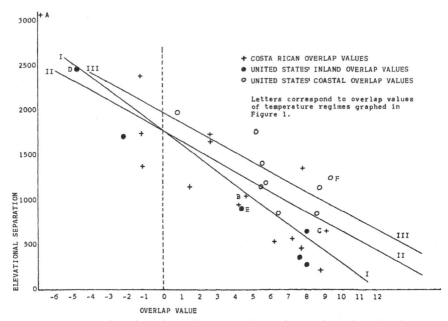

FIG. 2. Regression of overlap values for a 12-month period on elevational separation between the pair of temperature regimes for which the overlap value was calculated. Line I is that for 15 Costa Rican sites, line II is that for 15 United States' sites, and line III is that for the nine coastal sites included within the sites for line II.

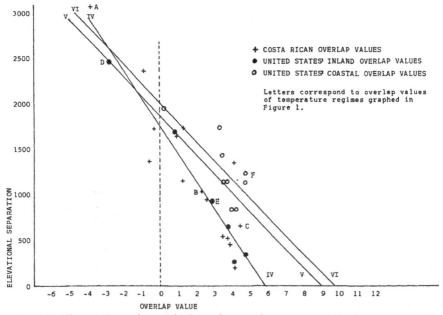

FIG. 3. The overlap values calculated for the six summer months for the temperate regimes (April through September) and the first six months of the rainy season for the tropical regimes (May through November). Line IV, Costa Rican sites; V, all United States sites; VI, nine U. S. coastal sites.

season. However, it should be noted that the removal of the winter months from the temperate data removes in absolute value much more of the variation in overlap from month to month than does the removal of the dry season months from the Costa Rican data.

4. If the six continental United States' records are removed, to make the relation to oceans more equivalent between the United States and Costa Rican sites, the 12 month regression line (line II) moves up the overlap scale to become line III (overlap value changes from 5.22 to 6.31), but lines II and III remain parallel (line II, $b = -114$; line III, $b = -110$; $t_{20 d.f.} = 0.1948$). The six-month regression line also moves following this manipulation but to a lesser extent (mean overlap changes from 3.335 to 3.650). From the distribution of the midcontinental United States' points in Fig. 2 and 3, it can be seen that, in respect to overlap for a given elevational separation, the Costa Rican sites have more in common with the midcontinental United States' sites than with the coastal ones. In other words, maximal differences between tropical and temperate overlaps are recorded when coastal areas of the United States are compared with Costa Rican transects. It should be emphasized at this point that virtually all of the Costa Rican records are coastal in the sense that no point in the country is more than about 125 miles from an ocean. In retrospect, it would seem advisable to remove this major source of variation in later comparisons of this type.

5. The variation of overlap value is great for any given elevational separation, and a given overlap value may be representative of a wide range of elevational separations in both the Costa Rican and United States data. In Fig. 3, based on six-month values, the variation in overlap for a given elevational separation is reduced. To compare any two barriers involving an elevational separation, it is clear from this that the actual amount of overlap between the upper and lower temperature regimes should be determined, since the same elevational separation on two different mountain ranges can yield quite different overlap values.

6. There is no obvious trend in change of amount of variation in overlap value along the various regression lines. For example, at an elevational separation of approximately 1700 m, the overlap values range from -2.25 to 5.40, and at 1150 m, they range from 1.50 to 8.75 (Fig. 2). This, coupled with the apparent linearity of the data, indicates that at one latitude the overlaps between temperature regimes lying at sea level and 500 m elevation should have about the same mean values and the same variances as overlaps between temperature regimes at 500 m and at 1000 m.

7. There is a point where the Costa Rican and United States' regression lines intersect (Fig. 2, zero overlap, 1780 m elevational separation; Fig. 3, -1.2 overlap, 2100 m elevational separation), and, thus, for these elevational separations, the overlap for the paired regimes are the same in the tropical and temperate example. It is notable that this point of overlap equality is at a larger elevational separation when the six-month records are considered. This indicates that the Costa Rican and United States' overlap values are even more dissimilar during the growing seasons.

8. The temperate and tropical regression lines become increasingly divergent as the elevational separation becomes less; the smaller the elevational separation, the greater will be the difference between a temperate and a tropical overlap value calculated for a pair of adjacent temperature regimes. It is not possible to compare the regression lines (I and II, IV and V) on the basis of population means since the slopes of the lines are not similar. However, in both Fig. 2 and 3, the temperate regression lines are above the Costa Rican lines, indicating that the more "tropical" the physical environment pattern, the more important are the smaller topographic features.

While there is great variation around the regression lines in Fig. 2 and 3, a major part of this variation could be removed if many weather records were available for a specific set of slopes of a major mountain range in Costa Rica, and another in the United States, both with similar exposures to an ocean and similar precipitation regimes. This statement is based on the observation that in the few cases where a long elevational transect could be broken up into component temperature regimes and an overlap value calculated between each pair of regimes, the points yield a straight line with almost no deviation from the line (e.g., Costa Rica: the overlaps of Puntarenas–Esparta, Puntarenas–Naranjo, Puntarenas–San Jose, Esparta–San Jose, and Naranjo–San Jose form a very straight line). A second source of variation, implied in the preceding paragraph, is that the points from different precipitation regimes tend to form different lines; but within a given regime, the points form very straight lines.

The minimum of the valley and the maximum on the pass do not occur at the same time during the 24 hour cycle. Thus it is that at any one point in time there is probably little or no overlap between two temperature regimes separated by more than 500 m. However, an organism is present for long periods of time and thus is subjected to all the temperature levels in its habitat. Thus the overlap as graphed in Fig. 1 and quantified in Fig. 2 and 3 is a measure of the amount of similarity between the temperature regimes experienced by an organism living in the valley bottom and then moving up and over the pass. If all the overlap patterns had the same shape as those of Costa Rica, the tropical and temperate values could be compared without qualification (as in Fig. 2). However, the manipulation used in Fig. 3 is in part justified by the fact that major periods of animal and plant activity are during the northern summer and tropical rainy season, and a purpose of this paper is to illustrate the relation between overlap and activities of organisms.

ACCLIMATION AND EVOLUTIONARY ADAPTATION

Throughout this discussion it is assumed that an organism is less likely to evolve mechanisms to survive at a given temperature if that temperature falls outside of the temperature regime of the organism's habitat than if it falls within it.

Allee et al. (1949, p. 538-539) in summarizing Payne's studies of cold hardiness in insects, has stated this in the following manner: "(1) degree of cold hardiness was [positively] correlated with seasonal periodicity of temperature, and (2) that degree of cold hardiness in a series of species, from a variety of habitats in terrestrial communities, was [positively] correlated with the normal seasonal fluctuation of temperature in that community or habitat in which a particular species was normally resident." In other words, the larger the usual variation around the mean environmental values, the higher the probability that an organism will survive a given deviation from that mean; this should apply to daily, as well as seasonal, predictability of deviations. This relation has been indicated in another manner by DuRant and Fox (1966) when discussing the effect of soil moisture on soil arthropods. Those in soil of consistently lower moisture content were more sensitive to small changes in soil moisture content than those in soils of consistently higher moisture content; i.e., the reaction of an organism to a change in the environment is dependent upon the relative as well as absolute values of that change.

It is reasonable to expect that an organism living within the relatively uniform tropical temperature regimes depicted in Fig. 1 (a, b, c) will more probably be acclimated and evolutionarily adapted to a narrower absolute range of temperatures than one which lives within the more highly fluctuating temperature ranges depicted in Fig. 1 (d, e, f). "Fluctuation" as used above applies to the variation in the monthly means across the 12-month period, the changes in the difference between the monthly means of the daily maxima and minima, and the variation in maxima and minima from day to day. This should be true even if the organism is in a resting stage during some part of the year and thus, by regulating its activity, it places itself in a more uniform environment during major activity periods. It seems likely that there will be residual ability to withstand temperatures outside of the usual habitat values at the times of activity, as physiological "by-products" of the mechanisms that allow survival during the times of inactivity.

The relation between relative uniformity of the normal habitat and the dispersal ability of the organism is well shown with weed or fugitive species of both plants and animals. These organisms customarily live in disturbed sites and have many mechanisms for survival under the physical extremes present at these places. While they have evolved great dispersal ability as a necessary part of the strategy of living in the temporary habitat of disturbed sites, mechanisms for living in this variable habitat are of obvious value in crossing the various temperature, etc., regimes necessary to find further disturbed sites.

Assuming the same amount of overlap between the temperature regimes of a valley bottom, and the pass above, at a tropical and at a temperate site with equal elevational separation, it is proposed that a tropical organism from the valley bottom is less likely to get over the pass than is a temperate organism, because the tropical organism has a higher probability of encountering temperatures to which it is neither acclimated nor evolutionarily adapted than does the temperate organism. In addition to the monthly changes exemplified in Fig. 1, this is due to the central fact that in the Costa Rican temperature regimes, for example, if the monthly mean maxima and minima are 28 and 19 C, respectively, the standard deviation of each of these means is normally less than 2 C, while the standard deviation of a similar pair of values for the United States' records in Fig. 2 and 3 would be 4 to 5 C.

However, as has been shown with the six-months values (Fig. 3) and indicated with the 12-month values (Fig. 2) for a given elevational difference in Costa Rica and in the United States, there is likely to be more overlap between the temperate adjacent temperature regimes than the tropical ones. In this case, the tropical organism has even less chance than the temperate one of getting over a pass.

This is why it is postulated that mountains are higher in the tropics figuratively speaking; they are harder to get over because, for a given elevational separation, the probability is lower in the tropics that a given temperature found at the higher elevation will fall within the temperature regime of the lower elevation than is the case in the temperate area. For example, an insect living in the forest around Puerto Viejo, Costa Rica (83 m elev.) is subjected to an annual absolute range of about 37 to 17 C; if the insect lives anywhere other than the upper surface of the canopy, the range is reduced by the insulating value of the vegetation. To move up to and over the adjacent pass at Vara Blanca, Costa Rica (1804 m elev.) it must pass through an area that is rarely, if ever, over 22 C but ranges down to 8 C (a similar regime is depicted in Fig. 1b). An insect living around Sacramento, California (8 m elev.) is subjected to an annual range of at least 46 to −9 C. To cross the Sierra Nevada at Blue Canyon, California (1760 m elev.), it must pass through an area that experiences annual ranges of about 31 to 8 C (a similar regime is depicted in Fig. 1e). The 12-month overlap value calculated for the Puerto Viejo–Vara Blanca transect is 2.648 and that for the Sacramento–Blue Canyon transect is 5.358. During the summer months, the temperature regime at Blue Canyon is almost completely contained within that of Sacramento. Thus the temperate elevational separation of 1758 m should not be nearly as inimical to animal movements as the tropical elevational spread of 1721 m.

DISCUSSION

The relation between the climatic uniformity of a habitat and the ability of the organisms living in that habitat to cross adjacent areas with different climatic regimes may indicate a general concept. In respect to temperature, valleys may figuratively be deeper to an organism living on the ridge top in the tropics than in a temperate area. In respect to rainfall patterns, one would expect the following situation. An organism living in an area with a uniform water supply (ground water or rainfall) should have more difficulty in crossing an adjacent desert than would an organism living in an area with a six-month dry season. Seasonal swamp inhabitants should be able to cross rivers more readily than those organisms which always live on the dry ground around the swamp. In other words, the greater the fluctuation of the environment in the habitat of the organism, the higher the probability that it will not encounter an unbearable combination of events in the adjacent different habitat that it is attempting to cross. Since the tropics are in general more uniform in relation to temperature, and often in respect to rainfall patterns, for a given habitat or site, it is expected that barriers involving gradients in temperature or rainfall are more effective in preventing dispersal in the tropics.

It is clear that the "tropics" are not a single phenomenon. Classical tropical climates and vegetation are at the intersections of the high ends of gradients of uniform solar energy and capture rainfall patterns. The classical tropical image is destroyed as one moves into seasonality in rainfall pattern and temperature, upward into colder elevations, laterally into areas of uniform dryness, or into areas of increased unpredictability in the absolute values and pattern of the physical environment. Thus it is that overlaps between temperature regimes in the tropics and temperate zones cannot be precisely compared, and the effect of rainfall patterns on temperature regimes is associated with much of the variation seen in Fig. 2 and 3.

Even if the "significant differences" in overlap values in Costa Rica and the United States are not truly representative of tropical and temperate temperature regime overlaps, the amount of overlap between a temperature regime at the pass and the temperature regime in the valley below are characteristcis of valley-mountain systems that should be considered in discussions of mountains as barriers between populations. Secondly, there is so much variation in the overlap value for a given elevational separation along both temperate and tropical transects that each pair of transects should be regarded as a special case in specific studies of barriers.

Since overlap values are not the same throughout the year over a specific elevational separation at a specific site, the evolutionary timing of the dispersal phases of organisms that use immigration across barriers as a strategy may be associated with the periods of greatest overlap. On the other hand, if segregation of populations is of selective value, as may well be the case if the dispersal forms are also the sexual forms, timing of production of dispersal forms may "move" away from the time of greatest overlap through evolutionary selection. In other words, if it is detrimental to the population to mate with other sexuals "coming over the mountain," then this can be avoided by production of sexuals at a time when other sexuals are not coming over the mountain (group selection is not being invoked here).

It is not intended that the idea of greater effectiveness of tropical barriers as compared with temperate ones of equal absolute magnitude be offered as an explanation of tropical species diversity. However, it is my intent to emphasize the concept that greater sensitivity to change is promoted by less frequent contact with that change.

With a few notable exceptions (e.g., Schulz' 1960 analysis of dry seasons in Suriname), investigators in tropical areas have generally ignored the problem that the usefulness of a given unit of scale in portraying constancy, variation, and variance, is directly related to the organism's sensitivity to the part of the environment being measured; while we are intuitively satisfied with monthly values in understanding the four temperate seasons, it is clear that monthly values are hardly adequate in a tropical area such as Costa Rica where eight seasons are customarily recognized (from the beginning of the rainy season: invierno, veranielo, canicula, invierno, temporales, chubascos, verano fresco, and verano caliente). That eight seasons are recognized in a supposedly more uniform climate underscores the topic of this paper. While the seasonal and daily variations in Costa Rica are smaller than those of most areas in the United States, they are simultaneously much more predictable; thus, the behavioral and developmental patterns of populations are more easily, in an evolutionary sense, associated with them than they are with unpredictable changes of equal magnitude. However, it is to be expected that more precise fitting of population activities to more predictable environmental conditions should lead to less ability to tolerate the different conditions encountered outside of the usual temporal and spatial habitat. This leads directly to the idea that the more predictable the environment, the smaller the change in that environment needs to be to serve as an immediate or long-term barrier to dispersal. This should be important in understanding the higher fidelity of tropical animals and plants to spatial and to temporal habitats which are set off by apparently minor differences in physical conditions (as compared with temperate habitats). This fidelity is experienced by anyone collecting in tropical areas, and has been documented (e.g., MacArthur, Recher, and Cody [1966] and included references, Janzen and Schoener [1967]). Such fidelity is obviously an important element of the structure of communities; and it is proposed that the increase in predictability of the physical and biotic environment, as one progresses from classical temperate environmental regimes to tropical ones, is causally related in a positive manner to the increase in species' fidelity to their habitats across the same progression. Further, this is a mutually reinforcing system whereby increased biotic fidelity leads to further biotic fidelity through the medium of interdependency of members of the same food chain.

ACKNOWLEDGMENTS

This paper was first written while teaching in the course entitled "Fundamentals of tropical biology: an ecological approach" under the auspices of the Organization for Tropical Studies during the summer of 1965 in Costa Rica. I am greatly indebted to L. Wolf and N. Scott for inspiring discussions on the subject. Student and faculty members of the Fundamentals courses in 1965 and 1966 have greatly assisted in criticism of the manuscript. F. J. Rohlf assisted in numerical treatment of the temperature data. The following persons have read the manuscript and contributed considerably by their comments: D. Johnston, J. Eagleman, and G. Orians.

LITERATURE CITED

Allee, W. C., A. E Emerson, O. Park, T. Park, and K. P. Schmidt. 1949. Principles of animal ecology. Saunders Co., Philadelphia. 837 p.

246 THE AMERICAN NATURALIST

Annuario Meteorologico. 1961. Servicio Meteorologico Nacional, Seccion Climatologia. San Jose, Costa Rica. 43 p.
_____. 1964. Servicio Meteorologico Nacional, Seccion Climatologia. San Jose, Costa Rica. 61 p.
DuRant, J. A., and R. C. Fox. 1966. Some arthropods of the forest floor in pine and hardwood forests in the South Carolina Piedmont region. Ann. Entomol. Soc. Amer. 59:202-207.
Janzen, D. H., and T. W. Schoener. 1967. Differences in insect abundance and diversity between wetter and drier sites during a tropical dry season. Ecology (in press).
MacArthur, R. H., H. Recher, and M. Cody. 1966. On the relation between habitat selection and species diversity. Amer. Natur. 100:319-332.
Marr, J. W. 1961. Ecosystems of the east slope of the Front Range in Colorado. University of Colorado Studies, Series in Biology No. 8, Univ. Colorado Press. 134 p.
Pianka, E. R. 1966. Latitudinal gradients in species diversity: a review of concepts. Amer. Natur. 100:33-46.
Schulz, J. P. 1960. Ecological studies on rain forest in northern Suriname. Verhandelingen der Koniklijke Nederlandse Akademie van Wetenschappen, Afd. Natuurkunde 53(1):1-267.
Simpson, G. G. 1964. Species density of North American recent mammals. Syst. Zool. 13:57-73.
U.S. Department of Commerce Weather Bureau. 1959. Climates of the States; Colorado (1959), California (1959), Oregon (1960), New Hampshire (1959), Idaho (1959), Nevada (1960). U. S. Government Printing Office.

APPENDIX

Localities of pairs of temperature regimes from which overlap values were calculated

Locality	Elevation of site (m)	Degrees N Latitude	Elevational separation	12 month overlap value	6 month overlap value
United States:					
(1) Sacramento, California	8	38° 35'	1173 m	5.712	3.774
Mount Shasta, California	1181	41° 19'			
(2) Los Angeles, California	100	34° 03'	1406 m	5.462	3.547
Sandburg, California	1506	34° 45'			
(3) Eureka, California	14	40° 48'	1167 m	5.616	3.528
Mount Shasta, California	1181	41° 19'			
(4) Roseburg, Oregon	168	43° 14'	1107 m	8.862	4.806
Sexton Summit, Oregon	1275	42° 37'			
(5) Fresno, California	110	36° 46'	1259 m	9.415	4.979
Bishop, California	1369	37° 22'			
(6) Sacramento, California	8	38° 35'	1752 m	5.358	3.460
Blue Canyon, California	1760	39° 17'			
(7) Pendleton, Oregon	497	45° 41'	853 m	6.662	4.300
Meacham, Oregon	1350	45° 30'			
(8) Concord, New Hampshire	113	43° 12'	1974 m	0.982	0.238
Mount Washington Observatory, New Hampshire	2087	44° 16'			
(9) Medford, Oregon	437	42° 22'	838 m	8.735	4.215
Sexton Summit, Oregon	1275	42° 37'			
(10) Las Vegas, Nevada	721	36° 05'	648 m	8.038	3.703
Bishop, California	1369	37° 22'			
(11) Reno, Nevada	1466	39° 30'	294 m	8.308	4.123
Blue Canyon, California	1760	39° 17'			
(12) Boise, Idaho	944	43° 34'	367 m	7.771	4.818
Idaho Falls, Idaho	1311	43° 32'			

248   THE AMERICAN NATURALIST

APPENDIX (Continued)

Locality	Elevation of site (m)	Degrees N Latitude	Elevational separation	12 month overlap value	6 month overlap value
(13) Station A-1, Ponderosa Pine Zone, 6 miles west of Boulder, Colorado Alpine Zone, Top of Front Range above Boulder, Colorado	2400 4100	40° 40°	1700 m	−2.291	−0.935
(14) Grand Junction, Colorado Station D-1, Alpine Zone, Top of Front Range above Boulder, Colorado	1616 4100	39° 07' 40°	2484 m	−4.770	−2.469
(15) Station A-1, Ponderosa Pine Zone, 6 miles west of Boulder, Colorado Station C-1, Douglas Fir Zone, 14 miles west of Boulder, Colorado	2400 3333	40° 40°	933 m	4.453	2.777
Costa Rica:					
(16) Cairo, Limon Prov. Cartago, Cartago Prov.	94 1435	10° 07' 9° 40'	1341 m	7.717	4.027
(17) Palmar Sur, Puntarenas Prov. Villa Mills, Puntarenas Prov.	16 3096	8° 57' 9° 34'	3080 m	−6.896	−3.971
(18) Puntarenas, Puntarenas Prov. Esparta, Puntarenas Prov.	3 208	9° 58' 9° 59'	205 m	8.779	4.192
(19) Palmar Sur, Puntarenas Prov. San Isidro del General, San Jose Prov.	16 703	8° 57' 9° 22'	687 m	9.218	4.596
(20) Cartago, Cartago Prov. Villa Mills, San Jose Prov.	1435 3096	9° 40' 9° 34'	1661 m	2.656	0.969
(21) Esparta, Puntarenas Prov. San Jose, San Jose Prov.	208 1172	9° 59' 9° 56'	964 m	4.292	2.655
(22) Puerto Viejo, Heredia Prov. Vara Blanca, Heredia Prov.	89 1814	10° 26' 10° 10'	1715 m	2.648	1.090
(23) Canas, Guanacaste Prov. Tilaran, Guanacaste Prov.	45 562	10° 25' 10° 28'	517 m	6.291	3.689

APPENDIX (Continued)

Location				
(24) Quebrada Azul, Alajuela Prov.	83	10° 24'	479 m	7.895	3.900
Tilaran, Guanacaste Prov.	562	10° 28'			
(25) Quebrada Azul, Alajuela Prov.	83	10° 24'	567 m	7.366	3.511
Villa Quesada, Alajuela Prov.	656	10° 17'			
(26) Puntarenas, Puntarenas Prov.	3	9° 58'	1039 m	4.668	2.287
Naranjo, Alajuela Prov.	1042	10° 06'			
(27) Quebrada Azul, Alajuela Prov.	83	10° 24'	1731 m	-1.016	-0.351
Vara Blanca, Heredia Prov.	1814	10° 10'			
(28) Puntarenas, Puntarenas Prov.	3	9° 58'	1169 m	1.513	1.179
San Jose, San Jose Prov.	1172	9° 56'			
(29) Puntarenas, Puntarenas Prov.	3	9° 58'	1377 m	-1.147	-0.590
Monteverde, Puntarenas Prov.	1380	10° 20'			
(30) San Isidro del General, San Jose Prov.	703	9° 22'	2393 m	-1.162	-0.987
Villa Mills, San Jose Prov.	3096	9° 34'			

Late Pleistocene History of Coniferous Woodland in the Mohave Desert

New evidence records pluvial expansion of the pinyon-juniper zone at the close of the Wisconsin glacial.

Philip V. Wells and Rainer Berger

The Mohave Desert lies midway between the winter-cold Great Basin and the subtropical Sonoran deserts, in a sensitive zone where the expression of past climatic change happens to be well preserved in the paleobotanical record. The environmental history of now-arid regions of southwestern North America is of major interest to evolutionary biologists, and to anthropologists currently probing evidence of Early Man; it is also relevant to the regional problem of water resources. However, firm knowledge of the former distribution of vegetation in the arid Southwest has heretofore been limited by the paucity of macroscopic fossil material in Pleistocene and younger sediments.

Until the last few years the meager Quaternary data contrasted with the relative abundance of Tertiary leaf-impression floras (1). Pleistocene floras were virtually unknown in the Mohave Desert region, with the notable exception of coarsely comminuted plant remains present in unique coprolite deposits of the extinct ground sloth *Nothrotherium shastense*. Sloth dung has been found in Gypsum Cave, east of Las Vegas, Nevada, and in Rampart and Muav caves in the lower Grand Canyon, east of Lake Mead, Arizona. Much of the fecal plant material was difficult to identify, but a number of recognizable plant species from sloth dung, collected at the surface in Rampart and Muav caves (elevation, about 500 meters), were similar to desert species now growing in the vicinity of the

Dr. Wells, associate professor of botany at the University of Kansas, Lawrence, is acting director, Botanical Garden, and visiting associate professor of botany, University of California, Berkeley. Dr. Berger heads the Isotope Laboratory, Institute of Geophysics and Planetary Physics, and is assistant professor of anthropology, University of California, Los Angeles.

caves, such as creosote bush (*Larrea*), saltbushes (*Atriplex*), cacti (*Opuntia*), and *Nolina* (2). The radiocarbon age of sloth dung from the surface of the Rampart Cave deposit has since been determined as 10,050 ± 350 years. Pollen analysis of the dung from the surface indicated a hot and dry climate like today's, but a sample from the 46-centimeter level (radiocarbon age, 12,050 ± 400 years) contained significantly more pine, juniper, and *Artemisia* pollens, suggesting cooler and somewhat moister conditions (3). Macrofossils preserved in sloth dung at Gypsum Cave (elevation, 610 meters) indicated that a decidedly different vegetation formerly grew there. The outstanding feature noted by Laudermilk and Munz (4) was the preeminent abundance of leaf fragments of the Joshua tree (*Yucca brevifolia*), which no longer grows in the vicinity of the cave but does occur at a somewhat higher elevation in Detrital Valley, some 70 kilometers to the southeast. Unfortunately, the material examined during this excellent early study has not yet been dated, but other dung samples collected in Gypsum Cave have yielded radiocarbon ages ranging from 8500 to 11,700 years (5). The sloth dung from Gypsum Cave dated by Libby (C-221; radiocarbon age, 10,455 ± 340 years) has been examined for fossil content by us. Macroscopic remains of woodland conifers are lacking, but leaf fibers and epidermis of the xerophytic family Agavaceae predominate, as in the undated samples of Laudermilk and Munz (4). A principal identifiable constituent in Libby's sample is *Agave utahensis* (mescal), a xerophyte with semisucculent fibrous leaves, which is now restricted to higher elevations of the desert.

Plant Fossils Preserved in Pleistocene Wood-Rat Middens

Widespread occurrence in the Southwest of extremely old wood-rat midden deposits, containing abundant well-preserved plant macrofossils, has recently come to light (6, 7). Radiocarbon ages of middens from the Mohave Desert of California and southern Nevada and from the Chihuahuan Desert of western Texas range from about 4000 years to more than 40,000 years (the limit of the method), therefore encompassing some of the latest glacial maxima of the Pleistocene and extending into postglacial time. Cricetine rodents of the genus *Neotoma*, known as wood rats, pack rats, or trade rats, range widely through the arid and semi-arid lands of North America. The outstanding generic behavioral trait, from which the common names derive, is the gathering of a great diversity of plant materials within a limited foraging range and the accumulation of these, together with excreta, in often-large middens, dens, or stick-houses containing small fibrous nests in which the rats reside (8). *Neotoma* middens are compacted refuse heaps of plant-food debris, fecal pellets, and dried urine, usually merging with the loosely assembled principal abodes, which are designated dens when situated in caves or rock shelters, or houses when constructed in relatively open situations. The bulk constituents of these fortifications are woody, often-armed, branches, spiny cactus joints or areoles, leaves, leafy twigs, fibrous bark, grass, fruits, and seeds. The herbivorous wood rats utilize an astonishing variety of plant species, and, although the dominant plants of the surrounding vegetation usually make up the bulk of their deposits, a rather detailed inventory of the local flora often accumulates. Relative to the small size of the rats, *Neotoma* deposits may attain a very large volume, often more than 1000 liters, and sometimes contain surprisingly large objects such as woody branches nearly 1 meter in length and the long bones of much larger animals. The hoarding instinct is frequently diverted to objects having no apparent food or constructional value, including human artifacts, for which the rat "trades" whatever it happens to have in its mouth at the time.

Pleistocene *Neotoma* deposits, which are strictly confined to caves and rock shelters, are chiefly of the compacted midden type but often contain den and nest material. Lustrous dark-brown or black masses of dried wood-rat urine are conspicuous veneers on many old (and some modern) deposits, or on the adjacent walls of the rock shelter. This conspicuous material has attracted some attention and been called "amberat" (9). The ancient deposits usually occupy a relatively dry and secluded space on a ledge or in a crevice or cavity, where the mummified plant materials have often remained excellently preserved for tens of thousands of years. Because of the rapid accumulation of plant debris by wood rats, the time span represented by any one period of constructional activity appears to be small relative to the total age of the deposit. The largest ancient deposit yet uncovered, which completely filled a small tunnel of a cave in the Chihuahuan Desert with several thousand liters of debris from a pluvial pinyon-juniper-oak woodland, contained about 75 centimeters of *Neotoma* deposition, with radiocarbon ages of 11,560 ± 140 and 12,550 ± 130 years near the top and bottom, respectively (10). However, different deposits provide samples of vegetation often widely different in age, extending over much of the range of the radiocarbon method (6, 7).

A total of 17 ancient wood-rat middens has been uncovered in the Mohave Desert in the northeastern (Frenchman Flat), north-central (Funeral Mountains), southwestern (Lucerne Valley), and southeastern (Turtle Mountains) sectors (Table 1; Fig. 1). All but one of the dated deposits (Newberry Cave, 7400 years old) contain records of former juniper or pinyon-juniper woodland vegetation, evidenced by consistent abundance of leafy twigs and seeds of *Juniperus osteosperma* (Utah juniper) and, locally, of leaves, cone scales, and seeds of *Pinus monophylla* (pinyon pine; Fig. 2), together with many shrubby species characteristic of woodland, semidesert, or even desert vegetation. The late-Pleistocene macrofossil record clearly indicates that at least the higher areas of the Mohave Desert, now occupied by a scanty desert or semidesert shrub vegetation, supported xerophilous woodlands of pygmy conifers during pluvial times. One of the remarkable features of the history revealed to date is the hitherto-unsuspected late persistence of woodland trees at elevations that are now treeless desert. Prevalence of woodland vegetation at moderately low elevations throughout the Mohave Desert as recently as about

9000 years ago is apparent, and woodland persisted locally at elevations now desert in southern Nevada about 8400 and 7800 years ago (Table 1). Although radiocarbon dates associated with major fluctuations of the largest pluvial bodies of water in the Great Basin indicate an episode of aridity about 11,000 years ago (11), a precipitation-temperature regime inadequate to maintain vast lakes in a relatively arid region may have been sufficient then to maintain established xerophilous woodlands at moderately low elevations. However, a

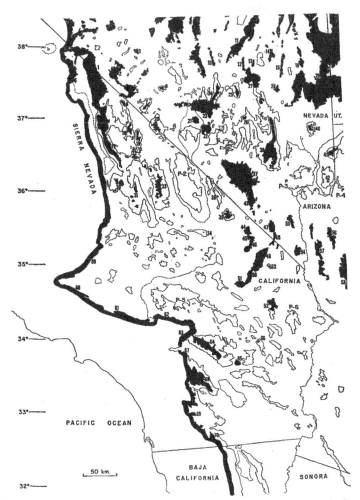

Fig. 1. Mountain ranges and playas of the Mohave, Colorado, and southern Great Basin deserts, showing in black the areas of higher elevation that now support woodland, forest, or other vegetation zones within the desert region to the east of the Sierra Nevada, Transverse, and Peninsular ranges. Mountain ranges are located by unbroken lines or by extent of existing woodland. Dashed lines show playas, which were often the sites of larger pluvial lakes, and principal streams. Symbols: P, Pleistocene wood-rat middens and coprolites of the ground sloth containing remains of former woodland or desert vegetation; P-1, Frenchman Flat series; P-2, Funeral Mountains; P-3, Gypsum Cave; P-4, Rampart and Muav caves; P-5, Negro Butte; P-6, Turtle Mountains. Other mountains cited are numbered as follows: P-1 (northernmost of four), Aysees Peak; P-1 (third from south), Ranger Mountains; P-1 (second from south), Mercury Ridge; P-1 (southernmost of four), Spotted Range; P-5 (north side), Ord Mountains; 33, Panamint Range; 34, Avawatz Range; 35, Kingston Range; 36, Nopah Range; 37, Spring Range; 44, Clark Mountain; 52, Old Woman Range; 61, San Gabriel Range; 62, San Bernadino Range; 66, Coxcomb Mountains.

Table 1. Fossil plant debris in, radiocarbon ages of, and elevations of wood-rat middens in the Mohave Desert of Nevada and California. Abbreviations: S, Spotted Range; A, Aysees Peak; M, Mercury Ridge; F, Funeral Range; R, Ranger Mountains; N, Negro Butte; T, Turtle Mountains. Symbols (relative abundance): +, low; ++, intermediate; +++, high (principal constituent).

Species, structures	Midden site, elevation (meters), and radiocarbon age (years)											
	S-1, 1830, 8420, ±100	S-2, 1550, 9450, ±90	A, 1525, 9320, ±300	M-1, 1390, 9000, ±250	M-2, 1280, 7800, ±150	M-3, 1250, 12,700, ±200	F, 1280, 11,600, ±160	R-1, 1130, 16,800, ±300	R-2, 1100, 10,100, ±160	N, 1070, 9140, ±140	T-1, 850, 19,500, ±380	T-2, 730, 13,900, ±200
Trees and arborescent shrubs												
Juniperus osteosperma, leafy twigs, seeds, wood	+++	+++	+++	++	+++	+++	+++	+++	+++	+++	+++	+++
Pinus monophylla, leaves, cones, seeds		++									++	+++
Acer glabrum, twigs with buds, samaras		+										
Cercocarpus ledifolius, leaves				+			+					
Cowania mexicana, leaves		++		+	+++							
Fraxinus anomala, twigs with buds, samaras							+					
Shrubs												
Artemisia nova, leaves	+	+	+	+			+		+	+		+
Atriplex canescens, fruits	+		+									
A. confertifolia, leaves, twigs, fruits	++	++						+	+++			
Ceanothus greggii, leaves		+										
Cercocarpus intricatus, leaves, calyces, achenes	++	++	+				+					+
Chamaebatiaria millefolia, twigs with buds, follicles		+										
Chrysothamnus sp., involucres		+							+			
Coleogyne ramosissima, achenes		+		+	+							
Encelia virginensis, achenes							+					
Ephedra viridis, twigs, seeds	++	++	++	+			+	+		++		+
Eriogonum microthecum, leaves		+							+	+		
Fallugia paradoxa, leaves, achenes		+	+	+	+		+					
Fendlerella utahensis, capsule		+										
Haplopappus cuneatus, leaves										+		
Hecastocleis shockleyi, leaves		+										
Lepidospartum latisquamum, involucres			+						+			
Petrophytum caespitosum, leaves, inflorescences, follicles		++							+			
Prunus fasciculata, leaves, drupes	++	++	++		+		+++					
Purshia glandulosa, leaves, fruits										+++		
Ribes montigenum, twigs with trifid spines or prickles		++					+		+	+		
Ribes cf. velutinum, twigs with spines											+	+
Senecio douglasii, involucres	+	+	+	+		+			+			
Sphaeralcea ambigua, leaves, fruits		+	+									
Symphoricarpos longiflorus, leaves, twigs, flowers, seeds	+	++	+	+	+		++	+	+			
Tetradymia canescens, involucres		+										
Agavaceae, Cactaceae (succulents)												
Agave utahensis, leaves, seeds	+	+										
Nolina parryi, leaves							+					
Yucca brevifolia, leaves, seeds	+	+	+							+		
Opuntia erinacea, areoles, fruits, seeds	++	+	+	+	+		+	+	++	+	++	+++
Grasses and forbs												
Oryzopsis hymenoides, fruits	+	++	+	+	+		+	+	+			+
Stipa arida, fruits		+										
S. speciosa, fruits	+	+	+	+					+			
Artemisia ludoviciana, leaves, twigs, flowers			++							+		
Amsinckia tessellata, nutlets		+										
Cryptantha confertiflora, nutlets	+											
C. flavoculata, nutlets		+										
Crypthantha sp., nutlets										+		
Penstemon palmeri, leaves, capsules		+	+	+		+						
Viguiera multiflora, achenes		++			+				+			

* Species no longer present in mountain range of midden site.

wood-rat midden having a radiocarbon age of 7400 ± 100 years, from Newberry Cave in the *Larrea* zone of the south-central Mohave Desert, shows an absence of woodland species and records the presence of creosote bush (*Larrea*), a warm-desert shrub, at about the onset of the period of maximum postglacial warmth (Hypsithermal time).

A Paleozonation of Vegetation

Evidence of an approximately synchronous zonal differentiation of vegetation in response to a gradient of elevation, on limestone in the northeastern Mohave Desert, has now become available. Briefly, the evidence consists of wood-rat middens at 1550, 1525, 1390, and 1100 meters with radiocarbon ages of about 9450, 9320, 9000, and 10,100 years, respectively, and ground-sloth coprolites at 610 and 530 meters with radiocarbon ages of 10,455 (also 8500 to 11,700) and 10,020 years, respectively. The deposit at 1550 meters in Spotted Range, near Frenchman Flat, Nevada, is the only midden in the sequence that contains remains of the more mesophytic pinyon-juniper woodland species; these include mountain maple (*Acer glabrum*), fernbush (*Chamaebatiaria millefolium*), and *Ceanothus greggii*, as well as *P. monophylla* and *J. osteosperma*. Closely associated with them in the midden are many relatively xerophytic, semidesert species (Table 1).

Similar combinations of species occur today at elevations of about 2200 meters in the highest mountains of the region. Despite the dramatic postglacial disappearance of the woodland conifers, the maple, and the more mesophytic shrubs, many of the species closely associated with them in the midden are now growing in the canyon at the midden site at this relatively high elevation (1550 meters), including small populations of characteristic woodland species such as a mountain mahogany (*Cercocarpus intricatus*), cliff rose (*Cowania mexicana*), Apache plume (*Fallugia paradoxa*), snowberry (*Symphoricarpos longiflorus*), and rock-spiraea (*Petrophytum caespitosum*). Also still present on open slopes, but in greater numbers, are the desert shrubs blackbrush (*Coleogyne*) and shadscale (*Atriplex confertifolia*).

The plant contents of the Aysees Peak deposit at 1525 meters also show a limited overlap of species with the existing canyon vegetation near the midden site, again including *Cercocarpus intricatus*. No trace of pinyon pine was detected in this midden, although leaves of the more mesophytic mountain mahogany, *C. ledifolius*, were present, and *J. osteosperma* was the principal constituent. On the other hand, the deposits at 1100 meters in Ranger Mountains contain a much more xerophytic assemblage paralleling the more arid aspect of existing vegetation: present in the middens are remains of juniper, the only characteristic woodland species, accompanied by great abundance of fruits, leaves, and spinose twigs of shadscale, a dominant shrub of existing cold-desert vegetation in the Great Basin. Absence of big sagebrush (*Artemisia tridentata*) and abundance of shadscale signify distinctly dry, though cool, conditions near the lower limit of woodland (*12*).

This roughly synchronous, late-Wisconsin zonation from relatively mesophytic pinyon-juniper-maple woodland to xerophytic juniper-shadscale semidesert occurred on limestone within an elevational span of 450 meters near Frenchman Flat basin, southern Nevada. A lower limit of elevation in this

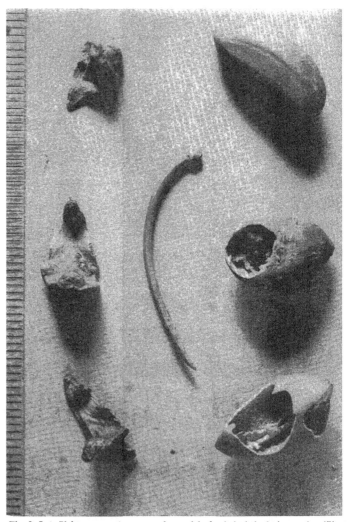

Fig. 2. Late Pleistocene seeds, cone scales, and leaf, of single-leaf pinyon pine (*Pinus monophylla* Torrey & Fremont), selected from a fossiliferous stratum having a radiocarbon age of 9450 ± 90 years in a wood-rat midden from the arid, unwooded Spotted Range of southern Nevada. Scale: millimeters.

drainage basin is, of course, imposed by the base level of 939 meters at Frenchman Playa, less than 200 meters below the lowest Pleistocene *Neotoma* deposit. However, a much lower range of elevation is available further south in Nevada and adjacent Arizona, where the dung deposits of the ground sloth in Gypsum and Rampart caves provide a record of former vegetation on limestone at 610 and 530 meters, respectively, during the time period spanned by some of the wood-rat middens in the Frenchman Flat area. The macrofossil record at Gypsum Cave was dominated by the Joshua tree and other Agavaceae in association with desert shrubs including creosote bush (*Larrea*), a xerophytic assemblage transitional to the warm-desert vegetation now occupying the site (*4*). This finding contrasts with the lowest midden record at Frenchman Flat, 500 meters higher and about 100 kilometers further north than Gypsum Cave; the evidence from the Flat suggests a transition from juniper woodland to shadscale, a cold-desert vegetation. At about the same time a decidedly warm-desert assemblage was accumulating in sloth dung at Rampart Cave, near the bottom of the lower Grand Canyon in Arizona, as indicated by fossils of creosote bush and other desert shrubs that still grow there.

Hence, in the northeastern Mohave Desert, during the time interval bracketed by radiocarbon ages of about 9000 and 10,000 years, juniper woodlands descended to an elevation of 1100 meters on limestone, some 600 meters below the present average lower limit of woodland on limestone, which lies at an elevation of about 1700 meters in the latitude of Frenchman Flat (*6*). But desert or semidesert shrubs coexisted with the woodland trees throughout much of the span of elevation corresponding to the pluvial lowering of the woodland zone, and the more meso-phytic phase of pinyon-juniper woodland (with pinyon pine, mountain maple, and other species requiring more water) was apparently confined to montane habitats at elevations above 1500 meters. Joshua trees, accompanied by desert shrubs, prevailed down to about 600 meters at Gypsum Cave, but only the shrubs of the existing warm-desert vegetation occurred at 530 meters near Rampart Cave.

In the southwestern sector of the Mohave Desert, another wood-rat midden dating from this time period, with a radiocarbon age of about 9140 years,

has been uncovered by E. Jaeger and one of us (P.V.W.) near Negro Butte, Lucerne Valley, California, at an elevation of 1070 meters and within the rain shadow of the lofty San Bernardino Mountains. Presence of abundant remains of *J. osteosperma* and absence of pinyon pine, despite a favorable granitic substratum, parallel the *Neotoma* record on limestone at about the same elevation near Frenchman Flat, some 250 kilometers to the northeast. An older *Neotoma* deposit (radiocarbon age, about 11,600 years) at an elevation of 1280 meters in the north-central part of the Mohave Desert, in the arid Funeral Range, California, provides additional macrofossil evidence (again great abundance of leafy twigs and seeds of *J. osteosperma*) of a widespread pluvial occurrence of xerophilous juniper woodlands, lacking pinyon pine, at moderately low elevations in certain sectors of the Mohave Desert. But the southeastern sector of the desert offers a seemingly anomalous record.

Two Pleistocene wood-rat middens have been found at 730 and 850 meters in the arid Turtle Mountains, California, with radiocarbon ages of about 13,900 and 19,500 years, respectively. Both contain abundant macrofossils of single-leaf pinyon pine (*P. monophylla*), as well as the formerly ubiquitous *J. osteosperma*. Woodland conifers, or even relict individuals of woodland shrubs, are now lacking in the low, arid Turtle Mountains (maximum elevation, 1289 meters). The apparent anomaly lies in the presence of pinyon pine at this extremely low elevation (730 meters) and latitude (34°24'N) during a time span when more-xerophytic juniper woodlands, lacking pinyon pine, prevailed at a higher elevation (about 1100 meters) and latitude (36°40'N) near Frenchman Flat, as evidenced by *Neotoma* deposits with radiocarbon ages ranging from 12,700 to 17,400 years (*6*). However, the paradox appears to be resolved by the astonishing fact that lower limits of elevation for existing pinyon-juniper woodlands are, in general, much lower in the southern than in the northern sector of the Mohave Desert and Great Basin. Woodlands of *P. monophylla* and *J. californica* (a close relative of *J. osteosperma*) thrive today at an elevation of 1200 meters in Old Woman Range (maximum elevation, 1620 meters), about 30 kilometers west of the Pleistocene *Neotoma* midden sites in Turtle Mountains. The lower limits of pinyon-juniper woodland in

Old Woman Range are in fact more than 500 meters *lower* than the average lower limits of woodland on the much higher (to 3631 meters) and more massive mountain ranges in the region of Frenchman Flat, 200 kilometers to the north.

Biogeographic Anomalies

Numerous anomalies, both present and past, concerning the altitudinal and latitudinal distribution of pinyon-juniper woodland in the Mohave Desert and Great Basin region have recently come to light, and their interpretation is of critical importance to an understanding of biogeography in the Southwest (*13*). The "Merriam effect," or inverse relation between mountain mass and elevations of vegetation zones (*14*), seems to be lacking or outweighed by other factors affecting the distribution of vegetation zones in this arid region. For example, the average lower limits of woodland, on mountains greatly differing in massiveness and maximum elevation, are approximately equal in the region of Frenchman Flat in southern Nevada. It is true that in certain deep canyons of the most massive range, namely Pine and Excelsior canyons in Spring Range, pinyon pine descends more than 400 meters below its average lower limits and ponderosa pine occurs fully 1000 meters below its usual lowest level (*15*); but these exceptional distributions clearly correlate with the extraordinary edaphic and microclimatic effects of deep canyons cut in porous sandstone and should not be misconstrued as examples of the "Merriam effect." On the other hand, there does appear to be a general relation between the average minimum elevation of existing woodland and latitude; data for 80 wooded mountain ranges in the Mohave and Colorado deserts and southern Great Basin are plotted in Fig. 3; the usual inverse relation between latitude and the elevation of montane vegetation zones seems to be reversed. A similar downward shift of vegetation zones with decreasing latitude occurs in eastern Arizona and Mexico (*16*), where decreasing distance from the source of moist Gulf air masses, yielding summer rain, may explain the anomaly. The Mohave Desert as a whole receives very little summer rain, although the incidence is significantly greater in the eastern sector (*17*).

However, numerous details of distribution of pinyon-juniper woodland in the Mohave Desert area suggest the existence of major local, as well as regional, orographic effects. For example, in the Avawatz-Granite Range, north of Barstow, woodland is absent from Granite Mountains (maximum elevation, 1680 meters) but present on the summit of Avawatz Mountain at 1890 meters. Nevertheless, Ord Mountains (maximum elevation, 1920 meters), situated south of Barstow, lack woodland entirely, possibly because they lie closer to the maximum intensity of rain shadow cast by the massive Transverse Ranges, particularly the San Bernardino–San Gabriel axis (maximum elevation, 3380 meters). Also, Funeral Range, with a maximum elevation of 2040 meters, now lacks woodland, whereas the less massive Nopah Range has small stands of woodland near the summit at 1940 meters; the high points of both ranges are carbonate rocks of Paleozoic age. Funeral Range lies just to the leeward of the lofty Panamint Range (maximum elevation, 3370 meters) and therefore may be in a more intense rain shadow than Nopah Range, which is 50 kilometers further to the southeast. Similarly, but on a grander scale, the vast rain shadow of the very massive Sierra Nevada (maximum elevation, 4410 meters), coupled with the great distance from air masses bearing summer rain, may cause the general high elevation of the woodland-vegetation zone observed in the northern Mohave Desert and Great Basin (Fig. 3).

Hence the Pleistocene wood-rat middens, at 730 and 850 meters on northward-facing canyon walls cut in porous tuffaceous agglomerate in the low, treeless Turtle Mountains (maximum elevation, 1289 meters), record a relatively large downward shift of the pinyon-juniper woodland zone in the southeastern sector of the Mohave Desert, paralleling the remarkably low minimum elevation of the existing woodland zone (about 1200 meters) on granite in the adjacent Old Woman Range (maximum elevation, 1620 meters) and in the less massive Coxcomb Mountains (maximum elevation, 1340 meters). However, the *Neotoma* deposits in Turtle Mountains, which record a pluvial lowering of the pinyon-juniper zone by 350 to 470 meters, do not necessarily establish a lower limit of woodland in this sector at that time (about 14,000 and 19,500 years ago). The abundance of fossil remains of *P.*

monophylla in the middens leads to the inference that more-xerophilous juniper woodlands, lacking pinyon pine, may have extended to still-lower elevations, as was true of the pluvial zonation recorded in the *Neotoma* middens of the northern Mohave Desert.

Magnitude of Pluvial Migration of Vegetation

In the present state of knowledge, it would be premature to draw a map purporting to show the maximum pluvial extent of pinyon-juniper woodland vegetation in the Mohave Desert area. It is sufficient to note that, when relatively low arid ranges, such as Turtle Mountains, formerly supported pinyon-juniper woodland, most of the other low arid mountains or drainage divides, not situated in especially intense rain shad-

ows cast by the more massive mountain ranges, were probably also wooded. The fossil evidence speaks for a former continuity of woodland, at least along the higher divides connecting most of the ranges, which implies more or less wooded corridors for the extensive pluvial migrations suggested by the disjunct distributions of existing woodland species (Fig. 1). But there is no macrofossil evidence of pluvial continuity for the more mesophytic coniferous-forest zone of ponderosa pine or white fir now occupying islands of relatively mesic environment on the highest mountains of the region, as in Spring, Sheep, Clark, and Kingston ranges (*18*). Indeed, general discontinuity of the highest zones is suggested by the uneven stocking of the isolated lofty mountains with mesophytic or boreal species of high montane habitats, and by the trend toward endemism (*19*). There is a parallel in

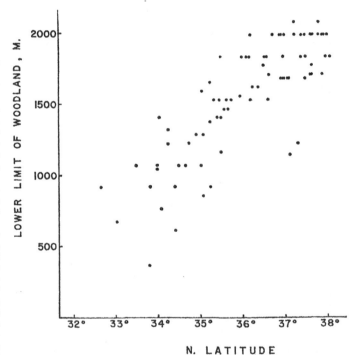

N. LATITUDE

Fig. 3. Lower limits of elevation (in meters) for the existing woodland-vegetation zone in relation to latitude of 80 wooded mountain ranges in the Mohave, Colorado, and southern Great Basin deserts. Prevailing elevations for lower limits of woodland were estimated on open slopes; exceptional downward extensions in canyons and on steep northward-facing slopes were not used. Some of the large variance probably reflects the east-west gradient of decreasing summer precipitation, and local effects of substratum and physiography—especially distance from orographic barriers producing rain shadows. The anomalous latitudinal trend may be related to the declining trend in average elevation, massiveness, or continuity of the principal orographic barriers from north to south along the western margin of the deserts.

the uneven pattern of distribution of montane species on the high peaks of the Chihuahuan Desert region, where a lack of pluvial continuity of range for mesophytic montane species has also been inferred from fossil evidence of xerophilous woodland trees and desert shrubs in the intervening lowlands even during full-glacial Wisconsin time (20).

Summary

Seventeen ancient wood-rat middens, ranging in radiocarbon age from 7400 to 19,500 years and to older than 40,000 years, have been uncovered in the northeastern, north-central, southeastern, and southwestern sectors of the Mohave Desert. Excellent preservation of macroscopic plant materials (including stems, buds, leaves, fruits, and seeds) enables identification of many plant species growing within the limited foraging range of the sedentary wood rat.

An approximately synchronous zonal differentiation of vegetation in response to a gradient of elevation on limestone in the northeastern Mohave Desert is apparent from the macrofossil evidence, preserved in wood-rat middens and ground-sloth coprolites, covering a time span bracketed by radiocarbon ages of about 9000 and 10,000 years. Xerophilous juniper woodlands descended to an elevation of 1100 meters, some 600 meters below the present lower limit of woodland (1700 meters) in the latitude of Frenchman Flat. But desert or semidesert shrubs coexisted with the woodland trees throughout much of the span of elevation corresponding to the pluvial lowering of the woodland zone, and the more mesophytic phase of pinyon-juniper woodland was evidently confined to montane habitats at elevations above 1500 meters. Joshua trees, accompanied by desert shrubs, prevailed down to about 600 meters at Gypsum Cave, Nevada, but only the shrubs of the existing warm-desert vegetation occurred at 530 meters near Rampart Cave, Arizona.

Pleistocene middens from the southeastern Mohave Desert record a relatively large downward shift of the pinyon-juniper woodland zone, paralleling the remarkably low minimum elevation of the existing woodland zone in that area. The macrofossil evidence speaks for former continuity of the many disjunct stands of woodland vegetation in the Mohave Desert region, at least along the higher divides connecting most of the ranges. However, there is no macrofossil evidence of pluvial continuity of range for the more mesophytic, montane, coniferous-forest zone of ponderosa pine or white fir now occupying islands of relatively mesic environment on the highest mountains of the region. On the contrary, the uneven stocking of the lofty mountains of the Mohave Desert with mesophytic or boreal species and the trend toward endemism suggest a long history of isolation.

Addendum: We have recently uncovered and dated three more late-Pleistocene *Neotoma* deposits, containing remains of woodland conifers, at elevations now desert in the northeastern Mohave Desert. A midden from Pintwater Cave in southern Nevada, at an elevation of 1280 meters on limestone, contains a relatively xerophytic semidesert assemblage, including remains of shadscale (*Atriplex confertifolia*) and rabbit brush (*Chrysothamnus*) but only a few leafy twigs of *J. osteosperma*, associated with a radiocarbon age of 16,400 ± 250 years (UCLA-1099). Also, pollen analysis of *Neotoma* feces of about the same age suggested semidesert vegetation dominated by *Atriplex*, with a relative percentage of *Juniperus* pollen of only 2.6 in a sample of 1500 grains (21). Ecologically the Pintwater Cave site is relatively xeric, since it is high on a steep, westward-facing, limestone scarp, with strong afternoon insolation; it now supports a very scanty warm-desert vegetation, but sparse woodland persists on westward-facing limestone scarps only 500 meters above the cave in the northern part of Pintwater Range (Fig. 1, No. 26).

Two ancient wood-rat middens at extremely low elevations (530 and 550 meters) in the low, arid North Muddy Mountains of southeastern Nevada, about 50 kilometers northeast of Gypsum Cave (Fig. 1, *P-3*), contain abundant remains of *J. osteosperma* and the semidesert shrubs *Purshia glandulosa* and *Ephedra viridis*, but there are no macroscopic remains of pinyon pine or other relatively less xerophytic woodland species in the fossil material examined. The radiocarbon ages of the 12,900 ± 180 years (UCLA-1218 and 1219). It is important to note that these sites are located in a narrow canyon cut in massive Mesozoic sandstone. On the same type of sandstone raised to much two deposits are 17,750 ± 200 and

higher elevations in Spring Range, southern Nevada, existing woodland descends to an elevation of 1180 meters, which is about 500 meters below the average lower limit of woodland on limestone in the same range. Woodland no longer exists in North Muddy Mountains, but the next-lower vegetation zone, dominated by *Coleogyne* (blackbrush), is present there above 1000 meters and permits correlation with the zonation in Spring Range. The lower limit of the *Coleogyne* zone on sandstone is at an elevation of about 1000 meters in both areas. Hence the *Neotoma* fossil record on sandstone in North Muddy Mountains documents a full-pluvial downward shift of xerophilous juniper woodland as much as 650 meters below the existing lower limit of woodland on sandstone, which parallels the 600-meter downward shift of juniper-dominated woodland on limestone that was previously established in the Frenchman Flat area (6).

References and Notes

1. D. I. Axelrod, *Carnegie Inst. Wash. Publ. 590* (1950), p. 215; *Botan. Rev.* 24, 431 (1958).
2. J. D. Laudermilk and P. A. Munz, *Carnegie Inst. Wash. Publ. 487* (1938), p. 273.
3. P. S. Martin, B. Sabels, D. Shutler, *Amer. J. Sci.* 259, 102 (1961); E. A. Olsen and W. S. Broecker, *Radiocarbon* 3, 141 (1961). The latter publication cites radiocarbon ages of 9900 ± 400 and 11,900 ± 500 years, respectively.
4. J. D. Laudermilk and P. A. Munz, *Carnegie Inst. Wash. Publ.* 453 (1934), p. 29. Leaf fragments of *Yucca* constituted fully 80 percent (by volume) of the sloth dung from Gypsum Cave; principal species was *Y. brevifolia* (Joshua tree), but *Y. baccata* Torr. and *Y. mohavensis* (*Y. schidigera* Roezl) were also identified. Presence of remains of the desert shrubs *Larrea divaricata* Cav., *Atriplex confertifolia* (T. & F.) Wats., *Ephedra nevadensis* Wats., and *Petalonyx* indicates a decidedly arid environment. Aside from the Joshua tree, only the remains of *Y. baccata*, *Agave utahensis* Engelm., and *Chrysothamnus* provided any definite evidence of a former lowering of vegetation zones near Gypsum Cave, although a few fiber cells or tracheids of a gymnosperm (genus undetermined) were found in the dung. The great abundance of *Y. brevifolia* and presence of *L. divaricata* imply a type of desert vegetation that occurs today at an elevation of about 800 meters on the slopes of Detrital Valley, Arizona, about 70 kilometers to the southeast of Gypsum Cave (elevation, 610 meters), proving a downward displacement of the existing Joshua-tree vegetation zone of less than 200 meters. Laudermilk and Munz compared their fossil assemblage at Gypsum Cave with the Joshua-tree zone at an elevation of 1500 meters at Clark Mountain, Nevada, about 70 kilometers to the southwest. Later study of the dung deposits in Rampart Cave, Arizona, uncovered three twigs and a seed of *Juniperus* 92 centimeters below the surface and 46 centimeters below a sample dated by radiocarbon at 12,050 ± 400 years old. *Juniperus* now grows about 400 meters above the cave (3).
5. J. R. Arnold and W. F. Libby, *Science* 113, 11 (1951); C. L. Hubbs, G. S. Bien, H. E. Suess, *Radiocarbon* 4, 204 (1962); 5, 254 (1963).
6. P. V. Wells and C. D. Jorgensen, *Science* 143, 1171 (1964).
7. P. V. Wells, *Bull. Ecol. Soc. Amer.* 45, 76 (1964); 46, 197 (1965); G. J. Fergusson and

W. F. Libby, *Radiocarbon* **4**, 109 (1962); **5**, 1 (1963); **6**, 318 (1964); R. Berger, G. J. Fergusson, W. F. Libby, *ibid.* **7**, 336 (1965); R. Berger and W. F. Libby, *ibid.* **8**, 467 (1966).

8. R. B. Finley, *Univ. Kans. Publ. Museum Nat. Hist.* **10**, 213 (1958); J. M. Lindsdale and L. P. Tevis, Jr., *The Dusky-footed Wood Rat* (Univ. of California Press, Berkeley, 1951), p. 57.

9. P. C. Orr, *Observations* (Western Speleol. Inst., 1957), No. 2, p. 1; R. Berger, W. S. Ting, W. F. Libby, in *Proc. Intern. Conf. Radiocarbon Tritium Dating 6th Pullman, Wash.* (1965), p. 731. This peculiar mineral-like substance was earlier designated "johnsonite" (E. R. Hall, personal communication).

10. P. V. Wells, *Science* **153**, 970 (1966). There appears to be at least one hiatus in this sequence.

11. W. S. Broecker, M. Ewing, B. C. Heezen, *Amer. J. Sci.* **258**, 429 (1960); W. S. Broecker and P. C. Orr, *Geol. Soc. Amer. Bull.* **69**, 1009 (1958). Firm evidence for a subsequent high stand of the pluvial lakes (Lahontan and Bonneville), beginning about 9500 years ago [W. S. Broecker and A. Kaufman, *Geol. Soc. Amer. Bull.* **76**, 537 (1965)], is in agreement with the wood-rat midden record of woodland vegetation (at elevations now desert) at that time.

12. The combination of open stands of juniper (*J. osteosperma*) with a low shrub synusia of shadscale (*Atriplex confertifolia*) and dwarf sagebrush (*Artemisia nova*) occurs today in a transitional zone between the more mesophytic pinyon-juniper woodland–sagebrush (*A. tridentata*) vegetation and the cold desert of treeless shadscale scrub in particularly arid sectors of the Great Basin, as in House Range, Utah.

13. An apparent anomaly, which seems to be particularly significant, is the contrast between interpretations of evidence derived from pollen analysis of sediments, in the light of macrofossil evidence of former vegetation of comparable age from the same region. High relative percentages of pine pollen (species unidentified), recorded in pluvial lake deposits at Tule Springs at an elevation of 700 meters in southern Nevada, are interpreted as implying a downward displacement of vegetation zones by as much as 1220 meters [P. J. Mehringer, *J. Ariz. Acad. Sci.* **3**, 186 (1965)]. On the other hand, the Pleistocene wood rat–midden evidence supports a downward displacement of about 600 meters in the Frenchman Flat area, which is 70 kilometers north of the Tule Springs site. However, the lower Pleistocene *Neotoma* deposits (at an elevation of about 1100 meters) contain abundant remains of juniper (*J. osteosperma*) and several species of desert shrubs, but no pine. The relative downward displacement recorded in Pleistocene wood-rat middens in Turtle Mountains is still less at 470 meters, but this is probably not a maximum figure for that area. As to time of occurrence, the Tule Springs date of 22,600 ± 550 years (UCLA-536, based on carbonate of mollusk shells) is on the further side of the classical Wisconsin glacial maximum (18,000 to 20,000 years ago) from a Frenchman Flat organic-carbon date of 17,450 ± 300 years (UCLA-555), but the

Turtle Mountains date of 19,500 ± 380 years (UCLA-1063) is clearly "full glacial." An alternative interpretation of the pollen samples at Tule Springs is that they integrated the pollen rain from a large area of watershed around the site of deposition. Because of the basin-and-range topography, steep gradients of elevation place relatively arid basin floors well within the range of dissemination of pollen of anemophilous conifers of the montane forest and woodland zones. The pollens of *Pinus* and *Abies* are equipped with bladder-like wings which increase the buoyancy in air or water, and *Pinus* produces notoriously large quantities of pollen. If the absolute quantity of pollen produced by vegetation immediately surrounding the depositional site were moderately low—as would be true, for example, of vegetation dominated by entomophilous species—high relative percentages of conifer pollen could have accumulated in the pluvial sediments, with no conifers at the Tule Springs site. The macrofossil evidence reported by us does support at least 600 meters of pluvial downward displacement of the pinyon-juniper woodland zone, and also records a great increase in the area occupied by conifers. Hence, the Tule Springs site, which now lies about 1000 meters below the average lower limits of existing woodland, was probably less than half that vertical distance below a source of conifer pollen during the Wisconsin pluvial. By the same reasoning, the lateral distance to the nearest woodland in Las Vegas Range would have been less than 8 kilometers at that time, compared to about 24 kilometers at present; and, of course, the total expanse of coniferous vegetation contributing pollen to the basin was greater during pluvial time. Now, it happens that the exceedingly dominant shrub of the lower woodland and upper desert zones, *Coleogyne ramosissima* (blackbrush), which extends in nearly pure stands about 400 meters below the existing lower limit of woodland, has entomophilous pollination. This xerophytic shrub appears in the fossil record in three late-Wisconsin *Neotoma* deposits in the Frenchman Flat area, and should have extended then at least 600 meters below its present lower limits at about 1300 meters in this latitude; this fact would place the Tule Springs site (at 703 meters) within the zone of dominance by *Coleogyne* during pluvial time. The scanty cohesive pollen of the entomophilous *Coleogyne*, if it entered the pollen record in significant amounts, would be drastically diluted by the abundant pollen rain from coniferous vegetation upslope from Tule Springs. Therefore the pollen preserved in pluvial sediments at the Tule Springs site may provide an integrated record of total pollen production over a large area of watershed, but a record weighted for anemophilous conifers. The likelihood that a weighted integration of pollen from two or more vegetation zones has entered the sedimentary record suggests that calculation of absolute downward displacement of vegetation zones on the basis of *relative* percentages of pollen recorded in pluvial sediments of desert basins is a complex and apparently unsolved problem. Nevertheless, pollen analysis of sediments in

desert basins has the advantage of providing a relatively continuous stratigraphic record, which may indeed sensitively portray in a generalized way the expansions and contractions of montane coniferous vegetation in response to changing climate over a wide but undetermined range of elevation.

14. C. H. Lowe, *J. Ariz. Acad. Sci.* **2**, 40 (1961); P. S. Martin, *The Last 10,000 Years* (Univ. of Arizona Press, Tucson, 1963), p. 9. However, F. Shreve [*Ecology* **3**, 271 (1922)] pointed out that in southeastern Arizona the lower limit of the ponderosa-pine zone is fully 550 meters higher on the more massive Pinaleno Range (maximum elevation, 2800 meters) than on the smaller Santa Catalina Range (maximum elevation, 2800 meters), which is similar in bedrock and topography but lower in base level.

15. I. W. Clokey, *Flora of the Charleston Mountains, Clark County, Nevada* (Univ. of California Press, Berkeley, 1951), p. 28.

16. J. T. Rothrock, *Report upon United States Geographical Surveys West of the One Hundredth Meridian* (Washington, D.C., 1878), vol. 6; P. S. Martin, *The Last 10,000 Years* (Univ. of Arizona Press, Tucson, 1963) p. 8.

17. J. Leighly, in *California and the Southwest*, C. M. Zierer, Ed. (Wiley, New York, 1956).

18. The coniferous-forest zone is defined by the dominance of yellow pine (*Pinus ponderosa* Laws.) or, in some instances, white fir (*Abies concolor* Lindl.); at higher elevations on lofty mountains, by dominance of bristle-cone pine (*P. aristata* Engelm.) and limber pine (*P. flexilis* James).

19. I. W. Clokey (*15*, p. 13) reports 31 endemic taxa of vascular plants from the isolated Spring Range, Nevada, which rises nearly 3000 meters above the floor of adjacent desert basins. An example of apparently uneven stocking is the distribution of *A. concolor* (white fir), *P. ponderosa*, *Acer glabrum* (mountain maple), and *Betula occidentalis* (western birch) among the high mountains of the Mohave Desert region. All four are present in Spring Range (maximum elevation, 3631 meters), but only the relatively mesophytic white fir and maple grow in the lower, more arid Clark and Kingston ranges (maximum elevations, 2430 and 2230 meters, respectively), where the hardy *P. ponderosa* is anomalously lacking. However, neither white fir nor ponderosa pine occur in the high Panamint Range (maximum elevation, 3370 meters), although *A. glabrum*, *B. occidentalis*, and other mesophytes of the ponderosa pine–white fir zone are present; and there is even a limited development in the Panamints of the next-higher vegetation zone of bristle-cone and limber pines (as in the more fully stocked Spring Range), a zone that is now totally lacking in the smaller Clark and Kingston ranges.

20. P. V. Wells, *Science* **153**, 970 (1966).

21. R. Berger, W. S. Ting, W. F. Libby, in *Proc. Intern. Conf. Radiocarbon Tritium Dating 6th Pullman, Wash.* (1965).

22. Supported by NSF grants GB-5002 and GP-1893. We thank D. I. Axelrod and W. F. Libby of the University of California, Los Angeles, and P. S. Martin and P. J. Mehringer of the University of Arizona for criticism.

The Late Quaternary vegetational history of the equatorial mountains

by J. R. Flenley

Until a few years ago, most people believed that the vegetation of equatorial mountains is essentially stable. The glaciations of temperate regions were believed to have been reflected in the tropics only as pluvials—periods of heavy precipitation—and even though many equatorial mountains were shown to have been glaciated to quite low altitudes, this was attributed more to increased precipitation than to reduced temperature. Altitudinal shifts of vegetation were believed not to exceed about 500 m at the most.

Since about 1960 several groups of enthusiasts have produced pollen diagrams purporting to show substantial changes in the vegetation of equatorial mountains during the Late Pleistocene. Such diagrams became available for sites in South America, East Africa and New Guinea. Even after the publication of this work, many scientists preferred to remain unconvinced. It was always possible to find a reason why any particular site should be unrepresentative of a larger area, and the palynological method was capable of only limited precision about the nature of the vegetational changes and the climatic shifts which were presumed to have caused them.

Some palynologists made extravagant claims and others retreated into excessive conservatism. Arguments grew as to what kinds of climatic change were indicated—particularly as to whether they were thermal or hydrologic changes. The non-specialist in this field can perhaps be forgiven for retreating in despair.

I believe this is too pessimistic a view. There comes a time in any subject when enough evidence accumulates for a dramatic change in orthodoxy to be appropriate, and I believe now is the time in this case. Further equatorial pollen diagrams have been published, using more advanced techniques, and the story from temperate and tropical regions has been much augmented by oxygen isotope measurements and other studies. With the greater general availability of dating, attempts can at last be made at worldwide climatic reconstruction. Although these are yet crude, I think they can no longer be completely ignored.

There has also been an important recent reevaluation of the importance of man in shaping the vegetation of equatorial mountains. It is true that certain areas of 'savannah' or 'grassland' had long been suspected of being man-made but, in general, the significance of man's effects on equatorial mountains was underestimated, or at least incorrectly estimated. This applies to both the areal and the temporal extent of human activity.

For these reasons a review of current developments seems appropriate and this has been essayed recently (Flenley, 1979). The present paper is an attempt at a further slight updating in a more succinct form. I have confined my attention to areas within roughly 10° north and south of the equator, but there is no special justification for this limit and I have occasionally mentioned places outside it. Altitudinally, my lower limit has generally been 1500 m, although sites lower than this may receive mention.

Many of the diagrams represent interpretations of fossil pollen data in terms of vegetation, rather than the raw pollen data. These interpretations are made by reference to modern pollen rain studies. I have omitted these studies for the sake of brevity; the reader who requires them should consult the fuller reference given above.

I   The present vegetation of equatorial mountains

Mountains occur in each of the three great equatorial regions: South America, Africa and Southeast Asia with New Guinea. For convenience the latter region will be referred to by the name botanists use for it, i.e. Malesia (cf. Flora Malesiana since 1964). This spelling was adopted to avoid confusion with the political state of Malaysia.

The vegetation on equatorial mountains, insofar as it remains undisturbed by man, is usually strongly differentiated altitudinally. For convenience, distinct zones have often been recognized and an attempt to show these is made in Figure 1.

In fact, the recognition of a zone is usually based at least partly on physiognomy and structure rather than just on the total floristic complement, and boundaries are usually derived entirely subjectively. Zones based purely on floristic information—the most generally useful in relation to palynology—often differ considerably from the structural-physiognomic zones, especially if they are derived by less subjective techniques. This was demonstrated at the specific level over a limited geographical and altitudinal range in New Guinea by Flenley (1969), using semi-objective computer techniques. At the generic level, Walker and Guppy (1976) extended this study over a much greater area and range of altitude in the same island, reaching similar conclusions. The first of these studies showed that an ordination, in which altitude was closely related to one of the major axes, might be at least as useful as a classification in considering the nature of the vegetation.

In general, as an observer climbs an equatorial mountain he encounters forest which becomes progressively more dwarf, less diverse in species, and more laden with epiphytic bryophytes. Eventually, often at 3200–3800 m, he emerges from the forest, through a transitional shrub zone, into a

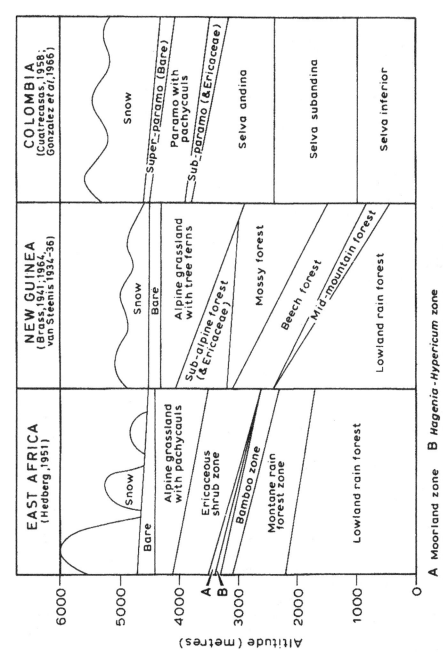

Figure 1 Altitudinal zonations of equatorial vegetation. *After Flenley (1967), Troll (1959) and other authors shown in the diagram*

A Moorland zone B *Hagenia - Hypericum* zone

'tropicalpine' vegetation of herbs and dwarf shrubs, many of which belong to genera of temperate affinities and have disjunct distributions. If the mountain is high enough, even this eventually gives way to tundra and finally to bare ground, before permanent snow is reached at about 4500–4700 m.

The precise reasons for this zonation have not been fully investigated. There is tacit assumption, but little proof, that mean annual air temperature is the controlling factor. Alternatives, such as soil temperature or ultra-violet radiation, are worthy of more attention than they have so far received.

The effects of man at the present time are widespread. Forest clearance for agriculture is concentrated at the lower altitudes but extends in densely populated areas up to 2700 m in New Guinea for cultivation of the sweet potato *(Ipomoea batatas)* and to over 3000 m in the Andes where the more hardy potato (*Solanum* spp.) has long been grown. It has recently been recognized, however, that the influence of man may go even higher than this. In many areas it has been shown that the tropicalpine and 'subalpine' vegetation has been modified by burning, and the forest limit shifted downhill by the same means. An example of this from New Guinea is provided by Wade and McVean (1969). The same phenomenon is wide-spread in Ethiopia (Leakey *et al.,* 1957), and in most other tropical montane areas.

II   The fossil pollen evidence

We now have, from the mountains of each of the three major equatorial regions, at least one pollen diagram covering the last 30 000 years, and several covering the last 10 000 years or more. It is necessary to treat each region independently because of the floristic differences between them.

1   *South America*

The evidence from South America has been ably reviewed by van der Hammen (1974) and what follows can be no more than a summary. The famous pollen diagram from the Sabana de Bogota, Colombia (van der Hammen and Gonzalez, 1960) covers perhaps a large part of Quaternary time and demonstrates multiple changes of vegetation. The Late Quaternary is, however, better covered in another fairly long pollen diagram. The Sabana de Bogota is one of a series of intermontane basins in the Cordillera Oriental of the Andes. The next basin to the north contains a lake, Laguna de Fuquene, which has also yielded a pollen diagram of substantial temporal coverage (van Geel and van der Hammen, 1973). This site is at 2580 m, slightly below the Sabana de Bogota. The pollen diagram from Fuquene is shown, greatly summarized, in Figure 2, and may be interpreted

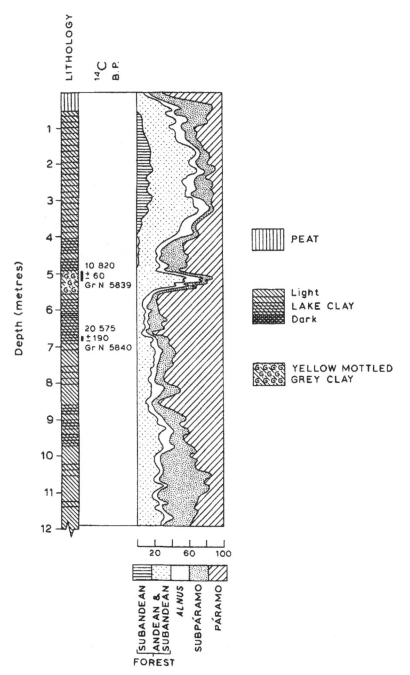

Figure 2 Summary pollen diagram from Laguna de Fuquene, Colombia, South America, altitude 2580 m. *After van Geel and van der Hammen (1973), van der Hammen (1974)*

as follows. About 30 000 years ago, the area was dominated by scrub vegetation similar to today's subparamo, the shrub-dominated zone at the forest limit. Shortly afterwards the grassy paramo, the tropicalpine vegetation of the Andes, started to dominate. This probably indicates colder conditions, which apparently culminated in the period 20 000–14 000 BP. These dates are strikingly similar to those for the last glacial maximum of temperate areas. The forest limit must have been about 1600 m lower than it is now, a difference which, at present lapse rates, would require mean annual temperature to have averaged about 10°C lower. The detailed composition of the pollen spectra also allows some conclusions about the hydrologic conditions at the time. An abundance of hydroseral elements such as Cyperaceae and Umbelliferae suggests that the lake level was lowered sufficiently to allow overgrowth of part of the lake bed by swamp. This could indicate arid conditions, perhaps more like those of the dry *puna* of Peru.

By 13 000 BP *Quercus* forest was invading the Fuquene plateau and its surrounding mountains, and the forest limit rose nearly to its present level. This almost certainly implies a rise of temperature, perhaps to only 2°C lower than today. Lake level apparently rose and inundated the swamp formed previously. In view of the rise in evaporation which would have resulted from the higher temperature, this must indicate a considerably higher precipitation.

Between 10 800 and 9500 BP there was an abrupt change in these trends. The forest limit went into reverse, declining to perhaps 800 m below its present-day altitude. A dwarf forest may have survived around the lake. The lake bottom again became available to colonizers. Climate must have been cold and dry once again.

From 9500 BP we have evidence for a rapid rise in forest limit, reaching present levels by 7500 BP when oak forest dominated. Climate must presumably have ameliorated rapidly at this time, and may have been warmer than the present, perhaps by 2°C, by 3000 BP.

At 3000 BP there are sudden changes in the pollen curves which are best explained by a slight decline in temperature to about its present level. Shortly after this there starts a decline of forest elements and an increase in grass pollen which is almost certainly the result of the destruction of forest by man.

One pollen diagram is of limited value without confirmatory evidence. There are now, however, several other pollen diagrams available from sites in Colombia, and each has been interpreted to yield a history of vegetation. Although the diagrams come from diverse environments, it seems worthwhile to try to correlate them and an attempt to do so is made in Figure 3. Here each pollen diagram, or rather the interpretation of each pollen diagram in terms of vegetational history, is shown as a horizontal bar passing through time. The vertical axis represents altitude. It now becomes possible to make a preliminary indication of the movements of the forest

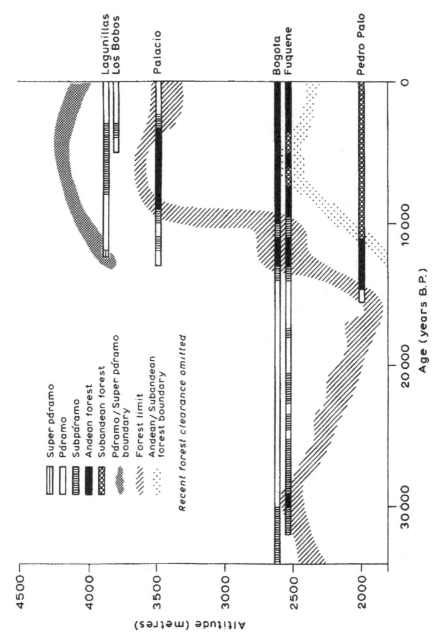

Figure 3   Summary diagram of Late Quaternary vegetational changes in the Colombian Andes. *After Flenley (1979)*

limit and of other altitudinal boundaries through time. The stippled bands representing these boundaries in Figure 3 are intended to give some indication of confidence limits, although it is emphasized that these are entirely subjectively chosen.

Several points emerge from this diagram. First, the ability of the pollen records to be correlated is remarkable. Second, the way in which the various vegetation boundaries appear to have moved up and down in harmony is also very striking. Of course, this degree of generalization obscures considerable variation in the composition of each vegetation type through time. Third, one cannot help but comment immediately on the general similarity of the forest limit curve to palaeo–temperature curves for many temperate areas. The low point about 20 000–15 000 BP, the early Holocene rise, even the Late Glacial uncertainty, are all there. Fourth, the magnitude of the maximum altitudinal shift shown is surprising. At something like 1600 m, this is far greater than expected by many biogeographers.

Figure 3 does nothing to confirm or refute the hypothesis that glacial times were arid in Colombia, as suggested by the Fuquene evidence: this must await further work.

Results of human activity have been omitted from Figure 3 for the sake of clarity, but in fact there is evidence of forest clearance in the last few thousand years in almost all the Colombian diagrams below 3000 m.

2   Africa

East Africa has provided us with a 30 000 year pollen diagram (Figure 4) from Sacred Lake at 2400 m in the forest on the slopes of Mount Kenya (Coetzee, 1967; Bakker and Coetzee, 1972). Unfortunately the diagram is not well dated and, as there is a stratigraphic change near the base, there is always the possibility that a part is missing, as pointed out by Livingstone (1975) in his excellent review of the Quaternary vegetation of Africa. Nevertheless, the diagram reveals a self-consistent vegetational history which may be summarized as follows.

At the time represented by the start of the diagram, about 33 000 BP the vegetation around the lake was probably shrub-dominated, resembling that of the Ericaceous Belt just above the forest limit today. The forest limit may have been not far below the lake, and after some time it began a slow rise, reaching the lake about 26 000 BP. From this time onwards there was a dramatic change. The forest limit became depressed far below the lake and shrubby vegetation, mixed with many grasses, dominated until 14 000 BP. Pollen of *Artemisia* was abundant at this time, possibly indicating dry conditions (Hamilton, 1972). For the next 3500 years aridity was maintained, but the forest limit slowly rose to just below the altitude of the lake.

At c. 10 500 BP another sharp change occurred. The forest limit advanced suddenly upwards, surrounding the lake by forest. This must have been, at

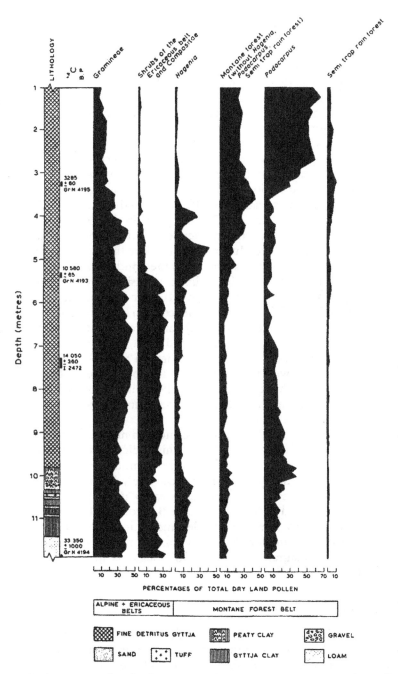

Figure 4 Summary pollen diagram from Sacred Lake on Mount Kenya, East Africa altitude 2400 m. Values are percentages of total dry land pollen. *After Coetzee (1967)*

least in part, a response to a rise in temperature. The species present show that the climate became moist, so hydrologic changes could also have been important in causing the vegetation to alter. Forest has persisted around the lake to the present day, although the composition of the forest has changed markedly, particularly through the immigration of *Podocarpus*, the present dominant taxon, since about 3000 BP. There is no major forest clearance shown around Sacred Lake although forest lower down the mountain has been totally destroyed.

Although East Africa is large and climatically diverse, an attempt has been made in Figure 5 to correlate the Sacred Lake record with others from the region. The early part of the Muchoya diagram, before 12 000 BP, is dated only by extrapolation, so that section must be regarded as uncertain in position. The general pattern is different from the South American one, but shows distinct similarities. The low point in the forest limit curve is somewhat later than the South American one, but this could be due to error in the extrapolation of the Muchoya dating. The time of the dramatic rise in forest limit is similar to that shown in the South American curve. The magnitude of the shift of the forest limit is considerably less than that for South America. Vegetational changes related to hydrologic variation are not shown in Figure 5. In fact, the evidence from highland pollen diagrams is not as clear as that from the East African Plateau, where Kendall (1969) showed Late Pleistocene desiccation, evidenced by pollen change and by indications of low water level in Lake Victoria at 1130 m. Possibly these extreme conditions were not expressed so forcibly in the cooler conditions of the mountains, although Livingstone (1967), in his pollen diagram from Mahoma Lake, believed he had better evidence in the Late Pleistocene for a dry period than for a cold period.

Man probably evolved in Africa and modern man has been there for at least 60 000 years (Clark, 1970). Despite this, indications of marked human effect on vegetation are not unduly early in montane Africa. There are pollen diagrams from Uganda which show likely forest clearance from c. 1000 BP. One of these is from Katenga swamp at 1980 m (Morrison and Hamilton, 1974), and another from Lake Bunyonyi, 14 km away, at 1950 m. A similar age for the onset of clearance is indicated in Livingstone's (1967) pollen diagram from Mahoma Lake on the Ruwenzori. The Lake Victoria pollen diagram (Kendall, 1969) suggests that clearance began about 3000 BP on the East African Plateau.

3   *Malesia*

Up to now all the available palynological evidence from above 1500 m comes out of New Guinea, although studies in Sumatra are under way. The New Guinea evidence consists of several pollen diagrams, although only one of these extends continuously back to 30 000 BP. This is the Sirunki diagram (Walker and Flenley, 1979) which is shown in Figure 6. It is

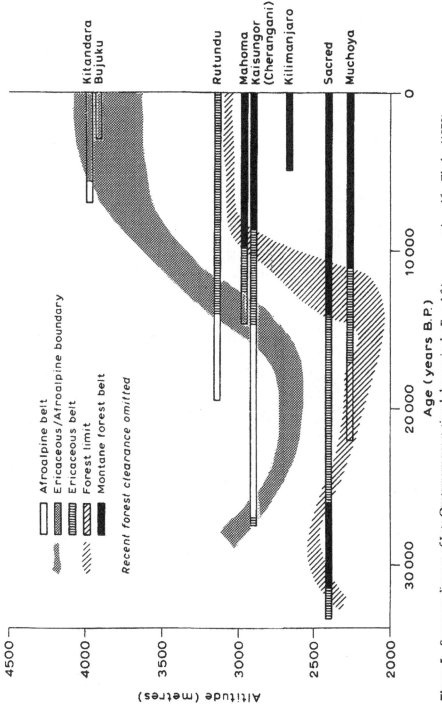

Figure 5   Summary diagram of Late Quaternary vegetational changes in the East African mountains. *After Flenley (1979)*

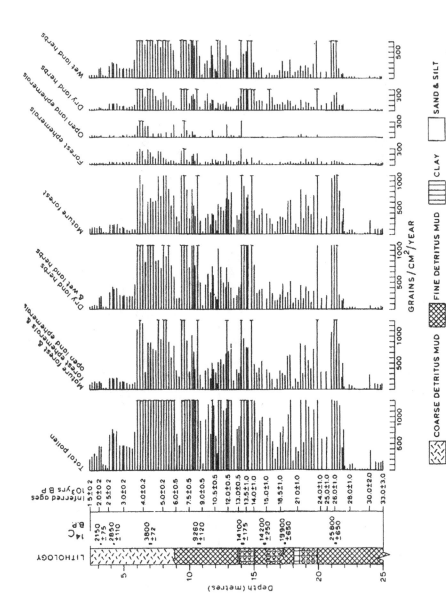

Figure 6　Summary pollen diagram from Sirunki Swamp, New Guinea, altitude 2500 m. Values are pollen deposition rates in grains per cm² per annum. *After Walker and Flenley (1979)*

different from the other diagrams in this paper in that the results are expressed in absolute terms, as pollen deposition rates (PDRs) in grains per cm² per annum. In addition, the number of radiocarbon dates was sufficient to permit the establishment of a chronology of 'Inferred Ages' (IA) based on isotopic, stratigraphic and other evidence. When interpreted by reference to modern values of PDR, this diagram suggests the following vegetational history for the Sirunki catchment surrounding at 2500 m the swamp from which the diagram was obtained. Between 33 000 and 27 000 IA the PDR is very low, indicating unforested conditions. For the following 2000 years, to 25 000 IA, PDR is high enough to indicate forest, although subalpine wetland herbs are also present. After this date forest apparently disappeared from the catchment until 14 000 IA, being replaced by a vegetation of dryland and wetland herbs. The forest limit must have lain well below 2500 m during much of this time. After 14 000 BP, forest enjoyed a brief incursion into the catchment, for perhaps c. 750 years, but was then replaced by tree-fern 'woodland', sedge swamps and grassland. Whether this change was under climatic control is uncertain. There is stratigraphic evidence for a rise in water level in the swamp, culminating about 13 000 IA. The 'postglacial' forest did not develop until about 9000–8000 IA, and was eventually dominated by *Nothofagus,* the southern beech genus. At about 5000 IA, but especially after 4300 IA, forest began to decline in favour of forest ephemerals and open land ephemerals, taxa which follow in the footprints of man today. These changes are therefore most readily explained in terms of forest clearance by man. The decline in total PDR from about 3800 IA is probably related to the overgrowth of the sampling site by swamp vegetation, and does not necessarily reflect changes in the vegetation of the catchment.

Several other pollen diagrams from the New Guinea highlands confirm this picture of the vegetational changes, especially in the later stages. This evidence is summarized in Figure 7. This diagram suggests movements of the forest limit which bear close comparison with those from South America (Figure 3), and to a lesser extent with those from Africa (Figure 5).

The evidence of forest clearance, which is deliberately omitted from Figure 7, is also strikingly confirmed and extended by other work. The Draepi site, at 1900 m on the edge of the densely inhabited Wahgi valley, shows evidence of grassland dominance from c. 5100 BP (Powell *et al.,* 1975), and archaeological excavation at Kuk Swamp in the same valley (Powell *et al.,* 1975; Golson and Hughes, 1976; Golson, 1977) shows evidence of modification of the swamp surface, presumably for agriculture, since 6000 BP and possibly 9000 BP. There is evidence that the pig, an introduced and domesticated animal, has been in New Guinea for at least 10 000 years (Bulmer, 1975).

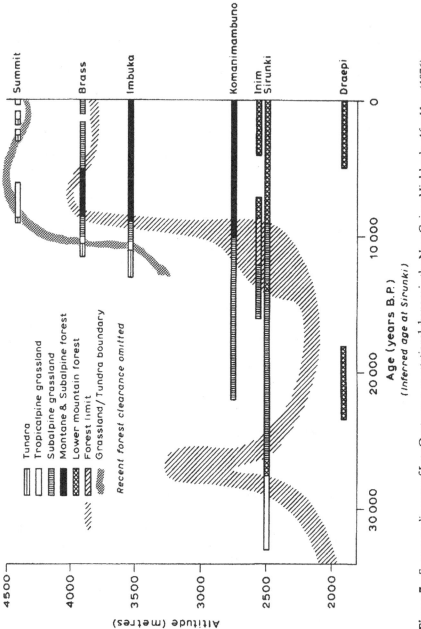

Figure 7 Summary diagram of Late Quaternary vegetational changes in the New Guinea Highlands. *After Hope (1976),* *Flenley (1979)*

502 *Vegetational history of the equatorial mountains*

III Discussion

Since this is an attempt to review the evidence for the vegetational history of equatorial montane areas in general, discussion will concentrate on areas of agreement between the different regions mentioned, while recognizing the individuality of each region and of each mountain or range of mountains. The similarities between the records from South America, Africa and New Guinea are rather striking. The suggested trend for the forest limit in all three areas shows that the limit was low before 30 000 years BP, shows a slight peak (of uncertain height and imprecise date) between 30 000 and 25 000 BP, reaches an especially low point about 18 000–15 000 BP, shows a steep climb between 14 000 and 9000 BP, with oscillations in some areas and considerable variation between sites, reaches modern altitudes or slightly above them by c. 7000 BP, and thereafter shows minor adjustment, if necessary, at 5000–3000 BP.

The most likely control on forest limit is mean annual temperature. It is therefore natural to compare this evidence with that for the deglaciation of equatorial mountains. A selection of published minimum dates for deglaciation of mountains between 10°N and 10°S is shown in Figure 8, where they are related to altitude. That these dates, coming as they do from widely separated corners of the earth, and from mountains under diverse climatic influences, should show much in the way of correlation is surprising. Yet there is some agreement between the records, and this suggests that worldwide climatic change was the likely overriding cause of deglaciation. Furthermore, the major tendency is for deglaciation at successively higher altitudes between 15 000 BP and 8000 BP, a trend closely similar to that followed by the forest limit. The simplest possible explanation is that both deglaciation and the rise of forest limit were the result of climatic change. This change correlates closely in time and magnitude with the climatic amelioration which occurred at the end of the last major glaciations in temperate regions of both hemispheres. There is now some pollen evidence that even the equatorial lowlands were affected by parallel changes (Flenley, 1979).

The conventional model of climatic change in the tropics was, until recently, the pluvial theory. Pluvials were envisaged as times of greatly increased available moisture, and were widely believed to coincide with the glacials of temperate regions. Temperature changes were believed to have been minor in the tropics. That temperatures should have changed sufficiently to shift the forest limit vertically by as much as 1500 m or more (equivalent to about 9°C variation in mean annual temperature at modern lapse rates) therefore demands reassessment of the pluvial theory. The palynological evidence gives some indication of changes in moisture availability. The Fuquene pollen diagram is compatible with a drier climate in the Late Pleistocene in montane South America, and the Mahoma diagram from montane Africa contains abundant *Myrica* and *Artemisia* which could

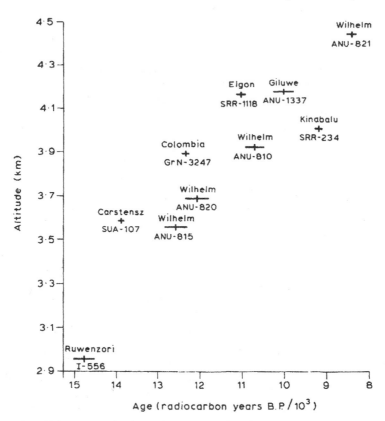

Figure 8   Minimum ages for the deglaciation of selected equatorial mountains at various altitudes. *After Livingstone (1962), Hope (1976), Hope and Peterson (1975; 1976), Gonzalez et al. (1966), Flenley and Morley (1978), Hamilton and Perrott (1978)*

lead to the same conclusion. The New Guinea highlands pollen evidence does not require any change in moisture availability in the past, but the stratigraphy at Sirunki suggests a rise in water level at about 13 000 IA. The Montane evidence is inadequate on its own for any firm conclusion but is not in conflict with lowland evidence, summarized by Flenley (1979), which supports the idea of the Late Pleistocene being drier than the present in many areas. The pluvial theory is therefore not supported; in fact, the evidence is on the whole contrary to it.

There are many facts in the pollen evidence which cannot be related to climatic change at all. Most of these cannot be seen in the summary diagrams shown here, but they are nonetheless important and can be seen in fuller diagrams (Flenley, 1979). The idea of a simple upward and downward shift of the forest limit and other vegetation boundaries is certainly a gross oversimplification. Three peculiarities stand out. First, the forests

504   *Vegetational history of the equatorial mountains*

seem rarely to have remained of uniform composition for long periods; the pollen evidence, where it permits such resolution, suggests continuous, perhaps sometimes even cyclical, change. This may be seen, for example, in the record of *Nothofagus* at Sirunki (Walker and Flenley, 1979). Second, when forest migrates into an area it does not do so as a unit: rather, the individual taxa arrive at their own pace. A good example of this is the late arrival of *Podocarpus* at Sacred Lake on Mount Kenya, where it appears several thousand years after many other montane forest taxa (Coetzee, 1967). Third, the grouping of taxa into forest types seems to have varied in the past. In New Guinea today there is a distinct upper mountain forest which is usually dominated by *Dacrycarpus* and occurs above the *Nothofagus* forests. At Sirunki and Inim there was little evidence of this vegetation type in the Pleistocene, although *Dacrycarpus* is known to be a reliable pollen producer. For this reason, upper mountain (= montane + subalpine) forest is shown on Figure 7 as appearing *de novo* at 10 000 BP. Possibly the taxa of this forest existed as rare components of the forest near the tree line during the Pleistocene, and only amalgamated to form a distinct vegetation type during the Holocene.

It is not to be expected that the evidence for forest clearance should correlate from one continent to another. What is interesting, however, is the pattern of relationship between the evidence of forest clearance in the mountains and the archaeological evidence for cultivation of crops. We have indications of cultivation in Africa for perhaps 3000 years (Clark, 1962), and for forest clearance since c. 1000 BP in the mountains. In South America there is evidence for cultivation (in the lowlands) since c. 7000 BP (Patterson, 1971; Moseley, 1975); the forest clearances in the highland diagrams date from c. 3000 BP. The evidence from New Guinea is for agriculture (in the highlands) from c. 6000 BP, or even 9000 BP, and for forest clearance from 5000 BP. In general then, palynology has recovered perhaps only half the length of the archaeological record. This is probably more a result of the low density of palynological sites than of any other cause.

IV Conclusions

1 *The fact of change*

A good deal of biogeography has been based on the idea of the immutability of equatorial vegetation, so it is worth drawing attention first to the fact that the pollen diagrams consistently demonstrate that this idea is incorrect. Equatorial vegetation, at least in the mountains, *has* changed markedly in the Late Quaternary.

2   The former vegetation

In general, although not in detail, the vegetation found at a given point in the Late Pleistocene was similar to that now found at a higher altitude in the same area. During the period 33 000–30 000 BP, forest limits were lower than at present by at least 700 m and perhaps much more. Between 30 000 and 27 500 BP the limits rose somewhat but then declined again, reaching their lowest levels of the last 30 000 years during the period 18 000–15 000 BP, when they lay at least 1000 m and possibly as much as 1700 m below present values. There is slight evidence from some areas that the vegetation at this time was of a type now found in drier areas. Between 15 000 and 8000 BP forest limits climbed to, or even above, their present levels. At some, but not all, sites there are indications of oscillation in the period 14 000–10 000 BP and there is considerable variation from site to site. Forest limits reached their present levels by c. 3000 BP.

There is evidence for forest clearance by man from c. 1000 BP in the African mountains, c. 3000 BP in The Andes and c. 5000 BP in the New Guinea mountains.

3   The former climate

The vegetational evidence strongly suggests marked climatic changes. The climate most different from that of the present, during the last 30 000 years, probably occurred during the period 18 000–15 000 BP. In New Guinea at that time the forest limit was c. 1700 m below its present value. The vegetation then at Sirunki was of a type now found at c. 2°C to c. 6°C mean annual temperature. This implies a reduction of mean annual temperature of c. 8°C to c. 12°C at Sirunki at the time (Walker and Flenley, 1979). Judging from Figures 3 and 5, the situation in South America may have been similar to that in New Guinea, but the African evidence is consistent with rather less climatic shift. Attempts have been made to reconstruct the former lapse rate in New Guinea (Walker and Flenley, 1979), but it is too early to generalize widely from this. There is some evidence of previously dry climates, especially during the time of low temperatures, but the evidence for this from the mountains remains fragmentary so far. There is certainly no palaeobotanical evidence of 'pluvials' coinciding with the 'glacials' of temperate regions.

4   Stability

The pollen diagrams show more or less continuous fluctuation in the composition of the vegetation in the last 30 000 years. Even the Holocene forests have existed for only a few tree generations. This argues against the long-held ideas about the stability of equatorial vegetation. The theory that

506   *Vegetational history of the equatorial mountains*

high diversity endows stability upon vegetation (MacArthur, 1955) is not supported.

5 *Individualistic behaviour*

The diagrams show that the vegetation types concerned did not usually migrate as holistic units but as individual taxa. This is evidence in favour of the individualistic concept of vegetation (Gleason, 1926) rather than the holistic view of it (Clements, 1916; Tansley, 1920).

6 *Climax theory*

Any vegetation which has existed for only a few generations of its component species can scarcely be regarded as 'climax'. It seems likely that the vegetational and climatic changes of the last 30 000 years are but the tail end of changes which have been continuous for the last two million years, with only a small percentage of time as warm as the present during that period. Vegetation has probably been continuously adjusting throughout, and the climax concept becomes inapplicable.

7 *Diversity*

Several explanations of the high species diversity of rain forest assume the stability of the forest (Ricklefs, 1973). It now seems unlikely that these theories can be maintained. It is true that the vegetational changes of the mountains may well have been softened in the lowlands, but there is now some evidence of Pleistocene vegetational changes even in the tropical lowlands (Flenley, 1979). Besides, the lower mountain forests, of whose instability we now have direct evidence, are themselves very diverse.

8 *Biographical distributions*

The idea that the disjunct occurrence of many montane species on numerous isolated peaks could be explained by vegetational movements resulting from climatic change goes back at least to Darwin (1859) and Wallace (1869). Clearly when forest limits were lower, tropicalpine species would have had an expanded distribution, perhaps incorporating many sites which could aid as 'stepping-stones' in distribution. While this cannot be a complete solution to the problems of montane disjunctions, it may simplify them.

University of Hull

V   References

Bakker, E. M. van Zinderen and **Coetzee, J. A.** 1972: A reappraisal of Late-Quaternary climatic evidence from tropical Africa. *Palaeoecology of Africa and of the Surrounding Islands and Antarctica* 7, 151–81.

Brass, L. J. 1941: The 1938–39 expedition to the Snow Mountains, Netherlands New Guinea. *Journal of the Arnold Arboretum* 22, 271–342.

1964: Results of the Archbold Expeditions no. 86. Summary of the Sixth Archbold Expedition to New Guinea (1959). *Bulletin of the American Museum of Natural History* 127, 145–216.

Bulmer, S. 1975: Settlement and economy in prehistoric Papua New Guinea: a review of the archaeological evidence. *Journal de la Société des Océanistes* 46, tome xxxi, 7–75.

Clark, J. D. 1962: The spread of food production in subSaharan Africa. *Journal of African History* 3, 211–28.

1970: *The prehistory of Africa.* London: Thames and Hudson. (302 pp.)

Clements, F. E. 1916: Plant succession: an analysis of the development of vegetation. *Publications of the Carnegie Institution no. 242.*

Climap Project Members 1976: The surface of the Ice-age Earth. *Science* 191, 1131–37.

Coetzee, J. A. 1967: Pollen analytical studies in East and Southern Africa. *Palaeoecology of Africa and of the Surrounding Islands and Antarctica* 3, 1–146.

Cuatrecasas, J. 1958: Aspectos de la vegetacion natural de Colombia. *Revista de la Academia colombiana de ciencias exactas, fisicas y naturales* 10, 221–64.

Darwin, C. 1859: *The origin of species by means of natural selection.* London: John Murray. (432 pp.)

Flenley, J. R. 1967: *The present and former vegetation of the Wabag region of New Guinea.* Australian National University, unpublished PhD thesis.

1969: The vegetation of the Wabag region. New Guinea Highlands: a numerical study. *Journal of Ecology* 57, 465–90.

1979: *The equatorial rain forest: a geological history.* London: Butterworth.

Flenley, J. R. and **Morley, R. J.** 1978: A minimum age for the deglaciation of Mt Kinabalu, East Malaysia. *Modern Quaternary Research in Southeast Asia* 4, 57–61.

Geel, B. van and **Hammen, T. van der** 1973: Upper Quaternary vegetational and climatic sequence of the Fuquene area (Eastern Cordillera, Colombia). *Palaeogeography, Palaeoclimatology, Palaeoecology* 14, 9–92.

Gleason, H. A. 1962: The individualistic concept of the plant association. *Bulletin of the Torrey Botanical Club* 53, 7–26.

Golson, J. 1977: No room at the top: agricultural intensification in the New Guinea Highlands. In Allen, J., Golson, J. and Jones, R., editors, *Sunda and Sahul: prehistoric study in southeast Asia, Melanesia and Australia,* London: Academic Press.

Golson, J. and **Hughes, P. J.** 1976: The appearance of plant and animal domestication in New Guinea. *Paper prepared for the ninth Congress, Union International des Sciences Préhistoriques et Protohistoriques, Nice, September 1976.*

Gonzalez, E., Hammen, T. van der and **Flint, R. F.** 1966: Late Quaternary

508   *Vegetational history of the·equatorial mountains*

glacial and vegetational sequence in Valle de Lagunillas, Sierra Nevada del Cocuy, Colombia. *Leidse geologische Mededelingen* 32, 157–82.

Hamilton, A. C. 1972: The interpretation of pollen diagrams from Highland Uganda. *Palaeoecology of Africa and of the Surrounding Islands and Antarctica* 7, 45–149.

Hamilton, A. C. and Perrott, A. 1978. Date of deglacierisation of Mount Elgon. *Nature* (London) 273, 49.

Hammen, T. van der 1974: The Pleistocene changes of vegetation and climate in tropical South America. *Journal of Biogeography* 1, 3–26.

Hammen, T. van der and Gonzalez, E. 1960: Upper Pleistocene and Holocene climate and vegetation of the 'Sabana de Bogota' (Colombia, South America). *Leidse geologische Mededelingen* 25, 261–315.

Hedberg, O. 1951: Vegetation belts of the East African Mountains. *Svensk botanisk Tidskrift* 45, 140–202.

Hope, G. E. 1976: The vegetational history of Mt Wilhelm, Papua New Guinea. *Journal of Ecology* 64, 627–63.

Hope, G. S. and Peterson, J. A. 1975: Glaciation and vegetation in the High New Guinea Mountains. *Bulletin of the Royal Society of New Zealand* 13, 155–62.

1976: Palaeoenvironments. In Hope, G. S., Peterson, J. A. and Radok, U., editors, *The equatorial glaciers of New Guinea*, Rotterdam: A. A. Balkema, 173–205. (244 pp.)

Kendall, R. L. 1969: An ecological history of the Lake Victoria Basin. *Ecological Monographs* 39, 121–76.

Leakey, C. L. A., Ballantine, W. J., Damte, A., Evans, I. M., Flenley, J. R., Hiller, R. G. and Lythgoe, J. N. 1957: *Interim Report, The Cambridge Botanical Expedition to Ethiopia, 1957*. (15 pp.)

Livingstone, D. A. 1962: Age of deglaciation in the Ruwenzori Range, Uganda. *Nature* (London) 194, 859–60.

1967: Postglacial vegetation of the Ruwenzori Mountains in equatorial Africa. *Ecological Monographs* 37, 25–52.

1975: Late Quaternary climatic change in Africa. *Annual Review of Ecology and Systematics* 6, 249–80.

MacArthur, R. H. 1955: Fluctuations of animal populations and a measure of community stability. *Ecology* 36, 533–36.

Morrison, M. E. S. and Hamilton, A. C. 1974: Vegetation and climate in the uplands of south-western Uganda during the Later Pleistocene Period. II: Forest clearance and other vegetational changes in the Rukiga Highlands during the past 8000 years. *Journal of Ecology* 62, 1–31.

Moseley, M. E. 1975: *The maritime foundations of Andean civilization*. Menlo Park, California: Cummings Publishing Co.

Patterson, T. C. 1971: The emergence of food production in Central Peru. In Struever, S., editor, *Prehistoric agriculture*, New York: Natural History Press, 81–207.

Powell, J. M., Kulunga, A., Moge, R., Pono, C., Zimike, F. and Golson, J. 1975: *Agricultural traditions of the Mount Hagen Area*. University of Papua New Guinea, Department of Geography, Occasional Paper no. 12. (67 pp. and 5 figures.)

Ricklefs, R. E. 1973: *Ecology.* London: Nelson. (861 pp.)

Steenis, C. G. G. J. van 1934–36: On the origin of the Malaysian mountain flora. *Bulletin du Jardin Botanique de Buitenzorg, Series III;* Part I, 13, 135–262; Part II, 13, 289–417; Part III, 14, 56–72.

Tansley, A. G. 1920: The classification of vegetation and the concept of development. *Journal of Ecology* 8, 118–49.

Troll, C. 1959: *Die tropischen Gebirge. Ihre dreidimensionale klimatische und pflanzengeographische Zonierung.* Bonn: Dummlers. (93 pp.)

Wade, L. K. and **McVean, D. N.** 1969: *Mt. Wilhelm Studies I: the alpine and subalpine vegetation.* Canberra: Australian National University, Research School of Pacific Studies, Department of Biogeography and Geomorphology, Publication BG/1. (225 pp.)

Walker, D. and **Flenley, J. R.** 1979: Late Quaternary vegetational history of the Enga District of Upland Papua New Guinea. *Philosophical Transactions of the Royal Society* B 286, 265–344.

Walker, D. and **Guppy, J. C.** 1976: Generic plant assemblages in the highland forests of Papua New Guinea. *Australian Journal of Ecology* 1, 203–212.

Wallace, A. R. 1869: *The Malay Archipelago.* London: Macmillan. (Volume I 478 pp.; Volume II 524 pp.)

The Discovery of America

The first Americans may have swept the Western
Hemisphere and decimated its fauna within 1000 years.

Paul S. Martin

America was the largest landmass
undiscovered by hominids before the
time of *Homo sapiens*. The Paleolithic
pioneers that crossed the Bering Bridge
out of Asia took a giant step. They
found a productive and unexploited
ecosystem of over 10^7 square miles
(2.6×10^7 square kilometers). As
Bordes has said (*1*), "There can be no
repetition of this until man lands on a
[habitable] planet belonging to another
star."

At some time toward the end of the
last ice age, big game hunters in Siberia
approached the Arctic Circle, moved
eastward across the Bering platform
into Alaska, and threaded a narrow
passage between the stagnant Cordilleran and Laurentian ice sheets. I propose that they spread southward ex-
plosively, briefly attaining a density
sufficiently large to overkill much of
their prey.

Overkill without Kill Sites

Pleistocene biologists wish to determine to within 1000 years at most the
time of the last occurrence of the dominant Late Pleistocene extinct mammals.
If one recognizes certain hazards of
"push-button" radiocarbon dating (*2*),
especially dates on bone itself, it appears that the disappearance of native American mammoths, mastodons,
ground sloths, horses, and camels coincided very closely with the first appearance of Stone Age hunters around
11,200 years ago (*3*).

Not all investigators accept this circumstance as decisive or even as adequately established. No predator-prey
model like Budyko's (*4*) on mammoth
extinction has been developed to show

The author is professor of geosciences, University of Arizona, Tucson 85721.

how the American megafauna might
have been removed by hunters (*5*).
Above all, prehistorians have been
troubled by the following paradox.

In temperate parts of Eurasia, large
numbers of Paleolithic artifacts have
been found in many associations with
bones of large mammals. Although the
evidence associating Stone Age hunters
and their prey is overwhelming, not
much extinction occurred there. Only
four late-glacial genera of large animals
were lost, namely, the mammoth
(*Mammuthus*), woolly rhinoceros (*Co-
elodonta*), giant deer (*Megaloceros*),
and musk-ox (*Ovibos*).

In contrast, the megafauna of the
New World, very rarely found associated with human artifacts in kill or
camp sites (*6*), was decimated. Of the
31 genera of large mammals (*7*) that
disappeared in North America at the
end of the last ice age, only the mammoth (*Mammuthus*) is found in unmistakable kill sites. The seven kill sites
listed by Haynes (*8*) lack the wealth
of cultural material, including art objects, associated with the Old World
mammoth in eastern Europe and the
Ukraine. It is not surprising that some
investigators discount overkill as a
major cause of the extinctions in
America.

But if the new human predators
found inexperienced prey, the scarcity
of kill sites may be explained. A rapid
rate of killing would wipe out the more
vulnerable prey before there was time
for the animals to learn defensive behavior, and thus the hunters would not
have needed to plan elaborate cliff
drives or to build clever traps. Extinction would have occurred before there
was opportunity for the burial of much
evidence by normal geological processes. Poor paleontological visibility
would be inevitable. In these terms, the

scarcity of kill sites on a landmass
which suffered major megafaunal losses
becomes a predictable condition of the
special circumstances which distinguish
a sudden invasion from more gradual
prehistoric cultural changes in situ.
Perhaps the only remarkable aspect of
New World archeology is that *any* kill
sites have been found (*9*).

Megafaunal Biomass

Bordes (*1*) and Haynes (*8*) believe
that the Stone Age hunters found
abundant game in America. Although
the fauna was diverse (*7*), no estimates
of the size of the Late Pleistocene game
herd have been attempted. I propose
two crude but independent methods of
estimating the biomass of the native
megafauna, both of which utilize present range-carrying capacity. In the first
method one projects estimates of the
biomass of large mammals in African
game parks to areas of comparable
range productivity in the New World.
The other method is based on the assumption that present managed livestock plus game populations in the
Americas would equal, and probably
exceed, the maximum herd size of the
Late Pleistocene.

Estimates of biomass in various African parks are shown in Table 1. The
drier parks such as Tarangire Game
Reserve, Kafue, Kagera, and others not
included in Table 1 such as Kruger
National Park, South Africa, and Tsavo
National Park, Kenya, support 10 to
20 animal units (*10*) per section (1.8
to 3.5 metric tons per square kilometer).
In the Americas during the Pleistocene,
similar values might be expected on
drier ranges (mean annual precipitation, 400 to 600 millimeters) dominated
by mammoth, horse, and camel. The
carrying capacity would have been
much less in the driest regions (annual
precipitation less than 200 millimeters).

African game parks supporting the
highest biomass, over 100 animal units
per section (over 18 metric tons per
square kilometer), occur in tall grass
savannas along the margin of wet tropical forest. The dominant species are
elephant, buffalo, and hippo. In the
tropical American savannas, along
coasts, and on floodplains of the temperate regions, one might expect a
similar biomass in the Pleistocene. The
dominant species were mastodons and
large edentates.

For the 3×10^6 sections of land

9 MARCH 1973

969

641

Table 1. Large mammal biomass in some African parks and game reserves [from Bourliere and Hadley (32)]. [Courtesy of Annual Reviews, Inc., Palo Alto, Calif.]

Location	Habitat	Number of species	Biomass (metric tons per square kilometer)	Biomass (animal units per square mile)
Tarangire Game Reserve, Tanzania	Open *Acacia* savanna	14	1.1	6
Kafue National Park, Zambia	Tree savanna	19	1.3	7
Kagera National Park, eastern Rwanda	*Acacia* savanna	12	3.3	18
Nairobi National Park, Kenya	Open savanna	17	5.7	32
Serengeti National Park, Tanzania	Open and *Acacia* savannas	15	6.3	36
Queen Elizabeth National Park, western Uganda	Open savanna and thickets	11	12	68
Queen Elizabeth National Park, western Uganda	Same habitat, overgrazed	11	27.8–31.5	158–179
Albert National Park, northern Kivu	Open savanna and thickets, overgrazed	11	23.6–24.8	134–141

$(7.8 \times 10^6$ square kilometers) in the unglaciated United States, I propose the following average stocking capacities, each covering roughly 10^6 sections: (i) savannas, forest openings, floodplains, and other highly to moderately productive habitats, 50 animal units per section, or 22.7×10^6 metric tons; (ii) arctic, boreal, semiarid, short grass ranges, and other low to moderately productive habitats, ten animal units per section, or 4.6×10^6 metric tons; (iii) closed canopy forest, extreme desert, barren rock, and other habitats unproductive for large herbivores, two animal units per section, or 0.9×10^6 metric tons. The total for North America north of Mexico is 62×10^6 animal units or 28.2×10^6 metric tons.

Turning from the African analogy to estimates based on current livestock plus game populations, one obtains a higher biomass. The United States alone supported 1.20×10^8 animal units (all types of livestock) in 1900 and 1.48×10^8 in 1945. Adding wild game, I project these values for the Western Hemisphere south of Canada to a total of 5.00×10^8 animal units, or 2.30×10^8 metric tons (11). Presumably this value, based on managed herds, exceeds the natural Pleistocene biomass.

A herd of 2.50×10^8 animal units during the Late Pleistocene would seem more realistic in terms of the African values. A hemispheric estimate of 10^8 animal units should be far too low for the primary plant productivity available to the native herbivores but would still be a sizable resource for the first Paleolithic hunters.

The alternate view, that the American large mammal biomass was in eclipse during the late glacial (12), cannot be tested quantitatively on the basis of fossils alone. Bones do not provide reliable estimates of past biomass (13). But the great numbers of mastodon, mammoth, extinct horse, camel, and bovid bones found in late-glacial sediments hardly suggest scarcity. Evidence that the Late Pleistocene megafauna was declining in numbers and diversity before 12,000 years ago, as Kurtén found in Late Paleolithic sites of the Old World (14), is lacking in the New World (11).

America's First Population Explosion

The *minimum* growth rate required to attain the estimated (A.D. 1500) population of the New World is negligible, 0.1 percent annually. Slow, imperceptible growth is what demographers are prone to project into the Paleolithic (15). They have no choice. Neither bones nor artifacts will reveal instantaneous rates of change. A century of maximum growth, followed by a year of massive mortality, would escape detection by archeologists.

It seems likely that, when entering a new and favorable habitat, any human population, whatever its economic base, would unavoidably explode, in the sense of Deevey [see (15)], with a force that exceeded ordinary restraint. The environment of the New World should have been particularly favorable. The hunters who conquered the frozen

tundra of eastern Siberia and western Alaska must have been delighted when they first detected milder climates as their route turned southward. Predation loss seems improbable (11). More important, the major hominid diseases endemic to the Old World tropics were unknown in the New World (16). Hunting accidents undoubtedly occurred, but presumably less often when New World elephants were at bay than in the case of the more wary and experienced mammoths of Eurasia.

When they reached the American heartland, the Stone Age hunters may have multiplied as rapidly as 3.4 percent annually, the rate Birdsell (17) reported for the settlement of Pitcairn Island and elsewhere. Anthropologists regard one person per square mile (0.4 person per square kilometer) as maximum for a preagricultural economy on its best hunting grounds. Had such a population density been attained throughout the Americas, it would represent a total population of 10^7 in the 10^7 square miles (2.6×10^7 square kilometers) outside of Canada and other glaciated regions. At a rate of population growth of 3.4 percent annually, or a doubling every 20 years, 340 (17 generations) would be the minimum time needed for a band of 100 invaders to saturate the hemisphere. Even at a rate of 1.4 percent annually, or a doubling every 50 years, saturation would require only 800 years. Presumably a population crash would soon follow the extinction of the megafauna (4).

One need not demand that a maximum growth rate was maintained for long, or that the New World Paleo-Indian population ever totaled 10^7 at any one time. Animal invaders expand along an advancing front (18). I propose that the human invasion of the Americas proceeded in the manner Caughley [see (18)] has reported for exotic mammals spreading through New Zealand (see Fig. 1). A high population density was concentrated only along the periphery. The advance of the hunters was determined partly by the abundance of fresh game within the front and partly by cultural limits to the rate of human migration. In a decade or less, the population of vulnerable large animals on the front would have been severely reduced or entirely obliterated. As the fauna vanished, the front swept on, while any remaining human population would have been driven to seeking new resources.

For the North American midcontinent, I assume the arrival near Edmonton of a band of 100 hunters (Fig. 2). If the average southward movement did not exceed 16 kilometers per year (*19*), 184 years would have been required to develop a population of 61,000, large enough to continue to expand southward at the required rate while maintaining the required frontal density of 0.4 person per square kilometer across an arc 160 kilometers deep. By then the front would have advanced southward 640 kilometers beyond Edmonton (Fig. 2).

Further expansion would be limited not by the maximum rate of population growth but by the assumed cultural limits to migration. A maximum population for North America would be about 600,000, with half that number on the front when it reached the Gulf of Mexico, 3300 kilometers south of Edmonton. The concordant radiocarbon ages Haynes finds among midcontinent man-mammoth sites (*8*) are conformable with the proposed rapid sweep of the hunters. Alternate solutions based on computer simulation are shown in Table 2.

Under the conditions of the model, the front reached Panama at 10,930 years ago. At this point a second slight lag ensued, imposed by the need to develop a broad front into South America after passage of the Panamanian bottleneck (Fig. 3). In this case, a larger initial population seems likely. Within about 130 years, a population growth rate of 3.4 percent annually would again begin to be limited by cultural restraints. By 10,500 years ago, 1000 years after the arrival of the hunters at Edmonton (1200 years after arrival in Alaska), Tierra del Fuego would be within view (Fig. 3).

Modeling Overkill

The impact of the hunters is best visualized if one considers a representative area on their front. If a sizable biomass, say, 50 animal units per square mile, were exposed for 10 years to hunters whose density is one person per square mile on the average, what removal rates would be necessary to reduce the fauna? The fraction of the standing crop of moose available annually to wolves on Isle Royale is 18 percent (*20*); mainly older and young animals are taken. For animals larger than moose, an annual removal rate of 20 percent of the biomass attributable

Table 2. Simulated values for New World population growth.* In example 1, 104 people reached Edmonton 11,500 years ago. They double in numbers every 20 years until limited by their southward migration rate, 16 kilometers per year. Population growth fills a sector of 90° and is concentrated along the arc ("front") through a depth of 160 kilometers. The density on the front is 0.4 person per square kilometer and behind the front is 0.04 person per square kilometer. Example 2 is the same as example 1 except that the population doubles every 30 years. Example 3 is the same as example 1 except that the migration rate is 8 kilometers per year. Example 4 is the same as example 1 except that the migration rate is 25 kilometers per year. Example 5 is the same as example 1 except that the front is 80 kilometers deep. Example 6 is the same as example 1 except that the front is 240 kilometers deep. Example 7 is the same as example 1 except that the front is 0.02 person per square kilometer.

Ex-ample	Filling the front		Point at which frontal advance is limited by migration rate			Front reaches the Gulf of Mexico	
	Time (years)	Population	Time (years)	Distance (km)	Population	Time (years)	Population
1	125	8,000	184	393	61,000	345	590,000
2	188	8,000	299	583	102,000	440	590,000
3	125	8,000	159	207	26,000	519	590,000
4	125	8,000	199	583	102,000	294	590,000
5	86	2,800	172	457	40,000	326	435,000
6	149	18,000	193	349	81,000	358	750,000
7	125	8,000	180	368	53,000	344	450,000

* Computer simulations programmed by D. P. Adam.

to all predators would very likely be lethal in a few years. Smaller animals (between 50 and 400 kilograms in adult body weight) reproduce at higher rates, but their vulnerability would increase if the hunters preyed less selectively than wolves, taking a higher percentage of adult females.

An annual removal of 30 percent of their biomass should exceed normal replacement by reproduction for all the mammals lost in the Late Pleistocene. If one person in four did all of the hunting, destroying one animal unit (450 kilograms) per week from an animal population on the front averaging 50 animal units per section, he would

eliminate 26 percent of the biomass in 1 year. Regions with a higher biomass (more animals), resembling the richest African game parks of today, would not have escaped if the density of hunters rose accordingly.

Provided that each carcass was carefully butchered and dried and all edible portions were ultimately consumed, the minimum caloric requirements for one person per section could have been met by the annual removal of only 5 percent of the assumed 50 animal units per section. But much more wasteful consumption is to be expected, especially if tempting, new prey were easily accessible (*11*). Wheat [see (*3*)] has

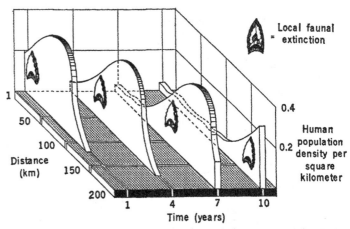

Local faunal = extinction

Human population density per square kilometer

Distance (km)

Time (years)

Fig. 1. Passage of the front showing theoretical changes in human population density. At any one point, the big game hunters and the extinct animals coexisted for no more than 10 years. Poor paleontological visibility of kill sites is thus inevitable.

reviewed the historic records of extraordinary meat consumption and occasional extreme waste among the Plains Indians.

Unless one insists on believing that Paleolithic invaders lost enthusiasm for the hunt and rapidly became vegetarians by choice as they moved south from Beringia, or that they knew and practiced a sophisticated, sustained yield harvest of their prey, one would have no difficulty in predicting the swift extermination of the more conspicuous native American large mammals. I do not discount the possibility of disruptive side effects, perhaps caused by the introduction of dogs and the destruction of habitat by man-made fires. But a very large biomass, even the 2.3×10^8 metric tons of domestic animals now ranging the continent, could be overkilled within 1000 years by a human population never exceeding 10^6. We need only assume that a relatively innocent prey was suddenly exposed to a new and thoroughly superior predator, a hunter who preferred killing and persisted in killing animals as long as they were available (21).

With the extinction of all but the smaller, solitary, and cryptic species, such as most cervids, it seems likely that a more normal predator-prey relationship would be established. Major cultural changes would begin. Not until the prey populations were extinct would the hunters be forced, by necessity, to learn more botany. Not until then would they need to readapt to the distribution of biomes in America in the manner Fitting (22) has proposed.

An explosive model will account for the scarcity of extinct animals associated with Paleo-Indian artifacts in obvious kill sites. The big game hunters achieved high population density only during those few years when their prey was abundant. Elaborate drives or traps were unnecessary.

Sudden overkill may explain the absence of cave paintings of extinct animals in the New World and the lack of ivory carvings such as those found in the mammoth hunter camps of the Don Basin. The big game was wiped out before there was an opportunity to portray the extinct species.

Finally, the model overcomes any objections that acceptable radiocarbon dates of around 10,500 years ago on artifacts from the southern tip of South America (23) require a crossing of the Bering platform thousands of years earlier (24).

As Birdsell (17) found in the case of Australia, it appears that prehistorians have overlooked the potential for a population explosion in what ecologists must regard as a uniquely favorable environment—the New World when first discovered. An outstanding difficulty remains, the question of "pre-Paleo-Indians" or "early-early man."

The Hunt for Early-Early Man

The population and overkill model I have proposed predicts that the chronology of extinction is as effective a guide to the timing of human invasion as the oldest artifacts themselves. According to Haynes (8), well-dated New World mammoth kill sites cluster tightly around 11,200 years ago. The population growth model presented here requires that the time of human entry into Alaska need be no older than 11,700 years ago to bring the hunters to Arizona by 11,200 years ago and to

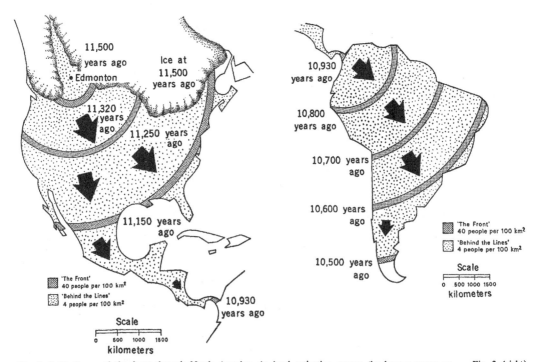

Fig. 2 (left). Sweep of the front through North America. As local extinction occurs, the hunter moves on.   Fig. 3 (right). Sweep of the front through South America. Local extinction accompanies passage of the front. (Figures 2 and 3 are not drawn to scale.)

the tip of South America by 10,500 years ago.

A growing number of claims and reviews of sites considered to be at least 13,000 years old or older, including some proposed to be over 20,000 years old, have appeared recently (25). The presence of people in the New World long before the big game hunters of 11,200 years ago seems all but conclusively established. Most prehistorians assume that the Americas were occupied by 15,000 years ago (26). However, questions of evidence loom.

An ephemeral or scarcely detectable invasion by or before 15,000 years ago implies slow population growth and a low population density. Few would claim that the putative early-early Americans were numerous, and Irwin-Williams [see (25)] concluded that they were scarce. A sizable hunt for new evidence of early-early Americans is under way. The more spectacular the claim, the more interest is generated in the announcement (27).

The nature of death assemblages, the subtleties of rebedding and redeposition, the uncertainty in diagnosing artifacts, and, especially, the limitations of various dating methods under ordinary field conditions are certain to generate difficulties even for the most careful investigator. Although replication or the critical verification of an original excavation assumes major significance, it is not often attempted.

In a notable exception, the reexcavation of Tule Springs, Nevada, a well-funded team of geologists, ecologists, and archeologists failed to verify the impressive claim of a 23,000-year-old human occupation (28). The oldest evidence of occupation that could be verified at Tule Springs occurred in depositional units considered to be between 11,000 and 13,000 years old (29).

Their research material has made behavioral scientists especially sensitive to interpreter and experimenter effects. According to Rosenthal, "Perhaps the greatest contribution of the skeptic, the disbeliever, in any given scientific observation is the likelihood that his anticipation, psychological climate, and even instrumentation may differ enough so that his observation will be a more independent one" (30). Site replication established the contemporaneity of ancient man and extinct fauna in the New World (31).

Replication is now needed if the early-early man sites are to be regarded seriously. The enthusiastic search and the growing number of claims can be viewed as destructive, not supportive, of the early-early man theory. At this point, the more unreplicated claims that are filed, the more likely that their authors may be victims of an experimenter, or, in this case, excavator, effect.

Begging each claim is an ecological paradox: If *Homo sapiens* was clever enough to master a technology that allowed him to penetrate the Arctic or the marine barriers standing in the way of discovery of the New World, why did he fail to exploit the highly productive ecosystem he found in warmer parts of this hemisphere? Why did he fail to leave a trail of evidence at least as obvious as the Mousterian, Gravettian, Solutrean, and other Middle and Upper Paleolithic cultures so abundant in Europe? My questions simply rephrase an objection voiced long ago by Hrdlička and revised by Graham and Heizer (31).

For the present, American archeologists can rest assured that ecological principles are not violated by evidence at the known validated sites. A brief moment of big game hunting, not only of mammoth but also of many other species, could have led to megafaunal extinction around 11,000 years ago and to major cultural readaptation in most of the hemisphere afterward. It is not necessary to postulate human invasion by or before 15,000 years ago.

Invasion by a slowly growing and chronically sparse population is not impossible. But it requires major ecological constraints that have yet to be identified in the American environment. Given the biology of the species, I can envision only one circumstance under which an ephemeral discovery of America might have occurred. It is that sometime before 12,000 years ago, the earliest early man came over the Bering Straits without early woman.

Summary

I propose a new scenario for the discovery of America. By analogy with other successful animal invasions, one may assume that the discovery of the New World triggered a human population explosion. The invading hunters attained their highest population density along a front that swept from Canada to the Gulf of Mexico in 350 years, and on to the tip of South America in roughly 1000 years. A sharp drop in human population soon followed as major prey animals declined to extinction.

Possible values for the model include an average frontal depth of 160 kilometers, an average population density of 0.4 person per square kilometer on the front and of 0.04 person per square kilometer behind the front, and an average rate of frontal advance of 16 kilometers per year. For the first two centuries the maximum rate of growth may have equaled the historic maximum of 3.4 percent annually. During the episode of faunal extinctions, the population of North America need not have exceeded 600,000 people at any one time.

The model generates a population sufficiently large to overkill a biomass of Pleistocene large animals averaging 9 metric tons per square kilometer (50 animal units per section) or 2.3×10^8 metric tons in the hemisphere. It requires that on the front one person in four destroy one animal unit (450 kilograms) per week, or 26 percent of the biomass of an average section in 1 year in any one region. Extinction would occur within a decade. There was insufficient time for the fauna to learn defensive behaviors, or for more than a few kill sites to be buried and preserved for the archeologist. Should the model survive future findings, it will mean that the extinction chronology of the Pleistocene megafauna can be used to map the spread of *Homo sapiens* throughout the New World.

References and Notes

1. F. Bordes, *The Old Stone Age* (McGraw-Hill, New York, 1968).
2. P. S. Martin, in *Pleistocene Extinctions, the Search for a Cause*, P. S. Martin and H. E. Wright, Jr., Eds. (Yale Univ. Press, New Haven, Conn., 1967), pp. 87–89.
3. Over the past two decades radiocarbon dates have been published which, if taken at face value, appear to show that mammoths, mastodons, ground sloths, and other common members of the extinct American megafauna lasted into the postglacial. Since my 1967 review (2), the following dates, all younger than 10,000 years old, have appeared: UCLA-1325 (fossil wood below Pleistocene mammal bones); ISGS-17 A,B,C (mastodon ivory and bone); OWU-224 A,B (gyttja with mastodon); A-806 A,C,D; A-874 C; A-876 C; A-195; A-584; A-619; A-536; I-2244; M-1764; M-1765 (bone apatite, acid-soluble organic matter, enamel, and other materials from mammoth bones); UGa-79 (sloth bone). For a complete description of field and laboratory treatment of the samples and for laboratory designations, see *Radiocarbon* 9–13 (1967–1971). In those cases in which stratigraphically associated charcoal dates were available, the bone dates were all significantly younger. They may be suspected of contamination by younger carbon. In another set of especially interesting cases, much younger organic material was found associated with mammoth bones (M-2361, 3310 ± 160 years, conifer cones), a mastodon rib (M-2436, 4470 ± 160 years, conifer log), and sloth dung (UCLA-1069, 2400 ± 60 years; UCLA-1223, 2900 ± 80 years on artifacts). On the basis of other information, the collectors could discount the associations as probably secondary. As long as sample selec-

tion is not foolproof and because bone contamination is not always avoidable, we may expect a steady increase in postglacial dates of the sort which misled me years ago [P. S. Martin, in *Zoogeography*, C. Hubbs, Ed. (Publication No. 51, AAAS, Washington, D.C., 1958), p. 397]. Admittedly, there is no theoretical reason why a herd of mastodons, horses, or ground sloths could not have survived in some small refuge until 8000 or even 4000 years ago. But in the past two decades, concordant stratigraphic, palynological, archeological, and radiocarbon evidence to demonstrate beyond doubt the postglacial survival of an extinct large mammal has been confined to extinct species of *Bison* [see S. T. Shay, *Publ. Minn. Hist. Soc.* (1971); J. B. Wheat, *Sci. Amer.* **216**, 44 (January 1967); *Amer. Antiquity* **37** (part 2) (No. 1) (1972); D. S. Dibble and D. Lorrain, *Tex. Mem. Mus. Misc. Pap. No. 1* (1968)]. No evidence of similar quality has been mustered to show that mammoths, mastodons, or any of the other 29 genera of extinct large mammals of North America were alive 10,000 years ago. The coincidence in time between massive extinction and the first arrival of big game hunters cannot be ignored.

4. M. I. Budyko, *Sov. Geogr. Rev. Transl.* **8** (No. 10), 783 (1967).
5. According to R. F. Flint [*Glacial and Quaternary Geology* (Wiley, New York, 1971), p. 778], "The argument most frequently advanced against the hypothesis of human agency is that in no territory was man sufficiently numerous to destroy the large numbers of animals that became extinct."
6. Apart from postglacial records of extinct species of *Bison*, very few kill sites have been discovered. J. J. Hester [in *Pleistocene Extinctions, the Search for a Cause*, P. S. Martin and H. E. Wright, Jr., Eds. (Yale Univ. Press, New Haven, Conn., 1967), p. 169], A. J. Jellinek (*ibid.*, p. 193), and G. S. Krantz [*Amer. Sci.* **58**, 164 (1970)] have all raised this point as a counterargument to overkill.
7. The North American megafauna that I believe disappeared at the time of the hunters includes the following genera: *Nothrotherium*, *Megalonyx*, *Eremotherium*, and *Paramylodon* (ground sloths); *Brachyostracon* and *Boreostracon* (glyptodonts); *Castoroides* (giant beaver); *Hydrochoerus* and *Neochoerus* (extinct capybaras); *Arctodus* and *Tremarctos* (bears); *Smilodon* and *Dinobastis* (saber-tooth cats); *Mammut* (mastodon); *Mammuthus* (mammoth); *Equus* (horse); *Tapirus* (tapir); *Platygonus* and *Mylohyus* (peccaries); *Camelops* and *Tanupolama* (camelids); *Cervalces* and *Sangamona* (cervids); *Capromeryx* and *Tetrameryx* (extinct pronghorns); *Bos* and *Saiga* (Asian antelope); and *Bootherium*, *Symbos*, *Euceratherium*, and *Preptoceras* (bovids).
8. C. V. Haynes, in *Pleistocene and Recent Environments of the Central Great Plains*, W. Dort, Jr., and J. K. Jones, Jr., Eds. (University of Kansas Department of Geology Special Publication No. 3, Lawrence, 1971), p. 77.
9. A. Dreimanis [*Ohio J. Sci.* **68**, 257 (1968)] estimates that there are more than 600 mastodon occurrences in northeastern North America. If we suppose that 500 of these were of late-glacial age and assign an equal probability of death, burial, and discovery to each in the time span from 10,500 to 15,500 years ago, then an average of ten may be expected for any given century and one for any decade. I assume that in any one region local extinction was swift. The elephants and their hunters were associated for no more than a decade. Even if the temporal overlap between the elephants and their hunters were as much as 100 years, and if half (five) of the finds represent animals killed by hunters, it is clear that the probability of the field evidence actually being detected and appreciated by the discoverers of the bones is small. Had the hypothetical mastodon kill sites been located on the uplands rather than in bogs or on lake shores, the probability of discovery becomes smaller still. My pessimistic appraisal should not deter those engaged in the search for more kill sites. It should refute the view that extinction by overkill would yield abundant fossil evidence.
10. One animal unit can be used as a standard for paleoecological comparison in the sense range managers have used it for comparing stocking rates under common use. One animal unit equals 1000 pounds (450 kilograms), or approximately the adult weight of one steer, one horse, one cow, four hogs, five sheep, or five deer. Some possible Pleistocene equivalents would be 0.2 mammoth (*Mammuthus columbi*), 0.3 mastodon (*Mastodon americanus*), 0.6 large camel (*Camelops*), one large horse (*Equus occidentalis*), one woodland musk-ox (*Symbos*), three woodland peccaries (*Mylohyus*), or ten cervicaprids (*Tetrameryx*).
11. P. S. Martin, in *Arctic and Alpine Research*, J. Ives and R. Barry, Eds. (Methuen, London, in press).
12. P. F. Wilkinson, in *Models in Archaeology*, D. L. Clarke, Ed. (Methuen, London, 1972).
13. R. D. Guthrie [*Amer. Midl. Natur.* **79**, 346 (1968)] concluded that mammoths constituted 3 to 11 percent of the megafauna in the rich Late Pleistocene deposits near Fairbanks, Alaska. Their size meant that they comprised 20 to 50 percent of the *relative* biomass. But fossil mammal deposits are obviously not randomly distributed. Within a deposit, the rates of bone deposition are unknown and apparently unknowable. There is no prospect of estimating accurately the size of a past population from its fossil bones.
14. B. Kurtén, *Acta Zool. Fenn.* **107**, 1 (1965). The late Würm fossil carnivores of Levant Caves reveal a shrinkage in range, decline in numbers, and reduction in body size.
15. F. Lorimer, in *The Determinants and Consequences of Population Trends* (Population Studies No. 17, United Nations, New York, 1953), pp. 5–20; A. Desmond, *Population Bull.* **18** (No. 1), 1 (February 1962); E. S. Deevey, Jr., *Sci. Amer.* **203**, 195 (September 1960). Deevey proposed an increase of 1.4 times in a 28-year generation, or 1.3 percent annually, as prehistoric man's best effort. C. V. Haynes [*Sci. Amer.* **214**, 104 (June 1966)] used this value to estimate the population growth of mammoth hunters. Much more rapid growth rates can be assumed.
16. Disease can be discounted. Microbiologists generally regard the Paleolithic as a healthy episode [R. Hare, in *Diseases in Antiquity*, D. Brothwell and A. T. Sandison, Eds. (Thomas, Springfield, Ill., 1967), pp. 115–131]. Their reasons are based less on detailed knowledge of skeletal pathologies or the scarcity of major parasites in prehistoric feces [G. F. Fry and J. G. Moore, *Science* **166**, 1620 (1969); R. F. Heizer and L. K. Napton, *ibid.* **165**, 563 (1969)] than on biological inference. Lacking closely related hosts, the New World held no major reservoir of hominid diseases. Cholera and African sleeping sickness never became established. American Indians suffered catastrophic losses in historic time, presumably through lack of prior exposure to Old World diseases such as smallpox and tuberculosis [H. F. Dobyns, *Curr. Anthropol.* **7**, 395 (1966)].
17. J. B. Birdsell, *Cold Spring Harbor Symp. Quant. Biol.* **22**, 47 (1957).
18. C. S. Elton, *The Ecology of Invasions by Animals and Plants* (Methuen, London, 1958). Introduced populations of the giant African snail, *Achatina fulica*, attain highest values at the time of establishment, declining rapidly after initial introduction [A. R. Mead, *The Giant African Snail* (Univ. of Chicago Press, Chicago, 1961)]. Exotic large mammals spreading through New Zealand attain peak population densities and maximum reproduction rates at the margin of their range [T. Riney, *Int. Union Conserv. Nature Publ.* (*n.s.*) *No. 4* (1964), p. 261; G. Caughley, *Ecology* **51**, 53 (1970)].
19. The proposed migration rate is well within the distance covered by groups of Zulus known to have moved from Natal to Lake Victoria (3000 kilometers) and halfway back in half a century [J. D. Clark, *The Prehistory of Southern Africa* (Penguin Books, Harmondsworth, Middlesex, England, 1959), p. 168].
20. P. A. Jordan, D. B. Botkin, M. L. Wolfe, *Ecology* **52**, 147 (1970). G. B. Schaller [*Natur. Hist.* **81**, 40 (1971)] says Serengeti predators remove roughly 10 percent of the prey biomass.
21. Even when most of their calories come from plants [see R. B. Lee, in *Man the Hunter*, R. B. Lee and I. Devore, Eds. (Aldine, Chicago, 1969)], men of modern nonagricultural tribes devote much time to the hunt. The arctic invaders of America had come through a region notably deficient in edible plants. As long as large mammals were flourishing, there was no need to devise new techniques of harvesting, storing, and preparing less familiar food. None of their artifacts suggests that the first American hunters also stalked the wild herbs, and none of their midden refuse suggests that the succeeding gatherers knew the extinct mammals.
22. J. E. Fitting, *Amer. Antiquity* **33**, 441 (1968).
23. J. B. Bird, *ibid.* **35**, 205 (1970).
24. M. Bates, *Where Winter Never Comes* (Scribner, New York, 1952); R. F. Black, *Arctic Anthropol.* **3**, 7 (1966); H. T. Irwin and H. M. Wormington, *Amer. Antiquity* **34**, 24 (1969).
25. A. L. Bryan, *Curr. Anthropol.* **10**, 339 (1969); unpublished data; C. S. Chard, *Man in Prehistory* (McGraw-Hill, New York, 1969); W. N. Irving, *Arctic Anthropol.* **8**, 68 (1971); C. Irwin-Williams, paper presented at the Conference on Pleistocene Man in Latin America, San Pedro de Atacoma, Chile, 1969; E. Lanning, *World Archaeol.* **2**, 90 (1970); T. F. Lynch, *Occas. Pap. Idaho Univ. State Mus. No. 21* (1967); —— and K. A. R. Kennedy, *Science* **169**, 1307 (1970); R. S. MacNeish, R. Berger, R. Protsch, *ibid.* **168**, 975 (1970); R. S. MacNeish, *Sci. Amer.* **224**, 36 (February 1971); H. Müller-Beck, *Science* **152**, 1191 (1966); P. C. Orr, *Prehistory of Santa Rosa Island* (Santa Barbara Museum of Natural History, Santa Barbara, Calif., 1968); B. E. Raemsch, *Yager Mus. Publ. Anthropol. Bull. No. 1* (1968); A. Stalker, *Amer. Antiquity* **34**, 428 (1969).
26. K. W. Butzer, *Environment and Archaeology: An Ecological Approach to Prehistory* (Aldine-Atherton, Chicago, ed. 2, 1971); J. M. Cruxent, in *Biomedical Challenges Presented by the American Indian* (Pan-American Health Organization Science Publication 165, Washington, D.C., 1968), pp. 11–16; J. E. Fitting, *The Paleo-Indian Occupation of the Holcombe Beach* (Michigan University Museum of Anthropology Anthropological Paper No. 27, Ann Arbor, 1966).
27. One of the boldest claims of great antiquity is that of L. S. B. Leakey, R. De E. Simpson, and T. Clements [*Science* **160**, 1022 (1968)] near Calico Hills, California. Flaked cherts have been reported in fan deposits first considered to be at least 40,000 years old, and later judged to be much older. A study based on a visitation to the quarry in October 1970 by 60 leading American geologists and archeologists sustained the view that the deposits are of great age. It failed to satisfy skeptics that the alleged artifacts were definitely man-made and in a cultural context [C. Behrens, *Sci. News* **99**, 98 (1971)].
28. M. R. Harrington and R. De E. Simpson, *Tule Springs, Nevada, with Other Evidences of Pleistocene Man in North America* (Southwest Museum Paper No. 18, Los Angeles, 1961).
29. C. V. Haynes, in *Pleistocene Studies in Southern Nevada*, H. M. Wormington and D. Ellis, Eds. (Nevada State Museum of Anthropology Paper No. 13, Reno, 1967); R. Shutler, Jr., *Current Anthropol.* **6**, 110 (1965).
30. R. Rosenthal, *Experimenter Effects in Behavioral Research* (Appleton-Century-Crofts, New York, 1966).
31. A. Hrdlička, "Early man in South America" [*Bur. Amer. Ethnol.* **52**, 4 (1912)]; J. A. Graham and R. F. Heizer, *Quaternaria* **9**, 225 (1968).
32. F. Bourliere and M. Hadley, *Annu. Rev. Ecol. Syst.* **1**, 138 (1970).
33. For comments and counterarguments I am grateful to D. P. Adam, J. B. Birdsell, A. L. Bryan, J. E. Guilday, E. W. Haury, C. V. Haynes, Jr., J. J. Hester, A. J. Jelinek, R. G. Klein, G. S. Krantz, L. S. Lieberman, E. H. Lindsay, Jr., D. I. Livingstone, A. Long, R. H. MacArthur, M. Martin, J. H. McAndrews, P. Miles, J. E. Mosimann, C. W. Ogston, B. Rippeteau, J. J. Saunders, G. G. Simpson, W. W. Taylor, N. T. Tessman, and most especially, R. F. Heizer. I thank D. P. Adam for programming the computer simulations in Table 2. Special thanks for editorial aid are due B. Fink. This study was supported in part by NSF grant 27406. Contribution No. 56, Department of Geosciences, University of Arizona, Tucson.

4

Revolutions in Historical Biogeography

Vicki A. Funk

As we have seen in previous parts of this volume, the distribution of a taxon is the result its biological history as well as the geologic and climatic history of the areas in which it has lived. As a result, historical biogeography comprises two parts: the history of life on earth and the history of earth itself. The methods and results of historical biogeography are central to studies in diverse fields ranging from systematics to ecology, and there are many opinions of exactly how one should attempt to reconstruct an accurate history of a group of organisms or of the parts of earth they inhabit. However, despite this diversity, one can identify certain points in the development of the field that have fundamentally changed the way biogeographers go about their work. Part 4 contains many of the classic papers that led to such changes and influenced the development of the field of historical biogeography.

Descriptive Biogeography

That different regions of the world are inhabited by distinct groups of plants and animals is a fact that has been known to naturalists at least since Buffon (1761; see part 1, paper 2). By the early 1800s this concept was generally accepted for a variety of organisms. In 1820 Candolle

recognized the identification of "botanical regions" as a problem of general interest: the works of most of the early biogeographers discussed in part 1 of this volume included explanations for plant and animal distributions ranging from Noah's Ark to centers of origin and land bridges. However, two men stand out for their contributions to historical biogeography because, unlike their predecessors and colleagues, they saw the need for developing a general method for explaining the differences among biogeographic regions. The earliest suggested methods for classifying biogeographic regions are probably those of a botanist, Joseph Dalton Hooker (1844–60) and an ornithologist, Philip Lutley Sclater (1858). Each had a question and a method to answer it.

As described by Nelson (1978), Sclater's question and method were:

Question: "What are the primary ontological divisions of the earth's surface?"
Method: "For each inquirer to take up the subject with which he is best acquainted, and to work out what he conceives to be the most natural divisions of the earth's surface from that alone. Such being done, we shall see how far the results correspond, and on combining the whole, may possibly arrive at a correct solution of the problem—to find the primary ontological divisions of the earth's surface."

Sclater used his impressive knowledge of distributions of birds to classify the world's biogeographic regions based on similarities and differences among their avifauna. He thought that if his methods were applied to other groups of organisms, then correspondence in patterns among groups would reveal the "primary ontological divisions."

Hooker (1844–60) sought to understand the origins of floras from around the world. In his treatments of the floras of lands of the southern hemisphere, he proposed a much more detailed method than that of Sclater to evaluate checklists and determine affinities of floras, a method that is still being used today. Hooker's *The Botany of the Antarctic Voyage* (1853, 209–23; see also part 1, paper 8), provides an example of what is probably the first stated method in what would become analytical biogeography, and an excellent evaluation is provided by LARS BRUNDIN (1966: 46–55; paper 36). Hooker's method was to place species from the areas in question into groups that summarized their total distribution. After examining the species in common among the floras of Australia, New Zealand, Tasmania, the Antarctic islands, South America, and South Africa, he came to the conclusion that the "bands of affinity" (species shared among areas) demonstrated that there had been a single center of evolution in the south that had been broken up by geological and climatic causes. As is discussed elsewhere in this volume, he believed that species had arrived to each of the continents in question by dispersal across land bridges postulated to have formerly existed between continents. Darwin, in contrast, preferred long-distance dispersal as an explanation for species distribution patterns (see Darwin 1859). Of all the early works in biogeography, Hooker's (1867) view of a "continuous extensive flora, . . . that once spread over a lager and more continuous tract of land" came the closest to getting it right. His method and conclusions are a "must read" for all students of biogeography.

Hooker had proponents, such as the well-known geologist Charles Lyell (1832) and the botanist Carl Skottsberg, who believed that the

Antarctic floral elements Hooker studied were derived from a common flora that inhabited a single area during an earlier period (see Skottsberg 1922, 1925). Unfortunately, and perhaps because of their penchant for proposing new land bridges for each vicariant pattern, Hooker and other so-called extensionists were overshadowed by Darwin and other proponents of "dispersal over a permanent geography." Darwin's colleagues believed that the earth existed first and that the plants and animals dispersed around it. While both sides of the dispersalist/extentionist debate had merit, evidence for transoceanic land bridges never surfaced, and it was not until the acceptance of the theory of continental drift that we fully understood the mechanism responsible for the patterns Hooker studied.

Evolutionary Biogeography

Through the works of such influential biogeographers as Wallace, Darwin, Matthew, Simpson, Mayr, Darlington and others, the concept of evolutionary biogeography took control of historical biogeography, and its dominance of the field lasted for more than 100 years. While Darwin spent most of his time on other topics concerning evolution, Wallace remained focused on biogeography, and he is extremely important to the history of the subject. A good example of his work is found in paper 12 (this volume), which contains Wallace's discussion of the distributions of animals. Many of the patterns he proposed are still studied today. He is probably best remembered by the biogeographic landmark often referred to as "Wallace's line." Brown and Lomolino (1998) provide an excellent summary of Wallace's work, and their table (reproduced here as table 1; see p. 650) lists his major principles. It is surprising how many of these are still in use today.

One idea that has existed for some time and which continues to play a role in biogeographic studies is that of a *center of origin*. This concept can be viewed in two ways, from the standpoint of a single taxon (e.g., mammals) or from the perspective of a biota (e.g., refugia). It has of course developed from Linnaeus and other early biogeographers who adapted the Noah's Ark concept of all life surviving the deluge and then dispersing from that one center to all regions of the globe. One of the more recent leaders in the centers of origin debate was Stanley A. Cain (1944) who proposed 13 criteria that could be used to determine such centers of origin. Cain's work was very influential with more recent practitioners of evolutionary biogeography such as Simpson, Mayr, and Darlington. Brown and Lomolino (1998) updated Cain's criteria and produced a useful table that summarizes his ideas (see table 2). Of course, most if not all of these criteria probably describe what has happened in one or more groups of organisms. The difficulty is in deciding which one applies to which group.

Philip J. Darlington Jr. (1965) was possibly the last of the evolutionary biogeographers to advocate dispersal over a fixed geology as the explanation for current patterns of distribution (see part 2 of this volume). Using Darlington's 1965 book as an example, Brundin (1966) explained the conceptual problems involved in evolutionary biogeography. What is interesting about all of these scientists is that, although they were proponents of evolutionary theory and they used it to explain distributions, none of them developed a method to use such relationships in their biogeographic studies. As a result, their explanations read much like Rudyard Kipling's *Just So Stories* (1902), although Kipling's illustrations were better.

The acceptance of the theory of continental drift and plate tectonics revolutionized the way most biogeographers studied the origin and spread of life on Earth. For some, it was an instant recognition of a solution to the age-old question of how various parts of the biota ended up in certain areas; for others, it was a long struggle, and there was much foot dragging. It wasn't until the 1960s that the theory of continental drift was generally accepted and the last of the distinguished opposition finally gave up (Darlington 1965). In a sense, Hooker had been right all along. There *had* been a uni-

form flora in the southern hemisphere, he just had the mechanism wrong: land bridges did not rise and fall across the great oceans, the continents moved.

Phylogenetic Biogeography

A shift from a largely static view of species distributions, with a reliance on dispersal events, to a more dynamic view of species and their distributions began in the 1950s and 60s. This more evenly balanced approach was rapidly adopted when combined with newly available information, especially that related to plate tectonics and paleoclimates. The shift also occurred because of new methodological approaches in the systematic description of species and the relationship between species and their geographic distributions. Foremost among the scientists contributing to this early impetus for change were Willi Hennig and Lars Burndin.

WILLI HENNIG first published his ideas on phylogenetic systematics in 1950, but it was not until the English version of his rewritten book, *Pylogenetic Systematics* (1966), that these ideas began to have a wider influence. His ideas about the delimitation of monophyletic groups (groups containing all of the descendents of a common ancestor) using only uniquely derived characters (apomorphies) had a profound effect on systematics and historical biogeography. In his books, Hennig described several methods for determining derived and primitive conditions for each trait, thus establishing the direction of evolutionary change in groups of characters and producing a branching diagram representing the phylogeny of the organisms in question. One way to determine the direction of evolution was "the chorological method" (Hennig 1966), which made use of the distribution of the taxa in question. The featured section of Hennig's discussion (paper 37), provided the first method for combining the phylogeny of a monophyletic group of organisms with the distribution of its members to trace the "progression in space." Although Sclater (1858) had alluded to the phenomenon that some "geographical divisions" have closer historical relations than oth-

Table 1
Biogeographic Principles Advocated by Alfred Russel Wallace

These conclusions are summarized from Wallace's writings, and have been verified many times by researchers in the twentieth century.

1. Distance by itself does not determine the degree of biogeographic affinity between two regions; widely separated areas may share many similar taxa at the generic or familial level, whereas those very close may show marked differences, even anomalous patterns.

2. Climate has a strong effect on the taxonomic similarity between two regions, but the relationship is not always linear.

3. Prerequisites for determining biogeographic patterns are detailed knowledge of all distributions of organisms throughout the world, a true and natural classification of organisms, acceptance of the theory of evolution, detailed knowledge of extinct forms, and knowledge of the ocean floor and stratigraphy to reconstruct past geological connections between landmasses.

4. The fossil record is positive evidence for past migrations of organisms.

5. The present biota of an area is strongly influenced by the last series of geological and climatic events; paleoclimatic studies are very important for analyzing extant distribution patterns.

6. Competition, predation, and other biotic factors play determining roles in the distribution, dispersal, and extinction of animals and plants.

7. Discontinuous ranges may come about through extinction in intermediate areas or through the patchiness of habitats.

8. Speciation may occur through geographic isolation of populations that subsequently become adapted to local climate and habitat.

---

ers, it was left to Hennig over a century later to propose a method for examining that phenomenon. In doing so he provided the earliest applications of what became known as *phylogenetic biogeography*.

Hennig believed that there were "close relations between the species and space" and that one could use the two together to determine the direction in which certain transformation series of characters needed to be read. He illustrated this by using the phylogenetic method (a comparison of the phylogeny of a group of species and the distribution of those species) to illustrate two types of *rassenkreisen* (extension of range accompanied by change) where "vicarying subspecies are arranged in approximately linear succession—in which extension evidently took place only in one main direc-

tion—and the other in which extension took place in all directions from a center of distribution" (1966: 135). Hennig also emphasized that the concept of space was not limited to *geographic space* but also could include a change in *ecological space* (i.e., phenology changes, habitat shifts) within the same geographic area. In all of his work, he used a model of dispersal from a point or center of origin. Hennig assumed that "phylogenetically homogeneous groups of higher rank also inhabit fundamentally continuous areas," but he tempered this with a very loose definition of continuous areas. Although both evolutionary biogeography and phylogenetic biogeography were based on a center of origin concept, they differed in that evolutionary biogeography had no scientifically rigorous method and developed around

Table 1
continued

9. Disjunctions of genera show greater antiquity than those of single species, and so forth for higher taxonomic categories.

10. Long-distance dispersal is not only possible, but is also the probable means of colonization of distant islands across ocean barriers; some taxa have a greater capacity to cross such barriers than others.

11. The distributions of organisms not adapted for long-distance dispersal are good evidence of past land connections.

12. In the absence of predation and competition, organisms on isolated landmasses may survive and diversify.

13. When two large landmasses are reunited after a long period of separation, extinctions may occur because many organisms will encounter new competitors.

14. The processes acting today may not be at the same intensity as in the past.

15. The islands of the world can be classified into three major biogeographic categories: continental islands recently set off from the mainland, continental islands that were separated from the mainland in relatively ancient times, and distant oceanic islands of volcanic and coralline origin. The biotas of each island type are intimately related to the island's origin.

16. Studies of island biotas are important because the relationships among distribution, speciation, and adaptation are easier to see and comprehend on islands.

17. To analyze the biota of any particular region, one must determine the distributions of its organisms beyond that region as well as the distributions of their closest relatives.

Source: Brown and Lomolino (1998: 28, box 2.1). Reproduced with permission.

descriptive enumeration and scenario building (see discussion in Ball 1975). In contrast, Hennig's method (albeit simplified), included the following:

1. Study a group and determine apomorphies (derived characters).
2. Produce a phylogeny.
3. Examine the phylogeny of a monophyletic group with respect to distribution of its members.
4. Reexamine the phylogeny using information from the distribution and perhaps make changes in the previously determined apomorphies.

Hennig was a great believer in *reciprocal illumination,* such that information from one method of study was fed back into the overall evaluation of the phylogeny, hence information from the distribution could be used to refine the phylogeny. Today we might call this "multiple converging lines of evidence."

Brundin (1966) devoted a section in his monograph to explaining and expanding on these methods, and to showing the importance of the sister-group in the reconstruction of biogeographic history (paper 36; see also part 2, paper 20). Brundin's work is especially important in that it is more readable than Hennig's. After its publication, it became the primary

Table 2

Cain's Criteria Used and Abused for Indicating Center of Origin of a Taxon

1. Location of greatest differentiation of a type (greatest number of species)

2. Location of dominance or greatest abundance of individuals (most successful area)

3. Location of synthetic or closely related forms (primitive and closely related forms)

4. Location of maximum size of individuals

5. Location of greatest productiveness and relative stability (of crops)

6. Continuity and convergence of lines of dispersal (lines of migration that converge on a single point)

7. Location of least dependence on a restricted habitat (generalist)

8. Continuity and directness of individual variation or modifications radiating from the center of origin along highways of dispersal (clines)

9. Direction indicated by geographic affinities (e.g., all Southern Hemisphere)

10. Direction indicated by the annual migration routes of birds

11. Direction indicated by seasonal appearance (i.e., seasonal preferences are historically conserved)

12. Increase in the number of dominant genes toward the centers of origin

13. Center indicated by the concentricity of progressive equiformal areas (i.e., numerous groups are concentrated in centers, and numbers decrease gradually outward)

Source: Brown and Lomolino (1998: 347, table 12.1), after Cain 1944.

way to introduce colleagues to phylogenetic systematics and to its use in biogeography. Because of this, Brundin's impact on biogeography has arguably been greater than Hennig's. Although Hennig is cited more frequently, it was Brundin who really developed the method and convinced other systematists that it worked, and it is to Brundin that many of us send our students for their first introduction to phylogenetic biogeography.

Hennig and Brundin differed somewhat in their concepts. Hennig did not look at numerous biogeographic patterns and search for commonality. Brundin did discuss groups of organisms and implied repeating patterns, but he tempered it by emphasizing the need for a phylogeny and a sister-group before one could evaluate the history of either groups or areas, and his comparison of patterns was largely descriptive. Hennig assumed that everything dispersed from a center of origin. Brundin (1966) was less caught up in dispersal, as is demonstrated by statements such as "The theory of continental drift provides a background fitting all demands raised by the nature of the transantarctic relationships, as displayed by the chironomid midges" (454). Clearly Brundin credited continental movement with having an overriding effect on current distributions. For more detailed descriptions and examples, Ross (1974), Cracraft (1975) and Ball (1975) have good applications of Hennig's biogeographic method.

In the midst of all of the discussion of evolutionary versus phylogenetic biogeography came GARRETH NELSON's article, "The Problem of Historical Biogeography" (1969; paper 38), which showed the beginning of a more rigorous way of thinking about distributions. Through the discussion of this simple method of determining ancestral areas, one can see the future of modern vicariance biogeography. But first we must pause and consider the other side of the story, Leon Croizat.

Panbiogeography

To some extent, LEON CROIZAT, a naturalized Venezuelan scientist, picked up where Sclater and Hooker left off. He tried to "combine the whole" as Sclater suggested, and worked mainly from the examination of numerous distributions as Hooker did. His work was completely independent of Hennig and he did not use a phylogeny as part of his analysis. Croizat believed that geography and life evolved together, and for 35 years (from 1947 to 1982), he wrote books and papers totaling approximately 9,000 pages, mainly focused on his method, *panbiogeography*, which he defined as "a consideration of vegetal and animal dispersal in time and space as one, correlated in addition with main geological concepts" (1958, 1:139n; for a list of Croizat's works, see Nelson 1973, Craw 1984).

Most of Croizat's work consisted of long monographs that were published over a 13-year period (1952, 1958, 1960, 1962). Because all of his books except the first were self-published, he did not benefit from the peer-review process. No doubt the lack of review contributes to the fact that these monographs are generally difficult to read: his writing style was verbose, he spent many pages lambasting one person after another, and he seemed to find it difficult to stay on a single topic for long. One of his books, *Panbiogeography* (1958), has a section in the introduction (xxi–xxxi) where he answers criticisms of his work: a highly instructive section for those who wish to understand Croizat. His book, *Space, Time, Form: The Biological Synthesis* (1964) is one of his

best, but even this one concludes with hundreds of pages attacking Darwin (135 pages), Fisher (45 pages), Catholicism (24 pages), and other people and ideas. To sample the flavor of Croizat and to expose the reader to the panbiogeographic method as well as his intemperate style, a few pages from *Space, Time, Form: The Biological Synthesis* (Croizat 1962) have been included here (see paper 39).

Although a concise explanation of Croizat's method is hard to find, Nelson (1973: 312) describes it as follows:

Croizat's method is to record on a map the ranges of the species of a given group, let us say a genus or family, and then to connect the ranges with a line to form a track. . . . Croizat found that if tracks . . . follow the same routes, they form a generalized track. . . . He concluded that a generalized track (1) is an empirical phenomenon and (2) unites continental areas that together are an estimate of the distribution of an ancestral biota. According to Croizat it is only with reference to a generalized track that individual tracks become worthy of interpretation."

Croizat also brought to biogeography the first attempt to examine the biota of an extremely large number of taxa. His generalized tracks do identify potentially interesting areas of endemism, and they alert us to possible similar histories among groups; however, they never include a phylogeny. The simple biogeographic method of Croizat stands in contrast to the ecological-dispersive, biogeographical views of Matthew, Simpson, and Darlington, who believed that improbable chance dispersal events of species to distant locations would become entirely probable given sufficient time, and consequently believed that dispersal was the primary mechanism responsible for disjunct distributions of closely related taxa.

In their paper entitled, "Centers of Origin and Related Concepts," LEON CROIZAT, GARETH NELSON, and DONN ERIC ROSEN (1974; paper 40) attempted to explain the method in a more concise manner. However, they described a modified panbiogeographic method by adding the concept of *monophyly* as a requirement for the taxa to be sampled. (Croizat later [1982] repudiated this collaboration; see the discussion below.) By 1974, Historical biogeography was primed for a major revolution in methods, the combination of Hennig and Brundin on one side, Croizat on the other, and Nelson and co-authors in the middle trying to synthesize the two.

Vicariance (Cladistic) Biogeography

In 1974, GARRETH NELSON broke with biogeographic practices based on dispersal and, in doing so, developed a new method for historical biogeography, one that combined Croizat's methods with Hennig's and Brundin's, under the "refutability" philosophy of Karl Popper, to form what became known as *vicariance* (later *cladistic*) *biogeography*. In his paper, "Historical Biogeography: An Alternative Formalization" (paper 41), Nelson (1974) stated the following:

The belief that speciation (vicariance) and dispersal (migration) must take place together may be due to the practice of drawing a phylogenetic tree on a map—with the implication that the tree has "grown" by dispersal as it were, from its root to the tip of its branch. . . . But I see no reason to assume that dispersal (migration) is a necessary or even common adjunct of the splitting (vicariance) of ancestral species. . . . I suggest, therefore, that for a given group the distribution of ancestral species can be estimated best by adding the descendant distributions. (555–56)

and

In summary, I reject as aprioristic all "clues" or "rules" used to resolve centers of origin and dispersal without reference to general patters of vicariance and sympatry (Croizat et al. 1974). . . . Unencumbered by aprioristic dispersal, historical biogeography is the discovery and interpretation, with reference to causal geographic factors, of the vicariance shown by the monophyletic groups resolved by phylogenetic ("cladistic") systematics." (557)

Here Nelson laid out the basic differences in the method of vicariance biogeography and those that came before. It was different from the phylogenetic biogeography method of Hennig and

Brundin, because it included the repeating patterns of Croizat, but different from Croizat's panbiogeography method in including phylogenies and monophyletic groups. Nelson had something new; however, he did not have any examples.

Donn Rosen (1975) and Norman Platnick (1976) used Nelson's method in respective studies of fishes and spiders. Both authors mention "testing" the hypotheses of biogeographic history against known geologic events, and both allow for dispersal if a certain distribution fails the test. Rosen included a clear description of the method; Platnick's presentation was easy to understand, but still the ideas were not fully formed. It took a collaboration by NORMAN I. PLATNICK and GARETH NELSON (1978)—"A Method of Analysis for Historical Biogeography" (paper 42)—to formulate the ideas in detail. In this featured paper, they departed from earlier work where dispersal is allowed and stated that the only testable hypotheses are those related to vicariance, and that because dispersal cannot be tested it is therefore ad hoc and cannot be invoked as an explanation. Cladograms were converted into area cladograms, which could be tested by comparing them against area cladograms from other groups and/or against cladograms reflecting the known geologic history of the areas in question. In 1978 and 1979, DONN E. ROSEN published what became one of the most analyzed data sets in historical biogeography, poeciliid fishes from the uplands of Guatemala. Rosen's (1978) featured paper, "Vicariant Patterns and Historical Explanations in Biogeography" (paper 43), was the first large-scale application of vicariance biogeography and as such is worthwhile examining in more detail.

In 1979, the American Museum of Natural History hosted a symposium called "Vicariance Biogeography: A Critique." A volume with the same name, containing papers from the symposium was published two years later (Nelson and Rosen 1981). Although this book is not often cited, it contains some important points in regard to methods of historical biogeography, and two of these are especially relevant to this discussion. Chapter 3, of that book contains an article by Brundin (1981), that contrasted Croizat's method with Hennig's; Slater and Platnick each replied to Brundin, and then Brundin rejoined. Their pointed exchange highlights the difference between phylogenetic biogeography and the direction that vicariance biogeography had taken. As Brundin stated in his rejoinder:

Dr. Platnick's criticism of my paper demonstrates that he is an adherent of the extreme type of vicariance biogeography in which there is no room for the study of dispersal, the existence of the deviation rule is denied, the application of the progression rule rejected, and the goal of the biologist is to discover how the geographical areas inhabited by endemic groups may be interrelated. Hence it is no wonder that there are several points of disagreement." (1981: 152)

He concluded:

Platnick's final and general charge—that I have used "a priori rules" to explain the history of groups—also lacks support. Indeed the rules of deviation and progression have not been simply applied but referred to so often because my own material is a perpetual reminder of the consequences of their action. Is not the blunt denial of their existence the expression of an aprioristic approach?" (158)

Thus, the break with the Hennig–Brundin phylogenetic method was complete. We next turn to Nelson's summary paper in the same volume (1981: 525) where he made a point often overlooked by some users of the vicariance biogeography method.

Vicariance biogeography is not an attempt to explain everything, or even the greater part of the geographical distribution of plants and animals, for its focus is allopatric differentiation as manifested by the phenomenon of endemism. It begins by asking. . . . Are areas of endemism interrelated among themselves in a way analogous to the interrelationships of the species of a certain group of organisms?

This statement makes it clear that Nelson does not think all patterns are the result of vicariance, but rather that these are the only ones that can be tested.

In 1982, Croizat severed his connections

with the method of vicariance biogeography and repudiated the use of phylogeny with his track method—a seemingly abrupt change of opinion that startled many biogeographers but not botanists. In this work Croizat said:

I would have pointed out . . . that [Nelson] was lightly taking on himself a particularly heavy burden of responsibility in trying to synthesize and extend the contributions of authors like Hennig and Croizat (Popper comes in strictly as window-dressing), who, to start with, have found their work incompatible. . . . Hennig is a dispersalist, who true to the style of his sect systematically overlooks Croizat as inconvenient and unanswerable; Croizat is a panbiogeographer whose thinking and work . . . is beyond comparison richer than the glittering generalities . . . constituting the contributions of Hennig. I do, of course reject as unwarranted Nelson's affirmation that the contributions of Hennig and Croizat are substantially similar. (1982: 295)

and

Naturally, I deeply resent that my life-work, Panbiogeography, has been dragged in with Hennigism to the very extent of publicly losing its identity under the improper designation of "Vicariance Biogeography." (296)

Croizat continues with an attempt to characterize differences between "vicariance," "vicariism," and "vicariance biogeography," terms which he considered to be hopelessly mixed up.

So, the proponents of vicariance biogeography rejected the phylogenetic method, and Croizat rejected the vicariance biogeography method. Bereft of either of its parent methods, vicariance biogeography did not disappear, but instead took on a life of its own and emerged from the ashes like a Phoenix. At that point it was a new method, distinct from either the phylogenetic or panbiogeographic methods, developed in detail by Nelson and Platnick in their book *Vicariance Biogeography* (1981). The text is somewhat difficult to read, but chapter 7 gives a thorough description of the "component analysis" first mentioned by Nelson in 1974, and chapter 8 summarizes the biogeographic results and makes clear their position on dispersal. Two "more general" explanations of the method were published by Nelson and Platnick (1980, 1984), and these provide useful summaries of their ideas. Other noteworthy publications that explain and sometimes alter the method include Wiley (1980, 1981, 1988), Humphries (1981), Wagner and Funk (1995), Crisci and Morrone (1995), and Crisci, Katinas, and Posadas (2000). Although vicariance biogeography became better known, the other methods have not vanished and they continue to have proponents. Perhaps following the lead of Croizat, the proponents of panbiogeography have been the most vocal (i.e., Craw, 1982, 1989; Craw, Grehan, and Heads 1999, reviewed in Mayden 1991).

Vicariance biogeography has continued to metamorphose. For one thing, the title was never the best one for the method (as was so dramatically explained in Croizat 1982) and it was replaced with "Cladistic Biogeography" by Humphries and Parenti (1986, 1999) whose books provide an excellent explanation and summary of the method as it is currently practiced. Nelson continued to develop cladistic biogeography, and he and his collaborators developed a method, called *three-item statement analysis*, that seeks to summarize most parsimoniously data which are coded as suites of informative, three-area cladograms (Nelson and Platnick 1991; Nelson and Ladiges 1991, 1995). This method is an outgrowth of the discussion on components in Nelson and Platnick (1981), but its roots can be found in Nelson's 1974 paper.

A natural outgrowth of cladistic and phylogenetic biogeography is the quest to determine the ancestral area for the groups under study (Nelson 1969). Building on the basic optimization technique often used in phylogenetic analysis (Farris 1970), Bremer (1992, 1995) and Ronquist (1994) have both developed methods based on the assumptions that areas that are positionally more basal in a cladogram, and areas that are represented on numerous branches of a cladogram are more likely parts of the ancestral area.

Naturally in this age of computers there have been several attempts to either appropri-ate existing programs for biogeographic purposes or to develop new ones to implement the explicit methods. In the former category, Brooks (1981, 1990) developed a method for reducing a tree to a data matrix that could then be analyzed using any tree-generating program; results were used for co-evolutionary (1981) and biogeographic studies (Brooks 1985; Brooks and McLennan 1991, 1993). In addition, many studies have used the technique of optimizing distributions on area cladograms (numerous examples in Wager and Funk 1995). In the latter category, programs have been developed to implement methods, notably by Page (1989), who used the basic ideas of component analysis developed by Nelson and Platnick in a program called COMPONENT that compares patterns (latest version 2.0-Page 1993), and by Nelson and Ladiges, who have developed programs to implement the three-item analysis method (TAS in 1991 and TASS in 1995).

Attempts to implement a method in biogeography began in the mid-1800s with "bands of affinities" and moved through evolutionary biogeography, phylogenetic biogeography, panbiogeography, and cladistic (vicariance) biogeography. These methods have met with varying degrees of success; in fact, all historical biogeography methods have a mixed record of answering the questions that intrigue us all. Perhaps this is because we continue to look for simple answers for some very complex patterns. One must not forget that areas, unlike most taxa, can have multiple and complex histories. The movement of organisms across the Earth's surface has not been under the same constraints as the evolution and diversification of species. As an example, one immediately thinks of such areas as India and Antarctica whose histories involve a complex array of geologic and climatic changes. While it is true that these events influenced the distribution of the different lineages that were present at the time, it is also true that different lineages were present at different periods, contributing to the complex legacies of the historical event. The complexity of the questions we ask in biogeography must be taken into account when we attempt to formulate the answers.

By most accounts, the cladistic method is the most rigorous, but as Nelson (1981) points out, the model of dispersal of a biota followed by vicariance cannot explain all distributions. If we have learned anything from our studies in historical biogeography it is that biogeographic reconstructions are influenced by the questions we ask and by the amount of information we have available. Just about every scenario proposed is true for some group of taxa. The difficult part is finding the method that is appropriate for the group or groups of interest. One thing is certain, we have a better chance of understanding biogeographic patterns when we have a phylogeny because it allows us to select a method that works best for the group. As Brundin (1966: 64) put it, "The study of phylogenetic relationships is time-consuming and difficult work, but there is no way around, and we are short of time. . . . Let us do it, with enthusiasm and humbleness and freedom from preconception."

From *Transantarctic Relationships and Their Significance, as Evidenced by Chironomid Midges*

Lars Brundin

II. REVIEW OF THE PROBLEM
OF TRANSANTARCTIC RELATIONSHIPS

Among the problems raised by the distribution of plants and animals in the southern hemisphere there is none which takes a more central position and is more stimulating to the imagination than the problem of transantarctic relationships. We have before us the broken circle of southern lands—southern South America, South Africa, Tasmania-Australia, New Zealand—separated by wide stretches of ocean, but populated by a biota containing numerous groups whose strongly disjunct elements are more closely related to one another than to any other group; and in the centre of the scene we are faced with the dormant Antarctic Continent, hiding its secrets beneath a mighty ice cap. The questions come thronging in upon our mind as to the meaning of these disjunct distribution patterns and the biogeographic and evolutionary role played by the Antarctic Continent.

JOSEPH HOOKER

More than 110 years have passed since the eminent botanist Sir Joseph Hooker outlined the above problem. The profound reality of the problem of transantarctic relationships was emphasized by the circumstance that Hooker could rely upon a wide first hand experience of the subantarctic and south temperate floras. In the years 1839–43, as an assistant surgeon, he took part in the Antarctic voyage of the two ships *Erebus* and *Terror* under the command of Captain Sir James Clark Ross. During that lengthy voyage, which in reality consisted of three different expeditions to the Antarctic areas with breaks between in lower southern latitudes, Hooker was enabled to study the floras of Tasmania, New South Wales, New Zealand (North Island), Auckland Islands, Campbell Island, Kerguelen, Cockburn Island, Falkland Islands, and Fuegia (Hermite Islands west of Cape Horn). The botanical results were presented in *Flora Antarctica* (1844–47), *Flora Novae-Zelandiae* (1853–55), and *Flora Tasmaniae* (1855–60). Although mainly systematic accounts these works have, however, introductory essays which are essentially phytogeographical.

Collating his results Hooker realized that he had discovered a circumpolar Antarctic flora. He was also deeply impressed by the sight of fossilized gymnosperm wood (*Araucaria, Podocarpus*) on the barren Kerguelen, and already in the second part of his *Flora Antarctica* (1847, pp. 210, 368) he made some allusions as to the possibility of more or less continuous transantarctic land connections between now isolated southern lands. That problem was treated more broadly in the Introductory Essay to *Flora Novae-Zelandiae* (1853). Since this essay has a classical position in the annals of biogeography, it seems appropriate to cite some of the relevant statements and conclusions.

Having stressed that the distribution of species has been effected by natural causes, but that these are not necessarily the same as those to which they are now exposed, Hooker makes the following general statement:

"Of all the branches of Botany there is none whose elucidation demands so much preparatory study, or so extensive an acquaintance with plants and their affinities, as that of their geographical distribution. Nothing is easier than to

46

explain away all obscure phenomena of dispersion by several speculations on the origin of species, so plausible that the superficial naturalist may accept any of them; and to test their soundness demands a comprehensive knowledge of facts, which moreover run great risk of distortion in the hands of those who do not know the value of the evidence they afford." (P. xix.)

The elucidation of the New Zealand flora brings up many instances difficult to account for, but all bearing upon the same point. They are arranged by Hooker as follows:

"1. Seventy-seven plants are common to the three great south temperate masses of land, Tasmania, New Zealand, and South America."

"2. Comparatively few of these are universally distributed species, the greater part being peculiar to the south temperate zone."

"3. There are upwards of 100 genera, subgenera, or other well-marked groups of plants entirely or nearly confined to New Zealand, Australia, and extra-tropical South America. These are represented by one or more species in two or more of these countries, and they thus effect a botanical relationship or affinity between them all, which every botanist appreciates."

"4. These three peculiarities are shared by all the islands in the south temperate zone (including even Tristan d'Acunha, though placed so close to Africa), between which islands the transportation of seeds is even more unlikely than between the larger masses of land."

"5. The plants of the Antarctic islands, which are equally natives of New Zealand, Tasmania, and Australia, are almost invariably found only on the lofty mountains of these countries." (Pp. xix–xx.)

Hooker finds, not least because of his own experience of the biology of the actual plants, that transoceanic migration is highly improbable as an explanation of the strongly discontinuous distribution patterns:

"To return to the distribution of existing species, I cannot think that those who, arguing for unlimited powers of migration in plants, think existing means ample for ubiquitous dispersion, sufficiently appreciate the difficulties in the way of the necessary transport. During my voyages amongst the Antarctic islands, I was led, by the constant recurrence of familiar plants in the most inaccessible spots, to reflect much on the subject of their possible transport; and the conviction was soon forced upon me, that, putting aside the almost insuperable obstacles to trans-oceanic migration between such islands as Fuegia and Kerguelen's Land, for instance (which have plants in common, not found elsewhere), there were such peculiarities in the plants so circumstanced, as rendered many of them the least likely of all to have availed themselves of what possible chances of transport there may have been. As species they were either not so abundant in individuals, or not prolific enough to have been the first to offer themselves for chance transport, or their seeds presented no facilities for migration, or were singularly perishable from feeble vitality, soft or brittle integuments, the presence of oil that soon became rancid, or from having a fleshy albumen that quickly decayed. Added to the fact that of all the plants in the respective floras of the Antarctic islands, those common to any two of them were the most unlikely of all to migrate, and that there were plenty of species possessing unusual facilities, which had not availed themselves of them, there was another important point, namely, the little chance there was of the seeds growing at all, after transport. Though thousands of seeds are annually shed in those bleak regions, few indeed vegetate, and of these fewer still arrive at maturity. There is no annual plant in Kerguelen's Land, and seedlings are extremely rare there; the seeds, if not eaten by birds, either rot on the ground or are washed away; and the conclusion is evident, that if such mortality attends them in their own island, the chances must be small indeed for a solitary individual, after being transported perhaps thousands of miles, to some spot where the available soil is pre-occupied." (Pp. xx–xix.)

"It was with these conclusions before me, that I was led to speculate on the possibility of the plants of the Southern Ocean being the remains of a flora that had once spread over a larger and more continuous tract of land than now exists in that ocean; and that the peculiar Antarctic genera and species may be vestiges of a flora characterized by the predominance of plants which are now scattered throughout the southern islands." (P. xxi.)

The last chapter of the Introductory Essay deals with the physiognomy and affinities of the New Zealand flora. Of special interest here is the discussion of plants which are common to New Zealand and other countries. About one-third of the flora belongs to this category, while more than two-thirds, or 26 genera and 507 species, are considered by Hooker to be "absolutely peculiar or endemic". The above-mentioned third of the flora is divided into five groups, for illustrating the relations of the plants to those of other countries:

"1. 193 species, or nearly one-fourth of the whole, are Australian.

2. 89 species, or nearly one-eight of the whole, are South American.

3. 77 species, or nearly one-tenth of the whole, are common to both the above.

4. 60 species, or nearly one-twelfth of the whole, are European.

5. 50 species, or nearly one-sixteenth of the whole, are Antarctic Islands', Fuegian, etc." (P. xxx.)

1. *Those of Australian affinity.* — Hooker's discussion of these plants has no direct bearing on the problem of transantarctic relationships, but is partly quoted here because it touches upon some remarkable negative facts illustrating the limitations of long distance dispersal.

"If the number of plants common to Australia and New Zealand is great, and quite unaccountable for by transport, the absence of certain very extensive groups of the former country is still more incompatible with the theory of extensive migration by oceanic or aerial currents. This absence is most conspicuous in the case of *Eucalypti*, and almost every other genus of *Myrtaceae*, of the whole immense genus of *Acacia*, and of its numerous Australian congeners, with the single exception of *Clianthus*, of which there are but two known species, one in Australia, and the other in New Zealand and Norfolk Island."

"The rarity of *Proteaceae*, *Rutaceae*, and *Stylideae*, and the absence of *Casuarina* and *Callitris*, of any *Goodeniae* but *G. littoralis* (equally found in South America), of *Tremandreae*, *Dilleniaceae*, and of various genera of *Monocotyledones*, admit of no explanation, consistent with migration over water having introduced more than a very few of the plants common to these tracts of land. Considering that *Eucalypti* form the most prevalent forest feature over the greater part of South and East Australia, rivalled by the *Leguminosae* alone, and that both these Orders (the latter especially) are admirably adapted constitutionally for transport, and that the species are not particularly local or scarce, and grow well wherever sown, the fact of their absence from New Zealand cannot be too strongly pressed on the attention of the botanical geographer, for it is the main cause of the difference between the floras of these two great masses of land being much greater than that between any two equally large contiguous ones on the face of the globe. If no theory of transport will account for these facts, still less will any of variation; for of the three genera of *Leguminosae* which do inhabit New Zealand, none favour such a theory; one, *Clianthus*, I have just mentioned; the second, *Edwardsia*, consists of one tree, identical with a Juan Fernandez and Chilian one, and unknown in New Holland; and the third genus (*Carmichaelia*) is quite peculiar, and consists of a few species feebly allied to some New Holland plants, but exceedingly different in structure from any of that extensive Natural Order." (P. xxxi.)

"2. *Species of South American affinity.*—The South American species in New Zealand amount to 89, or one-eighth: of these some are absolutely peculiar to the two countries, as *Myosurus aristatus*, two species of *Coriaria*, *Edwardsia grandiflora*, *Haloragis alata*, *Hydrocotyle Americana*, and *Veronica elliptica*. Of these the *Edwardsia* is by far the most striking case, from the size of the tree: it appears to have a much wider range in New Zealand than in Chili, and supposing it to have been transported between these countries, it is difficult to say which was the parent one; its affinities would, however, incline us to consider it amongst the aborigines of the former. It is by representative genera and species that the affinity of the New Zealand and South American floras is best shown, and this most conspicuously by *Fuschsia* and *Calceolaria*, two most remarkable genera, confined to these two countries, but by far the most abundant to the west of the Andes. Here again the amount of affinity is differently displayed by each; of the Calceolarias one is so closely allied to an American species, that I doubt the propriety of keeping them separate, while the other appears a very distinct species; the Fuchsias are both extremely peculiar, one of them being the only species that has no petals. Altogether there are 76 genera common to New Zealand and South America, and 17 of these are not found in Australia, or elsewhere in the Old World. It is curious that none of the latter belong to those peculiarly Arctic and north temperate genera mentioned in the note to p. xxiv, except *Caltha*, to a southern form of which, however, the New Zealand species belongs."

"3. *Plants common to New Zealand, Australia, and South America.*—Of the 77 plants common to these three countries, which include one-tenth of the flora of New Zealand, the majority are Grasses, 10; *Cyperaceae*, 7; moisture-loving Monocotyledons, 9; *Monochlamydeae*, 8; *Umbelliferae* and *Compositae* each 4; and fully 50 of the whole number are also found in Europe, and do not indicate any peculiar affinity between these three southern masses of land: of those that are not European, some are Antarctic plants found in mountainous districts of Australia and Tasmania, as *Oxalis Magellanica*. Of genera and species which, from their near affinity with one another, and marked distinction from any others, may be said to be represented in all three countries, the majority are Antarctic, and will be noticed under the fifth head." (Pp. xxxi–xxxii.)

We skip the fourth head, dealing with "European" plants in New Zealand, and go to the fifth, with the remark that the "Antarctic flora" in the sense of Hooker includes that of Tierra del Fuego, the Falklands, "with different islands east and south of them", Tristan d'Acunha, St. Paul's, Amsterdam and Kerguelen's Land, Lord Auckland's, Campbell's, and "other islands south and east of New Zealand".

"5. *Antarctic plants in New Zealand.*—Of these Antarctic plants, about 50 inhabit the mountains and southern extreme of New Zealand; a number which (as I have stated at p. 15) will probably be greatly increased by future discoveries. They may be geographically grouped as follows:—*a.* Those of general distribution, being common also to Europe, as *Callitriche, Montia, Cardamine hirsuta, Potentilla anserina, Epilobium tetragonum, Myriophyllum, Calystegia Soldanella* and *C. Sepium, Limosella,* many *Monochlamydeae,* and more *Monocotyledones.*—*b.* Those found also in Tasmania, and chiefly on its mountains, but not elsewhere; as *Oxalis Magellanica, Acaena,* some *Epilobia, Colobanthus, Scleranthus, Tillaea, Apium, Coprosma, Leptinella, Hierochloe antarctica,* etc."

"The botanical affinity between extra-tropical South America, the Antarctic islands, New Zealand, and Tasmania, is, however, much better indicated by the peculiar genera, by groups of those, or by individual species which, as it were, represent one another in two or more of these localities, and which give a peculiar botanical character to the flora of southern latitudes beyond latitude 35°."

"Of these genera, there are 50 which afford botanical characters in common, and give as decided a proof of close affinity in vegetation, as do the 50 identical species above mentioned. The most conspicuous of these genera common to all the above-named localities are, *Colobanthus, Drosera, Acaena, Gunnera, Oreomyrrhis, Leptinella, Lagenophora, Forstera, Pratia, Gaultheria, Gentiana, Euphrasia, Plantago, Drapetes, Fagus, Astelia, Juncus, Carpha, Chaetospora, Oreobolus, Uncinia, Carex,* and many Grasses, especially *Hierochloe, Alopecurus, Trisetum, Deyeuxia,* etc." (P. xxxiii.)

Hooker then gives a comparative table of plants comprising 228 species "which may be considered as representing one another (more or less remarkably) in two or all the three south temperate masses of land, *viz.* New Zealand (including Auckland and Campbell's Island), Australia (including Tasmania), and extra-tropical South America (including the Falkland Islands)". He arrives at the following general conclusion:

"Enough is here given to show that many of the peculiarities of each of the three great areas of land in the southern latitudes are representative ones, effecting a botanical relationship as strong as that which prevails throughout the lands within the Arctic and Northern Temperate zones, and which is not to be accounted for by any theory of transport or variation, but which is agreeable to the hypothesis of all being members of a once more extensive flora, which has been broken up by geological and climatic causes" (p. xxxvi).

In the Introductory Essay to *Flora Tasmaniae* (1860) Hooker compared and discussed the antiquity of the Northern and Southern floras. He underlined the presence of many signs of intimate connections between north and south, and as to the Old World he was struck with the appearance of "there being a continuous current of vegetation (if I may so fancifully express myself) from Scandinavia to Tasmania" (p. ciii). He did not believe, however, that the Southern Flora was a direct derivation of the Northern. "On the contrary, the many bands of affinity between the three southern Floras, the Antarctic, Australian, and South African, indicate that these may all have been members of one great vegetation, which may once have covered as large a southern area as the European now does a Northern. It is true that at some anterior time these two Floras may have had a common origin, but the period of this divergence antedates the creation of the principal existing generic forms of each." (Pp. civ–cv.)

In other words, Hooker felt convinced that there had been a centre of evolution in the south. We have right to use just these words, since Hooker was one of the first to accept the theory of evolution by means of natural selection, as set out by Darwin and Wallace in 1858 and 1859.

There is a vast amount of literature which has appeared on the subject after Hooker, and an exhaustive list would probably by far exceed 1000 titles. Indeed, Hooker's suggestion of the former existence of continuous land connections between the other southern continents by way of Antarctica has had a powerful effect on the imagination of scientists. But it is noteworthy that among several hundred authors which have expressed an opinion for or against Hooker, there are very few which have had any first hand experience of landscapes and biotas on more than one side of the Antarctic Continent; and most of the European and American opponents of Hooker have never visited the south temperate and subantarctic zones. It is strange, but it seems to be an open question if any of the authors discussing the problem of transantarctic relationships, except Skottsberg, has had field experience equivalent to that acquired by the pioneer himself in 1839–43.

The continuing biological exploration of the southern lands has meant a successive accumulation of new data also concerning the austral disjunctive groups, their number and distribution patterns. Since the first zoological evidence was brought forth by Rütimeyer (1867) and Hutton (1873) there has been an immense increase in our knowledge of intimate faunistic resemblances between the circumantarctic land areas. In several cases the proposed relationships will not stand a critical examination, it is true, but enough is left to enable us to state that there are numerous groups among the invertebrates whose isolated subgroups doubtless are more nearly related to each other than to any other group of the world fauna. It is also clear that most of these small animals are typical inhabitants of dense temperate rain forests and cool, fast-running streams. But there is still much to do, especially in temperate South America, and from my own experience I feel convinced that the known cases of transantarctic relationships among the invertebrate animals only form a fraction of the number really existing.

For several reasons I must refrain from giving a detailed review of the literature after Hooker and a survey of the plant and animal groups believed to display transantarctic relationships. The reader is referred to the excellent reviews given by Sarasin (1925), Harrison (1928), Wittman (1934), Du Rietz (1940), Thorne (1963), and del Corro (1964).

THE OPPONENTS OF HOOKER

Hooker had often discussed the biogeographical problems with his friend Charles Darwin who shared Hooker's idea of a Tertiary Antarctic Continent clothed with vegetation. However, touching upon the problem of austral distribution patterns Darwin (*The Origin of Species*, 1859; 1964, pp. 381–382) preferred another explanation than Hooker:

"I am inclined to look in the southern, as in the northern hemisphere, to a former and warmer period, before the commencement of the Glacial period, when the antarctic lands, now covered with ice, supported a highly peculiar and isolated flora. I suspect that before this flora was exterminated by the Glacial epoch, a few forms were widely dispersed to various points of the southern hemisphere by occasional means of transport (across the sea), and by the aid, as halting-places, of existing and now sunken islands, and perhaps at the commencement of the Glacial period, by icebergs."

Darwin was thus so little impressed by the circumantarctic occurrences of supposed elements of a Tertiary Antarctic flora that he, contrary to Hooker, did regard chance dispersal as an adequate explanation. He had, however, only a limited experience of his own to reply upon here (his expression "a few forms" is an obvious underrating of the scope of the problem), and it is fair to underline that the existence of the problem was demonstrated by Hooker 11 years after the voyage of the *Beagle* in 1831–36. But to Darwin the decisive point was, I think, his firm belief in the permanence of the main geographical features.

At an early stage there were thus divided opinions about the meaning of the transantarctic relationships and the history of the elements involved; and, as in all other cases of disjunctive occurrences separated by wide stretches of ocean, the biologists had to deal with the two usual main alternatives: long distance dispersal, or former land connections.

If we try to analyse why Hooker's idea of an Antarctic migration route "has been opposed as strongly as it has been supported" (Simpson 1940a), by participation on either side of many of the more prominent biologists of 11 decades, it will be apparent that the main reason for the conflicting opinions is the different approach of the two main camps. The opponents of Hooker, like Darwin, Wallace, Thiselton-Dyer, Engler, Matthew, and Simpson, have been firm believers in the permanence of the continents and oceans. Dana in his *Manual of Geology* (1863) and other orthodox geologists were the guarantors. But it is evident that the aprioristic confinement to a certain geological hypothesis means a great risk to a biologist trying to interpret the history in time and space of living beings, since on that assumption it will be very difficult to escape from using the geological hypothesis, more or less consciously, as a basic argument. The biological data have thus to fit a geological *a priori* pattern. As a consequence great efforts have been laid down

L. BRUNDIN, *Transantarctic relationships*   K. V. A. Handl. 11: 1

in order to demonstrate that the known distribution patterns can be satisfactorily explained on the basis of Dana's geological hypothesis, meaning that the assumption of great changes of the world map is an unnecessary complication. The confidence in the traffic capacity and climatic possibilities of an intermittent Beringian highway has been nearly unlimited and is comparable only with the reliance put on long distance dispersal as a major factor in the history of the biotas. The results of extensive investigations on means of dispersal, dispersal capacity, and on winds and ocean currents as agents in the dispersal of different organisms have been conceived as proofs of the soundness of the basic ideas.

There is something negative, sterile, and superficial involved in the above approach which offends a critical mind. Studying the literature one finds that the "Wallacean" biogeographer over and over again gets faced with puzzling cases of disjunction putting him into trouble. If a conjecture is offered, reference is often vaguely made to the former wide extension of groups which are now discontinuous. But that is nothing else than to suggest the principal problem. It is rarely realized that the reconstruction of the history of a group by a proper analysis of the mutual relationships of more or less isolated subgroups is the main task of the potential biogeographer. Other "baffling" cases of disjunctive distribution are often interpreted as the result of "some peculiar and exceptional means of dispersal" (Wallace). In connection with this it is argued, not least by recent writers, that even quite unbelievable things may happen if there is enough time. With the aid of some mathematics it is easy to show that there has been time enough for the realization of many unbelievable events. Several troubled biogeographers have found consolation and relief in the thought of a raft with a sufficient store of food and water put at the disposition of a pregnant female at the right moment. To me such an attitude is comparable with a confession of failure.

Reading the classic biogeographical works of Wallace (1876, 1880) one is struck by his daring approach to and sweeping generalizations of the great problems connected with the history of life since the middle Mesozoic. But such an attitude has to be conceived as fairly natural at a time so close to the breakthrough of the theory of evolution, at the coming into being of which Wallace himself had been one of the two giants. Much more striking is, after all, the circumstance that Wallace's manner of reasoning and treatment of the problems of causal biogeography still are so generally copied.

The general picture of the main trends in the history of the terrestrial biotas since the middle Mesozoic, as created by the opponents of Hooker, is a consequence of the acceptance of the permanence of the world map. The role of the northern continents as the great centres of origin, and the southern lands as receivers of a primarily northern biota, and, moreover, the wholly secondary importance of the transantarctic biotic connections, is stressed in the following statement of Wallace (1876; 1962, II, p. 159):

"If our views of the origin of the several regions are correct, it is clear that no mere binary division—into north and south, or into east and west—can be altogether satisfactory, since at the dawn of the Tertiary period we still find our six regions, or what may be termed the rudiments of them, already established. The north and south division truly represents the fact, that the great northern continents are the seat and birth-place of all the higher forms of life, while the southern continents have derived the greater part, if not the whole, of their vertebrate fauna from the north; but it implies the erroneous conclusion, that the chief southern lands—Australia and South America—are more closely related to each other than to the northern continent. The fact, however, is that the fauna of each has been derived, independently, and perhaps at very different times, from the north, with which they therefore have a true genetic relation; while any intercommunion between themselves has been comparatively recent and superficial, and has in no way modified the great features of animal life in each. The east and west division, represents—according to our views—a more fundamental diversity; since we find the northern continent itself so divided in the earliest Eocene, and even in Cretaceous times; while we have the strongest proof that South America was peopled from the Nearctic, and Australia and Africa from the Palaearctic region; hence, the Eastern and Western Hemispheres are the two great branches of the tree of life of our globe. But this division, taken by itself, would obscure the facts—firstly, of the close relation and parallelism of the Nearctic and Palaearctic regions, not only now but as far back as we can clearly trace them in the past; and, secondly, of the existing radical diversity of the Australian region from the rest of the Eastern Hemisphere."

The above conclusions of Wallace were arrived at on the basis of vertebrate distribution patterns, but they were widely accepted also by botanists. Some among the latter, realizing the difficulty of explaining the striking relationships between the austral floras by long distance dispersal, went still farther than Wallace and took their refuge in what Schröter (1913, 1934) called "the monoboreal relic hypothesis". The austral groups were looked upon as primitive groups driven southward through development in the northern continents of new, aggressive groups and then remaining definitely isolated: the southern lands are centres of preservation of old relic faunas and floras. In 1910 these lines of thought were formulated by the botanist Thiselton-Dyer in the following way: "The extraordinary congestion in species of the peninsulas of the Old World points to the long-continued action of a migration southwards. Each is in fact a *cul-de-sac* into which they have poured and from which there is no escape."

One of the most forceful defenders of the monoboreal relic hypothesis was the wellknown paleontologist W. D. Matthew. In his work *Climate and Evolution* (1915) he made an apparently strong case of his knowledge of the relationships and history of different mammalian groups and suggested that the southern groups have split off from the main line of evolution at an earlier date than the northern and that old and once widespread groups are likely to survive longer in the isolated land areas in the south.

Concealed behind the idea of a uni-directional stream of life from the north are not only aprioristic conceptions of the world geology, but also typological thinking and a general neglect of the principles of phylogenetic evolution. I refer to the foregoing chapter and to the discussion in the last chapter of the present paper.

In his look upon the major biogeographical problems and the history of the earth G. G. Simpson has always taken a position closely similar to those advocated by Wallace and Matthew. His interpretation of the problem of transantarctic relationships, set out in a special paper on the subject in 1940, is a confirmation of this. Simpson (l.c.) raises the following lucid question:

"Are there any known facts of the distribution of recent and fossil plants and animals that make it necessary or preferable to suppose that Antarctica was involved in a land-migration route between the other southern lands, or can all these facts be adequately and more probably explained by other routes for which there is more direct evidence?"

On the basis of a discussion of some examples of supposedly crucial evidence taken from each class among the vertebrates (galaxiid fishes, leptodactylid toads, meiolanian turtles, ratite birds, marsupials) Simpson arrives at the conclusion (l.c., p. 767) that a "definite answer" can now be made to his cited question summarizing the Antarctic migration problem: "There is no known biotic fact that demands an Antarctic land-migration route for its explanation and there is none that is more simply explained by that hypothesis than by any other."

But mere simplicity cannot be accepted as a proof of the correctness of the suggested explanation, and in spite of Simpson's declaration it was of course still an open question if the explanation performed by Wallace and his followers was more adequate and more probable than that of Hooker. Moreover, many opponents of Hooker have not realized that the actuality of his hypothesis would persist quite independently of the possible negative evidence delivered by some groups among the vertebrates. It was a serious misjudgment of Simpson to end his 1940 paper with the statement that "the Antarctic migration route hypothesis remains simply a hypothesis with no proper place in present scientific theory".

The adherents of Hooker

It was pointed out above that the main reason for the conflicting opinions is the different approach of the two camps. Starting from the conviction of the permanency of the main geographical features the opponents of Hooker generally looked upon the occurrence of related groups at the southern extremities of the widely separated southern lands as the result of successive independent dispersal over land from the north, often combined with secondary long distance dispersal in the south via intervening land (Antarctica, different islands). We can also trace a widespread tendency to depreciate the number of the above austral groups and the closeness of the relationships between the subgroups.

The adherents of Hooker, on the other hand, have been strongly impressed by the close relationships

L. BRUNDIN, *Transantarctic relationships*   K. V. A. Handl. 11: 1

displayed by the subgroups of the circumantarctic and amphiantarctic groups, by the many indications of them all representing persisting splinters of a once much larger austral biota, by the often striking primitiveness of many of the involved groups, and by several signs of dispersal northwards from an austral centre. This aspect and interpretation of the situation has mostly been arrived at through the direct study of representative groups, independently of geological hypotheses. Considering the widely differing opinions among the geologists as to the history of the earth since the end of the Paleozoic, the free, unfettered approach of Hooker and many of his followers must be considered sounder in principle than that of their opponents. In connection with their investigations of relationships and distribution patterns the Hookerians have found so many indications of orderly dispersal and independent extensive evolution in the south, and such weak parallelism between widespread disjunctive occurrence and fitness for long distance dispersal, that the existence of former direct connections between Antarctica and the surrounding southern lands seemed to be the most probable explanation. It may be added that these views were held perhaps predominantly by students (botanists, invertebrate zoologists) who did not consider the possible involvement of vertebrates in the discussed problems.

Land-bridge makers have always been looked upon with suspicion in certain quarters and the literature abounds in critical and more or less sarcastical comments upon indiscriminate bridge constructions which, if laid out on the same map, would indeed nearly fill the present oceans. It goes without saying that much excess has been committed, but this circumstance must of course not be allowed to conceal the fact that most land-bridge makers have arrived at their conclusions on purely biological grounds, as a consequence of objective efforts to interpret special patterns of relationship and distribution which did not seem to give any other choice.

No one has done more than Carl Skottsberg in order to penetrate the problem of transantarctic relationships. Basing his conclusions on wide field experience and extensive taxonomic studies of supposed Antarctic floral elements Skottsberg has enriched the debate with an authorship which extended over six decades and was highly appreciated even among his opponents. His analyses and discussions of the farflung outliers of an old Antarctic flora in different parts of the Pacific and elsewhere (1925, 1926, 1934a, b, c, 1936, etc.) and his extensive study of the Juan Fernandez flora (1956) have deepened the perspective in an essential way. Skottsberg was convinced that the Arcto-Tertiary flora of the northern hemisphere had its counterpart in an Antarcto-Tertiary flora and that both have played a prominent role in the history of life (1940). Already in 1915 (p. 141) he described the role of Antarctica as an old centre of evolution "from which animals and plants wandered north", thanks to the existence of land connections with South America and Australia–New Zealand. As to the possibility of a connection (via Kerguelen) between Antarctica and South Africa Skottsberg did remark: "So much seems to be certain that if this bridge did exist, separation took place early, long before the other Antarctic connections were broken off" (1956, p. 391). From his study of the composition and relationships of the Juan Fernandez flora he was led to the conclusion that it is a remnant of a continental austral biota formerly inhabiting a fore-land joining southern Chile and extending further northwards, parallel with South America, and not sunk until the newborn volcanic islands Masafuera and Masatierra, now reduced to ruins, had become a refuge for the ancient continental fauna and flora (1956, p. 379). Contrary to such phytogeographers as Wulff (1950) and Good (1953) Skottsberg never adopted Wegener-Du Toit's hypothesis of continental drift, mostly because it apparently did not help to solve the problem of oceanic island biotas.

The concept of Antarctica as an integrant part of an earlier unitary extensive centre of evolution and dispersal, forming an austral counterpart to the Laurasian centre, gained strong support through the investigations of austral fossil conifers performed by Florin (1940). On a broad basis of data he arrived at the conclusion "that the world's conifer vegetation has, from the Permian onwards, become steadily differentiated into two large separate phytogeographical dominions" kept apart by the Tethys sea. This view was further documented in his recent extensive review of the distribution of conifer and taxad genera in time and space (1963):

"Under the influence of great changes in the surface features of the earth and particularly in the climatic conditions, some interchanges have, it is true, taken place, but northern genera appear only rarely to have overstepped the equator to any considerable extent, while southern genera have done so during a geologically late period especially in the Malaysian-continental East Asian region. ... The degree of uniformity supposed to have characterized the middle Mesozoic floras in general has been overestimated. It did not amount to a total absence of climatic zonation, but merely to a general rise in temperature allowing vegetation of an ecological type now confined to regions much farther removed from the poles, to exist in high latitudes. All arguments put forward in favour of a northern origin of the whole southern conifer floras must fall to the ground in the light of our present knowledge of the distribution and affinities of their respective fossil members. Although the araucarias may have been an exception to the general rule, this has not yet been incontrovertibly proved. The supposed northern origin of the present southern assemblage of conifers is, finally, contradicted by the fact that most of its genera are entirely absent in northern lands, which they would be unlikely to be if that supposition were true." (Pp. 280–281.)

The results reached by Florin in his classical paper of 1940 have been confirmed by recent investigations of microfossils, especially of podocarps, from different parts of the southern hemisphere (Couper 1960, and others). Summarizing, it seems justified to state that the assembled knowledge of the history in time and space of the conifers stands out as the most important of the arguments which are at the disposal of the Hookerian camp. Next come tha data provided by the southern beeches.

The genus *Nothofagus*, comprising 40-odd species and now confined to temperate South America, south-eastern Australia, Tasmania, New Zealand, and mountain ridges in New Caledonia and New Guinea, is of special documentary importance because it forms a strongly disjunct austral group which is doubtless monophyletic and probably forms the sister group of a strictly boreal assemblage, the genus *Fagus*. The structures of the pollen grains do indicate that *Nothofagus*, living and fossil, is composed of 3 species groups, the *brassii* group, the *menziesii* group, and the *fusca* group; and it is an important point that all three groups have been or are still involved in transantarctic relationships. Macrofossils and the three pollen types are known from Upper Cretaceous-Lower Tertiary strata in the Graham Land section of West Antarctica. The mesothermal *brassii* group is now confined to New Guinea (16 species!) and New Caledonia. In New Zealand members of that group belonged to the dominant forest trees during most of the Tertiary, but they disappeared abruptly from the fossil record with the onset of the first Pleistocene glaciation. The dynamic history of the genus is further indicated by the fact that the southern half of Australia was widely covered with *Nothofagus* (and podocarp) forests during Eocene-Pliocene time. The absence of *Nothofagus* pollen in all investigated strata of northern South America, India, and Borneo, and its occurrence only during Pliocene-Pleistocene in the strata of New Guinea is conceived as indicating late dispersal from the south to the north in the latter sector and an austral origin of the genus (cf. Couper 1960 and Cranwell 1963). Cranwell (l.c., p. 388) alludes to the parallelism with the conifers and points out that *"Nothofagus* truly represents the angiosperm manifestation of Southern Hemisphere segregations".

Both Florin and Couper concluded that Antarctica once was much larger than at present and assumed land connections or at least proximity in the past between South America, Antarctica, and Australasia. They even agreed, for reasons which are not very clear, that Antarctica has functioned as a route of migration between the Australasian and South American regions rather than as a centre of evolution. In his recent paper of 1963 Florin stressed that fold-mountain belts along the continental margins probably have been the principal scene of the origin and diversification of the conifer and taxad genera and that the continental shields have been of secondary importance. In his fig. 68, showing the probable "main migration route across Antarctica in Mesozoic and presumably early Tertiary times", Florin has drawn a track from New Zealand via West Antarctica and the Scotia Arc to South America. Kerguelen is connected with the East Antarctic shield, but not with South Africa. Somewhat surprising is the circumstance that Tasmania and southeastern Australia mark the beginning of a track trending northwards, towards the Merauke Ridge of New Guinea, while there is no connection southwards (Florin's fig. 67).

Cranwell (l.c.) accepts the displacement hypothesis in the sense of Du Toit. She stresses that while fossil podocarps are known from South Africa, Madagascar, and Kerguelen, there is no trace of *Nothofagus* from

L. Brundin, *Transantarctic relationships*   K. V. A. Handl. 11: 1

these areas, which might mean that the southern beeches never lived there. If true, the absences may be "profoundly significant for estimating the time and nature of earlier land connections and severances in the Africa–Australasia–Antarctic–South America rhomboid" (p. 389).

The connection between bipolarity and austral disjunctive distribution was discussed in an important paper by Du Rietz (1940). He finds that there is very strong botanical evidence for former continuous land connections in the south and is convinced that the boreal and austral components of the bipolar groups once have been connected via continuous transtropical highland bridges, one American, one African, one Malaysian, and that these bridges were older than the mountain chains of the Alpine Orogen. Du Rietz shows by numerous examples that the single bridges have played a different role in the history of different bipolar plant groups.

Among the many cases of transantarctic relationships brought forward and discussed by invertebrate zoologists until 1960, I will only mention those performed by Jeannel on carabid and silphid beetles. Though he does not name the concept, Jeannel has made a fairly extensive use of the occurrence of geographical vicariism among groups of comparatively high rank. His "lignées phylétiques" are, further, based on a broad knowledge of evolutionary trends within the world fauna of the groups studied. In short, his concepts of relationships seem to be more reliable than those of many other authors, and his monographs of *Trechini* (1926–28), *Catopini* (1936), *Migadopini* (1938), and *Calosoma* (1940) are, therefore, well worth mentioning, likewise his *La genèse des faunes terrestres* (1942). Among the Coleoptera studied by him there are according to Jeannel several striking cases of primitive "paleoantarctic" groups with disjunct distribution in the south whose closest relatives turn out to be comparatively derivative groups of Laurasia. The situation indicates austral origin and diversification and early dispersal northwards of strongly progressive apomorph forms. An example of this is according to Jeannel the peculiar, strongly plesiomorph and endemic genera *Ceroglossus* of the South Andes and *Pamborus* of the Australian Alps on one hand, and the wellknown genus *Carabus* and related genera of Laurasia on the other. Jeannel believes that the ancestor of *Carabus* migrated from Australia to the Angara Continent during the end of the Jurassic when these land masses are supposed to have been connected.

Jeannel was just as much a believer in Wegener as others in Dana. This has been advanced as evidence against him; but the importance of his studies of phylogenetic relationships remains, independently of his special paleogeographic explanation.

Conclusive evidence and how to reach it

Dealing above with the cases presented by the adherents of Hooker I restricted myself to a few examples which seem to be better documented than most others. But even in these cases there is no conclusive evidence. We still do not know the real nature of the transantarctic relationships displayed, for example, by the austral conifers and beeches. As regards *Nothofagus* the significance of the structures of the pollen grains remains doubtful, and thus also the systematic structure of the much discussed *brassii*, *menziesii*, and *fusca* groups. It is in addition still an open question if *Nothofagus* is on the whole more primitive than *Fagus*.

However, have all available methods and principles of approach really been tried or sufficiently applied? The foregoing chapter on phylogenetic systematics and phylogenetic reasoning demonstrates that the answer is negative and that there is only one way out of the dilemma: to start the search for the sister group. It is of basic importance to realize that paleontology is not able to give direct information as to the phylogenetic relationships and history of a group. The reasons were given above. An analysis of the phylogenetic relationships of the recent groups by means of synapomorph characters is under all circumstances a necessary prerequisite for real advancement. But the establishment of a simple sister group relationship between, for example, one group in South America and another in New Zealand cannot be regarded as an argument for the former existence of an Antarctic centre of evolution or even an Antarctic route of migration, since in

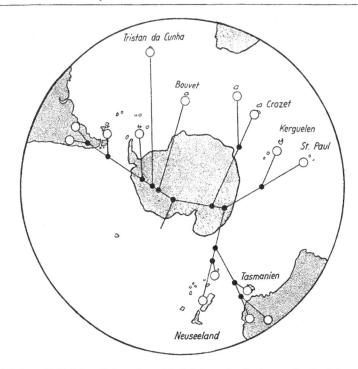

Fig. 12. Presence of phylogenetically intermediate species on islands lying on the direct connecting line between South America and Australia–New Zealand as argument for the earlier existence of direct migration routes between these areas. (From Hennig 1960, fig. 5.)

such a situation it might be argued that the bicentric occurrence of the two groups could as well be the result of independent dispersal from the north.

In his paper *Die Dipteren-Fauna von Neuseeland als systematisches und tiergeographisches Problem*, which is one of the most important contributions to the problem of transantarctic relationships, Hennig (1960) has stressed that, theoretically, there are three main patterns of such relationship which can be considered as strong arguments for a former more or less continuous migration route between Antarctica and other southern lands:

(1) The presence of phylogenetically intermediate species on islands lying on the direct connecting line between South America and Australia–New Zealand (fig. 12).

(2) The South American and Australasian groups display successive, comparatively apomorph grades of trend characters which are represented by more than two grades within a larger group enclosing the groups of South America and Australasia.

This argument is linked in with the phenomenon of parallelism between morphological and chorological progression (the progression rule of Hennig, 1950; cf. also p. 23 above). It implies that within the total distribution area of a group the species possessing the most primitive characters are found within the earliest, those with the most derivative characters within the latest occupied parts of the area. If the picture of the systematic design has remained undisturbed it is thus possible to infer the migrational course of a group from the distribution of species displaying certain patterns of primitive and derivative characters. In

L. Brundin, *Transantarctic relationships*   K. V. A. Handl. 11: 1

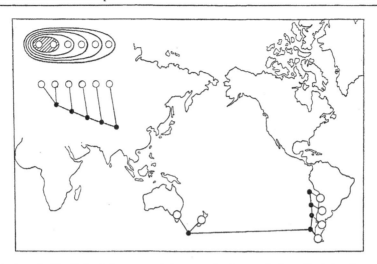

Fig. 13. Presence in Australia–New Zealand of a comparatively apomorph subgroup of a larger South American group as argument for the earlier existence of direct migration routes between these areas. (From Hennig 1960, fig. 8.)

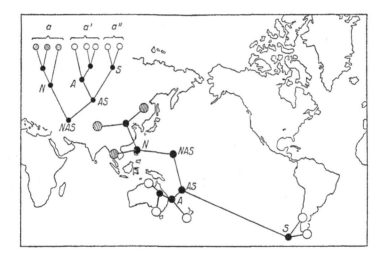

Fig. 14. Occurrence of three evolutionary grades of a character (or a group of characters) in the monophyletic subgroups of a group of Laurasia, Australia–New Zealand, and South America. Presence of the two apomorph grades in Australia–New Zealand and South America as argument for the earlier existence of direct migration routes between these areas. (From Hennig 1960, fig. 10.)

other words, if there is a larger monophyletic group in, for example, temperate South America and a small group in Tasmania which stands out as the apomorph sister group of a subgroup of the South American group, then it would be good reason to suppose not only that the Tasmanian group (or its ancestor) has arrived from South America, but also that Antarctica has functioned as a migration route (fig. 13).

Hennig believes, however, that a phylogenetic pattern of the type just mentioned is a rare phenomenon,

L. Brundin, *Transantarctic relationships*   K. V. A. Handl. 11: 1

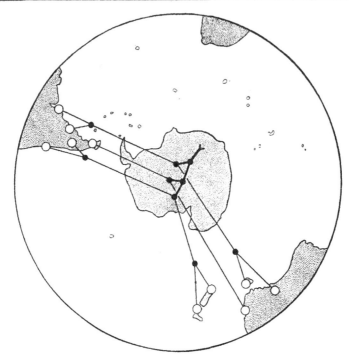

Fig. 15. Existence of sister group relationships between several groups occurring in New Zealand, Australia, and South America as argument for the earlier existence of direct migration routes between these areas. (From Hennig 1960, fig. 3.)

since the austral disjunct groups generally are too poor in species and possess distribution areas which are too limited. He presupposes therefore that it will be necessary also to consider groups which are present not only in South America and Australasia but also in other areas. The continued search for the sister group will sooner or later carry us to the northern continents, Laurasia; but the simple establishement of a sister group relationship between the austral aggregate and a boreal group cannot be a conclusive argument for a former direct connection via Antarctica. The situation would change considerably, however, if it could be shown that a character (or a group of characters) occurs as three different evolutionary grades in the three monophyletic subgroups of Laurasia, Australia–New Zealand, and South America, in such a way that, for example, characters of the Australia–New Zealand group are further developments of those of the Laurasian group and characters of the South American group are further developments of those of the Australia–New Zealand group (fig. 14).

(3) The presence on both sides of Antarctica of groups whose subgroups all are in their turn bicentric (fig. 15). The establishment of such complex transantarctic relationships within a monophyletic group would obviously be a very strong evidence that Antarctica has been at least a part of an austral centre of evolution.

The analysis performed by Hennig demonstrates that convincing cases of direct biotic connections via Antarctica must be based on the fulfilment of certain very rigid demands. Among Diptera there are no less than 50 groups which, because of their austral bicentric distribution, have been considered by leading specialists of the order as proofs or strong indications of former land connections via Antarctica. It is significant that Hennig, after a critical examination, is forced to state (l.c., p. 323) that the supposed system-

L. Brundin, *Transantarctic relationships*   K. V. A. Handl. 11: 1

atic evidence "has proved unable to yield for any of these groups a pattern of phylogenetic relationships which could be regarded as an argument for an antarctic evolution centre or even an antarctic migration route. This holds good also for those groups frequently cited as paradigms of the type of distribution under discussion and used as proof of a former antarctic connection." Hennig is of the opinion, however, that the possibility cannot be denied that a more intimate study of some dipteran groups might be able to yield the necessary patterns of relationship.

Reviewing the opinions of other authors Hennig ends his paper with a discussion of the problem which point of time can be fixed as *terminus post quem non* for the origin of animal groups distributed in Australia–New Zealand and South America. He states the boundary between Oligocene and Miocene as an "especially cautious solution" (l.c.), but prefers to regard the Eocene/Oligocene boundary as the lowest age of the actual groups. This problem will be discussed further in the last chapter of the present paper.

Surveying the matter we can state that until 1960 no one, owing to a striking deficiency in reasoning according to strict phylogenetic principles, had yet been able to produce any biotic data that made it necessary to suppose that Antarctica was a part of a continuous land-migration route between the other southern lands. There was even no conclusive evidence that Antarctica had been a centre of evolution. The meaning of many data was disputed and hypothesis stood against hypothesis. But in 1960 it was generally agreed, I think, that there had been some sort of faunal and floral interchange at least between South America and Australia–New Zealand, while the involvement of the South African sector was disputed. A majority believed in a former greater proximity of land between South America, Antarctica and the Australasian sector. Some authors preferred to assume intervening stretches of sea, at least between Antarctica and Australia–New Zealand, and had no difficulty in accepting accidental dispersal as an explanation of the disjunct distribution patterns, while others, impressed by the abundance of striking data, calculated with former continuous land, thereby preferring either land bridges or the Gondwanaland hypothesis (with inclusion of continental displacement). The presence of temperate forests during Upper Cretaceous–Paleogene in the Graham Land area of West Antarctica had been established and seemed to allow the conclusion that Antarctica enjoyed a temperate climate and was covered with beech and podocarp forests at least in the coastal sectors during those ages.

Regarding Hennig's three criteria it is important to note that the proper fulfilment of the demands raised can demonstrate the existence and extensiveness of austral biotic connections and suggest the role of Antarctica as an evolutionary centre, but that it cannot prove the former existence of continuous land and the nature of the continuity. This was admitted by Hennig, who did not enter upon the biological aspects connected with the latter problems.

When Hennig wrote his paper on the zoogeography of the New Zealand Diptera, Dr. J. Illies of Plön and the present writer were already deeply engaged in studies of two austral rheophil insect groups, stoneflies (Plecoptera) and chironomid midges (Diptera, Chironomidae) respectively. We had both accepted the principles of phylogenetic systematics, and as early as in 1960 Illies (1960a) was able to report that the family Eustheniidae, belonging to the strongly plesiomorph austral suborder Archiperlaria, affords an example of a monophyletic group fulfilling Hennig's third criterion, namely the display of complex transantarctic relationships. The subfamily *Stenoperlinae* of Eustheniidae occurs in Australia, New Zealand, and Chile. Its sister group is formed by the other two subfamilies of the family: *Thaumatoperlinae* is an apomorph group of SE Australia, while the third subfamily, *Eustheniinae*, is known from SE Australia, Tasmania, and Chile. The above paper was, however, preliminary and did not present phylogenetic arguments concerning the mutual relationships of the disjunct subgroups of Tasmania–Australia and New Zealand. But it meant certainly strong new evidence of the evolutionary importance of Antarctica; it likewise belongs to the picture that the cold-stenothermal Archiperlaria, as has been lucidly shown by Illies (1960a, b), do represent the most primitive living Plecoptera and, further, that plecopteran fossils are known from the Permian of both hemispheres (Siberia, Australia), and that it seems to be well established by Illies that the order arose in the south.

59

Further and stronger evidence of Antarctica as a vital part of a former evolutionary centre is furnished by the large plecopteran family *Gripopterygidae* (28 genera, 88 species) and the smaller, nearly related family *Austroperlidae*. In this case Illies (1963, 1965a) was able to show by concise arguments that the two exclusively austral families form a monophyletic aggregate comprising 5 groups; each of these displays transantarctic relationships. Unfortunately the phylogenetic reasoning is confined to the mutual relationships of those 5 major groups. It is stressed by Illies that few insects are less fitted for accidental long distance dispersal than the Plecoptera. Everyone with experience of the group will agree. The circumstance that plecopteran groups are involved in complex transantarctic relationships is, therefore, also good evidence of former continuous connections between the southern lands.

At the Fourth Pacific Science Congress in Honolulu (1961), on my way to the field work in the Southwest Pacific, I read a preliminary paper on limnic Diptera in their bearings on the problem of transantarctic faunal connections (Brundin 1963). Basing my arguments on the principles of phylogenetic systematics I was able to show that the austral genera *Parochlus*, *Podonomus*, *Podonomopsis*, and *"Podonomites"* (*Podochlus* of the present paper)—belonging to the chironomid subfamily Podonominae—display a very complex pattern of transantarctic relationships, where the disjunct subgroups of South Chile–Patagonia and New Zealand are formed by genera, species groups and even pairs of species, the whole answering to Hennig's third criterion. Another case, referring to the subfamily Orthocladiinae, illustrated the fulfilment of the demands raised by Hennig's second criterion.

The new data brought forward by Illies and Brundin confirmed the old idea of Hooker. The proven involvement of the Plecoptera in the problem of austral disjunctive animal groups sets the problem into a perspective of time that ranks it equal with that of the conifers. Considering, moreover, the insight into the history of the latter delivered by Florin, and the assembled knowledge of the austral beeches, there seemed to be little doubt that Antarctica has formed a part of an austral centre of evolution and diversification, now fragmented and largely destroyed, but with a history going back to the Permo–Carboniferous glaciations.

But such a general concept of a former austral centre of evolution is too vague and only forms a first step on the way to a deeper understanding of the real nature of the transantarctic relationships. The evidence of Illies and the present writer was incomplete. Our field experience of the involved austral biotopes was confined to temperate South America, and our material from the Australasian sector was too limited to enable us to discuss the mutual relationships of the components inhabiting Tasmania–Australia and New Zealand. We had also no possibility to touch upon the problem of the possible involvement of the South African biota in the history of the austral disjunct groups.

In order to be able to disentangle the main trends in the history of the splitted austral biota the investigator has to raise and try to answer some important questions:

(1) Do the structure of the phylogenetic relationships and the distribution of the isolated elements of different austral groups form some fixed patterns?

(2) If so, is there some connection between distribution and different relative age of the groups? In other words: is there some evolutionary indication of a time-table in the development of different types of disjunct distribution?

(3) If so, are there indications of causal connections between phylogenetic relationships, distribution patterns, and certain paleogeographic events or conditions?

It is evident that the answering of questions of the kind just outlined does presuppose a detailed and complete analysis of the involved sister group systems and thus the inclusion of the world fauna of the groups treated. Three major groups among the chironomid midges, all with disjunctive circumantarctic distribution and occurrence also in Southern Africa, will be analysed in the following chapter.

My paper of 1965 is a summary of the results set out in the present paper.

L. BRUNDIN, *Transantarctic relationships*   K. V. A. Handl. 11: 1

P. J. DARLINGTON, JR.

Just when writing this review I got into my hands a work by P. J. Darlington (1965) with the title: *Biogeography of the Southern End of the World. Distribution and history of far-southern life and land, with an assessment of continental drift.* Comprising 236 pages it is one of the largest papers which have as yet appeared on the problems of austral biogeography.

Darlington is the author of *Zoogeography: the geographical distribution of animals* (1957). Being published at such a late date this book arouses disappointment. Darlington is an entomologist, a specialist in carabid beetles; but in spite of that he preferred to discuss solely vertebrates, because he considered them the best animals for biogeographic work (which is debatable). When discussing relationships Darlington was as a consequence helpless in the hands of a long row of authorities. Fettered moreover by aprioristic belief in the permanence of the main geographical features and in the decisiveness of means of dispersal, all his endeavour to deepen the understanding of the history behind the great disjunctions was doomed to fail. The book illustrates in a conspicuous way all the weaknesses of the Wallacean type of approach, and it is not surprising that it became a highly appreciated target of Croizat (1958) in his *Panbiogeography*.

From Darlington's recent *Biogeography* it is evident that he has learned nothing from the blazing sermon of Croizat. The approach of the latter is not wholly sound, it is true, and his criticism of the Wallacean camp may be considered too violent, but his search for the truth has been frenetic, and it is only fair to stress that much of his message concerns everyone dealing with the history of life very deeply.

Darlingtons' general conclusions (1965, pp. 214–215) are as follows:

"I think, then, that at the time of the Permo-Carboniferous glaciation all the continents (except Antarctica) probably lay further south than now. Africa and South America may still have been joined together but the other southern continents were probably not united. Africa and South America separated not later than the Jurassic and probably earlier, too long ago for the union to effect the distribution of plants and animals now. And all the continents (except Antarctica) have moved northward since the Permo-Carboniferous, different continents moving independently. Water gaps have probably been narrower and dispersal has probably been easier in the past than now, but land bridges between southern continents are not indicated."

"Even when the land was not glaciated, the climate of the southern end of the world since the Permo-Carboniferous has probably always been cooler than tropical and also seasonal, and zonation of climate has probably always favored differentiation of a special southern cool-temperate biota. The biotic history of the southern end of the world seems in fact to have been the history of a continuously existing but continuously changing, climatically specialized, far-southern biota shared at least in part by all far-southern lands, including habitable parts of Antarctica. Beginning with near-contiguity of southern continents in the late Paleozoic, facilitating dispersal in the far south, the history has probably been one of gradual widening of gaps between continents and gradual lessening of dispersal and finally deterioration of climate and virtual cessation of dispersal across southern water gaps during the later Tertiary and Pleistocene."

"During the whole of the time in question, successive new groups of plants and animals have presumably been invading the southern end of the world from the tropics or across the tropics, and minor counterinvasions from the south northward may have occurred too. However, I see no evidence that any major groups of plants and animals have evolved on Antarctica and spread widely over the world from there. During this whole time dispersal around the southern end of the world has probably tended to be from west to east, since that has probably always been the direction of prevailing winds and currents."

Darlington compares the whole to a large tidal-river from the north towards "the southern end of the world". He ends the book with this: "Nevertheless I have oversimplified, because I have had to, and in doing so I may unintentionally have cut out essential facts or interpretations. Have I?"

The answer to that question is: Yes, indeed; but hardly unintentionally, because of the heavy burden of preconceptions involved in the considerations and conclusions throughout that book.

Apart from the allusions to the Permo-Carboniferous glaciations and the continental drift, the cited conclusions of Darlington could as well have been written, word for word, by Wallace in 1876 (cf. p. 51 above). The dominant idea of both is that of a stream of life from the north, feeding the small, cold pieces of land in the south with scattered fragments from the rich table of a tremendous creative centre in the north. The idea is old and very impressive, it is true, but does it mirror the truth? To Darlington that was precognition, as demonstrated by his method of approach and by all his discussions. The big news is that Darlington has accepted the hypothesis of continental displacement, but of course in a personal, toned down edition, with proper water gaps between the continents.

Darlington declares in his Preface that he will try to solve the problems connected with the history of far-southern life. But he has never tried to look upon the problems from the south, never made clear to himself what the task really implies, neither in regard to degree of difficulty and complexity, nor in regard to method of approach and freedom of anticipation. As a biogeographer he must know that the problems of great disjunctions still are much-debated and largely unsettled. Faced with the above unsolved problem of world-wide scope his only chance was to look upon every biogeographical and geological hypothesis with distrust and to start a self-dependent, detailed study of life itself. Since the history of life and the history of the phylogenies is by necessity the same thing, the task of the investigator should be self-evident: selection of at least one representative group and reconstruction of the involved sister group system by means of synapomorph characters, the rule of deviation, and the rule of geographical vicariism. But Darlington has never tried to use that master-key. All through the book he goes around the main problem, the phylogenetic relationships (a concept he has not even mentioned), at length discussing the exceptional importance of climate, the intimate connection between successful evolution and large tracts of land, and the dispersal possibilities, thanks to the strong westerly winds and currents, of the poor biotas of the far-southern *cul-de-sacs*.

Darlington's biotic evidence is in all essentials based on compilation. He stresses in the Introduction (p. 6) that "nothing has been settled", which is true. But the fact that he nevertheless prefers compilation to accumulation of new evidence by individual research can logically only mean that he believes that he, as a biogeographer, has a special predisposition to divine the truth where others have failed. This is freely admitted by Darlington (p. 184): "If there is such a thing as a professional in biogeography, I am one. I am therefore in a position to know the complexities and difficulties of the subject. They are many."

But is that remarkable confession security enough? Let us have a look at, for example, Darlington's discussion (pp. 152–156) of the role of Antarctica where he tries to show that that continent has not been a centre of evolution. His main argument here is *Nothofagus*, which according to him displays a typical immigrant pattern and not a relict pattern which the genus should do if it had evolved in Antarctica. His evidence is the circumstance that all three groups of the genus occur or have occurred in Australia, New Zealand, Antarctica, and South America. "The regularity of this pattern strongly suggests that *Nothofagus* has followed a path around the world in the far south and has not survived in three separate relict areas after evolution, diversification, and extinction on the Antarctic Continent. Of course the coast of Antarctica may have been a dispersal path ..."

The *Nothofagus* problem is, however, much more complex and difficult, and our knowledge in essential points much less complete than realized by the professional biogeographer. His thesis of a uni-directional movement of *Nothofagus* "around the world in the far south" (in Darlington's fig. 21 B marked as dispersal of the three main groups from Australia via New Zealand and Antarctica to South America) is a loose construction to say the least. No one has shown that the endemic subgroups of any of the three groups display successive apomorph grades of trend characters indicating any sort of uni-directional dispersal, still less that the most primitive species of all the three groups are concentrated in a special area. So little is still known about plesiomorph and apomorph characters in *Nothofagus* and *Fagus* (and Fagaceae in general) that no attempt has been made to reconstruct the pertinent sister group system. The *brassii*, *menziesii*, and *fusca* groups are only provisional from a phylogenetical point of view, their relative plesiomorphy, monophyletic status and mutual relationships still unsettled. To assume a similar history of evolution and dispersal for the three groups, as Darlington has done, is, therefore, wild speculation. What little is known speaks against him. The *fusca* group is represented by 4 species in New Zealand, 1 species in Tasmania, and by 5 species in Chile-Patagonia (Cookson & Pike 1955). The species *gunnii* of Tasmania is generally considered to be more closely related to *N. antarctica* and *N. pumilio* of South America than to any other living species of the *fusca* group. An important point is that those 3 species are characterized by deciduousness, a clearly derivative character which is a rare phenomenon among the southern forest trees. These things were discussed by Du Rietz at the symposium on the biology of the southern cold temperate zone held in London in 1959 (Du Rietz 1960, p. 501), and it is strange that the matter is passed in silence by Darlington who, moreover, attended that symposium. A strict phylogenetic discussion is not possible, but there seems to be reason to assume that the *gunnii-antarctica* subgroup is the apomorph sister group of a group formed by the rest of the *fusca* group, or a part of it. In other words, *within* the *fusca* group we are faced with the possibility of a phylogenetic connection between Tasmania and South America via Antarctica, independently of another connection between New Zealand, Antarctica and South America. It will be shown in the following chapters that this is a common pattern. There is in reality not a shadow of evidence against the assumption that Antarctica has taken a very active part in the evolutionary history of *Nothofagus* from, let us say, Lower Cretaceous to Paleogene.

Darlington believes of course that *Nothofagus* originated in the north and thinks, referring to a personal communication by Cranwell, that "we shall be near solving the problem of place of origin when we find pollen of *Nothofagus* and *Fagus* together ...." (p. 145). That the common ancestor of *Nothofagus* and *Fagus* lived in the north is possible, but it would be a serious mistake to believe that the very findings of typical *Nothofagus* pollen somewhere on the northern continents (such findings from the Upper Cretaceous in Kazakhstan and Western Siberia have been reported, Zaklinskaya 1964) would prove the northern origin of *Nothofagus*. Then in these, like in all other similar cases, it must at least be demonstrated that the fossil northern *Nothofagus* populations were more strongly plesiomorph than any known member of the southern aggregate, fossil or living. Moreover, if typical pollen grains of *Nothofagus* and *Fagus* are found together, such a situation, i.e. the establishment of sympatry, would be very strong evidence that the two genera have had a long history before that meeting, but not any evidence of a northern origin of *Nothofagus*. It is disappointing that Darlington does not realize that a phylogenetic analysis is an absolute prerequisite for a proper discussion. Only such an analysis can show what the structures of the pollen grains really tell and if they are sufficient for a reconstruction of the history of *Nothofagus*.

A strange thing, though in full accordance with his general attitude, is Darlington's treatment of his own special domain, the carabid beetles, an old, large insect group obviously displaying several examples of transantarctic relationships and bipolarity. Darlington has himself studied the austral carabids during extensive field work in Australia and Tasmania, and during several weeks of collecting in southernmost Chile, from Fuegia to Wellington Island; and he points out that he has studied carabid taxonomy for more than thirty years (p. 8). That sounds impressive, but one becomes suspicious when reading the following: "The profound effect of climatic (ecologic) factors on distribution of plants and animals in the southern cold-temperate zone must be seen to be fully appreciated. Seeing it—that is, seeing plant and animal distributions at the southern tip of South America in relation to climate—was the most important result of my recent visit" (p. 15). In passing: it was a serious mistake to restrict the South American field work to the southernmost tip of the continent. However, reading Darlington's discussion of the carabid groups *Migadopini*, *Broscini*, *Trechini*, and *Bembidion* and their disjunct components one is struck with the extent to which he has to rely on the results of others. Moreover, he not only does not deepen the discussion with any own data of phylogenetic significance, but leaves out such data performed by Jeannel, who probably has studied austral carabids more extensively than any other.

Let us have a look at, for example, the *Migadopini*. This tribe of prevailingly ground-living beetles is represented by 7 genera in southern South America, 4 genera in Tasmania–SE Australia, and by 4 genera in New Zealand with the Auckland Islands. All genera are endemic. Jeannel has stressed repeatedly the strong general plesiomorphy of the tribe (1938, 1942, 1949), but this relevant statement of the former leading specialist is simply passed over by Darlington, who argues (p. 36):

"Jeannel in his useful revision of the Migadopini (1938), considered the tribe wholly flightless (apterous) and thought it had originated on an ancient antarctic continent and spread from there. However, as usual in "antarctic" distributions, there are inconsistent details. First, the tribe is not wholly flightless. Both *Antarctonomus* in Chile and *Decogmus* in eastern Australia are strongly winged. Second, the tribe is not quite confined to *cold* areas in the south. The northernmost genera in both South America and Australia, although technically within the south temperate zone, occur at warm-temperate or subtropical locations: *Rhytidognathus* is known only from Uruguay, about 35° S; *Decogmus*, from Comboyne in northern New South Wales, at 31° 36′ S and less than 1000 m altitude. Third, existing genera of the tribe are extraordinarily diverse in form and characters, as if they are products of a complex ecologic as well as geographic radiation rather than of simple spread from an antarctic center. And fourth, the closest relatives of the tribe are probably the Elaphrini of the *north* temperate zone, although the relationship may not be very close (Jeannel 1938, Lindroth, personal communication, 1963). All this suggests that the ancestor of Migadopini was winged, that it may have lived in or dispersed through relatively warm climates, that the history of the tribe has been complex, and that a common ancestor of this tribe and the Elaphrini crossed the tropics a long time ago. These details do not disprove an antarctic origin of the Migadopini but do suggest other possibilities."

Here are some comments on the four points which according to Darlington are "inconsistent" with the assumption of an Antarctic origin of Migadopini.

(1) "The tribe is not wholly flightless." Interesting, but what is the inconsistency? Compare the results of the present investigation.

(2) "The tribe is not quite confined to *cold* areas in the south." A curious argument. Why should all subgroups of old austral disjunct groups necessarily be narrowly cold-adapted? The Migadopini pattern is perfectly normal.

(3) Is there any earthly reason to imagine that "spread from an antarctic center" must be simple? It is fatal preconception to say that diversity in form and characters is inconsistent with the assumption of austral (Antarctic) derivation of an old group.

(4) "The closest relatives of the tribe are probably the Elaphrini of the *north* temperate zone, although the relationships may not be very close." Darlington's line of thought, further marked by his italicizing of the word "north", is revealing. He is wholly unaware of the ever present rule of deviation and that difference cannot be a measure of phylogenetic relationship. He does not realize that his acceptance of Elaphrini as the closest relatives of Migadopini implies a suggestion that the two tribes are sister groups and, further, that the very character of Elaphrini of being northern is perfectly consistent with the view of Jeannel. Omitting the vital point as to the relative plesiomorphy of the two tribes, Darlington is unable to apply the phenomenon of parallelism between morphological and chorological progression.

The Migadopini were conceived by Jeannel as a strongly plesiomorph group that evolved in "Paleoantarctica" and that now displays an amphiantarctic relict pattern. Jeannel believed, further, (expressed in a more exact way) that the apomorph sister species of the migadopid ancestor migrated northwards and became the ancestor of Elaphrini. This conception, whether right or wrong, remains absolutely intact after the attack of Darlington.

Darlington has been largely unsuccessful because he has misunderstood the method of approach in biogeography. It is plainly meaningless to speculate in that wild way, without knowledge of the reliability and meaning of the basic data and by perpetual neglect of the principles of phylogenetic reasoning.

The study of phylogenetic relationships is time-consuming and difficult work, but there is no way around, and we are short of time. We have to realize that the principal answer as to the history of austral life most probably will be delivered by invertebrates and that much more of basic, strongly specialized field work is badly needed. But the utterly important virgin biotopes of the southern lands disappear very rapidly as a consequence of human agency. I am ending this review with the following words of C. F. A. Pantin (1960): "The study of biological relations in the southern hemisphere may in fact prove to be a key to understanding biological relations of the world generally; and though we may not have so many decades left in which to do what is needed, today it can still be done."

Let us do it, with enthusiasm and humbleness and freedom from preconception.

Works Cited

Brundin, L. 1963. Limnic Diptera in their bearings on the problem of trans antarctic faunal connections. In J. L. Gressitt, ed., *Pacific Basin Biogeography*, 425–34. Honolulu: Bishop Museum Press.

———. 1965. On the real nature of transantarctic relationships. *Evolution* 19: 496–505.

Cookson, I. C., and K. M. Pike. 1955. The pollen morphology of *Nothofagus* Bl. Subsection bipartitae Steen. *Austr. Jour. Bot.* 3: 197: 206.

Couper, R. A. 1960. Southern Hemisphere Mesozoic and Tertiary Podocarpaceae and Fagaceae and their palaeographic significance. In C. F. A. Pantin and Others, A discussion on the biology of the southern cold temperate zone. *Proc. Roy. Soc. London*, ser. B, 152: 491–500.

Cranwell, L. M. 1963. *Nothofagus*: Living and fossil. In J. L. Gressitt, ed., *Pacific Basin Biogeography*, 387–400. Honolulu: Bishop Museum Press.

Croizat, L. 1958. *Panbiogeography, or an Introductory Synthesis of Zoogeography, Phytogeography, and Geology;* with notes on evolution, systematics, ecology, anthropology, etc. 2 vols. Caracas: by the author.

Dana, J. D. 1863. *Manual of Geology*. Philadelphia: Bliss & Co.

Darlington, P. J., Jr. 1957. *Zoogeography: The Geographical Distribution of Animals*. New York: Wiley.

———. 1965. *Biogeography of the Souther End of the World: Distribution and History of Far-southern Life and Land, with an Assessment of Continental Drift.* Cambridge: Harvard University Press.

Darwin, C. 1859. *On the Origin of Species*. London: John Murray; facsimile of 1st ed., Harvard University Press, Cambridge, Mass.

del Corro, G. 1964. La Gondwania, el antiguo continente austral. *Publ. Ext. Cult. Didáct., Mus. Arg. Cienc. Nat. Buenos Aires* 12: 1–90.

Du Rietz, G. E. 1940. Problems of bipolar plant distribution. *Acta Phytogeogr. Suec.* 13: 215–82.

———. 1960. Remarks on the botany of the southern cold temperate zone. In C. F. A. Pantin and Others, A discussion on the biology of the southern cold temperate zone. *Proc. Roy. Soc. London*, ser. B, 152: 500–507.

Florin, R. 1940. The Tertiary fossil conifers in South Chile and their phytogeographical significance, with a review of the fossl conifers of southern lands. *Kungl. Sv. Vet. Akad. Handl.*, ser. 3, 19: 1–107.

———. 1963. The distribution of conifer and taxad genera in time and space. *Acta Hort. Berg.* 20 (4): 121–312.

Good, R. 1953. *The Geography of the Flowering Plants*. 2d ed. London: Longmans, Green, and Co.

Harrison, L. 1928. The composition and origin of the Australian fauna with special reference to the Wegener hypothesis. *Rep. Australas. Ass. Adv. Sci.* 18: 322–96.

Hennig, W. 1950. Grundzüge einer Theorie der phylogenetischen Systematik. *Forsch. Fortschr.* 25: 137–39.

———. 1960. Die Dipterenfauna von Neuseeland als systematisches und tiergeographisches Problem. *Beitr. Ent.* 10: 221–329.

Hooker, J. D. 1844–47. *The Botany of the Antarctic Voyage of H.M. Discovery Ships "Erebus" and "Terror" in the Years 1839–43*. Part 1: *Flora Antarctica*. 2 vols. London: Lovell Reeve.

———. 1853–55. *The Botany of the Antarctic Voyage ...* Part 2: *Flora Novae-Zelandiae*. 2 vols. London: Lovell Reeve.

———. 1855–60. *The Botany of the Antarctic Voyage ...* Part 3: *Flora Tasmaniae*. 2 vols. London: Reeve. (Introductory Essay, pp. 1–cxxviii, London 1860.)

Hutton, F. W. 1873. On the geographical relations of the New Zealand Fauna. *Proc. Nat. Acad. Sci.* 42 (2): 84–86.

Illies, J. 1960. Phylogenie und Verbreitungsgeschichte der Ordnung Plecoptera. *Verh. Deutsch. Zool. Ges.* Bonn/Rhein 1960: 384–394. Leipzig: Akadem. Verlagsges. Geest & Portig K.-G.

———. 1963. Revision der südamerikanischen Gripoterygidae (Plecoptera). *Mitt. Schweiz. Ent. Ges.* 36: 145–248.

———. 1965a. Verbreitungsgeschichte der Gripoterygiden (Plecoptera) in der südlichen Hemisphäre. *Proc. 12th Int. Congr. Ent.* London 1964: 464–68.

———. 1965b. Entstehung und Verbreitungsgeschichte einer Wasserinsektordnung (Plecoptera). *Limnologica* (Berlin), 3 (1): 1–10.

Jeannel, R. 1926–28. Monographie des Trechinae: Morphologie comparée et distribution géographique d'un groupe de Coléoptères. *L'Abeille* 32: 221–550 (1926); 33: 1–592 (1927); 35: 1–808 (1928).

———. 1936. Monographie des Catopides. *Mém. Mus. nat. Hist. Nat.* 1: 1–433.

———. 1938. Les Migadopides, une lignée subantarctique. *Rev. fr. D'Ent.* 5: 1–55.

———. 1940. Les Calosomes. *Mém. Mus. nat. Hist. Nat.* 13: 1–240.

———. 1942. *La genèse des faunes terrestres*. Paris: Presses Universitaires de France.

————. 1949. Les Insectes: Classification et Phylogénie; les insectes fossiles; evolution et géonémie. *Traité de Zoologie* 9: 1–110. Paris: Masson.

Matthew, W. D. 1915. Climate and evolution. New York Acad. Sci., *Annals* 24: 171–318.

Pantin, C. F. A. 1960. Introduction. In C. F. A. Pantin and Others, A discussion on the biology of the southern cold temperate zone. *Proc. Roy. Soc. London*, ser. B, 152: 431–33.

Rütimeyer, L. 1867. *Über die Herkunft unserer Thierwelt: Eine zoogeographische Skizze.* Basel: H. Georg's Verlagsbuchhandlung.

Sarasin, F. 1925. Über die Tiergeschichte der Länder des südwestlichen Pazifischen Ozeans auf Grund von Forschungen in Neu-Caledonien und auf den Loyalty-Inseln. In F. Sarasin and J. Roux, *Nova Caledonia* (A), 4: 1–177. Berlin: C. W. Kreidel's Verlag.

Schröter, C. 1913. *Geographie der Pflanzen.* 2. *Genetische Pflanzengeographie (Epiontologie).* Handwörterbuch d. Naturwiss. 4. Jena.

Simpson, G. G. 1940a. Antarctica as a faunal migration route. *Proc. 6th Pacific Sci. Congr. (1939):* 755–68.

Skottsberg, C. 1915. Notes on the relations between the floras of subantarctic America and New Zealand. *The Plant World* 18 (5): 129–42.

————. 1925. Juan Fernandez and Hawaii: a Phytogeographical discussion. *Bernice P. Bishop Museum Bull.* 16: 1–47.

————. 1928. Remarks on the relative independency of Pacific floras. *Proc. 3d Pan-Pacif. Sci. Congr. (1926):* 914–20.

————. 1934a. *Astelia,* an Antarctic-Pacific genus of Liliaceae. *Proc. 5th Pacif. Sci. Congr. (1933):* 3317–3323.

————. 1934b. Studies in the genus *Astelia. Kungl. Sv. Vet. Akad. Handl.*, n.s., 14 (2): 1–106.

————. 1934c. Le peuplement des îles pacifiques du Chili.

Société de Biogéographie 4. *Contribution a l'étude de peuplement zoologique et botanique des îles du Pacifique,* 271–80. Paris.

————. 1936. Antarctic plants in Polynesia. *Essays in Geobotany in Honor of William Albert Setchell,* 291–311. Berkeley: University of California Press.

————. 1940. Några drag av den antarktiska kontinentens biologiska historia. *Kgl. Norske Vidensk. Selsk. Skrifter* 12: 45–55.

————. 1956. Derivation of the flora and fauna of Juan Fernandez and Easter Island. In C. Skottsberg, ed., *The Natural History of Juan Fernandez and Easter Island,* vol. 1, part III (5): 193–438. Uppsala.

Thorne, R. F. 1963. Biotic distribution patterns in the tropical Pacific. In J. L. Gressitt, ed., *Pacific Basin Biogeography,* 311–50. Honolulu: Bishop Museum Press.

Wallace, A. R. 1876. *The Geographical Distribution of Animals,* with a study of the relations of living and extinct faunas as elucidating the past changes of the earth's surface. 2 vols. Reprinted ed., New York: Hafner Publishing Co.

————. 1880. *Island Life, or the Phenomena and Causes of Insular Faunas and Floras,* including a revision and attempted solution of the problem of geological climates. London: Macmillan.

Wittman, O. 1934. Die biogeographischen Beziehungen der Südkontinente: Die antarktischen Beziehungen. *Zoographica* 2: 246–304.

Wulff, E. V. 1950. *An introduction to Historical Plant Geography.* Waltham, Mass.: Chronica Botanica Co.

Zaklinskaya, E. D. 1964. On the relationships between Upper Cretaceous and Paleogene floras of Australia, New Zealand, and Eurasia, according to data from spore and pollen analysis. In Lucy M. Cranwell, ed., *Ancient Pacific Floras: The Pollen Story.* Honolulu: University of Hawaii Press.

Phylogenetic Systematics
Willi Hennig

were likewise brought about by phylogeny, but are only remotely connected with the form of the organisms. The reliability of phylogenetic relationships derived from certain structural relations can therefore be regarded as especially reliable if the distributional relations of the same organisms also show an orderly structure.

Naturally we can proceed equally well in the opposite direction and make the distributional relations themselves the starting point for deriving phylogenetic kinship relations. The reliability of the results can then be tested against the resulting holomorphological consequences. Thus the analysis of the distributional relations of organisms—the "chorological method"—is an important aid in taxonomic work.

The Chorological Method. In discussing taxonomy it was shown that there are close relations between the species and space. Every species originally occupies a certain area, and the breaking up of a species into several reproductive communities usually, if not always, is closely related to the dispersal of the species in space. Consequently the distribution of the closest reproductive communities in space could also be used as a criterion for the phylogenetic relationships between them (the "vicariance criterion"). It was recognized long ago that the spatial relationships between organisms can be used for determining phylogenetic relationships—and thus for solving the problems of phylogenetic systematics—in cases where the morphological method does not suffice.

In botany von Wettstein seems to have been the first to point explicitly to the importance of the geographical method for systematics. He used it in working over certain groups (e.g., *Euphrasia*). The use of the geographical method in zoology has become known particularly through the books of Rensch. It is almost exclusively restricted, however, to the lower taxa. Especially in the taxonomy of birds and mammals, the opinion is now rather general that vicarying reproductive communities are to be regarded as subspecies of a single species. The term *rassenkreislehre* is used for this method and all questions surrounding its principles of application and results. This fact alone indicates that geographic vicariance and the possibility of using it in phylogenetic systematics are regarded as a peculiarity limited to relationships among the lower taxa. This is not true at all. There are also relationships between distribution in space and the systematic classification of the higher taxa.

For the relationships between the chorological distribution and phylogenetic relationships in the higher taxa it may be taken as a ground rule that species groups belonging to a community of descent are restricted to unit areas that are to a certain extent unbroken. In the chorological method it plays about the same role as the statement that the more similar two groups are the more closely they are related phylogenetically does in morphological methods.

The example of the genus *Myennis* shows how this rule can be used in determining phylogenetic relationships. The dipteran groups known under the names Otitidae and Pterocallidae each include about 150 known species. These two groups are undoubtedly very closely related, and are not always easily distinguish-

able from one another. They have distinctly different ranges, however: the Otitidae are almost exclusively Holarctic, the Pterocallidae almost exclusively Neotropical. Until recently it appeared that the pterocallids were represented in the
Palaearctic by the genus *Myennis* (5 species). Neotropical and Palaearctic species were once united in this genus, but later study showed that the Palaearctic
species form a separate group not closely related to the Neotropical species
(which today comprise the genus *Neomycnnis*). The strange occurrence of a
pterocallid genus (*Myennis*) in the Palaearctic still remained. The presence of
Myennis far outside the range of all the other pterocallids led to a re-examination
of the genus, particularly since the great morphological similarity between the
pterocallids and otitids led to the suspicion that a false evaluation of the similarity relationships was responsible for the inclusion of the genus in the Pterocallidae.
In fact it turned out that the species of the genus *Myennis* differ from all other
pterocallids and agree with the Otitidae in a previously unobserved character,
the armament of the male genitalia. On the basis of this, *Myennis* was transferred
from the Pterocallidae to the Otitidae, in whose range it occurs.

The interpretation of Kiriakoff (1956) of the so-called subfamily Brephidiinae
of the lepidopteran family Lycaenidae rests on the same considerations. Kiriakoff
starts from the fact that America and Africa do not form a single area of distribution, and that "recent investigations have shown almost irrefutably that no
land connection between South America and Africa across the south Atlantic existed, at least during the Middle Mesozoic or later." He therefore considers it
improbable that the subfamily Brephidiinae is a monophyletic group. From
Kiriakoff's statements it seems that the correspondence between the American
and African species rests on symplesiomorphy. If this is true it could not be attributed to the action of Vavilov's law, as Kiriakoff assumes.

There is no need to give further examples. Any systematist revising a supposedly monophyletic group would check the affiliations of a species group that
occurs far outside the otherwise continuous range of the main group. He would
test particularly critically whether it really belonged in the group in question.
This he would do by investigating, by the criteria discussed above, whether the
morphological characters that had determined its systematic assignment must
actually be considered synapomorphous correspondences.

Within the continuous range of a monophyletic group it is also possible, as
noted above, under certain circumstances with certain transformation series of
characters, to determine the direction in which it must be read.

We have already pointed out several times that speciation apparently always
goes parallel with a progression in space. Perhaps this is expressed most conspicuously in the fact that the speciation process obviously goes through a stage
in which the daughter species are vicariant. This need not be vicariance in geographic space.

We can visualize two possibilities with respect to the relationship of the ranges
of the daughter species (or races at first) to the original range of the parent

species. In dichotomous speciation one of the two daughter species remains in the old range of the species, while the other inhabits the new area that has been acquired. The second possibility is that neither of the two daughter species remains in the original range because the group was forced by some kind of external circumstances to give it up. The question now arises as to what relationships there are between the "progression in space," which according to the above is connected with speciation, and the progression of the morphological characteristics of the daughter species.

We would expect a study of the geographic rassenkreise to give the most reliable answer to this question. For the purposes of the problem dealt with here we may distinguish between rassenkreisen in which the vicarying subspecies are arranged in approximately linear succession—in which extension evidently took place only in one main direction—and rassenkreisen in which extension took place in all directions from a center of distribution. In the first case (the so-called chains of races) we often find that one or several characters in each successive race are a further development of these characters in the preceding race, and that the direction of further development of the characters is the same. Consequently these are orthogenetic series, in which the direction of differentiation

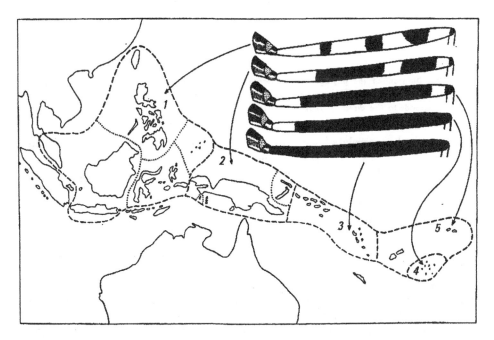

Figure 39. Pattern of the thigh in different subspecies of Mimegralla albimana (Diptera, Tylidae).
From top to bottom: 1, western group of races (albimana, sepsoides, galbula, palauensis); 2, New Guinea, Bismark archipelago, Key Islands (contraria, keiensis, striatofasciata); 3, Samoa (samoana); 4, Tonga Islands (tongana); 5, New Hebrides (extrema).

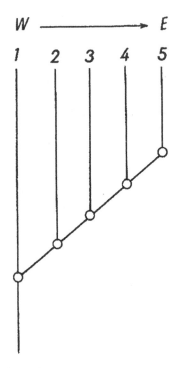

Figure 40. Diagram of the phylogenetic relationships between the five subspecies of *Mimegralla albimana* shown in Fig. 39. At the same time the stage series of increasing apomorphy in blackening of the thigh corresponds to the direction of spreading from west to east.

can scarcely be correlated with an approximately similar unidirectional change in living conditions. Examples of such unidirectional deviation are shown in Figs. 39, 40, and 41 for the dipteran genus *Mimegralla* and the reptile genus *Draco*.

Mimegralla albimana (Tylidae) breaks up into several subspecies in the area between the mainland of southeast Asia and the New Hebrides and Tonga Islands. These subspecies differ, among other things, in their leg markings, each subspecies to the east having darker leg markings than its neighbor to the west. This is clearly shown in Figs. 39 and 40.

In the flying lizards (*Draco*) there are two species (rassenkreise) distributed from the mainland of southeast Asia to the eastern limits of the Oriental region. In both rassenkreisen the most derived pattern on the flight membrane is found in the races farthest removed from the center of distribution of the rassenkreis (Fig. 41). In the rassenkreis *lineatus* the western races (*lineatus* from the Greater Sunda Islands, *beccarii* from southern Celebes) are characterized by primitive, relatively complete patterns on the flight membranes. The race occurring in north-

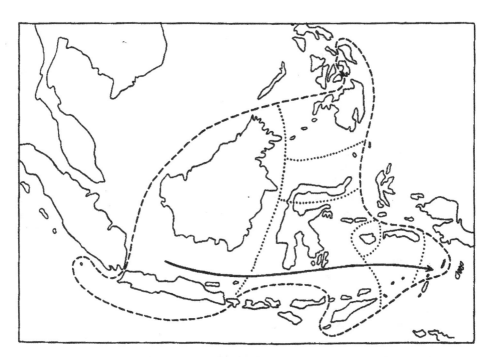

Figure 41. Range of distribution and supposed direction of spread of *Draco lineatus*. The general direction of spread corresponds to a stage series of increasing apomorphy in the pattern of the flight membrane (see text).

ern Celebes (*spilonotus*) has only vestiges of the pattern at the edge of the flight membrane closest to the body. In the most eastern race, *ochropterus* from the Key Islands, the flight membrane pattern is completely absent. The situation is even more striking in the other rassenkreis (*volans*). In the most easterly races (*boschmai* from the Lesser Sunda Islands, *reticulatus* from the Philippines) the partly very distinctly banded pattern of the western nominate form is fused into a reticulate pattern. It is very interesting that the forms from the Philippines and the Lesser Sunda Islands were until recently considered identical and were included under the same name. The two races apparently originated independently from the more western stem form, however, and so have nothing to do with each other in the immediate phylogenetic sense. Their nearly identical reticulate pattern is rather to be regarded as convergence (Hennig 1936a).

In all these examples the direction in which the species extended their ranges is known with adequate certainty: in all it was in general from west to east. This makes it possible to determine exactly the phylogenetic relationships between the individual subspecies. It must be assumed that each subspecies stands in a kinship relation of the first degree to all subspecies lying east of it, so that a hierarchy of degrees of phylogenetic relationship within the species, as shown in Fig. 40, is to be assumed. If the morphological differences are compared with this diagram of the degrees of phylogenetic relationship, it is evident that in each case the more easterly form shows a more advanced (more apomorphic) stage of development of the characters distinguishing them than does the neighboring form to the west.

But the concept of space in the systematics of the higher taxa need not be limited to geographic space any more than was the case in the differentiation of species (see p. 47). Rather, according to the first ground rule of the chorological method, we must demand that certain unbroken areas in "living space" be assumed for the higher taxa too. This in fact is usually the case. Everyone knows that most higher taxa, insofar as they are true phylogenetic units, have certain "ecological characters" in addition to the structural characters—just as species and subspecies do. In other words, they too occupy certain uniform areas of the living space.

This rule can also be used in the opposite direction in systematics. For example, we view with skepticism the suggestion—made most recently by Imms—that the Braulidae (bee lice) are closely related to the Chamaemyiidae; because all chamaemyiids are shield-louse parasites and therefore inhabit a peculiar and sharply characterized ecological niche in which *Braula* does not occur.

The occurrence of the higher taxa in certain ecological niches or zones is no doubt based, exactly as in species, on certain physiological, ethological, and in the narrow sense even morphological, peculiarities in their form. Consequently we may say that even in the higher taxa very definite dimensions of form (holomorphy) correspond to the different dimensions of the environment. Only the two or three dimensions of the environment that we assign to geographic space form an exception; no particular dimensions in the form of organisms corresponds to these. Although the division between ecological and geographic space is to a certain degree artificial and the concept of purely geographic space merely an abstraction, it must nevertheless be maintained—for reasons that have been discussed repeatedly—that the geographic dimensions of the environment are of particular importance to phylogenetic systematics. Fortunately in practice they can usually be separated sufficiently sharply from the other dimensions of the living space. Consequently in the following we will restrict ourselves essentially to them. We will never forget, however, that everything we say about the two or three geographic dimensions of the total environment also applies in the last analysis to the other dimensions, and thus to the total environment itself.

The rule that phylogenetically homogeneous groups of higher rank also inhabit fundamentally continuous areas should not be misunderstood to mean that such areas (a continent, for example) must be unbroken in the trivial sense. They may extend over broadly separated parts of different continents, and such cases are called "disjunctions." Darlington (1957), among others, has described the most important of them (Fig. 42). These are of particular importance in systematics because in a monophyletic group with disjunct distribution the partial areas are usually occupied by different partial groups. Species groups that arose from the stem species of a monophyletic group by one and the same splitting process may be called "sister groups." We can then show that there is often a vicariance relationship between monophyletic groups that represent such sister groups. We may find, for example, that the sister group of a Neotropical group is Holarctic or Australian (Fig. 42).

Up to now we have said that the task of phylogenetic systematics is to construct a system that contains only monophyletic groups, at the same time presenting all recognized monophyletic groups. Instead we may now say that its task is

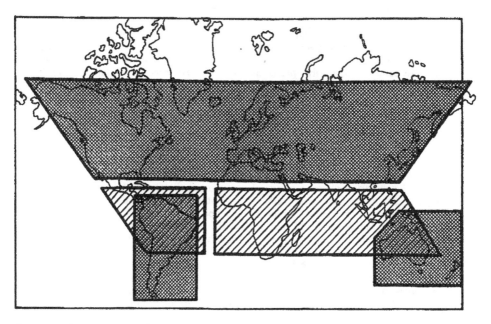

Figure 42. Some of the most important disjunctions (vicariance types of higher order).

to express all sister group relationships: every monophyletic group, together with its sister group (or groups), forms—and forms only with them—a monophyletic group of higher rank. Once a monophyletic group has been recognized, the next task of phylogenetic systematics is always to search for its sister group. The importance of the study of vicariance types is that it provides systematics with clues to the geographic region in which this sister group is to be found.

A little thought shows that sister groups must have the same absolute rank in a phyletic system. Consequently the study of vicariance types is also important in determining the absolute rank of systematic groups. This will be discussed further in the section devoted to this question.

Gareth J. Nelson
In Systematic Zoology, *volume 18*

Points of View

The Problem of Historical Biogeography

There are few branches of science that in their development have included more controversy than has historical biogeography. This is true to such an extent that even today there is no generally accepted methodology that enables biogeographers, when faced with the same data, to reach approximately the same answer to a given problem. Frequently, external authorities (e.g., geologists) are called in to resolve problems biogeographers initially have taken upon themselves (e.g., continental drift). Regrettably, theories of past geological events often have had such appeal that biogeographers simply lined up behind one or another alternative theory and interpreted their data accordingly.

Thus, it is necessary for those who discuss the historical geography of some group of organisms to give an account of methods of analysis. The following is a consideration of some problems of biogeography of particular interest to the writer, a consideration inspired largely by the recent discussions of Brundin (1966) and Hennig (1966).

It is generally agreed that speciation normally involves the splitting of an ancestral population into geographically isolated daughter populations. The simplest case might involve an ancestral species B being divided by some geographical barrier into two daughter species: species C in area x, and species D in area y (Fig. 1). Having some reason to believe that species C and D are each others closest, known Recent relatives, the biogeographer can consider the following hypotheses: the ancestral species B last occurred (1) only in area x; (2) only in area y; (3) in area xy; (4) in some area z.

Hypotheses similar to (1) and (2) often have been considered probable when one daughter species seems more primitive, has more individuals or has a larger distribution than the other; indeed, for these reasons, one daughter species often is treated as if it were the species ancestral to the other. Hypotheses similar to (3) have been considered probable when the geographical barrier separating the Recent species seems likely to have been the one initially isolating daughter populations of the ancestral species. Hypotheses similar to (4) have been considered as possible alternatives to those similar to (1), (2) and (3).

It may appear that the four types of hypotheses listed above in themselves are equally probable solutions to a geographical problem involving two, closely related allopatric species. But hypothesis (3) involves fewer assumptions, and the principle of parsimony dictates that (3) is preferable to the others, provided that the known data do not call for its rejection. Thus, on the basis of the Recent distribution of the daughter species C and D, one might predict that fossils attributable to species C might be found in area x, and fossils of species D in area y, back to the time of the ancestral species B, and that fossils of species B might be found in both areas x and y. Collection of such additional data would tend to confirm hypothesis (3). But finds of fossils of species C in area y or in an area z, fossils of species D in areas x or z, or fossils of species B in area z, might call for rejection of hypothesis (3), in favor of either (1), (2) or (4).

The hypothesis concerning the last occurrence of the ancestral species B, can be looked upon as one concerning the origin of species C and D. Thus, hypothesis (3) would be equivalent to one stating that species C originated in area x, and species D in area y.

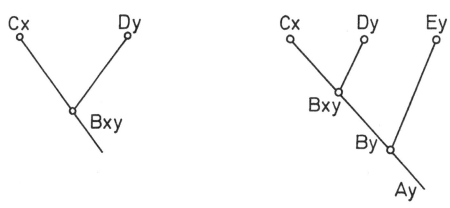

Fɪɢ. 1(left).—Two Recent species C and D, in areas x and y, with ancestral species B in both.

Fɪɢ. 2 (right).—Three Recent species, C, D and E in areas x and y, with derivation of ancestral species B from species A in area y.

Thus, data concerning Recent distributions can be adequate for the erection of parsimonious, testable hypotheses, and data concerning fossil distributions theoretically can be adequate to test such hypotheses. With actual data of this sort, the present biogeographic problem in the main could be said to be solved, at least that part of it concerning the origin of species C and D. But about the origin of species B, little would have been elucidated, beyond the parsimonious hypothesis that B probably originated in area xy.

Suppose, however, that there is in area y a third species E, which is the closest Recent relative of species C and D (Fig. 2). Regarding the last occurrence of ancestral species A (or the origin of species B and E), the biogeographer can consider the following hypotheses: the ancestral species A last occurred (1) only in area x; (2) only in area y; (3) in area xy; (4) in some area z. Of these hypotheses, (2) would be the most parsimonious, involving only one migration between areas, that of species B from area y into area x. Thus, on the basis of the Recent distributions of the species C, D and E, one might predict that fossils attributable to species B and E might be found in area y, back to the time of the ancestral species A,

and that fossils of species A also might be found in area y. Collection of such additional data would tend to confirm hypothesis (2). But finds of fossils of species E in area x or in some area z, fossils of species B in area z, or fossils of species A in areas x or z, might call for rejection of hypothesis (2), in favor of either (1), (3) or (4).

The significant point here is that the distribution of the Recent species E in area y, makes area y also the probable area of origin of species B, and leads to the replacement of the hypothesis that species B originated in area xy, by the more parsimonious hypothesis that species B originated in area y. Thus, data concerning Recent distributions can be adequate not only for erecting an initial hypothesis, but for improving it as more data become available. Provided that the phyletic relationships among the Recent species are correctly understood, the improved hypothesis not only is more parsimonious, but is more probable. Assuming the relationships shown in Fig. 3, who could doubt, e.g., that the occurrence of this lineage in area x is a secondary and relatively late one?

Suppose that there is a Recent species C, without a fossil record, living in area x and a

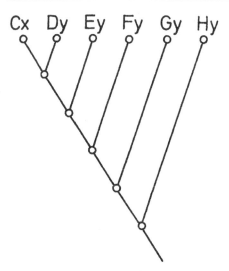

Fig. 3.—Six Recent species (C–H) in areas x and y, suggesting derivation of C from area y.

lation or species, that C evolved from †D and during its evolution moved from area y into x. If a given fossil could be demonstrated to have been a representative of a population ancestral to a Recent species, it might have some such significance. But this ancestor-descendant relationship strictly speaking cannot be demonstrated. Accordingly, if evidence is available suggesting a relationship between some fossil and a Recent species, their relationship can be assumed to involve only a common ancestor. Even if the fossil by chance was a member of an ancestral population, common ancestry would be no less true. It follows that fossil †D in area y distributionally would involve ancestral species B in area y, but no more than in area x, in which species C already is known to occur. The significance of this conclusion may better be grasped by allowing †D to survive in area y to the present day, giving an example similar to one already discussed (Fig. 1) and leading to the hypothesis that the common ancestor B last occurred in xy. Similarly, if in area y another fossil †E (not attributable either to C or †D but related to them) were known

similar fossil †D in area y (Fig. 4). With such data it often is assumed that the fossil is a representative of an ancestral popu-

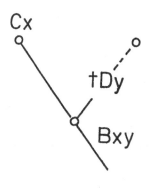

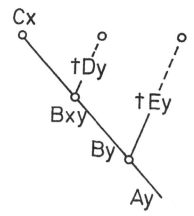

Fig. 4 (left).—One Recent species C in area x and a fossil †D in area y, with ancestral species B in both.

Fig. 5 (right).—One Recent species C in area x and two fossils †D and †E in area y, with derivation of ancestral species B from species A in area y.

(Fig. 5), it would have the same significance as if it, too, had survived to the present day, and lead to the hypothesis that the ancestral species A last occurred only in y (Fig. 2). For an historical, biogeographic analysis, it follows, therefore, that fossil distributions are no more nor less significant than Recent ones. It follows, also, that an hypothesis pertaining to the distribution of an ancestral species can be tested only by predicting where all probable descendants, both fossil and Recent, are to be found. It follows, finally, that the relative worth of alternative hypotheses can be judged only by the degree of parsimony shown by each. The above analysis of distribution of species of course can be applied to the distribution of monophyletic groups of any number of species, provided that the groups can be characterized geographically.

The abuse of information pertaining to fossil distributions has been flagrant in most discussions of vertebrate phylogeny and historical geography. Most often this abuse involves the assumption that some known fossil species or group is ancestral to a Recent one. In the writer's opinion this assumption is unjustifiable.

As here conceived, a biogeographic analysis implies, logically follows from, and at best can be no more reliable than, a prior phyletic analysis. Biogeographers, however, have not always recognized the interdependence of distribution and phylogeny, and often have been either unable or unwilling to come to grips with detailed problems of phyletic relationships. When phylogeny has been discussed in relation to distribution it almost always has been distorted by inappropriate assumptions, e.g., that one Recent species or group can be said to have given rise to another, that ancestral species can be recognized as such in the fossil record, that "primitive" species do not disperse as rapidly as "advanced" ones, etc.

It perhaps still is widely believed that historical biogeography has as its ultimate objective the elucidation of ancient land and water connections affecting the distribution of past biotas. As here conceived, the ob-

jective is a more limited one, concerning only the distribution of ancestral species. With an incomplete fossil record it is to be assumed that statements about distributions of ancestral species necessarily are hypothetical. In this respect they take on the hypothetical character of a morphotype (see e.g., Zangerl, 1948; Nelson, 1969).

It is a complicating fact that the distribution of organisms changes with time, either expanding or contracting in response to physical and biological factors of the environment. There is little likelihood, therefore, that Recent distributions are a very accurate mirror of past distributions of the same species or groups. In addition, Recent distributions of closely related species often are very complex, with ranges partially or completely overlapping. In such cases, very detailed geographic analyses become complicated or are rendered even impossible. However, a biogeographer is obliged to use all of the known distributional data to construct the most parsimonious hypothesis of earlier distributions. Historical biogeography deserves a place in science only to the extent that its methods, given the same distributional data, can lead to such an hypothesis, and produce agreement that in fact the hypothesis is the most parsimonious.

REFERENCES

BRUNDIN, L. 1966. Transantarctic relationships and their significance, as evidenced by chironomid midges with a monograph of the subfamilies Podonominae and Aphroteniinae and the austral Heptagyiae. K. Svenska VentenskAkad. Handl., ser. 4, vol. 11, no. 1.

HENNIG, W. 1966. Phylogenetic systematics. University of Illinois Press, Urbana.

NELSON, G. J. 1969. Origin and diversification of teleostrean fishes. Ann. New York Acad. Sci. (in press).

ZANGERL, R. 1948. The methods of comparative anatomy and its contribution to the study of evolution. Evolution, 2:351–374.

GARETH J. NELSON

Department of Ichthyology, American Museum of Natural History, New York, New York 10024.

From *Space, Time, Form: The Biological Synthesis*
Leon Croizat

CHAPTER I
On rudiments

A) On method

In the very beginnings of his history, Man saw that bolts from the high seared the clouds in time of heavy rain. He doubtless knew that lightning could set steppe and forest afire, and it did not escape his sight and hearing that the rumblings of thunder follow, do not precede a flash.

Thus did Man learn by immediate observation over 1 million years ago [1] what was in itself sufficient to tell him that a form of energy existed in nature greatly exceeding in its power the flame of the campfire and the impact of a club. Man could also know by then that lightning, the fire of a volcano, the flame from burnt wood stood bound by common properties. Thus did Man have at the very dawn of his ascent, by nothing more than trusting his senses, positive information based on direct observation and averages of performance that could be used to promote the immediate inception of a scientific age.

Why this age was on the contrary no less than about 1 million years late in dawning is a complex question on which a separate book may well be written. How this age arose is of course in itself a simple matter. It dawned when Man resolved to account by reason for what he had heretofore accepted without insisting on a positive interplay of natural causes and effects. Zeus had to be done away first of all as the Lord of Lightning, and much later some crude contraption had to be gotten together fit to release sparks at Man's will. Fear of the unknown, whimsical celestial and infernal powers had to leave the field clear to logic working through experimentation and disciplined imagination. In sum, the mind of Man had itself to change and to grow before it would stop reacting before Nature as but that of a child. Pertinent facts had been known for ages but attitudes had to mature and methods to be devised to interpret them with the purposeful rigour of laws.

---

[1] *Zinjanthropus*, a definite "hominid" from Tanganyka using tools, has been dated to about 1,750,000 years. See for an informative reference, Leakey, in Natl. Geogr. Mag. (Washington, U.S.A.) 118:420. 1960; op. cit., 120:564. 1961; and further, p. 579.

I

As heirs to this very long history, but here deprived of the possibility of leisurely dwelling upon its byways, we are fortunate at least to the extent of counting upon a few lines, penned some twenty centuries ago by a Roman philosopher imbued with Greek spirit, that bring to a head the whole of it. Writing of lightning, and the different attitude in its regard by the superstitious Etruscan and the positive Roman, Seneca said the following (Quaest. Nat. II, 32, 2): *"Hoc inter nos et Tuscos. . . . interest: nos putamus, quia nubes collisae sunt, fulmina emitti; ipsi existimant nubes collidi, ut fulmina emittantur (nam cum omnia ad deum referant, in ea opinione sunt, tamquam non, quia facta sunt, significent, sed quia significatura sunt, fiunt)."* [1]

Neither the Graeco-Roman nor the Etruscan explanation of lightning is correct as we know it today, and both nations argued accordingly outside of science. However, less than successful for the time being in point of sheer fact, the Graeco-Romans were beyond comparison more advanced than the Etruscans in their attitude toward natural phenomena. The latter theorized and believed by authority and rote in a purely religious frame of mind even though they were well schooled in precise observation and accurate reporting. The former did so no longer, and by a simple, quite radical change in approach they had devised a method leading through objective observation of the facts of nature to their eventual explanation as phenomena lying within the scope of human reason. By his "method", the Etruscan, son of a typically Oriental past, would forever go on believing. The Graeco-Roman, herald of the Western future, had by his own method completely altered on the contrary the relationships between Man and Nature. Before the latter, not the former, opened a field clear to explore, to reason, to meditate, eventually to explain, and so to advance free of limits. *A simple change in attitude thus did bring about a basic difference in methods, and spelled the difference between the unbound past and future.* Seneca did kill teleology.

The course of human events is fraught with unforeseeable contingencies, and liable to be altered by an interplay of causes and effects that the keenest mind cannot hope to plumb. No doubt, the Etruscans had preserved in their annals records fit to prove on the strength of past experience that decisions on matters of life and death taken to agree with divine omens, whether by lightning, hepatoscopy, the flight of birds, etc., are oftentimes equal to the shrewdest anticipations based on what Man calls reason. By

---

[1] In a rather free translation: This is how we, Graeco-Romans, do differ from Etruscans: we believe that lightning originates when cloud clash. They take it for granted that powers unseen cause cloud to clash in order thus to release lightning. As the Etruscans read in everything the will of the gods, they do not believe that natural happenings are meaningful in themselves but, rather, that they occur in order that divine powers may thereby reveal their intentions.

2

sheer law of probability, decisions ruled by superstitious formulae would prove right in about fifty per cent of the cases. Prodded by the natural sagacity of some servant of the Gods, even brute lightning may be reasonable in its portents in better than one case out of two. Indeed, so deeply tinged with stupidity is the run of history that a cogitative student of human events is drawn to wonder whether, in the end, the Etruscans were not wiser in trusting the Gods rather than their own free judgement. This question is still very pertinent at this hour.

Whatever the argument be in its philosophical implications — with these we may not be concerned here, of course — we will derive from what Seneca so beautifully expressed a lesson of immediate application to the whole of our enquiry. Understood as genuine learning rather than as a mere piling up of disjointed facts, science primarily rests upon approaches responsible for precise methods of enquiry. As a sheer matter of recorded observations, it is quite probable that the Etruscans knew much more of lightning than ever did the Romans [1], nor it is at all unlikely that the latter received their first explanation of it from the former. However, the Etruscan mind was pre-, even more anti-scientific in its approach, therefore method. Science only then is born when the facts of nature are perceived as such, and *explained away* on the basis of what they intrinsically contain in reference to averages secured by critical comparison. Nothing else avails, and the whole of the history of knowledge is starkly repetitious by a constant change of methods by which it became possible at each turn to see the old as if new, to dismiss outworn axioms for new ideas, in a word, *to advance*. By the application of proper method, Man can become an ever more intimate partner of Nature in her operations, which means that, in a way at least, Man turns nature into his servant. Whether with lightning or the records of geographic distribution of plants and animals the world over, the method that befits the will to know farther and deeper is the very same. The Etruscans did not know it, the Graeco-Romans started it at least on a straight course.

Thus warned by our distant elder, we will renounce throughout this book whatever is preconceived approach to the knowledge we seek. Instead of indorsing theories, authority, fiats, and to indulge accordingly in compilation of what others thought before us, we will but do our best freely to reason out certain manifestations of nature, exploring

---

[1] This is worth heavy underscoring. *An occasional error in point of sheer fact is by far less pernicious than the application of a mistaken method.* To illustrate: the first work published by Abel, one of the greatest mathematical geniuses of all time (Vera, 20 Matématicos Célebres 15 - 16. 1961) contained an error which — later corrected by Abel himself — originated the theorem that made Abel's name immortal. Abel's method, obviously, was the major consideration before and after the error in question. See also p. 684.

above all the factors of space and time that belong to their substances. We will make a positive effort to infer from what we are to deal with, what commonsense — doubtless the most powerful tool of scientific enquiry, today and in the past and future — is to advise case by case and overall. We cannot hope being fully successful, of course, and our explanations may for the time being be no better than the one which Romans had of lightning. We will, however, most certainly not think the like the Etruscans did, and so we will take nothing for granted on the strength of somebody else having said it, and everybody just now believing it. One thing is obvious: so long as the method we follow is in principle correct, matters of detail will fall into line without undue lag or difficulty. Science is perforce a collective undertaking, and what we will leave behind us undone or ill done will easily be rectified by those to follow in our wake if only we begin it right. The highest reward awaiting an enquirer is that he is corrected in short order in everything of detail, but not in point of method and approach. Chockfull of factual errors, the effort of Copernicus yet managed to make short work, because of its method and spirit, of fifteen centuries of dominance by that of Ptolemy.

B) On patterns of distribution, in general

Concrete examples are to furnish the subject of our enquiries throughout this book, and we will presently begin to work on one of them. To conform with the notion current today that the *geographic distribution* of plants obeys standards different from that of animals, I should of course be bound to tell the reader on the spot whether we are to deal with, e.g., a rose or a frog, giving for everything exact authority, citation, etc.

There is no reason to accept current notions for a standard, however. Had the Romans never begun to challenge by a shift of approach what the Etruscans had told them of lightning, the former would have forever remained quite as antiscientific as were the latter. It is accordingly eminently scientific to take absolutely nothing for granted. This precept and rule does apply first of all to what seem to be the foundations of a science. It is these foundations, as a matter of fact, that tend to escape critical reappraisal because everybody views them as sacred and certain. In the end, they invariably prove to be the exact contrary of what it was once thought; and it is by their being exposed as false or inadequate that science does progress. Its history teaches nothing else.

No naturalist ignores today that the biological evolution of the earth has called for a constant joint renewal of animals and plants. Mammals and angiospermous plants came to this world virtually together to replace dinosaurians and ancient forms of vegetation. When so clearly and finally associating the two in evolution *over space, in time, by form,* nature

4

would obviously not deal with either using standards particularly applying to only one group or individual.

Of course, if we deal with the anatomy of an oyster or the morphology
of a pine tree, we cannot act otherwise than as zoologists — indeed, con-
chologists — or as botanists, perhaps even as specialists on certain aspects of, e.g., coniferous life. However, we are here to enquire in another
capacity: we will deal with our subjects first and foremost as *biogeographers,* that is, essentially as *students of the effects of space and time on
the course of organic evolution.* A time will come, naturally, when we must
dovetail our findings with those of particular disciplines outside our
own, but this time is not yet, and whatever we may do just now will
answer our own needs, our own methods, our own aims as biogeographers.
If we are to use alien materials, we will do with them strictly as our own.

Let us, then, choose as our first example a certain group A, which is
subordinately divided into a number of consanguineous entities: a, b,
c, d, e, f, the geographic distribution of which is given by specialized
monographic work as follows: 1) *Aa* — Western North America (Baja
California to Washington (State)); 2) *Ab* — Western United States (California to Washington, inland to Western Colorado); 3) *Ac* — Southwestern United States (California); 4) *Ad* — Western United States
(Northern California); 5) *Ae* — Hawaii; 6) *Af* — Bolivia.

We will in the first place observe that the element *space* of this pattern
of *geographic distribution* (Fig. 1) would remain unaltered if *Aa, Ab, Ac*
and *Ad* were lumped together as a single taxonomic entity. In other words,
whether *Aa, Ab,* etc., are different *species,* or but *subspecies* or *varieties* of
one and the same *species* makes no difference at all, so — *as biogeographers* —
we cannot be concerned with the narrowly formal aspects of the case.
Common is indeed the occurrence when taxonomists of equal authority
and experience do not agree on specific or generic limits. The biogeographer will of course not take sides when beginning his work, even though
he may have good grounds at the end of it to formulate his own observations also on questions of pure classification. Space and time are basic
factors of form-making, and the biogeographer who competently interprets factors of the kind cannot remain for long ignorant of systematic
and taxonomic subjects. *He deals, as a matter of fact, with the substances
that underlie classification and mold out evolution,* and if he is at all competent in his own chosen field he cannot fail developing a critical, well
rounded appreciation of natural history in general. He may be only a specialist in biogeography, but he can surely not be a narrow-minded naturalist.

In a broad sense, the *geographic distribution* shown in Fig. 1 will of
course suffer no alteration if *Aa* is a taxon of the plains or of the moun-

5

tains, etc. Ecology is not biogeography much as the two science may, and indeed do, closely dovetail in the end.

The factor *time* we face will likewise suffer no change on account of the taxonomic rank of the entities considered. We may of course suppose that if we deal with *varieties* the chronology competent in the premises might not be exactly the same as if we handle *species*, the latter being possibly older than the former in development. However, as we are duly to learn (see, e.g., p. 217) when time will be for it, it is unwise to judge of what nature performs according to man-made, therefore academic standards. Trained as we are today, we are bound to assume that the *genus* comes first before the *species* because the *genus* is made up of *species*, and the whole is more important in the absolute than its parts. In reality, our training is less than successful: a *genus* could hardly exist independently from certain at least of its *species*, which but means that *genus* and *species* may be absolutely contemporary in origin. Let us then conclude that nothing can be taken for granted, and that *time* in evolution is not tantamount to *age* as expressed in terms of one or the other taxonomic group. The student of biogeographic distribution, and what follows from it, must *by all means* constantly be on his guard against standards of judgement that do not fit the essential requirements of the subject. *Life is virtually endless over space through time by form, and to it criteria of space, time, and form do necessarily apply of a kind that fits infinity rather more than its contrary.* On account of the intrinsic requirements of their subjects, the astronomer, geologist, and biogeographer can certainly not reckon by the watch and the calendar, the inch and the yard.

Finally, the factor *form* in the *geographic distribution* here before us will of course *technically* alter if we handle *varieties* or *species*. However, this will not interest the face of the pattern, therefore cannot be viewed as a primary biogeographic consideration, even if all taxonomists were for once agree as to the rank of the entities under enquiry. By ruling taxonomic considerations out of biogeography in principle I do not intend of course to divorce classification and biogeography as incompatible. Indeed, properly conducted enquiries into space and time are essential to successful taxonomy. The point is this: a biogeographer is primarily interested in *form* in general, as one of the members of the broad equation: *Evolution = Space + Time + Form.* He is not interested in questions of *form* in particular, to the extent for example of worrying as to certain technical characters of the tarsus of an insect making of it a good species or, perhaps, a good subspecies only, etc. The biogeographer is interested in *form-making*, that is, in the process responsible over space in time for the appearance of a certain taxonomic group at a certain point of the map, whether insect or spurge. Of course, a safe understanding of

6

form-making in general proves in the end to be a powerful coefficient of proper understanding of *form* in one or the other particular field of biology, whether entomology, botany, etc.

The objection may, and will be raised that we cannot be certain that the records of A are complete as we have them at this hour. For example: coming exploration may reveal the existence of a taxon *Ag* in, e.g., Peru bridging the gap between *Ad* and *Af*, therefore altering the face of the pattern of geographic distribution with which we deal. This is of course not impossible but it is a question, to begin with, whether the addition of *Ag* would be such as to detract from the cogency of the method of enquiry we follow, and as to alter materially the pattern in question. We may at any rate accept the scores in our hands as *random sampling* rather than as fully final. Naturally, random sampling may be subjected to proper consideration of limits by statistical comparison, and be accordingly used without serious danger of error in biogeography as in every other biological science. As we shall see, breaks in distribution between North and South America calling for gaps, or disconnections, as we will call them, of the order: Mexico/Peru, Mexico/Bolivia, Mexico/Chile, and the like, are so repetitious in the distribution of animals and plants as to be standard much sooner than unusual. In sum, whether the break we might face is Mexico/Peru, or Guatemala/Bolivia on the map of today, and by its political geography is indeed no biogeographic consideration at all.

Cutting these preliminaries short, we will now connect the stations of A by a line interesting them all. This line (Fig. 2; see p. 11) is what today most everybody is agreed to call a *track*. We will accept it as such, and for the moment at least understand the whole as a *graph of geographic distribution*. This graph is of course a proper analytical tool, whatever its true nature in the end. It tells us at a glance that a certain entity A, animal or plant that it may be, occurs in Western North America, Hawaii, and Bolivia, in the last two centers sharply disconnected away from the first. The *track* accordingly places in our hands the primary coordinates of A in *space*, and opens thus the way to an enquiry into factors of *time* and *form* material to considerations on *space*, everything of it against a general background of evolution.

Since these coordinates are essentially factual, the analysis to follow no longer needs theories to its support. By objectively comparing *tracks* in amounts sufficient to yield the averages of the behaviour of animals and plants in space through time in Bolivia, California, Hawaii, etc., etc., biogeography is quite competent to reach conclusions that owe nothing to preconceived guesses. In sum, thus understood and conducted, biogeographic analysis discards at the outset — because of a basic requirement of method — everything which does not strictly belong to the re-

7

cords of geographic distribution and to whatever these records may tell by averages of comparative performance. Naturally, to this very extent — on ground of method and principles — biogeography as we will practice it hardly can agree with the zoogeography and the phytogeography now current, and with the theories upon which they rest. The break is complete right at the start.

The objection may be anticipated, of course, that, as Fig. 1 and Fig. 2 implicitly admit, the connections laid down among different sectors of actual occupation are not factual, not at least as factual as are the recorded localities of occurrence. These connections, the objectors will say, are hypothetical, indeed personal with their authors; and it will finally be inferred from this prelude that the concept of "track" is essentially theoretical, not at all equivalent to that of a statistical graph as I incorrectly represent it to be.

My answer to this objection is that the proof of the pudding is in the eating, even in science. To believers in theoretical brands of "zoogeography" and "phytogeography" a "track" may seem very theoretical, as everything else of "sciences" of the kind, "migrations", "casual means", etc., etc. To the competent student of dispersal, however, the obvious conclusion is that a graph is a graph to begin with; and that only by putting many graphs to work in a steady play of comparisons and averages is the question eventually to be answered whether the graphs expressing "tracks" in Figs. 1 and 2 are illusory. Let us, then, continue our work using "tracks" for all they may return. If objectors are correct, we will learn it from what we are to find in the course of our enquiries, and since we owe nothing to any theory we will immediately agree to rectify our course to conform with the facts, whether with or against Darwin and all his followers. *At any rate, let us start doing things instead of but squabbling about theoretical postulates.* Some of the weightiest "objections" levelled by most learned professors and clerics against Copernicanism in the name of Aristotelism look today rather foolish when not entirely ridiculous. Since things and men cannot be very different today than they were some four centuries back — the tune is the very same if the tone be slightly different — the chances are that our "zoogeographers" and "phytogeographers" — being theorists to their marrows — do fire with their "objections" loud blanks rather than genuinely live ammunition.

According to the standards of "zoogeography" and "phytogeography" as now understood, the pattern of geographic distribution of Fig. 1 and Fig. 2 would by no means invite the thoughts we have just formulated. This pattern would on the contrary but invite questions such as these: 1) What are the "means of dispersal" that made possible for *Ae* to co-

8

lonize Hawaii, and for *Af* to invade Bolivia?; 2) Did these "means" act overland or oversea? In other words: was Hawaii reached over a "landbridge", by "stratospheric means of conveyance", by "rafts", by "chance colonizations", etc., etc.?; 3) It is possible to conclude anything at all about anything at all on the strength of a pattern of distribution which does not reveal whether the form disconnected in Bolivia is, or is not, the lone survivor of an once continuous range? What *if* new records should turn out, as they almost certainly *must*?; 4) What was, above all, the "center of origin" of A? Bolivia, Hawaii, Madagascar?; 5) What was the "species" the origin of which did determine the origin of A as a whole? What is "primitive" and "derivative" of the whole?; 6) What is the prevailing opinion among authorities on the subject of a distribution such as A's?

"Zoogeographers" and "phytogeographers" take these, and similar, questions as legitimate, and strive to answer them making reference for the purpose to the distribution of one, or at the most few selected cases, seen at any rate through the prism of a theory which, when not identically the same as (see p. 638) that advanced by Darwin a century ago, does in fact not differ from it in anything substantial.

It is easy to see that — legitimate or not that be questions of the kind — they can only be answered in reference to the averages of performance over space, in time, by form returned by a critical comparison among numerous patterns of geographic distribution. Even more: *questions had better not be moved until these averages are known.* There is no sense in looking for answers that the facts cannot return through no fault of their own, but simply because the questions are themselves inept. As a science of life over space, in time, by form — therefore, as a science implicitly of cosmic attainment — biogeography deals with patterns of geographic distribution in series the world over, that is, with $A + A^{n+1}$ cases, not only with case A in the hope of concluding from it something only valid for, e.g., the "zoogeography" of Hawaii or the "phytogeography" of Java. It is assuredly not nature operating through endless time and over space unbound that would pay attention of the geography of our maps, and respect claims such as these: I cannot be interested in the frogs of Java because I am a botanist specialized on the flora of Central America! Nature is not interested in our "specialties", and the very least duty she imposes upon the student of distribution is that he is a "specialist" on the entire distribution of plants and animals the world over. Self-evident as this duty is, still very few are the naturalists who would think today of it as normal. Most everybody in biology today mistakenly believes that concerns with distribution are sheer appendages to one or the other branch of classification.

9

Notions of so palpably thin and destructive a kind would not be current, of course, if they did not rest upon a long tradition essentially supported by uncritically accepted authority and compilation. As a matter of fact, these notions go straight back to Darwin's own "Origin of Species", 1859. Here for example is one of the Darwinian texts pertinent to our discussion (Origin of Species, Chapter XII, "Single Centers of supposed Creation"), as follows: "We are thus brought to the question which has been largely discussed by naturalists, namely, whether species have been created at one or more points of the earth's surface. Undoubtedly there are many cases of extreme difficulty in understanding how the same species could possibly have migrated from some one point to the several distant and isolated points, where now found. Nevertheless the simplicity of the view that each species was first produced within a single region captivates the mind. He who rejects it, rejects the *vera causa* of ordinary generation with subsequent migration, and calls in the agency of a miracle. . . . It seems to me, as it has to many other naturalists, that the view of each species having been produced in one area alone, and having subsequently migrated from that area as far as its powers of migration and subsistence under past and present conditions permitted, is the most probable. Undoubtedly many cases occur, in which we cannot explain how the same species could have passed from one point to the other. But the geographical and climatal changes which have certainly occurred within recent geological times, must have rendered discontinuous the formerly continuous range of many species. . . ".

Whether true or false, there is nothing of what Darwin thus states that rests upon a factual, comparative analysis of what life actually performs through space in time. The whole of it is speculative and assertive and cannot be received as even minimally supported. Apart from its intrinsic weakness from the standpoint of approach, this Darwinian text — *a text of enormous importance and influence, for with it began "zoogeography" and "phytogeography" as now still current* — induces the critically minded naturalist to wonder on the thoroughly justified basis that: 1) It does make no effort at squarely meeting fundamental issues of interrelationships among space, time, and form which are of the essence of a solid theory of evolution; 2) It rests its main conclusion on purely propagandistic affirmations concerning a specious "simplicity of view" which is the pretext for accepting without sustained consideration what does on the contrary demand consideration of the deepest; 3) It reverses the normal order of scientific methodology. *It is the exception that comes first before the assumed rule*, because so long as exceptions do challenge the rule — in a manner statistically relevant above all — the rule cannot be said to hold. In sum, it is overall but "Etruscan" reasoning based on the theory that

10

the "species" must have a "single centre of origin" attended by "migrations", and virtually the single argument tendered for so weighty an assumption is that its simplicity does captivate the mind. So do rest upon a like argument all the nostrums which, everywhere in life, captivate simple minds with their simplicity [1]. True simplicity is of course achieved only when rules and exceptions merge, *both*, within a simple explanation of either and both, but this is not a requirement that simple minds will appreciate for its ultimate — this time, genuine — simplicity. Shortcuts, whatever their merits, they like better, and Darwin perfectly understood, it seems, what he must offer others when being himself hazy in mind and counsel.

We are of course not interested in taking issue here with the Darwinian postulates and what has followed from them to this day [2]. The whole

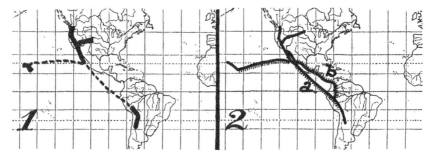

Fig. 1. *The geographic distribution of a certain group A.* Regions of actual occupation marked out by heavy lines indicatively connected by broken lines. See the main text for discussion.

Fig. 2. *The geographic distribution of a certain group A.* The sum of the sectors of actual occupation and of the "tracks" connecting them (hatched lines) yields a *graph of geographic distribution.* "Tracks" *a,b* are integrative rather than exclusive. See the main text for discussion.

[1] No naturalist better caught the weakness of the Darwinian argument in fewer words than did Agnes Arber there, where she wrote (see *Princ.* 1b: 1689): "The facile Darwinian way — so easy to understand, and therefore so fatally easy to accept".

[2] I would not be inclined to take up here in earnest the objection that the text I have just quoted can be contradicted by others scattered throughout the "Origin of Species". I am not unaware of this (see, e.g., p. 635), but the very fact that Darwinian oracles can be made to talk from both corners of the mouth is an indication of their finally dubious nature. Had Darwin, at any rate, spoken clearly and finally current "zoogeography" and "phytogeography" would not think and perform precisely along the lines of the text I have just brought to record. Right or wrong that I may be — and wrong I must be in better than a few cases, in matters of detail when not of method — I feel sure that I will not be easily quoted to different purposes in anything basic of my thinking. Nobody has the right of being ambiguous in fundamental matters, which does of course not mean that everybody must be omniscient in everything right at the start.

of it will be judged, amply even *if* only implicitly, through the whole of this book. Suffice therefore here to observe that, to judge from the face of the geographic distribution of A (see Fig. 1, Fig. 2), the quest for the "centre of origin" of the "species" responsible for the whole of A offers little assurance of concrete results. Assuming that, for instance, *A* is a genus, and that species *Af* of Bolivia looks "primitive", Bolivia and *Af* ought to be the fountainhead of the whole. However, it is altogether reasonable that the center of main massing, which lies on the Western United States on account of *Aa — Ad*, may owe its origins to ancestors no longer now in existence but much older than *Af*, therefore the Western United States may be the true "center of origin", not Bolivia. As to "migrations", the question must be, why great many animals and plants happen to be disconnected, on a clear statistical basis, between North and South America in the manner of *A*? Did the "migrations" run overland, and if they did why so many disconnections? If they did not, "migrations" from Mexico to Bolivia cannot be more challenging than "migrations" from Mexico to Hawaii whether overland or oversea; therefore, the issue of "means" becomes irrelevant on the basis of land or sea, landbridges vs. continental tracks, etc. It seems to be a well established fact that Hooker, who knew a marvellous amount of phytogeography over 125 years ago (*Princ.* 1b: 1282 ff.), had at least an inkling of these and like questions, and knowing he could not answer them — though in principle he already may; knowledge in his hands was sufficient to do so — steered clear of their coils. Not so others, who allowed themselves to be captived by simple viewpoints which, treasured without much justification from 1859 to this hour, have caused biological thinking a lag of a century at least.

Naturalists not aware of the basic importance that approach and method have for science may presently decide what seems to them best. I will for myself conclude right here that the actual and historical implications from what we have discussed do not favour on their very face Darwin's "geographic distribution". Darwin's approach to it could only lead — as its author had conceived it — to an unsatisfactory biogeographic method; a method based on half-baked axioms rather than upon a cogent analysis of factual interplays of space, time and form. It must right here be obvious that a false start is in itself enough to vitiate whatever follows in its wake. It will further be plain that problems on "center of origin" and the like call at all times for sharp reckonings of *space* and *form*, which invite on the rebound enquiries into *time* as their solvent. In sum, *biogeography cannot be extricated from evolution, and the other way around*, because, quite concretely speaking, a scientifically conducted analysis of the distribution of *A* does require a joint consideration of *form-making* and of *translation in space*, that is, must amount to a critical enquiry of the entire evolution of *A* within three points, Northwestern United States, Hawaii, and Bolivia. Darwin had at least an inkling of this because in one of his letters to Hooker written in 1845, he spoke of "Geographical Distribution" as: "That grand subject, that almost keystone of the laws of creation". This is eminently correct, but believers in its truth should by all means try at least to find the hole wherein the key can fit to open the laws of creation: simple simplicity will not map out this keyhole. [1]

On the basis of the conclusion that a genuinely scientific enquiry of *A*'s history through space in time by form requires joint understanding of *form-making* and *translation in space*, we may right here discriminate two concepts that are generally confused today, as follows: 1) *Geographic distribution* — As such, we will understand, as a simple fact of nature, the records of occurrence at different points of the map of the modern world of consanguineous entities forming a taxon. The *geographic distribution* of, e.g., A is given at a glance by Fig. 1 and Fig. 2; 2) *Dispersal* — As such, we will identify a coherent explanation of geographic distribution formulated in joint reference to *form-making* and *translation in space*. In sum, *geographic distribution* holds the record, *dispersal* interprets it. This means in regard to A for example, that Fig. 1 and Fig. 2, and what they stand for, are the beginning of the biogeography of A, by far not the end of it. The end will come according to a genuine science of dispersal (= (pan)biogeography) only when what Fig. 1 and Fig. 2 display shall have been satisfactorily accounted for by a method responsible not only in regard of the geographic distribution of A, but of that as well of the whole of life at large, plants and animals the world over. This requirement is not academic. It is as a matter of fact inconceivable that a scientific explanation valid on the basis of space and time in regard of the form-making of A can be untrue of other cases, A', A"...., whether in America or Africa, Hawaii or Bolivia.

Translation in space is obviously not flatly synonymous of *migration*, for, as we shall know (see p. 209), a certain taxon may have "migrated"

---

[1] Purely as a preliminary note: Darwin does not seem to have been very fortunate in the choice of the arguments he developed. He had clearcut perception, as we have just heard, of the basic importance of dispersal, but radically misconstrued it in his *opus magnum*. He had an understanding of broad directional streams of deployment ("laws of growth"), but he hardly did more than mentioning them. In sum: Darwin can in theory be given credit more or less fairly for a very great deal that he never did develop, but it is not always easy in practice to approve of what he did on the contrary choose to develop. Tragic is the circumstance that his warmest supporters have contributed to popularize what Darwin's thought contains that is not necessarily of the best, while failing to expound and to improve what this thought offers, now and then at least, that is genuinely noteworthy. Had not Darwin been ill served by his friends, there would be no reason today to oppose what his name stands for among naturalists. The subject here barely mentioned could furnish ample material for a separate book. See Chapter VIII.

by actual count about 10,000 miles (some 16,000 km) between Brazil and India without at the same time ever having left Brazil or India. In plain words: *Translation in space does not necessarily require migration.*

Entity A of which we have thus far spoken is a group of Dodders (*Cuscuta* sect. *Californicae*; *Man.* 220). The reader could of course not tell whether A is a botanical or zoological subject, and — as a student of dispersal — he knows now neither more nor less than he knew before being told what is A. I suggest accordingly that he stops taking for granted that the methods of zoogeography and phytogeography differ, as many naturalists today believe and teach. Should he not intend to accept my word for it, I would then advise that he repeats the scores analysing patterns of geographic distribution/dispersal drawn from plant and animal life of his choice, always thinking and acting primarily as a biogeographer — which is what he is here, and will be throughout this book — never as a botanist or as a zoologist. He may remain untroubled within the sphere of his specialty as either so long as he has no business with evolution, in its aspects particularly that interest space and time in form-making. As an evolutionist, his status shall of course alter. I say *evolutionist* in the clear understanding that, as it will be apparent from what we have discussed (see further, p. 481), as an *evolutionist* the reader will also necessarily be a *biogeographer. Form-making is unthinkable unless against a background of space and time which, in nature when not in the laboratory, is rather wide as to space and deep as to time, as everybody today well understands.*

As a matter of fact, what space and time do mean in regard of form-making cannot be properly understood in the laboratory, for it is only in nature that averages are compounded in a manner fit to yield laws and rules of broad significance. Naturally, to work with form-making over space through time in nature, better is needed than the Darwinian doctrine of "geographic distribution". Indeed, this doctrine is misleading, as we saw, because of a fundamental failure of approach and method.

To close these preliminaries, I would ask two simple questions: must a botanist and zoologist, whatever his particular "specialty", be interested in a proper method of biogeographic analysis? Can this method be found in the "zoogeography" and "phytogeography" now current?

To questions of the kind only one answer is possible, *yes* and *no*, respectively. *Yes*, as to the first question, because an understanding of the meaning of space, time, and form is essential to a precise appreciation of evolution and its processes, which will readily stand beyond dispute in reference to the numerous examples we are to discuss in the pages to follow. *No*, as to the second question, because the "zoogeography" and "phytogeography" now conventional rest upon a faulty appreciation of method historically traceable to Darwin's own "Origin of Species", 1859, and never rectified since as a whole.

C) On patterns of distribution, in particular

Convenience suggests that we continue working with plants for the moment, using a second example in which Western North America, Hawaii, and Bolivia are once again involved. As our work progresses, we will turn of course to animals. [1]

Labiatae Lepechinieae is a botanical tribe of which we have an excellent monograph (Epling, in Brittonia 6 : 352. 1948) by authority of the highest. It consists of two genera, the monotypic *Chaunostoma* (*mecistrandrum* in Mexico (Chiapas), and Southeastern Guatemala (Santa Rosa Dept.)), and the sizeable *Lepechinia* in 8 sections and 38 different species.

Of these 38 species, 34 are distributed in the New World between Central California and Central Chile with manifest stress on ranges toward the Eastern Pacific. Two sections, each with 2 species, score out as follows: A) Sect. *Thyrsiflorae* — 1) *Lepechinia nelsoni*: Mexico (Jalisco, Mexico (State), Guerrero); 2) *L. hastata*: Mexico (Baja California, Revilla Gigedo Archipelago: Socorro Island), Hawaii (Maui); B) Sect. *Campanulatae* — 3) *L. chamaedryoides*: Chile (Aconcagua, Concepción, Malleco, Valdivia); 4) *L. stellata*: Mascarene Islands southeast of Madagascar (La Réunion).

The pattern of distribution [2] of *Lepechinia* (Fig. 3) repeats the disconnection: Western North America — Hawaii displayed by *Cuscuta* sect. *Californicae* adding to score one more insular station (Revilla Gigedo). Mexico and the United States stand this time connected in pattern with Bolivia virtually without geographic break. In Panama alone is *Lepechinia* unreported. [3] Chile does tie with the Mascarenes.

A superficial student would of course readily conclude that the disconnection: Baja California/Hawaii, and Baja California/Revilla Gigedo was bridged by "casual dispersal", whatever the "means", and of course insist that these "means" excluded "landbridges", which are just now not *à la mode* in so called orthodox circles. He would add that *Lepechinia*

---

[1] The reader is to find examples of biogeographic analysis conducted without reference to their belonging to animals or plants in *Princ.* 1b: 1451 ff.

[2] We could call it also a *pattern of dispersal*, apparently synonymizing distribution to dispersal. This would not, however, alter the conceptual difference between the two. Biogeographically analysed, it will become a *pattern of dispersal*. Just now it may only be a *pattern of distribution*. The reader will readily be aware how and when one pattern passes into the other in these pages.

[3] *Lepechinia* avoids Panama the like do, e.g., rattle-snakes and certain birds (*Panbiog.* 1: 647, 791, etc.). Dispersal in Central America is rather less simple than currently imagined by authors who identify this neck-of-land as a "landbridge" between North and South America. See op. cit., Vol. 1 in general; further, *Princ.* 1b: 1659 ff.

CENTERS OF ORIGIN AND RELATED CONCEPTS

LEON CROIZAT,[1] GARETH NELSON AND DONN ERIC ROSEN

Abstract

Croizat, L., G. Nelson, and D. E. Rosen (Department of Ichthyology, The American Museum of Natural History, New York, New York 10024) 1974. Centers of origin and related concepts. Syst. Zool. 23:265–287.—The concept of center of origin in the Darwinian sense is often accepted and used as if it were a conceptual model necessary and fundamental to historical zoogeographical analysis. But in certain respects it is inconsistent with the principles of common ancestry and vicariance[2] (e.g., allopatric speciation), and its application to concrete examples of animal distribution generally yields ambiguous results. In the following pages we present a critique of the concept of center of origin, and outline an alternative conceptual model, involving generalized patterns of biotic distribution (generalized tracks). We assume that a given generalized track estimates an ancestral biota that, because of changing geography, has become subdivided into descendant biotas in localized areas. We assume that in such areas, more or less biotically isolated from one another by barriers to dispersal, the descendant biotas differentiate and produce more modern patterns of taxonomic diversity and distribution. We reject the Darwinian concept of center of origin and its corollary, dispersal of species, as a conceptual model of general applicability in historical biogeography. We admit the reality of dispersal and specify how examples of dispersal may be recognized with reference both to sympatry and to generalized tracks, but we suggest that on a global basis the general features of modern biotic distribution have been determined by subdivision of ancestral biotas in response to changing geography. [Biogeography; distribution; evolution.]

GENERALIZED TRACKS

1. Distributions (tracks)

a. The distribution (track) of a species or monophyletic group of organisms may coincide with the distributions (tracks) of other species and groups.

b. Coincident distributions involving monophyletic groups (coincident individual tracks) confirm the reality, and are components, of a general biotic distribution (generalized track).

c. The distribution of most species and of most monophyletic groups coincides with that of some other species or group and may, therefore, occupy part or all of some generalized track.

d. The most generalized tracks include the largest number of, and the most biologically diverse, groups of organisms both fossil and recent, and are, therefore, the most thoroughly confirmed.

2. Distributions (tracks) and biotas

a. All species are components of biotic systems (biotas) that tend to persist through time despite their more or less gradual change in distribution and species composition.

b. Modern biotas are descendants of one or more ancestral biotas that existed in the past.

c. Ancestral biotas subdivided (vicariated) in response to a changing geography, the history of which is, therefore, correlated with their subdivision and differentiation (vicariance).[2]

d. A generalized track estimates the biotic composition and geographical distribution of an ancestral biota before it subdivided (vicariated) into descendant biotas.

e. The components of one generalized track are geographically and biotically more closely related among themselves than they are to the components of some other generalized track. Therefore, descendant biotas (vicariants) resulting from the subdivision of an ancestral biota are biotically and geographically more closely related

---

[1] Present address: Apdo. 60262, Caracas, Venezuela (reprint requests should be addressed to Nelson or Rosen).

among themselves than they are to the subdivisions of some other ancestral biota.

3. Distributions (tracks) and dispersal

a. Some coincident distributions may include components of different generalized tracks, where generalized tracks, or one or more of their components, overlap.

b. Overlap of generalized tracks, or any of the components of different generalized tracks, reflects geographical overlap of different biotas due to dispersal.[3]

c. The occasional species, or group, whose distribution does not occupy part of a generalized track may have been distributed by chance dispersal.

d. Attempts to explain the distribution of individual plant and animal groups, based on their ecology and means of dispersal, may ignore and obscure existing generalized tracks and the ancestral biotas they represent.

If a given type of geographical distribution (individual track) recurs in group after group of organisms, the region delineated by the coincident distributions (generalized track) becomes statistically and, therefore, geographically significant, and invites explanation on a general level. The first step toward such generalization is to determine what major types of coincident distributions (generalized tracks) recur in the world biota, the number of individual tracks composing each, and the variety of organisms incorporated (Croizat, 1964:21). By this means ancestral biotas may be estimated and compared according to their geographical extent, and the number and diversity of their biotic elements. Interpreted as an ancestral biota, a generalized track serves as a constraining reference for interpretation of its individual elements. Thus a group of freshwater fishes (Galaxiidae), widely distributed in the temperate parts of the southern hemisphere, might by itself be interpreted, perhaps even credibly, as an example of transoceanic dispersal from some center of origin (McDowall, 1973a, 1973b). But when galaxiid distribution is compared with that of other southern hemisphere organisms, many of which have similar distributions but different means of dispersal, e.g., earthworms, freshwater crustaceans and their parasites, molluscs, and birds (Whitley, 1956), in addition to midges (Brundin, 1966) and plants of many different groups (e.g., Croizat, 1952; Good, 1964) a general problem is posed, concerning the original distribution and subsequent history of a pan-austral biota, of which the Galaxiidae might be only a small part (Rosen, 1974). Thus, an ancestral pan-austral biota, including ancestral Galaxiidae, might once have been geographically widespread, and later subdivided into local areas of evolution (Australia, New Zealand, New Caledonia, South America, and Africa) in relation to disruptive geological events.[4] Given these two alternative models (dispersal of Galaxiidae from a center of origin versus subdivision of a pan-austral biota), how might they be compared and evaluated on the basis of distributional data of organisms? The problem is one of different realities: the reality of long-distance dispersal versus the reality of a pan-austral biota. Can evidence be found that long-distance dispersal of Galaxiidae occurs, e.g., that galaxiids spawned in Australia were, or are, regularly distributed in the ocean from Australia to South America? Or, alternatively, can the pattern of galaxiid distribution be found in other groups for which long-distance dispersal is not known to occur; and if so, is the accumulation of individual tracks large and varied enough, as determined by comparison with worldwide standards, to justify the interpretation of galaxiid distribution as part of a subdivided ancestral biota?[5] If not—if the distribution of the Galaxiidae were unique, or unparalleled by a significant sample of the world biota, chance dispersal would be indicated (the "sweepstakes" dispersal of Simpson, 1940:152; see item 3c above). Indeed, the only conclusive evidence for chance dispersal may be the demonstration that a given distribution is unique, unparalleled by that of any other living organisms, and free—for other interpretation as it were —from the constraining reference of a generalized track. Unfortunately, few zoogeographers have ever envisioned, much less attempted explicitly, the necessary first step howard historical interpretation: the determination of the historical biotas as represented by generalized tracks. The bulk of historical zoogeography consists of repeated attempts (e.g., Darlington, 1957, 1965) to explain individual case after case of distribution with reference to the Darwinian concepts of center of origin and dispersal of species according to the means available to each—concepts that seem to us inapplicable on the general level of the historical biotas (as represented by generalized tracks). Centers of origin and means of dispersal only vary from species to species, and do not explain general patterns of biotic distribution (cf. Croizat, 1952, 1958a, 1960, 1964).

The significance of a generalized track may be demonstrated with reference not only to a subdivided continental biota, the parts of which are separated by oceans, as in the case of the austral biota; but also to a subdivided marine biota, the parts of which are separated by land. A simple and dramatic example is the subdivided amphi-American biota, the parts of which are now separated by the Panamanian isthmus. Of this biota Ekman (1953:30) wrote:

"In spite of the fact that the isthmus of Panama nowadays represents an unsurmountable barrier for sea animals, the tropical eastern Pacific and the tropical West Atlantic constitute nevertheless a faunistic unit. This emerges not from a one-sided consideration of the distribution of the species but becomes all the clearer if we consider the genera. This state of affairs . . . may be attributed to historical causes." "Some species, though only few in number, are found in identical development on both sides of Central America. In other cases Atlantic and Pacific species are to be found more closely related to each other than to other species. This suggests that they must have evolved from a common ancestor."

Ekman's concept (1953:31) embodies aspects of a generalized track:

"In determining the relationship of the various faunas to one another and the history of their distribution it is important that conclusions should be drawn not from a more or less subjective general impression of faunistic studies but as far as possible from numerical statements about the composition of the fauna which will permit statistical comparisons."

From assembled data of American warmwater crabs and echinoderms Ekman showed that 2% of the crab species and 18% of their genera, and 0.3% of the echinoderm species and 10% of their genera, occur in the Pacific and Atlantic. He concluded that the species common to both, whether endemic in American waters or more or less circumtropical, seem to be all ancient species that have changed little in their external morphology since the time of formation of the Panamanian isthmus. The shared genera Ekman collectively regarded as a good indication of the close relationship of the Atlantic and Pacific faunas as parts of a common American (amphi-American) thermophile biota. Other conspicuous amphi-American elements include molluscs, crustaceans, and fishes. As cited by Ekman, the numbers of vicariant pairs of species are impressive: 80 pairs of warmwater crabs, 40 pairs of echinoderms, 100+ pairs of fishes, and so on. The conclusion to be drawn from this evidence is that there is a generalized transisthmian track, thoroughly confirmed, which may be part of a larger circumtropical track; that hundreds, and probably thousands (Briggs, 1974), of ancestral species (an ancestral marine biota) once occupied a sea, and later a seaway connecting the Atlantic and Pacific basins; and that the populations of these ancestral species (biota) were subdivided by the gradual subdivision and final interruption of this seaway. There is no evidence of east- or west-bound traffic of dispersing species through the seaway, either before or after its interruption. But there is every indication of wholesale allopatric speciation (vicariance) that began at the time of interruption, and that, as measured by taxonomic differentiation, proceeded at different rates in different groups of organisms

(Rosenblatt, 1963). Like the austral biota, the amphi-American biota was initially recognized on the basis of biogeographical, not geological, data:

> "This conclusion of a former direct connection was reached by Günther [1869:398] before the geologists demonstrated an ancient channel across what is now Central America and his views were subsequently fully confirmed by geology, an example of the legitimacy of drawing in certain cases geophysical conclusions from purely zoogeographical premisses" (Ekman, 1953:36).

The important feature of the amphi-American biota for historical biogeography is that it demonstrates how comparable in detail may be the generalized tracks of a continental biota of which the parts are separated by water, and a marine biota of which the parts are separated by land. It demonstrates also how capricious may be proposals to explain, with reference to means of dispersal but without reference to generalized tracks, the presence of closely related forms, identical or not as to species, on each side of a barrier. It demonstrates finally how desirable may be the consideration that a given distribution (track) is potentially a member of a generalized biotic distribution (generalized track), as in the examples discussed below for the avian genus *Columba* (fig. 1) and the decapod genera *Typhlataya* and *Gecarcinus* (fig. 2).

The internal structure of a generalized track may be closely correlated with the historical subdivision of an ancestral biota. Among south temperate fishes now inhabiting the Gondwanian fragments, for example, closer ties are indicated among South America-New Zealand-Australia than any of these with Africa. Examples are the retropinnid osmeroids (smelts) of New Zealand-Australia; the galaxiids, with a single form in Africa and many interrelated forms in South America-New Zealand-Australia; and the petromyzonid (lamprey) *Geotria australis* of South America-New Zealand-Australia, but not Africa (Rosen, 1974). Following the initiative of Brundin (1966), Keast (1973:331; also Edmunds,

1972) analyzed the contemporary southern biota relative to the separation sequence of Africa, South America, Australia, and New Zealand, and observed that

> "The 'southern ends' of these continents contain a range of temperate, or cold temperate, biotas quite different from those of the sub-tropical or tropical regions to the north" and that there "is good evidence that the former have had long and continuous histories as southern, cool-adapted forms." Keast (1973:338) concluded that "The contemporary southern temperate biotas of South America and Australia are much more closely related than either is to the African one. Since this pattern is repeated in a wide variety of groups, of widely differing ecological requirements, there can be no doubt that this is a fundamental difference and is not an artifact resulting from secondary extinction in Africa. The biological data, hence, confirm the geological data that Africa separated off earlier from the Gondwana landmass."

What may be reflected in the history of the pan-austral biota is subdivision caused by large-scale geographical changes, i.e., the fragmentation of ancestral populations, and their subsequent isolation and differentiation, due to continental fracture and displacement. The existence of large-scale changes does not alter the fact that a given ancestral biota is also subject to subdivision through subtler changes, involving climate, elevation, flooding, erosion, glaciation, and other factors that lead to reproductive isolation of localized parts of the biota. Vicariance, therefore, is a more general phenomenon than continental fracture and displacement, even though vicariance may be well exemplified by those geological processes and their subsequent effects on the course of biotic evolution.

To the extent that vicariance underlies organic distribution, sympatry (range overlap) is evidence of subsequent dispersal (note 3). After observing sympatry, one may infer that dispersal has occurred; but the observation alone does not specify whether one member of the sympatric pair dispersed, and if one member dispersed, which one of the pair did; or whether both dispersed. For example, in the salamander genus *Plethodon*, widely distributed in the

United States, there are 28 unit taxa (definable, named populations), divided into two main groups, which are here assumed to be monophyletic (Highton, 1962, 1972; Brodie, 1970; Highton and Henry, 1970). One group includes eight western *Plethodon*; and the other, with two subgroups, 12 small and eight large eastern *Plethodon*. There is no sympatry between western and eastern *Plethodon*, but there is considerable sympatry among the members of each group, and among the members of the two eastern subgroups. Only a few of the 28 unit taxa are not sympatric with any other *Plethodon*. Within the western group, each of a number of cases of sympatry involves the widespread *P. dunni* or *P. vehiculum* as one of a sympatric pair. Possible interpretations include (1) that a widespread form dispersed into the more restricted ranges of the other unit taxa, or into the range of their common ancestor before it vicariated, and (2) that the unit taxa sympatric with *P. dunni* or *P. vehiculum* arose by vicariance from a widespread ancestral species that had already dispersed into the range of the population ancestral to *P. dunni* or *P. vehiculum*. Similar sympatric distributions occur among the 12 eastern small *Plethodon* and eight eastern large *Plethodon*.

We consider that two general factors are responsible for the complexity of the modern world biota: (1) a continuing temporal sequence of vicariance events, and (2) subsequent dispersal modifying earlier vicariant patterns. Without a history of vicariance, the modern world biota would consist of only one or a few species, most if not all of which would be sympatric. Without a history of dispersal, the modern world biota would consist of no or few sympatric species, although it might have become subdivided into numerous allopatric fragments (vicariants). Vicariance, therefore, produces geographical differentiation and multiplication of species, and dispersal produces sympatry and the possibility of interspecific interaction (competitive exclusion, ecological differentiation, extinction). We identify geological change as the general

causal principle of vicariance, but at present we are unable to identify a general causal principle of dispersal; we imagine that the causes of dispersal are as numerous as the species that have dispersed, although perhaps these causes may be grouped into several classes of phenomena, e.g., physiographic, hydrographic, climatological, biological, and such other factors that may on occasion act to open an environment to a species that previously found it closed to dispersal. We conclude, therefore, that historical biogeography, i.e., the study of the history of the world biota, is to be understood first in terms of the general patterns of vicariance displayed by the world biota. Sympatry (dispersal) means, after all, that a population has broken away from the original geographical constraints responsible for vicariance, and that the original vicariant pattern has, to some extent, become obscured as a result. Operationally, we consider that biogeographical investigation begins with the determination of general patterns of vicariance, and the determination of the geological changes that caused them.

CENTERS OF ORIGIN

"Every animal species originated from a few ancestors in a limited area; if a particular species is now found to be widespread, it must of necessity have reached parts of its present range at an earlier period" (Udvardy, 1969:7).

Applied to a species, the concept of center of origin is a "limited area" in which a "few ancestors" of a species may be supposed to have originated, and from which the species may be supposed to have dispersed to achieve its present distribution. Applied to a group of species, the concept is the "limited area" in which a "few ancestors" of the first existing species may be presumed to have originated. The other species of the group may be presumed to have been derived from the first either in the "limited area" of origin, or in some other area to which the first species dispersed.[6]

"The concepts of centers of origin and dispersal are deeply ingrained in biogeographic thought and supported by so much evidence in the best known cases that other concepts have received little attention. Yet in many specific cases the nagging questions of what the center was or whether there was a center do arise. What is commonly seen in the fossil record seems to suggest that evolution always occurred somewhere else" (Olson, 1971:738).

As early as 1901, Briquet (1901:65–66) stated that the concept of centers of origin ("le principe monotopique") had exerted an unfortunate influence ("a joué un rôle fâcheux") upon phytogeographical research. In his opinion the concept implied numerous assumptions of dubious validity that had to be accepted before a center of origin could be worked out for any given group (cf. Favarger and Küpfer, 1969; Croizat, 1971a). Cain (1943) reviewed the criteria by means of which centers of origin were recognized in phytogeographical studies (Croizat, 1964:595ff). According to Cain at least 13 different criteria had been advanced, not one of which was really reliable. He concluded that

"There seems to be only one conclusion possible, and it carries implications far beyond the scope of the present discussion of criteria of center of origin. The sciences of geobotany (plant geography, plant ecology, plant sociology) and geozoology carry a heavy burden of hypothesis and assumption which has resulted from an over-employment of deductive reasoning. What is most needed in these fields is a complete return to inductive reasoning (Raup, 1942) with assumptions reduced to a minimum and hypotheses based upon demonstrable facts and proposed only when necessary (Hultén, 1937). In many instances the assumptions arising from deductive reasoning have so thoroughly permeated the science of geography and have so long been a part of its warp and woof that students of the field can only with difficulty distinguish fact from fiction" (Cain, 1943:151).

Independently of Cain, Croizat came to the same conclusion and, accordingly, attempted to work out a more inductive approach to historical biogeography (bibliography in Nelson, 1973).

To some extent the conclusions of Briquet and Cain are echoed in the writings of zoologists, e.g., Kinsey (1936:58):

"C. C. Adams some years ago (1902) listed several criteria for the recognition of the center of origin of any taxonomic group; and while only scant argument for and no specific test of the principles was then presented, these criteria have found some approval and have been repeatedly quoted as usable means for finding what I do not believe ever existed."

Nevertheless, criteria for recognition of centers of origin are still listed as routine preliminaries for zoogeographical analysis (e.g., Erwin, 1970:184).

Despite its disadvantages, the quest for centers of origin continues to be a dominant theme of modern zoogeography. Mayr, for example, sorted the North American bird fauna into its presumed centers of origin (Mayr, 1946b:14–15; also, Cracraft, 1973). Of some 100 families he listed 29 as "unanalyzed" for reasons such as these:

"Most of the families of shore birds also are so widespread as to make it impossible to trace their origin." "Among the strictly terrestrial birds, there are eight families [Mayr listed only seven] that are so widespread or so evenly distributed as to make analysis difficult at the present time." "The evidence indicates that all of these families originated at such an early date (Eocene or Cretaceous) that subsequent shifts in distribution have obliterated most of the clues." Mayr nevertheless guessed that the "Caprimulgidae may well be of New World origin." He added that "The woodpeckers (Picidae) are represented about equally well in the Americas and the Oriental regions. They are rather poorly developed in Eurasia and Africa and are absent from the Australian region and from Madagascar. This pattern of distribution suggests a New World (but very early) origin for the family, although the fact that their nearest relatives, the wrynecks (Jyngidae), are exclusively Old World would seem to indicate the opposite."7 For the Hirundinidae he stated that "It is uncertain whether the family originated in South America . . . or whether the 'old-American' swallows are descendants of early invaders from Asia."

Mayr's attempt to resolve centers of origin thus caused him to abandon nearly 30% of the North American avian families as "unanalyzed"—and, presumably, unanalyzable beyond the ambiguous conclusions quoted above.

Darlington (1957:236) opened his chapter on bird distribution by stating that

"In some ways, birds are the best-known animals. Almost all existing species of them are probably known, some 8600 full species (Mayr 1946a; Mayr and Amadon 1951), plus thousands of geographical subspecies, and the distributions of many of the species are known in detail. Of all vertebrates, birds are the ones I know best myself. I have watched them almost all my life and have collected them in a small way in northern South America and Australia. I have had the benefit of many conversations about them with the late James L. Peters and with Ludlow Griscom and James C. Greenway, Jr., of the museum. staff. And Dr. Josselyn Van Tyne and Professor Ernst Mayr have read stages of the manuscript of this chapter and made useful criticisms of it; Professor Mayr has allowed me to use his carded references on bird geography. I have therefore had unusual opportunities. Nevertheless, I still find the distribution of birds very hard to understand. The present pattern is clear enough, though complex. But the processes that have produced the pattern— the evolution and dispersal of birds—are very difficult to trace and understand."

To us, Darlington's remarks convey the curious idea that, according to his method of analysis, the better the data the more difficult is their interpretation—even to the point of impossibility. In our opinion, a properly devised method of analysis should make short work of statistically optimum data such as those of the birds (cf. Croizat, 1958a, in which bird distribution is analyzed).

Analyzing the geography of *Columba*, Darlington (1957:272-273) stated that

"The one genus of pigeons common to the Old and New Worlds is *Columba* (from which domestic pigeons are derived). This genus is an example of ambiguity of numbers clues. It is nearly cosmopolitan. There are about 32 species of it native in the Old World and about 20 in the New, and the Old World species are more diversified, which suggest an Old World origin. But all 20 New World species occur in South and Central America and the West Indies. One of the Central American species extends into western North America north to southwestern Canada, but the genus is otherwise absent from the main part of North America, above southern Florida. There are about 14 species in temperate Eurasia and associated islands; 11 in the main part of Africa and closely associated islands, but none in Madagascar; 5 in the tropical Oriental Region etc.; and 2 in the Australian Region, but only one of them

reaches Australia proper, and only the eastern part of the continent; and none reaches New Zealand. Thus detailed, the numbers suggest a tropical American origin of *Columba*, dispersal to the Old World through the north (not by the existing western North American species but perhaps by an earlier one), and spread through the Old World from the north. The absence of the genus in Madagascar and the more remote part of Australia is consistent with this history. Alternatively, the genus may have originated in temperate Eurasia and radiated from there and then radiated secondarily in tropical America. Or (and I think this is most likely) it may have had a still more complex history."

In the above account, Darlington seems to hesitate among ambiguous clues, always grasping for an ever-elusive center of origin. His analysis of *Columba* tends to confirm the opinion of Fraipont and Leclerq (1932: 7) that the quest for a center of origin leads to an "effarante paléogéographie où les mers et les continents, les plantes et les animaux dansent, sur une terre épileptique, une ronde sans repos." Indeed, *Colomba* as conceived by Darlington seems to have flittered so restlessly between the Old World and the New that its wanderings through space and time are opaque to analysis.

Mayr and Phelps (1967) attempted to determine the centers of origin of the avifauna of the *cerros* and *mesas* of southeastern Venezuela (their "*Pantepui*") with reference to the Andes, the coastal cordillera of Venezuela, and the Brazilian shield. But it is equally reasonable to begin with the premise that the birds of *Pantepui* and of the cordilleras to the north and west had a common origin in a widespread fauna, that was later subdivided by the events of Tertiary geology into Andean, cordilleran, and pantepuian groups. Mayr and Phelps (1967: 293) considered this possibility (their "plateau theory"), but dismissed it for the reasons that *Pantepui*

"is geologically vastly more ancient than its bird fauna" and that "The irregular distribution within Pantepui, the different degrees of differentiation from the nearest relatives, and the various degrees of differentiation within Pantepui all contradict the assumption that the present fauna of Pantepui is the remnant of an old, formerly uniform plateau fauna."

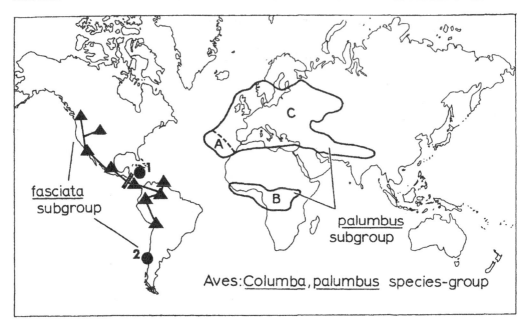

Fig. 1.—Distribution of *Columba* (Aves: Columbidae), species-group *palumbus* (partly after John-ston, 1962). The species-group includes two sub-groups: *palumbus* and *fasciata*. The distribution of the *palumbus* sub-group, in Eurasia and Africa, is circumscribed: A, Atlantic sector (Canary Islands, Madera, Azores) in which occur *C. palumbus maderensis*, *C. p. azorica*, *C. trocaz*, *C. bollii*, *C. junoniae*; B, range of *C. unicincta* (a classic west African taxon); C, range of *C. palumbus* outside the Atlantic sector. The distribution of the complex *C. fasciata/C. albilinea* (the former to the northwest of the double bar, Panama/Costa Rica, the latter to the southeast) is indicated by full triangles connected by a line (track); *C. caribaea* (Jamaica) by circle 1; and *C. araucana* (Chile and Argentina) by circle 2. The New World species form the *fasciata* sub-group.

Instead they (1967:297; cf. Haffer, 1970) favored the view that "the subtropical fauna of Pantepui is derived from that of other subtropical regions by 'island hopping'" (their "distance dispersal theory"; cf. Deignan, 1963:264: "so overwhelming is the evidence of sedentation in birds, especially in the tropics, that the whole hypothesis of dispersal of birds by island hopping must be suspect"). Mayr and Phelps (1967: 286) noted that the age of Pantepui has been variously estimated as Proterozoic to Eocene, but omitted any discussion of the age of the bird fauna. They stated, how-ever, that

"it is evident that there is a completely even gradation from endemic genera to species that have not even begun to develop endemic sub-species. Fewer than one-third (29 species) of the characteristic upper zonal element of Pan-

tepui (96 species) are endemic species. These facts are conclusive evidence for the continuity and long duration of the colonization of Pan-tepui" (1967:291).

Mayr and Phelps did not consider the pos-sibility that these "facts" might simply re-flect different evolutionary rates, rather than "chance colonization" (1967:298) by means of "long-distance flights" (1967:301).

The quest for a center of origin often leads to equivocal conclusions and conflict-ing opinion. For another example (fig. 1), the lone pigeon that occurs in southwestern Canada is *Columba fasciata*. This species is distributed from British Columbia to Trinidad and Argentina, and was once di-vided into two species: *C. albilinea* (South and Central America to Panama/Costa Rica); and *C. fasciata* (North America). According to Johnston (1962), it belongs to the *palumbus* species-group, which is itself divided into two sub-groups: a *palumbus* subgroup and a *fasciata* subgroup.[8] In the *palumbus* subgroup are included five species endemic to western Eurasia/northwestern Africa, Equatorial Africa, and particularly, the Atlantic islands (Madeira, Canaries, and Azores). The *fasciata* subgroup includes three species, all American. The attempt to determine the center of origin and subsequent dispersal of this group leads to a basic disagreement between those zoogeographers who would accept transatlantic relationships at their face value, and those who accept the Matthewian thesis that transatlantic relationships are impossible except via the Bering landbridge or through the agencies of remote chance. But how could one or another school of zoogeographers begin to evaluate the truth of the matter without first understanding the nature of the preconceptions that produce their disagreement? Moreover, neither school might consider the hypothesis that the ancestors of the *palumbus* group were already widely distributed—from which hypothesis it follows that the subsequent history of the species group involved neither a center of origin nor transoceanic (or transbering) dispersal.

The relation between centers of origin and the distribution of "primitive" and "advanced" taxa may, likewise, lead to conflict (Lutz, 1916). As an example, we refer to certain cetoniid Coleoptera (Wiebes, 1968: fig. 4; Croizat, 1971a:394, 397, fig. 2B): *Goliathus russus*, endemic to the Congo (Zaire) Basin, differs more sharply from the adjacent *G. fornasini* (Kenya, Tanzania [Usambara], Mozambique, etc.), *G. aureosparsus* (Nigeria), and *G. higginsi* (Ivory Coast) than these three species differ among themselves (Wiebes, 1968:30, mentioned "more examples of this phenomenon in other groups of African Cetoniidae," and Croizat noted numerous other cases in plants and animals, and discussed some as examples of "wing dispersal"). Some zoogeographers would assume that one species or group, e.g., the species (*G. russus*) in

the center of the assemblage, is relatively more primitive (or "plesiomorphous") than the remaining three, and that it should be assumed to indicate, or to occupy, the center of origin of the group as a whole; these zoogeographers assume that relatively primitive species are generally less apt to disperse than their relatively advanced (or "apomorphic") relatives (Hennig, 1966:232; Brundin, 1972). Other zoogeographers would assume that one species or group, e.g., the centrally located species, is advanced and that it indicates, or occupies, the center of origin of the group as a whole; these zoogeographers assume that relatively advanced species are generally less apt to disperse than their primitive relatives (Darlington, 1957:554–555, 1970). Other zoogeographers would assume that one species or group, especially if it were fossilized and demonstrably older than its relatives (the "right fossils in the right places" of Darlington, 1957:35) is actually ancestral to the other members of the group, and therefore reveals directly the center of origin (e.g., Simpson, 1940).[9] Still other zoogeographers might approach the problem with different sets of apriorisms. The conflict of opinion resulting from different apriorisms raises the question of the applicability of the concept, and even the existence, of a center of origin as envisioned in these discordant approaches. We would point out that, if a center of origin is imaginary, all of its corollaries are equally imaginary; and by an opportune choice of examples, anyone can "prove" whatever he wishes to "prove" about it.

CHARLES DARWIN

The question naturally arises, who first thought of the idea of center of origin? The idea seems to be very old, for Spanish clerics thought about the center of origin of the Indians they encountered in the New World shortly after 1492 A.D. (Croizat, 1960:1367–1372). The idea was one of the basic assumptions of Darwin's zoogeography:

"We are thus brought to the question which has been largely discussed by naturalists [e.g., Swainson, 1835], namely, whether species have been created at one or more points of the earth's surface. Undoubtedly there are very many cases of extreme difficulty, in understanding how the same species could possibly have migrated from some one point to the several distant and isolated points, where now found. Nevertheless the simplicity of the view that each species was first produced within a single region captivates the mind. He who rejects it, rejects the *vera causa* of ordinary generation with subsequent migration, and calls in the agency of a miracle" (1859:352); "Hence it seems to me, as it has to many other naturalists, that the view of each species having been produced in one area alone, and having subsequently migrated from that area as far as its powers of migration and subsistence under past and present conditions permitted, is the most probable" (1859:353); and "Whenever it is fully admitted, as I believe it will some day be, that each species has proceeded from a single birthplace, and when in the course of time we know something definite about the means of distribution, we shall be enabled to speculate with security on the former extension of the land. But I do not believe that it will ever be proved that within the recent period continents which are now quite separate, have been continuously, or almost continuously, united with each other, and with the many existing oceanic islands" (1859:357–358).[10]

It is apparent that Darwin was concerned with the origin of species, but he produced no factual evidence for his belief that species originate in centers from which they disperse, either actively or passively according to the means available to them, to points far and near in time and space. Darwin simply affirmed that his belief "captivates the mind," and that he who rejects it is guilty of invoking a miracle against the true cause of evolution.[11] Indeed, so overwhelmingly important to Darwin were means of dispersal that he believed knowledge of them would permit resolution of paleogeographic problems (but whatever the resolution, Darwin believed that continents and oceanic islands will never be found to have been in connection). In short, the zoogeography of Darwin is based on the preconditions of (a) centers of origin, (b) dispersal of spe-

cies according to available means, and (c) permanent continental outlines. To these, Wallace (1876) and, especially, Matthew (1915) added the thesis that dispersal proceeded from Holarctica to the rest of the earth,[12] and laid the foundations of the zoogeography of writers such as Simpson, Mayr, Darlington,[13] Schmidt, Hershkovitz, B. Patterson, MacArthur, and numerous others of the modern era (e.g., Sauer, 1969; Tobler et al., 1970).

Darwin's treatment of geographical distribution in the *Origin of Species* is less interesting than in the *Voyage of the Beagle*, at any rate in our opinion (cf. Ghiselin, 1969). There is no need to repeat here what has been detailed in many pages elsewhere (Croizat, 1964:592–706), but we would point out that Darwin in the *Voyage* was already aware of the general phenomenon of vicariance; with respect to three species of Galapagos "mocking birds" he stated that

"I examined many specimens in the different islands, and in each the respective kind was *alone* present. These birds agree in general plumage, structure, and habits; so that the different species replace each other in the economy of the different islands" (1839:475). He added that "it never occurred to me, that the productions of islands only a few miles apart, and placed under the same physical conditions, would be dissimilar" (1839:474). In a later edition he added, with reference to the Geospizinae, that "Seeing this gradation and diversity of structure in one small, intimately-related group of birds, one might really fancy that, from an original paucity of birds in this archipelago, one species had been taken and modified for different ends" (1846:148)

—a statement made without appeal to a center of origin and dispersal.[14] As for his interpretation of the biogeography of the Galapagos, it was incisive:

"It would be impossible for any one accustomed to the birds of Chile and La Plata to be placed on these islands, and not to feel convinced that he was, as far as the organic world was concerned, on American ground" (1839:474). In a later edition he expanded on this theme: "I have said that the Galapagos Archipelago might be called a satellite attached to America, but it should rather be called a group of satellites,

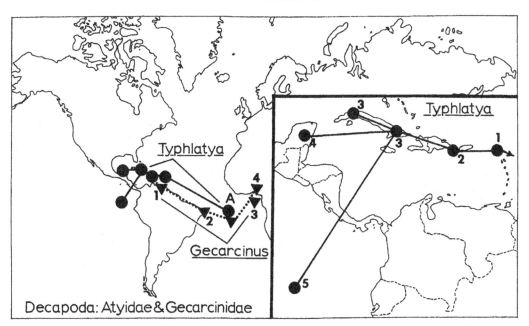

Fig. 2.—Distribution of *Typhlatya* (Decapoda: Atyidae) and *Gecarcinus lagostoma* (Decapoda: Gecarcinidae) (partly after Chace and Hobbs, 1969; Chace and Manning, 1972).

Main map.—The track for *Typhlatya* (see below for inset) is the solid line interconnecting circles. The track for *Gecarcinus lagostoma* is the dotted line interconnecting triangles. A, Ascension Island; 1, Trinidad; 2, Fernando de Noronha; 3, islands of the Gulf of Guinea; 4, Cameroon (cf. Croizat, 1968b; figs. 13–16, 22).

Inset.—The American stations of *Typhlatya*: 1, Barbuda (*T. monae*); 2, Mona (*T. monae*); 3, Cuba (Oriente, Pinar del Rio; *T. garciai*); 4, Yucatan (*T. pearsei*); 5, Galapagos (*T. galapagensis*); arrow to right of station 1 represents part of track between Barbuda and Ascension Island (A) as shown in main map.

physically similar, organically distinct, yet intimately related to each other, and all related in a marked, though much lesser degree, to the great American continent" (1846:172).

Darwin viewed the Galapagos, at least metaphorically, as a fragment of the Americas, isolated from that continental landmass long enough to evolve its own biota, but not long enough to have lost its American ties. Whatever Darwin might have speculated about centers of origin and dispersal does not detract from this basic view of a marine outpost of continental America that, for all we know of biotic distribution, carried with it from the mainland the ancestors of its actual biota of today.[15]

With reference to a fragment of rock collected on Ascension Island, Darwin remarked that it contained remains of "sili-

ceous-shielded, fresh-water infusoria, and no less than twenty-five different kinds of the siliceous tissue of plants, chiefly of grasses." He stated that

"we may feel sure that at some former epoch the climate and productions of Ascension were very different from what they now are. Where on the face of the earth can we find a spot on which close investigation will not discover signs of that endless cycle of change to which this earth has been, is, and will be subjected?" (1846:297).

The spirit of this passage ill agrees with the preconceptions of the *Origin*, many of which persist to the present day (center of origin, dispersal according to available means, and permanent continental outlines). But the fragment of rock and its contents agree well with recent data from Ascension

(Chace and Manning, 1972): the discovery of two endemic shrimps, one of which belongs to the genus *Typhlatya*, known also from Caribbean islands (Barbuda, Mona, and Cuba), Mexico (Yucatan), and—incredibly by the modern map (fig. 2)—Galapagos.[16] Similar distributions are shown by other groups, and sometimes even by single species (fig. 2).

On the Beagle's return to England, Darwin possessed at least the rudiments of the principles of modern systematics and biogeography. He understood vicariance clearly enough to visualize the Galapagos as a cluster of lesser areas of evolution.

Darwin was also certain that an obvious nexus bound the history of the earth with that of its inhabitants: plants and animals, past and present. He had evidence in his hands that "oceanic islands" such as Ascension had a very different biological and geological past.[17]

Upon his return to England, Darwin unfortunately, and perhaps tragically for biology, began building theories of "geographical distribution" based on concepts of species and their centers of origin and dispersal. In so doing Darwin avoided the general problem of vicariance (biological differentiation in time and space), as represented particularly by the material he himself collected during the voyage of the Beagle.

The majority of naturalists today accept concepts such as center of origin as foolproof fundamentals of biogeography without having much understanding of their history and real meaning, and without any awareness of the conflict in Darwin's own views (vicariance versus center of origin and dispersal).[18]

Having failed to dissect these concepts (center of origin, vicariance) to their core, contemporary zoogeographers founder in a self-created morass of chance hops; great capacities for, or mysterious means of, dispersal; rare accidents of over-sea transportation; small probabilities that with time become certainties; and other pseudo-explanations. Where such conceptual imprecision leads is exemplified by the statement that

> "the close relationship between the Old and New World members of the Pantropical element, whose ranges are now widely discontinuous, proves that . . . a faunal exchange must have taken place, and this places the zoogeographer in a real quandary. The customary solution for the problem is to ignore it" (Mayr, 1946b:36).[19]

But ignoring fundamental problems because they conflict with the principles of the geographical distribution of a Darwin or a Matthew is a form of bias repugnant to the spirit of science.

Nevertheless, Darwin did write that

> "it is obvious, that the several species of the same genus, though inhabiting the most distant quarters of the world, must originally have proceeded from the same source, as they have descended from the same progenitor" (1859: 351).

The principle of common ancestry implies vicariance, applicable to species and biota alike, meaning specifically that an ancestral species (or biota) with characters (or species) $a + b + c + d + \ldots n$ subdivides into an assemblage of species (or descendant biotas), each distinct by a different combination of characters (or species), e.g., $a + b + c' \ldots, d + e + f \ldots, a' + c + f \ldots$ This process of subdivision (vicariance) is geographical, involving particular geographical areas for species and biotas alike, such that the resulting species (or descendant biotas) vicariate, i.e., replace each other geographically without any of them migrating from one area to another. Vicariant patterns are generally shown by the various species within a genus, the subspecies within a species, and the varieties within a subspecies; and generally by higher taxa as well (e.g., Hoffstetter, 1973). But because the geographical subdivision of species into subspecies has occurred more recently within a given lineage than the formation of groups of species (e.g., genera) in that lineage, vicariance is generally more precise today for taxa of lower rank (notes 2 and 3).

The principles of common ancestry and vicariance do not require that species must ever have migrated from a center of origin in the Darwinian sense. But suppose we do interpret the above text by Darwin to refer to species originating in centers, spreading therefrom by means of dispersal often of the mysterious and unknowable kind. In an avian genus of some 10–20 species and subspecies, we might consequently imagine that one of the species or subspecies originated at some center, corresponding to a point on the map, and then dispersed to other centers corresponding to other points on the map, there to give rise to the other species and subspecies, which then dispersed to achieve their present respective geographical distributions. Such interpretation, which requires a center of origin and dispersal therefrom, is inconsistent with the principles of common ancestry and vicariance—the basis for the practice and philosophy of modern systematics since the days of Wagner (1889), Kleinschmidt (1926, 1930), and Rensch (1929; cf. Croizat, 1964: 177–216).

The discordance between these principles (common ancestry and vicariance) and the "geographical distribution" of Darwin (1859) and Matthew (1915), and those who took their cues from them, is stark and seemingly impossible to reconcile. It may be fortunate for biology that geophysics has finally managed to show how brittle are the foundations of the aprioristic "geographical distribution" of Darwin, Matthew, and others—at least with respect to belief in stable geography. But the biogeographical solution—as opposed to the geophysical—was outlined by Cain in 1943 and its principles as we see them today are (Croizat, 1973): to do away with aprioristic "theory" and the authority that supports such "theory"; to formulate explicit methods of statistical analysis (based on the concept of generalized tracks) that yield unambiguous and repeatable results; to reject affirmations lacking a demonstrably objective basis (centers of origin); to admit that ideas and beliefs have · a history; and, in the

search for that history, to be candid with students so that they may not wander in a world of make-believe and pretense—however reputable and orthodox that world might seem. In reality, science does have an orthodoxy of its own, demanding repeatable results and independent confirmation. No one well informed of the zoogeography of our times can have an illusion about its manifest disreputability.

ACKNOWLEDGMENTS

We are grateful to certain anonymous reviewers for their constructive criticism, and to Drs. J. W. Atz, L. Brundin, D. L. Hull, C. Patterson, and C. F. Thompson, who read and made suggestions for improvement of earlier drafts.

NOTES

[2] The terms *vicar, vicariad,* or *vicariant; vicarious* or *vicariant species;* and *vicariance, vicariism,* or *vicariation* have acquired a variety of meanings in biology (Cain, 1944:265–273; Schilder, 1956:90–92; Udvardy, 1969:192–194; Hennig, 1966; Lemée, 1967; Neill, 1969). On the one hand, they may designate ecologically similar but geographically separated species, for example of different land masses, as in a comparison between a marsupial and a placental; or they may designate ecologically different species living in the same area: in either case the two species involved may, or may not, be closely related. On the other hand, they may designate what Jordan (1908:75) termed *geminate species*—closely related species, usually very similar, that occupy adjacent (allopatric) areas separated by a barrier. This usage, apparently the original and most common one (Wagner, 1868:9, 1889:56; Hesse, 1924; Marcus, 1933; Geptner, 1936; Hesse et al., 1937; Cain, 1944; Ekman, 1953; Dansereau, 1957; Polunin, 1960; Good, 1964; Schmithüsen, 1968; Schmidt, 1969; Valentine, 1972), is similar to our own, and implies a common ancestry for the geminate pair. We view all of. the components (species and species clusters) of a monophyletic group as primarily (originally) allopatric. The separate components of the group are therefore, *vicariants,* and the historical process giving rise to them, *vicariance,* as embodied in the following premises and conclusions:

a. Allopatric species (*vicariants*) arise after barriers separate parts of a formerly continuous population, and thereby prevent gene exchange between them.

b. The existence of races or subspecies of a species that are separated by barriers (*vicariance*) means that a population has subdivided, or is subdividing, not that dispersal has occurred, or is occurring, across the barriers.

c. The earliest stages (races and subspecies) of differentiation (*vicariance*), separated by complete or incipient barriers to gene exchange, are entirely allopatric.

d. Sympatry between species of a monophyletic group implies dispersal of one or more species into the range(s) of the other(s) (note 3).

e. Allopatric speciation (*vicariance*) predominates over other forms of population differentiation; allopatry is the rule and sympatry the exception in present-day distributions of the species of a given monophyletic group.

f. *Vicariance* is, therefore, of primary importance in historical biogeography, and dispersal is a secondary phenomenon of biotic distribution.

[3] The general phenomenon of sympatry, including all cases of overlap (or coincidence) of distribution of unit taxa (definable, named populations) is itself evidence of dispersal. Sympatry is more prevalent among the members of more distantly related and, therefore, relatively older taxa, e.g., between a fish species and a crustacean species. We assume that most sympatry of this sort, between distantly related species, was caused by dispersal in the remote past, before the formation of the most recent ancestral biotas, as estimated by the generalized tracks displayed by the modern world biota. But sympatry (range overlap of reproductively isolated populations) is the same phenomenon at all taxonomic levels. For example, in the Middle American fish genus *Xiphophorus* there are 18 unit taxa grouped in two sections. Of nine examples of sympatry, five occur between members of different sections, and the other four examples occur between members of one section, in which the sympatric pairs are separated phylogenetically by two or more genealogical bifurcations. In the platyfish section of the genus, the distribution of unit taxa is completely allopatric (Rosen, 1960:fig. 4), a picture typical for groups displaying unaltered vicariance. In this connection, Sokal and Crovello (1970:148) point out that the criterion of reproductive isolation for the recognition of "biological species," for which the occurrence of sympatry is an ingredient, may be questioned on the grounds that the "well circumscribed biological species is *not* the rule but the exception." Indeed, acceptance of an allopatric speciation model is consistent with little or no sympatry at the lowest taxonomic levels. But regardless of the extent of broad sympatry (evidence of significant dispersal), we consider that vicariance underlies and antedates nearly all cases of sympatric distributions. Current practice in biogeography, however, involves an initial assumption of dispersal from a center of origin. But is it not reasonable that, before the causes and means of dispersal may be investigated in any specific case, evidence should first be found that dispersal has occurred?

[4] An ancestral pan-austral flora was recognized long ago by Hooker (1853:xxi; 1860:325–326): "I was led to speculate on the possibility of the plants of the Southern Ocean being the remains of a flora that had once spread over a larger and more continuous tract of land than now exists"; "the many bonds of affinity between the three southern floras, the Antarctic [which according to Hooker occurs in "Fuegia, the Falkands, and Lord Auckland's and Campbell's group, reappearing in the alps of New Zealand, Tasmania, and Australia"], Australian, and South African, indicate that these may all have been members of one great vegetation, which may once have covered as large a southern area as the European now does a northern. It is true that at some anterior time these two floras [southern and northern] may have had a common origin, but the period of their divergence antedates the creation of the principal existing generic forms of each." The recognition of a corresponding pan-austral fauna came later (Huxley, 1868; Hutton, 1873), but its existence was soon denied by Wallace (1876:159, probably following Darwin, see below): "The north and south division truly represents the fact, that the great northern continents are the seat and birth-place of all the higher forms of life, while the southern continents have derived the greater part, if not the whole, of their vertebrate fauna from the north; but it implies the erroneous conclusion, that the chief southern lands—Australia and South America—are more closely related to each other than to the northern continent. The fact, however, is that the fauna of each has been derived, independently, and perhaps at very different times, from the north, with which they therefore have a true genetic relation." The Darwin-Wallace influence was such that many later botanists rejected even Hooker's well founded notion of an ancestral pan-austral flora in favor of dispersal from northern centers of origin: "all the great assemblages of plants which we call floras seem to admit of being traced back at some time in their history to the northern hemisphere"; "The extraordinary congestion in species of the peninsulas of the Old World points to the long-continued action of a migration southwards. Each [peninsula] is in fact a *cul-de-sac* into which they [species] have poured and from which there is no escape"; "The theory of southward migration is the key to the interpretation of the geographical distribution of plants" (Thiselton-Dyer, 1878:441; 1909:311, 316). The view that the pan-austral biota is a mere artifact of independent dispersal from the northern hemisphere (e.g., Wallace and Thiselton-Dyer, 1885) was later termed the "monoboreal relic hypothesis" (Schröter, 1913:921) and

had much appeal earlier in this century (Matthew, 1915). The history of this period is well covered by Du Rietz (1940; also Wittman, 1934, 1935). Other authors viewed the pan-austral biota as an artifact due to chance dispersal over southern water gaps, a practice maintained for some years, especially by zoologists (e.g., Darlington, 1957, 1965; cf. Axelrod, 1952). Both views of the pan-austral biota as an artifact were historically based on a priori acceptance of stable geography: "Writers on geographical distribution now occupied themselves, to use Darwin's words, 'in sinking imaginary continents in a quite reckless manner' and in constructing land bridges in every convenient direction. They were brought back to the stern reality of fact when Dana in his *Manual of Geology* [1863:732, also 1847:92] first made the unexpected statement: 'The continents and oceans had their general outline or form defined in earliest times.' From this view Darwin, supported as he was from his own reflections, never deviated. Writing to Hooker in 1856 he said, 'you cannot imagine how earnestly I wish I could swallow continental extension, but I cannot.'" "Half a century has elapsed since Dana laid down his memorable principle [stable geography]. Time has strengthened and in no way diminished its force. But though adopted by Darwin and Wallace it is still ignored by those who prefer facile speculation to the sober contemplation of established facts" (Thiselton-Dyer, 1912:237, 239). Cf. Darlington (1957:22): "A synthesis of this sort—putting Wallace and Darwin together, so to speak, and adding geology and other things in reasonable proportions—ought to be the purpose of modern zoogeographers. Of some it is"; Raup (1942:328): "the effect of Darwinism upon the floristic view of plant geography was not so great as upon other views. The reasoning remained inductive in large measure, with conclusions growing slowly out of masses of fact which were sorted laboriously into patterns of coincidence and suspected actual relationship"; and Turrill (1953:226): "The modern tendency has been mainly against his [Darwin's] views in this respect, at least among botanists." Recent commentary about the history of the pan-austral biota may be found in Pantin et al. (1960), Gressitt (1963), Brundin (1966, 1970, 1972), Corro (1967, 1971), Valentine (1972), and Keast (1973).

[5] Assuming that galaxiids are Gondwanian, i.e., that their present distribution represents geographical isolation of fragments of one or two widespread species by fracture of the Gondwana landmass, one may be tempted to ask where the ancestral galaxiid species originated. Did it originate in some center, corresponding to a point on the map, and from there disperse across Gondwana? This question should be considered in relation to (1) the fact that the galaxiids have a sister group, the salmonids of Laurasia and (2) the possibility that the Gondwanian galaxiids and the Laurasian salmonids may be vicariants that formed in response to the initial fracture of Pangaea into Laurasia and Gondwana. If so, then the ancestral species common to both groups may already have been widespread over Pangaea, and the question of dispersal of the ancestral galaxiid species becomes unnecessary and, perhaps, irrelevant. We do not deny that, at some point or other, dispersal might have played a role in the formation of the ancient Pangaean distributions, of which we now have only the vicariant remnants; or that some dispersal, as indicated by sympatry between modern galaxiid species, has occurred since. But we see no need to assume that more dispersal occurred than is indicated by the evidence.

[a] The concept of "limited area" of origin is, of course, relative. Early Darwinians tended to assume that species originate from one or, at most, one pair of organisms, and that species generally have, consequently, a very small center of origin: "A new species recently come into existence would naturally, at least on any theory of evolution, have a limited range because it would have come into being at one locality and not have had time to extend its range"; "We must imagine each species setting out from its centre of origin and gradually extending itself by actual or passive migration right and left and in every possible direction from this focus" (Beddard, 1895:12–13; also Bartholomew et al., 1911:3). On this matter Darwin himself was equivocal: "It is also obvious that the individuals of the same species, though now inhabiting distant and isolated regions, must have proceeded from one spot, where their parents were first produced" (1859:351–352). "With those organic beings which never intercross . . . , all the individuals of each variety will have descended from a single parent. But in the majority of cases, namely, with all organisms which habitually unite for each birth . . . , the individuals of the species will have been kept nearly uniform by intercrossing; so that many individuals will have gone on simultaneously changing, and the whole amount of modification will not have been due, at each stage, to descent from a single parent" (1859:355–356). Despite the modern view (e.g., Fisher, 1930; Haldane, 1932; Dobzhansky, 1937; Waddington, 1939; Huxley, 1940, 1942) that, for sexually reproducing organisms at least, the unit of evolution is a population rather than an individual or pair, the early Darwinian view is still sometimes maintained: "If a species is strictly monophyletic, then all of its individuals are the descendants of one and the same ancestral plant and their total range, however extensive and peculiar it may be, must have grown by the processes of dissemination from the tiny area occupied by this ancestor" (Good, 1964:34). But even in the context of population biology, the process of "speciation" has often been viewed as necessarily beginning with a small population of

restricted distribution: "Never, since the days of the hypothesis of special creation, has it been maintained that a species originally arose over the whole of the area upon which it now occurs." "It is clear . . . that the large areas now occupied by many species must almost always, if not always, be due to spreading [dispersal] from others [areas] originally much smaller" (Willis, 1922:10–11; cf. Croizat, 1958b); "Of vital importance . . . is the determination of that initial territory whence . . . a species began its dispersal whereby it reached the present boundaries of its area." "There are no grounds for presuming that a new species will not extend its area beyond. the limits of the region of its origin. It will, without any doubt, begin to spread in all directions open to it, and the region of its origin will constitute the center of the area being formed" (Wulff, 1943:27); "*when* a new species evolves, it is almost invariably from a peripheral isolate [population]" (Mayr, 1963:513; also Takhtajan, 1969:27). For biogeography, the consequences of these views, embracing the concepts of center of origin and dispersal on an a priori basis, have been far-reaching, even though post-Darwinian authors often considered a center of origin to be relatively large, e.g., the "Holarctic centers" of Matthew (1915:172) and the "Old World tropics" of Darlington (1957:570–577; 1959b).

However large the size might have been imagined, the concepts were the same: "This I believe to be the type of pattern that would be shown by almost any form of life that had run its entire course from origin to extinction. A form appears in some center or 'cradle,' not an exact spot that could be marked with a monument but, say, a single biotic district or province. Thence it tends to spread steadily in all directions until it encounters insuperable barriers. After a time it begins to contract." "An excellent descriptive analogy is provided by the expansion and contraction of ice caps" (Simpson, 1940:144); "successive 'dominant forms of life' . . . rise and spread over the world, each dominant group competing with, destroying, and replacing older groups, then differentiating in different places until overwhelmed by the rise and spread of the next dominant group." "We know now that this process of evolution, spread, and replacement of successive dominant groups is the main process (infinitely more complex in detail than my description of it) that makes the main patterns in animal distribution" (Darlington, 1959a: 311; also 1957:552–556, 1959b:488; cf. Darwin, 1859:325–326 and e.g., Newbigin, 1948:9; Beaufort, 1951:2; Takhtajan, 1969:137; Banarescu, 1970:246; Laubenfels, 1970:21); "It is a basic tenet of zoogeography that an animal group arises in and spreads from a single area, its center of origin. For larger, more inclusive groups, as the more primitive members move out from the center of origin, successively more advanced forms evolve

in the center. As they in turn spread, they tend to eliminate the more primitive forms by competition. A large group that has been in existence for a long time typically shows a pattern of distribution in which the primitive species are located at the periphery of the range, in areas that the more advanced members have not yet reached or have reached only recently" (Goin and Goin, 1973:113). In contrast to these authors, we would separate the concept of vicariance from any and all a priori considerations of presumed population size and distributional extent of species at the time of their origin. We view vicariance as a phenomenon that may be displayed by populations of any size and geographical extent.

[7] Peters (1948:86) placed the wrynecks in a subfamily (Jynginae) of Picidae, but such placement would not make Mayr's interpretation any easier. However understood taxonomically, Picidae are a biogeographically interesting family, for the genus *Picumnus* is represented by about 25 species in South America (Schauensee, 1964:187) and one species in southeast Asia and Malaysia (Peters 1948:88–97)—totally isolated from its congeners.

[8] Johnston's taxonomy clarifies the transatlantic nature of the relationship between these two subgroups (Goodwin, 1959, is less clear on this point). The relationship seems transatlantic (fig. 1) also because (1) *Columba* in America may be secondarily distributed north of Mexico (Cracraft, 1973: 509); were the genus distributed transpacifically, one might expect it to be better represented in the United States—and Canada, as are generally the groups of plants and animals with transpacific distributions (Croizat, 1968a:236ff, figs. 29–30); (2) the considerable differentiation in the Atlantic sector, and the predominantly western distribution of *C. unicincta*, are common adjuncts of transatlantic relationships (Croizat, 1952, 1958a, 1960, 1964).

[9] "Here the paleontologist comes to the rescue. His discoveries are the historical documents of animal distribution" (Simpson, 1940:137–138; cf., e.g., Furon, 1958:41: "Il ne peut pas y avoir d'histoire biogéographique certaine sans paléontologie"); "It is extremely difficult, if not impossible, to reconstruct former distributions and colonization routes, if there is no fossil evidence (Mayr, 1952: 255); "In a really good fossil record the earliest, most primitive fossils of a group will be at the place of origin, and later and more derivative fossils will clearly show directions of movement" (Darlington, 1959a:314; 1959b:495). Exaggeration of the significance of paleontological data has been common, one might say even traditional, in biology since the time of Darwin; statements in the literature abound to the effect that "Verification of the actual history of a group . . . depends ultimately upon finding fossilized remains of its members" (Stahl, 1974:1). In contrast, we view the role of paleontology in historical biogeography as the same as its role in phylogenetic sys-

tematics, i.e., as an additional source of information for historical analysis (Schaeffer, Hecht, and Eldredge, 1972). Thus, paleontological data, if they reveal examples of sympatry or overlap of generalized tracks, might justify an inference of dispersal. In our opinion such an inference is neither more or less reasonable, justifiable, significant, certain, verified, or easily performed for paleontological than for neontological data.

[10] When writing about permanence of continents both Darwin and Matthew (1915:172) appealed to qualifying phrases such as "within the recent period" and "in later geological epochs" (cf. Croizat, 1971b). Modern writers have generally argued for continental stability only during the late Mesozoic and Caenozoic (e.g., Mayr, 1952; Hubbs, 1958), in the belief that the mammalian fossil record, as interpreted mainly by Simpson, was conclusive evidence in favor of stability (cf. McKenna, 1973). But for biogeography, as opposed to geophysics (for which Kasbeer, 1972), the important issue is not that the continents were, or were not, recently connected; for biogeography the issue is that the continents were, or were not, connected recently enough so that their subsequent separation significantly contributed to the vicariance displayed by the modern world biota. Traditionally, many persons considered the issue to involve only a choice between different means of dispersal (overland or oversea, respectively): "On a more theoretical level there has been a long-running . . . argument in southern biogeography. Many persons, like Hooker, have thought that dispersal must have occurred across land connections in the far south. Others, like Darwin, have postulated dispersal across far-southern water gaps" (Darlington, 1965:5). But the issue as we view it (see note 2 above) involves a choice between two basic explanatory principles (vicariance versus dispersal) on the basis of their relative generality: are the general patterns (generalized tracks) of modern biotic distribution due to vicariance (in response to a changing geography) or to dispersal (over a more or less stable geography)?

[11] "As Darwin's main problem was the origin of species, nature's way of making species by gradual changes from others previously existing, he had to dispose of the view, held universally, of the independent creation of each species and at the same time to insist upon a single centre of creation for each species; and in order to emphasize his main point, the theory of descent, he had to disallow convergent, or as they were then called, analogous forms. To appreciate the difficulty of his position we have to take the standpoint of fifty years ago, when the immutability of the species was an axiom and each was supposed to have been created within or over the geographical area which it now occupies. If he once admitted that a species could arise from many individuals instead of from one pair, there was no way of shutting the door against

the possibility that these individuals may have been so numerous that they occupied a very large district, even so large that it had become as discontinuous as the distribution of many a species actually is. Such a concession would at once be taken as an admission of multiple, independent, origin instead of descent in Darwin's sense" (Gadow, 1909:322; cf. Gadow, 1913:61).

[12] To a paleontologist specializing in mammalian faunas, the abundance and diversity of Eurasian and North American Tertiary fossils may suggest that life originated in Holarctica, radiating from there to the rest of the earth. But what happens when the field of mammalogy is left behind? Schmidt (1946:152) was forced to admit that "There is a general agreement of the South American, Australian, and African faunae in certain primitive elements, among which may be mentioned lung fishes; leptodactylid frogs; pleurodiran turtles; the more primitive groups of snakes and lizards; and the marsupials (absent from Africa). The list might be greatly extended among invertebrate groups." Schmidt nevertheless disposed of this conflict with the Holarctic theory by asserting simply that "These primitive faunae are probably the accumulated remnants of repeated dispersals in the late Paleozoic and early Mesozoic, i.e., from the 'Holarctic' fauna of those early ages."

[13] Darlington (1957:22), although crediting Matthew for having done "much to counteract the more irresponsible historical zoogeographers," credits Darwin and to some extent Wallace, rather than Matthew (Darlington, 1959a:313, 315; also 1959b:488–489, 1965:57–59; cf. Simpson, 1965: 53; Romer, 1973:345), with being "extraordinarily, almost incredibly, right about a hundred years ahead of his time," because Darwin made passing reference to " 'the more dominant forms, generated in the larger areas and more efficient workshops of the north,' and to 'The living waters . . . [that] have flowed with greater force from the north so as to have freely inundated the south' " (cf. Darwin, 1859:380, 382). According to Darlington, "Darwin was not guessing about these things. He presented evidence and reached correct conclusions. No one could have reached correct conclusions just by guessing. That he saw and understood all these things ["the fundamental concepts of evolutionary zoogeography"], which together are the whole heart of the subject, makes him pre-eminent in evolutionary zoogeography." Cf. Thiselton-Dyer (1909:308, 316): "If an observer were placed above a point in St. George's Channel from which one half of the globe was visible he would see the greatest possible quantity of land spread out in a sort of stellate figure. The maritime supremacy of the English race has perhaps flowed from the central position of its home. That such a disposition would facilitate a centrifugal migration of land organisms is at any rate obvious, and fluctuating conditions of climate operating

from the pole would supply an effective means of propulsion." "If, as is so often the case, the theory [of southward migration] now seems to be *à priori* inevitable, the historian of science will not omit to record that the first germ sprang from the brain of Darwin." "He was in more or less intimate touch with everyone who was working at it." "It is hardly an exaggeration to say that from the quiet of his study at Down he was founding and directing a wide-world school."

[14] Nevertheless, in the same edition Darwin stated that "The archipelago is a little world within itself, or, rather, a satellite attached to America, whence it has derived a few stray colonists, and has received the general character of its indigenous productions" (1846:145).

[15] Holden and Dietz (1972) place the historical beginnings of the Galapagos at about 40 million years ago. They suggest that the present-day Galapagos are simply the most recent (late Pliocene) islands of a volcanic chain with easterly components that subsided as new, more westerly islands emerged. They state (1972:269) that "The Galapagos Islands contain many endemic birds and bizarre animals which have required millions of years for their evolution in isolation. By our model, the modern Galapagos Islands may have inherited faunas from a whole series of ancestral 'Galapagos islands' which existed over a span of 40 m. y. Presumably the animals would have little difficulty negotiating the short span of water to a new volcanic island as an older extinct volcanic island drifted eastward and subsided beneath the sea (a subsiding 'stepping stone'), adding itself to the end of the Cocos and Carnegie ridges. To date, no guyots have been reported from either the Carnegie or Cocos chains, but this still is not conclusive evidence that these ridges were not subareal at some time in their history."

Holden and Dietz (1972; also Malfait and Dinkleman, 1972) discuss the history of the submarine Carnegie ridge that extends between the Galapagos and western South America. They point out that the history of the Galapagos is directly related to the history of the Panamanian isthmus, which is connected to the Galapagos by the undersea Cocos ridge. The region including these features they term the "Galapagos Gore." The gore encloses these two ridges, which bifurcate at the Galapagos and extend eastward. Together, the ridges form the sides of an isoceles triangle, with the Galapagos at the apex and the Panama fracture zone at the base. It is apparent, therefore, that the Panamanian isthmus formed over a considerable period of earth history, and was closely related to dynamic changes in land and water configurations of the entire region bounded by Central America, northwestern South America, and the Galapagos Islands. The formation of the isthmus was characterized by Nemeth and Libke (1972:19): "Uplift and downwarping throughout

the Oligocene and Miocene resulted in the first uninterrupted connection of Central and South America, completed during the Pliocene." This geophysical history and the zoogeographic suggestions of Holden and Dietz, who view the Galapagos as part of a Cocos-Carnegie ridge system that is subsiding in its eastern part, contrasts with the Darwinian view of the Galapagos as true "oceanic islands" without proximity, or any possible historical connections, to the mainland (cf. Croizat, 1958a:746–859, particularly figs. 105 and 110).

A review of the subject of "oceanic islands" is beyond the scope of this paper. But we would point out that the concept of "oceanic island"— an island that arose out of the ocean, far from any continent as judged by modern geography, and that must have been populated by means of chance dispersal—was developed by Darwin (1859:388–406) and, as an overpowering apriorism, has since influenced numerous biogeographers from Wallace (1881) and Guppy (1906) to Carlquist (1965) and MacArthur and Wilson (1967). We note, however, that Jeannel (1942:131) denied that in the Atlantic there were "oceanic islands" in the Darwinian sense (also note 16). Skottsberg (e.g., 1956), who did much to counter an aprioristic approach to Pacific island biogeography (see particularly Du Rietz, 1940:237–240), commented on the spirit of those times: "Guppy relied on fruit-eating doves as carriers of seeds from island to island in bygone times. As they are sedentary today he concluded that they had changed their habits: that once they had been great travellers" (Skottsberg, 1960:455); "When confronted with a peculiar insular flora like that found on many islands in the Pacific, our first thought invariably is: Where did it come from? and how did it get there? In our eagerness to answer these questions and our impatient desire to explain everything, we have tried to form theories before enough is known not only of the geology and physical geography of the Pacific, but even of the plants themselves, their taxonomy and geographical distribution. I am afraid that this is attacking the problem at the wrong end. It even may be worth while to ask why we always assume that everything there is in the Pacific must have come from some distant place. Nobody asks where the Chinese, or Malayan, or Brazilian floras came from. We are quite satisfied to believe that they have developed right where they are, that their early history goes back so far that it is useless, for the present at least, to ask any but general questions as to their origin" (Skottsberg, 1928:914); "How we are to get away from the controversy arising from the fact that some of the present strongholds of this [austral] flora are to be found on supposedly very young volcanic islands, I do not know. When the biologists ask for a little more land of greater age which has disappeared and become succeeded by volcanic chains, the geologists refuse them assistance. All evidence is contrary to the assumption that the present Pacific flora, with the exception of already widespread species of the seashore, is travelling from one island to another" (Skottsberg, 1928:917; see also note 16).

[16] The explanation of this Ascension-Caribbean bond given by Wilson (see below) agrees, from the geophysical side, with conclusions about how "new" islands and mountains manage to retain in their biota very "old" elements (Croizat, 1964: 247ff, 258, fig. 50): "An alternative explanation of the origin of the Ascension shrimps was proposed by J. Tuzo Wilson (in litt.): 'another possibility which I think much more likely and intriguing from your point of view is that Ascension is only the latest in a series of islands whose remains form scattered seamounts and ridges from Ascension Island to the Cameroons in one direction (the Guinea Rise) and in the other direction to the northeast corner of Brazil. The idea that I proposed in the Scientific American [Wilson, 1963] was that there had been a continuously active centre from the time that Recife separated from the Cameroons and that these two chevron-shaped ridges formed as a result of continuous volcanic action at the center now represented by Ascension Island. If that is so, it is just conceivable that forms of life might have survived on Ascension from the time when the Atlantic was very narrow and the forerunners of Ascension were in contact with Brazil and the Cameroons'" (Chace and Manning, 1972:6).

[17] Even though Darwin harbored questionable notions about "barriers to dispersal," as, for example when he stated that the Andes "have existed as a great barrier, since a period so remote that whole races of animals must subsequently have perished from the face of the earth" (1839:399), he could still have worked quite constructively from his own notes and observations. In a later edition this passage was modified to read "these mountains have existed as a great barrier since the present races of animals have appeared" (1846: 78). But by modern estimates the Andes are a relatively recent Tertiary feature. When they arose, the genera and species of passeriform birds, for example, were already modern enough to be assignable to extant families and genera (Howard, 1950). The Andes may accordingly have risen under the roots and feet of the immediate ancestors of the species and subspecies still living there today (Croizat, 1971a:383, fig. 1).

[18] The contrast between Darwin the keen observer and Darwin the casual theoretician has spawned an equivocal literature, in which Simpson (1949:268), for example, extols Darwin as "one of history's towering geniuses" and Himmelfarb (1959:viii), for example, views him as "limited intellectually and insensitive culturally" (cf. Croizat, 1964:592–706; Vorzimmer, 1970; Ghiselin,

1973; Hull, 1973). Accordingly, it is difficult to judge the history and present status of "Darwinism," and few naturalists seem willing to accept the chore.

[19] Cf. Darlington (1957:606–607): "I have tried to keep my mind open on this subject and have made a new beginning by trying once more (as I have done before) to see if I can find any real signs of drift in the present distribution of animals. I can find none." "Although I have made this trial as fairly as I could, I think the results were to be expected"; and (1959a:313): "I think all this can fairly be summarized by saying that Darwin considered the evidence he had and decided that as far back as he could see the main pattern of land had been the same as now, although many details had changed. Fifty-six years later Matthew, with much more evidence, reached the same conclusion, but saw farther back and in much more detail than Darwin could. And now, with still more evidence, we can see still farther back and in still more detail than Matthew could, but the conclusion is still the same. As far back as we can see, the distribution of animals and other evidence suggest a main pattern of land like the present one, in spite of all the details that have changed."

REFERENCES

ADAMS, C. C. 1902. Southeastern United States as a center of geographical distribution of flora and fauna. Biol. Bull. 3:115–131.

AXELROD, D. I. 1952. Variables affecting the probabilities of dispersal in geologic time. Bull. Amer. Mus. Nat. Hist. 99:177–188.

BANARESCU, P. 1970. Principii si probleme de zoogeografie. Editura Academie, Bucuresti.

BARTHOLOMEW, J. G., W. E. CLARKE, AND P. H. GRIMSHAW. 1911. Atlas of zoogeography. Bartholomew, London.

BEAUFORT, L. F., DE. 1951. Zoogeography of the land and inland waters. Sidgwick and Jackson, Ltd., London.

BEDDARD, F. E. 1895. A text-book of zoogeography. Cambridge University Press.

BRIGGS, J. C. 1974. Marine zoogeography. McGraw Hill Book Co., New York.

BRIQUET, J. 1901. Recherches sur la flore des montagnes de la Corse et ses origines. Ann. Conserv. Jard. Bot. Genève 5:12–119.

BRODIE, E. D., JR. 1970. Western salamanders of the genus Plethodon: systematics and geographic variation. Herpetologica 26:468–516.

BRUNDIN, L. 1966. Transantarctic relationships and their significance. K. Svenska Vetensk.-Akad. Handl., ser. 4, 11(1):1–472.

BRUNDIN, L. 1970. Antarctic land faunas and their history. In: Holdgate, M. W. (ed.), Antarctic ecology 1:42–53. Academic Press, London and New York.

BRUNDIN, L. 1972. Circum-Antarctic distribution patterns and continental drift. 17th Int. Zool. Congr., Theme 1:1–11.

CAIN, S. A. 1943. Criteria for the indication of center of origin in plant geographical studies. Torreya 43:132–154.

CAIN, S. A. 1944. Foundations of plant geography. Harper and Brothers, New York and London.

CARLQUIST, S. 1965. Island life. Natural History Press, Garden City.

CHACE, F. A., JR., AND H. H. HOBBS, JR. 1969. The freshwater and terrestrial decapod crustaceans of the West Indies with special reference to Dominica. Bull. U. S. Natl. Mus. 292:1–258.

CHACE, F. A., JR., AND R. B. MANNING. 1972. Two new caridean shrimps, one representing a new family, from marine pools on Ascension Island (Crustacea: Decapoda: Natantia). Smith. Contr. Zool. 131:1–18.

CORRO, G. DEL. 1967. El papel de la Antartandia en relacion con el poblamiento de las otras areas gondwanicas. Mus. Argen. Cien. Nat., Publ. Ext. Cult. Did. 15:1–39.

CORRO, G. DEL. 1971. Algunos ejemplos de distribucion gondwanica. Mus. Argen. Cien. Nat., Publ. Ext. Cult. Did. 17:1–16.

CRACRAFT, J. 1973. Continental drift, paleoclimatology, and the evolution and biogeography of birds. Jour. Zool., London, 169:455–545.

CROIZAT, L. 1952. Manual of phytogeography. W. Junk, The Hague.

CROIZAT, L. 1958a. Panbiogeography. Published by the author, Caracas.

CROIZAT, L. 1958b. An essay on the biogeographic thinking of J. C. Willis. Arch. Bot. Biogeog. Ital. 34:90–116.

CROIZAT, L. 1960. Principia botanica. Published by the author, Caracas.

CROIZAT, L. 1964. Space, time form: the biological synthesis. Published by the author, Caracas.

CROIZAT, L. 1968a. The biogeography of the tropical lands and islands east of Suez-Madagascar. Atti Ist. Bot. Lab. Crittogamico Univ. Pavia, ser. 6, 4:1–400.

CROIZAT, L. 1968b. Introduction raisonnée à la biogéographie de l'Afrique. Mem. Soc. Broteriana 20:1–451.

CROIZAT, L. 1971a. Polytopisme ou monotopisme? Le cas de *Viola parvula* Tin. et de plusieurs autres plantes et animaux. Bol. Soc. Broteriana, ser. 2, 45:379–433.

CROIZAT, L. 1971b. De la "pseudovicariance" et de la "disjonction illusoire." Anuar. Soc. Broteriana 37:113–140.

CROIZAT, L. 1973. La "panbiogeografia" in breve. Webbia 28:189–226.

DANA, J. D. 1847. A general review of the geological effects of the earth's cooling from a state of igneous fusion. Amer. Jour. Sci. Arts, ser. 2, 4:88–92.

DANA, J. D. 1863. Manual of geology. Theodore Bliss & Co., Philadelphia, and Trübner & Co., London.

DANSEREAU, P. 1957. Biogeography. An ecological perspective. Ronald Press, New York.

DARLINGTON, P. J., JR. 1957. Zoogeography: the geographical distribution of animals. John Wiley and Sons, Inc., New York.

DARLINGTON, P. J., JR.. 1959a. Darwin and zoogeography. Proc. Amer. Phil. Soc. 103:307–319.

DARLINGTON, P. J., JR. 1959b. Area, climate, and evolution. Evolution 13:488–510.

DARLINGTON, P. J., JR. 1965. Biogeography of the southern end of the world. Harvard University Press, Cambridge.

DARLINGTON, P. J., JR. 1970. A practical criticism of Hennig-Brundin "phylogenetic systematics" and Antarctic biogeography. Syst. Zool. 19:1–18.

DARWIN, C. 1839. Journal of researches into the geology and natural history of the various countries visited by H. M. S. Beagle. Henry Colburn, London.

DARWIN, C. 1846. Journal of researches into the natural history and geology of the countries visited during the voyage of H. M. S. Beagle. Vol. 2. Harper and Brothers, New York.

DARWIN, C. 1859. On the origin of species by means of natural selection. John Murray, London.

DEIGNAN, H. G. 1963. Birds in the tropical Pacific. In: Gressitt, J. L. (ed.), Pacific basin biogeography:263–269. Bishop Museum Press, Honolulu.

DOBZHANSKY, T. 1937. Genetics and the origin of species. Columbia University Press, New York.

DU RIETZ, G. E. 1940. Problems of bipolar plant distribution. Acta Phytogeogr. Suec. 13:215–282.

EDMUNDS, G. F. 1972. Biogeography and evolution of Ephemeroptera. Ann. Rev. Ent. 17:21–42.

EKMAN, S. 1953. Zoogeography of the sea. Sidgwick & Jackson, London.

ERWIN, T. L. 1970. A reclassification of bombardier beetles. Quest. Ent. 6:4–215.

FAVARGER, C., AND P. KÜPFER. 1969. Monotopisme ou polytopisme? Le cas du *Viola parvula* Tin. Bol. Soc. Broteriana, ser. 2, 43:315–331.

FISHER, R. A. 1930. The genetical theory of natural selection. Oxford University Press.

FRAIPONT, C., AND S. LECLERQ. 1932. L'évolution adaptations et mutations. Hermann et Cie., Paris.

FURON, R. 1958. Causes de la repartition des êtres vivants. Masson et Cie., Paris.

GADOW, H. 1909. Geographical distribution of

animals. *In*: Seward, A. C. (ed.), Darwin and modern science:319–336. Cambridge University Press, Cambridge.

GADOW, H. 1913. The wanderings of animals. Cambridge University Press, Cambridge.

GEPTNER, V. G. 1936. Obščaja zoogeografia. Moscow and Leningrad.

GHISELIN, M. T. 1969. The triumph of the Darwinian method. University of California Press, Berkeley and Los Angeles.

GHISELIN, M. T. 1973. Mr. Darwin's critics, old and new. Jour. Hist. Biol. 6:155–165.

GOIN, C. J., AND O. B. GOIN. 1973. Antarctica, isostacy, and the origin of frogs. Quart. Jour. Florida Acad. Sci. 35:113–129.

GOOD, R. 1964. The geography of flowering plants. Longmans, Green and Co., Ltd., London and Harlow.

GOODWIN, D. 1959. Taxonomy of the genus *Columba*. Bull. Brit. Mus. (Nat. Hist.), Zool. 6:1–23.

GRESSITT, J. L. (ed.). 1963. Pacific basin biogeography. Bishop Museum Press, Honolulu.

GÜNTHER, A. 1869. An account of the fishes of the states of Central America, based on collections made by Capt. J. M. Dow, F. Godman, Esq., and O. Salvin, Esq. Trans. Zool. Soc. London 6:377–494.

GUPPY, H. B. 1906. Observations of a naturalist in the Pacific between 1896 and 1899. Macmillan and Co., London and New York.

HAFFER, J. 1970. Entstehung und Ausbreitung nord-Andiner Berdvögel. Zool. Jb. Syst. 97:301–337.

HALDANE, J. B. S. 1932. The causes of evolution. Longmans, Green and Co., London, New York, Toronto.

HENNIG, W. 1966. Phylogenetic systematics. The University of Illinois Press, Urbana.

HESSE, R. 1924. Tiergeographie auf ökologischer Grundlage. Gustav Fischer, Jena.

HESSE, R., W. C. ALLEE, AND K. P. SCHMIDT. 1937. Ecological animal geography. John Wiley & Sons, Ltd., London.

HIGHTON, R. 1962. Revision of North American salamanders of the genus *Plethodon*. Bull. Florida St. Mus., Biol. Ser., 6:235–367.

HIGHTON, R. 1972. Distributional interactions among eastern North American salamanders of the genus *Plethodon*. Monogr. Virginia Polytech. Inst. St. Univ., Res. Div., 4:139–188.

HIGHTON, R., AND S. A. HENRY. 1970. Evolutionary interactions between species of North American salamanders of the genus *Plethodon*. Evol. Biol. 4:211–256.

HIMMELFARB, G. 1959. Darwin and the Darwinian revolution. Chatto and Windus, London.

HOFFSTETTER, R. 1973. Origine, compréhension et signification des taxons de rang supérieur: quelques enseignements tirés de l'histoire des mammifères. Ann. Paléont. 59:1–35.

HOLDEN, J. C., AND R. S. DIETZ. 1972. Galapagos gore, NazCoPac triple junction and Carnegie/Cocos ridges. Nature 235:266–269.

HOOKER, J. D. 1853. The botany of the Antarctic voyage of H. M. Ships Erebus and Terror in the years 1839–1843. Part II. Flora Novae-Zelandiae. Vol. 1. Lovell Reeve, London.

HOOKER, J. D. 1860. On the origin and distribution of species:—Introductory essay to the flora of Tasmania. Amer. Jour. Sci. Arts, ser. 2, 29: 1–25, 305–326.

HOWARD, H. 1950. Fossil evidence of avian evolution. Ibis 92:1–21.

HUBBS, C. L. (ED.). 1958. Zoogeography. Pub. Amer. Assoc. Adv. Sci. 51:1–509.

HULL, D. L. 1973. Darwin and his critics. Harvard University Press, Cambridge.

HULTÉN, E. 1937. Outline of the history of arctic and boreal biota during the Quartenary Period. Aktiebolaget Thule, Stockholm.

HUTTON, F. W. 1873. On the geographical relations of the New Zealand fauna. Trans. Proc. New Zealand Inst. 5:227–256.

HUXLEY, J. (ED.). 1940. The new systematics. Oxford University Press.

HUXLEY, J. 1942. Evolution. The modern synthesis. George Allen & Unwin, Ltd., London.

HUXLEY, T. H. 1868. On the classification and distribution of the Alectoromorphae and Heteromorphae. Proc. Zool. Soc. London 1868:294–319.

JEANNEL, R. 1942. La genèse des faunes terrestres. Presses Universitaires de France, Paris.

JOHNSTON, R. F. 1962. The taxonomy of pigeons. Condor 64:69–74.

JORDAN, D. S. 1908. The law of geminate species. Amer. Nat. 42:73–80.

KASBEER, T. 1972. Bibliography of continental drift and plate tectonics. Geol. Soc. Amer. Spec. Pap. 142:1–96.

KEAST, A. 1973. Contemporary biota and the separation sequence of the southern continents. *In*: Tarling, D. H., and S. K. Runcorn (eds.), Implications of continental drift to the earth sciences 1:309–343. Academic Press, London and New York.

KINSEY, A. C. 1936. The origin of higher categories in Cynips. Indiana Univ. Publ., Sci. Ser., 4:1–334.

KLEINSCHMIDT, O. 1926. Die Formenkreislehre und das Weltwerden des Lebens. Gebauer-Schwetschke Druckerei, Halle.

KLEINSCHMIDT, O. 1930. The formenkreis theory and the progress of the organic world. H. F. & G. Witherby, London.

LAUBENFELS, D. J., DE. 1970. A geography of plants and animals. W. C. Crown Co., Dubuque.

LEMÉE, G. 1967. Précis de biogéographie. Masson & Cie., Paris.

LUTZ, F. E. 1916. Faunal dispersal. Amer. Nat. 50:374–384.

MacArthur, R. H., and E. O. Wilson. 1967. The theory of island biogeography. Princeton University Press.

McDowall, R. M. 1973a. The status of the South African galaxiid (Pisces, Galaxiidae). Ann. Cape Prov. Mus. (Nat. Hist.) 9:91–101.

McDowall, R. M. 1973b. Zoogeography and taxonomy. Tuatara 20:88–96.

McKenna, M. C. 1973. Sweepstakes, filters, corridors, Noah's arks, and beached Viking funeral ships in palaeogeography. In: Tarling, D. H., and S. K. Runcorn (eds.), Implications of continental drift to the earth sciences 1:295–308. Academic Press, London and New York.

Malfait, B. T., and M. G. Dinkleman. 1972. Circum-Caribbean tectonic and igneous activity and the evolution of the Caribbean plate. Bull. Geol. Soc. Amer. 83:251–272.

Marcus, E. 1933. Tiergeographie. In: Klute, F. (ed.), Handbuch der geographischen Wissenschaft 2:81–166. Athenaion, Potsdam.

Matthew, W. D. 1915. Climate and evolution. Ann. N. Y. Acad. Sci. 24:171–318.

Mayr, E. 1946a. The number of species of birds. Auk 63:64–69.

Mayr, E. 1946b. History of the North American bird fauna. Wilson Bull. 58:3–41.

Mayr, E. (ed.). 1952. The problem of land connections across the South Atlantic, with special reference to the Mesozoic. Bull. Amer. Mus. Nat. Hist. 99:79–258.

Mayr, E. 1963. Animal species and evolution. Harvard University Press, Cambridge.

Mayr, E., and D. Amadon. 1951. A classification of recent birds. Amer. Mus. Novitates 1496:1–42.

Mayr, E., and W. H. Phelps, Jr. 1967. The origin of the bird fauna of the south Venezuelan highlands. Bull. Amer. Mus. Nat. Hist. 136:269–328 (Spanish translation: 1971. Bol. Soc. Ven. Cien. Nat. 29:309–401).

Neill, W. T. 1969. The geography of life. Columbia University Press, New York and London.

Nelson, G. 1973. Comments on Leon Croizat's biogeography. Syst. Zool. 22:312–320.

Nemeth, K., and M. Libke. 1972. Plate tectonics in the Caribbean. Compass 50:14–23.

Newbigin, M. I. 1948. Plant and animal geography. F. Dutton & Co., New York.

Olson, E. C. 1971. Vertebrate paleozoology. Wiley-Interscience, New York, London, Sydney, and Toronto.

Pantin, C. F. A., et al. 1960. A discussion on the biology of the southern cold temperate zone. Proc. Roy. Soc., London, 152B:429–682.

Peters, J. L. 1948. Check-list of birds of the world. Vol. 6. Harvard University Press, Cambridge.

Polunin, N. 1960. Introduction to plant geography and some related sciences. McGraw-Hill Book Co., New York, Toronto, London.

Raup, H. M. 1942. Trends in the development of geographic botany. Ann. Assoc. Amer. Geogr. 32:319–354.

Rensch, B. 1929. Das Prinzip geographischer Rassenkreise und das Problem der Artbildung. Gebrüder Borntraeger, Berlin.

Romer, A. S. 1973. Vertebrates and continental connections: an introduction. In: Tarling, D. H., and S. K. Runcorn (eds.), Implications of continental drift to the earth sciences 1:345–349. Academic Press, London and New York.

Rosen, D. E. 1960. Middle-American poeciliid fishes of the genus Xiphophorus. Bull. Florida State Mus. 5:57–242.

Rosen, D. E. 1974. The phylogeny and zoogeography of salmoniform fishes, and the relationships of Lepidogalaxias salamandroides. Bull. Amer. Mus. Nat. Hist. 153:265–326.

Rosenblatt, R. H. 1963. Some aspects of speciation in marine shore fishes. Syst. Assoc. Publ. 5:171–180.

Sauer, J. D. 1969. Oceanic islands and biogeographical theory: a review. Geogr. Rev. 59:582–593.

Schaeffer, B., M. K. Hecht, and N. Eldredge. 1972. Phylogeny and paleontology. Evol. Biol. 6:31–46.

Schauensee, R. M. de. 1964. The birds of Colombia and adjacent areas of South and Central America. Livingston Publ. Co., Narberth.

Schilder, F. A. 1956. Lehrbuch der allgemeinen Zoogeographie. Gustav Fischer, Jena.

Schmidt, G. 1969. Vegetationsgeographie auf ökologisch-soziologischer Grundlage. B. G. Teubner Verlagsgesellschaft, Leipzig.

Schmidt, K. P. 1946. On the zoogeography of the Holarctic Region. Copeia: 144–152.

Schmithüsen, J. 1968. Allgemeine vegetationsgeographie. W. De Gruyter & Co., Berlin.

Schröter, C. 1913. Genetische Pflanzengeographie. In: Korschelt, E., et al. (eds.), Handwörterbuch der Naturwissenschaften 4:907–942. Gustav Fischer, Jena.

Simpson, G. G. 1940. Mammals and land bridges. Jour. Wash. Acad. Sci. 30:137–163.

Simpson, G. G. 1949. The meaning of evolution. Yale University Press, New Haven and London.

Simpson, G. G. 1965. The geography of evolution. Chilton, Philadelphia and New York.

Skottsberg, C. 1928. Remarks on the relative independency of Pacific floras. Proc. Third Pan-Pac. Sci. Congr. 1:914–920.

Skottsberg, C. 1956. Derivation of the flora and fauna of Juan Fernandez and Easter Island. In: Skottsberg, C. (ed.), The natural history of Juan Fernandez and Easter Island 1:193–438. Almqvist & Wiksell, Uppsala.

Skottsberg, C. 1960. Remarks on the plant

geography of the southern cold temperate zone. Proc. Roy. Soc. London 152B:447–457.

SOKAL, R. R., AND T. J. CROVELLO. 1970. The biological species concept: a critical evaluation. Amer. Nat. 104:127–153.

STAHL, B. J. 1974. Vertebrate history: problems in evolution. McGraw-Hill Book Company, New York.

SWAINSON, W. 1835. A treatise on the geography and classification of animals. Longman, Rees, Orme, Brown, Green, & Longman and John Taylor, London.

TAKHTAJAN, A. 1969. Flowering plants origin and dispersal. Smithsonian Institution Press, Washington.

THISELTON-DYER, W. T. 1878. Lecture on plant distribution as a field for geographical research. Proc. Roy. Geog. Soc. 22:412–445.

THISELTON-DYER, W. T. 1909. Geographical distribution of plants. In: Seward, A. E. (ed.), Darwin and modern science:298–318. Cambridge University Press, Cambridge.

THISELTON-DYER, W. T. 1912. On the supposed Tertiary Antarctic continent. Jour. Acad. Nat. Sci. Phil., ser. 2, 15:237–239.

TOBLER, W. R., H. W. MIELKE, AND T. R. DETWYLER. 1970. Geobotanical distance between New Zealand and neighboring islands. BioSci. 20:537–542.

TURRILL, W. B. 1953. Pioneer plant geography. Lotsya 4:1–267.

UDVARDY, M. D. F. 1969. Dynamic zoogeography. Van Nostrand Reinhold Co., New York, Cincinnati, Toronto, London, Melbourne.

VALENTINE, D. H. (ED.). 1972. Taxonomy phytogeography and evolution. Academic Press, London and New York.

VORZIMMER, P. J. 1970. Charles Darwin: the years of controversy. Temple University Press, Philadelphia.

WADDINGTON, C. H. 1939. An introduction to modern genetics. Macmillan Co., New York.

WAGNER, M. 1868. Die Darwin'sche Theorie und das Migrationgesetz der Organismen. Duncker & Humblot, Leipzig.

WAGNER, M. 1889. Die Entstehung der Arten durch räumliche Sonderung. Benno Schwabe, Basel.

WALLACE, A. R. 1876. The geographical distribution of animals. Vols. 1–2. Macmillan and Co., London.

WALLACE, A. R. 1881. Island life. Harper & Brothers, New York.

WALLACE, A. R., AND W. T. THISELTON-DYER. 1885. The distribution of life, animal and vegetable, in space and time. Humboldt Lib. Pop. Sci. Lit. 6(64):227–274.

WHITLEY, G. P. 1956. The story of Galaxias. Australian Mus. Mag. 12:30–34.

WIEBES, J. T. 1968. Catalogue of the Coleoptera Cetoniidae in the Leiden Museum 1. Goliathus Lamarck, sensu lato. Zool. Meded. Rijkmus. Nat. Hist. Leiden 43:19–40.

WILLIS, J. C. 1922. Age and area. Cambridge University Press.

WILSON, J. T. 1963. Continental drift. Sci. Amer. 208:86–100.

WITTMAN, O. 1934. Die biogeographischen Beziehungen der Südkontinente. Zoogeographica 2:246–304.

WITTMAN, O. 1935. Die biogeographischen Beziehungen der Südkontinente. Teil 2. Die südatlantischen Beziehungen. Zoogeographica 3:27–65.

WULFF, E. V. 1943. An introduction to historical plant geography. Chronica Botanica Co., Waltham.

Manuscript received August, 1973
Revised December, 1973

Gareth J. Nelson
In Systematic Zoology, *volume 23*

Historical Biogeography: An Alternative Formalization

In an attempt to formalize the procedures of Hennig (1966) and Brundin (1966; also Ross, 1974), I argued that, given distributional data for a certain group of descendant species, ancestral distributions could be estimated and episodes of dispersal resolved (Nelson, 1969). Having considered the arguments of Croizat (1964 and other papers), I now believe that dispersal is not realistically resolvable by that formalization. Consider an example of an ancestral species (A) whose range extends over what will become two geographic areas, e.g., continental South America and Africa (Fig. 1). Suppose (1) that a geographic barrier appears within South America and subdivides ancestral species A into two descendant populations: A1 in southwestern South America, and A2 in eastern South America and Africa; (2) that another geographic barrier subsequently appears between South America and Africa, and subdivides population A2 into two descendant populations: A2a in eastern South America and A2b in Africa; and (3) that the three resulting descendant populations (A1, A2a, A2b) differentiate to the point where they might be recognized as species (or monophyletic taxa of whatever rank) and their kinship correctly interpreted. Given the distribution of the three descendant species (A1 in southwestern South America, A2a in eastern South America, A2b in Africa), an attempt to estimate ancestral distributions according to my previous formalization results in propositions (1) that ancestral species A was exclusively South American, (2) that ancestral species A2 occurred both in South America and Africa, and (3) that dispersal (range extension or migration) occurred from South America to Africa, after the splitting (vicariance) of ancestral species A, but before the splitting of ancestral species A2 (Fig. 2, above). With reference to the details of this hypothetical example,[1]

proposition (1) would be false, (2) would be true, and (3) would be false; in my understanding (at least some of) the procedures of Hennig (1966), Brundin (1966) and Ross (1974) would give the same results. I conclude, therefore, that my previous formalization—and the procedures of Hennig, Brundin, and Ross—is defective in resolving episodes of dispersal in cases where no dispersal occurred, and needs improvement to eliminate that defect.[2]

This conclusion breaks decisively with biogeographic practice that is based on a priori acceptance of a speciation model that requires, or at least encourages the resolution of, dispersal—even in cases where no dispersal has occurred: "Every species originally occupies a certain area, and the breaking up of a species into several reproductive communities usually, if not always, is closely related to the dispersal of the species in space" (Hennig, 1966:133); by "dispersal" Hennig clearly means expansion of range or migration. The belief that speciation (vicariance) and dispersal (migration) must take place together may be due to the practice of drawing a phylogenetic tree on a map—with the implication that the tree has "grown," by dispersal as it were, from its root to the tip of its branches: "An evolutionary tree like this can be laid on a map to show how evolving animals have moved over the earth" (Darlington, 1959:308; also Ross, 1973:179, 1974:214). But I see no reason to assume that dispersal (migration) is a necessary or even common adjunct of the splitting (vicariance) of an ancestral species (Croizat, Nelson, Rosen, 1974). Indeed, what is there, except the absence of a more reasonable alternative, to recommend the view that a formalization should be based on that apriorism? A more reasonable alternative has been available, albeit generally

---

[1] For a concrete example, see Chardon's (1967) consideration of the geographic history of catfishes, which he believes had their center of origin in South America (also Gosline, 1972).

[2] Resolving dispersal when none occurred might be termed a "Type-I error"; not resolving dispersal when dispersal did occur might be termed a "Type-II error." Type-I errors seem to be by far the more common in biogeography.

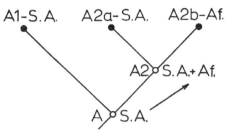

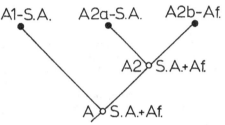

FIG. 2.—Three species (A1, A2a, A2b), their distributions (S.A., S. A., and Af., respectively), and alternative interpretations of the distributions of hypothetical common ancestral species (A, A2). Above, an interpretation that resolves dispersal (arrow, from S.A. to Af.), when no dispersal occurred. Below, an interpretation that does not erroneously resolve dispersal.

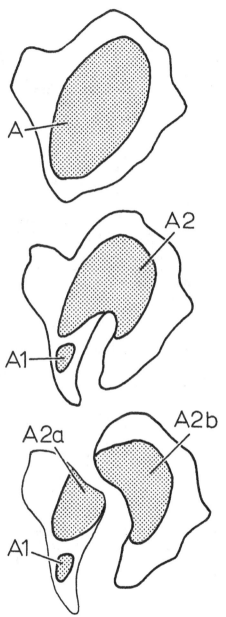

FIG. 1.—The splitting (vicariance event) of an ancestral species (A) into descendant species A1 and A2, and the subsequent splitting of A2 into descendant species A2a and A2b.

unrecognized, since the theory of hologenesis of Rosa (1918), and has been elaborated in great detail by Croizat in numerous publications during the period 1952–1974 (Nelson, 1973).

I suggest, therefore, that for a given group the distribution of ancestral species can be estimated best by adding the descendant distributions (Fig. 2, below). "Eliminating the unshared element" (Nelson, 1973:314) comes close to assuming that dispersal must occur with the speciation process—an assumption that may be rejected as too restrictive, unnecessary and unrealistic: estimation of ancestral distributions may be considered simply additive (hologenetic). If so, may episodes of dispersal—which no doubt have occurred—be resolved by more efficient means? In the manner of Croizat, I suggest that estimation of ancestral distributions must ultimately be reconciled with the history of the biota of which a given analyzed group of species is a part, and that the resolution of dispersal, as evidenced by the general phenomenon of sympatry, is more efficiently accomplished in the context of that synthetic approach (Croizat et al., 1974). I suggest, finally, that my previous formalization—and the procedures of Hennig, Brundin, and Ross—may still have merit, not in estimating ancestral distributions, but rather in estimating where barriers appeared, barriers that caused the splitting (vicariance) of ancestral species. In the above example, therefore, one may estimate that the barrier responsible for the splitting of ancestral species A appeared within South America, and that the barrier responsible for the splitting of ancestral species A2 appeared between South America and Africa.

Estimating where a specific barrier appeared is a step toward localizing and identifying a specific barrier as the cause of a specific vicariant pattern, as, for example, the formation of the Panamanian isthmus has been identified as the cause of the eastern Pacific-western Atlantic vicariance displayed by many shallow-water marine groups. If a specific barrier can be identified, and its time of formation determined (as e.g., Upper Oligocene as the time of formation of the Panamanian isthmus), its time of formation is an estimate of the time of origin of the specific vicariance occurring in relation to the barrier. Its time of formation is, also, an estimate of the absolute time of splitting of the evolutionary lineages displaying that specific vicariance (Hennig, 1948, 1966).

Estimating the time of splitting of lineages through a study of vicariance in relation to dated barriers is an alternative to the traditional approach through paleontology, which tends to underestimate the absolute age, or age of origin, of lineages. It is an alternative, also, to molecular and immunological approaches, which are based on the disputed assumption of constant rates of molecular evolution.

Two considerations emerge immediately from the vicariance approach: (1) that the various populations isolated by a specific barrier will evolve at different rates, such that their descendants, through the vagaries of taxonomic procedure, might be classed as different subspecies, species, subgenera, etc.,—all with the same absolute time of origin, and (2) that the known fossil record of the lineages displaying a particular vicariance will vary in completeness, such that some or most of the lineages actually dating from the time of formation of the barrier that caused their vicariance might be unrepresented in the known fossil record, and that few if any lineages would be represented from their time of origin. The above considerations suggest two conclusions: (1) that, as represented in traditional classification, discrepancy of taxonomic rank among lineages displaying a given vicariant pattern is evidence of different rates of evolution—not temporally different origins, and (2) that, as represented in the known fossil record, discrepancy in the known minimum ages among lineages displaying a given vicariant pattern is evidence of sampling variability of the fossil record—not temporally different origins.

In summary, I reject as aprioristic all "clues" or "rules" used to resolve centers of origin and dispersal without reference to general patterns of vicariance and sympatry (Croizat et al., 1974). With many others, I include as a rejectable apriorism Hennig's "Progression Rule" (Ashlock, 1974). Unencumbered by aprioristic dispersal, historical biogeography is the discovery and interpretation, with reference to causal geographic factors, of the vicariance shown by the monophyletic groups resolved by phylogenetic ("cladistic") systematics.

REFERENCES

Ashlock, P. D. 1974. The uses of cladistics. Ann. Rev. Ecol. Syst. 5:81–99.
Brundin, L. 1966. Transantarctic relationships and their significance. K. Svenska Vetensk.-Akad. Handl., ser. 4, 11(1):1–472.

CHARDON, M. 1967. Réflexions sur la dispersion des Ostariophysi à la lumière de recherches morphologiques nouvelles. Ann. Soc. Roy. Zool. Belgique 97:175–186.

CROIZAT, L. 1964. Space, time, form: the biological synthesis. Published by the author, Caracas.

CROIZAT, L., G. NELSON, AND D. E. ROSEN. 1974. Centers of origin and related concepts. Syst. Zool. 23:265–287.

DARLINGTON, P. J., JR. 1959. Darwin and zoogeography. Proc. Amer. Phil. Soc. 103:307–319.

GOSLINE, W. A. 1972. A reexamination of the similarities between the freshwater fishes of Africa and South America. 17th Int. Zool. Congr., Monaco, Theme 1, 12 pp.

HENNIG, W. 1948. Die Larvenformen der Dipteren. Akademie-Verlag, Berlin.

HENNIG, W. 1966. Phylogenetic systematics. University of Illinois Press, Urbana.

NELSON, G. J. 1969. The problem of historical biogeography. Syst. Zool. 18:243–246.

NELSON, G. 1973. Comments on Leon Croizat's biogeography. Syst. Zool. 22:312–320.

ROSA, D. 1918. Ologenesi. R. Bemporad and Figlio, Firenze.

ROSS, H. H. 1973. Evolution and phylogeny. Ann. Rev. Ent. 18:171–184.

ROSS, H. H. 1974. Biological systematics. Addison-Wesley Publishing Co., Inc., Reading, Menlo Park, London, Don Mills.

GARETH NELSON

Department of Ichthyology
The American Museum of Natural History
New York, New York 10024

A METHOD OF ANALYSIS FOR HISTORICAL BIOGEOGRAPHY

Norman I. Platnick and Gareth Nelson

Abstract

Platnick, N. I., and G. Nelson (Departments of Entomology and Ichthyology, The American Museum of Natural History, New York, New York 10024) 1978. A method of analysis for historical biogeography. Syst. Zool. 27:1–16.—Historical explanations of biotic distribution fall into two classes, dispersal explanations and vicariance explanations. Dispersal models explain disjunctions by dispersal across pre-existing barriers, vicariance models by the appearance of barriers fragmenting the ranges of ancestral species. Distributional data seem insufficient to resolve decisively either dispersal or vicariance as the cause of particular allopatric distribution patterns. When faced with such a pattern our first question should therefore be directed not to its cause, but to whether or not it conforms to a general pattern of relationships shown by taxa endemic to the areas occupied. Two-taxon statements are always compatible with a general pattern; three-taxon statements are therefore the most basic possible units of biogeographic (as well as phylogenetic) analysis. Analysis of three-taxon statements involves converting a hypothesis about the interrelationships of taxa (a cladogram indicating relative recency of common ancestry) to one concerning the interrelationships of areas (a cladogram indicating relative recency of common ancestral biotas). The generality of the area hypothesis may be tested by comparison with other groups endemic to the relevant areas. If the area hypothesis is corroborated as general, a statement of the relative recency of interconnections among areas is obtained, and evidence from historical geology may allow us to specify the nature of those interconnections and thereby the cause of those distributions that conform to the general pattern. Analysis of four-taxon statements indicates that the availability of structurally different patterns and of groups that can serve as adequate tests of the generality of those patterns increases with the addition of taxa to the hypothesis, and that neither extinction nor the failure of some groups to respond (by speciating) to given dispersal or vicariance events interferes with the analysis. [Biogeography; dispersal; vicariance.]

"Without waiting, passively, for repetitions to impress or impose regularities upon us, we actively try to impose regularities upon the world. We try to discover similarities in it, and to interpret it in terms of laws invented by us. Without waiting for premises we jump to conclusions. These may have to be discarded later, should observation show that they are wrong. ... Since there were logical reasons behind this procedure, I thought that ... scientific theories were not the digest of observations, but that they were inventions—conjectures boldly put forward for trial, to be eliminated if they clashed with observations; with observations which were rarely accidental but as a rule undertaken with the definite intention of testing a theory by obtaining, if possible, a decisive refutation" (K. R. Popper, 1965:46).

"To do science is to search for repeated patterns, not simply to accumulate facts, and to do the science of geographical ecology is to search for patterns of plant and animal life that can be put on a map. The person best equipped to do this is the naturalist .... But not all naturalists want to do science; many take refuge in nature's complexity as a justification to oppose any search for patterns" (R. H. MacArthur, 1972:1).

Historical biogeography is that field of inquiry that attempts to answer the question "Why are taxa distributed where they are today?" in terms of their history rather than exclusively in terms of their current ecology. Traditionally this question has been answered with dispersal explanations: 'Because they (or their ancestors) dispersed into the areas where they now occur.' Entire textbooks (for example, Darlington, 1957) provide answers of this sort for particular groups by first taking into account the known distributions of the fossil and recent members of those groups and then drawing dispersal routes that connect all the relevant localities. This type of answer could be called colonialistic biogeography (because it implies that taxa always originate in one area, their center of origin, and colonize other areas) or even vacuum biogeography (because areas are always assumed to have been originally devoid of the organisms that eventually dispersed there).

Croizat, Nelson, and Rosen (1974) presented an alternative answer: 'Because their ancestors originally occurred in the areas where they occur today, and the taxa now there evolved in place.' Those authors pointed out that under the model of allopatric speciation, related species represent isolated parts of a once joined ancestral population that has been divided by the appearance of some barrier (i.e., that has undergone vicariance). If the model of allopatric speciation is accepted, sympatry of related forms is evidence of dispersal, but allopatric and parapatric distributions of related forms can be explained without recourse to dispersal.

DISPERSAL AND VICARIANCE

Dispersal hypotheses are of two general types (illustrated, for example, by Darlington, 1957, figs. 67, 68), one of which is vicariance in disguise. In this first type of hypothesis, an ancestral species (A) enlarges its range through time and is then fragmented into two disjunct ranges, the populations of which differentiate through time, ultimately to form two allopatric species (B and C). Implied is some causal factor, the appearance of a barrier, responsible for the fragmentation of the range of the ancestral species. The reason why this example of "dispersal" really is vicariance is that the postulated dispersal takes place prior to the appearance of the barrier and prior to the fragmentation of the range of the ancestral species. The effect of the postulated dispersal is only the creation of primitive cosmopolitanism (a requirement of the vicariance model).

The second, classic dispersalist model postulates dispersal over a pre-existing barrier. In this case an ancestral species (A), by means of "accidental crossing of a barrier," expands its range and, in the process, simultaneously fragments its range. The effect of the postulated dispersal is immediate isolation and disjunction. The populations in the disjunct

areas subsequently differentiate into two allopatric species (B and C).

In both cases, the existence of a barrier is implied. In the vicariance model, dispersal, if it takes place at all, occurs in the absence of a barrier; in the dispersalist model, dispersal occurs across a barrier. The explanations offered by both models amount to a correlation between a particular disjunction and a particular barrier: according to the vicariance model, the disjunction and barrier are the same age; according to the dispersalist model the disjunction is younger than the barrier.

Both models allow the possibility that primitive cosmopolitanism may be achieved by an ancestral species that enlarges its range through the means of dispersal characteristic of the species. The models differ with respect to the causal factors invoked to explain disjunctions and, ultimately, allopatric differentiation. In the case of vicariance, disjunction is caused by the appearance of a barrier that fragments the range of an ancestral species; in the case of dispersal, disjunction is caused by dispersal of an ancestral species across a pre-existing barrier. The causal factors may thus be isolated: (1) Vicariance: the appearance of a barrier; (2) Dispersal: dispersal across a pre-existing barrier.

The vicariance model predicts that if we could find a single monophyletic group of organisms that (1) had a primitive cosmopolitan distribution (i.e., whose single ancestral species was worldwide in distribution), (2) had responded (by speciating) to every geological or ecological vicariance event that occurred after the origin of its single ancestral species, (3) had undergone no extinction, and (4) had undergone no dispersal, we could, by reconstructing the phylogenetic interrelationships of its members, arrive at a detailed description of the history in space of the ancestral biota of which the single ancestral species was a part. Since we would also have arrived at a detailed description of the history of the world from the time of the first speciation event within the group to the present, we could, by correlating the sequence of branching points thus reconstructed with the sequence of events indicated by studies in historical geology, arrive at a chronology of the biogeographic events.

That extant distribution patterns are diverse and do not all obviously correspond to each other in every detail is evidence that at least criteria 2–4 above do not always prevail in nature. Since under the model sympatry is evidence of dispersal, the fact that we find numerous sympatric taxa at any given locality is evidence that dispersal has occurred at least in direct proportion to the number of sympatric taxa. The fossil record provides ample evidence that extinction has occurred. Finally, the fact that within any biota some taxa are very widespread and others very localized is evidence that not all members of a biota need respond (by speciating) to every vicariance event.

Clearly, then, neither dispersal nor vicariance explanations can be discounted a priori as irrelevant for any particular group of organisms, and it might seem that the ideal method of biogeographic analysis would be one that allows us to choose objectively between these two types of explanations for particular groups.

TESTING BIOGEOGRAPHIC HYPOTHESES

Popper (1968) has presented the view that scientific explanations differ from non-scientific ones only by virtue of their falsifiability (i.e., that we must be able to test and potentially reject any explanation that is to be considered scientific). We cannot actually observe the history of organisms, but we can test historical explanations of their distribution by accepting one or more axioms that can serve as a bridge between distribution patterns and their historical causes. Two such axioms seem sufficient for this purpose. Because explanations of the history of groups in space are meaningful only if the groups actually had independent histories in time (i.e., if the groups are monophyletic), we must assume (1) that it is possible to construct accurate hypotheses of the genealogy of organisms (by the methods, for example, developed by Hennig, 1966, and justified as scientific under Popper's criterion by Wiley, 1975). Because we seek to explain how members of a single monophyletic group come to be found in some given set of areas, we must also assume (2) that there is a mechanism of evolution that results in the distribution of related forms in different areas (i.e., that we can apply a model of allopatric speciation to explain distribution patterns). A corollary of these two axioms is that the sympatric occurrence of two or more members of a single monophyletic group has been caused by the dispersal of one or more of those taxa into the area of sympatry. Thus, a question arises as to the cause of the distribution pattern of a group only when it contains allopatric elements (that could be the result of vicariance alone or of dispersal).

Because the vicariance and dispersal models differ with regard to the age of disjunctions and barriers, it might appear that a critical test between them could be made by investigating these ages. Both models appear to be open to falsification through statements made about the age of a particular disjunction and the age of a particular barrier. But the age of either one, in any particular case, is problematical. The minimum age of a taxon is that of the oldest known fossil attributed to it or to its most closely related sister-taxon (an age that always can be augmented by the discovery of an older fossil). The age of a barrier can be estimated through studies of historical geology. Both types of studies (paleontology and geology) are subject to wide margins of error. Both types of studies can, however, falsify a correlation between a particular disjunction and a particular barrier. The vicariance type of correlation is falsified if one of the disjunct taxa is shown to be older than the barrier; the dispersal type of correlation is falsified if one of the disjunct taxa is shown to be the same age as, or older than, the barrier.

Both types of correlation are potential-

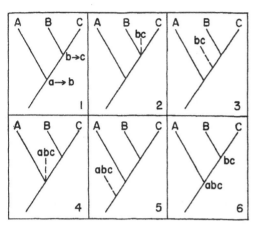

FIG. 1.—Three taxa (A, B, C) in three corresponding areas (a, b, c). 1.1. A dispersal explanation (arrows indicate direction of dispersal). 1.2–1.5. Widespread fossil taxa added. 1.6. A vicariance explanation.

ly falsified, therefore, through discovery of older fossils. But in each case, it is a particular correlation between a disjunction and a barrier, not the model itself, that is falsified. Both models include ad hoc protection from falsification through discovery of older fossils. The ad hoc principle is the rejection of a particular barrier (apparently of the wrong age to explain a disjunction) in favor of another (an older barrier, if one can be found; or if not found, postulated).

It is apparent that, in any particular case, neither model is exposed to a critical test by the dating of barriers and disjunctions, and that both models incorporate approximately the same ad hoc protection against falsification. In the absence of a critical test by the dating of barriers and disjunctions, we can proceed by adopting one model and attempting to refute other implications it may have that are not shared by the second model.

TESTING DISPERSAL HYPOTHESES

Assume that we have three allopatric taxa (A, B, and C) distributed in three corresponding areas (a, b, and c), and that we have tested and corroborated hypotheses that the three taxa form a mono-phyletic group (i.e., have a common ancestor unique to themselves) and that taxa B and C are more closely related to each other than either is to A (i.e., that B and C share a more recent common ancestor with each other than either does with A). We could construct a dispersal explanation like that shown in Fig. 1.1, to the effect that the common ancestor of the group was originally found only in area a and is represented there today by taxon A, that some members of that ancestral taxon dispersed to area b and subsequently speciated there, and that some members of this second taxon subsequently dispersed to area c and eventually speciated there. How might we test this hypothesis? To test any explanation, we must be able to deduce from it some prediction with which additional data can either agree or disagree. What can we deduce from explanation 1.1? Because the dispersal capabilities of these organisms may or may not be similar to those of other groups, we can make no prediction about what patterns other groups that occupy these areas might show. Because the postulated dispersal events involve only movements of ancestors of the three taxa, we can make no prediction about what the dispersal capabilities of the three extant taxa might be. However, it seems that we might be able to make some predictions about the distribution of fossil specimens of the group that we might find.

Since we have hypothesized that taxon C evolved within area c, we should presumably be able to find fossil specimens attributable to taxon C within area c if suitable deposits exist there. Suppose, however, that we find a fossil attributable to taxon C in area a; does this falsify our dispersal hypothesis? The presence of C in area a could be accounted for by yet another postulated dispersal that occurred after the speciation of C, and we could accommodate the additional data without abandoning the hypothesis in favor of a vicariance explanation. Suppose instead that we had found a fossil taxon in areas b and c that cannot be attributed to one of the extant taxa but that, because it both shares the synapomorphies uniting B and C and lacks their autapomorphies, has to be added to the cladogram as in Fig. 1.2. Our hypothesis requires that B and C had a common ancestor that for at least some period of time occurred in both areas (since the "founders" of taxon C did not speciate in transit), so the fossils are consistent with our explanation. Similarly, if the fossil taxon found in areas b and c shared only some of the synapomorphies of B and C and has therefore to be added to the cladogram as in Fig. 1.3, our hypothesis is still tenable as it does not specify the length of time for which the common ancestor of B and C may have occurred in area c before it speciated there. A similar fossil found only in areas a and c could also be accommodated in a dispersal hypothesis by postulating an initial dispersal from a to c and a second dispersal from c to b.

Suppose, however, that we found a single fossil taxon in areas a, b, and c that has to be added to the cladogram as shown in Figs. 1.4 or 1.5. Our hypothesis is that area c was populated by a taxon that had originated in area b (i.e., that had been isolated from a taxon in area a); thus, there should never have been a single taxon found in all three areas. Must we now abandon the dispersal hypothesis? It is possible that the presence of the fossil taxon in areas b and c represents an independent set of invasions into those areas from area a, and that the center of origin of the common ancestor of all four taxa was indeed only area a. If we accept this possibility, it appears that any distribution pattern whatsoever can be explained by dispersal if we are willing to postulate a sufficient number of separate dispersal events; this would mean that dispersal explanations can never be rejected and are therefore unscientific under Popper's criterion.

To prevent this untestability, we might adopt a methodological rule that requires us to minimize the number of parallel dispersals (dispersals from one given area to a second given area), in the same

way that we seek to minimize the number of parallel acquisitions of derived characters in cladograms by adopting a methodological rule that requires us to choose the most parsimonious hypothesis of relationships that will account for any given set of character distributions. Given this methodological rule, we could recognize that a dispersal explanation of patterns 1.4 or 1.5 requires parallel dispersals from area a to area b, and possibly to area c as well, and that we must therefore abandon the dispersal explanation in favor of a vicariance hypothesis (Fig. 1.6). In other words, given such data we would have to abandon the hypothesis that area a was the center of origin of the group in favor of a hypothesis of primitive cosmopolitanism (i.e., that the common ancestor of the group occurred in all the areas in which its descendants occur today). It should be noted that the methodological rule suggested here to render dispersal explanations testable is not one of parsimony in regard to the number of postulated dispersals (which would always lead to a vicariance explanation) but only in regard to the number of postulated parallel dispersals.

Clearly, then, a system in which we always adopt a dispersal explanation of allopatric patterns and use the discovery of plesiomorphic cosmopolitan fossil taxa to reject those explanations is possible. But is it sufficient for our desired purpose, to distinguish all cases of vicariance from cases of dispersal? Since there may be many groups whose distributions are due to vicariance alone but which may not be thus resolved because the relevant fossils are unavailable, the initial adoption of dispersal explanations may greatly overestimate the number of groups whose distributions are the result of dispersal.

TESTING VICARIANCE HYPOTHESES

Perhaps we should therefore choose a vicariance hypothesis as our initial explanation of allopatric distribution patterns. How might we test such a hypothesis for the same taxa and distributions consid-

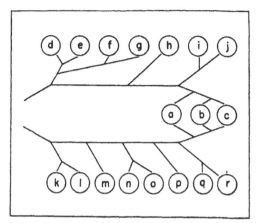

FIG. 2.—Sympatry of two incompatible general patterns in area abc.

ered above? From explanation 1.6 we can deduce that two vicariance events occurred, one of which divided area abc into two smaller areas (a and bc) and one of which subsequently subdivided area bc into two still smaller areas (b and c). If those geological or ecological events did occur during earth history, they would have affected other organisms living in area abc. Thus, if a vicariance event did divide area abc, we should be able to find other taxa living in area a that have their closest relatives in area bc, and other taxa living in area b that have their closest relatives in c. In other words, we can test explanation 1.6 by converting the cladogram of taxa A, B, and C (reflecting the relative recency of their common ancestry) into a cladogram of *areas* a, b, and c (reflecting the relative recency of their common ancestral biotas and the relative recency of the geological or ecological events involved). The converted cladogram thus states that areas b and c share a more recent common ancestral biota with each other than either of them do with area a, and that area bc was fragmented only after it was isolated from area a.

If we examine the interrelationships among taxa of other groups extant or known to be extinct in area abc and find that they correspond to the area clado-

gram, explanation 1.6 is corroborated. What if none of the other groups distributed in area abc shows that set of relationships (i.e., if the pattern is unique)? The cause of that unique pattern might be simply taxonomic error, but that possibility is subject to independent testing by cladistic character analysis and can be disregarded here. It is also possible that the unique pattern represents the only surviving component of a general pattern, but this ad hoc hypothesis requires parallel extinctions of all the other components of the general pattern to have taken place. Thus it can also be discounted (through the use of a methodological rule against parallel extinctions), allowing us to reject the vicariance explanation of the unique pattern in favor of some dispersal hypothesis.

Suppose further that upon subsequent investigation we find that the groups that do correspond to a general allopatric pattern do not also correspond in their higher-level relationships. Assume, for example, that we have detected a general pattern among the taxa distributed in area abc corresponding to explanation 1.6, but that the cladograms of some of the groups sharing that pattern relate taxa in area abc only to taxa in areas to the north (say to areas d through j, Fig. 2), and that the cladograms of other groups sharing that pattern relate taxa in area abc only to taxa in areas to the south (say to areas k through r, Fig. 2). In other words, we find that two larger and mutually incompatible general patterns share some elements in a limited area (abc), where they are sympatric. Here again, vicariance is not a sufficient explanation of the pattern; under our axioms, sympatry of elements of different general patterns is evidence of dispersal just as is sympatry of individual taxa.

How we might resolve the nature of the dispersal involved will be considered below; suffice it to say here (1) that the pattern shown in Fig. 2 requires not only a single case of dispersal but at least one case of biotic dispersal, i.e., of the dispersal of several elements of a biota into the same new area, promoting cosmopolitanism of taxa, (2) that the episode of biotic dispersal can be dated as having occurred before the division of area abc but after the division of either area hij (Fig. 2) or area pqr, or both, (3) that there are events in earth history (such as the fusion of India with mainland Asia or the appearance of the Panamanian isthmus) that provide opportunities for episodes of biotic dispersal, and (4) that the vicariance model requires episodes of cosmopolitanism to have occurred in order to account for present-day sympatry among members of many different monophyletic groups. For example, the vast number of patterns of large, worldwide groups whose subgroup interrelationships reflect the fragmentation of Pangaea all require episodes of biotic dispersal promoting cosmopolitanism to account for the (sympatric) Pangaean distributions of the single ancestral species of each of those groups.

Clearly, then, a system in which we always adopt a vicariance explanation of allopatric patterns and use the discovery of unique patterns or sympatry of general patterns to reject those explanations is also possible. But is it sufficient for our desired purpose, to distinguish all cases of vicariance from cases of dispersal? Since there may be many general allopatric patterns shared by groups whose distributions are due only to unidirectional and sequential ("steppingstone") dispersal like that shown in Fig. 1.1, but which may not be thus resolved because the sequence and destinations of the dispersal events have been the same for each component group of the pattern, the initial adoption of vicariance explanations may greatly overestimate the number of groups whose distributions are the result of vicariance.

PATTERNS OF DISTRIBUTION

If it is true that neither critical testing by the dating of barriers and disjunctions, nor the initial adoption of either dispersal or vicariance explanations provides a method sufficient in theory (much less in practice) to distinguish unambiguously instances of vicariance from instances of dispersal, what are the implications? Perhaps the question of whether a given allopatric distribution pattern is due to vicariance or dispersal is like the question of whether the differences we detect among organisms are due to anagenetic or cladogenetic change. The latter question seems impermeable to analysis because the results of both processes, anagenesis and cladogenesis, can be expressed only in one and the same way: a pattern of character-state distributions among different taxa, and because the empirical data available to us allow us to retrieve directly only the pattern and not its cause. Perhaps the first question is also impermeable to analysis because the results of both dispersal and vicariance can be expressed only in one and the same way: a pattern of taxon distributions among different areas, and because the empirical data available to us allow us to retrieve directly only the pattern and not its cause. If so, and if we do not wish to abandon the problem entirely, we must seek a way to answer the question indirectly.

If, as indicated above, a general allopatric pattern could be produced by vicariance or by sequential biotic dispersal, there is only one kind of information that can be obtained from distributional data, information about the relative recency of connections (common ancestral biotas) among different areas. If there is a general allopatric pattern corresponding to the cladogram in Fig. 1.1, areas b and c share a more recent connection than either of them does with area a. That connection might have involved, for example, an actual land connection (and vicariance) or merely changes in the relative interdistances of areas at various times in the past (and the resulting possibilities for biotic dispersal). In either case, of course, the general pattern of area interconnections tells us something about the history of those areas. Thus, distributional data seem sufficient to resolve a pattern of interconnections among areas that reflects their history, but not to specify the nature of those interconnections.

Questions regarding earth history and the interconnections of areas are open to several tests, however, of which biotic distribution is only one. Stratigraphy, paleomagnetism, geochemistry, and other similar sources of data contribute independent historical hypotheses. The possibility exists, therefore, of using distributional and geological data as reciprocally illuminating sources of evidence. Thus, having used biotic distribution to specify a pattern of interconnections among areas, we might in at least some cases be able to use data from historical geology to specify the nature of the interconnections themselves.

Take, for example, the case of the sympatric general patterns shown in Fig. 2. Because under our axioms sympatry of general patterns (as in area abc) must be due to biotic dispersal, we know that vicariance alone is not a sufficient explanation of the pattern of interconnections shown between area abc and other areas, but the biological data tell us nothing more specific about the nature of those interconnections. In what way can the history of area abc account for the patterns we observe? There seem to be four possibilities. It is possible that the dispersal occurred only within area abc; this could happen if area abc is actually a composite of two smaller areas (each belonging to one of the larger patterns) that have been joined together, and that are no longer discernible as separate areas because of an episode of biotic dispersal between the two after their merger. It is also possible that area abc is anciently a part of the southern area, and that a vicariance event resulted in a shift of area abc or a piece of northern land toward each other, permitting dispersal of parts of the northern biota into area abc (and possibly vice versa). Similarly, area abc could be anciently a part of the northern area with biotic dispersal from the south being facilitated by a vicariance event. Finally, it is possible that area abc represents new land that emerged between the northern and southern areas and was populated by biotic dispersal from both areas.

Given that correlations of the sequence of connections indicated by the two general biotic patterns with the sequence of connections indicated by geological data can allow us to date the episode of biotic dispersal within fairly precise limits, it seems likely that geological data would permit us to resolve the question of which of the four possible geological events was actually involved, and thereby allow us to specify the nature of the interconnections between area abc and the areas united by each general pattern. For an analysis of an actual area with a complex geological history belonging to at least four general patterns, see Rosen (1975).

Clearly, then, if resolution of the nature of interconnections among areas can only be accomplished through the use of independent data from historical geology, the first question we should ask when confronted with the allopatric distribution pattern of some group is not "Is this pattern the result of vicariance or dispersal?" but "Does this pattern correspond to a general pattern of area interconnections (and thus reflect the history of those areas) or not?". What is needed is a method of analysis that will allow us to determine whether two given distribution patterns correspond to each other or not, so that we can test a hypothesis that the pattern of relationships of areas indicated by one group is a general one. After a hypothesized general pattern is corroborated, we may be able to ascribe it to vicariance or dispersal by the use of independent evidence of earth history.

TWO-AREA PATTERNS

Given a pair of related taxa (A, B) in two separate areas (a, b), as in Fig. 3.1, we can hypothesize that this pattern of relationships conforms to a general pattern of interconnections of areas a and b, and test that hypothesis by finding other taxa endemic to area a and determining where their closest relatives occur. If we find another taxon in area a with its closest relative in area b, their relationships are obviously exactly congruent with our cladogram and the hypothesis is corroborated. What if we find another taxon in area a with its closest relative both in area b and somewhere else (i.e., in area bx)? Both sets of distributional data could belong to the same general pattern, since it is possible that our original taxon B primitively occurred in area bx and is now merely extinct in area x. In other words, although the two patterns are not congruent, they are compatible, and our hypothesized general pattern is not falsified.

Similarly, if we find another taxon endemic to area a with its closest relative not in area b but somewhere else (in area x, Fig. 3.2), the two patterns are again incongruent but compatible, since a group like that shown in Fig. 3.2 could have had a member in area b that is now merely extinct (as in Fig. 3.3). Since these cases exhaust the possibilities, it is apparent that no distribution pattern involving at least one of the two areas can be both incongruent and incompatible with a hypothesized two-taxon, two-area pattern. As a result, the most basic possible unit of analysis in biogeography, as in phylogenetics, must be a three-taxon statement. Just as any two organisms that we may choose will be related at some level, any two areas that we may choose will be connected at some level, and comparative analysis requires that we deal with at least three taxa and areas.

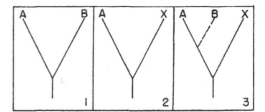

Fig. 3.—3.1, 3.2. Two apparently incompatible two-taxon patterns. 3.3. Pattern 3.2 rendered compatible by the addition on dashed line of an unknown extinct taxon.

ture of the specific barrier involved—a barrier effective for isolating flies may not also isolate birds) and cannot refute it, even if the single taxon is restricted to areas ab, ac, bc, a, b, or c (since it could always have occurred previously in the parts of the total area from which it is now missing and merely be extinct there now).

What about groups that have two taxa in area abc? Again, all such groups are compatible with the hypothesis and cannot refute it (since they could merely have failed to respond to one of the dispersal or vicariance events or now be extinct in one of the smaller areas). It might appear that a group with two taxa, one of which occurs in area ac and the other of which occurs in area b, would be incompatible with hypothesis 1.1. Thus, given that hypothesis of relationships and three areas situated as shown in the top left corner of Fig. 4, we could find cladogram 1 (of Fig. 4) if a group failed to respond to the first event and cladogram 2 (of Fig. 4) if a group failed to respond to the second event. We could not, however, find a cladogram in which a taxon AC in area ac is the sister group of a taxon B in area b, since an event isolating and disconnecting areas b and c must also disconnect area a from c. The difficulty here is that areas a, b, and c might be arranged so that each one is in contact with the others, as in Fig. 5. In that case, if a group failed to respond to the first event but did respond to the second, we might get single populations in area ab *or* ac, depending on whether area b or area c was isolated from

THREE-AREA PATTERNS

Given three taxa in three corresponding areas related as in Fig. 1.1, we may hypothesize that areas b and c are more recently connected to each other than either is to area a. What groups can test this hypothesis? Any group with only a single taxon in area abc is compatible with the hypothesis (since allopatric speciation requires the appearance of a barrier and any group may simply have failed to speciate in response to a dispersal or vicariance event because of the na-

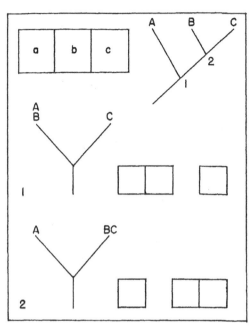

FIG. 4.—Three taxa (A, B, C) in three corresponding areas (a, b, c), with interrelationships as in top right cladogram and distributions as in top left map. Row 1, possible cladogram and area configuration if a test group failed to respond to event 1; row 2, possible cladogram and area configuration if a test group failed to respond to event 2.

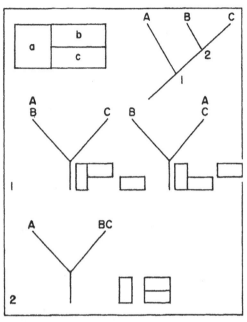

FIG. 5.—Taxa and relationships as in Fig. 4 but distributions as in top left map. Row 1, possible cladograms and area configurations if a test group failed to respond to event 1; row 2, possible cladogram and area configuration if a test group failed to respond to event 2.

area a by the same event that disconnected area b from area c (isolation of one of the areas from both of the others is of course necessary for allopatric speciation to occur). Since the spatial relationships shown by areas a, b, and c today need not be the same as they were in the past, we can never discount the possibility that their spatial arrangement was like that shown in Fig. 5, and two-taxon statements can never be used as tests.

Thus, the only groups that can serve as tests of the hypothesized general pattern are groups with three or more taxa in the total area abc. Further, such groups must have endemic taxa in each of the smaller areas a, b, and c, or we will only be able to obtain a one or two-area statement (all of which are compatible with the hypothesis and do not test it). Given groups with taxa endemic to each of the smaller areas, three relevant cladograms are possible:

either the taxa in areas b and c are each other's closest relatives, or those in areas a and c are closest relatives, or those in areas a and b are closest relatives.[1] If we find groups showing the first mentioned set of relationships, our hypothesis is corroborated, but if none of the available test groups shows that set of relationships we can reject the hypothesis that the distri-

[1] Trichotomous cladograms are of no significance as tests (or as initial hypotheses) unless the cladograms for all available test groups are trichotomous, in which case we may suspect that an event disconnected an area into three smaller areas simultaneously. Testing this hypothesis by biotic relationships seems impossible (because we have no way of distinguishing those cladograms reflecting actual trichotomies from those reflecting only our failure to find the relevant synapomorphies that would resolve a dichotomous cladogram), but it should be subject to independent testing by data from historical geology, which can either be in accordance with such a synchronous tripartite disconnection of the total area or not.

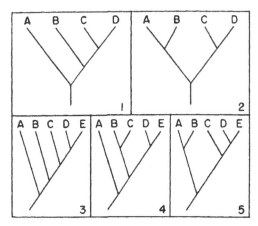

FIG. 6.—Top: two possible structurally different general patterns for four taxa in four corresponding areas. Bottom: three possible structurally different general patterns for five taxa in five corresponding areas.

bution of our original group reflects a general pattern of interrelationships of areas and their history. If our hypothesis is corroborated, we can then seek evidence in historical geology of the nature of the two connections involved and if sufficient data are available resolve vicariance or biotic dispersal as the cause of the general pattern, and also place the cladograms of the groups showing the general pattern on an actual time scale (as in Platnick, 1976, fig. 2).

FOUR-AREA PATTERNS

Since the only groups that provide adequate tests of the minimal three-taxon, three-area hypothesis are those for which it is also possible to construct three-taxon, three-area statements, it might be objected that compatibility analysis is not very practical because of the scarcity of groups that can serve as tests. The availability of test groups, however, increases each time we add another allopatric taxon (and thereby another area) to the hypothesis; the number of available structurally different general patterns also increases in such cases. Analysis of a four-taxon, four-area problem will serve to illustrate this as well as some ramifications of the

effects of extinction and the failure of groups to respond to given events.

Four allopatric taxa situated in four corresponding areas can show either of two structurally different patterns of relationship (Fig. 6, top); all other dichotomous four-taxon cladograms are convertible to pattern 6.1 by rotation at their nodes. We can adopt whichever pattern is shown by the group we wish to investigate, convert it into a hypothesis of area interconnections, and use other groups that have taxa endemic to those areas and for which dichotomous cladograms can be constructed to test the hypothesis.[2] Other groups with taxa endemic to each of the four areas (i.e., for which four-taxon, four-area cladograms can be drawn) obviously provide tests since their cladograms will either be exactly congruent with the hypothesis or not.

Groups with taxa endemic to only three of the four areas (as might happen if, for example, the group was now simply extinct in one of the smaller areas) and for which three-taxon, three-area statements can therefore be constructed also provide tests. For each of the two possible hypothesized general patterns, four of the 12 possible three-area cladograms (which are shown and numbered in Fig. 7) are compatible with the hypothesis but the other eight are not. Determination of compatibility involves separately omitting each of the four areas from the hypothesis and drawing the proper cladograms for the remaining three areas; only the cladograms numbered 1, 4, 7, and 10 (in Fig. 7) are compatible with hypothesis 6.1, and only cladograms 3, 6, 7, and 10 are compatible with hypothesis 6.2.

---

[2] Again cladograms incorporating polychotomies are of no significance as initial hypotheses or tests unless the cladograms for all available test groups are quadrichotomous or show the same pattern of trichotomous branching. In the latter case, three-taxon, three-area statements can serve as tests of the basal dichotomy of the hypothesized general pattern, since three of the possible dichotomous three-area cladograms (those showing the proper area as plesiomorphic to the other two) are compatible with the hypothesis but the other six are not.

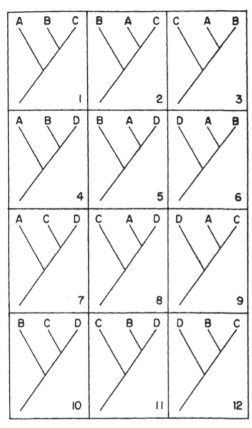

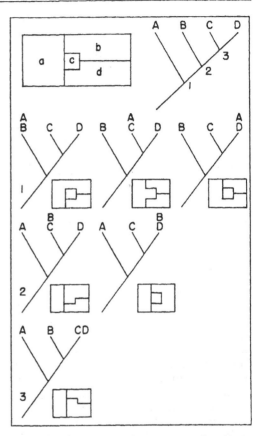

FIG. 7.—Twelve possible three-taxon, three-area cladograms.

FIG. 8.—Four taxa in four corresponding areas, with interrelationships as in top right cladogram and distributions as in top left map. Row 1, possible cladograms and area configurations if a test group failed to respond to event 1; row 2, possible cladograms and area configurations if a test group failed to respond to event 2; row 3, possible cladogram and area configuration if a test group failed to respond to event 3.

Similarly, groups with three taxa endemic to the four areas (as might happen if, for example, the group had simply failed to respond to one of the events) and for which three-taxon, four-area statements can therefore be constructed also provide tests. The compatibility between the possible hypotheses (Fig. 6, top) and the possible cladograms (shown and numbered in Fig. 9) is less obvious than in the last case. Here again we must assume that each of the four areas might be in contact with each of the others. Given a spatial arrangement of four areas and a hypothesis of their interconnections like those shown in Fig. 8, six cladograms are possible for groups that have failed to respond to one of the events. If a group failed to respond to the first event, the

population in area a could be shared with any of the other areas (depending on which areas the events that disconnected areas b, c, and d also isolated from area a), producing any of the cladograms in row 1 (of Fig. 8). If it was the second event that a group failed to respond to, the population in area b must be more closely related to those in areas c and d than to that in area a, but could be shared with either area c or area d (depending on which area the event that disconnected areas c and d also isolated from area b), producing either of the cladograms in row 2 (of Fig. 8). Finally, if it was the third event that a group failed to respond to, the cladogram in row 3 of Fig. 8 is the only possible result. Thus only the cladograms numbered 1, 4, 7, 10, 13, and 16 in Fig. 9 are compatible with hypothesis 6.1.

A similar analysis for hypothesis 6.2 is shown in Fig. 10. If a group failed to respond to the event disconnecting areas a and b, the cladogram in row 1 (of Fig. 10) is the only possible result; similarly, if it failed to respond to the event disconnecting areas c and d, the cladogram in row 2 is the only possible result. If a group failed to respond to the event disconnecting areas ab and cd, two alternatives appear. If the two later events were simultaneous, areas a and b must be separate, areas c and d must be separate, but the two pairs of areas need not be separate and could sort independently, producing either two-taxon cladogram shown in row 3 of Fig. 10. If the two later events did not occur simultaneously, each pair of areas could also be associated with only one member of the other pair. Thus, if the disconnection of areas a and b preceded that of areas c and d, the four cladograms in row 4 (of Fig. 10) could occur, and if the disconnection of areas c and d occurred first, the four cladograms in row 5 could be found. Thus 10 of the cladograms shown in Fig. 9 (numbers 1, 4, 6, 7, 8, 10, 12, 13, 14, and 18) are compatible with hypothesis 6.2 but the other eight are not.

Because two of the three possible two-taxon, four-area cladograms (those in row 3 of Fig. 10) are compatible with hypothesis 6.2, it might appear that such cladograms could serve as tests, but this is illusory, for the remaining two-taxon, four-area cladogram (showing AB as the sister group of CD) is also compatible with the hypothesis (and would be found if a group failed to respond to all but the first event).

If none of the available test groups (all those for which four-taxon, four-area cladograms, three-taxon, three-area clado-

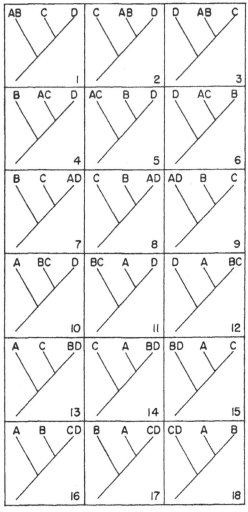

FIG. 9.—Eighteen possible three-taxon, four-area cladograms.

grams, or three-taxon, four-area cladograms can be constructed) show cladograms congruent or compatible with the hypothesis adopted, we can conclude that the distribution of our original group does not reflect a general pattern caused by the history of area interconnections.[3]

---

[3] It is of course possible that part of the four-taxon pattern (say one of the three-taxon patterns that could be abstracted from it) does correspond to a general pattern, and that only the remainder of it does not.

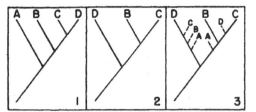

FIG. 11.—A four-area pattern (11.1), a three-area pattern incompatible with it (11.2), and the same three-area pattern rendered deceptively congruent by the additon on dashed lines of five unknown taxa (11.3).

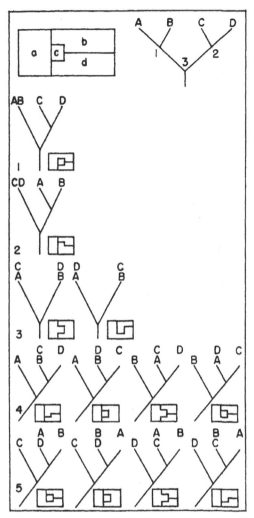

FIG. 10.—Taxa and distributions as in Fig. 8 with interrelationships as in top right cladogram. Possible cladograms and area configurations as in Fig. 8, with rows 3–5 possible if a test group failed to respond to event 3.

If, on the other hand, the pattern proves to be general, we can search for a correlated sequence of events affecting the relevant areas in the geological data on earth history.

It might be objected here that in this four-taxon analysis we have rejected patterns as incompatible that might be interpreted as congruent by postulating a number of unknown extinct taxa. A pat-tern, such as Fig. 11.2, rejected as incompatible with hypothesis 11.1 can be made to appear congruent by adding five unknown taxa (Fig. 11.3). If this is the case, could it not be argued that pattern 11.2 is the result of the same events as pattern 11.1, and that by rejecting pattern 11.2 as incongruent we may erroneously resolve it as a unique pattern or a component of a different general pattern? Examination of pattern 11.3 indicates that if this alternate explanation were correct, the common ancestors of the taxa on each side of the basal dichotomy would then *both* have been distributed in area abcd (i.e., they would have been completely sympatric). Since under our axioms sympatry must be the result of dispersal, the method of compatibility analysis outlined here would be correct in ruling out the events indicated by hypothesis 11.1 as a sufficient explanation for pattern 11.2.

LARGER PATTERNS

Analysis of larger allopatric patterns can be carried out in the same manner. For five allopatric taxa in corresponding areas there are three structurally different general patterns (Fig. 6, bottom), of which five-taxon, five-area cladograms, four-taxon, four-area cladograms, and three-taxon, three-area cladograms can serve as tests. Interestingly, however, four-taxon, five-area cladograms, three-taxon, four-area cladograms, and three-taxon, five-area cladograms are less severe tests. Since it is impossible to arrange five areas so that each one is in contact with all of the others (Gardner, 1976), at least one pair of the five areas must be unconnected. Further, since a priori knowledge of the spatial arrangement of areas during the past is unavailable, the calculated compatibility relationships will include some cladograms as compatible that cannot possibly be so (because they coalesce areas not originally in contact).

The point should be made here that it is not necessary or even expected that we find that the biota of one area belongs to only a single general pattern, or that if we find more than one, that only one pattern contributes information about the history of the areas involved. How the sympatric occurrence of parts of two or more incompatible general patterns might relate to the history of areas has been discussed above. We might also find that a large general pattern is represented in one of its component areas by two or more incongruent and incompatible smaller general patterns (one of which might, for example, connect areas b and c most recently, and the other areas a and b most recently). Such a situation would be expected if one of the small areas involved (say area a) was isolated at different points in time by different events, the earlier of which affected only parts of the biota (because of the specific barriers involved). Here again, independent data from historical geology might allow us to specify the nature of the disconnections involved.

IMPLICATIONS

If the only way to deal objectively with causal explanations of biotic distribution is to work with general patterns and the historical events that can be correlated with them, the implication is clear: we cannot justify the kinds of biogeographic analyses of particular groups commonly found relegated to the back pages of systematic revisions (analyses that automatically invoke dispersal to account for all distribution patterns, be they sympatric or allopatric, unique or general, and that are primarily concerned with drawing

scenarios of such dispersal). Systematists provide the basic data source of biogeography—statements of the distribution and relationships of monophyletic groups of organisms. It is incumbent on us as systematists to present our data as explicitly as possible (preferably through the use of maps and cladograms), but unless we are willing to consider more than single groups at a time, we cannot adequately analyze that data.

Geological data will be most readily usable when organized into cladogramlike statements based on the shared features of stratigraphy, paleomagnetism, or other data, that unite now disjunct areas; this has been done, for example, by McKenna (in Rosen, 1974, fig. 44).

APPLICATIONS

Two questions arise as to the applicability of the method detailed here. First, does it matter how we choose the areas with which we deal, or how widely separated they are? Clearly, any area with which we deal must be definable by the range of an endemic taxon of some rank, and must have at least one other endemic taxon of some rank that can be used to test the generality of the pattern of relationships shown by the first taxon and its relatives. However, the actual spatial separation of the areas chosen would seem to be irrelevant, since the answer arrived at if a hypothesized pattern is corroborated as general is a non-restrictive one: it can assert that two given areas share a more recent connection with each other than they do with a third area, but it cannot exclude the possibility that a more recently connected fourth area is also involved.

Second, is it necessary that we utilize only strictly monophyletic groups, or only groups for which complete hypotheses of the interrelationships of all their members are available? Certainly the groups that we use as tests are not all strictly monophyletic. Assume that we are investigating area abc and dealing with a monophyletic tribe containing three monophyletic genera, each of which has a dozen included species, and that genus A has only one species in area abc (in area a), genus B only one species in area abc (in b), and genus C only one species in area abc (in c). A group containing only the three mentioned species (and excluding the other 33) would be polyphyletic, but can still serve as a test of a three-area hypothesis for area abc. However, even though such test groups need not themselves be monophyletic (or even paraphyletic), we can select them only from larger groups previously established as monophyletic. Further, we must insist that test groups be minimally non-monophyletic in terms of available knowledge of taxa, their relationships, and their distribution (i.e., that we disregard no known members of the group under consideration that occur in the relevant areas), and we must accept the fact that new data on taxa, relationships, or distribution may result in changes in our conclusions. Such a proviso is of course required by any empirical method. When investigating given areas, we must choose from monophyletic groups all their members endemic to the areas of interest, but we can ignore all their members occurring outside those areas (although they must obviously be taken into consideration whenever the scope of the hypothesis is expanded to include the areas in which they occur). Thus, we need only to obtain hypotheses of relationship between and among those members of test groups that occur in the areas of interest to any particular problem.

CONCLUSIONS

It would appear that the method outlined here allows us to choose any set of three or more areas of the world that can each be delimited by the presence of two or more endemic taxa (of any rank) and, by comparing the patterns of interrelationships of the various groups with taxa endemic to those areas, test hypotheses of the interconnections of the areas themselves. Armed with corroborated hypotheses of the relative recency of the interconnections among areas, we can seek evidence in historical geology of the nature of those interconnections, and thereby resolve causal explanations of the distributions of those groups that share a general pattern of area interconnections. The limits of resolution of the method seem to be set only at the level of the smallest areas of the world possessing endemic taxa, and for which relevant geological data are available.

ACKNOWLEDGMENTS

We are grateful to Donn Rosen for his help with many aspects of this paper and for his willingness to endure numerous drafts of the manuscript. Useful comments were also received from Joel Cracraft, Niles Eldredge, Eugene Gaffney, and E. O. Wiley.

REFERENCES

CROIZAT, L., G. NELSON, AND D. E. ROSEN. 1974. Centers of origin and related concepts. Syst. Zool. 23:265–287.

DARLINGTON, P. J., JR. 1957. Zoogeography: The geographical distribution of animals. John Wiley & Sons, New York.

GARDNER, M. 1976. Snarks, Boojums and other conjectures related to the four-color-map theorem. Sci. Amer. 234:126–130.

HENNIG, W. 1966. Phylogenetic systematics. Univ. of Illinois Press, Urbana.

MACARTHUR, R. H. 1972. Geographical ecology: Patterns in the distribution of species. Harper & Row, New York.

PLATNICK, N. I. 1976. Drifting spiders or continents?: Vicariance biogeography of the spider subfamily Laroniinae (Araneae: Gnaphosidae). Syst. Zool. 25:101–109.

POPPER, K. R. 1965. Conjectures and refutations: The growth of scientific knowledge, second edition. Harper & Row, New York.

POPPER, K. R. 1968. The logic of scientific discovery, second English edition. Harper & Row, New York.

ROSEN, D. E. 1974. Phylogeny and zoogeography of salmoniform fishes. Bull. Amer. Mus. Nat. Hist. 153:265–326.

ROSEN, D. E. 1975. A vicariance model of Caribbean biogeography. Syst. Zool. 24:431–464.

WILEY, E. O. 1975. Karl R. Popper, systematics, and classification: A reply to Walter Bock and other evolutionary taxonomists. Syst. Zool. 24:233–243.

Manuscript submitted May 1977
Revised September 1977

VICARIANT PATTERNS AND HISTORICAL
EXPLANATION IN BIOGEOGRAPHY

DONN E. ROSEN

Abstract

Rosen, D. E. (Department of Ichthyology, American Museum of Natural History, New York, New York 10024) 1978. Vicariant patterns and historical explanation in biogeography. Syst. Zool. 27:159–188.—Geographic coincidence of animal and plant distributions to form recognizable patterns suggests that the separate components of the patterns are historically connected with each other and with geographic history. To seek evidence of these historical connections, cladograms of geographic areas, representing sequences of disruptive geologic, climatic, or geographic events, may be compared with biological cladograms, representing sequences of allopatric speciation events in relation to those geographic areas. Such comparisons, when they meet the minimum requirements of being among dichotomized three-taxon cladograms, can resolve similar or dissimilar historical factors; two-taxon statements do not distinguish between groups with different histories. Congruence of biological and geological area-cladograms at a high confidence level (such as congruence of a five-taxon clado-gram or four three-taxon cladograms with a geological cladogram, where the confidence level can be shown in cladistic theory to be 99%) means that specified events of paleogeography can be adopted as an explanation of the biological patterns. In such a cause and effect relationship, where the earth and its life are assumed to have evolved together, paleogeography is taken by logical necessity to be the independent variable and biological history, the dependent variable. Drawing a mathematical simile, the biological cladogram y (dependent variable), is a function of the geological cladogram x (independent variable), as in a simple regression of effect y on cause x where we are given no free choice as to which is the independent variable. Such a view implies that any specified sequence in earth history must coincide with some discoverable biological patterns; it does not imply a necessary converse that each biological pattern must coincide with some discoverable paleogeographic pattern, because some biological distributions might have resulted from stochastic processes (chance dispersal). Determining that all discoverable biological patterns conflict with a given corroborated or observed sequence of geologic, climatic, or geographic change (i.e., that y is not a function of x), in theory, therefore should falsify vicariance biogeography. Because dispersal biogeography presupposes stochastic processes, and any failure to meet the expectation of a postulated dispersal is explained by an additional dispersal, dispersal biogeography is immune to falsification. Without resort to paleontology or earth history, whether a given historical relationship implied by congruence of biological area-cladograms is the result of dispersal or vicariance can also be thought of in terms which minimize the number of necessary assumptions: did the sedentary organisms disperse with the vagile ones or did the vagile organisms vicariate with the sedentary ones? Cladistic congruence of a group of sedentary organisms with a group of vagile ones rejects dispersal for both. Hence, distributions of sedentary organisms have the potential to falsify dispersal theories as applied to vagile organisms, but distributions of vagile organisms cannot falsify vicariance theories as applied to sedentary ones. The problems that arise in various kinds of historical explanation are exemplified by several specific distributions of fishes and other organisms in North and Middle America and in the larger context of Pangaean history, and are discussed in relation to current species concepts. [Vicariance; species concepts; biocladistics; biohistory; geocladistics; geohistory; Neotropics; Gondwanaland.]

The patterns of spatial distribution attained by life on earth and the means by which these distributions were achieved are two essential concerns of biogeography. The first of these concerns refers to the manner in which the patterns of distribution are displayed on the world's geography. Such patterns may be analyzed in various ways, such as noting the numbers of species or groups per region and comparing them numerically or ecologically, or by noting the phylogenetic relationships of the components of one region with the components of another. Phylogenetic comparison of the components of various regions implies a search for historical connections between biotas in time and space—the pursuit which I identify as historical biogeography. The second concern of biogeography, how distributions were achieved, is a question of mechanisms or processes. When viewed from the perspectives of historical biogeography, the patterns have suggested to some persons that there are two processes that have molded biotas into their present configurations: large-scale dispersal to produce widespread ancestral biotas and later allopatric speciation events which have fragmented the ancestral biotas into their present highly subdivided states.

When patterns are viewed from the standpoints of their numerical species composition or ecological spectra, however, attention tends to be focused on the behavior of smaller biotic aggregations, usually at the population level, and on measuring small-scale changes in population structure over relatively short spans of time. Models of population structure and behavior, such as those of MacArthur and Wilson (1967), specify geographic dimensions that are believed to be in constant biological flux resulting in part from the emigration, immigration, and extinction of the individual species components. Ecological biogeography has tended, therefore, to make its practitioners view modern biotas mainly in terms of Quaternary history, to place emphasis on the study of organisms of high vagility, and to think of the distant past mainly in terms of what can be learned from paleoecology.

Ecological biogeography seems broadly to overlap the population biology of ecologists and may thus be in the act of divorcing itself from the objectives of historical biogeography. Vuilleumier (in press) may be quite right in seeing an unbridged gap between ecological (equilibrium) and historical (vicariance) biogeography: as the former is gradually merging with ecology, the latter is becoming increasingly integrated with the objectives of systematics and the historical aspects of earth science.

The sorting process that has been separating traditional biogeography into its ecological and historical ingredients, as its most substantive benefit, will provide opportunities for biologists to discover whether each ingredient can be imbued with its own body of principles and methods. It is my present opinion that equilibrium theory has the same relation to vicariance theory in biogeography that population genetics has to phylogenetic analysis in evolutionary biology. The relation is one of friendly but independent coexistence. A hope for the future reintegration of these divergent interests may be misguided, because current equilibrium theory, although claiming to "explain" how distribution patterns arise, is not designed to study major patterns of monophyletic groups of organisms in the context of long-term earth history. For this reason, I conclude that the present choice of methodology for biogeographers does not include equilibrium theory, but is a choice between the traditional approach with its rules for determining centers of origin and directions of dispersal and the vicariance approach which is attempting to discover the bases for uniting phylogenetic theory, concepts of distributional congruence, and theories of earth history (e.g., Platnick and Nelson, 1978).

The views on vicariance biogeography that I present below were derived directly from Croizat's (1958, 1962) concepts of distributional congruence and from Nelson's (MS.) ideas on component analysis in phylogenetic theory. The relation between these two kinds of ideas is that the components of cladograms (or phylogenetic trees), when viewed in the geographical context of the distributions of taxa within monophyletic groups, provide information for deciding whether distributions of several different monophyletic groups have some general significance greater than, or in addition to, empirically observable geographic coincidence. The greater general significance referred to concerns an estimate of whether congruence of distribution of

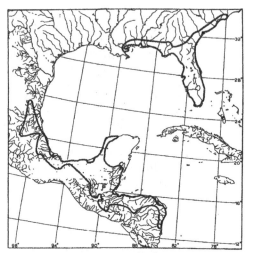

FIG. 1.—Distribution of the species of the poeciliid fish genus *Heterandria* in North and Middle America. The North American *H. formosa* is the sister group of an assemblage including all Middle American forms.

FIG. 2.—Distribution of the species of the poeciliid fish genus *Xiphophorus* in North and Middle America. The North American (Rio Grande) species pair, *X. couchianus* and *X. gordoni*, is the sister group of an assemblage including all Middle American forms.

two or more groups could have occurred by chance alone.

This relation extends Croizat's panbiogeographic method by integrating cladistic techniques. I shall attempt to illustrate how vicariance biogeography can be applied in specific cases and to suggest how the results of a vicariance analysis may be compared with theories of earth history.[1]

The plan of this paper is first to illustrate some general distribution patterns which coincide broadly in North and Middle America and which show similar disjunctions within the region. Two earlier thoughtful attempts to explain these

disjunctions are reviewed. Questions about the significance of these disjunctions are then framed in terms of vicariance theory, followed by a discussion of the relations among vicariant patterns, endemism and different kinds of cladistic statements drawn from elements of the North and Middle American biota. Cladistic statements with and without historical content are contrasted. Next the vicariant patterns are compared with the physical history of North and Middle America to illustrate the kinds of problems, and their solutions, that may be expected when selecting certain historical geologic or geographic events as explanations of the biological patterns. A distinction is, and must be, made between the original, or underlying, causes of biological patterns and the subsequent alterations of these patterns. A failure to make this distinction has led biogeographers to a temporally narrow view of biotic history. Because a temporally narrow view of biotic history focuses attention on taxa of low rank, and because some biogeographers have advocated that only the distributions of species-lev-

---

[1] A frequent complaint about vicariance biogeography (Keast, 1977:285; McDowall, 1978) is that the role of dispersal is ignored, considered irrelevant, or understated. I will therefore clear the decks of such notions by restating what has already been stated emphatically and explicitly before: 1) under the allopatric speciation model, sympatry is evidence of dispersal, 2) for the modern world to show evidences of a fragmented Mesozoic biota it is a necessary assumption that a large-scale dispersal of an ancestral biota, had first to have occurred on a predrift landscape, and 3) that, therefore, both local dispersal and dispersal of biotas (cosmopolitanism) are assumed in the vicariance paradigm (see Platnick, 1976).

FIG. 3.—Distribution of the poeciliid fishes of the *Gambusia affinis* species group in North and Middle America.

FIG. 5.—Distribution of the species of colubrid snakes, genus *Rhadinaea*, in North and Middle America. South American occurrences, to Ecuador, not shown.

el taxa form the data-base for biogeography, some consideration is also given to the use of concepts of subspecific, specific, and supraspecific taxa in biogeographic analysis. The final discussion, which incorporates ideas developed in the foregoing sections, describes a comparative cladistic method of analysis of biological distribution patterns and geologic or geographic patterns.

Although many groups of organisms are discussed to exemplify some particular point or method, primary attention is focused on two groups of tropical American fishes—poeciliid fishes of the genera *Heterandria* and *Xiphophorus*.

FIG. 4.—Partial distribution in North and Middle America of the lungless salamanders, family Plethodontidae. Western and northeastern North American and South American occurrences not shown.

FIG. 6.—Co-occurrences of a red-bellied snake, *Storeria occipitomaculata* (solid), flying squirrel, *Glaucomys volans* (dash), and barred owl, *Strix varia* (dash-dot).

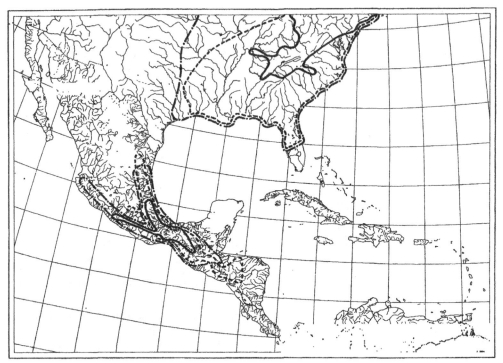

FIG. 7.—Co-occurrences of a pine, *Pinus strobus* (solid), sweet gum, *Liquidambar styraciflua* (dash), and blue beech, *Carpinus caroliniana* (dash-dot).

Poeciliids are tiny, viviparous, freshwater fishes such as the guppy. The two genera were selected because recent revisionary work has been done on both groups (Rosen, in press).

GENERAL PATTERNS AND HISTORICAL PERSPECTIVES

If the Rio Grande basin is taken as a southern geographic limit in North America, the distributions of both *Heterandria* (Fig. 1) and *Xiphophorus* (Fig. 2) show a disjunction between North and Middle American taxa north of the Rio Tamesí basin. The disjunction is especially striking in *Heterandria* where the North American representative is confined to the southeastern states.

A brief survey, based on combined zoological and botanical literature, showed that many other plant and animal distributions are also disjunct between North and Middle American occurrences, al-

though the exact location of the disjunction is somewhat variable. For example:

poeciliid fishes of the *Gambusia affinis* species group (Fig. 3), in part from Rosen and Bailey (1963);

plethodontid salamanders (Fig. 4), in part from Wake and Lynch (1976);

colubrid snakes of the genus *Rhadinaea* (Fig. 5), from Myers (1974);

Storeria occipitomaculata, the red-bellied snake (Fig. 6) (this and the following from Martin and Harrell, 1957);

Strix varia, the barred owl (Fig. 6);

Glaucomys volans, a flying squirrel (Fig. 6);

eight groups or species of mesophytic trees and shrubs and a root parasite (Figs. 7–9).

Apparently the distributions of *Heterandria* and *Xiphophorus* have some generality with respect to their southern lim-

FIG. 8.—Co-occurrences of the sour gum, *Nyssa sylvatica* (solid), and the species pairs of star anise, *Illicium floridanum* and *I. mexicanum* (dash), and yew, *Taxus floridana* and *T. globosa* (dash-dot).

FIG. 10.—Patterns produced by superimposed North and Middle American distributions of three groups of poeciliid fishes, *Heterandria, Xiphophorus*, and *Gambusia*.

its and the disjunction between Mexican and North American populations (Figs. 10, 11). The confinement of the North American representative of *Heterandria* to the southeast is also found in a few groups, but in most of those illustrated the North American distributions, al-

though eastern, are not confined to such low latitudes. A few groups also have a small western distribution west of the Rocky Mountains. Still other biogeographic literature pertaining to this region suggests that adding more distributions to the analysis would not appreciably change the percentage of

FIG. 9.—Co-occurrences of the beech, *Fagus grandifolia* (solid), and its root parasite, *Epifagus virginiana* (dash), and sugar maples of the *Acer saccharum* species group (dash-dot).

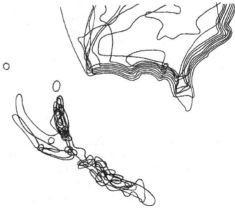

FIG. 11.—Patterns produced by superimposed North and Middle American distributions shown in Figs. 6 to 9.

groups with these somewhat different latitudinal restrictions in eastern North America. In other words, the figured distributions may fairly represent the diversity of North American distributions of extant groups that have eastern North American representatives with Mexican and Nuclear Central American vicariads.

Adding fossils to the analysis shows, however, that the North American representatives of those particular groups that are now tropical and subtropical were formerly both more numerous than now and also generally widespread in the eastern two thirds of North America. For example, crocodiles and giant land tortoises occurred as far north and west as South Dakota in early Pliocene times and in the late Pliocene they occurred at least as far north as Kansas; during interglacial times of the Pleistocene they were still present north to Kansas and Nebraska (Dott and Batten, 1976). Both crocodiles and tortoises did then occur also in the American tropics, including the Antillean islands, as crocodiles still do; crocodiles continue to occupy North America, but now only in the extreme southeast. Among the non-passerine birds formerly present in North America but now confined in the New World to the tropics are such groups as boobies, flamingos, chachalacas, parrots, and barbets, ranging in age from Middle Eocene to late Quaternary (Brodkorb, 1963, 1964, 1967, 1971). Among mammals, Floridian fossils belonging to taxa that are now confined to South America, Central America, Mexico, or the western Gulf States, or some combination of these, include three bats, a gopher, a capybara, a hognosed skunk, three cats, a tapir, several peccaries, and several camelids (data from Webb, 1974; Hall and Kelson, 1959).

Twenty years ago Martin and Harrell (1957) attempted to deal with the age of the disjunction between extant eastern North American and Middle American plant communities. They expressed concern with the proposal by various workers that the disjunction developed during late Pliocene–Pleistocene time—a proposal which requires the development of "a Pleistocene forest corridor, or at least a transitory spread of temperate forest, across the plains and isolated mountains of northern Mexico and south Texas," an area that today is arid grassland and thorn scrub. An additional requirement is that cooler and much wetter climatic conditions would have had to prevail to support a corridor of mixed mesophytic forest with beech, white pine, sugar maple, sour gum, sweet gum, evergreen magnolia, and so on. The alternative proposal is simply that the disjunction is much older, a possibility that is not unrealistic since the modern forest plants of this region were already present in the Eocene.

Martin and Harrell proposed that a rough test of these two phytogeographic views involves examining the fauna of the montane Mexican forests with the expectation of finding "unmistakable faunal evidence" of a Pleistocene forest corridor. They supposed that if the forest connection is more ancient, any residual faunal evidence "should accordingly be less obvious and at a higher taxonomic level." Their study of an isolated outpost of cloud forest in southwestern Tamaulipas showed that the number of vertebrate species with eastern North American vicariads was only 2% of the vertebrate fauna of their test station as compared with 29% of the flora with eastern North American vicariads. They concluded, therefore, that there is not "impressive faunal confirmation of a Pleistocene forest corridor to the northeast." Regarding the taxonomic level of Middle and North American faunal disjuncts, Martin and Harrell argued that the plethodontid salamanders are of special importance for this problem because "(1) a continuous forest corridor is necessary for plethodontid dispersal, (2) there is a closer relationship between the Plethodontidae in eastern and western United States, than between either of these and [those of] Middle America, and (3) the morphological development of the genera found south of the Mexican boundary requires the time of separation to antedate the

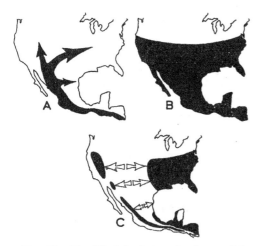

FIG. 12.—Simplified depiction of a theory of the Tertiary history of the Neotropical flora, derived from Axelrod (1975).

Pleistocene." This general line of argument is, of course, weak and partly flawed, especially the last part which assumes that plants have characteristically different rates of evolution from animals, that levels of taxonomic recognition by different taxonomists are directly comparable and significant, and that there is a known time frame for the evolution of the currently used taxonomic characters of plethodontid salamanders.

Martin and Harrell dealt with the climatic requirements of the Pleistocene-disjunction hypothesis by arguing that "it is not yet possible to correlate low latitude pluvial periods with glacial maxima, although this has been attempted repeatedly (Deevey 1953)," and they cited some data concerning glacial readvances of about 7,000 years BP to infer that, "in fact, the opposite may be true." Dott and Batten (1971:456) also disputed the necessity of such correlations by pointing out that mild-climate fossils may be found very near obvious glacial moraines. In the view of Dott and Batten "a really harsh climate did not grip most of the United States until the last (Wisconsin) glacial advance." Moreover, their map (p. 448) of Pleistocene paleogeog-

raphy at the time of the maximum glacial advance shows a hot, dry-climate flora occupying most of central and northeastern Mexico, southern and western Texas and the mountain states. These do not seem to me to be the conditions of a "cooler and a much wetter" climate required for a forest corridor across northeastern Mexico and southern Texas.

Axelrod (1975) dealt with the problem of this biotic disjunction based, to a large extent, on the paleontological evidence. The forest disjuncts considered by him include maple, hornbeam, hickory, redbud, dogwood, beech, witch hazel, holly, sweet gum, sour gum, sweet gale, ironwood, cherry, linden, 125 species of mosses (Crum, 1951) and other bryophytes and ferns, to mention only some of the plant vicariads.

Axelrod's argument concerns the known distribution of plant megafossils and fossil pollen representing plants closely related to or exactly like those now composing the temperate rain forests of Mexico and Central America. His argument is simple and of two parts. The tropical American rain forests are ancient and were already present in Mexico and Central America at the beginning of the Tertiary at a time when North America was vegetated at middle and low latitudes by a preponderance of tropical species. By the late Eocene the tropical element was reduced to insignificance and the forests were more or less typical of the Mexican temperate rain forests of today (Leopold and MacGinitie, 1972). In essence the evidence now available suggests that, in the Eocene, temperate rain forests dominated by a mixture of evergreen dicots and deciduous hardwoods were continuously distributed from the central United States to Central America along the present axis of the Sierra Madre Occidental (Fig. 12A, B). With drying conditions in the west during the middle Tertiary and the appearance of the Sierra Madre Oriental, the forest shifted to the east in Middle America as it did also in the United States (Fig. 12C). The drought that was moving eastward to northeastern

Mexico and southern Texas caused a floral replacement by xerophyllous woodland which was originally part of a low altitude subtropical forest that lived under a dry winter season. The non-tropical appearance of the Appalachians today was caused by the loss of tropical evergreen dicots following a late Tertiary deterioration in climate that left an impoverished, dominantly deciduous hardwood forest as almost sole reminder of a once more cosmopolitan neotropical flora. Hence, following Axelrod's arguments, there were two Tertiary events affecting the northern extension of the neotropical flora: a middle Tertiary period of increasing aridity in the southwest interrupting a once continuous North and Middle American neotropical forest and a later Tertiary (Miocene–Pliocene) deterioration in climate that eliminated most of the evergreen dicots from temperate eastern North America, leaving prominent signs of a more complete neotropical flora only in certain favorable locations on south-facing or coastal exposures and the Florida peninsula.

Now, whether one argues in favor of Martin and Harrell's, Axelrod's, or some other viewpoint, it is clear that there is a biological relationship between eastern North America and southeastern Mexico and northern Central America and that this relationship may have existed for some of the included taxa for a very long time (i.e., for at least 50–60 million years). One may ask, therefore, to what extent the different components of this inter-American biota have shared the same history and by what method this question might be answered most efficiently? What we are seeking is some statement concerning the included taxa of greater generality than their simple coincidence. That there exists a disjunction between North and Middle American vicariads is one such statement, but this statement (i.e., taxa in area A are related to taxa in area B), lacks complexity and contains no more information than that originally required to recognize the existence of the general problem. However,

if one of the two areas (area A, for example) is subdivided (i.e., a second disjunction is found), a more complex statement becomes possible: A' is more closely related to A" than either is to B. This three-taxon statement implies a historical component not contained in two-taxon statements, namely, that a speciation event (second disjunction) affecting the ancestor of A' and A" occurred only after the speciation event (first disjunction) that affected the ancestor of A and B.

VICARIANT PATTERNS, CLADISTIC SEQUENCES AND EARTH HISTORY

In order to learn if the distribution of *Heterandria* in the southeastern United States, eastern Mexico, and northern Central America (Fig. 1) has some generality with respect to at least two areas of disjunction, it was compared first with major disjunctions in freshwater fish distributions. The location of faunal disjunctions was derived from a matrix in which the occurrences of all known fish species (data from Meek, 1904) were plotted against river systems of the Atlantic versant of Mexico (Fig. 13). The rivers were ordered in north to south geographical sequence, the most northern being the Rio Grande and the most southern the Rio Papaloapan near the Isthmus of Tehuantepec. Such a matrix, which provides an immediate indication of regions of endemism, allows a more precise resolution of regions of disjunction than is possible by the graphic superimposition of distributions as shown in Figs. 10 and 11. Of 114 taxa shown in Fig. 13, 36 are endemic to the Rio Grande, none is endemic in the next two rivers south, 14 are endemic to the Rio Panuco basin, and 35 occur only in the rivers south of the Rio Panuco. Of the 36 in the Rio Grande, some are the same as, and others are most closely related to, species of the eastern United States (thus showing that the Rio Grande is part of a much larger region of endemism). Of the 35 taxa in six rivers south of the Rio Panuco some are the same as, and others are most closely re

TAXA	RIO GRANDE	RIO SAN FERNANDO	RIO SOTO LA MARINA	RIO PANUCO BASIN	RIO SAN FRANCISCO	LAS LAGUNAS	BOCA DEL RIO	RIO BLANCO	RIO OTAPA	RIO PAPALOAPAN
LEPISOSTEUS 1	x		x							
ICTALURUS 1	x		x	x						
ICTALURUS 2	x		x							
ICTALURUS 3	x	x	x							
ICTALURUS 4	x									
LEPTOPS 1	x									
CARPIODES 1	x		x	x						
CARPIODES 2	x									
CARPIODES 3	x	x	x							
PANTOSTEUS 1	x									
CATOSTOMUS 1	x									
MYZOSTOMA 1	x	x	x							
CAMPOSTOMA 1	x									
CAMPOSTOMA 2	x									
CAMPOSTOMA 3	x									
HYBOGNATHUS 1	x									
PIMELOCEPHALUS 1	x									
LEUCISCUS 1	x									
NOTEMOGONUS 1	x									
COCHLOGNATHUS 1	x									
NOTROPIS 1	x		x							
NOTROPIS 2	x									
NOTROPIS 3	x									
NOTROPIS 4	x									
NOTROPIS 5	x									
NOTROPIS 6	x	x								
NOTROPIS 7	x									
NOTROPIS 8	x									
PHENACOBIUS 1	x									
RHINICHTHYS 1	x									
HYBOPSIS 1	x	x								
COUSIUS 1	x									
ASTYANAX 1	x	x	x	x					x	
ANGUILLA 1	x		x							
DOROSOMA 1	x			x					x	
FUNDULUS 1	x	x	x							
FUNDULUS 2	x									
LUCANIA 1	x									
CYPRINODON 1	x			x						
CYPRINODON 2	x									
GAMBUSIA 1	x		x	x						
XIPHOPHORUS 1	x									
POECILIA 1	x	x	x	x	x	x	x		x	x
POECILIA 2	x									
POECILIA 3	x		x	x						
LEPOMIS 1	x									
LEPOMIS 2	x									
LEPOMIS 3	x									
LEPOMIS 4	x									
EUPOMOTIS 1	x									
MICROPTERUS 1	x	x	x							
ETHEOSTOMA 1	x									
ETHEOSTOMA 2	x									
ETHEOSTOMA 3	x									
APLODINOTUS 1	x		x							
CICHLASOMA 1	x									
CICHLOSOMA 2	x	x	x	x						
NEETROPLUS 1	x	x	x	x						
GOBIOMORUS 1	x		x	x	x		x			x
DORMITATOR 1	x						x	x		x
XIPHOPHORUS 2		x	x							
LEPISOSTEUS 2				x						
ICTALURUS 5				x			x			
ICTALURUS 6				x						
CARPIODES 4				x						
ALGANSEA 1				x						
HYBOGNATHUS 2				x						
AZTECULA 1				x						
NOTROPIS 9				x						
DOROSOMA 2				x						x
FUNDULUS 3				x						
GOODEA 1				x						
GOODEA 2				x						
POECILIA 4				x						
XIPHOPHORUS 3				x						
POMADASYS 1				x						x
CICHLASOMA 3				x						
CICHLASOMA 4				x						
AWAOUS 1				x	x		x			x
HETERANDRIA 1					x			x	x	x
XIPHOPHORUS 4					x			x	x	x
STRONGYLURA 1					x					x
CENTROPOMUS 1					x		x			x
CICHLASOMA 5				x	x	x				x
GOBIUS 1				x	x	x				
GOBIUS 2				x	x	x				
OPHISTERNON 1					x					
AGONOSTOMUS 1					x			x	x	
CICHLASOMA 6					x				x	x
CICHLASOMA 7					x			x	x	
ELEOTRIS 1				x	x					
GAMBUSIA 2						x			x	x
BELONESOX 1					x				x	x
SYPHOSTOMA 1					x					
RHAMDIA 1								x	x	x
RHAMDIA 2								x		x
ASTYANAX 2								x	x	x
ICTALURUS 7								x		
DOROSOMA 3									x	x
POECILIOPSIS 1									x	x
CICHLASOMA 8									x	x
ARIUS 1										x
ICTIOBUS 1										x
HYPHESSOBRYCON 1										x
RIVULUS 1										x
PRIAPELLA 1										x
XIPHOPHORUS 5										x
MENIDIA 1										x
POMADASYS 2										x
CICHLASOMA 9										x
CICHLASOMA 10										x
CICHLASOMA 11										x
CICHLASOMA 12										x
ACHIRUS 1										x

CONTINUED IN COLUMN 2

lated to, species in the remainder of southeastern Mexico and northern Central America. Three disjunct regions are thus defined (Fig. 14): (1) eastern North America southwest to the Rio Grande, (2) the Rio Panuco basin near Tampico, and (3) the rivers of southeastern Mexico and Central America south to the Great Lakes of Nicaragua; these three regions correspond with patterns of disjunction within *Heterandria* (Figs. 14, 16).

Comparing the geographical components of the *Heterandria* cladogram with the three areas of endemism shows that the principal dichotomy separates North American from Middle American taxa, or, put another way, the cladogram predicts a closer historical relationship between areas 2 and 3 than between either and area 1. The same pattern is present also in poeciliid fishes of the genus *Xiphophorus* (Figs. 14, 16) and box tortoises of the genus *Terrapene* (Fig. 14).[2] However, poeciliid fishes have their plesiomorph

FIG. 14.—Simplified depiction of the geography of principal regions of endemism, derived from the matrix in Fig. 13. Cladograms are described in text.

---

[2] That is, in *Heterandria, Xiphophorus,* and *Terrapene* Middle American taxa in areas 2 and 3 are more closely related to each other than either is to taxa in area 1. This may be seen readily from the more complete cladograms of *Heterandria* and *Xiphophorus* in Figs. 19 and 20. In the case of the forms of *Terrapene carolina,* the cladogram is constructed from data in Milstead (1969): within *T. carolina* seven subtaxa are recognized, five in North America and two in Middle America. In the two Middle American taxa, *yucatana* and *mexicana,* the carapace has a peculiar shape (posteriorly depressed and somewhat constricted), a shape unusual if not unique in box turtles. This form of the carapace is, therefore, presumably derived and is a synapomorphy uniting *yucatana* and *mexicana* with each other but not with any of the five North American taxa. *Terrapene c. mexicana* is from southwestern Tamaulipas, northeastern San Luis Potosi and northern Veracruz (area 2), and *yucatana* is from Campeche, Quintana Roo and Yucatan (area 3).

sister group in South America whereas the three consecutive sister groups to the entirely North and Middle American *Terrapene* (*Emydoidea, Emys* and *Clemmys,* in that order; Bramble, 1974) are confined to Laurasia and Africa north of the Atlas Mountains (a detached fragment of southern Europe) (Loveridge and Williams, 1957).

One implication of this difference in extralimital sister-group relationships but identity of vicariance patterns in the study area is that one or both groups of taxa dispersed into the study area and became sympatric prior to the relevant vicariance events and subsequent allopatric speciation.

Other fishes, for example, the alligator gar, show no recognized taxonomic differentiation within the region and are continuously distributed between at least

---

←

FIG. 13.—Matrix of the geographical occurrences of fish taxa (each numeral following a generic name represents a separate nominal species) for North America (Rio Grande only) and nine major freshwater areas of the Atlantic slope of Mexico (from Meek, 1904). Although there have been taxonomic additions, deletions and changes since Meek's formal survey (Darnell, 1962; Rosen and Bailey, 1963; Miller, 1976), the basic patterns of endemism revealed by the matrix remain substantially unaltered. A few names have been changed to correspond with current usage.

two of the three areas. Forms such as a gizzard shad exhibit the disjunct distributions of the endemic taxa, but show no apparent regional differentiation.

Why some groups show patterns of disjunction and others do not is related to what may be called a species' activity range—the geographic perimeter of occurrence of the individuals of a species during their diurnal, seasonal, or annual movements by active or passive means. For example, a vicariant event separating two once-connected streams will affect the fishes and aquatic invertebrates, but not necessarily the birds that feed on them. This problem might be generalized in the following three statements:

(1) When the geographical extent of an area exceeds the activity ranges of the species of the included taxa, a majority of the taxa should exhibit the same, or components of the same, patterns of vicariance.

(2) When the geographical extent of an area is exceeded by the activity ranges of the species of the included taxa, the taxa should exhibit no vicariant patterns.

(3) When the geographical extent of an area exceeds the activity ranges of the species of some of the included taxa, but not of others, only some of the taxa should exhibit vicariant patterns.

The activity ranges of some of the included taxa (e.g., gars) may in fact be found to exceed the geographical extent of the areas showing biogeographical disjunctions—for example, taxa with continuous distributions which span two or more regions of disjunction. For these taxa, no decision can be reached regarding their age in the study area since they are uninformative with regard to specifiable historical events.

For the included taxa that did respond in the regions of biogeographical disjunction, but nevertheless show no recognizable taxonomic differentiation in all three regions of endemism, one can only say that they too are cladistically uninformative with regard to the history of the region.

In the example where different groups have the same vicariant patterns, one might hope to associate each pattern with specific events of geographic history. A search for such associations must begin at some time level, and I will take the age of the oldest known fossil representatives of lineages in the modern biota (e.g., gars, lungless salamanders, boid snakes, crocodiles; see Romer, 1966) as a best initial estimate of where to begin the search. This is the Cretaceous.[3]

Since the Cretaceous, North America, Mexico and northern Central America have experienced a succession of major transformations of surface relief, climate, and coastline (Fig. 15). Mountain building in the far west during the Cretaceous and Paleocene was precursor to the Rocky Mountains of North America and the Sierra Madre Occidental of Middle America. Erosional reduction of these features was followed during mid-Tertiary times by an orogeny that uplifted and enlarged the western ranges and gave rise to the Sierra Madre Oriental in eastern Mexico. During the latter part of the Tertiary and continuing to the present day, volcanism contributed lesser ranges along the Pacific versant of Middle America and one that bisected Mexico latitudinally between Tampico and the Isthmus of Tehuantepec.

In eastern North America at the beginning of the Tertiary, the Appalachian chain was represented by an area of gentle relief, having been eroded down into a vast and low, undulating plain, the Schooley Peneplain. The Appalachians were later rejuvenated by a general uplift of the plain which caused renewed erosional activity. The present relief in this region is the result of persistent erosional sculpturing and rejuvenation throughout the Tertiary (Dott and Batten, 1976).

Climate, during Cretaceous and Paleocene times, was subtropical into Canadian latitudes, with no evidence, from

---

[3] Although a majority of the principal modern lineages of animals and plants are known from the Eocene, starting at that point would exclude those which are known to be older.

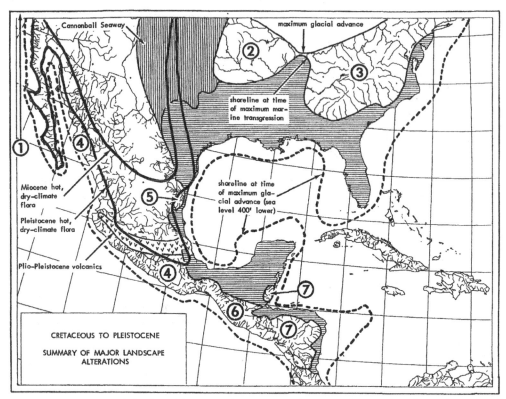

FIG. 15.—Summary of the broad features of Cretaceous-Pleistocene historical geology of North and Middle America. Lower Central America and West Indies omitted. North–south trending mountains not shown. Circled numbers represent areas remaining relatively stable.

geochemistry or the paleofloras, of significant seasonality. Beginning in the early Eocene, however, seasonally drier climate appeared in the western and central Cordilleras, conditions that gradually spread to most of the Cordilleras between Canada and northern Mexico by the mid-Tertiary and finally to the west coast and Great Plains and southward almost to the Isthmus of Tehuantepec in Mexico by the Pleistocene. As the drier conditions were developing in the west and extending gradually eastward toward the Great Plains in North America and toward the Sierra Madre Oriental in Mexico, there was a gradual southward recession of subtropical conditions, until, by about 40,000 years ago, the northern limit of the subtropics came to lie astride the central Florida peninsula. At the time of the maximum Pleistocene glacial advance, ice sheets, forming a V-shaped wedge, extended down to southern Illinois and paralleled the present courses of the Missouri and Ohio rivers (Dott and Batten, 1976).

Changes in the configuration of the coastline were also dramatic. In Cretaceous times, major portions of the Atlantic versant of North and Middle America were drowned (in North America, as far north as southern Illinois) and a shallow sea extended across North America from the Gulf to the Canadian Arctic—a sea which had retreated south to become the Cannonball Sea of Paleocene times. By the end of the Paleocene the Cannonball Sea had withdrawn farther south, leaving

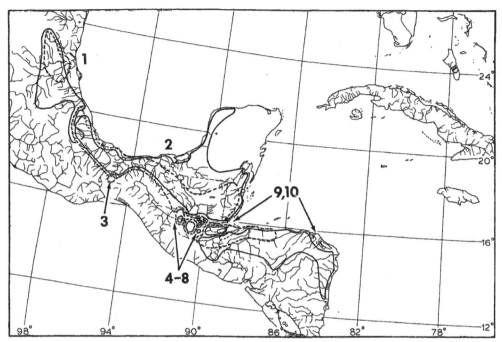

FIG. 16.—Co-occurrences of the Middle American species and recognizable populations of *Heterandria* (solid) and the swordtail species of *Xiphophorus* (dashed) within 10 subregions.

some dry land at middle latitudes. During the mid-Tertiary, Oligocene and Miocene times, the seaway gradually disappeared, there was a slight withdrawal of the marine transgression along the Gulf coast, and parts of the Florida and Yucatan peninsulas had emerged. By the end of the Tertiary, during the period of maximum glacial advance, the present coastline was fully exposed—and more, for the water locked up in the world's glaciers had contributed to a 400-foot lowering of sea level so that almost the entire continental shelf had become subaerial (Dott and Batten, 1976).

Thus, during the roughly 100 million year history just summarized, relatively few areas had remained both continually accessible to nonmarine life and relatively undisturbed by orogenic, volcanic or general tectonic activity: these relatively stable areas (Fig. 15) are 1) parts of extreme western North America, 2) the region of the Ozark Uplift, 3) parts of the

middle latitudes of eastern North America, including the Appalachians, 4) scattered areas of western and central Mexico, 5) the area just north of Tampico in eastern Mexico, 6) most of the base of the Yucatan Peninsula from the Isthmus of Tehuantepec eastward, and 7) the highlands of Belize, Honduras and northern Nicaragua on the Atlantic versant. These relatively stable areas are of special interest because they coincide with the generalized distribution patterns formed by the combined individual distribution patterns of many different components of the biota. To the extent that the different components of the biota shared a common history, their sequences of phylogenetic branching in relation to geography should also correspond. Conversely, a theory of the geographic history should predict the sequences of phylogenetic branching. The prediction should be in the form of a statement that *if* component groups of the biota had responded to two or more geographical vicariant events, the relative ages of these events should correspond with the branching sequence in their cladograms of relationships.

In relation to cladograms, such as those derived for *Heterandria, Xiphophorus*, or *Terrapene*, in which taxa in area 1 are the sister group of those in areas 2 and 3 (Fig. 14), an initial vicariant event is predicted for the region between the Rio Grande and the Rio Panuco basins and a subsequent vicariant event ·for the region between the Rio Panuco and southern Mexico. At the level of resolution of the geographic history presented above, the initial event could have been either a marine transgression of Cretaceous age or a climatologically induced shift in edaphic conditions (affecting the abundance of surface waters) of middle to late Tertiary times. The subsequent vicariant event providing opportunity for allopatric speciation south of the Panuco basin is attributed most simply to the Pliocene volcanics that bisected Mexico in this region. If the latter is accepted, it carries with it the implication that the initial vicariant event to the north of the Panuco basin had become effective in pre-Pliocene times or at least earlier than the Pliocene volcanics. Whether to accept as the effective agent Miocene edaphic conditions or a Cretaceous marine transgression can be resolved by another cladogram. For example, given the cladogram:

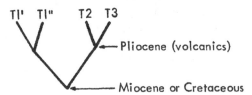

in which T1′ is a fossil taxon of older than Miocene age or in which T1′ and T1″ are both extant taxa associated with a pre-Miocene vicariance in area 1, the Miocene edaphic factors might be rejected in favor of the Cretaceous marine transgression. I have not attempted to resolve such a cladogram but I will provisionally

adopt the more conservative view that the initial vicariance in this instance is mid-Tertiary simply because the paleofloras associated with the region of disjunction are at present not known to be older than the Middle Eocene.

Assuming that each of the cladograms in Fig. 14 is correct, two historical explanations seem possible. One is that the ancestors of the poeciliids and box turtles had dispersed into the area prior to the Pliocene via coastal drainages that were still unaffected by eastward spreading aridity, and the other is that they were all already present. The discovery of a fossil *Heterandria, Xiphophorous*, or *Terrapene* in the Miocene or earlier deposits from the region south of the Rio Grande would cause us to reject the post-Miocene dispersal theory. The poeciliids, however, occur in the rivers that traverse the Cretaceous limestones of the Yucatan Peninsula and members of each genus are also found in the isolated intermontane, karst basins along the foothills of the northern Sierras in Guatemala—the northern limits of the stable land at the base of the Yucatan Peninsula. Living together with them in these karst basins are some endemics with their nearest relatives remote from the region (e.g., in the Great Lakes of Nicaragua and in southeastern South America), and this suggests a long period of isolation. If the age of isolation of the karst basins from the main river could be shown to be Miocene or earlier, this would also be sufficient to reject the post-Miocene dispersal theory. The time of isolation of these karst basins is not yet known and, I suspect, would be extremely difficult to document considering the plasticity of karst topography. A paleontological discovery is therefore likely to provide the most readily available test of the theory.

By way of a summary, the following points are offered:

1. A major biotic relationship exists across a disjunction between eastern North America and Middle America, implying a former biotic continuity.

2. Many of the earliest known representatives of this biota in North America are from Eocene deposits.

3. Many components of this widespread Tertiary tropical biota became extinct in temperate parts of North America during the latter half of the Tertiary and the Quaternary.

4. The major disjunction between eastern North American and Middle American elements in northeastern Mexico is a prominent and well-recognized feature of the modern remnants of this Tertiary biota.

5. By identifying areas of endemism (regions where populations have evolved in isolation), an additional area of disjunction (between the Rio Panuco of east-central Mexico and rivers to the south) was recognized. This finer subdivision of the biota permitted three-taxon cladistic statements that are capable of differentiating between groups with different histories. The cladistic components predict what patterns of interrelationship one might expect to find in other taxa, but impose some constraints on the time levels for certain allopatric speciation events. This contrasts with earlier attempts to deal with neotropical history such as those by Martin and Harrell and Axelrod which, by considering only the disjunction between North and Middle American elements, were limited to a kind of two-taxon problem that is incapable of resolving historical differences between different monophyletic groups included in both areas. One of the purposes of the discussion thus far is to show, by a simple example, how to avoid that limitation.

PROGENERATIVE AND EPIGENERATIVE ASPECTS OF DISTRIBUTION PATTERNS

In modern evolutionary theory, the vicariance history of every polytypic group is a history of allopatric speciation events, and this will be so whether the history is underlain by bouts of dispersal, in situ fragmentation of an ancestral species, or both. This process of diversification of an ancestral taxon into geograph-ically discrete descendent populations (cladogenesis) is estimated by constructing a character-state tree, or cladogram, which specifies a hierarchical scheme of relationships among the descendent taxa that may be interpreted as corresponding with a sequence of speciation events.[4] If the causal explanation of cladogenesis is sought in the interpretations of earth history, then one would hope to discover a sequence of geologic or geographic changes (vicariant events) that correspond in relative ages with the relative positions of taxa in a cladistic sequence and in geographic position with the geographic disjunctions between taxa. The events that were associated with the breakup of the ancestral taxon into descendent taxa are the original, or *progenerative*, causes of the vicariant pattern. Each vicariant pattern is subject to subsequent modification involving expansion, contraction, alteration in shape of the distributions or extinctions of descendent taxa (the vicariant segments) in response to changing ecology, climate, and so forth. These subsequent, or *epigenerative*, influences will, therefore, be highly correlated with the ecologic and physiographic conditions of the modern landscape.

The distinction between progenerative and epigenerative factors is not meant to imply that Quaternary or recent landscape changes and climatic fluctuations cannot cause speciation, but only that progenerative causes provide the opportunities for allopatric speciation and that epigenerative causes affect the boundaries and dimensions of vicariant segments. The importance of this distinction is that usually it isn't made. Müller (1973:185), for example, observed the strong, but expected, correlation of his

---

[4] I consider the character-state tree as a primary concept expressing perceived natural order, the cladogram as a derivative concept which expresses inferred relationships of taxa with the ordering data of the character-state tree stripped away, and the phylogeny as a further removed derivative concept which interprets the cladogram in an evolutionary context.

"dispersal centers" with Quaternary climate and vegetational fluctuations. He then concluded (p. 203) that "strong displacements of the biochores have taken place during the Quaternary period and these displacements led to the formation of refuges," and that "this is an essential reason for the richness of the tropical rain forest of Central and South America in species." Perhaps so, but to accept this conclusion we would also have to agree that the coincidence of a distributional center with an ecological regime is a sufficient correlation.

But let us suppose that cladistic sequences were worked out for several groups of taxa in several of Müller's "dispersal centers" and it happened that these sequences showed little or no congruence. Would this not suggest that these several groups of taxa had different histories of cladogenesis (allopatric speciation events)? And what then would be the significance of Müller's correlation? Or, suppose that their cladistic sequences were all perfectly congruent, but that the sequence of establishment of the refugia or "dispersal centers" was quite different. And, again, what then would be the significance of Müller's correlation? I submit that the observed correlation, under either of these hypothetical circumstances, could only be judged epigenerative. In turn, that judgment would mean that the progenerative causes (the vicariant events leading to allopatric speciation) lay elsewhere, in a more remote time. The problem with Müller's presentation, and with many others like it, is that Müller assembled, arranged and summarized a great deal of information in a useful and interesting fashion, but he stopped short of analysis. In place of analysis he substituted a scenario based on the assumption that he had actually found the progenerative factors to be part of Quaternary history. The kinds of correlations noted by Müller seem no more significant than the statement that fishes are found in water and that where the water goes, so go the fishes.

SPECIES CONCEPTS AND BIOGEOGRAPHY

To the extent that biogeography searches for patterns and their historical explanation, the choice of a species concept by the practicing biogeographer is crucially important. This is so only because conventional taxonomic practice regards the species as a fundamental evolutionary unit. A concept such as the biological species, however, appears to be inapplicable to observable nature, because it incorporates criteria that are generally undiscoverable. It requires that we identify noninterbreeding sympatric units (I have discussed the reproductive criterion elsewhere: Rosen, in press). But in practice what are these units—crabs and fishes, lions and zebras? Or, does the sympatry refer to more closely related groups and, if so, how closely related? Must they have a sister-group relationship to each other—that is, must they be each other's closest living relatives which, under the allopatric speciation model, are the least likely groups to occur sympatrically? Even if so, the sympatry of closest living relatives (the only nonarbitrary choice of sister-taxa) couldn't be shown to meet the requirements of the concept, because the 'closest living relatives' may be only the surviving plesiomorph and apomorph members of a formerly speciose group (e.g., man and the chimpanzee within the Homininae). The only way we can postulate that there were no extinctions of cladistically intermediate taxa is that the surviving forms have contiguous ranges which together estimate the range of their ancestral taxon, and that these allopatric distributions coincide with the distributions of other pairs of some groups of other plants and animals, i.e., are part of a general pattern that can be shown to have some historical integrity. In other words, information needed to satisfy the requirements of the 'biological species' concept for reproductive incompatability of sympatric sister-taxa is provided only by allopatric distributions!

What about the evolutionary species concept, which requires that the natural unit be an evolutionary lineage? Presumably this could refer to a population or several populations. A decision about how many populations are to be included depends entirely on the discovery and distribution of apomorphous character states among the populations. If two or more populations share some derived character they are lumped together because, only together, can they be defined by an apomorphous character. If subsequently each population is discovered to have a unique and recognizable apomorphous character, then each population can be recognized. The populations, both grouped and singly, are evolutionary lineages in the sense of being defined by uniquely derived character states, but the decision about whether the populations constitute one or several species has sometimes been, in practice, an arbitrary one dictated by whether a taxonomist has wanted to apply a polytypic species concept. It is the polytypic species concept (which is a logical necessity of the biological species concept) that has required an arbitrary decision about how many species to recognize within a cluster of diagnosable populations in order to avoid monotypic species—making the polytypic species operationally equivalent to the genus. When the choice has been made to confer species status on each unit (of one or more populations) that can be clearly diagnosed with apomorphous characters, the recognized species are those of everyday taxonomy which, when interpreted as lineages, are equivalent to evolutionary species. Thus, the evolutionary species appears to conform, in practice, with Regan's (1926) definition that a species is what a competent taxonomist says it is.

Both of the above species concepts have in common a requirement for the discovery of discontinuous variation; the type of variation which not only permits sharp definition of geographically isolated clusters of individuals (populations) but which also makes possible the adop-

tion of a stable nomenclatural system that requires a type specimen to conform with all the individuals in a reasonable sample of the population (or at least the assumption that it will so conform).

What about two or more geographically isolated clusters of individuals (populations) showing overlapping variation with statistically different modes? They are generally absorbed within the taxonomy of discontinuous variants (species), treated as subdivisions of species (subspecies or races), or ignored taxonomically.[5] This unequal treatment of populations showing overlapping and discontinuous variation involves an assumption that these two aspects of variability are fundamentally different and of different levels of significance from the standpoint of evolutionary history. What is the basis for this assumption? Both are the presumed results of genetic processes operating under conditions of geographic isolation. For the many kinds of organisms in a biota that has undergone geographic fragmentation, the conditions of isolation for all individuals will be the same although their responses to it may differ in accordance with their different genetic histories. Thus, some kinds of organisms will differentiate at what some taxonomists have called the generic or higher levels, some at the level of species, subspecies, race, or statistically recognizable population. The point is that *they have differentiated* and are potentially informative with respect to the relationships of their area to other areas, a quality that is independent of a taxonomist's particular bias concerning what amount of differentiation deserves species recognition.

I conclude, therefore, that a species, in the diverse applications of this idea, is a unit of taxonomic convenience, and that the population, in the sense of a geographically constrained group of individuals with some unique apomorphous characters, is the unit of evolutionary sig-

---

[5] This may be true mostly of vertebrate taxonomy, especially in ornithology and mammalogy.

nificance.[6] Some populations, whether recognized as species or not, are informative with respect to a history of geographic isolation because they have differentiated. Others are uninformative either because they haven't differentiated, because they are parts of larger populations that span two or more areas (are insensitive to existing barriers), or because taxonomists have thus far failed to detect the ways in which differentiation has occurred (physiologically, developmentally, behaviorally, etc.).

The problems that have beset biogeography by the casual application of species concepts are easily seen in center or origin-dispersal models of biotic history. Mayr and Phelps (1967), for example, have explained the occupation of the Pantepui highlands of northern South America by birds of different taxonomic rank as the result of a series of invasions of different ages, the oldest colonizations being represented by the taxa of highest rank. These authors have, of course, assumed that the bird taxa are evolving at a more or less constant rate so that differences in rank can accurately reflect differences in age. They also have assumed that certain degrees of difference are reliable indicators of certain taxonomic ranks, and they have assumed that the different taxonomic judgments of different taxonomists studying different bird groups are comparable. Presumably their 'colonization' theory was also influenced by prior acceptance of a center of origin-dispersal paradigm which neatly interlocked with their special uses of taxonomic rank.

But suppose Mayr and Phelps had considered the alternative theory that the different bird taxa of the Pantepui were the fragmented remains of an old occupation of these highlands? They would also presumably have allowed for different rates of differentiation of assorted character states in the different groups of birds. In fact, at some very general level such theories may be the only way we have of crudely estimating differences in evolutionary rates,[7] an option that is denied us by arbitrary application of the center of origin-dispersal model of biotic history. A choice between that and the vicariance theory could be made by asking which groups of birds, at whatever rank, yield congruent area cladograms by the method described below (pp. 178–182). Congruence would naturally suggest a single history for all such taxa, just as more than one pattern of congruence would suggest more than one history for the Pantepui bird fauna. I judge that such an approach is at least a concrete analytical beginning in the elucidation of what must be a complex historical problem. The alternative approach, used by dispersalists, has so many built-in apriorisms that it consists of little more than narrative invention. The absence in traditional biogeography of cladistic methodology and concepts of distributional congruence, as well as the application of doctrinal species concepts, seem to be largely responsible for the lack of rational choices between conflicting narratives.

The following, and final, section discusses a method for choosing between alternative theories of biogeographic history, using examples of two fish groups in Middle America and some groups of insects and fishes in the southern hemisphere.

CLADISTIC CONGRUENCE AND HISTORICAL EXPLANATION

In Middle America the two groups of poeciliids, *Heterandria* and *Xiphophorus*, show many detailed similarities in their general patterns of occurrence (Fig. 16). Both have species represented in and around the Rio Panuco basin (area 1

---

[6] If this view is accepted, it renders superfluous arguments about whether a 'biological species' is an individual or a class (see, e.g., Hull, 1976).

[7] A fossil, of course, provides only an estimate of how young its assigned lineage is. The rare continuous stratigraphic sequences of related fossils provide an estimate of the rate of change in that sequence, but are of no value in concepts of relative rates of change as would be provided by the many components of a single biota.

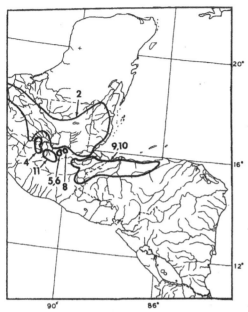

FIG. 17.—Distributions in southern Mexico and northern Central America of the swordtail species and recognizable populations of *Xiphophorus*. The samples from subregion 11 are intermediate between the taxon in subregions 4, 5, and 6 and that in subregion 2, and are provisionally interpreted (Rosen, in press) as intergrades.

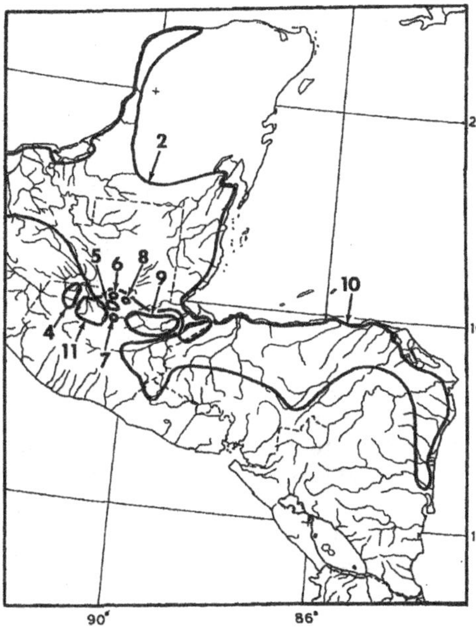

FIG. 18.—Distributions in southern Mexico and northern Central America of the species and recognizable populations of *Heterandria*. The samples from subregion 11 are intermediate between the taxon in subregions 4 and 5 and that in subregion 2, and are provisionally interpreted (Rosen, in press) as intergrades.

in Fig. 16) and both groups have a complex of species and recognizable populations southeast from the Panuco to Honduras and Nicaragua. Within the region south of the Panuco, each group has a widespread taxon or group of related taxa occupying most of the Atlantic versant southeast to Belize and eastern Guatemala (area 2), each has a form in the Rio Motagua basin and the rivers of coastal Honduras related to the widespread taxon (area 10), and each has a complex of species or populations ranged along the foothills of the Sierras in northern Guatemala (areas 4 to 8). Each group is also represented by an upland population in west-central Guatemala that is morphologically intermediate between a Sierran form and the widespread taxon (an area 11, not identified in Fig. 16). There are some dissimilarities, too. For example,

only the swordtail group of *Xiphophorus* is represented by a distinct endemic taxon in the headwaters of the Rio Coatzacoalcos basin of Mexico (area 3), and only *Heterandria* is represented by endemic species or differentiated populations in the Rio Polochic basin of Guatemala (area 9) and in certain isolated basins along the Sierras (areas 6 and 7, Figs. 17 and 18). Furthermore, *Heterandria* is, in general, of wider occurrence than the swordtail group of species south of the Rio Panuco basin, on the Yucatan Peninsula, in southeastern Guatemala, southern Honduras, and in extreme northeastern Nicaragua.

By replacing the named taxa or populations on cladograms derived for *Heterandria* and *Xiphophorus* (Figs. 19, 20) with numerals representing one or more of the 11 subregions, the cladograms of species or populations may be converted into cladograms of areas. The justification for this procedure is the underlying assumption that the history of all life has some generality with respect to the history of the earth's geography; hence, the search for general distribution patterns is equivalent to the search for related geographic areas. Inspection of the area cladograms for the two groups shows that *Heterandria* has a unique component in area 7 and that the two are incongruent with respect to three areas (3, 6 and 9). In relation to the question of whether each cladogram shows some generality with respect to the other, it is evident therefore that it is not with respect to area 7 (the unique component of *Heterandria*), or areas 3, 6 and 9. Deletion of unique area components and those which are incongruent because only one group has endemic taxa in certain areas (Fig. 21), shows that there is a residual congruence involving five main areas (1, 4–5, 10, 8 and 2). The significance of the identity of the two reduced area-cladograms is not the fact that they each includes the same areas, but that each includes the *same* areas in the *same* cladistic sequence.[8]

What is the probability that the geographically transformed and reduced cladogram of *Xiphophorus* will coincide with the transformed and reduced clado-

---

[8] One may, of course, ask questions about the significance of the unique or incongruent elements in the original cladograms. Assuming that the original, unreduced cladograms truthfully represent real phylogenies and that the unique or incongruent elements have no generality with respect to other as yet unanalyzed components of the Middle American biota, the incongruent elements would be most simply explained as dispersals. Because of the assumptions required, however, it is evident that such explanations have little scientific merit. In other words, such explanations are without significance at the level of general explanation. Mayr (1965), and others, have curiously continued to view biogeographic explanation almost wholly as a search for unique explanations of unique events. Mayr has written: "As a result of various historical forces, a fauna is composed of unequal elements, and no fauna can be fully understood until it is segregated into its elements and until one has succeeded in explaining the separate history of each of these elements."

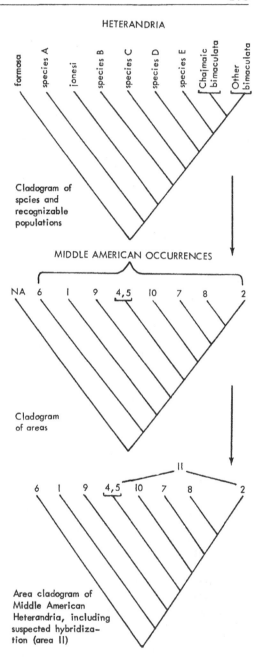

FIG. 19.—Conversion of cladogram of species and recognizable populations of *Heterandria* (derived from a character-state tree in Rosen, in press) into cladograms of areas. Species referred to here by letters (A to E) are described as new in Rosen (in press). Numbers refer to areas in Figs. 16 to 18; NA = North America.

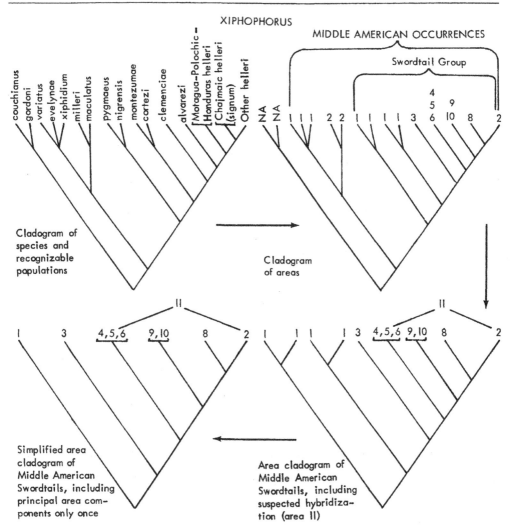

Fig. 20.—Conversion of cladogram of species and recognizable populations of *Xiphophorus* (derived from a character-state tree in Rosen, in press) into cladograms of areas. The first area cladogram is reduced to that part relevant to Middle American swordtails, which is in turn reduced to an area cladogram specifying only the cladistic sequence of principal area components (primary branch points). Numbers refer to areas in Figs. 16 to 18; NA = North America.

gram of *Heterandria* by chance alone? This question is related to the number of dichotomous configurations that are possible in groups containing different numbers of taxa (see Schlee, 1971). Going back for a moment to the simpler three-taxon system (Fig. 14), although only one pattern was found, there are three theoretically possible dichotomous area-cladograms (areas 1 and 2 versus area 3, 2 and 3 versus 1, and 1 and 3 versus 2). The probability that a second group sympatric with the first will duplicate the branching sequence of the first group is one out of three or 33% that such congruence will occur by chance alone. Such a relatively high probability (in fact, the highest P value possible in three-taxon

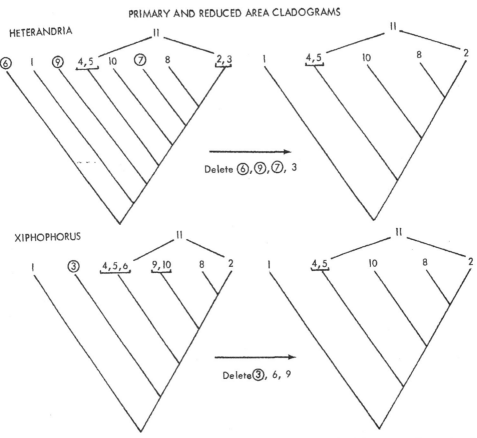

FIG. 21.—Method of reduction of primary cladograms (from Figs. 19 and 20) for *Heterandria* and *Xiphophorus* to produce cladograms representing residual congruence. Deleted unique or incongruent elements are shown below arrows. Circled numbers represent incongruent areas with endemic taxa in each genus.

comparisons) casts some doubt on the significance of finding two congruent three-taxon area-cladograms such as those for *Heterandria* and *Xiphophorus* in North and Middle America. But if a third congruent three-taxon area-cladogram is added to the system, for example the forms of *Terrapene carolina*, then the probability that two three-taxon cladograms will be congruent with the first one is one out of 3^2, or 11% probability that such congruence will occur by chance alone. For a five-taxon area-cladogram, such as those for *Heterandria* and the

swordtail species of *Xiphophorus* in Middle America (Fig. 21), the probability of duplicating a given dichotomous pattern by chance alone is one in 105 (the number of possible dichotomous cladograms for five taxa), or 0.9%. If the individuals in area 11 are correctly interpreted as hybrids (Figs. 19, 20), these putative hybrids, by representing a distinct historical event, could be treated as the equivalent of a sixth taxon in both *Heterandria* and *Xiphophorus*. The probability that the two five-taxon configurations with the sixth reticulate element

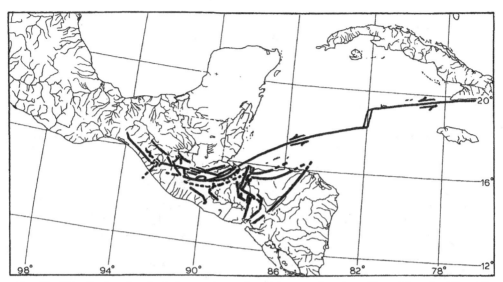

FIG. 22.—Distribution of major fault zones in southern Mexico and northern Central America. The strike-slip fault shown extending into the Caribbean (the Cayman Trench) has its landward extension in the Motagua and Polochic faults of Guatemala, and, according to one geophysical theory, represents the northern boundary of the Caribbean plate (see Muehlberger and Ritchie, 1975).

(area 11) related in the same manner to areas 4, 5, and 2 are congruent by chance alone is only one in 945, or 0.1%.

It appears, therefore, that the coincidence of *Heterandria* and *Xiphophorus* in Middle America can be said to be due to nonrandom factors that affected both equally in five, possibly six, subregions of Middle America (at the 99% confidence level), but that the coincidence of *Heterandria, Xiphophorus*, and *Terrapene* in North and Middle America can be said to be nonrandom only at the 89% confidence level. The discovery by random search within other monophyletic groups distributionally coincident in North and Middle America of another cladogram that coincides with those three would alter the estimate of nonrandomness to the 96% confidence level, and of still one more to the 99% level. This is merely another way of saying that the discovery of additional cladistically congruent distributions can only reduce the probability or likelihood that the observed coincidence has resulted from random historical factors.

What might the historical factors have been in Middle America, and on what basis might they be compared with the given biological distributions? The idea that biotic and historical geologic patterns can be compared requires two assumptions: that geologic or geographic change is the cause of biotic fragmentation and that the fragmentation of an ancestral population by the formation of barriers will result in heritable differences among the descendent populations. In such a cause and effect relationship, where the earth and its life are assumed to have evolved together, paleogeography is taken by logical necessity to be the independent variable and biological history, the dependent variable. Drawing a mathematical simile, a set of biological relationships y (dependent variable) is a function of a set of geological or geographic relationships x (independent variable), as in a simple regression of effect y on cause x where we are given no free choice as to which is the independent variable. Such a view implies that any specified sequence in earth history must coincide with some discoverable biological patterns; it does not imply a necessary converse that each biological pattern must coincide with some discoverable paleogeographic pattern, because some biological distributions might have resulted from stochastic processes (chance dispersal).

These assumptions do not require either that all components of a biota need respond to a given barrier or that heritable differences resulting from isolation need be recognizable by the taxonomist. Groups that either fail to respond to a barrier or have differentiated at an undetected level are simply uninformative with respect to the history of their area. The significance of endemism, then, is simply that some differentiation has occurred that can be viewed cladistically and therefore used to formulate statements about nested sets of distribution patterns. This is illustrated in the example where the Polochic and Motagua faults of eastern Guatemala (Fig. 22), elements of the tectonic history of the Caribbean region, are correlated with two detectable responses in *Heterandria* at the level of recognizable taxa and with one detectable response in *Xiphophorus* at the population level.

The extent to which detectable differentiation has proceeded is unimportant (e.g., recognizable taxa versus recognizable populations), but the *number of cases* of detectable allopatric differentiation is important since more kinds of differentiation make possible more complex cladograms which in turn make possible more predictions about the history of an area. Thus, *Heterandria* is more informative than *Xiphophorus* with respect to the historical distinctions between the Polochic and Motagua basins because only *Heterandria* has endemic species in both basins; furthermore, because of the plesiomorphic position of the species in area 9 relative to that in area 10 in the cladogram for *Heterandria* (Fig. 21), the branching pattern of the cladogram predicts that the events which isolated the Polochic basin preceded those which iso-

lated the Motagua basin and coastal Honduras drainages from northern Guatemala. In this regard *Xiphophorus* is simply uninformative, for, although its populations in northern Guatemala and those in the Polochic, Motagua, and Honduras are slightly differentiated from each other, no detectable differences have been found between those from the Motagua-Honduras region and from the Polochic.

The idea that a biological area-cladogram predicts something about the sequence of geological events in the region under study requires that geological data are capable of being arranged in nested sets or in the form of a cladogram of related areas now separated by discontinuities (such as water gaps, disrupted river channels, rifts along fault zones, mountains, etc.). Within an historical context, related geological areas are two or more areas that arose by the disruption or fragmentation of an ancestral one. The classic example of such a history is the disruption of Pangaea followed by continental drift, a cladistic representation of which is shown in Fig. 23 based on data from McKenna (1973), Rich (1975) and Ballance (1976). There is nothing fundamentally different between geological and biological cladograms since both incorporate the idea of nested sets of things that are grouped by special similarities (synapomorphies), such as the unique patterns of glacial ridges in the rocks of Gondwanian fragments or the uniquely matching contours of the margins of some of the fragments. Hence, in a geological cladogram, the "taxa" are the separate areas or physical features of the landscape which are united as sister-taxa by the common possession of some special property not known elsewhere, and the origin of the separate areas is inferred to have been the disruption of a formerly continuous area. The relative ages of the different branch points in such a cladogram would be determined by radiometric, paleomagnetic and other physical data pertaining to the actual or relative ages of the various disruptive geographic events (the phenomena that become vi-

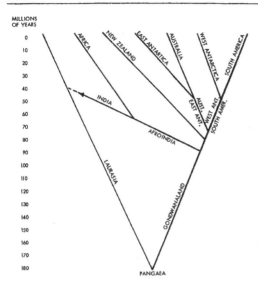

FIG. 23.—Cladistic representation of the history of the breakup of Pangaea greatly simplified, with an approximate time scale of the bifurcations (fragmentation events) and one conjunction (of India and Laurasia).

this may be reduced to the three-taxon statement, in the form of a biological area-cladogram, shown in Fig. 24B. The nature of any question about the generality of this distribution has therefore been specified with reference to three geographical regions. A comparison of reduced cladograms derived for the other two subfamilies treated by Brundin, the Aphroteniinae and Diamesinae, shows that they, too, have Australian and South American taxa more closely interrelated than are any of them to the taxa in Africa. Similarly, in galaxiid fishes of the subfamily Galaxiinae, Australian, and South American taxa may be more closely interrelated than are any of them to the single southern African species, *Galaxias zebratus*.[9] In order to compare these four coincident and cladistically congruent distributions with a cladistic representation of Gondwanian history, the Pangaean area-cladogram in Fig. 23 is reduced by the deletion of components that are unique with reference to a problem that specifies only Australia, South America and Africa. When the unique components of the unreduced Pangaean area-cladogram (Laurasia, India, New Zealand, and East and West Antarctica) are deleted, the remaining figure (Fig. 24A) is directly comparable with the biological cladograms and is seen to be congruent with them. Since the four biological area-cladograms are congruent and may be said to have resulted from nonrandom historical factors at the 96%

cariant events for the affected biota). The resulting geological cladogram representing related areas arranged according to their relative ages of origin can then be converted into an area-cladogram. As in the comparison of the two area-cladograms for *Heterandria* and *Xiphophorus*, the unique components of the geological area-cladogram are deleted so that the components of the geological area-cladogram are geographically comparable with those of a biological area-cladogram. The two kinds of reduced area cladograms are congruent to the extent that the component areas are interrelated in the same sequence.

The nature and method of comparison of biological and geological area-cladograms is most simply illustrated by reference to the Pangaean model of earth history (Fig. 23). We observe in Brundin's (1966) revision of chironomid midges that, within the subfamily Podonominae, Australian and South American taxa are more closely related to each other than any of them are to African taxa;

[9] This species, sometimes referred to a separate genus by itself or together with other non-African species, differs from all other galaxiines in the anterior extent of the dorsal fin insertion. Given the anterior position of the dorsal fin in the sister group of the galaxiines, the aplochitonines, and the equally advanced dorsal fin position in the sister group of both, the Salmonidae, it appears the *G. zebratus* has retained a primitive fin position and that all other galaxiines are united by the relatively more posterior location. Scott (1966) has suggested that the fin position of *zebratus* and certain other species results from a specific pattern of relative growth that causes a foreshortening of the anterior half of the body; although this may be so, such growth patterns have not been studied in a majority of galaxiine species or in the Aplochitoninae.

confidence level, and since they are congruent with a (presumably) corroborated cladistic representation of the relevant components of Gondwanian history, the ancient fragmentation of Australia, South America, and Africa can be taken as our best current estimate of the progenerative cause of the biological patterns.

McDowall (1978), in a lengthy effort to rebutt the view that galaxiid distribution patterns might have ancient roots, offered in evidence the widespread (South America, New Zealand, and Australia) *Galaxias maculatus* which, according to him, is taxonomically undifferentiated in its different outposts. If McDowall is correct in his taxonomic judgment, it appears that the most one can say about *Galaxias maculatus* is that it is uninformative with respect to the history of Australia, South America, and Africa and to the history of other galaxiids in those regions. Yet McDowall insisted that the distribution of *G. maculatus* and another widespread species freed him from any constraints to consider the possibility of an ancient origin for the group. What he claimed to be reasonable is that the distributions of these two widespread taxa permit one "to assume that early dispersal of galaxiid fishes could have been through the sea." Although McDowall emphasized the necessity of phylogenetic argument in biogeography, he provided none, but seemed intent, instead, on rescuing dispersal at the expense of an arguable position.

Congruence of a biological area-cladogram with a geological area-cladogram

→

FIG. 24.—Comparison of geological and biological area-cladograms in relation to four monophyletic animal groups in the southern hemisphere. Taken as an explanation of the generalized biological pattern, the geological cladogram specifies an age of about 85 my BP for the first dichotomy (between the taxa in Africa and Australia-South America) and about 70 my BP for the second (between the taxa in Australia and South America). Many other examples of these intercontinental relationships have been discussed by Keast (1973).

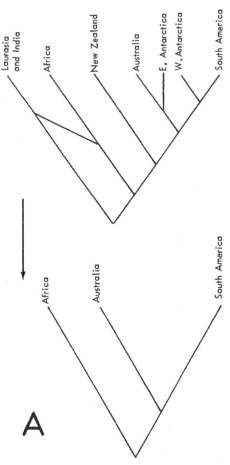

PRIMARY AND REDUCED GEOLOGICAL AREA CLADOGRAMS

A

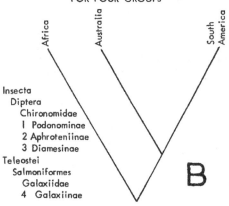

REDUCED BIOLOGICAL AREA CLADOGRAM FOR FOUR GROUPS

Insecta
 Diptera
  Chironomidae
   1 Podonominae
   2 Aphroteniinae
   3 Diamesinae
Teleostei
 Salmoniformes
  Galaxiidae
   4 Galaxiinae

B

does not test the reality of vicariance versus dispersal. However, it does provide a best estimate of a sequence of geological events implied by the geological relationships that might have been involved in the speciation events implied by the biological relationships. Hence, the probability that a five-term geological area-cladogram will correspond with a five-taxon biological area-cladogram by chance alone is one in 105 (about 1%); and, as in the Gondwanian example, above, the probability that four three-taxon biological area-cladograms will be congruent with a three-term geological area-cladogram by chance alone also is about 1%. A decision concerning the nature of this correspondence whether by dispersal or vicariance or some combination of the two, must be a parsimony decision concerned with minimizing the number of separate assumptions entailed by the different types of explanations. Interestingly, a parsimony decision only becomes possible when a given distribution pattern (a biological area-cladogram) can be shown to have some significant generality expressed as a low probability that a given pattern will have occurred in more than one monophyletic group by chance alone. McDowall looked with small favor on a remark by Croizat et al. (1974) that galaxiid distribution "might itself be interpreted, perhaps even credibly, as an example of transoceanic dispersal from some center of origin. But when galaxiid distribution is compared with that of other southern hemisphere organisms, many of which have similar distributions but different means of dispersal . . . a general problem is posed, concerning the original distribution and subsequent history of a pan-austral biota, of which the Galaxiidae might be only a small part. . . . The problem is one of different realities: the reality of long-distance dispersal versus the reality of a pan-austral biota." The clear implication and certainly the intent of that remark, which McDowall ignored in his own view of the problem, is that general patterns demand

general explanations. As noted above, cladograms congruent at a very high confidence level only specify that the groups involved have probably shared a common history. Whether that was a history of dispersal or vicariance seems to depend on other considerations. If, for example, a group of perching birds and a group of earthworms had congruent six-taxon area-cladograms (where $P = 0.1\%$ that the two were congruent by chance alone), would we assume that the earthworms dispersed with the birds or that the birds vicariated with the earthworms? I wonder which alternative McDowall would choose, and why.

A final, and most necessary, observation is that a geological area-cladogram neither tests nor in any way affects the generality of a biological area-cladogram. A geological area-cladogram differing from the biological pattern does not refute the pattern—it is simply irrelevant to it because it contains no explanatory information regarding the biological pattern. A geological area-cladogram that corresponds in part or whole with a biological pattern is simply adopted as an "explanation" of the pattern in the sense of providing a best current estimate of the historical factors that induced the biological pattern. A conflict of geological explanation could only arise, then, if, for the same geographic region, another series of geological relationships of entirely different chronology were shown to have the same cladistic structure.[10] With two congruent but allochronic geological area-cladograms for the same region, only appropriate fossil occurrences could resolve which one to accept as the relevant

---

[10] If, as logic demands, each set of historical relationships has its own unique properties due to its unique spacio-temporal coordinates, the correspondence between any two allochronic series of events would simply reflect the low resolution of that level of analysis. At the same time, it is the logical uniqueness of historical relationships that makes the comparison of generalized distribution patterns with geological or geographical patterns inherently reasonable, even if we now (and will continue to) face the severe practical problems of inadequate data.

explanation of a particular biogeographic history (e.g., by showing that a relevant part of the biological history is too old for one geological sequence but not for the other).

Returning now to the problem of the geographic history of *Heterandria* and *Ziphophorus* in Middle America: the region has been tectonically active since the end of the Mesozoic, and although a great deal has been written on the historical geology of this region, I am unable to obtain enough precise information to produce a cladistic statement that is relevant to the area extending from the Rio Panuco basin southeastward to Nicaragua. This does not mean that I consider a search for such historical explanation fruitless, but only that, as a nongeologist, I am unable to pick and choose among the varied and sometimes conflicting geological interpretations of the Tertiary history of Middle America.

On the other hand, I could abandon the pursuit of a cladistic synthesis of geographic history and simply try to find an event here and there that coincides with some biological disjunction: for example, the fault zones of northern Central America or the east–west oriented Pliocene volcanic zone south of the Rio Panuco basin in Mexico.

Or, I could renounce my responsibility altogether and simply assume that the biological patterns have been caused by Quaternary events, as Deevey (1949) said we should assume.

Or, I could do something else, which I am inclined to favor. This is to recognize that geology and biogeography are both parts of natural history and, if they represent the independent and dependent variables respectively in a cause and effect relationship, that they can be reciprocally illuminating. But for there to be reciprocal illumination between these two fields there must be a common language. This language must be the language of nested sets, i.e., hierarchical systems of sets and their subsets united by special similarities, i.e., synapomorphies. In short, taxonomists should be

encouraged to continue the current salutary trend to organize their data cladistically, and geologists should be encouraged to begin. At least until the geological data pertaining to Middle America are so ordered I am forced to draw the limited conclusion that the observed biological patterns have formed during a period spanning all or part of the last 80 million years.

ACKNOWLEDGMENTS

To my friend and colleague of many years, Gareth Nelson, I offer special thanks for the countless hours, adding up, I am certain, to months, of enjoyable and informative discussion of systematic and biogeographic theory. Appreciated criticism of the typescript was received from James Atz, Daniel Axelrod, Roger Batten, Niels Bonde, Sadie Coats, Leon Croizat, Jürgen Haffer, David Hull, Malcolm McKenna, Gareth Nelson, Lynne Parenti, Norman Platnick, and E. O. Wiley. Helpful discussions of Mexican fish distributions were offered by Robert R. Miller, and information concerning tropical American herpetology was supplied by Eugene Gaffney, Charles Myers, and Richard Zweifel.

REFERENCES

AXELROD, D. I. 1975. Evolution and biogeography of Madrean-Tethyan sclerophyll vegetation. Ann. Missouri Bot. Gard. 62:280–334.

BALLANCE, P. F. 1976. Evolution of the Upper Cenozoic magmatic arc and plate boundary in northern New Zealand. Earth and Planetary Sci. Lett. 28:356–370.

BRAMBLE, D. M. 1974. Emydid shell kinesis: biomechanics and evolution. Copeia 3:707–727.

BRODKORB, P. 1963. Fossil birds, Part 1. Bull. Florida State Mus. 7:179–293.

BRODKORB, P. 1964. Fossil birds, Part 2. Bull Florida State Mus. 8:195–355.

BRODKORB, P. 1967. Fossil birds. Part 3. Bull. Florida State Mus. 11:99–220.

BRODKORB, P. 1971. Fossil birds. Part 4. Bull. Florida State Mus. 15:163–266.

BRUNDIN, L. 1966. Transantarctic relationships and their significance. K. Svenska Vetensk.-Akad. Handl. Ser. 4, 11:1–472.

CROIZAT, L. 1958. Panbiogeography. Published by the author, Caracas.

CROIZAT, L. 1964. Space, time, form: the biological synthesis. Published by the author, Caracas.

CROIZAT, L., G. NELSON, AND D. E. ROSEN. 1974. Centers of origin and related concepts. Syst. Zool. 23:265–287.

CRUM, H. A. 1951. The Appalachian-Ozarkian element in the moss flora of Mexico with a check-list of all known Mexican mosses. Unpublished Ph.D. thesis, University of Michigan.

DARNELL, R. 1962. Fishes of the Rio Tamesi and

related coastal lagoons in east-central Mexico. Publ. Inst. Mar. Sci. 8:299–365.

DEEVEY, E. S., JR. 1949. Biogeography of the Pleistocene. Bull. Geol. Soc. Amer. 60:1315–1416.

DEEVEY, E. S., JR. 1953. Paleolimnology and climate. *In* Shapley, H. (ed.), Climatic change. Harvard University Press, Cambridge.

DOTT, R. H., JR., AND R. L. BATTEN. 1976. Evolution of the earth. McGraw-Hill, Inc., New York.

HALL, E. R., AND K. R. KELSON. 1959. The mammals of North America. 2 vols. The Ronald Press Co., New York.

HULL, D. L. 1976. Are species really individuals? Syst. Zool. 25:174–191.

KEAST, A. 1973. Contemporary biotas and the separation sequence of the southern continents. *In* Tarling, D. H., and S. K. Runcorn (eds.), Implications of continental drift to the earth sciences. Academic Press, London and New York. 1:309–343.

KEAST, J. A. 1977. Zoogeography and phylogeny: the theoretical background and methodology to the analysis of mammal and bird faunas. *In* Hecht, M. K., P. C. Goody, and B. M. Hecht (eds.), Major patterns in vertebrate evolution. Plenum Press, New York and London. pp. 249–312.

LEOPOLD, E. S., AND H. D. MACGINITIE. 1972. Development and affinities of Tertiary floras in the Rocky Mountains. *In* Graham, A. (ed.), Floristics and paleofloristics of Asia and eastern North America. Elsevier Publ. Co., Amsterdam. pp. 147–200.

LOVERIDGE, A., AND E. E. WILLIAMS. 1957. Revision of the African tortoises and turtles of the suborder Cryptodira. Bull. Mus. Comp. Zool. 115:163–557.

MACARTHUR, R. H., AND E. O. WILSON. 1967. The theory of island biogeography. Princeton University Press, Princeton.

MARTIN, P. S., AND B. E. HARRELL. 1957. The Pleistocene history of temperate biotas in Mexico and eastern United States. Ecology 38:468–480.

MAYR, E. 1965. What is a fauna? Zool. Jahrbuch., System. 92:473–486.

MAYR, E., AND W. PHELPS. 1967. The origin of the bird fauna of the south Venezuelan highlands. Bull. Amer. Mus. Nat. Hist. 136:269–328.

McDOWALL, R. M. 1978. Generalized tracks and dispersal in biogeography. Syst. Zool. 27:88–104.

McKENNA, M. C. 1973. Sweepstakes, filters, corridors, Noah's arks, and beached Viking funeral ships in paleogeography. *In* Tarling, D. H., and S. K. Runcorn (eds.), Implications of continental drift to the earth sciences. Academic Press, London and New York. 1:295–308.

MEEK, S. E. 1904. The fresh-water fishes of Mexico north of the Isthmus of Tehuantepec. Pub. Field Columbian Mus., Zool. Ser. 5:i–lxiii, 1–252.

MILLER, R. R. 1976. An evaluation of Seth E. Meek's contributions to Mexican ichthyology. Fieldiana Zool. 6:1–31.

MILSTEAD, W. W. 1969. Studies on the evolution of box turtles (genus *Terrapene*). Bull. Florida State Mus., Biol. Sci. 14:1–113.

MUEHLBERGER, W. R., AND A. W. RITCHIE. 1975. Caribbean-Americas plate boundary in Guatemala and southern Mexico as seen on *Skylab IV* orbital photography. Geology, May 1975:232–235.

MÜLLER, P. 1973. The dispersal centres of terrestrial vertebrates in the neotropical realm. Biogeographica 2:1–244.

MYERS, C. 1974. The systematics of *Rhadinaea* (Colubridae), a genus of New World snakes. Bull. Amer. Mus. Nat. Hist. 153:1–262.

NELSON, G. MS. Cladograms and trees.

PLATNICK, N. 1976. Concepts of dispersal in historical biogeography. Syst. Zool. 25:294–295.

PLATNICK, N. I., AND G. NELSON. 1978. A method of analysis for historical biogeography. Syst. Zool. 27:1–16.

REGAN, C. T. 1926. Organic evolution. Rept. British Assoc. Adv. Sci. 1925:75–86.

RICH, P. V. 1975. Antarctic dispersal routes, wandering continents, and the origin of Australia's non-passeriform avifauna. Mem. Nat. Mus. Victoria 36:63–126.

ROMER, A. S. 1966. Vertebrate paleontology. The University of Chicago Press, Chicago.

ROSEN, D. E. 1974. Phylogeny and zoogeography of salmoniform fishes and relationships of *Lepidogalaxias salamandroides*. Bull. Amer. Mus. Nat. Hist. 153:265–326.

ROSEN, D. E. In press. Fishes from the uplands and intermontane basins of Guatemala: revisionary studies and comparative geography. Bull. Amer. Mus. Nat. Hist.

ROSEN, D. E., AND R. M. BAILEY. 1963. The poeciliid fishes (Cyprinodontiformes), their structure, zoogeography, and systematics. Bull. Amer. Mus. Nat. Hist. 126:1–176.

SCHLEE, D. 1971. Die Rekonstruktion der Phylogenese mit Hennig's Prinzip. Aufsätze Red. senckenberg. naturforsch. Gesell. 20:1–62.

SCOTT, E. O. G. 1966. The genera of the Galaxiidae. Australian Zool. 18:244–258.

VUILLEUMIER, F. In press. Qu'est-ce que la biogéographie? Comptes Rendus de Séances de la Société de Biogéographie.

WAKE, D. B., AND J. F. LYNCH. 1976. The distribution, ecology, and evolutionary history of plethodontid salamanders in tropical America. Sci. Bull. Nat. Hist. Mus. Los Angeles Co. 25:1–65.

WEBB, S. D. 1974. Pleistocene mammals of Florida. University of Florida Press, Gainesville.

Manuscript received October 1977
Revised February 1978

5 Diversification

Lawrence R. Heaney and Geerat Vermeij

Diversification and the Interplay between Diversity and Geography

One of the central and most fundamental questions in the study of biological systems has been, "Why are there so many species?" In biology, as in other fields, such simple questions seem never to have simple answers, and the many attempts to answer this question demonstrate that it is perhaps the most challenging in biology. Indeed, phrased in this fashion, the question may not have an answer, and perhaps we should instead ask "How, when, and under what circumstances does species proliferation take place?" Regardless of the precise wording, the reasons for the existence of such an exuberant diversity of life on earth may be the greatest puzzle that has ever been faced by humankind (aside, perhaps, from the means to ensure its continued existence).

From the earliest origins of modern biology, issues of biodiversity and geography have been intertwined. Alexander Von Humboldt and Joseph Dalton Hooker, for example (see part 1 of this volume), based much of their writings on the diversity of plants in geographical context, often commenting on the existence of exceptionally high diversity within particular groups of organisms on particular archipelagos or in other isolated biotic regions. Indeed, much of the literature that documents patterns of biological diversity in time and space deals with its geographical correlates, and treats geography as playing an integral role in the evolution of biological diversity (e.g., Behrensmeyer et al. 1992, Ricklefs and Schluter 1993, Brown 1995, Rosenzweig 1995, Grant 1998, Magurran and May 1999). Even in the case of sympatric and parapatric speciation, in which geographic isolation is limited or absent, highly localized geographic circumstances are often said to play a role (Bush 1969, White 1978).

In part 5, we discuss key publications that address the interplay between biodiversity and geography: how geography has influenced the development and continued existence of biological species richness through the process of diversification. We view diversification as an outcome of speciation, and define *biological diversification* as "the increase in species richness within biological lineages over time." One might define diversification differently, to include the results of both speciation and colonization/invasion (see, e.g., Vermeij 1987), but here we follow the narrower definition because it deals with a single process. In this commentary, we discuss the mechanisms of speciation processes only to the extent that they are directly relevant. Similarly, we do not discuss the various means for determining phylogenetic relationships (cladistics, numerical taxonomy, etc.; but see above, part 4). Rather, this brief introduction focuses on key historical publications that have influenced the way biologists have interpreted the diversification of lineages in a geographic theater. The papers cited, and especially those papers or excerpts from books reprinted here, span roughly the period 1920 to 1978, but by no means represent a comprehensive review. They have been selected because they have had especially great impact, often serving as the spark to a long series of subsequent publications by a generation of biogeographers.

Diversification and the Modern Synthesis

The early classic papers in biogeography presented and discussed in part 1 show evidence of several stages of development. The earliest ones focused on documenting the distributions of organisms and beginning to assess diversity. Lacking an evolutionary framework, they often assumed static diversity and used typological definitions of species (*typology* refers to the traditional Greek philosophical belief in invariant "types," with variation seen as the product of error). Later studies in the mid- to late 1800s shifted the focus to the mechanisms of evolution by natural selection as a means to understand the development of species. Yet diversification, as a distinct topic, was rarely addressed, aside from simply remarking that it had taken place (e.g., Darwin's remarks on the tortoises of the Galapagos Islands).

By about 1900, the prevailing view of species diversification had begun to change in concert with the changing concept of what constitutes a species. Some of the earliest glimmerings came from the gradual acceptance of subspecies names to designate geographic variants of species; mammalogists, for example, commonly began to use such trinomials during the 1890s.

This resulted largely from the growing information about geographic variation among populations, and from the growing sense that typological species concepts were problematic. Some ornithologists soon took another step by developing a new definition of species; among the earliest, Erwin Stresemann (1919) stated that bird populations which ". . . have risen to species rank have become so different from each other physiologically that they . . . can come together again without interbreeding" (translation from Mayr 1982: 273). Stresemann and Rensch emphasized this definition throughout the 1920s and 1930s, and Dobzhansky (1937: 312) adopted it as well, bringing it firmly into mainstream biology; today it forms the basis for the *biological species concept*.

The rapid development and widespread acceptance of the biological species concept, with its emphasis on absence of interbreeding among populations as the primary criterion of species status, had profound impact on perceptions of both diversity and diversification. First, sweeping taxonomic revisions of birds, mammals, and some other groups reduced the number of recognized species. Under the new view, many former species were recognized only as subspecies or considered to be too trivial to recognize formally at all. Second, because virtually all discussions about species definitions and about the taxonomy of any given group emphasized geographically allopatric or parapatric populations, a geographic perspective became more pervasive. The means of diversification thus came to be viewed as one in which geographical heterogeneity and historical climatic changes combined to produce periods of restricted distribution and allopatric populations. This was then followed by periods of range expansion and, occasionally, sympatry among differentiated populations—populations which then met the definition of species. Thus, diversification came to be viewed in almost purely geographical terms. Further, because at this time (roughly the 1930s to the 1950s) the concept of continental drift was viewed skeptically by Europeans and rejected almost entirely by North Americans, diversification also came to be viewed as being associated either with climatic change (especially the ice ages of the Pleistocene), or with dispersal, including both gradualist, short-distance range expansion and rare, long-distance dispersal.

A burst of publications in the 1940s formalized this new view of the importance of geography in diversification. We begin with a section from the 1960 English edition of the book by BERNARD RENSCH entitled *Evolution above the Species Level* (see paper 44). This book was based largely on the author's studies from the 1920s and 1930s—work that had already been influential in Europe. Like several of the contributors to the newly emerging "modern synthesis" paradigm of evolution, Rensch was primarily an ornithologist, although he also studied land snails. Systematic collections of birds had been accumulating for many decades, and birds were among the first organisms for which reasonably comprehensive geographic samples were available for study. Rensch became intrigued early in his career with a geographically complex type of distribution among related taxa. He coined the term *rassenkreise* for this pattern (Rensch 1929), which literally means "race rings," but which he translated (1960: 23) as "mosaics of races." These cases involved situations in which closely related and morphologically distinctive, geographically dispersed populations often interbred and formed intermediate forms when they came into contact, but sometimes did not interbreed and remained distinct. Rensch viewed this as evidence of "race formation," a stage in the gradual development of reproductive isolation. He viewed reproductive isolation as being due to selective pressure for physiological adaptations to the climate of a particular geographic region, and ultimately to the incompatibility of animals with different physiologies.

Rensch (1960: 97) then went on to discuss "the causes of branching in the lines of descent," in the process coining the term *kladogenesis* (defined as "phyletic branching"). (Part of his discussion also referred to orthogenesis, a topic related but not central to diversification). He reviewed evidence for the previously postu-

lated "intense radiation of types, an 'explosive phase' in the early part of [the] phylogeny," concluding that this is typically "brought about by a temporary intensification of selection due to environmental changes, . . . or due to the colonization of new ranges with habitats unoccupied or inhabited by types inferior in competition" (112). In both cases, he emphasized climate, including both geographic and long-term temporal variation, as causing strong selection. In other words, Rensch argued for the importance of natural selection on populations that were responding to new climatic conditions and biotic interactions (including competition). Geographically distinctive populations eventually became reproductively isolated, thus increasing phyletic diversity. These new species then sometimes underwent secondary geographic expansions that resulted in another round of speciation events.

Because Rensch published his early papers in German, their impact in English-speaking countries was initially limited. However, his work was quoted in one of the most influential books defining the Modern Synthesis, *Genetics and the Origin of Species* by Theodosius Dobzhansky ([1937] 1951), who briefly summarized the issue of species diversification in this way:

A living species is seldom a single homogeneous population. Far more frequently species are aggregates of races, each race possessing its own complex of characters. The term "race" is used quite loosely to designate any subdivisions of species which consists of individuals having common hereditary traits. Races may be of very different orders. Formation of geographical races is probably the most usual method of differentiation of species both in animal and plant kingdoms (the zoological literature on this subject has been reviewed by Rensch, 1929), but a geographical race can occupy an area ranging from a few to hundreds of thousands of square miles. Geographical races are in turn subdivided into smaller secondary races (sometimes known as natio). Schmidt (1923) shows that populations of the small fish *Zoarces viviparus* inhabiting different Scandinavian fiords, or even parts of the same fiord, are distinct from each other in such characters as size, number of vertebrae, etc. The distinctions are hereditary.

Turesson (1922–31) finds that many plant species are split into numerous hereditary "ecotypes" adapted to living in definite ecological stations and found wherever the proper environment exists in the distribution region of the species. Coastal, dune, forest, swamp, alpine, and other ecotypes are distinguished." (47)

The bulk of Dobzhansky's aptly titled book deals in detail with the genetic basis for evolution, and especially for the influence of adaptation on the initial genetic differentiation between species. While much of the book is focused exclusively on genetics and the mechanisms of natural selection, Dobzhansky's view of the crucial importance of the geographic context is evident throughout the text. For example, in a section on "regularities in geographical variation" (165), he quoted Rensch at length for examples of eco-geographic rules, including Gloger's rule, concerning the relative abundance of melanin in populations from humid versus dry regions; Bergmann's rule, concerning a tendency for body size of "warm-blooded" animals to increase in colder climates; and Allen's rule, concerning the tendency for animals to have longer legs, tails, and ears in hot, dry areas than in cold, wet areas. In later writings, Dobzhansky argued for the linkage between speciation, environment, and geography even more strongly. From comparisons between the temperate zones and the tropics, he argued that "The diversity of organisms which live in a given territory is a function of the variety of available habitats. The richer and more diversified the environment becomes, the greater should be the multiformity of the inhabitants. And vice versa: diversity of the inhabitants signifies that the environment is rich in adaptive opportunities" (Dobzhansky 1950).

Another extremely influential volume appeared during this same period, *Systematics and the Origin of Species*, by ERNST MAYR (1942). In this book, which soon became one of the most frequently cited volumes in evolutionary biology, research by Stresemann, Rensch, and other European ornithologists (including Mayr himself, who had emigrated to the United States in the 1930s) was communi-cated to a broad English-speaking audience for the first time. While the emphasis of the book overall was on the biological species concept (and was, to some degree, biased by its strong reliance on information about birds and mammals, and by Mayr's own research on the birds of New Guinea and its offshore islands), one large chapter, reprinted here (paper 45), dealt with "the process of geographic speciation." Mayr was unambiguous: he viewed speciation as overwhelmingly and inextricably linked to geography, and saw diversification of lineages as the inevitable outcome. His figure 16 (p. 160) illustrating the "stages of speciation" became one of the most commonly reproduced figures in biology textbooks, and his definition of species was adopted widely.

Other authors writing at about the same time contributed to a solidification of "the Modern Synthesis." Julian Huxley's (1943) *Evolution: The Modern Synthesis* presented a broad summary of evidence for natural selection as the mechanism for speciation based on genetic variation. George Gaylord Simpson's (1944) *Tempo and Mode in Evolution* focused on rates of evolution and did not address issues of diversification directly, but, like Huxley, he saw phylogenesis (the increase in species richness within a given lineage over time) principally in terms of geographic differentiation. Similarly, G. Ledyard Stebbins's (1950) *Variation and Evolution in Plants* dealt predominantly with issues at the level of species: variation among populations, mechanisms of evolution, development of isolating barriers, and mechanisms of hybridization and polyploidy. Stebbins said little about factors associated with the increase in species richness within lineages. However, he, too very strongly emphasized the crucial importance of allopatry in the production of new species, stating that "Jordan's Law (Jordan 1905) usually holds [that] the nearest related species to any given species population is found, not in the same area or in a very different one, but in an adjacent geographic region or in a far distant one with similar climatic and ecological conditions" (Stebbins 1950: 238). In Stebbins's view, development of isolating mechanisms eventually permits sympatry: "Speciation, therefore, may be looked upon as the initial stage in the divergence of evolutionary lines which can enrich the earth's biota by coexisting in the same habitat" (195).

Conceptual Developments from 1945 to the 1970s

An influential book of a different kind from the broad syntheses by Rensch, Dobzhansky, Mayr, Huxley, Simpson, and Stebbins appeared in 1947: DAVID LACK's monograph, *Darwin's Finches* (see paper 46). This was perhaps the first book to focus on the evolution of diversity in a limited group of organisms that combined systematic, behavioral, and ecological perspectives. Darwin's finches have arguably remained the most famous example of an insular adaptive radiation, influencing the perspectives of generations of biologists on fundamental issues of how, when, and where evolution takes place (reinforced in recent years by further studies by Lack's student Peter Grant and his colleagues; e.g., Grant 1986, 1998). Reinforcing the views advanced by Streseman and Rensch, Lack stressed how speciation occurred and lineages proliferated as "forms differentiated in geographical isolation have later met and kept distinct" (1947: 132). In this instance, he argued that these cycles of colonization and isolation were frequently repeated, as a single ancestral population of finches gave rise to 14 species, with up to 10 of these occurring sympatrically on a single island, implying frequent dispersal between islands as the clade differentiated. He contrasted the Galapagos archipelago, with its many islands of varying degrees of isolation, from Cocos Island, located roughly 600 km northeast of the Galapagos, and the one place outside of the Galapagos where Darwin's finches occur: ". . . despite the length of time for which it has been there, despite the variety of foods and habitats which Cocos provides, and despite the almost complete absence of both food competitors and enemies, there is still

only one species of Darwin's finch on Cocos. But Cocos is a single island, not an archipelago, and so provides no opportunity for the differentiation of forms in geographical isolation" (133; but see Werner and Sherry 1987 for subsequent information on behavioral diversification of the Cocos finch).

The influence of geography on diversification was explored at much broader spatial and temporal scales in PHILIP J. DARLINGTON JR.'s, "Area, Climate, and Evolution" (1959; paper 47), which addresses the development of global terrestrial patterns of biodiversity as inferred from both the fossil record and current distributions. Darlington's paper is largely a critique of the views of Darwin (1859) and Matthew (1915), who argued that most organisms originated in temperate Eurasia and North America, and from there spread and diversified into the continental tropics and "peripheral areas." Following from Forster, Wallace, and other early biogeographers introduced in part 1 of this volume, Darlington asserted that maximum diversity occurred in the tropics (providing one of the first diagrams of a general latitudinal gradient in the process), and that *dominant* animals (taxa that are species-rich and widespread) could be examined together to determine if they had similar geographical histories. His analysis led him to conclude that most of these dominant groups originated in the Old World tropics, with the large area and the climatic stability of that land mass playing important roles in the evolutionary dynamics. In his view, dispersal and diversification were intertwined, especially over long periods of time.

Like many of his contemporaries, Darlington was deeply skeptical of the theory of continental drift until, in the 1960s, the evidence for plate tectonics finally became irrefutable. While his assumptions, and some of his conclusions, have subsequently been shown to be wrong, his writings summarized the views of a generation of biogeographers, and Darlington can reasonably be lauded for his insistence on making assumptions explicit, sticking to the available data, and seeking parsimonious explanations for general patterns. For example, as late as 1965, he spoke of the importance of understanding the role of climate in determining distribution patterns, and "against moving continents or making other fundamental changes in the world unless real evidence requires it" (Darlington 1965: 161).

The issue of diversification in island faunas returned to the fore with publication of Edward O. Wilson's paper, "The Nature of the Taxon Cycle in the Melanesian Ant Fauna" (1961; see also part 6, paper 53). Taking a quantitative and hypothesis-generating approach, Wilson depicted a dynamic system in which colonization by ants among isolated oceanic islands was followed by phyletic diversification, followed by further colonization, speciation, and extinction. In many respects, the origins of current island biogeography can be traced to this paper, as well as the origins of a continuing debate over the existence and interpretation of taxon cycles. The subsequent publications by MacArthur and Wilson (1963, 1967) had still greater impact, effectively dominating studies of organisms on islands for the next four decades (Whittaker 1998, Lomolino 2000a). There was, however, a subtle but significant shift in the emphasis of the MacArthur and Wilson publications from that of Wilson (1961): the later publications made very little mention of diversification, which had been a prominent part of the taxon cycle. Where Wilson made phylogenesis a core issue and presented his data in a long-term geological context, MacArthur and Wilson relegated phylogenesis to a few paragraphs or pages near the ends of their two seminal publications. Thus, after 1963 (with a few exceptions), island biogeography became an almost purely ecological topic, with colonization and extinction as the primary variables of interest, instead of being viewed as an evolutionary process involving diversification (see Heaney 2000, Whittaker 1998).

By the 1960s, studies of contemporary and fossil marine faunas became the focus for long-term analysis of diversification patterns (see, e.g., Fischer 1960, reprinted in part 8). One of the major contributions was the use of the fossil record to provide an historical perspec-

tive. In a series of papers culminating in the 1969, "Patterns of Taxonomic and Ecological Structure of the Shelf Benthos during Phanerozoic Times" (paper 48), JAMES W. VALENTINE argued that, while complete diversity at the species level could not be determined in the fossil record because of incomplete sampling, the global diversity of "well-skeletonized" marine shelf invertebrates at the family level was sufficiently well known for quantitative analyses. He concluded that the global diversity of these organisms shows clear and substantial variation over time, with apparent periods of rapid diversification and periods of rapid decline. In particular, relying on new compilations of an increasingly well-documented fossil record, he argued that previous evidence of massive extinction at the end of the Permian was well founded, and that this event was followed by rapid diversification that began in the Triassic and continued nearly unabated through to the Cenozoic and Recent periods. The post-Triassic increase in diversity, he stated, was due to both increasing specialization of species and to an increase in the number of distinct *provinces* (i.e., centers of endemism) associated with gradually intensifying latitudinal temperature gradients and fragmentation of shelf environments caused by continental drift.

Valentine's interpretations and conclusions were challenged in 1972 by DAVID M. RAUP's "Taxonomic Diversity during the Phanerozoic" (paper 49), in which the author argued that the number of families tallied by Valentine was highly correlated with the volume of sediments and areas of exposure from each of the relevant time periods, and thus was most readily explained as an artifact of biased sampling. Moreover, on the basis of computerized randomized sampling models (a newly developed procedure), Raup argued that Valentine's non-equilibrial description of continuously varying diversity levels could more accurately be viewed as an initial period of diversification and over-shoot, "followed by a decline to an equilibrium level" that then remained stable.

For 15 years, these two papers, with their very different approaches and conclusions, formed the basis for a remarkably productive series of exchanges that continues in modified form at the present time. Equilibrial views of diversity patterns were championed by Raup and his colleagues (e.g., Raup et al. 1973, who cited and were influenced by MacArthur and Wilson's 1967 monograph on equilibrium island biogeography), and apparent deviation from equilibrial values were ascribed to sampling bias in the fossil record. Non-equilibrial views were upheld by Valentine and some others (e.g., Bambach 1977; Valentine, Foin, and Peart 1978), who continued to offer evidence of temporal variation in diversity (as well as in rates of diversification and extinction), while acknowledging the existence of some potential for sampling bias. Simultaneously, other authors (e.g., Sepkoski 1976, 1978; Flessa and Sepkoski 1978), also drawing heavily on MacArthur and Wilson's conceptualizations, expanded a portion of Valentine's original (1969) argument that fragmentation of shelf areas by continental drift played a role. They showed that diversity was strongly correlated with continental shelf area at different points in time, and inferred that diversity was largely in equilibrium with habitat area, although they acknowledged some stochastic variation and periods of disequilibrium. By 1979, Raup had concluded that the late Permian extinction was far more massive than previously believed, and Knoll, Niklas, and Tiffney (1979) presented evidence that, while vascular plant diversity was correlated with the area of outcrop exposure and thus subject to sampling bias, there was strong evidence that stepwise increases in diversity accompanied major evolutionary innovations in plants during the Phanerozoic. Finally, in a remarkable example of compromise and synthesis, Sepkoski, Bambach, Raup, and Valentine (Sepkoski et al. 1981) concluded that careful interpretation of the fossil record could ameliorate the existing biases, and such analysis supported the view that substantial variation in diversity occurred during the Phanerozoic. Whether the data supported equilibrial or non-equilibrial conditions of diversity, however, was side-stepped by reference to succes-

sive "multiple equilibria"—and discussion of the point has continued (e.g., Miller 2000).

One of the most productive controversies regarding diversification of mainland biotas was sparked by JURGEN HAFFER's 1969 paper "Speciation in Amazonian Forest Birds" (paper 50). Returning to the question of why there are so many species in the tropics, especially in the fabled Amazon Basin, Haffer proposed a strikingly different hypothesis. He suggested that it was not the presence of a stable climate, as stated by generations of evolutionary biologists before him (e.g., Dobzhansky 1950, Darlington 1959), but rather the *instability* of the region's climate during the Pleistocene, that had induced alternating periods of isolation and contact of populations that in turn promoted diversification. Citing evidence from Moreau (1966) that the Amazonian Basin had repeatedly undergone periods of relative aridity during the many glacial maxima of the Pleistocene, he stated that "the repeated expansion and contraction of the forests as a result of climatic fluctuations during the Quaternary, [led] to repeated isolation and rejoining of forest animal populations" (Haffer 1969: 135). He further cited Moreau (1966) as providing evidence that "the speciation process in birds may be completed in 20,000 to 30,000 years or less, particularly in the tropics, where birds generally seem to occupy smaller niches than they do in cooler and less stable climates. . . . The above estimates are highly speculative, [but] if the order of magnitude is at least approximately correct, it indicates that the Tertiary ancestors of present Amazonian birds may have speciated *repeatedly* during the Quaternary" (Haffer 1969: 135).

Haffer's bold ideas stimulated a flurry of studies debating the number, location, and extent of rainforest refugia, as well as the locations of potential hybrid zones between refugia, the mechanisms of speciation (especially the question of parapatric speciation), and the implications of forest fragmentation and geographic range dynamics for conservation (Whitmore and Prance 1987, Haffer 1997). The current state of the debate, however, is perhaps the outcome least anticipated by Haffer in 1969: evidence is strong that most of the speciation events between sister-taxa in the Amazon Basin occurred in the very early Pleistocene or, more often, in the Pliocene (roughly 1.8 to as much as 5 million years ago), thus long predating the existence of the climatic and vegetation changes on which Haffer based his model (see Moritz et al. 2000, Bates 2001). While Haffer's model may be criticized for making a fundamentally incorrect assumption, it stimulated a great deal of research, and we know far more about diversity and diversification in the Amazon Basin (and other regions of lowland tropical rainforest) than we would without his thought-provoking hypothesis.

An entirely different perspective on the issue of diversification was presented in the same year as Haffer's seminal paper on diversification in Amazonian refugia. Guy L. Bush (1969), in "Sympatric Host Race Formation and Speciation in Frugivorous Flies of the genus *Rhagoletis* (Diptera, Tephritidae)" (paper 51), revived discussion of sympatric speciation. The paradigm of allopatric speciation had been virtually unquestioned for several decades, owing largely to the enormous influence of Ernst Mayr's (1942, 1963) views on sympatric speciation. Bush presented evidence that seasonality in the production of fruit by related species of trees provided an opportunity for rapid speciation by frugivorous flies, and, by extrapolation, by many other groups of insects and other invertebrates. By thus taking at least some of the geography out of the process of speciation, Bush and those who quickly followed him shifted much attention to the genetic, physiological, and behavioral consequences of seasonal reproduction and host-shifting (see, e.g., Duffy 1996, Via 2001).

Another historically influential publication by those advocating the importance of sympatric or nearly sympatric speciation was Michael White's (1978) *Modes of Speciation.* Relying heavily on studies of chromosomal variation conducted during the 1960s and 1970s (including many of his own), White concluded that "in spite of a rather woeful lack of

evidence as to the exact nature of the genetic processes involved, it seems impossible today to deny the reality of sympatric speciation, at least in many groups of insects" (260). His review, however, argued for the *existence* of sympatric speciation, not for its frequent occurrence. To the contrary, he argued strongly for the frequent occurrence of what he termed "stasipatric speciation," similar to parapatric speciation (described below), in which a chromosomal mutation, occurring within the boundaries of the ancestral species, causes greatly reduced fecundity in the heterozygous state. He stated,

The stasipatric model of chromosomal speciation was put forward by White, Blackith, Blackith, and Cheney (1967) and White (1968) in order to explain (1) the enormous number of cases in which closely related species of animals of limited vagility have visibly different karyotypes, and (2) the large number of instances in which the assumption of strictly allopatric speciation seems unreasonable, mainly because it would force us to assume that the parent species had a geographic range too small to be plausible. Essentially, the stasipatric model envisages a widespread species generating within its range daughter species characterized by chromosomal rearrangements that play a primary role in speciation because of the diminished fecundity or viability of the heterozygotes. The daughter species are assumed to gradually extend their range at the expense of the parent species, maintaining a narrow parapatric zone of overlap at the periphery of their distribution within which hybridization leads to the production of genetically inferior individuals (usually inferior because of irregularities at meiosis). (177)

This reduction in hybrid fitness serves as an incipient isolating mechanism that allows establishment of a new species within the boundaries of the ancestral species. This means of speciation could operate over very brief periods of time, and so was often cited, for example, as a potential mechanism for the type of diversification discussed by Haffer (1969). While White's data led him to view stasipatric speciation as often involving rather small areas and very brief periods of time, his approach also heavily emphasized the geographic aspect of speciation and the view that diversification leads to allopatric species, not to an increase in local (alpha) diversity (for more recent perspectives see Otte and Endler 1989; Rieseberg 2001).

John Endler (1977), in another highly influential book, entitled *Geographic Variation, Speciation, and Clines,* argued that clinal variation in environmental conditions may often lead to parapatric speciation, especially when an abrupt ecological transition (e.g., a rapid change in soil type associated with a change in vegetation) occurs within the region of initial clinal genetic variation. Further, he argued that the genetic effects of hybridization often cannot be distinguished from the genetic effects of incipient parapatric speciation due to divergent selection, and that many documented cases of presumed hybridization following secondary contact of formerly isolated populations may in fact represent primary parapatric speciation. Because of the nearly ubiquitous presence of clinal geographic variation in natural populations of animals and plants, Endler and others have speculated that much of the diversity of species may have been generated by this means. For example, he suggested that Haffer's (1969) data on Amazonian diversity and putative refugia support parapatric speciation as readily as they do Haffer's allopatric model (Endler 1977: 168–75). In concluding his discussion of the topic, Endler emphasized that the parapatric speciation models were highly consistent with rapid speciation of the type postulated by Haffer, and went on to state that:

. . . speciation is likely to be area-dependent not only under the allopatric model . . . but also under the parapatric model. Species distributed over broad geographic ranges are more likely to be split up by changes in geomorphology and climate than species with small ranges; on the other hand, species with larger geographic ranges are more likely to have greater habitat diversity . . . and more isolation-by-distance, hence are more likely to develop clines, subspecies, and parapatric speciation than species with smaller ranges. Repeated fragmentation and concomitant interruption or reduction in gene flow may accelerate the differentiation process but is not necessary for population differentiation and speciation. (174–75)

A final development during the 1970s involved an alternative approach to the equilibrium model of MacArthur and Wilson. This approach, most widely known as "vicariance biogeography," emerged from research programs on the phylogenetic relationships of organisms and newly emerging data on plate tectonics (see, e.g., Croizat, Nelson, and Rosen 1974 [paper 40], and discussion in part 4 of this volume). Vicariance biogeographers argued that much of the diversity of organisms must be due to disruption of gene flow when continents (or other areas) fragmented owing to geological forces. Among the early and most influential publications on this topic were Donn Rosen's (e.g., Rosen 1975).

From the 1970s to the 1990s, "equilibrium" and "vicariance" biogeography dominated much of biogeography. Unfortunately, during most of the period from 1970 to 2000, many "equilibrium biogeographers" continued to ignore deep history and the process of diversification, while many "vicariance biogeographers" ignored or rejected the possibility of colonization between islands and disregarded the impact of diversification on geographic patterns of species richness. The most recent trend in the further development of historical biogeography is the call for integration (or, speaking from the perspective of the history of biogeography, the re-integration) of the vicariance and equilibrium perspectives, and the development of methodologies that allow this (e.g., Heaney 2000; Lomolino 2000a; Schluter 2001; Whittaker 1998; Zink, Blackwell-Rago, and Ronquist 2000). As a result of both the recent increase in knowledge about the tectonic history of the earth and increased knowledge of phylogenetic relationships of organisms, a more complete history of the development of regional biotas is emerging (e.g., Hall and Holloway 1998).

Perhaps the most recent approach (or, at least, the one most readily identified) to biogeographical diversification was developed in the late 1980s and 1990s in response to the availability of DNA sequence data. New technology made possible the construction of detailed phylogenies of populations within species, and allowed the development of the new field of *phylogeography*, introduced and defined by John Avise and his colleagues (e.g., Avise et al. 1987). Like many of the earlier topics that captured the imagination of a generation of biogeographers, papers on this topic now dominate several journals, leading to greatly expanded data on the geographic and phylogenetic components of diversification (see, e.g., Arbogast and Kenagy 2001, Barraclough and Nee 2001). The usually explicit emphasis of phylogeographers on the geographical and geological circumstances of diversification appear, at this early date, to have potential to add greatly to our knowledge of pattern and process in diversification.

In preparing this part, we have come to the unexpected conclusion that diversification, as a topic in its own right, has been understudied. Many of the crucial insights have been developed as an adjunct to investigation of some other topic. During the 1940s and 50s, the classic period of the Modern Synthesis, most emphasis was on documenting the role of natural selection in evolution, and later on the mechanisms of the processes of speciation, with little direct attention to phylogenetic or geographic patterns in the diversity that was produced. Discussion of speciation in the Amazon Basin by Haffer (1969) and most of those who followed dealt with the geographic circumstances of speciation, but barely addressed the impact on patterns of species richness, and the approach was largely nonphylogenetic. Indeed, we are struck by the fact that much of the literature from about 1940 to about 1980 that dealt with diversity was nearly ahistorical, including neither phylogenesis nor explicit geological perspectives or fossil data; this has been especially true of equilibrium island biogeography. Vicariance biogeography emphatically brought in phylogenetics and geological history, but has largely ignored the impact of phylogenesis (as well as dispersal and extinction) on patterns of species richness, and the evaluation of evidence and development of conceptual models often proceeded in the absence of a fossil record. It should be borne in mind that diversity at any given place is a product of the cumulative effects of phylogenesis, extinction, and invasion/colonization, all of which are influenced by diverse geological, ecological, and evolutionary factors. Yet one could identify numerous examples of rather simplistic dichotomous treatments of issues (e.g., "history versus climate," "sympatric versus allopatric speciation," "dispersal versus vicariance") that have failed to address adequately the very real complexity of processes that have influenced diversification. Further, most studies of diversification have focused on terrestrial organisms, to the exclusion of both marine and freshwater organisms, and on vertebrates, to the usual exclusion of invertebrates and plants.

Looking from the foundations of biogeography to its frontiers, we believe that the study of diversification will continue to play a central role. Integration of many factors and perspectives are likely to be essential for key new insights about diversification (and diversity) to emerge, including the impact of scale (in time and space), ecology, genetics, and earth history, for biogeography is, at heart, a broadly and intrinsically integrative field.

From *Evolution above the Species Level*
Bernard Rensch

B. Geographic Races

In contrast to the cases of strictly historical race formation there is a quite different situation to be observed when the geographic races are studied. Here we need not search for suitable examples, as geographic variation can be observed in nearly every group, and there are even extremely large numbers of geographic variants, at least in those groups of terrestrial and fresh-water animals that present a great diversity of forms. (It should be noted that [23] usually the races that have been studied have differed in morphological traits only, and that the numerous physiological variations have not yet been dealt with sufficiently.)

In many cases it has proved useful to designate large polymorphic species comprising several geographic races as 'Rassenkreise' (mosaics of races; B. Rensch, 1926, 1929, 1934). This term was introduced because strongly differentiated members of a 'Rassenkreis' do not differ less from each other than do closely related 'good species' (in the old sense) without geographic variation, and because such races when getting into secondary contact either do not hybridize or produce a more or less infertile offspring. Nevertheless, 'Rassenkreise' and species differ only by degrees, and in most cases the terms are synonymous.

The systematic study of geographic race formation has not yet been completed in any animal group, but in several cases the situation is now fairly clear, as the following data will show. When Hartert and Steinbacher finished their well-known work (1938), 1,715 geographic races of the palearctic Oscines had been described (doubtful forms not included), which were grouped into 363 'Rassenkreise' (comprising from 2 to 31 races each, with an average of 4.7 per 'Rassenkreis'). There were 153 more 'species' in the old sense which could not be grouped into 'Rassenkreise' (8.2 percent of all forms mentioned), many of which, however, represent borderline cases between race and species. A similar situation is found on other continents.

Species of larger birds with a wider distribution often covering several continents show a considerably weaker geographic variation. So the five bird families of herons (Ardeidae), storks (Ciconiidae), ibises (Ibididae), bustards (Otididae), and cranes (Gruidae) of the palearctic region show a total of only 32 races in 20 species or 'Rassenkreise' (making an average of 1.6 races per 'Rassenkreis' if the extra-palearctic races are excluded) and 24 species without any geographic race differentiation. The decrease in geographic race formation in such larger birds is probably caused by their greater vagrancy. In the species of smaller birds, migrating and nonmigrating birds also show typical differences as to their number of geographic subspecies. The palearctic families of ravens (Corvidae), tree creepers (Certhiidae), nuthatches (Sittidae), titmice (paridae), wrens (Troglodytidae), and woodpeckers (picidae), most species of which are more or less sedentary, comprise a total of 586 races in 81 'Rassenkreise' and only 34 species without geographic variations. The migratory families of shrikes (Laniidae), flycatchers (Muscicapidae s. lat.), hedge sparrows (Accentoridae), swallows (Hirundinidae), wagtails (Motacillidae), and orioles (Oriolidae) present 550 races in 173 'Rassenkreise' besides 115 species not varying geographically. Thus, the average number of species per 'Rassenkreis' is 7.2 in the sedentary and 3.2 in the migratory bird families, and the percentage of geo-

Reprinted from Bernard Rensch, *Evolution above the Species Level* (New York: Columbia University Press, 1960), with permission of the publisher.

Page numbers in the original printed version are shown in brackets where the page begins. Gaps in pagination occur when figures occupying a full page in the original have been repositioned. The list of works cited was compiled for use in this volume.

graphically nonvarying species is 5.5 in the nonmigratory and 17.3 in the migratory form.

The European mammals have not yet been studied sufficiently as to their 'racial systematics'. Up to 1937, however, the critical list by Oekland [24] enumerated 399 mammalian races in 88 'Rassenkreise' (an average of 4.5 races per 'Rassenkreis'). Unfortunately Oekland did not mention the 'species' that showed no geographic variation. They amount to 90, which is 18 percent of all forms concerned. In studying Eurasiatic and Indo-Australian bats Tate (1941 a,b) applied the 'Rassenkreis' principle successfully (up to 18 races per 'Rassenkreis'). Mertens and Müller (1940), working on the European reptiles, could enumerate 205 geographic races forming 54 'Rassenkreise' (up to 24 races per 'Rassenkreis') and 50 nonvarying species (19.6 percent). In the European amphibians they could establish 20 'Rassenkreise' comprising 62 races. Twenty-one species (25.3 percent of all forms described) did not show geographic variation. (It is interesting to note that in 1928, when the first edition of the check list was published, 26.5 percent of the reptiles and 28.4 percent of the amphibian forms were considered to be species.)

In a zoogeographic survey of the European freshwater fishes, Berg (1933) classified 207 geographic races (besides numerous subraces and ecologic races) in 64 'Rassenkreise' and 186 species without geographic variation (47 percent of the total number of forms mentioned in the study, and 40 percent of this total if the said subraces are included).

Breuning's monograph (1932–6) on the ground beetles of the genus *Carabus* enumerates 366 geographic races in Europe (Caucasian and East Russian border countries excluded) in 62 'Rassenkreise' (comprising from 1 to 42 races each) and 27 (6.9 percent) geographically nonvarying species. (Besides the geographic races mentioned by Breuning there are many types referred to as 'natio' which can well be considered as genuine geographic races with relatively small distribution areas.) Inside the German borders (northern Alps included) there are 63 races in 22 'Raissenkreise' and only 3 nonvarying species (4.6 percent). Evidently, then, these ground beetles which have lost their ability to fly and show only a very limited vagrancy generally tend to geographic race formation. And if in Breuning's survey the Asiatic forms are nearly all considered as species, this is due only to our incomplete knowledge of the systematic and faunistic characters of these types.

Quite a similar situation prevails in many other insect groups. Numerous 'Rassenkreise' have already been established as the result of studies on beetles,

butterflies, Hymenoptera (especially bumblebees and ants), Orthoptera, and others. It is certain, however, that we are far from knowing enough about the geographic variation (except in bumblebees and in some groups of butterflies). The collections often do not include material that represents the series of specimens from all possible localities, and in some cases the specialist's interpretation points in a different direction from the principle of geographic 'Rassenkreise'. In some groups, such as the Drosophilae, genetically clear-cut races cannot be separated morphologically from each other (Dobzhansky, 1951). Therefore the extent to which speciation via the geographic variation occurred in the other insect groups cannot yet be defined precisely. Geographic variation in European butterflies cannot be surveyed as completely [25] as that in Indo-Australian forms, because in describing the latter types the authors, K. Jordan *et al.*, in the great work on butterflies by Seitz (1927) applied the principle of geographic 'Rassenkreise' systematically. Among the Papilionidae, Pieridae, and Danaidae they listed 2,268 geographic races in 412 'Rassenkreise' and 283 nonvarying species (11 percent of all forms mentioned). So we may now say that in nearly all insect groups the geographic race is probably the most frequent stage of formation of new species.

That the taxonomic method applied in studying a certain animal group is quite decisive for the evaluation of the systematic situation can clearly be seen from the revision of the Cypraeidae (Prosobranchia) by F. A. and M. Schilder (1939). Here formerly only a few 'variants' that could be considered as geographic races had been described. The monograph by Schilder and Schilder revealed 279 geographic races forming 84 'Rassenkreise', and only 77 species in the former sense (21 percent of all forms mentioned). Such a surprising change in the classification of an animal group of great geological age should inspire us to caution when evaluating other animal classes in which geographic variation has not yet been studied sufficiently, if at all.

Thus, it may not be superfluous to give a brief mention of those animal groups in which geographic 'Rassenkreise' have recently been established and in which formerly only species and nongeographic 'variants' had been described: Tunicata (Eisentraut, 1926: 3 'Rassenkreise' in *Ciona* and *Botryllus*), Echinodermata (Döderlein, 1902; Engel, 1934; Heding, 1942; Vasseur, 1952; Mayr, 1954), Cephalopoda (Adam, 1941: e.g. 4 races in *Sepia officinalis*), Brachiopoda (Helmcke, 1940), Phyllopoda (Wesenberg Lund, 1904–8, in Cladocera; Colosi, 1923, in *Apus cancriformis*), Copepoda (Ekman, 1917–20, in Limnocalanus; Baldi, 1941, even in pelagic Diaptomidae within the same lake; Tonolli, 1949, in diaptomids from high altitudes; Pirocchi, 1951, in cosmopolitan Copepoda and Cladocera; Kiefer, 1952, in many African and often pelagic species), Cumacea (C. Zimmer, 1930, 6 races in *Diastylis glabra*, 3 in *D. rathkei*), Amphipoda (d'Ancona, 1942, in Italian races of *Niphargus*), Hydracarina (Viets, 1926), Scorpionidae (Meise, 1933: 8 'Rassenkreise' in the genus *Rhopalurus*), Diplopoda (Karl, 1940), Trematoda (Erhardt, 1935), Turbellaria (Benazzi, 1945), Anthozoa (Pax, 1936: 4 European and American [pacific coast] races of *Metridium senile* form one 'Rassenkreis'; Jaworski, 1938, in *Actinia*; Frenzel, 1937, in *Alcyonium*), and salt-water Spongiae (W. Arndt, 1943: 17 'Rassenkreise').

In the plant kingdom geographic race formation is not a rare phenomenon, though systematists do not yet seem to pay sufficient attention to it and sometimes get into trouble, since the recognition of geographic speciation can be difficult because of hereditary ecologic race formation and hybridization. For further study, see Von Wettstein, 1898 (e.g. *Gentiana, Euphrasia*); B. Rensch, 1929, 1939; Du Rietz, 1930 (e.g. *Pinus, Picea, Silene, Celmisia*, etc.); Geyr von Schweppenburg, 1935 (*Larix, Abies*); W. Zimmermann, 1935 (*Pulsatilla*); Schwarz, 1936 (Pinus); Stebbins, 1950; Baker, 1953 (*Armeria maritima*).

Summing up, we may state that geographic race formation has been proved [26] to occur in nearly all animal groups and in several plant groups. It is evident that the number of examples studied recently and showing this mode of speciation is rapidly increasing, and that in many groups with an abundant variety of forms the taxonomic situation is ruled by this principle. This means that the large number of existing species may be considered to be mainly the result of geographic speciation.

Such a conclusion, of course, requires that the characteristics of the geographic races be hereditary. Modification and mutation often act in the same direction, giving similar phenotypic results (phenocopy), and they—in spite of their totally different causation—often seem to react to the same environmental conditions (because modification and mutation are the two possible ways by which the organisms adapt themselves to their environment). Therefore, a safe judgment of the possible heritability of the phenotype in situ cannot generally be given. Experimental tests on geographic races have been carried out in a relatively small number of cases, but these cases dealt with very different groups of animals. And all these cases showed clear cut genotypic differences in their essential racial characteristics. This has been proved in the North American rodents of the genus *Peromyscus* (Sumner, 1930, 1932; Dice, 1935; Svilha, 1935), in the European mouse *Arvicola scherman* (Müller-Böhme, 1935), and in the European hedgehogs (*Erinaceus europaeus* and *E. roumanicus*), which must be considered as strongly differing races (Herter, 1935). It was shown to be true in the European and American bison (*Bison europaeus* and *B. americanus;* the latter is a borderline case, see below) (Iwanow and Philipschenko, 1916), in red and silver foxes (*Vulpes*) (e.g. Demoll, 1930, p. 42), in the races of the African ostrich, *Struthio camelus,* (Duerden, 1919), in the races of the parrot, *Agapornis* (Duncker, 1929), in doves of the genus *Streptopelia* (Whitman, 1919), and in the races of the pheasant *Phasianus colchicus* (e.g. Cronau, 1902; Poll, 1911; Thomas and J. S. Huxley, 1927; and others: some races are still termed species). Additional examples are found in races of frogs (Porter, 1941; Moore, 1954) and of newts (Wolterstorff and Radovanovic, 1938; Callan and Spurway, 1951; Spurway, 1953); in races of the lady beetle *Epilachna* (K. Zimmermann, 1934, 1936); in bees and bumblebees (Armbruster, Nachtsheim, and Roemer, 1917); in flies, genus *Volucella* (Gabritschewsky, 1924); in the geographic races of butterflies, genus *Lymantria* (R. Goldschmidt, 1924–33), *Callimorpha* (R. Goldschmidt, 1924), *Spilosoma* (Federley, 1920; R. Goldschmidt, 1924), *Phragmatobia* (Seiler, 1925), *Colias* (Lorković, 1928), *Dicranura* (Federley, 1937: here the numbers of chromosomes vary in different races), *Leucodonta* (Suomalainen, 1941), *Zygaena* (Bovey, 1941), and *Pieris* (Petersen, 1947); in Diptera, genus *Drosophila* (in which the detection of racial differences is based on the pattern of the giant chromosomes, since in other structural and morphological details there are only slight differences: Dobzhansky, 1950); and finally, in races of the land snail, *Murella* (B. Rensch, 1937).

This short survey of the relevant literature could be considerably increased by carefully examining all entomological and game-management periodicals, [27] and those published by zoological gardens (for example, see Beninde, 1940, on hybridization in the red deer). Furthermore, one should mention the observations on the constancy of characteristics in different races kept under equal environmental conditions, as in all zoological gardens, where the strikingly different races of giraffes, zebras, lions, ostriches, cassowaries, and pheasants keep their respective differences through many generations. How far, however, modificatory alterations can proceed is shown, for instance, by the coregonoid fishes (*Coregonus*), which, if they have been transported from their native lake to another one, develop quite a different pattern of gill structure after a few generations. (Surbeck, 1920; Kreitmann, 1927; Thienemann, 1928).

Another proof of the heritability of geographic racial characteristics can be given if two races get into secondary contact under natural conditions (hybrid populations). This can be studied in central Europe in the hooded and the carrion crow (*Corvus corone corone* and *C. corone cornix*), and in the races of longtailed titmice (*Aegithalos c. caudatus* and *Ae. c. europaeus*); in southern Europe in the sparrows *Passer domesticus* and *Passer hispaniolensis;* in North America in two races of *Melospiza melodia* near San Francisco Bay (A. H. Miller, 1949), in two races of *Junco caniceps* in Arizona (A. H. Miller, 1941), and in other forms. Such cases are very numerous (for detailed study see also Meise, 1928, 1936; G. Dementiev, 1936, 1938; B. Rensch, 1936, 1945, pp. 42–3; Mayr, 1942; Johannsen, 1944). A typical feature of such regions with hybrid populations due to secondary contact is the extent of variation, which comprises the variability of both contacting races.

Finally, heritability of geographic characters can be considered to be fairly certain in the numerous cases where marked racial boundaries are not accompanied by any conspicuous environmental alterations, as is so often found where uniform races are distributed over a large area with strongly differing environmental conditions.

All such experiments and studies on populations have emphasized the fact that the characteristics used by systematists as typical features of geographic races generally have a genotypic background. Nevertheless, one should never forget that in all cases which have not been analyzed genetically one can deduce the possible heritability of characters only by analogy, although the probability may be fairly great. Concerning the problems dealt with in the present chapter we may state that geographic race formation based on genotypic variation can be a prelude to speciation.

In most cases geographic races differ from each other in several genes. By the formation of various geographic races, each of which is a more or less harmonious unit, a species becomes a very complicated genetic system capable of adapting in various directions, and thus possessing an enormous evolutionary plasticity (cf. Wright, 1939). It is apparently because of this plasticity that geographic variation became such a general phenomenon in speciation.

A further requirement for the assumption of geographic races as the most [28] frequent forerunner of new species is to prove that there are intermediate stages between geographic race and species. This is of importance because in some recent papers the opinion is favored that geographic variation should be considered as a secondary branching of species which originated due to other causes (see especially R. Goldschmidt, 1935, 1940, 1952; compare also Chapter 3, G of this book). Concerning this opinion, it should be noted that in geographic variation fundamentally the same characters which are typical of a species can be altered, and that not only phylogenetically nonessential features (such as differences in coloration, size, and proportion) are concerned. As I have pointed out in more detail (B. Rensch, 1929, 1934, 1943) many geographic races of mammals differ in skull structure, in dental formula, in the number of toes (races of the rodent *Dipodomys heermani* have four or five toes: Dahl, 1939), in relative size of organs, and in number of young. Bird races show different osteological structures, organ proportions, egg colors, structure of eggshells and pores, number of eggs, songs, and fixed hereditary requirements as to habitat and migration. Races of fishes differ in the number of vertebral and fin bones and their migration habits. Insect races can differ in the number of generations produced per year and (often strongly) in shape and structure of genital appendages and organs. Occasionally the same holds true in Mollusca, where the shape of the radula can also differ between the races, etc. (Rensch, 1943; Mayr, 1943).

Extreme members of geographic 'Rassenkreise' (species) may show such striking differences in the characteristics mentioned above that, judging them morphologically, one can consider them as 'good species'. (Hence, for a more precise classification the term 'Rassenkreis' is not superfluous.) It can be assumed that, in some cases, extreme races would behave like good species, i.e. that normally they would not hybridize if they should come into secondary contact in the same habitat, as is the case in the races *Parus m. major* and *P. m. wladiwostokensis* of the great tit in the Amur area (B. Rensch, 1934, 1945). This excellent example is again presented in a detailed map of distribution which was drawn according to the late revision of the 'Rassenkreis' by Delacour and Vaurie (1950). It can be seen from Figure 3 that in Persia the European-Asiatic forms with green

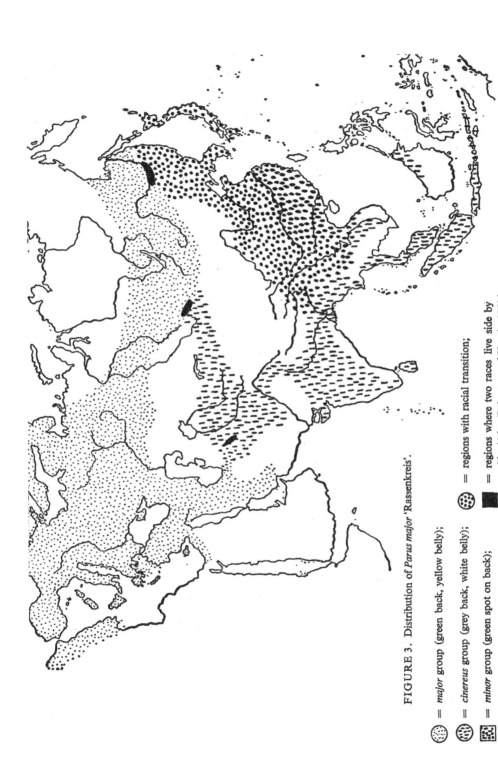

FIGURE 3. Distribution of *Parus major* 'Rassenkreis'.

◌ = *major* group (green back, yellow belly);

◉ = *cinereus* group (grey back, white belly);

▨ = *minor* group (green spot on back);

⊛ = regions with racial transition;

◼ = regions where two races live side by side. (After Delacour and Vaurie, 1950.)

backs and yellow bellies gradually shade into the South-Asiatic type with grey backs and white bellies (*intermedius*), which type turns into the intermediate form *commixtus* in China, and finally into the East-Asiatic races with pale green backs and white bellies. In the central Amur region (black on the map) one of the races concerned (*minor*) lives side by side as a 'good species' with the green-backed, yellow-bellied race of the *major* group, which probably came from the west in a secondary shifting of distribution. A similar behavior of two such races has been observed in some cases when the races *intermedius* and *bokharensis* or *turkestanicus* and *major* came into contact (black areas on the map).

The same type of non-mixing contact in some places only, as described above, has also been observed in the races *Pyrrhula p. pyrrhula* and *P. p. cineracea* of the bullfinch in the Sajan, Altai, and Kusnetzk Alatau [30] occasionally do hybridizations occur here, but a hybrid race lives in the Jenessei area: see Johannsen, 1944), in the warbler races *Phylloscopus trochiloides viridanus* and *Ph. tr. plumbeitarsus* of the upper Jenessei region (Ticehurst, 1938; Mayr, 1942), in the Mexican towhees, *Pipilo erythrophthalmus* and *P. ocai* (Sibley, 1954), in the mice *Peromyscus maniculatus artemisiae* and *P. m. osgoodi* on the eastern border of the Glacier National Park of Montana, in *P. m. austerus* and *P. m. areas* of the Puget Sound region in Washington, in *P. m. bairdii* and *P. m. gracilis* of Michigan (ecologic isolation, hybrids rare: Dice, 1931). According to Stegmann's investigations, the gulls *Larus argentatus* and *L.fuscus* living side by side as 'species' in a large part of Europe can be looked upon as races of a large circumpolar 'Rassenkreis', in which they represent the clear-cut and well-defined end-links overlapping in a certain part of their ring-shaped area (Europe) in secondary contact. Stresemann and N. W. Timoféeff-Ressovsky (1947) suggested classifying the nineteen dubious forms of this borderline case into three or four species (compare also Voipio, 1954). Finally, the 'Rassenkreis' of the butterfly *Junonia lavinia* in North and South America presents a clear racial organization, while in the West Indies the southern and northern types live side by side without mixing.

Besides such extreme racial differentiations within a 'Rassenkreis', there are many other borderline cases between geographic race and species. They are represented by the frequent examples which the systematist cannot clearly define as either a race or a species. Frequently forms are concerned in which the

morphologic divergence is so marked that one might just be able to term them species, but their distribution areas replace each other in various geographic regions. For instance, this is true in some island forms (see Figure 4), in some marine 'twin species' on both sides of the Panama region, or in closely related species and 'Rassenkreise' in the northern parts of the Old and New World, such as red deer and wapiti, European and American bison, stone marten and pine marten, European and North American badger, etc. I have designated such cases as 'Artenkreise' (mosaics of species) or geographic subgenera (B. Rensch, 1928, 1929, 1934), and more recently Mayr named them superspecies. In other cases, young species replacing each other occupy slightly overlapping areas but show hybrid populations only in some spots—similar to the above-mentioned races—whereas in other regions the forms live side by side without mixing. This is the case in the sparrows *Passer domesticus* and *P. hispaniolensis:* hybrids live only in Italy, Algeria, Corsica, and Crete; in the titmice *Porus caeruleus* and *P. cyaneus;* in the Egyptian desert snails *Eremina desertorum* and *E. hasselquisti;* and in the arboreal snails *Amphidromus contrarius* and *A. reflexilabris* of the Lesser Sunda Island of Timor.

Another and more advanced stage of such borderline cases can be seen in forms that are relatively young species and still fairly similar to each other, replacing each other in different geographic areas, but not showing any hybridization in the regions of secondary contact (examples: pairs of species which differentiated during glacial periods in the refuges of western and [31] southeastern Europe or western Asia and now live side by side in central Europe, such as the woodpeckers *Picus viridis* and *P. canus,* the nightingales *Luscinia megarhynchos* and *L. luscinia,* the ground beetles *Carabus purpurascens* and *C. violaceus,* etc. For further examples of such borderline cases, see B. Rensch, 1929, 1934, 1943). In the latter cases the physiologic distance probably is relatively small, and in captivity hybrids can be obtained, as is known from *Luscinia megarhynchos* × *L. luscinia.* Dice (1940) also published such an example: the rodents *Peromyscus leucopus* and *P. gossypianus* are very closely related but clearly separated species living in an area of secondary contact—Dismal Swamp, Virginia—without hybridization. In captivity, however, they can be hybridized and will yield fertile offspring. Hence we see that the important thing in speciation is not whether hybridization is possible, but whether it does or does

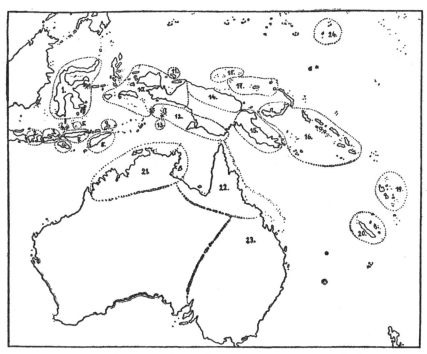

FIGURE 4. Range of races in the 'Rassenkreis' of the parrot *Trichoglossus ornatus*:
1. *T. o. ornatus*; 2. *mitchelli*; 3. *forsteri*; 4. *djampeanus*; 5. *stresemanni*; 6. *fortis*; 7. *weberi*;
8. *capistratus*; 9. *flavotectus*; 10. *haematodus*; 11. *rodenbergi*; 12. *nigrogularis*; 13. *brooki*; 14.
intermedius; 15. *micropteryx*; 16. *aberrans*; 17. *flavicans*; 18. *nesophilus*; 19. *massena*;
20. *deplanchii*; 21. *rubritorquis*; 22. *septentrionalis*; 23. *moluccanus*; 24. *rubiginosus*.

not take place freely in nature. From this point of view the borderline cases last mentioned above can really be considered species.

Moreover, there is evidence that the fertility of hybrids of extreme geographic races is decreased. From experiments on the hybrids of the strongly **[32]** different pheasant races *Phasianus colchicus formosanus* (Formosa) × *Ph. c. versicolor* (Japan) by Thomas and J. S. Huxley (1927), it is known that only 87 percent of the hybrid eggs were fertile, and only half the chickens reached maturity. Hybrids of the butterfly races *Smerinthus populi populi* (Europe) × *S. p. austauti* (northwest Africa) proved to be partly sterile in both sexes in F_1 and F_2 (Standfuss, 1909). In hybridizing markedly different geographic races of the moths *Lymantria dispar* or *Lasiocampa quercus,* the physiological disturbance becomes manifest in the appearance of intersexes (see R. Goldschmidt, 1920). Analyzing the hybrids of extreme newt races of the *Triturus cristatus* 'Rassenkreis,' Callan and Spurway (1951) brought forward the important evidence that chromosomal

conjugation during meiosis is severely disturbed, and that inversions sometimes occur. Studying hybrids of the newts *Pleurodeles waltl* and *P. hagenmülleri* (belonging to an 'Artenkreis' in Spain and Tunisia) Steiner (1942, 1945) found lethal segregants in the F_2 generation. Hybrids of geographic races of *Rana pipiens* from more distant regions show a retardation in rate of development and morphological defects. Hybridization between races from geographically extreme regions results in a high degree of hybrid inviability (Moore, 1946). Crosses of the frog races *Hyla aurea aurea* from New South Wales and *H. au. raniformis* from southwest Australia resulted in complete hybrid inviability. Crosses of the similarly distributed races of *Crinia signifera* gave more or less the same results (Moore, 1954). On the other hand, Breider (1936) found that fertile hybrids can be obtained from fish species (genus *Limia*) which replace each other geographically.

It is important that such borderline cases can be found in all animal groups showing geographic variation. This, however, is revealed only if the systematist concerned pays special attention to these intermediate forms and labels them accordingly, because terminologically they are counted as 'still being' geographic races or 'already being' good species. Working on the birds of the Lesser Sunda Islands of Lombok, Sumbawa, and Flores (1931), I could show 47 such borderline cases beside 160 definite 'Rassenkreise'. Stresemann (1931) when reviewing the white-eyes (Zosteropidae) found six borderline cases in addition to 22 'Rassenkreise' and 30 undivided species. Among the Fringillidae of western Siberia, Johannsen (1944) found six such cases ('species-groups') in addition to 38 'Rassenkreise' and species, and among the Corvidae there were three such examples in addition to eight 'Rassenkreise' and species. From these few examples it may be seen that borderline cases of geographic race and species are as frequent as could be expected in the course of a speciation which is due mainly to geographic isolation. (One should also study the' Artenkreise' of Tenebrionidae of the Sahara analyzed by Knoerzer, 1940.) There are further examples of borderline cases between race and 'good species' in the works by B. Rensch (1928, 1929, 1931, 1934 a,b, 1937, 1943); Meise (1928); Herre and Raviel (1939); Fitch (1940); Von Boetticher (1940, 1941, 1944); Ripley and Birckhead (1942); Mayr (1942); Hubbs (1943); Hubbs and R. R. Miller (1943); Hovanitz (1949); Nobis [33] (1949); De Lattin (1949); Herter (1954); Kauri (1954); E. O. Wilson (1955), etc.

Considering the geographic race formation as the most frequent and, in many animal groups, quite usual precursor of a species, we still have to ask whether Recent animal forms without geographic variation (species in the older sense) have passed stages of a *different* kind. A sure answer can rarely be given. It is possible that recent species are the only remains of former 'Rassenkreise'. Another means of evading the geographic variation—especially typical for geologically old species—may have been the capacity for modificatory adaptation to changing environmental conditions of the habitats. Many small fresh-water organisms capable of forming permanent cysts, statoblasts, spores or eggs, etc., and of modificatory changes in growth and structure, show such adaptive mechanisms. Such species may have been developed by historical alteration alone or by a combination of ecological and historical race formation.

Furthermore, it is important to ask which modes of selection are primarily effective in geographic race

formation. This can be answered fairly well by studying the parallelism between certain characters and some environmental factors. Thus, we may establish three evolutionary types of geographic races: (1) those in which the characters do not show any corresponding relations to their environment and may have emerged at random; (2) those in which the characters show a gradual succession of steps according to the geographic distribution, but in which selection due to environmental factors is unlikely; and (3) those in which the characters and certain environmental factors correspond and in which the characters probably resulted from natural selection. Let us have a brief look at these three types.

1. Undirected geographic variation can be observed mainly in those cases where the effect of isolation is stronger than that of selection, i.e. where 'mutation pressure' is more effective than 'selection pressure', or where the range of variation is reduced due to intense fluctuations of the population size or to the establishment of new populations by a few specimens. Hence, random race formation is especially frequent in polytypic species scattered over a range of islands with nearly the same environmental conditions (such as tropical islands). So the conspicuously colored birds, such as fruit doves, parrots, pittas, flower peekers (Dicaeidae), and so forth, vary quite remarkably as to the extent and patterns of colors, and there is no apparent correlation of these differences with any obvious environmental factor on the various isles. Thus, for instance, in the 'Rassenkreis' of the parrot *Trichoglossus ornatus* (*Tr. haematod,* Peters, 1937) ranging from Celebes and Bali via New Guinea and its surroundings to New Caledonia and Australia (Figure 4), we find the race *mitchelli* on the Lesser Sunda Islands with a dark-purple head, a red breast, and a purple-black belly with green lateral parts. On the neighboring Isle of Sumbawa (race *forsteri*) only the center of the belly shows dark-purple color, and the flanks are yellow. In the race *weberi* inhabiting [34] the Isle of Flores, red colors are totally lacking in the plumage, and the belly is yellow-green. On neighboring Timor we find the race *capistratus* with an orange-yellow breast and a blackish-green belly. On the Isle of Celebes, in the race *ornatus* the dark-purple color of the head is bordered by a red neck stripe with black scales, the red breast feathers show black edges, and the belly is dark green, so that this markedly different form is often considered to be a species. Probably all these different color patterns have the same pro-

tective function and camouflage the bird equally well when it feeds between red blossoms (*Erythrina* and others). So it is apparently of no importance at all whether a certain spot of the plumage is red, green, or yellow. Hence, it seems that in these cases either random mutation gave rise to new races without any auxiliary effect of selection, or the color variations are the by-product of pleiotropic genes, meaning that color is linked with a trait subject to certain factors of selection. In this context it is remarkable that in the one extreme race (*weberi*, see above) zooerythrin is totally lacking, while in the other extreme this red pigment is so intense that it bars all green and yellow shades. It is understandable that systematists consider this dark brown and red bird as a species of its own (*Tr. rubiginosus*), especially as it inhabits the peripheral regions of the *Trichoglossus* range, i.e. the Carolines (a borderline case). Quite instructive, too, is the synopsis of the color characters in the races of the shrike *Pachycephala pectoralis* of the Melanesian Archipelago, the races of *Monarcha castaneoventris* of the Solomon Islands, or the south-Asiatic races of *Microscelis leucocephalus*, given by Mayr (1932, 1942). In all these cases, the

groups. Thus, four races of the deer mouse *Peromyscus maniculatus* inhabiting British Columbia may be considered as the possible combinations of three different characters, two of which are linked:

1. Relatively short tail + relatively large skull + dark coloration = race of Bowen Island.
2. Relatively short tail + relatively large skull + light coloration = race of Satura Island.
3. Relatively long tail + relatively small skull + dark coloration = race on the continent.
4. Relatively long tail + relativey small skull + light coloration = race of Vancouver Island (Engels, 1936).

Mertens (1931), in a monograph on the 'Rassenkreis' of the lizard *Ablepharus boutonii*, comprising more than 40 races from Hawaii and Australia to East Africa and the Bonin Islands, concludes that this subspeciation occurred by random mutation and that the combination of 'bipolar' characteristics often resulted in certain extreme forms, so that there are giant types of 47–54 mm. in length and dwarfs of about 35–38 mm, melanic and pale [35] forms, spec-

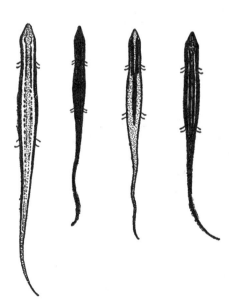

FIGURE 5. Geographic races of the lizard *Ablepharus boutonii* showing extreme variants due to random mutation. Left to right: giant race from the Isle of Juan de Nova (*caudarus*), dwarf race from Viti Levu (*eximius*), race from Lombok (*cursor*), race from an unknown locality, probably an East African island (*quinquetaeniatus*). (After Mertens.)

races originated from different combinations of alternatively varying characters.

Such transformations of racial characteristics, apparently not correlated at all with any environmental factor, may also be found in some other animal

imens with coarse or fine stripes (Figure 5), with a flat or a high head, with 30–34 lines of scales or with only 20–22 lines. None of these varying characters reveals any correlation whatever with environmental factors.

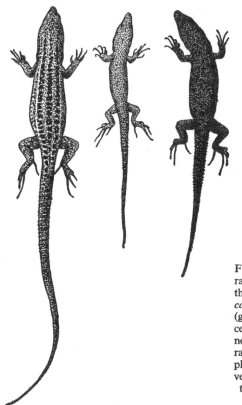

FIGURE 6. Three extreme races of *Lacerta pityusensis* from the Pityusic Islands. Left: Race *canensis* from Cana, east of Ibiza (green with brown lateral sides); center: race *grueni* from Trocados near Ibiza (sand-colored); right: race *maluguerorum* from Bleda plana near Ibiza (black with blue ventral side). (After M. Eisentraut; two-thirds natural size.)

Some of the bird mutants analyzed by Stresemann (1926) also show random color variations independent of environment. The same applies to the polytypic species of the ground beetle *Carabus monilis* ranging from France and England to southern Russia and southeastern Europe, and displaying quite a divergent surface structure of elytra in the various races. In *preyssleri* (Bohemia, Silesia, and northern Hungary) the surface of the *elytra* is practically smooth; in other races it shows a relief of chains and longitudinal stripes (primary, secondary, and tertiary intervals) of a more (race *mollilis*) or less (races *illigeri* from Bosnia, *semetrica* from Serbia, Banat) distinct character. This sculpture may also be variable (Rumanian race *kollari*), and so on.

Judging from the cases mentioned here, one might raise the objection that characters of only minor evolutionary importance are concerned. We should not forget, however, that quite frequently anatomical random differences also arise, capable of serving as essential steps in speciation. This may be seen in the genitalia of Carabidae and Sphingidae. In the latter family, K. Jordan (1905) found differences in the copulative organs in 131 out of 276 geographic races. In some cases, as in *Hyloicus pinastri*, no differences of coloration or wing venation are correlated with these differences of the genital appendages (K. Jordan, 1931). In this context mention should be made of the 'Arten-' and 'Rassenkreise' of the *Hipparchia semele* group that have been analyzed recently by De Lattin (1949). These forms, often closely resembling each other and possibly representing one single 'Artenkreis', differ markedly in the genital [36] structures of both sexes. Two sympatric forms of the northern and central Balkans (*H. semele danae* and *H. aristaeus senthes*) which did not differ in their wing patterns revealed marked differences in their genital structures.

We cannot properly estimate the share of such random variation, independent of factors of environment, in the formation of geographic races in some animal groups. All we can state is that apparently in colorful birds of the tropics this random race formation is more frequent than in birds of the temperate zones, and in polytypic species ranging from the colder regions to the tropics. In these latter forms, the race formation by characters parallel with some environmental factors is definitely more conspicuous than in tropical birds. This is true at least as regards coloration, relative length of certain parts of the plumage, and body size. In the ground beetles of the genus *Carabus* already referred to above, the racial variation seems independent of environment in types from similar climatic environment, e.g. from the Mediterranean countries. But in forms living in different climatic zones, many racial differences prove to be correlated with environment: length and proportions of legs and antennae, relative height of elytra, and coloration and sculpture of the integument to some extent, whereas the structures of the genital appendages are exempt (B. Rensch, 1934). It seems possible that such differences in the type of race formation also exist in mammals, fishes, butterflies, other groups of beetles, and some other animal groups. In other cases, too, the difference between characters dependent on or independent of the environment is rather striking. According to Hellmich (1934 *a,b*), in the lizards of the genus *Liolaemus* from Chile, geographic variation is correlated with environment as regards the proportions of tail and legs, the relative width of the body, the number of scales, the feeding habits (percentage of vegetable food), and the reproductive and escape instincts. There is no correlation with environment as far as the special arrangement of scales on the head or special sexual dimorphisms are concerned.

If very small areas, e.g. small islands, are invaded, usually the first population is small in number, and hence these few individuals cannot represent the whole gene pool of the species. Thus a separate race will originate even without any further processes of selection, as variability will be reduced and homozygosity increased in a very short time. Such splinter races frequently show certain extreme characters in size or pattern, characters which often did not show up at all in the larger former range, as, due to the continuous panmixia, the formation of extreme traits was invariably prevented there. Examples of such 'new' and extreme characters are provided by the types of the 'Rassenkreis' *Ablepharus boutonii* (see above) and the polytypic species of *Lacerta lilfordi* and *L. pityusensis* (Figure 6) inhabiting the small islands near the Balearic Islands. Some of these forms are extremely small or extremely large, especially quick or remarkably slow, conspicuously slender or definitely plump, and often extremely dark, if

not totally black, on the dorsal side (Eisentraut, 1929). In his thorough synopsis of these island variants, Eisentraut (1950) does not discuss the possibility that the extreme characters, with [37] the exception of the black color, were brought about by an increase of random homozygosity due to the small populations (Sewall Wright effect).

In lizards of small islands, melanism is a frequent phenomenon. But it has not been clarified whether this extreme coloration is only the result of a homozygosity in a small population, or whether it is the result of selection (pigmentation serving as a means of protection from irradiation: Kramer, 1949). Examples of such cases of island melanisms have been provided by T. Eimer (1874) and Kramer (1949): Faraglioni Rocks and some islets near Capri; by Mertens (1924): rocky islands of Kamik and Melisello near Lissa in the Adriatic Sea, the Isle of Filfalo near Malta and the Glenan Isles off Brittany; by Eisentraut (1925, 1930, 1950): Balearic Islands; by Kramer and Mertens (1938): some islets off the Istrian coast; and by Mertens (1932): Isola Madre in Lago Maggiore. Melanism may be supposed to be a means of protection from insolation, as there are further cases of island melanism in other animal groups. The terrestrial snails *Helix cincta* and *Otala vermiculata* on some islets off the Istrian coast are definitely melanic variants (La Figarola Grande near Rovigno: B. Rensch, 1928; Figure 7). In evaluating such examples of variation possibly correlated with

FIGURE 7. Above: left, *Helix cincta cincta* from the Dalmatian mainland; right, *H. c. melanotica* from the island La Figarola Grande. Below: left, *Otala verm. vermiculata* from the Dalmatian mainland; right, *O. v. figarolae* from La Figarola Grande. (After B. Rensch, 1928; two-thirds natural size).

environment, we should, however, account [38] for the fact that, due to the reduced gene pool of the small populations, certain characters, especially melanism, which so often is dominant, may be found quite frequently. In the same manner other parallel characters of related species may emerge in consequence of the reduced gene pool of small populations. In the polytypic species of the land snail *Murella muralis* from western Sicily, for instance, the 'normal' shells show a smooth surface and a more or less spherical shape, whereas in some races inhabiting places that were separate islets in former times (during the late Tertiary) the shells are flat and display solid ridges on their surface. Such alternative racial characters—spherical and smooth or flat and sculptured shells—have also evolved in the related polytypic species of *Tyrrheniberus villica* from Sardinia, *Rossmaessleria subscabriuscula* from northern Morocco, *Iberus gualterianus* from eastern Spain, *Eremina hasselquisti* from the Libyan desert, and in some forms of *Levantina* (B. Rensch, 1937. See Figure 68). Such limitations of original random variation and parallel subspeciation may easily be supposed to be evidence of definite evolutionary trends which, however, can hardly be true (compare Chapter 6, B III, on the rules of transspecific changes of structural types).

(2) Geographically graded variation may occur without selection by the successive loss of alleles in the course of migrations and expansions of small peripheral populations or even single individuals (compare the statements on population fluctuations on p. 9). The most direct consequence of such migrations

is a marked reduction of variability in the forms concerned. I could demonstrate this as highly probable in the terrestrial snail *Papuina wiegmanni* from New Britain of the Bismarck Archipelago (B. Rensch, 1939). Apparently this polytypic species spread from New Guinea, as the race *P. w. kubaryi* and some related races and species inhabit that island. The remotest part of New Britain (about one-third) has not yet been reached by the migrations, although the habitat as such would perfectly suit the snails. On the southern coast of New Britain facing New Guinea, we find a race with a relatively large range of variation (*P. w. wiegmanni*). Here the normal form with two black longitudinal stripes is found, but there are also variants in which these stripes are reduced to two cuneiform spots or are united to form one broad stripe. In other variants of this region the whole shell is covered by this black stripe, while in still others—with the normal pattern of the two stripes—the lip is [39] white instead of black or dark brown (Figure 8). The easternmost part of the southern coast, including the regions between Wide Bay and Cape Quoy, which obviously have been invaded last, is inhabited by only one variant with two normal bands (*P. w. disjuncta*). On the north coast the situation is quite similar: here the common type is *conjuncta*, in which the two dark bands usually unite with the dark lip. In the center of its range (near Talassea) this type is accompanied by a purely white variant with hyaline bands (frequency of this type: about 15 percent), and towards the eastern end of the range only one type is to be found (Figure 9).

FIGURE 8. Arboreal snail *Papuina wiegmanni* from New Britain. Above: three extreme variants of *P. wiegm. wiegmanni*; below: left, *P. w. disjuncta*; right, *P. w. conjuncta*. (After I. Rensch, 1934.)

FIGURE 9. Range of *Papuina wiegmanni* on New Britain.

▨ = Probably inhabited regions.

● = *P. wiegm. wiegmanni* with the full range of variability.

◉ = *P. w. conjuncta.*

◑ = *P.w. disjuncta.*

Arrows indicate probable direction of spreading. (After B. Rensch, 1939.)

As far as polygenic traits are concerned, successive loss of alleles in the course of continuous spreading may result in a geographically graded variation of a character. Such genotypic variation without any influence of selection was termed 'elimination' by Reinig (1938). Reinig presented various examples of such happenings and tried to prove that—especially in warm-blooded vertebrates—the rules of climatic variation (Bergmann's, Allen's, and Gloger's Rules) were the result of such 'elimination', as in these cases there is a successive gradation of a character, for instance, body size, with geographic latitude. For several reasons, however, I was convinced of the effectiveness of climatic selection in the formation of such gradations, which led to the establishment of the rules concerning the correlation of body size with climate. In a debate with Reinig (Rensch, 1938; Reinig, 1938, Rensch, 1939), I tried to demonstrate that several examples cited by Reinig were not relevant and that others were of doubtful validity. And there are cases in which the largest race will be found in the areas colonized last by the population concerned. Moreover, examples have been found providing ample proof of the effects [40] of climatic selection (see the next paragraph). According to my opinion, concluding the debate, only certain cases remain which are—for the time being—more reasonably comprehensible on the principles of elimination than of selection (compare also Henke, 1938).

To test the hypothesis of elimination, N. W. Timoféeff-Ressovsky (1940) suggested the special study of those cases in which the spreading occurred within historical times and in which its course can be traced fairly precisely. Thus, he studied the yellow-breasted bunting (*Emberiza aureola*) in its westward course in Russia during the last century. This species, however, does not support the hypothesis of elimination, as body size and similar measurements are not decreased in those specimens inhabiting the areas conquered last. On the contrary, there is a distinct increase of these measurements in the types from the 'youngest' areas of the range, and this is well in accord with Bergmann's Rule, and hence it may be regarded as the result of selection by climatic factors. I myself (1941) analyzed the spreading of the serin (*Serinus canaria serinus*) that has occurred since about 1800 and is still under way in central Europe. Here, too, there is no decrease of size and relevant measurements in the types inhabiting the recently occupied parts of the range (central, southern and southeastern Germany), but there is a slight

(though not statistically valid) increase in Franconia and the central Rhine areas (Figure 10).

In five cases of Fenno-Scandinavian butterflies, Petersen (1947) could demonstrate that in relatively young postglacial distributions no decrease of size is detectable and that there are geographic gradations of a different kind, thus contradicting Reinig's hypothesis of elimination.

Nor can the successive gradation of melanin pigmentation (Gloger's Rule) have been brought about by successive losses of alleles in the course of postglacial distributions, as it is impossible to comprehend why the coloration towards the northeast should tend to a gradual loss of brown pigments (phaeomelanins) and finally to a complete disappearance of the eumelanins, while towards the northwest of Europe (England) an increase of both types of melanins should occur. As the gradation is parallel and correlated with temperature and humidity, a much more reasonable interpretation is provided by the assumption that climatic selection was the important effect. There are further cases and climatic rules, such as the egg rule, that cannot be interpreted at all by the idea of elimination, as those bird races inhabiting young postglacial areas of their range in the temperate and cooler zones tend to lay more eggs per clutch than their southern relatives.

Quite frequently, races of warm-blooded animals inhabiting high altitudes show a larger body size than their relatives in closely neighboring valleys (for details, compare the findings on birds from western China and eastern Tibet by E. Schafer and Meyer de Schauensee, 1939). This fact, too, seems to contradict elimination, as according to Reinig's opinion a decrease of size was to be expected, and as on the short way from the valley to the higher elevation there could hardly be such a general effect of elimination. Hence, it is more [41] probable that there was a selection of size variants according to Bergmann's Rule. The variability of shape, color, and size of wings, and of genital appendages (vulva and aedeagus) in races of butterflies is decreased in some types of the high mountains, as stated by Eller (1936, 1937) in *Papilio machaon sikkimensis, P. m. ladakensis*, and others. But here, too, elimination can hardly be held responsible, as in the (spatially) short process of spreading from the valleys to the mountains, climatic ejection of preadapted variants or increased homozygosity due to the small populations of the high altitude habitats is more probable than loss of alleles in the course of elimination.

Elimination might be suspected in European

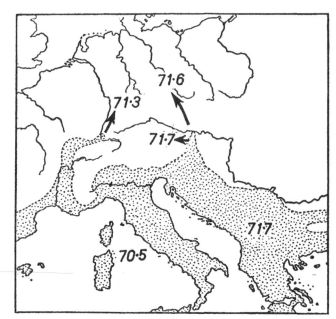

FIGURE 10. Average length of wings in populations of serins (♂).
Figures pertain to the populations from the countries of the Western
Mediterranean; from Franconia and the central Rhine regions; from
upper and lower Bavaria to upper Austria; and from Bohemia, Saxony to
western Prussia. Dotted: the original area inhabited before 1800. Arrows
indicate direction of spreading. (After B. Rensch, 1941.)

races of the ground beetles, genus *Carabus,* as the general variability decreases towards the northern part of the range, i.e. in places occupied last in the course of postglacial migration. (Tables 1 and 2). In some species, such as *C. ullrichi, C. auronitens, and C. nemoralis,* the number of individual color variants also decreases towards the north. But in this latter case we are not sure whether elimination is responsible, as the reduction of color variability is also correlated with climatic conditions (see below). In this context a statement by Voipio (1950) is of **[42]** interest. He demonstrates an increase of variability ('polymorphization') in warm-blooded animals, especially on the periphery of their range in northern Finland. Quite correctly the author attributes this phenomenon to the smaller size of the different northernmost populations.

[43] Summarizing our findings on the pro and contra of elimination, we may state that occasionally a successive gradation of a character may originate from a loss of alleles without any effects of selection, but that in most cases only the total range of variability will be reduced successively by elimination. This type of subspeciation is most likely in those cases where the differentiating traits do not depend

upon climatic conditions or similar factors of environment. Usually, however, 'elimination pressure proper' will be much less effective than selection pressure (see also Ford, 1949, p. 312).

(3) Very much commoner than the types of race formation just discussed is the formation of geographic races by selection, which can be recognized by the fact that an inherited character varies parallel with factors of the environment. It is quite obvious that in the course of spreading the preadapted variants will be favored by the new environment and that selection will act continuously upon any new mutations that may emerge after the population has settled in the new habitat. Frequently it is difficult to trace the effects of such processes of selection, as some effective environmental factors may be unknown or not fully known (e.g. certain enemies, diseases, difficulties of getting suitable food, and so on). It is much easier properly to evaluate the effect of inorganic factors of the environment, especially if they are varying gradually. In such cases the racial differences of the subspecies will also form gradients. J. S. Huxley's term, 'clines' (1939), referring to such gradients, has now come into general use in the literature (Kiriakoff, 1947; Braestrup, 1945).

The study of whole animal groups from this point of view leads to the establishment of certain rules of climatic character gradients. In each case, of course, the percentage of exceptions from the rule concerned must be taken into consideration (B. Rensch, 1929, and limbs (Allen's Rule, partly though not totally due to allometric shifts and hence a consequence of Bergmann's Rule), reduction of phaeomelanins and finally also of eumelanins (Gloger's Rule), increase of relative length of hair and of number of wool-

TABLE 1. Range of variability of absolute measurements in the ground beetles *Carabus cor. coriaceus* from central Europe and *C. cor. cerisyi* from Greece and Asia Minor in mm. and as percent of average measurements taken from 31 males. Decrease of variability indicated by −, increase by +. (After B. Rensch, 1943.)

	Range of variability in *Carabus cor. coriaceus*		Range of variability in *Carabus cor. cerisyi*	
	in mm.	as percent of average	in mm.	as percent of average
Width of head	0·9	26·5−	1·1	39·3
Length of elytra	4·3	19·4−	4·6	24·9
Width of abdomen	3·6	27·5+	2·6	23·8
Length of scutellum	1·6	25·0−	1·5	26·3
Width of scutellum	2·1	23·9+	1·7	21·3
Length of antennae	2·8	17·0−	4·2	26·9
Length of femur (1st leg)	1·9	26·8−	2·2	35·5
Length of tibia (1st leg)	1·4	26·9−	1·6	34·8
Length of tarsus (1st leg)	1·6	25·8−	1·5	25·9
Length of femur (3rd leg)	2·4	23·8−	3·1	34·9
Length of tibia (3rd leg)	2·7	25·2−	3·7	39·4
Length of tarsus (3rd leg)	2·2	23·7—	3·4	36·9

TABLE 2. Range of variability of absolute measurements in the ground beetles *Carabus auronitens auronitens* from central Europe and *C. aur. festivus* from southern France in mm. and as percent of average measurements taken from 18 males.

	Range of variability in *Carabus auronitens auronitens*		Range of variability in *Carabus aur. festivus*	
	in mm.	as percent of average	in mm.	as percent of average
Width of head	0·6	28·6+	0·4	20·0
Length of elytra	2·2	16·5−	3·1	24·6
Width of abdomen	1·5	18·1−	1·9	24·4
Length of scutellum	0·7	17·5−	0·9	22·5
Width of scutellum	0·8	13·8−	1·0	18·2
Length of antennae	1·9	15·2−	2·2	18·0
Length of femur (1st leg)	1·2	25·5−	1·3	27·1
Length of tibia (1st leg)	0·6	15·8−	0·7	19·5
Length of tarsus (1st leg)	0·6	13·6−	1·3	30·2
Length of femur (3rd leg)	1·2	17·9−	1·6	23·9
Length of tibia (3rd leg)	1·4	18·7−	2·0	28·2
Length of tarsus (3rd leg)	1·7	25·0−	2·2	31·9

1936, 1956). For instance, if the climate becomes cooler, the following changes (in the sense of a geographical gradation) will be found in warm-blooded vertebrates: increase of body size (Bergmann's Rule), shortening of relative lengths of tail, ears, bill, hairs of the underfur (hair rule). Later the wing tips will become more pointed and arm feathers relatively shortened (wing rule, called Rensch's Rule by Allee and Smith, 1951), the relative size of the heart, the pancreas, the liver, the kidneys, the stomach, and the intestine will be increased (B. Rensch, 1956), the number of eggs per clutch or young per litter will be greater, and the migratory instincts will be stronger, in the races of the cooler region (for further details concerning these rules, see B. Rensch, 1936, 1956). All these rules, of course, apply only to geographic variation in those polytypic species whose range covers various climatic zones, and this is especially true in the many examples provided by types inhabiting the temperate and cold zones. It is an important fact that geographic gradations will often be found in characters brought about by several genes or multiple [44] alleles, as is well known in the formation of gradations in coloration or size of body and organs.

vincing, as often a very close parallelism between body size and climate exists. In the willow titmouse (*Parus atricapillus*), for instance, I could demonstrate that the gradation of body size (inferred from measurements of wing length) is closely parallel to the January isotherms, but not to the average isotherms of the whole year (Figure 11). Apparently,

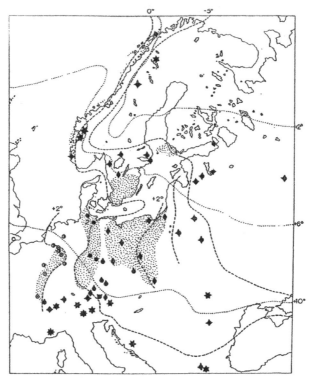

FIGURE 11. Distribution of size variants in the willow-titmouse *Parus atricapillus*. Average length of wings: white circle 57·0–57·9 mm.; circle with central dot 58·0–58·9 mm.; half black circles 59·0–59·9 mm.; black circles 60·0–60·9 mm.; circle with 1 wedge 61·0–61·9 mm.; circle with 2 wedges 62·0–62·9 mm.; circle with 3 wedges 63·0–63·9 mm.; and so forth. Dotted lines: annual isotherms; broken lines: January isotherms. Dotted areas: ranges of 58, 60, and 62 mm. populations. (After B. Rensch, 1939.)

The proof that graded differences of body size are due to selection must, of course, rely on more or less indirect evidence. But such evidence is rather con-

the minima of temperatures per year cause the extinction of the smallest variants, as the loss of body heat is greater in the small-sized variants in consequence of their relatively larger body surface, and as vital organs in the absolutely smaller body will [45] more easily be reached by the temperature. Moreover, the physiological functions of limbs and other peripheral organs not reaching the normal blood temperature will usually be more easily impaired in the smaller animals.

There is one more fact supporting the assumption that the minimum temperatures of the winter are a most important factor in size selection: European migratory birds not subjected to the harsh winters follow Bergman's Rule to a much smaller degree than do resident birds. The comparison of 21 relevant species of migratory birds having subspecies in central Europe as well as in northwest Africa (B. Rensch, 1939) revealed that there is an increase of size in only two central European races, whereas in the nonmigratory species 12 out of 22 showed a racial increase of body size in the central European races. The effect of temperature minima is also demonstrated by the following examples: in polytypic species of birds ranging from Morocco and Algeria to Madeira and the Canary Islands, in 12 out of 16 cases the continental races are larger than the island forms, in three races there is no size difference, and in only one case the Madeira Canary race is larger, but this race inhabits a higher altitude than its Algerian relatives. In both regions the average annual temperatures are now about equal, but the average minimum temperatures are definitely lower in northwest Africa (in the coastal regions of Tunisia and Algeria from −4.0° to +4.2°C., on the Canary Islands between 9.1° and 11.0°C.).

Climatic selection is likely to have occurred especially in all those cases where the larger races (contrary to what would be expected on the hypothesis of elimination) inhabit the areas of the range conquered last, and where the temperatures of these areas are lower than those of the central range. The races of the wren (*Troglodytes troglodytes*) show an ever-increasing size as we proceed from Scotland to more northern areas like the Hebrides, the Shetlands, the Faeroes, and Iceland (Figure 12). Birds that have spread from [46] tropical and subtropical zones and invaded Europe in postglacial times, such as the kingfisher (*Alcedo atthis*), the golden oriole (*Oriolus oriolus*), and the turtledove (*Streptopelia turtur*), have developed their largest races in Europe. Moreover in six polytypic species of birds of the Lesser Sunda Islands which may safely be considered as invaders from Java, Sumatra, and Indochina, the largest races are to be found in the peripheral parts of their range (B. Rensch, 1939).

Quite recently Scholander (1954) expressed doubts concerning the explanation of Bergmann's Rule by selection, but he did not discuss the possible extinction by temperature minima and he did not physiologically analyze subspecies or closely related species. The results of Winkel (1951), Salt (1952), and Saxena (1958) contradict Scholander's arguments.

As in the case of the warm-blooded animals, there are rules applying to the poikilothermic forms with regard to body size, relative size of organs, colorations, number of eggs per clutch, special characters in the cycle of development, and so forth. In many of these examples, however, the hereditary character of the traits concerned must still be proved. As far as body size is concerned, only the general rule may be established that size will decrease towards more rigorous regions. There are terrestrial snails, for instance, increasing in size from northern to southern Europe, and others showing a definite decrease in this direction (possibly those especially adapted to a cooler climate, such as *Eucanulus trochiformis, Vertigo alpestris*, etc.: B. Rensch, 1931).

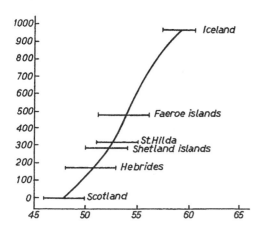

FIGURE 12. Increase of wing length towards the peripheral range in the wren *Troglodytes troglodytes*. Ordinate: distance from Grampian Mountains (northern Scotland) in km. Abscissa: wing length in mm. (After Salomonsen, 1933.)

In reptiles and amphibians, Mell (1928) and Schuster (1950) proved a similar correlation between size differences (probably genotypic) and gradations of temperature and humidity. Size gradations of limbs and tails in lizards seem to follow Allen's Rule only as a consequence of negatively allometric growth (Schuster, 1950). Wing length of Fenno-Scandinavian butterflies often decreases towards the northern regions (Petersen, 1947). In races of European ground beetles, size sometimes decreases towards the Mediterranean countries, as in *Carabus coriaceus*, *C. auronitens*, and others, but increases in *C. ullrichi*, for instance. Generally, in various ground beetles inhabiting the warmer countries, antennae and legs are relatively longer (Krumbiegel, 1936) which is not a consequence of allometric growth only (B. Rensch, 1943). There is a more metallic coloration in the Mediterranean races than in types from central and northern Europe. For climatic parallelism of the coloration of butterflies, the papers by Hovanitz (1941, 1950, 1953), Petersen (1947), and others should be studied. For correlation of climate and the number of generations per year, consult the studies by Mell (1943) and Petersen (1947). Climatic selection is also suggested by the close correlation between microclimatic factors and typical colorations in geographic races of bumblebees (Pittioni, 1941). Further rules may be drawn up regarding the number of vertebrae and fin bones in fishes, the number of offspring in several animal groups, physiologically caused optimum temperatures (Moore, 1949, on *Rana pipiens*: Winkel, 1951, on closely related birds; from different climatic zones), and so forth (compare also Margalef, 1955).

Works Cited

Adam, W. 1941. 'Cephalopoda. In: Res. Sci. Croisières "Mercator",' 3. *Mém. Mus. Hist. Nat. Belge*, 2nd sér., fasc. 21.

Allee, W. C., and K. P. Schmidt. 1951. *Ecological animal geography*. 2nd ed. New York and London.

Ancona, U. D'. 1942. 'I Niphargus Italiani. Tentativo di valutazione critica delle minori unità sistematiche.' *Mem. Ist. Ital. Speleol., Ser. Biol.*, Mem. IV. Trieste.

Armbruster, L., H. Nachtsheim, and T. Roemer. 1917. 'Die Hymenopteren als Studienobjekt azygoter Vererbungserscheinungen.' *Z. ind. Abstammungsund Vererbungslehre*, 17: 283–355.

———. 1943. 'Das "philippinische Elefantenohr" Spongia thienemanni n. sp. Zugleich ein Uberblick uber unsere bisherige Kenntnis des Vorkommens geographischer Rassen bei Meeresschwämmen.' *Arch. F. Hydrobiol.*, 30: 381–442.

Baker, H. G. 1953. 'Race formation and reproductive method in flowering plants.' *Symposia Soc. Exper. Biol.*, No VII, Evolution 144–45.

Baldi, E. 1941 'Différents mécanismes d'isolement au sein de populations de crustacés planktiques.' In: *Symposium sui fattori ecologici e genetici della speciazione negli animali. La Ricerca Scientif.*, 19, Suppl., 119–22.

Benazzi, M, 1945. 'Dendrocoelum lacteum verbanese: nuova razza del Lago Maggiore.' *Atti acad. d. Fisiocritici, Sez. Agr.*, 10: 31–3.

Beninde, J. 1940. 'Die Fremdblutkreuzung (sog. Blutauffrischung) beim deutschen Rotwild.' *Z. f. Jagdkde.*, Sonderheft. Neudamm u. Berlin.

Berg, E. S. 1933. 'Die bipolare Verbreitung der Organismen und die Eiszeit.' *Zoogeogr.*, 1: 449–84.

Boetticher, H. Von. 1940. 'Die Formenkreisbildung bei den Raken, Gattung *Coracias Linnaeus*.' *Senckenbergiana*, 22: 362–9.

———. 1941. 'Artenkreise oder Gattungen und Untergattungen.' *Z. f. Naturwiss.* (Halle), 94: 52–60.

———. 1944. 'Die Verwandtschaftsbeziehungen der afrikanischen Papageien (Poicephalus und Agapornis).' *Zool. Anz.*, 145: 10–27.

Bovey, P. 1941. 'Contribution à l'étude génétique et biogéographique de *Zygaena ephialtes* L.' *Rev. Suisse Zool.*, 48: 1–90.

Braestrup, F. W. 1945. 'Progressive clines.' *Nature*, 156: 337–9.

Breider, H. 1936. 'Eine Erbanalyse von Artmerkmalen geographisch vikariieren-der Arten der Gattung *Limia. Z. Vererbungs.*, 71: 441–99.

Breuning, S. 1932–6. 'Monographie der Gattung *Carabus* L.' *Bestimm. Tab. europ. Coleopt*. H. 104–10. Troppau.

Callan, H. G., and H. Spurway. 1951. 'A Study of meiosis in interracial hybrids of the newt *Triturus cristatus*.' *J. Genetics*, 50: 235–49.

Carl, J. 1940. 'Un "cercle de races" en miniature chez les diplopods de l'Inde méridionale.' *Arch. Sci. Phys.*, 22: 227–33.

Colosi, G. 1923. 'Note sopra alcuni eufillopodi. IV. Triops cancriformis e le sue forme.' *Atti Soc. Ital. Mus. Civ. Milano*, 62: 75–87.

Cronau, B.C. 1902. *Der Jagdfasan, seine Anverwandten und Kreuzungen*. Berlin.

Dale, F. H. 1939. 'Variability and environmental responses of the kangaroo rat, *Dipodomys heermanni saxatilis*.' *Amer. Midland Naturalist*, 22: 703–31.

Delacour, J., and C. Vaurie. 1950. 'Les mésanges charbonniéres (Revision de l'espéce *Parus major*).' *L'Oiseau et Rev. Franç. D'Ornithol.*, 20: 91–121.

Dementiev, G. P. 1936. 'Zur Frage der Grenzen der systematischen Kategorien Art und Unterart.' *Zool. J. Moskau*, 15: 82–95 (Russian, German Summary.).

Demoll, R. 1930. *Die Silberfuchszucht*. Munich.

Dice, L. R. 1931. 'The occurrence of two subspecies of the same species in the same area.' *J. Mammal.*, 12: 210–13.

———. 1935. 'A Study of racial hybrids in the deermouse *Peromyscus maniculatus*). *Contrib. Lab. Vert. Gen. Univ. Mich.*, No. 312.

———. 1940. 'Relationships between the wood mouse and the cotton mouse in Eastern Virginia.' *J. Mammal.*, 21: 14–23.

Doederlein, L. 1902. 'Über die Beziehungen nahevervandter "Thierformen" zueinander.' *Z. Morphol. Anthrop.*, 4: 394–442.

Duerden, J. E. 1919. 'Crossing the North African and South African. Ostrich. I.' *Genetics*, 8.

Duncker, H. 1929, 'Farbenvererbung bei Buntvögeln.' *Vögel ferner Länder*, 3.

Du Rietz, E. 1930. 'The fundamental units of biological taxonomy.' *Svensk Botan. Tidskr.*, 24: 333–428.

Eisentraut, M. 1926. 'Das geographische Prinzip in der Systematik der Ascidien.' *Zool. Anz.*, 66: 171–79.

———. 1929. 'Die Variation der balearischen Inseleidechse *Lacerta lilfordi* Gthr.' *Sitz. Ber. Ges. Naturforsch. Freude Berlin*, 24–36.

———. 1930. 'Beitrag zur Eidechsenfauna der pityusen und Columbreten.' *Mitt. Zool. Mus. Berln*, 16: 397–410.

———. 1950. *Die Eidechsen der spanischen Mittelmeerinseln und ihre Rassenaufspaltung im Lichte der Evolution*. Berlin.

Ekman, S. 1917–20. 'Studien über die marinen Relikte der nordeuropäischen Binnengewäser. VI.' *Int. Rev. Hydrobiol.*, 8: 477–528.

Eller, K. 1936. 'Die Rassen- und Artfrage in dem Formenkreis von *Papilio machaon* L., *Z. angew. Entom.*, 24: 145–49.

Engel, H. 1934 'Über Echinodermen aus der Nordsee und dem Nordatlantik.' *Zool. Anz.*, 107: 23–30.

Engels, W. L. 1936. 'An insular population of *Peromyscus maniculatus* subsp. with mixed racial characters.' *Amer. Midland Naturalist*, 17: 776–80.

Erhardt, A. 1935. 'Systematische und geographische Verbreitung der Gattung *Opishorchis*.' *Z. Parasitenkunde*, 8: 188–225.

Federley, H. 1920. 'Die Bedeutung der polymeren Faktoren für die Zeichnung der Lepiopteren.' *Hereditas*, 1.

———. 1937. 'Fusion zweier Chromosomen als Folge einer Kreuzung.' *Acta Soc. Fauna Flora Fennica*, 60: 685–95.

Fitch, H. S. 1940. 'A biogeographical study of the Ordinoides Artenkreis of garter snakes (Genus *Thamnophis*).' *Univ. Calif. Publ. Zool.*, 44 (No. 1): 1–150.

Ford, E. B. 1949. 'Early stages in allopatric speciation.' In: Jepsen, Simpson, Mayr, *Genetics, Paleontology, and Evolution*, 309–14. Princeton.

Frenzel, G. 1937. 'Die systematische Stellung des adriatischen Alcyonium.' *Notiz. Dtsch. Ital. Inst. Rovigno*, 2 (No. 6): 1–15.

Gabritschewsky, E. 1924. 'Farbenpolymorphismus and Vererbung mimetischer Varietäten der Fliege Volucella bombylans und anderer "hummelähnlicher" Zweiflügler.' *Z. ind. Abstamm. U. Vererbl.*, 32: 321–53.

Geyr von Schweppenburg, H. 1935 'Zur Systematik der Gattung *Larix*.' *Mitt. D. Dendrol. Ges.*, No. 47.

Goldschmidt, R. 1920. *Mechanismus und Physiologie der Geschlechtsbestimmung*. Berlin.

———. 1924. 'Erblichkeisstudien an Schmetterlingen IV.' *Z. ind. Abstamm. U. Vererbl.*, 34: 229–44.

———. 1924–33. 'Untersuchungen zur Genetik der geographischen Variation, I–VII.' *Roux' Arch.*, 101 (1924); 116 (1929); 126 (1932); 130 (1933).

———. 1935. 'Geographische Variation und Artbildung.' *Naturwiss.*, 23: 169–76.

———. 1940. *The Material Basis of Evolution*. New Haven.

———. 1952. 'Evolution, as viewed by one geneticist.' *Amer. Scientist*, 40: 84–135.

Hartert, E. 1910–38. *Die Vögel der paläarktischen Fauna*. With suppl. By E. Hartert and F. Steinbacher. 4 vols. Berlin.

Heding, S. 1942. 'Holothuroidea. Pt. II.' *Danish Ingolf-Exp.*, 4, pt. 3. Copenhagen.

Hellmich, W. 1934a. 'Die Eidechsen Chiles, insbesondere die Gattung Liolaemus.' *Abh. Bayr. Ak. Wiss., Math. Nat. Abt.*, N.F., Heft, 24.

———. 1934b. 'Zur näheren Analyse der geographischen Variabilität.' *Forsch. u. Fortschr.*, 10: 358–39.

Helmcke, J. G. 1940. 'Die Brachiopoden der Deutschen Tiefsee-Expedition.' *Wiss-Ergebn. Tiefsee-Exp. Validvia*, 24: 215–316.

Henke, K. 1938. Review of: W. F. Reinig, *Elimination und Selection. Biol. Zentrabl.*, 58: 553–5.

Herre, W., and F. Rawiel. 1939. 'Vergleichende Untersuchungen an Unken.' *Zool. Anz.*, 125: 290–99.

Herter, K. 1935. 'Ingelbastarde (*Erinaceus roumanicus* ♂ × *E. europaeus* ♀).' *Sitz.-Ber. Ges. Naturforsch. Freunde Berlin*, 188–21.

Herter, K., and W. R. Herter, 1954. 'Die Verbreitung der Kreuzkröte (*Bufo calamita* Laur.) und der Wechselkröte (*Bufo viridis* Laur.) in Europa.' *Zool. Beiträge* (Berlin), N.F. 1: 203–18.

Hovanitz, W. 1941. 'Parallel ecogenotypical color variation in butterflies.' *Ecology*, 22: 259–84.

———. 1950. 'The biology of Colias butterflies. II. Parallel geographical variation of dimorphic color phases in North American Species.' *Wasmann J. Biol.*, 8: 197–219.

———. 1953. ' Polymorphism and evolution.' *Sympos. Soc. Exper. Biol.*, No. 7, 238–53.

Hubbs, C. L. 1943. 'Criteria for subspecies, species, and genera, as determined by researches on fishes.' *Ann. New York Acad. Sci.*, 44: 109–21.

Hubbs, C. L., and R. R. Miller. 1943. 'Mass hybridization between two genera of Cyprinid fishes in the Mohawe Desert, California.' *Pap. Michigan Acad. Sci.*, 28: 343–78.

Huxley, J. S. 1939. 'Clines: An auxiliary method in taxonomy.' *BijDr. Dierk.*, editor, 27: 491–520.

Iwanoff, E., and J. Philiptschenko. 1916. 'Beschreibung von Hybriden zwischen Bison, Wisent und Hausrind.' *Z. ind. Abstamm. U. Vererbl.*, 16.

Jaworski, E. 1938. 'Untersuchungen über Rassenbildung bei Anthozoen.' *Thalassia*, 3: 1–57.

Johansen, H. 1944. 'Die Vogelfauna Westsiberiens. II.' *J.f. Ornith.*, 92: 1–105.

Jordan, K. 1905. 'Der Gegensatz zwischen geographischer und nicht geographischer Variation.' *Z. wiss. Zool.*, 83: 151–210.

Jordan, K., H. Fruhstorfer, et al. 1927. 'Die indoaustralischen Tagfalter.' In: Seitz, *Großschmetterlinge der Erde*, 9. Stuttgart.

Kauri, H. 1954. 'Über die systematische Stellung der europäischen grünen Frösche Rana esculenta L. und *R. ridibunda* Pall.' *Lunds Univ. Arsskr.*, N.F., Avd. 2, 50, No. 12.

Kiefer, F. 1952. 'Copepoda Calanoida und Cyclopoida.' In: *Explor. Parc. Nat. Albert, Mission H. Damas*, Fasc. 21. Brussels.

Kiriakoff, S. G. 1947 'Le cline, une nouvelle catégorie systématique intraspécifique.' *Ann. Soc. Entom. Belg.*, 83: 130–9.

Knoerzer, A. 1940. 'Der saharo-sindische Verbreitungstypus bei der ungeflügelten Tenebrioniden-Gattung Mesostena.' *Riv. Biol. Colon.*, 3: 1–133.

Kramer, G. 1949. 'Über Inselmelanismus bei Eidechsen.' *Z. induct. Abstamm. u. Vererbl.*, 83: 157–64.

Kramer, G., and R. Mertens 1938. 'Zur Rassenbildung bei West-instrianischen Inseleidechsen in Abhängigkeit von Isolierungsalter und Arealgröße.' *Arch. F. Naturgesch.*, N.F., 7: 189–234.

Kreitmann, L. 1927. 'L'acclimatisation du Lavaret de Bourget dans le Lac Léman et sa relation avec la systématique des Corégones.' *Verh. Int. Ver. Limnol.*, 4.

Lattin, G. De. 1949. 'Über die Artfrage in der *Hiparchia semele* L.-Gruppe.' *Entom. Z.*, 59, No. 15–17.

Lorkovič, Z. 1928. 'Analyse der Speziesbegriffe und der Variabilität der Spezies auf Grund von Unstersuchungen einiger Lipidopteren.' *Act. Soc. Sci. Nat. Croaticase*, 38: 1–64.

Margalef, R. 1955. 'Temperatura, dimensiones y evolución.' *Publ. Inst. Biol. Aplicada (Barcelona)*, 19.

Mayr, E. 1932. 'Birds collected during the Whitney South Sea expedition.' *Amer. Mus. Novit.*, 20: 1–22; 21: 1–23.

———. 1942. *Systematics and the origin of species.* New York.

———. 1954. 'Geographic speciation in tropical echinoids.' *Evolution*, 8: 1–18.

Meise, W. 1928. 'Rassenkreuzungen an den Arealgrenzen.' *Verh. Dtsch. Zool. Ges.*, 96–105.

———. 1933. 'Scorpiones (Norweg. Zool. Exp. Galapagos Isls. 1925).' *Meddel. Zool. Mus. Oslo*, No. 39, 25–43.

———. 1936. 'Zur Systematik und Verbreitungsgeschichte der Haus- und Weidensperlinge, *Passer domesticus* (L.) und *hispaniolensis* (T.).' *J. Ornith.*, 84: 631–72.

Mell, R. 1929. *Grundzüge zu einer Ökologie der chinesischen Reptilien und einer herpetologischen Tiergeographic Chinas.* Berlin and Leipzig.

———. 1943. 'Inventur und ökologisches Material zu einer Biologie der südchinesischen Pieriden.' *Zoologica*, 36, Lief. 6, Heft 100. Stuttgart.

Mertens, R. 1924. 'Ein Beitrag zur Kenntnis der melanotischen Inseleidechsen des Mittelmeeres.' *Pallasia*, 2: 40–52.

———. 1931. '*Ablepharus boutonii* (Desjardin) und seine geographische Variation.' *Zool. Jb., Abt. Syst.*, 61: 63–210.

———. 1932. 'Über duster gefärbte Inseldeidechsen des Lago Maggiore.' *Zool. Anz.*, 101: 106–11.

Mertens, R., and L. Müller. 1940. 'Die Amphibien und Reptilien Europas.' *Abh. senckenb. naturf. Ges.*, 451.

Miller, A. H. 1941. 'Speciation in the avian genus Junco.' *Univ. Calif. Publ. Zool.*, 44: 173–434.

———. 1949. 'Some concepts of hybridization and intergradation in wild populations of birds.' *Auk*, 66: 338–42.

Moore, J. A. 1946. 'Incipient intraspecific isolating mechanism in Rana pipiens.' *Genetics*, 31: 304–26.

———. 1949. 'Geographic variation of adaptive characters in Rana pipiens Schreber.' *Evolution*, 3: 1–24.

———. 1954. 'Geographic and genetic isolation in Australian Amphibia.' *Amer. Nat.*, 88: 65–74.

Müller-Böhme, H. 1935. 'Beiträge zur Anatomie, Morphologie und Biologie der "Großen Wühlmaus".' *Arb. Biol. Anstalt Berlin*, 21: 363–453.

Nobis, J. 1949. 'Vergleichende und experimentelle Untersuchungen an heimischen Schwanzlurchen.' *Zool. Jb., Bt. Anat.*, 70: 333–96.

Økland, F. 1937. 'Die geographischen Rassen der extramarinen Wirbeltiere Europas.' *Zoogeogr.*, 3: 389–484.

Pax, F. 1936. 'Anthozoa.' In: G. Grimpe and A. Remane, *Tierwelt der Nord- und Ostsee*, Lier. 30. Leipzig.

Peters, J. L. 1937. *Check-list of birds of the world.* Vol III. Cambridge.

Petersen, B. 1947. 'Die geographische Variation einiger Fennoskandischer Lepidopteren.' *Zool. BiDr. Uppsala*, 26: 329–521.

Pirocchi, B. 1951. 'Hochendemische Copepoden- und Cladoceren-Lokalformen im Karst.' *Arch. F. Hydrobiol.*, 45: 245–53.

Pittioni, B. 1941. 'Die Variabilität des *Bombus agrorum* F.

in Bulgarien.' *Mitt. Kgl. Naturwiss. Inst. Sofia*, 14: 238–311.

Poll, H. 1911. 'Über Vogelmischlinge.' *Verh. 5th Int. Ornith. Kongr.* (1910), 399–468.

Porter, K. R. 1941. 'Diploid and androgenetic haploid hybridization between two forms of *Rana pipiens* Schreber.' *Biol. Bull.*, 80: 238–64.

Reinig, W. F. 1938. *Elimination und Selektion*. Jena.

Rensch, B. 1928. 'Inselmelanismus bei Mollusken.' *Zool. Anz.*, 78: 1–4.

———. 1928. 'Grenzfälle von Rasse und Art.' *J. F. Ornith.*, 76: 222–31.

———. 1929. *Das Prinzip geographischer Rassenkreise und das Problem der Artbildung*. Berlin.

———. 1931. 'Die Vogelwelt der Kleinen Sunda-Inseln Lombok, Sumbawa und Flores.' *Mitt. Zool. Mus. Berlin*, 17: 451–637.

———. 1934. *Kurze Anwisung für zoologisch-systematische Studien*. Leipzig.

———. 1934. 'Die Molluskenfauna der Kleinen Sunda-Inseln.' III. *Zool. Jb., Abt. Syst.*, 65: 389–422.

———. 1936. 'Studien über klimatische Prallelität der Merkmalsausprägung bei Vögeln und Säugern.' *Arch. f. Naturgesch.*, N.F., 5: 317–63.

———. 1937. 'Untersuchungen über Rassenbildung und Erblichkeit von Rassenmerkmalen bei sizilischen Landschnecken.' *Z. ind. Abstamm. u. Vererbl.*, 72: 564–88.

———. 1938. 'Bestehen die Regeln klimatischer Parallelität bei der Merkmalsausprägung bon homöothermen Tieren zu Recht?' *Arch. f. Naturgesch.*, N.F., 7: 364–89.

———. 1939. 'Klimatische Auslese von Größenvarianten.' *Arch. f. Naturgesh.*, N.F., 8: 89–129.

———. 1939. 'Typen der Artbildung.' *Biol. Rev.*, 14: 180–222.

———. 1939. 'Über die Anwendungsmöglichkeit zoologisch-systematischer Prinzipien in der Botanik.' *Chronica Botanica*, 5: 46–9.

———. 1941. '"Elimination" oder Selektion bei der Birlitzausbreitung?' *Ornithol. Monatsber.*, 49: 94–104.

———. 1943. 'Studien über Korrelation und klimatische Parallelität der Rassenmerkmale von Carabus-Formen.' *Zool. Jb., Abt. Syst.*, 76: 103–70.

———. 1943. 'Die biologischen Beweismittel der Abstammungslehre.' In: Heberer, *Evolution der Organismen*, 57–85. Jena.

———. 1956. 'Relative Organmasse bei tropischen Warmblütern.' *Zool. Anz.*, 156: 106–24.

Ripley, S. D., and H. Birckhead. 1942. 'On the fruit pigeons of the *Ptilinopus pupuratus* group.' *Amer. Mus. Novitates*, No. 1192.

Salt, G. W. 1952. 'The relation of metabolism to climate and distribution in three finches of the genus *Carpodacus*.' *Ecol. Monogr.*, 22: 121–52.

Schäfer, E., and R. Meyer de Schauensee. 1939. 'Zoological results of the 2nd Dolan expedition to western China and eastern Tibet 1934–1936. 2. Birds.' *Proc. Acad. Nat. Hist. Philadelphia* (1938), 90: 185–260.

Schilder, M., and F. A. Schilder. 1939. 'Prodrome of a monography of living Cypraeidae.' *Proc. Malacol. Soc. London*, 23: 119–231.

Scholander, P. F. 1955. 'Evolution of climatic adaptation in homeotherms.' *Evolution*, 9: 15–26.

Schuster, O. 1950. 'Die klimaparallele Ausbildung der Körperproportionen bei Poikilothermen.' *Abh. Senckenberg. Naturforsch. Ges.*, Abh. 482. Frankfurt a. M.

Schwarz, O. 1936. 'Über die Systematik und Nomenklatur der europäischen Schwarzkiefern.' *Notizbl. Botan. Garten u. Mus. Berlin-Dahlem*, 13: 216–43.

Seiler, J. 1925. 'Zytologische Vererbungsstudien an Schmetterlingen.' *Arch. Vererb. Sozialanthrop.*, 1: 63–117.

Sibley, C. G. 1954. 'Hybridization in the red-eyed towhees of Mexico.' *Evolution*, 8: 252–90.

Spurway, H. 1953. 'Genetics of specific and subspecific differences in European newts.' *Sympos. Soc. Exper. Biol.*, No. VII, 200–37.

Standfuss, M. 1909. 'Hybridationsexperimente.' *Proc. 7th Int. Congre. Cambridge, Mass.*, 57–73.

Stebbins, G. L., jr. 1950. *Variation and evolution in plants*. New York.

Steiner, H. 1942. 'Bastardstudien bei Pleurodeles-Molchen. Letale Fehlentwicklung in der F_2-Generation bei artspezifischer Kreuzung.' *Arch. Jul.-Klaus-Stift. f. Vererbl. Sozialanthrop.*, 17: 428–32.

———. 1945. Über letale Fehlentwicklung der zweiten Nachkommenschaftsgeneration bei tierischen Artbastarden.' *Arch. Jul-Klaus-Stift. f. Vererb. Sozialanthrop.*, Erg. Bd., 20: 236–51.

Stresemann, E. 1926. 'Übersicht über die Mutationsstudien I–XXIV und ihre wichtigsten Ergebnisse.' *J. f. Ornith.*, 74: 377–85.

———. 1931. 'Die Zosteropiden der indoaustralischen Subregion.' *Mitt. Zool. Mus. Berlin*, 17: 201–38.

Stresemann, E., and N. W. Timoféeff-Ressovsky, 1947. 'Artentstehung in geographischen Formenkreisen. I. Der Formenkreis *Larus argentatus-cachin-nans-fuscus*.' *Biol. Zentralbl.*, 66: 57–76.

Sumner, F. B. 1930. 'Genetic and distributional studies of three subspecies of *Peromyscus*.' *J. Genetics*, 23: 277–376.

———. 1932. 'Genetic, distributional, and evolutionary studies of the subspecies of deer mice (*Peromyscus*).' *Bibliogr. Genet.*, 9: 1–106.

Suomalainen, E. 1941. 'Vererbungsstudien an der Schmetterlingsart *Leucodonta bicoloria*.' *Hereditas*, 27: 313–18.

Surbeck, G. 1920. 'Beitrag zur Kenntnis der Schweizerischen Coregonen.' *Festschrift Zschokke*, No. 15. Basel.

Svilha, A. 1935. 'Development and growth of the prairie deer mouse *Peromyscus maniculatus bairdii*.' *J. Mammal.*, 16: 109–16.

Tate, G. H. H. 1941. 'A review of the genus *Hipposideros*

with special reference to Indo-Australian species.' *Bull. Amer. Mus. Nat. Hist.*, 78: 353–93.

———. 1941. 'Review of *Myotis* of Eurasia.' *Bull. Amer. Mus. Nat. Hist.*, 78: 537–65.

Thienemann, A. 1928. 'Die Felchen des Laacher Sees.' *Zool. Anz.*, 75: 226–34.

Ticehurst, C. B. 1938. *A systematic review of the genus Phylloscopus.* London, British Museum.

Timoféeff-Ressovsky, N. W. 1940. 'Zur Frage über die "Eliminationsregel": Die geographische Größenvariabilität von *Emberiza aureola* Pall.' *J. f. Ornith.*, 88: 334–40.

Tonolli, V. 1949. 'Isolation and stability in populations of high altitude diaptomids.' *La Ricerca Scientif.*, 19, Suppl.

Vasseur, E. 1952. 'Geographic variation in the Norwegian sea-urchin, *Strongylocentrotus*.' *Evolution*, 6: 87–100.

Viets, K. 1926. 'Indische Wassermilben.' *Zool. Jb., Abt. Syst.*, 52: 369–94.

Voipio, P. 1950. 'Evolution at the population level with special reference to game animals and practical game management.' *Papers on Game Research*, 5. Helsinki.

———. 1954. 'Über die gelbfüssigen Silbermöven Nordwesteuropas.' *Acta Soc. Fauna Flora Fennica*, 71. (No. 1): 1–56.

Wesenberg-Lund, C. 1904–8. *Plankton investigations of the Danish lakes.* I and II. Copenhagen.

Wettstein, R. von. 1898. *Grundzüge der geographisch-morphologischen Pflanzen-systematik.* Jena.

Whitman, C. O. 1919. *Inheritance, fertility and the dominance of sex and color in hybrids of wild species of pigeons.*, ed. By O. Riddle. Washington.

Wilson, E. O. 1955. 'A monograph revision of the ant genus *Lasius*.' *Bull. Mus. Comp. Zool. Harvard Coll.*, 113: 1–205.

Winkel, K. 1951. 'Vergleichende Untersuchungen einiger physiologischer Konstanten bei Vögeln aus verschiedenen Klimazonen.' *Zool. Jb., Abt. Syst.*, 80: 256–76.

Woltersdorff, W., and M. Radovanovic. 1938. *Triturus alpestris reiseri* Wern. Und T. *alpestris* Laur., vergesellschaftet im Proskoško-See.' *Zool. Anz.*, 122: 23–30.

Wright, S. 1940. 'The statistical consequences of Mendelian heredity in relation to speciation.' In: Huxley, J.; *The New systematics*, 161–83. Oxford.

Zimmer, C. 1930. 'Untersuchungen an Diastyliden (Ordnung Cumacea).' *Mitt. Zool. Mus. Berlin*, 16: 583–658.

Zimmermann, K. 1934. 'Zur Genetik der geographischen Variabilität von *Epilachna chrysomelina* F.' *Entom. Beih. Berlin-Dahlem*, 1: 86–90.

———. 1936. 'Die geographischen Rassen von *Epilachna chrysomelina* F. und ihre Beiziehungen zu *Epilachna capensis* Thunbg. 'Z. ind. Abst.- u. Vererbl.*, 71: 527–37.

Zimmermann, W. 1935. 'Rassen- und Artbildung bei Wildpflanzen.' *Forsch. u. Fortschr.*, 11: 272–4.

From *Systematics and the Origin of Species*
Ernst Mayr

The Process of Geographic Speciation

The term speciation includes two processes: the development of diversity and the establishment of discontinuities between the diverging forms. To be sure, the two processes are correlated and frequently go hand in hand, but nevertheless they represent two rather different aspects of the course of evolution. The development of diversity, which is the more obvious of the two, has been discussed by us in detail in Chapters III and IV, under the heading of geographic variation. But variation and mutation alone do not necessarily produce new species. After all, it is quite thinkable that such variation might lead only to a single, interbreeding, immensely variable community of individuals. But this is not what we find in nature. What we find are groups of individuals that share certain characters, and that are more or less sharply segregated from other groups with different character combinations. These groups of individuals, these populations, races, or subspecies can be combined into species, and the latter into higher categories. This is a rough description of the situation as it occurs in nature. But we are not satisfied with mere description; we are interested in the dynamics of this process of speciation. Therefore, we want the answers to certain questions, such as: (1) Do species originate from individuals or from infraspecific units, and, if the latter, from which units? and (2) Is there any evidence for a broadening of intraspecific gaps, to the extent that they become interspecific gaps?

The answer to the first question is not simple, since it involves indirect proof. Even if no new species had ever developed under domestication or under other conditions of close observation, somebody might still insist that the spontaneous production of individuals representing **[155]** new species

was the usual process of species formation and that this had never been observed, merely because no interested observer had happened to be present when the new species first appeared. When De Vries described his first mutations, it seems that he was convinced that they demonstrated spontaneous species formation, and Lotsy insisted on this point even at a much later date. In the meantime the species concept has been clarified by the taxonomist, and we know now that species differ by so many genes that a simple mutation would, except for some cases in plants, never lead to the establishment of a new species. Goldschmidt, therefore, modified the simple De Vriesian concept and replaced it by the hypothesis of speciation through systemic mutations: "Species and the higher categories originate in single macroevolutionary steps as completely new genetic systems." To him a species is like a Roman mosaic, consisting of thousands of bits of marble. A systemic mutation would be like the simultaneous throwing out of all the many thousands of pieces of marble on a flat surface so that they would form a completely new and intelligible picture. To believe that this could actually happen is, as Dobzhansky has said in review of Goldschmidt's work, equivalent to "a belief in miracles." It seems to me not only that Goldschmidt did not prove his novel ideas, but also that the existing facts fit orthodox ideas on species formation so adequately that no reason exists for giving them up. This statement requires proof and there is perhaps no better way to introduce our arguments than to state briefly how we visualize the course of geographic species formation:

A new species develops if a population which has become geographically isolated from its parental species acquires during this period of isolation characters which promote or guarantee reproductive isolation when the external barriers break down.

Reprinted from Ernst Mayr, *Systematics and the Origin of Species* (New York: Columbia University Press, 1942), with the permission of the publisher and by kind permission of Ernst Mayr.
See the footnote on page 789.

This definition contains a number of postulates which we shall now discuss. To begin with, it involves the concept of the "incipient" species. Geographic speciation is thinkable only, if subspecies are incipient species. This, of course, does not mean that every subspecies will eventually develop into a good species. Far from it! All this statement implies is that every species that developed through geographic speciation had to pass through the subspecies stage. There is, naturally, a considerable infant mortality among subspecies and only a limited number reaches adulthood, or the full species stage. We shall see in Chapter IX under what conditions subspecies are most likely to be successful. At this point a few figures may be helpful. There are, in the entire world, approximately 8,500 species of living birds, with probably 35,000 recognizable subspecies. Apparently all the present orders of birds already existed at **[156]** the beginning of the Tertiary period, some fifty-five million years ago, and we can think of no reason why the number of species should have increased materially during the last ten or twenty million years. As some became extinct, others took their place. This replacement is apparently a rather slow process, since there is much evidence that most of the present species or the "lines" to which they belong have existed for considerable periods (Miller 1940). Occasionally a species succeeds in entering a previously unoccupied ecological niche. We are forced to the conclusion, on the basis of such considerations, that probably less than 10 or 15 percent of the existing subspecies of birds will both diverge sufficiently and survive long enough in isolation to become good species. The statement that subspecies are incipient species should therefore be amended to read: Some subspecies are incipient species, or subspecies are potentially incipient species. Furthermore, the isolated incipient species may consist of several subspecies or of a subspecifically as yet unmodified population.

We have called the theory of geographic speciation an orthodox theory, and this is correct when we realize how old it is and how widespread its acceptance. It had considerable support among thinking biologists, even long before Darwin. Leopold von Buch, for example, in a description of the fauna and flora of the Canary Islands (1825), writes as follows:

The individuals of a genus spread out over the continents, move to far distant places, form varieties (on account of differences of the localities, of the food, and the soil), which owing to their segregation [geographical

isolation] cannot interbreed with other varieties and thus be returned to the original main type. Finally these varieties become constant and turn into separate species. Later they may reach again the range of other varieties which have changed in a like manner, and the two will now no longer cross and thus they behave as "two very different species."

We can hardly improve on this statement, except for choosing a few different terms. The two points which von Buch makes, namely that geographic isolation was needed to permit the species difference to "become constant" and that proof of the species difference was given by their reproductive isolation, were, curiously enough, not recognized with the same clarity by later authors. Darwin, for example, was primarily interested in the development of the diversity which precedes species formation and he therefore neglected to explain the development and maintenance of discontinuities. M. Wagner seems to have been the first author to realize this gap in Darwin's argumentation, and it led him to propose in 1869, his "Migrationsgesetz der Organismen," which he later **[157]** called more correctly the "separation theory" (Wagner 1889). On the basis of his extensive collecting experiences in Asia, Africa, and America, Wagner emphasized the nonexistence of sympatric speciation and stated that "the formation of a real variety which Mr. Darwin considers as 'incipient species,' can succeed in nature only where some individuals cross the previous borders of their range and segregate themselves a long period from the other members of their species." Darwin himself, in a letter to Wagner, admitted later that he had overlooked the importance of this point.

The speciation process does not need to be completed during this isolation. Dobzhansky (1940, 1941a) has pointed out that selective mating in a zone of contact of two formerly separated incipient species (zone of secondary intergradation, p. 99) may play an important role. The two incipient species must be sufficiently distinct, so that the hybrid offspring of mixed matings has discordant (unbalanced) gene patterns; in other words, the individuals produced in such matings must have a reduced viability and survival value.

Let it be assumed that two incipient species, A and B, are in contact in a certain territory, and that mutations arise in either or in both species which make their carriers less likely to mate with the representatives of the other species. The nonmutant individuals of the species A which cross to B will produce a progeny which is, by hypothesis, inferior in viability to the pure species; the off-spring of the mutant individuals will have, other things being equal, a normal viability. Since the mutants breed only or mostly within the species, their progeny will be more numerous or more vigorous than that of the non-mutants. Consequently, natural selection will favor the spread and establishment of the mutant condition [Dobzhansky 1941a],

until only conspecific pairs are formed or, in other words, until complete discontinuity (a bridgeless gap) has developed between the two species. Dobzhansky presents a plausible case, and we agree that such a selective process may help to complete the establishment of discontinuity, in those cases in which some interbreeding has taken place between incipient species.

The question is, however, whether or not this is the only way by which reproductive isolation can be established. Naturalists, from L. von Buch down to our contemporaries, have always believed that good species can complete their development in isolation. They find an abundance of cases in nature which seem to permit no other interpretation. The most conclusive evidence is, of course, presented by the multiple invasion of islands by separate colonizing waves coming from the same parental stock. Let us, for example, take Norfolk Island, 780 miles **[158]** from the coast of Australia, and surely never in con-tinental connection with any of the surrounding is-land areas or continents. Among its scanty bird fauna (about 15 species of land birds) there are 3 spe-cies of *Zosterops: norfolkensis, tenuirostris,* and *al-bogularis,* which in the same order are progressively more different from their only close relative, *Zos-terops lateralis,* from the Australian mainland. The island is about 44 square kilometers in area and can easily harbor several thousands of pairs of each spe-cies. The following interpretation of this situation is obvious, and there seems no other interpretation nearly so convincing (Stresemann 1931). There were three waves of immigration. The first had already become specifically distinct when the second wave arrived. If the single or the two pairs of *Zosterops* which comprised the second colonization had hy-bridized with the more-than-thousand-pair popula-tion of the first wave, they would have been swamped out of existence within one or two generations. The second wave had developed into a separate species when the colonizing pair of the third wave appeared. The discontinuity between the three species could not have become established through a slow, selec-tive process, as described above by Dobzhansky. The bridgeless gap must have been there already, when the second and third set of colonists arrived; other-wise there would not be three species on the island. The distance of 780 miles between Norfolk and the mainland precludes the possibility of numerous at-tempts at colonization, by which an isolating mecha-nism could have been built up gradually through selection. The same explanation applies, *mutatis mutandis,* to all other cases of double or triple inva-sions (see p. 173). It is also the best interpretation of many other situations in which two related species now have partly overlapping ranges, owing to the breakdown of former barriers (p. 176).

Some geneticists endorse the viewpoint of the naturalist, that the accumulation of small genetic changes in isolated populations can lead in the course of time to a new integrated genetic system, of such difference that it thereby acquires all the characters of a new species, including reproductive isolation. S. Wright (1941c) describes this in the following words:

If isolation of any portion of a species becomes suffi-ciently complete, the continuity of the fabric is broken. The two populations may differ little if any at the time of separation but will drift ever farther apart, each car-rying its subspecies with it. The accumulation of genic, chromosomal and cytoplasmic differences tends to lead in the course of ages to intersterility or hybrid sterility, making irrevocable the initial merely geographic or ecologic isolation.

[159] For a more detailed discussion of the genetic aspects of the establishment of biological discontinu-ity, we refer to Muller (1940). We maintain, there-fore, that the discontinuity between species is due to their divergence (difference), both in regard to their cytogenetics and in regard to their crossability (eco-logical and ethological). The establishment of dis-continuity is closely associated with the process of divergence, and, to make a pun, one might say: "The establishment of discontinuity is a continuous pro-cess." In other words, the big gaps which we find be-tween species are preceded by the little gaps which we find between subspecies and by the still lesser gaps which we find between populations. Of course, if these populations are distributed as a complete *continuum,* there are no gaps. But with the least iso-lation, the first minor gaps will appear.

Stages of speciation.—That speciation is not an abrupt, but a gradual and continuous process is proven by the fact that we find in nature every imaginable level of speciation, ranging from an almost uniform species at one extreme to one in which isolated populations have diverged to such a degree that they can be considered equally well as separate, good species at the other extreme. I have tried in a recent paper (Mayr 1940a) to analyze this continuous process and to demonstrate its different phases by subdividing it into various stages. [See figure 16.] I am well aware that these divisions are somewhat artificial and that a polytypic species may be in different stages in different parts of its range at the same time. Still, this analysis is useful, as we shall see in the subsequent discussion. The classification of my 1940 paper has been somewhat modified, since I now realize that what I then called stage (1) is as much the final as the first stage of speciation. A species may have a small range because it is so new that it has had as yet no time to expand (Willis's age and area concept), or because it is adapted to a unique situation, or because it developed in a particularly isolated location (island, cave), or because it became extinct in the other parts of its range. A widespread species is more likely to represent the first stage of speciation than one with a narrowly restricted range.

There are many cases in nature which cannot be fitted very well into this scheme, but still it will be possible to take the entire number of species of a systematic group (let us say birds or butterflies) from one particular region and classify them according to the stage of speciation to which they belong. The resulting figures of such an analysis shed much light on the degree of speciation and, in particular, on the degree of geographic isolation in the respective region.

[161] To demonstrate the value of this method, which is applicable only to well-known groups, I have listed in Table 10 the passerine birds of three geographic regions, including the extraterritorial range of each species. Stage 5 was omitted and in stage 4 every uniform species with small range was included.

An analysis of this tabulation shows that stages 3 and 4, which indicate the final stages of evolution, are almost nonexistent when geography and geology favor continuous ranges, while stages 1 and 2a, indicating the early stages of evolution, reach a definite high in such continental areas. In contradistinction, we find that where geographic factors break up the species ranges to a high degree, as, for example, on an old tropical archipelago such as the Solomon Islands, a great number of the species are in the final

stages of evolution (3 and 4) and comparatively few in the early stages (1 and 2a). A student of speciation must study regions with continuous ranges as well as those with discontinuous ranges before he can generalize on the dynamics of the speciation process. To base all conclusions on the temperate zones of the large Old World and New World continents leads inevitably into error or to a very one-sided viewpoint, because these two regions are characterized by special conditions. Not only are there very few effective geographic barriers, but most of the populations are also comparatively young, because they occupied their present ranges only after the rather recent retreat of the ice. So far as I know, all workers who minimize the importance of geographic variation for species formation base this opinion **[162]** on work done in the Holarctic region. On the other hand, von Buch, Darwin, Wallace, and others derived their clear ideas on evolution from a study of both continental and insular species. Kinsey (1937b) has presented us with a particularly graphic description of the differences between these two types of species.

The Proof for Geographic Speciation

The conclusion of the taxonomist of birds, mammals, butterflies, and other well-known groups that geographic isolation is in most groups of animals one of the necessary conditions of speciation has not remained unchallenged. Rensch (1939a) recently has cited a whole list of books and papers by authors who deny the importance of geographic speciation (allopatric speciation). Goldschmidt (1940) devoted most of the first half of his book (183 pages) to a refutation of this thesis. It may be in order, therefore, to gather additional proof for the existence and importance of geographic speciation. But what is proof? Is it not sufficient to point out, as we have done, that the majority of well-isolated subspecies have all the characters of good species and are indeed considered to be such by the more conservative systematists? Is it not sufficient to show that subspecific characters are of exactly the same kind as specific characters? Is it not sufficient to point out that certain *Rassenkreise* clearly merge into each other?

All this evidence is highly indicative, but it may not be completely convincing. The fact that Goldschmidt and others, who know this evidence, deny the importance of geographic speciation makes it necessary to present additional proof. The evidence

Stage 1. A uniform species
with a large range

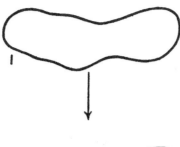

Followed by:
Process 1. Differentiation
into subspecies

Resulting in:
Stage 2. A geographically
variable species with a more
or less continuous array of
similar subspecies (2a all
subspecies are slight, 2b
some are pronounced)

Followed by:
Process 2. a) Isolating action
of geographic barriers between some of the popula-
tions;
also b) development of isolating mechanisms in the
isolated and differentiating
subspecies

Resulting in:
Stage 3. A geographically
variable species with many
subspecies completely isolated, particularly near the
borders of the range, and
some of them morphologi-
cally as different as good
species

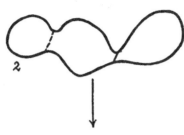

either   or

Followed by:
Process 3. Expansion of
range of such isolated populations into the territory of
the representative forms

Resulting in either
Stage 4. Noncrossing, that is,
new species with restricted
range
or

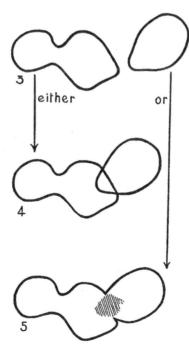

Stage 5. Interbreeding, that
is, the establishment of a
hybrid zone (zone of secondary intergradation)

Fig. 16. Stages of speciation.

TABLE 10

STAGES OF SPECIATION IN DIFFERENT GEOGRAPHICAL REGIONS

STAGE	MANCHURIA (CONTINUOUS RANGES)		NEW GUINEA REGION (PARTLY CONTINUOUS RANGES)		SOLOMON ISLANDS (DISCONTINUOUS RANGES)	
	SPECIES		SPECIES		SPECIES	
	Number	*Percent*	*Number*	*Percent*	*Number*	*Percent*
1	15	14.0 ⎫	21	7.2 ⎫	1	2 ⎫
2a	59	55.1 ⎭ 69	118	40.7 ⎭ 48	11	22 ⎭ 24
2b	30	28.0	84	29.0	12	24
3	1	1.0 ⎫	33	11.4 ⎫	17	34 ⎫
4	2	1.9 ⎭ 3	34	11.7 ⎭ 23	9	18 ⎭ 52

in favor of geographic speciation can be summarized as follows:

ALL DIFFERENCES BETWEEN SPECIES ARE SUBJECT TO GEO-GRAPHIC VARIATION; THERE IS NO DIFFERENCE OF KIND BETWEEN SPECIFIC AND SUBSPECIFIC CHARACTERS.

The discussion as to what characters are subject to geographic variation (Chapter III), proves this point so conclusively that nothing need be added. If species formation and subspecies formation were two fundamentally different processes, we should find that two different classes of characters were subject to variation in the two categories (geographic race and species). But this is not what we found in our analysis of geographically variable characters. Not a species character is known, be it morphological, physiological, or other, which is not subject to geographic variation. As a general rule it can be said that characters separating full [163] species tend to be more pronounced, and that there are often more differences between species than between subspecies. But this criterion breaks down completely in all the really doubtful cases, and many subspecies are characterized by more striking differences than some "good" species. It is therefore obvious that there is no "gap" between subspecies and species, as far as systematic characters are concerned.

Reduced Fertility between Geographic Races of One Species

We have already called attention to a number of mistakes in logic in the discussion of sterility as a species criterion (p. 119). Lack of interbreeding in nature between two forms of animals (in breeding condition) may be due to two different obstacles, sexual isolation and sterility. Even though these two factors are frequently correlated, they actually belong to two entirely different fields: ethology and cytogenetics. It has been proven again and again for birds and many other animals that several species can live side by side in nature without normally hybridizing, even though they are highly or completely fertile with one another in artificial crosses. It is therefore not to be expected that fertility should always be reduced between the geographic races of all the species. Still, the number of cases of partial sterility between geographically distant races of the same species is surprisingly high, particularly among insects. Rensch (1929: 93–94, 1933) has listed a number of them and new ones are being discovered and described continually. Very striking are the intersexes or otherwise less viable forms which occur if certain subspecies of *Lymantria dispar* are hybridized (Goldschmidt 1934). A considerable degree of sterility occurs also between some of the geographic races of many of the "wild" species of *Drosophila* (excluding *melanogaster* and other cosmopolitan "domestic" species). These data have been summarized by Spencer (1940) and Dobzhansky (1941a). Pictet (1937) found reduced fertility, in certain species of butterflies and moths, between populations that are not even separated into different subspecies. This is true for *Nemeophila plantaginis* and for the geographic and altitudinal races of *Lasiocampa quercus*. Neighboring races are highly fertile, but the wider their geographic separation, the more pronounced is the reduction of fertility, some races being almost or completely sterile. It does not require much imagination to picture what would happen, if such races should meet in nature. The presence of even partial sterility would speed up considerably the establishment of biological isolating mechanisms between the two incipient species. The most recent work on Drosophila proves that not only may fertility be reduced in geographic races of the same species, but also the sexual [164] attraction (Patterson 1942, Stalker 1942). This also proves that reproductive isolation (misogamy) is a by-product of genetic divergence.

The Border-line Cases

It is a logical postulate of our thesis (that species originate from geographic races) that we should find certain subspecies which have just about reached the threshold of the species. Such cases are usually called border-line cases, since it is impossible in these situations to decide whether the questionable form is "still" a subspecies or "already" a species; they are in the border zone between the two categories. There are a number of different situations which can be included with the border line cases. For example, a form may be a species on the basis of one species definition and a subspecies on the basis of another definition. Or certain forms may behave like subspecies of a single species in part of their range and like good species in other parts of their range. Border-line cases are by no means exceptional; in fact, we find a surprisingly high proportion of such situations in all the regions in which geographic or ecological conditions promote active speciation. Border-line cases may be classified as follows:

No criteria permit satisfactory distinction between species and isolated subspecies.—The ranges of two allopatric forms are often separated by a geographic gap, a form of distribution which is particularly common among island, mountain, and cave species, where there is, in all cases, a discontinuous distribution of the habitat. The taxonomist who adheres to a strictly morphological species definition is not particularly baffled by such cases. Every isolated form that is separate from its nearest relative by a clear-cut discontinuity of taxonomic characters is regarded by him as a good species. He is, in consequence, forced to admit as good species many isolated forms, which differ by very minute, but constant and unbridged differences. Such a proce-

dure is defensible, as long as we are merely interested in the pigeonholing of specimens in the correct collection cases. It becomes an absurdity when we view the species as a biological unit.

In the Atlantic Ocean, about one hundred miles east of Nova Scotia, there is a small, isolated land mass, Sable Island. On this island, and only on it, lives a sparrow, the Ipswich Sparrow (*Passerculus (sandwichensis) princeps*), which is unquestionably derived from the same stock as the Savanna Sparrow (*Passerculus sandwichensis*) of the mainland of North America. The differences are, however, very striking; the Sable Island bird is much larger (wing, ♂ 73.5–79.5, against 66–72 mm.), and of a distinctly different coloration (much more whitish) from the races [165] of the neighboring mainland. Even in the field the two forms can be told apart at a glance.

Some ornithologists hold that the Ipswich Sparrow is nothing but a subspecies of the Savanna Sparrow, whereas others insist that the morphological gap should be recognized as a species gap. Similar situations are encountered in nearly every well-worked taxonomic group.

It is of interest to find out how common such cases are. I have made an analysis of all the North American birds listed in the A. O. U. Checklist (1931), a work which is rather conservative in its taxonomic point of view. I have omitted only introduced species and the purely marine order Tubinares. In 374 genera there are 755 species with a total of 1,367 species and subspecies. At least 94 of the listed 755 full species of North American birds will be considered by some authors to be merely subspecies of other species. In other words 12.5 per cent of the species of North American birds have reached a very interesting taxonomic stage: They still show by their distribution and general similarity that they had been only recently geographical forms of some other species, but they have, in their isolation, developed morphological characters of such a degree of difference that the majority of authors now prefer to call them good species. Typical examples in the North American fauna are: Ipswich and Savanna Sparrow (*Passerculus*), Red-shafted and Yellow-shafted Flicker (*Colaptes*), Audubon's and Myrtle Warbler (*Dendroica*), the various species of the genera *Junco* and *Leucosticte*, etc. The majority of these forms are more or less isolated, either on the islands off the California coast or on the various mountain ranges of the Rocky Mountains or in the lowlands east and west of the Rocky Mountains. These "semi-species" comprise $12\frac{1}{2}$ per cent of the total of species in the rather continental fauna of North America. For a typically insular region, namely the Lesser

Sunda Islands, Rensch (1933) thinks that not less than 47 species are intermediate among a total of 160 species. I have analyzed the birds of the Solomon Islands and find that if we employ a narrow species concept there are 174 species of land and fresh water birds; if we, however, employ a wide species concept (= include within one species all geographical representatives) there are only 125 species. In other words, of 174 species there are 49 of intermediate status, that is 28.2 per cent [Mayr 1940a].

An even clearer impression can be gained if we analyze, in a similar manner, the entire bird fauna of a single island (Table 11).

groups is as follows: the gall wasps (*Cynipidae*) 76 percent, the salamanders of the family *Plethodontidae* 74 percent, the cave crickets of the genus *Ceuthophllus* 62 percent, the pond weeds of the genus *Potamogeton* 15 percent, the spiderworts of the genus *Tradescantia* 9 percent, and so forth. On the basis of these data, it seems as if animals tended more to the formation of localized, isolated forms than plants, although both *Tradescantia* and *Potamogeton* are rather "weedy" plants, and perhaps not typical for all plants.

These isolated forms, or "insular species," are excellent evidence in support of geographic speciation

TABLE 11

PERCENTAGE OF BORDER-LINE CASES ON ISLANDS IN THE PAPUAN REGION

NAME OF ISLAND	TOTAL NUMBER OF SPECIES	MAY BE CONSIDERED EITHER ENDEMIC SPECIES OR SUB-SPECIES OF MAIN-LAND SPECIES	BORDER-LINE FORMS, PERCENT OF TOTAL
Biak Island (+Numfor)	69	21	30
Rennell Island	34	7	20
Waigeu* (+Batanta)	71	3	4.2
Aru Islands*	72	2	2.8

* Passeres only.

The percentage of border-line cases depends on a number of factors (size, distance from mainland, and so forth) which will be treated in a later chapter. Only one of the birds of the British Islands, the red grouse (*Lagopus scoticus*), is a border-line case. It is by many authors considered to be a race of the continental *Lagopus lagopus*.

There is no doubt that similar conditions prevail in other animal groups and even in plants, but the taxonomy of most of these groups has not yet been clarified to the point where we can express the number of [166] border-line cases in actual percentage figures. Kinsey (1937b) divides species into two classes, continental and insular. A study of his data has convinced me that an ornithologist would call most of his "insular" species either subspecies or border-line cases. Kinsey's data are therefore of interest to us in this connection. According to him the percentage of insular species in various taxonomic

and, as such, welcome to the student of evolution; but, on the other hand, they are also very troublesome to the modern taxonomist, as far as their practical treatment is concerned. Our species definition included the statement: "A species consists of a group of populations which replace each other geographically which are potentially capable of interbreeding . . . where contact is prevented by geographical . . . barriers." The question remains, how can we determine which of these isolated forms are "potentially capable of interbreeding"? Unfortunately, there is no way of testing this in most cases, and we may as well admit that a decision is then possible only by inference. We must study other polytypic species of the same genus or of related genera and find out how different the subspecies can be [167] that are connected by intermediates, and, vice versa, how similar good sympatric species can be. This scale of differences is then used as a yardstick in the doubtful situations. And even after all these data have been given due consideration, the decision will often be, to a large extent, as Stresemann would say, "a matter of taste." Such arbitrary decisions have to be made in all modern taxonomic work. In a revision of the genus *Megapodius*, I proposed to unite, into one polytypic species, the seven species *nicobariensis, tenimberensis, reinwardt, freycinet, eremita, affinis,* and *layardi,* which had been recognized even by such progressive authors as Peters and Stresemann. The reasons were that I found not only that all these species were strictly allopatric, but also because the form (*macgillivrayi*) from the Louisiade Islands combined the characters of *reinwardt* and *eremita* and because members of a hybrid population between *affinis* and *eremita* (Dampier Island) were, by convergence, similar to *freycinet*. On the other hand, it was decided to retain as full species the forms *lapérouse* (Micronesia) and *pritchardi* (Niouafu, central Polynesia), which, although strictly allopatric, are separated from all other forms of the genus by very striking morphological as well as geographic gaps (Mayr 1938b). Meinertzhagen (1935: 765) lists *Alauda arvensis* and *A. gulgula, Apus apus* and *A. pallidus,* and *Riparia obsolta* and *R. rupestris* as typical border-line cases among birds.

Extreme morphological development of terminal subspecies.—The isolated forms of *Megapodius,* which are considered separate species, differ only in size, color, and proportions. But sometimes such isolated forms develop such a degree of difference that they might be considered different genera if they were judged only on morphological criteria. As a matter of fact, many of these forms have originally been described as separate genera, and their true systematic position has become clear only recently. That they are nothing but subspecies, or at best allopatric species, is particularly evident in cases in which the widely diverging species are the extreme ends of a long chain of intermediate subspecies.

The distribution map (Fig. 17) of the barking pigeon (*Ducula pacifica*) of Polynesia well illustrates the geographic conditions under which such extreme morphological development may occur. This species

has developed a form (*galeata*) on the Marquesas Islands which, on account of its peculiarly developed bill, was, until nine years ago, considered a good genus (*Serresius*).

Other genera that are based on morphologically distinct geographic forms are: in pigeons, *Oedirhinus* (of *Ptilinopus iozonus*) and *Chrysophaps* (of *Chrysoena luteovirens*); in kingfishers, *Todirhamphus* (of *Halcyon chloris*); in birds of paradise, *Taeniaparadisea* and *Astrarchia* **[169]** (of *Astrapia nigra*), *Schlegelia* (of *Diphyllodes magnificus*), *Uranornis* (of *Paradisaea apoda*); in drongos, *Dicranostephes* (of *Dicrurus bracteatus*); in rails, *Porphyriornis* (of *Gallinula chloropus*); in Passeres, *Galactodes* (of *Erythropygia*), *Conopoderas* (of *Acrocephalus*), *Pinarolestes* (of *Clytorhynchus*), *Papuorthonyx* (of *Orthonyx*), *Allocotops* (of *Melano cichla lugubris*); and so forth.

I could quote many other similar cases in which subspeciation, that is geographic variation, has actually brought about the formation of unquestionably new species of birds, Unfortunately, the systematics of most other groups of animals is not sufficiently well known to justify our drawing comparable conclusions, but Kinsey reports exactly the same situation in cynipid gall wasps (Kinsey 1930, 1936, 1937a, 1941).

The superspecies, a border-line situation.—Nobody will deny that all the strongly specialized allopatric forms which we have just listed are merely "glorified" geographic races, and it seems possible to combine groups of them into single species if one wanted to carry the principle of geographic representation to an extreme. This is just about what Kleinschmidt does in his *Formenkreise*. Rensch (1929, 1934) realizing that two rather different taxonomic concepts were hidden under the term *Formenkreise,* namely ordinary polytypic species and groups of allopatric species, proposed the term *Artenkreis* for the latter. I have suggested the replacement of this term, for more convenient international usage, by the term *superspecies*[2] since it is the supraspecific counterpart to the intraspecific unit, the subspecies (Mayr 1931a).

2. *Super,* beyond, is the counterpart of *sub,* below (or within); *supra,* above, is the counterpart of *infra,* below. The term supraspecies, used by several recent authors, seems to me to be an unfortunate combination. We can speak of supraspecific categories, as we speak of infraspecific factors, but as we use the term subspecies for a specieslike category that is below the species, we must use the term superspecies for a specieslike category that goes beyond the species. This corresponds to a similar usage of these prepositions in subgenus and supergenus, in subfamily and superfamily, and so forth.

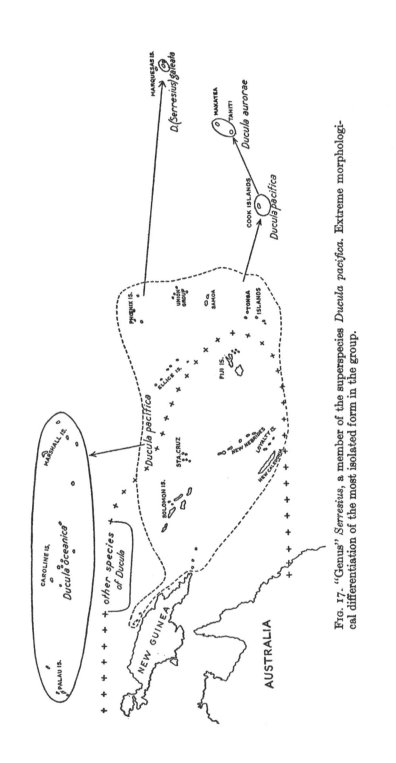

Fig. 17. "Genus" *Serresius*, a member of the superspecies *Ducula pacifica*. Extreme morphological differentiation of the most isolated form in the group.

A *superspecies consists of a monophyletic group of geographically representative (allopatric) species which are morphologically too distinct to be included in one species.* It is inconsequential whether the species of which the superspecies is composed are monotypic or whether some of them, break up into geographic races. The principal feature of the superspecies is that it presents, geographically, the picture of an ordinary polytypic species, but that morphologically these allopatric species are different to such a degree that reproductive isolation between them may be suspected. One of the most important aspects of the superspecies is that it is the highest category which can be delimited objectively, as is apparent **[170]** from the definition. Rensch has pointed out that the adoption of this concept affects in no way the nomenclature of the species which are involved and that no objections can be raised against it on this basis. On the other hand, it offers a number of considerable advantages in the preparation of faunal lists, in zoogeographic studies, and in discussions on speciation. Some critics (for example Meise 1938: 63) have proposed the elimination of the term superspecies by broadening the scope of the polytypic species to the point at which it includes all geographic representatives. If we go back to these Kleinschmidtian views, we shall have to include in one species the red and the yellow birds of paradise; we shall have to call all the *Astrapia*, all the *Parotia*, and all the juncos one species, to mention some avian examples. To call all these forms subspecies not only obscures their distinctness, but it violates even our species concept. There is some evidence that many of the species of which the superspecies are composed are reproductively isolated. On the other hand, it may also be called a mistake to list them merely as ordinary species, without combining them into superspecies, because this ignores an important relationship. The superspecies should be employed only in cases of strikingly different allopatric species. It would be an abuse of this concept if an author were to call every polytypic species, composed of insular and thus well-marked subspecies, a superspecies.

The members of a superspecies form a taxonomic and phylogenetic unit, all being descendants of one ancestral population. The recognition of the superspecies helps very materially to reduce the gap between subspecies and species, and, since every superspecies is a border-line case, it calls attention to these intermediate situations. The superspecies has its greatest practical importance in zoögeographic work. It is unwarranted to count the members of

superspecies as separate species, if we compare two faunas. For example, it is altogether misleading to say that Polynesia has more species of fruit doves (*Ptilinopus*) than New Guinea. Current check lists record 17 species from Polynesia and 11 species from the mainland of New Guinea, but there are only 3 superspecies in Polynesia as compared to 11 on New Guinea. The comparison of the number of superspecies indicates, therefore, much more accurately how rich the New Guinea fauna is in fruit doves than does a comparison of the number of species.

The ranges of several typical superspecies have been illustrated by me in a recent paper (Mayr 1940a, Figs. 2, 3, 4, 7). The superspecies *Zosterops rendovae* consists of three species: *Z. rendovae, Z. luteirostris,* and *Z. vellalavella;* the superspecies *Ducula pacifica* consists of the species *galeata, aurorae, pacifica, oceanica,* and perhaps *myristicivora,* and so forth. **[171]** The 10 species of the genus *Junco* which Miller recognizes in his recent revision of the genus (Miller 1941) comprise one extensive superspecies. In fact, it would seem proper to reduce the number of species of this superspecies to 3 or 4, since several of them interbreed freely in their zone of contact and would not pass as species on the basis of a biological species definition.

We have already encountered, in our discussion on the application of the polytypic species principle, many situations in which groups of closely related forms are best characterized as superspecies. Just a few cases from groups other than birds may be added. The Mediterranean lizard *Lacerta lepida* forms a superspecies with *L. atlantica* (eastern Canary Islands), *L. galloti:* (western Canary Islands), and *L. simonyi* (Gran Canaria) (Mertens 1928a). Most of the species of the genus *Orcula* seem to belong to one superspecies (St. Zimmermann 1932). Usinger (1941) lists superspecies in the hemipteran genus *Neseis* from Hawaii, and the 4 species of the Central American fish genus *Platypoecilus (couchianus, xiphidinus, variatus,* and *maculatus)* are another example, although these 4 "species" could equally well be considered subspecies of a polytypic species. The *Drosophila macrospina* group, as described by Patterson (1942), can also be cited as an illustration of a superspecies.

Superspecies are not exceptional cases. On the contrary, they comprise a regular and sometimes rather high percentage of every fauna. Rensch (1933: 29) has tabulated their occurrence among birds and snails and has found, as is to be expected, that they are particularly frequent where effective geographic barriers are present, that is, where insular ranges are involved. There are 17 superspecies (or 13.6 percent) among the 125 species of Solomon Island birds. Among the 9 widespread, mountain-inhabiting species of birds of paradise from New Guinea 3 are superspecies (33 percent). But even among continental species, the number of superspecies is rather high. Among the Palearctic Corvidae (crow family) there are 4 superspecies, in addition to 40 polytypic species; in the Palearctic starlings, one superspecies and 2 polytypic species; among the finchlike birds of the Palearctic there are 6 superspecies in addition to 70 polytypic species; and so forth (Rensch 1934: 50). At least a dozen superspecies of birds occur in North America (in the genera *Branta, Larus, Otus, Colaptes, Lanius, Dendroica, Leucosticte, Junco,* and so forth). The percentage of superspecies may be expected to be even higher among animals, which are more sedentary than birds. Among the European Clausiliidae (snails) there are no less than 10 superspecies, in addition to 16 polytypic species (Rensch, loc. cit.)

The superspecies is, of course, only a stage in the speciation scale, and **[172]** it is to be expected that it also has its border cases. The species pairs, *Acanthiza pusilla* and *ewingi, Tanysiptera hydrocharis* and *galatea, Ptilinopus dupetithouarsi* and *mercieri,* and *Lalage maculosa* and *sharpei,* which we shall discuss later in this chapter, must be mentioned here. Each of these pairs of species would be considered as belonging to the same superspecies if there was not some overlap of ranges. The superspecies is the stage at which the transition from allopatric to sympatric species is most likely to occur.

The indivisible gradient of the lower systematic categories.—We have stated repeatedly that every one of the lower systematic categories grades without a break into the next one: the local population into the subspecies, the subspecies into the monotypic species, the monotypic species into the polytypic species, the polytypic species into the superspecies, the superspecies into the species group. This does not mean that we find the entire graded series within every species group. It simply means that in the absence of definite criteria it is, in many cases, equally justifiable to consider certain isolated forms as subspecies or as species, to consider a variable species monotypic or to subdivide it into two or more geographic races, to consider well-characterized forms as subspecies of a polytypic species or to call them representative species.

In a revision of the neotropical snake genus

Dryadophis, Stuart (1941) recognizes 17 forms in 6 monotypic and 3 polytypic species, belonging to 4 species groups. Isolated forms which do not intergrade are considered full species. Many ornithologists would not recognize this criterion and, by considering some of the "species" subspecies, they would reduce the total number of species to 4 or 5 (*bifossatus, pulchriceps, pleei, amarali,* and *boddaerti,* with *heathii, melanolomus,* and *dorsalis* as subspecies). In the bunting genus *Junco* there are about seven or eight possible ways of delimiting the species, and none of the disagreeing authors can prove that his arrangement is more correct than the others. The presence of graded series and the absence of all decisive criteria makes it necessary to rely on subjective judgments. But the fact that so many geographic races stand on the border-line between subspecies and species is further proof of the importance of geographic speciation.

A similar gradient of categories may be observed, if we compare the degree of geographic variation of a number of related species in the same geographic region. Usinger (1941) describes very graphically such a situation among the hemiptera which have colonized the Hawaiian Islands. After arrival on the islands, these species

proceeded to diverge, and have now reached varying degrees of differentiation, the extent of which can not be determined without breeding experiments. **[173]** Thus the various species in the endemic genera fall into a series, ranging from (1) the widespread and variable *Oceanides nimbatus* Kirk., not yet broken up into distinguishable forms on the various islands, through (2) the scarcely differentiated *Neseis saundersianus* Kirk. to (3) the "polytypic species" (Huxley 1938) or "Rassenkreis" (Rensch 1929) *Neseis nitidus* White, which is structurally distinct but closely allied races on each island, then to (4) the "supra-species" (Huxley 1938) *Neseis hiloensis* Perkins, the Oahu form of which was unhesitatingly called a distinct species, until a connecting link was discovered on Molokai, and finally to (5) that which Huxley (1938) has called a "geographical subgenus" and Rensch (1929) has called an "Artenkreis," namely, the *Nescis mauiensis* Blackburn group which has diverged to such an extent that the Oahu and Kauai forms have attained the status of full species and had not even been recognized as belonging to this group previously.

A similar gradient of systematic categories was described by me for some islands and mountains of the Papuan Region (Mayr 1940a: 267).

Double Invasions

Oceanic islands are defined as all those islands that have received their fauna from other islands or from neighboring continents by transoceanic colonization, and not over land bridges (Mayr 1941b). The immigrants soon start to diverge from the original parent population (a process which is speeded up by the small size of most of these island populations) and if, after a sufficient time interval, a second set of immigrants arrives from the same source, the two waves of immigrants will behave like good species.

Simple cases.—Cases of double colonization are known from nearly every sufficiently isolated oceanic island, for example among birds, from Tenerife (*Fringilla teydea* and *F. coelebs canariensis*), western Canary Islands (*Columba laurivora* and *C. bollii*), Norfolk Island (*Zosterops albogularis, Z. tenuirostris, Z. lateralis norfolkiensis*), and Samoa (*Lalage maculosa* and *sharpei*). Double invasions also occur on continental islands, for example Ceylon (*Brachypternus erythronotus* and *B. benghalensis intermedius*), Luzon (*Pitta kochi* and *P. erythrogaster*), and Celebes (*Dicrurus montanus* and *D. hottentottus*). Isolated mountain peaks may present exactly the same phenomenon, since they act as distributional islands. *Dendrobiastes bonthaina* (together with *D. rufigula*) on the Pic of Bonthain (S. Celebes) and several of the endemic species of Mount Kina Balu (Borneo) may be explained in this manner. Willis (1940) has listed a number of endemic plant species on the mountains of Ceylon, for which the same manner of origin is probable.

Not a single case of double colonization is known to me from recent continental islands, such as Britain or Ireland, or from any oceanic island [174] (such as Biak or Rennell) which is situated close to a continent. The reason for this is obvious. Two species can develop from immigrant descendants of the same parent species only if the time interval between the first and the second colonization was sufficient to permit the earlier arrivals to develop sexual isolation. The new arrivals are simply absorbed by the earlier ones if this condition is not fulfilled. This is the reason why such twin species are absent from incompletely isolated islands.

Double colonizations of islands are, of course, not restricted to birds; it is only that the advanced condition of avian systematics makes their detection easier. I know of at least one well-analyzed case in butterflies. The common European Swallowtail (*Papilio machaon*) occurs as a single species in all of the Mediterranean countries. Two species of the *machaon* group are found only on the islands of Corsica and Sardinia, *Papilio machaon* subsp. and *Papilio hospiton* (Eller 1936: 79). *P. machaon* subsp. is closest to the Italian and southern French races of the *mediterraneus* group, and *P. hospiton* to the North African races (*saharae* group) of *P. machaon*. When the two sets of colonists met on Corsica and Sardinia, they had diverged sufficiently from each other not to interbreed. Some of the species of the hemipteran genus *Nysius* reported by Usinger (1941) seem also to belong here.

A particularly puzzling case is presented by the Tasmanian thornbill, which reveals how difficult it is to decide, purely on the basis of morphological criteria, whether an island form is a species or a subspecies (Mayr and Serventy 1938). On the island of Tasmania (AE) south of Australia, there are two very closely related species of *Acanthiza* (thornbill). One of these (A), *Acanthiza pusilla diemensi*, is very similar to the subspecies *Acanthiza pusilla pusilla* (B), of the mainland of Australia, opposite. The other species *Acanthiza ewingi* (E), which lives beside *diemensi* like a perfectly good species without any signs of interbreeding, is also fairly similar to B (*Acanthiza pusilla*) and clearly an earlier offshoot of B. However, E is as different from C (western Australia) as is B, but B and C are completely connected by intergradation and interbreeding. E is morphologically closer to B than is C, but since A also occurs on Tasmania, E cannot be considered a subspecies of B. There is no question that we would list E as a member of the species B . . C, if the second invasion (A) had not taken place on Tasmania and revealed the specific distinctness of E. This teaches us that analogy is a poor tool in analyzing these cases, and that in many of these border-line cases one guess is as good as another.

Archipelago speciation.—The chances of double invasions are particularly [175] favorable in archipelagos consisting of two or three good-sized islands. The representative subspecies which develop in isolation on these islands have a good chance to become in time so different that they can spread to the neighboring island without mixing. This probably explaim the presence of two species of related hummingbirds (*Eustephanus fernandensis* and *galeritus*) on Masatierra Island of the Juan Fernandez group; of two species of related doves on Hivaoa and Nukuhiva Islands in the Marquesas Islands (*Ptilino-*

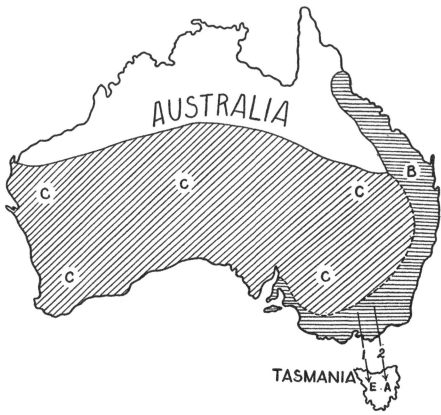

Fig. 18. Double invasion of Tasmania by *Acanthiza pusilla*. Completion of the speciation process proved by successful second colonization. A = *Acanthiza pusilla diemenensis*; B = *A. pusilla pusilla* group; C = *A. pusilla apicalis-albiventris* group; E = *Acanthiza ewingi*.

pus dupdithouarsi and *mercieri*); of two related flycatchers (*Mayrornis versicolor* and *M. lessoni orientalis*) on Ongea Levu, Fiji; and of two species of finches (*Nesospiza acunhae* and *Nesospiza wilkinsi*) on Nightingale Island, Tristan da Cunha.

These cases are forerunners of that amazing speciation which has **[176]** taken place on ancient archipelagos, such as Hawaii or the Galápagos Islands. The case of the Geospizidae on the Galápagos Islands has been excellently reviewed by D. Lack (1940a, 1942), while the more ancient and more complicated case of the Drepanididae on Hawaii has, so far, defied adequate analysis. Even richer species swarms than those of the Drepanididae have developed among the Hawaiian invertebrates.

The *Proterhinus* weevils with one hundred and fifty species, Cerambycids of the genera *Plagithmysus* and *Neoclytarhus*, Lygaeid bugs of the genera *Nysius, Ne-*

seis and *Oceanides* in the tribe Orsillini and a host of other genera in all the principal orders of Hawaiian insects, have developed unique branches of from six to over one hundred species. Each of these is a small phylogenetic world in itself. Here we find geographical replacement, well developed, with distinct forms on each separate island and often on each host [Usinger 1941].

The same is true for the Hawaiian snails and in particular for the genus *Achatinella*. There is no doubt that archipelago speciation presents some of the most instructive examples of geographic species formation.

Partial Distributional Overlap

Border invasions.—Another class of border-line situations is presented by cases in which two otherwise allopatric species show a slight overlap of ranges.

Particularly interesting are those cases in which the two representative species are so similar that they would probably be considered subspecies, if it was not for the existence of the area of overlap. Even so, the entire distributional picture indicates the former subspecific relationship. Cases of this sort are not frequent, because, aside from some ecological competition, there is no reason why the two species should remain largely allopatric after the biological isolating mechanisms have developed to the point of complete reproductive isolation. As soon as such a species moves back into the range of a sister species (stage 4, p. 160), it is likely to spread so fast that all traces of the original allopatric condition are soon wiped out. This is particularly true for all large genera (with numerous species). Cases of slight overlap, as we shall presently see, therefore indicate generally a rather recently completed establishment of discontinuity between species.

The overlap is usually due to the recent breakdown of a geographic or ecological barrier. *Tanysiptera galatea* now lives in South New Guinea side by side with *hydrocharis,* because the arm of the sea that had separated them previously (Fig. 15) recently dried up. *T. hydrocharis* lived on an island which connected the [177] Islands with the Oriomo River plateau. *T. galatea* was restricted to the mainland of New Guinea, and offshore islands. When the erosion debris of the rapidly rising central range of New Guinea filled the sea, *T. galatea* was enabled to intrude into the formerly isolated range of *hydrocharis,* but no interbreeding took place. A similar situation exists in a Venezuelan snake (Stuart 1941). *Dryadophis amarali* developed apparently from *pleei* stock during insular isolation on Tobago Island or on the Paria Peninsula. Recent geological events have led to an overlap of its range with that of *pleei,* but are no signs of interbreeding. A third situation of the same sort has been described in the case of a Florida dragon fly. The Florida species *Progomphus alachuensis* and the Cuban species *P. integer* developed in insular isolation from the eastern North American species *P. obscurus.*

The reunion of the central Florida Island (Pleistocene) with the mainland North America brought the ranges of *P. obscurus* and *P. alachuensis* intact with each other . . . and *P. obscurus* invaded north and north central Florida. In north-central Florida the invading species overlapped range of the endemic one, but remained ecologically distinct, inhabiting rivers and streams, leaving the lakes for the species already established [Byers 1940].

More frequent than the joining of an island with the mainland is range expansion due to the breakdown of ecological barriers in connection with climatic changes. The coming and the going of the ice during the Pleistocene age has been responsible for a great many such changes, of which very few have as yet been analyzed. It is not always clear whether the isolation was due to glaciation or occurred at an earlier date. Some authors, notably Salomonsen (1931), list a very high number of European species pairs as being due to Pleistocene separation; other authors hold that this separation was only in exceptional cases long enough to permit the development of interspecific gaps. The final decision cannot be reached until we know more details as to climate and plant distribution during Miocene and Pliocene. Until such time, cases discussed below will have to be treated with some reservation. There is an eastern and a western species in many genera of European birds. Stresemann (1919), who studied the distribution of the western Tree Creeper (*Certhia brachydactyla*) and the eastern Tree Creeper (*C. familiaris*), suggested that this peculiar pattern of distribution was of Pleistocene origin. When, at the height of glaciation, the Scandinavian and the Alpine ice caps approached each other in central Europe to within a distance of about 200 miles, they forced all European animal life into a southwestern (southern France, Spain) and southeastern (Balkans) refuge. During this period of isolation, the parental *Certhia* population developed specific differences and reproductive isolation and did not interbreed [178] when the ranges of the expanding species finally met. (But see Steinbacher 1927.) Today the two creepers occur side by side without interbreeding, in a broad zone which extends from northwestern Germany to the Alps. Salomonsen (1931) explains on the same basis similar species pairs in the avian genera *Hippolais* (*polyglotta* western, *icterina* eastern), *Luscinia* (*megarhyncha* western, *luscinia* eastern), and *Muscicapa* (*hypoleuca* western, and *albicollis* eastern). A number of parallel situations exist among European amphibia. The western toad, *Bombina variegata,* became a mountain form during the glacial separation, while the eastern toad, *Bombina bombina,* remained a lowland species. Range expansion after the retreat of the ice led to a considerable distributional overlap, but the two species remain effectively isolated, since they occur at different altitudinal levels. At a few localities there is an overlap of the altitudinal ranges, and it is in such places that intermediate (hybrid) individuals have been found. The two species can be hybridized in captivity without difficulty (Mertens 1928a, b). A similar overlap of ranges is shown in the case of two central European frogs, but the ecological factor which keeps the two species separate is, in this case, the breeding season. The western species (*Rana esculenta*) breeds from the end of May well into June, while the eastern species (*Rana ridibunda*) completes its breeding season before the end of May (Mertens 1928a). The two species of newts *Triturus cristatus* and *T. marmoratus* developed apparently during glacial separation. There is now a narrow zone of overlap in central France, in which a few hybrids with reduced fertility have been observed ("*T. blasii*"). A similar case in presented by *Triturus vulgaris* and *T. helveticus*. The present overlap between these two species is very considerable, comprising the British Isles and the region between eastern France and western Germany. *T. helveticus* prefers the mountains, *T. vulgaris* the lowlands, but both have been found in the same waters, where lack of sexual affinity prevents interbreeding (W. Herre 1936). The glaciers, which at the height of the Pleistocene era advanced into the Po basin from the southern foot of the Alps, separated very effectively a number of snail and insect populations, which lived on southern spurs or foothills of the Alps. These populations expanded when the ice retreated, and, even though they are still largely allopatric, there are now a number of places where two of such "forms" overlap without any signs of interbreeding. Good examples of this can be found in the work of Klemm (1939) on the snail genus *Pagodulina* and of St. Zimmermann (1932) on *Orcula.*

A few similar cases have also been described from North America, although isolation was not as long-continued and effective as in Europe, **[179]** where the formidable barrier of the Alps has been so important. The geographic ranges of the two mice *Peromyscus leucopus* and *gossypinus* are exclusive except for some areas of overlap in the Dismal Swamp of Virginia, in northern Alabama, and, more widely, in the lower Mississippi Valley. They occupy, in part, the same habitats, where their ranges overlap; but there is no evidence of any interbreeding in nature, except for two presumed hybrids reported from Alabama. The two species are very similar and fully fertile in the laboratory (Dice 1940b). Quite a number of similar cases have been described from North American snakes, of which I shall report only a few.

The polytypic species (or species group) *Crotalus atrox* (diamondback rattler) exhibits very clearly the effects of isolation during the height of the Pleistocene age (Gloyd 1940). The species became separated into three portions, one on the west coast of North America, which developed into *ruber;* a second one in Mexico, which developed into *atrox;* and a third one in Florida, which became *adamanteus.* Additional populations were isolated on the tip of Lower California (*lucasensis*) and on some islands near Lower California (*exsul* and *tortugensis*). After the retreat of the ice, the populations expanded northward, but the gap between *adamanteus* and *atrox* in the lower Mississippi Valley was never closed. However, *atrox* moved westward until it reached the border of the range of *ruber* in the western part of San Diego County, California. There are no hybrids or intergrades known from this district, but the forms (species?) seem to be ecologically separated. Another interesting case of speciation is presented by the species *Crotalus viridis* and *mitchellii*, which are very similar and the ranges of which are still largely exclusive. In southern California there is, however, a considerable area of overlap, without any signs of intergradation or hybridization. Among the North American bull snakes (*Pituophis*) there is an overlap of the ranges of the species of *catenifer* and *sayi*, which species indicate by their pattern of distribution that they were formerly subspecies of a single polytypic species (together with *melanoleucus*) (Stull 1940). The overlap results occasionally in a limited amount of hybridization without an actual breaking down of the species limits. This happens for example, where the moth *Platysamia cecropia* overlaps the ranges of the closely related species *nokomis, columbia,* and *gloveri* (Sweadner 1937).

All these cases have one feature in common, namely, that owing to range expansion two formerly allopatric forms begin to overlap and to prove thereby to be good species. If no overlap existed and if we had to classify these forms merely on the basis of their morphological distinctness, we would probably decide, in most cases, that they were subspecies. **[180]** But overlap without interbreeding shows that they have attained species rank.

Overlap of the terminal links of the same species.—The perfect demonstration of speciation is presented by the situation in which a chain of intergrading subspecies forms a loop or an overlapping circle, of which the terminal forms no longer interbreed, even though they coexist in the same localities. To be sure, such speciation by force of distance is much rarer than speciation by strict isolation, but at the same time these cases demonstrate species formation by geographic variation in the most perfect manner. One of the reasons why such cases have not

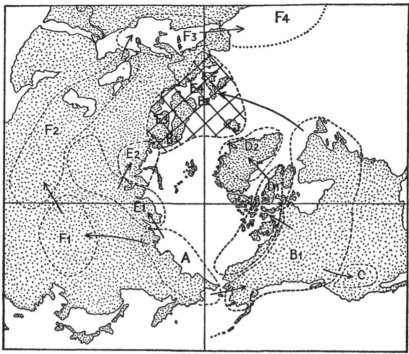

Fɪɢ. 19. Circumpolar projection of the ranges of the forms of the *Larus argentatus* group, showing overlap of the terminal links of a chain of races. A = *vegae;* Bɪ = *smithsonianus;* B2 = *argentatus;* B3 = *omissus;* C = *californicus;* Dɪ = *thayeri;* D2 = *leucopterus;* Eɪ = *heuglini;* E2 = *antelius;* E3 = *fuscus;* E4 = *graellsi;* Fɪ = *mongolicus;* F2 = *cachinnans;* F3 = *michahellis;* F4= *atlantis.—L. fuscus* (with *graellsi*) lives now beside *L. argentatus* (with *omissus*) like a good species. (From Mayr 1940a.)

been recorded in the literature more frequently is a purely psychological one. The puzzled systematist who comes across such cases is tempted to "simplify" them by making two species out of one ring, without frankly telling the facts. Overlapping rings are disturbing to the orderly mind **[182]** of the cataloguing systematist, but they are welcome to the student of speciation.

In birds such cases are rather frequent, even though the situation is generally more complex than can be indicated in the subsequent discussion. The Great Titmouse (*Parus major*), for example, was apparently split into at least three groups during the Pleistocene (Rensch 1933). As the three groups came together again after the retreat of the ice, they either formed broad or narrow hybridization zones or they expanded into the same area (upper Amur Valley), behaving like good species. We now have both *minor* and *major* in the Amur Region, without signs of intergradation or hybridization, although the two

"species" are connected via China-India-Persia through a completely linked chain. Similar cases are those of the *Larus argentatus* group in northwestern Europe (Fig. 19) and of the *Halcyon chloris* group in the Palau Islands (Fig. 20). Additional cases are those of *Zosterops* in the Lesser Sunda Islands, of *Lalage* in southern Celebes, and of the honey buzzards (*Pernis*) in the Philippines. A more detailed analysis of the relationship of the babblers *Eupetes caerulescens* and *nigricrissus* in the Wanggar district of New Guinea may also lead to similar conclusions.

The warbler *Phylloscopus trochiloides* (Fig. 21) has a wide distribution in Asia. It occurs over most of northern Asia and also on the mountains which surround the arid central-Asiatic plateau. Two forms (*viridanus* and *plumbeitarsus*) meet in the Altai Mountains (western Sayan and Uriankhai) without interbreeding. The two forms are connected by a gapless chain of inter grading subspecies: *obscuratus, trochiloides,* and *ludlowi*. The area of overlap is prob-

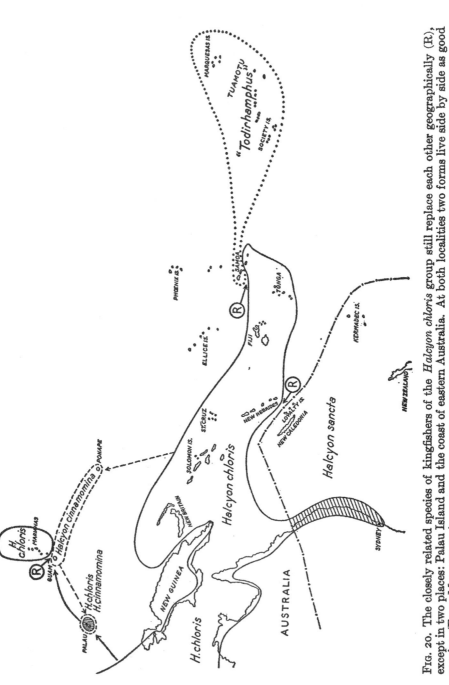

Fig. 20. The closely related species of kingfishers of the *Halcyon chloris* group still replace each other geographically (R), except in two places: Palau Island and the coast of eastern Australia. At both localities two forms live side by side as good species. (From Mayr 1940a.)

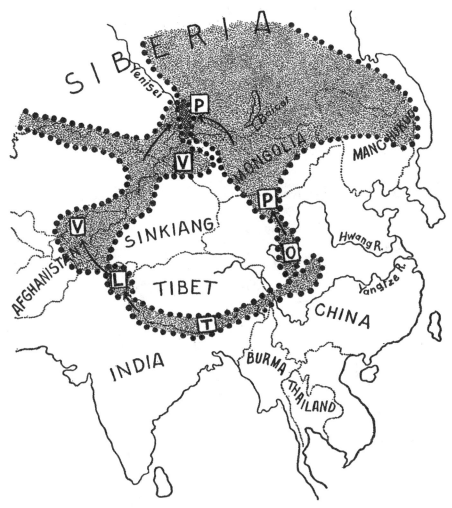

Fig. 21. Overlap of two terminal links in a ring of subspecies of the warbler *Phylloscopus trochiloides*. The subspecies of this ring are: V = *viridanus*; L = *ludlowi*; T = *trochiloides*; O = *obscuratus*; and P = *plumbeitarsus*. The overlap between *viridanus* and *plumbeitarsus* in the district between the western Sayan Mts. and the Yenisei River is indicated by cross-hatching. (From Ticehurst 1938a.)

ably rather recent—of post-Pleistocene origin (Ticehurst 1938a).

In the species *Phylloscopus collybita* there is another possible case of coexistence of two "subspecies" within the same area. The race *abietinus*, which came from northern Europe, meets in the western Caucasus the subspecies *lorenzii*, which came from the Himalayas and the western central Asiatic mountains. Specimens of both forms have been collected during the breeding season in the same localities, although, in the main, the ranges of the two forms exclude each other. Nothing is known about possible differences of song, habits, and habitat in the area of overlap, but there are some indications that *lorenzii* is largely an altitudinal representative of *abietinus* (Ticehurst, op. cit.: 42–52). The ring is closed in the Pamir-Altai region, through the forms *tristis* and *sindianus*. The case of the House Sparrow (*Passer domesticus*) and the Willow Sparrow (*P. hispaniolensis*) is slightly different (see p. 268), but agrees in one [183] respect, namely, that two forms may have reproductive isolation in part

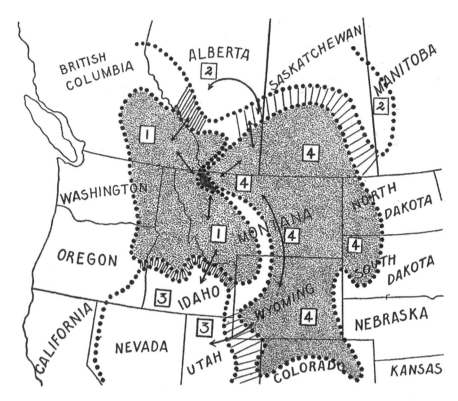

Fig. 22. Overlap (without interbreeding) of subspecies of the deermouse *Peromyscus maniculatus* in Glacier National Park. 1 = *artemisiae;* 2 = *arcticus;* 3 = *sonoriensis;* 4 = *osgoodi.* Habitat segregation is more or less maintained in the zone of overlap. (After Osgood 1909 and other sources.)

of their range and may be interbreeding in other parts.

The known number of cases of circular overlap in other animal groups is constantly increasing. The butterfly *Junonia lavinia* colonized the West Indies from South America and from North America (W. T. M. [184] Forbes 1928). The two colonizing lines met in Cuba, where they now live side by side without interbreeding. Goldschmidt's question (1940: 120): "Would they be able to mate and produce fertile offspring, if brought together?" is beside the point. The point is not what they would do under the artificial conditions of captivity, but what they do in nature. Many good species can be crossed in captivity, but that does not in the least weaken their status as good species. A very interesting overlapping circle of races exists in the mouse *Peromyscus maniculatus.* In Glacier National Park, Montana, a forest-inhabiting subspecies, *P. m. artemisiae,* meets a grassland race, *P. m. osgoodi,* with no evidence of interbreeding

(Murie 1933). The failure of the two subspecies to interbreed in the zone of overlap is only partly due to the differences in their ecological requirements, for at some places near the margins of their habitats [185] the two races live together without interbreeding. The two forms would undoubtedly be considered good species if the chain of intergrading races, now connecting them, were broken. (Fig. 22).

The evidence discussed by me on pages 162 to 185 is, it seems to me, conclusive proof for the existence of geographic speciation: If an isolated population of a species remains long enough in this isolation, it may acquire biological isolating mechanisms which permit it, after the breakdown of the isolating barrier, to exist as a separate species within the range of the parental species. The reproductive isolation, which originally was maintained by the extrinsic means of a geographic barrier, is being replaced during this isolation by intrinsic isolating barriers. One species has developed into two.

Works Cited

Buch, L. von. 1825. Physicalische Beschreibung der Canarischen Inseln, 132–133. Kgl. Akad. Wiss., Berlin.

Byers, C. F. 1940. A study of the dragon-flies of the genus *Progomphus,* etc. Proc. Florida Acad. Sci., 4: 19–86.

Dice, L. R. 1940b. Speciation in Peromyscus. Amer. Nat., 74: 289–298.

Dobzhansky, Th. 1940. Speciation as a stage in evolutionary divergence. Amer. Nat., 74: 312–321.

———. 1941a. Genetics and the origin of species. Revised edition. Columbia Univ. Press, New York.

Eller, K. 1936. Die Rassen von Papilo machaon. Abh. Bayer. Akad. Wiss., N.F., H. 36: 1–96.

Forbes, W. T. M. 1928. Variation in *Junonoia lavinia* (Lep., Nymphalidae). J. N. Y. Ent. Soc., 36: 306.

Gloyd, H. K. 1940. The rattlesnakes, genera Sistrurus and Crotalus. Chicago Acad. Sci., Spec. Publ., No. 4: 1–270.

Goldschmidt, R. 1934. Lymantria. Bibl. Genet., 11: 1–180.

———. 1940. The material basis of evolution. Yale Univ. Press, New Haven.

Herre, W. 1936. Über Rasse und Artbilduing. Studien an Salamandriden. Abh. Ber. Mus. Naturk. Magdeburg, 6: 193–221.

Kinsey, A. C. 1930. The gallwasp genus Cynips. Indiana Univ. Studies, 16: 1–577.

———. 1936. The Origin of Higher Categories in Cynips. Indiana Univ. Publ., Sci. Ser., No. 4: 1–334.

———. 1937a. Supra-specific variation in nature and in classification. From the viewpoint of zoology. Amer. Nat., 71: 206–222.

———. 1937b. An evolutionary analysis of insular and continental species. Proc. Nat. Acad. Sci., 23: 5–11.

———. 1941. Local populations in the gallwasp *Biorrhiza eburnea.* Genetics, 26: 158.

Klemm, W. 1939. Zur rassenmässigen Gliederung des Genus *Pagodulina* Clessin. Arch. Naturg., N.F., 8: 198–262.

Lack, David. 1940a. Evolution of the Galapogos Finches. Nature, 146: 324ff.

———. 1942. The Galapagos Finches (Geospizinae): A study in variation. Proc. Calif. Acad Sci. (in press).

Mayr, E. 1931a. Notes on *Halcyon chloris* and some of its subspecies. Amer. Mus. Novit., No. 469: 1–10.

———. 1938b. Notes on New Guinea birds. IV. Amer. Mus. Novit., No. 1006: 11.

———. 1940a. Speciation in phenomena in birds. Amer. Nat., 74: 249–278.

———. 1941b. The origin and the history of the bird fauna of Polynesia. Proc. Sixth Pac. Sci. Congr. (1939), 4: 197–216.

Mayr, E., and D. L. Serventy. 1938. A review of the genus *Acanthiza* Vigors and Horsfield. Emu, 38: 245–292.

Meinertzhagen, R. 1935. The races of *Larus argentatus* and *Larus fuscus,* etc. Ibis, (13) 5: 765.

Meise, W. 1938. Fortschritte der orithologischen Systematik seit 1920. Proc. Eighth Int. Orn. Congr. Oxford (1934): 49–189.

Mertens, R. 1928a. Über den Rassen- und Artenwandel auf Grund des Migrationsprinzipes, dargestellt an einigen Amphibien und Reptilien. Senckenbergiana, 10: 81–91.

———. 1928b. Zur Naturgeschichte der europäischen Unken. Zeitschr. Morph. Okol., 11: 613–628.

Miller, A. H. 1940. Climatic conditions of the pleistocene reflected by the ecological requirements of fossil birds. Proc. Sixth Pacific Sci. Congr., 1939, Geol.: 807–810.

———. 1941. Speciation in the avian genus *Junco.* Univ. Calif. Publ. Zool., 44: 173–434, 33 figs.

Muller, H. J. 1940. Bearings of the 'Drosophila' work on systematics. In Huxley, J., The New Systematics, 185–268.

Murie, A. 1933. The ecological relationship of two subspecies of *Peromyscus* in the Glacier Park Region, Montana. Univ. Mich. Occ. Pap. Mus. Zool., 270: 1–17.

Patterson, J. T. 1942. Drosophila and speciation. Science, 95: 153–159.

Rensch, B. 1929. Das Prinzip geographischer Rassenkreise und das Problem der Artbildung. Borntraeger Verl., Berlin.

———. 1933. Zool. Systematik und Artbildungsproblem. Verh. Dtsch. Zool. Ges., 1933: 19–83.

———. 1934. Kurze Anweisung für zoologisch-systematische Studien. Akademische Verlagsgesellschaft, Leipzig.

———. 1939a. Typen der Artbildung. Biol. Rev., 14: 180–222.

Salomonsen, F. 1931. Diluviale Isolation und Artbildung. Proc. VII. Int. Ornith. Congr. Amsterdam (1930): 413–438.

Spencer, W. P. 1940. Levels of divergence in Drosophila speciation. Amer. Nat., 74: 299–311.

Stalker, H. 1942. *Drosophila virilis-americana* Crosses. Genetics, 27: 238–257.

Steinbacher, F. 1927. Die Verbreitungsgebiete einiger europäischer Vogelarten als Ergebnis der geschichtlichen Entwicklung. J. Ornith., 75: 535–567.

Stresemann, E. 1919. Über die europäischen Baumläufer. Verh. Ornith. Ges. Bayern, 14: 39–74.

———. 1931. Die Zosteropiden der indo-australischen Region. Mitt. Zool. Mus. Berlin, 17: 201–238.

Stuart, L. C. 1941. A revision of the genus *Dryadophis*. Misc. Publ. Mus. Zool. Univ. Mich., No. 49: 1–106, 4 pls.

Stull, O. G. 1940. Variations and relationships in the snakes of the genus *Pituophis*. U.S. Nat. Mus., Bull. 175: 1–225.

Sweadner, W. R. 1937. Hybridisation and the phylogeny of the genus *Platysamia*. Ann. Carnegie Mus., 25: 163–242.

Ticehurst, Cl. B. 1938a. A Systematic Review of the Genus Phylloscopus. Brit. Mus. London.

Usinger, R. L. 1941. Problems of insect speciation in the Hawaiian Islands. Amer. Nat., 75: 251–263.

Wagner, M. 1889. Die Entstehung der Arten durch räumliche Sonderung. Benno Schwalbe, Basel.

Willis, J. C. 1940. The course of evolution by differentiation, etc. Univ. Press, Cambridge.

Wright, S. 1941c. The material basis of evolution. Sci. Monthly, 53: 165–170.

Zimmermann, S. 1932. Über die Verbreitung und die Formen des Genus *Orcula* Held in den Ostalpen. Arch. Naturg., N.F., 1: 1–56.

From *Darwin's Finches*
David Lack

Chapter XIII:
The Origin of Subspecies

How has it happened in the several [Galapagos] islands situated within sight of each other, having the same geological nature, the same height, climate, etc., that many of the immigrants should have been differently modified, though only in a small degree. This long appeared to me a great difficulty: but it arises in chief part from the deeply-seated error of considering the physical conditions of a country as the most important for its inhabitants; whereas it cannot be disputed that the nature of the other inhabitants with which each has to compete, is at least as important, and generally a far more important element of success. . . . When in former times an immigrant settled on any one or more of the islands, or when it subsequently spread from one island to another, it would undoubtedly be exposed to different conditions of life in the different islands, for it would have to compete with different sets of organisms. . . . If then it varied, natural selection would probably favour different varieties in the different islands.

Charles Darwin: The Origin Of Species, Ch. XII

Degree of Difference Shown
by Island Forms

The recognition of geographical variation in animals, like the composition of the Star-spangled Banner, came as a by-product of the otherwise unfortunate war of 1812. When replenishing his food supplies with Galapagos tortoises, Captain Porter of the U.S. frigate *Essex* noticed that the specimens from different islands were different in appearance. The point was again noticed by Mr. Lawson, the vice-governor of the colony on Charles in 1835, whose name is re-

membered because he happened to mention the matter to the visiting naturalist of the *Beagle*. The collections then made by Darwin established the existence of distinctive island forms not only in the Galapagos tortoise, but also in the mockingbird *Nesomimus*, in some of the plants such as *Scalesia*, and in some of the finches, though observations on the last group were obscured by his unfortunate mixing of specimens before he became aware of the peculiar state of affairs. Darwin's realization that a species may be represented by **[116]** different forms in different regions was one of the most important results of the voyage of the *Beagle,* since it led directly to his questioning the immutability of species.

The extent to which Darwin's finches are divided into island forms differs considerably in different species. Thus in the vegetarian tree-finch *Camarhynchus crassirostris,* the birds from different islands show few, if any, differences in either plumage, beak or size. Likewise, the three ground-finches *Geospiza magnirostris, G. fortis* and *G. fuliginosa* show no differences in plumage on different islands; but they differ fairly markedly in the average size of beak and wing. An opposite tendency is found in the warbler-finch *Certhidea,* in which the island forms differ markedly in colour of plumage, but less obviously in size of beak and wing. Greater differences occur in the sharp-beaked ground-finch *Geospiza difficilis,* in which there are three clearly defined subspecies differing from each other in plumage, beak, wing-length and ecological niche, while each of these races is in turn slightly different on the two islands on which it occurs. In the cactus ground-finch *G. scandens,* the James and Bindloe forms show differences as marked as those which separate some of the ground-finch species (compare the heads in Fig. 11 (iv), p. 68 and Fig. 12 (i), p. 69). Finally, the large cactus ground-finch *G. conirostris* differs so markedly from *G. scandens* that, though the two

Reprinted with the permission of Cambridge University Press.
  See the footnote on page 789.

forms replace each other geographically, it is doubtful whether they represent pronounced geographical forms of the same original species, or whether they have originated separately; in either case, the differences between them are so great that the birds are classified as separate species.

To sum up, the island forms of Darwin's finches show every stage of differentiation, from differences that are barely perceptible to others that are as marked as those which separate species. These differences are primarily in the size and shape of the beak, the size of the body, and the general shade and the amount of streaking of the female plumage. It is in just these same characters that the individual specimens of one form show differences from each other, and the island populations have presumably diverged from each other by an accumulation and restriction of such individual differences. It is now generally agreed that in animals the differences between geographical [117] races of the same species are hereditary, though in Darwin's finches, as in most other birds, experimental proof of this statement is as yet lacking.

Adaptive and Non-Adaptive Differences

Darwin accounted for the existence of distinctive island forms in the manner quoted at the head of this chapter. As he pointed out, physical and climatic conditions are very similar on the various Galapagos islands, so that there is no reason to think that the differences between island forms are correlated with differences in the physical conditions to which they are subjected. On the other hand, Darwin's postulate that differences in the nature of the competing species are important is abundantly borne out by the data in the latter part of Chapter VI. It was there shown that the small outlying Galapagos islands have a different complement of ground-finch species from the central islands; as a result some species occupy different ecological niches on different islands, in which case the island forms show corresponding differences in the shape of the beak.

There are many other variations between island forms which cannot be related either to differences in the nature of the competing species or to any other differences in the conditions on different islands. This applies to some of the variations in the colour of the female plumage discussed in Chapter IV, to differences in the proportion of fully plumaged males discussed in Chapter V, and to many of the differences in the size of beak and wing discussed in Chapter VII. Both the similarity of the island environments, and the absence of regular trends of variation in the finches, strongly suggest that many of the differences involved are without adaptive significance. As stated earlier, such a negative view is extremely difficult to establish with certainty. There may well be some cases in which an adaptive correlation exists, but has so far been overlooked. But it is extremely unlikely that such a correlation is present but has been overlooked in all of the many cases involved.

A similar problem is presented by the Galapagos reptiles. Many of the islands possess peculiar forms of the tortoise *Testudo*, the snake *Dromicus*, the lizard *Tropidurus* and the gecko *Phyllodactylus*. The ecological niches occupied by these animals seem [118] closely similar on different islands, and it is extremely difficult to believe that the differences in appearance between the island forms are in all cases adaptive. The same applies to many of the differences between the island forms of the Galapagos mockingbird *Nesomimus*, though, as already noted, the long beak of the Hood form is perhaps adaptive and correlated with its shorefeeding habits.

Unfortunately, the only experimental test of the above views is impracticable, namely, to interchange every single individual of each of two island forms, leaving the two islands and their faunae otherwise unchanged, and then to observe the evolution of each form in the environment of the other over a period of a few hundred, or a few thousand, years. On the view adopted here, there would in many cases be no particular tendency for the transferred island form to evolve the distinctive characters of the previous occupant. But in the absence of such an experiment, this view remains unverified.

Importance of Isolation

In continental birds, colour of plumage and average size of beak or wing often show a very gradual change (a cline) over an extensive region, followed by an abrupt change over a narrow region. The regions of rapid change form convenient boundaries at which to delimit geographical races or subspecies, and the narrowness of such zones has been attributed to the fact that hybrids between populations are usually at a disadvantage compared with either parent form (Huxley, 1942). The zones of rapid change tend to occur where intermixing of populations is restricted by geographical obstacles, but they sometimes occur in the absence of the latter, while at the other extreme some continental races are isolated by geographical barriers as completely as are island races. Many of the differences between continental races are undoubtedly adaptive, as mentioned for plumage in Chapter IV and for size in Chapter VII, but in other cases the differences seem as pointless as those which so often separate island races.

Some might claim that all the differences between geographical races of birds are really adaptive; that if any seem pointless, this is due merely to ignorance, and further study will reveal their [119] adaptive significance. But in this connection, a comparison between continental and island races of birds is

often provide very similar environments, so that one might expect any adaptive differences between island races to be much smaller than those separating continental races. Despite this, island races differ from each other much more strikingly than do continental races. This principle is illustrated by almost every bird species whose range includes both continental areas and islands, and the more isolated the island, the greater the degree of differentiation found among its land birds. The primary cause of geographical variation in birds would seem to be not adaptation, but isolation. This point is demonstrated for Darwin's finches in Table XVII (overleaf).

Table XVII and Fig. 22 show that the most iso-

TABLE XVII. ISOLATION AND ENDEMISM IN DARWIN'S FINCHES

Island	Degree of isolation	No. of resident species	Endemic subspecies not found on other islands	
			No.	%
Cocos	Very extreme	1	1	100
Culpepper and Wenman	Extreme	4	3	75
Hood	Marked	3	2	67
Tower	Marked	4	2	50
Chatham	Moderate	7	2½	36
Abingdon and Bindloe	Moderate	9	3	33
Charles	Moderate	8	2	25
Albemarle and Narborough	Small	10	2	20
Barrington	Small	7	1	14
James	Very slight	10	½	5
Jervis	Very slight	9	—	0
Indefatigable	Very slight	10	—	0
Duncan	Very slight	9	—	0

Notes. (i) The number of endemic forms is somewhat arbitrary and is based on Table IV, p. 20. If the classification by Swarth (1931) had been used, the same general correlation would have been apparent, but the percentage of endemic forms would have been higher.

(ii) The forms of *Geospiza scandens* were merged for purely practical reasons of nomenclature (see p. 20), and for assessing insular differentiation the Abingdon and Bindloe birds may be reckoned as distinct, and those from James and Chatham as half-differentiated.

(iii) If Darwin's specimens of *G. magnirostris* and *G. nebulosa* came from Charles (see pp. 22–3), then the figures for Charles should be 10 resident species, 4 endemic forms, or 40 per cent.

(iv) The degree of isolation is assessed from the map. In three cases it was considered best to group a pair of islands together, as they are near each other but distant from all other islands.

highly illuminating. Over large land areas the environment shows gradual but marked changes, so that there is every reason to expect continental races to differ from each other adaptively, and in fact they often do so. On the other hand, neighbouring islands

lated islands, namely, Cocos, Culpepper, Wenman, Tower and Hood, have a much higher proportion of peculiar forms than have the central Galapagos islands, while on moderately isolated islands, such as Abingdon, Bindloe, Chatham and Charles, there is

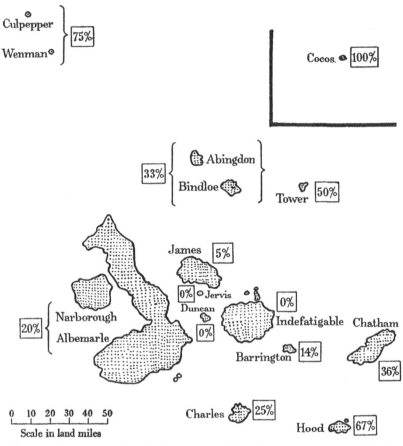

Fig. 22. Percentage of endemic forms of Darwin's finches on each island,
showing effect of isolation.

an intermediate condition, with proportionately fewer endemic forms than on the remote islands, but proportionately more than on the central islands. Hence in Darwin's finches there is a marked correlation between the degree of isolation and the tendency to produce peculiar forms. In some cases the peculiarities of the forms on the outlying islands are adaptive, being correlated with differences in ecological niche, while in other cases the differences seem unrelated to any possible environmental differences.

The taxonomic review by Swarth (1931) shows that the same principle holds for other Galapagos land birds. For instance, the ground-dove *Nesopelia galapagoensis,* found throughout the archipelago, is represented by a distinctive race on remote Culpepper and Wenman. The mockingbird *Nesomimus* is most distinctive on Hood, Charles and Chatham. The latter islands are more isolated from the bird

point of view than a map might suggest, since they lie to the south and south-east, in which direction bird dispersal is comparatively difficult as the trade [120] wind blows from this quarter. The vermilion flycatcher *Pyrocephalus* is also most distinctive on Chatham.

Similarly, Murphy (1938a) has correlated the degree of isolation and the degree of differentiation in the island forms of various Polynesian land birds, and has shown the important influence of the prevailing wind. Again, Perkins (1918) found that in Hawaii the most isolated islands had the most highly modified birds, and also the greatest number of peculiar insects. The same principle holds in other animals, particularly clear examples being given by Kramer and Mertens for the lizard *Lacerta sicula* on the Adriatic islands, by Kinsey for the gall wasp *Cynips* in North America, and by Reinig for the bumblebee [121] *Bombus* in Europe. These latter cases are summarized by Huxley (1942) and Mayr (1942), who also give numerous other examples of geographical variation in animals, the differences being in some cases adaptive, and in others apparently not.

It is at first sight curious that the bird populations of neighbouring islands could be sufficiently isolated from each other to permit the evolution of distinctive forms, since birds are one of the few groups of land animals capable of active dispersal from one island to another. Thus one of Darwin's finches could often reach an island inhabited by a different form by means of an hour's continuous flight, and these birds seem sufficiently [122] strong on the wing to fly such a distance with ease. Mayr (1942) has produced an even more striking case in the white-eye *Zosterops rendovae* of the Solomons, in which one island form could fly into the region inhabited by another form in under five minutes. As Mayr points out, though birds are capable of flying long distances, they tend to use their wings to stay in, or fly back to, their homes, for which reason bird populations are often more isolated from each other than might otherwise have been expected.

Evolution of Non-Adaptive Differences

The evidence considered above shows the fundamental importance of isolation in the differentiation of bird populations, and suggests that many of the differences between the forms are not correlated with differences in their environments. Writers on genetics have suggested several possible mechanisms for the origin of non-adaptive differences between populations. These views have been discussed in detail by Dobzhansky (1987), Huxley (1942) and others, so that only a brief summary is given here.

First, Muller (1940) considers that if two populations are isolated, then through chance alone some of the mutations occurring in one population will be different from those occurring in the other. Such differences lead in turn to further balancing mutations, so that with time the two populations become increasingly divergent. Except as noted in the next paragraph, a mutation must normally be advantageous in order to spread through a population, hence in one sense the differences which arise between the two populations are adaptive; but the point is that such new characters need not be related to possible differences between the environments of the two

forms, and they are evolved even if the two environments are identical.

Secondly, Sewall Wright (1940) has shown that if an island population is sufficiently small, of the order of several hundred individuals, then through accident alone slightly favourable mutations may become eliminated and slightly unfavourable mutations established. In this way two small isolated populations of a species may acquire genuinely nonadaptive differences.

Thirdly, a mutation tends to have more than one effect. If a [123] mutation occurred which had a favourable effect, it would tend to spread through a population, though it might carry with it other effects which were neutral in their influence.

Fourthly, one island population might differ from another because the island was originally colonized only by a very few individuals, and these happened not to be typical of the population from which they came. But, at least in birds, the importance of this factor has perhaps been exaggerated. As pointed out by Mayr (1942), bird species tend to have periods of expansion and rapid spread and other periods when they remain more or less static. Most colonization of new regions probably occurs in the periods of expansion, and at such times it seems likely that, if one individual can reach a new locality, quite a number of others will follow. This applies particularly in such regions as the Galapagos, where the distances between islands are comparatively short, in most cases less than 50 miles.

The way in which a bird species becomes established on new ground is being demonstrated at the present time by the colonization of England by the black redstart *Phoenicurus ochrurus*. The data of Witherby and Fitter (1942) show that this species has appeared independently in a number of widely separated localities, while its rate of increase seems greater than can be accounted for solely through the breeding of the existing English stock. Evidently there has been multiple colonization from the European continent, and further individuals are still continuing to arrive. It is therefore probable that the eventual English breeding population will be a typical sample of the stock from western Europe.

Exact counts are not available for the size of the populations of any of Darwin's finches, but rough estimates can be made from the known size of the islands, together with the density of the birds as assessed from field observations. A calculation of this nature suggests that on Daphne, which is only half a mile across, there can exist at one time only a few hundred individuals of the peculiar local form of *Geospiza fortis,* and the Crossman form of *G. fuliginosa* is probably represented by a similar total. Likewise each of the distinctive races inhabiting the islands of Culpepper and Wenman probably consists of only a few hundred, or at most a few thousand, individuals at any one time, and the **[124]** same holds for the Tower form of *G. conirostris.* Mayr (1942) cites several other island birds with populations of under a thousand individuals. In such cases the existence of marked and non-adaptive differences between island forms might be explained through Sewall Wright's views on the accidental elimination and fixation of hereditary factors in small populations.

But most other island forms of Darwin's finches consist of larger populations than those cited above. On Tower the observed density of *G. magnirostris,* *G. difficilis* and *Certhidea* suggests that there are several thousand individuals of each present there. The other Galapagos islands are considerably larger, and here most forms probably run to tens of thousands, and others to hundreds of thousands, of individuals. Few estimates are available for the size of the populations of other birds, but the great majority of geographical races almost certainly include at least ten thousand individuals alive at any one time, and many include a much greater number. For instance, some of the endemic British song-birds probably include over a million individuals, and some of the continental races even more.

Sewall Wright's views, on the accidental elimination and recombination of heredity characters are not applicable to populations consisting of more than a few hundred individuals. Hence they cannot be invoked to explain the differences between geographical races of birds except in a small minority of cases. It might, of course, be contended that Darwin's finches decrease periodically to numbers much smaller than those given above, but there is no evidence for this, and there are many other bird races whose populations almost certainly never fall as low as a thousand individuals. Perhaps the views of Muller, summarized earlier, are sufficient to account for the differences in appearance between these larger populations, but this subject requires much further study. To conclude, the evidence from Darwin's finches and other birds shows the great importance of geographical isolation in producing hereditary differences between populations, but there is still considerable doubt as to the way in which they are actually brought about.

[125]

Chapter XIV:
The Origin of Species

No clear line has as yet been drawn between species and sub-species . . . or, again, between sub-species and well-marked varieties, or between lesser varieties and individual differences. These differences blend into each other in an insensible series; and a series impresses the mind with an actual passage.

Charles Darwin: *The Origin of Species,* Ch. II

Incipient Species

The various specimens of Darwin's finches from any one island do not form a continuously graded series from large to small, thick-billed to thin-billed, or dark to pale. Instead they fall into distinct segregated groups, each group having a characteristic appearance, while the individuals of any one group do not normally interbreed with the members of any other. A similar state of affairs is found in the birds breeding in Britain, and for that matter in every other region of the world. The segregated groups are, of course, the units termed species, and their manner of origin has aroused discussion and controversy ever since the theory of evolution came to be accepted.

The apparent fixity of species is most striking, and provides the basis for systematic zoology. But with the full acceptance of the doctrine of evolution there has arisen a tendency among general biologists, though not among taxonomists, to underestimate the definite nature of species, and a corresponding tendency to exaggerate the frequency of intermediate forms. Charles Darwin and many after him are partly wrong when they assert that the determination of species is purely arbitrary. Provided the ornithologist keeps within a limited district, he usually is in no doubt as to which birds should be regarded as separate species, and the same holds in many other groups of animals. Difficulty over intermediate forms arises mainly when the naturalist compares related forms of birds or other animals from different districts, but then the difficulty immediately becomes considerable, a fact which provides the essential clue to the way in which new species originate.

Big evolutionary changes are normally achieved in a series of small steps, so that it is to be expected that the gaps between **[126]** species would come into existence gradually, in which case some of the intermediate stages ought to be visible. The closely related species of Darwin's finches differ from each other in beak, in size of body, in the shade and amount of streaking of the female plumage, and in the amount of black in the male plumage. It is in just these characters that island forms of the same species differ from each other and, as discussed in the last chapter, such island forms show every stage of divergence from differences that are barely perceptible to differences as marked as those which separate some of the species. Moreover, this is the only kind of incipient differentiation found among Darwin's finches. These facts strongly suggest that island forms are species in the making, and that new species have arisen when well-differentiated island forms have later met in the same region and kept distinct.

Camarhynchus on Charles

An instance of such a manner of origin is provided by the two species of large insectivorous tree-finch, *Camarhynchus psittacula* and *C. pauper,* which occur together on Charles. These are undoubtedly two separate species, differing in size of beak, winglength and shade of plumage. The differences between them are small, but constant and reliable, and each collected specimen can safely be allocated to one or the other type.

If *C. psittacula (sens. strict.)* did not occur on Charles, all the large insectivorous tree-finches could be included in one species *C. psittacula,* divided into four well-marked geographical races as follows: *psittacula (sens. strict.)* on the central islands of James, Indefatigable and Barrington, *habeli* to the north on Abingdon and Bindloe, *affinis* to the west on Albemarle and Narborough, and *pauper* to the south on Charles, as shown in Fig. 23. Of these four forms the Charles form *pauper* appears to be the more primitive, being much streaked and possessing the smallest and most finch-like beak. The Albemarle form *affinis* shows close resemblance to *pauper* both in plumage and beak, so links up with it. The central island form *psittacula (sens. strict.)* is less streaked, larger and with a more parrot-shaped beak, but links up with the Albemarle form through a population of intermediate type on the intervening island of Duncan. The [127] northern form *habeli* shows most resemblance to *psittacula (sens. strict.).*

This simple situation is complicated by the fact that Charles is inhabited not only by the form *pau-*per but also by the form *psittacula (sens. strict.),* the Charles individuals of the latter being indistinguishable from those found on the central islands. So far as known these two forms do not interbreed on Charles, and there [128] are no specimens intermediate between them in appearance. The facts suggest that Charles has been colonized by the large insectivorous tree-finch on two separate occasions. Originally, the island was inhabited by the form *pauper,* or by a form which later turned into *pauper,* while more recently it has been invaded from the north by the form *psittacula (sens. strict.).* Formerly, *pauper* and *psittacula (sens. strict.)* were geographical races of the same species, but by the time that they met on Charles they had become so different that they did not interbreed, and so they have become separate species.

Similarly, if the Albemarle race *affinis* were now to colonize Abingdon, where the form *habeli* lives, the two are so distinctive that they might keep separate, in which case they also would have to be classified as separate species. But such an invasion has not yet occurred, so that it is more convenient to consider *affinis* and *habeli* as races of the same species.

Species-Formation in Other Birds

Darwin's finches provide no further cases in which the origin of a new species can be traced in this way. But during the last ten years many similar examples have come to light in other birds. These have been fully reviewed by Mayr (1942), so that detailed comment is scarcely necessary here. A close parallel with the case of *Camarhynchus* on Charles is provided by the two species of thornbill *Acanthiza ewingii* and *A. pusilla* in Tasmania. On the Australian mainland the species *A. pusilla* is divided into a number of well-defined geographical races, each occupying a particular region. But on Tasmania two related forms, both obviously derived from *pusilla* stock, occur together in the same region without interbreeding. They must therefore be reckoned as separate species, though they differ no more in appearance than do some of the mainland races of *pusilla.* Presumably the original Tasmanian form *ewingii* had become so distinctive that, when a second invasion of *pusilla* took place from Australia, the two forms kept separate.

Mayr gives a number of other examples in which a remote island is inhabited by two closely related species, both evidently derived from the same main-

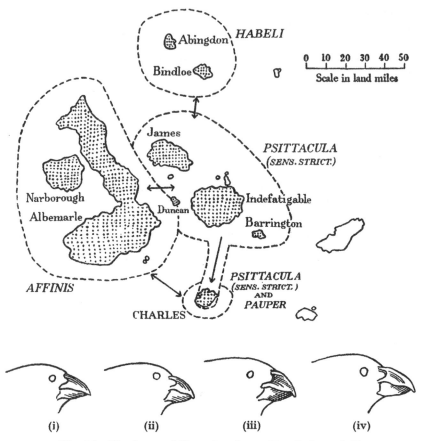

Fig. 23. The forms of *Camarhynchus psittacula* (*sens. lat.*).

Showing the existence of two forms on Charles, the earlier form (*C. pauper*) being related to the Albemarle form (*affinis*), and the later form (*psittacula, sens. strict.*) coming from the central islands. The two have met, but do not interbreed.

(i) *pauper* (Charles)   (ii) *affinis* (Albemarle)   (iii) *psittacula, sens. strict.,* (James)
(iv) *habeli* (Bindloe)
Heads ⅔ natural size (*after* Swarth).

land species, and which have presumably originated as a result of two separate colonizations of **[129]** the island by the mainland form. More comparable with the situation in Darwin's finches are several cases in which an island appears to have been colonized on two separate occasions from adjoining islands. This, for instance, probably accounts for the existence of two related species of the finch *Nesospiza* on the same islands in the Tristan da Cunha archipelago, for the two species of the flycatcher *Mayrornis* on Ongea Levu in the Fijis, for the two species of the fruit pigeon *Ptilinopus* in the Marquesas, and other cases.

There are also a number of comparable examples among continental birds. For instance, the herring gull *Larus argentatus* and the lesser black-backed gull *Larus fuscus* both breed in Britain, and they do not usually interbreed. But if they had not thus met in western Europe, they might have been classified as geographical races of the same species, as they are linked by a series of geographical forms extending across Europe, Asia and North America. By the time that this species had spread right round the world, the two end-forms had evidently become sufficiently different not to interbreed where they met. A similar situation is presented by the two species of great tit *Parus major* and *P. minor*. These species occur together in the Amur valley, where they keep distinct, but each is linked by a chain of geographical races with the same great-tit stock. Mayr gives other examples.

Prevention of Interbreeding

In birds generally, as in Darwin's finches, geographical forms show every stage of divergence, from differences which are barely perceptible to differences as marked as those which separate full species. From the evidence which has now accumulated, it is clear that the commonest method of species-formation in birds is through the meeting in the same region of two geographical forms which have become so different that they keep separate. The fundamental problem in the origin of species is not the origin of differences in appearance, since these arise at the level of the geographical race, but the origin of genetic segregation. The test of species-formation is whether, when two forms meet, they interbreed and merge, or whether they keep distinct.

As discussed in the last chapter, when two populations of the **[130]** same form are isolated from each other, differences gradually arise between them. Muller (1940) considers further that such differences inevitably lead to some degree of sterility between the individuals of the two populations. If this view is correct, there should sometimes be partial sterility between geographical races of the same species, and this has now been established in the case of a number of insects, as summarized by Mayr (1942). Similar evidence is not yet available in birds, as they are rather unsuitable for quantitative breeding experiments.

If the members of two well-differentiated races meet later in the same region, and if they are partially intersterile, or if their hybrid offspring are at a disadvantage, then those individuals which breed with members of their own kind tend to leave more offspring than those which interbreed with individuals of the other race. Hence even if genetical segregation between two races is not complete when they first meet, natural selection will tend to deepen the gap between them. Indeed, it has even been claimed that, owing to the disadvantages possessed by hybrids, selection will initiate intersterility between forms which meet in this way. However, the latter view is not certain.

The above considerations show that any factors which prevent the interbreeding of forms have survival value, hence the frequency with which specific recognition marks have been evolved in birds, as discussed in Chapter V. Darwin's finches are unusual in that the beak is used as a recognition mark, but as this is the most prominent racial and specific difference, it is not surprising that the birds should have evolved behaviour responses relating to it.

Barriers to interbreeding might also be provided by differences in breeding season or habitat. But the various species of Darwin's finches breed at the same season and most of them are not separated in habitat, so that these factors have little or no importance. The latter conclusion applies to birds generally. Most related species found in the same region breed at the same season, and though they commonly occupy different habitats, the degree of isolation thus provided is usually quite inadequate to ensure genetic isolation. In other birds, as in Darwin's finches, the primary factors which prevent attempts at interbreeding are psychological ones correlated with breeding behaviour.

[131] Darwin's statement quoted at the head of this chapter suggests that in animals every gradation exists between mere varieties and full species. Later knowledge has shown that, at least in birds, this statement is somewhat misleading. There is only one kind of variety, if such it should be called, which grades insensibly with the full species, namely the geographical race. In birds there is no other type of variety which can reasonably be termed a subspecies, indeed, the terms subspecies and geographical race have become synonymous. This strongly suggests that in birds the only regular method of species-formation is via races differentiated in geographical isolation. However, it has sometimes been claimed that, to produce the variety of species found in Darwin's finches, some quite peculiar method of evolution must have been involved. Even Rensch (1933), who was the first to advocate species-formation from geographical races as a widespread principle, was greatly puzzled by Darwin's finches, and considered that for them some different process must have operated.

The frequency with which closely related bird species occupy different habitats suggests that an alternative method of species formation is by ecological, instead of geographical, isolation. Since a bird tends to breed in the same type of habitat as that in which it was raised, it might have been expected that, where a species breeds in a variety of habitats, it would tend to become subdivided into populations each with a rather different habitat preference, and that with time this might result in the formation of new species. This is a plausible view and has been put forward by a number of writers, formerly including myself (1933). But there are two insuperable objections. First, the degree of isolation provided by differences in habitat is not usually at all complete, and the bird species which occupy separate habitats usually have numerous border zones where they come in contact with other species. To produce well-differentiated forms, complete isolation seems essential. Secondly, no cases are known in birds of incipient species in process of differentiation in adjoining habitats. All subspecies are isolated from each other geographically and, though they occasionally [132] differ in habitat as well, geographical isolation is the essential factor. These points are treated in further detail by Mayr (1942). Finally, the frequent existence of habitat differences between closely related bird species has a quite different explanation, consideration of which is postponed to the next chapter.

Another alternative has been suggested by Lowe (1930, 1936), who supposes that Darwin's finches represent the varied products of interbreeding between a small number of original forms, as has happened in certain 'species-swarms' in plants. The reasons for rejecting this view have been discussed in Chapter X. There is no evidence that hybridization has been of importance in species-formation in any group of birds.

Streseman (1936) is the only previous writer to suggest that species-formation has followed the same course in Darwin's finches as in other birds, i.e. that forms differentiated in geographical isolation have later met and kept distinct. With this conclusion, I fully agree. There is only one apparent case, in the large insectivorous tree-finches *Camarhynchus psittacula* and *C. pauper* on Charles. But such cases will rarely be apparent, since once a form has become firmly established in the range of another it will tend to spread rapidly right through that range, so that its place and means of origin quickly become obscured. The only type of incipient differentiation found in Darwin's finches is that shown by geographical races, and there is nothing to suggest that geographical isolation is not the essential preliminary to species-formation in this group. The existence of an unusually large number of similar species may be attributed first to the great length of time for which the finches have been in the Galapagos, secondly to the paucity of other land birds, and thirdly to the un-

usually favourable conditions provided by a group of oceanic islands, both for differentiation in temporary geographical isolation, and also for the subsequent meeting of forms after differentiation.

The primary importance of the geographical factor is strongly corroborated by the situation on Cocos Island. Here there occurs one, and only one, species of Darwin's finch, *Pinaroloxias inornata*. That it has been on Cocos a long time is suggested by the extent to which it differs from all the other species of Darwin's finches. Yet despite the length of time for which it has been [133] there, despite the variety of foods and habitats which Cocos provides, and despite the almost complete absence of both food competitors and enemies, there is still only one species of Darwin's finch on Cocos. But Cocos is a single island, not an archipelago, and so provides no opportunity for the differentiation of forms in geographical isolation.

Species-Formation in Other Organisms

That geographical isolation can lead to the origin of new species in groups other than birds is shown by examples given by Mayr (1942) and Huxley (1942) for mammals, reptiles, amphibia, molluscs and several groups of insects. Populations of organisms such as insects or land molluscs are sometimes isolated from each other in a much smaller space than are bird populations. It might therefore be better to replace the term geographical isolation by topographical isolation when considering animals generally.

The extent to which ecological isolation can lead to the formation of new species in other animals is still in doubt (Mayr, 1942; Huxley, 1942, modified in 1943). If an insect occurred on several types of food plant, or if a parasite had several host species, it seems possible that populations might become isolated, each restricted to a particular food plant or host species, thus allowing subspecific and eventually specific differentiation to take place. But this is not proven and, except in these rather special circumstances, it seems doubtful whether ecological differences without topographical isolation can provide a sufficient degree of segregation between populations. There are many groups of animals and plants in which the closely related species differ in their ecology, but it does not necessarily follow that ecological isolation preceded the formation of the species in question, as will become clear in the next chapter.

In certain types of plants big changes in the chromosomes can produce sterility between individuals, and it is possible that such changes can lead to the effective isolation of groups within the species and so to the formation of new species. Such purely genetic isolation may also have been important in a few animal groups but, as Huxley (1942) points out, in most animals purely genetic [134] isolation becomes important only secondarily, after initial differentiation in topographical isolation. A different type of genetic isolation is found in the few plants and the much smaller number of animals which reproduce only by asexual means, and such forms are often highly differentiated.

To conclude, in all organisms the isolation of populations is an essential preliminary to the origin of new species. In birds, geographical isolation is of primary importance, and isolation by genetic factors arises secondarily; ecological isolation is not known to initiate species-formation. Possibly these conclusions apply to many other groups of animals and also to plants, but definite conclusions cannot be reached until the systematics of other organisms is known as well as that of birds, and, in particular, the systematics of the subspecific units which are species in the making. [. . .]

[146]
Chapter XVI: Adaptive Radiation

Natural selection will always act according to the nature of the places which are either unoccupied or not perfectly occupied by other beings; and this will depend on infinitely complex relations. But as a general rule, the more diversified in structure the descendants from any one species can be rendered, the more places they will be enabled to seize on, and the more their modified progeny will be increased.

Charles Darwin: *The Origin of Species,* Ch. III

Species-Formation and Adaptive Evolution

Many of the differences between the subgenera and genera of Darwin's finches are undoubtedly adaptive. This applies particularly to their beaks, as de-

scribed in Chapter VI, but also to numerous other characters and habits. Thus the ground-finches hop about the ground, the insectivorous tree-finches are agile and tit-like among the branches, the woodpecker-finch climbs vertical trunks and inserts a cactus spine into crevices, the Cocos finch and the cactus ground-finch have bifid tongues, and *Certhidea* has the quick flitting movements of a warbler. Like a warbler, too, *Certhidea* repeatedly flicks the wings partly open when hopping about the bushes. The reason for this habit is not known, but it is found in no other of Darwin's finches, and its parallel evolution in *Certhidea* and in the true warblers presumably means that it has some significance in the lives of these birds.

Darwin's finches provide an example of an adaptive radiation, with seed-eaters, fruit-eaters, cactus-feeders, wood-borers and eaters of small insects. Some feed on the ground, others in the trees. Originally finch-like, they have become like tits, like woodpeckers and like warblers. There are, of course, gaps; for instance, none have taken to open country, none have become birds of prey, and none are aquatic. But the considerable similarity between all the species in regard to breeding habits, plumage and internal structure implies that their divergence has been [147] comparatively recent and rapid. Though scarcely in the same class with the astounding radiation of the Australian pouched mammals, Darwin's finches are sufficiently impressive.

Adaptive evolution, as now generally agreed, is brought about by the action of natural selection on random mutations. Those mutations which happen to be favourable spread through the population, and are repeatedly modified by further mutations, thus gradually building up the finely adapted structures and behaviour patterns which make up each living organism. This process is long-term and accumulative. At first sight it has nothing to do with the origin of new species, which is essentially the formation of gaps between populations. For this reason, and influenced by the apparent absence of adaptive differences between closely related species, some recent writers have claimed that the origin of species and long-term adaptive evolution are independent processes, and that the former has no important influence on the latter. This view I believe to be mistaken.

The point is illustrated by the sharp-beaked ground-finch *Geospiza difficilis* in the Galapagos. Probably in former times this bird bred in both the arid and the humid forest of the central islands, but it has since been driven from all except the humid zone by a newer species, the small ground-finch G. *fuliginosa* (see p. 28). Hence on the central islands G. *difficilis* can now become increasingly specialized for life in the humid forest. But this would not have been possible for it so long as it also bred in the arid zone, and is therefore a direct result of the appearance of the newer species G. *fuliginosa*.

Similar considerations apply wherever two species, originally geographical races of the same species, have become established in the same region. In most cases adaptive or ecological differences probably arose between the two forms during their period of geographical isolation before they met, but their meeting must almost inevitably have resulted in further restriction of their foods or habitats, and so must have tended to accelerate their adaptive specialization. For instance, the blue chaffinch of the Canary Islands probably occurred formerly in both pine and broad-leaved forest, but since the arrival of the second chaffinch species it has been confined to pine forest, after which, but not before, it could become specialized for the latter habitat. **[148]** Likewise the earlier and large-beaked species of the white-eye *Zosterops* on Lord Howe Island probably had a more varied diet before the arrival of the second and smaller species. The meeting of these two species in the same area may be expected to have led to a restriction of their diets, and thereafter to an increase in the size-difference between them.

The origin of new species, so far from being irrelevant, is therefore an active agent in promoting specialization, and in particular it is an essential precursor to an adaptive radiation. I consider that the adaptive radiation of Darwin's finches can have come about only through the repeated differentiation of geographical forms, which later met and became established in the same region, that this in turn led to subdivision of the food supply and habitats, and then to an increased restriction in ecology and specialization in structure of each form. On Cocos, where conditions are unsuitable for species-formation, there has likewise been no adaptive radiation among the land birds.

Various investigators have sought to explain the evolution of Darwin's finches through unusual genetic factors, such as an excessive inherent variability or frequent hybridization. But it is the persistence, rather than the origin, of new types which chiefly requires explanation, and in my opinion the peculiarities of the finches are primarily due to an unusual combination of geographical and ecological factors. The chief geographical factor has been the existence of a number of islands, which has provided favourable opportunities for the differentiation of forms in isolation and their subsequent meeting. The chief ecological factor has been the scarcity of other passerine birds, which has permitted an unusually great degree of ecological divergence between forms, and thus has allowed an unusually large number of related species to persist alongside each other without competition. Divergence may have been accelerated by the scarcity of predators (see p. 114), but this would seem, at most, very subsidiary.

Nature of the Genus in Song-Birds

The genus is essentially an artificial unit for the convenient cataloguing of similar species. However, in continental passerine birds, systemists have tended to attach particular importance to **[149]** morphological divergence, including beak differences, as a criterion for generic separation. On the continents, closely related passerine species do not usually differ greatly from each other in beak or other morphological characters. Hence the use of morphological characters for separating genera has tended to give a convenient number of species in each genus, and so has proved useful in practice.

On oceanic islands, a greater proportion of closely related passerine species differ prominently from each other in beak than is the case on the continents. This is not because beak differences are more frequent among insular than continental land birds—the reverse is usually the case—but because on oceanic islands the species which differ in beak are often very similar in other respects, and so tend to be placed in the same genus, whereas on the continents they are usually distinctive in other ways as well, and so tend to be placed in separate genera. Thus in Britain the small-beaked goldfinch, the medium-sized greenfinch, and the large-beaked hawfinch are related species living in similar habitats but eating mainly different foods. They therefore provide a close ecological parallel with the small, medium and large ground-finches of the Galapagos. But whereas the latter species differ in little except beak and are placed in the same genus *Geospiza*, the three British finches differ also in plumage, nesting habits and various other ways, and are placed in separate genera, *Carduelis*, *Chloris* and *Coccothraustes* respectively [fig. 25].

In applying generic names to oceanic land birds,

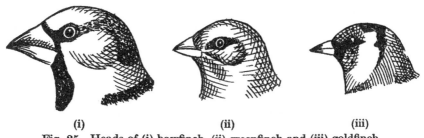

(i)   (ii)   (iii)

**Fig. 25. Heads of (i) hawfinch, (ii) greenfinch and (iii) goldfinch.
⅔ natural size (*after* ten Kate).**

the taxonomist is often in somewhat of a dilemma. Related species frequently differ so much in beak that, were they mainland birds, they would **[150]** unhesitatingly be placed in different genera. But in other respects, including plumage, they are often very similar, indicating close relationship. If every island land bird that differs markedly in beak is placed in a separate genus, many of the resulting genera contain only a single species, which not only multiplies names excessively, but obscures the close relationship of forms that differ in little except beak. The difficulty discussed in Chapter II regarding the number of genera of Darwin's finches is not merely a question of terminology, but reflects the fact that on oceanic islands the land birds are at an earlier stage of evolution than on the continents. On the continents the divergence into finches, warblers, tits and woodpeckers obviously took place in a much more distant past than the corresponding divergence among Darwin's finches in the Galapagos.

Birds on Other Oceanic Islands

An adaptive radiation like that of Darwin's finches is rarely found on other oceanic islands. For reasons already considered it was not to be expected on solitary islands, however remote. Such islands often have a single highly peculiar species, but in no case is there a group of related birds. The greatest number known is on Norfolk Island, 800 miles from Australia, where there are three species of white-eye *Zosterops*, differing conspicuously in size and in size of beak (Mathews, 1928; Stresemann, 1931). But these species have almost certainly evolved as a result of three separate colonizations of Norfolk Island from outside, either from the Australian mainland or from Lord Howe Island. There is no reason to think that they became differentiated on Norfolk Is-

land itself, which has an area of less than 20 square miles.

It is more surprising that no parallel with Darwin's finches is found on most oceanic archipelagos. For example, the Azores are volcanic islands and further from a continent than are the Galapagos. But their songbird inhabitants differ only subspecifically from those of Europe, and no endemic form has given rise to an adaptive radiation. The Azores have been colonized by a larger number of passerine birds than have the Galapagos, which suggests that they are more accessible. This is corroborated by the frequent records of European song-birds passing through **[151]** the Azores on migration, whereas only two species of mainland passerines have been recorded on migration in the Galapagos, the American barn swallow *Hirundo erythrogaster* and the bobolink *Dolichonyx oryzivorus* (Swarth, 1931). This may be correlated with the fact that the Azores are in an area of winds, whereas the Galapagos are in a region of calms. Though in the Galapagos the south-east trade wind blows for part of the year, it does not blow from the South American mainland but forms at sea, away from the land.

The land organisms of remote islands seem less efficient than continental types, perhaps because, owing to the greater number of both individuals and species on the continents, competition is more severe there. When a remote island receives new bird colonists from a mainland area, they often eliminate the original inhabitants. This is being demonstrated in New Zealand at the present time, where artificially introduced European birds are spreading rapidly at the expense of the native species. Similarly, Usinger (1941) reports that in the Hawaiian Islands many of the peculiar endemic insects are being eliminated by introduced continental types. This suggests that even if the Azores were at one time so isolated that an endemic group comparable with Darwin's finches evolved there, such peculiar forms would probably have been eliminated later, with the arrival of the newer and more efficient species which now frequent the islands. That the latter birds are only recent arrivals from the mainland is shown by the slight extent to which they differ from European forms.

Similar considerations probably account for the absence of local adaptive radiations among the land birds of the Canary Islands, the Cape Verde Islands, the Revilla Gigedos, and other archipelagos off Europe, Africa and America. Even the Polynesian archipelagos provide no parallel with Darwin's finches. They possess a variety of land birds, but these are derived from the lands to the west or from adjoining archipelagos. No Polynesian bird has diverged into a group of species within one archipelago, presumably because invasions of new and more efficient species from outside have occurred at too frequent intervals. Though the distances between some of the Polynesian archipelagos are considerable, all are in a region of winds.

[152] There are in the world only two archipelagos whose land birds show an adaptive radiation comparable with that of Darwin's finches. The three islands of the Tristan da Cunha group lie in the Atlantic some 2000 miles from South America and nearly as far from Africa. Aided by strong westerly winds, two South American passerine birds have colonized the islands, a thrush *Nesocichla* and a finch *Nesospiza*. The thrush is a single species and need not be considered further. But the finch has evolved into two species, one large and large-beaked, the other small and small-beaked, as shown in, Fig. 24 (p. 139). Both species occur together on Nightingale Island, where Hagen (1940) found that they differ in food and habitat. The smaller species is divided into three island races, differing in size of beak, but the Tristan form has become extinct. Two hundred miles away from Tristan lies Gough Island, and here occurs a related finch *Rowettia (Nesospiza) goughensis*, rather similar to *Nesospiza* in appearance but with a more elongated beak. This bird may have evolved from *Nesospiza* stock, though Lowe considers that it was independently derived from South America (Clarke, 1905; Lowe, 1923).

As Lowe points out, Tristan da Cunha is Galapagos in miniature. At the other extreme are the astonishing Drepaniidae, the sicklebills of Hawaii. The Hawaiian Islands are as remote in the Pacific as is Tristan da Cunha in the Atlantic, and only five passerine forms have succeeded in reaching them. Of these, the crow belongs to the mainland genus *Corvus*. The flycatcher *Chasiempis* is placed in a distinct genus, and there is one form on each island. The thrush *Phacornis* is also peculiar to Hawaii; there is one form on each island, while on Kauai occur two species which differ in size. The fourth colonist, a honeyeater, has diverged into two distinct genera, *Chaetoptila* and *Moho*. Finally, the fifth colonist, the sicklebill, has produced a multitude of forms far more diverse than Darwin's finches.

The ancestor of the sicklebills is considered to have been a finch, but it has given rise to some strikingly unfinchlike forms. A modern study of these birds has not been undertaken, but Rothschild (1893–1900) and Perkins (1903) divide them into as many as eighteen genera, some of which would probably be merged on modern standards. In most cases the genera are represented by distinctive island forms, formerly classified as [154] separate species, while in *Chlorodrepanis*, also in *Rhodacanthis*, two related species differing prominently in size are found on the same island. Some genera feed on both nectar and insects, some mainly on nectar, and others mainly on insects. Others eat fruit, others seeds, one feeds mainly on a native bean and on caterpillars, and another on nuts. Some of the associated modifications in beak are shown in Fig. 26. The most remarkable is in *Heterorhynchus*; This bird climbs trunks and branches like a wood pecker, and feeds on longicorn beetle larvae in the wood. It taps with its short lower mandible, and probes out the insects with its long decurved upper mandible. This method may be compared first with that of a true woodpecker, which excavates with its beak and probes with its long tongue; secondly with that of the Galapagos wood pecker-finch *Camarhynchus pallidus*, which trenches with its beak and probes with a cactus spine; and thirdly with that of the extinct New Zealand huia *Heterolocha acutirostris*, in which the male excavated with its short beak, the female probed with its long decurved beak, the pair to some extent co-operating (Buller, 1888).

Other Organisms on Islands

It is significant that the regions which have given rise to adaptive radiations among birds have produced a similar multiplicity of related species in various other land organisms. The most striking case in the Galapagos is that of the land mollusc *Nesiotus*, which has evolved into fifty-three species, or sixty-six geographical races (Dall and Oschner, *1928b*; Gulick, 1932). A much smaller radiation has occurred in the iguanas, of which the two Galapagos

Susan Williams-Ellis

Fig. 26. Adaptive radiation of Hawaiian sicklebills
(*after* Keulemanns).

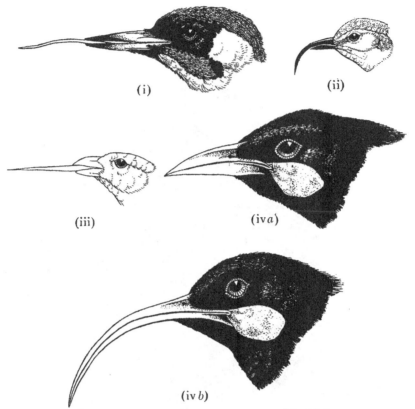

Fig. 27. Extraction of insects from wood by birds.

(i) European green woodpecker *Picus viridis* excavates with beak, probes with long tongue.

(ii) Hawaiian *Heterorhynchus* taps with short lower mandible, probes with long upper mandible. (Head *after* Keulemanns.)

(iii) Galapagos woodpecker-finch *Camarhynchus pallidus* trenches with beak, probes with cactus spine.

(iv) New Zealand huia *Heterolocha acutirostris*: (*a*) male excavated with short beak, (*b*) female probed with long beak. (Heads *after* Keulemanns.)

Note. Various other trunk-feeding birds either excavate only or probe only, but they do not do both.

Heads drawn ½ natural size.

genera probably evolved within the archipelago from one original colonist. The land iguana *Conolophus* lives on land and feeds primarily on cactus, while the marine iguana *Amblyrhynchus* lives on the shore and feeds primarily on marine green algae. Both iguanas show some differentiation into island forms.

In the other Galapagos reptiles, evolution has proceeded only to the stage of differentiation into well-marked geographical forms. The giant tortoise *Testudo* is divided into as many as fifteen geographical forms, five of which occur on separate parts of Albemarle, and there are seven or eight island forms of the lizard *Tropidurus*, the gecko *Phyllodactylus* and the snake [155] *Dromicus*[1] (Van Denburgh, 1912–14; Van Denburgh and Slevin, 1913).

1. Actually two species of the snake *Dromicus* are found on Indefatigable and Jervis. Van Denburgh considers that these did not evolve from each other within the Galapagos, but colonized the islands independently from South America.

Well-marked island forms, from four to nine in number, are also found in the Galapagos ants *Camponotus macilentus* and **[156]** *C. planus* (Wheeler, 1919), and in the four grasshoppers *Halmenus, Liparoscelis, Schistocerca* and *Sphingonotus* (McNeill, 1901; Snodgrass, 1902b; Hebard, 1920). It is curious that no Galapagos insect is recorded as having produced an adaptive radiation, with a multiplicity of species living on each island. Such radiations are common on other oceanic islands, so further collecting may possibly reveal their existence in the Galapagos.

One of the Galapagos plants has produced a radiation, the endemic helianthoid genus *Scalesia* having evolved into a group of trees and bushes consisting of six main species, each of which is in turn divided into a number of geographical forms (Howell, 1941). Groups of up to about a dozen endemic Galapagos species are also found in the shrubs and bushes of the genera *Telanthera* (Amarantaceae), *Acalypha* and *Euphorbia* (Euphorbiaceae), *Cordia* (Boraginaceae) and a few others. These genera are not endemic to the Galapagos, and botanists do not state whether the various species are the result of independent colonization from outside regions, or whether, as in *Scalesia*, a group of species has evolved within the archipelago from one original form. Some of the species are divided into island forms, a good example being *Euphorbia viminea*, with eight geographical forms (Robinson, 1902; Stewart, 1911).

In other groups of land organisms, as in birds, the most remarkable radiations are those which have occurred in the Hawaiian Islands. The most remarkable of all is that of the achatinellid land molluscs, of which there are nearly five hundred known forms, all apparently descended from one original form on the islands (Usinger, 1941).

The many striking radiations among Hawaiian insects have been described by Perkins (1913). They include three endemic subfamilies and four other genera of Coleoptera, five endemic genera of Lepidoptera, and from one to three endemic genera of Orthoptera, Odonata, Hemiptera, Neuroptera and Hymenoptera. Each of these groups presumably evolved from one original species, though each now contains at least thirty species, while in the *Proterhinus* weevils, the Anchomenini ground-beetles and the *Hyposmochoma* moths, there are well over a hundred related species in each group. These are the figures given by Perkins, and since he wrote, the number of known forms has been considerably **[157]** increased. In numerous cases an insect is represented by a different form on different islands, while when closely related species are found on the same island, they often occur on different food plants. Perkins concluded that geographical isolation has been the factor of primary importance in the origin of the insect species.

Radiations have also occurred among Hawaiian plants. Thus according to Gulick (1932), there are 146 species and varieties of Hawaiian lobelias, these having evolved from an extremely small number of original colonists. The survey by Hillebrand (1888) shows that there are several other genera of trees and shrubs endemic to Hawaii and represented in the islands today by between ten and thirty species, presumably as a result of local evolution within the archipelago.

In Polynesia, as already mentioned, no local radiations are found among the land birds, but they occur in some cases in the trees, insects and land molluscs (Buxton, 1938; Crampton, quoted by Huxley, 1942, etc.). In general, the insects, land molluscs and land plants have produced more local radiations on remote islands than have the land birds. It is of special interest to find such radiations on the solitary island of St Helena, which is in the Atlantic 1200 miles from West Africa and remote from all other islands. Wollaston (1877) described as many as ninety-one species of weevils endemic to St Helena, and five of the endemic genera have each given rise to between ten and fourteen species, presumably as a result of local evolution on St Helena. One endemic genus of land molluscs has also diverged into about a dozen species (Smith, 1892), and there seems to have been a small radiation among trees of the family Compositae (Melliss, 1875).

No single island, however remote, has given rise to an adaptive radiation of land birds. This is probably because an essential step in the formation of new bird species is the isolation of geographical forms. For the latter, a single small island does not suffice, and either an extensive land area or a number of small islands is required. That the insects and land molluscs have produced radiations on a solitary island, such as St Helena, suggests that populations of these organisms can be isolated topographically in a much smaller space than bird populations. This is corroborated by the work of Crampton, Gulick and others, **[158]** who have shown that in some of the Hawaiian and Polynesian land molluscs, each steep valley may have its own subspecies.

The comparative ease with which insects and land molluscs are isolated means that in these groups new species tend to arise more quickly than they do in birds. This is presumably one reason why in Hawaii and elsewhere the radiations among insects and land molluscs are more extensive than those of birds. But another, and probably more important, reason is that a much greater number of ecological niches are available for insects and land molluscs than for land birds. It is not only the origin, but also the persistence, of new species which requires explanation.

That on oceanic islands radiations have occurred not only among the land birds, but also among the insects, land molluscs and land plants, suggests that the evolutionary factors involved in these other groups are fundamentally the same as they are in Darwin's finches. This is one of the chief reasons why I hope this book may have been worth writing.

Works Cited

Buller, W. L. 1888. A History of the Birds of New Zealand. 2nd ed. 1, pp. 7, 10.

Buxton, P. A. 1938. The formation of species among insects in Samoa and other oceanic islands. *Proc. Linn. Soc. Lond.* 150: 264–7.

Clarke, W. E. 1905. Ornithological results of the Scottish National Antarctic Expedition. I. On the birds of Gough Island, South Atlantic Ocean. *Ibis*, pp. 255–8.

Dall, W. H., and W. H. Oschner. 1928b. Land shells of the Galapagos islands. *Proc. Calif. Acad. Sci.* (4), 17: 141–84.

Darwin, C. 1859. *On the Origin of Species by means of Natural Selection.* (Quotations are made from the second edition.)

Dobzhansky, Th. 1937. *Genetics and the Origin of Species.* Also 2nd ed. 1941.

Gulick, A. 1932. Biological peculiarities of oceanic islands. *Quart. Rev. Biol.* 7: 405.

Hagen, Y. 1940. In Christopherson, E. *Tristan da Cunha. The Lonely Isle.* Eng. Trs. R. L. Benham, pp. 96–99.

Hebard, M. 1920. Expedition of the California Academy of Sciences to the Galapagos Islands, 1905–6. XVII. Dermaptera and Orthoptera. *Proc. Calif. Acad. Sci.* (4) 2: 311–46.

Hillebrand, W. 1888. *Flora of the Hawaiian Islands.*

Howell, J. T. 1941. The Templeton Crocker Expedition of the California Academy of Sciences, 1932. No. 40. The genus *Scalesia. Proc. Calif. Acad. Sci.* (4), 22: 221–71.

Huxley, J. S. 1942. *Evolution: the Modern Synthesis.*

———. 1943. Evolution in action. (Review of Mayr, 1942.) *Nature, Lond.* 151: 347–8.

Lack, D. 1933. Habitat selection in birds. *J. Anim. Ecol.* 2: 239–62. See also (1937), *Brit. Birds* 31: 130–6 and (1940), *Brit. Birds* 34: 80–4.

Lowe, P. R. 1923. Notes on some land birds of the Tristan da Cunha group collected by the *Quest* expedition. *Ibis*, pp. 519–23.

———. 1930. Hybridisation in birds in its possible relation to the evolution of the species. *Bull. Brit. Orn. Cl.* 50: 22–9.

———. 1936. The finches of the Galapagos in relation to Darwin's conception of species. *Ibis*, pp. 310–21.

Mathews, G. M. 1928. *The Birds of Norfolk and Lord Howe Islands*, pp. 50–3.

Mayr, E. 1942. *Systematics and the Origin of Species.*

McNeill, J. 1901. Papers from the Hopkins-Stanford Galapagos Expedition, 1898–1899. IV. Entomological results (4). Orthoptera. *Proc. Wash. Acad. Sci.* 3: 487–506.

Melliss, J. C. 1875. *St Helena*, pp. 283–6.

Muller, H. J. 1940. Bearings of the *Drosophila* work on systematics. *The New Systematics* (ed. J. S. Huxley), pp. 185–268, esp. pp. 194–8.

Murphy, R. C. 1938a. The need of insular exploration as illustrated by birds, *Science* 88: 533–9.

Perkins, R. C. L. 1903. *Fauna Hawaiiensis*, vol. 1, pt. IV, Vertebrata: Aves, pp. 368–466.

———. 1913. *Fauna Hawaiiensis*, vol. 1, pt. VI, Introduction, being a review of the land-fauna of Hawaii, pp. xv–ccxxviii.

Robinson, B. L. 1902. Flora of the Galapagos Islands. *Proc. Amer. Acad. Arts Sci.* 38; contr. *Gray Herb.* 24: 77–269.

Rothschild, W. 1893–1900. *The Avifauna of Laysan and the neighboring Islands, with a complete History to date of the Birds of the Hawaiian Possessions.*

Smith, E. A. 1892. On the land-shells of St Helena. *Proc. Zool. Soc. Lond.* pp. 258–70.

Snodgrass, R. E. 1902b. *Schistocera, Sphingonotus* and *Halmenus.* Papers from the Hopkins-Stanford Galapagos Expedition, 1898–1899. VIII. Entomological results (7). *Proc. Wash. Acad. Sci.* 4: 411–55.

Stewart, A. 1911. Expedition of the California Academy of Sciences to the Galapagos Islands, 1905–1906. II. A botanical survey of the Galapagos Islands. *Proc. Calif. Acad. Sci.* (4), 1: 7–288.

Stresemann, E. 1931. Die Zosteropiden der indo-australischen Subregion. *Mitt. Zool. Mus. Berl.* 17: 201–38.

———. 1936. Zur Frage der Artbildung in der Gattung

Geospiza. Organn der Club Van Nederlandische Vo-gelkundigen 9, no. 1, pp. 13–21.

Swarth, H. S. 1931. The avifauna of the Galapagos Islands. *Occ. Pap. Calif. Acad. Sci.* 18.

Usinger, R. L. 1941. Problems of insect speciation in the Hawaiian Islands. *Amer. Nat.* 75: 251–63.

Van Denburgh, J. 1912a. Expedition of the California Academy of Sciences to the Galapagos Islands, 1905–1906. IV. The snakes of the Galapagos Islands. *Proc. Calif. Acad. Sci.* (4), 1: 323–74.

———. 1912b. Expedition of the California Academy of Sciences to the Galapagos Islands, 1905–1906. VI. The geckos of the Galapagos archipelago. *Proc. Calif. Acad. Sci.* (4), 1: 405–30.

———. 1914. Expedition of the California Academy of Sci-ences to the Galapagos Islands, 1905–1906. X. The gi-gantic land tortoises of the Galapagos archipelago. *Proc. Calif. Acad. Sci.* (4), 2: 203–374.

Van Denburgh, J., and J. R. Slevin. 1913. Expedition of the California Academy of Sciences to the Galapagos Is-lands, 1905–1906. IX. The Galapagoan lizards of the genus Tropidurus; with notes on the iguanas of the gen-era *Conolophus* and *Amblyrhynchus. Proc. Calif. Acad. Sci.* (4), 2: 133–202.

Wheeler, W. M. 1919. Expedition of the California Acad-emy of Sciences to the Galapagos Islands, 1905–1906. The ants of the Galapagos Islands. *Proc. Calif. Acad. Sci.* (4), 2: 259–97.

Wollaston, T. V. 1877. *Coleoptera Sanctae Helenae.*

AREA, CLIMATE, AND EVOLUTION

Philip J. Darlington, Jr.

Museum of Comparative Zoology, Harvard University

Received February 3, 1958

Introduction

This is an attempt to find what the geographical distribution of animals can tell about the evolution of dominant, successful animals.

Geographical distribution and evolution are very closely connected. Historically, animal distribution (the geographical patterns of it) provided some of the first, decisive evidence that convinced Darwin and Wallace of the fact of evolution (Darlington, 1959). In fact, evolution has made many of the patterns of animal distribution, and the patterns can still tell us new things about how evolution occurs. The particular question I shall ask and try to answer now is, where do dominant animals evolve: in large or small areas, and in warm or cold, stable or cyclical climates? The answer may tell not only where but also something about how dominant animals evolve.

Three kinds of evolution. Evolution can be divided (arbitrarily—it is an oversimplification) into three kinds (Darlington, 1948: 109; 1957: 565): speciation (which may or may not involve adaptation), adaptation to special environments, and general adaptation. General adaptation includes all the general improvements of structures and functions that increase animals' efficiency in many environments, that (for example) increase the speed of reactions or the efficiency of reproduction. It is adaptation to the general environment of the world, and it should lead to general dominance, to success over great areas and in many special environments.

Dominant animals are conspicuously successful ones. Dominant groups are usually numerous in individuals and numerous in species, often (eventually) diverse in adaptations, and often (eventu-ally) widely and continuously distributed in diverse habitats. Also, of course, they are relatively successful in competition. The history of dispersal of animals seems to be primarily the history of successions of dominant groups, which in turn evolve, spread over the world, compete with and destroy and replace older groups, and then differentiate in different places until overrun and replaced by succeeding groups. (This is, of course, another oversimplification—radiation of successive dominant groups is only the main pattern among many patterns of animal dispersal.) The question (asked in different terms above) is, where do the dominant animals evolve in relation to area and climate?

Darwin and Matthew. Darwin was the first important evolutionary zoogeographer. A hundred years ago, in *The Origin of Species* (1859: in the two chapters on geographical distribution and in parts of other chapters; see also Darlington, 1959) he showed clearly how the geographical distribution of animals is related to evolution. The next really important treatment of the subject was by Matthew (1915) in *Climate and Evolution.* Matthew was a distinguished paleontologist. His work on fossil mammals was extensive and very good, and his ideas about the permanence of continents and the histories of islands were right and deservedly influential. In the following pages I shall have to pick out and criticize his theory of the climatic control of evolution. When I do it, readers should remember that this theory is a very minor part of Matthew's work. To judge Matthew just by his climatic theory (if it turns out to be wrong) would be as unfair as to judge Wordsworth by selections

488

from *The Stuffed Owl,* an anthology of bad verse by good poets. Repeatedly in the following pages I shall go back to Darwin, compare his ideas with Matthew's, and compare the ideas of both men with current ideas as set forth in my recent *Zoogeography* (1957).

Patterns of dispersal. I shall assume for present purposes (but only for present purposes—matters like this will never be beyond re-examination) that the continents and climatic zones have been constant *in position* during the period under consideration, which is mainly the later Tertiary, Pleistocene, and Recent, but that the north temperate zone was colder than now at times in the Pleistocene and warmer than now before that. Darwin, Matthew, and (probably) most modern zoogeographers agree about this.

Both Darwin and Matthew thought that animal (and plant) dispersal follows a main pattern, that successive dominant groups usually evolve (become dominant) in certain places and move (spread) in certain directions. Darwin drew his evidence primarily from the behavior of plants but extended his conclusions to organic "productions," by which he probably meant to include animals. He detected world-wide movements from north to south and from continents to islands, and his explanation was (1859, 1950: 322), "I suspect that this preponderant migration from north to south is due to the greater extent of land in the north, and to the northern forms having existed in their own homes in greater numbers, and having consequently been advanced through natural selection and competition to a higher stage of perfection or dominating power than the southern forms." In other words Darwin thought that *large area* induces evolution of dominant plants and animals.

Matthew drew his evidence primarily from the distribution and fossil record of vertebrates, chiefly mammals. He too detected a general movement from north to south, but he thought it was due to evolution of dominant groups north of the tropics, during the dry, cool phases of climatic cycles. In other words Matthew thought that *northern cyclical climate* induces evolution of dominant animals.

When I considered this matter recently (1957), it seemed to me that much evidence pointed to the main Old World tropics (tropical Asia and Africa) as the usual place of origin of dominant groups of vertebrates, including warm-blooded ones, and that the dominant groups spread from there in all directions that the land allows: to South Africa; across the islands to Australia; northward, to North America, and thence to South America; and from all the continents to any islands that could be reached. And my explanation was Darwin's, sharpened and extended: that dominant animals usually evolve in large, favorable areas and move (spread) to smaller and/or less favorable areas. In other words I thought that *large area and favorable climate* induce evolution of dominant animals. By "favorable" I meant physically favorable to existence of life. I suppose the conditions of existence are most favorable in the tropics, less so in colder or less stable climates. It does not necessarily follow that dominant animals evolve in the physically most favorable climates, but I think that in fact they do. Matthew thought not. He thought (1915, 1939: 7) that dominant groups should evolve under "conditions . . . unfavorable to abundance of life and the ease with which animals could obtain a living."

Matthew's suggested pattern of dispersal (A) and mine (B) are compared in figure 1. If they were meaningless shapes, these patterns would not seem very different, but they are not meaningless. They imply different things about evolution. Pattern A implies that dominant animals evolve in response to northern cyclical climates; B, that they evolve in large areas and favorable, relatively stable climates. These two patterns would not be reconciled if tropical or subtropical climate extended into the north-temperate zone, as it formerly did. Matthew's thesis

490   PHILIP J. DARLINGTON

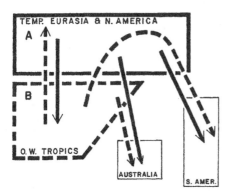

FIG. 1. Two suggested dispersal patterns: A (solid lines), with center in the north temperate zone (Matthew's pattern); B (broken lines), with center in the Old World tropics.

is that dominant vertebrates evolve in a zone above the tropics, when it is not tropical, and because it is not tropical.

Effect of area. Area—the relative sizes of different land masses—does affect the evolution and dispersal of dominant vertebrates, whatever the effect of climate may be. Vertebrates have moved back and forth over the world in very complex ways, but the sum of the movements has been from larger to smaller land masses. Evidence of this comes from many sources (Darlington, 1957: 545–547). For example, existing distributions suggest more movement of vertebrates from Asia to Australia than the reverse, more from the Old World to North America than the reverse, and more from North to South America than the reverse. Movements of vertebrates from continents to islands have been even more strongly directional: existing distributions show that many vertebrates from all the continents have reached islands across moderate water gaps, while very few have returned from islands to continents. In some cases the fossil record shows general (average) directions of movement, for example movement of mammals from Asia to North America during much of the Tertiary (Simpson, 1947) and from North to South America since the late Pliocene (Simpson, 1940; and see below, p. 491).

All this concerns natural dispersal, in the past.

There is additional evidence in the behavior of plants and animals dispersed by man. Darwin (1859, 1950: 286) says, "From the extraordinary manner in which European productions have recently spread over New Zealand, and have seized on places which must have been previously occupied, we may believe, if all the animals and plants of Great Britain were set free in New Zealand, that in the course of time a multitude of British forms would become thoroughly naturalized there and would exterminate many of the natives. On the other hand, from what we see now occurring in New Zealand, and from hardly a single inhabitant of the southern hemisphere having become wild in any part of Europe, we may doubt, if all the productions of New Zealand were set free in Great Britain, whether any considerable number would be enabled to seize on places now occupied by our native plants and animals. Under this point of view, the productions of Great Britain [i.e. of Eurasia] may be said to be higher [more dominant] than those of New Zealand. Yet the most skillful naturalist from an examination of the species of the two countries could not have foreseen this result."

This is a sample of a kind of evidence that tells much about the relative dominances of organisms of different parts of the world. Man, intentionally or by accident, has carried many plants and animals, especially weeds and insects but also some vertebrates including some mammals and birds, all over the world, back and forth in every direction. But man-carried species have taken hold in some directions more than others. Species from Eurasia have probably established themselves in North America more than the reverse (I am not quite sure about this). Species from Eurasia and North America have certainly established themselves in Australia and probably also in southern South America more than the reverse. And species from continents have established themselves on islands more than the reverse. This pattern follows area. The plants and animals of larger areas have usually established themselves in smaller areas, although there have been some reverse cases too. This—the behavior of living plants and animals—is good evidence that dominant forms usually evolve in large areas, although the matter is much more complicated than my summary of it and needs detailed study.

I have been asked whether movements of animals between different parts of the world are simply proportional to the areas or the faunas concerned. They are not.

Take, as a hypothetical example, two areas, one ten times the size of the other (10A and 1A), and suppose they have faunas limited by area to 2N and 1N species respectively (see p. 504 for reasons for selecting these figures), and suppose no extinction occurs in either fauna. Then there can be no exchange. The larger fauna has more to give, but the smaller fauna has as many species as its area allows and no more can be added.

Now suppose that extinction (but not evolution) occurs. If each fauna consists of dominant species (D, capable of spreading) and undominant ones (E, liable to extinction) in the same proportion, so that the faunas are $2(D + E)$ and $1(D + E)$ respectively, one unequal exchange can occur in which twice as many species (D's) spread from the large to the small fauna as the reverse, the additions to each fauna being balanced by extinction of E's (if there are enough E's in the smaller fauna). Then no further exchange can occur.

Now suppose that evolution as well as extinction occurs, more effectively in the 2N than in the 1N fauna. Then exchange will become unequal again, but the ratio of exchange will not be fixed. It will depend, partly, on time. If the two faunas are separated for a short time and then allowed to exchange, the exchange may be nearly equal. But if they are separated for a long time, during which evolution

steadily increases the relative dominance of the larger fauna, the eventual exchange may be very unequal or may even approach a one-way movement.

A very unequal exchange of mammals has actually occurred between North and South America, after the South American fauna had been isolated for sixty million years or more (Simpson, 1940). When the two continents became connected, perhaps a couple of million years ago, many North American mammals moved (extended) to South America. Some South American ones moved northward too, but they were relatively few and most of them later became extinct in North America. The net result is that, while recent invaders from North America now make up about half the mammal fauna of South America (Darlington, 1957: 333, table 7), only three species of South American mammals now exist in North America (excepting bats and a few terrestrial forms that reach the tropical southern edge of North America). Moreover there has been extensive replacement in South America. All the old South American (marsupial) carnivores have been replaced by invading carnivores. All the old South American hoofed mammals have been replaced by invading ungulates. And some other South American mammals have probably been replaced by invaders. But of the three South American species now in North America, two (the armadillo and porcupine) have specialized food habits and may have moved into unoccupied niches without replacing anything else. Only the oppossum is an unspecialized, presumably competitive invader!

This exchange has really been between the enormous area of Megagea, which includes North America, and the much smaller area of South America. Many of the groups of mammals that have invaded South America came originally (through North America) from the Old World. But no South American mammal has returned to the Old World. As between the Old World and South America,

492   PHILIP J. DARLINGTON

the exchange has, in fact, been a one-way movement.

Darwin suspected that large area induces evolution of dominant plants and animals (above, p. 489). Matthew apparently did not suspect it. Matthew's pattern of dispersal (fig. 1, A) runs mostly from large to small areas, but he was apparently unaware of Darwin's explanation, and it apparently did not occur to Matthew that area as well as climate might be involved in evolution of dominance. Matthew (1915, 1939: 140) does mention "wider area" as favoring "expansive evolution" (radiation or diversification of dominant groups), but not as favoring evolution of dominance. He thought that dominant animals were made by northern climatic cycles, and that they then diversified in and spread from large central areas in the north. Nevertheless, having noted Matthew's apparent lack of agreement (I think he did not so much disagree as fail to consider the possibility), I am going to assume as a working hypothesis justified by the evidence that area is an important factor (but not the only one) in evolution of dominance, and that, other things being equal, the most dominant animals usually evolve in the largest areas. This leaves climate to be considered.

Effect of climate. We now need a biologists' definition of "the tropics." Many definitions are possible and no simple one is entirely satisfactory, for the tropics vary in climate (wet to dry), vegetation (rain forest to desert), and animal life. The diversity of the tropics is well described by Aubert de la Rüe *et al.* (1957). However we do not need a full or precise definition, but one that will emphasize the significant differences between the tropics and the north temperate zone. For this purpose the tropics may be defined as the zone in which, compared to the temperate zone, temperatures are relatively uniformly warm by day and night and at all seasons (and have probably been relatively uniform for a long time in the past), and in which plants and animals are relatively numerous, diverse, and diversely adapted, reaching a maximum in number of species, diversity, and interadaptation in the rain forest—the tropics could be defined as the zone within which, when other conditions are suitable, warmth and stability of temperature permit development of what we call tropical rain forest with its associated complex fauna. This zone now lies mainly between the Tropics of Cancer and Capricorn, but, biologically, its boundaries are irregular and in part poorly defined, and the tropical zone was certainly wider than now and perhaps less well defined at times in the past. But exact boundaries are not important for present purposes. Both temperature and plant and animal life are irregularly graded from the equator to the arctic (fig. 2). At the tropical end of the gradient temperature is relatively warm and stable, and floras and faunas are large and complex; at the northern

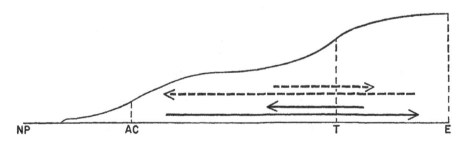

FIG. 2. Gradient of number of species in relation to climate, based (roughly) on number of mammals in eastern Asia. NP, North Pole; AC, Arctic Circle; T, Tropic of Cancer; E, Equator. Solid arrows suggest expected movements if dominant groups evolve mostly in the north temperate zone; broken arrows, if they evolve mostly in the tropics.

AREA, CLIMATE AND EVOLUTION   493

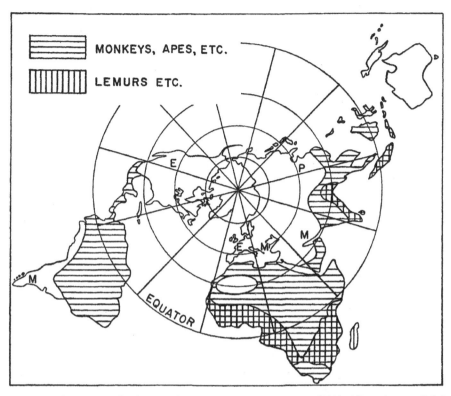

MONKEYS, APES, ETC.

LEMURS ETC.

E   P   M   M   E   M   EQUATOR

Fig. 3. Distribution of primates other than man, after Matthew (1939: 46), redrawn, slightly simplified. Letters show approximate localities of northernmost and southernmost fossil primates in: E, Eocene; M, Miocene; P, Pliocene.

end, temperature is colder and less stable (fluctuating daily, seasonally, and during recent geological time), and floras and faunas are relatively small and simple. The question is not the position of dividing lines but at which end of the gradient dominant animals usually evolve and in which direction they usually disperse. Stated differently, the question is whether evolution of dominant animals occurs more in simple faunas and unfavorable, unstable climates (as Matthew thought) or in complex faunas and favorable, relatively stable climates. ["Favorable" here refers to conditions of existence, not to factors affecting evolution.]

The gradient (fig. 2) is probably steeper now than it has usually been in the past, but a less steep gradient, reflecting a fundamental relation between cli-

mate and the size and complexity of floras and faunas, has presumably always existed. Climate has apparently tended to restrict exchange of mammals between Eurasia and North America at least since the later Eocene (Simpson, 1947: 220), which suggests that the gradient has been steep enough to have a marked effect on faunas at least since then. This is as far back as we need to go for present purposes.

Climate can limit animal distributions without involving evolution. Figure 3 is Matthew's map of the distribution of primates, other than man. Matthew considered this distribution radial from an Holarctic center, with primitive forms (lemurs etc.) in a "marginal position." But figure 4 is the distribution of the same primates on an orthographic pro-

494   PHILIP J. DARLINGTON

Fɪɢ. 4. Distribution of primates other than man, on an orthographic projection, with details brought up to date. Dotted lines are approximate limits of existing Prosimii (lemurs, tarsiers, etc.) ; broken lines, of monkeys etc.; solid lines, of anthropoid apes.

jection, with details brought up-to-date. On this map, it looks as if the effect of climate is primarily to limit distribution both northward and (in South America) southward. The lemurs etc. (Prosimii) in Africa and the Orient no longer seem marginal; they are within the limits of distribution of higher primates; they inhabit all the big, accessible forests of the Old World tropics; on the continents, in fact, existing Prosimii are no more marginal that the great apes, which Matthew thought represented the last prehuman cycle of dispersal from the north. The lemurs on Madagascar are marginal, outside the geographical limits of other primates, but this is presumably an effect of geographical isolation, not of dispersal with regard to climate. Fossils show that primates extended both farther north and farther south than now at times in the Tertiary, but that need mean only that limits were wider when climates were warmer. The record is too limited and too one-sided geographically to show whether or not particular primates were

in the Old World tropics at critical times in the Tertiary. Climate plainly sets general limits to primate distribution and probably affects many details of it. But does the distribution of primates (fig. 4) really show a pattern of successive radiations from the north, first of lemurs (dotted lines), then of monkeys (broken lines), and then of great apes (solid lines) ? I think a fair answer is that there certainly have been successive radiations of primates, that the present distribution of primates does not forbid northern origins, but that their distribution does not form a clear pattern of successive radiations from the north and therefore is not evidence that northern climatic cycles have controlled the evolution and dispersal of primates. This example suggests that Matthew did not sufficiently distinguish the present limiting effect of climate from its supposed effect on evolution in the past. What we want now is evidence of the effect of climate *on evolution.*

The Evidence

Where is the evidence? If area is a factor, the effect of climate on evolution should be looked for in places where area has comparatively little effect. Movements from the great continents to Australia or South America are probably influenced by relative areas and are not good evidence of effect of climate: patterns A and B of figure 1 agree that movement is from the great continents to Australia and South America. It is within the limits of the great continents, where A and B are different, that we should look for evidence of the effect of climate on evolution. What we want is evidence of whether dominant groups usually evolve in the north temperate zone or in the Old World tropics. And we want especially the evidence of warm-blooded mammals and birds, because they are the ones there is most doubt about, and because Matthew based his theory mainly on the mammals.

Evidence of the fossil record. The best evidence might be expected to come from fossils. Is the fossil record good enough in the north temperate zone and the Old World tropics to show directly in which place dominant groups have usually evolved or in which direction (north or south) they have usually moved? Mammals have the best fossil record of any animals. In parts of the north temperate zone it is very good, but in the Old World tropics the record even of mammals is very poor. Matthew knew this. He says (1915, 1939: 19), "In the Oriental region, we know nothing of the land life of the early Tertiary (except for a glimpse at the Eocene fauna of Burma), and in the later Tertiary we know only the life of its northern borders, . . ." and "In the Ethiopian region, we have but a single glimpse of the (early) Tertiary land fauna, and that is derived from Egypt, ..." We know a little more now than Matthew did, at least in Africa (Hopwood, 1954; Leakey *et al.*, 1955), but it

is still true that in the whole of the Old World tropics there is no record of mammals in the very early Tertiary and only a fragmentary one later. Consider what this means geographically. Suppose (this example is imaginary) that there were ten times as many mammals known fossil in the north temperate zone as in the Old World tropics and that everything else were equal, then, with incomplete samples and on the basis of chance alone, we should expect to find about ten groups known first in the north to one known first in the tropics. This would have nothing to do with place of origin or direction of movement, but it might be misinterpreted! The evidence of fossil mammals, then, is one-sided, very scanty in the tropics, and very difficult to interpret. The fossil record of birds is practically useless for present purposes.

Evidence of selected groups. Most of Matthew's evidence was presented as a series of histories of separate groups of mammals and other vertebrates. Each history was derived from the present distribution and fossil record of the group concerned. This should be the perfect method, if there is enough evidence, and if it is used properly. One of Matthew's histories, that of the primates other than man, has been discussed and criticized above (p. 493). In this case the method did not seem to work well. The distribution of primates looked radial on Matthew's map (fig. 3) but not (or not so much so) on a more world-like projection (fig. 4), and, although climate seemed to set limits to primate distribution, the distribution did not seem to form a clear pattern of successive waves of dispersal from the north. Moreover the fossil record of primates seemed too limited and too one-sided geographically to show places of origin and directions of movement within the main part of the Old World. The distribution of primates might be reconciled with a history of successive dispersals from the north but does not seem to give real evidence of it. This,

I think, is the best that can be said of most of Matthew's histories. They might be reconciled with radiation from the north but most of them seem to give no real evidence of it. Most of the histories (like that of the primates) are weakened by failure to distinguish the present limiting effect of climate from its supposed past effect on evolution and by failure to distinguish the effects of relative area and of geographical isolation from the effect of climate, as well as by failure to allow for the great gaps in the fossil record in the Old World tropics.

It is unnecessary to criticize all Matthew's other cases in detail, but I shall review the apparent histories of a few selected groups: man, elephants etc., cattle etc., horses, and murid rodents. Elephants and horses have been selected because they have exceptionally good fossil records. Man, cattle, and murids have been selected because they are probably the most recently dominant groups of mammals, most likely to give clear evidence of where and how dominant groups evolve. In studying the evolution of dominant animals we should, of course, give most attention to the most dominant groups.

Man. At first thought it seems as if man has recently risen to predominance in the north temperate zone. North temperate man now can atomize anyone else. But this is not what is meant here by dominance. Dominance is a biological matter, and the test of it is in the evolution and movements of populations. There are two pertinent questions to ask. Has the main course of man's evolution, the long rise to dominance during the last million years, been primarily in the Old World tropics or north of them? This question cannot be given a final answer but the evidence seems to favor the tropics. The distribution of a few human bones and many stone tools shows that man has been wide-spread in the Old World tropics probably throughout his history. We do not know how far northward he extended until recently, but ap-

parently neither he nor his immediate ancestors got far enough north to reach America until almost the end of the Pleistocene. The other question is, are human populations now moving from north temperate regions into the Old World tropics, or vice versa? I do not know the answer; certainly the owners of atomic and fusion bombs are not moving into the tropics of Africa and Asia in very large numbers. (Matthew thought that "the center of dispersal of the human race" was in Asia probably north of the Himalayas, but he oversimplified human history and did not know what is now known of the occurrence of man in Africa as well as tropical Asia in the Pleistocene.)

Elephants etc. Proboscideans—elephants, mastodonts, mammoths, etc.—first appear fossil in Egypt, in the Upper Eocene and Lower Oligocene, but had probably been evolving and diversifying in Africa for some time before that. Their history since then has been one of spectacular radiation from Africa or at least from the main part of the Old World. As new groups arose, they spread rapidly, and most of them sooner or later (from the Miocene to the Pleistocene) extended to North America, although a few groups did not. Probably at least a dozen groups reached North America, and eventually (in the late Pliocene and Pleistocene) three or four reached South America too. In the Pleistocene proboscideans were still numerous on all continents except Australia and in cold as well as hot climates, but in the short time since then they have been reduced to the two existing elephants, in different genera, in Africa and tropical Asia. (Matthew, without much evidence, questioned the African origin of proboscideans and thought they might have originated in southern Asia and later transferred their dispersal center northward, a "quite exceptional" history that did not fit his pattern.)

The geographical history of one group of proboscideans, the mastodonts, has been diagrammed by Simpson (1940: 143) (my fig. 5). Many details of this diagram

AREA, CLIMATE AND EVOLUTION 497

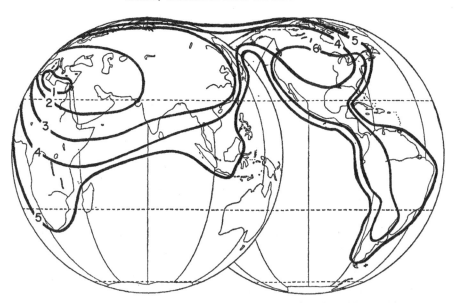

Fig. 5. Geographical history of mastodonts, diagrammatic, after Simpson (1940: 143), slightly simplified and transferred to an orthographic projection. Distributions are suggested in: 1, Oligocene; 2, earliest Miocene; 3, later Miocene; 4, Mio-Pliocene; 5, Plio-Pleistocene; 6, late Pleistocene. Many details are hypothetical (see text).

are necessarily doubtful. For example it is doubtful if mastodonts were ever confined to a small northern corner of Africa. It seems more likely that they arose in the main part of Africa, where there is no pertinent fossil record. Nevertheless the diagram does suggest the probable history of one group of elephant-like mammals. And it illustrates a kind of geographical history that may have occurred in other cases. In this case the mastodonts apparently began in Africa, then spread through the warmer part of Eurasia, and then spread far enough north to reach America. Other groups of animals may have followed the same main pattern, in some cases beginning in tropical Asia rather than in Africa, or they may have begun in the same way but never spread far enough north to reach America. Failure of a dominant Old World group to reach America suggests that it has evolved primarily in the warmer part of the Old World.

Cattle etc. The family Bovidae (cattle, antelopes, sheep, goats, etc.) is the most dominant existing family of hoofed mammals and one of the most recent of all mammal families to evolve. Bovids are now most numerous and diverse in Africa, less so in Eurasia (tropical and temperate), and very few in North America. Their fossil record begins in the Miocene and is extensive, but is probably very incomplete in Africa. Several tribes that are fossil in Europe and Asia are now confined to Africa, but this is evidence of withdrawals rather than of place of evolution. Simpson (1945: 157–162) lists 134 fossil or living genera of the family in the Old World. Only five of them have reached North America, where one additional genus is endemic. The present distribution (concentrated in Africa) and the fact that so few stocks have reached America (apparently none before the Pleistocene) suggest that the family has evolved, very complexly, mainly in the warmer part of the Old World. (Matthew thought that the centers of dispersal of antelopes etc. and of cattle have been in southwestern and southeastern Asia respectively, but he oversimplified bovid history and did not allow for the poverty of the fossil record in the tropics.)

Horses etc. Equids (horses, zebras, asses, and their ancestors) appear first in the Lower Eocene of Eurasia and North America. It is probable (but not absolutely certain) that the whole main line of further evolution of equids was in temperate North America. Their evolution was moderately complex (but much less so than that of some other families of mammals), with several successive radiations of genera. Several genera spread from North America to Eurasia, and at least one reached Africa in the Pliocene, but all the older genera that reached the Old World became extinct. Then most of the five or more Pliocene stocks became extinct even in North America; but one survived, radiated again, and produced *Equus* in North America (still in the Pliocene); and *Equus,* with numerous species, spread widely over the world, extending through Eurasia to Africa. (*Equus* and one or more other equids reached South America too in the late Pliocene or Pleistocene.) Finally, all equids became extinct in North (and South) America, leaving only the few existing species of *Equus* in temperate Eurasia and in Africa. This well documented history clearly follows Matthew's pattern of evolution in the north, although the invasions of the Old World tropics have not been extensive.

Murid rodents. Within their ecological limits, rodents are dominant mammals. The family Muridae (common Old World rats and mice) is probably the most recent family of rodents to become dominant. It includes the Black and Norway Rats and the House Mouse, which have spread over the world with man. However the fossil record of the family is so poor that its history has to be deduced almost entirely from its present distribution. Murids are numerous and diverse throughout the Old World tropics. Several different ones have reached Australia at different times since (perhaps) the late Miocene, which

suggests that the family has been dominant in tropical Asia since then. Murids occur also in temperate Eurasia but, in spite of their dominance, they have not reached America except as carried by man, which suggests that they have never been numerous in the north and that the family has evolved primarily in the warmer part of the Old World. (Matthew thought that "the myomorph families are evidently of Holarctic origin" but did not otherwise discuss the place of origin of the murids. He argued that the related cricetids are of northern origin but he did not consider the possibility that they may once have been world-wide and may have been replaced in the Old World tropics by murids.)

Summary of selected groups. The examples just discussed form two categories. The first includes dominant groups that seem to have evolved primarily in the Old World tropics or at least in the warmer part of the Old World: man, other primates (p. 493), proboscideans, bovids, and murid rodents. Of these groups only the proboscideans and perhaps the bovids have anything like an adequate fossil record. The other groups are placed here because their present distributions center on or include the Old World tropics and because they have not reached North America or have done so only infrequently or recently. Some other dominant existing mammals could be placed here for the same reasons: e.g. viverrids (civets, mongooses, etc.), pigs, and fruit bats.

The second category includes groups that have evolved in the north temperate zone or that seem to be more northern than tropical in evolution. Of the groups discussed, only the horses belong here. They have clearly evolved in the north and invaded the Old World tropics, but their invasions have not been extensive. Other families that might be placed here include camels and pronghorns, but they are primarily North American rather than north temperate and they are not known to have entered the Old World tropics at all. Bears and deer might be placed here because they are well represented in the north and absent in Africa, but we do not know how they have evolved and moved within Eurasia. Some additional smaller or more specialized families (moles, pikas, beavers, jumping mice, etc.) might be placed here, but they are hardly to be counted as dominant groups.

A third category might be made for widely distributed dominant or formerly dominant groups in which there is even less evidence of geographical origins, whether in the Old World tropics, the north temperate zone, or both. This category might include shrews, weasels, cats, dogs etc., rhinoceroses, rabbits, squirrels, cricetid rodents, and some insectivorous bats, as well as many non-dominant and less well known groups.

Although this summary does not include all mammals and is to some extent subjective, it does include the best known and most recently dominant families, which should give the best evidence of where and how dominant groups evolve. What general conclusions can fairly be drawn from this evidence?

If my presentation and interpretation of this evidence is even approximately correct, the groups that have evolved in the Old World tropics and spread northward are, if not more numerous, certainly larger and more dominant than those that have evolved in the north temperate zone. However some, usually smaller groups do seem to have evolved in the north and some of them, at least the horses and perhaps a few others, have or may have spread into the Old World tropics. This suggests an unequal exchange, with most dominant groups rising in the Old World tropics and spreading northward but with some countermovement too. This is what the evidence seems to show, and an exchange is also what would be expected. The great faunal movements that have been analyzed are exchanges rather than simple directional movements. For example, Simpson's (1947) analysis of movement of mammals across Bering Strait during the Tertiary shows a series of complex exchanges, each apparently with more movement from Asia to America than the reverse, but each with some countermovement too; and Simpson's (1940) analysis of recent movements of mammals between North and South America again shows a complex exchange, with (finally) most movement from north to south but with a little countermovement.

But suppose my presentation of the evidence of selected groups is not convincing. Most of the separate cases are less than conclusive, and readers may feel that all of them together do not justify my general conclusion. Can these cases then prove Matthew's thesis, that most dominant groups of mammals have evolved in the north and spread into the tropics? I think not. Some of the cases might be reconciled with a history of successive dispersals from the north, but most of them seem to give no clear evidence of it, and certainly not proof (p. 496 above). This amounts to saying that, if the cases do not prove that tropical origins are the rule, they do not prove that northern origins are the rule either. This is a compromise (which, I think, does less than justice to the evidence of tropical origins) but it is also an important conclusion, and readers should consider very carefully whether or not they agree with it. It is important because most of Matthew's evidence, which he thought proved that dominant animals evolve north of the tropics, came from the histories of separate groups of mammals. If this evidence does not prove what Matthew thought it proved, that leaves little to support his thesis of northern origins.

We have now to find what other evidence there is and what it seems to show.

Evidence of whole faunas. The distribution of dominant families in whole faunas (of mammals and birds) in the north temperate zone and the Old World tropics might give evidence of place of evolution of dominant groups. If dominant groups usually evolve in the north, replacing older groups there while the older groups survive in the tropics (as

500   PHILIP J. DARLINGTON

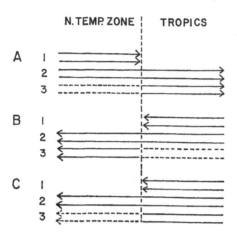

FIG. 6. Expected distributions of dominant families of mammals and birds in the north temperate zone and the Old World tropics: A, if successive dominant families evolve in the north and spread into the tropics, with replacements beginning in the north (Matthew's hypothetical pattern); B, if successive dominant families evolve in the tropics and spread northward, with replacements beginning in the tropics; and, C, if successive dominant families evolve in the tropics and spread northward, with replacements beginning in the north (the apparent actual pattern). Solid lines indicate present occurrence; broken lines, past occurrence. The numbers 1, 2, 3 indicate relative time, 1 being the first and 3 the last stage in the cycle of rise, spread, and beginning of replacement of dominant families.

Matthew thought), they should make a pattern of distribution something like figure 6A. But if the dominant groups usually evolve in the tropics, replacing older groups there while the older groups survive in the north, the pattern should be something like 6B. Arrowheads show the direction of movement in each case. If, as I have suggested, there is an unequal exchange rather than movement in a single direction, the patterns would be more complicated. We should therefore be prepared to find that no simple diagram fits the facts exactly, and if none does, we shall have to try to decide what diagram comes closest to fitting.

The existing mammal faunas of tropical Asia and of an approximately equal area of temperate eastern Asia differ in two ways (Darlington, 1957: 327–329). The tropical fauna is larger, containing nearly twice as many species as the temperate fauna; and the tropical fauna is more diverse, consisting of widely distributed families plus many additional families, while the temperate fauna consists of the same widely distributed families plus few additional ones. This is the general pattern of zonation of both mammals and birds: tropical faunas are large and diverse; temperate faunas are like tropical faunas from which much has been subtracted; and arctic faunas are like temperate faunas from which much more has been substracted (*op. cit.:* 341). The northern faunas have few important groups confined to them, except of water birds. This pattern is diagrammed in figure 6C. It is not 6A and not exactly 6B, but seems to be a modification of the latter. It would be formed if dominant mammals and birds usually evolve in and spread from the tropics, but if older groups usually begin their disappearance in the north, shrinking back into the tropics instead of forming a pattern (like 6B) of successive northward dispersals. This seems to be what does happen, among mammals and birds (*op. cit.:* 560–561). The proboscideans (above, p. 496) are an example.

Figure 6C is a simple model of a very complex situation, but as a static pattern, without the arrowheads, it does, I think, show the essential pattern of distribution of dominant families of mammals and birds in the Old World tropics and the north temperate zone now. It shows the relative richness and diversity of the tropical fauna and the absence of dominant families confined to the north temperate zone (see below). The arrowheads suggest how the pattern seems to have been formed.

I do not think Matthew's hypothetical pattern (6A) can be reconciled with the actual, existing pattern (6C, without the arrowheads), unless the existing pattern is abnormal, temporarily modified by special factors. Figure 6A puts too few dominant groups in the tropics and too many in the north temperate zone. Additional old (stage 3) groups could be added to 6A to give the required diversity in the tropics (if the tropical fauna is mainly a huge accumulation of relicts while dominant animals evolve mainly in the much smaller and simpler north temperate fauna), but the dominant northern (stage 1) groups required by 6A do not seem to exist.

Where are the dominant groups of mammals and birds that, according to Matthew, should be evolving in the north temperate zone but have not yet spread into the tropics? This is a critical question. I want to answer it fairly, but the answer is partly a matter of personal judgment. Dominant groups are conspicuously successful ones. Existing, primarily north temperate families and subfamilies of land mammals (Darlington, 1957: 327) are Soricinae (red-toothed shrews), Talpidae (moles), Ochotonidae (pikas), Castoridae (beavers), Microtinae (field mice etc.), Zapodidae (jumping mice), Aplodontidae (sewellels), Spalacidae (mole rats), Seleviniidae (related to dormice), and Antilocapridae (pronghorns). Of these, only the Soricinae and Microtinae seem to me conspicuously successful, and they do not seem to be enough to satisfy the requirements of figure 6A. Existing, primarily north temperate families and subfamilies of land birds (op. cit.: 255) are Tetraoninae (grouse), Prunellidae (hedge sparrows), Bombycillinae (waxwings), and Certhiidae (creepers). These four groups total only 39 species. I do not consider any of them conspicuously successful. (Some families of far northern water birds are more successful.)

Can Matthew's dominant northern groups have been temporarily eliminated by special factors? I think not. They have hardly been destroyed by man; man has eliminated some mammals and birds in northern regions but many have survived, and the survivors should include the most dominant, evolving groups. They have hardly been destroyed by Pleistocene climatic fluctuations; many animals sur-

vived the Pleistocene north of the tropics, and evolving dominant mammals and birds should have been particularly able to survive. In fact, according to Matthew's thesis, conditions north of the tropics should be especially favorable to evolution of dominant groups now, for we are in a period of continental emergence and markedly zonal climate (1915, 1939: 7). Then, where are the dominant northern groups?

What does Matthew say about this? Not in *Climate and Evolution* but thirteen years later (1928: 84–85) he said, "The outstanding feature of the Holarctic fauna is that it includes today all of the most progressive and highly advanced groups of animals. Its animals are very distinctly at the top of the tree in their various branches. They are not always the largest of their kind but they have the most perfected mechanisms, the most finished adaptations for their several modes of life." Matthew did not consider the existing situation unfavorable. He thought evolution is proceeding on his pattern now, and that the best animals are all in the north temperate zone now. When he made the statement just quoted, he probably had in mind everything that had gone into *Climate and Evolution*, but here is the evidence he chose to cite (*loc. cit.*): "Rabbits are brought to Australia and flourish amazingly, at the expense of the native rabbit-like animals. Dogs are brought [to Australia] and almost wipe out the native dog-like carnivora. The mongoose is brought to Jamaica, nearly exterminates the few native mammals and makes heavy inroads on the birds. In most cases the introduction of Holarctic animals and plants into the southern continents and oceanic islands has wrought havoc with the native flora and fauna. . . ." This is all true, but Matthew deduced too much from it. The Holarctic Region, as Matthew himself defined it (*loc. cit.*), is essentially the world north of the tropics. The examples cited all compare Holarctic animals only with those of small, isolated continents and islands. These examples (and many others like them) show that animals from the main part of the world are usually superior to those of smaller land masses, but they tell nothing about the relative dominances of north temperate and tropical animals within the main part of the world.

It seems to me that the pattern of distribution of dominant families in whole faunas of mammals and birds (fig. 6C) is inexplicable by Matthew's theory. He does not explain the pattern (he may not have known that it existed) but skips over it to compare Holarctic animals with those of smaller, isolated areas. But the pattern is explained if successive dominant families of mammals and birds have risen in the tropics and spread northward, and if some of them have later shrunk back into the tropics again. Exceptional groups have probably risen in the north and spread into the tropics, so that there has been an exchange. But figure 6C is close enough to the actual pattern of distribution of dominant families of mammals and birds in the main part of the Old World to suggest that the exchange has been very unequal, and that tropical origin and northward spread is the general rule.

Matthew's explanations. Matthew (1915, 1939: 7–8) thought that arid and cool phases of climate above the tropics would require animals to maintain themselves against inclemency of nature, scarcity of food, and variations of temperature, as well as against competition of rivals and attack of enemies, and that this would favor "greater activity and higher development of life" [i.e., general adaptation] as well as "special adaptations." He thought also that moist tropical climate, with abundant food, relatively constant temperature, and "larger percentage of carbonic acid and probably smaller percentage of oxygen in the atmosphere" would favor "sluggishness." And he thought that active, more highly developed northern animals would invade the tropics and compete with and replace sluggish tropical animals. These were *"a priori* deductions" to be tested against the evidence, especially against the record of

mammals. Matthew's evidence seems at best inconclusive, as I have said. Are there other reasons to think that his climatic theory is right? Should northern climates offer a more effective challenge to evolving animals than, say, mass of life in the tropics? I do not know. If so, should not the most dominant animals evolve where the challenge of climate is greatest, in the arctic or in deserts? They do not seem to do so. Are northern animals really more active, tropical ones sluggish? They do not seem to be. Matthew's explanations, then, do not seem to agree with observed facts.

THE WORKING HYPOTHESIS

The relation of evolution and dispersal to climate is a complex and difficult subject, and not everyone will agree with my conclusions about it. Therefore, instead of stating fixed conclusions, I shall set up a working hypothesis which seems justified by the evidence, and then shall try to find what it implies and how it agrees with observable facts and evolutionary theory.

It is part of the hypothesis, justified by evidence, that area is one important factor in evolution of dominance and that, other things being equal, the most dominant animals usually evolve in the largest areas (p. 492). To this may now be added that climate is another important factor. Taken at face value, the evidence, especially the apparent histories of the best known and most recently dominant families of mammals and the pattern of distribution of dominant families of mammals and birds in the main part of the Old World (fig. 6C), justifies the hypothesis that the most dominant groups usually evolve in the Old World tropics and spread into the north temperate zone. Countermovements occur, but they seem to be relatively unimportant.

If evolving, dominant mammals and birds rise in the Old World tropics and spread northward, they follow the apparent pattern of cold-blooded vertebrates. The most dominant existing fresh-water fishes (cyprinids) and frogs (*Rana*) and some dominant genera of lizards and snakes all extend from the Old World tropics far northward. Mammals and birds probably can spread northward more easily than cold-blooded vertebrates, but the pattern of movement seems the same. Why should it not be? Movement northward is, for all the groups, including the warm-blooded ones, from more favorable to less favorable climates and from more complex to simpler faunas (fig. 2). Why, then, should the warm-blooded groups reverse the pattern of evolution and movement of the cold-blooded ones?

Rising, spreading groups presumably continue to evolve, complexly, in the whole area they occupy. However, if the groups begin their rise in response to conditions in the Old World tropics, they would be expected usually to continue to evolve most effectively there, in continuing response to the same conditions. In this case, within each major group, successive dominant subgroups should rise in and spread from the Old World tropics.

EVOLUTIONARY IMPLICATIONS

Area, climate, and populations. Why should dominant animals evolve in the Old World tropics more than in other places? Apparently neither area alone nor climate alone can supply a satisfactory explanation. If area were involved, evolution should be as effective in the north temperate zone as in the Old World tropics. If climate alone were involved, evolution should be as effective in tropical America as in the Old World tropics. Apparently the effect of area and the effect of climate must be added together to make the Old World tropics a unique center of evolution of dominant vertebrates. The great area of Megagea (Africa + Eurasia + North America) apparently favors evolution of dominant groups. And so apparently does the warm, stable climate of the tropics. Where tropical climate is focused in a broad zone across Megagea, in the main part of the Old World, the effect of climate is added to that of area, making (apparently) op-

timum conditions for evolution of dominant animals.

How may area and climate exert effects on evolution?

Theoretically, warm, stable, tropical climate might accelerate evolution by increasing the rates of mutation or the numbers of generations of animals. But these effects are most likely to occur among cold-blooded animals and, I think, have not been observed among mammals and birds. Are there other ways in which both area and climate observably affect both cold-blooded and warm-blooded vertebrates so as to produce the requisite geographical pattern?

Area and climate might affect evolution through size of populations. Adaptation involves selection of advantageous mutations, and mutations occur in proportion to number of individuals in populations: the more individuals, the more (and sooner) the mutations, and the more rapid adaptation (including general adaptation) should be. If, therefore, large area and favorable climate increase the size of populations, that should accelerate general adaptation and the evolution of dominant animals. Or, area and climate might exert their effects through number of populations. If adaptation proceeds partly by selection of superior species—i.e., by selection of whole populations—it should be most rapid where populations are numerous as well as where they are large, for (other things being equal) the chance of superior populations appearing should be proportional to number of populations. Increase in number of populations might increase the force of selection, too. These are preliminary suggestions, to be considered in more detail in the following pages.

How are size of populations and number of populations related to area and climate?

Size of populations depends on area in some cases. Widely distributed populations on continents are obviously larger than populations confined to very small islands (even though the island populations may be denser), and this probably gives the continental populations a great advantage in adaptative evolution. But size of populations is not so simply correlated with climate. Large populations occur in some very adverse climates (hares and lemmings in the arctic are examples), and populations are not demonstrably larger in more favorable climates. In the tropics most animals seem to have small (sparse) populations. Some large populations do occur in the tropics, but whether they are larger than large populations in the north temperate zone is unknown. It is therefore impossible to say what role size of populations plays in the relation of evolution to climate. In earlier discussions (1948: 109–110; 1957: 565 ff.) I have probably overestimated its role.

Number of populations is correlated with both area and climate. To demonstrate the correlation with area we begin by comparing a very small island with a continent. If the island is small enough, it will have no vertebrates on it or perhaps a few small species but no large ones, while the continent will have many, diverse species. Then, comparing other islands of intermediate sizes, we find that there is no critical area below which faunas are limited and above which they are not. There is (other things being equal) an orderly relationship—the larger the area, the more numerous and more diverse the species in it. This relationship holds in a general way from the smallest island to the largest continent (Darlington, 1957: 482–483), although the effect of area on number of species is evidently modified by many other factors in particular cases.

The correlation between climate and number and diversity of species is similar. In the most adverse climates, say in the interior of Antarctica or on the Greenland ice cap, there are no vertebrates except passing birds. In less extreme parts of the arctic there are small numbers of species. In temperate, seasonal climates species are more numerous. And in warm, relatively stable, tropical climates they are still more numerous, and more diverse. There is (other things being equal) an orderly relationship here, like but more complex than the relationship of area to number and diversity of species— the warmer and more stable the climate the more and more diverse the species. This is true even of mammals (fig. 2) and birds, which are much more numerous (in species) and more diverse (in ancestry and relationships) in the tropics than elsewhere.

Area and climate, then, have similar effects on number of species—or of populations—and they reinforce each other. The largest and most diverse faunas—the most populations—are in the largest areas and the warmest, most stable climates.

The observed general relation of area and climate to number of populations is mapped diagrammatically in figure 7. The three principal land masses of the world are represented in proportion to their areas. Each mass is divided into squares according to its area: in Megagea (area 10A) the squares are double the dimensions of those in Australia (area 1A), and in South America (area 3A) the squares are intermediate. And within each mass the squares in the tropics are double the dimensions of those in the temperate zone. I have made the diagram in this way because multiplication of area by 10 does double number of species in some cases (e.g. species of reptiles and amphibians on some islands in the West Indies, Darlington, 1957: 483–484), and because number of species is sometimes approximately doubled in tropical as compared with temperate regions (e.g. species of mammals in tropical as compared with temperate eastern Asia, *op. cit.:* 328). As thus drawn, the sides of the squares are VERY ROUGHLY proportional to the numbers of populations to be expected in different places, if my assumptions are correct. Moreover if the complexity of interaction of populations increases exponentially with their number (as suggested by Prof. E. O. Wilson, in conversation), the areas rather than the sides of the squares may be VERY ROUGHLY proportional to the complexity (and force?) of selection in different places.

Arrows have been added to the diagram (fig. 7) to suggest VERY ROUGHLY the relative amounts of movement in different directions that might occur if area and climate determine numbers of populations, and if numbers of populations determine the effectiveness of adaptive evolution. The arrows make a world-wide pattern that agrees reasonably well with the apparent main pattern of dispersal of dominant vertebrates (fig. 1, B) (Darlington, 1957: 571).

This diagram (fig. 7) is extremely oversimplified. Rainfall is ignored, and the distribution of temperature is much simplified. Habitat is ignored; the area available to different animals in different regions must depend partly on whether they live in forests or steppes or deserts, and number of populations must vary with continuity or discontinuity of habitats; but habitat restrictions may be partly overcome by ecological radiations of dominant animals. Partial or intermittent barriers (e.g. Bering Strait) are ignored, although they must often reduce or divide populations. The structure of populations is ignored; it is assumed that each species is

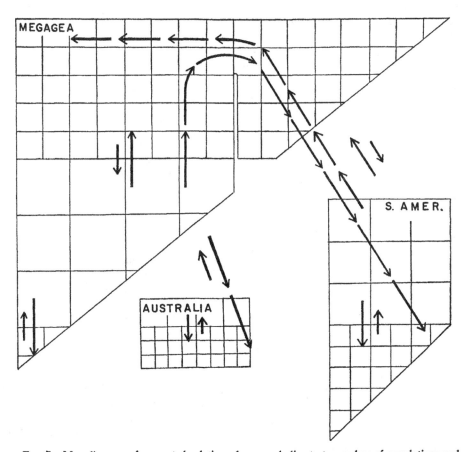

FIG. 7. Map-diagram of suggested relation of area and climate to number of populations and to evolution and dispersal of dominant vertebrates. See text for method of construction. Length of arrows is (very roughly) proportional to amount of movement expected in directions indicated.

one population, but some species in some places (especially in large areas or discontinuous habitats) probably consist of many separate subpopulations, and this probably affects evolution. And, perhaps most important of all, figure 7 ignores the effect of time. Under special conditions, when large faunas become isolated on small islands, time may reduce the number of species. More often, time probably increases the number. If, for example, there are more species of birds in a given area in tropical South America than in tropical Asia, this may be partly because species of birds have been multiplying in South America for a longer time without major replacements.

Nevertheless figure 7 gives a pattern which agrees in general with the apparent pattern of dispersal of dominant vertebrates and which is based on the observed general relation of number of species to area and climate. I can see no other way of deriving the requisite pattern from observed facts. I think, therefore, that the diagram is probably somewhere near the truth, in a very oversimplified way. At least it provides a further working hypothesis by which to examine general adaptation.

The diagram adds the effect of climate to the effect of whole area. This is necessary to make a pattern that conforms to the apparent pattern of vertebrate dispersal, with one main center, in the Old World tropics (p. 503). But the effect of climate can be added to the effect of whole area only if, besides much movement out of the tropics, there is some movement back into them from other parts of the whole area: the whole of Megagea must produce moderately dominant groups some of which then move into the tropics and are raised to predominance there. Counter-movements like that of the horses, then, are not only observed, and expected from what is known of the nature of complex faunal movements (p. 499), but apparently must occur to make the Old World tropics the main center of evolution of dominant vertebrates.

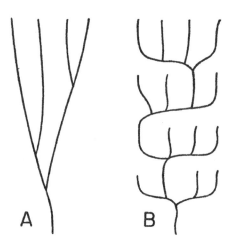

Fig. 8. Diagrams of evolution of four species: A, without selection of populations; B, with repeated selection and re-radiation of populations.

Populations, faunas, and evolution. Large populations probably have an adaptive advantage everywhere (p. 503). The diagram (fig. 7) implies and requires that large number of populations gives an additional, perhaps greater evolutionary advantage. Mass and diversity of populations might increase selective pressures on individuals within populations. Or populations might be selected as wholes, with continual radiations and reradiations of successive, selected, superior populations, extinction of others, continual replacement, and continual movement (spreading) of the superior, evolving populations (fig. 8). That dominant animals often do evolve in this way, with continual replacement and much movement, is suggested by the present distributions of many animals and by the fossil records of the best known groups. The history of man during the last few thousand years exemplifies evolution with continual replacement and much movement of populations.

Evolution by selection of whole populations may be more efficient than evolution by selection of individuals in single populations, for two reasons. First, although selection may proceed in many different directions in one population at one time (e.g. toward concealing coloration, better teeth, and increased intelligence) without the different selective progressions interfering much with each other, interference probably does occur in some cases. For example sexual selection may interfere with or oppose non-sexual selective processes and may produce, often in one sex only, characters that give some individuals an advantage in mating but that are otherwise disadvantageous to the species. In such cases there may be an opposition of interests between individual animals and the species they belong to: what is good (or selectively advantageous) for individuals may be bad for the species. This may be true of increase of size. Large size may give individuals an advantage in many ways but may impose an evolutionary disadvantage on species in the course of time (see below). Evolution of single species is therefore likely to be a compromise between what is good for individuals and what is good for the species, and evolution of single species is likely to be retarded or deflected accordingly. But selection among species—selection of whole populations—involves less compromise and is more likely to be simply selection of the most efficient animals. Selection of species (or whole populations) may therefore be more efficient than selection of individuals within species, so far as general adaptation—evolution of generally better animals—is concerned.

The second way in which selection of whole populations may be more efficient than selection of individuals in single populations depends on the nature of the evolutionary process. Evolution involves changes of gene systems by mutation, recombination, and selection. Recombination is very important. Its evolutionary advantage is evidently so great that it more than outweighs the wastefulness of sexual reproduction, which often halves the number of individuals that bear offspring but makes recombination of genes possible. The great importance of recombination is confirmed by mathematics.

Potential recombinations of genes give all ordinary species a great "store of variability" sufficient for varied and extensive evolutionary modifications without the occurrence of any new mutations (Fisher, 1930: 95–96; Wright, 1949). Only a comparatively small amount of recombination-variability is available in a single, random-breeding population. More becomes available if the population is divided into subpopulations which are partly isolated from each other but which interbreed occasionally. Such subpopulations may evolve divergently, by recombination of genes, and subpopulations may then be selected almost as wholes. The most favorable situation may well be where many widely distributed but very sparse populations occur together, each tending to form many small, partly isolated subpopulations. Under these conditions a maximum of recombination-variability should occur, and selection both of subpopulations and of whole populations can occur, with extinctions and replacements at both levels, and this should result in much more efficient use of recombination-variability, and more efficient adaptation, than can occur in single populations. These conditions may be approximated in large, widely distributed faunas in the tropics.

Under these conditions, the size of individual animals might have a profound effect on evolution. Among given animals during limited times large size may be advantageous in competition for mates and in other competition among individuals of one species and (less often ?) among different species, and size may increase progressively by selection. But increase of size increases the requirements of each individual and reduces the size and/or number of populations that can exist in a given area and climate. Moreover, since large vertebrates can move farther than small ones, increase of size of individuals may tend to make populations more uniform, less likely to become divided into partly isolated subpopulations. These changes toward smaller, fewer, less subdivided populations would presumably reduce the efficiency of adaptive evolution. And, eventually, the advantage of large size might be overcome by the more efficient evolution (general adaptation) of smaller animals. This might give a basis for evolution of successive groups, each group going through a cycle of radiation, increase of size, and then extinction of the large forms, followed by re-radiation of new, smaller forms, which in turn would increase in size, etc. Successions something like this, but irregular and very complex, seem to have occurred during the evolution of terrestrial vertebrates.

The relation of number and structure of populations to evolution can be approached in another way. Evolution presumably makes situations favorable to itself: effectively evolving populations are presumably selected, and combinations of populations favorable to evolution are presumably selected too. This kind of selection is complex and presumably requires much time and also a fairly stable climate, for catastrophic changes would destroy evolving combinations. We might therefore expect to find populations and combinations of populations favorable to effective evolution in the relatively stable tropics rather than the unstable north temperate zone. This is what we do seem to find, as I have tried to show in the preceding pages.

METHODS; RESTATEMENT OF THE HYPOTHESIS

I first wrote this paper as a sequence of evidence, arguments, and "proved" conclusions, but then it read as if I had taken an arbitrary position and were defending it. This was of course the case, and it inevitably still is the case to some extent. However I have tried to limit this aspect of the paper by partly changing the method of presentation. I have had to present facts and situations as they seem to me to be, and say what they seem to me to mean. But then, instead of trying to prove conclusions, I have set up a

working hypothesis based on evidence taken at face value, and I have tried to test it and explain it by means of observable facts and relationships, so far as possible. This method may help to reduce personal bias, may prepare the way for experimental and mathematical treatment of the observed relationships, and may lead toward the truth more surely than personal judgment would. The change in method has led to two significant changes in my original conclusions: to stressing unequal exchange rather than one-way movement between the Old World tropics and the north temperate zone, and to stressing number rather than size of populations in relation to evolution. The latter change brought my hypothesis more into line with the expectations of mathematical evolutionists.

Restated very briefly, the hypothesis is this. The most dominant vertebrates usually evolve in the Old World tropics and spread northward, but some countermovements do and must occur. Both area and climate contribute to making the Old World tropics a unique evolutionary center: moderately dominant groups evolve in the great area of Megagea, and are then raised to predominance in the Old World tropics. Both area and climate are observably correlated with number of populations: the larger the area and the more favorable the climate, the more and the more diverse the populations. Large number of populations may accelerate adaptation by selection of whole populations, which should tend to avoid individual-versus-species oppositions and make effective use of recombination variability. The situation in the tropics, of many, sparse, perhaps subdivided populations, may be, and logically ought to be, most favorable to efficient selection and adaptation, including general adaptation.

This hypothesis is very much oversimplified. It stresses only one (but perhaps the most important) of many factors that probably affect the evolution of different animals in different places.

AREA, CLIMATE AND EVOLUTION   509

POSTSCRIPT

G. E. Hutchinson's "Homage to Santa Rosalia or Why Are There So Many Kinds of Animals?" (1959, *American Naturalist*, 93: 145–159) and William L. Brown Jr.'s "General Adaptation and Evolution" (1959, *Systematic Zoology*, 7: 157–168), both concerned with some aspects of evolution dealt with here, appeared while this paper was in press. Also while this paper was in press I have reread "Evolution in the Tropics" (1950, *American Scientist*, 38: 209–221) by Theodosius Dobzhansky. He suggests that selection is controlled in the cold zone mainly by the physical environment and in the tropics mainly by the more complex biological environment, and that the latter favors evolution of new modes of life and more advanced types of organization. It would be interesting to see this idea developed in detail.

ACKNOWLEDGMENTS

I am indebted to Dr. William L. Brown Jr. and Prof. Edward O. Wilson for reading the manuscript of this paper and for a number of useful suggestions.

SUMMARY

(1) This paper is concerned with general adaptation, which is the kind of evolution that produces dominant animals (p. 488).

(2) Area is one important factor in evolution of dominance: the most dominant animals seem usually to evolve in the largest favorable areas (pp. 490–492). How climate affects evolution is the question (fig. 2).

(3) The apparent histories of the best known and most recently dominant groups of mammals and the distribution of dominant families of mammals and birds in the Old World tropics and the north temperate zone (fig. 6C) indicate that the most dominant warm-blooded (like cold-blooded) vertebrates usually evolve in the Old World tropics and spread northward, although some countermovements occur (pp. 495–502, summarized p. 502).

(4) Apparently neither area alone nor climate alone can make the Old World tropics a unique evolutionary center. The effects of area and climate must be added together (p. 503).

(5) Tropical climate might accelerate mutation or reproduction and thus accelerate evolution; but this is unlikely among warm-blooded animals. Large populations presumably have an advantage in evolution by mutation and selection of individuals; but, although large populations occur in large areas, they do not seem to be especially characteristic of tropical climate. Number of populations of both cold- and warm-blooded vertebrates is correlated with both area and climate: the larger the area and the warmer and more stable the climate, the more and more diverse the populations (pp. 503–504).

(6) If the relation of area and climate to number of populations is mapped diagrammatically (fig. 7), a world-wide pattern is formed, centered on the Old World tropics, which fits the apparent geographical pattern of evolution and dispersal of dominant vertebrates (fig. 1, B) (pp. 504–506).

(7) This suggests that effective evolution—general adaptation—is correlated with number of populations and occurs partly by selection of whole populations, with continual extinctions, replacements, and movements (spreadings) of evolving populations (fig. 8). There are indications that dominant groups do evolve in this way (p. 507).

(8) Selection of whole populations tends to avoid individual-versus-species oppositions and makes effective use of recombination-variability. The situation in the tropics—occurrence of many, sparse, perhaps subdivided populations—may be especially favorable to selection of populations and use of recombination-variability. Evolution ought to make situations favorable to itself in the stable tropics (pp. 507–508).

510   PHILIP J. DARLINGTON

(9) Conclusions are presented as a working hypothesis, justified by evidence taken at face value, based as far as possible on observable facts and relationships, and capable of experimental and mathematical testing.

LITERATURE CITED

AUBERT DE LA RÜE, E., F. BOURLIÈRE, AND J.-P. HARROY. 1957. The Tropics. New York, Alfred A. Knopf.

DARLINGTON, P. J., JR. 1948. The geographical distribution of cold-blooded vertebrates. Quart. Rev. Biol., 23: 1–26, 105–123.

——. 1957. Zoogeography, . . . New York, John Wiley; London, Chapman & Hall.

——. 1959. Darwin and zoogeography. Proc. American Phil. Soc., 103: 307–319.

DARWIN, C. (1859) 1950. On the Origin of Species . . . [reprint of the first ed.]. London, Watts & Co.

FISHER, R. A. 1930. The Genetical Theory of Natural Selection. Oxford, Clarendon Press.

HOPWOOD, A. T. 1954. Notes on the recent and fossil mammalian faunas of Africa. Proc. Linn. Soc. London, 165: 46–49.

LEAKEY, L. S. B., AND W. E. LE GROS CLARK. 1955. British-Kenya Miocene expeditions. Interim report. Nature, 175: 234.

MATTHEW, W. D. (1915) 1939. Climate and evolution [reprint]. Special pub. New York Acad. Sci., 1.

——. 1928. Outline and general principles of the history of life. Univ. California Syllabus Ser., No. 213.

SIMPSON, G. G. 1940. Mammals and land bridges. J. Washington Acad. Sci., 30: 137–163.

——. 1945. The principles of classification and a classification of mammals. Bull. American Mus. Nat. Hist., 85.

——. 1947. Evolution, interchange, and resemblance of the North American and Eurasian Cenozoic mammalian faunas. EVOLUTION, 1: 218–220.

WRIGHT, S. 1949. Adaptation and selection. (In) Jepsen et al., Genetics, Paleontology, and Evolution: 365–388. Oxford, Clarendon Press.

PATTERNS OF TAXONOMIC AND ECOLOGICAL STRUCTURE OF THE SHELF BENTHOS DURING PHANEROZOIC TIME

by JAMES W. VALENTINE

ABSTRACT. The taxonomic and ecological structure of the shelf biota are intimately related at the species–population levels. Early Paleozoic faunas contained relatively few species representing relatively many higher taxa, and ecosystems were relatively generalized. Medial and late Paleozoic faunas contained more species representing fewer higher taxa, and ecosystems were relatively specialized. This suggests that, as higher taxa became extinct, they were not replaced except at lower taxonomic levels; diversification was proceeding through increasing specialization. After Permo-Triassic extinctions, rediversification was chiefly confined to low taxonomic levels. Late Mesozoic and Cenozoic diversification at lower taxonomic levels has been remarkably great, resulting not only from increasing specialization at the population level but from a marked increase in provinciality due to rising latitudinal temperature gradients on the shelves and to the fragmentation and isolation of shelf environments by continental drift.

THIS paper examines the historical relationships between the ecological and taxonomic structures of the marine biosphere, and attempts to account in a general way for the patterns of their evolution. Each of these structures is hierarchic. The units composing the levels of the ecological hierarchy include individuals, populations, communities, and provinces, while the units composing the levels of the taxonomic hierarchy are such categories as species, genera, and families.

There has been relatively little theoretical discussion of the evolution of these structures for marine invertebrates, yet the geological record of skeletonized taxa of the shallow marine invertebrate benthos is longer and more complete than for any comparable group of organisms. This paper therefore deals with the rich and lengthy record of shallow marine environments.

This restriction to a specific group of communities has some special advantages. The diversity pattern for the world at large is obviously very much influenced by the deployment of organisms into new environments, such as the invasion of the terrestrial habitat by vertebrates. By restricting the data to a limited group of communities it may be possible to investigate the patterns of diversity changes within ecosystems.

Much of the structural evolution which the taxonomic and ecological hierarchies have undergone is a product of the diversification and extinction of species. There has been much discussion of the patterns of taxonomic diversifications and extinctions through geologic time, especially of higher taxa, and ecological relations are commonly invoked to account for these patterns, particularly for extinctions. The processes of diversification assumed herein are those of the synthetic theory of evolution based on Darwinian selection and upon modern genetic concepts. Speciation and the origin of higher taxa have been discussed from this viewpoint in a number of larger works (for example Huxley 1942; Mayr 1963; Rensch 1947; and Simpson 1953). As diversity rises, there must be a mechanism of accommodation of the new forms in ecological systems; such mechanisms are discussed by Klopfer (1962), MacArthur and Wilson (1967), and

[Palaeontology, Vol. 12, Part 4, 1969, pp. 684–709.]

Miller (1967), among others. Possible causes of extinction, and hypotheses of the processes of extinction that have operated to create the Permo-Triassic faunal change, have been reviewed by Rhodes (1967).

THE ECOLOGICAL STRUCTURE OF THE BIOSPHERE

The ecological hierarchy is regarded as being composed of the levels that are depicted in text-fig. 1. This paper is chiefly concerned with the functional aspects of the hierarchy,

GENETIC HIERARCHY		ECOLOGICAL HIERARCHY		
UNIT	COLLECTIVE	DESCRIPTIVE UNITS	FUNCTIONAL UNITS	LEVEL
$*^2$	$*^3$	Marine Shelf Biota	Shelf Realm of the Biosphere	High
$*^1$	$*^2$	Province	Provincial System	
Gene Pool	$*^1$	Community	Community System (Ecosystem)	
Genotype	Gene Pool	Population (Deme, Species)	Population System (Niche)	
Functional Genetic Unit	Genotype	Individual	Ontogenetic System	Low

$*^1$ *Collection of gene pools*
$*^2$ *Collection of gene pool collections*
$*^3$ *Collection of collected gene pool collections*

TEXT-FIG. 1. Some levels of organization in the ecological hierarchy employed in this paper (after Valentine 1968b).

that is, with the interacting systems of organisms and environments. From the highest level down, each functional system is composed of subsystems representing the systems of the next lower level. The lowest functional level in the figure, that of the individual, is certainly capable of further subdivision into sorts of functions of 'unit characters', each underpinned by a system of genes and its regulators. For the most part, however, the present discussion concerns population and higher levels.

It is convenient to consider the ecological units in terms of the environment with which they interact. Hutchinson (1957, 1967) has developed a formal conceptual model that treats the environment as a multi-dimensional region (see also Simpson 1944, 1953). Only an informal treatment, based on Hutchinson's model, is required here. If each separate environmental parameter is visualized as a single geometric dimension of this region, then all possible environments are represented by the resulting multi-dimensional space or hyperspace, which contains as many dimensions as there are possible environmental parameters. The space extends along each dimension to the physical limits of each parameter. It is assumed that this multi-dimensional environment model is standardized by having each axis allotted an arbitrary but permanent direction to form an

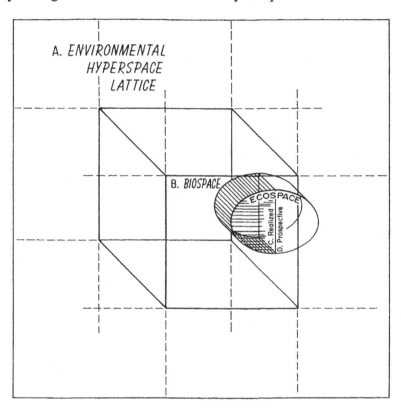

TEXT-FIG. 2. Highly diagrammatic representation of some aspects of environment–organism relations, visualized as a multi-dimensional space, of which each dimension is some environmental factor, physical or biotic. Each point within the lattice represents a unique combination of factors. Only three of the many dimensions are depicted. A, the total possible range of all environmental factors represented as a multi-dimensional lattice. B, the portion of the environment that actually exists on earth, the *biospace*; it is available for occupation by organisms. C and D, the region of environmental space that coincides with factors tolerated by an organism and that is bounded by its limits of tolerance—the *ecospace* of the organism. Only a portion of the ecospace is realized (C); the remainder is prospective ecospace (D) that may be inhabited if the environment fluctuates so as to include more of that portion of the lattice. The ecospace concept may be expanded to population, community, province or biosphere levels, and to species, genus, family, and higher taxonomic levels.

environmental hyperspace lattice, hereinafter called simply a lattice for brevity. Only a certain portion of the total possible lattice (the prospective lattice) actually represents conditions of the environment. This 'realized ecological hyperspace' may be called *biospace* (text-fig. 2), a term employed in a similar but less generalized sense by Doty (1957).

For any organism there is some more or less small volume (actually a hypervolume) within the lattice corresponding to the range of environmental conditions under which it may live. This functional hypervolume will be called the *ecospace* of that organism (text-fig. 2). Each population also has its own ecospace, which is the hypervolume of its niche within the lattice. Indeed, the ecological units at all levels have ecospaces. A community ecospace is the multi-dimensional model of its ecosystem, and a provincial ecospace is the model of the provincial system. Although the highest functional level standing above that of the provincial system is the level of the biosphere, the system of the shallow marine realm is being used here in its place as a matter of simplicity, and this realm has its own ecospace. The total ecospace that an organism or other ecological unit may utilize if it is physically available may be called the *prospective ecospace*, while the portion of the ecospace that actually overlaps with realized biospace may be called the *realized ecospace*. These terms are modelled on the discussion of Parr (1926) and Simpson (1944, 1953).

Dimensions of the lattice which have special properties are those that represent the *real* dimensions of space, in which discontinuities occur that permit the occupation of similar functional regions in different geographic regions (Miller 1967) and of time, in which the changing shapes and sizes of ecospaces and of biospace are perceived.

The structure of the ecological hierarchy may be illustrated by considering just one level, for example the community level. Community ecospace is composed of the ecospaces of all the niches of the component populations, and includes some dimensions that are not niche properties but are organizational properties of the ecosystem. The size of the community ecospace, measured by the number of dimensions occupied and the extent of occupation along each dimension, depends upon the sizes of the component niche ecospaces and to a small extent upon the organizational properties. Into a community ecospace of a given size, a relatively large number of small niches or a relatively small number of large niches may be packed. All niches in a community ecospace overlap to some degree, for all share a common tolerance for certain salinity ranges, for example, and for certain oxygen concentrations, and for other parameters. The more that niches overlap, other things being equal, the more populations that can be packed into a community ecospace of a given size (Klopfer 1962; Miller 1967).

Consider, then, a community (*A*) composed of relatively few populations that have very large niches that overlap only narrowly on the whole. The animals tend to be rather generalized feeders, so that energy flows in relatively broad streams through the trophic levels. This community, though of low diversity, may displace a large biospace in the lattice, that is, may have a large ecospace. Consider another community (*B*) composed of many populations of different species that tend to have very small niches which overlap broadly on the average. The animals are highly specialized with relatively narrow ranges of food sources, so that energy flows through the trophic levels in relatively discrete paths along chains of organisms that tend to be rather isolated owing to their high specialization. Energy flow is not like a stream but more like a shower that breaks up into numerous jets. A community of this sort, though rich in species, may displace no more biospace in the lattice than community (*A*) and may displace considerably less.

These communities have vastly different structures in the lattice, and yet it seems possible for one to evolve from the other. They may thus represent relatively early (*A*) and advanced (*B*) stages in the evolution of a community 'lineage' that has inhabited a similar biotope through its history. In this event the ecospaces of the two communities will approximately coincide, although the way in which each community biospace is occupied by niches is different. The community structure has evolved.

Structural states of ecological systems at other levels may be described in an analogous way. All the systems evolve by changes in the quality, relative proportions, and diversity of their subsystems (Valentine 1968b). Thus evolution of ecological systems need not involve organic evolution, but may result merely from the readjustment of existing populations in new patterns of association. However in the present discussion the chief interest lies in changes that *are* based upon organic evolution, upon changes in gene frequencies within populations that produce changes in niches, and upon the accommodation of the changed niches in ecosystem structures. Enough is now known of these processes to permit the construction of a provisional model of the diversification of ecosystems. But before proceeding to the model, it is appropriate to examine the main patterns of taxonomic structure during the Phanerozoic.

THE TAXONOMIC STRUCTURE OF THE BIOSPHERE

The taxonomic hierarchy is too well known to require any general remarks. For purposes of this paper only a few levels need be considered: phylum, class, and order, which will be called 'higher' taxonomic categories; and family, genus, and species, which will be called 'lower' categories. It is possible to visualize the ecospace of any genus as composed of the ecospaces displaced by all its component species, and the ecospace of a family as composed of all the generic ecospaces, and so on. Thus defined, the ecospace of a higher taxon displaces the actual regions of the lattice that have been occupied by the members of that taxon. Thus the taxonomic hierarchy possesses a precise structure at any time. This structure changes through time in well-defined patterns.

The main trends of evolution of the taxonomic structure may be characterized by considering the trends of diversity among higher and lower taxa through geologic time. The fossil record of diversity, however, is certainly biased. An important source of bias is the differential preservation of taxa. It seems possible to use the skeletonized taxa that are best represented as a sample, from which to attempt to generalize to the entire biota. The basic data from which generalizations will be attempted are the records of easily fossilized shallow benthonic taxa of nine phyla: Protozoa, Porifera, Archaeocyatha, Coelenterata, Ectoprocta, Brachiopoda, Mollusca, Arthropoda, and Echinodermata. The ranges of these phyla and of their taxa are taken chiefly from the *Treatise on Invertebrate Paleontology* (ed. Moore 1953–67), the *Fossil Record* (ed. Harland *et al.* 1967), and the Russian *Osnovy Paleontologii* (Orlov 1958–64).

As the assignment of groups of organisms to taxonomic categories involves a large element of subjectivity, it is fair to ask to what extent the trends in taxonomic diversity are real. In the first place, if one constructs a hierarchical classification of fossils that appear at different times, the average time of appearance of higher taxa will be earlier than that of lower, simply because some of the lower taxa appeared later than others, but none appeared earlier than the higher taxa to which they belong. The mode of first appearance should shift progressively towards the present at lower and lower taxonomic levels (Simpson 1953, pp. 237–9). Similarly, the mode of highest diversity will tend to shift towards the recent at progressively lower levels provided that the earlier taxa at each level persist or are replaced. These considerations account for such shifts in the mode of appearance and diversity in text-fig. 3.

Secondly, there is no doubt that the present data contain monographic artifacts (for

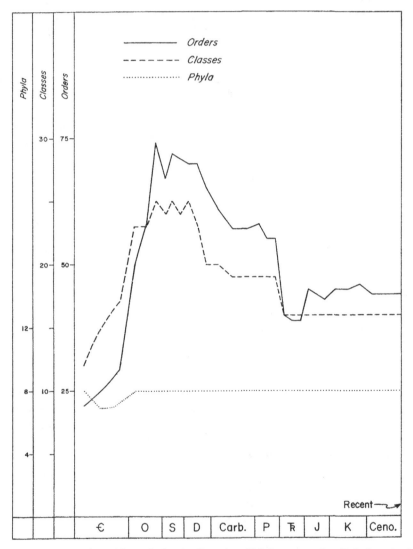

TEXT-FIG. 3. Stratigraphic variation in diversity of higher taxa of well-skeletonized marine shelf invertebrates. Data chiefly from Harland *et al.* (1967) and Moore (1953–7).

an instructive example see Williams 1957). However some of the main points to be discussed here concern relative diversities among the several taxonomic levels. Presumably, monographic artifacts would tend to appear at all levels, and relative diversities would be much less affected than absolute diversities. It is unlikely that there is a consistent monographic bias in the same direction among a majority of the taxa, and I therefore believe that the major trends are real.

Another important consideration is the extent to which trends among skeletonized taxa represent the biota as a whole. At present the non-skeletonized Invertebrata have the same biogeographic and synecological patterns as skeletonized groups (Lipps, *in press*), and there is no reason to expect that patterns of diversity of non-skeletonized taxa would follow different trends than the skeletonized ones. Furthermore Lipps has pointed out (pers. comm.) that the preserved groups are morphologically diverse and unrelated, yet they often exhibit similar patterns. Therefore it is assumed that major trends among skeletonized and non-skeletonized groups tend to be in phase.

Experience has shown clearly that the chances of preservation of an organism that does not possess a well-mineralized skeleton are exceedingly small. Indeed, the lack of a record of a taxon that does not have a relatively high probability of preservation can hardly be taken as proof that the taxon was not living at the time. And the probabilities of preservation cannot yet be specified even for taxa with highly mineralized skeletons, under many of the stratigraphic situations common in the geologic record. It is therefore difficult to assess the significance of negative records.

Most of the known phyla had appeared in the record by Cambrian time, although even among our sample of nine, one (Ectoprocta) does not appear until the Lower Ordovician. The phyla are well-differentiated and some contain relatively complex organisms when they first appear, so that a fairly long period of evolution can be assumed to have preceded their appearance in the record. However, it is possible to argue for many phyla that their final organization into the ground-plans that are now considered as characteristic may have only narrowly preceded their appearance in the record (Cloud 1949, 1968). It has been suggested that such a great evolutionary event may have been permitted by an increase in atmospheric oxygen past a critical level (Berkner and Marshall 1965). At any rate, it is likely that nearly all of the invertebrate phyla had become established before the Cambrian, and the relative timing of their appearance in the record may partly indicate the order in which they acquired hard parts or the chance occurrences of unusual preservations. Nicol (1966) and others have suggested that the acquisition of hard parts may have ensued as a result of widespread phyletic body-size increases.

Many of the nine phyla in the sample contain taxa that are not members of the shelf benthos or that have relatively low probabilities of preservation. Examples are the planktonic Scyphozoa and the soft-bodied Keratosa. Such taxa are excluded from the tallies. A few other taxa probably participated only partly in the benthonic ecosystems. An important example is the Ammonoidea. The effects of such taxa on the diversity curves are considered separately. Diversity graphs for phyla, classes and orders that seem to be chiefly members of the benthos are presented in text-fig. 3; diversities are classed by geological epochs, and therefore do not exactly represent the standing diversities at any given time. The diversity levels depicted in text-fig. 3 represent a balance between diversification and extinction, but the amount of taxonomic turnover that has occurred in any epoch cannot be inferred from the diversity levels. Text-fig. 4 depicts the numbers of appearances and disappearances (presumed to be extinctions) among the higher taxa per epoch.

From text-figs. 3 and 4 the following history of higher taxonomic diversity can be

VALENTINE: PATTERNS OF TAXONOMIC AND ECOLOGICAL STRUCTURE   691

inferred. The higher the category the earlier it tends to reach its maximum diversity. If the phyla were not all present throughout the Cambrian, at least they all appear by early Ordovician time, but the highest diversity is recorded in the Middle and Upper Cambrian. New classes continue to appear until the Lower Carboniferous, but the highest class diversity is recorded in the Middle and Upper Ordovician. Orders have

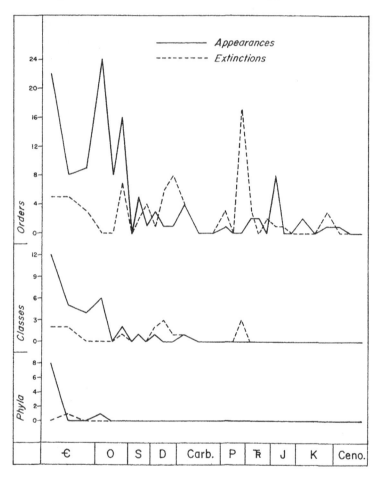

TEXT-FIG. 4. Appearances and extinctions of phyla, classes and orders of shelf benthos in the sample, classed by Epochs. Data as in text-fig. 3.

continued to appear until the end of the Cretaceous, but achieved their highest recorded diversity in medial Ordovician time. Ordinal diversification, however, was great during early Ordovician time, whereas the greatest class diversification was during the Cambrian (text-fig. 4).

It also appears that the higher the taxonomic category the less it has been ravaged by extinction, and the earlier extinction has stopped. Only one phylum (Archaeocyatha), which comprises 11% of the sample, disappears, although the sample is so small that this figure cannot be taken as very precise. However, 16 classes comprising 50%, and

75 orders comprising 64%, disappear. These extinct taxa are never fully replaced by other higher taxa, at least not from among the taxa that we are considering, so that the diversity of each higher taxonomic category has decreased, rather markedly in the cases of classes and orders, since the early Paleozoic. For the taxa in the sample, the extinction of phyla is complete by the end of Cambrian time, of classes by the end of Permian time, and of orders by the end of the Cretaceous.

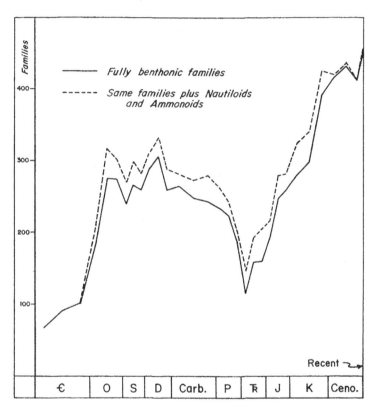

TEXT-FIG. 5. Stratigraphic variation in diversity of skeletonized families of shelf benthos belonging to phyla included in text-fig. 3. Data chiefly from Harland *et al.* (1967), Moore (1953–7), and Orlov (1958–64)

The lower taxa present a somewhat different pattern. Text-fig. 5 depicts the geological record of family diversity in the sample, again excluding unsuitable taxa such as the planktonic foraminiferal families, and text-fig. 6 depicts the record of appearances and disappearances. About 100 families are recorded by the end of the Cambrian, and 300 occur in the Upper Ordovician; diversity remains near 300 until the later Paleozoic, when it gradually falls off. A marked drop occurs in late Permian and early Triassic times. The pattern has until this point been not too unlike that of the higher taxa, with the mode of major diversity shifted towards the present. The Jurassic rise is even anticipated on the ordinal level (text-fig. 3). However it is in the great diversity rise of the Cretaceous and Cenozoic that the pattern of the families departs in a fundamental way

VALENTINE: PATTERNS OF TAXONOMIC AND ECOLOGICAL STRUCTURE   693

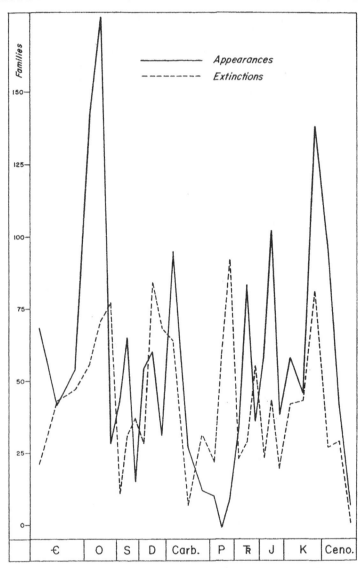

TEXT-FIG. 6. Appearances and extinctions of the families of shelf benthos in the sample, including Nautiloidea and Ammonoidea, classed by Epochs. Data as in text-fig. 5.

from the pattern of the higher taxa, for the higher taxa reach and maintain rather steady levels of diversity from Ordovician onwards for phyla, Lower Triassic onwards for classes, and lower Jurassic onwards for orders. The Cretaceous and Cenozoic diversification of families marks a major change in the evolutionary trends of taxonomic structure.

It can be seen from text-fig. 6 that the times of diversification of families tend to alternate with times of extinction. This pattern is well shown in a figure by Newell (1967, fig. 7) that is based upon a different taxonomic sample. The pattern certainly suggests that times favourable for diversification and those favourable for extinction were distinct, and that there is therefore a complementary relation between these processes. Newell suggests in effect that there have been extinctions to provide unoccupied biospace before there are major diversifications, and of course there must be many taxa resulting from diversification before there can be major extinctions. However, the data in text-figs. 5 and 6 do not entirely bear out this thesis. The very high extinction peaks in the Ordovician, Devonian, and Cretaceous are not accompanied by massive reductions in standing diversity. In fact, the early Ordovician and Cretaceous extinction highs are nearly hidden in text-fig. 5 owing to the great contemporary diversifications. Late Ordovician and Devonian extinction peaks reduce the diversity level somewhat, but only by about 6 and 13% respectively, because diversification is fairly high at these times. Only near the Permo-Triassic boundary, when diversification is exceedingly low (text-fig. 5), does the diversity level suffer a major decline of about 50%. The unusually low level of Permo-Triassic diversity is not unique because of the extinction peak alone, but because of the lack of a corresponding peak of diversification.

It is interesting to examine the patterns of family diversification and extinction within each of the higher taxa. Newell (1967) has presented graphs of family diversification among a number of higher taxa; I have prepared similar charts for diversification as well as for extinction of the families of higher taxa in the present sample which confirm the patterns he has presented. In general, however, the high rates of family diversification and of extinction within a higher taxon do not alternate in time but are highly correlated. For example, the Brachiopoda diversify strongly during the Ordovician and Devonian, but extinction peaks are found at these times also. Secondary levels of diversification during the Silurian and Lower Carboniferous correlate with secondary peaks of extinction. Only in the Permian is there a lack of correlation; the great extinction is not accompanied (nor is it followed) by diversification at the family level, but is accompanied by the lowest diversification rate known for brachiopods during the Phanerozoic.

Trilobites display a similar correlation. Cambrian diversification rises to a peak in the late Cambrian and falls off progressively during Ordovician epochs; extinction levels do precisely the same, except that they rise a bit in latest Ordovician. There is no following rise in diversity, however, although there is a secondary peak of extinction in medial and late Devonian time. Certainly there is no alternation of diversification and extinction. The same may be said of diversity patterns of the Porifera, the Echinoidea, and several of the Paleozoic echinoderm groups. The Foraminiferida, Anthozoa, and Gastropoda have more complex patterns, but there is no suggestion of alternating extinction and diversification except at the Permo-Triassic boundary. In the Ostracoda there is some indication that extinction follows diversification and not the reverse, and similar trends are found during parts of the record of other taxa. Even in the Ammonoidea and Nautiloidea peaks of diversification and extinction tend to correlate and certainly do not alternate.

The alternation of peaks of diversification and extinction of all families in the sample (text-fig. 6), then, are chiefly due to the alternation of high rates of family extinction of some higher taxa with high rates of diversification of different higher taxa (which is usually accompanied by a rise in extinction among these different taxa also). Except at the Permo-Triassic boundary, all this tends to be accomplished while diversity as a whole remains surprisingly stable considering the magnitude of extinction and diversification peaks. The major events that alter standing diversity on the family level are the diversification in Cambro-Ordovician times, the Permo-Triassic diversity low, and the Cretaceous–Cenozoic diversification (text-fig. 5).

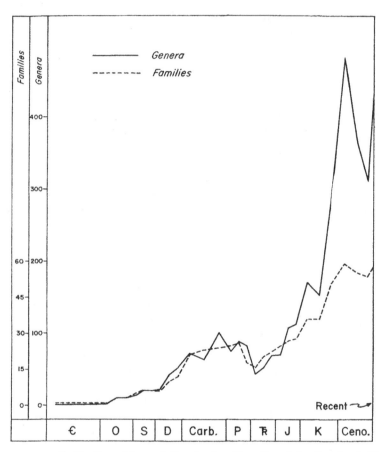

TEXT-FIG. 7. Stratigraphic variation in the diversity of benthonic shelf families (dashed line) and genera (solid line) of Foraminiferida, excepting the poorly skeletonized Allogromiina. Data from Loeblich and Tappan (1964).

It is possible to demonstrate that with the genera as with the families, there is a striking rise in numbers in the Cretaceous and Cenozoic. Text-fig. 7 depicts the diversity of families and of genera of benthonic Foraminiferida through geological time. This is one of the taxa that contributes strongly to the late Cretaceous and Cenozoic rise in family diversity. Throughout the Paleozoic there are, on the average, about 3–4 genera per family described. Across the Permo-Triassic boundary this ratio drops and then rises again in the Jurassic and early Cretaceous. In the late Cretaceous the genus/family ratio climbs to nearly 6, and in the early Cenozoic to over 8, where it stands at present after a late Cenozoic decline. The size of families is somewhat a matter of opinion. There is no special reason, however, to believe that the disproportionate Cretaceous–Cenozoic rise is a taxonomic artifact. The data are certainly subject to monographic and other biases, but have all been reviewed by the same team of authorities (Loeblich and Tappan 1964). It is interesting in this regard that the same trend can be inferred from the data charted by Henbest (1952) based chiefly on the work of Cushman (1948). If the trend is an artifact it is an enduring one. Incidentally, peaks of extinction of genera of foraminiferida correlate rather than alternate with peaks of diversification, just as is common among invertebrate families.

Another taxon that contributes heavily to the Cretaceous rise in family diversity is the Gastropoda, among which the same pattern of disproportionate generic diversification is present. There is no satisfactory recent review of all marine gastropod genera,

ORDERS	GEOLOGIC RANGE	FAMILIES	Genera & Subgenera	Genera – Subgenera/Family
Archaeogastropoda[1]	U. Cambrian–Recent[2]	82	1314	16.0
Mesogastropoda	U. Ord. (Caradocian)–Recent	75	1467	19.6
Opisthobranchia	L. Carb. (Visean)–Recent	9	240	26.6
Neogastropoda	L. Cretaceous (Albian)–Recent	20	1119	56.0

[1] Including Bellerophontacea.

[2] Possibly from L. Cambrian, depending upon ordinal assignment of early groups.

TEXT-FIG. 8. Families, genera, and subgenera of orders of shallow marine gastropods. The most advanced orders contain higher numbers of genera and subgenera, on the average, than primitive orders. Data from Taylor and Sohl (1962).

but Taylor and Sohl (1962) have published a census of gastropod genera and subgenera combined by family and higher taxa. It is possible to show that the more recently an order has appeared in the record, the more genera and subgenera per family it contains on the average (text-fig. 8). The Neogastropoda, which appear in the Lower Cretaceous (Albian or possibly earlier) and which diversify chiefly in the Upper Cretaceous and later, have 56 genera and subgenera per family on the average. This is 3½ times as many as the average of the Archaeogastropoda. Furthermore, even the groups of archaeogastropods with the largest living representatives, such as the trochaceans, tend to have differentiated strongly at the generic level in the Upper Cretaceous and Cenozoic. For the Trochacea, for example, there are 14 genera and subgenera recorded from the Lower Cretaceous, 32 from the Upper Cretaceous, and 66 from the Upper Cenozoic (data from Moore 1953–7). The other orders of Gastropoda are intermediate, both in time of appearance and in genus–subgenus/family ratios, between the Archaeogastropoda and the Neogastropoda. It follows from this situation that, for the shallow marine shelled Gastropoda, the late Cretaceous–Cenozoic rise in family diversity (from 37 families in the Lower Cretaceous to 66 in the Upper Cretaceous and to 83 or 84 in the Cenozoic) is disproportionately exaggerated on the generic level. This agrees with the data for Foraminiferida. Two other groups that contribute especially strongly to the rise in Upper Cretaceous and Cenozoic family diversity, the Ectoprocta and the

Echinoidea, appear to display similar trends (see Newell 1952 and the appropriate *Treatise* volumes).

In contrast to the preceding groups, the phylum Brachiopoda displayed its greatest familial diversity during the Paleozoic. It has not diversified during the Cretaceous–Cenozoic but maintained an average of about 12 or 13 families and about 50–65 genera

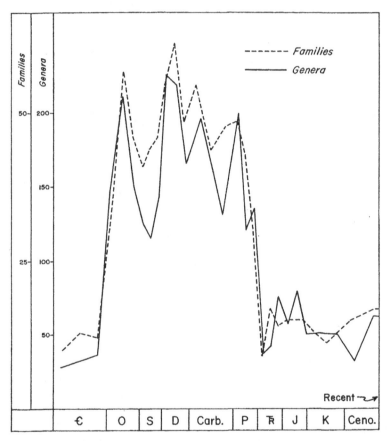

TEXT-FIG. 9. Stratigraphic variation in brachiopod diversity by families (broken line) and by genera (solid line). The genera/families ratio remains near 4. Genera after Williams (1965), and family data from Williams *et al.* (1965).

during this time (text-fig. 9). It is interesting, therefore, that the brachiopods have about the same genus/family ratio recorded for the Paleozoic as in the Mesozoic and Cenozoic —about 4 times as many genera as families throughout the Phanerozoic. Even during the greatest periods of diversification, the numbers of genera per family did not rise disproportionately, although generic turnover was significantly greater than family turnover. Judging from a comparison of the graphs of evolutionary rates among trilobite genera (Newell 1952) with family diversity trends, disproportionate generic diversity is not found among trilobites during their time of greatest family diversity either.

In summary, diversification on the family level during the Paleozoic and early Mesozoic seems to be accompanied chiefly by simple proportionate diversification on the generic level. Diversification on the family level during the late Cretaceous and Cenozoic seems to be accompanied by a disproportionately high diversification on the generic level.

It is impracticable to attempt a census on the species level for even a few higher taxa, and there are strong reasons for doubting the significance of fossil species counts in any event. It is necessary to approach species diversity at least in part from a theoretical point of view. Species evolve at greater frequencies than genera or families or higher taxa, and their standing diversities are therefore more volatile. The appearance of isolated habitats can produce swarms of closely related species, and the development of specialized communities may permit the development of a great number of species, not necessarily closely related, but endemic to the community. Reef communities, for example, appear to contribute large opportunities for both these types of speciation. A great number of specializations are possible on reefs, and thus large numbers of relatively specialized species from various phylogenetic backgrounds may appear. Reef tracts are also characterized by patchy and discontinuous distributions of reefs, and the isolation of outlying patches might often serve as a basis for speciation. Communities such as those on reefs that appear, endure long enough for a highly specialized biota to develop, and then disappear or become greatly reduced, can produce temporally localized but significantly large fluctuations in standing species diversity. If species diversification is great within such communities, generic diversity would also be enhanced. Since the species endemic to such communities are normally specialized, the average niche size of the shelf biota would be decreased while they flourish and increased when they wane.

At times the middle and upper Paleozoic record contains numerous reef associations and at these times it is likely that species diversity reaches disproportionately high levels, relative to families. It is expected that generic diversity might also rise disproportionately at these times. Although early and middle Permian reefs are widespread and contain probably the most specialized Brachiopoda recorded (Rudwick and Cowen 1968), a disproportionate generic diversity peak does not appear at that time in the available data (text-fig. 9). The description and evaluation of Permian reef biota is far from finished, however.

Another important way in which specific (and generic) diversity may be disproportionately multiplied is through a rise in provinciality. For example, theoretical considerations suggest that there are many more shelf species today than in the past, owing to the high degree of shallow-water provinciality at present (Valentine 1967, 1968a). This provinciality is both latitudinal, correlating with the great latitudinal temperature gradients at present, and longitudinal, owing to the presence of efficient biogeographic barriers of continents and ocean deeps. In the early Jurassic provinciality was not strongly developed. A Middle and Upper Jurassic Boreal fauna that contains endemic forms has been widely recognized (Neumayr 1883; Arkell 1956). Evidence has now been advanced to suggest that the Boreal fauna of the Jurassic signifies a low-salinity facies in a region of rather stable palaeogeography rather than a climatic province (Hallam 1969). Whatever its environmental basis, the appearance of the widespread fauna marks an increase in environmental heterogeneity on a sub-continental scale and a rise in species diversity. The general trend towards increasing provinciality in late Cretaceous and Cenozoic times must have greatly enhanced the numbers of species on the shelves (Valentine 1967, 1968a). Many genera which contain several species in a given province are now represented in different provinces by separate suites of species, so that the total numbers of their species are immense. Such is the case with species of *Nucula*, *Macoma*, and *Mactra* among the Bivalvia and *Conus*, *Calliostoma*, and *Fissurella* among the Gastropoda, to choose a few of the many examples. This situation must have been much less marked during times of low provinciality. For these reasons alone it is contended that the number of species in the shelf environment has increased disproportionately relative to the genera, especially during the Cretaceous and Cenozoic. Thus a species diversity curve would have about the same pattern as a generic diversity curve, but the peaks would be exaggerated (and the curve would be offset slightly towards the present). There are still other reasons, discussed below, for believing this pattern to be correct.

The major diversity trends in time among the fossil taxa are assumed to reflect real diversity trends among the ancient shelf biota, and they can be described in terms of the structure of the taxonomic hierarchy (text-fig. 10). In the earliest Paleozoic each phylum was represented by only a few classes, each class by relatively few orders, and so on down the hierarchy. By the close of Ordovician time, however, the average phylum was well differentiated into classes, and the average class into orders. Some phyla became extinct, but were not replaced by other phyla. After the Middle Ordovician the diversity of classes and of orders declined but the diversity of families rose, so that the structure became relatively more diversified among the lower taxa (text-fig. 10). It is possible that the average generic and specific diversity of the Upper Carboniferous indicated in text-fig. 10 is too low, owing to a disproportionate diversification at these lower levels that culminated during the early Permian. At about the Permo-Triassic boundary, both the numbers of classes and of orders were reduced by just less than half relative to their Middle Ordovician peaks, as were the families. The Permo-Triassic diversity low was most marked at lower taxonomic levels (contrast the familial and ordinal diversity decreases).

After the Permian the only gains in diversity registered among higher taxa are on the ordinal level. The numerous lost classes are not replaced, and even the rise in ordinal diversity is relatively small. The great climb in diversity at the familial level returns the hierarchy to a pattern commensurate with the early Upper Paleozoic pattern by Middle Jurassic time. Thereafter the number of families per higher taxon increases, especially during the Upper Cretaceous, and the diversities of lower taxa follow suit (text-fig. 10).

The so-called nekto-benthonic cephalopod groups Nautiloidea and Ammonoidea have not been included in the basic sample because of uncertainty as to the degree to which they participated in benthonic ecosytems. Nevertheless, some of them were surely regular members of a benthonic food chain. In text-fig. 5, the families of these cephalopod taxa are added to the families of the sample. The pattern of diversity is not much altered thereby; there was a slight increase in the steepness of the diversity rise from the Triassic to the Lower Cretaceous and the appearance of a rough plateau during the late Cretaceous and Cenozoic. The effect on the ordinal level is to emphasize the Ordovician rise in diversity and the mid- and upper-Paleozoic decline. Therefore, the exclusion of these Cephalopoda from the sample has a conservative effect insofar as the major trends are concerned, and in no way contributes to special conclusions.

Another major group which might have merited some representation in the figures is the fishes. At the family level their inclusion would raise the curve in text-fig. 5 to an even higher Devonian peak, and somewhat steepen the Mesozoic trend of rediversification (as

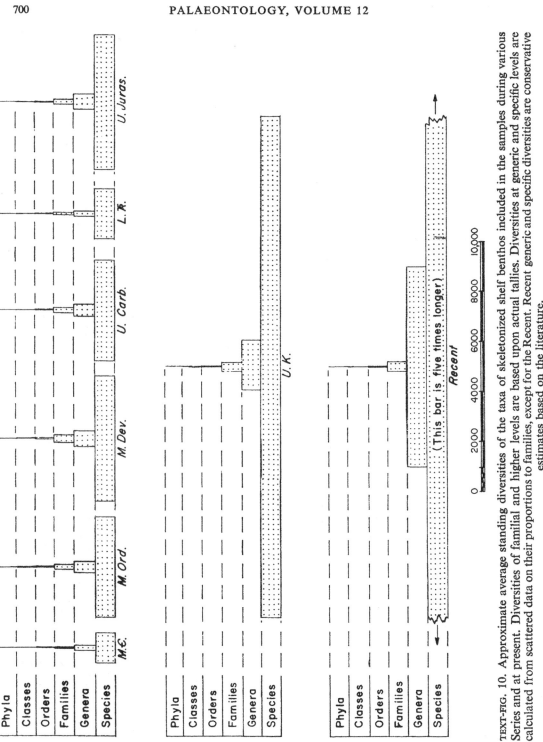

TEXT-FIG. 10. Approximate average standing diversities of the taxa of skeletonized shelf benthos included in the samples during various Series and at present. Diversities of familial and higher levels are based upon actual tallies. Diversities at generic and specific levels are calculated from scattered data on their proportions to families, except for the Recent. Recent generic and specific diversities are conservative estimates based on the literature.

would almost any additional taxon). At higher levels, they would raise ordinal diversity in medial Paleozoic and class diversity in early Paleozoic times, emphasizing graph patterns. Clearly, their exclusion does not affect the major diversity trends.

COADAPTATION OF ECOLOGICAL AND TAXONOMIC HIERARCHIES

The correlation of the structures of the two hierarchies under consideration is best approached at the species level. The fluctuations in species diversity, which have changed through a whole order of magnitude and more at times (Valentine 1967), naturally cause wide fluctuations in the numbers of populations in the ecological hierarchy, which must be accommodated in some manner.

In general there are three ways in which new species populations might be accommodated in ecological units (Klopfer 1962): (1) they may colonize parts of the environmental lattice that were previously unoccupied, in which case their niches represent an extension of ecospace in the ecological unit but average niche size is little affected; (2) space for their niches may be created in the lattice by shrinking one or several of the pre-existing niches, in which case the existing ecospace becomes more crowded by partitioning and average niche size decreases; or (3) parts of their niches may overlap with one or several pre-existing niches and thus crowd the lattice by overlap rather than by partitioning. The last sort of accommodation, by niche overlap, occurs hand in hand with one of the other sorts. The fitting of a new niche into the lattice may commonly involve all these sorts of accommodation. Theoretical or practical aspects of niche partitioning, which were touched on by Darwin (1859), have been considered in a modern perspective by a number of workers (for example, Bray 1958; Brown and Wilson 1956; Klopfer and MacArthur 1960, 1961; Klopfer 1962; McLaren 1963; MacArthur and Levins 1967; MacArthur and Wilson 1967; Hutchinson 1967; and Miller 1967). Yet data on variations in niche size and its relation to species diversity among marine shelf invertebrates is scanty. Marine research includes the work of Kohn (1959, 1966) on the gastropod *Conus* and Connell (1961) on some intertidal barnacles.

The most recent major rise inferred in species diversity from the late Cretaceous to the present seems to have involved an extension of ecospace, an invasion of parts of the lattice which were becoming newly realized, thus expanding the available environment. This increase in environmental heterogeneity in the shallow marine realm was evidently due partly to the cooling of shelf waters in high latitudes (Smith 1919; Durham 1950; Valentine 1967, 1968a), which permitted the rise of new biogeographic provinces in separate chains along north–south-trending coastlines. New provinces may also have been created by the drifting apart of some continents, progressively isolating, from about medial Cretaceous time, shelf regions that had previously been connected. This permitted an increase in endemism.

Thus this expansion of ecospace is envisaged as due primarily to two factors: (1) extension of the thermal factors and of numerous other parameters that are related to temperature, creating one set of biogeographic barriers; and (2) the creation of another set of barriers through the breaking up of formerly continuous or nearly continuous epicontinental seaways and continental shelves through continental drift (which in effect multiplies biospaces along dimensions of real space). The relative timing of these events is not yet clear, although there is a suggestion that longitudinal provinciality was strengthening in the late Cretaceous (Sohl 1961) while latitudinal provinciality was still weak. The Cretaceous–Cenozoic expansion of ecospace was fundamentally on the level of the biosphere and involved the rise of provinciality, which in turn permitted new communities based upon endemic populations to appear in each province. The increase in isolation among the populations living in separate provinces or separate communities permitted the formation of many new species.

The increase in generic diversity follows from the multiplication of species and their isolation in separate provinces or communities. Different but related species with similar morphological adaptations would arise in similar habitats in different provinces, forming associations such as Thorson's (1957) parallel bottom communities or becoming 'geminate' or twin species such as occur on opposite sides of the Isthmus of Panama (Ekman 1953). The increase in generic diversity would be proportionately less than in species diversity, owing to the multiplication of similar morphological types in distinct provinces that would be grouped as genera by taxonomic practice. The same principle seems to apply to the family level. A marked increase in the diversity and provinciality of genera would lead to a more or less modest increase in family diversity. That the present biogeographic pattern of diversity is similar at the familial, generic, and specific levels has been well documented for the Bivalvia by Stehli, McAlester, and Helsley (1967). Kurtén (1967) suggested that mammalian diversity is high at the ordinal level because of endemism arising through the isolating effects of continental drift.

At higher taxonomic levels on the marine shelves a different factor must be operating to control diversity, since higher taxa do not much participate in the Mesozoic–Cenozoic diversification. Perhaps this diversification has occurred too recently for evolution to have proceeded to the level of higher taxa. The pattern of ordinal diversity suggests that this may be the case. Yet class diversity has declined since the Ordovician. A better explanation may be that the available biospace of the epicontinental seas and shelves was nearly fully occupied since at least very early in the Phanerozoic, and most increases in taxonomic diversity had to be accomplished by ecospace partitioning and overlap (see Rhodes 1962, pp. 270–2; Nicol 1966). Thus only in times of unusual expansion of the marine shelf biospace, when the realized parts of the environmental lattice increased persistently, would significant diversity increases result simply from the invasion of new biospace. Ecospace partitioning involves a decrease in the niche sizes of populations, at least along dimensions where competition may occur (Miller 1967), and this is not a process that lends itself to the appearance of organisms with wholly new ground-plans or with major modifications thereof, such as are required for the development of higher taxa. It is instead a process suited more to the modification in detail of pre-existing morphological types, so as to accommodate to smaller ecospaces—in other words, a process suited to the increase of specialization.

By Cambrian time or shortly thereafter the ground-plans that are the hallmarks of the major invertebrate phyla had been established. Most of the species were by present standards primitive and functionally generalized; modal niche size was no doubt far larger than at present, though there certainly may have been some highly specialized forms. Evidently, diversification in the Cambrian and Ordovician led to the presence in late Ordovician time of a large number of higher taxa. This may have been partly due to an expansion of biospace and should have included an increase in resources. As much of the former marine shelf biospace may have been occupied by soft-bodied organisms, the expansion of skeletonized forms may have involved the appropriation of some resources that had formerly been utilized by soft-bodied groups. On the other hand, on the assumption that soft-bodied taxa should have responded to the same opportunities as skeletonized taxa, it seems even more likely that there was a concomitant diversification of soft-bodied lineages. These are debatable points, but it is clear that it cannot be assumed that Cambro–Ordovician radiation was occurring in vacant biospace. Vast regions of the lattice must have been occupied, and this may have served to channel the evolutionary pathways of diversifying lineages.

Certainly there was unusual extinction among higher taxa during this time. Higher taxa in the sample that appear in the Cambrian, but which are not known to have survived the Ordovician, include the phylum Archaeocyatha, the echinoderm classes Homostelea and Helicoplacoidea, the inarticulate brachiopod orders Obolellida, Paterinida, and Kutorginida, the monoplacophoran order Cambridoida, and the trilobite orders Redlichiida and Corynexochida. Numbers of other taxa that are poorly known also disappeared early, and may have represented higher taxonomic levels. They were not included in the sample because of their questionable status. These include some trilobitoids, some early echinoderm stocks, and early gastropod-like molluscan stocks. Perhaps most of these are functionally generalized forms, the morphological architecture of which proved unsuitable to the demands for specialization.

Biospace formerly occupied by populations of extinct lineages would soon be recolonized by populations of the lineages that remained, if such colonization did not actually precede and contribute to the extinction. This easily leads to the diversification of extant lower taxa, but not to the creation of taxa on a comparably high level. In the early Phanerozoic the large average size of the former ecospaces of extinct taxa provided ecological room, so to speak, which allowed the lineages that reoccupied vacated biospace a certain leeway for progressive morphological modification that could still lead to the establishment of a higher taxon. Later, when vacated biospace was to be in smaller parcels, opportunities for morphological modifications became limited, as during medial Paleozoic time, when the diversity of classes declined markedly (text-fig. 3). Although diversity of orders (text-fig. 3) and families (text-fig. 4) also declined during that time, the decline was less marked, and order/class and family/order ratios were both rising. This suggests that the extinction of a higher taxon was not usually accompanied by replacement at the level of the higher taxon but at a lower level. It also suggests that biospace was decreasing. In the Lower and Middle Permian the numbers of species and genera may well have been disproportionately higher than is suggested by the number of families, owing to the development of reef associations.

Towards the close of the Paleozoic the change in the taxonomic structure suggests a great reduction in the heterogeneity of the shelf environment. Provinciality was already low. It is not certain, therefore, whether a significant further reduction occurred or indeed was even possible in late Permian time. However the numbers of communities certainly decreased (for example, the late Paleozoic reefs disappear) and, although there are not well-documented field studies, it appears from faunal lists that the numbers of populations in the remaining communities also declined. Thus the ecological structure shrank at all, or nearly all, levels, suggesting a general decline in biospace. There are too few data on the ecological hierarchy of Permo-Triassic times, however, to support speculation on the precise causes of the extinctions on this basis. Rhodes (1967) has reviewed the major hypotheses of extinction and remarks that probably none of them alone would cause the sorts of changes found at the Permo-Triassic boundary.

Rediversification evidently began by medial Triassic time, unless the increase in family diversity then is an artifact. If it is real (and it includes the beginnings of Scleractinian radiation as well as the expansion of gastropod families), it suggests that generic and specific diversification was proceeding at even higher rates. Presumably biospace had expanded (or was expanding) once more and the newly realized parts of the environmental lattice were being recolonized. However, compared with Cambro–Ordovician lineages, Triassic lineages were rather specialized, with smaller modal niche sizes, so that the average colonizing lineage must have occupied a relatively smaller part of the lattice. The opportunities now presented for the formation of higher taxa could not be much exploited by the relatively specialized populations. There may be exceptions; the Scleractinia may have taken advantage of biospace vacated by Paleozoic coelenterate lineages to become skeletonized and reoccupy some of the same biospace. Finally, at some time in the Mesozoic a diversification involving the marked rise in provinciality discussed previously began to occur as well. It seems likely that the Jurassic and early Cretaceous diversity increases, which are not inconsiderable and which take place at a high rate (text-fig. 5), are partly owing to an increase of latitudinal provinciality due to cooling poles and to an increase of longitudinal provinciality due to the separation of some continental masses in Jurassic time. A thorough analysis of the biogeographic patterns of Mesozoic diversification is badly needed.

The Cretaceous–Cenozoic boundary is marked by extinctions of some benthonic marine groups (see Hancock 1967), but if there was any appreciable alteration in the taxonomic diversity structure at the family and higher levels it was of so short a duration that it does not appear in the present data. The structure of communities, provinces and of the entire shelf realm must have undergone qualitative changes, but this is a more or less continuous process on the broad scale we are considering. The Late Cretaceous rise in diversity extends unabated across the Cretaceous–Cenozoic boundary. Surely there was no large-scale reduction in biospace.

Two major modes of taxonomic diversification have been described, both proceeding at progressively lower taxonomic levels through time. The first involves a biospace that fluctuates about some size that does not vary much in time. Diversification at the population level at first proceeds by colonization of untenanted biospace, but soon must be accompanied by a progressive decrease in average niche size. Communities therefore become increasingly packed with more and more specialized populations and begin to fragment into portions, each of which has an energy flow that is partially independent of the others. The isolation and independence of these portions will increase with further specialization until they form ecosystems that are as independent as the original one from which they fragmented. Slighter and slighter environmental discontinuities will form community boundaries until the environmental mosaic of a given primitive community, relatively heterogeneous but occupied by primitive populations with large niches, is broken up into a number of smaller environments, each more homogeneous than the original and each occupied by more specialized populations. Provinces become packed with more and more communities that have progressively smaller ecospaces. The diversity of provinces is not so sensitive to this progressive specialization, although it may eventually be affected if the trend continues long enough. Any widespread partitioning of temperature or temperature-correlated parameters would lead to increased provinciality even in the absence of progressive changes in the latitudinal temperature gradient. If partitioning were to continue, smaller and smaller changes in thermal regimes would act to localize range end-points, and thus form provincial boundaries (Valentine 1966). In sum, in this mode the ecological structure evolves from lower levels towards higher.

The other mode of diversification involves a biospace that expands to create new environments or to add new dimensions (or at least to extend old dimensions) to old environments. New environments may be created, for example, through climatic changes, and old environments may be extended through the improvement of limiting factors, that is through the amelioration of conditions which tended to inhibit diversification.

From what is known and can be inferred of the hierarchies of the Paleozoic, diversification (at least after an initial radiation of skeletonized taxa and probably before) was proceeding chiefly in the first mode, from the bottom of the ecological hierarchy upwards, implying a relatively stable biospace. Rediversification following the Permo-Triassic extinction was probably in the second mode, involving an amelioration of factors that had inhibited diversification, and the Upper Cretaceous and Cenozoic diversification seems to have also been in the second mode, but involved the creation of new environments. Nevertheless there must have been continuing specialization and thus the first mode was also active in the taxonomic and ecological evolution, Processes that bring species that appeared during biospace expansion into sympatry with older lineages, such as 'species pumps' of various kinds (Valentine 1967, 1968a), may link these two modes into a single system of diversification. Finally, the more lineages that exist the greater the opportunity for large-scale diversification under appropriate circumstances in either mode. This factor is certainly at work in the disproportionate multiplication of lower taxa during Cretaceous and Cenozoic times.

A number of authors have suggested that extensive changes in sea level may control some of the diversification and extinction patterns (Newell 1952, 1956, 1963; Moore 1954; see Rhodes 1967 for other references). Widespread epicontinental seas, it is asserted, provide more inhabitable area for shelf invertebrates and therefore more opportunity for diversification, while regressions reduce the inhabitable area and thus the diversity. There is some theoretical support for this position in the species-area work of Preston (1962), Williams (1964), MacArthur and Wilson (1967), and others. Little work has been done in marine environments; the areas of ancient shelf seas, especially during regressive phases, are difficult to estimate (though see Ronov 1968); and the effects that could be expected in a biosphere of vastly different ecological and taxonomic structure and composition are largely uncertain. The problem is further complicated by facies differences between epicontinental seas and shelves bordered by open oceans. Although uncertainties in calculations must be great, preliminary estimates suggest that the species-area effect would have been far too small to account for major diversifications and extinctions by itself. Moreover, we live at present in a time of great continental emergence yet the shelves are richly diverse in lower taxa and in ecological units at all levels, precisely the opposite of the pattern of Permo-Triassic extinction. Indeed, a relative lowering of sea level must commonly result in the emergence of land barriers which isolate regions formerly connected and permit the rise of an endemic biota in each region. This would have the effect of increasing the total number of species in these regions.

Nevertheless, the elimination or rise of species resulting from shelf-area fluctuations would certainly contribute to diversity patterns, and further evaluation of this subject is clearly merited.

CONCLUSIONS: THE PROGRESSIVE CANALIZATION OF ECOSPACE

It is concluded that a major Phanerozoic trend among the invertebrate biota of the world's shelf and epicontinental seas has been towards more and more numerous units at all levels of the ecological hierarchy. This has been achieved partly by the progressive partitioning of ecospace into smaller functional regions, and partly by the invasion of previously unoccupied biospace. At the same time, the expansion and contraction of available environments has controlled strong but secondary trends of diversity. Present marine biospace is in fact unusually extensive, and the world's shelf seas are therefore unusually heterogeneous and support a large number of ecological units today. The relations of these trends to trends within the taxonomic structure of the benthonic invertebrates are intimate.

Assuming for the moment that evolutionary trends among marine benthonic invertebrates will continue and that biospace does not change much (a dim prospect in view of rising pollution), what might be predicted of the future structures of the ecological and taxonomic hierarchies? Speculation on this point may be of some value to underline the sort of process that is postulated to have gone before. Clearly, the trend towards specialization would further reduce the average niche sizes of species. It would be increasingly difficult for evolving lineages to depart much from their modal functions and morphologies, as biospace would become available only in increasingly smaller compartments. The amount of change necessary to produce a new family would be increasingly difficult to attain, and eventually no new families could appear. In fact, some families would become extinct so that familial diversity would decrease, and lineages from other families would fill any vacated biospace. After some time genera could no longer appear, for biospace would be packed too tightly to permit morphological variation even at that level, and generic diversity would decline for a while as some extinction, inevitably, occurred. Eventually, all the biospace would become filled with evolving lineages with an incredibly small modal niche size, each lineage constrained by the presence of all the others to evolve in only a narrow pathway directed by the trends of evolution of the entire biota, and of changes in the entire biosphere. Ecological units are now exceedingly small by today's standards, with virtually every few food-chains forming a separate community and every moderate topographic irregularity forming a provincial boundary. Canalization of ecospace is complete. The biosphere has become a splitter's paradise.

Although this extrapolation cannot be taken too seriously, it does point to some important consequences of ecospace partitioning. First, average species of the early Paleozoic, with their broad niches, may have had different patterns of morphological variation than the specialized species of today. Secondly, the occurrence in the early Paleozoic of numbers of unusual 'aberrant' higher taxa that contain few lower taxa is not necessarily due to a poor fossil record but is probably the natural consequence of adaptive strategies that prevailed in primitive ecosystems of low diversity. Finally, extinction of taxa of high diversity is less likely than extinction of taxa of low diversity, other

VALENTINE: PATTERNS OF TAXONOMIC AND ECOLOGICAL STRUCTURE   707

things being equal (Simpson 1953), simply because so many more lineages must disappear. Similarly, the markedly rising provinciality of the late Cretaceous and Cenozoic will tend to make the extinction of the newly diverse taxa that have representation in many provinces—a common situation even on the generic level—more difficult.

Acknowledgements. Gratitude for extensive discussion of the ideas presented herein is expressed to Dr. A. Hallam (Oxford University) and Professor A. L. McAlester (Yale University). The manuscript was carefully reviewed by Dr. W. S. McKerrow (Oxford University) and Professor R. Cowen, Professor J. H. Lipps, and Robert Rowland (University of California, Davis). All this attention resulted in much improvement. The manuscript was written during a Guggenheim Fellowship spent at the Department of Geology and Mineralogy, Oxford University, and at the Department of Geology and Geophysics, Yale University. The generosity of the Guggenheim Foundation and the hospitality of these departments is gratefully acknowledged.

REFERENCES

ARKELL, W. J. 1956. *Jurassic geology of the world.* Edinburgh.

BERKNER, C. V. and MARSHALL, C. C. 1965. Oxygen and evolution. *New Scientist,* **28**, 415–19.

BRAY, J. R. 1958. Notes towards an ecologic theory. *Ecology,* **39**, 770–6.

BROWN, W. L., JR. and WILSON, E. O. 1956. Character displacement. *Syst. Zool.* **5**, 49–64.

CLOUD, P. E. 1949. Some problems and patterns of evolution exemplified by fossil invertebrates. *Evolution, Lancaster, Pa.* **2**, 322–50.

—— 1968. Pre-metazoan evolution and the origin of the metazoa. *In* DRAKE, E. T. (ed.), *Evolution and environment.* 1–72. New Haven, Conn.

CONNELL, J. H. 1961. The influence of interspecific competition and other factors on the distribution of the barnacle *Chthamalus stellatus. Ecology,* **42**, 710–23.

CUSHMAN, J. A. 1948. *Foraminifera, their classification and economic use.* Cambridge, Mass.

DARWIN, C. R. 1859. *On the origin of species by means of natural selection.* London.

DOTY, M. S. 1957. Rocky intertidal surfaces. *In* HEDGPETH, J. W. (ed.), Treatise on marine ecology and paleoecology. *Mem. geol. Soc. Am.* **67**, 1, 535–85.

DURHAM, J. W. 1950. Cenozoic marine climates of the Pacific Coast. *Bull. geol. Soc. Am.* **61**, 1243–64.

EKMAN, S. 1953. *Zoogeography of the sea.* London.

HALLAM, A. 1969. Faunal realms and facies in the Jurassic. *Palaeontology,* **12**, 1–18.

HANCOCK, J. M. 1967. Some Cretaceous–Tertiary marine faunal changes. *In* HARLAND, W.B. *et al.* (eds.), *The fossil record,* 91–104. London (Geological Society).

HARLAND, W. B. *et al.* (eds.). 1967. *The fossil record.* London (Geological Society).

HENBEST, L. G. 1952. Significance of evolutionary explosions for diastrophic division of earth history—introduction to the symposium. *J. Paleont.* **52**, 299–318.

HUTCHINSON, G. E. 1957. Concluding remarks. *Cold Spring Harbor Symp. quant. Biol.* **22**, 415–27.

—— 1967. *A treatise on limnology, Volume 2. Introduction to lake biology and the limnoplankton.* New York.

HUXLEY, J. S. 1942. *Evolution, the modern synthesis.* London.

KLOPFER, P. M. 1962. *Behavioral aspects of ecology.* Englewood Cliffs, N.J.

—— and MACARTHUR, R. H. 1960. Niche size and faunal diversity. *Am. Nat.* **94**, 293–300.

—— —— 1961. On the causes of tropical species diversity: niche overlap. *Am. Nat.* **95**, 223–6.

KOHN, A. J. 1959. The ecology of *Conus* in Hawaii. *Ecol. Monogr.* **29**, 47–90.

—— 1966. Food specialization in *Conus* in Hawaii and California. *Ecology,* **47**, 1041–3.

KURTÉN, B. 1967. Continental drift and the palaeogeography of reptiles and mammals. *Soc. Scient. Fennica,* **31** (1), 1–8.

LEVINS, R. 1962. Theory of fitness in a heterogeneous environment. II. Developmental flexibility and niche selection. *Am. Nat.* **97**, 75–90.

LIPPS, J. H. (in press). Plankton evolution. *Evolution.*

708 PALAEONTOLOGY, VOLUME 12

LOEBLICH, A. R. and TAPPAN, HELEN. 1964. Protista 2, Sarcodina, chiefly 'Thecamoebians' and Fora-miniferida. *In* MOORE, R. C. (ed.), *Treatise on invertebrate paleontology, Part C.* Geol. Soc. Amer. and Univ. Kansas Press.

MACARTHUR, R. H. and LEVINS, R. 1964. Competition, habitat selection and character displacement in a patchy environment. *Proc. nat. Acad. Sci. U.S.* **51**, 1207–10.

—— 1967. The limiting similarity, convergence, and divergence of coexisting species. *Am. Nat.* **101**, 377–85.

—— and WILSON, E. O. 1967. *The theory of island biogeography.* Princeton, N.J.

MAYR, E. 1963. *Animal species and evolution.* Cambridge, Mass.

MCLAREN, I. A. 1963. Effects of temperature on growth of zooplankton, and the adaptive value of vertical migration. *J. Fish res. Bd. Canada*, **20**, 685–727.

MILLER, R. S. 1967. Pattern and process in competition. *In* CRAGG, J. B. (ed.), *Adv. Ecol. Res.* **4**, 1–74.

MOORE, R. C. (ed). 1953–7. *Treatise on invertebrate paleontology.* Geol. Soc. Amer. and Univ. Kansas Press.

NEUMAYR, M. 1883. Ueber klimatische Zonen während der Jura- und Kreidezeit. *Denkschr. Akad. Wiss., Wien, Math.-nat. Kl.* **18**, 277–310.

NEWELL, N. D. 1952. Periodicity in invertebrate evolution. *J. Paleont.* **26**, 371–85.

—— 1956. Catastrophism and the fossil record. *Evolution, Lancaster, Pa.* **10**, 97–101.

—— 1963. Crises in the history of life. *Scient. Am.* **208**, 76–92.

—— 1967. Revolutions in the history of life. *Spec. Pap. Geol. Soc. Am.* **89**, 63–91.

NICOL, D. 1966. Cope's rule and Precambrian and Cambrian invertebrates. *J. Paleont.* **40**, 1397–9.

ORLOV, Y. A. 1958–64. *Osnovy Paleontologii*, Akad. nauk SSSR., Moscow (in Russian).

PARR, A. E. 1926. Adaptiogenese und Phylogenese; zur Analyse der Anpassungserscheinungen und ihre Entstehung. *Abh. Theor. org. Entw.* **1**, 1–60.

PRESTON, F. W. 1962. The canonical distribution of commonness and rarity. *Ecology*, **43**, 185–215, 410–32.

RENSCH, B. 1947. *Neuere Probleme der Abstammungslehre.* Stuttgart.

RHODES, F. H. T. 1962. *The evolution of life.* Baltimore, Md.

—— 1967. Permo-Triassic extinction. *In* HARLAND, W. B., *et al.* (eds.), *The fossil record*, 57–76. London (Geological Society).

RONOV, A. B. 1968. Probable changes in the composition of sea water during the course of geological time. *Sedimentology*, **10**, 25–43.

RUDWICK, M. J. S. and COWEN, R. 1968. The functional morphology of some aberrant strophomenide brachiopods from the Permian of Sicily. *Bol. Soc. Paleont. Ital.* **6**, 113–76.

SIMPSON, G. G. 1944. *Tempo and mode in evolution.* New York.

—— 1953. *The major features of evolution.* New York.

SMITH, J. P. 1919. Climatic relations of the Tertiary and Quaternary faunas of the California region. *Proc. Calif. Acad. Sci.* (4) **9**, 123–73.

SOHL, N. F. 1961. Archaeogastropods, Mesogastropods, and stratigraphy of the Ripley, Owl Creek and Prairie Bluff Formations. *Prof. pap. U.S. geol. Surv.* **331A**, 151 pp.

STEHLI, F. G., MCALESTER, A. L., and HELSLEY, C. E. 1967. Taxonomic diversity of Recent bivalves and some implications for geology. *Bull. geol. Soc. Am.* **78**, 455–66.

TAYLOR, D. W. and SOHL, N. F. 1962. An outline of gastropod classification. *Malacologia*, **1**, 7–32.

THORSON, G. 1957. Bottom communities (sublittoral or shallow shelf). *In* HEDGPETH, J. W. (ed.), Treatise on marine ecology and paleoecology. *Mem. geol. Soc. Am.* **67** (1), 461–534.

VALENTINE, J. W. 1966. Numerical analysis of marine molluscan ranges on the extratropical north-eastern Pacific shelf. *Limnol. Oceanogr.* **11**, 198–211.

—— 1967. Influence of climatic fluctuations on species diversity within the Tethyan Provincial System. *In* ADAMS, C. G. and AGER, D. V. (eds.), *Syst. Ass. Pub.* **7**, 153–66.

—— 1968a. Climatic regulation of species diversification and extinction. *Bull. geol. Soc. Am.* **79**, 273–76.

—— 1968b. The evolution of ecological units above the population level. *J. Paleont.* **42**, 253–67.

—— 1969. Niche diversity and niche size patterns in marine fossils. *Ibid.* **43**, 905–15.

WILLIAMS, A. 1957. Evolutionary rates in brachiopods. *Geol. Mag.* **94**, 201–11.

VALENTINE: PATTERNS OF TAXONOMIC AND ECOLOGICAL STRUCTURE   709

WILLIAMS, A. 1965. Stratigraphic distribution. *In* MOORE, R. C. (ed.), *Treatise on invertebrate paleontology, Part H*. H237–50. Geol. Soc. Am. and Univ. Kansas Press.

—— *et al.* 1965. *In* MOORE, R. C. (ed.), *Treatise on invertebrate paleontology, Part H*. Geol. Soc. Am. and Univ. Kansas Press.

WILLIAMS, C. B. 1964. *Patterns in the balance of nature and related problems in quantitative ecology.* New York.

J. W. VALENTINE
Department of Geology
University of California
Davis, California, 95616
U.S.A.

Typescript received 8 April 1969

22 September 1972, Volume 177, Number 4054

SCIENCE

Taxonomic Diversity during the Phanerozoic

The increase in the number of marine species since the Paleozoic may be more apparent than real.

David M. Raup

The evolution of taxonomic diversity is receiving increasing attention among geologists. The immediate reason for this is that diversity data may have a direct bearing on problems of plate tectonics and continental drift. The tantalizing possibility exists that diversity may be a good indicator of past arrangements of continents or climatic belts, or both. Valentine (1, 2) has related temporal changes in fossil diversity to changes in climate and to the evolutionary consequences of continental drift. Stehli (3) and others have used spatial differences in diversity to interpret paleoclimates and paleolatitudes for single intervals of time.

Diversity information from the fossil record is also important because of its bearing on general models of organic evolution. Is the evolutionary process one that leads to an equilibrium or steady-state number of taxa, or should diversification be expected to continue almost indefinitely? Has equilibrium (or saturation) been attained in any habitats in the geologic past? If mass extinction has led to a significant reduction in diversity, what are the nature and rate of recovery? The answers to these and comparable questions depend in part on theoretical arguments, but their documentation must come ultimately from the fossil record itself.

The large-scale analysis of taxonomic

The author is professor of geology at the University of Rochester, Rochester, New York 14627.

diversity has been facilitated in the past few years by several important publications. The American *Treatise on Invertebrate Paleontology* (4) and the Russian *Osnovy Paleontologii* (5) are particularly valuable in having brought together vast amounts of taxonomic data with a minimum of inconsistency. Also, the British publication *The Fossil Record* (6) provides a useful synthesis of the geologic ranges of the higher taxa. This new literature, plus advances in data-processing technology, makes possible a more sophisticated study of diversity problems than has been possible heretofore (7).

Valentine (1, 2) used the newly published data to estimate temporal changes in diversity during the Phanerozoic, the geologic time since the end of the Precambrian. His conclusions were not dramatically different from those of earlier workers, but the breadth of documentation was far greater.

My purpose in this article is to investigate the nature of the diversity data to determine if more can be learned from it. In particular, I will examine the proposition that systematic biases exist in the raw data such that the actual diversity picture may be quite different from that afforded by a direct reading of the raw data. My study will be limited to the major groups of readily fossilizable marine invertebrates (as was Valentine's) and to changes in their worldwide diversity through time.

Traditional View

Figure 1 shows three histograms of taxonomic diversity for the Phanerozoic. The three sets of data differ somewhat in scope. Those of Valentine (1) and Newell (8) are principally tied to the family level, whereas Müller's (9) are numbers of genera. All three are limited mostly to the major groups of fossilizable marine invertebrates: Protozoa, Archaeocyatha, Porifera, Coelenterata, Bryozoa, Brachiopoda, Arthropoda, Mollusca, and Echinodermata, but Newell's data also include vertebrates. All three sets of data inevitably include some nonmarine and terrestrial taxa, but in none is this influence numerically significant.

The important fact is that all three show essentially the same picture and the one that has constituted the consensus for many years. The overall pattern is one of (i) a rapid rise in the number of taxa during the Cambrian and Early Ordovician, (ii) a maximum at about the Devonian, (iii) a slight but persistent decline to a minimum in the Early Triassic, and (iv) a rapid increase to an all-time high in diversity at the end of the Tertiary. Valentine (1, 2) has suggested that the rise in diversity at the species level in Mesozoic to Tertiary time was an exponential one, with the late Tertiary having up to 20 times more species than the average for the mid-Paleozoic. This rise would appear even greater if insects, land plants, and terrestrial vertebrates were considered. These are particularly "noticeable" groups, important to man, and the history of their diversity has influenced thinking on the general subject.

It should be emphasized that the Phanerozoic diversity pattern yielded by the published taxonomic data depends on the choice of taxonomic level. As Valentine has pointed out, diversities at the levels of phylum, class, and order have behaved very differently from those at the lower levels. The number of phyla has been essentially constant since the Ordovician, for example.

1065

Sedimentary Record and Diversity

It has been established that the general quality of the sedimentary rock record improves with proximity to the Recent (*10, 11*). That is, the younger parts of the record are represented by larger volumes of rock (per unit of time), and the amount of metamorphism, deformation, and cover by overlying rocks is generally less. This is usually interpreted as resulting from the fact that the younger rocks are closer to "the top of the stack" and that, being younger, they have had less chance to be destroyed by erosion, metamorphism, and the like.

Figure 1 includes a graphic display of Gregor's estimate (*12*) of change in the sedimentary record through the Phanerozoic. The vertical coordinate in the lower graph is what Gregor calls the "survival rate" and is expressed as cubic kilometers of sediment per year now known and dated stratigraphically. This shows, for example, that the Devonian is represented by about twice the volume of sediments

as the Cambrian (after adjustment for the relative durations of the periods). Gregor's survival data are comparable to estimates made on quite different bases by others (*10, 13*).

There is unquestionably a strong similarity between the patterns of taxonomic diversity at the genus and family levels and the pattern of sediment survival rate. This similarity suggests that changes in the quantity of the sedimentary record may cause changes in apparent diversity by introducing a sampling bias.

In spite of the fact that the patterns in Fig. 1 are correlated, a causal relationship is by no means demonstrated. Furthermore, the correspondence is not perfect, and both the diversity and sedimentary data are subject to many errors and uncertainties. The remainder of this article is devoted to a more detailed assessment of these relationships.

Gregor's data (Fig. 1) are estimates of survival rate for all sedimentary rocks, without distinction between marine and nonmarine. This detracts from the comparison with diversity because

the biologic data are nearly free of nonmarine elements. Also, with the exception of the interval from the Devonian through the Jurassic, Gregor's numbers are derived from estimates of maximum sediment thickness (*14*). This part of the data is suspect because of the logical problems involved in going from the maximum known thickness (in a local section) for a geologic system to the total volume of rock in that system (*11*). Furthermore, Gregor's rates are all sensitive to errors in estimates of the absolute time durations of the periods.

Thus, although there is little doubt about the general validity of Gregor's pattern, the inherent weaknesses prevent its use in more rigorous analysis.

By far the best data for sediment volumes are those published by Ronov (*15*). They are based on the results of a 10-year project of compiling lithological-paleogeographic maps and must be considered the most comprehensive data available. They are limited, however, to the Devonian-Jurassic interval. Ronov's data were used by Gregor where possible, but were modified by his calculation of survival rates. Ronov carefully distinguished between continental clastics, marine clastics, evaporites, marine and lagoonal carbonate rocks, and volcanics.

In Fig. 2, the taxonomic diversity data of Newell, Müller, and Valentine are compared with Ronov's estimates for the total volume of marine and lagoonal clastics and carbonates. Absolute time does not enter in because for each stratigraphic series total number of taxa and total sediment volumes are used. The diagram is thus free of most of the effects of errors in radiometric dating.

The correspondence between diversity and quantity of sediments is much stronger than indicated in Fig. 1. In particular, it should be noted that the Early Triassic diversity minimum coincides with a sediment minimum, which was not the case when Gregor's data were used. This is primarily because Gregor used Ronov's data for all sedimentary facies and because of the effect of Gregor's rate calculation.

It could be argued that the similarity between the patterns in Fig. 1 is due simply to a broad but independent increase in both sediment volume and diversity from the Cambrian through the Tertiary and that similarity in detail is quite accidental. Figure 2 largely denies this interpretation because the Carboniferous-Permian interval shows

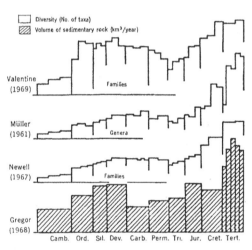

Fig. 1. Comparison of the number of taxa and the volume of sedimentary rock during the Phanerozoic. The diversity data are based mainly on well-skeletonized marine invertebrates (*1, 8, 9, 12*).

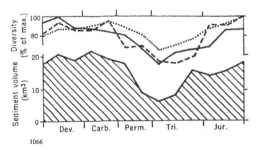

Fig. 2. Apparent taxonomic diversity compared with estimated volume of marine and lagoonal clastic and carbonate sediments. The diversity data are from Fig. 1. (Solid line) Valentine (*1*), (dotted line) Müller (*9*), (dashed line) Newell (*8*).

the reverse trend in both measures. Thus, although a causal relation is not proved, the empirical relation appears to be strong enough to justify further investigation.

There is no disagreement on the proposition that the number of taxa known from the fossil record is less than the number that actually lived. This stems simply from the fact that some taxa (particularly at the species level) are rarely or never preserved. The effect is most striking when late Tertiary diversity is compared with the diversity of living organisms. There is no evidence for widespread extinction in the late Tertiary yet most groups have much smaller Tertiary records than would be predicted from neontological data. Furthermore, it is agreed that some biologic groups show fossil diversities closer to their actual diversities than do other groups because of inherent differences in preservability. Crustaceans, for example, are clearly underrepresented as fossils when compared with brachiopods or bivalves. The real problem, however, in the present context, is to evaluate relative changes in diversity over time, using the fossil record as the only available measure.

Sampling Problems

Many fossil taxa remain to be discovered. At the species level, this number probably exceeds even the number that have been described, although this would vary greatly from group to group. The diversity problem is thus in the realm of sampling theory and can be attacked from a mathematical viewpoint.

Exploration for fossils is analogous to problems in probability theory known variously as cell occupancy and urn problems. Consider a wooden tray which is divided into small compartments or cells, and assume that small balls are thrown randomly at the tray in such a way that each ball falls into a cell, without being influenced by the position of the cell or whether it is already occupied. The first ball thrown will inevitably result in the occupancy of one cell. The second ball may fall in the same cell and thus not add to the number of cells occupied: The probability of this event will be greatest if the total number of cells is small. At some point, all the cells will be occupied by at least one ball, and the waiting time necessary to accomplish

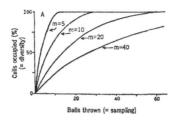

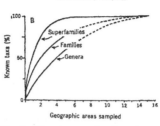

Fig. 3. Diversity as a function of sampling. (A) Illustration of cell occupancy problem. The average waiting time for cell occupancy varies with the number of cells to be occupied (m). (B) Effect of sampling on apparent diversity in fossil ammonoids of the *Meekoceras* zone (Triassic).

this (measured in number of balls thrown) will depend only on the number of cells in the tray.

As noted above, the waiting time for occupancy of one cell is equal to 1 (one ball thrown). It can be shown (16) that the average additional waiting time for occupancy of a second cell is:

$$\frac{m}{m-1}$$

where m is the total number of cells in the tray. The additional waiting time for the third occupancy is:

$$\frac{m}{m-2}$$

and so on. The total waiting time for complete cell occupancy then becomes:

$$m\left(\frac{1}{m}+\frac{1}{m-1}+\frac{1}{m-2}+\ldots+\frac{1}{2}+1\right)$$

Calculated curves for the expected waiting time for various values of m are shown in Fig. 3A.

The appropriate paleontological analogy is as follows: Let m be the total number of taxa available for discovery (thus, one cell equals one taxon), and let the balls thrown be the number of fossils found and identified or described. The first fossil discovered inevitably means recognition of one taxon. The second fossil may be the same or it may be from a second taxon (second

cell occupied). Groups with fewer subgroups will require less sampling to be completely discovered.

This reasoning can be applied directly to the influence of taxonomic level on observed diversity. In any fossiliferous rock unit, the number of families represented is inevitably equal to or greater than the number of phyla, the number of genera is equal to or greater than the number of families, and so on. Thus, much less sampling is required to find all or nearly all the phyla (low m) than the families or genera (higher m values). At any point in the sampling process, a larger percentage of the phyla will be known than of the lower taxa. In Fig. 3, the curves of low m are what would be expected for discovery of high taxa, and the curves of high m would be representative of lower taxa. It should be noted that as sampling progresses, the ratio of the numbers of lower to higher taxa (genera per family, for example) steadily increases.

Figure 3 also shows a paleontological analog of the calculated curves. It is based on published data for ammonoids of the *Meekoceras* zone (Lower Triassic) (17). The data include the known occurrences of 58 genera in 15 geographic assemblages around the world. The ammonoid data (Fig. 3B) show the relationship between sampling and apparent diversity. Sampling is in this case expressed as the number of sites or areas sampled and is analogous to the number of balls thrown in the cell occupancy problem. The number of taxa found at one site in the *Meekoceras* case depends, of course, on which site is used. China, for example, yields well over half the genera and about three-quarters of the families; at the other extreme, the assemblage from the Caucasus has only two of the genera. The curves in Fig. 3 are therefore based on average expectations. For each taxonomic level, some of the values could be calculated directly; other values were determined by simulation based on a random selection of the published distributional data. The remainder (dashed lines) were extrapolated.

The ammonoid example demonstrates that apparent diversity is severely controlled by (i) the extent of sampling and (ii) the taxonomic level. At the order level (Ammonoidea) any one of the 15 sites is sufficient to yield 100 percent of the known diversity. At the generic level it requires (in this case) an average of 5 sites to exceed 50 percent. It should be emphasized that the leveling off of the curves at 100 percent does not mean that the 15 sites yield all of the ammonoid diversity in the *Meekoceras* zone: New genera and new localities are still being found.

As noted above, an increase in sampling is accompanied by predictable increases in the apparent number of genera per family, and so forth, and the effect is seen in the *Meekoceras* zone data. That this is a general phenomenon was noted by Simpson (*18*) as follows: "Sampling at few, restricted localities certainly reveals a much higher percentage of the genera than of the species that existed at any one time."

The sampling problem need not be analyzed only in the context of geographic extent of collecting. The sampling axes of Fig. 3 could be replaced by various measures of the intensity of collecting or study (such as number of paleontologists or years of study) or by measures of the quality of the fossil or rock record (extent of outcrops, type of preservation, and even accessibility of outcrops). The fact that new taxa are constantly being defined or discovered means that the fossil record is still in a relatively early stage of sampling and thus may be represented by the steeper parts of the curves in Fig. 3.

Sources of Error in Diversity Data

In the following numbered sections I consider seven major sources of error that may affect any set of diversity data. All of them certainly have influenced published diversity data of the type shown in Fig. 1.

1) *Range charts.* When the objective of a diversity study is to estimate how many taxa lived during a given interval of geologic time, the primary source of information is usually a range chart drawn at the appropriate taxonomic level. If a family has a range from the base of the Silurian to the top of the Lower Devonian, for example, it is assumed that the family lived throughout the entire range. Thus, the family is registered for the Upper Silurian even though the Upper Silurian fossil record may not actually contain species of the family.

This procedure is valid biologically as a means of estimating actual diversity, but it does have the effect of overestimating "observed" diversity for relatively unfossiliferous intervals. In fact, an interval can be completely unfossiliferous yet still be credited with

having considerable fossil diversity. This source of error becomes important when one is assessing the biasing effect of low sediment volume, as in the Permian-Triassic of Fig. 2. In this instance, the drop in fossil content of Permian rocks may be greater than it appears from the range chart data.

More important, the use of range charts introduces a systematic, time-related bias, as follows. Many (or most) range charts are incomplete in that the true first and last occurrences have not yet been found. In fact, the fossil record may not even contain the first or last occurrences (due to nonpreservation). Ranges of taxa may be truncated at either end, but truncation at the older end (first occurrence) has a higher probability because the older rocks have a greater chance of nonexposure or destruction by erosion and metamorphism. This means that the Phanerozoic diversity data are inevitably biased toward an increase in observed diversity through time.

2) *Influence of "extant" records.* Cutbill and Funnel (*7*) have already noted the biasing effect of the fact that ranges of fossil taxa are generally said to include the Recent if the taxa have living representatives. A not uncommon example would be a living group which has only one fossil occurrence, let us say in the Jurassic. Its range would be listed as Jurassic-Recent which, again, is valid for many purposes but causes problems in the present context. If the group had the same sparse fossil record but had not survived to the Recent, its range would be given as simply Jurassic. Cutbill and Funnel concluded that truncation at the "last occurrence" end of a range through nondiscovery is less likely if the group has living representatives, and since younger rocks contain more extant forms, the late Mesozoic and Cenozoic diversity data are consistently biased toward larger diversity and fewer extinctions than older parts of the column.

3) *Durations of geologic time units.* Consider the effects of the durations of periods and epochs on the diversity data in Fig. 1. The horizontal axis in the diagram is roughly adjusted for relative durations—albeit with little justification in many cases—but the vertical axes showing numbers of taxa are not. The height of each bar on the histograms indicates the total taxa which are found anywhere in the system or series or which have ranges that include those rocks. All things being equal, a long time interval will show a

higher diversity than a short one. The effect of the bias is probably to overestimate diversity in the early Paleozoic, where period and epoch durations are generally greater (*7*). This bias thus operates in a direction opposite to that of the two discussed above.

Furthermore, the bias is not easily corrected. The calculation of a simple ratio, such as families per million years, is valid when working with, for example, extinction rates, but only makes matters worse in the present context, where "standing crop" is the objective.

4) *Monographic effects.* The effects of the quality and quantity of taxonomic activity on apparent diversity are well known. It has been noted, for example, that the peak number of brachiopod genera shifted from Devonian to Ordovician largely as a result of the publication of one monograph (*19*). It is interesting to note that the generic peak has since shifted back to the Devonian.

Some of the monographic effects stem from the stratigraphic distribution of taxonomic specialists and taxonomic and phylogenetic philosophy, and perhaps even from the geographic distribution of taxonomists. Fossiliferous rocks in western Europe and eastern North America are more likely to be fully studied and thus to show higher diversity than rocks in other parts of the world.

If monographic effects are randomly distributed among the major phyla and throughout the stratigraphic column, then the consequences for overall trends in Phanerozoic diversity are minimal. Whether this lack of systematic bias exists is difficult to prove. If more families and superfamilies have been defined in the lower Paleozoic than in other parts of the geologic time table, it is impossible to say whether the difference reflects a tendency of lower Paleozoic paleontologists to be quick to erect such taxa, or whether it results from different kinds of diversity and states of preservation. At the very least, the monographic factors make highly precise studies of diversity impossible.

One special type of monographic effect is surely time dependent. If a group of organisms has many living representatives, and if biologists have subdivided it into many higher taxa, fossil representatives of these higher taxa are more likely to be recognized than if living forms are absent. This says in effect that it is easier to recognize a fossil taxon as distinct if the classification has already been established on the basis of the more complete morphological information afforded by living species. This bias has the effect that diversity is underestimated in extinct groups relative to nonextinct groups. For example, the discovery in Japan of a bivalved gastropod led to the reassignment of its Eocene counterpart from the Bivalvia to the Gastropoda. This greatly extended the stratigraphic range of the gastropod order Sacoglossa and thus increased the apparent gastropod diversity of the Tertiary (20).

5) *Lagerstätten.* Our knowledge of the history of life would be very different were it not for the occasional instances of spectacular preservation of large assemblages (Lagerstätten). Individual formations such as the Solnhofen, the Burgess shale, and the Baltic Amber as well as unusually fossiliferous groups of rocks such as in Timor and Madagascar have significant effects on diversity curves. In some cases, the lack of Lagerstätten is also significant. For example, the observed diversity of insects during the Cretaceous is essentially zero, but this is presumably only an artifact resulting from the lack of the special conditions required for good insect preservation during that period.

The distribution of Lagerstätten through time does not appear to be systematic although they are probably more common in younger rocks. To the extent that this is true, there will be a bias toward high diversity in younger rocks. The greatest effect, however, is to add "noise" to the diversity data in much the same way that monographic bursts produce irregularity in diversity trends in the affected groups.

6) *Area-diversity relationships.* When a new geographic region is opened to exploration, new taxa are almost inevitably discovered. This is due in part to increased sampling, but it also results from the fact that taxa tend to be geographically restricted because of either climatic factors or barriers to dispersal. Also, diversity has been shown empirically to be area dependent (21).

Many instances of geographic effects could be cited. One example comes from Mortensen's tabulation of distributions of living cidarid echinoids, which shows that the 148 species and subspecies of the 27 genera are distributed among 18 geographic regions (22). Only one genus, *Eucidaris*, is found in as many as half the 18 regions,

and 63 percent of the genera are confined to fewer than four regions. No single region contains even one-third of the species. This is in spite of the fact that most cidarids have a free-swimming larval stage.

If the cidarid distribution is looked at in terms of the probable fossil record it will leave, the potential effect of geographic restriction becomes greater. The biogeography of living echinoids is based on a reasonably good sampling of three-quarters of the earth's surface—that is, the oceanic areas. In the fossil record, sampling is limited for all intents and purposes to one-quarter of the earth's surface (the continents and islands), and a significant part of that quarter has remained out of the marine realm by being emergent during most of the Phanerozoic. Thus, the paleontologists can examine only a small fraction of the ocean area for any point in the geologic past. If one were to look at only 5 percent of the present ocean area (or even 5 percent of the present continental shelf area), the apparent diversity in groups such as echinoids would be greatly reduced at all taxonomic levels. This is particularly true since, in most geologic systems, the bulk of the record is usually concentrated in a few areas—rather than being randomly scattered over the world.

The effect of biogeography on diversity is greatest at the species level and decreases upward in the taxonomic hierarchy. Most modern phyla have worldwide distributions but even so are missing in some large regions, mainly due to climatic factors. At the family level, endemism becomes much more common, although this varies greatly from group to group.

The net effect of the biogeographic factor in the present context is to make the observed fossil diversity dependent not only on the area of rock exposure but also on the nature of the world distribution of exposures. Relatively small exposures on several continents are likely to yield a higher overall diversity than the same total exposure concentrated on one continent.

Although Gilluly has demonstrated a clear increase in area of exposure through the Phanerozoic column (10), no studies have been made on the manner in which these rocks are distributed spatially. However, because the probability of finding older Phanerozoic rocks is less than that of finding younger ones (assuming equal time durations) it would seem reasonable that geographic cov-

erage improves toward the Recent. This should produce higher observed diversities in younger rocks.

7) *Sediment volume.* This article started with the empirical correlation between sediment volume per unit of time and diversity of major marine groups. It is clear from sampling considerations that more sedimentary record should produce more diversity. The correlation shown in Fig. 2 is thus quite plausibly a causal one. But the strength of the resulting bias depends on (i) the taxonomic level and (ii) the kinds of differences in sediment volume from one part of the column to another. A figure for sediment volume for one geologic system [such as used by Ronov (15)], may be higher than the figure for another geologic system for many reasons. Discontinuous sedimentation may mean that many short-lived taxa are not preserved, but the fossil record of longer-lived taxa, characteristic of families and orders, may not be much affected. Thus, for example, if the Paris Basin had twice the volume of sediments, species diversity would be higher but family diversity little if any different. If sediment volume figures are influenced by differences in area of sedimentation, then the biogeographic relationships discussed above become significant, even at high taxonomic levels.

Postdepositional destruction or covering of sediments is the most widely accepted explanation for the temporal trends in sediment volume. Such losses of record are likely to have a spotty geographic distribution. That is, loss of the sedimentary record from one or more whole regions is more likely than small-scale reductions in all areas. This suggests that loss of biogeographic coverage is the important factor for diversity and that the sediment volume bias is closely tied to the geographic bias discussed earlier.

Models for Phanerozoic Diversity

Figure 4 shows in generalized form Phanerozoic diversity patterns at several taxonomic levels for shelf invertebrates with well-developed skeletons. The illustration is a composite of several from Valentine (1) and one from Müller (9). Minor irregularities were removed in making the composite, and vertical scales were adjusted. Valentine based the species curve on inference, but all the others were drawn directly from observed diversities.

Valentine concluded that the patterns are a plausible result of a combination of the evolutionary process of diversification and certain events in the physical history of the Phanerozoic. The basic biologic process envisioned requires that diversification take place first at high taxonomic levels (phylum, class, order) and later at successively lower taxonomic levels. The number of phyla (not shown) reached a maximum during or before the Early Ordovician, classes and orders later in the Ordovician, families in the Devonian, and genera and species in the Carboniferous or earliest Permian. According to Valentine, the diversity of the higher taxa (except phyla) declined after the initial peaks because as high taxa became extinct, they were replaced not by equally distinct groups but rather by specialized lower taxa (genera and species) within the surviving groups.

Still following Valentine's interpretation, the Permian-Triassic mass extinctions sharply reduced the diversity at all levels, and this was followed by a dramatic rise in diversity at the family, genus, and species levels, leading to the present-day array. Valentine argues that the driving forces behind this Mesozoic-Cenozoic rediversification were (i) continental drift and (ii) an increase in latitudinal temperature gradients. The diversity increase would presumably have taken place anyway—but to a lesser degree—as a continuation of the trend to specialization that was interrupted by the Permian-Triassic extinctions.

Figure 4 and its interpretation represent, therefore, one model for Phanerozoic diversity. It is an appealing one in that it is based largely on a "face value" use of empirical data and because it is biologically and ecologically plausible.

The foregoing interpretations are subject to several problems. The patterns in Fig. 4 contain elements that are qualitatively those which would be predicted from the biases discussed in this paper, as follows:

1) If the quality or quantity of sampling increases through time, it is inevitable that the ratios of species to genera, genera to families, and so on, will also increase.

2) Time-dependent biases should produce a rise in diversity at the lower taxonomic levels as the Recent is approached. The post-Paleozoic increase in numbers of families, genera, and species seen in Fig. 4 may be due to this factor.

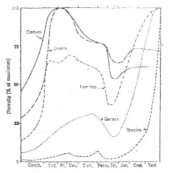

Fig. 4. Variation in apparent taxonomic diversity for several taxonomic levels of well-skeletonized marine invertebrates during the Phanerozoic.

3) Time-dependent biases should also shift any diversity peak toward the Recent (to the right in Fig. 4), and the amount of shift should be greatest at the lowest taxonomic levels. The fact, noted by Valentine (1), that diversities at lower taxonomic levels appear to have peaked after those at higher levels may actually be due to the effects of biases.

The last point deserves more consideration. From an evolutionary viewpoint, it is certainly plausible that diversity maxima for species and genera should occur after those for higher taxa in the same group. The question is whether the time lag is large enough and sufficiently universal to produce distinct offsets when diversities of several major animal groups are plotted together as in Fig. 4. If this were the case, periods of widespread extinction

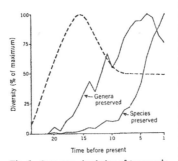

Fig. 5. Computer simulation of taxonomic diversity. The dashed line is a hypothetical diversity distribution before fossilization and is based on simulated ranges of 2000 species constituting 100 genera. The solid lines indicate the diversity trends after biases are applied to the range data.

should be followed by recognizable intervals of low diversity, during which rediversification takes place. But the fact is that most major extinctions are not followed by periods of low diversity. Lowered diversity must have occurred at such times, but it evidently did not last long enough to be noticeable on the time scale used here. Valentine points out that the Permian-Triassic extinction is the only one which is followed by a diversity drop. Figure 2 indicates that in that interval the diversity drop may be an artifact of sampling.

An alternate model for Phanerozoic diversity is suggested by the dashed line in Fig. 5 and consists of a diversity maximum followed by a decline to an equilibrium level. The time scale in Fig. 5 is arbitrary, but a mid-Paleozoic position for the maximum is implied: The curve was suggested by the curves in Fig. 4 for classes and orders (where effects of biases should be least). The alternative model makes no distinction between taxonomic levels and thus is meant to apply to all levels below phylum. Thus, the assumption has been made that the offset of diversity peaks caused by gradual diversification either is not large enough to be observed at this scale or is masked by noise resulting from the fact that many animal groups with different evolutionary histories are plotted together. The proposed model is, of course, valid only if the biases described in this article are quantitatively significant.

The plausibility of the alternative model was checked by a computer simulation. By using random numbers, hypothetical first and last occurrences were generated, and a range chart was constructed showing the distributions in time of 2000 hypothetical species (segregated into 100 genera). The dashed diversity curve in Fig. 5 was computed from the simulated range chart. The curve thus represents a hypothetical diversity pattern before biasing factors are applied.

Next, information was removed from the range chart by a random process designed to simulate the biasing factors. For each species, portions of the record were "destroyed," with the probability of destruction increasing back in time. Record losses occurring only inside a range had no effect. If, however, a loss included the beginning or end of the range, the range was shortened accordingly. In many cases, species were completely removed by this process. The Recent was made immune from these information losses to stimulate the biasing effect of "extant" records.

Finally, a new range chart was constructed from what was left after the information removals. The diversity curves computed from this are also shown in Fig. 5. Species diversity increases sharply toward the Recent whereas generic diversity shows a maximum, offset to the right of the original maximum. When genera are grouped into hypothetical families (not shown), the diversity maximum is offset to the right but not as far.

The simulation demonstrates that diversity patterns such as are observed in the fossil record can be produced by the application of known biases to quite different diversity data. The simulation does not, of course, prove the alternative model for Phanerozoic diversity because of our present ignorance of the actual impact of the biases. The simulation does suggest, however, that the model proposed in Fig. 5 is a plausible one for the Phanerozoic record of marine invertebrates.

The alternative model cannot be applied literally to land-dwelling forms because the exploitation of terrestrial habitats started much later in geologic time and may be still going on. The fossil record of terrestrial organisms is subject to the same biases, however, and so should be read with caution.

Summary

Apparent taxonomic diversity in the fossil record is influenced by several time-dependent biases. The effects of the biases are most significant at low taxonomic levels and in the younger rocks. It is likely that the apparent rise in numbers of families, genera, and species after the Paleozoic is due to these biases. For well-skeletonized marine invertebrates as a group, the observed diversity patterns are compatible with the proposition that taxonomic diversity was highest in the Paleozoic. There are undoubtedly other plausible models as well, depending on the weight given to each of the biases. Future research should therefore be concentrated on a quantitative assessment of the biases so that a corrected diversity pattern can be calculated from the fossil data. In the meantime, it would seem prudent to attach considerable uncertainty to the traditional view of Phanerozoic diversity.

References and Notes

1. J. W. Valentine, *Paleontology* 12, 684 (1969).
2. ———, *Bull. Geol. Soc. Amer.* 79, 273 (1968); *J. Paleontol.* 44, 410 (1970).
3. F. G. Stehli, in *Evolution and Environment*, E. T. Drake, Ed. (Yale Univ. Press, New Haven, Conn., 1968), p. 163.
4. R. C. Moore *et al.*, Eds., *Treatise on Invertebrate Paleontology* (Geological Society of America and Univ. of Kansas Press, Lawrence, 1953–1972).
5. Y. A. Orlov, *Osnovy Paleontologii* (Akademiia Nauk SSSR, Moscow, 1958–1964).
6. W. B. Harland *et al.*, Eds., *The Fossil Record* (Geological Society of London, London, 1967).
7. J. L. Cutbill and B. M. Funnel, *ibid*, p. 791.
8. N. D. Newell, in *Uniformity and Simplicity*, C. C. Albritton, Jr., Ed. (Geological Society of America, New York, 1967), p. 63.
9. A. H. Müller, *Grossabläufe der Stammesgeschichte* (G. Fischer, Jena, Germany, 1961).
10. J. Gilluly, *Bull. Geol. Soc. Amer.* 60, 561 (1949).
11. J. D. Hudson, in *The Phanerozoic Time-Scale*, W. B. Harland *et al.*, Eds. (Geological Society of London, London, 1964), p. 37.
12. C. B. Gregor, *Proc. Kon. Ned. Akad. Wetensch.* 71, 22 (1968).
13. See also the general discussion by R. M. Garrels and F. T. Mackenzie, *Evolution of Sedimentary Rocks* (Norton, New York, 1971).
14. The data on maximum thickness are mostly from A. Holmes, *Trans. Edinburgh Geol. Soc.* 17, 117 (1959); M. Kay, in *Crust of the Earth*, A. Poldervaart, Ed. (Geological Society of America, New York, 1955), p. 665.
15. A. B. Ronov, *Geokhimiya* 1959, 397 (1959), translated in *Geochemistry USSR* 1959, 493 (1959).
16. M. Dwass, *Probability* (Benjamin, New York, 1970).
17. B. Kummel and G. Steele, *J. Paleontol.* 36, 638 (1962). A few of the ammonoid occurrences were designated as doubtful; these were eliminated for the present purpose with the effect that one of the original 16 localities was eliminated.
18. G. G. Simpson, *The Major Features of Evolution* (Columbia Univ. Press, New York, 1953), p. 31. See also the excellent discussions of sampling problems and biases by J. W. Durham, *J. Paleontol.* 41, 559 (1967) and by G. G. Simpson, in *Evolution After Darwin*, S. Tax, Ed. (Univ. of Chicago Press, Chicago, 1960), vol. 1, pp. 117–180.
19. G. A. Cooper, *J. Paleontol.* 32, 1010 (1958); see also the general discussion of monographic effects in A. Williams, *Geol. Mag.* 94, 201 (1957).
20. L. R. Cox and W. J. Rees, *Nature* 185, 749 (1960).
21. F. E. Preston, *Ecology* 43, 185 (1962); N. D. Newell, *Amer. Mus. Nov.* 2465 (1971).
22. T. Mortensen, *A Monograph of the Echinodea* (Reitzel, Copenhagen, 1928), vol. 1.

Maize and Its Wild Relatives

Teosinte and *Tripsacum*, wild relatives of maize, figured prominently in the origin of maize.

H. Garrison Wilkes

The close relatives of maize, teosinte and the genus *Tripsacum*, have assumed increasing importance in the understanding of the evolution under domestication of the New World's most important plant food. *Tripsacum* hybridizes with maize under experimental conditions, and teosinte crosses with maize in its native habitat, Mexico and Central America. Much of the heterotic vigor of maize is attributed to introgressive hybridization from its closest relative, teosinte. Today, the maize crop is the single largest harvest in the United States and is the staple food for most of the inhabitants of Latin America.

Considering the importance of the hybridization of maize (*Zea mays* L.) with its wild relatives (Fig. 1) teosinte [*Z. mexicana* (Schrad.) O. Ktze.] (*1*), an annual grass looking very much like maize, and *Tripsacum* (*2*), perennial grasses quite distinct from maize in appearance, it is startling to realize how little is known about this phenomenon in the wild. Maize and teosinte are genetically compatible and hybridize freely with each other in places where the isolating mechanisms between the two have broken down, as in the Sierra Madre Occidental of northern Mexico, the Central Plateau and Valley of Mexico in central Mexico, and in Heuhuetenango of northern Guatemala. *Tripsacum* does not hybridize readily with maize in the field, but hybrids can be produced under experimental conditions. There is reason to be alarmed by the rapid extinction of these wild relatives in and around maize fields where teosinte is known to have hybridized with maize for at least three millennia. This extinction of the native populations of teosinte is disastrous from the standpoint of future introgression, since it

The author is a botanist in the department of biology at the University of Massachusetts, Boston 02116.

11 July 1969, Volume 165, Number 3889

SCIENCE

Speciation in Amazonian Forest Birds

Most species probably originated in forest refuges during dry climatic periods.

Jürgen Haffer

The richest forest fauna of the world is found in the tropical lowlands of central South America. This fauna inhabits the vast Amazonian forests from the base of the Andes in the west to the Atlantic coast in the east, and its range extends far to the north and south of the Amazon valley onto the Guianan and Brazilian shields, respectively (Fig. 1). Here I propose a historical explanation of the immense variety of the Amazonian forest bird fauna, postulating that, during several dry climatic periods of the Pleistocene and post-Pleistocene, the Amazonian forest was divided into a number of smaller forests which were isolated from each other by tracts of open, nonforest vegetation. The remaining forests served as "refuge areas" for numerous populations of forest animals, which deviated from one another during periods of geographic isolation. The isolated forests were again united during humid climatic periods when the intervening open country became once more forest-covered, permitting the refuge-area populations to extend their ranges. This rupturing and rejoining of the various forests in Amazonia probably was repeated several times during the Quaternary and led to a rapid differentiation of the Amazonian forest fauna in geologically very recent times.

The author is research geologist in the Field Research Laboratory of Mobil Research and Development Corporation, Dallas, Texas. He worked for several years in South America prior to his assignment in the United States.

This interpretation should be considered merely a working model based on a number of inferences. It may, however, serve for testing the distribution pattern of various groups of organisms. Much more concrete information on the climatic and vegetational history of Amazonia, as well as on the population structure and species relationships of Amazonian birds and other animals, is needed if one is to reconstruct the actual course of species formation in particular areas or in certain families.

Climatic Fluctuations during the Quaternary

The worldwide climatic fluctuations of the Pleistocene and post-Pleistocene severely influenced environmental conditions in the tropics. In the mountains, altitudinal temperature zones and life zones were repeatedly compressed and expanded vertically during cold and warm periods, respectively (1). At the same time the lowlands probably remained "tropical," but humid and dry climatic periods caused vast changes in the distribution of forest and nonforest vegetation. The present continuity of the Amazonian forest seems to be a rather recent and temporary stage in the vegetational history of South America (2). Geomorphological observations in southern Venezuela (3), lower Amazonia (4), central Brazil (5), and eastern Peru (6) indicate that, during

the Quaternary, arid climatic conditions repeatedly prevailed over large parts of Amazonia. During these periods dense forests probably survived in a number of rather small, humid pockets (7). Palynological studies in northern South America (8) also revealed repeated vegetational changes over large areas during the Pleistocene and post-Pleistocene. The absolute ages of the various humid and arid climatic phases, in particular the age of the last severe arid period, are not yet known. Moreover, correlation of the warm-dry and cool-humid periods of the low-latitude lowlands with the glacial and interglacial periods of the temperate regions remains a matter of controversy.

The immense importance of the Quaternary climatic fluctuations for the latest differentiation of tropical faunas has been recognized for some time (9). Stresemann and Grote (10) long ago emphasized the significance of humid and dry periods for the history of the fauna of central Africa and the East Indies. The extent of vegetational changes in Africa has been amply demonstrated in recent years by detailed palynologic studies (11). Moreau (12) analyzed the differentiation of African bird faunas in the light of the geological and climatic history of the African continent, in a very convincing interpretation. Similar zoogeographic analyses have been published on the bird faunas of Australia (13), Tasmania (14), and parts of the Old World tropics (15). But the significance of Quaternary climatic fluctuations for the differentiation of the forest faunas of tropical South America has hitherto received little attention.

Reconstruction of Forest Refuges in Amazonia

On the basis of the theory of geographic speciation (16) let us assume that most or all Amazonian forest species originated from small populations which were isolated from their parent population and deviated by selection and chance. Most of this differentiation probably took place in re-

131

stricted refuge areas. Thus our main problem becomes the reconstruction of the probable geographic location of the forest refuges. I have used the following criteria for a first tentative approach to this problem: (i) current inequalities of annual rainfall in Amazonia, and (ii) current distribution patterns of Amazonian birds, particularly members of superspecies.

Current rainfall maxima. Rainfall is not evenly distributed over the Amazonian lowlands (*17, 18*). There are three main centers of rainfall (Fig. 2) which receive over 2500 millimeters of rain per year. These areas usually are humid throughout the year, having no *pronounced* dry season. The largest of these rainfall centers comprises upper Amazonia from the Río Juruá and the upper Río Orinoco west to the base of the Andes, where the annual rainfall increases to 4000 to 5000 millimeters. The second of the three centers is the Madeira–upper Río Tapajós region, which is separated from upper Amazonia by a somewhat drier corridor between the Negro, Purús, and Juruá rivers. The third center comprises the southern Guianas and the area extending southeastward to the mouth of the Amazon. The very humid regions of western and central Amazonia are separated from the rainfall center near the Atlantic coast by a comparatively dry transverse zone extending in a northwest-southeast direction and crossing the lower Amazon around Óbidos and Santarém. Although most of this region is forested, numerous isolated savannas are found here. This comparatively dry belt has a very dry season, and the total annual rainfall is less than 2000 or even 1500 millimeters per year. This zone connects the open plains of central Venezuela with the unforested region of central and northeastern Brazil. The relatively dry areas north and south of the humid upper Amazonian lowlands extend their influence near the base of the Andes far to the southwest and northwest, respectively. In Fig. 2 this is obvious from the conspicuous southwestward bulging of the isohyets in eastern Colombia and from the characteristic course of the 2000-millimeter isohyet near the Andes in eastern Peru [where rather dry forests are found from Bolivia northwest to the middle Río Ucayali and Huallaga valleys (*19*)]. The foothill regions of the Andes in eastern Colombia and southeastern Peru and Bolivia represent narrow extensions of humid Amazonia which receive 2000 to 5000 millimeters of rain per year. In these regions the air is forced to rise, and it loses its moisture in the form of mist and frequent rain.

Reinke (*18*) has given a detailed climatologic interpretation of this rainfall pattern and has discussed the significance of the Andes and of the mountains in the interior of the Guianas for the location of current rainfall maxima.

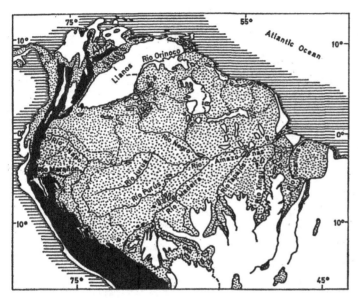

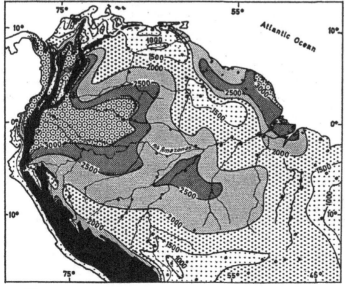

Fig. 1 (above left). Distribution of humid tropical lowland forests in central and northern South America. Forests surrounding savanna regions are mostly semi-deciduous. (Black areas) Andes mountains above 1000 meters. [Adapted from Hueck (*38*), Haffer (*39*), Aubréville (*40*), and Denevan (*41*)]

Fig. 2 (bottom left). Total annual rainfall (in millimeters) in northern and central South America. (Large dots) Weather stations; (black area) Andes mountains above 1000 meters. [Adapted from Reinke (*18*)]

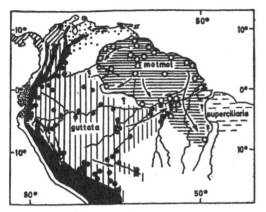 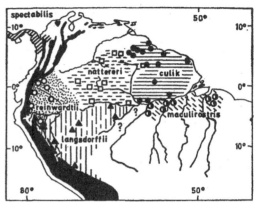

Fig. 3 (left). Distribution of the chachalaca, the *Ortalis motmot* superspecies, in Amazonia. Additional species occur north and south of the area shown. The trans-Andean forms are (stippled area) *O. erythroptera*, (hatched area) *O. garrula*, and (sparsely dotted area) *O. ruficauda*. [Adapted from Vaurie (*42*)]   Fig. 4 (right). Distribution of a toucanet, the *Selenidera maculirostris* superspecies, in Amazonia. Additional isolated populations of *S. maculirostris* occur in eastern Brazil. Hybridization between *S reinwardtii* and *S. langsdorffii* is known in northeastern Perú.

During arid periods the effective humidity in Amazonia was reduced by a reduction in rainfall or by a rise in temperature and an increase in evaporation, or by both. I assume that, during dry phases, rainfall in the areas of current rainfall maxima remained high enough to permit the continued growth of forests, while the forest probably disappeared from the intervening areas of lower rainfall. Since the major orographic features that cause the current inequalities in rainfall in Amazonia were present during most of the Pleistocene, possibly the basic rainfall pattern was fairly constant, though probably not entirely so, during the various climatic periods. For this reason I suggest that the main Amazonian forest refuges of arid climatic periods coincided with the present centers of high rainfall.

Current distribution patterns of Amazonian birds. Important indirect evidence concerning the possible geographic location of former forest refuges may also be obtained from the present distribution of localized Amazonian animal species, which apparently never extended their ranges far beyond their center of survival or area of origin. For Amazonian birds I distinguish the following "centers of distribution": (i) upper Amazonia from the base of the Andes east to the Río Negro and the Río Madeira; (ii) the Guianas west to the Río Negro and south to the Amazon; and (iii) lower Amazonia south of the Amazon from the lower Río Tapajós east to the Atlantic coast. Each of these areas is characterized by many distinc-

tive and morphologically rather isolated bird species which do not range beyond the *approximate* limits indicated above. An additional "center of distribution" is the region between the Río Madeira and the upper Río Tapajós, where several endemic species occur. The foothills of the Peruvian Andes and the forests of the upper Río Negro–Río Orinoco region represent other such centers.

A number of Amazonian forest birds of particular zoogeographic interest form allopatric (that is, mutually excluding) species assemblages which are designated superspecies (*16*). Members of different superspecies often have similar distributions. Many are restricted to the Guianas and adjacent territories, to parts of upper Amazonia, or to various portions of the lowlands south of the Amazon. Numerous examples could be cited from the Cracidae (*20*), the toucans, the antbirds, the cotingas, the manakins, and others. The distribution of two superspecies is shown in Figs. 3 and 4. The component species of the Amazonian superspecies probably originated in forest refuges from a common ancestor whose range was split into a number of isolated portions during arid periods. By comparing the ranges of localized forms we may derive important clues concerning the former location of refuge areas.

Location of forest refuges. Using the above indirect evidence derived from rainfall inequalities and from patterns of avian distribution, I have reconstructed and named the probable geographic location of various Quaternary forest

refuges in the lowlands of tropical South America (Fig. 5). The postulated refuges west of the Andes are as follows.

Chocó refuge, which comprised the central Pacific lowlands of Colombia (*21*).

Nechí refuge, on the northern slope and the foreland of the central and western Andes of Colombia (*21*).

Catatumbo refuge, on the eastern slope and base of the Serranía de Perijá (*21*).

There are six postulated refuges east of the Andes, as follows.

Napo refuge, which comprised mainly the lowlands of eastern Ecuador from the Andes to the Marañón River. This may have been the largest and ecologically the most varied forest refuge for a great number of Amazonian forest animals. It is named after the Río Napo in eastern Ecuador (*22*).

East Peruvian refuges. Several isolated lowland forests probably existed along the eastern base of the Peruvian Andes and, farther east, on the low mountains between the Río Ucayali and the Juruá-Purús drainage (*23*).

Madeira-Tapajós refuge, which comprised the lowlands between the middle Río Madeira and the upper Río Tapajós (*24*).

Imerí refuge—a small area around the Sierra Imerí and Cerro Neblina between the headwaters of the Río Orinoco and the upper Río Negro (*25*).

Guiana refuge, on the northern slope and foreland of the mountains of Guyana, Surinam, and Cayenne (*26*).

Belém refuge, in the region south of

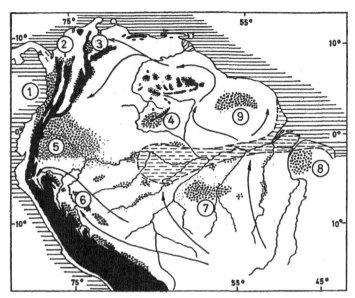

Fig. 5. Presumed forest refuges in central and northern South America during warm-dry climatic periods of the Pleistocene. The arrows indicate northward-advancing nonforest faunas of central Brazil. (1) Chocó refuge; (2) Nechí refuge; (3) Catatumbo refuge; (4) Imerí refuge; (5) Napo refuge; (6) East Peruvian refuges; (7) Madeira-Tapajós refuge; (8) Belém refuge; (9) Guiana refuge; (hatched area) interglacial Amazonian embayment (sea level raised by about 50 meters); (black areas) elevations above 1000 meters.

the mouth of the Amazon and west to the lower Río Tocantins (27).

There were probably additional smaller forests along the major river courses of Amazonia, on the slopes of isolated mountains, and in the extensive lowlands between the upper Río Madeira and the Marañón River. Even when palynological field data from Amazonia become available it will remain difficult to map the distribution of forest and nonforest vegetation at any given time under the constantly changing climatic conditions of the Pleistocene. The table mountains of southern Venezuela are today located in the dry transverse zone of Amazonia and may have supported no tropical, but only subtropical, forests on the higher slopes during arid climatic phases. These may have served as refuges for the montane forest fauna of the highlands of Southern Venezuela.

Since climatic fluctuations were much more pronounced during the Pleistocene than in post-Pleistocene time, it seems possible that the rupturing of the Amazonian forest was most marked during the arid periods of the Pleistocene. During the post-Pleistocene merely a separation of an upper Amazonian forest from lower Amazonian forests

may have resulted from the disappearance of forest growth in the dry transverse zone through the Obidos-Santarém region.

During these arid periods many nonforest animals of central Brazil probably advanced across the lower Amazon to reach the upper Río Branco valley and the nonforest regions of central Venezuela and eastern Colombia (28). Many relict populations of nonforest bird species still inhabit the isolated remnants of savannas and campos found in the forests on both sides of the lower Amazon (29). Nonforest regions probably also extended during dry periods northwestward from central Brazil and eastern Bolivia, near the Andes, through the Ucayali-Huallaga valleys, to connect with the arid upper Marañón valley. A number of Brazilian bird species even crossed the Andes, probably in the area of the low Porculla Pass west of the upper Marañón River, to reach the arid Pacific lowlands of Peru and southwestern Ecuador (28). Without palynological data from eastern Peru we have no means of knowing during which dry phase (or phases) of the Quarternary the postulated connection of the upper Marañón fauna and the Brazilian

nonforest fauna may have been established.

During humid periods the Amazonian forest probably was repeatedly connected with the forests of southeastern Brazil over the now unforested tableland of central Brazil. The connecting forests may not have been very extensive, but they probably made possible the exchange of numerous plants and animals (30, 31). Remnants of these forests are still preserved in small humid pockets and are inhabited by isolated populations of Amazonian animals.

Secondary Contact Zones

Upon the return of humid climatic conditions many forest refuge populations followed the expanding forests and often came in contact with sister populations of neighboring or even far-distant refuges. Because of variation in the rate of differentiation of different animal species, the populations that came in contact reached many different levels in the speciation process. Basically, we may distinguish the following situations.

1) Geographic overlap. The speciation process was completed during the period of geographic isolation—that is, the allies had attained reproductive isolation as well as ecologic compatibility. This resulted in sympatry and a more or less extensive overlap of the ranges occupied.

2) Geographic exclusion. The speciation process was not fully completed. Although reproductively isolated (and therefore treated taxonomically as species) the allies remained ecologically incompatible. This situation led to mutual exclusion presumably as a result of ecologic competition without hybridization along the zone of contact (31a).

3) Hybridization. The speciation process was not completed and the allies hybridized along the zone of contact. Hybridization may occur along a rather narrow belt, indicating that a certain degree of incompatibility of the gene pools had been reached before contact was established in ecologically more or less continuous forests (32). Hybridization over a broad zone may lead to the more or less complete fusion of the populations in contact.

The zones of presumably secondary contact of Amazonian forest birds are in areas between the postulated forest refuges—for example, north and south of the middle Amazon River and in the Huallaga-Ucayali region of eastern

Peru (Fig. 6). A number of representative bird species doubtless met along broad rivers, such as the Amazon or its large tributaries, which separate their ranges today. In these cases, expansion of the allies' range beyond the rivers appears to be inhibited by competition or by swamping of the occasional colonists which manage to get across the watercourse (*33*). A few forms were able to build up small populations on the opposite river bank. Most bird species eventually crossed the rivers in the course of extending their ranges, provided the opposite bank was not occupied by a close relative or, if it was, provided the two allies had acquired complete ecologic compatibility. Few species seem to be definitely halted solely by the river courses of Amazonia, especially a number of birds inhabiting the dark forest interior. However, many such bird species as well as arboreal and small terrestrial mammals probably surrounded the broad portions of the rivers by crossing the latter in the narrow middle and upper parts, in this way often extending their ranges far beyond the areas of origin or survival; again, provided they did not meet ecologically competing close relatives which extended their ranges from other refuge areas. Interspecific competition seems to be very important in limiting the range of numerous forest birds in tropical South America.

In summarizing, the rivers probably are not a causal factor of avian speciation in Amazonia (except perhaps in a few cases), but merely modified, or occasionally, limited, the dispersal of forest bird species after the latter had originated in forest refuges during dry climatic periods.

Conclusion and Summary

The Tertiary forest fauna of central South America inhabited comparatively restricted forests along marginal portions of the Guianan and Brazilian shields. Although several species of the present Amazonian forest bird fauna may represent direct descendants of Tertiary forms, most or all species seem to have undergone considerable evolutionary change during the Pleistocene. Several factors, in combination, probably caused this rather recent faunal differentiation in Amazonia: (i) the great expansion of dense forests onto the fully emerging Amazonian basin and into the lowlands around the rising

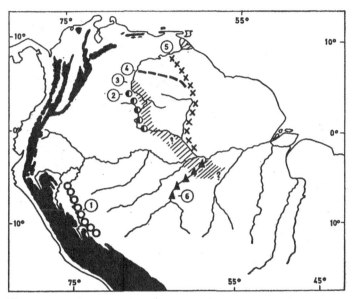

Fig. 6. Location of some secondary contact zones of Amazonian forest birds. (1, open circles) *Pipra chloromeros–P. rubrocapilla* and *Gymnopithys lunulata–G. salvini*; (2, half-solid circles) *Ortalis guttata–O. motmot* and *Gymnopithys leucaspis–G. rufigula*; (3, hatched area) *Ramphastos vitellinus culminatus–R. v. vitellinus*, north of the Amazon, and *R. v. culminatus–R. v. ariel*, south of the Amazon (this hybrid belt probably extends in a southeasterly direction beyond the Tapajós and Xingú rivers); (4, dashed line) *Pteroglossus pluricinctus–P. aracari*; (5, crosses) *Celeus grammicus–C. undatus* and *Tyranneutes stolzmanni–T. virescens*; (6, triangles) *Pteroglossus flavirostris mariae–P. bitorquatus*; (black areas) Andes mountains above 1000 meters.

northern Andes, at the end of the Tertiary; (ii) the repeated contraction and expansion of the forests as a result of climatic fluctuations during the Quaternary, leading to repeated isolation and rejoining of forest animal populations; (iii) the increased rate of extinction of animal forms. Because the populations of many tropical forest animals are small, a reduction in the size of their habitat must have drastically increased the chances of extinction of a number of forms, either within the forest refuges or through competition with newly evolved forms from other refuges upon the return of humid conditions.

Possibly the Amazonian forest fauna was not nearly so rich and diversified in the upper Tertiary as it is at present, mainly because of intensified speciation in the greatly enlarged forests and because of the fluctuating climate of the Pleistocene. This evolutionary boom may have been comparable to that of the montane fauna of the Andes during the Quaternary.

On the basis of evidence discussed by Moreau (*12*), it appears possible that, under favorable circumstances, the

speciation process in birds may be completed in 20,000 to 30,000 years or less, particularly in the tropics, where birds generally seem to occupy smaller niches than they do in cooler and less stable climates. This estimate refers mainly, though not exclusively, to passerine birds with a high reproductive rate and evolutionary potential. Under the same conditions speciation may take longer in larger birds, perhaps requiring on the order of a hundred thousand to several hundred thousand years. Factors such as the size of the refuge population and the degree of isolation of course influence the rate of speciation considerably. The above estimates are highly speculative, and the error involved may be very substantial. However, if the order of magnitude is at least approximately correct, it indicates that the Tertiary ancestors of present Amazonian birds may have speciated *repeatedly* during the Quaternary, and that many connecting links may have disappeared due to extinction. A similar assumption may also apply to the insect (*34*), amphibian, reptile (*35*), and mammal faunas (*36*) of Amazonia. Some of the more strongly differentiated species probably originated in early Pleistocene refuges, while most other species and semi-species may date back to the late Pleistocene or, in the case of the latter, to the post-Pleistocene only. In view of the length of the Tertiary period (60 million years), during which the Amazonian fauna probably evolved rather slowly under quite uniform environmental conditions, the Quaternary faunal differentiation in tropical South America during the last 1 to 2 million years is, geologically speaking, very "recent" and occurred rather "rapidly."

It follows from the foregoing discussion that the Quaternary history of tropical faunas was basically quite similar to that of the faunas of higher latitudes (37). In the temperate regions, as well as in the tropics, climatic fluctuations caused pronounced changes in the vegetation cover and led to the isolation of comparatively small populations in refuge areas. The presumably smaller niche size (and lower population density) of tropical relative to temperate-zone forest animals and the correspondingly higher rate of speciation in the tropics under conditions of large-scale climatic fluctuations may explain the rapid differentiation of tropical forest faunas during the Pleistocene.

References and Notes

1. During cold and warm periods the vertical temperature gradient probably was increased and decreased, respectively, relative to the present gradient (a change of approximately 0.5°C per 100-meter difference in elevation); see J. Haffer [*Amer. Museum Novitates No. 2294* (1967)] for details pertaining to South America. The repeated vertical displacement of the temperature zones led to frequent interruptions and rejoining of the animal populations along the mountain slopes, thereby causing a rapid differentiation of the montane faunas during the Pleistocene.
2. The main uplift of the Andes mountains did not take place until the upper Pliocene and lower Pleistocene [R. W. R. Rutland, J. E. Guest, R. L. Grasty, *Nature* **208**, 677 (1965); J. Haffer, *J. Ornithol.* **109**, 67 (1968)]. The rise of the Andes caused the climate to be very humid along the eastern base and foreland of the mountains. This was partly responsible for the vast expansion of dense forests onto the fully emerging Amazonian lowlands and northward and southward along the base of the rising Andes to Colombia and Bolivia, respectively. During the Tertiary, prior to the Andean uplift and the emergence of the Amazonian lowlands, forests probably had a rather restricted distribution along rivers and marginal lowlands of the elevated land areas north and south of the present Amazon valley.
3. H. F. Garner, *Rev. Geomorphol. Dynamique* **2**, 54 (1966); *Sci. American* **216**, 84 (1967).
4. A. N. Ab'Saber, *Bol. Soc. Brasileira Geol.* **6**, 41 (1957); *Notic. Geomorfol.* **1**, 24 (1958); A. Barbosa, *ibid.*, p. 87.
5. A. Cailleux and J. Tricart, *Compt. Rend. Soc. Biogeograph.* **293**, 7 (1957); J. J. Bigarella and G. O. de Andrade, *Geol. Soc. Amer. Spec. Paper 84* (1965), p. 433; M. M. Cole, *Geograph. J.* **126**, 166 (1960).
6. H. F. Garner, *Bull. Geol. Soc. Amer.* **70**, 1870 (1959).

7. The shrinkage of the humid lowland forests probably was more pronounced than it is shown to be on hypothetical vegetation maps by J. Hester [*Amer. Naturalist* **100**, 383 (1966)] and T. C. Patterson and E. P. Lanning [*Bol. Soc. Geograf. Lima* **86**, 8 (1967)]. The sweeping interpretation of Pleistocene vegetational changes in Amazonia by A. Aubréville [*Adansonia* **2**, 16 (1962)] appears to be unacceptable in view of the fact that climatic changes occurred simultaneously in the Northern and Southern hemispheres.
8. T. van der Hammen and E. Gonzalez, *Leidse Geol. Mededel.* **25**, 261 (1960); T. A. Wijmstra and T. van der Hammen, *ibid.* **38**, 71 (1966); T. A. Wijmstra, *ibid.* **39**, 261 (1967). Additional evidence of climatic fluctuations is available from currently arid western Peru; see E. P. Lanning, *Sci. Amer.* **213**, 68 (1965); *Peru before the Incas* (Prentice-Hall, Englewood, N.J., 1967).
9. P. J. Darlington, *Zoogeography* (Wiley, New York, 1957), pp. 586–88; E. Mayr, *Animal Species and Evolution* (Harvard Univ. Press, Cambridge, 1963), p. 372.
10. E. Stresemann and H. Grote, *Trans Intern. Congr. Ornithol. 6th, Copenhagen, 1926* (1929), p. 358; E. Stresemann, *J. Ornithol.* **87**, 409 (1939).
11. R. E. Moreau, *Proc. Zool. Soc. London* **141**, 395 (1963); E. M. van Zinderen Bakker, Ed., *Palaeoecology of Africa and of the Surrounding Islands and Antarctica* (Balkema, Cape Town, 1967), vols. 2 and 3.
12. R. E. Moreau, *The Bird Faunas of Africa and Its Islands* (Academic Press, New York, 1966).
13. A. Keast, *Bull. Museum Comp. Zool.* **123**, 305 (1961).
14. M. G. Ridpath and R. E. Moreau, *Ibis* **108**, 348 (1966).
15. P. Hall, *Bull. Brit. Museum Zool.* **10**, 105 (1963).
16. E. Mayr, *Systematics and the Origin of Species* (Columbia Univ. Press, New York, 1942); *Animal Species and Evolution* (Harvard Univ. Press, Cambridge, 1963).
17. K. Knoch, in *Handbuch der Klimatologie*, H. Köppen and C. Geiger, Eds. (Borntraeger, Berlin, 1930), vol. 2, pp. 68–95; M. Velloso, in *Geografia do Brasil*, A. Teixeira, Ed. (Rio de Janeiro, 1959), vol. 1, pp. 61–111.
18. R. Reinke, "Das Klima Amazoniens," dissertation, University of Tübingen (1962).
19. J. A. Tosi, *Inst. Interamer. Cienc. Agr. OEA, Bol. Tec. No. 5* (1960), with ecological map of Peru.
20. C. Vaurie, *Bull. Amer. Museum Nat. Hist.* **138**, 131 (1968).
21. J. Haffer [*Amer. Museum Novitates No. 2294* (1967); *Auk* **84**, 343 (1967)] has given details on this refuge as well as data on southern Central America.
22. Bird species which may have originated in this refuge include *Mitu salvini, Nothocrax urumutum, Gymnopithys leucaspis, Grallaria (Thamnocharis) dignissima, Metopothrix aurantiacus, Ancistrops strigilatus, Porphyrolaema porphyrolaema, Heterocercus aurantiivertex,* and *Todirostrum capitale.*
23. Bird species which may have originated in these refuges include *Pithys castanea, Gymnopithys lunulata* and *G. salvini, Formicarius rufifrons, Grallaria (Thamnocharis) eludens, Conioptilon mcilhennyi, Pipra chloromeros, Todirostrum albifacies,* and *Rhegmatorhina melanosticta.*
24. Bird species which may have originated in this refuge include *Neomorphus squamiger, Pyrrhura rhodogaster, Dendrocolaptes hoffmannsi, Myrmotherula sclateri, Rhegmatorhina hoffmannsi, Phlegopsis (Skutchia) borbae, ?Heterocercus linteatus, Pipra nattereri, Pipra vilasboasi* (isolated forest east of the Tapajós river), *Todirostrum senex,* and *Idioptilon aenigma.*
25. Bird species which may have originated in this refuge include *Mitu tomentosa, Selenidera nattereri, Herpsilochmus dorsimaculatus, Myrmotherula ambigua, Myrmeciza disjuncta, Myrmeciza pelzelni, Percnostola caurensis, Rhegmatorhina cristata, Pipra cornuta, Heterocercus flavivertex,* and *Cyanocorax heilprini.*
26. Bird species which may have originated in this refuge include *Ortalis motmot, Brotogeris chrysopterus, Pionopsitta caica, Pteroglossus aracari, Pteroglossus viridis, Selenidera culik, Ramphastos t. tucanus, R. v. vitellinus, Celeus undatus, Hylexetastes perrotii, Myrmotherula guttata, M. gutturalis,*

Gymnopithys rufigula, Xipholaena punicea, Iodopleura fusca, Pachyramphus surinamensis, Haematoderus militaris, Perissocephalus tricolor, Pipra serena, Tyranneutes virescens, Microcochlearius josephinae, Phylloscartes virescens, Euphonia cayennensis, and *Phaethornis malaris.*
27. Bird species which may have originated in this refuge include *Ortalis superciliaris, Pyrrhura perlata, Xipholaena lamellipennis, Selenidera gouldi, Ramphastos vitellinus ariel, ?Pteroglossus bitorquatus, Pipra iris,* and *Gymnostinops bifasciatus.*
28. J. Haffer, *Hornero (Buenos Aires)* **10**, 315 (1967).
29. This historic interpretation of the occurrence of nonforest birds in lower Amazonia contrasts with the earlier explanation given by E. Snethlage [*Bol. Museu Goeldi* **6**, 226 (1910); *J. Ornithol.* **61**, 469 (1913); *ibid.* **78**, 58 (1930)], who assumed that the nonforest birds reached their present stations by following the river valleys. The following facts strongly support the interpretation of a natural rather than a secondary (man-made) origin of the isolated savannas of lower Amazonia. (i) The soil of the savannas is a bleached sand (podzol type) in contrast to the lateritic brown loamy soil of the forests. The forest soil could not have been replaced completely by podzol in the short period since the supposed artificial clearing by man [H. Sioli, *Erdkunde* **10**, 100 (1956)]. (ii) The flora of the isolated campos is decidedly nonhylean and is similar to that of the *cerrado* of central Brazil [A. Ducke and G. A. Black, *Anais Acad. Brasil. Cienc.* **25**, 1 (1953); *Bol. Tec. Inst. Agr. Norte Belém* **29**, 50 (1955); K. Hueck, *Die Wälder Südamerikas* (Fischer, Stuttgart, 1966), pp. 18, 21, 23]. (iii) The fauna of the isolated campos must be comparatively old, as a number of endemic forms are present. Examples are the mockingbird, *Mimus s. saturninus*; the grassland finch, *Coryphaspiza melanotis marajoara*; and the snakes *Bothrops marajoensis* and *Crotalus durissus marajoensis*, the latter being restricted to Isla Marajó [P. Müller, *Die Herpetofauna der Insel von São Sebastião (Brasilien)* (Saarbrücker Zeitung, Saarbrücken, 1968), pp. 60–61].
30. L. Smith, *U.S. Nat. Museum Contrib. U.S. Nat. Herbarium* **35**, 222 (1962); P. Müller, *Die Herpetofauna der Insel von São Sebastião (Brasilien)* (Saarbrücker Zeitung, Saarbrücken, 1968), pp. 60–61.
31. P. E. Vanzolini, *Arquiv. Zool. (São Paulo)* **17**, 105 (1968).
31a. Several examples of this interesting situation have been discussed by J. Haffer [*Amer. Museum Novitates No. 2294* (1967); *Auk* **84**, 343 (1967)] and by C. Vaurie (20).
32. This situation is probably much more common among Amazonian animals than is recognized. Examples are found among toucans and other forest birds.
33. According to this view the rivers merely keep the representative species (which originated in distant forest refuges) geographically separated. By contrast, H. Sick [*Atlas Simp. Biota Amazônica* (1967), vol. 5, p. 517] recently postulated that the ancestors of many Amazonian forest birds "must have lived at a time when the area was not yet divided by large rivers as it is today." He assumed that the rivers later acted as effective barriers and caused the differentiation of the representative species on opposite banks. The effect of the river barriers may be restricted to variation at the subspecies level.
34. M. G. Emsley, *Zoologica* **50**, 244 (1965). Contrary to Emsley's views, I believe that the differentiation of the *Heliconius* butterfly species may be related to the Quaternary climatic history of tropical South America rather than to the Tertiary paleogeographic history of this region.
35. E. E. Williams and P. E. Vanzolini, *Papéis Avulsos Dept. Zool. (São Paulo)* **19**, 203 (1963); ———, in *Simp. sobre o Cerrado* (Univ. of São Paulo, São Paulo, 1963), p. 307; ———, *Atlas Simp. Biota Amazônica* (1967), vol. 5, p. 85; P. Müller, *Die Herpetofauna der Insel von São Sebastião (Brasilien)* (Saarbrücker Zeitung, Saarbrücken, 1968). These authors emphasized the importance of vegetational changes in Amazonia for the most recent differentiation of the neotropical reptile fauna. Direct evidence of a rapid rate of speciation in Brazilian reptiles

136

has been discussed recently by P. E. Vanzolini and A. N. Ab'Saber, *Papéis Avulsos Dept. Zool. (São Paulo)* **21**, 205 (1968).

36. P. Hershkovitz, *Proc. U.S. Nat. Museum* **98**, 323 (1949); *ibid.* **103**, 465 (1954); in *Ectoparasites of Panamá*, R. L. Wenzel and V. J. Tipton, Eds. (Field Museum of Natural History, Chicago, 1966), pp. 725–751; *Evolution* **22**, 556 (1968). The distributional history of the monkeys, tapirs, and rodents discussed in these articles may well be interpreted on the basis of Quaternary climatic and vegetational changes, many species probably having originated during the Pleistocene. B. Patterson and R. Pascual

[*Quart. Rev. Biol.* **43**, 440 (1968)] also assumed that a rapid differentiation at the species level, in some cases to the generic level, took place in South American mammals, particularly the rodents, during the Pleistocene.

37. G. de Lattin [*Grundriss der Zoogeographie* (Fischer, Stuttgart, 1967), pp. 327–329] summarized the Pleistocene history of the north temperate faunas.

38. K. Hueck, *Die Wälder Südamerikas* (Fischer, Stuttgart, 1966).

39. J. Haffer, *Amer. Museum Novitates No. 2294* (1967); *Auk* **84**, 343 (1967).

40. A. Aubréville, *Etude écologique des principales formations végétales du Brasil et contribution à la connaissance des forets de l'Amazonie brésilienne* (Centre Technique Forestier Tropicale, Nogent-sur-Marne, France, 1961), pp. 1–265.

41. W. M. Denevan, *Ibero Americana* **48**, 7 (1966).

42. I am grateful to Professor Ernst Mayr, Harvard University, for many helpful suggestions concerning the manuscript of this article. I also thank Eugene Eisenmann, American Museum of Natural History, New York, and Dr. François Vuilleumier, University of Massachusetts, Boston, for critical remarks on an earlier version.

Biosynthesis of Oligosaccharides and Polysaccharides in Plants

Mechanisms of enzymic synthesis of complex plant carbohydrates are reviewed.

W. Z. Hassid

Plants are the chief producers of carbohydrates in nature by the process of photosynthesis. Most forms of life which are unable to photosynthesize depend either directly or indirectly on the assimilation of carbon dioxide by plants. All the organic substances which arise from photosynthetic processes serve the other forms of life as starting materials for diverse metabolic functions. While there are many ways in which organic substances are decomposed, there is only one reaction, photosynthesis, which for millions of years has counterbalanced death and decomposition.

Monosaccharides are synthesized by green plants, starting with a carboxylation reaction in which D-ribulose-1,5-diphosphate serves as the acceptor of CO_2 for the formation of phosphoglyceric acid (1). By a subsequent series of enzymic reactions, a number of phosphorylated monosaccharide derivatives are produced in the photosynthetic carbon dioxide cycle. Some of these phosphorylated sugars, such as D-glucose-6-phosphate and D-fructose-6-phosphate, are hydrolyzed to free sugars, causing in some cases the accumulation of large concentrations of D-glucose and D-fructose in plants.

The phosphorylated monosaccharides produced in the photosynthetic carbon dioxide cycle are partially consumed in respiration with the production of energy which is utilized for the numerous metabolic reactions of the plants. They are also converted by a series of enzymic reactions to sugar nucleotides, chiefly UDP-D-glucose (2), and to other sugar nucleotides, such as UDP-D-galactose, GDP-D-glucose, and ADP-D-glucose (3). The sugar moieties of these nucleotides are interconverted by various specific epimerases, and serve as donors for the formation of the numerous oligosaccharides and polysaccharides (4).

A monosaccharide must be activated to enable the enzyme to transfer it to an acceptor for the synthesis of an oligosaccharide or to lengthen the chain by subsequent transfers for the formation of a polymer. From the thermodynamic point of view, nucleoside diphosphate sugars are superior donors for complex saccharide formation, because they have the higher negative free energy of hydrolysis ($\Delta F°$) than other glycosyl compounds (5). Uridine diphosphate-D-glucose has a relatively high negative $\Delta F°$ of hydrolysis of -7600 calories; although it has never been determined for other sugar nucleotides, it is assumed that the nucleoside

diphosphate sugars, containing bases other than uracil or sugars other than D-glucose, have approximately the same $\Delta F°$ values. The most important reaction for complex saccharide formation appears to involve sugar nucleotides.

Oligosaccharides

Sucrose, the most abundant oligosaccharide in higher plants, was first synthesized in vitro by a bacterial enzyme obtained from *Pseudomonas saccharophila* from α-D-glucose-1-phosphate and D-fructose (6). However, in this reaction the equilibrium favors the breakdown rather than the synthesis of sucrose. Various attempts by a number of investigators to synthesize sucrose by an enzyme from a plant source from α-D-glucose-1-phosphate and D-fructose failed. Leloir and his collaborators (7) later found that the donor of D-glucose for sucrose formation was not α-D-glucose-1-phosphate but the sugar nucleotide UDP-D-glucose. The synthesis takes place by two separate enzymes, one utilizing D-fructose and another D-fructose-6-phosphate as the acceptors according to the following two reactions

$$\text{UDP-D-glucose} + \text{D-fructose} \rightleftharpoons \text{sucrose} + \text{UDP} \quad (1)$$

and

$$\text{UDP-D-glucose} + \text{D-fructose-6-phosphate} \rightleftharpoons \text{sucrose phosphate} + \text{UDP} \quad (2)$$

The sucrose phosphate formed in reaction 2 is hydrolyzed by a phosphatase, resulting in the formation of free sucrose.

Since $\Delta F°$ for hydrolysis of UDP-D-glucose is about -7500 calories per mole, formation of the glycosyl bond of sucrose is favored. For most glycosides, synthesis from a nucleotide sugar precursor would proceed with a favorable free-energy change of about

The author is a member of the Department of Biochemistry, University of California, Berkeley 94720.

SYMPATRIC HOST RACE FORMATION AND SPECIATION IN FRUGIVOROUS FLIES OF THE GENUS *RHAGOLETIS* (DIPTERA, TEPHRITIDAE)

Guy L. Bush

Department of Zoology, University of Texas, Austin, Texas

Received July 10, 1968

The origin and evolution of species and host races[1] in certain phytophagous insect groups have long been a source of disagreement among evolutionary biologists. The rapid establishment of new host races by some stenophagous insects on introduced plants, as well as various other aspects of their biology and distribution, has led several biologists to suggest that new host races and species may arise sympatrically (Brues, 1924; Thorpe, 1930; Smith, 1941; Haldane, 1959; Alexander and Bigelow, 1960; Bush, 1966). Others regard geographic isolation as a prerequisite for speciation in all groups of sexually reproducing animals (for examples see Mayr, 1963).

The difficulty in resolving whether either one or the other, or both, modes of speciation may occur in these insects stems directly from the paucity of detailed studies of wild stenophagous insects. Our information on these insects has been derived primarily from studies made over the past hundred years on a few economically important plant pests. One such group that has received particular attention is the Holarctic and Neotropical genus *Rhagoletis* whose larvae feed within the developing fruits of many plant species such as cherries, blueberries, apples, currants, rose hips, walnuts, and tomatoes.

---

[1] For the purpose of this paper I restrict the term host race to a population of a species living on and showing a preference for a host which is different from the host or hosts of other populations of the same species. Host races represent a continuum between forms which freely interbreed to those that rarely exchange genes. The latter may approach the status of a species generally regarded as an interbreeding population reproductively isolated from all other such populations (Mayr, 1963).

The objective of this paper is to point out how the biological attributes of these flies may have permitted new forms to arise rapidly in the absence of geographical barriers to gene flow.

General Biological Characteristics of *Rhagoletis*

Before proceeding with a discussion of host race formation and speciation in *Rhagoletis*, it will be necessary to outline briefly certain pertinent biological characteristics of these flies and the unique differences which exist between the three species-groups discussed in this paper. A detailed analysis has been presented elsewhere (Bush, 1966).

Courtship behavior.—The bodies of most Tephritidae, including all species of *Rhagoletis*, are ornamented with brightly contrasting color patterns and their wings usually bear elaborate, frequently species specific, and in some cases sexually dimorphic patterns. Féron (1962), Tauber and Toschi (1965), Bush (1966), and others have shown that these distinctive body and wing patterns serve as visual releasers in courtship and agonistic displays. However, visual releasers are effective only at close range, and they are used after the flies have congregated on their host fruits which serve as the rendezvous for courtship, mating and later oviposition. Male *Rhagoletis* are territorial and set up and defend a territory on a single host fruit while awaiting the arrival of a female (Bush, 1966). Hence, host and mate selection are directly correlated. This feature has important implications in the sympatric host race formation of these flies which will be discussed later.

TABLE 1. *The hosts of the* pomonella, cingulata, *and* suavis *species groups.*

Species Group	Host Plant Family	Adults Reared From	No. Plant Spp. Infested
POMONELLA GROUP			
R. pomonella	Rosaceae	*Crataegus*	7
		Pyrus	3
		Cotoneaster	1
		Prunus	2
R. mendax	Ericacae	*Vaccinium*	9
		Gaylussacia	3
R. cornivora	Cornaceae	*Cornus*	3
R. zephyria	Caprifoliaceae	*Symphoricarpus*	1
CINGULATA GROUP			
R. cingulata	Rosaceae	*Prunus*	5
R. indifferens	Rosaceae	*Prunus*	5
R. osmanthi	Oleaceae	*Osmanthus*	1
R. chionanthi	Oleaceae	*Chionanthus*	1
SUAVIS GROUP			
R. suavis			
R. completa			
R. juglandis	Juglandaceae	*Juglans*	8
R. boycei			
R. zoqui			

Host selection.—*Rhagoletis* species, although not monophagous, have fairly narrow host preferences and restrict their attacks to plant species within a genus or closely related genera. No species whose biology is well documented infests the fruits of more than one plant family. However, the factors related to host selection in this genus and other members of the family are poorly understood. The studies by Currie (1932) on *Euaresta aequalis* Loew and by Féron (1962) on *Ceratitis capitata* strongly suggest that adults of these species initially locate their hosts by means of olfactory cues emanating from the plants. These authors, as well as Prokopy (1969) in his recent work on *R. pomonella*, have shown that once contact with the host has been made, visual and tactile cues become important in host recognition.

Diapause and emergence.—Larvae pupate about 2 to 5 inches below the soil surface where they pass the winter and complete development in the spring. Temperate zone species of *Rhagoletis* are usually univoltine with 60 to 90% of the adults emerging over a span of 2 to 4 weeks. This period corresponds with the maximum availability of host fruits in a suitable condition for oviposition. Diapause, although facultative under certain conditions, is normally broken by a period of low temperature (i.e., 40 F for 3 months). Some pupae may require winter chillings over 2 to 5 successive years before diapause is terminated.

Life span and mobility.—Under controlled laboratory conditions, *R. pomonella* adults may live an average of 60 to 70 days (Prokopy, 1968). However, it is doubtful that this species or other temperate climate *Rhagoletis* can survive for more than 20 to 30 days under natural field conditions (Porter, 1928). During this period they may travel over a mile (Barnes, 1959), while marked individuals of other tephritid species such as *Anastrepha ludens*, *Dacus dorsalis*, and *Ceratitis capitata* have frequently been found up to 15–20 miles from their release point (Christenson and Foote, 1960; Shaw et al., 1967). In addition, Prokopy (1969) has shown that *R. pomonella* females usually do not move to their host plant until they are ready to mate and

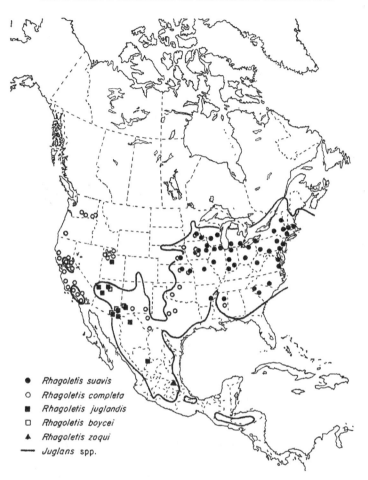

● Rhagoletis suavis
○ Rhagoletis completa
■ Rhagoletis juglandis
□ Rhagoletis boycei
▲ Rhagoletis zoqui
— Juglans spp.

FIG. 1. Distribution of the *suavis* species group and native species of its host plant genus *Juglans*. Cultivated species of *Juglans* infested by *R. completa* in California, Oregon, Washington and Utah have not been included. *R. completa* was introduced to California in about 1922.

oviposit. Thus, their movements are not restricted to their host plants except during the critical periods of oviposition and mate selection. The evidence therefore suggests that only major geographic or ecological barriers would be effective in limiting the distribution and dispersal of these flies.

BIOLOGICAL CHARACTERISTICS OF SPECIES GROUPS

Rhagoletis suavis *species group.*—Four of the five species in this North American species group have overlapping, parapatric,

or sympatric ranges with at least one other species in the group (Fig. 1). All infest the husks of any native or introduced species of walnut (*Juglans*) (Table 1), and thus appear to have similar if not identical host preferences. One species, *R. zoqui*, is known from only a single locality, but it is probably more widespread as walnuts are common throughout Mexico.

On the basis of genitalic structure, general habitus, karyotypes, and host preference, members of the *suavis* group form a rather distinct, closely related complex of

240   GUY L. BUSH

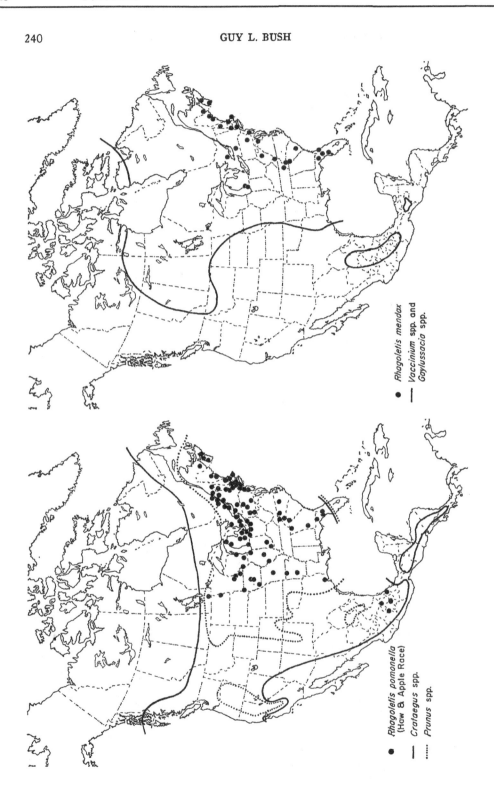

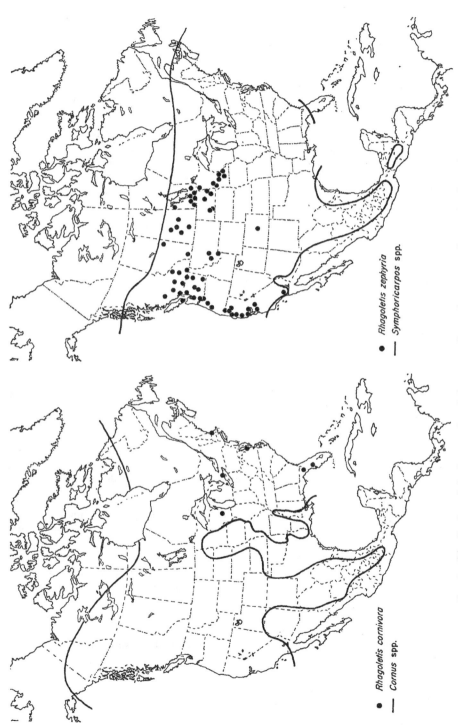

FIG. 2. Distribution of the four species of the *pomonella* species group and respective host plants.

- Rhagoletis zephyria
- Symphoricarpos spp.

- Rhagoletis cornivora
- Cornus spp.

242   GUY L. BUSH

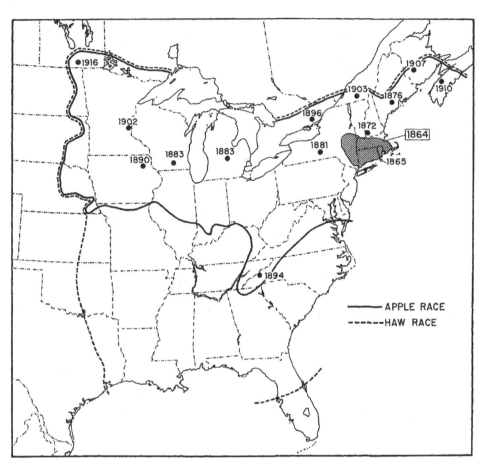

Fig. 3. Initial area of infestation, dispersal and current distribution of the apple and hawthorn races of *Rhagoletis pomonella*. Dates represent earliest known infestation in the state.

species within the genus *Rhagoletis* (Bush, 1966). However, they differ conspicuously from one another in their wing markings and by the fact that three species, *suavis*, *completa*, and *zoqui*, are sexually dimorphic for black and brown maculations on the thorax and abdomen. In the case of the remaining two species, which are broadly sympatric, *juglandis* is entirely yellow and *boycei* is completely black. These two species also have different emergence periods and different but somewhat overlapping altitudinal distributions (Bush, 1966).

Rhagoletis pomonella *species group.*— This group of four sibling species is restricted to the Nearctic region (Fig. 2) and

has no close relatives within the genus (Bush, 1966). The most striking feature of the *pomonella* group is the ecological diversity of its species. In contrast to the walnut-infesting *suavis* group which infests only a single host plant genus, speciation in the *pomonella* group has been accompanied by a shift to a radically new host family in every case (Table 1).

Until recently it was difficult to distinguish one species from another on morphological grounds alone, and most authors considered the four forms as host races of a single species. However, hybridization, oviposition choice experiments, and ecological studies carried out by four independent

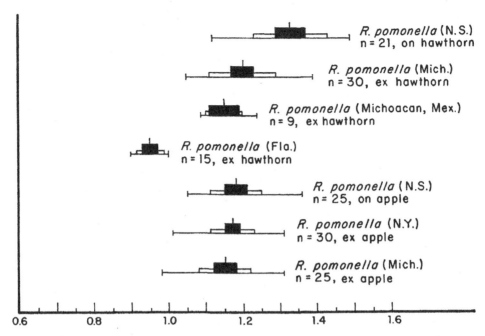

Fig. 4. Intraspecific variation in ovipositor length of the apple and hawthorn races of *Rhagoletis pomonella*. Horizontal line represents observed range; clear rectangular mark represents standard deviation; solid black rectangle indicates 95% confidence interval for the mean (vertical line).

groups in the 1930's (Lathrop and Nichols, 1931; McAlister and Anderson, 1935; Pickett, 1937; Pickett and Neary, 1940; Hall, 1938, 1943), and recent comparative serology studies (Simon, 1969) leave no doubt that the three sympatric eastern forms, *pomonella*, *mendax*, and *cornivora*, are reproductively isolated from one another and possess quite different biological characteristics. Although similar studies have not been made between the primarily western *zephyria* and the three eastern species, there is strong evidence that this form is also quite distinct; morphologically it is the most divergent of the four species.

Of particular interest in this species group is a new host race of *R. pomonella* which was established on introduced apples a little over 100 years ago from the original hawthorn infesting form. The first reports of *R. pomonella* attacking apples came from farmers in the Hudson River Valley, and from this region it spread rapidly to adjacent orchards in Massachusetts and Connecticut. The cross-shading in Figure 3 shows the approximate area of the original infestation as it was first noted in 1865 by Ward (1866, vide Illingworth, 1912). The exact time at which the first apple infestation occurred is not known. However, the widespread common occurrence of hawthorn and apples in the area, the economic importance of this fly, and its well documented mobility would preclude any lengthy period of isolation. It should be pointed out that native crab apples are not infested (possibly because they mature too late in the season and lack suitable nutritive qualities), and have been ruled out as the source of the apple race (Illingworth, 1912; O'Kane, 1914; Porter, 1928; Bush, 1966).

Today certain sympatric populations of the apple and hawthorn races have slight but significant differences in relative body size, number of postorbital bristles, and

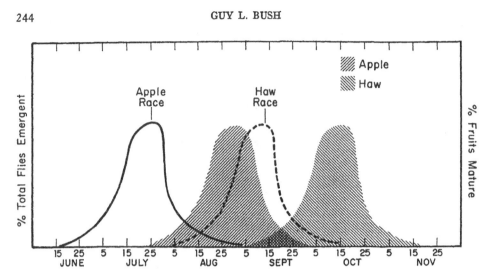

FIG. 5. Emergence period of the apple and hawthorn races of *Rhagoletis pomonella*. The cross hatched curve represents the approximate period of fruit maturation when larvae are leaving the fruit and pupating in the soil.

ovipositor length (Bush, 1966, and unpublished data). There are, for instance, no statistically significant differences in the ovipositor lengths between populations of the apple race over much of their range, yet quite significant differences exist between populations of the hawthorn race from various localities (Fig. 4). Thus, in the more peripheral regions, such as Nova Scotia, there is a marked and significant difference between sympatric populations of the two races. This is to be expected if the apple race is of recent origin and arose and spread from a small local population in the Hudson River Valley.

Another striking difference is apparent in the emergence times of these two races as first pointed out by Pickett and Neary, (1940). Both races emerge from the pupal stage at a time when their respective host fruits are in suitable condition for oviposition (Fig. 5). Accurate and well documented surveys have established the emergence period of the apple race in many areas of eastern North America, but little work has been done on the original hawthorn race. Data taken from reared specimens, reports in the literature, and an examination of herbarium specimens of *Crataegus*,

as well as my own field observations, indicate that the apple race emerges and begins oviposition approximately 4 to 5 weeks before the hawthorn race. The two populations are, therefore, allochronically isolated from one another, but the exact extent of overlap in emergence for any given area and how much, if any, gene flow occurs between them is yet to be established.

Another allochronically isolated race of *R. pomonella* also occurs in eastern North America on native and cultivated plums which ripen considerably earlier than either apples or haws, but little is known concerning its biology. It may prove to be yet another long established sibling species.

Rhagoletis cingulata *species group*.—The four sibling species in this group are strictly Nearctic in distribution. Two species, *R. cingulata* and *R. indifferens*, infest native and introduced cherries (Table 1) and are allopatrically isolated from one another in the eastern and western parts of North America (Fig. 6). Like *zephyria* in the *pomonella* group, western *indifferens* upon close examination is quite distinct from the three eastern forms which resemble each other closely. The other two species,

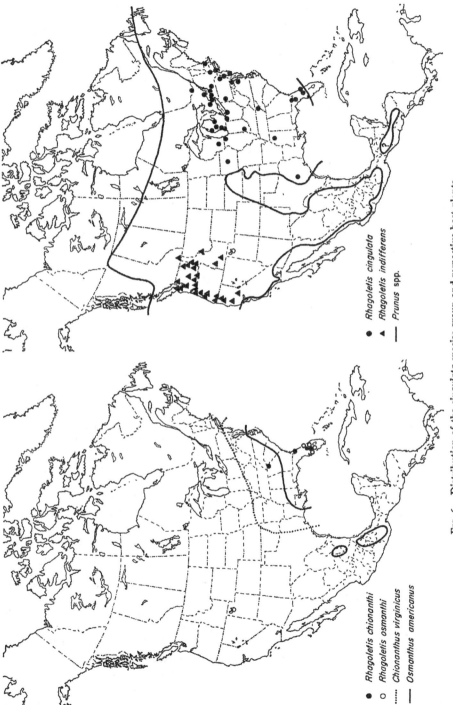

Fig. 6.   Distribution of the *cingulata* species group and respective host plants.

● *Rhagoletis cingulata*
▲ *Rhagoletis indifferens*
— *Prunus* spp.

● *Rhagoletis chionanthi*
○ *Rhagoletis osmanthi*
⋯ *Chionanthus virginicus*
— *Osmanthus americanus*

R. chionanthi and *R. osmanthi*, which are sympatric with *R. cingulata* in the Southeast (Fig. 6), infest the native olives, *Chionanthus virginicus* and *Osmanthus americanus* respectively (Table 1).

The principal native host of *R. indifferens* is *Prunus emarginata* (pin cherry). This fly was found infesting cultivated cherries (*P. avium* and others) in Oregon about 80 years ago (Jones, 1963, *in litt.*) soon after the first seedlings were introduced into the Northwest. Today there appear to be two races, one on *P. emarginata* whose fruits mature from late July to early September, and the other on cultivated cherries which mature in late June and July. There is about a two week overlap between the late maturing varieties of cultivated cherries and early maturing pin cherry.

California is the only area where *R. indifferens* still restricts its attacks to *P. emarginata*. A commercial cherry orchard in the Mount Shasta area, for example, may be completely surrounded by pin cherry and remain free from attack by this fly. Occasionally, however, late maturing cultivated cherries within a single orchard growing in the vicinity of infested *P. emarginata* may become infested. The California State Department of Agriculture systematically eradicates these newly formed and highly localized cultivated cherry fruit fly populations. This approach has thus far precluded the establishment of a population on commercial cherries in California.

There is another aspect of the California cherry fruit fly problem that must be mentioned. Domestic cherries in California for the most part are grown at considerably lower altitudes than the native pin cherry, but a variable altitudinal zone of overlap of aboutt 1000 feet exists between the native and cultivated cherries (Robinson, 1966, *in litt.*). The fruits in the majority of cultivated cherry orchards, therefore, are not only semichronically but also semigeographically isolated from those of *P. emarginata*.

Allochronic isolation between the two sympatric olive infesting species is even more pronounced. During a federal and state eradication campaign conducted against the Mediterranean fruit fly (*Ceratitis capitata*) in Florida in the late 1920's, extensive collections and rearings were made of many Tephritidae including the two native olive infesting *Rhagoletis* species. These records, compiled by Nicholson (1929–30), show that there is only one generation of each species a year. *Rhagoletis chionanthi* emerges in the summer when its host fruit is abundant, while *R. osmanthi* emerges during the winter months (Nov.–Feb.) when the fruit of its host plant matures.

Evolutionary Interpretation

Rhagoletis suavis species group.— McVaugh (1952), Manning (1957), and others have reported that there has been a continuous shift in the distribution of *Juglans* in southwestern United States and Mexico in response to climatic changes during the late Tertiary and particularly throughout the Pleistocene. Such shifts could have furnished ample opportunity for *Juglans*-infesting populations of the *suavis* group to become isolated and diverge genetically in response to local selection pressures. It is likely that once contact was established between previously isolated walnut fruit fly populations, intensive competition would be inevitable. Under these conditions selection would probably have favored adaptations which increased the frequency of homogamic matings and reduced competition.

The observed well-developed differences in characters associated with visual releasers involved in courtship are, therefore, of the type one would consider most likely to arise during the course of speciation in this species group. Thus, there appears to be little doubt that members of this group have speciated allopatrically.

Rhagoletis pomonella species group.— The suggestion has been made that few species of herbivorous insects are truly monophagous, and that at one time or another most will have a secondary host somewhere within their range. Mayr (1963) has proposed that a population may become adapted to the secondary host in an isolated locality when climatic or other changes lead to the extinction of the primary host. Such a shift could conceivably occur quite rapidly in temporarily isolated peripheral populations even before the primary host is entirely extinct. This form of geographic speciation could account for the three sympatric eastern sibling species in the *pomonella* group as well as the primarily allopatric sibling species pairs, such as *pomonella* and *zephyria*, which are associated with different host plants in eastern and western North America.

Although some form of allopatric speciation can be invoked to explain the origin of all the currently sympatric sibling species in this genus, other modes of speciation appear to be more plausible when the known biological characteristics of this group of flies are considered. The recent cases of host race formation in this genus under apparently sympatric conditions also support this conclusion.

I would therefore like to propose a model to account for the rapid formation of the apple race of *R. pomonella* and for the evolution of other sibling species in this species group. This model is based on the following assumptions:

1) Diapause and emergence times are ultimately under genetic control.

2) Initial orientation to and selection of a host plant is in response to a chemical cue.

3) Host selection has a genetic basis. In this case homozygous *AA* and heterozygous *Aa* individuals move preferentially to haws while homozygous *aa* flies move to apples.

4) *A* mutates to *a* at a locality where apples are available and a few homozygous *aa* individuals are eventually produced as a result of recombination.

5) Host plant and mate selection are positively correlated. Individuals move preferentially to their respective host plants depending on their genotype, and mating occurs on the host plant.

The initial stages of host selection may involve a response to a chemical cue, such as a terpene or other aromatic component produced by the plant. A minor change in the olfactory system of the fly could result in the recognition of a related but quite distinct compound produced by another species of plant. If the fruit of this plant has suitable nutritive qualities for larval development, and the plant furnishes the appropriate tactile and visual stimuli for oviposition, a small population could become established in a single generation.

The populations on apples may have been founded initially by only a few early emerging flies with *aa* genotypes which oviposited on late maturing apples. The initial infestation was probably followed by a rapid rise in population density accompanied by a spread of the apple race to other areas occupied by the new host. A negligible amount of introgression might have occurred each generation at low frequency within the area where the *a* allele originated as a result of *aa* individuals arising by recombination in the original hawthorn race. The effect of introgression would be diminished, however, as the apple race increased and spread.

There would also be a concurrent shift in the mean emergence time of the apple race to correspond with the period of optimum availability of the new host fruit, as flies with early emerging genotypes would find a greater number of suitable oviposition sites. Within a few generations two populations could become established, one adapted to the original hawthorn host, and a new early emerging allochronically or semichronically isolated population infesting apples.

The evolutionary future of these host races would depend on the interrelationship between gene flow and selection. If selection can successfully eliminate introgressing genes between the two populations, then eventually the races may diverge genetically to a point where they are completely reproductively isolated from one another. Otherwise, a stable *polytrophic* species might be established in which heterozygous individuals are maintained either with or without heterozygote superiority, depending on the intensity and type of selection (i.e., stabilizing or disruptive) (Smith, 1966).

Rhagoletis cingulata species group.—A more contemporary example of sympatric host race formation in *Rhagoletis* is to be found in certain populations of the western cherry fruit fly, *R. indifferens* Curran.

The details concerning the development of a host race on domestic cherries from the wild western cherry population are unknown. However, the recent appearance of new populations on domestic cherries in California where only the wild cherry is infested offers some evidence in support of a sympatric origin.

As previously mentioned, cultivated cherries mature earlier and for the most part grow at lower altitudes than wild cherries. Thus, the two are semichronically and semigeographically isolated. The same model proposed for the origin of the apple race of *R. pomonella* could equally well apply to *R. indifferens*. The newly established population on cultivated cherry could move down the mountain to lower altitudes, adjusting its emergence period to an earlier date to coincide with the time of optimum host availability. Some gene flow might continue to occur in the zone of overlap between early emerging wild cherry flies and late emerging cultivated cherry flies, but, as in the case of the apple and hawthorn races of *R. pomonella*, this would probably be negligible once large populations of the cultivated cherry fly became well established and semichronically and semigeographically isolated.

There are three alternative explanations for the origin of the two olive infesting species in this species group.

1) The periods of fruit maturation of sympatric *Osmanthus* and *Chionanthus*

may have overlapped considerably, or perhaps were synchronous in the past, and both were infested by the same species of fly. The fruiting time of *Osmanthus* may then have shifted in response to climatic changes, possibly occurring during the Pleistocene, and in the process slowly split the original olive infesting species into two distinct, allochronically isolated populations.

2) Alternatively, the establishment of a new host race on *Osmanthus* from a population infesting *Chionanthus* may have occurred after the fruiting time of the two host plants had diverged, following the pattern already discussed for the apple race of *R. pomonella* and the cherry fruit fly race of *R. indifferens*.

3) A third and more complicated model involves geographic isolation of the new host plants. Originally, the fruiting times of the two host plants may have been synchronous or broadly overlapping, and the fruits infested by a single species of *Rhagoletis*. At a later stage, each host plant and its population of flies became geographically isolated. This geographic isolation was accompanied by a shift in the fruiting period of *Osmanthus*. Once contact was reestablished, the two populations, although sympatric, have become allochronically isolated from one another.

FACTORS ENHANCING ISOLATION AND REDUCING GENE FLOW

There are several genetic and nongenetic factors that could further enhance the degree of isolation and reduce gene flow between two recently established sympatric host races.

Disruptive selection.—Levene (1953) has shown that when two or more niches are available, two or more alleles can be maintained in equilibrium without the heterozygote being superior to either of the homozygotes in any single niche. Under these conditions, disruptive rather than stabilizing selection (i.e., selection favoring extremes rather than average phenotypes) maintains the heterozygote in equilibrium. Levene's model does not take into account the possibility of individuals moving preferentially to niches for which they are best fitted, or that mating may occur within a niche.

These factors have been considered by Smith (1966) who not only confirmed Levene's model, but demonstrated that the establishment of a stable polymorphism without heterozygote superiority accompanied by disruptive selection on genes associated with habitat selection, pleiotropism, assortative mating and modifier genes, could lead to a drastic reduction or complete elimination of gene flow between extreme genotypes without geographic isolation.

Conditioning.—Another possibly important factor which has been given little attention in the past with respect to the evolution of host races and speciation is conditioning, where the animal learns or becomes accustomed to a particular host stimulus. Conditioning may in some instances further reduce gene flow between newly established host races and enhance to some degree the effects of allochronic isolation and disruptive selection. Recently Manning (1967), as well as Hershberger and Smith (1967), confirmed Thorpe's (1939) earlier findings that conditioning occurs in *Drosophila*. Host plant conditioning in the larvae of the tobacco worm, *Manduca sexta*, and the corn ear worm *Heliothus zea* (Jermy et al., 1968) has also been demonstrated. It is regretable that so little work of a similar nature has been done on phytophagous insects with narrow host requirements.

Magnitude of host shift.—A final consideration must also be given to the "magnitude" of the shift from one host to another. Shifts to closely related species of plants whose biochemistry is undoubtedly similar will not initially require a radical change in the genetic makeup of the new host race from that of the parental population. This type of host race formation can be expected to occur frequently among phytophagous insects and also can be the most ephemeral and least likely to develop

successfully into completely reproductively isolated species. The allochronic nature of the fruiting or flowering times of their host plants, conditioning, disruptive selection or some other ecological factors, either singly or in combination, may be the major barriers to gene flow which permit these populations to acquire and maintain distinct racial qualities in all or part of their ranges under sympatric conditions. A change in agricultural practices or climatic conditions could rapidly break down these barriers.

Host race formation resulting from shifts to radically different hosts, such as between plant subfamilies or families, is likely to occur much less frequently. The new host plant may differ considerably in nutritive, tactile, olfactory, and gustatory qualities from the original host. These differences could require a much more extensive alteration of the overall gene pool of the parental population if it is to adapt to the new host plant. Once established, such races can be expected to develop quite distinct gene pools rapidly in response to the more intense selection pressure brought about by different environmental conditions.

The evidence suggests, therefore, that slight alterations in genes involved in host plant selection of a few individuals within a population, coupled with the effects of disruptive selection and conditioning, could rapidly lead to the sympatric formation of new host races and species.

Although allopatric models of speciation can always be invoked to explain the origin of sympatric host races and sibling species of phytophagous insects, alternative and often simpler possibilities, such as models involving allochronic isolation and disruptive selection, seem feasible and warrant more careful consideration than they have been given in the past. We actually know far too little about the population biology, gene flow, or selection pressures acting on the six or seven hundred thousand plant-feeding and parasitic insects to make a blanket statement that in every case they have speciated only after periods of geo-

250   GUY L. BUSH

graphic isolation. The current tendency to discuss speciation only within the framework of complete geographic isolation may cloud the true picture of the factors acting to reduce gene flow during the course of divergence between natural populations in some groups.

Summary

Courtship and mating in univoltine frugivorous *Rhagoletis* species occur on the larval host plant. Thus, there is a direct correlation between mate and host selection. This characteristic, coupled with other biological attributes of the genus and evidence provided from studies on recently established host races, suggests that some members of certain groups of sibling species may have evolved sympatrically as a result of minor alterations in genes associated with host plant selection.

Other factors such as allochronic isolation on unrelated plants with different fruiting times, disruptive selection, conditioning, and semigeographic isolation, which might enhance the reproductive isolation between a recently established host race and its parent population, are discussed. It is concluded that such factors may considerably reduce gene flow between host races and lead to the rapid sympatric evolution of host races and sibling species.

Acknowledgments

I am indebted to Professors Paul R. Ehrlich, Howard E. Evans, and Daniel H. Janzen for their advice and criticisms of the concepts and examples discussed in this manuscript. This work was partially supported by a National Science Foundation Fellowship and by a grant from the National Institutes of General Medical Sciences GM 14600 and 15769.

Literature Cited

ALEXANDER, R. D., AND R. S. BIGELOW. 1960. Allochronic speciation in field crickets, and a new species *Acheta veletis*. Evolution 14: 334–346.

BARNES, M. M. 1959. Attractants for the walnut husk fly. J. Econ. Entomol. 51:686–689.

BRUES, C. T. 1924. The specificity of food plants in the evolution of phytophagous insects. Amer. Natur. 58:127–144.

BUSH, G. L. 1966. The taxonomy, cytology, and evolution of the genus *Rhagoletis* in North America (Diptera, Tephritidae). Bull. Mus. Comp. Zool. 134:431–562.

CHRISTENSON, L. D., AND R. H. FOOTE. 1960. Biology of fruit flies. Ann. Rev. Entomol. 5:171–192.

CURRIE, G. A. 1932. Oviposition stimuli of the burr-seed fly, *Euaresta aequalis* Loew, (Diptera: Trypetidae). Bull. Entomol. Res. 23: 191–203.

FÉRON, M. 1962. Le comportement de reproduction chez la mouche Méditerranéenne des fruit *Ceratitis capitata* Wied. (Diptera: Trypetidae). Comportement sexuel-comportment de ponte. University of Paris. Ph.D. Thesis. (Series A, No. 3868). No. d'ordre: 4719.

HALDANE, J. B. S. 1959. Natural selection, p. 101–149. *In*: Darwin's biological work, P. R. Bell, Ed., Cambridge Univ. Press, 343 p.

HALL, J. A. 1938. Further observations on the biology of the apple maggot, *Rhagoletis pomonella* (Walsh). Rep. Entomol. Soc. Ontario 69:53–58.

HALL, J. A. 1943. Notes on the dogwood fly, a race of *Rhagoletis pomonella* (Walsh). Canadian Entomol. 75:202.

HERSHBERGER, W. A., AND M. P. SMITH. 1967. Conditioning in *Drosophila melanogaster*. Anim. Behav. 15:259–262.

ILLINGWORTH, J. F. 1912. A study of the biology of the apple maggot (*Rhagoletis pomonella*), together with an investigation of methods of control. Cornell Univ. Agr. Exp. Sta. Bull. 324:129–187.

JERMY, T., F. E. HANSON, AND V. G. DETHIER. 1968. Induction of specific food preferences in lepidopterous larvae. Entomol. Exp. Appl. 11:203–211.

LATHROP, F. H., AND C. B. NICHOLS. 1931. The blueberry maggot from an ecological viewpoint. Ann. Amer. Entomol. Soc. 24:260–274.

LEVENE, H. 1953. Genetic equilibrium when more than one ecological niche is available. Amer. Natur. 87:331–333.

MANNING, A. 1967. Pre-imaginal conditioning in *Drosophila*. Nature 216:338–340.

MANNING, W. E. 1957. The genus *Juglans* in Mexico and Central America. J. Arnold Arb. 38:121–150.

MAYR, E. 1963. Animal species and evolution. Harvard Univ. Press. Cambridge, Massachusetts, 797 p.

McALISTER, L. C., AND W. H. ANDERSON. 1935. Insectary studies on the longevity and pre-oviposition period of the blueberry maggot and on cross-breeding with the apple maggot. J. Econ. Entomol. 28:675–678.

McVAUGH, R. 1952. Suggested phylogeny of

Prunus serotina and other wide ranging phylads in North America. Brittonia 7:317–346.

NICHOLSON, D. J. 1929–30. Report upon native *Trypetidae* of Florida. U. S. Dept. Agr. Bur. Entomol. (unpublished report).

O'KANE, W. C. 1914. The apple maggot. New Hampshire Exp. Sta. Bull., 171. 120 p.

PICKETT, A. D. 1937. Studies on the genus *Rhagoletis* (Trypetidae) with special reference to *Rhagoletis pomonella* (Walsh). Canadian J. Res. (D), 15:53–75.

PICKETT, A. D., AND M. E. NEARY. 1940. Further studies on *Rhagoletis pomonella* (Walsh). Sci. Agr. 20:551–556.

PORTER, B. A. 1928. The apple maggot. U. S. Dept. Agr. Tech. Bull. 66. 48 p.

PROKOPY, R. J. 1968. Influence of photoperiod, temperature and food on initiation of diapause in the apple maggot. Canadian Entomol. 100: 318–329.

PROKOPY, R. J. 1969. Visual responses of apple maggot flies, *Rhagoletis pomonella* (Walsh) (Diptera: Tephritidae): Orchard studies. Entomologia. (in press).

SHAW, J. C., M. SANCHEZ-RIVIELLO, L. M. SPESHAKOFF, G. TRUJILLO, AND F. LOPEZ D. 1967. Dispersal and migration of Tepa-sterilized Mexican fruit flies. J. Econ. Entomol. 60:992–994.

SIMON, J. P. 1969. Biochemical systematics of the *Rhagoletis pomonella* species group (Diptera, Tephritidae): Comparative serology using pupal antigens. Systematic Zoology (in press).

SMITH, H. S. 1941. Racial segregation in insect populations and its significance in applied entomology. J. Econ. Entomol. 34:1–12.

SMITH, J. M. 1966. Sympatric speciation. Amer. Natur. 100:637–650.

TAUBER, M. J., AND C. A. TOSCHI. 1965. Bionomics of *Euleia fratria* (Loew) (Diptera: Tephritidae). 1. Life history and mating behavior. Canadian J. Zool. 43:369–379.

THORPE, W. H. 1930. Biological races in insects and allied groups. Biol. Rev. 5:177–212.

THORPE, W. H. 1939. Further studies on preimaginal olfactory conditioning in insects. Roy. Soc. (London), Proc., B., 127:424–433.

6   The Importance of Islands

Robert J. Whittaker

Islands as Natural Laboratories

Biogeographers have often turned to the real world to find natural "experiments"—simplified systems in which key factors vary so that their effects can be isolated. Islands, being numerous, clearly defined entities, of varied geographical circumstances, and with tractable biotas, have been exploited to such effect that "island biogeography" is often identified as a distinct branch or subfield. In contrast to most island species, ideas from island biogeography have successfully colonized the continents, in application to the problem of fragmentation and loss of habitats therein, forming an important part of the foundations of conservation biology. Indeed, it is important to stress that the processes and mechanisms that have been invoked for islands are not unique to islands, or to obvious habitat islands, it is just that islands have provided the context for their isolation and analysis.

The papers featured in this part focus on a crucial phase in the development of the discipline from the 1950s through the mid-1970s.

They offer insights into the context in which particular ideas were put forward, and we can remind ourselves of what their authors actually wrote, and not what the trail of secondary citations might have us believe they said—or failed to say! They are introduced below in chronological sequence, and related to some of the papers featured in other parts of the volume. Indeed, it should be noted at the outset that island biogeography has deep roots within the discipline, as is illustrated in our excerpt (paper 3) from Johann Reinhold Forster's *Observations Made during a Voyage round the World*, dating from Cook's second voyage (1772–75). In the nineteenth century both Alfred Russel Wallace and Charles Darwin gained important insights from their observations of island systems (the Malay Archipelago and Galápagos islands, respectively). More recently, David Lack (see part 5) and Sherwin Carlquist (see part 2), made important contributions to evolutionary and biogeographic theory through their work on islands. Similarly, papers featured in this part were foundational to those by Jared Diamond, and by Ed

---

The title of part 6 is borrowed from MacArthur and Wilson (1967) who used it for the first chapter of their monograph.

Connor and Daniel Simberloff, which appear in part 7.

Diversity Patterns

Several of the featured articles in part 6 are concerned with understanding patterns in diversity, and especially relationships between species richness and environmental variables such as area and isolation. The 1921 paper "Species and Area," by OLOF ARRHENIUS (paper 52) is instructive for two reasons. First, because it demonstrates that quantitative theory in the subject was not first invented in the 1960s, and second, because of the academic debate in which it is rooted. Whereas the paper is most often cited today for its foundational role in understanding species-area patterns, the issues that most bothered Arrhenius were questions of phytosociology (i.e., organization of plant communities). At the time, the dominant view of natural vegetation in both Europe and North America was strongly deterministic, vegetation associations were viewed as natural units, and Frederic E. Clements (e.g., 1916) was developing the ideas of communities as "superorganisms": a misconception of vegetation most famously contested by Henry A. Gleason (e.g., 1926).

Arrhenius begins by pointing out that it is necessary to control area in order to compare species lists from different localities. There are two ways of dealing with this problem: one is to use standard-sized sample units, arguably preferable for certain tasks (Whittaker, Willis, and Field 2001), and the other is to "control" for area statistically. It is this second approach that Arrhenius developed, drawing from his analyses the observations that "the number of species increases continuously as the area increases" and that "the species in an association are distributed according to the laws of probability" (Arrhenius 1921: 99). He uses these observations to attack the notion of discrete associations, comprising "constant" species; rather, he writes ". . . the plant associations pass into each other quite continuously." In short, as a contribution to vegetation science this article adopts a line that might be termed "Gleasonian"—a rejection of the highly deterministic views dominant at the time in both Europe and North America.

The reason for the inclusion of the article, however, is for its foundational role in the literature on species-area relationships. Arrhenius wasn't the first to describe this fundamental relationship, but his paper was notable in describing (at least to a good first-order fit) the form of the relationship and in making it mathematically tractable. Arrhenius' demonstration of a geometric series depends on taking logarithms of both ordinate and abscissa, which is to say that there is a linear relationship between log species number and log area. Preston (1960, 1962) labeled this log-log plotting an *Arrhenius plot*. It is also termed the *power model* because it is described by the power function $S = CA^z$. This may be expressed using the formula $\log S = C + z (\log A)$, where S is the species number, C is another constant usually referred to as the intercept (although there is no true zero on a log-log plot), A is the area, and z is the slope in log-log space. Arrhenius (1921) demonstrated that although his formula worked well, there was variation in the constant for different vegetation types. He also recognized that it was important to establish the generality of his formula beyond the Swedish island systems he had analyzed. Subsequent work led to the insight that the slope (the z-value) of Arrhenius plots appeared to differ systematically between islands and non-isolated sample areas on continents (or within large islands), a difference MacArthur and Wilson (1963, 1967) seized on in developing their island theory (see also: Wilson 1961; Brown, 1971). Apart from this empirical relationship, Arrhenius offered the important insight, again later developed by Preston (e.g., 1962), that the species-area relation is essentially an emergent property of underlying species-abundance distributions (and see Hubbell 2001). Beyond this, however, he provided no insight into mechanisms that might be at work.

Biogeographical Processes

In the mid-1950s, Edward O. Wilson and William L. Brown, both at Harvard, made important contributions to ant taxonomy and to evolutionary theory. Their review paper on character displacement (Brown and Wilson 1956) provides a notable pointer to the way Wilson's ideas were developing at the time. In it Brown and Wilson not only named and characterized character displacement, but they set out to convince the reader of the importance of this mechanism by number of cases and weight of supporting evidence. They also focused, as Wilson did in his featured paper, on hypotheses concerning the relative importance of allopatric and sympatric phases in lineage development. In its emphasis on the importance of competition, Brown and Wilson (1956) joined a debate that has run back and forth across ecological studies since.

This collaboration led up to EDWARD O. WILSON's seminal island biogeographical analyses of the Melanesian ants, the first of which was published under the title "Adaptive Shift and Dispersal in a Tropical Ant Fauna" (1959). In this, and a subsequent paper, Wilson (1959, 1961) put forward an evolutionary-ecological model, termed in the second paper the *taxon cycle*. We reproduce the earlier paper here (paper 53), as it is the first description of the cycle. It is based on analyses of the Ponerinae, one of eight subfamilies of ants known from Melanesia. The 1961 paper extends the concept to other ant subfamilies, and adds additional detail, without significantly recasting the model. The opening paragraphs of the 1959 article indicate the ambition and vision of its author: ". . . The nature of general adaptation and dispersal mechanisms underlying major biotic movement is clearly one of the great problems of modern evolutionary theory. . . . There is a need for a "biogeography of the species," oriented with respect to the broad background of biogeographic theory but drawn at the species level and correlated with studies on ecology, speciation, and genetics" (122).

Wilson (1959) makes primary use of two data sources: (1) the updated taxonomy (on which he and W. L. Brown had been collaborating), and (2) distributional data, at two scales of analysis—the distribution among islands and the within-island use of habitats. Although the interpretation is compelling, only a few remarks concerning nesting site preferences and dispersal powers were offered in support of the dynamic evolutionary ecological model (see also Wilson 1961). The supporting papers published by Wilson in the period 1958–1961 describe the fieldwork he conducted in Melanesia in 1955 and 1956 and from which the synthetic model was drawn. They present a rather sparse picture of both the phylogenetic relationships and ecology of the ants. Typically, the "ecological notes" of particular species appear to be based on observations of one or a very few colonies and sometimes of just a handful of stray workers sampled on short forays into the islands in question (although more time was spent on the field work on New Guinea itself). The absence of a formal sampling design or statistical testing does not mean that the taxoncycle model is invalid, just that it provides a fairly bold interpretation of the very limited data.

Intriguingly, there seems to have been no attempt made by other authors to follow up Wilson's taxon cycle by experimental or other ecological studies of the ponerine ants, and relatively few scientists have studied the taxon cycle in other taxa. As a model, the taxon cycle involves some core concepts—competitive interactions and displacements—and repeated phases of dispersal/invasion from a center of origin. However, it has been considerably modified whenever it has been applied to other taxa and other archipelagos. One of the most prominent illustrations, by Robert E. Ricklefs and George W. Cox (1972, 1978), for Caribbean birds, has been the subject of some pointed criticisms (Pregill and Olson 1981) and it has taken a further twenty years and the application of modern phylogenetic analyses of gene sequences to deal with these questions and advance the understanding of taxon cycles in the Antillean avifauna (e.g., Ricklefs and Bermingham 2002). The generality of the cycle thus remains open to question, but it does, at the least, provide a heuristic framework for the evolution of insular biotas (Whittaker 1998). Moreover, Wilson's studies of Melanesian ants played a crucial role in the development of the MacArthur–Wilson dynamic equilibrium model and of Diamond's (1974, 1975a) later work on assembly of bird communities in the same region (see paper 65, in part 7 of this volume). For another study of insular evolution, see David Lack's (1947) paper on the evolutionary radiation of the Galápagos finches (paper 46, in part 5).

On the Dynamics of Species Richness

The "equilibrium," or "dynamic" theory of island biogeography, developed by ROBERT H. MACARTHUR and EDWARD O. WILSON, remains enormously influential throughout biogeography, ecology, and conservation biology. The first statement of the theory, in their 1963 article "An Equilibrium Theory of Insular Zoogeography" (paper 54), is less fully developed than the more frequently cited 1967 monograph, but the guts of the theory are all there. Hopefully, reading the 1963 paper will stimulate readers to examine the 1967 monograph: it remains a rich source of insight. The 1963 paper begins by making several observations about patterns of species diversity: that species-area curves (Arrhenius plots) vary in a fairly systematic fashion and that species richness declines with distance from source. They also argue against the more traditional explanation for impoverishment with distance, which held that time has been insufficient for remote islands to fill up (with the implication that over time the species richness of such islands will increase further). This "non-equilibrium" explanation is contrasted with MacArthur's and Wilson's big idea, that the pattern in species diversity represents a dynamic steady state due to the offsetting effects of immigration (influ-

enced by distance) and extinction (influenced by area).

While their model is highly simplified, they recognize within the paper that, for instance: (1) species vary in their likelihood of immigrating to distant islands and of dying out; (2) the size of the available species pool can vary, and; (3) factors such as climate and ecological succession do influence the capacity of an island to support species.

It is often forgotten that the first formulaic statement of the MacArthur–Wilson model recognizes that species can be added to an island not only by immigration but also by local speciation. As signified by publication of the 1963 paper in the journal *Evolution*, it was indeed intended as a contribution to evolutionary ecology as well as to ecological biogeography. In it the authors note: "the increase of s [species richness] by local speciation on single islands and exchange of autochthonous species between islands, probably becomes significant only in the oldest, largest, and most isolated archipelagoes, such as Hawaii and the Galápagos. Where it occurs, the exchange among the islands can be predicted . . . with individual islands in the archipelago serving as both source regions and recipient islands" (380).

They suggest that extinction rate will vary with the average life span of the taxon, the richness of the taxon, and the mean size of the species populations, these being necessarily related to island area. The model was thus offered as a general theory for all taxa on all islands. It was this generality, combined with the apparent testability of the model, that accounts for its immediate and lasting influence.

Although MacArthur and Wilson (1963) suggested several ways of testing the model, work undertaken since reveals that it is harder to test conclusively than they had hoped. A case in point is provided by the Krakatau bird data, the only species-time data set analyzed within their paper. As they themselves caution in the final paragraph of that section, successional changes in the Krakatau ecosystem seem to have been responsible for a phase of apparent equilibrial turnover during the 1920s, which was followed by an upward drift in species number over the following decades (Thornton 1996). Obtaining time series of sufficient quality and precision with which to test their theory is a nontrivial problem, as is shown for Krakatau plants by Whittaker, Field, and Partomihardjo (2000). Although spatial patterns have been used in evaluating the theory, they have similar problems, and in any case are less diagnostic in distinguishing the dynamic MacArthur–Wilson model from alternative hypotheses (see review in Whittaker 1998).

It was noted by MacArthur and Wilson (1963) that Preston (1962) had independently produced insights covering part of the theory of island biogeography. It has recently been pointed out that K. W. Dammerman's (1948) work on Krakatau faunal recolonization was also moving in the direction of a dynamic view of island biogeography (Thornton 1992), although only in broad descriptive terms. More remarkably, a mathematical version of the core hypothesis was set out in a long-ignored doctoral thesis written in 1948 by Eugene Gordon Munroe (Brown and Lomolino 1989). But although these and other authors may have sketched parts of the picture, it was MacArthur and Wilson who combined the basic ideas, in terms of processes and principles, in a compelling model that could be depicted in simple graphical form. They also had the insight to articulate, in their 1967 monograph, the potential of these ideas to change island biogeography from a rather descriptive tradition, to one that was far more dynamical, analytical, and experimental.

Although they recognized that modification of their simple initial colonization model might be needed, their thesis was clearly intended to apply even to remote archipelagos with high levels of endemicity (see MacArthur and Wilson 1967: 171–76). They write of the "radiation zone . . . in which intra-archipelago exchange of autochthonous species approaches or exceeds extra-archipelagic immigration toward the outer limits of the taxon's range," adding that it "is predicted as still another consequence of the equilibrium condition" (MacArthur and Wilson 1963: 386). They therefore regarded distant archipelagos, in which phylogenesis is a dominant signal, as providing special cases that are consistent with their general theory.

In practice, empirical evidence to support the notion that, on remote islands, speciation and extinction typically result in equilibrium is equivocal at best (Whittaker 1998, 2000). Indeed, the assumption that *most* islands are at or near equilibrium most of the time, also appears to be in doubt, and the MacArthur–Wilson model is increasingly seen as but one of an array of alternative hypotheses (see, e.g., Brown and Lomolino 1998, 2000; Heaney 2000; Lomolino 2000c). There is no doubting, however, the enormous stimulus that was provided by their theory, especially after publication of the longer version in 1967.

The most powerful recent illustration of this is provided by Stephen P. Hubbell's *The Unified Neutral Theory of Biodiversity and Biogeography* (2001), which appears to forge the links between population abundance distributions and species richness more convincingly than Arrhenius, Preston, and MacArthur and Wilson. Hubbell's theory is in part built upon MacArthur and Wilson's theory, a point he recognizes explicitly in his monograph.

Experimentation

Following the publication of MacArthur and Wilson (1963) and before their 1967 monograph appeared, Wilson had already begun to work on testing the theory experimentally in collaboration with his graduate student Daniel S. Simberloff (Simberloff 1969; Simberloff and Wilson 1969, 1970; Wilson and Simberloff 1969). One of the resulting papers (Simberloff and Wilson 1969) is reproduced in the volume *Foundations of Ecology* (Real and Brown 1991). Their approach was to manipulate a replicated set of small, simple "island" systems—mangrove clumps of 11 to 18 m in diameter and isolated by distances of 2 to 1,183 m. Their subject organisms were arthropods, which might be anticipated to respond rapidly to experimental manipulations. The experiments involved the elimination of all arthropods by fumigation, and the censusing of the islands before and at intervals following this "defaunation" event. Details of the "islands" are given in Wilson and Simberloff (1969).

"Experimental Zoogeography of Islands: A Two-Year Record of Colonization" (paper 55), by DANIEL S. SIMBERLOFF and EDWARD O. WILSON (1970), is noteworthy as being the most overtly experimental study reprinted in this volume. The results demonstrated a return to equilibrium, and then a fairly balanced turnover due to offsetting colonization and extinction events. Although seeming to provide strong support for the theory, the results in fact require some modification of the initial, simple MacArthur–Wilson model. First, the data show a slight overshoot from the anticipated equilibrium species number, followed by a fall, and then a gradual rise to a new, so-called assortative equilibrium. Second, the findings suggest convergence toward the original species composition; i.e., they point to some structure in the recolonization process, whereby "more highly co-adapted species sets find themselves by chance on an island and persist longer as sets" (Simberloff 1976: 576). A third, more problematic limitation is that these systems were chosen to be about the largest systems that might practically be experimentally manipulated in this way. They nonetheless are tiny (again, just 11 to 18 m in diameter) islets of uniform habitat. The extent to which results from such simple systems can be "scaled-up" to more complex islands and successional ecosystems remains to be established (e.g., see Bush and Whittaker 1991).

Simberloff (1976) provides a careful, critical evaluation of the mangrove islet defaunation experiments. He discusses at length the problems of documenting turnover, and the likelihood of *cryptoturnover* (turnover occurring between surveys, which therefore goes unobserved) and *pseudoturnover* (species being present throughout but being undetected in some surveys). He concluded that "most observed turnover on these small mangrove islands involved transients: that is, it is 'pseudo-

turnover'" (577). Pointing to the extreme difficulty of generating data of sufficient quality and precision to test the theory, his 1976 paper suggested caution in evaluating the MacArthur–Wilson theory. Subsequent empirical analyses of other systems demonstrate that turnover is often dominated by species that might be labeled "transient," "fugitive," or "ephemeral" (see, e.g., Schoener and Spiller 1987; Williamson 1989a, b). As turnover at equilibrium is central to the MacArthur–Wilson model, such observations reveal a serious weakness that has led to calls for reappraisal of the theory (e.g., Brown and Lomolino 2000; Heaney 2000).

Non-equilibrium Island Biogeography

MacArthur and Wilson implied that non-equilibrium islands could be found in nature, but they were not the dominant signal: most patterns of insular diversity were consistent with the equilibrium model. This argument stood in contrast to the previous emphasis on island impoverishment reflecting very slow processes of island-filling, a non-equilibrium explanation. Since the 1960s, the array of alternative non-equilibrium ideas has increased, or at least taken clearer shape, as exemplified first by JAMES H. BROWN's (1971) "Mammals on Mountaintops: Nonequilibrium Insular Biogeography" (paper 56).

Brown used sample areas from the Sierra Nevada "mainland" and compared their species-area relation to that of 17 mountain "islands." He found that there was no correlation between species richness and variables likely to affect rates of colonization (e.g., distance to the mainland), and inferred from this and other data that the boreal mammals have "an exceptionally low rate of immigration to isolated mountains." Brown's explanation for the present distributional patterns is essentially historical: the mountain-top faunas represent relics from the Pleistocene, when the climatic barriers that currently isolate these habitats were removed. Since then, the dominant process has been *attrition* ("relaxation") of the fauna through extinctions, which have been greater on small mountains than larger ones.

Despite the finding of a non-equilibrial pattern, the framework for the analysis—and the tools used—are those provided by MacArthur and Wilson, that is, they constitute analytical rather than "narrative" biogeography. Thus, for instance, Brown makes predictions as to which taxa (e.g., nonvolant mammals versus birds) are likely to be found to demonstrate non-equilibrium patterns, predictions which he tested in a later publication (Brown 1978). The paper also builds upon the MacArthur and Wilson model by identifying ecological structure in the form of attributes such as body size, trophic level, and habitat specialization, a development taken further for a different system in Diamond's subsequent papers (discussed below; see also Brown and Lomolino 1998: fig. 14.2).

Brown's (1971) inference of a slow attrition of species on the one hand and effectively zero immigration on the other, has recently been challenged by revised data suggesting that Great Basin mountain-top species richness versus "island" size relationships are quite weak, and by indications that cross-valley (i.e., cross-desert) dispersal of montane forest mammals may be more frequent than Brown's data suggested (Lawlor 1998, Grayson and Madsen 2000). It seems likely that the ability of these mountain mammal species to cross desert barriers varies as a function of body size, the span of the desert area, and the habitat requirements of the species: with woodland species being relatively mobile and cold-adapted subalpine species less likely to immigrate across intermountain habitats (Lomolino and Davis 1997, Lawlor 1998, Rickart 2001).

The inferred Great Basin non-equilibrium took the form of a slow dynamic, in which species numbers have been declining over thousands of years. Other islands may be non-equilibrial because their environmental carrying capacities are highly variable and have been perturbed by active disturbance regimes. In such cases, successional processes may lag behind changing environments, resulting in a more dynamic non-equilibrium, featuring relatively high rates of immigration (e.g., as postulated for Krakatau communities by Bush and Whittaker 1991). Another interesting dynamic non-equilibrium model put forward in the 1970s was Joseph Connell's (1978) *intermediate disturbance hypothesis*. His insight was that diversity should be maximal, not in places where disturbance was frequent and intense or in patches of very low disturbance (where competition would result in relatively few species dominating the site), but in sites of intermediate intensity or frequency of disturbance. This "patch dynamics" idea was not, of course, an island model, but it serves to illustrate the way in which ecology began to move on from a reliance on dynamic equilibrial models of diversity to consider dynamic non-equilibrial models.

Island Assembly Theory

JARED M. DIAMOND's (1974, 1975a) work on the avifauna of New Guinea and its surrounding islands shares the dynamic equilibrial footings of the MacArthur–Wilson theory but builds significantly on them through its emphasis on patterns of species composition and the mechanisms that may account for that structure. As a contribution to biogeography it has been dogged by controversy, but it nonetheless amounts to a comprehensive body of theory, known as "island assembly theory" (Whittaker 1998). The monographic paper, "Assembly of Species Communities" (Diamond 1975a) is too long to be reproduced in its entirety but part of that paper is featured in part 7 of this volume. Here we reprint a short article published in *Science* in 1974 and entitled "Colonization of Exploded Volcanic Islands by Birds: The Supertramp Strategy" (paper 57).

Diamond's 1974 paper builds once again on Arrhenius plots and on the MacArthur–Wilson theory. Thus, a species-area plot is used to demonstrate that islands disturbed within the previous two centuries are "under-saturated"; i.e., they hold fewer species than other islands of comparable area. Note also that there appears to be a breakpoint at about 1 ha: islands smaller than this show a very poor species-area relation, and those above a very tight fit, consistent with the notion of scale-dependence identified most recently by Lomolino (2000b) and Lomolino and Weiser (2001). Diamond (1974) wisely uses only the seven largest control islands to generate the "equilibrial" line in the Arrhenius plot. The paper focuses on the two exploded volcanoes (the tiny remnant island Ritter, and the larger Long island). Not only do they have fewer lowland bird species than expected as a function of area, they also have higher population densities than most islands sampled.

Diamond's explanation for these phenomena is both a successional and an evolutionary ecological model. It is founded on the "arrested" development of the vegetation of the defaunated islands, with Long island still having early-successional "open, savannah-like lowland forest" and Ritter even sparser vegetation. The response of the birds to successional processes on a time-scale of centuries is reflected not only in species numbers and densities of individuals, but also in the identities of the species. Here Diamond seizes on another important theoretical contribution of MacArthur and Wilson (1967), the notion of r-selected and K-selected species: crudely put, the contrast between pioneer, rapid breeding, fast dispersing, quick life-cycle species on the one hand, and late successional, slow maturing, limited dispersing, slow life-cycle species on the other. Diamond translates the K to r spectrum into three groups of species: "sedentary species," "tramps," and "supertramps." The supertramps, are less competitive, but have survived by rapidly dispersing to fill early successional habitat created by major disturbances, like volcanic eruptions. In time, as individual island systems mature, later successional species move in and out-compete early colonists, but meanwhile a new disturbance will have occurred somewhere else, allowing the supertramps to move on.

Diamond's model can thus be considered an attempt to import succession theory into the equilibrium theory, and simultaneously to focus on species compositional patterns more than species richness. The 1974 paper provides a plausible model, but is not without limitations, including a degree of seemingly post-hoc modification to accommodate contrary observations (e.g., regarding montane birds), and a considerable reliance on mechanisms of competition and resource use that are not directly tested. Reproduced in part 7 is the summary from the 1975a paper and the list of seven "assembly rules" that encapsulated the empirical findings as Diamond interpreted them (see paper 65, and also below). Diamond adopted the working hypothesis, subsequently challenged, that through diffuse competition, the component species of a community are assembled, and co-adjusted in their niches and abundances, so as to fit with each other and to resist invaders. The assembly rules are:

1. If one considers all the combinations that can be formed from a group of related species, only certain ones of these combinations exist in nature.
2. Permissible combinations resist invaders that would transform them into forbidden combinations.
3. A combination that is stable on a large or species-rich island may be unstable on a small or species-poor island.
4. On a small or species-poor island, a combination may resist invaders that would be incorporated on a larger or more species-rich island.
5. Some pairs of species never coexist, either by themselves or as part of a larger combination.
6. Some pairs of species that form an unstable combination by themselves may form part of a stable larger combination.
7. Conversely, some combinations that are composed entirely of stable sub-combinations are themselves unstable.

In his book chapter, Diamond (1975a) relied upon a variety of patterns and analyses including incidence functions, recolonization of defaunated islands, checkerboard distributions, and patterns of guild structure. The assembly rules are descriptions and interpretations of these emergent patterns.

As Gotelli explains in part 7 of this volume, Diamond's approach drew pointed criticism from Daniel Simberloff and colleagues (see Simberloff 1978; Connor and Simberloff 1979, 1983; for responses, see Diamond and Gilpin 1982; Gilpin and Diamond 1982). The problems appeared to be sufficiently troublesome, or intractable, as to hamper subsequent progress, and Diamond's approach has until recently failed to receive the attention that its intrinsic interest warrants (Whittaker 1998, Weiher and Keddy 1999). There were two key points of criticism. First, Diamond appeared to give too much emphasis (particularly in summary statements) to competition and too little to predation, dispersal, habitat differences, chance, and historical events. Second, using null models, his critics questioned the extent to which the distributional patterns departed from a random expectation. This, however, has proven a really taxing challenge, as it has become clear that the findings of null model analyses depend heavily on the biological and/or biogeographical assumptions that particular authors build into their "null" models (Gotelli 2001). The debate over Diamond's assembly rules continues (e.g., Sanderson, Moulton, and Selfridge 1998), and has reappeared in other guises, as the assembly rule concept has been applied to other systems, for example, desert rodents and communities of herbaceous plants (Weiher and Keddy 1999), and to the process of anthropogenic "disassembly" of native communities (Fox 1987, Mikkelson 1993, Lomolino and Perault 2000).

Applied Island Biogeography

JARED M. DIAMOND's 1975 paper "The Island Dilemma: Lessons of Modern Biogeographic Studies for the Design of Natural Reserves" (paper 58), is one of the foundational papers of conservation biology. In reading the paper one is struck by the authority with which the conclusions are offered in the Abstract, although the paper is very much a theoretical treatise. Basing his arguments on the MacArthur–Wilson model, Diamond argues that habitat fragmentation will produce predictable patterns of species loss as previously equilibrial patches become "supersaturated" and thereafter "relax" (i.e., shed species).

Once again, species-area relationships form a key part of the argument, with supporting quantitative evidence for relaxation coming largely from "landbridge islands." Analogous to Brown's (1971, 1978) studies of mountain-top biotas, the assumption is that before sea levels rose at the end of the Pleistocene and turned these land areas into islands, they would have shared close to the full quota of species now present on the mainland. The discrepancy between this assumed starting value and that derived from species-area plots for "control" areas is taken to represent the number of extinctions, with the speed of equilibration being seen as an inverse function of area. The possibilities that the observed discrepancies reflect the influence of variables other than area and isolation, or that biotas may not initially be equilibrial in terms of richness, are recognized but downplayed. The paper goes on to make fairly bold statements about our ability to predict not only total numbers of losses (the often cited 50 percent of species remaining in 10 percent of the area figure) but also rates of losses. Then, as now, such predictions appear to be derived largely from inferences based on studies of actual islands, rather than on studies of recorded losses from fragmented habitats within large landmasses.

Diamond recognized that species have differing degrees of mobility and, in comments on movements between patches, he seems to foreshadow the eventual blossoming of *metapopulation theory* (Hanski 1999) within conservation. Diamond noted other ecological features that influence the likelihood of persistence in reserves of different sizes, and captured some of this signal by the use of "incidence functions"—a static analysis of the distribution of a species across a system of isolates. He also acknowledged that the principles derived may have to be qualified, for example "by the statement that separate reserves in an inhomogeneous region may each favor the survival of a different group of species; and that even in a homogeneous region, separate reserves may save more species of a set of vicariant similar species, one of which would ultimately exclude the others from a single reserve" (144). This marks the beginning of the so-called SLOSS debate, which focused on the question of whether a Single Large or Several Small reserves were better for conservation (reviewed in Whittaker 1998). A careful reading of the paper will reveal not only the seeds of metapopulation ideas, but also of sources and sinks (Pulliam 1996): it thus encapsulates several of the key ideas that have occupied conservation biology over the final two decades of the twentieth century.

Human-Induced Extinction

The final paper in this part continues the theme of biodiversity loss and human-caused extinction. As indicated above, work undertaken since the 1960s has undermined MacArthur and Wilson's (1963, 1967) assumption that the biotas of remote oceanic islands are typically in equilibrium. Although far from the first indication of this, the 1982 article, "Fossil Birds from the Hawaiian Islands: Evidence for Wholesale Extinction by Man before Western Contact," by STORRS L. OLSON and HELEN F. JAMES (paper 59), was selected because it provides such a dramatic and unequivocal demonstration of anthropogenic non-equilibria. Olson and James were well aware of the implications of their findings for island theory, and focus on this issue in their concluding paragraph. One of the key features of the Hawaiian, and many other Pacific islands, is that the colonization of these land areas by humans occurred within the last few thousand years. This means that well-dated fossil (and subfossil) evidence can be used to establish the background, prehuman extinction rate, and then to document the impact of human colonization. This has proven rather more difficult and contentious on continents, as is shown by the continuing debate concerning

the Pleistocene "overkill" hypothesis, put forward for the Americas by Paul S. Martin (1973, featured in part 3 of this volume [paper 35]). On continents some element of doubt, or at least of dispute, remains concerning the timing of extinctions, the arrival of humans, and the influence of congruent climatically driven biological change (e.g., Flannery 1994, Choquenot and Bowman 1998, Diamond 2001). Island studies, on the other hand, clearly demonstrate that pre-industrial societies have been responsible for accelerated and often highly selective pulses of extinction (for recent syntheses of continental "overkill" see Brown and Lomolino 1998; for islands, see Milberg and Tyrberg 1993, Steadman 1997, and Whittaker 1998). We now know that as a direct result of human action during the last 400 years, island birds were 40 times more likely to become extinct than continental birds, and that 80 percent of documented animal extinctions were island species. This destruction of island biodiversity seems likely to continue as, for example, one in three threatened plant species are island endemics (see review in Whittaker 1998).

Closing Thoughts

The selection of articles featured here directs a great deal of attention to the work of Edward O. Wilson, Robert H. MacArthur, and of two of their associates, Daniel S. Simberloff and Jared M. Diamond. These articles share a key characteristic: they provide a dynamic interpretation of diversity patterns on islands. The ideas expressed in the taxon cycle paper, on an evolutionary time-frame, can be seen as feeding logically into the development of the ecological time-scale dynamic model of MacArthur and Wilson. Moreover, there are close parallels between Diamond's assembly rules ideas and Wilson's taxon cycle concept. Both seek to explain distributional patterns by means of competitive effects, habitat relationships and evolutionary considerations. This body of work has provided a huge stimulus to biogeography over the last three decades. Nonetheless, there is more to island biogeography than this small selection of featured authors. Our final featured article, by Olson and James, with its reliance on paleoecological evidence for species extinctions, provides a hint of the wider array of approaches and issues.

The unique characteristics of different kinds of organisms (their mobility and life cycle attributes) and the special features of different insular ecosystems may be crucial to the applicability of particular theories of island biogeography. It is patently unrealistic to view the ecosystems of the world as entirely equilibrial or non-equilibrial, or to build theory on the assumption that only one of these conditions may apply (Whittaker, Willis, and Field 2001). Considered in a spatial/temporal perspective, each of the theories or models considered herein would seem to have a domain of applicability, where the processes captured in the model are prominent, while elsewhere the effects are weak or overridden by other more prominent mechanisms (Whittaker 1998, 2000; Brown and Lomolino 2000; Heaney 2000).

If some of the debates within island biogeography have been heated, then the benefit has been the gradual development of more rigorous requirements for collecting and analyzing data and for evaluating alternative theories or hypotheses. It should be clear that it is as important to have good tools, methods, and data as it is to have good theory. I hope that this short collection of foundational contributions to island biogeography will stimulate the reader to think about the role of theory in biogeography, and about the central role of islands in developing and testing biogeographic theory.

SPECIES AND AREA

By OLOF ARRHENIUS.

(Stockholm, Sweden.)

Both for the plant-geographer and the ecologist it is of great importance to know if an area or a district is rich or poor from the point of view of vegetation. Often the lists of plants from different districts are quite incomparable owing to the different sizes of the areas investigated. The problem of comparing lists of flora from several districts of different size therefore is of great interest and many attempts have been made to solve the question. Most writers, however, have obtained no results except to confirm the well-known and obvious fact, that the larger the area taken the greater the number of species. In two recent papers (**1, 2**) I have tried to solve the question and have ventured to propose an empirical formula. The material sampled showed that this formula is correct for areas of such different sizes as square decimetres, square metres and hectares.

As it is of great interest to know if the formula only holds for complexes of associations, floral districts, etc., or if it is also valid for pure communities, the summer of 1920 was used for collecting material from several (altogether 13) associations of different types all lying in the islands of Stockholm as described in *Öcologische Studien*[1]. The results are tabulated in the following table.

In the left-hand column the area is given in square decimetres, the next column gives the mean values for the different areas as observed in the field, while the third column gives the values obtained by the formula. The last column gives the deviations between the calculated and observed values estimated as percentages of the former. It is easily seen that the values calculated and observed agree very well. Generally there is an increase in the deviation corresponding to increasing area. This depends on the fact that the values of the smaller areas are the average of a greater number of observations than those of the larger.

As it is shown that the formula holds for all the values obtained from these communities and as these examples are picked out quite by chance it can be stated as a general rule that it holds for every association and as a consequence of this also for complexes of associations, formations, large areas, districts, etc.

---

[1] In *Öcologische Studien*, p. 15, there is a misprint, the equation is written in this way:

$$y^{\log 9} = a \left(\frac{x}{b}\right)^{\log 0}. \text{ For this read } y^{\log 9} = a^{\log 9} \left(\frac{x}{b}\right)^{\log 0}.$$

Later (2) I simplified the formula to $\dfrac{y}{y_1} = \left(\dfrac{x}{x_1}\right)^n$ where x is the number of species growing on the area y, and x_1 that on y_1; n is a constant.

96 *Species and Area*

Table showing the differences between the observed and calculated numbers of species on areas of different sizes in 14 different communities.

obs. =observed number of species on each area.   calc. =calculated ditto.
diff. =difference expressed as percentage of calc.

Weed-association

area in dm²	obs.	calc.	diff.
1	1·4	1·4	0
2	2·0	1·9	5
4	2·6	2·6	0
8	3·7	3·5	6
16	5·2	4·8	8
32	6·8	6·6	3
64	8·8	9·0	2
128	10·3	10·2	1
256	14·5	16·9	14
300	16	18·2	12

Calluna-Pinus wood

area	obs.	calc.	diff.
1	0·9	0·9	0
2	1·2	1·1	9
4	1·5	1·4	7
8	2·0	1·8	11
16	2·2	2·2	0
32	2·3	2·8	17
64	2·5	3·5	28
100	3·0	4·0	25

Aira-Pinus wood

area	obs.	calc.	diff.
1	1·2	1·2	0
2	1·4	1·4	0
4	1·8	1·9	5
8	2·4	2·4	0
16	3·3	3·2	3
32	4·6	4·0	15
64	5·5	5·2	6
100	7	6·1	15

Herb-Pinus wood

area	obs.	calc.	diff.
1	4·8	4·8	0
2	7·0	6·7	4
4	9·8	9·4	4
8	14·3	13·1	9
16	18·9	18·5	2
32	23·0	25·8	11
64	27·0	33·0	18
100	33	41	19

Vaccinum vitis-Pinus wood

area	obs.	calc.	diff.
1	1·4	1·8	22
2	1·9	1·9	0
4	2·0	2·0	0
8	2·1	2·1	0
16	2·2	2·2	0
32	2·3	2·3	0
64	2·5	2·5	0
100	3	2·6	15

Arctostaphylos-Pinus wood

area	obs.	calc.	diff.
1	1·4	1·4	0
2	1·6	1·6	0
4	1·8	1·8	0
8	2·1	2·0	5
16	2·2	2·3	4
32	2·3	2·5	8
64	2·5	2·8	10
100	3	3	0

Myrtillus-Picea wood

area	obs.	calc.	diff.
1	1·9	1·9	0
2	2·6	2·5	4
4	3·5	3·2	9
8	4·5	4·1	9
16	5·1	5·2	2
32	6·0	6·7	11
64	6·5	8·7	25
100	7	9·9	30

Herb-Picea wood

area	obs.	calc.	diff.
1	2·5	2·5	0
2	3·6	3·5	3
4	5·4	5·0	8
8	7·6	7·1	7
16	10·2	9·9	2
32	12·7	14·0	9
64	16·5	19·9	17
100	18	24·8	23

Empetrum-moor

area	obs.	calc.	diff.
1	1·3	1·3	0
2	1·5	1·6	6
4	2·0	2·2	10
8	2·8	2·9	3
16	4·0	3·9	3
32	5·6	5·2	4
64	6·5	6·8	4
100	8·0	8·0	0

Herb-hill I

area	obs.	calc.	diff.
1	3·3	3·4	3
2	4·4	4·2	5
4	5·3	5·2	2
8	6·7	6·4	5
16	7·5	7·7	3
32	9·0	9·6	6
64	11·0	10·7	3
100	12	13·5	11

Herb-hill II

area	obs.	calc.	diff.
1	6·8	6·8	0
2	9·1	8·8	4
4	11·6	11·4	2
8	14·8	14·8	0
16	17·2	19·1	10
32	22·3	25·0	11
64	27	31·0	13
100	30	38	21

Shore-association I

area	obs.	calc.	diff.
1	2·2	2·2	0
2	2·8	2·8	0
4	3·6	3·6	0
8	4·8	4·7	2
16	6·2	6·0	3
32	8·3	7·7	8
64	9·5	9·8	3
100	12	11·6	3

Shore-association II

area	obs.	calc.	diff.
1	0·9	1·0	10
2	1·4	1·4	0
4	2·1	2·0	5
8	2·8	2·7	4
16	3·2	3·6	11
32	4·3	4·9	12
64	4·5	6·9	32
100	6	8·5	29

If the floral territories also were uniformly distributed over the earth's surface it would be possible to calculate the number of species growing on the earth by the aid of the formula, but as they are not no formula can be found which will make such a calculation possible.

Name of association	Constant (n)	Name of association	Constant (n)
Calluna-Pinus wood ...	2·9	Empetrum-moor	2·5
Aira-Pinus wood	2·8	Herb-hill	3·3
Arctostaphylos-Pinus wood	6·2	,, ...	2·6
Vaccinium-Pinus wood ...	12·5	Shore-association ...	2·8
Herb-Pinus wood ...	2·0	,, ,, ...	2·1
Myrtillus-Picea wood ...	2·7	Weed ,, ...	2·2
Herb-Picea wood	2·0		

In the formula there is a constant n. As it is of great interest to see if this constant varies from association to association the values of n for the associations examined are collected in above table, and the values are seen to vary from 2 to 12·5. In a recent paper (**2**) I thought I had found that the constant n does not vary. But there I was working almost entirely with complexes of associations and the n found was an average of values of n for the pure associations, so that the truth was obscured. This approximate formula is only an empirical one which is justified as long as it can be shown to be valid and it is therefore of great importance to ascertain this.

With this formula one is calculating the average number of species growing on the area. But why should there be a certain number of species on a specific area, in other words, what is the probability of finding a given number of species on the area chosen? There is a certain probability of finding each particular species on an area (lying in a certain association) and the sum of these probabilities is the probable number of species on the area.

If we know the absolute degree of frequency of a species (that is the number of individuals belonging to one species growing on a large area Y) we can calculate the probable occurrence on every area (y) which is smaller than Y.

If the probable occurrence of a given species on the area y is called a, and the probability of not finding it is $\left(1 - \frac{y}{Y}\right)^{n_1}$, then $a_1 = 1 - \left(1 - \frac{y}{Y}\right)^{n_1}$. In the case of another species with a frequency n_2, $a_2 = 1 - \left(1 - \frac{y}{Y}\right)^{n_2}$, etc. The probable number of species (A) on the area y is

$$A = a_1 + a_2 + a_3 + \ldots = 1 - \left(1 - \frac{y}{Y}\right)^{n_1} + 1 - \left(1 - \frac{y}{Y}\right)^{n_2} + 1 - \left(1 - \frac{y}{Y}\right)^{n_3} + \ldots.$$

If one calculates the sum of species according to this formula the results obtained can be compared with the number of species found by the field survey. The results are given in the last table.

The agreement between the values calculated and observed is very good. The deviations between the values obtained by the approximate formula and

98   *Species and Area*

the probability calculation are also not very great, but they do vary a good deal, and this must depend on the approximate formula holding in some cases and not in others. If the formula holds the number of species (A) must increase in a geometrical series.

Area in sq. dm.	Calluna-Pinus wood		Aira-Pinus wood		Vaccinium vitis-Pinus wood		Arctosta-phylos-Pinus wood		Herb-Pinus wood		Myrtillus-Picea wood	
	O.	P.	O.	P.	O.	P.	O.	P.	O.	P.	O.	P.
1	0·9	0·9	1·2	1·3	1·4	1·8	1·4	1·5	4·8	5·5	1·9	2·3
2	1·2	1·2	1·4	1·6	1·9	1·9	1·6	1·8	7·0	8·2	2·6	3·0
4	1·5	1·5	1·8	2·1	2·0	1·9	1·8	1·9	9·8	11·8	3·5	3·6
8	2·0	1·9	2·4	2·9	2·1	2·0	2·1	2·1	14·3	15·7	4·5	4·2
16	2·2	2·1	3·3	4·1	2·2	2·1	2·2	2·2	18·9	19·2	5·1	5·1
32	2·3	2·3	4·6	5·5	2·3	2·3	2·3	2·3	23·0	22·8	6·0	5·9
64	2·5	2·6	5·5	6·5	2·5	2·6	2·5	2·6	27·0	25·7	6·5	6·6
100	3·0	3·0	7·0	7·0	3·0	3·0	3·0	3·0	33·0	33·0	7·0	7·0

Area in sq. dm.	Herb-Picea wood		Empetrum-moor		Herb-hill I		Shore-association I		Shore-association II		Weed-association	
	O.	P.	O.	P.	O.	P.	O.	P.	O.	P.	O.	P.
1	2·5	2·5	1·3	1·2	3·3	3·6	2·2	2·2	0·9	0·9	1·4	1·4
2	3·6	3·7	1·5	1·6	4·4	4·5	2·8	2·8	1·4	1·4	2·0	1·9
4	5·4	5·6	2·0	2·2	5·3	5·5	3·6	3·6	2·1	2·1	2·6	2·7
8	7·6	8·0	2·8	3·2	6·7	6·6	4·8	4·8	2·8	2·7	3·7	4·0
16	10·2	10·5	4·0	4·6	7·5	8·1	6·2	6·7	3·2	3·7	5·2	5·9
32	12·7	13·4	5·6	5·8	9	9·8	8·3	7·1	4·3	4·0	6·8	8·0
64	16·5	16·5	6·5	7·1	11·0	11·7	9·5	8·8	4·5	4·0	8·8	9·6
100	18·0	18·0	8·0	8·0	12·0	12·0	12·0	12·0	6·0	6·0	—	—

O. = observed number of species.   P. = probable number of species.

As the relation y/Y varies from 0 to 1, the equation approximates to a geometrical series, as n_1, n_2, n_3, etc. approximate to 1. Expressed in words, the equation satisfies best the condition of a geometrical series when the species growing on the area have a low degree of frequency. It is seen from all the cases here cited that the distribution of species in plant associations follows the laws of probability.

This is of very great importance for the science of the organisation of plant associations, or synecology.

Several authors have worked on this subject in order to find the laws regulating plant associations, especially from the purely botanical standpoint. In a recent paper (3) this question is again taken up and analysed, and the authors give results which if true would make an epoch in the science.

For instance, they believe they have found that the association is a unit as well defined and limited as the species. The characteristics of the association are the *constants*, that is, species which will always be found in the association when a part of it larger than a certain area, the minimum area, is examined.

The authors try to find something more exact than the old words "leading species," etc. That there should be a certain skeleton of constants in every association is a very fascinating idea, which on the first view is supported by material taken from different associations.

But according to the probability calculation on page 97 we find that when there are on an average two individuals of a species on a plot the probability of finding it on every plot is 0·99, that is nearly 1. When the plant is found on every one of the small areas it is a constant according to the "Gesetze." But then every species belonging to the association can become a constant; it is only necessary to take the plots so large that the average number of individuals belonging to one species is about 2. Thus, one can say that the idea of a "constant" is too pretentious and rather a misleading substitute for the old terms. How the result is obtained is quite easily seen. The material used is collected from associations with one or two leading species and some rather rare ones. According to the laws of probability there must very soon, by increasing the area taken, be some species which become pseudo-constants and when very large areas are taken the other rarer species will also become "constants."

The number of species increases continuously as the area increases. A consequence of this will also be that there are very seldom any limits between different associations but the passage of one to another is quite continuous. The passage-belt between the associations may be very narrow but always exists.

The results may be concluded in the following summary. An approximate formula given in a former paper has been shown to hold within wide limits. Using this formula one is enabled to find a standard for the relative richness or poorness of a floral district.

The species in an association are distributed according to the laws of probability. The number of species increases continuously as the area increases, and the plant associations pass into each other quite continuously.

For valuable help in the field-work I must thank my wife. To Mr J. Östlind I am very much indebted for his kind help with the mathematical formulae.

LITERATURE CITED.

(1) **Arrhenius, O.** *Öcologische Studien in den Stockholmer Schären.* Stockholm, 1920.

(2) **Arrhenius, O.** "Distribution of the species over the area." *Medd. fr. K. Vet. Akad. Nobelinstitut*, **4**, nr. 7, 1920.

(3) **Du Rietz, G. E., Fries, Th. C. E., Osvald, H.** und **Tengwall, T. Å.** "Gesetze der Konstitution natürlicher Pflanzengesellschaften. Vetenskapl. o. prakt. unders. i Lappland." Upsala o. Stockholm, 1920.

ADAPTIVE SHIFT AND DISPERSAL IN A TROPICAL ANT FAUNA

EDWARD O. WILSON

Biological Laboratories, Harvard University

Received April 16, 1958

INTRODUCTION

The tropics have long been recognized as the site of maximum evolutionary activity on land. In tropical rain forests, evolution proceeds simultaneously in the largest number of species. A growing amount of evidence of diverse kinds also points to the rain forests, particularly those of continents and the large islands, as the "center" of evolution of a majority of major animal and plant groups, where these groups have diversified maximally in various stages of their phyletic history and out of which they have tended to spread into adjacent temperate and arid regions (Richards, 1952; Darlington, 1957).

According to the theory stressed by Darlington it is the Old World tropics specifically that form the principal evolutionary center of the vertebrates. Successful groups, achieving "general adaptation," send emigrant species out of the Old World tropics into temperate zones, where they may diversify secondarily and in time come to show zonation and radial dispersal of their own. From time to time faunal drift occurs across the region of the present-day Bering Straits. Movement across this barrier has been predominantly out of the Old World into the New. Once emigrant species reach the New World, further diversification may take place; this becomes especially probable if members of the group succeed in penetrating the Neotropical forests.

The nature of general adaptation and the dispersal mechanisms underlying major biotic movement is clearly one of the great problems of modern evolutionary theory. It appears that our knowledge has now reached the stage where finer analyses of these causal processes can and

should be undertaken. There is a need for a "biogeography of the species," oriented with respect to the broad background of biogeographic theory but drawn at the species level and correlated with studies on ecology, speciation, and genetics. In the present paper one such analysis is attempted in very preliminary form, dealing with the ponerine ants [1] of Melanesia. This fauna is unusually suitable for a study of the kind proposed. Melanesia, including New Guinea, has proven to be a peripheral or "recipient" zoogeographic and for ants, i.e., most of the present fauna has been derived ultimately from immigrations from southeastern Asia and Australia, while proportionately few Melanesian-centered groups have emigrated into these adjacent source areas. Hence dominant, successful groups can conceivably be distinguished in their early stages of expansion as they first enter Melanesia, and older resident elements can be studied to piece together details of the later history of invading groups. Theoretically, it should then be possible to characterize the early invading groups ecologically and to infer some of the attributes that have contributed to their successful dispersal, in short, the "attributes of success" associated with general adaptation.

PATTERNS OF SPECIATION

Taxonomic analysis of the Melanesian ponerine fauna has revealed a remarkably

[1] The subfamily Ponerinae is one of eight subfamilies of ants (Formicidae) known from Melanesia. The species known constitute approximately one-fifth of the described ant fauna. Revisionary work on the Melanesian ants has been conducted at Harvard University during the past three years and is now virtually complete (Wilson, 1957 et seq.; Brown, 1958; Willey and Brown, ms.).

EVOLUTION 13: 122–144. March, 1959.

122

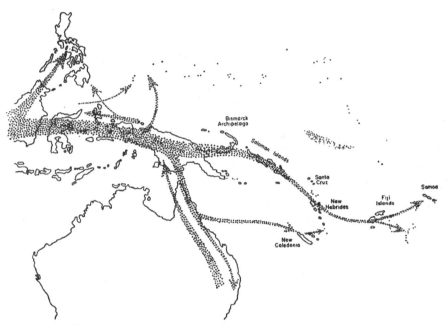

Fig. 1. Routes of dispersal in and out of Melanesia followed by the ponerine ants.

consistent order of zoogeographic patterns. These are summarized in the sections below.

(1) Most groups seem to invade New Guinea from southeastern Asia. Many of these also reach tropical Queensland, and a substantial part of the Queensland fauna is thus composed of Oriental stocks held in common with New Guinea. A few old Australian groups, lacking any apparent affinity with the Oriental fauna, also invade New Guinea. Occasionally, endemic Papuan groups move back in the direction of southeastern Asia but rarely if ever reach beyond the Moluccas and Philippines (fig. 1).

(2) From New Guinea a fraction of the invading stocks presses on to outer Melanesia. An ever-diminishing number reaches the Bismarck Archipelago, Solomon Islands, New Hebrides, and Fiji Islands. Progress along this route follows the classical "filter" effect that applies generally to the faunas of archipelagic chains with permanent water gaps. Some secondary radiation occurs in the various island groups, especially the Fiji Islands, but there is at present no concrete evidence indicating any reverse movement of these precinctive species back toward New Guinea (fig. 1, table 1).

(3) Because radiation has been more extensive on the Fiji Islands, the fauna of this archipelago appears to be larger than that of the adjacent New Hebrides. However, this is true only in the sense that the total number of known species is larger. When species groups are counted instead of species, the fauna of the New Hebrides proves to be somewhat the larger. Since the number of species groups corresponds more closely to the actual number of original immigrant species, this estimate is to be considered a truer measure of relative accessibility of archipelagos to the mainland fauna (fig. 11, table 1).

(4) New Caledonia is faunistically very distinct from the remainder of Melanesia. Despite its proximity to the New Hebrides and Fiji Islands, it does not appear to have received any of its ponerine fauna

124   E. O. WILSON

TABLE 1. *List of the species of Ponerinae known from Melanesia, broken into species groups.* Roman numerals after the names indicate evolutionary classification; lower case letters (a–f) indicate ecological distribution on New Guinea alone, referring specifically to the major habitats given in figure 9. Further explanation in text.

New Guinea	Bismarcks	Solomons	N. Hebrides	Fijis	Samoa
Amblyopone australis (I, f)		*A. australis* (I)	*A. australis* (I)		
		A. celata (III)			
Myopopone castanea (I, ef)	*M. castanea* (I)	*M. castanea* (I)			
Mystrium camillae (I)					
Prionopelta majus-cula (III)	*P. majuscula* (III)				
P. opaca (? II, ef)					
Platythyrea parallela (I, cdf)	*P. parallela* (I)	*P. parallela* (I)			*P. parallela* (I)
P. quadridenta (II)					
Rhytidoponera araneoides (I, bc)	*R. araneoides* (I)	*R. araneoides* (I)			
R. celtinodis (III)					
R. inops (II, def)					
R. gagates (III)	*R. nexa* (III)				
R. nexa (III)					
R. purpurea (II, f)					
R. abdominalis (III, f)					
R. aenescens (III)					
R. laciniosa (III, ef)					
R. rotundiceps (III, f)					
R. strigosa (III, ef)					
R. subcyanea (III, e)					
Gnamptogenys biroi (II, f)					
G. grammodes (II, f)					
G. macretes (II, f)					
G. major (II, f)					
G. cribrata (II, e)		*G. malaensis* (II)			
G. epinotalis (II, f)		*G. albiclava* (III)			
		G. crenaticeps (III)		*G. aterrima* (III)	
		G. lucida (III)			
Proceratium papua-num (II, f)				*P. relictum* (III)	
Discothyrea clavi-cornis (II, f)		*D. clavicornis* (II)			
Ponera biroi (III, ef)					
P. macradelphe (III, f)	*P. biroi* (III)	*P. biroi* (III)		*P. eutrepta* (III)	
P. punctiventris (III, f)		*P. sororcula* (III)		*P. turaga* (III)	
P. sororcula (III, cd)					
P. confinis (I, ef)	*P. confinis* (I)	*P. confinis* (I)	*P. confinis* (I)		
P. pallidula (II, ef)					

TROPICAL ANTS   125

TABLE 1—*Continued*

New Guinea	Bismarcks	Solomons	N. Hebrides	Fijis	Samoa
P. papuana (III)					
P. pruinosa (I, cdef)		*P. pruinosa* (I)	*P. pruinosa* (I)	*P. monticola* (III)	
P. sabronae (III, f)				*P. vitiensis* (III)	
P. clavicornis (I, def)					
P. elegantula (III, f)					
P. selenophora (III, e)		*P. clavicornis* (I)	*P. clavicornis* (I)	*P. colaensis* (III)	
P. syscena (III, f)					
P. xenagos (III, f)					
		P. gleadowi (I)			*P. gleadowi* (I)
P. tenella (III, f)					
P. huonica (III, f)					
P. petila (III, e)					
P. szaboi (III, e)	*P. ratardorum* (III)	*P. ratardorum* (III)	*P. ratardorum* (III)		
P. szentivanyi (III, e)					
P. tenuis (III, f)					
Brachyponera arcuata (I)		*B. croceicornis* (I)			
B. croceicornis (I, cdef)					
Mesoponera manni (I, f)		*M. manni* (I)			
M. papuana (II, e)					
Trachymesopus crassicornis (III, f)		*T. crassicornis* (III)			
		T. sheldoni (III)			
T. darwini (I)		*T. darwini* (I)	*T. darwini* (I)		
T. stigma (I, cdef)	*T. stigma* (I)	*T. stigma* (I)	*T. stigma* (I)	*T. stigma* (I)	*T. stigma* (I)
Ectomomyrmex aciculatus (II, e)		*E. acutus* (II)			
E. acutus (II)	*E. acutus* (II)	*E. aequalis* (II)			
E. exaratus (II, e)					
E. scobinus (II, e)					
E. simillimus (II, e)					
E. striatulus (II, ef)					
					E. insulanus (II)
Bothroponera incisus (?)					
B. obesus (?)					
Cryptopone butteli (I, ef)	*C. butteli* (I)				
C. fusciceps (II, ef)		*C. fusciceps* (II)			
C. testacea (I, e)		*C. testacea* (I)			
C. motschulskyi (III, def)					
Diacamma rugosum (I, cdef)					

126   E. O. WILSON

TABLE 1—*Continued*

New Guinea	Bismarcks	Solomons	N. Hebrides	Fijis	Samoa
Myopias cribriceps (III, e)					
M. delta (III, e)					
M. concava (III, ef)					
M. foveolata (III)					
M. gigas (III, e)					
M. julivora (III, e)					
M. levigata (III, f)					
M. loriai (III, f)					
M. media (III, f)					
M. ruthae (III, e)					
M. xiphias (III)					
M. latinoda (II, f)					
M. tenuis (?III, cdef)					
M. tylion (?III, ef)					
Leptogenys bituberculata (II, ef)				*L. foveopunctata* (II)	
				L. fugax (II)	
L. drepanon (II)			*L. hebrideana* (II)	*L. humiliata* (II)	
L. indagatrix (II, f)				*L. letilae* (II)	
L. papuana (II)				*L. navua* (II)	
L. triloba (III, f)				*L. vitiensis* (II)	
L. breviceps (II, ef)					
L. caeciliae (III, f)					
L. optica (III, ef)					
L. diminuta (I, cdef)		*L. diminuta* (I)			
L. nitens (II)	*L. diminuta* (I)	*L. oresbia* (II)			
L. purpurea (II, f)					
L. foreli (I)	*L. emeryi* (III)	*L. foreli* (I)	*L. foreli* (I)		
		L. truncata (III)			
L. keysseri (III, f)					
Odontomachus latissimus (II)					
O. linae (II, f)					
O. malignus (I, a)		*O. emeryi* (II)			
O. imperator (II)					
O. montanus (II)	*O. malignus* (I)	*O. malignus* (I)		*O. angulatus* (II)	
O. opaculus (II, f)					
O. papuanus (I, ef)					
O. rufithorax (II)					
O. saevissimus (I)					
O. tauerni (II)					
O. aciculatus (III)					
O. aeneus (III)	*O. simillimus* (I)	*O. simillimus* (I)	*O. simillimus* (I)	*O. simillimus* (I)	*O. simillimus* (I)
O. cephalotes (I, ef)					
O. simillimus (I, bcdf)					
O. nigriceps (III, e)					
O. testaceus (III, ef)	*O. tyrannicus* (III)				
O. tyrannicus (III, f)					
Anochetus cato (III, f)	*A. cato* (III)	*A. cato* (III)			
		A. isolatus (III)			

TABLE 1—*Continued*

New Guinea	Bismarcks	Solomons	N. Hebrides	Fijis	Samoa
A. chirichinii (III, ce)					
A. fricatus (III, e)					
A. graeffei (I, d)	*A. graeffei* (I)	*A. graeffei* (I)	*A. graeffei* (I)	*A. graeffei* (I)	*A. graeffei* (I)
A. variegatus (III)					
A. vesperus (III, d)					

from these islands. Rather, all endemic stocks appear to have come from tropical or subtropical Australia. This is true even in the cases of Oriental groups found both on New Caledonia and elsewhere in Melanesia; the New Caledonian species are in each instance most closely related to eastern Australian representatives of the invading Oriental group. Thus Oriental groups reaching New Caledonia, of which there are a goodly number, appear to have entered by way of eastern Australia. One species (*Amblyopone australis*) has spread from New Caledonia to the southern New Hebrides, thus providing a slender link between these two segments of the Melanesian fauna. The New Caledonian endemics show nearly every conceivable degree of evolutionary divergence from the cognate Australian species, from a clearly conspecific condition to a highly modified form approaching a distinct generic level (figs. 1, 4, 5, 6).

(5) Individual ponerine species and species-groups in Melanesia seem to show various steps in a sequence of expansion, diversification, and contraction. In fact, when the distributions and phylogenetic positions of all of the species are considered as a whole, the conclusion is almost inescapable that such a process is nearly universal in this group of ants. The steps are illustrated by the actual distributions of species shown in figures 2–8. These examples are admittedly selected, and the evolutionary history they are supposed to follow hypothetical. Yet virtually all of the patterns of the species groups fit somewhere within this unilateral zoogeographic classification, and the evolutionary interpretation proposed seems to be the most reasonable and simple one within the limits of the existing evidence. Three classes of species can be distinguished, each representing a major step in the historical sequence.

Stage-I species. These are the species apparently in the process of expansion, either from southeastern Asia or Australia into Melanesia, or out of Melanesia back into the primary source areas, or out of New Guinea through most of the remainder of Melanesia. Thus two categories of Stage-I species are hypothesized: *primary,* including Oriental- or Australian-based species invading Melanesia (as well as Oriental-based species invading New Guinea and then Australia by way of New Guinea); and *secondary,* including Papuan species that are pushing well out into outer Melanesia or invading southeastern Asia or Australia. Stage-I species generally have continuous ranges and show little geographic variation, although non-geographic variation may be considerable (fig. 2).

Stage-II species. These are species which have differentiated to species level in Melanesia, i.e., are Melanesian precinctives, but which belong to species groups centered outside Melanesia, either in southeastern Asia or Australia. They represent stocks which are relatively recent invaders but which have been resident sufficiently long to diverge to species level (Figs. 3, 4).

Stage-III species. These are Melanesian precinctives belonging to Melanesian-centered species groups. That such

128   E. O. WILSON

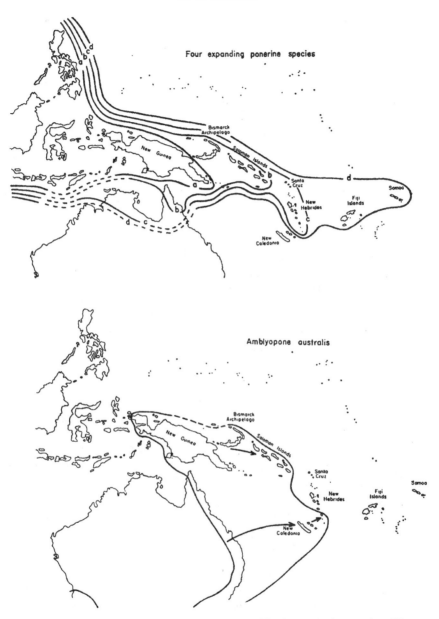

Fig. 2. Maximum known distributions of several Stage-I ponerine species. Upper: four unrelated species believed to represent successive stages of invasion from Asia; *a*, *Diacamma rugosum*; *b*, *Myopopone castanea*; *c*, *Trachymesopus darwini*; *d*, *T. stigma*. Lower: the range of *Amblyopone australis*, which has entered Melanesia from Australia in two places; a double entry is inferred from the fact that the Papuan-Solomons and New Caledonian-New Hebridean populations are marked by opposing trends in geographic variation.

In this and subsequent distribution maps the following qualifications should be noted. The precise limits of the species ranges in the Philippines, lesser Sundas, and lesser island groups such as the Louisiade and Manus Archipelagos are unknown. Dashed lines indi-

groups are ultimately Oriental or Australian in origin is suggested by the fact that groups in stages intermediate between II and III exist (e.g., groups of *Ponera selenophora* and *P. tenuis*). These have a few members with relict distributions outside of Melanesia that appear (within the framework of the present hypothesis) to be the last remnants of a contracting central (Asian or Australian) fauna (figs. 5–8).

Most Stage-III species have sufficient potency to continue diversifying, and some are able to expand out of Melanesia again, thus secondarily achieving Stage-I status (figs. 6, 7, 8, 12). Eventually, however, they probably enter upon a phase of irrevocable recession, their ranges contracting more and more until they can properly be labeled "relict" species (fig. 8). Presumably the next and final step for most is extinction, or extreme specialization, perhaps hastened by the inroads of freshly competing Stage-I and Stage-II species.

Ecological Attributes of the Expanding (Stage-I) Species

Using the zoogeographic criteria outlined in the preceding section, it is now possible to make a qualified distinction of those species that are in the early stages of expansion in or out of Melanesia. In the great majority of cases these belong to "dominant" Oriental- or Australian-centered species groups, i.e., groups that are relatively abundant, widespread, and diversified in the source areas. When ecological data gathered by the author during recent field work in New Guinea was analyzed, it was shown that Stage-I species in this area are characterized by (1) a proportionately greater concentration in marginal habitats, marginal in this case being defined specifically as open lowland forest, grassland, and littoral [2] (table 2, fig. 9) and (2) a greater ecological amplitude (table 3, fig. 10). Further, when the composition of the Melanesian fauna is broken down according to successive archipelagos, it was shown that a significantly larger proportion of Stage-I species occupies central Melanesia than is the case in New Guinea and the Fiji Islands (table 4, fig. 11). The evolutionary implications of these interesting correlations will now be considered.

Discussion

If the zoogeographic sequence postulated here is correct, it seems to lead to the conclusion that ponerine species invade New Guinea, and outer Melanesia through New Guinea, by way of the relatively sparsely populated marginal habitats. When occasional Melanesian-based species succeed in expanding out of New Guinea into the Philippines or Australia, they apparently leave by the same adaptive route. This ecological attribute is the most significant one evident within the limits of the existing data.

It should be mentioned that there does not seem to be any other peculiarity in mode of dispersal or nest-site preference in Stage-I species that can account for their current expansion. These species include *Diacamma rugosum* and *Leptogenys foreli*, which have wingless queens, as well as many species that are presumed

[2] The central habitats are the "inner" rain forest, including denser lowland forest and montane (mid mountain) forest up to approximately 1000 meters elevation. Marginal habitats also have the smallest ponerine faunas in terms of absolute numbers of species.

cate that the presence of the species anywhere in the island group under question is not known but believed likely. In most cases the precise limits of the ranges within larger archipelagos, such as the Solomon Islands and New Hebrides, are not known; where the entire archipelago is enclosed this can be taken certainly to mean only that the species occurs somewhere within the archipelago. Detailed locality data can be found in the revisionary basis of this article published elsewhere and cited in the bibliography.

130   E. O. WILSON

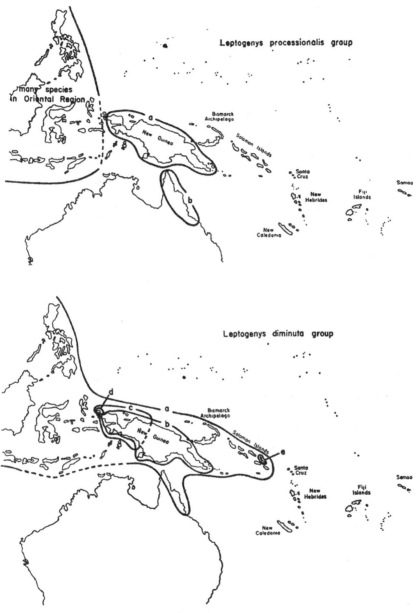

Fig. 3. Maximum known distributions of species in early Stage II. Upper: *a, Leptogenys breviceps; b, L. fallax*. Lower: *a, L. diminuta; b, L. purpurea; c, L. nitens; d, L. violacea; e, L. oresbia*.

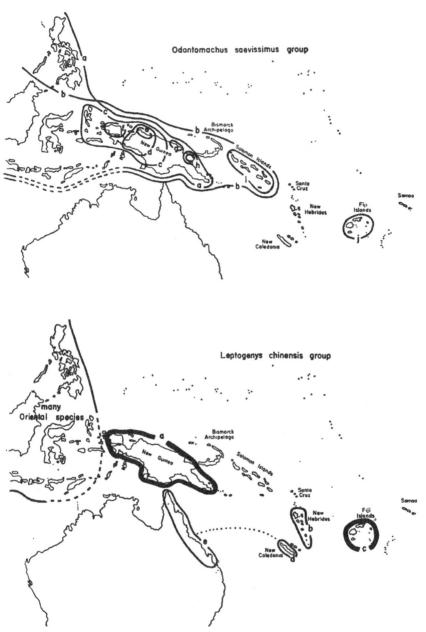

Fɪɢ. 4. Maximum known distribution of species in advanced Stage II. Upper: *a*, *Odontomachus rixosus*; *b*, *O. malignus*; *c*, *O. tauerni*; *d*, *O. rufithorax*; *e*, *O. linae*; *f*, *O. imperator*; *g*, *O. opaculus*; *h*, *O. latissimus* and *O. montanus*; *i*, *O. emeryi*; *j*, *O. angulatus*. Lower: *a*, *Leptogenys bituberculata; L. drepanon, L. indagatrix, L. papuana, L. triloba* (the precise limits of these species on New Guinea is not indicated); *b*, *L. hebridcana; c, L. foveopunctata; L. fugax, L. humiliata, L. letilae, L. navua, L. vitiensis; d, L. sagaris; e, L. anitae.* The dotted line suggests the close relationship of the New Caledonian *sagaris* and Australian *anitae*.

132   E. O. WILSON

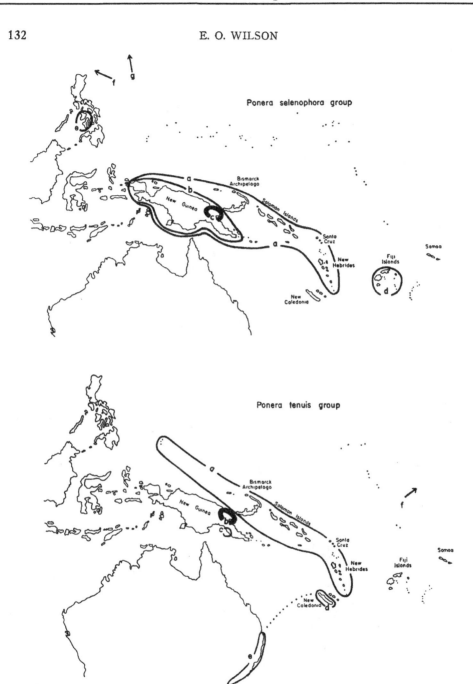

Fig. 5. Maximum known distribution of species in early Stage III, in which the Oriental source fauna has contracted and at present contains only a few relict species. Upper: *a, Ponera clavicornis* (ranked as a secondary Stage-I species); *b, P. selenophora; c, P. elegantula, P. syscena, P. xenagos; d, P. colaensis; e, P. oreas* (Philippines); *f, P. sinensis* (Hong Kong); *g, P. scabra* (Honshu). Lower: *a, P. ratardorum, b, P. huonica, P. petila, P. szaboi, P. tenuis; c, P. szentivanyi; d, P. caledonica; e, P. exedra* (Australia); *f, P. swezeyi* and *P. zwaluwenburgi* (Hawaii); *g, P. incerta* (Java). The dotted line suggests the close relationship between the New Caledonian *caledonica* and Australian *exedra*.

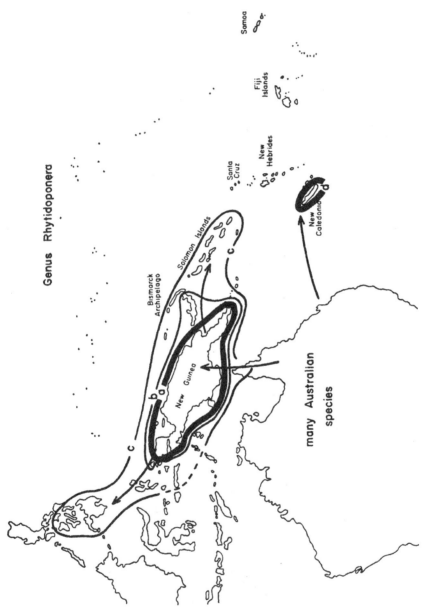

Fig. 6. The distribution of the genus *Rhytidoponera*, which is ultimately Australian in origin but which is differentiating and expanding secondarily out of New Guinea. Most of the Papuan species belong to Papuan-centered species groups and are classified in Stage III. *Rhytidoponera araneoides*, however, has expanded to the Philippines and Solomon Islands and is classified as Stage I (secondary). *a* represents the bulk of the Papuan species, which are confined to the New Guinea mainland; *b, R. nexa; c, R. araneoides; d,* six endemic New Caledonian species, whose closest affinities are with Australian stocks.

134   E. O. WILSON

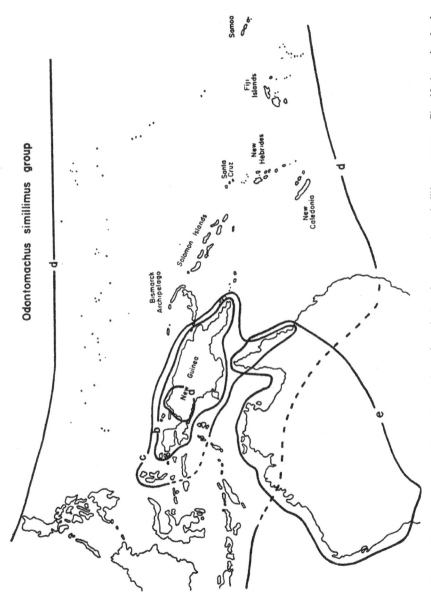

Fig. 7. Maximum known distributions of species of the *Odontomachus simillimus* group. Classified variously in Stage I (secondary) or Stage III, depending on their present ranges. *a*, *O. aeneus*; *b*, *O. aciculatus*; *c*, *O. cephalotes*; *d*, *O. cimillimus*; *e*, *O. ruficeps*. *O. simillimus* (=*O. haematodus auct.*) ranges over most of the Pacific and has probably been transported through a large part of this area by man.

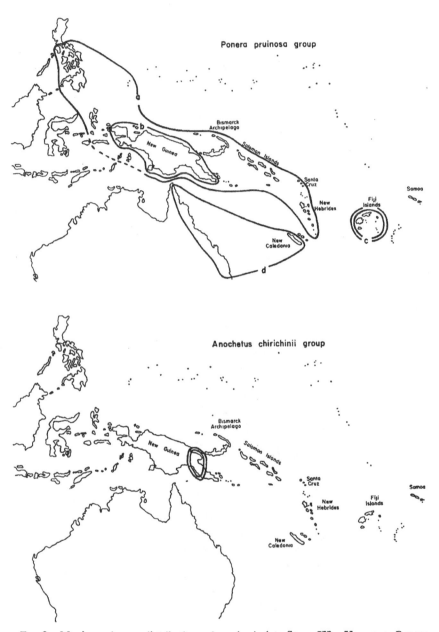

FIG. 8. Maximum known distributions of species in late Stage III. Upper: *a, Ponera pruinosa; b, P. sabronae; c, P. monticola* and *P. vitiensis; d, P. elliptica.* Lower: *Anochetus chirichinii* and *A. vesperus.* All of these species are classified in Stage III except *Ponera pruinosa,* which has expanded to outer Melanesia, the Philippines, and Palau. *P. elliptica* is not considered in the present analysis.

136   E. O. WILSON

TABLE 2. *Partition by evolutionary stage of the Ponerine fauna in six major New Guinea habitats*

Habitat	Stage I	Stage II	Stage III	Total	χ^2	P*
a. Littoral	1(*1.0*)**	—	—	1	0	
b. Savanna	2(*1.0*)	—	—	2		
c. Monsoon forest	8(*0.8*)	—	2(*0.20*)	10	0.41	
d. Lowland rain forest (marginal)	9(*0.69*)	1(*0.08*)	3(*0.23*)	13	0.81	
e. Lowland rain forest (interior)	12(*0.27*)	13(*0.29*)	20(*0.44*)	45	8.08	<0.02
f. Montane rain forest	15(*0.24*)	20(*0.32*)	27(*0.44*)	62	0.17	

* P indicates the probability that the evolutionary and ecological characteristics in two adjacent rows are independent. It is given here in the single instance where it is "significantly" small (<0.05).
** Italicized numbers in parentheses are the proportions with respect to evolutionary stage.

to conduct normal nuptial flights.[3] Stage-I includes a larger proportion of species nesting in open soil and beneath stones, as opposed to rotten wood and trees, than

do Stages II and III, so that it is not easy to account for the greater dispersal of Stage-I species by means of a differential rate of "rafting."

[3] The existence of a normal nuptial flight is no guarantee of a relatively high potential dispersal rate, since the organization of the flight itself may be of such a nature as to limit greatly this rate. In the Nearctic myrmicine species *Pheidole sitarches*, mating takes place on the ground directly beneath the swarms of males, and the queen does not attempt to fly afterward, so that the species range can be advanced within a generation only over the distance by which the males are able to form their swarms (Wilson, 1958c). If similar behavioral phenomena occur in Indo-Australian species, they can presumably normally cross water gaps only if the gaps are less than the distance over which the swarms can be formed.

TABLE 3. *Partition by evolutionary stage of species occurring in various numbers of major habitats in New Guinea* *

Number of habitats occupied	Stage I	Stage II	Stage III	Total
1	5(*0.09*)	19(*0.33*)	33(*0.58*)	57
2	6(*0.30*)	6(*0.30*)	8(*0.40*)	20
3	2(*0.40*)	2(*0.40*)	1(*0.20*)	5
4	6(*1.0*)	—	—	6

* On the basis of two classes, occupants of a single habitat vs. occupants of multiple (2, 3, 4) classes, $\chi^2 = 16.62$ and P = 0.001.

TABLE 4. *Partition by evolutionary stage of the species comprising the Ponerine faunas of various of the Melanesian Archipelagos and Samoa*

Place	Stage I	Stage II	Stage III	Total	χ^2	P
New Guinea Mainland*	24(*0.22*)	35(*0.31*)	53(*0.47*)	112	10.52	0.005
Bismarck Archipelago**	10(*0.56*)	1(*0.06*)	7(*0.38*)	18	1.67	
Solomon Islands**	18(*0.49*)	7(*0.19*)	12(*0.32*)	37	3.83	
New Hebrides**	9(*0.82*)	1(*0.09*)	1(*0.09*)	11	11.10	0.001–
Fiji Islands*	3(*0.18*)	7(*0.41*)	7(*0.41*)	17		0.002
Samoa**	5(*0.83*)	1(*0.17*)	—	6	8.69	0.01–0.02

* The New Guinea and Fijian faunas do not differ significantly from each other in proportionment by evolutionary stage; χ^2 is only 0.64.
** The faunas of central Melanesia (Bismarcks, Solomons, New Hebrides) and Samoa do not differ significantly from each other in proportionment by evolutionary stage; χ^2 between New Hebrides and Samoa is 0.87.

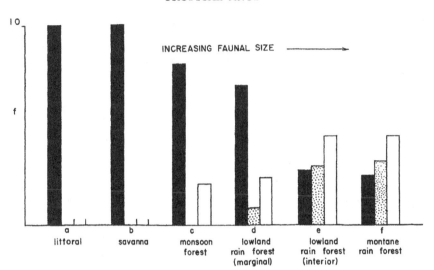

FIG. 9. Partition by evolutionary stage of the ponerine faunas of various major habitats in New Guinea. Black, Stage-I species; stippled, Stage II; blank, Stage III. See table 2.

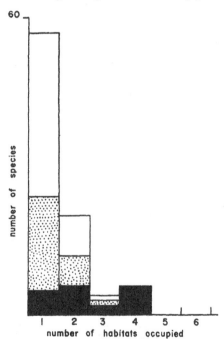

FIG. 10. Relationship of evolutionary stage and number of habitats occupied in the New Guinea fauna. Conventions as in figure 8. See table 3.

Is it possible that faunal movement occurs through marginal habitats simply because these habitats are more often nearer the coast? The present pattern of vegetation on New Guinea does not lend strong support to such a view. It is true that mountains, and montane rain forest, containing the largest number of ponerine species, are predominantly inland. But it is also generally true that where rainfall is sufficient to maintain lowland rain forest, this forest extends right to the coast, and where rainfall is not sufficient, grassland and monsoon forest extend far inland. It does not seem possible to invoke the slightly greater geographic accessibility of marginal habitats as the deciding factor in the marked preference of Stage-I species for these habitats.

Perhaps a stronger possibility to be considered is that the fauna of the marginal habitats is more prone to dispersal by wind and water, since it is more open to the action of wind currents and tends to be distributed differentially near waterways, e.g., "open aspect" rain forest predominates along streams in forested areas. There is at the present time insufficient evidence to test the validity of this supposition. If anything, two characteristics of Stage-I species, the common occurrence

138   E. O. WILSON

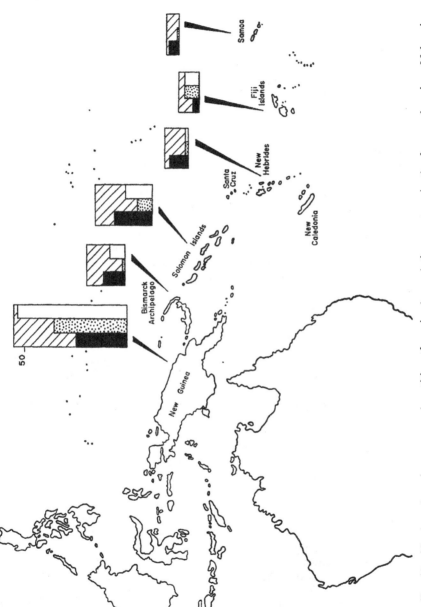

Fig. 11. Number of species groups and partition of species by evolutionary stage in the faunas of various Melanesian archipelagos. Hatched column represents the number of species groups; otherwise conventions are as in figure 9. The size of the Bismarck fauna as given is probably not indicative of its true relative size, since it is the most poorly explored; it is probably larger than that of the Solomons. See table 4.

of flightless queens and the common preference of soil as a nesting site, seem to weigh against regarding such an explanation too seriously at this time. Furthermore, there is some direct evidence, to be considered below, that the resident Stage-II and Stage-III species play a role in keeping Stage-I species out of the inner rain forest habitat.

It might also be argued that parts of the Celebes-Moluccan corridor, including some of Celebes itself and Buru, are drier and more "marginal" in their vegetation cover, hence posing a filter that has favored species adapted to marginal habitats. Again the evidence is not adequate to evaluate this possibility, and it may not be soluble until ecological studies are carried out in the Celebes and Moluccas. At least it is known that vast areas of these islands, including the entire island surfaces in some cases, are covered with dense rain forests, thus seeming to provide excellent stepping-stones that would allow a true rain forest fauna to cross over to New Guinea.

If the provisional interpretation adopted here is true, that different dispersal mechanisms and purely geographic phenomena are not sufficient to account for the marginal-habitat effect, one is inclined to turn to features of the biotic environment within Melanesia for a more complete explanation. There is likely significance in the fact that marginal habitats, containing as they do the smallest number of species, offer the least diversified biotic resistance. It is in fact the hypothesis advanced here that Stage-I species, by penetrating marginal habitats initially, are "flanking" the competition of the resident fauna rather than undertaking a "frontal assault" on it. The ant fauna of the inner rain forest does indeed present the appearance of a formidable closed "association." The species are very numerous (over 160 at the lower Busu River alone) and highly specialized to fill, in the aggregate, almost every imaginable niche available to ants. Moreover, the absolute composition of local faunas, as well as the

relative abundance of individual species, changes over distances as short as 12 kilometers, producing a "kaleidoscope effect" in the total faunal structure (Wilson, 1958d). In short, the potential competition offered by the resident inner rain forest fauna is very great at any given locality, and its overall effect may be greatly heightened by the spatially and temporally shifting nature of the total faunal structure, which would serve to thwart the extension of any local adaptation achieved by the first invading populations.

Now the question must be raised, if Stage-I species normally enter by way of marginal habitats, why has there not been a faunal buildup in these habitats, eventually closing them to further invasion? There is abundant evidence to indicate that species do not linger long in the marginal habitats. By the time they have differentiated and achieved Stage-II status, they usually have succeeded in penetrating the deep rain forest habitat.[4] Moreover, there is some indication that Stage-I species enter deep rain forest wherever competitive forms are relatively scarce. Thus *Rhytidoponera araneoides* is restricted to marginal habitats in New Guinea, but is apparently abundant in deep rain forest in the Solomon Islands, where the ponerine and myrmicine fauna is significantly smaller. On the Huon Peninsula of New Guinea, which has one of the largest and most diversified ponerine-myrmicine faunas of any comparable area in the world, *Odontomachus simillimus* (= *haematodus auct.*) is limited to marginal habitats. In the vicinity of the Brown River, Papua, with a local fauna only slightly larger than half that of local

---

[4] For the Ponerinae the most favored type of habitat in the New Guinea lowlands is what I have described elsewhere as "medium aspect" rain forest (Wilson, 1958d). Within this habitat most Stage-II and Stage-III ponerine species nest on the ground in rotting logs at the "zorapteran" and "passalid" stages of decomposition and prey on small arthropods. Microhabitat and nest-site choice is more variable in the mountain rain forests.

140 E. O. WILSON

faunas on the Huon Peninsula, *O. simillimus* is moderately abundant in primary rain forest as well as marginal habitats. On Espiritu Santo, New Hebrides, which has a depauperate fauna, *O. simillimus* is one of the dominant ants on the floor of primary rain forest.

At this point the possibility should be considered that the invasion of Melanesia by way of the marginal habitats is a recent, unique event and not a continuing historical process, as supposed. This would be true if the coming of man, or else extensive drying of parts of New Guinea during the Pleistocene independent of man, had been necessary to open marginal habitats sufficiently for the Stage-I species to invade. In the author's opinion it is highly unlikely that such a circumstance existed. Even if at various geologic periods there were no savanna or monsoon forest, there would always be extensive rain forest border, and within the rain forest abundant patches of open-aspect, second-growth vegetation resulting from the fall of large trees, stream erosion, and landslides. Anyone who has worked in primary tropical forest is

familiar with the extensiveness of these phenomena. Native cultivation on New Guinea has undoubtedly greatly increased the extent of marginal habitats, and along with it the size of Stage-I populations, but if its effects were to be obliterated overnight, there would still be sufficient marginal areas to support large Stage-I populations. The same is probably true if all savannas and monsoon forests present in New Guinea today were to be replaced by continuous rain forest; there would remain abundant patches of open-aspect vegetation within the forest.

For invading (Stage-I) species, entering New Guinea by the marginal habitat route, there is a strong selective pressure to penetrate the inner rain forest (see fig. 12). Apparently they must achieve this adaptive shift in a relatively short time or face extinction, either through unfavorable environmental changes or under the pressure of newly invading competitor species. In evidence is the fact that there are relatively few old endemic species in the marginal habitats and even fewer that appear to be confined to them. This situation poses something of a paradox,

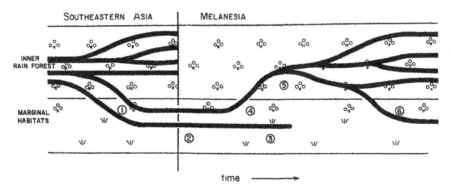

FIG. 12. Schematic representation of the hypothesized evolution of ponerine species groups in Melanesia, in this case tracing the history of groups derived ultimately from Asia. (1) Species or infraspecific populations adapt to marginal habitats in southeastern Asia, then cross the water gap to New Guinea and colonize marginal habitats there (2). In time these Stage-I species either become extinct (3) or invade the inner rain forest of Melanesia (4). Having successfully adapted to the inner rain forest, they diverge to species level (5), thus entering Stage II. As diversification continues in Melanesia, the source fauna in Asia may be contracting, so that in time the group as a whole becomes Melanesian-centered and its Melanesian species are classified as Stage III. A few of the New Guinea species may readapt to the marginal habitats (6) and expand back out of New Guinea, thus entering Stage I secondarily.

for it is true that the marginal habitats support (at least at present) large populations of ants. For the period of their tenure in the marginal habitats, the Stage-I species appear to be relatively abundant. What, then, causes their frequent extinction and the resultant selective pressure to shift into the deep rain forest habitat? Two causal phenomena can be hypothesized: (1) the environment of the marginal habitats, both physical and biotic, is more variable and capricious; (2) historically the marginal habitats have been limited in size and diversity of available niches, so that resident populations have tended to remain genetically relatively homogeneous and hence more vulnerable to the deleterious effects of rapid environmental change. (It should be noted that while the marginal habitats have fewer niches, it is thought that these niches tend to remain more open than is the case in rain forest, because of greater fluctuations in population size of resident species.)

Thus it is conceivable that the marginal habitats provide an opening for invading species while at the same time serving as an evolutionary trap for those that remain in residence. The islands of outer Melanesia may function in a similar fashion. As shown here, all of the archipelagos, with the single exception of the Fiji Islands, contain a higher proportion of Stage-I species than New Guinea. This youthful character of the fauna of central Melanesia may be due to the fact that these islands are geologically younger or less stable. But it may be also true that their smaller size and less diversified biota cause them to function also as evolutionary traps, "weakening" resident species and rendering them more susceptible to competition from subsequently invading species. The idea that oceanic islands do in fact have such a general effect on the evolution of endemic faunas has already been developed at some length by Mayr (1942, 1954) and Darlington (1957) and does not need further expression here.

Dobzhansky (1955) has recently summarized the population geneticist's view of the process of adaptation to diversified environments in the following rule: "Granted that genetic variability is an instrumentality whereby Mendelian populations master environmental diversity, one may expect that, other things being equal, populations which control a greater variety of ecologic niches will be more variable than those having a more limited hold on the environment." In a penetrating analysis extending over the past ten years, Dobzhansky and his colleagues have gone far to document just a phenomenon in the *Drosophila willistoni* group (da Cunha, Burla, and Dobzhansky, 1950; da Cunha and Dobzhansky, 1954; Birch and Battaglia, 1957; Dobzhansky, 1957; Townsend, 1958). Although the evidence is of a different and more limited kind, one is tempted to extend Dobzhansky's rule to include consideration of the evolution of the Melanesian ponerines. At first glance there appears to be a basic contradiction. If Stage-I species tend to be genetically impoverished, as suggested in the present paper, how is it that they occur in a greater array of major habitats than do old resident species? Would not greater ecological amplitude of this nature presuppose greater genetic diversification? The answer suggested by our much more extensive knowledge of *Drosophila willistoni* is that occurrence of a species or a population of a species in a wide range of major habitats (as defined here) does not of itself imply genetic diversity. If the habitats are marginal, or the distribution peripheral, the species (or population) may be relatively homogeneous. Of much greater importance are length of residence within the part of the range under consideration and the microecological diversification within the major habitats occupied. In the case of *Drosophila willistoni* it is the tropical forests and adjacent savannas of central Brazil that contain the largest number of inversion types and the highest percentage of inversion heterozygotes. The Brazilian forests appear to be the center of the range of the species and also that part of the range to which resident populations are maximally adapted. In the case of the Melanesian ponerines, most of the Stage-I species are derived from Oriental stocks that are probably primarily rain-forest dwellers. In the case of these that have reached Stage-I status by expanding out of New Guinea by way of the marginal habitats, it is known with certainty that they belong to groups that are otherwise adapted to the inner rain forest. In making an adaptive shift to marginal habitats, the pioneer populations of these stocks are entering a peripheral, generally less favorable area, and it is probable that their genetic variability is correspondingly diminished. Furthermore, as Mayr (1954) has duly emphasized in his elaboration of the "founder principle" in speciation, pioneer populations such as those crossing water gaps in Melanesia are certain to have diminished variability, and the consequently narrowed "genetic environment" may have an accelerating effect on subsequent evolution.

With the foregoing theoretical background in mind, it is now possible to construct a model of the sequence of genetic events that have occurred in the colonization of Melanesia by the ponerine ants. Five steps can be conceived.

1. Prior to its invasion of New Guinea, the species occupies a central range, or permanent breeding area, probably composed mostly of deep rain forest, and on occasion marginal areas as well, in sum making up what Brown (1957) has referred to as the "maximum range." Brown has suggested the following events as being typical for animal populations: "Within the maximum range, the populations of the species normally undergo successive expansions into the less favorable areas, alternating with contractions into more favorable refuges. The expansions and contractions are the sequelae of inevitable density fluctuations affecting all or part of the species at one time."

2. Evolutionary changes accompany the continuing process of expansion and contraction. Populations occupying marginal areas will tend to adapt to the differing habitats found there.

3. Populations that are adapted to marginal habitats are now candidates for the colonization of New Guinea. Queens and colony fragments from both rain-forest and marginal-habitat populations probably disseminate from time to time across the water gaps to New Guinea, but generally only those adapted to marginal habitats are able to establish pioneer populations. The basis of this selection is the more open nature of the resident endemic marginal-habitat populations on New Guinea, containing fewer species and offering overall less competition to invading forms.

4. Having become established in New Guinea, the pioneer populations tend to remain genetically homogeneous because of the "niche-poor" condition of the marginal habitats. Their future is rendered further hazardous by the more intense fluctuations that occur in the marginal environments. Hence there is a strong selective pressure on these populations to penetrate the inner rain forest habitat.

5. If penetration of the inner rain forest is successful, new evolutionary opportunities are realized, and the populations tend to diversify and speciate. In time Group-II status is reached. With the further passage of time, however, the phyletic line will ultimately tend to diminish, as new, better adapted stocks penetrate the inner rain forest from outside. Occasionally, a species will break out of the inner rain forest habitat secondarily and expand out of New Guinea in the same fashion that the ancestral species entered, by way of the marginal habitat route. Such species, however, rarely if ever are able to push back beyond Australia or the Philippines.

SUMMARY

1. The zoogeography of the Melanesian ponerine fauna is preliminarily analyzed. Most of the fauna has apparently been

derived ultimately from Oriental stocks entering by way of New Guinea; some invading species are able to spread beyond this island to Queensland and outer Melanesia. A smaller part of the fauna has been derived from old Australian stocks that have entered by way of New Guinea or New Caledonia. Faunal flow from New Guinea through outer Melanesia has been unidirectional, with an ever diminishing number of species groups found outward from the Bismarcks to the Fiji Islands. New Caledonia draws almost all of its fauna from eastern Australia and has engaged in very little direct faunal exchange with the remainder of Melanesia.

2. A cyclical pattern of expansion, diversification, and contraction is hypothesized to account for later evolutionary events following initial dispersal. Following invasion of Melanesia (Stage I, *primary*), the pioneer populations may then diverge to species level (Stage II) and further diversify. Eventually the source populations outside Melanesia tend to contract, leaving the species group as a whole peripheral and Melanesian-centered (Stage III). Endemic Melanesian species occasionally enter upon a secondary phase of expansion (Stage I, *secondary*) but are rarely if ever able to push beyond Australia or the Philippines.

3. Stage-I species are characterized on New Guinea by their greater concentration in "marginal" habitats, including open lowland forest, savanna, and littoral. The central habitats, including denser lowland forest and montane forest, contain significantly larger faunas as well as a higher percentage of Stage-II and Stage-III species. Stage-I species are also characterized by their individual occurrence in a greater range of major habitats. Finally, these species make up a significantly higher proportion of the faunas of the archipelagos of central Melanesia, including the Bismarck Archipelago, the Solomon Islands, and the New Hebrides.

4. On the basis of these data it is suggested that ponerine species normally invade New Guinea by way of the marginal habitats. Evolutionary opportunity is nevertheless limited in the marginal habitats, and there is a strong selective pressure favoring re-entry into the inner rain forest habitats. In general, Stages II and III, leading to the origin of the great bulk of the Melanesian fauna and its most distinctive endemic elements, are played out only in the inner rain forest.

Acknowledgments

The author wishes to acknowledge his gratitude to W. L. Brown, P. J. Darlington, T. Dobzhansky, A. E. Emerson, C. Lindroth, E. Mayr, and E. C. Zimmerman for reading the manuscript and offering many helpful criticisms and suggestions.

Literature Cited

Birch, L. C., and B. Battaglia. 1957. Selection in *Drosophila willistoni* in relation to food. Evolution, 11: 94–105.

Brown, W. L. 1957. Centrifugal speciation. Quart. Rev. Biol., 32: 247–277.

——. 1958. Contributions toward a reclassification of the Formicidae. II. Tribe Ectatommini (Hymenoptera). Bull. Mus. Comp. Zool. Harvard, 118: 175–362.

da Cunha, A. B., H. Burla, and Th. Dobzhansky. 1950. Adaptive chromosomal polymorphism in *Drosophila willistoni*. Evolution, 4: 212–235.

da Cunha, A. B., and Th. Dobzhansky. 1954. A further study of chromosomal polymorphism in *Drosophila willistoni* in its relation to the environment. Evolution, 8: 119–134.

Darlington, P. J. 1957. Zoogeography. Wiley.

Dobzhansky, Th. 1955. A review of some fundamental concepts and problems of population genetics. Cold Spring Harbor Symp. Quant. Biol., 20: 1–15.

——. 1957. Genetics of natural populations XXVI: chromosomal variability in island and continental populations of *Drosophila willistoni* from Central America and the West Indies. Evolution, 11: 280–293.

Mayr, E. 1942. Systematics and the origin of species. Columbia University Press.

——. 1954. Change of genetic environment and evolution. In Evolution as a process. Allen and Unwin.

Richards, P. W. 1952. The tropical rain forest. Cambridge University Press.

144 E. O. WILSON

TOWNSEND, J. I. 1958. Chromosomal polymorphism in Caribbean island populations of *Drosophila willistoni*. Proc. Nat. Acad. Sci., 44: 38–42.

WILSON, E. O. 1957. The *tenuis* and *selenophora* groups of the ant genus *Ponera* (Hymenoptera: Formicidae). Bull. Mus. Comp. Zool. Harvard, 116: 355–386.

——. 1958a. Studies on the ant fauna of Melanesia. I. The tribe Leptogenyini. II. The tribes Amblyoponini and Platythyreini. Bull. Mus. Comp. Zool. Harvard, 118: 101–153.

——. 1958b. Studies on the ant fauna of Melanesia. III. *Rhytidoponera* in western Melanesia and the Moluccas. IV. The tribe Ponerini. Bull. Mus. Comp. Zool. Harvard, 118: 303–371.

——. 1958c. Organization of a nuptial flight of the ant *Pheidole sitarches* Wheeler. Psyche, 64: 46–50.

——. 1958d. Patchy distributions of ant species in New Guinea rain forests. Psyche (In press.)

EVOLUTION

INTERNATIONAL JOURNAL OF ORGANIC EVOLUTION

PUBLISHED BY

THE SOCIETY FOR THE STUDY OF EVOLUTION

| Vol. 17 | DECEMBER, 1963 | No. 4 |

AN EQUILIBRIUM THEORY OF INSULAR ZOOGEOGRAPHY

ROBERT H. MACARTHUR[1] AND EDWARD O. WILSON[2]

Received March 1, 1963

THE FAUNA–AREA CURVE

As the area of sampling A increases in an ecologically uniform area, the number of plant and animal species s increases in an approximately logarithmic manner, or

$$s = bA^k, \qquad (1)$$

where $k < 1$, as shown most recently in in the detailed analysis of Preston (1962). The same relationship holds for islands, where, as one of us has noted (Wilson, 1961), the parameters b and k vary among taxa. Thus, in the ponerine ants of Melanesia and the Moluccas, k (which might be called the *faunal coefficient*) is approximately 0.5 where area is measured in square miles; in the Carabidae and herpetofauna of the Greater Antilles and associated islands, 0.3; in the land and freshwater birds of Indonesia, 0.4; and in the islands of the Sahul Shelf (New Guinea and environs), 0.5.

THE DISTANCE EFFECT IN PACIFIC BIRDS

The relation of number of land and freshwater bird species to area is very orderly in the closely grouped Sunda Is-

lands (fig. 1), but somewhat less so in the islands of Melanesia, Micronesia, and Polynesia taken together (fig. 2). The greater variance of the latter group is attributable primarily to one variable, distance between the islands. In particular, the distance effect can be illustrated by taking the distance from the primary faunal "source area" of Melanesia and relating it to faunal number in the following manner. From fig. 2, take the line connecting New Guinea and the nearby Kei Islands as a "saturation curve" (other lines would be adequate but less suitable to the purpose), calculate the predicted range of "saturation" values among "saturated" islands of varying area from the curve, then take calculated "percentage saturation" as $s_i \times 100/B_i$, where s_i is the real number of species on any island and B_i the saturation number for islands of that area. As shown in fig. 3, the percentage saturation is nicely correlated in an inverse manner with distance from New Guinea. This allows quantification of the rule expressed qualitatively by past authors (see Mayr, 1940) that island faunas become progressively "impoverished" with distance from the nearest land mass.

[1] Division of Biology, University of Pennsylvania, Philadelphia, Pennsylvania.
[2] Biological Laboratories, Harvard University, Cambridge, Massachusetts.

EVOLUTION **17**: 373–387. December, 1963   373

374   ROBERT H. MACARTHUR AND EDWARD O. WILSON

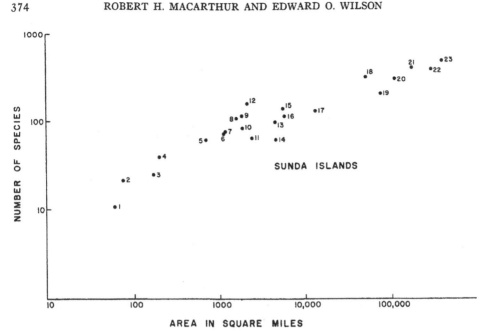

FIG. 1. The numbers of land and freshwater bird species on various islands of the Sunda group, together with the Philippines and New Guinea. The islands are grouped close to one another and to the Asian continent and Greater Sunda group, where most of the species live; and the distance effect is not apparent. (1) Christmas, (2) Bawean, (3) Engano, (4) Savu, (5) Simalur, (6) Alors, (7) Wetar, (8) Nias, (9) Lombok, (10) Billiton, (11) Mentawei, (12) Bali, (13) Sumba, (14) Bangka, (15) Flores, (16) Sumbawa, (17) Timor, (18) Java, (19) Celebes, (20) Philippines, (21) Sumatra, (22) Borneo, (23) New Guinea. Based on data from Delacour and Mayr (1946), Mayr (1940, 1944), Rensch (1936), and Stresemann (1934, 1939).

AN EQUILIBRIUM MODEL

The impoverishment of the species on remote islands is usually explained, if at all, in terms of the length of time species have been able to colonize and their chances of reaching the remote island in that time. According to this explanation, the number of species on islands grows with time and, given enough time, remote islands will have the same number of species as comparable islands nearer to the source of colonization. The following alternative explanation may often be nearer the truth. Fig. 4 shows how the number of new species entering an island may be balanced by the number of species becoming extinct on that island. The descending curve is the rate at which *new* species enter the island by colonization. This rate does indeed fall as the number

of species on the islands increases, because the chance that an immigrant be a new species, not already on the island, falls. Furthermore, the curve falls more steeply at first. This is a consequence of the fact that some species are commoner immigrants than others and that these rapid immigrants are likely, on typical islands, to be the first species present. When there are no species on the island ($N = 0$), the height of the curve represents the number of species arriving per unit of time. Thus the intercept, I, is the rate of immigration of species, new or already present, onto the island. The curve falls to zero at the point $N = P$ where all of the immigrating species are already present so that no new ones are arriving. P is thus the number of species in the "species pool" of immigrants. The shape of the rising curve in the same figure, which represents the

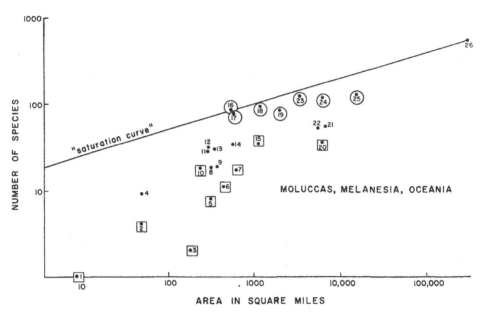

Fig. 2. The numbers of land and freshwater bird species on various islands of the Moluccas, Melanesia, Micronesia, and Polynesia. Here the archipelagoes are widely scattered, and the distance effect is apparent in the greater variance. Hawaii is included even though its fauna is derived mostly from the New World (Mayr, 1943). "Near" islands (less than 500 miles from New Guinea) are enclosed in circles, "far" islands (greater than 2,000 miles) in squares, and islands at intermediate distances are left unenclosed. The saturation curve is drawn through large and small islands at source of colonization. (1) Wake, (2) Henderson, (3) Line, (4) Kusaie, (5) Tuamotu, (6) Marquesas, (7) Society, (8) Ponape, (9) Marianas, (10) Tonga, (11) Carolines, (12) Palau, (13) Santa Cruz, (14) Rennell, (15) Samoa, (16) Kei, (17) Louisiade, (18) D'Entrecasteaux, (19) Tanimbar, (20) Hawaii, (21) Fiji, (22) New Hebrides, (23) Buru, (24) Ceram, (25) Solomons, (26) New Guinea. Based on data from Mayr (1933, 1940, 1943) and Greenway (1958).

rate at which species are becoming extinct on the island, can also be determined roughly. In case all of the species are equally likely to die out and this probability is independent of the number of other species present, the number of species becoming extinct in a unit of time is proportional to the number of species present, so that the curve would rise linearly with N. More realistically, some species die out more readily than others and the more species there are, the rarer each is, and hence an increased number of species increases the likelihood of any given species dying out. Under normal conditions both of these corrections would tend to increase the slope of the extinction curve for large values of N. (In the rare situation in which the species which enter most often as immigrants are the ones which die out most readily—presumably because the island is atypical so that species which are common elsewhere cannot survive well—the curve of extinction may have a steeper slope for small N.) If N is the number of species present at the start, then $E(N)/N$ is the fraction dying out, which can also be interpreted crudely as the probability that any given species will die out. Since this fraction cannot exceed 1, the extinction curve cannot rise higher than the straight line of a 45° angle rising from the origin of the coordinates.

It is clear that the rising and falling curves must intersect and we will denote by $\hat{s}$ the value of N for which the rate of immigration of new species is balanced by

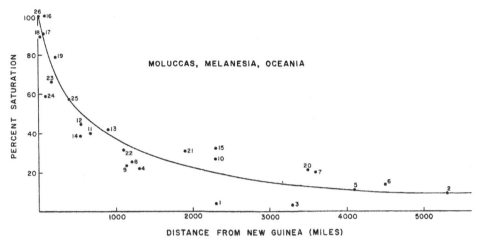

MOLUCCAS, MELANESIA, OCEANIA

DISTANCE FROM NEW GUINEA (MILES)

FIG. 3. Per cent saturation, based on the "saturation curve" of fig. 2, as a function of distance from New Guinea. The numbers refer to the same islands identified in the caption of fig. 2. Note that from equation (4) it is an oversimplification to take distances solely from New Guinea. The abscissa should give a more complex function of distances from all the surrounding islands, with the result that far islands would appear less "distant." But this representation expresses the distance effect adequately for the conclusions drawn.

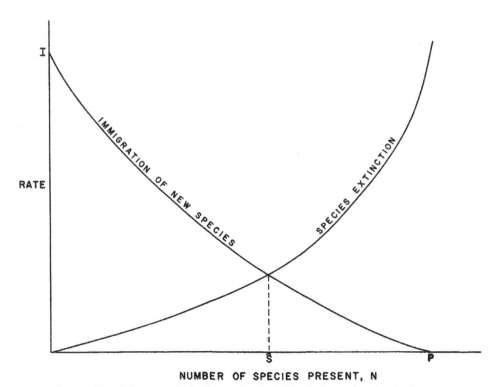

NUMBER OF SPECIES PRESENT, N

FIG. 4. Equilibrium model of a fauna of a single island. See explanation in the text.

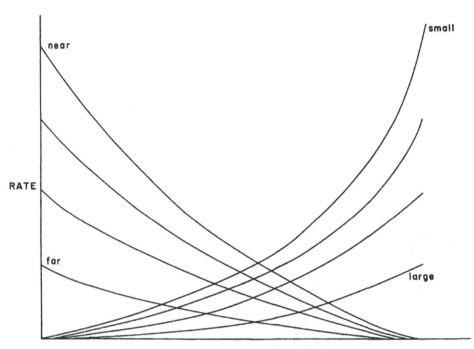

Fig. 5. Equilibrium model of faunas of several islands of varying distances from the source area and varying size. Note that the effect shown by the data of fig. 2, of faunas of far islands increasing with size more rapidly than those of near islands, is predicted by this model. Further explanation in text.

the rate of extinction. The number of species on the island will be stabilized at $\hat{s}$, for a glance at the figure shows that when N is greater than $\hat{s}$, extinction exceeds immigration of new species so that N decreases, and when N is less than $\hat{s}$, immigration of new species exceeds extinction so that N will increase. Therefore, in order to predict the number of species on an island we need only construct these two curves and see where they intersect. We shall make a somewhat oversimplified attempt to do this in later paragraphs. First, however, there are several interesting qualitative predictions which we can make without committing ourselves to any specific shape of the immigration and extinction curves.

A. An island which is farther from the source of colonization (or for any other reason has a smaller value of I) will,

other things being equal, have fewer species, because the immigration curve will be lower and hence intersect the mortality curve farther to the left (see fig. 5).

B. Reduction of the "species pool" of immigrants, P, will reduce the number of species on the island (for the same reason as in A).

C. If an island has smaller area, more severe climate (or for any other reason has a greater extinction rate), the mortality curve will rise and the number of species will decrease (see fig. 5).

D. If we have two islands with the same immigration curve but different extinction curves, any given species on the one with the higher extinction curve is more likely to die out, because $E(N)/N$ can be seen to be higher [$E(N)/N$ is the slope of the line joining the intersection point to the origin].

E. The number of species found on islands far from the source of colonization will grow more rapidly with island area than will the number on near islands. More precisely, if the area of the island is denoted by A, and $\hat{s}$ is the equilibrium number of species, then d^2s/dA^2 is greater for far islands than for near ones. This can be verified empirically by plotting points or by noticing that the change in the angle of intersection is greater for far islands.

F. The number of species on large islands decreases with distance from source of colonization faster than does the number of species on small islands. (This is merely another way of writing E and is verified similarly.)

Further, as will be shown later, the variance in $\hat{s}$ (due to randomness in immigrations and extinctions) will be lower than that expected if the "classical" explanation holds. In the classical explanation most of those species will be found which have at any time succeeded in immigrating. At least for distant islands this number would have an approximately Poisson distribution so that the variance would be approximately equal to the mean. Our model predicts a reduced variance, so that if the observed variance is significantly smaller than the mean for distant islands, it is evidence for the equilibrium explanation.

The evidence in fig. 2, relating to the insular bird faunas east of Weber's Line, is consistent with all of these predictions. To see this for the non-obvious prediction E, notice that a greater slope on this log-log plot corresponds to a greater second derivative, since A becomes sufficiently large.

THE FORM OF THE IMMIGRATION AND EXTINCTION CURVES

If the equilibrium model we have presented is correct, it should be possible eventually to derive some quantitative generalizations concerning rates of immigration and extinction. In the section to follow we have deduced an equilibrium equation which is adequate as a first ap-

proximation, in that it yields the general form of the empirically derived fauna–area curves without contradicting (for the moment) our intuitive ideas of the underlying biological processes. This attempt to produce a formal equation is subject to indefinite future improvements and does not affect the validity of the graphically derived equilibrium theory. We start with the statement that

$$\Delta s = M + G - D, \qquad (2)$$

where s is the number of species on an island, M is the number of species successfully immigrating to the island per year, G is the number of new species being added per year by local speciation (not including immigrant species that merely diverge to species level without multiplying), and D is the number of species dying out per year. At equilibrium,

$$M + G = D.$$

The immigration rate M must be determined by at least two independent values: (1) the rate at which propagules reach the island, which is dependent on the size of the island and its distance from the source of the propagules, as well as the nature of the source area, but not on the condition of the recipient island's fauna; and (2) as noted already, the number of species already resident on the island. Propagules are defined here as the minimum number of individuals of a given species needed to achieve colonization; a more exact explication is given in the Appendix. Consider first the source region. If it is climatically and faunistically similar to other potential source regions, the number of propagules passing beyond its shores per year is likely to be closely related to the size of the population of the taxon living on it, which in turn is approximately a linear function of its area. This notion is supported by the evidence from Indo-Australian ant zoogeography, which indicates that the ratio of faunal exchange is about equal to the ratio of the areas of the source regions (Wilson, 1961). On the other hand, the number of propagules reaching the recipient island probably varies linearly with the angle it subtends with reference to the center of the source region. Only near islands will vary much because of this factor. Finally, the number of propagules reaching the recipient island is most likely to be an exponential function of its distance from the source region. In the simplest case, if the probability that a given propagule ceases its overseas voyage (e.g., it falls into the sea and dies) at any given instant in time remains constant, then the fraction of propagules reaching a given distance fits an exponential holding-time distribution. If these assumptions are correct, the number of propagules reaching an island from a given source region per year can be approximated as

$$\alpha A_i \frac{\text{diam}_i\, e^{-\lambda d_i}}{2\pi d_i}, \qquad (3)$$

where A_i is the area of the source region, d_i is the mean distance between the source region and recipient island, diam_i is the diameter of the recipient island taken at a right angle to the direction of d_i, and α is a coefficient relating area to the number of propagules produced. More generally, where more than one source region is in position, the rate of propagule arrival would be

$$\frac{\alpha}{2\pi} \sum_i \frac{\text{diam}_i}{d_i} A_i e^{-\lambda d_i}, \qquad (4)$$

where the summation is of contributions from each of the ith source regions. Again, note that a propagule is defined as the minimum number of individuals required to achieve colonization.

Only a certain fraction of arriving propagules will add a new species to the fauna, however, because except for "empty" islands at least some ecological positions will be filled. As indicated in fig. 4, the rate of immigration (i.e., rate of propagule arrival times the fraction colonizing) declines to zero as the number of resident species (s) approaches the limit P. The curve relating the immigration rate to degree of unsaturation is probably a concave one, as indicated in fig. 4, for two

reasons: (1) the more abundant immigrants reach the island earlier, and (2) we would expect otherwise randomly arriving elements to settle into available positions according to a simple occupancy model where one and only one object is allowed to occupy each randomly placed position (Feller, 1958). These circumstances would result in the rate of successful occupation decelerating as positions are filled. While these are interesting subjects in themselves, a reasonable approximation is obtained if it is assumed that the rate of occupation is an inverse linear function of the number of occupied positions, or

$$\left(1 - \frac{s}{P}\right). \qquad (5)$$

Then

$$M = \frac{\alpha(1 - s/P)}{2\pi} \sum_i \frac{\text{diam}_i}{d_i} A_i e^{-\lambda d_i}. \qquad (6)$$

We know the immigration line in fig. 4 is not straight; to take this into account we must modify formula 5 by adding a term in s^2. However, this will not be necessary for our immediate purposes.

Now let us consider G, the rate of new productions on the island by local speciation. Note that this rate does not include the mere divergence of an island endemic to a specific level with reference to the stock species in the source area; that species is still counted as contributing to M, the immigration rate, no matter how far it evolves. Only new species generated from it and in addition to it are counted in G. First, consider an archipelago as a unit and the increase of s by divergence of species on the various islands to the level of allopatric species, i.e., the production of a local archipelagic superspecies. If this is the case, and no exchange of endemics is yet achieved among the islands of the archipelago, the number of species in the archipelago is limited to

$$\sum_{i=1}^{\infty} n_i \hat{s}_i, \qquad (7)$$

where n_i is the number of islands in the archipelago of ith area and $\hat{s}_i$ is the num-

380   ROBERT H. MACARTHUR AND EDWARD O. WILSON

ber of species occurring at equilibrium on islands of ith area. But the generation of allopatric species in superspecies does not multiply species on single islands or greatly change the fauna of the archipelago as a whole from the value predicted by the fauna–area curve, as can be readily seen in figs. 2 and 3. G, the increase of s by local speciation on single islands and exchange of autochthonous species between islands, probably becomes significant only in the oldest, largest, and most isolated archipelagoes, such as Hawaii and the Galápagos. Where it occurs, the exchange among the islands can be predicted from (6), with individual islands in the archipelago serving as both source regions and recipient islands. However, for most cases it is probably safe to omit G from the model, i.e., consider only source regions outside the archipelago, and hence

$$\Delta s = M - D. \tag{8}$$

The extinction rate D would seem intuitively to depend in some simple manner on (1) the mean size of the species populations, which in turn is determined by the size of the island and the number of species belonging to the taxon that occur on it; and (2) the yearly mortality rate of the organisms. Let us suppose that the probability of extinction of a species is merely the probability that all the individuals of a given species will die in one year. If the deaths of individuals are unrelated to each other and the population sizes of the species are equal and nonfluctuating,

$$D = sP^{N_r/s}, \tag{9}$$

where N_r is the total number of individuals in the taxon on the recipient island and P is their annual mortality rate. More realistically, the species of a taxon, such as the birds, vary in abundance in a manner approximating a Barton–Davis distribution (MacArthur, 1957) although the approximation is probably not good for a whole island. In s nonfluctuating species ordered according to their rank (K) in relative rareness,

$$D = \sum_{i=1}^{s} p^{(N_r/s)} \sum_{1=i}^{K} 1/(s-i+1). \tag{10}$$

This is still an oversimplification, if for no other reason than the fact that populations do fluctuate, and with increased fluctuation D will increase. However, both models, as well as elaborations of them to account for fluctuation, predict an exponential increase of D with restriction of island area. The increase of D which accompanies an increase in number of resident species is more complicated but is shown in fig. 4.

MODEL OF IMMIGRATION AND EXTINCTION PROCESS ON A SINGLE ISLAND

Let $P_s(t)$ be the probability that, at time t, our island has s species, λ_s be the rate of immigration of new species onto the island, when s are present, μ_s be the rate of extinction of species on the island when s are present; and λ_s and μ_s then represent the intersecting curves in fig. 4. This is a "birth and death process" only slightly different from the kind most familiar to mathematicians (cf. Feller, 1958, last chapter). By the rules of probability

$$P_s(t + h) = P_s(t)(1 - \lambda_s h - \mu_s h)$$
$$+ P_{s-1}(t)\lambda_{s-1}h$$
$$+ P_{s+1}(t)\mu_{s+1}h,$$

since to have s at time $t + h$ requires that at a short time preceding one of the following conditions held: (1) there were s and that no immigration or extinction took place, or (2) that there were $s - 1$ and one species immigrated, or (3) that there $s + 1$ and one species became extinct. We take h to be small enough that probabilities of two or more extinctions and/or immigrations can be ignored. Bringing $P_s(t)$ to the left-hand side, dividing by h, and passing to the limit as $h \to 0$

$$\frac{dP_s(t)}{dt} = -(\lambda_s + \mu_s)P_s(t) + \lambda_{s-1}P_{s-1}(t)$$
$$+ \mu_{s+1}P_{s+1}(t). \tag{11}$$

For this formula to be true in the case where $s = 0$, we must require that $\lambda_{-1} = 0$ and $\mu_0 = 0$. In principle we could solve

(11) for $P_s(t)$; for our purposes it is more useful to find the mean, $M(t)$, and the variance, $\mathrm{var}(t)$, of the number of species at time t. These can be estimated in nature by measuring the mean and variance in numbers of species on a series of islands of about the same distance and area and hence of the same λ_s and μ_s. To find the mean, $M(t)$, from (11) we multiply both sides of (11) by s and then sum from $s = 0$ to $s = \infty$. Since $\sum_{s=0}^{\infty} sP_s(t) = M(t)$, this gives us

$$\frac{dM(t)}{dt} = -\sum_{s=0}^{\infty} (\lambda_s + \mu_s)sP_s(t)$$
$$+ \sum_{s-1=0}^{\infty} \lambda_{s-1}[(s-1)+1]P_{s-1}(t)$$
$$+ \sum_{s+1=0}^{\infty} \mu_{s+1}[(s+1)-1]P_{s+1}(t). \cdot$$

(Here terms $\lambda_{-1}\cdot 0 \cdot P_{-1}(t) = 0$ and $\mu_0 \cdot (-1)P_0(t) = 0$ have been subtracted or added without altering values.) This reduces to

$$\frac{dM(t)}{dt} = \sum_{s=0}^{\infty} \lambda_s P_s(t) - \sum_{s=0}^{\infty} \mu_s P_s(t)$$
$$= \overline{\lambda_s(t)} - \overline{\mu_s(t)}. \qquad (12)$$

But, since λ_s and μ_s are, at least locally, approximately straight, the mean value of λ_s at time t is about equal to $\lambda_{M(t)}$ and similarly $\overline{\mu_s(t)} \sim \mu_{M(t)}$. Hence, approximately

$$\frac{dM(t)}{dt} = \lambda_{M(t)} - \mu_{M(t)}, \qquad (13)$$

or the expected number of species in Fig. 4 moves toward $\hat{s}$ at a rate equal to the difference in height of the immigration and extinction curves. In fact, if $d\mu/ds - d\lambda/ds$, evaluated near $s = \hat{s}$ is abbreviated by F, then, approximately $dM(t)/dt = F(\hat{s} - M(t))$ whose solution is $M(t) = \hat{s}(1 - e^{-Ft})$. Finally, we can compute the time required to reach 90% (say) of the saturation value $\hat{s}$ so that $M(t)/\hat{s} = 0.9$ or $e^{-Ft} = 0.1$. Therefore,

$$t = \frac{2.303}{F}. \qquad (13a)$$

A similar formula for the variance is obtained by multiplying both sides of (11) by $(s - M(t))^2$ and summing from $s = 0$ to $s = \infty$. As before, since $\mathrm{var}(t) = \sum_{s=0}^{\infty} (s - M(t))^2 P_s(t)$, this results in

$$\frac{d\,\mathrm{var}(t)}{dt}$$
$$= -\sum_{s=0}^{\infty} (\lambda_s + \mu_s)(s - M(t))^2 P_s(t)$$
$$+ \sum_{s-1=0}^{\infty} \lambda_{s-1}[(s-1-M(t))+1]^2 P_{s+1}(t)$$
$$+ \sum_{s+1=0}^{\infty} \mu_{s+1}[(s+1-M(t))-1]^2 P_{s+1}(t)$$
$$= 2\sum_{s=0}^{\infty} \lambda_s(s - M(t))P_s(t)$$
$$- 2\sum_{s=0}^{\infty} \mu_s(s - M(t))P_s(t)$$
$$+ \sum_{s=0}^{\infty} \lambda_s P_s(t) + \sum_{s=0}^{\infty} \mu_s P_s(t). \qquad (14)$$

Again we can simplify this by noting that the λ_s and μ_s curves are only slowly curving and hence in any local region are approximately straight. Hence, where derivatives are now evaluated near the point $s = M(t)$,

$$\lambda_s = \lambda_{M(t)} + [s - M(t)]\frac{d\lambda}{ds}$$
$$\mu_s = \mu_{M(t)} + [s - M(t)]\frac{d\mu}{ds}. \qquad (15)$$

Substituting (15) into (14) we get

$$\frac{d\,\mathrm{var}(t)}{dt}$$
$$= 2(\lambda_{M(t)} - \mu_{M(t)})\sum_{s=0}^{\infty} (s - M(t))P_s(t)$$
$$+ 2\left(\frac{d\lambda}{ds} - \frac{d\mu}{ds}\right)\sum_{s=0}^{\infty} (s - M(t))^2 P_s(t)$$
$$+ [\lambda_{m(t)} + \mu_{M(t)}]\sum_{s=0}^{\infty} P_s(t)$$
$$+ \left(\frac{d\lambda}{ds} + \frac{d\mu}{ds}\right)\sum_{s=0}^{\infty} (s - M(t))P_s(t),$$

which, since $\sum\limits_{s=0}^{\infty} P_s(t) = 1$ and

$$\sum (s - M(t))P_s(t) = M(t) - M(t) = 0,$$

becomes,

$$\frac{d \operatorname{var}(t)}{dt} = -2\left(\frac{d\mu}{ds} - \frac{d\lambda}{ds}\right)\operatorname{var}(t)$$
$$+ \lambda_{M(t)} + \mu_{M(t)} . \qquad (16)$$

This is readily solved for $\operatorname{var}(t)$:

$$\operatorname{var}(t)$$
$$= e^{-2[(d\mu/ds)-(d\lambda/ds)]t} \qquad (16a)$$
$$\times \int_0^t (\lambda_{M(t)} + \mu_{M(t)}) e^{2[(d\mu/ds)-(d\lambda/ds)]t} \, dt .$$

However, it is more instructive to compare mean and variance for the extreme situations of saturation and complete unsaturation, or equivalently of $t = $ near ∞ and $t = $ near zero.

At equilibrium, $\dfrac{d \operatorname{var}(t)}{dt} = 0$, so by (16) ·

$$\operatorname{var}(t) = \frac{\lambda_{\hat{s}} + \mu_{\hat{s}}}{2\left(\dfrac{d\mu}{ds} - \dfrac{d\lambda}{ds}\right)} . \qquad (17)$$

At equilibrium $\lambda_{\hat{s}} = \mu_{\hat{s}} = x$ say and we have already symbolized the difference of the derivatives at $s = \hat{s}$ by F (cf. eq. [13a]). Hence, at equilibrium

$$\operatorname{var} = \frac{X}{F} . \qquad (17a)$$

Now since μ_s has non-decreasing slope $X/s \leqslant d\mu/ds \big|_{s=\hat{s}}$ or $X \leqslant \hat{s}\, d\mu/ds \big|_{\hat{s}}$.

Therefore, variance $\leqslant \dfrac{\hat{s}\, d\mu/ds}{d\mu/ds - d\lambda/ds}$ or, at equilibrium

$$\frac{\text{variance}}{\text{mean}} \leqslant \frac{d\mu/ds}{d\mu/ds - d\lambda/ds} . \qquad (18)$$

In particular, if the extinction and immigration curves have slopes about equal in absolute value, (variance/mean) $\leqslant \frac{1}{2}$. On the other hand, when t is near zero, equation (16) shows that $\operatorname{var}(t) \sim \lambda_0 t$. Similarly, when t is near zero, equations (13) or (14) show that $M(t) \sim \lambda_0 t$. Hence, in a very unsaturated situation, approximately,

$$\frac{\text{variance}}{\text{mean}} = 1 . \qquad (19)$$

Therefore, we would expect the variance/mean to rise from somewhere around $\frac{1}{2}$ to 1, as we proceed from saturated islands to extremely unsaturated islands farthest from the source of colonization.

Finally, if the number of species dying out per year, X (at equilibrium), is known, we can estimate the time required to 90% saturation from equations (13a) and (17a):

$$\frac{2.303}{t} = \frac{X}{\text{variance}}$$
$$t = \frac{2.303 \text{ variance}}{X} = \frac{2.303 \text{ mean}}{2} \frac{\text{mean}}{X} . \qquad (19a)$$

The above model was developed independently from an equilibrium hypothesis just published by Preston (1962). After providing massive documentation of the subject that will be of valuable assistance to future biogeographers, Preston draws the following particular conclusion about continental versus insular biotas: "[The depauperate insular biotas] are not depauperate in any absolute sense. They have the correct number of species for their area, provided that each area is an isolate, but they have far fewer than do equal areas on a mainland, because a mainland area is merely a 'sample' and hence is greatly enriched in the Species/Individuals ratio." To illustrate, "in a sample, such as the breeding birds of a hundred acres, we get many species represented by a single pair. Such species would be marked for extinction with one or two seasons' failure of their nests were it not for the fact that such local extirpation can be made good from outside the 'quadrat,' which is not the case with the isolate." This point of view agrees with our own. However, the author apparently missed the precise distance effect and his model is consequently not predictive in the direction we are attempting. His model is, however, more accurate in its account of

TABLE 1. *Number of species of land and freshwater birds on Krakatau and Verlaten during three collection periods together with losses in the two intervals (from Dammerman, 1948)*

| | 1908 | | | 1919–1921 | | | 1932–1934 | | | Number "lost" | |
	Non-migrant	Migrant	Total	Non-migrant	Migrant	Total	Non-migrant	Migrant	Total	1908 to 1919–1921	1919–1921 to 1932–1934
Krakatau	13	0	13	27	4	31	27	3	30	2	5
Verlaten	1	0	1	27	2	29	29	5	34	0	2

relative abundance, corresponding to our equation (10).

THE CASE OF THE KRAKATAU FAUNAS

The data on the growth of the bird faunas of the Krakatau Islands, summarized by Dammerman (1948), provide a rare opportunity to test the foregoing model of the immigration and extinction process on a single island. As is well known, the island of Krakatau proper exploded in August, 1883, after a three-month period of repeated eruptions. Half of Krakatau disappeared entirely and the remainder, together with the neighboring islands of Verlaten and Lang, was buried beneath a layer of glowing hot pumice and ash from 30 to 60 meters thick. Almost certainly the entire flora and fauna were destroyed. The repopulation proceeded rapidly thereafter. Collections and sight records of birds, made mostly in 1908, 1919–1921, and 1932–1934, show that the number of species of land and freshwater birds on both Krakatau and Verlaten climbed rapidly between 1908 and 1919–1921 and did not alter significantly by 1932–1934 (see table 1). Further, the number of non-migrant land and freshwater species on both islands in 1919–1921 and 1932–1934, i.e., 27–29, fall very close to the extrapolated fauna–area curve of our fig. 1. Both lines of evidence suggest that the Krakatau faunas had approached equilibrium within only 25 to 36 years after the explosion.

Depending on the exact form of the immigration and extinction curves (see fig. 4), the ratio of variance to mean of numbers of species on similar islands at or near saturation can be expected to vary between about ¼ and ¾. If the slopes of the two curves are equal at the point of intersection, the ratio would be near ½. Then the variance of faunas of Krakatau-like islands (same area and isolation) can be expected to fall between 7 and 21 species. Applying this estimate to equation (19a) and taking t (the time required to reach 90% of the equilibrium number) as 30 years, X, the annual extinction rate, is estimated to lie between 0.5 and 1.6 species per year.

This estimate of annual extinction rate (and hence of the acquisition rate) in an equilibrium fauna is surprisingly high; it is of the magnitude of 2 to 6% of the standing fauna. Yet it seems to be supported by the collection data. On Krakatau proper, 5 non-migrant land and freshwater species recorded in 1919–1921 were not recorded in 1932–1934, but 5 other species were recorded for the first time in 1932–1934. On Verlaten 2 species were "lost" and 4 were "gained." This balance sheet cannot easily be dismissed as an artifact of collecting technique. Dammerman notes that during this period, "The most remarkable thing is that now for the first time true fly catchers, *Muscicapidae*, appeared on the islands, and that there were no less than four species: *Cyornis rufigastra, Gerygone modigliani, Alseonax latirostris* and *Zanthopygia narcissina*. The two last species are migratory and were therefore only accidental visitors, but the sudden appearance of the *Cyornis* species in great numbers is noteworthy. These birds, first observed in May 1929, had already colonized three islands and may now be called common there. Moreover the *Gerygone*, unmistakable from his gentle note and common along the coast and in the mangrove forest, is certainly a new acquisition." Extinctions are less susceptible of proof but the following evidence is suggestive. "On the other hand two species mentioned by Jacobson (1908) were not found in 1921 and have not been observed since, namely the small kingfisher *Alcedo coerulescens* and the familiar bulbul *Pycnonotus aurigaster*." Between 1919–1921 and 1932–1934 the conspicuous *Demiegretta s. sacra* and *Accipter* sp. were "lost," although these species may not have been truly established as breeding populations. But "the well-known greybacked shrike (*Lanius schach bentet*), a bird conspicuous in the open field, recorded in 1908 and found breeding in 1919, was not seen in 1933. Whether the species had really completely disappeared or only diminished so much in numbers that it was not noticed, the future must show." Future research on the Krakatau fauna would indeed be of great interest, in view of the very dynamic equilibrium suggested by the model we have presented. If the "losses" in the data represent true extinctions, the rate of extinction would be 0.2 to 0.4 species per year, closely approaching the predicted rate of 0.5 to 1.6. This must be regarded as a minimum figure, since it is likely that species could easily be lost and regained all in one 12-year period.

Such might be the situation in the early history of the equilibrium fauna. It is not possible to predict whether the rate of turnover would change through time. As other taxa reached saturation and more species of birds had a chance at colonization, it is conceivable that more "harmonic" species systems would accumulate within which the turnover rate would decline.

PREDICTION OF A "RADIATION ZONE"

On islands holding equilibrium faunas, the ratio of the number of species arriving from other islands in the same archipelago (G in equation no. 2) to the number arriving from outside the archipelago (M in no. 2) can be expected to increase with distance from the major extra-archipelagic source area. Where the archipelagoes are of approximately similar area and configuration, G/M should increase in an orderly fashion with distance. Note that G provides the best available measure of what is loosely referred to in the literature as adaptive radiation. Specifically, adaptive radiation takes place as species are generated within archipelagoes, disperse between islands, and, most importantly, accumulate on individual islands to form diversified associations of sympatric species. In equilibrium faunas, then, the following prediction is possible: adaptive radiation, measured by G/M, will increase with distance from the major source region and after corrections for area and climate, reach a maximum on archipelagoes and large islands located in a circular zone close to the outermost range of the taxon. This might be referred to as the "radiation zone" of taxa with equilibrium faunas. Many examples possibly conforming to such a rule can be cited: the birds of Hawaii and the Galápagos, the murid rodents of Luzon, the cyprinid fish of Mindanao, the frogs of the Seychelles, the gekkonid lizards of New Caledonia, the Drosophilidae of Hawaii, the ants of Fiji and New Caledonia, and many others (see especially in Darlington, 1957; and Zimmerman, 1948). But there are conspicuous exceptions: the frogs just reach New Zealand but have not radiated there; the same is true of the insectivores of the Greater Antilles, the terrestrial mammals of the Solomons, the snakes of Fiji, and the lizards of Fiji and Samoa. To say that the latter taxa have only recently reached the islands in question, or that they are not in equilibrium, would be a premature if not facile explanation. But it is worth considering as a working hypothesis.

ESTIMATING THE MEAN DISPERSAL DISTANCE

A possible application of the equilibrium model in the indirect estimation of the mean dispersal distance, or λ in equation

(3). Note that if similar parameters of dispersal occur within archipelagoes as well as between them,

$$\frac{G}{M} = \frac{A_1 \, \mathrm{diam}_1 \, d_2}{A_2 \, \mathrm{diam}_2 \, d_1} e^{\lambda(d_2 - d_1)}, \qquad (20)$$

and

$$\lambda = \ln \frac{A_2 \, \mathrm{diam}_2 \, d_1 \, G}{A_1 \, \mathrm{diam}_1 \, d_2 \, M} \bigg/ (d_2 - d_1), \qquad (21)$$

where, in a simple case, A_1, diam_1, and d_1 refer to the relation between the recipient island and some single major source island within the same archipelago; and A_2, diam_2, and d_2 refer to the relation between the recipient island and the major source region outside the archipelago.

Consider the case of the Geospizinae of the Galápagos. On the assumption that a single stock colonized the Galápagos (Lack, 1947), G/M for each island can be taken as equal to G, or the number of geospizine species. In particular, the peripherally located Chatham Island, with seven species, is worth evaluating. South America is the source of M and Indefatigable Island can probably be regarded as the principal source of G for Chatham. Given G/M as seven and assuming that the Geospizinae are in equilibrium, λ for the Geospizinae can be calculated from (21) as 0.018 mile. For birds as a whole, where G/M is approximately unity, λ is about 0.014 mile.

But there are at least three major sources of error in making an estimate in this way:

1. Whereas M is based from the start on propagules from an equilibrium fauna in South America, G increased gradually in the early history of the Galápagos through speciation of the Geospizinae on islands other than Chatham. Hence, G/M on Chatham is actually higher than the ratio of species drawn from the Galápagos to those drawn from outside the archipelago, which is our only way of computing G/M directly. Since λ increases with G/M, the estimates of λ given would be too low, if all other parameters were correct.

2. Most species of birds probably do not disperse according to a simple exponential holding-time distribution. Rather, they probably fly a single direction for considerable periods of time and cease flying at distances that can be approximated by the normal distribution. For this reason also, λ as estimated above would probably be too low.

3. We are using $\hat{S}_G/\hat{S}_M$ for G/M, which is only approximate.

These considerations lead us to believe that 0.01 mile can safely be set as the lower limit of λ for birds leaving the eastern South American coast. Using equation no. 12 in another case, we have attempted to calculate λ for birds moving through the Lesser Sunda chain of Indonesia. The Alor group was chosen as being conveniently located for the analysis, with Flores regarded as the principal source of western species and Timor as the principal source of eastern species. From the data of Mayr (1944) on the relationships of the Alor fauna, and assuming arbitrarily an exponential holding-time dispersal, λ can be calculated as approximately 0.3 mile. In this case the first source of error mentioned above with reference to the Galápagos fauna is removed but the second remains. Hence, the estimate is still probably a lower limit.

Of course these estimates are in themselves neither very surprising nor otherwise illuminating. We cite them primarily to show the possibilities of using zoogeographic data to set boundary conditions on population ecological phenomena that would otherwise be very difficult to assess.

Finally, while we believe the evidence favors the hypothesis that Indo-Australian insular bird faunas are at or near equilibrium, we do not intend to extend this conclusion carelessly to other taxa or even other bird faunas. Our purpose has been to deal with general equilibrium criteria, which might be applied to other faunas, together with some of the biological implications of the equilibrium condition.

386 ROBERT H. MACARTHUR AND EDWARD O. WILSON

SUMMARY

A graphical equilibrium model, balancing immigration and extinction rates of species, has been developed which appears fully consistent with the fauna–area curves and the distance effect seen in land and freshwater bird faunas of the Indo-Australian islands. The establishment of the equilibrium condition allows the development of a more precise zoogeographic theory than hitherto possible.

One new and non-obvious prediction can be made from the model which is immediately verifiable from existing data, that the number of species increases with area more rapidly on far islands than on near ones. Similarly, the number of species on large islands decreases with distance faster than does the number of species on small islands.

As groups of islands pass from the unsaturated to saturated conditions, the variance-to-mean ratio should change from unity to about one-half. When the faunal buildup reaches 90% of the equilibrium number, the extinction rate in species/year should equal 2.303 times the variance divided by the time (in years) required to reach the 90% level. The implications of this relation are discussed with reference to the Krakatau faunas, where the buildup rate is known.

A "radiation zone," in which the rate of intra-archipelagic exchange of autochthonous species approaches or exceeds extra-archipelagic immigration toward the outer limits of the taxon's range, is predicted as still another consequence of the equilibrium condition. This condition seems to be fulfilled by conventional information but cannot be rigorously tested with the existing data.

Where faunas are at or near equilibrium, it should be possible to devise indirect estimates of the actual immigration and extinction rates, as well as of the times required to reach equilibrium. It should also be possible to estimate the mean dispersal distance of propagules overseas from the zoogeographic data. Mathematical models have been constructed to these ends and certain applications suggested.

The main purpose of the paper is to express the criteria and implications of the equilibrium condition, without extending them for the present beyond the Indo-Australian bird faunas.

ACKNOWLEDGMENTS

We are grateful to Dr. W. H. Bossert, Prof. P. J. Darlington, Prof. E. Mayr, and Prof. G. G. Simpson for material aid and advice during the course of the study. Special acknowledgment must be made to the published works of K. W. Dammerman, E. Mayr, B. Rensch, and E. Stresemann, whose remarkably thorough faunistic data provided both the initial stimulus and the principal working material of our analysis. The work was supported by NSF Grant G-11575.

LITERATURE CITED

DAMMERMAN, K. W. 1948. The fauna of Krakatau 1883–1933. Verh. Kon. Ned. Akad. Wet. (Nat.), (2) **44**: 1–594.

DARLINGTON, P. J. 1957. Zoogeography. The geographical distribution of animals. Wiley.

DELACOUR, J., AND E. MAYR. 1946. Birds of the Philippines. Macmillan.

FELLER, W. 1958. An introduction to probability theory and its applications. Vol. 1, 2nd ed. Wiley.

GREENWAY, J. G. 1958. Extinct and vanishing birds of the world. Amer. Comm. International Wild Life Protection, Special Publ. No. 13.

LACK, D. 1947. Darwin's finches, an essay on the general biological theory of evolution. Cambridge University Press.

MACARTHUR, R. H. 1957. On the relative abundance of bird species. Proc. Nat. Acad. Sci. [U. S.], **43**: 293–294.

MAYR, E. 1933. Die Vogelwelt Polynesiens. Mitt. Zool. Mus. Berlin, **19**: 306–323.

——. 1940. The origin and history of the bird fauna of Polynesia. Proc. Sixth Pacific Sci. Congr., **4**: 197–216.

——. 1943. The zoogeographic position of the Hawaiian Islands. Condor, **45**: 45–48.

——. 1944. Wallace's Line in the light of recent zoogeographic studies. Quart. Rev. Biol., **19**: 1–14.

PRESTON, F. W. 1962. The canonical distribution of commonness and rarity: Parts I, II. Ecology, **43**: 185–215, 410–432.

RENSCH, B. 1936. Die Geschichte des Sundabogens. Borntraeger, Berlin.

STRESEMANN, E. 1934. "Aves." *In* Handb. Zool., W. Kukenthal, ed. Gruyter, Berlin.

———. 1939. Die Vögel von Celebes. J. für Ornithologie, **87**: 299–425.

WILSON, E. O. 1961. The nature of the taxon cycle in the Melanesian ant fauna. Amer. Nat., **95**: 169–193.

ZIMMERMAN, E. C. 1948. Insects of Hawaii. Vol. 1. Introduction. University of Hawaii Press.

APPENDIX: MEASUREMENT OF A PROPAGULE

A rudimentary account of how many immigrants are required to constitute a propagule may be constructed as follows. Let η be the average number of individuals next generation per individual this generation. Thus, for instance, if $\eta = 1.03$, the population is increasing at 3% interest rate.

Let us now suppose that the number of descendants per individual has a Poisson distribution. If it has not, due to small birth rate, the figures do not change appreciably. Then, due to chance alone, the population descended from immigrants may vanish. This subject is well known in probability theory as "Extinction probabilities in branching processes" (cf. Feller 1958, p. 274). The usual equation for the probability ζ of eventual extinction (Feller's equation 5.2 with $P(\zeta) = e^{-\eta(1-\zeta)}$, for a Poisson distribution), gives

$$\zeta = e^{-\eta(1-\zeta)}.$$

Solving this by trial and error for the

TABLE 2. *Relation of replacement rate (η) of immigrants to probability of extinction (ζ)*

η	1	1.01	1.1	1.385
ζ	1	0.98	0.825	0.5

probability of eventual extinction ζ, given a variety of values of η, we get the array shown in table 2. From this we can calculate how large a number of simultaneous immigrants would stand probability just one-half of becoming extinct during the initial stages of population growth following the introduction. In fact, if r pairs immigrate simultaneously, the probability that all will eventually be without descendants is ζ^r. Solving $\zeta^R = 0.5$ we find the number, R, of pairs of immigrants necessary to stand half a chance of not becoming extinct as given in table 3. From this it is clear that when η is 1, the propagule has infinite size, but that as η increases, the propagule size decreases rapidly, until, for a species which increases at 38.5% interest rate, one pair is sufficient to stand probability 1.2 of effecting a colonization. With sexual species which hunt for mates, η may be very nearly 1 initially.

TABLE 3. *Relation of replacement rate (η) to the number of pairs (R) of immigrants required to give the population a 50% chance of survival*

η	1	1.01	1.1	1.385
R	∞	34	3.6	1

EXPERIMENTAL ZOOGEOGRAPHY OF ISLANDS. A TWO-YEAR RECORD OF COLONIZATION[1]

Daniel S. Simberloff

Department of Biological Science, Florida State University, Tallahassee, Florida 32306

AND

Edward O. Wilson

Biological Laboratories, Harvard University, Cambridge, Massachusetts 02138

Abstract. In 1966–1967 the entire arthropod faunas of six small mangrove islands in the Florida Keys were removed by methyl bromide fumigation. In earlier articles we described the process of recolonization through the first year, during which the numbers of species in five of the six faunas rose to what appear to be noninteractive equilibria and then slumped slightly to interactive equilibria. The sixth, that of island E1, we believed to be climbing more slowly because of its greater distance from the source area. It had not reached the predefaunation (interactive) equilibrium by 1 year. Here we give the results of censuses taken at the end of the second year on the four islands in the group located in the lower Keys (E1, E2, E3, ST2). The numbers of species were found to have changed little from the previous year, providing further evidence that they are in equilibrium. Species immigrations and extinctions have continued at a high rate, and the species compositions on three of the four islands appear to be moving slowly in the direction of the original, predefaunation states.

In the first two articles of this series we described an experiment in which the entire arthropod faunas of six small mangrove islands in the Florida Keys were exterminated by methyl bromide fumigation and the process of recolonization was monitored thereafter by frequent censuses (Wilson and Simberloff 1969, Simberloff and Wilson 1969). The experiment was designed both to test and to extend certain aspects of species equilibrium theory (MacArthur and Wilson 1967, Simberloff 1969, Wilson 1969) and to observe the actual processes of immigration and extinction. At the end of the first year following defaunation, the numbers of species had reattained approximately the original (prefumigation) levels on five of the six islands. The most distant island (E1) supported the fewest species, just as it had prior to defaunation; it alone had not approached the original species number. Intermediate islands (E3, ST2) reattained intermediate numbers of species. The time-colonization curves appeared to have assumed the logarithmic forms predicted by basic equilibrium theory. Moreover, higher levels of species numbers were reached prior to the buildup of the populations belonging to the constituent species. These numbers then dipped slightly as the densities of the constituent populations approached the predefaunation levels. The decline occurred near the end of the first year following defaunation (see Fig. 1). We interpreted the first, higher levels to represent "noninteractive species equilibria," that is, species equilibria reached before the extinction rates could be greatly influenced by interspecific interactions such as

[1] Received April 23, 1970; accepted May 27, 1970.

predation and competition; the second, lower levels were considered to represent "interactive species equilibria," in which species interactions contributed significantly to species extinction rates. The colonizations had thus been followed to the mere beginnings of the interactive equilibria. It was clearly desirable to continue monitoring the islands to watch for subsequent developments.

Censuses at the end of the second year

Near the end of the second year following defaunation, we visited the islands again. Time permitted the examination of only the four islands in the Sugarloaf area of the lower Keys, namely E1, E2, E3, and ST2. We employed the same procedures for censusing as described in our earlier reports (Wilson and Simberloff 1969, Simberloff and Wilson 1969). The results are summarized in Figure 1 and in Tables 1 and 2.

From Figure 1 it can be seen that the estimated species numbers have not changed greatly since the end of the first year—and therefore since the period just before the entire original fauna was destroyed by fumigation. The equilibrial numbers continue to be an inverse function of the distance of the islands from the nearest mangrove islands and swamps that can serve as a source of fresh immigrants. Only E1 has failed to reattain fully the original predefaunation level. We do not know the reason for this exception. Our best guess is the following: basic equilibrium theory predicts that distant islands approach equilibrium more slowly than near islands, and it is possible that E1 is still in the process of climbing toward equilibrium. This purely statistical effect, caused

TABLE 1. The colonists of four experimental islands 2 years after defaunation

E1 March 8, 1969	E2 March 6-7, 1969	E3 March 9, 1969	ST2 March 10-11, 1969
			Embioptera *Diradius caribbeana*
	Orthoptera *Latiblattella* n. sp. *Latiblattella rehni* *Cyrtoxipha* sp. *Tafalisca lurida*	Orthoptera *Latiblattella* n. sp. *Cycloptilum* sp. *Tafalisca lurida*	Orthoptera *Latiblattella* n. sp. *Tafalisca lurida*
Coleoptera *Sapintus fulvipes* *Tricorynus* sp. *Cryptorhynchus* *minutissimus* *Pseudoacalles* sp.	Coleoptera *Tricorynus* sp. *Styloleptus biustus* *Chrysobothris* sp. *Pseudoacalles* sp. *Trischidias minutissima*	Coleoptera *Tricorynus* sp. *Pseudoacalles* sp.	Coleoptera *Tricorynus* sp. *Styloleptus* sp. *Pseudoacalles* sp.
	Thysanoptera *Barythrips sculpticauda*	Thysanoptera *Liothrips* sp. *Pseudothrips inequalis*	Thysanoptera *Haplothrips* sp. *Pseudothrips inequalis*
Corrodentia *Psocidus texanus* *Peripsocus* sp.	Corrodentia *Caecilius subflavus* aff. *Caecilius flavidus* aff. *Psocidus texanus* Hemiptera *Paraleyrodes* sp. *Pseudococcus* sp. Anthocoridae gen. sp. Gen. sp.	Corrodentia *Liposcelis bostrychophilus* *Psocidus texanus*	Corrodentia Liposcelidae gen. sp. *Caecilius subflavus* aff. *Psocidus texanus* Hemiptera Miridae (Phylinae) gen. sp.
Lepidoptera *Alarodia slossoniae* *Ecdytolopha* sp. *Bema ydda* *Phocides batabano* Diptera Hippoboscidae gen. sp.	Lepidoptera *Bema ydda* *Nemapogon* sp. *Phocides batabano*	Lepidoptera *Ecdytolopha* sp. *Phocides batabano*	Lepidoptera *Alarodia slossoniae* *Ecdytolopha* sp. *Nemapogon* sp.
Hymenoptera *Pseudomyrmex elongatus* *Crematogaster ashmeadi*	Hymenoptera Ichenumonoidea gen. sp. *Pseudomyrmex elongatus* *Xenomyrmex floridanus* *Crematogaster ashmeadi* *Paracryptocerus varians* *Paratrechina bourbonica* *Camponotus floridanus*	Hymenoptera *Calliephialtes* sp. *Pseudomyrmex elongatus* *Xenomyrmex floridanus* *Monomorium floricola* *Crematogaster ashmeadi* *Paratrechina bourbonica*	Hymenoptera *Scleroderma macrogaster* *Pseudomyrmex elongatus* *Crematogaster ashmeadi* *Paracryptocerus varians* *Tapinoma littorale* *Camponotus floridanus*
Araneae *Ayscha* sp.	Araneae *Ayscha* sp. *Hentzia grenada* Salticidae gen. sp. *Nephila clavipes* *Ariadna arthuri*	Araneae *Ayscha* sp. *Hentzia grenada*	Araneae *Ayscha* sp. *Hentzia palmarum* *Ariadna arthuri*
Acarina *Scheloribates* sp. Gen. sp.	Acarina Gen. sp. 1 Gen. sp. 2	Acarina *Scheloribates* sp. *Eupodes* sp. Gen. sp. Diplopoda *Lophoproctinus* sp.	Acarina Gen. sp. Diplopoda *Lophoproctinus* sp.
Total seen = 16 Estimated present = 18	Total seen = 34 Extimated present = 39	Total seen = 23 Estimated present = 28	Total seen = 26 Estimated present = 30

by a uniform lowering of invasion rates with increasing distance, may have been enhanced by abnormal population growth of the few early colonists (Simberloff 1969).

A comparison of the species lists in Table 1 with those provided from earlier censuses by Simberloff and Wilson (1969) shows that a high rate of species turnover occurred during the second year, resulting in a significant alteration of species composition, even though the species numbers remained nearly the same. In fact, 35–52% of the species making up the total lists from individual islands at the ends of the first and second years were encountered during both censuses (Table 2, section C).

A question of major significance that continued monitoring should answer is whether the species compositions will eventually converge toward those prevailing before defaunation. Wilson (1969) has suggested the general proposition that as new

936 DANIEL S. SIMBERLOFF AND EDWARD O. WILSON Ecology, Vol. 51, No. 5

TABLE 2. Percentages of species that were present at both of two given censuses on four of the experimental islands

Name of experimental island	A. Censuses: just before defaunation and one year later			B. Censuses: just before defaunation and two years later			C. Censuses: one and two years after defaunation		
	No. spp. in common	Total no. in both censuses	Per cent in common	No. spp. in common	Total no. in both censuses	Per cent in common	No. spp. in common	Total no. in both censuses	Per cent in common
E1.......	2	29	6.9%	5	26	19.2%	7	18	38.9%
E2.......	10	54	18.5%	13	51	25.5%	16	34	37.2%
E3.......	8	40	20.0%	7	35	20.0%	16	31	51.6%
ST2.......	11	37	29.7%	17	31	54.8%	12	34	35.3%

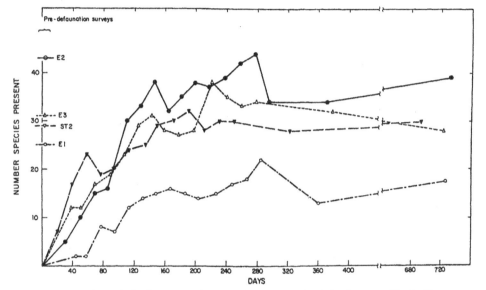

FIG. 1. The colonization curves of four small mangrove islands in the lower Florida Keys whose entire faunas, consisting almost solely of arthropods, were exterminated by methyl bromide fumigation. The figures shown are the estimated numbers of species present, which are the actual numbers seen plus a small fraction not seen but inferred to be present by the criteria utilized by Simberloff and Wilson (1969) and Simberloff (1969). The number of species in an inverse function of the distance of the island to the nearest source of immigrants. This effect was evident in the predefaunation censuses and was preserved when the faunas regained equilibrium after defaunation. Thus, the near island E2 has the most species, the distant island E1 the fewest, and the intermediate islands E3 and ST2 intermediate numbers of species.

combinations of species are generated by turnover, combinations of longer-lived species must eventually accumulate. Such species persist longer either because they are better adapted to the peculiar physical conditions of the local environment or else because they are able to coexist longer with the particular set of species among which they find themselves. Thus in time an "assortative equilibrium" will succeed the original "interactive equilibrium." Because the individual survival times of the resident species populations are greater on the average, while the total propagule invasion rates of the faunas remain about the same, the

numbers of species in an assortative equilibrium should be higher. It is also true that the numbers of possible assortative equilibria are fewer, and therefore under similar conditions the species compositions should converge to some degree. Our data are still too few to be conclusive, but they do suggest that the faunas of the experimental islands are drifting in the direction of the original compositions. By comparing section A and B in Table 2 it can be seen that on three of the islands the faunas were closer in composition to the predefaunation faunas at the end of the second year than they were at the end of the first year. In one of the four islands (E3) there was no change. We regard these trends as being suggestive only. Time should tell us with certainty whether convergence has really been occurring.

ACKNOWLEDGMENTS

We are happy to acknowledge once again the assistance of the following entomologists who identified most of the arthropod colonists: J. A. Beatty, G. W. Dekle, R. C. Froeschner, D. G. Kissinger, E. L. Mockford, E. S. Ross, L. M. Russell, R. L. Smiley, L. J. Stannard, T. J. Walker, S. L. Wood. Without their help this entire project would have been quite impossible.

LITERATURE CITED

MacArthur, R. H., and E. O. Wilson. 1967. The theory of island biogeography. Princeton University Press, xi + 203 p.

Simberloff, D. S. 1969. Experimental zoogeography of islands. A model for insular colonization. Ecology 50: 296–314.

Simberloff, D. S., and E. O. Wilson. 1969. Experimental zoogeography of islands. The colonization of empty islands. Ecology 50: 278–296.

Wilson, E. O. 1969. The species equilibrium. Brookhaven symposia in biology 22: 38–47.

Wilson, E. O., and D. S. Simberloff. 1969. Experimental zoogeography of islands. Defaunation and monitoring techniques. Ecology 50: 267–278.

Vol. 105, No. 945   The American Naturalist   September–October 1971

MAMMALS ON MOUNTAINTOPS: NONEQUILIBRIUM INSULAR BIOGEOGRAPHY

James H. Brown*

Department of Zoology, University of California, Los Angeles, California 90024

INTRODUCTION

MacArthur and Wilson (1963, 1967) have provided a theoretical model to account for variation in the diversity of species on islands. This model, which attributes the number of species on an island to an equilibrium between rates of recurrent extinction and colonization, appears to account for the distribution of most kinds of animals and plants on oceanic islands. However, in addition to these islands, there are many other kinds of analogous habitats. Obvious examples are caves, desert oases, sphagnum bogs, and the boreal habitats of temperate and tropical mountaintops. It is of interest to ask whether the variables which determine the number of species on oceanic islands have similar effects on the biotas of other isolated habitats. For example, it has recently been shown that aquatic arthropods in caves (Culver 1970) and Andean birds in isolated paramo habitats (Vuilleumier 1970) are distributed as predicted by the equilibrium model of MacArthur and Wilson.

The boreal mammals of the Great Basin of North America provide excellent material for testing the generality of the equilibrium model. Almost all of Nevada and adjacent areas of Utah and California are covered by a vast sea of sagebrush desert, interrupted at irregular intervals by isolated mountain ranges. The cool, mesic habitats characteristic of the higher elevations in these ranges contain an assemblage of mammalian species derived from the boreal faunas of the major mountain ranges to the east (Rocky Mountains) and west (Sierra Nevada). Thanks largely to the work of Hall (1946), Durrant (1952), and Grinnell (1933), mammalian distributions within the Great Basin are documented quite thoroughly. I have used the work of these authors and data accumulated during 3 summers of my own field work in the Great Basin to produce the following analysis.

I shall show that the diversity and distribution of small mammals on the montane islands cannot be explained in terms of an equilibrium between colonization and extinction. Boreal mammals reached all of the islands during the Pleistocene; since then there have been extinctions but no colonizations.

* Present address: Department of Biology, University of Utah, Salt Lake City, Utah 84112.

METHODS

The Montane Islands

Because mountains do not have discrete boundaries, islands were defined by operational criteria applied to topographic maps (U.S. Geological Survey maps of the states; scale 1:500,000). A mountain range was considered an island if it contained at least one peak higher than 10,000 feet and was isolated from all other highland areas by a valley at least 5 miles across below an elevation 7,500 feet. This altitude corresponds approximately to the lower border of montane piñon-juniper woodland. Application of the above criteria defined 17 islands (fig. 1; table 1) which lie in a sea, the Great Basin, between two mainlands, the Sierra Nevada to the west and the central mountains of Utah (a part of the Rocky Mountains) to the east. The sizes (areas) of the islands and the distances between islands and between islands and mainlands have been determined from the topographic maps.

The Boreal Mammals

As with the montane islands, the mammals restricted in range to the higher altitudes must be defined somewhat arbitrarily. I have selected those

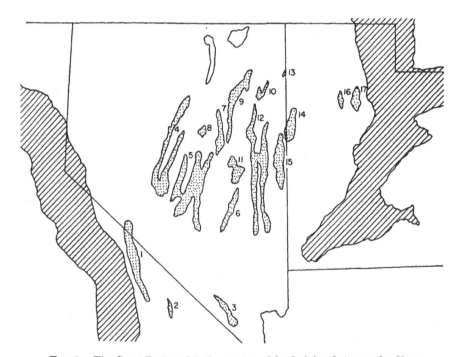

Fɪɢ. 1.—The Great Basin, with the montane islands lying between the Sierra Nevada (left) and Rocky Mountains (right). The shaded islands were used for the present analysis and are identified in table 1. The two unshaded islands were not used because they lie on the northern perimeter of the Great Basin and their faunas are poorly known.

TABLE 1
CHARACTERISTICS OF THE MOUNTAIN RANGES CONSIDERED AS ISLANDS

Mountain Range	Area above 7,500 Feet (Sq Miles)	Highest Peak (Ft)	Highest Pass* (Ft)	Nearest Island† (Miles)	Nearest Mainland (Miles)	Boreal Mammal Species (N)
1. White-Inyo	738	14,242	7,000	82	10	9
2. Panamint	47	11,045	5,500	19	52	1
3. Spring	125	11,918	3,500	108	125	3
4. Toiyabe	684	11,353	6,000	...	110	12
5. Toquima-Monitor	1,178	11,949	7,000	9	114	9
6. Grant	150	11,298	7,000	17	138	3
7. Diamond	159	10,614	7,000	7	190	4
8. Roberts Creek ...	52	10,133	7,000	22	216	4
9. Ruby	364	11,387	6,000	...	173	12
10. Spruce	49	10,262	6,500	12	156	3
11. White Pine	262	11,188	7,000	43	150	6
12. Schell Creek-Egan	1,020	11,883	7,000	11	114	7
13. Pilot	12	10,704	5,000	33	114	2
14. Deep Creek	223	12,101	7,000	9	104	6
15. Snake	417	13,063	7,000	76	89	8
16. Stansbury	56	11,031	6,000	4	39	3
17. Oquirrh	82	10,704	5,500	88	19	5

* Elevation of the highest pass separating the island from the mainland or from another island with more species.

† Distance to the nearest island with more species. No distance is given for the Toiyabe and Ruby Mountains because no other islands have more species.

species that occur only at high elevations in the Rockies and Sierra Nevada and are unlikely to be found below 7,500 feet at the latitudes of the Great Basin. I have excluded large carnivores and ungulates from the analysis because their distributions were drastically altered by human activity before accurate records were kept. I have also ignored the bats because their distributions are very poorly known and because their dispersal by flight introduces a completely new variable that could only complicate the present discussion. After these omissions, there remain 15 species of mammals (table 2) which occur in the Sierra Nevada and Rocky Mountains and on at least one of the montane islands of the Great Basin. All of these species occur in piñon-juniper, meadow, or riparian habitats. A striking feature of the mammalian faunas of the isolated peaks is the absence of those species which are restricted to dense forests of yellow pine, spruce, and fir in the Sierra Nevada and Rocky Mountains. None of these species, which include *Martes americana*, *Aplodontia rufa*, *Eutamias speciosus*, *Tamiasciurus hudsonicus*, *T. douglasi*, *Glaucomys sabrinus*, *Clethrionomys gapperi*, and *Lepus americanus*, are found on any of the isolated mountain ranges of the Great Basin, even though some of the large islands have large areas of apparently suitable habitat. However, the well-developed coniferous forests on the large islands have a good sample of the avian species characteristic of these habitats on the mainlands.

Records of occurrence are taken from the literature (Hall 1946; Hall and Kelson 1959; Durrant 1952; Durrant, Lee, and Hansen 1955; Grinnell 1933) and from my own observations which concentrated on the small mountain ranges. Undoubtedly, there are a few errors of omission that will

TABLE 2

CHARACTERISTICS AND DISTRIBUTION IN THE GREAT BASIN OF THE FIFTEEN SPECIES OF BOREAL MAMMALS CONSIDERED

SPECIES	BODY WEIGHT (g)	HABITAT AND DIET	1	2	3	4	5	6	7	8	9	10	11	12	13	14	15	16	17	TOTAL NO. ISLANDS INHABITED
*Sorex vagrans**	6.7	Carnivore	..	..	..	x	..	..	..	..	x	..	..	x	..	x	x	..	x	6
Sorex palustris	14	Carnivore	x	..	..	x	x	..	..	x	x	..	..	..	..	..	x	..	..	6
Mustela erminea	58	Carnivore	x	..	..	..	..	..	..	..	..	..	..	..	..	..	x	x	..	3
Marmota flaviventer	3,000	General herbivore	x	..	..	x	x	x	..	..	..	..	x	x	..	x	x	..	..	9
Spermophilus lateralis	147	General herbivore	x	..	x	x	x	x	x	..	x	x	x	x	x	x	x	x	..	13
Spermophilus beldingi	382	Meadow herbivore	..	..	..	x	x	..	..	..	x	..	..	..	..	..	..	..	..	3
Eutamias umbrinus†	57	General herbivore	x	..	x	x	x	x	x	x	x	x	x	x	..	x	x	x	x	14
Thomomys talpoides	102	Deep soil herbivore	x	..	..	x	x	..	x	x	x	x	x	x	..	..	..	..	x	8
Microtus longicaudus	47	General herbivore	x	..	x	x	x	x	x	x	x	x	x	x	x	x	x	x	x	12
Neotoma cinerea	317	General herbivore	x	x	x	x	x	x	..	x	x	x	x	x	x	x	x	x	..	14
Zapus princeps	33	Riparian herbivore	..	..	..	x	..	..	..	x	x	..	..	..	..	..	..	..	x	4
Ochotona princeps	121	Talus herbivore	x	x	..	..	..	..	..	..	x	..	..	..	..	..	..	..	x	4
Lepus townsendi	2,500	Meadow herbivore	..	..	..	..	..	..	..	..	..	..	..	..	..	..	..	..	..	1

* On island 17 the species is *Sorex obscurus*, a Rocky Mountain form, which forms a superspecies with *S. vagrans*.

† On island 3 the species is *Eutamias palmeri*, an endemic, which forms a superspecies with *E. umbrinus*.

be revealed by more intensive collecting, but these should not affect the present analysis significantly.

Island Area and Extinction

The number of mammalian species inhabiting a montane island is closely correlated with the area of the island (fig. 2). When both variables are plotted logarithmically, the data are well described ($r = .82$) by a straight line with a slope (z) of .43. Four comparable areas on the mainland (Sierra Nevada) have more species and the species-area curve has much less slope ($z = .12$). These areas can be considered as "saturated" with suitable species. These relationships are similar to those described for various groups of animals and plants on oceanic islands and continental mainlands (Mac-Arthur and Wilson 1967; Johnson, Mason, and Raven 1968) and for birds on tropical montane islands in South America (Vuilleumier 1970), except that the z value describing the data for mammals on montane islands is

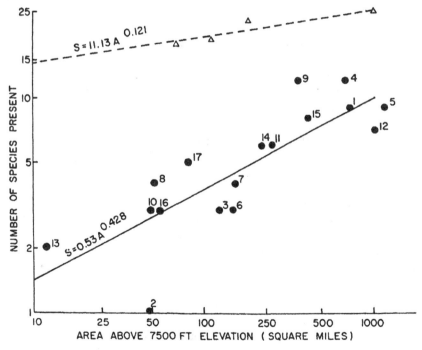

FIG. 2.—Double logarithmic plot of number of species of boreal mammals against area for the 17 montane islands (circles) listed in table 1. The solid line indicates the relationship between number of species (S) and area (A) fitted by least-squares regression. The dashed line represents the "saturation values" based on four areas (triangles) in the Sierra Nevada; data are from Hall (1946) and Grinnell (1933).

considerably higher than most of the values (.24–.34) previously obtained for insular biotas.

Local areas acquire species by immigration and lose them by extinction. In the absence of historical perturbations and speciation, the number of species inhabiting an area will represent an equilibrium between the opposing rates of extinction and immigration. The biotas of most continental areas and oceanic islands (including islands on the continental shelf) approximate such an equilibrium condition (Preston 1962a, 1962b; Mac-Arthur and Wilson 1963, 1967; Diamond 1969; Simberloff and Wilson 1969). Rates of extinction are dependent largely on population size and, in the absence of recurrent colonization, should be similar for comparable areas of islands or mainlands; to the extent that the absence of competitors results in greater population densities on islands, extinction rates should be slightly lower for islands than for comparable areas on mainlands. Islands, because of their isolation, have much lower rates of immigration (colonization) than local areas of comparable size within relatively homogeneous habitat on mainlands. The net result is that the slope of the species-area curve should be related to the degree of isolation of the islands concerned—the lower the rate of colonization, the higher the z value. The fact that the z value obtained in the present study is higher than the range observed for the biotas of oceanic islands (MacArthur and Wilson 1967) indicates that boreal mammals have an exceptionally low rate of immigration to isolated mountains.

The population size (and, hence, probability of extinction on an island) for a species is greatly affected by three variables—body size, trophic level, and habitat specialization. These variables affect the distribution of mammalian species on montane islands in the manner we would expect from considerations of their effects on population size: small mammals are found on more islands than large ones, herbivores are better represented than carnivores, and herbivores that can live in most montane habitats inhabit more islands than herbivores which can live only in restricted habitats (fig. 3). Those species which occur on only a few islands usually are found only on large islands (see tables 1 and 2). Thus, at least for mammals on montane islands, insular area not only affects the number of species but also can be used to predict quite reliably some ecological attributes of those species which are present.

Distance from the Mainland, Climatic Change, and Colonization

It is frequently observed that the number of species on an island is related to the proximity of the island to the nearest mainland as well as to the size of the island. This suggests that the probability of colonization of an island by new species from the mainland is proportional to the distance involved and that recurrent colonization of islands occurs at a rate sufficient to offset partially the extinction of species. Using this sort of reason-

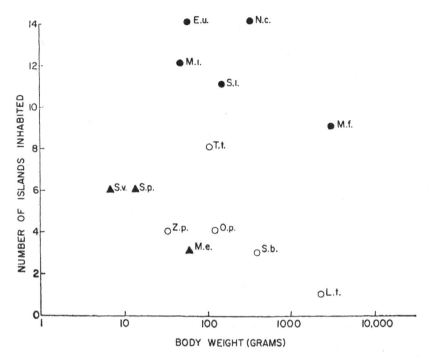

Fɪɢ. 3.—Frequency of occurrence on the montane islands of species of boreal mammals plotted against their body weight; shaded circles represent herbivores which are found in most habitats, unshaded circles indicate herbivores with specialized habitat requirements, triangles denote carnivores. The abbreviations are of species names which can be identified by reference to table 2.

ing, MacArthur and Wilson (1963) have proposed a model which represents the number of species on an island as a dynamic equilibrium between rates of colonization and extinction. The effect of recurrent colonization on species diversity is usually assessed by expressing the number of species on an island as a percentage of the number present in an area of the same size on the mainland and plotting this percentage saturation against the distance to the nearest mainland. When this is done for the montane mammals of the Great Basin, no relationship $(r = .005)$ is observed (fig. 4). This is in marked contrast to the inverse correlation predicted by the equilibrium model and observed for various groups of organisms on oceanic islands (MacArthur and Wilson 1967) and for birds on montane islands (Vuilleumier 1970).

Before concluding that recurrent colonization has not been a significant factor in determining the diversity of mammals on the montane islands, it is necessary to exclude three alternative explanations: (1) Differences in habitat between islands may be sufficiently great to obscure the distance effect (see Diamond 1969). The montane islands of the Great Basin have similar vegetation, climate, and geomorphology; but they differ somewhat in elevation, which may affect the amount and diversity of boreal habitats

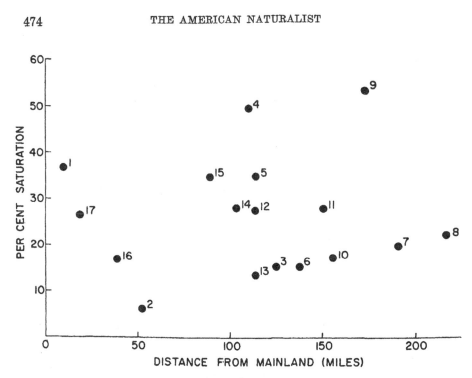

Fig. 4.—Percentage of faunal saturation plotted against distance from the Sierra Nevada or Rocky Mountains, whichever is nearer, for the 17 montane islands listed in table 1. Percentage of saturation is defined in the text.

in a manner different from insular area. (2) The islands may acquire most of their mammalian colonists from other islands (by a stepping-stone process) rather than directly from the nearest mainland, so that distance between islands might be an important variable affecting immigration. If this is so, the nearest island with more species is the most likely source of colonists if it is nearer than the closest mainland. (3) The effectiveness of the desert barriers surrounding the islands rather than the distance between mountains may be the most important variable influencing the rate of colonization. If this is true, the altitude of the highest pass should be the best measure available of the severity of the climatic and habitat barriers which separate the mountain ranges.

The possibility that any of these three explanations may account for some of the variability in insular-species diversity and provide evidence of the effects of immigration can be evaluated by subjecting the data in table 1 to stepwise multiple regression analysis. The results of such analysis using a linear model are shown in table 3. Again it is apparent that the number of mammalian species inhabiting an island is closely related to insular area, although the correlation is not as good as when a logarithmic model is used. Neither elevation of the highest peak nor any of the three variables (distance from nearest mainland, distance from nearest mountain with more species, elevation of highest intervening pass) which might be expected to

TABLE 3
INFLUENCE OF SEVERAL VARIABLES ON THE NUMBER OF SPECIES OF SMALL MAMMALS
INHABITING MONTANE ISLANDS IN THE GREAT BASIN ANALYZED BY
STEPWISE MULTIPLE REGRESSION USING LINEAR MODEL[a]

Variable	Contribution to R^2	F Value	Order Entered in Equation
Area	.49421	14.65*	1
Highest peak	.00006	0.32	2
Nearest mainland	.00042	0.83	3
Nearest island	.00002	0.16	4
Highest pass	.00000	0.04	5
Total R^2	.49471	...	...

[a] Data are from table 1.
* Significant at 1% level; the other F values are not significant.

influence immigration rate contributes significantly to the reduction of the remaining variability in the number of insular species. Apparently, at the present time the rate of colonization of the islands is effectively zero. The desert valleys of the Great Basin are virtually absolute barriers to dispersal by small boreal mammals.

I conclude that the boreal mammalian faunas of the montane islands of the Great Basin do not represent equilibria between recurrent rates of colonization and extinction. This is somewhat surprising, because conditions of dynamic equilibria with surprisingly high and measurable steady state turnover rates have been documented for some oceanic islands (see Mac-Arthur and Wilson 1967, on Krakatau; Simberloff and Wilson 1969, Diamond 1969). However, my explanation for the distribution of boreal mammals on the montane islands of the Great Basin—that all the islands were inhabited by a common pool of species at some time in the past and subsequent extinctions have reduced the number of species on individual islands to their present levels—is consistent with what we know of the paleoclimatic history of the Southwest and the way small mammals disperse over land.

At intervals during the Pleistocene, colder climates in the Southwest forced piñon-juniper woodland down to elevations approximately 2,000 feet below their present distribution (Wells and Berger 1967). This was sufficient to make piñon-juniper and associated streamside and meadow habitats contiguous across most of the Great Basin perhaps as recently as 8,000 years ago. The climatic and habitat barriers between the isolated peaks and the mainlands were removed for an entire set of species; these paleoclimatic and habitat shifts are sufficient to account for all 15 species of boreal mammals now found on the montane islands. Two lines of evidence indicate that the islands were colonized in this manner. First, the limited fossil evidence indicates that some of the boreal mammals that are now confined to only a few of the islands were more widely distributed in the late Pleistocene; for example, Wells and Jorgensen (1964) found fossils of *Marmota flaviventer* in the Spring Range in southern Nevada. Second, paleobotanical evidence indicates that the climate changes during the Pleistocene were not sufficient to connect the dense coniferous forests of yellow pine, spruce, and fir between the islands and the mainlands. Those mammalian species which are restricted to such forests in the Sierra Nevada and Rocky Mountains are absent from all of the isolated peaks in the Great Basin, even though there are large areas of suitable habitat on several of the larger islands. It is apparent that the islands have been colonized only during those periods in the past when the climatic and habitat barriers which now isolate them were temporarily abolished.

A few thousand feet of elevation, with the associated differences in climate and habitat, constitute a nearly absolute barrier to dispersal by small mammals (with the exception of bats). This is not surprising in view of the fact that small mammals must disperse over land on foot. It may take several days for an individual to travel only a few miles, and during this period it must obtain food and avoid predation in an unfamiliar habitat and survive the stresses imposed by a different climate.

GENERAL DISCUSSION

The mammalian faunas of the isolated mountains in the Great Basin do not represent equilibria between rates of extinction and colonization. This is shown by the unusually steep slope of the species-area curve and by the lack of correlation between the percentage saturation of the insular faunas and any of the variables (distance to mainland, distance between islands, and elevation of surrounding valleys) likely to influence the probability of immigration to an island. The diversity of mammalian species on the mountaintops must be explained by historical events (climatic changes during the Pleistocene) which temporarily abolished the isolation of the mountains and permitted their colonization by a group of mammalian species. Subsequent extinctions, related to island size, have reduced the faunas of each island but not sufficiently to restore equilibrium with the vanishingly small rates of colonization. Approximately 8,000 years after their isolation, the largest islands still have almost all of their original species of boreal mammals, and even on the smallest islands one or more species remain. This suggests that the rate of extinction is very low once the biota of the island has initially adjusted to its isolation; when colonization rates approach zero, extinction of all species is reached only in geological time spans.

Island biotas which do not represent equilibria between rates of extinction and colonization should be fairly common, particularly in the temperate zones where the climatic fluctuations of the Pleistocene have drastically altered the distribution of terrestrial and freshwater habitats. Some organisms are capable of crossing the barriers between isolated habitats, and these may be expected to have distributions predicted by the equilibrium model. Examples are the birds, bats, flying insects, and most kinds of plants on montane islands. However, other groups of organisms are unable to cross the barriers between insular habitats, and these should be distributed as relicts in a nonequilibrium pattern similar to the montane mammals described here. Likely examples are amphibians, reptiles, and large, nonflying arthropods on mountaintops and fish is isolated bodies of fresh water. The fish fauna of the isolated springs and streams of the Great Basin clearly does not represent an equilibrium situation; these habitats were colonized in the Pleistocene when there were aquatic connections between them, and extinctions have subsequently reduced the fauna of each isolate to its present composition (Hubbs and Miller 1948). Certainly, other nonequilibrium patterns of island distributions will be described. In those cases where environmental changes (often the result of human activity) have caused massive extinctions, the number of species on an island will be less than the equilibrium number.

SUMMARY

An analysis of the distribution of the small boreal mammals (excluding bats) on isolated mountaintops in the Great Basin led to the following conclusions:

1. The species-area curve is considerably steeper ($z = .43$) than the curves usually obtained for insular biotas.

2. There is no correlation between number of species of boreal mammals and variables which are likely to affect the probability of colonization, such as distance between island and mainland, distance between islands, and elevation of intervening passes. Apparently the present rate of immigration of boreal mammals to isolated mountains is effectively zero.

3. Paleontological evidence suggests that the mountains were colonized by a group of species during the Pleistocene when the climatic barriers that currently isolate them were abolished.

4. Subsequent to isolation of the mountains, extinctions have reduced the faunal diversity to present levels. Probability of extinction is inversely related to population size and, therefore, is influenced by body size, diet, and habitat. The rate of extinction has been low, and all of the islands still have one or more species of boreal mammals.

5. The mammalian faunas of the mountaintops are true relicts and do not represent equilibria between rates of colonization and extinction.

ACKNOWLEDGMENTS

My wife, Astrid, assisted with all aspects of the study. Dr. F. Vuilleumier made many helpful comments on the manuscript and performed the multiple regression analysis shown in table 3. Drs. G. A. Bartholomew, M. L. Cody, J. M. Diamond, and D. E. Landenberger kindly read and criticized the manuscript. The work was supported in part by a grant (GB 8765) from the National Science Foundation.

LITERATURE CITED

Culver, D. C. 1970. Analysis of simple cave communities. I. Caves as islands. Evolution 29:463–474.

Diamond, J. M. 1969. Avifaunal equilibria and species turnover rates on the Channel Islands of California. Nat. Acad. Sci. (U.S.), Proc. 64:57–63.

Durrant, S. D. 1952. Mammals of Utah, taxonomy and distribution. Univ. Kansas Publ. Mus. Natur. Hist. 6:1–549.

Durrant, S. D., M. R. Lee, and R. M. Hansen. 1955. Additional records and extensions of known ranges of mammals from Utah. Univ. Kansas Publ. Mus. Natur. Hist. 9:69–80.

Grinnell, J. 1933. Review of the recent mammal fauna of California. Univ. California Publ. Zool. 40:71–234.

Hall, E. R. 1946. Mammals of Nevada. Univ. California Press, Berkeley. 710 p.

Hall, E. R., and K. R. Kelson. 1959. The mammals of North America. Ronald, New York. 1083 p.

Hubbs, C. L., and R. R. Miller. 1948. Correlation between fish distribution and hydrographic history in the desert basins of western United States. *In* The Great Basin, with emphasis on glacial and postglacial times. Bull. Univ. Utah Biol. Ser. 38:20–166.

Johnson, M. P., L. G. Mason, and P. H. Raven. 1968. Ecological parameters and plant species diversity. Amer. Natur. 102:297–306.

MacArthur, R. H., and E. O. Wilson. 1963. An equilibrium theory of insular zoogeography. Evolution 17:373–387.

———. 1967. The theory of island biogeography. Princeton Univ. Press, Princeton, N.J. 203 p.

Preston, F. W. 1962a. The canonical distribution of commonness and rarity. I. Ecology 43:185–215.

———. 1962b. The canonical distribution of commonness and rarity. II. Ecology 43:410–432.

Simberloff, D. S., and E. O. Wilson. 1969. Experimental zoogeography of islands. The colonization of empty islands. Ecology 50:278–296.

Vuilleumier, F. 1970. Insular biogeography in continental regions. I. The northern Andes of South America. Amer. Natur. 104:373–388.

Wells, P. V., and R. Berger. 1967. Late Pleistocene history of coniferous woodland in the Mojave Desert. Science 155:1640–1647.

Wells, P. V., and C. D. Jorgensen. 1964. Pleistocene wood rat middens and climatic change in Mojave Desert: a record of juniper woodlands. Science 143:1171–1173.

Colonization of Exploded Volcanic Islands by Birds: The Supertramp Strategy

Abstract. *After volcanic explosions or tidal waves had defaunated several islands near New Guinea, bird species number rapidly returned to equilibrium on coral islets and rapidly returned to quasi-steady-state values limited by regrowth of vegetation in lowland forest of larger islands. However, reequilibration in montane forest has been limited by slow dispersal of the birds. Colonists have been drawn disproportionately from r-selected "supertramp" species, which maintain much higher population densities than do K-selected faunas, perhaps due to selection for resource overexploitation by the latter.*

The repopulation of Krakatau (Indonesia) by plants and animals, after a volcanic explosion destroyed all life, furnished a classic natural experiment in island colonization. A similar opportunity may be provided by the recolonization of Long and Ritter, two volcanic islands in Vitiaz and Dampier straits between the "source" islands of New Guinea and New Britain. On 13 March 1888 Ritter disintegrated in a cataclysmic explosion that removed more than 95 percent of its bulk and destroyed all vegetation and almost certainly all life on the remaining fragment (1, 2). Today Ritter supports pandanus trees up to 12 m high on its gentler slopes but is still bare in steeper areas, the regrowth of vegetation having been retarded by landslides, rain runoff, porous soil, and strong prevailing winds. Long was devastated about two centuries ago by an explosion that deposited a layer of ash up to 30 m thick or more (1, 3, 4). The forest in the lowlands of Long today is more open and savanna-like than on older volcanic islands, probably due to porous soil and the disappearance of streams during the prolonged annual dry season, and has remained in this arrested subclimax state for at least much of the 20th century. However, the montane forest above 900 m on Long is kept moist by standing cloud banks and is already similar in physiognomy to forest on the summits of older volcanic islands. Near Long and Ritter are "control" islands and "wave-defaunated" islands of varying sizes: the relatively undisturbed islands Umboi, Sakar, Tolokiwa, Crown, Malai, and Midi; and seven coral islets whose birds must have disappeared in a tidal wave set up by the Ritter explosion but which now support forest similar to that on remote islets. In 1972 I surveyed the birds of these 15 islands, examining all major types of habitat on each island from sea level to the summit. Since the survey party

consisted of six observers familiar with New Guinea birds and received much additional information from people living on the larger islands, our lists of presently resident bird species should be virtually complete. This report summarizes, for birds, the return to equilibrium species number, the types of colonizing species, and the striking contrast in population densities between the colonist avifaunas and older avifaunas.

In Fig. 1 the bird species numbers on the control islands are used to assess by comparison how far the defaunated islands have come toward equilibrium. For the seven larger control islands (represented by the seven closed circles on the right in Fig. 1) the number of lowland nonmarine bird species S increases very regularly with the island area A (in square kilometers) according to the empirical relation

$$S = 18.9 \, A^{0.18} \qquad (1)$$

A linear logarithmic relation between species number and area similarly describes distributions of most plant and animal groups on most other archipelagoes (5). Such a relation is interpreted as meaning that island species numbers represent an equilibrium between extinction and immigration, larger islands reaching equilibrium at more species because of larger populations, lower extinction rates, and greater habitat diversity (5–7). Examination of the spe-

cies and races involved shows that most bird populations on the Vitiaz-Dampier islands are more immediately derived from the source island of New Britain than from New Guinea, although New Guinea is 20 times larger and approximately as close. The reason is that many New Guinea species are sedentary and rarely cross water gaps, but all New Britain species were originally derived by overwater colonization from New Guinea and were thus selected from New Guinea's pool of superior colonists. The same factor may underlie the high immigration rates implicit in the low value of the area exponent in Eq. 1 [0.18, compared to 0.22 for satellite islands of New Guinea (8) and 0.24 to 0.30 for many other archipelagoes (5)].

Species numbers for the eight smaller islets ($A = 0.003$ to 0.07 km²) lie generally below the species-area relation defined by the larger islands and show much more scatter. These islets are comparable in size to one territory for even the commoner bird species, and many of the populations on the islets were found to consist of only a single bird pair. Since it is thus marginal whether even species with minimal territory requirements can occur on these islets at all, S is lower than would be the case if the species-area relation could be extrapolated to the left by infinite subdivision of bird individuals. The deviation of the observed S value below the regression line of Fig. 1 is significant at $P < .05$ for six islets and at $P < .01$ for four islets. In addition, S is subject to fluctuation because of the small number of species (only four on some islets) as well as the small number of individuals. The seven islets defaunated by tidal wave (Fig. 1, triangles) have S values similar to that of the noninundated control islet (Fig. 1, closed circle on the left) and may already have achieved equilibrium. This was to be expected because they have

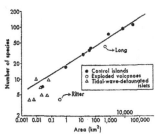

Fig. 1. Number of resident, nonmarine, lowland bird species, S, on Vitiaz-Dampier islands, plotted as a function of island area, A, on a double logarithmic scale. (Closed circles) Relatively undisturbed "control" islands (from left to right, Midi, Malai, Crown, Sakar, Tolokiwa, Umboi, New Ireland, and New Britain); (open circles) exploded volcanoes; (triangles) coral islets inundated by the Ritter tidal wave. The straight line $S = 18.9 \, A^{0.18}$ was fitted by least mean squares through points for the seven larger control islands.

already regained climax vegetation and because their tiny population sizes and hence high natural turnover rates restrict successful colonization to the most rapidly dispersing species. Rapid equilibration is unequivocally demonstrated by the fact that after experimental removals of birds from a small islet the bird species number returned to its original value within times as short as a few days and with initial colonization rates of approximately one species per hour.

Lowland species numbers on the two exploded volcanoes lie below the line for control islands. Ritter has only four species, far below the 16 species predicted for its area from Eq. 1 or the 17 species observed on a control island of similar size, Malai. No other ornithologically explored island of the tropical southwest Pacific deviates so far below the species-area relation for its archipelago as does Ritter. Long has only 43 species, considerably fewer than the 57 species predicted from Eq. 1 and nearly matched by the 40 species observed on the control island Tolokiwa, which is ten times smaller than Long. Although the deviation below the control curve (Eq. 1) is less striking for Long than for Ritter, the deviation for Long is still significant at $P < .001$. A priori, these deficits could be due either to slow colonization by birds or to the postexplosion forest not yet having regained the climax structure found on older volcanoes. The latter explanation is surely the correct one. The arrival at Ritter of species that could not establish resident populations, because of sparse vegetation and lack of wind shelter, was documented both by direct observation and by feathers at the plucking perch of a resident peregrine falcon. On Long, my party found 40 species in the same habitats where Coultas (4) had already found 37 species in 1933. The difference between 37 and 40 is insignificant, probably arose because Coultas was working alone rather than in a party of six, and suggests that the species number on Long scarcely changed from 1933 to 1972. The species turnover rate on Long, calculated as in table 1 in (6) from apparent immigration and extinction rates between 1933 and 1972, is 0.18 percent per year. This is even less than the turnover rate of 0.32 percent per year for a similar-sized, older volcano, Karkar (9), whose species number is at equilibrium (8, 10). Thus, by 1933, about a century and a half after the explosion, the bird species number on Long had built up to 75 percent of the equilibrium value, become "stuck" in this quasi-steady state, and ceased to turn over more rapidly than on old islands. Since the open, savanna-like lowland forest of Long should hold fewer species than the more structured rainforest of old islands (11), the arrested equilibration of the Long lowlands with bird species may be attributed to the arrested development of forest. The more marked deficit in species number on Ritter than on Long is consistent with Ritter's much sparser vegetation.

On mountainous southwest Pacific islands, each 1000 m of elevation enriches the avifauna by a number of montane species equal on the average to 8.9 percent of the species number at sea level (8, 10). For the control islands Crown, Tolokiwa, and Umboi, this formula predicts 2, 5, and 11 montane species, in good agreement with the 2, 4, and 9 observed, respectively. For Long, however, the actual number of montane species, 2, is far below the predicted 7, this deficit being greater than that for any other known mountainous island of the southwest Pacific. Unlike the lowland deficit, the montane deficit cannot be attributed to arrested vegetation, since the montane forest of Long is already structurally mature. Colonization by montane birds must be much slower than colonization by lowland birds and must be limited by dispersal of the birds themselves. In agreement with this conclusion, the montane birds of New Britain are much more distinct from their New Guinea relatives than are the lowland birds, suggesting lower rates of gene flow.

A remarkable finding of the surveys was the spectacularly high population densities of birds on Long and two other islands. This finding, which was qualitatively obvious within a short time of our landing on Long, was quantitated by mist-netting techniques (12, 13). Figure 2 shows that on most control islands the combined population densities of all bird species increase with the local number of species present, linearly in lowland forest and supralinearly in montane forest. Values for two Vitiaz-Dampier control islands,

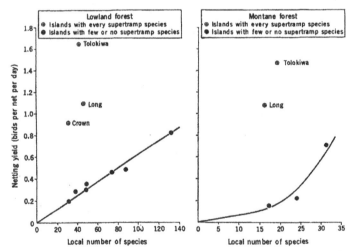

Fig. 2. Ordinate: rate of catching land birds under standardized conditions in mist nets on a particular island (expressed in birds caught per net per day), in lowland forest (left) or montane cloud forest (right). Abscissa: local number of species present on the island. If one compares the same type of habitat on different islands (but not if one compares different habitats), this catch rate is proportional to the combined population densities of all species present (12, 13). (Closed circles) Control islands with not more than three supertramp species [values for Umboi and Sakar are from this report; other values are from table 1 and figure 1 in (12)]; (crossed circles) the exploded volcano of Long and two control islands, which support all nine supertramp species. Note that, on islands with few supertramps, netting yields (and total population densities) increase with the local number of species, and that islands with all the supertramp species have much higher yields than islands with the same total number of species but few supertramps.

Umboi and Sakar (61 and 93 km from Long, respectively), fall on the curve defined by other New Guinea satellite islands and interpreted previously (12). On the exploded volcano of Long and the two nearest control islands Crown and Tolokiwa (12 and 37 km, respectively, from Long), population densities are 4 to 11 times those on more distant control islands with similar species numbers. This pattern could not be explained by postulating that these three islands share exceptionally fertile soil deposited as an ash blanket by the Long explosion, since ash fallout on Tolokiwa was slight (1).

What does set Long, Crown, and Tolokiwa apart is the unusual, shared species composition of their avifaunas. Bird species of the Bismarck Archipelago (New Britain and its neighbors, including the Vitiaz-Dampier group) may be divided according to their dispersal strategy into three somewhat arbitrary categories: (i) "sedentary species," confined to the larger islands; (ii) "tramps," present on not only the larger islands but also many smaller and more remote islands; and (iii) "supertramps," nine species which are confined as residents mainly to small islands and virtually absent as residents (although often recorded as vagrants) from the larger islands (14). In the terminology of MacArthur and Wilson (5), sedentary species represent the extreme of K selection (selection for competitive ability at the expense of dispersal ability), while supertramps represent the extreme of r selection (selection for dispersal ability and reproductive potential at the expense of competitive ability). Tramps comprise most species on all the Vitiaz-Dampier islands. On most islands, except for Long, Crown, and Tolokiwa, there are only a few supertramp species. These three islands are unique in the whole Bismarck Archipelago in supporting all nine supertramps.

These distributional patterns suggest the following interpretation of the supertramp strategy. Supertramps specialize in rapid breeding and overwater colonization, but they have paid a price for these adaptations and are excluded from most islands by competitors that can harvest resources more thoroughly and tolerate lower resource levels. Whenever the supertramps find islets too small for stable populations to persist for a long time, or else an empty island recently devastated by a tidal wave or volcanic explosion, they breed on a nearly year-round basis (15), fill the island, and generate new emigrants. By the time these populations have disappeared or been squeezed out by more efficient later arrivals, the supertramps have already ensured their survival as species by finding other transiently empty islands. With the explosion of a large island like Long, the supertramps "struck it rich" as first arrivals. For some of the supertramp species, the colonization of Long quadrupled the area inhabited by all populations of that species combined. The frequency with which I saw groups of land birds over mid-ocean around Long, or else leaving land and disappearing out of sight, suggests that Long is producing numerous emigrants and that these colonists have inundated parts of the former avifaunas of the two nearest islands, Crown and Tolokiwa. However, on islands more distant from Long, where the flood of colonists is lower, more efficient residents have been able to exclude the supertramps.

The excess population densities in the supertramp-rich avifaunas of Long, Crown, and Tolokiwa largely represent the abundances of the supertramp species themselves, even though they comprise only about one-quarter of the species present (16). At first it seems paradoxical that the supertramps maintain much higher population densities than the more efficient competitors which exclude them from older islands. However, there is increasing reason to suspect that "self-renewing resources can be exploited to the detriment of the predator's population, and this overexploitation will be a natural consequence of competition among the predator species" (17, p. 31; see pp. 56–57 for a simple model). That is, competition in a species-rich fauna selects for species that can reduce resource levels below the point where other species can survive, even though this diminishes the rate of resource production and hence the population density of harvesting species. For instance, K-selected species of insectivorous birds, by finding and catching insects more efficiently than do supertramps, may be depressing sustainable insect yields far below the level that exists on supertramp-rich islands. K-selected frugivores may eat fruits which are not yet full-sized and are more unripe than supertramps tolerate, even though this reduces the caloric value of the fruit crop. This "non-overexploitation" interpretation of high supertramp densities on the Vitiaz-Dampier islands is the mirror image of Elton's explanation (18), in terms of heavy predation pressure, for the low total abundance (despite high diversity) of insects in the most species-rich community on Earth, the neotropical rainforest.

JARED M. DIAMOND
Department of Physiology, School of Medicine, University of California, Los Angeles 90024

References and Notes

1. R. W. Johnson, G. A. M. Taylor, R. A. Davies, *Bur. Miner. Resour. Aust. Rec. 21* (1972) (unpublished).
2. Anonymous, *Nachr. Kaiser Wilhelms-Land 4*, 76 (1888); Anonymous, *Mitt. Forschungsreis. Geleprt. Dtsch. Schutzgeb. 4*, 59 (1891); R. R. Steinhäuser, *Westermanns Illus. Dtsch. Monatsh. 71*, 265 (1892); K. L. Hammer, *Die Geographische Verbreitung der Vulkanischen Gebilde und Erscheinungen* (Münchow Druckerei, Giessen, 1907), pp. 17–19.
3. J. M. Bassot and E. E. Ball, *Papua New Guinea Sci. Soc. Proc. 23*, 26 (1972).
4. W. F. Coultas, "Journal of the Whitney South Sea expedition" (1933) (manuscript), vol. 4.
5. R. H. MacArthur and E. O. Wilson, *The Theory of Island Biogeography* (Princeton Univ. Press, Princeton, N.J., 1967).
6. J. M. Diamond, *Proc. Natl. Acad. Sci. U.S.A. 64*, 57 (1969).
7. E. O. Wilson and D. S. Simberloff, *Ecology 50*, 267 (1969).
8. J. M. Diamond, *Proc. Natl. Acad. Sci. U.S.A. 69*, 3199 (1972).
9. ———, *ibid. 68*, 2742 (1971).
10. ———, *Science 179*, 768 (1973).
11. R. H. MacArthur, H. Recher, M. Cody, *Am. Nat. 100*, 319 (1966).
12. J. M. Diamond, *Proc. Natl. Acad. Sci. U.S.A. 67*, 1715 (1970).
13. R. H. MacArthur, J. M. Diamond, J. R. Karr, *Ecology 53*, 330 (1972).
14. The supertramps are *Ptilinopus solomonensis, Ducula pistrinaria, Macropygia mackinlayi, Halcyon chloris stresemanni, Monarcha cinerascens, Pachycephala melanura dehli, Myzomela sclateri, Myzomela pammelaena,* and *Zosterops griseotincta.* On Long, all of these species are widespread in virtually all habitats with bushes or trees, from sea level to the summit, and from gardens to savanna to cloud forest. The supertramps lack resident populations in the corresponding habitats of most species-rich islands, although they are often present on islets a few kilometers or less distant. Since vagrants, dispersing juveniles, or transient populations have by now been recorded for most supertramp species on the best-studied large islands such as New Britain and Umboi, the supertramps must be excluded on a permanent basis by competitors. In some cases the competitors can be identified from details of distributional patterns. For example, *Macropygia mackinlayi* is excluded by *M. nigrirostris* and certain combinations of other cuckoo doves, *Ducula pistrinaria* by *D. rubricera, Monarcha cinerascens* by certain combinations of six other flycatchers, *Pachycephala melanura dehli* by *P. pectoralis* or by certain combinations of other flycatchers, and *Myzomela sclateri* and *M. pammelaena* by certain combinations of other myzomelids and sunbirds.
15. Compare observations on year-round breeding by five of the supertramp species [O. Meyer, *J. Ornithol. 78*, 19 (1930)]. A single pair of the tramp flycatcher *Rhipidura leucophrys* has been observed to rear six broods within 6 months [R. Mackay, *The Birds of Port Moresby and District* (Nelson, Melbourne, 1970), p. 46].
16. More individuals were netted of the territorial supertramp species than of the nonterritorial ones, because the latter happen to be species that forage largely above the height

of the nets. Thus, wanderings of nonterritorial supertramps are unlikely to have significantly inflated the mist-net yields. Predators of birds are more diverse and far more abundant on Long than on control islands, so that supertramp abundance cannot be attributed to reduced predation pressure.

17. R. H. MacArthur, *Geographical Ecology* (Harper & Row, New York, 1972).
18. C. S. Elton, *J. Anim. Ecol.* **42**, 55 (1973).

19. I thank the National Geographic Society and the Sanford Trust of the American Museum of Natural History for support; E. E. Ball and R. W. Johnson for discussions of Long Island; and many New Guinea residents for making fieldwork possible.

25 June 1973; revised 29 October 1973

THE ISLAND DILEMMA: LESSONS OF MODERN BIOGEOGRAPHIC STUDIES FOR THE DESIGN OF NATURAL RESERVES

JARED M. DIAMOND

*Physiology Department, University of California Medical Center,
Los Angeles, California 90024, USA*

ABSTRACT

A system of natural reserves, each surrounded by altered habitat, resembles a system of islands from the point of view of species restricted to natural habitats. Recent advances in island biogeography may provide a detailed basis for understanding what to expect of such a system of reserves. The main conclusions are as follows:

The number of species that a reserve can hold at equilibrium is a function of its area and its isolation. Larger reserves, and reserves located close to other reserves, can hold more species.

If most of the area of a habitat is destroyed, and a fraction of the area is saved as a reserve, the reserve will initially contain more species than it can hold at equilibrium. The excess will gradually go extinct. The smaller the reserve, the higher will be the extinction rates. Estimates of these extinction rates for bird and mammal species have recently become available in a few cases.

Different species require different minimum areas to have a reasonable chance of survival.

Some geometric design principles are suggested in order to optimise the function of reserves in saving species.

INTRODUCTION

For terrestrial and freshwater plant and animal species, oceanic islands represent areas where the species can exist, surrounded by an area in which the species can survive poorly or not at all and which consequently represents a distributional barrier. Many situations that do not actually involve oceanic islands nevertheless possess the same distributional significance for many species. Thus, for alpine species a mountain top is a distributional 'island' surrounded by a 'sea' of lowlands;

129

Biol. Conserv. (7) (1975)— © Applied Science Publishers Ltd, England, 1975
Printed in Great Britain

130   JARED M. DIAMOND

for an aquatic species a lake or river is a distributional island surrounded by a sea of land; for a forest species a wooded tract is a distributional island surrounded by a sea of non-forest habitat; and for a species of the intertidal or shallow-water zones, these zones represent distributional islands compressed between seas of land and of deep water.

Original equilibrium situation | Situation at time when protec-tive measures go into force | Final equilibrium situation

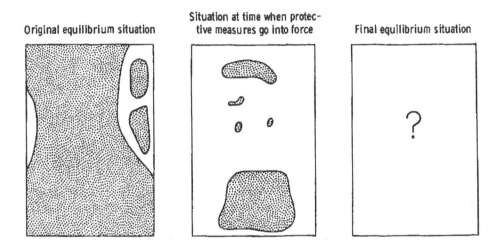

Fig. 1.   Illustration of why the problems posed by designing a system of natural reserves are similar to the problems of island biogeography. In the situation before the onset of accelerating habitat destruction by modern man, many natural habitats were present as continuous expanses covering large areas (indicated by shaded areas of sketch on left). Species characteristic of such habitats were similarly distributed over large, relatively continuous expanses. By the time that extensive habitat destruction has occurred and some of the remaining fragments are declared natural reserves, the total area occupied by the habitat and its characteristic species is much reduced (centre sketch). The area is also fragmented into isolated pieces. For many species, such distributions are unstable. Applying the lessons of modern island biogeography to these islands of natural habitat surrounded by a sea of disturbed habitat may help predict their future prospects.

Throughout the world today the areas occupied by many natural habitats, and the distributional areas of many species, are undergoing two types of change (Fig. 1). First, the total area occupied by natural habitats and by species adversely affected by man is shrinking, at the expense of area occupied by man-made habitats and by species benefited by man. Second, formerly continuous natural habitats and distributional ranges of man-intolerant species are being fragmented into disjunctive pieces. If one applies the island metaphor to natural habitats and to man-intolerable species, island areas are shrinking, and large islands are being broken into archipelagos of small islands. These processes have important practical consequences for the future of natural habitats and man-intolerant species (Preston, 1962; Willis, 1974; Diamond, 1972, 1973; Terborgh, in press, *a, b*; Wilson & Willis, in press). Ecologists and biogeographers are gaining increasing understanding of these processes as a result of the recent scientific revolution stemming from the work of MacArthur & Wilson (1963, 1967) and MacArthur (1972). In this paper I shall explore four implications of recent biogeographic work for conservation policies: (1) The ultimate *number* of species that a natural reserve will save is likely to be an increasing function of the reserve's area. (2) The *rate* at which species go extinct in a reserve is likely to be a decreasing function of the reserve's area. (3) The relation between reserved area and probability of a species' survival is characteristically different for different species. (4) Explicit suggestions can be made for the optimal geometric design of reserves.

HOW MANY SPECIES WILL SURVIVE?

Let us first examine the relation between reserve area and the number of species that the reserve can hold at equilibrium. As a practical illustration of this problem, consider the fact that we surely cannot save all the rain forest of the Amazon Basin. What fraction of Amazonia must be left as rain forest to guarantee the survival of half of Amazonia's plant and animal species, and how many species will actually survive if only 1% of Amazonia can be preserved as rain forest? Numerous model

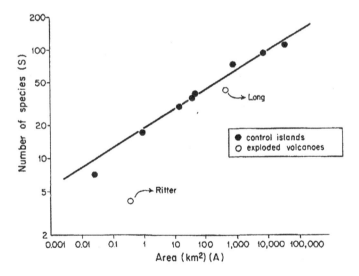

Fig. 2. Example of the relation between species number and island area in an archipelago. The ordinate is the number of resident, non-marine, lowland bird species (S) on the islands of Vitiaz and Dampier Straits near New Guinea in the south-west Pacific Ocean, plotted as a function of island area (A, in km^2) on a double logarithmic scale. The points ● represent relatively undisturbed islands. The straight line $S = 18.9A^{0.18}$ was fitted by least mean squares through the points for these islands. Note that species number increases regularly with island area. The two points ○ refer to Long and Ritter Islands, whose faunas were recently destroyed by volcanic explosions and which have not yet regained their equilibrium species number.

systems to suggest answers to these questions are provided by distributional studies of various plant or animal groups on various archipelagos throughout the world. If one compares islands of different size but with similar habitat and in the same archipelago, the number of species S on an island is usually found to increase with island area A in a double logarithmic relation:

$$S = S_0 A^z \tag{1}$$

where S_0 is a constant for a given species group in a given archipelago, and z usually assumes a value in the range 0.18-0.35 (Preston, 1962; MacArthur & Wilson, 1963, 1967; May, in press). A rough rule of thumb, corresponding to a z value of 0.30, is that a tenfold increase in island area means a twofold increase in the number of species. Figure 2 illustrates the species/area relation for the breeding land and freshwater bird species on the islands of the Bismarck Archipelago near New Guinea and shows that the number of bird species increases regularly with island area. If one compares islands of similar area but at different distances from the continent or large island that serves as the main source of colonisation, then one finds that the number of species on an island decreases with increasing distance. This feature is illustrated by Fig. 3, which shows that the number of bird species on

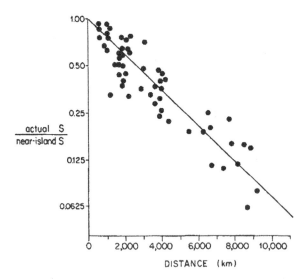

Fig. 3. Example of the relation between species number and island distance from the colonisation source in an island archipelago. The ordinate is the number of resident, non-marine, lowland bird species S on tropical south-west Pacific islands more than 500 km from New Guinea, divided by the number of species expected on an island of equivalent area less than 500 km from New Guinea. The expected near-island S was read off the species/area relation for such islands (Fig. 5). The abscissa is the island distance from New Guinea. Note that S decreases by a factor of 2 per 2600 km distance from New Guinea. (After Diamond, 1972.)

islands of the south-west Pacific decreases by a factor of 2 for each 2600 km of distance from New Guinea. For plants or animals with weaker powers of dispersal than birds, the fall-off in species number with distance is even more rapid.

Similar findings are obtained if, instead of oceanic islands, one compares habitat 'islands' within a continent or large island. For example, isolated as enclaves within the rain forest that covers most of New Guinea are two separate areas of savanna, which received most of their plant and animal species from Australia (Schodde & Calaby, 1972; Schodde & Hitchcock, 1972). The savanna which is larger and also closer to Australia supports twice as many savanna bird species as the smaller and more remote savanna (Fig. 4). Other examples are provided by mountains rising out of the 'sea' of lowlands, such as the isolated mountain ranges of Africa, South America, New Guinea and California. Thus, the number of bird species on each 'island' of alpine vegetation at high elevations in the northern Andes increases with area of alpine habitat and decreases with distance from the large alpine source area in the Andes of Ecuador (Vuilleumier, 1970).

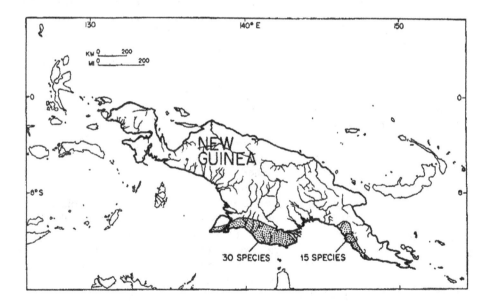

Fig. 4. Example of the relation between area of 'habitat islands' and the number of characteristic species they support. Most of New Guinea is covered by rain forest, but two separate areas on the south coast (shaded in the figure) support savanna woodland. The characteristic bird species of these savannas are mostly derived from Australia (the northern tips of Australia are just visible at the lower border of the figure). The so-called Trans-Fly savanna (left) not only has a larger area than the so-called Port Moresby savanna (right), but is also closer to the colonisation source of Australia. As a result, the Trans-Fly savanna supports twice as many bird species characteristic of savanna woodland (c. 30 compared with 15 species) as does the Port Moresby savanna.

Why is it that species number increases with increasing area of habitat but decreases with increasing isolation? In explanation of these findings, Preston (1962) and MacArthur & Wilson (1963, 1967) suggested that species number S on an island is set by (or approaches) an equilibrium between immigration rates and extinction rates. Species immigrate into an island as a result of dispersal of colonists from continents or other islands; the more remote the island, the lower is the immigration rate. Species established on an island run the risk of extinction due to fluctuation in population numbers; the smaller the island, the smaller is the population and the higher the extinction rate. Area also affects immigration and extinction rates in several other ways: through its relation to the regional magnitude of spatial and temporal variation in resources; by being correlated with the variety of available habitats as stressed by Lack (1973); and by being correlated with the number of 'hot spots', or sites of locally high utilisable resource production for a particular species (Diamond, in press). On a given island, extinction rates increase, and immigration rates decrease, with increasing S. The S value on an island in the steady state is the number at which immigration and extinction rates become equal. The larger and less isolated the island, the higher is the species number at which it should equilibrate.

The correctness of this interpretation has been established by several types of study. One has involved observing the increase in species number on an island whose fauna and/or flora have been destroyed. The most famous such study was provided by a 'natural experiment', the colonisation by birds of the vocanic island of Krakatoa after its fauna had been destroyed by an eruption in 1883 (Dammerman, 1948; see MacArthur & Wilson, 1967, pp. 43-51). Similar 'natural experiments' are provided by the birds of Long Island near New Guinea, whose fauna was destroyed by a volcanic eruption two centuries ago (*see* Fig. 2), and by the birds of seven coral islets in the Vitiaz-Dampier group near New Guinea, when a tidal wave destroyed the fauna in 1888 (Diamond, 1974). Simberloff & Wilson (1969) created an analogous 'artificial experiment' by fumigating several mangrove trees standing in the ocean off the coast of Florida and observing the recolonisation of these trees by arthropods. In all these studies, the number of species on the island returned within a relatively short time to the value appropriate to the island's area and isolation, confirming that this value really was an equilibrium value. Naturally, the rate of approach to equilibrium depends on the plant or animal group studied and the island's location: for example, successive surveys have shown the number of plant species on Krakatoa still to be rising and not yet to have reached equilibrium (Docters van Leeuwen, 1936; MacArthur & Wilson, 1967, p. 49).

Another type of test of the MacArthur-Wilson interpretation is provided by turnover studies at equilibrium. According to the MacArthur-Wilson interpretation, although the *number* of species on an island may remain near an equilibrium value, the *identities* of the species need not remain constant, because new species are continually immigrating and other species are going extinct. Estimates of immigration and extinction rates at equilibrium have been obtained by comparing surveys of an island in separate years. Such studies have been carried out for the birds of the Channel Islands off California (Diamond, 1969; Hunt & Hunt, 1974; Jones & Diamond, in press), Karkar Island off New Guinea (Diamond, 1971), Vuatom Island off New Britain (Diamond, in press), and Mona Island off Puerto Rico (Terborgh & Faaborg, 1973). All these studies found that a certain number of species present in the earlier survey had disappeared by the time of the later survey, but that a similar number of other species immigrated in the intervening years, so that the total number of species remained approximately constant unless there was a major. habitat disturbance. As expected from considering the risk of extinction in relation to population size, most of the populations that disappeared had initially consisted of few individuals. The turnover rates per year (immigration or extinction rates) observed in these studies have been in the order of 0.2-6% of the island's bird species for islands of 300-400 km² area.

Thus, the number of species that a reserve can 'hold' at equilibrium is likely to be set by a balance between immigration rates and extinction rates. The set-point will be at a larger number of species, the larger the reserve or the closer it is to a source of colonists:

1. If 90% of the area occupied by a habitat is converted by man into another habitat and the remaining 10% is saved as an undivided reserve, one might expect to save roughly about half of the species restricted to the preserved habitat type, while the populations of the remaining half of the species will eventually disappear from the reserve. It should be stressed explicitly that increased habitat diversity is part of the reason, but not the only one (cf. p. 134 for others), why larger areas hold more species. Thus, even if a reserve does include some of the type of habitat preferred by a threatened species, the species may still disappear because of population fluctuations, spatial or temporal variation in resources, and too few or too small 'hot spots'.

2. If one saves two reserves, the smaller reserve will retain fewer species if it is remote from the larger reserve than it would if it were near the larger reserve.

3. As the contrast increases between the preserved habitat types and the surrounding habitat types, or between the ecological requirements of a threatened species and the resources actually available in areas lying between reserves, the results of island biogeographic studies become increasingly relevant. The greater this contrast, the lower will be the population density of the threatened species in the area between reserves, and the lower will be the species' dispersal rate between the reserves. To some species the intervening area may be no barrier at all, while to other species it may be as much of a barrier as the ocean is to a flightless mammal.

HOW RAPIDLY WILL SPECIES GO EXTINCT ?

Suppose that 90% of a habitat is destroyed and the remaining 10% is saved as a faunal reserve. The reserve will initially support most, though not all, species restricted to the original expanse of habitat. (The actual proportion of the species present in such a portion of a larger habitat is discussed on pp. 9-10 and 16 of MacArthur & Wilson (1967).) However, we have just seen that at equilibrium the reserve will support only about half the species of the original expanse of habitat. Thus, at the time that the reserve is set aside, it will contain more species than its area can support at equilibrium as an island. Species will go extinct until the new equilibrium number is reached. Such a reserve will constitute the exact converse of an island which has had its fauna destroyed: equilibrium of species number will be approached from above, by an excess of extinction over immigration, rather than from below, by an excess of immigration over extinction. The important practical question thus arises: how rapidly will species number 'relax' to the new equilibrium value? If equilibrium times were of the order of millions of years, these extinctions would not be a matter of practical concern, whereas a reserve that lost half of its species in a decade would be unacceptable.

A natural experiment that permits one to assess 'relaxation rates' as a function of the reserve's area is provided by so-called land-bridge islands (Diamond, 1972, 1973). During the late Pleistocene, when much sea-water was locked up in glaciers, the ocean level was about 100-200 m lower than at present. Consequently, islands separated from continents or from larger islands by water less than 100 m deep formed part of the continents or larger islands, and shared the continental faunas and floras. Examples of such 'land-bridge islands' are Britain off Europe, Aru and other islands off New Guinea, Tasmania off Australia, Trinidad off South America, Borneo and Java off south-east Asia, and Fernando Po off Africa. When rising sea levels severed the land-bridges about 10,000 years ago, these land-bridge islands must have found themselves supersaturated; they initially supported a species-rich continental fauna rather than the smaller number of species appropriate to their area at equilibrium. Gradually, species must have been lost by an excess of extinctions over immigrations. Figure 5 illustrates how far the avifaunas of the satellite land-bridge islands of New Guinea have returned towards equilibrium in 10,000 years. The larger land-bridge islands, with areas of several hundred to several thousand km², still have more bird species than predicted for their area from the species/area relation based on islands at equilibrium, though they do have considerably fewer bird species than New Guinea itself. That is, the larger land-bridge islands have lost many but not all of their excess species in 10,000 years. However, land-bridge islands smaller than about 250 km² at present have the same number of bird species as similar-sized oceanic islands that never had a land-bridge. Thus, the smaller land-bridge islands have lost their entire excess of bird species in 10,000 years.

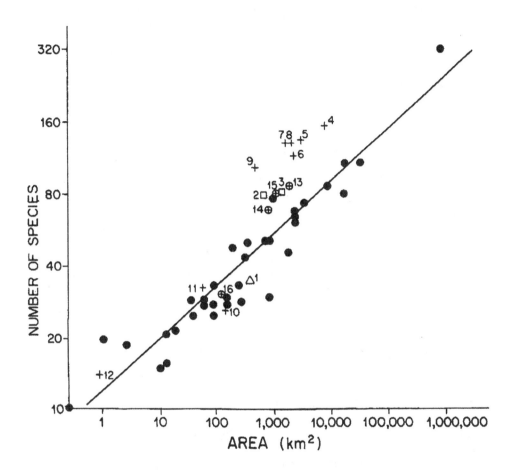

Fig. 5.   Example of how one can use land-bridge islands to estimate extinction rates in the faunas of natural reserves. The ordinate is the number of resident, non-marine, lowland bird species on New Guinea satellite islands, plotted as a function of island area on a double logarithmic scale. The points ● are islands which have not had a recent land-connection to New Guinea and whose avifaunas are presumed to be at equilibrium. The numbered point △ (1) refers to a recently exploded volcano whose avifauna has not yet returned to equilibrium; points + (4-12), to islands connected to New Guinea by land-bridges at times of lower sea-level 10,000 years ago; points ⊕ (13-16), to islands formerly connected by land-bridges to some other large island but not to New Guinea itself; and points □ (2-3), to islands that lie on a shallow shelf and had a much larger area at times of lower sea-level. Up to the time that the land-bridges were severed by rising sea-level, the New Guinea land-bridge islands (+, 4-12) must have supported nearly the full New Guinea quota of 325 lowland species (point in the upper right-hand corner). At present none of these land-bridge islands supports anything close to 325 species; the larger ones (+, 4-9) do, however, still have more species than expected at equilibrium (as given by points ● and the straight line); and the smaller ones (+, 10-12) already have about the number of species expected at equilibrium. The conclusion is that no land-bridge island has been able to hold more than half its initial number of species, but that the larger islands have been able to hold an excess of species for longer. The same conclusion follows from points ⊕ and □. (From Diamond, 1972.)

138   JARED M. DIAMOND

The re-equilibration of land-bridge islands is the resultant of the extinction rate E (in species/year) exceeding the immigration rate I (in species/year) until an equilibrium species number S_{eq} is attained. Both I and E depend on the instantaneous species number $S(t)$, where t represents time (in years). As a highly simplified model, let us assume constant coefficients K_i and K_e (in year^{-1}) of immigration and extinction, respectively:

$$E = K_e S(t) \qquad (2)$$

$$I = K_i[S^* - S(t)] \qquad (3)$$

where S^* is the mainland species pool, and $[S^* - S(t)]$ is the number of species in the pool not present on the island at time t, hence available as potential immigrants. At equilibrium, when $dS/dt = I - E = 0$, S_{eq} is given by

$$S_{eq} = K_i S^*/(K_i + K_e) \qquad (4)$$

If a land-bridge island initially (at $t = 0$) supports a species number $S(0)$ that exceeds S_{eq}, the rate at which $S(t)$ declines from $S(0)$ towards S_{eq} is obtained by integrating the differential equation

$$dS/dt = I - E = (K_i + K_e)[K_i S^*/(K_i + K_e) - S(t)]$$

with the boundary condition $S(t) = S(0)$ at $t = 0$, to obtain:

$$[S(t) - S_{eq}]/[S(0) - S_{eq}] = \exp(-t/t_r) \qquad (5)$$

The relaxation time t_r is the length of time required for the species excess $[S(t) - S_{eq}]$ to relax to $1/e$ or 36.8% of the initial excess $[S(0) - S_{eq}]$, where e is the base of natural logarithms. Relaxation is 90% complete after 2.303 relaxation times.

As an example of the use of this formula, consider the land-bridge island of Misol near New Guinea. At the time 10,000 years ago when it formed part of New Guinea, Misol must have supported nearly the full New Guinea lowlands fauna of 325 bird species. With an area of 2040 km^2, Misol should support only 65 species at equilibrium, by comparison with the species/area relation for islands that lacked land-bridges and are at equilibrium. The present species number on Misol is 135, much less than the initial value of 325 but still in excess of the final equilibrium value. Substituting $S(0) = 325$, $S(t) = 135$, $S_{eq} = 65$, $t = 10,000$ years into eqn. (5) yields a relaxation time of 7600 years for the avifauna of Misol.

Similar calculations have been carried out for other land-bridge islands formerly connected to New Guinea, for islands formerly connected to some other large satellite island but not to New Guinea itself, and for islands that lie on a shallow-water shelf and that formerly must have been much larger in area although without connection to a larger island. A similar analysis in a continental situation was made by Brown (1971), who studied distributions of small non-volant mammals in forests which are now isolated on the tops of mountains rising out of western North American desert basins but which were formerly connected by a continuous forest belt during times of cooler Pleistocene climates. Terborgh (in press, a, b) has made a similar analysis of the avifaunas of Caribbean islands and has dramatically confirmed the accuracy of his calculations by showing that they correctly predict the extinction rates observed within the present century on Barro Colorado Island (Willis, 1974). Both Terborgh's analyses of Caribbean birds and mine of New Guinea birds show that relaxation times increase with increasing island area. Both analyses also show that eqns. (2) and (3) are oversimplified: K_i actually increases with $S(t)$, and K_i decreases with $S(t)$.

Thus, the gradual decline of species number from a high initial value to a lower equilibrium value on land-bridge islands may furnish a model for what could happen when a fraction of an expanse of habitat is set aside as a reserve and the remaining habitat is destroyed. A small reserve not only will eventually contain few species but will also initially lose species at a high rate. For reserves of a few km², extinction rates of sedentary bird and mammal species unable to colonise from one reserve to another are so high as to be easily measurable in a few decades. Within a few thousand years even a reserve of 1000 km² will have lost most such species confined to the reserve habitat. These estimates assume that man's land-use practices do not grossly alter the preserved habitat. More rapid changes in species composition are likely to occur if sylviculture or other human use changes the habitat structure.

WHAT SPECIES WILL SURVIVE?

In the preceding pages we have considered the problem of survival from a statistical point of view: what fraction of its initial fauna will a reserve eventually save, and how rapidly will the remainder go extinct? We have not yet considered the survival probabilities of individual species. If each species had equal probabilities of survival, then it would be a viable conservation strategy to be satisfied with large numbers of small reserves. Each such vest-pocket reserve would lose most of its species before reaching equilibrium, but with enough reserves any given species would be likely to be among the survivors in at least one reserve. In this section we shall examine the flaw in this strategy: different species have very different area requirements for survival.

The survival problem needs to be considered from two points of view: the chance that a reserve where a species has gone extinct will be recolonised from another reserve, and the chance that a species will go extinct in an isolated reserve. Consider the former question first. Suppose that there are many small reserves. Suppose next that a given species is incapable of dispersing from one reserve to another across the intervening sea of unsuitable habitat. The isolated populations in each reserve run a finite risk of extinction. If there is no possibility of recolonisation, each extinction is irrevocable, and it is only a question of time before the last population of the species disappears. Suppose on the other hand that dispersal from one reserve to another is possible. Then, although a species temporarily goes extinct in one reserve, the species may have recolonised that reserve by the time it goes extinct in another reserve. If there are enough reserves or high enough recolonisation rates or low enough extinction rates, the chances of the species disappearing simultaneously from all reserves are low, and the long-term survival prospects are bright. Dispersal ability obviously differs enormously among plant and animal species. Flying animals tend to disperse better than non-flying ones; plants with wind-borne seeds tend to disperse better than plants with heavy nuts. The more sedentary the species, the more irrevocable is any local extinction, and the more difficult will it be to devise a successful conservation strategy. Thus, conservation problems will be most acute for slowly dispersing species in normally stable habitats, such as tropical rain forest. Even power of flight cannot be assumed to guarantee high dispersal ability. For instance, 134 of the 325 lowland bird species of New Guinea are absent from all oceanic islands more than a few km from New Guinea, and are confined to New Guinea plus islands with recent land-bridge connections to New Guinea. Similarly, many neotropical bird families with dozens of species have not even a single representative on a single New World island lacking a recent land-bridge to South or Central America; and not a single member of many large Asian bird families has been able to cross Wallace's Line separating the Sunda Shelf land-bridge islands from the oceanic islands of Indonesia. Such bird species have insuperable psychological barriers to crossing water gaps, and are generally characteristic of stable forest habitats. Thus, low recolonisation rates may mean either that a species *cannot* cross unsuitable habitats (a mountain forest rodent faced by a desert barrier), or that it *will not* cross unsuitable habitats (some tropical forest birds faced by a water gap).

Having seen that species vary in their ability to recolonise, let us now consider how species vary in extinction rates of local populations. The New Guinea land-bridge islands again offer a convenient test situation (Diamond, 1972, in press). Recall that these islands initially supported most of the New Guinea lowlands fauna, that the land-bridges were severed about 10,000 years ago, that 134 New Guinea lowlands bird species do not cross water gaps, and that any extinctions of populations of these species on the land-bridge islands cannot therefore have been reversed by recolonisation. Virtually all these species are now absent from all land-bridge islands smaller than 50 km², because extinction rates on small islands are so high that virtually no isolated population survives 10,000 years. However, these 134 species vary greatly today in their distribution on the seven larger (450-8000 km²) land-bridge islands. At the one extreme, some species, such as the frilled monarch flycatcher *(Monarcha telescophthalmus)*, have survived on all seven islands. At the other extreme, 32 species have disappeared from all seven islands, and must be especially prone to extinction in isolated populations. Most of these 32 species fit into one or more of three categories: birds whose initial populations must have numbered few individuals because of very large territory requirements (*e.g.* the New Guinea harpy eagle *(Harpyopsis novaeguineae)*); birds whose initial populations must have numbered few individuals because of specialised habitat requirements (*e.g.* the swamp rail *(Megacrex inepta)*); and birds which are dependent on seasonal or patchy food sources and normally go through drastic population fluctuations (*e.g.* fruit-eaters and flower-feeders).

Another natural experiment in differential extinction is provided by New Hanover, an island of 1200 km² in the Bismarck Archipelago near New Guinea. In the late Pleistocene, New Hanover was connected by a land-bridge to the larger island of New Ireland and must then have shared most of New Ireland's species. Today New Hanover has lost about 22% of New Ireland's species, a fractional loss that does not sound serious. However, among these lost species are 19 of the 26 New Ireland species confined to the larger Bismarck islands, including every endemic Bismarck species in this category. That is, New Hanover differentially lost those species most in need of protection. As a faunal reserve, New Hanover would rate as a disaster. Yet its area of 1200 km² is not small by the standards of many of the tropical rain forest parks that one can realistically hope for today.

As a further example of a natural experiment in differential extinction, consider the mammals isolated on mountain tops rising from North American desert basins, mentioned in the previous section. Like the bird species restricted to the New Guinea land-bridge islands, the isolated populations of these mammal species have been exposed to the risk of extinction for the past 10,000 years, without opportunity for recolonisation. Today, some of these mammal species are still present on most of the mountains, while other species have disappeared from all but a few mountains. The species with the highest extinction rates are those whose initial populations must have numbered few individuals: either because the species is a carnivore rather than a herbivore, or because it has specialised habitat requirements, or because it is a large animal (Brown, 1971).

A method of quantifying the survival prospects of a species is to determine its so-called incidence function (Diamond, in press). On islands of the New Guinea region one notes that some bird species occur only on the largest and most species-rich islands; other species also occur on medium-sized islands; and others also occur on small islands. To display these patterns graphically, one groups islands into classes containing similar numbers of bird species (*e.g.* 1-4, 5-9, 10-20, 21-35, 36-50, etc.); calculates the *incidence J* or fraction of the islands in a given class on which a particular species occurs; and plots incidence against the total species number S on the island (Fig. 6). Since S is closely correlated with area, in effect these graphs represent the probability that a species will occur on an island of a particular size.

142   JARED M. DIAMOND

For most species, J goes to zero for S values below some value characteristic of the particular species, meaning that there is no chance of survival on islands below a certain size. These incidence functions can be interpreted in terms of the biology of the particular species (*e.g.* its population density, reproductive strategy, and dispersal ability). From these incidence functions one can estimate what chance a certain species has of surviving on a reserve of a certain size.

Thus, different species have different probabilities of persisting on a reserve of a given size. These probabilities depend on the abundance of the species and the

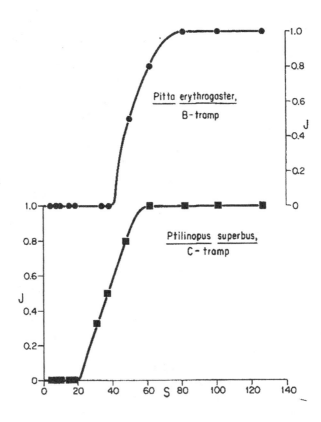

Fig. 6.   So-called incidence functions for two bird species of the Bismarck Archipelago near New Guinea. The incidence $J(S)$ is defined as the fraction of the islands with a given total number of bird species S that a given species occurs on. For example, the so-called B-tramp *Pitta erythrogaster* (•) is on all islands (*i.e.* $J = 1.0$) with $S > 80$, on about half of the islands ($J = 0.5$) with S around 55, and on no island ($J = 0$) with $S < 40$. Other bird species of the Bismarck Archipelago have different incidence functions: for example, the so-called C-tramp *Ptilinopus superbus* (▪) is on all islands with $S > 60$ and on many islands ($J = 0.3$-0.8) with $S = 30$-50. Since S is mainly a function of island area, the message is that each species requires some characteristic minimum area of island for it to have a reasonable chance of surviving.

magnitude of its population fluctuations, and also on its ability to recolonise a reserve on which it has once gone extinct. Even on reserves as large as 10,000 km², some species have negligible prospects of long-term survival. Such species would be doomed by a system of many small reserves, even if the aggregate area of the system were large.

WHAT DESIGN PRINCIPLES WILL MINIMISE EXTINCTION RATES IN NATURAL RESERVES?

In the preceding sections we have examined how the eventual number of species that a reserve can hold is related to area, how extinction rates are related to area, and how area-dependent survival prospects vary among species. Given this background information, let us finally consider what the designer of natural

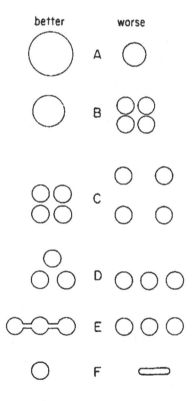

Fig. 7. Suggested geometric principles, derived from island biogeographic studies, for the design of natural reserves. In each of the six cases labelled A to F, species extinction rates will be lower for the reserve design on the left than for the reserve design on the right. See text for discussion.

reserves can do to minimise extinction rates (Diamond, 1972, 1973; Terborgh, in press, *a, b,* Wilson & Willis, in press). Figure 7 (modified from Wilson & Willis, in press) summarises a series of design principles, identified as A, B, C, D, E and F.

A large reserve is better than a small reserve (principle A), for two reasons: the large reserve can hold more species at equilibrium, and it will have lower extinction rates.

In practice, the area available for reserves must represent a compromise between competing social and political interests. Given a certain total area available for reserves in a homogeneous habitat, the reserve should generally be divided into as few disjunctive pieces as possible (principle B), for essentially the reasons underlying principle A. Many species that would have a good chance of surviving in a single large reserve would have their survival chances reduced if the same area were apportioned among several smaller reserves. Many species, especially those of tropical forests, are stopped by narrow dispersal barriers. For such species even a highway swath through a reserve could have the effect of converting one large island into two half-sized islands. Principle B needs to be qualified by the statement that separate reserves in an inhomogeneous region may each favour the survival of a different group of species; and that even in a homogeneous region, separate reserves may save more species of a set of vicariant similar species, one of which would ultimately exclude the others from a single reserve.

If the available area must be broken into several disjunctive reserves, then these reserves should be as close to each other as possible, if the habitat is homogeneous (principle C). Proximity will increase immigration rates between reserves, hence the probability that colonists from one reserve will reach another reserve where the population of the colonist species has gone extinct.

If there are several disjunctive reserves, these should ideally be grouped equidistant from each other rather than grouped linearly (principle D). An equidistant grouping means that populations from each reserve can readily recolonise, or be recolonised from, another reserve. In a linear arrangement, the terminal reserves are relatively remote from each other, reducing exchange of colonists.

If there are several disjunctive reserves, connecting them by strips of the protected habitat (Preston, 1962; Willis, 1974) may significantly improve their conservation function at little further cost in land withdrawn from development (principle E). This is because species of the protected habitat can then disperse between reserves without having to cross a sea of unsuitable habitat. Especially in the case of sedentary species with restricted habitat preferences, such as understorey rain forest species or some bird species of California oak woodland and chaparral, corridors between reserves may dramatically increase dispersal rates over what would otherwise be negligible values.

Any given reserve should be as nearly circular in shape as other considerations permit, to minimise dispersal distances within the reserve (principle F). If the

THE ISLAND DILEMMA 145

reserve is too elongate or has dead-end peninsulas, dispersal rates to outlying parts of the reserve from more central parts may be sufficiently low to perpetuate local extinctions by island-like effects.

ACKNOWLEDGEMENTS

Field work in the south-west Pacific was supported by the National Geographic Society, Explorers Club, American Philosophical Society, Chapman Fund and Sanford Trust of the American Museum of Natural History, and Alpha Helix New Guinea Program of the National Science Foundation.

REFERENCES

BROWN, J. H. (1971). Mammals on mountaintops: nonequilibrium insular biogeography. *Am. Nat.*, **105**, 467-78.
DAMMERMAN, K. W. (1948). The fauna of Krakatau 1883-1933. *Verh. Koninkl. Ned. Akad. Wetenschap. Afdel. Natuurk.*, **44**(2), 1-594.
DIAMOND, J. M. (1969). Avifaunal equilibria and species turnover rates on the Channel Islands of California. *Proc. natn. Acad. Sci. USA*, **64**, 57-63.
DIAMOND, J. M. (1971). Comparison of faunal equilibrium turnover rates on a tropical island and a temperate island. *Proc. natn. Acad. Sci. USA*, **68**, 2742-5.
DIAMOND, J. M. (1972). Biogeographic kinetics: estimation of relaxation times for avifaunas of southwest Pacific islands. *Proc. natn. Acad. Sci. USA*, **69**, 3199-203.
DIAMOND, J. M. (1973). Distributional ecology of New Guinea birds. *Science, N.Y.*, **179**, 759-69.
DIAMOND, J. M. (1974). Colonization of exploded volcanic islands by birds: the supertramp strategy. *Science, N.Y.*, **184**, 802-6.
DIAMOND, J. M. (in press). Incidence functions, assembly rules, and resource coupling of New Guinea bird communities. In *Ecological structure of species communities*, ed. by M. L. Cody & J. M. Diamond. Cambridge, Mass., Harvard University Press.
DOCTERS VAN LEEUWEN, W. M. (1936). Krakatau, 1833 to 1933. *Ann. Jard. Bot. Buitenzorg*, **56-57**, 1-506.
HUNT, G. J. Jr & HUNT, M. W. (1974). Trophic levels and turnover rates: the avifauna of Santa Barbara Island, California. *Condor*.
JONES, H. L. & DIAMOND, J. M. (in press). *Species equilibrium and turnover in the avifauna of the California Channel Islands*. Princeton, N.J., Princeton University Press.
LACK, D. (1973). The numbers and species of hummingbirds in the West Indies. *Evolution*, **27**, 326-7.
MACARTHUR, R. H. & WILSON, E. O. (1963). An equilibrium theory of insular zoogeography. *Evolution*, **17**, 373-87.
MACARTHUR, R. H. & WILSON, E. O. (1967). *The theory of island biogeography*. Princeton, N.J., Princeton University Press.
MACARTHUR, R. H. (1972). *Geographical ecology*. New York, Harper & Row.
MAY, R. M. (in press). Patterns of species abundance and diversity. In *Ecological structure of species communities*, ed. by M. L. Cody & J. M. Diamond. Cambridge, Mass., Harvard University Press.
PRESTON, F. W. (1962). The canonical distribution of commonness and rarity. *Ecology*, **43**, 185-215, 410-32.
SCHODDE, R. & CALABY, J. H. (1972). The biogeography of the Australo-Papuan bird and mammal faunas in relation to Torres Strait. In *Bridge and barrier: the natural and cultural history of Torres Strait*, ed. by D. Walker, 257-300. Canberra, Australian National University.
SCHODDE, R. & HITCHCOCK, W. B. (1972). Birds. In *Encyclopedia of Papua and New Guinea*, 1, ed. by P. A. Ryan, 67-86. Melbourne, Melbourne University Press.

146   JARED M. DIAMOND

SIMBERLOFF, D. S. & WILSON, E. O. (1969). Experimental zoogeography of islands: the colonization of empty islands. *Ecology*, **50**, 278-96.
TERBORGH, J. W. (in press, *a*). Faunal equilibria and the design of wildlife preserves. In *Trends in tropical ecology*. New York, Academic Press.
TERBORGH, J. W. (in press, *b*). Preservation of natural diversity: the problem of extinction prone species. *BioScience*.
TERBORGH, J. W. & FAABORG, J. (1973). Turnover and ecological release in the avifauna of Mona Island, Puerto Rico. *Auk*, **90**, 759-79.
VUILLEUMIER, F. (1970). Insular biogeography in continental regions. 1. The northern Andes of South America. *Am. Nat.*, **104**, 373-88.
WILLIS, E. O. (1974). Populations and local extinctions of birds on Barro Colorado Island, Panama. *Ecol. Monogr.*, **44**, 153-69.
WILSON, E. O. & WILLIS, E. O. (in press). Applied biogeography. In *Ecological structure of species communities*, ed. by M. L. Cody & J. M. Diamond. Cambridge, Mass., Harvard University Press.

References and Notes

1. A 50,000-man army, of the Persian general Cambyses, is reported by Herodotus (H. Carter, *The Histories of Herodotus* (Heritage, 1958), book 3, chapter 26] to have perished in a sand storm in 525 B.C. while trying to reach and destroy the temple of Jupiter Amun by crossing the Sand Sea presumably from Dakhla Oasis. In more recent times a camel caravan route from Kufra Oasis in Libya to Abu Mungar Well in Egypt crossed the southern end by winding its way through gaps in the longitudinal (seif) dunes.
2. G. Rohlfs, *Drei Monate in der Libyschen Wüste* (Theodor Fischer, Cassel, 1875).
3. P. A. Clayton and J. J. Spencer, *Mineral. Mag.* 23, 501 (1934); R. A. Bagnold, *Libyan Sands* (Hodder and Staughton, London, 1935), chap. 7; L. de Almasy, *Publications Speciale de la Societe Royale de Geographie d'Egypte* (Schindler, Cairo, 1936), p. 45.
4. R. A. Bagnold, *The Physics of Blown Sand and Desert Dunes* (Methuen, London, 1941).
5. The concept of a fluvial origin for the sand grains of the Western Desert has been stated by C. Squyres (personal communication); R. Said, in (6), p. 281; C. S. Breed, J. F. McCauley, M. J. Grolier, *J. Geophys. Res.*, in press; B. Issawi, *Geol. Soc. Am. Absts. Programs*, in press.
6. R. Wendorf and R. Schild, Eds., *Prehistory of the Eastern Sahara* (Academic Press, New York, 1980).
7. C. Squyres and S. E. Whitten, personal communication.
8. Originally called Libyan Desert silica glass, the chunks of clear pale greenish glass have eluded unequivocal explanation [L. J. Spencer, *Mineral. Mag.* 25, 425 (1939)]. My investigation of the Sand Sea began with a preliminary trip to the area 80 km west of the remote well of Ain Dalla (Fig. 1) in 1978 in an effort to find playa deposits within the Sand Sea while en route to the LDG area (C. V. Haynes, Jr., *Natl. Geogr. Soc. Res. Rep.*, 1978 Proj., in press). Although one playa deposit was believed to have been seen at a distance, it was not visited because of mechanical difficulties. It was with great pleasure, therefore, that I accepted an invitation to visit the southwestern edge of the Sand Sea in March 1981. The expedition, organized by B. Issawi, Geological Survey of Egypt, and R. F. Giegengack, University of Pennsylvania, consisted of specialists in aeolian processes, stratigraphy, physics, archeology, and Quaternary geochronology (R. F. Giegengack et al., in preparation). My function was to attempt to date the linear dunes by studying the stratigraphic relations of the playa deposits and artifacts described to me by R. F. Giegengack on the basis of a trip to the area in 1980 [R. F. Giegengack and J. R. Underwood, in *NASA Tech. Memo. 82385* (1980), pp. 314–316].
9. The streets between the linear dunes were given letter designations during the pioneering study of the LDG by P. A. Clayton and J. J. Spencer [*Mineral. Mag.* 23, 501 (1934)]. Dune ridge B-C is therefore between streets B and C.
10. Radiocarbon date A-2517 is on shell carbonate as are dates of 6290 ± 150 (A-2518) and 6270 ± 50 (A-2315) years ago, but the latter sample was large enough to allow analysis of the organic carbon to provide A-2516 after pyrolysis and hydrolysis. Because carbonates are subject to exchange with atmospheric CO₂, A-2516 is considered to be the most accurate of the three values closest to it in apparent age.
11. F. Wendorf, R. Schild, R. Said, C. V. Haynes, A. Gautier, M. Kobusiewicz, *Science* 193, 103 (1976); B. Issawi, *Ann. Geol. Surv. Egypt* 8, 295 (1978); F. Wendorf and R. Schild, (6); C. V. Haynes, Jr., in *ibid.*, p. 353. The term pluvial is used here in the general sense of a period of greater effective moisture for vegetation, soil development, and ground water recharge whether due to increased precipitation, decreased evaporation, or both.
12. J. W. Olsen, unpublished manuscript.
13. D. A. Roe, J. W. Olsen, J. R. Underwood, Jr., R. F. Giegengack, *Antiquity*, in press.
14. Absolute dates pertaining to the end of the Early Paleolithic are few and unknown for Egypt; 300,000 ± 100,000 years is a minimum estimate based on potassium-argon dates for the Middle Paleolithic in Ethiopia. See F. Wendorf, R. L. Laury, C. C. Albritton, R. Schild, C. V. Haynes, P. E. Damon, M. Shafiqullah, R. Scarborough, *Science* 187, 740 (1975); C. E. Stearns, *ibid.* 190, 809 (1975); R. G. Klein, *ibid.* 197, 115 (1977).
15. C. V. Haynes, Jr., in preparation; G. E. Wickens, *Boissiera* 24, 43 (1975); H. J. Pachur and G. Braun, in *Palaeoecology of Africa and the Sur-*
rounding Islands, E. M. Van Zinderen Bakker, Sr., and J. A. Coetzee, Eds. (Balkema, Rotterdam, 1980), p. 351.
16. C. V. Haynes, Jr., in (6), p. 353.
17. _____, *Geogr. J.* 146, 59 (1980).
18. Soil colors are indexed according to the Munsell system. Where specified, the first valuechroma ratio is for dry soil, the second for moist soil.
19. R. Said, in (6), p. 281.
20. The Qoz is an area of ancient dunes and red soils, between 10° and 16°N, stabilized by vegetation [A. Warren, *Z. Geomorphol. Suppl.* 10, 154 (1970)].
21. The recurrence interval is not known, but R. A. Bagnold [in *Biology of Deserts*, J. L. Cloudsley-Thompson, Ed. (Institute of Biology, London, 1954), p. 7] estimates 30 to 50 years. We are probably still a long way from learning the actual value because, as he points out, "it is against human nature to look conscientiously at an empty rain gauge for several years on end, by the time rain does come the gauge has probably been put to some other use, or the observer is elsewhere."
22. P. J. Mehringer, Jr., *Natl. Geogr. Soc. Res. Rep.*, 1976 Proj., in press; C. V. Haynes, P. J.
Mehringer, Jr., El S. A. Zaghloul, *Geogr. J.* 148, 437 (1979). Mehringer is currently analyzing fossil pollen samples from a Holocene sequence of laminated, organic lake beds at Selima Oasis, Sudan, and from lake sediment cores from Birqet Qarun, Faiyum depression, Egypt.
23. N. Wade, *Science* 186, 234 (1974); S. W. Matthews, *Natl. Geogr. Mag.* 150, 576 (1976).
24. Investigations supported by grants from the Smithsonian Foreign Currency Program (grant FC 10215300), the National Science Foundation (grant EAR-7926362), and the National Geographic Society (grant 2258) in cooperation with the Geological Survey of Egypt, and the Geological and Mineral Resources Department, Ministry of Energy and Mining, Sudan. Radiocarbon samples were pretreated by J. Quade and analyzed under the auspices of A. Long. Participation in the Libyan Desert Glass expedition was made possible by R. F. Giegengack and B. Issawi. Participation of P. J. Mehringer, Jr., is appreciated. Reading of preliminary versions of the manuscript by C. C. Albritton, Jr., R. F. Giegengack, and M. Grolier is appreciated.

16 November 1981; revised 1 April 1982

Fossil Birds from the Hawaiian Islands: Evidence for Wholesale Extinction by Man Before Western Contact

Abstract. *Thousands of fossil bird bones from the Hawaiian Islands collected since 1971 include remains of at least 39 species of land birds that are not known to have survived into the historic period; this more than doubles the number of endemic species of land birds previously known from the main islands. Bones were found in deposits of late Quaternary age; most are Holocene and many are contemporaneous with Polynesian culture. The loss of species of birds appears to be due to predation and destruction of lowland habitats by humans before the arrival of Europeans. Because the historically known fauna and flora of the Hawaiian Islands represent only a fraction of natural species diversity, biogeographical inferences about natural processes based only on historically known taxa may be misleading or incorrect.*

Since 1971, tens of thousands of fossil bird bones have been found in various geological settings on five of the main Hawaiian islands (*1*). At least 39 endemic species of land birds and one species of seabird are now known only from fossil remains (*2*); only three of these have been named previously (*3*). We have completed a general overview of the fossil deposits and their faunas (*1*), but systematic revisions and descriptions of new taxa are not completed (*4*). We now report on the role of Polynesians, who colonized the Hawaiian Islands by A.D. 600, and perhaps as early as A.D. 400 (*5*), in the disappearance of native birds.

The largest collections of fossil birds were found on the islands of Molokai, Oahu, and Kauai (*6*) (Fig. 1), and a few remains were found in lava tubes on Maui and Hawaii. Bones of prehistorically extinct birds (that is, extinct before Europeans arrived to keep written records, beginning in 1778) have also been recovered from archeological midden sites on Hawaii, Molokai, and Oahu.

The endemic species of land birds (*7*) that survived into the historic period on the main Hawaiian Islands include a goose, a hawk, a flightless rail, a crow,

two thrushes, a flycatcher, five honeyeaters, and 27 Hawaiian finches (Drepanidini, previously called "Hawaiian honeycreepers"). To these, the fossil record now contributes the following additional endemic taxa: at least seven species of geese (many of them flightless), two species of flightless ibises, a sea eagle (*Haliaeetus*), a small hawk (*Accipiter*), seven flightless rails (Rallidae), three species of owls belonging to an extinct genus, two large crows (*Corvus*), one honeyeater (*Chaetoptila*), and at least 15 Hawaiian finches (Drepanidini). Thus, the number of species of endemic land birds known for the main islands has been more than doubled by the fossil taxa. The number of colonizations by birds that are known to have produced endemic species in the main islands has likewise now been doubled (*1*).

In addition to providing evidence of the extinction of many species, the fossil record shows that numerous taxa with restricted ranges in the historic period were formerly more widely distributed. For instance, certain species that are known historically only from the Hawaiian Leeward Islands (*Pterodroma hypo-*

SCIENCE, VOL. 217, 13 AUGUST 1982

0036-8075/82/0813-0633$01.00/0 Copyright © 1982 AAAS

633

leuca, Psittirostra cantans, and *Psittirostra ultima*) or only from the island of Hawaii (*Branta sandvicensis, Buteo solitarius, Chaetoptila angustipluma, Psittirostra bailleui, Psittirostra kona, Psittirostra flaviceps*, and *Ciridops anna*) are represented in fossil deposits from other islands by the same or closely allied species.

As an indication of the extent of extinction, a combined total of only 33 island populations of endemic land birds were recorded from Molokai, Oahu, and Kauai during the historic period, whereas 74 populations are known from the same islands as fossils (Table 1). Only 22 (30 percent) of these fossil populations survived long enough to be recorded by ornithologists. In addition, at least five species of marine birds became extinct or were reduced in range prehistorically (*1*).

Extinction took varying proportions of different elements of the avifauna. Of the 24 endemic species of nonpasserine land birds now known from the main Hawaiian Islands, only three (12.5 percent) are definitely known to have survived into the historic period, whereas 62 percent of the species of passerines discovered so far survived. Of the 13 to 17 species of flightless birds that occur as fossils, only one small rail is known historically. Only one endemic species of raptorial bird (*Buteo solitarius*) now exists in the archipelago, whereas at least five species in three genera became extinct prehistorically, a situation that must alter assumptions concerning the role of predation in the evolution of the Hawaiian avifauna. Because the fossil record is incomplete, the figures on the extent of survivorship may actually be exaggerated.

Although at least one of the Hawaiian fossil deposits is of late Pleistocene age (*1, 8*), most of the important sites appear to be late Holocene. The major deposits

Table 1. Island areas and numbers of species of endemic land birds in the historic and fossil avifaunas of Oahu, Kauai, and Molokai. For comparison, the large island of Hawaii (10,464 km²) has only 23 historically known endemic species of land birds.

Island	Area (km²)	Endemic species of land birds	
		Historic	Fossil
Oahu	1536	11	32
Kauai	1422	13	21
Molokai	676	9	21

on Molokai and Kauai have yielded maximum radiometric ages ranging from 5145 ± 60 to 6740 ± 80 years before present (B.P.) (*9*), an indication that extinct species in these deposits survived any Pleistocene climatic perturbations that may have affected the Hawaiian Islands.

At least 12 species that are either extinct, or that were extirpated on the island where their bones were found, have been collected in prehistoric archeological sites; these provide evidence that prehistorically extinct species of birds persisted until Polynesians colonized the islands. Charcoal from a hearth in a large sinkhole at Barber's Point, Oahu, was associated with charred bones of extinct birds and yielded a radiocarbon age of 770 ± 70 years B.P. (*1*). Noncultural deposits at Barber's Point also provide evidence of the contemporaneity of prehistoric man and extinct birds. In these sites, the Pacific rat *Rattus exulans* and the adventive land snail *Lamellaxis* (*10*), both introduced by Polynesian colonists, are ubiquitous in the same stratigraphic levels that contain the greatest concentrations of bones of extinct birds. This evidence suggests that all of the 23 extinct populations of land birds from the Barber's Point de-

posits were present on Oahu when Polynesians first arrived.

The Polynesian residents may have been responsible for the disappearance of more than half the endemic avifauna of the Hawaiian Islands. We attribute the extinction that occurred to a combination of habitat destruction and predation. Flightless species, as well as ground-nesting land birds and burrowing seabirds, would have been particularly vulnerable to predation by humans and by the dogs, pigs, and rats that arrived with them. Predation, however, was probably not the principal factor in the prehistoric extinction of most Hawaiian birds. It is unlikely, for instance, that 29 extinct populations of small passerines succumbed to hunting pressure. A more plausible explanation for the disappearance of these and many other Hawaiian land birds is the clearing of lowland forest, primarily by fire, for agricultural purposes. Journals of early western voyagers to the islands, including those of James Cook, James King, and George Vancouver, record extensive deforestation and heavy cultivation of the lowlands, as well as the use of fire in clearing (*1*). Archeological research on prehistoric land use supports these early descriptions (*1*). Changes through time in the land snail fauna in the Barber's Point deposits on Oahu also reflect habitat alterations that took place in the prehistoric Polynesian period (*10*).

In the historic period, endemic Hawaiian forest birds have been reported mainly from the wet montane regions where native forest persisted. Yet evidence from the fossil deposits shows that many of these species once occurred, sometimes abundantly, in relatively dry regions near sea level. Early botanical surveys have shown that the drier lowland regions of the Hawaiian Islands once supported a distinctive forest vegetation with many endemic species of plants, although only scattered remnants of this flora were in existence when they were first described by botanists (*11*). Species of birds that were restricted entirely to such habitats would have become extinct. Wet montane forest was probably not the optimal habitat for many others, which perhaps accounts for the scarcity of certain species of Hawaiian birds throughout the historic period.

We should emphasize that the fossil record for the Hawaiian Islands is still incomplete. We have good fossil samples from only three of the main islands, and even these samples lack species that must have been present at the time of deposition (*1*). Fossil material from the

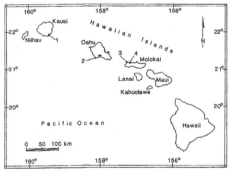

Fig. 1. The main Hawaiian Islands, showing the more important collecting localities for fossil birds: 1, Makawehi dunes, Kauai; 2, Barber's Point, Oahu; 3, Ilio Point, Molokai; and 4, Moomomi dunes, Molokai.

two largest islands, Hawaii and Maui, is scant, and there is as yet no way to assess changes in the avifaunas of these islands caused by prehistoric man, although three species of birds are known to have become extinct prehistorically on each. There is no fossil record from the islands of Lanai, Kahoolawe, or Niihau. No endemic species of land birds were ever recorded from the last two, although the absence of endemic birds cannot be a reflection of natural conditions. It is probable that the historically known avifauna represents only a third, or less, of the total number of endemic species of birds that were present in the Hawaiian Islands when man first arrived there.

These findings have implications for studies of island biogeography. The equilibrium theory of island biogeography (12), for example, was applied to the historically known avifauna of the Hawaiian Islands, with the results being congruent with the theory (13); the fossil record shows these results to be spurious, however (1). The assumption that the historically known biota of a prehistorically inhabited island contains an intact complement of species in a natural state of equilibrium is invalid for the Hawaiian Islands, and is most likely invalid for other islands as well.

Note added in proof: Much more extensive deposits of bird bones have very recently been found in lava tubes on Maui. Two or three species of geese, including flightless forms, are represented, along with other birds.

STORRS L. OLSON
HELEN F. JAMES

National Museum of Natural History, Smithsonian Institution, Washington, D.C. 20560

References and Notes

1. S. Olson and H. James, Smithsonian Contrib. Zool., in press.
2. We use the term "fossil" to refer both to fossil and "subfossil" bones, including those from archeological midden sites.
3. The three previously described fossil species are Geochen rhuax, a goose known from fragmentary remains from the island of Hawaii [A. Wetmore, Condor 45, 146 (1943)], Thambetochen chauliodous, a large flightless goose, and Apteribis glenos, a flightless ibis, both from Molokai [S. Olson and A. Wetmore, Proc. Biol. Soc. Wash. 89, 247 (1976)].
4. S. L. Olson and H. F. James, in preparation.
5. P. Kirch, Archaeol. Phys. Anthropol. Oceania 9, 110 (1974).
6. On Molokai and Kauai the major collecting localities are in calcareous dune sand. Those on Oahu are from a raised coral-algal reef replete with sinkholes and caverns containing abundant fossil birds.
7. We exclude taxa that are not endemic at the species level from the calculations; that is, freshwater birds that are only subspecifically distinct from mainland species, as well as the short-eared owl Asio flammeus, which appears to have colonized the archipelago subsequent to the arrival of man (1).
8. H. Stearns, Occas. Pap. Bernice Pauahi Bishop Mus. 24, 144 (1973).
9. Radiocarbon ages from dune deposits are based on three samples of land snail shells and one of crab claws (Smithsonian Radiation Biology Laboratory, Washington, D.C.).
10. P. Kirch and C. Christensen, "Nonmarine molluscs and paleoecology at Barber's Point, Oahu (unpublished report prepared for the U.S. Corps of Engineers; manuscript No. ARCH 14-115; copy deposited in Smithsonian Institution Libraries) (1981), pp. 242–286; C. Christensen and P. Kirch, Bull. Am. Malacol. Union 1981, 31 (1981).
11. J. Rock, The Indigenous Trees of the Hawaiian Islands ("Published under patronage," Honolulu, 1913).
12. R. H. MacArthur and E. O. Wilson, The Theory of Island Biogeography (Princeton Univ. Press, Princeton, N.J., 1967).
13. J. Juvik and A. Austring, J. Biogeogr. 6, 205 (1979).
14. We thank the many persons who aided our research and who are acknowledged in detail in (1); we also thank A. Kaeppler and D. W. Steadman for reading the manuscript.

7 January 1982; revised 8 April 1982

Androgens Alter the Tuning of Electroreceptors

Abstract. Weakly electric fish possess electroreceptors that are tuned to their individual electric organ discharge frequencies. One genus, Sternopygus, displays both ontogenetic and seasonal shifts in these frequencies, possibly because of endocrine influences. Systemic treatment with androgens lowers the discharge frequencies in these animals. Concomitant with these changes in electric organ discharge frequencies are decreases in electroreceptor best frequencies; hence the close match between discharge frequency and receptor tuning is maintained. These findings indicate that the tuning of electroreceptors is dynamic and that it parallels natural shifts in electric organ discharge frequency.

In communicatory and active sensory systems, motor outputs and sensory inputs are often matched so that sensory receptors are most sensitive to the frequency components that predominate in the output. For example, in different species of weakly electric fish, tuberous electroreceptors, which are modified hair cells, are most sensitive to the peak power of the species-specific electric organ discharge (EOD) used in electrolocation and communication (1, 2). Among the "wave" species, so called because their EOD is nearly sinusoidal, each animal discharges within a species-specific frequency band, with each individual

typically discharging at its own characteristic frequency within that band. These individual differences in discharge frequency are reflected in individual variations in electroreceptor tuning: the receptors of a given fish are closely tuned to its EOD frequency (1).

Despite the high stability of discharge frequencies in these fish (3), there may be changes over the lifetime of an individual. In the South American gymnotoid Sternopygus there is a sexual dimorphism in discharge frequencies whereby mature males discharge at lower frequencies than mature females (4). This difference apparently results from (i) the gradual divergence of male and female discharge frequencies from the intermediate discharge frequencies found in juveniles (4, 5) and (ii) a seasonal enhancement of the difference between male and female discharge frequencies (5). These shifts in EOD frequency may be under hormonal control, since treatment of fish

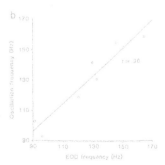

Fig. 1. Receptor oscillation characteristics in Sternopygus. (a) Average oscillation from a fish whose EOD frequency was 122 Hz, showing the initial stimulus artifact followed by five peaks. The signal was analog-to-digital converted at a sampling rate of 5 kHz; each point represents a bin width of 200 μsec. We used 512 stimulus presentations in obtaining this response. The number of bins between peaks and then to calculate the frequency of the oscillation. The frequencies, as calculated from the four interpeak periods, are shown above the oscillation. (b) Illustration of the close correspondence between individual EOD frequencies and oscillation frequencies for the eight animals used in the study (day 0).

0036-8075/82/0813-0635$01.00/0 Copyright © 1982 AAAS

635

7 Assembly Rules

Nicholas J. Gotelli

The search for simple "assembly rules" has dominated the literature in ecological biogeography for more than 50 years, and is still one of the most active areas of current research (Weiher and Keddy 1999, Gotelli and McCabe 2002). What, precisely, is meant by an assembly rule? The term would seem to imply a mechanism that determines the order in which species are added to a community (Drake 1990). Operationally, assembly rules have come to refer to the description of patterns seen among replicated sets of communities (Wilson 1999); mechanisms are secondarily inferred from the presence of these patterns. Because most community assembly rules are framed in terms of species interactions—especially interspecific competition—the study of assembly rules might seem to be more in the purview of ecology than of biogeography. However, the recognition of these patterns requires comparisons among sets of communities, usually at regional or continental scales. It is for this reason that the seminal papers on assembly rules emerged in the biogeographic literature.

Two antagonistic themes have dominated the study of assembly rules. The first is the description of biogeographic and community patterns from which assembly rules are first de-

rived. Four of the papers in this part are the genesis for assembly rules based on forbidden species combinations (MacArthur 1972, Diamond 1975a), species per genus (S/G) ratios (Elton 1946), and species nestedness (Darlington 1957). The second theme is that of testing for the statistical significance of the pattern against a null hypothesis of randomness (Gotelli 2001). The focus on statistical methodology has been important because it is rarely possible to "test" assembly rules with small-scale manipulative experiments. Therefore, the validity of assembly rules rests on our ability to distinguish them from patterns that might arise by random colonization or extinction of species. Papers by Connor and Simberloff (1979) and Williams (1947a) are classics in this literature on null model tests.

The readings by MacArthur (1972), Diamond (1975a), and Connor and Simberloff (1979) form a "successional sequence" in which MacArthur's (1972) local models of competitively structured communities were expanded to the biogeographic scale with assembly rules proposed by Diamond (1975a), and then challenged with null model tests by Connor and Simberloff (1979). Similarly, Williams's (1947a) paper is a statistical critique of Elton's (1946) assembly rules derived from the study of taxonomic ratios in local communities. Darlington's (1957) contribution on nestedness stands alone. The study of species nestedness did not gain popularity until the mid-1980s (Patterson and Atmar 1986, Patterson 1987), and even then Darlington's (1957) novel contribution was not recognized and appreciated until the 1990s (Lomolino 1996). All of these papers stimulated a long line of research—and controversy—that continues today. In contrast to other "classic" papers that may be of interest for purely historical reasons, the papers in this part continue to be actively cited and discussed by biogeographers today.

Nestedness of Island Biotas

Important research ideas are occasionally "lost" and then independently rediscovered in science. Mendel's studies of inheritance are the classic example from genetics. In biogeography, the MacArthur and Wilson (1963, 1967) equilibrium theory was independently discovered by Munroe (1948; see Brown and Lomolino 1989), and important elements of it are imbedded in Preston's (1962) treatment of species-abundance distributions.

The discussion of species nestedness patterns in PHILIP J. DARLINGTON JR.'s *Zoogeography: The Geographic Distribution of Animals* (1957; paper 60) qualifies as another example of an important idea that was published early on, lay nascent, was independently published (May 1978, Patterson and Atmar 1986), and not properly publicized until much later (Lomolino 1996). Consider a set of communities that differ in species richness. If the species composition of small communities are subsets of the larger communities, the assemblage is said to be "nested" in its distribution (Patterson and Atmar 1986). As with other assembly rules, a variety of null models can be used to test the hypothesis that the degree of nestedness differs from that expected by chance (Simberloff and Martin 1991).

Darlington (1957) first described the nestedness pattern and illustrated it for a hypothetical island biota. He believed that this pattern could arise through differential migration of species to near and distant islands:

Unless other factors are very unequal, animals dispersing from a continent to an archipelago may usually be expected to reach the nearest islands first and to spread to other islands across the narrowest water gaps. The resulting pattern of distribution should be orderly, with related forms occurring in series on adjacent islands along the routes of immigration. This is *an immigrant pattern.* (485)

Darlington then considered more complex scenarios, in which the patterns become modified because of dispersal rates and ongoing extinctions. He emphasized that his treatment was purely conjecture: "All that has just been said about immigrant and relict patterns, and mixed and modified ones, is preliminary and hypothetical. It remains to be seen whether the hypothetical patterns occur in fact" (488).

The study of species nestedness did not resume until May (1978) and Patterson and Atmar (1986) proposed that nested faunas could be produced through orderly extinction. Thus, the biogeographic assembly rule was that faunas "fall apart" in a repeatable, deterministic sequence; the role of immigration was not considered in this and subsequent studies of nestedness (Patterson 1987, 1990; Patterson and Brown 1991). In contrast, Darlington (1957) clearly recognized that both immigration and extinction will contribute to species occurrences on islands. However, Darlington believed that extinctions would not occur in a predictable sequence, and would therefore lead to a pattern that was not nested:

In contrast to this is the pattern that would be expected if a group of animals were once well represented on an archipelago and were then reduced in numbers and eliminated on some of the islands. . . . In this case the survivors would probably not form an orderly series on adjacent islands but would occur irregularly, partly on the largest and most favorable islands regardless of position and partly according to chance. This would be a simple *relict pattern.* (485)

Almost four decades would pass before Lomolino (1996) called attention to Darlington's ideas and introduced tests for assessing the relative importance of immigration and extinction processes in creating nestedness patterns.

Why were Darlington's ideas overlooked? First, they were presented in a few pages of a full-length book. Scholars who rely exclusively on computer searches and publication abstracts won't find gems such as this. Equally important is the fact that, by 1986, Darlington's ideas on the role of dispersal in biogeography had been discredited by the discovery of plate tectonics. Although the criticisms of Darlington's (1957) work are relevant to the distribution of continental biotas, they do not apply to oceanic biotas that have originated by long-distance dispersal. Biogeographers are currently rethinking their ideas on species nestedness; it remains to be seen how much of the nestedness patterns can be attributed to extinction processes, and how much to differential immigration (Lomolino 1996), as Darlington (1957) originally envisioned.

Taxonomic Ratios

CHARLES S. ELTON's paper, "Competition and the Structure of Ecological Communities" (1946; paper 61), qualifies as a classic in both biogeography and community ecology. In this paper, Elton developed a "working hypothesis" that the taxonomic diversity of small ecological communities was regulated by interspecific competition and resource limitation. Predating the current interest in meta-analysis (Gurevitch et al. 1992), Elton's (1946) careful compilation of taxonomic records from 55 animal communities in small, well-defined habitats formed a data base that was analyzed (and debated) by other researchers for decades to follow. The striking pattern that emerged from these data was the large number of genera that were represented by only a single species in local communities. In contrast, "regional" lists of species from the same habitat typically contained many different species in the same genus. Elton (1946) used the species/genus (S/G) ratio as a convenient way to summarize

these differences between local and regional communities. This biogeographic approach allowed Elton to establish that reduced S/G ratio was quite general for local communities and not idiosyncratic to a particular taxon or part of the world. The pattern of a reduced S/G ratio constitutes one of the first biogeographic assembly rules. His paper also popularized the use of taxonomic ratios to quantify biodiversity patterns for any hierarchical classification of organisms (individuals/species, species/family, etc.).

What was the explanation for the reduced S/G ratio in small local communities? Elton (1946) developed his hypothesis by drawing on previous ideas of Charles Darwin and Gyorgyi F. Gause. Darwin (1859) had pointed out that closely related species are typically much more similar in body size and morphology than distantly related species. Consequently, closely related species might be expected to compete more severely for limited resources than would distantly related species. Gause's (1934) work emphasized the importance of competitive exclusion, and the necessity for some degree of niche differentiation to allow species coexistence. Elton (1946) boldly proposed that competition for limited resources in small communities restricted the number of species within a genus that could coexist locally.

There are both precedents and antecedents to Elton's (1946) work. The precedent is a set of papers by European plant ecologists that analyzed *generic coefficients* (G/S ratios, expressed as percentages) in the context of gradients of ecological diversity (Jaccard 1901; Maillefer 1929; Palmgren 1929; Pólya 1930). However, this older work was unfortunately ignored (reviewed in Järvinen 1982), and it is Elton's (1946) paper which has had lasting impact on biogeography. The response to Elton's (1946) work was a fairly heated (for that time) controversy over the criteria he used for defining "local" communities (Moreau 1948, Bagenal 1951), the logic of inferring competition from taxonomic ratios (Hairston 1964), and, more significantly, an extended investigation of the statistical properties of taxonomic ratios that continues to this day (Williams 1947a, 1951, 1964; Simberloff 1970; Gotelli and Colwell 2001). Taxonomic ratios, now informed by statistical modeling, have enjoyed a more recent renaissance as tools for quantifying and estimating what remains of biodiversity in the face of global habitat change (Colwell and Coddington 1994).

Elton's (1946) analysis represented an important shift in emphasis from the "vertical," or trophic structure of communities, to the "horizontal" structure of coexisting species that use common resources. This perspective dominated the study of community ecology well into the 1980s (MacArthur 1972, Strong et al. 1984). Elton's (1946) use of a simple and intuitive numerical index to quantify community patterns across spatial scales, and his compilation of published data sets representing a broad geographic area constitute methodological advances that are hallmarks of modern biogeographic studies.

Elton's (1946) paper served a second purpose in that it elicited a response from CARRINGTON BONSOR WILLIAMS, which is also a landmark paper in biogeography, one of the first studies to pose an explicit null hypothesis. In his paper entitled, "The Generic Relations of Species in Small Ecological Communities" (paper 62), Williams (1947a) emphasized the statistical properties of the S/G ratio, and asked what the expected values of the ratio would be in a small, local community that was not structured by competition: "It is, however, most important to consider in detail what exactly happens when a selection of a relatively small number of species is made from a larger fauna or flora, without reference to their generic relations . . . as a true interpretation can only be made by comparing the observed data with the results of a selection of the same size made at random" (11). Whereas Elton (1946) had advocated comparisons of S/G ratios of small versus large communities, Williams (1947a) advocated comparisons of actual S/G ratios in small communities versus S/G ratios of null communities assembled by random draws of the same number species from larger

assemblages. Interestingly, Elton (1946) did not entirely ignore the statistical properties of the S/G ratio. He suggested—incorrectly, as it turned out—that if competitive structure was not present in small communities, the observed S/G ratio would resemble the S/G ratio of the regional biota from which it was derived. But, as Williams (1947a) carefully demonstrated, the S/G ratio is quite sensitive to the number of species in the assemblage, and will decrease in small communities even in the absence of competitive interactions.

Williams's (1947a) most salient contribution was in determining the expected S/G ratio for a random sample of species. He used two approaches. First, he assumed that the underlying species-genus distribution followed a *log-series* (a simple mathematical relationship that can be used to describe the relative abundance of species in a large collection of individuals). The model provided a good fit to many data sets, particularly those in which large numbers of individuals were sampled randomly, such as light trap catches of moths (Fisher et al. 1943). If one scales up hierarchically, and assumes a log-series distribution of species into genera, it is straightforward to derive mathematically the expected number of genera for a sample of a given number of species (Williams 1944, 1947b). To relax the assumption of the log-series distribution, Williams (1947a) also made random draws by hand using numbered cards, each one representing a species in the sample, just as Maillefer (1929) had done for taxonomic ratios of the Swiss flora almost two decades earlier. Later, Williams (1964) and Simberloff (1970) would use computers to carry out these random simulations.

Both the analytical solutions and simulation exercise gave similar results. The observed S/G ratios in small communities were remarkably similar to those predicted by the null model. When deviations did occur, they tended to be positive, with slightly more species per genus occurring than expected by chance. This is actually the opposite pattern to the one first described by Elton (1946), and the slight increase in S/G ratios in small communities relative to the null model predictions seems to be a general pattern (Simberloff 1970). Williams (1947a) pointed out that whereas competition would tend to reduce the S/G ratio in small communities, similar habitat affinities of congeners would tend to increase it, and his results suggested the balance tipped slightly toward the latter. In either case, Williams (1947a) emphasized that the observation of reduced S/G ratios in small communities could not be accepted as simple evidence in favor of competitive structuring.

What was the response to Williams's (1947a) critique? Bagenal (1951) and Moreau (1948) objected to the data sets that Williams analyzed, but Williams (1951) responded effectively by establishing the same patterns in Moreau's (1948) data. Hairston (1964) would later object to the entire procedure of trying to infer competition from taxonomic ratios, and instead advocated a retreat to small-scale experimental ecology as the only reliable source of evidence for competitive structuring.

Sadly, none of these exchanges highlighted the revolutionary approach that Williams (1947a) had taken by posing and testing an explicit null hypothesis in biogeography. In spite of Williams's thorough treatment of this subject (Williams 1947a, 1951, 1964), many investigators ignored (or did not understand) the statistical properties of taxonomic ratios and continued to interpret them as Elton (1946) had done (Grant 1966, Moreau 1966, MacArthur and Wilson 1967, Cook 1969). The issue was not really laid to rest until Simberloff (1970) reiterated Williams's (1947a) basic result and used computer simulations to unequivocally document the patterns for several island–mainland comparisons.

In a parallel, but largely independent, effort, ecologists developed the statistical technique of *rarefaction* (Sanders 1968; Hurlbert 1971; Heck, van Belle, and Simberloff 1975) to predict the expected number of species in sample of individuals drawn at random from a larger collection. The statistical issues are identical to those first raised by Williams (1947a) in the study of taxonomic ratios. Surprisingly, rar-

efaction and other sampling statistics continue to be neglected, and many recent analyses of biodiversity have failed to recognize the distinction between samples that are standardized by area and those that are standardized by the number of individuals collected (Gotelli and Colwell 2001). For both its historical significance as one of the first examples of a null model in biogeography, and its current relevance to the problem of estimating biodiversity, Williams (1947a) deserves to be read and cited.

Species Co-occurrence

A longstanding question for biogeographers, and community and landscape ecologists as well, is whether or not species are distributed randomly with respect to each other and geographic gradients in environmental factors. Few papers have proven more insightful in this regard than ROBERT HARDING WHITTAKER's (1967) "Gradient Analysis of Vegetation" (1967), which we feature here (paper 63). Whittaker and his colleagues developed gradients analysis as an approach to understand "variation of vegetation of a landscape in terms of gradients of variables on three levels—environmental factors, species populations and characteristics of communities" (Whittaker 1967: 207; see also Whittaker, 1956, 1960; Whittaker and Niering, 1964, 1965, 1968a and b, 1975). Many contemporary scientists share Whittaker's earlier assessment of the value of this approach, i.e., that "gradients analysis has changed the conception of vegetation as much as research on the genetic basis of variation and evolution has changed the concepts of plant species" (Whittaker 1967:207). Whittaker's 1967 paper also represents a classic illustration of how major advances in science result from the "complex interplay" between observation, theory and advances in methodology.

Whittaker's observations were based on field studies conducted in different regions of North America (the Great Smoky Mountains of Tennessee, The Siskiyou Mountains of Oregon and California, and the Santa Catalina Mountains of Arizona), but in each case he used a similar methodology. He first identified a gradient selected so as to keep all other variables (e.g., slope, exposure, aspect, parent rock, and soil) as similar as possible except for the one of interest (e.g., elevation). Then he laid out standardized, elevational transects and sampled quadrats (0.1 ha) in which the vegetation was counted and measured, summarized these data across similar sites, and used graphs to illustrate patterns of variation along geographic and environmental gradients.

Whittaker developed his gradient analysis largely to address fundamental questions about the nature of ecological communities. Ecologists had long been (and still are) debating the extent to which communities are coherent sets of coevolved species that tend to be distributed as a group over both space and time, as opposed to opportunistic associations, with the abundance and distribution of each species reflecting its capacity to disperse to and grow in the local environment. This question is often framed in terms of the debate between Frederic Clements (1916), who characterized communities as "superorganisms," and Henry A. Gleason (1917, 1926), who instead thought of communities as opportunistic associations between "individualistic" species. Whittaker's careful documentation of the distribution of plant species along primary gradients of elevation and secondary gradients of moisture, soil type, or other variables, strongly supported a Gleasonian concept. Most readers will be familiar with Whittaker's coenocline graphs of abundance, showing species distributed seemingly independently, rather than in groups, along the gradients.

Although Whittaker's gradients analysis strongly suggested that communities do not represent superorganism, many of his contemporaries continued to observe structure in ecological communities—nonrandom patterns in species occurrence that were hypothesized to result from interactions among species. During the 1960s and 1970s, interspecific competition had assumed the status of a paradigm in community ecology.

Key papers by G. E. Hutchinson and his student ROBERT H. MACARTHUR laid the theoreti-

cal framework and popularized the idea that local assemblages could be understood largely in the context of niche adjustment, resource utilization, and diffuse competition (Hutchinson 1957a, MacArthur and Levins 1967). MacArthur died at age 42, and his final synthesis, *Geographical Ecology* (1972) was written in just a few months following the diagnosis of terminal illness. In *Geographical Ecology*, MacArthur summarized his life's work and "scaled up" local processes of competition and resource utilization to explain species distributions at a biogeographic scale. The brief excerpt that we reprint here (paper 64; see also paper paper 71 in part 8 of this volume) gives a good flavor of MacArthur's approach.

MacArthur first reviewed his graphical 2-resource competition model in which two consumer species (X_1 and X_2) come to a competitive equilibrium when using two shared resources. This system can be invaded by a third species (X_3), which is a generalist and an efficient forager on both resources. The new equilibrium will either contain species X_1 and X_3 or X_2 and X_3. The important result is that competition controls not simply the number of species that coexist, but the particular combination of species, which is dictated by the degree of resource specialization. He then scaled up this model to consider two independently evolved faunas. In each fauna, species fill the entire niche space and diffuse competition makes each assemblage resistant to invasion. Yet, each assemblage utilizes similar resources in its own region. MacArthur speculated that this type of biotic integrity might be responsible for the maintenance of distinct regional biotas, and his book closes with a figure of Wallace's biogeographic provinces.

This excerpt typifies MacArthur's approach to ecology and biogeography. He often relied on intuitive graphical models, of the same sort that made his equilibrium theory (MacArthur and Wilson, 1963, 1967) so accessible and famous. Although a theoretical perspective is dominant, MacArthur (1972) skillfully linked natural history observations on avian distributions at the biogeographic scale with the qualitative predictions of his models. Finally, the writing is clean and straightforward, and MacArthur's (1972) text and ideas still seem fresh and relevant 30 years after they were written.

After MacArthur died, his colleagues organized a symposium in his honor. The resulting volume, *Ecology and Evolution of Communities* (Cody and Diamond 1975), included synthetic papers by MacArthur's collaborators and represents the apogee of the MacArthurian approach to community ecology. JARED M. DIAMOND's (1975a) "Assembly of Species Communities" is a seminal paper from this volume that forms the basis for modern ideas about community assembly rules. In this 101-page treatise, Diamond summarized decades of study of the distribution of 513 bird species on New Guinea and the satellite Bismarck Islands, and emphasized that islands with similar habitats do not always support the same species. Even when the same species occur on different islands, they do not always use the same microhabitats or resources. Diamond (1975a) greatly expanded MacArthur's ideas on the importance of unique species combinations as indicators of diffuse competition and niche adjustment. He emphasized that such patterns would only be realized in replicate communities in which dispersal limitations were not important.

Diamond's (1975a) long and complex treatise consists of two parts. In the first part, reprinted here (paper 65), Diamond proposed a series of broad, general assembly rules, based on detailed observations of four avian guilds. For a set of S species, there are $2^S - 1$ unique species combinations that can be formed. Diamond noted that not all of these combinations will be found in nature, and proposed that the missing combinations are forbidden by incidence, compatibility, and combination rules. The second and longer part of his text detailed the operation of these assembly rules in the Bismarck Archipelago. Diamond (1975a) introduced important concepts such as the incidence function, in which the frequency of occurrence of a species is plotted against the number of species in an assemblage, stating that "such a

graph can serve as a 'fingerprint' of the distributional strategy of a species" (353). He also pioneered the idea of "checkerboard distributions" in which two competing species never co-occur, but can be found in allopatry on different islands. Species distributional strategies, supersaturated island faunas, and habitat niche shifts also received detailed treatment. In many ways, this complex blend of data, theoretical concepts, and narrative treatment resembles the biogeographic narratives of Darlington, Wallace, Simpson, and other biogeographers who attempted to describe the "assembly" of continental avifaunas. Diamond (1975a) recognized the risks inherent in such an approach:

The power of [diffuse competition] is that, in principle, it can explain anything. Its heuristic weakness is that, if it is important at all, its operation is likely to be so complicated that its existence becomes difficult to establish and impossible to refute. Such a concept deserves to be greeted with skepticism until its importance can be documented. (348)

The detailed complexity of Diamond's narrative does not sit comfortably with reductionist hypothesis-testing, and Simberloff (1978) objected a few years later that "one is left with the uneasy feeling that the rules lack predictive power, in that all the data are required before any prediction can be made" (723). However, the "general" assembly rules that Diamond (1975a) outlined at the start of his paper have, indeed, proven to be testable, and it was Simberloff himself who first formulated operational tests of community assembly rules against statistical null hypotheses.

In their sharply worded response, "The Assembly of Species Communities: Chance or Competition?" (paper 66), EDWARD F. CONNOR and DANIEL S. SIMBERLOFF (1979) attacked the logic of Diamond's (1975a) general assembly rules and presented a pioneering null model analysis to determine the expected pattern of species co-occurrence in the absence of competitive interactions. This pair of papers sparked a controversy over assembly rules, null models, and the statistical analysis of presence-absence matrices that has continued unabated for more than 25 years (see, e.g., Diamond and Gilpin 1982; Stone and Roberts 1990; Manly 1995; Sanderson, Moulton, and Selfridge 1998; Gotelli and Entsminger 2001). For many biogeographers, these baroque exchanges are synonymous with null models.

Why was the null model analysis of assembly rules so controversial? After all, many of the same issues arose in the debate between Elton (1946) and Williams (1947a) over species/genus ratios and their interpretation. However, there were some important differences. First, whereas Williams's (1947a) tone was circumspect, Connor and Simberloff (1979: 1132) aggressively attacked the central tenets of Diamond's thesis: "We show that every assembly rule is either tautological, trivial, or a pattern expected were species distributed at random." Second, although Williams (1947a) clearly stated the null expectation for his analysis, he did not overtly champion the null model approach for biogeographic analysis. In contrast, Connor and Simberloff (1979) argued that explicit null hypotheses were essential for the establishment of biogeographic patterns. Their exposition added to a growing body of manifesto papers by the "Tallahassee Mafia" on null model analysis (Strong, Szyska, and Simberloff 1979; Strong 1980; Simberloff 1978; Simberloff 1980).

Connor and Simberloff's (1979) null model was a breakthrough for biogeographic analysis. Row and column sums of a binary presence-absence matrix were retained, which preserved differences among sites (column sums) and among species (row sums), but the matrix was filled randomly, so that co-occurrence patterns would presumably not reflect species interactions. This paper established protocols for so-called R and Q mode analyses (Simberloff and Connor 1979) that have become widely used in biogeography. It is interesting to see that the quantitative claim of Connor and Simberloff—random co-occurrence patterns—was not well-supported. Indeed, two of the three presence-absence matrices tested (West Indies birds and West Indies bats) were highly nonrandom by their own analyses, although Connor and Sim-

berloff chose to downplay this result. The Vanuatu avifauna matrix appeared to be random, but there were problems for that matrix in the implementation of the null model (Diamond and Gilpin 1982). Other authors suggested different statistical analyses, although many of these alternatives had statistical flaws of their own (Gotelli and Graves 1996). The most recent studies do confirm that the Vanuatu avifauna matrix is nonrandom, and that at least some of the co-occurrence patterns are consistent with the predictions of Diamond's (1975a) assembly rules model (Stone and Roberts 1990, Gotelli and Entsminger 2001). With some important modifications, the Connor and Simberloff (1979) procedure of preserving row and column totals in presence-absence matrices has good statistical properties for null model analysis (Gotelli, 2000).

Lost in this statistical crossfire were perhaps the most important contributions of Connor and Simberloff (1979). Whereas the significance of Williams (1947a) was overlooked at the time, Connor and Simberloff (1979) represented a profoundly different perspective in biogeography, and it led to an immediate and permanent change in how the science was conducted. Connor and Simberloff (1979) forced scientists to distinguish clearly between patterns in the data and hypotheses that might account for them (cf. Roughgarden 1983). For better or worse, after Connor and Simberloff (1979) it became increasingly difficult to publish complex narrative papers in biogeography without reference to an explicit, and usually statistical, null hypothesis. Although still somewhat controversial, null model analysis has become increasingly accepted as a useful tool in biogeography, and the publication of Connor and Simberloff (1979) paved the way towards such acceptance.

Ironically, null model analysis has now brought the interpretation of Diamond's (1975a) assembly rules model full circle. A meta-analysis of published presence-absence matrices confirms the predictions of Diamond's (1975a) general assembly rules: plant and animal communities exhibit fewer species combinations, more checkerboard species pairs, and less species co-occurrence than would be expected by chance (Gotelli and McCabe 2002). It remains to be seen how much of this pattern can be attributed to ecological forces of competition and niche segregation, and how much to biogeographic forces of dispersal and speciation. Nevertheless, the study of community assembly rules with null models is an active research front, and all of the "classic" papers in this part continue to inform and guide current biogeographic research.

From *Zoogeography: The Geographic Distribution of Animals*
Philip J. Darlington Jr.

TABLE 17. APPROXIMATE RELATION OF AREA TO NUMBER OF SPECIES OF AMPHIBIANS AND REPTILES ON CERTAIN WEST INDIAN ISLANDS (SEE TABLES 15 AND 16)

Approximate Area, sq. mi.	Approximate Number Species	(Actual Number Species)
40,000	80	(76–84)
4,000	40	(39–40)
(400	20)	—
40	10	(9)
4	5	(5)

ent places on single islands. More species occur *together* on large than on small islands. Of course, the effect of area is modified by many other things, including extent of high land and ecological diversity; and, on the other hand, area probably has other effects on island animals. In some cases limitation of area may limit the size of individual animals (but this effect is complex and probably not predictable—see Mertens 1934, pp. 120–123), and it probably also affects the structure of populations and the direction of evolution.

Effect of distance; immigrant and relict patterns

The effect of distance—of the width of water gaps—on the composition and patterns of distribution of island faunas is profound too. To understand it, it is necessary to understand something about the dispersal of land animals across water (see Darlington 1938).

The dispersal of terrestrial animals across water is often referred to as "accidental dispersal" or "random dispersal," but these are not good terms. The non-committal term "over-water dispersal" is preferable.

Dispersal of *individual* land animals over water is largely accidental (and accidents must occur also in dispersal over land), but in the course of time statistical probability comes into play and determines what sorts of animals cross water most often and what islands they most often reach. Even if there is only one chance in a million that a given individual of a species will get across a given water gap, out of many million individuals some may be almost sure to cross; and, other things being equal, they will be much more likely to cross narrow water gaps than wide ones. This is obvious, at least in a general way.

The effect of the width of water gaps is not just inversely proportional to distance or to the square of distance. It depends on rate of loss, on the death rate during dispersal. The death rate of most

Island patterns   485

terrestrial animals during dispersal across salt water is presumably high and presumably forms a geometric progression: if only one individual in a thousand survives the crossing of a hundred miles of sea, only one in a thousand of the remainder will be expected to survive a second hundred miles, etc. In this case, if there is one chance in a thousand that a given sort of animal will reach an island 100 miles out at sea, there is only one chance in a million that it will reach an island 200 miles out, and only one in a billion that it will reach one 300 miles out.

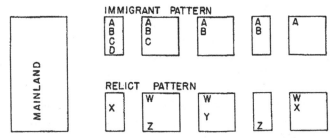

Fig. 57. Diagram to compare a simple immigrant pattern, formed by dispersal of several groups of animals from the mainland for different distances along a series of islands, and a relict pattern, formed by partial extinction of an old fauna formerly common to all the islands.

In nature, this matter is complicated by biological factors and by differences in the nature and age of different islands, the direction of winds and currents, and other things. But distance is basically important. Unless other factors are very unequal, animals dispersing from a continent to an archipelago may usually be expected to reach the nearest islands first and to spread to other islands across the narrowest water gaps. The resulting pattern of distribution should be orderly, with related forms occurring in series on adjacent islands along the routes of immigration. This is an *immigrant pattern.*

In contrast to this is the pattern that would be expected if a group of animals were once well represented on an archipelago and were then reduced in numbers and eliminated on some of the islands. This might happen if the archipelago were detached from a continent and had at first a continental fauna, much of which later became extinct. In this case the survivors would probably not form orderly series on adjacent islands but would occur irregularly, partly on the largest and most favorable islands regardless of position and partly according to chance. This would be a simple *relict pattern.*

Simple immigrant and relict patterns are contrasted in Figure 57.

Immigrant and relict patterns might be mixed. This could happen if new immigrants dispersed into an archipelago already occupied by a relict fauna, or if an old immigrant fauna were decimated by some process other than replacement from the mainland, for example by partial submergence of the islands. Replacement from the mainland should form a different sort of pattern, described below.

Immigrant patterns probably become modified in the course of time, and the pattern of modification probably varies with the rate of dispersal of the animals concerned. If the rate of dispersal is low, if few stocks reach even the nearer islands of an archipelago and if still fewer extend to the farther ones, little extinction may occur. In this case, the more diverse fauna and eventually (as the mainland fauna changes) the exclusive relict groups should accumulate on the nearer islands, and the farther ones should have less diverse faunas (descended from fewer ancestors) and no relict groups not represented on the nearer islands. But if the rate of dispersal is high, the nearer islands may become overpopulated, and new arrivals may replace some of the older animals. In this case, the nearer and farther islands should have equally diverse faunas (in proportion to area), but the farther ones will probably have the exclusive relicts, the last survivors of groups that have been replaced on the nearer islands as well as on the mainland. These different results may be reached in different groups of animals on the same archipelago. Under appropriate conditions, most terrestrial mammals and amphibians, which cross salt water with difficulty, should accumulate principally on the nearer islands of an archipelago, and their exclusive relicts should be there; other animals that disperse more freely should undergo more replacement on the nearer islands, and their exclusive relicts should be on the farther islands.

This can be illustrated by the simple case of two similar islands lying one beyond the other (Fig. 58). In this case, among animals whose rate of dispersal has been low, the most diverse fauna and most of the exclusive relicts should be on the near island. Among animals whose rate of dispersal has been high, the two islands should have about equally diverse faunas, but most of the exclusive relicts should be on the outer island. There might also be animals whose rate of dispersal and replacement would be intermediate and would produce an approximately equal number of exclusive relicts on each island; in this an immigrant pattern might simulate a relict one. But, if the islands are truly comparable in size and habitats and if enough stocks

Island patterns   487

are involved, it is unlikely that a relict pattern would come to simulate an immigrant one.

Of course these different patterns would probably be more or less modified by redispersals of old island animals and in other ways. Complications would be most likely to occur among the most actively dispersing animals. Animals with low powers of dispersal, indicated

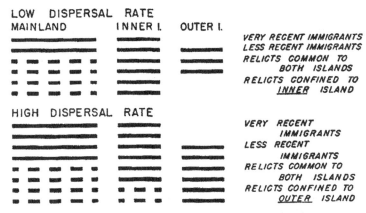

Fig. 58. Diagram of effect of dispersal rate on pattern of distribution on a series of two islands. Each island is limited (by area) to a fauna of six units. In the first case, rate of dispersal is low, the inner island is not overpopulated, and the outer island has received less than a maximum fauna; and, when groups become extinct on the mainland (broken lines), the exclusive relicts occur on the *inner* island. In the second case, rate of dispersal is high, the inner island has been overpopulated, and some older groups have become extinct there as well as on the mainland, and the outer island has received a full fauna; and the exclusive relicts occur on the *outer* island.

by simplicity of their immigrant patterns and by occurrence of localized relict-endemics mostly on islands nearest the mainland, should be the most dependable indicators of the past.

In considering the areas of islands and the widths of water gaps, the past as well as the present must be considered, especially the changes of sea level that occurred in the Pleistocene (p. 583). At times during the Pleistocene, when the ice on land reached its maximum, sea level everywhere was probably at least 100 meters lower than now. Such a lowering of the sea would increase the size of many islands and reduce the width of many water gaps. In fact it would join some islands, such as Sumatra, Java, and Borneo, to the mainland or to each other. On the other hand, when there was little or no ice on land, before the Pleistocene and perhaps at times during interglacial ages, sea level was probably 20 to 50 meters higher than now; this must have reduced the size of some islands, completely drowned others (*e.g.*, some of the Bahamas), and increased the width of some water gaps.

All that has just been said about immigrant and relict patterns, and mixed and modified ones, is preliminary and hypothetical. It remains to be seen whether the hypothetical patterns occur in fact.

The following accounts of island faunas are brief. Additional information about some of them is given in other connections (see page references), and many of the island animals are treated more fully in the lists of families (ends of Chapters 2 to 6), which should be referred to.

Work Cited

Darlington, P. J., Jr. 1938. The origin of the fauna of the Greater Antilles, with discussion of dispersal of animals over water and through the air. *Quarterly Review of Biology* 13:274–300.

[54]

COMPETITION AND THE STRUCTURE OF ECOLOGICAL COMMUNITIES

By CHARLES ELTON, *Bureau of Animal Population, Oxford University*

(With 1 Figure in the Text)

1. AN ANALYSIS OF SOME COMMUNITY SURVEYS

(a) General

If one peruses the lists of species recorded in various ecological surveys of clearly defined habitats, the thing that stands out is the high percentage of genera with only one species present. This is quite a different picture from a faunal list for a whole region or country, in which many large genera are to be found.

There are, of course, theoretical difficulties in deciding exactly what we mean by a clearly defined community or a major habitat, and also considerable practical difficulties during ecological survey work in the field in separating genuine inhabitants from accidental visitors, especially as some of the latter may play a real part in the life of the community. The following analysis is made with clear realization that all community surveys to a certain degree set arbitrary limits to the radiating connexions between species. Such partly arbitrary sections of the larger system of interspersed habitats with their communities will nevertheless show something of the typical structure, without supplying a complete story.

(b) Animal communities

Table 1 gives analyses of fifty-five ecological surveys of animal communities from an extremely wide range of habitats. In three instances some grouping has been adopted to give more reliable figures, which reduces the total to forty-nine units, distributed among twenty-one major types of habitat. The communities cover land, fresh-water, estuary and marine; Arctic, Subarctic, Temperate and one Tropical; free-living and parasitic; and mostly include a very large proportion of the groups of animals present in each habitat.

The percentage of genera with only one species present varies from 69 to 100%, but the greatest frequency is centred round 85%, while about three-quarters of the figures lie between 81 and 95% (Table 2). The corresponding percentage of the number of species belonging to genera in which only one species is present varies more widely, from 46 to 100%. The greatest frequency lies between 71 and

85%, and about three-quarters of the figures lie between 66 and 90% (Table 3).

Genera with four or more species present form a very small fraction of the whole—on the total figures, only 1·32%. In the fifty-five communities, only eleven recorded five or more species in the same genus; while there is only one instance of more than six (in no. 22). These facts are expressed in the figures for the average number of species per genus in each community, which in all instances lies between 1 and 2, the average for the whole lot being 1·38 (range 1·00–1·63). This is shown in another way in Fig. 1, which indicates something like a straight-line relationship between the number of species and number of genera in an animal community.

It is necessary to discuss the validity of the survey data a little, before considering the explanation of these relationships between genera and species:

(1) The range of habitats included is very wide, but it does not contain samples of the most complex habitats, particularly woodland, for the reason that no complete ecological surveys of them have yet been done. Such communities might prove to differ in their structure from those of the simpler kind. This point is discussed again in § 2 (b).

(2) There is really no such thing as a uniform habitat, since all habitats consist of interspersed mosaics of micro-habitats or are internally patchy in the distribution of population densities (as with plankton); and since they also are subject to variations in conditions caused by seasonal and other temporal changes. The habitat units chosen as samples have fairly uniform habitat patterns within them, and provide well-established ecosystems that have been studied fairly or very thoroughly by the surveyors.

(3) Few ecological surveys can be complete, yet many of those analysed are undoubtedly very nearly complete within the limitations of the collecting and recording methods used. These limitations usually affect whole groups of organisms, rather than genera within the same family or order, and so do not harm the present analysis. Thus a plankton net will collect all planktonic Crustacea but not any fish; the bottom sampling of benthos may ignore the micro-fauna; the log communities only give the invertebrates, not the

Table 1. *Analysis of genus/species relations in fifty-five animal communities*

General habitat	Community	1 (A)	2	3	4	5	6	+	Total no. of species analysed (B)	Total no. of genera analysed (C)	Average no. of species per genus (B/C)	% of genera with one species present (A/C)	% of 'single species' present (A/B)	Reference
Arctic fjaeldmark	1 a Frost-weathered rocky soil, Bear Island	24	2	1	—	—	—	—	31	27	—	—	—	Summerhayes & Elton, 1923, pp. 221–2, 245, 262–3
	1 b Rocky Lowland, Prince Charles Foreland, Spitsbergen	13	1	—	—	—	—	—	15	14	—	—	—	
	1 c Raised beaches, Klaas Billen Bay, West Spitsbergen	14	8	1	—	—	—	—	33	23	—	—	—	Summerhayes & Elton, 1928, pp. 216–17
	1 d Lowland, Reindeer Peninsula, West Spitsbergen	17	5	2	1	1	—	—	42	26	—	—	—	
	1 a–d Average	17	4	1	0·25	0·25	—	—	30·25	22·5	1·35	75	56	
Subarctic fjaeldmark	2 Rocky plateau, Akpatok Island, Ungava Bay, Canada	23	3	—	—	—	—	—	29	26	1·12	88	79	Davis, 1936, p. 321
Arctic heath	3 Cassiope tetragona heath, Wijde Bay, West Spitsbergen	21	3	1	—	—	—	—	30	25	1·20	84	70	Summerhayes & Elton, 1928, p. 235
Subarctic heath	4 Better vegetated slopes of ravines, Akpatok Island, Ungava Bay, Canada	26	5	1	—	—	—	—	39	32	1·22	81	67	Davis, 1936, p. 325
	5 Heath (excluding willow scrub), Godthaabsfjord, West Greenland	56	7	1	1	—	—	—	77	65	1·19	86	73	Longstaff, 1932, p. 122
Subarctic scrub	6 Willow (Salix glauca) scrub, Godthaabsfjord, West Greenland	34	6	2	2	—	—	—	60	44	1·36	77	57	Longstaff, 1932, p. 126
Temperate grassland	7 Invertebrates of soil and surface vegetation, meadow on clay, near Oxford, England	61	11	2	1	—	—	—	93	75	1·24	81	66	Ford, 1935, p. 198
Temperate woodland	Invertebrates in logs (complete succession, several years), Duke Forest, North Carolina:													Savely, 1939, pp. 377, 381
	8 Pinus taeda and echinata logs	92	8	1	—	—	—	—	111	101	1·10	91	83	
	9 Quercus alba, borealis (=rubra), velutina and stellata logs	118	6	2	—	—	—	—	136	126	1·08	94	87	
Subarctic bog	10 Eriophorum bog, Godthaabsfjord West Greenland	17	2	—	—	—	—	—	21	19	1·11	89	81	Longstaff, 1932, p. 131
	11 Bog and pool margins, plateau valley, Akpatok Island, Ungava Bay, Canada	26	3	—	—	—	—	—	32	29	1·10	90	81	Davis, 1936, p. 324
Temperate fresh-water pond	12 General fauna of 8 fresh-water ponds, Bardsey Island, north Wales	21	4	1	—	—	—	—	32	26	1·23	81	66	Pyefinch, 1937, p. 128
	13 Moorland hill pond, rich vegetation, Wales	61	8	1	1	0	1	—	90	72	1·25	85	68	Laurie, 1942, p. 172
Temperate lake benthos	14 Invertebrates, sublittoral, Windermere, English Lake District	38	6	0	1	1	—	—	59	46	1·28	83	64	Humphries, 1936, p. 32
	15 Ditto, profundal	6	1	0	0	1	—	—	13	8	1·63	75	46	Ditto

Table 1 (continued)

General habitat		Community	1 (A)	2	3	4	5	6	+	Total no. of species analysed (B)	Total no. of genera analysed (C)	Average no. of species per genus (B/C)	% of genera with one species present (A/C)	% of 'single species' present (A/B)	Reference
Temperate lake benthos	16	Invertebrates, sublittoral 10–40 m., Vättern (cold relict fresh-water lake), Sweden	44	8	1	3	1	1	—	86	58	1.51	76	51	Ekman, 1915, p. 373
Temperate lake, zooplankton	17	Lac de Bret, Switzerland, altitude 673 m.	22	3	1	—	—	—	—	31	26	1.19	85	71	Linder, 1904, p. 166
	18	Lake Washington, Washington, near sea-level	18	8	—	—	—	—	—	34	26	1.31	69	53	Scheffer & Robinson, 1939, p. 117
	19	Lake Michigan (Great Lakes):													Eddy, 1927, p. 212
		(a) 1887–8	21	8	1	—	—	—	—	40	30	—	70	52	
		(b) 1926–7	19	6	1	—	—	—	—	34	26	—	73	56	
		Average	20	7	1	—	—	—	—	37	28	1.32	71	54	
Temperate river	20	Invertebrates, River Wharfe, Yorkshire	82	7	6	3	1	1	—	131	99	1.44	83	63	Percival & Whitehead, 1930, p. 296
	21	Invertebrates and fish, River Lark, East Anglia	42	5	1	1	0	1	—	65	50	1.30	84	65	Butcher, Pentelow & Woodley, 1931, p. 103
	22	Invertebrates, fish and Amphibia, River Rheidol, Wales (recovering from lead pollution)	47	6	3	1	0	1	1 (9)	87	59	1.47	80	54	Laurie & Jones, 1938, p. 280
	23	Ditto, River Melindwr, Wales (Stations A–F, lead pollution)	46	4	3	0	1	—	—	68	54	1.26	85	68	Jones, 1940a, p. 193
	24	Ditto, River Dovey, Wales (no pollution)	81	16	5	—	—	—	—	128	102	1.25	79	63	Jones, 1941, p. 18
	25	Invertebrates, River Ystwyth, Wales (zinc pollution; no fish or molluscs)	43	4	1	0	1	—	—	59	49	1.20	88	73	Jones, 1940b, p. 374
Temperate estuary	26	Plankton, River Thames, Southend, 5 years	56	5	3	1	—	—	—	79	65	1.22	86	71	Wells, 1938, p. 116
	27	Benthos and a few plankton invertebrates, and the fish, River Tamar, Devonshire. (Upper salinity at stations, 0·06°/oo–25·3°/oo)	49	4	1	—	—	—	—	60	54	1.11	91	82	Percival, 1929, p. 95
Arctic marine drift-line	28	Intertidal and bottom benthos, River Tay, Scotland	89	5	2	—	—	—	—	105	96	1.19	93	85	Alexander, 1932, p. 37
	29	Invertebrates and birds, Reindeer Peninsula, West Spitsbergen	12	3	1	—	—	—	—	21	16	1.31	75	57	Summerhayes & Elton, 1928, p. 250
Subarctic marine drift-line	30	Invertebrates, Godthaabsfjord, West Greenland	19	2	—	—	—	—	—	23	21	1.10	90	83	Longstaff, 1932, p. 134
Subarctic marine intertidal	31	Invertebrates, Amerdloq Fjord, west Greenland:													Steven, 1938, p. 61
		(a) Rock	14	1	—	—	—	—	—	16	15	—	93	87	
		(b) Sand	5	—	—	—	—	—	—	5	5	—	100	100	
		(c) Mytilus beds	33	2	—	—	—	—	—	37	35	—	94	89	
		(a–c) Three types combined	33	2	—	—	—	—	—	37	35	1.06	94	89	

The left-hand margin groups the entries under the categories: *Temperate marine intertidal* (32–35), *Tropical marine intertidal* (36–37), *Temperate marine benthos* (38–41), *Temperate marine zoo-plankton*, and *Parasite faunas* (42–49).

Group	No.	Community	A	B	C	D	E	F	G	H	I	J	K	L	Reference
Temperate marine intertidal	32	Invertebrates, rocky shore, Plymouth, England (4 years after colonization began)	54	3	3	—	—	—	—	73	60	1·22	90	74	Moore & Sproston, 1940, p. 61
	33	Invertebrates and fish, sand shore, Port Erin Bay, Isle of Man	17	—	—	—	—	—	—	17	17	1·00	100	100	Pirrie, Bruce & Moore, 1932, p. 287
	34	Invertebrates and fish, rocky shore (excluding rock-pools), Bardsey Island, Wales	85	10	1	—	—	—	—	108	96	1·12	88	79	Pyefinch, 1943, p. 84
	35	Wharf piles, Wood's Hole, Massachusetts	74	8	3	—	—	—	—	99	85	1·17	87	75	Allee, 1923, pp. 213, 218
Tropical marine intertidal	36	Exposed rocks, ditto	50	4	3	—	—	—	—	67	57	1·18	88	75	Allee, 1923, pp. 213, 218
	37	Coral reef flat, Low Isles, Great Barrier Reef, Australia	50	5	3	—	—	—	—	69	58	1·19	86	72	Stephenson *et al.*, 1931, p. 44
Temperate marine benthos	38	Dogger Bank, North Sea, (*a*) 150 stations [Voyage 48]	38	9	—	—	—	—	—	56	47	1·19	81	68	Davis, 1923, p. 9
	39	Ditto, (*b*) smaller area (900 sq. miles), 100 stations [Voyages 22, 26, 29]	35	4	—	—	—	—	—	43	39	1·10	90	81	Ditto
	40	Ditto, (*c*) still smaller area (340 sq. miles), 189 stations [Voyage 39]	29	2	—	—	—	—	—	33	31	1·06	94	88	Ditto
	41	Shields area, Northumberland, North Sea [Stations 14–30]	39	3	—	—	—	—	—	45	42	1·07	93	87	Savage, 1926, Table 3
Temperate marine zoo-plankton · Parasite faunas	42	Endoparasites, salamander, *Desmognathus fuscus*, stream and swamp-stream margin in woodland, lower Piedmont, North Carolina	13	2	—	—	—	—	—	17	15	1·13	87	76	Rankin, 1937, p. 184
	43	Endoparasites, salamander, *Plethodon cinereus*, oak-hickory forest, above streams, Blue Ridge Mountains, North Carolina	9	1	—	—	—	—	—	11	10	1·10	90	82	Ditto
	44	Ectoparasites, wood-mouse, *Apodemus sylvaticus*, Bagley Wood, Oxford	17	1	—	1	—	—	—	23	19	1·21	89	74	Elton, Ford, Baker & Gardner, 1931, p. 706, etc.
	45	Nest fauna of wild house-mouse (*Mus musculus*) (parasites and other inhabitants), Isle of Lewis, Outer Hebrides	16	1	—	—	—	—	—	18	17	1·06	94	89	Ditto, p. 683, etc.
	46	"	13	2	—	—	—	—	—	17	15	1·13	87	76	Elton, 1934, p. 109
	47	Ecto- and endoparasites of brown rat (*Rattus norvegicus*), England	20	1	1	—	—	—	—	25	22	1·14	91	80	Balfour, 1922, p. 290
	48	Ecto- and endoparasites, steppe ground squirrel (*Citellus pygmaeus*), South-east Russia	23	2	1	—	—	—	—	30	26	1·15	88	77	Sassuchin & Tiflow, 1933, p. 438
	49	Ecto- and endoparasites, cotton-tail (*Sylvilagus floridanus mallurus*), Durham County, North Carolina	10	1	—	—	—	—	—	12	11	1·09	91	83	Harkema, 1936, p. 160
Totals (49 units)			1912	225	55	17·25	6·25	5	1	2666·25	2221·5	1·38	86	72	
%			86·07	10·13	2·48	0·78	0·28	0·22	0·04						

58   *Competition and the structure of ecological communities*

birds, etc. But within the groups collected analysis can be made, provided a high proportion of the genera have all their species identified.

(4) It is more important that a number of groups should have been collected completely and separated into reliable species, than that all groups should be recorded. A good many lists that were insufficiently broken down into species had to be omitted. I have, however, accepted certain surveys, mostly fresh-water benthos ones, that record 'species *a,·b*', etc., without actual Latin names; most of these being immature stages not yet correlated with known adult species. It will be realized, therefore, that the 'total number of species analysed' in Table 1 is seldom the total number present on the area, and is usually a little less than the total number given in the published surveys. Some of the detailed decisions that had to be made are relegated to an Appendix.

Table 2. *Frequency distribution of the percentages of genera with only one species present, in forty-nine animal communities*

Percentage	66–	71–	76–	81–	86–	91–	96–
No.	1	4	4	11	18	10	1

Table 3. *Frequency distribution of the percentages of 'single species' present, in forty-nine animal communities*

Percentage	46–	51–	56–	61–	66–	71–	76–	81–	86–	91–	96–
No.	1	4	3	4	7	9	6	9	5	0	1

(5) Although these communities are treated as if they are a random sample, they do include most of the reliable and fairly complete surveys known to me (except that, in order to retain some balance between different types of habitat, the proportion of European fresh-water surveys is not high). Also, although the percentages are all grouped together into one frequency table (Table 2), it will probably turn out, when enough surveys have accumulated, that some major habitats will show figures consistently higher or lower than the average. At present too few reliable surveys exist to decide whether such differences are really present in the ecosystems concerned, or whether they are inherent in the collecting methods or even in the taxonomic conventions for particular groups.

(c) *Plant communities*

Only a small sample analysis of plant communities is given here (Table 4), out of the very large published material that exists. No special selection was exercised in the choice of twenty-seven communities for analysis, provided they were complete and on clearly defined habitat areas and the lists were local rather than very large regional ones, except that the samples were intended to cover a wide range of conditions. It is sufficient to prove that the genus/ species relationships are similar to those in animal

communities: the frequencies summarized in Table 4 are remarkably similar to those in Table 2, e.g. the average percentage of genera with only one species present is 84 (range 63–96), compared with 86 for animal communities, and the average number of species per genus 1·22 (range 1·06–1·47), compared with 1·38 for animal communities. The almost exact correspondence of these averages may be partly a coincidence, but considering the very wide range of communities analysed, the resemblances are certainly remarkable and would lead one to suppose that there is some common principle operating both for plants and animals. The agreement is important also because most plant ecological surveys are more complete than animal ones, and because we know by direct evidence something of the direct competition that exists between plant species.

2. DISCUSSION AND WORKING HYPOTHESIS

(a) *Faunal statistics*

One possible explanation of the statistical relationships described above would be that the frequencies of species in genera simply reflect those of the fauna as a whole. For if, say, 86% of species in the British Isles belonged to genera of which only one species was present in this region, the figures in Tables 1 and 4 would be the record of a faunistic distribution, rather than of any peculiarity of homogeneous communities taken separately. Since insects form a high proportion of the fauna on land and in fresh water, we can take the British insect fauna as a test of this question, using the recently published 'Check list of British insects' (Kloet & Hincks, 1945), in which the numbers of established genera and species are summarized for each order (or in some cases, suborder). In Table 5 these thirty-one insect groups are arranged in ascending series of size, the Hemiptera being split into four, the Hymenoptera into six, and the Diptera into two subgroups. The frequencies for different groups are given in Table 6, with those of the animal communities for comparison. Whereas all the community figures for the number of species per genus lie in the frequency class 1·00– (actually, 1·00–1·63), only about 10% of those for the insect groups are in that class, and their percentages range from this to over 7·00. Although the greatest frequencies lie in the classes 2·00– to 4·00–, with a peak in 2·00–, the weight of the very large insect groups Hymenoptera, Lepidoptera, Diptera and Coleoptera (which together form 84% of the total British insect species and which all have high ratios) brings the average for the whole assemblage of insect groups to 4·23..

Since Kloet & Hincks seem to have inclined rather strongly towards the splitting of genera (i.e. calling subgenera genera) and the ecological surveys analysed here were done at earlier periods when generic splitting in most groups had gone less far, it may safely be stated that, *on the average*, for every species of insect present in a British animal community there are at least three or four others of the same genus

for the eleven largest groups, the percentages being given in the fourth column of Table 5. They range from 28 to 57 %, but most of them lie between 44 and 57 %, and the average for the whole lot is 50 %. This is the figure that we may compare with 86 % for the animal communities. The differences between particular communities and the fauna as a whole are evidently considerable in this respect, whether we consider the average number of species per genus or the percentage of genera with 'single species'. The difference is greater for the former figure than for

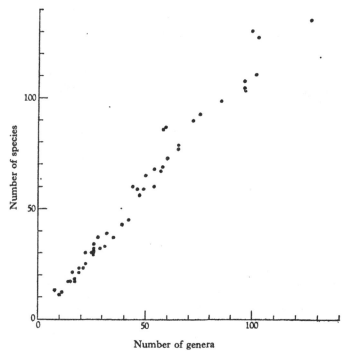

Number of genera

Fig. 1. Relation between the number of species and the number of genera present in forty-nine animal communities (from Table 1).

present in the country. But in the communities considered, the average number of species per genus was only 1·38, i.e. *on the average* every species only had two-fifths of another species living with it. This comparison is not quite satisfactory, because the insect statistics refer only to the insect groups of a fairly typical temperate continental island, while the community figures are derived from a very wide sample of varied animal groups from habitats in the Northern Hemisphere; but I think it illustrates a real difference that will be generally found to occur.

The Check List does not summarize the numbers of monospecific genera. I have therefore done this

the latter, owing to the presence in the general faunal lists of a great many large genera, with numbers far exceeding the usual limit of three or four found in the community lists.

It can still be said that the community statistics might be reflecting the general fauna picture for a smaller region within a country. I have considered making a further check by analysing the lists of county faunas, such as those published in the *Victoria County Histories*, but came to the conclusion that the comparison would probably be meaningless, because these lists are compiled over a very long period of time and do not necessarily describe the fauna of

Table 4. *Analysis of genus/species relations in twenty-seven plant communities*

General habitat		Community	No. of genera with the following nos. of species present							Total no. of species analysed (B)	Total no. of genera analysed (C)	Average no. of species per genus (B/C)	% of genera with one species present (A/C)	% of 'single species' present (A/B)	Reference
			1 (A)	2	3	4	5	6	+						
Arctic fjaeldmark	1	Frost-weathered rocky soil, Bear Island	16	4	1	1	—	—	—	31	22	1·41	73	52	Summerhayes & Elton, 1923, p. 220
	2	Rocky lowland (non-polygon soil), Prince Charles Foreland, Spitsbergen	24	5	3	—	—	—	—	43	32	1·34	75	56	Ditto, p. 244
Arctic heath	3	Cassiope tetragona heath, Wijde Bay, West Spitsbergen	28	7	0	1	1	—	—	51	37	1·38	76	55	Summerhayes & Elton, 1928, p. 233
Subarctic heath	4	Empetrum-Betula nana heath, Billefjordelv area, Finmark, Norway	26	3	2	1	—	—	—	42	32	1·28	81	62	Leach & Polunin, 1932, p. 420
Subarctic woodland	5	Drier Betula odorata forest, same area as (4)	24	1	1	—	—	—	—	29	26	1·12	92	83	Ditto, p. 417
Temperate grassland	6	Nardetum, Cader Idris, north Wales	20	3	—	—	—	—	—	26	23	1·13	87	77	Evans, 1932, p. 25
	7	Deschampsia flexuosa grass heath, Bunter sandstone, Nottinghamshire, England	42	10	1	0	1	—	—	70	54	1·30	78	60	Hopkinson, 1927, p. 159
	8	Primitive chalk grassland, Buriton, West Sussex, England	52	10	2	1	—	—	—	82	65	1·26	80	63	Tansley & Adamson, 1925, p. 185
Steppe	9	Avena desertorum meadow steppe, Kuznetsak District, Saratov, Russia	46	8	2	—	—	—	—	68	56	1·21	82	68	Keller, 1927, p. 220
Temperate scrub	10	Limestone heath-scrub, Ballyvaghan, Co. Clare, Ireland	25	0	1	—	—	—	—	28	26	1·08	96	89	Tansley, 1939, p. 474
Temperate carr	11	Open carr, Esthwaite Fens, Lake District, England	50	5	2	0	0	0	1(7)	73	58	1·26	86	68	Pearsall, 1918, p. 61, summarized by Tansley, 1939, p. 644
	12	Valley fenwoods, River Lark, East Anglia, England	35	5	3	1	—	—	—	58	44	1·32	80	60	Farrow, 1915, p. 226, summarized by Tansley, 1939, p. 467
	13	Coppiced alder carr, Cothill, Berkshire, England	38	4	0	1	—	—	—	50	43	1·16	90	76	Tansley, 1939, p. 469
	14	Alder carr, Wheatfen Broad, Norfolk, England	47	9	1	1	—	—	—	72	58	1·24	81	65	Ellis, 1935, summarized by Tansley, 1939, p. 461

Category	No.	Locality													Reference
Temperate woodland	15	Ashwood, Ling Ghyll, Yorkshire, England	50	8	2	—	—	—	—	72	60	1·20	83	69	Tansley, 1939, p. 433
		Highland oakwood, Glenmore, Scotland:													Ditto, p. 344
	16	(a) Portclair (*Quercus robur*)	38	4	—	—	—	—	—	46	42	1·10	90	83	
	17	(b) Loch Leven (*Q. sessiliflora*)	49	3	—	—	—	—	—	55	52	1·06	94	89	
	18	Birchwood, Cader Idris, north Wales, 800–1100 ft.	40	5	—	—	—	—	—	50	45	1·11	89	80	Evans, 1932, p. 20
	19	Mature beechwood (sere 1, stage 3), Singleton Forest, Sussex Downs, England	39	3	—	—	—	—	—	45	42	1·07	93	87	Watt, 1925, p. 50
	20	Red fir forest, Sierra Nevada, California	60	12	3	—	—	—	—	93	75	1·24	80	64	Oosting & Billings, 1943, p. 271
Temperate lake plankton	21	Windermere, England	18	1	0	1	—	—	—	24	20	1·20	90	75	Pearsall, 1932, p. 261
Temperate lake littoral	22	Phanerogams, Lake Mendota, Wisconsin	13	1	0	0	0	1	—	21	15	1·40	87	62	Denniston, 1921, p. 500
Subarctic marine intertidal	23	Amerdloq Fjord, West Greenland	10	1	—	—	—	—	—	12	11	1·09	91	83	Steven, 1938, p. 62
Temperate marine intertidal	24	Rocky shore, Plymouth, England (4 years after colonization)	12	1	—	—	—	—	—	14	13	1·08	92	86	Moore & Sproston, 1940, p. 319
Tropical marine intertidal	25	Coral reef flat, Low Isles, Great Barrier Reef, Australia	12	5	2	—	—	—	—	28	19	1·47	63	43	Stephenson *et al*, 1931, p. 45
Marine phytoplankton	26	Aberystwyth Harbour, Wales	18	1	1	1	—	—	—	23	20	1·15	90	78	Lloyd, 1925, p. 103
	27	Shields Area, Northumberland, North Sea [Station 26]	7	1	0	0	—	1	1	13	9	1·45	78	54	Savage, 1926
		Totals (27 units)	839	120	27	9	2	1	1	1219	999	1·22	84	69	
		%	83·99	12·01	2·70	0·90	0·20	0·10	0·10						

62   *Competition and the structure of ecological communities*

a region in one year or a short period of years; whereas the community surveys are done usually within a year or at most a few years. Nevertheless, the faunal list does indicate the species that either live or have attempted to live within the area, and it is notable that 54% of the British species of Hemiptera Heteroptera have been recorded from an area as small as the county of Oxfordshire (China, 1939) and

conclusions drawn here. But although the list of habitats is an extremely varied one and gives a very wide sampling, it is deficient (except for no. 7, Table 1) in one very important class of animal community, that of terrestrial habitats of temperate and tropical regions containing complex plant associations. The reason for this omission is, of course, the absence of sufficiently complete surveys hitherto.

Table 5. *Relation between the number of genera and of species in thirty-one British insect orders or suborders (established species), columns 1 and 2 from Kloet & Hincks (1945)*

Order	No. of species	No. of genera	No. of species per genus	% of genera with only one species present
Mecoptera	4	2	2·00	—
Megaloptera	6	3	2·00	—
Dermaptera	9	7	1·29	—
Protura	17	4	4·25	—
Strepsiptera	17	5	3·40	—
Thysanura	23	7	3·29	—
Plecoptera	32	15	.2·13	—
Orthoptera	38	27	1·41	—
Odonata	42	21	2·00	—
Ephemeroptera	46	19	2·42	—
Siphonaptera	47	24	1·96	
Neuroptera	54	18	3·00	
Psocoptera	68	33	2·06	
Hemiptera, Coccoidea	103	38	2·71	
Thysanoptera	183	42	4·35	
Trichoptera	188	70	2·69	—
Hymenoptera, Cynipoidea	228	49	4·65	—
Collembola	261	62	4·21	—
Anoplura (including Mallophaga)	286	73	3·92	—
Hemiptera, Aphidoidea	375	122	3·07	—
Hymenoptera, Symphyta	430	92	4·67	35
Hemiptera, Homoptera	434	92	4·71	46
Hemiptera, Heteroptera	499	221	2·26	57
Hymenoptera, Aculeata	531	139	2·82	44
Hymenoptera, Proctotrupoidea	613	93	6·60	28
Hymenoptera, Chalcidoidea	1564	214	7·31	51
Diptera, Cyclorrhapha	2072	652	3·18	53
Lepidoptera	2187	657	3·33	57
Hymenoptera, Ichneumonoidea	2825	485	5·82	44
Diptera, Orthorrhapha	3127	480	6·53	45
Coleoptera	3690	947	3·90	50
Total insects*	19999	4713	4·23	
		Average (from raw totals, 11 groups)		50

* Omitting 25 'Addenda' to the list.

that Waters (1929) found that at least 58% of the British species of micro-Lepidoptera had been recorded within seven to ten miles of Oxford.

(b) *Limitations of the community data*

It was pointed out in § 1 that the community surveys analysed are certainly incomplete in many respects, but that the features in which they themselves were lacking were not likely to invalidate the

On the whole, the list given in Table 1 contains animal communities in which the species live on a comparatively few different basic sources of food. This applies to log-dwelling herbivores, which depend on phloem, on fungi growing in the galleries made by the phloem-eaters, and on the wood (Savely, 1939); to soil animals; to fresh-water, estuarine and marine bottom detritus or plankton feeders; to intertidal animals dependent on plankton; to zooplankton itself; to drift-line animals eating decaying matter; and to blood-sucking ectoparasites. I do not mean that there is no ecological differentiation in the food habits of species living, say on phloem in logs or mud on the sea bottom; only that this type of community consists mainly of a few ecological groups each broadly drawing upon the same natural resources for its basic food, with of course the usual predator-parasite food cycle rising from it.

Even in those aquatic habitats that have a number of plant species (e.g. fresh-water benthic phanerogams, intertidal sea-weeds, and phytoplankton) few of the herbivores seem to be restricted to one or a few plant species. The same thing applies in general to Arctic and Subarctic terrestrial communities, where specialization of herbivores is rather exceptional.

mostly species pairs—also show strong ecological differences in habits, although they live within the same community. Lack's study (1944) of passerine birds has brought this point out especially clearly. He also gives many instances of genera whose species are split up between different major habitats. In other instances, although we have not yet got any direct evidence of different habits or tolerance ranges, quantitative survey shows one species of a pair to be much more abundant than another. Thus in his Dogger Bank samples Davis (1923, Voyage 48) got 1182 specimens of the lamellibranch *Spisula subtruncata*, and only four of *S. solida*. These appear, however, as a species pair in Table 1 of the present paper. Undoubtedly some other 'species pairs' will be due to chance immigration of one species not living in that habitat, but these (as also instances of

Table 6. *Frequency distributions of numbers of species per genus in (a) forty-nine animal communities (b) thirty-one British insect orders or suborders*

	1·00–	2·00–	3·00–	4·00–	5·00–	6·00–	7·00–
No. of communities	·49	—	—	—	—	—	—
% of communities	100	—	—	—	—	—	—
No. of orders or suborders	3	10	8	6	1	2	1
% of orders or suborders	9·7	32·3	25·8	19·4	3·2	6·4	3·2

The situation in highly organized terrestrial communities like heath, meadow, scrub and woodland is different. Here we find large numbers of monophagous species, especially among insects, attached to particular plant species. Since there may be commonly up to seventy species of plants in one association, the majority of which are phanerogams edible to some animals, the possibilities for ecological differentiation within the community are therefore much greater than in communities of the type so far surveyed with any completeness. Such terrestrial communities do not cover a larger area of the globe than ones of the simpler trophic type, but they do contain some of the most highly organized and complicated relationships known between species, and I wish to make it quite clear that the conclusions that follow should be treated as an approach to the more complex problem, through the evidence for simpler communities that has already been accumulated in the short time that animal ecology has been a science.

(c) Ecotypic differentiation

It is already well known that ecotypic differentiation occurs between many species of the same genus. What this community analysis shows is that the amount of differentiation is apparently very high, and that it is a prominent feature of all the communities for which we have sufficient knowledge to make the analysis. There is no doubt that some of the 14% of genera that have more than one species present—

temporary establishment) might also apply to the single species, and we cannot therefore make any statistical proviso from them. It can be concluded that the amount of ecotypic differentiation in genera is really very high in communities of the type we are considering here, and that it is the exception to find groups of species of the same genus occupying the same ecological niche on the same area or apparently doing so (as does, however, occur in genera like the lamellibranch *Pisidium* or the larvae of the black-fly *Simulium* in fresh water).

(d) Competition and community structure

Ecological research has discovered a good deal about the 'vertical' organization of animal communities. By 'vertical' I mean here, not vertical layering of the habitat, but the flow of matter and energy through different levels of consumption, as found in food chains, with their herbivore-plant, predator-prey, parasite-host and other relationships; fluctuating equilibrium between the stages in these chains; cover, making such equilibrium possible; daily and other activity rhythms causing alternating mass action of different components of the community in response to environmental cycles; and the ultimate limit (usually about five stages) set to the number of consumer levels by the size relations of animals and the pyramid of numbers. In this field of ecology it is possible to proceed with some general measure of agreement on the fundamental principles at work

64   *Competition and the structure of ecological communities*

(see Lindeman, 1942, who has restated the subject in a useful essay).

We also have a great deal of information, though none of it complete in respect to any single species, about the tolerance ranges, optima and preferenda that animals have in regard to various habitat factors like temperature, humidity and amount and quality of food—the ranges, etc. often varying with the sex and life history stage of the species and with the type of life process (growth, viability, reproduction, activity, etc.) studied. Here again, research is progressing along well-defined lines, though still very weak upon the fundamental problem of habitat selection in nature.

When we come to the 'horizontal' organization of animal communities, i.e. the dynamic relations between species of the same consumer level, we find that little is known except from very simplified laboratory experiments and from certain lines of *a priori* reasoning. The pros and cons of argument on this question were partly explored in the British Ecological Society's Symposium (21 March 1944) on 'The ecology of closely allied species' (Anon. 1944), at which the substance of the present paper was put forward for discussion. But that discussion was mainly concerned with closely allied forms, whereas I wish to consider now the relation between all species of the same genus. The statistics of Table 1, and the field lists on which they are based, really mean that these animal communities have, at each level of consumption (i.e. food-chain stages 1, 2, 3, 4,...) a certain number, from several to a score or more, of species that mainly belong to separate genera. These, as has been indicated, could be broken up into subgroups each drawing its food from a common pool, though not necessarily from exactly the same part of it or at the same time, or in the same way.

We simply do not understand exactly why populations of, say, a Pentatomid bug, a grasshopper, a moth caterpillar, a vole, a rabbit and an ungulate should be able to draw upon the same common resource (grassland vegetation) and yet remain in equilibrium at any rate sufficiently to form a stable animal community over long periods of years. In all communities the primary resources of plant or decaying matter (or with parasites, tissues or the food of the host) are split up in this way between a number of species, each of which is able to maintain, though with fluctuating equilibrium, its own share of the common natural resource. I think it has usually been assumed, by analogy with the specific food habits of monophagous or not very polyphagous insects, that the equilibrium is made possible by some specialized division of labour, and that the animals do not come into direct competition at all; or else that the amount of resources is generally sufficient to provide for all

the populations present because they are limited by factors other than food in the increase of their populations. The second idea is on the whole supported by the general evidence that animals do not normally become limited in numbers by starvation, and that the biomass of phanerogamic vegetation is far beyond that of animals dependent on it. However this may be, we know extremely little about the whole subject.

Darwin, in *The Origin of Species*, remarked that 'As species of the same genus have usually, though by no means invariably, some similarity in habits and constitution, and always in structure, the struggle will generally be more severe between species of the same genus, when they come into competition with each other, than between species of distinct genera. We see this in the recent extension over parts of the United States of one species of swallow having caused the decrease of another species. The recent increase of the missel-thrush in parts of Scotland has caused the decrease of the song-thrush. How frequently we hear of one species of rat taking the place of another species under the most different climates! In Russia the small Asiatic cockroach has everywhere driven before it its great congener. One species of charlock will supplant another, and so in other cases. We can dimly see why the competition should be most severe between allied forms, which fill nearly the same place in the economy of nature; but probably in no one case could we precisely say why one species has been victorious over another in the great battle of life.'

More recently, Gause (1934) and other laboratory workers have shown by experiments in controlled environments how one of two similar species of a genus introduced into a culture will prevail eventually over the other. That this type of competition is not confined to species of the same genus is well proved by the experiments of Crombie (1945, 1946) with grain insects, in which a beetle was able eventually to crowd out a moth, both having larvae living inside the grains of wheat; also in similar experiments with beetles by Park, Gregg & Lutherman (1941). Here competition was effective between members of two different orders of insects, and we have the type of equilibrium problem that is most commonly encountered in the field. The importance of the community analysis given in the present paper is that it confirms the general proposition that some (though not necessarily all) genera of the same consumer level that are capable of living in a particular habitat at all can coexist permanently on an area; whereas it is unusual, in the communities analysed, for species of the same genus to coexist there. We therefore arrive at some points for a working hypothesis to cover our present limited knowledge of competition in relation to basic community structure.

It should be stressed that 'competition' is here used not merely for direct antagonism or struggles for space, etc., but as an objective description (in the same way that 'natural selection' or the 'struggle for existence' are only shorthand terms) of the interplay of longevity and fertility factors of all kinds (known and unknown) favouring one species at the expense of another.

(1) Every habitat that supports a whole community of animal species contains one or more pools of natural resources available for the building up of animal populations, plant resources (alive or dead) being usually the most immediately important, and in all cases the ultimate source there or elsewhere.

(2) In habitats where there are not large numbers of terrestrial plant species suitable for food specialization by animals (i.e. most of those considered in Table 1), these resources are exploited for the greater part by genera with only one species present on that area of habitat, genera with more than two species present forming a small fraction only. Very large genera do not seem to be represented in full force at all. The main extra-specific ecological relations in such communities are between organisms with generic differences.

(3) We do know a little, experimentally and from field observations (especially on introduced species and their allies) about the effectiveness of competition between species of the same genus. We also know that similarly effective competition can occur between species of separate genera, or even orders. We do not at present know what maintains the state of equilibrium between the different genera actually found in the natural communities analysed, but must postulate that there is some ecological condition that buffers or cuts down the effectiveness of competition between species separated by generic characters. This problem is therefore seen to be the central problem in animal community structure, because it is the variety of species that can coexist at the primary consumer level, that makes possible the considerable complexity of the superstructure of secondary and other consumers.

(4) The comparative shortness of the lists in the community surveys analysed suggests that on any one area of a given habitat, there is in fact restriction upon the number of primary consumers that can coexist (see Elton, 1933), and that there is therefore also some real state of population competition or tension between the different primary consumers, just as there is known to be between primary producers in plant communities.

There is a point about competition that has perhaps not yet been brought out clearly in this discussion. It does not follow that because we find only one species of a certain genus living in a particular animal community on a particular area, that this is the only species of that genus that is capable of living in that habitat. In other words, we have to distinguish carefully between ability to live in a habitat and ability to live among a particular assemblage of other species present there. There may be lists of species, though usually short ones, of the same genus, all capable of contending with the habitat conditions (tundra, marine intertidal, etc.), but it may still be true that only one (or not often more than two) of them can coexist permanently on the same area of it. So we might not be surprised to find a given niche occupied in one area by one species and in another area of the same general habitat by another. Such differences in distribution might either be due to minor variations in the habitat that we have perhaps not yet detected, or they might be due to the process of competition that has been postulated.

The laboratory experiments of Park et al. (1941) illustrate the working of such a situation between two species of beetles, *Tribolium confusum* and *Gnathocerus cornutus* of the same family Tenebrionidae, competing for a common food supply of ground cereals and yeast, but each cannibalizing the early stages of the other species. Competition was well marked between these two species, but the end result depended partly upon the relative initial densities of the two forms. It was possible to get pure or almost pure cultures of either species developing from cultures that had been initially mixed. In nature we might expect to find a number of instances of this happening between species of the same genus, although all the species concerned might be found in some area or other of the general habitat under observation. In discussing the statistical picture of communities given in the present paper, it is therefore essential to remember that each survey was made on a relatively (though not always absolutely) small area of the total habitat.

Finally, something must be said about current trends in taxonomic methods, and especially about fashions for lumping or splitting genera, for these will have a good deal of influence in future handling of community statistics of the type we have been considering. Practically all the animal surveys in Table 1 were done in the 20 years 1921-40, and their nomenclature is a random sample of the taxonomic practices, fashions, advances and retreats, and equilibria of group specialists in various countries during that period and before it. No doubt a good deal of the differences in the frequency pictures for different surveys can be attributed to these variations in taxonomic treatment. Future comparisons will have to take into account the marked tendency for further splitting of genera in many groups of animals, and in some plants, with the reduction in average numbers of species per genus that this involves, and increase

66   *Competition and the structure of ecological communities*

in number (though not always necessarily in the percentage) of monotypic genera. I believe that further research on community structure will eventually give us some new, ecological, criteria for generic classification.

3. ACKNOWLEDGEMENTS

I would like to thank Captain Cyril Diver, Mr David Lack and Mr P. H. Leslie for stimulus and help, and some modifying ideas, given during discussions of these problems during the last two years.

4. SUMMARY

1. Analysis was made of the published ecological surveys of fifty-five animal (including some parasite) communities and twenty-seven plant communities from a wide range of habitats, and the frequencies of genera with different numbers of species tabulated. A rather constant and high percentage of genera with only one species present was found, the average being 86 % for animal and 84 % for plant communities. The corresponding average numbers of species per genus were 1·38 and 1·22.

2. These figures differ considerably from those of a faunal list for any large region, e.g. the percentage of genera with only one species present for eleven

large British insect groups is 50, and the average number of species per genus for all British insects is 4·23.

3. The difference in species/genus frequencies between ecological surveys of relatively small parts of any general habitat, and those for faunal lists from larger regions, is attributed to existing or historical effects of competition between species of the same genus, resulting in a strong tendency for the species of any genus to be distributed as ecotypes in different habitats, or if not, to be unable to coexist permanently on the same area of the same habitat.

4. These conclusions apply at present only to the list of communities hitherto surveyed with any completeness, which does not include a sufficient sample of terrestrial habitats like heath, meadow, scrub and woodland containing many plant species. The animal communities analysed are mostly ones in which the primary consumer species depend on only a few natural resources.

5. The ability of certain groups of species, mostly separated by generic characters, to exist together on the same area while drawing upon a common pool of resources, is one of the central unsolved problems in animal community structure and population dynamics.

APPENDIX

Special notes on the compilation of Table 1

Community no.

1–11	Insect parasites omitted from the table.
1–6	Immigrant adult aquatic flies are included because they are an integral part of the food supply of spiders on land. A part of the micro-fauna of mites and Collembola has probably been omitted from all these surveys, as it requires special methods of collection (see Hammer, 1944).
8–9	Mites omitted from the table.
11	Subgenera of *Spaniotoma* treated as separate genera, for uniformity with other surveys.
12	Entomostraca omitted from the table, also five records from an earlier survey.
14–15	All Chironomidae genera containing any species just marked 'sp.' or 'gr.' (for group) omitted from the table.
16	The depth taken for sublittoral is slightly arbitrary, as the actual limits vary in different parts

Community no.

	of a lake. Chironomidae and most Trichoptera omitted from the table.
19	Two independent surveys of Lake Michigan were done 40 years apart. There were some significant differences in species, but little in the genera present.
20	Chironomidae omitted from the table.
22–23	Diptera omitted from the table.
26	Fish omitted from the table.
35–36	Surveys covered 9 years.
38–40	This survey does not include all bottom-living fish.
41	Autotrophic flagellates omitted from the table.
45–47	Spirochaetes and bacteria omitted from the table.
48	Single records omitted from the table.

REFERENCES

Allee, W. C. (1923). 'Studies in marine ecology. III. Some physical factors related to the distribution of littoral invertebrates.' Biol. Bull. Wood's Hole, 44: 205–53.

Alexander, W. B. (1932). 'The natural history of the Firth of Tay.' Trans. Perthshire Soc. Nat. Sci. 9: 35–42.

Anon. (1944). 'British Ecological Society. Easter meeting 1944. Symposium on "The ecology of closely allied species".' J. Anim. Ecol. 13: 176–8; also in (1945) J. Ecol. 33: 115–16.

Balfour, A. (1922). 'Observations on wild rats in England, with an account of their ecto- and endoparasites.' Parasitology, 14: 282–98.

Butcher, R. W., Pentelow, F. T. K. & Woodley, J. W. A. (1931). 'An investigation of the River Lark and the effect of beet sugar pollutions.' Minist. Agric. Fish., Fish. Invest. Ser. 1, 3, No. 3: 1–112.

China, W. E. (1939). 'Hemiptera.' In B. M. Hobby, 'Zoology', in 'The Victoria History of the County of Oxford', 1: 69–77.

Crombie, A. C. (1945). 'On competition between different species of graminivorous insects.' Proc. Roy. Soc. B, 132: 362–95.

Crombie, A. C. (1946). 'Further experiments on insect competition.' Proc. Roy. Soc. B, 133: 76–109.

Darwin, C. (1859). 'On the origin of species by means of natural selection, or the preservation of favoured races in the struggle for life.' London. P. 76.

Davis, D. H. S. (1936). 'A reconnaissance of the fauna of Akpatok Island, Ungava Bay.' J. Anim. Ecol. 5: 319–32.

Davis, F. M. (1923). 'Quantitative studies on the fauna of the sea bottom. No. 1. Preliminary investigation of the Dogger Bank.' Minist. Agric. Fish., Fish. Invest. Ser. 2, 6, No. 2: 1–54.

Denniston, R. H. (1921). 'A survey of the larger aquatic plants of Lake Mendota.' Trans. Wis. Acad. Sci. Arts Lett. 20: 495–500.

Eddy, S. (1927). 'The plankton of Lake Michigan.' Bull. Ill. St. Nat. Hist. Surv. 17, Art. 4: 203–32.

Ekman, S. (1915). 'Die Bodenfauna des Vättern, qualitativ und quantitativ untersucht.' Int. Rev. Hydrobiol. 7: 146–204, 275–425.

Ellis, E. A. (1935). 'Wheatfen Broad, Surlingham.' Trans. Norfolk Norw. Soc. Nat. 13: 422–51.

Elton, C. (1933). 'The ecology of animals.' London, pp. 17–23.

Elton, C. (1934). 'Metazoan parasites from mice in the Isle of Lewis, Outer Hebrides.' Parasitology, 26: 107–11.

Elton, C., Ford, E. B., Baker, J. R. & Gardiner, A. D. (1931). 'The health and parasites of a wild mouse population.' Proc. Zool. Soc. Lond.: 657–721.

Evans, E. Price (1932). 'Cader Idris: a study of certain plant communities in south-west Merionethshire.' J. Ecol. 20: 1–52.

Farrow, E. P. (1915). 'On the ecology of the vegetation of Breckland. I. General description of Breckland and its vegetation.' J. Ecol. 3: 211–28.

Ford, J. (1935). 'The animal population of a meadow near Oxford.' J. Anim. Ecol. 4: 195–207.

Gause, G. F. (1934). 'The struggle for existence.' Baltimore.

Hammer, M. (1944). 'Studies on the Oribatids and Collemboles of Greenland.' Medd. Grønland, 141, No. 3: 1–210.

Harkema, R. (1936). 'The parasites of some North Carolina rodents.' Ecol. Monogr. 6: 153–232.

Hopkinson, J. W. (1927). 'Studies on the vegetation of Nottinghamshire. I. The ecology of the Bunter sandstone.' J. Ecol. 15: 130–71.

Humphries, C. F. (1936). 'An investigation of the profundal and sublittoral fauna of Windermere.' J. Anim. Ecol. 5: 29–52.

Jones, J. R. E. (1940a). 'The fauna of the River Melindwr, a lead-polluted tributary of the River Rheidol in North Cardiganshire, Wales.' J. Anim. Ecol. 9: 188–201.

Jones, J. R. E. (1940b). 'A study of the zinc-polluted river Ystwyth in north Cardiganshire, Wales.' Ann. Appl. Biol. 27: 368–78.

Jones, J. R. E. (1941). 'The fauna of the River Dovey, West Wales.' J. Anim. Ecol. 10: 12–24.

Keller, B. A. (1927). 'Distribution of vegetation on the plains of European Russia.' J. Ecol. 15: 189–233.

Kloet, G. S. & Hincks, W. D. (1945). 'A check list of British insects.' Stockport.

Lack, D. (1944). 'Ecological aspects of species-formation in passerine birds.' Ibis, 86: 260–86.

Laurie, E. M. O. (1942). 'The fauna of an upland pond and its inflowing stream at Ystumtuen, North Cardiganshire, Wales.' J. Anim. Ecol. 11: 165–81.

Laurie, R. & Jones, J. R. E. (1938). 'The faunistic recovery of a lead-polluted river in North Cardiganshire, Wales.' J. Anim. Ecol. 7: 272–89.

Leach, W. & Polunin, N. (1932). 'Observations on the vegetation of Finmark.' J. Ecol. 20: 416–30.

Lindeman, R. L. (1942). 'The trophic-dynamic aspect of ecology.' Ecology, 23: 399–418.

Linder, C. (1904). 'Étude de la faune pélagique du Lac de Bret.' Rev. Suisse Zool. 12: 149–258.

Lloyd, B. (1925). 'Marine phytoplankton of the Welsh coasts, with special reference to the vicinity of Aberystwyth.' J. Ecol. 13: 92–120.

Longstaff, T. G. (1932). 'An ecological reconnaissance in West Greenland.' J. Anim. Ecol. 1: 119–42.

Moore, H. B. & Sproston, N. G. (1940). 'Further observations on the colonization of a new rocky shore at Plymouth.' J. Anim. Ecol. 9: 319–27.

Oosting, H. J. & Billings, W. D. (1943). 'The red fir forest of the Sierra Nevada: *Abietum magnificae*.' Ecol. Monogr. 13: 261–74.

Park, T., Gregg, E. V. & Lutherman, C. Z. (1941). 'Studies in population physiology. X. Interspecific competition in populations of granary beetles.' Physiol. Zoöl. 14: 395–430.

Pearsall, W. H. (1918). 'The aquatic and marsh vegetations of Esthwaite Water.' J. Ecol. 6: 53–74.

Pearsall, W. H. (1932). 'Phytoplankton in the English Lakes. II. The composition of the phytoplankton in relation to dissolved substances.' J. Ecol. 20: 241–62.

Percival, E. (1929). 'A report on the fauna of the estuaries of the River Tamar and the River Lynher.' J. Mar. Biol. Ass. U.K. 16: 81–108.

Percival, E. & Whitehead, H. (1930). 'Biological survey of the River Wharfe. II. Report on the invertebrate fauna.' J. Ecol. 18: 286–302.

Pirrie, M. E., Bruce, J. R. & Moore, H. B. (1932). 'A quantitative study of the fauna of the sandy beach at Port Erin.' J. Mar. Biol. Ass. U.K. 18: 279–96.

Pyefinch, K. A. (1937). 'The fresh and brackish waters of Bardsey Island (North Wales): a chemical and faunistic survey.' J. Anim. Ecol. 6: 115–37.

Pyefinch, K. A. (1943). 'The intertidal ecology of Bardsey Island, North Wales, with special reference to the recolonization of rock surfaces, and the rock-pool environment.' J. Anim. Ecol. 12: 82–108.

Rankin, J. S. (1937). 'An ecological study of parasites of some North Carolina salamanders.' Ecol. Monogr. 7: 171–269.

Sassuchin, D. & Tiflow, W. (1933). 'Endo- und ecto-

68   *Competition and the structure of ecological communities*

parasiten des Steppenziesels (*Citellus pygmaeus* Pall.) im Süd-Osten RSFSR.' Z. Parasitenk. 5: 437–42.

Savage, R. E. (1926). 'The plankton of a herring ground.' Minist. Agric. Fish., Fish. Invest. Ser. 2, 9, No. 1: 1–35.

Savely, H. E. (1939). 'Ecological relations of certain animals in dead pine and oak logs.' Ecol. Monogr. 9: 321–85.

Scheffer, V. B. & Robinson, R. J. (1939). 'A limnological study of Lake Washington.' Ecol. Monogr. 9: 95–143.

Stephenson, T. A., Stephenson, A., Tandy, G. & Spender, M. (1931). 'The structure and ecology of Low Isles and other reefs.' Great Barrier Reef Expedition 1918–29, Sci. Rep. 3, No. 2: 1–112.

Steven, D. (1938). 'The shore fauna of Amerdloq Fjord, West Greenland.' J. Anim. Ecol. 7: 53–70.

Summerhayes, V. S. & Elton, C. S. (1923). 'Contributions to the ecology of Spitsbergen and Bear Island.' J. Ecol. 11: 214–86.

Summerhayes, V. S. & Elton, C. S. (1928). 'Further contributions to the ecology of Spitsbergen.' J. Ecol. 16: 193–268.

Tansley, A. G. (1939). 'The British islands and their vegetation.' Cambridge.

Tansley, A. G. & Adamson, R. S. (1925). 'Studies of the vegetation of English chalk. III. The chalk grasslands of the Hampshire-Sussex border.' J. Ecol. 13: 177–223.

Waters, E. G. R. (1929). ' A list of the Micro-Lepidoptera of the Oxford District.' Proc. Ashmol. Nat. Hist. Soc. for 1928 (Supplement): 1–72.

Watt, A. S. (1925). 'On the ecology of the British beechwoods with special reference to their regeneration. Part II. Sections II and III. The development and structure of beech communities on the Sussex Downs (*continued*).' J. Ecol. 13: 27–73.

Wells, A. L. (1938). 'Some notes on the plankton of the Thames Estuary.' J. Anim. Ecol. 7: 105–24.

THE GENERIC RELATIONS OF SPECIES IN SMALL ECOLOGICAL COMMUNITIES

By C. B. WILLIAMS, *Rothamsted Experimental Station, Harpenden*

1. INTRODUCTION

The problem of the importance of competition between two or more species in the same genus, in determining whether they can survive side by side in the same community, is of great ecological interest. The following is a statistical approach to the problem based largely on the fact that in most animal and plant communities the numbers of genera with one, two, three, etc., species appear to form a mathematical series very close to the 'logarithmic series'.

The species in any group of animals and plants that are found living side by side in any relatively small community have presumably been selected in the course of time from all the species in the surrounding areas which have been able to reach the smaller area in question. Those that survive are those which are capable of existing in the physical environment of the area and also in association with, or in competition with, the other members of the community.

Such a natural selection could conceivably be brought about under three different conditions: (1) without reference to the generic relations of the species, (2) more or less *against* species in the same genus, (3) more or less *in favour of* species in the same genus. Extremes of either (2) or (3) would result in only one species to each genus on the one hand, or in all the species in each genus on the other hand being represented. We know, however, that biologically, neither of these extremes is correct.

It is, however, most important to consider in detail what exactly happens when a selection of a relatively small number of species is made from a larger fauna or flora, without reference to their generic relations (no. (1) above), as a true interpretation can only be made by comparing the observed data with the results of a selection of the same size made at random. If the average number of species per genus in a small community is smaller than that expected by random sampling, then there is evidence of a selection against generically related species; if the number of species per genus is larger than would be expected by random sampling, then there is evidence of selection in favour of species in the same genus.

Before we can discuss the problem of sampling, it is, however, necessary to consider what is the general structure of the relative numbers of genera with 1, 2, 3, etc., species in any group of animals or plants.

2. GENERIC CLASSIFICATION AND THE LOGARITHMIC SERIES

I have shown recently (Williams, 1944) that in a number of classifications of particular families or orders of plants, insects and other animals, the number of genera with 1, with 2, with 3, etc., species forms a series which can be represented very closely by the logarithmic series.

This will be discussed more fully below, but for the moment it should be noted that if we know the number of species and the number of genera in any group, a logarithmic series of the genera with 1, 2, 3 species (which we will call n_1, n_2, n_3, etc.) can be calculated and this can be checked against the observed frequencies. If a close fit is found, it is evidence of the likelihood that the logarithmic series is a correct interpretation, and we can then use for further argument known properties of the logarithmic series.

I have shown, as stated above, that the logarithmic series gives a very close fit to published classifications of large groups, such as the flowering plants of Britain, the Mantidae of the world, the birds of Great Britain, and many other systematic and geographical groups. It is, however, important to know if the same principle is found in the classification of small communities. For this purpose some evidence brought forward recently by Elton (1946) is of great value.

Table 1 shows particulars of ten of his animal communities and three of his plant communities, together with the total (reduced to an average) of his 49 animal and of his 27 plant communities; with the number of genera with 1, 2 and 3 species in each case as given by Elton and as calculated by the logarithmic series.

The communities analysed here are all comparatively simple ones, because reliable surveys of more complex communities such as woodland do not yet exist; some of them, however, are quite large in area.

No one can deny the extremely close approximation of the observed and calculated figures in nearly all the examples and particularly in the averages. Thus in the average figure for 49 animal communities the calculated number of genera with one species is 38·1—the observed 39; in the average of 27 plant communities the calculated number of genera with one species is 30·7 and the observed 31·1. The fit for n_2 and n_3 is almost equally good.

Thus we have very strong evidence that the number of genera with 1, with 2 and with 3, etc., species is arranged in a logarithmic series in quite small or simple ecological communities as well as in large systematic groups and in large areas.

where the successive terms are the numbers of groups (in this case genera) with 1, with 2, with 3, etc., units (in this case species).

n_1 (which $= \alpha x$) is the number of genera with one species; x is a constant less than unity, α is a constant which we have called the Index of Diversity. If we know the number of species (N) and the number of genera (S) in a population which is arranged in a log series, we can calculate both n_1 and x, and

Table 1

Elton's community	No. of genera S	No. of species N	No. of genera with 1 species n_1	2 species n_2	3 species n_3	Index of diversity α
Animal						
9 Temperate woodland (logs)	126	136	118	6	2	
			116	*8·7*	*0·9*	*773·*
10 Subarctic bog	19	21	17	2	0	
			19·1	*1·6*	*0·2*	*94*
11 Subarctic bog	29	32	26	3	0	
			26·3	*2·2*	*0·25*	*155*
12 Temperate fresh-water pond	26	32	21	4	1	
			21·4	*3·5*	*0·8*	*65*
13 Temperate fresh-water pond	72	90	61	8	1	
			59	*11*	*2·2*	*171*
14 Temperate lake benthos	46	59	38	6	0	
			36·8	*6·9*	*1·7*	*98·1*
15 Temperate lake benthos	8	13	6	1	0	
			5	*1·8*	*0·7*	*8·3*
16 Temperate lake benthos	58	86	44	8	1	
			40	*10·8*	*3·9*	*74*
20 Temperate river	99	131	82	7	6	
			79·6	*15·6*	*4·1*	*203*
22 Temperate river	59	87	47	6	3	
			42	*10·8*	*3·7*	*81·5*
Average of 49 animal communities	45·3	54·41	39·0	4·6	1·12	
			38·1	*5·7*	*1·14*	*127*
Plant						
1 Arctic rocky soil	22	31	16	4	1	
			16·4	*3·9*	*1·2*	*35*
3 Arctic heath	37	51	28	7	0	
			27·8	*6·3*	*1·9*	*62*
4 Subarctic heath	32	42	26	3	0	
			26	*4·9*	*1·35*	*68*
Average of 27 plant communities	37·0	95·1	31·1	4·4	1	
			30·7	*4·9*	*1·05*	*96*

Data from Elton (1946). Numbers in col. 1 are those of communities given in his Tables 1 and 4.
In the figures for n_1, n_2 and n_3 the upper line is the observed number, the lower line (in italics) that calculated on the basis of the log series.

3. THE PROPERTIES OF THE LOGARITHMIC SERIES

The logarithmic series will be found fully discussed in Fisher, Corbet & Williams (1943), Williams (1944) and Williams (1947).

The series can be represented in two ways, either

$$n_1, \quad \frac{n_1 x}{2}, \quad \frac{n_1 x^2}{3}, - \frac{n_1 x^3}{4},$$

or

$$\alpha x, \quad \alpha \frac{x^2}{2}, \quad \alpha \frac{x^3}{3}, \quad \alpha \frac{x^4}{4},$$

hence the whole series, and also the Index of Diversity.

If a random sample is taken from a population which is arranged in a logarithmic series, a new logarithmic series is found in the sample, with a new n_1 and a new x. But for all samples of whatever size from one population the ratio of n_1 to x, which we have called 'α', is constant. In other words, this is a property of the population and not of one particular sample. It is a measure of the extent to which the species are grouped into genera, or the genera

C. B. WILLIAMS   13

divided into species. It is high when there are a large number of genera in relation to the number of species, and low when there are a small number of genera relative to the number of species, and it is independent of the size of the sample. It is thus a measure of the generic diversification of the species population, and for this reason we have called it the 'Index of Diversity'. It is, indeed, an index of exactly the property which is at present at issue. If a small sample of a larger population has the same Index of Diversity as the larger population, then it has been selected without reference to the generic relationships of the species. If the smaller sample has a larger Index of Diversity, it is evidence that there has been selection against species in the same genus. If it has a smaller Index of Diversity, it is

error is particularly high when the average number of species per genus is very low, which is usually the case in small communities.

An example of the application of the logarithmic series is as follows: In Bentham & Hooker's *British Flora*, 1906 edition there are 1251 species of plants classified into 479 genera. These are arranged reasonably closely to a logarithmic series and give an Index of Diversity = 284 (Williams, 1944, p. 30).

If samples of different numbers of species are taken from the above flora, *without any reference to generic relationships*, they must each have the same index of diversity; and from this and the number of species we can calculate that the expected number of genera would be, as shown in Table 2.

Table 2. *Calculated properties of samples of different numbers of species taken from Bentham & Hooker's 'British Flora' (1906 ed.), without any bias with respect to generic relations. All samples have an Index of Diversity of 284.*

No. of species	Expected no. of genera	Average no. of species per genus	Expected no. of genera with 1 species	% of genera with 1 species
N	S	N/S	n_1	$100\, n_1/S$
		Original population		
1251	479	2·61	231	48
		Samples		
1000	428	2·34	222	52
500	288	1·74	181	63
200	151	1·32	117	78
100	86	1·16	74	86
50	46	1·09	42·5	92
30	28·5	1·05	—	—
20	19·3	1·04	18·7	95
10	9·8	1·02	—	—

evidence that there has been selection in favour of species in the same genus.

The average number of species per genus (N/S), the proportion of genera with one species (n_1/S), and the proportion of species in genera with one species (n_1/N) are all dependent on the size of the sample (see Table 2); but the Index of Diversity is the same for all samples from the same population provided that they have been randomized for generic relationships.

It is, however, important to note for the present study that the error of estimation of the Index of Diversity increases rapidly as the sample gets smaller (see Fisher *et al.* 1943, pp. 53 and 56). Thus for populations with 1000 species in 100 genera the error (standard deviation) of α is about 6 %; but with 100 species in 70 genera it is 20 %, and with 10 species in 9 genera it is just under 100 %. The

It will be seen that the average number of species per genus steadily falls, and the percentage of genera with one species each steadily rises as the sample gets smaller. In a random sample of only 20 species of British flowering plants less than one genus with more than one species would be expected to be present.

It should be noted that if our biologically selected community contains as few as 30 species, the only possible high values for n_1 (since fractions do not exist in nature), are 30; 29; 28; 27; 26; 25. These give respectively Indexes equal to infinity; 420; 270; 195; and 150, etc. so that no community as small as this will give statistical data sufficient to distinguish between small changes in α, or small biases in favour of, or against, species in the same genus.

The data brought forward by Elton are not suitable for further inquiry along these lines as firstly

14   *Generic relations of species in small communities*

the numbers of species are in general too small, and secondly we have no available lists of the number of species and genera (and that is to say of the Index of Diversity) in the larger groups, floras or faunas, from which these communities have been selected by nature. What we require is a series of natural groups of different sizes selected by nature from one much larger group; all the species and genera in each series being of course in exactly the same classification.

For example, if we could take the butterflies of the world, the butterflies of England, the butterflies of an English county, the butterflies of a small area in this county, and the butterflies of a single ecological community in the small area (all in the same classification), then we would have a series of observed facts which could be analysed to see if the Index of Diversity in each sample indicated greater or less generic diversification as the samples got smaller.

It has not been possible to get such a perfect series of data, but in what follows I give the evidence I have been able to find that appears to be suitable for study.

The two Broadbalk Communities thus have—as would be expected from the small size of the sample —a lower average number of species per genus; but the Index of Diversity is in each case lower than the whole British flora, and the smaller sample has a lower value than the larger. Thus the evidence is that there is less generic diversification in the Broadbalk Wilderness flora, or a smaller number of genera than would be expected by a random sample including the same number of species.

In fact a random sample of 73 species from the British flora, as classified by Bentham & Hooker, would have about 65 genera instead of the 59 observed.

(b) Flowering plants of Scolthead Island, Norfolk

Chapman (1934) gives a list of the flowering plants of Scolthead Island, an island of sand-dunes which covers an area of about one and a half square miles.

His list, when altered to the nomenclature of Bentham & Hooker's *British Flora* (1906 ed.), gives the following results (Table 3).

Table 3

		Species per genus N/S	No. of genera with					
Species N	Genera S		1 species n_1	2 species n_2	3 species n_3	4 species n_4	5 species n_5	Index of Diversity
161	114	1·41	Obs. 82	22	7	1	2	
			Cal. 85·3	20·1	6·3	2·26	0·8	182

4. COMPARISONS OF SMALL WITH LARGE SERIES

(a) Plants on Broadbalk Wilderness, Rothamsted Experimental Station, Hertfordshire

Brenchley and Adam (1915) have published a list of the species of flowering plants that have been found on this piece of abandoned wheat field which covers an area of about half an acre. In their table on p. 198 (emended to agree with the nomenclature of Bentham & Hooker's *British Flora*, 1906 ed.), they mention 73 species which were found in four surveys carried out between 1867 and 1913. These are classified into 59 genera. Of these, 50 genera have one species; six have two species; one has three and one has four. This gives an average number of species per genus of 1·24; and an Index of Diversity of 147.

If only the plants observed in 1913 are considered, the numbers are 65 species in 52 genera (41 with one species, ten with two, one with three). The average number of species per genus is 1·25 and the Index of Diversity 134.

I have already published (Williams, 1944, p. 30) data on the whole classification of Bentham & Hooker's 1906 edition which gives an average number of species per genus of 2·61 and an Index of Diversity = 284.

It will be seen that the observed numbers conform very closely to a logarithmic series as checked by the calculated values of n_1, n_2, and n_3; and that the Index of Diversity is 182, as compared with 284 for the whole British flora.

In a purely random sample of 161 species from the British flora, irrespective of generic relations, we would have expected to find representatives of approximately 126 genera. Thus the observed number is smaller than the calculated, indicating a selection in favour of generically related species rather than against them.

(c) Flora of Park Grass Plots, Rothamstead Experimental Station, Hertfordshire

At Rothamsted there are a series of plots of grass, of areas varying from $\frac{1}{2}$ to $\frac{1}{8}$ acre, which have been manured in special ways for many years, but in which no further interference has been made in the natural flora which develops under such conditions of manuring and soil.

Brenchley (1924) has given a series of tables from which Table 4 is extracted; the classification has been altered slightly to agree with Bentham & Hooker's *British Flora* (1906 ed.).

C. B. WILLIAMS   15

Table 4

	No. of species N	No. of genera S	Average species per genus N/S	Index of Diversity
All plots all years	59	53	1·11	263
1919 Survey only				
Plot 3. Unmanured:				
Without lime	30	27	1·11	134
With lime	30	27	1·11	134
Plot 13. Farmyard manure:				
Without lime	20	18	1·11	89
With lime	23	21	1·10	111

The original Index of Diversity for the whole of Bentham & Hooker was 284 so that all the smaller floras have a smaller Index of Diversity. The error of estimation of α is however very high.

Plots 9 and 10, with sulphate of ammonia for many years, gave (with and without potash) from 9 to 14 species each in its own genus. It is not possible to calculate a finite value of α from such data. It is, however, interesting to note that in a random sample of ten species from Bentham & Hooker's *British Flora* (α = 284) one would expect 9·8 genera! So the observed figure of 10 cannot be taken as evidence of any extreme departure from randomization.

(d) Lepidoptera of Wicken Fen, Cambridgeshire

Farren (1936) gives a list of the Macrolepidoptera of Wicken Fen according to the classification of Meyrick's *Handbook of the British Lepidoptera* (1895). In the whole group 368 species are listed in 135 genera. This gives an average of 2·73 species per genus and an Index of Diversity of 76·4. The number of species in the same families for the whole of Great Britain, as listed in Meyrick's handbook, is 788 in 212 genera, which gives an average of 3·72 species per genus and an Index of Diversity of 96·5.

For the family Caradrinidae (Agrotidae) alone, there are 146 species in 29 genera for Wicken Fen (average per genus 5·04; Index of Diversity 10·8); and 273 species in 39 genera for the whole of Great Britain (average per genus 7·00; Index of Diversity 12·7). In the family Plusiidae there are 24 species in 12 genera for Wicken Fen (average 2·00; α = 9·3) and 54 species in 22 genera for the British Isles (average 2·45, α = 14·0).

So we see that for the whole Macrolepidoptera and also for two separate families, one large and one small, the Index of Diversity of the local fauna is smaller than that of the larger area. In other words, there are fewer genera in the local fauna than would be expected by a random selection, thus giving evidence of a selection in favour of generically related species.

(e) Coleoptera of Windsor Forest, Berkshire

Donisthorpe (1939) gives a list of 1825 species of Coleoptera found in Windsor Forest. They are grouped into 553 genera, according to the classification of Beare & Donisthorpe (1904). The list gives a good approximation to a logarithmic series. There is an average of 3·30 species per genus and an Index of Diversity of 278. The original classification of the Coleoptera of the British Isles gave 3268 species in 804 genera, with an average of 4·60 species per genus and an Index of Diversity of 341 (see Williams, 1944, p. 24). Thus the Coleoptera of Windsor Forest have, as expected from the smaller number of species, a lower average number of species per genus; but have also a smaller Index of Diversity, indicating a selection in favour of species of the same genus rather than against.

(f) Capsidae (= Miridae) (Heteroptera) of Hertfordshire

China (1943) gives a list of the Heteroptera of the British Isles. The family Capsidae (= Miridae) in this includes 186 species in 76 genera as shown in Table 5. The Index of Diversity is 48.

Table 5

No. of species N	No. of genera S	Average sp. per gen. N/S		No. of genera with				Index of Diversity
				1 species n_1	2 species n_2	3 species n_3	4 species n_4	5 species n_5	
			British						
186	76	2·45	Obs.	40	16	11	3	1	
			Cal.	38·1	15·2	8	4·8	—	48
			Hertfordshire						
127	60	2·12	Obs.	39	12	3	1	0	
			Cal.	32·5	12·1	6	—	—	43·7

16 *Generic relations of species in small communities*

Bedwell (1945) has given a list of the Hemiptera of Hertfordshire; the particulars of the Capsidae from this are given in the second half of Table 5. It will be seen that the Index of Diversity is slightly below that of the British fauna, but probably not outside the limits of error.

It shows, however, no evidence of selection against species of the same genus.

180 species in 33 genera in Hertfordshire (average 5·15, $\alpha = 13\cdot2$), and 290 species in 45 genera for the British Isles (average 6·44; $\alpha = 14\cdot9$). Thus the differences indicated that the smaller faunas have an Index of Diversity just equal to or smaller than the larger areas.

It is interesting to note that Meyrick wrote his first *Handbook* in 1895, and his 'Revised' edition in

Table 6

Plants	No. of species observed	No. of genera		Av. no. of species per genus		Index of Diversity	
		obs.	calc.	obs.	calc.	orig.	sample
Broadbalk: 4 surveys	73	59	65·0	1·24	1·12	284	147
1913 survey	65	52	58·5	1·25	1·11	284	1·34
Scolthead Island	161	114	126	141	1·28	284	182
Park Grass, all plots	59	53	55·9	1·11	1·055	284	263
Plot 3 (1919)	30	27	28·5	1·11	1·05	284	134
Plot 13 (1919)	20	18	19·3	1·11	1·04	284	89
Plot 13 (1919)	23	21	22·2	1·10	1·04	284	111
Insects							
Wicken Fen:							
Macrolepidoptera	368	135	151·6	2·73	2·43	96·5	76·4
Caradrinidae	146	29	32·1	5·04	4·55	12·7	10·8
Plusiidae	24	12	14·0	2·00	1·71	14·0	9·3
Windsor Forest:							
Coleoptera	1825	553	630·0	3·30	2·90	341·0	278
Hertfordshire:							
Miridae	127	60	62·1	2·12	2·05	48	43·7
Macrolepidoptera	561	186	188·4	3·02	2·98	99·6	99
Caradrinidae	180	33	38·31	5·15	4·70	14·9	13·2

Table 7

No. of species in sample	No. of genera in sample			Av. no. of genera in sample	Observed no. in natural samples of same
	1	2	3		
20	20	19	20	19·7	18
23	23	22	22	22·3	21
30	29	28	29	28·7	27
59	56	50	54	53·3	53
65	60	54	59	59·7	52
73	67	60	66	64·3	59
161	128	117	126	123·7	114

(g) *Lepidoptera of Hertfordshire*

Foster (1937) gives a list of the Lepidoptera recorded for Hertfordshire. The classification used is that of Meyrick's, *Revised handbook of British Lepidoptera* (1927). He enumerates 561 species in 186 genera. This gives an average of 3·02 species per genus and an Index of Diversity of about 99. In Meyrick's *Revised handbook* the figures for the same families for the whole British Isles are 806 species in 218 genera. This is an average of 3·7 species per genus and an Index of Diversity of 99·6. In the Family Caradrinidae (Agrotidae) alone, there are

1927. The first gives an Index of Diversity of 96·5, the second of 99·6; which shows how little his ideas on the scope of genera had changed in the 32 years between the two editions.

5. DISCUSSION

In Table 6 is a summary of the evidence brought forward in seven plant communities and seven animal communities. None of them is very small, as we have already explained that in the very small communities the chance of getting two species in one genus is too low to be the basis of discussion.

C. B. WILLIAMS   17

The table shows the number of genera observed in each natural sample and also the number of genera calculated on the assumption that it is a random sample from a larger flora or fauna which is arranged in a logarithmic series. It will be seen that in practically every case the observed number of genera is smaller than the calculated. It follows that the observed average number of species per genus is larger than that calculated on the assumption of a random selection from a logarithmic series, although much smaller than that in the larger area used for comparison.

These figures are calculated on the acceptance of the logarithmic series. If this is not accepted, it is still possible to make a practical demonstration that the same difference still exists between observed figures and a random sample. Bentham & Hooker's *British Flora* contains 1251 species. A series of numbers up to 1251 was typed on small cards and extremely well mixed. Three complete sets of random samples were then drawn from the complete set, each set being returned before the next set was selected. At intervals during the draw the number of genera represented in the species already selected was checked up. The results were as shown in Table 7. It will be seen again that in every case except one the observed natural number of genera is smaller than any of the three sets of mechanically randomized samples taken from the original flora. The amount of data at present available is insufficient for an overwhelming proof that the number of genera is really smaller, but at least it contains no evidence whatever that the number of genera is larger in a natural sample than in one selected absolutely independent of generic relationship, which would be expected if competition between species of the same genus was a major factor in determining survival.

From all this evidence it will be seen that a statistical treatment of the number of genera and species of different groups of animals or plants, in small and in large communities, indicates that the 'diversification' of species into genera is smaller in the small samples than in the larger. In other words, the smaller samples have fewer genera than would be expected in a random sample of the same number of species taken from the larger fauna or flora. This can only be interpreted as a natural selection—in the course of time—in favour of species in the same genus rather than against them.

It is possible to suggest reasons for this—for example, if one species in a genus is capable of survival in a given physical environment, it seems likely that other species in the same genus might be more likely to have a similar genetic make-up than species in another genus, and so might also have a good chance of survival.

If two or more species in the same genus are each capable of surviving in a particular ecological niche, the problem of their joint survival seems to be a question of the balance between the advantages of suitability to the physical environment and the disadvantages of the increased number of competitors with very similar habits—not only individuals of the same species, but other species very closely related.

There are undoubtedly increasing difficulties to a species when the numbers increase beyond a certain level and in some cases the individuals of two closely related species might act almost as individuals of the same species, and so bring all the difficulties of increased numbers (e.g. competition for available food) without any compensating advantages.

The evidence at our disposal, however, seems to indicate that on an average, the advantages of increased suitability to the environment outweighs the disadvantages of any possible intensification of competition due to close relationship. Such increased competition possibly, or even probably, exists, but its effect cannot be estimated by study of the present type, as it is undoubtedly overshadowed by other factors acting in a reverse direction.

6. SUMMARY

1. Evidence of conditions being more favourable or less favourable to species of the same genus, as compared with species on different genera (intra-generic versus inter-generic competition), can be found in the relative number of species and genera in small and in large natural communities of animals or plants.

2. It is, however, insufficient to show that the average number of species per genus is smaller in the smaller communities than in the larger as this is a mathematical result of taking a smaller sample from a larger group. It is necessary to show that the proportion of genera to species in the smaller communities is *smaller* or *larger* than would have been expected in a randomized sample of the same number of species, selected without reference to generic relationships from the larger fauna or flora.

3. Evidence had previously been brought forward to show that in large groups of animals or plants the number of genera with 1, 2, and with 3 and more species are closely represented by the mathematical 'logarithmic series'. New evidence is here given to show that the same order exists in the genera and species of quite small ecological communities.

4. The logarithmic series has several mathematical properties of biological interest, including the possibility of calculating, for any population or sample, a factor known as the 'Index of Diversity' which is common to all random samples from a single population. It is a measure of the extent to which the species are grouped into genera, and it is inde-

18 *Generic relations of species in small communities*

pendent of the size of the sample. If the Index of Diversity is high, there are many genera in relation to the number of species; if the Index is low, there are fewer genera in relation to the number of species.

5. It thus becomes possible to compare the Index of Diversity in natural small or simple ecological communities, with that of the larger population from which these have been selected by nature. This has been done for a number of cases, including both animal and plant communities, and in every case from which significant results can be obtained the

Index of Diversity in the small community is smaller than that of the larger fauna or flora.

6. The result therefore indicates that there are fewer genera in a small or simple community than would be expected in a sample of the same number of species selected at random—that is, independent of genera relationships—from the larger series. In other words, the evidence brought forward by this method indicates a selection by nature in favour of more than one species in the same genus rather than in favour of single species in different genera.

REFERENCES

Beare, T. H. & Donisthorpe, H. St J. K. (1904). 'Catalogue of British Coleoptera.' London.

Bedwell, E. C. (1945). 'The county distribution of the British Hemiptera-Heteroptera.' Ent. Mon. Mag. 81: 253–73.

Brenchley, W. E. (1924). 'Manuring of grass land for hay.' London.

Brenchley, W. E. & Adam, H. (1915). 'Recolonisation of cultivated land allowed to revert to natural conditions.' J. Ecol. 3: 193–210.

Chapman, V. J. (1934). 'Appendix II. Floral list.' In J. A. Steers 'Scolthead Island....' Cambridge. Pp. 229–34.

China, W. E. (1943). 'The generic names of the British Hemiptera-Heteroptera, with a check list of the British species.' In 'The generic names of British Insects,' Part 8: 217–316. London: R. Ent. Soc.

Donisthorpe, H. St J. K. (1939). 'A preliminary list of the Coleoptera of Windsor Forest.' London.

Elton, C. (1946). 'Competition and the structure of ecological communities.' J. Anim. Ecol. 15: 54–68.

Farren, W. (1936). 'A list of Lepidoptera of Wicken and the neighbouring Fens.' In J. S. Gardener, 'The natural history of Wicken Fen', Part III: 258–66. Cambridge.

Fisher, R. A., Corbet, A. S. & Williams, C. B. (1943). 'The relation between the number of species and the number of individuals in a random sample of an animal population.' J. Anim. Ecol. 12: 42–58.

Foster, A. H. (1937). 'A list of the Lepidoptera of Hertfordshire.' Trans. Herts. Nat. Hist. Soc. 20: 171–279.

Williams, C. B. (1944). 'Some applications of the logarithmic series and the Index of Diversity to ecological problems.' J. Ecol. 32: 1–44.

Williams, C. B. (1947). 'The logarithmic series and its application to biological problems.' J. Ecol. 34: 253–72.

Biol. Rev. (1967), **4** *pp. 207–264*

207

GRADIENT ANALYSIS OF VEGETATION*

By

R. H. WHITTAKER

*Department of Population and Environmental Biology,
University of California, Irvine*

(*Received 1 May 1966*)

CONTENTS

I. Introduction		207	(2) Matrices and quantitative classification		233
(1) The concept of gradient analysis		207	(3) Plexuses		239
(2) Brief history		209	(4) Early Wisconsin gradient analysis		241
(3) Preliminary definitions		211	(5) Wisconsin comparative ordination		243
II. Direct gradient analysis		212	(6) Factor analysis		248
(1) A transect along a single gradient		212	IV. Conclusion and Summary		254
(2) Ordination		214	(1) Choice of techniques		254
(3) Pattern analysis		220	(2) Concepts		254
(4) Hyperspaces and evolution		225	(3) Perspective		255
III. Indirect gradient analysis		231	V. References		256
(1) Similarity measurements		231			

I. INTRODUCTION

(1) *The concept of gradient analysis*

Gradient analysis is a research approach for study of spatial patterns of vegetation. It seeks to understand the structure and variation of the vegetation of a landscape in terms of gradients in space of variables on three levels—environmental factors, species populations and characteristics of communities. This article reviews gradient analysis both as a group of techniques for analyzing and describing vegetation and as a source of new theoretical understanding of natural communities. It may be fair to say that gradient analysis has changed the conception of vegetation as much as research on the genetic basis of variation and evolution has changed the concepts of plant species. In both cases the change involved shift of emphasis from classification of the objects of study to analysis of kinds and degrees of relationship among these objects.

Gradient analysis and classification are alternative approaches to the vegetation of a landscape. If, for example, one stands on a viewpoint in the Southern Appalachian Mountains in the autumn, one sees a complex and varicoloured mantle of vegetation covering the mountain topography. Different kinds of plant communities are marked

* A contribution from the ecology programme Biology Department, Brookhaven National Laboratory, Upton, Long Island, N.Y., under the auspices of the U.S. Atomic Energy Commission.

out in the pattern by differences in the forms and colours of their trees. There is a marked relationship between kinds of vegetation and kinds of topographic position in the landscape. In the valley at one's feet one may observe these plant communities: (a) a mixed and multicoloured broad-leaved 'cove forest' in the valley bottom, (b) a dark-evergreen hemlock forest on the moist, lower north-facing slope just above the cove forest, (c) oak forests marked by red and yellow foliage, on many open slopes above the hemlock forest, (d) more open oak heaths in which a layer of evergreen shrubs may be seen through the separated crowns of oak trees and which occupy most of the upper slopes of the valley, and (e) pine forest on the dry, upper south-facing slope, with pines forming an open canopy above an open shrub layer. Much the same sequence may be observed in other valleys of these mountains.

The student of vegetation seeks to construct systems of abstraction by which relationships in this mantle of vegetation may be comprehended. The traditional approach is through classification of plant communities into community types. The five kinds of vegetation given above constitute a system of community types by which much of the vegetation of a given elevation belt in the mountains may be classified. As regards the three levels of study, each community type may be characterized by: (a) environment—kinds of topographic positions or ranges of magnitudes of environmental factors in which communities of the type occur, (b) species populations—which species are usually present, in what numbers of individuals, in communities of the type, and (c) over-all community characteristics (such as structure of the vegetation, total numbers of species present, total mass of organic material and rate of production) which are shared, within some ranges of values, by communities of the type.

In gradient analysis as an alternative, the five vegetation types are treated as parts of a single continuum from cove forest to pine forest. Relations of the three levels of study now appear in a different light. (a) There exists an environmental gradient, along which many characteristics of soils and climates change, from moist valley to dry open slope. (b) Species populations are distributed, each according to its own physiological responses, along this gradient. (c) The different combinations of species along the gradient are recognized as community types by ecologists, and these community types are related to one another along gradients of community characteristics. The community types are the 'colours' which man recognizes in the vegetational spectrum (Brown & Curtis, 1952); but the spectrum may also be studied in terms of parallel (or otherwise related) gradients of environmental factors, species populations and community characteristics.

This article will develop in greater detail the meaning of gradient analysis in terms of techniques, concepts and theory. As is often the case, the method, as a broadly conceived research approach, has evolved by a complex interplay of suggestion and substantiation among techniques, concepts and theory. These need consequently to be treated in parallel, rather than as if one had simply resulted from the other. Development of gradient analysis will be considered through a series of stages or phases. In each phase a more detailed 'discussion' of techniques and results is followed by a 'conclusion' stating interpretation in terms of concepts and theoretic implications. It is hoped thus to serve the interests both of readers interested in details of method for possible research application and of those interested primarily in surveying the meaning of gradient analysis by way of the conclusions.

The phases will be arranged in two parallel sequences; for two, complementary, approaches must be distinguished within gradient analysis even though there has been extensive exchange of ideas between them and their results are convergent. In the first of these approaches vegetation samples are arranged and studied according to known magnitudes of (or indexes of position along) an environmental gradient which is accepted as a basis of the study. This approach, to which the term gradient analysis was originally applied (Whittaker, 1951), may be termed *direct gradient analysis*. In the other approach vegetation samples are compared with one another in terms of degrees of difference in species composition and on the basis of these degrees of difference are arranged along axes of variation. The axes may or may not correspond to environmental gradients; but if they do correspond, the approach to environmental gradients is indirect or inferential. The approach may consequently be termed *indirect gradient analysis*. My own work has been primarily in direct gradient analysis for the purpose of inquiry into the theory of vegetation structure and classification; extensive development of techniques of indirect gradient analysis has occurred in the School of Wisconsin of Curtis and his associates. The direct approach will be discussed first and the indirect approach second, but the reader should recognize that these two streams run parallel in time.

(2) *Brief history*

In earlier vegetation studies it was generally taken for granted that vegetation 'consisted' of the community types into which it was classified. These community types were assumed to be well-defined natural units which were part of the structure of vegetation (and not simply part of the structure of a classification) and which generally contacted one another along narrow boundaries called 'ecotones'. It was thought that research methods had, of necessity, to be based on these units of which vegetation consisted. So fully accepted was this theory of the structure of vegetation that it had no name as a theory; recently (Whittaker, 1956, 1962) it has been designated the 'community-unit theory'.

In American ecology the dissent from this theory was first effectively stated by Gleason (1926) in a paper on 'The individualistic concept of the plant association.' Gleason advanced two central ideas which may be restated as follows. (1) The principle of species individuality—each species is distributed in relation to the total range of environmental factors (including effects of other species) it encounters according to its own genetic structure, physiological characteristics and population dynamics. No two species are alike in these characteristics, consequently, with few exceptions, no two species have the same distributions. (2) The principle of community continuity—communities which occur along continuous environmental gradients usually intergrade continuously, with gradual changes in population levels of species along the gradient. Gleason's ideas met with intense opposition (Nichols, 1929; Clements, Weaver & Hanson, 1929). Gleason restated his views in another paper (1939), and later, when the course of time had begun to make the field ready for them, his ideas were supported by Cain (1947) and Mason (1947). During the two decades

R. H. WHITTAKER

between 1926 and 1947 these ideas were latent, uninvestigated and largely forgotten.

In the summer of 1947 a study was carried out in the Great Smoky Mountains of Tennessee which was designed to test the community-unit theory and individualistic hypothesis (Whittaker, 1948, 1951, 1956). Results supported Gleason's ideas. Vegetation was conceived as primarily a complex continuum of populations, rather than a mosaic of discontinuous units. The method of research which dealt with vegetation in terms of continuity and gradient relationships was termed 'gradient analysis.' These results were set forth in a thesis (Whittaker, 1948, see 1956), the logic and evidence of the test of the community-unit theory were developed in a paper (1951, see also 1956), and the findings extended to animal communities (1952).

During the same period research on the forests of Wisconsin was independently carried out by J. T. Curtis and his associates. Results, which strikingly paralleled my own in their demonstration of species individuality and community continuity, were also published in 1951 and 1952 (Curtis & McIntosh, 1951; Brown & Curtis, 1952). Work by Ellenberg (1948, 1950, 1952) in Germany during the same period used related methods and obtained similar results, though these were not applied to testing the individualistic hypothesis. Work on different bases led Matuszkiewicz (1947, 1948), Motyka (1947), Major (1951), Walter & Walter (1953), Goodall (1953a, 1954a, b) and Beard (1955) to approaches emphasizing continuity. As is so often the case, discoveries for which the time was ripe were made by a number of scientists working independently. As is so often the case too, others had anticipated the discoveries to a greater extent than was realized at the time.

The Americans had not known, until it appeared from a search of the European literature (Whittaker, 1953), that the statement and rejection of Gleason's ideas in the United States had been paralleled in France in statements of Lenoble (1927, 1928) and their rejection by phytosociologists (Allorge, 1927; Braun-Blanquet, 1928; Pavillard, 1928), and that much the same ideas had been expressed in Russia by Ramensky (1924). Ramensky's formulation of the two principles stated above and their implications for research was clearer than Gleason's, and his conception of the vegetational mantle was much like that developed by myself in the Great Smoky Mountains. Unlike Gleason and Lenoble, Ramensky (1930) went beyond argument to extensive research on species distribution and community relationships. It is Ramensky, rather than Gleason, Lenoble, Ellenberg, or the recent Americans, who should be recognized as the originator of gradient analysis.

History of schools of ecology and expressions of the community-unit theory have been reviewed in more detail elsewhere (Whittaker, 1962). Despite the prevalence of the community-unit theory, there were a number of other antecedents of gradient analysis; among them should be mentioned the notable studies in Iceland and Denmark by Hansen (1930, 1932), the work on ecological series in Russia by Keller (1925–26), Sukatschew (1928, 1932) and others, and studies of forest site-type series by Cajander and Ilvessalo (1921) and others.

Gradient analysis of vegetation   211

(3) *Preliminary definitions*

A number of terms may best be defined in advance.

A particular, limited area of vegetation which seems homogeneous—the area is limited so that there is no marked, progressive change within it toward a different kind of vegetation—is a *plant community*. The sum of the environmental factors—or, better, the pattern or gestalt of those interrelated factors—which affect the plants in that community constitute its *environmental complex* (Billings, 1952). Both plant community and environmental complex are part of a broader system comprising a natural community (of plants, animals and saprobes) and its environment. This open system of community-and-environment is termed an *ecosystem* (Tansley, 1935; Evans, 1956; Odum, 1959).

When information about a plant community (species present, kind of soil, etc.) is obtained, that information constitutes a vegetation *sample*. Usually a sample is taken from an area of specified size and shape; such a sample area is a *quadrat*. A collection of samples which represent the vegetation of a landscape or part of a landscape, and which are used for gradient analysis or classification, will be termed a *set*. Subsets, or smaller numbers of samples from the set which are grouped together for any purpose of gradient analysis or classification, will be termed *groups*. If an ecologist chooses to formulate a class concept defining a group, the group also represents a (lower level) *community type*. When samples of a group have their characteristics averaged, the resulting average or synthetic vegetation sample is a *composite sample*. Composite samples may serve to represent either community types or positions along a gradient in gradient analysis.

A vegetation sample consists of, at least, information on environment and a list of species present, usually with indications of their relative importance. *Occurrence* refers to the mere presence or absence of a species in the sample area. An *importance value* is some expression of the massiveness, conspicuousness, activity or interest of a given species in the community. A variety of importance values have been used for different kinds of organisms and research purposes, among them, for terrestrial communities (cf. Curtis & McIntosh, 1950), the following. *Density* is the number of individuals of a species per unit ground surface area (or other spatial measurement). *Coverage* is the percentage of ground-surface area in a community above which foliage of a given species occurs. *Basal area* is the area occupied by cross-sections of tree stems at 1·4 m. above the ground surface (or other basal measurement) per unit ground surface area. *Frequency* is the percentage of small subquadrats within a larger sample quadrat in which a given species has been observed. *Constancy* is the percentage of the larger sample quadrats of a group or set in which a given species has been observed. *Presence* is the percentage of samples in which a species occurs when the samples are not quadrats of the same size. Species *biomass* is the total mass (usually, dry weight of organic matter) of a species present at a given time per unit ground surface area. Species *production* is an expression of the amount of organic matter (dry weight) produced, or energy bound, per unit ground surface per unit time.

R. H. WHITTAKER

If an importance value for a given species is divided by the total of the same kind of importance values for all species in the sample, the resulting percentage is a *relative importance value*. If two or more kinds of importance values are added together (often in the form of relative importance values) the result is a *synthetic importance value*. The most important one, two or few species in a community or sample are termed its *dominants*. Their *dominance* implies that they have the highest value for some importance value in the community or that they are judged most important in community structure and function. *Character species* are species (which may or may not be important) whose distributions are centred in or largely restricted to a given community type; their relative restriction to it is termed *fidelity*.

II. DIRECT GRADIENT ANALYSIS

(1) *A transect along a single gradient*

(a) Discussion

In a first and simplest application of gradient analysis, vegetation samples are taken at equal intervals along an environmental gradient—for example, 50 m. elevation intervals up a long, even mountain slope. The samples will usually include measurements of plant populations in quadrats; I use samples combining counts of trees and shrubs in 0·1 hectare (20 × 50 m.) quadrats with counts of herbs in twenty-five 1 m. square subquadrats within the 0·1 ha quadrat, and coverage measurements. A field transect—that is, a series of such samples taken as one moves along a gradient in the field—can effectively show how densities of some major plant populations change along the gradient. Clearer results are possible when several samples represent each interval, and their population data are combined so that much of the sample-to-sample irregularity is averaged out. Five samples may, for example, be combined into a composite sample to represent each 100 m. interval along an elevation gradient. Most of my work in gradient analysis has been based on such composite transects. Field transects with one sample per interval were taken along specific elevation and topographic gradients to control the possibility that averaging groups of samples might blur population discontinuities, or otherwise alter apparent population relations from those existing in the field.

Results from a composite transect of the elevation gradient on southwest-facing slopes bearing pine forests in the Great Smoky Mountains are shown in Fig. 1 (Whittaker, 1948, 1956). Species populations do not have sharp boundaries along the gradient at points which might correspond to sharply defined physiological limits of tolerance. Each species has a central mode or peak of maximum density, and decreases in density gradually with increasing departure from this peak in both directions; the curves appear to be binomial or Gaussian in form (Gause, 1930; Whittaker, 1951, 1952; Brown & Curtis, 1952). As asserted by Ramensky (1924) and Gleason (1926), no two species have closely parallel distributions. No boundaries separate the three community types an ecologist is likely to distinguish along this gradient—*Pinus virginiana* forest at low, *Pinus rigida* heath at middle, and *Pinus pungens* heath at high elevations.

Gradient analysis of vegetation   213

The population curves and community gradients are responses to the 'elevation gradient', but elevation is a variable without relevance to the physiology of plants. Along the elevation gradient many factors of environmental complexes—temperature and growing season, precipitation and humidity, wind velocity, atmospheric pressure, evaporation—change concomitantly. The elevation gradient is a complex climatic gradient for which elevation itself is merely one useful index of relative position. Neither elevation nor any other particular gradient can be accepted, without experiment, as 'cause' of a population distribution. Rather, it is accepted that a complex-gradient of many characteristics of environment exists and population distributions are observed in relation to one another along this complex-gradient.

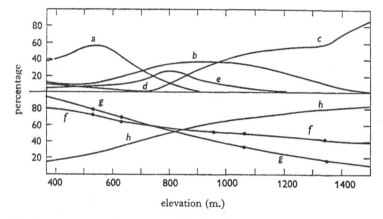

Fig. 1. Results from a composite transect of the elevation gradient, Great Smoky Mountains, Tennessee. Samples from dry, south-facing slopes bearing pine forests and pine heaths are grouped by 100 m. intervals. Above, species populations in percentages of stems over 1 cm. d.b.h. (diameter at breast height, 1·4 m. above the ground) in the sample (Whittaker, 1956): *a, Pinus virginiana; b, P. rigida; c, P. pungens; d, Quercus marilandica; e, Q. coccinea.* Below, trends in community characteristics: *f,* species diversity of all vascular plant strata in sample quadrats (in percentage of a maximum of forty-four for a cove forest sample, data of Whittaker, 1965); *g,* tree-stratum above-ground net annual production (in percentage of a maximum of 1200 g./m²./year for a cove forest sample, data of Whittaker, 1966); *h,* coverage of the predominantly ericaceous shrub stratum (percentages of ground surface above which foliage occurs, data of Whittaker, 1956). Transect data have been smoothed for all curves except *f* and *g.*

Many of the gradient relations observed must remain observations only—they are essentially correlations for which the context of interlinkage and causation is inadequately known. Most of them are neither simply coincidental nor causal; they are expressions, in two particular measured variables, of the many reciprocal influences and cyclic processes relating communities and environments in ecosystems. In some cases, however, when the correlation is strong and when there is reason to regard the magnitude of one gradient as functionally consequent on the magnitude of another, the two gradients may be regarded as 'cause' and 'effect' in a sense appropriate to this context (Whittaker, 1962). It is reasonable, for example, to regard the gradient of decreasing net production of Fig. 1 as an effect of the gradient of decreasing temperatures and lengths of growing season with increasing elevation.

214   R. H. Whittaker

(b) *Conclusion*

Results of broad significance emerge from so simple a technique as following changes in species populations and community characteristics along an environmental gradient:

(1) Populations of species along continuous environmental gradients typically form bell-shaped, binomial curves, with densities declining gradually to scarcity and absence on each side of a central peak (Fig. 1).

(2) Species are not organized into groups with parallel distributions along the gradient. Each species is distributed in its own way, according to its own population response to environmental factors that affect it (including effects of other organisms), as stated in Ramensky and Gleason's principle of species individuality.

(3) Because of the tapered form of population distributions of species, composition of communities changes continuously along environmental gradients (if the gradients are uninterrupted and the communities undisturbed). Community types are class concepts abstracted from the continuum of community variation along environmental gradients.

(4) Environments and communities together constitute ecosystems. Since a transect relates a community gradient to an environmental gradient, a transect is an approach to studying a gradient of ecosystems or, to use a term of Clements' (1936), an *ecocline*. The environmental aspect of the ecocline, a gradient of environmental complexes in which innumerable factors of environment vary together through space, may be termed a *complex-gradient* (Whittaker, 1956) in distinction from a factor-gradient for one measurable characteristic of environment.

(5) The corresponding community gradient may be termed a *coenocline* (Whittaker, 1960). Within the coenocline or whole-community gradient particular characteristics of communities change, forming gradients or *trends* of such community characteristics as coverage, productivity and species diversity, in which the adaptations of communities to changing conditions along the environmental gradient are expressed (Fig. 1).

(6) Study of transects thus permits the relating to one another of gradients on the three levels of study—environment (factor-gradients), species populations (binomial curves and their modifications), and communities (gradients of community composition and trends of community characteristics).

(a) *Discussion*   (2) *Ordination*

In a second phase of gradient analysis an environmental gradient may be accepted as given, but there is no satisfactory environmental index (such as elevation) by which samples may be arranged in sequence. In such cases the samples can usually be arranged by their own characteristics. The process of arranging samples (or species) in relation to one or more gradients or axes of variation is *ordination* (Goodall, 1954*b*, as a translation of Ramensky's *Ordnung*, in German versions of his articles, 1924, 1930).

Together with elevation, the 'topographic moisture gradient' has a major effect on distribution of plants in mountains. This gradient in a given valley may extend from a moist ravine bottom outward (at essentially the same elevation) to a north-facing slope

Gradient analysis of vegetation   215

and around the ridge to northwest-, west-, and southwest-facing slopes. Great contrasts in moisture conditions characterize the extremes of the gradient, the moist ravine and dry southwest slope. Although it is not feasible in most field-work to measure moisture factors of both soil and atmosphere and integrate them into expressions of position along the gradient, there are ways in which vegetation samples can be arranged:

(*a*) Topographic position may be used as a crude index of moisture conditions. Samples are arranged in a composite transect of which the intervals may be deeper ravines, shallower canyons, sheltered lower slopes and open slopes of varying exposures from northeast and north through northwest, east, and west, to south and southwest. Techniques for more exacting treatment of topographic moisture gradients are discussed by Loucks (1962).

(*b*) Species distributions in the composite topographic transect may be used for a second arrangement of the samples. Species are grouped by relative positions along the gradient; for the topographic moisture gradient the groups may be: (o) *mesics* (species with their population modes at or near the moist extreme), (1) *submesics* (centred in less moist situations than the preceding, but in the moister half of the transect), (2) *subxerics* (centred in the drier half of the transect but not near the dry extreme), and (3) *xerics* (populations centred at or near the dry extreme). The numbers given are applied as weights to data on species composition of samples. Thus a ravine forest sample might include 100 trees of which 60 are of mesic species, 30 of submesic, 10 of subxeric, and none of xeric species. Multiplied by weights and divided by the unweighted total number—$(60 \times 0 + 30 \times 1 + 10 \times 2 + 0 \times 3)/100 = 0.50$—a weighted average of species composition results, as an index of position along the gradient.

Weighted average techniques were developed independently by authors including, at least, Ellenberg (1948, 1950, 1952), Whittaker (1948, 1951, 1956), Curtis & McIntosh (1951), and Rowe (1956). I use two independent weightings—in forests one with weights applied to densities of tree stems and another with weights applied to frequencies of herbs and shrubs (Whittaker, 1960). When each sample has been thus twice weighted, samples as points are plotted against the two scales as axes of a scatter-figure (Fig. 2). The samples are now grouped, using the two weightings as checks on each other, along the oblique axis of the scatter figure so that five samples represent each interval of a composite transect.

(*c*) As a third means of arranging samples, (Whittaker, 1960; cf. Bray & Curtis, 1957) they may be compared with composite samples representing the extremes of the gradient. Composition is averaged for a group of samples from most mesic (moist) ravines and for a group of samples from most xeric (dry) southwest slopes, to form two composite end-point samples. Each sample in the full set is compared with both these end points by percentage similarity or some other measurement expressing the degree to which a sample differs from the end-point samples. (Meaning of such measurements will be discussed later.) Samples may now be arranged in sequence from those strongly related to the samples from the mesic extreme and with little in common with the xeric extreme, through various intermediate compositions, to samples strongly related to the xeric extreme with little in common with the mesic

216   R. H. WHITTAKER

extreme. The technique may sometimes be aided by use of a composite sample for the mid-point of the gradient (open east-facing slopes, or samples with equal similarity to the mesic and xeric end-point samples) (Whittaker, 1960, fig. 3).

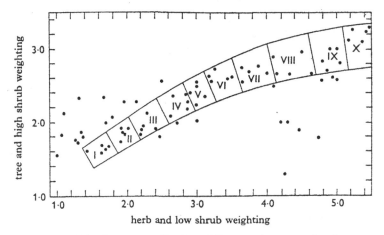

Fig. 2. A double weighted-average ordination of Sonoran desert samples from upper (low indexes) to lower (high indexes) valley plain or bajada below the Santa Catalina Mountains, Arizona (Whittaker & Niering, 1965). The scale on the ordinate applies weights to occurrences of tree and arborescent shrub species in o·1 ha quadrats, that on the abscissa to frequencies of herbs and shrubs in metre-square subquadrats. Points representing samples are grouped by fives into ten steps of a composite transect. Samples which are deviant in relation to this ecocline from upper to lower bajada lie to one side of the axis. Samples above the axis on the left are from less xeric upper bajada disturbed by grazing, those below the axis on the right are from desert washes.

The effectiveness of different techniques for ordinating the same set of samples has been measured (Whittaker 1960; Bray 1961; Loucks, 1962). A number of approaches are possible, among them:

(*a*) It is assumed that increased effectiveness of ordination will be expressed in reduced dispersion of species distributions through the intervals of the transects. The numbers of intervals through which species distributions extend are averaged for the transects being compared, using the same lists of species which are represented by sufficient measurements to be useful and occur in several but not all steps of the transect (Whittaker, 1960; cf. Bray, 1961).

(*b*) It is assumed that more effective ordination will be expressed in closer approach to binomial pattern of importance values along the gradient and hence greater consistency in the decline of importance values away from the peak or modal values. Differences between successive transect steps which decline in sequence away from the peak importance value are summed; from this sum are subtracted the amounts by which importance values violating these trends would have to be reduced to render them consistent with the trends. These values are averaged for species in the transects being compared (Bray, 1961; cf. Whittaker, 1960).

(*c*) It is assumed that more effective ordination will result in closer mean similarities of adjacent samples in the sequence into which samples have been ordinated. A statistical test, '*z*' test) comparing the means of similarities to given samples of the three samples closest to them in the sequence with the mean similarity value for all comparisons of samples in the set was applied by Loucks (1962).

The tests used by myself (1960) showed that weighted averages gave a more effective ordination than either arrangement by topographic position or comparison with end-point samples. Bray's (1961) tests did not show significant difference in effectiveness between weighted average and sample similarity ordinations. Both Bray's (1961) and Loucks's (1962) tests showed that vegetation characteristics (weighted averages or sample comparisons) give more effective ordinations than such factor-gradient magnitudes as light intensity and soil water-retaining capacity.

Ordination by characteristics of the vegetation itself is not only possible, it is in some cases more effective than ordination by any available measurement of environment. A normal research perspective suggests that one should first use measurements of environment to arrange samples along a gradient, and second study the vegetational response to the gradient as expressed in composition of those samples. It is a legitimate alternative, however, first to ordinate samples by measurements of the vegetation, and second to observe the way environmental factors change in the ecocline which the ordination represents. When an ordination based on species distributions is used to study species distributions the approach is circular (Whittaker, 1956). It is usually possible, however, to use a field transect or other independent data to establish that patterns of species distribution in the ordinated sequence of samples are not artifacts of ordination.

The 'topographic moisture gradient' may further illustrate the meaning of the ecocline to which these techniques are applied. The complex-gradient and coenocline are not merely parallel but coupled, for each contributes to the determination of the other. A topographic moisture gradient in mountains is occupied by vegetation. This vegetation contributes, along with weathering, to the development of soil characteristics which affect water penetration into the soil, subsurface flow, and availability to plants; above-ground vegetation affects, by transpiration, re-evaporation of rain from leaves and microclimatic effects on wind and humidity, the water available to the community and the evaporative stress on plant foliage in the community. Moisture conditions affecting plants are determined not merely by precipitation and topography, but by function of ecosystems which develop in relation to precipitation and topography. The bell-shaped distributions of plant populations represent the responses of populations not only to 'moisture' but to all factors of the ecocline, which (in relation to genetic structure and physiological characteristics of species populations) affect the dynamics of species populations. In studying the ecocline as a system of functionally interrelated gradients of environmental factors, species populations and community characteristics, it is reasonable to use for ordination those characteristics of either environments or communities which best combine effective expression of position in the ecocline with relative ease of measurement (Whittaker, 1954*b*).

Composition of plant communities often provides the most sensitive and most easily measured expression of position in the ecocline. The upper panel of Fig. 3 illustrates, for a larger number of species than Fig. 1, change in community composition along

218   R. H. WHITTAKER

an ecocline. For the continuum of changing proportions among species populations
we may use the term *compositional gradient* (Bray & Curtis, 1957). The compositional gradient is one of a number of possible abstractions from the coenocline (the
trends illustrated in Fig. 3 bottom panel, and the sequence of community types to
be discussed are others); among these the compositional gradient is most useful for
ordination.

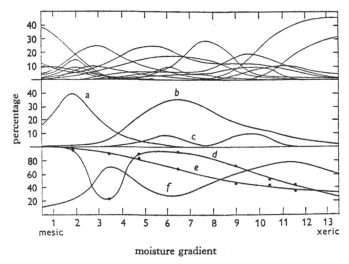

moisture gradient

Fig. 3. Results from composite transects of the topographic moisture gradient, Great Smoky
Mountains, Tennessee. Above, smoothed population curves of major tree species in percentage of stems over 1 cm. d.b.h. in samples, 460–760 m. elevation (Whittaker, 1951). Samples
were grouped in thirteen steps by weighted-average ordination. Middle, three types of population curves (Whittaker, 1956): *a*, more sharply peaked (*Tsuga canadensis*, 760–1070 m.);
b, more broadly dispersed (*Acer rubrum*, 760–1060 m., see also *Pinus rigida*, Fig. 1); *c*, bimodal
(*Quercus alba*, 460–760 m., see also Figs. 4 and 7). Bottom, trends in community characteristics,
970–1370 m. elevation: *d*, species diversity of all vascular plant strata in sample quadrats (in
percentage of a maximum of forty-four for a cove forest sample, data of Whittaker, 1965);
e, total above-ground net annual production (in percentage of a maximum of 1200 g./m.²/year
for a cove forest sample, data of Whittaker, 1966); *f*, shrub-stratum coverages (percentages of
ground surface above which foliage occurs, data of Whittaker, 1956). The peak in curve *f* and
dip in curve *d* at transect steps 3–4 reflect high coverage of *Rhododendron maximum* and low
species diversity in forests of *Tsuga canadensis*, the peak density of which is in the same transect
steps at that elevation.

Weighted averages use distributional relations of species to provide expressions of
relative positions along the compositional gradient. For this purpose the species must
be grouped and assigned numbers for weights. The groupings are clearly arbitrary; the
same species can as easily be grouped into three or five groups as into the four used
for the ordination on which Fig. 3 was based. Because the species groups are based
primarily on location of population modes along the gradient, they have been termed
commodal groups or *commodia* (Whittaker, 1956). Ellenberg's (1950) term, which will
be used here, is *ecological groups*. Weighted averages are probably the most useful
means of ordination by the vegetation itself, along a gradient accepted as given
(Whittaker, 1954*b*).

Gradient analysis of vegetation   219

The diversity of species distributions in Fig. 3 may be noted. Not only are the population peaks scattered along the gradient, the curves show a range of variation of forms (middle panel) from sharp-peaked binomial curves of relatively narrow amplitude, through broader, flatter, distributions of wide dispersion, to bimodal curves having two peaks along the gradient. Some of the widely distributed species show morphological gradients or clines, as evidence of genetic complexity of their populations along the gradient. Populations of wide-ranging forest species, like those of the prairies (McMillan, 1960, 1965), may include gradients of genetic composition and physiological characteristics in adaptation to the environmental gradients along which they occur. Other, bimodal populations show partial separation of ecotypes—sub-populations with different genetic, physiological and sometimes morphological characteristics, adapted to different environments—within the species population. Width and pattern of population along a gradient are in part expressions of genetic structure of the population (Whittaker, 1954*b*, 1956). Bimodal species reduce the effectiveness of both weighted averages and measurements of sample similarity for ordination, and species recognized to have bimodal distributions may be excluded from the computations.

Despite its continuity the coenocline can be divided into community types. The coenocline illustrated in Fig. 3 includes the five community types along the moisture gradient referred to in the Introduction. A sequence of community types along an environmental gradient is an *ecological series*, and this term and concept have a long and significant history in Finnish and Russian ecology (Whittaker, 1962) from their origin in work of Cajander (1903) and Keller (Dimo & Keller, 1907; Keller, 1925–26).

By means of ecological series, two major approaches to natural communities and their environments—classification and gradient analysis—may be co-ordinated. On the one hand results from a gradient analysis may be summarized in terms of community types familiar to ecologists. On the other hand, community types from a study based on classification may be arranged in sequence in relation to an environmental gradient. Thus the remarkable study of Hansen (1930) arranged community types along gradients of elevation and snow cover and observed the trends in life-form composition and other characteristics in these ecological series. Trends of forest production and other community characteristics in ecological series of forest site-types have been studied in both the Finnish school of Cajander (Cajander & Ilvessalo, 1921; Ilvessalo, 1922; Palmgren, 1928; Kujala, 1945) and the Russian of Sukatschew (1928, 1932; Sočava, 1927, Sambuk, 1930; Sokolowa, 1935). In the approach of Poore (1956, 1962; cf. Ratcliffe, 1959; Gimingham, 1961; McVean & Ratcliffe, 1962; Ramsay & DeLeeuw, 1965*a, b*) lower-level community types, termed *noda*, are conceived as reference points in the largely continuous, multidimensional pattern of vegetational variation. Relations of noda in the pattern are expressed in terms of ecological and successional series in relation to environmental and successional gradients. In these studies, community types (or the composite samples representing them) have been ordinated; and when the ways in which environmental gradients, species distributions and community trends relate the community types to one another are observed, the approach combines classification and gradient analysis.

(b) Conclusions

(1) Variations from the typical bionomial distribution along a gradient (Fig. 3) express genetic complexity of the populations. Some species show widely dispersed population curves and morphological evidence of genetic heterogeneity. Other species show partial or full separation of subpopulations or ecotypes with different adaptive centres. Results of a gradient analysis may thus be closely linked with genecology.

(2) These species populations form a complex, flowing population continuum along the environmental gradient. This gradient of populations, as distinguished from the community gradient or coenocline of which it is one aspect, may be termed a *compositional gradient*. The species occurring along an environmental gradient can be arbitrarily grouped into sets of species having their population centres or modes close together, these sets are *ecological groups*.

(3) When it is not feasible to arrange vegetation samples into a transect by measurements of environment, they may often be arranged by measurements applied to the vegetation itself. The arrangement of samples (or community types or species) in relation to one or more gradients is *ordination*. Most ordinations not based directly on measurements of environment use measurements expressing relative position in a compositional gradient to arrange the samples.

(4) A sequence of community types along an environmental gradient forms an *ecological series*. The community types may be either products of an approach through classification, or arbitrarily bounded segments of a coenocline. In either case the ecological series represents a means by which treatment in terms of both community types and the continuities by which these community types are related, results from classification and gradient analysis as different modes of abstraction, may be combined.

(3) *Pattern analysis*

(a) Discussion

Means are needed for analyzing communities in relation to two or more gradients. Much of the observed variation of vegetation on mountain landscapes is related to two complex-gradients, elevation and the topographic moisture gradient, and it is natural to treat these as two axes of analysis. The elevation gradient may be used as ordinate and the topographic moisture gradient as abscissa of charts on which population levels are plotted and outlined with contours (Fig. 4). Source of the background pattern of vegetation will be discussed. In two dimensions, distributional figures for species become binomial solids (Fig. 4, *Quercus prinus*), any transection of which cuts a binomial curve. These figures may be conceived as hills with peaks at the population optima for species, and slopes declining in all directions with increasing departure from these optima.

No two species have population figures which fully parallel one another, and the population modes are scattered in the pattern of environments and communities represented by the diagram—quite according to the principle of species individuality. Structure of the vegetation pattern may consequently be conceived in terms of some hundreds of these population figures superimposed, forming a complex population

Gradient analysis of vegetation   **221**

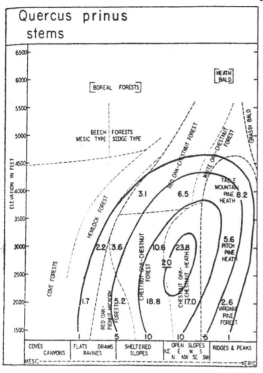

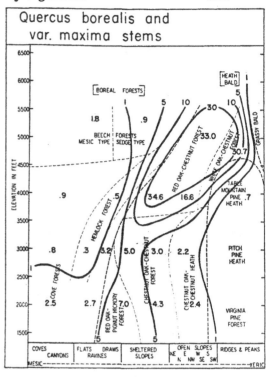

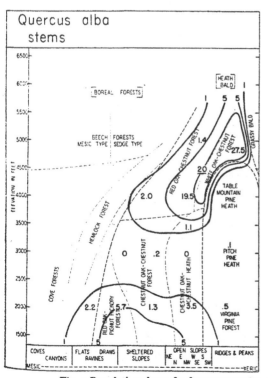

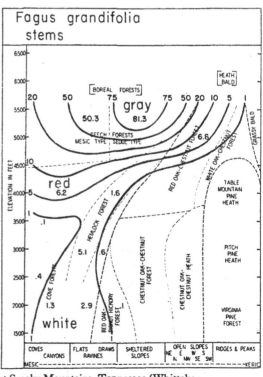

Fig. 4. Population charts for four tree species, Great Smoky Mountains, Tennessee (Whittaker, 1956). Data are percentages of tree stems over 1 cm. d.b.h. in composite samples of approximately 1000 stems.

continuum (Whittaker, 1951, 1956), any transection of which is a compositional gradient.

Genetic complexity appears in the figures for some species (Fig. 4). Two ecotypes, separated by elevations in which the species is nearly absent, appear within the population of *Quercus alba*. *Quercus borealis* shows an extended cline connecting a more mesic low-elevation population (*Q. borealis* var. *maxima*) with a less-mesic high-elevation population (*Q. borealis* var. *borealis*). The population of *Fagus grandifolia* includes three ecotypes, one of them ('gray' beech, Camp, 1951) dominant in the beach gap forests of high elevations, another ('white' beech) in beech–hemlock ravine forests of low elevations, and the third ('red' beech) a subordinate tree in cove forests between 1100 and 1400 m.

The ridges of these population figures are displaced toward the left from higher toward lower elevations. Such displacements express a familiar distributional relationship. When a species population is followed from a more humid to a drier climate, the centres and limits of species populations shift along the topographic moisture gradient toward its mesic end (Boyko, 1947; Billings, 1952; Walter & Walter, 1953; Whittaker, 1960). This shift toward the mesic may be observed both from higher elevations toward the drier climates of lower elevations within a mountain range (Whittaker, 1956, 1960; Whittaker & Niering, 1964, 1965) and along a geographic gradient from more humid to more arid climate (Walter & Walter, 1953; Whittaker, 1960). Similar displacements appear when vegetations on parent materials forming soils with widely different properties are compared within the same climate (Whittaker, 1954*a*, 1960).

As dominant species shift along the topographic gradient from one climate to another, the community types they characterize must shift. In the Siskiyou Mountains the response of one community type, the mixed evergreen forest (dominated by *Pseudotsuga menziesii* and broad-sclerophyll trees), to the climatic gradient from humid coastal climates inland to drier continental climates was observed (Whittaker, 1960). Along this climatic gradient the mixed evergreen forest shows continuous shift in topographic position, progressive increase to, followed by decrease from, maximum importance in relation to other community types, and gradual change in floristic composition. In the middle part of this climatic gradient mixed evergreen forest is prevailing climax type, but its prevalence gives way continuously into the area of redwood prevalence near the coast and that of oak woodland prevalence near the inner end of the range.

If vegetation is to be compared between different climates, effects of the topographic gradient within each of these climates must be controlled. There are, then, a number of possible approaches to comparison (Whittaker, 1956, 1960; Whittaker & Niering, 1965): (*a*) Coenoclines as units and population distributions in them are compared between different climates, elevations or parent materials. If these climates, elevations or parent materials form a sequence along a gradient, vegetation patterns are being approached through a transect of transects. (*b*) Topographic coenoclines may be treated in continuity with one another in a chart representing vegetation and species distributions in relation to two gradients (Fig. 4). (*c*) Coencoclines as expressions of climate may be compared in terms of average composition of the samples representing

Gradient analysis of vegetation   223

them, or composition of mid-point samples, or extent of change in community composition along the gradient. (*d*) Comparisons involving three gradients are more difficult, but it is possible to compare along one gradient whole vegetation patterns in relation to two other gradients. Fig. 5 represents two vegetation patterns along the parent-material gradient in the Siskiyou Mountains.

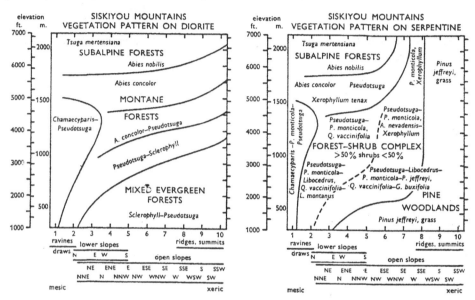

Fig. 5. Mosaic charts of mountain vegetation on quartz diorite (left) and peridotite and serpentine (right), Siskiyou Mountains, Oregon (Whittaker, 1960). The patterns are based on 270 samples on diorite, 160 samples or records on serpentine, plotted by elevation and topographic position; the lines mark out in the continuous patterns community types of the author's classification.

It is possible for someone intimately acquainted with a vegetation pattern to comprehend much of its design in terms of distributions of species populations, but it is not easy to convey this comprehension to others. Simpler means of abstraction are needed and these are likely to depend on more conventional use of community types as units. In one technique the vegetation samples are plotted, each as an individual point, on a chart with elevation and topographic position as axes. Each sample has been classified as a community type, or as intermediate to two community types. Boundaries are drawn around community types outlining the ranges of elevation and topographic positions they occupy and revealing the manners in which they relate to one another in the pattern as a whole. Fig. 5 and the background pattern of Fig. 4 represent such 'mosaic charts' (Whittaker, 1951, 1956, 1960; Whittaker & Niering, 1965; cf. Waring & Major, 1964). Alternative approaches (see also Gams, 1961) include the construction of compound ecological series, such as that of Sukatschew (1932) for Russian pine forests shown in Fig. 6 (cf. Sukatschew, 1928; Matuszkiewicz, 1947; Gams, 1961), schematic arrangements of community types without the labour of mosaic chart construction (Poore & McVean, 1957; Ratcliffe, 1959; Wace, 1961;

224   R. H. WHITTAKER

Johnson & Billings, 1962; Bliss, 1963; Hartl, 1963; Ellenberg, 1963; Florence, 1964), and the community-type plexus to be discussed below.

(b) Conclusion

The major concepts of gradient analysis developed above each undergoes transformation from unidimensional to multidimensional form when more than one gradient is considered:

(1) The ecocline or ecosystemic gradient becomes in more than one direction an *ecosystemic pattern* or *landscape pattern*. The complex-gradient in one becomes in several dimensions an *environmental pattern*. The coenocline in one becomes a *community pattern* in more than one direction.

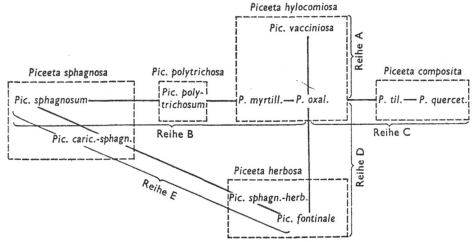

Fig. 6. A compound ecological series for Russian spruce forests (Sukatschew, 1932, fig. 34). Spruce forests with oxalis (*P. oxal.* = *Piceeta oxalidosum*) are the central or climax type, from these radiate series: series (Reihe) A toward drier, nutrient-poor soils; series B toward wetter soils with stagnant water and poor nutrient conditions, and bog forests; series C toward drier soils with more favourable nutrient content and forests mixed with oaks (*Piceeta quercetosum*); and series D toward wet soils with moving water; series E connects the latter with the bog forests.

(2) The binomial curves of species populations become, in two dimensions, *binomial solids*, modes of which represent adaptive peaks (in terms of population function, not simply of individual physiology) in relation to the pattern of ecosystemic environments that the population encounters (Fig. 4). Genecological complexity of some species populations appears in the form of two or more modes, partially or wholly separated from one another in relation to the environmental pattern.

(3) The compositional gradient in one dimension becomes a *complex population continuum* in more than one dimension. This population continuum may be conceived in terms of many binomial solids superimposed on one another in relation to the environmental pattern. Trends of community characteristics along gradients become patterns of these characteristics in relation to more than one gradient (Whittaker, 1952, 1956; Whittaker & Niering, 1965).

(4) The ecological series along one gradient becomes a *community-type pattern* when community types are arranged in relation to more than one gradient (Figs. 5, 6). Although such patterns represent high levels of abstraction at which most detail on species populations is lost sight of, they are a fruition of gradient analysis which can bring into communicable form some most important relations among environmental gradients, dominant species populations, and the community types they characterize.

(a) *Discussion*   (4) *Hyperspaces and evolution*

It may be appropriate now to state a theory of the population structure of vegetation based on community continuity and species individuality, and exceptions thereto.

One may choose to recognize in the landscape certain complex-gradients as major directions of variation of environments, gradients along which many factor-gradients vary and along which environmental complexes of particular points in the landscape are continuously related. These *n* complex-gradients for a landscape may serve as axes for an *n*-dimensional abstract space or hyperspace (cf. Goodall, 1963).

At a given place in the landscape a community develops through successional time. Through the course of succession species populations may be replaced by other species populations, and in most cases the environmental complex is increasingly modified by the communities developing in relation to that environmental complex. A succession is thus an ecocline of communities-and-environments changing through time. The succession leads to a self-maintaining community or climax, adapted to the conditions of the environmental complex (as modified) and to sustained utilization of the resources of environment in an ecosystem of relatively stable, steady-state function (Whittaker, 1953). Although the climax communities in two places with different environments may be less widely different than the successional communities which preceded them, their convergence is only partial. Differences in environment are expressed in differences in composition of climax communities; differences in environments along a complex-gradient are expressed in a climax coenocline. Differences of environments in the complex environmental pattern of the landscape are expressed in a pattern of climax communities (Whittaker, 1953), and the pattern of actual vegetation is further complicated by disturbance and successional communities.

Pattern and hyperspace concepts emphasize the continuity of vegetation, but actual vegetation is a complex mixture of continuity and discontinuity (Whittaker, 1956). Some of the discontinuities are from causes extrinsic to the plant communities, such as topographic discontinuities, contrast of adjacent soil parent materials, and differences in disturbance effects. In general, the more disturbed the vegetation is the more likely it is to appear discontinuous. Some relative discontinuities in vegetation are neither simply extrinsic nor intrinsic. A prairie–forest border, for example, may be made abrupt by effects of fires which are in part set by man but also are part of the normal environment of the fire-adapted grasslands. There are also some relative discontinuities which are intrinsic to vegetation; these have been inadequately studied but deserve brief discussion.

When cove forests of the Great Smoky Mountains are studied in elevation transects from 450 m. up to 1350 m., average composition of the forests changes gradually and

15

226   R. H. WHITTAKER

continuously. At elevations around 1350–1400 m., however, especially in south-facing canyons and concave slopes, the population density of beech (*Fagus grandifolia*) increases rapidly as one climbs (Fig. 7). Mesic deciduous forests of higher elevations are dominated by beech. The beech forests thus appear as a 'zone' of vegetation in marked contrast with the cove forests of all lower elevations, separated from them by a relatively abrupt transition (Whittaker, 1956). Similar results were obtained at a far point of the world from the Smokies, in transects from mixed forest into southern beech (*Nothofagus menziesii*) in New Zealand (Mark, 1963; Scott, Mark & Sanderson, 1964).

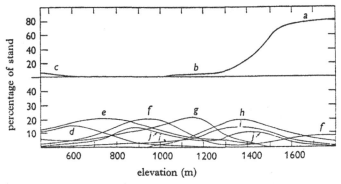

Fig. 7. Population curves in a composite elevation transect of mesic forests, Great Smoky Mountains, Tennessee (Whittaker, 1956). Above, 'plateau' distribution of *Fagus grandifolia*: *a*, grey; *b*, red, and *c*, white ecotypes (see also Fig. 4). The boundary of the grey beech population, 1350–1500 m., is sharper in some field transects in south-facing canyons (Whittaker, 1956, fig. 13), less sharp in transects in north-facing canyons. Below, other major tree species: *d*, *Acer rubrum*; *e*, *Tsuga canadensis*; *f*, *Halesia monticola* (bimodal); *g*, *Tilia heterophylla*; *h*, *Acer spicatum*; *i*, *Aesculus octandra*; and *j*, *Betula alleghaniensis* (bimodal). Data are percentage of stems over 1 cm. d.b.h. in composite samples at 100 m. elevation intervals.

Since all major tree species of middle-elevation cove forests in the Smokies extend upward into the beech forests, the partial discontinuity is not a product of competitive exclusion of other species by beech. In certain subalpine shrub communities in other mountains closed patches of the different species form a mosaic; and in each unit of the mosaic one species is strongly dominant and the other shrub species are largely excluded. Evidence from the broad overlap of species populations in general, however, is against the application to plant communities of the concept of sharp boundaries of competitive exclusion for certain animal species occupying closely similar niches (Gause, 1934; Hairston, 1951; Whittaker, 1965). As an alternative interpretation it may be suggested that one species, in this case *Fagus*, has a strong competitive advantage over other species in some range of the environmental gradient. Throughout that range of the gradient this species forms 60–90% of the canopy trees of the forests. Since it can never form more than 100% or some value below this, its population figure is flattened or vertically truncated, compared with the binomial distributions of other species. Such distributions have been termed 'plateau' distributions (Whittaker 1956).

Other observations affect the evaluation of plateau distributions as exceptions to

Gradient analysis of vegetation   227

vegetation continuity. (*a*) I have never observed plateau distributions in communities of mixed dominance (and a majority of natural communities are such). (*b*) Relative discontinuities of species populations were in the Smokies limited to the borders of the beech forests and the grassy balds, two communities of relatively 'special' or 'extreme' environments which form a small part of the vegetation pattern. (*c*) Studies of mountain vegetation in the western United States, which has been described in terms of the life-zones of Merriam (1898), indicate that plant populations in these zones are fully continuous (Whittaker, 1960; Whittaker & Niering, 1964, 1965). Data presented by Daubenmire (1966) suggest (though they do not show the forms of the population curves) that in eastern Washington *Artemisia tridentata* and *Festuca idahoensis* may form plateau distributions in vegetation in which other species are continuously distributed. In the study of Beschel & Webber (1962), continuity of vegetation was shown by distribution curves for dominant species and by statistical tests (rank comparison) in swamp forests in which belts dominated by different species could be recognized from the air.

The question of discontinuity may be differently stated in relation to hyperspaces. In the hyperspace of which complex-gradients are axes, positions of samples (or centres of species distributions) may be located by measurements of environmental variables or indexes chosen to represent each complex-gradient. Alternatively, positions may be located by relative similarities of sample composition in a compositional hyperspace of which compositional gradients are axes. One may then ask the following questions. (*a*) Do vegetation samples of a set form natural clusters, separated by space with relatively few samples, in compositional hyperspace? (The samples must be of adequate number, taken from the landscape by means of choice which exclude subjective preference toward under-representation of transitional samples). (*b*) Do species populations form natural clusters, relatively separate from other clusters, in environmental hyperspace (or compositional hyperspace, or a hyperspace of directions of correlation among species distributions)?

Statistical approaches to these questions are difficult and results of any technique bearing on them require cautious interpretation. In general, the scattering of samples and species in hyperspaces, in studies of indirect gradient analysis to be described, are in agreement with the continuity of sample composition and scattering of species centres along gradients observed in direct gradient analysis. Goodall (1954*b*) found that small samples from the Australian mallee formed two fairly distinct clusters representing the ridge and hollow communities. These clusters appeared to be due, however, to the fact that the ridge and hollow environments covered greater areas than intermediate environments. In the beech transect discussed from the Smokies, the beech forest samples form a distinct cluster when compared by percentage similarity (expressing their sharing of dominant species), but not when compared by coefficient of community (expressing relative similarity in total community composition). It is likely that further work will reveal more frequent partial clustering of samples for varied reasons—difference in relative area of habitat types and corresponding community types, extrinsic and sometimes intrinsic vegetation discontinuities, the manner in which strong dominance effects percentage similarity and related measurements,

228   R. H. WHITTAKER

and the tendency toward spottiness in coverage of a vegetation pattern by a limited sample set, which are reasons mostly consistent with vegetational continuity.

In studies of species association (e.g. Bray, 1956; Vries, 1953; Dagnelie, 1960; Looman, 1963; Ramsay & De Leeuw, 1964; Beals, 1965) species do not form clearly distinct clusters, though it is possible to use the weak clusters which are suggested as bases of ecological groups. A technique applied to transects (Whittaker, 1960) suggests weakly defined clusters of species at the mesic (ravine) end of certain gradients, but not of other gradients. The clusters are limited to wet-soil streamside species, and these species are dispersed in their distributional relations to elevation. Forest–grassland borders have high species diversities, with numbers of species and ecotypes centred in the transition ('edge effect' of Odum, 1959). Analysis of species distributions shows that species are differently distributed within the transition and in relation to other environments and communities (Whittaker, 1956; Bray, 1956, 1960, 1961; Patten, 1963). Such transitions are not lines along which communities meet but are themselves steeper community gradients.

Few indeed are the ecological generalizations free from exceptions and limitations. Available evidence suggests, however, that intrinsic relative discontinuities in vegetation, and sample and species clusters, are of restricted occurrence and implication as limitations on the principles of community continuity and species individuality. Some authors have speculated toward the idea that species evolve as multi-species associations and hence may evolve toward formation of natural clusters of associated species characterizing relatively discrete community types (Du Rietz, 1921; Allee *et al.* 1949; Dice, 1952; Goodall, 1963). It is thus necessary to ask the evolutionary meaning of the scattering of species centres along environmental gradients which is actually observed.

The question converges with that of how species evolve in relation to one another within the community. The position of the species in the community, its particular way of relating to other species, environment and space within the community, and seasonal and diurnal time, is its niche. If niche characteristics are assumed to be related to one another along gradients, then these gradients as axes define a niche hyperspace (Hutchinson, 1957). The principle of Gause (1934; Odum, 1959; Hardin, 1960; Wallace & Srb, 1964) states that no two species in a stable community can occupy the same niche, competing for the same environmental resources in the same part of intracommunity space at the same time. Species consequently evolve toward avoidance of competition by differentiation of niche, by such division of niche hyperspace that each species occupies (or is centred in) a different part of that hyperspace (Hutchinson, 1957; MacArthur, 1960; Whittaker, 1965). Patterns of relative importances of species—lognormal distributions and other forms—linking dominant with subordinate and rare species result from the manners in which niche space and environmental resources are divided among species (MacArthur, 1960; Whittaker, 1965). The community is an assemblage of niche-differentiated species which conceivably, if ecologists understood enough, might be ordinated in a niche hyperspace. The relative richness of a particular community in species per unit area is the community's *alpha* species-diversity (Whittaker, 1960; MacArthur, 1965).

Gradient analysis of vegetation   229

Species can avoid competition also by occupying different habitats; that is, different positions along environmental gradients and in environmental hyperspace. Implications of species interactions for their distributions do not really support the assumption that they should evolve toward natural clusters of species with closely similar distributions (Whittaker, 1962). Instead, an extension of the principle of Gause implies that species should evolve toward dispersion of their distributional centres in environmental hyperspace (Whittaker, 1965). Since the species which occur together are also niche-differentiated, their populations do not form boundaries of mutual exclusion but overlap freely. The structure of compositional gradients—broadly overlapping population distributions mostly of binomial form, with centres scattered along the environmental gradient—is thus a consequence of species evolution toward both niche and habitat diversification.

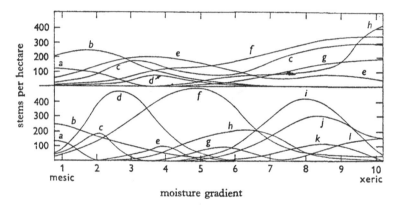

Fig. 8. Contrasts of beta diversities of vegetation along topographic moisture gradients. Above, moderately low beta diversity (less change in composition along the gradient, with widely dispersed population curves) at 460–470 m. elevation, Siskyou Mountains, Oregon (Whittaker, 1960). Below, high beta diversity (narrower population curves, greater change in composition along the gradient) at 1830–2140 m., Santa Catalina Mountains, Arizona (Whittaker & Niering, 1965). Half-change values expressing relative change in composition in the ten-step transects were 1·1 for the tree stratum of the Siskyou transect, 3·4 for that of the Catalina transect. Species, above, *a, Taxus brevifolia*; *b, Chamaecyparis lawsoniana*; *c, Castanopsis chrysophylla*; *d, Abies concolor*; *e, Pseudotsuga menziesii*; *f, Lithocarpus densiflora* (× 0·5); *g, Quercus chrysolepis*; *h, Arbutus menziesii*. Species, below, *a, Abies concolor*; *b, Quercus rugosa*; *c, Pseudotsuga menziesii*; *d, Pinus ponderosa*; *e, Arbutus arizonica*; *f, Quercus hypoleucoides* (× 0·5); *g, Pinus chihuahuana*; *h, Quercus arizonica*; *i, Arctostaphylos pringlei*; *j, Pinus cembroides*; *k, Garrya wrightii*; *l, Quercus emoryi*.

The implication of different degrees of species-habitat diversification for vegetational gradients is illustrated in Fig. 8. A larger number of species with narrower habitats (i.e. narrower distributional amplitudes or dispersions) are accommodated along the topographic moisture gradient in the lower coenocline. There is consequently a greater degree of floristic and compositional change along the gradient in the lower coenocline, a higher *beta* diversity (Whittaker, 1960, 1965). MacArthur (1965) has found that alpha diversities of bird communities are (for a given degree of structural diversity of the plant community) similar in temperate and tropical climates; but that

230   R. H. WHITTAKER

continental tropical bird communities show higher beta diversities. In studied cases of mountain vegetation in the western United States (Whittaker, 1960, 1965; Whittaker & Niering, 1965), both alpha and beta diversities, and consequently the landscape species-diversity, or *gamma* diversity resulting from both these, increase from maritime to continental climates.

Environmental and niche axes may be conceived as forming together a combined environment-and-niche hyperspace in which each species has its own distinctive position. A major trend of species evolution is toward reduction of competition by diversification in the positions of species in this hyperspace; from this trend result the gamma diversities of landscapes and, still more broadly, the richness in species of the living world. Some species have two or more population centres in this hyperspace; ecotypic populations thus considered are experiments in combinations of genetic possibility and adaptive opportunity, within the species. By gradual shift in genetic composition of populations and by evolution of ecotypic differentiation, species may change through evolutionary time their positions in the hyperspace and their distributional and niche relations to other species. Natural communities consequently do not evolve by a phylogeny of divaricate descent. Species may change their associative combinations with one another in evolutionary time (Mason, 1947; Whittaker, 1957), and the evolutionary relations of community types are reticulate.

(b) Conclusion

(1) From the environmental pattern of the landscape one may abstract an *environmental hyperspace*, axes of which are the *n* complex-gradients recognized in the landscape. From the landscape pattern one may also abstract a *compositional hyperspace*, with compositional gradients as axes; these axes may, but need not, correspond to those of the environmental hyperspace. Representations of compositional hyperspace provide the co-ordinate systems in relation to which samples (and species) are ordinated in indirect gradient analysis, as discussed in following sections.

(2) Species individuality and community continuity may be differently stated in relation to these hyperspaces. Centres of populations of species and their ecotypes are primarily scattered, rather than grouped into clusters in environmental hyperspace. Samples taken by objective procedures are primarily scattered, rather than grouped into clusters representing distinct community types, in compositional hyperspace. Exceptions to this continuity of sample composition occur, however.

(3) Species evolve toward niche differentiation, by which direct competition within the community is avoided. They evolve also toward habitat diversification, toward occupation of scattered positions in environmental hyperspace, so that plant species are in general not competing with one another in their population centres. The niche differentiation implies that species are partial competitors whose distributions may overlap broadly, forming the population continua along environmental gradients revealed by gradient analysis.

(4) From this evolution toward niche-and-habitat diversification there results the population structure of vegetation patterns. Numerous plant species, with population centres scattered along environmental gradients, each with binomial distributions

Gradient analysis of vegetation   231

broadly overlapping those of other species, freely and variously combine into communities which predominantly intergrade with one another, forming a complex and potentially continuous but variously interrupted population pattern. From this population structure of vegetation result the difficulties of classification which have vexed phytosociologists. Gradient analysis is an effective alternative approach to its investigation and understanding.

III. INDIRECT GRADIENT ANALYSIS
(1) *Similarity measurements*
(a) Discussion

The essential basis of indirect gradient analysis is the use of comparisons between samples, or between species, in such ways as to cause gradient relationships to emerge from the data. Measurements of relative similarity of sample composition, or of relative similarity of species distribution, are used to arrange samples or species along axes which may correspond to environmental gradients. The approach may consequently be described as *comparative ordination*. The underlying gradient relationships of species populations and environmental factors make the ordination possible. The ordination in turn makes possible the following of gradients of environments, species populations, and community characteristics along the axes of ordination.

There are many possible ways of expressing relative similarity of two community samples. These measurements have been reviewed in some detail by Dagnelie (1960; see also Greig-Smith, 1964; and citations of Goodall, 1962). Among them:

(*a*) The earliest and simplest measurement, the *coefficient of community* of Jaccard (1902), compares samples in terms of the number of species they share among the total number of species occurring in one or both, without regard to the importance values for species. $CC = c/(a+b-c)$, in which c is the number of species occurring in both samples, a the total number in one sample and b the total number in the other. A number of variants have been used, most widely among them Sørensen's (1948) $CC = 2c/(a+b)$.

(*b*) Other measurements, computed from importance values for species, express the relative similarity of samples in quantitative composition. Measurements of this sort have been independently devised and applied by a number of authors. A simplest version, which may be termed *percentage similarity* (Odum, 1950), sums the percentage of species composition shared by two samples in the form, $PS = 100 - 0.5 \Sigma |a-b| = \Sigma \min. (a, b)$, in which a and b are, for a given species, the percentages of importance values in samples A and B which that species comprises. This measurement was used (with quadrat frequencies as importance values) by Gleason (1920). It has been applied to samples of animal communities by Renkonen (1938), Agrell (1941), Odum (1950) Pearson (1963) and others; its computation and characteristics have been discussed (Whittaker, 1952, Whittaker & Fairbanks, 1958). The form, $PS = \sqrt{[\Sigma(a-b)^2]}$ is suggested by Vasilevich (1962) and Orloci (1966). Related measurements were used by Kulczyński (1928), Raabe (1952), Barkman (1958) and Morisita (1959; Ono, 1961).

(*c*) The Wisconsin variant of percentage similarity (Bray & Curtis, 1957) is based on a two-step conversion of importance values into percentages. First, in each horizontal

R. H. Whittaker

row (for a given kind of importance value for a given species in all samples in which it is represented) all values are converted to percentages of the maximum value in that row. Second, importance values are converted into percentages vertically; in each column the values become percentages of the total of the values (as already horizontally converted) in that column. Columns are now directly compared, with percentage similarities computed by $PS = \Sigma \min. (a, b)$, summing the smaller of the two percentages for species.

(d) Goodall (1953b), Hughes & Lindley (1955) and Groenewoud (1965a) have used the discriminant function (Fisher, 1936) and the related generalized distance (D^2) of Mahalanobis (1936) for sample comparison.

(e) Coefficients of correlation, or of rank comparisons, may be computed for the importance values of species in the two samples (Motomura, 1952; Beschel & Webber, 1962; Ghent, 1963).

Statistical tests of the probability that two samples are the same are of interest in some cases, but they are not appropriate to comparative ordination of samples from different communities. These samples are in fact different, and what is needed is an expression of the degree to which they differ. Such measurement is provided by coefficient of community, percentage similarity and their variants. The two measurements have complementary merits and limitations. Kontkanen (1950) and Whittaker (1960; Whittaker & Fairbanks, 1958) have found use of both measurements for the different evaluations they provide is often desirable. The Wisconsin variant of percentage similarity to some extent combines advantages of the two measurements. Quantitative data are taken advantage of, but the horizontal conversion to percentages has the further advantage of permitting different kinds of importance values, and data on minor, as well as major, species to be effectively used in the same computation.

Some results from direct gradient analysis may clarify the meaning of similarity measurements. Coefficients of community and percentage similarities have been computed for samples which are one unit, two units, three units, etc., apart along transects; the resulting curves are plotted in Fig. 9. Of these curves it may be observed: (a) In the upper parts of the curves the logarithms of similarity measurements decrease in proportion to distance along the environmental gradient. (b) Extrapolated back to zero distance along the gradient, the curves intersect the ordinate, not at 100% but at lesser values. Such must be the case, for two samples from the same community give similarity values below 100%. Such 'internal associations' computed for replicates of foliage insect samples (Whittaker, 1952) ranged from 50 to 80%; comparison of replicate vegetation samples (Bray & Curtis, 1957) gave a mean value of 82%. Extrapolations to zero units in vegetation transects have yielded values of 60–94% for different strata and different kinds of vegetation studied by the author. (c) In analogy to the half-life, a half-change unit based on reduction of sample similarity to one-half the internal association may be used as an expression of relative distance along compositional gradients (Whittaker, 1956, 1960; Whittaker & Niering, 1965). (d) Sooner or later the curves depart from the straight line and drop rapidly to zero similarity, as they must when distance along the gradient is great enough for no two species to be shared by the samples compared. (e) Slopes of these curves, and hence

Gradient analysis of vegetation   233

rates of change of community composition along the gradient, may be quite different for different fractions of the same communities along the same gradient.

The other face of the same problem is the measurement of relative distributional similarity of species. Sample similarity measurements compare vertical columns in tables in which the columns represent samples (or composite samples). Measures of distributional similarity of species (see also Cole, 1949; Goodall, 1952, 1962; Morisita, 1959; Dagnelie, 1960; Greig-Smith, 1964) compare horizontal rows, in which are recorded occurrences or importance values for species, in the same table. Approaches to such measurement include:

(*a*) Comparison of the actual number of samples in which the two species both occur with the number of joint occurrences to be expected from the product of their frequencies of occurrence (Forbes, 1907, 1925; Dice, 1945; Cole, 1949, 1957; Fager, 1957, 1963). Species association in this case becomes, most simply, $A = an/bc$ (Forbes, 1907); in more effective forms $A = (ad-bc)/[(a+b)(a+c)]$ and related equations of Cole (1949, applied by McIntosh, 1957; Omura & Hosokawa, 1959; Vasilevich, 1961; Cook & Hurst, 1962), or $A = a/\sqrt{(bc)} - 0.5\sqrt{b}$ (Fager & McGowan, 1963). In these, a is the number of samples containing both species, b is the number in which the first species occurs and c the number in which the second species occurs (so assigned that $b \leqslant c$), d is the number of samples in which neither species occurs and n is the total number of samples.

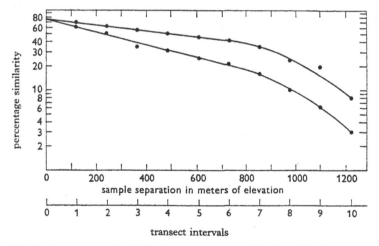

Fig. 9. Decline in mean sample similarity measurements with increasing sample separation along the elevation gradient in the Great Smoky Mountains (Whittaker, 1956, 1960, fig. 18). The upper curve is for coefficients of community, the lower for percentage similarities, on the same logarithmic scale on the ordinate. The abscissa indicates the separation of the samples being compared in terms of elevation in metres and transect intervals (of 122 m.).

(*b*) Statistical measures of departure of actual co-occurrence in samples from expected co-occurrence by chance, by application of χ^2 or fourfold point correlation coefficient to 2×2 contingency tables (Nash, 1950; Goodall, 1953*a*; Vries, 1953; Vries, Baretta & Hamming, 1954; Hopkins, 1957; Gounot, 1959; Welch, 1960; Agnew,

234   R. H. WHITTAKER

1961; Cook & Hurst, 1962; Dagnelie, 1962a; McDonough, 1963; Frei, 1963; Greig-Smith, 1952, 1964; Ramsay & De Leeuw, 1964; Beals, 1965; Mueggler, 1965; Yarranton, 1966).

(c) Application of coefficient of correlation, or preferably rank correlation, to the importance values for the two species in samples or columns where both occur (Iljinski & Poselskaja, 1929; Stewart & Keller, 1936; Tuomikoski, 1942; Dawson, 1951; Kershaw, 1961; Williamson, 1961a; Debauche, 1962; Dagnelie, 1962a).

(d) The simple percentages of samples in which both species occur together—'percentage co-occurrence' (Whittaker & Fairbanks, 1958). Two versions of percentage co-occurrence, analogous to the coefficient of community variants of Jaccard and Sørensen, are $PC = a/(b+c-a)$ (Agrell, 1945; Ellenberg, 1956; Whittaker & Fairbanks, 1958) and $PC = 2a/(b+c)$ (Bray, 1956) in both of which a is the number of samples in which both species occur, b is the number in which the first species occurs and c the number in which the second species occurs.

(e) To express separately the association of each of a pair of species with the other, the association index of Dice (1945) uses the ratio of the number of samples in which both species occur to the number in which the other species occurs, hence a/c for association of the first species with the second, and a/b for the second with the first (Gardner, 1951; Hegg, 1965).

(f) Measurement by the formula used for percentage similarity of samples, applied to horizontal rows of the table (when data in these rows have been converted to percentages of the horizontal sums)—'percentage similarity of distribution' (Whittaker & Fairbanks, 1958).

(g) Distance by which the modes of species populations are separated along a compositional gradient or in a compositional hyperspace (Bray & Curtis, 1957).

The statistical measurements (b) and (c), and measurements based on spatial distances between individual plants in a given community (Goodall, 1965), are appropriate for study of association between species in samples taken from a single community or from a narrow range of environments. Measurement of relative distributional similarity in samples from a wide range of environments and communities is a different problem. Since it is almost always the case that two species are differently distributed; the degree of improbability that their distributions are the same is not the question to be asked.

What is to be measured is the degree of overlap of species distributions (Whittaker & Fairbanks, 1958). Measures (d) and (e) provide simple expressions for this purpose. All these measures are, however, strongly affected by choice of samples; they express distributional relations of species in a set of samples, not in the field. Measurements in form (f) may be most appropriate in theory, but are also most difficult to obtain. There is, from the nature of the problem, no fully satisfactory measurement of distributional similarity of species in a set of samples from different communities. Despite their evident limitations both the non-statistical measures (a), (d) and (e) and, especially for narrower ranges of communities, the statistical measures (b) and (c) can in practice give useful indications of relative distributional similarity of species.

(b) Conclusion

(1) Most indirect ordinations are based on measurements of relative similarity of community samples or of species distributions in samples. Principal measures of sample similarity include the percentage of species shared by two samples (coefficient of community) and percentage of the quantitative composition of the samples that is shared (percentage similarity).

(2) These measures may be understood as approaches toward indication of *ecological distance*, the extent to which samples are distant from one another along compositional gradients, or in compositional hyperspace.

(3) Analogous measurements can be applied to distributional similarity of species. Statistical measurements through χ^2 or correlation are appropriate to study of distributional relations of species in a given community or a narrow range of environments. For study of species distribution in samples from a wider range of environments and communities, simpler expressions reflecting degree of distributional overlap of species populations are appropriate.

(4) The manner in which species populations relate to one another within communities and along environmental gradients limits the effectiveness of these measurements. Nevertheless, used with skill, perceptiveness of the ecological factors which affect them and consciousness of their limitations, they can give results of considerable scientific interest discussed in the following sections.

(2) *Matrices and quantitative classification*

(a) Discussion

The result of computing similarity values between all pairs of samples in a set is a matrix, such as illustrated in Table 1. In a matrix which has not been ordered (Table 1 has been ordered) the high and low similarity values are scattered. Much information of interest is present in the thicket of numbers, but techniques are needed by which significant relations may be made to emerge from the thicket.

A first step is the ordering of samples in the matrix to reveal groupings and gradient relations (Kulczyński, 1928; Renkonen, 1938; Matuszkiewicz, 1948; Guinochet, 1955; Guinochet & Casal, 1957; Bray 1956; Clausen, 1957b; Macfadyen, 1963; Balogh, 1958; Whittaker & Fairbanks, 1958; Faliński, 1960; Agrell, 1963). One sample which is in some sense 'extreme' for the set (it may have lowest total of its similarity values with all other samples) is chosen for first position. That sample which has highest similarity measurement with the first is listed second, that sample remaining which has highest similarity with the second is listed third, and so on. By this means (in some cases repeated with different choices of first sample) many, or all, of the highest similarity values are manoeuvred down to the oblique axis. The numbers in the matrix may be replaced with symbols, or degrees of darkening of squares, to form a 'trellis diagram' in which the arrangement of high similarity values is visually apparent.

In some cases an effective ordination results; for the samples have been arranged in sequence from the first sample chosen, representing one extreme of the gradient, to

Column headings (top, left to right):
Rock Lake, Deep Lake, Park Lake, Blue Lake, Falls Lake, Silver Lake, Clear pothole, Cow Lake, Moses Lake, Sprague Lake, Clear Lake, Yellowhead pond, Pothole no. 1, East Twin pond, Sunken pond, O'Sullivan pothole, Upper Crab pothole, South Tule Lake, North Tule Lake, Willow Lake, Granite Lake, Ewan pond, Lenore Lake, Three-Inch pond, Medical Lake, Marsh pond, Tree pond, Pear pond, Disappointment pond, Grass pond, North Rock pond, Ledge pond, Rock pond

Row headings (left, top to bottom):
Rock pond, Ledge pond, North Rock pond, Grass pond, Disappointment pond, Pear pond, Tree pond, Marsh pond, Medical Lake, Three-Inch pond, Lake Lenore, Ewan pothole, Granite Lake, Willow Lake, North Tule Lake, South Tule Lake, Upper Crab pothole, O'Sullivan pothole, Sunken pond, East Twin pond, Pothole no. 1, Yellowhead pond, Clear Lake, Sprague Lake, Moses Lake, Cow Lake, Clear Pothole, Silver Lake, Falls Lake, Blue Lake, Park Lake, Deep Lake, Rock Lake

Matrix values (each row, numbers as printed left → right):

Rock pond	30 39 39 39 39 39 39 39 39 39 39 · 19 9 · 2 1 · 1 · 8 6 6 · 3 31 · 49
Ledge pond	16 15 16 25 21 15 15 15 15 15 15 · 11 19 10 27 9 9 · 2 · 1 1 · 44 34 31 25 25 37
North Rock pond	12 12 12 12 12 12 12 12 12 12 12 · 12 9 · 2 1 · 69 75 75 69 72
Grass pond	75 75 91 94 97
Disappointment pond	75 91 94
Pear pond	6 6 6 6 6 6 6 6 6 6 6 · 6 6 · 2 1 · 75 75 97
Tree pond	7 6 7 9 9 6 6 6 6 6 6 · 3 3 · 9 6 · 2 1 1 · 75
Marsh pond	5 5 17 14 · 2 · 1 · 3 5
Medical Lake	28 28 28 · 10 10 · 3 13 13 13 13 · 22 87 87 88 87 87 87 · 90
Three-Inch pond	28 · 10 · 3 3 3 3 · 22 91 97 98 97 97 97
Lake Lenore	28 · 10 · 22 91 98 98 98 99
Ewan pothole	28 · 10 · 1 1 · 23 91 98 98 98
Granite Lake	1 1 1 1 1 · 28 · 10 · 3 · 1 1 · 23 91 98 98
Willow Lake	1 1 1 1 1 · 29 · 11 · 2 · 1 · 24 92 99
North Tule Lake	2 2 2 2 2 2 · 30 · 12 · 2 2 2 · 24 93
South Tule Lake	9 9 9 9 9 9 · 37 · 19 · 9 9 9 · 31
Upper Crab pothole	18 17 18 29 23 17 · 39 27 · 17 17 17
O'Sullivan pothole	1 1 · 12 6 · 46 46 58
Sunken pond	1 · 12 6 · 27 27 47
East Twin pond	20 20
Pothole no. 1	33
Yellowhead pond	30 30 56 58 67 67 67 67 67 67 66
Clear Lake	30 56 58 66 66 66 67 75 87
Sprague Lake	30 56 58 79 79 79 72 88
Moses Lake	30 56 58 81 81 81 82
Cow Lake	30 56 58 71 71 71
Clear Pothole	30 56 58 88 93
Silver Lake	31 56 59 94
Falls Lake	31 56 59
Blue Lake	41 64 97
Park Lake	44 62
Deep Lake	32
Rock Lake	

the sample least related to it, representing the other extreme (Bray, 1956; Dix & Butler, 1960). Such arrangement along a gradient is in part the case in Table 1, in which choice of a most oligotrophic lake as the first sample produces an approximate ordination of lakes of increasing solute content from oligotrophic, through eutrophic, to saline (Whittaker & Fairbanks, 1958). It is not necessarily the case, however, that the axis of an ordered matrix will represent a single environmental complex-gradient. Other treatment, beyond the ordered matrix, is often necessary to reveal adequately the directions of interrelation among samples.

The matrix may for, one thing, be used for classification as well as for ordination. In Table 1, groups of samples related by high similarity values have been boxed into triangles along the axis. Gaps between the groups on the axis may have no meaning more fundamental than that samples intermediate to these groups did not happen to be taken. The groups along the axis constitute community types which can be characterized in terms of species composition, environmental characteristics and levels of percentage similarities relating the member samples. It is thus possible to turn the techniques of similarity measurement and matrix treatment to the purposes of quantitative classification of communities.

Extensive work of a Polish school of ecology is based on such an approach. An earlier Polish ecologist, Kulczyński (1928) was one of the pioneers in application of similarity measurements and use of matrix ordering, ideas for which he gives credit to Czekanowski (1909). Application of these techniques led Motyka (1947) and Matuszkiewicz (1947, 1948) to early achievements in comparative ordination and rediscovery of the principles of species individuality and community continuity. A series of articles applied matrix techniques to the classification of Polish vegetation and related the units of classification to those of the school of Braun-Blanquet (Matuszkiewicz, 1950, 1952, 1955, 1958; Motyka, Dobrzański & Zawadzki, 1950; Motyka & Zawadzki, 1953; Izdebski, 1963 a, b; Faliński, 1958, 1960).

Sørensen (1948) also has used similarity measurements for quantitative classification of samples. The Sørensen and Jaccard coefficients of community have been used for classification by Dahl (1957) and others (Nordhagen, 1943; Evans & Dahl, 1955; Ellenberg, 1956; Hosokowa, Omura & Nishihara, 1957; Looman & Campbell, 1960). Barkman (1958) compared a number of similarity measurements for classification of epiphyte associations into higher units of the system of Braun-Blanquet. Groenewoud (1965 a) has used the D^2 measure of Mahalanobis (1936) to seek clustering of samples in his set. Dahl (1957, cf. Matuszkiewicz, 1948; Barkman, 1958; Ramsay, 1964) found coefficients of community useful aids to classification although it was not possible to specify particular magnitudes to define community-types on the different levels of a hierarchy.

The same techniques of similarity measurements and matrix treatment can be applied to ordination and classification of species. Bray (1956) computed percentage co-occurrence of species in samples from an ecocline between oak forest and prairie in Wisconsin. Co-occurrence values were compiled into a matrix, and the matrix was ordered to arrange species in sequence by relative positions along this ecocline. The species sequence was then divided into three ecological groups, and weights for these ecological groups were used for weighted average ordination of the samples. This sample ordination was strongly correlated with one independently obtained by ordering a matrix of percentage similarity measurements for samples.

A number of applications of species-association measurements to community classification are possible:

(a) Ecological groups of distributionally related species may be marked out in an ordered matrix of distributional similarity measurements or may be recognized from correlation analysis. Each of these (because distributional centres of its species are close together) should characterize a community type by occurrence of its species at relatively high importance values in samples from that type. In the system of Braun-Blanquet (1951) ecological groups become groups of character-species (and in some cases of differential-species) defining associations and other community types (Ellenberg, 1950, 1952; Gounot, 1959; Looman, 1963; Doing, 1963; Hegg, 1965). Quantitative definition of community types thus may be stated in terms of occurrence, or importance, of species of a given ecological group in samples. These community types may be classified into higher-level units by distributional similarity values linking member species of their ecological groups, by their sharing species of ecological groups of wider-ranging species, or by similarity measurements comparing composite samples for the community types.

(b) A contrasting premise may be used for definition of homogeneous community types. It may be assumed that, if an appropriate level of quadrat size for samples has been chosen, existence of correlations among species implies that populations of some species are responding in parallel to environmental factors affecting different samples of the set. Samples are consequently so grouped as to eliminate measured correlations among species within each group (Goodall, 1953a, 1954b; Greig-Smith, 1964). The resulting groups should be homogeneous communities or community types representing narrow spans of compositional gradients and presumably of environmental gradients (Goodall, 1953a; Hosokawa, 1955; Hosokawa et al. 1957; Rayson, 1957; Williams & Lambert, 1959; Looman, 1963). These homogeneous types may in turn be further grouped or classified.

(c) Williams & Lambert's (1959, 1960; Greig-Smith, 1964) 'association analysis' divides the sample set first by the presence or absence of that species which has the highest sum of association values with all other species. The resulting two groups may be in turn divided by the species having highest sums of association values with other species within them, and the resulting four groups similarly divided, and so on; a dichotomous hierarchy is thus generated. Williams & Lambert (1961; Lambert & Williams, 1962; Lambert & Dale, 1964; see also Gittins, 1965c) have applied related techniques to classification by sample similarities and to interrelation of results from species-association and sample-similarity treatments.

Quantitative techniques have had extensive recent development in the classification of organisms (Sokal & Sneath, 1963; Heywood & McNeill, 1964; Sokal, 1965, and others). Possible techniques for quantitative classification of 'individuals' with 'attributes' are closely parallel for taxonomy, where the individuals are organisms (or populations) with lists of characteristics or measured values of those characteristics and ecology, where the individuals are community samples (or composite samples) with lists of species or importance values for species. Degrees of discontinuity with one another vary widely among taxonomic groups, but community samples are taken from predominantly continuous vegetation patterns in which centres of species distributions are scattered. There are many ways of classifying a given set of samples, or a given set of species. Different similarity measurements may group the samples or species of a set differently and the same measurement applied to the different strata or fractions of the same communities may also group the samples for these communities differently. Quantitative techniques cannot solve the fundamental problems of classifying samples from continuously intergrading communities (Whittaker, 1962); but they have possibilities of interest—especially in the exploration and comparison of different ways of grouping samples that the ecologist's intuition alone cannot explore— which remain to be exploited.

(b) Conclusion

(1) Similarity values may be computed among samples of a set and compiled into a sample-similarity matrix, a table of values comparing each sample with every other sample. Values for relative similarity of distribution of the species occurring in a set of samples may be similarly compiled into a species-association matrix. The matrices embody much information about relations of species and samples to one another, but these relations are not easily apparent.

(2) A simplest means of rendering some of these relations apparent is the ordering of samples, or species, in the matrix so that high similarity values lie along the oblique axis of the matrix (Table 1). In some cases the samples (or species) may thus be arranged in a sequence representing a compositional gradient and an environmental complex-gradient. Such is the essence of *comparative ordination*—the use of similarity comparisons between samples or species to arrange them along a compositional axis. In many cases other techniques than ordering of the matrix are needed for effective ordination.

(3) One may also recognize and delimit groups of high similarity values along the axis of an ordered matrix (Table 1). A group of samples related by high similarity values then become a community type. A group of species related by high distributional similarity values are an ecological group and representation of species of this ecological group may be used to define a community type. A number of other techniques for quantitative classification of communities are possible.

(3) *Plexuses*

(a) Discussion

It is natural to represent distances by lines on a diagram. When a matrix includes a small number of samples, these may be arranged in a diagram called a plexus, in which lengths (or widths, or numbers, or distances without connecting lines) express relative similarity of samples (cf. Curtis, 1959, p. 492; Gimingham, 1961; Ono, 1961; Beals & Cope, 1964; Ramsay, 1964; Barkman, 1965; Ramsay & De Leeuw, 1965b). The Polish school of Matuszkiewicz (1947), Faliński (1960) *et al.* have published a number of such diagrams, which they term 'dendrites'. Thresholds may be subtracted from the similarity values or the values may be otherwise transformed. Even

240   R. H. WHITTAKER

after transformation, the similarity values will not in most cases fit on to a plane surface, and it is often impossible to use lines scaled by similarity values to connect a given sample with samples other than the 2, 3 or 4 with which it is most closely related.

As the number of samples increases, it becomes increasingly difficult to construct a sample plexus and to interpret the meaning of that which is constructed. Solution to the difficulty may involve these steps: (*a*) The samples are grouped into community types and importance values for species within each of these community types are averaged to obtain composite samples; (*b*) Similarity values are computed between the composite samples in all possible directions and are compiled into a second-order matrix; (*c*) High similarity values from this matrix are used to construct a second-order plexus relating community types, rather than individual samples. Such a second-order plexus (Fig. 10, left) is a community-type pattern, equivalent for indirect ordination (with community types arranged in relation to one another in a compositional hyperspace) to the mosaic chart and compound ecological series described above for direct ordination.

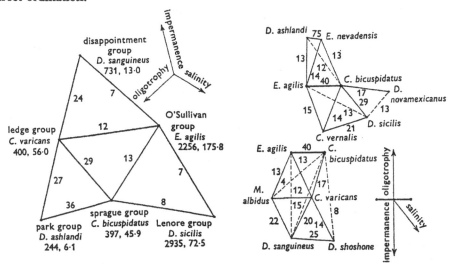

Fig. 10. A second-order (community-type) plexus and species-association plexus for the same copepod samples (Whittaker & Fairbanks, 1958, figs. 3 and 5). On the left the six community types bounded on the diagonal of Table 1 are arranged by percentage similarities (numbers on connecting lines) of composite samples. For each group are entered a characteristic species, mean total salt content of the water (parts per million, the first number), and average micro-crustacean summer standing crop (individuals per litre, the second number). On the right, copepod species are arranged by percentage co-occurrence values (numbers on connecting lines); lengths of the solid lines are scaled for these values. The heavy line connecting *Eucyclops agilis* and *Cyclops bicuspidatus* is a centre from which the remaining species radiate on three environmental axes—oligotrophy (*Diaptomus ashlandi* and *Epischura nevadensis*), salinity (*C. vernalis*, *D. sicilis*, and *D. novamexicanus*), and seasonal impermanence (*D. sanguineus* and *D. shoshone*).

It is a special value of a plexus, however, that it can sometimes be constructed and interpreted in circumstances which make the direct approach impossible. In favourable cases the plexus may arrange community types in such a way as to render apparent the design in relation to environments of the pattern of communities from which the samples were taken, Experience suggests, however, cautionary observations. The fallibility of similarity measurements, particularly when bimodal distributions are involved, implies the fallibility of plexuses constructed from them. The limitations of a two-dimensional surface imply that representation of complex relationships is often very difficult. Different (though related) plexuses may result from application of different similarity measurements to the same samples, and from application of the same similarity measurement to different strata or fractions of the same samples, as in the studies of Beals & Cope (1964) and Barkman (1965). Both the limitations of the approach and the real interest of its results in favourable circumstances should be recognized in balance with each other.

The procedures of sample similarity computation, matrix compilation and plexus construction can be turned through 90 degrees. Similarities of species distributions are then measured and values from the resulting matrix used to construct a species plexus. A number of authors have published plexuses for species association (Vries, 1953; Vries *et al.* 1954; McIntosh, 1957, 1962; Omura & Hosokawa, 1959; Welch, 1960; Agnew, 1961; Looman, 1963; Ramsay & De Leeuw, 1964; Groenewoud, 1965*b*; Beals, 1965; Yarranton, 1966).

When both community and species plexuses are formed from vertical and horizontal analyses of the same community table, they may produce figures which are convergent in their indication of environmental relations. Fig. 10 illustrates a relatively simple case in which the two plexuses can be effectively interpreted in relation to each other and three environmental axes. A more complex treatment was carried out by Hegg (1965) using an association measurement of Dice (1945) to arrange species in a plexus and in ecological groups characterizing lower and higher level community types.

(b) Conclusion

(1) Sample similarity values from a matrix may be used to construct a *plexus*, a reticulate diagram in which the lines connecting samples are, so far as possible, proportional to ecological distances between these samples. Plexuses can be constructed also on the basis of similarity values for community types (as represented in composite samples) and for species (from measurements of relative distributional similarity of species, Fig. 10).

(2) Various limitations effect these techniques, but in favourable circumstances they produce arrangements in which (*a*) ecological distances between samples, or types, or species, are represented in the spacing of these in the plexus, (*b*) the design of the plexus as a whole represents a compositional hyperspace, and (*c*) known environmental relations of the samples permit effective interpretation of that hyperspace in terms of environmental gradients.

(a) Discussion   (4) *Early Wisconsin gradient analysis*

The first major studies of the Wisconsin school analyzed upland forests from the southern (Curtis & McIntosh, 1951) and northern (Brown & Curtis, 1952) parts of the state. Samples were taken from all suitable forest tracts available, so far as possible, by the random-pairs technique of Cottam & Curtis (1949, 1956; Grieg-Smith, 1964), a technique suited to obtaining data from extensive forest stands on level terrain, in contrast to the concentrated, 0·1-hectare samples used by myself on mountain topography. The Wisconsin authors found that the samples did not simply fit into a number of distinct community-types, and they sought other means of analysis.

From the ninety-five southern samples, eighty were arranged by their 'leading dominants'—the tree species having the highest value of the Wisconsin synthetic importance value (the sum of relative frequency, relative density, and relative basal area). Four dominance types, each characterized by dominance of a single tree species, resulted. Importance values of species were averaged for the samples in each of these dominance types, to produce four composite samples. The samples from the northern part of the state were similarly classified into eleven dominance types. Different ways of arranging the four or eleven composite samples in tables were tried. In only one arrangement (and its mirror image) would importance values decrease progressively through the columns away from the peak values for the species. By study of the tables, weights were assigned to species, ranging from 1 for species most strongly concentrated at one side of the table to 10 for those most strongly concentrated at the other side. Weighted averages were computed, and the samples were regrouped by ranges of values of the weighted averages into steps of a composite transect.

These techniques of ordination are very close to those I have used (Whittaker, 1951, 1956, 1960), except that trial arrangements of dominance types ('method of leading dominants') replace field transects and topographic composite transects as the preliminary ordination from which species weights are derived. Success of the trial arrangement of dominance types depends both on unimodality of the populations of major species and on preponderance of the effects of one complex-gradient over those of others which would introduce multidimensional relations among the dominance types.

The tables for trial arrangement of dominance types and the composite transects by weighted averages both supported the ideas of vegetation continuity and species individuality (Curtis & McIntosh, 1951; Brown & Curtis, 1952; Curtis, 1955; McIntosh, 1958). Distributions of species populations were observed to be of binomial form (Brown & Curtis, 1952; McIntosh, 1958; Maycock & Curtis, 1960) as suggested by Gause (1930) and Whittaker (1951, 1952, 1956). Characteristics of soils were shown to vary, apparently continuously, along the gradient (Brown & Curtis, 1952; Maycock & Curtis, 1960). Distributions of understory plant species (Gilbert & Curtis, 1953) of mosses and lichens on bark (Hale, 1955; Culberson, 1955), of birds (Bond, 1957) and of soil microfungi (Tresner, Backus & Curtis, 1954; Christensen, Whittingham & Novak, 1962) were studied, and these were found to form continua paralleling that of the forest trees. A prairie continuum was investigated by Curtis (1955; Orpurt & Curtis, 1957; cf. Dix & Butler, 1960; Looman, 1963; Knight, 1965), using species weights derived from field observation of distributional relations of species to topography. Dix (1959 see also Horikawa & Itow, 1958; Itow, 1963), used weighted averages for indication of intensity of grazing effects and as expressions of ecological distance along a grazing ecocline, an approach related to that of Dyksterhuis (1948). Some of the early Wisconsin studies used measurements of similarity of species distribution (Gilbert & Curtis, 1953; Hale, 1955; Culberson, 1955) and sample similarity (Culberson, 1955) in addition to the weighted-average ordination. Additional applications of trial arrangement of dominance types and ordination by weighted averages include Habeck (1959a, b), Christensen, Clausen & Curtis (1959), Maycock & Curtis (1960), Bray (1960), Lindsey, Petty, Sterling & van Asdall (1961), Anderson (1963), Larsen (1965), and Buell, Langford, Davidson & Ohmann (1966).

To the extent that the characteristics of the vegetation gradient emerge from the trial arrangement of dominance types, rather than from field observation, the arrangement of samples in early studies of the Wisconsin school is indirect ordination. Some studies, however, are in fact direct ordination (e.g. Curtis, 1955; Maycock, 1963). Although details of techniques differ, my own work and early work of the Wisconsin school are closely parallel studies in gradient analysis, with similar methods leading to similar results on the gradient relations of environments, species populations and communities.

(b) Conclusion

(1) In early work of the school of Wisconsin, samples were ordinated by grouping the samples into dominance types and arranging composite samples for these dominance types into ecological series in which species importance values decline smoothly away from peak values. Weights are applied to species according to their relative positions in the ecological series and used to compute weighted averages and arrange samples in a composite transect.

(2) The Wisconsin transects gave results convergent with those from my work in direct gradient analysis, including (a) vegetational continuity, (b) binomial distributions of species populations, and (c) species individuality. Studies of gradient relations of environments and soils, trees and subordinate plants, animals, and saprobes, in upland forest continua of Wisconsin provided also an effective illustration of the concept of ecocline.

(a) Discussion   (5) *Wisconsin comparative ordination*

With the publication of Bray & Curtis (1957), the Wisconsin school turned in a new direction. Clausen (1957a) and Bond (1957) calculated percentage similarities (of forest herb and forest bird populations) for all samples in their sets compared with one another, chose least similar samples as end points of a gradient, and arranged other samples in sequence along the gradient by their relative similarities to the two end points. This approach, of indirect ordination by sample comparison, was elaborated by Bray & Curtis (1957) into a system by which several axes of ordination could be extracted from the sample comparisons. Comparative ordination has two significant advantages. First, it makes possible the ordination of samples without prior assumptions about environmental gradients; it will even in some cases lead to the recognition of gradient relationships which were not observed in the field. Second, it provides means of arranging a set of samples in relation to two, three, or more axes of variation, multidirectional treatment that may be possible with weighted averages (Ellenberg, 1950, 1952; Waring & Major, 1964) but is usually difficult. The technique is basically simple, but its details are less so; the essential design of the technique may be outlined first.

(a) Samples are taken from an area of vegetation, or range of communities, that is to be studied. (b) Characteristics of the samples which seem most useful for ordination are chosen. (c) Using these characteristics, sample similarity values are computed for every sample compared with every other sample and the similarity values are compiled in a matrix. (d) For each sample the total of its similarity values with all other samples is obtained. That sample which has the lowest similarity total is chosen as one end point of the first axis of ordination. That sample which has the lowest similarity value with the first sample chosen becomes the opposite end point of that axis. (e) All samples of the set are ordinated along the first axis on the basis of their relative similarities to the two end-point samples. (f) Samples in the intermediate range of this axis are observed and from them the pair of samples with lowest similarity to each other and the first end-point samples are chosen for end-point samples of a second axis. (g) All samples are subjected to a second ordination in relation to this axis, on the basis of their similarity values with the new pair of end-point samples. (h) Samples in the intermediate ranges of both the first and second axes are observed, and from them the pair of samples with lowest similarity to each other and the second end-point samples are chosen for end points of the third axis. (i) All samples are subjected to a third ordination on the basis of their similarities with the third pair of end-point samples.

The Wisconsin school has its own choice of techniques for the various stages of the method. (a) Forest samples are taken by random pairs or other plotless techniques. (b) The characteristics of forest samples chosen for ordination have included densities, basal areas and frequencies of larger trees, of saplings and of shrubs, and frequencies of herbs. (c) The Wisconsin 'index of similarity' is, as discussed above, a variant on percentage similarity in which importance values are converted to percentages first horizontally, then vertically. (e) A number of means of computing relative positions of samples between two end-point samples have been tried. Clausen (1957a, b; cf. Bond, 1957) and Maycock & Curtis (1960) used a simple algebraic expression of position, Bray & Curtis (1957) a geometric treatment of sample similarities. Beals (1960) subtracted similarities from a threshold of 85% and computed position along the axis (x) by a relation based on the Pythagorean theorem, $x = (L^2 + D_1^2 - D_2^2)/2L$, in which D_1 and D_2 are distances of a given sample from the end-point samples and L is the distance of the end-point samples from each other. Distance of the sample from this axis (hence along other axes of variation) may be estimated from $e = \sqrt{(D_1^2 - x^2)}$. Loucks (1962) used an angular transformation of percentage similarity values and computed positions by the formula of Beals (1960).

Some other variations may be mentioned. Bray (1956) and Swindale & Curtis (1957) used indirect ordination of species to derive species weights for weighted-average ordination of samples. Curtis (1959) brought twenty-eight community types of Wisconsin into an ordination by a matrix and plexus treatment, computing community-type similarities from presence values for ground-level species. Patterns of environmental characteristics, community characteristics and species populations were plotted in relation to the two main axes (temperature and moisture) of the three-dimensional plexus. Curtis's book (1959), culminating his work, is a masterly treatment of the vegetation of a state, in which ordination is a means to the purpose of organizing and presenting a wealth of information on Wisconsin plant communities.

Maycock & Curtis (1960) used weighted averages (with weights based on field observations of species along the moisture gradient) to select the end-point samples for the first ordination. The second and third axes were derived in the usual manner, by choice of stands of lowest similarity among those of intermediate positions on the axis or axes already analyzed. The approach thus combines direct ordination for an observed principal axis of vegetational variation with indirect ordination for secondary axes of less evident meaning.

Loucks (1962) compared a three-dimensional indirect ordination by sample similarity with an ordination by environmental scalars—synthetic scales derived by combining values for several environmental factors. Use of three scalars (for moisture conditions, soil nutrients and local climate) produced an ordination related to that from sample comparison, although the axes of the two ordinations were obliquely related to one another. Monk (1965) arranged samples by percentage similarities of their values for five soil variables, thus indirectly ordinating samples by environmental characteristics. Knight (1965) showed occurrence of marked trends in representation of different plant forms along the moisture gradient in a prairie. Samples could be ordinated by representation of plant forms and the resulting ordination was strongly related to that by representation of species. Gittins (1965 *a*, *b*; see also Beals, 1965) has shown that Bray & Curtis's system may be used in inverse form, ordinating species by their distributional association in samples, with results closely related to those of ordination by sample comparison. Other studies applying indirect comparative ordination following Bray & Curtis (1957) include Ayyad & Dix (1964), McIntosh & Hurley (1964), Gittins (1965 *a*, *c*), White (1965), Larsen (1965), and Swan & Dix (1966).

Two limitations of the Wisconsin method may be observed. First, in some circumstances the comparison with end-point samples is a less sensitive basis of ordination than weighted averages. Comparison with end points uses distributional information less efficiently, and significance of the similarity values may be reduced by unrecognized bimodalities of species populations. It is probable that, when a primary gradient is recognized in the field, effectiveness of the ordination may be increased by Maycock & Curtis's (1960) approach of direct, weighted-average ordination along this primary gradient, combined with indirect, comparative ordination along second and third gradients.

Second, the orientation of axes is determined by extreme samples (Austin & Orloci, 1966). The technique may give the impression that its arithmetic produces a more objective choice of primary axis than an observed gradient accepted for direct ordination. The objectivity is illusory, for the particular orientations of axes are determined by accidents of sample selection in the field. The effect of sample choice on the choice

R. H. WHITTAKER

of axis is not, however, as serious an objection as it might seem. In many cases the technique will select first end-point samples which reasonably indicate the major axis of vegetational variation. The fact that this principal axis may be obliquely related to environmental gradients as ordinarily accepted is not fundamentally objectionable. It is also possible to design techniques for determination of the first axis by the long-axis of variation in the full sample set, rather than by extreme stands (Orloci, 1966).

Beyond the question of oblique axes, indirect ordination will in some cases arrange samples along axes that do not correspond to environmental gradients at all. In the ordination by Beals & Cottam (1960) the central part of the community pattern was occupied by samples including successional species of *Populus* and *Betula*. Samples representing a moisture gradient were curved around these, with the end points of the gradient, the wettest and driest samples, close together on one side of the pattern. The arrangement has ecological meaning. It expresses both the compositional relationship of the samples dominated by widespread successional species with all other samples of the pattern, and the sharing of bimodal species by samples from the extremes of the moisture gradient (cf. Maycock, 1963)—extremes which may be alike in some qualities of environment other than position along the 'moisture gradient'. A Wisconsin ordination is a constructed expression of similarities in species composition of samples; correspondence of its axes to environmental gradients may be sought but not assumed. If, however, ordination itself is the objective, then a position of relativism may be taken on oblique axes and patterns like that of Beals & Cottam (1960). Any compositional axes will serve for ordination and express compositional relationships of samples, and there is no need to require of these axes any particular direction or relation to environmental gradients recognized by ecologists.

For visual representation of results the Wisconsin school has developed charts and models in which sample positions, magnitudes of environmental factors, species importance values and population centres, and community types and trends are shown in relation to the three axes of ordination. Fig. 11 (Ayyad & Dix, 1964; cf. Bray & Curtis, 1957, fig. 7) illustrates the scatter of samples in a three-dimensional hyperspace, with the axes taken in pairs for two-dimensional representation. Sizes of the circles at sample points indicate importance values for the two species. The figures thus represent also the three-dimensional equivalent of the binomial curve and solid, an 'atmospheric distribution' (Bray & Curtis, 1957) in which importance values decline in all directions away from the population centre for the species. Fig. 12 (Bray & Curtis, 1957) shows centres of species populations in relation to the three axes of sample ordination, as an alternative to a species-association plexus.

(b) Conclusion

(1) Like most other indirect techniques, Wisconsin comparative ordination is a system of extracting directions of variation from similarity measurements. Values of percentage similarity of samples are compiled into a matrix. The sample which has lowest total similarity with other samples becomes the first end-point sample, and the sample with lowest similarity to it becomes the second end point. Usually three pairs of end-point samples are chosen to define three axes of variation. The samples are

ordinated in relation to these axes by their similarities with the end-point samples. A number of other techniques have been used.

(2) The axes are compositional gradients, and similarity values are used as ecological distances to locate samples along these axes. The axes should also represent ecoclines of changing community characteristics in relation to environmental complex-gradients (which may include successional or disturbance gradients). Relations of axes to environments in the field will not always, however, be intelligible ones. The system of axes defines a compositional hyperspace which should represent (in a transformed version of different proportions) the pattern of environments and communities in the landscape from which the samples were taken. Relations of environmental factors, species populations and community characteristics to one another in the hyperspaces have been studied by various authors of the school.

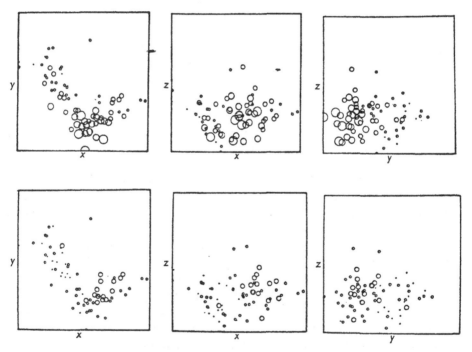

Fig. 11. Sample ordination and patterns of species importance in a three-dimensional hyperspace (Ayyad & Dix, 1964, fig. 2). Centres of circles locate samples in relation to the *y* (ordinate) and *x* (abscissa) axes on the left, the *z* and *x* axes in the middle, and the *z* and *y* axes on the right. Sizes of circles represent relative density values in these samples for two species, *Agropyron dasystachum* above, and *Koeleria cristata* below.

(3) Theoretical results are in accord with those of direct gradient analysis and the earlier Wisconsin research. The study of Bray & Curtis (1957) stated the concepts of compositional axis and atmospheric distribution (*n*-dimensional binomial solid) in application to species distributions. The strong point of the Wisconsin school's later work, however, is its versatile exploration of techniques of ordination.

248   R. H. WHITTAKER

(a) Discussion   (6) Factor analysis

The reader may have noted the resemblance of Wisconsin comparative ordination to the statistical techniques known as 'factor analysis'. Wisconsin ordination is, in fact, an approximation to factor analysis for ecological purposes, similar in design and objectives but based on simpler computations. Formal factor analysis (or principal component analysis) can also be applied to community samples, however, as shown by Goodall (1954a) and Dagnelie (1960, 1962b; see also Williamson, 1961b; Reyment, 1963; Colebrook, 1964; Groenewoud, 1965a; and Orloci, 1966). Ecologists interested in such applications may take advantage of the effective review by Greig-Smith (1964) and a discussion of notable lucidity by Dagnelie (1962b), to which the present account owes much. The computations are discussed by Thurstone (1947) and in other texts.

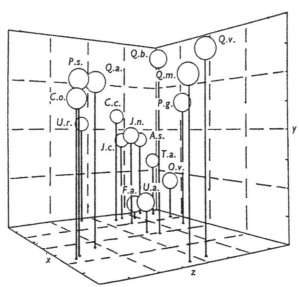

Fig. 12. An ordination of centres of species distribution in a three-dimensional hyperspace (Bray & Curtis, 1957, fig. 8). Tree species indicated by genus and species initials: *Acer saccharum, Carya cordiformis, C. ovata, Fraxinus americana, Juglans cinerea, J. nigra, Ostrya virginiana, Populus grandidentata, Prunus serotina, Quercus alba, Q. borealis, Q. macrocarpa, Q. velutina, Tilia americana, Ulmus americana, U. rubra.*

Measurements of distributional similarities of species in samples are first summarized in a matrix. Coefficients of correlation (comparing species distributions by importance values in samples) or coefficients of point correlation (comparing species distributions only by occurrence in samples) have been preferred to other similarity measurements for factor analysis. Values for distributional 'correlation' of each species with itself must be entered on the diagonal of the full matrix. The most appropriate of such values for factor analysis are estimated or computed 'communalities'. The matrix may be analyzed to extract from it underlying, initially unknown, 'factors' which, acting together in different combinations with different efficacies, determine the co-occurrence of species in samples.

Effects of 'factors' on species importance values may be expressed as

$$z_{ji} = a_{j1}F_{1i} + a_{j2}F_{2i} + a_{j3}F_{3i} \ldots + a_{jm}F_{mi} + a_{j}S_{ji}.$$

In this z_{ji} is the importance value of species j in sample i, F_{1i}, F_{2i}, $\ldots$, F_{mi} are the common 'factors' affecting more than one species and determining their importance values in samples, and S_{ji} is a special factor for that sample and species only. The coefficients a_{j1}, a_{j2}, $\ldots$, a_{jm} are 'loadings'. These express the correlations of the distribution of species j with the extracted common factors; but the loadings are interpreted as expressing the 'efficacy' of the factors in determing the species distribution, or the 'responsiveness' of the species' distribution to the factors. The analogy with an equation of multiple regression will be noted. The technique of factor analysis may in fact be conceived as solving first for coefficients (i.e. loadings) and second for values of unknown independent variables (i.e. extracted factors) when only the measurements of dependent variables (importance values) in samples are known. The solution must be obtained from analysis of the matrix of coefficients of correlation of the dependent variables with one another.

These coefficients of correlation may be represented by

$$r_{jk} = a_{j1}a_{k1} + a_{j2}a_{k2} + a_{j3}a_{k3} \ldots + a_{jm}a_{km}.$$

In this, r_{jk} is a coefficient of correlation for distributional similarity of species j and k. The magnitude of r_{jk} is taken to express the degree to which species j and k are alike in their distributional 'response' to combined effects of the factors F_1, F_2, $\ldots$, F_m, which correspond to the successive terms of the equation. Since values of r_{jk}, etc., are known, the matrix may be solved to obtain the loadings. An infinite number of solutions are possible. Factor analysis computations are designed to obtain a solution which is optimal in the sense of a maximum value for each successive column of loadings (for different species on a given factor, see bottom row of Table 2), hence maximum significance of a given factor in determining differences in species representation in samples. When the first factor has been extracted, a new matrix of residual coefficients of correlation, expressing the remaining effects of all other factors on species distribution and correlation, is computed. This matrix may be solved for a second factor, and a residual matrix may again be obtained and solved for a third factor, and so on until species correlations are judged sufficiently accounted for by the factors extracted.

Matrix analysis to this point yields loading values (Table 2). The right-hand column of the table (h_i^2) gives the sums of the squares of the loadings for each species. These values (communalities) consequently express the relative 'responsiveness' of species to the full set of factors extracted. The loadings may serve as co-ordinate values for ordinating species in a first (loading) hyperspace for which the extracted factors are axes. Such an ordination is illustrated in Fig. 13. Distance of a species' point from the origin on a given axis (its loading value for that factor) indicates the species' responsiveness to that factor. The shortest distance between the point and the origin is h,

250   R. H. WHITTAKER

the square root of the communality. Species near the origin are consequently those least correlated with, or least responsive to, the factors extracted.

The figure illustrates further steps of Dagnelie's (1960, 1962*b*) interpretive treatment. The circles include species known from field observation to have their populations centred in certain kinds of environments—dry poor soils (C), moist rich soils (A), and wetter soils (B). The species included are thus ecological groups possessing indicator value; they are also diagnostic-species groupings for the school of Braun-Blanquet and may be used to characterize community types. Circle (D) includes species of low communality which occur through the sample set with weak correlation with the factors; these are 'companions' in the terminology of phytosociology. Axes of the ordination may be rotated together, or separately rotated into oblique relation to each other, and new loading values may be calculated. The dashed lines in Fig. 13 are axes rotated to run through the three ecological groups; the new factor axes may then be identified with complex-gradients of soil characteristics.

Table 2. *Matrix of loadings on three factors for fourteen species of French beech forests* (*shortened from table 34, Dagnelie, 1960, with* a_j^2 *summed for these species only*)

Species	Loadings			h_i^2
	$a_{j\alpha}$	$a_{j\beta}$	$a_{j\gamma}$	
Fagus silvatica (h)	0·47	0·12	0·37	0·37
Sorbus aucuparia (h)	0·28	0·19	−0·38	0·26
Anemone nemorosa	0·48	−0·43	−0·17	0·44
Milium effusum	0·40	0·14	0·14	0·19
Festuca silvatica	0·64	−0·37	0·15	0·56
Luzula nemorosa	0·51	0·33	0·30	0·46
Oxalis acetosella	0·44	−0·10	0·43	0·38
Juncus sp.	0·11	0·46	0·27	0·29
Carex remota	0·24	0·28	0·22	0·19
Vaccinium myrtillus	−0·65	0·32	−0·19	0·56
Deschampsia flexuosa	−0·57	0·38	−0·18	0·50
Dicranum scoparium	−0·52	−0·18	0·27	0·37
Leucobryum glaucum	−0·48	−0·24	0·17	0·32
Hypnum cupressiforme	−0·29	−0·36	0·43	0·39
Σa_j^2	2·96	1·26	1·10	

Given loadings, values may be computed for the magnitudes of the extracted factor which, acting through the responses of the species, determine composition of each of the samples. Dagnelie (1960, pp. 140–6) describes a simplified approach to this computation. When these factor values have been computed, the position of each sample in relation to the factors is defined by them; and the samples may be ordinated in a second (factor-value) hyperspace (Fig. 14). As further steps, Dagnelie (1960, 1962*b*) has plotted species-importance values and forest-production values in relation to the factor-value hyperspace of Fig. 14, thus relating the extracted factors as presumed complex-gradients with gradients of species populations and community characteristics.

Techniques of factor analysis so far described are 'R' techniques, treating vegetation samples as individuals and species occurrences or importance values as attributes or tests. Like other techniques of indirect ordination, those of factor analysis can be

Gradient analysis of vegetation   251

turned through 90 degrees to treat species as 'individuals' and the samples in which they occur as 'attributes' in 'Q' techniques (Dagnelie, 1960). Dagnelie (1960, 1962b) has also applied factor analysis to nine attributes of environment of his beech samples. The matrix of correlation is in this case one of coefficients of correlation between factor gradients or other characteristics of sample environments; the extracted factors are common factors of environment which may be assumed to underlie the particular factor gradients.

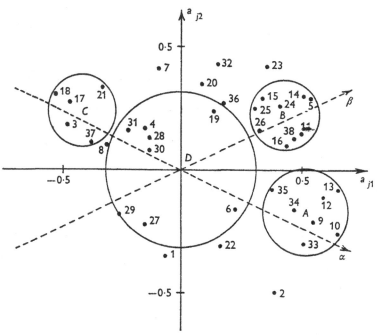

Fig. 13. An ordination of species (numbered points) in a loading hyperspace, from factor analysis of French beech forests (Dagnelie, 1962b, fig. 7). Circles enclose four ecological groups of species related to environmental factors and community types; dashed lines are axes rotated so that extracted factors will more nearly represent the environmental determinants of these groups.

As is the case with Wisconsin comparative ordination, extracted factors before rotation are as likely as not to be oblique in relation to environmental complex-gradients that ecologists recognize. Some extracted factors prove very difficult to interpret ecologically. It is probable that some of these are environmental effects which an ecologist is unable to recognize and that others are false factors resulting from the arithmetic of species correlations. Factor analysis is indeed not a simply 'objective' technique from the calculations of which will emerge the 'true' factors affecting a pattern of vegetation. The technique is demanding and it is unlikely to be profitable unless ecological skill, understanding and thoughtfulness are invested along with labour and computer time.

The most serious problems affecting factor analysis may result from the nature of compositional gradients. In psychological application, it may be assumed that

R. H. WHITTAKER

attributes (test scores) bear a relation to one another and the extracted factor that, if not strictly linear, is at least monotonic or unidirectional. The relation of environmental factor gradients to one another and an extracted common gradient is of analogous design, but the relationships of populations in a compositional gradient are not. Variation of a given attribute (species importance value) along the compositional gradient is normally curvilinear (the slope of the binomial curve), but will often also be bidirectional, sloping in opposite directions on the two sides of the mode. Slope of the importance value may reverse itself more than once along the gradient when the population distribution is bimodal.

If the samples are drawn from a narrow range of a compositional gradient, the species may form three distributional groups of about equal size. Species of two groups have their modes near or beyond the opposite limits of the range of the gradient sampled, and the importance values for species of these two groups decline in opposite directions along the gradient. Positive correlations within these groups and negative correlations between them are responsible for the extracted factor, in relation to which they have high loading values of opposite signs, and for the effectiveness of the ordination.

As samples are taken from a larger range of a gradient, a larger number of species will have their modes between the extremes of the gradient as sampled. The proportion of species with unidirectional slopes decreases in the total set of species, while the proportion with bidirectional slopes increases. Effectiveness of the unidirectional species for the factor analysis is decreased, especially within the central range of the gradient. Within this central range, positive correlations may be expected among species whose modes are close together and whose importance values decline together in both directions away from their modes (cf. Groenewoud, 1965 a). The implication of these correlations for factor analysis is unestablished, but unlikely to be favourable. It consequently seems that factor analysis is best suited for intensive studies of sample sets from narrower ranges of environments (Greig-Smith, 1964).

(b) Conclusion

(1) Mathematically more advanced techniques of factor analysis can be applied to vegetation samples, with results similar to those of Wisconsin comparative ordination. In this approach: (a) A matrix of distributional similarities of species is compiled. (b) The matrix is solved to yield one or more extracted 'factors' and 'loadings' of species expressing the degree to which their distributions are correlated with those factors. The loadings provide an ordination of species in a hyperspace of which the factors are axes (Fig. 13). (c) Factor values are computed for the samples. The factor values for a sample are estimates of magnitudes of the different extracted factors which (acting through the different loading values) determine species composition of the sample. Factor values permit ordination of samples in a hyperspace having the same extracted factors as axes, but different scales of magnitudes (factor values rather than loadings) along these axes (Fig. 14). (d) Values for environmental factor gradients, species populations and community characteristics for the samples can be plotted in the factor-value hyperspace and thus related to one another and the extracted factors.

Gradient analysis of vegetation   253

(*e*) Groups of species which are close together in the loading hyperspace, and groups of samples which are close together in the factor-value hyperspace, also provide a basis for vegetational classification if this is desired.

(2) These techniques: (*a*) are often laborious in relation to their research rewards; (*b*) produce extracted factors which require ecological understanding from other sources, and often rotation in the hyperspace, for their interpretation; (*c*) can probably in some cases produce extracted factors which are ecologically unidentifiable or even meaningless in relation to environmental gradients; (*d*) are subject to limitations resulting from the nature of species distributions and the limitations of correlations as means of expressing distributional relations of species; and (*e*) are increasingly affected by these limitations as the range of environments and communities represented in the sample set increases. The techniques consequently seem most appropriate for sample sets from narrower ranges of environment on which an investigator is willing to make heavy investment of effort and interpretation.

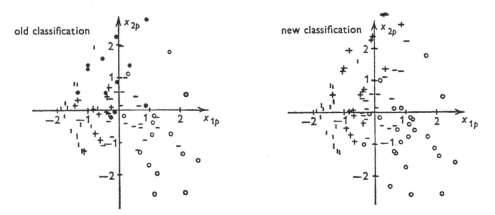

Fig. 14. An ordination of samples in a factor-value hyperspace, from factor analysis of French beech forests (Dagnelie, 1962, fig. 9). Samples have been classified into community types by a conventional classification by undergrowth, left, and by representation of the ecological groups of Fig. 13, right, as indicated by the symbols at sample points. Old classification: O, beech forest with *Festuca silvatica*; —, beech forest with *Luzula nemorosa* (fresh variants); +, beech forest with *L. nemorosa* (dry variant); ●, intermediate beech forest; |, beech forest with *Vaccinium myrtillus*. New classification by representation of ecological groups (see Fig. 13): O, groups A and B; —, groups B; +, groups B and C; | group C.

(3) The extracted factors may be regarded as compositional gradients. Factor analysis applied to natural communities thus converges with other techniques of indirect gradient analysis in meaning. An indirect gradient analysis should make possible: (*a*) a treatment causing the emergence from sample data of axes representing directions of variation in community composition, (*b*) use of these axes to ordinate samples and species in a hyperspace, (*c*) identification of environmental complex-gradients corresponding to these axes, in some cases, (*d*) observation of variations in environmental factors, species importance values and community characteristics through the hyperspace, and (*e*) comprehension, through the ordination and these observations, of the pattern of variation of the vegetation sampled.

R. H. WHITTAKER

IV. CONCLUSION AND SUMMARY

(1) *Choice of techniques*

Within gradient analysis as a method, two approaches may be distinguished—direct analysis in relation to an environmental gradient accepted as given and indirect analysis in which axes of variation are obtained from similarity measurements. Within each of these approaches a variety of techniques may be applied to different problems and purposes; choice among these techniques cannot be categorically stated. Rewards are often greatest, in fact, for the ecologist who experiments with more than one technique for clarifying the relationships in his sample set.

Advantages of direct and indirect techniques are evenly balanced in application to different circumstances. Direct gradient analysis has marked advantage in clarity and interpretability of its results, when conditions are favourable for the direct approach. In some circumstances, however, it is better not to accept environmental gradients as given, as a basis of ordination. Matrix and plexus techniques and Wisconsin comparative ordination can be applied to research circumstances and kinds of sample sets for which direct gradient analysis is inappropriate or ineffective. Direct gradient analysis is usually the approach of preference when environmental complex-gradients are recognized and samples can be taken to represent the gradients adequately in the sample set; indirect gradient analysis is the approach of preference in other circumstances.

As regards techniques of more, or less, mathematical complexity, varied purposes must be distinguished—the effort to understand a pattern of vegetation by the most direct and effective means, the testing of hypotheses on vegetation structure, and the exploratory application of mathematically advanced techniques to see what they may accomplish. All are well justified, but the first is the most common. Mathematical treatment has often the great merit of revealing relationships which are inherent in study material but otherwise unrecognizable. It is the case, however, and no fault of mathematically inclined ecologists, that the character of species distributions and compositional gradients imply diminishing returns from complex mathematical treatment of vegetation. Advances in theory of vegetation structure and classification have come predominantly from simpler techniques, from the results of which the basis of complex and indirect techniques may be understood. This is by no means to oppose mathematical treatment, but to balance against its interest and value the fact that for some kinds of vegetation research simpler and more direct methods, carried out with skill and perceptiveness, are more effective.

(2) *Concepts*

A natural community and its environment are linked together by cyclic and inter-determinant processes into an ecosystem. A landscape may be conceived as a pattern of ecosystems related to one another along environmental gradients. One may choose to distinguish, as aspects of the landscape pattern, an environmental pattern and a community pattern, which vary in parallel in the landscape space.

Along a single environmental gradient chosen for study, environments and com-

munities change together, forming an ecosystemic gradient or ecocline. The environmental part of the ecocline, comprising many gradients of particular environmental
factors, is a complex-gradient. The parallel gradient of communities is a coenocline,
and the coenocline includes various trends or gradients of particular characteristics of
communities. Species populations are generally distributed along complex-gradients
in the form of binomial curves, with their densities tapering on each side of the population optima. Form of the curve and location of the optimum express in part the
species' genecology—its adaptive centre and range of genetic differentiation as expressed in population distribution. Species evolve toward adaptive differences, as
indicated by different locations of their population optima, different niches within
communities, and the overlap of population distributions made possible by different
niches. Because of their scattered centres and distributional overlap, the species
populations along a complex-gradient form a population continuum, a compositional
gradient. A compositional gradient, and the coenocline of which it is one expression,
may be divided into segments, and the resulting sequence of community types is an
ecological series.

More than one ecocline may be recognized in a landscape pattern as directions of
variation of its environments and communities. Two or more complex-gradients as
axes define an environmental hyperspace, which may represent in abstract form much
of the environmental variation in the landscape. Two or more compositional gradients
as axes define a compositional hyperspace representing much of the variation in
vegetation in the landscape. In a hyperspace the binomial distributions of species
become binomial solids or atmospheric distributions. Groups of species whose
population centres are close together along a gradient or in the compositional hyperspace are ecological groups. Relative similarities of vegetation samples express their
ecological distances—their degrees of separation from one another along compositional
gradients or in compositional hyperspaces.

Representation of ecological groups and measurements of sample similarity provide
means by which samples can be located in the compositional hyperspace. The process
of locating or arranging samples, or species, or community types in relation to one
another along gradients or in a hyperspace is ordination. An ordination of community
types in relation to more than one gradient, a multidimensional ecological series, forms
a community-type pattern. The community types of the pattern are to be understood
as abstractions from the landscape pattern, interconnected by those gradients of
environments, species populations and community characteristics which gradient
analysis provides the means of studying in relation to one another.

(3) *Perspective*

The four preceding paragraphs state a system of interlinked concepts through which
the method and the relations of its various techniques to one another may be understood, and in which the principal theoretical contribution of gradient analysis is
embodied. This theoretical advance is the replacement of the conception of vegetation
as consisting of the units into which ecologists have classified that vegetation, units
often of unclear relation to one another, by the conception of vegetation as a pattern of

256   R. H. Whittaker

intergrading communities, a complex population continuum. Research techniques based on gradient relations and a multidimensional pattern, co-ordinate system, or hyperspace are appropriate to this conception.

Scientific knowledge may be viewed as a structure of relationships, and relationships of relationships, among unknown entities (Hutchinson, 1953). It is desired that the relationships should be, so far as possible, quantitatively expressed, experimentally established, tightly and coherently interrelated in deductive systems, and causal or predictive from lower levels of organization to higher. Results of gradient analysis cannot yet claim experimental verification, exact mathematical interlinkage, or predictive value except in a limited sense. Further understanding of the basis, in genetics and population dynamics, of observed patterns of species distribution must await further, more difficult research. Vegetation analysis is not one of the exact sciences; character, complexity, and relative looseness of the relationships studied imply that appropriate methods and feasible achievements differ widely from those of physical sciences (Whittaker, 1957, 1962; Slobodkin, 1965). It may be a fair measure of the contribution of gradient analysis that it has provided, despite these limitations, a clearer, deeper, and more coherent understanding of vegetation relationships, together with quantitative techniques appropriate to these relationships and a wide range of ecological problems.

V. REFERENCES

Agnew A. D. Q. (1961). The ecology of *Juncus effusus* L. in North Wales. *J. Ecol.* **49** 83–102.
Agrell I. (1941). Zur Ökologie der Collembolen. Untersuchungen im schwedischen Lappland. (Engl. summ.) *Opusc. ent.* suppl. 3 pp. 1–236.
Agrell I. (1945). The collemboles in nests of warmblooded animals with a method for sociological analysis. *Acta Univ. lund.* N.F., Avd. 2, **41**, (10), 1–19.
Agrell, I. (1963). A sociological analysis of soil Collembola. *Oikos* **14**, 237–47.
Allee, W. C., Emerson, A. E., Park, O., Park, T. & Schmidt, K. P. (1949). *Principles of animal ecology.* 837 pp. Philadelphia.
Allorge, P. (1927). (Comments following Lenoble, 1927.) *Bull. Soc. bot. Fr.* **73**, 892–3.
Anderson, D. J. (1963). The structure of some upland plant communities in Caernarvonshire. III. The continuum analysis. *J. Ecol.* **51**, 403–14.
Austin, M. P. & Orloci, L. (1966). Geometric models in ecology. II. An evaluation of some ordination techniques. *J. Ecol.* **54**, 217–27.
Ayyad, M. A. G. & Dix, R. L. (1964). An analysis of a vegetation-microenvironmental complex on prairie slopes in Saskatchewan. *Ecol. Monogr.* **34**, 421–42.
Balogh, J. (1958). *Lebensgemeinschaften der Landtiere, ihre Erforschung unter besonderer Berücksichtigung der zoozönologischen Arbeitsmethoden.* 560 pp. Budapest and Berlin.
Barkman, J. J. (1958). *Phytosociology and ecology of cryptogamic epiphytes.* 628 pp. Assen.
Barkman, J. J. (1965). Die Kryptogamenflora einiger Vegetationstypen in Drente und ihr Zusammenhang mit Boden und Mikroklima. (Engl. summ.) *Biosoziologie, Ber. Int. Symp. Stolzenau/Weser,* ed. R. Tüxen, 1960, pp. 157–71.
Beals, E. W. (1960). Forest bird communities in the Apostle Islands of Wisconsin. *Wilson Bull.* **72**, 156–81.
Beals, E. W. (1965). Species patterns in a Lebanese Poterietum. *Vegetatio* **13**, 69–87.
Beals, E. W. & Cope, J. B. (1964). Vegetation and soils in an eastern Indiana woods. *Ecology* **45**, 777–92.
Beals, E. W. & Cottam, G. (1960). The forest vegetation of the Apostle Islands, Wisconsin. *Ecology* **41**, 743–51.
Beard, J. S. (1955). The classification of tropical American vegetation types. *Ecology* **36**, 89–100.
Beschel, R. E. & Webber, P. J. (1962). Gradient analysis in swamp forests. *Nature, Lond.* **194**, 207–9.
Billings, W. D. (1952). The environmental complex in relation to plant growth and distribution. *Q. Rev. Biol.* **27**, 251–65.

Gradient analysis of vegetation   257

BLISS, L. C. (1963). Alpine plant communities of the Presidential Range, New Hampshire. *Ecology* 44, 678–97.

BOND, R. R. (1957). Ecological distribution of breeding birds in the upland forests of southern Wisconsin. *Ecol. Monogr.* 27, 351–84.

BOYKO, H. (1947). On the role of plants as quantitative climate indicators and the geo-ecological law of distribution. *J. Ecol.* 35, 138–57.

BRAUN-BLANQUET, J. (1928). À propos d'associations végétales. *Archs Bot. Bull. mens.* 2 (4), 67–68.

BRAUN-BLANQUET, J. (1951). *Pflanzensoziologie: Grundzüge der Vegetationskunde*, 2nd ed. 631 pp, 3rd ed., 1964, 865 pp. Wien.

BRAY, J. R. (1956). A study of mutual occurrence of plant species. *Ecology* 37, 21–8.

BRAY, J. R. (1960). The composition of savanna vegetation in Wisconsin. *Ecology* 41, 721–32.

BRAY, J. R. (1961). A test for estimating the relative informativeness of vegetation gradients. *J. Ecol.* 49, 631–42.

BRAY, J. R. & CURTIS, J. T. (1957). An ordination of the upland forest communities of southern Wisconsin. *Ecol. Monogr.* 27, 325–49.

BROWN, R. T. & CURTIS, J. T. (1952). The upland conifer-hardwood forests of northern Wisconsin. *Ecol. Monogr.* 22, 217–34.

BUELL, M. F., LANGFORD, A. N., DAVIDSON, D. W. & OHMANN, L. F. (1966). The upland forest continuum in northern New Jersey. *Ecology* 47, 416–32.

CAIN, S. A. (1947). Characteristics of natural areas and factors in their development. *Ecol. Monogr.* 17, 185–200.

CAJANDER, A. K. (1903, 1906). Beiträge zur Kenntniss der Vegetation der Alluvionen des nördlichen Eurasiens. I. Die Alluvionen des unteren Lena-Thales. *Acta Soc. Sci. fenn.* 32(1), 1–182.

CAJANDER, A. K. & ILVESSALO, Y. (1921). Über Waldtypen. II. *Acta for. fenn.* 20(1), 1–77.

CAMP, W. H. (1951). A biogeographic and paragenetic analysis of the American beech (*Fagus*). *Yb. Am. phil. Soc.* 1950, pp. 166–9.

CHRISTENSEN, E. M., CLAUSEN, J. J. & CURTIS, J. T. (1959). Phytosociology of the lowland forests of northern Wisconsin. *Am. Midl. Nat.* 62, 232–247.

CHRISTENSEN, M., WHITTINGHAM, W. F. & NOVAK, R. O. (1962). The soil microfungi of wet-mesic forests in southern Wisconsin. *Mycologia* 54, 374–88.

CLAUSEN, J. J. (1957a). A phytosociological ordination of the conifer swamps of Wisconsin. *Ecology* 38, 638–46.

CLAUSEN, J. J. (1957b). A comparison of some methods of establishing plant community patterns. *Bot. Tidsskr.* 53, 253–78.

CLEMENTS, F. E. (1936). Nature and structure of the climax. *J. Ecol.* 24, 252–84.

CLEMENTS, F. E., WEAVER, J. E. & HANSON, H. C. (1929). Plant competition: an analysis of community functions. *Publs Carnegie Instn* 398, 1–340.

COLE, L. C. (1949). The measurement of interspecific association. *Ecology* 30, 411–24.

COLE, L. C. (1957). The measurement of partial interspecific association. *Ecology* 38, 226–33.

COLEBROOK, J. M. (1964). Continuous plankton records: A principal component analysis of the geographical distribution of zooplankton. *Bull. mar. Ecol.* 6, 78–100.

COOK, C. W. & HURST, R. (1962). A quantitative measure of plant association on ranges in good and poor condition. *J. Range Mgmt* 15, 266–73.

COTTAM, G. & CURTIS, J. T. (1949). A method for making rapid surveys of woodlands by means of pairs of randomly selected trees. *Ecology* 30, 101–4.

COTTAM, G. & CURTIS, J. T. (1956). The use of distance measures in phytosociological sampling. *Ecology* 37, 451–60.

CULBERSON, W. L. (1955). The corticolous communities of lichens and bryophytes in the upland forests of northern Wisconsin. *Ecol. Monogr.* 25, 215–31.

CURTIS, J. T. (1955). A prairie continuum in Wisconsin. *Ecology* 36, 558–66.

CURTIS, J. T. (1959). *The vegetation of Wisconsin: An ordination of plant communities.* 657 pp. Madison, Wisc.

CURTIS, J. T. & MCINTOSH, R. P. (1950). The interrelations of certain analytic and synthetic phytosociological characters. *Ecology* 31, 434–55.

CURTIS, J. T. & MCINTOSH, R. P. (1951). An upland forest continuum in the prairie-forest border region of Wisconsin. *Ecology* 32, 476–96.

CZEKANOWSKI, J. (1909). Zur differential Diagnose der Neandertalgruppe. *KorrespBl. dt. Ges. Anthrop.* 40, 44–7 (*fide* Kulczyński, 1928).

DAGNELIE, P. (1960). Contribution à l'étude des communautés végétales par l'analyse factorielle. (Engl. summ.) *Bull. Serv. Carte phytogéogr.* B, 5, 7–71, 93–195.

DAGNELIE, P. (1962a). Étude statistique d'une pelouse à *Brachypodium ramosum*: les liaisons interspécifiques. *Bull. Serv. Carte phytogéogr.* B, 7, 85–97, 149–60.

258   R. H. WHITTAKER

DAGNELIE, P. (1962b). L'étude des communautés végétales par l'analyse des liaisons entre les espèces et les variables écologiques. 135 pp. Inst. Agron. de l'État, Gembloux.

DAHL, E. (1957). Rondane: Mountain vegetation in South Norway and its relation to the environment. Skr. norske Vidensk-Akad. Mat.-naturv. Kl. 1956, no. 3, pp. 1–374.

DAUBENMIRE, R. (1966). Vegetation: identification of typal communities. Science, N.Y. 151, 291–8.

DAWSON, G. W. P. (1951). A method for investigating the relationship between the distribution of individuals of different species in a plant community. Ecology 32, 332–4.

DEBAUCHE, H. R. (1962). The structural analysis of animal communities of the soil. Progress in soil zoology (ed. P. W. Murphy), pp. 10–25. London and Washington.

DICE, L. R. (1945). Measures of the amount of ecologic association between species. Ecology 26, 297–302.

DICE, L. R. (1952). Natural communities. 547 pp. Ann Arbor.

DIMO, N. A. & KELLER, B. A. (1907). In the semidesert regions. Soil and plant studies in the southern Tzaritzyn district of the Saratov province. Saratov. (In Russian.)

DIX, R. L. (1959). The influence of grazing on the thin-soil prairies of Wisconsin. Ecology 40, 36–49.

DIX, R. L. & BUTLER, J. E. (1960). A phytosociological study of a small prairie in Wisconsin. Ecology 41, 316–27.

DOING, H. (1963). Übersicht der floristischen Zusammensetzung, der Stuktur und der dynamischen Beziehungen niederländischer Wald- und Gebüschgesellschaften. (Engl. summ.) Meded. Landbouw-Hoogesch. Wageningen 63(2), 1–60.

DU RIETZ, G. E. (1921). Zur methodologischen Grundlage der modernen Pflanzensoziologie. 267 pp. Wien.

DYKSTERHUIS, E. J. (1948). Condition and management of range land based on quantitative ecology. J. Range Mgmt 2, 104–15.

ELLENBERG, H. (1948). Unkrautgesellschaften als Mass für den Säuregrad, die Verdichtung und andere Eigenschaften des Ackerbodens. Ber. Landtech. 4, 130–46.

ELLENBERG, H. (1950). Landwirtschaftliche Pflanzensoziologie. I. Unkrautgemeinschaften als Zeiger für Klima und Boden. 141 pp. Stuttgart.

ELLENBERG, H. (1952). Landwirtschaftliche Pflanzensoziologie. II. Wiesen und Weiden und ihre standörtliche Bewertung. 143 pp. Stuttgart.

ELLENBERG, H. (1956). Aufgaben und Methoden der Vegetationskunde. In Einführung in die Phytologie by H. Walter, vol. IV. Grundlagen der Vegetationsgliederung, Pt. 1, 136 pp. Stuttgart.

ELLENBERG, H. (1963). Vegetation Mitteleuropas mit den Alpen in kausaler, dynamischer and historischer Sicht. In Einführung in die Phytologie by H. Walter, vol. IV. Grundlagen der Vegetationsgliederung, Pt. 2, 943 pp. Stuttgart.

EVANS, F. C. (1956). Ecosystem as the basic unit in ecology. Science, N.Y. 123, 1127–8. Readings in ecology, pp. 166–7. Ed. E. J. Kormondy. Englewood Cliffs, N.J. 1965.

EVANS, F. C. & DAHL, E. (1955). The vegetational structure of an abandoned field in southeastern Michigan and its relation to environmental factors. Ecology 36, 685–706.

FAGER, E. W. (1957). Determination and analysis of recurrent groups. Ecology 38, 586–95.

FAGER, E. W. (1963). Communities of organisms. The sea, vol. II, pp. 415–37. Ed. M. N. Hill. New York.

FAGER, E. W. & McGOWAN, J. A. (1963). Zooplankton species groups in the North Pacific. Science, N.Y. 140, 453–60.

FALIŃSKI, J. B. (1958). Nomogramy i tablice wspołczynników podobieństwa między zdjęciami fitosocjologicznymi według wzoru Jaccarda i Steinhausa. (Engl. summ.) Acta Soc. bot. Pol. 27 115–30.

FALIŃSKI J. B. (1960). Zastosowanie taksonomii wrocławskiej do fitosocjologii. (Germ. summ.) Acta Soc. bot. Pol. 29 333–61.

FISHER, R. A. (1936). The use of multiple measurements in taxonomic problems. Ann. Eugen. 7, 179–188.

FLORENCE, R. C. (1964). Edaphic control of vegetational pattern in east coast forests. Proc. Linn. Soc. N.S.W. 89, 171–90.

FORBES, S. A. (1907). On the local distribution of certain Illinois fishes: an essay in statistical ecology. Bull. Ill. St. Lab. nat. Hist. 7, 273–303.

FORBES, S. A. (1925). Method of determining and measuring the associative relations of species. Science, N.Y. 61, 524.

Фрей Т. Е. А. (Frie, T. E. A.) (1963). О применении корреляционного коэффициента межвидовых отношений. Ботан. Журнал СССР. Bot. Zh. U.S.S.R. 48, 235–39. (Biol. Abstr. 45, 40701, 1964.)

GAMS, H. (1961). Erfassung und Darstellung mehrdimensionaler Verwandtschaftsbeziehungen von Sippen und Lebensgemeinschaften. Ber. geobot. Inst. ETH, Stiftg. Rübel, Zürich, 1960, 32, 96–115.

GARDNER, J. L. (1951). Vegetation of the creosotebush area of the Rio Grande valley in New Mexico. Ecol. Monogr. 21, 379–403.

GAUSE, G. F. (1930). Studies on the ecology of the Orthoptera. Ecology 11, 307–25.

Gradient analysis of vegetation   259

GAUSE, G. F. (1934). *The struggle for existence*. 163 pp. Baltimore.

GHENT, A. W. (1963). Kendall's 'tau' coefficient as an index of similarity in comparisons of plant or animal communities. *Can. Ent.* **95**, 568–75.

GILBERT, M. L. & CURTIS, J. T. (1953). Relation of the understory to the upland forest in the prairie-forest border region of Wisconsin. *Trans. Wis. Acad. Sci. Arts Lett.* **42**, 183–95.

GIMINGHAM, C. H. (1961). North European heath communities: a 'network of variation'. *J. Ecol.* **49**, 655–94.

GITTINS, R. (1965 a). Multivariate approaches to a limestone grassland community. I. A stand ordination. *J. Ecol.* **53**, 385–401.

GITTINS, R. (1965 b). Multivariate approaches to a limestone grassland community. II. A direct species ordination. *J. Ecol.* **53**, 403–9.

GITTINS, R. (1965 c). Multivariate approaches to a limestone grassland community. III. A comparative study of ordination and association-analysis. *J. Ecol.* **53**, 411–25.

GLEASON, H. A. (1920). Some applications of the quadrat method. *Bull. Torrey bot. Club* **47**, 21–33.

GLEASON, H. A. (1926). The individualistic concept of the plant association. *Bull. Torrey bot. Club* **53**, 7–26.

GLEASON, H. A. (1939). The individualistic concept of the plant association. *Am. Midl. Nat.* **21**, 92–110.

GOODALL, D. W. (1952). Quantitative aspects of plant distribution. *Biol. Rev.* **27**, 194–245.

GOODALL, D. W. (1953 a). Objective methods for the classification of vegetation. I. The use of positive interspecific correlation. *Aust. J. Bot.* **1**, 39–63.

GOODALL, D. W. (1953 b). Objective methods for the classification of vegetation. II. Fidelity and indicator value. *Aust. J. Bot.* **1**, 434–56.

GOODALL, D. W. (1954 a). Objective methods for the classification of vegetation. III. An essay in the use of factor analysis. *Aust. J. Bot.* **2**, 304–24.

GOODALL, D. W. (1954 b). Vegetational classification and vegetational continua. (Germ. summ.) *Angew. PflSoziol.*, Wein, Festschr. Aichinger, **1**, 168–82.

GOODALL, D. W. (1962). Bibliography of statistical plant sociology. *Excerpta bot.* B, **4**, 253–322.

GOODALL, D. W. (1963). The continuum and the individualistic association. (French summ.) *Vegetatio* **11**, 297–316.

GOODALL, D. W. (1965). Plot-less tests of interspecific association. *J. Ecol.* **53**, 197–210.

GOUNOT, M. (1959). L'exploitation mécanographique des relevés pour la recherche des groupes écologiques. *Bull. Serv. Carte phytogéogr.* B, **4**, 147–77.

GREIG-SMITH, P. (1952). Ecological observations on degraded and secondary forest in Trinidad, British West Indies. II. Structure of the communities. *J. Ecol.* **40**, 316–30.

GREIG-SMITH, P. (1964). *Quantitative plant ecology*, 2nd ed. 256 pp. London and Washington.

GROENEWOUD, H. van. (1965 a). Ordination and classification of Swiss and Canadian forests by various biometric and other methods. (Germ. summ.) *Ber. geobot. Inst. ETH, Stiftg. Rübel*, Zürich, 1964, **36**, 28–102.

GROENEWOUD, H. van. (1965 b). An analysis and classification of white spruce communities in relation to certain habitat features. *Can. J. Bot.* **43**, 1025–36.

GUINOCHET, M. (1955). *Logique et dynamique du peuplement végétale*. 143 pp. Paris.

GUINOCHET, M. & CASAL, P. (1957). Sur l'analyse différentielle de Czekanowski et son application à la phytosociologie. *Bull. Serv. Carte phytogéogr.* B, **2**, 25–33.

HABECK, J. R. (1959 a). A phytosociological study of the upland forest communities in the central Wisconsin sand plain area. *Trans. Wis. Acad. Sci. Arts Lett.* **48**, 31–48.

HABECK, J. R. (1959 b). A vegetational study of the central Wisconsin winter deer range. *J. Wildl. Mgmt* **23**, 273–8.

HAIRSTON, N. G. (1951). Interspecies competition and its probable influence upon the vertical distribution of Appalachian salamanders of the genus *Plethodon*. *Ecology* **32**, 266–74.

HALE, M. E., Jr. (1955). Phytosociology of corticolous cryptogams in the upland forests of southern Wisconsin. *Ecology* **36**, 45–63.

HANSEN, H. MØLHOLM (1930). Studies on the vegetation of Iceland. *The botany of Iceland*, vol. III, pt. I, no. 10. Ed. L. K. Rosenvinge and E. Warming, pp. 1–186. Copenhagen.

HANSEN, H. MØLHOLM. (1932). Nørholm Hede, en formationsstatistisk Vegetationsmonografi. (Engl. summ.). *K. danske Vidensk. Selsk. Skr.*, Naturv. Math. Afd. ser. 9, **3**(3), 99–196.

HARDIN, G. (1960). The competitive exclusion principle. *Science, N.Y.* **131**, 1292–7.

HARTL, H. (1963). Die Vegetation des Eisenhutes im Kärntner Nockgebiet. *Mitt naturw. Ver. Kärnten* **73/153**, 293–336.

HEGG, O. (1965). Untersuchungen zur Pflanzensoziologie und Ökologie im Naturschutzgebiet Hohgant (Berner Voralpen). (French and Engl. summs.). *Beitr. geobot, Landesaufn. Schweiz* **46**, 1–188.

HEYWOOD, V. H. & McNEILL, J. (eds) (1964). Phenetic and phylogenetic classification. *Publs Syst. Ass.* **6**, 1–164.

260   R. H. WHITTAKER

HOPKINS, B. (1957). Pattern in the plant community. *J. Ecol.* **45**, 451–63.

HORIKAWA, Y. & ITOW, S. (1958). The vegetational continuum and the plant indicators for disturbance in the grazing grassland. (Japan. with Engl. summ.) *Jap. J. Ecol.* **8**, 123–8.

HOSOKAWA, T. (1955). An introduction of 2 × 2 table methods into the studies of the structure of plant communities. (Japan. with Engl. summ.) *Jap. J. Ecol.* **5**, 58–62, 93–100; 150–3.

HOSOKAWA, T., OMURA, M. & NISHIHARA, Y. (1957). Grading and integration of epiphyte communities. *Jap. J. Ecol.* **7**, 93–8.

HUGHES, R. E. & LINDLEY, D. V. (1955). Application of biometric methods to problems of classification in ecology. *Nature, Lond.* **175**, 806–7.

HUTCHINSON, G. E. (1953). The concept of pattern in ecology. *Proc. Acad. nat. Sci. Philad.* **105**, 1–12. *Readings in population and community ecology,* pp. 2–13. Ed. W. E. Hazen. Philadelphia and London. 1964.

HUTCHINSON, G. E. (1957). Concluding remarks. *Cold Spring Harb. Symp. quant. Biol.* **22**, 415–27.

ИЛЬИНСКИЙ А. П. и Посельская М. А. (Iljinski, A. P. & Poselskaya, M. A.) (1929). К вопросу об ассоциированности растении. *Труд. по прикл. ботанике, генетике и селекции. Trud. prikl. Bot. genet. Selek.* **20**, 459–74. (Engl. summ.)

ILVESSALO, Y. (1922). Vegetationsstatistische Untersuchungen über die Waldtypen. *Acta for. fenn.* **20**(3), 1–73.

ITOW, S. (1963). Grassland vegetation in uplands of western Honshu, Japan. II. Succession and grazing indicators. *Jap. J. Bot.* **18**, 133–67.

IZDEBSKI, K. (1963a). Bory na Roztoczu Środkowym. (Engl. and Russ. summs.) *Annls Univ. Mariae Curie-Skłodowska,* C, 1962, **17**, 313–62.

IZDEBSKI, K. (1963b). Zbiorowiska leśne na Roztoczu Środkowym. Uogólnienie i uzupełnienie. (Engl. summ.) *Acta Soc. bot. Pol.* **32**, 349–74.

JACCARD, P. (1902). Lois de distribution florale dans la zone alpine. *Bull. Soc. vaud. Sci. nat.* **38**, 69–130.

JOHNSON, P. L. & BILLINGS, W. D. (1962). The alpine vegetation of the Beartooth Plateau in relation to cryopedogenic processes and patterns. *Ecol. Monogr.* **32**, 105–35.

KELLER, B. A. (1925–26). Die Vegetation auf den Salzböden der russischen Halbwüsten und Wüsten. *Z. Bot.* **18**, 113–37.

KERSHAW, K. A. (1961). Association and co-variance analysis of plant communities. *J. Ecol.* **49**, 643–54.

KNIGHT, D. H. (1965). A gradient analysis of Wisconsin prairie vegetation on the basis of plant structure and function. *Ecology* **46**, 744–7.

KONTKANEN, P. (1950). Quantitative and seasonal studies on the leafhopper fauna of the field stratum on open areas in North Karelia. (Finn. summ.) *Annls zool., Soc. Zool.-bot. fenn. Vanamo* **13**(8), 1–91.

KUJALA, V. (1945). Waldvegetationsuntersuchungen in Kanada mit besonderer Berücksichtigung der Anbaumöglichkeiten kanadischer Holzarten auf natürlichen Waldböden in Finnland. *Annls Acad. Sci. fenn.* A, 4 Biol. **7**, 1–434.

KULCZYŃSKI, S. (1928). Die Pflanzenassoziationen der Pieninen. *Bull. int. Acad. pol. Sci. Lett.,* Cl. Sci. Math. Nat., sér. B, 1927 (Suppl. 2), 57–203.

LAMBERT, J. M. & DALE, M. B. (1964). The use of statistics in phytosociology. *Adv. ecol. Res.* **2**, 59–99.

LAMBERT, J. M. & WILLIAMS W. T. (1962). Multivariate methods in plant ecology. IV. Nodal analysis. *J. Ecol.* **50**, 775–802.

LARSEN, J. A. (1965). The vegetation of the Ennadai Lake Area, N.W.T: Studies in subarctic and arctic bioclimatology. *Ecol. Monogr.* **35**, 37–59.

LENOBLE, F. (1926, 1927). À propos des associations végétales. *Bull. Soc. bot. Fr.* **73**, 873–93.

LENOBLE, F. (1928). Associations végétales et espèces. *Archs Bot. Bull. mens.* **2**(1), 1–14.

LINDSEY, A. A., PETTY, R. O., STERLING, D. K. & VAN ASDALL, W. (1961). Vegetation and environment along the Wabash and Tippecanoe Rivers. *Ecol. Monogr.* **31**, 105–56.

LOOMAN, J. (1963). Preliminary classification of grasslands in Saskatchewan. *Ecology* **44**, 15–29.

LOOMAN, J. & CAMPBELL, J. B. (1960). Adaptation of Sorensen's K (1948) for estimating unit affinities in prairie vegetation. *Ecology* **41**, 409–16.

LOUCKS, O. L. (1962). Ordinating forest communities by means of environmental scalars and phytosociological indices. *Ecol. Monogr.* **32**, 137–66.

MACARTHUR, R. H. (1960). On the relative abundance of species. *Am. Nat.* **94**, 25–36, and *Readings in population and community ecology,* pp. 307–18. Ed. W. E. Hazen. Philadelphia and London. 1964.

MACARTHUR, R. H. (1965). Patterns of species diversity. *Biol. Rev.* **40**, 510–33.

MACFADYEN, A. (1963). *Animal ecology, aims and methods,* 2nd, ed. 344 pp. London, New York and Toronto.

McDONOUGH, W. T. (1963). Interspecific associations among desert plants. *Am. Midl. Nat.* **70**, 291–9.

McINTOSH, R. P. (1957). The York Woods, a case history of forest succession in southern Wisconsin. *Ecology* **38**, 29–37.

McIntosh, R. P. (1958). Plant communities. *Science, N.Y.* **128**, 115–20.

McIntosh, R. P. (1962). Pattern in a forest community. *Ecology* **43**, 25–33.

McIntosh, R. P & Hurley, R. T. (1964). The spruce-fir forests of the Catskill Mountains. *Ecology* **45**, 314–26.

McMillan C. (1960). Ecotypes and community function. *Am. Nat.* **94** 245–55.

McMillan C. (1965). Grassland community fractions from central North America under simulated climates. *Am. J. Bot.* **52**, 109–16.

McVean, D. N. & Ratcliffe, D. A. (1962). *Plant communities of the Scottish Highlands,* 445 pp. London.

Mahalanobis, P. C.(1936). On the generalized distance in statistics. *Proc. natn. Inst. Sci. India* **2**, 49–55.

Major, J. (1951). A functional, factorial approach to plant ecology. *Ecology* **32**, 392–412.

Mark, A. F. (1963). Vegetation studies on Secretary Island, Fiordland. 3: The altitudinal gradient in forest composition, structure and regeneration. *N.Z. Jl Bot.* **1**, 188–202.

Mason, H. L. (1947). Evolution of certain floristic associations in western North America. *Ecol. Monogr.* **17**, 201–10.

Matuszkiewicz, A. (1955). Stanowisko systematyczne i tendencje rozwojowe dąbrów białowieskich. (French summ.). *Acta Soc. bot. Pol.* **24**, 459–94.

Matuszkiewicz, A. (1958). Materiały do fitosocjologicznej systematyki buczyn i pokrewnych zespołów (związek *Fagion*) w Polsce. (Germ. summ.). *Acta Soc. bot. Pol.* **27**, 675–725.

Matuszkiewicz, W. (1947). Zespoły leśne południowego Polesia. (Engl. summ.) *Annls Univ. Mariae Curie-Skłodowska,* E, **2**, 69–138.

Matuszkiewicz, W. (1948). Roślinność lasów okolik Lwowa. (Engl. summ.) *Annls Univ. Mariae Curie-Skłodowska,* C, **3**, 119–93.

Matuszkiewicz, W. (1950). Badania fitosocjologiczne nad lasami bukowymi w Sudetach. (Russ. and Engl. summs.) *Annls Univ. Mariae Curie-Skłodowska,* C (suppl.), **5**, 1–196.

Matuskiewicz, W. (1952). Zespoły leśne Białowieskiego Parky Narodowego. (Russ. and Germ. summs.). *Annls Univ. Mariae Curie-Skłodowska,* C (suppl.), **6**, 1–218.

Maycock, P. F. (1963). The phytosociology of the deciduous forests of extreme southern Ontario. *Can. J. Bot.* **41**, 379–438.

Maycock, P. F. & Curtis, J. T. (1960). The phytosociology of boreal conifer-hardwood forests of the Great Lakes region. *Ecol. Monogr.* **30**, 1–35.

Merriam, C. H. (1898). Life zones and crop zones of the United States. *Bull. U.S. biol. Surv.* **10**, 1–79

Monk, C. D. (1965). Southern mixed hardwood forest of northcentral Florida. *Ecol. Monogr.* **35**, 335–54.

Morisita, M. (1959). Measuring of interspecific association and similarity between communities. *Mem. Fac. Sci. Kyushu Univ.* E, **3**, 65–80.

Motomura, I. (1952). Comparison of communities based on correlation coefficients. (Japanese.) *Ecol. Rev.* **13**, 67–71 (*fide* Dagnelie, 1960).

Motyka, J. (1947). O zadaniach i metodach badań geobotanicznych. (French summ.). *Annls Univ. Mariae Curie-Skłodowska,* C (suppl.), **1**, 1–168.

Motyka, J., Dobrzański, B. & Zawadzki, S. (1950). Wstępne badania nad łąkami południowo-wschodniej Lubelszczyzny. (Russ. and Engl. summs.) *Annls Univ. Mariae Curie-Skłodowska,* E, **5**, 367–447.

Motyka, J. & Zawadzki, S. (1953). Badania nad łąkami w dolinie Huczwy koło Werbkowic. (Russ. and Engl. summs.) *Annls Univ. Mariae Curie-Skłodowska,* E, **8**, 167–231.

Mueggler, W. F. (1965). Ecology of seral shrub communities in the cedar-hemlock zone of northern Idaho. *Ecol. Monogr.* **35**, 165–85.

Nash, C. B. (1950). Associations between fish species in tributaries and shore waters of western Lake Erie. *Ecology* **31**, 561–6.

Nichols, G. E. (1929). Plant associations and their classification. *Proc. Int. Congr. Plant Sci.*, Ithaca, 1926, vol. 1, pp. 629–41.

Nordhagen, R. (1943). Sikilsdalen og Norges Fjellbeiter. En plantesosiologisk monografi. *Bergens Mus. Skr.* **22**, 1–607.

Odum, E. P. (1950). Bird populations of the Highlands (North Carolina) Plateau in relation to plant succession and avian invasion. *Ecology* **31**, 587–605, and *Readings in population and community ecology*, pp. 350–68. Ed. W. E. Hazen, Philadelphia and London. 1964.

Odum, E. P. (1959). *Fundamentals of ecology*, 2nd ed. 546 pp. Philadelphia and London.

Omura, M. & Hosokawa, T. (1959). On the detailed structure of a corticolous community analysed on the basis of interspecific association. *Mem. Fac. Sci. Kyushu Univ.* E, **3**, 51–63.

Ono, Y. (1961). An ecological study of the brachyuran community on Tomioka Bay, Amakusa, Kyushu. *Rec. oceanogr. Wks Japan,* spec. no. 5, pp. 199–210.

262   R. H. WHITTAKER

ORLOCI, L. (1966). Geometric models in ecology. I. The theory and application of some ordination methods. *J. Ecol.* **54**, 193–215.

ORPURT, P. A. & CURTIS, J. T. (1957). Soil microfungi in relation to the prairie continuum in Wisconsin. *Ecology* **38**, 628–37.

PALMGREN, P. (1928). Zur Synthese pflanzen- und tierökologischer Untersuchungen. *Acta zool. fenn.* **6** 1–51.

PATTEN, D. T. (1963). Vegetational pattern in relation to environments in the Madison Range, Montana. *Ecol. Monogr.* **33**, 375–406.

PAVILLARD, J. (1928). Espèces et associations. *Archs Bot. Bull. mens.* **2**(4), 68–72.

PEARSON, R. G. (1963). Coleopteran associations in the British Isles during the late Quaternary period. *Biol. Rev.* **38**, 334–63.

POORE, M. E. D. (1956). The use of phytosociological methods in ecological investigations. IV. General discussion of phytosociological problems. *J. Ecol.* **44**, 28–50.

POORE, M. E. D. (1962). The method of successive approximation in descriptive ecology. *Adv. ecol. Res.* **1**, 35–68.

POORE, M. E. D. & McVEAN, D. N. (1957). A new approach to Scottish mountain vegetation. *J. Ecol.* **45**, 401–39.

RAABE, E.-W. (1952). Über den 'Affinitätswert' in der Pflanzensoziologie. *Vegetatio* **4**, 53–68.

Раменский Л. Г. (Ramensky, L. G.) (1924). Управление по опытному делу Средне-Черноземной области наркомаземледелия. *Вестник опытного дела, Воронеж.* (Abstract *Bot. Zlb.* **7**, 1926, 453–5).

RAMENSKY, L. G. (1930). Zur Methodik der vergleichenden Bearbeitung und Ordnung von Pflanzenlisten und anderen Objekten, die durch mehrere, verschiedenartig wirkende Faktoren bestimmt werden. *Beitr. Biol. Pfl.* **18**, 269–304.

RAMSAY, D. McC. (1964). An analysis of Nigerian savanna. II. An alternative method of analysis and its application to the Gombe sandstone vegetation. *J. Ecol.* **52**, 457–66.

RAMSAY, D. McC. & DELEEUW, P. N. (1964). An analysis of Nigerian savanna. I. The survey area and the vegetation developed over Bima sandstone. *J. Ecol.* **52**, 233–54.

RAMSAY, D. McC. & DELEEUW, P. N. (1965a). An analysis of Nigerian savanna. III. The vegetation of the Middle Gongola region by soil parent materials. *J. Ecol.* **53**, 643–60.

RAMSAY, D. McC. & DELEEUW, P. N. (1965b). An analysis of Nigerian savanna. IV. Ordination of vegetation developed on different parent materials. *J. Ecol.* **53**, 661–77.

RATCLIFFE, D. A. (1959). The vegetation of the Carneddau, North Wales. I. Grasslands, heaths and bogs. *J. Ecol.* **47**, 371–413.

RAYSON, P. (1957). Dark Island heath (Ninety-Mile Plain, South Australia). II. The effects of micro-topography on climate, soils, and vegetation. *Aust. J. Bot.* **5**, 86–102.

RENKONEN, O. (1938). Statistisch-ökologische Untersuchungen über die terrestriche Käferwelt der finnischen Bruchmoore. (Finn. summ.) *Annls zool., Soc. Zool.-bot fenn. Vanamo* **6**(1), 1–231.

REYMENT, R. A. (1963). Multivariate analytical treatment of quantitative species associations: an example from palaeoecology. *J. Anim. Ecol.* **32**, 535–47.

ROWE, J. S. (1956). Uses of undergrowth plant species in forestry. *Ecology* **37**, 461–73.

Самбук Ф. В. (Sambuk, F. V.) (1930). Ботанико-географический очерк долины реки Печоры. *Труд. Ботан. Музея Акад. Наук СССР. Trud. bot. Mus. Akad. Nauk U.S.S.R.* (3) **22**, 49–155. (Germ. summ.)

SCOTT, G. A. M., MARK, A. F. & SANDERSON, F. R. (1964). Altitudinal variation in forest composition near Lake Hankinson, Fiordland. *N.Z. Jl Bot.* **2**, 310–23.

SLOBODKIN, L. B. (1965). On the present incompleteness of mathematical ecology. *Am. Scient.* **53**, 347–57.

Сочава В. Б. (Sočava, V. B.) (1927). Ботанический очерк лесов полярного Урала от р. Нельки до р. Хулги. *Труд. Ботан. Музея Акад. Наук СССР. Trud. bot. Mus. Akad. Nauk U.S.S.R.* (3) **21**, 1–78. (Germ. summ.)

SOKAL, R. R. (1965). Statistical methods in systematics. *Biol. Rev.* **40**, 337–91.

SOKAL, R. R. & SNEATH, P. H. A. (1963). *Principles of numerical taxonomy.* 359 pp. San Francisco and London.

Соколова Л. А. (Sokolowa, L. A.) (1935, 1937). Материалы к геоботаническому районированию Онего-Северодвинского водораздела и Онежского полуострова. *Труд. Ботан. Инстит. Акад. Наук СССР. Trud. bot. Inst. Akad. Nauk U.S.S.R.* (3)**2**, 9–80 (Germ. summ.)

SØRENSEN, T. A. (1948). A method of establishing groups of equal amplitude in plant sociology based on similarity of species content, and its application to analyses of the vegetation on Danish commons. *K. danske Vidensk. Selsk. Biol. Skr.* **5**(4), 1–34.

Gradient analysis of vegetation   263

STEWART, G. & KELLER, W. (1936). A correlation method for ecology as exemplified by studies of native desert vegetation. *Ecology* **17**, 500–14.

SUKACHEV, V. N. (1928). Principles of classification of the spruce communities of European Russia. *J. Ecol.* **16**, 1–18.

SUKATSCHEW, W. N. (1932). Die Untersuchung der Waldtypen des osteuropäischen Flachlandes. *Handb. biol. ArbMeth.* **11**, 6, 191–250.

SWAN, J. M. A. & DIX, R. L. (1966). The phytosociological structure of upland forest at Candle Lake, Saskatchewan. *J. Ecol.* **54**, 13–40.

SWINDALE D. N. & CURTIS J. T. (1957). Phytosociology of the larger submerged plants in Wisconsin lakes. *Ecology* **38**, 397–407.

TANSLEY, A. G. (1935). The use and abuse of vegetational concepts and terms. *Ecology* **16**, 284–307.

THURSTONE, L. L. (1947). *Multiple-factor analysis: A development and expansion of the vectors of mind.* 535 pp. Chicago.

TRESNER, H. D., BACKUS, M. P. & CURTIS, J. T. (1954). Soil microfungi in relation to the hardwood forest continuum in southern Wisconsin. *Mycologia* **46**, 314–33.

TUOMIKOSKI, R. (1942). Untersuchungen über die Vegetation der Bruchmoore in Ostfinnland. I. Zur Methodik der pflanzensoziologischen Systematik. *Annls bot., Soc. Zool.-bot. fenn. Vanamo* **17** (1), 1–203.

VASILEVICH, V. I. (1961). Association between species and the structure of a phytocoenosis. (In Russian.) *Dokl. Akad. Nauk SSSR* **139**, 1001–4. *Dokl. Bot. Sci. Sect. Translation* **139**, 133–5, 1962.)

VASILEVICH, V. I. (1962). The quantitative dimension of similarity between phytocoenoses. (In Russian.) *Problemy Bot.* **6**, 83–94. (*Biol. Abstr.* **45**, 76419, 1964.)

VRIES, D. M. de. (1953). Objective combinations of species. *Acta bot. neerl.* **1**, 497–9.

VRIES, D. M. de, BARETTA, J. P. & HAMMING, G. (1954). Constellation of frequent herbage plants, based on their correlation in occurrence. *Vegetatio* **5/6** 105–11.

WACE, N. M. (1961). The vegetation of Gough Island. *Ecol. Monogr.* **31**, 337–67.

WALLACE, B. & SRB, A. M. (1964). *Adaptation.* 2nd ed. 115 pp. Englewood Cliffs, N.J.

WALTER, H. & WALTER, E. (1953). Einige allgemeine Ergebnisse unserer Forschungsreise nach Südwestafrika 1952/53: Das Gesetz der relativen Standortskonstanz; das Wesen der Pflanzengemeinschaften. *Ber. dt. bot. Ges.* **66**, 228–36.

WARING, R. H. & MAJOR, J. (1964). Some vegetation of the California coastal redwood region in relation to gradients of moisture, nutrients, light, and temperature. *Ecol. Monogr.* **34**, 167–215.

WELCH, J. R. (1960). Observations on deciduous woodland in the Eastern Province of Tanganyika. *J. Ecol.* **48**, 557–73.

WHITE, K. L. (1965). Shrub-carrs of southeastern Wisconsin. *Ecology* **46**, 286–304.

WHITTAKER, R. H. (1948). *A vegetation analysis of the Great Smoky Mountains.* Ph.D. Thesis, University of Illinois, Urbana. 478 pp.

WHITTAKER, R. H. (1951). A criticism of the plant association and climatic climax concepts. *NW. Sci.* **25**, 17–31.

WHITTAKER, R. H. (1952). A study of summer foliage insect communities in the Great Smoky Mountains. *Ecol. Monogr.* **22**, 1–44.

WHITTAKER, R. H. (1953). A consideration of climax theory: The climax as a population and pattern. *Ecol. Monogr.* **23**, 41–78.

WHITTAKER, R. H. (1954a). The ecology of serpentine soils. IV. The vegetational response to serpentine soils. *Ecology* **35**, 275–88.

WHITTAKER, R. H. (1954b). Plant populations and the basis of plant indication. (Germ. summ.) *Angew. PflSoziol.*, Wien, Festschr. Aichinger, **1**, 183–206.

WHITTAKER, R. H. (1956). Vegetation of the Great Smoky Mountains. *Ecol. Monogr.* **26**, 1–80.

WHITTAKER, R. H. (1957). Recent evolution of ecological concepts in relation to the eastern forests of North America. *Am. J. Bot.* **44**, 197–206, and *Fifty years of botany*, pp. 340–58. Ed. W. C. Steere. New York. 1958.

WHITTAKER, R. H. (1960). Vegetation of the Siskiyou Mountains, Oregon and California. *Ecol. Monogr.* **30**, 279–338.

WHITTAKER, R. H. (1962). Classification of natural communities. *Bot. Rev.* **28**, 1–239.

WHITTAKER, R. H. (1965). Dominance and diversity in land plant communities. *Science, N.Y.* **147**, 250–60.

WHITTAKER, R. H. (1966). Forest dimensions and production in the Great Smoky Mountains. *Ecology* **47**, 103–21.

WHITTAKER, R. H. & FAIRBANKS, C. W. (1958). A study of plankton copepod communities in the Columbia Basin, southeastern Washington. *Ecology.* **39**, 46–65, and *Readings in population and community ecology*, pp. 369–88. Ed. W. E. Hazen. Philadelphia and London. 1964.

264   R. H. WHITTAKER

WHITTAKER, R. H. & NIERING, W. A. (1964). Vegetation of the Santa Catalina Mountains, Arizona. I. Ecological classification and distribution of species. *J. Ariz. Acad. Sci.* **3**, 9–34.

WHITTAKER, R. H. & NIERING, W. A. (1965). Vegetation of the Santa Catalina Mountains, Arizona: a gradient analysis of the south slope. *Ecology* **46**, 429–452.

WILLIAMS, W. T. & LAMBERT, J. M. (1959). Multivariate methods in plant ecology. I. Association-analysis in plant communities. *J. Ecol.* **47**, 83–101.

WILLIAMS, W. T. & LAMBERT, J. M. (1960). Multivariate methods in plant ecology. II. The use of an electronic digital computer for association-analysis. *J. Ecol.* **48**, 689–710.

WILLIAMS, W. T. & LAMBERT, J. M. (1961). Multivariate methods in plant ecology. III. Inverse association-analysis. *J. Ecol.* **49**, 717–29.

WILLIAMSON, M. H. (1961a). An ecological survey of a Scottish herring fishery. IV. Changes in the plankton during the period 1949 to 1959. *Bull. mar. Ecol.* **5**, 207–23.

WILLIAMSON, M. H. (1961b). A method for studying the relation of plankton variations to hydrography. *Bull. mar. Ecol.* **5**, 224–229.

YARRANTON, G. A. (1966). A plotless method of sampling vegetation. *J. Ecol.* **54**, 229–37.

From *Geographical Ecology: Patterns in the Distribution of Species*
Robert H. MacArthur

Alternate Stable Equilibria

History even leaves its mark on equilibria, although how long its influence will be felt is unknown. We have already seen (p. 91) how very hard it is for a second species to colonize an island containing a reasonably close competitor. In this sense whichever species arrives first is practically permanent, and the later arrival is virtually certain to remain missing. But this is not really stable. Given enough time, early species A will go extinct from some islands and B will certainly successfully invade some of them. By this time random processes will have erased most of the history.

There is another way in which history can leave a fairly permanent mark. To see this, we must return now to the graphical presentation of competition that appears in the appendix to Chapter 2 (pp. 46–56). Further, in what follows in the balance of the chapter, we must draw freely both on the theoretical results described in the Chapter 2 appendix and on the extension to this theory derived in Appendix 1 to Chapter 8.

Here we have two resources, probably quite similar, and three consumer species harvesting them. Two species, X_1 and X_2, are far better at harvesting one than the other resource, and in this context are specialists. The third consumer species, X_3, is not only equally good at using both resources, but in fact very good at utilizing both, so that its isocline lies inside the intersection of the first two (see Fig. 9-3). Then the two resource levels, A and B, are alternate stable equilibria, each resistant to invasion by the remaining species. If X_2 and X_3 are present, they maintain the resources at level A and the X_1 isocline lies outside A, so X_1 cannot invade; similarly, if X_1 and X_3 are present at equilibrium, they reduce the resources to level B and X_2 cannot invade. It looks as if history leaves an indelible trace on this kind of situation, but in fact this situation is usually vulnerable to another fate. To see this, we note that if the X_3 isocline passed outside of point C, there would not be alternate stable equilibria. Then the only stable equilibrium point would be C; either of the other intersections would be vulnerable to invasion by the third species and would end up at point C, with only consumers X_1 and X_2 present. Hence, the condition for alternate stable equilibria is that the generalist isocline lie inside the intersection, C, of the specialist isoclines. A glance at Fig.

From Robert H. MacArthur, *Geographical Ecology: Patterns in the Distributions of Species* (New York: Harper & Row, 1972). Reprinted by permission of Princeton University Press.

248

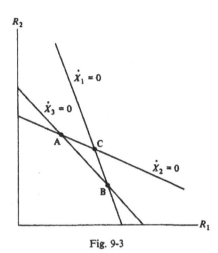

Fig. 9-3

Plotted as in Chapter 2, the isoclines $\dfrac{1}{X_1}\dfrac{dX_1}{dt}(=\dot{X}_1)=0,\ \dfrac{1}{X_2}\dfrac{dX_2}{dt}(=\dot{X}_2)=0,$ and $\dfrac{1}{X_3}\dfrac{dX_3}{dt}(=\dot{X}_3)=0.$ Since $\dot{X}_3=0$ lies inside the point C, X_3 can invade a community consisting of X_1 and X_2 and either eliminate X_2 and come to equilibrium with X_1 at point B or eliminate X_1 and come to equilibrium with X_2 at point A. Points A and B are alternate stable equilibria, each resistant to invasion by the remaining species.

2-13 shows that this condition is naturally met only when the resources are so similar that the consumers would have an evolutionary convergence. Thus, although we have exhibited the possibility of alternate equilibria resistant to ecological alterations, these equilibria are vulnerable to evolutionary alterations.

     We can, however, construct interesting whole communities each resistant to invasion by species from the other. These appear stable both in an ecological and an evolutionary sense. To do this we use the format and assumptions of Appendix 1 to Chapter 8 and of Fig. 8-16, where we found that those species would be present and with such abundance that their total utilization most closely matched the "useful production" along the whole resource spectrum. In Fig. 9-4 we show the simplest case of these alternate communities. The two areas have just slightly different useful production curves, P_j, but each has its production curve exactly matched by a sum of the utilizations of the separate species

*The Role of
History*

249

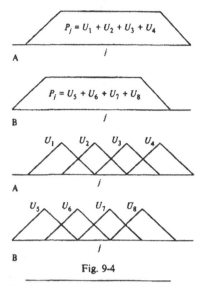

Fig. 9-4

Production curves P_j for regions A and B, and utilization curves U_i for species 1 to 4 of region A and species 5 to 8 of region B. The perfect matching of P_j by combinations of the four curves shows that each region is uninvadable by species of the other region. See the text for details.

and so each is uninvadable. However, the production curves overlap enough so that species 2 and 7 would each be able to live equally well in the other environment if there were no competitors. Their resources are equally present in both environments and yet neither can invade the other's environment. We can even go further and exhibit, slightly artificially, two alternate communities each occupying the same environment and each resistant to invasion by species from the other (see Fig. 9-5). This sort of situation is likely where two independently evolved faunas meet. In nature, it would not be quite as clear cut as in the figure because neither community would have its productions exactly matched by its utilizations and so each would be slightly, but only slightly, vulnerable to new invasions. Hence, where these two faunas meet we would expect each to maintain its integrity quite well, up to a very narrow zone of chaos where one is replaced by the other. This is diffuse competition in its purest sense. Let us consider a real case. A bird watcher in England or continental Eurasia finds warblers and flycatchers of many species breeding in the

250

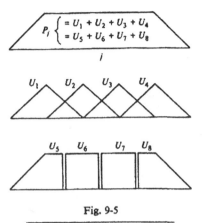

$$P_j \begin{cases} = U_1 + U_2 + U_3 + U_4 \\ = U_5 + U_6 + U_7 + U_8 \end{cases}$$

j

$U_1 \quad U_2 \quad U_3 \quad U_4$

$U_5 \quad U_6 \quad U_7 \quad U_8$

Fig. 9-5

Production and utilization curves as in Fig. 9-4 except that now we exhibit two alternate sets of species that utilize the same spectrum of resources, and yet each is resistant to invasion from the other region. See the text for details.

forests; so does the bird watcher in North America; and yet the New World warblers and flycatchers are unrelated to the Old World ones. The families have evolved in parallel to occupy similar environments. Yet few Old World warblers and no New World warbler or New or Old World flycatcher has ever really successfully invaded the other world. New World myrtle warblers (*Dendroica coronata*) enter slightly into Siberia and Old World Arctic warblers (*Phylloscopus borealis*) breed in northwestern Alaska, but each returns to winter with the rest of its species and neither has penetrated farther, to where there would be many competitors. We conclude that the difficulty of changing winter quarters, coupled with the diffuse competition from the other fauna, has prevented most exchanges of species. There has been some, but limited, exchange of less migratory species. Again, where the American bird species meet the tropical ones on the northeast coast of Mexico, there is a fairly abrupt transition. Few tropical species reach far beyond the Rio Grande, partly because they cannot stand the winters, and very few temperate species penetrate as far as Veracruz along the Mexican coast. In this case we must put most of the blame on diffuse competition.

This integrity of alternate communities provides the explanation for the patterns noticed by the earliest biogeographers. These

The Role of
History
———
251

men—Wallace and Sclater, for example—divided the world into biogeographic "realms," as in Fig. 9-6. The realms were Holarctic (divided into
Nearctic and Paleartic), Neotropical, Ethiopian, Oriental, and Australian.
The definition of realm was in part made possible by the independent
faunas evolving in the separate areas and exchanging relatively few species.
Their boundaries also match the great natural barriers to terrestrial
exchange such as the Sahara Desert, the Himalayan plateau, and waterways separating continents.

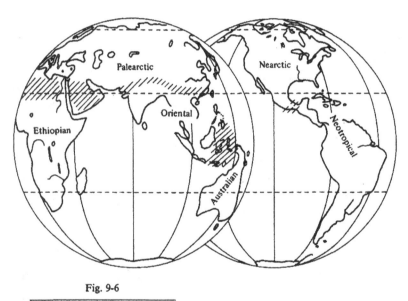

Fig. 9-6

The six continental faunal regions. Diagonal hatching shows approximate boundaries
and transition areas. (From Darlington, 1957.)

14 Assembly of Species Communities

Jared M. Diamond

Contents

Summary 342
Introduction 345
Statement of the problem 346
Background: the New Guinea
  biogeographic scene 349
Incidence functions 352
Relation between incidence functions
  and habitat requirements 361
Relation between incidence functions
  and island area 364
  Absence of suitable habitat 364
  Island size less than
    minimum territory requirement 364
  Seasonal or patchy food
    supply 364
  Population fluctuations 366
  "Hot spots" 369
Relation between incidence functions
  and dispersal 371
  Recolonization of defaunated
    islands 373
  Dispersal to Vuatom 373
  Supertramp dispersal 376
  Witnessed instances of
    over-water dispersal 376
  Rare dispersal events:
    gales and blooms 377
Relation between incidence functions
  and reproductive rates 378
The supertramp strategy versus
  the overexploitation ethic 380
The meaning of assembly rules 385
"Simple" checkerboard distributions:
  the dilemma of the empty squares 387

Assembly rules for the cuckoo-dove
  guild 393
Assembly of the gleaning flycatcher
  guild 400
Assembly of the myzomelid-sunbird
  guild 404
Assembly of the fruit-pigeon guild 405
Replicate communities on the
  same island 412
Communities at the habitat level 416
The origin of assembly rules 423
  Resource utilization 423
  Companions in starvation:
    a mechanism of coadjustment? 432
  Dispersal and assembly 437
  Transition probabilities 438
Unsolved problems 439
  Reconstructing incidence
    functions 440
  Applications to habitat
    communities, and to locally
    patchy communities 440
  Chance or predestination? 440
  Applications to conservation
    problems 442
Acknowledgments 442
References 443

Summary

This chapter explores the origin of differences in community structure, such as those between different islands of the same archipelago, between different localities on the same island, between different adjacent habitats, and between

Reprinted with the permission of the publisher from *Ecology and Evolution of Communiteis*, ed.
Martin L. Cody and Jared M. Diamond, pp. 342–49 (Cambridge, Mass.: Harvard University Press.
Copyright © 1975 by the President and Fellows of Harvard College.

different biogeographical regions. The working hypothesis is that, through diffuse competition, the component species of a community are selected, and coadjusted in their niches and abundances, so as to fit with each other and to resist invaders. Observations are derived from bird communities of New Guinea and its satellite islands, of which some are at, some above, and some below equilibrium in species number (S).

From exploration of numerous islands with various values of S, so-called incidence functions are constructed for individual species. These relate J, the incidence of occurrence of a particular species on islands of a certain S-class, to S. Species are classified according to their incidence functions into six categories: high-S species, confined to the most species-rich islands; A-, B-, C-, and D-tramps, present on the most species-rich islands and also on increasing numbers of increasingly more species-poor islands; and supertramps, confined to species-poor islands and absent from species-rich islands. Since different species have incidence functions of different shapes, the fauna of any real island is a very nonrandom subset of the total species pool.

The high-S category consists partly of endemic species of forest on large islands, partly of non-endemic species of scarce habitats often unrepresented or barely represented on smaller islands. Tramps, especially C- and D-tramps, are mostly nonendemic species characteristic of habitats that occur on virtually any island.

The dependence of incidence on area involves several factors, which vary from species to species: whether the required habitat of a species occurs on small islands; minimum territory size for species in which each pair maintains an exclusive territory; minimum year-round support area for species dependent on patchy or seasonal food supplies; population size in relation to short-term and long-term population fluctuations; and the role of "hot spots" (areas of locally-high utilizable resource production) in colonization and in recovery from population crashes.

Dispersal ability of species in different incidence categories has been assessed from data sources such as recolonization of islands defaunated by volcanic explosion or tidal wave, long-term records of vagrants, and direct observations of overwater colonization. Especially in the tropics, many bird species capable of strong flight refuse to cross water barriers of even a few miles. Dispersal rates are highest for supertramps and D-tramps, followed by C-tramps, B-tramps, and nonendemic A-tramps of scarce habitats. For high-S species, such dispersal as there is may be associated with rare population "blooms."

There is no obvious correlation between clutch size and incidence category. However, supertramps and D- and C-tramps have longer breeding seasons and raise more broods per year than do other species.

Supertramps have extraordinarily catholic and unspecialized habitat preferences, high reproductive potential, and high dispersal ability. They are competitively excluded from species-rich islands by "K-selected" species. However, faunas dominated by supertramps maintain population densities up to nine times *higher*. than those of K-selected faunas composed of the same number of species. Thus, the supertramp strategy may be contrasted with an inferred overexploitation ethic practised by high-S species, which are selected by competition to harvest early and overexploit. The high-S species thereby reduce resource levels below the point where other species can survive, even though this diminishes the rate of resource production and hence the population density of the harvesting species.

In a few instances, competition expresses itself in "simple" checkerboard distributions, by which species replace each other one-for-one. The frequent occurrence of "empty squares," however, shows that even these cases are complex. In the great majority of species groups or guilds, competitive exclusion involves so-called diffuse competition, i.e., the combined effects of several closely related species. Detailed examination of four guilds reveals the following types of assembly rules for species communities:

If one considers all the combinations that can be formed from a group of related species, only certain ones of these combinations exist in nature.

These permissible combinations resist invaders that would transform them into a forbidden combination.

A combination that is stable on a large or species-rich island may be unstable on a small or species-poor island.

On a small or species-poor island a combination may resist invaders that would be incorporated on a larger or more species-rich island.

Some pairs of species never coexist, either by themselves or as part of a larger combination.

Some pairs of species that form an unstable combination by themselves may form part of a stable larger combination.

Conversely, some combinations that are composed entirely of stable subcombinations are themselves unstable.

The forbidden combinations do not exist in nature because they would transgress one or more of three types of empirical rules: compatibility rules banning the coexistence of certain closely related species under any circumstances; incidence rules, implicit in incidence functions; and combination rules, which cannot be predicted from incidence functions.

Most of the evidence for these assembly rules is drawn from comparison of communities on different islands. However, examples are also drawn from communities at different localities, or in different habitats, or at different altitudes, or at different heights above the ground, on the same island. In some cases one can recognize simple effects of one-to-one competition. In other cases, one can recognize assembly rules describing more complex competitive effects and permitted combinations of several related species. In still more complex cases, competitive effects must be described by incidence functions relating the occurrence or niche limits of one species to diffuse competition from many other species. Thus, recognition of assembly rules may help us understand competitive effects on the spatial niche limits of a given species, and the puzzling tropical phenomenon of patchy distributions.

Much of the explanation for assembly rules has to do with competition for resources and with harvesting of resources by permitted combinations so as to minimize the unutilized resources available to support potential invaders. Communities are assembled through selection of colonists, adjustment of their abundances, and compression of their niches, in part so as to match the combined resource consumption curve of all the colonists to the resource production curve of the island. Members of permitted combinations must also be "companions in starvation"—i.e., must be similar in their tendencies to overexploit and in their tolerances for lowered resource levels, thereby starving less tolerant species off the island. Thus, consumer species form hierarchies with respect to exploitive strategy. The conditions under which overexploitation becomes a useful strategy for its practitioners are examined by loop analysis. Also relevant to the origin of assembly rules are two further factors: dispersal abilities, which permit only certain species to have a high incidence on small islands with high extinction rates; and transition probabilities, i.e., ease of assembling a species combination in one or a few steps from other permitted combinations.

Major unsolved problems include: the development of mathematical models for incidence functions; extensions to habitat communities and to locally patchy communities; the relative roles of chance and of predestination (i.e., detailed matches of different species combinations to slightly different local production curves) in the build-up of alternate communities; and applications to conservation problems.

Introduction

The understanding of alternate, stable, invasion-resistant communities of co-adjusted species poses a major current problem in ecology. Sets of such communities occur in similar habitats in different biogeographical regions, in similar habitats on different islands colonized from the same species pool, in similar habitats at different localities on the same large island or continent, and in different adjacent habitats. The theoretical basis for the existence of alternate stable communities was brilliantly explored by Robert MacArthur (1972) in *Geographical Ecology*. A conceptual framework is now available within which field observers can approach such unsolved problems as the following:

To what extent are the component species of a community mutually selected from a larger species pool so as to "fit" with each other?

Does the resulting community resist invasion? If so, how?

To what extent is the final species composition of a community uniquely specified by the properties of the physical environment, and to what extent does it depend on chance events (e.g., the question of which colonists arrive first, possibly also affecting which subsequent arrivals are compatible with the successful first colonists)?

The present chapter discusses such problems in the light of observations on bird communities of New Guinea satellite islands. It will be shown that (a) the probabilities or incidences of occurrence of particular species in a community bear

Jared M. Diamond   **346**

neat empirical relations to the total species number in the community; (b) these so-called incidence functions can be interpreted in terms of island area plus a species' habitat requirements, dispersal ability, birth and death schedule, exploitation strategy, and competitive relations; (c) the various species in a guild can coexist only in certain combinations; (d) these permitted combinations resist invaders that would result in forbidden combinations; and (e) lowering of resource levels by coadjusted constellations of species, to below the point where invaders can survive, may be an important mechanism of competitive exclusion.

Statement of the Problem

The structure of a species community may be described in terms of its species composition, together with the resource utilization, and distribution and abundance in space and time, of each component species. Comparison of different communities at any one of four levels generally reveals some differences in structure:

1. Differing but adjacent habitats differ in community structure, even though there may be no physical barriers preventing species of one habitat from invading another habitat (cf. Cody, Chapter 10).

2. Differences in community structure may exist between similar habitats in different areas of the same continent or large island, or even between similar habitats in areas that are in immediate contact and constitute artificially defined sections of a continuum. This phenomenon is es-pecially marked in the tropics. The result is often that tropical species are patchily distributed with respect to the available habitat. Figures 33–38 will present examples of these baffling distributional patterns.

3. Communities on similar islands colonized from the same species pool may differ. For example, the islands Sakar and Tolokiwa lie 29 miles apart in the Bismarck Sea near New Guinea, differ in area by only 13%, are geologically similar, support similar forest, have derived their birds from the same sources, and support similar numbers of lowland bird species (36 and 40, respectively). Yet Tolokiwa lacks three of the seven most abundant species of Sakar, Sakar lacks eight of the 15 most abundant species of Tolokiwa, and only 23 species are shared. In the Pearl Archipelago off Panama, MacArthur, Diamond, and Karr (1972) cite equally striking differences in bird species composition between Chitre and Contadora islands, which are only 1 mile apart. Furthermore, a species that is shared between similar islands may still occupy different habitats and have different abundances. For example, the fruit pigeon *Ptilinopus insolitus* is present both on Sakar and on Tolokiwa, but on Sakar it is widespread whereas on Tolokiwa it is confined to mid-montane forest. Its congener *Ptilinopus solomonensis* is present both on Sakar and on Tolokiwa and occupies similar habitats on the two islands, but is approximately six times more abundant on Tolokiwa than on Sakar.

4. The examples mentioned so far involve communities formed from the same species pool and lying within the same

biogeographic region or faunal province. Much larger differences are observed between more distant communities lying in different faunal provinces. For more than a century, from the time of Sclater and Wallace until the publication of *The Theory of Island Biogeography* by MacArthur and Wilson (1967), these differences formed the principal subject matter of biogeography. Although similar habitats in South America, Africa, and Australia may share few species in common, these communities may exhibit remarkably detailed convergent similarities in structure (Cody, Chapter 10; Karr and James, Chapter 11). The borders of the world's major faunal provinces are formed by present and past barriers to movement of organisms. These barriers have not served to eliminate colonization, but rather to reduce it to a level where great differences are maintained indefinitely between the communities on opposite sides of the barrier. If the communities did not possess some resistance to invasion, colonization across the barriers for millions of years would have smoothed many of the differences between even the major faunal provinces. Thus, the differences between the Australian Region and the Oriental Region present many of the same problems, albeit in more marked form, as the differences between Sakar and Tolokiwa islands in the Bismarck Sea.

These examples suggest (but do not prove) that the species in a community are somehow selected, and their niches and abundances somehow coadjusted, so that the community possesses some measure of "stability." Stability implies the existence of several different properties, some of which are easier to demonstrate than others. The most obvious thing we mean in describing a community as "stable" is that its present species composition is likely to persist with little change if there is no change in the physical environment. This property is easy to assess by comparing historical surveys with recent surveys. For instance, faunal surveys of a given New Guinea satellite island a century ago and today yield much more similar species compositions than do surveys of several different islands of similar size at the same time. The property of stable species composition suggests the existence of an additional property, namely, ability of a community to resist invasion by new species. This property is more difficult to document, because one needs much more than two faunal surveys at different times. A particular species may be absent from a particular island because the existing community prevents colonizing individuals of the new species from establishing themselves, or merely because colonizing individuals of the species may never reach the island at all. To document resistance to invasion requires sufficiently extensive observations so that arrivals of colonizing individuals, and their failures to establish stable populations, are detected. Finally, the property of resistance to invasion suggests a further property, which is still more difficult to document as well as to formulate, namely, that the existing community utilizes available resources in some optimal manner (MacArthur, 1970; MacArthur, 1972, pp. 231-234).

It seems likely that competition between species plays a key role in the integration of species communities. Real or potential

Jared M. Diamond

348

utilization of some of the same resources could be an obvious explanation for why similar species do not occur in the same community, unless their resource utilizations are somehow coadjusted. Numerous recent studies have provided clear-cut distributional evidence for competition between members of a pair of related species. These examples are valuable in documenting the existence of competition, but by themselves they do not account for much of the real world. Far more often, the presence or absence of a given species, and intercommunity variation in its abundance or spatial distribution, cannot be understood predominantly in terms of a correlated distribution of any single other species. It is then a logical extension of simple two-species distributional checkerboards to invoke "diffuse competition"—i.e., the complex situations resulting from the sum of competitive effects from many other somewhat similar species (Diamond, 1970a, p. 530; 1970b, pp. 1716–1717; MacArthur, 1972, pp. 43–46 and 249; Pianka, Chapter 12). The power of this concept is that, in principle, it can explain anything. Its heuristic weakness is that, if it is important at all, its operation is likely to be so complicated that its existence becomes difficult to establish and impossible to refute. Such a concept deserves to be greeted with skepticism until its importance can be documented. A profitable biogeographic approach to documenting diffuse competition would seem to be, first, to seek evidence whether variation in the incidence, niche, or abundance of a given species is correlated with variation in total species number; then, to

seek to trace out cases in which the distribution of a given species can be clearly related to the distribution of certain *combinations* of a few other species, yielding patterns that are analogous to two-species distributional checkerboards but more complex.

Such a test of the hypothesis of alternate, stable, invasion-resistant communities integrated by diffuse competition requires a field situation or experimental situation with the following properties: (a) a large number of communities that provide a similar physical environment and habitat structure; (b) a large species pool, varying fractions and combinations of which occur in the available communities; (c) availability of evidence that a species absent from a given community actually has had access, and that its absence is not simply due to a total lack of immigrants; (d) availability of evidence that the community does resist invasion, and that failure of attempted colonizations is not simply due to unsuitable habitat; (e) availability of cases in which a community has been displaced from equilibrium, so that relaxation towards equilibrium can be studied.

The avifauna of New Guinea and its satellite islands provides a favorable test situation. Considerable ecological and evolutionary information exists about the New Guinea species pool of 513 breeding nonmarine bird species. Surrounding New Guinea, and colonized by varying fractions of this species pool, are thousands of islands of varying sizes and at varying distances, providing numerous sets of replicate communities. Ornithological ex-

14 Assembly of Species Communities   **349**

Species

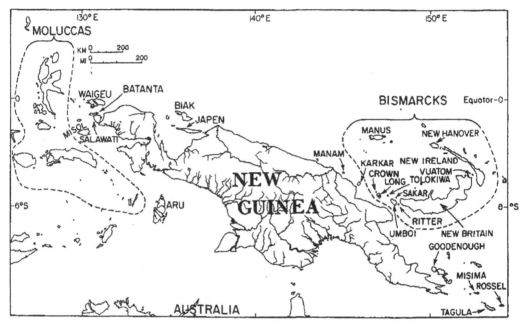

Figure 1 Map of the New Guinea region with names of some of the islands to be discussed.

ploration has been sufficiently intensive to provide not merely species lists but, for some islands, instances of successful and unsuccessful colonizations. Species numbers on some islands have been displaced above what would be their present value at equilibrium by Pleistocene episodes of lowered sea level, which joined some islands to New Guinea, joined other islands to each other, and expanded still other islands in area. Species numbers on other islands have been displaced below equilibrium by Krakatoa-like volcanic explosions or by tidal waves. Some species called supertramps are particularly useful in studying community integration, be-cause of their high colonization rates and sensitivity to competition. We shall see that the distributions of most species can be neatly related to total species number in a community; and that, in a few cases, it is possible to relate species distributions to diffuse competitive effects from specific combinations of related species.

Ecology, 60(6), 1979, pp. 1132–1140
© 1979 by the Ecological Society of America

THE ASSEMBLY OF SPECIES COMMUNITIES:
CHANCE OR COMPETITION?[1]

EDWARD F. CONNOR[2,3]
Department of Environmental Sciences, University of Virginia,
Charlottesville, Virginia 22903 USA

AND

DANIEL SIMBERLOFF
Department of Biological Sciences, Florida State University,
Tallahassee, Florida 32306 USA

Abstract. We challenge Diamond's (1975) idea that island species distributions are determined predominantly by competition as canonized by his "assembly rules." We show that every assembly rule is either tautological, trivial, or a pattern expected were species distributed at random. In order to demonstrate that competition is responsible for the joint distributions of species, one would have to falsify a null hypothesis stating that the distributions are generated by the species randomly and individually colonizing an archipelago.

Key words: assembly rules; bird communities; competition; exclusive distribution; island species distribution; species pairs.

In a widely cited new approach to the interpretation of biogeographic distributions, Diamond (1975) asserts that the assembly of bird communities manifests the following patterns:

a. "If one considers all the combinations that can be formed from a group of related species, only certain ones of these combinations exist in nature."

b. "Permissible combinations resist invaders that would transform them into forbidden combinations."

c. "A combination that is stable on a large or species-rich island may be unstable on a small or species-poor island."

d. "On a small or species-poor island, a combination may resist invaders that would be incorporated on a larger or more species-rich island."

e. "Some pairs of species never coexist, either by themselves or as a part of a larger combination."

f. "Some pairs of species that form an unstable combination by themselves may form part of a stable larger combination."

g. "Conversely, some combinations that are composed entirely of stable subcombinations are themselves unstable."

Examining data from 147 species of land birds distributed in various combinations over 50 islands in the Bismarck Archipelago near New Guinea, Diamond

concluded that these "assembly rules" can be explained by (1) interspecific competition for resources, (2) "overexploitation strategies" whereby certain "permissible combinations" of species together lower resources to a point such that other species are usually starved to extinction, (3) differences among species in dispersal rates, and (4) low transition probabilities between "permissible combinations" such that combinations A and B might both be "permissible," but transition from A to B might only be possible through combinations which are very unlikely, presumably for the three preceding reasons. This latter phenomenon is not unlike the genetic landscape of Wright (1967), with adaptive peaks separated by impassable low-fitness troughs.

We will show that every assembly rule is either a tautological consequence of the definitions employed, a trivial logical deduction from the stated circumstances, or a pattern which would largely be expected were species distributed randomly on the islands subject only to three constraints: (1) that each island has a given number of species, (2) that each species is found on a given number of islands, (3) and that each species is permitted to colonize islands constituting only a subset of island sizes. The last constraint is an acceptance, for the purpose of this paper, of Diamond's contention that each species has an "incidence function" of probabilities of being found on islands of given sizes. The allowable subset of island sizes constitutes the domain for which the values of the incidence function are nonzero. Diamond (1975), Diamond et al. (1976), and Diamond and Marshall (1977) all synonymize island size with the number of species on the island, a convention which we adopt here. We

[1] Manuscript received 8 June 1978; revised and accepted 30 January 1979.

[2] Order of authorship determined by a coin toss.

[3] Present address: Department of Zoology, South Parks Road, Oxford OX1 3PS, England.

regret we cannot use the same Bismarck data which Diamond first used, but its publication has been delayed by various unforseen complications (J. M. Diamond, *personal communication*, E. Mayr, *personal communication*). In lieu of these, we have used the New Hebridean bird data (Diamond and Marshall 1976), plus data for West Indies birds (Bond 1971) and bats (Baker and Genoways 1978) to examine the assembly rules.

Rules *a* and *e*

"If one considers all the combinations that can be formed from a group of related species, only certain ones of these combinations exist in nature."

"Some pairs of species never coexist, either by themselves or as part of a larger combination."

These rules are identical, except that rule *a* is restricted to related species and rule *e*, though unrestricted taxonomically, is concerned only with pairs of species. We deal with them together. Diamond's "proof" of rule *e* consists of five examples from the Bismarcks of exclusively distributed related species (his Figs. 20–24). Although he does not do so, it is possible for each example to calculate the probability of an arrangement as exclusive as that observed. For instance, the *Macropygia* doves (his Fig. 20) consist of two species, *M* and *N*, such that *M* is on 14 islands, *N* on 6 different ones, and 13 surveyed islands have neither. Presumably the remaining 17 islands have not been censused. The probability of an arrangement this exclusive over these 33 islands for randomly placed species with the same frequencies is:

$$\frac{\binom{33}{14}\binom{19}{6}}{\binom{33}{14}\binom{33}{6}} = .0245.$$

The terms in the numerator are the number of ways species *M* can be placed, and the number of ways species *N* can be placed on islands not occupied by species *M*, respectively. The terms in the denominator are simply the numbers of ways the two species can be placed irrespective of which islands are already occupied. Similarly, for the *Pachycephala* flycatcher example (his Fig. 21), species *P* is found on 11 islands, *D* on 18 different islands, and 21 islands have neither. The probability of such an exclusive arrangement for species placed randomly, subject only to their being on 11 and 18 islands, respectively, is:

$$\frac{\binom{50}{11}\binom{39}{18}}{\binom{50}{11}\binom{50}{18}} = .00345.$$

When one recalls that there are $\binom{141}{2} = 9870$ pairs

of birds in the Bismarcks, it is clear from the above probabilities that by chance alone certain species pairs would not occur together on any island. What one wants to know is how many such pairs, trios, etc. would be expected for randomly distributed birds, and how many such pairs, trios, etc. are actually observed.

It is impossible, lacking the data, to treat these rules for the Bismarcks, but for the New Hebrides and West Indies avifaunas and West Indies bats one would expect a large fraction of the species pairs and trios, whether related or not, not to coexist even were the species placed randomly within each archipelago subject only to three constraints:

i) For each island, there is a fixed number of species, namely, that which is observed.
ii) For each species, there is a fixed number of occurrences, namely, that which is observed.
iii) Each species is placed only on islands with species numbers in the range for islands which that species is, in fact, observed to inhabit. That is, the "incidence" range convention is maintained.

We simulated such a random placing 10 times, with the result that the total number of species occurrences was maintained, allocated as in nature among islands and among species. We then scanned each simulated arrangement for number of pairs not found anywhere (and number of trios for New Hebrides birds and West Indies bats), number of pairs (or trios) found on only one island, only two islands, only three islands, etc. Finally, we examined the actual arrangements. All analyses were performed with and without constraint (iii) "incidence functions". However, since relaxing incidence constraints does not affect the results, only the results including incidence constraints are presented. Details of the simulation are described in the Appendix.

For the New Hebrides birds (56 species on 28 islands), there are 1540 possible species pairs of which 63 are not found on any island. But by chance alone one would have expected 63.2 such pairs (SD = 2.9). Further, the entire distribution of number of species pairs vs. number of islands shared (Fig. 1) shows a close match to the random expectation; with the last two classes lumped so that expected number in each class is >5, $\chi^2 = 16.34$ (27 df), .95 > P > .90. Of the 27 720 trios of New Hebrides birds, 3070 do not coexist on any island, but the expected number of such trios is 3068.0 (SD = 105.1). For the entire distribution of number of species trios vs. number of islands shared, with the last three classes lumped so that expected number in each class is >5, $\chi^2 = 13.57$ (26 df), .98 > P > .95. If we restrict our attention, as in rule *a*, to birds within families (cf. Terborgh 1973), we find that there are 99 pairs in the New Hebrides that are confamilial in one or another of the 15 families, of

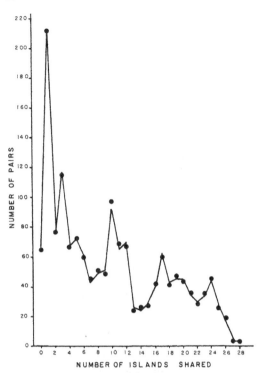

FIG. 1. Distribution of number of species pairs vs. number of islands shared for New Hebrides birds. Solid line is the expected distribution given the three constraints discussed in text. Dots are the observed values.

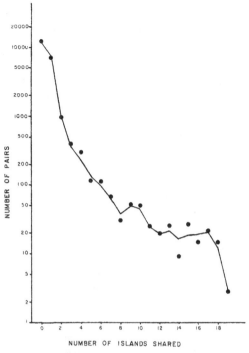

FIG. 2. Distribution of numbers of species pairs vs. number of islands shared for West Indies birds. Solid line is the expected distribution given the three constraints discussed in text. Dots are the observed values.

which only one is not found on any island; one would have expected 0.9 (SD = 0.3). For the entire distribution, lumping classes so that denominators are >5, we find $\chi^2 = .63$ (11 df), $P > .99$. For confamilial trios, we find that of 304 possible, seven are found nowhere in the archipelago, while one would have expected 6.4 (SD = 1.9) exclusive confamilial trios even had the birds been randomly distributed. For the entire distribution, with lumped classes as before, $\chi^2 = 1.04$ (17 df), $P > .99$. In a nutshell, there is nothing about the absence of certain species pairs or trios, related or not, in the New Hebrides that would not be expected were the birds randomly distributed over the islands as described above. Since there are so many possible sets of species, it is to be expected that a few sets are not found on any island; this does not imply that such sets are actively forbidden by any deterministic forces.

For West Indies birds (211 species on 19 islands), there are 22 155 pairs of which 12 757 are found on no island. But had the birds been randomly distributed on the islands as described above, one would have expected 12 448.1 (SD = 79.2) such exclusive pairs. For the entire distribution of number of species pairs vs. number of islands shared (Fig. 2), we find the observed

and expected values to be not nearly so close as in the New Hebrides: $\chi^2 = 66.18$ (18 df), $P < .01$. It is nevertheless clear from Fig. 2 that not only is the number of completely exclusive species pairs only slightly greater than expected for randomly distributed birds, but also there are no major anomalies in the degree of partial exclusivity which some species pairs achieve. Of 1029 pairs of birds which are confamilial in one or another of the 24 West Indian families, 621 are mutually exclusive. But a random arrangement would have 437.0 (SD = 18.3) such exclusive confamilial pairs; for the entire distribution, $\chi^2 = 271.44$ (17 df), $P < .01$. If we relax constraint (ii), that of specified total number of islands for each species, we find the fit to be much better, though observed and expected distributions still differ by a χ^2 test. Since there are 1 543 465 possible trios of West Indian birds, computing time limitations forbade our extending this analysis to cover trios.

Finally, for the West Indies bats (59 species on 25 islands) there are 1711 possible species pairs, of which 996 are exclusively distributed while a random arrangement would have produced 941.7 (SD = 11.6); for the entire distribution (Fig. 3), $\chi^2 = 15.26$ (6 df), $.02 > P > .01$. Of the 32 509 trios of West Indies bats,

27 397 do not coexist on any island, and the expected number of such trios is 26 965.7 (SD = 84.2). For the entire distribution of number of species trios vs. number of islands shared, with shared island classes 6 and 7, 8–11, and 12–25 lumped so that expected number in each class is >5, χ^2 = 91.96 (8 df), P < .01. Within the five families are 499 confamilial pairs, of which 325 are found on no single island while 208.6 (SD = 5.5) such pairs would have been expected. For the entire distribution, χ^2 = 183.99 (7 df), P < .01; as with the birds, relaxing constraint (ii) brings the observed and expected distributions much closer together. Of the 3850 confamilial trios, 3519 are mutually exclusive, while one would expect 2564.4 (SD = 47.2) given a random arrangement with all three constraints; for the entire distribution χ^2 = 851.15 (7 df), P < .01, with classes 6 and 7 lumped. Relaxing constraint (ii) yields a much better fit for confamilial trios, χ^2 = 10.29 (3 df), .025 > P > .01.

For the West Indies birds and bats, then, there are as many mutually exclusive species pairs as would have been expected had the species been randomly distributed on the islands subject only to constraints of incidence ranges (iii), some species being more widely distributed than others (ii), and some islands having more species than others (i). Although more of these exclusive pairs are of related species than chance alone would have dictated, one would have expected many exclusive related pairs even under the random hypothesis. For our three test biotas, we summarize the data on mutually exclusive pairs and trios in Table 1. It is clear that five examples of exclusive distribution, as Diamond (1975) presents, provide no support for assembly rules *a* and *e*.

The New Hebrides bird distribution fit the random hypothesis even more closely than the West Indies birds and bats for both related and unrelated species. We can suggest two possible reasons. First, our simulation placing birds on islands (at least without incidence range restriction) is analogous to randomly placing 0's and 1's in an $m \times n$ matrix (with m = number of species, n = number of islands) with row and col-

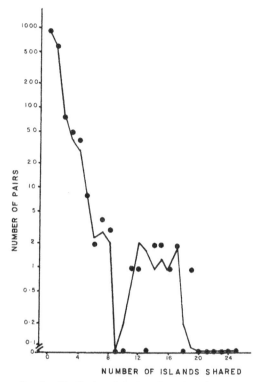

FIG. 3. Distribution of number of species pairs vs. number of islands shared for West Indies bats. Solid line is the expected distribution given the three constraints discussed in text. Dots are the observed values.

umn sums fixed. There are a limited number of such matrices, and since we would even allow row and column interchanges (which correspond, respectively, to exchanging species names or island names), the number of·different arrangements is even smaller. Exactly how many such arrangements there are, given a set of row and column sums, is an old, unsolved combina-

TABLE 1. Observed and expected exclusive species groups for several taxa.

Taxon	Group size	Confamilial	Total groups	Observed exclusive groups	Expected exclusive groups
New Hebrides birds	pair	No	1 540	63	63.2 (2.9)*
New Hebrides birds	pair	Yes	99	1	0.9 (0.3)
New Hebrides birds	trio	No	27 720	3 070	3 068.0 (105.1)
New Hebrides birds	trio	Yes	304	7	6.4 (1.9)
West Indies birds	pair	No	22 155	12 757	12 448.1 (79.2)
West Indies birds	pair	Yes	1 029	621	437.0 (18.3)
West Indies bats	pair	No	1 711	996	941.7 (11.6)
West Indies bats	pair	Yes	499	325	208.6 (5.5)
West Indies bats	trio	No	32 509	27 397	26 965.7 (84.2)
West Indies bats	trio	Yes	3 850	3 519	2 565.4 (47.2)

* Parenthetic values are standard deviations.

torics problem (N. Heerema, *personal communication*), but we noticed that all our West Indies random matrices looked very different, both from one another and from the matrix depicting the observed arrangement, while all the random New Hebrides matrices were similar, and similar to the observed matrix. We infer that for the species and island totals of the New Hebrides birds, there are very few possible arrangements even without the incidence ranges (which, incidentally, span the whole range of island sizes for 36 of the 56 New Hebrides species, and almost the whole range for most of the rest). Since these are all similar, it is not surprising that the observed distribution is not very different from a random one with respect to exclusive pairs and trios. But unless one is willing to ascribe to competition the facts that islands have different numbers of species and that species are found on different numbers of islands, the New Hebrides data still argue heavily against the claim that competition determines most aspects of the distribution of species on islands.

Second, where a distribution includes single-island endemics (as do two of Diamond's examples, *Myzomela* honeyeaters and *Zosterops* white-eyes) a statistical analysis like ours exaggerates the degree of biological exclusion, since the evolution of specific differences between two formerly conspecific populations would generate an exclusive pattern independent of competition. Such species should probably be excluded from the analysis, as Terborgh (1973) did for identical reasons in his examination of West Indian bird distributions. We used complete lists because we were uncertain which species pairs, trios, etc. represented recent cases of allopatric speciation. Much of the excessive exclusivity in West Indies birds and bats is "pseudo-exclusion" arising from either unsettled taxonomy, or the inclusion of superspecies. For example, Bond (1971) mentions that among the vireos, *Vireo modestus, V. crassirostris, V. griseus,* and *V. gundlachi* can be considered conspecific, as well as *V. altiloquus* and *V. magister.* Several other genera (*Contopus, Elaenia, Mimus, Quiscalus, Loxigilla, Melanerpes, Saurothera, Amazona,* and *Chlorostilbon*) contain two or more exclusive species, but it is likely that this represents allopatric speciation without subsequent reinvasion, rather than active competitive exclusion. Diamond (1975) provides a striking example of this problem in his discussion of *Zosterops* in the Bismarcks. Five of 12 species of white-eyes belong to a superspecies, and are by definition allopatric or parapatric in distribution. From a zoogeographic standpoint these are all one species. Yet he contends that this "checkerboard" distribution results from competitive exclusion, not geographic speciation without reinvasion. We conclude by observing that to the extent that such taxa are included in the analysis, they distort the results in the direction of increased number of observed exclusive pairs or trios (the 0-class in Figs.

1–3). It may be that such taxa are more common in the West Indies than in the New Hebrides.

RULE *b*

"Permissible combinations resist invaders that would transform them into forbidden combinations."

This is clearly a deduction from the definitions of "permissible combination" and "forbidden combination," plus Diamond's explanation of the assembly rules. Since Diamond defines a permissible combination to be one which exists somewhere in the archipelago and a forbidden one to be one which does not exist, and believes that forces 1–4 on page 1132 are the deterministic explanations for these observed combinations or absences, it follows that he believes the permissible combinations actively "resist" transformation into forbidden ones. Our discussion of rules *a, e,* and *g* makes it clear that the statistical distributions themselves do not demand an explanation of active resistance.

Nor is there compelling experimental evidence that any particular combination is actively forbidden by any force(s), or actively "resists" transformation. Diamond claims that five cases directly document active resistance of "forbidden" invaders, but examination of these suggests otherwise. Case 1 is that three cuckoo-doves (*A, N, R*) have been resident on New Britain during this century, while a fourth (*M*), "whose addition would create a combination forbidden by compatibility rules," is resident on islands 1.6 km away, but is only a vagrant on New Britain. But a "compatibility rule" is an ad hoc rationalization: "knowledge about species ecologies may suggest to us that a given pair . . . is incapable of coexistence . . . ." Diamond believes that "distributional information" can also lead us to infer a compatibility rule, but his reasoning is flawed here. He calculates the no-coexistence probability Z_{AB} of not finding coexistence on any of a set of n islands as the product of n terms of the form $1-J_A J_B$, where J_A and J_B for each island are the incidence probabilities for species A and B, respectively, on an island of that size. If the no-coexistence probability Z_{AB} is low, yet A and B are never seen together in the archipelago, incompatibility is inferred. But as our discussion of rules *a, e,* and *g* shows, there are a vast number of possible combinations of any size, so many that even for a large archipelago there will be so few islands that many combinations would not exist even were the species randomly distributed, whether or not constrained by incidence rules.

Now, the "combination forbidden by compatibility rules" in case 1 could be *MN, MNR, AMN,* or *AMNR,* since Diamond earlier has said that these four are all forbidden by compatibility rules. "The first two of these combinations are also forbidden by incidence rules," which simply means that in this set of 50 is-lands all islands that have *N* also have *A*, so that "any combinations containing *N* but not *A* . . . are forbidden" (page 394). Obviously he does not mean "*actively* forbidden" by the incidence rules! As for the compatibility rule which forbids these four combinations, we are left only with Diamond's assertion that their ecologies preclude their coexistence; he concedes that even his method of calculating no-coexistence probabilities does not support this particular compatibility rule (page 396). Whether *M* bred on New Britain in this century is not even clear; Diamond did not find it in 1969. Even if one concedes that it has not bred there in this century, where is the evidence of active resistance? So long as we accept that some species (including *M*) are not found on all islands, and that one can *conceive* of noncompetitive reasons for this, the absence per se of a species from an island is not evidence for active resistance by other species. Cases 3 and 4, that *N* has not bred on Vuatom and Karkar, can be explained similarly.

Case 2 is that species *A* and *M* are resident on Vua-tom, while *R* is believed not to have bred there this century in spite of its occasional presence there. It is said that the combination which would have resulted, *AMR*, is "forbidden by combination rules." Now, "combination rules" are defined strictly statistically, and are exactly the statistical part of the assembly rules. That is, when the no-coexistence probability of a combination, calculated as described above, is not so high that one would not have expected to see the combination, yet the combination is not seen in the archipelago, a "combination rule" of active exclusion is inferred. The flaw in the statistical reasoning has already been demonstrated, and once again there is no evidence of active resistance, unless one believes *in advance* that there can be *no* explanation of any species' failure to colonize all islands, except for competition. Case 5, that *M* has not bred on Umboi, is analogous to case 2.

Diamond's section title for these cases is "Historical proof of resistance to invasion . . . ," yet the proof seems always to consist of an aprioristic combination or compatibility rule which itself lacks proof. In no case is there evidence that active resistance occurs, unless one *begins* with the assumption that a distributional gap must be explained by active resistance, in which instance we have a tautology and why bother with the exercise of producing evidence?

Rule *c*

"A combination that is stable on a large or species-rich island may be unstable on a small or species-poor island."

First, since all Diamond's statistical analyses use species number as the operational definition of island size, we may omit "large" and "small" and restate

rule *c* thus: "A combination that is stable on a species-rich island may be unstable on a species-poor island." Nowhere are "stable" and "unstable" defined. If persistence is meant, no evidence is even attempted, so we presume that stable is here a synonym for "permissible" and unstable a synonym for "forbidden." We have in our discussion of rule *b* shown that there is no evidence for active resistance, and that permissible simply means exists somewhere in the archipelago (Diamond, page 344, appears also to define permissible thus), while forbidden means not found in the archipelago. So rule *c* becomes: "A combination which is found on species-rich islands may not be found on species-poor islands." Need we add that any species-rich island will contain far more combinations of all sizes than a species-poor island, so that one would expect by chance alone to find some combinations on species-rich islands which are found on no species-poor islands? And that by chance alone there should be many more of these than of combinations found on species-poor islands but not species-rich islands? For that matter, a certain number of combinations by chance alone ought not to be found on any island, species-rich or species-poor! We discuss this in our treatment of rules *a*, *e*, and *g*, but for now suffice it to say that rule *c* is a trivial consequence of the definitions of rich and poor plus the most elementary laws of combinatorial mathematics.

Rule *d*

"On a small or species-poor island, a combination may resist invaders that would be incorporated on a larger or more species-rich island."

This rule seems to be a composite of rules *b* and *c*, and thus both a tautology (since active resistance must be assumed, if one is to prove active resistance) and a direct consequence of the definitions of poor and rich. As discussed for rule *c*, one would expect by chance alone certain combinations to be found on species-rich but not species-poor islands. As discussed for rule *b*, there is no experimental or statistical evidence that any particular combination (on *any* size island) actively resists transformation to some other particular combination.

Rule *f*

"Some pairs of species that form an unstable combination by themselves may form part of a stable larger combination."

Suppose one had a pair of species, *A* and *B*, which he claimed exemplified this rule. That is, they are unstable by themselves, but stable as part of some larger combination (say, *ABC*). There are no islands with published faunal lists of only two species in the Bismarcks, New Hebrides, or West Indies, and for the Bismarcks just one island with only three species. This fact renders the "rule" an untestable proposition.

How is one to know that A and B are an unstable duo by themselves? Whenever they appear, they will have to be with some other species (say C), likely several other species. By its existence, the resultant larger combination (ABC) is defined as stable. So the rule boils down to: "Some pairs of species which are not found alone (but they couldn't be, since no island is small enough) *are* found together with other species." Or should it be, "Every pair of species which is found at all is found with other species"? And what has this to do with competition?

RULE g

"Conversely, some combinations that are composed entirely of stable subcombinations are themselves unstable."

It turns out that if the species were distributed randomly, subject to the three constraints described in our discussion of rule e, one would *expect* a number of species trios *not* to exist on any island even though each of the species pairs contained in the trio is found on one or more islands. For the New Hebrides birds, 27 720 trios are possible, of which 3070 do not occur on any island. Of these 3070 noncoexisting trios, 169 have all three of their component pairs existing on at least one island. For the random simulation in our test of rule e, the expected number of exclusive trios is 3068.0 (sd = 105.1), and one would expect 162.9 (sd = 7.2) of these to have each of their three component pairs represented on at least one island. For the West Indies bats, 32 509 trios are possible, of which 27 397 do not occur on any island. Of these 27 397 noncoexisting trios, eight have all their component pairs on at least one island, while one would expect 130.8 (sd = 33.9). Clearly, for the West Indies bats, few trios are composed entirely of stable subcombinations. The simulated and real data could as well be examined for species quartets and their component trios, etc., but the computer bookkeeping is expensive and we trust that our point is made.

CODA

We have shown that at least one of the assembly rules is untestable, three are tautological consequences of definition plus elementary laws of probability, and the remaining three describe situations which would for the most part be found even if species were randomly distributed on islands. Clearly the assembly rules do not compel us to posit interspecific competition as a major organizing force for avian communities. That such an all-encompassing theory should be built on so little evidence invites an examination of the procedures used in its construction, and one point stands out. At no time was a parsimonious null hypothesis framed and tested. Instead of asking what biogeographical distributions would arise were *no* biological forces acting to produce them other than

dispersal differences among species, and whether observed distributions differ from these, Diamond (1975) assumed competition to be the primary determinant and then sought post facto to rationalize the observed data in the light of this assumption. As Popper (1963) points out, it is easy to find confirmatory evidence for most reasonable hypotheses, but science progresses by a different route: by posing testable hypotheses and then attempting to falsify them.

Further, at least two other attempts to use biogeographic data to demonstrate that interspecific competition structures island communities suffer from the same defect: failure to pose and to test a null hypothesis (Simberloff 1979a). First, that species/genus ratios on islands are lower than the mainland has been invoked as evidence for intense island competition (Grant 1966) in spite of Williams' prior demonstration (1951) that such a situation would obtain even for random subsets of any mainland species pool (Simberloff 1970). Second, Schoener (1965), Grant (1968), and Abbott et al. (1977) claim that size ratios of "adjacent" species in a size ranking are greater on islands than mainland, and greater on small islands than large ones, because of more intense interspecific competition on small islands. But the proper null hypothesis for such a study is that random subsets of a mainland pool would produce these trends, and a test of the null hypothesis for several archipelagoes provides no cause for rejection (Strong et al. 1979).

All this is not to say that species *are* randomly distributed on islands, or that interspecific competition does not occur. Rather, statistical·tests of properly posed null hypotheses will not easily detect such competition, since it must be embedded in a mass of noncompetitively produced distributional data. Instead, one must make a strong argument for competitive exclusion via observed active replacement of one species by another (Simberloff 1978 reviews several cases), experiment, or very detailed autecological study.

ACKNOWLEDGMENTS

We thank J. Diamond for providing us with the New Hebrides bird list, R. Baker for a preprint of his West Indies bat paper, R. McIntosh, F. Vuilleumier, and the Florida State University Ecology group for reading the manuscript, the Florida State University Computing Center for computer time, and James W. Beever III for preparing the figures.

LITERATURE CITED

Abbott, I., L. K. Abbott, and P. R. Grant. 1977. Comparative ecology of Galapagos ground finches (*Geospiza* Gould): evaluation of the importance of floristic diversity and interspecific competition. Ecological Monographs 47:151–184.

Baker, R. J., and H. H. Genoways. 1978. Zoogeography of Antillean bats. Pages 53–97 *in* F. B. Gill, editor. Zoogeography in the Caribbean: the 1975 Leidy Medal Symposium. Academy of Natural Sciences of Philadelphia, Philadelphia, Pennsylvania, USA.

Bond, J. 1971. Birds of the West Indies. Collins, London, United Kingdom.

Diamond, J. M. 1975. Assembly of species communities. Pages 342–444 *in* M. L. Cody and J. M. Diamond, editors. Ecology and evolution of communities. Harvard University Press, Cambridge, Massachusetts, USA.

Diamond, J. M., M. E. Gilpin, and E. Mayr. 1976. Species-distance relation for birds of the Solomon Archipelago, and the paradox of the great speciators. Proceedings of the National Academy of Sciences of the United States of America 73:2160–2164.

Diamond, J. M., and A. G. Marshall. 1976. Origin of the New Hebridian avifauna. Emu **76**:187–200.

Diamond, J. M. and A. G. Marshall. 1977. Distributional ecology of New Hebridian birds: a species kaleidoscope. Journal of Animal Ecology 46:703–727.

Grant, P. R. 1966. Ecological compatibility of bird species on islands. American Naturalist **100**:451–462.

———. 1968. Bill size, body size, and the ecological adaptations of bird species to competitive situations on islands. Systematic Zoology 17:319–333.

Popper, K. R. 1963. Conjectures and refutations: the growth of scientific knowledge. Harper and Row, New York, New York, USA.

Schoener, T. W. 1965. The evolution of bill size differences among sympatric congeneric species of birds. Evolution 19:189–213.

Simberloff, D. S. 1970. Taxonomic diversity of island biotas. Evolution 24:23–47.

———. 1978. Using island biogeographic distributions to determine if colonization is stochastic. American Naturalist 112:713–726.

———. 1979, *in press*. Dynamic equilibrium island biogeography: the second stage. Proceedings of the 17th International Ornithological Congress, June 1978, West Berlin, Germany.

Strong, D. R., L. A. Szyska, and D. S. Simberloff. 1979. Tests of community-wide character displacement against a null hypothesis. Evolution 33:897–913.

Terborgh, J. 1973. Chance, habitat, and dispersal in the distribution of birds in the West Indies. Evolution 27:338–349.

Williams, C. B. 1951. Intra-generic competition as illustrated by Moreau's records of East African birds. Journal of Animal Ecology 20:246–253.

Wright, S. 1967. "Surfaces" of selective value. Proceedings of the National Academy of Sciences of the United States of America 58:165–172.

APPENDIX

To examine rules *a*, *e*, and *g* we generated values of expected number of species pairs or trios sharing 0, 1, 2, . . . N islands, for confamilial and nonconfamilial pairs and trios, as well as expected numbers of "unstable" trios with 0, 1, or 2 stable component pairs. The simulation algorithm to produce these expected values had two major parts; one to fill randomly a 0–1 matrix constrained by restrictions (i) row sums, (ii) column sums, and (iii) incidence ranges, and a second to inspect and to count the actual as well as expected values of the above-mentioned statistics.

The constraints (i, ii, and iii) to the randomly constructed matrix were determined by inspecting the actual matrix for a particular archipelago. Hence, the column sums (island species numbers), and row sums (species occurrences) for the random matrix are exactly equal to those for the actual matrix. The incidence ranges are as nearly equal to those in the actual matrix as we could make them (see below). Given these fixed constraints the random matrix can then be constructed.

Envision a matrix of 0's and 1's where each row is a species and each column an island. Presence is indicated by a 1 and absence by a 0. To construct an "expected" matrix given

that species are distributed at random with the matrix constrained as described above, the matrix is first sorted such that row sums (N_j) decrease from top to bottom and column sums (S_i) increase from left to right. Proceeding species by species (row by row), each species *j* is placed sequentially on N_j islands chosen by generating a sequence of random numbers. Before each species is placed on an island, a check is performed to determine if the species has already been placed on that island, or if that island has received its full complement if S_i species. If either of these conditions is met, another island is chosen on which to place the species. This procedure is repeated until species *j* has been placed on N_j islands. At this point the procedure is repeated for the next species, etc., performing the same checks to insure that each species *j* is placed on N_j islands and that after placing all species, each island *i* has S_i species.

When incidence constraints are used the selection of an island on which to place species *j* is limited to the range of island sizes (as measured by S_i) on which species *j* occurs in the actual matrix. As the simulation proceeds species by species, more and more islands are filled to their values of S_i. Occasionally the situation arises where the *j*th species cannot be placed on N_j islands all falling within its incidence range. In this event the incidence range constraint is expanded by one island in each direction (smaller and larger), and the random selection procedure repeated. The incidence range is continually expanded until species *j* can be placed on N_j islands.

Once the random matrix is constructed, it may be inspected in the same manner as the actual data set, to enumerate values of the statistics of interest. To count the number of species pairs sharing 0, 1, 2, . . . N islands, each pair of rows (species) is scanned for the number of positive matches (1's in the same column in both rows). To generate this distribution for trios, one need only to repeat this procedure for groups of three rows at a time. Obviously, larger groups could be examined in a similar manner, but even for a moderately large archipelago (e.g., the New Hebrides birds, 56 species ×28 islands) the number of computations necessary to perform this counting sequence rapidly becomes prohibitive (for pairs, 43 120 computations; for trios, 776 160; for quartets, 10 284 120 etc.). To compute expected values for these statistics, the entire sequence, matrix construction and counting, is repeated a number of times and the means and variances computed from this pool of values.

To determine the number of "unstable" trios that have 0, 1, or 2 "stable" component pairs, an "expected" matrix is again randomly constructed and all groups of three rows (species) having 0 positive matches are inspected. For each of these trios of rows all three component pairs are examined to determine how many (0, 1, or 2) have at least one positive match. These results are summed for all groups of three rows to generate the distribution of number of "unstable" trios with 0, 1, or 2 "stable" component pairs. Expected values and their variances are again computed from a number of repeated runs of the same simulation with different random number sequences.

The success of the matrix construction component of this simulation depends on the topology of the actual matrix from which row and column constraints are derived. If the N_j are mostly small (many species distributed on only a few islands, with few species widely distributed), the simulation can easily fill the matrix randomly subject to the specified constraints (e.g., West Indies birds and bats). However, if many N_j are large (species widely distributed) the simulation "hangs up" without completely filling the matrix according to the specified constraints. What happens is that since the matrix rows are sorted from large to small values of N_j, widely distributed species are placed early in the simulation, when each column (island) has few 1's (presences of species). Usually however, midway through the matrix-filling process a species is encountered that must be placed on, say, 10 islands, but of the, say, 28 islands (columns) in the archipelago, 19 have already been "filled up" (received S_i species) by the simulation. Hence, the expected matrix cannot be completed. We have determined that the reason for this hang up is not the simulation itself, but simply that when most N_j are large the population of non-equivalent matrices is very small (by nonequivalent we mean that the matrices are not derivable from the actual matrix by interchanges of the rows and/or columns).

This problem arose when we attempted to construct the New Hebrides bird matrix according to the simulation. To circumvent this problem we developed an alternative algorithm to insure that the random matrices generated for the New Hebrides birds were in fact all nonequivalent. This procedure involves placing the actual New Hebrides bird matrix in a canonical form. We did so by putting the row and column sums in echelon form. This involved first placing rows in decreasing row sum order from top to bottom, and then column sums in decreasing order from left to right. Among the tied row sums, the row translating to the largest binary number was placed first, etc., and this was repeated for all groups of tied rows. After the rows were sorted, tied columns were

treated similarly. When in canonical form, the full matrix was then scanned for submatrices of the form:

			N_j
	1	0	1
	0	1	1
S_i	1	1	

although they need not be this closely spaced. When located, these submatrices were switched so that the 1's and 0's appear on the opposite diagonal:

			N_j
	0	1	1
	1	0	1
S_i	1	1	

This allows the row and column sums to be maintained since the changes exactly balance. After several of these changes are made, the new matrix was placed in canonical form (as described above). If the actual and new matrices when placed in canonical form are not completely identical then they are nonequivalent and can be used to compute the statistics of interest using the counting part of the simulation. A number of matrices were thus created to generate means and variances of the statistics for the New Hebrides birds.

8 Gradients in Species Diversity: Why Are There So Many Species in the Tropics?

James H. Brown and Dov F. Sax

The richness of species is not uniformly distributed over the earth but instead is concentrated in the tropics and some other regions and habitats, especially rainforests, coral reefs and mountainous regions. For more than two centuries, Western scientists have been aware of this fundamental manifestation of nature and endeavored to explain it. In so doing, they have learned much about geographic gradients in diversity of species and communities, of variation in patterns of abundance and geographic distribution, and perhaps most importantly of the mechanisms that may cause these patterns across disparate spatial and temporal scales.

Early Accounts of Tropical Biodiversity

It would be hazardous to attempt to trace the earliest history of writings about geographic variation in species diversity. It is clear, however, that the species richness of the tropics made an indelible impression on the European naturalists who participated in voyages of exploration in the eighteenth and nineteenth centuries. This is seen, for example, in the writings of Joseph Banks and Johann Reinhold Forster (see part 1 of this volume), two botanists who circumnavigated the world with Captain James Cook.

Perhaps the most diagnostic early commentary is by Alexander von Humboldt, who is often viewed as the "father of phytogeography." Humboldt spent the years 1799 to 1804 in the New World tropics, including an extensive time in the Amazon Basin. His account, first published in 1807, translated in 1850, and reprinted in Hawkins (2001: 470), speaks eloquently for itself.

Over a great portion of the earth, therefore, only those organic forms that are capable of full development, which have the property of resisting the considerable abstraction of heat, or those which, destitute of leaf organs, can sustain a protracted interruption of their vital functions. Thus, the nearer we approach the tropics, the greater the increase in the variety of structure, grace of form, and mixture of colors, as also in perpetual youth and vigor of organic life.

One of the many avid readers of Humoldt's work was Charles Darwin, who was certainly one of the most influential naturalists of the nineteenth century, but who spent comparatively little time in the tropics. The only tropical continental habitats that he saw at first hand were the Atlantic rain forests of eastern Brazil, where the *Beagle* stopped in 1832 (Darwin 1839). Darwin was only 23 years of age and near the beginning of his life-altering voyage. His accounts focus on the grandeur and lushness of the forests, and the peculiar features of some of the organisms that he found, observing for example that, "the number of minute and obscurely-coloured beetles is exceedingly great" (38). He also made the following comparison:

In England any person fond of natural history enjoys in his walks a great advantage, by always having something to attract his attention; but in these fertile climates [i.e., Brazil], teeming with life, the attractions are so numerous, that he is scarcely able to walk at all. (29)

Among nineteenth-century naturalists, it was Alfred Russel Wallace who gave the clearest descriptions and most thoughtful discussions of the biological richness of the tropics. Wallace spent many years in the field, much of it in the spectacularly diverse tropical forests of Indonesia, which he called the "Malay Archipelago." There he collected thousands of specimens and documented biogeographic distributions of hundreds of plants and animals. Building upon this research, he published many books and essays. The following quotes from *Tropical Nature and Other Essays* (1878), provide just two examples of how Wallace, more than a century ago, was already grappling with patterns and processes that have preoccupied biogeographers ever since.

Atmospheric conditions are much more important to the growth of plants than any others. Their severest struggle for existence is against climate. As we approach towards regions of polar cold or desert aridity the variety of groups and species regularly diminishes; more and more are unable to sustain the extreme climatal conditions, till at last we find only a few specially organized forms which are able to maintain their existence. In the extreme north, pine or birch trees; in the desert, a few palms and prickly shrubs or aromatic herbs alone survive. In the

equable equatorial zone there is no such struggle against climate. Every form of vegetation has become alike adapted to its genial heat and ample moisture, which has probably changed little even throughout geological periods; and the never-ceasing struggle for existence between the various species in the same area has resulted in a nice balance of organic forces, which give the advantage, now to one, now to another, species, and prevents any one type of vegetation from monopolizing territory to the exclusion of the rest. (66)

The equatorial zone, in short, exhibits to us the result of a comparatively continuous and unchecked development of organic forms; while in the temperate regions, there have been a series of periodical checks and extinctions of a more or less disastrous nature, necessitating the commencement of the work of development in certain lines over and over again. In the one, evolution has had a fair chance; in the other it has had countless difficulties thrown in its way. The equatorial regions are then, as regards their past and present life history, a more ancient world than that represented by the temperate zones, a world in which the laws which have governed the progressive development of life have operated with comparatively little check for countless ages, and have resulted in those infinitely varied and beautiful forms. (123)

Wallace not only hypothesized that the causes of the latitudinal variation in species diversity were ultimately climatic, but also realized the need to understand the generation and maintenance of diversity in an historical and evolutionary context. Further, his conclusions, which may have been influenced by Dana's views on the factors limiting geographic ranges (see part 1), clearly anticipated later writings by Dobzhansky, MacArthur, and others who suggested that abiotic factors tended to determine the polar boundaries of geographic ranges whereas biotic interactions set the equatorial boundaries. Undoubtedly, those interested in the history of the study of biodiversity will be rewarded for delving much more deeply into Wallace's writings. His impressive grasp of both pattern and process is all the more remarkable, considering that much of the limited information available to him came from his own experience in the field.

Advances Triggered by the Modern Synthesis, and the Rise of Evolutionary Biology

Interest in global patterns of species diversity waned in the late nineteenth and early twentieth centuries, at least as reflected in the English language literature. American naturalists, including C. Hart Merriam, Henry C. Cowles, Joseph Grinnell, Forrest Shreve, Henry A. Gleason, Fredrick Clements, and Victor Shelford focused their field studies on temperate North American systems to address ecological questions about species distributions and community composition. In Great Britain, the emphasis of the most influential naturalists, notably Arthur G. Tansley, Charles S. Elton, and Alex S. Watt, focused primarily on more local, ecological questions—although, of course, Elton made important contributions to biogeography through his work on invasive species (see part 3) and assembly rules (see part 7).

The situation changed dramatically after World War II, with the solidification and extension of the Modern Synthesis in evolutionary biology. Before the war, Ronald Fisher, Sewall Wright, J. B. S. Haldane, and others had characterized the genetic mechanisms of evolutionary change that were missing or incorrect in Darwin's original formulation of natural selection. Theodosius Dobzhansky's *Genetics and the Origin of Species*, published in 1937, provided perhaps the most comprehensive discussion of the New Synthesis by combining the theoretical perspectives of Fisher, Wright, and Haldane with supporting empirical studies by himself and others like F. B. Sumner, L. R. Dice, and Frank Blair. By the 1950s, evolutionary biologists were extending the scope of this synthesis to provide an adaptive evolutionary perspective on longstanding questions in biogeography, ecology, paleontology, and other disciplines.

The first selection featured in this part is THEODOSIUS DOBZHANSKY's 1950 essay, "Evolution in the Tropics" (paper 67). Dobzhansky spent a lot of time in the American tropics, most of it in Brazil. In this paper, written for a general audience, Dobzhansky began by contrasting the "subdued . . . almost inhibited" nature of life forms in the temperate zones with those in the tropics, which are "exuberant, luxurious, flashy, often even gaudy, full of daring and abandon" (209). He went on to describe the central tenet of his essay that "any differences between tropical and temperate organisms must be the outcome of differences in evolutionary patterns."

Before attempting to account for these differences, Dobzhansky first documented tropical species richness and distribution. He noted that not only are there many species in the tropics, but that most of the species are rare. He presented data for two, 1 ha plots of forest near Belem, Brazil, one of which contained 423 trees of 87 species, with 33 species represented by only a single individual. Dobzhansky used these data to plot one of the first rank-abundance distributions, and then provided a major insight by characterizing latitudinal variation in species diversity as a gradient, not just a difference between zones, and by noting that this gradient is imperfect: "The progressive increase in diversity of species from the Arctic toward the equator is apparent in general, even though there are irregularities in this increase . . ." (214). He also summarized his extensive data on genetic diversity of *Drosophila*, showing that the number of chromosome polymorphisms, both within and between populations, is much greater in the tropics.

After demonstrating these features of tropical diversity, Dobzhansky explained how evolutionary processes could produce them. He argued that the dominant evolutionary pressures at high latitudes are abiotic, being principally due to temperature, and he suggested that abiotic constraints on evolution, and the periodic loss and gain of habitat at high latitudes due to glaciation, favor evolution of species that can adapt quickly to take advantage of these conditions. However, he believed that adaptation to immediate abiotic stress is likely to sacrifice future evolutionary plasticity, and so that these populations "become stranded in these evolutionary blind alleys" (219). In contrast, tropical organisms experience little abiotic stress, but instead are subjected to more severe biotic interactions: "Where physical conditions are easy, interrelationships between competing and symbiotic species become the paramount adaptive problem. The fact that physically mild environments are as a rule inhabited by many species makes these interrelationships very complex" (220).

It is this last point that most clearly lays important foundations for subsequent ideas about species diversity. Dobzhansky anticipated two seminal concepts that were developed more explicitly by subsequent authors. One is the asymmetry of abiotic and biotic limiting factors as a function of latitude, with physical conditions being more important at high latitudes and biological interactions more powerful in the tropics. This asymmetry figures prominently in Robert H. MacArthur's (1972) explanation for geographic patterns of species diversity (see below). The other concept that Dobzhansky anticipated is that of the "Red Queen." This idea, developed by Van Valen (1973), is based on the character in Louis Carroll's *Through the Looking Glass,* who said "it takes all the running you can do, to keep in the same place." The idea is that each species must continually evolve and coevolve if it is to persist, because the other organisms in its environment are continually evolving. Dobzhansky recognized that the high species diversity in the tropics is in large part owing to the evolutionary consequences of the numerous biotic interactions.

The 1960s saw the publication of several important papers that documented, quantitatively and in detail, geographic variation in species richness in different kinds of organisms and different biogeographic regions and environmental settings. These publications were based on the biotic inventories and geographic range maps that were becoming increasingly well documented in the literature. We have reprinted two of the most influential papers here.

The first, a major synthesis by ALFRED G. FISCHER, is "Latitudinal Variations in Organic Diversity" (1960; paper 68). In the first part of

his paper, Fischer compiled data on latitudinal patterns of diversity in many different taxonomic groups of both terrestrial and marine organisms. Fisher's data presentation may seem quaintly primitive by today's standards, and his statistical analysis was nonexistent. Nevertheless, his paper documented the generality of the latitudinal gradient of increasing species richness from poles to equator more cogently than any previous study. Ever since Fischer, it has been clear that most taxonomic groups follow this pattern, and only a few exceptional groups are most diverse outside the tropics. Fischer also clearly articulated that the mechanisms responsible for latitudinal gradients should operate at smaller biogeographic and local scales. Therefore, gradients in richness should be observed from, "hot lowlands into cold and variable mountain areas, from humid rainforest into deserts, from the warm shallow waters into the frigid ocean depths, and from normal warm-water areas into belts of cold upwelling" (68).

In the second part of his paper, Fischer evaluated evolutionary and ecological hypotheses to explain the general pattern. He presented a graphical model of factors affecting diversification over evolutionary time. These ideas, with their emphasis on the "immaturity" of high-latitude biotas that Fischer attributed to long-term, large-scale climatic fluctuations, may now seem somewhat antiquated. But they anticipated subsequent efforts to erect and evaluate alternative historical and ecological hypotheses by Pianka (1966), Brown and Gibson (1983), Rohde (1992), Rosenzweig (1995), Brown and Lomolino (1998), Sax (2001) and others. Finally, one of Fischer's most prognostic contributions was his speculation about the nature of species extinction, and whether it occurred as a result of gradual range contraction or wholesale collapse.

The other featured paper based on data compiled from the literature is by GEORGE GAYLORD SIMPSON (1964; paper 69). Simpson is best known for his many books and papers that extended the Modern Synthesis to interpret the fossil record of life on earth. This paper, on geographic variation in species richness of North American mammals, was different but also extremely influential. Simpson developed an innovative method of using geographic range maps to tally the number of species occurring in grid cells of a standardized area (in this case 150 miles on a side), and then depicting the resulting geographic pattern as *isopleths,* or contours of increasing diversity. This method has been widely used in subsequent studies, such as Stehli's (1968) of planktonic foraminifera, Cook's (1969) of birds, Kiester's (1971) of reptiles and amphibians, Badgley and Fox's (2000) of mammals, and many others (e.g., Rahbek and Graves 2001).

The other major contribution of Simpson's paper is the explicit demonstration that the latitudinal pattern of increasing species toward the tropics, although pronounced, is complicated by the influence of other factors. Simpson shows that the latitudinal pattern is far from a smooth gradient. He suggests possible reasons for the varying rate of change at different latitudes, but most of these are specific to mammals and therefore of limited general interest. More generally, he anticipates many subsequent studies by calling attention to the effects of peninsulas and mountains. He notes that diversity declines with distance along several peninsulas, including Florida, Baja California, and Yucatan. This appears to be one of the first clear documentations of the so-called *peninsula effect,* which has subsequently been documented for many taxonomic groups on many different peninsulas. The effect has usually been studied from the perspective of island biogeography theory (see Brown and Lomolino 1998).

Simpson (1964) also called attention to the higher diversity of mammals in mountainous regions (e.g., the Appalachians, Rocky Mountains, Cascades and Pacific Coast Ranges) than in the regions of low topographic relief (e.g., the Great Plains, Mississippi Valley, and Atlantic Coast Plain). A similar pattern has been found in nearly all subsequent studies of species diversity in terrestrial organisms in North America. Simpson rightly attributed this pattern largely

to spatial heterogeneity: the juxtaposition of habitats found at different elevations allows species with different ecological requirements to come into close proximity, so that the ranges of many species are included in a single grid cell. Simpson writes, "On the principle that higher altitudes are ecologically similar to higher latitudes and that higher latitudes generally have lower species densities, one might expect just the opposite of what is found, a *decrease* of species densities in mountains instead of an increase. . . . the increase of species densities here so obvious is connected with the high relief and not with mountains or altitude *per se*" (69). We will return to this issue of spatial scale when we discuss Robert MacArthur's (1972) attempt to synthesize the different patterns and processes of geographic variation in species diversity. The situation is a bit more complicated, however, than simply including diverse habitats and multiple adjacent, but nonoverlapping geographic ranges in large grid cells in regions of high topographic relief. We will consider elevational patterns of species richness again when we discuss the paper of Whittaker and Niering (1975).

As implied above, the 1950s and 1960s saw greatly increased attention to geographic patterns of species diversity, stimulated in large part by efforts to extend the Modern Synthesis to address questions in ecology and biogeography. The strong evolutionary emphasis is apparent in the selections by Dobzhansky, Fischer, and Simpson and other contemporary works. Note also, however, the strong ecological emphasis, especially in the paper by Simpson, who was after all one of the major architects of the Modern Synthesis. Simpson downplayed the role of evolutionary processes, especially the then-current suggestions that higher rates of evolution in the tropics overwhelm the influence of abiotic and biotic factors that might otherwise limit coexistence of species. In this respect, his work had much in common with the more ecological perspectives of contemporaries such as Hutchinson (1957a and b, 1959), Klopfer (1959), Klopfer and MacArthur (1960), Williams (1964), and MacArthur (1965).

Mechanistic Hypotheses and Synthetic Efforts

By the late 1960s, owing in large part to the quantitative data compilations of Fischer (1960), Simpson (1964), Stehli (1968), Sanders (1968) and others, its was clear that nearly all major taxonomic groups are most diverse in the tropics (see Brown and Lomolino 1998 for discussion of the few exceptions). It remained, however, to explain this near-universal pattern. Thus, Simpson in 1964 writes "The long-known and unquestionable phenomenon of tropical diversity has never been generally and adequately explained" (64). The efforts of evolutionary biologists, ecologists, and biogeographers to explain the pattern began with the works of Elton (1958), Hutchinson (1959, reprinted in *Foundations of Ecology*), Klopfer (1959), Fischer (1960), Klopfer and MacArthur (1960), Connell and Orias (1964), Simpson (1964), Williams (1964), MacArthur (1965) and others. These efforts proliferated and, indeed, they have continued right up until the present. Simpson's comment quoted above is as relevant today as it was when it was written in 1964.

The last three selections in this part illustrate different approaches taken by some of the most eminent evolutionary ecologists of the twentieth century. Paper 70, by ERIC R. PIANKA (1966), lists and evaluates "six more or less distinct hypotheses: (a) the time theory, (b) the theory of spatial heterogeneity, (c) the competition hypothesis, (d) the predation hypothesis, (e) the theory of climatic stability, and (f) the productivity hypothesis" (34). We will not consider these hypotheses individually here. It is sufficient to note that these and other possible explanatory factors have continued to be discussed and assigned varying importance by subsequent authors right up until the present (e.g., MacArthur 1965; Terborgh 1973; Schall and Pianka 1978; Brown 1981, 1988; Brown and Gibson 1983; Currie 1991; Rohde 1992; Rosenzweig 1995; and Brown and Lomolino 1998).

It is important, however, to note that Pianka's treatment has a distinctly ecological emphasis. He initially raised the *time theory*, or the hypothesis that the latitudinal gradient reflects the historical legacy of Pleistocene glaciation and related climatic perturbations at high latitudes. But then he invoked empirical studies of islands (see part 6 in this volume) and MacArthur and Wilson's (1963, 1967) equilibrium theory of island biogeography to suggest that species diversity may usually reflect a steady-state response of the biota to current environmental conditions. Pianka then went on to explore the possible influence of the five ecological factors listed above. The special focus on competition, predation, and niche characteristics may seem somewhat outdated today, but it reflected the powerful influence of the theoretical studies of Hutchinson, MacArthur, Preston (1962; reprinted in *Foundations of Ecology*), Williams (1964) and others on the biogeography and evolutionary biology of the time. Pianka also pointed to the seminal experimental studies of Connell (1961) and Paine (1966; both reprinted in *Foundations of Ecology*). On the one hand, he recognized the limitations of such studies, which usually manipulated only one or a small number of ecological variables on spatial and temporal scales much smaller than the geographic patterns. On the other hand, he called attention to the difficulty of relying exclusively on nonmanipulative comparative geographic studies, where many of the relevant variables were intercorrelated in ways that do not allow for unambiguous tests of hypotheses. He echoed the experimental emphasis of much of the ecology of the 1960s and 1970s by concluding "Unambiguous demonstration of causality can only be attained by experimental manipulation of the independent variables in the system" (Pianka 1966: 43).

Although Pianka was heavily influenced by ROBERT H. MACARTHUR (who had been his postdoctoral mentor), MacArthur's own final discourse on the problem of geographic variation in species diversity had a very different flavor from Pianka's. We have reprinted all twenty-six pages of chapter 7, "Patterns of Species Diversity," from MacArthur's book *Geo-*

graphical Ecology (paper 71). This book was published in 1972, the same year as MacArthur's death from cancer at the age of 42. Chapter 7 summarizes many of the characteristics of MacArthur's work that made it so influential: a focus on big, important conceptual questions; clear writing that spoke to a broad audience of ecologists, evolutionary biologists, and biogeographers; and simple mathematical and graphical models that placed the questions in a compelling rigorous theoretical framework. MacArthur knew that he was dying, and was writing a final exposition of his views on evolution, ecology and biogeography. Nevertheless, *Geographical Ecology*, like his earlier publications, was very much an effort to convey his current thinking. To a contemporary reader, it may seem somewhat dated and may fail to convey the magnitude of MacArthur's influence on ecology and biogeography, which began in the 1950s and lasted well into the 1990s (see Brown 1999).

Chapter 7 begins with a theoretical treatment that develops a simple graphical and mathematical model. This part, like much of MacArthur's theoretical work, places a large emphasis on competition. Subsequent sections of the chapter focus on empirical studies by other investigators that MacArthur interprets in terms of his model. Several things are apparent. First, MacArthur largely ignored the influences of mutualism, predation, herbivory, parasitism, and disease (but note the brief final section on "The Effect of Predators"). Second, he did not attempt a thorough survey of the relevant literature, but instead emphasized work of friends and associates. Third, although MacArthur supported his ideas with examples from "Habitats Differing only . . ." (or primarily) in structure, climatic stability, or productivity (177–85), he realized the limitations of such comparisons. In subsequent sections on "Habitats Differing in Structure, Stability, and Productivity" (185) and on "Matters of Environmental Scale" (186), MacArthur confronted not only the problems of complex intercorrelations among the variables, but also the problems of extrapolating uncritically

across divergent scales, where the relative roles of different processes may vary. His views on the "hierarchical structure" of nature laid the groundwork for subsequent examinations of the importance of scale by Peter Kareiva, Tim Allen, Thomas Hoekstra and other ecologists and biogeographers (e.g., Allen and Hoekstra 1992; Ives, Kareiva, and Perry 1993).

We have also reprinted a small section on "Tropical Richness of Species" from chapter 8 of MacArthur's *Geographical Ecology*. In this section, he presented empirical data from several authors (many of them cited above) and provided his own perspective on some of the possible causes, including freezing temperatures, seasonality, productivity, niche breadth and overlap in relation to competition and predation. MacArthur presented additional views on tropical species richness in two other chapters (which are not reprinted here). In "Species Distributions" (chapter 6), he modified Dobzhansky's ideas about the importance of biotic interactions in the tropics to suggest that, as a general rule, the polar boundaries of species ranges are determined primarily by abiotic factors that cause physiological stress (such as freezing temperatures), whereas equatorial range boundaries are set primarily by "diffuse" interactions with many other species. In "The Role of History" (chapter 9), he tried to reconcile his general worldview that most ecological patterns reflect a biota in approximate equilibrium with their environment, with seeming compelling examples of long time lags and other legacies of past events. The overall message is that MacArthur was an exceptionally observant and clear-thinking ecologist who had enormous influence during and for decades after his own lifetime. Now, after three decades, this influence appears to be waning, as evidenced by the fact that MacArthur is no longer the most frequently cited ecologist. But MacArthur's promotion of geographical ecology and his focus on big, pervasive patterns of biodiversity played a major role in elucidating the shortcomings of his own theories and in stimulating the new directions of research. It is impossible to understand the development of biogeography and ecology in the second half of the twentieth century without appreciating the unique role of Robert MacArthur.

The final selection of this part, "Vegetation of the Santa Catalina Mountains, Arizona, V: Biomass, Production, and Diversity along the Elevation Gradient," by ROBERT HARDING WHITTAKER and WILLIAM A. NIERING (1975; paper 72), is quite different from all of the preceding ones (see also Whittaker and Niering, 1964, 1965, 1968a and b). Whittaker was perhaps the preeminent plant ecologist of the late twentieth century. He is probably best known for developing *coenoclines* and *gradient analysis*, a way of quantifying the distribution and abundance of species across environmental gradients, especially the elevational gradients on mountains. He performed three major studies, in the Great Smokey Mountains of Tennessee and North Carolina, the Siskiyou Mountains of California and Oregon, and the one included here in the Santa Catalina Mountains of Arizona (see also C. H. Merriam's 1890 paper, featured in part 1). These studies resulted in many papers and review articles (e.g., Whittaker 1956, 1960, 1967) and a book, *Communities and Ecosystems*, published in 1970. Because Whittaker's gradient analysis proved fundamental to much of biogeography, community ecology, and landscape ecology, we featured his 1967 paper, "Gradient Analysis of Vegetation," in part 7 of this volume.

In addition to his work on gradient analysis, Whittaker also developed a theoretical structure for characterizing variation in diversity within and across communities at multiple spatial scales. Building upon the work of Fisher, Corbet, and Williams (1943), he popularized the term *alpha diversity*, which describes the richness of individual communities. He used the term *beta diversity* to describe the relative turnover of species composition among these communities. He introduced the term *gamma diversity* to characterize the species-richness of larger geographic regions. These concepts of alpha, beta, and gamma diversity have become the standard vocabulary for comparing biodi-versity across different scales (e.g., Ricklefs and Schluter 1993; Brown and Lomolino 1998).

Whittaker and Niering (1975) focus on somewhat different questions, namely how emergent properties of vegetation, such as productivity, biomass, and species diversity, vary along a gradient. Our interest is primarily on geographic gradients in species diversity, but other gradients are also of interest in evaluating mechanistic hypotheses. Similarities and differences in patterns of diversity along other gradients, where abiotic and biotic factors vary in different ways, can help sort out the correlation-causation dilemma mentioned by Pianka to identify mechanistic processes. Whittaker and Niering's paper is of particular interest because gradients of elevation are often at least superficially similar to those of latitude. Other studies of species diversity along elevational gradients documented relatively gradual declines from the warm wet lowlands to the perennially cold highlands, seemingly similar to the decline in terrestrial species richness from the equator to the poles (e.g., Yoda 1967; Kikkawa and Williams 1971; Terborgh 1977). Yet numerous systems such as desert mountains, and montane biotas in other regions are different. Shreve's (1915) earlier work in the same Santa Catalina Mountains noted that plant diversity peaked at intermediate elevations, not in the hot and arid lowlands or the cold and moist highlands (Brown 2001; Heaney 2001; Lomolino 2001).

Whittaker and Niering (1975: 783) quantify this pattern, showing that species diversity was highest at about 1,300 meters (upper panel of fig. 6). They also show that productivity and biomass had hump-shaped distributions, but peaked nearer the upper end of the gradient, at about 2,500 meters (figs. 2 and 3). Note, however, that if the Santa Catalina Mountains were high enough to have a tree line and alpine tundra vegetation, biomass and productivity would continue to decline to permanently frozen environments above 3,000 meters. The authors' discussion is very brief and barely mentions possible mechanisms likely to account for the hump-shaped pattern of diversity. They do, however, briefly consider the relationship of diversity to productivity, and comment that "figures 3 and 6 do not support such a correlation." Interestingly, numerous, more recent studies have shown that species diversity also peaks at intermediate elevations on tropical mountains, where the lowlands are not obviously stressfully arid and less productive (e.g., Rahbek 1997; Brown 2001, and included references).

Advances since 1975

The pace of research has continued to increase since the mid-1970s. We can identify two important trends. First, there has been an enormous increase in the quantity and quality of data. Numerous new studies provided much more precise information on the distributions of species and the patterns of species richness. So there are now high-quality data for many different taxonomic groups in many different habitats, including gradients not only of latitude and elevation, but also of aridity, salinity, and other variables in terrestrial environments, depth in oceans and lakes, size or order in streams, height in the intertidal, and so on. Further, computer-aided advances, such as large electronic databases, spatial statistics, and Geographic Information Systems (GIS), have facilitated the compilation and analysis of this information. We simply know much more about the patterns of variation in species richness than we did 25 years ago. Second, there have been many efforts to present and evaluate alternative mechanisms for the patterns. This has led to a proliferation of hypotheses and to much debate about their relative merits. In 1966 Pianka listed six possible causes for the latitudinal pattern, but by 1992 Rohde had listed nearly thirty. While some of the latter are specific to particular environmental settings (e.g., forests) or specific taxonomic or functional groups of organisms (e.g., epiphytes), several additional and potentially general mechanisms have been proposed. These include: (1) the area of different habitat (or biome) types,

suggested by Terborgh (1973) and Rosenzweig (1992, 1995), a hypothesis largely based on the fact that species-area relationships universally have positive slopes (see part 6, on islands) and that there is more land area at lower latitudes and elevations; and (2) geographic variation in temperature, which was suggested by Rohde (1992) to influence diversity by affecting the speed of evolution.

In addition, other latitudinal patterns have been described and suggested to be related to variation in species richness. For example, Hubbell (1979, 2001), building on the work of Dobzhansky (1950), has emphasized the latitudinal variation in rank-abundance distributions within communities, which vary consistently such that tropical habitats typically contain many rare species, whereas temperate and arctic environments tend to have a few common ones. Perhaps one of the most intriguing patterns is that known as *Rapoport's Rule.* Stevens (1989), building upon the work of Rapoport (1982), showed that in many taxa the area or latitudinal extent of the geographic range is correlated positively with latitude and negatively with species richness. Stevens suggested that this pattern of varying range size could explain the latitudinal gradient in richness, but both the generality of this pattern and its potential affects upon the latitudinal gradient in richness have been debated (e.g., Rohde, Heap, and Heap 1993; Roy, Jablonski, and Valentine 1994; Gaston, Blackburn, and Spicer 1998; Taylor and Gaines, 1999; see also Colwell and Hurtt, 1994). Nevertheless, Stevens's (1989) paper has played an important role in revitalizing efforts to explain the latitudinal gradient of species richness, and has led to a large body of research that has been extremely valuable in describing and understanding biogeographical variation in species' distributions.

Finally, some of the most intriguing work to be done in recent years, and potentially some of the most promising, has examined geographic gradients in diversity for groups or systems that might not be expected to show such patterns. The first of these is the pioneering work of Rex et al. (1993), who investigated the abyssal plains and deep-sea trenches of the ocean, where to many investigators surprise, they found pronounced latitudinal gradients in species richness despite the apparent near-constancy of abiotic conditions at such great depths. Since the publication of these findings, much additional work has been done to corroborate the common nature of this pattern in the deep sea (e.g., Culver and Buzas 2000; Rex, Stuart, and Coyne 2000). Understanding the mechanisms responsible for these patterns, and relating them to comparative studies in nearshore environments, and in terrestrial systems, should provide fertile ground for research in coming years. A second set of unexpected patterns has been explored by Rejmánek (1996) and Sax (2001). They show that latitudinal gradients in richness and geographical range size are found in some groups of non-native species within their naturalized continents. Often these patterns appear to be similar to those observed with native taxa. Careful comparisons of exotic and native biotas should be useful in discriminating between the importance of evolutionary and ecological mechanisms. On a final note, it is important to recognize that even though the geographic patterns of diversity may hold across many taxonomic groups, habitat types, and geographic regions, no single mechanistic hypothesis may be sufficient to account for them. An approach that incorporates multiple, interacting causative factors may ultimately provide the best explanation for these patterns.

Latitudinal and elevational gradients in diversity have been a principal focus of much biogeographic research and inquiry from the age of exploration to the present. While general agreement on a set of causal mechanisms has proven elusive, the scientific gains of this endeavor have nonetheless been enormous. These efforts have lead to a greater understanding of the geographic distribution and diversity of species, communities and ecosystems; i.e., to more accurate descriptions of spatial and temporal patterns of biodiversity on the Earth. They have also allowed more insightful examinations of variation in the size of geographic ranges, variation in community structure across geographic scales, and diversity gradients along peninsulas and at depths within the oceans. Ultimately, this accumulating wealth of information on biogeographic gradients has lead to the development of more comprehensive and, we think, more insightful theories of the local and geographic scale processes influencing biological diversity.

AMERICAN SCIENTIST

SPRING ISSUE APRIL 1950

EVOLUTION IN THE TROPICS

By THEODOSIUS DOBZHANSKY

Columbia University

BECOMING acquainted with tropical nature is, before all else, a great esthetic experience. Plants and animals of temperate lands seem to us somehow easy to live with, and this is not only because many of them are long familiar. Their style is for the most part subdued, delicate, often almost inhibited. Many of them are subtly beautiful; others are plain; few are flamboyant. In contrast, tropical life seems to have flung all restraints to the winds. It is exuberant, luxurious, flashy, often even gaudy, full of daring and abandon, but first and foremost enormously tense and powerful. Watching the curved, arched, contorted, spirally wound, and triumphantly vertical stems and trunks of trees and lianas in forests of Rio Negro and the Amazon, it often occurred to me that modern art has missed a most bountiful source of inspiration. The variety of lines and forms in tropical forests surely exceeds what all surrealists together have been able to dream of, and many of these lines and forms are endowed with dynamism and with biological meaningfulness that are lacking, so far as I am able to perceive, in the creations exhibited in museums of modern art.

Tropical rainforest impresses even a casual observer by the enormity of the mass of protoplasm arising from its soil. The foliage of the trees makes a green canopy high above the ground. Lianas, epiphytes, relatively scarce undergrowth of low trees and shrubs, and, finally, many fungi and algae form several layers of vegetational cover. Of course, tropical lands are not all overgrown with impenetrable forests and not all teeming with strange-looking beasts. One of the most perfect deserts in the world lies between the equator and the Tropic of Capricorn, in Peru and northern Chile. Large areas of the Amazon and Orinoco watersheds, both south and north of the equator, are savannas, some of them curiously akin to southern Arizona and Sonora in type of landscape. But regardless of the mass of living matter per unit area, tropical life is impressive in its endless variety and exuberance.

Since the animals and plants which exist in the world are products of the evolutionary development of living matter, any differences between tropical and temperate organisms must be the outcome of differences in evolutionary patterns. What causes have brought about the greater richness and variety of the tropical faunas and floras, compared to faunas and floras of temperate and, especially, of cold lands? How does life in tropical environments influence the evolutionary potentialities of the inhabitants? Should the tropical zone be regarded as an evolutionary cradle of new types of organization which sends out migrants to colonize the extratropical world? Or do the tropics serve as sanctuary for evolutionary old age where organisms that were widespread in the geological past survive as relics? These and related problems have never been approached from the standpoint of modern conceptions of the mechanism of evolutionary process. Temperate faunas and floras, and species domesticated by or associated with man, have supplied, up to now, practically all the material for studies on population genetics and genetical ecology.

Classical theories of evolution fall into two broad groups. Some assume that evolutionary changes are autogenetic, i.e., directed somehow from within the organism. Others look for environmental agencies that bring forth evolutionary changes. Although two eminent French biologists, Cuénot and Vandel, have recently espoused autogenesis, autogenetic theories have so far proved sterile as guides in scientific inquiry. Ascribing arbitrary powers to imaginary forces with fancy names like "perfecting urge," "combining ability," "telefinalism," etc., does not go beyond circular reasoning.

Environmentalist theories stem from Lamarck and from Darwin. Lamarck and psycholamarckists saw in exertion to master the environment the principal source of change in animals. Induction of changes in the body and the germ cells by direct action of physical agencies is the basis of mechanolamarckism. The organism is molded by external factors. (A blend of psycholamarckist and mechanolamarckist notions, the latter borrowed chiefly from Herbert Spencer, has been offered by Lysenko as "progressive," "Michurinist," and "Marxist" biology. Advances of genetics have made Lamarckist theories untenable. There is not only no experimental verification of the basic assumptions of Lamarckism, but the known facts about the mechanics of transmission of heredity make these assumptions, to say the least, far-fetched.

Important developments and many changes have taken place in Darwinism since the publication of *The Origin of Species* in 1859. The essentials of the modern view are that the mutation process furnishes the raw materials of evolution; that the sexual process, of which Mendelian segregation is a corollary, produces countless gene patterns; that the possessors of some gene patterns have greater fitness than the possessors of other patterns, in available environments; that natural selection increases the frequency of the superior, and fails to perpetuate the adaptively inferior, gene patterns; and that groups of gene combinations of proved adaptive worth become segregated into closed genetic systems called species.

The role of environment in evolution is more subtle than was realized in the past. The organism does not suffer passively changes produced

Evolution in the Tropics   211

by external agents. In the production of mutations, environment acts as a trigger mechanism, but it is, of course, decisive in natural selection. However, natural selection does not "change" the organism; it merely provides the opportunity for the organism to react to changes in the environment by adaptive transformations. The reactions may or may

Fig. 1. Interior of equatorial rainforest near Belem, Brazil. The "terra firme" association. (Courtesy of Mr. Otto Penner, of the Instituto Agronomico do Norte, Belem do Pará.)

not occur, depending upon the availability of genetic materials supplied by the mutation and recombination processes.

Diversity of Species in the Tropics

Gause pointed out in 1934 that two or more species with similar ways of life can not coexist indefinitely in the same habitat, because one of them will inevitably prove more efficient than the others and will crowd out and eliminate its competitors. This "Gause principle" is a fruitful working hypothesis in studies on evolutionary patterns in tropical and temperate climates. The diversity of organisms which live in a given territory is a function of the variety of available habitats. The richer and more diversified the environment becomes, the greater should be the multiformity of the inhabitants. And vice versa: diversity of the inhabitants signifies that the environment is rich in adaptive opportunities.

Now, the greater diversity of living beings found in the tropical compared to the temperate and cold zones is the outstanding difference which strikes the observer. This is most apparent when tropical and temperate forests are compared. In temperate and cold countries, the forest which grows on a given type of terrain usually consists of masses of individuals of a few, or even of a single, species of tree, with only an admixture of some less common tree species and a limited assortment of shrubs and grasses in the undergrowth. The vernacular as well as the scientific designation of the temperate forest associations usually refers to these dominant species ("pine forest," "oak woodland," etc.). The forests of northern plains may, in fact, be monotonously uniform. In the forested belt of western Siberia one may ride for hundreds of miles through birch forests interrupted only by some meadows and bogs. Mountain forests are usually more diversified than those of the plains. Yet in the splendid forest of the Transition Zone of Sierra Nevada of California one rarely finds more than half a dozen tree species growing together. By contrast tropical forests, even those growing on so perfect a plain as that stretching on either side of the Amazon, contain a multitude of species, often with no single species being clearly dominant. Dr. G. Black, of the Instituto Agronomico do Norte in Belem, Brazil, made, in cooperation with Dr. Pavan and the writer, counts of individuals and species of trees 10 or more centimeters in diameter at chest height on one-hectare plots (100 × 100 meters) near Belem. On such a plot in a periodically inundated (igapó) forest 60 species were found among 564 trees. The numbers of species represented by various numbers of individuals on this plot were as follows:

Individuals	1	2	3	4	5	6	7	8	9
Species	22	9	7	2	2	2	4	2	1

Individuals	10	14	15	16	21	29	33	41	241
Species	1	1	1	1	1	1	1	1	1

The commonest species was the assaí palm (*Euterpe oleracea*), of

Evolution in the Tropics   213

which 241 individuals were found in the hectare plot, but as many as 22 species were represented by single individuals. On a plot of similar size, only a mile away but on higher ground (terra firma), 87 species were found among 423 trees. The numbers of individuals per species were as follows:

Individuals	1	2	3	4	5	6	7	9	12	17	20	25	37	49
Species	33	15	15	3	4	2	3	4	1	2	1	1	2	1

Here the commonest species was represented by only 49 individuals and as many as 33 species were found as single individuals. This high frequency of species represented by single individuals means that if

Fig. 2. The wet ground ("igapo") forest near Belem, Brazil. (Courtesy of Mr. Otto Penner, of the Instituto Agronomico do Norte, Belem do Pará.)

we had studied other hectare plots contiguous to the one actually examined, many new species would have been found. It is probable that each of the plant associations which we have sampled in the vicinity of Belem contains many more than 100 tree species, and, incidentally, only a few species occur in both associations. Similar results were obtained by Davis and Richards in the forests of British Guiana and by Beard in Trinidad.

The numbers of breeding species of birds recorded in the literature for territories in various latitudes, from arctic North America to equatorial Brazil, are as follows (data kindly supplied by Dr. E. Mayr):

Territory	Number of Species	Authority
Greenland	56	F. Salomonsen
Peninsular Labrador	81	H. S. Peters

Newfoundland	118	H. S. Peters
New York	195	K. C. Parkes
Florida	143	S. A. Grimes
Guatemala	469	L. Griscom
Panama	1100	L. Griscom
Colombia	1395	R. M. de Schauensee
Venezuela	1148	W. H. Phelps, Jr.
Lower Amazonia	738	L. Griscom

The numbers of recorded species of snakes are as follows (data obtained through the courtesy of Dr. C. M. Bogert):

Territory	Number of Species	Authority
Canada	22	Mills
United States	126	Steineger and Barbour
Mexico	293	Taylor and Smith
Brazil	210	Amaral

The progressive increase in diversity of species from the Arctic toward the equator is apparent in general, even though there are irregularities in this increase, resulting from such factors as how the different territories compare in size, how uniform or varied they are ecologically and topographically, and how intensively their faunas have been studied. There can be no doubt that, for most groups of organisms, tropical environments support a greater diversity of species than do temperate- or cold-zone environments.

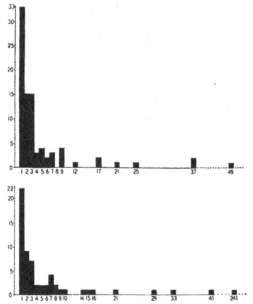

FIG. 3. Numbers of species of trees (ordinates) represented by different numbers of individuals (abscissae) on 1-hectare (100 × 100 meters) plots of equatorial rainforest near Belem, Brazil. "Terra firme" association (above) and "igapo" association (below). Only trees 10 or more cm. at chest height counted. (*Data of Black, Dobzhansky, and Pavan.*)

In order to survive and to leave progeny, every organism must be adapted to its physical and biotic environments. The former includes temperature, rainfall, soil, and other physical variables, while the latter is composed of all the organisms that live in the same neighborhood. A diversified biotic environment influences the evolutionary patterns of the inhabitants in several ways. The greater the diversity of inhabitants in a territory, the more adaptive opportunities exist in it. A tropical forest with its numer-

Evolution in the Tropics 215

ous tree species supports many species of insects, each feeding on a single or on several species of plants. On the other hand, the greater the number of competing species in a territory, the fewer become the habitats open for occupancy by each of these species. In the absence of competition a species tends to fill all the habitats that it can make use of; abundant opportunity favors adaptive versatility. When competing species are present, each of them is forced to withdraw to those habitats for which it is best adapted and in which it has a net advantage in survival. The presence of many competitors, in biological evolution as well as in human affairs, can be met most successfully by specialization. The diversity of habitats and the diversity of inhabitants which are so characteristic of tropical environments are conflicting forces, the interaction of which will determine the evolutionary fates of tropical organisms.

Chromosomal Polymorphism in Drosophila

Adaptive versatility is most easily attained by a species' becoming adaptively polymorphic, i.e., consisting of two or more types, each possessing high fitness in a certain range of environments. One of the most highly polymorphic species in existence is *Homo sapiens,* and it is this diversity of human nature which has engendered cultural growth and has permitted man to draw his existence from all sorts of environments all over the world. Now, the adaptive polymorphism of the human species is conditioned, on the cultural level, chiefly by the ability to become trained and educated to perform different activities. In species other than man, adaptive polymorphism is attained chiefly through genetic diversification—formation of a group of genetically different types with different habitat preferences.

In *Drosophila* flies, organisms best suited for this type of study, adaptive polymorphism takes its chief form in diversification of chromosome structure. In some American and European species of these flies, natural populations are mixtures of several interbreeding chromosome types which differ in so-called inverted sections. These chromosomal types have been shown, both observationally and experimentally, to have different environmental optima. The situation in tropical species has been studied in various bioclimatic regions in Brazil by a group consisting of Drs. A. Brito da Cunha, A. Dreyfus, C. Pavan, and E. N. Pereira of the University of São Paulo, A. G. L. Cavalcanti and C. Malogolowkin of the University of Rio de Janeiro, A. R. Cordeiro of the University of Rio Grande do Sul, M. Wedel of the University of Buenos Aires, H. Burla of the University of Zürich, N. P. Dobzhansky, and the writer. During the school year 1948-1949, the work of this group was supported by grants from the University of São Paulo, the Rockefeller Foundation, and the Carnegie Institution of Washington.

The commonest species in Brazil is *Drosophila willistoni.* It is also the adaptively most versatile species, since it has been found in every one of the 35 localities in various bioclimatic regions of Brazil in which collection was made. Significantly enough, this species, taken as a whole, shows not only the greatest chromosomal polymorphism among Brazilian species but the greatest so far known anywhere. The species *Drosophila nebulosa* and *Drosophila paulistorum* are also very common, but somewhat more specialized than *Drosophila willistoni*. *Drosophila nebulosa* is at its best in the savanna environments where dry seasons alternate with rainy ones, and *Drosophila paulistorum*, conversely, in superhumid tropical climates. The species are rich in chromosomal polymorphism, but not so rich as *Drosophila willistoni* in this respect. We have also examined some less common and biotically more specialized species, all of which showed much less or no chromosomal polymorphism.

The comparison of chromosomal polymorphism in *Drosophila willistoni* in different bioclimatic zones of Brazil proved to be even more interesting. The greatest diversity is found in those parts of the valleys of the Amazon and its tributaries where *Drosophila willistoni* is the dominant species. The exuberant rainforests and savannas of the Amazon basin are remarkable in the rich diversity of their floras and faunas; furthermore, the Amazon basin appears to be the geographical center of the distribution of the species under consideration, where it has captured the greatest variety of habitats. Yet, wherever in this region *Drosophila willistoni* surrenders its dominance to competing species, the chromosomal polymorphism is sharply reduced. This has been found to happen in the forested zone of the territory of Rio Branco, near Belem, in the state of Pará, and in the savanna of Marajó Island. In the first and the second of the regions named, *Drosophila paulistorum*, and in the third region, *Drosophila tropicalis*, reduce *Drosophila willistoni* to the status of a relatively rare species. In the peculiar desert-like region of northeastern Brazil, called "caatingas," *Drosophila willistoni* reaches its limit of environmental tolerance; *Drosophila nebulosa* seems to be the only species which still flourishes in this highly rigorous environment. The chromosomal variability of *Drosophila willistoni* is much reduced on the caatingas. It is also reduced in southern Brazil where *Drosophila willistoni* approaches the southern limit of its distribution, and is, presumably, losing its grip on the habitats.

Any organism which lives in a temperate or a cold climate is exposed at different periods of its life cycle or in different generations to sharply different environments. The evolutionary implications of nature's annually recurrent drama of life, death, and resurrection have not been sufficiently appreciated. In order to survive and reproduce, any species must be at least tolerably well adapted to every one of the environments which it regularly meets. No matter how favored a strain may be in summer, it will be eliminated if it is unable to survive winters, and vice versa. Faced with the need of being adapted to diverse environments, the organism may be unable to attain maximum efficiency in any one of them. Changeable environments put the highest premium on versatility rather than on perfection in adaptation.

Evolution in the Tropics   217

Adaptive Versatility in the Tropics

The widespread opinion that seasonal changes are absent in the tropics is a misapprehension. Seasonal variations in temperature and in duration and intensity of sunlight are, of course, smaller in the tropical than in the extratropical zones. However, the limiting factor for life in the tropics is often water rather than temperature. Some tropical climates, for example those of the caatingas of northeastern Brazil, have variations of such an intensity in the availability of water that plants and animals pass through yearly cycles of dry and wet environments which entail biotic changes probably no less serious than those brought about by the alternation of winter and summer in temperate lands. Absence of drastic seasonal changes in tropical environments is evidently a relative matter. In Belem, at the opening of the Amazon Valley to the Atlantic Ocean, the mean temperature of the warmest month, 26.2° C., is only 1.3° C. higher than that of the coolest month, and the highest temperature ever recorded, 35.1° C., is only 16.6° C. higher than the all-time low, which was 18.5° C. The wettest month has 458 mm. of precipitation, and the driest has 86 mm., which is still sufficient to prevent the vegetation from suffering from drought.

It might seem that the inhabitants of the relatively invariant tropical climates should be free from the necessity of being genetically adapted to a multitude of environments, and hence that evolution in the tropics would tend toward perfection and specialization, rather than to adaptive versatility. This is not the case, however. The climate of Espirito Santo Island, in the tropical Pacific, is seasonally one of the most constant in the world as far as temperature and humidity are concerned. Nevertheless, Baker and Harrison found that native plants have definite flowering and fruiting seasons, and animals have cycles of breeding activity in this climate. Our observations in Brazil show that populations of *Drosophila* flies undergo expansions or contractions from month to month, as well as changes in the relative abundance of different species. The magnitude and speed of these changes are quite comparable to those which occur in California, for example, or in the eastern part of the United States. Such pulsations have been observed even in the rainforests of Belem, about 1½° latitude south of the equator. They are caused mainly by seasonal variations in the availability of different kinds of fruits which are preferred by different species of *Drosophila*. Despite the apparent climatic uniformity, the biotic environment of tropical rainforests is by no means constant in time.

This writer observed several years ago that certain populations of the fly *Drosophila pseudoobscura* which live in the mountains of California undergo seasonal changes in the relative frequencies of chromosomal types. Dubinin and Sidorov found similar changes in the Russian *Drosophila funebris*. What happens is that some of the chromosomal types of these flies possess highest adaptive values in summer, and other types in winter or in spring environments. Natural selection augments the frequency of favorable types and reduces the frequency of unfavorable types. The populations thus react to changes in their environment by adaptive modifications. This is one of the rare occasions when evolutionary changes taking place in nature under the

FIG. 4. Giant trees and lianas in the "terra firme" rainforest near Belem. (Courtesy of Mr. Otto Penner, of the Instituto Agronomico do Norte, Belem do Pará.)

influence of natural selection can actually be observed in the process of happening. We may add that some of these changes have also been

Evolution in the Tropics   219

reproduced in laboratory experiments in which artificial populations of the species concerned were kept in special "population cages."

We have observed populations of *Drosophila willistoni* in three localities in southern Brazil for approximately one year. One of the localities, situated in the coastal rainforests south of São Paulo, has a rather uniform superhumid tropical climate. Periodic sampling of these populations has disclosed alterations in the incidence of chromosomal types similar in character to those observed in California and in Russian fly species. Adaptive alterations which keep living species attuned to their changing environments occur in tropical as well as in temperate-zone organisms. This constant evolutionary turmoil, so to speak, precludes evolutionary stagnation and rigidity of the adaptive structure of tropical and of temperate species equally.

It is nevertheless true that tropical environments are more constant than temperate ones, in a geological sense. Major portions of the present temperate and cold zones of the globe underwent drastic climatic and biotic changes owing to the Pleistocene glaciation. The present floras and faunas of the territories that were covered by Pleistocene ice are composed almost entirely of newcomers. The territories adjacent to the glaciated areas have passed through more or less radical climatic upheavals. Although the bioclimatic history of the tropical continents is still very little known, it is fair to say that their environments suffered less change.

The repeated expansions and contractions of the continental ice sheets, and the alternation of arid and pluvial climates in broad belts of land bordering on the ice, made large territories rather suddenly (in the geological sense) available for occupation by species that could evolve the necessary adaptations in the shortest possible time. This has not simply increased the rates of evolution, but often has favored types of changes which can be characterized collectively as evolutionary opportunisms. Such changes have the effect of conferring on the organism a temporary adaptive advantage at the price of loss or limitation of evolutionary plasticity for further change. Here belong the various forms of deterioration of sexuality observed in so many species of temperate- and cold-zone floras. An apomictic or an asexual species, with an unbalanced chromosome number and a heterosis preserved by loss of normal meiotic behavior of the chromosomes, may be highly successsful for a time but its evolutionary possibilities in the future are more limited than those of sexual and cross-fertilizing relatives. Polyploidy is also a form of evolutionary opportunism in so far as it produces at least a temporary loss of genetic variability. Although some plant species native in the tropics have also become stranded in these evolutionary blind alleys, the incidence of such species is higher in and near the regions which were glaciated.

Evolutionary Importance of Biotic Environment

The contradictory epithets of "El Dorado" and "Green Hell," so often used in descriptions of tropical lands, really epitomize the two aspects of tropical environments. The process of adaptation for life in temperate and especially in cold zones consists, for man as well as for other organisms, primarily in coping with the physical environment and in securing food. Not so in the tropics. Here little protection against winter cold and inclement weather is needed. In the rainforests, the amount of moisture is sufficient at any time to prevent the inhabitants from suffering from desiccation. Relatively little effort is necessary for man to secure food, and it seems that the amount of food is less often a limiting factor for the growth of populations of tropical animals than it is in the extratropical zones. But the biological environment in the tropics is likely to be harsh and exacting. Man must beware that his blood does not become infected with malarial plasmodia, his intestines with hookworms, and his skin with a variety of parasites always ready to pounce on him and rob him of his vitality if not of life itself. The tremendous intensity of the competition for space among plants in tropical forests can be felt even by a casual observer. The apparent scarcity, concealment, and shyness of most tropical animals attest to the same fact of extremely keen competition among the inhabitants.

Now, the processes of natural selection which arise from encounters between living things and physical forces in their environment are different from those which stem from competition within a complex community of organisms. The struggle for existence in habitats in which harsh physical conditions are the limiting factors is likely to have a rather passive character as far as the organism is concerned. Physical factors, such as excessive cold or drought, often destroy great masses of living beings, the destruction being largely fortuitous with respect to the individual traits of the victims and the survivors, except for traits directly involved in resistance to the particular factors. As pointed out by Schmalhausen, indiscriminate destruction is countered chiefly by development of increased fertility and acceleration of development and reproduction, and does not lead to important evolutionary advances. Physically harsh environments, such as arctic tundras or high alpine zones of mountain ranges, are inhabited by few species of organisms. The success of these species in colonizing such environments is due simply to the ability to withstand low temperatures or to develop and reproduce during the short growing season.

Where physical conditions are easy, interrelationships between competing and symbiotic species become the paramount adaptive problem. The fact that physically mild environments are as a rule inhabited by many species makes these interrelationships very complex. This is probably the case in most tropical communities. The effectiveness of natural selection is by no means proportional to the severity of the struggle for existence, as has so often been implied, especially by some early Darwinists. On the contrary, selection is most effective when, instead of more or less random destruction of masses of organisms, the survival and elimination acquire a differential character. Individuals that survive and reproduce are mostly those that possess combinations of

Evolution in the Tropics   221

traits which make them attuned to the manifold reciprocal dependences in the organic community. Natural selection becomes a creative process which may lead to emergence of new modes of life and of more advanced types of organization.

The role of environment in evolution may best be described by stating that the environment provides "challenges" to which the organisms "respond" by adaptive changes. The words "challenge" and "response" are borrowed from Arnold Toynbee's analysis of human cultural evolution, although not necessarily with the philosophical implications given to the terms by this author. Tropical environments provide more evolutionary challenges than do the environments of temperate and cold lands. Furthermore, the challenges of the latter arise largely from physical agencies, to which organisms respond by relatively simple physiological modifications and, often, by escaping into evolutionary blind alleys. The challenges of tropical environments stem chiefly from the intricate mutual relationships among the inhabitants. These challenges require creative responses, analogous to inventions on the human level. Such creative responses constitute progressive evolution.

REFERENCES

1. Anonymous. Normais Climatologicas. Serviço da Meteorologia, Ministerio da Agricultura. Rio de Janeiro, 1941.
2. Baker, J. R. The seasons in a tropical forest. Part 7. *Jour. Linn. Soc. London 41*, 248-258, 1947.
3. Beard, J. S. The natural vegetation of Trinidad. *Oxford Forestry Mem. 20*, 1-155, 1946.
4. Black, G. A., Dobzhansky, Th., and Pavan, C. Some attempts to estimate the species diversity and population density of trees in Brazilian forests. *Bot. Gaz.*, 1950. (In press.)
5. da Cunha, A. B., Burla, H., and Dobzhansky, Th. Adaptive chromosomal polymorphism in *Drosophila willistoni. Evolution*, 1950. (In press.)
6. Davis, T. A. V., and Richards, P. W. The vegetation of Moraballi Creek, British Guiana. *J. Ecol. 22*, 106-155, 1934.
7. Dobzhansky, Th. Observations and experiments on natural selection in *Drosophila. Proc. 8 Internat. Cong. Genetics, Hereditas Suppl.*, 210-224, 1949.
8. Dobzhansky, Th., and Pavan, C. Local and seasonal variations in relative frequencies of species of *Drosophila* in Brazil. *Jour. Animal. Ecol.*, 1950. (In press.)
9. Dubinin, N. P., and Tiniakov, G. G. Inversion gradients and natural selection in ecological races of *Drosophila funebris. Genetics 31*, 537-545, 1946.
10. Gause, G. F. The struggle for existence. Baltimore, 1934.
11. Lack, D. Darwin's finches. Cambridge, 1947.
12. Patterson, J. T. The Drosophilidae of the Southwest. *Univ. Texas Publ. 4313*, 7-216, 1943.
13. Schmalhausen, I. I. Factors of evolution. Philadelphia, 1949.
14. Stebbins, G. L. Variation and evolution in plants, New York, 1950. (In press.)
15. Toynbee, A. J. A study of history. New York and London, 1947.
16. Vandel, A. L'homme et l'évolution. Paris, 1949.
17. Vavilov, N. I. Studies on the origin of cultivated plants. Leningrad, 1926.

LATITUDINAL VARIATIONS IN ORGANIC DIVERSITY

Alfred G. Fischer

Princeton University, Princeton, N. J.

Received July 1, 1959

"Animal life is, on the whole, far more abundant and varied within the tropics than in any other part of the globe, and a great number of peculiar groups are found there which never extend into temperate regions. Endless eccentricities of form and extreme richness of color are its most prominent features, and these are manifested in the highest degree in those equatorial lands where the vegetation acquires its greatest beauty and its fullest development." Thus wrote A. R. Wallace (1878). His remarks apply equally to the vegetable and animal kingdoms, to the terrestrial realm and to at least the surficial parts of the oceans. Surely this correlation of floral and faunal diversity with latitude is one of the most imposing biogeographic features on earth. On the one hand, its existence poses large-scale problems in evolution. On the other, it offers a potential tool to the geologist-paleontologist who attempts to wring patterns of earth history out of the fossil record.

The following pages serve to review some previously known latitudinal gradients in organic diversity, to describe quantitatively gradients in molluscan diversity along North American shores, and to inquire into the origin of these patterns.

Some Examples of Diversity Gradients

Terrestrial gradients

The overwhelming variety of trees in tropical rain forests has impressed northern travellers, from the early navigators on, and offers a striking contrast to the solid stands of timber to be found in the austral and boreal regions. This floral diversity gradient is widely recognized by botanists.

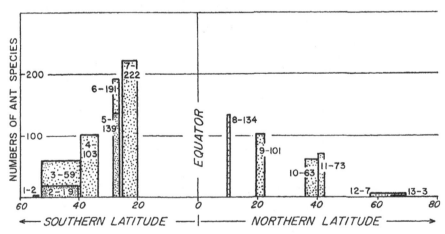

Fig. 1. Northern and southern diversity gradients in ants. 1—Tierra del Fuego; 2—Patagonia, humid western side; 3—Patagonia, as a whole; 4—Buenos Aires, Argentina; 5—Tucuman, Arg.; 6—Missiones, Arg.; 7—Sao Paulo, Brazil; 8—Trinidad; 9—Cuba; 10—Utah, USA; 11—Iowa, USA; 12—Alaska as a whole; 13—Alaska, arctic part. After Kusnezov (1957).

ORGANIC DIVERSITY   65

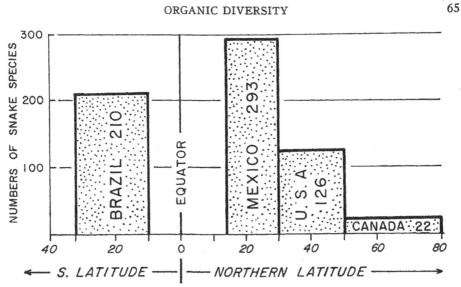

FIG. 2. Diversity gradient in American snakes. After Dobzhansky (1950, credited to Bogert).

Kusnezov (1957) compared ant diversity in areas of different latitudes, and his findings for the Americas are summarized in figure 1. In a stimulating paper dealing with the general problem of tropical diversity, Dobzhansky (1950) published figures on diversity gradients in American snakes (fig. 2) and birds (fig. 3). Distribution of snake species within Argentina follows a striking di-

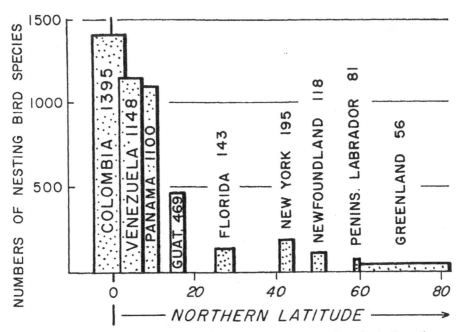

FIG. 3. Diversity gradient in nesting birds, from equator to Greenland. Data from Dobzhansky (1950, credited to Mayr).

66   ALFRED G. FISCHER

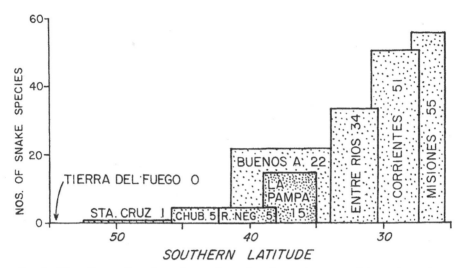

Fig. 4. Diversity gradient shown by numbers of snake species reported from the various provinces of Argentina. Data from Série (1936).

versity gradient described by Série (1936) (fig. 4). Bourlière (1957) shows comparisons of animal diversity between France and Barro Colorado Island in Panama.

Darlington (1957) does not deal with this problem as such, but points out repeatedly that amphibians, reptiles, birds and mammals are more diversified in the tropics than in higher latitudes; while the abundant figures provided by him are not readily plotted in diversity gradients, they illustrate the general principle.

Marine gradients

Any beachcomber who strays from northern shores into the tropics is im-

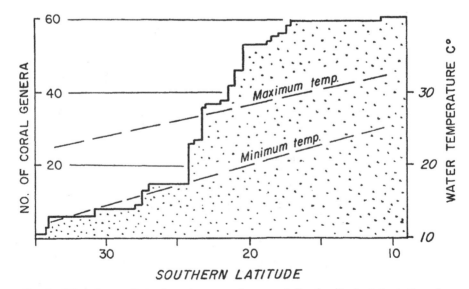

Fig. 5. Diversity gradient of coral genera along great Barrier Reef of Australia, after Wells (1956).

ORGANIC DIVERSITY 67

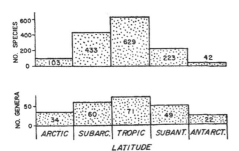

Fig. 6. Specific and generic diversity gradients in tunicates. After Hartmeyer (1909).

pressed with the greater variety of shells cast up on the shores there. Among general text and reference books the phenomenon has been noted, among others, in Hesse (1924) and Hesse, Allee, and Schmidt (1937), as well as by Thorson (1957).

Few if any groups of animals illustrate the principle more vividly than do the corals, with their bewildering variety of reef-building types in suitable tropical habitats, contrasted with a handful of solitary or at best bank-forming species in cold waters. Wells (1956) has studied the distribution of coral genera along the Great Barrier Reef of Australia, and his findings are summarized in figure 5. Some 60 genera of corals coexist in the northern part of this great reef belt, around lat. 9° S, and from this maximum the number dwindles to a single genus at the south end (35° S).

Hartmeyer (1919) illustrated similar diversity gradients in tunicates (fig. 6). Thorson (1952, 1957) shows similar gradients for amphipods, nudibranchs, and crabs (fig. 7), and Brodskij (1959) has illustrated the same principle in pelagic calanid crustaceans of the surficial water masses (fig. 8).

On the other hand, Thorson has shown that such gradients are not universal. As shown in figure 7, they are not exhibited by such burrowing groups as soft-bottom ophiuroids, holothuroids, cephalaspids and cumacids. These groups show little diversity anywhere. This difference between epifaunal or pelagic and infaunal

animals is particularly well shown by Thorson's comparison of prosobranch gastropods (fig. 9). Taking prosobranchs as a whole, and comparing areas of equal size, tropical faunas show approximately five times as many prosobranch species as do arctic ones. Throughout this range, the great majority of prosobranchs is epifaunal. However, the prosobranch family Naticidae is largely restricted to infaunal soft-bottom dwellers, and the naticids show only slightly more diversity in the tropics than in high latitudes.

Some groups of organisms are more diverse in the temperate latitudes than in the tropics. This appears to be true for certain groups of algae. No actual figures on diversity are available to me, but from Tilden (1937) one would gather that

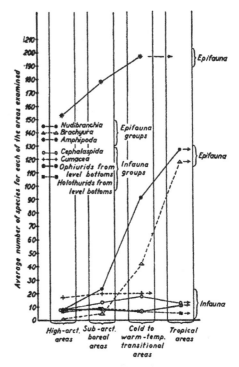

Fig. 7. Average number of species of different groups of bottom invertebrates from equal-sized coastal areas in different latitudes. Only depths from the shore to 300 m have been regarded, and all pelagic and parasitic species have been neglected. From Thorson (1952 and 1957).

68   ALFRED G. FISCHER

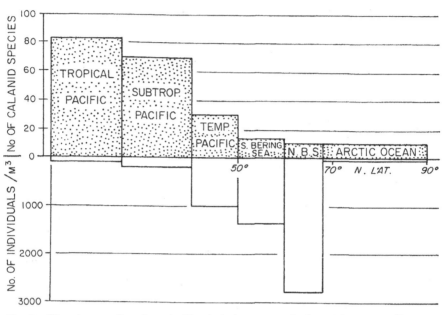

FIG. 8. Diversity gradient in calanids (pelagic crustacea) from the upper 50 meters. Abundance of individuals per cubic meter (lower graph) shows a strikingly divergent pattern. After Brodskij (1959).

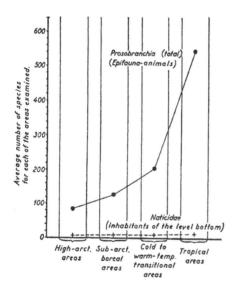

FIG. 9. Average number of prosobranch species from equally large coastal areas in different latitudes. Whereas the prosobranchs as a whole, a mainly epifaunal group, show a striking diversity gradient, the family Naticidae, mainly infaunal, does not. From Thorson (1952 and 1957).

whereas the green and the bluegreen algae are most diversely developed in the tropics, the red algae and kelps reach their acmes in the temperate zones.

Surely many similar examples of distribution, showing no correlation of diversity and latitude, or even "reversed" gradients from high diversity in temperate latitudes to little diversity in the tropics, could be found in various branches of the plant and animal kingdoms. They appear to be, however, no more than exceptions which prove the rule.

THE RELATION OF GRADIENTS TO CLIMATE

Within the tropical belt, similar gradients mark the passage from hot lowlands into cold and variable mountain areas, from the humid rainforest into deserts, from the warm shallow waters into the frigid ocean depths, and from normal warm-water areas into belts of cold upwelling. It is evident that diversity gradients are related to gradients in environmental factors—temperature, humidity, etc., whether developed on a regional or on a local scale.

On land we speak of the combinations of such physical factors as *weather*, and of the long-period weather patterns as *climate*. These concepts apply equally well to the underwater world, where there are both uniform and seasonal climates, and where, like on land, weather fluctuations may be gradual or catastrophic. Temperatures change, the place of winds is taken by waves and currents, and instead of moisture fluctuations there are changes in salinity (with rather similar physiologic effects). Suspended particles take the place of cloud cover in limiting penetration of solar energy, and excessive sedimentation replaces dust storms and ash falls. We may therefore extend the familiar terms to the world of water, and may speak of *atmospheric weather and climate* on the one hand, and *hydrospheric weather and climate* on the other. Hedgpeth (1957) speaks of *hydrographic* climate in this sense, but to me the adjective *hydrospheric* seems more apt and more consistent.

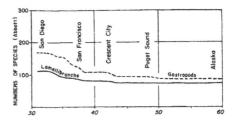

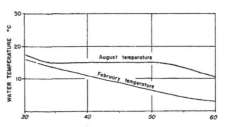

Fig. 11. Comparison of molluscan diversity and marine climate along west coast of Canada and United States. Data from Abbott (1954) and Sverdrup *et al.* (1942).

From a regional standpoint, temperature is the main factor in the climatic control of marine plant and animal distribution. This matter has been considered by Hutchins (1947) and is reviewed by Hedgpeth (1957b). The individuals which compose a population or a species can exist only within certain temperature limits; and within this viability range, there lies a further restricted temperature range outside of which reproduction is impossible.

The diversity gradients which have been been observed lend themselves to the following generalization or rule:

The diversity of biotas, on land and in the sea, is greatest in climates of relatively high and constant temperatures, such as those found over much of the tropics, and decreases progressively into the fluctuating and the cold climates normally associated with the higher latitudes.

MOLLUSCAN FAUNAS ALONG THE COASTS OF NORTH AMERICA

Climatic gradients. The east and west coasts of North America show distinct climatic gradients in respect to tempera-

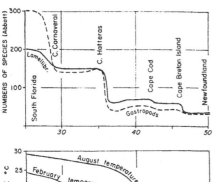

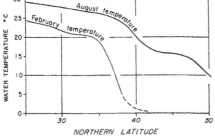

Fig. 10. Comparison of molluscan diversity and marine climate along east coast of Canada and United States. Data from Abbott (1954) and Sverdrup *et al.* (1942).

70   ALFRED G. FISCHER

ture, as illustrated in figures 10–11. Along the east coast this gradient is a very sharp one, for here the climatic differences to be expected as an expression of latitude are reinforced by oceanic circulation, which brings tropical waters northward to Cape Hatteras and arctic waters south to Cape Cod. On the West Coast, the opposite situation prevails; here the gradient is a very gentle one, for the circulation modifies the gradient to be expected from latitudinal differences: the northern part of the coastline is warmed by the Alaska current, while the southern shores are cooled by the upwelling California current.

Data desirable. A study relating the diversity patterns of marine organisms to climatic gradients such as these should ideally be based on data which include:

(1) reasonably detailed and evenly distributed systematic treatment of the chosen group along the entire belt studied, so that monographic highs resulting from patchy collecting or overzealous taxonomic splitting are avoided.

(2) restriction to shallow waters, so as to exclude the deeper forms which are under the influence of climates somewhat different from those expressed by the surface isotherms.

(3) accurate data on the geographic ranges of the species or genera chosen as the units of study.

Data used. The full measure of such information probably is not available for any group of organisms along the North American shoreline; yet, for many groups existing data are probably sufficient for a rough approximation which serves to delineate the basic pattern.

Abbott's *American Seashells* (1954) offers a source for such a compilation on mollusks. This magnificent volume, primarily a guide to shallow-water shells, deals with some 1,500 of the more than 6,000 species of mollusks described to date from North American waters. It provides range data. In emphasizing the shallow-water forms, it largely avoids species belonging to deep-water climates. In excluding the rare forms, it avoids the pit-

falls of poorly known ranges and monographic highs. The intense activity of shell collectors in Florida has perhaps introduced some bias, but probably not enough to distort the pattern seriously. Though many ranges given extend on into the Caribbean Island arc and into Mexican waters, Abbott's selection in these outlying regions may not be as representative of the actual fauna as it is off the U. S. and Canadian coasts, and therefore the compilations here made do not extend beyond the tip of Florida and the Coronado Islands. All in all, Abbott's selection within these limits appears to be an adequate index to molluscan diversity, and therefore worth the effort of compilation and presentation.

Results. Figures 10–15 show the diversity gradients in mollusks along the East and West Coast of North America, as compiled from Abbott. Snails and clams are considered separately. Among the snails, the naked forms, pelagic species and pyramidellids were not included. No

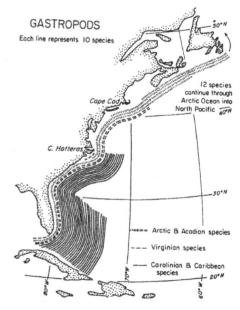

Fig. 12. Diversity gradient of gastropods along eastern coast of United States and Canada. Each line stands for ten species. Data from Abbott (1954).

ORGANIC DIVERSITY 71

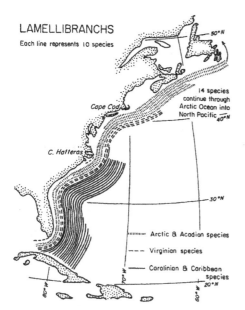

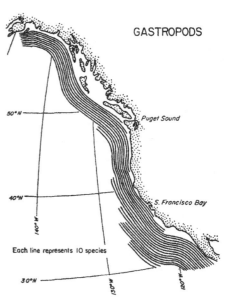

FIG. 13. Diversity gradient of lamellibranchs along eastern coast of United States and Canada. Each line stands for ten species. Data from Abbott (1954).

FIG. 14. Diversity gradient in gastropods along west coast of Canada and United States. Each line stands for ten species. Data from Abbott (1954).

compilations were made for the Amphineura, Scaphopoda and Cephalopoda.

The patterns found lend themselves to the following generalizations:

(1) Each case conforms to the rule of correlation between biotic diversity and climate.

(2) The gastropods (mainly epifaunal) illustrate this rule better than the lamellibranchs (which include large numbers of burrowers).

(3) There appears to be no particularly close correlation of the numbers of species with the pattern of seasonal high (August) temperatures, nor with that of seasonally low (February) temperatures, nor with a mean between these: All of these may have some effect, and in addition, the amount of seasonal variation appears to be important, for both graphs show a marked rise in the number of species associated with a convergence of August and February temperatures. No doubt other climatic factors not plotted are of importance; regions strongly influenced

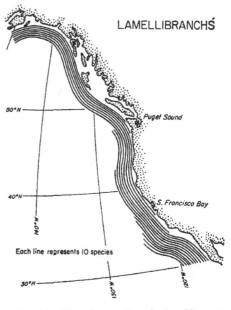

FIG. 15. Diversity gradient in lamellibranchs along west coast of Canada and United States. Each line stands for ten species. Data from Abbott (1954).

by currents are particularly subject to annual or occasional temperature changes of great magnitude and short duration, which would not find expression on normal isotherm maps, and which must have a catastrophic effect upon marine life (see below).

ORIGIN OF DIVERSITY GRADIENTS

Biogeographic patterns are the results of two closely interwoven processes: the evolution of organisms, and the evolution of their habitats. Both of these are time-rate problems: Organisms, climates, and the landscape itself are ceaselessly fluctuating, and the extent to which the status quo is altered depends (1) on the direction in which they change, (2) the rate at which they change, and (3) the time-span under consideration.

A model of biotic evolution

Let us consider the change of a fauna by means of a simple model, figure 16. A given fauna, F-1, occupies a certain area at a certain time. It contains 15 species (if you prefer, 150, or 1,500). Some millions of years later the same area is occupied by fauna F-2, numbering 20 species 6 of which are in common with F-1. Figure 16 shows graphically how the change has taken place: Of the original 15 species, 4 were exterminated without having left descendants; 6 survived unchanged; 12 are newly evolved species, of which we may arbitrarily consider 5 to be direct lineal descendants of a similar

number of species in F-1, and 7 to be "side-branches"; and 2 immigrated from other regions. It is evident that the difference between F-1 and F-2 depends (1) on the rates at which species are evolved, exterminated, and immigrated, and (2) on the duration of the timespan between F-1 and F-2.

Using this model, we may now inquire into the factors which bring about the addition and subtraction of species, and the ways in which these changes may be influenced by climate. In this inquiry, we shall neglect the immigration factor, to concentrate on (1) *speciation*, the addition of species to a biota by evolutionary processes, and (2) *extinction*.

Speciation

Addition of new species (other than immigrants) to a biota is dependent on two factors: (a) the evolutionary potential of existing species, as determined by mutation rates and other factors to be reviewed below, and (b) the rate at which mutants will be selected—something which might be termed environmental receptivity.

Evolutionary potential

The evolutionary potential of any one species would seem to depend on two factors: its genetic variability, and the length of its generations.

Genetic variability involves a number of more or less distinct qualities. One of these is the mutation rate, which may be defined as the ratio of mutant to normal

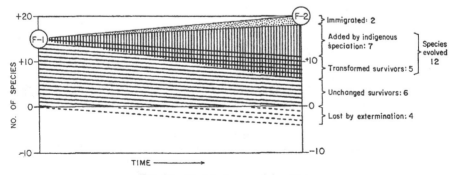

FIG. 16. Model of an evolving biota.

offspring. One might expect this to be related to the complexity of the genetic mechanism—surely the more elaborate the system, the greater the chances for mishaps, alterations and modifications within it—i.e., the greater the chance for mutations. Another is the extent to which mutations are retained and perpetuated within the population gene pool; this factor, also, may be dependent largely upon the complexity of the genetic system. Population size is a third factor: Just as the corner grocery store lacks variety as compared with a supermarket, so very small populations lack genetic variety (and therefore evolutionary potential) as compared with larger ones.

None of these factors appear to bear any relation to climate, and we may therefore incline to discount them as sources of diversity gradients related to climate.

A special aspect of population size, worth a short digression, is the number of offspring produced. Offhand one might think that an organism capable of producing some millions of young per year (such as the oyster) would enjoy a considerable advantage in evolutionary potential over the much less prolific cephalopods, or over certain mammals reproducing at less than one millionth that rate. But this apparent advantage is largely spurious: the production of vast numbers of progeny in oysters is the price paid by a benthonic animal for a defenseless pelagic larval stage, during which the vast majority of the offspring are destroyed in a nonselective fashion. This subject has been discussed by Thorson (1952), who has shown that at the height of the oyster breeding season a single medium-sized *Mytilus* will strain 100,000 oyster larvae in a period of 24 hours. Only a minute fraction of the oyster offspring ever reaches the stage which mutations or gene combinations for a modified filter system or circulatory system can be tested.

Generation length is another factor in evolutionary potential, since the generation is the unit link in the history of any life strain, and new mutants and new gene

combinations arise from one to the next. Simpson (1953, pp. 129–132) admits this, but on empirical evidence concludes that the genetic evolutionary potential of organisms is but rarely approached by actual evolutionary sequences. He contrasts the slow evolution of short-generation opossums with the rapid evolution of long-generation elephants, but admits that shortness of generation may be a factor in the rapid evolution of pathogenic bacteria. The development of DDT resistance in certain insects may be another example of rapid evolution facilitated by shortness of generations.

The question which here concerns us is the problem of variations in generation-length with environment: obviously, if species in general mature more rapidly in the tropics than in the higher latitudes, then tropical generations will succeed each other more rapidly; and this might then endow tropical organisms with a somewhat greater evolutionary potential. But observations on maturation rates and generation length show that there is no simple relation between these factors and environment (or latitude). On the one hand, the metabolism of some species seems to vary directly with temperature: Bourlière (1957) points out that the butterfly *Danaus chrysippus* matures in one year in North America and in 23 days in the Philippines, and the beetle *Crioceris asperagi* matures in Germany in one year, whereas *C. subpolita* matures in Java in 25–31 days. Thorson (1952) has reviewed this subject for marine organisms. Some observations on *Ostrea, Mytilus,* and *Balanus* show amazingly high maturation rates in the tropics. *Hydroides norvegica* becomes sexually mature 9 days after attachment in Madras, 4 months after attachment in England.

On the other hand, many species and genera appear to be "temperature-adapted," i.e. their tropical representatives and their cold-water forms live at the same rates, and show no significant variation in maturation rate and generation length. Thorson lists observations to this effect on *Littorina,* decapod crustacea, and echinoids. Furthermore, contrary to widespread opinion, many tropical plants and animals do not reproduce continuously, but have definite breeding periods. Bünning (1956) has pointed out that these may be very long—the sexual reproductive cycle of bamboos, for example, is measured in decades. Nor are the tropical mammals characterized by unusually rapid succession of generations. At this stage, then, it appears that *evolutionary advantage of tropical over high-latitude organisms due to different generation length is at best a minor factor.*

Environmental receptivity

Ecologic niches. A given physical environment provides a variety of possible ways for organisms to make a living, and the organisms themselves greatly multiply the number of these *ecologic niches,* in which properly adapted species can prosper and procreate.

Some workers believe that the main cause for differences in diversity lies in the number of ecological niches available. Thus Bourlière (1957), in writing about animals, states: "It is, however, in the abundance and variety of suitable habitats that one must seek the principle cause of richness of tropical faunas. The environments in which they live is in fact infinitely more diverse than that of temperate latitudes, and many of the habitats of the warm regions have no counterpart in other parts of the world."

The development of major physiographic characters—mountain ranges, alluvial plains etc.—occurs in temperate as well as in tropical regions; certainly there is in the tropics no more diversity of landforms than in other parts of the earth. The related edaphic and microclimatologic features show, if anything, a greater variety within temperate landscapes than within tropical ones: a case in point is the great variety of soil types in the north-temperate regions. Another is the uniform distribution of solar energy in the tropics in contrast to its directional character in the

higher latitudes: in the tropics the sun alternately shines from the north and south, as anyone there who tries to grow plants on the "sunny" or the "shady" side of the house will find out in the course of the year. Hence, the striking vegetation contrasts between the north and south flanks of hills, which we tend to take for granted, are not to be found in lower latitudes. Even the change in seasons increases the number of physical niches.

Among the various ecological features of the tropics, it is the organic ones which provide such a rich range of habitats, as discussed by Bünning (1956). Bourlière, in illustrating the richness of tropical environments, cites the variety of animal habitats provided by that most complex of all plant associations, the tropical rainforest. But this accounts neither for the astonishing diversity of tropical plant life, nor for the marine animal diversity gradients which are not associated with such highly organized plant communities.

Differences in selection. Since Darwin's time, the process of selection, resulting from the struggle for survival, has been considered to be one of the main springs of the evolutionary process. And, since biotic diversity is the result of evolution, we may well ask whether differences in the manner of natural selection may partly account for biotas of different diversity.

Schmalhausen (1949) and Dobzhansky, (1950) among others, believe that this is the case.

The size to which a given population may grow depends on its rate of reproduction, and on the pattern according to which individuals are killed off (I use the term pattern rather than rate, for the rate may vary from one stage of life history to another). The nature of the survivors will determine the course of evolution. Two things are therefore of great importance: (1) the selectivity (as opposed to randomness) of killing (or, survival); (2) the direction or directions in which variation is encouraged and accepted by the environment.

Selective factors may be divided arbitrarily into two groups: organic and in-organic. Both operate everywhere. At all places, some organisms die because of inclement weather; and others die because of lack of food, or because they are eaten by others, or because they fall prey to disease. But the relative balance between these two sets of factors varies from place to place.

Darwin, Schmalhausen (1949) and Dobzhansky (1950) all call attention to one aspect of this matter: In the high latitudes, one is struck by the havoc wrought among organisms by the physical environment: the changing seasons, and the thereon superimposed catastrophes of weather take a large toll. And, as these investigators have pointed out, the selection accomplished by winter storms, sum-mer droughts etc. is rather crude: mainly such disasters result in random killing of large numbers of individuals. To be sure, organic selective factors are also at work here, since individuals compete with each other, prey and are preyed upon, and succumb to parasites and diseases. And these factors appear to be more selective, in that they involve intense individual competition, and put a premium on in-dividual excellence. But the point is that *much of the killing in high latitudes is done by the less selective inorganic forces.*

In the tropics, on the other hand, the physical environment is more benign to most organisms, and the highly selective interorganic struggle for existence is more apparent. The struggle of plants for light in the rain forest, and the shyness of tropical mammals are dramatic expres-sions thereof. Dobzhansky believes that this difference may be an important factor in the differential evolution of high-lati-tude versus tropical biotas.

But the matter is really somewhat more involved than this. The extensive random killing in the high latitudes is largely offset by higher rates of reproduction (Schmalhausen, 1949). And the competi-tion between organisms in the tropics is so apparent because of such a large num-ber of species present. As in the case of

the ecologic niches, we are in danger of circular reasoning: is the diversity of tropical biotas due to more intense com-petition, or is the more intense competi-tion an outcome of the more diverse biota?

One thing appears fairly clear, although it seems not to have been emphasized: In a biota lacking diversity, a great deal of the competition is between members of the same species. This effect is intensi-fied if these species produce vast numbers of progeny, as many of the high-latitude plants and animals do. *In the high latitudes, organic selection results to a great extent from competition of an in-dividual with other members of his spe-cies; in the tropics, it tends to be much more a competition of one individual against the members of other species.* This difference must have some effects on the patterns of evolution; to some extent it is a result rather than a cause of dif-ferences in diversity, but in part it is linked to the numbers of offspring, and may therefore have had some effect in the differentiation of high-latitude and tropical faunas and floras.

There is, however, a much more direct difference in selection: the tropical tem-peratures (continental and shallow-water) are nearer the mid-point of the tempera-ture range which protoplasm can endure than are the high-latitude temperatures. Furthermore, the tropical temperatures show less seasonal variation (as has al-ready been mentioned). As a result, tropical conditions permit a wider range of physiologic and structural variation than do high-latitude conditions, which seasonally approach the lower limits at which life can be maintained, and at other times may range into temperatures comparable to those of the tropics. Thus *the tropical environment is receptive to wider latitudes of physiologic variation— to a wider range of mutations—than is the temperate and polar environment.* And this, it seems to me, may be one of the most important factors in the equation which has produced the biogeographic pattern under discussion.

76   ALFRED G. FISCHER

Extinction

In our model (fig. 16) the diversity of
F-2 is determined not only by the rate
at which species are added, but also by
the rate at which they have disappeared.
They vanished in two ways: by evolving
into new species, and by becoming exterminated. In a way these two modes of
disappearance are more alike than their
separation on the model would suggest,
inasmuch as the transformation of one
species into another (a into b) involves
the preferred killing-off of individuals of
type a. For practical purposes here we
may, however, restrict our attention to
the outright eradication of species without
descendants.

The question is: how does extinction
vary with climate; and do old species
survive more readily in the tropics than
in the higher latitudes?

Extinction of a species can be a sudden
process, as demonstrated by the passenger
pigeon. On the other hand extinction
of a species or a larger taxon can be a very
gradual kind of process. This may involve the gradual restriction of members
to a smaller and smaller area, where the
form may persist (and may be very prominent) for some time as a relict. Or it
may come about by the gradual thinning
of population density throughout its range,
until the spacing between individuals interferes with reproduction.

The mixed and diffuse population patterns of the tropical rainforest and the
low density of given species populations
suggest that in the tropics extinction may
come about chiefly as a result of this
gradual dilution. The patchy population
patterns and the prevalence of relict population in the higher latitudes suggest that
here the relict pattern dominates. In the
former case continued organic competition
is likely to strike the final blow; relict
patches, on the other hand, are particularly
vulnerable to the accidents of weather
mentioned before. Which extinction mechanism is the more efficient, and what role,
if any, such differences have played in the

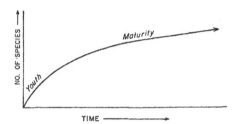

FIG. 17. The curve of biotic maturity.

development of diversity gradients, under
conditions of constant climate, remains to
be seen.

The time factor and biotic maturity

So far we have considered only relative
rates of speciation and extinction as related to climate. We have not and shall
not here deal with absolute rates. Neither
have we considered (1) relative changes
in rates through time, and (2) the evolution of habitats through time.

The maturity model

Again, let us refer to a model, figure
17, simple at the expense of being somewhat remote from reality. We assume a
given physical setting without organic inhabitants, and then introduce a limited
number of immigrants to start a biota.
Most of us would surely agree that in
this kind of a setting the availability of
many unfilled ecologic niches would lead
to a rapid evolution of species and a
corresponding diversification of the biota:
the diversity curve rises in a steep gradient. To some degree this is a self-gen-
erating process: as plants evolve in a
first approximation to fill all of the available physical niches, they provide the
basis for a first wave of animal evolution.
Secondary refinement will lead to the
evolution of many species more closely
adapted to specific subniches, and these
will largely replace their more generalized
ancestors; and so the process may go on,
almost *ad infinitum*. From the simple
dependence of plant on soil, air, sunlight
and water, and of animals on plants, there
will arise commensalism and symbiosis of

plant-with-plant, animal-with-plant, animal-with-animal, and even 3-way relationships. At this level of complexity we reach the stage shown by tropical coral reefs and rain forests. By now the curve of diversity increase has probably flattened out, as the rate of extinction of old species approaches the rate at which new ones are evolved, but I see no reason why the rise could not continue even beyond the richest biotas existing today.

We may speak of the different parts of this curve as youthful and mature, and accordingly of *degree of biotic maturity*.

From what has gone before, we may conclude that such a curve would rise somewhat faster in the tropics than in the temperate latitudes, and would rise slowest in the polar regions (fig. 18). Yet, the basic patterns should remain the same: evolution within each may be expected to lead from simple biotas to more and more diversified ones, as a result of increasingly specialized adaptation and of more elaborate interrelations with other organisms.

We now face two possibilities: either the biotas of the major climatic belts we have discussed have been evolving for the same length of time, and the differences in their diversity are simply a result of different rates of diversification (slope of the maturity curve) as shown in figure 18. Or the time factor itself is involved as well: some biotas have attained higher diversification (maturity) because they have been evolving steadily over greater periods of time; others are truly immature,

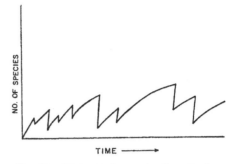

Fig. 19. History of organic diversity in an area subjected to climatic fluctuations during geologic time.

having originated much later or having been set back by periodic decimations (fig. 19). So long as we know so little of the quantitative aspects of evolution, the biotic patterns alone are not likely to give us a definitive answer to this question.

Climates through geologic time

If there is any one theme which runs through the geological record, it is that of ceaseless change.

Given a spheroidal globe, rotating on an inclined axis and circling the sun in a certain orbit, we may be sure that the solar energy received in the equatorial regions was always markedly greater than that received at the poles, and that the reception of this energy at the equator varied but little through the year, whereas it was subjected to seasonal variations in the polar areas. But beyond these basic invariables there lies a host of variable factors. Some of these are known to have operated, others are at the present stage of knowledge no more than possibilities and probabilities.

The variation is in the rate and pattern of redistribution of this energy. The rising of air in the tropical belt, its descent at the horse latitudes and in the polar regions, and the turbulent mixing in between greatly affect weather and climate —as does to a lesser extent the circulation of water in the oceans. Both are to a very large extent governed by the distribution of land and sea. This distribution is less

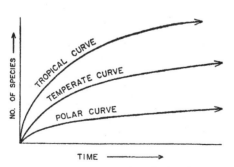

Fig. 18. Comparison of tropical and other maturity curves.

important in the tropical belt, which has probably always extended over a mixture of land and sea in which the sea was dominant, than it is in the small polar regions, which have at times been dominantly continental, at others dominantly marine.

Land areas have their surface layer heated quickly to comparatively high temperatures by incident sunlight. Downward conduction of heat is slow, but loss to the atmosphere is rapid, and thus little of the energy becomes stored. As a result we experience the extremes of "continental climates." In water covered areas, the solar energy is only partly absorbed at the surface, and heats deeper layers as well. Partly because of this, partly because of the much greater specific heat of water, and partly because of convection, the water mass stores the solar energy at lower temperatures and in a form which is gradually released to the atmosphere. As a result we speak of the ameliorating climatic effects of water bodies. In oceanic areas water circulation itself introduces an additional ameliorating factor, which tends to keep the circum-polar oceans warmer in winter than the circum-polar landmasses.

Four polar models. To visualize possible climatic extremes we may consider four situations; (1) a pole in the midst of an oceanic area; (2) a pole in the midst of a large continental area; (3) a pole in a limited continental area surrounded by seas (the present South Pole); and (4) a pole in a sea surrounded by continental areas (the present North Pole).

In the first model, the polar regions would be of moderate temperature in summer and in winter, maintained during the latter from the vast thermal reservoir of the oceans. A polar climate with extensive ice, such as we know today, would not exist. Neither would there be the giant struggle of the polar and the equatorial air masses, which stamps the present "temperate" region with their catastrophic character. Instead, the temperate regions around this polar ocean area might show climates like those of

present-day New Zealand, Tasmania, and coastal Alaska.

Model 2 would provide the opposite extreme. The polar landmass would be subject to extensive heating in the summer and to extremely frigid winter weather. Great polar air masses would make extensive inroads into the subtropical fringe during the winter. The "temperate" areas would thus be even less temperate than those we know. Ice sheets might accumulate in the central area, but the summer heat and lack of moisture would make widespread glaciation improbable.

Models 3 and 4 are available for study. In model 3, the polar landmass is utterly frigid, but the surrounding oceanic belt (famous for its violent storms) takes the brunt of the atmospheric mixing, and beyond them lies a truly temperate zone, protected from the catastrophic inroads of polar air masses.

In model 4, the polar ocean helps to ameliorate the climatic extremes of the polar area and adjacent regions, but provides a source of moisture which encourages the development of glaciers, and may thus cause a widespread development of what we normally think of as polar conditions. The history of the northern hemisphere during the last million years bears witness to this.

The history of the polar regions. Present polar and near-polar continental areas have in the past been widely covered by seas, just as have other continental regions. The existence of these seas is recorded in the form of marine sedimentary rocks. Thus we need no further proof that extensive changes of land and sea have taken place in the polar regions, and that profound climatic effects must have been felt in the polar and temperate regions.

But, beyond this, different lines of evidence suggest strongly that the poles and the earth's crust have shifted with respect to each other. The fossil record shows biotic patterns difficult to reconcile with the present geographic grid: for example, the lower Paleozoic faunas of South

America are very limited in variety (immature?), while exceedingly rich faunas and coral reefs thrived from central North America to the Arctic and to Scandinavia. The patterns of ancient glaciations, while still far from well understood, are also difficult to reconcile with the present pole positions. Remanent magnetism in sediments and lava flows offers a possible clue to pole positions of the past, and work to date suggests extensive polar shifting.

Such polar wandering might have come about either by shifting of the earth's crust over the underlying "mantle." or by displacement of the globe as a whole relative to its axis of rotation, or by extensive displacements of oceans and continents relative to each other and to the earth's axis. The work on remanent magnetism and the current discoveries of extensive wrench faulting in the earth's crust offer support for such theories.

Conclusions. It is concluded from this that widespread climatic changes have been a normal feature of earth evolution. The broad tropical belt, with its mixture of lands and seas, has probably been the most constant feature of the earth, although its position may have shifted. Polar climates of the present type have probably existed at some times and not at others. The temperate zones have undergone severe changes involving great expansions and contractions (as during the Pleistocene), changes from truly temperate to rigorous and subject to seasonal catastrophes, and possibly extensive shifting over the globe.

From these considerations it appears probable that the tropical marine and continental biotas of today are the products of a long and relatively undisturbed evolutionary history, and are truly mature, while the polar and temperate biotas have experienced a turbulent history of mass extinctions and gradual re-evolution. At times they have been more mature than at present. The great Tertiary mammalian faunas of North America, and the rich marine fauna of the East Coast Miocene, may be examples of comparatively mature

faunas adjusted to a truly temperate climate in the temperate zone. They were decimated by the more rigorous and catastrophic climates which may have had their beginnings in the Pliocene and reached their climax in the Pleistocene glacial stages. Our present North American temperate biotas probably represent a mixture of polar types and the most hardy survivors of the pre-Pleistocene temperate biotas, undergoing rapid evolution in a comparatively immature stage of the curve (fig. 17). This evolutionary surge is not likely to be carried very far: So long as the North Pole remains in the Arctic Ocean, a kind of self-induced oscillation involving freezing and thawing of the arctic seas, waxing and waning of ice caps, and rise and fall of sea level (Ewing and Donn 1956, 1958) may continue to bring alternate glacial advances and retreats, and may effectively prevent the northern temperate biotas from becoming mature.

Local diversity gradients

We may now return to glance at the local climatic diversity gradients, only mentioned above. The tropical rain forest and the warm tropical sea are normal and enduring features of the tropical belt. Mountain ranges, rainshadow deserts and areas of cold upwelling are geologically transient phenomena; their biotas are either lately evolved in place, or have been subjected to drastic migrations in the not-too-distant geologic past. They are therefore immature as compared to the normal tropical biotas.

GENERAL CONCLUSIONS

The diversity of biotas, on land and in the sea, is greatest in climates of relatively high and constant temperatures, such as those found over much of the tropics, and decreases progressively into the fluctuating and the cold climates associated with the higher latitudes. It is proposed that this pattern results largely from the following causes:

80   ALFRED G. FISCHER

(1) Biotas in the warm, humid tropics are likely to evolve and diversify more rapidly than those in the higher latitudes, mainly because of a more constant (favorable) normal environment, and relative freedom from climatic disasters.

(2) Biotic diversity is a product of evolution, and is therefore dependent upon the length of time through which a given biota has developed in an uninterrupted fashion. The low-level tropics represent that part of the globe which has been least affected by climatic fluctuations. Such factors as marine transgressions and regressions in the polar regions, possible polar wandering, and oscillations of glacial polar conditions have caused extensive fluctuations in polar and "temperate" climates, and have thereby profoundly disturbed the normal course of evolutionary diversification. The tropical coral reef and rainforest biotas are considered as examples of mature biotic evolution, whereas the biotas of the regions covered by Pleistocene ice sheets are prime examples of "immature" relicts of the more mature temperate Tertiary faunas and floras. We thus return to A. R. Wallace's words (1878): "The equatorial zone, in short, exhibits to us the result of a comparatively continuous and unchecked development of organic forms; while in the temperate regions there have been a series of periodical checks and extinctions of a more or less disastrous nature, necessitating the commencement of the work of development in certain lines over and over again. In the one, evolution has had a fair chance; in the other, it has had countless difficulties thrown in its way. The equatorial regions are then, as regards their past and present life history, a more ancient world than that represented by the temperate zones, a world in which the laws which have governed the progressive development of life have operated with comparatively little check for countless ages, and have resulted in those wonderful eccentricities of structure, of function, and of instinct—that rich variety of colour, and that nicely balanced harmony of relations which delight and astonish us in the animal productions of all tropical countries."

ACKNOWLEDGEMENTS

This paper grew out of a discussion with Professor Erling Dorf, who wished for material to illustrate latitudinal diversity gradients in his classes. To him, to Professors G. L. Jepsen, F. B. Van Houten and C. S. Pittendrigh, and to Mr. W. Zimmerman I am indebted for stimulating discussions, and for help with the literature and manuscript.

LITERATURE CITED

ABBOTT, P. T. 1954. American Seashells. Van Nostrand, New York, xiv + 541 pp.

AUBERT DE LA RÜE, E., F. BOURLIÈRE AND J. P. HARROY. 1957. The Tropics. Knopf, New York, 208 pp.

BOURLIÈRE, F. 1957. See AUBERT DE LA RÜE et al.

BRODSKIJ, A. K. 1959. Leben in der Tiefe des Polarbeckens. Naturwissenschaftliche Rundschau, 12: 52–56.

BÜNNING, E. 1956. Der tropische Regenwald. Verständliche Wissenschaft, vol. 56. Springer, Berlin-Göttingen-Heidelberg, 118 pp.

DARLINGTON, P. J. 1957. Zoogeography. John Wiley & Sons, New York, 675 pp.

DOBZHANSKY, T. 1950. Evolution in the tropics. Amer. Scientist, 38: 208–221.

EWING, M., AND W. L. DONN. 1956. A theory of ice ages. Science, 123: 1061–1066.

————. 1958. A theory of ice ages. II. Science, 125: 1159–1162.

HARTMEYER, R. 1911. Tunicata, ch. XVII, Die geographische Verbreitung. In Bronn's Klassen und Ordnungen des Tier-Reichs, III, Suppl.: 1498–1726.

HEDGPETH, J. W. 1957 a. Classification of marine environments. In Treatise on Marine Ecology and Paleonecology (ed. Ladd), Geol. Soc. Amer. Mem. 67, chap. 1.

————. 1957 b. Marine biogeography. Ibid, pp. 359–382.

HESSE, R. 1924. Tiergeographie auf ökologischer Grundlage. Fischer, Jena, xii + 613 pp.

————, W. C. ALEE, AND K. P. SCHMIDT. 1937. Ecological Animal Geography. John Wiley & Sons, New York, xiv + 597 pp.

HUTCHINS, L. W. 1947. The bases for temperature zonation in geographical distribution. Ecological Monogr. 17: 325–335.

KUSNEZOV, M. 1957. Numbers of species of ants in faunae of different latitudes. EVOLUTION, 11: 298–299.

SCHMALHAUSEN, I. I.   1949.   Factors in Evolution. Blakiston, Philadelphia, 327 pp.

SÉRIE, P.   1936.   Distribucion geográfica de los ofidios argentinos. Obra cincuentenario Museo de la Plata, 2: 33–61.

SVERDRUP, H. U., M. W. JOHNSON, AND R. H. FLEMING. 1942. The Oceans. Prentice-Hall, New York, 1087 pp.

THORSON, G.   1952.   Zur jetzigen Lage der marinen Bodentier-Ökologie.   Verh.   de Deutsch. Zool. Ges. 1951, Zool. Anzeiger Supplbd. 16: 276–327.

———. 1957. Bottom communities (sublittoral or shallow shelf). In Treatise on Marine Ecology and Paleoecology (ed. Ladd). Geol. Soc. Amer. Mem. 67: 461–534.

TILDEN, J. E.   1937.   The Algae and Their Life Relations. Milford, London, 550 pp.

WALLACE, A. R.   1878.   Tropical Nature and Others Essays. MacMillan, London & New York, xiii + 356 pp.

WELLS, J. W.   1955.   A survey of the distribution of reef coral genera in the Great Barrier Reef region. Reports of the Great Barrier Reef Committee, IV, pt. 2: 21–29.

Species Density of North American Recent Mammals

GEORGE GAYLORD SIMPSON

Introduction

This the first of several planned studies involving a quantification of data on the zoogeography of recent continental mammals in North America. This study deals with species density, that is, with the numbers of species present in different parts of the continent, the over-all pattern formed by those varying densities, and possible factors influencing the pattern.

Data were gathered and tabulated by Miss Marie V. V. Williams. The figures were drawn by Miss Myra L. Smith, with support from grant No. 7660/2 from the National Science Foundation.

Basic Data

The basic data for this and subsequent analyses consist of records of species of recent mammals present in quadrates of equal area covering continental North America. The method was to take a map on Lambert's azimuthal equal-area projection (Goode's series, by Henry M. Leppard, University of Chicago Press) and to superimpose a rectangular grid making quadrates 150 miles on each side (22,500 square miles each) not oriented with respect to physiographic features or other known zoogeographic factors. A list of species of recent mammals was compiled and the presence or absence of each species noted for each quadrate. The species lists and distributions were based on Hall and Kelson (1959), with modifications mentioned below. Only continuous mainland areas and species occurring on them were included. Insular distributions have their own special interest, but the fact that they are special would make their combination with mainland data confusing. The region covered is all of North America, including Central America to the Panama–Colombia border. The total number of quadrates is 453.

Quadrates of this size are suitable for some kinds of zoogeographical analysis and not for others. For almost any purpose using quadrates it is desirable that most or all species should occur in more than one and preferably in many quadrates, and that sets a rough upper limit of useful area. Although there are a few apparent exceptions for species the validity or extent of which is not in fact established, species of noninsular mammals in North America, at least, have areal distributions definitely and for the most part greatly in excess of 22,500 square miles. It is further necessary that quadrates be small enough for adequate resolution of the phenomena being investigated. Quadrates 500 miles on a side, for instance, probably would not resolve important zoogeographical changes adequately, even on a continental scale. For certain organisms and certain types of ecological studies, resolution may require quadrates as small as a square meter, or even smaller. Localized zoogeographic studies of vertebrates might utilize quadrates on the order of one to a hundred square miles, but that is too small to be practicable (at present) or desirable for studies of continental magnitude. For such small quadrates, actual presence or absence of a species would in many instances be variable, ephemeral, accidental, due to highly local factors, or indeterminate. A quadrate would not really represent the whole and long-term fauna of an area. Moreover, and equally important, our distributional data for species as a whole over the continent are certainly not accurate within 10 miles.

Although the choice of 150-mile quadrates was somewhat arbitrary, the order of magnitude was determined by the preceding considerations. For species density, at least, the subject of the present paper, the results seem to endorse the choice empirically. Quadrates much smaller than 150 miles on a side would have obscured parts of the over-all pattern by local, more or less random fluctuations and imperfections of the data. Quadrates much larger would have lacked significant resolution. After the fact, the results do suggest that if data sufficiently precise had been available, quadrates somewhat smaller might have given desirable increase of resolution without undue loss of pattern. It is, however, dubious whether that refinement would repay much increase in the labor of tabulation.

The data, as modified from the compilation by Hall and Kelson, have a number of known and not wholly corrigible deficiencies. Application of more meaningful criteria for biological species has markedly reduced the number of nominal species of North American mammals in recent years, but the number listed by Hall and Kelson is still certainly too large, as those authors recognize. I have gone beyond them, for instance and most obviously in eliminating almost all the ridiculously defined "species" and "genera" of bears that they still list. (Only three species and one genus, as *reasonably* defined, are present in North America.) In a number of other instances, also, I have united species that Hall and Kelson list as separate, but I have done so only when a student of the group has stated that the synonymy is probable, at least. There are numerous other cases, especially in groups not recently revised, in which the distributions and stated diagnoses make synonymy extremely likely on the face of things. I have not, however, eliminated such probable synonyms on my own authority. My list thus is also surely still too long, although somewhat shorter than that of Hall and Kelson. There is, nevertheless, no reason to believe that unrecognized

synonymy is more common in one *region* than in another. The sort of general patterns and conclusions here discussed are therefore probably not significantly falsified by this factor. Man and introduced mammals have not been included in my data. As noted above, entirely insular species are also omitted. The total number of species on my list is 670.

Most of the distributional patterns have now been worked over in considerable detail by several generations of mammalogists, and Hall and Kelson also checked these in numerous and large collections. There are, of course, exceptions, for instance, in some Central American species known only from one or a few scattered localities, but in general the distributions are probably now quite accurate at the scale of resolution implicit in this study and within methodological limitations. I have modified Hall and Kelson's data on areal distributions only by elimination of a few extensions obviously recent and under human influence, and of course by uniting distributions of species which they separate and I unite.

There are limitations in all the usual distribution maps for species, and these are not always understood. Such maps, like Hall and Kelson's, usually draw a smoothed limiting line around peripheral collecting records and indicate continuous distribution within that limit. Actually, of course, the limits are highly variable in time, and even if perfectly accurate for some one time (an impossibility in itself) they would not be quite the same at any other time. Moreover, the limits indicated for different species are not necessarily, or even usually, contemporaneous. An extreme but otherwise characteristic example is that of the bison, whose distribution has varied enormously and continually ever since they reached North America in the Pleistocene. The present, 1964, distribution obviously has little zoogeographical significance. The nominal distribution here used for the bison is that of latest prehistoric or earliest historic times, if indeed it is even roughly accurate for any *one* time. That is certainly not contemporaneous with other distributions, most of which are for twentieth-century dates.

The implication of continuous distribution within figured limits is also often or usually false. At most it could presumably mean that the whole area was covered by ranges of individuals of the species, which must rarely if ever be literally true. In fact, individuals of a stated species may be entirely absent over the greater part of its mapped distribution, for example, animals in a discontinuously distributed environmental niche or in a single faunal zone among several in the region. It is well known to field workers that at any one locality they can never hope to observe or collect all the species whose mapped distributional ranges cover that point.

These shortcomings must, of course, be borne in mind in any interpretation of the data. In some particulars they do seem to deprive analytical details of reliable significance. They do not, however, seem seriously to affect over-all patterns and trends for the continent. As regards species density in particular, the fact that a definite and understandable pattern does emerge is empirical evidence that the data are adequate and that their interpretation is significant in spite of some coarseness of resolution and uncertainty of detail.

The voluminous basic data (quadrate grid, faunal list, species distributions by quadrates, and other data in later studies) are not published but may be consulted on personal inquiry to the author.

Species Density Pattern

The species density pattern was established by counting species in each quadrate, entering the figures on the map, and contouring by isograms. The result is shown in Figure 1. The contour interval (change in number of species from one isogram to the next) is five for Canada and the United States. Near the Mexican border and southward it was found that species numbers change so rapidly that a contour interval of less than ten species was impractical on this scale. That difference is of some interest in itself. Since the absolute numbers of species are also higher in Mexico and Central America, the rate of change in terms of percentage of species present tends to be more nearly the same, although some difference persists. In other words, the density slope (change in number of species per unit of distance) tends to be a function (but not a simple linear function) of the absolute density (number of species in any one quadrate).

That is also visible in Figure 2. In Alaska and Canada the average species density is about 50. The slopes, all positive in a southerly direction and accelerating in that direction, average about one species per 40 miles. In the United States the average density is somewhere between 90 and 95 species, and the slopes (in both directions) average about one species per 25 miles. In Mexico and Central America the average density is about 130, and the slopes, in both directions but predominantly positive toward the south, average about one species per 20 miles. (These approximate figures apply only to the section given in Figure 2, along line A–A′ of Figure 1.) It is, however, evident that the relationship is loose and irregular.

The most obvious over-all feature of the map is a marked increase in species density from north to south, or equatorward. Another, more localized set of features seems to be correlated with topography, as it involves relative highs in mountain and plateau regions and relative lows on plains and in basins. A third possible but dubious regularity is what seems, as a first approximation, to be a tendency for species densities to increase from the coast inland. Also dubious but possible is some trend for increase from east to west. A striking special feature, related primarily to topography, is the occurrence of some lines of extremely abrupt change in density—the lines I have called "fronts" and indicated by a special symbol in Figure 1. Finally, there is an evident tendency for large peninsulas to have fewer species than otherwise similar

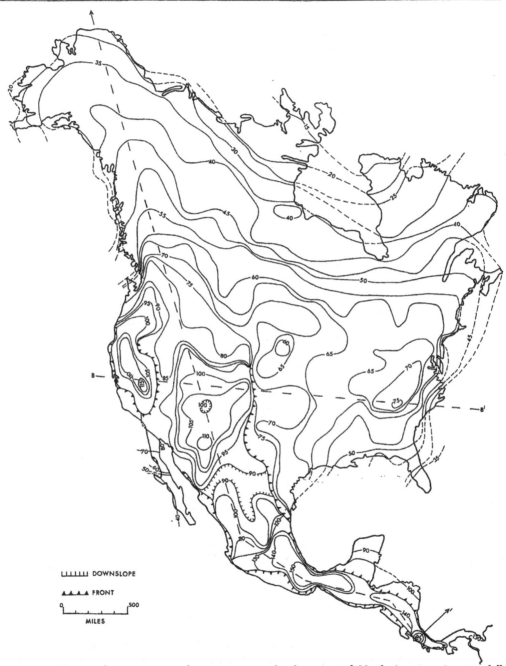

Fig. 1. Species density contours for recent mammals of continental North America. As more fully explained in the text, the contour lines are isograms for numbers of continental (nonmarine and noninsular) species in quadrates 150 miles square. The interval between isograms is 5 species for the northern and central parts of the map, approximately to the United States–Mexican border, and is 10 species south of there, through Mexico and Central America. The "fronts" are lines of exceptionally rapid change that are multiples of the contour interval for the given region (also see text). Indication of the downslope side of a contour is given in two areas only where this is not obvious at first sight.

inland regions. Each of these apparent major features or regularities will be discussed in sequence.

In addition to these apparent widespread tendencies, there are numerous minor curves and quirks in the contours, of the sort considered anomalies or "noise" in another recent study involving contouring of density of taxa (Stehli and Helsley, 1963). It would, of course, be unsound to qualify variations as "noise" simply because they fail to agree with a hypothesis, such as that of a polar-equatorial gradient in density of taxa. Some of the unresolved irregularities are doubtless due to deficiencies and inaccuracies of data, in which case they could reasonably but allegorically be considered noise in the sense of not containing correct information about an objective natural relationship or condition. I believe, however, that in our case (and perhaps also in that of Stehli and Helsley) most or all of these apparent irregularities do in fact contain information that is below the clear resolving power of the grid or that we are simply too ignorant to interpret. In other words, they are probably systematically caused by local conditions unknown to us or not evident at this scale.

The North–South Gradient

Mammalian species densities vary enormously in continental North America, from 13 per quadrate in one of its northernmost areas (Melville Peninsula, latitude 70°) to 163 in one of the southernmost (Costa Rica, latitude 10°). Between these extremes there is a clear but interrupted and irregular gradient. In the colder parts of the continent, from the northernmost continental land approximately to the Canadian–United States border, this gradient is fairly regular. South of that Arctic–Cold Temperate zonal gradient, more or less between latitudes 50° and 30°, that trend is absent. In 18 quadrates spanning the continent approximately along the 45th parallel, the average species density is 72. In 15 quadrates similarly arrayed along the 30th parallel, it is 68. That might, indeed, suggest a feeble trend *op-*

posite to that farther north, but from further examination it appears that there simply is no general north–south trend in this area. That is certain as regards any trend significant in the data as here tabulated. This area, roughly the United States, has a definite but quite irregular pattern dominated by other trends and local factors and devoid of the latitudinal gradient generally expected for such data.

To the south of that area, mainly in Mexico and roughly between latitudes 30° and 20° (i.e., in the subtropical–high tropical zone), the pattern is also quite irregular and reflects other factors as well, but there is again a definite and strong north–south gradient of increase in density. Still farther south, through Central America, the continental land is so narrow, is so abruptly varied in physiography, climate, and vegetation, and changes so little in latitude that any over-all trend that might exist is concealed by local perturbations. All one can say is that a north–south trend cannot be seen in our data for this region.

Figure 2 is a profile along line A–A' on the map given in Figure 1. It runs from extreme northwestern Alaska, an area of very low although not minimal species density, to the quadrate of maximal density, in Costa Rica. The striking over-all increase equatorward is evident, and so are the Arctic–Cold Temperate zonal gradient through Canada, the much steeper Subtropical–Tropical zonal gradient in Central Mexico, and the absence of a clear trend in Central America. The irregularities in Central America and in the United States (Midtemperate subzone) correlate with topography and confuse or conceal any possible underlying latitudinal trend. The section in Figure 3 follows the 100th meridian, along which there are few marked topographic irregularities north of Mexico. The trend for increase from the Arctic down to the 50th parallel, or somewhat beyond, is again evident, but here it is clear that there is no such trend on southward to beyond the 30th parallel. It is impossible to make a similar check in Central America,

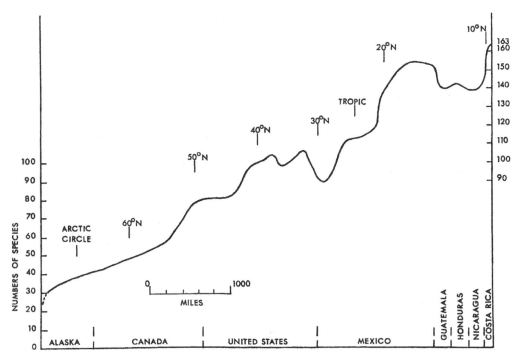

Fig. 2. Species densities along the line A–A' of Figure 1. Numbers of species are those contoured in Figure 1, with smoothed curves between contours filled in freehand. The line approximately follows the western and (to the south) central mountain axis of the continent.

where there are no extended possible section lines without great topographic irregularity.

Some details as to just what is going on as faunas change from north to south and east to west are deferred for later study. Here is given only brief mention of some points that could bear on attempts to explain the latitudinal gradient. Gradients for higher taxa tend to be similar to those for species, but the generic, familial, and ordinal gradients are generally flatter, both in absolute numbers and in terms of percentage change. Individual orders and families frequently do not follow the overall trends. For example, the orders Insectivora and Artiodactyla and the rodent families Geomyidae and Heteromyidae reach maximum diversities at various latitudes down to the Tropic of Cancer and become *less* diverse more to the south. The groups of which this is true are without exception old in the Holarctic region and newcomers in the Neotropical region, if they reach it at all. Some old Holarctic groups (e.g., Vespertilionidae, Sciuridae, Cricetidae), however, do maintain or even increase their diversity in the present tropics. Almost all of these basically northern groups have a definite increase in diversity from north to south in Canada, even though the trend may not be maintained through the United States and may be reversed in Central America.

The increase in tropical diversity as compared with the Temperate zone does involve some old northern groups, but it is primarily due to old southern groups, most of which even now do not extend far from the tropics: opossums, phyllostomatid bats, monkeys, edentates, and caviomorph rodents. Thus the numerical expansion of taxa of mammals in the tropics is not simply an inflation of numbers of species within

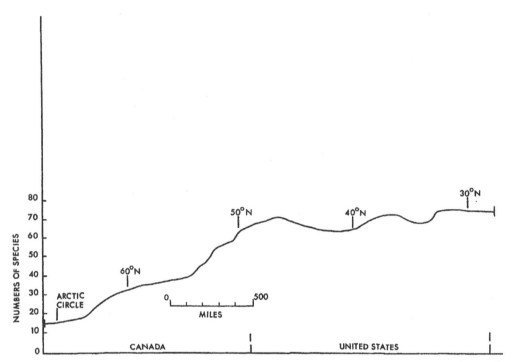

FIG. 3.   Species densities from the Arctic to the Mexican border along the 100th meridian.   The densities are those contoured in Figure 1 and the section is constructed as in Figure 2.

more or less the same higher categories but marks definite changes in over-all taxonomic composition and ecological characteristics of the faunas.

What actually exists as regards species densities in these mammals is not a single polar–equatorial gradient. There is first a northern gradient that is largely (but not altogether) a simple expansion of members of species within faunas of similar basic familial and ordinal composition. This reaches an apparent equilibrium in mid- to warm-temperate regions, where there is no further north–south trend in diversity. In the temperate–tropical transition there is again—and here much more steeply—an increase in species densities. Here this reflects an extensive (but not total) change in faunal type as regards both higher taxonomic categories and ecological make-up. Thus, what is to be explained is not a gradient but two gradients of different

kinds and the absence of gradient between them.

Discussion of Latitudinal Gradients

That many groups of organisms are more diverse in the tropics, or that there are latitudinal (polar–equatorial) gradients in species density, is one of the oldest generalizations of biogeography. It was already noted by Wallace (1878) and has since been found to hold for a great variety of plants, invertebrates, and vertebrates and for both continental and marine faunas. (See especially Fischer 1960; and in addition to his references: Hutchinson, 1959; Klopfer and MacArthur, 1960 and 1961; MacArthur and MacArthur, 1961; Klopfer, 1962; I am also indebted to a 1963 term paper for Biology 246 at Harvard University by Thomas W. Schoener.) There are, to be sure, a few groups that do not have tropical diversity or that may even have an opposite gradient (e.g., Thorson, 1957; as noted above this is also true of some mammalian taxa up to ordinal rank although not of mammals as a whole). Nevertheless, there is no doubt that tropical diversity is the rule within broad groups of organisms (higher taxa) almost throughout the realm of life and in the great majority of habitats. It has also long been known in a general way and by counts of taxa for broad faunal regions (e.g., Darlington, 1957; Hall and Kelson, 1959) that this rule applies to continental mammals, although I do not know of any previous attempt to follow their trends in detail.

There are, in fact, few data for any group that permit actual measurement of a trend, and fewer still that show details of trend patterns and interactions. Although it is based on extremely few taxa and exiguous data in general, special interest attaches to a recent study by Stehli and Helsley (1963) in which they have contoured densities of taxa in the same way as has been done here (independently, as it happens, Figure 1 of this paper having been roughed out before their paper was published). They attempt to locate the north pole in the Permian period on the assumption that species density in certain marine brachiopods then had a polar–equatorial gradient.

The long-known and unquestionable phenomenon of tropical diversity has never been generally and adequately explained (in my opinion, at least). The present data may make some contribution toward that end, but they also fail to suggest a full explanation. Major lines of previous attempts at explanations were summarized by Fischer (1960). Some are simply irrelevant. For example, it has often been suggested that evolutionary change is faster in the tropics for various reasons, such as increased competition or shortness of generations. In fact, it is not demonstrated that such factors would or do cause more rapid evolution in the tropics, and in any event they would be irrelevant. What is to be explained is the coexistence of more numerous species in the tropics, and that is not at all the same

thing as accelerated evolutionary change, nor is it a necessary or, indeed, probable result of such change. Other proposed explanations may be logically impeccable but are too lacking in specificity and concrete evidence to be really satisfactory. For example, Dobzhansky (1950) suggests that biotic competition is greater in the tropics, which thus present more evolutionary challenges, and that this leads to greater specialization of each species and so to accommodation of more species in a given area. That is a reasonable deduction from the fact that there are indeed more species but perhaps is not a useful and testable explanation of the given fact.

Still other explanations are proximate and *ad hoc*, without providing for the generality of the phenomenon. It has often been pointed out that the great species density of plants in the tropics permits if it does not induce a corresponding diversity of animal life. In reasoned, quantitative terms, MacArthur and MacArthur (1961) have shown that vegetational diversity is a good predictor of bird species diversity, but that it does not seem to be a fully sufficient explanation of the diversity of tropical birds. In any case the vegetational diversity would still have to be explained, so that the chain of causation could at best be followed back only one step. Still it is a useful observation that increase in numbers of taxa of *any* group in an integrated community would increase the number of potential niches and hence of species density capacity for other groups. In this way a circular or feedback reaction could readily increase species densities for many or most kinds of organisms in the community as a secondary result of an original cause that affected only one of them directly.

Fischer's (1960) own explanation may or may not apply to some groups in some circumstances, but it cannot apply to North American mammals, which, in fact, provide evidence against it. He concludes, first, that biotas evolve and diversify more rapidly in the tropics because of the more constant and favorable environment and, second, that the tropics being less subject to climatic changes have developed a mature fauna with characteristic high species densities, while in polar and temperate regions biotic diversity has been adversely affected by climatic disasters and has not yet had time to reach maturity since the last of these (the Pleistocene glaciation). The first point seems to me largely irrelevant, for reasons already suggested, and in any case it is hard to see why uniformity of climate would, in itself, accelerate speciation; one might expect quite the opposite.

Fischer's second and principal point seems to imply, although this is not quite explicit, that different climatic zones of sufficiently long, continuous existence would have biotas with equal species densities. (That would be the case unless the rates of faunal maturation were extremely different in the different zones, which would in turn require an explanation even more difficult and not given.) In spite of some misleading popularizations, there is no evidence that the earth has ever had a pantropical climate. On the contrary, it is practically certain that there has always been some zonation of climates. There is strong, explicit evidence that the Warm Temperate zone, at least, has existed since the Eocene (Axelrod, 1952), which must surely be more than long enough for biotic maturity in terms of numbers of species. (Among mammals, no species or genera and few families date from the Eocene.) The climate has moved geographically, but so slowly that associated biotas surely could and did follow it. On Fischer's theory, one might therefore expect the present Warm Temperate zone to have species densities comparable to the tropics, but that is decidedly not true of North American mammals. In fact, it is also untrue of most of the groups for which Fischer has gathered data.

If Fischer's premise, that only the Tropical zone is old enough for biotic maturity, were granted, the steepest gradient in species densities would be expected in the Temperate zone. That may be true for some organisms (there are few really clear data on such gradients), but it is again flatly false for North American mammals, which have low gradients or none at all in this zone. Fischer's theory might have some bearing on the existing Temperate–Arctic zonal gradient, in the sense that adaptation of many temperate groups to colder climates might require more time than has been available. Even here, however, I cannot believe that a mature arctic biota, after no matter how long a time, would be as diverse as either a temperate or a tropical biota.

In their discussion of tropical diversity, Klopfer and MacArthur (1960) have argued that nonpasserine birds, being more primitive than passerines, are less plastic in behavior and evolution and hence have narrow niche size. Further, it is supposed that temperate climates require more plasticity or breadth of niches than tropical ones. Data are given to suggest that nonpasserines may, with great irregularity, be somewhat more abundant relative to passerines in the warmer than in the colder parts of the United States and Mexico. The conclusion is that tropical niches for birds are narrower than those in the Temperate zone. The data refer to numbers of individuals, not directly to numbers of species. Assumptions taken as premises are entirely unsupported, and some, at least, seem improbable. The conclusion seems to me not to follow. In any case, I see no way of applying this approach to the mammals and so need not argue the matter here.

In a later note, Klopfer and MacArthur (1961) suggest that there is more overlap of niches among tropical birds and that therefore more species can coexist in the tropics. This seems to contradict their previous conclusion that niches are narrower in the tropics, and the statement that the *exclusive* parts of the niches are narrower does not seem fully to resolve the anomaly, nor yet to provide meaningful explanation for greater species numbers. The evidence provided for the existence of greater niche overlap is that, in four genera of birds with sympatric tropical (Central

American) species, the ratio of greatest to least bill length within the genus ranges from 1.11 to 1.00 (mean 1.04), whereas Hutchinson (1959) had previously given 1.28 as an average value for a number of sympatric species of birds. It is argued that the lower this ratio, the more similar the food habits of the birds concerned, and therefore the more overlap in their niches. That postulate is open to question, but in any event the data are inadequate or misleading. Some of MacArthur's (1958) own earlier figures give ratios for sympatric species (of *Dendroica*) in Maine lower (mean 1.01) than the average for Klopfer and MacArthur's few tropical examples, and a rather broad sampling of such data shows no latitudinal trends. (I am here indebted to the term paper by Schoener, previously mentioned.) In fact, in most of the available examples the stated ratio is so near 1.00 as to suggest either that there is no difference in niches or (much more probably) that this figure does not measure the actual difference and is not pertinent to the problem.

Another possible approach to this question can be applied to our data for mammals. It is probable and indeed in many cases is demonstrable that congeneric mammals are more similar in habits and ecological requirements than species of different genera. It is therefore quite likely, at least, that a fauna with numerous congeneric species has smaller niches, on the average, than one with few such species. Not all species occurring together in single quadrates of our data are strictly sympatric, but most of them are. Thus the ratio of species to genera in a quadrate might be a rough but, as an average and within limits, significant inverse indication of niche size. If niches are smaller in the tropics than elsewhere, these considerations lead to the prediction that the ratio of species to genera will on an average be larger in the tropics. For a sequence of 13 quadrates from the 50th parallel to the Tropic of Cancer, the mean value of that ratio is 1.57 and for a continuation of the sequence through 10 quadrates below the Tropic the mean value is 1.59. The difference obviously is not significant but happens to be in the direction opposite to that predicted. (These figures are for genera recognized by Hall and Kelson, which in my opinion are unduly split in some groups; the splitting is not greater on one side or the other of the Tropic, however, and so the comparison is valid.)

The greatly increased numbers of species in the tropics suggest that niches there might be smaller, and most naturalists who have worked there seem to have the subjective impression that this is so. The results reviewed above do not disprove that relationship, but they show that if it is true neither Klopfer and MacArthur nor I have found a way to demonstrate or measure it. It should also be noted that greater species density does not *necessarily* require smaller niches.

Klopfer and MacArthur predict that sharp changes in avifaunal diversity will occur at the boundaries of regions of maximum climatic stability, which are approximately those of tropical climates. The prediction for birds is fulfilled by mammals as regards the northern boundary of the region of maximum climatic stability in the Americas. Since the prediction is not specific or exclusive to Klopfer and MacArthur's theory of niche size, its fulfillment does not necessarily tend to confirm their theory. The boundary in question is a line of maximum or crucial climatic change, and the prediction would therefore be made from *any* theory that relates species density to climate. As regards the mammals, it is obvious enough that change in faunal type should tend to approximate this boundary, which is one of marked ecological change in many respects from the soil upward. For example, mammals living mainly or solely on fruits (other than seeds and nuts) are abundant south of the boundary and practically absent north of it, for fairly obvious reasons. That does not explain why the tropical fauna, qualitatively distinctive in taxonomic and ecological make-up, should also be characterized by higher *numbers* of taxa.

Since there are, in fact, more tropical species and since by definition each species occupies or defines a niche, it follows that:

1. Either there are more niches in the tropics, or
2. The niches of the tropics are more fully occupied, or
3. There is multiple occupation or overlap of niches in the tropics to a greater extent than elsewhere.

That is not explanatory, being simply a restatement of the observed phenomenon in language apt for its discussion, but it does suggest possible lines of explanation.

For mammals, at least, there certainly are many more niches in the tropics. That is most evident from the great diversity of plants, providing a greater variety of animal foods, available for longer periods, and permitting more specialization in food habits. Tropical uplands also provide habitats ecologically somewhat similar to those of the Temperate zone, whereas nowhere in the Temperate zone are tropical habitats simulated. Thus the tropics in general have, to some extent, both tropical and temperate niches and the Temperate zone has temperate niches only. Temperate animals must have coinciding seasonal cycles, whereas many tropical regions permit staggering of specific cycles and hence more niches. These and other factors clearly do give mammals more niches in the tropics, and this is probably sufficient to account for their tropical species density. However, the most obvious element here, diversity of vegetation, is only a proximate explanation, and I cannot carry it further.

The suggestion that tropical niches are more fully occupied could be substantiated only by showing that there are unoccupied niches elsewhere. We recognize niches by the species in them, so that a count of empty niches may be almost as abstract as a count of angels on the point of a needle. In any event, although this is a theoretical possibility, I see no good way to demonstrate or evaluate it, and it does not seem necessary for explanation of the phenomenon.

A long known, frequently rediscovered, and variously named principle (Gause's, exclusion, ecological incompatibility, etc., etc.) is that multiple, sympatric occupation of a niche cannot long endure. Students of tropical biotas, especially forests and marine faunas, must sometimes feel that the principle requires blind faith, but as regards mammals I see no serious reason to doubt it. The present tropical North. (i.e., Central) American fauna arose from the mingling of previously separate northern and southern faunas. In the most active phases of that interchange, mostly earliest to about middle Pleistocene, it is likely that there was, indeed, some multiple occupation of niches and that this was greater in the Tropical than in the other zones. However, there has since been a marked reduction in total numbers of taxa (throughout the Americas), and in the course of this it is highly probable that multiple niche occupation or significant overlap has been greatly reduced or completely eliminated. Indeed that is probably why the numbers of taxa have decreased.

The principle is valid for sympatric species only, and there can, in principle, be identity of niches for fully allopatric species. Small population ranges and numerous barriers against the spread and sympatry of related populations would therefore tend to increase density of species in a region as a whole. It will be suggested below that this is a factor in the increase of species densities in regions of high topographic relief. I do not, however, know of any evidence that it is more general or more effective in the tropics. It probably has no bearing on tropical diversity beyond the fact that when the two factors (tropical and topographic) coincide, maximum densities occur, as happens in Costa Rica in our example.

The most important single element in the Tropical–Temperate zonal transition is probably the frequent occurrence of killing winter frosts north of that transition. That has many concomitants and effects, such as the fact that more temperate than tropical angiosperms are annual or deciduous and that there is great reduction in diversity and quantity of winter food. Once that sort of ecological situation is established, no marked further effects would be expected from increasing severity of winters within rather broad limits, as long as the growing and (for most animals) breeding or infantile season remains several months long. Such considerations make it seem natural enough that no broadly climatically zoned changes in species density of mammals occur between the tropical transition and more or less the 50th parallel.

Northward from that parallel there is no abrupt change either faunally or climatically. There is, however, a shortening of the period between frosts, which becomes crucial more or less at that latitude. That has such concomitant ecological changes as the attenuation of numbers of perennial angiosperms and progressive reduction of variety and quantity of plant food in general. Adaptive requirements become more and more rigorous, population fluctuations and hazards of extinction greater, niches available to actually existing populations with geographical access fewer. The northward decrease in numbers of taxa, here an attenuation more than a change in faunal type, seems readily understandable in such terms. More explicit connections of faunal drop-outs with particular and measurable ecological conditions will doubtless be made when taxon frequency has been studied for a longer time and in more detail. For instance, the northward attenuation and final disappearance of the subterranean rodents, Geomyidae, can be correlated with temperature changes and other physical conditions of soils.

Topographic Factors

Even a glance at Figure 1 shows that there is certainly a relationship between topography and species density, although the correlation is neither perfect nor simple.

The species high for eastern North America (75) occurs precisely in the same quadrate as the highest eastern mountains and maximum relief. There is a clear although not very steep or high species isogram positive rather closely following the Appalachians from Pennsylvania to Georgia. (There is, however, no evident elevation of species density in the Adirondacks or the New England mountains.) In western United States the species high (115) is also precisely in the same quadrate as the highest mountains and maximum relief. Here, too, the whole mountain zone from British Columbia down through Washington, Oregon, and California is rather closely approximated by a species contour high.

The Rocky Mountains also evidently have a correlation with species density, but this is not quite as simple as for the Appalachians and the Pacific coastal and subcoastal mountains. In Colorado there is an extremely abrupt rise westward in species density, from 70 to 107 species in adjacent quadrates. In Figure 1 and Figure 5 this appears in eastern Colorado, seemingly east of the mountain front, but that is a failure of resolution in our method and data. This extreme rise is necessarily contoured near the quadrate boundaries. The western quadrate, with the specific high, extends about 150 miles west of the mapped contour front and reaches the mountain front and zone of maximum relief. Examination of individual species distributions confirms the inference that the most rapid rise in species numbers is indeed in the region of most rapid topographic change.

West of the Rocky Mountains proper, the high extends without marked change over most of the Colorado Plateau and parts of the Great Basin. Although not continuously mountainous, areas on the order of magnitude of our quadrates here do generally have high relief comparable to that in similar areas in the Rocky Mountains. A distinct trough in the species contours extending north and south through Nevada corresponds approximately with a region of lower average relief than those east and west of it. West of that comes another very abrupt rise, which, although again not precisely located by the loose resolution of our method, surely corresponds with the extreme relief along the eastern front of the Sierra Nevada.

Northward into Canada, the Rocky Mountain high in species merges into that of the coastal and subcoastal ranges, as is also true of the topographic highs. This northern extension is clear and definite all the way up into central Alaska, although it becomes much less strong and sharp north of Colorado. In the northern areas, the species highs are much less in proportion to the topographic relief. This can probably be ascribed in part, at least, to interaction with the north–south gradient, which markedly reduces all species numbers northward through Alaska and Canada but has no evident effect in the United States. Through Wyoming and Montana the change from plains to mountains is not so abrupt or so single-fronted as in Colorado and New Mexico, and the species density change is also less sharply localized.

The north–south density front of Rocky Mountain species continues from southeastern New Mexico south–southeastward into Mexico and to the Gulf Coast in Tamaulipas. It is here still following a rather abrupt topographic change from the low-relief plains east and northeast of the Pecos and lower Rio Grande to high relief areas west and southwest of those rivers.

In northern Mexico there is a region of lower relief between the eastern and western Sierra Madre, where there is also lower species density. South of this comes the steep species density gradient into the high tropics referred to previously. From this gradient zone on southeastward throughout Central America there are axial mountains with high relief, with correspondingly high axial species densities. These culminate in Costa Rica, with 163 species in a quadrate that is at the same time one of the southernmost (extending below 10° latitude) and one of highest relief (well over 12,000 feet in a single quadrate). Density fluctuations

occur along the axis, but these are not obviously related to topography. They may correlate with local features not resolved at this scale. The broad regions of relatively low relief in Honduras and the Yucatan peninsula have low species counts for these latitudes. (Instead of or in addition to the topographic effect, this may involve the peninsular effect discussed below.)

It is thus clear that species density has a strong correlation with topography. In the United States, where the north–south gradient is absent, the topographic gradients account for most major features of the pattern of species density. Northward and southward, where there are latitudinal gradients, these are additive with topographic gradients, the two accounting for most of the pattern.

On the principle that higher altitudes are ecologically similar to higher latitudes and that higher latitudes generally have lower species densities, one might expect just the opposite of what is found, a *decrease* of species densities in mountains instead of an increase. That may be true for some groups in detail and within single mountain ranges or local systems, although the few data known to me are inadequate to support that as a generalization. It is highly probable that the broad-scale topographic increase of species densities here so obvious is connected with high relief and not with mountains or with altitudes *per se*. It has long been recognized in a rather vague and general way that greater relief must tend to increase variety of ecological conditions and hence of ecological niches in a region. The present data substantiate that, illustrate it in objective, numerical terms, and demonstrate that the influence on faunal patterns is more important and widespread than has been commonly recognized.

In some western areas, a single quadrate on our scale may cover as many as four major life zones (for example, Upper Sonoran, Transition, Canadian, Hudsonian). In detail, such a region or any of comparable relief has an environmental complexity almost incomparably greater than in areas of low relief such as, for instance, the Kansas plains. It is therefore evident that higher relief does mean more environmental niches and so the possibility of higher species density, which is indeed observed. Nevertheless, it is rather surprising that the Rocky Mountains, where there are quadrates with relief up to about 9,000 feet, have only about 150 per cent as many species per quadrate as the adjacent high plains, with local relief a few hundred feet at most. It may be that greater altitude tends to decrease numbers of species and that this somewhat counteracts the effect of greater environmental diversity, but that is speculative at present.

Another somewhat speculative possibility is that regions of higher relief may frequently or on an average promote geographic isolation of populations and so increase the likelihood of speciation and consequently the number of species in such a region. Regions of topographic irregularity do often have discontinuous areas in different life and altitudinal zones, as well as other discontinuities. Hence, such regions have isolated demes that may readily develop and in some cases clearly have developed into new species. Some such influence, with alternatively or additionally some immigration of species into the area, is necessary if the more numerous niches in areas of high relief are in fact to be occupied by higher numbers of discrete species. Evolution of *continuous* populations *in situ* will not in itself increase species density no matter how varied topography and ecology may be.

Correlation of species density with topographic relief does not mean that there is always or predominantly a direct causal connection. The relationship must be largely indirect through all the elements that increase numbers of niches and numbers of isolated populations tending to speciate. Proximate causes must vary from region to region and include many factors such as topographic details (isolated peaks, topographic barriers, etc.), climate, and vegetation.

Species Density Fronts

In western United States and through Central America there are lines along which changes in species numbers are so abrupt and great that they cannot practically be represented by separate contours with the intervals and at the scale here used. The change in species numbers between adjacent quadrates is here several times the selected contour interval (5 in the United States, 10 in Central America). These lines, called "fronts" and given a special symbol in Figure 1 for convenience, are not qualitatively different from other density gradients but simply represent extremely steep gradients. They are in every instance topographically correlated, coinciding (within limits of resolution of the data) with abrupt changes in relief. At no point does a north–south or climatic gradient reach this degree of steepness. The central Mexican temperate–tropical gradient most nearly approaches it.

East–West Gradients

East–west sections anywhere north of central Mexico all show a pronounced lateral asymmetry, with average species densities decidedly higher to the west than to the east, as exemplified in Figures 4–5. (From central Mexico southward, no such tendency can be seen, but there are no east–west lines long enough to involve it there.) The asymmetry is influenced by topographic factors, the west having higher relief in general. In the Canadian section (Fig. 5), the species high in Alberta and British Colombia corresponds with the Rocky Mountains. The species high in Manitoba is also in an area of relief somewhat greater than immediately to the west or east, but the rise in species density here does seem more than would be expected from topography alone. Even the slighter rise in eastern Quebec approximates a local increase in topographic relief. Nevertheless,

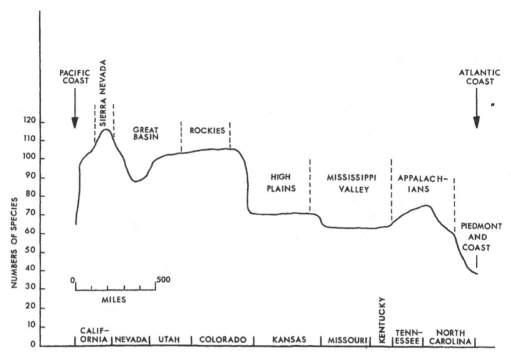

FIG. 4. Species densities from the Pacific to the Atlantic along line B–B' of Figure 1. Section constructed as in Figures 2 and 3. The section is a straight line on the projection used, approximately through the western and eastern species density highs in the United States. Because of the nature of the map projection, the compass bearing of the section is not constant.

the rather steady rise from central Quebec westward through Ontario is not clearly related to topography.

In Figure 4 the rises and falls in species density correspond closely with increases and decreases in topographic relief, as previously noted. Nevertheless, the species densities seem to be greater in the center and west than in the east for equivalent amounts of relief. The western high plains have species densities as high as those of the Appalachians even though the relief is much less in the former.

It is thus quite possible but not fully established that there is an east–west gradient in species density superimposed on and not clearly separable from the topographically correlated gradients. If the east–west gradients exists, it would be correlated roughly and *negatively* with mean annual precipitation. That is probably con-

trary to subjective expectations of most ecologists, who have sometimes assumed that drier climates are more rigorous and that more rigorous environments will have fewer species. There is a possible explanation in terms of topography plus precipitation. Relief in a more arid region does in general involve greater variety in climates, microclimates, and hence more generally in ecological niches. It is quite apparent that environments are more varied in, say, a southwestern desert range with 3,000–4,000 feet of relief than in an Appalachian area of equal relief. It is, however, hard to see how any such factor could account for the relatively high species densities (average about 71 per quadrate) in high plains areas of *low* relief in the United States west of the 20-inch isohyet and east of the mountain front. A possibility is that increased speciation in the mountainous areas has af-

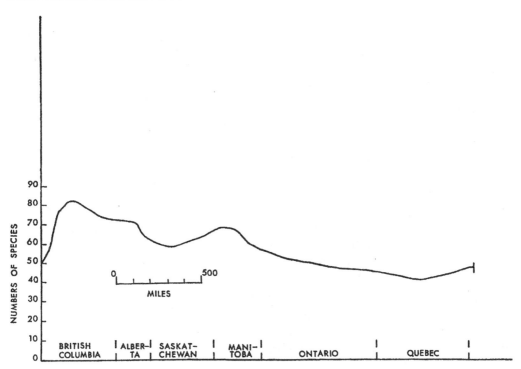

Fig. 5. Species densities from the Strait of Georgia (British Columbia) to the St. Lawrence River through southern Canada along the 50th parallel. Sections constructed as in Figures 2–4.

fected adjacent plains by subsequent expansion of originally montane species. It is of course known that montane species do sometimes spread to plains, and vice versa, but whether this does generally increase density of species has not been investigated.

However that may be, the species density pattern gives no evidence of a positive correlation with precipitation in equivalent latitudes but is definitely opposed to the existence of such a relationship.

Shore–Inland Gradients

As mapped in Figure 1, the eastern and western sides of the United States, Mexico, and Central America show a marked increase of species densities from the shores inland, and although less marked, this is also true of the shores of western Canada and Alaska. It is also true of the Arctic shores, but there is no separation of this from the broader north–south trend. A

tendency for increasing species density inland is not evident for the Atlantic shore of Canada or the Gulf Coast of the United States.

Increase in the immediate vicinity of the shore line is probably in part an artifact of method. Marginal quadrates extend across shore lines and include smaller or larger areas of sea. Their land areas are therefore smaller than for inland quadrates, and it is probable that on that account alone they will have, on an average, somewhat smaller numbers of continental mammals. Apart from that point, where there are steep shore–inland gradients in numbers of species, there are also mountains at or near the shore. Along the Gulf Coast, where there are no mountains near the shore, the species gradient is absent. The increase inland thus could well be and probably is largely or entirely topographical in nature. In view of these considerations, it seems unlikely that there is any real shore–inland gradient as an analytically separable factor.

Peninsulas

North America has the following more or less strongly defined peninsulas large enough for resolution by our data: Alaska (Aleutian), Seward, Boothia, Melville, Nova Scotia, Florida, Yucatan, and Baja California. On all of these, species densities are definitely lower than the average for nonpeninsular areas in the same latitude and of similar topographic relief. This is doubtless due in part to the artifact previously mentioned in connection with coastal densities: no quadrate on these peninsulas is entirely on land, thus the pertinent areas of the peninsular quadrates are decidedly smaller than for inland quadrates.

That cannot, however, wholly account for the peninsular lowering of species densities. Without regard for quadrates, the total number of species in peninsular Florida, for example, is definitely less than for any *equal* area in nonpeninsular parts of the continent of anywhere near the same latitude and including regions (in southern Texas, for instance) of equally low relief. It is therefore highly probable that peninsular areas are underinhabited in terms of available niches occupied by distinct species. Why that should be so is far from clear. One possible hypothesis is that restricted peninsular areas, generally permitting extensive spread in one direction only, have limited population sizes and continuities in some species to a degree making them more liable to (at least) local extinction or preventing long-continued coloniza-

tion of the area. A comparison of species ranges, population densities and structures, and related factors in peninsular and nonpeninsular species should cast some light on this problem.

REFERENCES

AXELROD, D. I. 1952. A theory of angiosperm evolution. Evolution, 6:29–60.

DARLINGTON, P. J. 1957. Zoogeography. Wiley, New York.

DOBZHANSKY, TH. 1950. Evolution in the tropics. Amer. Scientist, 38:209–221.

FISCHER, A. G. 1960. Latitudinal variations in organic diversity. Evolution, 14:64–81. [With citations to earlier literature not repeated here.]

HALL, E. R., and K. R. KELSON. 1959. The mammals of North America. Ronald Press, New York.

HUTCHINSON, G. E. 1959. Homage to Santa Rosalia, or why are there so many kinds of animals? Amer. Nat., 93:145–159.

KLOPFER, P. H. 1962. Behavioral aspects of ecology. Prentice-Hall, Englewood Cliffs.

KLOPFER, P. H., and R. H. MACARTHUR. 1960. Niche size and faunal diversity. Amer. Nat., 94:293–300.

———. 1961. On the causes of tropical species diversity: Niche overlap. Amer. Nat., 95:223–226.

MACARTHUR, R. A. 1958. Population ecology of some warblers of northeastern coniferous forests. Ecology, 39:599–619.

MACARTHUR, R. A., and J. W. MACARTHUR. 1961. On bird species diversity. Ecology, 42:594–598.

STEHLI, F. G., and C. E. HELSLEY. 1963. Paleontologic technique for defining ancient pole positions. Science, 142:1057–1059.

THORSON, G. 1957. Bottom communities (sublittoral or shallow shelf). Geo. Soc. Amer., Mem. 67:461–534.

WALLACE, A. R. 1878. Tropical nature and other essays. Macmillan, London and New York.

GEORGE GAYLORD SIMPSON is Alexander Agassiz Professor of Vertebrate Paleontology in the Museum of Comparative Zoology, Harvard University, Cambridge 38, Massachusetts.

Vol. 100, No. 910   The American Naturalist   January–February, 1966

LATITUDINAL GRADIENTS IN SPECIES DIVERSITY:
A REVIEW OF CONCEPTS

Eric R. Pianka

Department of Zoology, University of Washington, Seattle, Washington*

INTRODUCTION: DIVERSITY INDICES

The simplest index of diversity is the total number of species, usually of a specific taxon under investigation, inhabiting a particular area. Since this index does not take into account differing abundances of species, divergent communities may show similar "diversities." Because of this, more sophisticated measures have been proposed which weight the contributions of species according to their relative abundances. As early as 1922 Gleason described and discussed the now well known "species-area" curve (Gleason, 1922, 1925). Later, Fisher, Corbet, and Williams (1943) proposed an index, alpha, discussed in detail by C. B. Williams (1964), which can be shown to approximate Gleason's "exponential ratio" (H. S. Horn, personal communication). Margalef (1958) has also used a modification of this index "d," in phytoplankton diversity studies, as well as several other indices (Margalef, 1957). The most recent, and currently widely used diversity index, is the information theory measure, H, derived by Shannon (1948). This index, $-\Sigma p_i \log p_i$, in which p_i represents the proportion of the total in the i-th category, has been used to quantify the "dispersion" of the distribution of entities with no ordered sequence, such as species in a community, alphabetic letters on a page, etc. Unfortunately, there has as yet been little discussion of the application of statistical procedures to this quantity. However, even without statistical embellishments, H has been a useful and productive tool (Crowell, 1961, 1962; MacArthur, 1955, 1964; MacArthur and MacArthur, 1961; Margalef, 1957, 1958; Paine, 1963; and Patten, 1962).

The choice of the index used in any particular investigation depends on several factors, especially the difficulty of appraisal of species abundances, but also on the degree to which relative abundances shift during the period of study, and for many purposes the simplest index, the number of species present, may be the most useful measure of local or regional diversity. This index weights rare and common species equally, and is the logical measure of diversity in situations with many rare, but regular, species (such as desert lizard faunas, Pianka, in preparation).

THE PROBLEM: SPECIES DIVERSITY GRADIENTS

Latitudinal gradients in species diversity have been recognized for nearly a century, but only recently have some of these polar-equatorial

*Present address: Department of Biology, Princeton University, P.O. Box 704, Princeton, New Jersey.

trends been discussed in any detail (Darlington, 1959; Fischer, 1960; Simpson, 1964; Terent'ev, 1963). A few groups, such as the marine infauna (Thorson, 1957), and some fresh water invertebrates and phytoplankton appear not to follow this pattern, but many plant and animal taxa display latitudinal gradients. A phenomenon as widespread as this may have a general explanation, knowledge of which would be of considerable utility in making predictions about the operation of natural selection upon community organization. Because of the global scope of the problem, however, it has usually been impossible for a single worker to study a complete species diversity gradient.

Approaches to the study of diversity gradients have so far been mainly of two types, the method of gross geographic lumping with comparison of total species lists for a group (Simpson, 1964; Terent'ev, 1963), and the approach by synecological studies on a smaller scale, comparing the diversity of a taxon through many different habitats (MacArthur and MacArthur, 1961; MacArthur, 1964, and in press). Terent'ev and Simpson used the number of species as indices of diversity, and MacArthur and MacArthur used Shannon's information theory formula to calculate indices of faunal and environmental diversity. Simpson (1964) points out that diversity gradients indicated by the method of gross geographic lumping have two components, one due to the number of habitats sampled by a given quadrate (and thus to the topographic relief) and another component due to ecological changes of some kind. Low latitude regions have more kinds of habitats, (i.e., Costa Rica has a whole range of habitats from low altitude tropical to middle altitude temperate to high altitude boreal habitats; whereas regions of higher latitude progressively lose some of these habitats) and therefore the presence of more species there is neither surprising nor theoretically very interesting. The question of basic ecological interest is that of the second component of diversity—namely, what are the factors that allow ecological co-existence of more species at low latitudes? Ecological data relating to species diversity gradients are scant, and no one has yet attempted the logical step of merging the synecological with an autecological approach.

Despite the handicap of insufficient ecological data, or perhaps because of it, theorization and speculation as to the possible causes of diversity gradients has been frequent and varied (Connell and Orias, 1964; Darlington, 1957, 1959; Dobzhansky, 1950; Dunbar, 1960; Fischer, 1960; Hutchinson, 1959; Klopfer, 1959, 1962; Klopfer and MacArthur, 1960, 1961; MacArthur, 1964, and in press; Paine, in press; and C. B. Williams, 1964). These efforts have produced six more or less distinct hypotheses: (a) the time theory, (b) the theory of spatial heterogeneity, (c) the competition hypothesis, (d) the predation hypothesis, (e) the theory of climatic stability, and (f) the productivity hypothesis. It is instructive to consider each of these hypotheses separately, attempting to suggest possible tests and observations for each, even though only one pair represent mutually exclusive alternatives, and thus several of the proposed mechanisms of control of diversity could be operating simultaneously in a given situation.

In the following discussion, ecological and evolutionary saturation are defined as the ecological and evolutionary upper limits to the number of species supported by a given habitat. The assumption of ecological saturation is implicit in ecological studies of species diversity gradients, an assumption without which the study of such gradients must be made in terms of the history of the area. There is reasonable evidence that the majority of habitats are ecologically saturated (Elton, 1958; MacArthur, in press).

The time theory

Proposed chiefly by zoogeographers and paleontologists, the theory of the "history of geological disturbances" assumes that all communities tend to diversify in time, and that older communities therefore have more species than younger ones. (The evidence behind this assumption is scanty, and the assumption may or may not be valid.) Temperate regions are considered to be impoverished due to recent glaciations and other disturbances (Fischer, 1960). It is useful to distinguish between ecological and evolutionary processes as subcategories of the theory. Ecological processes would be applicable to those circumstances where a species exists which can fill a particular position in the environment; but this species has not yet had time enough to ,disperse into the relatively newly opened habitat space. Evolutionary processes apply to longer time spans, to those cases where a newly opened habitat is not yet utilized, but will be occupied given time enough for speciation and the evolution of an appropriate organism.

Tests of the ecological and evolutionary time theories are by necessity indirect, but several authors have suggested possibilities for assessing the importance of evolutionary time as a control of species diversity. Simpson (1964) has argued that the warm temperate regions have had a long undisturbed history (from the Eocene to the present), long enough to become both ecologically and evolutionarily saturated, and that since there are fewer species in this zone than there are in the tropics, other factors must be invoked to explain the difference between tropical and temperate diversities. Beyond this, he reasons that if the time theory were correct, the steepest gradient in species diversities should occur in the recently glaciated temperate zone. Since, for North American mammals, at least, this zone shows a fairly flat diversity profile, there is some evidence against the evolutionary time theory. Simpson (1964) emphasizes that temperate zones have probably been in existence as long as have tropical ones. Newell (1962) stresses that temperate areas at intermediate latitudes were probably not eliminated during the glacial periods, but were simply shifted laterally along with their floras and faunas, and that, if this is the case, they have had as long a time to adapt as have the non-glaciated areas. R. H. MacArthur (personal communication) has suggested that possibilities exist for a test of the effects of glaciation, by comparisons of areas in the glaciated north temperate with their non-glaciated southern temperate counterparts. However, the evolutionary time theory is not readily amenable to conclusive tests, and will probably remain more or less unevaluated for some time.

The evidence relating to the ecological time theory, summarized in detail by Elton (1958), and discussed by Deevey (1949), indicates that most continental habitats are ecologically saturated. Only in those cases where barriers to dispersal are pronounced can the ecological time theory be of importance in determining species diversity. Islands have sometimes been considered cases of historical accident in which maximal utilization of the biotope is often achieved by pronounced behavioral modifications of those species which have managed to inhabit them (Crowell, 1961, 1962; Lack, 1947). More recently, several theories for equilibrium in insular zoo-geography have been proposed, and data given which shows strong dependence of species composition on island size, distance from "source" areas, and time available for colonization (Hamilton, Barth, and Rubinoff, 1964; MacArthur and Wilson, 1963; Preston, 1962). These papers indicate that predictable patterns of species diversity occur even on islands, and further lessen the probable importance of the ecological time theory.

The theory of spatial heterogeneity

Proponents of this hypothesis claim that there might be a general increase in environmental complexity as one proceeds towards the tropics. The more heterogeneous and complex the physical environment becomes, the more complex and diverse the plant and animal communities supported by that environment. Again it is useful to distinguish two subcategories of this theory, one on a macro-, the other on a micro- scale. The first is the factor Simpson (1964) calls topographic relief, discussed in more detail by Miller (1958). The factor of topographic relief is especially interesting in the study of speciation, and has been much discussed in books and symposia on that subject (Blair, 1961; Mayr, 1942, 1957, 1963). The component of total diversity due to topographic relief has been mentioned earlier in this paper.

In contrast to topographic relief, micro-spatial heterogeneity is on a local scale, with the size of the environmental elements corresponding roughly to the size of the organisms populating the region. Elements of the environmental complex in this class might be soil particle size, rocks and boulders, karst topography, or if one is considering the animals in a habitat, the pattern and complexity of the vegetation. Environmental heterogeneity of the micro-spatial type has been little studied by zoologists, and is more interesting to the ecologist than to the student of speciation (considerations of sympatric speciation processes will, however, involve micro-spatial attributes of the environment). It should be noted here that in the only study which relates species diversity of a taxon to environmental diversity in a quantitative way, the environmental diversity is of the micro-spatial type (MacArthur and MacArthur, 1961; MacArthur, 1964). These authors demonstrate that foliage height diversity is a good predictor of bird species diversity, and that knowledge of plant species diversity does not improve the estimate. Further tests of the theory of environmental complexity will probably follow similar lines, although it would be useful to consider alternative ways of examining this hypothesis.

Spatial heterogeneity has several shortcomings when applied to the explanation of global diversity patterns. The component of total diversity due to topographic relief and number of habitats (macro-spatial heterogeneity) certainly increases towards the tropics, but does not offer an explanation for diversity gradients within a given habitat-type. Vegetative spatial heterogeneity is clearly dependent on other factors and explanation of animal species diversity in terms of vegetative complexity at best puts the question of the control of diversity back to the control of vegetative diversity. Since there is no reason to suppose that micro-spatial heterogeneity of the physical environment changes with latitude, the theory of micro-spatial heterogeneity seems to explain only local diversity. Ultimately, resolution into independent variables will require consideration of non-biotic factors such as climate which change more or less continuously from pole to pole (see section on the climatic stability hypothesis).

The competition hypothesis

Advocated by Dobzhansky (1950) and C. B. Williams (1964), this idea is that natural selection in the temperate zones is controlled mainly by the exigencies of the physical environment, whereas biological competition becomes a more important component of evolution in the tropics. Because of this there is' greater restriction to food types and habitat requirements in the tropics, and more species can co-exist in the unit habitat space. Competition for resources is *keener* and niches 'smaller' in more diverse communities. Dobzhansky emphasizes that natural selection takes a different course in the tropics, because catastrophic indiscriminant mortality factors (density-independent), such as drought and cold, seldom occur there. He notes that catastrophic mortality usually causes selection for increased fecundity and/or accelerated development and reproduction, rather than selection for competitive ability and interactions with other species. Dobzhansky predicts that tropical species will be more highly evolved and possess finer adaptations than will temperate species, due to their more directed mortality and the increased importance of competitive interactions. No statement has been given as to exactly why competition might be more important in the tropics, but the hypothesis is testable in its present form.

Because the predation hypothesis predicts very nearly the opposite mechanisms of control of diversity than does the competition hypothesis, I will briefly outline the predation hypothesis before proceeding with discussion. These two hypotheses are almost mutually exclusive alternatives, and the same tests, by and large, apply to both.

The predation hypothesis

It has been claimed that there are more predators (and/or parasites) in the tropics, and that these hold down individual prey populations enough to lower the level of competition between and among them (Paine, in press). This lowered level of competition then allows the addition and co-existence of new intermediate prey types, which in turn support new predators in the system, etc. The mechanism can apply to both evolutionary and dispersal additions of new species into the community. Paine (1966) argues that the upper limits on the process are set by productivity factors, which will here be considered separately.

According to this hypothesis, competition among prey organisms is *less* intense in the tropics than in temperate areas. Thus, a test between these two hypotheses is possible, provided that the intensity of competition can be measured. Several approaches to the quantification of competition might find application here (Connell, 1961a, 1961b; Elton, 1946; Kohn, 1959; MacArthur, 1958; Moreau, 1948). Also, if the predation hypothesis holds, community structure should shift along a diversity gradient, with an increase in the proportion of predatory species as the communities become more diverse. Evidence for such a shift in trophic structure along a diversity gradient is given by Grice and Hart (1962). These authors present data showing that the proportion of predatory species in the marine zooplankton increases along a latitudinal diversity gradient. A similar shift in community structure accompanies a terrestrial species diversity gradient in the deserts of western North America (Pianka, in preparation). Fryer (1959, 1965) has argued that predation enhances migration and speciation, thereby resulting in increased species diversity, in some African lake fishes. As will be pointed out in a later section, demonstration that species have either finer or more overlapping habitat requirements in the tropics could be used to support three of the six hypotheses, and is therefore not a powerful distinguishing tool.

The theory of climatic stability

According to this hypothesis, restated by Klopfer (1959), regions with stable climates allow the evolution of finer specializations and adaptations than do areas with more erratic climatic regimes, because of the relative constancy of resources. This also results in "smaller niches" and more species occupying the unit habitat space. Another way of stating this principle in terms of the organism, rather than the environment, is that, in order to persist and successfully exploit an environment, a species must have behavioral flexibility which is roughly inversely proportional to the predictability of the environment (J. Verner, personal communication). In recent years the theory of climatic stability has become a favorite for explaining the generality of latitudinal gradients in species diversity, but has as yet remained untested. Rainfall and temperature can be shown to vary less in the tropics than in temperate zones, but rigorous correlation with faunal diversity, let alone demonstration of causal connection, has not yet been possible. It should be realized that climatic factors could well determine directly floral and/or vegetative complexity, while being only indirectly related to the faunal diversity of the area.

Evidence that tropical species have more restricted habitat requirements than temperate species would support the competition hypothesis, the predation hypothesis, and the theory of climatic stability. Klopfer and Mac-

Arthur (1960) have attempted to test the hypothesis that "niches" are "smaller" in the tropics by comparing the proportion of passerine birds to non-passerines along a latitudinal gradient. Their thesis is that the non-passerines, possessing a more stereotyped behavior, are better adapted to exploit the more constant tropical environment than are the passerines, whose more plastic behavior allows them to inhabit less predictable habitats. Klopfer (in press, in preparation) has compared the degree of behavioral stereotypy in temperate and tropical birds, and tentatively concludes that "while tropical species are in fact 'stereotyped,' this is more likely an effect rather than a cause of their greater diversity."

An interesting variation on this theme is that of increased "niche overlap" in more diverse communities (Klopfer and MacArthur, 1961). Klopfer and MacArthur attempted to test this idea by comparing the ratios of bill lengths in congeners among several sympatric bird species in Panama and Costa Rica. Simpson (1964) notes that this ratio may come as close to 1.00 in temperate birds as it did in Klopfer and MacArthur's tropical species, but fails to realize that morphological character displacement is expected only in species occupying the same space (Brown and Wilson, 1956; Hutchinson, 1959; Klopfer and MacArthur, 1961; MacArthur, in press). Ratios of culmen lengths may often approach unity in species such as *Dendroica* which clearly divide up the biotope space (MacArthur, 1958), and thus demonstrate behavioral, rather than morphological character displacement. However, it is apparent that even if the comparison were valid, the test would not distinguish between "smaller niches" and increased "niche overlap" (C. C. Smith, personal communication). It is difficult to devise tests which will distinguish between these alternatives, and perhaps none can be suggested until "niche" has been operationally defined. Use of some of the dimensions of Hutchinson's (1957) multidimensional niche may allow partial testing between these alternatives, as in the work of Kohn (1959) on *Conus* in Hawaii.

Increased overlap in selected dimensions of the niche can imply either increased, decreased, or constant competition; the first if the overlapping resources are in short supply, the second if the overlapping resources are so abundant that sharing of them is only slightly detrimental to each species, and the third if independent environmental factors (such as more predictable production) allow increased sharing of the same amount of resource. Hence, because data supporting the niche overlap idea has ambiguous competitive interpretations, it does not distinguish between three (or four) hypotheses either.

Tests distinguishing between the theory of climatic stability and the competition hypothesis are especially difficult to devise, as there is considerable overlap between the two, and indeed, they are usually mixed when either is suggested. This similarity makes it all the more important to evaluate the importance of each, and in keeping with the rest of this paper they will be considered separately and at least two possible distinguishing tests suggested.

According to the theory of climatic stability, a unit of habitat will support the same number of individuals in the tropics and temperate regions, but since each of the species may be rarer (without becoming extinct) in the tropics, there can be more of them. The competition hypothesis implies that more individuals occupy the same habitat space, or else competition would not be increased. Considering a fixed areal dimension as a unit of habitat, abundance data from the tropics generally suggest that the number of individuals is relatively similar from temperate to tropics and therefore support the theory of climatic stability (Klopfer and MacArthur, 1960; Skutch, 1954). Another way in which these two theories might be separated is by examining the intensity of competition occurring along an increasing diversity gradient; if the level of competition remained constant, or decreased along the gradient, the prediction of an increased proportion of predatory species could be used to separate the predation hypothesis from the theory of climatic stability.

The productivity hypothesis

The most recent and most complete statement of this hypothesis is that of Connell and Orias (1964). They blend this hypothesis with the theory of climatic stability, distinguishing between the energetic cost of maintenance and the energy left for growth and reproduction. Their synthesis also includes aspects of the theory of spatial heterogeneity, and reasonably explains latitudinal trends in diversity, but the productivity hypothesis will be considered here in its "pure" form.

The productivity hypothesis states that greater production results in greater diversity, everything else being equal. Since it is patently impossible to hold everything else equal, the hypothesis can only be tested in crude or indirect ways. Experimental manipulation of nutrient levels in freshwater lakes, for instance, might provide a possible test. Such enrichments have often been made, both intentionally and accidentally (sewage), and quantitative data have been taken on the response of the biota. The data needed for calculating diversities probably exist, and it would be interesting to see such calculations performed. Qualitative indications are that enrichment usually causes an impoverished fauna (Patrick, 1949; L. G. Williams, 1964).

If productivity were of overwhelming importance in the regulation of species diversity, one would expect a correlation despite uncontrolled extraneous variables. Only one such correlation is known to me (Patten, Mulford, and Warinner, 1963), and in fact there may often be an inverse relation between species diversity and abundance or standing crop (which should usually be positively correlated with production) (Hohn, 1961; Hulburt, 1963; Yount, 1956; L. G. Williams, 1964; and my own observations on desert lizards). Those who would claim that the above studies are on non-equilibrium populations and thus not applicable to the problem at hand, would do well to search for data from "equilibrium" conditions which are relevant to the productivity hypothesis.

A common modification of the productivity hypothesis which has been claimed to be of importance in regulating species diversity is the notion of increased temporal heterogeneity in the tropics. The main argument is that the longer season of tropical regions allows the component species to partition the environment temporally as well as spatially, thereby permitting the coexistence of more species (MacArthur, in press). This notion has been rephrased by Paine (in press) who argues that the "stability of primary production" is a major determinant of the species diversity of a community. Paine integrates the predation hypothesis with this idea to form a sort of synergistic system controlling diversity. This hypothesis is also a blend of the stability and productivity theories, but in this case, the mixing suggests new observations, and a new mechanism of control of diversity than does either hypothesis alone. The mechanism for the regulation of species diversity by stability of primary production may be similar to the mechanism suggested for climatic stability, except that in this case, plants may buffer climatic variability by utilizing their own homeostatic adaptations and storage capacities to increase the stability of primary production.

These notions can be tested by analyses, such as that of MacArthur (1964), of the length of breeding seasons, but there are other ways of examining them as well. Thus, comparisons of the division of the day (or night) and season into discrete activity periods by different animal species might elucidate latitudinal trends. Another possible angle of approach is by means of the "stability of primary production," which can be measured directly and examined for latitudinal trends. Unfortunately, there are all too few reliable measures of primary production, let alone the variability in this quantity, and at this point it is difficult to assess the stability of primary production along a latitudinal gradient. The necessary data are simple enough in theory, but in practice a single determination of primary productivity is tedious (especially in terrestrial habitats). An indirect possibility for testing the hypothesis exists, however, for arid regions, where primary productivity is strongly positively correlated with precipitation (Pearson, 1965; Walter, 1939, 1955, 1962). In this environment, the amount and variability of precipitation can be used to estimate the amount and variability of primary production. Preliminary analysis of weather and lizard data for the deserts of western North America shows no correlation with either the average amount or the variability of rainfall and the number of lizard species (Pianka, in preparation).

Since clutch size is closely related to these ideas of increased temporal heterogeneity and stability of primary production, it may be profitably considered here. The fact of reduced clutch size in tropical birds (Skutch, 1954) has been discussed as a possible factor allowing the coexistence of more species in the tropics (MacArthur, in press). MacArthur argues that by lowering its clutch size, a species reduces its total energy requirements and is therefore able to survive in less productive areas which were formerly marginal habitats. He reasons that such reductions in total energy requirements will also allow the existence of more species, when the total amount of energy available is held constant. Apart from the problems of population replacement raised by these theoretical arguments, there are other reasons for doubting the importance of reduced clutch size as a determinant of increased tropical diversity. For instance, it is highly possible that tropical habitats never achieve food densities as high as those usual further north, because of their greater species diversity and the fact that most of the breeding birds do not migrate. In contrast, great blooms of production characterize the temperate regions, and most of the breeding birds are migrants. Thus it may be energetically impossible for tropical birds to raise as many young as can be supported in the more productive northern areas (that is, more productive on a short term basis, during the short growing season) (Orians, personal communication). Support of this notion comes from the large territory sizes of many tropical bird species (Skutch, 1954), which suggests that food may be scarce. If this is indeed the case, the smaller clutches of tropical birds would be a result, rather than a cause of, the greater diversity in the tropics. These criticisms of the reduced clutch size hypothesis are, however, in themselves largely theoretical, and it will be worthwhile to examine clutch sizes of other taxa along various diversity gradients. Clutch sizes of desert lizards vary latitudinally, but whether or not the largest clutches are from the south depends on the species concerned (Pianka, in preparation).

CONCLUSIONS

Obviously, there is room for considerable overlap between these different hypotheses, and several may be acting in concert or in series in any particular situation. Because of the preliminary state of knowledge on the subject of species diversity, for the sake of clarity, and in order to suggest tests of the various hypotheses, it is useful first to consider and assess each of the components of control of diversity in isolation, before attempting various mixtures. Once the relative importance of each factor has been assessed for many different diversity gradients, an attempt may be made to merge them. In general, the compounding of hypotheses is to be avoided, unless such blending suggests new tests not applicable to the isolated theories. As more and more parameters are included, the more complex hypothesis tends to "answer" all cases and becomes less and less testable and useful.

A fact often overlooked is that most of the hypotheses can be either supported or rejected by appropriate observations on a limited scale; any species diversity gradient might be a suitable study system. If the broader geographical gradients are found to be qualitatively different from local diversity patterns, this in itself would be interesting, and understanding the difference would ultimately require thorough knowledge of the control of local species diversity.

Finally, since ecologists can seldom structure their experiments except by their choice of observations and measurements, the natural system usually sets the bounds within which they must work. The basic technique of descriptive science is correlation, and it is well to keep in mind that correlation does not necessarily mean causation. This is especially true in the study of latitudinal gradients in species diversity, where many different factors vary along the gradient in a fashion similar to the taxon studied, and spurious correlations may be frequent. For these reasons all significant correlations must be carefully examined and attempts made to understand the mechanisms and causal connections (if any) between variates. Unambiguous demonstration of causality can only be attained by experimental manipulation of the independent variables in the system.

SUMMARY

The six major hypotheses of the control of species diversity are restated, examined, and some possible tests suggested. Although several of these mechanisms could be operating simultaneously, it is instructive to consider them separately, as this can serve to clarify our thinking, as well as assist in the choice of the best test situations for future examination.

ACKNOWLEDGMENTS

Many fruitful discussions preceded this effort, particularly those with H. S. Horn, G. H. Orians, R. T. Paine, and C. C. Smith. Drs. A. J. Kohn, R. H. MacArthur, G. H. Orians, and R. T. Paine have read the manuscript and made many valuable suggestions. I wish to acknowledge J. F. Waters for translating Terent'ev's paper. The work was supported by the National Institutes of Health, predoctoral fellowship number 5-F1-GM-16,447-01 to -03.

LITERATURE CITED

Blair, W. F. [Editor]. 1961. Vertebrate speciation. Univ. Texas Press, Austin, Texas.

Brown, W. L., Jr., and E. O. Wilson. 1956. Character displacement. Syst. Zool. 5: 49-64.

Connell, J. H. 1961a. Effects of competition, predation by *Thais lapillus*, and other factors on natural populations of the barnacle *Balanus balanoides*. Ecol. Monogr. 31: 61-106.

———. 1961b. The influence of interspecific competition and other factors on the distribution of the barnacle *Chthamalus stellatus*. Ecology 42(4): 710-723.

Connell, J. H., and E. Orias. 1964. The ecological regulation of species diversity. Amer. Natur. 98: 399-414.

Crowell, K. 1961. The effects of reduced competition in birds. Proc. Nat. Acad. Sci. 47: 240-243.

———. 1962. Reduced interspecific competition among the birds of Bermuda. Ecology 43: 75-88.

Darlington, P. J., Jr. 1957. Zoogeography; the geographical distribution of animals. John Wiley & Sons, Inc., New York and London.

———. 1959. Area, climate, and evolution. Evolution 13: 488-510.

Deevey, E. S., Jr. 1949. Biogeography of the Pleistocene. Bull. Geol. Soc. Amer. 60: 1315-1416.

Dobzhansky, T. 1950. Evolution in the tropics. Amer. Sci. 38: 209-221.

44   THE AMERICAN NATURALIST

Dunbar, M. J. 1960. The evolution of stability in marine environments. Natural selection at the level of the ecosystem. Amer. Natur. 94: 129-136.

Elton, C. S. 1946. Competition and the structure of ecological communities. J. Anim. Ecol. 15: 54-68.

———. 1958. The ecology of invasions by animals and plants. Meuthen, London.

Fischer, A. G. 1960. Latitudinal variation in organic diversity. Evolution 14: 64-81.

Fisher, R. A., A. S. Corbet, and C. B. Williams. 1943. The relation between the number of species and the number of individuals in a random sample of an animal population. J. Anim. Ecol. 12: 42-58.

Fryer, G. 1959. Some aspects of evolution in Lake Nyasa. Evolution 13: 440-451.

———. 1965. Predation and its effects on migration and speciation in African fishes: A comment. Proc. Zool. Soc. London 144: 301-322.

Gleason, H. A. 1922. On the relation between species and area. Ecology 3: 158-162.

———. 1925. Species and area. Ecology 6: 66-74.

Grice, G. D., and A. D. Hart. 1962. The abundance, seasonal occurrence and distribution of the epizooplankton between New York and Bermuda. Ecol. Monogr. 32: 287-309.

Hamilton, T. H., R. H. Barth, Jr., and I. Rubinoff. 1964. The environmental control of insular variation in bird species abundance. Proc. Nat. Acad. Sci. 52: 132-140.

Hohn, M. H. 1961. The relationship between species diversity and population density in diatom populations from Silver Springs, Florida. Trans. Amer. Microscop. Soc. 80: 140-165.

Hulburt, E. M. 1963. The diversity of phytoplanktonic populations in oceanic, coastal, and esturine regions. J. Marine Res. 21: 81-93.

Hutchinson, G. E. 1957. Concluding remarks. Cold Spring Harbor Symp. Quant. Biol. 22: 415-427.

———. 1959. Homage to Santa Rosalia, or why are there so many kinds of animals? Amer. Natur. 93: 145-159.

Klopfer, P. H. 1959. Environmental determinants of faunal diversity. Amer. Natur. 93: 337-342.

———. 1962. Behavioral aspects of ecology. Prentice-Hall, Englewood Cliffs, N. J.

Klopfer, P. H., and R. H. MacArthur. 1960. Niche size and faunal diversity. Amer. Natur. 94: 293-300.

———. 1961. On the causes of tropical species diversity: niche overlap. Amer. Natur. 95: 223-226.

Kohn, A. J. 1959. The ecology of *Conus* in Hawaii. Ecol. Monogr. 29: 47-90.

Lack, D. 1947. Darwin's finches. Cambridge Univ. Press, Cambridge, England. Reprinted 1961 by Harper and Brothers, New York.

MacArthur, R. H. 1955. Fluctuations of animal populations, and a measure of community stability. Ecology 36: 553-536.

———. 1958. Population ecology of some warblers of north-eastern coniferous forests. Ecology 39: 599-619.

————. 1964. Environmental factors affecting bird species diversity. Amer. Natur. 98: 387–398.

————. 1965. Patterns of species diversity. Biol. Rev. (In press).

MacArthur, R. H., and J. W. MacArthur. 1961. On bird species diversity. Ecology 42: 594–598.

MacArthur, R. H., and E. O. Wilson. 1963. An equilibrium theory of insular zoogeography. Evolution 17: 373–387.

Margalef, D. R. 1957. Information theory in ecology. Gen. Syst. 3: 37–71. Reprinted 1958.

————. 1958. Temporal succession and spatial heterogeneity in phytoplankton. In Perspectives in marine biology. A. Buzzati-Traverso [ed.], Univ. California Press, Berkeley.

Mayr, E. 1942. Systematics and the origin of species. Columbia Univ. Press, New York. Reprinted 1964 by Dover Publications, Inc., New York.

————. 1957. [Editor], The species problem. Amer. Ass. Advance. Sci., Publ. No. 50.

————. 1963. Animal species and evolution. The Belknap Press of Harvard Univ. Press, Cambridge, Mass.

Miller, A. H. 1958. Ecologic factors that accelerate formation of races and species in terrestrial vertebrates. Evolution 10: 262–277.

Moreau, R. E. 1948. Ecological isolation in a rich tropical avifauna. J. Anim. Ecol. 17: 113–126.

Newell, N. D. 1962. Paleontological gaps and geochronology. J. Paleontol. 36: 592–610.

Paine, R. T. 1963. Trophic relationships of eight sympatric predatory gastropods. Ecology 44: 63–73.

————. 1966. Food web complexity and species diversity. Amer. Natur. 100: 65–75.

Patrick, Ruth. 1949. A proposed biological measure of stream conditions, based on a survey of the Conestoga Basin, Lancaster County, Pennsylvania. Proc. Acad. of Natur. Sci. Philadelphia 101: 277–341.

Patten, B. C. 1962. Species diversity in net phytoplankton of Raritan Bay. J. Marine Res. 20: 57–75.

Patten, B. C., R. A. Mulford, and J. E. Warinner. 1963. An annual phytoplankton cycle in the lower Chesapeake Bay. Chesapeake Sci. 4: 1–20.

Pearson, L. C. 1965. Primary production in grazed and ungrazed desert communities of eastern Idaho. Ecology 46(3); 278–286.

Preston, F. W. 1962. The canonical distribution of commonness and rarity. Part I: Ecology 43: 185–215. Part II: Ecology 43: 410–431.

Schoener, T. W. 1965. The evolution of bill size differences among sympatric congeneric species of birds. Evolution 19: 189–213.

Shannon, C. E. 1948. The mathematical theory of communication. In C. E. Shannon and W. Weaver, The mathematical theory of communication. Univ. Illinois Press, Urbana.

Simpson, G. G. 1964. Species density of North American recent mammals. Syst. Zool. 13: 57–73.

46   THE AMERICAN NATURALIST

Skutch, A. F. 1954. Life histories of Central American birds. Vols. I and II. Cooper Ornithological Society Pacific Coast Avifauna Numbers 31 and 34.

Terent'ev, P. V. 1963. Opyt primeneniya analiza variansy k kachestvennomu bogatstvu fauny nazemnykh pozvonochnyk. Vestnik Leningradsk Univ. Ser. Biol. 18(21: 4); 19–26. English abstract in Biol. Abstr. 80822 (45).

Thorson, G. 1957. Bottom Communities (sublittoral or shallow shelf). In H. S. Ladd [ed.], Treatise on marine ecology and paleoecology. Geol. Soc. Amer. Mem. 67: 461–534.

Walter, H. 1939. Grasland, Savanne und Busch der arideren Teile Afikas in ihrer ökologischen Bedingtheit. Jahrbucher für wissenschaftliche Botanik 87: 750–860.

———. 1955. Le facteur eau dans les regiones arides et sa signification pour l'organisation de la vegetation dans les contrees sub-tropicales, p. 27–39. In Colloques Internationaux du Centre National de la Recherche Scientifique, Vol. 59; Les Divisions Ecologiques du Monde. Centre National de la Recherche Scientifique, Paris. 236 p.

———. 1962. Die Vegetation der Erde in ökologischer Betrachtung. Veb Gustav Fischer Verlag Jena. Jena, Germany.

Williams, C. B. 1964. Patterns in the balance of nature. Academic Press, New York and London.

Williams, L. G. 1964. Possible relationships between plankton-diatom species numbers and water-quality estimates. Ecology 45: 809–823.

Yount, J. L. 1956. Factors that control species numbers in Silver Springs, Florida. Limnol. Oceanogr. 1: 286–295.

From *Geographical Ecology: Patterns in the Distribution of Species*
Robert H. MacArthur

Patterns of Species
Diversity

7

There are more species of intertidal invertebrates on the coast of Washington than on the coast of New England, more species of birds breeding, and also more wintering, in forests than in fields, more species of diatoms in unpolluted streams than in polluted ones, more species of trees in eastern North America than in Europe, and more flies of the family Drosophilidae on Hawaii than anywhere else. There is an even more dramatic difference in the number of species in the tropics than in the temperate (the discussion of which is reserved for the next chapter). Will the explanation of these facts degenerate into a tedious set of case histories, or is there some common pattern running through them all?

A very brief review of explanations that naturalists have proposed will help put the discussion in perspective. One explanation that has been offered is (a) that there are more species where there are more opportunities for speciation and that the presence of many species simply reflects the head start that some areas have over others. Another is (b) that many species occur where fewer hazards have occurred, and that areas with few species have lost species through catastrophes of history. Others explain that (c) there are more species where competitors can safely be packed closely and that the numbers of both species and competitors will not increase with time. Climate, too, has been used to explain diversity; some claim that there are more species where (d) the climate is benign, others where (e) the climate is more stable. It has been suggested that (f) there are more species where the environment is complex and therefore more readily subdivided. Others suggest (g) that there are more species where the environment is more productive. Finally, the abundance of predators has been cited because (h) heavy predation puts a low ceiling on the abundances of separate species, thus allowing more species to fit in, and (i) predators sweep an area clean, leaving it ripe for recolonization by different species.

Some of these explanations are almost meaninglessly vague (what is a benign environment?), and nearly every explanation has

From Robert H. MacArthur, *Geographical Ecology: Patterns in the Distributions of Species* (New York: Harper & Row, 1972), chapters 7 and 8. Reprinted by permission of Princeton University Press.

170
---

been proposed with some particular organisms in mind; but most are plausible and could under some conditions alter the numbers of species. How can we make order out of such chaos? Here is where theory is useful; it can exhibit the roles played by each factor in combination with the others.

Theory of Species Diversity

We will give a very elementary theory relating diversity to the various ingredients by which it can be measured, and we will use the theory to reconcile the various proposals. In its most elementary form we use three ingredients. (1) We need a coordinate representing the range of resource subdivided among the species. For example, we picture height above the ground as this coordinate; we suppose food is available in the foliage from ground level up to the top canopy at 100 ft. Hence we would have a resource line 100 units long. If the canopy were only 40 ft above the ground, the resource line would be only 40 units long. In general, we call it R units long. In another situation foragers might subdivide food by size. In this case food size would be the resource coordinate. If the smallest food were 1 mm long and the largest were 75 mm long, then foods would vary from 1 to 75 mm in length, so $R = 75 - 1 = 74$ units. (2) We need a shorter line for the part of this resource coordinate used by a given species. If a bird feeding in the forest with the 100-ft-high canopy uses only the portion between 5 and 25 ft, we can represent that utilization by a bar (of length $U = 25 - 5 = 20$) between 5 and 25 on the resource axis. (3) We assume there are enough species in combination to utilize every height of resource and that adjacent ones overlap as represented in Fig. 7-1. We split the overlaps, as shown in the figure, so that the segments between the dashed vertical lines sum to the total resource R. The number of species N is, of course, the number of segments between vertical lines so $R = \sum_{i=1}^{N} H_i =$ $N \cdot \frac{1}{N} \sum_{i=1}^{N} H_i = N\bar{H}$ where $\bar{H}$ is the mean distance between vertical lines. Therefore, $N = \frac{R}{\bar{H}} = \frac{R}{\bar{U}} \cdot \frac{\bar{U}}{\bar{H}}$. Since the U's are composed of an H part and O parts, we can write $\sum_{i=1}^{N} U_i = \sum_{i=1}^{N} H_i + 2 \sum_{i=1}^{N-1} O_{i, i+1}$, and dividing by N,

Patterns of Species
Diversity

171

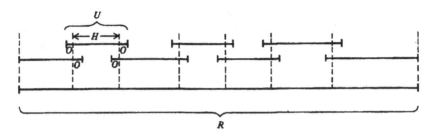

Fig. 7-1

The length R of a spectrum of resources, the utilization, U, per species, and the utilization overlaps. The overlaps are divided into two O's of equal length. See the text for definitions of H and O and for a discussion.

$U = H + 2\bar{O}$. (Notice, however, that $\bar{O}$ is not the ordinary average of the lengths of the overlap segments; it is the average, per species, of these lengths, and the end species only overlap with one species while the middle species overlap with 2. Hence, our $\bar{O}$ is slightly less than the strict average of the separate O's.) Substituting this into $N = \dfrac{R}{U}\dfrac{U}{H}$ we find

$N = \dfrac{R}{U} \cdot \dfrac{H + 2\bar{O}}{H} = \dfrac{R}{U}\left(1 + 2\dfrac{\bar{O}}{H}\right)$. This is just arithmetic and contains no biology; it simply relates the lengths of the overlapping U segments to the line R they fill and the necessary overlaps. It describes the obvious fact that you cannot increase the number N of U segments (i.e., of species) without increasing the relative overlap $\dfrac{\bar{O}}{H}$ or decreasing U or increasing R.

This same picture applies if the resource is subdivided in more than one dimension. If, for instance, both resource height and size are subdivided, R is a rectangle with length the range of resource heights, and width the range of resource sizes. Each species utilization is also a rectangle and we have enough overlapping rectangles to cover R. Now there are more neighboring species, 4 or more, depending upon the arrangement of rectangles, so we expect

$$N = \frac{R}{U}\left(1 + C\frac{\bar{O}}{H}\right) \tag{1}$$

where C is some number measuring numbers of neighbors.

172

This formula is necessarily correct; it does not assume the importance of competition or predation or history or anything else. It is crude in one way, however. Resources are not uniformly distributed at all heights, and the range, R, of heights is thus a crude measure of the diversity of available resources. Similarly, utilizations are not uniform but are concentrated on some resources more than on others, different species have different abundances, and the competition coefficient α is a more suitable measure of overlap than $\frac{\bar{O}}{\bar{H}}$. Alpha is defined precisely in the Chapter 2 appendix; roughly, it is the harm done by an individual to a competing species compared to the harm done by a member of the competing species to its own population. It is pleasant, and a little remarkable, to find that a strictly analogous formula holds when we substitute more practical measures of diversity of the production of resources and of utilization. As before (p. 113), we use the formula $\frac{1}{\sum_i p_i^2}$ as a measure of diversity. If the p_i are the proportions of the total resources in height interval i in the tree, then $\frac{1}{\sum_i p_i^2}$ is called the diversity of resources, D_R; if p_i is the proportion of a species' resource utilization in this height interval, i, then $\frac{1}{\sum_i p_i^2}$ is the diversity of utilization, D_U; if p_i is the proportion, of all of the individuals, that belongs to the ith species, then $\frac{1}{\sum_i p_i^2}$ is the species diversity, D_S.

Then it is true that

$$D_S = \frac{D_R}{D_U} \cdot \lambda \tag{2}$$

and λ (called the Rayleigh ratio of the α matrix) is not far from $1 + C\bar{\alpha}$ where $\bar{\alpha}$ is the mean competition coefficient and C again measures the number of neighbors and depends upon the number of subdivided dimensions. This is technically somewhat better than Eq. (1), but its intellectual content is the same; hence, we postpone its lengthy proof to the first appendix to the chapter.

What is the subdivided resource continuum? Is it really height above the ground or food size, or both, or neither? Only the naturalist familiar with a group of organisms is prepared to guess R or D_R. He can often test his guess by the tidiness of the results, as we shall see. Equation (1) or (2) is correct even if he uses the wrong measure of the resource continuum, for as we said before, they contain no biology. But the biologist who guesses the right resource coordinates for his species gets a special bonus: the R, or D_R, is then independent of the overlap $\frac{\bar{O}}{\bar{H}}$, or α; the overlap may then measure the "true" overlap experienced by the species. Suppose, for example, that birds are really subdividing food by both size and height in the tree, but that the biologist is guessing that only height in the tree is relevant. Accordingly, he observes a large amount of feeding height overlap, an amount which is actually greater than exists since, although many species are feeding at the same height, they are eating foods of different sizes and thus reducing their overlap. So far so good. Now suppose the biologist compares a second habitat of about the same foliage height but with a smaller range of food sizes. He then records the same R but finds fewer species and less overlap. Here he blames a change in the species overlap for the reduction in the numbers of species, while if he had guessed the entire appropriate food spectrum, he would have observed the same overlap but would have found reduced R in the second habitat. This would be a useful finding because then the overlap could be viewed as a biological property of the species and R as a measure of the environment; but these interpretations of overlap and R hold only if the biologist has guessed correctly.

In what follows, we use Eqs. (1) and (2) to interpret the various species diversity patterns.

History or Equilibrium:
   The Principle of
   "Equal Opportunity"

Some problems in numbers of species truly seem to be outcomes of a capricious history. That is, they are interpretable in historical terms and not in terms of the machinery controlling species diversity.

174

In short, our Eqs. (1) and (2), although applicable, are irrelevant in cases where history is paramount.

For example, the fossil records suggest that Europe and North America had roughly equal numbers of tree species prior to the Pleistocene glaciations. During the glaciations Europe lost a large fraction of tree species while North America lost virtually none, and the time since then has been insufficient for Europe to reacquire the species it formerly had. Hence, Europe is impoverished in trees for historical reasons. Presumably, the actual machinery involved was that the advancing glaciers forced the trees to move south ahead of them. In eastern North America where the mountains are in a north-south chain and form no barrier, the retreating tree species were always able to continue south to areas where climates were suitable. In Europe where the Alps and the Mediterranean lie east-west and therefore form barriers to southward movement, warm-loving tree species being pushed south by glaciers were squeezed in a vanishing warm zone between cold glaciers and cold mountains or cold glaciers and the sea. These trees found no refuge and went extinct. History, pure and simple, appears to be the only explanation for the difference between European and American tree diversities.

However, if we study the birds breeding in these very forests we get different results. If we compare a few acres of forest in Europe with similar acreage in North America, we find about 20 species in both habitats; and we get equivalent results if we contrast other habitats in Europe with similar habitats in North America. A small field in both places would have only a handful of bird species breeding, but a bushy field would have more, and a forest of many layers would have even more. Hence, in the comparison of similar habitats in North America and Europe, tree species diversity is explained by history; bird species diversity, which does not differ between habitats of similar structure but between habitats of differing structure, is explained by those habitat differences. Of course, birds are much more mobile than trees, and history has apparently left little trace on the comparison of numbers of bird species between Europe and North America.

Thus, for birds at least, we begin to look for explanations in terms of how many bird species a habitat will "hold." A forest, for some reason that we wish to investigate, will "hold" more bird species than a field. This is not to imply that both are "full" and cannot possibly

*Patterns of Species
Diversity*

———

175

receive another. Rather, it is more like a gas that equalizes its pressure
between two connected vessels, the larger of which automatically equili-
brates with more gas. This analogy can be made more precise with what may
be called the principle of "equal opportunity." We picture two habitats,
A and B, connected by a corridor permitting free migration back and
forth (Fig. 7-2). We also assume that the species can rapidly and easily
adapt to either habitat upon migration, but that changes are necessary,
and that a single species can seldom simultaneously occupy both. Finally,
suppose there is a large enough total number of species so that when they
are all together, they compete rather severely. Then the benefit of an
A → B migration might fall off as shown in Fig. 7-2, where p is the proportion of species in habitat B such that both habitats are equally good as far
as a potential migrant is concerned. When B has fraction p of the species,
both habitats offer equal opportunity for further colonization and the exchange stops. As further species enter the area, both habitats may acquire
new species but the balance is always maintained between the two, provided the entry of new species is slow compared to the shifting between
habitats.

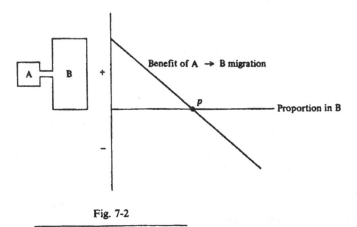

Fig. 7-2

———

The principle of equal opportunity. A and B are two different habitats of unequal size,
joined by a corridor as shown on the left. On the right is portrayed the benefit of A-to-B
migration as a function of the proportion of the species in habitat B. When this line
drops below zero, B → A migration is favored. Where the line crosses at proportion p,
the habitats have equal opportunity for further colonization.

176

In the case where the only differences between the habitats are in the lengths of the R lines, and possibly in the overlap between the species utilizations, we can be perfectly precise about what equal opportunity means. Then the species K values would be the same in either habitat, so the species will prefer the one with less competition. Equal opportunity will mean equal $\dfrac{\bar{O}}{\bar{H}}$ or equal α. Thus, two equally productive habitats that might be expected to have equal K's for species will be expected each to contain that number of species that will maintain an equal α in the two habitats. If one habitat temporarily has smaller α, there will be a net flow of species entering this habitat, restoring the equality of the α's. When different resources have different productions, then K's as well as α's will vary and it is not so easy to say just when the migrations will cease. However, there will still be a time, and a proportion p in habitat B, at which further migration will cease or balance because the habitats have equal opportunity. In such cases, nearby habitats should reveal a pattern of species diversity, a pattern describable in terms of Eqs. (1) and (2), or something similar.

There is another kind of historical pattern that we can avoid if we wish. Hawaii has more flies of the family Drosophilidae than anyplace else, perhaps because it has fewer flies of other kinds and fewer other ecologically similar insects. The clever entomologist by plotting the total number of fly-like insects might well find Hawaii equivalent to other tropical areas (or perhaps poorer because it is an island). To make this clearer we recall (from p. 174) that by plotting total breeding bird species, differences between habitats became regular and quite independent of history. If we had considered only flycatchers of the family Muscicapidae, the differences would have been purely historical since those flycatchers are confined to the Old World and appear to be replaced by the tyrant flycatchers of the family Tyrannidae in the New World. If we plotted all flycatching birds, the patterns might reappear; but it would probably be safer to plot all birds, or some other larger category.

At the other extreme people often suggest we should explain the diversity of all living things, not just of trees, or birds, or butterflies. But to suggest this is not only a little masochistic because the counting job would be virtually impossible—it also misses the point. We are looking for general patterns, which we can hope to explain. There are many of these if we confine our attention to birds or butterflies, but no one has

Patterns of Species
Diversity

———

177

ever claimed to find a diversity pattern in which birds plus butterflies made more sense than either one alone.

Hence, we use our naturalist's judgment to pick groups large enough for history to have played a minimal role but small enough so that patterns remain clear.

Habitats Differing Only in Structure

We have seen that if the biologist is clever, he measures R so that it reflects the environment and U so that it is a measure of the species' abilities determined as in Chapter 3 on economics, and overlap $\frac{\bar{O}}{\bar{H}}$ becomes a true measure of the species' interactions. It is very plausible then (recall the principle of equal opportunity) that nearby habitats would be occupied by species of the same U and same mean overlap $\frac{\bar{O}}{\bar{H}}$, and differ only in their R values. Comparing the numbers of species in such habitats should reveal that the number of species is proportional to the range of resources (Eq. (1)), or, using Eq. (2), that the diversity of species is proportional to the diversity of resources.

Cases of this sort have been explored in birds, first by R. and J. MacArthur (1961). These authors used a slightly different diversity measure, so we here (Fig. 7-3) replot some old plus some new data to show breeding bird species diversity, D_S (calculated in homogeneous census areas large enough to hold about 25 pairs), against "foliage height diversity," which seemed the best estimate of the real D_R. For calculating foliage height diversity, $\frac{1}{\sum_i p_i^2}$, p_1 was the proportion of the total foliage in the layer of herbaceous vegetation, p_2 the proportion of foliage in the layer of bushes and understory, and p_3 the proportion in the canopy. The general linear arrangement of the points suggests that R variations— variations in arrangement of foliage layers—are responsible for most of the variation in bird species diversity. This is not proof that foliage height is the only resource subdivided; there surely are others, such as food size,

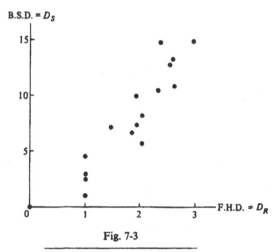

Fig. 7-3

Bird species diversity, B.S.D., is plotted against foliage height diversity, F.H.D. (the number of equally used layers of foliage). F.H.D. is a measure of the length of R.

but apparently nearby habitats differed in foliage height and not in food size spectrum, so the differences in numbers of bird species were controlled by the differences in foliage profile. The MacArthurs also showed that in eastern deciduous forests knowledge of the number of plant species was irrelevant to the prediction of the number of bird species. Thus, a pure red maple forest of a given foliage profile would have as many breeding bird species as a mixed forest with the same profile.

This result does not prove that bird species recognize their habitat by profile alone; many may actually require a particular tree species for food or nest site. Rather it is the *number* of species that depends upon the profile, and if one disappears because some tree species disappears, another replaces it.

In future studies some better measure of the range or diversity of resources will doubtless be devised, perhaps based on direct sampling of the food. Such studies may improve the clarity of the relation between the number of species and the range of resource.

Structure of the habitat may affect more than just R. Cody (1968) showed that in dense wet grassland of medium height, bird species are hard pressed to subdivide food. Different species end up selecting patches of slightly different density or wetness. Thus, Leconte's sparrows were in the wettest fields. But when he studied taller grass fields, the birds had used a new dimension. Not only do some species select denser or wetter areas, but some feed on the ground and others from the grass tops. There is now a vertical as well as a horizontal dimension of their subdivision. In terms of Eq. (1) or (2), the constant C has increased in the tall grass areas, and consequently more species could be present. Although they are less well worked out, there must be innumerable cases where added geometrical structure gives added dimensionality to the species subdivision and hence increases C.

Patrick (1968) compared the diatom floras developing independently in separate boxes filled with water from the same uniform source. Thus she tested for the role of accident in species diversity. In fact she found that in the absence of variation in R or production and under replicated conditions the diatom floras were very similar. Over 95% of the individuals were of species shared in common among the boxes, and other measures of diversity were very similar also. Hence, diversities are similar when conditions are similar.

Patrick (1963) also contrasted the numbers of species of diatoms in unpolluted rivers of different chemistry. Remarkably enough, provided the rivers were of complex structure the numbers of species (but not their names) were remarkably uniform. Relatively unstructured springs and other simple habitats had fewer species however.

Habitats Differing in
Climatic Stability

There are two ways in which the fluctuations of climate would affect Eqs. (1) and (2). First, a species adapting to a varying environment must have a large U, whereas in a constant environment U can be smaller. For example, in a wet tropical environment fruit is available throughout the year and many birds can, and do, specialize on it. In more seasonal places, fruits are only sometimes available and a resident species must enlarge its range of utilization, U, until it includes some items for every season. Since U appears in the denominator, increase in climatic fluctuations causing increasing U will decrease the number of species. For the same reasons competing species in a fluctuating environment tolerate

---
180
---

less overlap (Chapter 2), so that we may also have reduced $\dfrac{\bar{O}}{\bar{H}}$ in a fluctuating environment. For this reason, too, we expect fewer species, although this reason applies only to competitors.

　　The best account of the role of environmental stability is due to Sanders (1968, see also Sanders 1969, Slobodkin and Sanders, 1969). Since his technique is interesting in its own right we discuss it with some care. Most ecologists interested in diversity have wanted a number or a few numbers describing aspects of diversity (e.g., one number for number of species and another for degree of dominance of common species over rare; others have used the statistics of a log-normal distribution fitted to relative abundance data, etc.). They then use these numbers in regression analyses or in graphs to get numerical relations between diversity and measurable aspects of the environment. Sanders, however, describes diversity with a function, not just one or two numbers. In this respect, of course, he includes more information in his measure of diversity, but he does sacrifice the ability to plot diversity numerically on a graph and contents himself with qualitative comparisons.

　　Basically Sanders plots a curve of the expected number of species in random subsamples of different sizes. If there are, say, 1000 individuals in his total sample, he calculates (by a crude but effective approximation) the expected number of species in each size of subsample and plots these as in Fig. 7-4. Each curve represents a "species-individual" function calculated from a single sample of bivalves and polychaetes from marine benthic sediments. Specifically, suppose there are n_i ($i = 1, \ldots, N$) individuals in the ith of the N species and that $\sum_{i=1}^{N} n_i = M$, the total number of individuals. Sanders then plots the point (M, N) on the graph. To get the interpolated, "rarefaction" points connecting this to zero, Sanders proceeds as in the following example. To get the value of the curve over, say, 25 individuals, Sanders lets $S = \dfrac{25}{M}$ be the fraction of the total individuals in the subsample of 25. Any species i for which $Sn_i \geqslant 1$ he assumes is automatically in the subsample. Let $R = \sum n_j$, summed over all those j such that $Sn_j < 1$, be the number of individuals in species that are too rare to be guaranteed present in the subsample. Then Sanders wants a fraction S of these individuals from rare species in the subsample.

Patterns of Species
Diversity

181

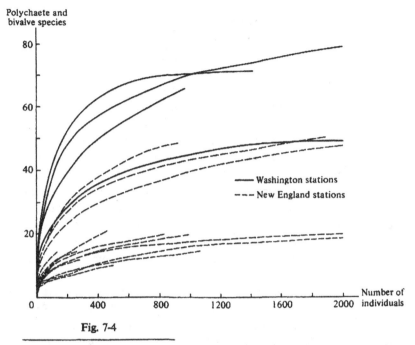

Fig. 7-4

Comparison of diversity values for the Friday Harbor, Washington, and southern New England series of stations. See the text for a discussion. (From Sanders, 1969.)

He assumes they are found one individual per species so he assumes the subsample has SR rare species. Hence, in total, the number of species in the subsample is SR plus the number of species with $Sn_i \geqslant 1$. This is a rough approximation to the expected number of species in a random sample of 25 individuals.

The trouble with treating the curve itself as diversity is that one can't in general call one curve "greater than" another curve the way one can call one real number greater than another. However, in most cases Sanders' curves don't cross. Then one curve is uniformly above another and it is reasonable to say that the former represents an area of greater diversity.

One final warning: Sanders' curves are not the same as the species-individual curves one would get by counting progressively larger real samples. Instead, they are roughly the species-individual curves one would get by stirring up all of the individuals in the whole area into a random homogeneous mixture and then counting from progressively increasing samples. In other words, any microenvironmental heterogeneity of the mud in Sanders' samples is overlooked in his measure, and one area with more heterogeneity would produce a higher diversity curve than a similar area with less heterogeneity. Of course, his technique of rarefaction can be applied, with appropriate modifications, to make a curve out of any diversity measure: Simply substitute "diversity" for "number of species" on the vertical ordinate axis of the graph. Figures 7-4 and 7-5 show some of Sanders' results. The first figure compares the diversities of stations near Friday Harbor, Washington, with its rich and constant conditions, with diversities on the more variable New England coast. With the exception of Friday Harbor curve 1, all the Friday Harbor curves are higher than all the New England ones, and we infer the diversities are greater where the ocean is more constant and productive. Curve 1, in fact, is the shallowest water station at Friday Harbor and hence the most variable. So far, Sanders' results could be either due to greater productivity or to reduced temporal fluctuations in Friday Harbor. But his next graph, Fig. 7-5, compares shallow water and deep water off the New England coast. Clearly the deep-water stations have greater diversity. Here the deep waters are less seasonal and less productive. The greater diversity can then only be due to the reduced seasonality or other temporal fluctuations in deep water. Sanders reports that the shallow-water stations have up to 23°C seasonal temperature change whereas waters as deep as 300 meters have only a 5°C seasonal change. Thus the reduced seasonality is the one factor he always finds associated with greater diversity. It seems inescapable that the reduced seasonality has caused an increased diversity. But in the absence of more numerical data we cannot rule out a subsidiary effect of productivity. To rule out productivity by his methods, Sanders would need to compare areas with identical seasonalities but very different productivities.

Sanders does not attribute all of diversity to climatic stability. He also says areas that subject their organisms to reduced "stress" have greater diversity. Unfortunately, though seasonality has some measurable objectivity, stress is harder to specify. Is reduced productivity a form of stress? If so, Sanders implies productivity affects diversity, but he seems to deny this. In any case his data are of extraordinary interest and we return to them briefly in Chapter 8.

Patterns of Species
Diversity

———

183

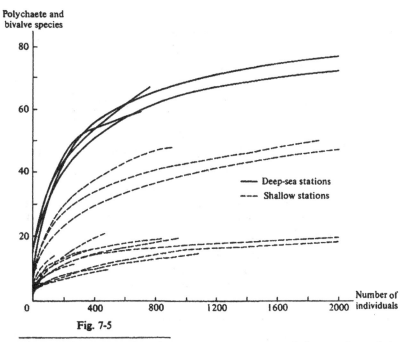

Fig. 7-5

Comparison of diversity values for the continental shelf and slope stations of the southern New England transect. (From Sanders, 1969.)

Habitats Differing
in Productivity

The knowledge that two habitats differ in their total productivity is not sufficient information to allow us to predict whether one will have more or fewer species. If every resource has uniformly increased production in one habitat, that habitat should have greater R and reduced U; the R will be greater because resources that were formerly too scarce to form an adequate diet and therefore not counted as part of R now allow a species to survive and are counted. U will be less because, as we proved in Chapter 3, species should have a more specialized diet where food is denser. Since food is often denser where production is greater, we conclude U will often be reduced. Both growth in R and reduction in U will cause an increased number of species, according to the

---
184

equations, so production clearly can affect the number of species. In extreme cases it obviously does: where there is zero production, there must be no species. Thus, we have theoretical reason to expect areas of uniformly increased production to support more species.

In practice this has never been adequately tested, but there is a general correlation that looks convincing. Where there is ocean upwelling on the west sides of continents, productivity is marvellously high and diversity is great. Tide pools in California have far more species than those in New England. Similarly, tropical wet forests and coral reefs are without doubt both productive and rich in species. The trouble is, however, that obviously more than just productivity changes between California tide pools and New England ones, or between tropics and temperate. For example, the more productive area is also more stable in time. The water in California tide pools may fluctuate less in temperature and salinity, so that its extra species can also be partially explained as in the section on habitats differing in climatic stability. It would, in fact, be misleading to try to explain the increase in terms of either stability or productivity alone. Without doubt they act hand in hand. However, for this reason the productivity effect has never been properly documented.

J. Brown of the University of Utah (pers. comm.) has discovered one case in which productivity may control species diversity. He finds that the species diversity of the heteromyid rodents (kangaroo rats and pocket mice) that occupy patches of desert in Nevada is greater where the rainfall is greater, even though there is no conspicuous difference in vegetation in these rainier patches. Brown conjectures that the heavier rainfall increases the seed production of the plants. The rodents divide resources partly by seed size and partly also by foraging position; some feed under bushes and others feed in the open. Hence it is plausible that the increased seed production would allow enough seeds to fall in the open to make that way of life feasible, whereas the open-feeding rodents would be unable to make a living where seed production is low.

On the other hand, there are many cases where an increase in total productivity actually reduces the number of species. Patrick, Hohn, and Wallace (1954) have compared the numbers and diversity of diatom species on the same river just above and just below where organic pollution was pouring in. The polluted waters were more productive in the sense that they supported more diatom individuals and more rapid photosynthesis, but there were fewer species and lower species diversity. In the polluted waters there were few, but exceedingly common, species; in unpolluted waters, many, but less common, species.

There are several ways of viewing such results. They make use of the fact that the complete spectrum of nutrients is not being enriched by the pollutants, so that some nutrients become, relative to others, very abundant. Diatoms consuming mainly these nutrient resources will become very abundant and will do proportionately more damage to their competing species, which consume mainly the unenriched nutrients. Thus the species will only coexist in the polluted environment if their overlaps, $\frac{\bar{O}}{\bar{H}}$, are relatively small. A greater overlap would have been tolerated in the unpolluted stream. Hence increased production, if concentrated on some resources, can reduce $\frac{\bar{O}}{\bar{H}}$ and thereby decrease species diversity.

Another, supplementary explanation for reduced species diversity in polluted areas is that if R is not reduced, at least D_R is, for D_R builds in the relative abundance of the resources and is interpreted as the number of equally dense resources. When one nutrient becomes excessively common, it reduces D_R, and for this reason reduces species diversity.

Habitats Differing in Structure,
    Stability, and Productivity

Most pairs of randomly selected habitats will differ in more than one respect; often they will differ in structure, environmental stability, and in productivity. Then all of the ingredients of Eqs. (1) and (2) may be altered, conceivably in ways that would cause opposing effects upon the numbers of species. In this case, to understand changes in the numbers of species there is no alternative but to measure R, U, C, and $\frac{\bar{O}}{\bar{H}}$ and combine them according to the equations. Since people have seldom even measured single components in these equations, it is no wonder no progress has been made in comparing habitats that differ in many ways simultaneously.

186

Matters of
Environmental Scale

A real environment has a hierarchical structure. That is to say, it is like a checkerboard of habitats, each square of which has, on closer examination, its own checkerboard structure of component subhabitats. And even the tiny squares of these component checkerboards are revealed as themselves checkerboards, and so on. All environments have this kind of complexity, but not all have equal amounts of it. In an orchard or creosote bush (*Larrea divaricata*) desert there is not much large scale variation; the big squares of our checkerboard are all identical, if their sides are, say, about 100 meters long. In contrast, 100-meter-squares laid on a field being recolonized by trees would be very different from each other. Some would contain almost solid patches of forest, others would still be grassland, and yet others would be filled with bushes and vines. Our problem is this: Over what size square should we count the number of species? Are the patterns of species diversity more orderly and more accessible if we choose small squares, medium squares, or very large ones? Three fairly obvious points help us to answer.

First, if our squares are too small, they will hold a very poor sample of individuals and species. A square in which we sample only three individuals cannot have more than three species. We must adjust the census technique or choose a square large enough so that the sample is big enough to reveal many species if many are present.

Second, if we choose very large squares, of the size, say, of 2000 miles or kilometers on a side, then we are involved in areas big enough for speciation. Two such squares would include most of eastern United States and most of western United States. The eastern square would have relatively little topographic diversity and hence little opportunity for the geographic isolation used in speciation. Even the Appalachian Mountains are essentially a continuous range, so that there are very few superspecies with different parts of the range containing semispecies. The west is very different. There are innumerable mountain barriers and mountain islands and hence there are innumerable superspecies and species groups. Figure 7-6 shows the numbers of North American chipmunk species of the genera *Tamias* and *Eutamias*, compiled from maps in Hall and Kelson (1959). Even in squares of the size shown on the map, topo-

*Patterns of Species
Diversity*

187

graphic diversity has an evident effect. The east has 1 or 2 species and the west has 14, but these western species are almost always in distinct habitats. There can be no doubt that history, the history of species formation, is revealed even in squares no bigger than those on the map.

Third, if competition is the cause of most patterns of species diversity, then the square that best revealed these patterns would be one small enough so that its species would be coexisting. Hence, we hope for patterns in relatively homogeneous habitats of a size just large enough to hold an adequate sample of species. In squares of this size we have reason to hope the traces of history will have been erased. This is why homogeneous habitats, even in the west, usually or almost always have at most a single chipmunk species.

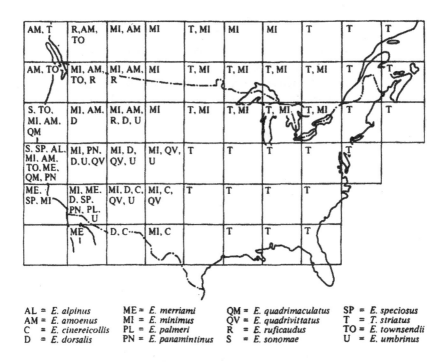

AL	= *E. alpinus*	ME	= *E. merriami*	QM	= *E. quadrimaculatus*	SP	= *E. speciosus*
AM	= *E. amoenus*	MI	= *E. minimus*	QV	= *E. quadrivittatus*	T	= *T. striatus*
C	= *E. cinereicollis*	PL	= *E. palmeri*	R	= *E. ruficaudus*	TO	= *E. townsendii*
D	= *E. dorsalis*	PN	= *E. panamintinus*	S	= *E. sonomae*	U	= *E. umbrinus*

Fig. 7-6

Ranges of chipmunk species showing that areas of the size of the squares have many more species in western United States than in eastern United States.

188

If we have large squares within which there are many species for historical reasons but small component squares with just the small number of species that can coexist, then we are sure that the component squares must have different species. In this case, the number of species grows rapidly with area sampled, from the number of coexisting species to the total in the large area. An area like the east with very little geographic diversity will have its number of species grow less fast with area sampled (Fig. 7-7).

It is hard to reverse this argument and infer from the growth in number of species with area how finely the species have subdivided their habitat. This is difficult because the topographic diversity is confounded with the species subdivision. The species-area curve will be steep if either the area is topographically diverse or the species have subdivided by very subtle criteria. Thus, if we wish to know how subtle the species habitat subdivision is, we must use other methods. Basically, we should compare the difference between the censuses with the difference between habitats. The steepness of this curve would reveal the subtlety of habitat subdivision alone, independent of topographic diversity. What remains is to work out how to measure species and habitat difference.

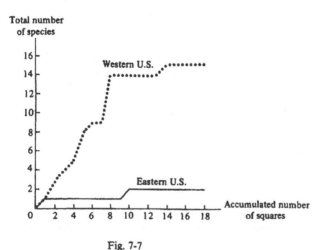

Fig. 7-7

Species-area curves for the chipmunks of Fig. 7-6 The curves were constructed by beginning at the bottom left in the map and working upward, recording new species as they appeared.

Patterns of Species
Diversity

$\overline{189}$

There are many suitable measures of the difference between the species of two habitats and of the difference in structure (Horn, 1966). If P and Q are two habitats being compared, and if p_i and q_i are the fractional abundances of the ith species in habitats P and Q, respectively, then $\dfrac{p_i + q_i}{2}$ is the fraction composed of the ith species in a mixture containing equal-sized samples of both habitats, P and Q. By our former reasoning (p. 113) $\dfrac{1}{\sum\limits_i \left(\dfrac{p_i + q_i}{2}\right)^2} = \dfrac{4}{\sum (p_i + q_i)^2}$ is the number of equally common species in the combined census of both habitats together. If we divide this by $\dfrac{1}{\sum\limits_i p_i^2}$, the number of equally common species in habitat P, we get the multiple of the number of species in habitat P that P plus Q contain. To be impartial, we also divide it by $\dfrac{1}{\sum\limits_i q_i^2}$ and average the resulting multiples. The result is the average multiple, M, of the number of species in separate habitats that the combined census contains:

$$M = \frac{1}{2}\left[\frac{\dfrac{1}{\sum\left(\dfrac{p_i+q_i}{2}\right)^2}}{\dfrac{1}{\sum p_i^2}} + \frac{\dfrac{1}{\sum\left(\dfrac{p_i+q_i}{2}\right)^2}}{\dfrac{1}{\sum q_i^2}}\right] = \frac{1}{2}\left[\frac{4\sum p_i^2 + 4\sum q_i^2}{\sum p_i^2 + 2\sum p_i q_i + \sum q_i^2}\right]$$

$$= \frac{2}{1 + \dfrac{2\sum p_i q_i}{\sum p_i^2 + \sum q_i^2}} = \frac{2}{1 + ov}$$

where ov is a well-known measure of the overlap between the communities. This may seem complicated, but it is really simple. If P has 4 equally common species and Q has 4 more, all different, then there are 8 species in all and $M = \frac{8}{4} = 2$. This is the largest value M can take; the combined census is only double the separate ones if there are no species in common. Suppose P has 4 and Q has 4, but now 2 are shared; then $M = \frac{6}{4} = 1.5$, indicating that the combined census has 1.5 times as many species as the

190

separate ones. Finally, if the 4 species in P and Q are all shared, then M takes on its minimum value $\frac{4}{4} = 1$. The elaborate formula for M is only different in that it takes into account the possibility that not all species are equally common. Suppose, for example, that P has three species of relative abundance $p_1 = 0.25$, $p_2 = 0.25$, $p_3 = 0.50$, and $p_4 = 0$ and that Q has two species of abundances $q_1 = 0$, $q_2 = 0$, $q_3 = 0.50$, and $q_4 = 0.50$; only species 3 is shared. Then $\sum p_i^2 = (0.25)^2 + (0.25)^2 + (0.50)^2 = 0.374$; $\sum q_i^2 = (0.5)^2 + (0.5)^2 = 0.5$; and $2 \sum p_i q_i = 2p_3 q_3 = 0.5$; so overlap, ov, equals $\dfrac{0.5}{0.874} = 0.57$. Hence, $M = \dfrac{2}{1.57} = 1.27$. In other words, the combined census has on the average 1.27 times as many "equally common species" as does P or Q.

MacArthur and Recher (MacArthur, 1965) compared pairs of bird censuses in Puerto Rico and also pairs of censuses in the United States. Each pair produces a point in Fig. 7-8, in which differences in bird census and differences in layering of the habitats are compared. (What they call *BS diff* (bird species difference) is essentially the logarithm to the base e of our M for differences in bird species, and what they call *FH diff* (foliage height difference) is essentially the $\log_e$ of our M value for layer densities in which p_1 = proportion of herbaceous foliage, p_2 is proportion of bushes, and p_3 is proportion of canopy over 25 ft. $\log_e 2 = 0.693$,

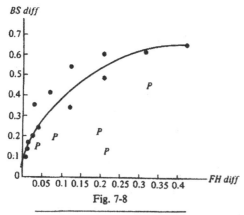

Fig. 7-8

Difference in bird species composition (*BS diff*) between census areas differing in foliage profile is plotted against foliage profile difference (*FH diff*), showing that in Puerto Rico (*P* points) the same difference in habitat causes much less change in bird species than on the U.S. mainland (● points). (From MacArthur, 1965.)

which is the maximum value on the scales of the figure and $\log_e 1 = 0$, which is the minimum.) Clearly the same difference in profiles causes a larger difference in bird species on mainland United States than in Puerto Rico. In other words, different habitat squares in Puerto Rico are more likely to contain the same species than those in the United States. M has been called the "between habitat diversity," and the preceding statement can be rephrased by saying, Puerto Rico has smaller bird "between habitat diversity" than the United States. We already know island species have expanded habitats. This is an exact documentation of the fact.

MacArthur, Recher, and Cody (1966) attempted to extend these results to the tropics with a new set of points of Panama comparisons. But these results were less satisfactory. Their censuses and their measures of habitat appear to have been inadequate.

Pianka (1969), comparing the rich lizard fauna of the Australian deserts to the poorer one of American deserts, concluded that Australian lizards have subdivided habitats more finely. One would infer that a lizard species-area curve would grow faster in Australia. But on top of this greater between-habitat effect, Pianka believes Australia also has a greater environmental heterogeneity—in other words, a larger R in Eq. (1).

The general conclusion of this section is that there are areas too small, and areas too large, to show clear diversity patterns, but that for the proper intermediate census area, the patterns are clear. Comparisons of pairs of these small census areas reveal the average multiple M, by which their combined census exceeds the separate ones. By using this, we can reconstruct how many species compound environments would have.

The Effect of Predators

If abundant predators prevent any species from becoming common, the entire picture changes. Resources are no longer of any concern and our Eqs. (1) and (2) are irrelevant. More correctly, resources are still a concern, but their manner of subdivision is irrelevant. J. Connell (pers. comm.) and Janzen (1970) have both shown that tropical tree seedlings seldom grow under their parent tree. This is no accident of seed fall;

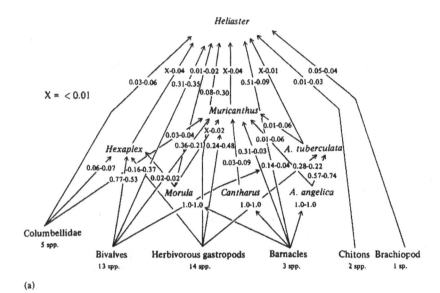

(a)

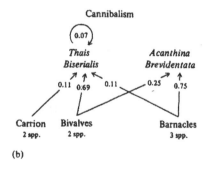

(b)

Fig. 7-9

(a) The complicated food web of a predator-rich community in the Gulf of California (genus A is *Acanthina*). Numbers on the left are fractions of numbers of food items in the predator's diet; numbers on the right are fractions of calories in the predator's diet. (b) A simpler food web in Costa Rica without a secondary carnivore. Numbers are fractions of numbers in the predator's diet. (From Paine, 1966.)

most must fall under the parent. But predators seem to gather where the seeds are commonest and there eat them. The result is that each kind of seedling grows more often under other species of canopy trees than under its own species. In itself, this does not allow a great tree species diversity, but a slight modification would. What we need is a ceiling on the abundance of each species. If predators gathered to collect the seeds only of common trees they would produce such a ceiling; any tree commoner than some ceiling abundance would be reduced by the predators. Alternately, if root parasites, fungi, or other diseases were acting as predators, nearby trees of the same species would be eliminated. Only isolated trees, far from sources of contagion, would be safe. In this way, too, an upper limit to the species' abundance would be set. Now, if we have ten tree species, each with maximum abundance of 1 per hectare, their canopies simply will not fill the space and there will be ground and sunlight for new trees to invade. They cannot be of the old species, so our community is vulnerable to invasion by new species. In this way there can be more species where predation is severe. This is result 3 on page 32.

      This result is not automatic: Where predation is severe it is not inevitable that there will be more species. An indiscriminate predator would not focus on the common prey and reduce them to below a ceiling; neither would an indiscriminate predator preserve trees isolated from contagion. In fact Patrick (1970) has described how predation by the snail *Physa heterostropha* decreases diatom species diversity by selectively leaving the common diatom *Cocconeis placentula*. What is required is most easily achieved by species-specific predators, such as diseases and parasites. A predator with a " search image " that switches to whichever prey is commonest could put a ceiling on abundance and increase the number of species, but it will only have this effect if the ceiling it puts on total abundance is below the ceiling that resources would set. These ceilings are usually interacting: the predator ceiling is lower where resources are scarce because consumers are ill fed and poorer at escaping, and the resource ceiling is lower where predation is severe because it is more dangerous to search for food. What is relevant to an increased number of species is an independent predator ceiling low enough so that the community is vulnerable to invasion.

      Paine (1966) has demonstrated another way that predation increases species diversity. When a starfish (*Pisaster*) feeds in an

_____
194

intertidal zone, it cleans a swath free from the mussels or barnacles that would otherwise outcompete the other species and dominate the community. In these swaths other species can maintain a sort of refuge. The intertidal community is then a mosaic of different succession stages. When Paine removed the predators, the number of other species fell from 15 to 8. Figure 7-9 shows how communities with more predators tend to be richer in species.

Predation certainly provides an alternative regulation of species diversity to that offered by competition. Probably the best evaluation is that both are sometimes important and that which one dominates depends upon the structure of the environment. Some environments are easy to search and are predator vulnerable; in these, predators doubtless exert a control over species diversity. In other areas prey species are hard to find and predators probably have little effect. Looking back over our results, the success of Eqs. (1) and (2) in environments differing in R and in stability is a suggestion that for these species groups, predation is not of importance.

210

this gap is due to insufficient observation; rather more people look for birds in Panama than in northern Saskatchewan! The third is the cardinal (*Cardinalis cardinalis*), which is missing from deserts and rivers of west Texas and eastern New Mexico although it occupies those habitats in Arizona and is found throughout eastern United States. It is probably no accident that the pyrruloxia (*Cardinalis sinuata*) is found primarily in the gap of the cardinal range although they are sometimes found together in Arizona. In summary, there seem to be very few range gaps found among American birds of the kind that seems quite common in the tropics, and we conclude tropical birds have patchier distributions.

Essentially four explanations have been proposed for these patchy distributions: (1) In history the gap areas contained no suitable habitats and only recently have such habitats existed. The species have not yet had time to recolonize. (2) Competition, usually diffuse competition, prevents the species from persisting in the gap areas. (3) The habitats in the gap areas are not really suitable; only our ignorance makes us think the species' habitats are continuous. (4) The patches are remnants of a once continuous distribution of a species going extinct. Explanation (3) is not really independent of the competition explanation, (2). The more competitors there are, the subtler the properties of the habitat where a species can persist. Hence (2) and (3) may be the same explanation. Even the historical explanation, (1), could interact with competition. In the absence of competitors recolonization may be much more rapid than in their presence. However, it would be hard to claim that the tropics have had a more disturbed last million years than the temperate zones that were invaded at least four times by continental glaciers. Hence the likely explanation of the greater tropical patchiness seems to be that competitors are packed more closely, causing a species to persist only where very subtle and particular conditions are met. See the chapter appendix for further discussion of tight packing of species and patchy geographic distributions.

Tropical Richness of Species

One of the most conspicuous facts about the tropics is their diversity of life. In many areas there are well over 100 tree species, many so rare that there may be only one specimen in 100 hectares. The numbers of insect species are not known; often the species have never been named, but naturalists are aware of a great tropical diversity. Schoener and Janzen (1968), with 2000 sweeps of a net, collected 545 species in a wet lowland tropical forest. Two thousand sweeps in a mixed Massachusetts forest yielded between 360 and 410 species for each month. Darlington showed the number of ant species at various south latitudes in South America (Table 8-1). The insects not only come in many species but also in many sizes and forms. The gradient in numbers of bird species is best illustrated by Fig. 8-8, which shows how the numbers of breeding land bird species grow toward the tropics. There is every reason to think that most other terrestrial taxa of large enough size show similar patterns. Thus there is a great tropical increase in amphibians, although salamanders as a subgroup of amphibians may be richer in the temperate. Marine faunas too seem to be richer in the tropics. Certainly coral reefs are famous for their richness in species and in their forms. Figure 8-9 shows the number of mollusc species at various stations along the Atlantic coast, following Fischer (1960). Figure 8-10 shows the pattern for planktonic Foraminifera following Stehli (1968). Certainly tropical richness in species is a very common phenomenon.

There are two exceptions to great tropical diversities that are intriguing. First, tropical mountaintops have fewer species than tropical lowlands. We saw a detailed example of this in Fig. 5-19, where we explained it as partially an island effect. But even in continuous mountain chains, the highlands are impoverished and tend to be occupied by a rather temperate-like fauna. Figure 8-11 from Kikkawa and Williams (1971) shows data for the whole of New Guinea. To give another instance,

Table 8-1
Decrease in Number of Species of
Ants Southward in South America

Area	Approximate S latitude	Number of species
São Paulo, Brazil	20°–25°	222
Misiones, Argentina	26°–28°	191
Tucumán, Argentina	26°–28°	139
Buenos Aires, Argentina	33°–39°	103
Patagonia as a whole	39°–52°	59
Patagonia, humid west	40°–52°	19
Tierra del Fuego	43°–55°	2

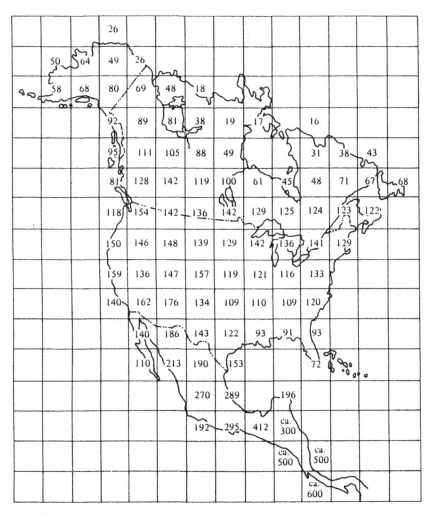

Fig. 8-8

Number of breeding land bird species in different parts of North America, from various sources. (From MacArthur, 1969, after MacArthur and Wilson.)

Comparisons of
Temperate and Tropics

213

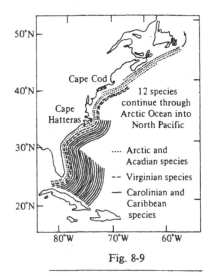

Fig. 8-9

Diversity gradient of gastropods along the eastern coast of the United States and Canada. Each line stands for ten species. (From Fischer, 1960, after Abbott.)

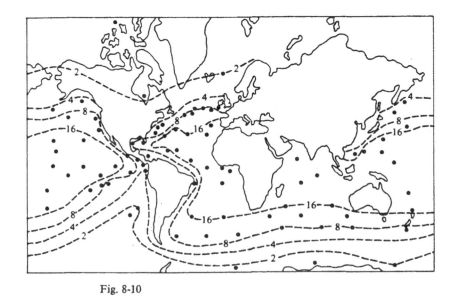

Fig. 8-10

Contoured raw diversity for Recent species of planktonic Foraminifera. (From Stehli, 1968.)

214

Central American mountains have lost, or nearly lost, many typical tropical lowland bird families such as antbirds (Formicariidae), puffbirds (Bucconidae), and manakins (Pipridae) and tend instead to be relatively rich in a few species of warblers (Parulidae), vireos (Vireonidae), finches (Fringillidae), and hummingbirds (Trochilidae), all of which are prominent in the temperate zones. But the mountains are not replicas of temperate zones! Although the mean annual temperature and the short height of the trees may be temperate, there is no winter. In fact seasonal temperate variation may be less on the tropical mountains than in the lowlands, due to a perpetual cloud of moisture. However, it does freeze on the mountains, and insects that dislike freezing weather have no summer, when they can count on no frosts, or winter, when the frosts are predictably concentrated. Productivity is doubtless lower and the range of resources shorter than in the lowlands. And even wet tropical mountains may have spells of dry weather when the wind makes everything crisp with lack of moisture. This gives a kind of seasonality to tropical mountains, but not an exact counterpart of the temperate seasonality. In this environment there are few species.

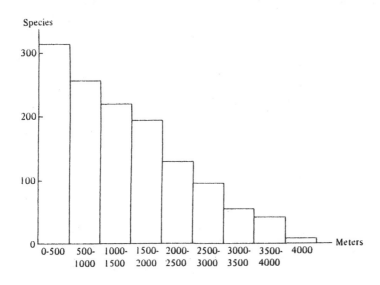

Fig. 8-11

Number of bird species in each altitudinal zone on New Guinea. (From Kikkawa and Williams, 1971.)

Comparisons of
Temperate and Tropics

215

The other chief exception to the rule of increased tropical diversity occurs in freshwater. Patrick (1966) compared the numbers of species of major groups of organisms found in hard waters of the upper Amazon basin (Rio Tulumayo and Quebrada de Puente Perez) with waters of comparable chemistry in temperate North America (Ottawa River and Potomac River). Table 8-2 from Patrick shows that the temperate waters are fully as rich as those of the Amazon basin. The tropical waters were not more productive than the temperate ones and were perhaps even more seasonal because the seasonal tropical rains caused periodic flooding of the rivers.

We saw in Chapter 7, on species diversity, that it is possible for one large area to have twice as many species as another without any of its component habitats having more species. The explanation would lie in a greater faunal difference between habitats in the richer area. But this is not the sole explanation of high tropical diversities as is shown in Fig. 8-12. The temperate data are typical of temperate forests nearly everywhere in the United States; the tropical data are probably moderately reliable for large areas although a few new species will be added with further observation. The small tropical areas are from Barro Colorado Island in Gatun Lake in the Canal Zone. Gatun Lake was made when the Panama Canal was constructed, so Barro Colorado Island is well under a century old. Even so, it seems to be poor in species compared to the adjacent mainland, and the 5-acre census may also be an underestimate. In any case, even small tropical areas of 5 acres are at least two

Table 8-2
Comparison of Species Numbers in
Temperate and Tropical Waters
(From Patrick, 1966)

	Rio Tulumayo	Quebrada de Puente Perez	Ottawa River 1955–1956	Potomac River
Algae	73[a]	62	69	103
Protozoa	33	40	47	68
Lower invertebrates[b]	5	6	8–15	27
Insects	104	78	51–64	104
Fish	26	22	17–28	28

The first two rivers are in Peru; the second two, in North America.
[a] These numbers represent established numbers of taxa and do not include taxa represented by less than six specimens.
[b] Excluding rotifers.

$\overline{216}$

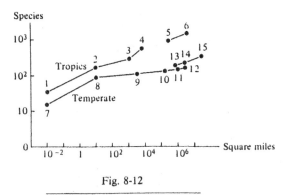

Fig. 8-12

Number of breeding land bird species plotted against the area in square miles for tropical and temperate areas. 1, 5-acre census on Barro Colorado Island; 2, Barro Colorado Island (6 sq miles); 3, Panama Canal Zone; 4, Republic of Panama; 5, Ecuador; 6, Colombia plus Ecuador plus Peru; 7, 5-acre Vermont census; 8, 6 sq miles in southern Vermont; 9, southern Vermont; 10, New England; 11, northeastern United States and adjacent Canada; 12, eastern United States and Canada; 13, Texas; 14, western United States; 15, North America (north of Mexico). The points for western United States indicate the effect of greater topographic diversity. (From various sources, after MacArthur, 1969.)

and one-half times as rich in bird species as temperate areas. These species live so close together that they are coexisting by some device. In the terminology of the last chapter, either the spectrum of resources, R, is greater in the tropics; the utilization per species, U, is less in the tropics; the overlap between species, $\dfrac{\overline{O}}{\overline{H}}$, is greater in the tropics; or the dimensionality of the environment, C, is greater in the tropics. Any of these can cause coexistence of the increased number of species that has been observed in the tropics. The embarrassment is that all are likely to be true, at least in some places! We have already seen that R is greater and probably also C. The reduced seasonality will allow a smaller U (p. 202) and larger overlap (Chapter 2, and appendix to this chapter); and the greater patchiness of tropical species distributions is evidence that overlap really is greater.

The data of Sanders (1968) are of considerable interest in this context. (His methods were discussed in detail on pages 180–182, to which the reader can turn for interpretation of the graphs.) Figure 8-13 shows Sanders' comparisons of tropical and temperate marine benthic

Comparisons of
Temperate and Tropics

217

polychaete and bivalve diversities. Sanders as before interprets these patterns as consequences of reduced tropical seasonality. The Bay of Bengal site was too deep to be affected by seasonal, monsoon freshening of the water and he inferred that it was very stable and hence diverse. This certainly seems a reasonable inference but as before we may question whether other factors also influence the diversity of his samples. Do we know whether there is an increased spectrum of resources in the tropical benthic sediments? What about productivity? and structural complexity? Lacking this knowledge we are not in a position to infer that the tropical climatic stability is the only factor, but it surely seems to be a primary one.

We have also seen that more intense predation can increase the diversity of prey species. Is predation more intense in the tropics? This is a complicated question. It appears that predation is more intense on nestling birds but perhaps not on adults. That it is very intense on nestling birds is the consensus of naturalists who have studied nesting

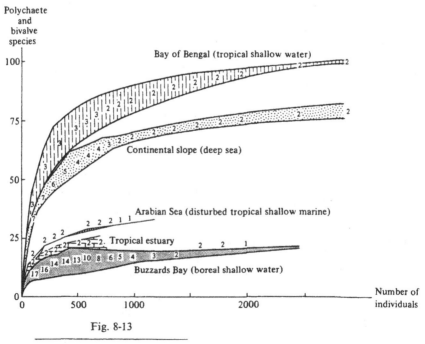

Fig. 8-13

Range for diversity values found for a number of regions in the temperate and the tropics. (From Sanders, 1968.)

218

success, or rather failure, in the tropics (e.g., Skutch, 1954, 1960, 1969). Data in which the same species is compared in the temperate and the tropics were kindly provided by G. Orians as follows:

	Costa Rica	Washington
Nests fledging at least one young	20	225
Nests fledging no young	73	274
Total nests	93	499

This doesn't quite give an estimate of mortality since the number of eggs per nest is less in the tropics, but predators usually ruin all of the eggs in a nest, so the nests from which no young fledged are usually those that were attacked by predators. We conclude that nest predation, at least for red-winged blackbirds (*Agelaius phoeniceus*), is more intense in the tropics. On the other hand adult mortality is far less. We can prove this for birds and can suggest why it is likely for many other organisms. Snow (1962) made an intensive study of the tropical black and white manakin (*Manacus manacus*) in Trinidad. He put colored aluminum bands on the legs of adult males and observed 89% of them the following year. Since some birds might have left the study area, 89% is a minimum estimate of survival. Karr (1971) banded many species over 2 years in Panama and recaptured similarly high percentages, again providing minimum estimates of survival. For the temperate we can provide a better estimate than that. The number of recaptures after 2 years divided by the number after 1 year is much closer to a true survival rate. Farner (1955) reviewed methods and results for temperate birds. For temperate passerine birds, such survival estimates averaged about 50%. These are far smaller than the tropical recapture rate. Hence there seems little doubt that adult birds in the tropics have much greater success in surviving than those in the temperate. In fact, the half-life of a bird with 0.9 survival is that number C such that $0.9^C = 0.5$. This $C = 6.6$, so half of the tropical birds are alive after 6.6 years. Half of the temperate birds are dead after somewhere between 1 and 2 years. These mortalities are not all due to predation, of course. Storms and the hazards of migration take a large toll in temperate regions. But the low adult mortality in the tropics makes it hard to claim predation is intense there, at least on adults. There seems to be no evidence that nest predators switch to whichever species is common and regulate populations in that way. This requirement for predator-controlled tropical species diversity in birds is therefore not met, and we tentatively discard this explanation. We only do so for birds, however; there is abundant evidence, as we mentioned, for trees and marine intertidal organisms. Janzen (1970) gives evidence that tropical tree species are kept sparse by herbivores who thus allow more species to fit in.

Finally, there is yet another explanation of tropical diversity that has been proposed by Fischer (1960) among others. According to this view the tropics had a head start in accumulating species either because more speciated there or fewer went extinct; given enough time the temperate species may increase until there are eventually as many in the temperate as the tropics have now. At first sight this explanation looks incompatible with the others; either the tropics are full or they are not. If they are full, history is irrelevant; if they are not full, competition is irrelevant. Now, for species of given morphology and physiology there may be such a thing as a full community (see chapter appendix 1), but "full" would apply only to small communities of coexisting species. The number of species in a large area of many habitats could easily keep increasing by increasing the faunal differences between habitats. It is significant that Fischer was describing the richness of whole faunas covering wide areas, so we have sufficient reason for history to have been important. Even if the component habitats are not full, however, history need not be the only relevant factor. We saw that a principle of equal opportunity for further colonization should prevail, producing a balance between the numbers of species in adjacent habitats. This balance is understandable in terms of competition even if, through history, both habitats are becoming enriched. The critical question, whether the number of species is increasing as the historical explanation would suggest, remains unsolved. Figure 8-14 from Simpson (1965) suggests the contrary, as does Brodkorb's (1971) estimate that the number of bird species has, if anything, declined since Miocene times.

Tropical Clutch Sizes

There is an interesting unexplained pattern in birth rates, or at least clutch sizes—the number of eggs per nest. Most small temperate

Works Cited

Brodkorb, P. 1971. Origin and evolution of birds. In D. S. Farner and J. R. King [ed.] *Avian Biology*. New York: Academic.

Cody, M. L. 1968. On the methods of resource division in grassland bird communities. *Amer. Natur.* 102: 107–147.

Farner, D. S. 1955. Birdbanding in the study of population dynamics. In A. Wolfson [ed.] *Recent Studies in Avian Biology*. Urbana: University of Illinois Press.

Fischer, A. 1960. Latitudinal variations in organic diversity. *Evolution* 14(1): 64–81.

Hall, E. R., and Kelson, K. R. 1959. The mammals of North America. New York: Ronald.

Horn, H. S. 1966. Measurement of overlap in comparative ecological studies. *Amer. Natur.* 100: 419–424.

Janzen, D. H. 1970. Herbivores and the number of tree species in tropical forests. *Amer. Natur.* 104: 501–529.

Karr, James R. 1971. Structure of avian communities in selected Panama and Illinois habitats. *Ecol. Monogr.* 41(3): 207–233.

Kikkawa, J., and W. T. Williams. 1971. Altitudinal distribution of land birds in New Guinea. *Search* 2: 64–69.

MacArthur, R. 1965. Patterns of species diversity. *Biol. Rev.* 40: 510–533.

———. 1969. Patterns of communities in the tropics. *Biol. J. Linnaean Soc.* 1: 19–30.

MacArthur, R., and J. MacArthur. 1961. On bird species diversity. *Ecology* 42: 594–598.

MacArthur, R., H. Recher, and M. Cody, 1966. On the relation between habitat selection and species diversity. *Amer. Natur.* 100: 319–332.

Paine, R. T. 1966. Food web complexity and species diversity. *Amer. Natur.* 100: 65–76.

Patrick, R. 1963. The structure of diatom communities under varying ecological conditions. *Ann. N.Y. Acad. Sci.* 108: 353–358.

———. 1966. *The Catherwood Foundation Peruvian Amazon Expedition*. Monographs of Academy of Natural Sciences of Philadelphia.

———. 1968. The structure of diatom communities in similar ecological conditions. *Amer. Natur.* 102: 173–184.

———. 1970. Benthic stream communities. *Amer. Sci.* 58: 546–549.

Patrick, R., M. Hohn, and J. Wallace. 1954. A new method of determining the pattern of the diatom flora. *Notulae Natura* 259. Academy of Natural Sciences of Philadelphia.

Pianka, E. 1969. Habitat specificity, speciation and species density in Australian desert lizards. *Ecology* 50: 498–502.

Sanders, H. L. 1968. Marine benthic diversity: A comparative study. *Amer. Natur.* 102: 243–282.

———. 1969. Benthic marine diversity and the stability time hypothesis. In *Diversity and Stability in Ecological Systems*. Brookhaven Symp. Biol. 22.

Schoener, T. W. and D. Janzen. 1968. Notes on environmental determinants of tropical versus temperate insect size patterns. *Amer. Natur.* 102: 207–224.

Simpson, G. G. 1965. The geography of evolution. Philadelphia: Chilton.

Skutch, A. F. 1954. *Life Histories of Central American Birds*, I. Pacific Coast Avifauna 31. Berkeley: Cooper Ornithological Society.

———. 1960. *Life Histories of Central American Birds*, II. Pacific Coast Avifauna 34. Berkeley: Cooper Ornithological Society.

———. 1969. *Life Histories of Central American Birds*, III. Pacific Coast Avifauna 35. Berkeley: Cooper Ornithological Society.

Slobodkin, L., and H. Sanders. 1969. On the contribution of environmental predictability to species diversity. In *Diversity and Stability in Ecological Systems*. Brookhaven Symp. Biol. 22.

Snow, D. W. 1962. A field study of the black and white manakin (*Manacus manacus*) in Trinidad. *Zoologica* 47: 65–104.

Stehli, F. G. 1968. Taxonomic diversity gradients in pole location: The recent model. In *Evolution and Environment*. Peabody Museum Centennial Symposium. New Haven: Yale University Press.

Ecology (1975) **56**: pp. 771–790

VEGETATION OF THE SANTA CATALINA MOUNTAINS, ARIZONA.
V. BIOMASS, PRODUCTION, AND DIVERSITY ALONG
THE ELEVATION GRADIENT[1]

R. H. WHITTAKER

Ecology and Systematics, Cornell University, Ithaca, New York 14850 USA

AND

W. A. NIERING

Botany Department, Connecticut College, New London, Connecticut 06320 USA

Abstract. Measurements were taken in 15 communities along the elevation gradient from fir forest at high elevations, through pine forest, woodlands, and desert grassland, to deserts at low elevations in the Santa Catalina Mountains, Arizona, and in a *Cercocarpus* shrubland on limestone. Eight small-tree and shrub species of woodlands and deserts were subjected to dimension analysis by the Brookhaven system. Aboveground biomass decreased along the elevation gradient from 36–79 dry kg/m² in fir and Douglas-fir forest to 0.26–0.43 kg/m² in the desert grassland and two desert samples. Net aboveground primary productivity similarly decreased from 1,050–1,150 g/m²·yr in mesic high-elevation forests to 92–140 g/m²·yr in desert grassland and deserts. Both biomass and production show a two-slope relation to elevation (and, probably, to precipitation), with a steeper decrease from the high-elevation forests to the mid-elevation woodlands, and a less steep decrease from dry woodlands through desert grassland into desert. The two groups of communities at higher vs. lower elevations also show different relations of leaf area index and chlorophyll to elevation and to productivity. The two groups may represent different adaptive patterns: surface-limiting, with low productivity in relation to precipitation but high production efficiency in relation to surface in the more arid lower elevations, vs. surface-abundant, with high productivity relative to precipitation based on high community surface area, but lower production efficiency in relation to this area, in the more humid higher elevations. Vascular plant species diversity shows no simple relation to productivity, but decreases from high-elevation fir forests to the pine forests, increases from these to the open woodlands, and decreases from dry woodlands through the desert grassland and mountain slope desert to the lower bajada (creosotebush) desert.

Key words: Arizona; biomass; desert; diversity; elevation; forest; grassland; productivity; Santa Catalina Mountains; woodland.

INTRODUCTION

The Santa Catalina Mountains are a range with strong Mexican affinities, northeast of the city of Tucson in southeastern Arizona. The south slope of the range bears an uninterrupted vegetational gradient from subalpine forest through woodlands and grasslands to desert. This gradient, described by Shreve (1915) and Whittaker and Niering (1964, 1965) is unique in the Southwest for its wide range of communities on largely consistent parent materials (Catalina gneiss and granite, and the bajada materials derived from these). On the north side of the mountains a vegetation pattern from fir forest to desert grassland occurs on a more complex mosaic of parent materials and includes distinctive communities on limestone among which mountain mahogany (*Cercocarpus breviflorus*) shrubland is most extensive (Whittaker and Niering 1968a, b). For this study production samples were taken in 14 climax vegetation types on the south slope of the range, and in a successional stand of aspen and a *Cercocarpus* scrub

on limestone. Some major species of the woodlands and deserts were subjected to dimension analysis by the Brookhaven system (Whittaker and Woodwell 1968, 1969, 1971). Primary purposes of the study were (1) to obtain measurements of aboveground net primary productivity and biomass for kinds of communities—woodlands and semideserts especially —for which few data are available, and (2) to observe interrelations of biomass, production, leaf area and chlorophyll, and species diversity along the extended physiognomic gradient from fir forest to desert.

BASIS OF THE STUDY

The elevation gradient

Elevations range from 2,766 m at the summit of Mt. Lemmon to 850–980 m at the southwestern base of the range near Tucson, and down the desert plain or bajada to 730 m at Tucson. The gradient of vegetation extends from subalpine fir forests near the summit of Mt. Lemmon through montane fir forest and pine forest, pine–oak forest, pine–oak woodland, pygmy conifer–oak scrub, open oak wood-

[1] Manuscript received 1 June 1973; accepted 19 December 1974.

772 R. H. WHITTAKER AND W. A. NIERING Ecology, Vol. 56, No. 4

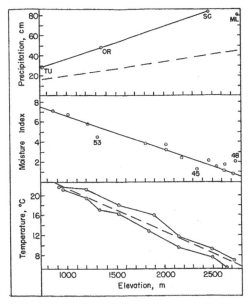

FIG. 1. Elevation and climate, Santa Catalina Mountains.

Top: Mean annual precipitation in cm at four stations: Tucson (TU) and Oracle (OR) (Sellers 1960, McDonald 1956), Soldiers Camp (SC) (6-yr record, Mallery 1936), and Mt. Lemmon (ML) (3-yr record, Weather Bureau 1952–54). The solid line is a trend ($y = 3.0 + 34.2x$, with x in 1,000 m) for the first three points. The dashed line is the trend line for mean daily precipitation in mm/day times 100, from the network of rain gauges of Battan and Green (1971); these values times 1.8 approximate the slope for year-round precipitation.

Middle: Weighted-average moisture indices for the vegetation samples of this study. Three stations lie off the trend line ($y = 10 - 3.4x$, with x in 1,000 m) because of topographic position—48 (a summit SSW-slope), 45 (a ravine), and 53 (a lower NNW-slope).

Bottom: Mean annual soil temperatures at 20-cm depth on south-facing slopes on the south side of the range (upper series) and north-facing slopes on the north side of the range (lower series), from Whittaker et al. (1968). The regression line is $y = 30.42 - 8.90x$, with x in 1,000 m. Short-term data of Shreve (1915) gave a mean decrease of 7.5°C per 1,000 m elevation.

land, and desert grassland to spinose–suffrutescent Sonoran semidesert on the mountain slopes, while on the desert plain or bajada below the mountains the desert gradation continues through the paloverde–bursage semidesert to the creosotebush desert of the lower bajada.

Changes in climate with elevation in the Santa Catalina Mountains were described by Shreve (1915), and available climatic data were summarized by Whittaker and Niering (1965). Changes in soils with elevation were described for the Catalinas by Whit-

taker et al. (1968) and for the nearby Pinaleno Mountains by Martin and Fletcher (1943). The top panel of Fig. 1 shows together the mean annual precipitation in four weather stations at different elevations, and a trend line for summer precipitation in a network of rain gauges. The data are not compelling, but suggest a linear increase in precipitation from 730 to 2,400 m elevation. Because the production samples were taken from different kinds of topographic positions, elevation is only a crude expression of their probable moisture relationships. A moisture index providing a better expression of these relationships was based on the intensive study of species distributions by Whittaker and Niering (1964, 1965). Plant species were classified by their relations to the moisture gradient in topographic transects for elevation belts from 0 (most mesic and highest elevation) to 8 (most xeric and lowest elevation). These species classes, or ecological groups, are applied as weights to the composition of vegetation samples to obtain weighted-average indices of the relative positions of samples along the gradient (Whittaker and Niering 1965). Index values for the production samples are given in Table 1-B and are plotted in relation to elevation in the middle panel of Fig. 1, where they show a linear trend.

The bottom panel of Fig. 1 gives the trend of mean annual soil temperatures at 20 cm depth, measured at 2-wk intervals from September 1962 to August 1963. At 10 of the stations, 5 pairs on each side of the range between 1,200 and 2,140 m, mean annual soil temperatures were obtained for both north-facing and south-facing slopes; the latter averaged 3.35°C warmer (Whittaker et al. 1968). Mean monthly temperatures for January and July, and mean annual temperatures, are 10.0°, 30.1°, and 19.6°C at Tucson, 7.7°, 26.5°, and 16.7°C at Oracle (1,370 m elevation) (Sellers 1960). Soil trends from high elevation to low included decreasing litter cover and organic content of soils, decreasing nitrogen content and carbon : nitrogen ratios, increasing pH, and increasing contents of Ca, Mg, and K (Whittaker et al. 1968). Although elevation is a complex-gradient of many climatic and edaphic factors, moisture conditions are considered the principal variable affecting vegetation structure and productivity along the gradient from fir forest to desert. Community characteristics are plotted in relation to both elevation and the moisture index in some of the figures that follow.

Communities sampled

Elevations, topographic positions, and other characteristics of the samples are summarized in Table 1. Samples 44–57 were taken to represent the major kinds of communities along the elevation gradient on the south slope of the mountains. Samples 44–55

Summer 1975   ELEVATION GRADIENT IN ARIZONA MOUNTAINS   773

TABLE 1. Summarized characteristics of productivity samples, Santa Catalina Mountains

Sample number and characteristic	44	45	46	47	48	49	50	51	52	53	54	55	56	57	58	59
	Abies lasiocarpa subalpine forest	*Abies concolor* ravine forest	*Pseudotsuga menziesii–Abies concolor* forest	*Pseudotsuga menziesii* forest	*Pinus ponderosa–Pinus strobiformis* forest	*Pinus ponderosa* forest	*Pinus ponderosa–Quercus hypoleucoides* forest	*Pinus chihuahuana–Quercus arizonica* woodland	Pygmy conifer–oak scrub	Open oak woodland	*Bouteloua curtipendula* desert grassland	Spinose-suffrutescent desert scrub	*Cercidium microphyllum–Franseria deltoidea* desert scrub	*Larrea divaricata* desert scrub	*Cercocarpus breviflorus* shrubland	*Populus tremuloides* successional forest
A. Sample number	44	45	46	47	48	49	50	51	52	53	54	55	56	57	58	59
B. Environment[a]																
Elevation (m)	2,720	2,340	2,640	2,650	2,740	2,470	2,180	2,040	2,040	1,310	1,220	1,020	870	760	1,810	2,550
Exposure	NNE	Rav.	NW	NNW	SSW	WSW	SW	SSE	W	NNW	SSW	SSE	SSE	W	ESE	E
Inclination (degrees)	25		27	29	11	8	15	17	13	26	27	22	3	4	33	27
Moisture index[b]	0.89	1.26	1.07	1.82	2.05	2.16	2.54	3.16	3.67	4.52	5.80	6.76	7.03	7.50	3.74	1.46
C. Cover and light																
Individual-point cover (%)																
Conifers	90	76	80	114	104	100	74	32	6	8						34
Broadleaf trees	10	44	2				42	54	4	34	6	20	19	13		74
Shrubs	8	18	8	6	10	10	2	6	52	32	30	16	15		33	14
Herbs	<1	4	6	<1	10	10		4	4	4[c]	40	9	<1	<1	42	6
Thallophytes		10										18[c]				
Rock										10	7	6	5			
Light penetration (%)																
Through trees	7.3	12.8	5.6	4.5	37.9	21.5	41.1	42.7	91.2	95	98	88.2	87.9	85.5	81.5	17.2
shrubs and seedlings[d]	7.2	12.6	5.1	4.3	34.4	20.6	36.8	34.2	25.5			76.7	62.4	85.5	49.7	15.8
herbs	7.2	9.7	4.7	4.3	32.1	19.9	36.3	34.0	25.4	32.4[d]	32.8	73.5	62.4			15.5
Log light absorption	1.141	1.013	1.341	1.361	.693	.702	.441	.468	.430	.285	.484	.135	.205	.068	.293	.802
Leaf area index (m²/m²)	14.7	15.5	16.7	15.5	7.6	5.9	4.7	3.7	2.0	1.76	1.58	.94	.59	.60	1.40	6.4
Chlorophyll (g/m²)	5.8	6.2	7.0	6.1	2.0	1.7	1.8	1.8	1.0	.80	.75	.50	.34	.34	.85	3.0
D. Stand characteristics																
Stems/0.1 ha,[e] trees	59	151	40	34	270	110	128	278	57	19	2					235
shrubs	16		87	38				1	16		64	387[f]	146[f]	4,950[e]	393	71
Basal area (m²/ha),[e] trees	57.8	58.6	118.1	70.5	39.4	46.3	34.9	26.0	4.32	4.01	.01					31.6
shrubs	.3	1.4	1.4	.26				.02	.58		.21	10.9	2.96	4.15[e]	2.68	.38
Bark/stem basal area (%)	13.4	20.7	23.2	23.7	24.8	27.7	24.6	21.2	19.2	22.3						15.2
Weighted mean height (m)	33.5	25.5	27.9	27.6	12.8	18.4	15.2	7.5	2.7	5.3	2.5	3.4	3.1	1.7	3.7	16.1
Conic stem surface (m²/m²)	.78	.59	.72	.58	.36	.41	.30	.19	.02	.02	.0045			.09	.05	.51
Parabolic stem volume (m³/ha)	837	746	1666	980	253	425	265	98	6.6	10.7	.28	18.5	4.5	3.5	5.0	218
Mean radial increment (mm/yr)	.78	1.06	.95	.61	.59	.52	.38	.38	.28	.36						1.14
Basal area increment	.416	.530	.499	.243	.373	.275	.187	.238	.044	.026						.885
Estimated volume increment (cm³/m²·yr)	567	563	593	311	183	180	100	66	5.5	6.3						539
Weighted mean age (yr)	106	124	321	252	93	142	150	101	115	117						34

[a] Soils data are given in Whittaker et al. (1968, Tables I and II).
[b] Weighted-average index of Whittaker and Niering (1964, 1965); high values are xeric.
[c] *Selaginella.*
[d] Below stratum of shrubs and tree seedlings. In samples 53 and 54 the principal shrubs (*Agave schottii, Calliandra eriophylla*) are part of the herb stratum; penetration was consequently measured for the combined herb-shrub stratum.
[e] Stems over 1 cm at breast height (except sample 57, all stems at 10 cm above ground level).
[f] Arborescent shrubs and small trees together.

and sample·59 are all on the Catalina granite–gneiss soils; samples 56 and 57 are on coarse upper, and fine lower bajada deposits derived from the granite–gneiss slopes of the Catalinas. Data on soils of the samples are given by Whittaker et al. (1968); further information on species composition characteristic of the types of communities is given by Whittaker and Niering (1964, 1965). Probably all vegetation above the desert has been subject to ground-fires, at least, in the past. Two of the forest samples (44 and 48) may have less than climax biomass because of past crown fires, and one sample (59) was a successional stand of aspen (*Populus tremuloides*) resulting from a fire. The range has been subject also to grazing by cattle; but the south slope, where it is ascended by the Mt. Lemmon highway, has been protected against grazing since 1947. All the production samples are considered to represent climax or near-climax communities except for the aspen stand. In this statement "climax" does not refer to climatic climax in the sense of Clements (1936), but to steady-state natural communities that are self-maintaining in their particular climatic, topographic, and edaphic environments, and that are generally characterized by near-equality of gross primary productivity and total community respiration and by maximum biomass for successions in their particular environments (Whittaker 1953, 1974). Lightning-induced fires are part of the natural environment of all communities between the deserts and the fir forests (Whittaker and Niering 1965), and we believe that past grazing has not caused significant present departure from the climax condition in these communities. The samples were taken along the Mt. Lemmon highway or on the upper slopes of Mt. Lemmon, unless otherwise indicated.

Subalpine fir forest (sample 44).—*Abies lasiocarpa* is strongly dominant, making up 85% of stem volume in the sample, with *Pseudotsuga menziesii* as second species with 10% of volume. The undergrowth is very sparse; *Jamesia americana* was the principal shrub, *Pyrola virens* and *P. secunda* the principal herbs. The bulk of the trees were in size classes 25–50 cm dbh (and 100–130 yr of age), and there were few young trees. The stand occupies a limited area of convex, north-facing slope shortly below the summit of Mt. Lemmon.

Mesic ravine fir forest (sample 45).—This sample, in Marshall Gulch, is dominated by *Abies concolor* and *P. menziesii* (55% and 20% of volume), with several percent volume each for *Pinus strobiformis*, *Acer grandidentatum*, and *Alnus obtusa*. *Rubus neomexicanus* and *Symphoricarpos oreophilus* are the principal shrubs, *Glyceria elata*, *Pteridium aquilinum*, and *Bromus richardsonii* the major species in a rich herb stratum. Ages of trees were mostly 50–145 yr at breast height; it is likely that the largest and oldest trees were cut from the stand some time in the past.

North-slope, montane fir forest (sample 46).—*Pseudotsuga menziesii* is dominant (79% of volume), with *A.*

concolor as second species (17%). The canopy trees were 90–150 cm dbh, 25–45 m tall, and 150–190 yr old at breast height. The undergrowth is sparse, though not so meager as in sample 44; *S. oreophilus* is the principal shrub, *Thalictrum fendleri* and *B. richardsonii* the principal herbs.

Drier montane fir forest (sample 47).—*Pseudotsuga menziesii* is dominant (88%) with *A. concolor* (7%) and *Pinus strobiformis* (5%); the canopy trees were 50–95 cm dbh, 25–37 m tall, and 200–250 yr old. *Jamesia americana* is the principal shrub, *B. richardsonii* and *Carex* sp. the principal species in the meager herb stratum.

High-elevation pine forest (sample 48).—*Pinus ponderosa* makes up 68% and *P. strobiformis* 32% of stand volume; *P. menziesii* is present in the stand. There are few shrubs, but *Muhlenbergia virescens* and *Pteridium aquilinum* are the major species in an herb stratum of fair coverage (10%). Canopy stems were 25–50 cm dbh, 12–17 m tall, and 70–120 yr old; and in contrast to sample 44 the stand includes many younger trees. Maximum ages of trees suggest these may have replaced a forest destroyed by fire somewhat more than 130 yr ago.

Lower elevation pine forest (sample 49).—*Pinus ponderosa* is strongly dominant (99% of volume vs. 1% for *P. strobiformis*); the canopy trees were 40–60 cm dbh, 15–25 m tall, and 100–180 yr old at breast height. Undergrowth composition resembled that of sample 48.

Pine–oak forest (sample 50).—*Pinus ponderosa* makes up 95% of stand volume, *Q. hypoleucoides*·4%, and *Arbutus arizonica* less than 1%. The canopy pines were 25–45 cm dbh, 15–20 m tall, and 120–170 yr old; a few larger pines are present. The oaks (*Quercus hypoleucoides*) form a small-tree stratum beneath the pine canopy; the larger oaks were 10–25 cm dbh, 4.5–7.0 m tall, and 30–50 yr old. The undergrowth is sparse, with *Rhamnus californica* the principal shrub, *M. virescens* and *Commandra pallida* the principal herbs.

Pine–oak woodland (sample 51).—The stand is smaller and more open than the preceding, with dominance shared among two pine and two oak species—*Pinus chihuahuana* 45%, *P. ponderosa* 10%, *Q. hypoleucoides* 25%, *Q. arizonica* 18% of volume; *A. arizonica* and *Juniperus deppeana* are present. The rosette shrubs *Yucca schottii* and *Nolina microcarpa* and the sclerophyll *Arctostaphylos pungens* are the principal shrub species; *Muhlenbergia emersleyi* is strongly dominant in the herb stratum. The canopy pines were 25–40 cm dbh, 9–15 m tall, and 100–150 yr old; the oaks 10–20 cm dbh, 3–6 m tall, and 30–50 yr old.

Pygmy conifer–oak scrub (sample 52).—This physiognomically distinctive community has a very open stratum of *Pinus cembroides* (Mexican pinyon pine, 69% of volume), *Juniperus deppeana* (alligator juniper, 16%), and *Q. hypoleucoides* (silverleaf oak, 4%) above a denser but still open stratum of sclerophyll shrubs (*Arctostaphylos pringlei*, *A. pungens*, *Garrya wrightii*) and rosette shrubs (*Y. schottii*, *N. microcarpa*, *Agave palmeri*, and *Dasylirion wheeleri*). The trees are small; the "canopy" trees were mostly 12–20 cm dbh, 2–4 m tall, and 70–120 yr old (but some individual junipers were more than 200 yr old). Basal area and stem volume are in a range appropriate to a shrub community (cf. Whittaker 1963), and the radial wood increments of the trees are lower than in any other forest or woodland sampled. Herb coverage is very low, with *M. emersleyi* the principal species.

Open oak woodland (sample 53).—Scattered evergreen oaks are of low coverage (8%) and small volume (56% in *Q. oblongifolia*, 44% in *Q. emoryi*), above shrub and herb strata of moderate coverage. The larger oaks were 25–35 cm dbh, and 5–7 m tall; their ages could not be read. The principal shrub species are *G. wrightii*, *Selloa glutinosa*, *Dalea pulchra*, *Baccharis thesioides*, and the rosette shrubs *N. microcarpa*, *Y. schottii*, and *Dasylirion wheeleri*; cacti (*Opuntia phaeacantha* and *O. spinosior*) first appear as a significant part of the shrub stratum in this sample. The herb stratum is strongly dominated by a low rosette shrub (*Agave schottii*) that from a distance appears to be part of the grass cover; the other principal species of the herb stratum are the grasses *M. emersleyi*, *Andropogon cirratus*, *Bouteloua curtipendula*, and *Aristida orcuttiana*, the xeric fern *Bommeria hispida*, and *Gnaphalium wrightii*.

Desert grassland (sample 54).—*Fouquieria splendens* contributed the small basal area and volume of the "tree" stratum; *Prosopis juliflora* and *Vauquelinia californica* were present on the slope. The principal shrubs were *Calliandra eriophylla*, *Carlowrightia arizonica*, *Haplopappus laricifolius*, and *Jatropha cardiophylla*. *Bouteloua curtipendula* was the dominant grass, with *B. filiformis*, *Aristida ternipes*, *Heteropogon contortus*, *Trichachne californica*, and *Muhlenbergia porteri* other principal herb species.

Spinose–suffrutescent Sonoran semidesert of the lower mountain slopes (sample 55).—In the complex physiognomy of this type the open "canopy" is formed by spinose trees and arborescent spinose shrubs—*Carnegiea gigantea* (giant cactus or saguaro), *Cercidium microphyllum* (paloverde), *Fouquieria splendens* (ocotillo), and *Prosopis juliflora* (mesquite). With these occur lower spinose shrubs (*Acacia greggii*, *Opuntia phaeacantha*, and other cacti), low shrubs of which *Encelia farinosa* (brittlebush) and *Calliandra eriophylla* are most important, perennial forbs (*Boerhaavia gracillima*, *Euphorbia melanadenia*, and *Allionia incarnata*), and grasses (*Muhlenbergia porteri* and *Aristida ternipes*). Annual herbs also are conspicuous during rainy seasons. Community structure and dynamics are further described by Niering et al. (1963).

Paloverde–bursage (Cercidium microphyllum–Franseria deltoidea) *semidesert of the upper bajada (sample 56), sampled near Campbell Avenue, Tucson.*—The two dominants form distinct strata, 2–4 m and about 0.3 m tall; other major species are *Fouquieria splendens*, *Opuntia phaeacantha* and *O. fulgida*, *Mammilaria microcarpa* and *Echinocereus fendleri*, *Jatropha cardiophylla* and *Janusia gracilis*, and *Calliandra eriophylla* and *Carlowrightia arizonica*. Herb coverage was very low, with *Bouteloua aristidoides* the principal species.

Creosotebush (Larrea divaricata) *desert of the lower bajada (sample 57), sampled near the Tanque Verde-Sabino Canyon Road near Tucson.*—Larrea was more strongly dominant than in many creosotebush deserts (cf. Rickard 1963, Lowe 1964, Shreve 1964); the stands below the Catalinas lack *Franseria dumosa*. Density of the *Larrea* shrubs (205/0.1 ha) was near the high extreme for creosotebush desert; for other stands (with *Larrea* in most cases mixed with other species) Woodell

et al. (1969) gave a *Larrea* density range of 6–54/0.1 ha and Barbour (1969) 3–187/0.1 ha. Small coverage was contributed by *Ephedra nevadensis* and the semishrubs *Psilostrophe cooperi* and *Zinnia pumila*. Herb coverage in this stand, which had probably been grazed in the past, was minute.

Mountain mahogany (Cercocarpus breviflorus) *scrub on limestone (sample 58), sampled on Marble Mountain above the Oracle Road on the north slope of the Santa Catalina Mountains.*—The shrubs form an open cover (33%), 1.5–2.5 m tall, above a well-developed grass and forb layer. The community thus resembles a woodland in miniature, and on limestone replaces the oak and pine–oak woodlands of corresponding elevations and topographic positions on acid rocks (Whittaker and Niering 1968a, b). Shrubs other than the dominant *Cercocarpus* include *N. microcarpa*, *A. palmeri*, *G. wrightii*, and *Sphaeralcea fendleri*. In the herb stratum *B. curtipendula* is dominant; other species are *Viguiera dentata*, *Hymenothrix wrightii*, and *Verbena neomexicanus*.

Successional aspen (Populus tremuloides) *forest (sample 59).*—Trees other than *Populus* (which has 72% of stem volume) are *Robinia neomexicana* (7.5%), *Salix scouleri* (6.5%), and young *Pseudotsuga menziesii* (10%) and *A. concolor* (2.5%). The *Populus* were mostly 15–25 cm dbh, 12–18 m tall, and 36–65 yr old. *Jamesia americana* is the principal shrub species, *Pteridium aquilinum* and *Bromus richardsonii* the principal herb species. The stand is assumed to have replaced a fir forest following fire, and to be developing toward climax composition similar to that of sample 47.

Species subjected to dimension analysis

The Brookhaven system of dimension analysis (Whittaker and Woodwell 1968, 1971) was applied to eight of the major woody species of the woodlands, semideserts, and creosotebush desert.

Quercus hypoleucoides (silverleaf oak) is an evergreen oak with *Salix*-like leaves of upper-middle elevation woodlands. This, along with other evergreen oaks of the range, is a small tree species with relatively heavy branches; few of the stems exceed 15 cm dbh and 5–6 m, very few reach 30 cm dbh and 8–9 m height. Most twigs bear current and 1-yr-old leaves, and few 2-yr-old leaves; the estimate of older leaf growth of 10.4% of aboveground net production was based on increase in mean dry weight of 1-yr-old leaves compared with current (midsummer) leaves on the sample trees. Wood rings of the oaks are interpretable at higher elevations (the trees were from sample 50), but are increasingly difficult toward lower elevations.

Pinus cembroides (Mexican pinyon pine) is on the whole a smaller tree than the oak, growing in somewhat drier environments. The pinyons are irregular in form, with heavy branches; very few of them exceeded 15 cm dbh and 3–4 m height. Needles are persistent for 5–6 or rarely 7 yr; as in *Q. hypoleucoides*, older leaf growth was inferred from the curve of increased needle weight from current to 1- and 2-yr-old leaves. The wood rings are fairly clear, but light false rings alternating with darker true annual rings are frequent.

Cercocarpus breviflorus (mountain mahogany) is a deciduous, arborescent shrub that in this area occurs

776   R. H. WHITTAKER AND W. A. NIERING   Ecology, Vol. 56, No. 4

primarily on 'limestone. The shrubs analyzed, taken from sample 58, ranged 1.6–5.2 cm dbh, up to 4.0 m in height.

Arctostaphylos pringlei and *A. pungens* are evergreen manzanitas with thick, sclerophyllous leaves; both contrast with the preceding three species in having very thin bark. *Arctostaphylos pringlei* is arborescent and similar to *Cercocarpus* in size; both *Arctostaphylos*, however, tend to have several stems from a common root crown, the *Cercocarpus* a single main stem per shrub. Basal diameters of the sample shoots of *A. pringlei* were 2.4–7.4 cm, heights 0.8–1.9 m; those for *A. pungens* were 1.3–3.1 cm, and 0.5–1.1 m. In both species many of the stems of older shoots are partly dead; in some only a strip of living bark and wood runs up one side of the stem. High dispersions from the regression equations for these species result from the variable fractions of living stem and branch wood supporting live foliage, in shoots of a given size.

Cercidium microphyllum (paloverde) is a many-stemmed, arborescent, leguminous desert shrub with compound leaves of minute leaflets; the leaves are produced in, and shed between, the rainy seasons. The bark is green, with stomata and a palisade-like chlorenchyma (Scott 1935); and the large area of bark was found to photosynthesize an amount comparable to that of the leaves in *Cercidium floridum* (Adams et al. 1967, Adams and Strain 1969). The seasonal timing of leaf production and radial stem growth are largely independent (Turner 1963). The twigs are tapered green spines, and branching occurs as new such spiny twigs are formed. In manner of growth the whole shrub is thus a multiply-branched and compounded spine. Because the shrub is branched to near ground level, branches cut near ground level were taken as "shoots" for dimension analysis, and the basal diameters of the short main stem and of the branches from it were measured instead of diameter at breast height. Mature shrubs in the spinose–suffrutescent desert (sample 55) from which the analysis plants were taken and the paloverde–bursage desert (56) had basal diameters of 10–25 cm, heights of 2.0–3.5 m. Wood rings of *Cercidium* and other desert shrubs are interpretable (Shreve 1911), but often not with assurance.

Larrea divaricata (creosotebush) is the most important plant species of the North American warm deserts. Numerous slender stems rise from the root crown; their branches end in green twigs bearing bifid, evergreen leaves that lack spongy tissue but are not otherwise evidently xeromorphic (Runyon 1934). As the twig elongates, several pairs of leaflets are produced in a year's growth. The ratio of fruit to current twig and leaf production in a Mojave desert was 0.154 (Soholt 1973). The plant is evergreen and has been shown to be continuously photosynthetic at a low rate during dry seasons (Strain and Chase 1966, Strain 1969, Oechel et al. 1972). During severe drought some, but not all, leaves are shed and some shoots die back to ground level. The stems in the sample ranged up to 2 cm, and a few to 3 cm in diameter at 10 cm above ground level; shrub heights were mostly 0.5–1.5 m, a few to 2.3 m. The wood rings were interpreted as annual; on the basis of them the shrubs in this sample add 7–10 leaf-nodes to the stem axis per year. Despite its remarkable adaptation to desert climates, creosotebush is unremarkable in its dry weight and growth distributions and other relationships in Table 2.

Fouquieria splendens (ocotillo) consists aboveground of numerous (usually 5–20) long, nearly cylindrical, spine-armed wands that extend in varied directions and angles to the horizontal from the root crown at the ground surface. The stems taper gradually from basal diameters and ages of 2.0–2.5 cm and 20–25 yr in older stems, to 0.5–0.7 cm in current twigs. The stems sampled from the spinose–suffrutescent desert (sample 55) were 0.6–3.6 m long; some longer and taller stems (over 4 m) occur in the desert grassland (sample 54). During the rains soft, ephemeral leaves 1–2 cm long are put out along much or the whole length of the stem; the leaves are lost when the rains end (Cannon 1905). Two or three sets of leaves may thus be produced in a year; the leaf production in Table 2 assumes annual leaf production twice that sampled in the summer rainy season. Annual segments are clearly defined in all but the oldest stems and can be clearly related to wood rings. The bark includes a thick chlorenchyma (Scott 1932), and bark dry weight and growth exceed wood dry weight and growth in all but the old stems. The bark chlorenchyma is photosynthetic during favorable periods, apparently not during drought (Mooney and Strain 1964). Assuming no branches (though a few stems of *F. splendens* are branched) and two leaf growths per year, then stem wood, stem bark, and leaves each include about one-third of aboveground growth with a smaller, unmeasured fraction in flowers and fruits.

Two of the desert low shrubs or semishrubs, *Encelia farinosa* (brittlebush) and *Franseria deltoidea* (bursage), were sampled on a less intensive basis. Fifteen individual shrubs were obtained with root crowns, and the aboveground shoots were treated as sample branches. Both species average about 0.3 m tall in the spinose–suffrutescent desert (sample 55) from which *Encelia* was taken, and the paloverde–bursage desert (56) from which *Franseria* was taken. Drought adaptation of *Encelia* is discussed by Shreve (1964), Strain and Chase (1966), Cunningham and Strain (1969a), and Strain (1969); the species is able to increase light utilization for photosynthesis up to the highest light intensities (Cunningham and Strain 1969b). The analyses of these shrubs suggest that the ratio of "branch" growth (of aboveground perennial wood and bark) to "clipping" growth (of current twigs with leaves) is about 25:75% in *Encelia* and 21:79% in *Franseria*.

Measurement and estimation techniques

The production samples were based on the 0.1-ha quadrats used also by Whittaker (1963, 1966) and Whittaker and Woodwell (1969). In each stand all trees were measured and recorded by diameter at breast height and species and all shrubs were tallied by species in a rectangle 10 m on each side of a 50-m tape. Heights were measured and increment borings taken from all trees or (in the denser forest stands) all trees above 30 cm dbh and some of the trees, as samples by size classes, below that diameter. Woody stems over 1.5 cm dbh including those of arborescent shrubs were tallied with the trees; those under 1.5 cm dbh were tallied and clipped as shrubs. The arborescent shrubs of the semideserts were treated as trees, and heights and diameters were measured for the larger cacti (*Carnegiea, Opuntia,*

Ferocactus). Undergrowth clippings were taken from twenty 0.5 × 2.0 m subquadrats, random numbers of meters out from points at 5-m intervals along the central tape. The aboveground growth of herbs, and current twigs with leaves of shrubs, were clipped in the subquadrats, bagged by species, and weighed fresh and ovendry. Minor herb and shrub species missed in the subquadrats were clipped in the full 0.1-ha quadrat. In the low-elevation stands clippings were taken of both summer herbs and shrub growth (1964) and spring herbs (1965). Light penetration through the tree, shrub, and herb strata was measured at 50 points along the central tape with a Weston sunlight illuminometer, and plant-individual point coverage was recorded for these sample points. In the nonforest samples light penetration and coverage were read also along the 50-m borders of the quadrats, hence for 150 points in all. We adapted the sample to the creosotebush desert by measuring diameters, heights, and numbers of stems for the shrubs in the 0.1 ha, and measuring stem diameters at 10 cm above ground level for all shoots of 20 random shrubs within the plot.

The eight woody species of the woodlands and deserts just described were subjected to intensive aboveground dimension analysis by the procedures of Whittaker (1961, 1962) and Whittaker and Woodwell (1968). Trees of two species (*Quercus hypoleucoides* and *Pinus cembroides*) were felled, and a tape was laid along the stem from apex to base. Branches were tallied from the tip downward by age and basal diameter, and five sample branches per tree were taken for more detailed measurement of length, current twig number, and fresh and dry weights of live wood and bark, deadwood and bark, current twigs with leaves, older leaves by ages, and fruits. The stems were cut into segments or logs, and for each of these length, diameters, bark thickness, and fresh weight were recorded, and a 10-cm basal disc was taken for measurement of fresh and dry weight of wood and of bark, wood mean diameter and bark mean thickness, age, and mean radial wood increment for the last 10 or the last 5 yr. The shoots of shrubs were treated in the same way, except that in the smaller shrubs stem segments were weighed fresh and dry and used for the measurements obtained from discs of the trees. Fruits were collected from all branches of a given plant, as well as the sample branches; and special twig samples were taken for measurement of leaf weights, areas, and chlorophyll content, and dry weight distribution between twigs, petioles, and leaf blades. The dimension analysis samples normally included 15 shoots of each species, but 10 each were taken from *Arctostaphylos pringlei* and *A. pungens*.

The data from these sample plants were submitted to the computer program for dimension analysis developed at Brookhaven National Laboratory (Whittaker and Woodwell 1968, 1969, 1971). Regressions are calculated for the sample branches, relating the weights of branch fractions to branch basal diameter, and the weight of branch wood and bark to branch age. The regressions are used to calculate, from the tally of branch diameters and ages, the dry weights of wood and bark, current twigs with leaves, older leaves, and current annual growth of wood and bark (Whittaker 1965a) and older leaves (Whittaker and Garfine 1962), for all branches of each plant. The wood increment readings on discs or stem segments are used to calculate mean annual dry-weight growth of stem wood for the last 5 or 10 yr, and to estimate growth of stem bark from the ratio of current annual growth to mass for the wood, times the dry weight of the bark, for each stem segment. These and other calculations yield as results: (1) volume of stem wood and stem bark, (2) dry weight of stem wood, stem bark, branch wood and bark, current twigs and leaves, and older leaves, (3) annual dry-weight growth of stem wood, stem bark, branch wood and bark, current twigs and leaves, older leaves, and fruits, and (4) surface area of stem wood, stem bark, branch bark (Whittaker and Woodwell 1967), and leaves. The program then computes, for the 15 or 10 plant individuals, regressions of these as dependent variables on one or more of the independent variables: (1) dbh, or basal diameter, (2) parabolic volume estimate, one-half basal area times plant height, (3) conic surface estimate, one-half basal circumference times plant height, and (4) estimated volume increment, one-half wood area increment at breast height (or 10 cm) times plant height.

In the final stages of the program these regressions are used to compute the various dependent variables for the trees (or shrubs) recorded in the sample quadrats; and the results are summed by species in each quadrat. Each dependent variable is computed from two or more regressions, using different independent variables, and the more appropriate regression is used. In most cases regressions on parabolic volume have been used to estimate biomass values and growth of current twigs with leaves, whereas regressions on estimated volume increment have been used for other production estimates (cf. Whittaker and Woodwell 1969). Since only a few species were subjected to dimension analysis, regressions thought most appropriate have been applied to other species. (The regressions for *Quercus hypoleucoides* have been applied to other oaks, those for *Pinus cembroides* to *Juniperus deppeana*, those for *Arctostaphylos pringlei* to *Garrya wrightii*, etc.) For lack of regressions for the larger trees in the Catalinas, regressions from the southeastern forests

TABLE 2. Mean dimensions of trees and shrubs for dimension analysis, Santa Catalina Mountains, Arizona

Dimension	Quercus hypoleucoides	Pinus cembroides	Cercocarpus breviflorus	Arctostaphylos pringlei	Arctostaphylos pungens	Cercidium microphyllum	Larrea divaricata	Fouquieria splendens
A. Mean shoot dimensions								
Number of shoots measured	15	15	15	10	10	15	18	11
Diameter at breast height (cm)	7.65	8.43						1.49
Diameter at ground level (cm)			3.26	4.45	2.10	4.52	1.13	
Height (m)	4.81	3.65	2.58	1.32	.71	2.00	1.31	2.70
Age (yr)	52.3	107	18.9	54.8	27	25	12	21
Dry weight total (g)	24,645	26,932	1,482	1,406	205	2,054	148	383
Net production (dry g/yr)	1,844	1,482	269	153	51.4	422	45.8	78.8
Bark thickness (mm)	5.10	6.16	1.89	.38	.23	1.21	.56	2.99
Wood radial increment (mm/yr)	.69	.39	.64	.32	.32	.69	.33	.35
Biomass accumulation ratio	13.4	18.1	5.5	9.2	4.0	4.7	3.2	4.9
B. Volume (cm³)								
Parabolic volume estimate	21,444	19.356	1,641	1,246	139.4	2,242	114.6	312
True stem volume	24,014	25,424	1,305	1,049	87.6	1,280	95.5	653
Wood volume	17,854	19,247	980	1,014	83.4	1,120	79.6	328
Estimated volume increment	401	200	95.0	39.2	7.59	111	8.88	16.2
True volume increment	527	288	89.3	35.3	5.55	96	9.61	37.1
C. Surface (cm²)								
Conic surface estimate	7,037	5,764	1,510	972	243	1,583	286	707
Parabolic surface estimate	10,766	9,390	2,914	5,532	392	4,067	734	
Stem bark surface	9,231	8,811	1,643	1,032	186	1,418	313	1,415
Branch bark surface	71,960	122,322	33,467	3,252	1,507			
D. Shoot dry weight distribution, % in								
Stem wood	46.4	40.8	47.0	48.7	25.7	35.3	45.1	47.0
Stem bark	16.9	16.1	13.5	2.0	1.4	5.0	10.1	48.9
Branch wood and bark	24.3	30.7	29.4	30.5	30.6	56.9	28.4	
Current twigs and leaves	5.7	3.9	10.1	3.2	14.9	2.8	16.4	4.1
Older leaves	6.8	8.5		15.6	27.5			
E. Aboveground net production distribution, % in								
Stem wood	20.0	12.0	24.8	15.4	7.1	14.8	16.4	31.4
Stem bark	4.5	3.4	4.6	.5	.3	1.4	2.5	32.0
Branch wood and bark	27.9	25.1	30.3	22.7	16.5	48.7	21.2	
Current twigs and leaves	37.2	49.9	40.3	27.7	59.8	35.1	59.8	36.6
Older leaf growth	10.4	9.6		33.7	16.3			

were used: *Pinus echinata* (Whittaker et al. 1963) for pines, and *Picea rubens* biomass regressions (Shanks and Clebsch 1962) for the firs. Estimates of production and biomass are affected by high dispersions and by two sources of systematic error: a tendency to underestimation consequent on the logarithmic calculations (Crow 1971, Baskerville 1972, Beauchamp and Olson 1973) and a tendency to overestimate values for the largest individuals and consequently for samples (Ogawa et al. 1965, Whittaker and Woodwell 1968). Because these errors are of opposite direction, corrections for the logarithmic calculations have not been used here.

The production estimates for the forest trees were based primarily on calculations from estimated volume increments as described by Whittaker (1966). Production estimates for woodland trees and all arborescent shrubs were based on dimension analysis regressions. Estimates of production of smaller true shrubs used the clipping weights of current twigs with their leaves, times ratios of aboveground pro-

duction to clipping production from the dimension analyses. Estimates for suffrutescents or semishrubs used the clipping measurements times the ratios given for *Encelia farinosa* and *Franseria deltoidea*. Estimates for *Opuntia* were based on counts of new segments in the quadrats times mean dry weights per segment; production values for other cacti were based on volume and mass calculations, and division of the masses by age estimates. Production of herb species was based directly on quadrat clippings during the peak of summer growth and, in the woodlands and lower elevation communities, quadrat clippings taken also during winter or spring herb growth.

Many of the tree and shrub estimates are based on relationships for a different species, or the same species in a different sample. Leaf, twig, and chlorophyll data are limited to a few species. Annual wood rings are difficult to interpret in the woodland and desert samples, and very difficult or impossible in the open oak woodland; some of the growth rates

TABLE 3. Regressions for trees and shrubs of woodlands and desert, Santa Catalina Mountains. Regressions are in the forms: (linear) $y = a + bx$, and (logarithmic) $\log_{10} y = A + B \log_{10} x$. Coefficients of correlation are given as r; estimates of relative error in the forms e (SE of estimate divided by mean y) for linear regressions and E (antilog of SE of estimate) for logarithmic regressions. Stem diameter is measured at breast height (1.35 m) in the first two species, 10 cm above ground level in the remaining species

Regression	*Quercus hypoleucoides*	*Pinus cembroides*	*Cercocarpus breviflorus*	*Arctostaphylos* spp.	*Cercidium microphyllum*	*Larrea divaricata*
A. Whole-shoot regressions on log basal diameter, x (cm)						
	at breast height		at 10 cm above ground level			
Log, shoot height, y (cm)						
A	2.2776	2.1132	2.0346	1.6311	1.8473	2.0523
B	0.4875	0.5022	0.7444	0.7374	0.6882	1.0430
r	0.970	0.914	0.914	0.894	0.814	0.930
E	1.114	1.163	1.182	1.200	1.228	1.252
Log, stem volume, y (cm^3)						
A	2.5912	2.6454	1.5623	0.8687	1.2546	1.5675
B	1.8593	1.7586	2.7091	3.1586	2.6280	3.0012
r	0.997	0.990	0.984	0.956	0.978	0.984
E	1.131	1.185	1.279	1.608	1.262	1.343
Log, stem wood volume, y (cm^3)						
A	2.4449	2.3966	1.4239	0.8325	1.1919	1.4850
B	1.8702	1.8730	2.7358	3.1890	2.6367	2.9710
r	0.997	0.992	0.984	0.956	0.978	0.982
E	1.130	1.182	1.292	1.618	1.266	1.366
Log, stem surface, y (cm^2)						
A	2.9700	2.8903	2.2767	1.5397	1.9611	2.2771
B	1.1104	1.1200	1.7369	2.2032	1.7291	2.0089
r	0.997	0.992	0.975	0.931	0.940	0.968
E	1.084	1.105	1.222	1.532	1.302	1.330
Log, branch surface, y (cm^2)						
A	3.1942	3.4677	3.4642	2.6358		
B	1.7069	1.5843	1.9169	2.2032		
r	0.970	0.949	0.961	0.931		
E	1.455	1.430	1.323	1.852		
Log, stem dry weight, y (g)						
A	2.3819	2.4141	1.5139	0.6883	1.1167	1.5165
B	1.8933	1.7719	2.5992	3.1259	2.6682	2.9719
r	0.998	0.991	0.987	0.970	0.956	0.980
E	1.109	1.178	1.238	1.469	1.406	1.392
Log, stem wood dry weight, y (g)						
A	2.2409	2.1746	1.3932	0.6610	1.0957	1.4292
B	1.9010	1.8800	2.6215	3.1402	2.5855	2.9544
r	0.998	0.991	0.986	0.970	0.964	0.977
E	1.121	1.184	1.252	1.470	1.348	1.420
Log, stem bark dry weight, y (g)						
A	1.8256	2.0898	0.8977	−0.1068	0.2773	0.6745
B	1.8673	1.5015	2.5208	2.0857	2.5121	3.0113
r	0.997	0.984	0.986	0.949	0.935	0.848
E	1.146	1.201	1.243	1.470	1.490	2.082
Log, branch wood and bark dry weight, y (g)						
A	1.9187	2.2107	1.4406	0.9153	1.8203	1.1623
B	1.9353	1.6915	2.0754	2.4709	1.7653	3.0843
r	0.969	0.946	0.961	0.857	0.918	0.820
E	1.545	1.483	1.357	2.081	1.376	2.579
Log, older leaf dry weight, y (g)						
A	2.0387	2.1485		1.2795		
B	0.9376	1.0529		1.3195		
r	0.974	0.879		0.828		
E	1.212	1.475		1.552		
Log, aboveground dry weight, y (g)						
A	2.6775	2.7593	1.8794	1.1492	1.8979	1.8712
B	1.7728	1.6563	2.2999	2.7761	2.0835	2.5281
r	0.994	0.978	0.991	0.967	0.953	0.966
E	1.186	1.271	1.173	1.432	1.318	1.449
Log, stem wood production, y (g/yr)						
A	1.5251	1.0092	0.8158	−0.2136	0.4110	0.6827
B	1.1486	1.2540	1.8878	2.2273	2.0575	2.2367
r	0.964	0.903	0.968	0.886	0.912	0.972
E	1.322	1.500	1.285	1.776	1.474	1.345
Log, stem bark production, y (g/yr)						
A	0.8571	0.6104	−0.9288	−1.4757	−0.6919	−1.1104
B	1.1772	1.1029	1.8991	1.8358	2.1558	1.8991
r	0.977	0.934	0.971	0.868	0.901	0.953
E	1.255	1.332	1.270	1.680	1.545	1.392
Log, branch wood and bark production, y (g/yr)						
A	1.5927	1.3438	1.0467	0.3470	1.6134	0.8247
B	1.2556	1.2702	1.5989	1.6837	0.9954	1.4772
r	0.958	0.925	0.957	0.815	0.719	0.914
E	1.392	1.425	1.280	1.803	1.495	1.430
Log, current twig and leaf production, y (g/yr)						
A	1.9218	1.7755	1.2413	0.6931[a]	1.1344	1.2807
B	0.9988	1.1159	1.4461	1.1612	1.5400	1.5486
r	0.975	0.962	0.958	0.687[a]	0.923	0.932
E	1.222	1.239	1.248	1.640	1.307	1.390
Log, aboveground net production, y (g/yr)						
A	2.2929	2.0234	1.5532	1.2220	1.6974	1.4947
B	1.0784	1.1825	1.5935	1.0580	1.3722	1.6238
r	0.981	0.964	0.979	0.736	0.867	0.949
E	1.208	1.250	1.183	1.616	1.391	1.345
B. Regressions on parabolic volume, x (cm^3)						
Linear, stem volume, y (cm^3)						
a	4817.6	5099.9	76.03	69.58	120.91	2.098
b	0.8952	1.0501	0.7483	0.7198	0.5172	0.8144
r	0.979	0.991	0.996	0.932	0.982	0.989
e	0.267	0.186	0.098	0.465	0.193	0.190
Linear, stem wood volume, y (cm^3)						
a	3175.7	3021.3	74.69	62.28	120.93	0.9652
b	0.6845	0.8381	0.5514	0.7020	0.4457	0.6857
r	0.978	0.990	0.995	0.933	0.979	0.988
e	0.283	0.210	0.111	0.465	0.208	0.195
Linear, stem dry weight, y (g)						
a	2969.7	3249.9	105.39	4.515	147.71	4.316
b	0.6185	0.6227	0.5321	0.5286	0.3948	0.6952
r	0.977	0.987	0.995	0.963	0.905	0.981
e	0.284	0.219	0.110	0.379	0.452	0.238
Linear, branch wood and bark dry weight, y (g)						
a	1382.1	−558.0	138.47	−57.19	561.6	−6.663
b	0.2620	0.5076	0.1590	0.5185	0.2169	0.4933
r	0.872	0.981	0.895	0.946	0.826	0.918
e	0.719	0.360	0.395	0.554	0.356	0.594
Log, stem dry weight, y (g)						
A	0.9667	1.2261	−0.02677	−0.7200	−0.2692	−0.0978
B	0.7586	0.7026	0.9465	1.1456	0.9782	0.9805
r	0.996	0.987	0.994	0.982	0.958	0.993
E	1.161	1.216	1.163	1.348	1.396	1.220
Log, branch wood and bark dry weight, y (g)						
A	0.4806	1.0918	0.2408	−0.2291	0.9448	−0.2061
B	0.7732	0.6668	0.7452	0.9181	0.6339	0.8946
r	0.964	0.937	0.954	0.879	0.901	0.922
E	1.595	1.530	1.392	1.968	1.418	1.900
Log, current twig and leaf dry weight, y (g)						
A	1.1795	1.0325	0.4023	0.5130[a]	0.3583	0.4568
B	0.3991	0.4411	0.5203	.4353	0.5570	0.5005
r	0.970	0.956	0.952	.659	0.913	0.925
E	1.244	1.262	1.264	1.633	1.330	1.413
Log, aboveground dry weight, y (g)						
A	1.3557	1.6532	0.5280	−0.1071	0.8249	0.5084
B	0.7095	0.6556	0.8334	1.0197	0.7609	0.8277
r	0.991	0.973	0.992	0.981	0.952	0.971
E	1.237	1.307	1.156	1.313	1.325	1.408

TABLE 3. Continued

Regression	*Quercus hypoleucoides*	*Pinus cembroides*	*Cercocarpus breviflorus*	*Arctostaphylos* spp.	*Cercidium microphyllum*	*Larrea divaricata*
C. Regressions on estimated volume increment, x (cm^3/yr)						
Linear, stem wood production, y (g/yr)						
a	45.59	34.76	15.08	2.037	8.745	0.4631
b	0.7877	0.6922	0.5790	0.4970	0.5657	0.9058
r	0.976	0.988	0.977	0.947	0.935	0.979
e	0.159	0.160	0.158	0.452	0.328	0.208
Linear, stem bark production, y (g/yr)						
a	7.488	16.02	0.2141	0.2134	0.8811	0.0263
b	0.1888	0.1519	0.0116	0.0060	0.0533	0.0106
r	0.950	0.937	0.920	0.754	0.875	0.931
e	0.241	0.317	0.350	0.542	0.474	0.322
Linear, branch wood and bark production, y (g/yr)						
a	−40.96	−55.67	18.76	6.100	135.91	2.017
b	1.5626	2.419	0.6761	0.6743	0.5253	0.8060
r	0.854	0.918	0.974	0.943	0.593	0.933
e	0.529	0.635	0.178	0.398	0.403	0.315
Linear, current twig and leaf production, y (g/yr)						
a	91.22	193.9	43.54	23.87	74.37	4.502
b	1.4100	2.524	0.6262	0.2701	0.6698	2.517
r	0.881	0.909	0.883	0.663	0.841	0.968
e	0.376	0.431	0.301	0.370	0.318	0.261
Log, stem wood production, y (g/yr)						
A	0.6603	0.4915	0.1352	−0.1784	−0.1403	0.04582
B	0.7383	0.7719	0.8802	0.9421	0.9716	0.9272
r	0.991	0.985	0.984	0.880	0.925	0.980
E	1.152	1.175	1.191	1.801	1.431	1.285
Log, stem bark production, y (g/yr)						
A	0.01565	0.2803	−1.6000	−1.3568	−1.1842	−1.6398
B	0.7377	0.6174	0.8779	0.6922	0.9727	0.7706
r	0.979	0.927	0.979	0.768	0.873	0.941
E	1.243	1.353	1.225	1.949	1.628	1.447
Log, branch wood and bark production, y (g/yr)						
A	0.7478	1.0635	0.5004	0.3374	1.4593	0.4162
B	0.7644	0.6621	0.7285	0.7461	0.4103	0.5946
r	0.933	0.855	0.952	0.848	0.637	0.896
E	1.516	1.625	1.299	1.715	1.562	1.482
Log, current twig and leaf production, y (g/yr)						
A	1.2485	1.5704	0.7840	0.9024	0.8724	0.8594
B	0.6086	0.5615	0.6382	0.3860	0.6474	0.6131
r	0.950	0.858	0.922	0.840	0.834	0.898
E	1.325	1.500	1.347	1.552	1.469	1.493
Log, aboveground net production, y (g/yr)						
A	1.5484	1.7514	1.0152	1.1633	1.4348	1.0479
B	0.6645	0.6219	0.7224	0.5182	0.5923	0.6503
r	0.967	0.898	0.968	0.846	0.804	0.924
E	1.282	1.445	1.230	1.459	1.482	1.430
D. Regressions on conic surface estimate, x (cm^2)						
Linear, stem bark surface, y (cm^2)						
a	1797.0	1840.9	96.84	9.5674	58.35	−10.69
b	1.0564	1.2092	1.0239	0.9867	0.9329	1.1320
r	0.991	0.992	0.997	0.936	0.986	0.978
e	0.105	0.104	0.058	0.321	0.137	0.210
Log, stem bark surface, y (cm^2)						
A	1.1309	1.1950	0.04642	−0.7875	−0.1507	0.06522
B	0.7444	0.7382	0.9987	1.2720	1.0303	0.9838
r	0.997	0.992	0.996	0.954	0.984	0.988
E	1.077	1.100	1.089	1.416	1.145	1.193
Log, branch bark surface, y (cm^2)						
A	0.4097	1.1459	1.1281	1.2725		
B	1.1325	1.0228	1.0611	0.7401		
r	0.961	0.931	0.945	0.740		
E	1.536	1.516	1.396	1.803		
E. Branch regressions on branch basal diameter, x (cm)						
Log, branch wood and bark dry weight, y (g)						
A	−1.0935	−1.4973	−3.1248	−2.3090[a]	−1.4986	−2.5087
B	2.7050	2.7286	2.8165	2.3101	2.0522	2.4447
r	0.986	0.956	0.951	0.944[a]	0.836	0.931
Log, current twig and leaf dry weight, y (g)						
A	−0.08156	−0.7110	−2.0078		−1.9952	−1.2658
B	1.5285	1.5824	1.9897		1.6697	1.4499
r	0.904	0.750	0.883		0.700	0.745
Log, branch bark surface, y (cm^2)						
A	1.7696	0.9108	0.2230	1.4269[a]		
B	1.7240	2.5672	2.6137	1.5444		
r	0.880	0.796	0.854	0.794		
F. Branch regressions on branch age, x (yrs)						
Log, branch wood and bark dry weight, y (g)						
A	−1.1407	−2.3257	0.0619	−0.3548[a]	0.5555	−0.3067
B	2.6928	2.7885	2.4633	2.3823	2.3386	2.4365
r	0.904	0.868	0.788	0.912	0.846	0.944
Log, current twig and leaf dry weight, y (g)						
A		−0.7887	0.4115		−0.1270	0.07948
B		1.3409	1.5541		1.7017	1.3823
r		0.562	0.654		0.628	0.751

[a] *A. pringlei.*

must be considered estimates, not measurements. It may be taken for granted that these growth rates, and clipping productions for herbs and low shrubs, vary significantly between years of drought and those of higher rainfall. The study should be considered a series of careful estimates of aboveground biomass and production from detailed information on the quadrats, but not systematic measurement such as applied at Brookhaven (Whittaker and Woodwell 1968, 1969) and Hubbard Brook, New Hampshire (Whittaker et al. 1974). No root data were obtained in this study. Estimates of likely root-shoot ratios from the literature (Whittaker 1962, 1966, Chew and Chew 1965, Rodin and Bazilevich 1967, Whittaker and Marks 1975) have been used to suggest

780 R. H. WHITTAKER AND W. A. NIERING Ecology, Vol. 56, No. 4

total, as well as aboveground, productivity and biomass in Table 4 and some of the figures.

RESULTS

Sample plant means and regressions

Table 2 summarizes mean dimensions, including dry weight mass and annual growth and the distribution of these in aboveground tissues, for the sets of sample plants. All values are arithmetic means, to permit comparisons with the similar summaries of Whittaker (1962), Whittaker and Woodwell (1968), Whittaker et al. (1974), and Andersson (1970, 1971).

Some of the regressions from the dimension analysis are summarized in Table 3. Because of their close similarity of form, the 10 sample shoots each of *Arctostaphylos pringlei* and *A. pungens* were combined to compute the *Arctostaphylos* regressions given. Dispersions from the regression lines are indicated both as r (coefficient of correlation) and E (estimate of relative error). E is the antilog of the standard error of estimate for a logarithmic regression (Whittaker and Woodwell 1968, cf. Furneval 1961, Attiwill 1966, Bunce 1968). It consequently expresses the expected range of values for y (to include 68% of individuals, assuming lognormal distribution) for a given value of x as a factor by which y is to be multiplied and divided. An E of 1.10 thus implies an expected range of values of 1.10y to y/1.10, for a given value of the independent variable.

The regressions are generally similar to those given for shrubs and small trees in the Brookhaven forest (Whittaker and Woodwell 1968) and in the studies of Andersson (1970, 1971) and Reiners (1972). *Quercus hypoleucoides* and *Pinus cembroides* in the Catalinas are stockier and heavier-branched than *Q. alba* and *P. rigida* at Brookhaven and the *Quercus* studied by Andersson (1970, 1971) and Reiners (1972). Correspondingly the slope constants (B) for stem surface, volume, and weight and other variables with a height component are lower in the Catalina trees. Slope constants for growth of stem wood and bark, branches, and twigs with leaves also are lower, reflecting the slower growth of the trees in the more arid climate of the Catalina woodland. The slope constants for the shrubs are generally higher and more consistent with those for the Brookhaven shrubs and trees.

Biomass

Table 4-B summarizes the community biomass estimates by plant fractions and strata. The total aboveground biomass values for fir forest samples 44, 45, and 47, 360–440 t/ha (dry metric tons per hectare, equals 36–44 kg/m²), are in the range of

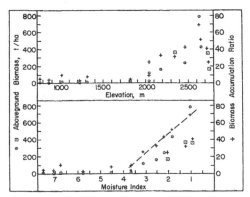

FIG. 2. Aboveground biomass (circles and squares, left ordinate) and biomass accumulation ratios (crosses and right ordinate) in relation to elevation and moisture index in the Santa Catalina Mountains. Biomass accumulation ratio (Table 4-B) is biomass/net annual production of vegetation, both in dry weights aboveground; moisture index (Table 1-B) is a weighted-average expression of position along the moisture gradient from subalpine fir forest (< 1.0) to desert (> 6.0). The three samples represented by squares are probably immature in biomass because of a fire about 130–150 yr ago (samples 44 and 48) or past selective cutting (45). A visual trend line has been fitted to the biomass accumulation ratios of the fully mature stands in the lower panel.

some of the climax forests of favorable environments in the Smokies (Whittaker 1966). Sample 46, in contrast, exceeds in biomass any of the southern Appalachian forests; with allowance for roots its aboveground value of 790 t/ha should become approximately 920 t/ha. The sample was taken from a particularly good stand of large trees by Santa Catalina standards; but it is not a large forest compared with those of the Pacific Northwest and the California coastal redwoods. The four pine forests and woodlands (samples 48–51) have total aboveground biomasses of 114–250 t/ha and compare with the pine forests and high-elevation fir forests (130–210 t/ha) in the Smokies. The low biomass of the aspen sample 59 (and a biomass accumulation ratio of 11.8) reflects its youth; these values are comparable with those for other young deciduous forests (Whittaker and Woodwell 1969, Andersson 1970, 1971, Duvigneaud et al. 1971, Reichle 1973).

The remaining woodland, grassland, and desert samples (52–58) have biomasses between 3 and 19 t/ha. The higher values compare with some of those for shrub communities in the southern Appalachians (heath balds, with biomasses of 11–110 t/ha, Whittaker 1963) and the lower values with those for other grasslands and deserts; comparable aboveground

TABLE 4. Summary of biomass and production estimates and diversity measurements for samples, Santa Catalina Mountains

Sample number and characteristic	44 Abies lasiocarpa subalpine forest	45 Abies concolor Abies lasiocarpa forest	46 Pseudotsuga menziesii– Abies concolor forest	47 Pseudotsuga menziesii forest	48 Pinus ponderosa– Pinus strobiformis forest	49 Pinus ponderosa forest	50 Pinus ponderosa– Quercus hypoleucoides forest	51 Pinus chihuahuana– Quercus arizonica woodland	52 Pygmy conifer– oak scrub	53 Open oak woodland	54 Bouteloua curtipendula grassland	55 Spinose– suffrutescent desert scrub	56 Cercidium microphyllum– Franseria deltoidea desert scrub	57 Larrea divaricata desert scrub	58 Cercocarpus breviflorus shrubland	59 Populus tremuloides successional forest
A. Sample number	44	45	46	47	48	49	50	51	52	53	54	55	56	57	58	59
B. Biomass (dry t/ha)																
Trees[a]— Stems	290	270	681	363	126	213	134	79	9.06	6.13	0.56	7.90	2.35	2.54[a]	4.89[a]	103
Branches	50	74	82	57	28	30	23	27.9	4.81	2.87	.04	4.00	1.35	1.33	1.84	16
Foliage	16.2	16.7	20	17	7.3	6.8	5.4	6.6	1.43	.40	.04	.18	.08	.42	.47	5.2
Total	356.2	360.7	783	437	161.3	249.8	162.4	113.5	15.30	9.40	.64	12.08	3.78	4.29	7.20	124.6
Shrubs[b]— Stems	.4	.012	4.0	.34	.05	.014	.20	.04	1.02	.04	.22	.14	.16	.0004	.03	.70
Branches	.2	.007	2.3	.22	.03	.005	.12	.01	.94	.09	.87	.64	.87	.002	.05	.24
Foliage	.04	.002	.46	.03	.02	.006	.04	.12	1.45	1.45	.42	.07	.11	.0001	.34	.04
Total	.64	.021	6.76	.59	.10	.025	.36	.17	3.41	1.61	1.51	.85	1.14	.0025	.42	.98
Herbs	tr.	.108	.006	.002	.039	.045	.004	.034	.031	.188	.483	.112	.004	tr.	.397	.030
Lichen and moss		.036	.045	.007	.028			.002	.044	.020	tr.					
Selaginella																
Total, above ground	357	361	790	438	161	250	163	114	18.8	11.22	2.63	13.10	3.92	4.29	8.02	126
Total, above and below[c]	420	420	920	520	190	300	200	150	30	17	21	21	21	6	14	200
Biomass accumulation ratio[d]	41.1	32.1	69.0	52.2	26.0	43.1	32.8	25.6	10.1	7.5	1.90	10.2	3.72	4.7	4.33	12.0
C. Net productivity (g/m²/yr)																
Trees[a]— Stem wood	275	395	340	227	138	136	129	75	6.6	13.9	4.1	24.7	12.0	16.7	32.3[a]	325
Stem bark	48	85	113	67	35	27	17	20	1.7	3.2	4.3	3.7	3.6	2.4	6.1	68
Branch wood and bark	130	164	177	139	100	97	72	90	12.5	22.1	1.4	29.9	20.0	18.7	37.1	217
Foliage	365	412	395	360	309	290	255	235	40.1	27.5	6.0	23.0	13.6	53.7	45.5	375
Fruit, flower, etc.	42	54	50	40	30	25	20	15	4.0	5.0	0.5	2.0	2.0	.2	4.0	45
Total	860	1,110	1,075	833	612	575	491	435	64.9	71.7	16.3	83.3	51.2	91.7	125.0	1,030
Shrubs[b]— Stems	2.1	.08	21	1.7	.4	.38	.6	.7	11.8	2.0	25.0	13.0	26.6	.01	.5	5.8
Branch wood and bark	2.2	.06	23	1.8	.3	.10	.7	.1	20.0	32.8	33.4	17.0	22.4	.02	1.0	5.9
Foliage	4.0	.16	46	3.4	1.0	.20	3.1	4.0	80.2	11.6	4.0	3.3	4.5	.03	9.5	6.6
Fruit, flower, etc.	0.2	.01	3	.2	.1	.04	.1	.6	5.0	3.4	62.4	3.3	53.6	.06	3.2	.5
Total	8.5	.31	93	7.1	1.8	4.2	4.4	6.7	117.0	47.3	62.4	33.3	53.6	.06	14.2	17.8
Herbs	.01	.31	.6	.17	1.1	.5	4.4	4.0	3.6	29.6	59.8	12.0	.2	.003	46.0	3.3
Total, above ground	869	1,123	1,146	840	618	580	496	446	186	149	139	129	105	92	185	1,051
Total, above and below[e]	1,020	1,300	1,340	1,000	740	700	620	580	350	300	280	210	170	140	330	1,220
Aboveground prod./leaf area (g/m³)	59.0	72.5	68.6	54.2	81.3	98.4	105.4	120.1	93.0	84.7	87.6	137	178	153	132	165
Aboveground prod./chlorophyll (g/g)	150	181	164	138	309	341	275	248	186	186	185	257	310	270	218	351
D. Diversity																
Species numbers[e]— Trees	6	9	6	3	4	5	5	8	4	3		30	27	4	15	7
Shrubs	3	5	2	1	1		4	6	10	23	26	11	6	2	22	
Herbs (20 m²)	6	30	8	6	8	3	3	4	6	32	20	24	23	8		8
Winter annuals											4	3	1	6		
Other herbs																
Sample total	15	44	16	10	13	8	12	18	20	58	46	41	33	18	37	18
Indices— Simpson C	.71	.27	.54	.83	.48	.86	.55	.35	.23	.21	.14	.26	.39	.99	.53	.36
Shannon-Wiener H'	.262	.667	.411	.168	.354	.143	.321	.553	.738	.856	1.066	.924	.543	.003	.478	.605
Equitability E_c	2.29	7.98	3.04	2.28	2.61	2.23	2.81	4.43	4.53	9.84	11.46	9.19	6.17	1.21	7.66	3.12
Equitability E_c	1.62	8.59	2.56	1.93	2.06	1.74	2.55	3.85	4.21	11.76	11.90	9.54	7.75	0.92	9.32	2.79

[a] "Trees" includes small trees and arborescent shrubs (Cercidium, Prosopis, Fouquieria, Carnegiea) in desert samples 55 and 56, and the canopy shrubs in samples 57 and 58.

[b] "Shrubs" include subordinate true shrubs and all semishrubs in samples dominated by shrubs (55–58). Rosette-shrubs and succulents (Opuntia, etc.) are responsible for some of the differences in distribution of biomass and production by tissues.

[c] No root data available; belowground estimates are based on plausible root/shoot ratios for different growth-forms.

[d] Biomass/net annual production, both as aboveground dry weight for all vascular plants.

[e] Trees, shrubs, and winter annual species in 0.1 ha. Perennial herbs and summer annuals are given both for the twenty 1-m²-subquadrats, and (as "other herbs") for the remainder of the 0.1-ha quadrat. Sample total includes trees and shrubs in 0.1 ha and perennial herbs and summer annuals in 20 m².

782 R. H. WHITTAKER AND W. A. NIERING Ecology, Vol. 56, No. 4

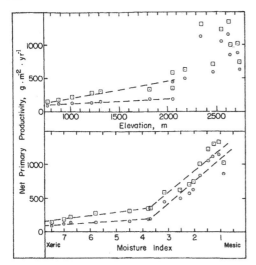

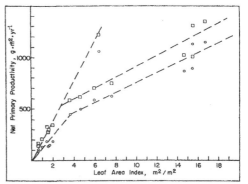

FIG. 4. Net primary productivity vs. leaf area index for communities in the Santa Catalina Mountains. Circles are aboveground productivity, squares estimates for aboveground and belowground productivity for the same samples. Visual trend lines suggest different slopes for deserts and desert grassland, with leaf area indices less than 3, and evergreen woodlands and forests, with leaf area indices of 4–16. The sample with a leaf area index of 6.5 is a high-elevation deciduous, successional forest (*Populus tremuloides*).

FIG. 3. Net primary productivity vs. elevation and moisture index in the Santa Catalina Mountains. Circles are aboveground productivity, squares estimates for aboveground and belowground productivity of the same samples. The moisture index (Table 1-B) is a weighted average of community composition, based on ecological groups for species responses to topographic moisture gradients and elevation.

values from Eurasia (Rodin and Bazilevich 1967) are as follows: arid steppes 0.3–3.0, *Artemisia* deserts 0.53–2.6, and shrub tundra 4.9 t/ha.

Figure 2 shows the trends of biomass and biomass accumulation ratios in relation to elevation and moisture index. The trend of biomass decrease toward lower elevations and increasing drought is evident, though the samples are scattered. The study in the Smokies (Whittaker 1966) gave a mean decrease in biomass of 230 t/ha with each gain of 1,000 m elevation. The increasing aridity toward lower elevations in the Catalinas produces a reverse biomass trend of about 590 t/ha increase per 1,000 m elevation gain within the forest and woodland zones (samples 44–52 only). No trend is evident for the lower elevation samples.

Biomass accumulation ratios (BAR, the biomass present/net annual production) express the delayed decomposition and accumulation of persistent, and particularly woody, tissues in terrestrial communities. In forests the ratios correlate with age of the dominants. BAR ranges in the Great Smoky Mountains were 2.5–11 for shrublands (heath balds), 10–20 in young forests, 20–30 in forests of intermediate stature, and 41–52 in mature cove forests (Whittaker 1966). The ratios in the Catalinas are consistent with these, except for the BAR of 69 in sample 46, exceeding that of any eastern forest sampled. In

the forest and woodland zones the trend of decrease of BAR toward lower elevations generally parallels the trend for biomass. Apart from the desert grassland (sample 54, BAR = 1.9), the BAR values of lower elevation samples 52–58 are in the range of expected values for shrublands and show no elevation trend. Both biomass and BAR are less scattered in relation to moisture index, in the lower part of Fig. 2, than in relation to elevation.

Production

Three of the forest samples (45, 46, 59, Table 4-C) have net primary productivities in the range suggested as normal for climax temperate forests of favorable environments (Whittaker 1966): 1,000–1,200 g/m² · yr aboveground, 1,200–1,500 g/m² · yr aboveground and belowground. The two samples at highest elevations, a fir and pine forest (44 and 48), are less productive. It is not evident what aspects of their environments (on opposite sides of the summit of Mt. Lemmon) may make these stands less productive than the other fir forests, or the spruce forests in the Great Smoky Mountains. Productivities of the pine forests and woodlands (48–51) range downward from 446–618 g/m² · yr, aboveground. Pine and oak heaths in the Smokies have similar values (419–578 g/m² · yr), but the low-elevation pine forests there were more productive (875–983 g/m² · yr).

The remaining open woodland, grassland, and desert productivities in the Catalinas, 92–186 g/m² · yr aboveground, are conspicuously low

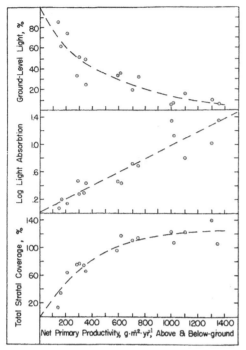

FIG. 5. Light and cover relationships to primary productivity in the Santa Catalina Mountains. Top: ground-level light intensity in per cent of incident sunlight; middle: $\log_{10}$ light absorption (logarithm of incident sunlight intensity, minus mean of logarithm of light intensity at ground level); bottom: total individual-point coverage (number of plant individuals of all strata with foliage above 100 points in the community).

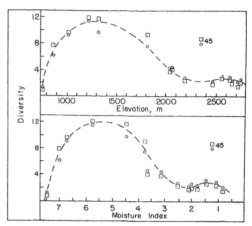

FIG. 6. Vascular plant diversity in relation to elevation and moisture index in the Santa Catalina Mountains. Equitability indices, mean number of species per logarithmic cycle (Whittaker 1972), are plotted $E_c = S/\log$ $p_1 - \log p_s$), circles, and $E'_c = S/4 \sqrt{\Sigma(\log p_i - \log \bar{p})^2/S}$, squares. $S =$ number of species in samples, and p_i aboveground net primary productivities—p_1 of the most productive and p_s of the least productive species in the sample, $\bar{p}$ the geometric mean for the sample; logarithms are to base 10. The curve for species diversity or richness, S, has the same form. Peak diversity occurs in the desert grassland and open oak woodland 1,220 and 1,310 m elevation, a lower secondary peak in fir forests at 2,500–2,700 m. Sample 45 is a ravine forest with diversity departing from the elevation trend.

compared with both forests and the closed heath balds (380–590 g/m² · yr) of the Smokies. The productivities of the open woodlands (samples 52, 53, and 58) are low compared with some other woodlands for which data are available (Whittaker and Woodwell 1969). Productivity of the desert grassland (54) is close to that of the spinose-suffrutescent semidesert (55). Sample 54 seems near the arid limit for grasslands; it may be compared with the dry steppes of Rodin and Bazilevich (1967), the short grass plains of Weaver (1924, 160 g/m² · yr), and African dry grasslands producing 90–170 g/m² · yr aboveground (Walter 1939, 1964). Chew and Chew (1965) estimated 130 g/m² · yr aboveground for a *Larrea* desert in Arizona. Their stand was in a somewhat less arid area and may have replaced desert grassland in consequence of grazing. Soholt (1973) measured aboveground production in a Mojave creosotebush desert, California, as 23.6 g/m² · yr in annual herbs and 7.0 g/m² · yr in leaves, twigs, and fruits of shrubs. The above-

ground total (estimating shrub stem and branch production as 4.1 g/m² · yr on the basis of the ratio of these to twig and leaf production in Table 2-E) would be 37.7 g/m² · yr. The four Arizona desert samples (Chew and Chew's and our samples 55–57) with net productivities of 92–130 g/m² · yr aboveground represent the mesophytic margin of southwestern desert environments, and much lower productivities occur in more arid areas (Lavrenko et al. 1955, Rodin and Bazilevich 1967).

The *Cercocarpus* shrubland on limestone (sample 58) is in a class with the pygmy conifer–oak scrub (52) and open oak woodland (53) in productivity. Sample 58, however, is at a higher elevation than the open oak woodland, and the pygmy conifer–oak scrub does not occur on the north side of the range where *Cercocarpus* scrub appears on limestone. The vegetation patterns of limestone and granite on the north slope (Whittaker and Niering 1968b) would predict a denser, mesic phase of the oak woodland or a pine–oak woodland for the *Cercocarpus* site if it were on granite. The limestone soil thus implies reduction of productivity, as well as a more xeric moisture index, compared with a corresponding site

784 R. H. WHITTAKER AND W. A. NIERING Ecology, Vol. 56, No. 4

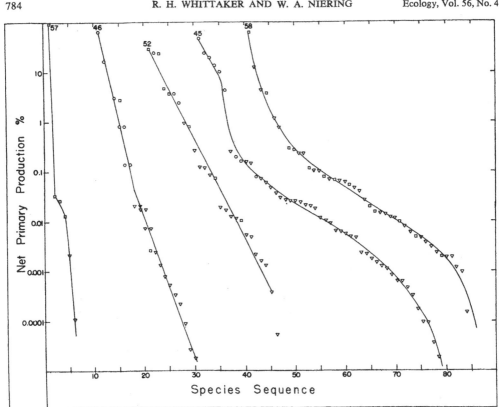

FIG. 7. Dominance-diversity curves for five samples from the Santa Catalina Mountains. The ordinate is the logarithm of aboveground net primary production for species, expressed as per cent of the totals for each sample. Species are arranged in sequence from the most to the least productive, by the scale on the abscissa. Four of the curves are displaced to the right to avoid overlap, and their first species are consequently at positions 11, 21, 31, and 41 on the abscissa. Circles are tree species, squares shrub species, and triangles herb species. Communities sampled: 57 a creosotebush (*Larrea divaricata*) desert; 46 a montane fir forest of *Pseudotsuga menziesii* and *Abies concolor*; 52 a pygmy conifer–oak scrub with *Pinus cembroides, Juniperus deppeana, Quercus hypoleucoides*, and *Arctostaphylos* spp.; 45 a ravine forest dominated by *Abies concolor*; and 58 a middle-elevation shrubland dominated by *Cercocarpus breviflorus* on limestone.

on an acid soil. The limestone is fissured and may permit rapid drainage of rainwater.

Figure 3 shows a trend of productivity increase with elevation up to 2,000 m, but at higher elevations the values are scattered. The moisture indices so transpose the samples as to produce a recognizable, though irregular, trend at higher elevations. Different slopes of production increase with elevation are suggested for more arid and more humid environments in Fig. 3. The visual trend lines give increases of about 25 g/m² · yr aboveground per moisture index unit for the more arid, and 320 g/m² · yr for the more humid series. Figure 1 suggests that a moisture index unit may correspond to about 10 cm precipitation difference. On this basis net aboveground production would increase by about 2.5 g/m² · yr

per cm of precipitation in the more arid, 32 g/m² · yr per cm in the more humid series. For dry grasslands Walter (1939, 1964) obtained a linear increase of aboveground productivity with rainfall of about 10 g/m² · yr per cm of precipitation between 10 and 55 cm.

Leaf area and chlorophyll

Estimates of leaf area index and chlorophyll per unit ground area are given in Table 1-C, ratios of aboveground productivity to these in Table 4-C. The surface area and chlorophyll estimates for fir forests, with evergreen needles persistent over 5 yr, are high in comparison with most literature values (Art and Marks 1971). The surface area and chlorophyll values for the lower elevation communities are underestimates in that photosynthetic surface and chlorophyll in stems and branches were estimated only for the cacti. In the lower elevation communities photosynthesis occurs in stems and branches of some shrub species during the dry season, when the leaves have been lost. The chlorophyll and photosynthetic surface estimates for leaves (and cacti) only are thus low for the rainy, but high for the dry season. The ratios of aboveground net productivity to leaf area and chlorophyll are low in the fir forests (54–72 g/m², 138–181 g/g), and consistently higher in the open woodland, grassland, and desert communities (85–165 g/m², 185–350 g/g).

The relationship is shown in another form in Fig. 4. Two slopes of total net primary productivity in relation to leaf area index are suggested: a steeper slope (of about 190 g/m²·yr increase per LAI unit) for the lower elevation communities, and a less steep slope (of about 50 g/m²·yr increase per LAI unit) for the higher elevation forests. A similar relation applies to production/chlorophyll ratios. The only deciduous community of a humid environment sampled, the aspen stand (59), falls on the line for the lower elevation communities.

Three other production relations—to ground-level light, log light absorption, and total stratal coverage—are shown in Fig. 5. Ground-level light percent (the geometric mean of light intensity at ground level, as a percent of incident sunlight) decreases as productivity and, in correlation with this, biomass and coverage, increase. Productivity in relation to log light absorption ($\log_{10}$ of incident sunlight minus $\log_{10}$ of ground-level sunlight) is complementary to the preceding, but is linear. In both cases the points are scattered in relation to the trend, and light penetration or absorption seems not effective as an index of production (Whittaker 1966). The curve of total stratal coverage on production relates to Fig. 4, though based on a different expression of community coverage. Total stratal coverage is the sum of individual-point coverages (number of plant individuals with foliage above 100 points) for the tree, shrub, and herb strata.

Growth-forms and diversity

Relations of vascular plant species diversity to elevation are shown in Table 4-D as several indices: S = number of tree and shrub species in 0.1-ha plus herb species in 25-m² quadrats (additional herb species in the 0.1 ha are indicated separately); the Simpson index, $C = \sum_s p_i^2$; the Shannon-Wiener index, $H' = -\sum_s p_i \log_{10} p_i$; and mean species per log cycle, E_c and E'_c (Fig. 6). In these p_i is relative productivity (percent in a given species of the total aboveground net productivity for a sample). The Simpson (1949) index expresses relative concentra-

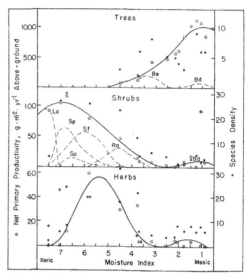

FIG. 8. Net primary productivity and species diversity of strata in relation to the moisture index, Santa Catalina Mountains. The circles are net primary productivity aboveground in g/m²/yr, left ordinate; the solid points are numbers of species per 0.1 ha (or, for herbs, 25 m²), right ordinate. Smoothed curves for productivities of fractions of strata are indicated by dashed lines, without datum points. Top: Bd is broadleaf deciduous and Be broadleaf evergreen trees; the remaining trees are needleleaf evergreen. (*Cercidium* and other small trees of the desert are here treated as shrubs.) Middle: Bd is broadleaf deciduous, Ro rosette shrubs, Sf suffrutescent semishrubs, Su succulents, and Sp spinose shrubs; La is *Larrea divaricata*. Bottom: the solid points are numbers of perennial and summer annual species in 25 m², the crosses numbers of winter annual species in 0.1 ha.

tion of dominance, H' primarily expresses equitability (Auclair and Goff 1971, Whittaker 1972), and E_c is an equitability measure (Whittaker 1972). The alternative form E'_c gives results paralleling E_c.

The fir forest sample 45 has a rich streamside flora that makes it diverse beyond the other high-elevation forests. Apart from this sample, the pattern of species diversity as represented by E_c and S is an increase from the high-elevation fir forests to the open oak woodland and desert grassland of lower elevations, followed by a decrease through the three deserts to a minimum in *Larrea* sample 57 (Fig. 6). If the samples are arranged by moisture index, the pattern is further complicated by a decrease from mesic fir (45, 46, 59) to more xeric pine (47–49) high-elevation forests, followed by the more conspicuous increase from the latter into the woodlands (51–53). The Shannon-Wiener index and the one complement of the Simpson index largely parallel the pattern for S and E_c; exception to this parallelism involve contrasts in relative dominance.

786 R. H. WHITTAKER AND W. A. NIERING Ecology, Vol. 56, No. 4

Dominance-diversity curves, using aboveground net productivity as the importance value, are shown for five communities in Fig. 7. Sample 57, the *Larrea* desert, has extreme dominance by one species and low species diversity. Three subordinate shrubs and semishrubs (*Ephedra nevadensis*, *Psilostrophe cooperi*, and *Zinnia pumila*) make up the hump in the middle of the curve, and two herbs with trivial productivities (*Tridens pulchellus* and *Pectis papposa*) the bottom points. In sample 46, as in all other forest samples of this study, the trees suggest a straight line approaching a geometric series (Whittaker 1965*b*, 1972). In sample 46 the herb stratum also suggests a geometric slope; in sample 45, in contrast, the rich herb stratum is sigmoid and suggestive of a lognormal distribution. Sample 52, a community of intermediate species diversity, suggests a geometric slope and resembles the curve for a pine forest in the Great Smoky Mountains (Whittaker 1965*b*, sample 10). Sample 58 represents a pattern more typical of woodlands, combining strong dominance with moderately high species diversity and sigmoid form. The remaining samples range from approximately geometric slopes (44, 47, 48, 49, 59) to sigmoid curves suggesting sparse lognormal distributions (50, 51, 53, 54, 55).

Correlation of diversity with productivity has been suggested (Connell and Orias 1964, MacArthur 1969). Figures 3 and 6 do not support such a correlation. The vegetation gradient from high to low elevation in the Catalinas is a physiognomic continuum, within which the various growth-forms have their peak importances in this sequence: needle-leaf evergreen trees, broadleaf-evergreen trees, rosette shrubs, grasses, semishrubs, and spinose shrubs (Whittaker and Niering 1965). Relations of diversity and productivity can be examined for these growth-forms as components of plant communities, but these relations are complex (Fig. 8). Consideration of growth-forms suggests these observations: (1) Within the tree stratum, no correlation of species number with productivity is in evidence. (2) Among perennial herbs both diversity and productivity are bimodal (with minima in the pine forests), and the two measures appear loosely correlated. Annual herbs are concentrated in the open and xeric communities, increasing in diversity and productivity from woodlands through desert grassland to less extreme deserts, but decreasing from these to the more extreme *Larrea* desert. (3) A loose correlation of diversity and productivity is suggested for the shrub stratum, when all shrub growth-forms are grouped together. (Certain arborescent shrubs forming part of the community canopy have been grouped with the tree stratum in Fig. 8—*Arctostaphylos* in sample 52, *Cercocarpus* in sample 58. Arborescent, spinose

plants of the desert—*Carnegiea, Cercidium, Fouquieria*—have, in contrast, been grouped with other spinose shrubs.) Within the shrub stratum the different growth-forms have separate centers of maximum diversity and productivity. (4) For the strata of dominants—whether trees, shrubs, or grasses—there is no correlation of diversity and productivity where these strata are dominant. For subordinate strata and growth-forms diversity and productivity are in general loosely correlated. (5) In temperate vegetation the tree stratum is in general less rich in species than are the herb and shrub strata, and the herb and shrub strata are on the average less rich beneath a closed forest canopy than in open woodland and grassland communities.

DISCUSSION

The relations of diversity to productivity do not encourage sweeping generalizations (cf. Whittaker 1965*b*, 1969, 1972). This study is in accord with others in indicating, however, that the highest temperate vascular plant species diversities are not in the most productive, closed forests but in less productive, open communities of intermediate environments—certain woodlands, grasslands, and shrublands. For the elevation gradient in the Catalinas the relation of diversity to moisture is combined with the relation to temperature, from which increasing diversity toward lower elevations would be expected. The high diversity of the spinose–suffrutescent semidesert may reflect the fact that this is a warm-temperate, near-subtropical community.

The results on productivity and biomass are generally in accord with those from the Great Smoky Mountains (Whittaker 1966) in suggesting characteristic ranges of these for most climax temperate forests and woodlands: for forests net productivity of 600–1,200 g/m²·yr aboveground and 700–1,500 g/m²·yr total, and biomass of 200–500 t/ha aboveground and 250–600 t/ha total; for woodlands net productivity of 150–600 g/m²·yr aboveground and 250–700 g/m²·yr total, and biomass of 20–200 t/ha aboveground and 30–250 t/ha total. Biomass of a mature Douglas-fir forest was higher, and that of an open oak woodland lower, than the ranges given. The transition from semidesert to one of the intermediate communities (grassland or woodland) appears to occur at about 150 g/m²·yr aboveground, probably 250 g/m²·yr total net primary productivity. Aboveground semidesert biomasses (4–13 t/ha in this study) exceeded those of desert grassland (2.6 t/ha in one sample of this study). Mountain-slope limestones in this area support vegetation that is Chihuahuan in floristic affinities and is more xeric in composition and structure than that on granite (Whittaker and Niering 1968*b*).

Many sites (as defined by elevation and topographic position) that support *Cercocarpus* shrubland on limestone support pine–oak woodland on granite. The contrast of these two types in biomass and production is represented by samples 58 and 51, which are roughly, though not closely, comparable in site: 8 t/ha and 185 g/m$^2\cdot$yr aboveground for *Cercocarpus* shrubland, 114 t/ha and 446 g/m$^2\cdot$yr for pine–oak woodland.

The most interesting result from the study may be the evidence of change in the character of productive relationships along the moisture gradient. In Fig. 2–4 different slopes for these relationships apply to the forest communities of more humid, and the woodland, grassland, and desert communities of more arid environments. The data are not adequate to assure that this is an abrupt change of slope, as distinguished from a curve of changing relationships toward higher elevations. To the extent a change of slope can be located, however, it is not at the desert border but at the transition from open to denser woodland, hence at the point along the gradient where trees become the primary basis of community productivity. As shown in Fig. 1, the limited climatic data suggest a linear relation of precipitation to elevation. One should be cautious in relating the productivities of this study to probable climate as implied by elevation. Accepting the evidence as it stands, however, the dual slopes in Fig. 2–4 suggest as an hypothesis different patterns of community response to the moisture gradient:

1) The communities of arid environments are surface-limiting, with transpiring surfaces minimized but with high productive efficiency of those surfaces made possible by their exposure to relatively full sunlight (and, in some species, such special photosynthetic adaptations as C_4 and crassulacean acid metabolism).

2) The evergreen forests of more humid environments are surface-abundant; with sufficient moisture they have much higher leaf area indices, but the productive efficiency per unit leaf surface is lower. The possibility of supporting high leaf surface areas permits, however, more rapid increase in productivity per unit of available moisture from woodlands into forests than in the surface-limiting communities.

3) With still greater moisture availability in humid forests, factors other than moisture (nutrient turnover, balance of respiring and photosynthetic tissue, light absorption) may become limiting, producing the 1,200–1,500 g/m$^2\cdot$yr range for mesophytic climax temperate forests. (Some young and floodplain temperate forests exceed this range.)

The two slopes of the Catalina productivity measurements (Fig. 3) in relation to probable precipitation values, and the assumption of a plateau

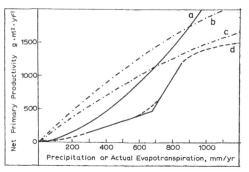

Fig. 9. Four interpretations of the relation of net primary productivity (dry g/m^2/yr) to precipitation and actual evapotranspiration. *a*: The curve fitted by Rosenzweig (1968) for aboveground net primary productivity of forests and shrublands in relation to actual evapotranspiration (mm/yr), with the form NPP = 1.66 log$_{10}$ AE − 1.66. *b*: The curve fitted by Lieth (1974) for total net primary productivity of diverse kinds of communities in relation to actual evapotranspiration, with the form NPP = 3,000 $(1 - e^{-0.0009695\,(AE-20)})$. *c*: The curve fitted by Lieth (1973, 1974) for total net primary productivity in relation to mean annual precipitation (mm/yr), with the form NPP = 3,000/$(1 + e^{1.315-0.119\,MAP})$. The nearly linear lower part of this curve is the ratio obtained by Walter (1939, 1964) for dry grasslands: NPP = MAP (g/m^2/yr aboveground, mm/yr), hence total NPP is about 2× MAP for the same units. *d*: A hand-drawn curve relating the two slopes of the Santa Catalina total net primary productivity estimates (Fig. 3) to probable mean annual precipitation, and adding to these a third, upper slope for limitation of climax temperate forest productivity at around 1,500 g/m^2/yr.

of productivity at precipitations higher than those reached in the Catalinas, have been combined in curve *d* of Fig. 9. The hollow lower part of the curve resembles the logarithmic fit of net productivity to evapotranspiration by Rosenzweig (1968), rather than the convex curves fitted to mean productivity in relation to precipitation and evapotranspiration by Lieth (1973, 1974). The curve may fall below Rosenzweig's (for primarily cooler temperate communities) because in a hot, dry climate the net production efficiency of a given amount of precipitation and evapotranspiration may be less (as a result of increased respiration and evaporative stress) than in a cooler climate. The response of productivity to combinations of moisture and temperature may be complex: for surface-abundant communities of humid environments higher net productivity for a given amount of precipitation should occur at higher temperatures; for surface-limiting communities of arid environments lower net productivity for a given amount of precipitation should occur at higher temperatures. (The latter statement excludes arctic-alpine communities in which productivity is limited by temperature and growing season.)

788   R. H. WHITTAKER AND W. A. NIERING   Ecology, Vol. 56, No. 4

The available data for evapotranspiration in Southwestern mountains indicate that actual evapotranspiration should be the same as precipitation at lower elevations in the Catalinas, up to the woodlands and precipitation values of 400–500 mm. For the cooler climates toward higher elevations evapotranspiration should remain roughly constant at values between 400 and 500 mm (Thornthwaite and Mather 1957, Buol 1964, Mather 1964). It is in this range of higher elevations with relatively constant actual evapotranspiration indices that the steep increase in productivity with increasing precipitation shown in Fig. 3 and 9 occurs. In this area the Thornthwaite actual evapotranspiration values seem not more useful than precipitation itself as a variable to which productivity may be related. Figure 9 suggests that the response of community productivity to climate is complex and nonlinear (comparing related communities in a given area, rather than Lieth's world averages), and therefore more interesting than has been recognized.

ACKNOWLEDGMENTS

Research supported in part by a grant to Brooklyn College from the National Science Foundation for "A study of southwestern mountain vegetation." Computer analysis of data was carried out at Brookhaven National Laboratory as part of a program in forest production with G. M. Woodwell, under the auspices of the U.S. Atomic Energy Commission.

LITERATURE CITED

Adams, M. S., and B. R. Strain. 1969. Seasonal photosynthetic rates in stems of *Cercidium floridum* Benth. Photosynthetica 3:55–62.

Adams, M. S., B. R. Strain, and I. P. Ting. 1967. Photosynthesis in chlorophyllous stem tissue and leaves of *Cercidium floridum*: Accumulation and distribution of ¹⁴C from ¹⁴CO₂. Plant Physiol. 42:1797–1799.

Andersson, F. 1970. Ecological studies in a Scanian woodland and meadow area, southern Sweden. II. Plant biomass, primary production and turnover of organic matter. Bot. Not. 123:8–51.

———. 1971. Methods and preliminary results of estimation of biomass and primary production in a south Swedish mixed deciduous woodland (French summ.), p. 281–288. *In* P. Duvigneaud [ed.] Productivity of forest ecosystems. Proc. Brussels Symp. 1969. UNESCO, Paris.

Art, H. W., and P. L. Marks. 1971. A summary table of biomass and net annual primary production in forest production ecosystems of the world, p. 1–32. *In* H. E. Young [ed.] Forest biomass studies. Univ. Maine, Orono.

Attiwill, P. M. 1966. A method for estimating crown weight in *Eucalyptus*, and some implications of relationships between crown weight and stem diameter. Ecology 47:795–804.

Auclair, A. N., and F. G. Goff. 1971. Diversity relations of upland forests in the western Great Lakes area. Am. Nat. 105:499–528.

Barbour, M. G. 1969. Age and space distribution of the desert shrub *Larrea divaricata*. Ecology 50:679–685.

Baskerville, O. L. 1972. Use of the logarithmic equation in the estimation of plant biomass. Can. J. For. Res. 2:49–53.

Battan, L. J., and C. R. Green. 1971. Summer rainfall over the Santa Catalina Mountains. Univ. Ariz., Inst. Atmos. Phys., Tech. Rep. 22:1–3.

Beauchamp, J. J., and J. S. Olson. 1973. Correction for bias in regression estimates after logarithmic transformation. Ecology 54:1403–1407.

Bunce, R. G. H. 1968. Biomass and production of trees in a mixed deciduous woodland. I. Girth and height as parameters for the estimation of tree dry weight. J. Ecol. 56:759–775.

Buol, S. W. 1964. Calculated actual and potential evapotranspiration in Arizona. Univ. Ariz. Agric. Exp. Sta., Tech. Bull. 162:1–48.

Cannon, W. A. 1905. On the transpiration of *Fouquieria splendens*. Bull. Torrey Bot. Club 32:397–414.

Chew, R. M., and A. E. Chew. 1965 The primary productivity of a desert-shrub (*Larrea tridentata*) community. Ecol. Monogr. 35:355–375.

Clements, F. E. 1936. Nature and structure of the climax. J. Ecol. 24:252–284.

Connell, J. H., and E. Orias. 1964. The ecological regulation of species diversity. Am. Nat. 98:399–414.

Crow, T. R. 1971. Estimation of biomass in even-aged stands—regression and "mean tree" techniques, p. 35–48. *In* H. E. Young [ed.] Forest biomass studies. Univ. Maine, Orono.

Cunningham, G. L., and B. R. Strain. 1969a. An ecological significance of seasonal leaf variability in a desert shrub. Ecology 50:400–408.

Cunningham, G. L., and B. R. Strain. 1969b. Irradiance and productivity in a desert shrub. Photosynthetica 3:69–71.

Duvigneaud, P., P. Kestemont, and P. Ambroes. 1971. Productivité primaire des forêts temperées d'essences feuillues caducifoliées en Europe occidentale (Engl. summ.), p. 259–270. *In* P. Duvigneaud [ed.] Productivity of forest ecosystems. Proc. Brussels Symp. 1969. UNESCO, Paris.

Furneval, G. M. 1961. An index for comparing equations used in constructing volume tables. For. Sci. 7:337–341.

Lavrenko, E. M., V. N. Andreev, and V. L. Leontief. 1955. Profile of the productivity of natural above ground vegetation of the U.S.S.R. from the tundra to the deserts (in Russian). Bot. Zhur. S.S.S.R. 40:415–419.

Lieth, H. 1973. Primary production: Terrestrial ecosystems. Human Ecol. 1:303–332.

———. 1975. Modeling the primary productivity of the world. *In* H. Lieth & R. H. Whittaker [ed.] Primary productivity of the biosphere. Springer, New York. (In press.)

Lowe, C. H. 1964. Arizona landscapes and habitats, p. 1–132. *In* C. H. Lowe [ed.] The vertebrates of Arizona. Univ. Arizona, Tucson.

MacArthur, R. H. 1969. Patterns of communities in the tropics. Biol. J. Linn. Soc. Lond. 1:19–30.

Mallery, T. D. 1936. Rainfall records for the Sonoran Desert. Ecology 17:110–121, 212–215.

Martin, W. P., and J. E. Fletcher. 1943. Vertical zonation of great soil groups on Mt. Graham, Arizona, as correlated with climate, vegetation, and profile characteristics. Univ. Ariz. Agric. Exp. Sta., Tech. Bull. 99:89–153.

Mather, J. R. 1964. Average climatic water balance

data of the continents. VII. United States. Publ. Climatol., Thornthwaite Lab. Climatol., Centerton, New Jersey 17:415–615.

McDonald, J. E. 1956. Variability of precipitation in an arid region: A survey of characteristics for Arizona. Univ. Ariz., Inst. Atmos. Phys., Tech. Rep. 1:1–88.

Mooney, H. A., and B. R. Strain. 1964. Bark photosynthesis in ocotillo. Madroño 17:230–233.

Niering, W. A., R. H. Whittaker, and C. H. Lowe, Jr. 1963. The saguaro: A population in relation to environment. Science 142:15–23.

Oechel, W. C., B. R. Strain, and W. R. Odening. 1972. Tissue water potential, photosynthesis, ^{14}C-labeled photosynthate utilization, and growth in the desert shrub *Larrea divaricata* Cav. Ecol. Monogr. 42:127–141.

Ogawa, H., K. Yoda, and T. Kira. 1965. Comparative ecological studies on three main types of forest vegetation in Thailand. II. Plant biomass. Nat. Life Southeast Asia 4:49–80.

Reichle, D. E., B. E. Dinger, N. T. Edwards, W. F. Harris, and P. Sollins. 1973. Carbon flow and storage in a forest ecosystem. *In* G. M. Woodwell [ed.] Carbon and the biosphere. Brookhaven Symp. Biol. 24:345–365.

Reiners, W. A. 1972. Structure and energetics of three Minnesota forests. Ecol. Monogr. 42:71–94.

Rickard, W. H. 1963. Vegetational analyses in a creosote bush community and their radioecologic implications, p. 39–44. *In* V. Schultz and A. W. Klement, Jr. [ed.] Radioecology. Proc. 1st Natl. Symp. Radioecol., Fort Collins, 1961. Reinhold and Am. Inst. Biol. Sci., Washington, D.C.

Rodin, L. E., and N. I. Bazilevich. 1967. Production and mineral cycling in terrestrial vegetation. Oliver & Boyd, Edinburgh and London. 288 p.

Rosenzweig, M. L. 1968. Net primary productivity of terrestrial communities: Prediction from climatological data. Am. Nat. 102:67–74.

Runyon, E. H. 1934. The organization of the creosote bush with respect to drought. Ecology 15:128–138.

Scott, F. M. 1932. Some features of the anatomy of *Fouquieria splendens*. Am. J. Bot. 19:673–678.

———. 1935. The anatomy of *Cercidium torreyanum* and *Parkinsonia microphylla*. Madroño 3:33–31.

Sellers, W. D. 1960. Arizona climate. Univ. Arizona, Tucson. 60 p. + tables.

Shanks, R. E., and E. E. C. Clebsch. 1962. Computer programs for the estimation of forest stand weight and mineral pool. Ecology 43:339–341.

Shreve, E. B. 1924. Factors governing seasonal changes in transpiration of *Encelia farinosa*. Bot. Gaz. 77:432–439.

Shreve, F. 1911. Establishment behavior of the palo verde. Plant World 14:289–296.

———. 1915. The vegetation of a desert mountain range as conditioned by climatic factors. Carnegie Inst. Wash. Publ. 217:1–112.

———. 1964. Vegetation of the Sonoran Desert, p. 1–186. *In* F. Shreve and I. L. Wiggins [ed.] Vegetation and flora of the Sonoran Desert. Stanford Univ. Press, Stanford, California.

Simpson, E. H. 1949. Measurement of diversity. Nature 163:688.

Soholt, L. F. 1973. Consumption of primary production by a population of kangaroo rats (*Dipodomys merriami*) in the Mohave Desert. Ecol. Mongr. 43:357–376.

Strain, B. R. 1969. Seasonal adaptations in photosynthesis and respiration in four desert shrubs growing in situ. Ecology 50:511–513.

Strain, B. R., and V. C. Chase. 1966. Effect of past and prevailing temperatures on the carbon dioxide exchange capacities of some woody desert perennials. Ecology 47:1043–1045.

Thornthwaite, C. W., and J. R. Mather. 1957. Instructions and tables for computing potential evapotranspiration and the water balance. Publ. Climatol., Thornthwaite Lab. Climatol., Centerton, New Jersey 10:181–311.

Turner, R. M. 1963. Growth in four species of Sonoran Desert trees. Ecology 44:760–765.

Walter, H. 1939. Grasland, Savanne und Busch der ariden Teile Afrikas in ihrer ökologischen Bedingtheit. Jahrb. Wissensch. Bot. 87:750–860.

———. 1964. Die Vegetation der Erde in ökologischer Betrachtung. Vol. 1. Die gemässigten und subtropischen Zonen. Fischer, Jena.

Weather Bureau. 1952–54. Climatological data, Arizona. Annual summaries. U.S. Dep. Commerce, Weather Bureau, Washington, D.C.

Weaver, J. E. 1924. Plant production as a measure of environment. J. Ecol. 12:205–237.

Whittaker, R. H. 1953. A consideration of climax theory: The climax as a population and pattern. Ecol. Monogr. 23:41–78.

———. 1961. Estimation of net primary production of forest and shrub communities. Ecology 42:177–180.

———. 1962. Net production relations of shrubs in the Great Smoky Mountains. Ecology 43:357–377.

———. 1963. Net production of heath balds and forest heaths in the Great Smoky Mountains. Ecology 44:176–182.

———. 1965a. Branch dimensions and estimation of branch production. Ecology 46:365–370.

———. 1965b. Dominance and diversity in land plant communities. Science 147:250–260.

———. 1966. Forest dimensions and production in the Great Smoky Mountains. Ecology 47:103–121.

———. 1969. Evolution of diversity in plant communities. Brookhaven Symp. Biol. 22:178–196.

———. 1972. Evolution and measurement of species diversity. Taxon 21:213–251.

———. 1974. Climax concepts and recognition. *In* R. Knapp [ed.] Vegetation dynamics. Handb. Veg. Sci. 8:137–154.

Whittaker, R. H., F. H. Bormann, G. E. Likens, and T. G. Siccama. 1974. The Hubbard Brook ecosystem study: Forest biomass and production. Ecol. Monogr. 44:233–254.

Whittaker, R. H., S. W. Buol, W. A. Niering, and Y. H. Havens. 1968. A soil and vegetation pattern in the Santa Catalina Mountains, Arizona. Soil Sci. 105:440–450.

Whittaker, R. H., N. Cohen, and J. S. Olson. 1963. Net production relations of three tree species at Oak Ridge, Tennessee. Ecology 44:806–810.

Whittaker, R. H., and V. Garfine. 1962. Leaf characteristics and chlorophyll in relation to exposure and production in *Rhododendron maximum*. Ecology 43:120–125.

Whittaker, R. H., and P. L. Marks. 1975. Measurement of net primary productivity on land. *In* H. Lieth and R. H. Whittaker [ed.] Primary productivity of the biosphere. Springer, New York. (*In press.*)

Whittaker, R. H., and W. A. Niering. 1964. Vegetation of the Santa Catalina Mountains, Arizona. I. Ecological classification and distribution of species. J. Ariz. Acad. Sci. 3:9–34.

Whittaker, R. H., and W. A. Niering. 1965. Vegetation of the Santa Catalina Mountains, Arizona. II. A gradient analysis of the south slope. Ecology 46: 429–452.

Whittaker, R. H., and W. A. Niering. 1968a. Vegetation of the Santa Catalina Mountains, Arizona. III. Species distribution and floristic relations on the north slope. J. Ariz. Acad. Sci. 5:3–21.

Whittaker, R. H., and W. A. Niering. 1968b. Vegetation of the Santa Catalina Mountains, Arizona. IV. Limestone and acid soils. J. Ecol. 56:523–544.

Whittaker, R. H., and G. M. Woodwell. 1967. Surface area relations of woody plants and forest communities. Am. J. Bot. 54:931–939.

Whittaker, R. H., and G. M. Woodwell. 1968. Dimension and production relations of trees and shrubs in the Brookhaven forest, New York. J. Ecol. 56:1–25.

Whittaker, R. H., and G. M. Woodwell. 1969. Structure, production and diversity of the oak–pine forest at Brookhaven, New York. J. Ecol. 57:157–174.

Whittaker, R. H., and G. M. Woodwell. 1971. Measurement of net primary production of forests (French summ.), p. 159–175. In P. Duvigeneaud [ed.] Productivity of forest ecosystems. Proc. Brussells Symp. 1969. UNESCO, Paris.

Woodell, S. R. J., H. A. Mooney, and A. J. Hill. 1969. The behaviour of Larrea divaricata (creosote bush) in response to rainfall in California. J. Ecol. 57:37–44.

References

Allen, T. F. H., and T. W. Hoekstra. 1992. *Toward a Unified Ecology*. New York: Columbia University Press.

Ammerman, A. J., and L. L. Cavalli-Sforza. 1971. Measuring the rate of spread of early farming in Europe. *Man* 6: 674–88.

Anker, P. 2001. *Imperial Ecology*. Cambridge, MA: Harvard University Press.

Arbogast, B., and G. J. Kenagy. 2001. Comparative phylogeography as an integrative approach to historical biogeography. *Journal of Biogeography* 28: 819–25.

Arrhenius, O. 1921. Species and area. *Journal of Ecology*, 9: 95–99.

Avise, J. C., J. Arnold, R. M. Ball, Jr., E. Bermingham, T. Lamb, J. E. Neigel, C. A. Reeb, and N. C. Saunders. 1987. Intraspecific phylogeography: The mitochondrial bridge between population genetics and systematics. *Annual Review of Ecology and Systematics* 18: 489–522.

Axelrod, D. I. 1967. Quaternary extinctions of large mammals. University of California Publications in Geological Science 74: 1–42.

Badgley, C., and D. L. Fox. 2000. Ecological biogeography of North American mammals: Species density and ecological structure in relation to environmental gradients. *Journal of Biogeography* 27: 1437–67.

Bagenal, T. B. 1951. A note on the papers of Elton and Williams on the generic relations of species in small ecological communities. *Journal of Animal Ecology* 20: 242–45.

Bailey, R. G. 1996. *Ecosystem Geography*. New York: Springer.

Ball, I. R. 1975. Nature and formulation of biogeographic hypotheses. *Systematic Zoology* 24: 407–30.

Bambach, R. K. 1977. Species richness in marine benthic habitats through the Phanerozoic. *Paleobiology* 3: 152–67.

Barraclough, T. G., and S. Nee. 2001. Phylogenetics and speciation. *Trends in Ecology and Evolution* 16: 391–99.

Bates, J. M. 2001. Avian diversification in Amazonia: Evidence for historical complexity and a vicariance model for a basic pattern of diversification. Pp. 119–138 in *Diversidade Biológica e Cultural da Amazônia*, ed. I. Vierra, M. A. D'Incao, J. M. Cardoso da silva, and D.

Oren. Belém, Pará, Brazil: Museu Paraense Emilio Goeldi.

Behrensmeyer, A. K, J. D. Damuth, W. A. DiMichele, R. Potts, H. Sues, and S. L. Wing. 1992. *Terrestrial Ecosystems through Time: Evolutionary Paleoecology of Terrestrial Plants and Animals*. Chicago: University of Chicago Press.

Berggren, W. A., and C. D. Hollister. 1977. Plate Tectonics and Palaeocirculation—Commotion in the ocean. *Tectonophysics* 38: 11–48.

Betancourt, J. L., T. R. Van Devender, and P. S. Martin, 1990. *Packrat Middens: The Last 40,000 years of Biotic Change*. Tucson: University of Arizona Press.

Bremer, K. 1992. Ancestral areas: A cladistic reinterpretation of the center of origin concept. *Systematic Biology* 41: 436–45.

———. 1995. Ancestral areas: Optimization and probability. *Systematic Biology* 44: 255–59.

Briggs, J. C. 1974. *Marine Zoogeography*. New York: McGraw-Hill.

———. 1995. *Global Biogeography*. Amsterdam: Elsevier.

Brooks, D. R. 1981. Hennig's parasitological method: A proposed solution. *Systematic Zoology* 30: 229–49.

———. 1985. Historical ecology: A new approach to studying the evolution of ecological associations. *Annals of the Missouri Botanical Garden* 72: 660–80.

———. 1990. Parsimony analysis in historical biogeography and coevolution: Methodological and theoretical update. *Systematic Zoology* 39: 14–30.

Brooks, D. R., and D. A. McLennan. 1991. *Phylogeny, Ecology and Behavior: A Research Program in Comparative Biology*. Chicago: University of Chicago Press.

———. 1993. *Parascript: Parasites and the Language of Evolution*. Washington, DC: Smithsonian Institution Press.

Brown, J. H. 1971. Mammals on mountaintops: Nonequilibrium insular biogeography. *The American Naturalist*, 105: 467–78.

———. 1978. The theory of insular biogeography and the distribution of boreal birds and mammals. *Great Basin Naturalist Memoirs* 2: 209–27.

———. 1981. Two decades of homage to Santa Rosalia:

Toward a general theory of diversity. *American Zoologist* 21: 877–88.

———. 1988. Species diversity. Pp. 57–89 in *Analytical Biogeography,* ed. A. A. Myers and P. S. Giller. London: Chapman and Hall.

———. 1995. *Macroecology.* Chicago: University of Chicago Press.

———. 1999. The legacy of Robert MacArthur: From geographical ecology to macroecology. *Journal of Mammalogy* 80: 333–44.

———. 2001. Mammals on mountainsides: Elevational patterns of diversity. *Global Ecology and Biogeography* 10: 101–9.

Brown, J. H., and A. C. Gibson. 1983. *Biogeography.* St. Louis: Mosby.

Brown, J. H., and M. V. Lomolino. 1989. Independent discovery of the equilibrium theory of island biogeography. *Ecology* 70: 1954–57.

———. 1998. *Biogeography.* 2d ed. Sunderland, MA: Sinauer.

———. 2000. Concluding remarks: Historical perspective and the future of island biogeography theory. *Global Ecology and Biogeography* 9: 87–92.

Brown, W. L., Jr., and E. O. Wilson, E. O. 1956. Character displacement. *Systematic Zoology* 7: 49–64.

Browne, J. 1983. *The Secular Ark.* New Haven: Yale University Press.

Brundin, L. 1966. Transantarctic relationships and their significance as evidenced by midges. *Kungliga Svenska Vetenskapsakademiens Handlingar,* ser. 4, 11: 1–472.

———. 1972. Phylogenetics and biogeography. *Systematic Zoology* 21: 69–79.

———. 1981. Croizat's Panbiogeography versus Phylogenetic Biogeography. Pp. 94–138, 151–158 in *Vicariance Biogeography: A Critique,* ed. G. Nelson and D. E. Rosen. New York: Columbia University Press.

———. 1988. Phylogenetic biogeography. Pp. 343–70 in *Analytical Biogeography,* ed. A. A. Myers and P. S. Giller. London: Chapman and Hall.

Bryson, R. A. 1966. Air masses, streamlines, and the boreal forests. *Geographical Bulletin* 8: 228–69.

Buffon, G. L. L., Compte de. 1761. *Histoire naturelle general.* Paris: Imprimerie Royale.

———. 1791. *Natural History, General and Particular.* Translated into English by W. Smellie. London: W. Strahan and T. Cadell.

Bush, G. L. 1969. Sympatric host race formation and speciation in frugivorous flies of the genus *Rhagoletis. Evolution* 23: 237–51.

Bush, M. B., and R. J. Whittaker. 1991. Krakatau: Colonization patterns and hierarchies. *Journal of Biogeography* 18: 341–56.

Bussing, W. A. 1985. Patterns of distribution of the Central American Icthyofauna. Pp. 453–73 in *The Great American Interchange,* ed. G. G. Stehli and S. D. Webb. New York: Plenum Press.

Cain, S. A. 1944. *Foundations of Plant Geography.* New York: Harper and Brothers.

Candolle, A. P. de. 1820. Essai elementaire de geographie botanique. In *Dictionnaire des sciences naturelles,* vol. 18. Strasbourg: Flevrault.

Carlquist, S. 1966. The biota of long-distance dispersal. I: Principles of dispersal and evolution. *The Quarterly Review of Biology* 4: 247–70.

———. 1981. Chance dispersal. *American Scientist* 69: 509–16.

CBDMS [Committee on Biological Diversity in Marine Systems]. 1995. *Understanding Marine Biodiversity. A Research Agenda for the Nation.* Washington, DC: National Academy Press.

Channell, R., and M. V. Lomolino. 2000. Dynamic biogeography and conservation of endangered species. *Nature* 403: 84–86.

Choquenot, D., and Bowman, D. M. J. S. 1998. Marsupial megafauna, aborigines and the overkill hypothesis: Application of predator-prey models to the question of Pleistocene extinction in Australia. *Global Ecology and Biogeography Letters* 7: 167–180.

Clements, F. E. 1916. *Plant Succession: An Analysis of the Development of Vegetation.* Publication no. 242. Washington, DC: Carnegie Institution of Washington.

Cliff, A. D., P. Haggett, J. D. Ord, and G. R. Versey, 1981. *Spatial Diffusion.* Cambridge: Cambridge University Press.

Cody, M. L., and J. M. Diamond, eds. 1975. *Ecology and Evolution of Communities.* Cambridge, MA: Belknap Press.

Colwell, R. K., and J. A. Coddington. 1994. Estimating terrestrial biodiversity through extrapolation. *Philosophical Transactions of the Royal Society of London,* B 345: 101–18.

Colwell, R. K., and G. C. Hurtt. 1994. Non-biological gradients in species richness and a spurious Rapoport effect. *The American Naturalist* 144: 570–595.

Connell, J. H. 1961. The influence of interspecific competition and other factors on the distribution of the barnacle *Chthamalus stellatus. Ecology* 42: 710–723.

———. 1978. Diversity in tropical rain forests and coral reefs. *Science* 199: 1302–10.

Connell, J. H., and E. Orias. 1964. The ecological regulation of species diversity. *The American Naturalist* 98: 399–414.

Connor, E. F., and D. Simberloff. 1979. The assembly of species communities: Chance or competition? *Ecology* 60: 1132–40.

———. 1983. Interspecific competition and species co-occurrence patterns on islands: Null models and the evaluation of evidence. *Oikos* 41: 455–65.

Cook, R. E. 1969. Variation in species density of North American birds. *Systematic Zoology* 18: 63–84.

Coope, G. R. 1975. Mid-Weichselian climatic changes in Western Europe, re-interpreted from coleopteran assemblages. Pp. 101–8 in *Quaternary Studies*, ed. R. P. Suggate and M. M. Cresswell. Wellington, NZ: International Union for Quaternary Research (INQUA).

Cox, C. B., I. N. Healey, and P. D. Moore. 1973. *Biogeography: An Ecological and Evolutionary Approach*. Oxford: Blackwell.

Cox, C. B., and P. D. Moore. 1985. *Biogeography: An Ecological and Evolutionary Approach*. 4th ed. Oxford: Blackwell.

Cracraft, J. 1975. Historical biogeography and earth history: perspectives for a future synthesis. *Annals of the Missouri Botanical Garden* 62: 227–50.

Craw, R. C. 1982. Phylogenetics, areas, geology and the biogeography of Croizat: A radical view. *Systematic Zoology* 31: 304–16.

———. 1984. Leon Croizat's biogeographic work: A personal appreciation. *Tuatara* 27: 8–13.

———. 1989. Panbiogeography special issue. *New Zealand Journal of Zoology* 16: 471–815.

Craw, R. C., J. R. Grehan, and M. J. Heads. 1999. *Panbiogeographphy: Tracking the History of Life*. Oxford Biogeography Series, no. 11. Oxford: Oxford University Press.

Crisci, J. V., L. Katinas, and P. Posadas. 2000. Introduccion a la teoria y practica de la biogeografia historica. Buenos Aires: Sociedad Argentina de Botanica.

Crisci, J. V., and J. J. Morrone. 1995. Historical biogeography: Introduction to methods. *Annual Review of Ecology and Systematics* 26: 373–401.

Croizat, L. 1952. *Manual of Phytogeography*. The Hague: W. Junk.

———. 1958. *Panbiogeography*. 2 vols. Caracas: By the author.

———. 1960. *Principia Botanica*. 2 vols. Caracas: By the author.

———. 1962. *Space, Time, Form: The Biological Synthesis*. Caracas: By the author.

———. 1982. Vicariance, vicariism, panbiogeography, 'vicariance biogeography,' etc. a clarification. *Systematic Zoology* 31: 291–304.

Croizat, L., G. Nelson, and D. E. Rosen. 1974. Centers of origin and related concepts. *Systematic Zoology* 23: 265–87.

Culver, S. J., and M. A. Buzas. 2000. Global latitudinal species diversity gradient in deep-sea benthic foraminifera. *Deep-Sea Research*. Part I, Oceanographic Research Papers, 47: 259–75.

Currie, D. J. 1991. Energy and large-scale patterns of animal- and plant-species richness. *The American Naturalist* 137: 27–49.

Curtis, J. T., and R. P. McIntosh. 1951. An upland forest continuum in the prairie-forest border region of Wisconsin. *Ecology* 32: 476–496.

Cushing, D. H. 1982. Climate and fisheries. London: Academic Press.

Dammerman, K. W. 1948. The fauna of Krakatau, 1883–

1933. *Verhandelingen der Koninklijke Nederlandse Akademie Van Wetenschappen Afdeling Natuurkunde*, 2d ser., 44: 1–594. Amsterdam: N. V. Noord-Hollandsche Uitgevers Maatschappij.

Dana, J. D. 1837. *A System of Mineralogy*. New Haven: Durrie and Peck, Herrick and Noyes.

———. 1848. *Manual of Mineralogy*. New Haven: Durrie and Peck.

———. 1853. On an isothermal oceanic chart illustrating the geographical distribution of marine animals. *American Journal of Science* 66: 153–57.

Darlington, P. J., Jr. 1957. *Zoogeography: The Geographic Distribution of Animals*. New York: Wiley.

———. 1959. Area, climate, and evolution. *Evolution* 13: 488–510.

———. 1965. *Biogeography of the Southern End of the World*. Cambridge, MA: Harvard University Press.

Darwin, C. 1839. *Journal of the Researches into the Geology and Natural History of Various Countries Visited by H.M.S. Beagle, under the Command of Captain Fitzroy, R.N. from 1832 to 1836*. London: Henry Colburn.

———. 1859. *On the Origin of Species by Means of Natural Selection, or The Preservation of Favoured Races in the Struggle for Life*. London: John Murray.

Davis, M. B. 1976. Pleistocene biogeography of temperate deciduous forests. *Geoscience and Man* 13: 13–26.

De Vries, H. 1901–5. *Die Mutationstheorie*. Leipzig: Von Veit.

Dexter, F. H., T. Banks, and T. Webb 1987. Modelling Holocene changes in the location and abundance of beech populations in eastern North America. *Review of Palaeobotany and Palynology* 50: 273–90.

Diamond, J. M. 1974. Colonization of exploded volcanic islands by birds: The supertramp strategy. *Science* 184: 803–6.

———. 1975a. Assembly of species communities. Pp. 342–44 in *Ecology and Evolution of Communities*, ed. M. L. Cody and J. M. Diamond. Cambridge, MA: Harvard University Press.

———. 1975b. The Island Dilemma: Lessons of modern biogeographic studies for the design of nature reserves. *Biological Conservation* 7: 129–146.

———. 2001. Australia's last giants. *Nature* 411: 755–57.

Diamond, J. M., and Gilpin, M. E. 1982. Examination of the "null" model of Connor and Simberloff for species co-occurrences on islands. *Oecologia* 52: 64–74.

Dobzhansky, T. [1937] 1951. *Genetics and the Origin of Species*. 3d ed. New York: Columbia University Press.

———. 1950. Evolution in the tropics. *American Scientist* 38: 209–21.

Docters van Leeuwen, W. M. 1936. Krakatau, 1883–1933. *Annales du jardin botanique de Buitenzorg* 46–47: 1–506.

Drake, J. A. 1990. Communities as assembled structures: Do rules govern pattern? *Trends in Ecology and Evolution* 5: 159–64.

Duffy, J. E. 1996. Resource-associated population subdivision in symbiotic coral-reef shrimps. *Evolution* 50: 360–73.

Ekman, S. 1935. *Tiergeographie des Meeres.* Leipzig: Akademische Verlagsgesellschaft.

———. 1953. *Zoogeography of the Sea.* London: Sidgwick and Jackson.

Elton, C. S. 1946. Competition and the structure of ecological communities. *Journal of Animal Ecology* 15: 54–68.

———. 1958. *The Ecology of Invasions by Animals and Plants.* London: Methuen.

Endler, J. A. 1977. *Geographic Variation, Speciation, and Clines.* Monographs in Population Biology 10. Princeton: Princeton University Press.

Engel, S. R., K. M. Hogan, J. F. Taylor, and S. K. Davis. 1998. Molecular systematics and paleobiogeography of the South American sigmodontine rodents. *Molecular Biology and Evolution* 15: 35–49.

Farris, S. 1970. Methods for computing Wagner Trees. *Systematic Zoology* 19: 83–92.

Fischer, A. G. 1960. Latitudinal variation in organic diversity. *Evolution* 14: 64–81.

Fisher, R. A. 1937. The wave of advance of advantageous genes. *Annals of Eugenics* 7: 355–69.

Fisher, R. A., A. S. Corbet, and C. B. Williams. 1943. The relation between the number of species and the number of individuals in a random sample of an animal population. *Journal of Animal Ecology* 12: 42–58.

Flannery, T. F. 1994. *The Future Eaters: An Ecological History of the Australasian Lands and People.* Sydney: Reed Books.

———. 2001. *The Eternal Frontier: Ecological History of North America and Its Peoples.* New York: Atlantic Monthly Press, 2001.

Flenley, J. R. 1979. The Late Quaternary vegetation history of the equatorial mountains. *Progress in Physical Geography* 3: 488–509.

Flessa, K. W. 1975. Area, continental drift and mammalian diversity. *Paleobiology* 1: 189–94.

Flessa, K. W., and J. J. Sepkoski, Jr. 1978. On the relationship between Phanerozoic diversity and changes in habitable area. *Paleobiology* 4: 359–366.

Forbes, E. 1844. Report on the Mollusca and Radiata of the Aegean Sea. Pp 130–93 in *Reports of the British Association of Science 1843.* London.

———. 1846. On the connection between the distribution of the existing flora and fauna of the British Isles and the geological changes which have affected their area, especially during epoch of the Northern Drift. *Memoirs of the Geological Society of Great Britain* 1: 336–42. London.

———. Forbes, E. 1856. Map of the distribution of marine life. In *The Physical Atlas of Natural Phenomena*, ed. A. K. Johnston. Philadelphia: Lea and Blanchard.

———. 1859. *The Natural History of European Seas.* London: John Van Voorst.

Forster, J. R. 1778. *Observations Made During a Voyage Round the World, on Geography, Natural History and Ethnic Philosophy.* London: G. Robinson. Reprint: *Observations Made during a Voyage round the World,* ed. N. Thomas, H. Guest, and M. Dettelbach, with a linguistics appendix by K. H. Rensch. Honolulu: University of Hawai'i Press, 1966.

Fox, B. J. 1987. Species assembly and the evolution of community structure. *Evolutionary Ecology* 1: 201–13.

Fox, H. M. 1929. Cambridge expeditions to the Suez Canal: Summary of results. *Transactions of the Zoological Society, London* 22/6, 843–63.

Fritts, H. C. 1976. *Tree Rings and Climate.* London: Academic Press.

Gaston, K. J., T. M. Blackburn, and J. I. Spicer. 1998. Rapoport's rule: Time for an epitaph? *Trends in Ecology and Evolution* 13: 70–74.

Gause, G. F. 1934. *The Struggle for Existence.* Baltimore: Williams and Wilkins.

Gilpin, M. E., and J. M. Diamond. 1982. Factors contributing to non-randomness in species co-occurrences on islands. *Oecologia* 52: 75–84.

Gleason, H. A. 1917. The structure and development of the plant association. *Bulletin of the Torrey Botanical Club* 44: 463–81.

———. 1926. The individualistic concept of the plant association. *Bulletin of the Torrey Botanical Club* 53: 7–26.

Good, R. A. 1931. A theory of plant geography. *New Phytology* 30: 149–71.

Gotelli, N. J. 2000. Null model analysis of species co-occurrence patterns. *Ecology* 81: 2606–21.

———. 2001. Research frontiers in null model analysis. *Global Ecology and Biogeography* 10: 337–43.

Gotelli, N. J., and R. K. Colwell. 2001. Quantifying biodiversity: Procedures and pitfalls in the measurement and comparison of species richness. *Ecology Letters* 4: 379–91.

Gotelli, N. J., and G. L. Entsminger. 2001. Swap and fill algorithms in null model analysis: Rethinking the Knight's Tour. *Oecologia* 129: 281–291.

Gotelli, N. J., and G. R. Graves. 1996. *Null Models in Ecology.* Washington, DC: Smithsonian Institution Press.

Gotelli, N. J., and D. J. McCabe. 2002. Species co-occurrence: A meta-analysis of J. M. Diamond's assembly rules model. *Ecology* 83: 2091–96.

Graham, R. W. 1986. Response of mammalian communities to environmental changes during the Late Quaternary. Pp. 300–313 in *Community Ecology,* ed. J. M. Diamond and T. J. Case. New York: Harper and Row.

Graham, R. W., E. L. Lundelius, Jr., M. A. Graham, E. K. Schroeder, R. S. Toomey III, E. Anderson, A. D. Barnosky, J. A. Burns, C. S. Churcher, C. K. Grayson, R. D. Guthrie, C. R. Harrington, G. T. Jefferson, L. D. Martin, H. G. McDonald, R. E. Morlan, H. A. Semken Jr., S. D. Webb, L. Werdelin and M. C. Wilson. 1996. Spatial response of mammals to late-Quaternary environmental fluctuations. *Science* 272: 1601–6.

Grant, P. R. 1966. Ecological compatibility of bird species on islands. *The American Naturalist* 100: 451–62.

———. 1986. *Ecology and Evolution of Darwin's Finches.* Princeton: Princeton University Press.

———, ed. 1998. *Evolution on Islands.* Oxford: Oxford University Press.

Gray, A. 1848. *Manual of the Botany of the Northern United States, from New England to Wisconsin and South to Ohio and Pennsylvania Inclusive.* New York: American Book Co.

Gray, A. 1856, 1857. Statistics of the flora of the northern United States. *American Journal of Science and Arts,* ser. 2, nos. 22: 204–32; 23: 62–84, 369–403.

———. 1859. Diagnostic characters of new species of phanerogamous plants collected in Japan by Charles Wright, Botanist of the U.S. North Pacific Exploring Expedition. With observations upon the relations of the Japanese flora to that of North America, and other parts of the Northern Temperate Zone. *Memoirs of the American Academy of Arts* 6: 377–452.

———. 1876. *Darwiniana: Essays and Reviews Pertaining to Darwinism.* New York: D. Appleton and Co.

Grayson, D. K., and D. B. Madsen. 2000. Biogeographic implications of recent low-elevation recolonization by Neotoma cinerea in the Great Basin. *Journal of Mammalogy* 81: 1100–1105.

Grinnell, J. 1922. The role of the accidental. *Auk* 39: 373–80.

Grinnell, J. 1943. *Joseph Grinnell's Philosophy of Nature: Selected Writings of a Western Naturalist.* Berkeley and Los Angeles: University of California Press.

Guilday, J. E. 1967. Differential extinction during Late Pleistocene and Recent times. Pp. 121–40 in *Pleistocene Extinctions: The Search for a Cause,* ed. P. S. Martin and H. E. Wright Jr. New Haven: Yale University Press.

Gurevitch, J., L. L. Morrow, A. Wallace, and J. S. Walsh. 1992. A meta-analysis of field experiments on competition. *The American Naturalist* 140: 539–72.

Haeckel, E. [1876] 1925. *The History of Creation, or the Development of the Earth and Its Inhabitants by the Action of Natural Causes,* trans. E. Ray Lankester. 2 vols. New York: D. Appleton.

Haffer, J. 1969. Speciation in Amazonian forest birds. *Science* 165: 131–37.

———. 1997. Alternative models of vertebrate speciation in Amazonia: An overview. *Biodiversity and Conservation* 6: 451–77.

Hägerstrand, T. 1967. *Innovation Diffusion as a Spatial Process.* Chicago: University of Chicago Press.

Hairston, N. G. 1964. Studies on the organization of animal communities. *Journal of Animal Ecology* 33: 227–39.

Hall, R., and J. D. Holloway. 1998. *Biogeography and Geological Evolution of S E Asia.* Leiden: Backhuys.

Hallam, A. 1967. The bearing of certain palaeogeographic data on continental drift. *Palaeogeography, Palaeoclimatology, and Palaeoecology* 3: 201–24.

Hanski, I. 1999. *Metapopulation Ecology.* Oxford: Oxford University Press.

Hawkins, B. A. 2001. Ecology's oldest pattern? *Trends in Ecology and Evolution* 16: 470.

Heaney, L. R. 2000. Dynamic disequilibrium: A long-term, large-scale perspective on the equilibrium model of island biogeography. *Global Ecology and Biogeography* 9: 59–74.

———. 2001. Small mammal diversity along elevational gradients in the Phillipines: An assessment of patterns and hypotheses. *Global Ecology and Biogeography* 10: 15–40.

Heck, K. L., Jr., G. van Belle, and D. Simberloff. 1975. Explicit calculation of the rarefaction diversity measurement and the determination of sufficient sample size. *Ecology* 56: 1459–61.

Hengeveld, R. 1990. *Dynamic Biogeography.* Cambridge: Cambridge University Press.

Hengeveld, R., and J. Haeck, 1981. The distribution of abundance. II. Models and implications. *Proc. Kon. Akad. Wetensch.* C84: 257–84.

Hengeveld, R., and G. H. Walter, 1999. The two coexisting ecological paradigms. *Acta Biotheoretica* 47: 141–70.

Hennig, W. 1966. *Phylogenetic Systematics.* Urbana: University of Illinois Press.

Holloway, J. D., and N. Jardine, 1968. Two approaches to zoogeography: A study based on the distributions of butterflies, birds and bats in the Indo-Australian area. *Proceedings of the Linnaean Society of London* 179: 153–88.

Hooker, J. D. 1844–60. *The Botany of the Antarctic Voyage of H.M. Discovery Ships "Erebus" and "Terror" in the Years 1839–1843, Under the Command of Captain Sir James Clark Ross.* London: Reeve Brothers. [Part 1: *Flora Antarctica,* 2 vols. (1844–47); part 2: *Flora Novae Zelandiae,* 2 vols. (1853–55); part 3: *Flora Tasmaniae,* 2 vols. (1855–60).]

———. 1861. Outline of the distributions of Arctic plants. *Transactions of the Linnaean Society of London* 23: 251–348.

———. 1867. Lecture on Insular Floras. London. Delivered before the British Association for the Advancement of Science at Nottingham, August 27, 1866. *Gardener's Chronicle.*

Hubbell, S. P. 1979. Tree dispersion, abundance, and diversity in a dry tropical forest. *Science* 203: 1299–1309.

———. 2001. *The Unified Neutral Theory of Biodiversity and Biogeography.* Princeton: Princeton University Press.

Hultén, E. 1937. *Outline of the History of Arctic and Boreal Biota during the Quaternary Period: Their Evolution during and after the Glacial Period as Indicated by the Equiformal Progressive Areas of Present Plant Species.* Stockholm: J. Cramer. [Repr.: Codicote, Herts.: Wheldon and Wesley/New York: Stechert-Hafner Service Agency, 1972.]

Humboldt, A. von 1805. *Essai sur la Geographie des Plantes; Accompagne d'un Tableau Physique des Regions Equinoxiales.* Paris: Levrault. Facsimile edition by the Society for the Bibliography of Natural History, Sherborn Fund Facsimiles No. 1 (1959).

———. 1815. *Personal Narrative of Travels to the Equinoctial Regions of America, during the years 1799–1804.* London: H.G. Bohn.

Humphries, C. J. 1981. Biogeographical methods and the southern beeches (Fagaceae: *Nothofagus*). Pp. 177–207 in *Advances in Cladistics: Proceedings of the First Meeting of the Willi Hennig Society,* ed. V. A. Funk and D. R. Brooks. New York: The New York Botanical Garden.

Humphries, C. J., and L. R. Parenti. 1986. *Cladistic Biogeography.* Oxford Monographs on Biogeography no. 2. Oxford: Clarendon.

———. 1999. *Cladistic Biogeography: Interpreting Patterns of Plant and Animal Distributions.* 2d ed. Oxford: Oxford University Press.

Huntley, B. 1980. Europe. Pp. 341–83 in *Vegetation History,* ed. B. Huntley and T. Webb. Dordrecht: Kluwer.

Hurlbert, S. H. 1971. The nonconcept of species diversity: A critique and alternative parameters. *Ecology* 52: 577–585.

Hutchinson, G. E. 1957a. *A Treatise on Limnology,* vol. 1: *Geography, Physics, and Chemistry.* New York: John Wiley and Sons.

———. 1957b. Concluding remarks. *Cold Spring Harbor Symposia on Quantitative Biology* 22: 415–27.

———. 1959. Homage to Santa Rosalia, or why are there so many kinds of animals? *The American Naturalist* 93: 145–59.

Huxley, J. S. 1943. *Evolution, the Modern Synthesis.* London: Allen and Unwin.

Ihering, H. von. 1900. The history of the Neotropical region. *Science* 12: 857–64.

———. 1927. *Die Geschichte des Atlantischen Ozeans.* Jena: Gustav Fischer.

Imbrie, J., and K. P. Imbrie. 1979. *Ice Ages, Solving the Mystery.* Short Hills, NJ: Enslow.

Ives, A. R., P. M. Kareiva and R. Perry. 1993. Response of a predator to variation in prey density at three hierarchical scales: Lady beetles feeding on aphids. *Ecology* 74: 1929–38.

Jablonski, D., Flessa, K. W., and Valentine, J. W. 1985. Biogeography and paleobiology. *Paleobiology* 11: 75–90.

Jaccard, P. 1901. Étude comparative de la distribution florale dans une portion des Alpes et du Jura. *Bulletin de la Société Vaudoise de la science naturelle* 37: 547–79.

Janzen, D. H. 1967. Why mountain passes are higher in the tropics. *The American Naturalist* 101: 233–49.

Järvinen, O. 1982. Species-to-genus ratios in biogeography: A historical note. *Journal of Biogeography* 9: 363–70.

Johnston, A. K. 1856. *The Physical Atlas of Natural Phenomena.* Edinburgh: William Blackwood and Sons.

Jokiel, P. L., 1990. Long distance dispersal by rafting: Reemergence of an old hypothesis. *Endeavour* (Oxford), 14: 66–73.

Jordan, D. S. 1905. The origin of species through isolation. *Science* 22:545–62.

Kiester, A. R. 1971. Species density of North American amphibians and reptiles. *Systematic Zoology* 20: 127–37.

Kikkawa, J., and E. E. Williams. 1971. Altitudinal distribution of land birds in New Guinea. *Search* 2: 64–69.

Kircher, A. 1675. *Arca Noe.* Amsterdam: Waesberg.

Klages, K. H. W. 1942. *Ecological Crop Geography.* New York: Macmillan.

Klopfer, P. H. 1959. Environmental determinants of faunal diversity. *The American Naturalist* 93: 337–42.

Klopfer, P. H., and R. H. MacArthur. 1960. Niche size and faunal diversity. *The American Naturalist* 94: 293–300.

Knoll, A. H., K. J. Niklas, and B. H. Tiffney. 1979. Phanerozoic land-plant diversity in North America. *Science* 206: 1400–1402.

Lack, D. 1947. *Darwin's Finches.* Cambridge: Cambridge University Press.

Lamb, H. H. 1972. *Climate: Present, Past and Future,* vol. 1: *Fundamentals and Climate Now.* London: Methuen.

———. 1977. *Climate: Present, Past and Future,* vol. 2: *Climatic History and the Future.* London: Methuen.

Lawlor, T. E. 1998. Biogeography of great mammals: Paradigm lost? *Journal of Mammalogy* 79: 1111–30.

Lessios, H. A. 1998. The first stage of speciation as seen in organisms separated by the Isthmus of Panama. Pp. 186–201 in *Endless Forms: Species and Speciation,* ed. D. J. Howard and S. H. Berlocher. New York: Oxford University Press.

Lincoln, G. A. 1975. Bird counts either side of Wallace's Line. *J. Zool. Lond.* 177: 349–61.

Linnaeus, C. 1781. On the increase of the habitable earth. *Amoenitates Academicae* 2: 17–27. Translation by F. J. Brandt.

Livingston, B. E., and F. Shreve. 1921. *The Distribution of Vegetation in the United States, as Related to Climatic Conditions.* Washington DC: Carnegie Institute of Washington.

Lomolino, M. V. 1996. Investigating causality of nestedness of insular communities: selective immigrations or extinctions? *Journal of Biogeography* 23: 699–703.

———. 2000a. A call for a new paradigm of island biogeography. *Global Ecology and Biogeography* 9: 1–6.

———. 2000b. Ecology's most general, yet protean pattern: The species-area relationship. *Journal of Biogeography,* 27: 17–26.

———. 2000c. A species-based theory of insular zoogeography. *Global Ecology and Biogeography,* 9: 39–58.

———. 2001. Elevational gradients of species-density: Historical and prospective views. *Global Ecology and Biogeography* 10: 3–14.

Lomolino, M. V., and R. Channell. 1995. Splendid isolation: Patterns of range collapse in endangered mammals. *Journal of Mammalogy* 76: 335–47.

Lomolino, M. V., and R. Davis. 1997. Biogeography scale and biodiversity of mountain forest mammals of western North America. *Global Ecology and Biogeography Letters,* 6: 57–76.

Lomolino, M. V., and D. R. Perault. 2000. Assembly and disassembly of mammal communities in a fragmented temperate rainforest. *Ecology* 81: 1517–32.

Lomolino, M. V., and M. D. Weiser. 2001. Towards a more general species-area relationship: Diversity on all islands, great and small. *Journal of Biogeography* 28: 431–45.

Lyell, C. [1832] 1991. *Principles of Geology,* vol. 2. Chicago: University of Chicago Press.

MacArthur, R. H. 1955. Fluctuations of animal populations and the measure of community stability. *Ecology* 36: 533–36.

———. 1965. Patterns of species diversity. *Biological Review* 40: 510–33.

———. 1972. *Geographical Ecology: Patterns in the Distribution of Species.* New York: Harper and Row.

MacArthur, R. H., and R. Levins. 1967. The limiting similarity, convergence, and divergence of coexisting species. *The American Naturalist* 101: 377–85.

MacArthur, R. H., and E. O. Wilson. 1963. An equilibrium theory of insular zoogeography. *Evolution* 17: 373–87.

———. 1967. *The Theory of Island Biogeography.* Monographs in Population Biology 1: 1–203. Princeton: Princeton University Press.

MacPhee, R. D. E., and P. A. Marx. 1997. The 40,000-year plague: Humans, hyperdisease, and first-contact extinctions. Pp. 169–217 in *Human Impact and Natural Change in Madagascar,* ed. S. Goodman and B. D. Patterson. Washington, DC: Smithsonian Institution Press.

Magurran, A. E., and R. M. May, eds. 1999. *Evolution of Biological Diversity.* Oxford: Oxford University Press.

Maillefer, A. 1929. Le Coefficient générique de P. Jacard et sa signification. *Mémoires de la Société Vaudoise des Sciences Naturelles* 3: 113–83.

Manly, B. F. J. 1995. A note on the analysis of species co-occurrences. *Ecology* 76: 1109–15.

Marshall, L. G., S. D. Webb, J. J. Sepkoski, Jr., and D. M. Raup. 1982. Mammalian evolution and the Great American Interchange. *Science* 215, 1351–57.

Martin, P. S. 1973. The discovery of America. *Science* 179: 969–74.

Martin, P. S., and H. E. Wright Jr., eds. 1967. *Pleistocene Extinctions: The Search for a Cause.* New Haven: Yale University Press.

Matthew, W. D. 1915. Climate and evolution. *Annals of the New York Academy of Science* 24, art. 6: 201–10. [Repr.: Special Publications of the New York Academy of Sciences, vol. 1; 2d ed., rev. and enlarged, 1939.]

May, R. M. 1978. The evolution of ecological systems. *Scientific American* 239: 160–75.

Mayden, R. L. 1991. The wilderness of panbiogeography: A synthesis of space, time and form? Review of *Panbiogeography* special issue, *New Zealand Journal of Zoology* 16, no. 4. *Systematic Zoology* 40: 503–519.

Mayr, E. 1942. *Systematics and the Origin of Species.* New York: Columbia University Press.

———. 1963. *Animal Species and Evolution.* Cambridge, MA: Belknap Press.

———. 1982. *The Growth of Biological Thought.* Cambridge, MA: Belknap Press.

Meltzer, D. J. 1997. Monte Verde and the Pleistocene peopling of the Americas. *Science* 276: 754.

Merriam, C. H. 1890. Results of a biological survey of the San Francisco Mountain region and the desert of the Little Colorado, Arizona. U.S. Department of Agriculture, *North American Fauna,* no. 3. Washington, DC: Government Printing Office.

———. 1894. Laws of temperature control of the geographic distribution of terrestrial animals and plants. *National Geographic* 6: 229–38.

Meusel, H. 1943. *Vergleichende Arealkunde.* 2 vols. Berlin-Zellendorf: Borntraeger.

Mikkelson, G. M. 1993. How do food webs fall apart? A study of changes in trophic structure during relaxation on habitat fragments. *Oikos* 67: 539–47.

Milberg, P., and T. Tyberg. 1993. Naïve birds and noble savages: A review of man-caused prehistoric extinctions of islands birds. *Ecography,* 16: 229–50.

Miller, A. I. 2000. Conversations about Phanerozoic global diversity. Pp. 53–73 in *Deep Time: Paleobiology's Perspective,* ed. D. H. Erwin and S. L. Wing. *Paleobiology* 26, supplement.

Mills, E. L. 1984. A view of Edward Forbes, naturalist. *Archives of Natural History* 11: 365–93.

Mollison, D. 1977. Spatial contact models for ecological and epidemic spread. *Journal of the Royal Statistical Society* B39: 283–326.

Moreau, R. E. 1948. Ecological isolation in a rich tropical avifauna. *Journal of Animal Ecology* 17: 113–26.

———. 1966. *The Bird Faunas of Africa and Its Islands.* New York: Academic Press.

Moritz, C., J. L. Patton, C. J. Schneider, and T. B. Smith. 2000. Diversification of rainforest faunas: An integrated molecular approach. *Annual Review of Systematics and Ecology* 31: 533–63.

Munroe, E. G. 1948. The geographical distribution of butterflies in the West Indies. Ph.D. diss., Cornell University.

Myers, A. A. 1991. How did Hawaii accumulate its biota? A test from the Amphipoda. *Global Ecology and Biogeography Letters* 1: 24–29

———. 1994. Biogeographic patterns in shallow-water marine systems and the controlling processes at different scales. Pp. 547–74 in *Aquatic Ecology, Scale, Pattern and Process,* ed. P. S. Giller, A. G. Hildrew, and D. G. Raffaelli. Proceedings of the 34th Symposium of the British Ecological Society. Oxford: Blackwell Scientific Publications.

———. 1997. Biogeographic barriers and the development of marine biodiversity. *Estuarine, Coastal and Shelf Science* 44: 241–48.

Myers, A. A., and Giller, P. S., eds. 1988. *Analytical Biogeography: An Integrated Approach to the study of Animal and Plant Distributions.* London: Chapman and Hall.

Nelson, G. 1969. The problem of historical biogeography. *Systematic Zoology* 18: 243–46.

———. 1973. Comments on Leon Croizat's Biogeography. *Systematic Zoology* 22: 312–20.

———. 1974. Historical biogeography: An alternative formalization. *Systematic Zoology* 23: 555–58.

———. 1978. From Candolle to Croizat: Comments on the history of biogeography. *Journal of the History of Biology* 11: 269–305.

———. 1981. Summary. Pp. 524–37 in *Vicariance Biogeography: A Critique,* ed. G. Nelson and D. E. Rosen. New York: Columbia University Press.

Nelson, G., and P. Y. Ladiges. 1995. TASS version 2.0. Three area subtrees. New York and Melbourne: Published by the authors.

———. 1991. Three-area statements: Standard assumptions for biogeographic analysis. *Systematic Zoology* 40: 470–85.

Nelson, G., and N. Platnick. 1980. A vicariance approach to historical biogeography. *Bioscience* 30: 339–43.

———. 1981. *Systematics and Biogeography: Cladistics and Vicariance.* New York: Columbia University Press.

———. 1984. *Biogeography.* Carolina Biology Readers no. 119. Burlington, NC: Carolina Biological Supply Company.

———. 1991. Three-taxon statements: A more precise use of parsimony? *Cladistics* 7: 351–66.

Nelson, G., and D. E. Rosen, eds. 1981. *Vicariance Biogeography: A Critique.* New York: Columbia University Press.

Nur, A., and Z. Ben-Avraham. 1982. Displaced terranes and mountain building. Pp 73–84 in *Mountain Building Processes,* ed. K. J. Hsü. London: Academic Press.

Okubo, A. 1980. *Diffusion and Ecological Problems: Mathematical Models.* Berlin: Springer.

Olson, S. L., and H. F. James. 1982. Fossil birds from the Hawaiian Islands: Evidence for wholesale extinction by man before Western contact. *Science* 217: 633–35.

———. 1984. The role of Polynesians in the extinction of the avifauna of the Hawaiian Islands. Pp. 768–80 in *Quarternary Extinctions,* ed. P. S. Martin and R. G. Klein. Tucson: University of Arizona Press.

Otte, D., and J. A. Endler, eds. 1989. *Speciation and Its Consequences.* Sunderland, MA: Sinauer.

Page, R. D. M. 1989. Quantitative cladistic biogeography: constructing and comparing area cladograms. *Systematic Zoology* 37: 254–70.

———. 1993. *COMPONENT, version 2.0.* London: The Natural History Museum.

Paine, R. T. 1966. Food web complexity and species diversity. *The American Naturalist* 100: 65–76.

Palmgren, A. 1929. Die Artenzahl als pflanzengeographischer Charakter sowie die Zufall und die sekuläre Landhebung als pflanzengeographische Faktoren. Ein pflanzengeographischer Entwurf, basiert auf Material aus dem åländischen Schärenarchipel. *Acta Botanica Fennica* 1: 1–143.

Patterson, B. D. 1987. The principle of nested subsets and its implications for biological conservation. *Conservation Biology* 1: 323–34.

———. 1990. On the temporal development of nested subject patterns of species composition. *Oikos* 59: 330–42.

Patterson, B. D., and W. Atmar. 1986. Nested subsets and the structure of insular mammalian faunas and archipelagos. *Biological Journal of the Linnaean Society* 28: 65–82.

Patterson, B. D., and J. H. Brown. 1991. Regionally nested patterns of species composition in granivorous rodent assemblages. *Journal of Biogeography* 18: 395–402.

Pianka, E. R. 1966. Latitudinal gradients in species diversity: A review of concepts. *The American Naturalist* 100: 33–46.

Platnick, N. 1976. Drifting spiders or continents? Vicariance biogeography of the spider subfamily Laroniinae (Araneae: Gnaphosidae). *Systematic Zoology* 25: 101–9.

———. 1981. Discussion. Pp. 144–50 in *Vicariance Biogeography: A Critique,* ed. G. Nelson and D. E. Rosen. New York: Columbia University Press.

Platnick, N., and G. Nelson. 1978. A method of analysis for historical biogeography. *Systematic Zoology* 27: 1–16.

Pólya, G. 1930. Eine Wahrscheinlichkeitsaufgabe in der Pflanzensoziologie. *Vierteljahrsschrift der Naturforschenden Gesellschaft in Zürich* 75: 211–19.

Por, F. D. 1971. One hundred years of Suez Canal: A century of Lessepsian migration, Retrospect and viewpoints. *Systematic Zoology* 10: 138–59.

Pregill, G. K., and S. L. Olson. 1981. Zoogeography of West Indian vertebrates in relation to Pleistocene climate cycles. *Annual Review of Ecology and Systematics* 12: 75–98.

Premoli, A. C., T. Kitzberger, and T. T. Veblen, 2000. Isozyme variation and recent biogeographical history of the long-lived conifer *Fitzroya cupressoides. Journal of Biogeography* 27: 251–60.

Preston, F. W. 1960. Time and space and the variation of species. *Ecology* 41: 785–90.

———. 1962. The canonical distribution of commonness and rarity, parts 1 and 2. *Ecology* 43: 185–215, 410–32.

Pulliam, H. R. 1996. Sources and sinks: Empirical evidence and population consequences. Pp. 45–69 in *Population Dynamics in Ecological Space and Time,* ed. O. E. Rhodes Jr., R. K. Chesser, and M. H. Smith. Chicago: University of Chicago Press.

Rahbek, C. 1997. The relationship among area, elevation and regional species richness in neotropical birds. *The American Naturalist* 149: 875–902.

Rahbek, C., and G. R. Graves. 2001. Multiscale assessment of patterns of avian species richness. *Proceedings of the National Academy of Sciences, USA* 98: 4534–4539.

Rainey, R. C. 1963. Meteorology and the migration of desert locusts. *Anti-Locust Mem.* 7: 1–115.

Ramensky, L. G. [1924] 1965. *Basic Regularities of Vegetation Cover and Their Study* [excerpt, trans. from Russian]. Pp. 151–52 in *Readings in Ecology*, ed. E. J. Kormondy. Englewood Cliffs, NJ: Prentice Hall.

Rapoport, E. H. 1982. *Areography: Geographical Strategies of Species*, trans. B. Drausal. New York: Pergamon Press.

Raup, D. M. 1972. Taxonomic diversity during the Phanerozoic. *Science* 177: 1065–71.

———. 1979. Size of the Permo-Triassic bottleneck and its evolutionary implications. *Science* 206: 217–18.

Raup, D. M., S. J. Gould, T. J. M. Schopf, and D. S. Simberloff. 1973. Stochastic models of phylogeny and the evolution of diversity. *Journal of Geology* 81: 525–42.

Raxworthy, C. J., M. R. Forstner, and R. A. Nussbaum. 2002. Chameleon radiation by oceanic dispersal. *Nature* 415: 784–86.

Real, L. A., and Brown, J. H., eds. 1991. *Foundations of Ecology*. Chicago: University of Chicago Press.

Rehbock, P. F. 1984. Edward Forbes (1815–1854): An annotated list of published and unpublished writings. *Archives of Natural History* 9: 171–218.

Rejmánek, M. 1996. Species richness and resistance to invasions. Pp. 153–72 in *Biodiversity and Ecosystem Processes in Tropical Forests*, ed. G. H. Orians, R. Dirzo, and J. H. Cushman. Ecological Studies, no. 122. New York: Springer-Verlag.

Rensch, B. 1929. *Das Prinzip geographischer Rassenkreise und das Problem der Artbildung*. Berlin: Borntraeger.

Rensch, B. 1960. *Evolution above the Species Level*. New York: Columbia University Press.

Rex, M. A., C. T. Stuart, and G. Coyne. 2000. Latitudinal gradients of species richness in the deep-sea benthos of the North Atlantic. *Proceedings of the National Academy of Sciences, USA* 97: 4082–85.

Rex, M. A., C. T. Stuart, R. R. Hessler, J. A. Allen, H. L. Sanders and G. D. F. Wilson. 1993. Global-scale patterns of species diversity in the deep sea benthos. *Nature* 365: 636–39.

Rickart, E. A. 2001. Elevational diversity gradients, biogeography and the structure of montane mammal communities in the intermountain region of North America. *Global Ecology and Biogeography* 10: 77–100.

Ricklefs, R. E., and Cox, G. W. 1972. Taxon cycles in the West Indian avifauna. *The American Naturalist* 106: 195–219.

———. 1978. Stage of taxon cycle, habitat distribution, and population density in the avifauna of the West Indies. *The American Naturalist* 112: 875–95.

Ricklefs, R. E., and E. Bermingham. 2002. The concept of the taxon cycle in biogeography. *Global Ecology and Biogeography* 11: 353–61.

Ricklefs, R. E., and D. Schluter, eds. 1993. *Species Diversity in Ecological Communities: Historical and Geographical Perspectives*. Chicago: University of Chicago Press.

Rieseberg, L. H. 2001. Chromosomal rearrangements and speciation. *Trends in Ecology and Evolution* 16: 351–58.

Rohde, K. 1992. Latitudinal gradients in species diversity: The search for the primary cause. *Oikos* 65: 514–27.

Rohde, K., M. Heap, and D. Heap. 1993. Rapoport's rule does not apply to marine teleosts and cannot explain latitudinal gradients in species richness. *The American Naturalist* 142: 1–16.

Ronquist, F. 1994. Ancestral areas and parsimony. *Systematic Biology* 43: 267–274.

Rosen, B. R. 1984. Reef and coral biogeography and climate through the late Cainozoic: Just islands in the sun or a critical pattern of islands? Pp. 201–62 in *Fossils and Climate*, ed. P. J. Brenchley. *Geological Journal* Special Issue 11.

Rosen, D. E. 1975. A vicariance model of Caribbean biogeography. *Systematic Zoology* 24: 431–64.

Rosen, D. E. 1978. Vicariant patterns and historical explanations in biogeography. Systematic Zoology 27: 159–188.

———. 1979. Fishes from the uplands and intermontane basins of Guatemala: Revisionary studies and comparative geography. *Bulletin of the American Museum of Natural History* 162: 267–376.

Rosenzweig, M. L. 1992. Species diversity gradients: We know more and less than we thought. Journal of Mammalogy 73: 715–730.

Rosenzweig, M. L. 1995. *Species Diversity in Time and Space*. Cambridge: Cambridge University Press.

Ross, H. H. 1974. *Biological Systematics*. Reading, PA: Addison-Wesley.

Rotondo, G. M., V. G. Springer, G. A. J. Scott, and S. O. Schlanger. 1981. Plate movement and island integration: A possible mechanism in the formation of endemic biotas, with special reference to the Hawaiian islands. *Systematic Zoology* 30(1): 12–21

Roughgarden, J. 1983. Competition and theory in community ecology. *The American Naturalist* 122: 583–601.

Roy, K., D. Jablonski, and J. W. Valentine. 1994. Eastern Pacific molluscan provinces and latitudinal diversity gradient: No evidence for Rapoport's rule. *Proceedings of the National Academy of Sciences, U.S.A.* 91: 8871–74.

Sagar, G. R., and J. L. Harper. 1964. Biological flora of the British Isles. *Plantago major* L., *P. media* L. and *P. lanceolata* L. *Journal of Ecology* 52: 189–221.

Sanders, H. L. 1968. Marine benthic diversity: A comparative study. *The American Naturalist* 102: 243–82.

Sanderson, J. G., M. P. Moulton, and R. G. Selfridge. 1998. Null matrices and the analysis of species co-occurrences. *Oecologia* 116: 275–83.

Sax, D. F. 2001. Latitudinal gradients and geographic ranges of exotic species: Implications for biogeography. *Journal of Biogeography* 28: 139–50.

Schall, J. J., and E. R. Pianka. 1978. Geographical trends in numbers of species. *Science* 201: 679–86.

Schlanger, S. O., and I. Premoli-Silva. 1981. Tectonic, volcanic and palaeographic implications of redeposited reef fauna of late Cretaceous and Tertiary age from the Naura Basin and the Line Islands. Pp. 817–28 in *Initial Reports of the Deep Sea Drilling Program*, ed. R. L. Larson and S. O. Schlanger. Washington, DC: U.S. Government Printing Office.

Schluter, D. 2001. Ecology and the origin of species. *Trends in Ecology and Evolution* 16: 372–80.

Schoener, T. W., and D. A. Spiller. 1987. High population persistence in a system with high turnover. *Nature* 330: 474–77.

Sclater, P. L. 1858. On the general geographical distribution of the members of the class Aves. *Journal of the Linnean Society of London, Zoology* 2: 130–45.

Sepkoski, J. J., Jr. 1976. Species diversity in the Phanerozoic: Species-area effects. *Paleobiology* 2: 298–303.

———. 1978. A kinetic analysis of Phanerozoic taxonomic diversity, I: Analysis of marine orders. *Paleobiology* 4: 223–51.

Sepkoski, J. J., Jr., R. K. Bambach, D. M. Raup, and J. W. Valentine. 1981. Phanerozoic marine diversity and the fossil record. *Nature* 293: 435–437.

Shreve, F. 1915. *The Vegetation of a Desert Mountain Range as Conditioned by Climatic Factors.* Washington, DC: Carnegie Institution of Washington.

Simberloff, D. S. 1969. Experimental zoogeography of islands: A model for insular colonization. *Ecology* 50: 296–314.

———. 1970. Taxonomic diversity of island biotas. *Evolution* 24: 23–47.

———. 1976. Species turnover and equilibrium island biogeography. *Science* 194: 572–78.

———. 1978. Using island biogeographic distributions to determine if colonization is stochastic. *The American Naturalist* 112: 713–26.

———. 1980. A succession of paradigms in ecology: Essentialism to materialism and probabilism. *Synthese* 43: 3–39.

Simberloff, D., and E. F. Connor. 1979. Q-mode and R-mode analyses of biogeographic distributions: Null hypotheses based on random colonization. Pp. 123–38 in *Contemporary Quantitative Ecology and Related Ecometrics*, ed. G. P. Patil and M. L. Rosenzweig. Fairland, MD: International Cooperative Publishing House.

Simberloff, D., and J.-L. Martin. 1991. Nestedness of insular avifaunas: Simple summary statistics masking complex species patterns. *Ornis Fennica* 68: 178–92.

Simberloff, D. S., and E. O. Wilson. 1969. Experimental zoogeography of islands: The colonization of empty islands. *Ecology* 50: 278–96.

———. 1970. Experimental zoogeography of islands: A two-year record of colonization. *Ecology* 51: 934–37.

Simpson, G. G. 1940. *Mammals and Land Bridges.* Publication no. 30, 137–63. Washington, DC: National Academy of Sciences.

———. 1943. Mammals and the nature of continents. *American Journal of Science* 241: 1–31.

———. 1944. *Tempo and Mode in Evolution.* New York: Columbia University Press.

———. 1946. Tertiary land bridges. *Transactions of the New York Academy of Sciences,* ser. 2, 8: 255–58

———. 1952. Probabilities of dispersal in geologic time. *Bulletin of the American Museum of Natural History* 99: 163–76.

———. 1964. Species density of North American recent mammals. *Systematic Zoology* 13: 57–73.

———. 1977. Too many lines: The limits of the Oriental and Australian zoogeographic region. *Proceedings of the American Philosophical Society* 121: 107–20.

Skellam, J. G. 1951. Random dispersal in theoretical populations. *Biometrika* 38: 196–218.

Skottsberg, C. 1922. *Natural History of Juan Fernandez and Easter Island,* vol. 2: The phanerogams of the Juan Fernandez Islands. Uppsala: Almqvist and Wiksells.

———. 1925. Juan Fernandez and Hawaii: A phytogeographical discussion. *Bishop Museum Bulletin* 16: 1–47

Slaughter, B. H. 1967. Animal ranges as a cue to Late-Pleistocene extinctions. Pp. 155–67 in *Pleistocene Extinctions: The Search for a Cause*, ed. P. S. Marion and H. E. Wright Jr. New Haven: Yale University Press.

Snider-Pellegrini, A., 1858. *La Création et ses Mystères Dévoilees.* Paris: A. Franck.

Springer, V. G. 1982. *Pacific Plate Biogeography with Special Reference to Shore Fishes.* Smithsonian Contributions to Zoology, no. 367, 1–182. Washington, DC: Smithsonian Institution Press.

Steadman, D. W. 1993. Biogeography of Tongan birds before and after human contact. *Proceedings of the National Academy of Sciences* 90: 818–22.

———. 1995. Prehistoric extinctions of Pacific island birds: Biodiversity meets zooarchaeology. *Science* 267: 1123–31.

———. 1997. Human caused extinction of birds. Pp. 139–61 in *Biodiversity II: Understanding and Protecting our Biological Resources*, ed. M. L. Reaka-Kudla, D. E. Wilson, and E. O. Wilson. Washington, DC: Joseph Henry Press.

Stebbins, G. L., Jr. 1950. *Variation and Evolution in Plants.* New York: Columbia University Press.

Stehli, F. G. 1968. Taxonomic diversity gradients in pole locations: The recent model. Pp. 163–227 in *Evolution and Environment*, E. T. Drake. Peabody Museum Centennial Symposium. New Haven: Yale Unversity Press.

Stehli, F. G., and S. D. Webb, eds. 1985. *The Great American Biotic Interchange.* New York: Plenum Press.

Stevens, G. C. 1989. The latitudinal gradients in geographic range: How so many species coexist in the tropics. *The American Naturalist* 132: 240–56.

Stone, L., and A. Roberts. 1990. The checkerboard score and species distributions. *Oecologia* 85: 74–79.

Stott, P. 1981. *Historical Plant Geography: An Introduction.* London: Allen and Unwin.

Stresemann, E. 1919. Uber die europaischen Baumlaufer. *Verh. Ornithol. Ges. Bayern* 14: 39–74.

Strong, D. R., Jr. 1980. Null hypotheses in ecology. *Synthese* 43: 271–85.

Strong, D. R., Jr., D. Simberloff, L. G. Abele, and A. B. Thistle, eds. 1984. *Ecological Communities: Conceptual Issues and the Evidence*. Princeton: Princeton University Press.

Strong, D. R., Jr., L. A. Szyska, and D. S. Simberloff. 1979. Tests of community-wide character displacement against null hypotheses. *Evolution* 33: 897–913.

Taylor, P. H., and S. D. Gaines. 1999. Can Rapoport's rule be rescued? Modeling causes of the latitudinal gradient in species richness. *Ecology* 80: 2474–82.

Terborgh, J. 1973. On the notion of favorableness in plant ecology. *The American Naturalist* 107: 481–501.

———. 1977. Bird species diversity on an Andean elevational gradient. *Ecology* 58: 1007–19.

Thomas, C. D., E. J. Bodsworth, R. J. Wilson, A. D. Simmons, Z. G. Davies, M. Musche, and L. Conradt. 2001. Ecological and evolutionary processes at expanding range margins. *Nature* 411: 597–81.

Thorne, R. F., 1963. Biotic distribution patterns in the tropical Pacific. Pp. 311–54 in *Pacific Basin Biogeography*, ed. J. J. Gresit. Honolulu: Bishop Museum Press.

Thornton, I. W. B. 1992. K. W. Dammerman: Fore-runner of island equilibrium theory. *Global Ecology and Biogeography Letters*, 2: 145–48.

———. 1996. *Krakatau: The Destruction and Reassembly of an Island Ecosystem*. Cambridge, MA: Harvard University Press.

Udvardy, M. D. F, 1969. *Dynamic Zoogeography, with Special Reference to Land Animals*. New York: Van Nostrand Reinhold.

Van Balgooy, M. M. J. 1971. Plant-geography of the Pacific. Supplement, *Blumea* 6: 1–222.

Valentine, J. W. 1969. Patterns of taxonomic and ecological structure of the shelf benthos during Phanerozoic time. *Paleontology* 12: 684–709.

Valentine, J. W., T. C. Foin, and D. Peart. 1978. A provincial model of Phanerozoic marine diversity. *Paleobiology* 4: 55–66.

Van den Bosch, F., R. Hengeveld, and J. A. J. Metz. 1992. Analysing the velocity of range expansion. *J. Biogeogr.* 19: 135–50.

Van den Bosch, F., J. A. J. Metz and O. Diekmann 1990. The velocity of population expansion. *J. Math. Biol.* 28: 529–65.

Van der Plank, J. E. 1963. *Plant Diseases: Epidemics and Control*. New York: Academic Press.

Van Valen, L. 1973. A new evolutionary law. *Evolutionary Theory* 1: 1–33.

Vanzolini, P. E., and W. R. Heyer. 1985. The American herptofauna and the interchange. Pp. 475–83 in *The Great American Interchange*, ed. F. G. Stehli and S. D. Webb. New York: Plenum Press.

Vermeij, G. J. 1978. *Biogeography and Adaptation: Patterns of Marine Life*. Cambridge, MA: Harvard University Press.

———. 1987. *Evolution and Escalation: An Ecological History of Life*. Princeton: Princeton University Press.

Via, S. 2001. Sympatric speciation in animals: The ugly duckling grows up. *Trends in Ecology and Evolution* 16: 381–90.

Vuilleumier, F. 1985. Fossil and recent avifaunas and the Interamerican interchange. Pp. 387–424 in *The Great American Biotic Interchange*, ed. F. G. Stehli and S. D. Webb. New York: Plenum Press.

Wagner, W. L., and V. A. Funk, eds. 1995. *Hawaiian Biogeography: Evolution on a Hot Spot Archipelago*. Washington, DC: Smithsonian Institution Press.

Walker, D., and J. R. Flenley 1979. Late Quaternary vegetational history of the Enga District of Upland Papua New Guinea. *Philos. Trans. Roy. Soc. B* 286: 265–344.

Wallace, A. R. 1876. *The Geographical Distribution of Animals*. 2 vols. London: Macmillan.

———. 1878. *Tropical Nature and Other Essays*. New York: Macmillan.

———. 1880. *Island Life: Or, the Phenomena and Causes of Insular Faunas and Floras*. London: Macmillan.

———. 1910. *The World of Life: A Manifestation of Creative Power, Directive Mind and Ultimate Purpose*. London: Chapman and Hall.

Walter, G. H., and R. Hengeveld. 2000. The structure of the two ecological paradigms. *Acta Biotheor.* 48: 15–46.

Webb, S. D. 1991. Ecogeography and the Great American Interchange. *Paleobiology* 17: 266–280.

Wegener, A. L. 1912, 1929. "Die Enstehung der Kontinente" *Dr. A. Petermanns Mitteilungen aus Justus Perthes' Geographischer Anstart year 58* (April), 185–95, (May), 253–56, (June), 305–9.

———. [1924] 1929. *The Origin of Continents and Oceans*, trans. J. G. A. Skerl. 4th ed. London: Methuen.

Weiher, E., and P. Keddy, eds. 1999. *Ecological Assembly Rules: Perspectives, Advances, Retreats*. Cambridge: Cambridge University Press.

Wells, P. V., and R. Berger. 1967. Late Pleistocene history of coniferous woodland in the Mohave Desert. *Science* 155: 1640–47.

Wells, P. V., and C. D. Jorgensen, 1964. Pleistocene woodland middens and climatic change in Mohave Desert: A record of Juniper woodlands. *Science* 143: 1171–74.

Werner, T. K., and T. W. Sherry. 1987. Behavioural feeding specialization in *Pinaroloxias inornata*, the "Darwin's Finch" of Cocos Island, Costa Rica. *Proceedings of the National Academy of Sciences USA* 84: 5506–10.

White, M. J. D. 1978. *Modes of Speciation*. San Francisco: W. H. Freeman.

Whitmore, T. C., ed. 1981. *Wallace's Line and Plate Tectonics*. Oxford: Oxford University Press.

Whitmore, T. C., and G. T. Prance, eds. 1987. *Biogeography and Quaternary History on Tropical America*. Oxford: Oxford University Press.

Whittaker, R. H. 1956. Vegetation of the Great Smoky Mountains. *Ecological Monographs* 22: 1–44.

———. 1960. Vegetation of the Siskiyou Mountains, Oregon and California. *Ecological Monographs* 30: 279–338.

———. 1967. Gradient analysis of vegetation. *Biological Reviews* 42: 207–264.

———. [1970] 1975. *Communities and Ecosystems.* 2d ed. New York: Macmillan.

Whittaker, R. H., and W. A. Niering. 1964. Vegetation of the Santa Catalina Mountains, Arizona, I: Ecological classification and distributions of species. *Journal of the Arizona Academy of Science* 3: 9–34.

———. 1965. Vegetation of the Santa Catalina Mountains, Arizona: A gradient analysis of the south slope. *Ecology* 46: 429–52.

———. 1968a. Vegetation of the Santa Catalina Mountains, Arizona, III: Species distribution and floristic relations on the north slope. *Journal of the Arizona Academy of Science* 5: 3–21.

———. 1968b. Vegetation of the Santa Catalina Mountains, Arizona, IV: Limestone and acid soils. *Journal of Ecology* 56: 523–44.

———. 1975. Vegetation of the Santa Catalina Mountains, Arizona, V: Biomass, production and diversity along the elevation gradient. *Ecology* 56: 771–90.

Whittaker, R. J. 1998. Island Biogeography: Ecology, Evolution, and Conservation. Oxford: Oxford University Press.

———. 2000. Scale, succession and complexity in island biogeography: Are we asking the right questions? *Global Ecology and Biogeography* 9: 75–85.

Whittaker, R. J., M. B. Bush, and K. Richards. 1989. Plant recolonisation and vegetation succession on the Krakatau islands, Indonesia. Ecological Monographs, 59, 59–123.

Whittaker, R. J., R. Field, and T. Partomihardjo. 2000. How to go extinct: Lessons from the lost plants of Krakatau. *Journal of Biogeography* 27: 1049–64.

Whittaker, R. J., K. J. Willis, and R. Field, 2001. Scale and species richness: Towards a general, hierarchical theory of species diversity. *Journal of Biogeography* 28: 453–70.

Wiley, E. O. 1980. Phylogenetic systematics and vicariance biogeography. *Systematic Botany* 5: 194–220.

———. 1981. *Phylogenetics: The Theory and Practice of Phylogenetic Systematics.* John Wiley and Sons, New York.

———. 1988. Vicariance biogeography. *Annual Review of Ecology and Systematics* 19: 513–42.

Williams, C. B. 1944. Some applications of the logarithmic series and the Index of Diversity to ecological problems. *Journal of Ecology* 32: 1–44.

———. 1947a. The generic relations of species in small ecological communities. *Journal of Animal Ecology* 16: 11–18.

———. 1947b. The logarithmic series and its application to biological problems. *Journal of Ecology* 34: 253–72.

———. 1951. Intra-generic competition as illustrated by Moreau's records of East African bird communities. *Journal of Animal Ecology* 20: 246–53.

———. 1964. *Patterns in the Balance of Nature.* New York: Academic Press.

Williamson, M. H. 1989a. The MacArthur and Wilson theory today: True but trivial. *Journal of Biogeography* 16: 3–4.

———. 1989b. Natural extinction on islands. *Philosophical Transactions of the Royal Society of London,* B 325: 457–68.

Willis, J. C. 1922. *Age and Area.* Cambridge: Cambridge University Press.

Wilsie, C. P. 1962. *Crop Adaptation and Distribution.* San Francisco: W. H. Freeman.

Wilson, E. O. 1959. Adaptive shift and dispersal in a tropical ant fauna. *Evolution* 13: 122–44.

———. 1961. The nature of the taxon cycle in the Melanesian ant fauna. *The American Naturalist* 95: 169–93.

Wilson, E. O., and D. S. Simberloff. 1969. Experimental zoogeography of islands. Defaunation and monitoring techniques. *Ecology* 50: 267–78.

Wilson, J. B. 1999. Assembly rules in plant communities. Pp. 130–64 in *Ecological Assembly Rules: Perspectives, Advances, Retreats,* ed. E. Weiher and P. Keddy. Cambridge: Cambridge University Press.

Wilson, J. T. 1963. Evidence from islands on the spreading of ocean floors. *Nature* 197: 536–38

Wolda, H. 1978. Seasonal fluctuations in rainfall, food and abundance of tropical insects. *Journal of Animal Ecology* 47: 369–81.

Wulff, E. V. [1932] 1943. *An Introduction to Historical Plant Geography,* trans. Elizabeth Brissenden. Waltham, Massachusetts: Chronica Botanica Co.

Yoda, K. 1967. A preliminary survey of the forest vegetation of eastern Nepal, II: General description, structure and floristic composition of sample plots chosen from different vegetation zones. *Journal of the College of Art and Science* 5: 99–140.

Zimmerman, E. C. 1948. *Insects of Hawaii,* vol. 1: Introduction. Honolulu: University of Hawaii Press.

Zink, R. M., R. C. Blackwell-Rago, and F. Ronquist. 2000. The shifting roles of dispersal and vicariance in biogeography. *Proceedings of the Royal Society of London,* B 267: 497–503.

Contributors

John C. Briggs
82-651 Skyview Lane
Indian Palms C.C.
Indio, California 92201
USA

James H. Brown
Biology Department
University of New Mexico
Albuquerque, New Mexico 87131
USA

V. A. Funk
Department of Botany, MRC 166
Smithsonian Institution
P.O. Box 37012
Washington DC 20013-7012
USA

Paul S. Giller
Faculty of Science Office
Kane Building
University College Cork
College Road
Cork
Ireland

Nicholas J. Gotelli
Department of Biology
209 Marsh Life Science Bldg,
University of Vermont
Burlington, Vermont 05405
USA

Lawrence R. Heaney
Division of Mammals
The Field Museum
1400 S. Lake Shore Drive
Chicago, IL 60605
USA

Robert Hengeveld
ALTERRA
Droevendaalsesteeg 3
P.O. Box 47
6800 AA Wageningen
The Netherlands

Christopher J. Humphries
Department of Botany
The Natural History Museum
Cromwell Road
London SW7 5BD
United Kingdom

Mark V. Lomolino
240 Illick Hall
SUNY-ESF
1 Forestry Drive
Syracuse, NY 13210
USA

Alan Myers
Department of Zoology and Animal Ecology
University College Cork
Lee Maltings, Prospect Row
Cork
Ireland

Brett R. Riddle
Department of Biological Sciences
University of Nevada, Las Vegas
4505 Maryland Parkway
Las Vegas, NV 89154-4004
USA

Dov F. Sax
Department of Ecology, Evolution and Marine
  Biology
University of California, Santa Barbara
Santa Barbara, CA 93106
USA

Geerat J. Vermeij
Department of Geology
University of California, Davis
1 Shields Avenue
Davis, CA 95616
USA

Dr Robert J. Whittaker
School of Geography and the Environment
University of Oxford
Mansfield Rd
Oxford, OX1 3TB
United Kingdom

Index

adaptation, 451, 651, 782, 933, 1147
adaptive radiation, 275, 783
allopatric, 655, 780, 786–88, 933
Amazon, 7, 785, 786, 788, 1145
ants, 784, 933
apomorphies, 649, 651
areas of endemism, 653, 655
assembly, 937–40, 1027–29, 1033, 1034, 1147
austral, 269
Australian, 10, 451
avifauna, 648, 934, 937, 1034

bands of affinities, 656
barrier(s), 3, 10, 267, 270, 271, 273, 275, 449–52, 651, 782, 936, 937
beetle(s), 274, 1146
Berger, Rainer, 453
biodiversity, 779, 783, 935, 940, 1029–31, 1145, 1146, 1151, 1154
biogeographic region, 6, 9, 451, 647, 648, 1148
biological diversity, xix, 273, 779, 1154
biological species concept, 780, 782
bird(s), ornithology, 9, 12, 272, 274, 275, 450–51, 452, 648, 652, 780–82, 785, 933, 934, 937, 938, 940, 1033, 1034, 1149
botany, 12
Brundin, Lars, 2, 269, 274, 275, 648, 649, 651, 652, 654, 655, 657
Buffon, Georges-Louis Leclerc, compte de, 6, 8, 9, 10, 11, 12, 647
Bush, Guy, 779, 786, 936, 937

Candolle, Augustin de, 7, 10, 11, 647
Carlquist, Sherwin, 271, 272, 931
center of origin (creation), 6, 8, 12, 13, 268, 271, 272, 273, 274, 449, 450, 451, 647, 648, 650, 652, 654
character displacement, 933
checkerboard distribution, 939, 1033
chorology, chorological method, 11
cladistics, 269, 452, 654, 656, 657, 780
cladogenesis, 781
cladogram, 656

climate, 6, 7, 8, 12, 267, 452, 454, 650, 781, 784, 785, 787, 788, 934, 1146
climax, 454
colonization, 270, 271, 275, 453, 455, 651, 779, 781, 784, 787, 788, 935, 936, 940, 1027
community assembly, 454, 1027, 1033, 1035
community theory, 454
competition, 275, 651, 781, 933, 937–39, 1027, 1029–33, 1035, 1150, 1151
competitive exclusion, 1029
Connor, Edward, 932, 939, 1027, 1033, 1034
conservation, xix, 452, 455, 785, 931, 934, 939
continental drift, 6, 12, 13, 268, 269, 273, 274, 648, 649, 652, 780, 783, 784, 785
coral, 8, 9, 1145
corridor, 272, 275
cosmopolitan, 271
Croizat, Leon, 12, 274, 275, 653, 654, 655, 656, 787
Cuvier, Georges, 6

Darlington, Philip, Jr., 268, 272, 274, 648, 649, 653, 783, 785, 1027, 1028, 1029, 1033
Darwin, Charles, 2, 7, 8, 9, 10, 13, 271, 272, 455, 648, 653, 780, 783, 931, 1029, 1145, 1147
desert, 271, 452, 937, 939, 1146, 1148
Diamond, Jared, 454, 931, 934, 937, 938, 939, 940, 1027, 1033, 1034
disjunction, 13, 267–69, 271, 273–75, 653
dispersal, 9, 10, 12, 267–76, 449–54, 648, 649–55, 657, 781, 783, 788, 933, 937, 939, 1028, 1029, 1033, 1035
diversification, xix, 10, 11, 13, 656, 779–88, 1148
diversity, 1, 7, 9, 267, 455, 647, 779–88, 932, 934, 936, 937, 940, 1029, 1145–50, 1153, 1154
Dobzhansky, Theodosius, 9, 780, 781, 782, 783, 785, 1146, 1147, 1148, 1149, 1151, 1153

ecological biogeography, 7, 451, 453, 934, 1027
ecology, xix, 1, 2, 6, 7, 11, 13, 275, 450, 452, 454, 788, 933, 934, 937, 1027, 1029–33, 1147, 1149, 1150, 1151
Ekman, Sven, 12
elevation, 6, 7, 8, 9, 12, 452, 453, 1149, 1153, 1154

Elton, Charles, 271, 276, 451, 452, 455, 1027, 1029, 1030, 1031, 1034, 1147, 1150
endemic, 9, 13, 270, 274, 655, 931
endemism, 269, 270, 655, 784
equiformal progressive area, 450, 652
equilibrium, equilibria, 2, 275, 451, 454, 784, 785, 787, 788, 934–38, 940, 1028, 1032, 1150, 1151
equilibrium theory, 2, 275, 934, 938, 1028, 1032, 1150
Ethiopia, 271
eustasis, eustasy, 267, 270
evolution, evolutionary, xix, 6, 9–11, 13, 268–72, 274–76, 450–52, 455, 648–50, 652, 656, 779–85, 788, 931, 933, 934, 938, 940, 1146–51, 1153, 1154
exotic, 5, 271, 452, 1154
extinction, 267–70, 274, 275, 455, 650, 651, 784, 785, 788, 934, 935–37, 939–41, 1027–29, 1146, 1148

filter, 272, 274
Fischer, Alfred, 784, 1148, 1149, 1150
fish(es), 270, 654, 781
Flenley, John R., 452, 453
Forbes, Edward, 8, 9, 10
Forster, Johann Reinhold, 6, 7, 12, 783, 1145
fossil(s), 11, 12, 268, 453, 650, 783–85, 788, 940, 1148
founder principle, 270
fragmentation, 273
function(s), 938, 939, 1033

Galapagos, 10, 780, 783
geographic range, 449–51, 453, 455, 785–87, 1146, 1148, 1149, 1153, 1154
glacier(s), glacial, 275, 453, 785
global warming, 271, 450
Gray, Asa, 9, 10
Grinnell, Joseph, 449, 450, 451, 1147

habitat, habitation, 3, 6, 7, 10, 12, 271, 273, 275, 276, 449, 450, 452, 650, 652–83, 785, 787, 931, 933, 936–40, 1029–31, 1033, 1145–47, 1149, 1153, 1154
Haeckel, Ernst Heinrich, 11
Haffer, Jürgen, 785, 786, 787, 788
Hallam, Tony, 2, 273, 274
Hawaii, 272, 934
Hennig, W., 2, 269, 649, 650, 651, 652, 653, 654, 655
historical biogeography, 8, 11, 267, 268, 275, 647–49, 653, 654, 656, 657, 787
Holarctic, 12, 450, 452
Holloway, J. D., 451, 787
Holocene, 453, 454
Hooker, Joseph Dalton, 2, 6, 7, 8, 9, 10, 11, 13, 268, 269, 272, 274, 647, 648, 649, 653, 779
humans, 452, 454, 455, 940
Humboldt, Alexander von, 2, 6, 7, 10, 12, 779, 1145

Ihering, Herman von, 11
immigration, 275, 934, 935, 936, 937, 1028, 1029
individualism, 3, 454, 455

Indo-Pacific, 269, 275
Interchange (biotic), 274, 275
intermediate disturbance hypothesis, 937
intertidal, 1153
invasion, invasions, 12, 451, 452, 455, 779, 788, 933, 1032
island biogeography, 2, 784, 785, 788, 931, 934, 935, 936, 939, 940, 941, 1149, 1150
island(s), 2, 6, 7, 10, 11, 270–72, 276, 451, 648, 651, 782–85, 787, 788, 931–41, 1028, 1031, 1033, 1149, 1150, 1153
isobar, 7
isocryme, 8
isolation, 267, 269, 271, 275, 650, 779, 781, 783, 785, 787, 931, 932, 939
isotherm, 7
Isthmus of Panama, 269

James, Helen, xix, 1, 8, 455, 784, 936, 940, 941, 1145
Janzen, Daniel H., 452
Jardine, N., 451

Kircher, Athanasius, 5
Krakatau, 271, 934, 935

Lack, David, 2, 783, 931, 934
land bridge, 8, 9, 11, 13, 268, 269, 272–76, 647–49
latitiude, 7, 8, 9, 12, 783, 784, 1146–50, 1153, 1154
Latreille, Pierre, 6
Lessepsian migration, 275
Linnaeus, Carolus, 5, 6, 7, 8, 11, 271, 453, 649
Lyell, Charles, 6, 9, 10, 11, 13, 272, 648

MacArthur, Robert H., 2, 9, 275, 452, 454, 784, 785, 787, 931, 932, 934, 935, 936, 937, 938, 939, 940, 1027, 1028, 1030, 1031, 1032, 1033, 1146, 1148, 1149, 1150, 1151
magneticm magnetism, 7, 268
mammals, mammalogy, 6, 9, 12, 273, 275, 451, 453, 648, 780, 782, 936, 937, 1149
marine, 2, 8, 9, 12, 268–73, 275, 784, 788, 1148
Marshall, L. G., 275
Martin, Paul, 455, 940, 1028
Matthew, William Diller, 12, 272, 274, 275, 648, 653, 783
Mayr, Ernst, 2, 648, 649, 780, 782, 783, 786
Merriam, Clinton Hart, 6, 12, 1147
midden(s), 453
mid-oceanic ridge, 268
migration, 10, 12, 272, 275, 452, 652, 654, 1028
modern synthesis, 781, 788
Mount Ararat, 6
Mount Chimborazo, 7
mountain, mountainous, montane, 6, 7, 8, 271, 452, 453, 936, 937, 938, 1145, 1148, 1149

natural selection, 10, 13, 780–82, 788, 1147
Nearctic, 271, 274, 452
Nelson, Gareth, 2, 5, 647, 653, 654, 655, 656, 657
Neotropical, 271, 274

nestedness, 1027–29
niches, 452, 785, 938
noble savage, 455
null model, null communities, 939, 1027, 1028, 1030, 1031, 1033, 1034

oceanographic events, 267
Olson, Stors, 455, 933, 940, 941
Oriental, 11, 271, 451
overkill, 455, 940

Palearctic, 452
paleobiology, 2
paleomagnetism, 12
paleontology, 2, 1147
palynology, pollen, 453
panbiogeography, 274, 653, 654, 656
parapatric, 779, 780, 785, 786, 787
parasite(s), parasitism, 1151
Permian extinction, 785
phylogeny, phylogenetics, 2, 11, 269, 649, 650, 651, 652, 653, 654, 655, 656, 657, 780, 781, 787, 788, 933
phylogeography, 788
phytosociology, 932
Pianka, Eric, 1148, 1150, 1153
plankton, planktonic, 270, 1149
plants, 2, 5, 6, 7, 9, 10, 13, 267, 268, 272, 274, 450, 452–55, 647, 648, 650, 655, 779, 781, 785, 787, 788, 932, 935, 939, 940, 1029, 1034, 1146
plate tectonics, 2, 12, 13, 269, 270, 272, 274, 276, 452, 649, 783, 787, 1029
Pleistocene, 450, 453–55, 781, 785, 786, 936, 939, 940, 1150
Por, F. D., 275
power function, power model, 932
predation, 650, 651, 939, 1150, 1151
progression rule, 6, 649, 655
provinces, provinciality, 7, 8, 9, 270, 454, 784, 1032

Quaternary, 450, 453–55, 785

range, 267–72, 275, 449–55, 650, 780, 786, 935, 1148, 1151, 1153, 1154. See also geographic range
range contraction, 271, 1148
range expansion, 268, 269, 271, 272, 276, 450, 452–54, 780
rassenkreise, rassenkreisen, 650, 781
Raup, David, 784, 785
recolonization, 935, 936, 938
Red Queen, 1148
refuge, refugia, 271, 648, 785–87
Rensch, Bernard, 780, 781, 782, 783
Rosen, Donn Eric, 2, 268, 653, 654, 787

Sclater, Philip Lutley, 9, 647, 648, 649, 653
sea-floor spreading, 268
sea level, 8, 11, 267, 272
seamounts, 272
Simberloff, Daniel, 932, 935, 936, 939, 940, 1027, 1028, 1030, 1031, 1033, 1034
Simpson, George Gaylord, 2, 272, 273, 274, 451, 648, 649, 653, 782, 783, 1033, 1148, 1149, 1150
speciation, 275, 450, 651, 654, 779, 781–88, 933–35, 1035
species area, 7, 932, 934, 936, 937, 939, 1153
species isolation, 7
species richness, 779, 782, 787, 788, 932, 934–38, 1145, 1147, 1148, 1149, 1151, 1153, 1154
station, 7, 782
stepping-stone, 272
stratigraphy, 650
successional, 934, 936, 937, 938, 1027
Suez Canal, 275
supersaturated fauna, 275
supertramp, 937
sweepstakes, 272
sympatric, 779, 786, 788, 933
systematics, systematist, 2, 5, 649, 652, 654

taxon cycle, 784, 933, 940
taxonomic ratio, 1028, 1029, 1030, 1031
tectonics. See plate tectonics
Tethys, 269
trees, 273, 453, 654, 656, 786, 1146, 1147
turnover, 275, 453, 454, 934, 936
typological, typology, 780

vicariance, vicariism, 6, 9, 13, 267–69, 273, 274, 276, 452, 648, 653–57, 787, 788, 940

Wallace, Alfred Russel, 2, 7, 9, 10, 13, 269, 271, 274, 451, 648, 650, 783, 931, 1032, 1033, 1146
Wallace's Line, Wallacea, 11
Wegener, Alfred Lothar, 6, 12, 13, 268, 269
Wells, Philip V., 453
Whittaker, Robert Harding, xx, 272, 454, 784, 787, 931, 932, 934, 935, 936, 937, 939, 940, 941, 1149
Williams, Carrington Bonsor, 1027, 1030, 1031, 1034, 1149, 1150
Wilson, Edward O., 2, 268, 273, 275, 452, 784, 785, 787, 931, 932, 933, 934, 935, 936, 937, 938, 939, 940, 1027, 1028, 1031, 1032, 1150
woodrat, 453
Wulff, E. V., 13, 451, 455

zoogeography, zoogeographers, 2, 12